# Biomedical Photonics

## HANDBOOK

# Advisory Board

# Biomedical Photonics

# HANDBOOK

Editor-in-Chief

**Tuan Vo-Dinh**

Oak Ridge National Laboratory
Oak Ridge, Tennessee

CRC PRESS

Boca Raton London New York Washington, D.C.

**Cover Art**: *Field of Lights*, oil painting by Kim-Chi Le Vo-Dinh. Reproduced with permission of the artist.

**Library of Congress Cataloging-in-Publication Data**

Biomedical photonics handbook / edited by Tuan Vo-Dinh.
p. cm.
Includes bibliographical references and index.
ISBN 0-8493-1116-0
1. Optoelectronic devices—Handbooks, manuals, etc. 2. Biosensors—Handbooks, manuals, etc. 3. Diagnositic imaging—Handbooks, manuals, etc. 4. Imaging systems in medicine—Handbooks, manuals, etc. I. Vo-Dinh, Tuan.

R857.06 B573 2002
610′.28—dc21 2002034914



Direct all inquiries to CRC Press LLC, 2000 N.W. Corporate Blvd., Boca Raton, Florida 33431.



International Standard Book Number 0-8493-1116-0
Library of Congress Card Number 2002034914
Printed in the United States of America 1 2 3 4 5 6 7 8 9 0
Printed on acid-free paper

*Inspired by the love and*
*infinite patience of*
*my wife, Kim-Chi, and*
*my daughter, Jade*

*This book is dedicated to the*
*memory of my parents,*
*Vo Dinh Kinh and*
*Dang Thi Dinh*

# Preface

The *Biomedical Photonics Handbook* is intended to serve as an authoritative reference source for a broad audience involved in the research, teaching, learning, and practice of medical technologies. Biomedical photonics is defined as the science that harnesses light and other forms of radiant energy to solve problems arising in medicine and biology. This research field has recently experienced an explosive growth due to the noninvasive or minimally invasive nature and cost-effectiveness of photonic modalities in medical diagnostics and therapy.

The field of biomedical photonics did not emerge as a well-defined, single research discipline like chemistry, physics, or biology. Its development and growth have been shaped by the convergence of three scientific and technological revolutions of the 20th century: the quantum theory revolution, the technology revolution, and the genomics revolution.

The quantum theory of atomic phenomena provides a fundamental framework for molecular biology and genetics because of its unique understanding of electrons, atoms, molecules, and light itself. Out of this new scientific framework emerged the discovery of the structure of DNA, the molecular nature of cell machinery, and the genetic cause of diseases, all of which form the basis of molecular medicine. The formulation of quantum theory not only gave birth to the field of molecular spectroscopy but also led to the development of a powerful set of photonics tools — lasers, scanning tunneling microscopes, near-field nanoprobes — for exploring nature and understanding the cause of disease at the fundamental level.

Advances in technology also played, and continue to play, an essential role in the development of biomedical photonics. The invention of the laser was an important milestone; the laser is now the light source most widely used to excite tissues for disease diagnosis as well as to irradiate tumors for tissue removal in interventional surgery ("optical scalpels"). The microchip is another important technological development that has significantly accelerated the evolution of biomedical photonics. Although the laser has provided a new technology for excitation, the miniaturization and mass production of integrated circuits, sensor devices, and their associated electronic circuitry made possible by the microchip has radically transformed the ways in which detection and imaging of molecules, tissues, and organs can be performed *in vivo* and *ex vivo*.

Recently, nanotechnology, which involves research on materials and species at length scales between 1 to 100 nm, has been revolutionizing important areas in biomedical photonics, especially diagnostics and therapy at the molecular and cellular levels. The combination of photonics and nanotechnology has already led to a new generation of devices for probing the cell machinery and elucidating intimate life processes occurring at the molecular level heretofore invisible to human inquiry. This will open the possibility of detecting and manipulating atoms and molecules using nanodevices, which have the potential for a wide variety of medical uses at the cellular level. The marriage of electronics, biomaterials, and photonics is expected to revolutionize many areas of medicine in the 21st century.

A wide variety of biomedical photonic technologies have already been developed for clinical monitoring of early disease states or physiological parameters such as blood pressure, blood chemistry, pH, temperature, and the presence of pathological organisms or biochemical species of clinical importance. Advanced optical concepts using various spectroscopic modalities (e.g., fluorescence, scattering, reflection, and optical coherence tomography) are emerging in the important area of functional imaging. Many photonic technologies originally developed for other applications (e.g., lasers and sensor systems in defense, energy, and aerospace) have now found important uses in medical applications. From the brain to the sinuses to the abdomen, precision navigation and tracking techniques are critical to position medical instruments precisely within the three-dimensional surgical space. For instance, optical stereotactic systems are being developed for brain surgery and flexible micronavigation devices engineered for medical laser ablation treatments.

With completion of the sequencing of the human genome, one of the greatest impacts of genomics and proteomics is the establishment of an entirely new approach to biomedical research. With whole-genome sequences and new automated, high-throughput systems, photonic technologies such as biochips and microarrays can address biological and medical problems systematically and on a large scale in a massively parallel manner. They provide the tools to study how tens of thousands of genes and proteins work together in interconnected networks to orchestrate the chemistry of life. Specific genes have been deciphered and linked to numerous diseases and disorders, including breast cancer, muscle disease, deafness, and blindness. Furthermore, advanced biophotonics has contributed dramatically to the field of diagnostics, therapy, and drug discovery in the post-genomic area. Genomics and proteomics present the drug discovery community with a wealth of new potential targets.

Biomedical photonics can provide tools capable of identifying specific subsets of genes, encoded within the human genome, which can cause development of disease. Photonic techniques based on molecular probes are being developed to identify the molecular alterations that distinguish a diseased cell from a normal cell. Such technologies will ultimately aid in characterizing and predicting the pathologic behavior of that diseased cell, as well as the cell's responsiveness to drug treatment. Information from the human genome project will one day make personal, molecular medicine an exciting reality.

This 1800-page handbook presents the most recent scientific and technological advances in biomedical photonics, as well as their practical applications, in a single source. The book represents the work of 150 scientists, engineers, and clinicians. Each of the 65 chapters provides introductory material with an overview of the topic of interest as well as a collection of published data with an extensive list of references for further details. The chapters are grouped in seven sections followed by an appendix of spectroscopic databases:

I. *Photonics and Tissue Optics* contains three introductory chapters on the fundamental optical properties of tissue, light–tissue interactions, and theoretical models for optical imaging.
II. *Photonic Devices* deals with basic instrumentation and hardware systems and contains chapters on lasers and excitation sources, basic optical instrumentation, optical fibers and waveguides, and spectroscopic systems.
III. *Photonic Detection and Imaging Techniques* deals with methodologies and contains nine chapters on various detection techniques and systems (lifetime imaging, microscopy, near-field detection, optical coherence tomography, interferometry, Doppler imaging, light scattering, and thermal imaging).
IV. *Biomedical Diagnostics I* contains ten chapters describing *in vitro* diagnostics (glucose diagnostics, *in vitro* instrumentation, biosensors, capillary electrophoresis, and flow cytometry) and *in vivo* diagnostics (functional imaging and photon migration spectroscopy). Two chapters describe specific techniques and applications of two important and mature technologies — x-ray diagnostics and optical pumping in magnetic resonance imaging.

V. *Biomedical Diagnostics II: Optical Biopsy* is composed of eight chapters mainly devoted to novel optical techniques for cancer diagnostics, often referred to as "optical biopsy" (fluorescence, scattering, reflectance, Raman, infrared, optoacoustics, and ultrasonically modulated optical imaging).
VI. *Interventional and Treatment Techniques* discusses in 15 chapters photodynamic therapy and various laser-based treatment techniques that are applied to various organs and disease endpoints (dermatology, pulmonary, neurosurgery, ophthalmology, otolaryngology, urology, gastroenterology, and dentistry).
VII. *Advanced Biophotonics for Genomics, Proteomics, and Medicine* examines the most recent advances in methods and instrumentation for biomedical and biotechnology applications. This section contains 14 chapters on emerging photonic technologies (e.g., biochips, nanosensors, quantum dots, molecular probes, bioluminescent reporters, optical tweezers) being developed for gene expression research, gene diagnostics, protein profiling, and molecular biology investigations for the "new medicine."
VIII. *Appendix* provides a comprehensive, single-chapter compilation of useful information on spectroscopic data of biologically and medically relevant species for more than 1000 compounds and systems.

The goal of this handbook is to provide a comprehensive forum that integrates interdisciplinary research and development of interest to scientists, engineers, manufacturers, teachers, students, and clinical providers. The handbook is designed to present the most recent advances in instrumentation and methods as well as clinical applications in important areas of biomedical photonics. Because light is rapidly becoming an important diagnostic tool and a powerful weapon in the armory of the modern physician, it is our hope that this handbook will stimulate a greater appreciation of the usefulness, efficiency, and potential of photonics in medicine.

**Tuan Vo-Dinh**
**Oak Ridge National Laboratory**
**Oak Ridge, Tennessee**
**February 2003**

# Acknowledgments

The completion of this work was made possible with the assistance of many friends and colleagues. It is a great pleasure for me to acknowledge, with deep gratitude, the contributions of the 150 contributors to the 65 chapters of this 1800-page handbook. I am indebted to the members of the Scientific Advisory Board, Drs. Mitchel S. Berger, Britton Chance, Thomas J. Dougherty, Daniel L. Farkas, James G. Fujimoto, Warren S. Grundfest, Leroy Hood, Joseph R. Lakowicz, Vladilen S. Letokhov, Praveen N. Mathur, Nitish V. Thakor, and Tony Wilson. Their thoughtful suggestions and useful advice in the planning phase of the project were important in achieving the breadth and depth of this handbook.

I wish to thank many scientists and co-workers at the Oak Ridge National Laboratory (ORNL) and many colleagues in academia for their kind help with reading and commenting on various chapters of the manuscript. I wish to thank my administrative assistant, Julia B. Cooper, at ORNL for her dedicated, efficient, and timely assistance over many years. The kind assistance of Dr. David Packer, the cheerful help of Debbie Vescio, and the tireless and dedicated afforts of Christine Andreasen at CRC Press are much appreciated. I greatly appreciate the collaboration and friendship of many colleagues, including Drs. Mitchel S. Berger, Martin Holland, Geoff Ling, Praveen N. Mathur, G. Wayne Morrison, Bergein F. Overholt, Masoud Panjehpour, and William P. Wiesmann, with whom I have had the opportunity to initiate various biomedical projects. My gratitude is extended to Dr. Urs P. Wild at the ETH in Zurich, Switzerland and Dr. James D. Winefordner at the University of Florida for kind advice in my early research in photonics. I would like to thank Drs. Barry A. Berven, W. Frank Harris, Lee L. Riedinger, and William J. Madia at ORNL and Drs. Michael Viola and Dean Cole at the Department of Energy (DOE) for their continued support throughout this undertaking.

I gratefully acknowledge the support of the DOE Office of Biological and Environmental Research, the National Institutes of Health, the Army Medical Research and Materiel Command, the Department of Justice, the Federal Bureau of Investigation, the Office of Naval Research, and the Environmental Protection Agency.

The completion of this work was made possible with the encouragement, love, and inspiration of my wife, Kim-Chi, and my daughter, Jade.

***Tuan Vo-Dinh***

# Editor-in-Chief

**Tuan Vo-Dinh, Ph.D.,** is a Corporate Fellow, Group Leader of the Advanced Biomedical Science and Technology Group, and Director of the Center for Advanced Biomedical Photonics, Oak Ridge National Laboratory (ORNL), Oak Ridge, Tennessee. He is also an Adjunct Professor at the University of California, San Francisco and the University of Tennessee, Knoxville. A native of Vietnam and a naturalized U.S. citizen, Dr. Vo-Dinh completed his high school education in Saigon (now Ho Chi Minh City). He continued his studies in Europe, where he received a Ph.D. in biophysical chemistry in 1975 from the Swiss Federal Institute of Technology (known as the ETH) in Zurich, Switzerland. His research interests have focused on the development of advanced technologies for the protection of the environment and the improvement of human health. His research activities involve laser spectroscopy, molecular imaging, medical diagnostics, cancer detection, chemical sensors, biosensors, nanosensors, and biochips.

Dr. Vo-Dinh has published more than 300 peer-reviewed scientific papers. He is an author of a textbook on spectroscopy and the editor of four books. He holds more than 26 patents, 5 of which have been licensed to environmental and biotech companies for commercial development. Dr. Vo-Dinh is a fellow of the American Institute of Chemists and of SPIE, the International Society for Optical Engineering. He serves on editorial boards of various international journals on molecular spectroscopy, analytical chemistry, biomedical optics, and medical diagnostics. He also serves the scientific community through his participation in a wide range of governmental and industrial boards and advisory committees.

Dr. Vo-Dinh has received six R&D 100 Awards for Most Technologically Significant Advance in Research and Development for his pioneering research and inventions of various innovative technologies, including a chemical dosimeter (1981), an antibody biosensor (1987), the SERODS optical data storage system (1992), a spot test for environmental pollutants (1994), the SERS gene probe technology for DNA detection (1996) and the multifunctional biochip for medical diagnostics and pathogen detection (1999). He received the Gold Medal Award, Society for Applied Spectroscopy (1988); the Languedoc-Roussillon Award (France; 1989), the Scientist of the Year Award, Oak Ridge National Laboratory (1992); the Thomas Jefferson Award, Martin Marietta Corporation (1992); two Awards for Excellence in Technology Transfer, Federal Laboratory Consortium (1995, 1986); the Inventor of the Year Award, the Tennessee Inventors Association (1996); and the Lockheed Martin Technology Commercialization Award (1998). In 1997, Dr. Vo-Dinh was awarded the Exceptional Services Award for distinguished contribution to a Healthy Citizenry from the U.S. Department of Energy.

# Editor-in-Chief

# Contributors


**Detlef F. Albrecht**
Medical College of Ohio
Toledo, Ohio,

**Leonardo Allain**
Oak Ridge National Laboratory
Oak Ridge, Tennessee

**R. Rox Anderson**
Wellman Laboratories of Photomedicine
Harvard Medical School
Boston, Massachusetts

**George Angheloiu**
The Cleveland Clinic Foundation
Cleveland, Ohio

**Vadim Backman**
Northwestern University
Evanston, Illinois

**Gregory Bearman**
California Institute of Technology
Pasadena, California

**Terry Beck**
TriLink BioTechnologies, Inc.
San Diego, California

**Moshe Ben-David**
Tel-Aviv University
Tel-Aviv, Israel

**Mitchel S. Berger**
University of California, San Francisco
San Francisco, California

**Rohit Bhargava**
Laboratory of Chemical Physics
National Institute of Diabetes and Digestive and Kidney Diseases
National Institutes of Health
Bethesda, Maryland

**Irving J. Bigio**
Boston University
Boston, Massachusett

**Devin K. Binder**
University of California, San Francisco
San Francisco, California

**David A. Boas**
Harvard Medical School
Massachusetts General Hospital
Athinoula A. Martinos Center for Biomedical Imaging
Charlestown, Massachusetts

**Claude Boccara**
Ecole Supérieure de Physique et Chimie Industrielles
Paris, France

**Darryl J. Bornhop**
Texas Tech University
Lubbock, Texas

**Nada N. Boustany**
Johns Hopkins University School of Medicine
Baltimore, Maryland

**Stephen G. Bown**
National Medical Laser Centre
Royal Free and University College Medical School
London, U.K.

**David Boyer**
Ophthalmology Research Laboratories,
Cedars-Sinai Medical Center
The Retina Vitreous Associates Medical Group
Doheny Eye Institute, Keck-USC School
of Medicine
Los Angeles, California

**Murphy Brasuel**
University of Michigan
Ann Arbor, Michigan

**Mark E. Brezinski**
Brigham and Women's Hospital
Boston, Massachusetts

**Gavin M. Briggs**
National Medical Laser Centre
Royal Free and University College Medical School
London, U.K.

**Richard D. Bucholz**
St. Louis University Health Sciences Center
St. Louis, Missouri

**Albert E. Cerussi**
Beckman Laser Institute
University of California, Irvine
Irvine, California

**Warren C.W. Chan**
University of Toronto
Toronto, Canada

**Thomas G. Chu**
The Retina Vitreous Associates Medical Group
Doheny Eye Institute, Keck-USC School
of Medicine
Los Angeles, California

**Gerald E. Cohn**
Cyber Tech Applied Science, Inc.
Evanston, Illinois

**Christopher H. Contag**
Stanford University School of Medicine
Stanford, California

**Pamela R. Contag**
Xenogen Corporation
Alameda, California
and
Stanford University School of Medicine
Stanford, California

**Gerard L. Coté**
Texas A&M University
College Station, Texas

**Maureen T. Cronin**
Genomic Health, Inc.
Redwood City, California

**Brian M. Cullum**
University of Maryland Baltimore County
Baltimore, Maryland

**Ramachandra Dasari**
Massachusetts Institute of Technology
Cambridge, Massachusetts

**Abby E. Deans**
New York University
New York, New York

**Volker Deckert**
Institut für Angewandte Photophysik
Dresden, Germany

**Richard Domanik**
Xomix, Ltd.
Chicago, Illinois

**Chen Y. Dong**
Massachusetts Institute of Technology
Cambridge, Massachusetts

**Thomas J. Dougherty**
Roswell Park Cancer Institute
Buffalo, New York

**Rose Du**
University of California, San Francisco
San Francisco, California

**Arnaud Dubois**
Ecole Supérieure de Physique
et Chimie Industrielles
Paris, France

**Gabriela Dumitrascu**
University of New Orleans
New Orleans, Louisiana

**Xiaohong Fang**
University of Florida
Gainesville, Florida

**Michael S. Feld**
Massachusetts Institute of Technology
Cambridge, Massachusetts

**Benoît C. Forget**
Ecole Supérieure de Physique
et Chimie Industrielles
Paris, France

**Daniel Fried**
University of California, San Francisco
San Francisco, California

**James G. Fujimoto**
Massachusetts Institute of Technology
Cambridge, Massachusetts

**Israel Gannot**
Tel Aviv University
Tel Aviv, Israel

**Irene Georgakoudi**
Massachusetts Institute of Technology
Cambridge, Massachusetts

**S. Douglass Gilman**
University of Tennessee
Knoxville, Tennessee

**Anuradha Godavarty**
Texas A&M University
College Station, Texas

**Sandra O. Gollnick**
Roswell Park Cancer Institute
Buffalo, New York

**Jay P. Gore**
Purdue University
West Lafayette, Indiana

**Guy D. Griffin**
Oak Ridge National Laboratory
Oak Ridge, Tenessee

**Abigail S. Haka**
Massachusetts Institute of Technology
Cambridge, Massachusetts

**Barbara W. Henderson**
Roswell Park Cancer Institute
Buffalo, New York

**Petr Herman**
Charles University
Prague, Czech Republic

**Richard Hogrefe**
TriLink BioTechnologies, Inc.
San Diego, California

**Jessica P. Houston**
Texas A&M University
College Station, Texas

**Ramesh Jagannathan**
Oak Ridge National Laboratory
Oak Ridge, Tennessee

**Alexander A. Karabutov**
Moscow State University
Moscow, Russia

**Tiina I. Karu**
Institute of Laser and Information Technologies
Russian Academy of Sciences
Troitsk, Moscow Region, Russian Federation

**Paul M. Kasili**
Oak Ridge National Laboratory
Oak Ridge, Tennessee

**Shannon Kelley**
University of Florida
Gainesville, Florida

**Raoul Kopelman**
University of Michigan
Ann Arbor, Michigan

**Eddy Kuwana**
Texas A&M University
College Station, Texas

**Joseph R. Lakowicz**
University of Maryland
Baltimore, Maryland

**Suzanne J. Lassiter**
Louisiana State University
Baton Rouge, Louisiana

**Keith A. Laycock**
St. Louis University Health Sciences Center
St. Louis, Missouri

**Andrew C. Lee**
National Medical Laser Centre
Royal Free and University College Medical School
London, U.K.

**Vladilen S. Letokhov**
Institute of Spectroscopy
Russian Academy of Sciences
Troitsk, Moscow Region, Russian Federation

**Richard Levenson**
Cambridge Research & Instrumentation
Woburn, Massachusetts

**Sandrine Lévêque-Fort**
Université Paris Sud
Orsay Cedex, France

**Ira W. Levin**
Laboratory of Chemical Physics
National Institute of Diabetes and Digestive and Kidney Diseases
National Institutes of Health
Bethesda, Maryland

**Julia G. Levy**
QLT Inc.
Vancouver, Canada

**John T. Leyland II**
Medical College of Ohio
Toledo, Ohio

**Jianwei Jeffrey Li**
University of Florida
Gainesville, Florida

**Jun Li**
Hunan University
Changsha, Hunan, People's Republic of China

**Kai Licha**
Schering AG
Berlin, Germany

**Hai-Jui Lin**
University of Maryland
Baltimore, Maryland

**Hong Liu**
University of Oklahoma
Norman, Oklahoma

**Hua Lou**
University of Florida
Gainesville, Florida

**Ezra Maguen**
Ophthalmology Research Laboratories,
Cedars-Sinai Medical Center
The Jules Stein Eye Institute,
UCLA School of Medicine
Los Angeles, California

**Anita Mahadevan-Jansen**
Vanderbilt University
Nashville, Tennessee

**Francis Mandy**
Health Canada
Ottawa, Canada

**Elaine S. Mansfield**
Stanford University School of Medicine
Stanford, California

**Barry R. Masters**
University of Bern
Bern, Switzerland

**Anubhav N. Mathur**
Indiana University Medical Center
Indianapolis, Indiana

**Praveen N. Mathur**
Indiana University Medical Center
Indianapolis, Indiana

**Karen McNally-Heintzelman**
Rose-Hulman Institute of Technology
Terre Haute, Indiana

**Roger J. McNichols**
BioTex, Inc.
Houston, Texas

**Joel Mobley**
Oak Ridge National Laboratory
Oak Ridge, Tennessee

**Eric Monson**
University of Michigan
Ann Arbor, Michigan

**Jason T. Motz**
Massachusetts Institute of Technology
Cambridge, Massachusetts

**Judith R. Mourant**
Los Alamos National Laboratory
Los Alamos, New Mexico

**Markus Müller**
Massachusetts Institute of Technology
Cambridge, Massachusetts

**Thuvan Nguyen**
University of New Orleans
New Orleans, Louisiana

**Shuming Nie**
Georgia Tech and Emory University
Atlanta, Georgia

**Gert E. Nilsson**
Linköping University
Linköping, Sweden

**Thomas Nishino**
University of Texas Medical Branch
Galveston, Texas

**Stephen J. Norton**
Geophex, Ltd.
Raleigh, North Carolina

**Alexander A. Oraevsky**
Fairway Medical Technologies
Houston, Texas

**Bergein F. Overholt**
Thompson Cancer Survival Center
Knoxville, Tennessee

**Clyde V. Owens**
Louisiana State University
Baton Rouge, Louisiana

**Ravindra K. Pandey**
Roswell Park Cancer Insititute
Buffalo, New York

**Masoud Panjehpour**
Thompson Cancer Survival Center
Knoxville, Tennessee

**Martin A. Philbert**
University of Michigan
Ann Arbor, Michigan

**Lionel Pottier**
Ecole Supérieure de Physique
et Chimie Industrielles
Paris, France

**Francois Ramaz**
Ecole Supérieure de Physique
et Chimie Industrielles
Paris, France

**Diether Recktenwald**
BD Biosciences
San Jose, California

**Lou Reinisch**
University of Canterbury
Christchurch, New Zealand

**Nitsa Rosenzweig**
Xavier University of New Orleans
New Orleans, Louisiana

**Zeev Rosenzweig**
University of New Orleans
New Orleans, Louisiana

**Ranadhir Roy**
Texas A&M University
College Station, Texas

**E. Göran Salerud**
Linköping University
Linköping, Sweden

**Meic H. Schmidt**
University of California, San Francisco
San Francisco, California

**Sheldon Schuster**
University of Florida
Gainesville, Florida

**Juliette Selb**
Ecole Supérieure de Physique
et Chimie Industrielles
Paris, France

**Steven Selman**
Medical College of Ohio
Toledo, Ohio

**Michael J. Sepaniak**
University of Tennessee
Knoxville, Tennessee

**Eva M. Sevick-Muraka**
Texas A&M University
College Station, Texas

**Sharat Singh**
ACLARA Biosciences, Inc.
Mountain View, California

**Peter T.C. So**
Massachusetts Institute of Technology
Cambridge, Massachusetts

**Joon Myong Song**
Oak Ridge National Laboratory
Oak Ridge, Tennessee

**Steven A. Soper**
Louisiana State University
Baton Rouge, Louisiana

**David L. Stokes**
Oak Ridge National Laboratory
Oak Ridge, Tennessee

**Dimitra N. Stratis-Cullum**
U.S. Army Research Laboratory
Adelphi, Maryland

**N.O. Tomas Strömberg**
Linköping University
Linköping, Sweden

**Kittisak Suthamjariya**
Wellman Laboratories of Photomedicine
Harvard Medical School
Boston, Massachusetts

**Weihong Tan**
University of Florida
Gainesville, Florida

**Andrew Taylor**
University of Surrey
Guildford, Surrey, U.K.

**Nitish V. Thakor**
Johns Hopkins University School of Medicine
Baltimore, Maryland

**Alan B. Thompson**
Texas A&M University
College Station, Texas

**Bruce J. Tromberg**
Beckman Laser Institute
University of California, Irvine
Irvine, California

**Valery V. Tuchin**
Saratov State University
Saratov, Russian Federation

**Rudi Varro**
BD Biosciences
San Jose, California

**Pierre M. Viallet**
University of Perpignan
Perpignan, France

**Marie Vicens**
University of Florida
Gainesville, Florida

**Carmen Virgos**
ACLARA Biosciences, Inc.
Mountain View, California

**Tuan Vo-Dinh**
Oak Ridge National Laboratory
Oak Ridge, Tennessee

**Emanuel Waddell**
Louisiana State University
Baton Rouge, Louisiana

**Kemin Wang**
Hunan University
Changsha, Hunan, People's Republic of China
and
University of Florida
Gainesville, Florida

**Youxiang Wang**
Turnerdesigns
Sunnyvale, California

**Karin Wårdell**
Linköping University
Linköping, Sweden

**Ashley J. Welch**
University of Texas at Austin
Austin, Texas

**M. Wendy Williams**
Oak Ridge National Laboratory
Oak Ridge, Tennessee

**Stephen J. Williams**
ACLARA Biosciences, Inc.
Mountain View, California

**Tony Wilson**
University of Oxford
Oxford, U.K.

**Xizeng Wu**
University of Alabama at Birmingham
Birmingham, Alabama

**Lisa S. Xu**
Purdue University
West Lafayette, Indiana

**Yichuan Xu**
Louisiana State University
Baton Rouge, Louisiana

**Kenji Yasuda**
University of Tokyo
Komaba, Meguro-ku, Tokyo, Japan

**Arjun G. Yodh**
University of Pennsylvania
Philadelphia, Pennsylvania

**Dmitry A. Zimnyakov**
Saratov State University
Saratov, Russian Federation

# Contents

## VIII Appendix

# 1

# Biomedical Photonics: A Revolution at the Interface of Science and Technology




Tuan Vo-Dinh
*Oak Ridge National Laboratory*
*Oak Ridge, Tennessee*


## 1.1 Introduction

Throughout human history, light has played an important role in medicine. In prehistoric times, the healing power of light was often attributed to mythological, religious, and supernatural powers. The history of light therapy dates back to the ancient Egyptians, Hindus, Romans, and Greeks, all of whom created temples to worship the therapeutic powers of light, especially sunlight, for healing the body as well as the mind and the soul. In Hindu mythology, Dhanvantar, originally a sun god, is physician of the gods and a teacher of healing arts to humans. In Greek mythology, Apollo, the god of healing, who taught medicine to man, is also called the sun god or the "god of light." These and other mythological figures are testaments to humankind's recognition of the healing power of light since the dawn of time.

The contribution of light to medicine has evolved throughout human history with the advent of science and technology. In the 17th century, the invention of the microscope by Dutch investigators was critical to the development of biological and biomedical research for the next 200 years. Cell theory emerged in the 1830s, when German scientists M. J. Schleiden and Theodor Schwann looked into their microscopes and identified the cell as the basic unit of plant and animal tissue and metabolism.[1] The microscope provided the central observation tool for a new style of research, out of which emerged the germ theory of disease, developed by Robert Koch and Louis Pasteur in the 1870s (Figure 1.1).

In the fall of 1895, the German physicist Wilhelm Roentgen, working with a standard piece of laboratory equipment, discovered a new type of radiation: the x-ray. This discovery extended the range of electromagnetic radiation well beyond its conventional limits. Furthermore, this discovery led to the development of a powerful new technique that uses x-rays to look into the intact body for disease diagnosis.

**FIGURE 1.1 (Color figure follows p. 28-30.)** Louis Pasteur used the microscope, which provided the central observation tool for a new style of research and out of which emerged the germ theory of disease. Albert Edelfelt's *Louis Pasteur* (1865), Musée d'Orsay, Paris, France. (©Réunion des Musées Nationaux/Art Resource, New York. Reproduced with permission of Artists Rights Society, New York.)

These examples are just some of the numerous cases where scientific discoveries and technological advances in photonics have opened new horizons to medicine and provided critical tools to investigate molecules, analyze tissue, and diagnose diseases.

## 1.2 Biomedical Photonics: A Definition

The field of biomedical photonics is often not well defined because it is a relatively new field that has emerged from research conducted at the interface of the physical and biological sciences and engineering. Therefore, it is useful to provide a definition here.

A related term that has commonly been used is "biomedical optics." Let us examine the similarity and difference between biomedical photonics and biomedical optics. By general definition, the field of optics involves "optical" light or "visible" light, which is a particular type of electromagnetic radiation that can be seen and sensed by the human eye. On the other hand, the field of photonics, which involves photons, the quanta of energy in the entire spectrum of electromagnetic radiation, is broader than the field of optics. We tend to think of optical radiation as "light," but the rainbow of colors that make up optical or visible light is just a very small part of a much broader range of the energy range of the photon.

Photonics includes optical and nonoptical technologies that deal with electromagnetic radiation, which is the energy propagated through space between electric and magnetic fields. The electromagnetic spectrum is the extent of that energy, ranging from cosmic rays, gamma rays, and x-rays throughout ultraviolet, visible, infrared, microwave, and radio frequency energy.

Biomedical photonics, therefore, may be defined as science and technology that use the entire range of electromagnetic radiation beyond visible light for medical applications. This field involves generating and harnessing light and other forms of radiant energy whose quantum unit is the photon. The science includes the use of light absorption, emission, transmission, scattering, amplification, and detection and uses a wide variety of methods and technologies, such as lasers and other light sources, fiber optics, electro-optical instrumentation, sophisticated microelectromechanical systems, and nanosystems, for medical applications. The range of applications of biomedical photonics extends from medical diagnostics to therapy and disease prevention.

# 1.3 Scientific and Technological Revolutions Shaping Biomedical Photonics

The field of biomedical photonics did not emerge as a well-defined, single research discipline like chemistry, physics, or biology. Its development and growth have been shaped by the convergence of three scientific and technological revolutions of the 20th century (Figure 1.2):

- The quantum theory revolution (1900–1950s)
- The technology revolution (1940s–1950s)
- The genomics revolution (1950s–2000)

Technological progress is usually represented as an "S curve," rising slowly at first, then more and more rapidly until it approaches natural limits, and then tending to level off to reach theoretical limits. Some examples of scientific discoveries and technological achievements are shown in Figure 1.2. The following sections discuss the three revolutions of the 20th century that have shaped the growth and development of biomedical photonics.

## 1.3.1 The Quantum Theory Revolution: A Historic Evolution of the Concept of Light

The field of photonics has significantly benefited from the development of quantum theory. With the advent of this theory, scientific fields such as molecular spectroscopy and photonic technologies (such as lasers, optical biopsy, optical tweezers, and near-field probes) have provided powerful tools to diagnose diseases noninvasively, interrogate the cell at the molecular level, and fight diseases at the gene level. The quantum theory of atomic phenomena provides a fundamental framework for molecular biology and genetics because of its unique understanding of electrons, atoms, and molecules, and light itself. Out of

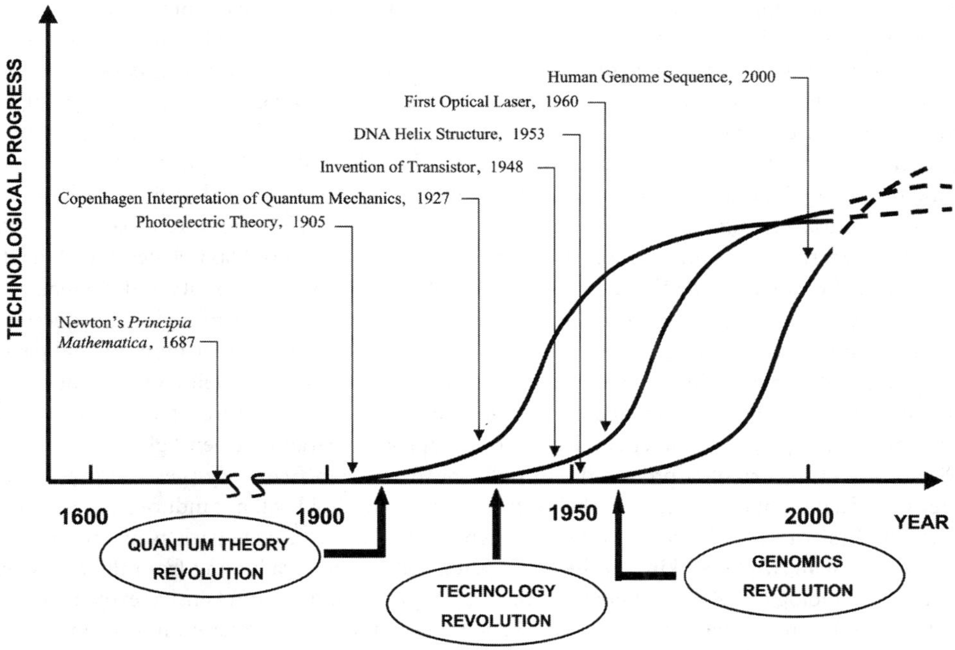


**FIGURE 1.2** Three revolutions of the 20th century shaped biomedical photonics. Technological progress is usually represented as an "S curve," rising slowly at first, then more and more rapidly until it approaches natural limits, and then tending to level off to reach theoretical limits.

this new scientific framework emerged the discovery of the DNA structure, the molecular nature of cell machinery, and the genetic cause of diseases, all of which form the basis of molecular medicine.

Quantum theory was one of the most amazing discoveries of the 20th century and brought about a monumental paradigm shift. The concept of a paradigm was proposed by Thomas Kuhn in his seminal book *The Structure of Scientific Revolutions*.[2] According to Kuhn's thesis, scientists perform their investigation within the framework of a collective background of shared assumptions, which make up a paradigm. During any given period, the scientific community in a particular field of research has a prevailing paradigm that shapes, defines, and directs research activities in that field. People often become attached to their paradigms until a paradigm shift occurs when a major revolutionary discovery triggers a drastic change in beliefs, like the dramatic upheavals occurring in revolutions.

A dramatic paradigm shift occurred in the 17th century with René Descartes' mechanistic philosophy and Isaac Newton's scientific revolution. Descartes, a French mathematician and philosopher, established the firm belief in scientific knowledge that forms the basis of Cartesian philosophy and the worldview derived from it. According to Bertrand Russell, such a shift in thought "had not happened since Aristotle .... There is a freshness about [Descartes'] work that is not to be found in any eminent previous philosopher since Plato."[3] For Descartes, the essence of human nature lies in thought, and all things conceived clearly and distinctly are true. Descartes' rational philosophy integrated a complete mechanistic interpretation of physics, biology, psychology, and medicine. His celebrated statement *Cogito, ergo sum* ("I think, therefore I am") has profoundly influenced Western civilization.

With Newton, a new agenda for scientific research in optics, mechanics, astronomy, and a wide variety of other fields was born. Newton's *Principia Mathematica Philosophica Naturalis*, a series of three books completed in 1687, laid the foundation for our understanding of the underlying physics of the world, which shaped the history of science and remained the main paradigm for the classical worldview for over two centuries. Newton's work concluded the intellectual quest that extended back through Galileo, Kepler, and Copernicus and, ultimately, back to Aristotle.

Since 300 B.C., Aristotle's work — which encompassed logic, physics, cosmology, psychology, natural history, anatomy, metaphysics, ethics, and aesthetics — had represented the culmination of the Hellenic Enlightenment Age and the source of science and higher learning for the following 2000 years. In the Aristotelian worldview, light was not considered among one of the four basic elements — air, earth, fire, and water — that made up the physical universe. In Newton's work, by contrast, light plays an important role in a series of three books called *Opticks*, which describes in detail a wide variety of light phenomena (such as the refraction of light, the nature of white light colors, thin-film phenomena) and optical instruments (such as the microscope and the telescope).[2] Newton performed groundbreaking experiments using optical instruments demonstrating that light was actually a mixture of colors by using a glass prism to separate the colors. In 1865, the British physicist James Clerk Maxwell developed the theory of light propagation by unifying the theories that describe the forces of electricity and magnetism.

Then a series of unexpected discoveries concerning the nature of light brought into question the underlying reality of the Newtonian worldview and set the stage for yet another monumental paradigm shift: the 20th century revolution in quantum physics launched by Albert Einstein. A phenomenon called the "photoelectric effect" raised intriguing questions about the exact nature of light. Discovered by Heinrich Hertz, the photoelectric effect dealt with an apparent paradox: when light irradiates certain materials, an electric current is produced, but only above a certain frequency (i.e., energy). Increasing the intensity of light that has a frequency below the requisite threshold will not induce a current. In 1901, the German physicist Max Planck suggested that light came only as discrete packets of energy.

However, Einstein, while working in the Swiss patent office after graduating from the Swiss Federal Institute of Technology in Zurich (known as the ETH), provided a comprehensive explanation of the photoelectric effect in a paper published in 1905 and launched the field of quantum mechanics. Einstein called the particles of light *quanta* (after the Latin *quantus* for "how much"); hence, the origin of the term "quantum theory." He showed that light consists neither of continuous waves, nor of small, hard particles. Instead, it exists as bundles of wave energy called photons. Each photon has an energy that corresponds to the frequency of the waves in the bundle. The higher the frequency (the bluer the color),

**FIGURE 1.3** Albert Einstein launched the field of quantum mechanics, which provides the theoretical foundation for molecular spectroscopy and photonics. Einstein called the particles of light quanta (after the Latin *quantus* for "how much"); hence, the origin of the term "quantum theory." Einstein showed that light consists neither of continuous waves nor of small, hard particles. Instead, it exists as bundles of wave energy called photons. (Photo from Bildarchiv ETH-Bibliothek, Zurich, Switzerland. With permission.)

the greater the energy carried by that bundle. Einstein's Nobel Prize was awarded in 1921, not for his theories on relativity (the Nobel Committee thought them too speculative at the time), but for his quantum theory on the photoelectric effect (Figure 1.3).

Einstein published another extraordinary paper in 1905 that drastically redirected modern physics. This paper dealt with special relativity, or the physics of bodies moving uniformly relative to one another. Here again, light took a central role as the ultimate reference element having the highest speed in the universe. By postulating that nothing can move faster than light, Einstein reformulated Newtonian mechanics, which contains no such limitation. This is the heart of his celebrated formula $E = mc^2$, which equates mass ($m$) and energy ($E$), and makes the speed of light ($c$) a constant in the equation. Einstein's theory of relativity shattered the classical worldview based on the three-dimensional space of classical Euclidean geometry, Newtonian mechanics, and the concept of absolute space and time. In Einstein's worldview, the universe has no privileged frames of references, no master clock, and no absolute time, because the velocity of light is the ultimate limit in speed, constant in all directions.

Acceptance of quantum theory was further reinforced by research conducted by Ernest Rutherford and Niels Bohr, using radioactive emission as a tool to investigate the structure of the atom. Essentially, quantum theory replaced the mechanical model of the atom with one where atoms and all material objects are not sharply defined entities in the physical world but rather "fuzzy clouds" with a dual wave–particle nature. According to quantum theory, all objects, even the subatomic particles (electrons, protons, neutrons in the nucleus), are entities that have a dual aspect. Light and matter exhibit this duality, sometimes behaving as electromagnetic waves, sometimes as particles called photons. The existence of any molecule or object can be defined by a "probability wave," which predicts the likelihood of finding the object at a particular place within specific limits. The mathematical formulations developed by Werner Heisenberg, Erwin Schrodinger, and Paul Dirac from 1926 through 1933 firmly established the theoretical foundation of the quantum theory.

The rules that govern the subatomic world of blurry particles and nuclear forces are called quantum mechanics. In quantum mechanics, the state of a molecule is described by a wavefunction $\psi$, a function of the position and spins of all electrons and nuclei and the presence of any external field. The probability of finding the molecule at a particular position or spin is represented by the square of the amplitude of wavefunction ψ. In other words, at the subatomic level, matter does not exist with certainty at definite places, but rather shows "probabilities to exist." Even Einstein could hardly accept the fundamental nature of probability in the quantum concept of reality, saying "God does not play dice with the universe."

By the 1930s, with such bizarre quantum entities as "spin" formulated by Wolfgang Pauli at the ETH (the Pauli exclusion principle), the worldview of quantum theory had completed its paradigm shift. Following a historic meeting in 1927, scientists acknowledged that a complete understanding of physical reality lies beyond the capabilities of rational thought, a conclusion known as the "Copenhagen Interpretation of Quantum Mechanics."[4] The quantum worldview implies that the structure of matter is often not mechanical or visible, and that the reality of the world cannot be explained by the physical perceptions of the human senses. Coming into full circle, a seed of this new scientific concept of the 21st century rejoined an important aspect of Plato's philosophy in 400 B.C. that referred to the concrete objects of the visible world as imperfect copies of the forms of which they "partake." Similarities between quantum physics and many metaphysical concepts in Eastern philosophies and Western religions have been a topic of great interest.[5]

Quantum theory also had a profound effect on many fields beyond science, such as art. It is noteworthy that modern art, with its seemingly strange distortions of visual reality, also appeared in the 1930s. It was no coincidence that, during the quantum revolution in science, Cubist and Surrealist art abolished realistic shapes referenced in fixed space and fixed time. Pablo Picasso's renderings of the human face with their multifaceted perspectives often reflected the dual nature of reality (Figure 1.4); Salvador Dali's vision of melting clocks evoked the elasticity of a relativistic time. Henry Moore's sculptures reshaping the physical form of the human body marked a departure from geometric forms (Figure 1.5) and Joan Miró's paintings often instilled a feeling of cosmic interrelationship of space–time of the new physics.

**FIGURE 1.4 (Color figure follows p. 28-30.)** Quantum theory also had a profound effect on many fields beyond science, such as art. It was no coincidence that, during the quantum revolution in science, Cubist and Surrealist art abolished realistic shapes referenced in fixed space and fixed time. Pablo Picasso's renderings of the human face with multifaceted perspectives often reflected the dual nature of reality. *Portrait of Dora Maar* (1937), Musée Picasso, Paris, France. (©Giraudon/Art Resource, New York. Reproduced with permission of Artists Rights Society, New York.)

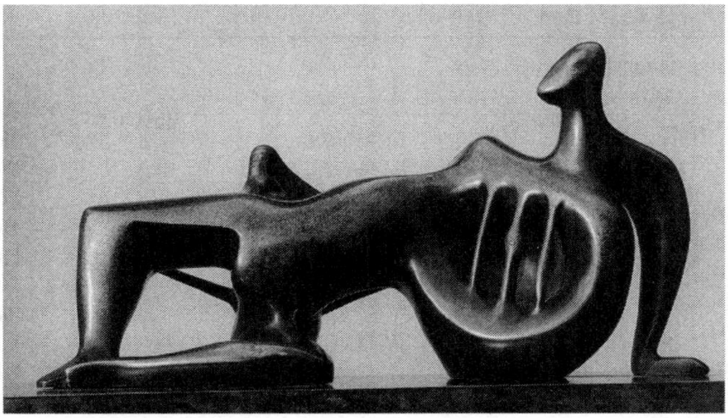

**FIGURE 1.5 (Color figure follows p. 28-30.)** It is noteworthy that modern art, with its seemingly strange distortions of visual reality, also appeared in the epoch of quantum theory. Henry Moore's sculptures reshaping the physical form of the human body marked a departure from geometric forms. *Reclining Figure* (1931), Dallas Museum of Art, Dallas, Texas. (From Kosinski, D., *Henry Moore: Sculpting the 20th Century*, Yale University Press, New Haven, CT, 2001, p. 91. Reprinted with permission from The Henry Moore Foundation.)

The Euclidean geometry of physical reality was further shattered by the creative movement of abstract art, which surged in the early 1910s with Wassili Kandinsky and Piet Mondrian and culminated in the 1950s with Jackson Pollock and Mark Rothko of the so-called New York School of Abstract Expressionism, where defined shapes and forms were replaced by shades, colors, and lines of fields — perhaps the visual equivalents of molecular waveforms, probability clouds, and energy fields of the new physics. The interdependence of science, technology, and art was well expressed in 1948 by Pollock in the avant-garde review *Possibilities*: "It seems to me that modern painting cannot express our area (the airplane, the atomic bomb, the radio) through forms inherited from the Renaissance and from any other culture of the past. Each area finds its own technique."[6]

At the same time that Niels Bohr introduced his theory of complementarity on the dual wave–particle nature of light, the Swiss psychologist and philosopher Carl Jung proposed his theory of synchronicity, an "acausal connecting" phenomenon where an event in the physical world coincides meaningfully with a psychological state of mind. Jung associated synchronistic experiences with the relativity of space and time and a degree of unconsciousness. Synchronicity reflected an almost mystical connection between the personal psyche and the material world, which are essentially different forms of energy.[7,8] Complementarity and synchronicity are concepts belonging to different fields, but both concepts, with their nature of duality, reflected the 20th century *Zeitgeist* of the quantum reality in the psyche and the physical world.

Quantum theory revolutionized molecular spectroscopy, an important field of photonics research. Molecular spectroscopy, which deals with the study of the interaction of light with matter, has been a cornerstone of the renaissance in biomedical photonics since the mid-1950s, mainly because of the foundations of quantum theory. *How do we learn about the molecules that make up the cell, the tissues, and the organs when most of them are too small to be seen even with the most powerful microscopes*? There is no simple answer to this question, but a large amount of information has come from various techniques of molecular spectroscopy.

Classical concepts of the world provide no simple way to explain the interaction of light with matter. The most satisfactory and complete description of the absorption and emission of light by matter is based on time-dependent wave mechanics. It analyzes what happens to the wavefunctions of molecules in the presence of light. Light is a rapidly oscillating electromagnetic field. Matter (including living species) is made up of molecules, which contain distributions of charges and spins that give matter its electrical and magnetic properties. These distributions are altered when a molecule is exposed to light.

The discovery of quantum theory has not only given birth to the new field of molecular spectroscopy, but also led to the development of a powerful set of photonics tools for exploring nature and understanding the cause of disease at the fundamental level. In essence, our current knowledge of how molecules

**TABLE 1.1** Spectroscopic Techniques for Biological and Biomedical Applications

| Spectral Region* | Wavelength* (cm) | Energy* (kcal mol$^{-1}$) | Techniques | Properties |
|---|---|---|---|---|
| γ-Ray | $10^{-11}$ | $3 \times 10^{8}$ | Mössbauer | Nucleus properties |
| X-ray | $10^{-8}$ | $3 \times 10^{5}$ | X-ray diffraction/scattering | Molecular structure |
| UV | $10^{-5}$ | $3 \times 10^{2}$ | UV absorption | Electronic states |
| Visible | $6 \times 10^{-5}$ | $5 \times 10^{3}$ | Visible absorption | Electronic states |
| | | | Luminescence | Electronic states |
| Infrared | $10^{-3}$ | $3 \times 10^{0}$ | IR absorption | Molecular vibrations |
| | | | IR emission | Electronic states |
| Microwave | $10^{-1}$ | $3 \times 10^{-2}$ | Microwave | Rotations of molecules |
| | $10^{0}$ | $10^{-3}$ | Electron paramagnetic resonance | Nuclear spin |
| Radio-frequency | 10 | $3 \times 10^{-4}$ | Nuclear magnetic resonance | Nuclear spin |

*Approximate values

bind together, how the building blocks of DNA cause a cell to grow, and how disease progresses on the molecular level has its fundamental basis in quantum theory.

With molecular spectroscopy, light can be used in many different ways to analyze complex biological systems and understand nature at the molecular level. Light at a certain wavelength $\lambda$ (or frequency $\nu = c/\lambda$) is used to irradiate the sample (e.g., a bodily fluid, a tissue, or an organ) in a process called excitation. Then some of the properties of the light that emerges from the sample are measured and analyzed. Some analyses deal with the fraction of the incident radiation absorbed by a sample; the techniques involved are called absorption spectroscopies (e.g., ultraviolet, visible, infrared absorption techniques). Other analyses examine the incident radiation dissipated and reflected back from the samples (elastic scattering techniques). Alternatively, it is possible to measure the light emitted and scattered by a sample, that occurs at wavelengths different from the excitation wavelength: in this case, the techniques involved are fluorescence, phosphorescence, Raman scattering, and inelastic scattering. Other specialized techniques can also be used to detect specific properties of emitted light (circular dichroism, polarization, lifetime, etc.).

The range of wavelengths used in molecular spectroscopy to study biological molecules is quite extensive. Molecular spectroscopic techniques have led to the development of a wide variety of practical techniques for minimally invasive monitoring of disease. For example, Britton Chance and co-workers have developed and used near-infrared absorption techniques to monitor physiological processes and brain function noninvasively.[9] Today, a wide variety of molecular spectroscopic techniques, including fluorescence, Raman scattering, and bioluminescence, are being developed for cancer diagnosis, disease monitoring, and drug discovery.[10–25] Table 1.1 summarizes the different types of spectroscopies and the wavelength of the electromagnetic radiation.

## 1.3.2 The Technology Revolution

In Western civilization science has often been associated with fundamental research and theoretical studies, whereas technology has often been viewed as originating from applied and experimental studies. Although the value of applied science was recognized very early by the mathematician Heron, who founded the first College of Technology in Alexandria in 105 B.C.,[26] science and technology have remained largely separate since the early times of Hellenic Greece. Aristotelian tradition held that the laws that govern the universe could be understood by pure thought, without requiring any experimental observation. In the 17th century, science and technology started to become interdependent. Galileo Galilei became a pivotal figure in the scientific revolution with his experimental observation of the movement of bodies by rolling balls of different weights down a slope. He also improved an important optical instrument, the telescope, which led to his revolutionary discoveries in astronomy.

The development and use of the microscope provided another revealing example of the interdependence between science and technology. It is very clear that it was indeed the discovery of a practical instrument that led to fundamental discoveries supporting the germ theory of diseases two centuries later. The 17th century witnessed the rise and spread of experimental science. This interdependence began to emerge when people believed that science should be useful and applied. Francis Bacon advocated his empirical, inductive method of investigation and envisioned a role for experiments designed to test nature and verify a hypothesis using an inductive process.

Descartes proposed a completely mechanical view of the world, a physical universe of objects that moves as a clockwork according to the laws of physics and the principles of geometry; this represented a radical departure from the more abstract worldview that dated back to Aristotle.[1] Father of the rationalistic mathematical philosophy, where rational thought and reason are the most important sources and tests of truth, Descartes was also influential in advocating what he called "practical philosophy" and the idea that knowledge should be applied "for the general good of all men." Opposing Cartesian rationalism based on pure thought and reason, British philosopher David Hume formulated his philosophy of empiricism, which regards empirical observation by the senses as the only reliable source of knowledge.

Shortly thereafter, the German philosopher Immanuel Kant built a bridge between pure thought and sensory perception and reconciled the philosophical divide between rationalists and empiricists (the so-called Battle of Descartes vs. Hume). In Kantian philosophy, the structure of consciousness, through an activity of thought called "synthesis," turns appearances observed by the senses into objects and perceptions, without which there would be nothing. Kant found roles for both the empiricist and the rationalist elements by including a rational element in his theory of knowledge (the 12 rational concepts of the understanding) with the publication in 1781 of his influential work *The Critique of Pure Reason*.[27]

In the 19th century the interdependence between science and technology became more important, culminating in the important technological discoveries of electricity and electromagnetic induction by Michael Faraday in 1831, the first electric telegraph in 1837 by Charles Wheaton, and the creation of the incandescent light bulb in 1879 by Thomas Edison in New Jersey and Joseph Swan in England. The discovery of the light bulb foreshadowed the next revolution in photonics in the 20th century. The following sections discuss three important examples of technological developments that have greatly affected the field of biomedical photonics.

#### 1.3.2.1 The Laser

The invention of the laser was an important milestone in the development of biomedical photonics. The laser is becoming the most widely used light source to excite tissues for disease diagnosis as well as to irradiate tumors for tissue removal in interventional treatment. The word "laser" is an acronym for "light amplification by stimulated emission of radiation." Einstein, who postulated photons and the phenomenon of stimulated emission, can be considered the grandfather of the laser. After Arthur Schawlow and Charles Townes published their paper on the possibility of laser action in the infrared and visible spectrum,[28] it was not long before many researchers began seriously considering practical devices. The first successful optical laser, constructed by Maiman in 1960, consisted of a ruby crystal surrounded by a helicoidal flash tube enclosed within a polished aluminum cylindrical cavity cooled by forced air.[29] Light was amplified (laser action) within a ruby cylinder, which formed a Fabry–Perot cavity by optically polishing the ends to be parallel to within a third of a wavelength of light. Each end was coated with evaporated silver; one end was made less reflective to allow some radiation to escape as the laser beam.

Lasers are now used as excitation light sources in disease diagnostics and as optical scalpels in interventional surgery. Ideal light sources because of their monochromaticity and intensity, lasers can be coupled to optical fibers inserted into endoscope for *in vivo* diagnosis of diseases (Figure 1.6). They also have the advantages of increased precision and reduced rates of infection and bleeding. Computers are used to control the intensity and direction of the laser beam, thus reducing human error. Nowadays lasers are commonly used to perform surgery on the skin and can be used to remove wrinkles, tattoos, birthmarks, tumors, and warts. Other types of growths can also be removed. Lasers are used to treat eye conditions; in some individuals, vision problems can be corrected with laser surgery. Lasers can help treat some forms of glaucoma and eye problems related to diabetes and are being incorporated into

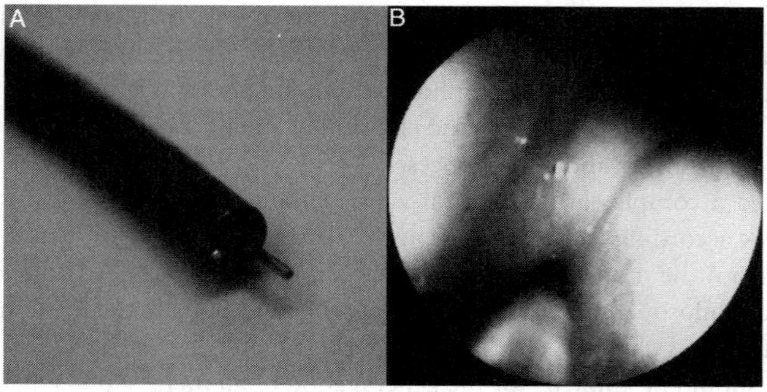


**FIGURE 1.6** Fiber-optic probes are used for *in vivo* laser-induced diagnostics of cancer. Laser-induced fluorescence has been used for gastrointestinal (GI) endoscopy examinations to directly diagnose cancer of patients without requiring physical biopsy. The LIF measurement was completed in approximately 0.6 s for each tissue site. The fiber-optic probe was inserted into the biopsy channel of an endoscope (A). The fiber-optic probe lightly touched the surface of the GI tissue being monitored (B).

surgical procedures for other parts of the body as well. These include the heart, prostate gland, and throat. Lasers are also used to open clogged arteries and remove blockages caused by tumors. Knives and scalpels may be completely eliminated one day.

### 1.3.2.2 The Microchip

The advent of the microchip is another important development that has significantly affected the evolution of biomedical photonics. Whereas the discovery of the laser has provided a new technology for excitation, the miniaturization and mass production of sensor devices and their associated electronic circuitry has radically transformed the ways in which detection and imaging of molecules, tissues, and organs can be performed *in vivo* and *ex vivo*.

Microchip technology comes from the development and widespread use of large-scale integrated circuits, consisting of hundreds of thousands of components packed onto a single tiny chip that can be mass-produced for a few cents each (Figure 1.7). This technology has enabled fabrication of microelectronic circuitries (microchips) and photonic detectors such as photodiode arrays (PDA), charge-coupled device (CCD) cameras, and complementary metal oxide silicon (CMOS) sensor arrays in large numbers at sufficiently low costs to open a mass market and permit the widespread use of these devices in biomedical spectroscopy and molecular imaging. The miniaturization and evolution of integrated circuit technology continues to exemplify a phenomenon known as "Moore's law." The observation made by Intel Corporation founder Gordon Moore in 1965 was that the number of components on the most complex integrated circuit chip would double each year for the next 10 years. Moore's law has come to refer to the continued chip size reduction and exponential decrease in the cost per function that can be achieved on an integrated circuit.

The miniaturization of high-density optical sensor arrays is critical to the development of innovative high-resolution imaging methods at the cellular or molecular scales capable of identifying and characterizing premalignant abnormalities or other early cellular changes. Photonic imaging detectors provide novel solutions for *in vivo* microscopic imaging sensors or microscopic implanted devices with high spatial contrast and temporal resolution.

During the past 10 years there has been an explosion of research in biomedical photonics, resulting in scores of publications, conventions, and manufacturers offering new products in the field. Sensor miniaturization has enabled significant advances in medical imaging technologies over the last 25 years in such areas as magnetic resonance imaging (MRI), computed tomography (CT), nuclear medicine, and optical and ultrasound imaging of diseases. The development of multichannel sensor technologies has led to the development of novel photonic imaging technologies that exploit current knowledge of the genetic and molecular bases of important diseases such as cancer. These molecular biological discoveries have great implications for prevention, detection, and targeted therapy.

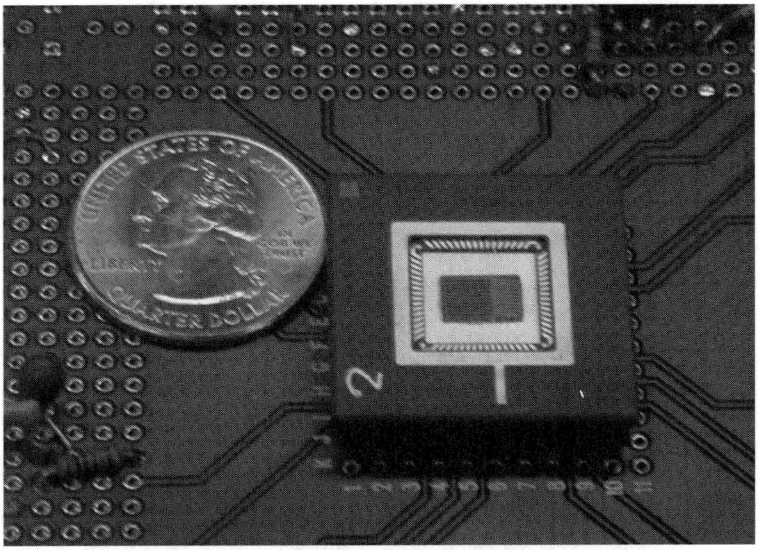

**FIGURE 1.7** Modern optical sensor array microchips are fabricated using CMOS technology. Microchip technology has enabled the fabrication of microelectronic circuitries (microchips) and photonic detectors such as PDA, CCD cameras, and CMOS sensor arrays in large numbers at sufficiently low costs to open a mass market and permit the widespread use of these devices in biomedical spectroscopy, molecular imaging, and clinical diagnostics. (Photo courtesy of Oak Ridge National Laboratory, Oak Ridge, Tennessee.)

### 1.3.2.3 Nanotechnology

Recently, nanotechnology, which involves research on and development of materials and species at length scales between 1 to 100 nm, has been revolutionizing important areas in biomedical photonics, especially diagnostics and therapy at the molecular and cellular level. The combination of molecular nanotechnology and photonics opens the possibility of detecting and manipulating atoms and molecules using nanodevices that have potential for a wide variety of medical uses at the cellular level.

Today, the amount of research in biomedical science and engineering at the molecular level is growing exponentially because of the availability of new investigative nanotools. These new analytical tools are capable of probing the nanometer world and will make it possible to characterize the chemical and mechanical properties of cells, discover novel phenomena and processes, and provide science with a wide range of tools, materials, devices, and systems with unique characteristics. The marriage of electronics, biomaterials, and photonics is expected to revolutionize many areas of medicine in the 21st century. The futuristic vision of nanorobots moving through bloodstreams armed with antibody-based nanoprobes and nanolaser beams that recognize and kill cancer cells might some day no longer be the "stuff of dream."

The combination of photonics and nanotechnology has already led to a new generation of devices for probing the cell machinery and elucidating intimate life processes occurring at the molecular level previously invisible to human inquiry. Tracking biochemical processes within intracellular environments can now be performed *in vivo* with the use of fluorescent molecular probes (Figure 1.8) and nanosensors (Figure 1.9). With powerful microscopic tools using near-field optics, scientists are now able to explore the biochemical processes and submicroscopic structures of living cells at unprecedented resolutions. It is now possible to develop nanocarriers for targeted delivery of drugs that have their shells conjugated with antibodies for targeting antigens and fluorescent chromophores for *in vivo* tracking.

The possibility of fabricating nanoscale components has recently led to the development of devices and techniques that can measure fundamental parameters at the molecular level. With "optical tweezer" techniques, small particles may be trapped by radiation pressure in the focal volume of a high-intensity, focused beam of light. This technique, also called "optical trapping," may be used to move small cells or subcellular organelles around at will by the use of a guided, focused beam.[30] For example, a bead coated

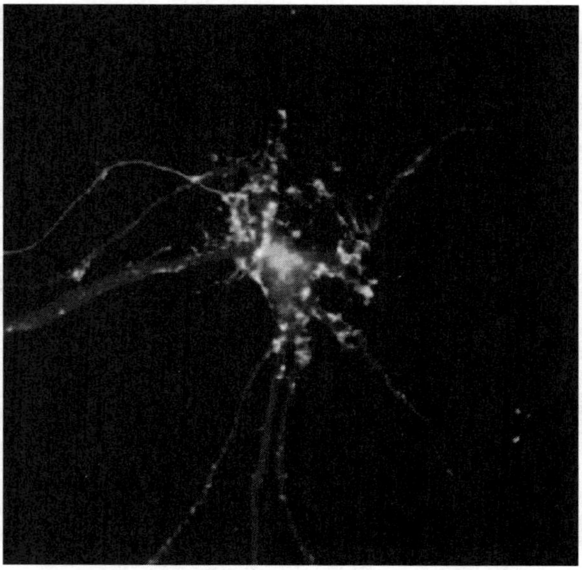

**FIGURE 1.8** (**Color figure follows p. 28-30.**) Biochemical processes in nerve cells can be tracked using green fluorescent protein. Tracking biochemical processes can now be performed with the use of fluorescent probes. For example, there is great interest in understanding the origin and movement of neurotrophic factors such as brain-derived neurotrophic factor (BDNF) between nerve cells. This figure illustrates the use of BDNF tagged with green fluorescent protein to follow synaptic transport from axons to neurons to postsynaptic cells. (From Berman, D.E. and Dudai, Y.K., *Science*, 291, 2417, 2002. With permission.)

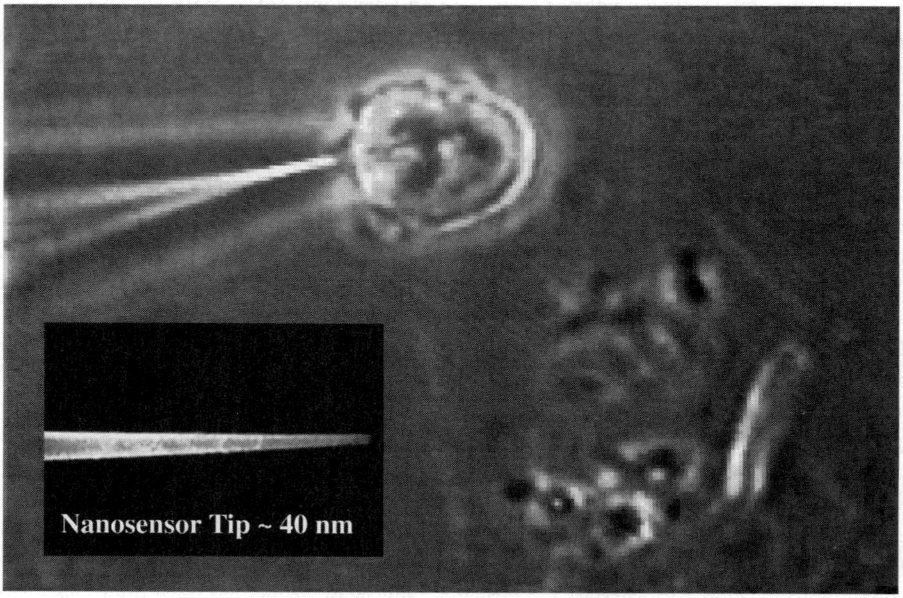


**FIGURE 1.9** A fiber-optic nanosensor with antibody probe is used to detect biochemicals in a single cell. The combination of photonics and nanotechnology has led to a new generation of devices for probing the cell machinery and elucidating intimate life processes occurring at the molecular level that were previously invisible to human inquiry. The insert (lower left) shows a scanning electron photograph of a nanofiber with a 40-nm diameter. The small size of the probe allowed manipulation of the nanoprobe at specific locations within a single cell. (Adapted from Vo-Dinh, T. et al., *Nat. Biotechnol.*, 18, 76, 2000.)

with an immobilized, caged bioactive probe could be inserted into a tissue or even a cell and moved around to a strategic location by an optical trapping system. The cage could then be photolyzed by multiphoton uncaging in order to release and activate the bioactive probe.

Optical tweezers can also be used to determine precisely the mechanical properties of single molecules of collagen, an important tissue component and a critical factor in diagnosing cancer and the aging process.[31] The optical tweezer method uses the momentum of focused laser beams to hold and stretch single collagen molecules bound to polystyrene beads. The collagen molecules are stretched through the beads using the optical laser tweezer system, and the deformation of the bound collagen molecules is measured as the relative displacement of the microbeads, which are examined by optical microscopy. Ingenious optical trapping systems have also been used to measure the force exerted by individual motor proteins.[32]

A recent advance in the field of biosensors has been the development of optical nanobiosensors, which have dimensions on the nanometer (nm) size scale. Typical tip diameters of the optical fibers used in these sensors range between 20 and 100 nm. Using these nanobiosensors, it has become possible to probe individual chemical species in specific locations throughout a living cell. Figure 1.9 shows a photograph of a fiber-optic nanoprobe with antibodies targeted to benzopyrene tetrol (BPT) and designed to detect BPT in a single cell.[33] An important advantage of the optical sensing modality is the capability of measuring biological parameters in a noninvasive or minimally invasive manner due to the very small size of the nanoprobe. Following measurements using the nanobiosensor, cells have been shown to survive and undergo mitosis.[34]

These photonic technologies are just some examples of a new generation of nanophotonic tools that have the potential to drastically change fundamental understanding of the life process. They could ultimately lead to development of new modalities of early diagnostics and medical treatment and prevention beyond the cellular level to that of individual organelles and even DNA, the building block of life.

### 1.3.3 The Genomics Revolution

James Watson and Francis Crick's publication of the helix structure of DNA in 1953 can be considered the first landmark achievement that launched the Genomics Revolution of the 21st century.[35] Almost 50 years after this landmark discovery, the completion of the sequencing of the human genome marked the second major achievement in the area of molecular genetics. The draft of the published sequence encompasses 90% of the human genome's euchromatic portion, which contains the most genes. Figure 1.10 illustrates the scientific progress achieved in genomics and proteomics research since the discovery of the DNA structure.

It is useful to get an overview of this remarkable 21st century achievement in molecular biology, known as the Human Genome Project. This project traces its roots to an initiative in the U.S. Department of Energy (DOE). In 1986, DOE announced its Human Genome Initiative, believing that precise knowledge of a reference human genome sequence would be critical to its missions to pursue a deeper understanding of the potential health and environmental risks posed by energy production and use. Shortly thereafter, DOE and the National Institutes of Health (NIH) teamed up to develop a plan for a joint Human Genome Project (HGP) that officially began in 1990.

From the beginning of the HGP project, photonics technologies provided the critical tools for accelerating the DNA sequencing process. In 1986, Leroy Hood and co-workers described a new technique for DNA sequencing whereby four fluorescent dyes were attached to the DNA instead of using radioactive labels and reading the x-ray films.[36] This photonic detection scheme — which allowed the DNA to run down only a single lane and be illuminated by a laser and the result from four fluorescent labels to be read into a computer — has been the workhorse of the sequencing efforts.

In June 2000, scientists announced biology's most spectacular achievement: the generation of a working draft sequence of the entire human genome.[37] In constructing the working draft, the 16 genome sequencing centers produced over 22.1 billion bases of raw sequence data, comprising overlapping fragments totaling 3.9 billion bases and providing sevenfold coverage (sequenced seven times) of the human genome. One

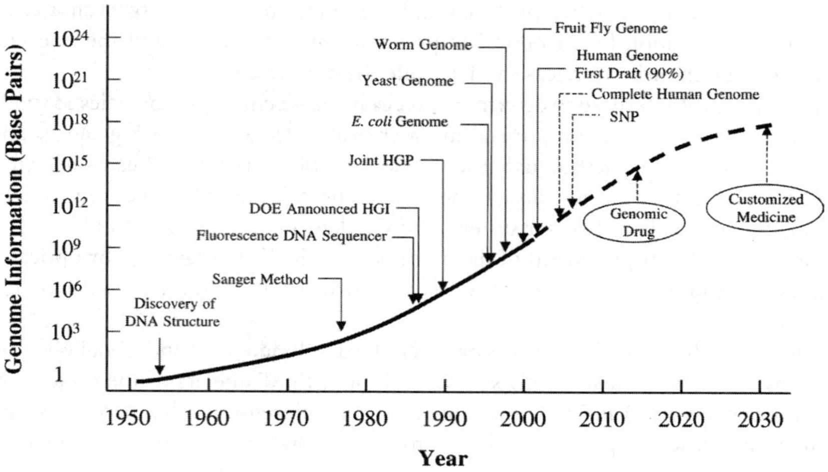


**FIGURE 1.10** Advances in the genomics revolution since the discovery of the DNA helix structure have resulted in new information, new approaches, and new technologies for the drug discovery process. Potential applications of this knowledge are numerous and include drug discovery and customized medicine.

of the greatest impacts of knowing the genome sequence is the establishment of an entirely new approach to biomedical research. In the past, researchers studied one gene or a few genes at a time. With whole-genome sequences and new automated, high-throughput technologies, they can address biological and medical problems systematically and on a large scale in a massively parallel manner. They are now able to investigate a large number of genes in a genome, or various gene products in a particular tissue or organ or tumor. They can also study how tens of thousands of genes and proteins work together in interconnected networks to orchestrate the chemistry of life. Specific genes have been pinpointed and associated with numerous diseases and disorders, including breast cancer, muscle disease, deafness, and blindness.

Potential applications of this knowledge are numerous and include drug discovery and customized medicine. Advanced photonics has contributed dramatically to the field of drug discovery in the post-genomic area. Genomics and proteomics present the drug discovery community with a wealth of new potential targets. In the pregenomics area, the basic approach to discovering new drugs mainly involved research efforts focused on individual drug targets. Nonsystematic methods were generally used to find these potential targets. Phenotype analyses were performed by examining the differential expression of some proteins in diseased vs. normal tissue to identify proteins associated with a specific disease process. When a possibly useful compound was successfully synthesized, it was evaluated using time-consuming animal studies and, later, human clinical tests.

The advent of genomics has provided new information, new approaches, and new technologies for the drug discovery process. By using the ability to predict all the possible protein-coding regions in experimental systems, biomedical studies and analyses can be expanded beyond the most abundant and best characterized proteins of the cells to discover new drug targets. Most drugs today are based on about 500 molecular targets; knowledge of the genes involved in diseases, disease pathways, and drug-response sites will lead to the discovery of thousands of new targets. The increasing number of potential targets requires new instrumental approaches for rapid development of new drugs. One area of great interest is the use of high-throughput cell-based assays, which have the potential to provide a fundamental platform, the living cell, to characterize, analyze, and screen a drug target *in situ*.

Combined with high-throughput techniques, fluorescence technologies based on a wide variety of reporter gene assays, ion channel probes and fluorescent probes have provided powerful tools for cell-based assays for drug discovery. An exciting new advance in fluorescent probes for biological studies has been the development of naturally fluorescent proteins for use as fluorescent probes. The jellyfish *Aequorea victoria* produces a naturally fluorescent protein known as green fluorescent protein (GFP). Active research is under way on the applications of GFP-based assays, which could allow analysis of intracellular signaling pathways using libraries of cell lines engineered to report on key cellular processes.

Another important advance in photonic technologies that has contributed to the dramatic growth of cell-based assays is the development of advanced imaging systems that have the combined capability of high-resolution, high-throughput, and multispectral detection of fluorescent reporters. These reporters allow the development of live cell assays with the ability for *in vivo* sensing of individual biological responses across cell populations, tracking the transport of biological species within intracellular environments, and monitoring multiple responses from the same cell.[38]

Cellular biosensors[39] and extrinsic cellular sensors based on site-specific labeling of recombinant proteins[40,41] are also important photonic tools recently developed for drug discovery. Advanced photonic techniques using time-resolved and phase-resolved detection, polarization, and lifetime measurements further extend the usefulness of cell-based assays. Novel classes of labels using inorganic fluorophors based on quantum dots or surface-enhanced Raman scattering (SERS) labels provide unique possibilities for multiplex assays.[42] Today, single-molecule detection techniques using various photonics modalities provide the ultimate tools to elucidate cellular processes at the molecular level.

## 1.4 Conclusion

As we enter the new millennium, it appears that we are witnessing a new paradigm shift in the contribution of science and technology to the development of human knowledge and the advancement of civilization. In past centuries a discovery in basic science usually led to new technology development. Today, however, technology often creates new devices and instruments that drive scientific discoveries. As we create the tools to peer at the stars, look into the inner world of atoms, or decipher the genetic code of human cells in order to understand and ultimately eliminate the cause of diseases, we are witnessing a shift from a "science-driven" process to a "technology-driven" process in the evolution of human knowledge. Neither science nor technology holds a privileged role in the quest for knowledge. Rather, as the yin and yang of life, science and technology will contribute equally to human development and societal progress.

Biomedical photonics is a relatively new field at the interface of science and technology that has the potential to revolutionize medicine as we know it. The rapid development of laser and imaging technology has yielded powerful tools for the study of disease on all scales, from single molecules to tissue materials and whole organs. These emerging techniques find immediate applications in biological and medical research. For example, laser microscopes permit spectroscopic and force measurements on single protein molecules and laser sources provide access to molecular dynamics and structure to perform "optical biopsy" noninvasively and almost instantaneously, while optical coherence tomography allows visualization of tissue and organs. Using genetic promoters to drive luciferase expression, bioluminescence methods can generate molecular light switches, which serve as functional indicator lights reporting cellular conditions and responses in the living animal. This technique could allow rapid assessment of and response to the effects of antitumor drugs, antibiotics, or antiviral drugs.

Developments in quantum chemistry, molecular genetics, and high-speed computers have created unparalleled capabilities for understanding complex biological systems. Information from the human genome project will make molecular medicine an exciting reality. Current research has indicated that many diseases such as cancer occur as the result of the gradual buildup of genetic changes in single cells. Biomedical photonics can provide tools capable of identifying specific subsets of genes encoded within the human genome that can cause the development of cancer. Photonic techniques are being developed to identify the molecular alterations that distinguish a cancer cell from a normal cell. Such technologies will ultimately aid in characterizing and predicting the pathologic behavior of that cancer cell, as well as the cell's responsiveness to drug treatment.

Biomedical photonics technologies can definitely bring a bright future to biomedical research because they are capable of yielding the critical information bridging molecular structure and physiological function, which is the most important process in understanding, treatment, and prevention of disease.

As this handbook explores in the following sections, it is our hope that light is rapidly finding its place as an important diagnostic tool and a powerful weapon in the arsenal of the modern physician.

## Acknowledgments

This work was sponsored by the U.S. Department of Energy (DOE) Office of Biological and Environmental Research, under contract DEAC05-00OR22725 with UT-Battelle, LLC, and by the National Institutes of Health (RO1 CA88787-01).

### References

1. McClellan, J.E., III and Dorn, H., *Science and Technology in World History*, Johns Hopkins University Press, London, 1999.
2. Kuhn, T.S., *The Structure of Scientific Revolutions*, University of Chicago Press, Chicago, 1962.
3. Russell, B., *A History of Western Philosophy*, Simon & Schuster, New York, 1972, p. 559.
4. Zukav, G., *The Dancing Wu Li Masters*, Bantam Books, New York, 1980.
5. Capra, F., *The Tao of Physics*, Shambhala, Berkeley, CA, 1975.
6. Ferrier, J.L., *Art of the 20th Century*, Chene-Hachette, Paris, 1999, p. 476.
7. Jung, C.G., *L'Homme a la Decouverte de son Ame*, Petite Bibliotheque Payot, Paris, 1962.
8. Shlain, L., *Arts and Physics*, Quill Publishers, New York, 1991.
9. Chance, B., Leigh, J.S., Miyake, H., Smith, D.S., Nioka, S., Greenfeld, R., Finander, M., Kaufmann, K., Levy, W., Young, M., Cohen, P., Yoshioka, H., and Boretsky, R., Comparison of time-resolved and time-unresolved measurements of deoxyhemoglobin in brain, *Proc. Natl. Acad. Sci. U.S.A.*, 85(14), 4971, 1988.
10. Andersson-Engels, S.J., Johansson, J., Svanberg, K., and Svanberg, S., Fluorescence imaging and point measurements of tissue — applications to the demarcation of malignant tumors and atherosclerotic lesions from normal tissue, *Photochem. Photobiol.*, 53(6), 807, 1991.
11. Vo-Dinh, T., Panjehpour, M., Overholt, B.F., Farris, C., Buckley, F.P., and Sneed, R., *In-vivo* cancer-diagnosis of the esophagus using differential normalized fluorescence (DNF) indexes, *Lasers Surg. Med.*, 16(1), 41, 1995.
12. Panjehpour, M., Overholt, B.F., Vo-Dinh, T., Haggitt, R.C., Edwards, D.H., and Buckley, F.P., Endoscopic fluorescence detection of high-grade dysplasia in Barrett's esophagus, *Gastroenterology*, 111(1), 93, 1996.
13. Alfano, R.R., Tata, D.B., Cordero, J., Tomashefsky, P., Longo, F.W., and Alfano, M.A., Laser-induced fluorescence spectroscopy from native cancerous and normal tissue, *IEEE J. Quantum Electron.*, 20(12), 1507, 1984.
14. Richards-Kortum, R. and Sevick-Muraca, E., Fluorescence spectroscopy for tissue diagnosis, *Ann. Rev. Phys. Chem.*, 47, 555, 1996.
15. Lam, S. and Palcic, B., Autofluorescence bronchoscopy in the detection of squamous metaplasia and dysplasia in current and former smokers, *J. Natl. Cancer Inst.*, 91(6), 561, 1999.
16. Wagnieres, G.A., Studzinski, A.P., and VandenBergh, H.E., Endoscopic fluorescence imaging system for simultaneous visual examination and photodetection of cancers, *Rev. Sci. Instr.*, 68(1), 203, 1997.
17. Manoharan, R., Shafer, K., Perelman, L., Wu, J., Chen, K., Deinum, G., Fitzmaurice, M., Myles, J., Crowe, J., Dasari, R.R., and Feld, M.S., Raman spectroscopy and fluorescence photon migration for breast cancer diagnosis and imaging, *Photochem. Photobiol.*, 67(1), 15, 1998.
18. Mahadevan-Jansen, A., Mitchell, M. F., Ramanujam, N., Malpica, A., Thomsen, S., Utzinger, U., and Richards-Kortum, R., Near-infrared Raman spectroscopy for *in vitro* detection of cervical precancers, *Photochem. Photobiol.*, 68(1), 123, 1998.
19. Fujimoto, J.G., Boppart, S.A., Tearney, G.J., Bouma, B.E., Pitris, C., and Brezinski, M.E., High resolution *in vivo* intra-arterial imaging with optical coherence tomography, *Heart*, 82, 128, 1999.

20. Boas, D.A., Oleary, M.A., Chance, B., and Yodh, A.G., Detection and characterization of optical inhomogeneities with diffuse photon density waves: a signal-to-noise analysis, *Appl. Opt.*, 36(1), 75, 1997.
21. Bigio, I.J, Bown, S.G., Briggs, G., Kelley, C., Lakhani, S., Pickard, D., Ripley, P.M., Rose, I.G., and Saunders, C., Diagnosis of breast cancer using elastic-scattering spectroscopy: preliminary clinical results, *J. Biomed. Opt.*, 5, 221, 2000.
22. Contag, C.H., Jenkins, D., Contag, F.R., and Negrin, R.S., Use of reporter genes for optical measurements of neoplastic disease *in vivo, Neoplasia*, 2, 41, 2000.
23. Mourant, J.R., Freyer, J.P., Hielscher, A.H., Eick, A.A., Shen, D., and Johnson, T.M., Spectroscopic diagnosis of bladder cancer with elastic light scattering, *Lasers Surg. Med.*, 17(4), 350, 1995.
24. Mycek, M.A., Schomacker, K.T., and Nishioka, N.S., Colonic polyp differentiation using time-resolved autofluorescence spectroscopy, *Gastrointest. Endosc.*, 48(4), 390, 1998.
25. Farkas, D.L. and Becker, D., Applications of spectral imaging: detection and analysis of human melanoma and its precursors, *Pigment Cell Res.*, 14(1), 2, 2001.
26. Grun, B., *The Timetables of History*, Touchstone Books, New York, 1982.
27. Lavine, T.Z., *From Socrates to Sartre: The Philosophic Quest*, Bantam Books, New York, 1989.
28. Bertolotti, M., *Masers and Lasers, An Historical Approach*, Adam Hilger, Bristol, U.K., 1983.
29. Maiman, T.H., Did Maiman really invent the Ruby-Laser — reply, *Laser Focus World*, 27(3), 25, 1991.
30. Ashkin, A., Dziedzic, J.M., and Yamane, T., Optical trapping and manipulation of single cells using infrared laser beam, *Nature*, 330, 769, 1987.
31. Luo, Z.P., Bolander, M.E., and An, K.N., A method for determination of stiffness of collagen molecules, *Biochem. Biophys. Res. Commun.*, 232, 251, 1997.
32. Kojima, H., Muto, E., Higuchi, H., and Yanagido, T., Mechanics of single kinesin molecules measured by optical trapping nanometry, *Biophys. J.*, 73(4), 2012, 1997.
33. Vo-Dinh, T., Alarie, J. P., Cullum, B., and Griffin, G.D., Antibody-based nanoprobe for measurement of a fluorescent analyte in a single cell, *Nat. Biotechnol.*, 18(7), 764, 2000.
34. Cullum, B. and Vo-Dinh, T., The development of optical nanosensors for biological measurements, *Trends Biotechnol.*, 18, 388, 2000.
35. Watson, J.D. and Crick, F.H.C., Molecular structure of DNA, *Nature*, 171, 737, 1953.
36. Smith, L.M., Sanders, J.Z., Kaiser, R.J., Hughes, P., Dodd, C., Connell, D.R., Heiner, C., Kent, S.B.H., and Hood, L.E., Fluorescence detection in automated DNA-sequence analysis, *Nature*, 321, 674, 1986.
37. Venter, J.C., Adams, M.D., Myers, E.W., Li, P.W., Mural, R.J., Sutton, G.G., Smith, H.O., Yandell, M., Evans, C.A., Holt, R.A., Gocayne, J.D., Amanatides, P., Ballew, R.M., Huson, D.H., Wortman, J.R., Zhang, Q., Kodira, C.D., Zheng, X.Q.H., Chen, L., Skupski, M., Subramanian, G., Thomas, P.D., Zhang, J.H., Miklos, G.L.G., Nelson, C., Broder, S., Clark, A.G., Nadeau, C., McKusick, V.A., Zinder, N., Levine, A.J., Roberts, R.J., Simon, M., Slayman, C., Hunkapiller, M., Bolanos, R., Delcher, A., Dew, I., Fasulo, D., Flanigan, M., Florea, L., Halpern, A., Hannenhalli, S., Kravitz, S., Levy, S., Mobarry, C., Reinert, K., Remington, K., Abu-Threideh, J., Beasley, E., Biddick, K., Bonazzi, V., Brandon, R., Cargill, M., Chandramouliswaran, I., Charlab, R., Chaturvedi, K., Deng, Z.M., Di Francesco, V., Dunn, P., Eilbeck, K., Evangelista, C., Gabrielian, A.E., Gan, W., Ge, W.M., Gong, F.C., Gu, Z.P., Guan, P., Heiman, T.J., Higgins, M.E., Ji, R.R., Ke, Z.X., Ketchum, K.A., Lai, Z.W., Lei, Y.D., Li, Z.Y., Li, J.Y., Liang, Y., Lin, X.Y., Lu, F., Merkulov, G.V., Milshina, N., Moore, H.M., Naik, A.K., Narayan, V.A., Neelam, B., Nusskern, D., Rusch, D.B., Salzberg, S., Shao, W., Shue, B.X., Sun, J.T., Wang, Z.Y., Wang, A.H., Wang, X., Wang, J., Wei, M.H., Wides, R., Xiao, C.L., Yan, C.H., et al., The sequence of the human genome, *Science*, 291(5507), 1304, 2001.
38. Thomas, N., Cell-based assays — seeing the light, *Drug Discovery World*, 3, 25, 2001.
39. Durick, K. and Negulescu, P., Cellular biosensors for drug discovery, *Biosensors Bioelectron.*, 16, 587, 2001.

40. Nakanishi, J., Nakajima, T., Sato, M., Ozawa, T., Tohda, K., and Umezawa, Y., Imaging of conformational changes of proteins with a new environment-sensitive fluorescent probe designed for site specific labeling of recombinant proteins in live cells, *Anal. Chem.*, 73, 2920, 2001.
41. Whitney, M., Rockenstein, E., Cantin, G., Knapp, T., Zlokarnik, G., Sanders, P., Durick, K., Craig, F.F., and Negulescu, P.A., A genome-wide functional assay of signal transduction in living mammalian cells, *Nat. Biotechnol.*, 16, 1329, 1998.
42. Vo-Dinh, T., Stokes, D.L., Griffin, G.D., Volkan, M., Kim, U.J., and Simon, M.I., Surface-enhanced Raman scattering (SERS) method and instrumentation for genomics and biomedical analysis, *J. Raman Spectrosc.*, 30(9), 785, 1998.

# I

# Photonics and Tissue Optics

# 2

# Optical Properties of Tissue



Joel Mobley
*Oak Ridge National Laboratory*
*Oak Ridge, Tennessee*

Tuan Vo-Dinh
*Oak Ridge National Laboratory*
*Oak Ridge, Tennessee*

## 2.1 Introduction

The electromagnetic spectrum provides a diverse set of photonic tools for probing, manipulating, and interacting with biological systems. A great variety of electromagnetic phenomena are used in biomedicine to detect and treat disease and to advance scientific knowledge. The focus of this chapter is the propagation of "light" in tissues. The main purpose is to provide an introduction to the linear optical properties of human tissues from a conceptual viewpoint, emphasizing how these properties fit into the description of light transport. The term *light* can have various meanings, but here it is used to refer to the portion of the electromagnetic spectrum with vacuum wavelengths $\lambda_{vac}$ in the range of 1 μm to 100 nm (or in the frequency $\nu = c/\lambda$, $3 \times 10^{14}$ to $3 \times 10^{15}$ Hz, where $c$ is the speed of light). This spectral range includes the near infrared (NIR), the visible, and the ultraviolet (UV) A, B, and C bands, and encompasses the so-called therapeutic (or diagnostic) window, which is of great importance in biomedical photonics (Figure 2.1).

As discussed in Chapter 1, the quantum description of light in terms of photons is essential for understanding the exchange of energy and momentum between light and matter and provides the theoretical basis for molecular spectroscopy. The term *photonics,* however, is used to refer to all electromagnetic wave phenomena, whether or not the quantum nature of light (photon) is a necessary component of its description. There is a parallel with the use of the term *electronics,* which deals with electrical circuit phenomena; in many applications the electric current flowing in a device can be thought of as a

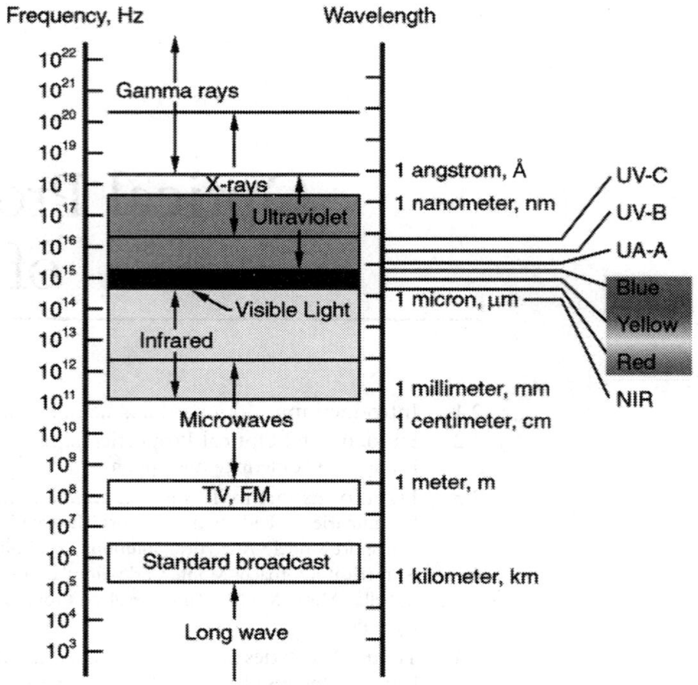


**FIGURE 2.1** The electromagnetic spectrum with the region discussed in this chapter expanded on the right. This includes the UV-C (100 to 280 nm), UV-B (280 to 315 nm), and UV-A (315 to 400 nm) bands, the visible region (400 to 760 nm), and a portion of the IR-A band (760 to 1400 nm). The therapeutic window (see Figure 2.16) extends from 600 nm (orange part of the visible) to 1300 nm (IR-A band).

continuous flow of charge, and the quantum nature of the electron is not required for understanding circuit behavior.

In physics, phenomena are often described from either "classical" or "quantum" viewpoints. The term *classical* refers to theories that do not use the concepts of quantum mechanics. In classical theory, light is considered to be an oscillating electromagnetic (EM) field that can have a continuous range of energies. In the quantum model, light waves consist of packets (i.e., "quanta") of energy called photons. Each photon has an energy proportional to the frequency of the EM wave. Quantum theory introduces the idea that light and matter exchange energy as photons, fundamental packets of energy, $E = h\nu = hc/\lambda$, where $h$ is Planck's constant. The description of light propagation in tissues incorporates both views: a classical picture is used to define the dynamics of light transport mathematically (e.g., to calculate scattering cross sections) and the photon concept is introduced in an *ad hoc* manner when necessary (e.g., to account for molecular transition processes, such as absorption, luminescence, and Raman scattering).

Because multiple scattering effects are important features of light propagation in tissues, the direct application of EM theory to the tissue optics problem is complicated. In place of the direct EM approach, a model known as radiation transport (RT) theory is used that explicitly ignores wave phenomena such as polarization and interference, and instead follows only the transport of light energy in the medium. As will be discussed later, the RT model does, however, implicitly incorporate elements of the classical and quantum descriptions of light. This model is not specific to light and has other important applications in areas such as neutron transport and thermodynamics.

In the electromagnetic description, light consists of oscillating electric and magnetic waves (Figure 2.2). Two important parameters of electromagnetic waves are phase and polarization. The *phase* is a general property of a wave and gives rise to important effects such as interference and diffraction. *Polarization*

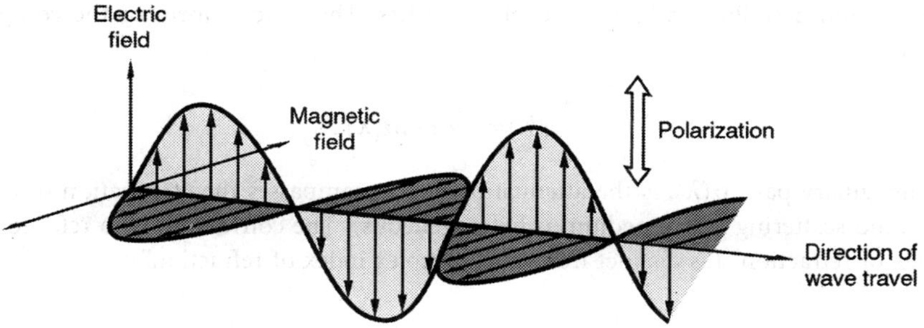


**FIGURE 2.2** Schematic diagram of an electromagnetic wave, which consists of the symbiotic oscillations of electric and magnetic fields. The polarization denotes the directions of the electric field vectors in the wave. In the case shown, the polarization is linear and in the vertical direction.

refers to the orientation of the electric field vector of the EM wave. Two waves must have some overlap in their polarizations for interference to occur. In free space, the polarization is perpendicular to the direction of propagation. Since RT theory deals with the transport of light energy, it explicitly ignores the phase and polarization of the EM waves, although it does use wave properties implicitly through the material parameters included in the fundamental equations. The validity of RT theory for strongly scattering optical media has been established empirically, and the theoretical link between RT theory and the fundamental laws of electromagnetism has also been demonstrated;[1] nevertheless, some attempt will be made to distinguish between phenomena that are inherently electromagnetic in nature vs. those that are part of the empirical RT model.

This chapter provides an introduction to fundamental photophysical concepts and basic theoretical models for describing light propagation in biological tissues. We examine in some detail RT theory, including the absorption-dominant and scattering-dominant limits of the problem, and the Monte Carlo method for numerically performing practical transport simulations. The chapter also provides a description of the Kubelka–Munk method. Finally, we provide a representative listing of optical properties of human tissues, including refractive indices, and scattering and absorption coefficients.

## 2.2 Fundamental Optical Properties

In this section we discuss three photophysical processes that affect light propagation in biological tissue: refraction, scattering, and absorption. These processes can be quantified by the following parameters:

- Index of refraction, $n(\lambda)$
- Scattering cross section, $\sigma_s$
- Differential scattering cross section, $\frac{d\sigma_s}{d\Omega}$
- Absorption cross section, $\sigma_a$

These basic properties, from which the inputs to RT theory are ultimately built, are defined in the following sections. For the purposes of this chapter, it is sufficient to assume that the media considered in this treatment are isotropic systems; this assumption will simplify the description of the fundamental properties while not compromising the essential features of the phenomenology.

### 2.2.1 Refraction

#### 2.2.1.1 Index of Refraction

The index of refraction is a fundamental property of homogeneous media. This index can also be defined for heterogeneous materials, as will be discussed later in Section 2.3. For a homogeneous medium, the index of refraction describes its linear optical properties. The general form of the complex index of refraction is

$$\tilde{n}(\lambda) = n(\lambda) - i\alpha(\lambda) \,. \tag{2.1}$$

where the imaginary part, $\alpha(\lambda)$, is the attenuation and encompasses the attenuation of a wave due to absorption (and scattering if the medium is heterogeneous). The convention is to refer to the real part as the "index of refraction" (as distinct from the "complex index of refraction"):

$$\mathrm{Re}\left[\tilde{n}(\lambda)\right] = n(\lambda) \,. \tag{2.2}$$

The real part of the refractive index is defined in terms of the phase velocity of light in the medium,

$$c_m(\lambda) = \frac{c}{n(\lambda)} \,, \tag{2.3}$$

where $c = 2.998 \times 10^8$ m/s is the speed of light in vacuum and $n_{\mathrm{vac}} = 1$. The wavelength of light in the medium $\lambda_m$ is given in terms of the vacuum wavelength $\lambda$ as

$$\lambda_m = \frac{\lambda}{n(\lambda)} \,. \tag{2.4}$$

Even though the phase speed and the wavelength of light depend on the refractive index, the wave frequency,

$$\nu = \frac{c}{\lambda} = \frac{c_m}{\lambda_m} \,,$$

and the size of its photon energy, $E = h\nu$, are always the same as in the vacuum. Moreover, although the wavelength in a material is different from its vacuum value, the prevailing convention is to use the quantity $\lambda$ to refer to the vacuum wavelength. When the actual wavelength in a material is required, this wavelength is denoted by the addition of a subscript (e.g., $\lambda_m$).

#### 2.2.1.2 Reflection and Refraction at an Interface

When a light wave propagating in a material with a given refractive index encounters a boundary with a second material with a different index, the path of the light is redirected. If the width of the light beam (or its wavelength) is small compared to the boundary or its curvature, reflection and refractive transmission result. This situation is illustrated in Figure 2.3. The amount of light reflected by and transmitted through a boundary depends on the refractive indices of the two materials, the angle of incidence, and the polarization of the incoming wave. The relation between the angle of incidence and the angle of refraction for the transmitted light is given by Snell's law:

$$\sin\theta_2 = \frac{n_1}{n_2}\sin\theta_1 \,. \tag{2.5}$$

For normal incidence onto a planar boundary, the fraction of the incident energy that is transmitted across the interface is given by

$$T = \frac{4n_1 n_2}{\left(n_1 + n_2\right)^2} \,. \tag{2.6}$$

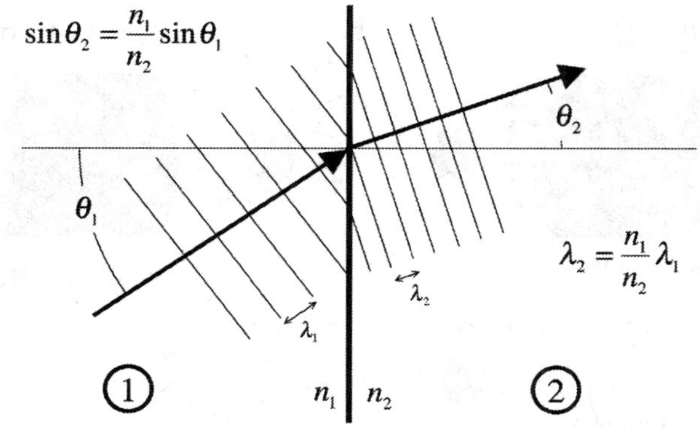


**FIGURE 2.3** Light incident on a planar boundary undergoes refraction. The refraction angle $\theta_2$ is given in terms of the angle of incidence $\theta_1$ and the refractive indices, $n_1$ and $n_2$, of the two media. The angle of reflection (not shown) is equal to the angle of incidence.

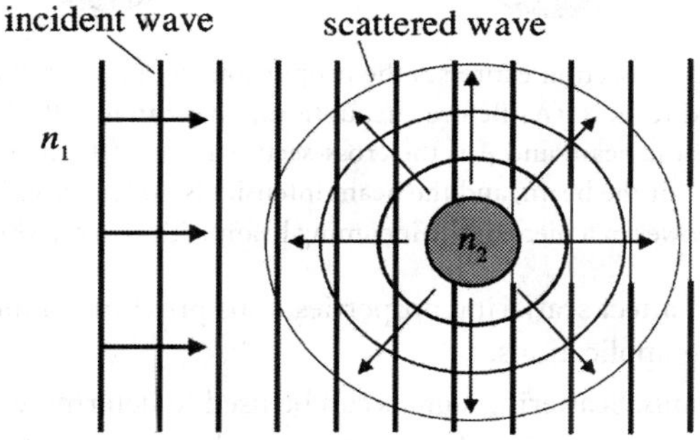


**FIGURE 2.4** Light incident on a localized particle embedded in a medium with a different refractive index will be partially scattered.

The fraction of the incident energy that gets reflected is

$$R = 1 - T = \frac{(n_1 - n_2)^2}{(n_1 + n_2)^2}. \tag{2.7}$$

This reflection from the interface, also known as the Fresnel reflection, is distinct from the diffuse reflectance to be discussed in Section 2.3, which in our context has a subsurface origin. (More general expressions describing transmission, reflection, and refraction phenomena can be found in Reference 2.)

## 2.2.2 Scattering

### 2.2.2.1 Scattering at a Localized Inclusion

When the guest material occupies only a localized region within the host material, scattering will occur. For example, a source of scattering in tissues is the mismatch in the refractive indices between subcellular organelles and the surrounding cytoplasm. In this case, some of the incident light is redirected over a range of angles relative to the scattering particle (Figure 2.4).

In biomedical photonics, scattering processes are important in both diagnostic and therapeutic applications:

- *Diagnostic applications:* Scattering depends on the size, morphology, and structure of the components in tissues (e.g., lipid membranes, nuclei, collagen fibers). Variations in these components

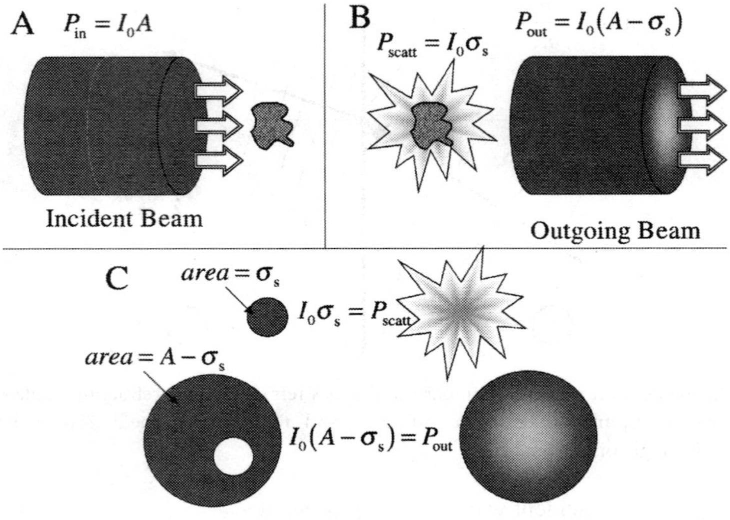


**FIGURE 2.5** The scattering cross section expresses the proportionality between the intensity of incident wave and the amount of power scattered from it. (A) Before encountering the scatterer, the beam has a uniform power $P_{in} = I_0 A$ where $I_0$ is the intensity of the beam and $A$ is the cross-sectional area. (B) After encountering the particle, some of the energy gets scattered out of the beam, and the beam intensity is no longer uniform. (C) The amount of power scattered is equivalent to the power in a piece of the incoming beam with area $\sigma_s$, which is the scattering cross section.

due to disease would affect scattering properties, thus providing a means for diagnostic purposes, especially in imaging applications.

- *Therapeutic applications:* Scattering signals can be used to determine optimal light dosimetry (e.g., during laser-based treatment procedures) and provide useful feedback during therapy.

Scattering is most simply described by considering the incident light as a plane wave (i.e., a wave of uniform amplitude in any plane perpendicular to the direction of propagation, at least over a size scale larger than the scattering particle). In principle, given the refractive indices of the two materials and the size and shape of the scatterer, the scattered radiation can be calculated. Scattering is quantified by the scattering cross section. For a monochromatic plane wave that has a given intensity (i.e., power per unit area) $I_0$ encountering the scattering object, some amount of power $P_{scatt}$ gets spatially redirected (i.e., scattered). The ratio of the power (i.e., energy per second) scattered out of the plane wave to the incident intensity (energy per second per area) is the *scattering cross section*,

$$\sigma_s(\hat{s}) = \frac{P_{scatt}}{I_0}, \tag{2.8}$$

where $\hat{s}$ is the propagation direction of the plane wave relative to the scatterer (Figure 2.5). The scattering cross section has units of area and is equivalent to the area that an object would have to "cut out" from the uniform plane wave in order to remove the observed amount of scattered power, $P_{scatt}$. The concept of cross section is also used for absorption where the power absorbed is proportional to the incident intensity. Note that the cross section is not the projected geometric area of an object as a glass sphere and a steel sphere of the same size will have different cross sections. It is merely a convenient way to quantify the scattering strength of an object.

Scattering depends on the polarization of the incoming wave, but we can consider Equation 2.8 to be an average over orthogonal polarization states. The angular distribution of the scattered radiation is given by the differential cross section (Figure 2.6),

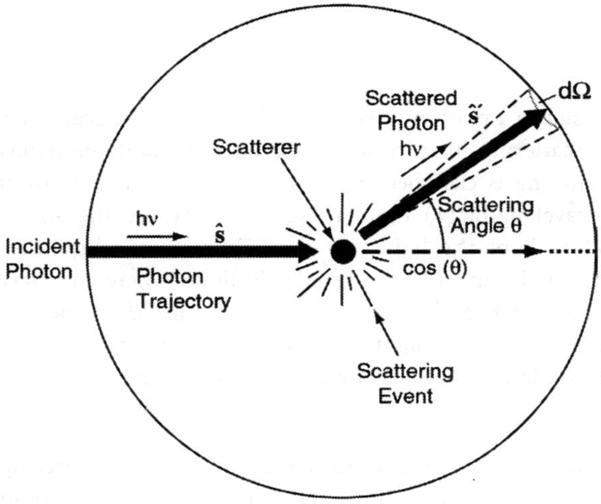


**FIGURE 2.6** The differential scattering cross section expresses the angular distribution of the scattered light relative to the incident light. The incident photon travels along the direction ŝ and the scattered photon exits in the ŝ′ direction.

$$\frac{d\sigma_s}{d\Omega}(\hat{s},\hat{s}'), \tag{2.9}$$

where $\hat{s}'$ defines the axis of a cone of solid angle $d\Omega$ originating at the scatterer.

For our purposes, we will assume that the scattering cross section is independent of the relative orientation of the incident light and the scatterer. This is equivalent to assuming that the object is spherically symmetric, a rough but adequate approximation valid within the context of the RT model. Under this approximation, the scattering cross section at a given wavelength is independent of the relative orientation of the particle and incident light,

$$\sigma_s(\hat{s}) = \sigma_s, \tag{2.10}$$

and the differential cross section depends only on the relative orientation of the incident and scattering directions, thus permitting us to write the differential cross section as a function of the cosine of the angle between $\hat{s}$ and $\hat{s}'$,

$$\frac{d\sigma_s}{d\Omega}(\hat{s},\hat{s}') = \frac{d\sigma_s}{d\Omega}(\hat{s}\cdot\hat{s}'). \tag{2.11}$$

(In a more general context, $\frac{d\sigma_s}{d\Omega}$ depends upon the polarization states of the incoming and outgoing waves, as with $\sigma_s$.)

A medium containing a uniform distribution of identical scatterers is characterized by the *scattering coefficient*,

$$\mu_s = \rho\sigma_s, \tag{2.12}$$

where $\rho$ is the number density of scatterers. The scattering coefficient is essentially the cross-sectional area for scattering per unit volume of medium. The *scattering mean free path*

$$l_s = \frac{1}{\mu_s} \tag{2.13}$$

represents the average distance a photon travels between consecutive scattering events.

In biological tissues, scattering interactions are often dominant mechanisms affecting light propagation. In practice, scattering is classified into three categories defined by the size of the scattering object relative to the wavelength: (1) the Rayleigh limit, where the size of the scatterer is small compared to the wavelength of the light; (2) the Mie regime, where the size of the scatterer is comparable to the wavelength; and (3) the geometric limit, where the wavelength is much smaller than the scatterer. The Rayleigh and Mie regimes are discussed further in the following sections. The geometric optics limit, where the light–scatterer interaction is described by rays that refract and reflect at the obstacles boundaries, is essentially the case discussed in Section 2.2.1.

#### 2.2.2.2 Rayleigh Limit

Scattering of light by tissue structures much smaller than the photon wavelength involves the so-called *Rayleigh limit.* Those structures include cellular components such as membranes and cell subcompartments (see Figure 2.7A), and extracellular components such as the banded ultrastructure of collagen fibrils. The most important implication of the small size-to-wavelength ratio is that, at any moment in time, the scatter sees only a spatially uniform electric field in the surrounding host. In the classical description (i.e., light as electromagnetic waves, not photons), this condition gives rise to a dipole moment in the scatterer, a slight and spatially simple redistribution of the charges in the scatterer. This dipole moment oscillates in time with the frequency of the incident field and as a consequence gives off dipole radiation. For a spherical particle of radius $a$, the differential cross section in the Rayleigh limit is

$$\frac{d\sigma_s}{d\Omega} = 8\pi^4 n_m^4 \left( \frac{n_s^2 - n_m^2}{n_s^2 + 2n_m^2} \right)^2 \frac{a^6}{\lambda^4} \left(1 + \cos^2 \theta\right), \tag{2.14}$$

where $\theta$ is the angle between the direction of the incoming wave and the outgoing direction of interest, and $n_m$ and $n_s$ are the refractive indices of the host medium and the scatterer, respectively (Figure 2.7B). This expression is averaged over polarization states of the incoming and scattered waves, as is appropriate for use in the RT theory.

In the quantum picture, Rayleigh scattering is called "elastic" because the energy of the scattered photon is the same as that of the incident photon. A related process is Raman scattering, which, in contrast, is "inelastic." In the Raman process, the frequencies of scattered photons are shifted from the incident frequency by amounts that are characteristic of molecular transitions, usually between vibrational energy states. The term *inelastic* refers to the fact that the scattered photon either loses energy to (Stokes) or gains energy from (anti-Stokes) the molecule. Raman scattering is extremely weak compared to the elastic Rayleigh scattering. (A rough approximation is 1 Raman-shifted photon for every $10^6$ Rayleigh photons.) Raman scattering is outside the scope of this chapter.

#### 2.2.2.3 Mie Regime

Scattering of light by spherical objects is described by *Mie theory,* which in principle is applicable at any size-to-wavelength ratio. In the intermediate size-to-wavelength ratio range, where the Rayleigh and geometric approximations are not valid, Mie theory is used. Thus this region where the scatterer size and wavelength are comparable is sometimes labled the *Mie regime.* Various cellular structures, such as mitochondria and nuclei (Figure 2.7A), and extracellular components like collagen fibers have sizes on the order of hundreds of nanometers to a few microns, and are comparable in dimension to the photon wavelengths generally used in biomedical applications (0.5 to 1 μm). Even though these structures are not necessarily spherical, their scattering behavior can be modeled reasonably well by Mie theory applied to spheres of comparable size and refractive index. Because the scatterer is on the order of the wavelength,

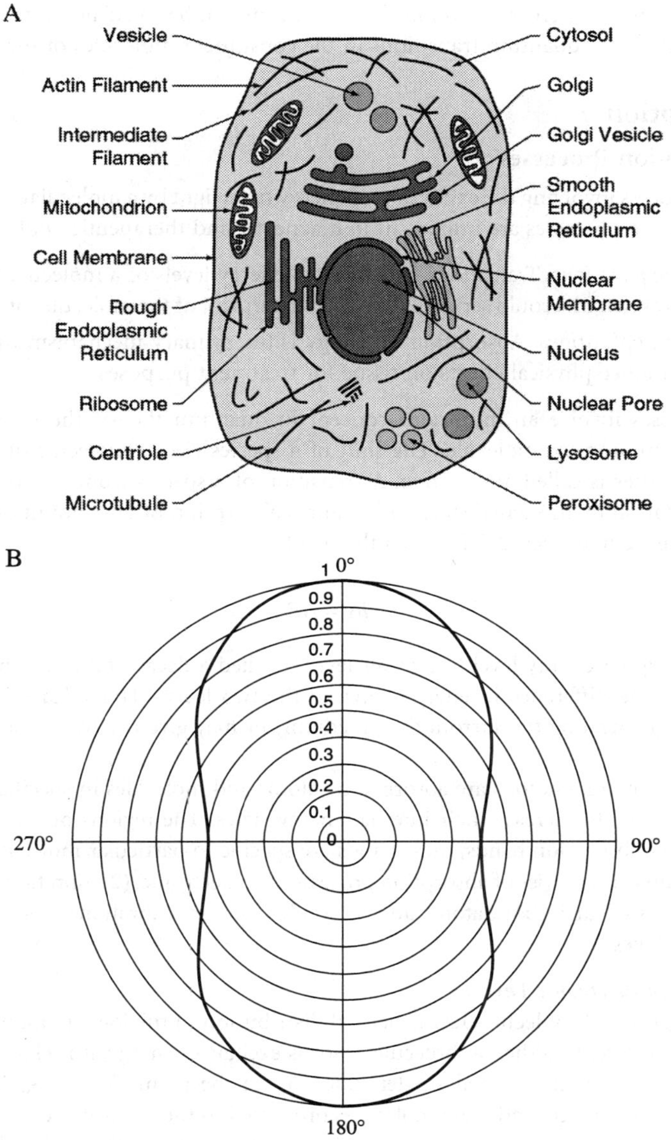


**FIGURE 2.7** (A) Cellular structures that act as scatterers in tissue. The organelles themselves scatter light in the Mie regime while the organelle and cellular membranes can act as Rayleigh scatterers. (B) Angular dependence of Rayleigh scattering. The distribution of unpolarized Rayleigh scattered light as a function of angle ($\theta = 0°$ is the forward direction).

it experiences a more complex field in the space around it at any moment in time, and thus the motion of charges in the scatterer in response to the field is also more complex. This results in a more complex angular dependence for the scattered light relative to the Rayleigh approximation. There can also be resonances and nulls due to the (classical) constructive and destructive interferences of the fields set up in the scatterer. The exact scattering patterns exhibited by the various biologic structures will of course depend on the detailed shape of the particle. The Mie scattering theory was specifically developed for spherical particles, although the label "Mie regime" is commonly (but loosely) used to refer to scattering from particles of arbitrary shape whose dimensions are on the order of or greater than the wavelength.

As in the Rayleigh limit, the cross section in the Mie regime can be calculated using the classical model, but may be modulated by quantum transitions in the constituent molecules of the scatterer.

## 2.2.3 Absorption

### 2.2.3.1 Absorption Processes

Absorption is a process involving the extraction of energy from light by a molecular species. In biomedical photonics, absorption processes are important in diagnostic and therapeutic applications:

- *Diagnostic applications:* Transitions between two energy levels of a molecule that are well defined at specific wavelengths could serve as a spectral fingerprint of the molecule for diagnostic purposes.
- *Therapeutic applications:* Absorption of energy is the primary mechanism that allows light from a laser to produce physical effects on tissue for treatment purposes.

Absorption processes involve an important concept in quantum theory, the *energy level*, which is a quantum state of an atom or molecule. The shift of a species (i.e., a molecule or an atom) from one energy level to another is called a *transition*. A transition of a species from a lower to a higher energy level involves excitation to an excited state and requires absorption of an amount of photon energy, $h\nu$, equal to the difference in energy, $\Delta E$, between the two levels:

$$h\nu = \Delta E \; . \tag{2.15}$$

A drop from a higher energy level to a lower level is called a decay and is accompanied by a release of energy equal to the difference in energy between the two levels. This release of energy may occur without radiation (by heating the surrounding medium) or may give rise to emission of a photon (e.g., luminescence).

In the quantum picture, photons are absorbed by atoms and molecules in specific transitions, and the photons' energy is used to increase their internal energy states. The regions of the spectrum where this occurs are known as absorption bands; these bands are specific to particular molecular/atomic species. In general, there are three basic types of absorption processes: (1) electronic, (2) vibrational, and (3) rotational. Electronic transitions occur in both atoms and molecules, whereas vibrational and rotational transitions occur only in molecules.

***Absorption between Electronic Levels***

At equilibrium, a group of molecules has a thermal distribution in the lowest vibrational and rotational levels of the ground state, $\mathbf{S}_o$. When a molecule absorbs excitation energy, it is elevated from $\mathbf{S}_o$ to some vibrational level of one of the excited singlet states, $\mathbf{S}_n$, in the manifold $\mathbf{S}_1 \ldots \mathbf{S}_n$. The intensity of the absorption (i.e., fraction of ground-state molecules promoted to the electronic excited state) depends on the intensity of the excitation radiation (i.e., number of photons) and the probability of the transition with photons of the particular energy being used. A term often used to characterize the intensity of an absorption (or an induced emission) band is the oscillator strength, $f$, which may be defined from the integrated absorption spectrum by the relationship

$$f = 4.315 \times 10^{-9} \int \varepsilon_\nu d\nu \, , \tag{2.16}$$

where $\varepsilon_\nu$ is the molar extinction coefficient at the frequency $\nu$.

Oscillator strengths of unity or near unity correspond to strongly allowed transitions, whereas lower values of $f$ are indicative of the smaller transition dipole matrix elements of forbidden transitions. Depending on the types of species, electronic transitions have energies corresponding to photons from the UV through the IR regions of the spectrum.

***Absorption Involving Vibrational Levels***

Vibrational levels characterize the different states of vibration of the atoms in a molecule. The vibrations in various degrees of freedom are quantized, giving rise to a series of vibrational levels for each vibrational pattern of a molecule. A vibrational transition occurs when a molecule shifts from one vibrational state to another. Typically, vibrational transitions correspond to the energies of photons from the IR.

***Absorption Involving Rotational Levels***

Rotational levels represent the different states of rotation of a molecule. These quantized energy levels correspond to photon energies from the far IR to submillimeter wavelengths.

The electronic energy levels of molecules are associated with molecular orbitals that describe the distribution of the electrons (i.e., electron clouds) in the molecule. When a molecule undergoes an electronic transition, an electron is transferred from one molecular orbital to another. However, there are cases where the excitation is considered to be localized to a particular bond or groups of atoms. Several types of molecular orbitals are involved in biological compounds: $\pi$ bonding orbitals, $\pi^*$ non-bonding orbitals, and *n* nonbonding (lone-pair) orbitals. The $\pi^*$ nonbonding orbital is less stable and has a higher energy than the $\pi$ bonding orbital. An electronic transition involving the transfer of an electron from a $\pi$ orbital to a $\pi^*$ orbital is called a $\pi\pi^*$ transition. Alternatively, the transfer of an electron from an *n* orbital to a $\pi^*$ orbital is called an $n\pi^*$ transition. In transition metal complexes (e.g., metal porphyrins), the electronic energy levels may also be described by molecular orbitals formed between the metal and ligands. The term *chromophore* refers to the part of the molecule that gives rise to the electronic transition of interest.

The probability of transition between different states or energy levels is governed by complex quantum mechanical rules that depend on the chemical structure, size, and symmetry of the molecules. Some transitions are said to be "allowed," which means that they are very likely to occur, thus giving rise to very strong absorption bands. Others are "forbidden" transitions, which are very unlikely to occur, producing very weak absorption bands.

### 2.2.3.2 Absorption Cross Section and Coefficient

For a localized absorber, the absorption cross section $\sigma_a$ can be defined in the same manner as for scattering, i.e.,

$$\sigma_a = \frac{P_{\text{abs}}}{I_0}, \tag{2.17}$$

where $P_{\text{abs}}$ is the amount of power absorbed out of an initially uniform plane wave of intensity (power per unit area) $I_0$. As previously with $\sigma_s$, we make the approximation that the cross section is independent of the relative orientation of the impinging light and the absorber. A medium with a uniform distribution of identical absorbing particles can be characterized by the *absorption coefficient*

$$\mu_a = \rho\sigma_a, \tag{2.18}$$

where $\rho$ is the number density of absorbers. The reciprocal,

$$l_a = \frac{1}{\mu_a}, \tag{2.19}$$

is the *absorption mean free path,* or *absorption length,* and represents the average distance a photon travels before being absorbed. For a medium, the absorption coefficient can be defined by the following relation:

$$dI = -\mu_a I dz, \tag{2.20}$$

where $dI$ is the differential change of intensity of a collimated light beam traversing an infinitesimal path $dz$ through a homogenous medium with absorption coefficient $\mu_a$. Integrating over a thickness $z$ yields the Beer–Lambert law:

$$I = I_0 \exp[-\mu_a z], \tag{2.21}$$

which can also be expressed as

$$I = I_0 \exp[-\varepsilon_\lambda a z] \tag{2.22}$$

where $\varepsilon_\lambda$ is the *molar extension coefficient* [$cm^2 \cdot mol^{-1}$] at wavelength $\lambda$, $a$ [$mol \cdot cm^{-3}$] is the molar concentration of the absorption species, and $z$ is the thickness [cm]. The molar extension coefficient $\varepsilon_\lambda$ (which is equal to $\mu_a/a$) is a measure of the "absorbing power" of the species.

A quantity commonly used is the *transmission,* $T$, defined as the ratio of transmitted intensity $I$ to incident intensity $I_0$:

$$T = I/I_0 . \tag{2.23}$$

The *attenuation*, also called *absorbance* (A) or *optical density* (OD), of an attenuating medium is given by

$$A = OD = \log_{10}(I_0/I) = -\log_{10}(T) . \tag{2.24}$$

The variation of $\varepsilon_\lambda$ with wavelength constitutes an absorption spectrum. Alternatively, a plot of $A$ vs. wavelength is also another way of presenting the absorption properties.

Once absorbed by a molecular species, the light energy can be dissipated optically by emitting a photon or nonradiatively by exchanging kinetic energy with other internal degrees of freedom of the absorbing species or external species (e.g., "heating" the medium). The quantum description of the absorption and emission processes (e.g., fluorescence, phosphorescence) of molecules is further discussed in Section 28.2 of Chapter 28 of this handbook. The most common situation is a combination of the two processes, where a small amount of the absorbed energy is dissipated nonradiatively and the majority is emitted as a photon in the transition back down to the ground state. This process of emission, known as luminescence, is further broken down into fluorescence (prompt emission from a singlet state) and phosphorescence (delayed emission from a triplet state). In fluorescence the delay between absorption and emission is typically on the order of nanoseconds, whereas in phosphorescence, emission follows much later (milliseconds and longer). As far as photon transport in tissues is concerned, fluorescence from excited singlet states is a much more common phenomenon, although triplet states are often involved in photodynamic therapy.

Under special conditions, the absorption process can also occur by the "simultaneous" extraction of two or more photons. This multiphoton absorption has important application in biomedical optics and is discussed in Chapter 11 of this handbook. Multiphoton absorption is a nonlinear process and is thus beyond the scope of our survey of the linear optical properties that are of primary importance in photon transport. Photons that are emitted in wavelength-changing processes such as fluorescence emission and Raman scattering are considered further in other chapters of this handbook.

## 2.3 Light Transport in Tissue

As media for light propagation, most human tissues are considered turbid (i.e., "cloudy" or opaque). Turbid tissues are heterogeneous structures and correspondingly have spatial variations in their optical properties. The spatial variation and density of these fluctuations make these tissues strong scatterers of light. In the absence of absorption, a significant fraction of the photons launched into these tissues are scattered multiple times, giving rise to a diffuse and largely incoherent field of penetrating light. As optical media, these tissues have been successfully modeled as two-component systems: (1) randomly positioned scattering and absorbing particles (2) embedded in a homogeneous continuum. In spite of the complexities of actual tissues, this simple two-component model has proven to give a satisfactory description of optical transport in tissue for many cases of interest. In this section, we examine RT solutions in limiting cases and also discuss Monte Carlo and Kubelka–Munk approaches to the light transport problem. At the end, we discuss some aspects of short-pulse time-resolved propagation.

## 2.3.1 Preliminaries to Radiation Transport Theory

### 2.3.1.1 Coherent and Incoherent Light

Light propagating in tissues can be classified into two types: coherent and incoherent. *Coherence* refers to the ability of a light field to maintain a definite (nonrandom) phase relationship in space and time and exhibit stable interference effects. An example of this interference is the speckle pattern produced when coherent light, such as a laser beam, is reflected from a surface. Speckle patterns occur because of the interference of a large number of elementary waves that arises when coherent light is reflected from a rough surface or when coherent light passes through a scattering medium. On the other hand, an *incoherent* source exhibits random phase patterns both temporally and spatially and thus is not capable of exhibiting stable or observable interference effects. An example of an incoherent source is an incandescent light bulb, as in a flashlight: unlike laser light, a flashlight beam reflected off a surface exhibits no speckle patterns. Speckle-related phenomena and measurement methods are discussed in Chapters 3 and 14 of this handbook.

Coherence is often a matter of degree because the same light source may exhibit observable interference effects in some situations and not in others. In the optical regime, coherent sources are necessarily monochromatic, because interferences between light of different wavelengths vary too rapidly to be meaningful. In tissues, light that is scattered multiple times can exhibit coherent and incoherent properties. Within the context of RT theory, the scattered fields are taken to be completely incoherent. This simplification is adequate for much of the phenomena of interest here. However, the scattered fields can also exhibit coherent effects; these are discussed in a later section of this chapter in the context of short-duration light pulses.

The following example illustrates the essential nature of coherence. In electromagnetic theory, the equations are linear in the electric and magnetic field amplitudes. This linear condition implies that the field at a point in a linear medium is the summation (superposition) of the field vectors contributed from each individual light source. For example, consider the total electric field at a point due to two scatterers of light (Figure 2.8),

$$\mathbf{E}_{\text{total}}(\mathbf{r},t) = \mathbf{E}_1(\mathbf{r},t) + \mathbf{E}_2(\mathbf{r},t), \tag{2.25}$$

where $\mathbf{E}_1(\mathbf{r},t)$ and $\mathbf{E}_2(\mathbf{r},t)$ are the electric field vectors of the two scattered waves. (To keep things simple, we will ignore the incident field and consider the two scatterers to be the only sources of light in the medium.) To calculate the energy $U$ associated with the scattered light field requires taking the square of its amplitude,

$$\begin{aligned} U(\mathbf{r}) &= \varepsilon \mathbf{E}_{\text{total}}(\mathbf{r}) \cdot \mathbf{E}_{\text{total}}(\mathbf{r}) = \varepsilon\left[\mathbf{E}_1^2(\mathbf{r}) + \mathbf{E}_2^2(\mathbf{r}) + 2\mathbf{E}_1(\mathbf{r}) \cdot \mathbf{E}_2(\mathbf{r})\right] \\ &= U_1(\mathbf{r}) + U_2(\mathbf{r}) + 2\varepsilon \mathbf{E}_1(\mathbf{r}) \cdot \mathbf{E}_2(\mathbf{r}) \end{aligned}, \tag{2.26}$$

where $\varepsilon$ is the *permittivity* of the medium. (Note that the permittivity $\varepsilon$ used in this section is distinct from the molar extension coefficient $\varepsilon_\lambda$ defined in Equation 2.22.)

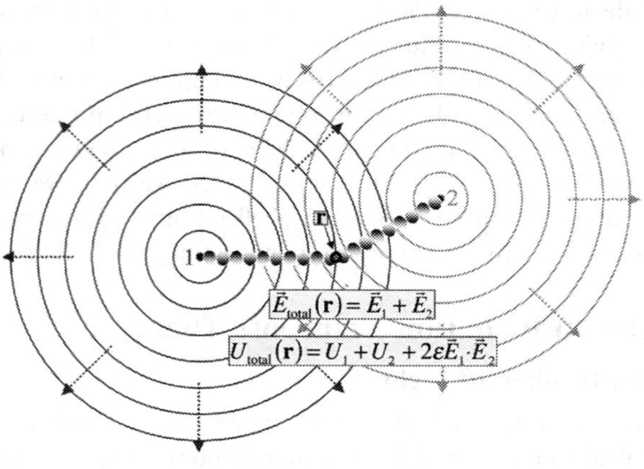


**FIGURE 2.8** Superposition of EM waves. The EM field $\mathbf{E}_{total}$ at point **r** is the vector sum of the two fields from the respective sources, 1 and 2. The energy $U_{total}$ at point **r** due to these two fields includes an interference term to account for constructive or destructive interference.

The energy expressed in Equation 2.26 is not just the addition of the separate energies contributed by the two respective sources but includes a cross term, $2\mathbf{E}_1 \cdot \mathbf{E}_2$, that accounts for the interference of the two sources. If the two electric fields have overlapping polarizations ($|\mathbf{E}_1 \cdot \mathbf{E}_2| > 0$), they can interfere constructively ($\mathbf{E}_1 \cdot \mathbf{E}_2 > 0$) or destructively ($\mathbf{E}_1 \cdot \mathbf{E}_2 < 0$) at point **r**. Therefore, the total energy at **r** can be greater than or less than the sum of the energies of the individual sources acting alone. Over all of space, however, the constructive and destructive interferences will balance out so that energy is conserved. Consider now the energy averaged over a region of space:

$$U_{avg} = \varepsilon \overline{\mathbf{E}_{total} \cdot \mathbf{E}_{total}} = \varepsilon \left[ \overline{\mathbf{E}_1^2} + \overline{\mathbf{E}_2^2} + 2\overline{\mathbf{E}_1 \cdot \mathbf{E}_2} \right]. \tag{2.27}$$

If the region is large enough, the interference contribution to the energy averages away,

$$U_{avg} = \varepsilon \left[ \overline{\mathbf{E}_1^2} + \overline{\mathbf{E}_2^2} \right] = \overline{U_1} + \overline{U_2}, \tag{2.28}$$

and

$$\overline{\mathbf{E}_1 \cdot \mathbf{E}_2} = 0. \tag{2.29}$$

For a greater number of scatterers, the energy at a point can be calculated in the same way, by first adding up the fields due to each scatterer,

$$\mathbf{E}_{total} = \sum_{j=1}^{N} \mathbf{E}_j, \tag{2.30}$$

and then squaring the total,

$$\begin{aligned} U = \varepsilon \mathbf{E}_{total} \cdot \mathbf{E}_{total} &= \varepsilon \sum_{j=1}^{N} \mathbf{E}_j \cdot \sum_{j=1}^{N} \mathbf{E}_j = \varepsilon \sum_{j=1}^{N} \mathbf{E}_j^2 + \varepsilon \sum_{j=1}^{N} \sum_{\substack{m=1 \\ m \neq j}}^{N} \mathbf{E}_j \cdot \mathbf{E}_m \\ &= \sum_{j=1}^{N} U_j + \varepsilon \sum_{j=1}^{N} \sum_{\substack{m=1 \\ m \neq j}}^{N} \mathbf{E}_j \cdot \mathbf{E}_m \end{aligned} \tag{2.31}$$

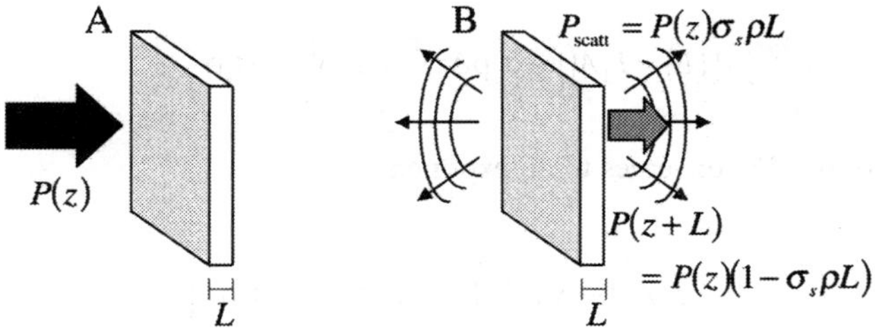


**FIGURE 2.9** Multiple scattering. Schematic depiction of the interaction of a planar light beam with a thin slab of scattering material. The material contains ρ scatterers per unit volume, each with cross section $\sigma_s$. (A) Before the wave enters the slab, it contains power $P(z)$. (B) After passing through the slab, the power scattered out of the wave is $\sigma_s \rho L$, where $L$ is the thickness of the slab.

The energy involves two types of terms: (1) the energy contributed from each individual source and (2) the interference contributions (which are the manifestation of the coherence of light field). Again, by averaging over a sufficiently large region around the point, the interferences will average to zero:

$$\overline{\sum_{j=1}^{N}\sum_{\substack{m=1\\m\neq j}}^{N} \mathrm{E}_j \cdot \mathrm{E}_m} \approx 0. \tag{2.32}$$

The size of the region necessary for the interferences to become zero depends on how the scatterers are distributed and on the nature of the scattering process. By considering only the spatially averaged energy, the wave nature of the light can be ignored and one need track only the movement of energy through the system. This can be a simpler task because the energies are scalar quantities without phase and polarization. In tissue media, which is modeled as a random collection of identical scattering particles, multiple scattering effects render the field largely incoherent, and the interference terms will be small relative to the average energy. Radiation transport theory will provide the equations that describe how this averaged light energy moves through such media.

### 2.3.1.2 Multiple Scattering

The systems of interest in this chapter are modeled as distributions of localized scatterers in a uniform background. Multiple scattering is an important phenomenon in these systems. Using a simple model one can illustrate the origin of these effects and their relation between single and multiple scattering. Consider a medium with a uniform distribution of scatterers. Each scatterer has a cross section $\sigma_s$, and the number density of scatterers is given by ρ. The medium can be broken up into slabs of thickness, $\Delta z$, and the field incident onto the first slab is a plane wave with intensity $I_0$ (Figure 2.9). The power incident on a local region of the layer with cross-sectional area $A$ is $I_0 A$. The power scattered out of the incident plane wave as it crosses the layer is given by

$$I_0 \sigma_s \rho A \Delta z = I_0 \mu_s A \Delta z = I_0 \sigma_s N_{\text{layer}}, \tag{2.33}$$

where $N_{\text{layer}}$ is the number of scatters encountered in the layer. After passing through the layer, the power remaining in the plane wave is

$$P_c(0+\Delta z) = I_0 A - I_0 \sigma_s \rho A \Delta z = I_0 A(1-\sigma_s \rho \Delta z). \tag{2.34}$$

(When $\sigma_s \rho \Delta z = 1$ for a layer, the effect saturates, so further increases in this quantity beyond unity have no physical effect.) Each successive layer decreases the intensity of the incident field by an addition factor of $(1 - \sigma_s \rho \Delta z)$. If the total number of layers is Γ, the total distance traveled is $L = \Gamma \Delta z$. The power remaining after propagating through this length, $L$, is

$$P_c(L) = I_0 A(1-\sigma_s\rho\Delta z)^\Gamma = I_0 A\left(1-\sigma_s\rho\frac{L}{\Gamma}\right)^\Gamma . \tag{2.35}$$

As $\Gamma$ increases, Equation 2.35 converges to an exponential:

$$I_0 A\left(1-\sigma_s\rho\frac{L}{\Gamma}\right)^\Gamma \to I_0 A\exp[-\sigma_s\rho L] . \tag{2.36}$$

Because there is no absorption, the total power in the scattered field given by

$$P_{\text{scatt}}^{\text{total}} = I_c(0)A - I_c(L)A = I_0 A(1-\exp[-\sigma_s\rho L]) = I_0 A\left(1-\exp\left[-\frac{\sigma_s}{A}N\right]\right) \tag{2.37}$$

where $N$ is the total number of scatterers encountered in the medium.

Using the power series expansion for the exponential, the term in parentheses can be written as

$$1-\exp\left[-\frac{\sigma_s}{A}N\right] = -\sum_{m=1}^{\infty}\frac{\left(-\frac{\sigma_s}{A}N\right)^m}{m!} = \frac{\sigma_s}{A}N - \frac{1}{2}\frac{\sigma_s^2}{A^2}N^2 + \frac{1}{6}\frac{\sigma_s^3}{A^3}N^3 + \ldots \tag{2.38}$$

The quantity $\sigma_s\rho L$ is proportional to the number of scatterers encountered along the entire path from 0 to $L$. Since $\sigma_s\rho L = \frac{\sigma_s}{A}N$ when the quantity $\sigma_s\rho L << 1$, the scattered power can be approximated by

$$P_{\text{total}}^{\text{scatt}} \cong N I_0 \sigma_s , \tag{2.39}$$

and the power is therefore proportional to the number of scatterers. Physically, this is equivalent to saying that the scattered power is entirely due to waves that have scattered only once. The contributions from waves that have scattered more than once are small enough to be safely ignored in this limit. However, when the approximation

$$1-\exp[-\sigma_s\rho L] \approx \sigma_s\rho L$$

breaks down, the scattered power has a more complicated nonlinear relationship with the number of scatterers encountered, and thus the power contained in multiply scattered waves is significant. So a criterion for defining a system as a single scattering medium is

$$\sigma_s\rho L << 1 , \tag{2.40}$$

or, in terms of the scattering coefficient, as

$$\mu_s L << 1 . \tag{2.41}$$

Even when the cross section is small if the scatterers are sufficiently dense and/or the propagation distance is sufficiently long, multiple-scattering effects can be important. Conversely, particles with large cross sections may not give rise to significant multiple-scattering effects if the concentration of the particles is low or if the path length is short. The significance of the distinction between single- and multiple-scattering processes is that the single-scattering case can be handled in a much more straightforward

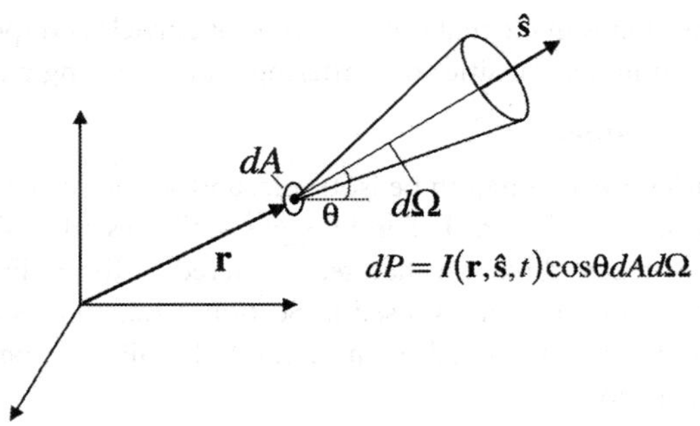


**FIGURE 2.10** The light power passing through a surface element *da* inside a cone of solid angle $d\Omega$ directed at an angle $\theta$ to the surface normal *n* is proportional to the specific intensity $I(\mathbf{r}, \hat{s}, t)$.

manner. The multiple-scattering effects provide the mechanism for "decohering" the scattered field and ultimately justify the use of RT theory, in which coherence processes are not explicitly taken into account.

## 2.3.2 The Radiation Transport Model

### 2.3.2.1 Basic Parameters

The multiple scattering and associated decoherence effects in turbid media effectively suppress the wave nature of light; therefore, instead of tracking light waves, one can track only the average energy they contain. The flow of light energy through the medium is described by RT theory, which explicitly disregards wave interference effects. The fundamental quantity that replaces the electromagnetic field in RT model is the specific intensity, $I(\mathbf{r}, \hat{s}, t)$. The following relation defines the specific intensity:

$$dP = I\left(\mathbf{r}, \hat{s}, t\right) d\omega da, \tag{2.42}$$

where $dP$ is the light power at time $t$ and at point $\mathbf{r}$ directed in a cone of solid angle $d\omega$ oriented in the direction defined by the unit vector $\hat{s}$, from a surface area $da$ normal to $\hat{s}$ (Figure 2.10). Thus, $I(\mathbf{r}, \hat{s}, t)$ is the light power per unit area per unit solid angle. Since the energy in a monochromatic light field is proportional to the number of photons the field contains, the specific intensity is representative of the number of photons per second passing through point $\mathbf{r}$ within the solid angle cone.

The medium through which the light energy propagates is characterized by three parameters:

1. The absorption coefficient, $\mu_a$
2. The scattering coefficient, $\mu_s$
3. The scattering "phase" function (SPF), $p(\hat{s} \cdot \hat{s}')$

(The function $p(\hat{s} \cdot \hat{s}')$ does not have anything to do with the phase of propagating waves; the "phase" label is mainly the legacy of an earlier notation.) The coefficients $\mu_a$ and $\mu_s$ were defined earlier in Equations 2.18 and 2.12. The attenuation coefficient $\mu_t$ combines absorption and scattering effects into a single quantity given as

$$\mu_t = \mu_a + \mu_s \,. \tag{2.43}$$

The total mean free path is defined through the total attenuation coefficient as

$$l_t = \frac{1}{\mu_a + \mu_s} = \frac{1}{\dfrac{1}{l_a} + \dfrac{1}{l_s}} \,. \tag{2.44}$$

In the RT model, the assumption is made that only one type of particle is responsible for both absorption and scattering and thus the number densities of scattering and absorbing particles are the same.

#### 2.3.2.2 Scattering Phase Function

In the RT model, the particles are assumed to be isotropic, and so the scattering phase function, or SPF, is expressed as a function of $\hat{s} \cdot \hat{s}' = \cos\theta$. The SPF, $p(\hat{s} \cdot \hat{s}')$, describes the fraction of light energy incident on a scatterer from the $\hat{s}'$ direction that gets scattered in the $\hat{s}$ direction. (Note that this is a reversal of the earlier convention for $\hat{s}$ and $\hat{s}'$ used in Section 2.2.2. Since $\hat{s} \cdot \hat{s}' = \hat{s}' \cdot \hat{s}$, this change in convention has no bearing on the functional form of the SPF.) SPF can be expressed in terms of the differential scattering cross section,

$$p(\hat{s} \cdot \hat{s}') = \frac{4\pi}{\sigma_s + \sigma_a} \frac{d\sigma_s}{d\Omega}(\hat{s} \cdot \hat{s}'). \tag{2.45}$$

The integral of the SPF over all the solid angle yields the albedo $W_0$,

$$\begin{aligned} W_0 \equiv \frac{1}{4\pi} \int_{4\pi} p(\hat{s} \cdot \hat{s}') d\Omega' &= \frac{\sigma_s}{\sigma_a + \sigma_s} \\ &= \frac{\mu_s}{\mu_a + \mu_s} \end{aligned}, \tag{2.46}$$

which is the fraction of the total cross section that is due to scattering. (It is important to mention here that the choice of normalization for $p(\hat{s} \cdot \hat{s}')$ varies among different treatments of RT theory.* To keep the outcomes the same among the different normalization schemes requires modification of one of the terms involving $p(\hat{s} \cdot \hat{s}')$ in the RT equations, Equations 2.49 to 2.51, discussed in the next section.)

Another constant of interest is the cosine-weighted average of the scattering, commonly known as the *average cosine of scatter*:

$$\begin{aligned} g \equiv \frac{\int_{4\pi} p(\hat{s} \cdot \hat{s}') \hat{s} \cdot \hat{s}' d\Omega'}{\int_{4\pi} p(\hat{s} \cdot \hat{s}') d\Omega'} &= \frac{1}{4\pi W_0} \int_{4\pi} p(\hat{s} \cdot \hat{s}') \hat{s} \cdot \hat{s}' d\Omega' \\ &= \frac{1}{2W_0} \int_{4\pi} p(\cos\theta) \cos\theta \sin\theta d\theta. \end{aligned} \tag{2.47}$$

The parameter $g$ is a measure of scatter retained in the forward direction following a scattering event. For a Rayleigh scatterer, the SPF varies as $1 + \cos^2\theta = 1 + (\hat{s} \cdot \hat{s}')^2$, and its average cosine of scatter, $g$, is zero. This is because backward and forward scattering are equally likely. A scatterer with $g > 0$ is more likely to forward-scatter a photon, while a negative $g$ indicates a preference for backward scattering. Most of the scattering processes of interest in tissue optics are in the Mie limit, and the SPF can be difficult to calculate without detailed knowledge of the system. An approximate SPF often used is the Henyey–Greenstein function[3] (Figure 2.11):

$$p_{HG}(\cos\theta) = \frac{4\pi\sigma_s}{\sigma_a + \sigma_s} \frac{1 - g^2}{(1 + g^2 - 2g\cos\theta)^{\frac{3}{2}}}. \tag{2.48}$$

*Investigators in biomedical optics use a variety of different units, notations, and normalization procedures. Despite the different notations, the photophysical phenomena described are similar, and the final results of the derivations are the same.

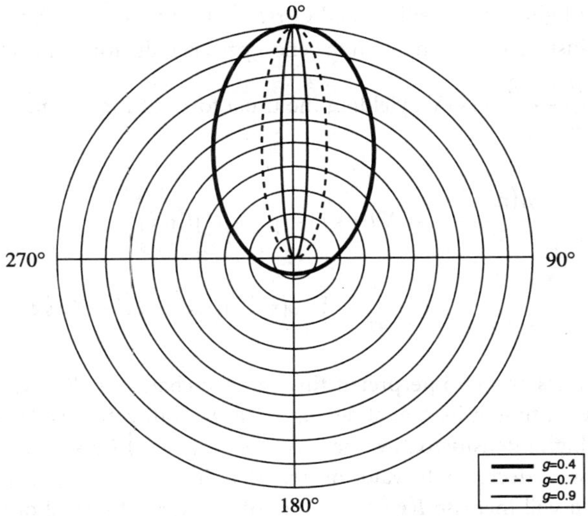


**FIGURE 2.11** The angular dependence of the Henyey–Greenstein scattering phase function for $g$ = 0.4, 0.7, and 0.9. Each of the patterns has been normalized to unity. When properly scaled, the maximum value of $p_{HG}(\theta)$ for $g$ = 0.9 is 49 times that of the $g$ = 0.4 case and 10 times that of the $g$ = 0.7 case.

Henyey and Greenstein have developed this expression to describe the angular dependence of light scattering by small particles, which they used to investigate diffuse radiation from interstellar dust clouds in the galaxy.[3] This function is convenient to use because it is parameterized by the average cosine of scatter, $g$. For tissues, $g$ ranges from 0.4 to 0.99. Such values indicate that scattering is strongly forward-peaked. For example, for $g$ = 0.6, almost 90% of the scattered energy is within 90° of the forward (i.e., $\theta$ = 0) direction; for $g$ = 0.99, 90% of the energy is scattered to within 5° of the forward direction.

The material constants can be calculated using electromagnetic wave theory, provided the properties of the particles (e.g., shape, size, refractive index, number density) and background medium are known. So even though the light transport in the RT model explicitly ignores the wave nature of light, electromagnetic wave phenomena are implicitly included through these material parameters.

### 2.3.2.3 Radiation Transport Equation

The fundamental equation describing light propagation in the RT model is the radiation transport equation (also known as the Boltzmann equation), which describes the fundamental dynamics of the specific intensity. Light is effectively treated as a collection of localized, incoherent photons. (Polarization properties can be considered, but we will avoid this complication.) Consider a small packet of light energy defined by its position, $\mathbf{r}(t)$, and direction of propagation $\hat{\mathbf{s}}$. Following along with it over the time interval $dt$ as it propagates in space, the packet loses energy due to absorption and scattering out of $\hat{\mathbf{s}}$, but also gains energy from light scattered into the $\hat{\mathbf{s}}$-directed packet from other directions and from any local source of the light at $\mathbf{r}(t)$. These processes are quantified by the integro-differential equation known as the radiation transport relation:

$$\begin{aligned}\frac{1}{c_m}\frac{d}{dt}I(\mathbf{r}(t),\hat{\mathbf{s}},t) &= -(\mu_a+\mu_s)I(\mathbf{r}(t),\hat{\mathbf{s}},t)\\ &\quad+\frac{\mu_a+\mu_s}{4\pi}\int p(\hat{\mathbf{s}}\cdot\hat{\mathbf{s}}')I(\mathbf{r}(t'),\hat{\mathbf{s}}',t)d\Omega'+Q(\mathbf{r}(t),\hat{\mathbf{s}},t),\end{aligned} \tag{2.49}$$

where $c_m$ is the speed of light in the medium and $Q(\mathbf{r}(t),\hat{s},t)$ is the source term. If the coordinate system is fixed to the medium instead of moving with the energy packet, the total time derivative can be reduced to partial derivatives, $\frac{d}{dt} \to \frac{\partial}{\partial t} + c\hat{s}\cdot\vec{\nabla}$, yielding the usual form of the radiation transport equation:

$$\frac{1}{c_m}\frac{\partial I(\mathbf{r},\hat{s},t)}{\partial t} = -\hat{s}\cdot\vec{\nabla}I(\mathbf{r},\hat{s},t) - (\mu_a + \mu_s)I(\mathbf{r},\hat{s},t) + \frac{\mu_a + \mu_s}{4\pi}\int_{4\pi} p(\hat{s}\cdot\hat{s}')I(\mathbf{r},\hat{s}',t)d\Omega' + Q(\mathbf{r},\hat{s},t)\ . \tag{2.50}$$

In this form, the dynamics can be interpreted through the change in $I(\mathbf{r}, \hat{s},t)$ with time. The specific intensity will increase with time if its spatial derivative in the $\hat{s}$ direction is decreasing — that is, it will "flow" from regions of high intensity to low intensity. The $-(\mu_a + \mu_s)\ I(\mathbf{r},\hat{s},t)$ term on the right-hand side of Equation 2.50 will always decrease the value of $I(\mathbf{r}, \hat{s},t)$, as it accounts for the scattering and absorption losses. The integral term will increase $I(\mathbf{r}, \hat{s},t)$ because of scattering from all other directions into the $\hat{s}$ cone, as will any source of light $Q(\mathbf{r},\hat{s},t)$ at $\mathbf{r}$. (Note that the integral term will have a different prefactor if an alternate normalization scheme for SPF is chosen, as mentioned in the previous section.)

An important simplification of the RT equation is the steady-state form. The term "steady state" implies that $\frac{\partial I(\mathbf{r},\hat{s},t)}{\partial t} = 0$; physically, this means that losses and gains are balanced such that the specific intensity at any point in the medium does not change with time. The steady-state limit arises out of the time-resolved case when the light source has been switched on and has illuminated the sample long enough for the light levels to reach equilibrium. Strictly, this occurs when the source light has had time to reverberate through the tissue layers (i.e., reflect from each boundary) an infinite number of times. In practice, a large but finite number of reflections are sufficient to reach the steady-state as eventually attenuation and transmission losses render the higher-order reflections negligible. The steady state is applicable to many practical situations, even when the light source is pulsed or modulated. For example, consider a uniform light pulse that is 1 μs in duration incident from air on a 1-cm-thick sample with an index of refraction of 1.33 with no absorbers or scatterers. One can show that the amount of light energy transmitted through the sample in the steady-state limit can be accounted for to better than one part in a million by just three reverberations of the initial light field through the sample. Over this 1-μs time interval, light could reverberate through this sample more than 10,000 times. Thus for almost the entire duration of the pulse, the problem could be considered to be in the steady-state limit. The steady-state RT equation in a source-free region is

$$\hat{s}\cdot\vec{\nabla}I(\mathbf{r},\hat{s}) = -(\mu_a + \mu_s)I(\mathbf{r},\hat{s}) + \frac{\mu_a + \mu_s}{4\pi}\int_{4\pi} p(\hat{s}\cdot\hat{s}')I(\mathbf{r},\hat{s}')d\Omega'\ . \tag{2.51}$$

### 2.3.3 Analytical Solutions for Limiting Cases

Although conceptually straightforward, the direct analytical solution of the RT equation for many problems of interest is difficult. Most of the complications arise from dealing with the boundaries at tissue interfaces, and geometric aspects of the tissues and light sources. The following discusses some of the important concepts regarding the light distributions and describes limiting cases of the RT model. The discussion will mainly be concerned with light fields propagating in a semi-infinite medium (i.e., cases in which the back wall of the tissue is far enough away that it has no influence on the distribution of light) in order to illustrate some basic properties of the diffuse field in various situations.

#### 2.3.3.1 Incident and Diffuse Light

In the model considered here, the specific intensity is zero until light energy is provided by an external source (e.g., a laser) or an embedded source (e.g., fiber optic). The intensity in the medium can be expressed as

$$I(\mathbf{r},\hat{\mathbf{s}},t) = I_c(\mathbf{r},\hat{\mathbf{s}},t) + I_d(\mathbf{r},\hat{\mathbf{s}},t), \tag{2.52}$$

where $I_c(\mathbf{r},\hat{\mathbf{s}},t)$ represents the unscattered component of the light energy and $I_d(\mathbf{r},\hat{\mathbf{s}},t)$ represents the scattered or diffuse component. If the light source is a coherent one, $I_c(\mathbf{r},\hat{\mathbf{s}},t)$ is the coherent field in the tissue. Even though coherence has no special implications in the RT model, it provides a convenient means for distinguishing between the field component that only loses energy from scattering, $I_c(\mathbf{r},\hat{\mathbf{s}},t)$, and the field component that receives and retains the energy from all orders of scattering, $I_d(\mathbf{r},\hat{\mathbf{s}},t)$. The single RT equation can be expressed as two separate equations: one involving only the coherent light,

$$\frac{1}{c_m}\frac{\partial I_c(\mathbf{r},\hat{\mathbf{s}},t)}{\partial t} + \hat{\mathbf{s}}\cdot\vec{\nabla} I_c(\mathbf{r},\hat{\mathbf{s}},t) = -(\mu_a+\mu_s)I_c(\mathbf{r},\hat{\mathbf{s}},t), \tag{2.53}$$

and the second involving both light components (coherent and incoherent),

$$\begin{aligned}
\frac{1}{c_m}\frac{\partial I_d(\mathbf{r},\hat{\mathbf{s}},t)}{\partial t} &+ \hat{\mathbf{s}}\cdot\vec{\nabla} I_d(\mathbf{r},\hat{\mathbf{s}},t) = \\
&-(\mu_a+\mu_s)I_d(\mathbf{r},\hat{\mathbf{s}},t) \\
&+\frac{\mu_a+\mu_s}{4\pi}\int_{4\pi} p(\hat{\mathbf{s}},\hat{\mathbf{s}}')I_d(\mathbf{r},\hat{\mathbf{s}}',t)\,d\Omega' \\
&+\frac{\mu_a+\mu_s}{4\pi}\int_{4\pi} p(\hat{\mathbf{s}},\hat{\mathbf{s}}')I_c(\mathbf{r},\hat{\mathbf{s}}',t)\,d\Omega'.
\end{aligned} \tag{2.54}$$

The last term in Equation 2.54 can be written as a source term for the diffuse component equation,

$$Q_c(\mathbf{r},\hat{\mathbf{s}},t) \equiv \frac{\mu_a+\mu_s}{4\pi}\int_{4\pi} p(\hat{\mathbf{s}},\hat{\mathbf{s}}')I_c(\mathbf{r},\hat{\mathbf{s}}',t)\,d\Omega'. \tag{2.55}$$

If the incident field is a plane wave, the coherent field solution to Equation 2.51 takes the form of the Beer–Lambert law (as shown earlier in Equation 2.21). For example, a planar source of intensity $I_0$ launched into a semi-infinite medium normal to its surface ($\hat{\mathbf{s}}_1 = \hat{z}$ and $z$ is the depth into the tissue) takes the form

$$\begin{aligned}
I_c(z,t) &= f(c_m t - z)I_c(0)\exp[-(\mu_a+\mu_s)z] \\
&= I_0(1-R^2)f(c_m t - z)\exp[-(\mu_a+\mu_s)z],
\end{aligned} \tag{2.56}$$

where $f(c_m t - z)$ is a positive definite, differentiable function describing the time envelope of the light pulse, valid for $z > 0$ and $t > 0$. In the steady state, the plane wave solution is simply

$$I_c(z) = I_0(1-R^2)\exp[-(\mu_a+\mu_s)z]. \tag{2.57}$$

Note that the intensity of a plane wave relates to the specific intensity as

$$I(z,t) = \int I(\mathbf{r},\hat{\mathbf{s}},t)\delta(\hat{\mathbf{s}} - \hat{z})d\Omega . \tag{2.58}$$

The behavior of the coherent field is straightforward in most cases, and its main role in tissue optics is to serve as a source for the diffuse field. The study of light transport in tissues means tracking the distribution of the diffuse field since in most situations it is the penetrating component of the light. Thus the determination of the specific intensity of the diffuse field $I_d$ is the principal goal. The diffuse light gives rise to important phenomena such as diffuse reflectance, which arises from the subsurface scattering of light. A portion of the light launched into tissue will be scattered back across the surface as it is redirected by multiple scattering events. This light is thus distinct from surface reflections and is largely incoherent as its source is the subsurface diffuse field. This diffusely "reflected" light contains information about the bulk of the medium because it has collectively sampled an extended volume of the tissue. Changes in the diffuse reflectance over time can be used to track the dynamics of certain absorbing species in tissue, such as oxygenated hemoglobin. (Note that in other contexts "diffuse reflectance" also refers to reflection from a rough surface or interface.)

To introduce some of the important aspects of the diffuse and coherent light, we will examine approximate solutions for the two limiting cases — $\mu_a >> \mu_s$ (absorption-dominant limit) and $\mu_s >> \mu_a$ (scattering-dominant limit). We will also briefly discuss Monte Carlo calculations and the propagation of short pulses of light.

### 2.3.3.2 Absorption-Dominant Limit

In the case where absorption of light is the dominant process, the light does not penetrate the tissue deeply enough for significant scattering to occur, and the diffuse field will remain small compared to the coherent one. Using Equation 2.54, one can find an approximate solution for $I_d$ by considering only the contributions of singly scattered light from the coherent field, encompassed in the term

$$\frac{\mu_a + \mu_s}{4\pi} \int_{4\pi} p(\hat{\mathbf{s}},\hat{\mathbf{s}}')I_c(\mathbf{r},\hat{\mathbf{s}}')d\Omega'$$

and ignoring the contributions of rescattered light, from the term

$$\frac{\mu_a + \mu_s}{4\pi} \int_{4\pi} p(\hat{\mathbf{s}},\hat{\mathbf{s}}')I_d(\mathbf{r},\hat{\mathbf{s}}')d\Omega' .$$

In the steady state, using this single-scattering approximation, Equation 2.54 becomes

$$\hat{\mathbf{s}} \cdot \vec{\nabla} I_d(\mathbf{r},\hat{\mathbf{s}}) = -(\mu_a + \mu_s)I_d(\mathbf{r},\hat{\mathbf{s}}) + \frac{\mu_a + \mu_s}{4\pi} \int_{4\pi} p(\hat{\mathbf{s}},\hat{\mathbf{s}}')I_c(\mathbf{r},\hat{\mathbf{s}}')d\Omega' . \tag{2.59}$$

The transport equation along a straight-line path of length $s$ parallel to the direction $\hat{\mathbf{s}}$ is

$$\frac{d}{ds} I_d(\mathbf{r},\hat{\mathbf{s}}) = -(\mu_a + \mu_s)I_d(\mathbf{r},\hat{\mathbf{s}}) + \frac{\mu_a + \mu_s}{4\pi} \int_{4\pi} p(\hat{\mathbf{s}},\hat{\mathbf{s}}')I_c(\mathbf{r},\hat{\mathbf{s}}')d\Omega' , \tag{2.60}$$

where $\mathbf{r}$ points to the path that begins at the point $\mathbf{r}_0$. This is a differential equation of the type $\frac{dy}{ds} + P(s)y = Q(s)$, which has the general solution

$$y(s) = Ce^{-\int_{s_0}^{s} P(w)dw} + e^{-\int_{s_0}^{s} P(w)dw} \int_{s_0}^{s} e^{-\int_{s_0}^{s'} P(v)dv} Q(s')ds' , \tag{2.61}$$

where $C$ is a constant determined by the boundary conditions. Integrating along the path from 0 to $s$, the diffuse field is

$$I_d(s,\hat{s}) = C\exp[-\mu_t s] + \exp[-\mu_t s]\frac{\mu_t}{4\pi}\int_0^s \exp[\mu_t s_1]\iint_{4\pi} p(\hat{s},\hat{s}')I_c(r_1,\hat{s}')d\Omega' ds_1 \,. \tag{2.62}$$

Now we take the surface of the scattering medium to be at the plane $\mathbf{r}\cdot\hat{z} = 0$, which includes the point $\mathbf{r}_0$ ($s = 0$), and assume that the coherent light in the medium is a plane wave traveling in the $\hat{z}$ direction:

$$I_c(\mathbf{r},\hat{s}) = I_0 \exp[-\mu_t(\mathbf{r}-\mathbf{r}_0)\cdot\hat{z}]\delta(\hat{s}-\hat{z}) \,. \tag{2.63}$$

The solution is then

$$\begin{aligned} I_d(s,\hat{s}) &= C\exp[-\mu_t s] + \exp[-\mu_t s]\frac{\mu_t}{4\pi}\int_0^s \exp[\mu_t s_1]\iint_{4\pi} p(\hat{s},\hat{s}')I_c(r_1,\hat{s}')d\Omega' ds_1 \\ &= C\exp[-\mu_t s] + \exp[-\mu_t s]\frac{\mu_t I_0}{4\pi} p(\hat{s}\cdot\hat{z})\int_0^s \exp[\mu_t s_1(1-\cos\theta)]ds_1 \\ &= C\exp[-\mu_t s] + \frac{I_0 p(\cos\theta)}{4\pi}\frac{\exp[-\mu_t s\cos\theta] - \exp[-\mu_t s]}{1-\cos\theta}, \end{aligned} \tag{2.64}$$

except for $\hat{s} = \hat{z}$ $(\cos\theta = 1)$, where it is

$$= C\exp[-\mu_t z] + \frac{I_0 p(1)}{4\pi}\mu_t z\exp[-\mu_t z] \,. \tag{2.65}$$

Along the path from 0 to $s$, only the forward-scattered energy from the plane wave can contribute. Thus, the calculated value of $I_d$ is only the forward-directed portion ($0 \le \theta < \pi/2$) of total diffuse field. So $I_d$ of Equations 2.64 and 2.65 is now denoted $I_d^{(+)}$. With the nature of the solution properly interpreted, the constant of integration is simple to determine. The forward-directed diffuse field vanishes at the front surface (because there is nothing behind it to scatter into it), so $C = 0$. The forward diffuse field is then

$$I_d^{(+)}(s,\hat{s}) = I_0\frac{p(\cos\theta)}{4\pi}\frac{\exp[-\mu_t s\cos\theta] - \exp[-\mu_t s]}{1-\cos\theta}\left(0<\theta<\frac{\pi}{2}\right), \tag{2.66}$$

and

$$I_d^{(+)}(z,\hat{z}) = I_0\frac{p(1)}{4\pi}\mu_t z\exp[-\mu_t z](\theta = 0) \,. \tag{2.67}$$

To solve for the diffuse field in the backscatter directions, $I_d^{(-)}$, the path must extend for the distance from $s$ to the back wall of the sample, $s_b$. Only the plane wave components ahead of $I_d^{(-)}(s)$ (i.e., further along the path than the depth $s$) can scatter in the backward direction and contribute to the $\pi/2 < \theta < \pi$ portion of the diffuse field. At the end of the path, the sample will be assumed to be sufficiently thick such that back wall effects (reflections) will not contribute to the diffuse fields. The general solution of Equation 2.61 now becomes

$$y(s) = Ce^{-\int_s^{s_b} P(s')ds'} + e^{-\int_s^{s_b} P(s')ds'}\int_s^{s_b} e^{-\int_{s'}^{s_b} P(s'')ds''} Q(s')ds' \,, \tag{2.68}$$

and $I_d^{(-)}$ is

$$I_d^{(-)}(s,\hat{s}) = C\exp\left[-\mu_t(s_b - s)\right] + I_0 \frac{p(\cos\theta)}{4\pi} \frac{\exp\left[-\mu_t s|\cos\theta|\right] - \exp\left[-\mu_t s_b\left(1+|\cos\theta|\right)+\mu_t s\right]}{1+|\cos\theta|} \quad (2.69)$$

for $\pi/2 < \theta \le \pi$, and for this range of angles using $\cos\theta = -|\cos\theta|$.

In the limit that $s_b >> s$ and $\mu_t s_b >> 1$, $I_d^{(-)}$ becomes

$$I_d^{(-)}(s,\hat{s}) = I_0 \frac{p(\cos\theta)}{4\pi} \frac{\exp\left[-\mu_t s|\cos\theta|\right]}{1+|\cos\theta|}. \quad (2.70)$$

Substituting the Henyey–Greenstein SPF for $p(\cos\theta)$, the total diffuse field as a function of depth $z = s\cos\theta$ (Figure 2.12) is

$$I_d(z,\cos\theta) = I_0 \frac{\mu_s}{\mu_t} \begin{cases} \dfrac{1-g^2}{\left(1+g^2-2g\cos\theta\right)^{\frac{3}{2}}} \dfrac{\exp\left[-\mu_t z\right]-\exp\left[-\mu_t z/\cos\theta\right]}{1-\cos\theta} & 0<\theta<\dfrac{\pi}{2} \\ \dfrac{1-g^2}{(1-g)^3}\mu_t z\exp\left[-\mu_t z\right] & \theta = 0 \\ \dfrac{1-g^2}{\left(1+g^2-2g\cos\theta\right)^{\frac{3}{2}}} \dfrac{\exp\left[-\mu_t z\right]}{1+|\cos\theta|} & \dfrac{\pi}{2}<\theta<\pi \end{cases}. \quad (2.71)$$

These fields are maximized at depths of

$$z_{max} = \begin{cases} -\dfrac{1}{\mu_t}\dfrac{\cos\theta\ln\left[\cos\theta\right]}{1-\cos\theta} & 0<\theta<\dfrac{\pi}{2} \\ \dfrac{1}{\mu_t} & \theta=0 \\ 0 & \dfrac{\pi}{2}<\theta<\pi \end{cases}. \quad (2.72)$$

With the Henyey–Greenstein SPF, one can see that the small diffuse field approximation may be violated for the forward fields when $g$ is close to 1. For example, at $g = 0.8$, in the $\theta = 0$ case

$$\frac{1-g^2}{(1-g)^3} = 45$$

and for $g = 0.95$

$$\frac{1-g^2}{(1-g)^3} = 780.$$

Thus the degree of anisotropy of the scattering should be taken into consideration before invoking the small albedo limit for a given problem.

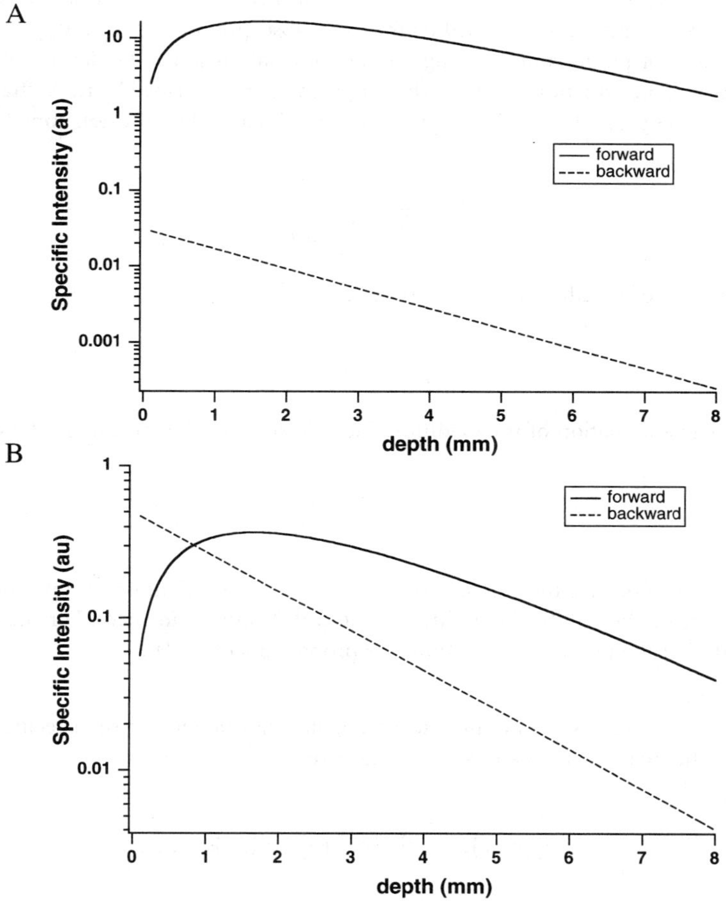


**FIGURE 2.12.** The forward ($\theta = 0°$) and backscattered ($\theta = 180°$) diffuse light in the absorption-dominant limit ($\mu_a >> \mu_s$) for $\mu_t \cong \mu_a = 0.6$ mm$^{-1}$. (A) $g = 0.8$ and (B) $g = 0.0$ (isotropic scattering).

### 2.3.3.3 Scattering-Dominant Limit: the "Diffusion Approximation"

When the absorption is sufficiently low to permit significant penetration of light into the tissue, scattering is the dominant transport process. This scattering-dominant limit, known as the *diffusion limit,* is important because photons are able to move through the tissue, although the strong scattering disperses the light in a random fashion. In the diffusion process, the particles (in our case photons) moving through a medium do so in a series of steps of random length and direction (i.e., random walk). Each step begins with a scattering event and is equally likely to be taken in any direction. This isotropic scattering is described by the reduced scattering coefficient $\mu_s'$, which is related to the previously defined anisotropic scattering parameters as follows:

$$\mu_s' = (1-g)\mu_s \,. \tag{2.73}$$

In an average sense, this relationship equates the number of anisotropic scattering steps, given by $m = 1/1 - g$, with one isotropic scattering event. For example, in a medium characterized by $g = 0.75$, it takes an average of four scattering events for a population of photons to disperse isotropically. In tissues, one encounters values of $g$ from 0.4 to >0.99, which results in isotropic dispersion of photons after as few as

2 steps to over 100 steps. The photons can also be absorbed as they propagate, and the absorption properties of the medium are encompassed by the usual absorption coefficient, $\mu_a$.

When the number of photons undergoing the random walk is large, the density of photons can be described as a continuous function in space whose dynamics are described by the diffusion equation. In the diffusion limit, the properties of the medium are contained in the diffusion constant,

$$D = \frac{c_m}{3\left(\mu_a + (1-g)\mu_s\right)}, \tag{2.74}$$

which has dimensions of length squared over time. The quantity

$$\mu_t' \equiv \mu_a + (1-g)\mu_s \tag{2.75}$$

is called the transfer attenuation of the medium. The diffusion coefficient can also be written as

$$D = \frac{1}{3} c_m l_t', \tag{2.76}$$

where $l_t' = 1/\mu_t'$ is the effective mean free path. The diffusion coefficient enters into the solutions for $I_d$ in factors that describe the width of the diffuse photon distribution in space. For the examples shown below, the "widths" of the photon distributions are proportional to $\sqrt{D}$.

***Diffusion Equation***

When the scattering processes are dominant, the angular dependence of the specific intensity is well approximated by the first-order expansion in the unit vector $\hat{\mathbf{s}}$,

$$I_d(\mathbf{r},\hat{\mathbf{s}},t) \cong \frac{1}{4\pi}\Phi_d(\mathbf{r},t) + \frac{3}{4\pi}F_d(\mathbf{r},t)\hat{\mathbf{s}}_f \cdot \hat{\mathbf{s}}, \tag{2.77}$$

where

$$\Phi_d(\mathbf{r},t) = \int_{4\pi} I_d(\mathbf{r},\hat{\mathbf{s}},t)\,d\Omega \tag{2.78}$$

is the total intensity at point $\mathbf{r}$ (also called the *fluence rate*), and

$$\mathbf{F}_d(\mathbf{r},t) = F_d(\mathbf{r},t)\hat{\mathbf{s}}_f = \int_{4\pi} I_d(\mathbf{r},\hat{\mathbf{s}},t)\hat{\mathbf{s}}\,d\Omega \tag{2.79}$$

is the net intensity vector. The total intensity can also be written as $\Phi_d(\mathbf{r},t) = h\nu c_m \eta_d(\mathbf{r},t)$, where $\eta_d(\mathbf{r},t)$ is the photon density. Similarly, the net intensity vector is proportional to the photon current density, $\mathbf{J}(\mathbf{r},t) = c_m\mathbf{F}_d(\mathbf{r},t)$. Integrating all the terms in the RT equation over the entire $4\pi$ of solid angle results in a new relation expressed in terms of $\Phi_d(\mathbf{r},t)$ and $\mathbf{F}_d(\mathbf{r},t)$:

$$\frac{1}{c_m}\frac{\partial}{\partial t}\Phi_d(\mathbf{r},t) + \vec{\nabla}\cdot\mathbf{F}_d(\mathbf{r},t) = -\mu_a\Phi_d(\mathbf{r},t) + Q_c + Q_s, \tag{2.80}$$

where $Q_c$ and $Q_s$ represent source terms due to the coherent field and local sources. This equation gives us an expression of the divergence of $\mathbf{F}_d$ in terms of the total intensity $\Phi_d$.

The essential step in obtaining Equation 2.80 from Equation 2.54 is the calculation of the integral over the SPF:

$$\frac{\mu_a+\mu_s}{4\pi}\iint_{4\pi} p(\hat{s},\hat{s}')I_d(\mathbf{r},\hat{s}',t)d\Omega' d\Omega =$$
$$\frac{\mu_a+\mu_s}{4\pi}\int_{4\pi}\left[\int_{4\pi} p(\hat{s},\hat{s}')d\Omega\right]I_d(\mathbf{r},\hat{s}',t)d\Omega' = \tag{2.81}$$
$$\mu_s\int_{4\pi} I_d(\mathbf{r},\hat{s}',t)d\Omega' = \mu_s\Phi_d(\mathbf{r},t).$$

This term then cancels out the scattering loss contribution on the right-hand side of the original RT equation (Equation 2.54). The next step is to define a second relation between $\mathbf{F}_d$ and $\mathbf{\Phi}_d$, and to use this new relation to eliminate $\mathbf{F}_d$. The lowest-order approximation is obtained by using Fick's law,

$$c_m\mathbf{F}_d(\mathbf{r},t) = -D\vec{\nabla}\Phi_d(\mathbf{r},t). \tag{2.82}$$

Fick's law asserts that the net photon current is proportional to the gradient of the photon density, or, in other words, when the photon density varies in space, there is a net flow of photons from high density to low density regions along the path of steepest descent. Substituting Equation 2.82 into Equation 2.80 yields the diffusion equation:

$$\frac{\partial}{\partial t}\Phi_d(\mathbf{r},t) = D\nabla^2\Phi_d(\mathbf{r},t) - \mu_a c_m\Phi_d(\mathbf{r},t) + Q_c + Q_s. \tag{2.83}$$

A higher-order approximation for the diffusion problem can be derived from the RT equation in which the constants *D*, $\mu'_{tr}$, and *g* arise naturally in the process. However, the resulting relation, known as the telegrapher's equation, has not been found to offer any significant advantages over the diffusion equation in tissue optics.[4]

Analytical solutions to the diffusion equation in practical situations can be quite complicated due to the finite sizes of the tissues and the related boundary effects. In the following section, we will examine solutions in some specific, ideal cases in order to demonstrate the nature of diffuse fields. The idealized results given below can often be used as starting points for the more complex situations that occur in solving practical problems.

***Time-Dependent Solutions***

In the following situation, time-dependent solutions of Equation 2.83 are given for the case in which material boundaries are far enough away to be neglected. A source is used to inject light at $t = 0$ but is removed immediately after and for all later times ($Q_c, Q_s = 0$, $t > 0$).

Following an injection of *N* photons into the medium at point $\mathbf{r}_0$, the solution of Equation 2.83 for $t > 0$ takes the form

$$\Phi_d(\mathbf{r},t) = h\nu c_m \frac{N\exp(-\mu_a c_m t)}{(4\pi D t)^{\frac{3}{2}}}\exp\left(-\frac{r^2}{4Dt}\right), \tag{2.84}$$

where $|r = \mathbf{r} - \mathbf{r}_0|$ (Figure 2.13). The spatial distribution of the light is Gaussian, with a time-varying $1/e$ radius of $r_{1/e} = \sqrt{4Dt}$. The numerator of the first factor, $N\exp(-\mu_a c_m t)$, contains the total number of photons that exist at time *t*, which is ever decreasing in time because of absorption. The denominator can be considered to be a volumetric normalization factor because it contains the factor $r_{1/e}^3 = (4Dt)^{3/2}$ is thus proportional to the volume of the sphere of radius $r_{1/e}$. At any time, and almost 90% of the photons will lie within the sphere of radius $r_{1/e}$. The photon density itself falls to half of its maximum inside of $r_{1/e}$:

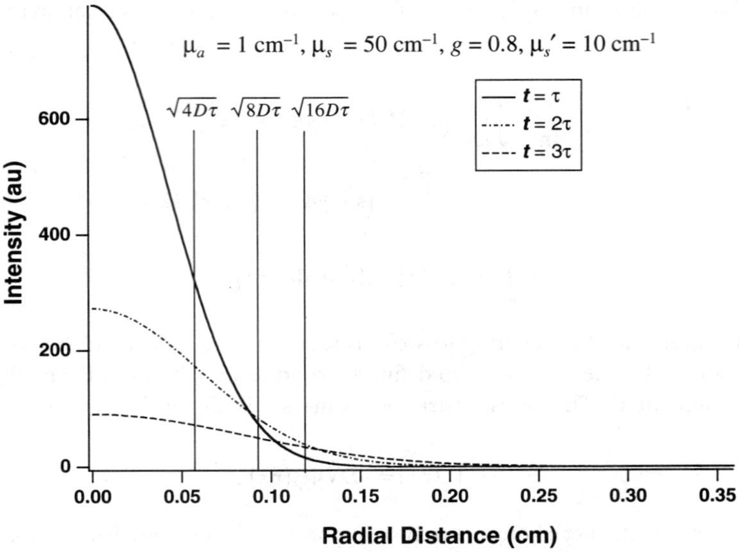


**FIGURE 2.13** The photon density in the time-dependent diffusion process due to the point injection of photons at $t = 0$. The three curves represent the solution for time $t = \tau$, $t = 2\tau$, and $t = 4\tau$.

$$r_{1/2} = 2\sqrt{\ln 2}\sqrt{Dt} = 1.67\sqrt{Dt} \, . \tag{2.85}$$

Toward the center of the distribution, the local photon density is decreasing with time, while farther out, the density is increasing with time. The boundary between the two regions (growth boundary) occurs at the radius:

$$r_g = \sqrt{6Dt\left(1 + \frac{2}{3}\mu_a c_m t\right)} \, . \tag{2.86}$$

For $r = |\mathbf{r} - \mathbf{r}_0| < r_g$, the photon density is decreasing with time, while for $r > r_g$ the photon density is increasing with time. This radius defines the boundary between the sphere from which energy is escaping and the surrounding region that sees a net increase in energy with time.

Similar solutions are available for other source geometries. For a line source of photons injected at $t = 0$, the solution is

$$h\nu c_m \frac{N \exp(-\mu_a c_m t)}{4\pi Dt} \exp\left(-\frac{(\rho - \rho_0)^2}{4Dt}\right), \tag{2.87}$$

where $\rho - \rho_0$ is the perpendicular distance from the source line to a point in the medium and $N$ now represents the number of photons emitted per unit of length. Here, the distribution has cylindrical symmetry, and 63% of the energy is within the cylinder of radius $\rho_{1/e} = \sqrt{4Dt}$ . The boundary between increasing and decreasing density occurs at

$$\rho_g = \sqrt{4Dt(1 + \mu_a c_m t)} \, . \tag{2.88}$$

For a planar source the solution is

$$hvc_m \frac{N \exp(-\mu_a c_m t)}{(4\pi Dt)^{1/2}} \exp\left(-\frac{(z-z_0)^2}{4Dt}\right), \tag{2.89}$$

where $N$ is now interpreted as the number of photons injected per unit area, and $z - z_0$ is the distance from the observation point to the plane. The boundary between increasing and decreasing density is planar and occurs at a distance of

$$z_g = \sqrt{2Dt(1+2\mu_a c_m t)} \tag{2.90}$$

in both directions normal to the source plane.

***Steady-State Solutions***

In the steady-state regime in a source free region, the diffusion equation becomes

$$\nabla^2 \Phi_d(\mathbf{r}) - \kappa_d^2 \Phi_d(\mathbf{r}) = 0, \tag{2.91}$$

where

$$\frac{1}{\kappa_d} = \sqrt{\frac{D}{\mu_a c_m}} = \sqrt{\frac{1}{3\mu_a(\mu_a + (1-g)\mu_s)}} \tag{2.92}$$

is the diffusion length. A point source of light at $r = 0$, injecting photons at a rate to maintain the steady state, has a photon density away from the source of

$$\frac{\Phi_d(\mathbf{r})}{hvc_m} = \frac{N\kappa_d^2}{4\pi} \frac{\exp[-\kappa_d r]}{r}. \tag{2.93}$$

In this case about 25% of the photons are inside the sphere of radius $1/\kappa_d$. A sphere of radius $4/\kappa_d$ contains 90% of the photons (see Figure 2.14).

For a line source the photon density is

$$\frac{\Phi_d(\mathbf{r})}{hvc_m} = \frac{N\kappa_d^2}{2\pi} K_0(\kappa_d \rho), \tag{2.94}$$

where $K_0(\kappa_d \rho)$ is the zero-order-modified Bessel function of the second kind. For $\kappa_d \rho << 1$,

$$K_0(\kappa_d \rho) \sim \ln \frac{1}{\kappa_d \rho},$$

while, for large $\kappa_d \rho$,

$$K_0(\kappa_d \rho) \sim \frac{\exp[-\kappa_d \rho]}{\sqrt{\kappa_d \rho}}.$$

A cylinder of radius $1/\kappa_d$ contains 44% of the photons and a cylinder of radius $3.2/\kappa_d$ holds 90%.

For a planar source, we have

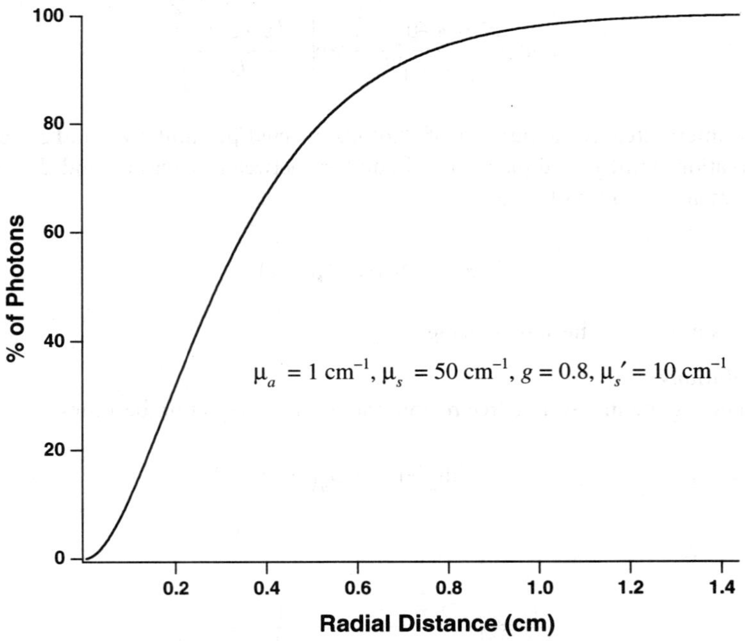


**FIGURE 2.14** The percentage of photons inside a spherical volume, plotted as a function of the radius of the sphere, for the point source solution of the steady-state diffusion problem.

$$\Phi_d(\mathbf{r}) = \frac{N\kappa_d}{2}\exp\left[-\kappa_d |z|\right]. \tag{2.95}$$

Here, 63% of the photons are within $1/\kappa_d$ of either side of the plane, and 90% are within $2.3/\kappa_d$.

Technically, these three solutions are the Green's functions for the steady-state diffusion equation

$$D\nabla^2\Phi_d^G(\mathbf{r},\mathbf{r}') - \mu_a c_m \Phi_d^G(\mathbf{r},\mathbf{r}') = -\delta(\mathbf{r}-\mathbf{r}') \tag{2.96}$$

for the various geometries (spherical, cylindrical, and planar). Because of their link to the Dirac delta function, they are considered distributions (i.e., generalized functions) as opposed to ordinary point functions. These solutions (Equations 2.93 through 2.95) can be used within the context of Green's theorem to solve problems involving finite sources and known boundary conditions.

### 2.3.4 Numerical Approach: Monte Carlo Simulations

Many problems of practical interest often involve a variety of light sources, multiple tissue types, and complex geometries. Analytical solutions for realistic scenarios are complicated at best, if even possible. These more realistic cases are instead solved with numerical techniques. For RT problems, the most widely used approach is the Monte Carlo (MC) method. The term *Monte Carlo* refers to a broad class of methods that employ random numbers in the course of solving a problem. Among the first large-scale applications of MC methods were RT problems involving neutron transport; thus, the MC approach is well suited to problems involving light transport in tissue because they are based on the same underlying framework of RT theory. The algorithms for implementing the basic elements of an MC simulation are straightforward. The technique thus possesses a great deal of flexibility and is widely applicable to practical transport problems. The MC approach can be used for any albedo and SPF.

The name refers to Monte Carlo, Monaco, where the primary attractions are casinos that offer games of chance. The random behavior in games of chance is similar to how MC simulation selects variable values at random to simulate a model. When one rolls a die, one knows that a number from 1 to 6 will come up, but not which number for any particular roll. The same phenomenon occurs with photon trajectories propagating inside tissues, which exhibit an uncertain value for any particular time or event (e.g., scattering, absorption). The simulation is based on the random walks that photons make as they travel through tissue, chosen by statistically sampling the probability distributions for step size and angular deflection per scattering event. As the number of photons grows, the net distribution of all the photon paths yields an increasingly accurate approximation to the RT problem.

In an MC approach to photon transport, single photons are traced through the sample step by step, and the distribution of light in the system is built from these single-photon trajectories. The parameters of each step (length, direction, "weight," etc.) are calculated using functions whose arguments are random numbers. As the number of photons in the simulation grows toward infinity, the MC prediction for the light distribution approaches an exact solution of the RT equation. The actual number of photons necessary for a realistic result (i.e., a good statistical sample) depends on the specifics of the simulation and which quantities one desires to determine. As few as 3000 may be adequate for determining diffuse reflectance from a sample, while more than 100,000 may be required for a complex three-dimensional simulation.[5]

As an example of how step parameters are calculated by this method, we will illustrate how the step size is determined. The exponential decrease in intensity with depth as expressed in Beer's law,

$$\frac{I(L)}{I_0} = p_L = \exp\left[-(\mu_a + \mu_s)L\right], \tag{2.97}$$

can be interpreted in the MC context as the probability $p_L$ $(0 \le p_L \le 1)$ that a photon will propagate a distance $L$ without being scattered or absorbed. This relation is then inverted to express $L$ in terms of the probability,

$$L = -\frac{\ln p}{\mu_a + \mu_s}, \tag{2.98}$$

and the probability is then treated as a random number. For each step, an $L$ is calculated by randomly selecting a number $p$ from a uniform distribution

$$f(p) = \begin{cases} 1 & 0 < p \le 1 \\ 0 & \text{otherwise} \end{cases}.$$

The expectation value (i.e., average) of $\ln p$ over the distribution $f(p)$ is –1; so, as the number of random samplings grows, the resulting mean free path will converge to $\bar{L} = 1/\mu_a + \mu_s$, as one would expect.

Monte Carlo methods have been used for various transport processes as well as for emission in tissues. For example, modeling laser-induced fluorescence involves an expression for the fraction of fluorescence emitted per unit depth that escapes from the medium. Accurate expressions for fluence rate and escape function for the one-dimensional case based upon MC simulation results have been reported.[6]

### 2.3.5 Kubelka–Munk Model

For optically inhomogeneous media, the expression for quantitative analysis can be derived from an equation developed for optical studies on surfaces with diffuse reflectance.[7] Schuster first conceived the continuum model and derived the basic differential equations for diffuse scattering materials to describe the behavior in a foggy atmosphere of radiation from a star.[8] Schuster's model described two diffuse fluxes traveling in opposite directions through the gas, contributing backscattered flux to each other. Kubelka and Munk further improved the model conceived by Schuster and derived differential equations similar to those developed by Schuster except for the emission terms.[9] Later, Kubelka developed a set of useful equations that provided the foundations for many quantitative studies of absorption, scattering, and luminescence processes in diffuse scattering media. These equations are often known as the Kubelka–Munk (K–M) equations:[10]

$$-di/dx = -(S + K)i + Sj, \tag{2.99}$$

$$dj/dx = -(S + K)j + Si, \tag{2.100}$$

where $i$ is the intensity of light propagating inside the sample in the forward (transmitted) direction, $j$ is the intensity of light propagating in the backscattered direction, $S$ is the scatter coefficient per unit thickness, $K$ is the absorption coefficient per unit thickness, and $x$ is the distance from the nonilluminated side.

The K–M equations simply describe the fact that the light beam traveling in the transmitted direction ($i$) decreases in intensity due to the absorption ($K$) and scattering ($S$) processes, and gains intensity from the scattering process that occurs in the beam coming from the other direction ($j$).

The basic assumptions in the K–M model are that the sample is a planar, homogeneous, and ideal diffuser illuminated on one side with diffuse monochromatic light. This model also assumes that the reflection and absorption processes occur at infinitesimal distances and are constant over the area under illumination and over the thickness of the sample. It is also implicitly assumed in the one-dimensional approximation of the K–M model that the reflected light beam is normal to the sample surface.

The general solutions in $i$ and $j$ for the K–M equations are given by

$$i = A \sinh bSx - B \cosh bSx, \tag{2.101}$$

$$j = (aA - bB)\sinh bSx - (aB - bA)\cosh bSx, \tag{2.102}$$

where

$$a = 1 + K/S \tag{2.103}$$

$$b = (a^2 - 1)^{1/2}. \tag{2.104}$$

The arbitrary constants $A$ and $B$ can be eliminated using the following conditions:

$$i = I_0 T\ ; \quad j = 0 \quad \text{for } x = 0 \tag{2.105}$$

$$i = I_0\ ; \quad j = I_0 R \quad \text{for } x = X\ . \tag{2.106}$$

The transmittance $T$ and reflectance $R$ are then given by

$$T = b(a \sinh bSX + b \cosh bSX)^{-1}, \tag{2.107}$$

$$\mathrm{R} = \sinh bSX(a \sinh bSX + b \cosh bSX)^{-1}, \tag{2.108}$$

where $X$ is the thickness of the sample.

The solution for $T$ and $R$ can be used to determine the coefficient $SX$.

For a scattering medium that does not absorb light (i.e., when $K = 0$) the transmission $T$ and reflectance $R$ become

$$T_0 = 1(SX + 1)^{-1}, \tag{2.109}$$

$$R_0 = SX(SX + 1)^{-1}. \tag{2.110}$$

The value of $SX$ can be determined using the above equations. In general $T_0$ can be derived from measurement of the absorbance $A_0 = \log 1/T_0$.

There has been a great interest in using the K–M model for investigating optical properties of diffuse media.[11–29] A simple modification of the two-flux K–M model for the solution of one-dimensional problems to the theory of radiation transfer in biological tissues and media was reported. The results were compared with the exact solution and experimental data for the substantiation of the applicability of the model to the problems of noninvasive laser diagnostics in medicine.[14] A method utilizing the K–M theory was developed to allow the quantification of tissue reflectance spectra to study *in vivo* kinetic changes in the oxygen saturation of hemoglobin and myoglobin.[16] The K–M model was used for investigating quantitative optical biopsy of tissue *ex vivo*.[19] Later models use four fluxes that, in addition to the diffuse forward and backward fluxes, also involve collimated forward and backward fluxes. Multiple-flux models provide data in good agreement with experimental results; however, the applicability of the K–M models is limited to simple slab geometries.

## 2.3.6 Time-Resolved Propagation of Light Pulses

When a light pulse is launched into a slab of tissue, the light that exits the sample can be broken into coherent and incoherent fields, as discussed previously. However, in contrast to the approximate RT model, some of the multiply scattered light retains phase coherence with the unscattered portion. In the forward direction, the interference of the coherent scatter and the unscattered light will effectively alter the refractive index of the medium. The coherent light is also called the "ballistic" component because it follows a straight path through the tissue. In the backscatter direction, there will be coherent components due to reflections from tissue boundaries and scattering. The remaining light will be diffuse. In the forward direction, the first part of the diffuse component to emerge consists of "snake" photons, so called because their paths approximate a straight line as they forwardscatter through the medium. If the detector used to observe the exiting light has sufficient time resolution, these photons can be isolated from the later-arriving components, consisting of a cloud of photons that have followed winding and varied paths through the slab.

The ballistic photons are the first to exit the tissue, and their mostly predictable paths (usually straight lines, although refraction may occur in multilayered samples) make them useful for absorption mapping and tomographic imaging (see Figure 2.15). The snake photons can also be used for mapping and tomography because their paths are nearly straight. Snake photons have the advantage over ballistic photons in scattering-dominant regimes because the ballistic component is more strongly attenuated as the tissue thickness increases. Most of the remaining multiple-scattering photons take random paths through the tissue; these paths can only be characterized statistically. Any information regarding local variations in the tissue (e.g., embedded tumors) carried by these later photons is dispersed and may be diluted in the cloud of diffuse light.

The specific conditions under which an experimental technique is considered time resolved depends upon the application. For example, if the aim is to be able to separate the ballistic, snake and diffuse components in a through-transmission experiment across a 1-cm sample, then the pulse width of light should be very short, typically on the order of ps ($10^{-12}$ s) or less.

### 2.3.6.1 Effective Index of Refraction

Since some portion of the scattered field, even if multiply scattered, can maintain phase coherence with the initial field, the coherent portion of the scattered field will interfere with the original unscattered light. This interference manifests itself as a phase shift, and hence a time delay for time-resolved pulses. This implies that the speed of light in the medium, and hence the effective index of refraction, is affected by the scatterers. For a slab of tissue with no embedded scatterers, the time a photon takes to traverse the tissue is

$$\tau_0 = \frac{n_m}{c} L , \tag{2.111}$$

where $L$ is the width of the tissue sample and $n_m$ is the index of refraction for the medium. When scatterers are present, the time of propagation becomes

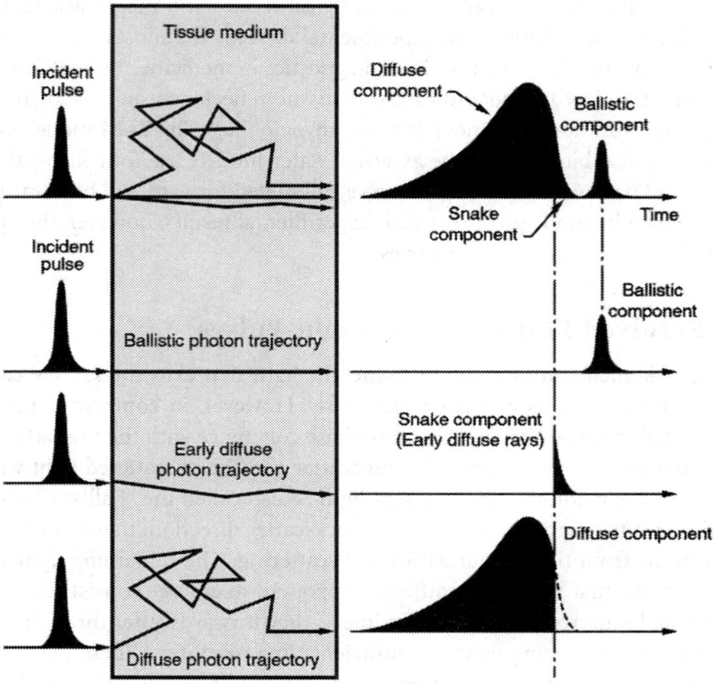


**FIGURE 2.15** Paths of ballistic, snake, and diffusion photons propagating through tissue. For a short pulse of light transmitted through tissue, the ballistic component emerges first, followed by the diffuse (multiply scattered) component. The early part of the diffuse component is known as the "snake" component because the paths of these photons approximate straight lines through the tissue.

$$\tau_0 = \frac{n_{\text{eff}}}{c} L , \tag{2.112}$$

where $n_{\text{eff}}$ is the *effective refractive index* of the two-component medium. The question then is how to calculate $n_{\text{eff}}$. A commonly used approximation is to compute the effective index of refraction as the volume-weighted average of its constituents. For our two-component medium, the effective index is then

$$n_{\text{eff}} = (1 - V) n_m + V n_{sc} , \tag{2.113}$$

where $V$ is the volume fraction of the scatterers embedded in the medium. This can also be written as

$$n_{\text{eff}} = n_m (1 + V\delta) , \tag{2.114}$$

where $n_{sc}/n_m = 1 + \delta$ and $\delta$ is the fractional contrast between the two indices. This volume-weighted approximation, however, does not take into account the coherent interference effects of the scattered field on the unscattered field. This interference will delay (or advance) the light over the medium alone, and is consistent with the volume-weighted approach only in limiting cases. However, the volume-weighted average approach is likely an adequate approximation in many cases of interest.

There are several theoretical formulations of the problem of defining the speed of ballistic transport across a multiply scattering medium. One such relation is that of Waterman and Truell,[30] who extended the earlier results of Foldy and Lax. The interference of the coherent scatter and the incident field gives rise to the following relation for the effective index of refraction of the medium:

$$n_{eff}^2 = n_m^2\left(\left(1+\frac{\rho\lambda^2}{2\pi n_m^2}f(0)\right)^2 - \left(\frac{\rho\lambda^2}{2\pi n_m^2}f(\pi)\right)^2\right), \tag{2.115}$$

where $f(\theta)$ is the scattering amplitude and is related to the differential cross section as

$$|f(\theta)|^2 = \frac{d\sigma}{d\Omega}(\theta). \tag{2.116}$$

For Rayleigh scatterers, using $|f(\theta)| = |f(\pi)|$ and Equation 2.14, the relation yields

$$\begin{aligned} n_{eff}^2 &= n_m^2\left(1+\frac{\rho\lambda^2}{\pi n_m^2}f(0)\right) \\ &= n_m^2\left(1+3V\frac{n_{sc}^2-n_m^2}{n_{sc}^2+2n_m^2}\right), \end{aligned} \tag{2.117}$$

where $V = \rho 4/3\pi a^3$ is the volume fraction of scatterers in the medium. Taking the square root,

$$n_{\text{eff}} = n_m\sqrt{1+2V\delta\frac{1+\delta/2}{1+2\delta/3+\delta^2/3}}. \tag{2.118}$$

For small $V\delta$, this can be approximated to lowest order in $\delta$,

$$n_{\text{eff}} = n_m(1+V\delta), \tag{2.119}$$

which is equivalent to the volume-weighted result. This result is essentially the single-scattering approximation to the problem, as it is linear in the volume fraction (and hence in the number) of scatterers. The question is then how well the approximate relation performs over realistic ranges of $\delta$ and $V$. Refractive indices for tissue constituents range from 1.33 (water) to above 1.6 (melanin particles), so 0.2 could be considered a relatively large value for $\delta$. A comparison of Equation 2.114 with Equation 2.118 for a volume fraction of 25% shows the two expressions to be the same to better than 1% for $\delta$ of up to 0.36. At a volume fraction of 10%, the approximation is good to 1% for $\delta$ up to 0.6. For Rayleigh scatterers, using reasonable values for $V$ and $\delta$, the volume-weighted approach can be accurate to at least the 1% level, if not better.

Another issue for pulse propagation is dispersion, the change in the pulse shape for the ballistic component. In this case, the notion of a single velocity is somewhat ill defined. As many as five different velocities (e.g., phase, group, signal, energy, etc.) have been defined for pulse transmission through dispersive media. We will consider two of these: phase velocity and group velocity. Phase velocity is defined as

$$c_m(\nu) = \frac{c_0}{n(\nu)} \tag{2.120}$$

and is the speed points of constant phase (e.g., the crests of the wave) for each individual wavelength component in the pulse. This can also be written as $1/(k_m/\omega)$, where $\omega = 2\pi\nu$ and $k_m = 2\pi/\lambda_m$. The speed of the envelope can be difficult to define because it changes shape in the material. A reasonable approximation to the speed of the pulse envelope is given by the group velocity, which is defined as

$$c_m^{gr} = \frac{1}{dk_m/d\omega} \tag{2.121}$$

$$= \frac{c_0}{n(\lambda)} \frac{1}{1 - \frac{\lambda}{n(\lambda)} \frac{d}{d\lambda} n(\lambda)}. \tag{2.122}$$

One can use this relation to define the group index of refraction,

$$n_g(\lambda) = n(\lambda)\left(1 - \frac{\lambda}{n(\lambda)} \frac{d}{d\lambda} n(\lambda)\right). \tag{2.123}$$

Variations in absorption and scattering in a material will give rise to index of refraction changes as they are fundamentally linked by causality. For cases where $n(\lambda)$ is known as a function of wavelength, the group index may be a more accurate estimate of the pulse speed for pulses whose dominant wavelength component is $\lambda$.

## 2.4 Tissue Properties

### 2.4.1 Refractive Indices

The refractive index for a medium determines the speed of light in the medium, and changes in the refractive index, either continuous or abrupt (i.e., at a boundary), give rise to scattering, refraction, and reflection. Since tissues are heterogeneous in composition, one may need to know the refractive indices for the various tissue constituents (e.g., to calculate scattering properties) or an "averaged" value for the tissue as a whole (e.g., the time delay of ballistic photons). The effective refractive index of a tissue is often approximated as the volume-weighted average of the values of its constituents, a reasonable estimate in many cases (see the discussion in Section 2.3.6.1).

Water makes up a significant portion of most tissues, and its index ($n = 1.33$) represents the minimum value for fluids and soft-tissue constituents. Among the other soft-tissue components, melanin particles, found mostly in the epidermal layer of the skin (see Section 2.4.3.2 below), are at the high end of the refractive index scale, with reported values above 1.6. Indices of refraction for whole tissues themselves — including parts of the brain, aorta, lung, stomach, kidney, and bladder — fall in the range of 1.36 to 1.40. Extracellular fluids and intracellular cytoplasm fall in the range of 1.35 to 1.38. The index of refraction for fatty tissues is around 1.45. The membranes that enclose cells and subcellular organelles are composed largely of lipids, and the index mismatch between the cytoplasm and these lipid structures is the origin of much of the scattering in cellular tissues. For hard tissues, the index for tooth enamel has been measured as 1.62 within the visible spectrum. Specific values for various bones in the body are not widely reported. The refractive indices of various tissues and tissue components are listed in Table 2.1.

### 2.4.2 Scattering Properties

As discussed previously, scattering occurs where there is a spatial variation in the refractive index, either continuous or abrupt (e.g., due to localized particles). In cellular media, the important scatterers are the subcellular organelles. The size range exhibited by these organelles includes the wavelengths of the therapeutic window, as their dimensions run from <100 nm to 6 μm. As scatterers, most of these structures fall in the Mie regime, exhibiting highly anisotropic forward-directed scattering patterns.

The mitochondria, the site of respiration and energy production in cells, are the dominant scatterers among the organelles. These structures are roughly cylindrical and vary in size depending on cell type, with characteristic dimensions of 0.5 to 2 μm. The mitochondria, in addition to being enclosed in a lipid

**TABLE 2.1** Index of Refraction for Various Tissues, Tissue Components, and Biofluids

| Tissue | Description | Wavelength (nm) | Index of Refraction | Ref. |
|---|---|---|---|---|
| Skin | Human stratum corneum | 400–700 | 1.55 | 40 |
| | Rat | 456–1064 | 1.42 | 41 |
| | Mouse | 456–1064 | 1.4 | 41 |
| Human brain | Gray matter | 456–1064 | 1.36 | 41 |
| | White matter | 456–1064 | 1.38 | 41 |
| | White and gray | 456–1064 | 1.37 | 41 |
| Human aorta | Normal intima | 456–1064 | 1.39 | 41 |
| | Normal media | 456–1064 | 1.38 | 41 |
| | Normal adventitia | 456–1064 | 1.36 | 41 |
| | Calcified intima | 456–1064 | 1.39 | 41 |
| | Calcified media | 456–1064 | 1.53 | 41 |
| Human bladder | Mucous | 456–1064 | 1.37 | 41 |
| | Wall | 456–1064 | 1.4 | 41 |
| | Integral | 456–1064 | 1.38 | 41 |
| Human colon | Muscle | 456–1064 | 1.36 | 41 |
| | Submucous | 456–1064 | 1.36 | 41 |
| | Mucous | 456–1064 | 1.38 | 41 |
| | Integral | 456–1064 | 1.36 | 41 |
| Human esophagus | Mucous | 456–1064 | 1.37 | 41 |
| Human fat | Subcutaneous | 456–1064 | 1.44 | 41 |
| | Abdominal | 456–1064 | 1.46 | 41 |
| Bovine fat | | 633 | 1.455 | 42 |
| Human heart | Trabecula | 456–1064 | 1.4 | 41 |
| | Myocardial | 456–1064 | 1.38 | 41 |
| Human femoral vein | | 456–1064 | 1.39 | 41 |
| Kidney | Human | 456–1064 | 1.37 | 41 |
| | Human | 633 | 1.417 | 42 |
| | Canine | 633 | 1.4 | 42 |
| | Porcine | 633 | 1.39 | 42 |
| | Bovine | 633 | 1.39 | 42 |
| Liver | Human | 456–1064 | 1.38 | 42 |
| | Human | 633 | 1.367 | 41 |
| | Canine | 633 | 1.38 | 42 |
| | Porcine | 633 | 1.39 | 42 |
| | Bovine | 633 | 1.39 | 42 |
| Lung | Human | 456–1064 | 1.38 | 41 |
| | Canine | 633 | 1.38 | 42 |
| | Porcine | 633 | 1.38 | 42 |
| Muscle | Human | 456–1064 | 1.37 | 41 |
| | Canine | 633 | 1.4 | 42 |
| | Bovine | 633 | 1.41 | 42 |
| Spleen | Human | 456–1064 | 1.37 | 41 |
| | Canine | 633 | 1.4 | 42 |
| | Porcine | 633 | 1.4 | 42 |
| Human stomach | Muscle | 456–1064 | 1.39 | 41 |
| | Mucous | 456–1064 | 1.38 | 41 |
| | Integral | 456–1064 | 1.38 | 41 |
| Human cerebrospinal fluid | | 400–700 | 1.335 | 40 |
| Human blood | | 633 | 1.4 | 42 |
| Cytoplasm | | 400–700 | 1.350–1.367 | 40 |
| Human eye | Aqueous humor | 400–700 | 1.336 | 40 |
| Cornea | Integral | 400–700 | 1.376 | 40 |
| | Fibrils | 400–700 | 1.47 | 40 |
| | Ground substance | 400–700 | 1.35 | 40 |
| Lens | Surface | 400–700 | 1.386 | 40 |
| | Center | 400–700 | 1.406 | 40 |

TABLE 2.1 Index of Refraction for Various Tissues, Tissue Components, and Biofluids (continued)

| Tissue | Description | Wavelength (nm) | Index of Refraction | Ref. |
|---|---|---|---|---|
| | Sclera | 442–1064 | 1.47–1.36 | 41 |
| | Vitreous humor | 400–700 | 1.336 | 40 |
| | Tears | 400–700 | 1.3361–1.3379 | 40 |
| Human tooth | Enamel | 220 | 1.73 | 40 |
| | Enamel | 400–700 | 1.62 | 40 |
| | Apatite | 400–700 | >1.623 | 40 |
| Human hair shaft | Black | 850 | 1.59(0.08) | 43 |
| | Red | 850 | 1.56(0.01) | 43 |
| Tooth | Enamel | 842–858 | 1.65 | 44 |
| | Dentin | 842–858 | 1.54 | 44 |
| Nail | | 842–858 | 1.51 | 44 |

membrane, also have lipid folds running inside them. This structure gives these organelles a high optical contrast to the surrounding cytoplasm and produces the observed strong scattering effects.

The largest of the cellular organelles is the cell nucleus, with a diameter in the 4- to 6-μm range. The endoplasmic reticulum and Golgi apparatus are some of the larger organelles. These two organelles also pinch off small submicron vesicles (spherical membrane-enclosed containers) that transport their various molecular products through the cell. These smaller vesicle-type organelles include lysosomes and peri-oxisomes that are 0.25 to 0.5 μm in diameter.

The cells vary in shape and size among the different tissue types with dimensions of a few microns and larger. An isolated cell can be a strong scatterer, but within a tissue the scattering is largely subcellular in origin. In skin, melaninosomes are important scattering structures ranging in size from 100 nm to 2 μm. These species contain melanin particles, which are strung together as beads. (Melanin is discussed further in Section 2.4.3.2.)

In blood, the disk-shaped red cells (erythrocytes) are the strongest scatterers. The erythrocyte disk is ~2 μm thick with a diameter of 7 to 9 μm. The scattering properties of blood are dependent on the hematocrit (volume fraction of red cells) and its degree of agglomeration.

Support tissues, consisting of cells and extracellular proteins such as elastin and collagen, provide mechanical strength and durability. The scattering properties of these tissues arise from the small-scale inhomogeneities and the large-scale variations in the structures they form. The characteristic sizes of the small-scale variations are subwavelength, and the scattering is of the Rayleigh type. For example, collagen fibrils have a banded structure that produces Rayleigh scattering and has a periodicity of 70 nm, on the order of 10 times smaller than the wavelengths in the diagnostic window.[31] The scattering coefficients ($\mu_s$) and average cosines of scatter ($g$) for various tissues are given in Table 2.2.

## 2.4.3 Absorption Properties

### 2.4.3.1 The Therapeutic Window

The ability of light to penetrate tissues depends on how strongly the tissues absorb light. Within the spectral range known as the therapeutic window (or diagnostic window), most tissues are sufficiently weak absorbers to permit significant penetration of light. This window extends from 600 to 1300 nm, from the orange region of the visible spectrum into the NIR. At the short-wavelength end, the window is bound by the absorption of hemoglobin, in both its oxygenated and deoxygenated forms. The absorption of oxygenated hemoglobin increases approximately two orders of magnitude as the wavelength shortens in the region around 600 nm. At shorter wavelengths many more absorbing biomolecules become important, including DNA and the amino acids tryptophan and tyrosine. At the IR end of the window, penetration is limited by the absorption properties of water. Within the therapeutic window, scattering is dominant over absorption, and so the propagating light becomes diffuse, although

**TABLE 2.2** Scattering Parameters for the Various Tissues, Tissue Components, and Biofluids

| Tissue | Description | Wavelength (nm) | $\mu_s$ (1/cm) | Uncert. (1/cm) | $\mu_s'$ (1/cm) | Uncert. (1/cm) | *g* | Uncert. | Notes | Ref. |
|---|---|---|---|---|---|---|---|---|---|---|
| Skin, *in vitro* | Stratum corneum | 250 | 2600 | | | | 0.9 | | | 45, 46* |
| | Stratum corneum | 308 | 2400 | | | | 0.9 | | | 45, 46* |
| | Stratum corneum | 337 | 2300 | | | | 0.9 | | | 45, 46* |
| | Stratum corneum | 351 | 2200 | | | | 0.9 | | | 45, 46* |
| | Stratum corneum | 400 | 2000 | | | | 0.9 | | | 45, 46* |
| | Epidermis | 250 | 2000 | | | | 0.69 | | | 45, 46* |
| | Epidermis | 308 | 1400 | | | | 0.71 | | | 45, 46* |
| | Epidermis | 337 | 1200 | | | | 0.72 | | | 45, 46* |
| | Epidermis | 351 | 1100 | | | | 0.72 | | | 45, 46* |
| | Epidermis | 415 | 800 | | | | 0.74 | | | 45, 46* |
| | Epidermis | 488 | 600 | | | | 0.76 | | | 45, 46* |
| | Epidermis | 514 | 600 | | | | 0.77 | | | 45, 46* |
| | Epidermis | 585 | 470 | | | | 0.79 | | | 45, 46* |
| | Epidermis | 633 | 450 | | | | 0.8 | | | 45, 46* |
| | Epidermis | 800 | 420 | | | | 0.85 | | | 45, 46* |
| | Dermis | 250 | 833 | | | | 0.69 | | | 45, 46* |
| | Dermis | 308 | 583 | | | | 0.71 | | | 45, 46* |
| | Dermis | 337 | 500 | | | | 0.72 | | | 45, 46* |
| | Dermis | 351 | 458 | | | | 0.72 | | | 45, 46* |
| | Dermis | 415 | 320 | | | | 0.74 | | | 45, 46* |
| | Dermis | 488 | 250 | | | | 0.76 | | | 45, 46* |
| | Dermis | 514 | 250 | | | | 0.77 | | | 45, 46* |
| | Dermis | 585 | 196 | | | | 0.79 | | | 45, 46* |
| | Dermis | 633 | 187.5 | | | | 0.8 | | | 45, 46* |
| | Dermis | 800 | 175 | | | | 0.85 | | | 45, 46* |
| | Skin:blood | 517 | 468 | | | | 0.995 | | | 47 |
| | Skin:blood | 585 | 467 | | | | 0.995 | | | 47 |
| | Skin:blood | 590 | 466 | | | | 0.995 | | | 47 |
| | Skin:blood | 595 | 465 | | | | 0.995 | | | 47 |
| | Skin:blood | 600 | 464 | | | | 0.995 | | | 47 |
| | Dermis (leg) | 635 | 244 | 21 | 78 | | 0.68 | | | 48 |
| | Dermis | 749 | | | 23.1 | 0.75 | | | | 49 |
| | Dermis | 789 | | | 22.8 | 1.29 | | | | 49 |
| | Dermis | 836 | | | 15.9 | 2.16 | | | | 49 |
| | Dermis | 633 | | | 11.64 | | 0.97 | | | 50 |

**TABLE 2.2** Scattering Parameters for the Various Tissues, Tissue Components, and Biofluids (continued)

| Tissue | Description | Wavelength (nm) | $\mu_s$ (1/cm) | Uncert. (1/cm) | $\mu_s'$ (1/cm) | Uncert. (1/cm) | $g$ | Uncert. | Notes | Ref. |
|---|---|---|---|---|---|---|---|---|---|---|
| | Dermis | 700 | | | 21.3 | 3.7 | | | | 51 |
| | Dermis | 633 | | | 23.8 | 3.3 | | | | 51 |
| | Dermis | 633 | | | 50.2 | | | | | 50 |
| | Skin and underlying tissues, including vein wall (leg) | 633 | 70.7 | | 11.4 | | 0.8 | | | 52 |
| | Caucasian male skin ($n$ = 3) | 500 | | | 50 | | | | | 53 |
| | Caucasian male skin ($n$ = 3) | 810 | | | 15.8 | | | | | 53 |
| | Cascasian male skin ($n$ = 3), ext. pressure 9.8 kPa | 500 | | | 167.4 | | | | | 53 |
| | Cascasian male skin ($n$ = 3), ext. pressure 9.8 kPa | 810 | | | 52.7 | | | | | 53 |
| | Cascasian male skin ($n$ = 3), ext. pressure 98.1 kPa | 500 | | | 156.7 | | | | | 53 |
| | Cascasian male skin ($n$ = 3), ext. pressure 98.1 kPa | 810 | | | 53.7 | | | | | 53 |
| | Caucasian female skin ($n$ = 3) | 500 | | | 23.9 | | | | | 53 |
| | Caucasian female skin ($n$ = 3) | 810 | | | 8.2 | | | | | 53 |
| | Caucasian female skin ($n$ = 3), ext. pressure 9.8 kPa | 500 | | | 31.5 | | | | | 53 |
| | Caucasian female skin ($n$ = 3), ext. pressure 9.8 kPa | 810 | | | 11.3 | | | | | 53 |
| | Caucasian female skin ($n$ = 3), ext. pressure 98.1 kPa | 500 | | | 40.2 | | | | | 53 |
| | Caucasian female skin ($n$ = 3), ext. pressure 98.1 kPa | 810 | | | 13.1 | | | | | 53 |
| | Hispanic male skin ($n$ = 3) | 500 | | | 24.2 | | | | | 53 |
| | Hispanic male skin ($n$ = 3) | 810 | | | 7.5 | | | | | 53 |
| | Hispanic male skin ($n$ = 3), ext. pressure 9.8 kPa | 500 | | | 37.6 | | | | | 53 |
| | Hispanic male skin ($n$ = 3), ext. pressure 9.8 kPa | 810 | | | 11.4 | | | | | 53 |
| | Hispanic male skin ($n$ = 3), ext. pressure 98.1 kPa | 500 | | | 40.4 | | | | | 53 |
| | Hispanic male skin ($n$ = 3), ext. pressure 98.1 kPa | 810 | | | 10.2 | | | | | 53 |

| | | | | | | |
|---|---|---|---|---|---|---|
| | Forearm:fat | 633 | 12 | | *ex vivo* | 54 |
| | Forearm:muscle | 633 | 5.3 | | *ex vivo* | 54 |
| Skin, *in vivo* | Skin (dermis) | 660 | 9–14.5 | | | 51 |
| | Skin | 633 | 32 | | | 55 |
| | Skin | 700 | 28.7 | | | 55 |
| | Skin (0–1 mm) | 633 | 16.2 | | | 47 |
| | Skin (1–2 mm) | 633 | 12 | | | 47 |
| | Skin (>2 mm) | 633 | 5.3 | | | 47 |
| | Forearm:epidermis | 633 | 17.5 | | | 54 |
| | Forearm:dermis | 633 | 17.5 | | | 54 |
| | Forearm (5 subjects, 14 measurements) | 800 | 6.8 | 0.8 | | 49 |
| | Arm | 633 | 9.08 | 0.05 | | 55 |
| | Arm | 660 | 8.68 | 0.05 | | 55 |
| | Arm | 700 | 8.14 | 0.05 | | 55 |
| | Foot sole | 633 | 11.17 | 0.09 | | 55 |
| | Foot sole | 660 | 10.45 | 0.09 | | 55 |
| | Foot sole | 700 | 9.52 | 0.08 | | 55 |
| | Forehead | 633 | 16.72 | 0.09 | | 55 |
| | Forehead | 660 | 16.16 | 0.08 | | 55 |
| | Forehead | 700 | 15.38 | 0.06 | | 55 |
| Brain, *in vitro* | White matter (female, 32 yrs. old, 24 h post mortem) | 415 | 24 | | | 46, 56* |
| | White matter (female, 32 yrs. old, 24 h post mortem) | 488 | 60 | | | 46, 56* |
| | White matter (female, 32 yrs. old, 24 h post mortem) | 630 | 32 | | | 46, 56* |
| | White matter (female, 32 yrs. old, 24 h post mortem) | 800–1100 | 40–20 | | | 46, 56* |
| | White matter (female, 63 yrs. old, 30 h post mortem) | 488 | 25 | | | 46, 56* |
| | White matter (female, 63 yrs. old, 30 h post mortem) | 633 | 22 | | | 46, 56* |
| | White matter (female, 63 yrs. old, 30 h post mortem) | 800–1100 | 20–10 | | | 46, 56* |
| | Gray matter (male, 71 yrs. old, 24 h post mortem) | 514 | 85 | | | 46, 56* |

**TABLE 2.2** Scattering Parameters for the Various Tissues, Tissue Components, and Biofluids (continued)

| Tissue | Description | Wavelength (nm) | $\mu_s$ (1/cm) | Uncert. (1/cm) | $\mu_s'$ (1/cm) | Uncert. (1/cm) | $g$ | Uncert. | Notes | Ref. |
|---|---|---|---|---|---|---|---|---|---|---|
| | Gray matter (male, 71 yrs. old, 24 h post mortem) | 585 | | | 63 | | | | | 46, 56* |
| | Gray matter (male, 71 yrs. old, 24 h post mortem) | 630 | | | 52 | | | | | 46, 56* |
| | Gray matter (male, 71 yrs. old, 24 h post mortem) | 800–1100 | | | 45–20 | | | | | 46, 56* |
| | Glioma (male, 65 yrs. old, 4 h post mortem) | 415 | | | 6 | | | | | 46, 56* |
| | Glioma (male, 65 yrs. old, 4 h post mortem) | 488 | | | 3 | | | | | 46, 56* |
| | Glioma (male, 65 yrs. old, 4 h post mortem) | 630 | | | 3 | | | | | 46, 56* |
| | Glioma (male, 65 yrs. old, 4 h post mortem) | 800–1100 | | | >1–2 | | | | | 46, 56* |
| | Melanoma (male, 71 yrs. old, 24 h post mortem) | 585 | | | 158 | | | | | 46, 56* |
| | Melanoma (male, 71 yrs. old, 24 h post mortem) | 630 | | | 75 | | | | | 46, 56* |
| | Melanoma (male, 71 yrs. old, 24 h post mortem) | 800 | | | 40 | | | | | 46, 56* |
| | Melanoma (male, 71 yrs. old, 24 h post mortem) | 900 | | | 30 | | | | | 46, 56* |
| | Melanoma (male, 71 yrs. old, 24 h post mortem) | 1100 | | | 25 | | | | | 46, 56* |
| | White matter | 633 | 532 | 41 | 91 | 5 | 0.82 | 0.01 | Freshly resected slabs | 46 |
| | White matter | 1064 | 469 | 34 | 60.3 | 2.5 | 0.87 | 0.007 | Freshly resected slabs | 46 |
| | Gray matter | 633 | 354 | 37 | 20.6 | 2 | 0.94 | 0.004 | Freshly resected slabs | 46 |
| | Gray matter | 1064 | 134 | 14 | 11.8 | 9 | 0.9 | 0.007 | Freshly resected slabs | 46 |
| | Adult head: scalp and skull | 800 | | | 20 | | | | | 57 |
| | Adult head: cerebrospinal fluid | 800 | | | 0.1 | | | | | 57 |
| | Adult head: gray matter | 800 | | | 25 | | | | | 57 |
| | Adult head: white matter | 800 | | | 60 | | | | | 57 |
| | White matter ($n$ = 7) | 360 | 402 | 91.9 | | | 0.702 | 0.093 | | 58 |
| | White matter ($n$ = 7) | 640 | 408.2 | 88.5 | | | 0.84 | 0.05 | | 58 |
| | White matter ($n$ = 7) | 860 | 353.1 | 68.1 | | | 0.871 | 0.028 | | 58 |

| | | | | | | | | |
|---|---|---|---|---|---|---|---|---|
| White matter ($n = 7$) | 1060 | 299.5 | 70.1 | | 0.889 | 0.01 | | 58 |
| White matter | 850 | 140 | 14 | | 0.95 | 0.02 | | 59 |
| White matter | 1064 | 110 | 11 | | 0.95 | 0.02 | | 59 |
| White matter | 1064 | 513 | 2.6 | | 0.96 | | | 58 |
| White matter, coagulated ($n = 7$) | 360 | 604.2 | 131.5 | | 0.8 | 0.089 | Coag. 2 h 80°C | 58 |
| White matter, coagulated ($n = 7$) | 860 | 417 | 272.5 | | 0.922 | 0.025 | Coag. 2 h 80°C | 58 |
| White matter, coagulated ($n = 7$) | 1060 | 363.3 | 226.8 | | 0.93 | 0.015 | Coag. 2 h 80°C | 58 |
| White matter, coagulated | 850 | 170 | 10.2 | | 0.94 | 0.02 | Homogenized tissue, coag. 75°C | 59 |
| White matter, coagulated | 1064 | 130 | 9.1 | | 0.93 | 0.02 | Homogenized tissue, coag. 75°C | 59 |
| Gray matter ($n = 7$) | 360 | 141.3 | 42.6 | | 0.818 | 0.093 | | 58 |
| Gray matter ($n = 7$) | 640 | 90.1 | 32.5 | | 0.89 | 0.04 | | 58 |
| Gray matter ($n = 7$) | 1060 | 56.8 | 18 | | 0.9 | 0.05 | | 58 |
| Gray matter | 1064 | 267 | | | 0.95 | | | 58 |
| Gray matter, coagulated ($n = 7$) | 360 | 426 | 122 | | 0.868 | 0.031 | Coag. 2 h 80°C | 58 |
| Gray matter, coagulated ($n = 7$) | 740 | | | | | | Coag. 2 h 80°C | 58 |
| Gray matter, coagulated ($n = 7$) | 1100 | 179.8 | 32.6 | | 0.954 | 0.001 | Coag. 2 h 80°C | 58 |
| Brain tumor, astrocytoma grade III WHO | 400 | 84 | | | 0.9 | | | 60 |
| Brain tumor, astrocytoma grade III WHO | 633 | 67 | 8 | | 0.883 | 0.011 | | 60 |
| Brain tumor, astrocytoma grade III WHO | 700 | 50 | | | 0.88 | | | 60 |
| Brain tumor, astrocytoma grade III WHO | 800 | 50 | | | 0.88 | | | 60 |
| White matter ($n = 7$) | 450 | 420 | | 92.4 | 0.78 | | <48 h Post mortem | 61 |
| White matter ($n = 7$) | 510 | 426 | | 80.94 | 0.81 | | <48 h Post mortem | 61 |
| White matter ($n = 7$) | 630 | 409 | | 65.44 | 0.84 | | <48 h Post mortem | 61 |
| White matter ($n = 7$) | 670 | 401 | | 60.15 | 0.85 | | <48 h Post mortem | 61 |
| White matter ($n = 7$) | 850 | 342 | | 41 | 0.88 | | <48 h Post mortem | 61 |
| White matter ($n = 7$) | 1064 | 296 | | 32.56 | 0.89 | | <48 h Post mortem | 61 |
| Gray matter ($n = 7$) | 450 | 117 | | 14.04 | 0.88 | | <48 h Post mortem | 61 |
| Gray matter ($n = 7$) | 510 | 106 | | 12.72 | 0.88 | | <48 h Post mortem | 61 |
| Gray matter ($n = 7$) | 630 | 90 | | 9.9 | 0.89 | | <48 h Post mortem | 61 |
| Gray matter ($n = 7$) | 670 | 84 | | 8.4 | 0.9 | | <48 h Post mortem | 61 |

**TABLE 2.2** Scattering Parameters for the Various Tissues, Tissue Components, and Biofluids (continued)

| Tissue | Description | Wavelength (nm) | $\mu_s$ (1/cm) | Uncert. (1/cm) | $\mu_s'$ (1/cm) | Uncert. (1/cm) | $g$ | Uncert. | Notes | Ref. |
|---|---|---|---|---|---|---|---|---|---|---|
| | Gray matter ($n$ = 7) | 1064 | 57 | | 5.7 | | 0.9 | | <48 h Post mortem | 61 |
| | White matter | 456 | 923 | | 73.84 | | 0.92 | | | 62 |
| | White matter | 514 | 1045 | | 73.15 | | 0.93 | | | 62 |
| | White matter | 630 | 386 | | 54.04 | | 0.86 | | | 62 |
| | White matter | 675 | 436 | | 56.68 | | 0.87 | | | 62 |
| | White matter | 1064 | 513 | | 25.65 | | 0.95 | | | 62 |
| | Gray matter | 456 | 686 | | 34.3 | | 0.95 | | | 62 |
| | Gray matter | 514 | 578 | | 17.34 | | 0.97 | | | 62 |
| | Gray matter | 630 | 473 | | 33.11 | | 0.93 | | | 62 |
| | Gray matter | 675 | 364 | | 32.76 | | 0.91 | | | 62 |
| | Gray matter | 1064 | 267 | | 10.68 | | 0.96 | | | 62 |
| | Gray matter ($n$ = 7) | 400 | 128.5 | 18.4 | | | 0.87 | 0.02 | <48 h Post mortem | 61* |
| | Gray matter ($n$ = 7) | 500 | 109.9 | 13.0 | | | 0.88 | 0.01 | <48 h Post mortem | 61* |
| | Gray matter ($n$ = 7) | 600 | 94.1 | 13.5 | | | 0.89 | 0.02 | <48 h Post mortem | 61* |
| | Gray matter ($n$ = 7) | 700 | 84.1 | 12.0 | | | 0.90 | 0.02 | <48 h Post mortem | 61* |
| | Gray matter ($n$ = 7) | 800 | 77.0 | 11.0 | | | 0.90 | 0.02 | <48 h Post mortem | 61* |
| | Gray matter ($n$ = 7) | 900 | 67.3 | 9.6 | | | 0.90 | 0.02 | <48 h Post mortem | 61* |
| | Gray matter ($n$ = 7) | 1000 | 61.6 | 5.7 | | | 0.90 | 0.02 | <48 h Post mortem | 61* |
| | Gray matter ($n$ = 7) | 1100 | 55.1 | 6.5 | | | 0.90 | 0.02 | <48 h Post mortem | 61* |
| | Gray matter, coagulated ($n$ = 7) | 400 | 258.6 | 18.8 | | | 0.78 | 0.04 | Coag. 2 h 80°C | 61* |
| | Gray matter, coagulated ($n$ = 7) | 500 | 326.5 | 7.7 | | | 0.85 | 0.03 | Coag. 2 h 80°C | 61* |
| | Gray matter, coagulated ($n$ = 7) | 600 | 319.0 | 15.2 | | | 0.87 | 0.03 | Coag. 2 h 80°C | 61* |
| | Gray matter, coagulated ($n$ = 7) | 700 | 319.0 | 7.5 | | | 0.88 | 0.03 | Coag. 2 h 80°C | 61* |
| | Gray matter, coagulated ($n$ = 7) | 800 | 252.7 | 18.3 | | | 0.87 | 0.02 | Coag. 2 h 80°C | 61* |
| | Gray matter, coagulated ($n$ = 7) | 900 | 214.6 | 10.3 | | | 0.87 | 0.02 | Coag. 2 h 80°C | 61* |
| | Gray matter, coagulated ($n$ = 7) | 1000 | 191.0 | 18.7 | | | 0.88 | 0.03 | Coag. 2 h 80°C | 61* |
| | Gray matter, coagulated ($n$ = 7) | 1100 | 186.6 | 13.5 | | | 0.88 | 0.03 | Coag. 2 h 80°C | 61* |
| | White matter ($n$ = 7) | 400 | 413.5 | 21.4 | | | 0.75 | 0.03 | <48 h Post mortem | 61* |
| | White matter ($n$ = 7) | 500 | 413.5 | 43.9 | | | 0.80 | 0.02 | <48 h Post mortem | 61* |
| | White matter ($n$ = 7) | 600 | 413.5 | 21.4 | | | 0.83 | 0.02 | <48 h Post mortem | 61* |
| | White matter ($n$ = 7) | 700 | 393.1 | 30.9 | | | 0.85 | 0.02 | <48 h Post mortem | 61* |
| | White matter ($n$ = 7) | 800 | 364.5 | 28.6 | | | 0.87 | 0.01 | <48 h Post mortem | 61* |
| | White matter ($n$ = 7) | 900 | 329.5 | 35.0 | | | 0.88 | 0.01 | <48 h Post mortem | 61* |
| | White matter ($n$ = 7) | 1000 | 305.4 | 15.9 | | | 0.88 | 0.01 | <48 h Post mortem | 61* |
| | White matter ($n$ = 7) | 1100 | 283.2 | 22.2 | | | 0.88 | 0.004 | <48 h Post mortem | 61* |

| | | | | | | | |
|---|---|---|---|---|---|---|---|
| White matter, coagulated ($n = 7$) | 410 | 568.7 | 111.9 | | 0.83 | 0.03 | Coag. 2 h 80°C | 61* |
| White matter, coagulated ($n = 7$) | 510 | 513.2 | 116.9 | | 0.87 | 0.02 | Coag. 2 h 80°C | 61* |
| White matter, coagulated ($n = 7$) | 610 | 500.2 | 129.9 | | 0.90 | 0.02 | Coag. 2 h 80°C | 61* |
| White matter, coagulated ($n = 7$) | 710 | 475.2 | 108.3 | | 0.91 | 0.01 | Coag. 2 h 80°C | 61* |
| White matter, coagulated ($n = 7$) | 810 | 440.0 | 114.3 | | 0.92 | 0.01 | Coag. 2 h 80°C | 61* |
| White matter, coagulated ($n = 7$) | 910 | 407.4 | 92.8 | | 0.93 | 0.01 | Coag. 2 h 80°C | 61* |
| White matter, coagulated ($n = 7$) | 1010 | 367.7 | 95.5 | | 0.93 | 0.004 | Coag. 2 h 80°C | 61* |
| White matter, coagulated ($n = 7$) | 1100 | 358.4 | 81.6 | | 0.93 | 0.001 | Coag. 2 h 80°C | 61* |
| Astrocytoma (tumor, WHO grade II, $n = 4$) | 400 | 198.4 | 55.6 | | 0.88 | 0.05 | Excised from patients | 61* |
| Astrocytoma (tumor, WHO grade II, $n = 4$) | 490 | 158.5 | 53.7 | | 0.93 | 0.03 | Excised from patients | 61* |
| Astrocytoma (tumor, WHO grade II, $n = 4$) | 600 | 132.4 | 49.0 | | 0.96 | 0.02 | Excised from patients | 61* |
| Astrocytoma (tumor, WHO grade II, $n = 4$) | 700 | 113.2 | 41.8 | | 0.96 | 0.02 | Excised from patients | 61* |
| Astrocytoma (tumor, WHO grade II, $n = 4$) | 800 | 96.7 | 41.8 | | 0.96 | 0.01 | Excised from patients | 61* |
| Astrocytoma (tumor, WHO grade II, $n = 4$) | 900 | 86.4 | 34.6 | | 0.96 | 0.01 | Excised from patients | 61* |
| Astrocytoma (tumor, WHO grade II, $n = 4$) | 1000 | 79.0 | 34.2 | | 0.96 | 0.004 | Excised from patients | 61* |
| Astrocytoma (tumor, WHO grade II, $n = 4$) | 1100 | 73.8 | 29.6 | | 0.96 | 0.004 | Excised from patients | 61* |
| Cerebellum ($n = 7$) | 400 | 276.7 | 19.1 | | 0.80 | 0.03 | <48 h Post mortem | 61* |
| Cerebellum ($n = 7$) | 500 | 277.5 | 32.6 | | 0.85 | 0.02 | <48 h Post mortem | 61* |
| Cerebellum ($n = 7$) | 600 | 272.1 | 12.3 | | 0.87 | 0.02 | <48 h Post mortem | 61* |
| Cerebellum ($n = 7$) | 700 | 266.8 | 12.1 | | 0.89 | 0.01 | <48 h Post mortem | 61* |
| Cerebellum ($n = 7$) | 800 | 250.3 | 17.2 | | 0.90 | 0.01 | <48 h Post mortem | 61* |
| Cerebellum ($n = 7$) | 900 | 229.6 | 15.8 | | 0.90 | 0.01 | <48 h Post mortem | 61* |
| Cerebellum ($n = 7$) | 1000 | 215.4 | 14.7 | | 0.90 | 0.005 | <48 h Post mortem | 61* |
| Cerebellum ($n = 7$) | 1100 | 202.1 | 13.9 | | 0.90 | 0.01 | <48 h Post mortem | 61* |
| Cerebellum, coagulated ($n = 7$) | 400 | 560.0 | 25.5 | | 0.61 | 0.01 | Coag. 2 h 80°C | 61* |
| Cerebellum, coagulated ($n = 7$) | 500 | 512.2 | 47.8 | | 0.77 | 0.02 | Coag. 2 h 80°C | 61* |
| Cerebellum, coagulated ($n = 7$) | 600 | 458.2 | 65.6 | | 0.78 | 0.01 | Coag. 2 h 80°C | 61* |
| Cerebellum, coagulated ($n = 7$) | 700 | 489.9 | 70.1 | | 0.85 | 0.01 | Coag. 2 h 80°C | 61* |
| Cerebellum, coagulated ($n = 7$) | 800 | 458.2 | 54.0 | | 0.87 | 0.02 | Coag. 2 h 80°C | 61* |
| Cerebellum, coagulated ($n = 7$) | 900 | 458.2 | 65.6 | | 0.89 | 0.02 | Coag. 2 h 80°C | 61* |
| Cerebellum, coagulated ($n = 7$) | 1000 | 419.1 | 49.4 | | 0.90 | 0.03 | Coag. 2 h 80°C | 61* |

**TABLE 2.2** Scattering Parameters for the Various Tissues, Tissue Components, and Biofluids (continued)

| Tissue | Description | Wavelength (nm) | $\mu_s$ (1/cm) | Uncert. (1/cm) | $\mu_s'$ (1/cm) | Uncert. (1/cm) | $g$ | Uncert. | Notes | Ref. |
|---|---|---|---|---|---|---|---|---|---|---|
| | Cerebellum, coagulated ($n$ = 7) | 1100 | 428.5 | 40.0 | | | 0.91 | 0.03 | Coag. 2 h 80°C | 61* |
| | Meningioma (tumor, $n$ = 6) | 410 | 197.4 | 19.8 | | | 0.88 | 0.02 | Excised from patients | 61* |
| | Meningioma (tumor, $n$ = 6) | 490 | 188.2 | 18.8 | | | 0.93 | 0.01 | Excised from patients | 61* |
| | Meningioma (tumor, $n$ = 6) | 590 | 171.1 | 12.7 | | | 0.95 | 0.01 | Excised from patients | 61* |
| | Meningioma (tumor, $n$ = 6) | 690 | 155.5 | 15.6 | | | 0.95 | 0.01 | Excised from patients | 61* |
| | Meningioma (tumor, $n$ = 6) | 790 | 141.3 | 14.2 | | | 0.96 | 0.01 | Excised from patients | 61* |
| | Meningioma (tumor, $n$ = 6) | 910 | 125.4 | 9.4 | | | 0.95 | 0.004 | Excised from patients | 61* |
| | Meningioma (tumor, $n$ = 6) | 990 | 119.6 | 8.9 | | | 0.95 | 0.01 | Excised from patients | 61* |
| | Meningioma (tumor, $n$ = 6) | 1100 | 116.8 | 8.6 | | | 0.97 | 0.003 | Excised from patients | 61* |
| | Pons ($n$ = 7) | 400 | 163.5 | 15.3 | | | 0.89 | 0.02 | <48 h Post mortem | 61* |
| | Pons ($n$ = 7) | 500 | 133.7 | 19.2 | | | 0.91 | 0.01 | <48 h Post mortem | 61* |
| | Pons ($n$ = 7) | 600 | 109.4 | 18.5 | | | 0.91 | 0.01 | <48 h Post mortem | 61* |
| | Pons ($n$ = 7) | 700 | 93.5 | 20.9 | | | 0.91 | 0.01 | <48 h Post mortem | 61* |
| | Pons ($n$ = 7) | 800 | 83.6 | 21.0 | | | 0.91 | 0.01 | <48 h Post mortem | 61* |
| | Pons ($n$ = 7) | 900 | 74.8 | 18.7 | | | 0.92 | 0.01 | <48 h Post mortem | 61* |
| | Pons ($n$ = 7) | 1000 | 69.9 | 17.5 | | | 0.91 | 0.01 | <48 h Post mortem | 61* |
| | Pons ($n$ = 7) | 1100 | 64.0 | 17.8 | | | 0.92 | 0.01 | <48 h Post mortem | 61* |
| | Pons, coagulated ($n$ = 7) | 410 | 685.7 | 63.7 | | | 0.85 | 0.02 | Coag. 2 h 80°C | 61* |
| | Pons, coagulated ($n$ = 7) | 510 | 627.5 | 73.6 | | | 0.89 | 0.005 | Coag. 2 h 80°C | 61* |
| | Pons, coagulated ($n$ = 7) | 710 | 402.5 | 67.7 | | | 0.89 | 0.004 | Coag. 2 h 80°C | 61* |
| | Pons, coagulated ($n$ = 7) | 810 | 329.7 | 55.4 | | | 0.89 | 0.005 | Coag. 2 h 80°C | 61* |
| | Pons, coagulated ($n$ = 7) | 910 | 276.0 | 46.4 | | | 0.88 | 0.004 | Coag. 2 h 80°C | 61* |
| | Pons, coagulated ($n$ = 7) | 1010 | 241.6 | 34.4 | | | 0.88 | 0.004 | Coag. 2 h 80°C | 61* |
| | Pons, coagulated ($n$ = 7) | 1100 | 221.1 | 31.5 | | | 0.88 | 0.004 | Coag. 2 h 80°C | 61* |
| | Thalamus ($n$ = 7) | 410 | 146.7 | 49.4 | | | 0.86 | 0.03 | <48 h Post mortem | 61* |
| | Thalamus ($n$ = 7) | 510 | 188.0 | 31.9 | | | 0.87 | 0.03 | <48 h Post mortem | 61* |
| | Thalamus ($n$ = 7) | 610 | 176.3 | 34.5 | | | 0.88 | 0.02 | <48 h Post mortem | 61* |
| | Thalamus ($n$ = 7) | 710 | 169.0 | 28.7 | | | 0.89 | 0.03 | <48 h Post mortem | 61* |
| | Thalamus ($n$ = 7) | 810 | 158.5 | 35.3 | | | 0.89 | 0.02 | <48 h Post mortem | 61* |
| | Thalamus ($n$ = 7) | 910 | 155.4 | 22.3 | | | 0.90 | 0.02 | <48 h Post mortem | 61* |
| | Thalamus ($n$ = 7) | 1010 | 139.3 | 34.9 | | | 0.90 | 0.02 | <48 h Post mortem | 61* |
| | Thalamus ($n$ = 7) | 1100 | 146.0 | 36.6 | | | 0.91 | 0.02 | <48 h Post mortem | 61* |
| | Thalamus, coagulated ($n$ = 7) | 400 | 391.1 | 56.1 | | | 0.83 | 0.04 | Coag. 2 h 80°C | 61* |
| | Thalamus, coagulated ($n$ = 7) | 500 | 399.9 | 67.7 | | | 0.90 | 0.01 | Coag. 2 h 80°C | 61* |
| | Thalamus, coagulated ($n$ = 7) | 600 | 365.7 | 43.2 | | | 0.92 | 0.01 | Coag. 2 h 80°C | 61* |

| | | | | | | | | | | |
|---|---|---|---|---|---|---|---|---|---|---|
| | Thalamus, coagulated ($n$ = 7) | 700 | 327.0 | 30.6 | | | 0.92 | 0.01 | Coag. 2 h 80°C | 61* |
| | Thalamus, coagulated ($n$ = 7) | 800 | 286.0 | 33.8 | | | 0.93 | 0.01 | Coag. 2 h 80°C | 61* |
| | Thalamus, coagulated ($n$ = 7) | 900 | 267.4 | 31.6 | | | 0.93 | 0.01 | Coag. 2 h 80°C | 61* |
| | Thalamus, coagulated ($n$ = 7) | 1000 | 233.8 | 39.7 | | | 0.93 | 0.01 | Coag. 2 h 80°C | 61* |
| | Thalamus, coagulated ($n$ = 7) | 1100 | 223.6 | 32.1 | | | 0.94 | 0.01 | Coag. 2 h 80°C | 61* |
| Brain, *in vivo* | Cortex, frontal lobe | 674 | | | 10 | 0.5 | | | | 63 |
| | Cortex, frontal lobe | 811 | | | 9.1 | 0.5 | | | | 63 |
| | Cortex, frontal lobe | 849 | | | 9.2 | 0.5 | | | | 63 |
| | Cortex, frontal lobe | 956 | | | 8.9 | 0.5 | | | | 63 |
| | Cortex, frontal lobe | 674 | | | 10 | 0.5 | | | | 63 |
| | Cortex, frontal lobe | 811 | | | 8.2 | 0.5 | | | | 63 |
| | Cortex, frontal lobe | 849 | | | 8.2 | 0.5 | | | | 63 |
| | Cortex, frontal lobe | 956 | | | 8.2 | 0.5 | | | | 63 |
| | Astrocytoma of optic nerve | 674 | | | 12.5 | 1 | | | | 63 |
| | Astrocytoma of optic nerve | 811 | | | 9.5 | 1 | | | | 63 |
| | Astrocytoma of optic nerve | 849 | | | 7.6 | 1 | | | | 63 |
| | Astrocytoma of optic nerve | 956 | | | 7.3 | 1 | | | | 63 |
| | Normal optic nerve | 674 | | | 17.5 | 2 | | | | 63 |
| | Normal optic nerve | 849 | | | 16 | 2 | | | | 63 |
| | Normal optic nerve | 956 | | | 15.2 | 2 | | | | 63 |
| | Skull | 674 | | | 9 | 1 | | | | 63 |
| | Skull | 849 | | | 9 | 1 | | | | 63 |
| | Skull | 956 | | | 8.5 | 1 | | | | 63 |
| | Cerebellar white matter | 674 | | | 13.5 | 1 | | | | 63 |
| | Cerebellar white matter | 849 | | | 8.5 | 1 | | | | 63 |
| | Cerebellar white matter | 956 | | | 7.8 | 1 | | | | 63 |
| | Medulloblastoma | 674 | | | 14 | 1 | | | | 63 |
| | Medulloblastoma | 849 | | | 10.7 | 1 | | | | 63 |
| | Medulloblastoma | 956 | | | 4 | 1 | | | | 63 |
| | Cerebellar white matter with scar tissue | 674 | | | 6.5 | 0.5 | | | | 63 |
| | Cerebellar white matter with scar tissue | 849 | | | 8 | 0.5 | | | | 63 |
| | Normal cortex, temporal and frontal lobe | 674 | | | 10 | 1 | 0.92 | | Measurements during surgery | 64 |
| | Normal cortex, temporal and frontal lobe | 849 | | | 9.2 | 1 | 0.92 | | Measurements during surgery | 64 |

**TABLE 2.2** Scattering Parameters for the Various Tissues, Tissue Components, and Biofluids (continued)

| Tissue | Description | Wavelength (nm) | $\mu_s$ (1/cm) | Uncert. (1/cm) | $\mu_s'$ (1/cm) | Uncert. (1/cm) | $g$ | Uncert. | Notes | Ref. |
|---|---|---|---|---|---|---|---|---|---|---|
| | Normal cortex, temporal and frontal lobe | 956 | | | 8.5 | 1 | 0.92 | | Measurements during surgery | 64 |
| | Normal optic nerve | 674 | | | 18 | 1 | 0.92 | | Measurements during surgery | 64 |
| | Normal optic nerve | 849 | | | 17 | 1 | 0.92 | | Measurements during surgery | 64 |
| | Normal optic nerve | 956 | | | 16 | 1 | 0.92 | | Measurements during surgery | 64 |
| | Astrocytoma of optical nerve | 674 | | | 14 | 1 | 0.92 | | Measurements during surgery | 64 |
| | Astrocytoma of optical nerve | 849 | | | 8.5 | 1 | 0.92 | | Measurements during surgery | 64 |
| | Astrocytoma of optical nerve | 956 | | | 8.5 | 1 | 0.92 | | Measurements during surgery | 64 |
| Female breast, *in vitro* | Fatty normal ($n$ = 23) | 749 | 8.48 | 3.43 | | | | | Kept in saline, 37°C | 49 |
| | Fatty normal ($n$ = 23) | 789 | 7.67 | 2.57 | | | | | Kept in saline, 37°C | 49 |
| | Fatty normal ($n$ = 23) | 836 | 7.27 | 2.4 | | | | | Kept in saline, 37°C | 49 |
| | Fibrous normal ($n$ = 35) | 749 | 9.75 | 2.27 | | | | | Kept in saline, 37°C | 49 |
| | Fibrous normal ($n$ = 35) | 789 | 8.94 | 2.45 | | | | | Kept in saline, 37°C | 49 |
| | Fibrous normal ($n$ = 35) | 836 | 8.1 | 2.21 | | | | | Kept in saline, 37°C | 49 |
| | Infiltrating carcinoma ($n$ = 48) | 749 | 10.91 | 5.59 | | | | | Kept in saline, 37°C | 49 |
| | Infiltrating carcinoma ($n$ = 48) | 789 | 10.12 | 5.05 | | | | | Kept in saline, 37°C | 49 |
| | Infiltrating carcinoma ($n$ = 48) | 836 | 9.1 | 4.54 | | | | | Kept in saline, 37°C | 49 |
| | Mucinous carcinoma ($n$ = 3) | 749 | | | 6.15 | 2.44 | | | Kept in saline, 37°C | 49 |
| | Mucinous carcinoma ($n$ = 3) | 789 | | | 5.09 | 2.42 | | | Kept in saline, 37°C | 49 |
| | Mucinous carcinoma ($n$ = 3) | 836 | | | 4.78 | 3.67 | | | Kept in saline, 37°C | 49 |
| | Ductal carcinoma, *in situ* ($n$ = 5) | 749 | | | 13.1 | 2.85 | | | Kept in saline, 37°C | 49 |
| | Ductal carcinoma, *in situ* ($n$ = 5) | 789 | | | 12.21 | 2.45 | | | Kept in saline, 37°C | 49 |
| | Ductal carcinoma, *in situ* ($n$ = 5) | 836 | | | 10.46 | 2.65 | | | Kept in saline, 37°C | 49 |
| | Grandular tissue ($n$ = 3) | 540 | | | 24.4 | 5.8 | | | Homogenized tissue | 65 |
| | Grandular tissue ($n$ = 3) | 700 | | | 14.2 | 3 | | | Homogenized tissue | 65 |
| | Grandular tissue ($n$ = 3) | 900 | | | 9.9 | 2 | | | Homogenized tissue | 65 |
| | Fatty tissue ($n$ = 7) | 540 | | | 10.3 | 1.9 | | | Homogenized tissue | 65 |

| | | | | | | | | | | |
|---|---|---|---|---|---|---|---|---|---|---|
| | Fatty tissue ($n$ = 7) | 700 | | | 8.6 | 1.3 | | | Homogenized tissue | 65 |
| | Fatty tissue ($n$ = 7) | 900 | | | 7.9 | 1.1 | | | Homogenized tissue | 65 |
| | Fibrocystic tissue ($n$ = 8) | 540 | | | 21.7 | 3.3 | | | Homogenized tissue | 65 |
| | Fibrocystic tissue ($n$ = 8) | 700 | | | 13.4 | 1.9 | | | Homogenized tissue | 65 |
| | Fibrocystic tissue ($n$ = 8) | 900 | | | 9.5 | 1.7 | | | Homogenized tissue | 65 |
| | Fibroadenoma ($n$ = 6) | 540 | | | 11.1 | 3 | | | Homogenized tissue | 65 |
| | Fibroadenoma ($n$ = 6) | 700 | | | 7.2 | 1.7 | | | Homogenized tissue | 65 |
| | Fibroadenoma ($n$ = 6) | 900 | | | 5.3 | 1.4 | | | Homogenized tissue | 65 |
| | Carcinoma ($n$ = 9) | 540 | | | 19 | 5.1 | | | Homogenized tissue | 65 |
| | Carcinoma ($n$ = 9) | 700 | | | 11.8 | 3.1 | | | Homogenized tissue | 65 |
| | Carcinoma ($n$ = 9) | 900 | | | 8.9 | 2.6 | | | Homogenized tissue | 65 |
| | Fatty tissue | 700 | | | 13 | 5 | 0.95 | 0.02 | | 66 |
| | Fibroglandular tissue | 700 | | | 12 | 5 | 0.92 | 0.03 | | 66 |
| | Carcinoma (central part) | 700 | | | 18 | 5 | 0.88 | 0.03 | | 66 |
| | Fatty tissue | 625 | | | 14.3 | 2.1 | | | | 67 |
| | Benign tumor | 625 | | | 3.8 | 0.3 | | | | 67 |
| Female breast, *in vivo* | Normal (30 Japanese women averaged for all ages) | 753 | | | 8.9 | 1.3 | | | Time-domain technique | 68 |
| | Normal (6 subjects, 26–43 years old) | 800 | | | 7.2–13.5 | | | | Time-domain technique | 69 |
| | Breast cancer (5 patients) | 630 | | | 9.41 | 7.35 | | | Relapsed cancer, HPD (72 h) | 70 |
| Aorta, *in vitro* | Post mortem, 6 h normal coagulated | 308 | | | 77 | | | | | 46 |
| | Post mortem, 6 h normal coagulated | 308 | | | 270 | | | | | 46 |
| | Post mortem, 6 h fibrous plaque | 308 | | | 81 | | | | | 46 |
| | Post mortem, 6 h fibrous plaque coagulated | 308 | | | 272 | | | | | 46 |
| | Normal | 1064 | 239 | 45 | 23.9 | | 0.9 | | | 46 |
| | Coagulated | 1064 | 293 | 73 | 29.3 | | 0.9 | | | 46 |
| | Fibro-fatty | 355 | | | 64.9 | | | | | 46 |
| | Fibro-fatty | 532 | | | 24.8 | | | | | 46 |
| | Fibro-fatty | 1064 | | | 7.7 | | | | | 46 |
| | | 633 | 316 | | 41 | | 0.87 | | | 46 |
| | | 1064 | 239 | | 23.9 | | 0.9 | | | 46 |
| | | 1064 | | | 22.4 | | | | | 46 |
| | | 1320 | 233 | | 23.3 | | 0.9 | | | 46 |
| | | 1320 | | | 17.8 | | | | | 46 |
| | | 470 | | | 42.6 | 6 | | | | 46 |

**TABLE 2.2** Scattering Parameters for the Various Tissues, Tissue Components, and Biofluids (continued)

| Tissue | Description | Wavelength (nm) | $\mu_s$ (1/cm) | Uncert. (1/cm) | $\mu_s'$ (1/cm) | Uncert. (1/cm) | $g$ | Uncert. | Notes | Ref. |
|---|---|---|---|---|---|---|---|---|---|---|
| | | 476 | | | 41.9 | 5.9 | | | | 46 |
| | | 488 | | | 39.9 | 5.6 | | | | 46 |
| | | 514.5 | | | 36.9 | 5.4 | | | | 46 |
| | | 580 | | | 31.1 | 4.9 | | | | 46 |
| | | 600 | | | 29.6 | 4.7 | | | | 46 |
| | | 633 | | | 27.4 | 4.4 | | | | 46 |
| | | 1064 | | | 15.5 | 2.8 | | | | 46 |
| | Intima | 476 | 237 | | 45 | | 0.81 | | | 29 |
| | Intima | 580 | 183 | | 34.8 | | 0.81 | | | 29 |
| | Intima | 600 | 178 | | 33.8 | | 0.81 | | | 29 |
| | Intima | 633 | 171 | | 25.7 | | 0.85 | | | 29 |
| | Intima | 1064 | 165 | | | | 0.97 | | | 41 |
| | Media | 476 | 410 | | 45.1 | | 0.89 | | | 29 |
| | Media | 580 | 331 | | 33.1 | | 0.9 | | | 29 |
| | Media | 600 | 323 | | 35.5 | | 0.89 | | | 29 |
| | Media | 633 | 310 | | 31 | | 0.9 | | | 29 |
| | Media | 1064 | 634 | | | | 0.96 | | | 41 |
| | Adventitia | 476 | 267 | | 69.4 | | 0.74 | | | 29,71,72 |
| | Adventitia | 580 | 217 | | 49.9 | | 0.77 | | | 29 |
| | Adventitia | 600 | 211 | | 46.4 | | 0.78 | | | 29, 71, 72 |
| | Adventitia | 633 | 195 | | 37.1 | | 0.81 | | | 29 |
| | Adventitia | 1064 | 484 | | | | 0.97 | | | 41 |
| Aorta, *in vivo* | Advanced fibrous atheroma | 355 | | | 72.1 | 7.2 | | | Photoacoustic method | 73 |
| | Advanced fibrous atheroma | 532 | | | 36.5 | 5.5 | | | Photoacoustic method | 73 |
| | Advanced fibrous atheroma | 1064 | | | 4.85 | 2.4 | | | Photoacoustic method | 73 |
| Blood | Oxygenated, Hct = 0.41 | 665 | 1246 | | 6.11 | | 0.995 | | | 29 |
| | Oxygenated, Hct = 0.41 | 685 | 1413 | | | | 0.99 | | | 29 |
| | Oxygenated, Hct = 0.41 | 960 | 505 | | 3.84 | | 0.992 | | | 29 |
| | Deoxygenated, Hct = 0.41 | 665 | 509 | | 2.49 | | 0.995 | | | 29 |
| | Deoxygenated, Hct = 0.41 | 960 | 668 | | 5.08 | | 0.992 | | | 29 |
| | Oxygenation >98%, Hct = 0.45–0.46 | 633 | 644.7 | | | | 0.982 | | | 74 |
| | Oxygenation >98%, Hct = 0.45–0.46 | 710 | 737 | 75 | | | 0.986 | 0.006 | | 74 |
| | Oxygenation >98%, Hct = 0.45–0.46 | 765 | 725 | 75 | | | 0.991 | 0.002 | | 74 |
| | Oxygenation >98%, Hct = 0.45–0.46 | 810 | 690 | 80 | | | 0.989 | 0.002 | | 74 |

| | | | | | | | | | |
|---|---|---|---|---|---|---|---|---|---|
| | Oxygenation >98%, Hct = 0.45–0.46 | 865 | 649 | 25 | | | 0.99 | 0.001 | 74 |
| | Oxygenation >98%, Hct = 0.45–0.46 | 910 | 649 | 25 | | | 0.992 | 0.002 | 74 |
| | Oxygenation >98%, Hct = 0.45–0.46 | 965 | 650 | 25 | | | 0.991 | 0.001 | 74 |
| | Oxygenation >98%, Hct = 0.45–0.46 | 1010 | 645 | 25 | | | 0.992 | 0.001 | 74 |
| | Oxygenation >98%, Hct = 0.45–0.46 | 1065 | 645 | 25 | | | 0.992 | 0.001 | 74 |
| | Oxygenation >98%, Hct = 0.45–0.46 | 1110 | 630 | 20 | | | 0.993 | 0.001 | 74 |
| | Oxygenation >98%, Hct = 0.45–0.46 | 1165 | 655 | 15 | | | 0.993 | 0.001 | 74 |
| | Oxygenation >98%, Hct = 0.45–0.46 | 1210 | 654 | 20 | | | 0.995 | 0.001 | 74 |
| | Oxygenation >98%, Hct = 0.38 | 633 | 400 | 30 | 1.7 | 1.2 | 0.971 | 0.001 | 74 |
| | Oxygenation >98%, Hct = 0.38 | 633 | 4130 | 170 | 12.4 | 0.9 | | | 74 |
| | Oxygenation >98%, Hct = 0.38 | 633 | 2390 | 160 | 9.1 | 0.8 | 0.9962 | 0.0001 | 74 |
| | Oxygenated (98%) flowing blood, Hct = 0.1 | 633 | 773 | 5 | 4.6 | | 0.994 | 0.01 | 75 |
| | Oxygenated (98%) flowing blood, Hct = 0.2 | 633 | 800 | | 8.8 | | 0.989 | | 75 |
| | Oxygenated (98%) flowing blood, Hct = 0.3 | 633 | 890 | | 12.5 | | 0.986 | | 75 |
| | Oxygenated (98%) flowing blood, Hct = 0.4 | 633 | 850 | | 16.8 | | 0.98 | | 75 |
| | Oxygenated (98%) flowing blood, Hct = 0.5 | 633 | 840 | | 20.2 | | 0.976 | | 75 |
| | Oxygenated (98%) flowing blood, Hct = 0.6 | 633 | 840 | | 22.7 | | 0.973 | | 75 |
| | Oxygenated (98%) flowing blood, Hct = 0.7 | 633 | 950 | | 22.8 | | 0.976 | | 75 |
| | Flowing blood, 24.5°C | 633 | | | 18–20 | | | | 76 |
| | Flowing blood, 24.5–42°C | 633 | | | 18–17 | | | | 76 |
| | Flowing blood, 47°C | 633 | | | 21 | | | | 76 |
| | Flowing blood, 54.3°C | 633 | | | 17 | | | | 76 |
| Misc. tissues, *in vitro* | Lung | 515 | 356 | 39 | | | | | 48 |
| | Lung | 635 | 324 | 46 | 81 | | 0.75 | | 48 |
| | Lung | 1064 | 39 | | | | 0.91 | | 41 |
| | Muscle | 515 | 530 | 44 | | | | | 48 |
| | Muscle | 1064 | 215 | | | | 0.96 | | 41 |
| | Meniscus | 360 | | | 108 | | | | 46 |
| | Meniscus | 400 | | | 67 | | | | 46 |
| | Meniscus | 488 | | | 30 | | | | 46 |

**TABLE 2.2** Scattering Parameters for the Various Tissues, Tissue Components, and Biofluids (continued)

| Tissue | Description | Wavelength (nm) | $\mu_s$ (1/cm) | Uncert. (1/cm) | $\mu_s'$ (1/cm) | Uncert. (1/cm) | $g$ | Uncert. | Notes | Ref. |
|---|---|---|---|---|---|---|---|---|---|---|
| | Meniscus | 514 | | | 26 | | | | | 46 |
| | Meniscus | 630 | | | 11 | | | | | 46 |
| | Meniscus | 800 | | | 5.1 | | | | | 46 |
| | Meniscus | 1064 | | | 2.6 | | | | | 46 |
| | Uterus | 635 | 394 | 91 | 122 | | 0.69 | | | 48 |
| | Bladder, integral | 633 | 88 | | 3.52 | | 0.96 | | Excised, kept in saline | 29 |
| | Bladder, integral | 633 | 29.3 | | 2.64 | | 0.91 | | | 41 |
| | Bladder, mucous | 1064 | 7.5 | | | | 0.85 | | | 41 |
| | Bladder, wall | 1064 | 54.3 | | | | 0.85 | | | 41 |
| | Bladder, integral | 1064 | 116 | | | | 0.9 | | | 41 |
| | Heart, endocardial | 1060 | 136 | | | | 0.97 | | Excised, kept in saline | 29 |
| | Heart, epicardial | 1060 | 167 | | | | 0.98 | | | 29 |
| | Heart, myocardial | 1060 | 177.5 | | | | 0.96 | | | 41 |
| | Heart, epicardlial | 1060 | 127.1 | | | | 0.93 | | | 41 |
| | Heart, aneurysm | 1060 | 137 | | | | 0.98 | | | 41 |
| | Heart, trabecula | 1064 | 424 | | | | 0.97 | | | 41 |
| | Heart, myocardial | 1064 | 324 | | | | 0.96 | | | 41 |
| | Heart, myocardial | 1060 | | | 4.48 | | | | | 77 |
| | Kidney, pars conv. | 1064 | 72 | | | | 0.86 | | | 41 |
| | Kidney, medulla ren | 1064 | 77 | | | | 0.87 | | | 41 |
| | Femoral vein | 1064 | 487 | | | | 0.97 | | | 41 |
| | Liver | 515 | 285 | 20 | | | | | | 48 |
| | Liver | 630 | 414 | | | | 0.95 | | | 77 |
| | Liver | 635 | 313 | 136 | 100 | | 0.68 | | | 48 |
| | Liver | 1064 | 356 | | | | 0.95 | | | 48 |
| | Colon, muscle | 1064 | 238 | | | | 0.93 | | | 41 |
| | Colon, submucous | 1064 | 117 | | | | 0.91 | | | 41 |
| | Colon, mucous | 1064 | 39 | | | | 0.91 | | | 41 |
| | Colon, integral | 1064 | 261 | | | | 0.94 | | | 41 |
| | Esophagus | 633 | | | 12 | | | | | 46 |
| | Esophagus(mucous) | 1064 | 83 | | | | 0.86 | | | 41 |
| | Fat, subcutaneous | 1064 | 29 | | | | 0.91 | | | 41 |
| | Fat, abdominal | 1064 | 37 | | | | 0.91 | | | 41 |
| | Prostate (0.5–3 h post mortem) | 850 | 100 | 20 | | | 0.94 | 0.02 | Shock frozen | 41 |
| | Prostate (0.5–3 h post mortem) | 980 | 90 | 20 | | | 0.95 | 0.02 | Shock frozen | 41 |

| | | | | | | | | | | |
|---|---|---|---|---|---|---|---|---|---|---|
| | Prostate (0.5–3 h post mortem) | 1064 | 80 | 20 | | | 0.95 | 0.02 | Shock frozen | 41 |
| | Prostate coagulated (0.5–3 h post mortem) | 850 | 230 | 30 | | | 0.94 | 0.02 | Coag. water bath 75°C, 10 min | 41 |
| | Prostate coagulated (0.5–3 h post mortem) | 980 | 190 | 30 | | | 0.95 | 0.02 | Coag. water bath 75°C, 10 min | 41 |
| | Prostate coagulated (0.5–3 h post mortem) | 1064 | 180 | 30 | | | 0.95 | 0.02 | Coag. water bath 75°C, 10 min | 41 |
| | Prostate normal (freshly excised) | 1064 | 47 | 13 | 0.64 | | 0.862 | | | 46 |
| | Prostate coagulated | 1064 | 80 | 12 | 1.12 | | 0.861 | | Coag. water bath 70°C, 10 min | 46 |
| | Spleen | 1064 | 137 | | | | 0.9 | | | 41 |
| | Stomach, muscle | 1064 | 29.5 | | | | 0.87 | | | 41 |
| | Stomach, mucous | 1064 | 732 | | | | 0.91 | | | 41 |
| | Stomach, integral | 1064 | 128 | | | | 0.91 | | | 41 |
| | Sclera | 650 | | | 25 | | | | | 78 |
| | Tooth, dentin | 633 | 1200 | | 672 | | 0.44 | | | 79 |
| | Tooth, enamel | 633 | 1.1 | | | | | | | 79 |
| *In vivo* | Calf (11 subjects, 14 meas.) | 800 | | | 9.4 | 0.7 | | | | 80 |

*Note:* Values are given for the scattering coefficient $\mu_s$, the average cosine of scatter $g$, and the reduced scattering coefficient $\mu_s' = \mu_s(1 - g)$.

* Data taken from plots.

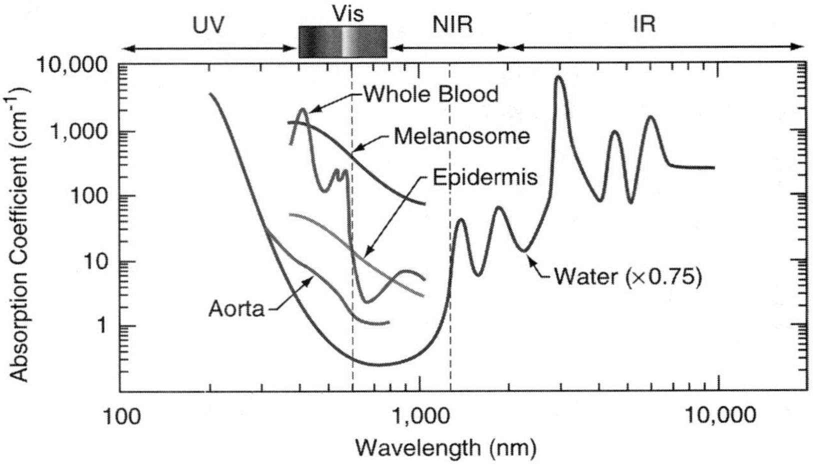


**FIGURE 2.16** **(Color figure follows p. 28-30.)** The "therapeutic window" (region between the dashed lines) in tissue, which runs from 600 to 1400 nm, and the spectra of several important absorbers in this range.

not necessarily entering into the diffusion limit. Figure 2.16 shows a diagram of the therapeutic window of tissue. The absorption properties of various components of tissues are discussed below.

#### 2.4.3.2 Absorption Properties of Tissue Components

Absorption in tissues is a function of molecular composition. Figure 2.17 shows the chemical structures of some molecules of biological and biomedical interest. Molecules absorb photons when the photon's energy matches an interval between internal energy states, and the transition between quantum states obeys the selection rules for the species. At the short-wavelength (higher photon energy) end of the spectrum, these transitions are electronic. Some of the important absorbers in the UV include DNA, the aromatic amino acids (tryptophan and tyrosine), proteins, melanins, and porphyrins (which include hemoglobin, myoglobin, vitamin B12, and cytochrome *c*) (Figure 2.17B). Figure 2.18 shows the absorption spectra of Fe(III) cytochrome *c* at various pH values from 6.8 to 1.7.[33]

*Nucleic acids.* DNA has an absorption peak at 258 nm ($\varepsilon = 6.6 \times 10^3$ cm$^2 \cdot$ mol$^{-1}$), but its absorption falls four orders of magnitude as the wavelength increases to 300 nm, and it has no significant activity in the visible and NIR. Figure 2.19 shows the absorption spectrum of DNA from *Escherichia coli* in the native form at 25°C (solid curve) and as an enzymic digest of nucleotides (dashed curve).[32] RNA exhibits similar absorption properties, having a maximum absorption at 258 nm ($\varepsilon = 14.9 \times 10^3$ cm$^2 \cdot$ mol$^{-1}$). The absorption of nucleic acids arise from $n - \pi^*$ and $\pi - \pi^*$ transitions. The absorption spectra of the constituent purine and pyrimidine bases occur between 200 and 300 nm. The absorption maxima ($\lambda_{max}$) and extinction coefficients ($\varepsilon$) of several purine and pyrimidine bases and their derivatives are shown in Table 2.3.

*Amino acids and proteins.* A number of amino acid side chains — including His, Arg, Glu, Gln, Asn, and Asp — have transitions around 210 nm. These transitions cannot usually be observed in proteins because they are overlapped by the much stronger absorption bands associated with amide backbone groups, which have strong absorption around 230 nm. The amino acids tryptophan, tyrosine, and phenylalanine contain an aromatic ring, which distinguishes them from the other amino acids. They are strong absorbers for $\lambda < 300$ nm and have local peaks in the 260- to 280-nm range and even larger peaks at shorter wavelengths (Figure 2.20).[33] The useful UV range for proteins is at wavelengths longer than 230 nm, where absorption bands from the aromatic side chains of phenylamine ($\lambda_{max} = 257$ nm; $\varepsilon = 0.2 \times 10^{-3}$ cm$^2 \cdot$ mol$^{-1}$), tryptophan ($\lambda_{max} = 280$ nm; $\varepsilon = 5.6 \times 10^3$ cm$^2 \cdot$ mol$^{-1}$), and tyrosine ($\lambda_{max} = 274$ nm; $\varepsilon = 1.4 \times 10^3$ cm$^2 \cdot$ mol$^{-1}$)[34] occur (Figure 2.20A). Because tryptophan has the most intense absorbance, it is often used as the basis for protein concentration measurements. Chemical modification of tryptophan

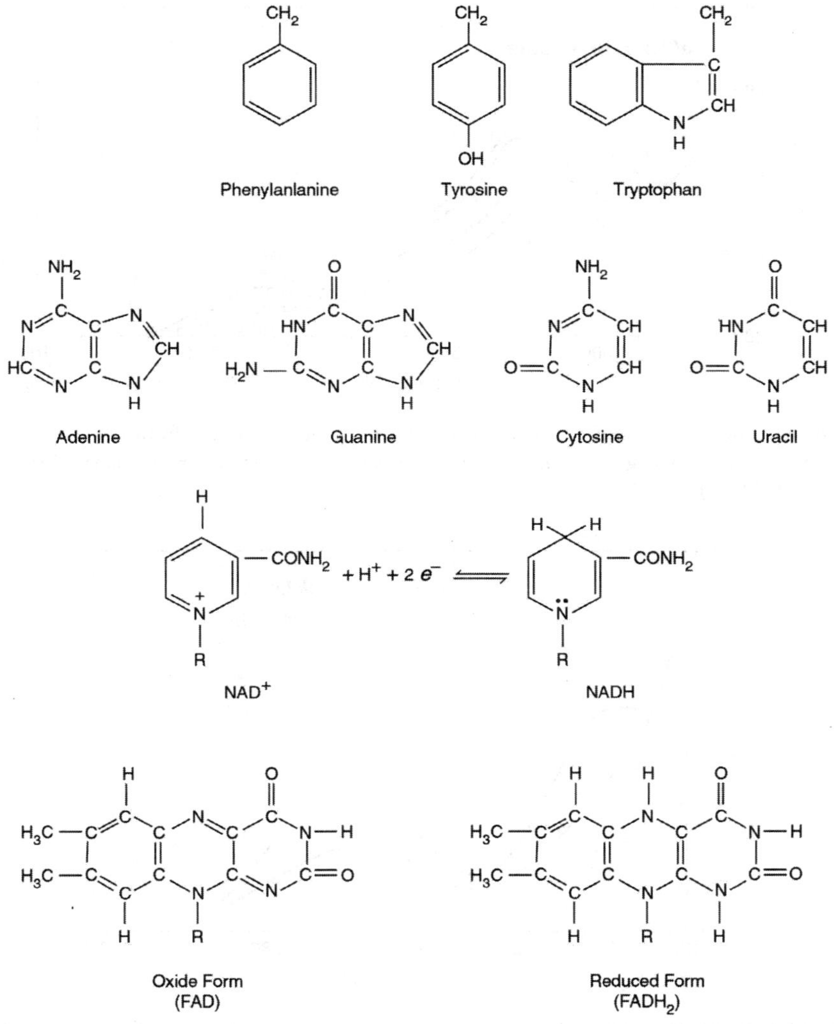


**FIGURE 2.17** Chemical structures of various species of biological interest.

or tyrosine can alter their absorption spectra. NADH absorbs at 340 nm ($\varepsilon = 6.2 \times 10^6 \cdot mol^{-1}$) and at 260 nm ($\varepsilon = 14 \times 10^6 \cdot mol^{-1}$), while NAD absorbs only at 260 nm ($\varepsilon = 18 \times 10^6 \cdot mol^{-1}$). Measurements of NADH absorption can be used to monitor a reaction occurring in biological fluids and tissues. Figure 2.20B shows the absorption spectra of other constituents in biological tissues.

*Skin.* The skin covers the entire surface of our bodies and protects internal organs from the harsh elements of the environment; it is usually described as consisting of three layers. The outer layer is the epidermis. Below is the dermis, which is made up of skin appendages such as hair follicles and sweat glands, surrounded by fibrous supporting tissue and collagen. Finally, the underlying layer, called the subcutaneous tissue, contains fat-producing cells and fibrous tissue. The outermost part of the epidermis is the corneal layer, consisting of millions of dead skin cells, which are shed and continuously replaced by living cells (epidermal keratinocytes) that originate from the basal layer. Melanocytes, which are among the basal cells of the epidermis, produce melanin, the protective pigment responsible for skin color. Melanocytes are stimulated by sunlight and therefore produce melanin to protect the skin against harmful UV radiation.

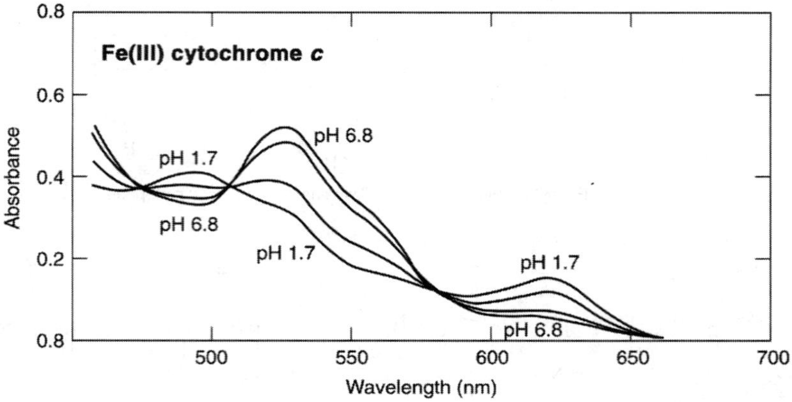


**FIGURE 2.18** Absorption spectra of Fe(III) cytochrome *c* at various pH values from 6.8 to 1.7. (Adapted from Tinoco, I., Jr., Sauer, K., and Wang, J., *Physical Chemistry: Principles and Applications in Biological Sciences*, Prentice-Hall, Englewood Cliffs, NJ, 1985.)

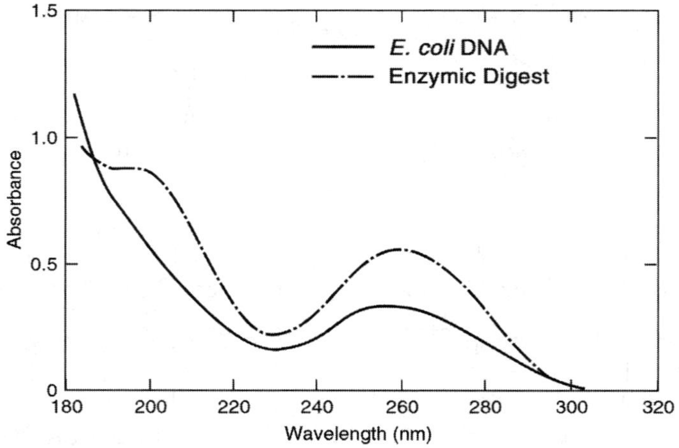


**FIGURE 2.19** Absorption spectrum of DNA from *E. coli* in the native form at 25°C (solid curve) and as an enzymic digest of nucleotides (dashed curve). (Adapted from Voet, D. et al., *Biopolymers*, 1, 93, 1963.)

**TABLE 2.3** Absorption Maxima and Extinction Coefficients of Several Purine and Pyrimidine Bases and Their Derivatives

| | Absorption Maximum ($\lambda_{max}$) (nm) | Extinction Coefficient ($\varepsilon$) ($cm^2 \cdot mol^{-1}$) |
|---|---|---|
| Adenine | 260.5 | $13.4 \times 10^3$ |
| Adenosine | 259.5 | $14.9 \times 10^3$ |
| Guanine | 275 | $8.1 \times 10^3$ |
| Guanosine | 276 | $9.0 \times 10^3$ |
| Cytidine | 271 | $9.1 \times 10^3$ |
| Cytosine | 267 | $6.1 \times 10^3$ |
| Uracil | 259.5 | $8.2 \times 10^3$ |
| Uridine | 261.1 | $10.1 \times 10^3$ |
| Thymine | 264.5 | $7.9 \times 10^3$ |
| Thymidine | 267 | $9.7 \times 10^3$ |

*Source:* Data from Campbell, I.D. and Dwek, R.A., *Biological Spectroscopy*, Benjamin Cummings, Menlo Park, CA, 1984.

A

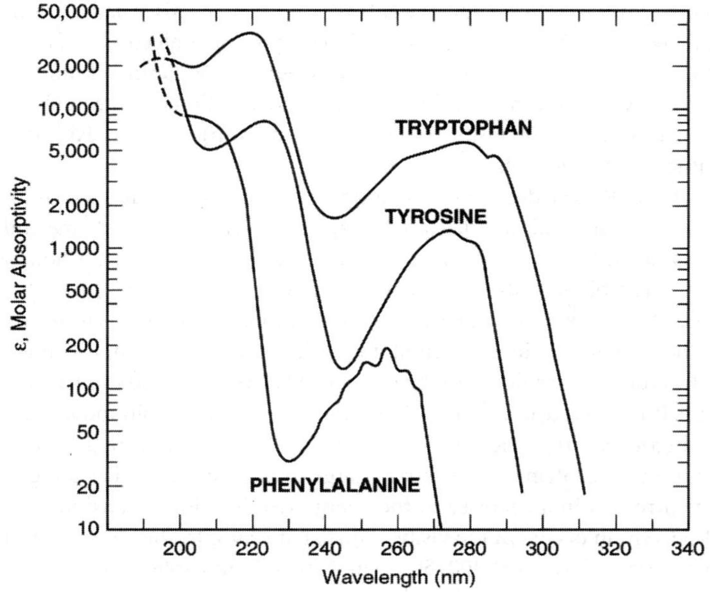


B

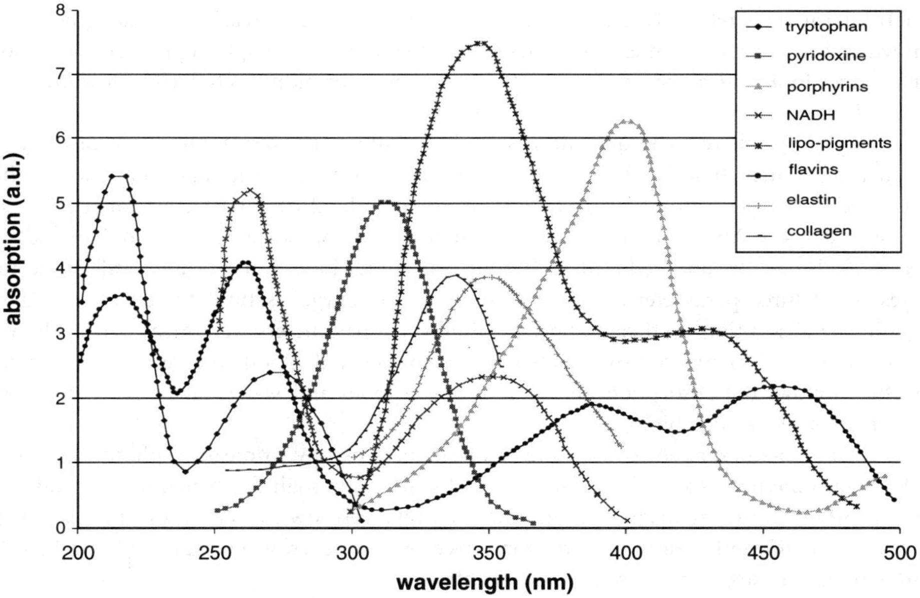


**FIGURE 2.20** (A) Absorption spectra of tryptophan, tyrosine, and phenylalanine. (Adapted from Tinoco, I., Jr. et al., *Physical Chemistry: Principles and Applications in Biological Sciences*, Prentice-Hall, Englewood Cliffs, NJ, 1985.) (B) Absorption spectra of various constituents in tissue.

*Melanins.* Melanins are a group of polymers that provide for pigmentation and are the dominant chromophores in the skin.[35] In most people the absorption of epidermis is usually dominated by melanin absorption. Melanin is a polymer built by condensation of tyrosine molecules and has a broad absorption spectrum exhibiting stronger absorption at shorter wavelengths. Melanin is found in the melanosome, a membranous particle whose internal membranes are studded with many melanin granules, particulates of about 10 to 30 nm in size. In the visible range (400 to 700 nm), melanin exhibits strong absorption, which decreases at longer wavelengths.

*Blood and hemoglobin.* Knowledge about the optical properties $\mu_a$, $\mu_s$, and $g$ of human blood plays an important role for many diagnostic and therapeutic applications in laser medicine and medical diagnostics. Hemoglobin is a constituent in biological tissue that exhibits several absorption bands. Hemoglobin molecules within the red blood cells (erythrocytes) carry 97% of the oxygen in the blood, while the remaining 3% is dissolved in the plasma. Each hemoglobin molecule contains four iron-containing heme groups as well as the protein globin. Oxygen atoms easily bind to the iron, causing the hemoglobin to assume different structures. Hemoglobin in the oxygenated state is referred to as oxyhemoglobin ($HbO_2$); in the reduced state, it is called deoxyhemoglobin (Hb). If the hemoglobin molecule is bound to carbon monoxide, then it becomes carboxyhemoglobin (HbCO). The porphyrin ring system is mainly responsible for the color in heme protein. This relatively large molecular system has a high degree of electron delocalization, which results in a decrease in the energy required for a transition to occur. As a result, the absorption of porphyrin occurs in the visible range. For example, the absorption spectrum of Fe(III) hemoprotein exhibits several bands at 400, 500, 530, 590, 625, and 1000 nm.[34]

The absorption spectrum of hemoglobin changes when binding occurs. Oxygenated hemoglobin is a strong absorber up to 600 nm; then its absorption drops off very steeply, by almost two orders of magnitude, and remains low. The absorption of deoxygenated hemoglobin, however, does not drop dramatically; it stays relatively high, although it decreases with increasing wavelengths. The isobestic point where the two spectra intersect occurs at about 800 nm.[36] The optical properties of blood depend strongly on physiological parameters such as oxygen saturation, osmolarity, flow conditions, and hematocrit.

The integrating sphere technique and inverse MC simulations were applied to measure $\mu_a$, $\mu_s$, and $g$ of circulating human blood.[37] At 633 nm the optical properties of human blood with a hematocrit of 10% and an oxygen saturation of 98% were found to be $2.10 \pm 0.02$ cm$^{-1}$ for $\mu_a$, $773 \pm 0.5$ cm$^{-1}$ for $\mu_s$, and $0.994 \pm 0.001$ for the $g$ factor. An increase of the hematocrit up to 50% led to a linear increase of absorption and reduced scattering. Variations in osmolarity and wall shear rate led to changes of all three parameters, while variations in the oxygen saturation led only to a significant change in the absorption coefficient. Spectra of all three parameters were measured in the wavelength range of 400 to 2500 nm for oxygenated and deoxygenated blood; the results showed that blood absorption follows the absorption behavior of hemoglobin and water. Figure 2.21 shows the extinction coefficient of hemoglobin.[36]

Chance and co-workers employed time-resolved spectroscopy to measure photon path lengths and thereby determine hemoglobin concentration in tissue.[38] Although deoxyhemoglobin and melanin are strong absorbers inside the diagnostic window, their relatively low concentrations in tissues explain why they do not significantly affect transmission processes. However, when light strikes a blood vessel, it encounters the full strong absorption of whole blood.

*Water.* At the NIR end of the window, the dominant absorber is water because of transitions between its vibrational energy states. The absorption of water in the spectral range of 700 to 1100 nm is shown in Figure 2.22. Note that even within the water transmission window (UV to 930 nm) where the absorption of water is rather low, water still contributes significantly to the overall attenuation of tissue because its concentration is very high in biological tissue.

*Skull.* The skull has been studied in the therapeutic window, and its absorption is quite low (0.02 to 0.05 mm$^{-1}$) compared with that of soft tissue.[39]

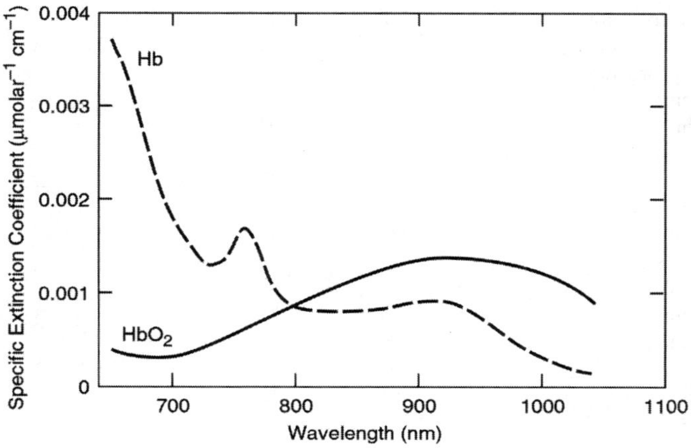


**FIGURE 2.21** Extinction coefficient of hemoglobin. (Adapted from Delpy, D. and Cope, M., *Philos. Trans. R. Soc. Lond. B*, 352, 1997.)

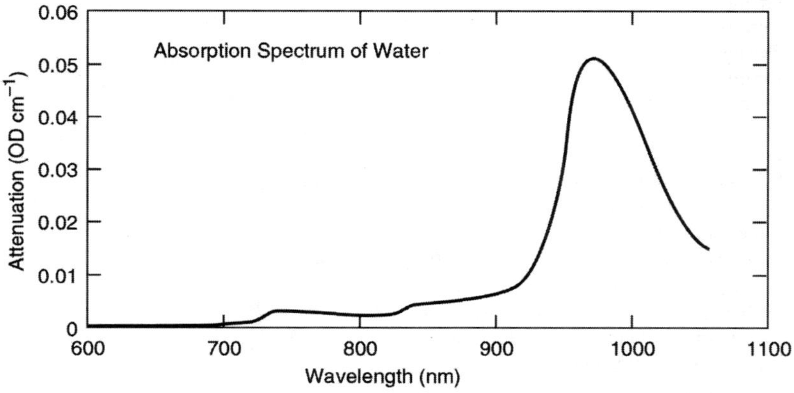


**FIGURE 2.22** Absorption of water.

**TABLE 2.4** Absorption Coefficient $\mu_a$ for Various Tissues, Tissue Components, and Biofluids

| Tissue | Description | Wavelength (nm) | $\mu_a$ (1/cm) | Uncert. (1/cm) | Notes | Ref. |
|---|---|---|---|---|---|---|
| Skin, *in vitro* | Stratum corneum | 193 | 6000 | | | 29 |
| | Stratum corneum | 250 | 1150 | | | 45, 46* |
| | Stratum corneum | 308 | 600 | | | 45, 46* |
| | Stratum corneum | 337 | 330 | | | 45, 46* |
| | Stratum corneum | 351 | 300 | | | 45, 46* |
| | Stratum corneum | 400 | 230 | | | 45, 46* |
| | Epidermis | 250 | 1000 | | | 45, 46* |
| | Epidermis | 308 | 300 | | | 45, 46* |
| | Epidermis | 337 | 120 | | | 45, 46* |
| | Epidermis | 351 | 100 | | | 45, 46* |
| | Epidermis | 415 | 66 | | | 45, 46* |
| | Epidermis | 488 | 50 | | | 45, 46* |
| | Epidermis | 514 | 44 | | | 45, 46* |
| | Epidermis | 585 | 36 | | | 45, 46* |
| | Epidermis | 633 | 35 | | | 45, 46* |
| | Epidermis | 800 | 40 | | | 45, 46* |
| | Dermis | 250 | 35 | | | 45, 46* |
| | Dermis | 308 | 12 | | | 45, 46* |
| | Dermis | 337 | 8.2 | | | 45, 46* |
| | Dermis | 351 | 7 | | | 45, 46* |
| | Dermis | 415 | 4.7 | | | 45, 46* |
| | Dermis | 488 | 3.5 | | | 45, 46* |
| | Dermis | 514 | 3 | | | 45, 46* |
| | Dermis | 585 | 3 | | | 45, 46* |
| | Dermis | 633 | 2.7 | | | 45, 46* |
| | Dermis | 800 | 2.3 | | | 45, 46* |
| | Skin:blood | 517 | 354 | | | 47 |
| | Skin:blood | 585 | 191 | | | 47 |
| | Skin:blood | 590 | 69 | | | 47 |
| | Skin:blood | 595 | 43 | | | 47 |
| | Skin:blood | 600 | 25 | | | 47 |
| | Dermis (leg) | 635 | 1.8 | 0.2 | | 48 |
| | Dermis | 749 | 0.24 | 0.19 | | 49 |
| | Dermis | 789 | 0.75 | 0.06 | | 49 |
| | Dermis | 836 | 0.98 | 0.15 | | 49 |
| | Dermis | 633 | <10 | | | 50 |
| | Dermis | 700 | 2.7 | 1 | | 51 |
| | Dermis | 633 | 1.9 | 0.6 | | 51 |
| | Dermis | 633 | 1.5 | | | 50 |
| | Skin and underlying tissues, including vein wall (leg) | 633 | 3.1 | | | 52 |
| | Caucasian male skin ($n = 3$) | 500 | 5.1 | | | 53 |
| | Caucasian male skin ($n = 3$) | 810 | 0.26 | | | 53 |
| | Caucasian male skin ($n = 3$), external pressure 9.8 kPa | 500 | 15.3 | | | 53 |
| | Caucasian male skin ($n = 3$), external pressure 9.8 kPa | 810 | 0.63 | | | 53 |
| | Caucasian male skin ($n = 3$), external pressure 98.1 kPa | 500 | 13.6 | | | 53 |
| | Caucasian male skin ($n = 3$), external pressure 98.1 kPa | 810 | 0.57 | | | 53 |
| | Caucasian female skin ($n = 3$) | 500 | 5.2 | | | 53 |
| | Caucasian female skin ($n = 3$) | 810 | 0.97 | | | 53 |
| | Caucasian female skin ($n = 3$), external pressure 9.8 kPa | 500 | 7.4 | | | 53 |

**TABLE 2.4** Absorption Coefficient $\mu_a$ for Various Tissues, Tissue Components, and Biofluids (continued)

| Tissue | Description | Wavelength (nm) | $\mu_a$ (1/cm) | Uncert. (1/cm) | Notes | Ref. |
|---|---|---|---|---|---|---|
| | Caucasian female skin ($n$ = 3), external pressure 9.8 kPa | 810 | 1.4 | | | 53 |
| | Caucasian female skin ($n$ = 3), external pressure 98.1 kPa | 500 | 10 | | | 53 |
| | Caucasian female skin ($n$ = 3), external pressure 98.1 kPa | 810 | 1.7 | | | 53 |
| | Hispanic male skin ($n$ = 3) | 500 | 3.8 | | | 53 |
| | Hispanic male skin ($n$ = 3) | 810 | 0.87 | | | 53 |
| | Hispanic male skin ($n$ = 3), external pressure 9.8 kPa | 500 | 5.1 | | | 53 |
| | Hispanic male skin ($n$ = 3), external pressure 9.8 kPa | 810 | 0.93 | | | 53 |
| | Hispanic male skin ($n$ = 3), external pressure 98.1 kPa | 500 | 6.2 | | | 53 |
| | Hispanic male skin ($n$ = 3), external pressure 98.1 kPa | 810 | 0.87 | | | 53 |
| | Forearm:fat | 633 | 0.026 | | | 54 |
| | Forearm:muscle | 633 | 0.96 | | | 54 |
| Skin, *in vivo* | Skin (dermis) | 660 | 0.07–0.2 | | | 51 |
| | Skin | 633 | 0.62 | | | 55 |
| | Skin | 700 | 0.38 | | | 55 |
| | Skin (0–1 mm) | 633 | 0.67 | | | 47 |
| | Skin (1–2 mm) | 633 | 0.026 | | | 47 |
| | Skin (>2 mm) | 633 | 0.96 | | | 47 |
| | Forearm:epidermis | 633 | 8 | | | 54 |
| | Forearm:dermis | 633 | 0.15 | | | 54 |
| | Forearm (5 subjects, 14 measurements) | 800 | 0.23 | 0.04 | | 49 |
| | Arm | 633 | 0.17 | 0.01 | | 55 |
| | Arm | 660 | 0.128 | 0.005 | | 55 |
| | Arm | 700 | 0.09 | 0.002 | | 55 |
| | Foot sole | 633 | 0.072 | 0.002 | | 55 |
| | Foot sole | 660 | 0.053 | 0.003 | | 55 |
| | Foot sole | 700 | 0.037 | 0.002 | | 55 |
| | Forehead | 633 | 0.09 | 0.009 | | 55 |
| | Forehead | 660 | 0.052 | 0.003 | | 55 |
| | Forehead | 700 | 0.024 | 0.002 | | 55 |
| Brain/head, *in vitro* | White matter (female, 32 yrs. old, 24 h post mortem) | 415 | 2.1 | | | 46, 56* |
| | White matter (female, 32 yrs. old, 24 h post mortem) | 488 | 1 | | | 46, 56* |
| | White matter (female, 32 yrs. old, 24 h post mortem) | 630 | 0.2 | | | 46, 56* |
| | White matter (female, 32 yrs. old, 24 h post mortem) | 800–1100 | 0.2–0.3 | | | 46, 56* |
| | White matter (female, 63 yrs. old, 30 h post mortem) | 488 | 2.7 | | | 46, 56* |
| | White matter (female, 63 yrs. old, 30 h post mortem) | 633 | 0.9 | | | 46, 56* |
| | White matter (female, 63 yrs. old, 30 h post mortem) | 800–1100 | 1.0–1.5 | | | 46, 56* |
| | Gray matter (male, 71 yrs. old, 24 h post mortem) | 514 | 19.5 | | | 46, 56* |
| | Gray matter (male, 71 yrs. old, 24 h post mortem) | 585 | 14.5 | | | 46, 56* |

**TABLE 2.4** Absorption Coefficient $\mu_a$ for Various Tissues, Tissue Components, and Biofluids (continued)

| Tissue | Description | Wavelength (nm) | $\mu_a$ (1/cm) | Uncert. (1/cm) | Notes | Ref. |
|---|---|---|---|---|---|---|
| | Gray matter (male, 71 yrs. old, 24 h post mortem) | 630 | 4.3 | | | 46, 56* |
| | Gray matter (male, 71 yrs. old, 24 h post mortem) | 800–1100 | ~1.0 | | | 46, 56* |
| | Glioma (male, 65 yrs. old, 4 h post mortem) | 415 | 16.6 | | | 46, 56* |
| | Glioma (male, 65 yrs. old, 4 h post mortem) | 488 | 12.5 | | | 46, 56* |
| | Glioma (male, 65 yrs. old, 4 h post mortem) | 630 | 3 | | | 46, 56* |
| | Glioma (male, 65 yrs. old, 4 h post mortem) | 800–1100 | 1 | | | 46, 56* |
| | Melanoma (male, 71 yrs. old, 24 h post mortem) | 585 | 2 | | | 46, 56* |
| | Melanoma (male, 71 yrs. old, 24 h post mortem) | 630 | 20 | | | 46, 56* |
| | Melanoma (male, 71 yrs. old, 24 h post mortem) | 800 | 8 | | | 46, 56* |
| | Melanoma (male, 71 yrs. old, 24 h post mortem) | 900 | 4 | | | 46, 56* |
| | Melanoma (male, 71 yrs. old, 24 h post mortem) | 1100 | 2 | | | 46, 56* |
| | White matter | 633 | 2.2 | 2 | Freshly resected slabs | 46 |
| | White matter | 1064 | 3.2 | 4 | Freshly resected slabs | 46 |
| | Gray matter | 633 | 2.7 | 2 | Freshly resected slabs | 46 |
| | Gray matter | 1064 | 5 | 5 | Freshly resected slabs | 46 |
| | Adult head: scalp and skull | 800 | 0.4 | | | 57 |
| | Adult head: cerebrospinal fluid | 800 | 0.01 | | | 57 |
| | Adult head: gray matter | 800 | 0.25 | | | 57 |
| | Adult head: white matter | 800 | 0.05 | | | 57 |
| | White matter ($n$ = 7) | 360 | 2.53 | 0.55 | | 58 |
| | White matter ($n$ = 7) | 640 | 0.8 | | | 58 |
| | White matter ($n$ = 7) | 860 | 0.97 | 0.4 | | 58 |
| | White matter ($n$ = 7) | 1060 | 1.08 | 0.51 | | 58 |
| | White matter | 850 | 0.8 | 0.16 | | 59 |
| | White matter | 1064 | 0.4 | 0.08 | | 59 |
| | White matter | 1064 | 1.6 | | | 58 |
| | White matter, coagulated ($n$ = 7) | 360 | 8.3 | 3.7 | Coag. 2 h 80°C | 58 |
| | White matter, coagulated ($n$ = 7) | 860 | 1.7 | 1.3 | Coag. 2 h 80°C | 58 |
| | White matter, coagulated ($n$ = 7) | 1060 | 2.15 | 1.3 | Coag. 2 h 80°C | 58 |
| | White matter, coagulated | 850 | 0.9 | 0.2 | Homogenized tissue, coag. 75°C | 59 |
| | White matter, coagulated | 1064 | 0.5 | 0.1 | Homogenized tissue, coag. 75°C | 59 |
| | Gray matter ($n$ = 7) | 360 | 3.33 | 2.2 | | 58 |
| | Gray matter ($n$ = 7) | 640 | 0.17 | 0.3 | | 58 |
| | Gray matter ($n$ = 7) | 1060 | 0.56 | 0.7 | | 58 |
| | Gray matter | 1064 | 1.9 | | | 58 |
| | Gray matter, coagulated ($n$ = 7) | 360 | 9.39 | 1.7 | Coag. 2 h 80°C | 58 |
| | Gray matter, coagulated ($n$ = 7) | 740 | 0.45 | 0.3 | Coag. 2 h 80°C | 58 |
| | Gray matter, coagulated ($n$ = 7) | 1100 | 1 | 0.5 | Coag. 2 h 80°C | 58 |
| | Brain tumor, astrocytoma grade III WHO | 400 | 10 | | | 60 |
| | Brain tumor, astrocytoma grade III WHO | 633 | 6.3 | 1.6 | | 60 |

**TABLE 2.4** Absorption Coefficient $\mu_a$ for Various Tissues, Tissue Components, and Biofluids (continued)

| Tissue | Description | Wavelength (nm) | $\mu_a$ (1/cm) | Uncert. (1/cm) | Notes | Ref. |
|---|---|---|---|---|---|---|
| | Brain tumor, astrocytoma grade III WHO | 700 | 4 | | | 60 |
| | Brain tumor, astrocytoma grade III WHO | 800 | 3 | | | 60 |
| | White matter ($n$ = 7) | 450 | 1.4 | | <48 h Post mortem | 61 |
| | White matter ($n$ = 7) | 510 | 1 | | <48 h Post mortem | 61 |
| | White matter ($n$ = 7) | 630 | 0.8 | | <48 h Post mortem | 61 |
| | White matter ($n$ = 7) | 670 | 0.7 | | <48 h Post mortem | 61 |
| | White matter ($n$ = 7) | 850 | 1 | | <48 h Post mortem | 61 |
| | White matter ($n$ = 7) | 1064 | 1 | | <48 h Post mortem | 61 |
| | Gray matter ($n$ = 7) | 450 | 0.7 | | <48 h Post mortem | 61 |
| | Gray matter ($n$ = 7) | 510 | 0.4 | | <48 h Post mortem | 61 |
| | Gray matter ($n$ = 7) | 630 | 0.2 | | <48 h Post mortem | 61 |
| | Gray matter ($n$ = 7) | 670 | 0.2 | | <48 h Post mortem | 61 |
| | Gray matter ($n$ = 7) | 1064 | 0.5 | | <48 h Post mortem | 61 |
| | White matter | 456 | 8.1 | | | 62 |
| | White matter | 514 | 5 | | | 62 |
| | White matter | 630 | 1.5 | | | 62 |
| | White matter | 675 | 0.7 | | | 62 |
| | White matter | 1064 | 1.6 | | | 62 |
| | Gray matter | 456 | 9 | | | 62 |
| | Gray matter | 514 | 11.7 | | | 62 |
| | Gray matter | 630 | 1.4 | | | 62 |
| | Gray matter | 675 | 0.6 | | | 62 |
| | Gray matter | 1064 | 1.9 | | | 62 |
| | Gray matter ($n$ = 7) | 400 | 2.6 | 0.6 | <48 h Post mortem | 61* |
| | Gray matter ($n$ = 7) | 500 | 0.5 | 0.2 | <48 h Post mortem | 61* |
| | Gray matter ($n$ = 7) | 600 | 0.3 | 0.1 | <48 h Post mortem | 61* |
| | Gray matter ($n$ = 7) | 700 | 0.2 | 0.1 | <48 h Post mortem | 61* |
| | Gray matter ($n$ = 7) | 800 | 0.2 | 0.1 | <48 h Post mortem | 61* |
| | Gray matter ($n$ = 7) | 900 | 0.3 | 0.2 | <48 h Post mortem | 61* |
| | Gray matter ($n$ = 7) | 1000 | 0.6 | 0.3 | <48 h Post mortem | 61* |
| | Gray matter ($n$ = 7) | 1100 | 0.5 | 0.3 | <48 h Post mortem | 61* |
| | Gray matter, coagulated ($n$ = 7) | 400 | 7.5 | 0.4 | Coag. 2 h 80°C | 61* |
| | Gray matter, coagulated ($n$ = 7) | 500 | 1.8 | 0.2 | Coag.: 2 h 80°C | 61* |
| | Gray matter, coagulated ($n$ = 7) | 600 | 0.7 | 0.0 | Coag. 2 h 80°C | 61* |
| | Gray matter, coagulated ($n$ = 7) | 700 | 0.7 | 0.1 | Coag. 2 h 80°C | 61* |
| | Gray matter, coagulated ($n$ = 7) | 800 | 0.8 | 0.1 | Coag. 2 h 80°C | 61* |
| | Gray matter, coagulated ($n$ = 7) | 900 | 0.9 | 0.1 | Coag. 2 h 80°C | 61* |
| | Gray matter, coagulated ($n$ = 7) | 1000 | 1.4 | 0.2 | Coag. 2 h 80°C | 61* |
| | Gray matter, coagulated ($n$ = 7) | 1100 | 1.5 | 0.2 | Coag. 2 h 80°C | 61* |
| | White matter ($n$ = 7) | 400 | 3.1 | 0.2 | <48 h Post mortem | 61* |
| | White matter ($n$ = 7) | 500 | 0.9 | 0.1 | <48 h Post mortem | 61* |
| | White matter ($n$ = 7) | 600 | 0.8 | 0.1 | <48 h Post mortem | 61* |
| | White matter ($n$ = 7) | 700 | 0.8 | 0.1 | <48 h Post mortem | 61* |
| | White matter ($n$ = 7) | 800 | 0.9 | 0.1 | <48 h Post mortem | 61* |
| | White matter ($n$ = 7) | 900 | 1.0 | 0.1 | <48 h Post mortem | 61* |
| | White matter ($n$ = 7) | 1000 | 1.2 | 0.2 | <48 h Post mortem | 61* |
| | White matter ($n$ = 7) | 1100 | 1.0 | 0.2 | <48 h Post mortem | 61* |
| | White matter, coagulated ($n$ = 7) | 410 | 8.7 | 1.7 | Coag. 2 h 80°C | 61* |
| | White matter, coagulated ($n$ = 7) | 510 | 2.9 | 0.6 | Coag. 2 h 80°C | 61* |
| | White matter, coagulated ($n$ = 7) | 610 | 1.7 | 0.4 | Coag. 2 h 80°C | 61* |
| | White matter, coagulated ($n$ = 7) | 710 | 1.4 | 0.5 | Coag. 2 h 80°C | 61* |
| | White matter, coagulated ($n$ = 7) | 810 | 1.5 | 0.5 | Coag. 2 h 80°C | 61* |
| | White matter, coagulated ($n$ = 7) | 910 | 1.7 | 0.6 | Coag. 2 h 80°C | 61* |
| | White matter, coagulated ($n$ = 7) | 1010 | 1.9 | 0.6 | Coag. 2 h 80°C | 61* |

**TABLE 2.4** Absorption Coefficient $\mu_a$ for Various Tissues, Tissue Components, and Biofluids (continued)

| Tissue | Description | Wavelength (nm) | $\mu_a$ (1/cm) | Uncert. (1/cm) | Notes | Ref. |
|---|---|---|---|---|---|---|
| | White matter, coagulated ($n = 7$) | 1100 | 2.4 | 0.5 | Coag. 2 h 80°C | 61* |
| | Astrocytoma (tumor, WHO grade II, $n = 4$) | 400 | 18.8 | 11.3 | Excised from patients | 61* |
| | Astrocytoma (tumor, WHO grade II, $n = 4$) | 490 | 2.5 | 0.9 | Excised from patients | 61* |
| | Astrocytoma (tumor, WHO grade II, $n = 4$) | 600 | 1.2 | 0.7 | Excised from patients | 61* |
| | Astrocytoma (tumor, WHO grade II, $n = 4$) | 700 | 0.5 | 0.3 | Excised from patients | 61* |
| | Astrocytoma (tumor, WHO grade II, $n = 4$) | 800 | 0.7 | 0.2 | Excised from patients | 61* |
| | Astrocytoma (tumor, WHO grade II, $n = 4$) | 900 | 0.3 | 0.2 | Excised from patients | 61* |
| | Astrocytoma (tumor, WHO grade II, $n = 4$) | 1000 | 0.5 | 0.3 | Excised from patients | 61* |
| | Astrocytoma (tumor, WHO grade II, $n = 4$) | 1100 | 0.6 | 0.2 | Excised from patients | 61* |
| | Cerebellum ($n = 7$) | 400 | 4.7 | 0.8 | <48 h Post mortem | 61* |
| | Cerebellum ($n = 7$) | 500 | 1.4 | 0.2 | <48 h Post mortem | 61* |
| | Cerebellum ($n = 7$) | 600 | 0.8 | 0.2 | <48 h Post mortem | 61* |
| | Cerebellum ($n = 7$) | 700 | 0.6 | 0.1 | <48 h Post mortem | 61* |
| | Cerebellum ($n = 7$) | 800 | 0.6 | 0.1 | <48 h Post mortem | 61* |
| | Cerebellum ($n = 7$) | 900 | 0.7 | 0.1 | <48 h Post mortem | 61* |
| | Cerebellum ($n = 7$) | 1000 | 0.8 | 0.1 | <48 h Post mortem | 61* |
| | Cerebellum ($n = 7$) | 1100 | 0.7 | 0.1 | <48 h Post mortem | 61* |
| | Cerebellum, coagulated ($n = 7$) | 400 | 19.3 | 7.7 | Coag. 2 h 80°C | 61* |
| | Cerebellum, coagulated ($n = 7$) | 500 | 5.1 | 1.7 | Coag. 2 h 80°C | 61* |
| | Cerebellum, coagulated ($n = 7$) | 600 | 2.9 | 1.4 | Coag. 2 h 80°C | 61* |
| | Cerebellum, coagulated ($n = 7$) | 700 | 1.7 | 0.4 | Coag. 2 h 80°C | 61* |
| | Cerebellum, coagulated ($n = 7$) | 800 | 1.1 | 0.2 | Coag. 2 h 80°C | 61* |
| | Cerebellum, coagulated ($n = 7$) | 900 | 1.1 | 0.3 | Coag. 2 h 80°C | 61* |
| | Cerebellum, coagulated ($n = 7$) | 1000 | 1.0 | 0.4 | Coag. 2 h 80°C | 61* |
| | Cerebellum, coagulated ($n = 7$) | 1100 | 1.1 | 0.5 | Coag. 2 h 80°C | 61* |
| | Meningioma (tumor, $n = 6$) | 410 | 4.1 | 0.5 | Excised from patients | 61* |
| | Meningioma (tumor, $n = 6$) | 490 | 1.3 | 0.2 | Excised from patients | 61* |
| | Meningioma (tumor, $n = 6$) | 590 | 0.7 | 0.2 | Excised from patients | 61* |
| | Meningioma (tumor, $n = 6$) | 690 | 0.3 | 0.1 | Excised from patients | 61* |
| | Meningioma (tumor, $n = 6$) | 790 | 0.2 | 0.1 | Excised from patients | 61* |
| | Meningioma (tumor, $n = 6$) | 910 | 0.2 | 0.1 | Excised from patients | 61* |
| | Meningioma (tumor, $n = 6$) | 990 | 0.4 | 0.2 | Excised from patients | 61* |
| | Meningioma (tumor, $n = 6$) | 1100 | 0.6 | 0.2 | Excised from patients | 61* |
| | Pons ($n = 7$) | 400 | 3.1 | 0.7 | <48 h Post mortem | 61* |
| | Pons ($n = 7$) | 500 | 0.9 | 0.3 | <48 h Post mortem | 61* |
| | Pons ($n = 7$) | 600 | 0.6 | 0.2 | <48 h Post mortem | 61* |
| | Pons ($n = 7$) | 700 | 0.5 | 0.2 | <48 h Post mortem | 61* |
| | Pons ($n = 7$) | 800 | 0.6 | 0.3 | <48 h Post mortem | 61* |
| | Pons ($n = 7$) | 900 | 0.7 | 0.3 | <48 h Post mortem | 61* |
| | Pons ($n = 7$) | 1000 | 1.0 | 0.4 | <48 h Post mortem | 61* |
| | Pons ($n = 7$) | 1100 | 0.9 | 0.4 | <48 h Post mortem | 61* |
| | Pons, coagulated ($n = 7$) | 410 | 17.2 | 1.6 | Coag. 2 h 80°C | 61* |
| | Pons, coagulated ($n = 7$) | 510 | 8.5 | 0.8 | Coag. 2 h 80°C | 61* |
| | Pons, coagulated ($n = 7$) | 610 | 7.7 | 0.5 | Coag. 2 h 80°C | 61* |
| | Pons, coagulated ($n = 7$) | 710 | 6.9 | 0.6 | Coag. 2 h 80°C | 61* |
| | Pons, coagulated ($n = 7$) | 810 | 6.5 | 0.6 | Coag. 2 h 80°C | 61* |
| | Pons, coagulated ($n = 7$) | 910 | 5.9 | 1.0 | Coag. 2 h 80°C | 61* |
| | Pons, coagulated ($n = 7$) | 1010 | 5.7 | 1.0 | Coag. 2 h 80°C | 61* |

**TABLE 2.4** Absorption Coefficient $\mu_a$ for Various Tissues, Tissue Components, and Biofluids (continued)

| Tissue | Description | Wavelength (nm) | $\mu_a$ (1/cm) | Uncert. (1/cm) | Notes | Ref. |
|---|---|---|---|---|---|---|
| | Pons, coagulated ($n$ = 7) | 1100 | 6.5 | 0.9 | Coag. 2 h 80°C | 61* |
| | Thalamus ($n$ = 7) | 410 | 3.2 | 1.0 | <48 h Post mortem | 61* |
| | Thalamus ($n$ = 7) | 510 | 0.9 | 0.3 | <48 h Post mortem | 61* |
| | Thalamus ($n$ = 7) | 610 | 0.6 | 0.2 | <48 h Post mortem | 61* |
| | Thalamus ($n$ = 7) | 710 | 0.5 | 0.3 | <48 h Post mortem | 61* |
| | Thalamus ($n$ = 7) | 810 | 0.7 | 0.3 | <48 h Post mortem | 61* |
| | Thalamus ($n$ = 7) | 910 | 0.7 | 0.3 | <48 h Post mortem | 61* |
| | Thalamus ($n$ = 7) | 1010 | 0.8 | 0.3 | <48 h Post mortem | 61* |
| | Thalamus ($n$ = 7) | 1100 | 0.8 | 0.3 | <48 h Post mortem | 61* |
| | Thalamus, coagulated ($n$ = 7) | 400 | 15.0 | 3.3 | Coag. 2 h 80°C | 61* |
| | Thalamus, coagulated ($n$ = 7) | 500 | 4.2 | 0.9 | Coag. 2 h 80°C | 61* |
| | Thalamus, coagulated ($n$ = 7) | 600 | 1.6 | 0.6 | Coag. 2 h 80°C | 61* |
| | Thalamus, coagulated ($n$ = 7) | 700 | 1.4 | 0.3 | Coag. 2 h 80°C | 61* |
| | Thalamus, coagulated ($n$ = 7) | 800 | 1.1 | 0.3 | Coag. 2 h 80°C | 61* |
| | Thalamus, coagulated ($n$ = 7) | 900 | 1.1 | 0.3 | Coag. 2 h 80°C | 61* |
| | Thalamus, coagulated ($n$ = 7) | 1000 | 1.4 | 0.4 | Coag. 2 h 80°C | 61* |
| | Thalamus, coagulated ($n$ = 7) | 1100 | 1.5 | 0.4 | Coag. 2 h 80°C | 61* |
| Brain/head, *in vivo* | Cortex, frontal lobe | 674 | <0.2 | 0.1 | | 63 |
| | Cortex, frontal lobe | 811 | <0.1 | 0.1 | | 63 |
| | Cortex, frontal lobe | 849 | <0.1 | 0.1 | | 63 |
| | Cortex, frontal lobe | 956 | 0.15 | 0.1 | | 63 |
| | Cortex, frontal lobe | 674 | 0.2 | 0.1 | | 63 |
| | Cortex, frontal lobe | 811 | 0.2 | 0.1 | | 63 |
| | Cortex, frontal lobe | 849 | <0.1 | 0.1 | | 63 |
| | Cortex, frontal lobe | 956 | 0.25 | 0.1 | | 63 |
| | Astrocytoma of optic nerve | 674 | 1.4 | 0.3 | | 63 |
| | Astrocytoma of optic nerve | 811 | 1.2 | 0.3 | | 63 |
| | Astrocytoma of optic nerve | 849 | 0.9 | 0.3 | | 63 |
| | Astrocytoma of optic nerve | 956 | 1.5 | 0.3 | | 63 |
| | Normal optic nerve | 674 | 0.6 | 0.3 | | 63 |
| | Normal optic nerve | 849 | 0.8 | 0.3 | | 63 |
| | Normal optic nerve | 956 | 0.7 | 0.3 | | 63 |
| | Skull | 674 | 0.5 | 0.2 | | 63 |
| | Skull | 849 | 0.5 | 0.2 | | 63 |
| | Skull | 956 | 0.5 | 0.2 | | 63 |
| | Cerebellar white matter | 674 | 2.5 | 0.5 | | 63 |
| | Cerebellar white matter | 849 | 0.95 | 0.2 | | 63 |
| | Cerebellar white matter | 956 | 0.9 | 0.2 | | 63 |
| | Medulloblastoma | 674 | 2.6 | 0.5 | | 63 |
| | Medulloblastoma | 849 | 1 | 0.2 | | 63 |
| | Medulloblastoma | 956 | 0.75 | 0.2 | | 63 |
| | Cerebellar white matter with scar tissue | 674 | <0.2 | | | 63 |
| | Cerebellar white matter with scar tissue | 849 | <0.2 | | | 63 |
| | Normal cortex, temporal and frontal lobe | 674 | >0.2 | | Measurements during surgery | 64 |
| | Normal cortex, temporal and frontal lobe | 849 | >0.2 | | Measurements during surgery | 64 |
| | Normal cortex, temporal and frontal lobe | 956 | >0.2 | | Measurements during surgery | 64 |
| | Normal optic nerve | 674 | 0.6 | 0.25 | Measurements during surgery | 64 |
| | Normal optic nerve | 849 | 0.75 | 0.25 | Measurements during surgery | 64 |

**TABLE 2.4** Absorption Coefficient $\mu_a$ for Various Tissues, Tissue Components, and Biofluids (continued)

| Tissue | Description | Wavelength (nm) | $\mu_a$ (1/cm) | Uncert. (1/cm) | Notes | Ref. |
|---|---|---|---|---|---|---|
| | Normal optic nerve | 956 | 0.65 | 0.25 | Measurements during surgery | 64 |
| | Astrocytoma of optical nerve | 674 | 1.6 | 1 | Measurements during surgery | 64 |
| | Astrocytoma of optical nerve | 849 | 1.1 | 1 | Measurements during surgery | 64 |
| | Astrocytoma of optical nerve | 950 | 1.8 | 1 | Measurements during surgery | 64 |
| Female breast, *in vitro* | Fatty normal (*n* = 23, excised) | 749 | 0.18 | 0.16 | Kept in saline, 37°C | 49 |
| | Fatty normal (*n* = 23, excised) | 789 | 0.08 | 0.1 | Kept in saline, 37°C | 49 |
| | Fatty normal (*n* = 23, excised) | 836 | 0.11 | 0.1 | Kept in saline, 37°C | 49 |
| | Fibrous normal (*n* = 35, excised) | 749 | 0.13 | 0.19 | Kept in saline, 37°C | 49 |
| | Fibrous normal (*n* = 35, excised) | 789 | 0.06 | 0.12 | Kept in saline, 37°C | 49 |
| | Fibrous normal (*n* = 35, excised) | 836 | 0.05 | 0.08 | Kept in saline, 37°C | 49 |
| | Infiltrating carcinoma (*n* = 48, excised) | 749 | 0.15 | 0.14 | Kept in saline, 37°C | 49 |
| | Infiltrating carcinoma (*n* = 48, excised) | 789 | 0.04 | 0.08 | Kept in saline, 37°C | 49 |
| | Infiltrating carcinoma (*n* = 48, excised) | 836 | 0.1 | 0.19 | Kept in saline, 37°C | 49 |
| | Mucinous carcinoma (*n* = 3, excised) | 749 | 0.26 | 0.2 | Kept in saline, 37°C | 49 |
| | Ductal carcinoma, *in situ* (*n* = 5, excised) | 749 | 0.076 | 0.068 | Kept in saline, 37°C | 49 |
| | Ductal carcinoma, *in situ* (*n* = 5, excised) | 836 | 0.039 | 0.068 | Kept in saline, 37°C | 49 |
| | Grandular tissue (*n* = 3) | 540 | 3.58 | 1.56 | Homogenized tissue | 65 |
| | Grandular tissue (*n* = 3) | 700 | 0.47 | 0.11 | Homogenized tissue | 65 |
| | Grandular tissue (*n* = 3) | 900 | 0.62 | 0.05 | Homogenized tissue | 65 |
| | Fatty tissue (*n* = 7) | 540 | 2.27 | 0.57 | Homogenized tissue | 65 |
| | Fatty tissue (*n* = 7) | 700 | 0.7 | 0.08 | Homogenized tissue | 65 |
| | Fatty tissue (*n* = 7) | 900 | 0.75 | 0.08 | Homogenized tissue | 65 |
| | Fibrocystic tissue (*n* = 8) | 540 | 1.64 | 0.66 | Homogenized tissue | 65 |
| | Fibrocystic tissue (*n* = 8) | 700 | 0.22 | 0.09 | Homogenized tissue | 65 |
| | Fibrocystic tissue (*n* = 8) | 900 | 0.27 | 0.11 | Homogenized tissue | 65 |
| | Fibroadenoma (*n* = 6) | 540 | 4.38 | 3.14 | Homogenized tissue | 65 |
| | Fibroadenoma (*n* = 6) | 700 | 0.52 | 0.47 | Homogenized tissue | 65 |
| | Fibroadenoma (*n* = 6) | 900 | 0.72 | 0.53 | Homogenized tissue | 65 |
| | Carcinoma (*n* = 9) | 540 | 3.07 | 0.99 | Homogenized tissue | 65 |
| | Carcinoma (*n* = 9) | 700 | 0.45 | 0.12 | Homogenized tissue | 65 |
| | Carcinoma (*n* = 9) | 900 | 0.5 | 0.15 | Homogenized tissue | 65 |
| | Carcinoma | 580 | 4.5 | 0.8 | | 66 |
| | Healthy tissue adjacent to carcinoma | 580 | 2.6 | 1.1 | | 66 |
| | Healthy tissue adjacent to carcinoma | 850 | 0.3 | 0.2 | | 66 |
| | Healthy tissue adjacent to carcinoma | 1300 | 0.8 | 0.6 | | 66 |
| | Fatty tissue | 625 | 0.06 | 0.02 | | 67 |
| | Benign tumor | 625 | 0.33 | 0.06 | | 67 |
| Female breast, *in vivo* | Normal (30 Japanese women averaged for all ages) | 753 | 0.046 | 0.014 | Time-domain technique | 68 |
| | Normal (6 subjects, 26–43 yrs. old) | 800 | 0.017–0.045 | | Time-domain technique | 69 |

**TABLE 2.4** Absorption Coefficient $\mu_a$ for Various Tissues, Tissue Components, and Biofluids (continued)

| Tissue | Description | Wavelength (nm) | $\mu_a$ (1/cm) | Uncert. (1/cm) | Notes | Ref. |
|---|---|---|---|---|---|---|
| | Normal (6 subjects) | 580 | 0.7 | 0.12 | Measurements of transmission | 66 |
| | Normal (6 subjects) | 780 | 0.23 | 0.02 | Measurements of transmission | 66 |
| | Normal (6 subjects) | 850 | 0.27 | 0.03 | Measurements of transmission | 66 |
| | Breast cancer (5 patients) | 630 | 0.305 | 0.16 | Relapsed cancer, HPD (72 h) | 70 |
| Aorta, *in vitro* | Post mortem, 6 h normal coagulated | 308 | 33 | | | 46 |
| | Post mortem, 6 h normal coagulated | 308 | 44 | | | 46 |
| | Post mortem, 6 h fibrous plaque | 308 | 24 | | | 46 |
| | Post mortem, 6 h fibrous plaque coagulated | 308 | 34 | | | 46 |
| | Normal | 1064 | 0.53 | 0.09 | | 46 |
| | Coagulated | 1064 | 0.46 | 0.18 | | 46 |
| | Fibro-fatty | 355 | 17.7 | | | 46 |
| | Fibro-fatty | 532 | 3.6 | | | 46 |
| | Fibro-fatty | 1064 | 0.09 | | | 46 |
| | | 633 | 0.52 | | | 46 |
| | | 1064 | 0.5 | | | 46 |
| | | 1064 | 0.7 | | | 46 |
| | | 1320 | 2.2 | | | 46 |
| | | 1320 | 4.3 | | | 46 |
| | | 470 | 5.3 | 0.9 | | 46 |
| | | 476 | 5.1 | 0.9 | | 46 |
| | | 488 | 4.5 | 0.9 | | 46 |
| | | 514.5 | 3.7 | 0.9 | | 46 |
| | | 580 | 2.8 | 0.9 | | 46 |
| | | 600 | 2.6 | 0.9 | | 46 |
| | | 633 | 2.6 | 0.9 | | 46 |
| | | 1064 | 2.7 | 0.5 | | 46 |
| | Intima | 476 | 14.8 | | | 29 |
| | Intima | 580 | 8.9 | | | 29 |
| | Intima | 600 | 4 | | | 29 |
| | Intima | 633 | 3.6 | | | 29 |
| | Intima | 1064 | 2.3 | | | 41 |
| | Media | 476 | 7.3 | | | 29 |
| | Media | 580 | 4.8 | | | 29 |
| | Media | 600 | 2.5 | | | 29 |
| | Media | 633 | 2.3 | | | 29 |
| | Media | 1064 | 1 | | | 41 |
| | Adventitia | 476 | 18.1 | | | 29, 71, 72 |
| | Adventitia | 580 | 11.3 | | | 29 |
| | Adventitia | 600 | 6.1 | | | 29, 71, 72 |
| | Adventitia | 633 | 5.8 | | | 29 |
| | Adventitia | 1064 | 2 | | | 41 |
| Aorta, *in vivo* | Advanced fibrous atheroma | 355 | 16.5 | 1.7 | Photoacoustic method | 73 |
| | Advanced fibrous atheroma | 532 | 3.53 | 0.53 | Photoacoustic method | 73 |
| | Advanced fibrous atheroma | 1064 | 0.15 | 0.07 | Photoacoustic method | 73 |
| Blood | Oxygenated, Hct = 0.41 | 665 | 1.3 | | | 29 |
| | Oxygenated, Hct = 0.41 | 685 | 2.65 | | | 29 |
| | Oxygenated, Hct = 0.41 | 960 | 2.84 | | | 29 |
| | Deoxygenated, Hct = 0.41 | 665 | 4.87 | | | 29 |

**TABLE 2.4** Absorption Coefficient $\mu_a$ for Various Tissues, Tissue Components, and Biofluids (continued)

| Tissue | Description | Wavelength (nm) | $\mu_a$ (1/cm) | Uncert. (1/cm) | Notes | Ref. |
|---|---|---|---|---|---|---|
| | Deoxygenated, Hct = 0.41 | 960 | 1.68 | | | 29 |
| | Oxygenation >98%, Hct = 0.45–0.46 | 633 | 15.5 | | | 74 |
| | Oxygenation >98%, Hct = 0.45–0.46 | 710 | 4 | 0.8 | | 74 |
| | Oxygenation >98%, Hct = 0.45–0.46 | 765 | 5.3 | 0.6 | | 74 |
| | Oxygenation >98%, Hct = 0.45–0.46 | 810 | 6.5 | 0.5 | | 74 |
| | Oxygenation >98%, Hct = 0.45–0.46 | 865 | 7.2 | 0.3 | | 74 |
| | Oxygenation >98%, Hct = 0.45–0.46 | 910 | 8.9 | 0.4 | | 74 |
| | Oxygenation >98%, Hct = 0.45–0.46 | 965 | 9.3 | 0.6 | | 74 |
| | Oxygenation >98%, Hct = 0.45–0.46 | 1010 | 8.3 | 0.4 | | 74 |
| | Oxygenation >98%, Hct = 0.45–0.46 | 1065 | 5.6 | 0.3 | | 74 |
| | Oxygenation >98%, Hct = 0.45–0.46 | 1110 | 4.2 | 0.3 | | 74 |
| | Oxygenation >98%, Hct = 0.45–0.46 | 1165 | 4.1 | 0.7 | | 74 |
| | Oxygenation >98%, Hct = 0.45–0.46 | 1210 | 5.5 | 0.5 | | 74 |
| | Oxygenation >98%, Hct = 0.38 | 633 | 15.2 | 0.6 | | 74 |
| | Oxygenation >98%, Hct = 0.38 | 633 | 16.1 | 0.6 | | 74 |
| | Oxygenation >98%, Hct = 0.38 | 633 | 16.3 | 0.5 | | 74 |
| | Oxygenated (98%) flowing blood, Hct = 0.1 | 633 | 2.1 | 0.02 | | 75 |
| | Oxygenated (98%) flowing blood, Hct = 0.2 | 633 | 4 | | | 75 |
| | Oxygenated (98%) flowing blood, Hct = 0.3 | 633 | 5.4 | | | 75 |
| | Oxygenated (98%) flowing blood, Hct = 0.4 | 633 | 6.6 | | | 75 |
| | Oxygenated (98%) flowing blood, Hct = 0.5 | 633 | 7.7 | | | 75 |
| | Oxygenated (98%) flowing blood, Hct = 0.6 | 633 | 10 | | | 75 |
| | Oxygenated (98%) flowing blood, Hct = 0.7 | 633 | 12.2 | | | 75 |
| | Flowing blood, 24.5°C | 633 | 2.9–3.5 | | | 76 |
| | Flowing blood, 24.5–42°C | 633 | 3.0–4.0 | | | 76 |
| | Flowing blood, 47°C | 633 | 4.5 | | | 76 |
| | Flowing blood, 54.3°C | 633 | 6.3 | | | 76 |
| Misc. tissues, *in vitro* | Lung | 515 | 25.5 | 3 | | 48 |
| | Lung | 635 | 8.1 | 2.8 | | 48 |
| | Lung | 1064 | 2.8 | | | 41 |
| | Muscle | 515 | 11.2 | 1.8 | | 48 |
| | Muscle | 1064 | 2 | | | 41 |
| | Meniscus | 360 | 13 | | | 46 |
| | Meniscus | 400 | 4.6 | | | 46 |
| | Meniscus | 488 | 1 | | | 46 |
| | Meniscus | 514 | 0.73 | | | 46 |

**TABLE 2.4** Absorption Coefficient $\mu_a$ for Various Tissues, Tissue Components, and Biofluids (continued)

| Tissue | Description | Wavelength (nm) | $\mu_a$ (1/cm) | Uncert. (1/cm) | Notes | Ref. |
|---|---|---|---|---|---|---|
| | Meniscus | 630 | 0.36 | | | 46 |
| | Meniscus | 800 | 0.52 | | | 46 |
| | Meniscus | 1064 | 0.34 | | | 46 |
| | Uterus | 635 | 0.35 | 0.1 | Frozen sections | 48 |
| | Bladder, integral | 633 | 1.4 | | Excised, kept in saline | 29 |
| | Bladder, integral | 633 | 1.4 | | | 41 |
| | Bladder, mucous | 1064 | 0.7 | | | 41 |
| | Bladder, wall | 1064 | 0.9 | | | 41 |
| | Bladder, integral | 1064 | 0.4 | | | 41 |
| | Heart, endocardial | 1060 | 0.07 | | Excised, kept in saline | 29 |
| | Heart, epicardial | 1060 | 0.35 | | | 29 |
| | Heart, myocardial | 1060 | 0.3 | | | 41 |
| | Heart, epicardgial | 1060 | 0.21 | | | 41 |
| | Heart, aneurysm | 1060 | 0.4 | | | 41 |
| | Heart, trabecula | 1064 | 1.4 | | | 41 |
| | Heart, myocardial | 1064 | 1.4 | | | 41 |
| | Heart, myocardial | 1060 | 0.52 | | | 77 |
| | Kidney, pars conv. | 1064 | 2.4 | | | 41 |
| | Kidney, medulla ren | 1064 | 2.1 | | | 41 |
| | Femoral vein | 1064 | 3.2 | | | 41 |
| | Liver | 515 | 18.9 | 1.7 | | 48 |
| | Liver | 630 | 3.2 | | | 77 |
| | Liver | 635 | 2.3 | 1 | | 48 |
| | Liver | 1064 | 0.7 | | | 48 |
| | Colon, muscle | 1064 | 3.3 | | | 41 |
| | Colon, submucous | 1064 | 2.3 | | | 41 |
| | Colon, mucous | 1064 | 2.7 | | | 41 |
| | Colon, integral | 1064 | 0.4 | | | 41 |
| | Esophagus | 633 | 0.4 | | | 46 |
| | Esophagus(mucous) | 1064 | 1.1 | | | 41 |
| | Fat:subcutaneous | 1064 | 2.6 | | | 41 |
| | Fat:abdominal | 1064 | 3 | | | 41 |
| | Prostate (0.5–3 h post mortem) | 850 | 0.6 | 0.2 | Shock frozen | 41 |
| | Prostate (0.5–3 h post mortem) | 980 | 0.4 | 0.2 | Shock frozen | 41 |
| | Prostate (0.5–3 h post mortem) | 1064 | 0.3 | 0.2 | Shock frozen | 41 |
| | Prostate coagulated (0.5–3 h post mortem) | 850 | 7 | 0.2 | Coag. water bath 75°C, 10 min | 41 |
| | Prostate coagulated (0.5–3 h post mortem) | 980 | 5 | 0.2 | Coag. water bath 75°C, 10 min | 41 |
| | Prostate coagulated (0.5–3 h post mortem) | 1064 | 4 | 0.2 | Coag. water bath 75°C, 10 min | 41 |
| | Prostate normal (freshly excised) | 1064 | 1.5 | 0.2 | | 46 |
| | Prostate coagulated | 1064 | 0.8 | 0.2 | Coag. water bath 70°C, 10 min | 46 |
| | Spleen | 1064 | 6 | | | 41 |
| | Stomach, muscle | 1064 | 3.3 | | | 41 |
| | Stomach, mucous | 1064 | 2.8 | | | 41 |
| | Stomach, integral | 1064 | 0.8 | | | 41 |
| | Sclera | 650 | 0.08 | | | 78 |
| | Tooth, dentin | 633 | 6 | | | 79 |
| | Tooth, enamel | 633 | 0.97 | | | 79 |
| | Gallstones, porcinement | 351 | 102 | 16 | | 29 |
| | Gallstones, porcinement | 488 | 179 | 28 | | 29 |
| | Gallstones, porcinement | 580 | 125 | 29 | | 29 |
| | Gallstones, porcinement | 630 | 85 | 11 | | 29 |
| | Gallstones, porcinement | 1060 | 121 | 12 | | 29 |

**TABLE 2.4** Absorption Coefficient $\mu_a$ for Various Tissues, Tissue Components, and Biofluids (continued)

| Tissue | Description | Wavelength (nm) | $\mu_a$ (1/cm) | Uncert. (1/cm) | Notes | Ref. |
|---|---|---|---|---|---|---|
| Misc. tissues, *in vivo* | Calf (11 subs, 14 meas.) | 800 | 0.17 | 0.05 | | 80 |

* Data taken from plots.

## 2.5 Conclusion

This chapter emphasizes the basic physics of light propagation in tissues with the goal of providing a fundamental framework for more in-depth study. The discussions are aimed at providing an introduction to and description of basic concepts and equations related to photon transport in biological tissues. The following chapters give detailed treatments of the wide range of phenomena that fall within the broad umbrella of biomedical photonics. A further description of specific theoretical methods for image reconstruction such as diffraction tomography is provided in Chapter 4. Various light phenomena in tissue are described in Chapter 3. Experimental techniques and clinical applications of light transport are described in Chapters 21, 22, and 33.

## 2.6 Summary

Across the spectrum, a great variety of electromagnetic phenomena are used in biomedicine to detect and treat disease, and to advance scientific knowledge. In principle, the propagation of light in tissue can be described using fundamental electromagnetic theory; however, the direct application of EM wave theory in tissue optics is not simple because of the complexity of tissue as a host medium for light propagation. Instead, the problem can be simplified by ignoring wave phenomena such as polarization and interferences, and particle properties such as inelastic collisions. This model, known as the *radiation transport (RT) theory,* is a general description of the flow of energy or particles in a system.

Several properties provide the basis on which the inputs to RT theory are built:

Index of refraction, $n(\lambda)$
Scattering cross section, $\sigma_s$
Differential scattering cross section, $d\sigma_s/d\Omega$
Absorption cross section, $\sigma_a$

- The *index of refraction* is a fundamental property of homogeneous media. The real part of the refractive index, $n(\lambda)$, can be defined in terms of the phase velocity of light in the medium,

$$c_m(\lambda) = \frac{c}{n(\lambda)}.$$

- Refractive index mismatches on a macroscopic scale (e.g., between soft tissue and bones, skin, and skull) give rise to refraction. On the other hand, refractive index mismatches at microscopic scales (e.g., cell membrane boundaries, organelles) give rise to scattering of light in biological tissue.
- The ratio of the power scattered out of a plane wave to the incident intensity is the *scattering cross section,*

$$\sigma_s(\hat{s}) = \frac{P_{scatt}}{I_0},$$

where $\hat{s}$ is the propagation direction of the plane wave relative to the scatterer.

- The *scattering coefficient* $\mu_s$ [cm$^{-1}$] can be expressed in terms of *particle density* $\rho$ [cm$^{-3}$] and scattering cross section $\sigma_s$ [cm$^2$] as

$$\mu_s = \rho\sigma_s.$$

- Scattering of light by tissue structures much smaller than the photon wavelength involves the so-called *Rayleigh limit*. Those structures include many components of the cell such as membranes, cell subcompartments, etc. The Rayleigh scattered light is equally distributed among the forward and rearward hemispheres, with peaks in the forward and backward directions.
- Scattering of light by spherical objects is described by *Mie theory*, which in principle is applicable at any size-to-wavelength ratio. In the intermediate size-to-wavelength range, where the Rayleigh and geometric approximations are not valid, Mie theory is used. Thus this region where the scatterer size and wavelength are comparable is sometimes labeled the *Mie regime*. Various cellular structures and extracellular components have sizes from hundreds of nanometers to a few microns and are comparable in dimension to the photon wavelengths of the therapeutic window. Even though these structures are not necessarily spherical, their scattering behavior can be modeled reasonably well by Mie theory applied to spheres of comparable size and refractive index. The scattering from tissue in this regime is strongly peaked in the forward direction.
- Absorption processes involve an important concept in quantum theory, the *energy level*, which is a quantum state of an atom and molecule. The energy of a particular level is determined relative to the ground level, or lowest possible energy level (zero energy). The shift of a species (e.g., molecule or atom) from one energy level to another is called a *transition*. A transition of a species from a lower to a higher energy level involves excitation to an excited state and requires absorption of an amount of photon energy, $h\nu$, equal to the difference in energy, $\Delta E$, between the two levels:

$$h\nu = \Delta E.$$

- For a localized absorber, the *absorption cross section* $\sigma_a$ can be defined in the same manner as for scattering:

$$\sigma_a = \frac{P_{abs}}{I_0},$$

where $P_{abs}$ is the amount of power absorbed out of an initially uniform plane wave of intensity (power per unit area)$I_0$.
- The *absorption coefficient* $\mu_a$ [cm$^{-1}$] can be defined in terms of particle density $\rho$ [cm$^{-3}$] and absorption cross section $\sigma_a$ [cm$^2$]

$$\mu_a = \rho\sigma_a.$$

- The fundamental equation that describes light propagation in the RT model is the *radiation transport (RT) equation*, which describes the fundamental dynamics of the specific intensity. The steady-state RT equation in a source-free region is

$$\hat{s}\cdot\vec{\nabla}I(\mathbf{r},\hat{s}) = -(\mu_a+\mu_s)I(\mathbf{r},\hat{s}) + \frac{\mu_a+\mu_s}{4\pi}\int_{4\pi} p(\hat{s}\cdot\hat{s}')I(\mathbf{r},\hat{s}')d\Omega',$$

where $I(\mathbf{r}, \hat{s}, t)$ is the specific intensity [light power per unit area per unit solid angle] at point **r** directed in a cone of solid angle oriented in the direction defined by the unit vector $\hat{s}$.
- The scattering phase function (SPF), $p(\hat{s}\cdot\hat{s}')$, describes the fraction of light energy incident on a scatterer from the $\hat{s}'$ direction that is scattered out into the $\hat{s}$ direction.

- Although conceptually straightforward, the direct analytical solution of the RT equation for many problems of interest is difficult. Most of the complications arise from dealing with the boundaries at tissue interfaces and geometric aspects of the tissues and light sources. The RT equation can be solved using analytical approaches based on deterministic approximations. Two common analytical approaches involve the *diffusion approximation* and the *Kubelka–Munk* model. In the numerical approach, the *Monte Carlo* method is often used to calculate photon transport through tissue.
- The ability of light to penetrate tissues depends on how strongly the tissues absorb light. Within the spectral range known as the *therapeutic window* (also called the *diagnostic window*), most tissues are sufficiently weak absorbers to permit significant penetration of light. This window extends from 600 to 1300 nm, from the orange region of the visible spectrum into the N/R.

## Acknowledgments

This work was sponsored by the U.S. Department of Energy (DOE) Office of Biological and Environmental Research, under contract DEAC05-00OR22725 with UT-Battelle, LLC, and by the National Institutes of Health (RO1 CA88787-01).

### References

1. Ishimaru, A., *Wave Propagation and Scattering in Random Media*, Vol. 1 and 2, Academic Press, New York, 1978.
2. Hecht, E., *Optics*, Addison-Wesley, Reading, MA, 1987.
3. Henyey, L. and Greenstein, J.L., Diffuse radiation in the galaxy, *Astrophys. J.*, 93, 70, 1941.
4. Das, B.B., Liu, F., and Alfano, R.R., Time-resolved fluorescence and photon migration studies in biomedical and model random media, *Rep. Prog. Phys.*, 60, 227, 1997.
5. Wang, L.H., Jacques, S.L., and Zheng, L.Q., MCML – Monte-Carlo modeling of light transport in multilayered tissues, *Comput. Methods Programs Biomed.*, 47, 131, 1995.
6. Gardner, C.M., Jacques, S.L., and Welch, A.J., Light transport in tissue: accurate expressions for one-dimensional fluence rate and escape function based upon Monte Carlo simulation, *Lasers Surg. Med.*, 18, 129, 1996.
7. Vo-Dinh, T., *Room Temperature Phosphorimetry for Chemical Analysis*, Wiley, New York, 1984.
8. Schuster, A., Radiation through foggy atmosphere, *Astrophys. J.*, 21, 1, 1905.
9. Kubelka, P. and Munk, F., Ein Beitrag zur Optik der Farbanstriche, *Z. Tech. Physik*, 12, 593, 1931.
10. Kubelka, P., New contributions to the optics of intensely light-scattering materials. Part I, *J. Opt. Soc. Am.*, 38, 448, 1948.
11. Ragain, J.C. and Johnston, W.M., Accuracy of Kubelka–Munk reflectance theory applied to human dentin and enamel, *J. Dent. Res.*, 80, 449, 2001.
12. Philips-Invernizzi, B., Dupont, D., and Caze, C., Bibliographical review for reflectance of diffusing media, *Opt. Eng.*, 40, 1082, 2001.
13. Tsuchikawa, S. and Tsutsumi, S., Analytical characterization of reflected and transmitted light from cellular structural material for the parallel beam of NIR incident light, *Appl. Spectrosc.*, 53, 1033, 1999.
14. Rogatkin, D.A., Development of the two-flux Kubelka–Munk model for solution of 1D problems of light propagation in scattering biological tissues and media, *Opt. Spectrosc.*, 87, 101, 1999.
15. Molenaar, R., ten Bosch, J.J., and Zijp, J.R., Determination of Kubelka–Munk scattering and absorption coefficients by diffuse illumination, *Appl. Opt.*, 38,2068, 1999.
16. Hoffmann, J., Lubbers, D.W., and Heise, H.M., Applicability of the Kubelka–Munk theory for the evaluation of reflectance spectra demonstrated for haemoglobin-free perfused heart tissue, *Phys. Med. Biol.*, 43, 3571, 1998.
17. Ragain, J.C. and Johnston, W.M., Agreement of measured reflectance with Kubelka–Munk theory for human enamel and dentin, *J. Dent. Res.*, 77, 1433, 1998.

18. Vargas, W.E. and Niklasson, G.A., Applicability conditions of the Kubelka–Munk theory, *Appl. Opt.*, 36, 5580, 1997.
19. Beuthan, J., Minet, O., and Muller, G., Quantitative optical biopsy of liver tissue *ex vivo*, *IEEE J. Sel. Top. Quantum Electron.*, 2, 906, 1996.
20. Rundlof, M. and Bristow, J.A., A note concerning the interaction between light scattering and light absorption in the application of the Kubelka–Munk equations, *J. Pulp Paper Sci.*, 23, J220, 1997.
21. Koukoulas, A.A. and Jordan, B.D., Effect of strong absorption on the Kubelka–Munk scattering coefficient, *J. Pulp Paper Sci.*, 23, J224, 1997.
22. Bjorn, L.O., Light propagation in biological materials and natural waters, *Sci. Mar.*, 60, 9, 1996.
23. Loyalka, S.K. and Riggs, C.A., Inverse problem in diffuse-reflectance spectroscopy — accuracy of the Kubelka–Munk equations, *Appl. Spectrosc.*, 49, 1107, 1995.
24. Christy, A.A., Kvalheim, O.M., and Velapoldi, R.A., Quantitative analysis in diffuse-reflectance spectrometry — a modified Kubelka–Munk equation, *Vib. Spectrosc.*, 9, 19, 1995.
25. Waters, D.N., Raman-spectroscopy of powders — effects of light-absorption and scattering. Part A molecular and biomolecular spectroscopy, *Spectrochim. Acta A Mol. Biomol. Spectrosc.*, 50, 1833, 1994.
26. Durkin, A.J. et al., Relation between fluorescence spectra of dilute and turbid samples, *Appl. Opt.*, 33, 414, 1994.
27. Agati, G. et al., Quantum yield and skin filtering effects on the formation rate of lumirubin, *J. Photochem. Photobiol. B-Biol.*, 18, 197, 1993.
28. Mandelis, A. and Grossman, J.P., Perturbation theoretical approach to the generalized Kubelka–Munk problem in nonhomogeneous optical media, *Appl. Spectrosc.*, 46, 737, 1992.
29. Cheong, W.F., Prahl, S.A., and Welch, A.J., A review of the optical properties of biological tissues, *IEEE J. Quantum Electron.*, 26, 2166, 1990.
30. Waterman, P.C. and Truell, R., Multiple scattering of waves, *J. Math. Phys.*, 2, 512, 1961.
31. Saidi, I.S., Jacques, S.L., and Tittel, F.K., Mie and Rayleigh modeling of visible-light scattering in neonatal skin, *Appl. Opt.*, 34, 7410, 1995.
32. Voet, D., Gratzer, W., Cox, R., and Doty, P., Absorption spectra of nucleotides, polynucleotides, and nucleic acids in the far ultraviolet, *Biopolymers*, 1, 93, 1963.
33. Tinoco, I., Jr., Sauer, K., and Wang, J., *Physical Chemistry: Principles and Applications in Biological Sciences*, Prentice-Hall, Englewood Cliffs, NJ, 1985.
34. Campbell, I. and Dwek, R., *Biological Spectroscopy*, Benjamin Cummings, Menlo Park, CA, 1984.
35. Prota, G., D'Ischia, M., and Napolitano, A., The chemistry of melanins and related metabolites, in *The Pigmentary System*, Normund, J. et al., Eds., Oxford University Press, Oxford, 1988.
36. Roggan, A. et al., Optical properties of circulating human blood in the wavelength range 400–2500 nm, *J. Biomed. Opt.*, 4, 36, 1999.
37. Delpy, D.T. and Cope, M., Quantification in tissue near-infrared spectroscopy, *Philos. Trans. R. Soc. Lond. B*, 352, 649, 1997.
38. Chance, B. et al., Comparison of time-resolved and time-unresolved measurements of deoxyhemoglobin in brain, *Proc. Natl. Acad. Sci. U.S.A.*, 85, 4971, 1988.
39. Firbank, M. et al., Measurement of the optical-properties of the skull in the wavelength range 650–950 nm, *Phys. Med. Biol.*, 38, 503, 1993.
40. Duck, F.A., *Physical Properties of Tissue: A Comprehensive Reference Book,* Academic Press, San Diego, 1990.
41. Muller, G. and Roggan, A., Eds., *Laser-Induced Interstitial Thermotherapy,* Vol. PM25, SPIE, Bellingham, WA, 1995.
42. Bolin, F.P., Preuss, L.E., Taylor, R.C., and Ference, R.J., Refractive index of some mammalian tissues using a fiber optic cladding method, *Appl. Opt.*, 28, 2297, 1989.
43. Wang, X.J., Milner, T.E., Dhond, R.P., Sorin, W.V., Newton, S.A., and Nelson, J.S., Characterization of human scalp hairs by optical low-coherence reflectometry, *Opt. Lett.*, 20(6), 524, 1995.

44. Ohmi, M., Ohnishi, Y., Yoden, K., and Haruna, M., *In vitro* simultaneous measurement of refractive index and thickness of biological tissue by the low coherence interferometry, *IEEE Trans. Biomed. Eng.*, 47(9), 1266, 2000.
45. van Gemert, M.J.C., Nelson, J.S., Jacques, S.L., Sterenborg, H.J.C.M., and Star, W.M., Skin optics, *IEEE Trans. Biomed. Eng.*, 36(12), 1146, 1989.
46. Tuchin, V.V., *Tissue Optics — Light Scattering Methods and Instruments for Medical Diagnosis*, Tutorial Texts in Optical Engineering, Vol. TT38, SPIE, Bellingham, WA, 2000.
47. Kienle, A., Lilge, L., and Patterson, M.S., Investigations of multilayered tissue with *in vivo* reflectance measurements, *Proc. SPIE*, 2326, 212, 1994.
48. Marchesini, R., Bertoni, A., Andreola, S., Melloni, E., and Sichirollo, A.E., Extinction and absorption coefficients and scattering phase functions of human tissues *in vitro*, *Appl. Opt.*, 28(12), 2318, 1989.
49. Troy, T.L., Page, D.L., and Sevick-Muraca, E.M., Optical properties of normal and diseased breast tissues: prognosis for optical mammography, *J. Biomed. Opt.*, 1(3), 342, 1996.
50. Simpson, C.R., Kohl, M., Essenpreis, M., and Cope, M., Near-infrared optical properties of *exo vivo* human skin and subcutaneous tissues measured using the Monte Carlo inversion technique, *Phys. Med. Biol.*, 43, 2465, 1998.
51. Graaf, R., Dassel, A.C.M., Koelink, M.H., Demul, F.F.M., Aarnoudse, J.G., and Zijlstra, W.G., Optical properties of human dermis *in vitro* and *in vivo*, *Appl. Opt.*, 32(4), 435, 1993.
52. Tuchin, V.V., Fundamentals of low-intensity laser radiation interaction with biotissues: dosimetry and diagnostical aspects, *Bull. Russ. Acad. Sci. Phys. Ser.*, 59(6), 120, 1995.
53. Chan, E.K., Sorg, B., Protsenko, D., O'Neil, M., Motamedi, M., and Welch, A.J., Effects of compression on soft tissue optical properties, *IEEE J. Sel. Top. Quantum Electron.*, 2(4), 943, 1996.
54. Kienle, A. and Hibst, R., A new optimal wavelength for treatment of port wine stains? *Phys. Med. Biol.*, 40, 1559, 1995.
55. Doornbos, R.M.P., Lang, R., Aalders, M.C., Cross, F.W., and Sterenborg, H.J.C.M., The determination of *in vivo* human tissue optical properties and absolute chromophore concentrations using spatially resolved steady-state diffuse reflectance spectroscopy, *Phys. Med. Biol.*, 44, 967, 1999.
56. Sterenborg, H.J.C.M., van Gemert, M.J.C., Kamphorst, W., Wolbers, J.G., and Hogervorst, W., The spectral dependence of the optical properties of human brain, *Lasers Med. Sci.*, 4, 221, 1989.
57. Tuchin, V.V., Podbielska, H., and Hitzenberger, C.K., Special section on coherence domain optical methods in biomedical science and clinics, *J. Biomed. Opt.*, 4(1), 94, 1999.
58. Schwarzmaier, H.-J., Yaroslavsky, A.N., Yaroslavsky, I.V., Goldbach, T., Kahn, T., Ulrich, F., Schulze, P.C., and Schober, R., Optical properties of native and coagulated human brain structures, *Proc. SPIE*, 2970, 492, 1997.
59. Roggan, A., Minet, O., Schroder, C., and Muller, G., The determination of optical tissue properties with double integrating sphere technique and Monte Carlo simulations, *Proc. SPIE*, 2100, 42, 1994.
60. Willmann, S., Terenji, A., Yaroslavsky, I.V., Kahn, T., Hering, P., and Schwarzmaier, H.-J., Determination of the optical properties of a human brain tumor using a new microspectrophotometric technique, *Proc. SPIE*, 3598, 233, 1999.
61. Yaroslavsky, A.N., Schulze, P.C., Yaroslavsky, I.V., Schober, R., Ulrich, F., and Schwarzmaier, H.-J., Optical properties of selected native and coagulated human brain tissues *in vitro* in the visible and near infrared spectral range, *Phys. Med. Biol.*, 47, 2059, 2002.
62. Gottschalk, W., Ein Messverfahren zur Bestimmung der optischen Parameter biologischer Gevebe *in vitro*, Doctoral thesis, University of Karlsruhe, Karlsruhe, Germany, 1992.
63. Bevilacqua, F., Piguet, D., Marquet, P., Gross, J.D., Tromberg, B.J., and Depeursinge, C., *In vivo* local determination of tissue optical properties: applications to human brain, *Appl. Opt.*, 38, 4939, 1999.
64. Bevilacqua, F., Piguet, D., Marquet, P., Gross, J.D., Tromberg, B.J., and Depeursinge, C.D., *In vivo* local determination of tissue optical properties, *Proc. SPIE*, 3194, 262, 1997.

65. Peters, V.G., Wyman, D.R., Patterson, M.S., and Frank, G.L., Optical properties of normal and diseased human tissues in the visible and near infrared, *Phys. Med. Biol.*, 35, 1317, 1990.
66. Key, H., Davies, E.R., Jackson, P.C., and Wells, P.N.T., Optical attenuation characteristics of breast tissues at visible and near-infrared wavelengths, *Phys. Med. Biol.*, 36, 579, 1991.
67. Zhadin, N.N. and Alfano, R.R., Correction of the internal absorption effect in fluorescence emission and excitation spectra from absorbing and highly scattering media: theory and experiment, *J. Biomed. Opt.*, 3(2), 171, 1998.
68. Suzuki, K., Yamashita, Y., Ohta, K., Kaneko, M., Yoshida, M., and Chance, B., Quantitative measurement of optical parameters in normal breast using time-resolved spectroscopy: *in vivo* results of 30 Japanese women, *J. Biomed. Opt.*, 1(3), 330, 1996.
69. Heusmann, H., Kolzer, J., and Mitic, G., Characterization of female breast *in vivo* by time resolved and spectroscopic measurements in near-infrared spectroscopy, *J. Biomed. Opt.*, 1(4), 425, 1996.
70. Driver, I., Lowdell, C.P., and Ash, D.V., *In vivo* measurement of the optical interaction coefficients of human tumors at 630 nm, *Phys. Med. Biol.*, 36, 805, 1991.
71. Keijzer, M., Jacques, S.L., Prahl, S.A., and Welch, A.J., Light distributions in artery tissue — Monte-Carlo simulations for finite-diameter laser beams, *Lasers Surg. Med.*, 9(2), 148, 1989.
72. Keijzer, M., Richards-Kortum, R.R., Jacques, S.L., and Feld, M.S., Fluorescence spectroscopy of turbid media — autofluorescence of the human aorta, *Appl. Opt.*, 28(20), 4286, 1989.
73. Oraevsky, A.A., Jacques, S.L., and Tittel, F.K., Measurement of tissue optical properties by time-resolved detection of laser-induced transient stress, *Appl. Opt.*, 36(1), 402, 1997.
74. Yaroslavsky, A.N., Yaroslavsky, I.V., Goldbach, T., and Schwarzmaier, H.-J., The optical properties of blood in the near infrared spectra range, *Proc. SPIE*, 2678, 314, 1996.
75. Roggan, A., Friebel, M., Dorschel, K., Hahn, A., and Mueller, G., Optical properties of circulating human blood in the wavelength range 400–2500 nm, *J. Biomed. Opt.*, 4(1), 36, 1999.
76. Nilsson, A.M.K., Lucassen, G.W., Verkruysse, W., Andersson-Engels, S., and van Gemert, M.J.C., Changes in optical properties of human whole blood *in vitro* due to slow heating, *Photochem. Photobiol.*, 65(2), 366, 1997.
77. Hourdakis, C.J. and Perris, A.A., Monte Carlo estimation of tissue optical properties for use in laser dosimetry, *Phys. Med. Biol.*, 40, 351, 1995.
78. Svaasand, L.O., Tromberg, B.J., Haskell, R.C., Tsay, T.T., and Berns, M.W., Tissue characterization and imaging using photon density waves, *Opt. Eng.*, 32(2), 258, 1993.
79. Zijp, J.R. and ten Bosch, J.J., Angular dependence of He-Ne laser light scattering by bovine and human dentine, *Arch. Oral Biol.*, 36(4), 283, 1991.
80. Matcher, S.J., Cope, M., and Delpy, D.T., *In vivo* measurements of the wavelength dependence of tissue scattering coefficients between 760 and 900 nm measured with time-resolved spectroscopy, *Appl. Opt.*, 36(1), 386, 1997.

# 3

# Light–Tissue Interactions




Valery V. Tuchin
*Saratov State University*
*Saratov, Russian Federation*


## 3.1 Introduction

Two large classes of biological media can be considered with regard to light interactions with biological tissues and fluids: (1) strongly scattering (opaque) like skin, brain, vessel walls, eye sclera, blood, and lymph and (2) weakly scattering (transparent) like cornea, crystalline lens, vitreous humor, and aqueous humor of the front chamber of eye.[1–7] Light interactions with tissues of the first class can be described in the model of multiple scattering of scalar or vector waves in a randomly nonuniform absorbing medium. For tissues of the second class, interactions can be described in the model of single (or low-step) scattering of an ordered isotropic or anisotropic medium with closely packed scatterers with absorbing centers.

The transparency of tissues reaches its maximum in the near-infrared (NIR), which is associated with the fact that living tissues do not contain strong intrinsic chromophores that would absorb radiation within this spectral range.[1–7] Light penetrates into a tissue over several centimeters, which is important for the transillumination of thick human organs (brain, breast, etc.). However, tissues are characterized by strong scattering of NIR radiation, which prevents obtaining clear images of localized inhomogeneities arising in tissues due to various pathologies, e.g., tumor formation, local increase in blood volume caused by a hemorrhage, or the growth of microvessels. Therefore, special attention in optical tomography and spectroscopy is focused on the development of methods for the selection of image-carrying

0-8493-1116-0/03/$0.00+$1.50

photons or detection of photons providing the information concerning optical parameters of the scattering medium.

Methods of noninvasive optical diagnosis and spectroscopy of tissues are concerned with two radiation regimes: continuous wave (CW) and time-resolved.[1–7] The time-resolved interactions are realized by means of exposure of a tissue to short laser pulses and subsequent recording of scattered broadened pulses (time-domain method) or by irradiation with modulated light and recording the depth of modulation of scattered light intensity and the corresponding phase shift at modulation frequencies (frequency-domain or phase method). The time-resolved regimes are based on the excitation of the photon-density wave spectrum in a strongly scattering medium, which can be described in the framework of the nonstationary radiation transfer theory (RTT). The CW regime is described by the stationary RTT.

Many optical medical technologies employ laser radiation and fiber optics; therefore, light coherence is very important for the analysis of the interaction of light with tissues and cell ensembles.[2–5,7–12] This problem can be considered in terms of the loss of coherence due to the scattering of light in a randomly nonuniform medium with multiple scattering or the change in the statistics of speckles in the scattered field. The coherence of light is of fundamental importance for the selection of photons that have experienced no or a small number of scattering events, as well as for the generation of speckle-modulated fields from scattering phase objects with single and multiple scattering. Such approaches are important for coherent tomography, diffractometry, holography, photon-correlation spectroscopy, laser Doppler anemometry, and speckle interferometry of tissues and biological flows. The use of optical sources with a short coherence length opens up new opportunities in coherent interferometry and tomography of tissues, organs, and blood flows.

The vector nature of light waves is important for light–tissue interaction because, in a scattering medium, it is manifested as polarization ability of an initially unpolarized incident light or the change in the character of polarization state of an initially polarized light propagating in a medium. Similar to coherence properties of a light beam reflected from or transmitted through a biological object, polarization properties of light can be employed as a selector of photons coming from different depths in an object.[3,5,7]

Quasi-elastic light scattering (QELS) as applied to monitoring of dynamic systems (chaotic or directed movements of tissue components or cells) is based mainly on the correlation or spectral analysis of the temporal fluctuations of the scattered light intensity.[3,5,7–10] QELS spectroscopy (also known as light-beating spectroscopy or correlation spectroscopy) is widely used for various biomedical applications, particularly for blood or lymph flow measurement and cataract diagnostics. Diffusion wave spectroscopy (DWS) is available for the study of optically thick tissue when multiple scattering prevails and photon migration (diffusion) within tissue is important for the character of intensity fluctuations.

Raman scattering is the basis for Raman vibrational spectroscopy. It is a great tool to study the structure and dynamic function of biological molecules and has been used extensively for monitoring and diagnosis of diseases such as cataracts, artherosclerotic lesions in coronary arteries, precancerous and cancerous lesions in human soft tissues, and bone and teeth pathologies.[13–15]

Light-induced fluorescence is also a powerful noninvasive method for tissue pathology recognition and monitoring.[7,17–19] Autofluorescence, fluorescence of introduced markers, and time-resolved, laser-scan, and multiphoton fluorescence have been used to study human tissues and cells *in situ*.

Light-induced thermal effects on tissue are important for diagnostics, therapy, and surgery.[5,7,12,20–25] The optothermal spectroscopy, based on detecting time-dependent heat generation induced in a tissue by pulsed- or intensity-modulated optical radiation, is widely used in biomedicine. Among optothermal methods, optoacoustic (OA) and photoacoustic (PA) techniques are of great importance. They allow one to estimate tissue optical, thermal, and acoustic properties that depend on peculiarities of tissue structure.

For thermal phototherapy and surgery, much higher light intensities are used. Controllable temperature rise and thermal or thermomechanical damage (coagulation, vaporization, vacuolization, pyrolysis, ablation) of a tissue are important in that case.[12,21–25]

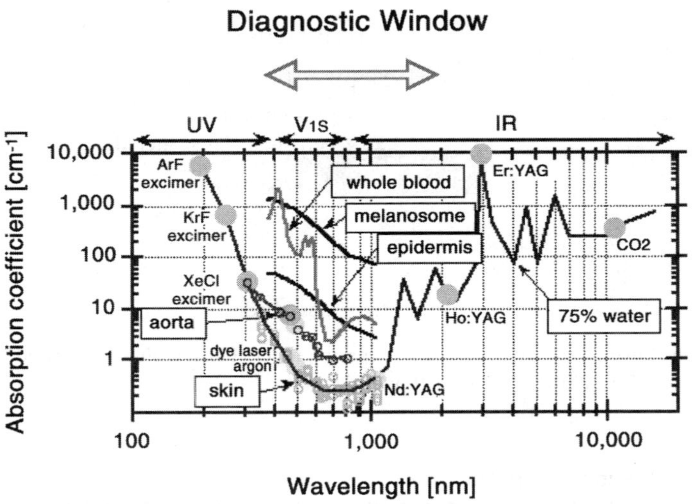


**FIGURE 3.1** Water, some tissues (aorta, skin), and tissue component (whole blood, melanosome, epidermis) absorption spectra. Wavelengths of lasers widely used in laser therapy and surgery are also shown (ArF, KrF, and XeCl excimer lasers; dye laser; argon laser; solid-state Nd: YAG, Ho:YAG, and Er:YAG lasers). (From Jacques, S.L. et al., *Proc. SPIE*, 4001, 14, 2000. With permission.)

## 3.2 Light Interactions with a Strongly Scattering Tissue

### 3.2.1 Continuous Wave (CW) Light

Biological tissues are optically inhomogeneous and absorbing media whose average refractive index is higher than that of air. This is responsible for the partial reflection of the radiation at the tissue–air interface (Fresnel reflection), while the remaining part penetrates the tissue. Multiple scattering and absorption are responsible for light beam broadening and eventual decay as it travels through a tissue, whereas bulk scattering is a major cause of the dispersion of a large fraction of radiation in the backward direction. Cellular organelles such as mitochondria are the main scatterers in various tissues.[1–7]

Absorbed light is converted to heat or radiated in the form of fluorescence; it is also consumed in photobiochemical reactions. The absorption spectrum depends on the type of predominant absorption centers and water content of tissues (see Figure 3.1). Absolute values of absorption coefficients for typical tissues lie in the range of $10^{-2}$ to $10^{4}$ $cm^{-1}$.[1–7] In the ultraviolet (UV) and infrared (IR) ($\lambda \geq 2$ μm) spectral regions, light is readily absorbed, which accounts for the small contribution of scattering and inability of radiation to penetrate deep into tissues (only across one or more cell layers). In the wavelength range of 600 to 1600 nm, scattering prevails over absorption; the intensity of the reflected radiation increases to 35 to 70% of the total incident light (due to backscattering).

Light interaction with multilayer and multicomponent tissues is a very complicated process (see Figure 3.2). For example, for skin, the horny layer (stratum corneum) reflects about 5 to 7% of the incident light. A collimated light beam is transformed to a diffuse one by microscopic inhomogeneities at the air–horny layer interface. A major part of the reflected light results from backscattering in different skin layers (stratum corneum, epidermis, dermis, blood, and fat). The absorption of diffuse light by skin pigments is a measure of bilirubin and melanin content and hemoglobin saturation with oxygen. These characteristics are widely used in the diagnosis of various diseases. Certain phototherapeutic and diagnostic modalities take advantage of ready transdermal penetration of visible and NIR light inside the body in the wavelength region corresponding to the so-called therapeutic or diagnostic window (600 to 1600 nm).

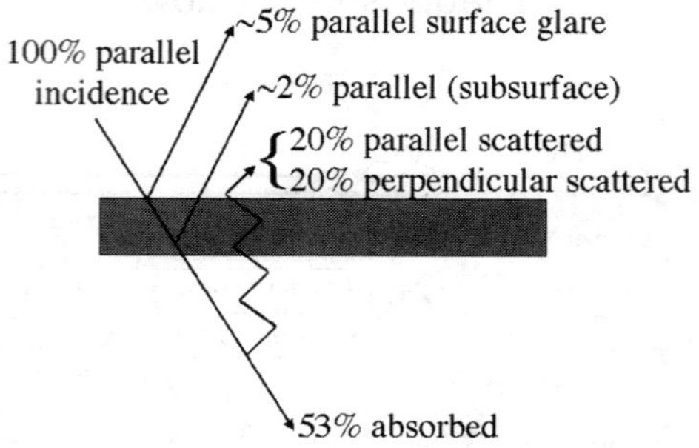


**FIGURE 3.2** Simplified two-layer scattering and absorption model of skin at linear polarized incident light. (From Jacques, S.L. et al., *Proc. SPIE*, 4001, 94, 2000. With permission.)

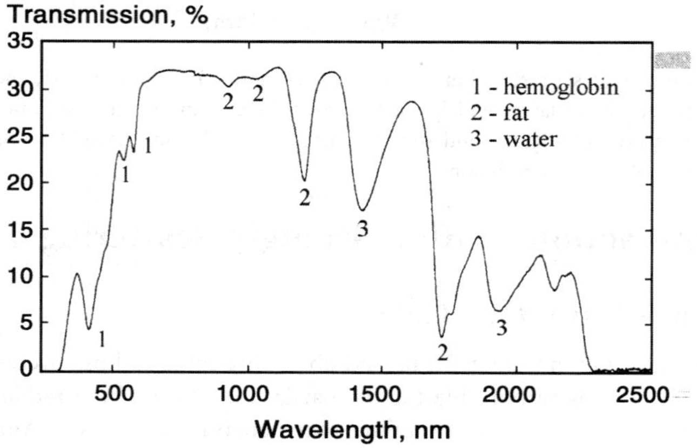


**FIGURE 3.3** Transmission spectrum of a 3-mm-thick slab of female breast tissue. A spectrometer with an integrating sphere was used. The contributions of absorption bands of the tissue components (hemoglobin = 1, fat = 2, and water = 3) are indicated. (From Marks, F.A., *Proc. SPIE*, 1641, 227, 1992. With permission.)

Another example of heterogeneous multicomponent tissue is a female breast. The absorption bands of hemoglobin, fat, and water are clearly seen in the *in vitro* measured spectrum of a 3-mm slab of breast tissue presented in Figure 3.3.[27] There is a wide window between 700 and 1100 nm, and narrow ones at about 1300 and 1600 nm where the lowest percentage of light is attenuated.

A collimated (laser) beam is attenuated in a tissue layer of thickness $d$ in accordance to the exponential law (Beer–Lambert law):

$$I(d) = (1 - R_F)\, I_0 \exp(-\mu_t d), \tag{3.1}$$

where $I(d)$ is the intensity of transmitted light measured using a distant photodetector with a small aperture (on-line or collimated transmittance), W/cm$^2$; $R_F$ is the coefficient of Fresnel reflection; at the normal beam incidence, $R_F = [(n - 1)/(n + 1)]^2$; $n$ is the relative mean refractive index of tissue and surrounding media; $I_0$ is the incident light intensity, W/cm$^2$; $\mu_t = \mu_a + \mu_s$ is the extinction coefficient (interaction or total attenuation coefficient), 1/cm; $\mu_a$ is the absorption coefficient, 1/cm; $\mu_s$ is the scattering coefficient, 1/cm. The mean free path length between two interactions is denoted by

$$l_{ph} = \mu_t^{-1}. \tag{3.2}$$

To analyze light propagation under multiple scattering conditions, it is assumed that absorbing and scattering centers are uniformly distributed across the tissue. Visible and NIR radiation is normally subject to anisotropic scattering characterized by a well-apparent direction of a scattered photon.

A sufficiently strict mathematical description of CW light propagation in a scattering medium is possible in the framework of the stationary RTT. This theory is valid for an ensemble of scatterers located far from one another and has been successfully used to work out some practical aspects of tissue optics. The main stationary equation of RTT for monochromatic light has the form:[1–7]

$$\frac{\partial I(\bar{r},\bar{s})}{\partial s} = -\mu_t I(\bar{r},\bar{s}) + \frac{\mu_s}{4\pi}\int 4\pi I(\bar{r},\bar{s}')p(\bar{s},\bar{s}')d\Omega', \tag{3.3}$$

where $I(\bar{r},\bar{s})$ is the radiance (or specific intensity) – average power flux density at a point $\bar{r}$ in the given direction $\bar{s}$, (W/cm$^2$sr); $p(\bar{s},\bar{s}')$ is the scattering phase function, 1/sr; and $d\Omega$ is the unit solid angle about the direction $\bar{s}'$, sr. It is assumed that no radiation sources are inside the medium.

To characterize the relation of scattering and absorption properties of a tissue, a parameter such as albedo is usually introduced $\Lambda = \mu_s/\mu_t$. The albedo ranges from zero for a completely absorbing medium to unity for a completely scattering medium.

The phase function $p(\bar{s},\bar{s}')$ describes the scattering properties of the medium and is in fact the probability density function for scattering in the direction $\bar{s}'$ of a photon traveling in the direction $\bar{s}$; in other words, it characterizes an elementary scattering act. If scattering is symmetric relative to the direction of the incident wave, then the phase function depends only on the scattering angle $\theta$ (angle between directions $\bar{s}$ and $\bar{s}'$), i.e., in practice, the phase function is usually well approximated with the aid of the postulated Henyey–Greenstein function:[1–7]

$$p(\bar{s},\bar{s}') = p(\theta) = \frac{1}{4\pi}\cdot\frac{1-g^2}{\left(1+g^2-2g\cos\theta\right)^{3/2}}, \tag{3.4}$$

where $g$ is the scattering anisotropy parameter (mean cosine of the scattering angle $\theta$). The value of $g$ varies in the range from 0 to 1: $g = 0$ corresponds to isotropic (Rayleigh) scattering and $g = 1$ to total forward scattering (Mie scattering at large particles).[28–31]

The integro-differential equation (Equation 3.1) is frequently simplified by representing the solution in the form of spherical harmonics. Such simplification leads to a system of $(N + 1)^2$ connected differential partial derivative equations known as the $P_N$ approximation. This system is reducible to a single differential equation of order $(N + 1)$. For example, four connected differential equations reducible to a single diffusion-type equation are necessary for $N = 1$.[1–7] The photon diffusion coefficient, cm$^2$/c,

$$D = c/3(\mu_s' + \mu_a), \tag{3.5}$$

and the reduced (transport) scattering coefficient, 1/cm,

$$\mu_s' = (1 - g)\mu_s, \tag{3.6}$$

are the major parameters of the diffusion equation; here $c$ is the velocity of light in the medium.

The transport mean free path (MFP) of a photon (cm) is defined as

$$\mathrm{lt} = (\mu_s' + \mu_a)^{-1}. \tag{3.7}$$

It is worthwhile to note that the transport mean free path in a medium with anisotropic single scattering significantly exceeds the mean free path in a medium with isotropic single scattering $l_t >> l_{\mathrm{ph}}$. The transport MFP $l_t$ is the distance over which the photon loses its initial direction.

### 3.2.2 Polarized Light

It is a common belief that the randomness of tissue structure results in fast depolarization of light propagating in tissues; therefore, polarization effects are usually ignored. However, in certain tissues (transparent tissues such as eye tissues, cellular monolayers, mucous membrane, superficial skin layers, etc.) the degree of polarization of transmitted or reflected light remains measurable even when the tissue has considerable thickness. In such a situation, the information about the structure of tissues and cell ensembles can be extracted from the registered depolarization degree of initially polarized light, the polarization state transformation, or appearance of a polarized component in the scattered light.[3,5,7,26,31–36] With regard to practical implications, polarization techniques are believed to give rise to simplified schemes of optical medical tomography compared with time-resolved methods and also provide additional information about the structure of tissues.

Polarization refers to the pattern described by the electric field vector as a function of time at a fixed point in space. When the electric field vector oscillates in a single fixed plane all along the beam, the light is said to be linearly polarized. This linearly polarized wave can be resolved into components parallel $E_{\|}$ and perpendicular $E_{\perp}$ to the scattering plane. If the plane of the electric field rotates, the light is said to be elliptically polarized, because the electric field vector traces out an ellipse at a fixed point in space as a function of time. If the ellipse happens to be a circle, the light is said to be circularly polarized. The connection between phase and polarization can be understood as follows: circularly polarized light consists of equal amounts of linear, mutually orthogonal polarized components where they oscillate exactly 90° out of phase. In general, light of arbitrary elliptical polarization consists of unequal amplitudes of linearly polarized components and the electric fields for the two polarizations oscillate at the same frequency but have some constant phase difference.

Light of arbitrary polarization can be represented by four numbers known as the Stokes parameters: *I, Q, U,* and *V*. *I* refers to the intensity of the light; the parameters *Q, U,* and *V* represent the extent of horizontal liner, 45° linear, and circular polarization, respectively.[31] In terms of the electric field components, the Stokes parameters are given by

$$\begin{aligned} I &= \langle E_{\|} E_{\|}^{*} + E_{\perp}E_{\perp}^{*}\rangle, \\ Q &= \langle E_{\|} E_{\|}^{*} - E_{\perp}E_{\perp}^{*}\rangle, \\ U &= \langle E_{\|} E_{\perp} + E_{\perp}E_{\|}^{*}\rangle, \\ V &= \langle E_{\|} E_{\perp}^{*} - E_{\perp}E_{\|}^{*}\rangle, \end{aligned} \quad (3.8)$$

and the irradiance or intensity of light by

$$I^2 = Q^2 + U^2 + V^2. \quad (3.9)$$

The polarization state of the scattered light in the far zone is described by the Stokes vector connected with the Stokes vector of the incident light

$$\vec{I}_s = M \cdot \vec{I}_i, \quad (3.10)$$

where *M* is the normalized 4 × 4 scattering matrix (intensity or Mueller's matrix):

$$M = \begin{bmatrix} M_{11} & M_{12} & M_{13} & M_{14} \\ M_{21} & M_{22} & M_{23} & M_{24} \\ M_{31} & M_{32} & M_{33} & M_{34} \\ M_{41} & M_{42} & M_{43} & M_{44} \end{bmatrix}, \quad (3.11)$$

and $\vec{I}_i$ is the Stokes vector of incident light:

$$\vec{I}_{i,s} = \begin{bmatrix} I_{i,s} \\ Q_{i,s} \\ U_{i,s} \\ V_{i,s} \end{bmatrix}. \tag{3.12}$$

The degree of linear polarization of scattered light is defined as

$$P_L = (I_{\parallel} - I_{\perp})/(I_{\parallel} + I_{\perp}) = [Q_s^2 + U_s^2]^{1/2}/\mathrm{I}_s, \tag{3.13}$$

and that of circular polarization as

$$P_C = V_s/I_s. \tag{3.14}$$

Elements of the light scattering matrix (LSM) depend on the scattering angle θ, the wavelength, and geometrical and optical parameters of the scatterers. There are only 7 independent elements (of 16) in the scattering matrix of a single particle with fixed orientation and nine relations that connect the others together. For scattering by a collection of randomly oriented scatterers, there are ten independent parameters.

$M_{11}$ is what is measured when the incident light is unpolarized, the scattering angle dependence of which is the phase function of the scattered light. It provides only a fraction of the information theoretically available from scattering experiments. $M_{11}$ is much less sensitive to chirality and long-range structure than some of the other matrix elements.[31] $M_{12}$ refers to a degree of linear polarization of the scattered light; $M_{22}$ displays the ratio of depolarized light to the total scattered light (a good measure of scatterers' nonsphericity); $M_{34}$ displays the transformation of 45° obliquely polarized incident light to circularly polarized scattered light (uniquely characteristic for different biological systems); the difference between $M_{33}$ and $M_{44}$ is a good measure of scatterers' nonsphericity.

If a particle is small with respect to the wavelength of the incident light, its scattering can be described as if it were a single dipole. The so-called Rayleigh theory is applicable under the condition that $m(2\pi a/\lambda) << 1$, where $m$ is the relative refractive index of the scatterers, $(2\pi a/\lambda)$ is the size parameter, $a$ is the radius of the particle, and λ is the wavelength of the incident light in a medium.[31] For the NIR light and typical biological scatterers with a refractive index referring to the ground matter $m$ = 1.05 to 1.1, the maximum particle radius is about 12 to 14 nm for the Rayleigh theory to remain valid. For this theory the scattered irradiance is inversely proportional to $\lambda^4$ and increases because $a^6$: the angular distribution of the scattered light is isotropic.

The Rayleigh-Gans or Rayleigh-Debye theory addresses the problem of calculating the scattering by a special class of arbitrary shaped particles. It requires $|m - 1| << 1$ and $(2\pi a'/\lambda)\,|m - 1| << 1$, where $a'$ is the largest dimension of the particle.[32] These conditions mean that the electric field inside the particle must be close to that of the incident field and the particle can be viewed as a collection of independent dipoles exposed to the same incident field. A biological cell might be modeled as a sphere of cytoplasm with a higher refractive index ($\bar{n}$ = 1.37) relative to that of the surrounding water medium ($\bar{n}$ = 1.35); then $m$ = 1.015 and, for the NIR light, this theory will be valid for the particle dimension up to $a'$ = 850 to 950 nm. This approximation has been applied extensively to calculations of light scattering from suspensions of bacteria.[32] It can be applicable for describing light scattering from cell components (mitochondria, lysosomes, peroxisomes, etc.) in tissues due to their small dimensions and refraction.

For the description of the region near the forward direction scattering caused by large particles (of order of 10 μm), the Fraunhofer diffraction approximation is useful.[32] According to this theory the scattered light has the same polarization as that of the incident light and the scatter pattern is independent of the refractive index of the object. For the small scattering angles, Fraunhofer diffraction approximation can accurately represent the change in irradiance as a function of particle size. That is why this approach is applicable in laser flow cytometry. The structures of the biological cell such as cell membrane, nuclear texture, and granules in the cytoplasm can be represented by variations in optical density.

Mie or Lorenz–Mie scattering theory is an exact solution of Maxwell's electromagnetic field equations for a homogeneous sphere.[31,32] In the general case, light scattered at a particle becomes elliptically polarized. For spherically symmetric particles of an optically inactive material, the Mueller scattering matrix is given by[31]

$$M = \begin{vmatrix} M_{11} & M_{12} & 0 & 0 \\ M_{12} & M_{11} & 0 & 0 \\ 0 & 0 & M_{33} & M_{34} \\ 0 & 0 & -M_{34} & M_{33} \end{vmatrix} . \quad (3.15)$$

Mie theory has been extended to arbitrary coated spheres and to arbitrary cylinders.[32] In Mie theory the electromagnetic fields of the incident, internal, and scattered waves are each expanded in a series.[32] A linear transformation can be made between the fields in each of the regions. This approach can also be used for nonspherical objects such as spheroids. The linear transformation is called the transition matrix (T-matrix), which for spherical particles is diagonal.

For a system of small, spatially uncorrelated particles, the degree of linear ($i = L$) and circular ($i = C$) polarization in the far region of the initially polarized (linearly or circularly) light transmitted through a layer of thickness $d$ is defined by the relation[34]

$$P_i \cong \frac{2d}{l_s} \sinh(l_s/\xi_i) \cdot \exp(-d/\xi_i), \quad (3.16)$$

where $l_s = 1/\mu_s$ is the scattering length; $\xi_i = (\zeta \cdot l_s/3)^{0.5}$ is the characteristic depolarization length for a layer of scatterers, $d >> \xi_i$, $\zeta_L = l_s/[\ln(10/7)]$, $\zeta_C = l_s/(\ln 2)$.

As can be seen from Equation 3.16, the characteristic depolarization length for linearly polarized light in tissues that can be represented as ensembles of Rayleigh particles is approximately 1.4 times greater than the corresponding depolarization length for circularly polarized light. One can employ Equation 3.16 to assess the depolarization of light propagating through an ensemble of large-scale spherical particles whose sizes are comparable with the wavelength of incident light (Mie scattering). For this purpose, one should replace $l_s$ by the transport length $l_t \cong 1/\mu_s'$ and take into account the dependence on the size of scatterers in $\zeta_L$ and $\zeta_C$. With the growth in the size of scatterers, the ratio $\xi_L/\zeta_C$ changes. It decreases from ~1.4 down to ~0.5 as $2\pi a/\lambda$ increases from 0 up to ~4, where $a$ is the size of scatterers and $\lambda$ is the wavelength of the light in the medium, and remains virtually constant at the level of 0.5 when $2\pi a/\lambda$ grows from ~4 to 15.

The Mueller matrix for the backscattering geometry was obtained by solving a radiative transfer equation with appropriate boundary conditions.[33] Analysis of this matrix structure showed that its form coincides with the single scattering matrix for optically active spherical scatterers. Thus, different tissues or the same tissues in various pathological or functional states should display different responses to the probing with linearly and circularly polarized light. This effect can be employed in optical medical tomography and for determining optical and spectroscopic parameters of tissues. As follows from Equation 3.16, the depolarization length in tissues should be close to the mean transport path length $l_t$ of a photon because this length characterizes the distance within which the direction of light propagation and, consequently, the polarization plane of linearly polarized light, become totally random after many sequential scattering events.

Since the length $l_t$ is determined by the parameter $g$ characterizing the anisotropy of scattering, the depolarization length should also substantially depend on this parameter. Indeed, the experimental data of Bicout et al.[34] demonstrate that the depolarization length $l_p$ of linearly polarized light, which is defined as the length within which the ratio $I_{\|}/I_{\perp}$ decreases to 2, displays such a dependence. The ratio mentioned above varied from 300 to 1, depending on the thickness of a sample and the type of a tissue (see Figure 3.4). These measurements were performed within a narrow solid angle (~$10^{-4}$ sr) in the direction of the

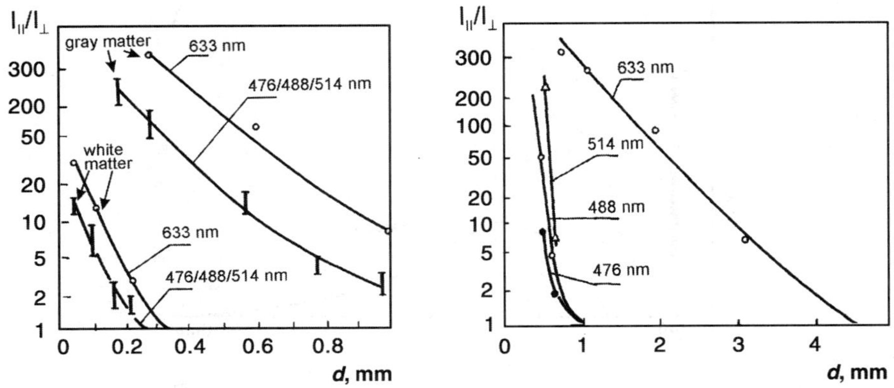


**FIGURE 3.4** Tissue polarization properties. Dependence of the depolarization degree ($I_{\|}/I_{\perp}$) of laser radiation (He-Ne laser, $\lambda = 633$ nm and Ar laser, $\lambda = 476/488/514$ nm) on the penetration depth ($d$) for brain tissue (gray and white matter) and whole blood (low hematocrit). Measurements were performed within a small solid angle ($10^{-4}$ sr) along the axis of a laser beam 1 mm in diameter. A strong influence of fluorescence was seen for blood irradiated by Ar laser.

incident laser beam. The values of $l_p$ considerably differed for white substance of brain and a tissue of cerebral cortex: 0.23 and 1.3 mm for $\lambda = 633$ nm, respectively. Whole blood is characterized by a considerable depolarization length (about 4 mm) at $\lambda = 633$ nm; this is indicative of dependence on the parameter $g$, whose value for blood exceeds the values of this parameter for tissues of many other types and can be estimated as 0.982 to 0.999.[5,7,10]

In contrast to depolarization, the attenuation of collimated light is determined by the total attenuation coefficient $\mu_t$ (see Equation 3.1). For many tissues, $\mu_t$ is much greater than $\mu_s'$. Therefore, in certain situations, it is impossible to detect pure ballistic photons (photons that do not experience scattering), but forwardscattered photons retain their initial polarization and can be used for imaging purpose. Ostermeyer et al.[36] experimentally demonstrated that laser radiation retains linear polarization on the level of $P_L \leq 0.1$ within $2.5l_t$. Specifically, for skin irradiated in the red and NIR ranges, $\mu_a \cong 0.4$ cm$^{-1}$, $\mu_s' \cong 20$ cm$^{-1}$, and $l_t \cong 0.48$ mm.

Consequently, light propagating in skin can retain linear polarization within the length of about 1.2 mm. Such an optical path in a tissue corresponds to a delay time on the order of 5.3 ps, which provides an opportunity to produce polarization images of macro-inhomogeneities in a tissue with a spatial resolution equivalent to the spatial resolution that can be achieved with the selection of photons by means of more sophisticated time-resolved techniques. In addition, polarization imaging makes it possible to eliminate specular reflection from the surface of a tissue (see Figure 3.2), which allows one to apply this technique for the imaging of microvessels in facile skin.[3] Polarization images can see skin subsurface textural changes and allow one to erase melanin from the image.[26]

Polarization imaging is a new direction in tissue optics.[5,7,26,33] The registration of two-dimensional polarization patterns for the backscattering of a polarized incident narrow laser beam is the base for this technique. Major informative images can be received using the backscattering Mueller matrix approach. To determine each of the 16 experimental matrix elements, a total of 16 images should be taken at various combinations of input and output polarization states.

## 3.2.3 Short Light Pulses

When probing the plane-parallel layer of a scattering medium with an ultrashort light pulse, the transmitted pulse consists of a ballistic (coherent) component, a group of photons having zigzag trajectories, and a highly intensive diffuse component (see Figure 3.5a).[1–7,17,23,37] Both unscattered photons and photons undergoing forward-directed, single-step scattering contribute to the intensity of the ballistic component (comprised by photons traveling straight along the light beam). This component (not shown in Figure

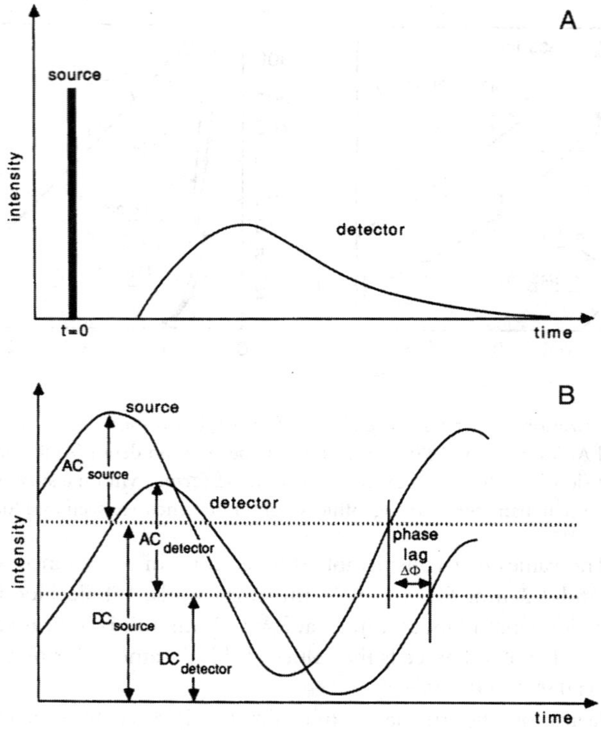


**FIGURE 3.5** Schematic representation of the time evolution of the light intensity measured in response to a very narrow light pulse (A) and a sinusoidally intensity-modulated light (B) transversing an arbitrary distance in a scattering and absorbing medium. (From Fishkin, J.B. and Gratton, E., *J. Opt, Soc. Am. A*, 10, 127, 1993. With permission.)

3.5a) is subject to exponential attenuation with increasing sample thickness. This accounts for the limited utility of ballistic photons for practical diagnostic purposes in medicine.

The group of snake photons with zigzag trajectories includes photons that experienced only a few collisions each. They propagate along trajectories, which only slightly deviate from the direction of the incident beam and form the first-arriving part of the diffuse component. These photons carry information about the optical properties of the random medium.

The diffuse component is very broad and intense because it contains the bulk of incident photons after they have participated in many scattering acts and therefore migrate in different directions and have different path lengths. Moreover, the diffuse component carries information about the optical properties of the scattering medium, and its deformation may reflect the presence of local inhomogeneities in the medium. The resolution obtained by this method at a high light-gathering power is much lower than in the method measuring straight-passing photons. Two principal probing schemes are conceivable — one recording transmitted photons and the other taking advantage of their backscattering.

The time-dependent reflectance is defined as[38,39]

$$R(r_{sd},t) = \frac{z_0}{(4\pi D)^{3/2}} t^{-5/2} \exp\left(-\frac{r_{sd}^2 + z_0^2}{2Dt}\right) \exp(-\mu_a ct), \tag{3.17}$$

where $t$ is the time; $z_0 = (\mu_s')^{-1}$, and $D$ is the photon diffusion coefficient, $cm^2/c$ (see Equation 3.5).

An important advantage of the pulse method is its applicability to *in vivo* study due to the possibility of the separate evaluation of $\mu_a$ and $\mu_s'$ using a single measurement in the backscattering or transillumination regimes.

### 3.2.4 Diffuse Photon-Density Waves

When probing the plane-parallel layer of a scattering medium with an intensity modulated light, the modulation depth of scattered light intensity $m_U \equiv AC_{detector}/DC_{detector}$ (see Figure 3.5B) and the corresponding phase shift relative to the incident light modulation phase $\Delta\Phi$ (phase lag) can be measured.[1–7,40] In applications to tissue spectroscopy and tomography compared with pulse measurements, this method is simpler and more reliable in terms of data interpretation and noise immunity. These happen because amplitude modulation is measured at low peak powers, slow rise time; therefore, smaller bandwidths than the pulse measurements are needed. The current measuring schemes are based on heterodyning of optical and transformed signals.[40]

The development of the theory underlying this method resulted in the discovery of a new type of wave: photon-density waves or progressively decaying waves of intensity.[1–7,40–42] Microscopically, individual photons make random migrations in a scattering medium, but collectively they form a photon-density wave at a modulation frequency ω that moves away from a radiation source (see Figure 3.5B). Photon-density waves possess typical wave properties, e.g., they undergo refraction, diffraction, interference, dispersion, and attenuation.

In strongly scattering media with weak absorption far from the walls and a source or a receiver of radiation, the light distribution may be regarded as a decaying diffusion process described by the time-dependent diffusion equation for photon density. For a point light source with harmonic intensity modulation at frequency $\omega = 2\pi\nu$ placed at the point $\bar{r} = 0$, an alternating component (AC) of intensity is a going-away spherical wave with its center at the point $\bar{r} = 0$, which oscillates at a modulation frequency with modulation depth

$$m_U(\bar{r},\omega) = m_I \exp(\bar{r}\sqrt{D/c\mu_a})\exp(-\bar{r}\sqrt{\omega/2D}), \tag{3.18}$$

and undergoes a phase shift relative to the phase value at point $\bar{r} = 0$ equal to

$$\Delta\Phi(\bar{r},\omega) = \bar{r}(\omega/2D)^{0.5}, \tag{3.19}$$

where $m_I$ is the intensity modulation depth of the incident light, $D = c/3(\mu_s' + \mu_a)$.

The length of a photon-density wave, $\Lambda_\Phi$, and its phase velocity, $V_\Phi$, are defined by

$$\Lambda_\Phi^2 = 8\pi^2 D/\omega \quad \text{and} \quad V_\Phi^2 = 2D\omega. \tag{3.20}$$

Measuring $m_U(\bar{r},\omega)$, $\Delta\Phi(\bar{r},\omega)$ allows one to determine the transport scattering coefficient $\mu_s'$ and the absorption coefficient $\mu_a$ separately and to evaluate the spatial distribution of these parameters.

Keeping medical applications in mind, we can easily estimate that, for $\omega/2\pi = 500$ MHz, $\mu_s' = 15$ cm$^{-1}$, $\mu_a = 0.035$ cm$^{-1}$, and $c = (3 \times 10^{10}/1.33)$ cm/s, the wavelength is $\Lambda_\Phi \cong 5.0$ cm and the phase velocity is $V_\Phi \cong 1.77 \times 10^9$ cm/s.

## 3.3 Optothermal Interactions

### 3.3.1 Temperature Rise and Tissue Damage

When photons traveling in tissue are absorbed, heat is generated. The generated heat, described by the heat source term $S$ at a point **r**, is proportional to the fluence rate of light $\phi(\mathbf{r})$ (mW/cm$^2$) and absorption coefficient $\mu_a(\mathbf{r})$ in this point[3,21–23,43]

$$S(\mathbf{r}) = \mu_a(\mathbf{r})\phi(\mathbf{r}). \tag{3.21}$$

The traditional bioheat equation originated from the energy balance describes the change in tissue temperature over time at point **r** in the tissue:

$$\rho\tilde{n}\,\frac{\partial T(\mathbf{r},t)}{\partial t} = \nabla\left[k_m \nabla T(\mathbf{r},t)\right] + S(\mathbf{r}) + \rho c w\left(T_a - T_v\right), \tag{3.22}$$

where ρ is the tissue density (g/cm³), $c$ is the tissue specific heat (mJ/g·°C), $T(\mathbf{r},t)$ is the tissue temperature (°C) at time $t$, $k_m$ is the thermal conductivity (mW/cm·°C), $S(\mathbf{r})$ is the heat source term (mW/cm³), $w$ is the tissue perfusion rate (g/cm³·s), $T_a$ is the inlet arterial temperature (°C), and $T_v$ is the outlet venous temperature (°C) — all at point $S(\mathbf{r})$ in the tissue. In this equation convection, radiation, vaporization, and metabolic heat effects are not accounted for because of their negligible effect in many practical cases. The source term is assumed to be stationary over the time interval of heating. The first term to the right of the equation describes any heat conduction (typically away from point **r**), and the source term accounts for heat generation due to photon absorption. In most cases of light (laser)–tissue interaction, the heat transfer caused by perfusion (last term) is negligible.

Initial and boundary conditions must be accounted for in order to solve this equation. The initial condition is the tissue temperature at $t = 0$ and the boundary conditions depend on tissue structure and geometry of light heating. Methods of solving the bioheat equation can be found in References 3 and 21 through 24.

Damage to a tissue results when it is exposed to a high temperature for a long time period.[3,21–24,43] The damage function is expressed in terms of an Arrhenius integral:

$$\Omega(\tau) = \ln\left(\frac{C(0)}{C(\tau)}\right) = A\int_0^{\tau} e^{-\frac{E_a}{RT(t)}}\,dt\ , \tag{3.23}$$

where τ is the total heating time (s); $C(0)$ is the original concentration of undamaged tissue; $C(\tau)$ is the remaining concentration of undamaged tissue after time τ; $A$ is an empirically determined constant (s$^{-1}$); $E_a$ is an empirically determined activation energy barrier (J/mol); $R$ is the universal gas constant (8.32 J/mol · K); and $T$ is the absolute temperature (K).

In noninvasive optical diagnostic and some photochemical applications of light one must keep the tissue below the damaging temperature, so-called the critical temperature $T_{crit}$. This temperature is defined as the temperature where the damage accumulation rate, $d\Omega/dt$, is equal to 1.0:[43]

$$T_{crit} = \frac{E_a}{R\ \ln(A)}. \tag{3.24}$$

The constants $A$ and $E_a$ can be calculated on the basis of experimental data when tissue is exposed to a constant temperature.[22] For example, for pig skin, $A = 3.1 \times 10^{98}$ and $E_a = 6.28 \times 10^5$, which gives $T_{crit} = 59.7$°C.

With CW light sources, due to the increase in temperature difference between the irradiation and the surrounding tissue, conduction of heat away from the light absorption point and into surrounding tissue increases. Depending on the light energy, large tissue volumes may be damaged or loss of heat at the target tissue component may be expected. For pulsed light, little heat is usually lost during the pulse duration because light absorption is a fast process while heat conduction is relatively slow; therefore, more precise tissue damage is possible.

The following forms of irreversible tissue damage are expected as tissue temperature rises past $T_{crit}$: coagulation (denaturization of cellular and tissue proteins) is the basis for tissue welding; vaporization (tissue dehydration and vapor bubble formation [vacuolization], $T \geq 100$°C) is the basis for tissue mechanical destruction; pyrolysis ($T \geq 350$ to 450°C). Vaporization, vacuolization, and pyrolysis combine to produce thermal ablation — the basis of laser surgical tissue removal.

The disadvantage of thermal ablation with CW light sources is undesirable damage to surrounding tissue via its coagulation. Pulsed light can deliver sufficient energy in each pulse to ablate tissue, but in a short enough time that the tissue is removed before any heat is transferred to the surrounding tissue. To achieve precise tissue cutting, lasers with a very short penetration depth, like excimer ArF laser, are used (see Figure 3.1).

As condensed matter, tissue can show any of noncoherent or coherent effects at laser irradiation.[44] Linear noncoherent effects exist within a wide area of pulse duration and intensities; for long pulses of 1 s the intensity should not exceed 10 W/cm$^2$; for shorter pulses of $10^{-9}$ s the intensity can be up to $10^9$ W/cm$^2$. Multiphoton processes may exist at relatively low pulse duration ($10^{-9}$ to $10^{-12}$ s) at intensities of $10^9$ to $10^{12}$ W/cm$^2$ and rather low light energies, not higher than 0.1 J/cm$^2$. The linear and nonlinear coherent effects may be induced only by a very short pulse with time duration comparable with relaxation time of biological molecules, $\tau \leq 10^{-13}$.

### 3.3.2 Optothermal and Optoacoustic Effects

The time-dependent heat generated in a tissue via interaction with pulsed or intensity-modulated optical radiation is known as optothermal (OT) effect.[5,20,45] Such interaction also induces a number of thermoelastic effects; in a tissue in particular, it causes generation of acoustic waves. Detection of acoustic waves is the basis for optoacoustic (or photoacoustic) methods. The informative features of this method allow one to estimate tissue thermal, optical, and acoustical properties that depend on tissue structure peculiarities. Two main modes can be used for excitation of tissue thermal response: (1) a pulse of light excites the sample and the signal is detected in the time domain with a fast detector attached to a wide-band amplifier (signal averaging and gating techniques are used to increase the signal-to-noise ratio) or (2) an intensity-modulated light source, a low-frequency transducer, and phase-sensitive detection for noise suppression are provided.

In every case the thermal waves generated by the heat release result in several effects that have given rise to various techniques: optoacoustics (OA) or photoacoustics (PA), optothermal radiometry (OTR) or photothermal radiometry (PTR), photorefractive techniques, etc.[5,20,45] (see Figure 3.6).

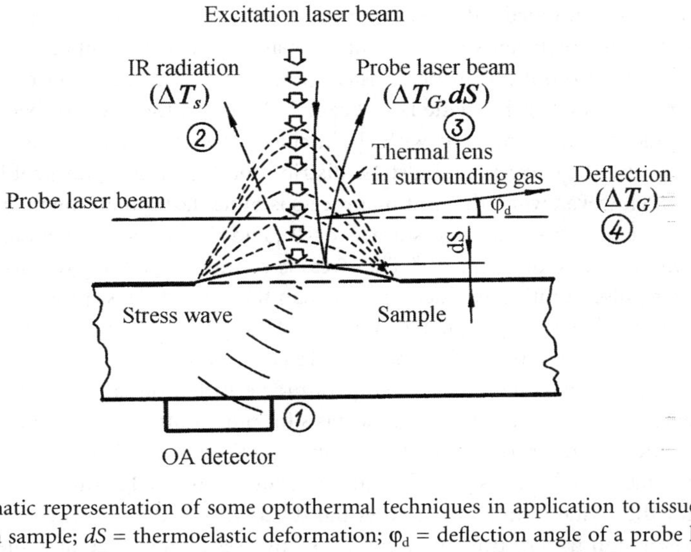


**FIGURE 3.6** Schematic representation of some optothermal techniques in application to tissue study. $\Delta T_s$ = temperature change of a sample; $dS$ = thermoelastic deformation; $\varphi_d$ = deflection angle of a probe laser beam; 1 = OA technique; 2 = OTR technique; 3 = thermal lens technique; 4 = deflection technique. (From Tuchin, V.V., *Laser and Fiber Optics in Biomedical Science*, Saratov University Press, Saratov, Russian Federation, 1998. With permission.)

The term optoacoustics (OA) refers primarily to the time-resolved technique utilizing pulsed lasers and measuring profiles of pressure in tissue; photoacoustics (PA) describes primarily spectroscopic experiments with CW modulated light and a PA cell.

When a laser beam falls down to the sample surface and the wavelength is tuned to an absorption line of the tissue component of interest, the optical energy is absorbed by the target component and most of the energy transforms to heat. The time-dependent heating leads to all of the earlier mentioned thermal and thermoelastic effects. In OA or PA techniques, a microphone or piezoelectric transducer in acoustic contact with the sample is used as a detector to measure the amplitude or phase of the resultant acoustic wave. In OTR technique, distant IR detectors and array cameras are employed for the sample surface temperature estimation and its imaging. The intensity of the signals obtained with any of the OT or OA techniques depends on the amount of energy absorbed and transformed into heat and on thermoelastic properties of the sample and its surrounding. When nonradiative relaxation is the main process in a light beam decay, and extinction is not very high, $\mu_a d << 1$ ($d$ is the length of a cylinder within the sample occupied by a pulse laser beam), the absorbed pulse energy induces the local temperature rise defined by

$$\Delta T \cong E\mu_a d / C_p V \rho, \tag{3.25}$$

where $C_P$ is the specific heat capacity for a constant pressure; $V = \pi w^2 d$ is the illuminated volume; $w$ is the laser beam radius; $\rho$ is the medium density. Assuming an adiabatic expanding of illuminated volume on heat with a constant pressure, one can calculate the change of the volume $\Delta V$. This expansion induces a wave propagating in the radial direction with the sound speed, $v_a$. The corresponding change of pressure $\Delta p$ is proportional to the amplitude of mechanical oscillations:

$$\Delta p \sim (f_a/w)(\beta v_a/C_P)E\mu_a, \tag{3.26}$$

where $\beta$ is the coefficient of volumetric expansion and $f_a$ is the frequency of the acoustic wave.

Equations 3.25 and 3.26 provide the basis for various OT and OA techniques. The information about the absorption coefficient $\mu_a$ at the definite wavelength can be received from direct measurements of the temperature change $\Delta T$ (optical calorimetry), volume change $\Delta V$ (optogeometric technique), or pressure change $\Delta p$ (OA and PA techniques). For a highly scattering tissue, measurements of the stress-wave profile and amplitude should be combined with measurements of the total diffuse reflectance in order to extract absorption and scattering coefficients of the sample separately. The absorption coefficient in a turbid medium can be estimated from the acoustic transient profile only if the subsurface irradiance is known.

For turbid media irradiated with a wide laser beam (>0.1 mm), the effect of backscattering causes a higher subsurface fluence rate compared with the incident laser fluence.[5] Therefore, the $z$-axial light distribution in tissue and the corresponding stress distribution have a complex profile with a maximum at a subsurface layer. However, when the heating process is much faster than the medium expansion, the stress amplitude adjacent to the irradiated surface, $\delta p(0)$, and the stress exponential tail into the depth of tissue sample, $\delta p(z)$, can be determined.[46] The stress is confined temporarily during laser heat deposition when the laser pulse duration is much shorter than the time of stress propagation across the light penetration depth in the tissue sample. Such conditions of temporal pressure confinement in the volume of irradiated tissue allow for the most efficient pressure generation.

The pulse laser heating of a tissue causes its temperature perturbations and corresponding modulation of its own thermal (infrared) radiation. This is the basis for pulse optothermal radiometry (OTR).[5,20,46,47] The maximum of intensity of thermal radiation of living objects falls at the wavelength range close to 10 μm. A detailed analysis of OTR signal formation requires knowledge on the internal temperature distribution within the tissue sample, tissue thermal diffusivity, and its absorption coefficients at the excitation $\mu_a$ and emission $\mu_a'$ (10 μm) wavelengths. Conversely, knowledge of some of the mentioned parameters allows one signal to reconstruct, for example, the depth distribution of $\mu_a$ on the basis of measured OTR.[47]

The surface radiometric signal $S_r(t)$ at any time $t$ is the sum of the contributions from all depths in the tissue at time $t$. The radiation from deeper depths is attenuated by the infrared absorption of the sample before reaching the detector. Because the initial surface temperature is known, the temperature distribution into the sample depth can be extracted from $S_r(t)$ measurement.

OTR, OA, and PA transient techniques provide a convenient means for *in vitro* or even *in vivo* and *in situ* monitoring of human skin properties (optical absorption, thermal properties, water content) and surface concentrations of topically applied substances (drugs and sunscreen diffusion). The main difficulty of the PA method in the case of *in vivo* measurements is the requirement for a closed sample cell that can guide the acoustic signal efficiently from sample to microphone. The use of pulsed OA and OTR techniques is more appropriate for *in vivo* and *in situ* measurements.

## 3.4 Refractive Index and Controlling of Light Interactions with Tissues

Refractive index in tissues is of great importance for light–tissue interaction. Most tissues have refractive indices for visible light in the 1.335 to 1.62 range (e.g., 1.55 in the stratum corneum, 1.62 in the enamel, and 1.386 at the lens surface).[5,48] It is worthwhile to note that *in vitro* and *in vivo* measured values of refractive indices may differ significantly. For example, the refractive index in rat mesenteric tissue *in vitro* was found to be 1.52 compared to only 1.38 *in vivo*. This difference can be accounted for by the decreased refractivity of ground matter $n_0$ due to impaired hydration.

The mean refractive index $\bar{n}$ of a tissue is defined by the refractive indices of its scattering centers material $n_s$ and ground matter $n_0$:[5]

$$\bar{n} = c_s n_s + (1 - c_s) n_0 , \tag{3.27}$$

where $c_s$ is the volume fraction of the scatterers.

The $n_s/n_0 \equiv m$ ratio determines the scattering coefficient. For example, in a simple monodisperse model of scattering dielectric spheres:[49]

$$\mu_s' = 3.28 \pi a^2 \rho_s (2\pi a/\lambda)^{0.37} (m-1)^{2.09} , \tag{3.28}$$

where $a$ is the sphere radius and $\rho_s$ is the volume density of the spheres. Equation 3.28 is valid for noninteracting Mie scatterers, $g > 0.9$; $5 < 2\pi a/\lambda < 50$; $1 < m < 1.1$.

It follows from Equation 3.28 that even a 5% change in the refractive index of the ground matter ($n_0 = 1.35 \rightarrow 1.42$), when that of the scattering centers is $n_s = 1.47$, will cause a sevenfold decrease of $\mu_s'$. Therefore, matching the refractive index of the scatterers and ground material allows one to reduce tissue scattering considerably. This phenomenon is very useful for improving measurement conditions for optical tomograpy and for obtaining more precise spectroscopic information from the depth of a tissue.

Optical parameters of a tissue, in particular refractive index, are known to depend on water content. The refractive index of water over a broad wavelength range of 0.2 to 200 μm has been reported.[48] The following relation was shown to be valid for the visible and NIR wavelength range ($\lambda$ in nm):[50]

$$n_{H_2O} = 1.3199 + 6878/\lambda^2 - 1.132 \times 10^9/\lambda^4 + 1.11 \times 10^{14}/\lambda^6 . \tag{3.29}$$

For different parts of a biological cell, values of refractive index in the NIR can be estimated as follows: extracellular fluid, $\bar{n}$ = 1.35 to 1.36; cytoplasm, 1.360 to 1.375; cell membrane, 1.46; nucleus, 1.38 to 1.41; mitochondria and organelles, 1.38 to 1.41; melanin, 1.6 to 1.7.[5,51] Scattering arises from a mismatch in refractive index of the components that make up the cell. When cells are surrounded by other cells or tissue structures of similar index, certain organelles become the important scatterers for tissues. For instance, the nucleus is a significant scatterer because it is often the largest organelle in the cell and its size increases relative to the rest of the cell throughout neoplastic progression.

Mitochondria (500 to 1500 nm in diameter), lysosomes (500 nm), and peroxisomes (500 nm) are very important scatterers whose size relative to the wavelength of light suggests that they must provide a significant contribution to the backscattering. Melanin granular, traditionally thought of as an absorber, must be considered an important scatterer because of its size and high refractive index.[51] Structures consisting of membrane layers such as the endoplasmic reticulum or Golgi apparatus may prove significant because they contain index fluctuations of high spatial frequency and amplitude. Besides cell components, tissue fibrous structures such as collagen and elastin must be considered important scatterers.

Refractivity measurements in a number of strongly scattering tissues at 633 nm performed with a fiber-optic refractometer have shown that fatty tissue has the largest refractive index (1.455) followed by kidney (1.418), muscular tissue (1.410), and then blood and spleen (1.400).[52] The lowest refractive indices were found in lungs and liver (1.380 and 1.368, respectively). There is a tendency to decreasing refractive indices with increasing light wavelength from 390 to 700 nm (for example, for bovine muscle in the limits of 1.42 to 1.39).

It is possible to achieve a marked impairment of scattering by matching the refractive indices of scattering centers and ground matter by means of intratissue administration of the appropriate chemical agents. Experimental optical clearing in human sclera in the visible wavelength range induced by administration of verografin, trazograph, glucose, propylene glycol, polyethylene glycol, and other solutions has been described.[5,7,53–57] Osmotic phenomena appear to be involved when optical properties of tissues are modulated by sugar, alcohol, glycerol, and electrolyte solutions.

Experimental studies on optical clearing of normal and pathological skin and its components (epidermis and dermis) and the management of reflectance and transmittance spectra using glycerol, glycerol–water solutions, glucose, sunscreen creams, cosmetic lotions, gels, and pharmaceutical products have been carried out.[5,7]

A marked clearing effect through rat[55] and human[56] skin and the rabbit sclera[7] occurred for an *in vivo* tissue within a few minutes of topical application or intratissue injection of glycerol, glucose, verografin, or trazograph. *In vivo* reflectance spectra of the human skin at inter-skin injection of 40% glucose are shown in Figure 3.7. Skin is well protected for penetration of any agent by the stratum corneum, and

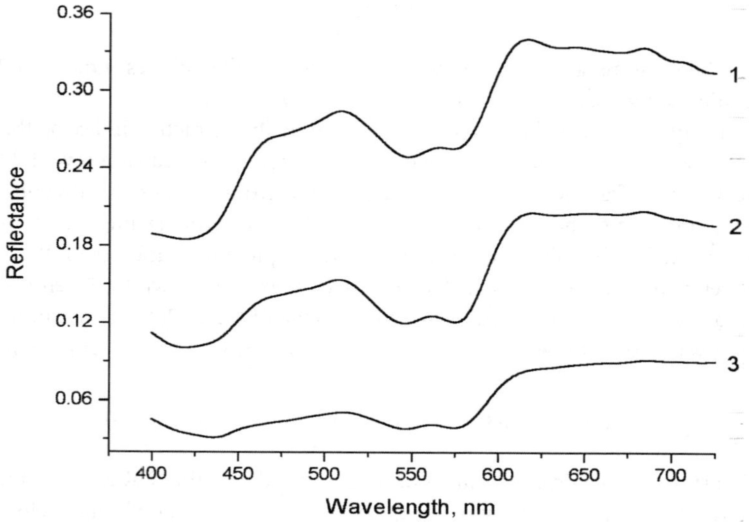


**FIGURE 3.7** Reflectance spectra measured before (1) and at 23 min (2) and 60 min (3) after interskin injection of 0.1 ml of 40% glucose into internal side of the forearm of the male volunteer. (From Tuchin, V.V. et al., *J. Tech. Phys. Lett.*, 27(6), 489, 2001. With permission.)

reflectance of the skin is defined mostly by dermis; therefore, the intradermis injection was applied. The fiber-optic photodiode array spectrometer providing about 0.9-mm-deep measurements was used.

The concept that index matching could improve the optical penetration depth of whole blood is proved experimentally in *in vitro* studies using optical coherence tomography.[58,59] For example, for whole blood twice diluted by saline, adding 6.5% of glycerol increases the optical penetration up to 117%; optical clearing up to 150% was achieved mostly due to refractive index matching for the high-molecular-weight dextran.[59]

## 3.5 Fluorescence

### 3.5.1 Fundamentals and Methods

Fluorescence arises upon light absorption and is related to an electronic transition from the excited state to the ground state of a molecule. In the case of thin samples, e.g., cell monolayers or biopsies with a few micrometers in diameter, fluorescence intensity $I_F$ is proportional to the concentration $c$ and the fluorescence quantum yield $\eta$ of the absorbing molecules.[18,60] In scattering media, the path lengths of scattered and unscattered photons within the sample are different, and should be accounted for.[18] However, in rather homogeneous thin samples, the linearity between $I_F$, $c$, and $\eta$ is still fulfilled.

At excitation biological objects by ultraviolet light ($\lambda \leq 300$ nm), fluorescence of proteins as well as of nucleic acids can be observed. Fluorescence quantum yields of all nucleic acid constituents, however, are around $10^{-4}$ to $10^{-5}$, corresponding to lifetimes of the excited states in the picosecond time range. Autofluorescence (AF) of proteins is associated with amino acids such as tryptophan, tyrosin, and phenylalanine with absorption maxima at 280, 275, and 257 nm, respectively, and emission maxima between 280 nm (phenylalanine) and 350 nm (tryptophan). Usually the protein spectrum is dominated by tryptophan. Fluorescence from collagen or elastin using excitation between 300 and 400 nm shows broad emission bands between 400 and 600 nm with maxima around 400, 430, and 460 nm. In particular, fluorescence of collagen and elastin can be used to distinguish various types of tissues, e.g., epithelial and connective tissue.[17,18,60–63]

The reduced form of coenzyme nicotinamide adenine dinucleotide (NADH) is excited selectively in a wavelength range between 330 and 370 nm. NADH is most concentrated within mitochondria where it is oxidized within the respiratory chain located within the inner mitochondrial membrane and its fluorescence is an appropriate parameter for detection of ischemic or neoplastic tissues.[60] Fluorescence of free and protein-bound NADH has been shown to be sensitive on oxygen concentration. Flavin mononucleotide (FMN) and dinucleotide (FAD) with excitation maxima around 380 and 450 nm have also been reported to contribute to intrinsic cellular fluorescence.[60]

Porphyrin molecules, e.g., protoporphyrin, coproporphyrin, uroporphyrin, or hematoporphyrin, occur within the pathway of biosynthesis of hemoglobin, myoglobin, and cytochromes.[60] Abnormalities in heme synthesis, occurring in the cases of porphyrias and some hemolytic diseases, may enhance the porphyrin level within tissues considerably. Several bacteria, e.g., *Propionibacterium acnes* or bacteria within dental caries lesions, accumulate considerable amounts of protoporphyrin.[60] Therefore, acne or caries detection based on measurements of intrinsic fluorescence appears to be a promising method.

At present, various exogenous fluorescing dyes can be applied for probing cell anatomy and cell physiology.[60] In humans, such dyes as fluorescein and indocyanine green are in use for fluorescence angiography or blood volume determination. Fluorescence spectra often give detailed information on fluorescent molecules, their conformation, binding sites, and interaction within cells and tissues. Fluorescence intensity can be measured as a function of the emission wavelength or of the excitation wavelength. The fluorescence emission spectrum $I_F(\lambda)$ is specific for any fluorophore and is commonly used in fluorescence diagnostics.

For many biomedical applications, an optical multichannel analyzer (OMA; a diode array or a CCD camera) as a detector of emission radiation is preferable because spectra can be recorded very rapidly and repeatedly with sequences in the millisecond range. Fluorescence spectrometers for *in vivo* diagnostics are commonly based on fiber optic systems.[7] The excitation light of a lamp or a laser is guided to the tissue (e.g., some specific organ) via fiber using appropriate optical filters. Fluorescence spectra are usually measured via the same fiber or via a second fiber or fiber bundle in close proximity to the excitation fiber.

Now available are various comprehensive and powerful fluorescence spectroscopies, such as microspectrofluorimetry, polarization anisotropy, time-resolved with pulse excitation and frequency domain, time-gated, total internal reflection fluorescence spectroscopy and microscopy, fluorescence resonant energy transfer method, confocal laser scanning microscopy, and their combinations.[18,60] These methods allow one to provide:[60]

- Three-dimensional topography of specimens measured in the reflection mode for morphological studies of biological samples
- High-resolution microscopy measured in the transmission mode
- Three-dimensional fluorescence detection of cellular structures and fluorescence bleaching kinetics
- Time-resolved fluorescence kinetics
- Studies of motions of cellular structures
- Time-gated imaging in order to select specific fluorescent molecules or molecular interactions
- Fluorescence lifetime imaging
- Spectrally resolved imaging

Principles of optical clinical chemistry based on measuring of changes of fluorescence intensity, wavelength, polarization anisotropy, and lifetime are described in Lakowicz.[18] Various fluorescence techniques of selective oxygen sensing and blood glucose and blood gases detection are available.[18]

### 3.5.2 Multiphoton Fluorescence

A new direction in laser spectroscopy of biological objects involves multiphoton (two- or three-photon) fluorescence scanning microscopy, which makes it possible to image functional states of an object or, in combination with autocorrelation analysis of the fluorescence signal, determine the intercellular motility in small volumes.[19] The two-photon technique employs ballistic and scattered photons at the wavelength of the second harmonic of the incident radiation coming to a wide-aperture photodetector exactly from the focal area of the excitation beam. A unique advantage of two-photon microscopy is the possibility of investigating three-dimensional distributions of chromophores excited with ultraviolet radiation in thick samples.

Such an investigation becomes possible because chromophores can be excited (e.g., at the wavelength of 350 nm) with laser radiation whose wavelength falls within the range (700 nm) where a tissue has a high transparency. This radiation can reach deeply lying layers and produces less damage in tissues. Fluorescence emission in this case occurs in the visible range (>400 nm) and emerges from a tissue comparatively easily and reaches a photodetector, which registers only the legitimate signal from the focal volume without any extraneous background. Investigations of tissues and cells by means of two-photon microscopy are characterized by the following typical parameters of laser systems: the wavelength ranges from 700 to 960 nm, the pulse duration is on the order of 150 fs, the pulse repetition rate is 76 to 80 MHz, and the mean power is less than 10 mW. Such parameters can be achieved with mode-locked dye lasers pumped by an Nd:YAG laser or with titanium sapphire lasers pumped by an argon laser. Diode-pumped solid-state lasers also hold much promise for the purposes of two-photon microscopy.

## 3.6 Vibrational Energy States Excitation

Mid-infrared (MIR) and Raman spectroscopies use light-excited vibrational energy states in molecules to obtain information about the molecular composition, molecular structures, and molecular interactions in a sample.[13–16,64] In MIR spectroscopy infrared light from a broad band source (usually 2.5 to 25 μm or 4000 to 400 $cm^{-1}$) is directly absorbed to excite the molecules to higher vibrational states. In a Raman scattering event, light is inelastically scattered by a molecule when a small amount of energy is transferred from the photon to the molecule (or vice versa). This leads to excitation of the molecule, usually from its lowest vibrational energy level in the electronic ground state to a higher vibrational state. The energy difference between the incident and scattered photon is expressed in a wavenumber shift ($cm^{-1}$). Some vibrations can be excited by Raman and MIR processes; others can only be excited by a Raman scattering or by MIR absorption. Both techniques enable recording of high-quality spectra in relatively short acquisition times (30 to 60 s).

The MIR and Raman spectroscopy techniques are successfully applied in various areas of clinical studies, such as cancerous tissue examination, the mineralization process of bone and teeth tissue monitoring, glucose sensing in blood, noninvasive diagnosis of skin lesions on benign or malignant cells, and monitoring of treatments and topically applied substances (e.g., drugs, cosmetics, moisturizers, etc.) in skin.[13–16,64]

Raman spectroscopy is widely used in biological studies, ranging from studies of purified biological compounds to investigations at the level of single cells. At present, combinations of spectroscopic techniques such as MIR and Raman with microscopic imaging techniques are explored to map molecular distributions at specific vibrational frequencies on samples to locally characterize tissues or cells.[64] Chemical imaging will become more and more important in clinical diagnosis.

## 3.7 Light Interaction with Eye Tissues

Healthy tissues of the anterior of the human eye chamber, e.g., the cornea and lens, are highly transparent for visible light due to their ordered structure and the absence of strongly absorbing chromophors.[5,7] Scattering is an important feature of light propagation in eye tissues. The size of the scatterers and the distance between them are smaller than or comparable with the wavelength of visible light; the relative refractive index of the scattering matter is equally small (soft particles). Typical eye tissue models are long, round dielectric cylinders (corneal and scleral collagen fibers) or spherical particles (lens protein structures) having refractive index $n_s$ and orderly (transparent cornea and lens) or quasi-orderly (sclera, opaque lens) distributed in the isotropic base matter with refractive index $n_0 < n_s$. Light scattering analysis in eye tissues often is possible using a single-scattering model owing to the small scattering cross section (soft particles).

In case of disordered (randomly distributed) scatterers, the resultant field intensity is the total intensity of fields scattered by individual scatterers (fibril or particles). For ordered scatterers, fields, rather than intensities, should be summed to take into account effects of interference arising in the presence of the near order of scatterers. In the integral form, the scattering indicatrix for symmetric scattering of particles with pair correlation is described by the expression:

$$I(\theta) = I_0(\theta)\left\{1+\rho_s\int_0^{\infty}[g(r)-1]\exp[i(\bar{s}'-\bar{s})\bar{r}]d^3r\right\} = I_0(\theta)F_{\text{int}}, \tag{3.30}$$

where $I_0(\theta)$ is the scattering indicatrix of an isolated scatterer; $\theta$ is the scattering angle; $\rho_s$ is the mean density of scatterers; $g(r)$ is the radial distribution function of scattering centers (local to average density ratio for scattering centers, for noninteracting centers $g(r) \rightarrow 1$); $\bar{s}$ and $\bar{s}'$ are unit vectors for incident and scattered waves, respectively; $\bar{r}$ is the radius-vector and $d^3r$ is the volume of a scatterer; and $F_{\text{int}}$ takes into account the effects of interference.

Equation 3.30 is valid for a monodisperse system of scatterers and can be used provided a scattering indicatrix for a single particle $I_0(\theta)$ is available from computations based on the Mie theory and corresponding approximate relations.

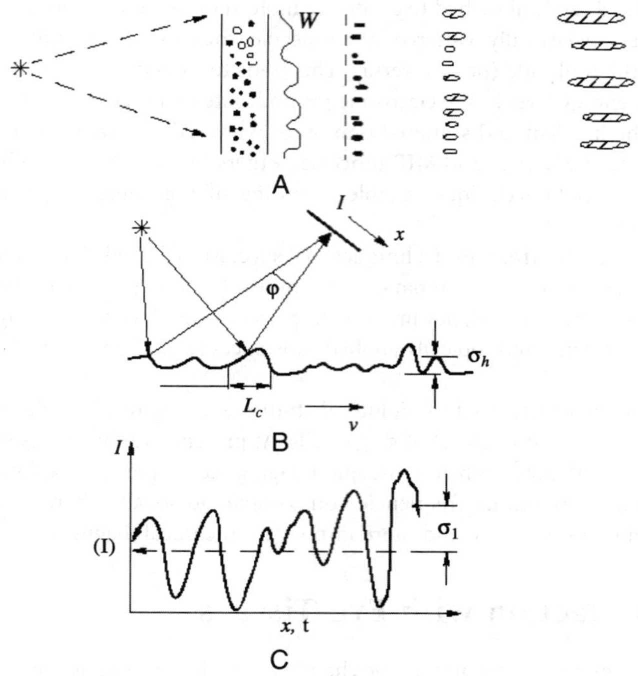


FIGURE 3.8 Formation and propagation of speckles (A), observation of speckles (B), and intensity modulation (C).

## 3.8 Formation of Speckles

Speckle structures are produced as a result of interference of a large number of elementary waves with random phases that arise when coherent light is reflected from a rough surface or when coherent light passes through a scattering medium.[3,7,65–67] Generally, there are two types of speckles: subjective speckles that are produced in the image space of an optical system (including an eye) and objective speckles, formed in a free space and usually observed on a screen placed at a certain distance from an object. Because the majority of tissues are optically nonuniform, irradiation of such objects with coherent light always gives rise to the appearance of speckle structures, which distort the results of measurements and, consequently, should be eliminated in some way, or provide new information concerning the structure and the motion of a tissue and its components.

Figure 3.8 schematically illustrates the principles of the formation and propagation of speckles. The average size of a speckle in the far-field zone is estimated as

$$d_{av} \sim \lambda/\varphi, \tag{3.31}$$

where $\lambda$ is the wavelength and $\varphi$ is the angle of observation.

Displacement of the observation point over a screen ($x$) or the scanning of a laser beam over an object with a certain velocity $v$ (or an equivalent motion of the object itself with respect to the laser beam) under conditions when the observation point remains stationary gives rise to spatial or temporal fluctuations of the intensity of the scattered field. These fluctuations are characterized by the mean value of the intensity $\langle I \rangle$ and the standard deviation $\sigma_I$ (see Figure 3.8B). The object itself is characterized by the standard deviation $\sigma_h$ of the altitudes (depths) of inhomogeneities and the correlation length $L_c$ of these inhomogeneities (random relief).

Because many tissues and cells are phase objects, the propagation of coherent beams in tissues can be described within the framework of the model of a random phase screen. Ideal conditions for formation of speckles, when completely developed speckles arise, can be formulated as: coherent light irradiates a diffusive surface (or a transparency) characterized by Gaussian variations of optical length $\Delta L = \Delta(nh)$ and standard deviation of relief variations, $\sigma_L >> \lambda$; the coherence length of light and the sizes of the scattering area considerably exceed the differences in optical paths due to the surface relief, and many scattering centers contribute to the resulting speckle pattern.

Statistical properties of speckles can be divided into statistics of the first and second orders. Statistics of the first order describes the properties of speckle fields at each point. Such a description usually employs the intensity probability density distribution function $p(I)$ and the contrast

$$V_I = \sigma_I / \langle I \rangle, \qquad \sigma_I^2 = \langle I^2 \rangle - \langle I \rangle^2, \tag{3.32}$$

where $\langle I \rangle$ and $\sigma_I^2$ are the mean intensity and the variance of the intensity fluctuations, respectively. In certain cases, statistical moments of higher orders are employed.

For ideal conditions, when the complex amplitude of scattered light has Gaussian statistics, the contrast is $V_I = 1$ (developed speckles), and the intensity probability distribution is represented by a negative exponential function:

$$p(I) = \langle I \rangle^{-1} \exp\{-I / \langle I \rangle\}. \tag{3.33}$$

Thus, the most probable intensity value in the corresponding speckle pattern is equal to zero, i.e., destructive interference occurs with the highest probability.

Partially developed speckle fields are characterized by a contrast $V_I < 1$. The contrast may be lower due to a uniform coherent or incoherent background added to the speckle field. For phase objects with $\sigma_\phi^2 >> 1$ and a small number of scatterers $N = w/L_\phi$ contributing to the field at a certain point in the observation plane, the contrast of the speckle pattern is greater than unity. The statistics of the speckle field in this case is non-Gaussian and nonuniform (i.e., the statistic parameters depend on the observation angle).

The specific features of the diffraction of focused laser beams from moving phase screens underlie speckle methods of structure diagnostics and monitoring of motion parameters of tissue, blood, and lymph.

Statistics of the second order shows how fast the intensity changes from point to point in the speckle pattern, i.e., characterizes the size and the distribution of speckle sizes in the pattern. Statistics of the second order is usually described in terms of the autocorrelation function of intensity fluctuations:

$$g_2(\Delta\xi) = \langle I(\xi + \Delta\xi) I(\xi) \rangle, \tag{3.34}$$

and its Fourier transform, representing the power spectrum of a random process; $\xi \equiv x$ or $t$ is the spatial or temporal variable; and $\Delta\xi$ is the change in variable. Angular brackets $\langle\ \rangle$ in Equation 3.34 stand for the averaging over an ensemble or over time.

In the elementary case, when reflected light in developed speckle structures retains linear polarization, intensity distribution at the output of a dual-beam interferometer can be written as[5]

$$I(r,t) = I_r(r) + I_s(r) + 2\left[I_r(r) I_s(r)\right]^{1/2} \left|\gamma_{11}(\Delta t)\right| \cos\{\Delta\Phi_I(r) + \Delta\Psi_I(r) + \Delta\Phi_I(t)\}, \tag{3.35}$$

where $I_r(r)$ and $I_s(r)$ are intensity distributions of the reference and signal fields, respectively; $r$ is the transverse spatial coordinate; $\gamma_{11}(\Delta t)$ is the degree of temporal coherence of light; $\Delta\Psi_I(r)$ is the deterministic phase difference of the interfering waves; $\Delta\Psi_I(r) = \Psi_{Ir}(r) - \Phi_{Is}(r)$ is the random phase difference; and $\Delta\Phi_I(t)$ is the time-dependent phase difference related to the motion of an object.

In the absence of speckle modulation, the deterministic phase difference $\Delta\Psi(r)$ governs the formation of regular interference fringes. On the average, the output signal of a speckle interferometer reaches its maximum when the interfering fields are phase-matched ($\Delta\Psi(r)$ = const within the aperture of the detector), focused laser beams are used (speckles with maximum sizes are produced), and a detector with a maximum area is employed.

## 3.9 Dynamic Light Scattering

### 3.9.1 Quasi-Elastic Light Scattering

Quasi-elastic light scattering (QELS) spectroscopy, photon-correlation spectroscopy, spectroscopy of intensity fluctuations, and Doppler spectroscopy are synonymous terms associated with the dynamic scattering of light, which underlies a noninvasive method for studying the dynamics of particles on a comparatively large timescale.[5] The implementation of the single-scattering mode and the use of coherent light sources are of fundamental importance in this case. The spatial scale of testing a colloid structure (an ensemble of biological particles) is determined by the inverse of the wave vector $|\bar{s}|^{-1}$:

$$|\bar{s}| = (4\pi n/\lambda_0)\sin(\theta/2), \tag{3.36}$$

where $n$ is the refractive index and $\theta$ is the angle of scattering. With allowance for self-beating due to the photomixing of the electric components of the scattered field, the intensity autocorrelation function (AF) can be measured $g_2(\tau) = \langle I(t)\, I(t+\tau)\rangle$. For Gaussian statistics, this AF is related to the first-order AF by the Siegert formula:

$$g_2(\tau) = A\left[1 + \beta_{sb}\,|g_1(\tau)|^2\right], \tag{3.37}$$

where $\tau$ is the delay time; $A = \langle i\rangle^2$ is the square of the mean value of the photocurrent, or the base line of the AF; $\beta_{sb}$ is the parameter of self-beating efficiency, $\beta_{sb} \approx 1$; and $g_1(\tau) = \exp(-\Gamma_T \tau)$ is the normalized AF of the optical field for a monodisperse system of Brownian particles, $\Gamma = |\bar{s}|^2 D_T$ is the relaxation parameter and $D_T = k_B T/6\pi\eta r_h$ is the coefficient of translation diffusion, $k_B$ is the Boltzmann constant, $T$ is the absolute temperature, $\eta$ is the absolute viscosity of the medium, and $r_h$ is the hydrodynamic radius of a particle. Many biological systems are characterized by a bimodal distribution of diffusion coefficients, when fast diffusion ($D_{\mathrm{Tf}}$) can be separated from slow diffusion ($D_{\mathrm{Ts}}$) related to the aggregation of particles. The goal of QELS spectroscopy is to reconstruct the distribution of scattering particles in sizes, which is necessary for the diagnosis or monitoring of a disease.

Description of the principles and characteristics of the homodyne and heterodyne photon-correlation spectrometers, the so-called laser Doppler anemometers (LDAs), differential LDA schemes, and laser Doppler microscopes (LDMs) and review of medical applications, mainly the investigation of eye tissues (cataract diagnosis), investigation of hemodynamics in isolated vessels (vessels of eye fundus or any other vessels) with the use of fiber-optic catheters, and blood microcirculation in tissues can be found in References 5, 7, and 10.

### 3.9.2 Diffusion Wave Spectroscopy

Diffusion wave spectroscopy (DWS) is a new class of technique in the field of dynamic light scattering related to the investigation of the dynamics of particles within very short time intervals.[5–8,41,68,69] A fundamental difference of this method from QELS is that it is applicable to the case of dense media with multiple scattering, which is critical for tissues. In contrast to the case of single scattering, the AF of the field $g_1(\tau)$ is sensitive to the motion of a particle on the length scale on the order of $[L/l_t]^{-1/2}$, which is generally much less than $\lambda$, because $L >> l_t$ ($L$ is the total mean photon path length and $l_t$ is the transport length of a photon, $l_t \approx 1/\mu_s'$). Thus, DWS AFs decay much faster than AFs employed in QELS.

Experimental implementation of DWS is very simple. A measuring system should provide irradiation of an object under study by a CW laser beam and measurement of intensity fluctuations of the scattered radiation within a single speckle with the use of a single-mode receiving fiber, photomultiplier, photon-counting system, and a fast digital correlator working in a nanosecond range. The possibilities of the DWS technique for medical applications have been demonstrated for blood flow monitoring in the human forearm. The AF slope is the indicative parameter for determination of blood flow velocity. The normalized AF of field fluctuations can be represented in terms of two components related to the Brownian and directed motion of scatterers (erythrocytes):

$$g_1(\tau) \approx \exp\left\{-2\left[\tau/\tau_B + (\tau/\tau_s)^2\right]L/l_t\right\}, \tag{3.38}$$

where $\tau_B^{-1} \equiv \Gamma_T$ is defined in Equation 3.31, $\tau_S^{-1} \cong 0.18 G_V |\bar{s}| l_t$ characterizes the directed flow, and $G_V$ is the gradient of the flow rate. It allows one to express the slope of the AF in terms of the diffusion coefficient (characterizes blood microcirculation) and the gradient of the directed velocity of blood.

## Acknowledgment

This chapter was written as a part of grants CRDF REC-006 and "Leading Scientific Schools" 00-15-96667 of the Russian Basic Research Foundation.

### References


1. Chance, B., Optical method, *Annu. Rev. Biophys. Biophys. Chem.*, 20, 1, 1991.
2. Mueller, G., Chance, B., and Alfano, R., Eds., *Medical Optical Tomography: Functional Imaging and Monitoring*, IS11, SPIE Press, Bellingham, WA, 1993.
3. Tuchin, V.V., Ed., *Selected Papers on Tissue Optics Applications in Medical Diagnostics and Therapy*, MS 102, SPIE Press, Bellingham, WA, 1994.
4. Minet, O., Mueller, G., and Beuthan, J., Eds., *Selected Papers on Optical Tomography, Fundamentals and Applications in Medicine*, MS 147, SPIE Press, Bellingham, WA, 1998.
5. Tuchin, V.V., *Tissue Optics: Light Scattering Methods and Instruments for Medical Diagnosis*, SPIE Tutorial Texts in Optical Engineering, TT38, SPIE Press, Bellingham, 2000.
6. Benaron, D., Bigio, I., Sevick-Muraca, E., and Yodh, A.G., Eds., Special issue honoring Professor Britton Chance, *J. Biomed. Opt.*, 5, 115, 269, 2000.
7. Tuchin, V.V., Ed., *Handbook of Optical Biomedical Diagnostics*, PM107, SPIE Press, Bellingham, WA, 2002.
8. Tuchin, V.V., Coherence-domain methods in tissue and cell optics, *Laser Phys.*, 8, 807, 1998.
9. Masters, B., Ed., *Selected Papers on Optical Low-Coherence Reflectometry and Tomography*, MS 165, SPIE Press, Bellingham, 2002.
10. Priezzhev, A.V. and Asakura, T., Eds., Special section on optical diagnostics of biological fluids, *J. Biomed. Opt.*, 4, 35, 1999.
11. Tuchin, V.V., Podbielska, H., and Hitzenberger, C.K., Eds., Special section on coherence domain optical methods in biomedical science and clinics, *J. Biomed. Opt.*, 4, 94, 1999.
12. Katzir, A., *Lasers and Optical Fibers in Medicine*, Academic Press, San Diego, 1993.
13. Ozaki, Y., Medical application of Raman spectroscopy, *Appl. Spectrosc. Rev.*, 24, 259, 1988.
14. Mahadevan-Jansen, A. and Richards-Kortum, R., Raman spectroscopy for detection of cancers and precancers, *J. Biomed. Opt.*, 1, 31, 1996.
15. Morris, M.D., Ed., Special section on biomedical applications of vibrational spectroscopic imaging, *J. Biomed. Opt.*, 4, 6, 1999.

16. Carden, A. and Morris, M.D., Application of vibration spectroscopy to the study of mineralized tissues (review), *J. Biomed. Opt.*, 5, 259, 2000.
17. Das, B.B., Liu, F., and Alfano, R.R., Time-resolved fluorescence and photon migration studies in biomedical and random media, *Rep. Prog. Phys.*, 60, 227, 1997.
18. Lakowicz, J.R., *Principles of Fluorescence Spectroscopy*, 2nd ed., Kluwer Academic, New York, 1999.
19. Denk, W., Two–photon excitation in functional biological imaging, *J. Biomed. Opt.*, 1, 296, 1996.
20. Braslavsky, S.E. and Heihoff, K., Photothermal methods, in *Handbook of Organic Photochemistry*, Scaiano, J.C., Ed., CRC Press, Boca Raton, FL, 1989.
21. Müller, G. and Roggan, A., Eds., *Laser-Induced Interstitial Thermotherapy*, SPIE Press, Bellingham, WA, 1995.
22. Welch, A.J. and van Gemert, M.J.C., Eds., *Optical-Thermal Response of Laser Irradiated Tissue*, Plenum Press, New York, 1995.
23. Niemz, H., *Laser–Tissue Interactions: Fundamentals and Applications*, Springer-Verlag, Berlin, 1996.
24. Jacques, S.L., Ed., *Laser-Tissue Interaction: Photochemical, Photothermal, Photomechanical I–XIII*, *Proc. SPIE*, 1990–2002.
25. Vij, D.R. and Mahesh, K., Eds., *Lasers in Medicine*, Kluwer, Boston, 2002.
26. Jacques, S.L., Lee, K., and Roman, J., Scattering of polarized light by biological tissues, *Proc. SPIE*, 4001, 14, 2000.
27. Marks, F.A., Optical determination of the hemoglobin oxygenation state of breast biopsies and human breast cancer xenografts in unde mice, *Proc. SPIE*, 1641, 227, 1992.
28. Ishimaru, A., *Wave Propagation and Scattering in Random Media*, Academic Press, New York, 1978.
29. Van de Hulst, H.C., *Multiple Light Scattering*, Vol.1, Academic Press, New York, 1980.
30. Van de Hulst, H.C., *Light Scattering by Small Particles*, Dover, New York, 1981.
31. Bohren, C.F. and Huffman, D.R., *Absorption and Scattering of Light by Small Particles*, Wiley, New York, 1983.
32. Salzmann, G.C., Singham, S.B., Johnston, R.G., and Bohren, C.F., Light scattering and cytometry, in *Flow Cytometry and Sorting*, 2nd ed., Melamed, M.R., Lindmo, T., and Mendelsohn, M.L., Eds., Wiley-Liss, New York, 1990, p. 81.
33. Racovic, M.J., Kattavar G.W., Mehrubeoglu, M., Cameron, B.D., Wang, L.V., Rasteger, S., and Coté, G.L., Light backscattering polarization patterns from turbid media: theory and experiment, *Appl. Opt.*, 38, 3399, 1999.
34. Bicout, D., Brosseau, C., Martinez, A.S., and Schmitt, J.M., Depolarization of multiply scattering waves by spherical diffusers: influence of the size parameter, *Phys. Rev. E.*, 49, 1767, 1994.
35. Svaasand, L.O. and Gomer, Ch.J., Optics of tissue, in *Dosimetry of Laser Radiation in Medicine and Biology*, SPIE Press, Bellingham, WA, 1989, p. 114.
36. Ostermeyer, M.R., Stephens, D.V., Wang, L., and Jacques, S.L., Nearfield polarization effects on light propagation in random media, *OSA TOPS*, 3, 20, 1996.
37. Fishkin, J.B. and Gratton, E., Propagation of photon-density waves in strongly scattering media containing an absorbing semi-infinite plane bounded by a straight edge, *J. Opt. Soc. Am. A*, 10, 127, 1993.
38. Patterson, M.S., Chance, B., and Wilson, B.C., Time-resolved reflectance and transmittance for the noninvasive measurement of tissue optical properties, *Appl. Opt.*, 28, 2331, 1989.
39. Jacques, S.L., Time-resolved reflectance spectroscopy in turbid tissues, *IEEE Trans. Biomed. Eng.*, 36, 1155, 1989.
40. Chance, B., Cope, M., Gratton, E., Ramanujam, N., and Tromberg, B., Phase measurement of light absorption and scatter in human tissue, *Rev. Sci. Instrum.*, 698, 3457, 1998.
41. Yodh, A.G. and Chance, B., Spectroscopy and imaging with diffusing light, *Phys. Today*, 48, 34, 1995.
42. Schmitt, J.M., Knuettel, A., and Knutson, J.R., Interference of diffusive light waves, *J. Opt. Soc. Am. A*, 9, 1832, 1992.
43. Wright, C.H.G, Barrett, S.F., and Welch, A.J., Laser-tissue interaction, in *Lasers in Medicine*, Vij, D.R. and Mahesh, K., Eds., Kluwer, Boston, 2002.

44. Letokhov, V.S., Laser biology and medicine, *Nature*, 316 (6026), 325, 1985.
45. Tuchin, V.V., *Lasers and Fiber Optics in Biomedical Science*, Saratov University Press, Saratov, Russian Federation, 1998.
46. Oraevsky, A.A., Jacques, S.J., and Tittel, F.K., Measurement of tissue optical properties by time-resolved detection of laser-induced transient stress, *Appl. Opt.*, 36, 402, 1997.
47. Sathyam, U.S. and Prahl, S.A., Limitations in measurement of subsurface temperatures using pulsed photothermal radiometry, *J. Biomed. Opt.*, 2, 251, 1997.
48. Duck, F.A., *Physical Properties of Tissue: A Comprehensive Reference Book*, Academic, London, 1990.
49. Graaff, R., Aarnoudse, J.G., Zijp, J.R., et al., Reduced light scattering properties for mixtures of spherical particles: a simple approximation derived from Mie calculations, *Appl. Opt.*, 31, 1370, 1992.
50. Kohl, M., Essenpreis, M., and Cope, M., The influence of glucose concentration upon the transport of light in tissue-simulating phantoms, *Phys. Med. Biol.*, 40, 1267, 1995.
51. Drezek, R., Dunn, A., and Richards-Kortum, R., Light scattering from cells: finite-difference time-domain simulations and goniometric measurements, *Appl. Opt.*, 38, 3651, 1999.
52. Bolin, F.P., Preuss, L.E., Taylor, R.C., and Ference, R.J., Refractive index of some mammalian tissues using a fiber optic cladding method, *Appl. Opt.*, 28, 2297, 1989.
53. Tuchin, V.V., Maksimova I.L., Zimnyakov, D.A., et al., Light propagation in tissues with controlled optical properties, *J. Biomed. Opt.*, 2, 304, 1997.
54. Liu, H., Beauvoit, B., Kimura, M., and Chance, B., Dependence of tissue optical properties on solute induced changes in refractive index and osmolarity, *J. Biomed. Opt.*, 1, 200, 1996.
55. Vargas, G., Chan, E.K., Barton, J.K., Rylander, H.G., III, and Welch, A.J., Use of an agent to reduce scattering in skin, *Laser Surg. Med.*, 24, 138, 1999.
56. Tuchin, V.V., Bashkatov, A.N., Genina, E.A., Sinichkin, Yu.P., and Lakodina, N.A., *In vivo* study of the human skin clearing dynamics, *J. Tech. Phys. Lett.*, 27(6), 489, 2001.
57. Wang, R.K., Tuchin, V.V., Xu, X., and Elder, J.B., Concurrent enhancement of imaging depth and contrast for optical coherence tomography by hyperosmotic agents, *J. Opt. Soc. Am. B*, 18, 948, 2001.
58. Brezinski, M., Saunders, K., Jesser, C., Li, X., and Fujimoto, J., Index matching to improve OCT imaging through blood, *Circulation*, 103, 1999, 2001.
59. Tuchin, V.V., Xu, X., and Wang, R.K., Dynamic optical coherence tomography in optical clearing, sedimentation and aggregation study of immersed blood, *Appl. Opt.*, 41(1), 258, 2002.
60. Schneckenburger, H., Steiner, R., Strauss, W., Stock, K., and Sailer, R., Fluorescence technologies in biomedical diagnostics, in *Optical Biomedical Diagnostics*, PM107, Tuchin, V.V., Ed., SPIE Press, Bellingham, WA, 2002, chap. 15.
61. Sinichkin, Yu.P., Kollias, N., Zonios, G., Utz, S.R., and Tuchin, V.V., Reflectance and fluorescence spectroscopy of human skin *in vivo*, in *Handbook of Optical Biomedical Diagnostics*, PM107, Tuchin, V.V., Ed., SPIE Press, Bellingham, WA, 2002, chap. 13.
62. Sterenborg, H.J.C.M., Motamedi, M., Wagner, R.F., Duvic, J.R.M., Thomsen, S., and Jacques, S.L., *In vivo* fluorescence spectroscopy and imaging of human skin tumors, *Lasers Med. Sci.*, 9, 344, 1994.
63. Zeng, H., MacAulay, C., McLean, D.I, and Palcic, B., Spectroscopic and microscopic characteristics of human skin autofluorescence emission, *Photochem. Photobiol.*, 61, 645, 1995.
64. Lucassen, G.W., Caspers, P.J., and Puppels, G.J., Infrared and Raman spectroscopy of human skin *in vivo*, in *Handbook of Optical Biomedical Diagnostics*, PM107, Tuchin, V.V., Ed., SPIE Press, Bellingham, WA, 2002, chap. 14.
65. Dainty, J.C., Ed., *Laser Speckle and Related Phenomena*, 2nd ed., Springer-Verlag, New York, 1984.
66. Aizu, Y. and Asakura, T., Bio-speckle phenomena and their application to the evaluation of blood flow, *Opt. Laser Technol.*, 23, 205, 1991.
67. Briers, J.D., Laser Doppler and time-varying speckle: a reconciliation, *J. Opt. Soc. Am. A*, 13, 345, 1996.

68. Boas, D.A., Campbell, L.E., and Yodh, A.G., Scattering and imaging with diffusing temporal field correlations, *Phys. Rev. Lett.*, 75, 1855, 1995.
69. Zimnyakov, D.A., Tuchin, V.V., and Yodh, A.G., Characteristic scales of optical field depolarization and decorrelation for multiple scattering media and tissues, *J. Biomed. Opt.*, 4, 157, 1999.

# 4

# Theoretical Models and Algorithms in Optical Diffusion Tomography




Stephen J. Norton
*Geophex, Ltd.*
*Raleigh, North Carolina*

Tuan Vo-Dinh
*Oak Ridge National Laboratory*
*Oak Ridge, Tennessee*


## 4.1 Introduction

The potential of optical tomography as a new diagnostic tool has stimulated considerable interest in the last 10 years.[1–4] Although limited to about the first 5 cm of the body, the technique offers several advantages often not available in established imaging modalities, such as ultrasound, x-ray computed tomography, and magnetic resonance imaging.[5,6] These benefits include nonionizing radiation, relatively inexpensive instrumentation, and the potential for functional (i.e., spectroscopic) imaging of optical tissue properties. In functional imaging, light at specific wavelengths can be used to excite specific biological molecules of interest, such as NAD, NADH, tryptophan, and hemoglobin, in order to provide real-time, *in vivo* information on the functional status of tissues and organs (e.g., pH, tissue oxygenation, glucose, dysplasia, tumor, etc.). Luminescence techniques based on bioluminescence allow monitoring gene expression *in vivo* in animals for drug discovery investigations.

The use of light for diagnostics of deep tissue still presents great challenges. Electromagnetic radiation in the UV and visible wavelength range is strongly absorbed by biological species in tissue. Due to absorption by tissue chromophores, the intensity of light decreases rapidly as it penetrates deep inside tissue. At all wavelengths, scattering of light occurs strongly in tissue; however, in the near-infrared "optical window," where hemoglobin absorbs weakly (600 to 900 nm), multiple scattering in tissue dominates absorption, and penetration can be substantial, sometimes reaching 5 cm or more.

As light propagates in an optically dense medium such as tissue, it rapidly loses its coherence and, as a result, light intensity (or photon flux density or radiance) is usually the observable quantity. To a reasonable approximation, the radiance can be shown to obey a diffusion equation, which, when the source is modulated, reduces to a Helmholtz equation with a complex wave number. With a harmonically modulated source, the radiance oscillates in space and in time; this oscillation is termed a diffuse photon density wave (DPDW). Such waves, although highly damped, are diffracted and scattered by optical inhomogeneities within the tissue. A number of authors have reported theoretical and experimental investigations of these waves.[7–28]

This chapter describes several approaches to optical tomography based on theoretical models of light propagation and scattering in tissue. We examine in some detail three optical tomography algorithms. The first employs Fourier methods to reconstruct an image from time-harmonic DPDW data recorded at one modulation frequency. This method can be regarded as a generalization of diffraction tomography developed originally for acoustical imaging. The second and third algorithms are iterative and attempt to minimize an error norm in the frequency and time domains, respectively. An overview is also given of other algorithmic approaches that have appeared in the literature in the last decade. Experimental techniques, instrumentation, and clinical applications of optical tomography are described in Chapters 9 (lifetime-based imaging), 21 (functional imaging with diffusing light), 22 (photon migration spectroscopy frequency-domain techniques), and 33 (near-infrared fluorescence imaging and spectroscopy in random media and tissues) of this handbook.

Here, we focus on imaging algorithms for reconstructing spatial maps of optical properties of tissue by exploiting models of the interaction of light with tissue. Two basic measurement methodologies are used for imaging: (1) frequency-domain methods that employ harmonically modulated photon density-waves, and phase-resolved detection, which measures the phase shift of the photon density waves,[17,18,20,21,23,29–45] and (2) "time-resolved" methods that use pulsed excitation and gated detection to examine the response of tissue to a short pulse of incident light.[46–60] Figure 4.1 shows a schematic diagram of the phase-resolved and time-resolved techniques. "Continuous-wave" imaging methods, which employ steady-state illumination, may also be classified as frequency-domain, since these correspond to the zero frequency limit.[61,62]

The simplest time-resolved imaging method exploits the fact that the first-arrival photons propagate along paths that have not deviated significantly from the line joining the source and receiver. Such photons are referred to as "ballistic photons." Photons that are scattered, predominantly in the forward direction, are sometimes referred to as "snake-like photons," as illustrated in Figure 4.2. In the case of ballistic photons, scattering is neglected, and conventional tomographic algorithms based on straight-line propagation are, in principle, applicable. Because of the need for ultrafast laser pulses for excitation

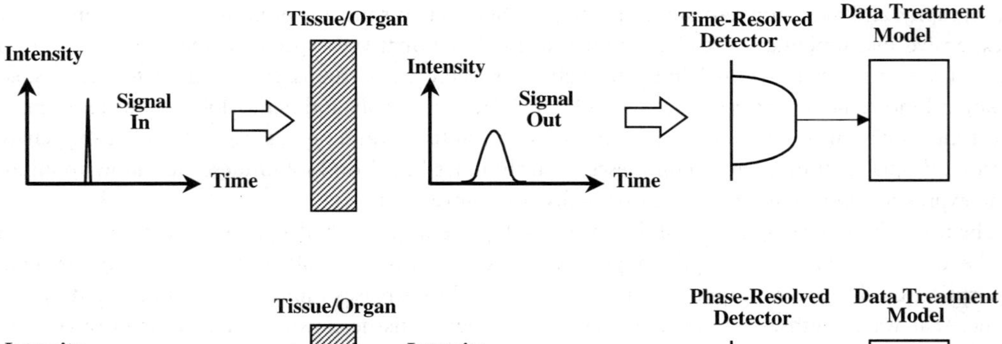

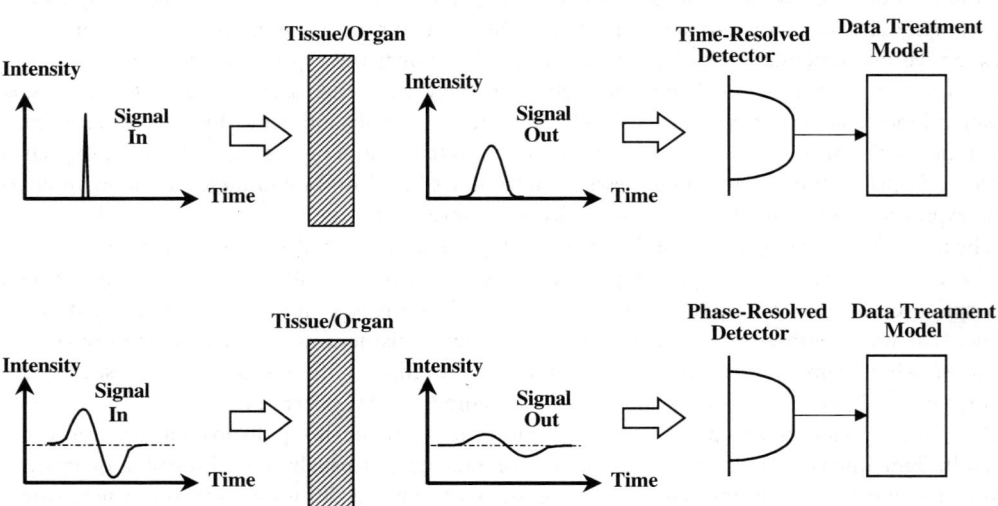


**FIGURE 4.1** Schematic diagram of the time-resolved (top) and phase-resolved (bottom) techniques.

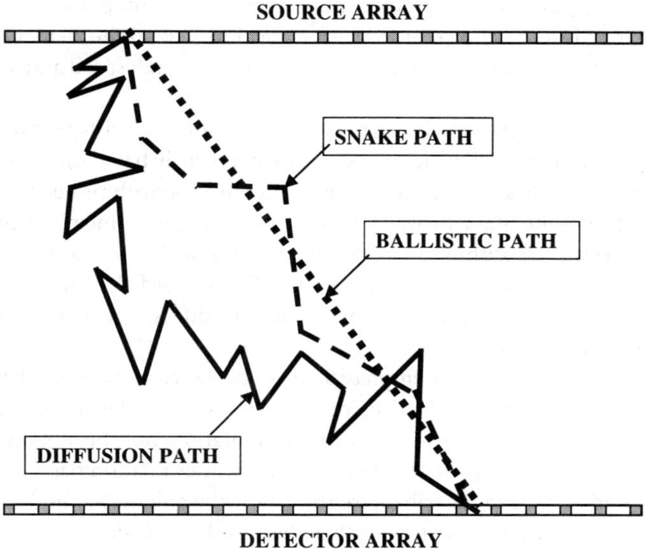


**FIGURE 4.2** Schematic diagram of various photon trajectories through tissue medium.

(femtosecond and picosecond time frames[63]), the very precise timing used to gate the first-arrival photons is technically challenging. In this regard, harmonic modulation of an intensity source is less demanding. Moreover, beyond a few centimeters, the relative number of ballistic photons becomes exceedingly small, so that ballistic measurements rapidly become signal-to-noise limited.

Frequency domain methods offer better signal-to-noise performance because more photons are available for measurement, and synchronous detection using phase-resolved measurements can be employed to enhance the signal. In time-resolved methods, the "boxcar" or a pulse-averaging scheme is often used to improve the signal-to-noise ratio. Modern time-resolved methods typically examine the entire transient response (rather than merely the ballistic photons) and attempt to fit this response to a model based, for example, on the time-dependent diffusion equation. In this manner, scattered photons are also accounted for, and time-domain methods that employ rapid averaging of the transient signals can achieve signal-to-noise ratios competitive with frequency-domain methods. In this chapter, we describe in detail two frequency-domain imaging algorithms, followed by a time-domain imaging algorithm applicable to time-resolved measurements.

We focus on two particular formulations of the inverse problem: analytical and iterative. Here, the term "inverse problem" refers to the task of deriving the spatial distribution of optical tissue properties from the detected scattered photons. The analytical approaches are noniterative, although we distinguish them from linear inversion schemes that are purely numerical, i.e., methods that solve a linear system of equations directly. These analytical approaches have largely been inspired by diffraction tomography (DT), which, in its conventional formulation, is a wave-equation-based inversion scheme.[64–67] In particular, DT provides an explicit solution to a linearized integral equation using Fourier inversion methods.

Traditional DT begins with the Helmholtz equation with a real wave number, but our starting point is the diffusion equation; the latter reduces to a Helmholtz equation with a complex wave number when the source is harmonically modulated. In this case, the required generalization of diffraction tomography resembles a Fourier-Laplace inversion. Markel and Schotland[44] have formulated the problem in this way, which requires an analytic continuation of the data in the Fourier domain. We describe a related approach that provides an explicit inversion formula,[36] as does Markel and Schotland's method, but does not invoke an inverse Laplace transform and more explicitly defines the process of analytic continuation.

Another category of algorithms attempts to minimize iteratively a global error norm that quantifies the difference between measured and predicted data. Typically, the error norm is the mean-square error, although other norms are possible. The predicted observations are computed using a forward solution to the transport problem, but the choice of forward algorithm (e.g., finite element, finite difference, or an integral equation solution) is not relevant for our purposes, although speed and accuracy are obviously desirable traits.

Iterative approaches are computationally expensive but have some advantages over analytical solutions. For example, analytical solutions, or Fourier methods, are necessarily based on a linearization procedure, which requires a weak-scattering assumption. Nonlinear iterative algorithms need make no such assumption because these methods iterate a forward algorithm, which, in principle, need not be subject to approximations. The forward algorithm can be designed to deal with voids and complicated boundaries, as well as arbitrary source and detector-array geometries. The forward algorithm may, for example, solve the full integro-differential transport equation, rather than the diffusion equation, which is generally the starting point for DT-based methods.

In investigating iterative algorithms, we emphasize the importance of the adjoint method in computing the functional gradient (or Fréchet derivative) of the error norm. The numerical evaluation of this gradient is generally the most demanding part of the computation, but the adjoint method renders this computation vastly more efficient than a "brute force" gradient evaluation based, for example, on finite differences. In the second algorithm described, we show how the adjoint method can be applied directly to the integro-differential transport equation, rather than to the diffusion equation, which is the more common approach.

## 4.2 Photon Transport in Tissue

In this section, we review some of the basic equations obeyed by photon density waves and describe some of their properties. Most imaging algorithms begin with an equation derived from the Boltzmann transport equation, whose time-dependent form is given by[68,69]

$$\frac{1}{c}\frac{\partial\psi(\mathbf{r},t,\mathbf{s})}{\partial t}+\mathbf{s}\cdot\nabla\psi(\mathbf{r},t,\mathbf{s})+\sigma_t(\mathbf{r})\psi(\mathbf{r},t,\mathbf{s})=\sigma_s(\mathbf{r})\int\psi(\mathbf{r},t,\mathbf{s}')f(\mathbf{s}\cdot\mathbf{s}')d^2s'+q(\mathbf{r},t,\mathbf{s}), \tag{4.1}$$

where $\psi(\mathbf{r},t,\mathbf{s})$ is the time-dependent radiance (or photon flux density) in the direction $\mathbf{s}$ at the point $\mathbf{r}$; $c$ is the velocity of light in the medium and $\sigma_t(\mathbf{r}) = \sigma_s(\mathbf{r}) + \sigma_a(\mathbf{r})$, where $\sigma_s(\mathbf{r})$ and $\sigma_a(\mathbf{r})$ are the spatially dependent scattering and absorption cross sections; $f(\mathbf{s}\cdot\mathbf{s}')$ is the phase function that denotes the probability of an incident photon from direction $\mathbf{s}$ being scattered into the direction $\mathbf{s}'$; and $q(\mathbf{r},t,\mathbf{s})$ is the source flux.

The diffusion approximation results on performing a spherical harmonic expansion of the radiance in Equation 4.1 and retaining the lowest-order nontrivial terms in the expansion. Defining the photon density as

$$\psi(\mathbf{r},t)=\int\psi(\mathbf{r},t,\mathbf{s})d^2s, \tag{4.2}$$

the diffusion approximation can be shown to give[68]

$$\frac{1}{c}\frac{\partial\psi(\mathbf{r},t)}{\partial t}-\nabla\cdot\left[D(\mathbf{r})\nabla\psi(\mathbf{r},t)\right]+\sigma_a(\mathbf{r})\psi(\mathbf{r},t)=Q(\mathbf{r},t), \tag{4.3}$$

where $D(\mathbf{r}) = 1/3[\sigma_a(\mathbf{r}) + \sigma_s'(\mathbf{r})]$ is the diffusion coefficient; $\sigma_s'(\mathbf{r}) = (1 - g)\ \sigma_s\ (\mathbf{r})$ is the reduced scattering coefficient; $g$ is the mean cosine of the scattering angle; and $Q(\mathbf{r},t)$ is the source density function. If we now assume that the sources in Equations 4.1 and 4.3 are harmonically modulated at frequency $\omega$ so that $q(\mathbf{r},t,\mathbf{s}) = q(\mathbf{r},\mathbf{s})\exp(-i\omega t)$ and $Q(\mathbf{r},t) = Q(\mathbf{r})\exp(-i\omega t)$, then, writing $\psi\ (\mathbf{r},t,\mathbf{s}) = \psi(\mathbf{r},\mathbf{s})\exp(-i\omega t)$, Equation 4.1 becomes

$$s\cdot\nabla\psi(r,s)+\left[\sigma_t(r)-\frac{i\omega}{c}\right]\psi(r,s)=\sigma_s(r)\int\psi(r,s')f(s\cdot s')d^2s'+q(r,s), \tag{4.4}$$

and, writing $\psi(r,t) = \psi(r)\exp(-i\omega t)$, Equation 4.3 reduces to

$$\nabla\cdot\left[D(r)\nabla\psi(r)\right]+\left[\frac{i\omega}{c}-\sigma_a(r)\right]\psi(r)=-Q(r). \tag{4.5}$$

Note that the radiance $\psi(r,s)$ in Equation 4.4 and the photon density $\psi(r)$ in Equation 4.5 are now complex quantities, since in-phase and quadrature components of the signals can be measured, corresponding to the real and imaginary parts of $\psi(r,s)$ and $\psi(r)$.

Now write $D(r) = D_0 + D_1(r)$ and $\sigma_a(r) = \sigma_0 + \sigma_1(r)$, where $D_1(r)$ and $\sigma_1(r)$ will denote perturbations assumed to be small compared to the background values $D_0$ and $\sigma_0$ if the equations are ultimately to be linearized. Then, substituting into Equation 4.5 gives

$$\nabla^2\psi(r)+k_0^2\psi(r)=-\frac{1}{D_0}\nabla\cdot\left[D_1(r)\nabla\psi(r)\right]+\frac{\sigma_1(r)}{D_0}-S(r), \tag{4.6}$$

where $S(r) = Q(r)/D_0$ and the wave number is given by

$$k_0=\left[\frac{i\omega}{cD_0}-\frac{\sigma_0}{D_0}\right]^{1/2} \tag{4.7}$$

By definition, the wavelength and the phase velocity of the photon density wave $\psi(r)$ are given, respectively, by[11] $\lambda = 2\pi/k_r$ and $v_p = \omega/k_r$, where $k_r$ is the real part of Equation 4.7. The "skin depth," or depth to which the photon density falls to $e^{-1}$ of its incident value, is given by $\delta = 1/k_i$ where $k_i$ is the imaginary part of Equation 4.7. If, for example, we choose a modulation frequency such that $\omega >> \sigma_a c$, then Equation 4.7 simplifies to $k_0=\sqrt{i\omega/cD_0}=(1+i)\sqrt{\omega/2cD_0}$, in which case $\lambda=2\pi\sqrt{2cD_0/\omega}$, $v_p=\sqrt{2cD_0\omega}$, and $\delta=\sqrt{2cD_0/\omega}$.

## 4.3 Optical Diffusion Tomography

### 4.3.1 Classes of Inversion Algorithms

The reconstruction of an image of optical properties requires the solution to an inverse problem. One can classify inversion algorithms as one of three types:

1. Algorithms based on linear numerical inversion methods
2. Algorithms based on Fourier-based inversion methods
3. Algorithms based on nonlinear iterative methods

The starting point for the first two methods is typically the diffusion equation, which is then transformed to an integral equation via Green's theorem (known as the Lippmann-Schwinger equation), linearized to first order in the inhomogeneity (the Born approximation), and then solved. In the first class of algorithms, a process of discretization converts the integral equation into a system of linear equations, which are then numerically solved using some type of regularized inversion scheme. Examples of this approach are singular-value decomposition and various iterative procedures such as an algebraic reconstruction technique or a conjugate gradient algorithm using, for example, a Tikhonov regularization procedure.[29,33,42,70]

In the second class of algorithms, one solves the linearized integral equation using Fourier methods. Diffraction tomography falls into this class; in this case, the integral equation is manipulated into a form from which the Fourier transform of the unknown optical inhomogeneity can be recovered. An inverse Fourier transform then yields the image. An aesthetically pleasing feature of Fourier-based methods is that they sometimes yield explicit inversion formulas, although the derivation of these formulas generally requires that the source and detection surfaces have simple geometries, such as lines or circles for 2-D problems and planes or spheres for 3-D problems.[66,71]

These inversion formulas are computationally efficient because the inversion has been performed analytically, whereas a purely numerical solution requires the equivalent of the inversion of a large matrix. Although numerical inversion algorithms allow greater freedom in formulating regularization schemes, regularization is still possible in Fourier-based methods, usually in the form of a low-pass, spatial-frequency filter. This filter attenuates the higher spatial-frequency components to reduce noise sensitivity.

Imaging algorithms in the third category are based on nonlinear optimization techniques, in which a global error norm, such as the mean-square error, is iteratively minimized. This method is the most general because, in principle, no approximations need be made for the benefit of analytical tractability. Here, the unknown inhomogeneity is sought that best predicts the data, subject perhaps to additional *a priori* constraints. At each iteration, the forward solution is computed on the basis of the current estimate of the unknown inhomogeneity and compared to the measurements. Normally, gradient descent methods are employed to minimize the mean-square error, although other approaches, such as simulated annealing or evolution algorithms,[72] are possible.

Gaudette and coworkers[42] recently conducted a comprehensive review of the first class of algorithms, linear numerical schemes. In this chapter, we shall focus on the second and third classes of algorithms, Fourier methods and nonlinear iterative methods. We will discuss examples of each of these approaches from our research to illustrate both techniques.

### 4.3.2 Analytical and Quasi-Analytical Methods

In the category of Fourier-based methods we include diffraction-tomography methods, as well as back-projection and angular spectrum techniques.

Inspired by x-ray tomography, several algorithms based on the concept of filtering and back-projection have been proposed. Walker et al.[73] proposed a particularly simple back-projection algorithm in which the data, after filtering, are back-projected along the straight lines joining each source and detector. Colak et al.[31] conducted a comprehensive investigation of back-projection methods using different filtering schemes. Such methods are rapid and robust but do not take into account the diffraction or scattering of the photon density waves.

By use of an angular spectrum decomposition of the scattered waves, Li et al.[21] and Durduran et al.[74] examined other algorithms that rigorously take into account DPDW diffraction. This approach is well suited for data acquisition over planar geometries because a spatial Fourier transform is performed over a plane (or line in 2-D imaging). The angular spectrum analysis decomposes the DPDW field into spatial frequency components, each of which propagates in different directions and attenuates at different rates. This method is particularly convenient for examining how spatial resolution changes with depth. The authors used the angular spectrum formulation to develop a quasi-analytical reconstruction algorithm based on filtering and back-propagation of the DPDW field.

A closely related approach developed by a number of authors begins by linearizing the Green's integral and then employing a plane-wave expansion of the Green's function.[39,75] This is a first step toward a conventional diffraction tomography formulation, except that in this case, the problem is complicated by the complex wave number. Along these lines, Matson[32,37,45] developed a projection slice theorem that relates the Fourier transform of the data to that of the object. However, the theorem in the form presented is not suitable for reconstructing the object directly via Fourier inversion because the complex wave number gives rise to complex spatial-frequency variables.

Norton and Vo-Dinh[36] showed how an explicit solution can be derived by a process of analytical continuation that makes the spatial-frequency variables real, after which a conventional inverse Fourier transform can be performed to reconstruct the object. This approach is described in more detail below. A method somewhat related to that of Norton and Vo-Dinh was developed by Markel and Schotland,[44] who obtained a solution to the Born integral equation based on a Fourier-Laplace inversion scheme. Their solution also implies analytically continued data. Other authors have also developed formal solutions for continuous-wave imaging (i.e., in the zero frequency limit) based on a Laplace-transform inversion.[61,62] Finally, Chen et al.[38] developed an approximate explicit solution by using a stationary-phase approximation to the Green's integral.

### 4.3.3 Nonlinear Iterative Methods

Nonlinear iterative approaches attempt to find the inhomogeneity that best predicts the data.[40,57,76,77] In these algorithms, regularization schemes are relatively easy to implement — for example, by attaching a penalty function to the error functional, which can be designed to emphasize objects that are smooth or have minimum norm or maximum entropy. Also, *a priori* constraints can be enforced at each iteration. Formulating the method in a maximum-likelihood or a Bayesian framework allows noise statistics to be taken explicitly into account.[78,79] Typically, the image is decomposed into pixels, each of which is described by one or more parameters, such as the diffusion or absorption coefficient at that point. A pixel basis is not necessary, however; other parameterizations are possible, such as the shape of the anomaly.[43,80]

An essential ingredient in most of these schemes is the computation of the functional gradient (or Fréchet derivative) of the mean-square error.[81] A brute-force approach to evaluating the gradient could be performed by computing finite differences. That is, if $N$ unknowns were characterizing the inhomogeneity (e.g., $N$ pixels in the image), then a minimum of $N$ forward problems would be required to compute the $N$ derivatives using finite differences. With the adjoint method, only two forward problems need to be solved: one forward problem and its adjoint.[81–86] In computing the functional gradient using the adjoint method, most published algorithms have employed the adjoint of the diffusion equation. However, Norton[83] and Dorn[85] employ the adjoint of the integro-differential transport equation, which is generally a more accurate model of DPDW propagation and scattering. We examine this method below in a simple example of a continuous-wave inverse problem for reconstructing the pixels of the scattering coefficient in a 2-D object. We also show how this approach can be applied to transient data.

## 4.4 Algorithms for Imaging

### 4.4.1 An Explicit Solution Based on Diffraction Tomography

We present in this section an analytical solution to the inverse diffusion problem using the methods of diffraction tomography.[36] As noted, DT is a wave-equation-based inversion procedure whose starting point is a Helmholtz equation with a real wave number. As we shall see, however, the complex wave number in the Helmholtz equation (Equation 4.7) that arises from the diffusion equation can be dealt with by analytical continuation of the data in the Fourier domain; then, the conventional DT methodology can be applied.

For simplicity, we assume that the frequency $\omega$ is large compared to $\mu_a c$; although this assumption is not necessary, it simplifies the algebra. In this case, Equation 4.7 becomes simply $k_0 = \sqrt{i\omega/cD_0}$. We also assume that scattering dominates absorption, so that the parameter to be imaged is the diffusion coefficient $D(\boldsymbol{r})$ or, equivalently, the perturbation $D_1(\boldsymbol{r})$. Then Equation 4.6 reduces to

$$\nabla^2\psi(\boldsymbol{r}) + k_0^2\psi(\boldsymbol{r}) = -\frac{1}{D_0}\nabla\cdot\left[D_1(\boldsymbol{r})\nabla\psi(\boldsymbol{r})\right] - S(\boldsymbol{r}). \tag{4.8}$$

Equation 4.8 can be converted to an integral equation using standard techniques, giving

$$\psi(\boldsymbol{r}) = \psi_i(\boldsymbol{r}) + \frac{1}{D_0}\int \nabla\cdot\left[D_1(\boldsymbol{r}')\nabla\psi(\boldsymbol{r}')\right]g(\boldsymbol{r}|\boldsymbol{r}')d^3\boldsymbol{r}', \tag{4.9}$$

where $g(\boldsymbol{r}\,|\,\boldsymbol{r}')$ is a Green's function and

$$\psi_i(\boldsymbol{r}) = \int S(\boldsymbol{r}')g(\boldsymbol{r}|\boldsymbol{r}')d^3\boldsymbol{r} \tag{4.10}$$

is the incident field. The Green's function is defined by

$$\nabla^2 g(\boldsymbol{r}|\boldsymbol{r}') + k_0^2 g(\boldsymbol{r}|\boldsymbol{r}') = -\delta(\boldsymbol{r}-\boldsymbol{r}').$$

If the source can be approximated as a point at $\boldsymbol{r}_s$ then letting $S(\boldsymbol{r}) = \delta(\boldsymbol{r} - \boldsymbol{r}_s)$, Equation 4.10 reduces to $\psi_i(\boldsymbol{r}) = g(\boldsymbol{r}\,|\,\boldsymbol{r}_s)$. In the Born approximation, the scattered field, represented by the integral on the right-hand side of Equation 4.9, is assumed to be much smaller than the incident field, $\psi_i(\boldsymbol{r})$. Then, to the first-order approximation in the inhomogeneity $D_1(\boldsymbol{r})$, we replace $\psi(\boldsymbol{r})$ by $\psi_i(\boldsymbol{r})$ in the integrand of Equation 4.9. Defining the scattered field by $\psi_s = \psi - \psi_i$ and setting $\boldsymbol{r} = \boldsymbol{r}_d$ in Equation 4.9, where $\boldsymbol{r}_d$ is the detection point, we obtain

$$\psi_s(\boldsymbol{r}_s,\boldsymbol{r}_d) = \frac{1}{D_0}\int \nabla\cdot\left[D_1(\boldsymbol{r})\nabla g(\boldsymbol{r}_s|\boldsymbol{r})\right]g(\boldsymbol{r}_d|\boldsymbol{r})d^3\boldsymbol{r}. \tag{4.11}$$

Here, we have substituted $\psi_i(\boldsymbol{r}) = g(\boldsymbol{r}\,|\,\boldsymbol{r}_s) = g(\boldsymbol{r}_s\,|\,\boldsymbol{r})$ in the integrand of Equation 4.9. Abbreviating $g_s = g(\boldsymbol{r}_d\,|\,r)$ and $g_d = g(\boldsymbol{r}_d\,|\,\boldsymbol{r})$, this integral can be rewritten using an integration by parts with the aid of the identity $\nabla\cdot[D_1\nabla g_s]g_d = \nabla\cdot[D_1 g_d \nabla g_s] - D_1\nabla g_d\cdot\nabla g_s$. When this is substituted into Equation 4.11, the volume integral of $\nabla\cdot[D_1 g_d \nabla g_s]$ can be converted to a surface integral over the domain of integration with the aid of the divergence theorem. This integral will vanish, since we can define $D_1(\boldsymbol{r}) = 0$ on this surface. Then Equation 4.11 becomes

$$\psi_s(\boldsymbol{r}_s,\boldsymbol{r}_d) = -\frac{1}{D_0}\int D_1(\boldsymbol{r})\nabla g(\boldsymbol{r}_s|\boldsymbol{r})\cdot\nabla g(\boldsymbol{r}_d|\boldsymbol{r})d^3\boldsymbol{r}. \tag{4.12}$$

For simplicity, at this point we consider a 2-D problem, although the theory can be readily generalized to 3-D. We then write $\boldsymbol{r} = (x,z)$, with all quantities assumed to be independent of the third dimension, $y$. For brevity we will refer to the sources as "points" lying along a line array, but for a true 2-D problem each "point" source is really a line source in the direction $y$. A similar statement can be made for the "point" detectors.

Consider now the 2-D geometry shown in Figure 4.3, in which the source point is confined to a line with coordinates $\boldsymbol{r}_s = (x_s,0)$ and the detector point is confined to a line with coordinates $\boldsymbol{r}_d = (x_d,L)$ parallel to, and at a distance $L$ from, the source line. We now employ the following plane-wave expansions of the 2-D Green's functions:

$$g(\boldsymbol{r}|\boldsymbol{r}_s) = \frac{i}{4\pi}\int_{-\infty}^{\infty}\frac{dk_x}{\gamma}\exp\left[ik_x(x-x_s)\right]\exp(i\gamma z) \tag{4.13a}$$

$$g(\boldsymbol{r}|\boldsymbol{r}_d) = \frac{i}{4\pi}\int_{-\infty}^{\infty}\frac{dk_x}{\gamma}\exp\left[ik_x(x-x_d)\right]\exp\left[i\gamma(L-z)\right], \tag{4.13b}$$

where

$$\gamma = \sqrt{k_0^2 - k_x^2}. \tag{4.13c}$$

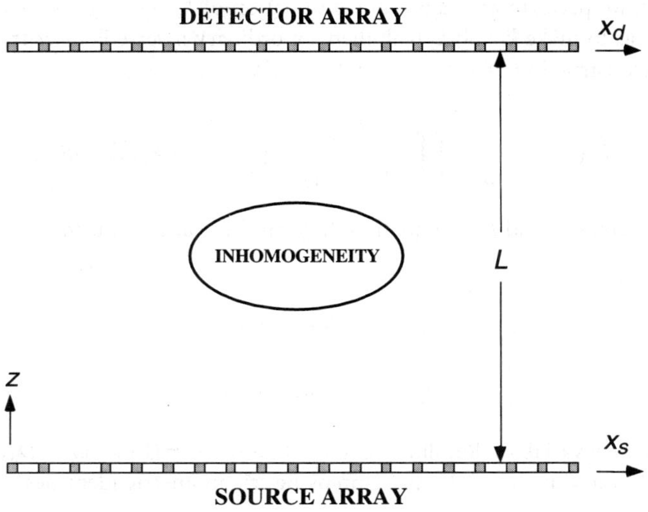


**FIGURE 4.3** Simplified 2-D geometry of source-detector system. The source point is confined to a line with coordinates $\mathbf{r}_s = (x_s,0)$ and the detector point is confined to a line with coordinates $\mathbf{r}_d = (x_d,L)$. See text for discussion.

We now compute the 2-D Fourier transform of Equation 4.12 with respect to $x_s$ and $x_d$, that is,

$$\tilde{\psi}_s(k_s,k_d) \equiv \int\int_{-\infty}^{\infty} \psi_s(\mathbf{r}_s,\mathbf{r}_d)\exp\left[i(k_d x_d + k_s x_s)\right]dx_s dx_d \tag{4.14}$$

Equations 4.13a and 4.13b are substituted into Equation 4.12 and the result into Equation 4.14. After interchanging the orders of integration and some manipulation, we obtain

$$\tilde{\psi}_s(k_s,k_d) = C(k_s,k_d)\int_0^{L_x} dx \int_0^{L_z} dz\ D_1(x,z)\exp\left[i(k_s + k_d)x + i(\gamma_s - \gamma_d)z\right], \tag{4.15}$$

where $\gamma_s \equiv \sqrt{k_o^2 - k_s^2}$ and $\gamma_d \equiv \sqrt{k_o^2 - k_d^2}$

$$C(k_s,k_d) \equiv -\frac{1}{4D_0}\frac{k_s k_d - \gamma_s\gamma_d}{\gamma_s\gamma_d}\exp\left[i\gamma_d L\right] \tag{4.16}$$

Note that in Equation 4.15 we have limited the nonzero domain of $D_1(x,z)$ to the rectangular regions $0 < x < L_x$ and $0 < z < L_z$. Note also that Equation 4.15 is in the form of a 2-D Fourier transform of the inhomogeneity $D_1(x,z)$, which can be written

$$\tilde{\psi}_s(k_s,k_d) = C(k_s,k_d)\int_0^{L_x} dx \int_0^{L_z} dz\, D_1(x,z)\exp\left[i(k_x x + k_z z)\right], \tag{4.17}$$

with transform variables defined by

$$k_x \equiv k_s + k_d, \tag{4.18a}$$

$$k_z \equiv \gamma_s - \gamma_d = \sqrt{k_0^2 - k_s^2} - \sqrt{k_0^2 - k_d^2}. \tag{4.18b}$$

For Equation 4.17 to assume the form of an ordinary 2-D Fourier transform, the transform variables $k_x$ and $k_z$ must be real; however, as written, $k_z$ in Equation 4.18b is complex, because $k_0^2 = i\omega/cD_0$. To invert Equation 4.17, we perform an analytical continuation of $k_s$ and $k_d$ into their respective complex planes in such a way as to make $k_x$ and $k_z$ real; then, an ordinary inverse Fourier transform of Equation 4.17 will yield the reconstruction of $D_1(x,z)$, denoted $\hat{D}_1(x,z)$:

$$\hat{D}_1(x,z)=\frac{1}{(2\pi)^2}\iint_{-\infty}^{\infty}\frac{\tilde{\psi}_s(k_s,k_d)}{C(k_s,k_d)}\exp\left[-i(k_x x+k_z z)\right]dk_x dk_z \ . \tag{4.19}$$

The transform variables $k_x$ and $k_z$ in Equation 4.18 can be made real using the following change of variables:

$$k_s \equiv k_0 \sin(p-q+iv)\ , \tag{4.20a}$$

$$k_d \equiv k_0 \sin(p+q+iv)\ , \tag{4.20b}$$

where $p$, $q$, and $v$ are new variables. Recalling that $k_0 \equiv \sqrt{i\omega / cD_0} = (1+i)\sqrt{\omega / 2cD_0}$, substituting Equations 4.20a and 4.20b into Equation 4.18 and employing trigonometric identities, gives

$$k_x = \sqrt{2\omega/cD_0}\,(1+i)\sin(p+iv)\cos q\ , \tag{4.21a}$$

$$k_z = \sqrt{2\omega/cD_0}\,(1+i)\sin(p+iv)\sin q\ . \tag{4.21b}$$

We now choose the variables $p$ and $v$ such that the imaginary part of $(1 + i)\sin(p + iv)$ vanishes. This gives the relation

$$\tan p = -\tan v\ . \tag{4.22}$$

With this condition, Equations 4.21a and 4.21b become purely real. Employing Equation 4.22 to eliminate $v$ in Equations 4.21a and 4.21b, we obtain

$$k_x = f(p)\cos q\ , \tag{4.23a}$$

$$k_z = f(p)\sin q\ , \tag{4.23b}$$

where $f(p) \equiv \sqrt{2\omega / cD_0}\,\sin 2p / \sqrt{\cos 2p}$.

Note that by forcing $k_x$ and $k_z$ to be real in Equation 4.19, we have made $k_s$ and $k_d$ complex. The implications of this will be explained shortly. In evaluating Equation 4.19, it is convenient to change the variables of integration from $(k_x,k_x)$ to $(p,q)$ since, in the integrand of Equation 4.19,

$$\tilde{\psi}_s(k_s,k_d)=\tilde{\psi}_s\left[k_s(p,q),k_d(p,q)\right]\equiv\tilde{\psi}_s(p,q)\ ,$$

and similarly,

$$C(k_s,k_d)=C\left[k_s(p,q),k_d(p,q)\right]\equiv C(p,q)\ ,$$

where the relations $k_s(p,q)$ and $k_d(p,q)$ are defined through Equations 4.21 and 4.22. From Equation 4.23, the entire Fourier plane $-\infty < k_x < \infty$ and $-\infty < k_z < \infty$ maps into the domain $0 \le p < \pi/4$ and $0 \le q < 2\pi$. Then, on changing variables to $(p,q)$, Equation 4.19 becomes

$$\hat{D}_1(x,z)=\frac{1}{(2\pi)^2}\int_0^{\pi/4}dp\, J(p)A(p)\int_0^{2\pi}dq\frac{\tilde{\psi}_s(k_s,k_d)}{C(k_s,k_d)}\exp\left[-if(p)(x\cos q+\sin q)\right]\ , \tag{4.24}$$

where $J(p)$ is the Jacobian of the transformation from $(k_x,k_x)$ to $(p,q)$, which can be shown to be $J(p) = (2\omega/cD_0)(1 + \cos^2 2p)\sin 2p/\cos^2 2p$ and $A(p)$ has been inserted to serve as a spectral apodization function, or lowpass filter, that forces the integrand to zero for large $p$. The simplest choice for $A(p)$ is $A(p) = 1$ for $p < p_{max}$ and $A(p) = 0$ for $p > p_{max}$ The function $A(p)$ is necessary to compensate for any exponential growth in the analytically continued data $\tilde{\psi}_s(k_s,k_d)$ caused by noise. The lowpass filter $A(p)$ may be regarded as a form of regularization.

We conclude by discussing how measurements can be analytically continued. This is, in principle, performed by substituting the complex variables $(k_s,k_d)$, defined by Equations 4.20 and 4.22, into the Fourier relation (Equation 4.14) and performing the integration. This will in general give rise to growing exponentials in the factor $\exp[i(k_d,x_d + k_d,x_d)]$, which, in the absence of noise, will be compensated for by the dying exponentials in the data $\tilde{\psi}_s(k_s,k_d)$. If we limit the domain of integration in Equation 4.14 (which will always be the case for finite-length source and detector arrays), the integral (Equation 4.14) will remain bounded. In this process, however, noise in the data will be amplified relative to the signal, which is a reflection of the ill-conditioning of the inverse problem. As noted, the purpose of the function $A(p)$ in Equation 4.24 is to reduce this noise at the expense of some loss of resolution.

To summarize, given the measurements $\psi_s(\boldsymbol{r}_s,\boldsymbol{r}_d)$, the reconstruction procedure requires the following steps. First, the modified Fourier transform of $\psi_s(\boldsymbol{r}_s,\boldsymbol{r}_d)$, defined by Equation 4.14, is evaluated using the complex transform variables $k_s$ and $k_d$ as defined by Equations 4.20 and 4.22. This result is then substituted into Equation 4.24 to obtain the image. Figure 4.4 shows a simulated reconstruction of two point inhomogeneities with no noise, 2% noise, and 5% noise added to the simulated data. Further details on the approach described in this section, with further discussion of the simulation, can be found in Reference 36.

## 4.4.2 Nonlinear Iterative Algorithm: Frequency Domain Data

The nonlinear iterative approach involves attempting to find a set of parameters that best predicts the data. For simplicity, we assume that scattering dominates absorption (i.e., $\sigma_s >> \sigma_a$), so that the scattering cross section $\sigma_s(\boldsymbol{r})$ is the parameter of interest. Then the transport equation (Equation 4.4) becomes

$$\boldsymbol{s}\cdot\nabla\psi(\boldsymbol{r},\boldsymbol{s}) + \left[\sigma_s(\boldsymbol{r}) - \frac{i\omega}{c}\right]\psi(\boldsymbol{r},\boldsymbol{s}) = \sigma_s(\boldsymbol{r})\int\psi(\boldsymbol{r},\boldsymbol{s}')f(\boldsymbol{s}\cdot\boldsymbol{s}')d^2\boldsymbol{s}' + q(\boldsymbol{r},\boldsymbol{s}). \quad (4.25)$$

The adjoint equation to Equation 4.25 is

$$-\boldsymbol{s}\cdot\nabla\psi^+(\boldsymbol{r},\boldsymbol{s}) + \left[\sigma_s(\boldsymbol{r}) - \frac{i\omega}{c}\right]\psi^+(\boldsymbol{r},\boldsymbol{s}) = \sigma_s(\boldsymbol{r})\int\psi^+(\boldsymbol{r},\boldsymbol{s}')f(\boldsymbol{s}\cdot\boldsymbol{s}')d^2\boldsymbol{s}' + q^+(\boldsymbol{r},\boldsymbol{s}), \quad (4.26)$$

where the adjoint flux $\psi^+(\boldsymbol{r},\boldsymbol{s})$ is defined as the solution to Equation 4.26 given the adjoint source $q^+(\boldsymbol{r},\boldsymbol{s})$, which will be defined. Equations 4.25 and 4.26 can be abbreviated as

$$L\{\psi\} = q \quad (4.27)$$

and

$$L^+\{\psi^+\} = q^+, \quad (4.28)$$

where $L$ and $L^+$ are the linear forward and adjoint operators defined by Equations 4.25 and 4.26. One can then show that[87]

$$\iint\psi L^+\{\psi^+\}d^3\boldsymbol{r}\,d^2\boldsymbol{s} = \iint\psi^+ L\{\psi\}d^3\boldsymbol{r}\,d^2\boldsymbol{s}. \quad (4.29)$$

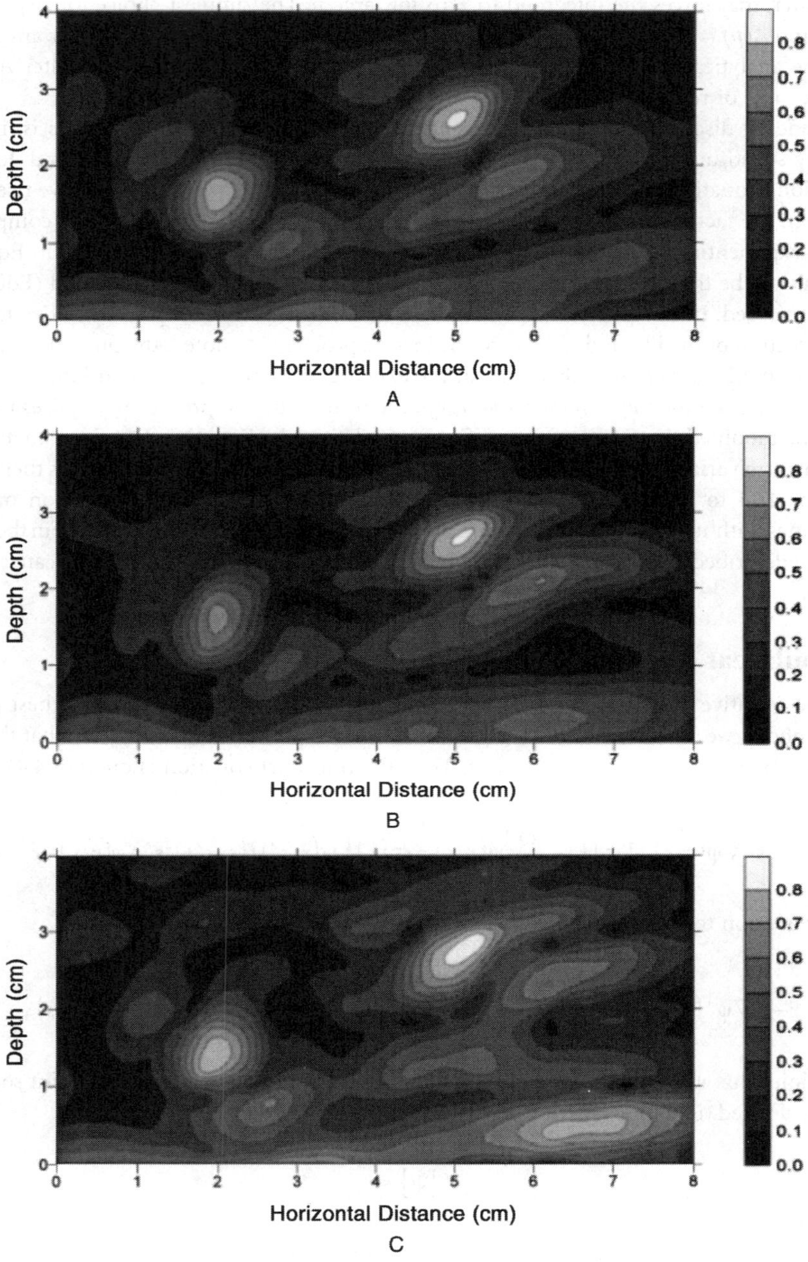


**FIGURE 4.4** Simulated reconstruction of two-point inhomogeneities using theoretical algorithms with (A) no noise, (B) 2% noise, and (C) 5% noise added to the simulated data. (From Norton, S.J. and Vo-Dinh, J., *Opt. Soc. Am.*, A15, 2670, 1998. With permission.)

We define the error norm expressing the difference between the predicted flux, $\psi(\boldsymbol{r,s})$, and the observed flux, $\psi^{(obs)}(\boldsymbol{r,s})$, as

$$E = \frac{1}{2}\int\int D(\boldsymbol{r},\boldsymbol{s})\left|\psi(\boldsymbol{r},\boldsymbol{s}) - \psi^{(obs)}(\boldsymbol{r},\boldsymbol{s})\right|^2 d^2\boldsymbol{r}d^2\boldsymbol{s}\,, \tag{4.30}$$

where $D(\boldsymbol{r,s})$ is a function that restricts the domain $(\boldsymbol{r,s})$ to "detector space"; we refer to $D(\boldsymbol{r,s})$ as the detector response function. For example, suppose we have $N$ point detectors at $\boldsymbol{r}_j$, $j = 1, \ldots, N$.

Further, suppose that each detector has an angular response $R_j(\boldsymbol{s})$ at point $\boldsymbol{r}_j$. Then

$$D(\boldsymbol{r},\boldsymbol{s}) = \sum_{j=1}^{N} R_j(\boldsymbol{s})\delta^{(3)}(\boldsymbol{r} - \boldsymbol{r}_j)\,, \tag{4.31}$$

where $\delta^{(3)}(\boldsymbol{r})$is the 3-D Dirac delta function. Substitution of Equation 4.31 into Equation 4.30 gives the error norm

$$E = \frac{1}{2}\sum_{j=1}^{N}\int d^2\boldsymbol{s}\, R_j(\boldsymbol{s})\left|\psi(\boldsymbol{r}_j,\boldsymbol{s}) - \psi^{(obs)}(\boldsymbol{r}_j,\boldsymbol{s})\right|^2 . \tag{4.32}$$

Our objective is to find the scattering cross section $\sigma_s(\boldsymbol{r})$ that minimizes $E$.

Now a small variation $\delta\sigma_s(\boldsymbol{r})$ in the cross section $\sigma_s(\boldsymbol{r})$ will give rise to a variation $\delta\psi(\boldsymbol{r,s})$ in the flux and a corresponding variation $\delta E$ in the error $E$ defined by Equation 4.30:

$$\delta E = \mathrm{Re}\int\int D\left(\psi - \psi^{(obs)}\right)^* \delta\psi\, d^3\boldsymbol{r}\, d^2\boldsymbol{s}, \tag{4.33}$$

where Re means real part and * denotes the complex conjugate. The corresponding variation in the transport equation (Equation 4.27) will be

$$\delta L\{\psi\} + L\{\delta\psi\} = \delta q\,, \tag{4.34}$$

because $L$ is a linear operator. Normally, the source $q$ is fixed, so that $\delta q = 0$ and Equation 4.34 gives

$$L\{\delta\psi\} = -\delta L\{\psi\}\,. \tag{4.35}$$

We wish to eliminate the unknown flux variation, $\delta\psi$, from Equation 4.33 with the aid of Equation 4.35 to express $\delta E$ explicitly in terms of the operator variation $\delta L$, which, in turn, can be related to the variation in the scattering cross section, $\delta\sigma_s$ through Equation 4.25, as we show next.

To accomplish this, define the adjoint source $q^+ \equiv D(\psi - \psi^{(obs)})^*$ so that Equation 4.28 becomes

$$L^+\{\psi^+\} \equiv D\left(\psi - \psi^{(obs)}\right)^*. \tag{4.36}$$

Substituting Equation 4.36 into Equation 4.33 and using Equation 4.29 gives

$$\delta E = \mathrm{Re}\int\int \delta\psi L^+\{\psi^+\} d^3\boldsymbol{r}\, d^2\boldsymbol{s} = \mathrm{Re}\int\int \psi^+ L\{\delta\psi\} d^3\boldsymbol{r}\, d^2\boldsymbol{s}\,. \tag{4.37}$$

Substituting Equation 4.35 into the second integral finally results in

$$\delta E = -\mathrm{Re}\int\int \psi^{+}\delta L\{\psi\} d^3r\, d^2s\,. \tag{4.38}$$

Now, variation of the linear transport operator $L$ defined in Equation 4.25 gives

$$\delta L\{\psi\} = \delta\sigma_s\psi - \delta\sigma_s\int \psi f\, d^2s'\,. \tag{4.39}$$

Substituting Equation 4.39 into Equation 4.38 and interchanging orders of integration gives

$$\delta E = \int\left[\nabla_s E(\mathbf{r})\right]\delta\sigma_s(\mathbf{r})d^3\mathbf{r}\,, \tag{4.40}$$

where

$$\nabla_s E(\mathbf{r}) \equiv \mathrm{Re}\int\int \psi^{+}(\mathbf{r},\mathbf{s})\left[\psi(\mathbf{r},\mathbf{s}) - \int \psi(\mathbf{r},\mathbf{s}')f(\mathbf{s}\cdot\mathbf{s}')d^2s'\right]d^2\mathbf{s}\,. \tag{4.41}$$

Here, $\nabla_s E(\mathbf{r})$ is the functional gradient (or Fréchet derivative) of the mean-square error, $E$, computed with respect to the scattering cross section $\sigma_s(\mathbf{r})$. Equation 4.40 may be regarded as the continuous-space analogue of a directional derivative, where the integral is analogous to a dot product between the functional gradient, $\nabla_s E(\mathbf{r})$, and the infinite-dimensional "vector" $\delta\sigma_s(\mathbf{r})$. If $\sigma_s(\mathbf{r})$ depends on a finite number of discrete parameters, Equation 4.40 reduces to the usual finite-dimensional gradient, as shown below. Equation 4.41 may also be interpreted as the mean-square error "sensitivity" in the sense that it tells us how sensitive a change in the error functional, $E$, is to a change in $\sigma_s(\mathbf{r})$ at the point $\mathbf{r}$.

Once the functional gradient has been computed, any convenient gradient descent algorithm may be employed to drive the error norm (Equation 4.32) to a minimum. Examples of descent algorithms are the method of steepest descent, the conjugate-gradient method, or various quasi-Newton methods.[88,89] The point is that any of these descent algorithms requires the gradient, which traditionally has been the most expensive part of the computation. For example, the conjugate-gradient algorithm updates the previous estimate of $\sigma_s(\mathbf{r})$, denoted $\sigma_s^{(n-1)}(\mathbf{r})$, as follows:[88,89]

$$\sigma_s^{(n)}(\mathbf{r}) = \sigma_s^{(n-1)}(\mathbf{r}) + \alpha_n f_n(\mathbf{r})$$

where

$$f_n(\mathbf{r}) = -\nabla_s^{(n-1)}E(\mathbf{r}) + \beta_n f_{n-1}(\mathbf{r})$$

In the latter equation, the gradient $\nabla_s^{(n-1)}E(\mathbf{r})$ is computed on the basis of the estimate $\sigma_s^{(n-1)}(\mathbf{r})$, and $\alpha_n$ and $\beta_n$ are step-size parameters updated at each iteration, with the initial condition $\beta_0 \equiv 0$.

### 4.4.2.1 Finite Number of Parameters

The finite-dimensional case can also be treated using the preceding results. Suppose, for example, that we represent the spatially varying cross section $\sigma_s(\mathbf{r})$ using a finite number of pixels at the points $\mathbf{r}_p$, $p =$ 1, …, $P$ More generally, we can expand $\sigma_s(\mathbf{r})$ in a set of $P$ basis functions, as follows:

$$\sigma_s(\mathbf{r};\mathbf{a}) = \sum_{p=1}^{P} a_p\varphi_p(\mathbf{r}), \tag{4.42}$$

where $\mathbf{a} = \{a_1, \ldots, a_P\}$ is a set of coefficients to be determined. The variation $\delta\sigma_s(\mathbf{r})$ can then be written

$$\delta\sigma_s(\boldsymbol{r};\boldsymbol{a}) = \sum_{p=1}^{P} \varphi_p(\boldsymbol{r})\delta a_p . \tag{4.43}$$

Substituting Equation 4.43 into Equation 4.40 gives

$$\delta E = \sum_{p=1}^{P} \left[ \int [\nabla_s E(\boldsymbol{r})] \varphi_p(\boldsymbol{r}) d^3\boldsymbol{r} \right] \delta a_p \tag{4.44}$$

However,

$$\delta E = \sum_{p=1}^{P} \frac{\partial E}{\partial a_p} \delta a_p \tag{4.45}$$

Comparing Equations 4.44 and 4.45, we obtain the components of the finite-dimensional gradient of the norm $E$:

$$\frac{\partial E}{\partial a_p} = \int [\nabla_s E(\boldsymbol{r})] \varphi_p(\boldsymbol{r}) d^3\boldsymbol{r} \tag{4.46}$$

where $\nabla_s E(\boldsymbol{r})$ is defined by Equation 4.41.

#### 4.4.2.2 Adding a Regularization Term

In general, the inverse transport problem is ill posed, and consequently ill conditioned, because of the diffuse nature of the photon migration in tissue; thus, a regularization scheme is needed to stabilize the inversion in the presence of noisy data. One approach involves adding a penalty term to the error norm (Equation 4.30) designed, for example, to penalize large variations in the unknown $\sigma_s(\boldsymbol{r})$ or to emphasize smoother solutions. Regularization of this kind not only helps mitigate ill-conditioning but also serves to force uniqueness in an under-determined problem. Equation 4.30 is then modified to read

$$E = \frac{1}{2} \int\int\int D \left| \psi - \psi^{(obs)} \right|^2 d^3\boldsymbol{r} d^2\boldsymbol{s} + \frac{1}{2} \lambda \int C(\sigma_s) d^3\boldsymbol{r} , \tag{4.47}$$

where $C(\sigma_s)$ is a penalty function and $\lambda$ is a parameter that controls the relative weighting of the two terms in Equation 4.47. Possible choices of $C(\sigma_s)$ are

$$C(\sigma_s) = \sigma_s(\boldsymbol{r})^2 , \tag{4.48a}$$

$$C(\sigma_s) = \nabla\sigma_s(\boldsymbol{r}) \cdot \nabla\sigma_s(\boldsymbol{r}), \tag{4.48b}$$

$$C(\sigma_s) = -\sigma_s(\boldsymbol{r}) \ln \sigma_s(\boldsymbol{r}). \tag{4.48c}$$

Working through the same derivation as above and letting $\nabla_s E(\boldsymbol{r})$ denote the Fréchet derivative defined by Equation 4.41, the augmented norm (Equation 4.47) gives the following new Fréchet derivatives corresponding, respectively, to the preceding three penalty functions:

$$\nabla_s^{(1)} E(\boldsymbol{r}) = \nabla_s E(\boldsymbol{r}) + \lambda\sigma_s(\boldsymbol{r}), \tag{4.49a}$$

$$\nabla_s^{(2)} E(\boldsymbol{r}) = \nabla_s E(\boldsymbol{r}) - \lambda\nabla^2\sigma_s(\boldsymbol{r}), \tag{4.49b}$$

$$\nabla_s^{(3)} E(\boldsymbol{r}) = \nabla_s E(\boldsymbol{r}) + \lambda\left[1 - \ln \sigma_s(\boldsymbol{r})\right]. \tag{4.49c}$$

#### 4.4.2.3 Example: Two-Dimensional Imaging of a Scattering Cross Section

To illustrate how the preceding theory can be implemented, consider a simple 2-D continuous wave (steady-state) imaging problem. We divide a square imaging domain into $N \times N$ square pixels, each with an unknown scattering cross section denoted by $\sigma_{mn}$, with $m,n = 1, \ldots, N$. We wish to determine the $N^2$ unknowns $\sigma_{mn}$ based on scattering observations on the boundary of the square domain. We consider a very simple discrete transport model for purposes of illustration; in this model, a photon incident on a pixel can be scattered in just four directions — up, down, left, and right — where each direction is described, respectively, by the unit vectors $\boldsymbol{s}_1$, $\boldsymbol{s}_2$, $\boldsymbol{s}_3$, $\boldsymbol{s}_4$. Denote the photon flux moving in the direction $\boldsymbol{s}_j$ in pixel $(m,n)$ by $\psi_{mn}(\boldsymbol{s}_j)$. This flux then obeys the following discrete transport equation:

$$\boldsymbol{s}_i \cdot \nabla_d \psi_{mn}(\boldsymbol{s}_i) + \sigma_{mn}\psi_{mn}(\boldsymbol{s}_i) = \sigma_{mn}\sum_{j=1}^{4} f(\boldsymbol{s}_i \cdot \boldsymbol{s}_j)\psi_{mn}(\boldsymbol{s}_j) + q_{mn}(\boldsymbol{s}_i) \tag{4.50}$$

for $i = 1, \ldots, 4$. This is the discretized version of Equation 4.25 in the zero-frequency limit ($\omega = 0$). Here $\nabla_d$ is a difference operator defined below, $f(\boldsymbol{s}_i \cdot \boldsymbol{s}_j)$ is a known scattering law, and $q_{mn}(\boldsymbol{s}_i)$is a source term, also assumed known. If we assume that the pixels are of unit dimensions ($\Delta x = \Delta y = 1$), then we have for the four directions:

$$\begin{aligned}
\boldsymbol{s}_1 \cdot \nabla_d \psi_{mn}(\boldsymbol{s}_1) &= \psi_{mn}(\boldsymbol{s}_1) - \psi_{m-1,n}(\boldsymbol{s}_1)\\
\boldsymbol{s}_2 \cdot \nabla_d \psi_{mn}(\boldsymbol{s}_2) &= \psi_{mn}(\boldsymbol{s}_2) - \psi_{m+1,n}(\boldsymbol{s}_2)\\
\boldsymbol{s}_3 \cdot \nabla_d \psi_{mn}(\boldsymbol{s}_3) &= \psi_{mn}(\boldsymbol{s}_3) - \psi_{m,n-1}(\boldsymbol{s}_3)\\
\boldsymbol{s}_4 \cdot \nabla_d \psi_{mn}(\boldsymbol{s}_4) &= \psi_{mn}(\boldsymbol{s}_4) - \psi_{m,n+1}(\boldsymbol{s}_4)
\end{aligned}$$

This is a reasonable approximation to the spatial derivatives, assuming that the flux is slowly varying on the scale of a pixel. More realistic transport models obviously exist but can be treated similarly. The forward problem can be defined as the task of computing the fluxes $\psi_{mn}(\boldsymbol{s}_i)$, given the set $\{\sigma_{mn}\}$; a simple forward algorithm can be devised by iterating Equation 4.50. Our objective is to minimize the following mean-square error with respect to $\{\sigma_{mn}\}$:

$$E = \frac{1}{2}\sum_{m,n=1}^{N}\sum_{i=1}^{4} D_{mn}(\boldsymbol{s}_i)\left[\psi_{mn}(\boldsymbol{s}_i) - \psi_{mn}^{(obs)}(\boldsymbol{s}_i)\right]^2, \tag{4.51}$$

in which the detector response function is defined by $D_{mn}(\boldsymbol{s}_i) \equiv 1$ for the observed flux on the boundary of the domain, and $D_{mn}(\boldsymbol{s}_i) \equiv 0$ otherwise.

We wish to show in this section how, using the adjoint method, we can compute the partial derivatives of the mean-square error $\partial E/\partial \sigma_{mn}$ in terms of known quantities. Defining for brevity $F_{ij} \equiv f(\boldsymbol{s}_i \cdot \boldsymbol{s}_j) - \delta_{ij}$, where $\delta_{ij}$ is the Kronecker delta, we can write Equation 4.50 in the more compact form

$$\boldsymbol{s}_i \cdot \nabla_d \psi_{mn}(\boldsymbol{s}_i) - \sigma_{mn}\sum_{j=1}^{4} F_{ij}\psi_{mn}(\boldsymbol{s}_j) = q_{mn}(\boldsymbol{s}_i). \tag{4.52}$$

The adjoint of Equation 4.52 is

$$-\mathbf{s}_i \cdot \nabla_d^+ \psi_{mn}^+(\mathbf{s}_i) - \sigma_{mn} \sum_{j=1}^{4} F_{ij} \psi_{mn}^+(\mathbf{s}_j) = q_{mn}^+(\mathbf{s}_i), \tag{4.53}$$

where the adjoint source, $q^+_{mn}(\mathbf{s}_i)$, is to be defined, and

$$\mathbf{s}_1 \cdot \nabla_d^+ \psi_{mn}^+(\mathbf{s}_1) = \psi_{mn}^+(\mathbf{s}_1) - \psi_{m+1,n}^+(\mathbf{s}_1),$$
$$\mathbf{s}_2 \cdot \nabla_d^+ \psi_{mn}^+(\mathbf{s}_2) = \psi_{mn}^+(\mathbf{s}_2) - \psi_{m-1,n}^+(\mathbf{s}_2),$$
$$\mathbf{s}_3 \cdot \nabla_d^+ \psi_{mn}^+(\mathbf{s}_3) = \psi_{mn}^+(\mathbf{s}_3) - \psi_{m,n+1}^+(\mathbf{s}_3),$$
$$\mathbf{s}_4 \cdot \nabla_d^+ \psi_{mn}^+(\mathbf{s}_4) = \psi_{mn}^+(\mathbf{s}_4) - \psi_{m,n-1}^+(\mathbf{s}_4).$$

If we abbreviate Equations 4.52 and 4.53 by

$$L_{mn}\left\{\psi_{mn}(\mathbf{s}_i)\right\} = q_{mn}(\mathbf{s}_i),$$
$$L_{mn}^+\left\{\psi_{mn}^+(\mathbf{s}_i)\right\} = q_{mn}^+(\mathbf{s}_i),$$

then the adjoint relation analogous to Equation 4.29 is

$$\sum_{m,n=1}^{N} \sum_{i=1}^{4} \psi_{mn}(\mathbf{s}_i) L_{mn}^+\left\{\psi_{mn}^+(\mathbf{s}_i)\right\} = \sum_{m,n}^{N} \sum_{i=1}^{4} \psi_{mn}^+(\mathbf{s}_i) L_{mn}\left\{\psi_{mn}(\mathbf{s}_i)\right\}.$$

We now define the adjoint source as

$$q_{mn}^+(\mathbf{s}_i) \equiv D_{mn}(\mathbf{s}_i)\left[\psi_{mn}(\mathbf{s}_i) - \psi_{mn}^{(obs)}(\mathbf{s}_i)\right].$$

If we follow the same procedure as described in the previous section, we obtain the following simple result for the derivatives of the mean-square error:

$$\frac{\partial E}{\partial \sigma_{pq}} = \sum_{i=1}^{4} \sum_{j=1}^{4} F_{ij} \psi_{pq}^+(\mathbf{s}_i) \psi_{pq}(\mathbf{s}_j), \tag{4.54}$$

where $\psi_{pq}(\mathbf{s}_i)$ and $\psi^+_{pq}(\mathbf{s}_i)$ are the solutions to Equations 4.52 and 4.53 computed on the basis of the current estimate of $\{\sigma_{mn}\}$. A simulation was performed by Norton[83] using these formulas to reconstruct a 64-pixel-square region with the scattering law $f(\mathbf{s}_i \cdot \mathbf{s}_j) = 0.1$ for $i \neq j$ and $f(\mathbf{s}_i \cdot \mathbf{s}_i) = 0.7$.

These formulas can be easily generalized to account for multiple sources and multiple modulation frequencies. If the modulation frequencies can be made sufficiently high so that the DPDW "skin depth" $\delta$ is comparable to the size of the region of interest, then multifrequency data can be shown to improve the conditioning of the inverse problem.[90]

### 4.4.3 Nonlinear Iterative Algorithm: Time-Resolved Data

As noted, when recording transient data, it is not advantageous to restrict ourselves to the relatively small number of first-arrival photons. Arridge and coworkers,[58,60,91] for example, have advocated measurement of the first few moments of the temporal response (e.g., the first moment being the mean arrival time); these measurements can then be employed in an iterative descent algorithm to minimize an appropriate error norm, similar to the scheme described earlier in this chapter. In this section, we describe a time-domain iterative algorithm designed to minimize an error norm that weights the entire transient waveform. For simplicity, we assume that the diffusion approximation holds, in which case the photon flux density, $\psi(\boldsymbol{r},t)$, obeys the time-dependent diffusion equation given by Equation 4.3:

$$\frac{1}{c}\frac{\partial\psi(\boldsymbol{r},t)}{\partial t}-\nabla\cdot\left[D(\boldsymbol{r})\nabla\psi(\boldsymbol{r},t)\right]+\sigma_a(\boldsymbol{r})\psi(\boldsymbol{r},t)=Q(\boldsymbol{r},t), \tag{4.55}$$

subject to the initial condition $\psi(\boldsymbol{r},0) = 0$. Although we employ the diffusion model (Equation 4.55) to illustrate our procedure, a similar time-domain treatment can be carried out with the time-dependent transport equation as a starting point.[85]

Suppose the region of interest is illuminated by a light pulse, which diffuses through the tissue and is detected at $N$ point detectors at $\boldsymbol{r}$, $j = 1, \ldots, N$; the pulse is observed over the interval of time $[0,T]$ and the observations are denoted by $\psi^{(obs)}(\boldsymbol{r}_j,t)$. The optical parameters, $D(\boldsymbol{r})$ and $\sigma_a(\boldsymbol{r})$, that minimize the time-integrated mean-square error are then sought:

$$E=\frac{1}{2}\sum_{j=1}^{N}\int_0^T w_j(t)\left[\psi(\boldsymbol{r}_j,t)-\psi^{(obj)}(\boldsymbol{r}_j,t)\right]^2 dt\,. \tag{4.56}$$

Here, the integration is over the observation interval $T$, and $w_j(t)$ is an arbitrary temporal weighting function. To compute the Fréchet derivative of Equation 4.56, we employ the adjoint diffusion equation, given by

$$-\frac{1}{c}\frac{\partial\psi^+(\boldsymbol{r},t)}{\partial t}-\nabla\cdot\left[D(\boldsymbol{r})\nabla\psi^+(\boldsymbol{r},t)\right]+\sigma_a(\boldsymbol{r})\psi^+(\boldsymbol{r},t)=Q^+(\boldsymbol{r},t), \tag{4.57}$$

where the adjoint source is defined by

$$Q^+(\boldsymbol{r},t)=\sum_{j=1}^{N}w_j(t)\left[\psi(\boldsymbol{r}_j,t)-\psi^{(obs)}(\boldsymbol{r}_j,t)\right]\delta(\boldsymbol{r}-\boldsymbol{r}_j)\,. \tag{4.58}$$

We require that the adjoint field obey the terminal condition $\psi^+(\boldsymbol{r},T) = 0$. The reason for this will become apparent shortly. We now vary Equation 4.56, giving

$$\delta E=\sum_{j=1}^{N}\int_0^T w_j(t)\left[\psi(\boldsymbol{r}_j,t)-\psi^{(obs)}(\boldsymbol{r}_j,t)\right]\delta\psi(\boldsymbol{r}_j,t)dt\,. \tag{4.59}$$

In view of the definition (Equation 4.58) of the adjoint source, we can write Equation 4.59 as

$$\delta E=\int d^3\boldsymbol{r}\int_0^T dt\,Q^+(\boldsymbol{r},t)\delta\psi(\boldsymbol{r},t), \tag{4.60}$$

where the volume integral includes the spatial domain of interest. Next, substituting the left-hand side of Equation 4.57 into Equation 4.60 and integrating by parts, once with respect to time and twice with respect to space,[81] we obtain

$$\delta E=\int d^3\boldsymbol{r}\int_0^T dt\left\{\frac{1}{c}\frac{\partial\delta\psi(\boldsymbol{r},t)}{\partial t}-\nabla\cdot\left[D(\boldsymbol{r})\nabla\delta\psi(\boldsymbol{r},t)\right]+\sigma_a(\boldsymbol{r})\delta\psi(\boldsymbol{r},t)\right\}\psi^+(\boldsymbol{r},t)\,. \tag{4.61}$$

When integrating by parts with respect to time, we find that the integrated part vanishes on account of the initial condition $\delta\psi(\boldsymbol{r},0) = 0$ and the terminal condition $\psi^{+}(\boldsymbol{r},T) = 0$. Similarly, we assume that appropriate homogeneous boundary conditions hold on the surface of the domain of integration, which ensures that the boundary terms vanish when integrating by parts with respect to space. Next, we vary Equation 4.55 to obtain

$$\frac{1}{c}\frac{\partial\delta\psi(\boldsymbol{r},t)}{\partial t}-\nabla\cdot\left[\delta D(\boldsymbol{r})\nabla\psi(\boldsymbol{r},t)\right]-\nabla\cdot\left[D(\boldsymbol{r})\nabla\delta\psi(\boldsymbol{r},t)\right]+\delta\sigma_a(\boldsymbol{r})\psi(\boldsymbol{r},t)+\sigma_a(\boldsymbol{r})\delta\psi(\boldsymbol{r},t)=0$$

or

$$\frac{1}{c}\frac{\partial\delta\psi(\boldsymbol{r},t)}{\partial t}-\nabla\cdot\left[D(\boldsymbol{r})\nabla\delta\psi(\boldsymbol{r},t)\right]+\sigma_a(\boldsymbol{r})\delta\psi(\boldsymbol{r},t)=\nabla\cdot\left[\delta D(\boldsymbol{r})\nabla\psi(\boldsymbol{r},t)\right]-\delta\sigma_a(\boldsymbol{r})\psi(\boldsymbol{r},t)\,.$$

Substituting this into Equation 4.61 gives

$$\delta E=\int d^3\boldsymbol{r}\int_0^T dt\left\{\nabla\cdot\left[\delta D(\boldsymbol{r})\nabla\psi(\boldsymbol{r},t)\right]-\delta\sigma_a(\boldsymbol{r})\psi(\boldsymbol{r},t)\right\}\psi^{+}(\boldsymbol{r},t)$$

The first term can be integrated once more by parts with respect to $\boldsymbol{r}$ to yield

$$\begin{aligned}\delta E&=-\int d^3\boldsymbol{r}\int_0^T dt\,\nabla\psi(\boldsymbol{r},t)\cdot\nabla\psi^{+}(\boldsymbol{r},t)\delta D(\boldsymbol{r})-\int d^3\boldsymbol{r}\int_0^T dt\,\psi(\boldsymbol{r},t)\psi^{+}(\boldsymbol{r},t)\delta\sigma_a(\boldsymbol{r})\\&=\int d^3\boldsymbol{r}\,\nabla_D E(\boldsymbol{r})\delta D(\boldsymbol{r})+\int d^3\boldsymbol{r}\,\nabla_a E(\boldsymbol{r})\delta\sigma_a(\boldsymbol{r}),\end{aligned}$$

where the Fréchet derivatives for $D(\boldsymbol{r})$ and $\sigma_a(\boldsymbol{r})$ are, respectively,

$$\nabla_D E(\boldsymbol{r})=-\int_0^T dt\,\nabla\psi(\boldsymbol{r},t)\cdot\nabla\psi^{+}(\boldsymbol{r},t), \tag{4.62}$$

$$\nabla_a E(\boldsymbol{r})=-\int_0^T dt\,\psi(\boldsymbol{r},t)\psi^{+}(\boldsymbol{r},t). \tag{4.63}$$

When these functional derivatives are used to update the parameters $D(\boldsymbol{r})$ and $\sigma_a(\boldsymbol{r})$ in a descent algorithm of one's choice, the time-integrated error functional (Equation 4.56) can be driven to a minimum. In a manner similar to that of the previous section, a penalty term can also be added to Equation 4.56 to regularize the solution.

As noted, the time-dependent residual in Equation 4.56 can be appropriately weighted through the selection of the weighting function $w_j(t)$ to control the signal-to-noise ratio. Arridge[68] argues that employing the logarithm of $\psi^{(obs)}(\boldsymbol{r},t)$ may be advantageous. The above approach can be carried out in the latter case as well, after the field Equations 4.55 and 4.57 are modified by substituting $\psi(\boldsymbol{r},t) = \exp[\gamma(\boldsymbol{r},t)]$, which defines new field equations in $\gamma(\boldsymbol{r},t)$. After this logarithmic transformation, the Fréchet derivatives of $D(\boldsymbol{r})$ and $\sigma_a(\boldsymbol{r})$ can be derived in the same way as above. The interesting question of what constitutes sufficient data to reconstruct $D(\boldsymbol{r})$ and $\sigma_a(\boldsymbol{r})$ simultaneously is beyond the scope of our discussion, but has been discussed by Arridge.[92]

## 4.5 Conclusion

This chapter has examined in some detail three optical tomography algorithms. The first of these employs Fourier methods to reconstruct an image from time-harmonic DPDW data recorded at one modulation frequency. This approach may be regarded as a modification of conventional diffraction tomography. The other algorithms employ a gradient descent algorithm to minimize a global error norm measuring the difference between recorded and predicted data. This approach is very general and can accommodate either time- or frequency-domain measurements.

Whether the algorithm is iterative or Fourier-based, the inverse problem will inevitably be ill conditioned because of the diffuse nature of the multiply scattered light or, equivalently, because of the smoothing effect of the forward transport operator. This implies that any inversion procedure that attempts to "undo" this smoothing will amplify noise in the data and will, as a result, require some form of regularization to mitigate the noise amplification. A variety of such schemes exists, but the price paid for regularization is some loss of spatial resolution. In this regard, it is important to note that resolution is ultimately noise-limited, since the resolution typically achieved is well beyond the "diffraction limit" — that is, less than the wavelength of a photon density wave. Moreover, the size of the region probed may sometimes be less than a single DPDW wavelength, implying operation well within the near field.

In addition to reviewing two fundamental methods of image reconstruction (diffraction tomography and the iterative minimization of an error norm), we have tried to provide a representative sampling of the literature in this area. A more comprehensive review of time- and frequency-domain techniques was given recently by Arridge.[68] We have not referenced many papers that have addressed related but more specialized topics, such as forward algorithms, more accurate photon transport models, fluorescence imaging and spectroscopic methods, effects of boundaries and unusual measurement geometries, and clinical studies. Also, in the interests of space, we have, with a few exceptions, emphasized references in archival journals rather than conference proceedings, although a great deal of work has been published in the latter (particularly the SPIE proceedings[2]). The number of research papers in optical tomography has grown rapidly in the last 10 years, and one has every reason to expect that continued technical and algorithmic progress will ultimately yield systems of significant clinical value.

## Acknowledgments

This work was sponsored by the National Institutes of Health (Grant R01 CA88787-01) and by the U.S. Department of Energy, Office of Biological and Environmental Research, under contract DEAC05-00OR22725 with UT-Battelle, LLC.

### References

1. Singer, J.R., Grunbaum, F.A., Kohn, P., Zubelli, J.P., Image reconstruction of the interior of bodies that diffuse radiation, *Science,* 248, 990, 1990.
2. Chance, B. et al., Eds., SPIE Proceedings: Photon propagation in tissues, Vol. 2626, 1995; Photon propagation in tissues II, Vol. 2925, 1996; Photon propagation in tissues III, Vol. 3194, 1997; Photon propagation in tissues IV, Vol. 3566, 1998; Optical tomography and spectroscopy of tissue, Vol. 2389, 1995; Optical tomography and spectroscopy of tissue II, Vol. 2979, 1997; Optical tomography and spectroscopy III, Vol. 3597, 1999; Optical tomography and spectroscopy IV, Vol. 4250, 2001.
3. Minet, O., Muller, G., and Beuthan, J., Eds., *Selected Papers on Optical Tomography, Fundamentals and Applications*, SPIE Press, Bellingham, WA, 1998.
4. Tuchin, V., *Tissue Optics,* SPIE Press, Bellingham, WA, 2000.
5. Hebden, J.C., Arridge, S.R., and Delpy, D.T., Optical imaging in medicine. I. Experimental techniques, *Phys. Med. Biol.*, 42, 825, 1997.
6. Arridge, S.R. and Hebden, J.C., Optical imaging in medicine: II. Modeling and reconstruction, *Phys. Med. Biol.*, 42, 841, 1997.
7. Ishimaru, A., Diffusion of light in turbid material, *Appl. Opt.*, 28, 2210, 1989.
8. Jacques, S.L., Time resolved propagation of ultrashort laser pulses within turbid tissues, *Appl. Opt.*, 28, 2223, 1989.
9. Profio, A.E., Light transport in tissue, *Appl. Opt.*, 28, 2216, 1989.

10. Arridge, S., Schweiger, M., Hiraoka, M., and Delpy, D.T., A finite element approach for modeling photon transport in tissue, *Med. Phys.*, 20, 299, 1993.
11. Tromberg, B.J., Svaasand, L.O., Tsay, T., and Haskell, R.C., Properties of photon density waves in multiple-scattering media, *Appl. Opt.*, 32, 607, 1993.
12. Knuttel, A., Schmitt, J.M., and Knutson, J.R., Spatial localization of absorbing bodies by interfering diffuse photon-density waves, *Appl. Opt.*, 32, 381, 1993.
13. Boas, D.A., O'Leary, M.A., Chance, B., and Yodh, A.G., Scattering of diffuse photon density waves by spherical inhomogeneities within turbid media: analytical solution and applications, *Proc. Natl. Acad. Sci. U.S.A.*, 91, 4887, 1994.
14. Feng, S., Zeng, F., and Chance, B., Photon migration in the presence of a single defect: a perturbation analysis, *Appl. Opt.*, 34, 3826, 1995.
15. Yodh, A.G. and Chance, B., Spectroscopy and imaging with diffusing light, *Phys. Today*, 48, 34, 1995.
16. Boas, D.A. and Yodh, A.G., Spatially varying dynamical properties of turbid media probed with diffusing temporal light correlation, *J. Opt. Soc. Am.*, A14, 192, 1997.
17. Boas, D.A., O'Leary, M.A., Chance, B., and Yodh, A.G., Detection and characterization of optical inhomogeneities with diffuse photon density waves: a signal-to-noise analysis, *Appl. Opt.*, 36, 75, 1997.
18. Boas, D.A., O'Leary, M.A., Chance, B., and Yodh, A.G., Scattering and wavelength transduction of diffuse density waves, *Phys. Rev. E.*, 47, R2999, 1993.
19. Ostermeyer, M.R. and Jacques, S.L., Perturbation theory for diffuse light transport in complex biological tissues, *J. Opt. Soc. Am.*, A14, 255, 1997.
20. Boas, D.A., O'Leary, M.A., Chance, B., and Yodh, A.G., Detection and characterization of optical inhomogeneities with diffuse photon desity waves: a signal-to-noise analysis, *Appl. Opt.*, 36, 75, 1997.
21. Li, X.D., Durduran, T., Yodh, A.G., Chance, B., and Pattanayak, D.N., Diffraction tomography for biochemical imaging with diffuse-photon density waves, *Opt. Lett.*, 22, 573, 1997.
22. Furutsu, K., Theory of a fixed scatterer embedded in a turbid medium, *J. Opt. Soc. Am.*, A15, 1371, 1998.
23. Ripoll, J., Nieto-Vesperinas, M., and Carminati, R., Spatial resolution of diffuse photon density waves, *J. Opt. Soc. Am.*, A16, 1466, 1999.
24. Sevick-Muraca, E.M., Lopez, G., Troy, T.L., Reynolds, J.S., and Hutchinson, C.L., Fluorescence and absorption contrast mechanisms for biomedical opical imaging using frequency-domain techniques, *Photochem. Photobiol.*, 66, 55, 1997.
25. Jacques, S.L., Roman, J.R., and Lee, K., Imaging superficial tissues with polarized light, *Lasers Surg. Med.*, 26, 119, 2000.
26. Jacques, S.L., Ramanujam, N., Vishnoi, G., Choe, R., and Chance, B., Modeling photon transport in transabdominal fetal oximetry, *J. Biomed. Opt.*, 5, 277, 2000.
27. Lee, J. and Sevick-Muraca, E.M., Three-dimensional fluorescence enhanced optical tomography using referenced frequency-domain photon migration measurements at emission and excitation wavelengths, *J. Opt. Soc. Am.*, A19, 759, 2002.
28. Sun, Z.G. and Sevick-Muraca, E.M., Investigation of particle interactions in dense colloidal suspensions using frequency domain photon migration: bidisperse systems, *Langmuir*, 18, 1091, 2002.
29. O'Leary, M.A., Boas, D.A., Chance, B., and Yodh, A.G., Experimental images of heterogeneous turbid media by frequency-domain diffusing-photon tomography, *Opt. Lett.*, 20, 426, 1995.
30. Reynolds, J.S., Przadka, A., Yeung, S.P., and Webb, K.J., Optical diffusion imaging: a comparative numerical and experimental study, *Appl. Opt.*, 35, 3671, 1996.
31. Colak, S.B., Papaioannou, D.G., 't Hooft, G.W., van der Mark, M.B., Schomberg, H., Paasschens, J.C.J., Melissen, J.B.M. and van Asten, N.A.A.J., Tomographic image reconstruction from optical projections in light-diffusing media, *Appl. Opt.*, 36, 180, 1997.
32. Matson, C.L., A diffraction tomographic model of the forward problem using diffuse photon density waves, *Opt. Express*, 1, 6, 1997.

33. Yao, Y., Wang, Y., Pei, Y., Wenwu, Z., and Barbour, R.L., Frequency-domain optical imaging of absorption and scattering distributions by a Born iterative method, *J. Opt. Soc. Am.*, A14, 325, 1997.
34. Zhu, W., Wang, Y., Yao, Y., Chang, J., Graber, H.L., and Barbour, R.L., Iterative total least-squares image reconstruction algorithm for optical tomography by the conjugate gradient method, *J. Opt. Soc. Am.*, A14, 799, 1997.
35. Lasocki, D.L., Matson, C.L., and Collins, P.J., Analysis of forward scattering of diffuse photon-density waves in turbid media: a diffraction tomography approach to an analytic solution, *Opt. Lett.*, 23, 558, 1998.
36. Norton, S.J. and Vo-Dinh, T., Diffraction tomographic imaging with photon density waves: an explicit solution, *J. Opt. Soc. Am.*, A15, 2670, 1998.
37. Matson, C.L. and Liu, H., Analysis of the forward problem with diffuse photon density waves in turbid media by use of a diffraction tomography model, *J. Opt. Soc. Am.*, A16, 455, 1999.
38. Chen, B., Stamnes, J.J., and Stamnes, K., Reconstruction algorithm for diffraction tomography of diffuse photon density waves in a random medium, *Pure Appl. Opt.*, 7, 1161, 1998.
39. Matson, C.L. and Liu, H., Backpropagation in turbid media, *J. Opt. Soc. Am.*, A16, 1254, 1999.
40. Ye, J.C., Webb, K.J., Millane, R.P. and Downar, T.J., Modified distorted Born iterative method with an approximate Fréchet derivative for optical diffusion tomography, *J. Opt. Soc. Am.*, A16, 1814, 1999.
41. Braunstein, M. and Levine, R.Y., Three-dimensional tomographic reconstruction of an absorptive perturbation with diffuse photon density waves, *J. Opt. Soc. Am.*, A17, 11, 2000.
42. Gaudette, R.J., Brooks, D.H., DiMarzio, C.A., Kilmer, M.E., Miller, E.L., Gaudette, T., and Boas, D.A., A comparison study of linear reconstruction techniques for diffuse optical tomographic imaging of absorption coefficient, *Phys. Med. Biol.*, 45, 1051, 2000.
43. Kilmer, M.E., Miller, E.L., Boas, D., and Brooks, D., A shape-based reconstruction technique for DPDW data, *Opt. Express*, 7, 481, 2000.
44. Markel, V.A. and Schotland, J.C., Inverse problem in optical diffusion tomography. I. Fourier-Laplace inversion formulas, *J. Opt. Soc. Am.*, A18, 1336, 2001.
45. Matson, C.L., Diffraction tomography for turbid media, in *Advances in Imaging and Electron Physics*, Vol. 124, Hawkes, P., Ed., Academic Press, New York, 2002.
46. Chance, B., Leigh, J.S., Miyake, H., Smith, D.S., Nioka, S., Greenfeld, R., Finander, M., Kaufmann, K., Levy, W., Young, M., Cohen, P., Yoshioka, H., and Boretsky, R., Comparison of time-resolved and -unresolved measurements of deoxyhemoglobin in grain, *Proc. Natl. Acad. Sci. U.S.A.*, 85, 4971, 1988.
47. Patterson, M.S., Chance, B., and Wilson, B.C., Time resolved reflectance and transmittance for the non-invasive measurement of tissue optical properties, *Appl. Opt.*, 28, 2331, 1989.
48. Benaron, D.A. and Stevenson, D.K., Optical time-of-flight and absorbance imaging of biologic media, *Science*, 259, 1463, 1993.
49. Das, B.B., Yoo, K.M., and Alfano, R.R., Ultrafast time-gated imaging in thick tissues — a step towards optical mammography. *Opt. Lett.*, 18, 1092, 1993.
50. Hee, M.R., Izatt, J.A., Swanson, E.A., and Fujimoto, J.G., Femtosecond transillumination tomography in thick issues, *Opt. Lett.*, 18, 1107, 1993.
51. Das, B.B., Barbour, R.L., Graber, H.L., Chang, J., Zevallos, M., Liu, F., and Alfano, R.R., Analysis of time-resolved data for tomographic image reconstruction of opaque phantoms and finite absorbers in diffuse media, *Proc. SPIE*, 2389(Part 1), 16, 1995.
52. Chang, J., Zhu, W., Wang, Y., Graber, H.L., and Barbour, R.L., Regularized progressive expansion algorithm for recovery of scattering media from time-resolved data, *J. Opt. Soc. Am.*, 14, 306, 1997.
53. Grosenick, D., Wabnitz, H., and Rinneberg, H., Time-resolved imaging of solid phantoms for optical mammography, *Appl. Opt.*, 36, 221, 1997.
54. Winn, J.N., Perelman, L.T., Chen, K., Wu, J., Dasari, R.R. and Feld, M.S., Distribution of the paths of early-arriving photons traversing a turbid medium, *Appl. Opt.*, 34, 8085, 1998.

55. Cai, W., Gayen, S.K., Xu, M., Zevallos, M., Alrubaiee, M., Lax, M., and Alfano, R.R., Optical tomographic image reconstruction from ultrafast time-sliced transmission measurements, *Appl. Opt.*, 39, 4237, 1999.
56. Morin, M., Mailloux, A., Painchaud, Y., and Beaudry, P., Time-resolved transmission through homogeneous scattering media: time-response effects, *Appl. Opt.*, 38, 3681, 1999.
57. Hielscher, A.H., Klose, A.D., and Hanson, K.M., Gradient-based iterative image reconstruction scheme for time-resolved optical tomography, *IEEE Trans. Med. Imag.*, 18, 262, 1999.
58. Schmidt, F.E.W., Hebden, J.C., Hillman, M.C., Fry, M.E., Schweiger, M., Dehghani, H., Delphy, D.T., and Arridge, S.R., Multiple-slice imaging of a tissue-equivalent phantom by use of time-resolved optical tomography, *Appl. Opt.*, 39, 3380, 2000.
59. Gao, F., Poulet, P., and Yamada, Y., Simultaneous mapping of absorption and scattering coefficients from a three-dimensional model of time-resolved optical tomography, *Appl. Opt.*, 39, 5898, 2000.
60. Hebden, J.C., Veenstra, H., Dehghani, H., Hillman, E.M.C., Schweiger, M., Arridge, S.R., and Delpy, D.T., Three-dimensional time-resolved optical tomography of a conical breast phantom, *Appl. Opt.*, 40, 3278, 2001.
61. Schotland, J.C., Continuous-wave diffusion imaging, *J. Opt. Soc. Am.*, A14, 275, 1997.
62. Cheng, X. and Boas, D.A., Diffuse optical reflection tomography with continuous-wave illumination, *Opt. Express*, 3, 118, 1998.
63. Alfano, R.R., Govindjee, W. Y. R., Becher, B., and Ebrey, T. G., Picosecond kinetics of fluorescence from chromophore of purple membrane-protein of Halobacterium-Halobium, *Biophys. J.*, 16, 541, 1976.
64. Devaney, A.J., Inversion formula for inverse scattering within the Born approximation, *Opt. Lett.*, 7, 111, 1982.
65. Devaney, A.J., A filtered backpropagation algorithm for diffraction tomography, *Ultrason. Imaging*, 4, 336, 1982.
66. Devaney, A.J. and Beylkin, G., Diffraction tomography using arbitrary transmitter and receiver surfaces, *Ultrason. Imaging*, 6, 181, 1984.
67. Devaney, A.J., Reconstructive tomography with diffracting wave fields, *Inverse Probl.*, 2, 161, 1986.
68. Arridge, S.R., Optical tomography in medical imaging, *Inverse Probl.*, 15, R41, 1999.
69. Case, K.M. and Zweifel, P.F., *Linear Transport Theory*, Addison-Wesley, Reading, MA, 1967.
70. Pogue, B., McBride, T., Prewitt, J., Osterberg, U., and Paulsen, K., Spatially variant regularization improves diffuse optical tomography, *Appl. Opt.*, 38, 2950, 1999.
71. Norton, S.J. and Linzer, M., Ultrasonic reflectivity imaging in three dimensions: exact inverse scattering solutions for plane, cylindrical and spherical apertures, *IEEE Trans. Biomed. Eng.*, BME-28, 202, 1981.
72. Hielscher, A., Klose, A., and Beuthan, J., Evolution strategies for optical tomographic characterization of homogeneous media, *Opt. Express*, 7, 507, 2000.
73. Walker, S.A., Fantini, S., and Gratton, E., Image reconstruction from frequency-domain optical measurements in highly scattering media, *Appl. Opt.*, 36, 170, 1997.
74. Durduran, T., Culver, J.P., Holboke, M.J., Li, X.D., Zubkov, L., Chance, B., Pattanayak, D.N., and Yodh, A.G., Algorithms for 3D localization and imaging using near-field diffraction tomography with diffuse light, *Opt. Express*, 4, 247, 1999.
75. Pattanayak, D.N. and Yodh, A.G., Diffuse optical 3D-slice imaging of bounded turbid media using a new integro-differential equation, *Opt. Express*, 4, 231, 1999.
76. Bluestone, A.Y., Abdoulaev, G., Schmitz, C.H., Barbour, R.L. and Hielscher, A.H., Three-dimensional optical tomography of hemodynamics in the human head, *Opt. Express*, 9, 272, 2001.
77. Roy, R. and Sevick-Muraca, E.M., A numerical study of gradient-based nonlinear optimization methods for contrast enhanced optical tomography, *Opt. Express*, 9, 49, 2001.
78. Ye, J.C., Webb, K.J., and Bouman, C.A., Optical diffusion tomography by iterative coordinate-descent optimization in a Bayesian framework, *J. Opt. Soc. Am.*, A16, 2400, 1999.

79. Eppstein, M.J., Dougherty, D.E., Troy, T.L., and Sevick-Muraca, E.M., Biomedical optical tomography using dynamic parameterization and Bayesian conditioning on photon migration measurements, *Appl. Opt.*, 38, 2138, 1999.
80. Kolehmainen, V., Vauhkonen, M., Kaipio, J.P., and Arridge, S.R., Recovery of piecewise constant coefficients in optical diffusion tomography, *Opt. Express*, 7, 468, 2000.
81. Norton, S.J., Iterative inverse-scattering algorithms: methods for computing the Fréchet derivative, *J. Acoust. Soc. Am.*, 106, 2653, 1999.
82. Norton, S.J. and Bowler, J.R., Theory of eddy current inversion, *J. Appl. Phys.*, 73, 501, 1993.
83. Norton, S.J., A general nonlinear inverse transport algorithm using forward and adjoint flux computations, *IEEE Trans. Nucl. Sci. Eng.*, NS-44, 153, 1997.
84. Arridge, S.R. and Schweiger, M., A gradient-based optimization scheme for optical tomography, *Opt. Express,* 2, 213, 1998.
85. Dorn, O., Scattering and absorption transport sensitivity functions for optical tomography, *Opt. Express*, 7, 492, 2000.
86. Dorn, O., A transport-backtransport method for optical tomography, *Inverse Probl.*, 14, 1107, 1998.
87. Bell, G.I. and Glasstone, S., *Nuclear Reactor Theory*, Van Nostrand Reinhold, New York, 1970.
88. Gill, P.E., Murry, W., and Wright M.H., *Practical Optimization*, Academic Press, New York, 1981.
89. Luenberger, D.G., *Linear and Nonlinear Programming*, 2nd ed., Addison-Wesley, Menlo Park, CA, 1984.
90. Norton, S.J., Electromagnetic induction imaging, Oak Ridge National Laboratory Report No. K/NSP-315, Oak Ridge National Laboratory, Oak Ridge, TN, August 1995.
91. Schweiger, M. and Arridge, S.R., Comparison of two- and three-dimensional reconstruction methods in optical tomography, *Appl. Opt.*, 37, 7419, 1998.
92. Arridge, S.R. and Lionheart, W.R.B., Nonuniqueness in diffusion-based optical tomography, *Opt. Lett.*, 23, 882, 1998.

# II

# Photonic Devices

# 5

# Laser Light in Biomedicine and the Life Sciences: From the Present to the Future

Vladilen S. Letokhov
*Institute of Spectroscopy*
*Russian Academy of Sciences*
*Troitsk, Moscow Region,*
*Russian Federation*



## 5.1 Introduction

The progress made in various fields of science and technology, especially in the development of technical and instrumental tools, has always led, is now leading, and will undoubtedly continue to lead to new, useful, and sometimes revolutionary applications in medicine. As medicine penetrates to ever deeper organizational levels (cellular and molecular) in the human organism (levels that are closely similar in all living creatures), it evolves into biomedicine and becomes more closely connected with all the life sciences. Figure 5.1 depicts the "biomedical tools tree," illustrating the applications of various tools and methods in biomedicine, from its origin in the Middle Ages to genetic engineering at present and into the future. Biomedical photonics, the subject of this volume, embraces a wide range of laser light applications in biomedical diagnostics at various levels of the organism, in therapy, and, finally, in surgery. In many applications laser light is exceptionally efficient and well matched to present-day advanced technology and to the human body — the object of all these applications.

The human body is a uniquely organized functional system including biomolecular, cellular, and tissue structures measuring from a fraction of a nanometer to a few meters, i.e., ranging over ten orders of magnitude. Figure 5.2 illustrates the spatial spectrum of various human structures, from individual atoms and molecules to organs and the body as a whole. The entirety of this spectrum is as yet unknown (except for its extreme sections), but one can assume that it features a fractal dependence like that of the cardiovascular system.[1] Throughout this huge range of sizes, the mutual interaction of all the structures is integrated and matched.

0-8493-1116-0/03/$0.00+$1.50

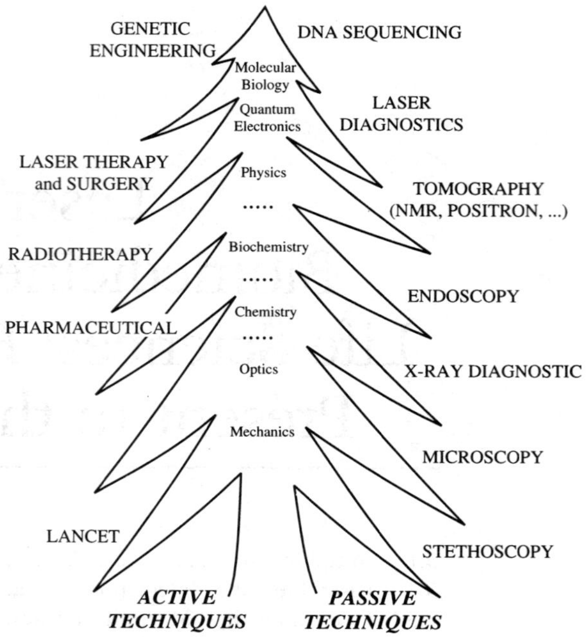


**FIGURE 5.1** Biomedical "tools tree."

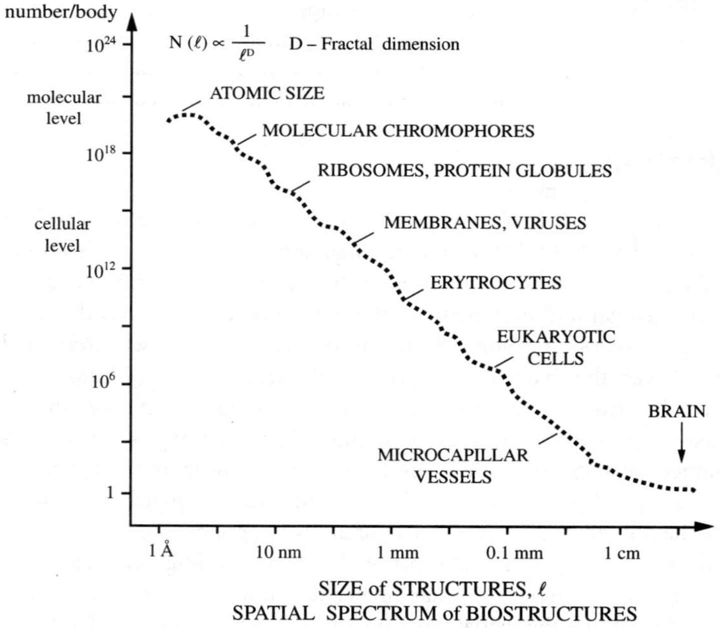


**FIGURE 5.2** Presumable spatial spectral distribution of nonhomogeneities in biotissue.

Researchers in one of the major avenues of research in modern technology, namely, nanotechnology, are endeavoring to create nanostructures the size of a molecule and organize their interaction. Naturally, nanotechnology here is faced with enormous difficulties associated with the organization of an addressable mode of interaction of nanoelements with one another and with trying to create a highly organized conglomerate of nanostructures such as a molecular computer, a robot, etc. This problem seems more involved than the development of the nanostructures themselves. Yet, in living organisms, from a cell to the human being, this problem is solved quite easily and efficiently, probably owing to their adequate hierarchic organization. Perhaps this fact explains why laser light proves an effective diagnostic, therapeutic, and surgical means at all organizational levels of the human body: biomolecules, cells, organs, and the body as a whole. The capabilities of lasers here seem inexhaustible.

Coherent laser light can be localized within a region measuring a few fractions of a micron so that it can irradiate intracellular structures, and nonlocalized to cover a region a few meters in size, with its coherence remaining unimpaired. Laser light supplies energy in a noncontact manner, i.e., without the need to use electrodes that cause mechanical damage to biostructures. Of special value is the possibility of conveying light via the optical information window of man, i.e., the human eye.

Laser light is monochromatic and can be made to have any wavelength, i.e., the supply of laser-light energy is spectral-selective, and so its action on biomolecules can be as selective as are their absorption or light-action spectra. Laser light can supply energy of any magnitude necessary, and its irradiation intensity can be varied over a wide range (from a few $mW/cm^2$ to $10^{11}$ $W/cm^2$) by varying the energy and duration of the laser pulse (from a few femtoseconds to a few seconds, minutes, and more).

The combination of these unique capabilities, absent in any other type of radiation, makes the laser a universal and efficient biomedical tool.[2–5]

## 5.2 Laser–Biomatter Interaction

Laser light interacts with living matter at various biological organization levels: (1) the biomolecule, (2) subcellular structures, (3) cells, (4) biotissues, (5) organs, and (6) the whole organism. All these structures fall within different sections of the spatial spectrum (Figure 5.2).

All types of laser–biomatter interaction can be categorized as (1) resonance processes, involving the absorption of laser light and the resultant excitation of biomolecules, and (2) nonresonance processes, involving the scattering of laser light by molecular vibrations or by refractive index irregularities in the biotissues exposed or its reflection from boundaries between media differing in refractive index (interfaces).

When a biomolecule absorbs laser light, the expenditure of the excitation energy can follow various pathways, depending on the type of excitation (electronic or vibrational) of the molecules, its surroundings, and radiation intensity (Figure 5.3). Some absorbed energy degrades to heat in a radiationless manner (for example, vibrational excitation by IR light degrades to heat in a picosecond timescale) and some of it may be transferred to the neighboring molecules. The electronically excited molecules may take part in chemical reactions or they may be raised to higher quantum states, provided the light intensity is high enough (ultraviolet pulses), and then be involved in photochemical processes. Almost all these excitation pathways are now successfully utilized in laser biomedicine: in laser fluorescence diagnostics, laser photodynamic therapy, laser phototherapy, and laser thermal surgery. The sum of the quantum yields attained in particular channels is equal to unity. Depending on the conditions of irradiation and the state of the medium, the quantum yield in any desired channel can be varied with a view to improving the efficiency of its utilization.

Laser radiation parameters (wavelength, spectral bandwidth, intensity, duration) can be varied over very wide limits, thus making it possible to implement various types of light–biomatter interaction (linear and nonlinear, single- and multiple-photon, coherent and noncoherent, thermal and nonthermal, etc.). Thus one can induce various effects in biotissues (photochemical modification, thermal destruction, explosive ablation, optical breakdown, shock pressure waves, photodisruption, etc.), as illustrated in Figure 5.4.

Figure 5.5 illustrates various nonresonant light-scattering processes occurring at a molecular and a macroscopic level. In addition to Raman light scattering by molecular vibrations, important biomedical applications are the processes of scattering light by macroscopic particles that increase the photon pathway in the exposed biotissue and give rise to a Doppler light-frequency shift by moving particles (Doppler flowmetry),[6] and the gradient (dipole) force due to the transfer of photon momentum $h\omega/c$) to the scattering particle (the physical basis of optical tweezers[7]).

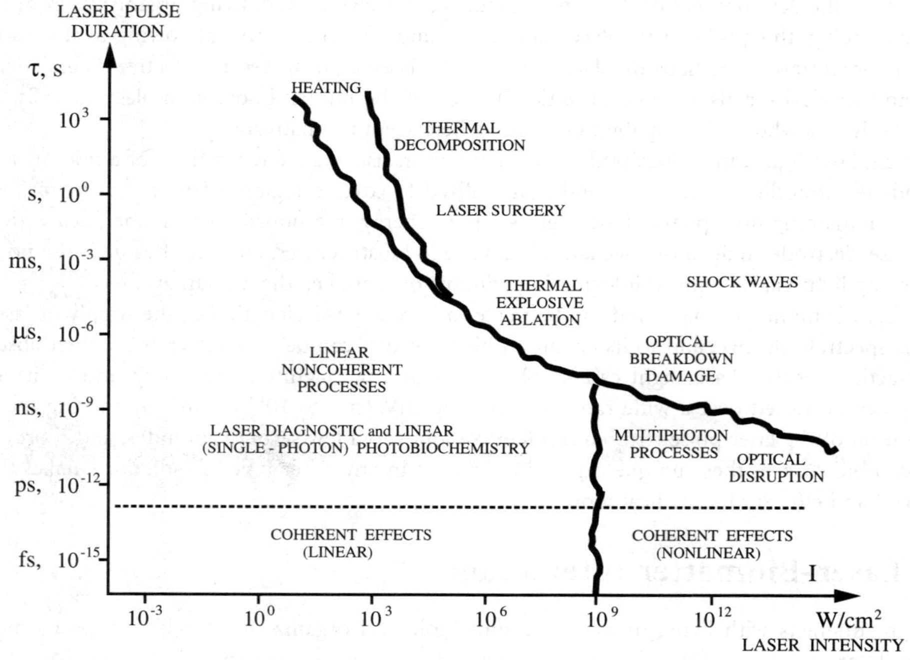


**FIGURE 5.3** Pathways of use of laser excitation.

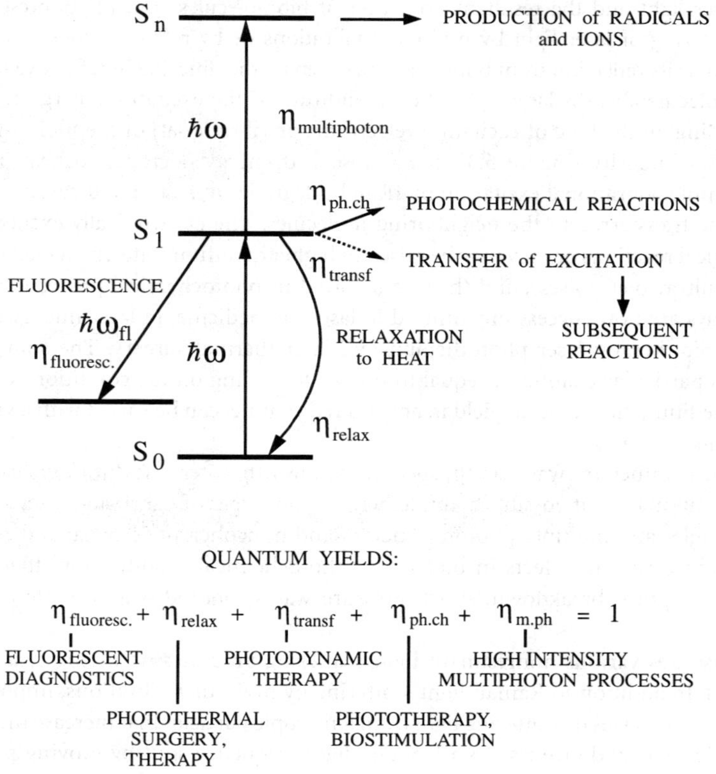


**FIGURE 5.4** Various regimes of laser pulse interaction with biotissue.

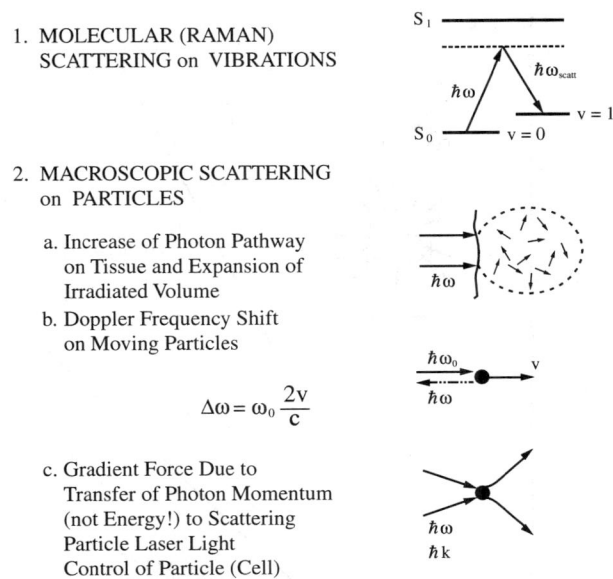

1. MOLECULAR (RAMAN) SCATTERING on VIBRATIONS
2. MACROSCOPIC SCATTERING on PARTICLES
   a. Increase of Photon Pathway on Tissue and Expansion of Irradiated Volume
   b. Doppler Frequency Shift on Moving Particles

$$\Delta\omega = \omega_0 \frac{2v}{c}$$

   c. Gradient Force Due to Transfer of Photon Momentum (not Energy!) to Scattering Particle Laser Light Control of Particle (Cell) Motion (Optical Tweezers)

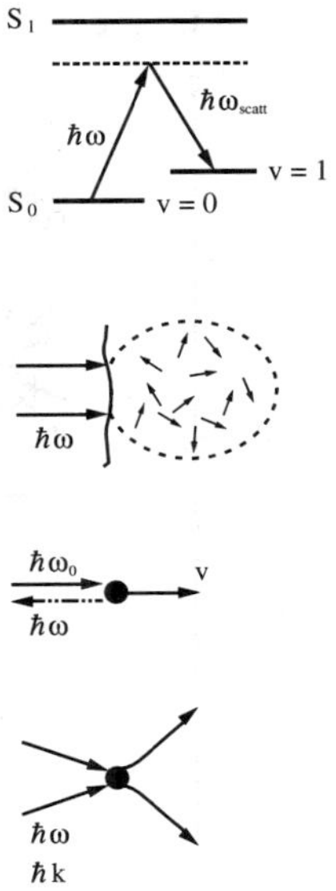


**FIGURE 5.5** Effects of nonresonant scattering interaction of laser light with biotissue.

When considering the interaction between laser radiation and biotissues, one should take into account (and possibly use) the inhomogeneous character of the various biological, chemical, and physical (optical, acoustical, electrical, etc.) properties of tissues. The characteristic size of these inhomogeneities varies over very wide limits, from the size of individual molecules to that of blood cells and organs (Figure 5.2).

The spatial inhomogeneity of the optical characteristics of biotissue (specifically its absorption characteristics) leads to a nonuniform distribution of the heat released as a result of absorbing pulsed laser radiation. The following three characteristic times are important in this case: $\tau_p$ — the laser pulse duration; $\tau_c = a/\upsilon_s$ — the time it takes for sound to propagate with a velocity of $\upsilon_s$ beyond an inhomogeneity of size $a$, i.e., the relaxation time of the pressure resulting from the pulsed local heating; and $\tau_{td} = a^2 k\chi$ — the time it takes for heat to propagate by diffusion beyond the inhomogeneity ($k$ is a numerical coefficient depending on the shape of the inhomogeneity and $\chi$ is the thermal diffusivity), i.e., the relaxation time of pulsed heating. Figure 5.6 shows regions of enhanced pressure, $\Delta P$ (pressure confinement), and enhanced temperature, $\Delta T$ (temperature confinement), in the case of microsphere of radius $R$ ($R = a$, $k = 27$) for pulses differing in duration. Both these effects are important in the irradiation of real biotissues, and should not only be taken into consideration, but also be used to advantage.

## 5.3 Laser Biomedical Macrodiagnostics

All laser biomedical diagnostic techniques can conveniently be subdivided into two classes: (1) macrodiagnostics of various objects (tomography, topography, detection of pulsation, blood flow measurements, etc.) and (2) micro(spectro)diagnostics at a molecular level. Macrodiagnostic techniques are discussed below, and microdiagnostic techniques are discussed in the following section.

Some biomedically oriented macrodiagnostic techniques are already well developed and are used in clinics. One may cite as an example the Doppler flowmetry of blood flow in microcapillary blood vessels, based on the Doppler frequency shift of the light backscattered by moving particles. This principle is at the root of a whole class of instruments used for biomedical and other purposes. Of interest are the capabilities of laser light combined with laser-beam scanning in topographic studies of the human body, remote sensing of heartbeats, etc. In principle, all these methods may give rise to a new approach to the intelligent characterization of the human body, including the dynamics of the processes occurring

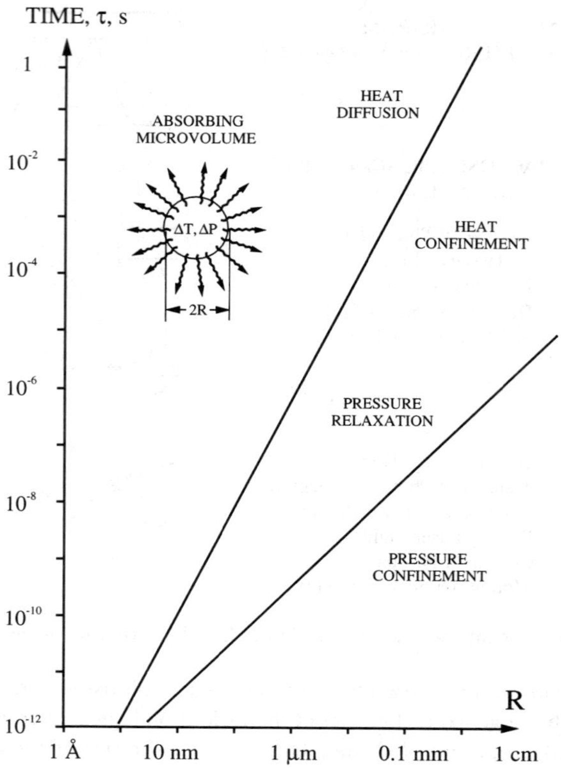


**FIGURE 5.6** Regions of homogeneous heating, local heating without local rise of pressure, and local heating and local rise of pressure.

there (blood flows, breathing, heartbeats, etc.) and their interplay. The combination of laser-optical and computer technologies will prove useful here, with the latter including CD ROM recording and storage of physiological data on individual patients in various periods of their life, various stages of illness, and so on.

But the central problem is the development of methods for laser diagnostics of inhomogeneities (tomography) in biotissue, which is an absorbent and highly scattering medium, and therefore a very difficult object for optical techniques. The tomography of biotissue is important for medical diagnostics, laser therapy, and surgery. To choose the optimal regime and correct laser irradiation dose, it is necessary to know the light absorption and scattering values of the tissue. Moreover, it is important to know the spatial distribution of the light absorption and scattering inhomogeneities over the area being irradiated. For example, determination of the location, size, and depth of tissue injuries, blood vessels, inclusions, pigments, etc. is of real importance.

Diagnostic methods for light absorption inhomogeneities in turbid media are widely discussed in the literature: these methods include photon migration,[8] time-resolved photon migration,[9–12] optical coherence tomography,[13–15] and others. The optical properties of tissue can be measured by such methods.[16,17] They prove most effective in transparent media and are mainly sensitive to refractive index inhomogeneities. The most exciting results have been obtained by using optical coherence tomography[15] for eye diagnostic purposes. Here, the lateral and longitudinal spatial resolution (in tissue depth) is limited to the wavelength of light, i.e., to a size much smaller than that of cells.

One of the promising techniques for measuring absorption inhomogeneities in turbid media is the optoacoustic method. It is based on the excitation of thermal or acoustic waves (thermooptical excitation) through absorption of time-modulated or space-modulated laser radiation (see, for example, References

18 through 20). The excitation of acoustic waves (optoacoustic signals) occurs as a result of transient nonuniform thermal expansion of the exposed area. The advantage of the optoacoustic method is the direct dependence of the amplitude of the excited wave on the light-absorption coefficient. Light-absorption measurements by means of time-resolved laser optoacoustic tomography can be taken in real time. Therefore, this method may prove useful in medical diagnostics.

Optical catheters are now successfully used for medical diagnostic purposes. The development of microminiature photonic spectral sensors and the advancement of nanotechnology will lead to the creation of self-contained, intelligent optical sensors. These sensors will be inserted in a controlled way into the human body via all the channels available, especially the highly branched system of blood vessels, transmitting information not only via the optical fiber but also by radio channels. This approach is fairly promising because it is based on the rapid progress of semiconductor microlasers operating within various spectral regions at very low power.

## 5.4 Spectral Biomedical Microdiagnostics

Spectral diagnostics at a molecular level makes use of a wide variety of laser spectroscopy methods (see, for example, References 16 and 17) to obtain extremely important information on tissue properties (spectral and time resolution, sensitivity, selectivity, and spatial resolution). The following discussion lists the most important performance characteristics of opticospectral diagnostics. The order of discussion is rather arbitrary because in actual application there is usually the need for a combination of several of these capabilities. For example, high sensitivity is a key parameter with any method.

### 5.4.1 Spectral Resolution

The well-developed laser spectroscopy techniques[21] are capable of practically any required spectral resolution, from a few megahertz to a few nanometers. But in medical diagnostics, which almost always is concerned with species in condensed media, spectral resolution is usually limited because of their relatively wide absorption and fluorescence spectral bands. A moderately high spectral resolution of a few nanometers is quite sufficient for the spectroscopy of physiologically important chemical species in biotissues, for example, in the measurements of natural endogeneous molecules in blood and tissues ($O_2$ concentration, pH level, glucose concentration, etc.). The noninvasive photonic spectral technique is an almost ideal, noninvasive way to monitor a number of important physiological parameters. One can anticipate the development of photonic spectral instruments for personal use, as well as "diagnostics cafés" (similar to Internet cafés) for online measurement of the patient's physiological parameters, saving the patient the time and expense of a clinic visit.

A higher spectral resolution is necessary to analyze the molecular composition of the gas-phase products of metabolism (breathing test) and, on this basis, to diagnose certain diseases and to conduct drug and alcohol tests. Recent progress in the development of semiconductor quantum-cascade lasers operating in the IR range at room temperature[22,23] has opened up exciting possibilities for developing photonic tools for diagnosing gaseous media in medical practice, as well as photonic tools for private use.

### 5.4.2 Time Resolution

The development of short- and ultrashort-pulse lasers has led to the improvement of time resolution from a few nanoseconds to a few picoseconds and then to a few femtoseconds. Ultrafast laser spectroscopy methods have contributed greatly to understanding fundamental processes in biological systems (see References 24 and 25). The list of intrinsically fast processes in biological systems is very long, including energy and electron transport in photosynthesis and reaction centers, and proton transfer and isomerization in the vision pigment and bacterial membranes. The biomedical diagnostic capabilities of present-day, simple solid-state femtosecond lasers[26] are practically unlimited. The main avenue of inquiry here is the search for biomolecular characteristics that require a high time resolution to be determined.

Femtosecond spectroscopy of the primary photochemical processes (femtochemistry[27]) is likely to provide insight into the primary chemical processes involving the most important biomolecules (DNA, RNA, proteins) and will be an open field for many investigations in the years to come.

### 5.4.3 Sensitivity

Laser spectroscopic techniques possess an ultrahigh sensitivity, and today they enable one to detect individual atoms (laser resonance photoionization spectroscopy[28]) and molecules (fluorescence spectroscopy with laser excitation[29]).

The detection of single atoms is of interest in determining trace elements in biological specimens. For example, the resonance photoionization method has been used to measure trace amounts of aluminum in blood.[30] Existing analysis techniques — the present-day mass spectrometric techniques, for example — meet practically all the requirements. Highly sensitive methods of detecting trace elements are important for understanding the pathways of microelements in the environment, food products, and human body.

Single molecules can be detected directly in condensed media (*in situ*), especially when combined with laser confocal microscopy, which allows one to irradiate a small volume containing a small number of fluorescing molecules, even a single such molecule.[31,32] In the case of two-photon microscopy, laser radiation can be made to scan a three-dimensional space, an especially convenient capability in studies at the cellular and intracellular levels.[33] Molecules with low fluorescence yield can be detected through use of chemical labeling with specially selected fluorescent molecular dyes; this method is promising for detecting nucleotides in DNA sequencing.[34]

### 5.4.4 Selectivity

The wide and frequently overlapping absorption and fluorescence spectral bands of biomolecules materially limit the selectivity of all spectral methods for biomolecules in their native condensed medium. Therefore, when a higher selectivity is required, one should combine spectral methods with other (nonoptical spectroscopy) methods.

The chemical labeling of nucleotides with fluorescent dye molecules in DNA sequencing[34] is an effective method for improving detection selectivity because the excitation spectra of the nucleotides coincide for all practical purposes. The preliminary chromatographic separation of a mixture of biomolecules, followed by the spectral-selective detection of their laser-induced fluorescence, is an effective method for improving selectivity at a very high sensitivity.

The fluorescence spectra of molecules may coincide, but the difference in fluorescence decay time between them makes it possible to use time-gating detection of the fluorescence of the desired molecule. Such a time selection substantially improves the contrast, especially in the fluorescent imaging of biotissues.[16] Where the time selection technique is applicable, it can be used *in vivo* and *in situ*, i.e., in actual human diagnostics.[16]

Combining various selection methods with opticospectral selection techniques substantially extends the field of application of laser optical methods.

### 5.4.5 Spatial Resolution

Laser methods are applicable practically across the whole of the spatial spectrum range of biostructures presented in Figure 5.2. The two-dimensional (2D) lateral resolution is simplest to achieve. In the range down to a fraction of a micron, various optical microscopic techniques whose capabilities have been substantially extended with the advent of lasers, particularly scanning confocal microscopy.[35] Femtosecond laser pulses cause the two-photon excitation of the fluorescing molecules of interest without causing any photodamage to them. When the femtosecond pulses are focused with a microscope, the excitation of the molecules occurs predominantly at the focal point of the microscope; this forms the basis of three-dimensional, two-photon microscopy.[33,36–38] All these methods in principle are capable of ensuring a spatial resolution of up to a fraction of the optical wavelength, which is sufficient for studying intracellular structures.

The next steps forward are investigations with spatial resolutions from a few tens to a few hundred nanometers. Such resolutions are quite realizable by use of scanning near-field optical microscopy (SNOM).[39–41] SNOM methods can be used only for 2D imaging because they are capable only of spatial resolution of the surface of biostructures.

Of much interest is the development of methods to obtain direct information on the molecular structure of individual biomolecules, especially DNA and proteins. Solving this challenging biophotonics problem can lead to the development of tabletop DNA sequencers and ultramicroscopy tools for uncrystallizable protein. Such a possibility has arisen[42] with the advent of modern femtosecond lasers capable of producing coherent light pulses in the x-ray region of the spectrum.[43,44]

## 5.5 Laser Therapy

The existing methods of laser therapy can be subdivided into two classes: (1) low-intensity laser therapy techniques and (2) photodynamic laser therapy techniques.

Low-intensity laser therapy is based on the excitation of endogenous chromophores in biotissues that absorb radiation in certain visible and near-IR spectral bands. This method requires monochromatic light of relatively low intensity (a few fractions of a $W/cm^2$) that causes no average heating of the exposed biotissue. The difference between laser and nonlaser (light-emitting diode, or LED) light sources is in their depth of coherence, $l_{coh}$, determined by the light spectrum width $\Delta\nu(cm^{-1})$:$l_{coh} \simeq 1/(\Delta\nu)$. Figure 5.7 illustrates the difference between the coherence volumes of the exposed biotissue.

In the coherence volume, as a result of random interference in the course of the scattering of light, a spatially nonuniform light intensity distribution (speckles)[45] is formed, which causes a spatially nonuniform deposition of the absorbed light energy. This, in turn, leads to a spatially nonuniform distribution of the photobiochemical reaction rates and local temperature variations. Although the primary mechanisms of the phototherapeutic effect are still to be understood,[46] the effectiveness of treatment by low-intensity laser therapy has been proved by numerous clinical trials conducted in a number of countries.[47] What is important is that existing semiconductor lasers and LEDs are compact, reliable, and cheap, which makes for widespread use not only at clinics, but also in private practice.

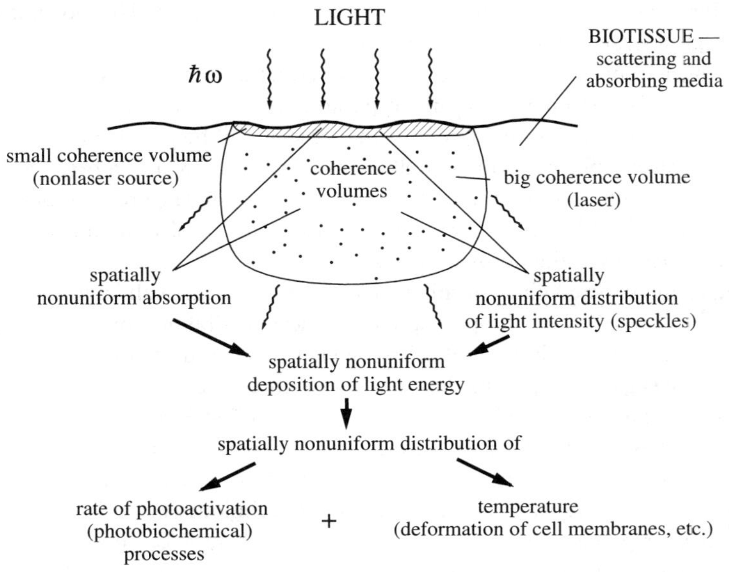


**FIGURE 5.7** Coherence volumes for irradiation of biotissue by monochromatic optical beam.

Photodynamic therapy is based on various types of photochemical reactions involving endogenous chromophores. In this field, the existing lasers will be replaced by solid-state lasers pumped by an array of laser diodes. The progress of high-power semiconductor lasers in the blue range of the spectrum is so rapid that they will also be used in clinical setups convenient to operate.

The development of photothermotherapy — a potential method based on the microlocal biotissue transient heating effect due to the nonradiative relaxation of the absorbed laser radiation in the tissue — seems quite feasible. This method would involve a nondestructive local transient being heated by short-pulse laser radiation.[49] At least three mechanisms are responsible for the transient local heating of biotissue exposed to laser radiation.

First, it is well known that when a chromophore absorbs a photon, some excitation energy relaxes in a radiationless manner, and this leads to an increase in the temperature of the chromophore and its immediate surroundings. This was observed to occur, for example, during excitation of the heme in hemoglobin and in hemoglobin itself.[48] Following the absorption of a visible photon, the heme temperature rose to 500 K. The transfer of vibrational energy to the immediate surroundings caused a rapid cooling (within 2 to 20 ps). For this reason, the effect of local heating of chromophores is important on a picosecond time scale and perhaps may play some part in biomolecular processes.

The second mechanism relates to the nonuniform spatial distribution of the laser radiation intensity. For example, when coherent light propagates through an optically inhomogeneous, scattering biotissue, random interferences in the tissue inevitably give rise to speckle structures,[45] with irregularities characteristically measuring up to $\lambda/2$. If the given effect — for example, local heating — is shorter than the correlation time of the speckle pattern fluctuations (which depends on the internal motions in the exposed medium), this speckle-like distribution of laser intensity distribution will be fairly substantial and perhaps play some role in low-intensity laser therapy.

Third, the absorption coefficient of the tissue at a certain radiation wavelength varies over sufficiently wide limits throughout the tissue volume. A simple example is the enhanced absorption of hemoglobin in blood that is distributed over the bulk of the tissue by vessels with a great variety of diameters and lengths (fractal structure with a dimension of $D = 2.7$).[1] The spatial variations of the tissue absorption are especially great in the visible and near-IR regions of the spectrum, but are smoothed out in the UV and IR regions because of the strong absorption by all organic molecules in the UV region and by water in the IR region. The magnitude of the local transient temperature rise is not very high, although it is much higher than that of time- and volume-averaged heating.[49] The gradient of the local temperature rise can be very high (Figure 5.8), for example, $\partial T/\partial r \sim 10^6$ degrees/cm.[49] This possible effect can play a role for thin biological structures (such as membranes).

Of course, stronger transient heating causes photothermal damage. This effect can also be used for photothermal therapy, for example, by means of photothermal sensitizers (metallo derivatives of porphyrines and porphyrinoid compounds, etc.). In that case, it will be possible to cause selective photothermal damage to macromolecules or subcelluar organelles to treat tumors.[50] Because the spatial spectrum of absorption inhomogeneities in biotissues is very wide (Figure 5.2), properly selecting the duration and wavelength of the laser pulses used can cause the local transient heating of the desired micro- and macrostructures in the exposed biotissue.

Pulsed heating of the targeted kinds of microvolumes in biotissues with laser pulses of certain wavelengths, durations, and intensities will open up entirely new possibilities for photothermotherapy. This is because the local transient overheating of microvolumes may be substantial ($\Delta T = 1$ to 100°C), while the time- and space-averaged macroscopic heating of the entire irradiated region is much lower. Also, by varying the laser radiation wavelength, one can affect the pulsed heating of the desired biologically important microvolumes, the relationship between the locally absorbing regions, and the wavelength. Locally measured pulsed heating of targeted microvolumes will have a material effect on biochemical processes, and so will form the basis for new therapeutic techniques based in principle on the use of tunable, ultrashort laser pulses. However, the actual types of therapy can only be determined through experimental investigations into local absorption regions, the nature and size of those tissues, and the

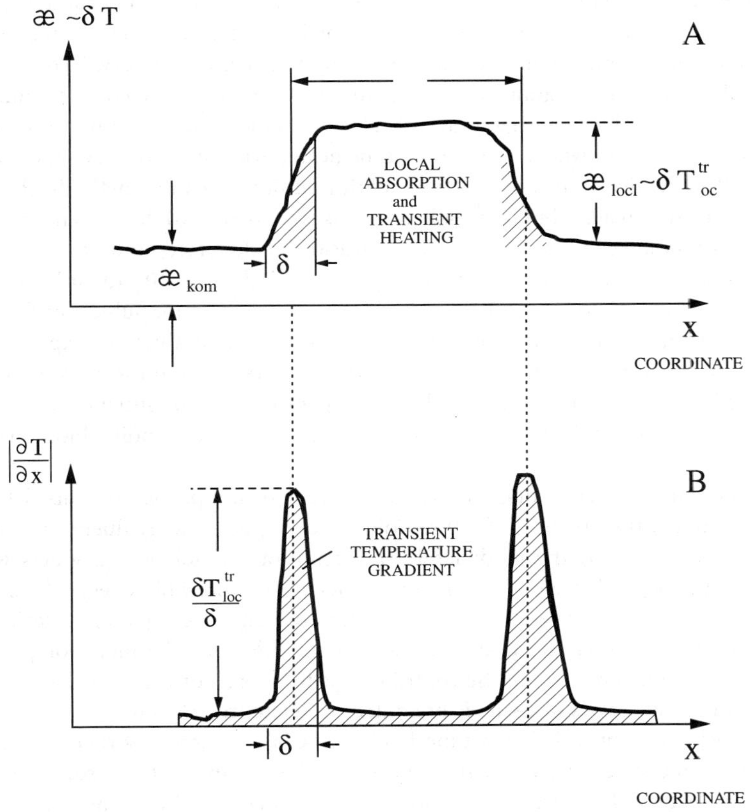


**FIGURE 5.8** Spatial temperature gradient (A) produced at the boundary of a local absorption microvolume in the case of energy deposition by short laser pulses (B).

characteristics of pulsed overheating and temperature gradients produced. All this should be the aim of special systematic studies in the near future.

## 5.6 Laser Surgery

Laser surgery uses three types of laser–biotissue interactions, which require different laser radiation parameters and so are carried out by means of entirely different types of lasers. These three types of interaction are arbitrarily referred to as photocoagulation, photoablation, and photodisruption. In fact, each of these modes of interaction comprises a fairly complex sequence of processes that are the subject of special investigations.

When a biotissue is exposed to a continuous wave (CW) radiation that it absorbs well, the tissue temperature rises, and that temperature increase causes, consecutively, enzymes to denature and membranes to loosen (40 to 45°C), coagulation and necrosis (60°C), drying out (100°C), carbonization (150°C), and, finally, pyrolysis and vaporization (above 300°C). This process is usually associated with photocoagulation, for it is important in stopping profuse bleeding when performing surgical operations on blood-containing organs. Here lies the advantage of the photocoagulative laser surgery technique over the standard surgical methods. The shortcoming of this technique is the thermal injury caused to adjacent biotissue regions as a result of heat conduction, which can transfer heat a few millimeters over tens of seconds (Figure 5.6).

By using pulsed laser radiation of certain wavelength, one can selectively heat local, restricted areas with elevated absorption (this is achieved when operating in the heat confinement region of Figure 5.6). This selective photothermolysis technique is at the root of precise microsurgery.[51] The effect of local laser heating of spatially nonuniform (granulated) biotissues has long been discussed in connection with the problem of pulsed-laser-induced retinal injuries.[52] This raises the following important question: What is the maximum temperature to which biotissues can be heated without risking destruction? It is essential that this temperature be tied to irradiation time. With traditional heating methods, short-term heating conditions are difficult to realize, but laser radiation makes it possible to deposit energy in a small tissue volume that cools rapidly as a result of thermal diffusion. It therefore becomes possible to study the limits of permissible local tissue overheating during time periods from $10^{-3}$ to $10^{-12}$ s for the previously discussed nondestructive photothermotherapy. This obviously will be the subject of future studies.

Also, high-power pulsed laser radiation can help effect the pulsed ablation (expositive evaporation) of the surface of soft and hard biotissues.[53] This process occurs in the pressure confinement region of Figure 5.6, where heating is strong enough and is accompanied by a substantial pulsed pressure rise. The mechanism and conditions of the laser ablation of various materials, including biotissues, are discussed in Reference 54.

The main phenomena experimentally observed to occur in the process of pulsed laser ablation of biological tissues are (1) the presence of a threshold incident pulse energy fluence for ablation, (2) the formation of shock waves, (3) the gasdynamic sputtering of the ablation products with supersonic velocities, and (4) the dependence of the ablation efficiency and threshold energy fluence on the attenuation coefficient of the tissue. These phenomena can, in principle, be explained within the framework of any of the mechanisms mentioned above: photochemical,[53] photothermal,[54] or photomechanical[55] mechanisms. The relationship between the contributions from each of these mechanisms for particular irradiation conditions and types of tissue is presented in References 56 and 57.

With the photochemical pulsed ablation mechanism, the laser energy absorbed by the tissue biomolecules excites electronic states in biopolymers (proteins) that lie above the dissociation and ionization limits of these molecules. The direct dissociation of molecular bonds, or their mediate dissociation through reactions with ions or radicals formed from highly excited electronic states, leads to splitting of long polymer chains into short fragments. The numerous bond breaks cause pressure to rise materially inside the irradiated tissue, which causes the molecular fragments to escape from the tissue. The threshold dissociation and ionization energies $h\omega_{d,i}$ range between approximately 6 and 8 eV. The corresponding laser wavelengths fall within the UV region of the spectrum. Therefore, any appreciable quantum yields of photochemical reactions in proteins and other biopolymers can be expected only at irradiation wavelengths $\lambda$ shorter than 200 nm.[58] The role of photochemical processes in the ablation of biotissues by pulses with a wavelength of $\lambda > 250$ nm can be considered negligible.

The necessary condition for photothermal explosion is to make the tissue substance in the irradiated tissue volume go to an overheated metastable state within a time interval shorter than that required for heat to diffuse from this volume, with the temperature of the tissue rising high enough for an intense formation within the tissue of gas-phase bubbles in a fluctuating manner. Within the framework of this model, the pulsed laser ablation process can be subdivided into three stages: optophysical, thermophysical, and gasdynamic.[59]

During the first stage, the laser radiation is absorbed by the tissue, and subsequently, the excitation energy is converted into heat. Depending on the wavelength of the laser radiation, the absorbing chromophores in the tissue are either biomolecules (mainly proteins) or water. In both cases, water serves as a reservoir for the thermal energy into which the laser-radiation energy is converted. As the irradiated tissue volume is heated so that the tissue temperature rises above the threshold value, the optophysical stage of the ablation processes comes to an end. The second stage, thermophysical ablation, consists in the formation of gas-phase bubbles in the overheated tissue volume. The process is accompanied by the rise of pressure in this volume and the resultant shock wave propagates over the heated tissue layer faster than the metastable overheated liquid phase is formed. The thermophysical stage is followed by the gasdynamic stage. During this last stage of the pulsed laser ablation process, the overheated tissue volume is ejected, and the ablation products undergo gasdynamic sputtering with supersonic velocities.

The experimental magnitudes of the energy fluence threshold for laser ablation are a factor 5 to 12 lower than the energy fluence required for heating the biotissue to the boiling point. This paradox may be explained by the spatially nonuniform heating of the biotissue by the laser pulse. At wavelengths shorter than 1.4 μm, laser radiation can be absorbed by microchromophore centers in the tissue, whose volume is materially smaller than the volume of the total irradiated tissue. The explosive boiling of such overheated microcenters and the ensuing mechanical removal of the entire irradiated tissue bulk sharply reduces the volume energy density $E_{thr}$ necessary for the ablation process to occur.[59]

Another approach to explaining the experimentally observed pulsed laser ablation parameters is by assuming the action of a photomechanical stress mechanism. To understand this mechanism, let us compare the pressures produced in the heated tissue volume by a long and a short laser pulse. We will compare the pulse duration $\tau_p$ with the characteristic hydrodynamic expansion time of the heated tissue.[56] Let the pulse duration $\tau_p$ be long enough in comparison with the time $\tau_{exp}$ so that it is equal to the time it takes for sound to propagate to the depth of light penetration into the tissue (pressure confinement region in Figure 5.6).

In this case, while a given tissue layer is heated by the laser pulse, the increased pressure (compression) pulse resulting from the thermal expansion of the irradiated tissue propagates over the irradiated tissue volume, followed by a negative-pressure depression pulse. Thus, the pressure rise in the irradiated tissue volume due to the gradual heating of the tissue is compensated for by the depression produced behind the sound wavefront. Esenaliev et al.[60] have carried out quantitative studies of the absolute pressure values of the acoustical and shock waves generated and propagating in a biotissue under pulsed ($\tau_p$ = 50 ns) UV ($\lambda$ = 308 nm) laser irradiation (below and above the ablation threshold). These authors found that powerful (several hundred bars of pressure) acoustic compression and depression short pulses were generated in the biotissue.

The spatially nonuniform heating of biotissues by laser radiation gives rise to local overheated microregions in which microexplosions are induced.[57,59] This phenomenon explains quite naturally the presence of vapor and microparticles in the ejected products, as well as the low laser energy fluence marking the ablation threshold. The specific energy deposition at the ablation threshold may be approximately an order of magnitude lower than that necessary to heat the tissue uniformly up to the explosive evaporation temperature.[57] This is evidence that only 10% of the bulk of the tissue suffers evaporation at the ablation threshold; the rest of the ejected mass is in the form of microparticles. The local overheating of microregions may result from the spatial variation of the absorbed energy due to the nonuniformity of the absorptivity of the absorbing medium and the laser energy fluence.[61,62]

To effect explosive ablation of absorbing biotissues, pulsed lasers whose pulse duration is much shorter than the time it takes for heat to be withdrawn from the exposed zone are used. The laser ablation process enables one to perform precision surgical operations only where the necessary light penetration depth is shallow enough. For example, the radiation produced by the ArF excimer laser at a wavelength of 193 nm can penetrate the optically transparent cornea to a depth of only a few microns, because proteins strongly absorb radiation in the region of 200 nm.

However, progress made in the field of ultrashort-pulse lasers, especially solid-state femtosecond lasers, has made it possible to perform precision surgical operations on biotissues featuring a low linear absorption at the laser radiation wavelength. This achievement is possible because of the high intensity of focused laser pulses, which results in the multiple-photon absorption (Figure 5.3) usually observed in the form of "optical breakdown."[63] In this case, plasma generation, shock-wave formation, and disruption take place, followed by the ejection of the heated biotissue region. What is important is that at a high radiation intensity the multiple-photon absorption coefficient is much higher than its linear absorption counterpart, so that the threshold energy fluence here drops to a few fractions of a $J/cm^2$, compared with 10 to 100 $J/cm^2$ in the case of nanosecond-pulse irradiation at the same wavelength.

Thus, moderate requirements for femtosecond lasers are quite acceptable today because they can operate at high pulse repetition frequency and good efficiency. Moreover, the possibility of pumping such lasers by radiation from a laser diode array makes them fairly reliable and suitable for use in clinical practice. One can anticipate considerable progress in this area of precision laser surgery, and not only in corneal corrective surgery.

Apart from direct medical applications, laser surgery can also be made very effective at the cellular level by focusing the laser radiation used (microirradiation).[64] It seems possible to increase the extent of localization of laser radiation so as to concentrate it within a submicron-size region by simultaneously employing near-field optics and femtosecond-pulse techniques. Optical fiber nanotips can be used for this purpose. Two important facts will be utilized here. First, the multiple-photon absorption of radiation in biotissues is much higher than in the transparent material of the optical nanotip, and so the delivery of the necessary laser-pulse energy fluence can be ensured. Second, the distance to which heat can be transferred in the course of a subpicosecond time is no more than 1 nm (Figure 5.6). One can foresee the development along these lines of femtosecond nanolocal "molecular laser surgery."

## 5.7 Conclusion

Concluding this brief analysis of some future trends in laser and photonics biomedicine, which is a supplement to my review of 15 years ago,[65] I clearly see the rapid development and fruitfulness of this field of laser application. In terms of its speed of development and usefulness, this field can probably only be compared with laser fiber science and telecommunications technology. Both of these areas will have a great impact on the further development of human society. Incidentally, when the laser first appeared, neither of these fields was considered a potentially serious application — one more example of how subjective and unreliable all sorts of predictions can be, even in science and technology.

### References


1. Mandelbrot, B., *The Fractal Geometry of Nature*, W.H. Freeman, New York, 1983.
2. Goldman, L., Ed. *The Biomedical Laser*, Springer-Verlag, New York, 1981.
3. Katzir, A., *Lasers and Optical Fibers in Medicine*, Academic Press, San Diego, 1993.
4. Simunovic, Z., Ed., *Lasers in Medicine and Dentistry*, Vitgraf, Rijeka (Croatia), 2000.
5. Waynant, R.W., Ed., *Lasers in Medicine*, CRC Press, Boca Raton, FL, 2001.
6. Shepard, A.P. and Öberg, V.A., Eds., *Laser Doppler Blood Flowmetry*, Kluwer Academic, Boston, 1990.
7. Ashkin, A., Dziedzic, J.M., and Yamane, T., Optical trapping and manipulation of single cells using infrared laser beams, *Nature*, 330, 769, 1987.
8. Chance, B., Ed., *Photon Migration in Tissue*, Plenum Press, New York, 1989.
9. Patterson, M.S., Chance, B., and Wilson, B.C., Time-resolved reflectance and transmitance for the noninvasive measurement of tissue optical properties, *Appl. Opt.*, 28, 2331, 1989.
10. Jacques, S.L., Time resolved reflectance spectroscopy of turbid tissue, *IEEE Trans. Biomed. Eng.*, 36, 1155, 1989.
11. Patterson, M.S., Moulton, J.D., Wilson, B.C., Berndt, K.W., and Lakowicz, J.R., Frequency domain reflectance for the determination of the scattering and absorption properties of tissue, *Appl. Opt.*, 30, 4474, 1991.
12. Arridge, S.R., Cope, M., and Delpy, D.T., The theoretical basis for the determination of optical pathlengths in tissue: temporal and frequency analysis, *Phys. Med. Biol.*, 37, 1531, 1992.
13. Hee, M.R., Izatt, J.A., Jacobson, J.M., Fujimoto, J.G., and Swanson, E.A., Femtosecond transillumination optical coherence tomography, *Opt. Lett.*, 18, 950, 1993.
14. Terney, G.J., Brezinski, M.E., Bouma, B.E., Boppart, S.A., Pitris, C., Southern, J.F., and Fujimoto, J.G., *In vivo* endoscopy optical biopsy with optical coherence tomography, *Science*, 1997, 276, 2037.
15. Huang, D., Swanson, E.A., Lin, C.P., Schuman, J.S., Stinson, W.G., Chang, W., Hee, M.R., Flotte, T., Gregory, K., Puliafito, C.A., and Fujimoto, J.G., Optical coherence tomography, *Science*, 254, 1178, 1991.
16. Svanberg, S., Tissue diagnostics using lasers, in *Lasers in Medicine*, Waynant, R.W., Ed., CRC Press, Boca Raton, FL, 2001, p. 135.

17. Tuchin, V.V., Ed. *Selected Paper on Tissue Optics Applications in Medical Diagnostics and Therapy*, SPIE Milestone Series, Vol. MS102, SPIE Optical Engineering Press, Washington, D.C., 1994.
18. Zharov, V.P. and Letokhov, V.S., *Laser Opto-Acoustical Spectroscopy*, Springer-Verlag, Berlin, 1986.
19. Tam, A.C., Applications of photoacoustic sensing techniques, *Rev. Mod. Phys.*, 58, 381, 1986.
20. Karabutov, A.A., Podymova, N.B., and Letokhov, V.S., Time-resolved laser optoacoustical tomography of inhomogeneous media, *Appl. Phys.*, B63, 545, 1996.
21. Svanberg, S., *Atomic and Molecular Spectroscopy: Basic Aspects and Practical Applications*, 3rd ed., Springer-Verlag, Berlin, 2001.
22. Kosterev, A.A., Curl, R.F., and Tittel, F.K., Chemical sensors using quantum cascade lasers, *Laser Phys.*, 11, 39, 2001.
23. Farst, J., Capasso, F., Sivco, D.L., Sirtori, C., Hutchinson, A.L., and Cho, A.Y., Quantum cascade laser, *Science*, 264, 553, 1994.
24. Alfano, R., Ed., *Biological Events Probed by Ultra-Fast Laser Spectroscopy*, Academic Press, New York, 1982.
25. Letokhov, V.S., Ed., *Laser Picosecond Spectroscopy and Photochemistry of Biomolecules*, Adam Hilger, Bristol, U.K., 1987.
26. Spielmann, C., Curley, P.F., Brabec, T., and Krausz, F., Ultrabroadband femtosecond lasers, *IEEE J. Quantum Electron.*, 30, 1100, 1994.
27. Zewail, A., Femtochemistry: atomic-scale dynamics of the chemical bond, *J. Phys. Chem.*, A104, 5660, 2000.
28. Letokhov, V.S., *Laser Photoionization Spectroscopy*, Academic Press, San Diego, 1987.
29. Weiss, S., Fluorescent spectroscopy of single biomolecules, *Science*, 283, 1676, 1999.
30. Bekov, G.I. and Letokhov, V.S., Laser photoionization spectral analysis, in *Laser Analytical Spectrochemistry*, Letokhov, V.S., Ed., Adam Hilger, Bristol, U.K., 1985, p. 98.
31. Nie, S.M., Chiu, D.T., and Zare, R.N., Probing individual molecules with confocal fluorescence microscopy, *Science*, 266, 1018, 1994.
32. Moerner, W.E. and Oritt, M., Illuminating single molecules in condensed matter, *Science*, 283, 1670, 1999.
33. Denk, W., Strickler, J.H., and Webb, W.W., Two-photon laser scanning fluorescence microscopy, *Science*, 248, 73, 1990.
34. Keller, R.A., Ambrose, W.P., Goodwin, P.M., Jett, J.H., Martin, J.C., and Wu, M., Single-molecule fluorescence analysis in solution, *Appl. Spectrosc.*, 50, 12A, 1996.
35. Wilson, T. and Sheppard, C.J.R., *Theory and Practice of Scanning Optical Microscopy*, Academic Press, London, 1984.
36. Basche, Th., Moerner, W.E., Orrit, M., and Wild, U.P., Eds., *Single-Molecule Optical Detection, Imaging and Spectroscopy*, VCH Publishers, Weinheim, Germany, 1996.
37. Mertz, J., Xu, C., and Webb, W.W., Single-molecule detection by two-photon-excited fluorescence, *Opt. Lett.*, 20, 2532, 1995.
38. Sonnleitner, M., Schültz, G.J., and Schmidt, Th., Imaging individual molecules by two-photon excitation, *Chem. Phys. Lett.*, 300, 221, 1999.
39. Betzig, E. and Chichester, R.J., Single molecule observed by near-field scanning optical microscopy, *Science*, 262, 1422, 1993.
40. Ohtsu, M., Ed., *Near-Field Nano/Atom Optics and Technology*, Springer-Verlag, Berlin, 1998.
41. Gimzewski, J.K. and Joachin, C., Nanoscale science of single molecules using local probes, *Science*, 283, 1683, 1999.
42. Letokhov, V.S., The possibility of ESCA microscopy with laser femtosecond EUV x-ray pulses, *JETP Lett.*, 74, 464, 2001.
43. Rundquist, A., Durfee, C.G., III, Chang, Z., Herke, C., Backus, S., Murnane, M.M., and Kapteyn, H.C., Phase-matched generation of coherent soft x-rays, *Science*, 280, 1412, 1998.
44. Schnürer, M., Cheng, Z., Hentschel, M., Krausz, F., Wilhein, T., Hambach, D., Schmahl, G., Drescher, M., Lim, Y., and Heinzmann, H., Few-cycle-driven XUV laser harmonics: generation and focusing, *Appl. Phys.*, B70, S227, 2000.

45. Dainty, J.C., Ed., *Laser Speckle and Related Phenomena*, Springer-Verlag, Berlin, 1984.
46. Karu, T., *The Science of Low Power Laser Therapy*, Gordon & Breach, London, 1998.
47. Turner, J. and Hode, L., *Low Level Laser Therapy — Clinical Practice and Scientific Background*, Prima Books, Stockholm, 1999.
48. Hochstrasser, R.M., Biological applications of ultrafast laser methods, *Ber. Bunsenges. Phys. Chem.*, 93, 239, 1989.
49. Letokhov, V.S., Effects of transient local heating of spatially and spectrally heterogeneous biotissue by short laser pulses, *Nuov. Cim.*, 13D, 939, 1991.
50. Jori, G. and Spikes, J.D., Photothermal sensitizers: possible use in tumor therapy, *Photochem. Photobiol. B: Biol.*, 6, 93, 1990.
51. Anderson, R.R. and Parrish, J.A., Selective photothermolysis: precise microsurgery by selective absorption of pulsed radiation, *Science*, 220, 524, 1983.
52. Hansen, W.P. and Fine, S., Melanin granule models for pulsed laser induced retinal injury, *Appl. Opt.*, 7, 155, 1968.
53. Srinivasan, R., Ablation of polymers and biological tissue, *Science*, 234, 559, 1986.
54. Oraevsky, A.A. and Letokhov, V.S., Pulsed laser ablation of athero-sclerotic plaque in blood vessels, in *The Practise of Interventional Cardiology*, Vogel, J.H.K. and King, S.B., III, Eds., Mosby, St. Louis, 1992, p. 203.
55. Dingus, R.S. and Scammon, R.J., Grüneisen-stress induced ablation of biological tissue, *Proc. SPIE*, 1427, 45, 1992.
56. Miller, J.C. and Haglund, R.F., Jr., Eds., *Laser Ablation: Mechanisms and Applications*, Springer-Verlag, Berlin, 1991.
57. Oraevsky, A.A., Esenaliev, R.O., and Letokhov, V.S., Pulsed laser ablation of biological tissue: review of mechanisms, in *Laser Ablation: Mechanisms and Applications*, Miller, J.C. and Haglund, R.F., Eds., Springer-Verlag, Berlin, 1991.
58. Petit, G.H., The physics of ultraviolet laser ablation, in *Lasers in Medicine*, Waynant, R.W., Ed., CRC Press, Boca Raton, FL, 2001, p. 109.
59. Esenaliev, R.O., Oraevsky, A.A., and Letokhov, V.S., Laser ablation of atherosclerotic blood vessel tissue under various irradiation conditions, *IEEE Biomed. Eng.*, 36, 1188, 1989.
60. Esenaliev, R.O., Oraevsky, A.A., Letokhov, V.S., Karabutov, A.A., and Malinsky, T.V., Sutides of acoustical and shock waves in the pulsed laser ablation of biotissue, *Lasers Surg. Med.*, 13, 470, 1993.
61. Golovlev, V.V., Esenaliev, R.O., and Letokhov, V.S., Ablation of an optically homogeneous absorbing medium by scattered pulsed laser radiation, *Appl. Phys.*, B57, 451, 1993.
62. Esenaliev, R.O., Karabutov, A.A., Podymova, N.B., and Letokhov, V.S., Laser ablation of aqueous solutions with spatially homogeneous and heterogeneous absorption, *Appl. Phys.*, B59, 73, 1994.
63. Kuper, S. and Stuke, M., Femtosecond UV excimer laser ablation, *Appl. Phys.*, B44, 199, 1987.
64. Berns, M.W., *Biological Microirradiation*, Prentice-Hall, London, 1974.
65. Letokhov, V.S., Laser biology and medicine, *Nature*, 316, 325, 1985.

# 6

# Basic Instrumentation in Photonics




Tuan Vo-Dinh

*Oak Ridge National Laboratory*
*Oak Ridge, Tennessee*


This chapter presents an overview of the basic instrumentation used in photonic applications. The chapter is intended to provide an introduction to the basic setups, apparatus, and system components for readers from other research fields who wish to become further acquainted with biomedical photonics. Only basic devices and components are described in this chapter. More advanced instrumentation and specialized systems for specific applications are described in detail in Chapters 7 through 9 of this handbook.

## 6.1 Basic Spectrometer

The detection and analysis of the optical signal require a spectrophotometer. This instrument is commercially available from many manufacturers, or it may be built using standard components. For routine analytical work, it is more convenient to use an available commercial instrument. For research investigations and special applications, it is sometimes necessary to build an apparatus to specific performance requirements.

### 6.1.1 Basic Apparatus

This section describes how various basic components can be combined to develop instrumental setups for different types of spectroscopic measurements. The common feature of spectroscopic measurements is that that they all measure spectroscopic properties related to the molecular composition and structure of biochemical species in the sample of interest. There are several types of spectroscopic measurements: absorption, scattering (elastic and inelastic), and emission. A typical spectroscopic experiment that allows us to analyze complex biological systems is conceptually simple. Light at a certain wavelength $\lambda$ (or frequency $\nu = c/\lambda$) is used to irradiate a sample of interest. This process is called "excitation." Then some of the properties of the light that emerges from the sample are measured and analyzed.

Some properties deal with the fraction of the incident radiation absorbed by the sample; the techniques involved are called absorption spectroscopy, e.g., ultraviolet (UV), visible, and infrared (IR) absorption

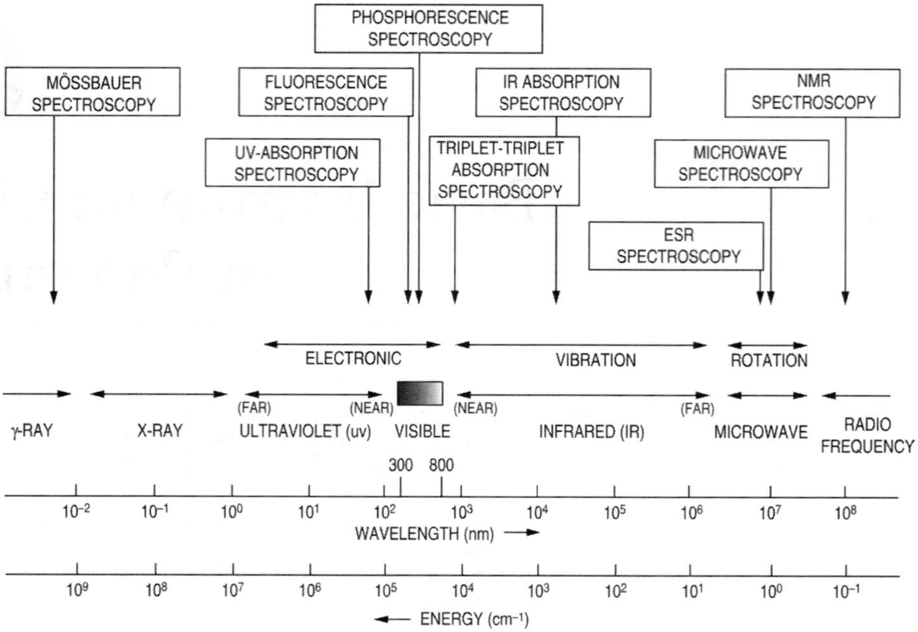


**FIGURE 6.1** Various types of spectroscopies and related spectral ranges in the electromagnetic spectrum.

techniques. Other properties are related to the incident radiation dissipated and reflected back from the samples (elastic scattering techniques). Alternatively, one can measure the light emitted and scattered by the sample, processes that occur at wavelengths different from the excitation wavelength; the techniques involved are fluorescence, phosphorescence, and inelastic scattering (Raman scattering). Other specialized techniques can be used to detect specific properties of the emitted light, such as its degree of polarization and decay times.

The range of wavelengths used in various types of molecular spectroscopy to study biological molecules is quite extensive (Figure 6.1). Table 6.1 summarizes the different types of molecular spectroscopy and the associated wavelength ranges of the electromagnetic radiation.

A basic spectrophotometer generally consists of the following components:

1. An excitation light source
2. Dispersive devices (optical filters, monochromators, or polychromators)
3. A sample (usually in a compartment with a sample holder)
4. A photometric detector (equipped with a read-out device)

Successful application of photonic methods requires considerable attention to experimental details and a good understanding of the instrumentation. The recorded spectra (absorption, emission, or excitation) represent the photon emission rate or power recorded at each wavelength, over a wavelength interval determined by the slit widths, and dispersion of the monochromator. There are many manufacturers of spectrometers, each offering several models with different performance characteristics and different options. Basic instrument components are also commercially available. An investigator may assemble off-the-shelf components for his or her particular applications. The basic components can be adapted to design the instrument for each type of spectroscopic measurement.

## 6.1.2 Instrument for Absorption Measurements

Figure 6.2A shows a schematic arrangement of a typical instrument setup for absorption measurements. The collimated output of a light source is focused on the entrance slit of an excitation monochromator

**TABLE 6.1** Spectroscopic Techniques for Biological and Biomedical Applications

| Spectral Region* | Wavelength* (cm) | Energy* (kcal mol$^{-1}$) | Techniques | Properties |
|---|---|---|---|---|
| γ-Ray | $10^{-11}$ | $3 \times 10^{8}$ | Mössbauer | Nucleus properties |
| X-ray | $10^{-8}$ | $3 \times 10^{5}$ | X-ray diffraction/scattering | Molecular structure |
| UV | $10^{-5}$ | $3 \times 10^{2}$ | UV absorption | Electronic states |
| Visible | $6 \times 10^{-5}$ | $5 \times 10^{3}$ | Visible absorption | Electronic states |
| | | | Luminescence | Electronic states |
| Infrared | $10^{-3}$ | $3 \times 10^{0}$ | IR absorption | Molecular vibrations |
| | | | IR emission | Electronic states |
| Microwave | $10^{-1}$ | $3 \times 10^{-2}$ | Microwave | Rotations of molecules |
| | $10^{0}$ | $10^{-3}$ | Electron paramagnetic resonance | Nuclear spin |
| Radio-frequency | 10 | $3 \times 10^{-4}$ | Nuclear magnetic resonance | Nuclear spin |

*Approximate values

for wavelength scanning. The output of the excitation monochromator is directed to the sample inside the sample compartment. The light transmitted by the sample is collected through appropriate optics and focused onto a detector. This simple instrumental setup is often used in a single-beam absorption spectrometer. Double-beam instruments include a reference beam, which is used to correct spectral fluctuations in the lamp automatically in order to reduce electronic drift and lamp warm-up periods.

### 6.1.3 Instrument for Scattering Measurements

The elastic scattering (ES) technique involves detection of the backscattering of a broadband light source irradiating the sample of interest. Figure 6.2B shows a typical instrument setup that can be used for elastic scattering measurements. A spectrometer records the backscattered light at various wavelengths and produces a spectrum that is dependent on sample structure, as well as chromophore constituents. In general, the sample is illuminated with the excitation light, which is selected with a dispersive element and then directed to a specific point location (e.g., via an optical fiber) of the sample. The scattered light is measured at the same wavelength as the excitation wavelength.

With inelastic scattering measurements, one measures the scattered light from the sample in a spectral region different from the excitation wavelength. In this case, the basic setup is similar to the ES setup but has an additional dispersive element to analyze the scattering emission from the samples (Figure 6.2C).

### 6.1.4 Instrument for Emission Measurements

Figure 6.2D shows a schematic arrangement of a typical spectrometer for emission (e.g., fluorescence, phosphorescence) measurements. The excitation light source is usually a laser or high-intensity xenon arc lamp. The collimated output of the light source is focused on the entrance slit of an excitation monochromator whose output is directed to the sample. When a laser is used as the excitation source, the excitation monochromator is not required. The emission from the sample is collected through appropriate optics and focused onto the entrance slit of an emission monochromator. The excitation beam and the emission beam are usually focused at right angles for minimum interference from scattered light.

There are three basic classes of spectrophotometers: filter instruments, monochromator instruments, and multichannel devices. The first type of device uses optical filters, whereas the latter two systems use prisms or gratings as dispersive elements. The more expensive grating spectrometers are more versatile than filter instruments and can be used for either applied or basic studies.

Selection of an instrument appropriate to specific needs requires careful examination of many factors, including cost. The three major features to consider are the intensity of the excitation light source, the resolution and throughput of the monochromators, and the sensitivity of the detector. Although it is desirable to have an intense excitation light source, the cost-benefit ratio should be taken into consideration.

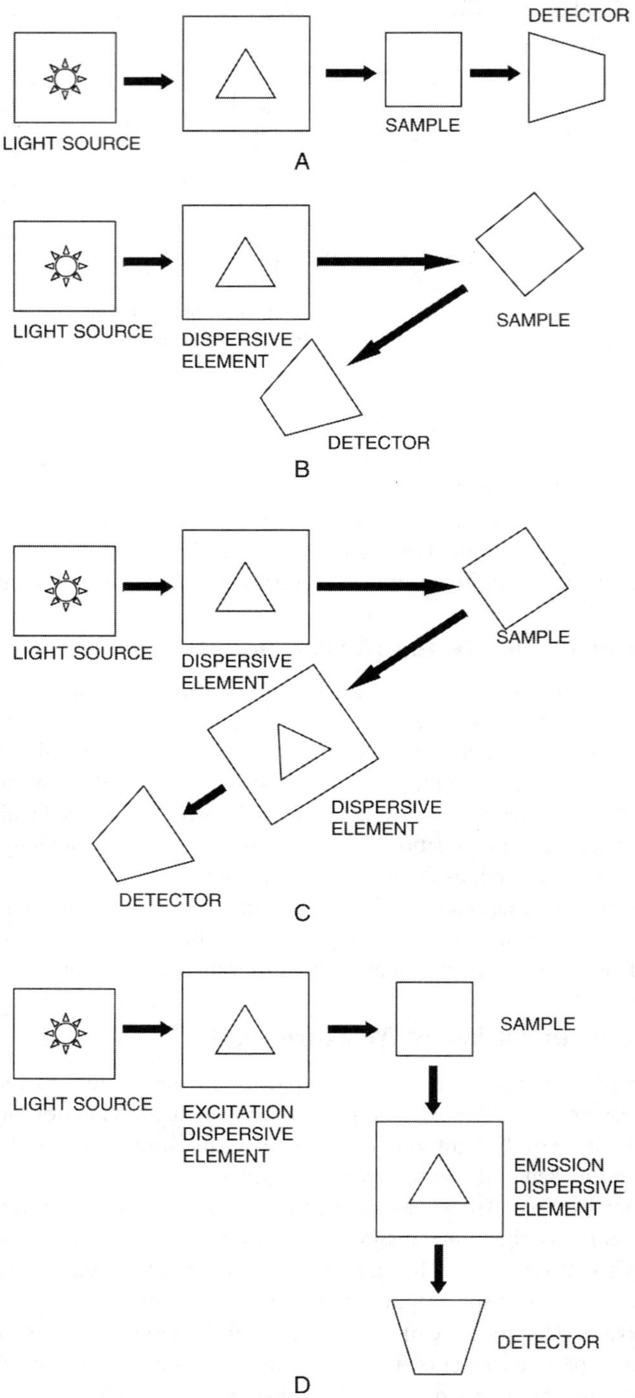


**FIGURE 6.2** Various instrumental setups: (A) absorption; (B) elastic scattering; (C) inelastic scattering; (D) emission.

For most fluorescence measurements at room temperatures, the spatial resolution need not be excessively high because the bandwidths of fluorescence spectra are usually larger than 5 nm. A monochromator with low spectral dispersion, however, would allow use of larger slit widths and consequently provide for higher radiance throughput. On the other hand, IR absorption and Raman instruments require very high spectral resolution (1 nm or less) in order to resolve the much narrower vibrational bands spectrally.

The spectrometer used for phosphorescence measurements is essentially the same as that used for fluorescence measurements. The only additional equipment required is a phosphoroscope attachment for isolating the long-lived phosphorescence from shorter-lived emissions (fluorescence, excitation scattered light). Frequently, a fluorescence spectrometer (fluorimeter) can be modified easily or equipped with a phosphoroscope to perform measurements.

The sensitivity of the spectrometer is strongly dependent on the choice of the detector, such as the photomultiplier (PM) tube or the charge-coupled device (CCD). The relative sensitivity of the PM tube or CCD can be found in data sheets from the manufacturer.

In addition to these factors, the choice of a spectrometer also depends on other features that are more difficult to characterize, such as the reliability of electronic components, the repeatability and accuracy of scanning mechanisms, and the stray-light rejection ability of the monochromators.

A cost-benefit evaluation must be made in the choice between a single-beam and a double-beam spectrometer. In the more expensive double-beam instruments, spectral fluctuations in lamp output are automatically corrected, reducing electronic drift and lamp warm-up periods. However, the performance of a single-beam instrument is adequate for most applications.

## 6.2 Instrumental Components: General Considerations

### 6.2.1 Excitation Light Sources

UV light is generally used for excitation in many spectroscopic measurements. These UV sources may be classified into two categories, namely, line or continuum type, and can be used in a continuous wave (CW) mode or in a pulsed mode. The line sources provide sharp spectral lines, whereas continuum sources exhibit a broadband emission.

#### 6.2.1.1 High-Pressure Arc Lamps

High-pressure arc lamps are the most commonly used radiation sources. These lamps produce an intense quasi-continuum radiation ranging from the UV (<200 nm) to the near-infrared (NIR) (>1000 nm) with only a few broad bands at approximately 450 to 500 nm. The lamps consist of two tungsten electrodes in a quartz envelope containing gases under high pressure, for example, xenon (Xe), mercury, an Xe–mercury mixture. Lamps of this type are commercially available in a wide range of input power from a few watts to several kilowatts.

The high-pressure mercury arc lamp is similar to the high-pressure Xe lamp in appearance and performance. The spectral output of mercury lamps is of a line type, whereas that of the Xe lamps is of a continuum type. If excitation can be carried out at only one wavelength or a few fixed wavelengths of the mercury emission lines, the mercury lamp is probably the most effective radiation source. The Xe lamp, however, is more commonly used because it provides a smoother spectral profile more suitable for conducting excitation spectra measurements.

The Xe arc lamp is the most versatile light source for steady-state spectrometers and has found widespread use. This lamp provides a relatively continuous light output from 250 to 700 nm. Xe arc lamps emit a continuum of light as a result of the recombination of electrons with the ionized Xe atoms. Complete separation of the electrons from the atoms yields the continuous emission. These lamps are available with an ellipsoidal reflector as part of the lamp. Parabolic reflectors in some commercially available Xe lamps collect a large solid angle of light and provide a collimated output.

The operation of high-pressure arc lamps requires special care and handling, such as reduction of the excitation stray light with a good monochromator, use of a highly regulated DC power supply, and removal of the heat generated by the lamp output in the IR range. A warm-up period is also necessary to minimize arc wandering, because of a tendency for the arc to change its location inside the lamp envelope during the first half hour of operation. This arc wandering effect may cause sudden variations in the observed intensity, especially when the image of the arc is focused into a small slit aperture.

Extreme care should be exercised when inspecting high-pressure arc lamps. They may explode when dropped or bumped because they are filled with gases at high pressures (~5 atm at ambient temperature and 20 to 30 atm at operating temperature conditions). It is recommended that special leather gloves, safety glasses, and protective headgear be used whenever the lamp housing is opened. One should not look directly at an operating Xe lamp; the extreme brightness will damage the retina, and the UV light can damage the cornea. With some older lamps, proper ventilation or use of deozonators is required to remove the ozone produced by the UV radiation of the lamp. Nowadays, many available Xe lamps are considered ozone-free because their operation does not generate ozone in the surrounding environment.

#### 6.2.1.2 Low-Pressure Vapor Lamps

Low- (or medium-) pressure mercury vapor lamps are often used as line sources. They are simple to use, require little power, and offer intense UV radiation concentrated in a few lines (e.g., 253.7 nm, 365.0/265.5/366.3 nm multiplet). The mercury vapor lamps are widely used in simple filter-type spectrometers because of their low cost, intense emission characteristics, and good stability. The lamps do not need a complex power supply system and provide excellent reference light sources for calibration of spectrometers.

#### 6.2.1.3 Incandescent Lamps

The tungsten filament incandescent lamp is the simplest continuum source. This type of incandescent lamp exhibits a smooth, continuous spectral profile determined by the blackbody radiation characteristics given by Planck's equation:

$$S_\lambda = \frac{5.8967\lambda^{-5} E_\lambda}{\exp(14388/\lambda T)^{-1}}$$

where $S_\lambda$ is the spectral radiance ($W - \text{cm}^{-2} - \text{sr}^{-1} - \text{nm}^{-1}$), $\lambda$ is the wavelength (nm), $T$ is the temperature (K), and $E_\lambda$ is the spectral emissivity of the filament material (dimensionless).

Since incandescent lamps usually have low UV output, they are seldom used as excitation sources, especially for luminescence measurements where samples absorb the UV. Their smooth spectral profile, however, makes them very suitable for intensity calibration procedures.

Standard incandescent lamps with calibration data provided by the National Institute of Standards and Technology (NIST) are readily available commercially. Intensity calibration data are available for the spectral range from 250 nm to 2.5 μm.

#### 6.2.1.4 Solid-State Light Sources

Light-emitting diodes (LEDs) are solid-state light sources that provide output over a wide range of wavelengths. These devices require little power and generate little heat. One can use a few LEDs to cover a spectral range from 400 to 700 nm. LEDs are practical light sources for many low-power photonic applications and can be amplitude-modulated up to hundreds of megahertz. Another type of solid-state light source is the solid-state laser, described in the next section.

#### 6.2.1.5 Lasers

Although conventional light sources have primarily been used for absorption analyses, lasers are increasingly used in luminescence and Raman measurements. The advantages offered by lasers as excitation sources include the following:

1. Monochromaticity
2. High degree of collimation

3. High intensity
4. Phase coherence
5. Short pulse duration (with pulsed lasers)
6. Polarized radiation

Selection of laser excitation sources is determined by the wavelengths that can be matched to the absorption band of the compounds to be analyzed in order to take advantage of maximum absorption. If time-resolved measurements are performed, the pulse width of the laser is an important factor to consider. The intensity of a laser is very high at (or even near) the laser emission lines and, therefore, often interferes with the lower intensity of the emission or scattering signal being measured. A number of devices, such as a spike filter or a single monochromator, may be used to reject the Rayleigh scattered light. Notch filters, which consist of crystalline arrays of polystyrene spheres, exhibit very high efficiency rejection of laser lines.

Many publications have described the principle and applications of lasers in detail.[1–6] Some general properties of various types of lasers are detailed in the following section.

#### 6.2.1.5.1 *General Properties of Lasers*

The laser process is a stimulated emission following population inversion between different electronic levels of an active medium. Laser operation occurs inside a resonant optical cavity filled with an active medium that can be a gas, solid crystal, semiconductor, or dye solution.

Some lasers are operated in the CW mode; others are pulsed. There are several methods for producing very short pulses. The simplest is to induce a population inversion for only short periods of time using pulsed electrical discharges. Other pulsing methods include Q-switching, cavity dumping, and mode locking. The Q-switching method consists of first decreasing the reflectivity or the quality, Q, of the laser cavity and then suddenly switching the Q value back to a relatively high value. Active Q-switching can be done by shutter devices, such as an electro-optic Kerr cell,[6] a Pockels cell (a device that provides a modulation linearly proportional to the applied electric field), or simply a rotating mirror.

Passive Q-switching uses a saturated absorber such as gas or a dye. Cavity dumping is a pulse-forming technique for laser materials having excited-state lifetimes too short for Q-switching operation. Another method that produces subnanosecond pulses is the mode-locking technique. This technique involves locking the phase of the electromagnetic longitudinal modes inside the cavity. The repetition rate of mode-locked pulses, determined by the speed of light and twice the length of the optical cavity, is too high for phosphorescence measurements. For a 1-m cavity, the time interval between two pulses is 6.6 ns. Even with a cavity design based on a folded-path mirror with a total length up to $10^3$ m, the laser still produces pulses at 5.5-μs intervals.

#### 6.2.1.5.2 *Gas Lasers*

The active medium of a gas laser is either an atomic (e.g., helium–neon), molecular (e.g., carbon dioxide, hydrogen cyanide, rare gas halide, water vapor), or ionized (e.g., argon, krypton, xenon, helium–cadmium, helium–selenium) gas. Output powers of CW gas lasers range from milliwatts to kilowatts. Peak powers of pulsed gas lasers can reach several megawatts.

A compact, low-power gas laser often used for alignment and for calibration is the helium–neon laser. The output of this laser ranges from less than a milliwatt to several tens of milliwatts at 632.8 nm. Certain lasers also emit at 1152 nm.

Two commonly used gas lasers are the argon ion laser and the krypton ion laser. These rare gas lasers emit a series of lines from the near-UV to the IR region and have CW output ranging from tens of milliwatts to tens of watts. An ion laser that uses a metal vapor as the active medium is the helium–cadmium laser. It provides a CW output in the milliwatt range in the blue or UV regions. A CW gas laser with high power (up to several hundred watts) at 10 μm is the carbon dioxide laser.

A very practical and widely used laser is the nitrogen laser. It operates only in the pulsed mode. This laser produces subnanosecond to tens-of-nanosecond pulses in the near-UV region (337 nm). The energy per pulse ranges from a few microjoules to a few joules.

**TABLE 6.2** Types and Spectral Characteristics of CW and Pulsed Commercial Gas Lasers

| Active Medium | Output Wavelength (nm) | CW Gas Laser | Pulsed Gas Laser |
|---|---|---|---|
| Argon | 330–514 | X | X |
| Argon fluoride | 193 | | X |
| Argon–krypton | 450–1,090 | X | |
| Barium | 1,500 | | X |
| Carbon dioxide | 9,000–11,000 | X | X |
| Carbon monoxide | 5,000–7,000 | X | X |
| Copper | 510.5 – 578.2 | | X |
| Deuterium cyanide | $190 \times 10^3 - 195 \times 10^3$ | X | |
| Deuterium fluoride | 3,600–4,000 | X | X |
| Fluoride | 624–780 | | X |
| Gold | 627.8 | | X |
| Helium–cadmium | 325 – 442 – 441 | X | |
| Helium–neon | 632.8 – 1,152 – 3,391 | X | |
| Helium–selenium | 460–653 | X | |
| Helium–xenon | 3,805 | X | X |
| Hydrogen cyanide | $311 \times 10^3 - 337 \times 10^3$ | X | X |
| Hydrogen fluoride | 2,600–3,000 | X | X |
| Krypton | 330–790 | X | X |
| Lead | | | X |
| Methanol | $118.8 \times 10^3$ | X | X |
| Methyl fluoride | $190 \times 10^3 - 596 \times 10^3$ | X | X |
| Nitrogen | 337 | | X |
| Water | $(18 - 78 - 118) \times 10^3$ | X | X |
| Xenon bromide | 282 | | X |
| Xenon chloride | 318 | | X |
| Xenon fluoride | 351 – 353 – 354 | | X |

Pulsed output in the UV region is also obtained with excimer lasers, in which the active medium is an excimer, a molecule that is stable only in the excited state and dissociates immediately after it emits light (e.g., krypton fluoride, xenon fluoride, xenon chloride, or argon fluoride).

In chemical lasers, the active medium, such as hydrogen fluoride or deuterium fluoride, is produced by a chemical reaction induced by electrical discharges. Table 6.2 summarizes various characteristics of several CW and pulsed gas lasers.

#### 6.2.1.5.3 *Solid-State Lasers*

Solid-state lasers can be the pulsed type, such as the chromium-doped ruby laser with principal output at 694 nm, or the CW type, such as the YAG (yttrium aluminum garnet) laser, with principal output at 1064 nm. Solid-state lasers are pumped optically. Optical pumping of pulsed lasers is usually achieved with an Xe flash tube. Pulsed outputs range from tenths of a joule to tens of joules. The repetition rate of these pulsed lasers is relatively low (0.1 Hz to tens of hertz). The CW YAG laser is generally used in the pulsed mode following mode-locking, cavity-dumping, or Q-switching operations. The CW output of YAG lasers is between a tenth of a watt and several watts.

Other solid-state lasers include glass lasers that operate only in a pulsed mode like the ruby lasers, and the F-center lasers, which provide a tunable CW output in the 2000- to 3000-nm spectral range.

#### 6.2.1.5.4 *Semiconductor Lasers*

One important family of semiconductor lasers is diode lasers. In a semiconductor diode laser, the front and back surfaces of the diode form the resonant cavity that generates light emission when electrons and holes recombine at the *p–n* junction. When driven by a low current, the diode emits incoherent light. Above an input current threshold, the diode begins to emit coherent laser light. The small size of semiconductor lasers allows efficient packaging. They can be operated on battery power and therefore are useful for remote operation. High-power diode lasers are generally not tunable; however, many wavelengths of interest are available for medical applications. One type of diode laser has a tunable output: the lead salt type. The other type has a fixed-wavelength output: the gallium arsenide (GaAs) or gallium-aluminum (GaAl) type. The lead salt lasers emit at 2000 to 3000 nm and are used for high-resolution spectroscopy in the IR region. GaAs and GaAl lasers are usually employed in communications utilizing fiber-optic cables or in information processing. Diode lasers can be amplitude modulated to several gigahertz. Titanium-sapphire lasers, which typically produce 100-fs pulses, are often used for ultrafast measurements or multiphoton excitation.

#### *6.2.1.5.5 Tunable Dye Lasers*

A dye laser consists of an organic dye solution optically pumped by a light source (flash lamp, nitrogen laser, or ion laser). Dye lasers offer many properties that make them close to ideal light sources. Dye lasers can be tuned across a large spectral range from the UV to the NIR with a series of different dyes. The type of output desired and the absorption properties of the dye determine the choice of the pump source. Pulsed sources usually provide peak powers at levels sufficiently high for laser action in most dyes when the energy per pulse is low.

#### *6.2.1.5.6 Tunable Lasers with Optical Parametric Oscillators*

Optical parametric oscillators (OPO) have extended the laser radiation from visible to the IR and have opened a new area for spectroscopists. With OPO systems, no changing of dye solutions is needed. In a process similar to harmonic generation, the operation of an OPO is based on the nonlinear response of a medium to a driving field (the pump laser beam) to convert photons of one wavelength to photons of other, longer wavelengths. Specifically, in the so-called parametric process, a nonlinear medium (usually a crystal) converts the high-energy photon (the pump wave) into two lower-energy photons (the signal and idler waves).

The exact wavelengths of the signal and idler are determined by the angle of the pump wave vector relative to the crystal axis. Energy can be transferred efficiently to the parametric waves if all three waves (pump, signal, idler) are traveling at the same velocity. Under most circumstances, the variation of index of refraction with crystal angle and wavelength allows this "phase matching" condition to be met only for a single set of wavelengths for a given crystal angle and pump wavelength. Thus as the crystal rotates, different wavelengths of light are produced.[7]

When the crystal is contained in a resonant cavity, feedback causes gain in the parametric waves in a process similar to buildup in a laser cavity. Thus, light output at the resonated wavelength (and the simultaneously produced other parametric wavelength) occurs. The cavity can be singly resonant at either the signal or idler wavelength, or it can be doubly resonant at both wavelengths.

## 6.2.2 Optical Fibers and Dispersive Devices

### 6.2.2.1 Optical Filters

The simplest dispersive element is the optical filter, which can be used when variation of the excitation or emission wavelength is not needed. In this case the filter is simply a single-wavelength selector device. Filters may be used to select the excitation light (excitation filters) or remove the scattered excitation light from the emission (emission filters). It is often desirable to use optical filters in addition to monochromators. In general, excitation filters with a narrow spectral bandpass are used to provide selectivity and should be able to withstand high-intensity light and, in some cases, high temperatures. Conversely, emission filters usually have a broad bandpass and are generally employed to provide maximum light throughput.

Filters may be classified into three main categories: neutral, cutoff, and bandpass. Neutral density filters, which have a nearly constant transmission over a wide spectral range, are generally used to attenuate the light equally at all wavelengths. For example, one can use neutral density filters to strongly decrease luminescence signals and intense excitation light, or to adjust or match the intensity of two signals.

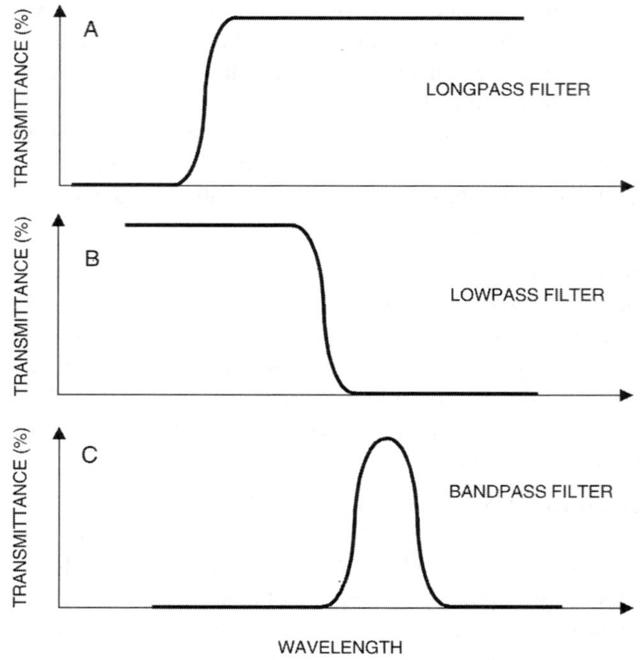


**FIGURE 6.3** Schematic diagram of transmission curves of various types of filters: (A) longpass filter; (B) lowpass filter; (C) bandpass filter and notch filter.

Cutoff filters, which have a sharp transmission cutoff, are generally used to remove undesired radiation, such as stray light or second-order diffraction light. Longpass (or highpass) filters are often used to reject scattered light from the excitation and to transmit the emission of interest (e.g., fluorescence). Similarly, shortpass (or lowpass) filters are used to transmit light at the shorter wavelengths and remove light occurring at wavelengths longer than the spectral range of interest. Notch filters are used to remove light within a specific bandwidth of interest. They are often used to remove the interfering laser excitation light from the emission light (e.g., Raman or fluorescence).

Bandpass filters generally serve to isolate a particular spectral region of interest for excitation or detection. Great care should be taken in the selection of bandpass filters to avoid the possibility of emission from the filter itself. When illuminated with UV light, some filters exhibit luminescence that can interfere with the sample emission. For this reason, it is usually desirable to position the filter farther away from the sample, rather than close to it. Figure 6.3 shows schematic diagrams of transmission curves for various types of filters: longpass (A), shortpass (B), and bandpass filters (C). Figure 6.4 depicts typical transmission curves of holographic and dielectric notch filters.

Filters can also be classified according to their operating principle: absorption, interference, or birefringence. Absorption filters include neutral density, cutoff, and wide bandpass filters. They may be fashioned from a variety of materials, including colored glass, gelatin, chemical solution, gas, sapphire, and alkali–halide materials. Colored glass filters are the most practical and least expensive. The spectral characteristics of the coloring inorganic species determine the transmission properties of the filters. Gelatin containing a dye is also used in absorption filters. Gelatin filters are thermally less stable than glass filters and should be used only as emission filters. Optical cells filled with a specific chemical solution or gas may be used as filters in special cases, but they are not convenient for general use.

Interference filters are generally of the narrow bandpass (10 nm) or broadband (50 to 80 nm) type. Carefully controlled thicknesses of vacuum-deposited layers of dielectric and metallic layers are arranged so that light outside the spectral bandpass is eliminated by destructive interference, whereas light within the bandpass is transmitted by constructive interference. Some interference filters have linearly variable

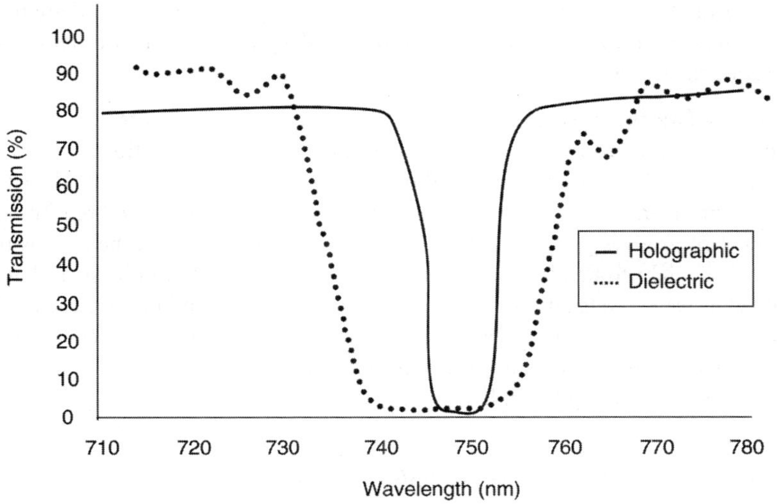


**FIGURE 6.4** Typical transmission curves of holographic and dielectric notch filters.

transmittance across the substrate; because these filters are angle sensitive, great care should be taken when mounting them in an optical system. Interference filters, which transmit typically 30% of the intensity in the UV and 60% in the visible range, are generally used as excitation filters because of their narrow bandpass (from 50 to 1 nm). Heat from the excitation light beam is usually not a problem because radiation that is not transmitted is rejected by reflection. Interference filters covering a broad spectral range from the UV to the IR are commercially available from many suppliers.

Dichroic filters are substrates coated with thin films to produce the desired transmission and reflection properties across the spectral range of interest. They are often used as color filters (both additive and subtractive) or as devices designed to transmit light at certain wavelengths and reflect light at other wavelength ranges. Dichroic filters are angle sensitive, so care should be taken to mount them properly in optical systems.

Birefringence filters, which are constructed with polarizers and retardation plates, may be used in polarization measurements. These filters are generally made of stretched polymer films, the absorption properties of which determine the characteristics of the filters. They usually have a bandwidth similar to that of interference filters.

### 6.2.2.2 Monochromators

Continuous wavelength selection is performed with monochromators, used to disperse polychromatic or white light into the various colors. The performance specifications of a monochromator are characterized by the spectral dispersion, the efficiency, and the stray light levels. The spectral dispersion is usually expressed in nanometers per millimeter; the slit width is expressed in millimeters. Low stray light and high efficiency are desired qualities in selecting a monochromator. The two types of monochromators are prism and grating monochromators.

#### *6.2.2.2.1 Prism Monochromators*

In prism monochromators, light dispersion is due to the change of the refractive index of the prism material with the wavelength of the incident light. The angular dispersion $D$ is given by

$$D = \frac{d\theta}{d\lambda} = \frac{dn}{d\lambda}\frac{d\theta}{dn}, \tag{6.1}$$

where $\theta$ is the angular deviation, $n$ is the refractive index of the prism materials, and $\lambda$ is the wavelength of the light source.

Prism monochromators usually produce less stray light than grating devices and are free from overlap from multiple orders, but they are less convenient to use than grating monochromators because of their nonlinear scanning dispersion.

#### 6.2.2.2.2 Grating Monochromators

Most spectrometers are now equipped with monochromators having diffraction gratings. Gratings comprise a large number of lines, or grooves, ruled on a highly polished surface. The density and shape of these grooves determine the characteristics of a grating. Energy throughput and resolution increase with increasing number of grooves per millimeter. The width of a groove should be approximately equal to the wavelength of the light to be dispersed. The shape of the groove should be such that the maximum amount of light at a given wavelength is concentrated at only one specific angle for each order. The design and construction of the grating also determine the other properties of the monochromator, for example, reflectivity (or radiance throughout) and stray light rejection.

The general diffraction grating formula is given by

$$\sin\theta + \sin\theta' = k\frac{\lambda}{d} = kn\lambda , \tag{6.2}$$

where $\theta$ is the angle of incidence, $n$ is the number of grooves per unit length, $d$ is the groove spacing, and $k$ is the dispersion order (see Figure 6.5A).

As shown in this formula, the grating disperses light because $\theta'$ is dependent on $\lambda$. A spectrum is obtained for each value of $k$. As a consequence, gratings also produce second- and higher-order dispersions that may overlap the lower-order emission of interest. One simple procedure to overcome this problem is to place a cutoff filter at the exit slit of the grating monochromator so that only one order is transmitted. The first-order dispersion ($k = 1$), which produces the highest throughput, is generally used for emission measurements. Higher orders, however, result in higher resolution. At the zero order ($k = 0$), there is no dispersion because all wavelengths are reflected at the same angle.

Most gratings used in modern spectrometers are of the reflection type ($\theta = \theta'$). In this case, the observation is in the direction of illumination (Littrow configuration). The grating formula then becomes

$$\sin\theta = k\frac{\lambda}{d} . \tag{6.3}$$

This relation shows that a spectral scan, which is linear with respect to the wavelength $\lambda$, can easily be produced with a simple sine-bar mechanism.

The scanning mechanisms of almost all commercial grating monochromators are readily made linear with respect to wavelength. If one prefers to record a spectrum on a linear wave-number ($cm^{-1}$) scale, a mechanical device that generates a reciprocal function may be attached to the monochromator. This latter device is available commercially, but it is used mostly in Raman spectrometers.

Three interrelated factors — spectral dispersion, resolving power, and throughput — should be considered in optimizing the conditions for a given experimental situation. The spectral dispersion, $D_S$, is defined by

$$D_S = \frac{d\lambda}{dl} = \frac{1}{l_f}\frac{d\Phi^{-1}}{d\lambda} = \frac{1}{l_f} = d' , \tag{6.4}$$

where $d'$ is the distance measured across the slit, $l_f$ is the focal length of the lens or mirror, $\Phi$ is the angle of deviation (or diffraction), and $d\Phi/d\lambda$ is the rate of change of $\Phi$ with wavelength.

This equation indicates that a constant spectral dispersion and resolution are readily provided when the slit width $d$ is kept constant. No elaborate slit program is required (as in prism monochromators) to keep the spectral resolution constant over a spectral scan.

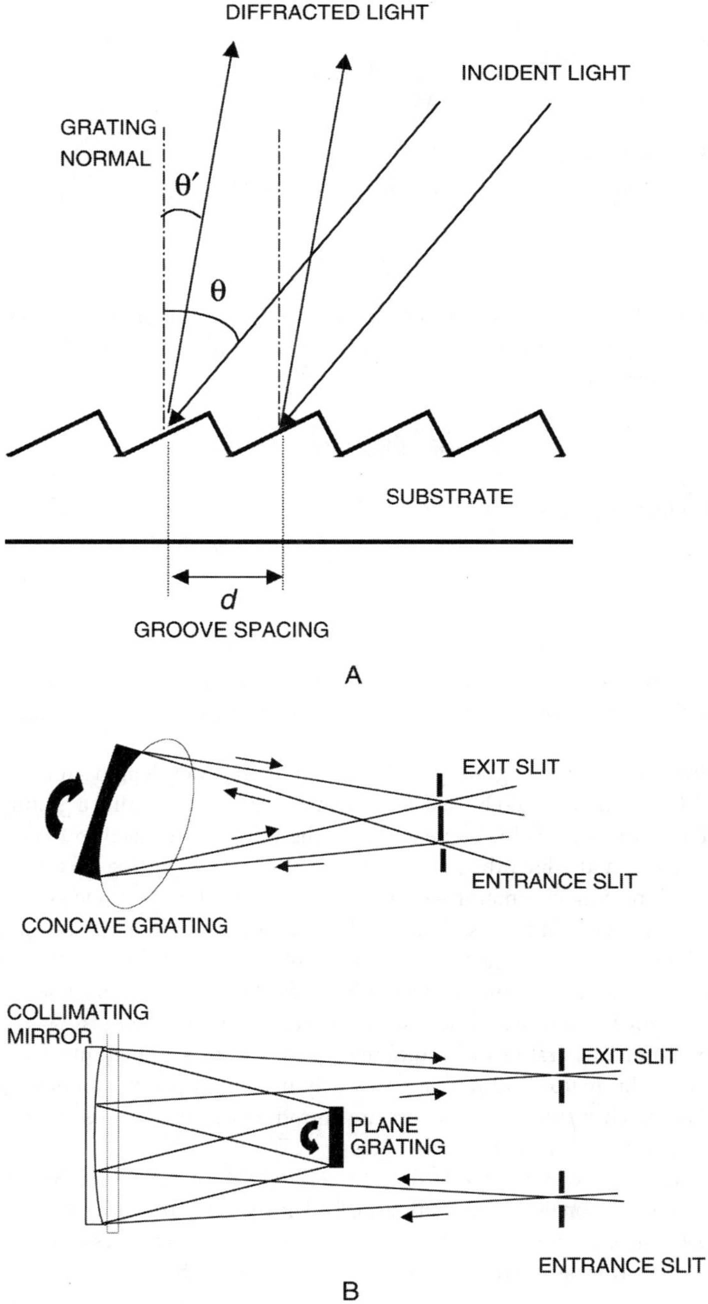


**FIGURE 6.5** (A) The grating rule ($\theta$ = angle of incidence; $\theta'$ = angle of diffraction; $d$ = groove spacing) and (B) optical configuration of planar grating and concave grating.

The resolving power $R$, which denotes the ability of a grating monochromator to separate adjacent emission lines, is defined by

$$R = \frac{\lambda}{\Delta\lambda} = \frac{w}{\lambda}(\sin\theta + \sin\theta'), \tag{6.5}$$

where $w$ is the width of the grating.

By combining Equations 6.2 and 6.5, one obtains (for $k = 1$)

$$R = wn. \tag{6.6}$$

This relation shows that the resolving power is directly proportional to the grating dimension (width) and to the groove density. For example, a grating with 1200 grooves/mm and 100-mm width would have a resolving power $R$ given by

$$R = wn = 100 \times 1200 = 120{,}000. \tag{6.7}$$

The spectral bandpass $B_S$ is given by

$$B_S = \frac{d'}{l_f D_S}. \tag{6.8}$$

This relation indicates that a smaller slit width would yield a smaller bandpass and provide better resolution. Larger slit widths, however, yield increased signal levels and therefore higher signal-to-noise (S/N) values.

The transmission efficiency of a grating is dependent on the wavelength and on the design of the grating. Two other factors that should be taken into consideration in selecting a grating are the blaze and the blaze angle. The blaze angle is the angle at which the grooves are ruled in the grating surface, and the blaze is the wavelength at which the maximum efficiency of the grating is concentrated. In general, one selects an excitation monochromator with high efficiency in the UV range and an emission monochromator with high efficiency in the visible spectral range. For example, a grating blazed at 300 nm may be chosen for excitation, whereas a grating blazed at 500 nm may be used for emission because the fluorescence of most organic compounds occurs between 400 and 600 nm. Use of gratings with an improper blaze angle may be a source of low sensitivity for many compounds.

In many monochromators, gratings with different blazes can be snapped into and out of the grating mount. Grating monochromators are now more widely used than prism monochromators because of their higher resolution. These monochromators are convenient to use because they produce a spectral dispersion that is linear with respect to wavelength.

The light throughput for various standard gratings varies between 50 and 90% and increases with groove density. For conventionally ruled gratings, this density has a practical limit above which it may produce substantial stray light and "ghosts." Ghosts are spurious spectral lines caused by periodic imperfections in the gratings. Stray light is defined as any light that passes through the monochromator outside the monochromator bandpass. For Raman measurements (and in some degree fluorescence measurements) that involve very low signal levels in the presence of an intense excitation source, the stray light level of the monochromator is perhaps the most critical parameter.

The advent of lasers and holography in the early 1960s made possible the production of holographic gratings with high throughput and low stray light. Monochromators may have planar or concave gratings. Planar gratings are usually produced mechanically and may contain imperfections in some of the grooves.

Concave gratings are usually produced by holographic and photoresist methods. Because they are produced optically, holographic gratings have fewer imperfections than mechanically produced (i.e., ruled) gratings. As a result of their optical configuration, a concave grating functions as the diffracting and the focusing element, resulting in one instead of three reflecting surfaces (Figure 6.5B). Fewer reflecting surfaces generally result in increased efficiency.

#### 6.2.2.3 Tunable Filters

Tunable filters are special types of dispersive devices. These devices, such as acousto-optic tunable filters (AOTFs) and liquid crystal tunable filters (LCTFs), allow the investigator rapidly to record an image at various wavelengths. An AOTF is a solid-state optical bandpass filter that can be tuned to various wavelengths within microseconds by varying the frequency of the acoustic wave propagating through the medium. The solid-state nature of the AOTF provides a high-throughput (90% diffraction efficiency) dispersive element with no moving parts, thus increasing the ruggedness of the instrumentation.

AOTF devices consist of a piezoelectric transducer bonded to a birefringent crystal. The transducer is exited by a radio frequency (rf) (50 to 200 MHz) and generates acoustic waves in a birefringent crystal. Those waves establish a periodic modulation of the index of refraction via the elasto-optic effect. Under proper conditions, the AOTF will diffract part of the incident light within a narrow frequency range. Only light that enters the crystal so that its angle to the normal of the face of the crystal is within a certain range can be diffracted by the Bragg grating.

In a non-collinear AOTF, the diffracted beam is separated from the undiffracted beam by a diffraction angle. The undiffracted beam exits the crystal at an angle equal to the incident light beam, while the diffracted beam exits the AOTF at a small angle with respect to the original beam. A detector can be placed at a distance so that the diffracted light can be monitored while the undiffracted light does not irradiate the detector. In addition, when the incident beam is linearly polarized and aligned with the crystal axis, the polarization of the diffracted beam is rotated 90° with respect to the undiffracted beam. This can provide a second means to separate the diffracted and undiffracted beams.

LCTFs use electrically controlled liquid crystal elements that transmit a certain wavelength band while being relatively opaque to others. The LCTF is based on a Lyot filter, a device constructed of a number of static optical stages, each consisting of a birefringence retarder (quartz for LCTFs) sandwiched between two parallel polarizers. A stack of stages function together to pass a single narrow wavelength band. Tunability is provided by the partial alignment of the liquid crystals along an applied electric field between the polarizers; the stronger the field, the more the alignment and the greater the increase in retardance. Tuning times for randomly accessing wavelengths depend on the liquid crystal material used and the number of stages in the filter. At the moment, commercial devices use nematic components that result in tuning times of approximately 50 to 75 ms.

The operating principles of AOTFs and LCTFs and their use in imaging are described in detail in Chapter 8 of this handbook.

### 6.2.3 Optical Fibers

Optical fibers and waveguides are described in detail in Chapter 7 of this handbook. This section briefly mentions only the basic and salient features of optical fibers. A widely used component that provides an optical link between a spectroscopic instrument and a remotely located sample is the optical fiber. The rapid growth of fiber optic sensing has paralleled the commercial availability of low-attenuation optical fibers. In many applications, the optical fibers comprise a core made with an optically transparent material (e.g., glass, quartz, or polymer) with a certain refractive index, $n_1$, surrounded by a cladding made with another material (e.g., quartz or plastic) having another refractive index, $n_2$. Light transmission is based on total internal reflection as depicted in Figure 6.6. Light rays that impinge on the core–cladding interface at an angle equal to or greater than the critical angle $\theta_c$, as determined by Snell's law, can be transmitted along the fiber by total internal reflection. Accordingly, an important fiber characteristic, the half acceptance angle $\beta$, is given by the equation:

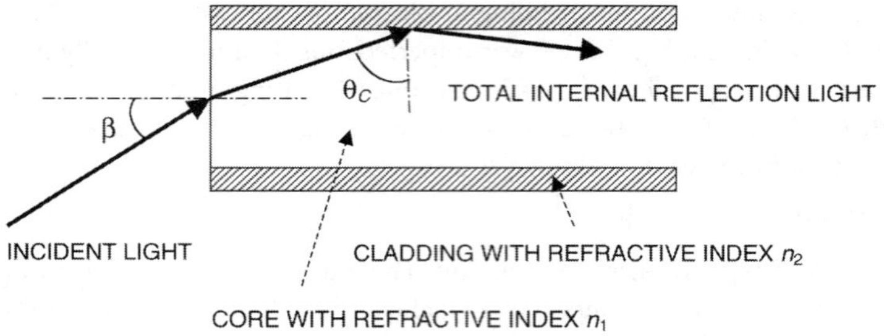


**FIGURE 6.6** Total internal reflection in an optical fiber.

$$\sin\beta = \left(n_1^2 - n_2^2\right)^{1/2} / n_0,$$

where $n_0$ is the refractive index of the medium in which the end of the fiber resides.

Low attenuation over desired spectral regions and a large β value are desirable fiber characteristics. Fibers with large values of β permit the coupling of a large amount of excitation radiation from the source at the incident end of the fiber and the efficient collection of fluorescence signal at the sensing end of the fiber.

Optical fibers can be used to transmit the excitation light to a sample and transmit the signal (reflected or scattered light) from the sample to the detector. Several possible optical fiber configurations can perform these measurements: single-fiber system, bifurcated fiber system, and dual-fiber system (Figure 6.7). In Figure 6.7A, a single fiber is used to transmit the excitation beam from a light source to the sample and transmit the emission from the sample to the detector. A dichroic filter is used to transmit the light at the excitation wavelength and reflect light at the emission wavelength. In Figure 6.7B, a bifurcated fiber is used, one end transmitting the excitation light to the sample, and the other end transmitting the sample emission light to the detector. In Figure 6.7C, two separate parallel fibers are used, one fiber transmitting excitation light and the other transmitting the emission light. In Figure 6.7D, two perpendicular fibers are used. The angle between the fibers can be varied and optimized in order to optimize the overlap of the excitation and detection volumes and to minimize scattered light.

### 6.2.4 Polarizers

Light can be described as consisting of two transverse electric and magnetic fields that are perpendicular to the direction of light propagation and to each other. The direction of the electric vector is used to describe the polarization of light. Most incoherent light sources consist of a large number of atomic and molecular emitters. The rays from such sources, which have electric fields with no preferred orientation, are called unpolarized. On the other hand, if a light beam (e.g., from a laser) consists of rays where the electric field vectors are oriented preferentially in the same direction, the beam is said to be linearly polarized.

Polarizers transmit light that has its electric vector aligned with the polarization axis, and block light that is rotated 90°. A commonly used device is the Glan–Thompson polarizer. This device consists of a calcite prism, a birefringent material, where the refractive index is different along each optical axis of the crystal. In general, the polarization characteristics of a crystal are determined by the crystal structure and electron bonding. Calcite is widely used as a polarizing material because of its excellent transmission well into the UV range and large difference in the two index-of-refraction values. Glan–Thompson polarizers exhibit high extension coefficients (near $10^6$) and high acceptance angle (10 to 15°).

Other types of polarizers include film polarizers composed of materials that absorb light polarized in one direction more strongly than light of the orthogonal polarization. These are thin films of a stretched polymer (e.g., polyvinyl alcohol) that transmit the light polarized in one direction and absorb the light polarized in another direction. The film sheets are stretched to orient and align the molecules. Polarizer

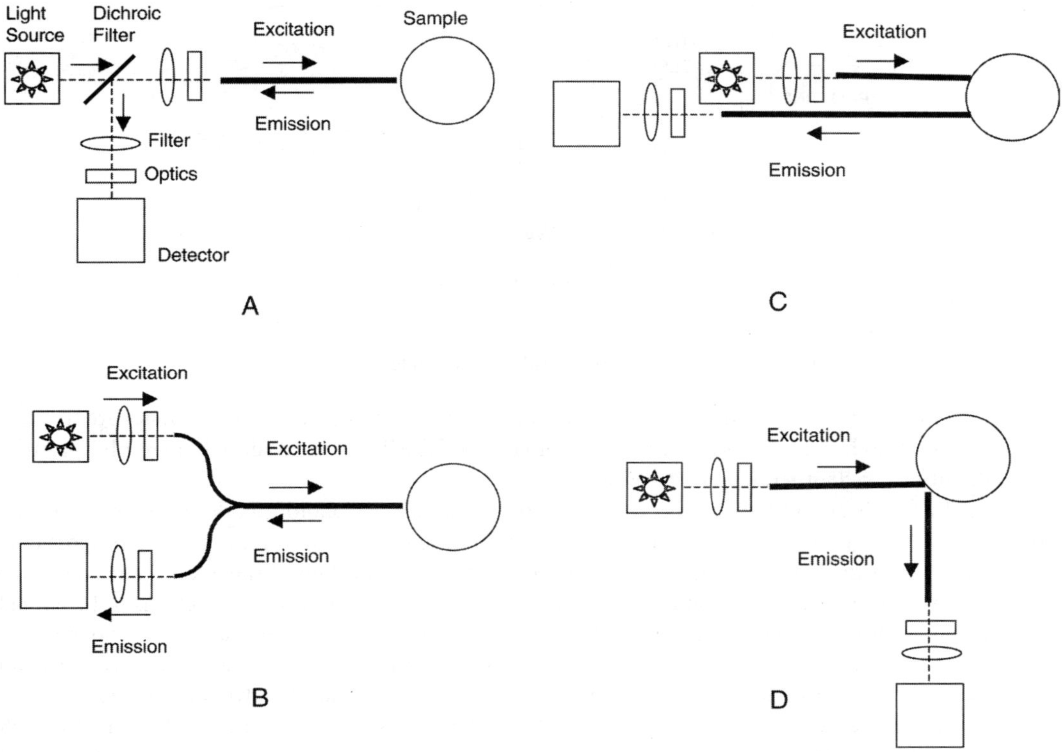


**FIGURE 6.7** Various configurations using optical fibers: (A) single-fiber system; (B) bifurcated-fiber system; (C) parallel dual-fiber system; (D) dual-fiber system.

films are not suitable for use with intense light beams (e.g., lasers) because they absorb light and therefore can be easily damaged. Other types of polarizers, such as the Wollaston polarizers and Rochon prism polarizers, split the unpolarized light into two beams, which then must be spatially selected.

## 6.2.5 Detectors

Selection of a suitable detector is one of the most critical steps in the development of a spectrometer. Detectors for electromagnetic radiation can be classified into photoemissive, semiconductor, and thermal types. Photoemissive detectors are generally used in optical measurements. These devices include PM tubes, photodiodes, and imaging tubes. The PM tubes are the most commonly used because they are the most sensitive detectors for the visible and near-UV regions. There are two types of detectors, single-channel detectors and multichannel. Multichannel detectors include one-dimensional and two-dimensional detector arrays. Traditionally, spectroscopy has involved using a scanning monochromator and a single-element detector (e.g., photodiode, PM). With monochromators, only one spectral resolution element, or channel, can be monitored at a time. Detectors that permit the recording of the entire spectrum simultaneously, thus providing the multiplex advantage, are known as multichannel detectors. With multichannel systems, a complete spectrum can be recorded in the same time it takes to record one wavelength point with a scanning system.

### 6.2.5.1 Single-Channel Detectors

#### *6.2.5.1.1 Photomultipliers*

PMs are widely used for their high spectral sensitivity, wide operating range, low cost, and relatively simple electronics. A PM coupled with a monochromator is the simplest and most commonly used

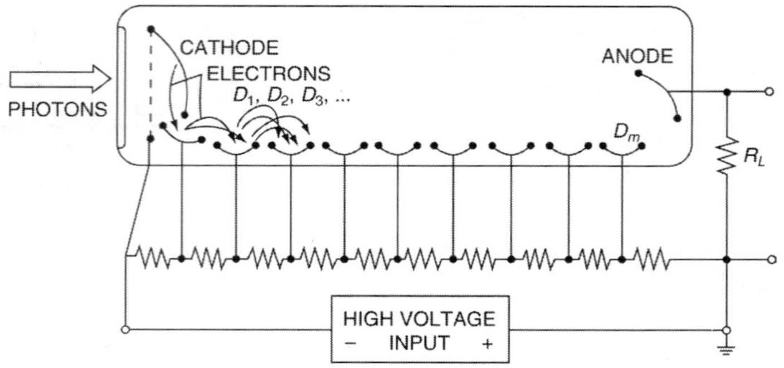


**FIGURE 6.8** Schematic diagram of a photomultiplier tube ($D_i$, dynode $i$; $R_L$, load resistor).

single-channel spectrometer. The spectral resolution and the sensitivity of the spectrometer are strongly dependent on the characteristics of the monochromator and the PM tube. Modern holographic gratings generally offer excellent stray light rejection.

The PM is a vacuum tube containing a highly sensitive surface, the photocathode, generally made of a metal oxide. The PM tube operates as a current source. The intensity of the current output is proportional to the intensity of the light striking the photocathode, which is a thin film of metal on the inside of the PM window. Figure 6.8 is a schematic diagram of a PM tube. The photons cause electrons to be ejected from the photocathode surface as the result of the photoelectric effect. The photocathode is held at a high negative potential, typically –1000 to –2000 V. The photoelectrons then cascade down the dynode chain in increasing numbers as a result of secondary emission, which occurs at each dynode surface.

Each dynode is an electrode coated with a material that emits several secondary electrons for each incoming electron. The electrons are accelerated down the dynode chain by a potential on the order of 100 V between two consecutive dynodes. The secondary electrons ejected from the first dynode are then accelerated onto the next dynode, where they induce further emission of secondary electrons. The secondary emission occurring at every successive dynode stage can provide an overall internal amplification factor of up to $10^8$. The total number of dynodes is usually between 7 and 12, and the total high voltage between the first and last dynode is commonly between 0.7 and 1.5 kV. The total current collected at the anode is proportional to the incoming photon flux and is linear over many orders of magnitude.

The photocathode materials and the transmission characteristics of the window materials determine the spectral response of the photocathode. Only a photon with an energy greater than the work function of the cathode material results in the release of a photoelectron from the cathode. Photocathode material determines the long-wavelength cutoff, and the window material determines the short-wavelength cutoff. Figure 6.9 depicts examples of some typical cathode sensitivity responses.

The PM is equivalent to a current power supply, since all secondary electrons ejected from the dynodes are replaced from the power supply via the dynode resistor chain. The number of dynodes determines the amplification gain of the PMT, and the gain per dynode stage is determined by the interdynode potential. The values of the dynode resistors are selected so that the current through the dynode resistor chain is about 100 times the maximum anode current. This design ensures that the anode potential remains constant during measurement. The manufacturers provide designs of typical dynode circuitry.

One limiting factor of the PM is the dark current that may be due to leakage current (imperfect insulation), ionization of residual gases inside the tube by electrons, and thermionic emission from the cathode or the dynodes. Proper preparation of PM connections significantly reduces the dark current in PM tubes. Figure 6.10 shows the five most common dynode configurations. The PM response time depends basically on the number of dynode stages and the dynode configurations; the line-focused configuration generally exhibits the fastest time response (<3 ns). In general, the capacitance between the anode and other electrodes determines the rise time of the PM tube.

Cooling the PM tube reduces the thermionic component of the dark current. With phototubes having GaAs cathodes, cooling produces a falloff in sensitivity in the red region but increases the sensitivity over

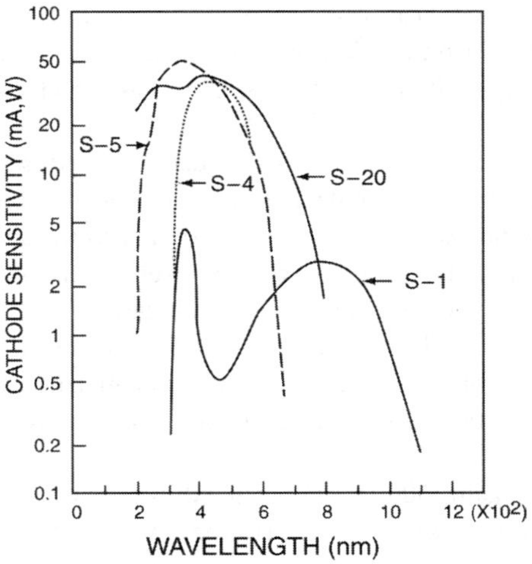


**FIGURE 6.9** Examples of spectral responses curves of photomultipliers.

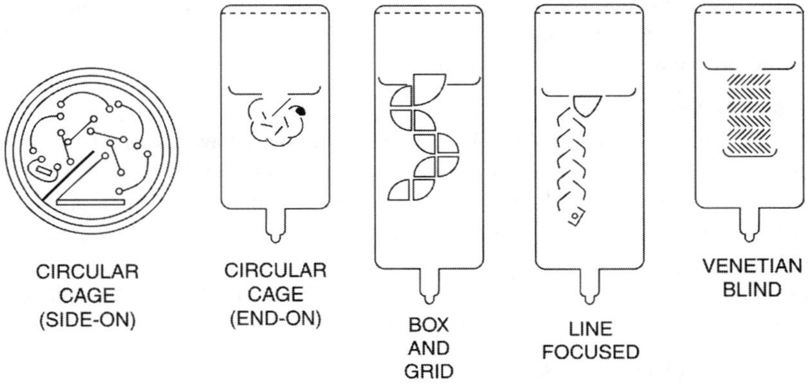


**FIGURE 6.10** Typical dynode configurations of photomultipliers.

the rest of the visible spectrum. PMs require various accessories such as housing assemblies, power supplies, current stabilization devices, and magnetic shielding. Cooling equipment is usually optional. Many suppliers of PM tubes also provide the accessories and frequently offer the complete detection system in one package.

Microchannel plates (MCPs) are special PM tubes that contain numerous holes, microchannels (instead of a dynode chain), which are lined with the secondary emissive dynode material. In an MCP, the electrons are amplified as they drop down the voltage gradient across the MCP. MCP devices have very fast time responses and are used in the most demanding high-speed, time-resolved measurements.

#### *6.2.5.1.2 Photodiode and Avalanche Photodiode*

A simple detector is a diode, a semiconductor device that produces, as a result of the absorption of photons, a photovoltage or free carriers that support the conduction of photocurrent. Photodiodes (PDs) are used for the detection of optical communication signals and for the conversion of optical power to electrical power. However, the lack of high amplification limits their usefulness for high-sensitivity measurements.

An avalanche photodiode (APD) is a PD that uses avalanche breakdown to achieve internal multiplication of photocurrent. The APD is usually a silicon-based semiconductor containing a *pn* junction consisting of a positively doped *p* region and a negatively doped *n* region sandwiching an area of neutral charge termed the depletion region. These diodes provide gain by the generation of electron–hole (e–h) pairs from an energetic electron that creates an "avalanche" of electrons in the substrate. Photons entering the diode first pass through the silicon dioxide layer and then through the *n* and *p* layers before entering the depletion region, where they excite free electrons and holes, which then migrate to the cathode and anode, respectively.

When a semiconductor diode has a reverse bias (voltage) applied, and the crystal junction between the *p* and *n* layers is illuminated, then a current will flow in proportion to the number of photons incident upon the junction. Avalanche diodes are very similar in design to the silicon *p–i–n* diode; however, the depletion layer in an APD is relatively thin, resulting in a very steep localized electrical field across the narrow junction. In operation, very high reverse-bias voltages (up to 2500 V) are applied across the device. As the bias voltage is increased, electrons generated in the *p* layer continue to increase in energy as they undergo multiple collisions with the crystalline silicon lattice. This "avalanche" of electrons eventually results in electron multiplication analogous to the process occurring in one of the dynodes of a PM tube.

APDs are capable of modest gain (500 to 1000) but exhibit substantial dark current, which increases markedly as the bias voltage is increased. They are compact and immune to magnetic fields, require low currents, are difficult to overload, and have a high quantum efficiency that can reach 90%. Usually, APDs are run with a reverse bias that is less than the breakdown value. In this mode, they have a gain of 200 to 300, which is too low for photon counting. However, APDs operated in the Geiger mode (with a reverse bias slightly greater than the breakdown value) can give similar gains to those of a PM, around $10^6$, and can be used for single-photon counting of low-level signals. The development of APDs as high-performance photodetectors is currently an active field of research.

#### 6.2.5.1.3 *Hybrid Detectors*

Hybrid PM (HPM) tubes and hybrid PDs (HPDs) are devices that have the potential for use in single-photon counting applications. These devices have a photocathode similar to that of conventional PM tubes. Following the absorption of a photon, an electron may be emitted from the photocathode. However, unlike in PM tubes, the photoelectrons are amplified in a solid-state structure by an avalanche process, resulting in an electrical pulse. In an HPD, the electron is accelerated in a strong electric field toward a silicon sensor, which represents the anode. The electron is stopped in the depleted silicon, where a large number of e–h pairs are created, depending on its kinetic energy (10 to 20 keV).

### 6.2.5.2 Multichannel Detectors

A well-known example of a multichannel detector is the photographic plate. Other multichannel detectors include the diode array, the CCD, or the charge-injection device. In multichannel spectrometers, the detector is placed at the focal plane of a polychromator, which is a monochromator with the exit slit removed. As a result, the entire emission dispersed at all wavelengths within the polychromator is detected simultaneously. The simultaneous detection of all of the dispersed emission using *n* spectral resolution elements reduces the measurement time by a factor of *n* in the case of a signal-to-noise-limited measurement, or improves the S/N ratio by a factor of *n* in the case of a time-limited measurement.

#### 6.2.5.2.1 *Vidicons*

The vidicon is a device used in the 1970s that is essentially a television-type device comprising an array of microscopic photosensitive diode junctions grown upon a single-silicon-crystal wafer. These diode junctions form individual microelements used to detect simultaneously the radiation dispersed by the polychromator. Prior to detection, each diode is charged to a preset reversed-bias potential by a fast-scanning electron beam. The electromagnetic radiation that strikes the diode causes depletion of the charge in each diode. A current proportional to the incident radiation recharges the diode to the preset potential when the electron beam scans again, thus providing information on the intensity of the incident light that has struck the diode. Consequently, an almost instantaneous recording of the spectral intensity at each diode is generated in the computer memory of the vidicon detector. Vidicons are capable of integrating radiation intensity over multiple scanning cycles because of their charge storage capabilities.

Sensitivity can be further enhanced by incorporating an image-intensification section in front of the vidicon to yield a silicon intensified target vidicon.

#### *6.2.5.2.2 Photodiode Array*

Another multichannel detector is the photodiode array (PDA), which also utilizes PDs as detection elements. Recording the signal is performed with direct on-chip circuitry rather than with a scanning electron beam. The electronic switches needed to read from individual PDs are built right onto the chip. One diode is read at a time, and the analog signal read out through each electronic switch correlates with the amount of light intensity impinging on the PD. Most multichannel PDAs are silicon based, operating from 180 to 1100 nm. PDA systems with InGaAs detectors can be used in the NIR spectral range (800 to 1700 nm). Signal amplification is achieved by a microchannel plate image intensifier. Gated detection down to 5-ns time resolution can be performed with the intensified diode array.

#### *6.2.5.2.3 Charge-Coupled Device*

More recently, multichannel devices such as CCDs are increasingly used as a result of their high quantum yield, two-dimensional imaging capability, and very low dark current. CCDs represent a great advance in detector instrumentation from the UV to NIR spectral range. They are very widely used because of their two-dimensionality and unique combination of sensitivity, speed, low noise, ruggedness, and durability in a relatively compact package. Because of their widespread use in spectroscopy, this section provides a detailed description of the operating principle of CCDs and the instrumental factors involved in the selection of CCDs for specific applications.

CCDs are one- or two-dimensional arrays of silicon PDs with metal-oxide semiconductor architectures. The detector arrays consist of individual detector elements, called pixels, which are defined by capacitors, called gates. These capacitors are charged by electrons generated by the light impinging onto the CCD. Silicon exhibits an energy gap of 1.14 eV. Incoming photons with energy greater than this can excite valence electrons into the conduction band, thus creating e–h pairs. The average lifetime for these carriers is 100 μs. After this time, the e–h pair will recombine. Photons with energy of from 1.14 to 5 eV generate single e–h pairs. Photons with energy of >5 eV produce multiple pairs. A 10-eV photon will produce three e–h pairs, on average, for every incident photon. Soft x-ray photons can generate thousands of signal electrons, making it possible for a CCD to detect single photons. For use as an IR imager, a CCD must be made of another material, like germanium (band gap 0.55 eV).

The current sources (e–h pairs) produced are localized in small areas, an array of capacitors, called pixels. Common two-dimensional CCD chips have 512 × 512 or 1024 × 1024 pixels. The charge accumulates in proportion to the light intensity impinging onto the pixel. A CCD sensor provides only one serial output: the readout register through which each capacitor can be discharged (each pixel can be read). A differential voltage is applied across each gate to perform charge transfer. The photogenerated charge is moved to the readout register by a series of parallel shifts, sequentially transferring charge from one pixel to the next within a column, until the charge finally collects in the readout register.

The charge from a row of pixels can be binned before readout to improve the S/N value. Furthermore, the dark count of CCDs is very low, especially when the detector is cooled. A CCD array can accumulate charge generated by photoelectrons almost noiselessly. However, CCD noise is produced in the act of commutating the charge out to a charge detector. Readout noise also tends to increase with increasing readout speeds; typically, the best CCD camera systems currently available give around five electrons of readout noise per pixel.

CCDs have several structures, including front-illuminated, back-illuminated, and open-electrode structures (Figure 6.11). In front-illuminated CCDs, incident photons have to penetrate a polysilicon electrode before reaching the depletion region. In back-illuminated or back-thinned CCDs, the substrate is polished and thinned to remove most of the bulk silicon substrate. Because illumination occurs from the back, the polysilicon on the front does not affect the quantum efficiency (QE) of the detector. CCDs are currently the detectors with the highest QEs. Typically, a CCD array has a QE of around 40%, but back-thinning can increase the QE to around 80% at 600 nm. Back-thinned CCDs are usually coated

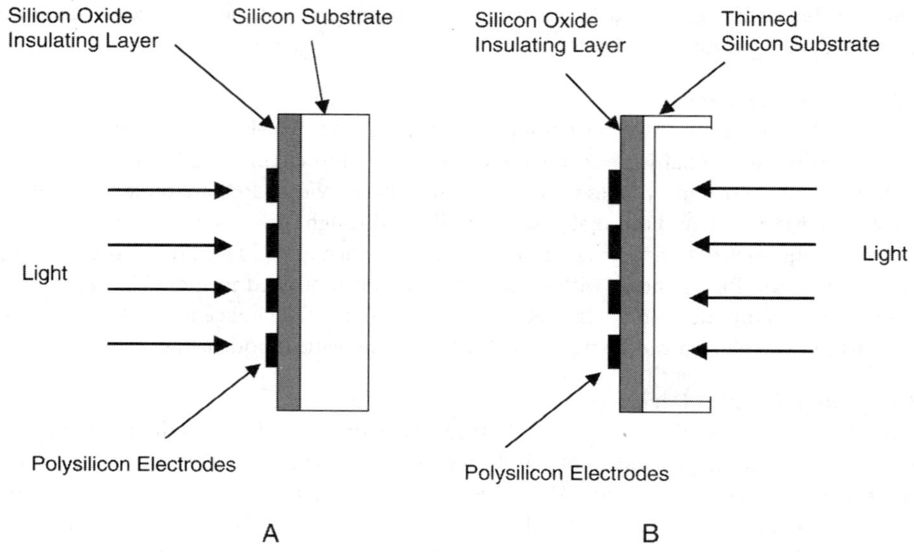


**FIGURE 6.11** Structure of front-illuminated (A) and back-illuminated (B) CCDs.

with an antireflection material for enhanced response in the UV or the NIR region. Often reflections from boundaries of the back-thinned devices form constructive and destructive interference patterns, referred to as the "etaloning effect." The etaloning effect often causes an undesired oscillation superimposed on the spectrum at wavelengths longer than 650 nm.

The transmittance of the electrode depends on its thickness. Because the polysilicon electrode material does not transmit at below 400 nm, some versions of front-illuminated UV CCDs have the detector coated with a phosphor, which converts UV radiation to green light. These UV CCDs can provide a 10 to 15% QE response in the 120- to 450-nm spectral range. The coating is selected so that it does not degrade the visible and NIR response of the detector. In open-electrode CCDs, the central area of the electrode is etched to expose the underlying photosensitive silicon. These types of CCDs exhibit QEs of 30% or greater in the UV.

Cooling CCDs is important because the operating temperature of the detector determines the dark-current noise. In general, the dark-current level is reduced 50% for every 9°C drop in temperature. Two common cooling methods are liquid-nitrogen cooling and thermoelectric cooling using multistage Peltier devices. Liquid-nitrogen cooling provides the lowest operating temperatures, achieving typical dark-current levels of one to three electrons per pixel per hour at –133°C. Four-stage thermoelectrically cooled CCDs have typical operating temperatures of –73°C with typical noise levels of one to two electrons per pixel per minute. Thermoelectric cooling systems are often a good choice because they can offer good performance, approaching that of liquid-nitrogen-cooled systems, and provide the convenience of uninterrupted operation. Research spectrometers generally require cooled CCDs to detect low-level signals, but conventional consumer-application CCDs are uncooled because of the high light levels and the very low bit depth of the analyzing electronics.

CCDs are often used as steady-state detectors. Because they have relatively slow time response, an intensifier stage with fast gating capability is used for time-resolved measurements. Intensified charge-coupled devices (ICCDs) can have nanosecond gate times.

Selection of CCDs is based on the spectral range of interest, resolution, and expected optical signal level. No single system can satisfy all of the possible spectral and S/N requirements in spectroscopy. The user should carefully analyze needs and evaluate the tradeoffs involved in selecting a detector type, structure, and cooling method. Several considerations in the selection of a CCD for various applications are summarized in Table 6.3.[8]

**TABLE 6.3** Various Types of Charge-Coupled Devices and Photonics Applications

| Light Level | Applications | CCD Chip Structure | Cooling Type |
|---|---|---|---|
| High | Absorption<br>Transmission<br>Reflection<br>Elastic scattering | Front-illuminated | Thermoelectric (two-stage) |
| Medium | Analytical luminescence<br>Analytical Raman | Front-illuminated<br>Open-electrode | Thermoelectric (four-stage) |
| Low | Research luminescence<br>Research Raman | Back-illuminated | Liquid nitrogen |

#### 6.2.5.2.4 Other Solid-State Detectors

Other popular types of solid-state detectors include charge-injection devices (CIDs) and active-pixel sensors (APSs). CIDs have pixels composed of two metal-oxide semiconductor gates that overlap and share the same row and column electrodes. The APS consists of a photodiode, a reset transistor, and a row-select transistor. These devices can be highly integrated and can be manufactured using complementary metal-oxide semiconductor (CMOS) technologies.

#### 6.2.5.2.5 CMOS Array

PDs and phototransistors are single-channel solid-state detectors. With the silicon fabrication process, the integrated electro-optic system on integrated circuit (IC) microchips, photodetector elements with amplifier circuitry, has led to the development of a new generation of multichannel (one- and two-dimensional) detector arrays based on the CMOS technology.

CMOS is the semiconductor technology used in the transistors integrated into most computer microchips. Semiconductors are made of silicon and germanium, materials that "sort of" conduct electricity, but not enthusiastically. Areas of these materials are "doped" by adding impurities, which supply extra electrons with a negative charge (*N*-type transistors) or of positive charge carriers (*P*-type transistors). In CMOS technology, both kinds of transistors are used in a complementary way to form a current gate that creates an effective means of electrical control. CMOS transistors use almost no power when they are not needed.

The use of IC systems based on the CMOS technology has led to the development of extremely low-cost diagnostic sensor chips for medical applications. In CMOS, both *N*-type and *P*-type transistors are used to realize logic functions. Today, CMOS technology is the dominant semiconductor technology for microprocessors, memories, and application-specific ICs (ASICs). The main advantage of CMOS is the low power dissipation. Power is dissipated only if the circuit actually switches. This design allows integration of increased numbers of CMOS gates on an IC, resulting in much better performance.

Using the standard CMOS process allows production of PDs and phototransistors, as well as numerous other types of analog and digital circuitry, in a single IC chip. This feature is the main advantage of the CMOS technology compared with other detector technologies such as CCDs and CIDs. The PDs are produced using the *n*-well structure generally used to make resistors or as the body material for transistors. The capability of large-scale production using low-cost IC technology is an important advantage. The assembly process of various components is made simple by integration of several elements on a single chip. For medical applications, this cost advantage will allow the development of extremely low-cost, disposable biochips that can be used for in-home medical diagnosis of diseases without sending samples to a laboratory for analysis.[9–14] The use of CMOS sensor arrays for medical biochips is further discussed in Chapter 51 of this handbook.

#### 6.2.5.2.6 Streak Cameras

Streak cameras are detectors that can provide very fast time gating with temporal resolution from several picoseconds to hundreds of femtoseconds.[15] Streak cameras operate by dispersing photoelectrons across an imaging screen at high speed using deflection plates within the detector. These cameras can simultaneously provide measurements of time decay at different wavelengths. In general, they have a faster response time than that of microchannel plate PM tubes but exhibit relatively low S/N values and low dynamic ranges (three orders of magnitude).

## 6.2.6 Detection Methods

### 6.2.6.1 Direct Current Technique

The analog or DC method is the oldest and simplest detection technique. Because the anode is basically a current generator, a current meter may measure the anode current. It is common to measure the output by amplifying the voltage across a load resistor, typically between $10^5$ and $10^8$ ohms. The noise in the DC amplification signal is commonly reduced by a resistor-capacitor (RC) low-frequency band-pass filter. Too slow a response in the RC filter, however, necessitates very slow scanning speeds and long measurement times. A typical DC detection system involves the use of a picoammeter. The time constant for such an instrument at the highest intensity scale is a few seconds. A zero-suppression device is generally provided to compensate for dark current and zero drift.

### 6.2.6.2 Alternating Current Technique

The alternating current (AC) method consists of modulating the excitation signal using electronic or mechanical means. An AC amplifier is then tuned to the modulating frequency for detection. This technique rejects the noise outside the amplifier bandpass, eliminating the direct-current (DC) dark current. A further improvement is based on the technique of synchronous (or lock-in) detection. Stability is improved because any drift or instability of the modulating system is locked in by the amplifier. However, synchronous detection cannot eliminate some types of noise ("l/*f* noise," multiplicative noise) that are amplified along with the signal.

In certain situations, it is desirable to obtain spectral information in digital form to allow data processing for correction or smoothing of spectra. One simple method to obtain data from the DC anode current uses a voltage-to-frequency converter.[16] This method is equivalent to using an analog-to-digital converter and counting the output pulses from the converter. A variety of voltage-to-frequency converters is at present commercially available.

### 6.2.6.3 Digital Photon Counting Technique

A truly digital technique that counts discrete photon pulses is the single-photon counting technique. Unlike the more conventional analog detection method, the single-photon counting signal output is digital in nature, producing discrete pulses of charge. The digital technique has proved to have several advantages over the analog method, especially for low-level signal detection. This method uses high-gain phototubes and high-speed electronics for detecting individual photon pulses.[19]

Unlike the previously described methods, in which the signal is analog, the PM signal output here is digital in nature. In the digital technique, discrete pulses of charge are produced at the anode, with the number proportional to the number of photons incident on the photocathode. This technique has proved to have several advantages over the conventional DC method, especially for low-level light detection.[17–19] First, the digital data can be processed directly in a manner suitable for further computer data treatment. The processing of information by digital circuitry is less susceptible to long-term drifts, which usually limit analog systems; this feature permits measurement of extremely low radiation flux. The method also makes optimization of the S/N ratio possible by discriminating against PM dark current.

With an ideal detection device, one would expect the pulses resulting from single photoelectrons to have exactly the same pulse height. Actually, the output pulses have a certain height distribution due to thermionic emission of higher dynodes and spurious pulses generated by the PM. Electrons that do not originate at the photocathode, but from farther down the dynode chain, are the main sources of PM noise or dark current. These undesired electron pulses have a height distribution that generally differs from that of the photoelectrons. For example, thermionic electrons have a pulse height distribution lower than that of photoelectrons, whereas other sources of noise — such as cosmic ray muons, after-pulsing, and radioactive contamination of tube materials — generally produce pulses of higher amplitude. Most of these unwanted pulses can be eliminated by discriminatory units that select pulses within the range of height levels expected for true photosignals of interest.[20]

Whereas digital photon counting techniques are very useful for low-level signal measurements, they may not be suitable for high-intensity conditions because of pulse pile-up phenomena. If two pulses arrive at the cathode at the same time or are closely spaced in time, they could be counted as a single pulse, resulting in an inaccurate count rate. The anode pulse width corresponding to a single photon for a typical PM is approximately 5 ns. This limits the maximum frequency response of the PM to 200 MHz for a periodic signal. For random events, the count rate needs to be about 100-fold less to avoid pulse pile-up. Therefore, the single-photon counting technique should be employed only for photon count rates of less than 2 MHz.

### 6.2.6.4 Time-Resolved and Phase-Resolved Detection Methods

Various detection techniques can be designed to measure optical signals from a wide variety of spectroscopic processes, including absorption, fluorescence, phosphorescence, elastic scattering, Raman scattering, etc. One important parameter of these signals is the lifetime of the radiation. The lifetimes of various processes are:

- Absorption: instantaneous with excitation
- Fluorescence: $10^{-10}$ to $10^{-8}$ s
- Phosphorescence: $10^{-6}$ to $10^{-3}$ s
- Scattering: almost instantaneous with excitation

Two methods of measuring signal that allow differentiation of lifetimes involve time-resolved and phase-resolved detection schemes.[21] These methods also improve the S/N values by differentiating the actual signal of interest from the background noise (DC signal). The use of time-resolved and phase-resolved techniques in imaging is described in Chapter 9 of this handbook.

#### *6.2.6.4.1 Time-Resolved Detection*

In the time-resolved method, a pulse excitation source is used. The width of the excitation is generally much shorter than the emission process of interest, i.e., much shorter than the lifetime (decay time) of the samples. If one desires to measure the lifetime, then the time-dependent intensity is measured following the excitation pulse, and the decay time $\tau$ is calculated from the slope of a plot of log $I(t)$ vs. $t$, or from the time at which the intensity decreases to $1/e$ of the initial intensity value $I(t = 0)$. To measure the emission intensity free from the excitation pulse intensity, one can gate the detection after a delay time when the excitation pulse has decreased to zero (Figure 6.12A). Different compounds having different decay times can be differentiated by using different delay times ($dT$) and gate times ($\Delta T$) (Figure 6.12B). An important source of noise in many measurement situations is the DC noise from the background. Improvement in the S/N ratio can be achieved by using multiple excitation pulses and applying the "boxcar" method by integrating the emission signal after each pulse (Figure 6.12B).

#### *6.2.6.4.2 Phase-Resolved Detection*

Another method that can differentiate the lifetimes involves the phase-resolved techniques (often referred to as the frequency domain techniques). In this case, the sample is excited with intensity-modulated light. The intensity of the incident light changes with a very high frequency ($\omega = 2\pi f$, where $f$ is the frequency in hertz) compared with the reciprocal of the decay time $\tau$. Following excitation with a modulated signal, the emission is also intensity-modulated at the same modulation frequency. However, because the emission from the sample follows a decay profile, there is a certain delay in the emission relative to the excitation (Figure 6.13). This delay is measured as a phase-shift ($\phi$), which can be used to calculate the decay time. Note that at each modulation frequency $\overline{\omega}$, the delay is described as the phase shift $\phi_\omega$, which increases from 0 to 90° with increasing modulation frequency $\overline{\omega}$.

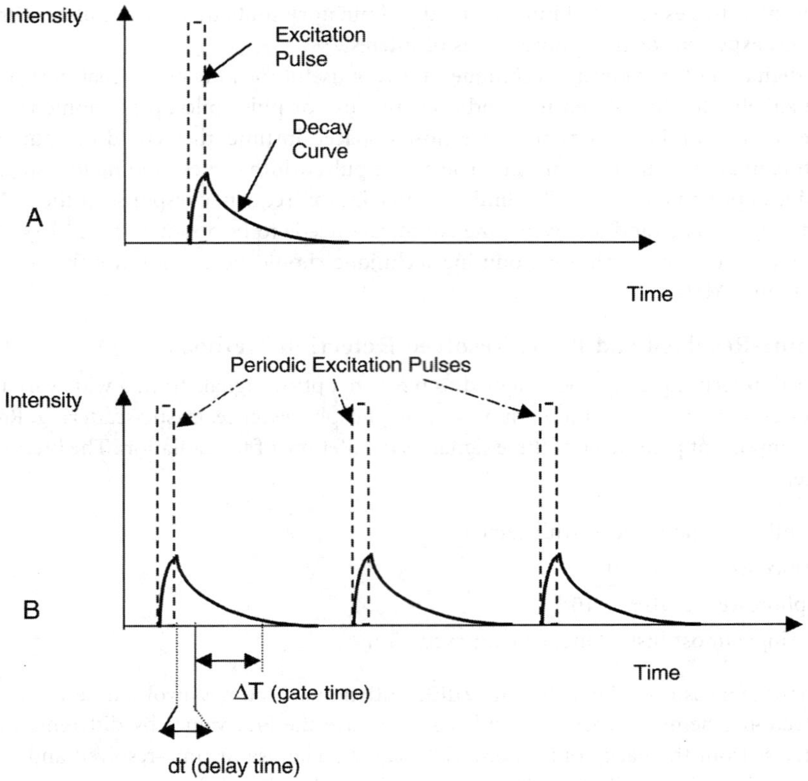


**FIGURE 6.12** Time-resolved detection: (A) excitation and decay profiles; (B) gated detection technique.

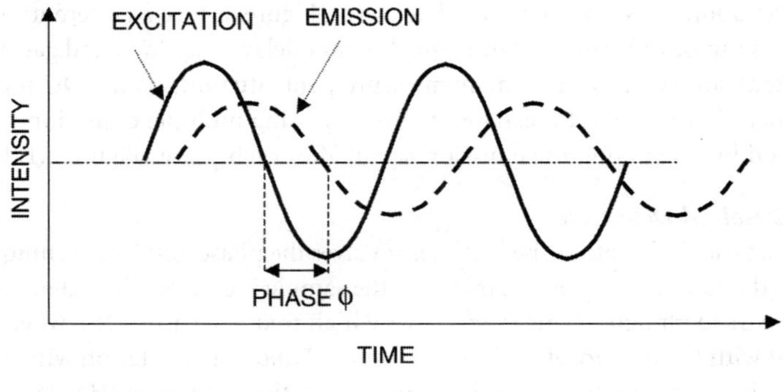


**FIGURE 6.13** Phase-resolved detection.

The finite time response of the sample also results in demodulation of the emission by a factor $m_{\varpi}$. This factor decreases from 1.0 to 0 with increasing modulation frequency. At low frequency, the emission closely follows the excitation. Hence, the phase angle is near 0 and the modulation is near 1. As the modulation frequency is increased, the finite lifetime of the emission prevents the emission from closely following the excitation. This results in a phase delay of the emission and a decrease in the peak-to-peak amplitude of the modulated excitation (Figure 6.13).

The shape of the frequency response is determined by the number of decay times displayed by the sample. If the decay is a single exponential, the frequency is simple. One can use the phase angle or modulation at any frequency to calculate the lifetime. For single-exponential decay, the phase and modulation are related to the decay time (τ) by

$$\tan\phi_{\varpi} = \varpi\tau$$

and

$$m_{\omega} = \left(1+\omega^2\tau^2\right)^{-1/2}.$$

Therefore, one can differentiate various emissions having different decay times by selecting the phase shift optimized to the decay time of interest. This method is referred to as phase-resolved detection.

### 6.2.6.5 Multispectral Imaging

Photonics diagnostic technologies can be broadly classified into two categories: (1) spectroscopic diagnostics and (2) optical imaging, often referred to as optical tomography. Spectroscopic diagnostic techniques are generally used to obtain an entire spectrum of a single tissue site within a wavelength region of interest. These techniques are often referred to as point-measurement methods. On the other hand, optical imaging methods are aimed at recording a two-dimensional image of an area of the sample of interest at one specific wavelength. A third category, which combines the two modalities, is often referred to as multispectral imaging or hyper-spectral imaging.

Spectral imaging represents a hybrid modality for optical diagnostics that obtains spectroscopic information and renders it in image form. In principle, almost any spectroscopic method can also be combined with imaging. Some of these techniques use computer-based image processing in combination with microscopy. These techniques have had a major impact on research and are being implemented in cytological diagnostics. There is also considerable future potential for direct clinical applications.

Instrumentation for multispectral imaging is described in detail in Chapter 8 of this handbook. Briefly, the concept of multispectral imaging is schematically illustrated in Figure 6.14. With conventional imaging, the optical emission from every pixel of an image can be recorded, but only at a specific wavelength or spectral bandpass. With conventional spectroscopy, the signal at every wavelength within a spectral range can be recorded, but for only a single analyte spot. The multispectral concept combines these two recording modalities and allows recording of the entire emission for every pixel on the entire image in the field of view with the use of a rapid-scanning solid-state devices, such as the acousto-optic tunable filter or a liquid crystal tunable filter. The multispectral imaging approach provides a "data cube" of spectral information of the entire image at each wavelength of interest (Figure 6.15).

## 6.3 Conclusion

Photonics instrumentation includes a wide variety of basic devices available to the investigator for many applications. A simple instrumental setup can be easily assembled using off-the-shelf components. More sophisticated devices can be designed for special research requirements. Selection of components or a complete system is based on the spectral range of interest, resolution, and expected optical signal level. No single system can satisfy all of the possible spectral and S/N requirements in spectroscopy. The user

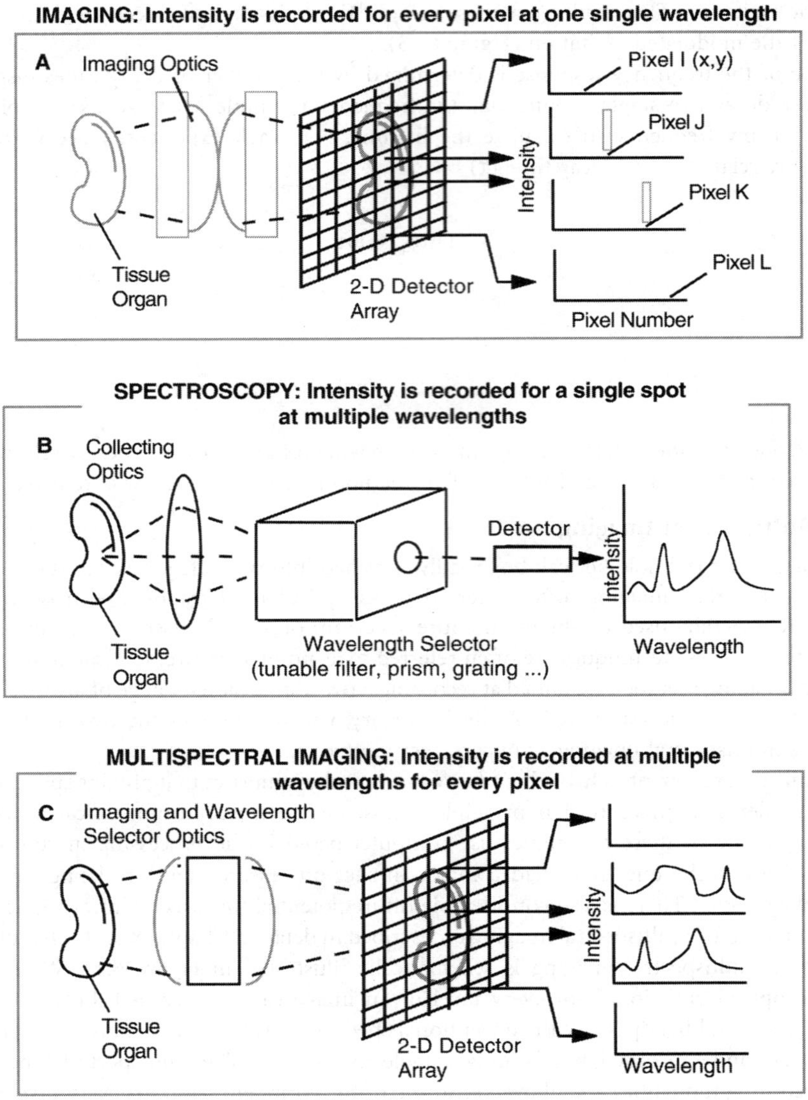


**FIGURE 6.14** Principle of multispectral imaging. With conventional spectroscopy, the signal at every wavelength within a spectral range can be recorded, but for only a single analyte spot. The multispectral concept combines these two recording modalities and allows recording of the entire emission for every pixel on the entire image in the field of view.

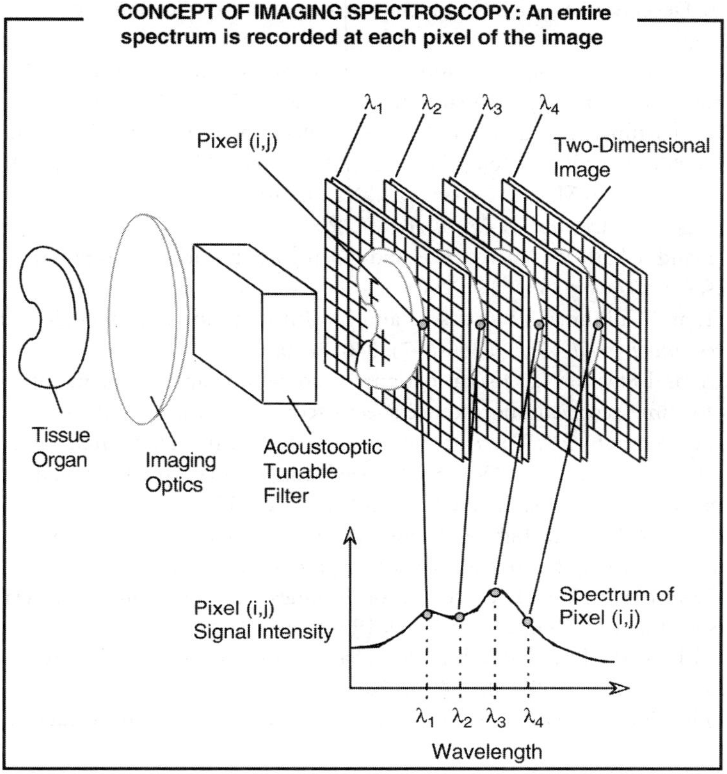


**FIGURE 6.15** Multispectral data cube. The multispectral imaging approach provides a "data cube" of spectral information of the entire image at each wavelength of interest.

should carefully analyze his or her needs and evaluate the tradeoffs involved in selecting a detector type, structure, and cooling method.

## Acknowledgments

This work was sponsored by the U.S. Department of Energy (DOE) Office of Biological and Environmental Research, under Contract DEAC05-00OR22725 with UT-Battelle, LLC, by the National Institutes of Health (RO1 CA88787-01), the Department of the Army (MIPR 6F5UFM6699), and by the Department of Justice, Federal Bureau of Investigation (NO. 2051-K041-A1).

## References

1. Schawlow, A.L., Laser spectroscopy of atoms and molecules, *Science*, 202(4364), 141, 1978.
2. Omenetto, N., Ed., *Analytical Laser Spectroscopy*, John Wiley & Sons, New York, 1979.
3. Kaminov, I.R. and Siegman, A.I., *Laser Devices and Applications*, IEEE Press, New York, 1973.
4. Kliger, D.S., *Ultrasensitive Laser Spectroscopy*, Academic Press, New York, 1983.
5. West, M.A., *Lasers in Chemistry*, Elsevier, New York, 1977.
6. Svelto, O., *Principles of Lasers*, 4th ed., Plenum Press, New York, 1998.
7. Orr, B.J., Optical parametric oscillators, in *Tunable Laser Applications*, Duarte, F.J., Ed., Marcel Dekker, New York, 1995.
8. Gilchrist, J.R., Choosing a scientific CCD detector for spectroscopy, *Photonics Spectra*, 36(3), 83, 2002.

9. Vo-Dinh, T., Development of a DNA biochip: principle and applications, *Sensors Actuators B Chem.*, B51, 52, 1999.
10. Vo-Dinh, T., Alarie, J.P., Isola, N., Landis, D., Wintenberg, A.L., and Ericson, M.N., DNA biochip using a phototransistor integrated circuit, *Anal. Chem.*, 71(2), 358, 1999.
11. Stokes, D.L., Griffin, G.D., and Vo-Dinh, T., Detection of *E. coli* using a microfluidics-based antibody biochip detection system, *Fresenius J. Anal. Chem.*, 369(3–4), 295, 2001.
12. Vo-Dinh, T., Cullum, B.M., and Stokes, D.L., Nanosensors and biochips: frontiers in biomolecular diagnostics, *Sensors Actuators B Chem.*, B74, 2, 2001.
13. Vo-Dinh, T. and Askari, M., Microarrays and biochips: applications and potential in genomics and proteomics, *Curr. Genomics,* 2, 399, 2001.
14. Vo-Dinh, T. and Cullum, B., Biosensors and biochips: advances in biological and medical diagnostics, *Fresenius J. Anal. Chem.*, 366(6–7), 540, 2000.
15. Wiessner, A. and Staerk, H., Optical design considerations and performance of a spectro-streak apparatus for time-resolved fluorescence spectroscopy, *Rev. Sci. Instrum.*, 64(12), 3430, 1993.
16. Ingle, J.D. and Crough, S.R., *Spectrochemical Analysis*, Prentice-Hall, Englewood Cliffs, NJ, 1988.
17. Woodruff, T.A. and Malmstad, H.V., High-speed charge-to-count data domain converter for analytical measurement systems, *Anal. Chem.*, 46(9), 1162, 1974.
18. Vo-Dinh, T. and Wild, U.P., High resolution luminescence spectrometer. 1. Simultaneous recording of total luminescence and phosphorescence, *Appl. Opt.*, 12(6), 1286, 1973.
19. Vo-Dinh, T. and Wild, U.P., High resolution luminescence spectrometer. 2. Data treatment and corrected spectra, *Appl. Opt.*, 13(12), 2899, 1974.
20. Gustafson, T.L., Lytle, F.E., and Tobias, R.S., Sampled photon-counting with multilevel discrimination, *Rev. Sci. Instrum.*, 49(11), 1549, 1978.
21. Lakowicz, J.R., *Principles of Fluorescence Spectroscopy*, 2nd ed., Kluwer Academic, New York, 1999.

# 7

# Optical Fibers and Waveguides for Medical Applications


Israel Gannot
*Tel-Aviv University*
*Tel-Aviv, Israel*

Moshe Ben-David
*Tel-Aviv University*
*Tel-Aviv, Israel*




## 7.1 Introduction

Lasers are used in medical applications for a wide range of the electromagnetic spectrum: x-ray, ultraviolet (UV), the visible, the near-infrared, and ending, at this time, in the mid-infrared. This wide range of wavelengths must be transmitted from the source (laser) to the target tissue by a flexible device that enables easy manipulation of the laser beam in a medical setting.

At the beginning of laser use in medicine, energy was transmitted by an articulated arm, which was a set of tubes connected by joints and reflecting mirrors to three freedom ranks (Figure 7.1). With time, this setup was miniaturized and made more precise, and the drift of the laser beam was reduced, but the device is still cumbersome and limited to external use.

Surgical procedures are usually open procedures; however, with the introduction of the endoscope, the trend is toward noninvasive or minimally invasive procedures. This practice reduces hospitalization time, causes less postoperative pain and discomfort, and results in fewer complications. Surgical tools are inserted through existing openings of the body or through minor cuts to the skin. The development of the endoscope (especially the flexible one; see Figure 7.2 for schematic drawing) was enabled by optical fibers, which deliver the light into the body, and by the coherent optical bundle, which delivers the image from the body cavity outside to the eyepiece or monitor. The working channel is used to insert surgical microtools and an optical fiber that can deliver laser radiation into the body.


0-8493-1116-0/03/$0.00+$1.50

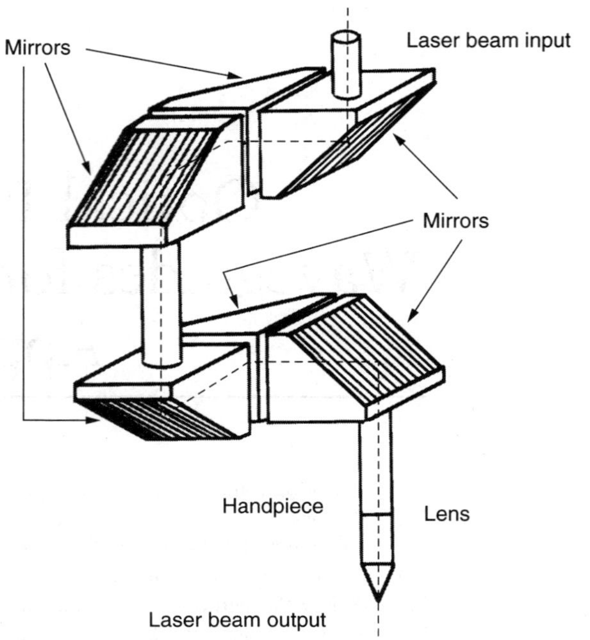


**FIGURE 7.1** Laser-articulated arm.

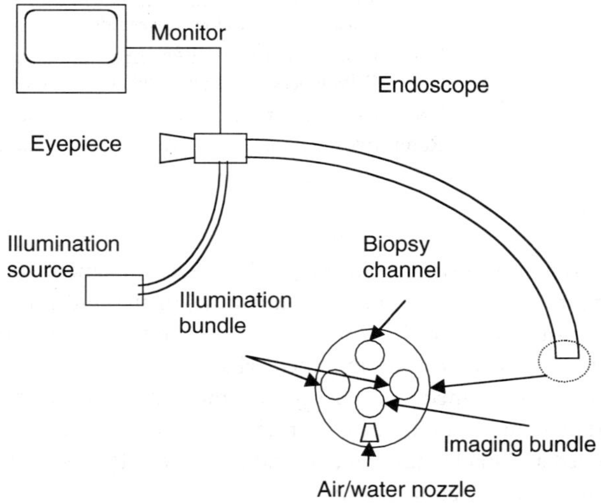


**FIGURE 7.2** Schematic drawing of an endoscope.

Optical fibers, which can transmit specific laser wavelengths, will make minimally invasive procedures even more common. They can also transmit information from tissue back to a detector. When tissue is heated it radiates; the radiation wavelength is a function of the tissue temperature (Planck curves). This information can be transmitted by an IR transmitting fiber, and the temperature of the tissue can be measured. One such use is a feedback mechanism for tissue welding/soldering.[1] Developments in biomedical optics and photon migration in tissue have made it possible to measure specific tissue properties through transmission and reflection. Spectroscopy of tissue can be made through evanescent waves at fibers with no cladding.[2] Fluorescence signals from tissue (either endogenous fluorophores or exogenous materials present in tissue) can be transmitted to detectors[3] or fluorescence cameras through coherent

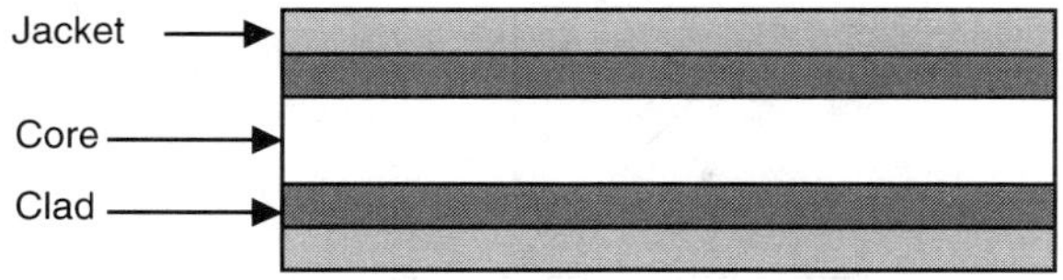


**FIGURE 7.3** General structure of a solid-core fiber.

optical bundles. These signals help diagnose changes in tissue, tissue transformations, and the presence of tumors. Measurements can be taken in a minimally invasive way using endoscopes. Diagnosis is immediate and less invasive, and response to treatment can be monitored.

The following sections examine these issues:

- Theory of laser radiation propagation in optical fiber/waveguide
- Attenuation mechanisms
- Fiber and waveguide structure
- Transmission properties
- Distal tips
- Coupling devices
- Materials for making fibers and waveguides

## 7.2 Theory

Two types of optical fibers are used in medical applications: (1) the conventional solid-core fibers such as silica and (2) hollow waveguides. Each type of optical fiber guides radiation according to different physical phenomena. Solid-core fibers guide optical radiation by using total internal reflection, while most of the hollow waveguides guide optical radiation by pure reflection.

### 7.2.1 Solid-Core Optical Fibers

#### 7.2.1.1 Fiber Basics

The general structure of a solid-core optical fiber (Figure 7.3) consists of a core, a clad, and a jacket. The optical radiation is guided through total internal reflection; hence, the core index of refraction is slightly higher than that of the clad. Solid-core fibers can be characterized by several criteria, the most common of which is index profile. According to this criterion there are two types of solid core fibers: (1) step index fiber, in which the refractive index profile of the core is a step function, and (2) graded index fiber, in which the refractive index of the core depends on the core radius. Fibers may also be single or multimode. Multimode fibers are the most commonly used type of fiber in medical applications.

#### 7.2.1.2 Ray Theory

Ray theory is the simplest way to understand the guiding mechanism of a solid-core fiber. It is applicable as long as the wavelength is much smaller than the fiber core diameter ($\lambda >> a$). When a ray impinges on a surface between two materials, it changes its angle of propagation (Figure 7.4) according to Snell's law,

$$n_1 \sin\theta_1 = n_2 \sin\theta_2 \ , \tag{7.1}$$

where $n$ is the material index of refraction and $\theta$ is the angle of propagation.

As can be seen from Snell's law, in an angle of incidence, $\theta_{1c}$, the angle of refraction equals 90°. Rays that impinge the surface at angles above $\theta_{1c}$ will totally reflect from the surface. These rays will be guided through the fiber while other rays will be lost and decay in the clad region (Figure 7.5). Once the

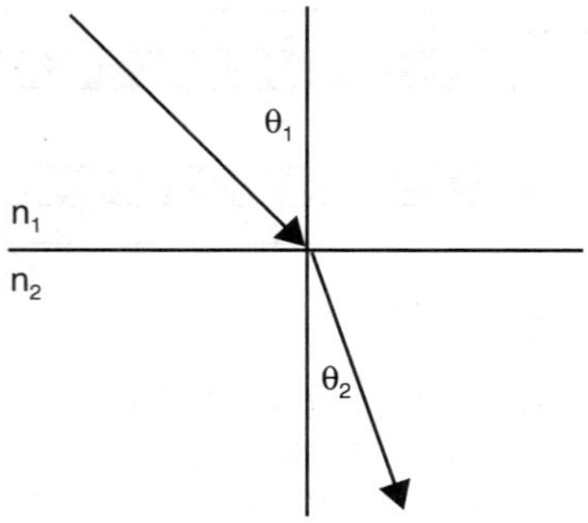


FIGURE 7.4 Snell's law.

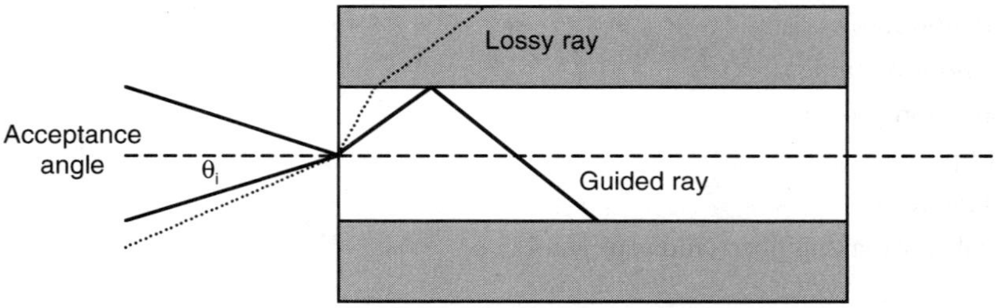


FIGURE 7.5 Guided ray and lossy ray.

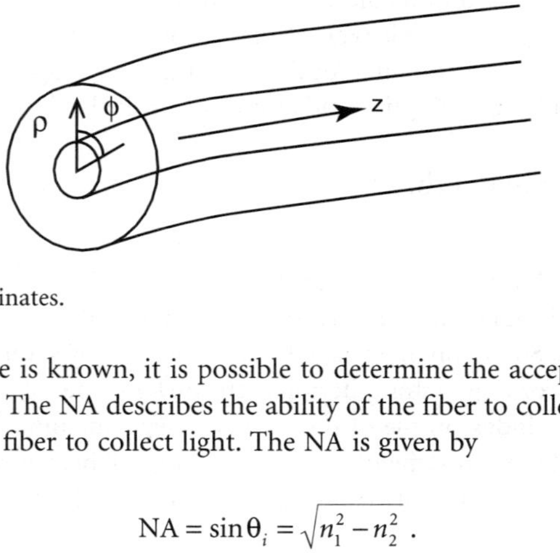


FIGURE 7.6 Cylindrical coordinates.

maximum angle of incidence is known, it is possible to determine the acceptance angle of the fiber or its numerical aperture (NA). The NA describes the ability of the fiber to collect light. The larger the NA, the greater the ability of the fiber to collect light. The NA is given by

$$\mathrm{NA} = \sin\theta_i = \sqrt{n_1^2 - n_2^2}\,. \tag{7.2}$$

### 7.2.1.3 Mode Propagation in Solid-Core Optical Fibers[4]

An alternative means of ray analysis is to solve the Maxwell equation with the appropriate boundary condition. The wave equations are given by

$$\nabla^2 E - \frac{1}{c^2}\frac{\partial^2 E}{\partial t^2} = 0; \quad \nabla^2 H - \frac{1}{c^2}\frac{\partial^2 H}{\partial t^2} = 0; \qquad \text{where } c^2 = \frac{1}{\mu\varepsilon}. \tag{7.3}$$

This is a representation of a single-frequency wave and phase velocity. The general wave motion can be represented by $\sin(wt - kz)$ or $\exp[i(\omega t - kz)]$, where $\omega$ is the *temporal frequency* and $k$ is the *spatial frequency* (or *propagation vector, wave vector, wave number*). The temporal frequency denotes the number of repetitions of a wave per unit time while the spatial frequency denotes the number of repetitions of a wave per unit distance; $\omega/k$ (or $\beta$) is known as the *phase velocity*, the velocity of any constant-phase point of a wave.

The wave equation in cylindrical coordinates (Figure 7.6) is given by

$$\frac{\partial^2}{\partial\rho^2}E_z+\frac{1}{\rho}\frac{\partial}{\partial\rho}E_z+\frac{1}{\rho^2}\frac{\partial^2}{\partial\phi^2}E_z+\frac{\partial^2}{\partial z^2}E_z-n^2k_0^2E_z=0 \tag{7.4}$$

for a step index fiber

$$n=\begin{cases} n_1 & \rho\le a \\ n_2 & \rho>a \end{cases}.$$

It is possible to solve the wave equation by separating the variables according to the coordinates

$$E_z(\rho,\phi,z)=F(\rho)\Phi(\phi)Z(z), \tag{7.5}$$

which yields

$$\frac{Z''}{Z}=\beta^2,\Rightarrow Z=e^{i\beta z}, \tag{7.6}$$

$$\frac{1}{\rho^2}\frac{\Phi''}{\Phi}=\frac{m^2}{\rho^2},\Phi=e^{im\phi}, \tag{7.7}$$

and

$$\rho^2\frac{\partial^2F}{\partial\rho^2}+\rho\frac{\partial F}{\partial\rho}+\left(k^2\rho^2-m^2\right)F=0. \tag{7.8}$$

The famous Bessel functions are the solution to the latter equation. Because we want to confine the energy only to the core, all the energy that propagates in the radial direction through the clad must decay with the distance $\rho$. Thus, in the core the solution is

$$F(\rho)=AJ_m(k\rho)+A'Y_m(k\rho)\quad \rho<a \tag{7.9}$$

where

$$k^2=n_1^2k_0^2-\beta^2 \tag{7.10}$$

and in the clad

$$F(\rho)=CK_m(\gamma\rho)+C'I_m(\gamma\rho)\quad \rho>a. \tag{7.11}$$

In the clad, the energy decreases exponentially with the distance, with a factor $\gamma$ given by

$$\gamma^2=\beta^2-n_2^2k_0^2. \tag{7.12}$$

It is possible to simplify the equations by setting

$$\begin{array}{lll} \rho \to \infty & F(\rho) \to 0 & C' = 0 \\ \rho \to 0 & F(\rho) \to \text{finite} & A' = 0 \end{array} \tag{7.13}$$

and thus obtaining

$$E_z = \begin{cases} AJ_m(k\rho)e^{im\phi}e^{i\beta z} & \rho \le a \\ CK_m(\gamma)e^{im\phi}e^{i\beta z} & p > a \end{cases} \tag{7.14}$$

and

$$H_z = \begin{cases} BJ_m(k\rho)e^{i\phi}e^{i\beta z} & \rho \ge a \\ DK_m(\rho)e^{i\phi}e^{i\beta z} & \rho > a\,. \end{cases} \tag{7.15}$$

#### 7.2.1.4 Attenuation Mechanisms in Solid-Core Fibers

There are four attenuation mechanisms in solid-core fibers. The mechanisms and their causes are listed in Table 7.1.

#### 7.2.1.5 Reflection

In solid-core fibers the difference in the index of refraction between the core and the entrance and exit surfaces causes backward reflection of some of the radiation. The amount of energy reflected depends on the difference between the index of refraction and the angle of incidence. If the ray impinges perpendicularly to the boundary the refection coefficient is given by

$$R = \left( \frac{n_1 - n_2}{n_1 + n_2} \right). \tag{7.16}$$

#### 7.2.1.6 Scattering

Scattering in solid-core fibers can be divided into two categories: intrinsic scattering and extrinsic scattering. Mie, Rayleigh, Brillouin, and Raman are the four intrinsic scattering mechanisms. Mie and Rayleigh scattering are caused by inhomogeneity in the index of refraction, granulation of the fiber material, or geometric faults in the fiber. These mechanisms cause elastic scattering, i.e., the wavelength of the radiation does not change in the process. The attenuation is proportional to $\lambda^{-4}$ and is the lower limit to the losses in the fiber.

Brillouin scattering is caused by interaction of the photon with acoustic phonons. Raman scattering is caused by interaction of the photon with acoustic photons, which changes the wavelength of the radiation.

Extrinsic scattering is caused by impurities in the fiber materials, faults during manufacturing, and the like. It is possible to decrease the effects of this type of scattering by closely controlling the manufacturing process of the fiber.

#### 7.2.1.7 Absorption

When radiation propagates through a certain material, some of the energy is transferred to heat — a phenomenon known as *absorption*. All absorption mechanisms involve moving particles (molecules or electrons) from one energy level to another. The absorption process must satisfy the following condition:

$$\lambda = \frac{hc}{\Delta E}, \tag{7.17}$$

where $\Delta E$ is the difference between the energy levels.

**TABLE 7.1** Loss Mechanisms in Fibers

| Attenuation Mechanism | Cause |
|---|---|
| Reflection (Fresnel losses) | Difference in index of refraction at the entrance and exit surfaces of the fiber |
| Scattering | Intrinsic — Mie, Rayleigh, Brillouin, Raman |
| | Extrinsic — impurities and defects |
| Absorption | Intrinsic — material |
| | Extrinsic — impurities |
| Radiation | Mode coupling |

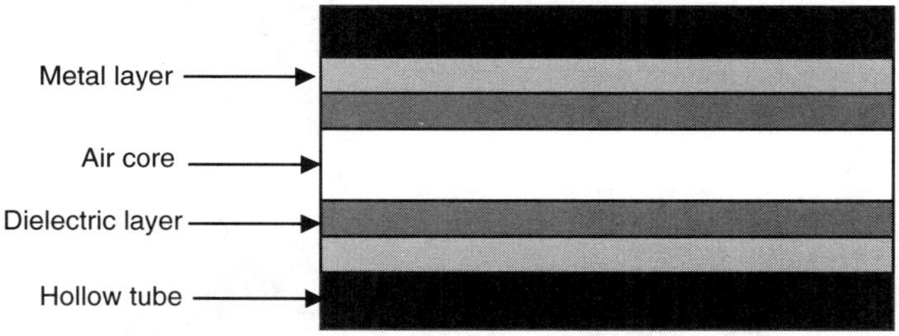


**FIGURE 7.7** General structure of a hollow waveguide.

Intrinsic absorption depends on the material and determines the range of wavelengths that the material transmits. Electronic absorption occurs at the UV range, while multiphonon absorption occurs at longer wavelengths when the wavelength corresponds to an integer multiplication of the resonance frequency. Extrinsic absorption is caused by microscopic faults in the fibers, irregularities in the core materials, and impurities.

#### 7.2.1.8 Radiation

Radiation losses are due to inappropriate coupling between the fiber and the laser beam, fiber bending, and intermodal interaction. Inappropriate coupling causes some of the rays to violate the total reflection condition and thus not to propagate through the fiber. Bending causes the rays to change their angle of incidence, and some may leave the fiber. Bending also causes coupling between low modes of propagation to higher ones. Since low modes are less attenuated, the attenuation of the fiber increases.

### 7.2.2 Hollow Waveguides[5,6]

#### 7.2.2.1 Hollow Waveguides Basics

The general structure of a hollow waveguide (Figure 7.7) consists of a hollow tube internally coated with a metal layer and a thin dielectric layer. (The dielectric layer is optional.) The optical radiation is guided through reflection from the inner layers. Hollow waveguides are characterized by guiding mechanism of the radiation. There are two types of hollow waveguides: (1) leaky waveguides, which have the same structure as that shown in Figure 7.7, and (2) attenuated total reflection (ATR) waveguides, which are made of hollow tubes. The tube index of refraction is lower than that at a region of wavelength; hence, its guiding mechanism is similar to that of solid-core fibers.

#### 7.2.2.2 Attenuation Mechanisms in Hollow Waveguides

The main attenuation mechanisms in hollow waveguides are reflection from a thin layer and scattering due to surface roughness. The first mechanism determines the wavelength that will be transmitted through the waveguide and how much it will be attenuated. In this case, the reflection coefficients are given by Fresnel equations.

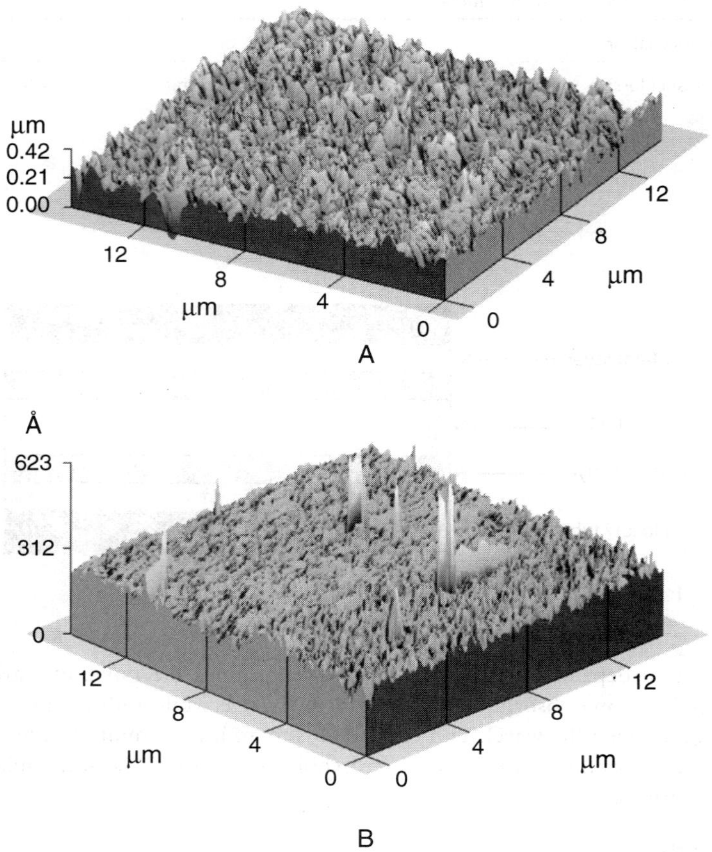


**FIGURE 7.8** AFM measurements for glass waveguide's Ag layer (A) and AgI layer (B).

The second mechanism, scattering, is caused by the roughness of the hollow tube and the deposited layers. Figure 7.8 shows the surface of Ag and AgI layers deposited on a glass tube. The measurement was made using an atomic force microscope (AFM). Measurements of the surface roughness and the height distribution enable us to compute the scattering coefficient[7] and the influence of different parameters on the waveguide's attenuation. The main influence of scattering on the ray propagation is to change the ray's propagation angle and energy distribution. Some of the energy is reflected backward and is lost; the angle of incidence of the other part is changed (usually decreased).

### 7.2.2.3 Ray Theory

One advantage of hollow waveguides is their ability to be "tuned" to almost any wavelength. The tuning is done by coating the metal layer with a thin dielectric layer, whose thickness corresponds to the desired wavelength. This can be seen from the following equation for the reflection coefficient from a thin layer:

$$r_{\text{tot}} = \frac{r_2 + r_1 e^{-2i\delta}}{1 + r_1 r_2 e^{-2i\delta}}, \tag{7.18}$$

where $r_{1/2}$ are Fresnel coefficients from a layer and $\delta$ is a phase shift that determines the wavelength and is given by

$$\delta = \frac{2\pi}{\lambda} n_1 d \cos(\phi_1) \tag{7.19}$$

To analyze the influence of the waveguide's geometrical parameters, let us calculate the distance that a ray passes between two hits on the waveguide wall; $z_i$ can be calculated using Figure 7.9 and is given by

$$z_i = 2r\tan(\phi_i), \tag{7.20}$$

where $r$ is the waveguide radius. The number of times the ray impinges on the waveguide wall of length $l$, $p_i$, is given by

$$p_i = \mathrm{int}\left(\frac{l}{z_i}\right) = \mathrm{int}\left(\frac{l}{2r\tan(\phi_i)}\right). \tag{7.21}$$

The total reflection coefficient of the ray, the multiplication of all the reflections it passed on the way, is given by

$$R_{\text{total}}(\phi_i) = R(\phi_i)^{p_i}. \tag{7.22}$$

The transmission, $T$, and the attenuation, $A$, are given by

$$T_i = \frac{I_{i,\text{out}}}{I_{i,\text{in}}} \qquad A_i = -10\log(T) \tag{7.23}$$

More explicitly, the attenuation may be written as

$$A_i = -10\left[\mathrm{int}\left(\frac{1}{4r\tan(\phi_i)}\right)\log(R(\phi_i))\right]. \tag{7.24}$$

From the last equation one can learn about the dependence of the waveguide's attenuation on its physical dimensions. It can be seen that, when the waveguide's length increases or its inner radius decreases, the attenuation increases.

As the preceding analysis shows, ray models explain the working principles of fibers and waveguides in a simple manner. A more complicated way is to solve Maxwell equations with the appropriate boundary

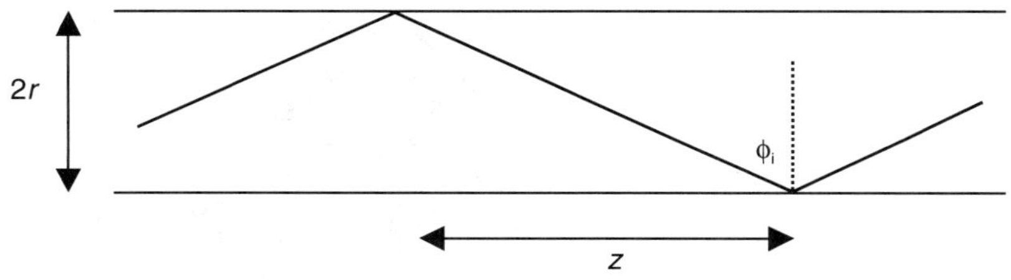


**FIGURE 7.9** Representative ray in a waveguide.

conditions. This may provide additional knowledge about the different modes that may propagate through the waveguide.

## 7.3 Multilayer Waveguides

One way to overcome the attenuation mechanisms in solid-core fibers and hollow waveguides is to develop a new type of fiber, a multilayer one. The general structure of a multilayer fiber is shown in Figure 7.10. A multilayer waveguide is made of an alternating structure of two dielectric materials. Such a structure is also known as a photonic crystal. There are two ways to analyze such a structure. The first way[8] is to use the symmetry properties of the system and quantum mechanics techniques to solve the Maxwell equation. This approach leads to an equation similar to the Bloch equation in solid-state physics. According to Bloch theory, forbidden band gaps exist through which photons (electrons in solid-state physics) with certain energies cannot propagate. In this case the photons are "perfectly" reflected; thus, such a device can have zero attenuation.

The second method is to solve the Maxwell equation with the appropriate boundary conditions.[9] Consider the linearly polarized wave shown in Figure 7.11, impinging on a thin dielectric film between two semi-infinite transparent media. Each wave $E_{rI}$, $E'_{rII}$, $E_{tII}$, and so forth, represents the result of all possible waves traveling in that direction, at that point in the medium. The summation process is therefore built in. The boundary conditions require that the tangential components of the electric field, **E**, and the magnetic field, **H**, be continuous across the boundaries.

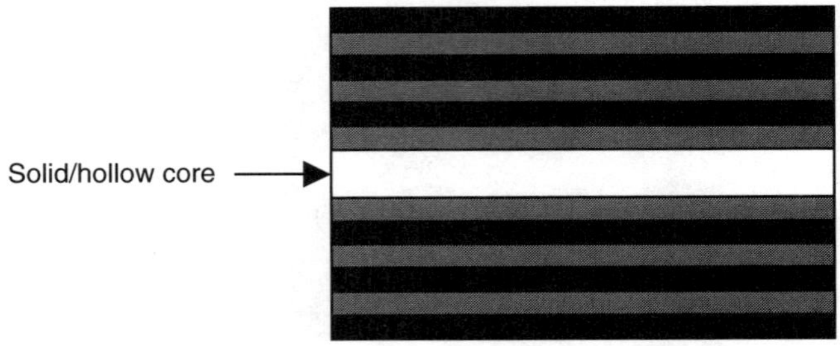


**FIGURE 7.10** Schematic drawing of a multilayer waveguide.

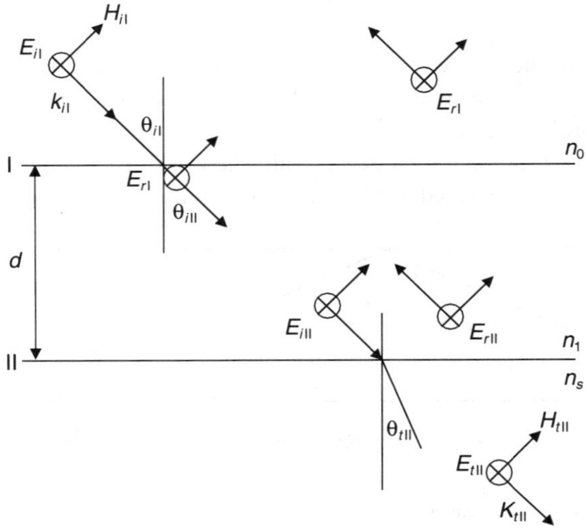


**FIGURE 7.11** Fields at the boundaries.

By applying the boundary conditions and using the Maxwell equation, it is possible to relate the fields on each side of the boundary. The relations are given by

$$\begin{bmatrix} E_{\mathrm{I}} \\ H_{\mathrm{I}} \end{bmatrix} = \begin{bmatrix} \cos(k_0 h) & \left(i\sin(k_0 h)\right)/Y_1 \\ Y_1 i\sin(k_0 h) & \cos(k_0 h) \end{bmatrix} \begin{bmatrix} E_{\mathrm{II}} \\ H_{\mathrm{II}} \end{bmatrix} \tag{7.25}$$

or

$$\begin{bmatrix} E_{\mathrm{I}} \\ H_{\mathrm{I}} \end{bmatrix} = M_I \begin{bmatrix} E_{\mathrm{II}} \\ H_{\mathrm{II}} \end{bmatrix}, \tag{7.26}$$

where

$$Y_1 \equiv \sqrt{\frac{\varepsilon_0}{\mu_0}}\, n_1 \cos\theta_{i\mathrm{II}} \tag{7.27}$$

or

$$Y_1 \equiv \sqrt{\frac{\varepsilon_0}{\mu_0}}\, n_1 \Big/ \cos\theta_{i\mathrm{II}} \tag{7.28}$$

for TE polarization and TM polarization, respectively. The same equations hold for the second and third boundaries. It follows, therefore, that if two overlaying films are deposited on the substrate, there will be three boundaries or interfaces, and now:

$$\begin{bmatrix} E_{\mathrm{II}} \\ H_{\mathrm{II}} \end{bmatrix} = M_{\mathrm{II}} \begin{bmatrix} E_{\mathrm{III}} \\ H_{\mathrm{III}} \end{bmatrix}. \tag{7.29}$$

Multiplying both sides of this expression by $M_I$, we obtain

$$\begin{bmatrix} E_{\mathrm{I}} \\ H_{\mathrm{I}} \end{bmatrix} = M_{\mathrm{I}} M_{\mathrm{II}} \begin{bmatrix} E_{\mathrm{III}} \\ H_{\mathrm{III}} \end{bmatrix}. \tag{7.30}$$

In general, if $p$ is the number of layers, each with a particular value of $n$ and $h$, then the first and the last boundaries are related by

$$\begin{bmatrix} E_{\mathrm{I}} \\ H_{\mathrm{I}} \end{bmatrix} = M_{\mathrm{I}} M_{\mathrm{II}} \ldots M_p \begin{bmatrix} E_{p+1} \\ H_{p+1} \end{bmatrix}. \tag{7.31}$$

The characteristic matrix of the entire system is the result of the product (in the proper sequence) of the individual $2 \times 2$ matrices, that is,

$$M = M_{\mathrm{I}} M_{\mathrm{II}} \ldots M_p = \begin{bmatrix} m_{11} & m_{12} \\ m_{21} & m_{22} \end{bmatrix} \tag{7.32}$$

From the above equations we can derive the reflection and transmission coefficients. Defining $Y_0$ and $Y_s$ for the medium and substrate, the reflection and transmission coefficients are given by

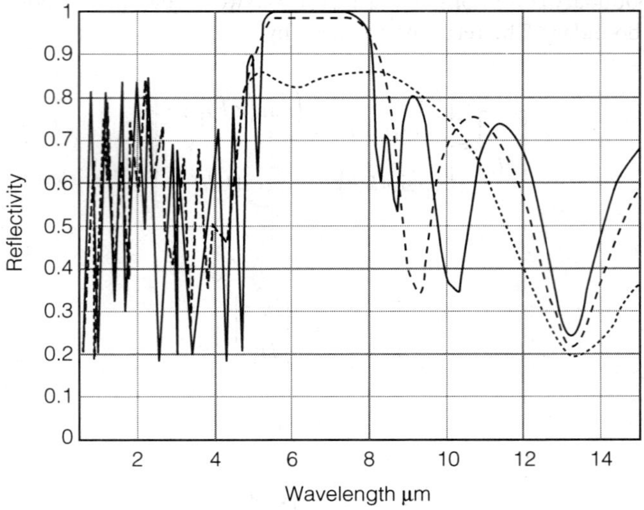


**FIGURE 7.12** Transmission as a function of wavelength (6-μm optimized waveguide).

$$r = \frac{Y_0 m_{11} + Y_0 Y_s m_{12} - m_{21} - Y_s m_{22}}{Y_0 m_{11} + Y_0 Y_s m_{12} + m_{21} + Y_s m_{22}} \tag{7.33}$$

and

$$t = \frac{2Y_0}{Y_0 m_{11} + Y_0 Y_s m_{12} + m_{21} + Y_s m_{22}} \tag{7.34}$$

To find $r$ or $t$ for any configuration of films, we need only compute the characteristic matrices for each film, multiply them, and then substitute the resulting matrix elements into the above equations.

As an example, let us design a multilayer mirror for a laser with a wavelength of 6 μm. We use two dielectric materials that have indices of refraction far apart. Such materials could be germanium ($n = 4$) and zinc selenide ($n = 2.4$). Figure 7.12 shows the reflectivity of a multilayer film made of a different number of pairs of Ge and ZnSe as a function of wavelength for a 0° angle of incidence. As the figure shows, the more pairs, the sharper the region where the reflectivity is perfect. Attempts to create photonic crystal fibers have been made by several groups in the United States, Denmark, England, Australia, and Israel.[10–12]

## 7.4 X-Ray Waveguides

X-ray radiation has been used for imaging in medicine for more than a century. X-ray laser sources have made new x-ray applications available.[13] However, x-ray radiation is difficult to manipulate because it is very energetic. The conventional way to focus x-ray radiation is coherently by using Fresnel zone plates and Bragg Fresnel lenses or incoherently by using bent crystal optics and coated fibers. The main drawback to these methods is the large spot size, which is unsuitable for most applications (x-ray spectroscopy and microscopy), particularly noninvasive medical treatments, which require waveguides.

To overcome the drawbacks of focusing lenses and crystal, Feng and co-workers[14] suggested a new device composed of a law density layer, such as carbon, surrounded by two high-density layers, such as nickel (Figure 7.13). They showed experimentally and theoretically that such a device may produce an

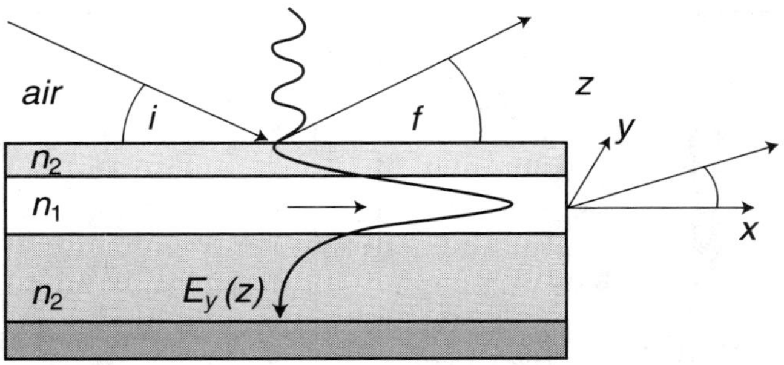


**FIGURE 7.13** Single-layer x-ray waveguide. (Courtesy of Prof. Salditt's group, Saarbrücken, Germany.)

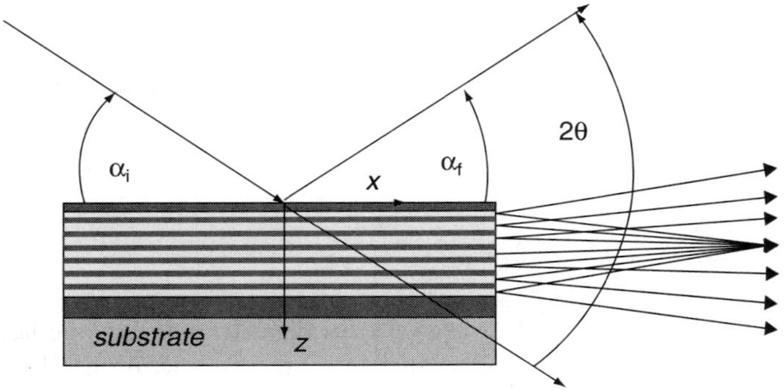


**FIGURE 7.14** Multilayer x-ray waveguide. (Courtesy of Prof. Salditt's group, Saarbrücken, Germany.)

x-ray beam with defined beam shape, divergence, and coherence, which corresponds to the geometrical properties of the device.

Grazing angle incidence couples the x-ray beam to the device. Because only a discrete set of angles can propagate through the device, the output should be coherent and its spot size should correspond to the low-density layer thickness. Using such a device, Feng et al. produced an x-ray beam with spot size of less than 100 Å. However, a single layer structure also has drawbacks. Beam shape control is difficult because it corresponds to the geometrical structure that can produce only a single spot. Coupling efficiency is very limited, even poor, because a single layer has a very small angular acceptance.

Pfeiffer et al.[15,16] suggested a multilayer structure to overcome the single layer structure drawbacks. The new structure (Figure 7.14) consists of one or more guiding layers and enables generation of more complicated beam shapes (Figure 7.15). This new concept enables new ways to perform x-ray spectroscopy and holography.

## 7.5 Coupling Devices

The laser beam can be coupled to the fiber in several ways. The first and most straightforward way is to use standard lenses. Other and more sophisticated ways involve different types of tapers.[17,18]

Coupling the laser beam into the fiber using an ordinary lens is simple. All one has to do is to ensure that the focused laser spot size is less than the fiber's core diameter, that the NA of the lens is suitable to that of the fiber, and that the optical axis of the lens and the fiber are aligned.

The laser spot size at the entrance to the waveguide is given by

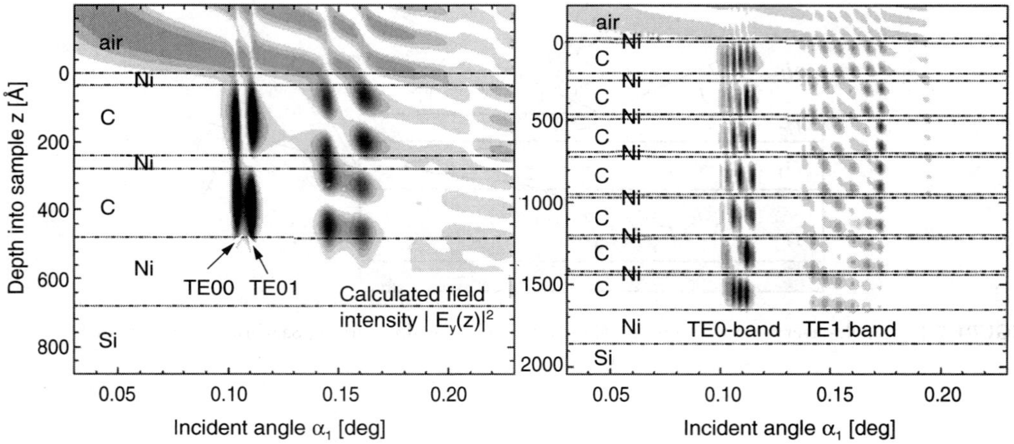


**FIGURE 7.15** Field distribution of different multilayer x-ray waveguide. (Courtesy of Prof. Salditt's group, Saarbrücken, Germany.)

$$\omega_0 = \frac{1.9\lambda f}{D} \tag{7.35}$$

where $f$ is the focal length of the coupling lens, $D$ is its diameter, and $\lambda$ is the laser wavelength. The maximum angle (i.e., the angle that determines the NA) at the entrance of the fiber is given by

$$\tan(\varphi) = \frac{D}{f} \tag{7.36}$$

If the laser spot size is larger than the core diameter of the fiber, then some of the energy will enter the fiber clad and be lost. Furthermore, if the laser beam is energetic enough it can cause damage to the fiber. If the NA of the focal end is greater than that of the fiber, then some of the rays will not be guided through the fiber, i.e., they would be lossy rays.

An alternative way to couple a laser beam into a fiber is to use a taper — a device that is attached to the fiber (see Figure 7.16). A taper increases the angle propagation through the waveguide and thus decreases the attenuation. Furthermore, because the taper changes the angle of propagation through the fiber it also functions as a higher mode filter or coupler, i.e., it couples higher modes of propagation to lower modes of propagation.

Let us look at a laser beam with a divergence angle, $\theta$, and a fiber with an NA that corresponds to a certain angle, $\phi$, as depicted in Figure 7.16. We want to find the angle $\beta$ of the taper that would enable us to couple the laser beam into the fiber successfully, i.e., a ray that emerges from the laser with an angle equal to $\theta$ will propagate through the fiber.

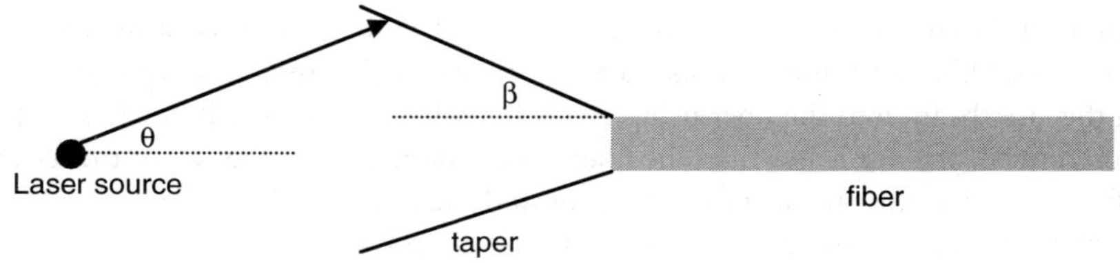


**FIGURE 7.16** Tapering.

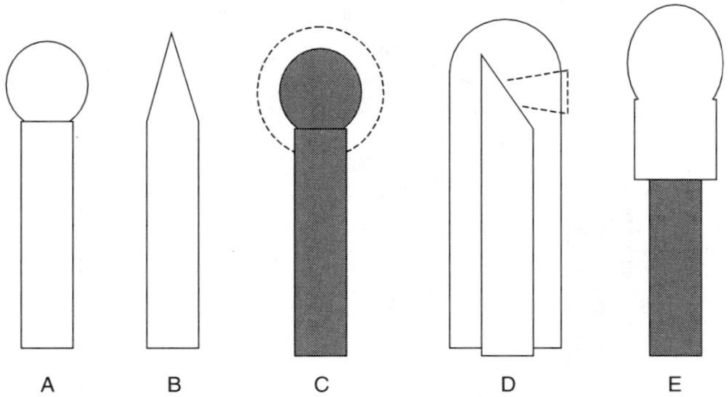


**FIGURE 7.17** Types of fiber tips: (A) ball shape; (B) tapered; (C) diffusing probe; (D) side firing; (E) metal probe.

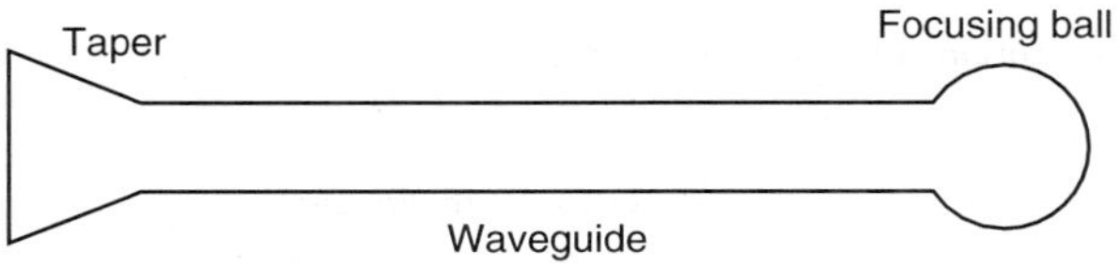


**FIGURE 7.18** Waveguide with input taper and ball shape distal tip.

A simple geometric calculation shows that the angle β equals

$$\beta = \frac{\phi - \theta}{2}. \quad (7.37)$$

## 7.6 Distal Tips

Although coupling the laser beam into the fiber is very important, in many medical applications, controlling the beam shape at the fiber's exit is equally important. Laser beam manipulation at the end of the fiber is done using different types of tips (see Figure 7.17). The tips can be made by polishing the fiber materials (in the case of solid-core fiber) or from different materials adhered to the fiber.

The different tips play different roles in the various types of medical application. A few examples are described here (see Figure 7.17). Tip C is used in photodynamic therapy (PDT)[19] application where radiation should diffuse to the surrounding area. Tip D is used in side firing of the laser beam to ablate tissue in the esophagus or colon. Tip E is used in thermotherapy.[20]

We now examine the change in beam shape when an input taper and a ball-shape distal tip are added to a straight waveguide. We performed ray tracing on a straight waveguide and drew the beam shape of the emerging radiation at 20 cm after waveguide output. The result is shown in Figure 7.19. We have added an input taper and a ball shape distal tip and recalculated the beamshape in the new conditions. The results are shown in Figure 7.20.

The focusing of the beam is easily observed. Another advantage of the ball is that is seals the waveguide and enables work in liquid medium, such as in body cavities.

## 7.7 Materials for Fabrication of Optical Fibers and Waveguides

### 7.7.1 Silica Fibers[21–24]

Silica fibers constitute the backbone of optical communication and are manufactured by many companies. The first silica fibers, manufactured by Corning in the mid-1970s, had an attenuation of about 20 dB/km. The manufacturing process had steadily improved, and attenuation has almost reached its theoretical

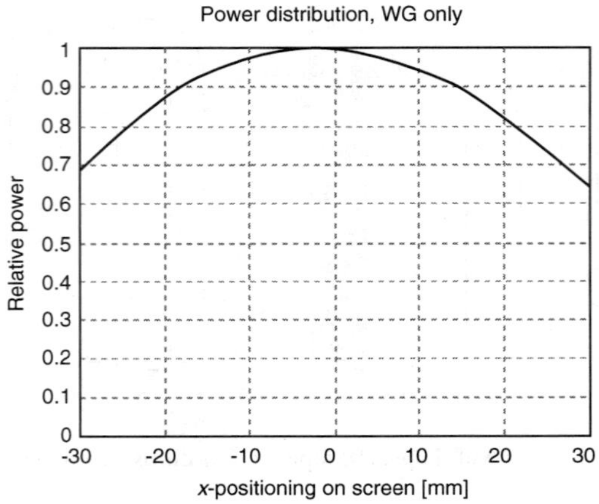


**FIGURE 7.19** Energy distribution at the output of a straight waveguide.

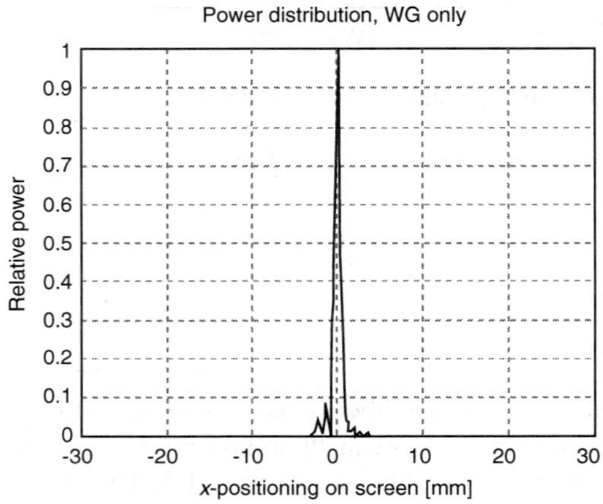


**FIGURE 7.20** Beam shape at the a distance of 20 cm after the addition of the input taper and distal tip.

limit. (The typical attenuation of a silica fiber is about 0.2 dB/km.) The spectral range of the fibers begins at the UV region (200 nm) and ends at the NIR (2000 nm). These fibers are useful for telecommunication, medical applications, power delivery, and remote sensing.

Silica fibers are solid-core fibers. Their basic structure is core clad; the core is made of silica ($SiO_2$) and the clad is made of silica doped with other elements such as germanium (Ge), boron (B), fluoride (F), etc. Doping the clad is needed in order to change its index of refraction. The fiber is coated with a plastic material (polyimide) that protects it and improves its bending capabilities. Depending on the core diameter, the fiber can transmit either single-mode or multimode radiation. The fiber can be drawn at any length and its transmission characteristics are easily analyzed using conventional ray-tracing models.

It is possible to change the transmission characteristics by manufacturing multicomponent silica glasses, in which silica is less than 75%/w, or by high silica content fibers, in which silica comprises more than 90%. It was hoped that these types of fibers would have better transmission capabilities; however, impurities in these fibers increase scattering and high OH concentration increases absorption, thus increasing attenuation. Furthermore, multicomponent glasses have intrinsically lower strength and radiation hardness, while high-content silica fibers are very sensitive to moisture. These drawbacks limit their applications.

TABLE 7.2 Types of Hollow Waveguides

| Tube | Metal Layer | Dielectric Layer | Group |
|---|---|---|---|
| Teflon, polyimide, fused silica, glass | Ag | AgI | Croitoru et al. |
| Nickel, plastic, glass, fused silica | Ag, Ni, Al | Si, ZnS, Ge, polymers | Miyagi et al. |
| Plastic, glass, fused silica | Ag | AgI | Harrington et al. |
| Silver | Ag | AgI | Morrow et al. |
| Stainless steel | Ag, Al, Au | ZnSe, $PbF_2$ | Laakman et al. |

The OH concentration within the silica determines the transparency of the fiber. Although low OH concentration is needed for transmission in the IR region, a higher concentration contributes to a large transmission in the UV.

### 7.7.2 Hollow Waveguides[25–39]

Hollow waveguides were initially developed for IR radiation delivery, but they may be adapted to almost any wavelength. The basic structure of a hollow waveguide is shown in Figure 7.6. A hollow waveguide consists of a hollow tube (glass, fused silica, plastic, etc.), internally coated with a metal layer (silver, nickel, gold, etc.) and a dielectric layer (AgI, Se, Ge, polymers).

Hollow waveguides are made from nontoxic materials. They have low attenuation (less than 1 dB at the IR region) and deliver high laser power (1 kW of $CO_2$ laser was demonstrated). They may be tuned to any desired wavelength by adapting the thickness of the dielectric layer and choosing the appropriate tube and inner layers. These characteristics explain why hollow waveguides are suitable for many medical applications.

Over the years most research on hollow waveguides concentrated on developing hollow waveguides for the mid-IR region because of the lack of materials and fibers with desirable characteristics needed for medical applications. Until now, different hollow waveguides have been developed for the IR region, especially for wavelengths of interest for medical applications.

Several research groups around the world have investigated hollow waveguides. Table 7.2 shows different hollow waveguides for different lasers. Their attenuation of a straight waveguide varies between 0.2 and 1 dB/m, depending on tube roughness and laser wavelength.

Another type of hollow waveguide is attenuated total reflection (ATR). These waveguides exploit the anomalous dispersion phenomena. Anomalous dispersion means that, at a certain wavelength, the material has an index of refraction of less than one. In this case, the guiding mechanism of the ATR waveguide is similar to that of a solid-core fiber.

## 7.8 Fibers for the IR Region

Silica fibers are suitable for a wide range of wavelengths from the UV to the NIR. Their manufacturing process is well controlled; hence, their low attenuation. This type of fiber is suitable up to 2000 nm, beyond which there is a need for new materials and fibers. For the past three decades intensive research has been conducted to develop optical fibers for the mid- and far-IR regions. One solution, hollow waveguides, was described in Section 7.7.2.

### 7.8.1 Glass Fibers

There are several glass-based fibers for the IR region. Silica-based fibers were discussed in Section 7.7.1. Other types are fluoride-based glass and chalcogenide fibers.

#### 7.8.1.1 Fluoride-Based Glass[40–42]

Fluoride glass fibers are based on $ZrF_4$. They offer many advantages over silica-based glasses. Their theoretical attenuation is about 0.01 dB/km, which is an order of magnitude lower than that of silica.

TABLE 7.3 Attenuation of Chalcogenide Fibers

| Chalcogenide Glass | Attenuation [dB/m] | Wavelength [μm] |
|---|---|---|
| Selenide | 10 | 10.6 |
| Sulfide | 0.3 | 2.4 |
| Telluride | 1.5 | 10.6 |
| | 4 | 7 |

This is due to a shift to longer wavelength in phonon absorption. They also posses a negative thermo-optic coefficient and are good hosts for rare earth elements. Such characteristics make fluoride glass fibers good candidates for optical fiber communications, laser amplifier, and chemical sensing applications.

Over the years different compounds of fluoride glass were investigated. The first compounds were composed of $BeF_2$. These compounds are very stable and resistant to damage. They are transparent in a wide region of wavelengths including the UV. However Be is a very toxic element and the hygroscopic nature of the compound make it difficult to handle, thus eliminating potential medical applications.

Other fluoride compounds are heavy metal fluoride glasses that include fluorozirconate ($ZrF_4$). Such compounds are ZB — $ZrF_4$–$BaF_2$, ZBT, and ZBL, which are ZB that include $ThF_4$ and $LaF_3$, respectively. Others are ZBLA–$ZrF_4$–$BaF_2$–$LaF_3$–$AlF_3$ and ZABLAN — $ZrF_4$–$BaF_2$–$LaF_3$–$AlF_3$–NaF. The main disadvantage of ZB is its instability. Adding other components to the compound increases its stability.

All fluoride glasses are transparent from the UV to the mid-IR and some of them even to the far-IR. However, it is difficult to achieve the ultralow loss that is predicted, especially due to bubble formation during fiber drawing, microcrystalization, and material contamination. Typical fluoride fiber loss is on the order of 10 dB/km.

#### 7.8.1.2 Chalcogenide Fibers[43–45]

Chalcogenide glass is formed when elements from groups 4B and 5B, such as As, Ge, Pb, Ga, Si, etc., are melted into silica glass. There are three groups of chalcogenide glass: sulfide, selenide, and telluride. The transparency of the glass depends on the glass composition. Sulfide glass is transparent in the visible region, while incorporating selenide or telluride into the glass shifts the transparency into the IR region.

Mechanical properties also depend on the glass composition. Although the mechanical strength and thermal stability are lower than those of silica glass they are sufficient for fiber drawing and typical fiber applications. The attenuation of chalcogenide glass also depends on the glass compound. Table 7.3 summarizes the attenuation of the different kinds of chalcogenide fibers.

The main drawback of these fibers is their poisonous nature. This characteristic makes them difficult to handle and limits their use in medical applications, unless protective coating is added.

## 7.8.2 Crystalline Fibers

About 100 crystalline materials are transparent in the infrared region. Most of these materials have similar optical characteristics but very different mechanical properties. Although transparency in a desired wavelength region is imperative, the mechanical strength and durability of the fibers cannot be overlooked.

#### 7.8.2.1 Single-Crystal Fibers

Single-crystal materials have many advantages. It is possible to obtain a very pure material, thus overcoming scattering from impurities in the fiber material. Most of the materials have a high melting point, which makes them suitable for laser power delivery. They are also chemically inert, and thus can operate in harsh environments. An example of such a material is sapphire.

These materials also have drawbacks. Unlike silica fibers that can be drawn to considerable length, single-crystal fibers have to be grown from a melt; thus, long fibers are difficult to manufacture. Some of the crystals are poisonous and brittle.

- **Sapphire fibers.**[46–48] Sapphire is transparent up to 3.5 μm. These fibers are grown at very low rates (about 1 mm/min). They are made as core-only fibers and have a core diameter between 0.18 and 0.55 mm. Their attenuation is 2 dB/m at 2.93 μm and maximum power delivery is 600 mJ. Sapphire fibers have very good mechanical properties. They have a high melting point and mechanical strength. They are chemically inert, insoluble in water, and biocompatible. These properties make them good candidates for medical applications.
- **AgBr.**[49] These fibers are manufactured by a pulling method, in which the material is heated and then pulled through a nozzle, or by pressing the material through a nozzle. Fibers of up to 2 m have been manufactured using these methods. The attenuation of such fibers is 1 dB/m at 10.6 μm. AgBr fiber may transmit up to 4 W of $CO_2$ laser.

#### 7.8.2.2 Polycrystalline Fibers

Polycrystalline fibers made of TlBr-TlI (KRS-5), KCl, AgCl, NaCl, and other compounds of heavy metals and halides are attractive because of their low theoretical attenuation. However, the attenuation of the first fibers manufactured was much higher and most of the early research concentrated on finding the origin of that loss.

Although transparency at the desired wavelength region is imperative, polycrystalline fibers must have additional attributes. The material must be deformed plastically in order to enable manufacturing of sufficiently long enough fibers. Furthermore, the crystal must be solid in nature and optically isotropic to avoid intrinsic scattering. Finally, the recrystallization of the material after fiber manufacturing must not impair the fiber optical characteristics. Potential materials for fiber manufacturing are thallium halides and silver halides, which are made using the extrusion method.

- **Thallium halides.**[50–52] Thallium halides have low theoretical attenuation (about 6.5 dB/km). However, the measured attenuation is much larger, between 120 and 350 dB/km. Scattering caused by material impurities, surface imperfection, grain boundaries, and dislocation lines is the main reason for the large attenuation. These fibers suffer from aging effects. They are soluble in water and sensitive to UV light.
- **Silver halide.**[53–55] Silver halide fibers are made of $AgCl_xBr_{1-x}$. They have very low attenuation at 10.6 μm to 0.15 dB/m. The main reason for the attenuation is bulk scattering, and intrinsic and extrinsic absorption.

### 7.8.3 Liquid-Core Fibers[56–57]

Liquid-core fibers are hollow tubes filled with liquid that is transparent in the IR region. The only advantage of these fibers over solid-core fibers is the possibility of obtaining a purer liquid than solid material. The liquids used are $C_2Cl_4$, which has a very high attenuation of about 100 dB/m at 3.39 μm, and $CCl_4$, with an attenuation of about 4 dB/m at 2.94 μm.

## 7.9 Conclusions

Fibers and waveguides play a major role in the application of lasers in medicine. A great variety of materials are used for laser transmission. Each fiber or waveguide can be used in some applications, but none can be used in all applications. Fibers need to be improved further for better laser transmission, wider range of spectrum, better beam shape at the output of the fibers, higher energy transmission, and ultrashort energetic pulses.

The ultimate optical fiber or waveguide for medical applications has yet to be developed.

## References

1. Cilesiz, I. and Katzir, A., Thermal-feedback-controlled coagulation of egg white by the $CO_2$ laser, *Appl. Opt.*, 40(19), 3268, 2001.

2. Gotshal, Y., Simhi, R., Sela, B.-A., and Katzir, A., Blood diagnostics using fiberoptic evanescent wave spectroscopy and neural networks analysis, *Sensors Actuators B Chem.*, B42(3), 157, 1997.
3. Gannot, I., Gannot, G., Garashi, A., Gandjbakhche, A., Buchner, A., and Keisari, Y., Laser activated fluorescence measurements and morphological features — an *in vivo* study of clearance time of FITC tagged cell markers, *J. Biomed. Opt.*, 7(1), 14, 2002.
4. Kawano, K. and Kitoh, T., *Introduction to Optical Waveguide Analysis*, John Wiley & Sons, New York, 2001.
5. Ben-David, M., Inberg, A., Gannot, I., and Croitoru, N., The effect of scattering on the transmission of IR radiation through hollow waveguides, *J. Optoelectron. Adv. Mater.*, 3, 23, 1999.
6. Inberg, A., Ben-David, M., Oksman, M., Katzir, A., and Croitoru, N., Theoretical and experimental studies of infrared radiation propagation in hollow waveguides, *Opt. Eng.*, 39(5), 1384, 2000.
7. Beckman, P. and Spizzichino, A., *The Scattering of Electromagnetic Waves from Rough Surfaces*, Pergamon Press, New York, 1963.
8. Joannopoulos, J.D., Meade, R.D., and Winn, J.N., *Photonic Crystals — Molding the Flow of Light*, Princeton University Press, Princeton, NJ, 1995, p. 3.
9. Gannot, I., Ben-David, M., Inberg, A., Croitoru, N., and Katzir, A., Mid-IR optimized multi-layer hollow waveguides, *BIOS 2001*, 4253, 11, 2001.
10. Hart, S.D., Maskaly, G.R., Temelkuran, B., Prideaux, P.H., Joannopoulos, J.D., and Fink, Y., External reflection from omnidirectional dielectric mirror fibers, *Science*, 296, 510, 2002.
11. Bjarklev, A., Broeng, J., Libori, S.E., and Knudsen, E., Photonic crystal fibers: a variety of applications, *Optical Fibers and Sensors for Medical Applications II*, Gannot, I., Ed., *Proc. SPIE*, 4616, 73, 2002.
12. Knight, J., Broeng, J., Birks, T.A., and Russell, P., Photonic band gap guidance in optical fibers, *Science*, 282, 1476, 1998.
13. Carroll, F. and Brau, C., Vanderbilt MFEL program, X-ray project, http://www.vanderbilt.edu/fel/Xray/default.html.
14. Feng, Y.P., Sinha, S.K., Deckman, H.W., Hastings, J.B., and Siddons, D.P., X-ray flux enhancement in thin-film waveguides using resonant beam couplers, *Phys. Rev. Lett.*, 71(4), 537, 1993.
15. Pfeiffer, F., X-ray waveguides, Diploma thesis, University of Munich, Munich, Germany, July 1999.
16. Pfeiffer, F., Salditt, T., Høghøj, P., Anderson, I., and Schell, N., X-ray waveguides with multiple guiding layers, *Phys. Rev. B*, 62, 16939, 2000.
17. Yaegashi, M., Matsuura, Y., and Miyagi, M., Hollow-tapered launching coupler for Er:YAG lasers, *Rev. Laser Eng.*, 28(8), 516, 2000.
18. Ilev, I. and Waynant, R., Uncoated hollow taper as a simple optical funnel for laser delivery, *Rev. Sci. Instrum.*, 70, 3840, 1999.
19. Marynissen, J.P., Jansen, H., and Star, W.M., Treatment system for whole bladder wall photodynamic therapy with *in vivo* monitoring and control of light dose rate and dose, *J. Urol.*, 142(5), 1351, 1989.
20. Tong, L., Zhu, D., Luo, Q., and Hong, D., A laser pumped Nd($^{3+}$)-doped YAG fiber-optic thermal tip for laser thermotherapy, *Lasers Surg. Med.*, 30(1), 67, 2002.
21. Sanghera, J.S. and Aggarwal, I.D., *Infrared Fiber Optics*, CRC Press, Boca Raton, FL, 1998.
22. Greek, L.S., Schulze, H.G., Blades, M.W., Haynes, C.A., Klein, K.-F., and Turner, R.F.B., Fiber-optic probes with improved excitation and collection efficiency for deep-UV Raman and resonance Raman spectroscopy, *Appl. Opt.*, 37, 170, 1998.
23. Karlitschek, P., Hillrichs, G., and Klein, K.-F., Influence of hydrogen on the colour center formation in optical fibers induced by pulsed UV-laser radiation. 2. All-silica fibers with low-OH undoped core, *Opt. Commun.*, 155(4–6), 386, 1998.
24. Klein, K.F., Arndt, K.F., Hillrichs, R., Ruetting, G., Veidemanis, M., Dreiskemper, M., Clarkin, R., Nelson, J.P., and Gary, W., UV fibers for applications below 200 nm, *Optical Fibers and Sensors for Medical Applications*, Gannot, I., Ed., *Proc. SPIE*, 4253, 42, 2001.

25. Miyagi, M. and Matsuura, Y., Delivery of $F_2$-excimer laser light by aluminum hollow fibers, *Opt. Express*, 6(13), 257, 2000.
26. Matsuura, Y., Yamamoto, T., and Miyagi, M., Hollow fiber delivery of F2-excimer laser light, *Optical Fibers and Sensors for Medical Applications*, Gannot, I., Ed., *Proc. SPIE,* 4253, 37, 2001.
27. Gannot, I., Schruener, S., Dror, J., Inberg, A., Ertl, T., Tschepe, J., Muller, G.J., and Croitoru, N., Flexible waveguides for Er:YAG laser radiation delivery, *IEEE Trans. Biomed. Eng.*, 42, 967, 1995.
28. Gannot, I., Inberg, A., Croitoru, N., and Waynant, R.W., Flexible waveguides for free electron laser radiation transmission, *Appl. Opt.*, 36(25), 6289, 1997.
29. Miyagi, M., Harada, K., Aizawa, Y., and Kawakami, S., Transmission properties of dielectric-coated metallic waveguides for infrared transmission, *Infrared Optical Materials and Fibers III, Proc. SPIE,* 484, 1984.
30. Matsuura, Y., Miyagi, M., and Hongo, A., Dielectric-coated metallic hollow waveguide for 3 μm Er:YAG, 5 μm CO, and 10.6 μm $CO_2$ laser light transmission, *Appl. Opt.*, 29, 2213, 1990.
31. Harrington, J., A review of IR transmitting hollow waveguides, *Fibers Integ. Opt.*, 19, 211, 2000.
32. Bhardwaj, P., Gregory, O.J., Morrow, C., Gu, G., and Burbank, K., Performance of a dielectric coated monolithic hollow metallic waveguide, *Mater. Lett.*, 16, 150, 1993.
33. Laakman, K.D., Hollow Waveguides, U.S. patent 4,652,083, 1985.
34. Laakman, K.D. and Levy, M.B., U.S. patent 5,500,944, 1991.
35. Garmire, E., Hollow metal waveguides with rectangular cross section for high power transmission, *Infrared Optical Materials and Fibers III, Proc. SPIE,* 484, 1984.
36. Haidaka, T., Morikawa, T., and Shimada, J., Hollow core oxide glass cladding optical fibers for middle infrared region, *J. Appl. Phys.*, 52, 4467, 1981.
37. Falciai, R., Gireni, G., and Scheggi, A.M., Oxide glass hollow fibers for $CO_2$ laser radiation transmission, *Novel Optical Fiber Techniques for Medical Application, Proc. SPIE,* 494, 1984.
38. Gregory, C.C. and Harrington, J.A., Attenuation, modal, polarization properties of $n < 1$ hollow dielectric waveguides, *Appl. Opt.*, 32, 5302, 1993.
39. Nubling, R. and Harrington, J.A., Hollow waveguide delivery systems for high power industrial $CO_2$ lasers, *Appl. Opt.*, 34, 372, 1996.
40. Fujiura, K., Nishida, Y., Kanamori, T., Terunuma, Y., Hoshino, K., Nakagawa, K., Ohishi, Y., and Sudo, S., Reliability of rare-earth-doped fluoride fibers for optical fiber amplifier application, *IEEE Photonics Technol. Lett.*, 10, 946, 1998.
41. Peizhen, D., Ruihua, L., Po, Z., Haobing, W., and Fuxi, Y., Defects and scattering loss in fluoride fibers: a review, *J. Non-Crystalline Solids*, 140(1–3), 307, 1992.
42. Busse, L.E. and Aggarwal, I.D., Design parameters for multimode fluoride fibers: effects of coating and bending on loss, *Opt. Fiber Mater. Process. Symp., Mater. Res. Soc.*, 177, Pittsburgh, PA, 1990.
43. Moon, J.A. and Schaafsma, D.T., Chalcogenide fibers: an overview of selected applications, *Fiber Integ. Opt.*, 19(3), 201, 2000.
44. Busse, L.E., Moon, J.A., Sanghera, J.S., and Aggarwal, I.D., Mid-IR high power transmission through chalcogenide fibers: current results and future challenges, *Proc. SPIE*, 2966, 553, 1997.
45. Hocde, S., Loreal, O., Sire, O., Turlin, B., Boussard-Pledel, C., Le Coq, D., Bureau, B., Fonteneau, G., Pigeon, C., Leroyer, P., and Lucas, J., Biological tissue infrared analysis by chalcogenide glass optical fiber spectroscopy, *Biomonitoring Endosc. Technol., Proc. SPIE*, 4158, 49, 2000.
46. Merberg, G.N. and Harrington, J.A., Optical and mechanical properties of single-crystal sapphire optical fibers, *Appl. Opt.*, 32, 3201, 1993.
47. Merberg, G.N., Current status of infrared fiber optics for medical laser power delivery, *Lasers Surg. Med.*, 13(5), 572, 1993.
48. Waynant, R.W., Oshry, S., and Fink, M., Infrared measurements of sapphire fibers for medical applications, *Appl. Opt.*, 32, 390, 1993.
49. Bridges, T.J., Hasiak, S., and Strand, A.R., Single crystal AgBr infrared optical fibers, *Opt. Lett.*, 5, 85, 1980.

50. Ikedo, M., Watari, M., Tateshi, F., and Ishiwatwri, H., Preparation and characterization of the TLBr-TlI fiber for a high power $CO_2$ laser beam, *J. Appl. Phys.*, 60, 3035, 1986.
51. Artjushenko, V.G., Butvina, L.N., Vojteskhosky, V.V., Dianov, E.M., and Kolesnikov, J.G., Mechanism of optical losses in polycrystalline KRS-% fibers, *J. Lightwave Technol.*, 4, 461, 1986.
52. Saito, M., Takizawa, M., and Miyagi, M., Optical and mechanical properties of infrared fibers, *J. Lightwave Technol.*, 6, 233, 1988.
53. Alimpiev, S.S., Artjushenko, V.G., Butvina, L.N., Vartapetov, S.K., Dianov, E.M., Kolesnikov, Yu.G., Konov, V.I., Nabatov, A.O., Nikiforov, S.M., and Mirakjan, M.M., Polycrystalline, IR fibers for laser scalpels, *Int. J. Optoelectron.*, 3(4), 333, 1988.
54. Saar, A., Barkay, N., Moser, F., Schnitzer, I., Levite, A., and Katzir, A., Optical and mechanical properties of silver halide fibers, *Proc. SPIE*, 843, 98, 1987.
55. Saar, A. and Katzir, A., Intrinsic losses in mixed silver halide fibers, *Proc. SPIE*, 1048, 24, 1989.
56. Takahashi, H., Sugimoto, I., Takabayashi, T., and Yoshida, S., Optical transmission loss of liquid-core silica fibers in the infrared region, *Opt. Commun.*, 53, 164, 1985.
57. Klein, S., Meister, J., Diemer, S., Jung, R., Fuss, W., and Hering, P., High-power laser waveguide with a circulating liquid core for IR applications in *Specialty Fiber Optics for Biomedical and Industrial Applications*, Katzir, A. and Harrington, J.A., Eds., *Proc. SPIE*, 2977, 155, 1997.

# 8

# Biological Imaging Spectroscopy

Gregory Bearman
*California Institute of Technology*
*Pasadena, California*

Richard Levenson
*Cambridge Research and Instrumentation*
*Woburn, Massachusetts*



## 8.1 Introduction

Improved detectors, new electro-optical devices, and vastly improved computational power for data analysis have fueled the recent interest in combining biology and spectroscopy. This chapter examines three aspects of biomedical imaging spectroscopy:

1. Data acquisition: What instruments are available for acquiring an image cube and what are the performance trade-offs involved in choosing one over the other?
2. Data analysis: What are some of the approaches for examining very large and multivariate data sets? We shall see that the remote sensing community, focused primarily on geology, has many tools that can be applied to biomedical data.
3. Applications: Which current research areas in biology and medicine can exploit the power of imaging spectroscopy?

It is reasonable to wonder what distinguishes spectral imaging from standard red, green, blue (RGB), full-color imaging. After all, our computer monitors tell us they can display 16.7 million colors — surely that should be enough. To answer this question, we need to understand the difference between "color" and spectral content. Light is composed of photons with different energies. Although we can think of higher-energy (shorter wavelength) photons as "blue" and less energetic (longer wavelength) photons as "red," these color attributes are an artifact of the human visual system. In fact, no simple relationship exists between wavelength content of light and the color we actually perceive. This is (in part) because our eyes (and conventional color film and color digital cameras) allocate visible light, no matter how

0-8493-1116-0/03/$0.00+$1.50

spectrally complex, into only about three different color bins: red, green, and blue. Light with completely different spectral content can have precisely the same RGB coordinates, a phenomenon known as metamerism. For example, red light and green light can combine to form yellow. If we see a yellow object, we cannot tell if the color is spectrally pure (as it would be if it were created by a prism or rainbow) or if it arose from a mixture of red and green.

Researchers have used human color vision to interpret images since the first microscope was developed. Although we perceive three spectral bands and cover a relatively narrow range, the human eye is quite sensitive to subtle color differences within that range. When exogenous dyes were used to color cellular structures or molecules differentially, the interpretation still relied on color vision, and more recently, on electronic color cameras. The addition of fluorescent dyes to the microscopist's tool kit began to push the limit of color vision, electronic or otherwise. The standard detection tool kit of fluorescence microscopy is an array of dichroic mirrors, filter cubes, and other filters designed to separate multiple colored probes in absorption or in fluorescence emission. Increasing the number of probes, as biologists want to do, creates so much spectral overlap that filtering cannot separate the probes; i.e., color images of fluorescent probes that differ only slightly spectrally appear the same. In that case, we need to use some sort of spectroscopy.

Spectroscopy usually uses single point detectors that cannot easily sample large areas or small areas at high resolution. On the other hand, imaging spectroscopy can spectrally image large areas, combining the function of a camera (recording spatial information) with that of a spectrometer. These devices can measure the spectral content of light at every point in the image: a $1000 \times 1000$ pixel sensor provides 1 million individual spectra. Once a spectral stack is acquired, mathematical approaches ranging from simple to very sophisticated can be used for analysis. Analysis of fluorescence microscopy uses spectral signatures to match each pixel with one of the known probes used in the experiment. Imaging spectroscopy tells us *what* is *where*.

Once properly calibrated, these images can be used to obtain corrected spectrum for each image pixel, which can then be used to identify components in the target. For the geologist, imaging spectroscopy yields compositional maps of geologic sites to show *which* minerals are *where*[1] or to determine the composition of the rain forest canopy.[2] It can detect agricultural pests,[3] drought stress, or fertilizer application levels. Spectral imaging has uses in industrial process control, in detection of ordinarily invisible bruising in fruit, in assessing the viability of transplanted organs,[4,5] in uncovering forgeries, etc. Finally, modifications in existing designs and novel approaches have made spectral imaging easy to accomplish with a microscope; this combination has promising applications in surgical pathology and molecular biology. Fluorescent dyes, which have recently become available, will increase the usefulness of spectral imaging in a variety of areas, including high-throughput screening, genomics, and clinical diagnostics.

## 8.2 Spectral Image Cubes

Simply put, an imaging spectrometer acquires the spectrum of each pixel in a two-dimensional spatial scene. As shown in Figure 8.1, the easiest way to think of such a scheme is as band sequential imaging, in which multiple images of the same scene at different wavelengths are acquired. A key point is that the spectra be sampled densely enough to reassemble a spectrum (commensurate with the needs for analysis). A remote sensing instrument may take hundreds more images over the visible to near-infrared (NIR) range. There are many technological means of obtaining these data, and this chapter presents a catalog of current technologies. The images are typically stacked in a computer, from the lowest wavelength to the highest, to create an image cube of the data set. The spectrum of a selected pixel is obtained by skewering it in its third dimension, wavelength, as the inset in Figure 8.1 shows. Although there are many ways of acquiring and storing the data, this representation is band-sequential (often termed BSQ), which the images are stacked like a deck of cards, and resembles a cube with sides $x$, $y$, and $\lambda$ (wavelength). Even if the data are acquired in some other fashion, they can be reconfigured into this mode. Two other data modes are band-interleaved-pixel (BIP) and band-interleaved-line (BIL) modes. In BIP, the spectra of successive pixels are stored sequentially. This is advantageous for computation because the spectrum

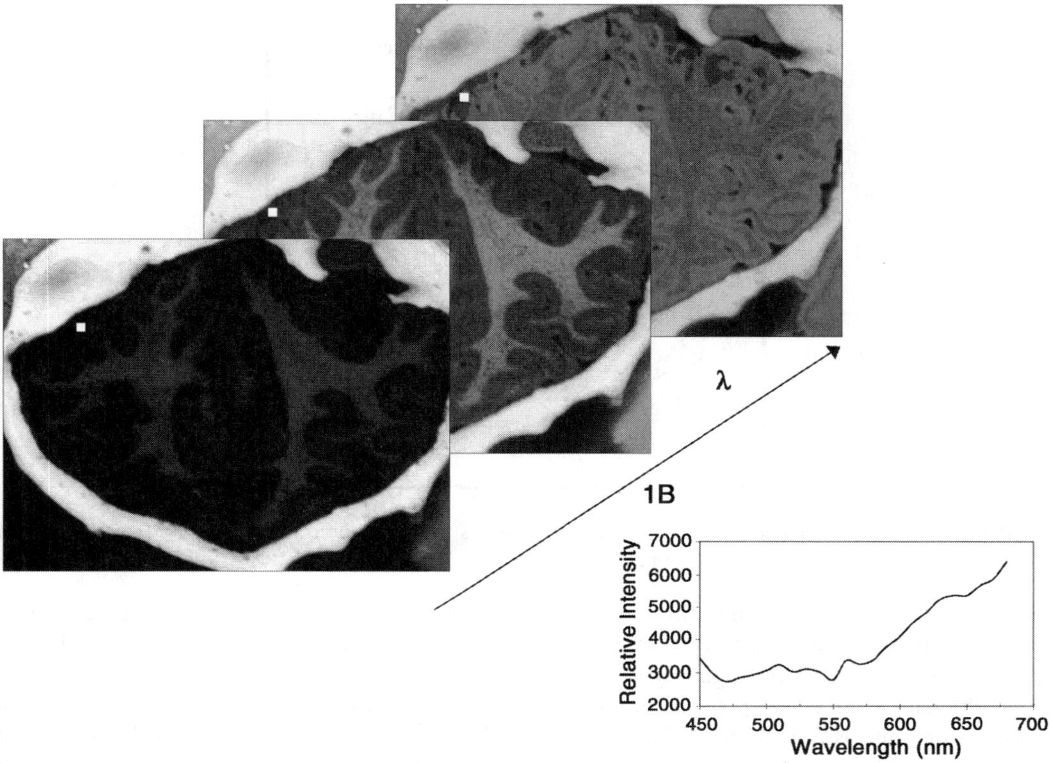


**FIGURE 8.1** Basics of imaging spectroscopy. Multiple images of the same scene are acquired at many different wavelengths, as schematically shown. The spectrum of any pixel in the image can be obtained by plotting signal against wavelength over the spectral range available. The images here are from a human brain, imaged after being frozen. The white areas on the side are the frozen matrix that maintains the brain's shape and provide some hydration.

of each pixel can be read directly, as opposed to band-sequential data, where one must read in the entire cube to calculate a spectrum of any given pixel.

## 8.3 Instruments

Before describing specific instruments, it is worthwhile to compare spectral imaging with what can be accomplished using standard imaging systems based on conventional RGB sensors. Because most such systems rely on single-chip cameras, color images can be acquired in a single exposure, typically at near-video rate. In contrast, most spectral systems require a series of exposures, so improvements in the quality or utility of the data collected should be large enough to justify the potential penalties in cost, and throughput and data acquisition time.

For example, although earlier systems for automated or assisted immunohistochemistry quantitation — a relatively simple problem in color analysis — used grayscale cameras and two or more color filters somewhere in the light path, recent approaches exploit RGB cameras and analytical strategies of varying levels of sophistication and complexity. With automatic thresholding operations, Ruifrok[6] was able to differentiate between a DAB (brown) stain alone, DAB plus hematoxylin (blue), and hematoxylin alone. More recently, this group has shown that conversion of RGB images into optical density units allowed for more accurate discrimination. However, RGB sensors have intrinsically broad and overlapping regions of spectral sensitivity for their three color channels; this adversely affects unmixing accuracy, especially when separation of similar chromogens is attempted. Thus, for example, a dense brown stain can generate

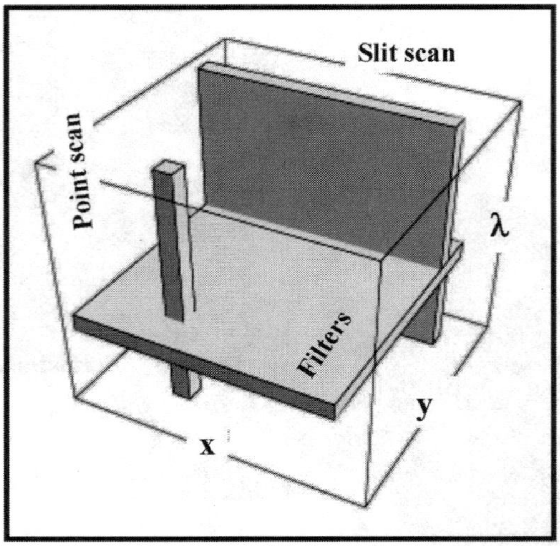


**FIGURE 8.2** Schematic illustration of the various approaches requiring spectral image cubes. In each case, it is shown how a technique slices through an $(x,y,\lambda)$ data set.

a signal in the postanalysis red channel.[7] While it may be possible to unmix red, brown, and blue using only three input images, the optimal wavelengths and bandwidths will differ from the broad channels provided by standard RGB imaging systems.

Additional technical and practical problems are associated with conventional color imaging. First, many color cameras use a charge-coupled device (CCD) that produces a color-encoded analog signal that is digitized by a computer video board into R, G, and B pixel intensities The fidelity and consistency of such a system can be variable. Section-to-section variability, along with interactions with camera controls such as automatic gain control, can induce fluctuations in the image quality. Because the color of a stained object is a product of the stain's transmittance and the camera's spectral response, it is possible that the camera could sense dyes differing in spectral properties similarly and thus be indistinguishable. Finally, the spatial resolution of single-chip color CCD cameras is typically lower than that of monochrome cameras with the same pixel count because of the color mask and interpolation routines that merge information from three or more pixels when determining RGB and intensity values.

True imaging spectrometers, in some fashion, acquire a three-dimensional data set, spatial (two dimensional) and wavelength as the third dimension. The approaches for instruments traditionally involved scanning one of the dimensions, either acquiring a complete spectrum for each pixel (or line of pixels) at a single shot and then spatially scanning through the scene or, alternatively, taking in the complete scene in a single exposure, and then stepping through wavelengths to complete the data cube. Typically, the light emerging from the imaged object is filtered for spectral content, but it is also possible to control the spectral content of the illumination. Recently, other instruments have been developed that acquire spectral and spatial information in a single exposure, although with some trade-offs.

Figure 8.2 shows an image cube and how different cuts through the data illustrate the different approaches. Some of the terminology comes from the origins of imaging spectroscopy, which involved performing remote sensing from a moving platform. For example, the whiskbroom imaging spectrometer is one in which a single point is scanned perpendicularly (cross track) to the direction of motion. The spectrum of each pixel is acquired with a spectrometer and the data are taken a spectrum at a time, pixel by pixel, along a line. The name comes from the resemblance of the path of the scanned pixels to that of a whiskbroom in action. Similarly, a pushbroom spectrometer images a slit onto a focal plane array; the spatial dimension occupies one axis of the array and the spectrum for each pixel is spread out perpendicularly to it. A complete image is acquired one line at a time as the slit is scanned in the direction of motion. In biological imaging, point-scanning and slit-scanning confocal microscopes use similar image collection geometries, respectively. In addition to these techniques, which collect spectral data directly, other modalities can require mathematical processing of the intermediate data.

We divide the discussion of instrument types into four general types:

1. Spectral scanning: These use electro-optical devices such as liquid crystal or acousto-optic tunable filters, as well as filter wheels, and project complete images onto CCDs or other focal plane arrays. Controlling the spectral content of the illumination source rather than filtering the remitted light can also achieve spectral scanning.
2. Spatial scanning: These use either pushbroom or whiskbroom configurations with prisms, gratings, or beam-splitters to create spectral discrimination.
3. Interferometric: These typically (but not always) acquire a two-dimensional image and scan optical path differences in some manner to obtain a complete interferogram at each pixel; the data need to be mathematically converted into spectra in wavelength space.
4. Other approaches: These include instruments such as the computed tomographic imaging spectrometer and a polarization-dependent rotogram device.

## 8.3.1 Spectral Scanning Instruments

These instruments are easy to understand and have very simple optics. They consist of imaging optics, a tunable filter of some sort for spectral selection (or a tunable light source), and a camera. Because the components can be in-line or folded, such systems can be made rather compact, suitable for mounting on microscopes or other instruments. The tunable filter can be a mechanical filter wheel, a linear variable filter, or an electro-optical filter that can be tuned electronically.

### 8.3.1.1 Fixed Filters

The simplest implementation of an imaging spectrometer incorporates a filter wheel equipped with a set of fixed bandpass filters in a rotating mount. A variant often necessary for fluorescence imaging would substitute a set of filter cubes (combinations of dichroic mirrors, excitation, and emission barrier filters) for a simple filter wheel. For applications in which a relatively small number of pre-set and invariant wavelengths are needed, this can be a useful technique. For example, Speicher et al.[8] have demonstrated fluorescence-based spectral imaging with a filter system generating a combinatorial library of 27 colors, enough to paint all the human chromosomes. Furthermore, compared with other approaches, a filter wheel can be relatively inexpensive and is also quite light efficient (although the latter consideration is not straightforward and can depend on the degree of spectral cross talk tolerated between channels). These instruments have limitations:

1. They lack spectral flexibility because only a relatively small number of wavelength choices are available in any one configuration. One could make the filter holder larger to accommodate more filters, but this would increase the size and expense commensurately.
2. The performance of the filters can change unpredictably over time due to aging.
3. Switching speeds can be low.
4. Moving parts create noise and vibration.
5. There can be image registration problems due to misalignment of filters.

### 8.3.1.2 Linear Variable Filters

A linear variable filter can also act as a spectral filtering element for an imaging spectrometer. For such filters, the transmission varies linearly along the filter; at any wavelength, $\lambda_o$ (or position along the filter), the local transmission is a bandpass filter with a width that is a fixed fraction of $\lambda_o$. That width is 1 to 1.5%, depending on the filter, so a typical bandwidth is ~4 to 10 nm over the visible spectrum. One vendor, OCLI, has marketed a spectrometer without a grating, using a linear filter directly on top of a linear CCD detector array. Similar versions are known as circular variable filters (CVF); the transmission changes with rotational angle of the filter.

An optical system with a beam waist can use a linear or circular variable filter to create an imaging spectrometer by inserting the filter at the location of the minimum spot size. Because the filter's transmission is spatially dependent, a large spot size would give a large and spatially varying bandwidth, so the filter is located at a beam waist to reduce the resultant spectral smearing. In this mode, the filter acts like a filter wheel with a large number of filters. Images are acquired at each wavelength and filter translated or rotated to the next wavelength. Kairos Scientific (www.kairos-scientific.com) has developed a system using a circular variable filter that mounts onto a microscope. Surface Optics Corp. (San Diego, CA) has developed an innovative variant based on TDI (time-domain integration) that reads out an imaging array row by row, synchronized to the motion of a spinning CVF (to avoid the problem of spectral smearing). In conjunction with algorithms implemented in hardware, their instrument is capable of acquiring and processing 30 image cubes per second.

#### 8.3.1.3 Tunable Filters

As the name implies, these devices can tune their spectral passband electronically, and without moving parts. Advantages include quiet and vibration-free operation, switching speed, spectral selectivity, spectral purity, and flexibility. Such filters need to meet several important criteria:

1. Because the entire image is being filtered, the filter wavelength needs to be constant over the entire image or meet some lower limit for edge effects.
2. Introduction of the filter into the optical path cannot introduce (significant) image distortion.
3. The tuning time must be commensurate with the dynamics of the experiment.
4. Out-of-band rejection must be sufficiently good that dim in-band signals are not contaminated by out-of-band intrusions.[9]

*Liquid crystal tunable filters (LCTFs)*: LCTFs use electrically controlled liquid crystal elements that transmit a certain wavelength band while being relatively opaque to others. The rejection of the unselected wavelengths, without further manipulation, is about $10^4$:1.[10] The bandpass can be as narrow as 1 nm or even less, and the spectral range with a single device can range from 400 to 720 nm in the visible. The LCTF is based on a Lyot filter — a device constructed of a number of static optical stages, each consisting of a birefringence retarder (quartz for LCTFs) sandwiched between two parallel polarizers. A stack of stages function together to pass a single narrow wavelength band. As the incident linearly polarized light traverses the retarder, it is divided into two rays, the ordinary and extraordinary, that have different optical paths given by

$$\Gamma(\lambda) = 2\pi^* \Delta d/\lambda,$$

where $\Delta$ is the birefringence and $d$ is the thickness. After transmission through the retarder, only those wavelengths of light in phase are transmitted by the polarizer and passed on to the next filter stage. The transmission of a stage is

$$T(\lambda) = \cos^2[\Gamma(\lambda)/2],$$

as illustrated in Figure 8.3. The overlap of these continuously varying transmission curves determines which wavelengths are passed by the filter stack as a whole. To introduce tunability, a liquid crystal layer is added to each stage, as shown in Figure 8.3A, that creates minor changes in retardance affecting the position along the spectrum where the curves constructively interact. Tunability is provided by the partial alignment of the liquid crystals along an applied electric field between the polarizers; the stronger the field, the more the alignment and the greater the increase in retardance. Tuning times for randomly accessing wavelengths depend on the liquid crystal material used and the number of stages in the filter. At the moment, commercial devices use nematic components that result in tuning times of approximately 50 to 75 ms.

Polarizers introduce some restrictions into the operating range of an LCTF filter. Plastic sheet polarizers function below ~730 nm and Polarcor glass polarizers are usable from ~630 to 1700 nm. In practical terms,

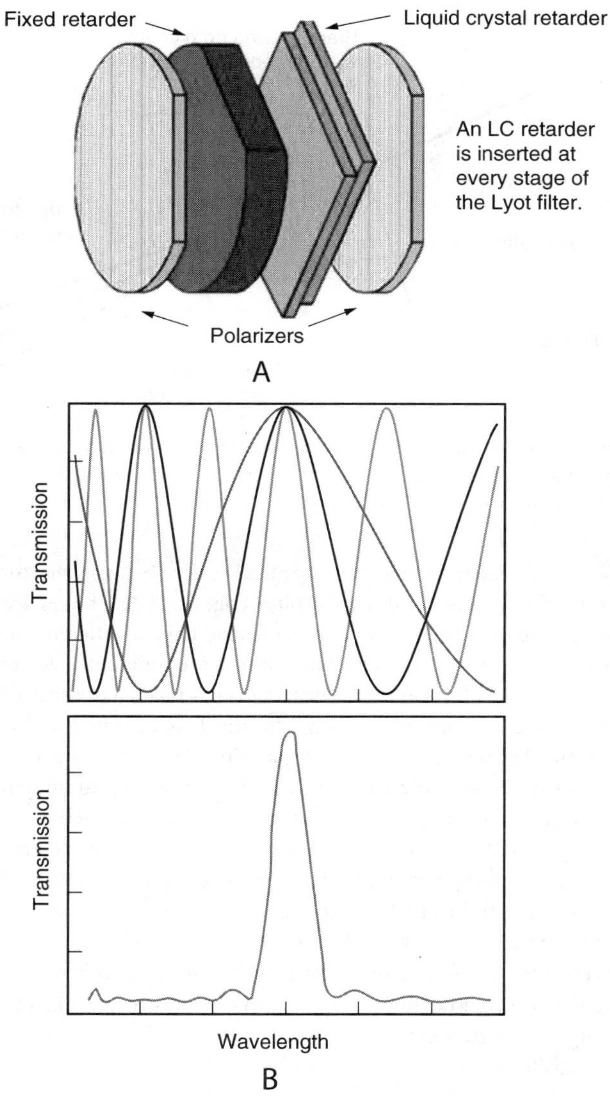


**FIGURE 8.3** (A) Operation of a liquid crystal tunable filter. Each successive stage of the filter is used to help cancel unwanted transmission of the previous stages. (B) For a multistage filter, the final transmission is at a fixed peak. Tunability is introduced via the liquid crystal layer in (A), the voltage of which tunes the retardance of each stage. This tunability is used to move the band of constructive interference to the desired wavelength.

the visible to NIR range can be covered from 420 to 1800 nm using three devices available from Cambridge Research and Instrumentation (CRI, Inc.), which address the following spectral regions: 400 to 720 nm, 650 to 1100 nm, and 850 to 1800 nm; each device covers about 1 octave (twofold spectral range).

Although the position of the bandpass is actively tunable, its width is fixed and depends on the construction of the device. A typical bandpass in the visible is ~10 nm, which is wavelength dependent ($\propto\lambda^2$; 10 nm at 550 nm grows to 16 nm at 700 nm). Because the bandpass is related to the number of stages in the device, any bandpass can be designed and fabricated, from 16 $cm^{-1}$ to 50 nm. The 16 $cm^{-1}$ device has been used for Raman imaging spectroscopy.[11] The devices are rather spectrally flat over a relatively large aperture (38 mm). Like the AOTF (acousto-optic tunable filter), the LCTF is polarization sensitive, which reduces the transmission by half, unless optical means are provided to harvest both polarization states.

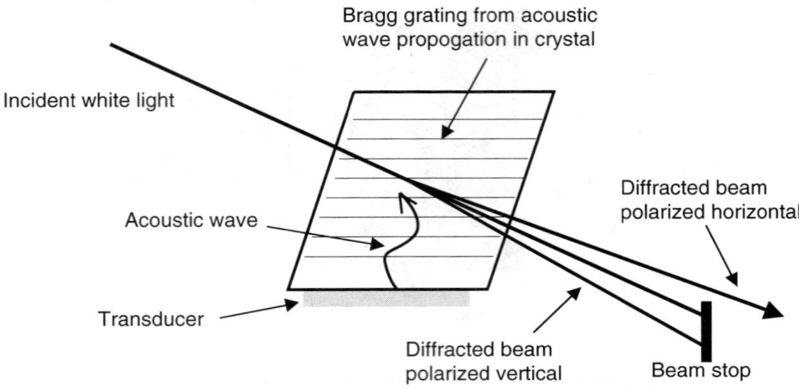


**FIGURE 8.4** Operation of an acousto-optic tunable filter. An acoustic wave, launched into a crystal by a transducer, produces a Bragg grating that diffracts the incident light. There are three output beams, a undiffracted beam and two monochromatic ones, one polarized vertically and the other horizontally.

LCTFs work best in a collimated or telecentric optical space, because the maximum f-number that provides an off-axis shift of less than 2 nm at the filter edge is ~2.5. The device operation depends on interference effects, so photons that are significantly off-axis have a different optical path than on-axis photons, creating edge effects. However, because many optics are inherently slower than f/2.5, they can be used before optical elements. One of the authors (Bearman) has taken a number of image cubes of remote scenes with an LCTF mounted in front of a Nikon 135-mm lens operated at f/4, as have others. Similar arrangements are also available commercially from Opto-Knowledge Systems, Inc. (www.oksi.com).

The LCTF approach has been used to create image cubes for biological imaging (see below), confocal microscopy,[12] and agriculture, and imaging archaeological documents such as the Dead Sea Scrolls.[13]

*Acousto-optic tunable filter (AOTF):* An AOTF uses the interaction between a crystal lattice and an acoustic wave to diffract an incoming beam into a fixed wavelength, as shown in Figure 8.4. As an applied acoustic wave propagates through the crystal, it creates a grating by alternately compressing and relaxing the lattice. Those density changes create a local index of refraction changes that acts like a transmission diffraction grating, except that it diffracts one wavelength at a time, so it behaves as a tunable filter. In practice, the undiffracted zero-order beam is stopped with a beam stop and the monochromatic diffracted beam is available. Changing the frequency (and wavelength) of the acoustic wave changes the grating spacing, adjusting the wavelength of the diffracted beam. In addition, if multiple RF frequencies are launched into the crystal, then combinations of frequencies can be diffracted simultaneously; in this, this filter is more flexible than LCTFs, which generate only a single bandpass at a time.

For visible wavelengths in a tellurium oxide crystal, the applied acoustic wave is RF and can be switched very quickly (typically in less than 50 μs) compared to other technologies. Unlike an LCTF in which the bandwidth is fixed by the design and construction, an AOTF can vary the bandwidth by using closely spaced RFs simultaneously. There are several standard problems with AOTFs, some of which have been successfully addressed: blurred images and poor out-of-band rejection ($<10^{-3}$). The acoustic wave spreads as it propagates through the crystal so diffracted rays leave at a variety of angles, resulting in blurred images. Use of a compensating prism[14] has significantly improved resolution. Narrowing the acceptance angle and attending to details of crystal fabrication can also overcome image blur and shift, albeit at a cost in light throughput.[15]

Both the AOTF and LCTF imaging spectrometers share an important attribute: they make it easy to get very good signal-to-noise spectra, due to the band-sequential nature of their operation. When spanning a wavelength range, say 400 to 720 nm, the sample may have a considerable variation in reflectivity or transmission over that range. In addition, at the blue end of the spectrum CCD sensitivity declines, as does the illumination intensity of many laboratory light sources. As a result, typically less signal is in the blue relative to the red or green part of the spectrum. However, that can be compensated for by longer integration times at the wavelengths with reduced signal, something not possible with many other devices. In fact, the ideal way to operate a BSQ imaging spectrometer is to set a pixel data target value and integrate at each wavelength as long as necessary to obtain that value, maintaining the signal-to-noise ratio (SNR) at each wavelength. In that case, the *model* raw data spectrum of the target pixel would be a straight line, with the real data contained in the varying integration times for each wavelength. This is a major advantage, especially when there are no restrictions on the data acquisition time.

Spectral leakage can contaminate the acquired spectra. One advantage of the LCTF is a high rejection ratio for out-of-band transmission ($10^{-3}$ to $10^{-5}$), critical for recovering spectra that can be compared with those from other laboratories or with standard spectral libraries. LCTFs can be fabricated with larger apertures than AOTFs, although that is not an issue for integration into microscopes, which do not need the large aperture. On the other hand, their major drawback is longer tuning time relative to an AOTF: ~30 to 50 ms vs. microseconds for the AOTF. A switching mode for LCTFs is somewhat faster, around 20 ms, but that is for a limited palette of perhaps three to four wavelengths. For situations in which the integration time is photon-limited and the exposure time is ≥50 ms, the tuning time of either device becomes less of a bottleneck for data acquisition. In fact, it is usually the camera data transfer rate that dominates acquisition time when light is ample.

## 8.3.2 Spatial Scanning Systems

### 8.3.2.1 Pushbroom

Lightform, Inc.[16] has developed a pushbroom imaging spectrometer designed to mount to a C-mount camera port. It collects a slit image from the object onto a two-dimensional camera in which the spatial information is displayed along one axis and wavelength information along the other. Wavelength dispersion is provided by a prism. This approach is well suited for scanning gels or searching an object for specific spectral features because the entire spectrum of each pixel in the slit image is available in real time. Because gels are too large to image easily, they can be mechanically scanned with this system. With this approach, the user does not have to collect an entire image cube, but can assemble an image that records hits only for the spectra of interest. If a spectral feature is a known, that feature can be identified in each spectral line scan in real time and used to assemble a classified image without acquiring the entire spectral cube for the whole image field.

### 8.3.2.2 Interferometers

Rather than scanning in wavelength space, one can also scan in optical-path-difference space and capture an interferogram for each pixel, which is then inverted via the fast Fourier transform (FFT) algorithm to obtain an image cube in wavelength space. Several devices have been developed and one is available commercially. Although seemingly different from instruments that acquire sequential wavelength images, many of the interferometric devices are similar in spirit and suffer from similar problems. Like the BSQ imagers, interferometric imaging spectrometers also require acquisition of many images — sometimes, an order of magnitude more images. For so many images, the data acquisition time may become limited by camera image transfer time.

Applied Spectral Imaging of Israel was perhaps the first company to make a commercially available imaging spectrometer. The device is a common-path Sagnac interferometer in which the interferogram is spread out over a two-dimensional sensor.[17] An optical element changes the optical path difference (OPD) in stepwise fashion, while a CCD (or other technology) focal plane array captures the resulting interference pattern at each step. Because the interferogram moves with each OPD image, object motion is a challenge for this instrument. If the object moves and the images are corrected by re-registering the spatial content to compensate, any errors will show up as incorrect interferograms and propagate into the spectra after inversion. The ASI instrument has been used for cytogenetics[18] and cell pathology,[19] to name a few applications.

Itoh[20] has developed another interferometric device that uses a tilted and wedged lens array and mirrors to acquire all the necessary images at different OPD *simultaneously* on a two-dimensional imager. Itoh has demonstrated imaging of rapidly moving objects with this approach, a laser ablation plume, and rotating (1800 RPM) targets. Because all the multiple images must be acquired on a single detector, there is a trade-off between image size and spectral resolution.

One problem with interferometers is that of the center burst (OPD = 0), which is quite bright relative to the rest of the fringes. The detector is an imager, so the integration time or illumination intensity must be reduced sufficiently to avoid saturation (or blooming) for pixels at the center burst, thereby reducing the fringe contrast farther out in the interferogram. The reduced fringe contrast decreases the signal and results in increased noise in the image cube in wavelength space.

Considerable discussion has taken place in the literature about the relative photon efficiency of interferometers compared to scanning instruments. Although on the surface the interferometer appears to have a substantial advantage over other approaches due to the fact that it collects all the spectral information simultaneously,[21] several papers[22,23] have argued that, for real instruments with read-noise and other noise sources, this advantage disappears in most imaging regimes. Furthermore, in the spectrally sparse scenes typical of fluorescence imaging, in which signals occupy only a fraction of the total spectral range, the ability of tunable filters to capture images only at informative wavelengths improves their performance relative to interferometer-based approaches that need to collect all wavelengths, informative or not.

## 8.3.3 Other Approaches

### 8.3.3.1 Rotogram

Microcosm, Inc. (Columbia, MD) recently developed an imaging spectrometer using a new spectral imaging technique based on the phenomenon of the dispersion of optical rotation.[24] Figure 8.5 schematically represents the optical layout of the HSI (hyperspectral imager) with polarizers at the input and output of an optically active rotating medium. The polarization plane of linearly polarized light is rotated during propagation through the optically active rotator element so that the rotation angle, $\varphi(\lambda)$, of the plane of polarization of the output light depends on the wavelength of the light and the path length through the optically active rotator element. The path length through the active medium can be varied,

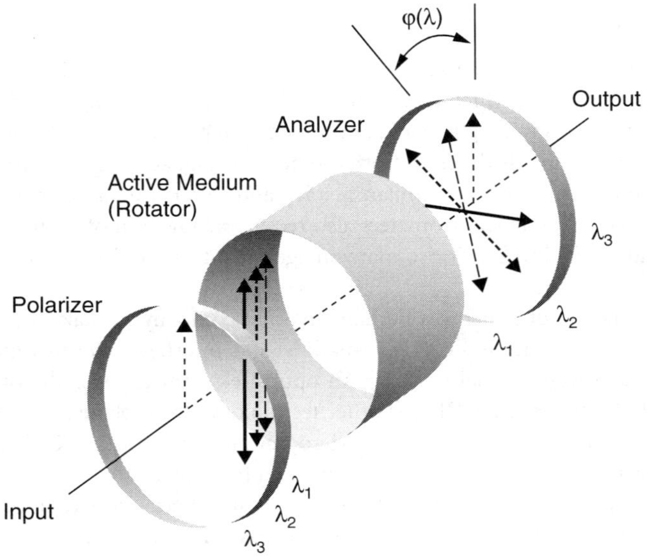


**FIGURE 8.5** Operation of the microcosm imaging spectrometer.

thus incrementally increasing the rotation of each wavelength component present. After passing through the output polarization analyzer, an intensity function is measured for each incremental polarization path length. This arrangement permits the wavelength-dependent polarization rotation to be uniform over a large two-dimensional aperture so a CCD is a suitable detector. The intensity function resulting from a stack of images can be analyzed by mathematical methods at any pixel in the image to yield a highly accurate spectrum at each point in the image. As with several data acquisition methods, some computation is required for reconstructing the image cube.

Like the LCTF or the AOTF, the transmission is polarization dependent. The unit will transmit 40 to 45% of the incoming unpolarized light depending upon the rotator material. If the incoming light is polarized, then the throughput efficiency can reach 85 to 90%. The HSI can have a large clear aperture that provides excellent light coupling through the device for the use of wide-field spectral imaging. It does not induce any beam deflection or image movement and can be used in-line with almost any imaging system. In a different configuration, the HSI is suitable for use with a point-scanning device such as laser scanning microscopes or, alternatively, the classical pushbroom and whiskbroom configurations.

#### 8.3.3.2 Computed Tomographic Imaging Spectrometer (CTIS)

A recent approach to imaging spectroscopy is tomographic imaging, as illustrated in Figure 8.6. With this technique, diffractive optics disperse the spectral and spatial information of each pixel onto an imaging sensor; an image cube in wavelength space is reconstructed from a *single* image. Because it turns out that the mathematics of the reconstruction is the same as tomographic imaging, such devices are known as computed tomographic imaging spectrometers (CTIS). Originally proposed by several researchers[25,26] in the early 1990s, they have been further developed by Volin et al.[27]

A diffractive optical element operating at multiple orders creates an array of images on the sensor. Development of techniques for fabricating the grating with e-beam lithography has been the main driver in development of this instrument.[28] It is important to note that each image is not simply composed of single wavelengths; information is multiplexed over the entire array. Figure 8.6 shows how the diffractive disperser distributes the spectrum of a single pixel. Note the zero-order image, which can be used for focusing — a difficult task for many spectral imagers. A calibration matrix is necessary to perform the reconstruction; it is obtained by measuring the location on the image plane of pixels in the object plane at different wavelengths with a movable fiber optic coupled to a monochromator. A CTIS can operate over a large wavelength range, easily from 400 to 800 nm, and, with the proper detector, can operate in the IR or UV. The data from a single image can be reconstructed in a variety of ways to adjust image size and wavelength bands. For example, an image can be reconstructed with 128 × 128 pixels with 20 bands or 64 × 64 pixels and 32 bands, using the *same* data set. The only difference between the two reconstructions is the calibration matrix.

A major advantage of the CTIS is speed. Because it takes a single image that contains all the spectral/spatial data, it can be run at video rates,[27] assuming sufficient light and a high-speed camera (since a large pixel array is typically required by this method). This potential speed makes it suitable for studies such as endoscopy and rapid processes that other instruments cannot handle. Alternatively, it is useful for collecting ratiometric data because all wavelengths are acquired simultaneously. A major issue with spectral imagers has always been bandwidth — they tend to produce enormous amounts of data that present downlink or transmission problems. For example, a satellite hyperspectral imager can produce hundreds of gigabytes of data a day. In the same vein, a remotely sited or operated imaging spectrometer can easily present significant bandwidth demands for data transmission in a power-limited environment (power = bandwidth for telecommunications).

CTIS devices have been used for some real-world imaging studies; Descour et al. have demonstrated ratiometric pH-imaging with standard probes using a CTIS,[29] while Ford et al.[30] used one for toxicology studies. In both cases, the device allowed capture of the entire spectrum of fluorescent probes at once.

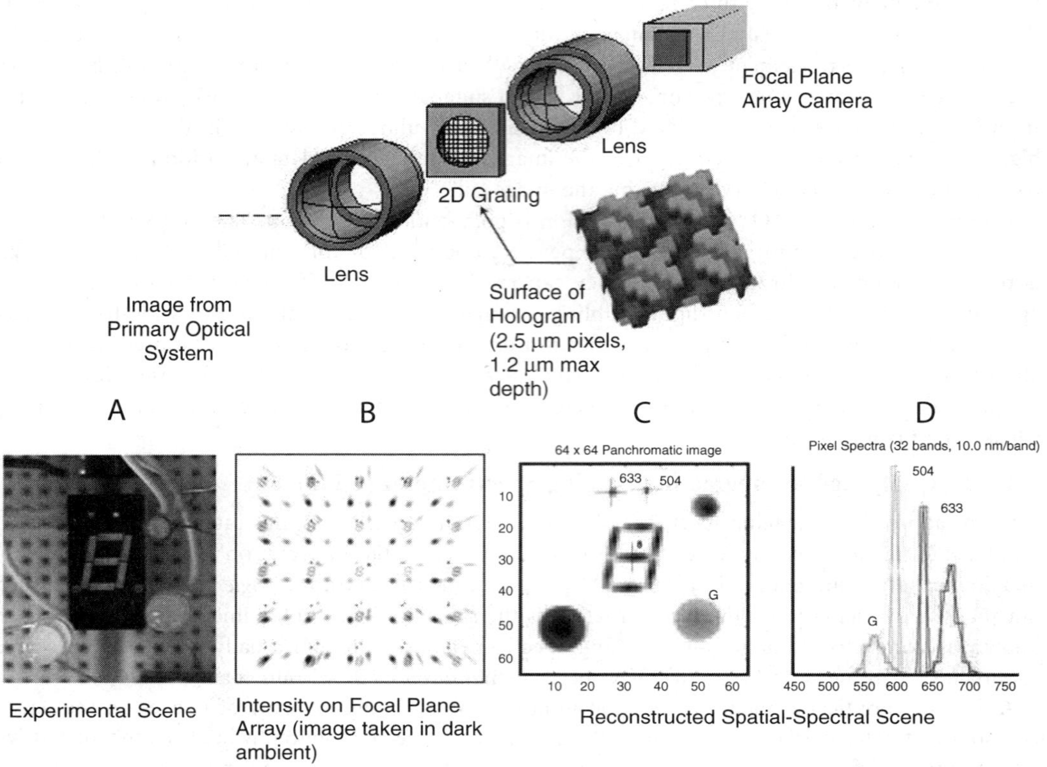


**FIGURE 8.6** Computed tomographic imaging spectrometer. A diffractive grating, written by e-beam lithography, multiplexes the spatial and spectral information of a single pixel onto many sensor pixels. (A) The observed scene, consisting of three color LEDs, a HeNe laser spot, and an eight-segment indicator. (B) The resulting image from the focal plane. There is a zero-order image (C) that can be used for focusing as well as providing an initial starting point for the reconstruction of the image cube. (D) Recovered spectra.

### 8.3.3.3 Hadamard Transform Imaging Spectroscopy

Hadamard transforms have been used for spectroscopy for some time[31] and have been adapted to fluorescence microscopy. The fabrication of large-format liquid crystal spatial modulators has made this application possible because they can create the Hadamard masks rapidly and with no moving parts. In a series of papers, Jovin and colleagues have developed this implementation of imaging spectroscopy and microscopy.[32]

Like the CTIS or an interferometer, the Hadamard transform spectrometer requires computation to reconstruct the image cube in wavelength space. However, it also requires a large number of images; for example, Hanley reports acquiring 511 images in ~11 min with 5 min of computation to transform the data. Increasing optical efficiency can clearly reduce some of the data time but, like the interferometer, the basic nature of the device requires many images.

### 8.3.3.4 Fiber-Optic Image Compression

Several groups have developed fiber-optic-based systems that acquire spectral image cubes using a combination of a spectrograph, two-dimensional detector, and a custom fiber bundle. The fiber bundle has a two-dimensional input but the output is reordered into a line, which then serves as the input for a spectrograph. The spectrograph output is a two-dimensional image, containing the spatial pixel in one direction while the other dimension is the spectrum of that pixel. Myrick has constructed such a system[33] and used it to image laser ablation plumes.[34] This is, in fact, the same data format as that of a pushbroom imaging spectrometer, except that the input now represents a scrambled two-dimensional image, rather than a line imaged across an object. Similarly, Amotz has used a fiber-optic compression system for Raman imaging with a 61-bundle (pixel) image.[35] One limitation of these systems is image size; the number of pixels in the image is limited to the number of rows in the detector. Myrick reports on a $17 \times 32$ image that uses a detector with 544 elements in the spatial dimension.

#### 8.3.3.5 Spectral Source

For microscopy brightfield applications, it is possible to accomplish spectral imaging by tuning the illumination light as well as filtering or otherwise analyzing the remitted light. Monochromators (usually relying on diffraction gratings and white light sources) are available for such purposes. However, as the name implies, monochromators provide illumination consisting only of one spectral band of light at a time. Nevertheless, it is possible to create more flexible sources that can produce illumination of any desired pure wavelength (like a monochromator) or any selected mixture of pure wavelengths simultaneously, with white light output an easy option.[36] The resulting images can be collected by a high-resolution gray-scale CCD camera and interpreted using appropriate algorithms and displays. This can be used to create a complete spectral image cube for a sample by taking sequential images while illuminating with a series of pure wavelengths, with greater ease and economy than by means of devices on the imaging path, such as tunable filters or interferometers.

An advantage over tunable filters in some applications is that contamination of individual bands by out-of-band light is minimal. Furthermore, using software approaches such as projection pursuit vectors or principal components to define specific illuminants, the sample can be illuminated with a precisely controlled mixture of wavelengths so that the image presented to the detector is a linear superposition of the sample properties at many wavelengths. Thus, spectral discrimination that would previously have required the collection of complete spectral cubes might require acquisition of as few as one to three matched spectral images per field. Data acquisition is simplified and, because spectral processing is performed optically rather than computationally, acquisition and analysis times are greatly reduced.

#### 8.3.3.6 Multispectral Confocal Microscopy

Much biology today uses confocal microscopy as a major tool to provide high-resolution three-dimensional imaging of cells and tissue. Considerable effort has gone into developing optics and software to provide diffraction-limited imaging in commercial instruments. Similar effort has been expended on fluorescent probes to illuminate cellular activity. Can we apply imaging spectroscopy to confocal microscopy? The answer is yes, but with some limitations.

Laser scanning confocal microscopes (LSCM) raster a laser spot across the object, obtaining a full image a point at a time, similar to the way that a whiskbroom imaging spectrometer operates. Because a pinhole present in the optical path to provide confocality attenuates the signal considerably, a PMT is typically used as a detector to provide sufficient gain and reasonably short pixel dwell time. To obtain an emission image cube of a fluorescently labeled sample, there are two options. Because the image is already being scanned, if one can somehow filter emission wavelength prior to detection, one can assemble an image cube. This was done recently by inserting an LCTF in front of the PMT in an LSCM and stepping through the emission wavelength range.[12] One could also use an AOTF as the filtering element. While serviceable, this is not a practical method, because it requires multiple scans at different wavelengths to acquire a full image cube. Aside from the significantly increased time for assembling a z-stack or a time series, the repeated scans can cause excessive photobleaching.

For an LCSM, the best data collection scheme would be to acquire the entire spectrum from each spatial pixel as it is scanned, making it in effect a whiskbroom imaging spectrometer. In this approach, the instrument collects an entire image cube in the same time as a single spatial scan. Zeiss has recently introduced a spectral imager that obtains the spectrum for each pixel as the LCSM scans through the image. Due to scanning nature of the data collection and the need for gain, it would be largely unrewarding to adapt other spectroscopic techniques to an LCSM.

However, one could imagine adapting an interferometric or tunable filter device to a confocal microscope that uses a Nipkow disk because these present an entire image to a CCD. Thus, any spectroscopic device that can be interposed before a CCD camera should be compatible with this or similar confocal designs. However, any method that requires a lot of images may pose practical problems when conducting real experiments due to the possibility of photobleaching and the time involved in acquiring an $(x,y,z,\lambda,t)$ image stack.

## 8.4 Data Analysis

### 8.4.1 Image Analysis

A number of freestanding spectral analysis tools are available commercially or as free-ware. These include ENVI™ and MultiSpec™ from Research System International.[37] Other programs can be assembled by the sophisticated user with resources such as the statistics, chemometrics, and image analysis toolboxes available in such packages as MatLab™, supplemented with researcher-generated MatLab-compatible algorithms downloadable from various Internet sites. Still others are available bundled with commercially available spectral imaging hardware (OKSI, Kairos Scientific, Spectral Dimensions, ChemIcon, etc.). These software tools will not be described further, except for some aspects that are touched on in the following discussion.

One of the appeals of developing spectral imaging systems or applications is the richness of the data sets comprising spatial and spectral information, which invites the use of intriguing analytical tools. Indeed, many of the algorithms, such as automatic clustering tools, developed for use with genomics data sets (such as the huge expression arrays) are applicable to spectral cubes, with the proviso that these methods do not encompass any of the spatial content to be found in the images. Methods attempting to link spatial and spectral data are under development.[38] Another thread in current investigations is determining how to select the minimum number of wavelengths needed to accomplish specific tasks.[39] It may seem intuitive that more spectral data and higher spectral resolution may provide increased analytical precision; however, this is often not the case. Many wavelengths may be "uninformative," and their inclusion in the data set merely adds noise. This consideration is partially related to the so-called "curse of dimensionality," which also deals with consequences of the huge internal volumes of the hyperspheres that can be used to represent high-dimensional data sets.[40] (This is more of a problem in remote sensing, in which data sets can contain images at hundreds of wavelengths.)

For the relatively simple problems posed by imaging in the visible range, and where the targets may be simply defined fluorescent dyes or chromogens, one may be able to lower the number of wavelengths acquired to approximate the number of distinct species sought in the image. Thus analytical techniques can be used not only to work with the data sets but also to shape how they are collected. At the limit, spectral flexibility can be used simply to provide a capability to select one or more specific wavelengths for the purposes of increasing contrast or enhancing the utility of straightforward image-analysis tools. In a recent publication, Ornberg et al.[41] describe using a tunable filter to identify optimal wavelengths for separating signal from background in samples stained with a single chromogen plus background stain. Using a simple processing routine, the authors were able to collect and analyze two to three images per minute.

### 8.4.2 Analysis of Spectral Images

Assuming that more than a couple of wavelengths have been collected, the task of analysis usually involves classification, unmixing, or both. Classification involves assigning each pixel to one or more spectrally defined classes (or to an "unclassified" class). Classification is equivalent to spectral segmentation; it is an *exclusive* operation in which a pixel or object is assigned to a *single* class using one or more of a variety of metrics. On the other hand, when pixels can be or are composed of more than one spectral class, as is often encountered when multiplexed protein or nucleic acid probes are used, then the pixels must be "unmixed," yielding estimates of the proportion of each class present. Overall, the steps involved typically consist of:

1. Detection or selection of appropriate spectra for subsequent analysis
2. Spectral classification or pixel unmixing

#### 8.4.2.1 Pixel Classification

There are several approaches to classifying pixels in a spectral image. The minimum squared error method compares the spectra at each pixel in the image with a set of reference spectra, choosing the most "similar" and using a least-squares (Euclidean distance) criterion. This metric compares spectral means; other distance-metrics such as Mahalanobis distance[42] can be used that are sensitive to higher-order statistics such as class variances. Related approaches convert spectra into *n*-dimensional vectors and the angles between such vectors can be used as measures of similarity.[43] Determining which spectra to use for the classification procedure is not always straightforward. In simple cases, the reference spectra can be selected from obvious structures in the image (foci of cancer vs. normal cells, for example) or from established spectral libraries. Alternatively, informative spectra can be extracted using statistical analysis methods, such as principal component analysis (PCA) or clustering methods.[44] Instead of using a classified pseudo-color display, spectral similarity can also be illustrated by mapping the degree of similarity using gray-scale intensity. This operation can reveal otherwise unapparent morphological details.[45]

#### 8.4.2.2 Pixel-Unmixing

Spectral classification methods are suitable for images in which no pure spectral components are likely to exist, such as in histologically stained samples. In other types of images, such as those generated by immunofluorescence or *in situ* hybridization procedures, multiple distinct spectral signals may co-exist in a single pixel to form the detected signal. Spectrally mixed pixels result when objects cannot be resolved either at an object boundary (spatial mixture), when more than one object is located along the optical path (depth mixture), or when multiple probes are co-localized within a pixel. In fluorescence, due to the additive nature of the light signal, the observed spectrum is a linear mixture of the component spectra, weighted by the amount of each probe. A linear combination algorithm can be used to unmix the summed signal arising from the pure spectral components in order to recover the weighting coefficients. Given an appropriate set of standards, the algorithm can quantitate the absolute amount of each label present.[46]

In contrast to fluorescence images, imaging multiplexed samples (such as immunohistochemistry studies) in brightfield must take into account the behavior of absorbing chromophores that, rather than being additive, subtract signal from the transmitted light in a nonlinear fashion. Conversion of the brightfield image from transmittance to optical density (OD), a straightforward mathematical procedure, permits the use of the same linear unmixing algorithms that work with fluorescence.

**Automated end-member detection:** How does one select which spectra to use for unmixing? In many cases, this is easy. In standard immunofluorescence studies, the spectra of the fluorophores, imaged one at a time, can be stored in a spectral library and used to unmix images in which multiple fluorophores are present. But what if one does not have pure spectral species with which to work, for example, if a single, multiply labeled image is available, or if, to change applications, one is trying to analyze a remote scene about which little *a priori* knowledge is available? Some tools can identify the pure spectral species present in an image, without *a priori* knowledge by deconstructing the spectral content into its presumed components. ENVI™ provides a tool based on convex hull analysis that considers spectral end members (the pure species) to occupy the periphery of a data cloud, all of whose mixed species must fall within, rather than on, the surface. The cloud (in which each pixel's location is determined by its spectral content in *n*-dimensional space) is rotated randomly and projected onto a hyperplane. Pixels that repeatedly end up on the periphery after multiple projections are considered end members and can be used to unmix the image. This procedure can be quite time-consuming. Another specialized utility, N-FINDR,[47] uses an analytical approach rather than multiple projections and accomplishes the same task quite efficiently.

**Dimensionality reduction and automated cluster analysis:** Spectral data, as noted above, can be expressed as points in hyperspace. Spectrally similar pixels will cluster together and algorithms, some of which are similar to those used for analyzing genetic expression arrays, can be used to identify such clusters that might represent meaningful bases for spectral classification.[48,49] Frequently, such analysis is either impossible or inefficient when all wavelengths are included in the data set. Because a great deal of covariance exists in typical data sets (i.e., the intensity at one wavelength predicts to a high degree the intensity at neighboring wavelengths), the number of dimensions needed to express the actual information content in a data set is often far less than the number of dimensions in the data set.

PPCA is one of a family of statistical tools that can identify the most informative combinations of wavelengths (by rotating the basis vectors of the original data set) and segregate signal from noise (with some major limitations). Typically, the dimensionality of a 25-wavelength image cube of a standard histology sample can be reduced to three or four dimensions (composed of linear combinations of many of the original wavelengths) while preserving virtually all of the spectral information. Clustering algorithms can then readily work on such a reduced data set to identify meaningful spectral clusters, although some techniques, such as support vector machines,[50] are designed to use the original full feature space. A large variety of published and proprietary clustering methods, including iterative, analytical, neural net, fuzzy logic, and genetic algorithm-based approaches have been developed; however, their descriptions and virtues are beyond the scope of this review. A number of these tools are available as part of the software resources identified at the beginning of this section.

**Combined spectral and spatial analysis:** All the tools described here are designed to work only on the spectral content of the data cube. Remarkably, the pixels could be randomly scrambled and, if their associated spectra were preserved, analysis of the resulting scrambled images by the purely spectral-based algorithms would be unaffected. Obviously, a more powerful approach would somehow combine the rich spatial information present in the images with the spectral data. This is an evolving field with ongoing attempts to adapt remote-sensing expertise to problems in biomedical imaging.

## 8.5 Applications

Spectral imaging holds promise for a number of areas. This section concentrates on applications involving microscopy and visual light, while touching on applications in other areas. In microscopy, the goal can be variously the spectral measurements of natural chromophores or environmentally sensitive indicator molecules (imaging spectroscopy), the detection and discrimination of multiple analytes (multiplex imaging), and the analysis of complex scenes (spectral segmentation and morphometry); these functions can be combined. Microscopy can thus serve as a bridge between the morphological (the traditional strength of pathology) and the molecular.

### 8.5.1 Imaging Spectroscopy

Conceptually, the most straightforward application of spectral imaging involves the simple acquisition of spectra from naturally occurring or adventitious chromophores within a sample. Potential uses in biomedicine include the characterization of different melanin moieties in normal skin, dysplastic and malignant pigmented lesions, discrimination between oxy- and deoxyhemoglobin, or the study of any pigments of interest in biological or nonbiological samples. Comparison between the acquired spectra and preexisting spectral libraries can be used to aid in the identification of specific species. An example of oxygenation-based studies of ischemic regions in a pig heart perfusion model is shown in Figure 8.7, which demonstrates application of macroscopic optics and a spectral range encompassing the near-IR.

Another use of spectral imaging in which the acquired spectrum has intrinsic importance is the detection of spectral shifts in (typically fluorescent) indicator dyes. Ion-sensitive dyes that shift their emission maxima in response to changing ion concentrations are well known but are not as frequently used as dyes that change their excitation profile, in part because it has been easier to switch rapidly between excitation wavelengths than to do the same on the emission side. Emission-responsive dyes, such

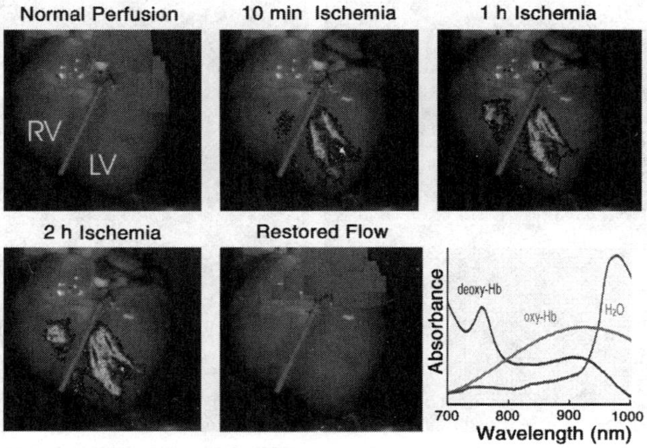


**FIGURE 8.7** **(Color figure follows p. 28-30.)** Near-infrared spectroscopy can detect variations in tissue oxygen levels by means of hemoglobin absorption peaks (lower right). Wavelength control was achieved using a liquid crystal tunable filter in front of a CCD. Pseudocolor highlights left ventricle deoxygenation during occlusion of the artery that normally supplies it. (Images courtesy of Henry Mantsch, National Research Council, Winnipeg, Canada.)

as Indo-1, SNARF-1, Acridine Orange, and Nile Red, can be excited at a single wavelength and their emission behavior monitored using high resolution or spectroscopy or by detecting intensities at only two or (perhaps) more specific wavelength ranges. An example using propidium iodide, which is sensitive to the relative proportion of DNA and RNA in a specimen, is shown in Figure 8.8, which compares three samples of yeast under three different experimental conditions. Spectral shifts, highlighted using PCA, identify yeast in each class. Such small spectral shifts are easily separated by data from an image cube, but are not separable by filters without significant cross talk.

For ratio-based ion-sensitive imaging approaches, one would ideally wish to monitor emission at a number of wavelengths simultaneously, rather than sequentially, in order to obtain an instantaneous pixel-by-pixel measure of ion concentration. Although LCTFs can be configured to switch between wavelengths with a switching time of 1 to 5 ms, and AOTFs in around 30 μs, these still represent serial measurements. Simultaneous measures can be achieved by using the CTIS approach described earlier, or by using beam-splitters and interference filters to direct light with the desired wavelengths to one or more detectors in parallel. A commercial device that sends up to four images at different wavelengths simultaneously to a single detector is available from Optical Insights. To our knowledge, a comparison of the light efficiency and signal-to-noise capabilities of these different approaches has not yet been done.

### 8.5.2 Multiplex Imaging, Including Immunohistochemistry and *in Situ* Hybridizations

Spectral imaging on an analytic level facilitates multiprobe detection techniques for proteins, RNA, and DNA. Histochemical, immunohistochemical, immunofluorescent, and fluorescent molecular probes bind specifically to intra- or extracellular components and can be visualized with fluorescence or brightfield (transmission) optics. Ideally, one would like to apply more than one specific probe at a time.

### 8.5.3 Spectral Karyotyping

Pioneering work in spectral karyotyping (SKY) using combinatorial labeling of metaphase chromosomes[51] allowed nonambiguous identification of 27 chromosomes or chromosome pairs. Applied Spectral Imaging has commercialized this approach, which has demonstrated considerable clinical utility when applied to difficult cytogenetic problems (Figure 8.9). Similar approaches using multiple fixed filter sets (M-FISH) have been described by David Ward and colleagues.[8]

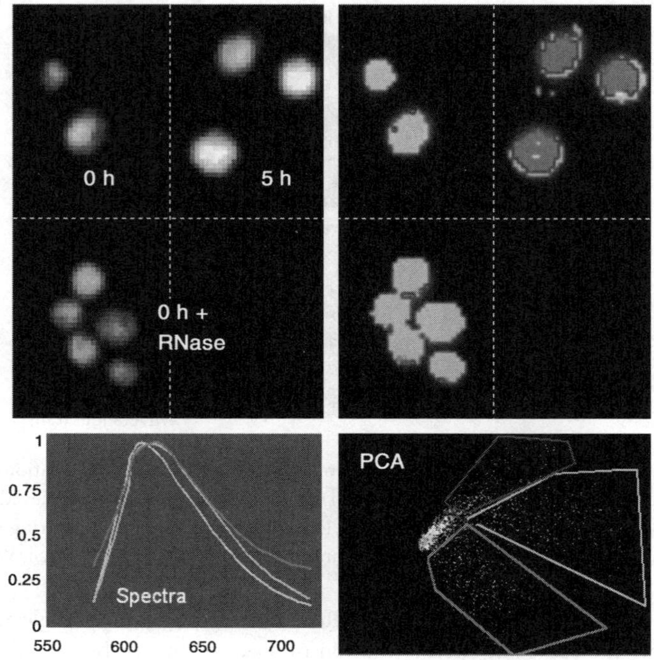


**FIGURE 8.8 (Color figure follows p. 28-30.)** Spectral imaging and image segmentation of yeast cells stained with propidium iodide (PI). Top left: Composite showing three panels of PI-stained yeast cells imaged under three conditions: immediately after transfer of yeast to new medium, after 5 h of culture, and after 5 h of culture plus the addition of RNAse. Bottom left: Spectra from yeast in each culture condition obtained by imaging using a liquid crystal tunable filter. Small differences in peak position and shoulder configuration are visible. Bottom right: Scatter-plot of the spectral image after principal components analysis (PCA). Three clusters are circled and the pixels contained in each cluster pseudocolored. Top right: Result of mapping pseudocolored PCA clusters back to the original image, resulting in robust segmentation. (Analysis: R. Levenson, CRI, Inc.)

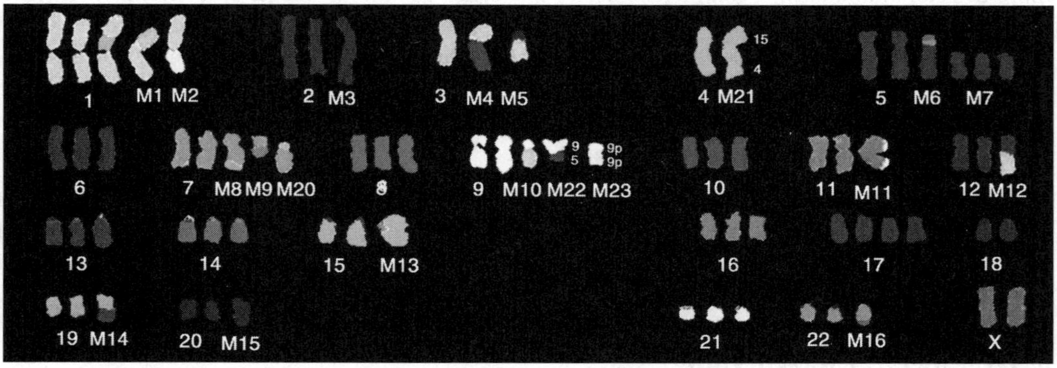


**FIGURE 8.9 (Color figure follows p. 28-30.)** Comprehensive cytogenetic analysis of a metaphase spread from a child (CK1) with dysmorphic features and developmental delay resembling an 18q-syndrome. Spectral karyotyping (SKY) was performed on a metaphase spread. The multicolor hybridization clearly reveals an aberrant chromosome (arrow) that contains chromosomes 18 (red) and X (dark green) material. The G-banding interpretation of a normal male. (From Schrock, E. et al., *Hum. Genet.*, 101, 255, 1997. With permission.)

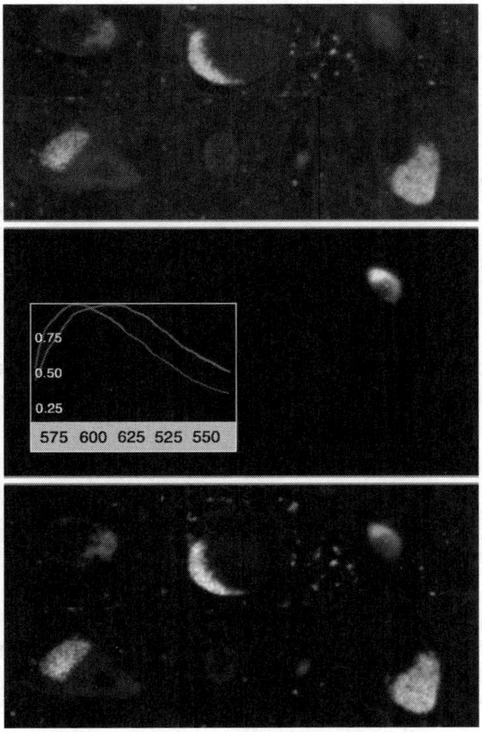


**FIGURE 8.10 (Color figure follows p. 28-30.)** Removal of autofluorescence using pixel unmixing. Top panel: Composite of six neurons in formalin-fixed, paraffin-embedded human brain stained with anti-GDNF labeled with Cy2. The bulk of the fluorescent signal is autofluorescence. Center panel: The single positively stained neuron is separated from the abundant autofluorescence using pixel unmixing (unmixing spectra shown in the insert). Bottom panel: Cy2 signal (in green) overlain on top of the autofluorescence signal (in gray). (Samples courtesy of Neelima Chauhan and George Siegel; analysis: R. Levenson, CRI, Inc.)

## 8.5.4 Immunofluorescence

Multiprobe immunophenotyping has become widely used in evaluation of hematological malignancies, with four or even as high as eight fluorescent signals discriminated with sophisticated flow cytometry instruments.[52] The imaging approach to molecular characterization improves on flow cytometry in its ability to visualize the cells under study directly, to localize (and co-localize) cellular features, to count discrete objects on a per-cell basis, and to allow correlation with tissue microarchitecture in tissue sections. Using more than three or, at most, four labels simultaneously in the absence of spectral imaging tools is currently difficult because of the problem of spectral overlap: it is not easy to prevent signal from one dye "leaking" into the spectral channel of another. The problem becomes intractable for conventional interference filter sets as the number of dyes is increased.[53]

Using spectral imaging, seven labels have been successfully discriminated.[54] Similar approaches can be used with multiple cell-compartment dyes. One problem with immunofluorescence is the interference of autofluorescence, which can be particularly troubling when formalin-fixed tissues are examined. Other troubling specimens include many plant samples, insects, and *Caenorhabditis elegans* nematodes. One solution to autofluorescence difficulties is to shift the excitation and emission wavelengths into the red and far-red, where autofluorescence is much less intense. If that is not possible, then spectral imaging can be used to separate the unwanted autofluorescence signal from that of specific fluorescent dyes (Figure 8.10). In this approach, autofluorescence is treated as another spectral feature, as if it were a fluorescent probe.

### 8.5.5 Immunohistochemistry

More popular in clinical applications than immunofluorescence, immunohistochemistry (IHC) is widely used clinically for the detection of diagnostically or prognostically significant molecules in or on cells. In the past two decades the technique has become central to the practice of oncologic pathology[55] because it can distinguish between look-alike lesions (mesothelioma vs. carcinoma, for example), or divine the cellular lineage of extremely undifferentiated neoplasms (lymphoma vs. other so-called "small blue cell" tumors). It can also be used to highlight the presence of otherwise easily overlooked microscopic foci of tumor, such as micrometastases lurking in lymph nodes, and can be used to measure quantitatively the levels of diagnostically or prognostically important markers such as estrogen- and progesterone-receptors, Her2-neu, p53, ki-67, and a host of others.[56]

Under some clinical circumstances, and often in research situations, double- or triple-staining single slides with different chromophore-coupled antibodies may be desirable. Triple-staining procedures are not often performed because of technical difficulties; however, with the advent of programmable staining systems, complex staining protocols may become less of a hindrance. Despite the nonlinear effects of enzyme amplification, immunohistochemistry can be made quantitative, if precautions are taken.[57,58] The major problem is the difficulty in visually determining where and to what extent the different stains may physically overlap when co-expression of two or more analytes may be in the same cellular compartment. Spectral imaging can overcome this difficulty, even in the presence of considerable spectral overlap with the chromogens. Figure 8.11 demonstrates spectral unmixing of a triple-stained breast cancer sample. This specimen was probed with an antiprogesterone receptor (PR) immunostain coupled to a brown chromogen (DAB) and an antiestrogen receptor (ER) immunostain coupled to a red chromogen (Fast Red); all nuclei were counterstained with a fairly dark hematoxylin wash. The RGB image reveals how difficult it is to determine by eye which cells are expressing PR, which are expressing ER, and which are expressing both. After converting the image to OD and using previously determined spectra for linear unmixing, separate images demonstrating localization of the PR, ER, and hematoxylin stains are shown.

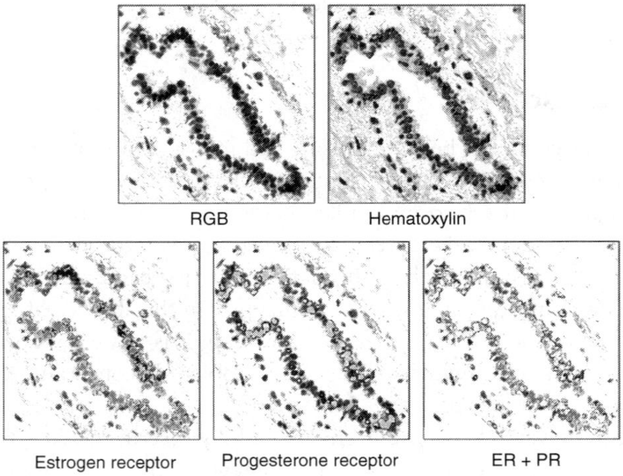


**FIGURE 8.11** **(Color figure follows p. 28-30.)** Multicolor immunohistochemistry and spectral unmixing. Top left: RGB image of a cluster of breast cancer cells stained for the presence of estrogen receptor (ER, red), progesterone receptor (PR, brown); nuclei are counterstained with hematoxylin (blue). It is difficult to determine how much of each antigen is present in the cancer cell nuclei. After collecting a spectral stack, the signals corresponding to nucleus, ER, and PR are spectrally unmixed and shown in separate images. Bottom right shows where ER and PR are co-expressed (yellow signal). (Sample courtesy of DAKO, Inc.; analysis: R. Levenson, CRI, Inc.)

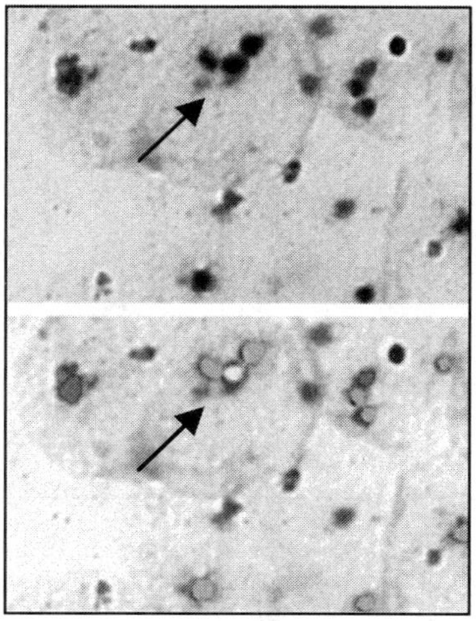

**FIGURE 8.12 (Color figure follows p. 28-30.)** Spectral unmixing of transmission *in situ* hybridization (TRISH). Nuclei of cytospun bladder carcinoma cells probed for three chromosome centromeres. Detection was performed using DAB, New Fuchsin, and TMB as chromogens. Lower panel: Spectral unmixing reveals overlap of brown and red signals (pseudocolored as green and red, respectively). The arrow points to a yellow spot representing the overlap. (Sample courtesy of Anton Hopman, University of Maastricht, Maastricht, the Netherlands; analysis: R. Levenson, CRI, Inc.)

## 8.5.6 FISH and TRISH

*In situ* hybridization (ISH) has proved to be an invaluable molecular tool in research and diagnosis and has enabled major strides in the fields of gene structure and expression at the level of individual cells and in complex tissues. To date, the vast majority of ISH applications have relied on fluorescence readout systems because of their sensitivity, spatial resolution, relative simplicity, and easy adaptation to multicolor and quantitative methods. As noted earlier, it can be difficult to use conventional filter sets to image multiple fluors simultaneously. With spectral imaging, it is possible to visualize six or more probes simultaneously, although similar feats can be accomplished using multiple filter cube sets and cross-talk correction (Larry Morrison, personal communication).

As noted earlier,[22,23] speed and signal-to-noise issues with the various approaches have aroused some degree of controversy. In any event, FISH-based techniques have proved to be somewhat problematic in the clinical arena. Drawbacks include the disadvantage that most fluorescent signals fade upon exposure to light and during storage, interference by autofluorescence (which can be severe in formaldehyde-fixed tissues), and the cost of the microscopic and imaging equipment needed (not to mention the inconvenience of dimming the lights around the imaging station). In addition, it is difficult to combine FISH with routine histopathological stains that can reveal the morphological context of the images.

Some of these difficulties have recently been overcome with the development of nonfading, brightfield detection methods for ISH signals.[59–61] Signals can readily be detected in tissue sections, which can also be counterstained with hematoxylin or other general histology stains. Finally, brightfield, or transmission-ISH (TRISH), can be combined with immunohistochemistry to provide truly multiparameter molecular characterization. An example of spectrally unmixed three-color TRISH is shown in Figure 8.12, which includes an example of spectrally resolving physically overlapped centromeric chromosome probes.

## 8.5.7 Spectral Segmentation and Morphometry

Prostate cancer cells can be spectrally detected in images of prostate biopsy tissue stained with hematoxylin and eosin. This capability could be useful, for example, in automated screening of prostate "chips" removed

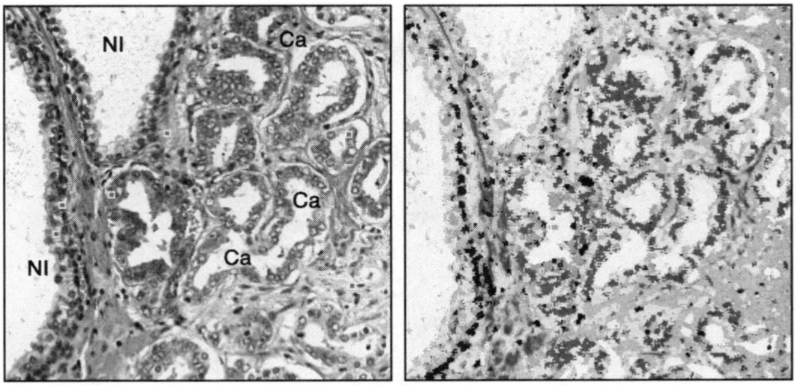


**FIGURE 8.13** **(Color figure follows p. 28-30.)** Spectral segmentation of hematoxylin- and eosin-stained prostate cancer specimen. Left: RGB image of prostate cancer (Ca) and normal (Nl) glands. The normal glands are lined with a double cell layer consisting of epithelial and basal cells; the cancerous glands have a single cell layer. Right: Result of spectral segmentation using spectra chosen manually from representative pixels in the image and a minimum square error classification algorithm. The two cell layers (pseudocolored green and blue) are spectrally distinct from one another and from the cancer cells (pseudocolored red). (Analysis: R. Levenson, CRI, Inc.)

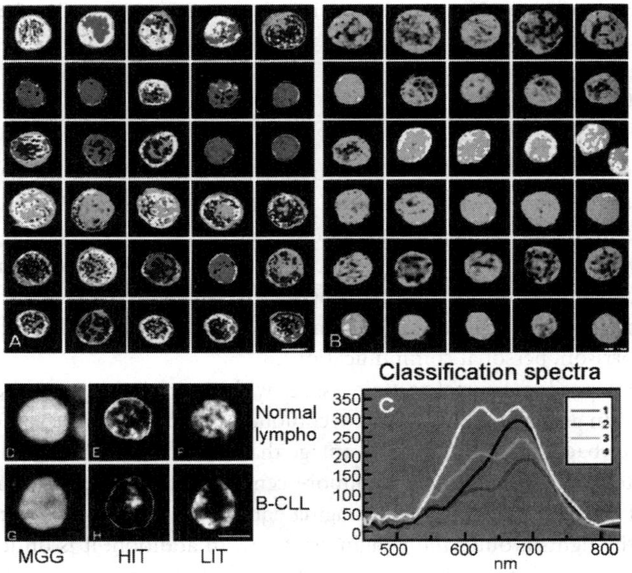


**FIGURE 8.14** **(Color figure follows p. 28-30.)** Spectral classification and morphological analysis. Normal lymphocytes are compared to small B-cell lymphocytic leukemia cells, both stained with Giemsa. By eye these are virtually indistinguishable. Spectral classification, using the spectra shown in the lower right panel, reveals spectral differences in content and spatial distribution of the spectral features (top right and left panels). Lower left: Spectral similarity mapping algorithms indicate more clearly the differences in distribution of spectral features in normal vs. lymphocytic leukemia cells. (From Malik, Z. et al., *J. Histochem. Cytochem.*, 46, 1113, 1998. With permission.)

for benign prostatic hyperplasia. Large volumes of tissue must be examined in a search for potentially tiny foci of clinically unsuspected cancer. Figure 8.13 demonstrates that it is possible to separate malignant and normal epithelial cells spectrally and to detect basal cells as well. (These are a second cell layer found in normal prostate glands but absent in cancer.) The segmentation is not perfect. Some of the imperfections (such as isolated misclassified pixels) can be suppressed using image-processing techniques. However, the limitations in the present case include use of the relatively unsophisticated minimum square error classification algorithm. More generally, it is likely that the stains hematoxylin and eosin — convenient, ubiquitous, and used for generations — may not be the optimal choice for spectral analysis of tissue.

Another demonstration of spectral classification (Figure 8.14) is provided by Malik and colleagues, who used spectral characteristics to distinguish between morphologically similar circulating B-cell lymphocytic leukemia cells and normal lymphocytes.[62] These authors also showed how spectral tools can be used to highlight morphological features that can then be used to further characterize cells or tissues (Figure 8.14).

## 8.6 Conclusion

Fueled by rapid advances in instrumentation, software and algorithmic developments, novel dyes, and chromogens and improvements in sample processing, and stimulated by the genomics revolution and a need to increase throughput and multiplexing capabilities, spectral imaging is poised to make an ever-increasing contribution to biomedicine and related arts.

## Acknowledgment

This research was carried out in part at the Jet Propulsion Laboratory, California Institute of Technology, under a contract with the National Aeronautics and Space Administration.

### References

1. Kruse, F.A., Analysis of AVIRIS data for the northern Death Valley region, California/Nevada, *Proc. 2nd AVIRIS Workshop*, JPL publication 90-54, Pasadena, 1990.
2. Johonson, L.F., Baret, F., and Peterson, D., Oregon transect: comparison of leaf-level with canopy-level and modeled reflectance, *Summaries of the 3rd Annu. JPL Airborne Geosci. Workshop*, JPL publication 92-14, Vol. 1, Pasadena, CA, 1992, p. 113.
3. Fitzgerald, G.J., Maas, S.J., and DeTar, W.R., Early detection of spider mites in cotton using multispectral remote sensing, *Beltwide Cotton Conf.*, Orlando, FL, 1999.
4. Abdulrauf, B.M., Stranc, M.F., Sowa, M.G., Germscheid, S.L., and Mantsch, H.H., A novel approach in the evaluation of flap failure using near-IR spectroscopy and imaging, *Plast. Reconstr. Surg.*, 8, 68, 2000.
5. Sowa, M.G., Payette, J.R., Hewko, M.D., and Mantsch, H.H., Visible–near-infrared multispectral imaging of the rat dorsal skin flap, *J. Biomed. Opt.*, 4, 474, 1999.
6. Ruifrok, A.C., Quantification of immunohistochemical staining by color translation and automated thresholding, *Anal. Quant. Cytol. Histol.*, 19, 107, 1997
7. Ruifrok, A.C. and Johnston, D.A., Quantification of histochemical staining by color deconvolution, *Anal. Quant. Cytol. Histol.*, 4, 291, 2001.
8. Speicher, M.R., Ballard, S.G., and Ward, D.C., Karyotyping human chromosomes by combinatorial multi-fluor FISH, *Nat. Genet.*, 4, 368, 1996.
9. Gat, N., Imaging spectroscopy using tunable filters: a review, *Proc. SPIE*, 4056, 50, 2000.
10. Morris, H.R., Hoyt, C.C., and Treado, P.J., Imaging spectrometers for fluorescence and Raman microscopy — acousto-optic and liquid-crystal tunable filters, *Appl. Spectrosc.*, 48, 857, 1994.
11. Morris, H.R., Hoyt, C.C., and Treado, P., Liquid crystal tunable filter Raman chemical imaging, *Appl. Spectrosc.*, 50, 805, 1996.
12. Landsford, R., Bearman, G.H., and Fraser, S., Resolution of multiple green fluorescent protein color variants and dyes using two-photon microscopy and imaging spectroscopy, *J. Biomed. Opt.*, 6, 311, 2001.
13. Bearman, G.H. and Spiro, S.I., Archeological applications of advanced imaging techniques, *Biblical Archaeol.*, 59, 56, 1996.

14. Wachman, E.S., Niu, W., and Farkas, D.L., Imaging acousto-optic tunable filter with 0.35 micrometer spatial resolution, *Appl. Opt.*, 35, 5220, 1996.
15. Lou Denes, CMRI, personal communication, 2001.
16. www.lightforminc.com.
17. Garini, Y., Katzir, N., Cabib, D., Buckwald, B., Soenksen, D., and Malik, Z., Spectral bio-imaging, in *Fluorescence Imaging Spectroscopy and Microscopy, Chemical Analysis Series,* Vol. 137, John Wiley & Sons, New York, 1996.
18. Schröck, E., du Manoir, S., Veldman, T., Schoell, D., Wienberg, J., Ferguson-Smith, M.A., Ning, Y., Ledbetter, D.H., Bar-Am, I., Soenksen, D., Garini, Y., and Ried, T., Multicolor spectral karyotyping of human chromosomes, *Science*, 26, 494, 1996.
19. Malik, Z. et al., Fourier transform multipixel spectroscopy for quantitative cytology, *J. Microsc.*, 182, 133, 1996.
20. Itoh, K., Watanabe, W., and Masuda, Y., Parallelisms in interferometric fast spectral imaging, *Proc. SPIE,* 3261, 278, 1998.
21. Garini, Y., Gil, A., Bar-Am, I., Cabib, D., and Katzir, N., Signal to noise analysis of multiple color fluorescence imaging microscopy, *Cytometry*, 35, 214, 1999.
22. Castleman, K.R., Eils, R., Morrison, L., Piper, J., Saracoglu, K., Schulze, M.A., and Speicher, M.R., Classification accuracy in multiple color fluorescence imaging microscopy, *Cytometry,* 41, 139, 2000.
23. Miller, P. and Harvey, A., Signal-to-noise analysis of various imaging systems, *Proc. SPIE*, 4259, 16, 2001.
24. Herman P., Malak H., Moore, W.E., and Vecer, J., Compact hyperspectral imager (HSI) for low light applications, *Proc. SPIE,* 4259, 8, 2001.
25. Okamato, T. and Yamaguchi, I., Simultaneous acquisition of spectral image information, *Opt. Lett.*, 16, 1277, 1991.
26. Descour, M., Nonscanning imaging spectrometry, Ph.D. dissertation, University of Arizona, Tucson, 1994.
27. Volin, C.E., Ford, B.K., Descour, M.R., Wilson, D.W., Maker, P.M., and Bearman, G.H., High speed spectral imager for imaging transient fluorescence phenomena, *Appl. Opt.*, 37, 8112, 1998.
28. Wilson, D., Maker, P., and Muller, R., Binary optic reflection grating for an imaging spectrometer, *Diffractive Holographic Opt. Technol.*, Vol. III, SPIE 2689, 1996, p. 255.
29. Ford, B.K., Volin, C.E., Murphy, S.M., Lynch, R.M., and Descour, M.R., Computed tomography-based spectral imaging for fluorescence microscopy, *Biophys. J.*, 80, 986, 2001.
30. de la Iglesia, F., Haskins, J., Farkas, D., and Bearman, G., Coherent multi-probes and quantitative spectroscopic multimode microscopy for the study of simultaneous intracellular events, *Int. Soc. Anal. Cytol. 20th Congr.*, Montpellier, June 2000.
31. Treado, P.J. and Morris, M.D., Multichannel Hadamard transform Raman microscopy, *Appl. Spectrosc.*, 44, 1, 1990.
32. Hanley, Q.S., Verveer, P.J., and Jovin, T.M., Spectral imaging in a programmable array microscope by Hadamard transform fluorescence spectroscopy, *Appl. Spectrosc.*, 53, 1, 1999.
33. Nelson, P.M. and Myrick, M.L., Fabrication and evaluation of a dimension-reduction fiber-optic system for chemical imaging applications, *Rev. Sci. I*, 70, 2836, 1999.
34. Nelson, P.M. and Myrick, M.L., Single-frame chemical imaging: dimension reduction fiber-otpic array improvements and application to laser-induced breakdown spectroscopy, *Appl. Spectrosc.*, 53, 751, 1999.
35. Ma, J. and Ben-Amotz, D., Rapid micro-imaging using fiber-bundle image compression, *Appl. Spectrosc.*, 51, 1845, 1997.
36. Miller, P.J. and Levenson, R., Beyond image cubes: an agile lamp for practical 100% photon-efficient spectral imaging, *Proc. SPIE*, 4259, 1, 2001.
37. http://www.ece.purdue.edu/~biehl/MultiSpec/.

38. Perkins, S.J., Theiler, J., Brumby, S.P., Harvey, N.R., Porter, R.B., Szymanski, J.J., and Bloch, J.J., GENIE: a hybrid genetic algorithm for feature classification in multispectral images, *Proc. SPIE,* 4120, 52, 2000.
39. de Wolf, G. and van Vliet, L.J., Design of a four channel spectral analyzer to resolve linear combinations of two fluorescent spectra, *Proc. SPIE,* 3920, 21, 2000.
40. Jimenez, L.O. and Landgrebe, D.A., Supervised classification in high dimensional space: geometric, statistical, and asymptotical properties of multivariate data, *IEEE Trans. Syst. Man Cybernetics, Part C: Appl. Rev.,* 28, 39, 1998.
41. Ornberg, R.L., Woerner, B.M., and Edwards, D.A., Analysis of stained objects in histological sections by spectral imaging and differential absorption, *J. Histochem. Cytochem.,* 47, 1307, 1999.
42. Mark, H.L. and Tunnell, D., Qualitative near-infrared reflectance analysis using Mahalanobis distances, *Anal. Chem.,* 57, 1449, 1985.
43. Kruse, F.A., Lefdoff, A.B., Boardman, J.W., Heidebrecht, K.B., Shapiro, A.T., Barloon, J.P., and Goetz, A.F., The spectral image processing system (SIPS) — interactive visualization and analysis of imaging spectrometer data, *Remote Sensing Environ.,* 44, 145, 1993.
44. Harsanyi, J.C. and Chang, C.I., Hyperspectral image classification and dimensionality reduction: an orthogonal subspace projection approach. *IEEE Trans. Geosci. Remote Sensing,* 32, 779, 1994.
45. Garini, Y., Katzir, N., Cabib, D., Buckwald, R.A., Soenksen, D.G., and Malik, Z., Spectral bioimaging, in *Fluorescence Imaging Spectroscopy and Microscopy,* Wang, X.F. and Herman, B., Eds., John Wiley & Sons, New York, 1996.
46. Farkas, D.L., Du, C., Fisher, G.W., Lau, C., Niu, W., Wachman, E.S., and Levenson, R.M., Noninvasive image acquisition and advanced processing in optical bioimaging, *Comput. Med. Imaging Graph.,* 22, 89, 1998.
47. Winter, M.E., Fast autonomous spectral end-member determination in hyperspectral data, *Proc. 13th Int. Conf. Appl. Geol. Remote Sensing,* Vancouver, B.C., Canada, 2, 337, 1999.
48. Landgrebe, D.A., Information extraction principles and methods for multispectral and hyperspectral image data, in *Information Processing for Remote Sensing,* Chen, C.H., Ed., World Scientific, River Edge, NJ, 2000.
49. Mansfield, J.R., Sowa, M.G., Payette, J.R., Abdulrauf, B., Stranc, M.F., and Mantsch, H.H., Tissue viability by multispectral near infrared imaging: a fuzzy C-means clustering analysis, *IEEE Trans. Med. Imaging,* 17, 1011, 1998.
50. Perkins, S., Harvey, N.R., Brumby, S.P., and Lacker, K., Support vector machines for broad area feature extraction in remotely sensed images, *Proc. SPIE,* 4381, 268, 2001.
51. Ried, T., Tumor cytogenetics revisited: comparative genomic hybridization and spectral karyotyping, *J. Mol. Med.,* 75, 801, 1997.
52. Roederer, M., De Rosa, S., Gerstein, R., Anderson, M., Bigos, M., Stovel, R., Nozaki, T., Parks, D., and Herzenberg, L., Eight-color, 10-parameter flow cytometry to elucidate complex leukocyte heterogeneity, *Cytometry,* 29, 328, 1997.
53. Brelje, T.C., Wessendorf, M.W., and Sorenson, R.L., Multicolor laser scanning confocal immunofluorescence microscopy: practical application and limitations, in *Cell Biological Applications of Confocal Microscopy,* Matsumoto, B., Ed., Vol. 38, Academic Press, San Diego, 1993, p. 97.
54. Tsurui, H., Nishimura, H., Hattori, S., Hirose, S., Okumura, K., and Shirai, T., Seven-color fluorescence imaging of tissue samples based on Fourier spectroscopy and singular value decomposition, *J. Histochem. Cytochem.,* 48, 653, 2000.
55. Taylor, C.R. and Cote, R.J., Immunohistochemical markers of prognostic value in surgical pathology, *Histol. Histopathol.,* 12, 1039, 1997.
56. Albonico, G., Querzoli, P., Ferretti, S., Magri, E., and Nenci, I., Biophenotypes of breast carcinoma *in situ* defined by image analysis of biological parameters, *Pathol. Res. Pract.,* 192, 117, 1996.
57. Zhou, R., Parker, D.L., and Hammond, E.H., Quantitative peroxidase-antiperoxidase complex-substrate mass determination in tissue sections by a dual wavelength method, *Anal. Quant. Cytol. Histol.,* 14, 73, 1992.

58. Fritz, P., Wu, X., Tuczek, H., Multhaupt, H., and Schwarzmann, P., Quantitation in immunohistochemistry. A research method or a diagnostic tool in surgical pathology? *Pathologica,* 87, 300, 1995.
59. Speel, E.J., Ramaekers, F.C., and Hopman, A.H., Cytochemical detection systems for *in situ* hybridization, and the combination with immunocytochemistry: "Who is still afraid of red, green and blue?" *Histochem. J.*, 27, 833, 1996.
60. Hopman, A.H., Claessen, S., and Speel, E.J., Multi-colour brightfield *in situ* hybridisation on tissue section, *Histochem. Cell Biol.,* 108, 291, 1998.
61. Speel, E.J., Hopman, A.H., and Komminoth, P., Amplification methods to increase the sensitivity of *in situ* hybridization: play card(s), *J. Histochem. Cytochem.*, 47, 281, 1999.
62. Malik, Z., Rothmann, C., Cycowitz, T., Cycowitz, Z.J., and Cohen, A.M., Spectral morphometric characterization of B-CLL cells versus normal small lymphocytes, *J. Histochem. Cytochem.*, 46, 1113, 1998.

# III

# Photonic Detection and Imaging Techniques

# 9

# Lifetime-Based Imaging

Petr Herman
*Charles University*
*Prague, Czech Republic*

Hai-Jui Lin
*University of Maryland*
*Baltimore, Maryland*

Joseph R. Lakowicz
*University of Maryland*
*Baltimore, Maryland*



## 9.1 Introduction

Since the first report on the fluorescence phenomenon in 1845,[1] interest in fluorescence has rapidly increased. During the last two decades, fluorescence spectroscopy—in particular, time-resolved fluorescence—has become recognized as an important research tool in the biological sciences.[2–5] Time-resolved fluorescence is of interest because the molecular information content is greater than for steady-state measurements. However, the information content of time-resolved fluorescence is usually not accessible with imaging techniques such as fluorescence microscopy. The situation has changed during the last decade as remarkable developments in time-resolved fluorescence facilitated the transfer of time-resolved fluorescence from solution spectroscopy to the field of lifetime-based sensing and imaging.[6–8]

Fluorescence emission is a radiative process that occurs on the nanosecond timescale for most fluorophores. Normally, fluorescence is a first-order kinetic process and the intensity decay obeys the exponential law. The fluorescence lifetime $\tau$ characterizes the average amount of time that a molecule spends in the excited state following absorption of a photon. The importance of time-resolved fluorescence for imaging can be understood from the fact that unlike intensity, the fluorescence lifetime is largely independent of the probe concentration, photobleaching, and lightpath. All these parameters are extremely difficult to control during microscopic cellular experiments. The difficulties can be overcome by fluorescence ratiometric measurements; however, the number of fluorescent ratiometric probes is much smaller than the class of lifetime sensors.

Lifetime-based imaging is an experimental technique in which characteristics of the fluorescence decay are measured at each spatially resolvable location within a fluorescent image. This allows generation of image contrast based on the lifetime. The concept of the lifetime contrast and its advantages for imaging applications are demonstrated in Figure 9.1. This figure shows two adjacent capillary tubes containing

0-8493-1116-0/03/$0.00+$1.50

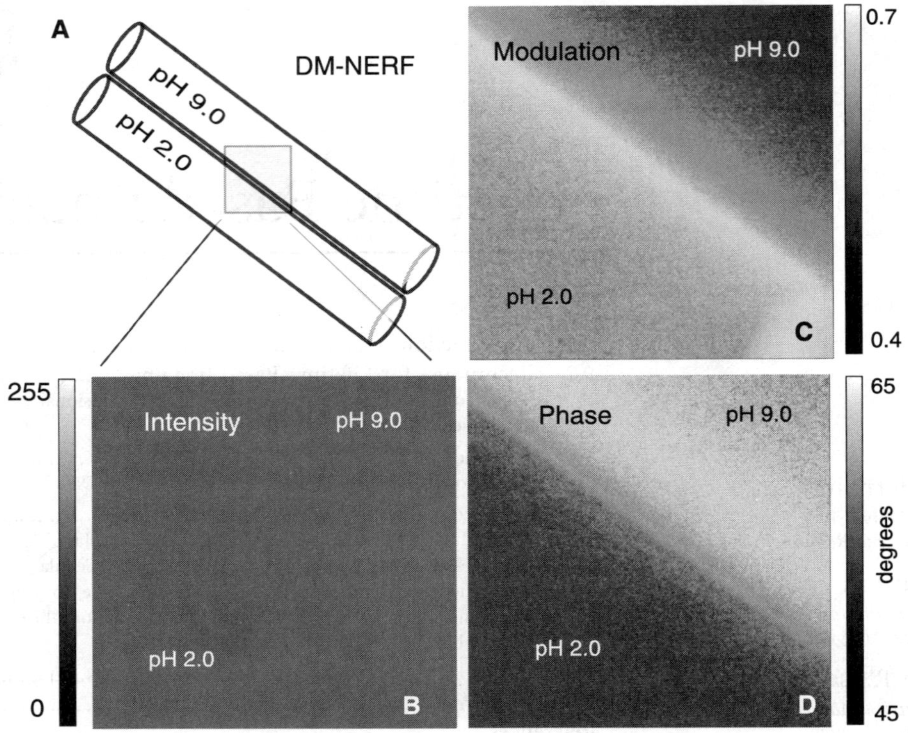


**FIGURE 9.1** Demonstration of lifetime contrast. (A) Two capillaries filled with a buffer containing the pH-sensitive lifetime probe DM-NERF. The pH was adjusted to 9.0 and 2.0 in the upper and lower capillaries, respectively. Intensities in both capillaries were adjusted to the same level. (B) Fluorescence intensity image does not show any contrast. (C) Modulation image. (D) Phase image. Images (C) and (D) exhibit significant contrast because modulation and phase shift depend on fluorescence lifetime (see Equation 9.9).

the pH-sensitive fluorophore DM-NERF, but at two different pH values of 2 and 9. The steady-state intensities were identical (lower left). Differences between the two capillaries are seen in the phase angle and modulation images (right panels), reflecting these different lifetimes. Hence, lifetime imaging can reveal spatial differences in the sample even when the steady-state intensities are identical.

For simplicity we will not distinguish between fluorescence and phosphorescence lifetime imaging because the concepts are the same; only the timescales differ. Unless specifically noted, the same applies for fluorescence lifetime imaging (FLI) and fluorescence lifetime imaging microscopy (FLIM), where the main difference between the techniques is an optical system for imaging of macroscopic and microscopic samples, respectively.

FLIM has evolved as a part of a broader discipline called fluorescence lifetime microscopy, as shown in Figure 9.2. Fluorescence lifetime microscopy also includes nonimaging applications where fluorescence decays are acquired from one or several localized areas of the sample.[9–18] Due to the potential of accessing intensity-independent lifetime information from subcellular volumes and due to a broad availability of lifetime probes, FLIM has gained popularity in cellular biology, biophysics, and biomedical sciences. Fluorescence lifetime microscopy and lifetime-based imaging have been discussed in several review articles.[6,7,19–22]

## 9.2 Techniques for Lifetime-Based Imaging

Spatially resolved measurements of fluorescence lifetimes can be accomplished by several means. Generally we have two types of time-resolved fluorescence: time-domain (TD) and frequency-domain (FD)

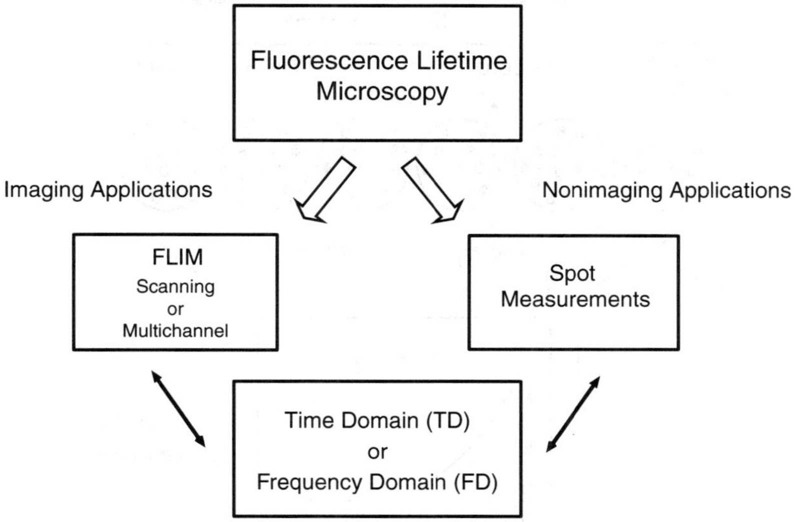


**FIGURE 9.2** Fluorescence lifetime microscopy. Measurements may be done at a single point or as imaging. Both TD and FD techniques are used.

measurements. In principle, the two methods are equivalent. They yield the same kind of information about the examined object, and they are related to each other by the Fourier transform. In practice, however, the methods use different instrumentation. The experimental data have a different appearance and the data analysis is formally different. Depending on detailed experimental conditions, the TD lifetime measurements can gain comparative advantage over the FD method and vice versa. However, it is mostly a personal preference or availability of instrumentation that determines whether the TD or FD approach is used.

Another aspect of FLI is the imaging modality. Images can be acquired pixel by pixel with a single channel detector in a sequential scanning mode or line by line with an array detector in a line scanning or push-broom configuration. The scanning imaging techniques were developed for conventional and spectral imaging and can be applied for lifetime imaging as well. The FLIM images can also be acquired in a wide-field (WF) multichannel regime when lifetime information is simultaneously recorded from the whole field of view (from all pixels) by a two-dimensional detector. This detector is typically an intensified (CCD) camera even though other spatially sensitive detectors such as a multiple-anode photomultiplier (PMT) suitable for lifetime-based measurements have been reported.[23–25] In the following paragraphs we discuss in more detail different approaches to fluorescence lifetime microscopy and lifetime-based imaging.

## 9.2.1 Time Domain

In time-domain measurement, the fluorophore is excited by a short pulse of light and a time course of the fluorescence response is recorded (Figure 9.3). The rate of photoemission is proportional to the number of excited fluorophores. The fluorescence intensity decay $I(t)$ can be a monoexponential function. Typically, the intensity decay is a multiexponential function,

$$I(t) = \sum_i a_i \cdot e^{-t/\tau_i} \tag{9.1}$$

where $\tau_i$ are lifetimes and $a_i$ are the corresponding amplitudes. For a mixture of fluorophores, each displaying a single exponential decay, the decay times will be those of the individual fluorophores and the amplitudes will be related to the intensity and concentration of each species.

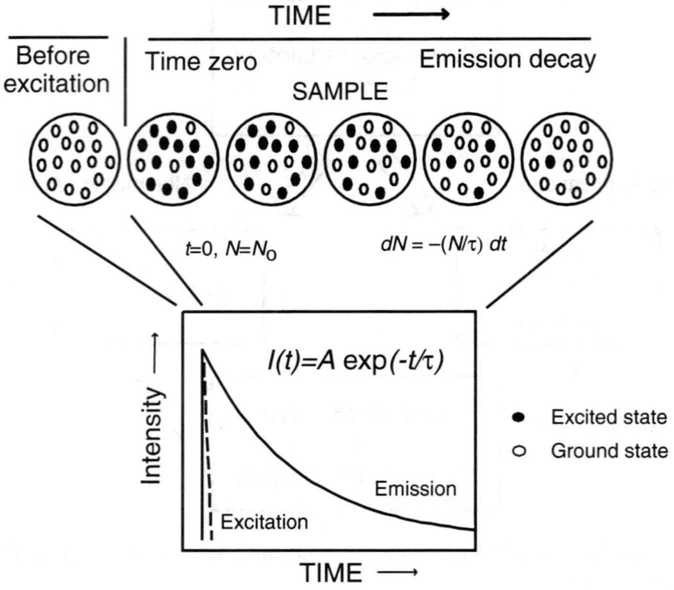


**FIGURE 9.3** Concept of the emission lifetime. $N$ = number of molecules in the excited state; $\tau$ = emission lifetime.

Several TD approaches have been used to measure fluorescence decays. They include single-shot experiments when the full fluorescence decay is acquired from a single-excitation pulse with fast digitizers or streak cameras,[26] repetitive stroboscopic sampling, gated boxcar detection, and optical pump-probe methods,[27] the widely used time-correlated single-photon counting (TCSPC),[28] which is a dominant method in nonimaging TD fluorescence spectroscopy. A number of these methods were adopted for lifetime-based imaging in a wide variety of instrumental implementations.

### 9.2.1.1 Time-Correlated Single-Photon Counting

A great deal of literature is available on the subject of TCSPC. Extensive description and discussion of the method can be found in the monograph by O'Connor and Philips.[28] For purposes of comparison with other lifetime imaging methods, it is useful to reiterate some properties and limitations of TCSPC.

The principle of TCSPC is shown in Figure 9.4. The method is based on repetitive measurements of the time difference between the excitation flash and a subsequent detection of the first emission event. The time difference is measured electronically by a time-to-amplitude converter (TAC), which outputs electrical pulses with amplitudes proportional to the time difference between the start and stop pulses generated by detectors in the excitation and emission channels. The output of the TAC is digitized and sorted by the multichannel analyzer (MCA), where a probability histogram of the count number vs. time channels is built up by sampling a large number of excitation-emission events.

If the emission intensity is low enough so that no more than one photon is detected per start–stop cycle, the histogram exactly reflects time evolution of the fluorescence intensity decay $I(t)$. In the opposite case the TAC does not register the second arriving photon and the measured intensity decay is biased to shorter times by the pile-up effect. To avoid pulse pile up, the detection rate has to be limited so that detection of the two photons per excitation pulse is unlikely. An acquisition rate lower than 1 to 2 photons per 100 excitation pulses is considered a practical limit for elimination of the pile-up effect.[28] Even though the pile-up can be to some extent corrected mathematically[29–32] or suppressed by electronic discriminating the multipulse events,[33,34] the safest way to avoid pile-up problems is to use low counting rates.

The upper limit on the data acquisition rate places practical limitation on the two-dimensional lifetime mapping where thousands of fluorescence decays have to be accumulated in order to obtain a lifetime

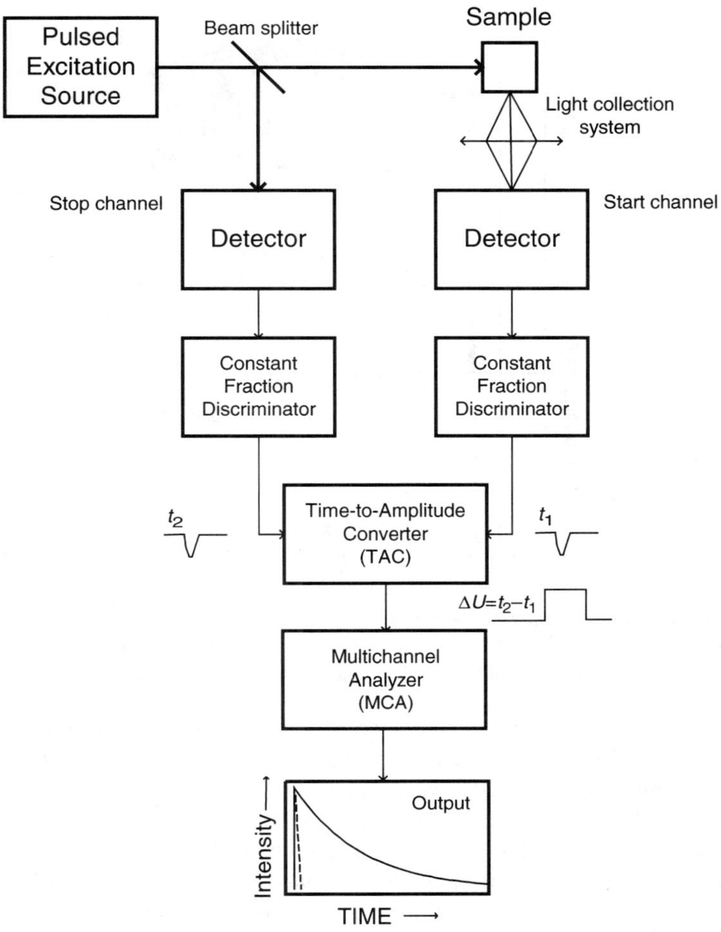


**FIGURE 9.4** Schematic diagram of a TCSPC instrument.

image. As an example, let us assume a $512 \times 512$-pixel image and an excitation repetition rate of 80 MHz. To eliminate pile-up artifacts, the data collection rate should not exceed 1% of the excitation frequency, i.e., 800 kHz. For cuvette experiments, it is not unusual to collect $10^6$ counts/decay in order to have data with a good signal-to-noise ratio (SNR) for multicomponent decay analysis. In FLI we will be more likely to acquire only $10^3$ counts/decay in a single pixel. This value is close to the limit where lifetime information starts to be hidden in the noise and further reduction of the counts per pixel might not be practical. Under these rather favorable conditions the acquisition of the $512 \times 512$-pixel image would take more than 5 min.

In our example we used the laser repetition frequency of 80 MHz with light pulses separated by 12.5 ns. The time window between pulses allows measurements of only lifetimes up to 2 to 3 ns because fluorescence should substantially diminish before the next pulse arrives. For longer lifetimes, the excitation frequency has to be reduced and the acquisition time proportionally increases. It is evident that the TCSPC is impractical for FLI experiments with fluorophores exhibiting microsecond and longer lifetimes because image acquisition time would exceed several days.

Despite the acquisition rate limitation, TCSPC has single-photon sensitivity, excellent time resolution, and a wide dynamic range. This is why the technique was employed with single- and two-photon excitation for fluorescence lifetime microscopy or FLI by a number of researchers.[14,16,25,35–42] Applications

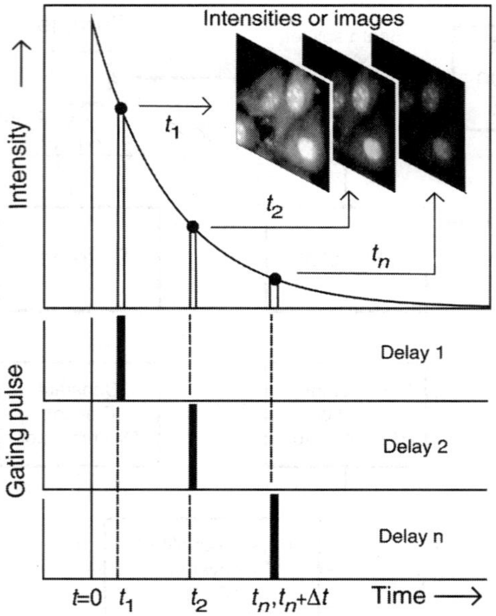


**FIGURE 9.5** Gated detection. During each excitation cycle the detector is turned on for a period of time $\Delta t$ and intensity is measured. Depending on the imaging mode, the image is acquired simultaneously or pixel by pixel for each time delay. The delay is changed until the whole emission decay is sampled.

range from time-resolved fluorescence measurements from single or multiple locations within a microscopic sample,[14,16] confocal scanning implementations,[36, 37,41,43] and three-dimensional imaging[40] to near-field scanning optical microscopy (NSOM) with fluorescence lifetime as a contrast mechanism.[44] Willemsen et al.[37] combined TCSPC-based lifetime imaging with atomic force microscopy in one scanning device. TCSPC has also been used for lifetime imaging in a scanning DNA sequencer.[39]

Methods have been developed to increase the data collection rate above the TCSPC limit. One such method uses parallel data acquisition with multichannel position-sensitive detectors. The detector could be a multianode PMT[23,25] placed in the image plane with outputs from the individual anodes routed to the multiplexed TCSPC detection system. In this system the count rate can be increased to a value close to 0.37,[45] which is a considerable improvement over single-channel TCSPC. Even though PMTs with up to 96 anodes have been reported,[46] the imaging capability and spatial resolution of FLIM instruments with the multianode PMTs is sufficient only for low spatial-resolution imaging.

### 9.2.1.2 Multichannel Photon Counting

In order to overcome limitations of TCSPC and increase the data collection rate required for two-dimensional lifetime scanning, multichannel photon counting has been developed for lifetime microscopy.[15] The method utilizes simultaneous digital sampling of all photons in the emission burst while preserving their arrival-time information. Because the method does not use TAC for time measurements, the detection rate can exceed the 1% limit of TCSPC, which substantially reduces measurement time compared to TCSPC. Modest time resolution in the nanosecond range, however, limits applicability of the technique for short-lived fluorophores.

### 9.2.1.3 Sampling Methods

Unlike the time-resolved cuvette experiments mostly performed by TCSPC, the field of TD lifetime imaging is dominated by sampling methods with gated detection. The principle of the most commonly used boxcar approach[27,47] can be understood from Figure 9.5. The sample is excited by a periodic train of light pulses and the signal is recorded within a defined time interval $\Delta t$ after each excitation pulse. More specifically, the detector is turned off during the fluorescence decay except for a brief period of

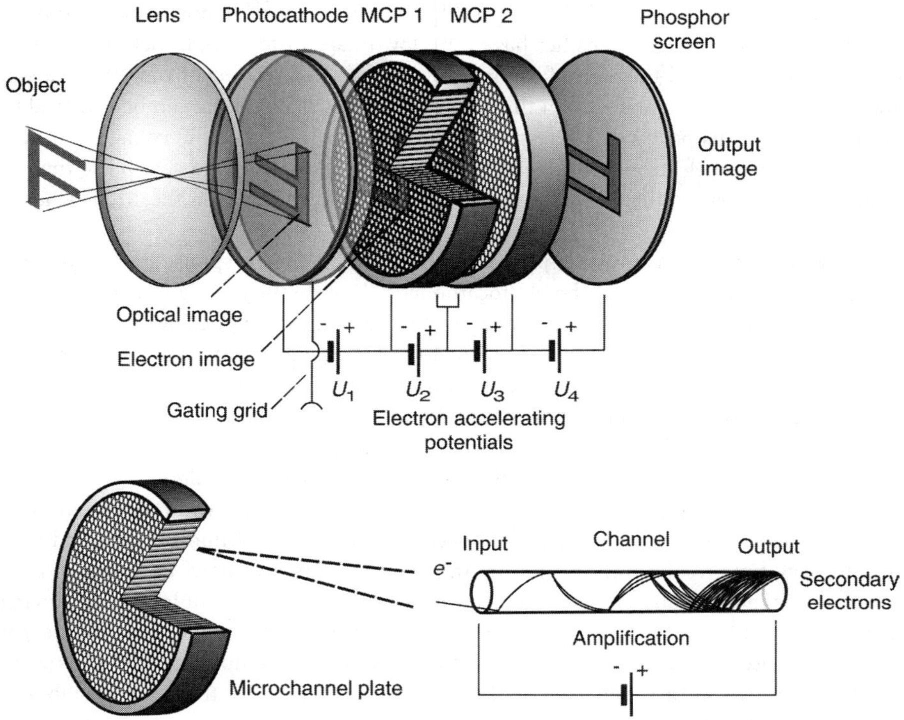


**FIGURE 9.6** Schematic of a two-stage image intensifier. An image of the observed object is projected on the photocathode. Emitted primary photoelectrons are accelerated and enter a pair of the microchannel plates where the number of electrons is amplified. Accelerated secondary electrons impinge on the phosphor screen where the bright optical image of the object is generated. The gating grid allows the gain of the intensifier to be turned ON and OFF, or continuous modulation of the gain.

time $\Delta t$. The fluorescence intensity in the time window of $t$ and $t + \Delta t$ is repetitively accumulated until a desired S/N ratio is reached. The process is repeated for a set of delays $t_i$, until the whole fluorescence decay is sampled. It is straightforward to implement the boxcar method for nonimaging time-resolved microscopy.[12,13] The boxcar method has been used also for scanning confocal FLIM with single-photon[48–52] or two-photon excitation[53–55] with a gated PMT as a detector. In such experiments full images are sequentially acquired for each delay $t_i$ and fluorescence decays are constructed at each pixel as a function of the pixel intensity vs. $t_i$.

**FLIM with gated image intensifier.** The full advantage of gated detection in lifetime-based imaging was realized by the introduction of a gated image intensifier. The whole image can then be sampled at once without need of pixel-by-pixel scanning. For a given S/N ratio and image size, simultaneous sampling of all pixels brings tremendous improvement of the data acquisition time with respect to the serial scanning methods. This property allowed fast lifetime mapping to the WF imaging.[56–76]

An image intensifier is an optoelectronic imaging device that amplifies brightness of an input light pattern on the photocathode while preserving the pattern. The device comprises a photocathode, one or several microchannel plates, and a phosphorescent screen where the amplified optical image is formed. The screen is typically observed by a CCD camera for digital quantification of the image. The concept of the image intensifier is illustrated in Figure 9.6. Primary photoelectrons generated on the photocathode by an incident light are accelerated and enter microchannel plate (MCP) consisting of a large number of metal-lined capillaries. The opposite ends of the capillaries are held at the electron accelerating potentials and their inner walls are covered by an electron emissive material. The accelerated photoelectrons therefore generate an avalanche of secondary electrons, which subsequently form a bright fluorescence spot on the phosphor screen.

Modern intensifiers exhibit practically no crosstalk between adjacent channels and all components are proximity focused, so the image intensifier has a very low image distortion. Depending on the number of multichannel plates MCPs, the gain of the intensifier can typically range from $10^3$ for a single-stage intensifier up to $10^7$ for an intensifier with three MCPs in series. Gain gating can be achieved by biasing photocathode several volts positive with respect to the MCP (by reversing the voltage $U_1$ in Figure 9.6) in order to prevent photoelectrons from being collected on the MCP.[77] The gate pulse then drives the photocathode more negative than the MCP and the photoelectrons enter the MCP where they are amplified.

Alternatively, the gating can be accomplished with improved time response by biasing an auxiliary gating grid positioned on the back of the photocathode.[57] Shutter ratios as high as $10^9$ to $10^{12}$ have been reported.[77] Recent gated intensifiers can operate at high repetition frequencies with gating pulses as narrow as several hundreds of picoseconds.[78–80] Small-diameter gated intensifiers are becoming available with an exposure time as short as 50 ps. These intensifiers, however, can only operate with a very low duty cycle with repetition frequencies up to several kHz.[79,80]

Although gating methods do not possess a dynamic range or the time resolution of the TCSPC with a microchannel PMT, they offer a number of important advantages for lifetime-based imaging. The boxcar acquisition method is highly configurable. The number and time-width of the individual gates can be arbitrarily changed in order to match the decay characteristics of fluorophores and balance data acquisition time vs. time resolution. The minimum number of time windows necessary to calculate the lifetime of monoexponentially decaying emission is two.[56] Such a double-gate approach to FLI experiments[48,50,51] greatly reduces data acquisition and processing time while discarding a part of the information otherwise accessible by multigate methods. Nevertheless, double-gate FLI is very efficient in generating a lifetime contrast even for complex decays. The lifetime contrast can be calibrated against known samples, which consequently allows extraction of biologically relevant information.

Unlike the TCSPC, gating can be efficiently used with low repetition rate lasers. Also, multiple photons detected per laser flash are not a limitation of the technique, which can thus be used for measurements of lifetimes in the microsecond and millisecond time ranges. An important property of gating methods is the capability for background suppression. When gating is used with longer-lived fluorophores, scattered light, reflections, autofluorescence, and short-lived fluorescence, which usually decrease image contrast in conventional fluorescence microscopy, can be easily rejected.[64,71] The normally off detector is turned on at the moment when the intensity of the short-lived components has diminished and only long-lived fluorescence is acquired. For very long-lived emission, such a gating scheme has also been accomplished without expensive electronics by attaching two phase-locked choppers to a microscope.[81]

**Gatable CCD cameras.** Lifetime imaging with long-lived millisecond or microsecond probes can be accomplished without expensive gatable image intensifiers. Khait et al.[82] reported a fluorescence lifetime microscope with a millisecond temporal resolution equipped with a free-running externally synchronized CCD camera acquiring data with a speed of 33 images/second. A similar approach has been used for oxygen imaging with a millisecond phosphorescent sensor Green 2W.[83] Phosphorescence decays were sampled at preset times after the excitation flash by an externally triggered CCD camera working with an exposure time of 2.5 ms. Phosphorescence lifetime maps and a distribution of $pO_2$ in tissues were constructed by this method.

During the past decade, fast, directly gatable CCD cameras have become available.[84–86] The technology utilizes a semiconductor-based electronic shutter integrated into the structure of the CCD chip, which allows for short optical exposure times. Shutter rise and fall times shorter than 55 ns and opened/closed ratios as high as $10^4$ to $10^5$ have been reported.[84] The directly gatable cameras have been used for microsecond lifetime imaging of luminescent ruthenium(II) complexes.[85] Decay times shorter than 100 ns have been measured with the directly gatable cameras.[86]

#### 9.2.1.4 Spatially Sensitive Multichannel Plate Detectors

The photoelectron emitted from the photocathode and amplified by the MCP creates a localized charge on the anode centered at positions corresponding to the positions of the incident photon. This property

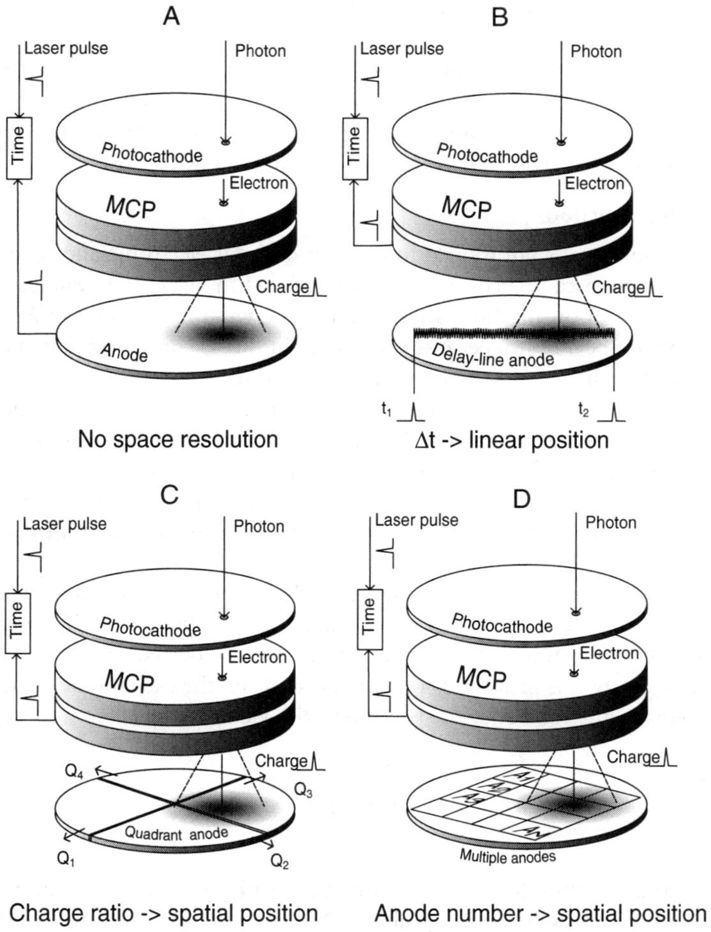


**FIGURE 9.7** Principle of position-sensitive detectors for time-resolved spectroscopy. (A) Conventional MCP-PMT without space resolution. (B) Delay line anode MCP-PMT with linear position resolution. The position information is derived from the difference in pulse arrival times to the terminals of the delay-line anode. (C) Quadrant anode MCP PMT with two-dimensional spatial resolution. The position of the incident photon is calculated from the ratio of charges collected by the individual quadrants $Q_1 - Q_4$ of the anode. (D) Multiple-anode MCP-PMT. The spatial position correlates with the electrical signal on the individual, spatially separated anodes. (Modified from Kemnitz, K. et al., *J. Fluoresc.*, 7, 93, 1997.)

provides the means to record the spatial and temporal information about incident photons simultaneously and to use these fast detectors for lifetime-based imaging. The temporal information is typically measured by TCSPC electronics. Several strategies for retrieving the spatial information of the incident photons are illustrated in Figure 9.7.

The spatial information can be accessed by replacing a conventional disk anode (Figure 9.7A) with a resistive (R) or a delay-line (DL) anode (Figure 9. 7B), quadrant anode (QA) (Figure 9.7C), or by a set of individual anodes (Figure 9.7D). While the DL-MCP-PMTs were used to encode a linear position across the photocathode,[35,87] the resistive[38] or quadrant anode[88] MCP-PMTs provide continuous two-dimensional position information. The detectors with continuously coded spatial location are not multichannel devices for simultaneous signal acquisition from multiple locations, because they can register only one detection event at a time. If multiple photons simultaneously strike the photocathode at different places, the photons are registered as one detection event and an incorrect location is assigned by electronics. In order to alleviate such cases the detection rates have to be limited to an extent as described earlier for TCSPC.

From this point of view a lifetime mapping with such detectors is slow and can be compared with a sequential random scanning of the sample. Due to the recent advance in the image intensifier technology, which offers better spatial resolution, lower image distortion, and faster data acquisition, the importance of the position-sensitive detectors for lifetime-based imaging has decreased.

#### 9.2.1.5 Multipulse Methods

Increasing the time resolution is not limited to creation of faster electronic and optoelectronic devices. The excited state population can be controlled by intense light pulses. The availability of femtosecond lasers offers time resolution on the time-scale of the pulse width. Nonlinear multipulse methods are especially suitable for confocal FLIM, because the laser light flux is concentrated in a diffraction-limited spot where a threshold for nonlinear phenomena is easier to reach.

One multipulse optical technique implemented on a scanning confocal microscope is a double-pulse fluorescence lifetime imaging (DPFLIM).[89,90] The method is based on the excitation of fluorophores with ultra-short laser pulses, which populate the excited state to near the saturation value. The method does not require electronic gating of the detector, but rather relies on the optical pumping of the excited state of fluorophores. Because the time resolution depends mainly on the duration of the excitation pulses, the DPFLIM can have subpicosecond time resolution.

The principle of the method is schematically depicted in Figure 9.8. The sample is illuminated with intense single pulses saturating the first excited state and the time-integrated fluorescence intensity $I_1$ is measured by a conventional detector. Then the measurement is repeated with saturating pulse doublets and again the total intensity $I_2$ is measured. The two pulses from the excitation doublet are separated by the time delay $\Delta t$. Finally, the ratio $R$ of the two intensities yields the fluorescence lifetime $\tau$:[89]

$$R = \frac{I_2}{I_1} = 2 - \exp(-\Delta t / \tau) \tag{9.2}$$

From Figure 9.8 and Equation 9.2, it can be seen that, with increasing separation of the pulses, the ratio $R$ increases until it reaches the limiting value of 2 for $\Delta t >> \tau$. By calculation of $R$ at each pixel, the lifetime contrast is created and a fluorescence lifetime map can be constructed.

The DPFLIM was first used with a confocal microscope where saturation of the excited state can be reached without problem. With some modification, the DPFLIM is usable even with weaker nonsaturating illumination[89] and has a potential to be applied to conventional WF-FLIM.

Additional examples of multipulse nonlinear methods potentially useful for lifetime-based imaging have been outlined in Reference 47 and in the review of Lakowicz,[21] where manipulation of the excited state population by light quenching has been proposed.

### 9.2.2 Frequency Domain

The history of FD fluorometry begins in 1926 when Gaviola performed his first fluorescence lifetime measurement.[91] During the years the FD methodology rapidly evolved[47] until in the last decades of the 20th century it was adopted for lifetime microscopy and lifetime-based imaging.[92,93] The principle of phase and modulation measurement is schematically depicted in Figure 9.9 for a single point in the image. In FD lifetime imaging, the investigated object is exposed to the excitation light $E(t)$, harmonically modulated at an angular frequency of $\omega = 2\pi f$ with a modulation degree of $m_E = a/A$:

$$E(t) = E_0 \cdot [1 + m_E \sin(\omega t)] \tag{9.3}$$

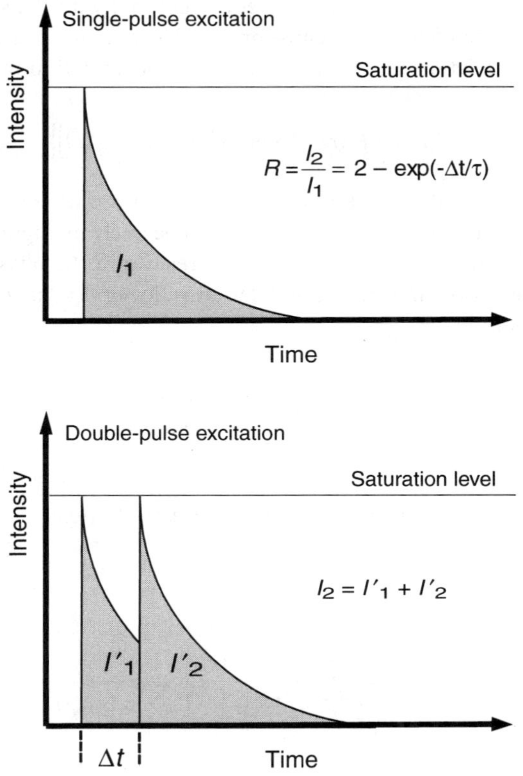


**FIGURE 9.8** Principle of the optical double-pulse lifetime determination used for a confocal FLIM. (Figure was Redrawn from Muller, M. et al., *J. Microsc. (Oxford)*, 171, 177, 1995.)

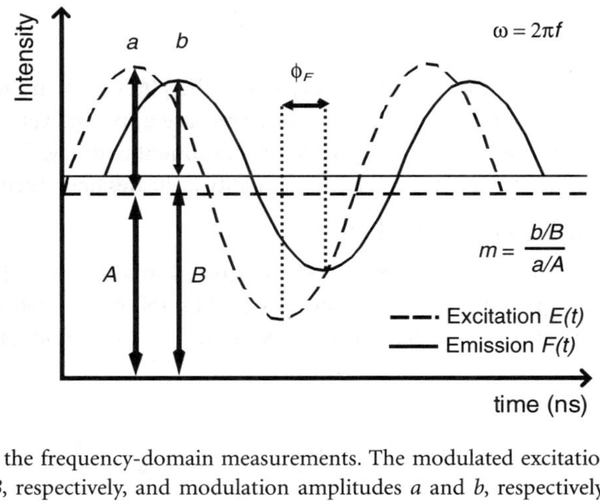


**FIGURE 9.9** Concept of the frequency-domain measurements. The modulated excitation and emission light have DC components $A$ and $B$, respectively, and modulation amplitudes $a$ and $b$, respectively. The emission exhibits a phase lag $\phi_F$ with respect to the excitation light and a demodulation ratio $m$.

Such periodic excitation causes fluorophores to emit light $F(t)$ with the same modulation frequency $\omega$. Due to a finite fluorescence lifetime, the emission is delayed, exhibiting a spatially dependent phase lag $\phi_F(r)$ and decreased modulation $m_F(r) = b/B$, relative to the excitation radiation:

$$F(r,t) = F_0(r) \cdot \{1 + m_F(r)\sin[\omega t - \phi_F(r)]\} \tag{9.4}$$

The spatial coordinate $r$ maps the position within the object from where the emission is observed. For simplicity we set the excitation phase to zero because it can be freely manipulated, by insertion of a delay line to the excitation path, and all phases are measured relative to the phase angle. It has been shown that the fluorescence phase $\phi_F(r)$ and the demodulation ratio $m(r) = m_F(r)/m_E(r)$ depend on lifetime components of the emission:[94,95]

$$\phi_F(\omega,r) = \arctan\left(\frac{N(\omega,r)}{D(\omega,r)}\right) \tag{9.5}$$

$$m(\omega,r) = \frac{m_F(\omega,r)}{m_E(\omega,r)} = \sqrt{N^2(\omega,r) + D^2(\omega,r)} \tag{9.6}$$

where

$$N(\omega,r) = \left[\sum_i \frac{a_i(r)\omega\tau_i^2(r)}{1+\omega^2\tau_i^2(r)}\right] \cdot \frac{1}{\sum_i a_i(r)\tau_i(r)} \tag{9.7}$$

$$D(\omega,r) = \left[\sum_i \frac{a_i(r)\tau_i(r)}{1+\omega^2\tau_i^2(r)}\right] \cdot \frac{1}{\sum_i a_i(r)\tau_i(r)} \tag{9.8}$$

For the simplest case of a monoexponential decay, the lifetime $\tau(r)$ at any location of the sample can be directly calculated from Equations 9.5 through 9.8:

$$\tau_\phi(r) = \frac{1}{\omega}\arctan(\phi_F(r)), \qquad \tau_m(r) = \frac{1}{\omega}\cdot\sqrt{\frac{1}{m^2(r)} - 1} \tag{9.9}$$

and $\tau = \tau_\phi = \tau_m$. Equation 9.9 can also be used to calculate the phase ($\tau_\phi$) and modulation ($\tau_m$) lifetime for more complex decays. The resulting lifetimes, however, are apparent values representing complex mixing of different lifetime components.[47,96,97] For a multiexponential decay, the phase lifetime is shorter than the modulation lifetime, $\tau_\phi < \tau < \tau_m$, and such inequality indicates heterogeneous fluorescence decay.

#### 9.2.2.1 Homodyne and Heterodyne FLIM

Theory of the phase and modulation FLIM has been discussed in a number of papers.[98–103] Briefly, the fluorescent image $F(r,t)$ (Equation 9.4) of an object excited by the modulated light is projected on the photocathode of the image intensifier that has the electronic gain $G(t)$ modulated at the RF frequency $\omega + \Delta\omega$:

$$G(t) = G_0 \cdot \{1 + m_D \sin[(\omega + \Delta\omega)\cdot t - \phi_D(r)]\} \tag{9.10}$$

Inside the intensifier, frequency mixing occurs as the primary photocurrents modulated at the frequency $\omega$ interact with the electronic gain modulated at the frequency $\omega + \Delta\omega$. Due to the long time constant of the phosphor screen caused by the millisecond phosphor lifetime, all high-frequency components become time averaged and we observe an amplified image with intensity slowly varying with the frequency $\Delta\omega$:

$$I(r,t) = I_0(r)\cdot\left\{1+\frac{m_D\cdot m_F(r)}{2}\cos\left[\Delta\omega t-\Delta\phi(r)\right]\right\} \tag{9.11}$$

where

$$\Delta\phi(r) = \phi_F(r) - \phi_D \tag{9.12}$$

The phase and modulation information is retained after the frequency mixing and can be extracted from Equations 9.11 and 9.12. The method is called homodyne and heterodyne for $\Delta\omega = 0$ and $\Delta\omega \neq 0$, respectively. Both methods rapidly evolved when image intensifiers suitable for high-frequency gain modulation became available.[98,104]

The homodyne method is the dominating method used in FD WF-FLIM.[7,98–100,102,103,105–124] The principle of the method is schematically shown in Figure 9.10 with the time-domain counterpart. The generic

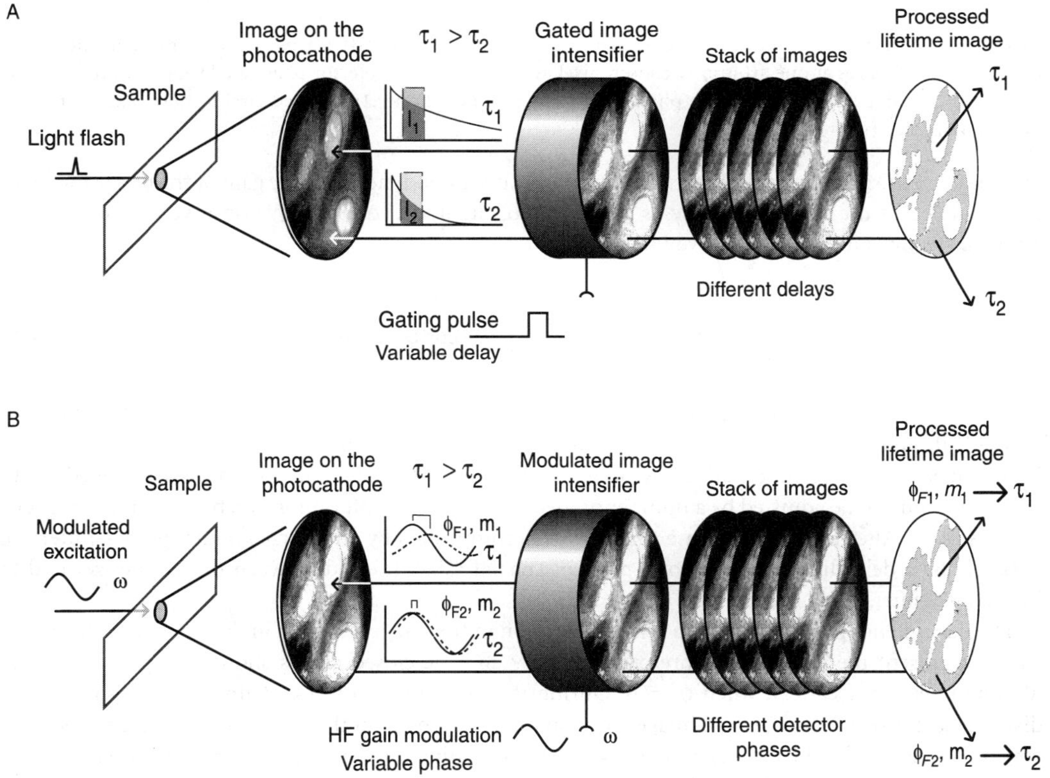


**FIGURE 9.10** Schematic comparison of the time-domain (A) and the frequency-domain (homodyne) (B) methods for wide-field FLI. In the time domain, images are acquired for several delays of the gating pulse with respect to the excitation flash. Since the gain of the intensifier is on only during the presence of the gating pulse, fluorescence decays at different locations of the sample are simultaneously sampled and the lifetime image can be constructed on a pixel-by-pixel basis from the measured stack of the images. In the frequency domain homodyne, FLIM, the gain of the image intensifier is modulated at the same frequency as the excitation light. As a result, we observe a steady-state phase-sensitive image on the phosphor screen of the intensifier. After the stack of images is acquired at detector phases distributed between 0 and $2\pi$, the modulation $m$ and phase lag $\phi_F$ are extracted and a lifetime image is generated.

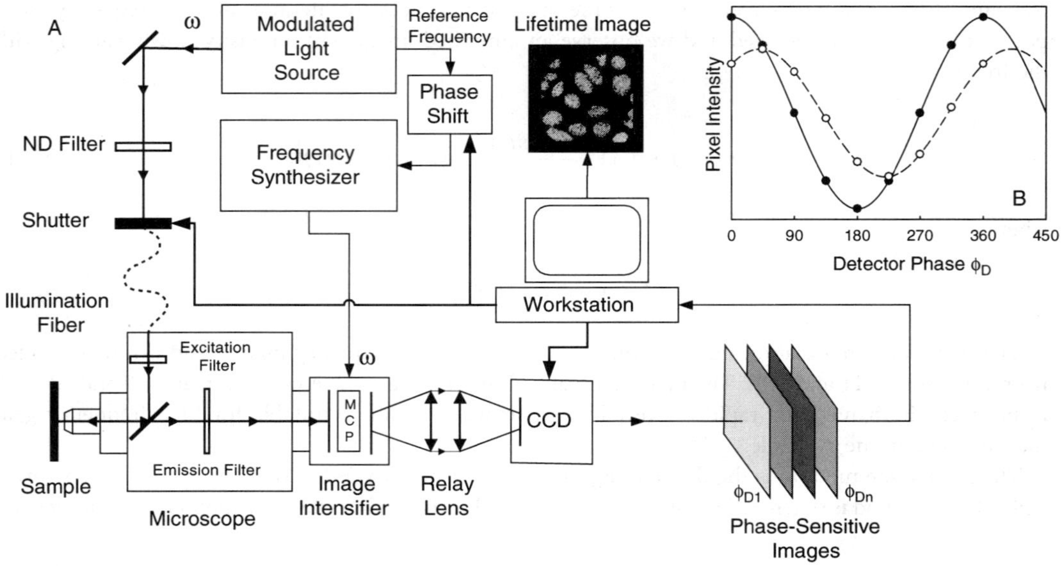


**FIGURE 9.11** Homodyne FLIM. (A) Scheme of a generic instrument. (B) Intensity profile through the stack of phase-sensitive images at one arbitrarily chosen pixel as a function of a detector phase angle. Open circles represent intensities of a fluorescent signal with phase lag $\phi_F > 0$ (see Equation 9.13); closed circles show signal for a zero-lifetime reference (scatter) when $\phi_F = 0$.

scheme of an FD instrument is shown in Figure 9.11. In a homodyne FLIM the gain of the image intensifier is modulated at the same frequency as the excitation light, i.e., $\Delta\omega = 0$. We observe on the phosphor screen a steady-state phase-sensitive image:

$$I(r) = I_0(r) \cdot \left\{ 1 + \frac{m_D \cdot m_F(r)}{2} \cos\left[\phi_F(r) - \phi_D\right] \right\} \tag{9.13}$$

The intensity $I(r)$ at any location of the phosphor screen depends on the cosine of the difference between fluorescence and detector phase angles and on the lifetime through the phase angle $\phi_F(r)$, Equation 9.5, and the modulation factor $m_F(r)$, Equation 9.6. The value of $\phi_D$ is controlled by the apparatus and can be adjusted by a number of means. The phase-shift option can be within the frequency synthesizer.[100] Alternatively the phase shift can be accomplished by an external digital phase-shifter[110] or passively by a delay line consisting of a piece of coaxial cable inserted between the synthesizer and the image intensifier.[98]

The dependence of $I(r)$ on $\phi_D$ for a selected location in the image is shown in Figure 9.11B. The opened circles represent the measured fluorescent intensity and the closed circles show the signal for a zero-lifetime reference (scatter), when $\phi_F = 0$. During the experiment a stack of images is acquired for $\phi_D$ distributed between 0 and $2\pi$. The images contain information about the $\phi_F(r)$ and $m_F(r)$ at each position, which can be extracted by Fourier analysis or by performing a linear least-squares fit for each pixel in the image.[99,100] Then the lifetime image can be constructed or one can simply use the phase angle or modulation images.

Heterodyne detection uses cross-correlation mixing for translation of the high-frequency fluorescence modulation to the frequency region of several hertz. When used with the image intensifier, an image with periodically oscillating intensity at the frequency $\Delta\omega$ is observed on the phosphor screen. The time-dependent image can be sampled by a CCD camera. This seems to be a disadvantage because camera exposure times much shorter than $1/\Delta\omega$ are required to sample the oscillations properly, which decreases the duty cycle for data acquisition. This is a disadvantage because measurements of dim samples may require longer integration times, and because the longer illumination times may result in excessive photobleaching.

To overcome this drawback, a method combining high-frequency heterodyne mixing on the image intensifier with low-frequency optical-homodyne image acquisition has been developed.[101,104,125] The heterodyne signal on the phosphor screen was recorded during a fraction of the heterodyne cycle by a boxcar method. This resulted in a steady-state signal on the phosphor screen integrated by the camera. The heterodyne cycle was sampled by changing delay of the acquisition window.

The heterodyne detection method does not require as strict a high-frequency phase stability as the homodyne counterpart. Synchronization on the low cross-correlation frequency is much easier to achieve. However, with modern synthesizers phase stability is not a problem. Due to the boxcar integration when only a slice of the heterodyne cycle is used for data accumulation, the acquisition times are longer with heterodyne than with homodyne detection. This disadvantage, together with more complex instrumentation, has caused lower use of the heterodyne WF-FLIM compared to the homodyne counterpart. Nevertheless, the heterodyne method of the lifetime determination has been successfully implemented in a scanning microscope.[126]

#### 9.2.2.2 Optical Methods

Dong et al. have reported a unique optical cross-correlation method implemented on a scanning fluorescence lifetime microscope.[127] Cross-correlation frequency mixing was accomplished by periodic quenching of the fluorescence by stimulated emission. The method utilizes the harmonic content[128–130] of the two pulse trains from two lasers running at repetition frequencies offset by the $\Delta\omega$ and focused in the sample. One of the laser beams excites fluorescence and the wavelength of the other is tuned to quench the emission. At the place of the beam overlap a cross-correlation fluorescence signal at multiple harmonics of the frequency $\Delta\omega$ is created. The low-frequency fluorescence signal is detected by a standard PMT; the harmonics are electronically isolated and analyzed for demodulation and phase-shift.

Optical heterodyning has two main advantages. The low-frequency fluorescence signal implies that the FLIM can be carried out without using fast photodetectors. Lifetime measurements up to 6.7 GHz have been reported with a conventional PMT. Overlap of the pump-and-probe lasers also results in an axial sectioning since the out-of-focus signal is not cross correlated and is rejected by the electronic filters. Because of the wavelength used in the one-photon pump-and-probe process, the spatial resolution of the technique is comparable to confocal microscopy.

#### 9.2.2.3 FLIM with Lock-In Amplifiers

Homodyne and heterodyne techniques are not the only detection methods for the FD FLI. Reports have been published where lifetime-based imaging was carried out with lock-in amplifiers (LIA) in place of the detection electronics.[131,132] The LIAs are devices sensitive to frequency and phase of the input signal and work directly with the high-frequency signal from the detector. This makes lifetime microscopes simpler; however, due to a final bandwidth of the LIAs, their usage has been limited to a frequency range up to several hundred MHz. Since LIAs for massively parallel data processing are not available, the LIAs have been utilized with scanning lifetime microscopes only. Phase sensitivity of the LIAs has been used mainly for selective lifetime suppression in the intensity-modulated multiple wavelength scanning (IMS) technique.[131,133] Recently, dual-phase LIAs have been demonstrated to be usable for lifetime mapping.[132] The technique has been used for lifetime-based pH imaging with fluorescence probe SNAFL-2.[134]

#### 9.2.2.4 Multifrequency FLIM

Unlike solution spectroscopy in which multifrequency instruments have been used for a long time,[135,136] until recently, FD FLI has been limited to a single modulation frequency. The main reason was a relatively long time required for sequential acquisition of the multifrequency FLI data, which resulted in photobleaching and photodamage of the samples. Single frequency imaging allowed generation of lifetime contrast and calculation of apparent modulation and phase lifetime images (Equation 9.9). However, this is a limitation when taking into account that fluorescence kinetics are rarely pure exponentials in a biological environment. Moreover, the heterogeneity of the decay is often the information of interest

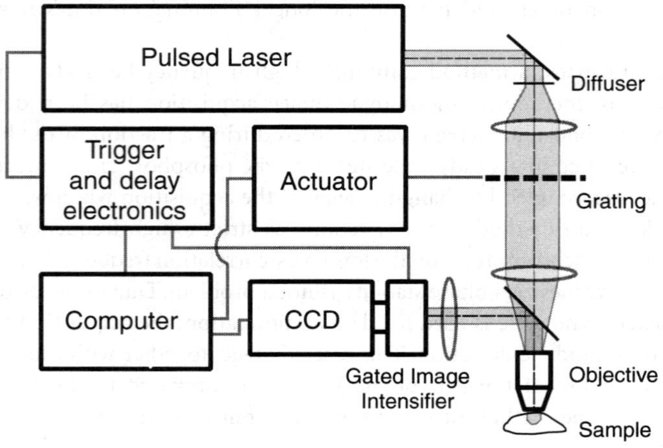


FIGURE 9.12 Scheme of a lifetime microscope with a structured illumination. A grid pattern produced by a grating in the illumination lightpath is projected on the sample. Three sets of time-gated images are taken at three positions of the grid and a deblurred stack of intensity images is calculated. To obtain a lifetime map, the resulting stack is processed by standard methods used with the time-gated FLIM. (Modified from Cole, M.J. et al., *Opt. Lett.*, 25, 1361, 2000.)

when looking for different fluorescence species within a cell. With the single-frequency FLIM, the information content of heterogeneous fluorescence decays remains inaccessible because resolution of $N$ lifetimes needs at least $N$ modulation frequencies.[137]

Fortunately, a solution has been found for rapid multiharmonic data acquisition. The multiharmonic method utilizes the harmonic content of the excitation pulse train[128–130] and the harmonic content of the pulse-modulated gain of a detector for simultaneous data acquisition at multiple modulation frequencies. The technique was initially applied for cuvette experiments[138–141] and single-spot heterodyne frequency-domain microscopy.[9] Recently, the multiharmonic approach has been implemented for the homodyne WF multifrequency FLIM (mfFLIM).[142] Simultaneous multiharmonic detection is achieved by mixing frequencies from the harmonic content of the emission with corresponding harmonics from the pulse-modulated gain of the image intensifier. The resulting optical image on the phosphor screen is formed as a sum of phase- and modulation-dependent images corresponding to the different harmonic frequencies. The phase and modulation is extracted at each harmonic frequency by Fourier analysis.

Compared to the serial frequency acquisition, the main advantage of the multiharmonic approach is a shorter data acquisition time and a higher modulation depth.[142] The reduction of the data acquisition time is not as high as it would correspond to the number of simultaneously acquired frequencies. The reason is a lower duty cycle of the pulse-modulated detector. Performance of the mfFLIM has been demonstrated by the lifetime resolution of green fluorescent protein variants co-expressed in live cells.

### 9.2.3 Three-Dimensional Wide-Field FLIM

Wide-field microscopy suffers from the lack of sectioning capability, which is an inherent property of confocal and multiphoton imaging systems. The reason is a mixing of the signal from out-of-focus planes, which blurs the intensity image and decreases contrast and spatial definition of observed structures. WF-FLIM suffers from the same problem. At any spatial location, the measured lifetime is an intensity-weighted average of lifetime contributions arising from out-of-focus planes and the lifetime contrast is compromised. Improvement of the spatial and lifetime resolution of the WF-FLIM has been the subject of several studies.[75,103,122,143,144]

One method for obtaining the optical sectioning with the WF lifetime microscope is depicted in Figure 9.12. The sectioning capability was achieved by implementation of a structured illumination,[145,146] when a grid pattern produced by a grating in the illumination lightpath is projected on the sample. Then three

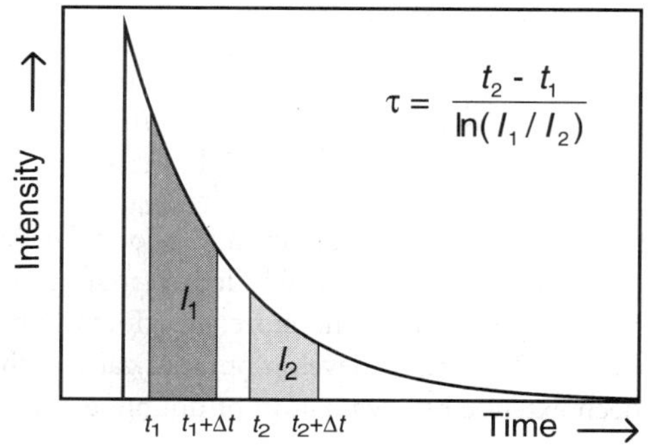


**FIGURE 9.13** Principle of the two-gate method for the rapid calculation of the emission lifetime. The formula for calculation of the lifetime is valid for a monoexponentially decaying emission.

sets of the time-gated images are taken for three positions of the grid pattern. Calculations then yield a deblurred lifetime map.[75]

Enhancement of the three-dimensional lifetime resolution of WF-FLIM can be achieved with multifocal multiphoton microscopy (MMM).[147] Multiphoton excitation and three-dimensional sectioning is accomplished by insertion of the spinning microlens disk to the illumination path of the conventional WF microscope. This causes a near-IR excitation light from the femtosecond laser to be transformed into an array of fast-scanning foci in the object plane where multiphoton excitation occurs. A WF-FLIM instrument based on the MMM technique and gated image intensifier has been recently reported.[143]

If one knows the point spread function (PSF), the three-dimensional temporal and spatial resolution of the WF-FLIM can be increased by a numerical deconvolution.[103,122] Squire et al.[103] performed deconvolution with the measured PSF in the Fourier space before construction of lifetime images. Alternatively, the FLIM data were first globally analyzed in order to separate fluorophore populations associated with different lifetimes. This allowed the separated image stacks to be individually corrected for photobleaching before the deconvolution.[122] In both cases significant enhancement of the spatial and temporal resolution was achieved.

## 9.3 Specifics of FLIM Data Analysis

Several methods have been proposed to analyze rapidly the substantial amount of data generated by lifetime imaging systems. Typically, this reduces a stack of images to a few images that contain spatial mapping of decay parameters, e.g., lifetimes and associated amplitudes. Going even a step farther, we usually want to obtain a single chemical image carrying biologically relevant information. Depending on the fluorescent lifetime sensor, it can be a spatial distribution of analyte concentration, two-dimensional map of a gas pressure, and, eventually, a spatial map of distances when FLIM is used for fluorescence resonance energy transfer (FRET) experiments.

### 9.3.1 Fast Two-Gate Analysis

Time-domain lifetime mapping does not require lengthy iterative fitting of a large number of points. Fluorescence lifetime maps can be calculated from two intensity images, $I_1(r)$ and $I_2(r)$, acquired in two distinct time windows.[148] Assuming windows of the equal width of $\Delta t$ delayed by $t_1$ and $t_2$ relative to the excitation pulse (Figure 9.13) we can write:

$$I_1(r) = \int_{t_1}^{t_1+\Delta t} a(r)\cdot\exp[-t/\tau(r)]dt, \qquad I_2(r) = \int_{t_2}^{t_2+\Delta t} a(r)\cdot\exp[-t/\tau(r)]dt \tag{9.14}$$

After integration the lifetime can be expressed as:

$$\tau(r) = \frac{t_2 - t_1}{\ln\left[I_1(r)/I_2(r)\right]} \tag{9.15}$$

Equation 9.15 holds true for monoexponential decays and allows for very fast computation of the lifetime maps. To obtain good accuracy of the measured lifetime map, prior knowledge of the lifetime is essential for selecting optimal gating windows. Therefore, an effort has been made to evaluate errors associated with the gating technique[149,150] and to develop optimal gating schemes.[151] This rapid lifetime determination technique has been extended to evaluation of double-exponential decays.[152]

### 9.3.2 Global Analysis of FLIM Data

Analysis of FLI data is a problem of the same scale as the analysis of $10^5$ to $10^6$ of time-resolved cuvette experiments. Besides the computational expenses, the number of data points on the time or frequency axis in the time or frequency domain, respectively, has to be limited in order to collect data before irreversible processes damage the sample. A smaller number of data points and lower S/N ratio, compared to values typically obtained for cuvette measurements, make reliable multicomponent lifetime analysis from a single pixel questionable. One can increase the S/N ratio by grouping several pixels together; however, the price is a deterioration of the spatial resolution. An important possibility in FLI data analysis is to use prior knowledge about the sample, which leads to a dramatic improvement of computational speed and fit quality.[117,142]

One of the approaches uses the global-analysis algorithm,[94,95,153–156] which analyzes all pixels simultaneously while constraining the component lifetimes in all pixels to be equal.[117] This is an acceptable assumption for many FLI experiments where a spatially distributed mixture of fluorescence species is analyzed. Such an approach leads to a significant reduction of the number of fitted parameters and to improved fit reliability. It has been shown that the FD global analysis of multiple pixels can yield twice the number of lifetimes that could be recovered with ordinary single-point measurements.[117] An important consequence is a possibility of recovering two lifetimes from a single-frequency FLI experiment. Examples and comparison of the single- and multifrequency global fit of FLIM data have been demonstrated on two co-expressed variants of the green fluorescence protein (GFP).[117]

A special case of the described method is called a lifetime invariant fit.[117,142] When lifetimes are known *a priori*, their values can be fixed during the fit and only amplitudes or intensity fractions are calculated in each pixel. In this case the fitting process evaluates only linear parameters, which are usually less correlated and can be calculated significantly faster than the nonlinear lifetime parameters. The global approach with variable lifetimes is a more general approach, but the high speed of the data analysis could be of prime importance in real-time clinical applications.[102,110]

The global fit and the lifetime invariant fit are not limited to the analysis of two-dimensional FLIM data. The concept of invariant lifetimes has been extended to the whole sample volume and used with the WF-FLIM for the three-dimensional deconvolution of fluorescent populations associated with phosphorylation states of the epidermal growth factor receptor ErbB1 tagged with GFP.[122]

## 9.4 Selected FLIM Applications

Biological and biomedical applications of the FLIM technique have been treated in several recent reviews.[5,8,22,157] The technique has been successfully applied in cell biology, clinical diagnostics, and analytical chemistry for determination of spatial ion and metabolite distributions, monitoring of interactions between cellular components by fluorescence resonance energy transfer (FRET), and for detection of abnormal tissues. In the following sections we briefly outline a few selected biological applications.

### 9.4.1 Intracellular Lifetime-Based pH Imaging and Ion Mapping

Cellular functions critically depend on the maintenance of a proper intracellular environment. Changes of pH and ion concentrations can trigger signalling pathways, activate expression of genes, or influence cellular morphology. A number of lifetime probes suitable for imaging of pH distribution in cytosol and cellular compartments have been characterized[158–160] and used for imaging of living systems.[48,53,134,161–163] Lifetime-based pH imaging has often been found to be more convenient than conventional ratiometric methods due to a simpler pH calibration[48,161] and possibly to distinguish between fractions of bound and free probe.[162] This allows for elimination of pH estimation errors caused by binding of the probe to cytosolic constituents, which is difficult to achieve with ratiometric imaging. The potential of the lifetime-based imaging with a multiphoton excitation has been demonstrated on microbial biofilms when pH gradients up to the biofilm depth of 140 μm were resolved with the carboxyfluorescein as the pH indicator.[53]

Lifetime-based detection has been applied for imaging of cellular ions as well. Spatial calcium distributions measured with nonratiometric lifetime probes have been reported from our laboratory;[100,164,165] concentrations of $Na^+$ ions in HeLa cells have been monitored with sodium green and phase/modulation fluorometry.[11] Optical sensors for a number of other biologically critical metal ions, e.g., Zn(II), Co(II), Cd(II), Cu(II), and Ni(II) have been systematically characterized[166–168] and can be used for lifetime-based imaging.

Similarly to lifetime-based pH indicators, the intracellular lifetime response of the ion-selective fluorophores is typically different from the response in a buffer. Differences are caused mainly by interactions of probes with cellular components or due to a mismatch of intracellular viscosity, pH, and temperature.[10,65,169] Therefore, a careful *in vivo* calibration is usually necessary for reliable estimation of intracellular ion concentrations.

### 9.4.2 Lifetime-Resolved Imaging of Cellular Processes

Cellular FLIM utilizing environmental sensitivity of the emission lifetime provides more information than conventional intensity-based imaging methods. The two-photon FLIM has been used to monitor the macrophage-mediated phagocytosis with the fluorescein-BSA conjugate as an exogenous fluorescent antigen.[170] Due to intracellular proteolysis, the initial subnanosecond lifetime of the conjugate increased with time to a value close to 3.0 ns, providing a monitor of the cellular proteolytic activity.

Cells contain a variety of intrinsic fluorophores. One of them, NADH, can provide information about the oxidation-reduction status of cells and tissues. Lifetime images of the free and protein-bound NADH autofluorescence were first reported by Lakowicz et al.[105] Later, the FLIM of the NAD(P)H/NAD(P) pairs revealed pathological effects induced by the exposure of cells to strong illumination pulses.[126,171]

Imaging of nuclear DNA is another example of the cellular FLIM. Cell cycle-dependent topology of DNA has been studied with the probe YOYO-1, which exhibits a base-pair dependent lifetime.[120] Cells have also been studied by polarized time-resolved fluorescence. Intracellular anisotropy decays were used to map membrane fluidity and the spatial distribution of the cytoplasmic viscosity.[12–14] The measured microviscosities served for assessment of the spatial organization of the cytoplasm and membranes.

### 9.4.3 Cellular Interactions Determined by the FRET-FLIM

Due to the inverse 6th power dependence of energy transfer efficiency on the distance between donor and acceptor molecules,[172] FRET is a powerful strategy to detect interactions, associations, and proximity of cellular components.[5,8] Pixel-by-pixel microscopic maps of FRET efficiency can be monitored by fluorescence lifetime imaging of the donor lifetime thus eliminating problems of intensity-based measurements.

Applications of the FRET-FLIM in cellular biology include monitoring phosphorylation, oligomerization and internalization of epidermal growth factor receptors,[114,173,174] studies related to signal transduction and imaging of the activated protein kinase C and its association with downstream effectors,[112,113,125] and, eventually, topological organization of nuclear DNA.[116,119] Time-resolved FRET microscopy has also been employed to study vesicular transportation of the cholera toxin by following the energy transfer between toxin subunits.[125]

A novel labeling strategy for FRET experiments involved tagging donor-acceptor pairs to the corresponding monoclonal antibody of the object proteins. Recently, genetically encoded fluorophores based on the GFP and its spectral variants represent an exciting possibility for *in vivo* protein labeling.[175] Today FRET experiments can take full advantage of the fusion proteins constructed as GFP-chimeras.[157]

### 9.4.4 Tissue Imaging and Clinical Applications

*In vivo* applications of lifetime-resolved imaging at the tissue level are emerging. Compared to conventional fluorescence imaging that uses intensity and spectral information as an indication of a tissue abnormality, lifetime-based imaging provides additional physiochemical diagnostic information. Time-gated imaging has been used for diagnosis of bones and teeth in order to develop diagnostic methods for bone fractures and early caries, respectively.[73,74] The diagnostic potential of lifetime-based imaging has been demonstrated by distinguishing lifetime differences between intact tooth enamel and a white-spot lesion area.[176] Additionally, multiphoton excitation has been employed for generation of three-dimensional lifetime images of human skin autofluorescence.[177] Lifetime distributions from the near-surface skin area up to a depth of ~200 μm revealed variations in metabolic states of the cells and indicated a possibility of using lifetime-based imaging for dermatological diagnosis.

For an early clinical diagnosis of the tumor tissue in hollow organs, an endoscope equipped with a real-time, lifetime-based contrast microscopy has been developed.[110] Preliminary results include endoscopic lifetime images of the excised bladder, where autofluorescence of the flavin molecules has been recorded. Finally, the frequency-domain photon migration technique has been applied to tissue-simulating phantoms in order to externally image fluorescence properties inside the tissue volume for possible disease detection.[178]

### 9.4.5 Lifetime Imaging with Long-Lived Fluorophores

The best-known application of the long-lifetime fluorophores is for oxygen imaging. Emission of many Ru(II) complexes and Pt(II) porphyrines is dynamically quenched by oxygen and can be effectively used as a lifetime oxygen sensor. Two-dimensional oxygen fluxes and partial pressure maps of surface oxygen have been monitored using such long-lived fluorophores, as well as the TD and FD techniques.[109,179–181] The long lifetimes enable adaptation of inexpensive light-emitting diode (LED) light sources and allow for suppression of the short-lived background by proper gating of the detection. Planar chemical sensors based on the long-lived Ru(II) complexes with lifetimes responding to biologically relevant agents such as $CO_2$, $K^+$, $Cl^-$, $NH_4^+$, and glucose have been integrated on microplate arrays for possible clinical applications.[86,182]

## 9.5 Outlook

Much progress has been made in technology and data analysis since lifetime-based imaging was first reported. Over the years, the potential of lifetime-based imaging has become well recognized and unquestionable. It seems evident that further progress of the technique is tightly linked not only to technological advances but also to advances of probe chemistry and to introduction of new concepts in fluorescence spectroscopy.

To answer the question of where the lifetime-based imaging is heading, one must identify current limitations of the technique and define application-driven goals. Real-world experience could be the starting point: (1) there have never been enough photons for imaging; (2) there is a need for high spatial resolution required by biological applications; (3) there is a demand for faster data acquisition and analysis to study dynamic processes. It seems evident that to satisfy all three requirements simultaneously is not an easy task and compromises must be made. Our vision of the future FLIM instrument is an affordable real-time apparatus with a three-dimensional lifetime imaging capability and the possibility of multiphoton excitation. The instrument will probably combine lifetime, spectral, and conventional intensity imagers in one multifunctional device for simultaneous multidimensional imaging.

The quality and accuracy of lifetime imaging are determined by the S/N ratio of the data. This ratio can be improved mainly by acquisition of more photons per pixel. Due to the requirement for real-time operation, longer data integration time is unacceptable. Stronger excitation is also problematic due to photobleaching and sample destruction. The solution could be an introduction of brighter probes. Here the concept of fluorescence enhancement near the surface of metallic particles could play a significant role.[183,184] Because the enhancement of fluorescence intensity is caused by faster radiative rate, accompanied by decrease in fluorescence lifetime, future instruments utilizing the intensity-enhanced probes should operate with a picosecond lifetime resolution.

An effort has already been made to build a real-time FLIM apparatus suitable for clinical use and for monitoring dynamic biological events.[102,110,115] Overall data acquisition rate favors WF multichannel imaging systems with an image intensifier as compared to single-channel scanning imagers. A lack of sectioning capability of WF lifetime imagers, however, has been neutralizing this comparative advantage. At present, this drawback seems to be disappearing and technologies utilizing multifocal multiphoton microscopy[143] or structured illumination[75,145,146] have enabled efficient three-dimensional fluorescence lifetime imaging with the WF imagers. In the future, we expect to see fast three-dimensional fluorescence lifetime imagers suitable for imaging turbid biological tissues[178] or for lifetime tomography.

Clinical and biological lifetime-based imaging of turbid samples will require not only a sectioning capability but also a high rejection of scattered light and suppression of autofluorescence. Two-photon excited long-lived long-wavelength fluorophores based on energy transfer[185,186] or long-lived metal–ligand complexes[187,188] will be useful with gated detection. Suppression of the prompt and short-lived emission components is no longer limited to TD gated methods; it also was recently introduced to FD fluorometry.[189,190]

It is reasonable to expect that price of future fluorescence lifetime imaging instruments will drop, since expensive pulsed laser systems could be replaced by pulsed or HF-modulated laser diodes or LEDs for routine applications.[17]

In this chapter, we were unable to discuss all possible modalities and applications of FLIM. It is obvious that each implementation of lifetime-based imaging has specific advantages and disadvantages and a chosen approach will always be dictated by the application. We conclude that lifetime-based imaging today is a mature technique that is emerging from research laboratories. Nevertheless, more powerful, less-expensive, and easier-to-operate apparatus for lifetime-based imaging is still a future goal.[6]

## Acknowledgment

This work was supported by the National Institutes of Health, National Center for Research Resources, RR-08119.

### References

1. Herschel, Sir J.F.W., On a case of superficial colour presented by homogeneous liquid internally colourless, *Phil. Trans. R. Soc. London*, 135, 143, 1845.
2. Blake, R., Cellular screening assays using fluorescence microscopy, *Curr. Opin. Pharmacol.*, 1, 533, 2001.
3. Turconi, S., Bingham, R., Haupts, U., and Pope, A., Developments in fluorescence lifetime-based analysis for ultra-HTS, *Drug Discovery Today*, Suppl. 6, 27, 2001.
4. D'Auria, S. and Lakowicz, J.R., Enzyme fluorescence as a sensing tool: new perspectives in biotechnology, *Curr. Opin. Biotechnol.*, 12, 99, 2001.
5. Wouters, F.S., Verveer, P.J., and Bastiaens, P.I., Imaging biochemistry inside cells, *Trends Cell Biol.*, 11, 203, 2001.

6. Wang, X.F., Periasamy, A., Herman, B., and Coleman, D.M., Fluorescence lifetime imaging microscopy (FLIM) — instrumentation and applications, *Crit. Rev. Anal. Chem.*, 23, 369, 1992.
7. Lakowicz, J.R., Fluorescence lifetime sensing creates cellular images, *Laser Focus World*, 60, 1992.
8. Bastiaens, P.I. and Squire, A., Fluorescence lifetime imaging microscopy: spatial resolution of biochemical processes in the cell, *Trends Cell Biol.*, 9, 48, 1999.
9. Verkman, A.S., Armijo, M., and Fushimi, K., Construction and evaluation of a frequency-domain epifluorescence microscope for lifetime and anisotropy decay measurements in subcellular domains, *Biophys. Chem.*, 40, 117, 1991.
10. Despa, S., Steels, P., and Ameloot, M., Fluorescence lifetime microscopy of the sodium indicator sodium-binding benzofuran isophthalate in HeLa cells, *Anal. Biochem.*, 280, 227, 2000.
11. Despa, S., Vecer, J., Steels, P., and Ameloot, M., Fluorescence lifetime microscopy of the $Na^+$ indicator sodium green in HeLa cells, *Anal. Biochem.*, 281, 159, 2000.
12. Dix, J.A. and Verkman, A.S., Pyrene eximer mapping in cultured fibroblasts by ratio imaging and time-resolved microscopy, *Biochemistry*, 29, 1949, 1990.
13. Dix, J.A. and Verkman, A.S., Mapping of fluorescence anisotropy in living cells by ratio imaging — application to cytoplasmic viscosity, *Biophys. J.*, 57, 231, 1990.
14. Srivastava, A. and Krishnamoorthy, G., Cell type and spatial location dependence of cytoplasmic viscosity measured by time-resolved fluorescence microscopy, *Arch. Biochem. Biophys.*, 340, 159, 1997.
15. Wang, X.F., Kitajima, S., Uchida, T., Coleman, D.M., and Minami, S., Time-resolved fluorescence microscopy using multichannel photon-counting, *Appl. Spectrosc.*, 44, 25, 1990.
16. Keating, S.M. and Wensel, T.G., Nanosecond fluorescence microscopy — emission kinetics of Fura-2 in single cells, *Biophys. J.*, 59, 186, 1991.
17. Herman, P., Maliwal, B.P., Lin, H.J., and Lakowicz, J.R., Frequency-domain fluorescence microscopy with the LED as a light source, *J. Microsc. (Oxford)*, 203, 176, 2001.
18. Ambroz, M., MacRobert, A.J., Morgan, J., Rumbles, G., Foley, M.S., and Phillips, D., Time-resolved fluorescence spectroscopy and intracellular imaging of disulphonated aluminium phthalocyanine, *J. Photochem. Photobiol. B*, 22, 105, 1994.
19. Morgan, C.G. and Mitchell, A.C., Fluorescence lifetime imaging: an emerging technique in fluorescence microscopy, *Chromosome Res.*, 4, 261, 1996.
20. Periasamy, A., Wang, X.F., Wodnick, P., Gordon, G.W., Kwon, S., Diliberto, P.A., and Herman, B., High-speed fluorescence microscopy: lifetime imaging in biological sciences, *J. Microsc. Soc. Am.*, 1, 13, 1995.
21. Lakowicz, J.R., Emerging applications of fluorescence spectroscopy to cellular imaging: lifetime imaging, metal-ligand probes, multi-photon excitation and light quenching, *Scanning Microsc.*, Suppl. 10, 213, 1996.
22. Tadrous, P.J., Methods for imaging the structure and function of living tissues and cells: 2. Fluorescence lifetime imaging, *J. Pathol.*, 191, 229, 2000.
23. Bonushkin, Y., Dworkin, L., Hauser, J., Kim, C., Lindgren, M., Scott, A., and Appollinari, G., Tests of a third generation multianode phototube, *Nucl. Instrum. Methods Phys. Res. Sect. A*, 381, 349, 1996.
24. Salomon, M. and Williams, S.S.A., A multi-anode photomultiplier with position sensitivity, *Nucl. Instrum. Methods Phys. Res. Sect. A*, 241, 210, 1985.
25. McLoskey, D., Birch, D.J.S., Sanderson, A., Suhling, K., Welch, E., and Hicks, P.J., Multiplexed single-photon counting. 1. A time-correlated fluorescence lifetime camera, *Rev. Sci. Instrum.*, 67, 2228, 1996.
26. Vecer, J., Herman, P., and Beranek, J., Low-temperature luminescence spectra and fluorescence lifetimes of polycytidylic acid in polyalcoholic glasses, *Photochem. Photobiol.*, 57, 792, 1993.
27. Demas, J.N., *Excited State Lifetime Measurements*, Academic Press, New York, 1983.
28. O' Connor, D.V. and Philips, D., *Time-Correlated Single Photon Counting*, Academic Press, London, 1984.

29. Coates, P.B., The correction for photon pile-up in the measurement of radiative lifetimes, *J. Phys. E (Series 2)*, 1, 878, 1968.
30. Selinger, B.K. and Harris, C.M., The pile-up problem in the pulse fluorometry, in *Time-Resolved Fluorescence Spectroscopy in Biochemistry and Biology*, Cundall, R.B. and Dale, R.E., Eds., Plenum Publishing, New York, 1983, p. 115.
31. Harris, C.M. and Selinger, B.K., Single-photon decay spectroscopy: II. The pileup problem, *Aust. J. Chem.*, 32, 2111, 1979.
32. Williamson, J.A., Kendalltobias, M.W., Buhl, M., and Seibert, M., Statistical evaluation of dead time effects and pulse pileup in fast photon-counting — introduction of the sequential model, *Anal. Chem.*, 60, 2198, 1988.
33. Meltzer, R.S. and Wood, R.M., Nanosecond time-resolved spectroscopy with a pulsed tunable dye laser and single photon time correlation, *Appl. Opt.*, 16, 1432, 1977.
34. Schuyler, R. and Isenberg, I., A monophoton fluorometer with energy discrimination, *Rev. Sci. Instrum.*, 42, 813, 1970.
35. Kemnitz, K., Pfeifer, L., Paul, R., and Coppey-Moisan, M., Novel detectors for fluorescence imaging on the picosecond time scale, *J. Fluoresc.*, 7, 93, 1997.
36. Sasaki, K., Koshioka, M., and Masuhara, H., Three-dimensional space-resolved and time-resolved fluorescence spectroscopy, *Appl. Spectrosc.*, 45, 1041, 1991.
37. Willemsen, O.H., Noordman, O.F. J., Segering, F.B., Ruiter, A.G.T., Moers, M.H.P., and Vanhulst, N.F., Fluorescence lifetime contrast combined with probe microscopy, in *Optics at the Nanometer Scale*, Nieto-Vesperinas, M. and Garcia, N., Eds., IBM, the Netherlands, 1996, p. 223.
38. Charbonneau, S., Allard, L.B., Young, J.F., Dyck, G., and Kyle, B.J., Two-dimensional time-resolved imaging with 100-ps resolution using a resistive anode photomultiplier tube, *Rev. Sci. Instrum.*, 63, 5315, 1992.
39. Lassiter, S.J., Stryjewski, W., Legendre, B.L., Jr., Erdmann, R., Wahl, M., Wurm, J., Peterson, R., Middendorf, L., and Soper, S.A., Time-resolved fluorescence imaging of slab gels for lifetime base-calling in DNA sequencing applications, *Anal. Chem.*, 72, 5373, 2000.
40. Schonle, A., Glatz, M., and Hell, S.W., Four-dimensional multiphoton microscopy with time-correlated single-photon counting, *Appl. Opt.*, 39, 6306, 2000.
41. Barzda, V., de Grauw, C.J., Vroom, J., Kleima, F.J., van Grondelle, R., van Amerongen, H., and Gerritsen, H.C., Fluorescence lifetime heterogeneity in aggregates of LHCII revealed by time-resolved microscopy, *Biophys. J.*, 81, 538, 2001.
42. Tinnefeld, P., Herten, D.P., and Sauer, M., Photophysical dynamics of single molecules studied by spectrally resolved fluorescence lifetime imaging microscopy (SFLIM), *J. Phys. Chem. A*, 105, 7989, 2001.
43. Böhmer, M., Pampaloni, F., Wahl, M., Rahn, H.J., Erdmann, R., and Enderlein, J., Time-resolved confocal scanning device for ultrasensitive fluorescence detection, *Rev. Sci. Instrum.*, 72, 4145, 2001.
44. Kwak, E.S., Kang, T.J., and Vanden Bout, D.A., Fluorescence lifetime imaging with near-field scanning optical microscopy, *Anal. Chem.*, 73, 3257, 2001.
45. Suhling, K., McLoskey, D., and Birch, D.J.S., Multiplexed single-photon counting. II. The statistical theory of time-correlated measurements, *Rev. Sci. Instrum.*, 67, 2238, 1996.
46. Howorth, J.R., Ferguson, I., and Wilcox, D.A., Developments in microchannel plate photomultipliers, in *Proc. SPIE, Advances in Fluorescence Sensing Technology* II, Lakowicz, J.R., Ed., 1995, p. 356.
47. Lakowicz, J.R., *Principles of Fluorescence Spectroscopy,* 2nd ed., Kluwer Academic/Plenum Press, New York, 1999.
48. Sanders, R., Draaijer, A., Gerritsen, H.C., Houpt, P.M., and Levine, Y.K., Quantitative pH imaging in cells using confocal fluorescence lifetime imaging microscopy, *Anal. Biochem.*, 227, 302, 1995.
49. Sanders, R., van Zandvoort, M., Draaijer, A., Levine, Y.K., and Gerritsen, H.C., Confocal fluorescence lifetime imaging of chlorophyll molecules in polymer matrices, *Photochem. Photobiol.*, 64, 817, 1996.

50. Buurman, E.P., Sanders, R., Draaijer, A., Gerritsen, H.C., Vanveen, J.J.F., Houpt, P.M., and Levine, Y.K., Fluorescence lifetime imaging using a confocal laser scanning microscope, *Scanning*, 14, 155, 1992.
51. Gerritsen, H.C., Sanders, R., Draaijer, A., Ince, C., and Levine, Y.K., Fluorescence lifetime imaging of oxygen in living cells, *J. Fluoresc.*, 7, 11, 1997.
52. Roorda, R.D., Ribes, A.C., Damaskinos, S., Dixon, A.E., and Menzel, E.R., A scanning beam time-resolved imaging system for fingerprint detection, *J. Forensic Sci.*, 45, 563, 2000.
53. Vroom, J.M., De Grauw, K.J., Gerritsen, H.C., Bradshaw, D.J., Marsh, P.D., Watson, G.K., Birmingham, J.J., and Allison, C., Depth penetration and detection of pH gradients in biofilms by two-photon excitation microscopy, *Appl. Environ. Microbiol.*, 65, 3502, 1999.
54. de Grauw, C.J. and Gerritsen, H.C., Multiple time-gate module for fluorescence lifetime imaging, *Appl. Spectrosc.*, 55, 670, 2001.
55. Sytsma, J., Vroom, J. M., De Grauw, C.J., and Gerritsen, H.C., Time-gated fluorescence lifetime imaging and microvolume spectroscopy using two-photon excitation, *J. Microsc.(Oxford)*, 191, 39, 1998.
56. Wang, X.F., Uchida, T., Coleman, D.M., and Minami, S., A two-dimensional fluorescence lifetime imaging-system using a gated image intensifier, *Appl. Spectrosc.*, 45, 360, 1991.
57. Oida, T., Sako, Y., and Kusumi, A., Fluorescence lifetime imaging microscopy (flimscopy). Methodology development and application to studies of endosome fusion in single cells, *Biophys. J.*, 64, 676, 1993.
58. Cubeddu, R., Canti, G., Taroni, P., and Valentini, G., Time-gated fluorescence imaging for the diagnosis of tumors in a murine model, *Photochem. Photobiol.*, 57, 480, 1993.
59. Cubeddu, R., Taroni, P., and Valentini, G., Time-gated imaging-system for tumor-diagnosis, *Opt. Eng.*, 32, 320, 1993.
60. Schneckenburger, H., Konig, K., Dienersberger, T., and Hahn, R., Time-gated microscopic imaging and spectroscopy in medical diagnosis and photobiology, *Opt. Eng.*, 33, 2600, 1994.
61. Periasamy, A., Wodnicki, P., Wang, X.F., Kwon, S., Gordon, G.W., and Herman, B., Time-resolved fluorescence lifetime imaging microscopy using a picosecond pulsed tunable dye laser system, *Rev. Sci. Instrum.*, 67, 3722, 1996.
62. Hennink, E.J., deHaas, R., Verwoerd, N.P., and Tanke, H.J., Evaluation of a time-resolved fluorescence microscope using a phosphorescent Pt-porphine model system, *Cytometry*, 24, 312, 1996.
63. Cubeddu, R., Canti, G., Pifferi, A., Taroni, P., and Valentini, G., Fluorescence lifetime imaging of experimental tumors in hematoporphyrin derivative-sensitized mice, *Photochem. Photobiol.*, 66, 229, 1997.
64. Dowling, K., Hyde, S.C.W., Dainty, J.C., French, P.M.W., and Hares, J.D., Two-dimensional fluorescence lifetime imaging using a time-gated image intensifier, *Opt. Commun.*, 135, 27, 1997.
65. Herman, B., Wodnicky, P., Kwon, S., Periasamy, A., Gordon, G.W., Mahajan, N., and Wang, X.F., Recent developments in Monitoring calcium and protein interactions in cells using fluorescence lifetime microscopy, *J. Fluoresc.*, 7, 85, 1997.
66. Schneckenburger, H., Gschwend, M.H., Strauss, W.S.L., Sailer, R., and Steiner, R., Time-gated microscopic energy tranfer measurements for probing mitochondrial metabolism, *J. Fluoresc.*, 7, 9, 1997.
67. Scully, A.D., Ostler, R.B., Phillips, D., O'Neill, P., Townsend, K.M.S., Parker, A.W., and MacRobert, A.J., Application of fluorescence lifetime imaging microscopy to the investigation of intracellular PDT mechanisms, *Bioimaging*, 5, 9, 1997.
68. Dowling, K., Dayel, M.J., Lever, M.J., French, P.M.W., Hares, J.D., and Dymoke-Bradshaw, A.K.L., Fluorescence lifetime imaging with picosecond resolution for biomedical applications, *Opt. Lett.*, 23, 810, 1998.
69. Dowling, K., Dayel, M.J., Hyde, S.C. W., Dainty, J. C., French, P.M. W., Vourdas, P., Lever, M.J., Dymoke-Bradshaw, A.K. L., Hares, J. D., and Kellett, P.A., Whole-field fluorescence lifetime imaging with picosecond resolution using ultrafast 10-kHz solid-state amplifier technology, *IEEE J. Sel. Top. Quantum Electron.*, 4, 370, 1998.

70. Konig, K., Boehme, S., Leclerc, N., and Ahuja, R., Time-gated autofluorescence microscopy of motile green microalga in an optical trap, *Cell Mol. Biol. (Noisy-le-grand)*, 44, 763, 1998.
71. Vereb, G., Jares-Erijman, E., Selvin, P.R., and Jovin, T.M., Temporally and spectrally resolved imaging microscopy of lanthanide chelates, *Biophys. J.*, 74, 2210, 1998.
72. Dowling, K., Dayel, M.J., Hyde, S.C.W., French, P.M.W., Lever, M.J., Hares, J.D., and Dymoke-Bradshaw, A.K.L., High resolution time-domain fluorescence lifetime imaging for biomedical applications, *J. Mod. Opt.*, 46, 199, 1999.
73. Konig, K., Schneckenburger, H., and Hibst, R., Time-gated *in vivo* autofluorescence imaging of dental caries, *Cell. Mol. Biol. (Noisy-le-grand)*, 45, 233, 1999.
74. Zevallos, M.E., Gayen, S.K., Das, B.B., Alrubaiee, M., and Alfano, R.R., Picosecond electronic time-gated imaging of bones in tissues, *IEEE J. Sel. Top. Quantum Electron.*, 5, 916, 1999.
75. Cole, M.J., Siegel, J., Webb, S.E. D., Jones, R., Dowling, K., French, P.M. W., Lever, M.J., Sucharov, L.O.D., Neil, M.A.A., Juskaitis, R., and Wilson, T., Whole-field optically sectioned fluorescence lifetime imaging, *Opt. Lett.*, 25, 1361, 2000.
76. Lee, K.C., Siegel, J., Webb, S.E., Leveque-Fort, S., Cole, M.J., Jones, R., Dowling, K., Lever, M.J., and French, P.M., Application of the stretched exponential function to fluorescence lifetime imaging, *Biophys. J.*, 81, 1265, 2001.
77. Ushida, K., Nakayama, T., Nakazawa, T., Hamanoue, K., Nagamura, T., Mugishima, A., and Sakimukai, S., Implementation of an image intensifier coupled with a linear position-sensitive detector for measurements of absorption and emission-spectra from the nanosecond to millisecond time regime, *Rev. Sci. Instrum.*, 60, 617, 1989.
78. Image intensifiers, Hamamatsu Photonics K.K., Electron Tube Center, Document TII 001E01, 2000.
79. Kentech Instruments Ltd., Gated image intensifiers — specifications, 2001.
80. LaVision GmbH, ICCD camera systems — specifications, 2001.
81. Marriott, G., Clegg, R.M., Arndt-Jovin, D.J., and Jovin, T.M., Time resolved imaging microscopy. Phosphorescence and delayed fluorescence imaging, *Biophys. J.*, 60, 1374, 1991.
82. Khait, O., Smirnov, S., and Tran, C.D., Multispectral imaging microscope with millisecond time resolution, *Anal. Chem.*, 73, 732, 2001.
83. Vinogradov, S.A., Lo, L.W., Jenkins, W.T., Evans, S.M., Koch, C., and Wilson, D.F., Noninvasive imaging of the distribution in oxygen in tissue *in vivo* using near-infrared phosphors, *Biophys. J.*, 70, 1609, 1996.
84. Reich, R.K., Mountain, R.W., McGonagle, W.H., Huang, J.C.M., Twichell, J.C., Kosicki, B.B., and Savoye, E.D., Integrated electronic shutter for back-illuminated charge-oupled-devices, *IEEE Trans. Electron. Devices*, 40, 1231, 1993.
85. Hartmann, P. and Ziegler, W., Lifetime imaging of luminescent oxygen sensors based on all-solid-state technology, *Anal. Chem.*, 68, 4512, 1996.
86. Liebsch, G., Klimant, I., Frank, B., Holst, G., and Wolfbeis, O.S., Luminescence lifetime imaging of oxygen, pH, and carbon dioxide distribution using optical sensors, *Appl. Spectrosc.*, 54, 548, 2000.
87. Ainbund, M.R., Buevich, O.E., Kamalov, V.F., Menshikov, G.A., and Toleutaev, B.N., Simultaneous spectral and temporal resolution in a single photon-counting technique, *Rev. Sci. Instrum.*, 63, 3274, 1992.
88. Arzhantsev, S. Yu., Ainbund, M.R., Chikischev, A. Yu., Koroteev, N.I., Shkurinov, A.P., Toleutaev, B.N., Turbin, E.V., Lehmann, A., Pfeifer, L., Fink, F., and Kemnitz, K., Picosecond fluorescence lifetime imaging microscopy at 1mm space- and 10 ps time-resolution: 50 × 50 ch MCP-PMT with quadrant anode, presented at Int. Symp. 2nd Conf. Fluorescence Spectrosc. Fluorescence Probes, Prague, 1997.
89. Muller, M., Ghauharali, R., Visscher, K., and Brakenhoff, G., Double-pulse fluorescence lifetime imaging in confocal microscopy, *J. Microsc. (Oxford)*, 177, 171, 1995.
90. Buist, A.H., Müller, M, Gijsbers, E.J., Brakenhoff, G.J., Sosnowski, T.S., Norris, T.B., and Squier, J., Double-pulse fluorescence lifetime measurements, *J. Microsc.*, 186, 212, 1997.

91. Gaviola, Z., Ein Fluorometer, Apparat zur Messung von Fluoreszenzabklingungszeiten, *Z. Phys.*, 42, 853, 1926.
92. Murray, J.G., Cundall, R.B., Morgan, C.G., Evans, G.B., and Lewis, C., A single-photon-counting Fourier-transform microfluorometer, *J. Phys. E Sci. Instrum.*, 19, 349, 1986.
93. Wang, X.F., Uchida, T., and Minami, S., A fluorescence lifetime distribution measurement system based on phase-resolved detection using an image dissector tube, *Appl. Spectrosc.*, 43, 840, 1989.
94. Gratton, E., Limkeman, M., Lakowicz, J.R., Maliwal, B.P., Cherek, H., and Laczko, G., Resolution of mixtures of fluorophores using variable-frequency phase and modulation data, *Biophys. J.*, 46, 479, 1984.
95. Lakowicz, J.R., Laczko, G., Cherek, H., Gratton, E., and Limkeman, M., Analysis of fluorescence decay kinetics from variable-frequency phase shift and modulation data, *Biophys. J.*, 46, 463, 1984.
96. Kilin, S.F., The duration of photo- and radioluminescence of organic compounds, *Opt. Spectrosc.*, 12, 414, 1962.
97. Spencer, R.D. and Weber, G., Measurement of subnanosecond fluorescence lifetimes with a cross-correlation phase fluorometer, *Ann. N.Y. Acad. Sci.*, 158, 361, 1969.
98. Lakowicz, J.R. and Berndt, K.W., Lifetime-selective fluorescence imaging using an rf phase-sensitive camera, *Rev. Sci. Instrum.*, 62, 1727, 1991.
99. Szmacinski, H., Lakowicz, J.R., and Johnson, M.L., Fluorescence lifetime imaging microscopy: homodyne technique using high-speed gated image intensifier, *Methods Enzymol.*, 240, 723, 1994.
100. Lakowicz, J.R., Szmacinski, H., Nowaczyk, K., Lederer, W.J., Kirby, M.S., and Johnson, M.L., Fluorescence lifetime imaging of intracellular calcium in COS cells using Quin-2, *Cell Calcium*, 15, 7, 1994.
101. Gadella, T.W.J., Jovin, T.M., and Clegg, R.M., Fluorescence lifetime imaging microscopy (FLIM) — spatial-resolution of microstructures on the nanosecond time-scale, *Biophys. Chem.*, 48, 221, 1993.
102. Schneider, P.C. and Clegg, R.M., Rapid acquisition, analysis, and display of fluorescence lifetime-resolved images for real-time applications, *Rev. Sci. Instrum.*, 68, 4107, 1997.
103. Squire, A. and Bastiaens, P.I., Three-dimensional image restoration in fluorescence lifetime imaging microscopy, *J. Microsc.*, 193, 36, 1999.
104. Clegg, R.M., Feddersen, B., Gratton, E., and Jovin, T.M., Time resolved imaging fluorescence miscoscopy, in *Proc. SPIE, Time-Resolved Laser Spectroscopy* in *Biochemistry* III, Lakowicz, J.R., Ed., 1992, p. 448.
105. Lakowicz, J. R., Szmacinski, H., Nowaczyk, K., and Johnson, M.L., Fluorescence lifetime imaging of free and protein-bound NADH, *Proc. Natl. Acad. Sci. U.S.A.*, 89, 1271, 1992.
106. Szmacinski, H. and Lakowicz, J.R., Fluorescence lifetime-based sensing and imaging, *Sensors Actuators B Chem.*, 29, 16, 1995.
107. Mizeret, J., Wagnieres, G., Studzinski, A., Shangguan, C., and van den Bergh, H., Endoscopic tissue fluorescence life-time imaging by frequency domain light-induced fluorescence, in *Proc. SPIE, Optical Biopsies*, Cubeddu, R., Mordon, S.R., and Svanberg, K., Eds., 1995, p. 40.
108. Gadella, B.M., Vanhoek, A., and Visser, A.J.W.G., Construction and characterization of a frequency-domain fluorescence lifetime imaging microscopy system, *J. Fluoresc.*, 7, 35, 1997.
109. Hartmann, P., Ziegler, W., Holst, G., and Lubbers, D.W., Oxygen flux fluorescence lifetime imaging, *Sensors Actuators B Chem.*, 38, 110, 1997.
110. Wagnières, G., Mizeret, J., Studzinski, A., and van den Bergh, H. Frequency-domain fluorescence lifetime imaging for endoscopic clinical cancer photodetection: apparatus design and preliminary results, *J. Fluoresc.*, 7, 75, 1997.
111. Pepperkok, R., Squire, A., Geley, S., and Bastiaens, P.I., Simultaneous detection of multiple green fluorescent proteins in live cells by fluorescence lifetime imaging microscopy, *Curr. Biol.*, 9, 269, 1999.
112. Ng, T., Squire, A., Hansra, G., Bornancin, F., Prevostel, C., Hanby, A., Harris, W., Barnes, D., Schmidt, S., Mellor, H., Bastiaens, P.I., and Parker, P.J., Imaging protein kinase C-alpha activation in cells, *Science*, 283, 2085, 1999.

113. Ng, T., Shima, D., Squire, A., Bastiaens, P.I., Gschmeissner, S., Humphries, M.J., and Parker, P.J., PKC-alpha regulates beta1 integrin-dependent cell motility through association and control of integrin traffic, *EMBO J.*, 18, 3909, 1999.
114. Wouters, F.S. and Bastiaens, P.I., Fluorescence lifetime imaging of receptor tyrosine kinase activity in cells, *Curr. Biol.*, 9, 1127, 1999.
115. Holub, O., Seufferheld, M.J., Gohlke, C., Govindjee, and Clegg, R.M., Fluorescence lifetime imaging (FLI) in real-time — a new technique in photosynthesis research, *Photosynthetica*, 38, 581, 2000.
116. Murata, S., Herman, P., Lin, H.J., and Lakowicz, J.R., Fluorescence lifetime imaging of nuclear DNA: effect of fluorescence resonance energy transfer, *Cytometry*, 41, 178, 2000.
117. Verveer, P.J., Squire, A., and Bastiaens, P.I., Global analysis of fluorescence lifetime imaging microscopy data, *Biophys. J.*, 78, 2127, 2000.
118. Verveer, P.J., Wouters, F.S., Reynolds, A.R., and Bastiaens, P.I., Quantitative imaging of lateral ErbB1 receptor signal propagation in the plasma membrane, *Science*, 290, 1567, 2000.
119. Murata, S., Herman, P., and Lakowicz, J.R., Texture analysis of fluorescence lifetime images of nuclear DNA with effect of fluorescence resonance energy transfer, *Cytometry*, 43, 94, 2001.
120. Murata, S., Herman, P., and Lakowicz, J.R., Texture analysis of fluorescence intensity and lifetime images af AT-and GC-rich regions in nuclei, *J. Histochem. Cytochem.*, 49, 1443, 2001.
121. Ng, T., Parsons, M., Hughes, W.E., Monypenny, J., Zicha, D., Gautreau, A., Arpin, M., Gschmeissner, S., Verveer, P.J., Bastiaens, P.I., and Parker, P.J., Ezrin is a downstream effector of trafficking PKC-integrin complexes involved in the control of cell motility, *EMBO J.*, 20, 2723, 2001.
122. Verveer, P.J., Squire, A., and Bastiaens, P.I., Improved spatial discrimination of protein reaction states in cells by global analysis and deconvolution of fluorescence lifetime imaging microscopy data, *J. Microsc.*, 202, 451, 2001.
123. Tertoolen, L.G., Blanchetot, C., Jiang, G., Overvoorde, J., Gadella, T.W., Jr., Hunter, T., and Hertog, J.D., Dimerization of receptor protein-tyrosine phosphatase alpha in living cells, *BMC Cell Biol.*, 2, 8, 2001.
124. Harpur, A.G., Wouters, F.S., and Bastiaens, P.I., Imaging FRET between spectrally similar GFP molecules in single cells, *Nat. Biotechnol.*, 19, 167, 2001.
125. Bastiaens, P.I. and Jovin, T.M., Microspectroscopic imaging tracks the intracellular processing of a signal transduction protein: fluorescent-labeled protein kinase C beta I, *Proc. Natl. Acad. Sci. U.S.A.*, 93, 8407, 1996.
126. Konig, K., So, P.T., Mantulin, W.W., Tromberg, B.J., and Gratton, E., Two-photon excited lifetime imaging of autofluorescence in cells during UVA and NIR photostress, *J. Microsc.*, 183, 197, 1996.
127. Dong, C.Y., So, P.T., French, T., and Gratton, E., Fluorescence lifetime imaging by asynchronous pump-probe microscopy, *Biophys. J.*, 69, 2234, 1995.
128. Merkelo, H.S., Hartman, S.R., Mar, T., Singhal, G.S., and Govindjee, Mode-locked lasers: measurements of very fast radiative decay in fluorescent systems, *Science*, 164, 301, 1969.
129. Gratton, E. and Delgado, R.L., Use of synchrotron radiation for the measurement of fluorescence lifetimes with subpicosecond resolution, *Rev. Sci. Instrum.*, 50, 789, 1979.
130. Gratton, E. and Lopez-Delgado, R., Measuring fluorescence decay times by phase-shift and modulation using the high harmonic content of pulsed light sources, *Nuovo Cimento B*, 56, 110, 1980.
131. Carlsson, K. and Liljeborg, A., Confocal fluorescence microscopy using spectral and lifetime information to simultaneously record four fluorophores with high channel separation, *J. Microsc. (Oxford)*, 185, 37, 1997.
132. Carlsson, K. and Liljeborg, A., Simultaneous confocal lifetime imaging of multiple fluorophores using the intensity-modulated multiple-wavelength scanning (IMS) technique, *J. Microsc.*, 191, 119, 1998.
133. Aslund, N. and Carlsson, K., Confocal scanning microfluorometry of dual-labeled specimens using 2 excitation wavelengths and lock-in detection technique, *Micron*, 24, 603, 1993.

134. Carlsson, K., Liljeborg, A., Andersson, R.M., and Brismar, H., Confocal pH imaging of microscopic specimens using fluorescence lifetimes and phase fluorometry: influence of parameter choice on system performance, *J. Microsc.*, 199, 106, 2000.
135. Gratton, E. and Limkeman, M., A continuously variable frequency cross-correlation phase fluorometer with picosecond resolution, *Biophys. J.*, 44, 315, 1983.
136. Lakowicz, J.R. and Maliwal, B.P., Construction and performance of a variable-frequency phase-modulation fluorometer, *Biophys. Chem.*, 21, 61, 1985.
137. Weber, G., Resolution of the fluorescence lifetimes in a heterogeneous system by phase and modulation measurements, *J. Phys. Chem.*, 85, 949, 1981.
138. Feddersen, B.A., Piston, D.W., and Gratton, E., Digital parallel acquisition in frequency-domain fluorimetry, *Rev. Sci. Instrum.*, 60, 2929, 1989.
139. Mitchell, G.W. and Swift, K., 48000 MHF: a dual-domain Fourier transform fluorescence lifetime spectrofluorometer, in *Proc. SPIE, Time-Resolved Laser Spectroscopy in Biochemistry* II, Lakowicz, J.R., Ed., 1990, p. 270.
140. Mitchell, G.W., Picosecond multiharmonic Fourier fluorometer, U.S. patent 4,937,457, 1990.
141. Watkins, A.N., Ingersoll, C.M., Baker, G.A., and Bright, F.V., A parallel multiharmonic frequency-domain fluorometer for measuring excited-state decay kinetics following one-, two-, or three-photon excitation, *Anal. Chem.*, 70, 3384, 1998.
142. Squire, A., Verveer, P.J., and Bastiaens, P.I., Multiple frequency fluorescence lifetime imaging microscopy, *J. Microsc.*, 197, 136, 2000.
143. Straub, M. and Hell, S.W., Fluorescence lifetime three-dimensional microscopy with picosecond precision using a multifocal multiphoton microscope, *Appl. Phys. Lett.*, 73, 1769, 1998.
144. Siegel, J., Elson, D.S., Webb, S.E.D., Parsons-Karavassilis, D., Leveque-Fort, S., Cole, M.J., Lever, M.J., French, P.M.W., Neil, M.A.A., Juskaitis, R., Sucharov, L.O., and Wilson, T., Whole-field five-dimensional fluorescence microscopy combining lifetime and spectral resolution with optical sectioning, *Opt. Lett.*, 26, 1338, 2001.
145. Neil, M.A.A., Juskaitis, R., and Wilson, T., Method of obtaining optical sectioning by using structured light in a conventional microscope, *Opt. Lett.*, 22, 1905, 1997.
146. Neil, M.A.A., Squire, A., Juskaitis, R., Bastiaens, P.I.H., and Wilson, T., Wide-field optically sectioning fluorescence microscopy with laser illumination, *J. Microsc. (Oxford)*, 197, 1, 2000.
147. Bewersdorf, J., Pick, R., and Hell, S.W., Multifocal multiphoton microscopy, *Opt. Lett.*, 23, 655, 1998.
148. Woods, R.J., Scypinski, S., Love, L.J.C., and Ashworth, H.A., Transient digitizer for the determination of microsecond luminescence lifetimes, *Anal. Chem.*, 56, 1395, 1984.
149. Ballew, R.M. and Demas, J.N., An error analysis of the rapid lifetime determination method for the evaluation of single exponential decays, *Anal. Chem.*, 61, 30, 1989.
150. Waters, P.D. and Burns, D.H., Optimized gated detection for lifetime measurement over a wide-range of single exponential decays, *Appl. Spectrosc.*, 47, 111, 1993.
151. Chan, S.P., Fuller, Z.J., Demas, J. N., and DeGraff, B.A., Optimized gating scheme for rapid lifetime determinations of single-exponential luminescence lifetimes, *Anal. Chem.*, 73, 4486, 2001.
152. Sharman, K.K., Periasamy, A., Ashworth, H., Demas, J.N., and Snow, N.H., Error analysis of the rapid lifetime determination method for double-exponential decays and new windowing schemes, *Anal. Chem.*, 71, 947, 1999.
153. Beechem, J. M., Knutson, J.R., Ross, J.B.A., Turner, B.W., and Brand, L., Global resolution of heterogeneous decay by phase modulation fluorometry — mixtures and proteins, *Biochemistry*, 22, 6054, 1983.
154. Knutson, J.R., Beechem, J.M., and Brand, L., Simultaneous analysis of multiple fluorescence decay curves — a global approach, *Chem. Phys. Lett.*, 102, 501, 1983.
155. Beechem, J.M., Ameloot, M., and Brand, L., Global and target analysis of complex decay phenomena, *Anal. Instrum.*, 14, 379, 1985.

156. Beechem, J.M. and Brand, L., Global analysis of fluorescence decay — applications to some unusual experimental and theoretical-studies, *Photochem. Photobiol.*, 44, 323, 1986.
157. Bastiaens, P.I. and Pepperkok, R., Observing proteins in their natural habitat: the living cell, *Trends Biochem. Sci.*, 25, 631, 2000.
158. Szmacinski, H. and Lakowicz, J.R., Optical measurements of pH using fluorescence lifetimes and phase-modulation fluorometry, *Anal. Chem.*, 65, 1668, 1993.
159. Lin, H.-J., Szmacinski, H., and Lakowicz, J.R., Lifetime-based pH sensors: indicators for acidic environments, *Anal. Biochem.*, 269, 162, 1999.
160. Lin, H.-J., Herman, P., Kang, J.-S., and Lakowicz, J.R., Fluorescence lifetime characterizaiton of novel low pH probes, *Anal. Biochem.*, 294, 118, 2001.
161. Sanders, R., Gerritsen, H.C., Draaijer, A., Houpt, P.M., Van Veen, S.J.F., and Levine, Y.K., Confocal fluorescence lifetime imaging of pH in single cells, in Time-Resolved Laser Spectroscopy in *Proc. SPIE, Biochemistry* IV, Lakowicz, J.R., Ed., 1994, p. 56.
162. Srivastava, A. and Krishnamoorthy, G., *Time-resolved fluorescence microscopy* could correct for probe binding while estimating intracellular pH, *Anal. Biochem.*, 249, 140, 1997.
163. Andersson, R.M., Carlsson, K., Liljeborg, A., and Brismar, H., Characterization of probe binding and comparison of its influence on fluorescence lifetime of two pH-sensitive benzo[c]xanthene dyes using intensity-modulated multiple-wavelength scanning technique, *Anal. Biochem.*, 283, 104, 2000.
164. Lakowicz, J.R., Szmacinski, H., Nowaczyk, K., and Johnson, M.L., Fluorescence lifetime imaging of calcium using Quin-2, *Cell Calcium*, 13, 131, 1992.
165. Lakowicz, J.R., Szmacinski, H., and Johnson, M.L., Calcium imaging using fluorescence lifetime and long-wavelength probes, *J. Fluoresc.*, 2, 47, 1992.
166. Thompson, R.B., Jr., Whetsell, W.O., Maliwal, B.R., Fierke, C.A., and Frederickson, C.J., Fluorescence microscopy of stimulated Zn(II) release from organotypic cultures of mammalian hippocampus using a carbonic anhydrase-based biosensor system, *J. Neurosci. Methods*, 96, 35, 2000.
167. Thompson, R.B., Maliwal, B.R., and Fierke, C.A., selectivity and sensitivity of fluorescence lifetime-based metal ion biosensing using a carbonic anhydrase transducer, *Anal. Biochem.*, 267, 185, 1999.
168. Birch, D.J.S., Holmes, A.S., and Darbyshire, M., Intelligent sensor for metal ions based on fluorescence resonance energy transfer, *Meas. Sci. Technol.*, 6, 243, 1995.
169. Oliver, A.E., Baker, G.A., Fugate, R.D., Tablin, F., and Crowe, J.H., Effects of temperature on calcium-sensitive fluorescence probes, *Biophys. J.*, 78, 2116, 2000.
170. French, T., So, P.T., Weaver, D.J., Jr., Coelho-Sampaio, T., Gratton, E., Voss, E.W., Jr., and Carrero, J., Two-photon fluorescence lifetime imaging microscopy of macrophage-mediated antigen processing, *J. Microsc.*, 185, 339, 1997.
171. Paul, R.J., Oxygen concentration and the oxidation-reduction state of yeast: determination of free/bound NADH and flavins by time-resolved spectroscopy, *Naturwissenschaften*, 83, 32, 1996.
172. Förster, T., Zwischenmolekulare Energiewanderung und Fluoreszenz, *Ann. Phys.*, 2, 57, 1948.
173. Gadella, T.W., Jr. and Jovin, T.M., Oligomerization of epidermal growth factor receptors on A431 cells studied by time-resolved fluorescence imaging microscopy. A stereochemical model for tyrosine kinase receptor activation, *J. Cell Biol.*, 129, 1543, 1995.
174. Martin-Fernandez, M.L, Tobin, M.J., Clarke, D.T., Gregory, C.M., and Jones, G.R., A high sensitivity time-resolved microfluorimeter for real-time cell biology, *Rev. Sci. Instrum.*, 3716, 1996.
175. Pollok, B. and Heim, R., Using GFP in FRET-based applications, *Trends Cell Biol.*, 9, 57, 1999.
176. Birmingham, J.J., Frequency-domain lifetime imaging methods at Unilever Research, *J. Fluoresc.*, 7, 45, 1997.
177. Masters, B.R., So, P.T., and Gratton, E., Multiphoton excitation fluorescence microscopy and spectroscopy of *in vivo* human skin, *Biophys. J.*, 72, 2405, 1997.
178. Sevick-Muraca, E.M., Reynolds, J. S., Troy, T.L., Lopez, G., and Paithankar, D.Y., Fluorescence lifetime spectroscopic imaging with measurements of photon migration, *Ann. N.Y. Acad. Sci.*, 838, 46, 1998.

179. Holst, G., Kohls, O., Klimant, I., Konig, B., Kuhl, M., and Richter, T., A modular luminescence lifetime imaging system for mapping oxygen distribution in biological samples, *Sensors Actuators B Chem.*, 51, 163, 1998.
180. Holst, G. and Grunwald, B., Luminescence lifetime imaging with transparent oxygen optodes, *Sens. Actuators B Chem.*, 74, 78, 2001.
181. Morgan, C.G., Mitchell, A.G., Murray, J. G., and Wall, E.J., New approach to lifetime-resolved luminescence imaging, *J. Fluoresc.*, 7, 65, 1997.
182. Wolfbeis, O.S.I., Klimart, W.T, Huber, C., Kosch, U., Krause, C., Neurauter, G., and Dürkop, A., Set of luminescence decay time based chemical sensors for clinical applications, *Sensors Actuators B Chem.*, 51, 17, 1998.
183. Lakowicz, J.R., Gryczynski, I., Shen, Y.B., Malicka, J., and Gryczynski, Z., Intensified fluorescence, *Photon. Spectrosc.*, 35, 96, 2001.
184. Lakowicz, J.R., Shen, B., Gryczynski, Z., D'Auria, S., and Gryczynski, I., Intrinsic fluorescence from DNA can be enhanced by metallic particles, *Biochem. Biophys. Res. Commun.*, 286, 875, 2001.
185. Lakowicz, J.R., Piszczek, G., and Kang, J.S., On the possibility of long-wavelength long-lifetime high-quantum-yield luminophores, *Anal. Biochem.*, 288, 62, 2001.
186. Maliwal, B.P., Gryczynski, Z., and Lakowicz, J.R., Long-wavelength long-lifetime luminophores, *Anal. Chem.*, 73, 4277, 2001.
187. Terpetschnig, E., Szmacinski, H., Malak, H., and Lakowicz, J.R., Metal-ligand complexes as a new class of long-lived fluorophores for protein hydrodynamics, *Biophys. J.*, 68, 342, 1995.
188. Terpetschnig, E., Szmacinski, H., and Lakowicz, J.R., Long-lifetime metal-ligand complexes as probes in biophysics and clinical chemistry, *Methods Enzymol.*, 278, 295, 1997.
189. Lakowicz, J.R., Gryczynski, I., Gryczynski, Z., and Johnson, M.L., Background suppression in frequency-domain fluorometry, *Anal. Biochem.*, 277, 74, 2000.
190. Herman, P., Maliwal, B.P., and Lakowicz, J.R., Frequency-domain fluorometry enhanced by real time background suppression, *Biophys. J.*, 80, 367A, 2001.

# 10

# Confocal Microscopy



Tony Wilson
*University of Oxford*
*Oxford, U.K.*

## 10.1 Introduction

It is probably fair to say that the development and wide commercial availability of the confocal microscope have been one of the most significant advances in light microscopy in the recent past. The main reason for the popularity of these instruments derives from their ability to permit the structure of thick specimens of biological tissue to be investigated in three dimensions by resorting to a scanning approach together with a novel (confocal) optical system.

The traditional wide-field conventional microscope is a parallel processing system that images the entire object field simultaneously. This is quite a severe requirement for the optical components, but we can relax this requirement if we no longer try to image the whole object at once. The limit of this relaxation is to require an image of only one object point at a time. In this case, all that we ask of the optics is to provide a good image of one point. The price that we pay is that we must scan in order to build up an image of the entire field. The answer to the question whether this price is worth paying will, to some extent, depend on the application in question.

A typical arrangement of a scanning confocal optical microscope is shown in Figure 10.1 in which the system is built around a host conventional microscope. The essential components are some form of mechanism for scanning the light beam (usually from a laser) relative to the specimen and appropriate photodetectors to collect the reflected or transmitted light.[1] Most of the early systems were analog in nature; however, it is now universal, thanks to the serial nature of the image formation, to use a computer to drive the microscope and to collect, process, and display the image.

In the beam scanning confocal configuration of Figure 10.1 the scanning is typically achieved by using vibrating galvanometer-type mirrors or acousto-optic beam deflectors. The use of the latter gives the possibility of TV-rate scanning, whereas vibrating mirrors are often relatively slow when imaging an extended region of the specimen, although significantly higher scanning speeds are achievable over smaller scan regions. Note that other approaches to scanning may be implemented, such as specimen scanning and lens scanning. These methods, although not generally available commercially, do have advantage in certain specialized applications.[1,2] This chapter is necessarily limited in length, but much additional material may be found in other sources[1–7] in which a detailed list of references may be found.

0-8493-1116-0/03/$0.00+$1.50

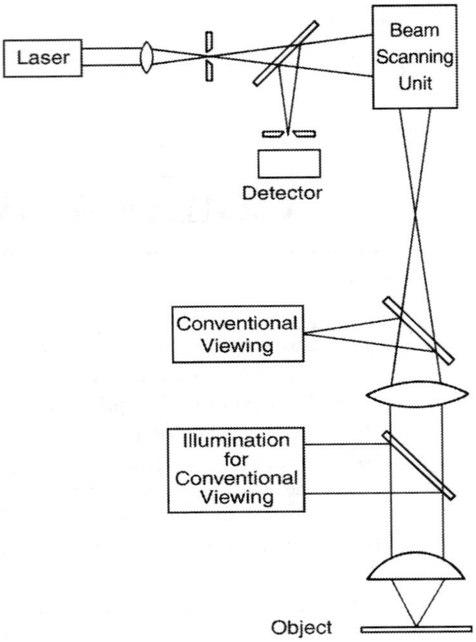


**FIGURE 10.1** Schematic diagram of a confocal microscope.

## 10.2 Image Formation in Scanning Microscopes

We will not discuss the fine detail of the optical properties of confocal systems because this is already widely available in the literature. The essence is shown in Figure 10.1, where the confocal optical system consists simply of a point source of light that is then used to probe a single point on the specimen. The strength of the reflected or fluorescence radiation from the single object point is then measured via a point, pinhole detector. The confocal — point source and point detector — optical system therefore merely produces an "image" of a single object point and hence some form of scanning is necessary to produce an image of an extended region of the specimen. However, the use of single-point illumination and single-point detection results in novel imaging capabilities that offer significant advantages over those possessed by conventional wide-field optical microscopes. In essence, these are enhanced lateral resolution and, perhaps more importantly, a unique depth discrimination or optical sectioning property. It is this latter property that leads to the ability to obtain three-dimensional images of volume specimens.

The improvement in lateral resolution may at first seem implausible. However, it can be explained simply by a principle given by Lukosz,[8] which states, in essence, that resolution can be increased at the expense of field of view. The field of view can then be increased by scanning. One way of taking advantage of Lukosz's principle is to place a very small aperture extremely close to the object. The resolution is now determined by the size of the aperture rather than the radiation. In the confocal microscope we do not use a physical aperture in the focal plane but, rather, use the back-projected image of a point detector in conjunction with the focused point source. Figure 10.2 indicates the improvement in lateral resolution that may be achieved.

The confocal principle, which was first described by Minsky,[9] was introduced in an attempt to obtain an image of a slice within a thick specimen that was free from the distracting presence of out-of-focus information from surrounding planes. The confocal optical system fulfils this requirement; its inherent optical sectioning or depth discrimination property has become the major motivation for using confocal microscopes, and is the basis of many of the novel imaging modes of these instruments.

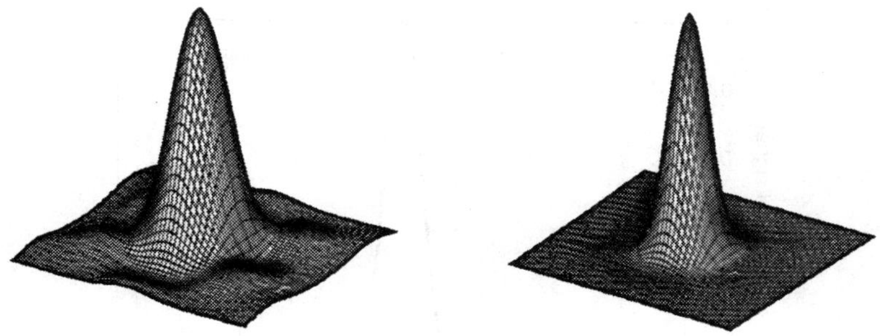


**FIGURE 10.2** The point spread functions of conventional and confocal microscopes showing the improvement in lateral resolution that may be obtained in the confocal case.

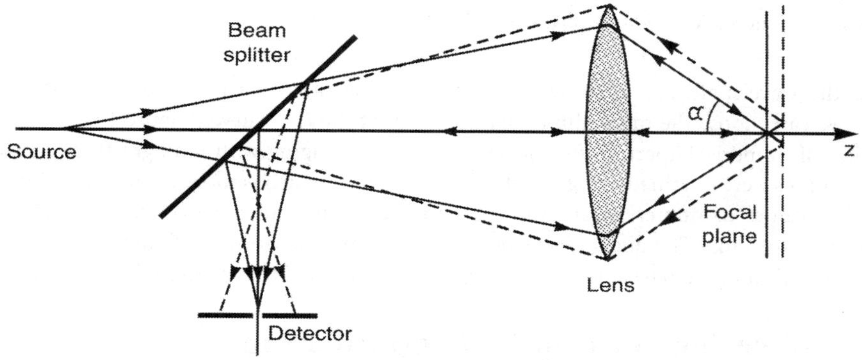


**FIGURE 10.3** The origin of the optical sectioning or depth discrimination property of the confocal optical system.

The origin of the depth discrimination property may be understood very easily from Figure 10.3, where we show a reflection-mode confocal microscope and consider the imaging of a specimen with a rough surface. The full lines show the optical path when an object feature lies in the focal plane of the lens. At a later scan position, the object surface is supposed to be located in the plane of the vertical dashed line. In this case, simple ray tracing shows that the light reflected back to the detector pinhole arrives as a defocused blur, only the central portion of which is detected, and contributes to the image. In this way the system discriminates against features that do not lie within the focal region of the lens.

A very simple method of demonstrating the effect and giving a measure of its strength is to scan a perfect reflector axially through focus and measure the detected signal strength. Figure 10.4 shows a typical response. These responses are frequently termed the $V(z)$ by analogy with a similar technique in scanning acoustic microscopy, although the correspondence is not perfect. A simple paraxial theory models this response as:

$$I(u)=\left[\frac{\sin(u/2)}{u/2}\right]^2 \tag{10.1}$$

where $u$ is a normalized axial coordinate related to real axial distance, $z$, via:

$$u=\frac{8\pi}{\lambda}nz\sin^2(\alpha/2) \tag{10.2}$$

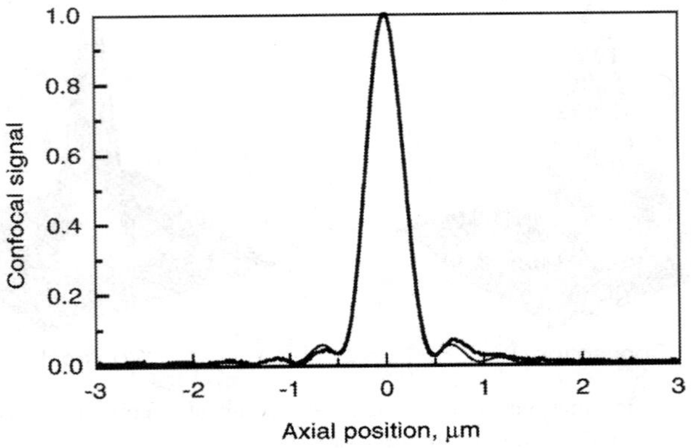


**FIGURE 10.4** The variation in detected signal as a plane reflector is scanned axially through focus. The measurement was taken with a 1.3 numerical aperture objective and 633-nm radiation.

where λ is the wavelength and n sin α the numerical aperture. As a measure of the strength of the sectioning, we can choose the full width at half intensity of the $I(u)$ curves. Figure 10.5 shows this value as a function of numerical aperture for the specific case of imaging with red light from a helium neon laser. These curves were obtained using a high aperture theory, which is more reliable than Equation 10.1 at the highest values of numerical aperture. We note, of course, that these numerical values refer to nonfluorescence imaging. The qualitative explanation of optical sectioning, of course, carries over to the fluorescence case, but the actual value of the optical sectioning strength is different in the fluorescence case.

## 10.3 Applications of Depth Discrimination

This property is one of the major reasons for the popularity of confocal microscopes, so it is worthwhile, at this point, to review briefly some of the novel imaging techniques that have become available with confocal microscopy.

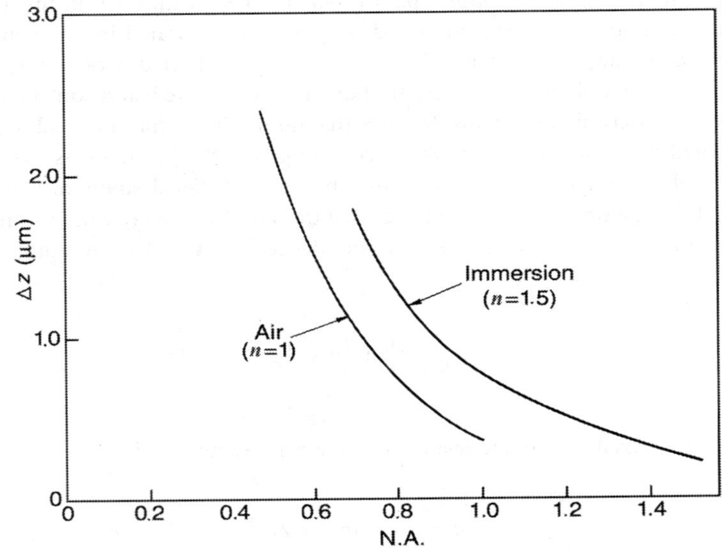


**FIGURE 10.5** The optical sectioning width as a function of numerical aperture. The curves are for red light (0.6328 μm wavelength). $\Delta z$ is the full width at the half-intensity points of the curves of $I(u)$ against $u$.

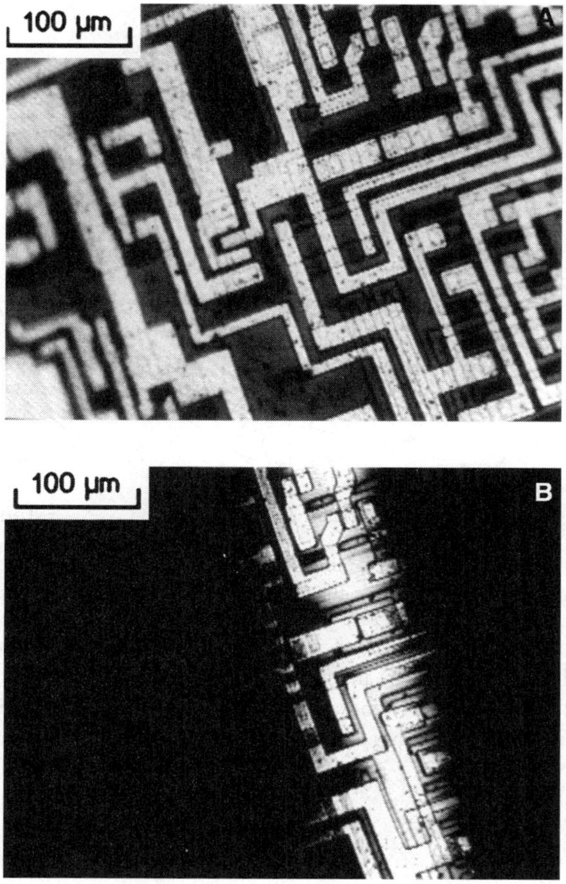


**FIGURE 10.6** (A) Conventional scanning microscope image of a tilted microcircuit: the parts of the object outside the focal plane appear blurred. (B) Confocal image of the same microcircuit: only the part of the specimen within the focal region is imaged strongly.

Figure 10.6 illustrates the essential effect: Figure 10.6A shows a conventional image of a planar microcircuit that has deliberately been mounted with its normal at an angle to the optic axis. We see that only one portion of the circuit, running diagonally, is in focus. Figure 10.6B shows the corresponding confocal image: here the discrimination against detail outside the focal plane is clear. The areas that were out of focus in Figure 10.6A have been rejected. Furthermore, the confocal image appears to be in focus throughout the visible band, which illustrates that the sectioning property is stronger than the depth of focus.

This suggests that, if we try to image a thick translucent specimen, we can arrange, by the choice of our focal position, to image detail exclusively from one specific region. In essence, we can section the specimen optically without resorting to mechanical means. Figure 10.7 shows an idealized schematic of the process. The portion of the beehive-shaped object that we see is determined by the focus position. In this way, it is possible to take a through-focus series of images and obtain data about the three-dimensional structure of the specimen. If we represent the volume image by $I(x,y,z)$, then, ideally, by focusing at a position $z = z_1$, we obtain the image $I(x,y,z_1)$. This, of course, is not strictly true in practice because the optical section is not infinitely thin.

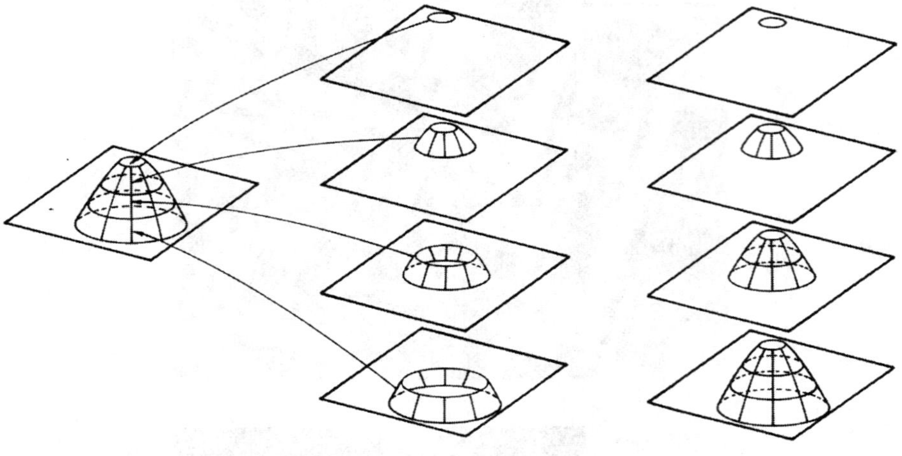


**FIGURE 10.7** An idealization of the optical sectioning property showing the ability to obtain a through-focus series of images, which may then be used to reconstruct the original volume object at high resolution.

It is clear that the confocal microscope allows us to form high-resolution images with a depth of focus sufficiently small that all the detail that is imaged appears in focus. This suggests immediately that we can extend the depth of focus of the microscope by adding together (integrating) the images taken at different focal settings without sacrificing the lateral resolution. Mathematically, this extended-focus image is given by:

$$I_{EF} = I(x, y, z)dz \tag{10.3}$$

As an alternative to the extended-focus method, we can form an auto-focus image by scanning the object axially and, instead of integrating, selecting the focus at each picture point by recording the maximum in the detected signal. Mathematically, this might be written:

$$I_{AF}(x, y) = I(x, y, z_{max}) \tag{10.4}$$

where $z_{max}$ corresponds to the focus setting giving the maximum signal. The images obtained are somewhat similar to the extended focus and, again, substantial increases in depth of focus may be obtained. We can go one step further in this case and turn the microscope into a noncontacting surface profilometer. Here we simply display $z_{max}$.

It is clear by now that the confocal method gives us a convenient tool for studying three-dimensional structures in general. We essentially record the image as a series of slices and play it back in any desired fashion. Naturally, in practice it is not as simple as this, but we can, for example, display the data as an $x$-$z$ image rather than an $x$-$y$ image. This is somewhat similar to viewing the specimen from the side. As another example, we might choose to recombine the data as stereo pairs by introducing a slight lateral offset to each image slice as we add them up. If we do this twice, with an offset to the left in one case and the right in another, we obtain, very simply, stereo pairs. Mathematically, we form images of the form:

$$\int I(x \pm \gamma z, y, z)dz \tag{10.5}$$

where $\gamma$ is a constant. In practice it may not be necessary to introduce offsets in both directions to obtain an adequate stereo view.

All that we have said so far on these techniques has been by way of simplified introduction. In particular we have not presented any fluorescence images, because these will be dealt with adequately later. The key point is that, in both bright-field and fluorescence modes, the confocal principle permits the imaging of specimens in three dimensions. Of course, the situation is more involved than we have implied. A thorough knowledge of the image formation process, together with the effects of lens aberrations and absorption, is necessary before accurate data manipulation can take place.

In conclusion to this section, it is important to emphasize that the confocal microscope does not produce three-dimensional images. The opposite is true: it essentially produces very high-quality two-dimensional images of a (thin) slice within a thick specimen. A three-dimensional rendering of the entire volume specimen may then be generated by suitably combining a number of these two-dimensional image slices from a through-focus series of images.

## 10.4 Fluorescence Microscopy

We now turn our attention to confocal fluorescence microscopy because this is the imaging mode usually employed in biological applications. Although we have introduced the confocal microscope in terms of bright-field imaging, the comments concerning the origin of the optical sectioning, etc. carry over directly to the fluorescence case. However, the numerical values describing the strength of the optical sectioning are, of course, different and we will return to this point later.

If we assume that the fluorescence in the object destroys the coherence of the illuminating radiation and produces an incoherent fluorescent field proportional to the intensity of the incident radiation, $I(v,u)$, then we may write the effective intensity point spread function, which describes image formation in the incoherent confocal fluorescence microscope, as

$$I(v,u)\, I\left( {}^{v}\!/_{\beta}, \frac{u}{\beta} \right) \tag{10.6}$$

where the optical coordinates $u$ and $v$ are defined relative to the primary radiation and

$$\beta = {}^{\lambda_2}\!/_{\lambda_1}$$

is the ratio of the fluorescence to the primary wavelength. We note that

$$v = {}^{2\pi}\!/_{\lambda_1}\, r n \sin\alpha$$

where $r$ denotes the actual radial distance.

This suggests that the imaging performance depends on the value of $\beta$. In order to illustrate this, Figure 10.8 shows the variation in detected signal strength as a perfect fluorescent sheet through focus. This serves to characterize the strength of the optical sectioning in fluorescence microscopy in the same way that the mirror was used in the bright-field case. We note that the half width of these curves is essentially proportional to $\beta$ and so for optimum sectioning the wavelength ratio should be as close to unity as possible.

We have just discussed what we might call one-photon fluorescence microscopy in the sense that a fluorophore is excited by a single photon of a particular wavelength. It then returns to the ground state and emits a photon at the (slightly longer) fluorescence wavelength. It is this radiation that is detected via the confocal pinhole. Recently, however, much interest has centered on two-photon excitation fluorescence microscopy.[10,11] This process relies on the simultaneous absorption of two, longer wavelength

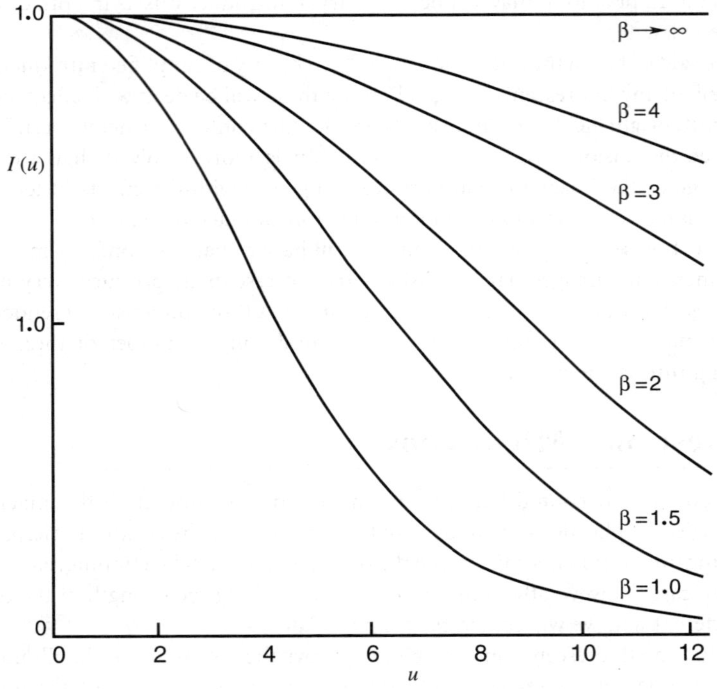


**FIGURE 10.8** The detected signal as a perfect planar object is scanned axially through focus for a variety of fluorescent wavelengths. If we measure the sectioning by the halfwidth of these curves, the strength of the sectioning is essentially proportional to β.

photons, following which a single fluorescence photon is emitted. The excitation wavelength is typically twice that used in the one-photon case.

The beauty of the two-photon approach lies in the quadratic dependence of the fluorescence intensity on the intensity of the illumination. This leads to fluorescence emission, which is always confined to the region of focus. In other words, the system possesses an inherent optical sectioning property. Other benefits of two-photon fluorescence over the single-photon case include the use of red or infrared lasers to excite ultraviolet dyes, confinement of photo-bleaching to the focal region (the region of excitation), and the reduced effects of scattering and greater penetration. However, it should also be remembered that, compared with single photon excitation, the fluorescence yield of many fluorescent dyes under two-photon excitation is relatively low.[12]

In order to make a theoretical comparison between the one- and two-photon modalities, we shall assume that the *emission* wavelength $\lambda_{em}$ is the same irrespective of the mode of excitation. Since the wavelength required for single-photon excitation is generally shorter than the emission wavelength, we may write it as $\gamma\,\lambda_{em}$ where $\gamma < 1$. Because two-photon excitation requires the simultaneous absorption of two photons of half the energy, we will assume that the excitation wavelength in this case may be written as $2\gamma\,\lambda_{em}$, which has been shown to be a reasonable approximation for many dyes.[13]

If we now introduce optical coordinates $u$ and $v$ normalized in terms of $\lambda_{em}$ we may write the effective point spread functions in the one-photon confocal and two-photon case as

$$I_{1p-conf} = I\left(\frac{v}{\gamma}, \frac{u}{\gamma}\right) I(v,u) \tag{10.7}$$

and

$$I_{2p} = I^2\left(\frac{v}{2\gamma}, \frac{u}{2\gamma}\right) \tag{10.8}$$

respectively. We note that, although a pinhole is not usually employed in two-photon microscopy, it is perfectly possible to include one if necessary. In this confocal two-photon geometry the effective point spread function becomes

$$I_{2p\text{-}conf} = I^2\left(\frac{v}{2\gamma}, \frac{u}{2\gamma}\right)I(v,u) \tag{10.9}$$

If we now look at Equations 10.7 and 10.8 in the $\gamma = 1$ limit, we find

$$I_{1p\text{-}conf} = I^2(v,u) \tag{10.10}$$

and

$$I_{2p} = I^2\left(\frac{v}{2}, \frac{u}{2}\right) \tag{10.11}$$

We now see that, because of the longer excitation wavelength used in two-photon microscopy, the effective point spread function is twice as large as that of the one-photon confocal in both the lateral and axial directions. The situation is somewhat improved in the confocal two-photon case, but it is worth remembering that the advantages of two-photon excitation microscopy are accompanied by a reduction in optical performance compared to the single-photon case.

From the practical point of view, the two-photon approach has certain very important advantages over the one-photon excitation in terms of image contrast when imaging through scattering media apart from the greater depth of penetration afforded by the longer wavelength excitation. In a single-photon confocal case, it is quite possible that the desired fluorescence radiation from the focal plane may be scattered after generation in such a way that it is not detected through the confocal pinhole. Further, since the fluorescence is generated throughout the entire focal volume, it is also possible that undesired fluorescence radiation, which was not generated within the focal region, may be scattered so as to be detected through the confocal pinhole.

In either case this leads to a reduction in image contrast. The situation is, however, completely different in the two-photon case. Here the fluorescence is generated only in the focal region and not throughout the focal volume. Furthermore, because all the fluorescence is detected via a large area detector — no pinhole is involved — it is not so important if further scattering events take place. This leads to high-contrast images, which are less sensitive to scattering. This is particularly important for specimens that are much more scattering at the fluorescence ($\lambda/2$) wavelength than the excitation ($\lambda$) wavelength.

## 10.5 Optical Architectures

In addition to the generic architecture of Figure 10.1, a number of alternative implementations of confocal microscopes have been recently developed in order to reduce the alignment tolerances of the architecture of Figure 10.1 as well as to increase image acquisition speed. Because great care is required to ensure that the illumination and detector pinholes lie in equivalent positions in the optical system, a reciprocal geometry confocal system has been developed that uses a single pinhole to launch light into the microscope as well as to detect the returning confocal image signal. A development on this is to replace the physical pinholes with a single mode optical fiber. However, these systems are essentially developments of the traditional confocal architecture. We will now describe two recent attempts to develop new confocal architectures. The first system is based on the pinhole source/detector concept, whereas the second approach describes a simple modification to a conventional microscope so as to permit optically sectioned images to be obtained.

### 10.5.1 The Aperture Mask System

The question of system alignment and image acquisition speed have already been answered to a certain extent by Petrán and his colleagues, who originally developed the tandem scanning (confocal) microscope;[14] this was later reconfigured into a one-sided disk embodiment (see, for example, Corle and Kino[4]).

The main component of the system is a disk containing many pinholes. Each pinhole acts as the illumination and the detection pinhole, so the system acts rather like a large number of parallel, reciprocal-geometry, confocal microscopes, each imaging a specific point on the object. However, because it is important that no light from a particular confocal system enter an adjacent system, i.e., no cross talk between adjacent confocal systems, the pinholes in the disk must be placed far apart — typically ten pinhole diameters apart — which has two immediate consequences. First, only a small amount — typically 1% — of the available light is used for imaging and, second, the wide spacing of the pinholes means that the object is only sparsely probed. In order to probe, and hence image, the whole object, it is necessary to arrange the pinhole apertures in a series of Archimedean spirals and to rotate the disk. These systems have been developed and are capable of producing high-quality images without the need to use laser illumination in real time with TV rate as well as higher imaging speeds.

In order to make greater use of the available light in this approach we must, inevitably, place the pinholes closer together. This, of course, leads to cross talk between the neighboring confocal systems and thus a method must be devised to prevent this. In order to do this, we replace the Nipkow disk of the tandem scanning microscope with an aperture mask consisting of many pinholes placed as close together as possible. This aperture mask has the property that any of its pinholes can be opened and closed independently of the others in any desired time sequence. This might be achieved, for example, by using a liquid crystal spatial light modulator. Because we require no cross talk between the many parallel confocal systems, it is necessary to use a sequence of openings and closings of each pinhole completely uncorrelated with the openings and closings of all the other pinholes.

Many such ortho-normal sequences are in the literature. However, they all require the use of positive and negative numbers and, unfortunately, we cannot have negative intensity of light! The pinhole is either open, which corresponds to 1, or closed, which corresponds to 0. No position can correspond to –1. The way out of the dilemma is to not obtain the confocal signal directly. In order to use a particular ortho-normal sequence, $b_i(t)$, of plus and minus ones for the *i*th pinhole, we must add a constant shift to the desired sequence in order to make a sequence of positive numbers that can be encoded in terms of pinhole opening and closing.

Thus we encode each of the pinhole openings and closings as $(1 + b_i(t))/2$, which will correspond to open (1) when $b_i(t) = 1$ and to close (0) when $b_i(t) = -1$. The effect of adding the constant offset to the desired sequence is to produce a composite image that will be partly confocal due to the $b_i(t)$ terms and partly conventional due to the constant term. The method of operation is now clear. We first take an image with the pinholes encoded as we have just discussed and so obtain a composite conventional plus confocal image. We then switch all the pinholes to the open state to obtain a conventional image. It is then a simple matter to subtract the two images in real time in a computer to produce the confocal image.

Although this approach may be implemented using a liquid crystal spatial light modulator, it is cheaper and simpler merely to impress the correlation codes photolithographically on a disk and rotate it so that the transmissivity at any picture point varies according to the desired ortho-normal sequence. A blank sector may be used to provide the conventional image. If this approach is adopted, then all that is required is to replace the single-sided Nipkow disk of the tandem scanning microscope with a suitably encoded aperture disk, as shown in Figure 10.9.[15] Figure 10.10 shows a through-focus series of images of a fly's eye; Figure 10.11 shows three-dimensional representation of the fly's eye obtained with this white light real-time confocal system.

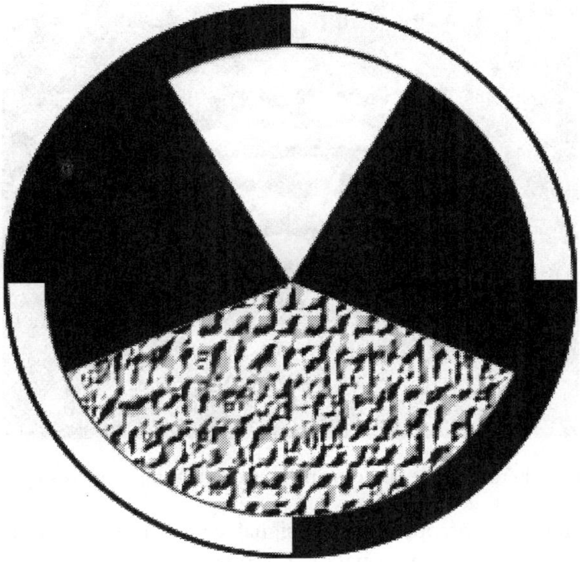

**FIGURE 10.9** A typical aperture mask. When light passes through the encoded region of the disk, a composite conventional and confocal image is obtained. Light passing through the unobstructed sector provides a conventional image.

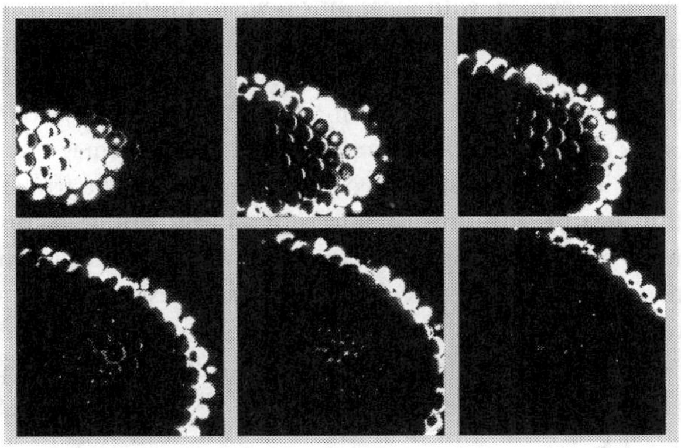

**FIGURE 10.10** A through-focus series of images of a portion of a fly's eye. Each optical section was recorded in real time using standard nonlaser illumination.

## 10.5.2 The Use of Structured Illumination to Achieve Optical Sectioning

The motivation for the development we have just described was to produce a light-efficient, real-time, three-dimensional imaging system. The approach was to start with the traditional confocal microscope design — point source and point detector — and to engineer a massively parallel, light-efficient three-dimensional imaging system (see also Liang et al.[16] and Hanley et al.[17]).

An alternative approach is to realize that the conventional light microscope already possesses many desirable properties: real-time image capture, standard illumination, ease of alignment, etc. However it does not produce optically sectioned images in the sense usually understood in confocal microscopy. In order to see how this deficiency may be corrected via a simple modification of the illumination system, let us look at the theory of image formation in a conventional fluorescence microscope and ask in what way the image changes as the microscope is defocused.

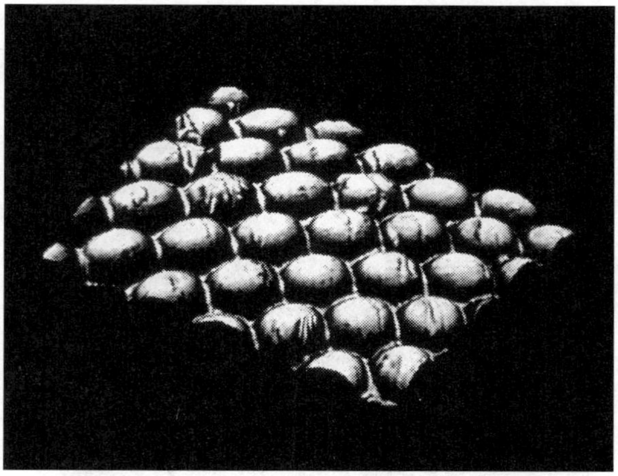

FIGURE 10.11 A computer generated three-dimensional representation of the fly's eye.

We know that in a confocal microscope the image signal from all object features attenuates with defocus and that this does not happen in a conventional microscope. However, when we look closely at the image formation process we find that it is only the zero spatial frequency (constant) component that does not change with defocus (Figure 10.12); all other spatial frequencies actually do attenuate with defocus to a greater or lesser extent. Figure 10.13 shows the image of a single spatial frequency one-dimensional bar pattern object for increasing degrees of defocus. When the specimen is imaged in focus, a good image of the bar pattern is obtained. However, with increasing defocus the image becomes progressively poorer and weaker until it eventually disappears, leaving a uniform gray level. This suggests a simple way to perform optical sectioning in a conventional microscope.

If we simply modify the illumination path of the microscope so as to project onto the object the image of a one-dimensional, single spatial frequency fringe pattern, then the image we see through the microscope will consist of a sharp image of those parts of the object where the fringe pattern is in focus, but an out-of-focus blurred image of the rest of the object. In order to obtain an optically sectioned image, it is necessary to remove the blurred out-of-focus portion as well the fringe pattern from the in-focus optical section. There are many ways to do this — one of the simplest involves simple processing of three images taken at three different spatial positions of the fringe pattern. The out-of-focus regions remain fairly constant between these images and the relative spatial shift of the fringe pattern allows the three

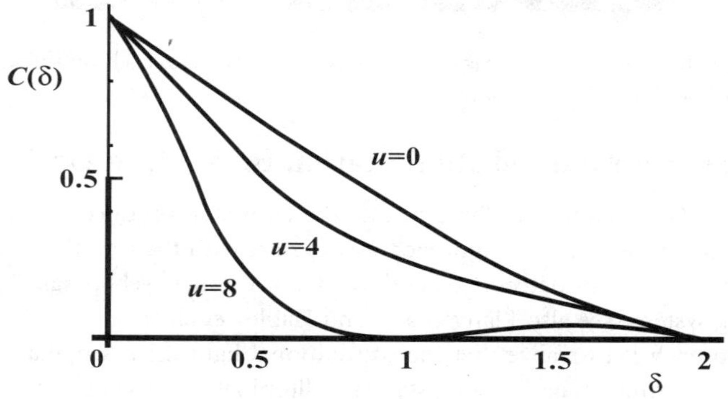


FIGURE 10.12 The transfer function for a conventional fluorescence microscope for a number of values of defocus (measured in optical units). The response for all non-zero spatial frequencies decays with defocus.

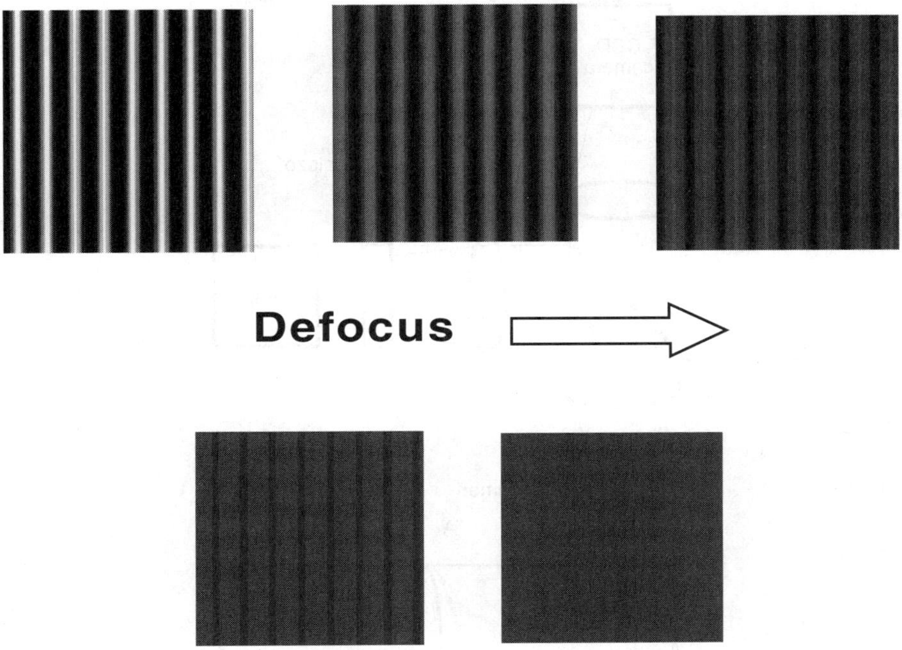


**FIGURE 10.13** The image of a one-dimensional single spatial frequency bar pattern for varying degrees of defocus. For sufficiently large values of defocus, the bar pattern is not imaged at all.

images to be combined in such a way as to remove the fringes. This permits us to retrieve an optically sectioned image as well as a conventional image in real time.[18]

The approach involves processing three conventional microscope images, so the image formation is fundamentally different from that of the confocal microscope. However, the depth discrimination or optical sectioning strength is very similar and this approach, which requires very minimal modifications to the instrument, has been used to produce high-quality three-dimensional images of volume objects directly comparable to those obtained with confocal microscopes. A schematic of the optical arrangement is shown in Figure 10.14A together with experimentally obtained axial responses in Figure 10.14B. These responses are substantially similar to those obtained in the true confocal case.

As an example of the kinds of images that can be obtained with this type of microscope, Figure 10.15 shows two images of a spiracle of a head louse. The first is an auto-focus image of greatly extended depth of field constructed from a through-focus series of images. The second is a conventional image taken at mid-focus. The dramatic increase in depth of field is clear when compared with a mid-focus conventional image. These images were taken using a standard microscope illuminator as light source. Indeed, the system is so light efficient that good quality optical sections of transistor specimens have been obtained using simply a candle as light source.

Imaging using fluorescence light is also possible using this technique. However, an alternative approach that does not require a physical grid is possible if a laser is used as the light source. In this system, the laser illumination is split into two beams that are allowed to interfere at an angle in the fluorescent specimen. This has the effect of directly "writing" a one-dimensional fringe pattern in the specimen. Spatial shift of the fringe pattern is achieved by varying the phase of one of the interfering beams. As before, three images are taken from which the optically sectioned image and the conventional image may be obtained.[19] The beauty of this approach is that no imaging optics are required at the illuminating wavelength and system alignment is trivial.

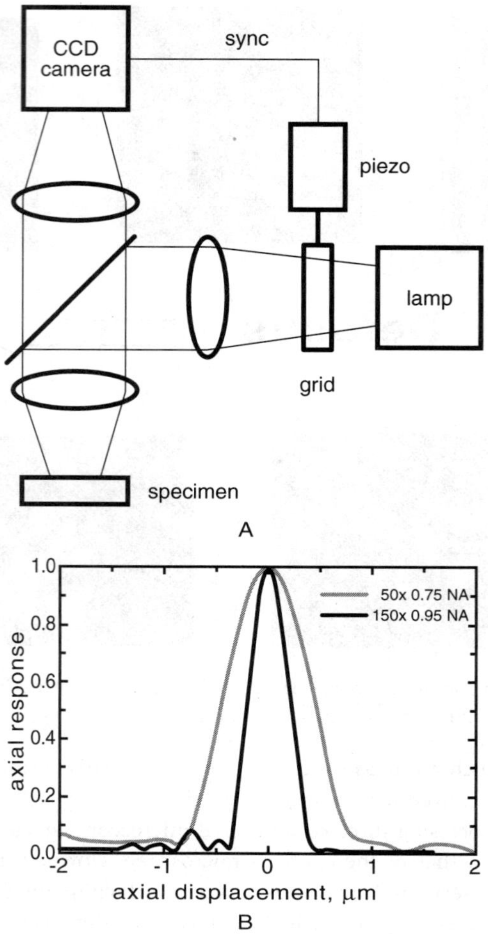


**FIGURE 10.14** (A) Optical system of the structured illumination microscope together with experimentally obtained axial responses in (B), which confirm the optical sectioning ability of the instrument.

## 10.6 Abberation Correction

As we have seen, the most important feature of the confocal microscope is its optical sectioning property, which permits volume structures to be rendered in three dimensions via a suitable stack of through-focus images. However, in order to achieve the highest performance, it is necessary that the optical resolution be the same at all depths within the optical section. Unfortunately, there are fundamental reasons why this cannot be achieved in general. One is that high aperture microscope objectives are often designed to give optimum performance when imaging features located just below the cover glass. Another is more fundamental and is due to specimen-induced refractive index mismatches. These could be caused by a variety of reasons, such as the use of an oil immersion objective to image into a watery specimen, or may be due to refractive index inhomogeneities within the specimen. In many cases they cannot be removed by system design and a new approach must be taken. An adaptive method of correction in which the aberrations are measured using a wavefront sensor and removed with an adaptive wavefront corrector, such as a deformable mirror, is required.

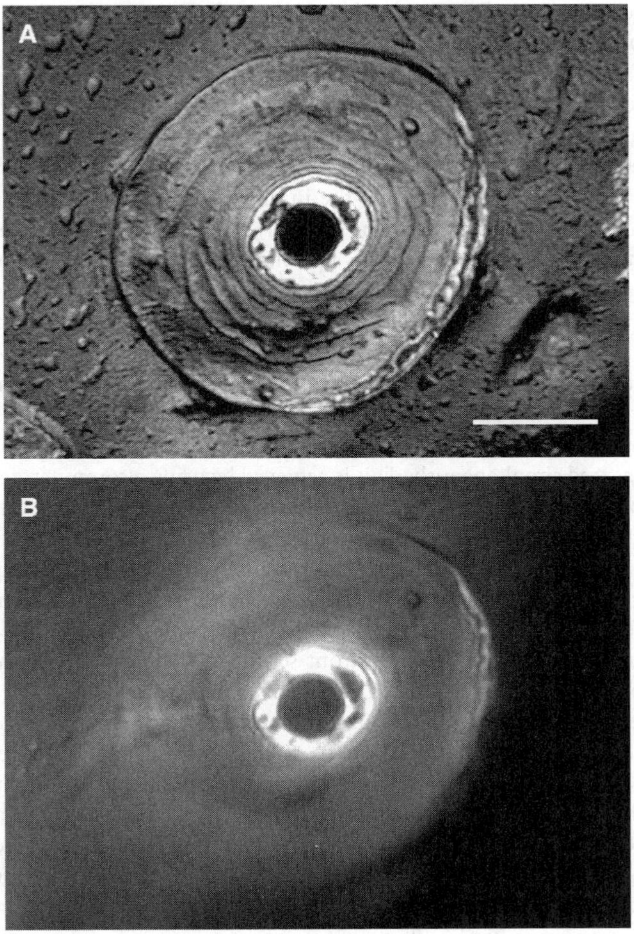


**FIGURE 10.15** Two images of the region around the spiracle of a head louse. As in the previous figure, (A) is an autofocus image and (B) shows a mid-focus conventional image. Scale bar = 10 μm.

A number of methods of wavefront sensing have already been developed for astronomical applications. Unfortunately these approaches are not necessarily appropriate for microscopy. Further, because the optical sectioning property of the confocal system limits the region of the specimen to be imaged to that lying close to the focal plane, it is clearly desirable that the wavefront sensor possess a similar optical sectioning property.

A very powerful method of describing a particular wavefront aberration is in terms of a superposition of aberration modes such as those described by the Zernike circle polynomials.[20] Mathematically speaking, the wavefront aberration is represented by a truncated series of Zernike polynomials (modes). These polynomials, which are two-dimensional functions, are orthogonal over the unit circle and hence are particularly useful in optics where apertures and lenses are typically circular in cross section. Their real power, however, lies in the fact that real-world aberrations can often be described by only a few low-order Zernike modes. We have shown, for example, that the aberrations introduced when focusing into a biological specimen are dominated by a few lower order modes.[21] It is therefore attractive to consider basing an adaptive system on Zernike polynomials since it will be possible to simplify the overall system design because the Zernike modes of interest could be measured directly. Equally importantly, because only a small number of modes need to be measured, the complexity of the wavefront sensor would be reduced.

The operation of the modal wavefront sensor is based upon the concept of "wavefront biasing." This process involves the combination of certain amounts of the Zernike aberration mode being measured (the "bias mode") with the input wavefront. Conceptually, this can be done by including appropriately shaped or etched glass plates in the optical beam path, although in practice other methods are used. First, the beam containing the input wavefront passes through a beamsplitter to create two identical beams. In the first beam, a positive amount of the bias mode is added to the wavefront, which is then focused by a lens onto a pinhole. In the second beam, an equal but opposite negative amount of the bias mode is added to the wavefront. This is again focused onto a pinhole. Behind each pinhole lies a photodetector that measures the optical power passing through the pinhole. The output signal of the sensor, which is taken to be the difference between the two photodetector signals, is found to be sensitive to the amount of the particular "bias" Zernike mode present in the input wavefront.[22]

One method of implementation of this wavefront sensor permits the entire Zernike modal content of a wavefront to be measured simultaneously. This is achieved by using a computer-generated binary phase hologram, which produces a diffraction pattern consisting of a number of spots created with the appropriate bias aberrations. A pair of spots in the diffraction pattern corresponds to the two spots produced in the positively and negatively biased paths for a given bias mode. An array of pinholes and photodetectors is positioned behind the diffraction pattern and, as before, the output signal for each mode is taken as difference between the detector signals from oppositely biased spots.

Another implementation of the modal wavefront sensor involves the sequential application of bias modes using an adaptive element. Adaptive optics systems use some form of adaptive wavefront shaping element in order to remove unwanted aberrations from the wavefront. Suitable devices include membrane or bimorph deformable mirrors and liquid crystal spatial light modulators. Effectively, the device flattens the aberrated wavefront by adding an equal but opposite aberration to it. Naturally, a device with such capabilities can also be used to apply the bias aberrations in a wavefront sensor sequentially. The sensor in this case comprises simply the biasing element, one lens, and a single pinhole/photodetector. The positive and negative bias aberrations for each mode are applied in turn and the corresponding detector signals are measured. Because the adaptive element applying the bias modes can be the same device used for correction of the wavefront, the resulting adaptive optics system can operate without a separate wavefront sensor. Considering also that only a single photodetector is required, this facilitates the design of a simple, low-cost adaptive optics system.

The practicability of this new approach has been demonstrated in a confocal microscope and has been used to measure and correct aberrations in the imaging of biological specimens.[23] Figure 10.16 shows the uncorrected and corrected *x-y* and *x-z* images of a specimen of fluorescently labeled mouse intestine.

## 10.7 Summary

We have discussed the origin of the optical sectioning property in the confocal microscope in order to introduce the range of imaging modes to which this unique form of microscope leads. A range of optical architectures has also been described. By far the most universal is that shown in Figure 10.1 where a confocal module is integrated around a conventional optical microscope. Other, more recent real-time implementations have also been described. A number of practical aspects of confocal microscopy have not been discussed. This is because they are readily available elsewhere, such as advice on the correct choice of detector pinhole size, or because they are still the focus of active research, such as the development of new contrast mechanisms for achieving enhanced three-dimensional resolution such as the stimulated emission depletion method (STED)[24] or 4-Pi.[25] These are still under development but offer great promise.

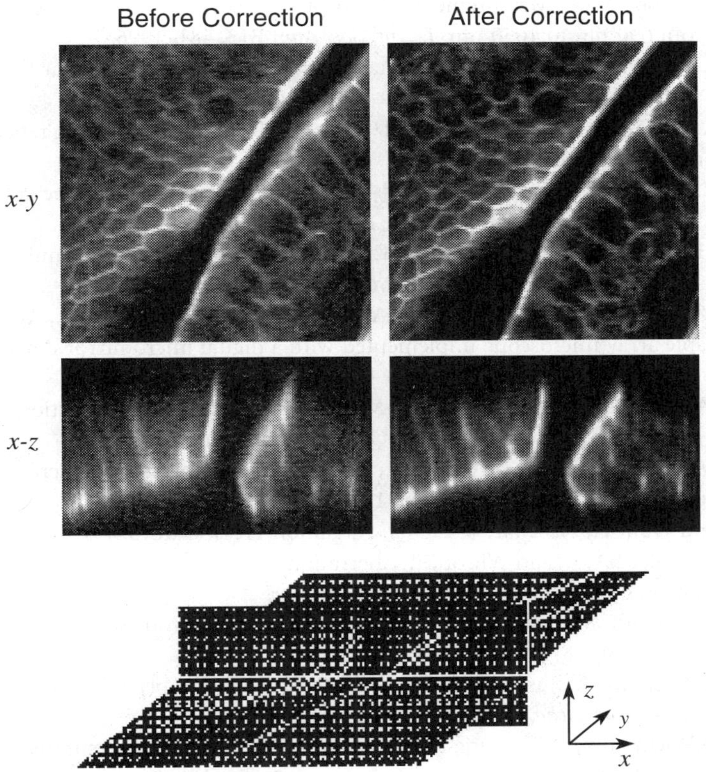


**FIGURE 10.16** Confocal microscope scans before and after aberration correction. *x-y* (lateral) and *x-z* (axial) scans of a fluorescently labeled section of mouse intestine specimen are shown. The schematic beneath illustrates the relative three-dimensional orientation of the *x-y* and *x-z* scans. The image dimensions in the *x* and *y* directions are 80 μm; the *z* dimension is 15 μm.

## References


1. Wilson, T. and Sheppard, C.J.R., *Theory and Practice of Scanning Optical Microscopy,* Academic Press, London, 1984.
2. Wilson, T., Ed., *Confocal Microscopy,* Academic Press, London, 1990.
3. Pawley, J.B., Ed., *Handbook of Biological Confocal Microscopy,* Plenum Press, New York, 1995.
4. Corle, T.R. and Kino, G.K., *Confocal Scanning Optical Microscopy and Related Imaging Systems*, Academic Press, New York, 1996.
5. Gu, M., *Principles of Three-Dimensional Imaging in Confocal Microscopes*, World Scientific, Singapore, 1996.
6. Masters, B.R., *Confocal Microscopy,* SPIE, Bellingham, WA, 1996.
7. Diaspro, A., *Confocal and Two-Photon Microscopy,* Wiley-Liss, New York, 2002.
8. Lukosz, W., Optical systems with resolving power exceeding the classical limit, *J. Opt. Soc. Am.*, 56, 1463, 1966.
9. Minsky, M., Microscopy Apparatus, U.S. patent 3,013,467, Dec. 19, 1961 (filed Nov. 7, 1957).
10. Denk, W., Strikler, J.H., and Webb, W.W., Two-photon laser scanning fluorescence microscopy, *Science*, 248, 73, 1990.
11. Göppert-Mayer, M., Über Elementarakte mit zwei Quantensprungen, *Ann. Phys.* (Leipzig), 5, 273, 1931.

12. Xu, C. and Webb, W.W., Measurement of two-photon cross-sections of molecular fluorophores with data from 690 nm to 1050 nm, *J. Opt. Soc. Am.*, B13, 481, 1996.
13. Xu, C. and Webb, W.W., Multiphoton excitation cross-sections of molecular fluorophores, *Bioimaging*, 4, 198, 1996.
14. Petran, M., Hadravsky, M., Egger, M.D. and Galambos, R., Tandem-scanning reflected-light microscopy, *J. Opt. Soc. Am.*, 58, 661, 1968.
15. Juskaitis, R., Wilson, T., Neil, M.A.A., and Kozubek, M., Efficient real-time confocal microscopy with white light sources, *Nature*, 383, 804, 1996.
16. Liang, M.H., Stehr, R.L., and Krause, A.W., Confocal pattern period in multi-aperture confocal imaging systems with coherent illumination, *Opt. Lett.*, 22, 751, 1997.
17. Hanley, Q.S., Verveer, P.J., Gemkow, M.J., Arndt-Jovin, D., and Jovin, T.M., An optical sectioning programmable array microscope implemented with a digital micromirror device, *J. Microsc.*, 196, 317, 1999.
18. Neil, M.A.A., Juskaitis, R., and Wilson, T., Method of obtaining optical sectioning by using structured light in a conventional microscope, *Opt. Lett.*, 22, 1905, 1997.
19. Neil, M.A.A., Juskaitis, R., and Wilson, T., Real time 3D fluorescence microscopy by two beam interference illumination, *Opt. Commun.*, 153, 1, 1998.
20. Born, M. and Wolf, E., *Principles of Optics*, Pergamon Press, Oxford, 1975.
21. Booth, M.J., Neil, M.A.A., and Wilson, T., Aberration correction for confocal imaging in refractive-index-mismatched media, *J. Microsc.*, 192, 90, 1998.
22. Neil, M.A.A., Booth, M.J., and Wilson, T., New modal wave-front sensor: a theoretical analysis, *J. Opt. Soc. Am.*, 17, 1098, 2000.
23. Booth, M.J, Neil, M.A.A., Juskaitis, R. and Wilson, T., Adaptive aberration correction in a confocal microscope, *Proc. Nat. Acad. Sci. U.S.A.*, 99, 5788, 2002.
24. Hell, S. and Wichmann, J., Breaking the diffraction resolution limit by stimulated emission: stimulated-emission-depeletion microscopy, *Opt. Lett.*, 19, 870, 1994.
25. Hell, S. and Stelzer, E.H.K., Fundamental resolution improvement with a 4Pi-confocal fluorescence microscope using two-photon excitation, *Opt. Commun.*, 93, 277, 1992.

# 11

# Two-Photon Excitation Fluorescence Microscopy




Peter T. C. So
*Massachusetts Institute of Technology*
*Cambridge, Massachusetts*

Chen Y. Dong
*Massachusetts Institute of Technology*
*Cambridge, Massachusetts*

Barry R. Masters
*University of Bern*
*Bern, Switzerland*


## 11.1 Introduction

Maria Göppert-Mayer first predicted the possibility of multiphoton excitation process in her doctoral dissertation presented in Göttingen.[1] Experimental verification of multiphoton processes was not realized until 1963 when Kaiser and Garret first observed two-photon excitation of $CaF_2$:$Eu^{2+}$ fluorescence.[2] Later, Kaiser and Garret further demonstrated two-photon excitation fluorescence from organic molecules. Three-photon excitation was subsequently reported by Singh and Bradley.[3] Multiphoton excitation of biochemical molecules provides complementary spectroscopic information to standard one-photon studies.[4–7]

The utilization of nonlinear optical processes to provide image contrast for microscopic studies was realized by pioneers such as Freund, Hellwarth, and Chirstensen.[8] Nonlinear optical imaging having inherent three-dimensioned (3D) resolution was first suggested by the Oxford group including Sheppard, Kompfner, Gannaway, and Wilson.[9,10] They predicted that optical emissions that have quadratic or higher-order dependence on the excitation power will be confined to the focal plane of the microscope objective. The utilization of this principle for 3D imaging was realized in the seminal work of Denk, Strickler, and Webb.[11] Denk and co-workers further demonstrated the importance of this approach for limiting photodamage and for extending penetration depth in biological specimens. This work revolutionized


0-8493-1116-0/03/$0.00+$1.50

the application of 3D microscopy in biomedical imaging. The applications of other nonlinear optical mechanisms in biological imaging, such as second and third harmonic generation, sum frequencies generation, and coherent anti-Stokes Raman scattering, soon followed.[12–14] This chapter focuses on multiphoton fluorescence microscopy, one of the most widely used nonlinear microscopy approaches, and its applications in biology and medicine.

The subject of multiphoton microscopy has been covered in a number of other excellent reviews providing different perspectives on this subject.[15–18] This chapter presents a discussion of the basic principle of multiphoton excitation and image formation, a review of two-photon instrumentation, and fluorescent probe choices. Because multiphoton microscopy is finding its most important applications in tissue imaging, a discussion on the strengths and limitations of this technology for deep tissue imaging is provided. This chapter concludes with a review of successful tissue imaging applications using two-photon microscopy.

## 11.2 Basic Principles of Multiphoton Excitation and Image Formation

### 11.2.1 The Physics of Multiphoton Excitation

The theory of two-photon excitation was predicted by Göppert-Mayer in 1931.[1] The basic physics of this phenomenon has also been described in a number of standard quantum mechanical texts.[19] Fluorescence excitation is an interaction between the fluorophore and an excitation electromagnetic field and is described by a time-dependent Schroedinger equation where the Hamiltonian contains an electric dipole interaction term: $\vec{E}_\gamma \cdot \vec{r}$, where $\vec{E}_\gamma$ is the electric field vector of the photons and $\vec{r}$ is the position operator. This equation can be solved by perturbation theory. The first-order solution corresponds to the one-photon excitation with transition probability $P$.

$$P \sim \left| \langle f | \vec{E}_\gamma \cdot \vec{r} | i \rangle \right|^2, \tag{11.1}$$

where $|i\rangle$ denotes the ground electronic state and $|f\rangle$ denotes the excited electronic state. The $n$-photon transitions are represented by the $n$th order solutions. In the case of two-photon excitation, the transition probability between the molecular ground state $|i\rangle$ and the excited state $|f\rangle$ is represented by

$$P \sim \left| \sum_m \frac{\langle f | \vec{E}_\gamma \cdot \vec{r} | m \rangle \langle m | \vec{E}_\gamma \cdot \vec{r} | i \rangle}{\varepsilon_\gamma - \varepsilon_m} \right|^2, \tag{11.2}$$

where $\varepsilon_\gamma$ is the photonic energy associated with the electric field vector $\vec{E}_\gamma$, and the summation is over all intermediate states $m$ with energy $\varepsilon_m$.

Therefore, two-photon excitation of molecules is a nonlinear process involving the absorption of two photons whose combined energy is sufficient to induce a molecular transition to an excited electronic state. Conventional one-photon technique uses ultraviolet or visible light to excite fluorescent molecules of interest. Excitation occurs when the absorbed photon energy matches the energy gap between the ground and excited state. The same transition can be excited by a two-photon process where two less energetic photons are simultaneously absorbed. Quantum mechanically, the first photon excites the molecule to a virtual intermediate state, and the molecule is eventually brought to the final excited state by the absorption of a second photon. A comparison of one- and two-photon absorption is shown in Figure 11.1. In the absence of major vibrational perturbations, inductive effects, or solvent relaxation, the probability of the transition depends on the magnitude of the overlap integrals between the ground,

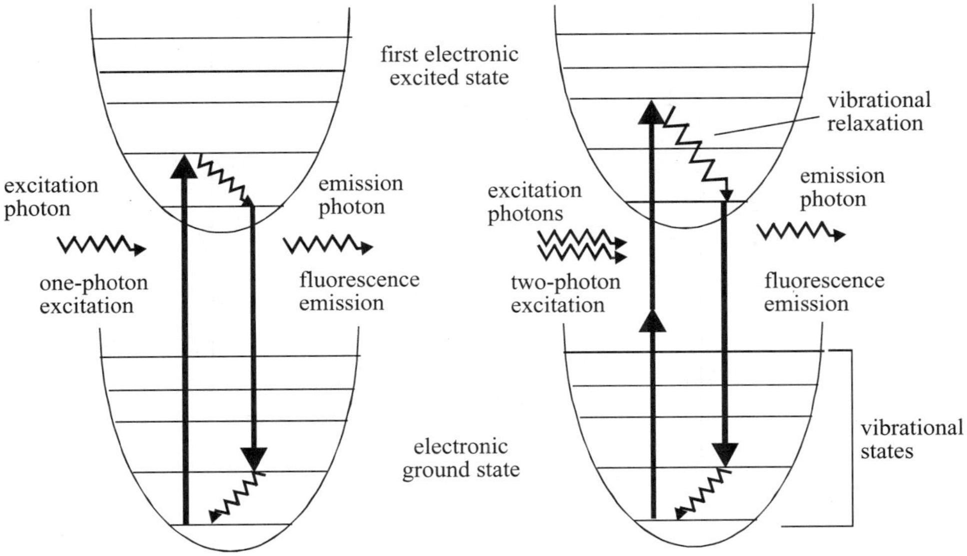


**FIGURE 11.1** Jablonski diagrams for one-photon (left) and two-photon (right) excitation. Excitations occur between the ground state and the vibrational levels of the first electronic excited state. After either excitation process, the fluorophore relaxes to the lowest energy level of the first excited electronic states via vibrational processes. The subsequent fluorescence emission process for both relaxation modes is the same.

intermediate, and excited states of a particular fluorophore. Choosing fluorophores with higher multiphoton cross sections allows more efficient excitation with lower photon dose.

A molecular state can be described by an eigenfunction with a definite parity. For a molecular state with even parity, there is no change in the sign of the wavefunction when there are sign changes in the spatial coordinates. For a molecular state with odd parity, the sign of the wavefunction changes when the sign changes in the spatial coordinates. Because the dipole operator has odd parity (i.e., absorbing one photon changes the parity of the state), the one-photon transition couples initial and final states with opposite parity. On the other hand, the two-photon moment only allows transition between two states with the same parity.[19,20] In general, an $n$-photon process couples states with equal parity, if $n$ is even, and opposite parity, if $n$ is odd.

Finally, it should be noted that while the absorption cross sections for one-, two-, and multiphoton excitation of a given fluorophore are different, the molecule rapidly relaxes to the same vibrational level in the excited electronic state. The excited state residence time (fluorescence lifetime) and the fluorescence decay processes depend only on the molecular structure and its microenvironment. Therefore, the fluorescence quantum yield and emission spectra are independent of the initial excitation process.

## 11.2.2 Imaging Properties of Two-Photon Microscopy

One of the most important attributes of two-photon microscopy is its inherent 3D sectioning capability. The sectioning capability of this method originates from the quadratic and higher-order dependence of the fluorescence signal upon the excitation intensity distribution.

Consider the intensity distribution at the focal point of a high numerical aperture objective. The spatial profile of the diffraction-limited focus for an objective with numerical aperture, NA = sin(α), is

$$I(u,v) = \left| 2\int_0^1 J_0(v\rho) e^{-\frac{i}{2}u\rho^2} \rho \, d\rho \right|^2, \tag{11.3}$$

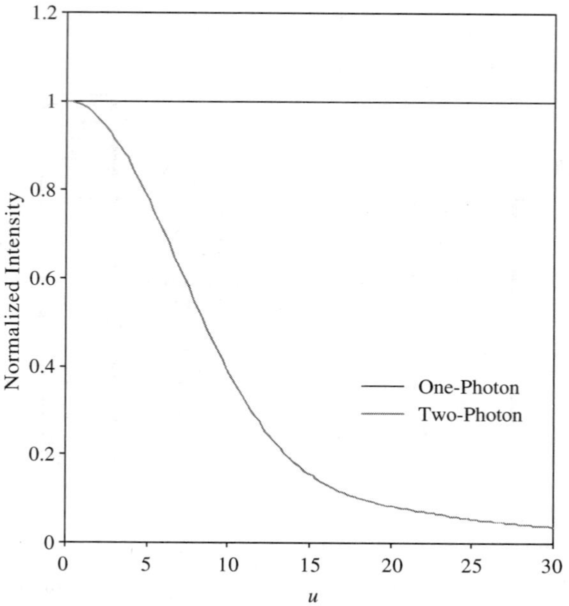


**FIGURE 11.2** Total fluorescence generated at a given *z*-plane is calculated. This quantity is plotted as a function of its distance from the focal plane. For one-photon excitation, equal fluorescence intensity is observed in all planes and there is no depth discrimination. For two-photon excitation, the integrated intensity decreases rapidly away from the focal plane. In this figure, *u* is a dimensionless optical axial coordinate as conventionally defined in Wilson, T. and Sheppard, C.J.R., *Theory and Practice of Scanning Optical Microscopy*, Academic Press, New York, 1984.

where $J_0$ is the 0th-order Bessel function, $\lambda$ is the wavelength of the excitation light, $u = 4k\sin^2(\alpha/2)z$ and $v = k\sin(\alpha)r$ are the respective dimensionless axial and radial coordinates normalized to wavenumber $k = 2\pi/\lambda$.[21,22] The point-spread function (PSF) of the fluorescence signal is a quadratic function of the excitation profile, which is just $I^2(u/2,v/2)$. In contrast, one-photon fluorescence PSF has a functional form of $I(u,v)$. For *n*-photon excitation, the fluorescence intensity profile is in general $I^n(u/n,v/n)$.

The difference of the fluorescence PSF resulting from one- and two-photon excitation is profound. The contribution of fluorescence signal from each axial plane can be computed for one- and two-photon excitation and is shown in Figure 11.2. Assuming negligible attenuation, the total fluorescence generated is equal at each axial plane for one-photon microscopy. In contrast, two-photon fluorescence falls off quickly from the focal plane resulting in the localization of the signal to the focal region. For higher-photon processes, the localization property is even more pronounced.

Qualitatively, the localization of multiphoton excitation can be understood by realizing that fluorescence generation is only appreciable at a region of high spatial photon density at the focal point of the microscope objective. The localization of the multiphoton excitation can be further appreciated by examining the dimensions of fluorescence PSF. Figure 11.3 illustrates the radial and axial PSF for two-photon microscopy at 960 nm. The two-photon photointeraction volume is on the order of 0.1 fl.

The axial depth discrimination of multiphoton microscopy greatly improves image contrast as compared with wide-field fluorescence microscopy, especially for thick specimens. This localization further reduces the photointeraction volume and greatly reduces photobleaching and photodamage in thick sample. Further, since photointeraction occurs in a femtoliter size volume, localized photochemical reactions can be initiated.[23]

An important advantage of multiphoton excitation is the use of infrared light to excite fluorophores with one-photon excitation in the ultraviolet and blue–green spectral range. For one-photon excitation, the excitation spectral band overlaps the fluorescence emission band. Excitation light cannot be eliminated

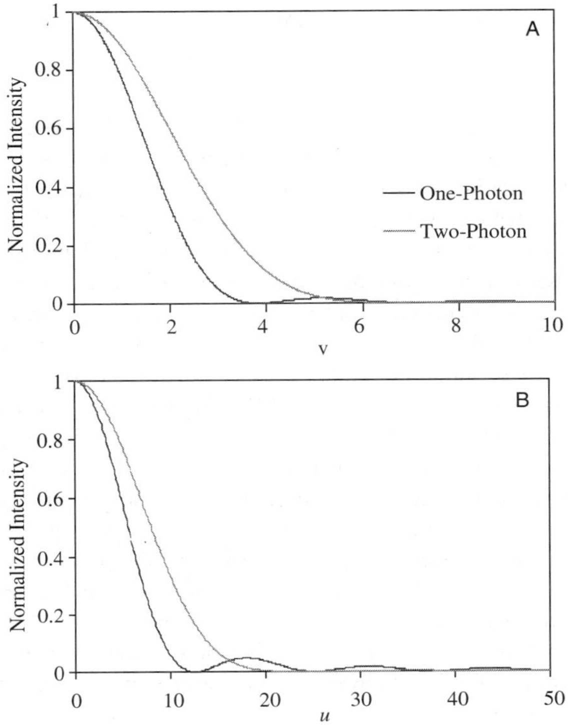


**FIGURE 11.3** A comparison of the one- and two-photon PSF in the (A) radial and (B) axial directions. In these figures, $v$ and $u$ are dimensionless optical coordinates along radial and axial directions as conventionally defined in Wilson, T. and Sheppard, C.J.R., *Theory and Practice of Scanning Optical Microscopy*, Academic Press, New York, 1984.)

without losing a fraction of fluorescence photons; microscope sensitivity is reduced. For two-photon excitation, the excitation spectrum is completely separated from the emission spectrum and very efficient filters can be used to eliminate the excitation with a minimal loss of the fluorescence. For weak fluorescence specimen in one-photon excitation, noise originated from Raman scattering is Stokes-shifted into the emission band and can sometimes obscure the fluorescence signal. In multiphoton excitation, the Raman signal is red-shifted from the excitation light and is further removed from the emission band. Hyper-Raman and hyper-Rayleigh scattering can still occur in the emission band for multiphoton excitation but these processes are typically weak.[24]

# 11.3 Experimental Considerations of Multiphoton Microscopy

## 11.3.1 Instrument Design of Multiphoton Microscopy

Multiphoton excitation efficiency increases with the spatial and temporal density of the excitation photons. Spatial localization of photons is relatively straightforward by using high-numerical-aperture optics. Temporal localization of photons is comparatively more difficult and was not efficiently achieved until the advent of femtosecond pulsed lasers. The typical cross section for two-photon absorption is on the order of 1 GM, which is equivalent to $10^{-50}$ $cm^4$ s, a unit adopted in honor of Göppert-Mayer. The need for temporal localization of photons using pulsed laser can be observed by considering the absorption efficiency of a two-photon probe with cross section, $\delta$. Two-photon absorption efficiency can be measured by $n_a$, the number of photons absorbed per fluorophore per pulse:

$$n_a \approx \frac{p_0^{\,2}\delta}{\tau_p f_p^{\,2}}\left(\frac{(\mathrm{NA})^2}{2\hbar c\lambda}\right)^2, \tag{11.4}$$

where $\tau_p$ is the pulse duration, $\lambda$ is the excitation wavelength, $P_0$ is the average laser intensity, $f_p$ is the laser's repetition rate, NA is the numerical aperture of the focusing objective, $\nu$ is Planck's constant, and $c$ is the speed of light.[11] Equation 11.4 shows that, for the same average laser power and repetition frequency, the excitation probability is increased by increasing the NA of the focusing lens, which increases spatial localization of the laser excitation. The absorption efficiency further improves linearly by reducing the pulse width of the laser indicating the importance of temporal localization.

For multiphoton excitation, femtosecond, picosecond, and continuous wave laser sources have been used. The most commonly used laser for multiphoton microscopy is femtosecond titanium-sapphire lasers. These lasers characteristically produce 100-fs pulse train at 100 MHz repetition rate. The tuning range of Ti:Sapphire systems extends from 680 to 1050 nm. Cr:LiSAF and pulse compressed Nd:YLF lasers are some of the other femtosecond lasers used in multiphoton microscopy.[25]

Multiphoton excitation can also be generated using picosecond light sources, although at a lower excitation efficiency. Two-photon excitation has been achieved using mode-locked Nd:YAG (~100 ps), picosecond Ti:Sapphire lasers, and pulsed dye lasers (~1 ps). Two-photon excitation using continuous wave lasers has also been demonstrated using ArKr lasers and Nd:YAG lasers.[26]

Comparing different laser sources, for fixed $\delta$, $P_0$, and NA, Equation 11.4 implies that the difference in excitation efficiency per unit time for pulsed and continuous wave lasers for a two-photon process is $\sqrt{f_p \tau_p}$. This factor is typically 300 for femtosecond Ti:Sapphire laser; two-photon excitation of a fluorophore that typically requires 1 mW of power using a femtosecond Ti:Sapphire laser will require 300 mW using a continuous wave laser.

The generalization of Equation 11.4 to the multiphoton case can be expressed as

$$n_a \propto \frac{1}{\tau_p^{\,n-1}}. \tag{11.5}$$

One can conclude that multiphoton excitation becomes increasingly difficult for nonfemtosecond light sources with increasing order of excitation process.

Because multiphoton excitation generates signal from a single point, the production of 3D images requires raster scanning of this excitation volume in 3D (Figure 11.4). In a typical multiphoton microscope, *x-y* raster scanning uses galvanometer-driven scanner. After beam power control and pulse width compensation optics, the *x-y* scanner deflects the excitation light into a fluorescence microscope.

A typical multiphoton microscope uses epi-luminescence geometry. The scan lens is positioned such that the *x-y* scanner is at its eye-point while the field aperture plane is at its focal point. For a telecentric microscope, this arrangement ensures that angular scanning of the excitation light is converted to linear translation of the objective focal point. A tube lens is positioned to re-collimate the excitation light directed toward the infinity-corrected objective via a dichroic mirror. The scan lens and the tube lens function together as a beam expander that overfills the back aperture of the objective lens to ensure that excitation light is diffraction limited. Typically, high numerical aperture objectives are used to maximize excitation efficiency. An objective positioner translates the focal point axially for 3D imaging.

In the emission path, the fluorescence emission is collected by the imaging objective and is transmitted through the dichroic mirror. Additional barrier filters are used to further attenuate the scattered excitation light and to select the emission band of interest. The fluorescence signal is subsequently directed toward the detector system. A number of photodetectors have been used in two-photon microscope including photomultiplier tubes (PMT), avalanche photodiodes, and charge coupled device (CCD) cameras. PMTs are the most common implementation because they are robust and low cost, and have a large active area and relatively good sensitivity. PMT and avalanche photodiode systems further allow the use of ultralow

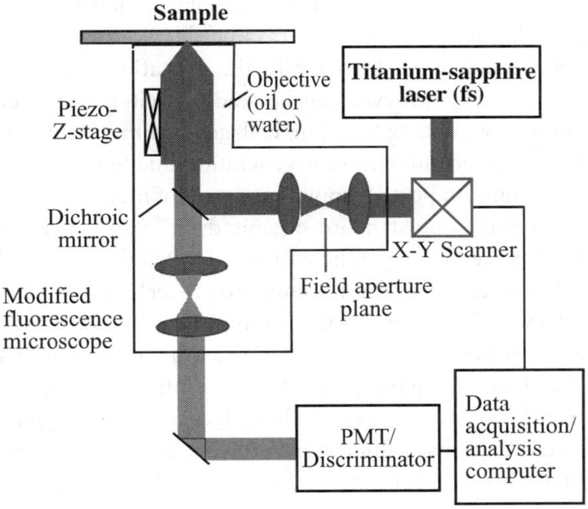


**FIGURE 11.4** Schematic of a typical two-photon microscope.

noise and high-sensitivity, single-photon counting electronic circuitry. Electronic signal from the detector is converted to digital signal and is recorded by the data acquisition computer. The control computer further permits archival of the image data and 3D rendering of these images.

### 11.3.2 Two-Photon Fluorescent Probes and Their Biological Applications

The design of a multiphoton imaging experiment requires not only high-sensitivity optical instrumentation but also specific and efficient fluorophores. Most fluorophores can be excited in two-photon mode at approximately twice their one-photon absorption wavelength; similarly, $n$-photon excitation of fluorophores typically occurs at approximately $n$ times their one-photon excitation wavelength. However, one- and multiphoton absorption processes actually have very different quantum mechanical selection rules. For example, a fluorophore's two-photon excitation spectrum scaled to half the wavelength is in general not equivalent to its one-photon excitation spectrum. Determining the multiphoton absorption spectra of fluorophores is required to optimize multiphoton imaging. Significant progress in this area has been made.[27–29]

Multiphoton imaging can be based on both endogenous and exogenous fluorophores. On the one hand, endogenous fluorophores make multiphoton imaging easier by alleviating the difficulty of *in vivo* labeling of biological specimens, especially in thick tissues. On the other hand, multiphoton imaging based on endogenous fluorophores is difficult because endogenous probes typically have low extinction coefficients and low quantum efficiencies.

Considering imaging based on endogenous probes, most proteins are fluorescent due to the presence of tryptophan and tyrosine. Two-photon induced fluorescence from tryptophan and tyrosine in proteins has been studied extensively.[30–32] Although these fluorophores are occasionally used for multiphoton imaging, the application of these amino acid probes is not common due to the lack of efficient laser source for two-photon excitation of these probes. Endogenous β-nicotinamide-adenine dinucleotide phosphate (NAD(P)H) is the most common endogenous fluorophore used for *in vivo* cellular imaging.[32] The production of NAD(P)H is associated with the cellular metabolism and the intracellular redox state.[33] Flavoproteins are another endogenous marker in cellular cytoplasm and are present almost exclusively in the mitochondria. The multiphoton excitation spectrum of NAD(P)H and flavin mononucleotide (FMN) has been determined.[28]

Endogenous fluorophores are also present in the extracellular matrix; the primary fluorophores are collagen and elastin. Tropocollagen and collagen fibers typically have absorption spectra similar to amino acids; their fluorescent emission is typically weak in the excitation wavelength range of Ti:Sapphire lasers.[34–36] Elastin fiber fluorescence, however, can be readily excited and visualized using two-photon excitation.[37–39] While fluorescence imaging based on collagen is difficult, recent studies have shown that collagen can be imaged based on second harmonic generation due to its chiral crystalline structure.[40–46]

Tissue imaging is one of the most important application areas of multiphoton microscopy. While cellular systems can be effectively labeled using standard organic dyes,[27–29] effective uniform labeling of tissues with exogenous organic probes is extremely difficult due to limited diffusion and differential partitioning in 3D tissue components. The invention of fluorescence protein technology presents a powerful molecular biology solution to this dilemma.[47] By transfecting or inflecting cells and tissues with fluorescent protein vectors, one can monitor the activation of a particular gene, the trafficking of specific proteins, and the changes in local cellular biochemical environment. After the introduction of green fluorescent proteins, fluorescence proteins of a wide range of colors including blue, cyan, yellow, and red have been created.[48,49] Two-photon absorption spectra for many fluorescent proteins have been determined.[50,51]

While many existing one-photon fluorophores can be used for multiphoton imaging, they are not necessarily optimized. Optimizing two-photon absorption properties in fluorophores has two important consequences. The development of efficient fluorophores can reduce the excitation laser intensity required for imaging, and thus can reduce system cost and specimen photodamage. Alternatively, with high two-photon cross sections, significant excitation can be achieved with the more economical continuous wave lasers, thus reducing the cost of two-photon systems that typically use femtosecond Ti:Sapphire lasers. Recent searches for molecules with high two-photon absorption cross sections have led to the synthesis of molecules with two-photon cross sections over 1000 GM.[52]

## 11.4 Optimization of Multiphoton Microscopy for Deep Tissue Imaging

Deep tissue imaging is an area in which multiphoton microscopy has found the most important and diverse applications. An understanding of how tissue optics affects light transmission and image formation properties is essential to optimize the use of multiphoton microscopy for deep tissue imaging. It is also important to consider photodamage mechanisms in tissues. Finally, some of the most promising deep tissue imaging applications of multiphoton microscopy are reviewed here.

### 11.4.1 Effect of Tissue Optical Properties on Multiphoton Microscopy Efficiency and Image Formation

On one level, light propagation through tissues can be characterized by multiple scattering of light in a homogeneous medium. In the multiple scattering regime, the modeling light propagation requires a knowledge of the scattering and absorption coefficients of the tissue. The scattering coefficient measures the propensity of the tissue to deflect photons from their path. The absorption coefficient measures the propensity of the tissue to absorb photons. To quantify scattering, in addition to the scattering coefficient, it is also important to quantify the directional change of the photon after scattering. The *g*-factor measures the cosine of the average angular change of the photon direction after scattering. Assuming small scattering and absorption coefficients, the light attenuation through a thickness of tissue can be modeled as an exponential function:

$$I(z) = I_0 e^{-[(1-g)\mu_s+\mu_a]z}, \quad (11.6)$$

where $I(z)$ is the intensity at depth $z$, $I_0$ is the intensity at tissue surface, $\mu_s$ is the scattering coefficient, $\mu_a$ is the absorption coefficient, and $g$ is the *g*-factor. The scattering coefficient, absorption coefficient, and the *g*-factor for typical tissues are 20 to 200 $mm^{-1}$, 0.1 to 1 $mm^{-1}$, and 0.7 to 0.9, respectively. Since the fluorescence signal depends quadratically on the excitation power, the fluorescence signal has a mean free path of 25 to 250 μm.

In homogeneous media, these phenomenological parameters fairly well characterize the microscopic light–tissue interactions. However, real tissues are not homogeneous and mesoscopic variations in tissue optical properties can also have dramatic effects in light propagation. For example, the dermal system consists of multiple structural layers, each with very distinct scattering coefficients, absorption coefficients, and *g*-factors. Further, these layers also have very different indices of refraction. An understanding of light propagation in tissues will require knowing the geometries of these mesoscopic inhomogeneities.

Effective multiphoton imaging of deep tissue requires efficient transmission of excitation light to the focal volume inside thick tissues and the efficient detection of the fluorescence signal generated. The imaging depth of two-photon microscopy can be limited by the efficacy of delivering femtosecond laser pulses into tissues. Assuming that the image point spread function is invariant with depth, excitation light penetration is limited by power delivery and pulse broadening. In terms of power delivery, average laser power as a function of depth is an exponential function of depth and is governed by Equation 11.6. Similar to the attenuation of the excitation light, the fluorescence signal generated is also reduced due to scattering and absorption effects. In fact, because scattering and absorption efficiencies are decreasing functions of wavelength, one expects attenuation of the fluorescence signal by the surrounding tissue to be more severe than the exciting light.

Compared with confocal microscopy, multiphoton microscopy is more efficient in signal detection inside multiple scattering medium. Because the mean free path of visible photons is less than 100 μm in typical tissues, the emitted photon will encounter one or more scattering events in thick tissues. However, many of these scattered photons can be collected because the *g*-factors of most tissue are close to unity (primarily forward scattering). In confocal microscopy, these scattered photons are lost at the confocal aperture. In two-photon microscopy, as long as the trajectories of these scattered photons remain within the collection solid angle of the objective lens, these photons are retained using a large area detector such as a PMT tube. Because these scattered photons have more divergent light paths, detection efficiency can be improved by increasing the solid angles of all optics in the emission light path and the physical size of the detector.[53]

From Equation 11.5, one concludes that efficient multiphoton excitation requires ultrafast laser pulses. Laser pulses from Ti:Sapphire lasers, which are typically on the order of 150 fs, can be broadened in the sample, reducing the multiphoton excitation efficiency. This degradation is particularly important for higher-order multiphoton processes. Laser pulse dispersion in microscope objectives is significant and pulse width can easily broaden to beyond 500 fs. The effect of pulse broadening in the objective can be negated by the use of dispersion compensation optics to maintain a short pulse width. However, further pulse dispersion may occur in the tissues due to scattering and index of refraction heterogeneity. The severity of this broadening effect as femtosecond pulses travel through tissues has not been extensively studied.

In addition to the reduction in excitation and detection efficiency, light propagation through tissue may also result in contrast and resolution losses. While signal strength reduction can be compensated for by increasing laser power, as long as photodamage and photobleaching are not too severe, contrast and resolution loss can result in irreversible loss of image information content. The major factors that can be responsible for resolution degradation are the scattering of the excitation and emission light, and spherical aberration. Spherical aberration is caused by mismatch in the indices of refraction between the sample and the immersion medium of the objective.

Scattering of the excitation and emission photons from the femtoliter excitation volume may lead to a degradation of the image resolution. To quantify this factor, Dunn et al.[54] pioneered the study of the scattering effects on resolution degradation. Using Monte Carlo simulations, they showed that the loss of signal intensity was due more to the scattering of the excitation photons than to the emission

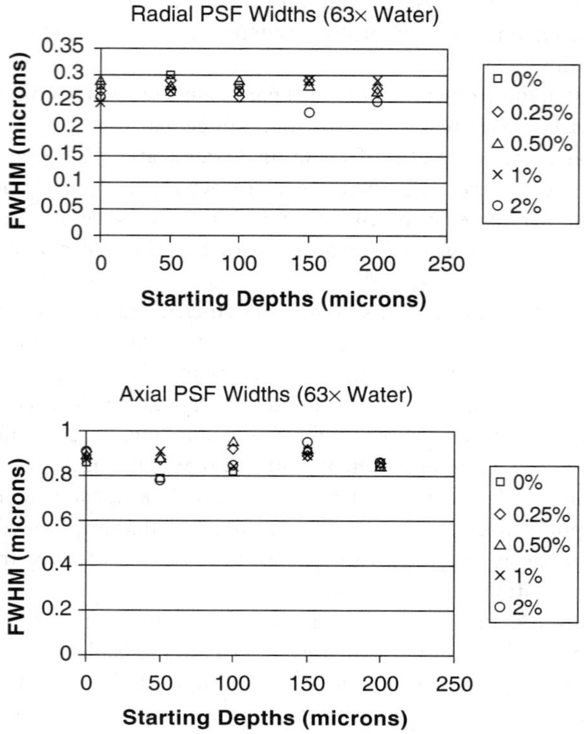


**FIGURE 11.5** Averaged widths of the radial and axial PSFs measured in multiple scattering media with different scattering coefficients containing Liposyn III at concentrations of 0, 0.25, 0.5, 1, and 2%. Data were taken with a Zeiss 63X, water-immersion, C-Apochromat objective (NA 1.2).

photons. More importantly, this group further found that the signal strength is the limiting factor in two-photon microscope image quality. Their simulation showed that resolution did not change as the image depth was increased as measured by the full-width-at-half-maximum (FWHM) of the image PSF. Their simulation was further supported by experimental measurements and that is in agreement with Centonze et al.[55] Dong and co-workers (manuscript in preparation) further measure two-photon microscopy PSF directly by imaging 100-nm spheres in agarose gels with varying scattering coefficient. It is found that scattering coefficients up to physiological range have minimal or no effect on the lateral and the axial FWHM of the PSF for both oil and water immersion objectives (Figure 11.5).

Considering the effect of spherical aberration, Dong et al. showed that the water immersion objective suffers no significant PSF degradation in thick samples. On the other hand, the oil immersion objective suffered significant degradation due to the spherical aberration introduced by index of refraction mismatch. The lateral and axial FWHM were broadened. One may conclude that, for typical tissue thickness on the order of a few hundred microns, resolution loss due to scattering is insignificant, but index refraction mismatch can be a major factor. The use of water-immersion objectives is preferred in biological tissues in general, given better index matching.

However, Dong et al. also studied the performance of oil- vs. water-immersion objectives for imaging more optically heterogeneous biological tissues such as excised human skin. Human skin is an optically heterogeneous layered structure with significantly different index of refraction in each layer. It is found that there is significant spherical aberration in the imaging of dermal structures for both oil- and water-immersion objectives. These objectives perform equivalently because neither can match the varying index of refraction in the skin and both suffer from spherical aberration. For optically heterogeneous samples, it is best to choose an immersion media that best matches the average index of the sample. For dermal structure, a glycerine-immersion objective may potentially outperform oil- or water-immersion lenses.

### 11.4.2 Photodamage Mechanisms in Tissues

Multiphoton microscopy is particularly suitable for *in vivo* tissue imaging due to its minimization of specimen photodamage. The region of photodamage is restricted to within the axial depth of the PSF. For specimens with thickness comparable to the thickness of the PSF, the use of multiphoton imaging does not significantly improve specimen viability. However, the decrease in photodamage is substantial for 3D imaging of thick tissues. Photodamage is approximately reduced by a factor equal to the ratio of the sample thickness to the axial dimension of the PSF. For 3D imaging of a 100-μm-thick specimen using multiphoton microscopy, the specimen is exposed to a light dosage about 100 times less than confocal microscopy.

A number of embryology studies demonstrate well the minimal invasive nature of multiphoton microscopy. Long-term monitoring of *Caenorhabditis elegans* and hamster embryos using confocal microscopy have not been successful due to photodamage-induced developmental arrest. However, these developing embryos can be repeatedly imaged using multiphoton microscopy for hours without appreciable damage.[56–58] Further, hamster embryos that were reimplanted after imaging experiments were successfully developed into normal adults.

One should note that multiphoton microscopy can still cause considerable photodamage in the focal volume where photochemical interaction occurs. Multiphoton photodamage proceeds through three main mechanisms:

1. Oxidative photodamage can be caused by two- or higher-photon excitation of endogenous and exogenous fluorophores similar to ultraviolet exposure. Tissue fluorophores act as photosensitizers in the photo-oxidative process.[59,60] Photoactivation of these fluorophores results in the formation of reactive oxygen species that trigger the subsequent biochemical damage cascade in cells. Current studies found that the degree of photodamage follows a quadratic dependence on excitation power indicating that the two-photon process is the primary damage mechanism.[61–65] Experiments have also been performed to measure the effect of laser pulse width on cell viability. Results indicate that the degree of photodamage is proportional to two-photon excited fluorescence generation independent of pulse width; therefore, using shorter pulse width for more efficient two-photon excitation also produces greater photodamage.[62,64] Flavin-containing oxidases have been identified as one of the primary endogenous targets for photodamage.[65]
2. One- and multiphoton absorption of high-power infrared radiation can also produce thermal damage. The thermal effect resulting from two-photon absorption by water has been estimated to be on the order of 1 mK for typical excitation power.[16,66] Therefore, heating damage due to multiphoton water absorption is insignificant. However, there can be appreciable heating due to one-photon absorption in the presence of a strong infrared absorber such as melanin;[67,68] thermal damage has been observed in the basal layer of human skin.[69]
3. Photodamage may also be caused by mechanisms resulting from the high peak power of the femtosecond laser pulses. There are indications that dielectric breakdown occasionally occurs.[61]

### 11.4.3 Tissue-Level Applications of Two-Photon Microscopy

Two-photon microscopy is well suited for imaging in highly scattering specimens. A recent comparison study has convincingly demonstrated that two-photon microscopy is a superior method for this application.[55]

Multiphoton tissue imaging has been successfully applied to study physiology of many tissue types such as the cornea structure of rabbit eyes,[69,70] the light-induced calcium signals in salamander retina,[71] the toxin effect on human intestinal mucosa,[72] and the metabolic processes of pancreatic islets.[73,74] The combination of multiphoton imaging and magnetic resonance imaging was used to investigate the

**FIGURE 11.6 (Color figure follows p. 28-30.)** Three-dimensional reconstructed two-photon images of dermal structures in a mouse ear tissue specimen. The three images show distinct structural layers: (A) epidermal keratinocytes; (B) basal cells; and (C) collagen/elastin fibers.

heterogeneous microscopic structure of mammalian tongue tissue to obtain spatial information on two different length scales.[75]

Today, multiphoton microscopy is widely used in three areas: neurobiology,[76] embryology, and dermatology. In neurobiology studies, multiphoton microscopy has been applied to study the neuron structure and function in intact brain slices.[77] The role of calcium signaling in dendritic spine function,[78–86] neuronal plasticity and the associating cellular morphological changes,[87] and hemodynamics in rat neocortex.[88] Hyman and co-workers further applied this technology to study the formation of β-amyloid plaques associated with Alzheimer's disease.[89,90] In embryology studies, multiphoton imaging has been used to examine calcium passage during sperm–egg fusion,[91] the origin of bilateral axis in sea urchin embryos,[92] cell fusion events in *C. elegans* hypodermis,[56,57] and hamster embryo development.[58]

Multiphoton microscopy has been used extensively in dermatology studies. Masters et al.[93] employed multiphotons to image the autofluorescence of *in vivo* human skin down to a depth of 200 μm. Cellular strata in the epidermis, including the corneum, spinosum, and basal layer, can be clearly resolved based on NAD(P)H fluorescence in the cytoplasm. NAD(P)H fluorescence is seen to be more pronounced in the basal layer due to its relatively higher metabolic activity. Below the epidermis, the dermal structure is visible due to elastin fluorescence and second harmonic signal generated from the collagen matrix (Figure 11.6). Recently, two-photon microscopy has been further applied to study the transport properties of the skin,[94,95] facilitating the development of transdermal drug delivery technology.

## 11.5 Conclusion

Multiphoton fluorescence microscopy is becoming one of the most important recent inventions in biological imaging. This technology enables noninvasive study of biological specimens in three dimensions with submicron resolution. Two-photon excitation of fluorophores results from the simultaneous absorption of two photons. This excitation process has a number of advantages such as reduced specimen photodamage and enhanced penetration depth. It also produces higher-contrast images and is a novel method to trigger localized photochemical reactions. As the technology of multiphoton microscopy reaches maturity and becomes commercially available, the range of applications in biology and medicine is rapidly increasing. The suitability of multiphoton microscopy for studying the physiology of a variety of tissues has been clearly demonstrated. In the near future, other promising emerging applications of multiphoton microscopy include the use of multiphoton excitation for single molecular analysis for pharmaceutical drug discovery and the use of multiphoton processes for nanosurgery and nanoprocessing in cells and tissues.[96,97]

## Acknowledgments

P.T.C. So acknowledges kind support from National Science Foundation MCB-9604382, National Institutes of Health R29GM56486-01, and American Cancer Society RPG-98-058-01-CCE. C.Y. Dong acknowledges fellowship support from National Institutes of Health 5F32CA75736-02.

### References

1. Göppert-Mayer, M., Über Elementarakte mit zwei Quantensprungen, *Ann. Phys.* (Leipzig), 5, 273, 1931.
2. Kaiser, W. and Garrett, C.G.B., Two-photon excitation in $CaF_2$:$Eu^{2+}$, *Phys. Rev. Lett.*, 7, 229, 1961.
3. Singh, S. and Bradley, L.T., Three-photon absorption in naphthalene crystals by laser excitation, *Phys. Rev. Lett.*, 12, 162, 1964.
4. McClain, W.M., Excited state symmetry assignment through polarized two-photon absorption studies in fluids, *J. Chem. Phys.*, 55, 2789, 1971.
5. Friedrich, D.M. and McClain, W.M., Two-photon molecular electronic spectroscopy, *Annu. Rev. Phys. Chem.*, 31, 559, 1980.
6. Friedrich, D.M., Two-photon molecular spectroscopy, *J. Chem. Educ.*, 59, 472, 1982.
7. Birge, R.R., Two-photon spectroscopy of protein-bound fluorophores, *Acc. Chem. Res.*, 19, 138, 1986.
8. Hellwarth, R. and Christensen, P., Nonlinear optical microscopic examination of structures in polycrystaline ZnSe, *Opt. Commun.*, 12, 318, 1974.
9. Sheppard, C.J.R., Kompfner, R., Gannaway, J., and Walsh, D., The scanning harmonic optical microscope, in *IEEE/OSA Conference on Laser Engineering and Applications*, Washington, D.C., 1977, p. 100D.
10. Gannaway, J.N. and Sheppard, C.J.R., Second harmonic imaging in the scanning optical microscope, *Opt. Quantum Electron.*, 10, 435, 1978.
11. Denk, W., Strickler, J.H., and Webb, W.W., Two-photon laser scanning fluorescence microscopy, *Science,* 248, 73, 1990.
12. Barad, Y., Eisenberg, H., Horowitz, M., and Silberberg, Y., Nonlinear scanning laser microscopy by third harmonic generation, *Appl. Phys. Lett.*, 70, 922, 1997.
13. Squier, J.A., Muller, M., Brakenhoff, G.J., and Wilson, K.R., Third harmonic generation microscopy, *Opt. Express,* 3, 315, 1998.
14. Campagnola, P.J., Wei, M.-D., Lewis, A., and Loew, L.M., High-resolution nonlinear optical imaging of live cells by second harmonic generation, *Biophys. J.*, 77, 3341, 1999.
15. Williams, R.M., Piston, D.W., and Webb, W.W., Two-photon molecular excitation provides intrinsic 3-dimensional resolution for laser-based microscopy and microphotochemistry, *FASEB J.*, 8, 804, 1994.
16. Denk, W.J., Piston, D.W., and Webb, W.W., Two-photon molecular excitation laser-scanning microscopy, in *Handbook of Biological Confocal Microscopy,* 2nd ed., Pawley, J.B., Ed., Plenum Press, New York, 1995, p. 445.
17. Hell, S.W., Nonlinear optical microscopy, *Bioimaging,* 4, 121, 1996.
18. So, P.T., Dong, C.Y., Masters, B.R., and Berland, K.M., Two-photon excitation fluorescence microscopy, *Annu. Rev. Biomed. Eng.*, 2, 399, 2000.
19. Baym, G., *Lectures on Quantum Mechanics,* Benjamin Cummings, Menlo Park, CA, 1973.
20. Callis, P.R., The theory of two-photon induced fluorescence anisotropy, in *Nonlinear and Two-Photon-Induced Fluorescence,* Lakowicz, J., Ed., Plenum Press, New York, 1997, p. 1.
21. Sheppard, C.J.R. and Gu, M., Image formation in two-photon fluorescence microscope, *Optik,* 86, 104, 1990.

22. Gu, M. and Sheppard, C.J.R., Comparison of three-dimensional imaging properties between two-photon and single-photon fluorescence microscopy, *J. Microsc.*, 177, 128, 1995.
23. Denk, W., Two-photon scanning photochemical microscopy: mapping ligand-gated ion channel distributions, *Proc. Natl. Acad. Sci. U.S.A.*, 91, 6629, 1994.
24. Xu, C., Shear, J.B., and Webb, W.W., Hyper-Rayleigh and hyper-Raman scattering background of liquid water in two-photon excited fluorescence detection, *Anal. Chem.*, 69, 1285, 1997.
25. Wokosin, D.L., Centonze, V.E., White, J., Armstrong, D., Robertson, G., and Ferguson, A.I., All-solid-state ultrafast lasers facilitate multiphoton excitation fluorescence imaging, *IEEE J. Sel. Top. Quantum Electron.*, 2, 1051, 1996.
26. Hell, S.W., Booth, M., and Wilms, S., Two-photon near- and far-field fluorescence microscopy with continuous-wave excitation, *Opt. Lett.*, 23, 1238, 1998.
27. Xu, C., Guild, J., Webb, W.W., and Denk, W., Determination of absolute two-photon excitation cross sections by *in situ* second-order autocorrelation, *Opt. Lett.*, 20, 2372, 1995.
28. Xu, C. and Webb, W.W., Measurement of two-photon excitation cross sections of molecular fluorophores with data from 690 to 1050 nm, *J. Opt. Soc. Am.*, B13, 481, 1996.
29. Xu, C., Zipfel, W., Shear, J.B., Williams, R.M., and Webb, W.W., Multiphoton fluorescence excitation: new spectral windows for biological nonlinear microscopy, *Proc. Natl. Acad. Sci. U.S.A.*, 93, 10763, 1996.
30. Lakowicz, J.R. and Gryczynski, I., Tryptophan fluorescence intensity and anisotropy decays of human serum albumin resulting from one-photon and two-photon excitation, *Biophys. Chem.*, 45, 1, 1992.
31. Lakowicz, J.R., Kierdaszuk, B., Callis, P., Malak, H., and Gryczynski, I., Fluorescence anisotropy of tyrosine using one- and two-photon excitation, *Biophys. Chem.*, 56, 263, 1995.
32. Kierdaszuk, B., Malak, H., Gryczynski, I., Callis, P., and Lakowicz, J.R., Fluorescence of reduced nicotinamides using one- and two-photon excitation, *Biophys. Chem.*, 62, 1, 1996.
33. Masters, B. and Chance, B., Redox confocal imaging: intrinsic fluorescent probes of cellular metabolism, in *Fluorescent and Luminescent Probes for Biological Activity*, Mason, W.T., Ed., Academic Press, London, 1999, p. 361.
34. LaBella, F.S. and Gerald, P., Structure of collagen from human tendon as influence by age and sex, *J. Gerontol.*, 20, 54, 1965.
35. Dabbous, M.K., Inter- and intramolecular cross-linking in tyrosinase-treated tropocollagen, *J. Biol. Chem.*, 241, 5307, 1966.
36. Hoerman, K.C. and Balekjian, A.Y., Some quantum aspects of collagen, *Fed. Proc.*, 25, 1016, 1966.
37. LaBella, F.S., Studies on the soluble products released from purified elastic fibers by pancreatic elastase, *Arch. Biochem. Biophys.*, 93, 72, 1961.
38. Thomas, J., Elsden, D.F., and Partridge, S.M., Degradation products from elastin, *Nature*, 200, 651, 1963.
39. LaBella, F.S. and Lindsay, W.G., The structure of human aortic elastin as influenced by age, *J. Gerontol.*, 18, 111, 1963.
40. Stoller, P., Kim, B.M., Rubenchik, A.M., Reiser, K.M., and Da Silva, L.B., Polarization-dependent optical second-harmonic imaging of a rat-tail tendon, *J. Biomed. Opt.*, 7, 205, 2002.
41. Theodossiou, T., Rapti, G.S., Hovhannisyan, V., Georgiou, E., Politopoulos, K., and Yova, D., Thermally induced irreversible conformational changes in collagen probed by optical second harmonic generation and laser-induced fluorescence, *Lasers Med. Sci.*, 17, 34, 2002.
42. Campagnola, P.J., Clark, H.A., Mohler, W.A., Lewis, A., and Loew, L.M., Second-harmonic imaging microscopy of living cells, *J. Biomed. Opt.*, 6, 277, 2001.
43. Theodossiou, T., Georgiou, E., Hovhannisyan, V., and Yova, D., Visual observation of infrared laser speckle patterns at half their fundamental wavelength, *Lasers Med. Sci.*, 16, 34, 2001.
44. Kim, B.M., Eichler, J., Reiser, K.M., Rubenchik, A.M., and Da Silva, L.B., Collagen structure and nonlinear susceptibility: effects of heat, glycation, and enzymatic cleavage on second harmonic signal intensity, *Lasers Surg. Med.*, 27, 329, 2000.

45. Freund, I., Deutsch, M., and Sprecher, A., Connective tissue polarity. Optical second-harmonic microscopy, crossed-beam summation, and small-angle scattering in rat-tail tendon, *Biophys. J.*, 50, 693, 1986.
46. Roth, S. and Freund, I., Optical second-harmonic scattering in rat-tail tendon, *Biopolymers*, 20, 1271, 1981.
47. Chalfie, M., Tu, Y., Euskirchen, G., Ward, W.W., and Prasher, D.C., Green fluorescent protein as a marker for gene expression, *Science*, 263, 802, 1994.
48. Tsien, R.Y., The green fluorescent protein, *Annu. Rev. Biochem.*, 67, 509, 1998.
49. Baird, G.S., Zacharias, D.A., and Tsien, R.Y., Circular permutation and receptor insertion within green fluorescent proteins, *Proc. Natl. Acad. Sci. U.S.A.*, 96, 11241, 1999.
50. Niswender, K.D., Blackman, S.M., Rohde, L., Magnuson, M.A., and Piston, D.W., Quantitative imaging of green fluorescent protein in cultured cells: comparison of microscopic techniques, use in fusion proteins and detection limits, *J. Microsc.*, 180, 109, 1995.
51. Potter, S.M., Wang, C.M., Garrity, P.A., and Fraser, S.E., Intravital imaging of green fluorescent protein using two-photon laser-scanning microscopy, *Gene*, 173, 25, 1996.
52. Albota, M., Beljonne, D., Bredas, J.-L., Ehrlich, J. E., Fu, J.-Y., Heikal, A.A., Hess, S.E., Kogej, T., Levin, M.D., Marder, S.R., McCord-Maughon, D., Perry, J.W., Rockel, H., Rumi, M., Subramaniam, G., Webb, W., Wu, X.-L., and Xu, C., Design of organic molecules with large two-photon absorption cross sections, *Science*, 281, 1653, 1998.
53. Oheim, M., Beaurepaire, E., Chaigneau, E., Mertz, J., and Charpak, S., Two-photon microscopy in brain tissue: parameters influencing the imaging depth, *J. Neurosci. Methods*, 111, 29, 2001.
54. Dunn, A.K., Wallace, V.P., Coleno, M., Berns, M.W., and Tromberg, B.J., Influence of optical properties on two-photon fluorescence imaging in turbid samples, *Appl. Opt.*, 39, 1194, 2000.
55. Centonze, V.E. and White, J.G., Multiphoton excitation provides optical sections from deeper within scattering specimens than confocal imaging, *Biophys. J.*, 75, 2015, 1998.
56. Mohler, W.A., Simske, J.S., Williams-Masson, E.M., Hardin, J.D., and White, J.G., Dynamics and ultrastructure of developmental cell fusions in the *Caenorhabditis elegans* hypodermis, *Curr. Biol.*, 8, 1087, 1998.
57. Mohler, W.A. and White, J.G., Stereo-4-D reconstruction and animation from living fluorescent specimens, *Biotechniques*, 24, 1006, 1012, 1998.
58. Squirrell, J.M., Wokosin, D.L., White, J.G., and Bavister, B.D., Long-term two-photon fluorescence imaging of mammalian embryos without compromising viability, *Nat. Biotechnol.*, 17, 763, 1999.
59. Keyse, S.M. and Tyrrell, R.M., Induction of the heme oxygenase gene in human skin fibroblasts by hydrogen peroxide and UVA 365 nm radiation: evidence for the involvement of the hydroxyl radical, *Carcinogenesis*, 11, 787, 1990.
60. Tyrrell, R.M. and Keyse, S.M., New trends in photobiology. The interaction of UVA radiation with cultured cells, *J. Photochem. Photobiol.*, B4, 349, 1990.
61. Konig, K., So, P.T.C., Mantulin, W.W., Tromberg, B.J., and Gratton, E., Two-photon excited lifetime imaging of autofluorescence in cells during UVA and NIR photostress, *J. Microsc.*, 183, 197, 1996.
62. Konig, K., Becker, T.W., Fischer, P., Riemann, I., and Halbhuber, K.-J., Pulse-length dependence of cellular response to intense near-infrared laser pulses in multiphoton microscopes, *Opt. Lett.*, 24, 113, 1999.
63. Sako, Y., Sekihata, A., Yanagisawa, Y., Yamamoto, M., Shimada, Y., Ozaki, K., and Kusumi, A., Comparison of two-photon excitation laser scanning microscopy with UV-confocal laser scanning microscopy in three-dimensional calcium imaging using the fluorescence indicator Indo-1, *J. Microsc.*, 185, 9, 1997.
64. Koester, H.J., Baur, D., Uhl, R., and Hell, S.W., $Ca^{2+}$ fluorescence imaging with pico- and femtosecond two-photon excitation: signal and photodamage, *Biophys. J.*, 77, 2226, 1999.
65. Hockberger, P.E., Skimina, T.A., Centonze, V.E., Lavin, C., Chu, S., Dadras, S., Reddy, J.K., and White, J.G., Activation of flavin-containing oxidases underlies light-induced production of $H_2O_2$ in mammalian cells, *Proc. Natl. Acad. Sci. U.S.A.*, 96, 6255, 1999.

66. Schonle, A. and Hell, S.W., Heating by absorption in the focus of an objective lens, *Opt. Lett.*, 23, 5, 325, 1998.
67. Jacques, S.L., McAuliffe, D.J., Blank, I.H., and Parrish, J.A., Controlled removal of human stratum corneum by pulsed laser, *J. Invest. Dermatol.*, 88, 88, 1987.
68. Pustovalov, V.K., Initiation of explosive boiling and optical breakdown as a result of the action of laser pulses on melanosome in pigmented biotissues, *Kvantovaya Elektron.*, 22, 1091, 1995.
69. Buehler, C., Kim, K.H., Dong, C.Y., Masters, B.R., and So, P.T.C., Innovations in two-photon deep tissue microscopy, *IEEE Eng. Med. Biol. Mag.*, 18, 23, 1999.
70. Piston, D.W., Masters, B.R., and Webb, W.W., Three-dimensionally resolved NAD(P)H cellular metabolic redox imaging of the *in situ* cornea with two-photon excitation laser scanning microscopy, *J. Microsc.*, 178, 20, 1995.
71. Denk, W. and Detwiler, P.B., Optical recording of light-evoked calcium signals in the functionally intact retina, *Proc. Natl. Acad. Sci. U.S.A.*, 96, 7035, 1999.
72. Riegler, M., Castagliuolo, I., So, P.T., Lotz, M., Wang, C., Wlk, M., Sogukoglu, T., Cosentini, E., Bischof, G., Hamilton, G., Teleky, B., Wenzl, E., Matthews, J.B., and Pothoulakis, C., Effects of substance P on human colonic mucosa *in vitro*, *Am. J. Physiol.*, 276, G1473, 1999.
73. Bennett, B.D., Jetton, T.L., Ying, G., Magnuson, M.A., and Piston, D.W., Quantitative subcellular imaging of glucose metabolism within intact pancreatic islets, *J. Biol. Chem.*, 271, 3647, 1996.
74. Piston, D.W., Knobel, S.M., Postic, C., Shelton, K.D., and Magnuson, M.A., Adenovirus-mediated knockout of a conditional glucokinase gene in isolated pancreatic islets reveals an essential role for proximal metabolic coupling events in glucose-stimulated insulin secretion, *J. Biol. Chem.*, 274, 1000, 1999.
75. Napadow, V.J., Chen, Q., Mai, V., So, P.T.C., and Gilbert, R.J., Quantitative analysis of three-dimensional-resolved fiber architecture in heterogeneous skeletal muscle tissue using NMR and optical imaging methods, *Biophys. J.*, 80(6), 2968, 2001.
76. Fetcho, J.R. and O'Malley, D.M., Imaging neuronal networks in behaving animals, *Curr. Opin. Neurobiol.*, 7, 832, 1997.
77. Denk, W., Delaney, K.R., Gelperin, A., Kleinfeld, D., Strowbridge, B.W., Tank, D.W., and Yuste, R., Anatomical and functional imaging of neurons using 2-photon laser scanning microscopy, *J. Neurosci. Methods*, 54, 151, 1994.
78. Yuste, R. and Denk, W., Dendritic spines as basic functional units of neuronal integration, *Nature*, 375, 682, 1995.
79. Yuste, R., Majewska, A., Cash, S.S., and Denk, W., Mechanisms of calcium influx into hippocampal spines: heterogeneity among spines, coincidence detection by NMDA receptors, and optical quantal analysis, *J. Neurosci.*, 19, 1976, 1999.
80. Denk, W., Sugimori, M., and Llinas, R., Two types of calcium response limited to single spines in cerebellar Purkinje cells, *Proc. Natl. Acad. Sci. U.S.A.*, 92, 8279, 1995.
81. Svoboda, K., Denk, W., Kleinfeld, D., and Tank, D.W., *In vivo* dendritic calcium dynamics in neocortical pyramidal neurons, *Nature*, 385, 161, 1997.
82. Svoboda, K., Helmchen, F., Denk, W., and Tank, D.W., Spread of dendritic excitation in layer 2/3 pyramidal neurons in rat barrel cortex *in vivo*, *Nat. Neurosci.*, 2, 65, 1999.
83. Helmchen, F., Svoboda, K., Denk, W., and Tank, D.W., *In vivo* dendritic calcium dynamics in deep-layer cortical pyramidal neurons, *Nat. Neurosci.*, 2, 989, 1999.
84. Shi, S.H., Hayashi, Y., Petralia, R.S., Zaman, S.H., Wenthold, R.J., Svoboda, K., and Malinow, R., Rapid spine delivery and redistribution of AMPA receptors after synaptic NMDA receptor activation, *Science*, 284, 1811, 1999.
85. Mainen, Z.F., Malinow, R., and Svoboda, K., Synaptic calcium transients in single spines indicate that NMDA receptors are not saturated, *Nature*, 399, 151, 1999.
86. Maletic-Savatic, M., Malinow, R., and Svoboda, K., Rapid dendritic morphogenesis in CA1 hippocampal dendrites induced by synaptic activity, *Science*, 283, 1923, 1999.

87. Engert, F. and Bonhoeffer, T., Dendritic spine changes associated with hippocampal long-term synaptic plasticity, *Nature*, 399, 66, 1999.
88. Kleinfeld, D., Mitra, P.P., Helmchen, F., and Denk, W., Fluctuations and stimulus-induced changes in blood flow observed in individual capillaries in layers 2 through 4 of rat neocortex, *Proc. Natl. Acad. Sci. U.S.A.*, 95, 15741, 1998.
89. Christie, R.H., Bacskai, B.J., Zipfel, W.R., Williams, R.M., Kajdasz, S.T., Webb, W.W., and Hyman, B.T., Growth arrest of individual senile plaques in a model of Alzheimer's disease observed by *in vivo* multiphoton microscopy, *J. Neurosci.*, 21, 858, 2001.
90. Bacskai, B.J., Kajdasz, S.T., Christie, R.H., Carter, C., Games, D., Seubert, P., Schenk, D., and Hyman, B.T., Imaging of amyloid-beta deposits in brains of living mice permits direct observation of clearance of plaques with immunotherapy, *Nat. Med.*, 7, 369, 2001.
91. Jones, K.T., Soeller, C., and Cannell, M.B., The passage of $Ca^{2+}$ and fluorescent markers between the sperm and egg after fusion in the mouse, *Development*, 125, 4627, 1998.
92. Summers, R.G., Piston, D.W., Harris, K.M., and Morrill, J.B., The orientation of first cleavage in the sea urchin embryo, *Lytechinus variegatus*, does not specify the axes of bilateral symmetry, *Dev. Biol.*, 175, 177, 1996.
93. Masters, B.R., So, P.T., and Gratton, E., Multiphoton excitation fluorescence microscopy and spectroscopy of *in vivo* human skin, *Biophys. J.*, 72, 2405, 1997.
94. Yu, B., Dong, C.Y., So, P.T.C., Blankschtein, D., and Langer, R., *In vitro* visualization and quantification of oleic acid induced changes in transdermal transport using two-photon fluorescence microscopy, *J. Invest. Dermatol.*, 117, 16, 2001.
95. Grewal, B.S., Naik, A., Irwin, W.J., Gooris, G., de Grauw, C.J., Gerritsen, H.G., and Bouwstra, J.A., Transdermal macromolecular delivery: real-time visualization of iontophoretic and chemically enhanced transport using two-photon excitation microscopy, *Pharm. Res.*, 17, 788, 2000.
96. Tirlapur, U.K. and Konig, K., Femtosecond near-infrared laser pulse induced strand breaks in mammalian cells, *Cell Mol. Biol.* Noisy-le-Grand, 47 Online Pub, OL131-4, 2001.
97. Konig, K., Multiphoton microscopy in life sciences, *J. Microsc.*, 200, 83, 2000.
98. Wilson, T. and Sheppard, C.J.R., *Theory and Practice of Scanning Optical Microscopy*, Academic Press, New York, 1984.

# 12

# Near-Field Imaging in Biological and Biomedical Applications




Volker Deckert
*Institut für Angewandte Photophysik*
*Dresden, Germany*


## 12.1 Introduction

Optical microscopy is a standard technique in biology and medicine because it is convenient and easy to use and allows rapid assessment of problems at an early stage. Ability to view the sample is very important; all techniques other than optical microscopy must devise "tricks" to visualize sample information content. Samples ranging from tissue to single cells like bacteria can be investigated in a wide variety of environments, temperatures, and pressures. Sample preparation is straightforward and depends primarily on what information is desired. Because investigation is typically noninvasive, samples are not destroyed and can be reused in other applications.

From the beginning, a major challenge has been the need to improve spatial resolution. Since the late 19th century instrument resolution has come close to the so-called diffraction limit.[1] This is a fundamental restriction that limits the resolution of any optical instrument based on diffraction. Essentially, the diffraction limit uses the greatest possible collection angle and the shortest possible wavelength. Simply put, blue light is better than red light and allows a spatial resolution of 300 nm and smaller. Using an electron microscope increases the resolution; the energy (wavelength) is very high (short) and a resolution down to 1 nm can be achieved. Elaborate sample preparation is a drawback; experiments require an extensive vacuum system. This method is not suitable for environments with great variations and high energy used often destroys the sample.


0-8493-1116-0/03/$0.00+$1.50

Another problem often encountered in biological samples is transparent specimens that show only weak contrasts. In the 20th century, contrast was improved by using the phase information of the image or special illumination schemes. One of the most important achievements was the introduction of fluorescent labeling of the specimen, which provided a means of distinguishing distinct cell compartments.

In 1928 the Irish scientist E. Synge proposed a totally different approach to imaging that scanned very small apertures across a surface.[2] It was almost 50 years before this approach was experimentally tested in the microwave region by Ash and Nichols.[3] A decade later two research groups independently demonstrated use of the technique for optical frequencies.[4,5]

Near-field optical microscopy uses different instrumentation based on scanning probe microscopies, such as scanning tunneling microscopy (STM) or atomic force microscopy (AFM).[6,7] STM/AFM and NSOM are similar in two ways. First, image formation is based on step-by-step recording. Second, the electronic feedback mechanism that moves the subwavelength aperture across the sample is essentially the same. The acronyms used for the technique, NSOM for near-field scanning optical microscopy and SNOM for scanning near-field microscopy, refer to exactly the same technique. These feedback mechanisms have a very useful side effect unique to near-field microscopy. Because the tip–sample distance remains constant, the feedback signal directly yields a topographic image of the sample. Optical and topographical images are therefore measured simultaneously and can be directly correlated. In the last 10 years improvements have resulted in better stability and handling. As a result, near-field microscopy has become a powerful tool for a large variety of applications in physics, chemistry, and biology.[8–12]

Despite the differences in image formation mechanism, the basics of standard (far-field) microscopy and near-field optical microscopy are very similar. In both techniques a change in refractive index or in absorption can be identified. Consequently, many of the established techniques in microscopy can be adapted to the near-field optical technique. Fluorescence labeling has been especially successful in biological applications and can be used to increase spatial resolution to less than 50 nm.

This chapter provides a brief introduction to near-field optics (NFO); however, the main focus is on state-of-the-art applications for investigating biological samples. Many of the current investigations focus on model systems under precisely controlled conditions; however, the first investigations have already been made on living cells. Recently, new techniques related to the original idea of NSOM have been used successfully to investigate problems in life science. Some examples will be introduced here.

## 12.2 Near-Field Optical Microscopy

### 12.2.1 Basic Principles of Near-Field Optical Microscopy

A detailed review of the theory and physics of near-field imaging is not appropriate here. The reader is referred to several excellent publications and reviews.[13–17] The focus here is on basic principles and the instrumentation required to conduct a near-field optical experiment.

Image formation in a classic (far-field) microscope is achieved by specimen illumination using a monochromatic plane wave. The object scatters the light in a characteristic way. The light is then collected by transmission or reflection and focused on a detector. The lens is several wavelengths away from the object, in the optical far-field. High spatial frequencies correspond to the fine details of the specimen and generate Fourier components of the field that decay exponentially along the normal object.[18] Such frequencies cannot be collected by the lens. This effect is the well-known Abbé limit of diffraction, $\Delta x = 0.61 \cdot \lambda/NA$,[1,19] where $\Delta x$ corresponds to the smallest resolvable distance between two points, $\lambda$ is the wavelength of the light, and $NA$ is the numerical aperture of the microscope objective or the lens. If a confocal setup is used instead of wide-field illumination, the resolution increases slightly;[20] multiphoton techniques can also improve the resolution.

In contrast to far-field microscopy in which the light source is confined by a lens, in near-field optical microscopy the light source is confined by a metal aperture. Within a short distance beyond the screen, the size of the illuminated spot is limited only by the dimensions of the aperture. This area is the so-called optical near-field. If such a small light source is scanned above a surface and the distance between aperture and sample is in this near-field region, all scattering or absorption phenomena must originate from that small illumination spot. Consequently, aperture size and distance determine the resolution.

## 12.2.2 Instrumentation

### 12.2.2.1 General Considerations

The key development for practical application of near-field optics was the fabrication of a functional subwavelength aperture. Instead of a hole in a planar metal screen, an aperture was formed at the apex of a pointed glass tip by coating it with a metal.[4,5,21–24] This design ensured an easy way to approach the sample and keep it in close proximity to the surface, while still satisfying the optical requirement for the aperture size. Figure 12.1 shows some of the most important setups using aperture tips. Figure 12.1A shows the most common setup. The sample is illuminated through the aperture and the light is collected by reflection or transmission by standard optical techniques (see, for example, References 4, 5, 11, 21–23, 12, and 25). Figure 12.1B shows an example in which the light is also collected back through the aperture in the illumination–collection mode.[24] Figure 12.1C shows a more specialized arrangement that uses a different illumination by evanescent waves.[26–29] Figure 12.1D illustrates a relatively new development

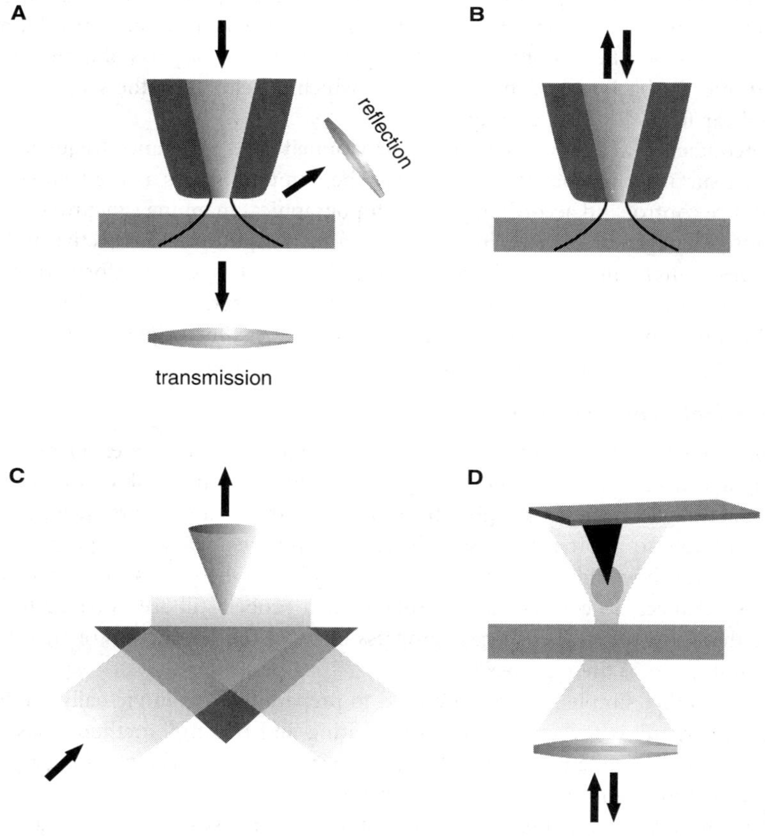


**FIGURE 12.1** Different modes for near-field scanning optical microscopy: (A) illumination through the aperture, detection reflection or transmission via standard optics; (B) illumination–collection mode, both via the near-field aperture; (C) photon scanning tunneling microscope (PSTM); (D) apertureless or probe-enhanced techniques. The illuminated tip enhances the field and serves as a subwavelength light source.

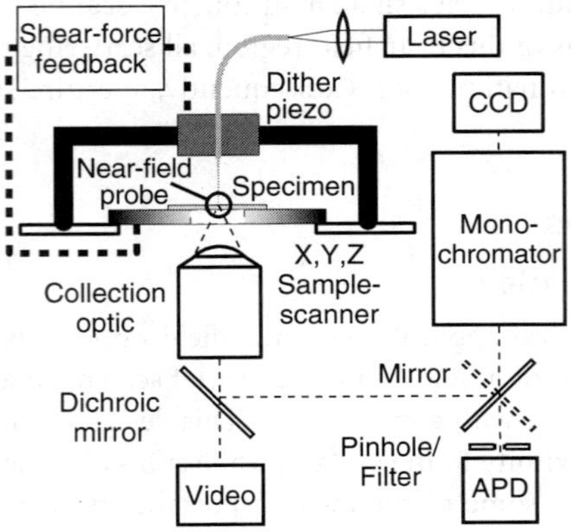


**FIGURE 12.2** Standard experimental setup for transmission of NSOM. For specific analytic requirements different detection schemes can be applied.

without an aperture. Here, the light source is generated at the tip apex and can therefore be even smaller than any aperture.[30–37] This technique will be discussed in more detail in Section 12.4.2.

The most commonly used setup is the design depicted in Figure 12.1A. The entire instrument is built around the core shown. It can be described as a microscope in which the standard illumination source has been replaced by a near-field probe. Figure 12.2 sketches the major parts of a modern NSOM. Laser light is coupled into the back end of the fiber probe, which is held above the sample at a distance of 5 to 15 nm by a shear-force feedback arrangement.

This is a noncontact AFM mode where the tip is vibrated at its resonance frequency.[38,39] As soon as the tip "feels" the surface this vibration is damped. The damping serves as the feedback signal for the tip–sample distance control and accordingly allows topographical mapping synchronously to the optical signal acquisition. The light is collected in transmission by an appropriate objective and, depending on the experimental requirements, simply imaged onto a detector. If spectral information is also desired, the collected light is dispersed by a spectrograph and then imaged onto a multichannel detector. To maintain the alignment of near-field probe and collection optics the sample is scanned underneath the tip; hence, the image is built up, point by point.

### 12.2.2.2 Near-Field Optical Probes

The fabrication of near-field optical probes is a crucial prerequisite for the experiment. Aperture size, efficiency of light transmission, and the damage threshold are considerabions in probe fabrication. Because large taper angles result in a higher transmission,[40] the goal is to produce probes with a large angle that are still able to approach the sample surface and yield a useful topographical image. The straightforward approach would be a microfabrication similar to the process used to manufacture AFM cantilevers. However, because commercial microfabricated probes still are not readily available, many groups prepare these tips themselves. Recent progress made in the production of such tips may lead to commercial manufacture of these probes.[41–47]

Fortunately, two rather simple methods are used to prepare the tips in virtually every lab. The most frequently used technique is the melt-drawn, or "heating and pulling," method. Glass fiber is heated locally using a laser or a filament and the fiber is then pulled apart. The resulting tip shapes depend largely on the temperature and the timing of the procedure.[48,49]

The second method, based on chemical etching of glass fibers, is often called "Turner's method."[50,51] The tip formation occurs at the meniscus between hydrofluoric acid and an organic overlayer. Tips generated in this way generally show considerably larger angles and therefore the transmission is higher than for melt-drawn tips.[52,53] However, the parameters are difficult to control and the quality is less reproducible. Recently, a variation of the standard etching scheme was proposed in which the taper is

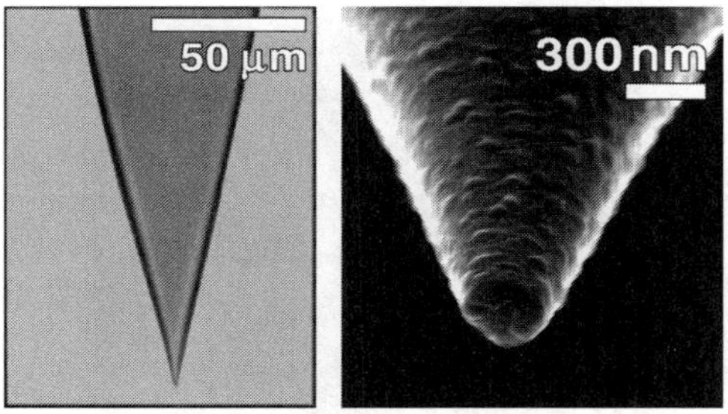


**FIGURE 12.3** Optical (left) and scanning electron microscope image (right) of a "tube-etched" near-field probe.

formed inside the polymer cladding of the glass fibers.[54,55] This so-called "tube-etching" preserves the advantages of the etching; however, it is much more reproducible compared to the original Turner etching.

The metal coating needed to form the aperture is relatively easy to achieve. The glass fiber tip is rotated along its main axis while the metal is evaporated. The arrangement is such that the metal vapor cannot reach the very end of the tip, but only the sides of the taper. Using this kind of shading, the aperture size can be selected by varying the angle between the rotating tip and the evaporation source. Figure 12.3 shows an example of a tube-etched near-field probe and a zoom into the aperture region.

Finally, the near-field tip must be mounted into the scanning probe head. Here the dithering necessary for the feedback loop will be applied by piezoactuators or quartz tuning forks.[38,39,56,57]

# 12.3 Biological Applications of Near-Field Optical Microscopy

## 12.3.1 Practical Considerations

The main reason for using a near-field microscope in biological applications is the need for spatial resolution beyond the limits of a typical microscope. Another advantage is the synchronous sampling of the sample morphology and its optical characteristics. If one of these two considerations cannot be addressed using another, simpler method, then NSOM is the researcher's only choice. NSOM is only 20 years old and still not as developed as standard optical microscopy; hence, the technique is not yet used routinely. The scanning probe part of the instrument, the delicate probes, and the very long acquisition times are aspects still unfamiliar to scientists accustomed to confocal microscopes.

These problems are especially obvious when the method is applied to biological problems, where the samples and their environments differ considerably from those of surface science. Soft samples and scanning in liquids are still challenges peculiar to scanning probe microscopy. As the following discussion of biological and medical applications shows, model systems often must be considered before investigation can commence. Finally, we show that NSOM is now used to investigate living cells.

## 12.3.2 Investigation of Cell Material

### 12.3.2.1 Near-Field Fluorescence Microscopy

#### *12.3.2.1.1 Stained Cell Tissue*

Betzig and co-workers were among the first to investigate biological tissue using an NSOM.[58] Thin tissue sections from the hippocampus region of monkey brain were investigated in the standard transmission mode (see Figure 12.4.) From the toluidine blue stained section, different features can be clearly identified in Figure 12.4: pyramidal cell dendrites (D), nucleus (N), glial cell (Gc), and myelinated axon (M). As indicated by the scale bar, the spatial resolution is clearly better than that of a conventional microscope. An electron microscope would yield an even better resolution, but sample preparation for an NSOM is much simpler and the measurement can be done more quickly with an NSOM.

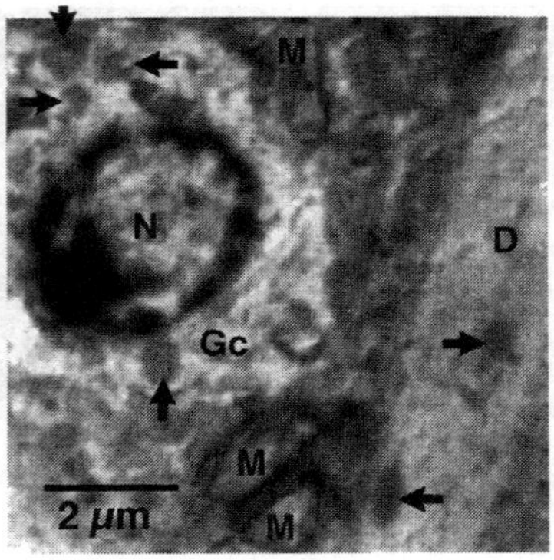


**FIGURE 12.4** Illumination mode NSOM image taken from a thin-tissue section of the hippocampus region of the monkey brain. (From Betzig, E. and Trautman, J.K., *Science*, 257, 189, 1992. Copyright 2002, American Association for the Advancement of Science. Reprinted with permission.)

#### *12.3.2.1.2 Actin Filaments*

Similar experiments attempt to distinguish single actin fibers in the cyto skeleton.[59,60] Single filaments are too big for confocal microscopy and they cannot be readily accessed by force microscopy because they are below the cell membrane. Figure 12.5 illustrates one way to approach that problem. Here MDCK cells (Madin-Darby canine kidney) were grown on a silica slide using glutaraldehyde and lysed so that the interior of the cell walls became accessible.[61] The actin filaments were stained selectively with rhodamine labeled phalloidin. Figure 12.5A shows the NSOM fluorescence image and Figure 12.5B shows the topography image. In contrast to the fluorescence image, the topography image shows no clear structure inside the cell. This rules out topographic coupling, which often obscures near-field images and may give rise to incorrect or misleading results.[62] A line plot across the fluorescent image visualizes the spatial resolution of the measurement. In this case the so-called 10:90 step resolution is around 100 nm.

Recent experiments by Doyle et al. on fixed or living Glial cells involve a novel feedback scheme that relies on the detected fluorescence.[60] By measuring the signal at different distances from the specimen, Doyle et al. distinguished between the pure near-field and the ubiquitous far-field contributions.[62] Such experiments would then allow measurement of at least small distances through the plasma membrane. The achieved spatial resolution that has been extracted from the data is already on the order of 50 nm. Although no further conclusions have been drawn from the data, these experiments provide a useful methodology for soft samples such as cells.

#### *12.3.2.1.3 NSOM inside Cells*

Correlation of topographic and optical information is usually considered a unique advantage of NSOM. In the case of cell investigation, it can also be considered a great disadvantage because the interior of the cells cannot be accessed. Recently, Lei and co-workers[63] approached the mitochondrial membrane of a living breast carcinoma cell by penetrating the outer cell membrane. The cells were treated with JC-1, a dye that changes conformation — and hence its emission characteristics — when it is aggregated. This high concentration state occurs particularly in the mitochondrial membrane and thus distinguishes this membrane from other cell compartments.

To access the mitochondria, the surface topography of the cell was first determined, using the shear-force feedback in the aqueous environment of the setup. Next, the plasma membrane of the cell was penetrated by the NSOM probe and three fluorescence spectra were recorded at different distances from the mitochondrium. Figure 12.6 schematically shows this penetration of the cell membrane and the corresponding near-field fluorescence spectra, respectively. When the probe is close to the mitochondrial membrane a clear change in the two emission peaks can be detected. The relative peak height change between positions 1 and 2 is negligible and the overall intensity changes are due to concentration

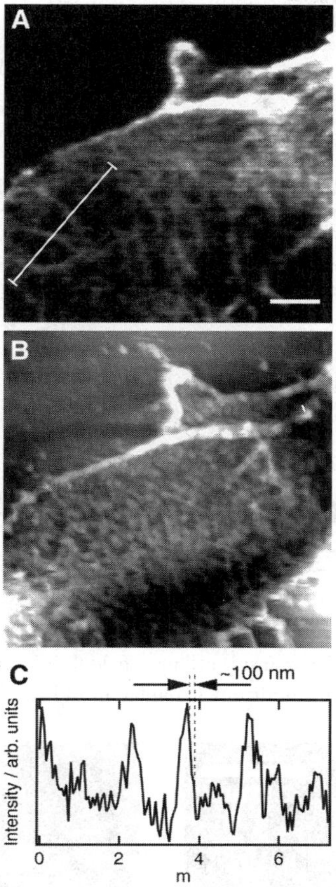


**FIGURE 12.5** (A) Reflection NSOM image of labeled actin filaments in Madin-Darby canine kidney (MDCK) cells; (B) corresponding topography image; (C) line scan through the section marked in (A).

fluctuations. However, the ratio between the two emission peaks at position 3 changed significantly, which indicates that the probe is very close to the mitochondrial membrane.

The main emphasis here is on the height resolution, which is not generally considered in normal NSOM. Lateral scanning inside the cell is not possible with this technique because the tip would drag the entire cell. Therefore, high spatial resolution imaging in the *x* and *y* directions is not feasible.

#### *12.3.2.1.4 In Vitro Chemical Imaging of Tobacco Mosaic Virus*

An intriguing advantage of near-field optical microscopy is the ability to visualize virus particles that cannot be investigated at all by standard microscopy techniques. Keller and co-workers were the first to apply NSOM to virus samples.[64] They used a special force feedback setup, in which the sample is placed on an AFM cantilever underneath the actual near-field aperture.[65] Their specific arrangement allows a more sensitive feedback for the distance control between sample and near-field probe. This is important because of the previously discussed difficulties of measuring directly in liquids.

Figure 12.7 shows the results of this investigation. Figures 12.7A and 12.7B show the topography and near-field optical image, respectively. An interesting labeling technique was used to determine the direction of the virus. The tobacco mosaic virus (TMV) was labeled with a metatope monoclonal antibody (M) that binds only to the 5′ end of the virus. A biotin labeled polyclonal antibody (P) was attached to the monoclonal antibody. Finally, an avidin labeled latex bead (~30 nm diameter) was bound to the remaining biotin sites. The composition can be seen from Figure 12.7D. The presence of bovine serum albumin (BSA) prevents aggregation of the virus and is needed to obtain single virus particles. This kind of label

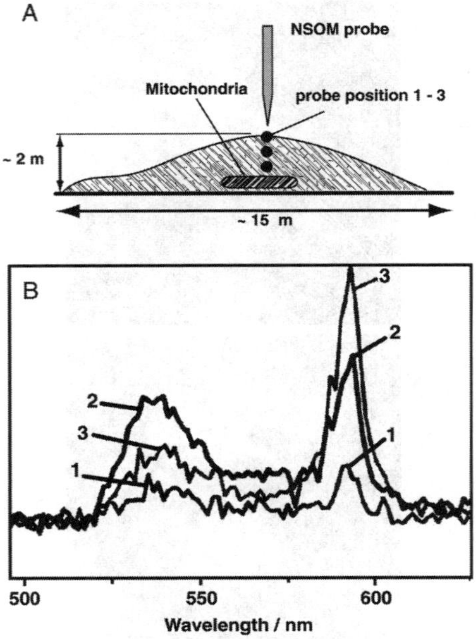


**FIGURE 12.6** (A) Experimental setup for an inside cell near-field fluorescence experiment. (B) Fluorescence spectra taken at different distances from the mitochondrium. (Adapted from Lei, F.H. et al., *Appl. Phys. Lett.*, 29, 2489, 2001.)

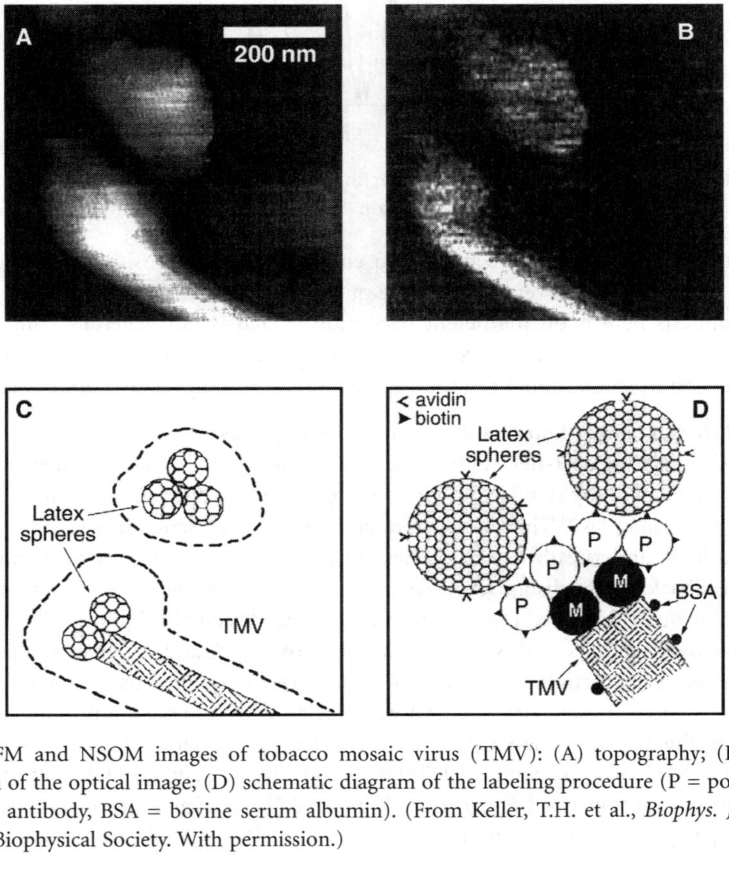


**FIGURE 12.7** AFM and NSOM images of tobacco mosaic virus (TMV): (A) topography; (B) NSOM image; (C) interpretation of the optical image; (D) schematic diagram of the labeling procedure (P = polyclonal antibody, M = monoclonal antibody, BSA = bovine serum albumin). (From Keller, T.H. et al., *Biophys. J.*, 74, 2076, 1998. Copyright 2002, Biophysical Society. With permission.)

can be detected by either topographic or optical measurements, as shown in Figure 12.7. The sketch in Figure 12.7C shows the results from the raw images more clearly. Because a near-field optical image is always a convolution between the object size and the aperture dimensions, it is possible to extract the object features if the size and shape of the near-field source are known. This is also valid for the topographic image.

A remarkable fact that can be deduced from the images is that the optical resolution in this experiment was better than the topographic resolution. In principle, even without the extensive labeling, the features of the virus could be easily detected with a resolution of ~60 nm; however, only the label provides direct information about the virus.

A major drawback of this method is the elaborate sample preparation on an AFM cantilever. Nevertheless, this method offers a means to control environmental conditions in a way not possible with other methods.

#### *12.3.2.1.5 Single Green Fluorescing Proteins*

The detection of intact single molecules is the ultimate goal of microscopy. In 1993, Betzig and Chichester[66] were the first to observe individual fluorescing molecules at room temperature, using a near-field optical microscope, thus establishing single-molecule detection as a new branch of science.[67–72] Although, in principle, single-molecule sensitivity can be achieved readily by standard microscopy techniques and appropriate dilution, near-field optical techniques offer some distinct advantages:[73] the small detection volume, which allows localization to a few nm; the sensitivity toward the molecular orientation in all three dimensions; and, perhaps most important of all, the correlation to surface morphology resulting from the simultaneous detection of topography and optical signal.

As an example, the near-field fluorescence and the corresponding topography images of single green fluorescing proteins (GFP) are shown in Figure 12.8.[73] This molecule has attracted considerable attention because it can be produced by many proteins without changing the biological function. An extensive exogenous staining procedure is obsolete in many cases.[74,75] The individual proteins were adsorbed on freshly cleaned hydrophilic glass.

The image shows a 2 × 2-μm scan of two distinct sample areas, each with corresponding fluorescence and topographic information. The most obvious feature is the irregular shape of the spots in the fluorescence image. This is a direct indication of single-molecule behavior, the so-called on/off switching observed very early by researchers.[76] The origin of the photodynamic is not yet fully understood.[77,78] The fluorescent events can be correlated with topographical features that, with a height of 2 to 5 nm and a width of ~10 nm, correspond well to the known size of 4 nm of the barrel-shaped GFP.

Interestingly, virtually all the fluorescent spots also can be found in the topography, whereas not all the morphological features fluoresce. Subsequent measurements revealed that most of the features are indeed proteins; however, they light up only from time to time. This indicates that only a few molecules are "active" at a given time and that most of the time the GFP is in a so-called dark state.[79] The structure

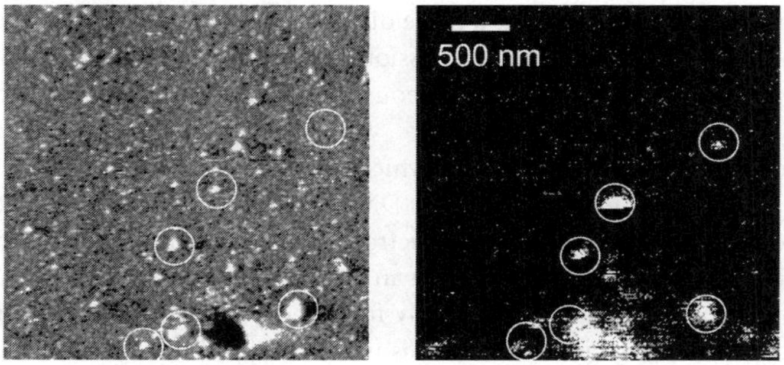


**FIGURE 12.8** NSOM and topography images (2 × 2 μm) of individual S65T GFPs. Right: Fluorescence image showing abrupt changes in the emission of isolated spots (circles). Left: The corresponding topography shows individual GFPs 2 to 4 nm high. (Courtesy of Niek van Hulst, University of Twente, the Netherlands.)

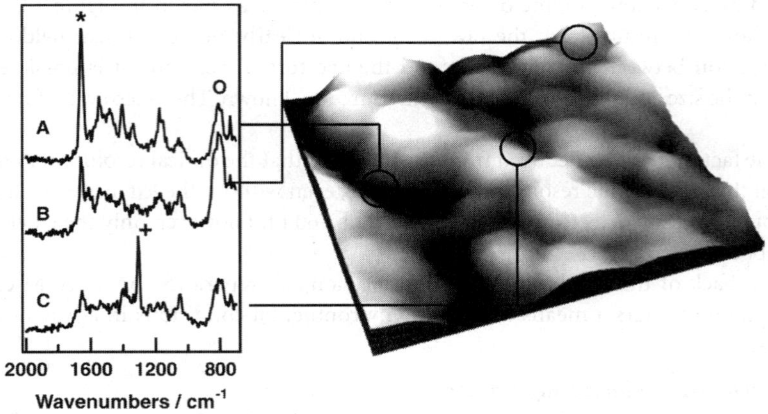


**FIGURE 12.9** Shear-force topography of a SERS substrate covered with brilliant cresyl blue labeled DNA fragments. The Raman spectra on the left correspond to the specified local positions on the substrate.

and biological nature of this dark state are still under investigation; nevertheless, these factor affect the sensitivity of the single-molecule detection because the off-time can last for several seconds and the probability of detecting a single GFP molecule is only 5%.

### 12.3.2.2 Near-Field Raman Spectroscopy of Labeled DNA

Raman spectroscopy has the potential to identify any material by its characteristic molecular vibration without the need for further labeling and, hence, in many cases without any further sample preparation. Raman microscopy is fast becoming a standard technique to characterize surfaces directly.[80] The major disadvantage of standard Raman spectroscopy is the poor scattering cross section. Its efficiency is about ten orders of magnitude lower than fluorescence spectroscopy. This makes routine investigations impossible for many analytical problems. In biology, nevertheless, many structural problems of proteins have been solved by using resonance Raman scattering.[81] Here, the efficiency is much better because the molecules or certain fragments of the molecule are resonantly excited.[82] This results in an enhancement of several orders of magnitude.

Even greater signal enhancements can be gained by the implementation of surface-enhanced Raman scattering (SERS).[83–86] Using certain rough metal substrates, enhancement factors up to $10^{10}$ and even higher have been reported, resulting in single-molecule sensitivity comparable to fluorescence, as discussed previously.[87,88] Kneipp and co-workers reported on the detection of single DNA base molecules using SERS.[88] At present, the still lower light emission of SERS compared with fluorescence restricts the use of near-field optical techniques. To utilize the enhancement it is helpful to label the substance of interest with an especially active SERS compound. This may seem like a more difficult way of performing fluorescence labeling, but the advantage here is the narrowness of the emission lines in Raman scattering compared to those in fluorescence. This narrowness allows the simultaneous detection of several labels with just one excitation line.

This kind of labeling is shown in Figure 12.9, which shows the rough topography of a SERS substrate covered with brilliant cresyl blue (BCB) stained DNA fragments.[89,90] Near-field Raman spectra were recorded simultaneously and some example spectra are shown. The change in the spectral pattern indicates a change in sample composition without an elaborate band assignment. The central part of the right substrate shown in Figure 12.9 differs clearly from the outer regions. The signal-to-noise ratio is not sufficient for further investigation. Because this was an atypical result, all but two prominent bands can be used for analysis. The only conclusion that can be drawn from the data is that the results are definitely not due to the dye label, and therefore a change in the sample composition occurred. The nature of this change remains, however, unclear.

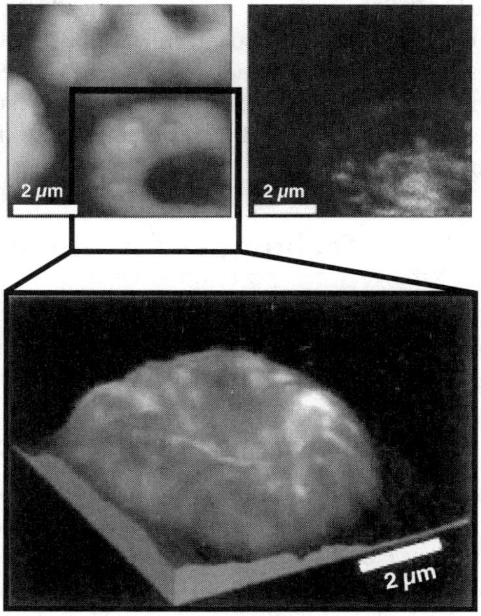


**FIGURE 12.10** Fluorescence and topography images of malaria-infected cells.

The combination of SERS and near-field optical techniques is promising because of the ease of distinguishing many samples simultaneously. However, there are still drawbacks, the most important of which is the long acquisition time of about 1 min per position. Therefore, the specific investigation of selected regions of interest appears to be more promising than a complete imaging.

## 12.3.3 Model Cell Membranes

An important issue in biology is the probing of cell membranes. Typical questions concern the organization and orientation of the membrane or the location of specific molecules such as proteins. Because of their softness, many membranes tend to fluctuate, and this requires a very sensitive feedback scheme. Consequently, many researchers choose Langmuir-Blodgett mono- and bilayers that are easier to control and serve as a model system.[91–93] The capabilities of NSOM to gain information on phase changes or changes resulting from additives such as cholesterol or small peptides have been demonstrated by several research groups.[94–99] One very interesting approach is the detection of the three-dimensional dipole orientation of single fluorophores to determine the structure and phase composition of a membrane.[66,100] Depending on environmental conditions, these orientations change; in combination with simultaneous topographic measurements, the correlation between structure and morphology can be deduced.

Some groups have begun to look at cell membranes directly.[101–104] Figure 12.10, for example, shows the fluorescence and topography images of malaria-infected cells.[105] Here the mapping of malarial protein in the erythrocyte membrane is shown. A blood sample was reacted with antibodies against a specific malaria protein (*Plasmodium falciparum* histidine-rich protein) and then stained with tetramethylrhodamine. The topography image clearly shows the red blood cells, but cannot distinguish between infected and healthy cells. The fluorescence image to the upper right clearly shows that only the cell in the lower right is infected. In the superposition of fluorescence and topography images, the distribution of the antibodies on the erythrocyte membrane can be visualized quite clearly. The investigators claim the resolution to be around 100 nm — well below the far-field control experiment. A further near-field dual-color experiment (not shown) proved that different malarial proteins interact differently with certain blood proteins. The high resolution of the near-field experiment is necessary to distinguish the local origins of the labeled proteins.

The progress in the feedback stability and resolution in such membrane investigations is very promising. In contrast to many other experiments, the resolving power of the instrument is a major issue here, because the concentration in these experiments cannot simply be diluted to make it convenient for standard microscopy. Furthermore, because the light field does not really extend into the cell, background fluorescence is a minor issue in the near-field. Finally, only NSOM offers the advantage of synchronous topography and light detection.

## 12.4 Special Near-Field Techniques for Biological Applications

This section provides an overview of some promising but not yet established techniques and approaches to obtain near-field optical resolution. Only time will tell whether these methods can compete with the methods currently used.

### 12.4.1 Fluorescence Resonance Energy Transfer

Fluorescence resonance energy transfer (FRET) is a method that uses the distance dependence of nonradiative energy transfer from an excited donor molecule to an acceptor molecule. The transfer efficiency is given by a sixth-power dependence on the intermolecular separation:

$$E = 1\Big/\left(1+\left(R/R_0\right)^6\right),$$

where $R$ is the distance between donor and acceptor and $R_0$ is the distance at which 50% of the energy is transferred. Consequently, this interaction is sensitive even to angstrom scale distance changes.[106–108] Experimentally, a pair of fluorescing molecules is needed, only one of which (the donor) can be excited directly with the excitation source. The acceptor dye only emits light if it is in close proximity to an excited donor. In an experimental setup one has to separate donor and acceptor fluorescence spectrally. A red shifted fluorescence then means that a donor resonantly transferred its energy to an acceptor. The relevant distance is on the order of a few nanometers, which is ideal for monitoring conformation changes in proteins or events in biomembranes.[106,108–112]

Discussing the details of this technique is beyond the scope of this chapter; we outline only the results achieved in biology using a near-field aperture. In a simple scheme, a near-field microscope can be applied within a dual color mode scan (as already mentioned) to find resonant pairs simply with higher resolution.[113,114] Sekatskii and Letokhov[115] proposed an experimental setup that, in principle, could result in a resolution of a few nanometers. In their setup, the Förster pair is composed of an acceptor attached to the near-field aperture, whereas the donor is on the sample surface. The distance between donor and acceptor can then be controlled by the tip sample feedback regulation. Vickery and Dunn[116] applied this technique to the near-field investigation of Langmuir-Blodgett films and observed an improvement in resolution; however, the results using this method are still at an early stage and no biosamples have been investigated.

### 12.4.2 "Apertureless" Near-Field Microscopy

The main restriction of near-field optical microscopy is the aperture. Although an aperture is necessary to achieve the resolution, it also imposes serious limitations. A different approach is the use of an active light source with subwavelength dimensions. Zenhausern and co-workers were the first to report that an externally illuminated sharp tip can serve as such a light source.[31,117] These investigators used an elaborate setup to distinguish the minute differences of light intensity when the vibrating tip was either close to or away from the sample. Sanchez and co-workers utilized the strongly enhanced electric field at a metal tip to induce two-photon fluorescence.[36] The spatial resolution on j-aggregates was about 20 nm, very difficult to achieve with standard NSOM. The problem of background fluorescence signal was avoided by using two-photon excitation. In general, the signal was spectrally far from the fluorescence excited with one photon; furthermore, without the enhancing tip, no two-photon fluorescence signal at all was observed.

Another approach that involves field enhancement with metal was suggested by Wessel in 1985.[118] He proposed that a sharp metal tip brought very close to the external sample surface would enhance the Raman signal in the same way as do the well-known rough metal surfaces used for SERS. Interestingly, a similar but much weaker enhancement effect exists for infrared spectroscopy; here, Knoll and Keilmann were the first to observe a vibrational signal that was enhanced by the presence of a metal tip.[35] The first reports on probe-enhanced Raman scattering followed shortly thereafter;[37,119,120] the enhancement factors reported were on the order of several thousand compared to the background Raman signal.[37] This is still far from the maximum enhancement factors achieved with rough surfaces or even colloids;[87,121] thus, there is room for improvement. Thus far, apertureless Raman or infrared spectroscopy have not been used on biological samples; however, there are no indications that the method would fail. All the advantages of Raman spectroscopy are maintained and the disadvantage of low scattering cross sections can be overcome. Thus, apertureless or probe-enhanced techniques may well be the future of high spatial resolution spectroscopy.

### 12.4.3 Multiphoton Near-Field Microscopy

In the previous section a type of multiphoton near-field microscopy was mentioned.[36] However, it is also possible to excite two- or three-photon processes with a standard near-field probe. Pulsed[122] and continuous-wave (CW)[123] near-field experiments on cell material have been reported. The advantage of this approach, the high signal-to-background level, is identical to the two-photon probe enhanced technique mentioned in the previous section. The detected signal is blue shifted with respect to the excitation source, so that normal fluorescence is spectrally far away and can easily be filtered. In addition, a resolution enhancement occurs because the two-photon energy is the same as one single UV-photon of half the wavelength. Using a UV source directly can be disadvantageous because of absorption effects and possible destruction of the specimen. The advantage of using a longer wavelength laser source in combination with two-photon processes is therefore twofold. Very good results, with respect to the signal level, have been reported using two- and three-photon excitation on polytene chromosomes.[122,123] The spatial resolution in these first experiments was below 200 nm but further improvement appears possible.

The application of this technique appears straightforward, at least for the CW technique. In cases where UV-dyes are the only choice, this method is a good alternative to direct excitation. For dye labels that absorb in the visible region the availability of an affordable laser source for two-photon excitation may be a problem.

### 12.4.4 Nonoptical Near-Field Microscopy

A unique scheme of observing biological specimens has been proposed by Kawata and coworkers.[124,125] The specimens are on top of a photosensitive copolymer. Instead of raster scanning the sample, the authors illuminate the whole area of interest with a short laser pulse. After the illumination the sample is washed and the polymer is then scanned with a standard AFM. Depending on the optical properties of the specimen and the polymer, a different corrugation is detected and the specimen is scanned indirectly. Using multiple illumination, even moving cells can be detected, as demonstrated by Kawata et al.[125]

This method will probably not become a general tool for detecting subwavelength features because it cannot correlate topographical features with molecular contrast. Nevertheless, it is a surprisingly simple and straightforward technique that allows observation of dynamic phenomena on a rather large scale with high resolution.

## 12.5 Outlook and Conclusions

Since the first experiments about 20 years ago, near-field optics has been applied in many research fields. Particularly in biological and biomedical applications, the possibility of observing features below 100 nm is very tempting and it is not surprising that many NSOM applications are used in biology-related material. The most successful near-field application to date is fluorescence NSOM. A "down scaling" from standard microscopy rather than a new technique, it combines high lateral optical resolution with the synchronous detection of the sample topography. This combination is a unique advantage of near-field microscopy. Similar to microscopy, the sensitivity of the fluorescence NSOM is down to the single-molecule level, despite the notoriously low efficiency of the near-field probes.

Other spectroscopic techniques such as infrared or Raman are much more limited by restrictions imposed by the aperture. Collection of near-field Raman images, for example, still takes a matter of hours. The emergence of probe-enhanced or "apertureless" techniques might well change that situation. The implementation and testing of these new methods is still at an early stage. The enhancement factors found in macroscopic experiments and theoretically predicted have not yet been demonstrated in the near-field. Basic research in these fields is still required.

Compared to normal microscopy the market share of NSOM today is quite small. It will probably remain a specialized technique for some time to come. Interestingly, the development of new techniques in standard microscopy (for example, the 4Pi microscope[126] and the solid immersion lens microscope[127]) that improve resolution, for the first time after almost a century, correspond to the rise of near-field techniques. Coincidence or not, developments in the two techniques will likely improve the applications of both in the future.

### References


1. Abbé, E., Beiträge zur Theorie des Mikroskops und der mikroscopischen Wahrnehmung, *Arch. Mikroskop.*, 9, 413, 1873.
2. Synge, E.H., A suggested method for extending microscopic resolution into the ultra-microscopic region, *Phil. Mag.*, 6, 356, 1928.
3. Ash, E.A. and Nicholls, G., Super-resolution aperture scanning microscope, *Nature*, 237, 510, 1972.
4. Pohl, D.W., Denk, W., and Lanz, M., Optical stethoscopy: image recording with resolution 1/20, *Appl. Phys. Lett.*, 44, 651, 1984.
5. Lewis, A., Isaacson, M., Harootunian, A., et al., Development of a 500 A spatial resolution light microscope i. Light is efficiently transmitted through 1/16 diameter apertures, *Ultramicroscopy*, 13, 227, 1984.
6. Binnig, G., Quate, C.F., and Gerber, C., Atomic force microscope, *Phys. Rev. Lett.*, 56, 930, 1986.
7. Binnig, G. and Rohrer, H., Scanning tunneling microscopy, *Helv. Phys. Acta*, 55, 726, 1982.
8. Pohl, D.W., Ed., *Advances in Optical and Electron Microscopy*, Bd. 12, Academic Press, New York, 1991, p. 243.
9. Heinzelmann, H. and Pohl, D.W., Scanning near-field optical microscopy, *Appl. Phys. A*, 59, 89, 1994.
10. Girard, C. and Dereux, A., Near-field optics theories, *Rep. Prog. Phys.*, 59, 657, 1996.
11. Dunn, R.C., Near-field scanning optical microscopy, *Chem. Rev.*, 99, 2891, 1999.
12. Barbara, P.F., Adams, D.M., and O'Connor, D.B., Characterization of organic thin film materials with near-field scanning optical microscopy (NSOM), *Annu. Rev. Mater. Sci.*, 29, 433, 1999.
13. Fillard, J.P., *Near-Field Optics and Nanoscopy*, World Scientific, Singapore, 1996.
14. Paesler, M.A. and Moyer, P.J., *Near-Field Optics: Theory, Instrumentation, and Applications*, John Wiley & Sons, New York, 1996.
15. Ohtsu, M. and Hori, H., *Near-Field Nano-Optics: From Basic Principles to Nano-Fabrication and Nano-Photonics*, Kluwer Academic, New York, 1999.
16. Ohtsu, M., *Near-Field Nano/Atom Optics and Technology*, Springer-Verlag, Tokyo, 1998.

17. Hecht, B., Sick, B., Wild, U.P., et al., Scanning near-field optical microscopy with aperture probes: fundamentals and applications, *J. Chem. Phys.*, 112, 7761, 2000.
18. Goodman, J.W., *Introduction to Fourier Optics*, McGraw-Hill, New York, 1968.
19. Born, M. and Wolf, E., *Principles of Optics*, 6th ed., Cambridge University Press, Cambridge, U.K., 1980.
20. Wilson, T. and Sheppard, C.J.R., *Theory and Practice of Scanning Optical Microscopy*, Academic Press, London, 1984.
21. Dürig, U., Pohl, D.W., and Rohner, F., Near-field optical-scanning microscopy, *J. Appl. Phys.*, 59, 3318, 1986.
22. Pohl, D., U.S. patent 4,604,520, optical near-field microscope, 1986.
23. Harootunian, A., Betzig, E., Isaacson, M., et al., Super-resolution fluorescence near-field scanning optical microscopy, *Appl. Phys. Lett.*, 49, 674, 1986.
24. Betzig, E., Isaacson, M., and Lewis, A., Collecting mode near-field scanning optical microscopy, *Appl. Phys. Lett.*, 51, 2088, 1987.
25. Betzig, E. and Trautman, J. K., Near-field optics: microscopy spectroscopy and surface modification beyond the diffraction limit, *Science*, 257, 189, 1992.
26. Courjon, D., Sarayeddine, K., and Spajer, M., Scanning tunneling optical microscopy, *Opt. Commun.*, 71, 23, 1989.
27. Reddick, R.C., Warmack, R.J., and Ferrell, T.L., New form of scanning optical microscopy, *Phys. Rev. B*, 39, 767, 1989.
28. Marti, O., Bielefeldt, H., Hecht, B., et al., Near-field optical measurement of the surface plasmon field, *Opt. Commun.*, 96, 4, 1993.
29. Krenn, J.R., Dereux, A., Weeber, J.C., et al., Squeezing the optical near-field zone by plasmon coupling of metallic nanoparticles, *Phys. Rev. Lett.*, 82, 2590, 1999.
30. Fischer, U.C. and Pohl, D.W., Observation of single-particle plasmons by near-field optical microscopy, *Phys. Rev. Lett.*, 62, 458, 1989.
31. Zenhausern, F., Martin, Y., and Wickramasinghe, H.K., Scanning interferometric apertureless microscopy: optical imaging at 10 angstrom resolution, *Science*, 269, 1083, 1995.
32. Inouye, Y. and Kawata, S., A scanning near-field optical microscope having scanning electron tunnelling microscope capability using a single metallic probe tip, *J. Microsc.*, 178, 14, 1995.
33. Girard, C., Martin, O.J.F., and Dereux, A., Molecular lifetime changes induced by nanometer scale optical fields, *Phys. Rev. Lett.*, 75, 3098, 1995.
34. Koglin, J., Fischer, U.C., and Fuchs, H., Scanning near-field optical microscopy with a tetrahedral tip at a resolution of 6 nm, *J. Biomed. Opt.*, 1, 75, 1996.
35. Knoll, B. and Keilmann, F., Near-field probing of vibrational absorption for chemical microscopy, *Nature*, 399, 134, 1999.
36. Sanchez, E.J., Novotny, L., and Sunney Xie, X., Near-field fluorescence microscopy based on two-photon excitation with metal tips, *Phys. Rev. Lett.*, 82, 4014, 1999.
37. Stockle, R.M., Suh, Y.D., Deckert, V., et al., Nanoscale chemical analysis by tip-enhanced Raman spectroscopy, *Chem. Phys. Lett.*, 318, 131, 2000.
38. Toledo Crow, R., Yang, P.C., Chen, Y., et al., Near-field differential scanning optical microscope with atomic force regulation, *Appl. Phys. Lett.*, 60, 2957, 1992.
39. Betzig, E., Finn, P.L., and Weiner, J.S., Combined shear force and near-field scanning optical microscopy, *Appl. Phys. Lett.*, 60, 2484, 1992.
40. Novotny, L., Pohl, D.W., and Hecht, B., Scanning near-field optical probe with ultrasmall spot size, *Opt. Lett.*, 20, 970, 1995.
41. Heinzelmann, H., Freyland, J.M., Eckert, R., et al., Towards better scanning near-field optical microscopy probes-progress and new developments, *J. Microsc.*, 194, 2, 1999.
42. Kim, Y.J., Kurihara, K., Suzuki, K., et al., Fabrication of micro-pyramidal probe array with aperture for near-field optical memory applications, *Jpn. J. Appl. Phys. Part 1*, 39, 1538, 2000.

43. Minh, P.N., Ono, T., and Esashi, M., Microfabrication of miniature aperture at the apex of $SiO_2$ tip on silicon cantilever for near-field scanning optical microscopy, *Sensors Actuators A,* 80, 163, 2000.
44. Noell, W., Abraham, M., Ehrfeld, W., et al., Microfabrication of new sensors for scanning probe microscopy, *J. Micromech. Microeng.*, 8, 111, 1998.
45. Schurmann, G., Indermuhle, P.F., Staufer, U., et al., Micromachined SPM probes with sub-100 nm features at tip apex, *Surf. Interface Anal.*, 27, 5, 1999.
46. Young Joo, K., Kurihara, K., Suzuki, K., et al., Fabrication of micro-pyramidal probe array with aperture for near-field optical memory applications, *Jpn. J. Appl. Phys.*, 39, 1538, 2000.
47. Münster, S., Werner, S., Mihalcea, C., et al., Novel micromachined cantilever sensors for scanning near-field optical microscopy, *J. Microsc.*, 186, 17, 1997.
48. Betzig, E., Trautman, J.K., Harris, T.D., et al., Breaking the diffraction barrier: optical microscopy on a nanometric scale, *Science,* 251, 1468, 1991.
49. Valaskovic, G.A., Holton, M., and Morrison, G.H., Parameter control, characterization, and optimization in the fabrication of optical fiber near-field probes, *Appl. Optics,* 34, 1215, 1995.
50. Turner, D.R., U.S. patent 4,469,554, etch procedure for optical fibers (AT&T Bell Laboratories, Murray Hill, New Jersey), 1983.
51. Hoffmann, P., Dutoit, B., and Salathe, R.P., Comparison of mechanically drawn and protection layer chemically etched optical fiber tips, *Ultramicroscopy,* 61, 1, 1995.
52. Zeisel, D., Nettesheim, S., Dutoit, B., et al., Pulsed laser-induced desorption and optical imaging on a nanometer scale with scanning near-field microscopy using chemically etched fiber tips, *Appl. Phys. Lett.*, 68, 2491, 1996.
53. Yatsui, T., Kourogi, M., and Ohtsu, M., Increasing throughput of a near-field optical fiber probe over 1000 times by the use of a triple-tapered structure, *Appl. Phys. Lett.*, 73, 2090, 1998.
54. Lambelet, P., Sayah, A., Pfeffer, M., et al., Chemically etched fiber tips for near-field optical microscopy: a process for smoother tips, *Appl. Opt.*, 37, 7289, 1998.
55. Stockle, R.M., Schaller, N., Deckert, V., et al., Brighter near-field optical probes by means of improving the optical destruction threshold, *J. Microsc. (Oxford),* 194(Part 2–3), 378, 1999.
56. Karrai, K. and Grober, R.D., Piezo-electric tuning fork tip-sample distance control for near field optical microscopes, *Ultramicroscopy,* 61, 1, 1995.
57. Ruiter, A.G.T., Veerman, J.A., van der Werf, K.O., et al., Dynamic behavior of tuning fork shear-force feedback, *Appl. Phys. Lett.*, 71, 28, 1997.
58. Betzig, E., in *Near-Field Optics*, Pohl, D. and Courjon, D., Eds., Kluwer Academic Publishers, Arc-et-Senans, France, 1993, p. 7.
59. Zenobi, R. and Deckert, V., Scanning near-field optical microscopy and spectroscopy as a tool for chemical analysis, *Angew. Chem. Int. Ed.*, 39, 1747, 2000.
60. Doyle, R.T., Szulzcewski, M.J., and Haydon, P.G., Extraction of near-field fluorescence from composite signals to provide high resolution images of glial cells, *Biophys. J.*, 80, 2477, 2001.
61. Ziegler, R., Vickier, A., Kernen, P., et al., Preparation of basal cell membranes for scanning probe microscopy, *Microsc. Res. Tech.*, 1, 1997.
62. Hecht, B., Bielefeld, H., Inouye, Y., et al., Facts and artifacts in near-field optical microscopy, *J. Appl. Phys.*, 81, 2492, 1997.
63. Lei, F.H., Shang, G.Y., Troyon, M., et al., Nanospectrofluorometry inside single living cell by scanning near-field optical microscopy, *Appl. Phys. Lett.*, 29, 2489, 2001.
64. Keller, T.H., Rayment, T. and Klenerman, D., Optical chemical imaging of tobacco mosaic virus in solution at 60-nm resolution, *Biophys. J.*, 74, 2076, 1998.
65. Keller, T.H., Rayment, T., Klenerman, D., et al., Scanning near-field optical microscopy in reflection mode imaging in liquid, *Rev. Sci. Instrum.*, 68, 1448, 1997.
66. Betzig, E. and Chichester, R.J., Single molecules observed by near-field scanning optical microscopy, *Science,* 262, 1422, 1993.

67. Moerner, W.E., High-resolution optical spectroscopy of single molecules, *Acc. Chem. Res.*, 26, 563, 1996.
68. Moerner, W.E. and Orrit, M., Illuminating single molecules in condensed matter, *Science*, 283, 1670, 1999.
69. Xie, X.S., Single-molecule spectroscopy and dynamics at room temperature, *Acc. Chem. Res.*, 26, 598, 1996.
70. Xie, X.S. and Trautman, J.K., Optical studies of single molecules at room temperature, *Annu. Rev. Phys. Chem.*, 49, 441, 1998.
71. Basché, T., Moerner, W.E., Orrit, M., and Wild, U.P., Eds., *Single Molecule Optical Detection, Imaging, and Spectroscopy*, VCH, Weinheim, Germany, 1997.
72. Goodwin, P.M., Ambrose, W.P., and Keller, R.A., Single molecule detection in liquids by laser-induced fluorescence, *Acc. Chem. Res.*, 26, 607, 1996.
73. van Hulst, N.F., Veerman, J. A., Garcia Parajo, M.F., et al., Analysis of individual (macro)molecules and proteins using near-field optics, *J. Chem. Phys.*, 112, 7799, 2000.
74. Chalfie, M. and Kain, S., John Wiley & Sons, New York, 1998.
75. Cubitt, A.B., Heim, R., Adams, S.R., et al., Understanding, improving and using green fluorescent proteins, *TIBS*, 20, 448, 1995.
76. Dickson, R.M., Cubitt, A.B., Tsien, R.Y., et al., Improved green fluorescence, *Nature*, 388, 355, 1997.
77. Moerner, W.E., Petermann, E.J.G., Brasselet, S., et al., *Cytometry*, Optical methods for exploring dynamics of single copies of green fluorescent protein, 36, 232, 1999.
78. Garcia-Parajo, M.F., Veerman, J. A., Segers-Nolten, G.M.J., et al., Visualising individual green fluorescent proteins with a near field optical microscope, *Cytometry*, 36, 239, 1999.
79. Garcia-Parajo, M.F., Segers-Nolten, G.M. J., Veerman, J.-A., et al., Real-time light-driven dynamics of the fluorescence emission in single green fluorescent protein molecules, *Proc. Natl. Acad. Sci. U.S.A.*, 97, 7237, 2000.
80. Manfait, M. and Nabiev, I., Raman microscopy applications to medicine, in *Raman Microscopy*, Corset, J., Ed., Academic Press, London, 1996, 379.
81. Carey, P.R., *Biochemical Applications of Raman and Resonance Raman Spectroscopies*, Academic Press, New York, 1982.
82. Asher, S.A., UV resonance raman spectroscopy for analytical, physical, and biophysical chemistry, *Anal. Chem.*, 65, 201A, 1993.
83. Chang, R.K. and Furtak, T.E., *Surface-Enhanced Raman Scattering*, Plenum Press, New York, 1982.
84. Moskovits, M., Surface-enhanced spectroscopy, *Rev. of Mod. Phys.*, 57, 783, 1985.
85. Kerker, M., in *Selected Papers on Surface Enhanced Raman Scattering*, Thomson, B.J., Ed., Milestone Series 10, SPIE, Bellingham,WA, 1990.
86. Vo-Dinh, T., Stokes, D.L., Griffin, G.D., et al., Surface-enhanced Raman scattering (SERS) method and instrumentation for genomics and biomedical analysis, *J. Raman Spectrosc.*, 30, 785, 1999.
87. Nie, S. and Emory, S.R., Probing single molecules and single nanoparticles by surface-enhanced Raman scattering, *Science*, 275, 1102, 1997.
88. Kneipp, K., Kneipp, H., Kartha, B., et al., Detection and identification of a single DNA base molecule using surface-enhanced Raman scattering (SERS), *Phys. Rev. B*, 57, R6281, 1998.
89. Deckert, V., Zeisel, D., Zenobi, R., et al., Near-field surface enhanced Raman imaging of dye-labeled DNA with 100 nm resolution, *Anal. Chem.*, 70, 2646, 1998.
90. Zeisel, D., Deckert, V., Zenobi, R., et al., Near-field surface-enhanced Raman spectroscopy of dye molecules adsorbed on silver island films, *Chem. Phys. Lett.*, 283, 381, 1998.
91. Gennis, R.B., *Biomembranes: Molecular Structure and Function*, Springer-Verlag, New York, 1989.
92. McConnel, H.M., Structures and transitions in lipid monolayers at the air-water interface, *Annu. Rev. Phys. Chem.*, 42, 171, 191.
93. Möhwald, H., Phospholipid and phospholipid-protein monolayers at the air/water interface, *Annu. Rev. Phys. Chem.*, 41, 441, 1990.

94. Hwang, J., Tamm, L.K., Böhm, C., et al., Nanoscale complexity of phospholipid monolayers investigated by near-field scanning optical microscopy, *Science,* 270, 610, 1995.
95. Tamm, L.K., Bohm, C., Yang, J., et al., Nanostructure of supported phospholipid monolayers and bilayers by scanning probe microscopy, *Thin Solid Films,* 285, 813, 1996.
96. Moers, M.H.P., Gaub, H.E., and van Hulst, N.F., *Langmuir,* 10, 2774, 1994.
97. Shiku, H. and Dunn, R.C., Direct observation of DPPC phase domain motion on mica surfaces under conditions of high relative humidity, *J. Phys. Chem. B,* 102, 3791, 1998.
98. Hollars, C.W. and Dunn, R.C., Submicron fluorescence, topology, and compliance measurements of phase-separated lipid monolayers using tapping-mode near-field scanning optical microscopy, *J. Phys. Chem. B,* 101, 6313, 1997.
99. Hollars, C.W. and Dunn, R.C., Submicron structure in L-alpha-dipalmitoylphosphatidylcholine monolayers and bilayers probed with confocal, atomic force, and near-field microscopy, *Biophys. J.,* 75, 342, 1998.
100. Hollars, C.W. and Dunn, R.C., Probing single molecule orientations in model lipid membranes with near-field scanning optical microscopy, *J. Chem. Phys.,* 112, 7822, 2000.
101. Hwang, J., Gheber, L.A., Margolis, L., et al., Domains in cell plasma membranes investigated by near-field scanning optical microscopy, *Biophys. J.,* 74, 2184, 1998.
102. Enderle, T., Ha, T., Chemla, D.S., et al., Near-field fluorescence microscopy of cells, *Ultramicroscopy,* 71, 303, 1998.
103. Nagy, P., Jenei, A., Kirsch, A.K., et al., Activation-dependent clustering of the erbB2 receptor tyrosine kinase detected by scanning near-field optical microscopy, *J. Cell Sci.,* 112, 1733, 1999.
104. Shinkarev, V.P., Brunner, R., and Wraight, C.A., Application of near-field scanning optical microscopy in photosynthesis research, *Photosynth. Res.,* 61, 181, 1999.
105. Enderle, T., Ha, T., Ogletree, D.F., et al., Membrane specific mapping and colocalization of malarial and host skeletal proteins in the *Plasmodium falciparum* infected erythrocyte by dual-color near-field scanning optical microscopy, *Proc. Natl. Acad. Sci. U.S.A.,* 94, 520, 1997.
106. Stryer, L., Fluorescent energy transfer as a spectroscopic ruler, *Annu. Rev. Biochem.,* 47, 819, 1978.
107. Lakowicz, J.R., *Principles of Fluorescence Microscopy,* Plenum Press, New York, 1983.
108. van der Meer, B.W., Coker, G., and Simon Chen, S.-Y., *Resonance Energy Transfer Theory and Data,* VCH Publishers, New York, 1994.
109. Chen, Q. and Lentz, B.R., Fluorescence resonant energy transfer study of shape changes in membrane-bound bovine prothrombine and meizothrombine, *Biochemistry,* 36, 4701, 1997.
110. Dos Remedios, C.G. and Moens, P.D.J., Fluorescence resonance energy transfer is a reliable ruler for measuring structural changes in proteins, *J. Struct. Biol.,* 115, 175, 1995.
111. Gonzales, J.E. and Tsien, R.Y., Voltage sensing by fluorescence resonance energy transfer in single cells, *Biophys. J.,* 69, 1272, 1995.
112. Miawaki, A., Llopis, J., Heim, R., et al., Fluorescent indicators for $Ca^{+}$ based on green fluorescent proteins and calmodulin, *Nature,* 388, 882, 1997.
113. Ha, T., Enderle, T., Ogletree, D.F., et al., Probing the interaction between two single molecules: fluorescence resonance energy transfer between a single donor and a single acceptor, *Proc. Natl. Acad. Sci. U.S.A.,* 93, 6264, 1996.
114. Ha, T., Ting, A.Y., Liang, J.W., et al., Single-molecule fluorescence spectroscopy of enzyme conformational dynamics and cleavage mechanism, *Proc. Natl. Acad. Sci. U.S.A.,* 96, 893, 1999.
115. Sekatskii, S.K. and Letokhov, V.S., Nanometer-resolution scanning optical microscope with resonance excitation of the fluorescence of the samples from a single-atom excited center, *JETP Lett.,* 63, 319, 1996.
116. Vickery, S.A. and Dunn, R.C., Scanning near-field fluorescence resonance energy transfer microscopy, *Biophys. J.,* 76, 1812, 1999.
117. Zenhausern, F., O'Boyle, M.P., and Wickramasinghe, H.K., Apertureless near-field optical microscope, *Appl. Phys. Lett.,* 65, 1623, 1994.
118. Wessel, J., Surface-enhanced optical microscopy, *J. Opt. Soc. Am. B,* 2, 1538, 1985.

119. Anderson, M.S., Locally enhanced Raman spectroscopy with an atomic force microscope, *Appl. Phys. Lett.*, 76, 3130, 2000.
120. Hayazawa, N., Inouye, Y., Sekkat, Z., et al., Metallized tip amplification of near-field Raman scattering, *Opt. Commun.*, 183, 333, 2000.
121. Kneipp, K., Wang, Y., Kneipp, H., et al., Single molecule detection using surface-enhanced Raman Scattering (SERS), *Phys. Rev. Lett.*, 78, 1667, 1997.
122. Jenei, A., Kirsch, A.K., Subramaniam, V., et al., Picosecond multiphoton scanning near-field optical microscopy, *Biophys. J.*, 76, 1092, 1999.
123. Kirsch, A.K., Subramaniam, V., Striker, G., et al., Continuous wave two-photon scanning near-field optical microscopy, *Biophys. J.*, 75, 1513, 1998.
124. Kawata, Y., Egami, C., Nakamura, O., et al., Nonoptically probing near-field microscopy, *Opt. Commun.*, 161, 1, 1999.
125. Kawata, Y., Murakami, M., Egami, C., et al., Nonoptically probing near-field microscopy for the observation of biological living cells, *Appl. Phys. Lett.*, 78, 2247, 2001.
126. Hell, S.W., Lindek, S., Cremer, C., et al., Measurement of the 4Pi-confocal point spread function proves 75 nm axial resolution, *Appl. Phys. Lett.*, 64, 1335, 1994.
127. Kino, G.S. and Mansfield, S.M., Solid immersion microscope for real-time near field imaging, *AIP Conf. Proc.*, 241, 61, 1991.

# 13

# Optical Coherence Tomography Imaging




James G. Fujimoto
*Massachusetts Institute of Technology*
*Cambridge, Massachusetts*

Mark E. Brezinski
*Brigham and Women's Hospital*
*Boston, Massachusetts*


## 13.1 Introduction

Optical coherence tomography (OCT) is a fundamentally new type of optical imaging modality. OCT performs high-resolution, cross-sectional tomographic imaging of the internal microstructure in materials and biological systems by measuring backscattered or backreflected light.[1–3] Image resolutions of 1 to 15 μm can be achieved — one to two orders of magnitude higher than conventional ultrasound. Imaging can be performed *in situ* and nondestructively. High-speed, real-time imaging is possible with acquisition rates of several frames per second. OCT enables "optical biopsy," the imaging of tissue structure or pathology on resolution scales approaching that of histopathology, with imaging performed *in situ* and in real time, without the need to excise specimens and process them as in standard excisional biopsy and histopathology.

OCT has a number of technological features that make it attractive for a broad range of applications. OCT technology is fiber-optically based and uses components developed for the telecommunications industry. Thus, it takes advantage of a well-established technology base. OCT can be interfaced with a wide range of imaging delivery systems and imaging probes. Image information is generated in electronic form that is amenable to image processing and analysis as well as electronic transmission, storage, and retrieval. Finally, OCT systems can be engineered to be compact and low cost, suitable for applications in research, manufacturing, or the clinic.


0-8493-1116-0/03/$0.00+$1.50

OCT is a powerful imaging technology in medicine because it enables the real-time, *in situ* visualization of tissue microstructure without the need to remove and process a specimen excisionally. The concept of nonexcisional "optical biopsy" performed by OCT and the ability to visualize tissue morphology in real time under operator guidance can be used for diagnostic imaging and to guide intervention. OCT promises to be useful in three general types of clinical situations:

1. Where standard excisional biopsy is hazardous or impossible. Examples include imaging the eye, coronary arteries, or nervous tissues.
2. Where standard excisional biopsy suffers from sampling errors. Biopsy followed by histopathology is the standard for cancer diagnosis; however, if the biopsy misses the lesion, then a false negative result is obtained. OCT might be used to guide biopsy to reducing sampling error or may ultimately provide diagnostic information directly.
3. For guidance of interventional procedures. The ability to see beneath the surface of tissue enables the assessment and guidance of microsurgical procedures such as vessel and nerve anastomosis or the guidance of procedures such as stent placement or atherectomy in interventional cardiology. Coupled with catheter, endoscopic, or laparoscopic delivery, OCT promises to have a powerful impact on many medical applications, ranging from improving the screening and diagnosis of neoplasia to enabling new microsurgical and minimally invasive surgical procedures. This chapter provides an overview of OCT technology and applications.

## 13.2 Principles of Operation of Optical Coherence Tomography

OCT imaging is analogous to ultrasound B mode imaging except that it uses light instead of sound. OCT images are generated by performing axial measurements of the echo time delay and magnitude of reflected light at different transverse positions, as shown in Figure 13.1. OCT images are a two- or three-dimensional data set that represents differences in optical backscattering or backreflection in a cross-sectional plane or volume of a tissue or material. Because of the analogy between OCT and ultrasound, it is helpful to consider the factors that govern OCT imaging compared to ultrasound imaging.[4–7] In ultrasound, a high-frequency sound wave is launched into the material or tissue being imaged using an ultrasonic probe transducer. The sound wave travels into the material or tissue and is reflected or backscattered from internal structures having different acoustic properties. The time behavior or echo structure of the

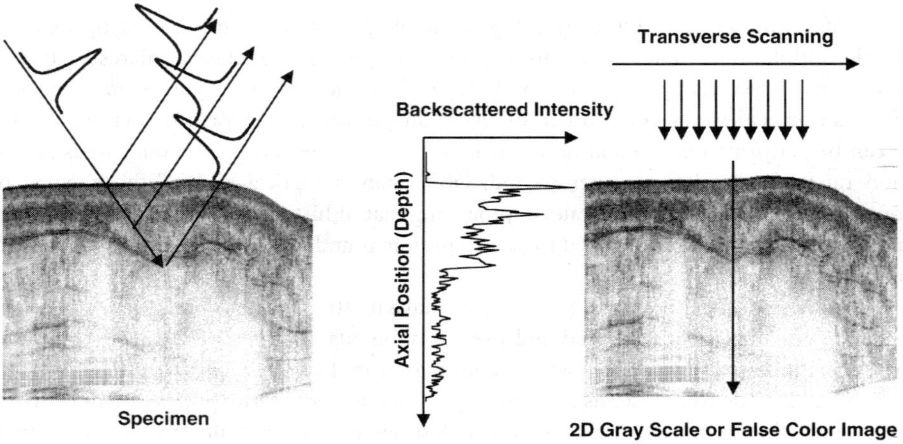


**FIGURE 13.1** OCT performs imaging by measuring the echo time delay and magnitude of backreflected or backscattering light. Cross-sectional images are constructed by performing axial measurements of the echo time delay and magnitude of backscattered or backreflected light at different transverse positions. This results in a two-dimensional data set that represents the backscattering in a cross-sectional plane of the material or tissue being imaged.

reflected sound waves is detected by the ultrasonic probe and the ranges and dimensions of internal structures determined from the echo delay. This principle is also similar to that used in radar.

In OCT, measurements of distance and microstructure are performed by directing a light beam onto the material or tissue and measuring the backreflecting and backscattering light from internal microstructural features as shown in Figure 13.1.[1] For conceptual purposes, it is possible to visualize the operation of OCT by thinking of the light beam as composed of short optical pulses. However, although OCT may be performed using short pulse light, most OCT systems operate using continuous wave, short coherence length light.

The principal difference between ultrasound and optical imaging is the fact that the velocity of propagation of light is approximately a million times faster than the velocity of sound. Because distances within the material or tissue are determined by measuring the "echo" time delay of backreflected or backscattered light waves, distance measurement using light requires ultrafast time resolution. For example, the measurement of a structure with a resolution on the 10-μm scale, which is typical in OCT, corresponds to a time resolution of approximately 30 fs. Direct electronic detection is not possible on this timescale. Thus, OCT measurements of echo time delay are based on correlation techniques that compare the backscattered or backreflected light signal to reference light traveling a known pathlength. Although OCT imaging is analogous to ultrasound, the core technology of OCT is quite different.

Finally, since ultrasound imaging depends on sound waves, it requires direct contact with the material or tissue being imaged, or immersion in a liquid or other medium to facilitate the transmission of sound waves. In contrast, optical imaging techniques such as OCT rely on the use of light rather than sound and can be performed without physical contact to the structure being imaged and without the need for a special transducing medium. In applications such as ophthalmology, the ability to perform noncontact imaging is important for patient comfort during examination. In endoscopy or bronchoscopy, the ability to image without a sound transducing medium can be a powerful advantage because contact with or occlusion of the lumen is not required.

### 13.2.1 Measuring Ultrafast Optical Echoes

The concept of using high-speed optical gating to perform imaging in biological tissues was first proposed by M. Duguay almost 30 years ago.[8,9] Duguay developed an ultrafast optical shutter to photograph light in flight based on the Kerr effect. The Kerr shutter can achieve picosecond or femtosecond resolutions and operates by using an intense light pulse to induce transient birefringence (the Kerr effect) in an optical medium placed between crossed polarizers. The principal disadvantage of the high-speed optical Kerr shutter is that it requires high-intensity laser pulses to induce the Kerr effect and operate the shutter. Duguay pointed out that optical scattering limits imaging biological tissues and that a high-speed shutter could be used to gate out unwanted scattered light, thereby detecting light echoes from internal structures at different depths in the tissue. Such a technology could be used to see through tissues and noninvasively image their internal structure.

An alternate approach to performing high-speed gating is to use nonlinear processes such as harmonic generation or parametric conversion. In this technique, short pulses are used to illuminate the object or specimen being imaged, and the backscattered or backreflected light is upconverted or parametrically converted with a reference pulse in a nonlinear optical crystal.[10,11] The time resolution is determined by the pulse duration and the sensitivity is determined by the conversion efficiency of the nonlinear process. Optical time-of-flight ranging measurements were first demonstrated in biological tissues using short pulses and a nonlinear intensity autocorrelation technique to measure corneal thickness and the depth of the stratum corneum and epidermis.[12] Dynamic ranges of $10^6$ or higher can be achieved, but these are insufficient for imaging scattering tissues where the attenuation from scattering can be $10^{-10}$. Nonlinear cross correlation does not require intensity pulses as high as the Kerr shutter, but still requires the use of short pulses. It is also important to note that this technique detects the intensity (rather than the field) of the backscattered or backreflected light.

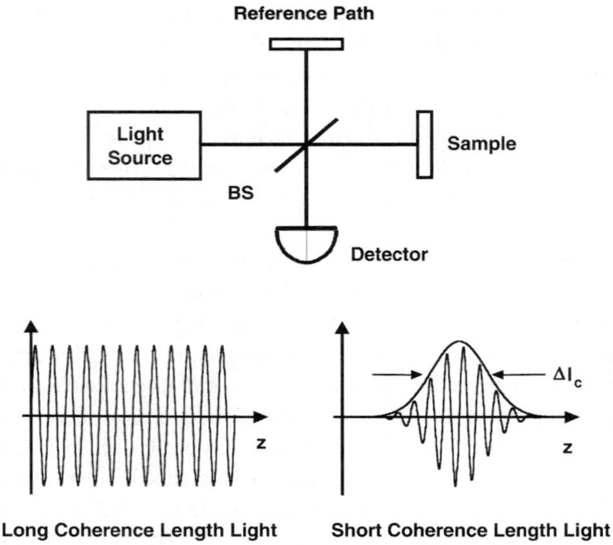


**FIGURE 13.2** The echo time delay and magnitude of backreflected or backscattering light can be measured using interferometry. The most common detection method is based on a Michelson interferometer with a scanning reference delay arm. Backreflected or backscattered light from the object being imaged is correlated with light that travels a known reference path delay.

OCT uses interferometric detection and correlation methods to measure the echo time delay of backreflected or backscattered light with high dynamic range and high sensitivity. These techniques are analogous to coherent optical detection in optical communications. OCT is based on a classic optical measurement technique known as low coherence interferometry or white light interferometry, first described by Sir Isaac Newton. More recently, low coherence interferometry has been used to characterize optical echoes and backscattering in optical fibers and waveguide devices.[13–15] The first biological application of low coherence interferometry was in ophthalmologic biometry for measurement of eye length.[16] Since then, related versions of this technique have been developed for noninvasive high-precision and high-resolution biometry.[17,18] High-resolution measurements of corneal thickness *in vivo* have been demonstrated using low coherence interferometry.[19]

Low coherence interferometry measures the field of the optical beam rather than its intensity. Figure 13.2 shows a schematic diagram of a simple Michelson type interferometer. The incident optical wave into the interferometer is directed onto a partially reflecting mirror or beamsplitter, which yields two beams, one of which acts as a reference beam and the other as a measurement or signal beam. The beams travel given distances in the two arms of the interferometer. The reference beam $E_r(t)$ is reflected from a reference mirror while the measurement or signal beam $E_s(t)$ is reflected from the biological specimen or tissue to be imaged. The beams then recombine and interfere at the beamsplitter. The output of the interferometer is a sum of the fields from the reference beam reflected from the reference mirror and the measurement or signal beam reflected from the specimen or tissue. The detector measures the intensity of the output optical beam that is proportional to the square of the electromagnetic field. If the distance that light travels in the reference path of the interferometer is $l_R$ and the distance that light travels in the measurement path, reflected from the specimen, is $l_S$, then the intensity of the output from the interferometer will oscillate as a function of the difference in distance $\Delta l = l_r - l_s$ because of interference effects. The intensity as a function of $E_s$, $E_r$, and $\Delta l$ is given as

$$I_o(t) \sim (1/4)|E_r|^2 + (1/4)|E_s|^2 + (1/2)E_r E_s \cos\{2(2\pi/\lambda)\Delta l\}.$$

If the position of the reference mirror is varied, then the pathlength that the optical beam travels in the reference arm changes. Interference effects will be observed in the intensity detected by the detector if the relative pathlength is changed by scanning the reference mirror.

If the light is highly coherent (narrow linewidth or bandwidth) with a long coherence length, then interference oscillations will be observed for a wide range of relative pathlength delays of the reference and measurement arms. For applications in optical ranging or optical coherence tomography, it is necessary to measure absolute distance and dimensions of structures precisely within the material or tissue. In this case, light with a short coherence length (broad bandwidth) is used. Low coherence light can be characterized as having statistical discontinuities in phase over a distance known as the coherence length ($l_c$). The coherence length is inversely proportional to the frequency bandwidth.

When low coherence light is used, interference is only observed when the pathlengths of the reference and measurement arms are matched to within the coherence length of the light. If the pathlengths differ by more than the coherence length, then the electromagnetic fields from the two beams are not correlated, and there is no interference. The interferometer measures the field autocorrelation of the light. For the purposes of ranging and imaging, the coherence length of the light determines the resolution with which optical range or distance can be measured. The magnitude and echo time delays of reflected light can be measured by scanning the reference mirror delay and demodulating the interferometer interference signal.

### 13.2.2 Resolution and Sensitivity of Optical Coherence Tomography

In contrast to conventional microscopy, in OCT the mechanisms that govern the axial and transverse image resolution are decoupled. The axial resolution in OCT imaging is determined by the coherence length of the light source, and high axial resolution can be achieved independent of the beam focusing conditions. The coherence length is the spatial width of the field autocorrelation produced by the interferometer. The envelope of the field autocorrelation is equivalent to the Fourier transform of the power spectrum. Thus, the width of the autocorrelation function, or the axial resolution, is inversely proportional to the width of the power spectrum. For a source with a Gaussian spectral distribution, the axial resolution $\Delta z$ is given as

$$\Delta z = (2 \ln 2/\pi)(\lambda^2/\Delta\lambda),$$

where $\Delta z$ and $\Delta\lambda$ are the full-widths-at-half-maximum of the autocorrelation function and power spectrum, respectively, and $\lambda$ is the source center wavelength. Broad bandwidth optical sources are required to achieve high axial resolution.

The transverse resolution in OCT imaging is the same as for conventional optical microscopy and is determined by the focusing properties of an optical beam. The minimum spot size to which an optical beam may be focused is inversely proportional to the numerical aperture or the angle of focus of the beam. The transverse resolution is given as

$$\Delta x = (4\lambda/\pi)(f/d),$$

where $d$ is the spot size on the objective lens and $f$ is its focal length. High transverse resolution can be obtained by using a large numerical aperture and focusing the beam to a small spot size. The transverse resolution is also related to the depth of focus or confocal parameter $b$, which is $2z_R$, two times the Rayleigh range:

$$b = 2z_R = \pi\Delta x^2/2\lambda .$$

Thus, increasing the transverse resolution produces a decrease in the depth of focus, similar to that which occurs in conventional microscopy.

OCT can achieve high detection sensitivity because interferometry is equivalent to optical heterodyne detection and measures the field rather than the intensity of light. Because the beam from the reference mirror can have a large amplitude, the weak electric field of the signal beam from the specimen is multiplied by a large field, thereby increasing the magnitude of the interference output of the interferometer. The interferometer thus produces heterodyne gain for weak optical signals.

In most OCT embodiments the reference mirror is scanned at a constant velocity $v$ resulting in a Doppler shift of the reflected light. This produces a modulation of the interference signal at the Doppler beat frequency $f_D = 2v/\lambda$, where $v$ is the reference mirror velocity. By electronically bandpass filtering and demodulating the photodetector signal, echoes from different reflecting surfaces in the biological specimen may be detected. The signal-to-noise ratio (SNR) performance can be calculated using well-established mathematical models and is given as

$$\mathrm{SNR} = 10\log\left(\eta P/2h\nu\mathrm{NEB}\right),$$

where $\eta$ is the detector quantum efficiency, $h\nu$ is the photon energy, $P$ is the signal power, and NEB is the noise equivalent bandwidth of the electronic filter used to demodulate or detect the signal. This expression implies that the SNR scales as the detected power divided by the noise equivalent bandwidth of the detection. High image acquisition speeds or higher image resolutions require higher optical powers for a given SNR.

The performance of optical coherence tomography systems varies widely according to their design and data acquisition speed requirements. However, for typical measurement parameters, sensitivities to reflected signals in the range of –90 to –100 dB can be achieved, corresponding to the detection of backreflected or backscattered signals that are $10^{-9}$ or $10^{-10}$ of the incident optical power.

## 13.2.3 Image Generation in Optical Coherence Tomography

Optical coherence tomography cross-sectional imaging is achieved by performing successive axial measurements of backreflected or backscattered light at different transverse positions.[1] Figure 13.1 schematically illustrates one example of how optical coherence tomography is performed. A two-dimensional cross-sectional imaging is acquired by performing successive rapid axial measurements of optical backscattering or backreflection profiles at different transverse positions by scanning the incident optical beam. The result is a two-dimensional data set where each trace represents the magnitude of reflection or backscattering of the optical beam as a function of depth in the tissue.

OCT image data are acquired by computer and displayed as a two-dimensional gray scale or false color image. Figure 13.3 shows an example of a tomographic image of the retina. The image displayed consists of 200 pixels (horizontal) by 500 pixels (vertical). The vertical direction corresponds to the direction of the incident optical beam and the axial data sets. The backscattered signal ranges from approximately –50 dB, the maximum signal, to –100 dB, the sensitivity limit. Because the signal varies over five orders of magnitude, it is convenient to use the log of the signal to display the image. This expands the dynamic range of the display but results in compression of relative variations in signal.

The previous images of the retina are examples of imaging media that have weak reflections. Figure 13.4 shows an OCT image of the human bronchus *in vitro* as an example of an image in highly scattering tissue. In a highly scattering material or tissue, light is rapidly attenuated with propagation depth resulting in a gradation of signal in the image. Also, significant levels of speckle noise or other noise can arise from the microstructural features of the material or specimen. For these reasons it is more common to display OCT images in highly scattering materials or tissues using a gray scale.

The detection sensitivity determines the imaging performance of optical coherence tomography for different applications. When OCT imaging is performed in very weakly scattering media, the imaging depth is not strongly limited, because there is very little attenuation of the incident beam. Instead, the sensitivity of the OCT detection establishes a limit on the smallest signals that can be detected. For example, in ophthalmic imaging, high sensitivity is essential in order to image structures, such as the

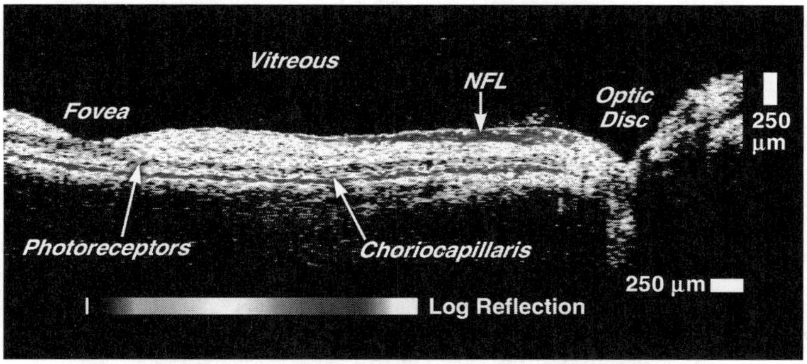


**FIGURE 13.3** False color OCT image in low scattering tissue. Example of an OCT image of the normal retina *in vivo* in a human subject. The image was performed at 800 nm and has a 10-μm axial resolution. The image consists of 200 axial scans and shows the differentiation of retinal layers possible using false color display. The log of the backscattering or backreflected signal is mapped to a false color scale. The retinal pigment epithelium, choroid, and retinal nerve fiber layers are visible as highly backscattering red layers. (From Hee, M.A. et al., *Arch. Ophthalmol.*, 113, 325, 1995. With permission.)

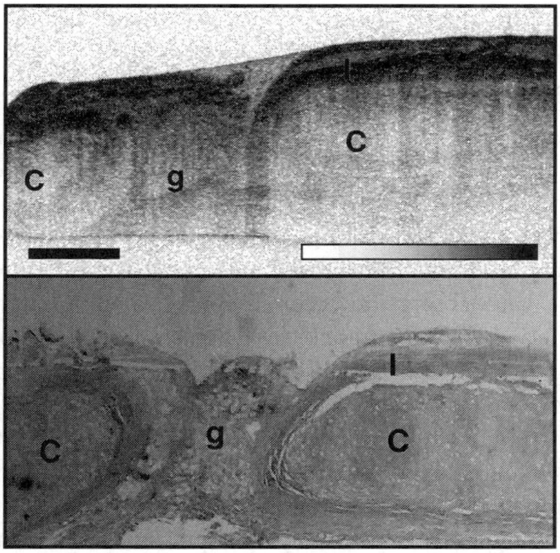


**FIGURE 13.4** Gray scale OCT image in scattering tissue. Example of an OCT image of the human secondary bronchus *in vitro* and corresponding histology. Cartilage, lamina propria, and glands can be visualized. OCT was able to image through the entire thickness of the bronchus. This image was performed at 1300-nm wavelengths and illustrates features common in many scattering tissues. These images show strong attenuation of the signal with depth as well as speckle noise; hence, they are more prone to display artifacts than OCT images in weakly scattering tissues and are often displayed using a log gray scale. The bar represents 500 μm. (From Pitris, C. et al., *Am. J. Respir. Crit. Care Med.*, 157, 1640, 1998. With permission.)

retina, that are nominally transparent and have very low backscattering. In ophthalmic applications, image contrast arises because of differences in the backscattering properties of different tissues. Structures such as the retinal pigment epithelium (RPE) or retina nerve fiber layer can be differentiated from other structures because of their different scattering amplitudes. Typical retinal images have signal levels between –50 to –100 dB of the incident signal. For retinal imaging, the ANSI standards govern the maximum permissible light exposure and set limits for the sensitivity of OCT imaging and OCT imaging speeds.[20,21]

In contrast, OCT imaging in highly scattering media also requires high sensitivity and dynamic range. In this case, the detection sensitivity determines the maximum depth to which imaging can be performed.

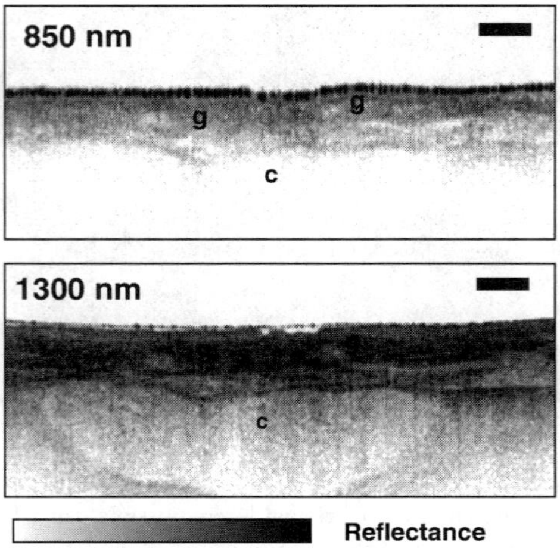


**FIGURE 13.5** Wavelength dependence of OCT imaging depth. OCT images of human epiglottis *in vitro* performed with light sources at 850 and 1300 nm. Attenuation from scattering is reduced using longer wavelengths. Superficial glands (g) can be visualized at both wavelengths but underlying structures such as cartilage (c) can be imaged only with 1300-nm light. With a 100-dB detection sensitivity, image penetration depths of 2 to 3 mm are achieved in most scattering tissues. The bar represents 500 μm. (From Brezinski, M.E. et al., *Circulation*, 93, 1206, 1996. With permission.)

Most biological tissues are highly scattering and OCT imaging in tissues other than the eye became possible with recognition that using longer optical wavelengths, 1300 nm compared to 800 nm, can reduce scattering and increase image penetration depths.[18,22–24] Figure 13.5 shows an example of OCT imaging in a human epiglottis *in vitro* comparing 800- to 1300-nm imaging wavelengths. The dominant absorbers in most tissues are melanin and hemoglobin, which have absorption in the visible and near-infrared wavelength range. Water absorption becomes appreciable for wavelengths approaching 1.9 to 2 μm. In most tissues, scattering at near-infrared wavelengths is one to two orders of magnitude higher than absorption. Scattering decreases for longer wavelengths, so that OCT image penetration increases.[22] For example, if a tissue has a scattering coefficient in the range of $\sim 40\ \text{cm}^{-1}$ at 1300 nm, then the round-trip attenuation from scattering alone from a depth of 3 mm is $e^{-24} \sim 4 \times 10^{-11}$. Thus, if the detection sensitivity is –100 dB, backscattered or backreflected signals are attenuated to the sensitivity limit when the image depth is 2 to 3 mm. Because the attenuation is exponential with depth, increasing the detection sensitivity would not appreciably increase the imaging depth. Image penetration depths of 2 to 3 mm are possible in most scattering tissues.

## 13.3 Optical Coherence Tomography Technology and Systems

Optical coherence tomography has the advantage that it can be implemented using compact fiber-optic components and integrated with a wide range of medical instruments. Figure 13.6 shows a schematic of an OCT system using a fiber-optic Michelson-type interferometer. A low coherence light source is coupled into the interferometer, and the interference at the output is detected with a photodiode. One arm of the interferometer emits a beam that is directed and scanned on the sample being imaged, while the other arm of the interferometer is a reference arm with a scanning delay.

OCT systems can be considered from a modular viewpoint in terms of integrated hardware and software modules with various functionalities. Figure 13.7 shows a schematic of the different modules in an OCT imaging system. The OCT system can be divided into the imaging engine, low coherence light source, beam delivery and probes, computer control, and image processing.

In general, the imaging engine can be any optical detection device that performs high-resolution and high-sensitivity detection of backreflected or backscattered optical echoes. Most OCT systems have

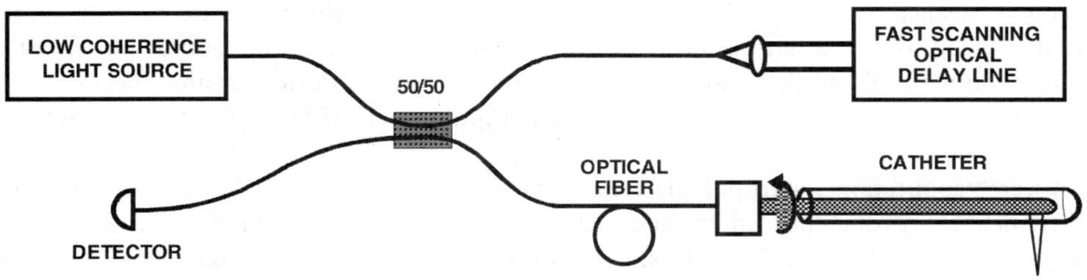


**FIGURE 13.6** Schematic showing of an OCT instrument using a fiber-optic Michelson interferometer with low coherence light source. One arm of the interferometer is integrated with an imaging probe, and the other arm has a scanning delay line. The system shown is configured for catheter/endoscope imaging. The fiber-optic implementation allows OCT to be integrated with a wide range of instruments.

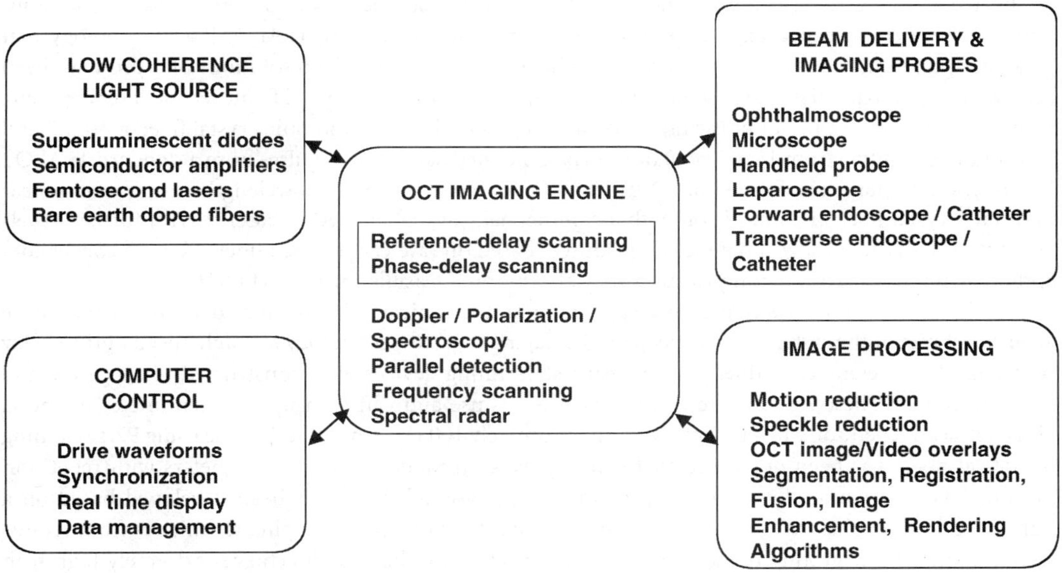


**FIGURE 13.7** OCT from a systems viewpoint. OCT is a modular technology consisting of hardware and software. The major modules in the OCT system include the imaging engine, light source, delivery system and optical probes, computer control, and image processing.

employed a reference delay scanning interferometer using a low coherence light source. There are many different embodiments of the interferometer and imaging engine for specific applications such as polarization diversity (insensitive) imaging, polarization sensitive imaging, and Doppler flow imaging. Doppler flow imaging has been performed using imaging engines that directly detect the interferometric output rather than demodulating the interference fringes.[25–27] Polarization sensitive detection techniques have been demonstrated using a dual channel interferometer.[28–30] These techniques permit imaging of the birefringence properties of structures. Conversely, polarization diversity or polarization insensitive interferometers can also be built using well-established detection methods from coherent heterodyne optical detection. Finally, other imaging engine approaches have been demonstrated that are based on spectral analysis of broadband light sources as well as tunable narrow linewidth sources.[31] These imaging engines have the advantage that they do not require scanning an optical delay, and thus no moving parts are required.

The short coherence length light source determines the resolution of the OCT system. For clinical applications, compact superluminescent diodes or semiconductor-based light sources have been used. Although these sources have not yet reached the performance levels of research systems, they are compact and robust enough to be used in the clinical environment. Ophthalmic OCT systems have employed commercially available compact GaAs superluminescent diodes, which operate near 800 nm with bandwidths of 20 to 30 nm and achieve axial resolutions of ~10 μm with output powers of a few milliwatts. Commercially available sources based on semiconductor devices or amplifiers operating at 1.3 μm can achieve axial resolutions of ~15 μm with output powers of 15 to 20 mW, sufficient for real-time OCT imaging.

For research applications, short pulse lasers are powerful light sources for OCT imaging because they have extremely short coherence lengths and high output powers, enabling high-resolution, high-speed imaging. Studies have been performed using a femtosecond $Cr^{4+}$:Forsterite laser. This laser produces output powers of greater than 100 mW with ultrashort pulses of less than 100 fs at wavelengths near 1300 nm. By using nonlinear self-phase modulation in optical fibers, the spectral output can be broadened to produce bandwidths sufficient to achieve axial resolutions of 5 to 10 μm.[32] Image acquisition speeds of several frames per second are achieved with SNRs of 100 dB using incident powers in the 5 to 10 mW range.

Short pulse Ti:$Al_2O_3$ lasers operating near 800 nm have also been used to achieve high resolutions. Early studies demonstrated axial resolutions of 4 μm.[33] Recently, using Ti:$Al_2O_3$ laser technology that generates pulse durations of ~5 fs and bandwidths of over 300 nm, axial resolutions of 1 μm have been achieved.[34] Broadband light can also be generated using standard commercial femtosecond lasers by using continuum generation in nonlinear fibers. Using a high nonlinearity photonic crystal fiber, a broadband continuum extending from 1400 to 400 nm can be generated using 100-fs pulses from a standard Ti:$Al_2O_3$. Recently, OCT image resolutions of ~3 μm were achieved at 1300-nm wavelengths using a nonlinear photonic crystal fiber in conjunction with a commercially available Ti:$Al_2O_3$ laser.[35] With further development, other types of low coherence light sources based on rare earth doped fibers, Raman conversion, and other nonlinear conversion processes may also become feasible for clinical OCT systems.

In OCT systems that use reference delay scanning, the optical delay scanner determines the image acquisition rate of the system. High-speed OCT imaging requires technology for high-speed optical delay scanning of the reference pathlength. The earliest scanning devices were constructed using galvos and retroreflectors.[36] These systems are simple and easy to use and can scan up to a centimeter or more. However, scan repetition rates are limited to approximately 100 Hz. A novel technique using PZT scanning of optical fibers has been developed that can achieve scan ranges of a few millimeters with repetition rates of 500 Hz or more.[37,38] A high-speed scanning phase delay line has been developed based on a diffraction grating phase control device.[39] This device is based on pulse shaping techniques from femtosecond optics.[40] The grating phase control scanner is attractive because it achieves extremely high scan speeds and also permits independent control of the phase and group velocity of the scanning. Imaging can be performed at several frames per second.[41]

OCT beam delivery and optical probes can be designed for many applications. Because OCT technology is fiber-optic based, it can be easily integrated with many standard optical diagnostic instruments including instruments for internal body imaging. OCT has been integrated with slit lamps and fundus cameras for ophthalmic imaging of the retina.[21,42] In these instruments, the OCT beam is scanned using a pair of galvonometric beam steering mirrors and relay imaged onto the retina. The ophthalmic OCT imaging instrument is a relatively complex design because it must allow the retina (fundus) to be viewed *en face* while showing the position of the tomographic scanning beam as well as the resulting OCT tomographic image. The ability to register OCT images with *en face* features and pathology is an important consideration for many applications. Similar principles apply for the design of OCT using low numerical aperture microscopes, which have been used for imaging developmental biology specimens *in vivo* as well as for surgical imaging applications.[43–45]

Other beam delivery systems include forward imaging devices that perform one- or two-dimensional beam scanning. Rigid laparoscopes are based on relay imaging using Hopkins-type relay lenses or graded index rod lenses. OCT can be readily integrated with laparoscopes to permit internal body OCT imaging with a simultaneous *en face* view of the scan registration.[37,46] Handheld imaging probes, devices that resemble light pens and use piezoelectric or galvonometric beam scanning, have also been demonstrated.[46,47] Handheld

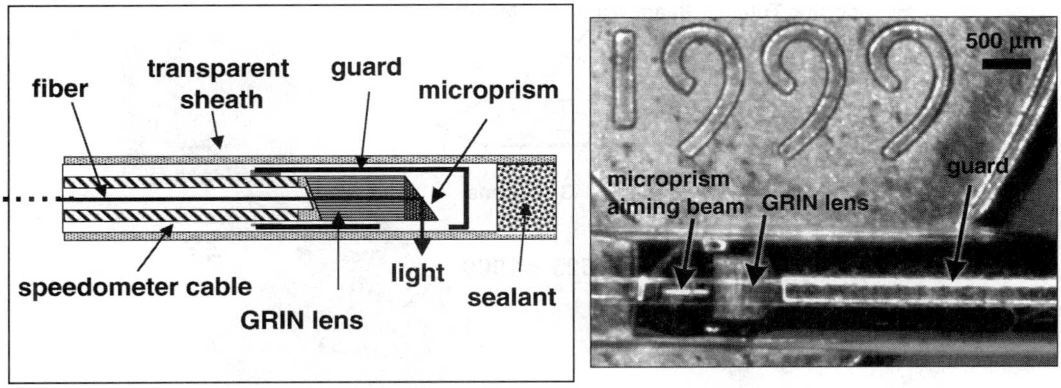


**FIGURE 13.8** Miniature fiber-optic OCT catheter probe. This catheter is designed for transverse scanning OCT imaging. A single-mode fiber is contained in a rotating flexible speedometer cable. The distal end has beam delivery optics that direct and focus the beam at 90° from the axis of the catheter. The rotating cable and optics are enclosed in a protective plastic sheath. The diameter of the catheter is 2.9 F or 1 mm. The catheter is shown on a penny for scale.

probes can be used in open field surgical situations to permit the clinician to view the subsurface structure of tissue by aiming the probe at the desired structure. These devices can also be integrated with conventional scalpels or laser surgical devices to permit simultaneous, real-time viewing as tissue is being resected.

Another class of imaging probes includes flexible miniature devices for internal body imaging. Because OCT beam delivery is performed by fiber-optics, small-diameter scanning endoscopes and catheters can be developed.[48] The first clinical studies of internal body imaging were performed using a novel forward scanning flexible probe that could be used in conjunction with a standard endoscope, bronchoscope, or trocar.[37,49] This device used a miniature magnetic scanner to produce a one-dimensional forward beam scanning in a 1.5- to 2-mm-diameter probe.

Transverse scanning OCT catheters/endoscopes have also been developed and demonstrated.[41,48,50] Figure 13.8 shows an example of a transverse scanning OCT catheter/endoscope. This device has a single-mode optical fiber encased in a hollow rotating torque cable coupled to a distal lens and a microprism that directs the OCT beam radially outward from the catheter. The cable and distal optics are encased in a transparent housing. The OCT beam is scanned by rotating the cable to perform imaging in a radar-like pattern, cross sectionally through vessels or hollow organs. Imaging may also be performed in a longitudinal plane by push–pull movement of the fiber-optic cable assembly. The catheter/endoscope shown in Figure 13.8 has a diameter of 2.9 French or 1 mm, comparable to the size of a standard intravascular ultrasound catheter. The development of even smaller diameter OCT catheters/endoscopes is possible. These types of OCT imaging probes are small enough to fit in the accessory port of a standard endoscope or bronchoscope and permit an *en face* view of the area being scanned simultaneously with OCT imaging. Alternately, catheters/endoscopes can be used independently, for example, in intravascular imaging.

Even smaller needle imaging devices can be developed.[51] Recently, a 27-gauge OCT imaging needle has been developed for performing OCT imaging in solid tissues or tumors, as shown in Figure 13.9. Following the concept of acupuncture needles, these devices can be inserted into solid tissues with minimal trauma. Since imaging can be performed to depths of 2 to 3 mm in most tissues, image needles enable imaging a cylindrical volume of tissue with a radius of 2 to 3 mm or a diameter of 4 to 6 mm and with arbitrary length. Figure 13.9 shows a protoype of an OCT imaging needle constructed using an optical fiber, graded index lens, and microprism as well as an example of *in vivo* OCT imaging the hamster leg muscle, an example of solid tissue. Ultimately, OCT imaging devices could be integrated with excisional biopsy devices such as needle or core biopsy devices to enable imaging of the tissue to be excised, thereby reducing sampling errors.

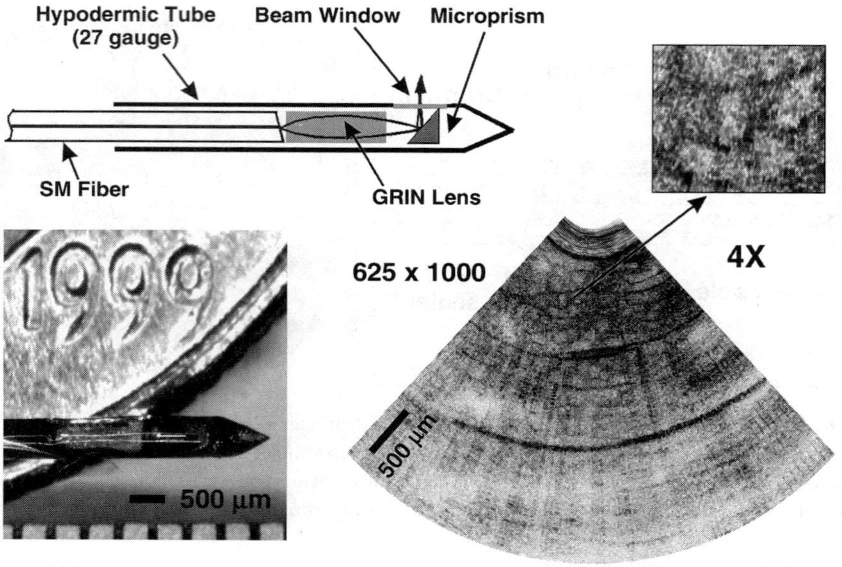


**FIGURE 13.9** OCT imaging needle. This needle consists of a single mode optical fiber with a GRIN lens and a micoprism used to focus and deflect the beam. A thin-wall, 27-gauge, stainless steel hypodermic tube houses the optical fiber and distal optics. A photo of a prototype OCT needle on a U.S. dime shows the size of the needle. Also shown is an OCT image of a hamster leg muscle *in vivo* acquired with the needle. Muscle fascicles are clearly discerned. The needle enables imaging of a 4- to 5-mm-diameter cylindrical region in solid tissues. The scale bar is 500 µm. (From Li, X.D. et al., *Opt. Lett.*, 25, 1520, 2001. With permission.)

## 13.4 Applications of Optical Coherence Tomography

Optical coherence tomography — performing two-dimensional measurements of optical backscattering of backreflection to image internal cross-sectional microstructure — was demonstrated in 1991.[1] OCT imaging was first demonstrated *in vitro* in the human retina and in atherosclerotic plaque as examples of imaging in transparent, weakly scattering media and highly scattering media. Since that time, numerous applications of OCT for both materials and medical applications have emerged. The performance and capabilities of OCT imaging technology have also improved dramatically. The next sections describe different examples of general areas of application for OCT.

### 13.4.1 Optical Coherence Tomography Imaging in Ophthalmology

Optical coherence tomography was first applied for imaging in the eye and, to date, OCT has had the largest clinical impact in ophthalmology. The first *in vivo* tomograms of the human optic disc and macula were demonstrated in 1993.[20,52] Numerous clinical studies have been performed in the last few years. OCT enables the noncontact, noninvasive imaging of the anterior eye, including the cornea, iris, and lens, as well as imaging of morphologic features of the human retina including the fovea and optic disc.[21,42,53]

Figure 13.3 showed an example of an OCT image of the normal retina of a human subject, which provides a cross-sectional view of the retina with unprecedented high resolution. Although the retina is almost transparent and has extremely low optical backscattering, the high sensitivity of OCT allows extremely small backscattering from features such as inter-retinal layers and the vitreal–retinal interface to be imaged. The retinal pigment epithelium, which contains melanin, and the choroid, which is highly vascular, are visible as highly scattering structures in the OCT image. Because typical OCT retinal images have a resolution of 10 µm, there can be residual motion of the patient's eye during the measurement. However, image processing algorithms can be used to measure the axial motion of the eye and correct for motion artifacts.[20]

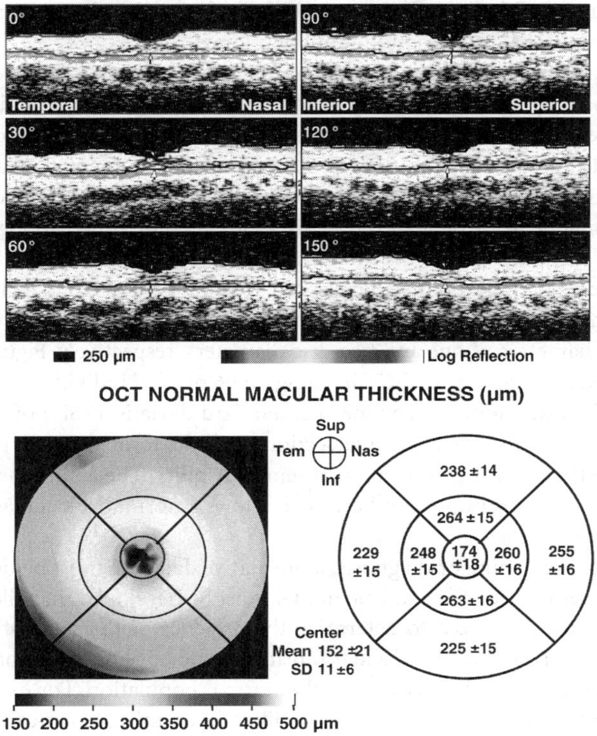


**FIGURE 13.10** Topographical mapping of retinal thickness as an example of image processing and display. The topographical representation is constructed by performing six OCT tomograms at different angles through the fovea. The tomograms are image-processed to segment the retinal thickness and the thickness values linearly interpolated over the macular region. The retinal thickness is displayed using a false color scale. Quantitative data of retinal thickness in nine different zones are also shown. These values are from a normative database and are displayed as mean and standard deviation. Quantitative data are more useful for screening or detection purposes. (From Hee, M.R. et al., *Ophthalmology*, 105, 360, 1998. With permission.)

OCT has been demonstrated for the detection and monitoring of a variety of macular diseases,[53] including macular edema,[54] macular holes,[55] central serous chorioretinopathy,[56] and age-related macular degeneration and choroidal neovascularization.[57] The retinal nerve fiber layer thickness, a predictor for early glaucoma, can be quantified in normal and glaucomatous eyes and correlated with conventional measurements of the optic nerve structure and function.[58–60] In addition, OCT has been applied for the evaluation of choroidal tumors,[61] congenital optic disc pits,[62] and argon laser lesions.[63] Retinal OCT images have been compared to histology,[64] and OCT has been used to investigate retinal degeneration in transgenic animal models.[65]

OCT is especially promising for the diagnosis and monitoring of diseases such as glaucoma or macular edema associated with diabetic retinopathy because it can provide quantitative information on retinal pathology that is a measure of disease progression. Images can be analyzed quantitatively and processed using intelligent algorithms to extract features such as retinal or retinal nerve fiber layer thickness.[66] Mapping and display techniques have been developed to represent the tomographic data in alternate forms, such as thickness maps, in order to aid interpretation.

As an example, consider the measurement of retinal thickness, shown in Figure 13.10. Measurement of retinal thickness in the macula is important in patients with diabetes who are prone to macular edema and diabetic retinopathy. The utility of OCT in clinical practice depends on the ability of the physician to accurately and quickly interpret the OCT results in the context of conventional clinical examination. Standard ophthalmic examination techniques give an *en face* view of the retina. Therefore, OCT topographic methods of displaying retinal and nerve fiber layer thickness, which could be directly compared with standard view of the retina during ophthalmic examination, were developed.[66]

Figure 13.10 shows an example of an OCT topographic map of retinal thickness. The thickness map is constructed by performing six standard OCT scans at varying angular orientations through the fovea. The images are then segmented to detect the retinal thickness along the OCT tomograms. The retinal thickness is then linearly interpolated over the macular region and represented in false color. The diameter of the topographic map is approximately 6 mm. Retinal thickness values between 0 and 500 μm are displayed by colors ranging from blue to red. This yields a color-coded topographic map of retinal thickness that displays the retinal thickness in an *en face* view.

For quantitative interpretation, the macula was divided into nine regions including a central disc of 500 μm diameter (denoted as the foveal region), and an inner and outer ring, each divided into four quadrants, with outer diameters of one and two disc diameters, respectively. Figure 13.10 also shows the average retinal thickness measured from OCT imaging on 96 normal individuals with statistics including mean values and standard deviations.[66] The mean ± standard deviation (SD) of central foveal thickness was also recorded for the six A-scans at the intersection of all the tomograms in the central macula. The SD provides a simple estimate of the measurement reproducibility. The ability to reduce image information to numerical information is important because it allows a normative database to be developed and statistics to be calculated.

In order to be used for screening or diagnosis, a normative database must be developed and extensive cross-sectional studies performed. A diagnostic model must be developed that allows the OCT image or numerical information to be analyzed to determine the presence or the stage of a given disease. These diagnostic criteria must perform with sufficient sensitivity and specificity. In many cases, screening or diagnosis is complicated by natural variations in the normal population. For example, normal subjects lose retinal nerve fibers with age, so retinal nerve fiber layer thickness measurements must be age adjusted.[67] Ongoing clinical studies are addressing the viability of using OCT for screening and diagnosis of diabetic macular edema and glaucoma, two of the leading causes of blindness.

OCT has the potential to detect and diagnose early stages of disease before physical symptoms and irreversible loss of vision occur. Furthermore, repeated imaging can easily be performed to track disease progression or monitor therapy. Many thousands of patients have been examined to date, and numerous clinical studies are currently under way in many research and clinical groups internationally. The technology was transferred to industry and introduced commercially for ophthalmic diagnostics in 1996 (Zeiss Humphrey Systems, Dublin, CA); a third-generation ophthalmic instrument was introduced in 2001. With continued development and investigation, we expect OCT ultimately to become a standard ophthalmic diagnostic.

### 13.4.2 Optical Coherence Tomography and Optical Biopsy

With recent research advances, OCT imaging of optically scattering, nontransparent tissues has become possible, thus enabling a wide variety of applications in internal medicine and internal body imaging. One of the techniques that enabled OCT imaging in nontransparent tissues was imaging at longer wavelengths where optical scattering is reduced.[18,23,24] Using 1.3-μm wavelengths, image penetration depths of 2 to 3 mm can be achieved in most tissues. This imaging depth is shallow compared to ultrasound, but is comparable to the depth over which many biopsies are performed. In addition, many diagnostically important changes of tissue morphology occur at the epithelial surfaces of organ lumens. The capability to perform "optical biopsy" — the *in situ*, real-time imaging of tissue morphology — could be important in a variety of clinical scenarios, including:

1. Assessing tissue pathology in situations where conventional excisional biopsy is hazardous or impossible
2. Guiding conventional biopsy to reduce false negative rates from sampling errors
3. Guiding surgical or microsurgical intervention

**FIGURE 13.11** OCT image showing atherosclerotic plaque *in vitro* and corresponding histology. This shows an example of OCT imaging in tissues where standard excisional biopsy is not possible. The plaque is highly calcified with a low lipid content and a thin intimal cap layer is observed. The high resolution of OCT can resolve small structures, such as the thin intimal layer, that are associated with unstable plaques. The bar is 500 μm. (From Brezinski, M.E., *Circulation*, 93, 1206, 1996. With permission.)

### 13.4.3 Imaging Where Excisional Biopsy Is Hazardous or Impossible

One class of applications where OCT could be especially powerful is where conventional excisional biopsy is hazardous or impossible. In ophthalmology, biopsy of the retina is not possible and OCT can provide high-resolution images of retinal structure that cannot be obtained using any other *in situ* imaging technique. Another situation where biopsy is not possible is in assessing atherosclerotic plaque morphology in the coronary arteries.[24,68] Recent research has demonstrated that most myocardial infarctions result from the rupture of small to moderately sized cholesterol-laden coronary artery plaques followed by thrombosis and vessel occlusion.[69–72] The plaques at highest risk for rupture are those that have a structurally weak fibrous cap.[73] These plaque morphologies are difficult to detect by conventional radiologic techniques and their microstructural features cannot be determined. Identifying high-risk unstable plaques and patients at risk for myocardial infarction is important because of the high percentage of infarctions that result in sudden death.[74] OCT could be a powerful diagnostic tool for intravascular imaging in risk stratification as well as guidance of interventional procedures such as atherectomy or stenting.

Imaging studies of arterial lesions *in vitro* have been performed to investigate the correlation of OCT and histology.[24,68,75] Figure 13.11 shows an example of an unstable plaque morphology from a human abdominal aorta specimen with corresponding histology. OCT imaging was performed at 1300-nm wavelength with an axial resolution of ~16 μm. The optical scattering properties of lipid, adipose tissue and calcified plaque are different and provide contrast between different structures and plaque morphologies. The OCT image and histology show a small intimal layer covering a large atherosclerotic plaque that is heavily calcified and has a relatively low lipid content. The thin intimal wall increases the likelihood of plaque rupture. Currently, these lesions can only be accurately diagnosed by postmortem histology; therefore, interventional techniques cannot be used effectively to treat them. The ability to identify structural details such as the presence and thickness of intimal caps using catheter-based OCT imaging techniques could lead to significant improvements in patient outcome.

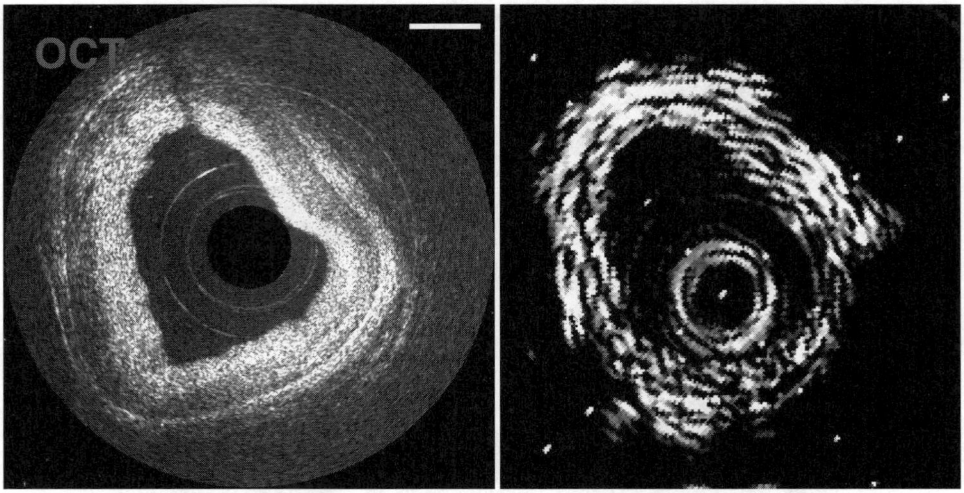


**FIGURE 13.12** OCT catheter-based image (left) vs. IVUS (right) in a human artery *in vitro*. OCT imaging was performed with a prototype OCT catheter. The OCT image enables the differentiation of the intima, media, and adventitia of the artery. Intimal hyperplasia is evident in the OCT image. (From Tearney, G.J. et al., *Circulation*, 94, 3013, 1996. With permission.)

Studies have been performed that compare IVUS and OCT imaging *in vitro* and demonstrate the ability of OCT to identify clinically relevant pathology.[68,76] Figure 13.12 shows a comparison of an OCT catheter image vs. an intravascular ultrasound (IVUS) in a human artery *in vitro. In vivo* OCT arterial imaging has also been performed in New Zealand white rabbits and porcine coronary arteries using a catheter-based imaging system.[50,77] For *in vivo* imaging, optical scattering from blood limits OCT imaging penetration depths so that saline flushing at low rates is required for imaging. These effects depend on vessel diameter as well as other factors, and additional investigation is required. Clinical studies of intravascular OCT imaging human subjects are just beginning.

## 13.4.4 Detecting Early Neoplastic Changes

Another major class of OCT imaging applications is in situations where conventional excisional biopsy has unacceptably high false negative rates due to sampling errors, important for cancer diagnosis. OCT can resolve changes in architectural morphology that are associated with many early neoplasias. Studies have been performed *in vitro* to correlate OCT images with histology for pathologies of the gastrointestinal, biliary, female reproductive, pulmonary, and urinary tracts.[78–87] One of the key issues for the use of OCT imaging in the detection of early neoplastic changes is the ability to differentiate clinically relevant pathologies with a given image resolution. Many early neoplastic changes are manifested by disruption of the normal glandular organization or architectural morphology of tissues. These changes can be imaged using standard 10 to 15 μm resolution OCT systems.

Figure 13.13 shows an example of an *in vitro* OCT image of normal colon and carcinoma. The OCT image of the colon epithelium shows normal glandular organization associated with columnar epithelial structure.[46,87] The mucosa and muscularis mucosa can be differentiated due to the different backscattering characteristics within each layer. Architectural morphology such as crypts or glands within the mucosa can be seen. The OCT image of carcinoma shows disruption of architectural morphology or glandular organization with dilation and distortion of crypt structures.

Standard OCT has image resolutions of 10 to 15 μm, which are sufficient to visualize tissue architectural morphology. OCT can also be used to assess other features such as the integrity of the basement membrane. However, the resolution of standard OCT is not sufficient for cellular level imaging and visualization of nuclear features. Thus, cancer diagnosis would be difficult using OCT images alone, and

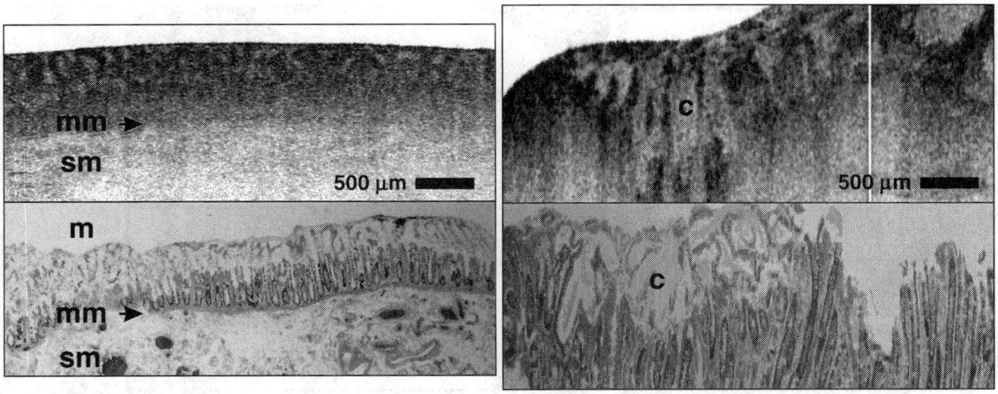


**FIGURE 13.13** OCT images of gastrointestinal tissues *in vitro* showing normal colon and carcinoma. OCT can image differences in architectural morphology or glandular organization that are associated with neoplasia. The normal colon shows columnar epithelial morphology with crypt structures visible. The image of carcinoma shows a loss of normal epithelial structure with disruption and disorganization of the crypts. Architectural morphology such as this can be imaged with standard OCT resolutions of 10 to 15 μm. (From Tearney, G.J. et al., *Am. J. Gastroenterol.*, 92, 1800, 1997. With permission.)

it is more feasible to apply OCT to guide conventional excisional biopsy. Ultrahigh-resolution OCT with resolutions of 1 to 5 μm enables cellular level imaging and could expand the range of neoplastic changes that can be imaged.[35,88,89] The imaging depth of OCT is 2 to 3 mm, less than that of ultrasound. However, for diseases that originate from or involve the mucosa, submucosa, and muscular layers, imaging the microscopic structure of small lesions is well within the range of OCT.

Conventional excisional biopsy often suffers from high false negative rates because the biopsy process relies on sampling tissue and the diseased tissues can easily be missed. OCT could be used to identify suspect lesions and to guide excisional biopsy and reduce the false negative rates. This would reduce the number of costly biopsies; at the same time, clinical diagnoses could be made using excisional biopsy and histopathology, which is the established clinical standard. In the future, after detailed clinical studies and more extensive statistical data are available, it may be possible to use OCT directly for the diagnosis of certain types of neoplasias or for determining the grade of dysplasias. For example, OCT might be used to determine whether there is submucosal vs. mucosal involvement in early gastric cancer, a criterion used to determine the therapy. OCT imaging has the advantage of providing a real-time diagnostic that could enable the integration of diagnosis and treatment.

The development of miniature, flexible OCT imaging probes and high-speed OCT imaging systems enables OCT imaging of internal organ systems. Imaging the pulmonary, gastrointestinal, and urinary tracts and arterial imaging have been demonstrated in New Zealand white rabbits.[41,50] A short-pulse $Cr^{4+}$:Forsterite laser was used as the light source to achieve high resolution with real-time image acquisition rates.[32] The laser output was near 1.3 μm with a spectral bandwidth of 75 nm (FWHM), to yield an axial resolution of 10 μm. A sensitivity of 110 dB was achieved with 5- to 10-mW power incident on the tissue. In order to achieve high-speed imaging at several frames per second, a novel high-speed phase scanning optical delay line was used.[39]

Figure 13.14 shows an example of a catheter/endoscope OCT image of the gastrointestinal tract in the rabbit. The image shown was obtained with a 512-pixel lateral resolution at 4 frames per second in order to optimize lateral sampling. The two-dimensional image data were displayed using a polar coordinate transformation and inverse gray scale. As shown in Figure 13.14, the mucosa of the esophagus was readily identifiable because of its low optical backscattering compared with the submucosa. The ability to differentiate mucusal vs. submucosal invasion of early cancers can be an important criterion for determining therapy in early cancers.

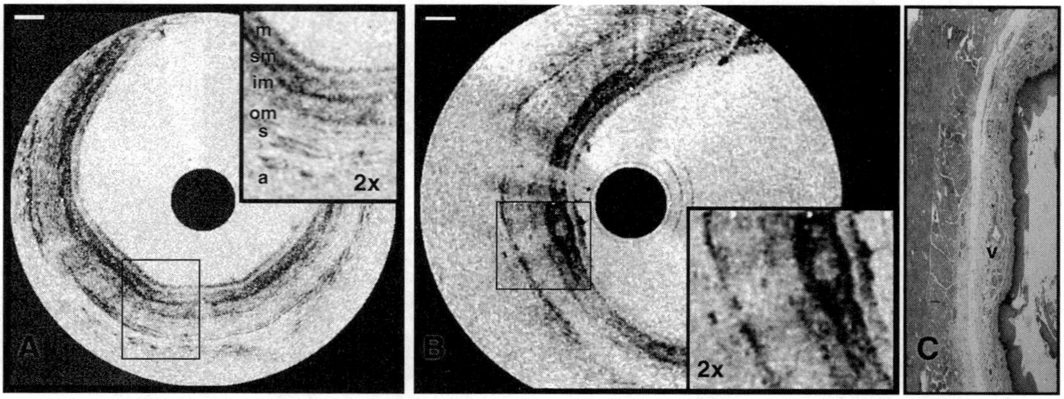


**FIGURE 13.14** Endoscopic OCT imaging of the rabbit esophagus *in vivo*. (A) OCT enables visualization of the esophageal layers of the rabbit including the mucosa (m), the submucosa (sm), the inner muscular layer (im), the outer muscular layer (om), the serosa (s), and the adipose and vascular supportive tissues (a). (B) A blood vessel (v) can be seen within the submucosa of the esophagus. (C) Corresponding histology for (B). The scale bars are 500 μm. (From Tearney, G.J. et al., *Science*, 276, 2037, 1997. With permission.)

Endoscopic OCT imaging in humans can be performed by introducing OCT imaging probes into the accessory port of standard endoscopes. Figure 13.15 shows an example of endoscopic OCT imaging of the human esophagus.[90] Imaging was performed with 13-μm resolution at 1.3-μm wavelengths. These images show representative linear scan OCT images, endoscopic images, and biopsy histology of normal squamous epithelium compared to Barrett's esophagus, a metaplastic condition associated with increased cancer risk. The OCT image (4 × 2.5 mm, 512 × 256 pixels) of normal epithelium illustrates the epithelium, lamina propria, muscularis mucosa, submucosa, and muscularis propria. The OCT image

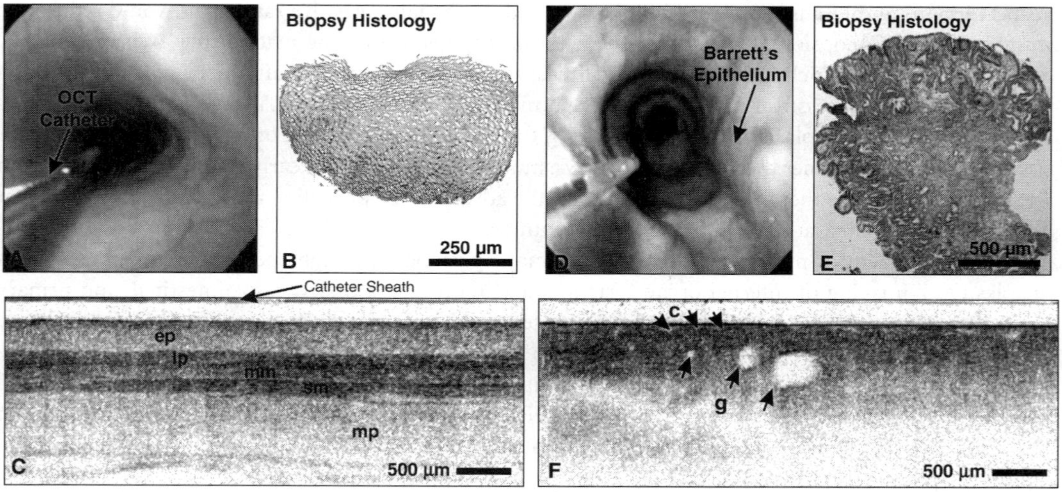


**FIGURE 13.15** Clinical endoscopic OCT imaging of the human esophagus. Images were performed using a linear (push–pull) scanning OCT probe introduced through the working port of a standard endoscope. (A) Endoscopic image of normal region. (B) Biopsy histology of normal squamous epithelium. (C) OCT image of normal squamous epithelium with relatively uniform and distinct layered structures. (D) Endoscopic image of region showing pathology (finger-like projection of Barrett's epithelium). (E) Biopsy histology of Barrett's esophagus showing characteristic specialized columnar epithelium. (F) OCT image of Barrett's epithelium with disruptions of layered morphology due to multiple crypt- and gland-like structures (arrows). (From Li, X.D. et al., *Endoscopy*, 32, 921, 2000. With permission.)

of Barrett's metaplasia shows loss of normal structure and replacement of squamous epithelium with columnar epithelium and glandular structures, which are characteristic of Barrett's metaplasia.

OCT imaging studies in patients have been performed in many different organ systems, including the gastrointestinal, urinary, pulmonary, and female reproductive tracts.[37,49,90–94] These studies demonstrate the feasibility of performing clinical OCT imaging and suggest a broad range of future clinical applications. Additional systematic studies that compare OCT imaging with histological or other diagnostic endpoints are still needed.

## 13.4.5 Guiding Surgical Intervention

Another important class of applications for OCT is guiding surgical intervention. The ability to see beneath the surface of tissue in real time can guide surgery near sensitive structures such as vessels or nerves as well as assist in microsurgical procedures.[44,45,95] Optical instruments such as surgical microscopes are routinely used to magnify tissue to prevent iatrogenic injury and to guide delicate surgical techniques. OCT can be easily integrated with surgical microscopes. Handheld OCT surgical probes and laparoscopes have also been demonstrated.[46]

One example of a surgical application for OCT is the repair or anastomosis of small vessels and nerves following traumatic injury. A technique capable of real-time, subsurface, three-dimensional, micron-scale imaging would permit the intraoperative monitoring of microsurgical procedures, giving immediate feedback to the surgeon that could improve outcome and enable difficult procedures. Studies have been performed *in vitro* that demonstrate the use of OCT imaging for diagnostically assessing microsurgical procedures using an OCT microscope.[45]

Figure 13.16 shows OCT images of an arterial anastomosis of a rabbit inguinal artery, demonstrating the ability of OCT to assess internal structure and luminal patency. An artery segment was bisected cross sectionally with a scalpel and then reanastomosed using a No. 10-0 nylon suture with a 50-μm diameter needle in a continuous suture. Cross-sectional OCT images (2.2 × 2.2 mm, 250 × 600 pixel) of a 1-mm-diameter rabbit inguinal artery are shown. Figures 13.16A through 13.16D were acquired transversely at different positions through the anastomosis. The images of the ends of the artery clearly show arterial morphology corresponding to the intimal, medial, and adventitial layers of the elastic artery. The image

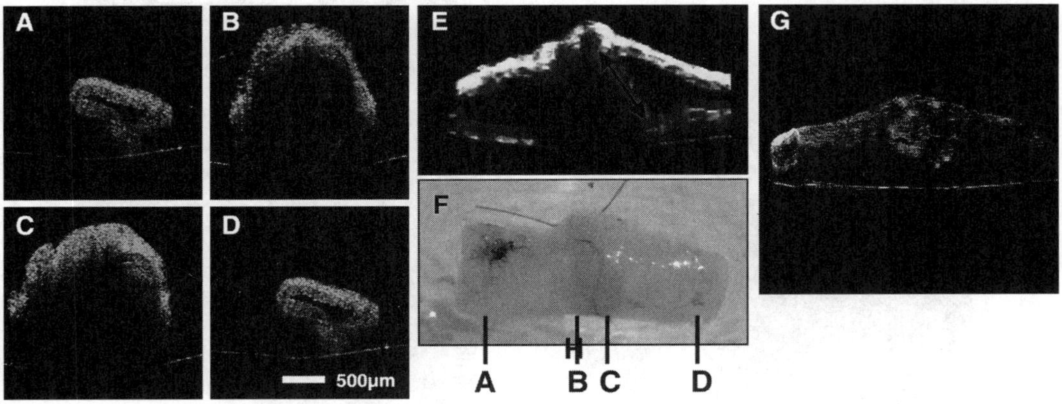


**FIGURE 13.16** OCT images of an anastomosis in a rabbit artery as an example of microsurgical imaging. The 1-mm-diameter rabbit artery was anastomosed with a continuous suture as seen in the digital image (F). The labels indicate the planes from which corresponding cross-sectional OCT images were acquired (A, D). Opposite ends of the anastomosis show the multilayered structure of the artery with a patent lumen. (B) Partially obstructed lumen and the presence of a thrombogenic flap. (C) Fully obstructed portion of the anastomosis site. (E) Longitudinal section virtual view constructed from three-dimensional image data sets shows the obstruction (o). (G) Three-dimensional data set. (From Boppart, S.A. et al., *Radiology*, 208, 81, 1998. With permission.)

from the site of the anastomosis shows that the lumen was obstructed by a tissue flap. By assembling a series of cross-sectional two-dimensional images, a three-dimensional data set was produced as shown in Figure 13.16G. Arbitrary virtual image planes (Figure 13.16E) can be constructed from the three-dimensional data set and corresponding virtual cross sections displayed.

Because OCT imaging can be performed in real time at high speeds, it can be integrated directly with surgery. The feasibility of using high-speed OCT imaging to image the dynamics of surgical laser ablation has been investigated.[96] These studies show examples of image rendering and reconstruction techniques as well as real-time imaging. The use of OCT to monitor ablative therapy in real time could enable more precise control of laser delivery and reduction in iatrogenic injury.

Argon laser ablation was performed in five different *ex vivo* rat organs to assess OCT imaging performance and variations between tissue types. These studies were performed using a commercially available low-coherence light source with a wavelength at 1.3 μm and an axial resolution of 18 μm. The signal-to-noise ratio was 115 dB using 5 mW of incident power on the specimen. For high-speed image acquisition, the size of the region imaged was 3 × 3 mm and 256 × 256 pixels with an acquisition rate of 8 frames per second or 125 ms for each image. Images were displayed on a computer monitor and simultaneously recorded to video. The argon beam was aligned and centered within the OCT imaging plane. Specimens were obtained from Sprague–Dawley rats after euthanasia, including brain, liver, kidney, lung, and rectus abdominis muscle. A three-dimensional, micron-precision, computer-controlled stage was used to position the specimens under the OCT imaging beam and was used to acquire three-dimensional data sets.

Figure 13.17 shows an example of a three-dimensional projection of a laser ablation crater in a rat rectus abdominis muscle. The laser exposure was 3 W of argon laser power focused to a 0.8-mm-diameter spot for 10 s. The OCT images show the ablation zone viewed from an external and an internal virtual viewpoint. Because image data are acquired over a three-dimensional volume, arbitrary cross-sectional tomographic planes can be synthesized. Figure 13.18 shows a series of virtual *en face* images at various depths through the ablation crater. OCT imaging of laser ablation can also be performed in real time. Figure 13.19 shows an example of real-time OCT imaging during laser ablation of a bovine muscle

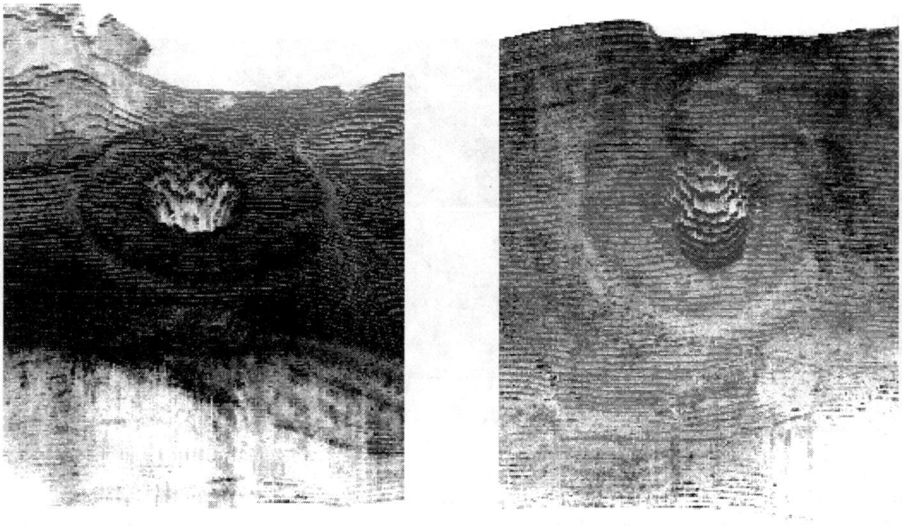

**FIGURE 13.17** Three-dimensional OCT projection image of a laser ablation crater demonstrating the ability to reconstruct OCT images with different perspective views. The lesion was produced in rat rectus abdominis muscle by a 3-W, 10-s argon laser exposure. The projection view illustrates a central ablation crater and concentric zones of thermal injury. (From Boppart, S.A. et al., *J. Surg. Res.*, 82, 275, 1999. With permission.)

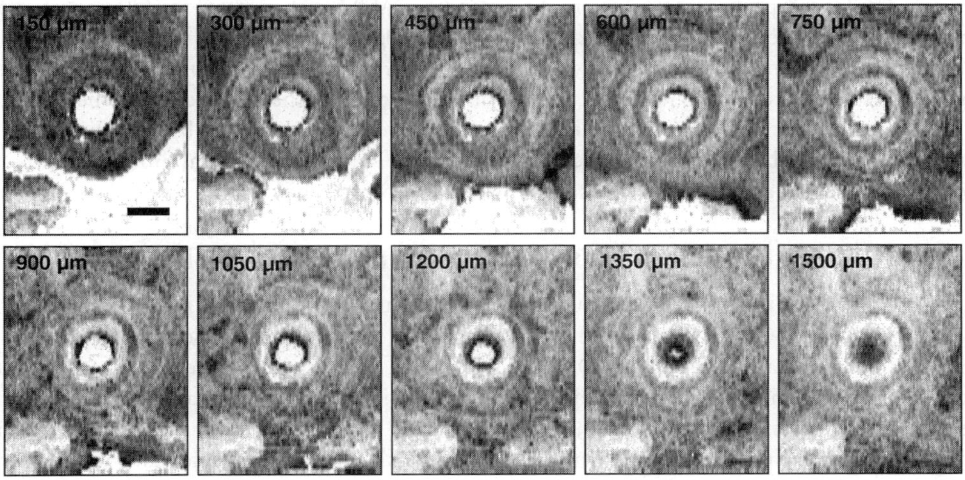


**FIGURE 13.18** *En face* images through an ablation crater. These images demonstrate the ability to render different planes of view using a three-dimensional OCT data set. *En face* sections at varying depths illustrate concentric zones of damage. The numbers show the depth below surface. The bar represents 1 mm. (From Boppart, S.A. et al., *J. Surg. Res.*, 82, 275, 1999. With permission.)

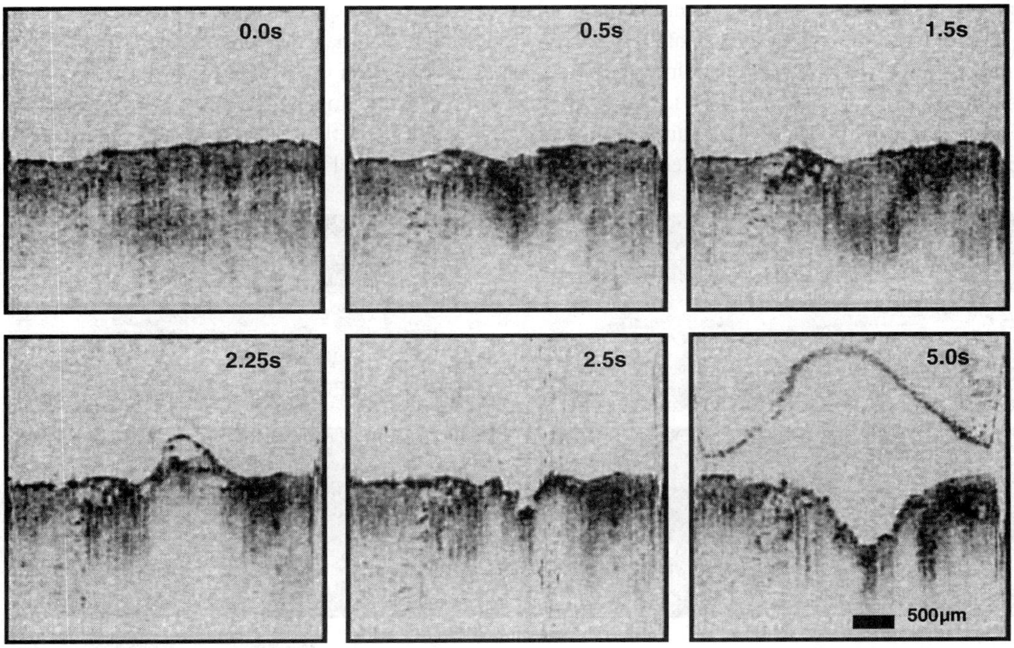


**FIGURE 13.19** Real-time OCT imaging of laser ablation showing a series of OCT images taken at 8 frames per second during argon laser exposure of beef muscle. The exposure was 1 W with a 0.8-mm spot diameter on tissue. The times indicated are in seconds after exposure is initiated. At 2.25 s, the formation of a blister at the surface is observed. At 2.5 s, the blister explodes and a crater develops (5 s). These images show the ability of OCT to perform high-speed, real-time imaging during interventional procedures. (From Boppart, S.A. et al., *J. Surg. Res.*, 82, 275, 1999. With permission.)

specimen. The laser exposure was 2 W from the argon laser. At 0.5 s, changes in optical properties of the tissue are observed due to heating, and explosive ablation begins at 2.25 s.

These examples demonstrate the ability of OCT to perform real-time imaging, as well as the use of image processing and rendering techniques to provide a more comprehensive view of tissue microstructure. Optical coherence tomography should improve intraoperative diagnostics by providing high-resolution, real-time, subsurface, cross-sectional imaging. Three-dimensional imaging and image reconstruction permit assessment of the spatial orientation of morphology.

### 13.4.6 Ultrahigh-Resolution Optical Coherence Tomography

The development of ultrahigh-resolution OCT is also an area of active research. As discussed previously, the axial resolution of OCT is determined by the coherence length of the light source used for imaging. Light sources for ultrahigh-resolution OCT imaging must have a short coherence length or broad bandwidth, but also must be single spatial mode to enable interferometry. In addition, since the SNR depends on the incident power, light sources with average powers of several milliwatts are typically necessary to achieve real-time imaging. One approach for achieving high resolution is to use short-pulse, femtosecond solid-state laser light sources.[34,43,97]

State-of-the-art femtosecond $Ti:Al_2O_3$ lasers can now directly generate pulse durations of ~5 fs.[89,98] These pulse durations correspond to only two optical cycles and have bandwidths of up to 400 nm centered around 800 nm. These high-performance lasers have been made possible through the development of double-chirped mirror technology, which yields extremely wide bandwidths and also compensates for higher-order dispersion in the laser. Unlike ultrafast femtosecond time-resolved measurements, where special care must be exercised to maintain the short pulse duration, OCT measurements depend on field correlations rather than intensity correlations. Thus, dispersion in the reference and signal paths of the interferometer must be precisely matched, but need not be equal to zero.

Figure 13.20 shows an example of an ultrahigh-resolution OCT image of the normal retina of a human subject compared to standard resolution OCT.[89] The axial resolution is 3 μm and is by limited chromatic aberration in the eye. Comparison to standard resolution OCT imaging with 10 μm resolution shows that ultrahigh-resolution OCT significantly improves image quality and the ability to resolve internal retinal architectural morphology. It also enables visualization of the foveal and optic disc contour, as well as

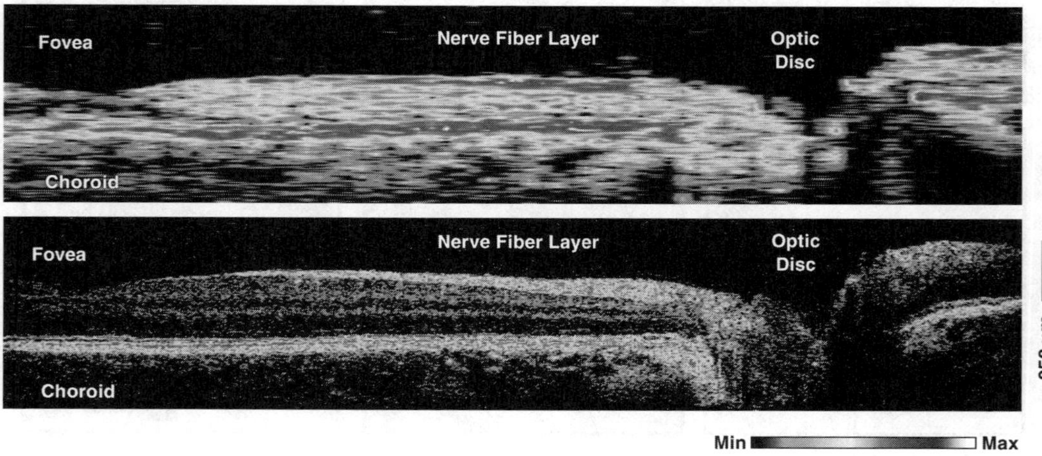


**FIGURE 13.20** Ultrahigh- vs. standard-resolution OCT images of the normal human retina. The ultrahigh-resolution image has a 3-μm axial resolution compared to the standard image with a 10-μm axial resolution. The images have been expanded by a factor of two in the axial direction to permit better visualization. Ultrahigh-resolution OCT enables the architectural morphology of the retina to be imaged noninvasively and in real time. (From Drexler, W. et al., *Nat. Med.*, 7, 502, 2001. With permission.)

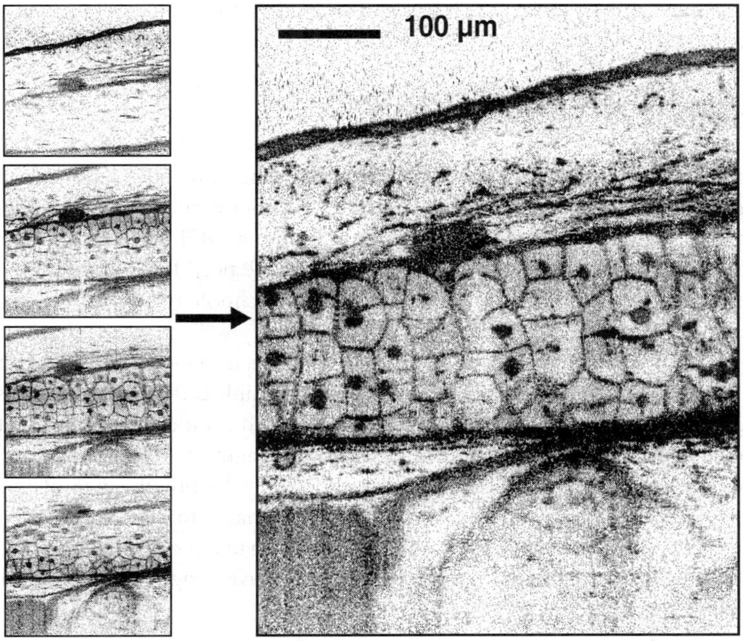


**FIGURE 13.21** Ultrahigh-resolution cellular imaging. *In vivo* subcellular level resolution 1 × 3 μm (axial × transverse) 1700 × 1000 pixel image of an African frog (*Xenopus laevis*) tadpole. Images were recorded with different depths of focus and fused to construct the image shown. Multiple mesenchymal cells of various sizes and nuclear-to-cytoplasmic ratios and intracellular morphology, as well as the mitosis of cell pairs, are clearly shown. The bar is 100 μm. (From Drexler, W. et al., *Opt. Lett.*, 24, 1221, 1999. With permission.)

internal architectural morphology of the retina and choroid that is not resolvable with conventional resolution OCT. The retinal nerve fiber layer (NFL) is clearly differentiated and the variation of its thickness toward the optic disc can be observed. In ophthalmology, improving the resolution should allow more precise morphmetric measurements of retinal features such as retinal thickness and retinal nerve fiber layer thickness, which are relevant for the detection of diseases such as macular edema and glaucoma.

Cellular level OCT imaging can also be performed.[34,88] Figure 13.21 shows an example of ultrahigh-resolution OCT imaging of a *Xenopus laevis* tadpole (African frog) *in vivo*. Imaging was performed with ~1 μm axial and 3 μm transverse resolution; subcellular features are visible at these resolutions. Depth of field limitations may be overcome by using a novel technique from ultrasound known as C mode imaging. Multiple images are acquired with the focus set to different depths within the specimen. Each image is in focus over a depth range comparable to the confocal parameter. The in-focus regions from each of the images are selected and fused together to form a single image, which has a greatly extended depth of field.

The fused image covers an area of 0.75 × 0.5 mm and consists of 1700 × 1000 pixels. Because of the small focal spot size, the confocal parameter was only 40 μm; however, using image fusion of eight images enabled imaging over a depth of 750 μm. Because the axial and transverse resolutions are extremely fine, images have large pixel densities. These images can be beyond the resolution of a standard computer monitor and need to be viewed with panning and zooming.

In developmental biology, the ability to image subcellular structure can be an important tool for studying mitotic activity and cell migration, which occur during development. Ultrahigh-resolution imaging would also extend the diagnostic capabilities of OCT for early neoplasias. Standard OCT image resolutions enable the imaging of architectural morphology on the 10- to 15-μm scale and can identify many early neoplastic changes. The extension of imaging to the cellular and subcellular level would not only enhance the spectrum of early neoplasias and dysplasias that could be imaged, but should also improve sensitivity and specificity.

## 13.5 Summary

OCT is a fundamentally new imaging modality with rapidly emerging applications spanning a range of fields. OCT can perform micron-scale imaging of internal microstructure in materials and biological tissues *in situ* and in real time. For medical imaging applications, OCT can function as a type of optical biopsy; image information is available immediately without the need for excision and histological processing of a specimen. The development of high-speed OCT technology permits real-time imaging with high pixel densities.

A wide range of OCT imaging platforms and probes has been developed, including ophthalmoscopes, microscopes, laparoscopes, handheld probes, and miniature, flexible catheters/endoscopes. These imaging probes can be used separately or in conjunction with other medical imaging instruments such as endoscopes and bronchoscopes; they can permit internal body imaging in a wide range of organ systems. There are numerous research applications of this technology in a broad range of fields. In addition, advances in the technology are continuing. More research remains to be done, and numerous clinical studies must be performed in order to determine where OCT can play a role in clinical medicine. However, the unique capabilities of OCT imaging suggest that it will have a significant impact on fundamental research as well as on medicine.

## Acknowledgments

Our research and development in optical coherence tomography would not have been possible without the long-term collaboration and support of a talented multidisciplinary research team. The invaluable contributions of Dr. Joel Schuman and Dr. Carmen Puliafito, from the New England Eye Center and Tufts University School of Medicine, Eric Swanson from Sycamore Networks, Dr. Jacques Van Dam from Stanford University School of Medicine, Dr. Annekathryn Goodman from Massachusetts General Hosptial, Dr. Hiroshi Mashimo from the Veterans Administration Medical Centers, and Dr. Scott Martin from the Brigham and Women's Hospital are greatly appreciated. Postdoctoral associates, M.D./Ph.D. students, and Ph.D. students, including Dr. Stephen Boppart, Dr. Brett Bouma, Dr. Stephane Bourquin, Dr. Christian Chudoba, Dr. Wolfgang Drexler, Dr. Ravi Ghanta, Dr. Ingmar Hartl, Dr. Michael Hee, Dr. Juergen Herrmann, Pei-Lin Hsiung, Dr. David Huang, Christine Jesser, Tony Ko, Dr. Xingde Li, Nirlep Patel, Dr. Constantinos Pitris, Kathleen Saunders, and Dr. Gary Tearney, have made invaluable contributions.

This research was supported in part by the National Institutes of Health, contracts NIH-1-RO1-CA75289-04 and NIH-1-RO1-EY11289-15, the Medical Free Electron Laser Program, F49620-01-1-0186, the Air Force Office of Scientific Research, contract F49620-98-01-0084, and National Science Foundation contract ECS-0119452.

### References

1. Huang, D., Swanson, E.A., Lin, C.P., Schuman, J.S., Stinson, W.G., Chang, W., Hee, M.R., Flotte, T., Gregory, K., Puliafito, C.A., and Fujimoto, J.G., Optical coherence tomography, *Science*, 254, 1178, 1991.
2. Brezinski, M.E. and Fujimoto, J.G., Optical coherence tomography: high resolution imaging in nontransparent tissue, *IEEE J. Sel. Top. Quantum Electron.*, 5, 1185, 1999.
3. Fujimoto, J.G., Pitris, C., Boppart, S.A., and Brezinski, M.E., Optical coherence tomography: an emerging technology for biomedical imaging and optical biopsy, *Neoplasia*, 2, 9, 2000.
4. Kremkau, F.W., *Diagnostic Ultrasound: Principles, Instrumentation, and Exercises*, 2nd ed., Grune & Stratton, Philadelphia, 1984.

5. Fish, P., *Physics and Instrumentation of Diagnostic Medical Ultrasound,* John Wiley & Sons, New York, 1990.
6. Kremkau, F.W., *Doppler Ultrasound: Principles and Instruments,* W.B. Saunders, Philadelphia, 1990.
7. Ziebel, W.J. and Sohaey, R., *Introduction to Ultrasound,* W.B. Saunders, Philadelphia, 1998.
8. Duguay, M.A., Light photographed in flight, *Am. Sci.*, 59, 551, 1971.
9. Duguay, M.A. and Mattick, A.T., Ultrahigh speed photography of picosecond light pulses and echoes, *Appl. Opt.*, 10, 2162, 1971.
10. Bruckner, A.P., Picosecond light scattering measurements of cataract microstructure, *Appl. Opt.* 17, 3177, 1978.
11. Park, H., Chodorow, M., and Kompfner, R., High resolution optical ranging system, *Appl. Opt.*, 20, 2389, 1981.
12. Fujimoto, J.G., De Silvestri, S., Ippen, E.P., Puliafito, C.A., Margolis, R., and Oseroff, A., Femtosecond optical ranging in biological systems, *Opt. Lett.*, 11, 150, 1986.
13. Youngquist, R.C., Carr, S., and Davies, D.E.N., Optical coherence-domain reflectometry: a new optical evaluation technique, *Opt. Lett.*, 12, 158, 1987.
14. Takada, K., Yokohama, I., Chida, K., and Noda, J., New measurement system for fault location in optical waveguide devices based on an interferometric technique, *Appl. Opt.*, 26, 1603, 1987.
15. Gilgen, H.H., Novak, R.P., Salathe, R.P., Hodel, W., and Beaud, P., Submillimeter optical reflectometry, *IEEE J. Lightwave Technol.*, 7, 1225, 1989.
16. Fercher, A.F., Mengedoht, K., and Werner, W., Eye-length measurement by interferometry with partially coherent light, *Opt. Lett.*, 13, 1867, 1988.
17. Clivaz, X., Marquis-Weible, F., Salathe, R.P., Novak, R.P., and Gilgen, H.H., High-resolution reflectometry in biological tissues, *Opt. Lett.*, 17, 4, 1992.
18. Schmitt, J.M., Knuttel, A., and Bonner, R.F., Measurement of optical-properties of biological tissues by low-coherence reflectometry, *Appl. Opt.*, 32, 6032, 1993.
19. Huang, D., Wang, J., Lin, C.P., Puliafito, C.A., and Fujimoto, J.G., Micron-resolution ranging of cornea and anterior chamber by optical reflectometry, *Lasers Surg. Med.*, 11, 419, 1991.
20. Swanson, E.A., Izatt, J.A., Hee, M.R., Huang, D., Lin, C.P., Schuman, J.S., Puliafito, C.A., and Fujimoto, J.G., *In vivo* retinal imaging by optical coherence tomography, *Opt. Lett.*, 18, 1864, 1993.
21. Hee, M.R., Izatt, J.A., Swanson, E.A., Huang, D., Lin, C.P., Schuman, J.S., Puliafito, C.A., and Fujimoto, J.G., Optical coherence tomography of the human retina, *Arch. Ophthalmol.*, 113, 325, 1995.
22. Parsa, P., Jacques, S.L., and Nishioka, N.S., Optical properties of rat liver between 350 and 2200 nm, *Appl. Opt.*, 28, 2325, 1989.
23. Fujimoto, J.G., Brezinski, M.E., Tearney, G.J., Boppart, S.A., Bouma, B., Hee, M.R., Southern, J.F., and Swanson, E.A., Optical biopsy and imaging using optical coherence tomography, *Nat. Med.*, 1, 970, 1995.
24. Brezinski, M.E., Tearney, G.J., Bouma, B.E., Izatt, J.A., Hee, M.R., Swanson, E.A., Southern, J.F., and Fujimoto, J.G., Optical coherence tomography for optical biopsy: properties and demonstration of vascular pathology, *Circulation*, 93, 1206, 1996.
25. Izatt, J.A., Kulkami, M.D., Yazdanfar, S., Barton, J.K., and Welch, A.J., *In vivo* bidirectional color Doppler flow imaging of picoliter blood volumes using optical coherence tomography, *Opt. Lett.*, 22, 1439, 1997.
26. Chen, Z., Milner, T.E., Srinivas, S., Wang, X., Malekafzali, A., van Gemert, M.J.C., and Nelson, J.S., Noninvasive imaging of *in vivo* blood flow velocity using optical Doppler tomography, *Opt. Lett.*, 22, 1119, 1997.
27. Chen, Z., Milner, T.E., Dave, D., and Nelson, J.S., Optical Doppler tomographic imaging of fluid flow velocity in highly scattering media, *Opt. Lett.*, 22, 64, 1997.
28. Hee, M.R., Huang, D., Swanson, E.A., and Fujimoto, J.G., Polarization-sensitive low-coherence reflectometer for birefringence characterization and ranging, *J. Opt. Soc. Am. B*, 9, 903, 1992.

29. de Boer, J.F., Milner, T.E., van Gemert, M.J.C., and Nelson, J.S., Two-dimensional birefringence imaging in biological tissue by polarization-sensitive optical coherence tomography, *Opt. Lett.*, 22, 9346, 1997.
30. Everett, M.J., Schoenenberger, K., Colston, B.W., and Da Silva, L.B., Birefringence characterization of biological tissue by use of optical coherence tomography, *Opt. Lett.*, 23, 228, 1998.
31. Hausler, G. and Lindner, MW., Coherent radar and spectral radar — new tools for dermatological diagnosis, *J. Biomed. Opt.*, 3, 21, 1998.
32. Bouma, B.E., Tearney, G.J., Bilinsky, I.P., and Golubovic, B., Self phase modulated Kerr-lens mode locked Cr:forsterite laser source for optical coherent tomography, *Opt. Lett.*, 21, 1839, 1996.
33. Bouma, B., Tearney, G.J., Boppart, S.A., Hee, M.R., Brezinski, M.E., and Fujimoto, J.G., High-resolution optical coherence tomographic imaging using a mode-locked Ti-$Al_2O_3$ laser source, *Opt. Lett.*, 20, 1486, 1995.
34. Drexler, W., Morgner, U., Kaertner, F.X., Pitris, C., Boppart, S.A., Li, X.D., Ippen, E.P., and Fujimoto, J.G., *In vivo* ultrahigh resolution optical coherence tomography, *Opt. Lett.*, 24, 1221, 1999.
35. Hartl, I., Li, X.D., Chudoba, C., Ghanta, R., Ko, T., Fujimoto, J.G., Ranka, J.K., Windeler, R.S., and Stentz, A.J., Ultrahigh resolution optical coherence tomography using continuum generation in an air-silica microstructure optical fiber, *Opt. Lett.*, 26, 608, 2001.
36. Swanson, E.A., Huang, D., Hee, M.R., Fujimoto, J.G., Lin, C.P., and Puliafito, C.A., High-speed optical coherence domain reflectometry, *Opt. Lett.*, 17, 151, 1992.
37. Sergeev, A.M., Gelikonov, V.M., Gelikonov, G.V., Feldchtein, F.I., Kuranov, R.V., Gladkova, N.D., Shakhova, N.M., Snopova, L.B., Shakov, A.V., Kuznetzova, I.A., Denisenko, A.N., Pochinko, V.V., Chumakov, Y.P., and Streltzova, O.S., *In vivo* endoscopic OCT imaging of precancer and cancer states of human mucosa, *Opt. Express*, 1, 432, 1997.
38. Tearney, G.J., Bouma, B.E., Boppart, S.A., Golubovic, B., Swanson, E.A., and Fujimoto, J.G., Rapid acquisition of *in vivo* biological images by use of optical coherence tomography, *Opt. Lett.*, 21, 1408, 1996.
39. Tearney, G.J., Bouma, B.E., and Fujimoto, J.G., High-speed phase- and group-delay scanning with a grating-based phase control delay line, *Opt. Lett.*, 22, 1811, 1997.
40. Weiner, A.M., Heritage, J.P., and Kirschner, E.M., High resolution femtosecond pulse shaping, *J. Opt. Soc. Am. B*, 5, 1563, 1986.
41. Tearney, G.J., Brezinski, M.E., Bouma, B.E., Boppart, S.A., Pitris, C., Southern, J.F., and Fujimoto, J.G., *In vivo* endoscopic optical biopsy with optical coherence tomography, *Science*, 276, 2037, 1997.
42. Puliafito, C.A., Hee, M.R., Schuman, J.S., and Fujimoto, J.G., *Optical Coherence Tomography of Ocular Diseases*, Slack Inc., Thorofare, NJ, 1996.
43. Boppart, S.A., Brezinski, M.E., Bouma, B., Tearney, G.J., Swanson, E.A., and Fujimoto, J.G., *In-vivo* imaging of developing morphology using optical coherence tomography, *Mol. Biol. Cell*, 6, 662, 1995.
44. Brezinski, M.E., Tearney, G.J., Boppart, S.A., Swanson, E.A., Southern, J.F., and Fujimoto, J.G., Optical biopsy with optical coherence tomography, feasibility for surgical diagnostics, *J. Surg. Res.*, 71, 32, 1997.
45. Boppart, S.A., Bouma, B.E., Pitris, C., Southern, J.F., Brezinski, M.E., and Fujimoto, J.G., Intraoperative assessment of microsurgery with three-dimensional optical coherence tomography, *Radiology*, 208, 81, 1998.
46. Boppart, S.A., Bouma, B.E., Pitris, C., Tearney, G.J., Fujimoto, J.G., and Brezinski, M.E., Forward-imaging instruments for optical coherence tomography, *Opt. Lett.*, 22, 1618, 1997.
47. Feldchtein, F.I., Gelikonov, G.V., Gelikonov, V.M., Iksanov, R.R., Kuranov, R.V., Sergeev, A.M., Gladkova, N.D., Ourutina, M.N., Warren, J.A., and Reitze, D.H., *In vivo* OCT imaging of hard and soft tissue of the oral cavity, *Opt. Express*, 3, 239, 1998.
48. Tearney, G.J., Boppart, S.A., Bouma, B.E., Brezinski, M.E., Weissman, N.J., Southern, J.F., and Fujimoto, J.G., Scanning single mode catheter/endoscope for optical coherence tomography, *Opt. Lett.*, 21, 543, 1996.

49. Feldchtein, F.I., Gelikonov, G.V., Gelikonov, V.M., Kuranov, R.V., Sergeev, A.M., Gladkova, N.D., Shakhov, A.V., Shakhova, N.M., Snopova, L.B., Terent'eva, A.B., Zagainova, E.V., Chumakov, Y.P., and Kuznetzova, I.A., Endoscopic applications of optical coherence tomography, *Opt. Express*, 3, 257, 1998.
50. Fujimoto, J.G., Boppart, S.A., Tearney, G.J., Bouma, B.E., Pitris, C., and Brezinski, M.E., High resolution *in vivo* intra-arterial imaging with optical coherence tomography, *Heart*, 82, 128, 1999.
51. Li, X.D., Chudoba, C., Ko, T., Pitris, C., and Fujimoto, J.G., Imaging needle for optical coherence tomography, *Opt. Lett.*, 25, 1520, 2001.
52. Fercher, A.F., Hitzenberger, C., Drexler, W., Kamp, G., and Sattmann, H., *In vivo* optical coherence tomography, *Am. J. Ophthalmol.*, 116, 113, 1993.
53. Puliafito, C.A., Hee, M.R., Lin, C.P., Reichel, E., Schuman, J.S., Duker, J.S., Izatt, J.A., Swanson, E.A., and Fujimoto, J.G., Imaging of macular diseases with optical coherence tomography, *Ophthalmology*, 102, 217, 1995.
54. Hee, M.R., Puliafito, C.A., Wong, C., Duker, J.S., Reichel, E., Rutledge, B., Schuman, J.S., Swanson, E.A., and Fujimoto, J.G., Quantitative assessment of macular edema with optical coherence tomography, *Arch. Ophthalmol.*, 113, 1019, 1995.
55. Hee, M.R., Puliafito, C.A., Wong, C., Duker, J.S., Reichel, E., Schuman, J.S., Swanson, E.A., and Fujimoto, J.G., Optical coherence tomography of macular holes, *Ophthalmology*, 102, 748, 1995.
56. Hee, M.R., Puliafito, C.A., Wong, C., Reichel, E., Duker, J.S., Schuman, J.S., Swanson, E.A., and Fujimoto, J.G., Optical coherence tomography of central serous chorioretinopathy, *Am. J. Ophthalmol.*, 120, 65, 1995.
57. Hee, M.R., Baumal, M.R., Puliafito, C.A., Duker, J.S., Reichel, E., Wilkins, J.R., Coker, J.G., Schuman, J.S., Swanson, E.A., and Fujimoto, J.G., Optical coherence tomography of age-related macular degeneration and choroidal neovascularization, *Ophthalmology*, 103, 1260, 1996.
58. Schuman, J.S., Hee, M.R., Arya, A.V., Pedut-Kloizman, T., Puliafito, C.A., Fujimoto, J.G., and Swanson, E.A., Optical coherence tomography: a new tool for glaucoma diagnosis, *Curr. Opinions Ophthalmol.*, 6, 89, 1995.
59. Schuman, J.S., Hee, M.R., Puliafito, C.A., Wong, C., Pedut-Kloizman, T., Lin, C.P., Hertzmark, E., Izatt, J.A., Swanson, E.A., and Fujimoto, J.G., Quantification of nerve fiber layer thickness in normal and glaucomatous eyes using optical coherence tomography, *Arch. Ophthalmol.*, 113, 586, 1995.
60. Schuman, J.S. and Noecker, R.J., Imaging of the optic nerve head and nerve fiber layer in glaucoma, *Ophthalmol. Clin. North Am.*, 8, 259, 1995.
61. Schaudig, U., Hassenstein, A., Bernd, A., Walter, A., and Richard, G., Limitations of imaging choroidal tumors *in vivo* by optical coherence tomography, *Graefe's Arch. Clin. Exp. Ophthalmol.*, 236, 588, 1998.
62. Krivoy, D., Gentile, R., Liebmann, J.M., Stegman, Z., Rosen, R., Walsh, J.B., and Ritch, R., Imaging congenital optic disc pits and associated maculopathy using optical coherence tomography, *Arch. Ophthalmol.*, 114, 165, 1996.
63. Toth, C.A., Birngruber, R., Boppart, S.A., Hee, M.R., Fujimoto, J.G., DiCarlo, C.D., Swanson, E.A., Cain, C.P., Narayan, D.G., Noojin, G.D., and Roach, W.P., Argon laser retinal lesions evaluated *in vivo* by optical coherence tomography, *Am. J. Ophthalmol.*, 123, 188, 1997.
64. Toth, C.A., Narayan, D.G., Boppart, S.A., Hee, M.R., Fujimoto, J.G., Birngruber, R., Cain, C.P., DiCarlo, C.D., and Roach, W.P., A comparison of retinal morphology viewed by optical coherence tomography and by light microscopy, *Arch. Ophthalmol.*, 115, 1425, 1997.
65. Huang, Y., Cideciyan, A.V., Papastergiou, G.I., Banin, E., Semple-Rowland, S.L., Milam, A.H., and Jacobson, S.G., Relation of optical coherence tomography to microanatomy in normal and rd chickens, *Invest. Ophthalmol. Vis. Sci.*, 39, 2405, 1998.
66. Hee, M.R., Puliafito, C.A., Duker, J.S., Reichel, E., Coker, J.G., Wilkins, J.R., Schuman, J.S., Swanson, E.A., and Fujimoto, J.G., Topography of diabetic macular edema with optical coherence tomography, *Ophthalmology*, 105, 360, 1998.

67. Schuman, J.S., Pedut-Kloizman, T., Hertzmark, E., Hee, M.R., Wilkins, J.R., Coker, J.G., Puliafito, C.A., Fujimoto, J.G., and Swanson, E.A., Reproducibility of nerve fiber layer thickness measurements using optical coherence tomography, *Ophthalmology*, 103, 1889, 1996.
68. Brezinski, M.E., Tearney, G.J., Weissman, N.J., Boppart, S.A., Bouma, B.E., Hee, M.R., Weyman, A.E., Swanson, E.A., Southern, J.F., and Fujimoto, J.G., Assessing atherosclerotic plaque morphology: comparison of optical coherence tomography and high frequency intravascular ultrasound, *Br. Heart J.*, 77, 397, 1997.
69. Falk, E., Plaque rupture with severe pre-existing stenosis precipitating coronary thrombosis, characteristics of coronary atherosclerotic plaques underlying fatal occlusive thrombi, *Br. Heart J.*, 50, 127, 1983.
70. Davies, M.J. and Thomas, A.C., Plaque fissuring — the cause of acute myocardial infarction, sudden ischemic death, and crescendo angina, *Br. Heart J.*, 53, 363, 1983.
71. Richardson, P.D., Davies, M.J., and Born, G.V.R., Influence of plaque configuration and stress distribution on fissuring of coronary atherosclerotic plaques, *Lancet*, 1, 941, 1989.
72. Fuster, V., Badimon, L., Badimon, J.J., and Chesebro, J.H., Mechanisms of disease — The pathogenesis of coronary-artery disease and the acute coronary syndromes. 1., *N. Engl. J. Med.*, 326(4), 242, 1992.
73. Loree, H.M. and Lee, R., Stress analysis of unstable plaque, *Circ. Res.*, 71, 850, 1992.
74. Gillum, R.F., Sudden coronary death in the United States; 1980–1985, *Circulation*, 79, 756, 1989.
75. Tearney, G.J., Brezinski, M.E., Boppart, S.A., Bouma, B.E., Weissman, N., Southern, J.F., Swanson, E.A., and Fujimoto, J.G., Catheter-based optical imaging of a human coronary artery, *Circulation*, 94, 3013, 1996.
76. Patwari, P., Weissman, N.J., Boppart, S.A., Jesser, C., Stamper, D., Fujimoto, J.G., and Brezinski, M.E., Assessment of coronary plaque with optical coherence tomography and high-frequency ultrasound, *Am. J. Cardiol.*, 85, 641, 2000.
77. Tearney, G.J., Jang, I.K., Kang, D.H., Aretz, H.T., Houser, S.L., Brady, T.J., Schlendorf, K., Shishkov, M., and Bouma, B.E., Porcine coronary imaging *in vivo* by optical coherence tomography, *Acta Cardiol.*, 55, 233, 2000.
78. Izatt, J.A., Kulkarni, M.D., Hsing-Wen, W., Kobayashi, K., and Sivak, M.V., Jr., Optical coherence tomography and microscopy in gastrointestinal tissues, *IEEE J. Sel. Top. Quantum Electron.*, 2, 1017, 1996.
79. Tearney, G.J., Brezinski, M.E., Southern, J.F., Bouma, B.E., Boppart, S.A., and Fujimoto, J.G., Optical biopsy in human gastrointestinal tissue using optical coherence tomography, *Am. J. Gastroenterol.*, 92, 1800, 1997.
80. Tearney, G.J., Brezinski, M.E., Southern, J.F., Bouma, B.E., Boppart, S.A., and Fujimoto, J.G., Optical biopsy in human urologic tissue using optical coherence tomography, *J. Urol.*, 157, 1915, 1997.
81. Kobayashi, K., Izatt, J.A., Kulkarni, M.D., Willis, J., and Sivak, M.V., High-resolution cross-sectional imaging of the gastrointestinal tract using optical coherence tomography: preliminary results, *Gastrointest. Endosc.*, 47, 515, 1998.
82. Tearney, G.J., Brezinski, M.E., Southern, J.F., Bouma, B.E., Boppart, S.A., and Fujimoto, J.G., Optical biopsy in human pancreatobiliary tissue using optical coherence tomography, *Dig. Dis. Sci.*, 43, 1193, 1998.
83. Pitris, C., Brezinski, M.E., Bouma, B.E., Tearney, G.J., Southern, J.F., and Fujimoto, J.G., High resolution imaging of the upper respiratory tract with optical coherence tomography — a feasibility study, *Am. J. Respir. Crit. Care Med.*, 157, 1640, 1998.
84. Pitris, C., Goodman, A., Boppart, S.A., Libus, J.J., Fujimoto, J.G., and Brezinski, M.E., High-resolution imaging of gynecologic neoplasms using optical coherence tomography, *Obstet. Gynecol.*, 93, 135, 1999.

85. Jesser, C.A., Boppart, C.A., Pitris, C., Stamper, D.L., Nielsen, G.P., Brezinski, M.E., and Fujimoto, J.G., High resolution imaging of transitional cell carcinoma with optical coherence tomography: feasibility for the evaluation of bladder pathology, *Br. J. Radiol.*, 72, 1170, 1999.
86. Boppart, S.A., Goodman, A., Libus, J., Pitris, C., Jesser, C.A., Brezinski, M.E., and Fujimoto, J.G., High resolution imaging of endometriosis and ovarian carcinoma with optical coherence tomography: feasibility for laparoscopic-based imaging, *Br. J. Obstet. Gynaecol.*, 106, 1071, 1999.
87. Pitris, C., Jesser, C., Boppart, S.A., Stamper, D., Brezinski, M.E., and Fujimoto, J.G., Feasibility of optical coherence tomography for high resolution imaging of human gastrointestinal tract malignancies, *J. Gastroenterol.*, 35, 87, 2000.
88. Boppart, S.A., Bouma, B.E., Pitris, C., Southern, J.F., Brezinski, M.E., and Fujimoto, J.G., *In vivo* cellular optical coherence tomography imaging, *Nat. Med.*, 4, 861, 1998.
89. Drexler, W., Morgner, U., Ghanta, R.K., Kärtner, F.X., Schuman, J.S., and Fujimoto, J.G., Ultrahigh-resolution ophthalmic optical coherence tomography, *Nat. Med.*, 7, 502, 2001.
90. Li, X.D., Boppart, S.A., Van Dam, J., Mashimo, H., Mutinga, M., Drexler, W., Klein, M., Pitris, C., Krinsky, M.L., Brezinski, M.E., and Fujimoto, J.G., Optical coherence tomography: advanced technology for the endoscopic imaging of Barrett's esophagus, *Endoscopy*, 32, 921, 2000.
91. Bouma, B.E. and Tearney, G.J., Power-efficient nonreciprocal interferometer and linear-scanning fiber-optic catheter for optical coherence tomography, *Opt. Lett.*, 24, 531, 1999.
92. Rollins, A.M., Ung-Arunyawee, R., Chak, A., Wong, C.K., Kobayashi, K., Sivak, M.V., and Izatt, J.A., Real-time *in vivo* imaging of human gastrointestinal ultrastructure by use of endoscopic optical coherence tomography with a novel efficient interferometer design, *Opt. Lett.*, 24, 1358, 1999.
93. Sivak, M.V., Jr., Kobayashi, K., Izatt, J.A., Rollins, A.M., Ung-Runyawee, R., Chak, A., Wong, R.C., Isenberg, G.A., and Willis, J., High-resolution endoscopic imaging of the GI tract using optical coherence tomography, *Gastrointest. Endosc.*, 51, 474, 2000.
94. Brand, S., Ponero, J.M., Bouma, B.E., Tearney, G.J., Compton, C.C., and Nishioka, N.S., Optical coherence tomography in the gastrointestinal tract, *Endoscopy*, 32, 796, 2000.
95. Boppart, S.A., Brezinski, M.E., Pitris, C., and Fujimoto, J.G., Optical coherence tomography for neurosurgical imaging of human intracortical melanoma, *Neurosurgery*, 43, 834, 1998.
96. Boppart, S.A., Herrmann, J., Pitris, C., Stamper, D.L., Brezinski, M.E., and Fujimoto, J.G., High-resolution optical coherence tomography-guided laser ablation of surgical tissue, *J. Surg. Res.*, 82, 275, 1999.
97. Boppart, S.A., Bouma, B.E., Brezinski, M.E., Tearney, G.J., and Fujimoto, J.G., Imaging developing neural morphology using optical coherence tomography, *J. Neurosci. Methods*, 70, 65, 1996.
98. Morgner, U., Kärtner, F.X., Cho, S.H., Chen, Y., Haus, H.A., Fujimoto, J.G., Ippen, E.P., Scheuer, V., Angelow, G., and Tschudi, T., Sub-two-cycle pulses from a Kerr-lens mode-locked Ti:sapphire laser, *Opt. Lett.*, 24, 411, 1999.

# 14

# Speckle Correlometry




Dmitry A. Zimnyakov
*Saratov State University*
*Saratov, Russian Federation*

Valery V. Tuchin
*Saratov State University*
*Saratov, Russian Federation*


## 14.1 Introduction

Scattering of coherent light by spatially inhomogeneous disordered media and rough surfaces leads to the formation of stochastic spatial distributions of the scattered light intensity, or *speckle* patterns. These patterns have a specific granular structure that results from the random interference of a great number of partial waves scattered by bulk or surface inhomogeneities (see Figure 14.1). The stochastic nature of scattering systems causes random phase shifts between these waves and results in statistical properties of the amplitude, phase, and intensity of speckle-modulated scattered fields.

Historically, attempts to understand how properties of observed speckles are related to the structural characteristics of a scattering system and observation conditions began with pioneer studies of laser light–matter interaction. The classic works dedicated to analysis of the statistical properties of laser speckles were made by Goodman, Dainty, Pedersen, Jakeman, Pusey, Asakura, Ioshimura, and many other researchers in the 1960s, 1970s, and early 1980s (see, for example, References 1 through 14). In these works, the concepts of speckle-pattern formation were based in general on the scalar approach for description of the random interference of partial waves scattered by the small-scale inhomogeneities of a scattering medium. Such an assumption leads to certain limitations in description of speckle formation by various scattering systems (especially in the case of bulk multiple scattering, where a significant change in the polarization state of propagating light occurs) but, despite these restrictions, the theory of speckle-pattern formation based on the scalar wave approach has allowed a rigid interpretation or prediction of a large number of experimentally observed phenomena. Moreover, the scalar theory can be applied for interpretation of the statistical properties of multiply scattered speckle patterns for certain observation conditions (in particular, with use of the polarization discrimination of scattered field partial components).


0-8493-1116-0/03/$0.00+$1.50

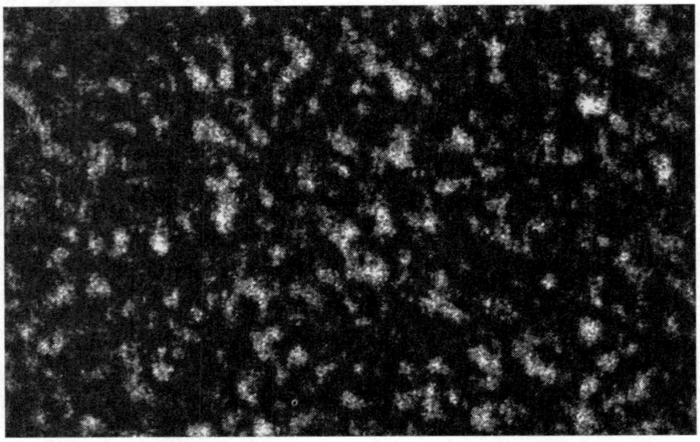

**FIGURE 14.1** Speckle pattern induced in the case of laser beam scattering by a layer of human epidermis (*in vitro* sample).

Beginning in the mid-1980s, further development of speckle science was related to the establishment of certain analogies between classical waves propagating in disordered media and quantum-mechanical behavior of electrons forced by potential with random spatial fluctuations. The pioneering studies of interference effects were conducted by Golubentsev, Stephen, John, MacKintosh and other researchers[15–18] in the area of light propagation in random multiply scattering media. The classic examples of these effects are the manifestation of the weak localization of light in the form of coherent backscattering[19–21] or the existence of long-range correlations of scattered stochastic optical fields.[22,23] The fundamental phenomena accompanying coherent light propagation in disordered and weakly ordered systems and related to the statistical properties of multiply scattered speckles are now the object of significant research interest from many research groups. These studies resulted in modern diffusing light technologies,[24–28] widely applied in material science, biology, and medicine for probing and imaging optically dense scattering media.

Speckle patterns form when coherent light is scattered by living tissue or *in vitro* tissue samples and is sometimes termed biospeckles, or biospeckle phenomena. However, there is no noticeable difference between the general properties of biospeckles and the properties of speckles induced when light is scattered by a weakly ordered nonbiological system characterized by the same optical properties as the scattering tissue. The possibility of obtaining information about structure and dynamic properties of tissues by using statistical or correlation analysis of biospeckles for certain illumination and detection conditions has initiated an abundance of theoretical and experimental works dedicated to various applications of speckle technologies in biomedicine. Various applications of speckle methods in medical diagnostics (especially in the case of bioflow monitoring) are now possible, due to tremendous work by Stern, Fercher, Briers, Asakura, Aizu, Ruth, Nilsson, and many other researchers in the developing theoretical and experimental bases for biospeckle technologies.[29–42]

This chapter presents a brief review of the basic principles of speckle technologies in the context of potential applications for medical diagnostics. Typical examples of such applications are also discussed.

## 14.2 Statistical Properties of Speckles: Basic Principles and Results

### 14.2.1 First-Order Speckle Statistics

Speckle patterns result from the random interference of a great number of statistically independent waves scattered by inhomogeneities inside the scattering volume. Therefore, they may be characterized by amplitude, phase, and intensity, the values of which vary randomly from one detection point to another in the scattered optical field. The probability of determining these parameters for an arbitrarily chosen detection point within the given ranges of their possible values is characterized by the corresponding probability density functions. Knowledge of the probability density distributions of scattered-field characteristics for a given detector position allows one to evaluate the statistical moments, such as mean values and variances, for these characteristics and thus obtain the description of the first-order statistics of the observed speckle pattern. Only the intensity fluctuations can be analyzed in a conventional speckle experiment carried out without any specific interference technique. Thus, the study of the first-order statistics of speckle intensity for known illumination and observation conditions can be considered the basis for certain types of speckle technologies.

The simplest and best-known case of the first-order statistics of speckles corresponds to so-called "fully developed speckles" and can be obtained in terms of the scalar wave approach.[1–4] Let us consider a stochastic field induced by single scattering of a plane monochromatic wave by an ensemble of $N$ statistically independent scatters. For example, if an incident collimated laser beam undergoes random spatial modulation of phase (e.g., by transmittance through a thin, inhomogeneous nonabsorbing layer in which the thickness or refractive index varies from point to point in a stochastic manner and in which the characteristic length of these variations is much larger than the wavelength used*), the number of independent scatters inside the illuminated area can be estimated as $N \approx (D/d)^2$, where $D$ is the illuminating beam aperture and $d$ is the correlation length of the phase fluctuations of boundary field associated with the spatial distribution of the complex amplitude of the transmitted light immediately behind the scattering object. In this case, the scattered field in an arbitrarily chosen point within the diffraction zone can be considered as the sum of partial waves scattered by statistically independent scattering sites:

$$E(\bar{r},t) = \sum_{i=1}^{N} E_i \exp\left[j\left(\omega t - \phi_i(\bar{r})\right)\right], \tag{14.1}$$

where $\omega$ is the frequency of incident light and statistical properties of the ensemble $\phi_i(\bar{r})$ of phase shifts for each partial wave are determined by structural characteristics of the scattering object and observation conditions. In the case of a large $N$ and

$$\sigma_\phi^2 = \left\langle \left(\phi_i - \langle\phi\rangle\right)^2 \right\rangle >> 1,$$

the mean value of the scattered field amplitude is very close to zero and the induced speckle pattern is classified as the fully developed one. For such speckles, the probability density distributions of the real and imaginary parts of the scattered-field complex amplitude have a Gaussian form derived from the central limit theorem.[44] The probability density of the field phase is characterized by a distribution that is very close to the uniform one. Correspondingly, the intensity probability density for such fully developed speckles has the well-known negative exponential form:

$$\rho(I) = \frac{1}{\langle I\rangle} \exp\left(-\frac{I}{\langle I\rangle}\right). \tag{14.2}$$

A manifestation of such exponential statistics is the high visibility of fully developed speckle patterns. This property is manifested as the unit value of the speckle contrast, which is estimated as the ratio of the standard deviation of speckle intensity fluctuations to the mean value of speckle intensity:

$$C = \sigma_I / \langle I\rangle = 1 .$$

*This scatter model is usually determined as the random phase screen (RPS) and, with certain limitations, allows the description of scattered field formation in terms of scalar diffraction theory (see Reference 43).

Another intrinsic property of intensity fluctuations for fully developed speckle patterns is the relation between high-order and first-order statistical moments of speckle intensity,

$$\frac{\langle I^n \rangle}{\langle I \rangle^n} = n!\,, \tag{14.3}$$

which can easily be obtained by calculating the statistical moments

$$\langle I^n \rangle = \int_0^\infty I^n \rho(I) di$$

using the probability density given by Equation 14.2. In the case of fully developed speckle patterns, the first-order statistics of speckle-intensity fluctuations are independent of the structural properties of the scattering object and thus cannot be used for its characterization.

This situation dramatically changes with transition from the fully developed speckle patterns to the partially developed speckles. This transition can be caused by a change in scattering or illumination and observation conditions. In particular, a decrease in the effective number of scattering sites $N$ causes the divergence of the speckle-amplitude statistics from the Gaussian statistics; this divergence leads to formation of partially developed speckles in the diffraction zone. If deep phase modulation of an incident beam by a scattering system takes place $(\sigma_\phi^2 >> 1)$, the contrast of partially developed speckles may significantly exceed 1; in many cases the dependence of $C$ on the effective number of scattering sites may be approximated by the following relation:

$$C \approx \sqrt{1 + K/N}\,, \tag{14.4}$$

where the dimensionless parameter $K$ is determined by properties of a scattering system. Analytical forms of $K$ for various scattering systems with small numbers of $N$ were reviewed by Jakeman.[7] In particular, for a scattering system with the Poisson-distributed number of identical scatters into a scattering volume, the parameter $K$ is equal to 1. An ensemble of identical spherical particles that move into and out of a scattering volume can be considered as an example of such a scattering system.

Another type of scattering system that causes non-Gaussian speckles is a scattering surface consisting of equal-sized facets with statistically independent slopes. The speckle patterns producted by this scattering system are characterized by the following dependence of normalized second-order moment-of-intensity fluctuations on system parameters:

$$\langle I^2 \rangle / \langle I \rangle^2 = 2\left(1 - N^{-1}\right) + \left[k^2 \xi^2 / 4\pi P(\theta)\right] N^{-1}\,, \tag{14.5}$$

where $k$ is the wave number of light, $N$ is the number of facets inside the illuminated area, $\xi$ is the facet radius, and $P(\theta)$ is the probability of finding a facet facing the detector. It is easy to see that this expression may be reduced to a form similar to Equation 14.4 with $K$ depending on $k$, $\xi$, and $P(\theta)$.

If a weak phase modulation of the illuminating beam occurs $(\sigma_\phi^2 \le 1)$, then the dependence of the speckle contrast on $N$ has a nonmonotonic character with the expressed maximum. The value of $C$ falls to 0 with $N$ diminishing to 0. In the case of a large, increasing $N$, the contrast value asymptotically approaches magnitudes that depend on the standard deviation $\sigma_\phi$ of phase fluctuations and is less than 1. In particular, theoretical analysis of similar behavior of first-order speckle statistics was given by Escamilla[45] for the case of a random phase screen with a finite correlation length of phase fluctuations and that is illuminated by the bounded monochromatic collimated beam.

Similar effects can be obtained in the case of illumination of a random phase-modulating object by a plane wave if the scattered field is observed in the near-field zone.[43] Figure 14.2 illustrates the typical behavior of the speckle contrast in the Fresnel diffraction zone for these conditions. If the illuminating

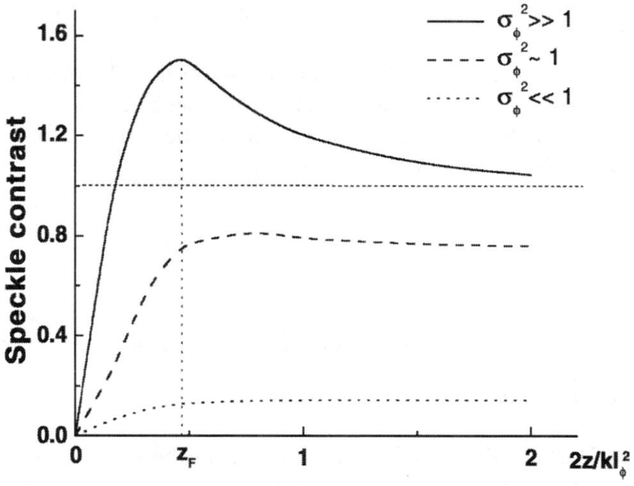


**FIGURE 14.2** Typical behavior of the speckle contrast in the near-field diffraction zone in the dependence on the depth of incident wave phase modulation; $l_\phi$ is the correlation length of the boundary field phase fluctuations; $z$ is the distance between scatter and observation plane.

wave undergoes strong phase modulation (phase fluctuations of the boundary field are characterized by $\sigma_\phi > 1$), the dependencies of $C$ on the detection point position $z$ behind a scattering object exhibit the nonmonotonic behavior with expressed maxima located at certain values of $z_F$ and asymptotic decay to 1 with increasing $z$. Scatter parameters such as $\sigma_\phi$ and correlation length of boundary field phase fluctuations can be estimated by measuring the values of $z_F$ and $C_F$.[43]

Sharp increase of the speckle contrast in the vicinity of $z_F$ can be explained by a manifestation of the "focusing" effect. Scattering of a plane wave by an inhomogeneous medium causes the formation of a boundary field with wavefront inhomogeneities; the development of these phase inhomogeneities in the course of scattered field propagation at relatively short distances from the object leads to the appearance of a random network of caustics formed within the focusing zone behind the scattering object by structural inhomogeneities as the set of stochastically distributed small-sized lenses. For $z > z_F$, the contributing fields propagating from different inhomogeneities will overlap; this leads to interference effects. Finally, for $z > z_F$, the fully developed speckle pattern appears.

Non-Gaussian statistics of the scattered field amplitude fluctuations in the case of a small number of scatters are manifested in the peculiarities of the probability density distributions of speckle-intensity fluctuations such as, for example, the appearance of probability density bimodality for certain illumination and detection conditions.[46]

One specific case is speckle formation in the diffraction or near-field zone caused by random scatterers that exhibit the self-similarity property[47] at different length scales (fractal-like scatterers). In particular, if speckles appear when an illuminating beam is scattered by a rough surface with fractal properties, two different regimes can be considered:[48]

1. Scattering by the surfaces with fractal distribution of the relative heights measured from the baseline $D_h(\bar{r}) = \left\langle \{h(0) - h(\bar{r})\}^2 \right\rangle \sim |\bar{r}|^\nu$, where the exponent $\nu$ of the surface height structure function $D_h(\bar{r})$ is related to a specific parameter of the fractal surface, such as its fractal dimension;[46] this relation can be written as $\nu = 2(2 - D)$. Theoretical estimations of the second-order statistical moment of scattered light intensity fluctuations for a scattering model made for the far diffraction zone (i.e., at distances from the scattering object that significantly exceed the ratio $W^2/\lambda$, where $W$ is the characteristic size of illuminated area and $\lambda$ is the wavelength used) as well as for the near diffraction zone (i.e., at distances that are comparable or less, $W^2/\lambda$) gives the value of

$\langle I^2\rangle/\langle I\rangle^2$ close to 2, as for conventional cases of Gaussian speckles (see, for example, References 1 through 4).

2. Scattering by subfractal surfaces, which are characterized by fractal distributions of the local surface slope. In this case, formation of a scattered field can be described in terms of ray optics, and simple consideration leads to the power-law dependence of the normalized second-order statistical moment of scattered intensity on the illuminated area:

$$\langle I^2\rangle/\langle I\rangle^2 \sim W^{2-\nu}, \tag{14.6}$$

where $\nu$ is the exponent of the structure function $D_s(\bar{r}) = \left\langle \{s(0) - s(\bar{r})\}^2 \right\rangle^{\nu}$, which characterizes the properties of distribution of the surface local slope $s(\bar{r})$.

Statistical analysis of partially developed speckle patterns in the near or far diffraction zone can be applied in comparative study of the morphological differences between diseased tissues and healthy tissues. Figure 14.3 illustrates the typical shapes of histograms that characterize intensity probability distributions for partially developed speckle patterns induced by probe-light scattering in thin layers of healthy and diseased (psoriasis) human epidermis.[49]

Experiments with *in vitro* epidermis samples prepared as skin strips were carried out with the focused beam of an He-Ne laser; speckle patterns were caused by random interference of the forward-scattered partial components and were analyzed in the diffraction zone behind the sample. Intensity fluctuations at the detection point were induced by the lateral scanning of the probed sample with respect to the illuminating beam. The characteristic changes in epidermal structure caused by the progress of psoriasis (e.g., the appearance of parakeratosis focuses, desquamation, and some other phenomena) are manifested as the noticeable distortion of intensity histograms and, as a result, a decrease of the contrast value for observed speckle patterns.

### 14.2.2 Second-Order Speckle Statistics

The statistical properties of a scattered field can also be characterized by simultaneous analysis of the correlation of the complex amplitude values for two spatially separated observation points and for

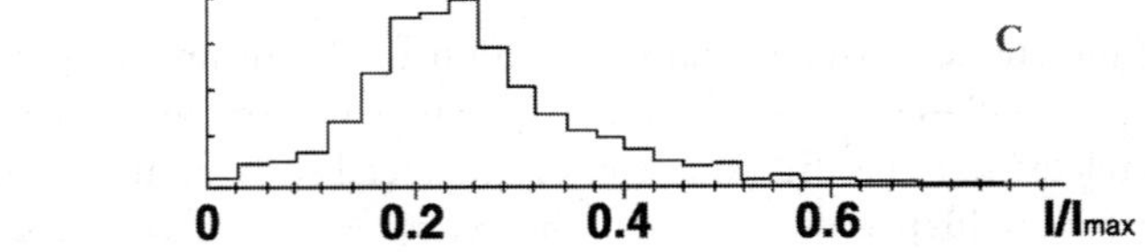

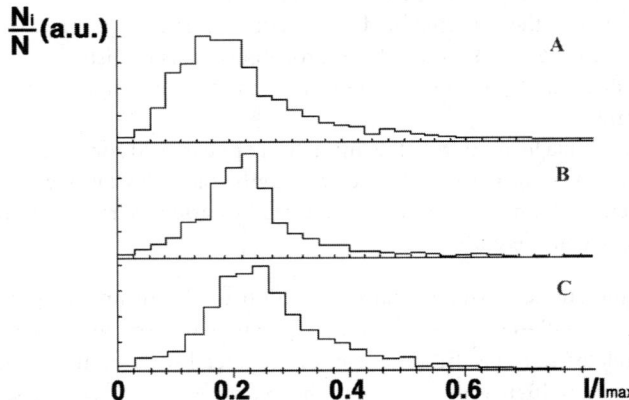


**FIGURE 14.3** Histograms of intensity distributions for partially developed speckle patterns induced by focused laser beam scattering in healthy (A) and diseased (B, C, psoriasis) human epidermis layers. The thickness of the layers is about 30 to 40 μm. The diameter of the light spot on the sample surface is approximately 100 μm.

different moments of time. In this way, the spatial–temporal correlation function of scattered-field fluctuations is introduced as follows:

$$G_1(\bar{r}_1,t_1,\bar{r}_2,t_2)=\langle E_1E_2^*\rangle=\langle E(\bar{r}_1,t_1)E^*(\bar{r}_2,t_2)\rangle, \tag{14.7}$$

where the symbol * denotes complex conjugation. For many cases, the spatial–temporal fluctuations of scattered-field amplitude can be considered the stationary random fields; this leads to the following form of $G_1(\bar{r}_1,t_1,\bar{r}_2,t_2)$:

$$G_1(\bar{r}_1,t_1,\bar{r}_2,t_2)=G_1(\bar{r}_1,\bar{r}_2,t_1-t_2)=G_1(\bar{r}_1,\bar{r}_2,\tau). \tag{14.8}$$

In a similar manner, the spatial-temporal correlation function of scattered light intensity fluctuations can be introduced:

$$G_2(\bar{r}_1,\bar{r}_2,\tau)=\langle I_1I_2\rangle=\langle I(\bar{r}_1,t)I(\bar{r}_2,t+\Delta\tau)\rangle. \tag{14.9}$$

Moreover, for statistically homogeneous speckle patterns, the field and intensity correlation functions depend only on $\Delta\bar{r}=\bar{r}_2-\bar{r}_1$:

$$G_1(\bar{r}_1,\bar{r}_2,\tau)=G_1(\Delta\bar{r},\tau);G_2(\bar{r}_1,\bar{r}_2,\tau)=G_2(\Delta\bar{r},\tau). \tag{14.10}$$

If a scattered optical field is characterized by the Gaussian statistics of a complex amplitude that has zero mean value, then the normalized correlation functions of amplitude and intensity fluctuations $g_2(\Delta\bar{r},\tau)=\langle I_1I_2\rangle/\langle I_1\rangle\langle I_2\rangle$ and, correspondingly, $g_1(\Delta\bar{r},\tau)=\langle E_1E_2^*\rangle/\left(\left(|E_1|^2\right)\left(|E_2|^2\right)\right)^{1/2}$ are related to each other as follows:[9,43]

$$g_2(\Delta\bar{r},\tau)=1+|g_1(\Delta\bar{r},\tau)|^2. \tag{14.11}$$

This is the well-known Siegert relation.

Three types of correlation measurements are typically considered:

- Correlations between spatially separated detectors measured with zero time delay; this type of measurement gives information about the typical size of a feature of a speckle intensity pattern that can be considered the average speckle size in the detection plane.
- Temporal intensity correlations measured at a single detection point; this type of measurement is the most typical for many practical applications and is therefore an object of further consideration.
- Simultaneous measurement of spatial-temporal correlations provided by changes in the positions of two detectors and by changes in the time delay $\tau$.

Usually, the correlation analysis of speckle intensity fluctuations is used to study a dynamic behavior of scattering media as well as their structural characteristics; in the latter case, the dependence of speckle intensity on time may be caused by controllable motion (e.g., lateral displacement) of the probed medium with respect to the illuminating laser beam. The relationship between the dynamic properties of the scattering medium and time-dependent behavior of speckle-intensity fluctuations sufficiently depends on observation conditions. In particular, three specific zones of observation in scattered optical fields induced by probe light scattering by inhomogeneous objects with properties of the random phase screen can be mentioned:

- The near-field region in vicinity of the scattering object ($z << z_F$); in this case, each point of the observed speckle pattern is associated with a certain region of the phase front of the boundary field emerging from the scattering object. For the case of strong phase modulation of a probe beam by the scattering object, the average size of this region is significantly less than the phase correlation length that characterizes the scattering object as the random phase screen. Thus we should expect the temporal evolution of the observed speckle pattern to reflect the time-dependent dynamics of local fluctuations of the boundary field phase. In its turn, this dynamic is determined by local motions of scattering sites.
- The "focusing" region with $z \sim z_F$; for this detection condition, the intensity in each point of the speckle pattern is the result of superposition of partial waves coming from local areas inside the region of the boundary field phase front with a characteristic dimension of the order of $\xi$. Local changes in time-dependent spatial distributions of speckle intensity depend on the relative motions of spatially separated regions of the scattering object. Therefore, the relationship between the time-dependent correlation decay of intensity fluctuations and the dynamic properties of a scattering system has a more complicated form than in the case of near-field detection.
- The far field region with $z >> z_F$; in this case, each point of the detection plane receives all contributions coming from the whole scattering region of the probed object.

The influence of detection conditions on the correlation decay of speckle intensity fluctuations is evident for speckle patterns that result from laser light scattering by moving a spatially heterogeneous "frozen" object that exhibits no mutual motions of local scattering sites and that displaces with respect to the illuminating beam as a whole. In this case, the specific property of the time-dependent spatial distributions of speckle intensity is the existence of two different types of speckle pattern evolution during the observation time; these are associated with two types of speckle motions: "boiling" motion and "translational" motion.

In the common case, the dynamic speckles translate while changing their structure and a decorrelation of intensity fluctuations occurs because of the mutual effect of the displacement and deformation of speckles. The difference between these types of speckle pattern evolution, which is intuitively understandable, can be explained in terms of an approach suggested by Yoshimura[13] and Yoshimura et al.[14] In particular, for a scattering object with properties of the random phase screen characterized by the Gaussian statistics of boundary field phase fluctuations and by the Gaussian-like form of their spatial correlation function, the normalized spatial–temporal correlation function of speckle intensity fluctuations can be expressed as

$$g_2(\Delta\bar{r},\tau) = 1 + \exp\left(-\frac{|\Delta\bar{r}|^2}{r_s^2} + \frac{\tau_d^2}{\tau_s^2}\right)\exp\left(-\frac{(\tau-\tau_d)^2}{\tau_s^2}\right), \tag{14.12}$$

where the parameters $r_s$, $\tau_s$, and $\tau_d$ are determined by the illumination and detection conditions as well as by the scattering object velocity. If intensity fluctuations are detected with two spatially separated detectors with zero time delay, then the parameter $r_s$ can be associated with the average size of the speckles. In order to characterize speckle motion, two additional parameters, such as the correlation distance, $r_c$, and translation distance, $r_T$, can be introduced. With used notations and following Yoshimura's considerations, these parameters may be introduced as:

$$r_c = \left(1/r_s^2 - \tau_c^2 V_s^2/r_s^4\right)^{-1/2};\ r_T = V_s\left(1/\tau_c^2 - V_s^2/r_s^2\right)^{-1/2}, \tag{14.13}$$

where $V_s$ is the speckle velocity. If the absolute value of ratio $r_T/r_s$ is less than 1, then the boiling motion is dominant.

In the opposite case of $|r_T/r_T| > 1$, we can observe the more expressed translation motion of speckles. In the particular case of Gaussian beam illumination of in-plane moving scatter and free-space formation of a speckle field, this consideration leads to the relation

$$r_T/r_s = \{(l+z)l/a + a\}/z\,, \tag{14.14}$$

where $z$ is the distance between the scattering object and the observation plane, $l$ is the distance between the object and the waist plane, and $a$ is the Rayleigh range parameter of illuminating beam evaluated as $\pi w_0^2/\lambda$, where $w_0$ is the waist radius.

It is easy to see that if $l = 0$ (the moving object is positioned in the waist plane), then speckles observed near the scattering object ($z < a$) exhibit the dominating translation motion; in contrast, far-zone speckles are characterized by the boiling-like motion with a vanishing translation component.

The most typical speckle-correlation measurement applied to study the dynamic behavior of various scattering systems (e.g., living tissues) is single-point correlation analysis, or measurement of the time-dependent correlation decay of speckle intensity fluctuations. The basic principles of such measurements are discussed in the next section.

## 14.3 Temporal Correlation Analysis of Speckle Intensity Fluctuations as the Tool for Scattering Media Diagnostics

### 14.3.1 Single-Scattering Systems

For a single-scattering system consisting of moving scattering particles, the scattered field in the detection point can be expressed using a simple modification of Equation 14.1:

$$E(t) = \exp(j\omega t)\sum_{i=1}^{N}|E_i(t)|\exp\{-j\phi_i(t)\}\,, \tag{14.15}$$

where the dependence of phase for each partial contribution on time is caused by scatter motions and therefore causes fluctuations of the scattered field amplitude in the detection point. Thus, the temporal correlation function of scattered field fluctuations in the detection point can be written as follows:

$$G_1(\tau) = \langle E(t)E^*(t+\tau)\rangle = \left\langle \exp(-j\omega\tau)\sum_{i=1}^{N}\sum_{i'=1}^{N}|E_i(t)||E_{i'}(t+\tau)|\exp[-j\{\phi_i(t) - \phi_{i'}(t+\tau)\}]\right\rangle.$$

By using simplifying assumptions about the statistical independence of partial contributions scattered by different scattering sites and slow varying amplitudes of scattered waves, $|E_i(t) \approx E_i(t+\tau)|$, the temporal correlation function can be expressed as (see Reference 9):

$$G_1(\tau) \approx \exp(-j\omega\tau)N\left\langle |E_i(t)|^2 \exp\{j\Delta\phi_i(\tau)\}\right\rangle. \tag{14.16}$$

Here we consider motions of scattering sites as the stationary and ergodic random process causing the increments of phase of interfering partial waves that depend only on the time delay value $\tau$. For the zero time delay, the value of $G_1(0)$ is equal to the mean intensity of scattered light in the detection point: $G_1(0) = \langle I\rangle \approx N\langle |E_i|^2\rangle$. For simplicity, without loss of generality, we can assume the equality of contributions from each partial scattered wave and express the temporal correlation function as:

$$G_1(\tau) \approx \langle I\rangle \exp(-j\omega\tau)\langle \exp\{j\Delta\phi(\tau)\}\rangle. \tag{14.17}$$

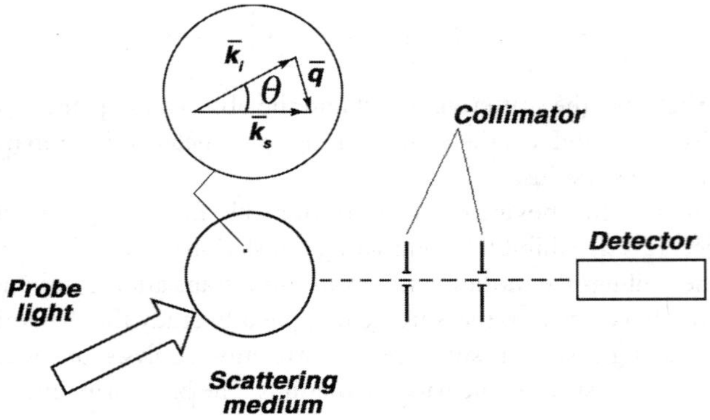


**FIGURE 14.4** Typical scheme of the dynamic light scattering experiment: $\bar{k}_i$ = the wave vector of the incident wave; $\bar{k}_s$ = the wave vector of the detected scattered wave; $\theta$ = the scattering angle.

Phase increment $\Delta\phi_i(\tau)$ for each partial wave can be expressed as $\Delta\phi_i(\tau)=\bar{q}_i\Delta\bar{r}_i(\tau)$, where $\bar{q}_i$ is the momentum transfer for each scattering event and $\Delta\bar{r}_i(\tau)$ describes the displacement of $i$-th scattering site for the observation time $\tau$. For the observation geometry presented in Figure 14.4, the momentum transfer module $|\bar{q}_i|$ for each scattering event can be presented as $|\bar{q}_i|\approx|\bar{q}|=(4\pi/\lambda)\sin(\theta/2)$, where $\theta$ is the scattering angle.

Finally, taking into account the property of a complex exponent with an argument as a combination of three independent random variables with the same statistical properties, the temporal correlation function of field fluctuations for single scattered dynamic speckles can be written as follows:

$$G_1(\tau)\approx\langle I\rangle\exp(-j\omega\tau)\exp\left\{-\left(4\pi\sin(\theta/2)/\lambda\right)^2\Delta\bar{r}^2(\tau)/6\right\}, \tag{14.18}$$

where the displacement variance for scattering sites measured with the given time delay is determined by the dynamic properties of the scattering system. In particular, for ensembles of Brownian particles (this model is usual for description of the dynamical behavior of various biological systems), the value of $\Delta\bar{r}^2(\tau)$ is equal to $6D\tau$ ($D$ is the self-diffusion coefficient of scattering particles).

In conventional schemes of scattering experiments with no use of a reference beam for detection of scattered light, the measurable parameter of a scattered optical field is the instantaneous value of speckle intensity in the detection point. The detected intensity fluctuations are usually characterized by the intensity correlation function $G_2(\tau)$. For Gaussian scattered fields with zero mean amplitude, the normalized correlation function of intensity fluctuations $G_2(\tau) = G_2(\tau)/G_2(0)$ is related to the normalized correlation function $G_1(\tau)$ by the previously presented Siegert relation (see Equation 14.11):

$$g_2(\tau)=1+\beta|g_1(\tau)|^2, \tag{14.19}$$

where the parameter $\beta$ depends on detection conditions. For ideal conditions (i.e., when the detector aperture is significantly smaller than the speckle size), $\beta = 1$.

Thus, for Brownian scattering systems the normalized correlation function of intensity fluctuations has the exponential form:

$$g_2(\tau)\approx1+\exp\left\{-2\left(4\pi\sin(\theta/2)/\lambda\right)^2D\tau\right\}, \tag{14.20}$$

and the corresponding power spectrum has the Lorentzian form. The value of $\tau_0 = (4\pi^2D/\lambda^2)^{-1} = (k^2D)^{-1}$ is usually considered the correlation time of the field fluctuations. One of the traditional approaches to measure the properties of scattering particles (in particular, their diffusion coefficient) is the estimation of the correlation time of speckle intensity fluctuations for given scattering angle and wavelength of the probe light.

A specific case involves coherent light scattering by moving phase scatters with a fractal-like structure. In this case, the temporal intensity fluctuations detected in the fixed point of the far diffraction zone have the properties of the one-dimensional random fractal process and the corresponding fractal dimension of such a process is directly related to the fractal dimension of the scattering object (e.g., a fractal-like surface). The form of corresponding relationship depends on illumination conditions. In particular, use of a sharply focused illuminating beam causes increase of the fractal dimension of detected speckle intensity fluctuations in comparison with the fractal dimension of scattering object (the effect of "stochastization" of speckle intensity fluctuations).[50]

### 14.3.2 Multiple-Scattering Systems

The formation of multiply scattered dynamic speckles is a significantly more interesting case and is important for practical applications. The discrete scattering model (Equation 14.15) can be modified in order to describe dynamic speckle formation due to multiple light scattering in disordered systems; such an approach was discussed in the classic work by Maret and Wolf.[51] In this case, each partial contribution is considered as the result of sequence of $n$ scattering events:

$$E_k(t) = \exp(j\omega t)\prod_{i=1}^{N_k} E_i \exp\left(-j\bar{q}_i\bar{r}_i(t)\right),$$

and the total scattered field in the detection point is expressed as follows:

$$E(t) = \sum_k E_k(t).$$

In further analysis, the *single-path* correlation function of field fluctuations is introduced as

$$\begin{aligned} G_2^k(\tau) &= \left\langle E_k(t)E_k^*(t+\tau)\right\rangle \approx \exp(-j\omega\tau)\left\langle \prod_{i=1}^{N_k} |a_i|^2 \exp\left\{j\bar{q}_i\Delta\bar{r}_i(\tau)\right\}\right\rangle \\ &\approx \exp(-j\omega\tau)\left\langle |a_i|^2\right\rangle \exp\left\{-\left\langle\bar{q}^2\right\rangle\left\langle\Delta\bar{r}^2(\tau)\right\rangle N_k/6\right\}. \end{aligned} \tag{14.21}$$

For the discussed case, the mean value of $\bar{q}^2$ estimated for a sequence of scattering events can be expressed as $\left\langle\bar{q}^2\right\rangle = 2k^2 l/l^*$, where $l$ is the scattering mean free path and $l^*$ is the transport mean free path for the scattering medium. The number of scattering events for each partial contribution can be expressed as $N_k \approx s_k/l$, where $s_k$ is the corresponding propagation path for $k$-th partial component inside a scattering medium. Thus, the single-path correlation function has the following form:

$$G_1^k(\tau) \approx \exp(-j\omega\tau)\left\langle |a|^2\right\rangle \exp\left\{-k^2\left\langle\Delta\bar{r}^2(\tau)\right\rangle s_k/3l^*\right\}. \tag{14.22}$$

The temporal correlation function of field fluctuations in the detection point $G_1(\tau) = \left\langle E(t)E^*(t+\tau)\right\rangle$ can be obtained by the statistical summation of the single-path correlation functions over the ensemble of partial contributions:

$$G_1(\tau) \approx \sum_k P(k) G_1^k(\tau),$$

where $P(k)$ are the statistical weights characterizing contributions of partial components to formation of a scattered field in a detection point. This expression may be modified for multiple scattering systems characterized by the continuous distribution of optical paths $s$ by integration over the range of all possible values of $s$:

$$G_1(\tau) \approx \int_0^{\infty} \exp\left(-\frac{k^2 \langle \Delta \bar{r}^2(\tau) \rangle s}{3l^*}\right) \tilde{\rho}(s) ds \,, \qquad (14.23)$$

where $\tilde{\rho}(s)$ is the probability density of optical paths that obeys the following normalization condition: $\int_0^{\infty} \tilde{\rho}(s) ds = \langle I \rangle$. The normalized temporal correlation function can be introduced as $g_1(\tau) = \int_0^{\infty} \exp(-k^2 \langle \Delta \bar{r}^2(\tau) \rangle s / 3l^*) \rho(s) dx$ by using the following normalization condition: $\rho(s) = \tilde{\rho}(s) / \int_0^{\infty} \tilde{\rho}(s) ds$. In particular, for Brownian scattering systems, the argument of exponential kernel on the right-hand side of Equation 14.23 has the well-known form $2\tau s/\tau_0 l^*$, where $\tau_0$ is the previously introduced correlation time of function of scattered field fluctuations for a single-scattering regime.

If a localized light source is used for illumination of a dynamic scattering system, then the correlation characteristics of scattered field fluctuations (e.g., the correlation time) strongly depend on observation conditions because of sensitivity of these characteristics to changes in path statistics characterized by variations of the pathlength density $\rho(s)$. This gives the opportunity for probing various spatially inhomogeneous dynamic media by the spatially separated light source and detector (for instance, by the probe system consisting of light-delivering and light-collecting optical fibers). Various examples of such probing technology are illustrated in Figure 14.5.

In this case, the probed scattering system consists of two layers — the superficial "static" layer consisting of motionless scatters, and underlying "dynamic" or "modulating" layer with moving scattering sites. Burned tissue with shallow necrotic layer with no blood microcirculation is a good example of a real scattering system with similar geometry. Below this layer, the living tissue is characterized by a sufficient level of blood microcirculation. For small source-detector separations, the depth of light penetration into the probed volume is small compared with the thickness of "static" layer; this scattering geometry causes the formation of dynamic speckles with large values of the correlation time. Increase of the penetration depth due to enlargement of the distance between source and detector fibers leads to faster decay of the field temporal correlation.

Potential applications of this technique were demonstrated by Boas and Yodh for burn depth diagnostics.[52] In their experiment, an Ar-ion laser was used as an illumination source. Laser light was delivered to the probed tissue by a multimode optical fiber. Probe light scattered in the backward direction was collected by an eight-channel fiber-optic probe. The input tips of each single-mode light-collecting fiber were fixed along the detection line with 1.5 mm of separation between them. Output tips were connected with a photodetector unit through the 8 × 1 fiber-optic switcher. A photomultiplier tube (PMT) in the photon-counting mode was used as photodetector. Each sequence of photocounts was processed by digital correlator. This construction allowed the investigators to provide correlation analysis of speckle-intensity fluctuations for various distances between source and detector without mechanical scanning.

*In vivo* pig skin with thermally induced burn diseases of different levels was used to study the influence of the burned tissue thickness on parameters of normalized temporal correlation function $g_2(\tau)$ of intensity fluctuations at a given detection point. Burn diseases were induced by applying a heated metal block to the skin surface. Diseases of various levels with significantly differing thicknesses of necrotic layers can be obtained by varying the provocation time. For experimental conditions used in the study, the necrotic layer thickness was varied from ≈ 100 μm (shallow burn) to ≈ 1000 μm (deep burn, average level of disease).

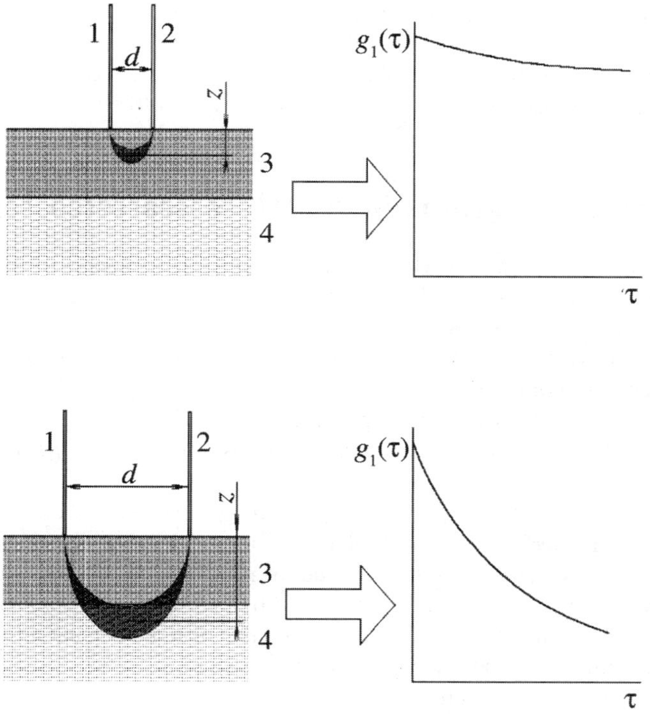


**FIGURE 14.5** Location of the underlying dynamic layer (4) using measurements of the correlation decay of speckle amplitude fluctuations. 1 = light-delivering fiber; 2 = light-collecting fiber; 3 = superficial static layer. $g_1(\tau)$ is the normalized temporal correlation function of the scattered-field fluctuations.

The parameter most sensitive to changes of the burned tissue structure was found to be the slope of $\ln\{g_2(\sqrt{\tau})\}$ dependencies obtained for different source-detector separations. For instance, this parameter slowly increases with an increase in source-detector separation up to the value of the order of the necrotic layer thickness. After the "critical" penetration depth comparable with the layer thickness is reached, the dramatic change in slope value occurs.

Correlation analysis of speckle intensity fluctuations associated with object scanning by the probe (for instance, as shown in Figure 14.5) can be applied for location and imaging of dynamic macro-inhomogeneities in multiple-scattering media. This possibility was demonstrated by Boas et al.[26] in experiments with a phantom scattering object such as a resin cylinder (a static scatterer) with dynamic inclusions (a spherical cavity filled with a colloidal suspension). Titanium dioxide particles were added to resin in order to improve its scattering properties. Both scattering substances (the mixture of resin and $TiO_2$ particles and the colloidal suspension) were carefully prepared in order to provide equal values for the transport mean free path length. A set of data collected for different positions of a light source and a detector around the surface of the resin cylinder was used for reconstruction of the object image. The image reconstruction algorithm was based on the concept of field correlation transfer in turbid media.[53] Under these experimental conditions, this concept leads to a correlation diffusion equation similar to the light diffusion equation; the reconstructed image of the phantom object adequately characterizes the position, form, and dynamic properties of the spherical inclusion.

The possibility of recognizing various types of dynamic inhomogeneities hidden in a multiply scattering dynamic medium with use of correlation analysis of speckle intensity fluctuations was shown by Heckmeier et al.[54] They studied the dynamic light-scattering response of a multiply scattering system consisting of a Brownian medium with an embedded glass capillary in which the regular flow of a scattering substance (a suspension of polystyrene beads in water) was provided. The dependence of the decay of the temporal correlation of speckle intensity fluctuations on capillary position inside the Brownian scattering medium and flow rate was studied theoretically and experimentally. Regular flow was found to be reliably located at the depths on the order of the transport mean free path length.

Some applications of speckle correlometry techniques to *in vitro* tissue diagnostics are discussed in References 55 through 59.

## 14.4 Angular Correlations of Multiply Scattered Light

The existence of long-range angular correlations, a fundamental property of optical fields multiply scattered by random media, was considered in terms of "angular memory" effect by Feng et al.[22] Berkovits and Feng discussed[60] the possibility of using this effect as the physical basis for tomographic imaging of optically dense disordered media. The relations between angular correlations of multiply scattered coherent light and optical properties of scattering media for the transmittance mode of light propagation were studied theoretically and experimentally by Hoover and co-workers.[61] In this study, the potential of using the angular correlation analysis for disordered scattering media characterization was investigated.

Also, Tuchin et al.[62] considered an original approach to this problem, based on the influence of angular correlation decay on an interference of optical fields induced by two illuminating coherent beams incoming in the scattering medium at different angles of incidence. In this case, the probed medium is illuminated by a spatially modulated laser beam formed by overlapping the two collimated beams. The spatial modulation of the resulting illuminating beam has the form of a regular interference pattern with the fringe spacing determined by the angle between the overlapping beams. In the absence of scattering, the angular spectra of incident beams have the δ-like forms; the appearance of scattering causes broadening of these angular spectra and decay in the interference pattern contrast of the outgoing spatially modulated beam. Analysis of the interference pattern contrast for the outgoing beam and its dependence on the distance between the scatter and the observation plane and interference fringe period allows one to characterize the scattering properties of the probed medium.

## 14.5 Use of Time-Varying Speckle Contrast Analysis for Tissue Functional Diagnostics and Visualization

Analysis of speckle intensity fluctuations can be provided by acquisition of time sequences in the fixed detection point with further data processing. Another approach can be based on the statistical analysis of spatial fluctuations of time-averaged dynamic speckle patterns. Evaluation of the parameters of time-dependent intensity fluctuations (e.g., the correlation time) by use of this technique is possible in the case of ergodic dynamic speckle patterns. In this case, ensemble statistical averaging leads to the same results as temporal statistical averaging. Usually, statistical analysis of spatial fluctuations of speckle images captured with given exposure time can be carried out in the image or diffraction plane.

Two typical examples of biomedical applications of this technique can be considered; one is laser speckle contrast analysis, or the LASCA technique, developed by Fercher and Briers[33]as the tool for microcirculation analysis in upper layers of *in vivo* human tissues. The typical scheme of the LASCA device is very simple, as shown in Figure 14.6. The surface of probed tissue is illuminated by the broad collimated laser beam. A speckle-modulated image of the tissue surface is captured by a charge-coupled device (CCD) camera with given exposure time. In the case of expressed blood microcirculation in the upper layer of the probed tissue, the speckle pattern, which modulates the surface image, has a dynamic character due to random Doppler shifts of partial contributions scattered by moving erythrocytes. Temporal fluctuations of modulating speckles in the image plane cause the blurred captured image with a decayed value of contrast.

The contrast value for the time-averaged image depends on exposure time and falls when exposure increases. The following expression can be obtained for time-dependent contrast:

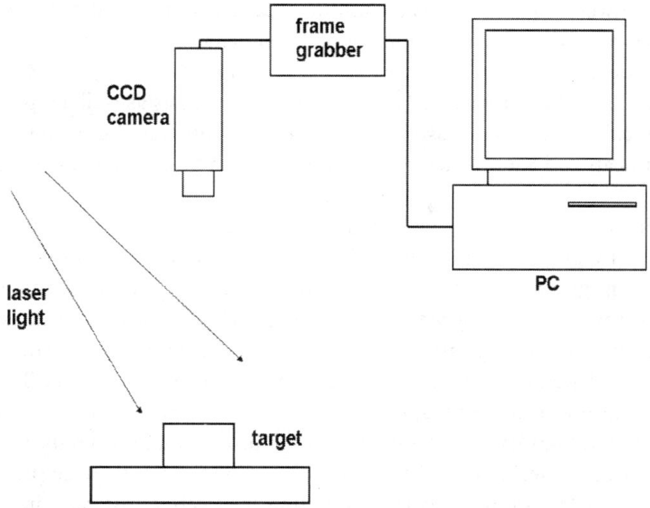


**FIGURE 14.6** Basic setup for LASCA.

$$C(T) = \left( \frac{1}{T} \int_0^T \tilde{g}_2(\tau)\, d\tau \right)^{1/2}, \tag{14.24}$$

where $\tilde{g}_2(\tau) = \left(G_2(\tau) - \langle I \rangle^2\right) / \langle I \rangle^2$ and $T$ is the exposure time. In particular, if the intensity correlation function exhibits the exponential decay: $\tilde{G}_2(\tau) = \langle I \rangle^2 \exp(-2\tau/\tau_c)$, where the correlation time $\tau_c = 1/ak\nu$ depends on the mean velocity $\nu$ of erythrocytes, wavenumber $k$ of probe light and scattering properties of tissue described by the factor $a$, then the time-dependent speckle contrast and the correlation time are related through

$$C(T) = \left[ \frac{\tau_c}{2T} \left\{ 1 - \exp\left(-2T/\tau_c\right) \right\} \right]^{1/2}. \tag{14.25}$$

If the analyzed tissue has the layered structure with significantly differing blood perfusion levels from layer to layer (e.g., human skin), then the average velocity, or mobility, of erythrocytes, which is evaluated by use of the LASCA technique, typically characterizes some average level of mobility for tissue volume with the thickness of the order of the depth of probe light penetration. This is the main disadvantage of speckle contrast analysis with broad collimated beam illumination, when a modulating speckle pattern is induced by random interference of partial components of a backscattered optical field coming from different depths of probed tissue.

In the case in which significant differences exist in the local perfusion level for various zones of the probed tissue area, the spatially inhomogeneous dynamic speckle pattern modulates the image of the tissue surface. For such an image, local estimations of exposure-dependent contrast can be used for characterization of the differences of average perfusion level from one zone of imaged surface to another. Estimated local values of the average mobility of erythrocytes can be used for reconstruction of "microcirculation maps." In the course of reconstruction of these maps, each pixel of the probed tissue surface is imaged by use of the color or gray-level scale related to the range of estimated erythrocyte mobility levels.

Briers et al.[63] demonstrated the potential of this technique for functional tissue imaging; more recent results in this field were reported by Dunn et al.[64] In this study, the LASCA technique was applied for dynamic imaging of cerebral blood flow (CBF). By illuminating the cortex with laser light and imaging the resulting speckle pattern, relative CBF images with tens of microns of spatial resolution and milliseconds of temporal resolution were obtained. This technique is easy to implement and can be used to monitor the spatial and temporal evolution of cerebral blood flow changes with high resolution in studies of cerebral pathophysiology.

The possibilities of using time-integrated speckle contrast analysis as well as speckle correlometry for studying the mechanical properties of tissue samples were demonstrated in recent works by Jacques and Kirkpatrick[65] and Duncan and Kirkpatrick.[66] This approach can be classified as the "speckle elastography of tissues."

Axial resolution in the case of time-dependent speckle contrast analysis in the image plane can be significantly improved by the use of a localized light source such as focused laser beam. Used for study of macroscopically homogeneous scattering media with thick slab geometry, this illumination scheme causes formation of statistically heterogeneous dynamic speckle patterns across the surface of the probed medium. This is because the average path of probe light propagation from the source points to the detection point strongly depends on the spatial separation between these points. The increasing distance from source and detection points causes spectral broadening of speckle intensity fluctuations from the center of the backscattered light spot on the surface of the probed medium to its edge, due to an increase in the average number of scattering events for partial contributions with large propagation paths.

The study of spatial distributions of correlation time for such statistically inhomogeneous speckle patterns allows one to obtain information not only about dynamic properties of the scattering medium, but also about conditions of probe light propagation inside the scattering volume. In particular, if the probe light travels in a layered scattering medium characterized by significant differences in dynamic properties from layer to layer, the analysis of time-averaged speckle-modulated images of the illuminated surface in the part of spatial distributions of local estimates of the time-dependent speckle contrast makes it possible to characterize the geometry of scattering system. A good example is the study of speckle contrast analysis potential for burn-depth diagnostics carried out by Sadhwani et al.[67] In this experiment, they used a burned tissue phantom composed of two layers — the static layer as the phantom of a superficial necrotic tissue layer and a modulating layer as the phantom of living tissue with an expressed microcirculation of blood (Figure 14.7).

Such scattering geometry in the case of a localized probe source leads to formation of two specific zones inside the backscattered light spot. Surrounding the source light spot, the first is characterized by the presence of a modulating static speckle pattern induced by scattering of probe light inside the static layer. The influence of partial contributions that reach the modulating layer and come back is negligibly small. The second zone, or area of dynamic speckle modulation, is located around the first. If the surface of the phantom scattering system is imaged with exposures significantly exceeding the typical value of correlation time of speckle intensity fluctuations, image for the second zone blurs with contrast value approaching 0. Thus, it is possible to estimate the thickness of the static ("necrotic") layer by measuring the inner radius of the blurred zone.

The following phantom object was used for approbation of this technique. Teflon films of various thicknesses (from 0.13 to 1.3 mm) were applied as static layers. A lipid solution (Intralipid-10%) was used as the modulating medium. An He-Ne laser operating at 633 nm, used as a source of probe light, was focused to an 80-μm diameter spot on the Teflon film surface. The image of the illuminated surface was captured by CCD camera with a Nikkor 50 mm f/1.8AF lens. The lens aperture diameter was set to minimum value (f/22) to maximize the average speckle size in the image plane. A polarizer was used between the CCD lens and the target in order to minimize the influence on image formation of specular reflection from the target surface.

Experiments with the burned tissue model have shown the strong correlation between the diameter of the nonblurred speckle pattern and the thickness of static layer. Because of the relative simplicity of the design, the inexpensive arrangement, and unsophisticated image-processing algorithm, this technique seems to be an adequate prospective tool for clinical applications.

The important question for various methods to monitor blood microcirculation on the basis of statistical, correlation, or spectral analysis of dynamic speckles is the influence of the detected light component induced by scattering of probe light by motionless scatterers existing in the scattering volume

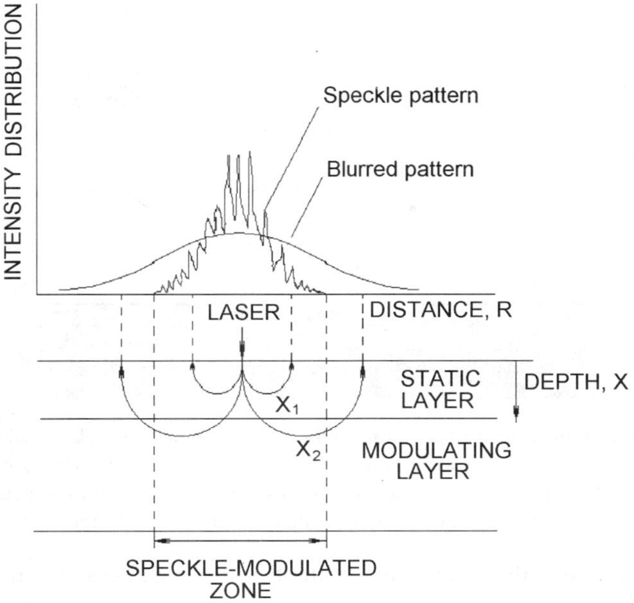


**FIGURE 14.7** Formation of time-integrated speckle pattern image in the case of backward multiple scattering of the probe laser light by a two-layered medium. Photons that travel to depth $x_1$ in the static layer produce a speckle pattern upon returning to the surface, whereas photons that travel to depth $x_2$ in the modulated layer produce a time-averaged blurred pattern at the surface. The size of the speckle-modulated zone is determined by the thickness of the static layer.

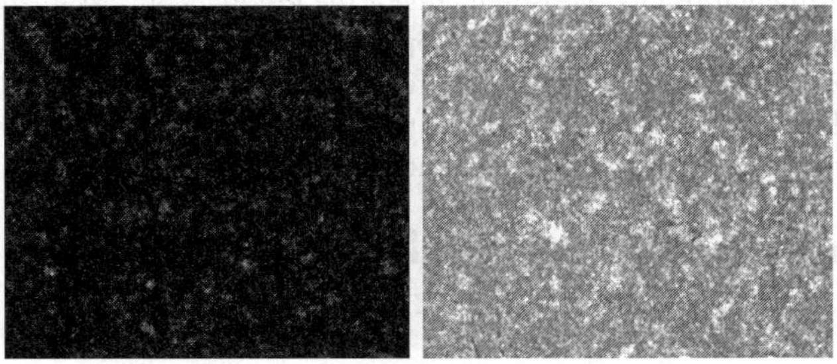

**FIGURE 14.8** Time-integrated speckle patterns recorded with differing exposure times (object under study is the nail bed of a 44-year-old male). Left: 8-ms exposure time; right: 25-ms exposure time.

of probed tissue. For instance, this influence will be significant in the case of laser diagnostics of blood microcirculation in the dermal layer. In the case of blood flow monitoring by means of speckle contrast analysis (e.g., the LASCA technique), the presence of a static component in the image-modulating speckle pattern will cause a noticeable value of the residual speckle contrast even for large values of exposure time — values that significantly exceed the correlation time of intensity fluctuations for the dynamic component. This effect is illustrated by Figure 14.8, which presents two time-averaged dynamic speckle patterns induced by laser light scattering in human forearm skin.

The residual contrast depends on the ratio of mean intensity values of the fluctuating component and the static component of an observed speckle pattern, and on the fraction of Doppler-shifted partial contributions $f$ in the scattered optical field. The simple analytical model that describes this ratio in the dependence on $f$ for the case of single-point detection of fluctuating speckle intensity was considered by

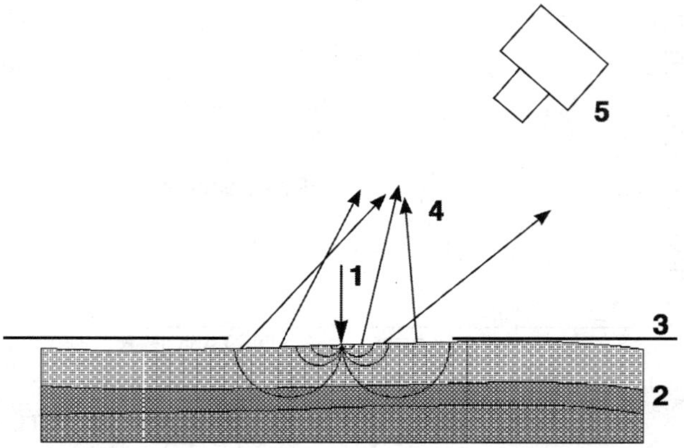


**FIGURE 14.9** Use of spatial filtering of a backscattered optical field in the object plane makes it possible to improve the spatial resolution of the speckle contrast technique. 1 = localized light source (focused laser beam); 2 = tissue under study; 3 = filtering diaphragm; 4 = scattered light; 5 = CCD camera without imaging lens. (Detection of scattered light in the diffraction regime allows one to exclude the saturation of captured speckle pattern.)

Serov et al.[68] They found that the modulation depth of fluctuating intensity in the detection point can be expressed as

$$\frac{\delta_I^2}{\langle I \rangle^2} = \frac{1}{N} f(2-f), \tag{14.26}$$

where $N$ is the number of speckles inside the detector aperture.

Having been verified for the case of a layered scattering model, this expression has shown excellent agreement with experimental results. Measurements of the residual contrast for *in vivo* tissues in combination with spatial filtration of a backscattered optical field by use of ring-like diaphragms offer the opportunity to analyze depth distributions of erythrocyte concentrations in probed tissue. The basic idea of this technique is illustrated by Figure 14.9. By using the circular and ring-like diaphragms of different radii, one can select the partial contributions of light penetrating into different depths of the tissue layer.

## 14.6 Imaging of Scattering Media with Use of Partially Coherent Speckles

If a scattering medium is illuminated by partially coherent light, only partial waves with pathlength differences less than the coherence length of the illuminating light will contribute to the formation of a random interference pattern. Such discrimination of partial contributions will lead to decay of the contrast of observed speckle patterns with an increase in the spectral width of the illumination source. This effect is clearly observed in the case of multiply scattering disordered media illuminated by a broadband light source (e.g., a superluminescent diode). Qualitative analysis of the physical picture of scattered field formation due to stochastic interference of partial waves, each of them characterized by the random value of the propagation path $s_i$, leads us to the following condition for the visibility of an observed speckle pattern: if the mean value of the pathlength difference $\Delta s = s_i - s_j$ for two arbitrarily chosen partial contributions $i$ and $j$ is significantly less than the coherence length $l_c$ of used light source: $\langle \Delta s \rangle = \int_0^\infty \Delta s \rho(\Delta s) d(\Delta s) << l_c$ [$\rho(\Delta s)$] is the probability density function of the pathlength difference), then the partial coherence of the illuminating beam does not strongly influence the visibility of the speckle pattern. In the opposite case of $\langle \Delta s \rangle >> l_c$, the speckle modulation of scattered field is suppressed and speckle contrast falls to zero.

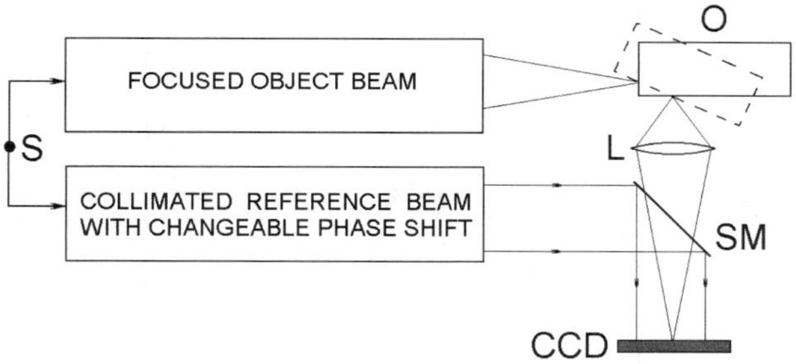


**FIGURE 14.10** Schematic of low-coherence optical system for a scattering media probing with use of the *photon horizon* detection technique. S is the low-coherent light source, O is the object under study, L is the imaging lens, and SM is the semitransparent mirror.

The strong influence on speckle visibility of pathlength statistics, determined by the propagation conditions for partially coherent light in a multiple-scattering medium, allows one to consider the measurements of the contrast of partially coherent speckles as a technique for inhomogeneous media probing and imaging. The presence of scattering or absorbing inhomogeneities in a probed medium between the light source and detector area will distort the pathlength distribution inside the scattering volume; consequently, they will appear as changes in the contrast of speckles observed in the detection area in comparison with the homogeneous medium with the same background optical properties.

In particular, Thompson et al.[69,70] considered one version of this technique. They measured the contrast of speckles that result from propagation of laser light in the multiple-scattering slab with a strongly scattering or absorbing inclusion, because the position of the inclusion varied with respect to illuminating beam and detector (a CCD camera). A speckle-modulated image of the backward surface of the illuminated slab was formed by a lens; therefore, the observed speckles were subjective ones with the average size depending on the aperture size of the imaging lens. A polarization filter was used to improve the visibility of the observed speckles by selection of the linearly polarized component of the scattered field. The experimental results have shown that it is possible to use partially coherent speckle contrast analysis for probing and imaging macroscopically heterogeneous scattering media with the appropriately chosen spectral bandwidth for the illumination source.

The original version of an imaging technique based on observation of spatial distributions of partially coherent speckle contrast was reported by Hausler et al.[71] The schematic of the experimental setup is presented in Figure 14.10. In this case, adjustment of the reference arm of the interferometer (e.g., a Mach-Zehnder interferometer) allows one to find a region on the tissue surface where partial components of a scattered field with a certain propagation path will emerge after propagation through a scattering volume. In order to measure the speckle contrast, subtraction of two sequential images of a scattering object is used. In this case, the second image is obtained with the reference phase shifted by $\pi$.

By using this technique, the region of interest, which can be interpreted as the position of the photon horizon for the given time delay, is imaged as a border of speckle-modulated area on the image of the object surface (Figure 14.11). Varying the time delay by changing the pathlength difference between the reference and object arms of interferometer, one can study light propagation in the probed medium based on its optical properties.

## 14.7 Summary

Various speckle technologies based on analysis of decay of statistical moments of spatial, temporal, or angular fluctuations of scattered optical fields are adequately powerful tools for diagnostics and visualization of structure and dynamic properties of multiply-scattering, weakly ordered media such as human and animal tissues.

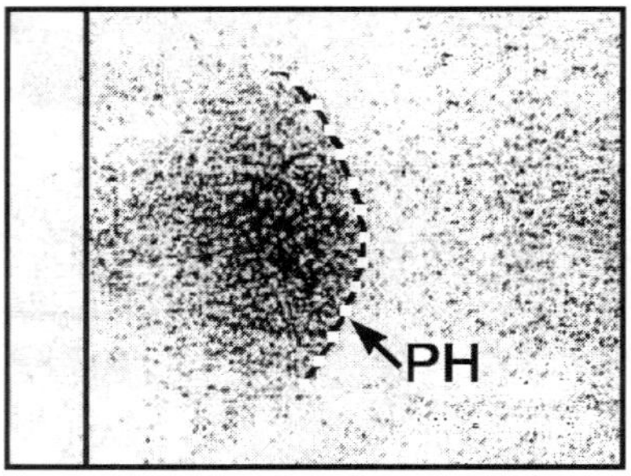


**FIGURE 14.11** Visualization of the *photon horizon* (PH) with use of the speckle contrast imaging technique. The vertical line in the left part of the figure shows the front surface of the object, which is illuminated by a low-coherent beam. The speckle-modulated image is formed by light emerging from the side surface of sample.

## Acknowledgments

The work on this chapter was supported by grants 01-02-17493 and 00-15-96667 of the Russian Foundation for Basic Research, and by grant REC-006 of the Civil Research and Development Foundation for the Independent States of the Former Soviet Union.

### References

1. Goodman, J.W., Some fundamental properties of speckles, *J. Opt. Soc. Am.*, 66, 1145, 1976.
2. Goodman, J.W., Dependence of image speckle contrast on surface roughness, *Opt. Commun.*, 14, 324, 1975.
3. Goodman, J.W., Statistical properties of laser speckle patterns, in *Laser Speckle and Related Phenomena,* 2nd ed., Dainty, J.C., Ed., Springer-Verlag, Heidelberg, 1984, p. 9.
4. Goodman, J.W., *Statistical Optics,* Wiley, New York, 1985.
5. Dainty, J.C., Ed., *Laser Speckle and Related Phenomena, Topics in Applied Physics,* Vol. 9, Springer-Verlag, Heidelberg, 1984.
6. Pedersen, H.M., Theory of speckle dependence on surface roughness, *J. Opt. Soc. Am.*, 66, 1204, 1976.
7. Jakeman, E., Speckle statistics with a small number of scatterers, *Opt. Eng.*, 23, 453, 1984.
8. Pusey, P.N., Photon correlation study of laser speckle produced by a moving rough surface, *J. Phys. D,* 9, 1399, 1976.
9. Cummins, H.Z. and Pike, E.R., Eds., *Photon Correlation and Light-Beating Spectroscopy,* NATO Advanced Study Institute Series B: Physics, Plenum Press, New York, 1974.
10. Jakeman, E. and Welford, W.T., Speckle statistics in imaging systems, *Opt. Commun.*, 21, 72, 1977.
11. Outsubo, J. and Asakura, T., Statistical properties of speckle intensity variations in the diffraction field under illumination of coherent light, *Opt. Commun.*, 14, 30, 1975.
12. Uozumi, J. and Asakura, T., First-order intensity and phase statistics of Gaussian speckles produced in the diffraction region, *Appl. Opt.,* 20, 1454, 1981.
13. Yoshimura, T., Statistical properties of dynamic speckles, *J. Opt. Soc. Am. A*, 3, 1032, 1986.
14. Yoshimura T., Nakagawa K., and Wakabayashi, N., Rotational and boiling motion of speckles in a two-lens imaging system, *J. Opt. Soc. Am. A*, 3, 1018, 1986.
15. Golubentsev, A.A., On the suppression of the interference effects under multiple scattering of light, *Zh. Eksp. Teor. Fiz.*, 86, 47, 1984 (in Russian).
16. Stephen, M.J., Temporal fluctuations in wave propagation in random media, *Phys. Rev. B*, 37, 1, 1988.
17. MacKintosh, F.C. and John, S., Diffusing-wave spectroscopy and multiple scattering of light in correlated random media, *Phys. Rev. B*, 40, 2382, 1989.

18. John, S., Localization of light, *Phys. Today*, May 1991, p. 32.
19. Van Albada, M.P. and Lagendijk, A., Observation of weak localization of light in a random medium, *Phys. Rev. Lett.*, 55, 2692, 1985.
20. Wolf, P.-E. and Maret, G., Weak localization and coherent backscattering of photons in disordered media. *Phys. Rev. Lett.*, 55, 2696, 1985.
21. Akkermans, E., Wolf, P.E., Maynard, R., and Maret, G., Theoretical study of the coherent backscattering of light by disordered media, *J. Phys. France*, 49, 77, 1988.
22. Feng, S., Kane, C., Lee, P.A., and Stone, A.D., Correlations and fluctuations of coherent wave transmission through disordered media, *Phys. Rev. Lett.*, 61, 834, 1988.
23. Freund, I., Rosenbluh, M., and Feng, S., Memory effects in propagation of optical waves through disordered media, *Phys. Rev. Lett.*, 61, 2328, 1988.
24. Pine, D.J., Weitz, D.A., Chaikin, P.M., and Herbolzheimer, E., Diffusing-wave spectroscopy, *Phys. Rev. Lett.*, 60, 1134, 1988.
25. Boas, D.A., Campbell, L.E., and Yodh, A.G., Scattering and imaging with diffusing temporal field correlations, *Phys. Rev. Lett.*, 75, 1855, 1995.
26. Yodh, A.G., Georgiades, N., and Pine, D.J., Diffusing-wave interferometry, *Opt. Commun.*, 83, 56, 1991.
27. Kao, M.H., Yodh, A.G., and Pine, D.J., Observation of Brownian motion on the time scale of hydrodynamic interactions, *Phys. Rev. Lett.*, 70, 242, 1993.
28. Boas, D.A., Bizheva, K.K., and Siegel, A.M., Using dynamic low-coherence interferometry to image Brownian motion within highly scattering media, *Opt. Lett.*, 23, 319, 1998.
29. Stern, M.D., *In vivo* evaluation of microcirculation by coherent light scattering, *Nature* (London), 254, 56, 1975.
30. Stern, M.D., Laser Doppler velocimetry in blood and multiply scattering fluids: theory, *Appl. Opt.*, 28, 1968, 1985.
31. Briers, J.D., Wavelength dependence of intensity fluctuations in laser speckle patterns from biological specimens, *Opt. Commun.*, 13, 324, 1975.
32. Fercher, A.F., Velocity measurement by first-order statistics of time-differentiated laser speckles, *Opt. Commun.*, 33, 129, 1980.
33. Fercher, A.F. and Briers, J.D., Flow visualization by means of single-exposure speckle photography, *Opt. Commun.*, 37, 326, 1981.
34. Fujii, H., Asakura, T., Nohira, K., et al., Blood flow observed by time-varying laser speckle, *Opt. Lett.*, 10, 104, 1985.
35. Fercher, A.F., Peukert, M., and Roth, E., Visualization and measurement of retinal blood flow by means of laser speckle photography, *Opt. Eng.*, 25, 731, 1986.
36. Aizu, Y., Ogino, K., Koyama, T., et al., Evaluation of retinal blood flow using laser speckle, *J. Jpn. Soc. Laser Med.*, 8, 89, 1987.
37. Fujii, H., Nohira, K., Yamamoto, Y., Ikawa, H., and Ohura, T., Evaluation of blood flow by laser speckle image sensing, *Appl. Opt.*, 26, 5321, 1987.
38. Ruth, B., Non-contact blood flow determination using a laser speckle method, *Opt. Laser Technol.*, 20, 309, 1988.
39. Aizu, Y. and Asakura, T., Bio-speckle phenomena and their application to the evaluation of blood flow, *Opt. Laser Technol.*, 23, 205, 1991.
40. Nilsson, G., Jakobsson, A., and Wårdell, K., Tissue perfusion monitoring and imaging by coherent light scattering, *Proc. SPIE*, 1524, 90, 1991.
41. Briers, J.D. and Webster, S., Quasi real-time digital version of single-exposure speckle photography for full-field monitoring or flow fields, *Opt. Commun.*, 116, 36, 1995.
42. Dacosta, G., Optical remote sensing of heartbeats, *Opt. Commun.*, 117, 395, 1995.
43. Rhytov, S.M., Kravtsov, U.A., and Tatarsky, V.I., *Introduction to Statistical Radiophysics*, Part 2, *Random Fields*, Nauka Publishers, Moscow, 1978.

44. Korn, G.A. and Korn, T.M., *Mathematical Handbook for Scientists and Engineers,* McGraw-Hill, New York, 1968.
45. Escamilla, H.M., Speckle contrast from weak diffusers with a small number of correlation areas, *Opt. Acta*, 25, 777, 1978.
46. Zimnyakov, D.A. and Tuchin, V.V., About two-modality of intensity distributions of speckle fields for large-scale phase scatterers, *J. Tech. Phys. Lett.*, 21, 10, 1995.
47. Feder, J., *Fractals*, Plenum Press, New York, 1988.
48. Jakeman, E., Scattering by fractals, in *Fractals in Physics*, Pietronero, L. and Tozatti, E., Eds., North-Holland, Amsterdam, 1986, p. 82.
49. Zimnyakov, D.A., Tuchin, V.V., and Utts, S.R., A study of statistical properties of partially developed speckle fields as applied to the diagnostics of structural changes in human skin, *Opt. Spectrosc.*, 76, 838, 1994.
50. Zimnyakov, D.A. and Tuchin, V.V., Fractality of speckle intensity fluctuations, *Appl. Opt.*, 35, 3328, 1996.
51. Maret, G. and Wolf, P.E., Multiple light scattering from disordered media. The effect of Brownian motions of scatterers, *Z. Phys. B*, 65, 409, 1987.
52. Boas, D.A. and Yodh, A.G., Spatially varying dynamical properties of turbid media probed with diffusing temporal light correlation, *J. Opt. Soc. Am. A*, 14, 192, 1997.
53. Ackerson, B.J., Dougherty, R.L., Reguigui, N.M., and Nobbman, U., Correlation transfer: application of radiative transfer solution methods to photon correlation problems, *J. Thermophys. Heat Trans.*, 6, 577, 1992.
54. Heckmeier, M., Skipetrov, S.E., Maret, G., and Maynard, R., Imaging of dynamic heterogeneities in multiple-scattering media, *J. Opt. Soc. Am. A*, 14, 185, 1997.
55. Zimnyakov, D.A., Tuchin, V.V., and Mishin, A.A., Spatial speckle correlometry in applications to tissue structure monitoring, *Appl. Opt.*, 36, 5594, 1997.
56. Zimnyakov, D.A., Tuchin, V.V., and Yodh, A.G., Characteristic scales of optical field depolarization and decorrelation for multiple scattering media and tissues, *J. Biomed. Opt.*, 4, 157, 1999.
57. Zimnyakov, D.A., Maksimova, I.L., and Tuchin, V.V., Controlling of tissue optical properties. II. Coherent methods of tissue structure study, *Opt. Spectrosc.*, 88, 1026, 2000.
58. Zimnyakov, D.A. and Tuchin, V.V., Laser tomography, in *Lasers in Medicine*, Vij, D.R., Ed., Kluwer Academic, Dordrecht, the Netherlands, 2002.
59. Tuchin, V.V., Ed., *Handbook on Optical Biomedical Diagnostics,* Vol. PM107, SPIE Press, Bellingham, WA, 2002.
60. Berkovits, R. and Feng, S., Theory of speckle-pattern tomography in multiple-scattering media, *Phys. Rev. Lett.,* 65, 3120, 1990.
61. Hoover, B.G., Deslauriers, L., Grannell, S.M., Ahmed, R.E., Dilworth, D.S., Althey, B.D., and Leith, E.N., Correlations among angular wave component amplitudes in elastic multiple-scattering random media, *Phys. Rev. E,* 65, 026614-1, 2002.
62. Tuchin, V.V., Ryabukho, V.P., Zimnyakov, D.A., Lobachev, M.I., Lyakin, D.V., Radchenko, E.Yu., Chaussky, A.A., and Konstantinov, K.V., Tissue structure and blood microcirculation monitoring by speckle interferometry and full-field correlometry, *Proc. SPIE,* 4251, 148, 2001.
63. Briers, J.D. and Webster, S., Laser speckle contrast analysis (LASCA): a non-scanning, full-field technique for monitoring capillary blood flow, *J. Biomed. Opt.,* 1, 174, 1996.
64. Dunn, A.K., Bolay, H., Moskowitz, M.A., and Boas, D.A., Dynamic imaging of celebral blood flow using laser speckle, *J. Cereb. Flow Metab.,* 21, 195, 2001.
65. Jacques, S.L. and Kirkpatrick, S.J., Acoustically modulated speckle imaging of biological tissues, *Opt. Lett.*, 23, 879, 1998.
66. Duncan, D.D. and Kirkpatrick, S.J., Processing algorithms for tracking speckle shifts in optical elastography of biological tissues, *J. Biomed. Opt.*, 6, 418, 2001.
67. Sadhwani, A., Schomacker, K.T., Tearney, G.T., and Nishioka, N., Determination of Teflon thickness with laser speckle. I. Potential for burn depth diagnosis, *Appl. Opt.*, 35, 5727, 1996.

68. Serov, A., Steenbergen, W., and de Mul, F., A method for estimation of the fraction of Doppler-shifted photons in light scattered by mixture of moving and stationary scatterers, *Proc. SPIE*, 4001, 178, 2000.
69. Thompson, C.A., Webb, K.J., and Weiner, A.M., Diffuse media characterization using laser speckle, *Appl. Opt.*, 36, 3726, 1997.
70. Thompson, C.A., Webb, K.J., and Weiner, A.M., Imaging in scattering media by use of laser speckle, *J. Opt. Soc. Am. A*, 14, 2269, 1997.
71. Hausler, G., Herrmann, J.M., Kummer, R., and Lindner, M.V., Observation of light propagation in volume scatterers with $10^{11}$-fold slow motion, *Opt. Lett.*, 21, 1087, 1996.

# 15

# Laser Doppler Perfusion Monitoring and Imaging

Gert E. Nilsson
*Linköping University*
*Linköping, Sweden*

E. Göran Salerud
*Linköping University*
*Linköping, Sweden*

N.O. Tomas Strömberg
*Linköping University*
*Linköping, Sweden*

Karin Wårdell
*Linköping University*
*Linköping, Sweden*



## 15.1 Introduction

Microcirculation — including capillaries, small arteries (arterioles), small veins (venules), and shunting vessels (arteriovenous anastomosis) — comprises the blood vessels of the most peripheral part of the vascular tree.[1] In the skin, the microvascular network is composed of different compartments, each with a different anatomy and function. The top layer of the skin, the epidermis, is avascular, while the dermal papillae host the capillaries responsible for the exchange of oxygen and metabolites with the surrounding tissue. Therefore, the blood perfusion through the capillaries is frequently referred to as the nutritive blood flow. Although this nutritive blood flow is low during resting conditions, it is of vital importance for maintaining the minute but important metabolic requirements of the skin. In the deeper dermal structures, the arterioles, venules, and shunting vessels reside. The main role of these vessels is to feed and drain the capillary network and to promote the maintenance of an adequate body temperature by dissipating heat to the environment through the modulation of shunting vessel blood flow.

Microvascular blood perfusion possesses temporal fluctuations and spatial variability. Temporal fluctuations may be either rhythmic, as in the case of vasomotion,[2] or more stochastic[3] and are of either local or central origin; spatial variability originates in the anatomical heterogeneity of the microvascular network.[4] Perfusion of blood in the skin is influenced by external factors such as heat and topically applied vasoactive substances and by the intake of beverages and drugs, as well as by smoking and mental stimuli. In addition, microvascular blood perfusion is regulated via the autonomous nervous system and through vasoctive agents released by the endocrine glands into the bloodstream.

A number of common diseases — including various inflammatory conditions, allergic reactions, and tumors — influence perfusion of blood in the skin. In addition, impaired circulation in association with diabetes and peripheral vascular disease may lead to the formation of ulcers and ultimately to tissue necrosis. Consequently, the perfusion of blood in the skin as well as in other tissues needs to be investigated by the use of noninvasive methods with a minimal influence on the parameters under study. In addition,

0-8493-1116-0/03/$0.00+$1.50

such methods should preferably record the perfusion in real time and be applicable in experimental and in clinical settings. In order to evaluate the dynamics of blood perfusion and its spatial variability over a given tissue surface, methods that enable the recording of both of these microvascular aspects need to be employed.

Many methods for the assessment of microvascular blood perfusion have been presented in the literature, including photoelectric plethysmography,[5,6] thermal[7] and radioisotope[8,9] clearance, orthogonal polarization spectral imaging,[10,11] and video-photometric capillaroscopy.[12,13] None of those methods, however, has fulfilled the requirements of noninvasiveness and applicability to a large number of tissues as well as offering the possibility of recording the tissue perfusion continuously or visualizing the results in terms of two-dimensional perfusion maps. Laser Doppler flowmetry (LDF) has emerged in the absence of any other method to fulfill those requirements.

The first measurements of microvascular blood flow employing the Doppler shift of monochromatic light (named after the Austrian scientist Johan Christian Doppler[14]) were reported by Riva et al.[15] in 1972; they studied blood cell velocity in the rabbit eye. Some years later the blood velocity of exposed microvessels was investigated by use of a laser Doppler microscope.[16,17] *In vivo* evaluation of skin microcirculation by the use of coherent light scattering was first demonstrated by M.D. Stern in 1975.[18] In his setup, in which an He-Ne laser beam illuminates the skin, a portion of the laser light is scattered by moving blood cells as well as static tissue structures, while the remaining light is scattered by static tissue structures alone. If the backscattered Doppler-broadened light is brought to the surface of a photodetector, the speckle pattern appearing on the detector surface fluctuates because of the optical mixing of waves of different frequencies. These fluctuations are manifested as audiofrequency components in the photocurrent superimposed on a steady baseline. Analysis of the photocurrent reveals a clear shift of the spectral content toward lower frequencies following the occlusion of the brachial artery by use of a pressure cuff placed around the upper arm and with the laser beam aimed at the fingertip. This fundamental discovery was soon implemented in a portable fiber-optic-based laser Doppler flowmeter,[19] which provided results comparable with those obtained by the $^{133}$Xe-clearance technique for assessment of tissue blood perfusion.

As this early prototype was used more extensively, laser mode interference noise proved to be a major limitation of the system. To overcome this problem, a differential detector system was introduced[20] that rejected the common mode modal interference signal while amplifying the sum of the uncorrelated blood perfusion–related signals from each detector. A signal processor using the first un-normalized moment of the photocurrent as a continuous output signal was introduced,[21] and a theory describing quasi-elastic light scattering in perfused tissues was proposed by Bonner and Nossal.[22] This signal-processing algorithm still remains the most favorable way of producing a laser Doppler flowmeter output signal that scales linearly with the average speed and concentration of moving blood cells within the scattering volume. In the context of LDF, this quantity is generally referred to as the perfusion.[23] Although LDF does not selectively record the movement of red blood cells (RBCs), it is frequently assumed that the RBCs represent the majority of the blood-borne objects in the undisturbed microcirculation.

As LDF was commercialized in the early 1980s, laser Doppler perfusion monitoring (LDPM) became available to a wide range of researchers and clinicians. The ease of use of this new technology quickly created an extensive interest in recording microvascular perfusion in a wide variety of disciplines, and to date several thousands of publications in the scientific literature cite its use.[24–26]

One of the most remarkable features of microvascular blood perfusion is its substantial temporal and spatial variability. Consequently, the comparatively small sampling volume (about 1 mm$^3$) of fiber-optic-based LDPM instruments constitutes one of the main limitations of this technology: with the spatial variability in tissue blood perfusion, gross differences in perfusion readings may appear at recordings from adjacent sites. This disadvantage of LDPM triggered the development of multichannel instruments[27] and the laser Doppler perfusion imaging (LDPI) technology at the end of the 1980s.[28–30] In LDPI, an airborne laser beam successively probes a large number of tissue volumes, and a data set of perfusion values is generated from which a color-coded perfusion map is compiled. The commercialization phase of the LDPI-technology started in the early 1990s, and several hundred publications have already verified its usefulness, particularly in dermatology,[31] wound healing,[32] and burn treatment.[33]

LDPM and LDPI constitute a versatile pair of related noninvasive medical technologies that facilitate the study of temporal and spatial variability of tissue blood perfusion. Both technologies are easy to use, but the design of a study and interpretation of the results require establishment of a well-prepared protocol and an understanding of microcirculation as well as the factors that influence it. Consequently, guidelines have been developed to assist the safe and reliable use of LDPM[34] and LDPI.[35]

## 15.2 Theory

Coherent light directed toward a tissue will be scattered by moving objects and by static tissue structures as the photons migrate through the tissue in a random pattern. Scattering in moving objects (predominantly RBCs) changes the frequency of the light according to the Doppler principle, while light scattered in static structures alone remains unshifted in frequency. If the remitted light is detected by a photodetector, optical mixing of light shifted and unshifted in frequency will result in a stochastic photocurrent, the power spectral density of which depends on the number of RBCs and their shape and velocity distribution within the scattering volume. In the following section we present a theory[36] that relates the photocurrent power spectrum $P(\omega)$ to the properties of the blood cells in the illuminated volume. More precisely, we will show that the quantity $\int \omega P(\omega) d\omega$ scales with the product of the RBC concentration and average velocity, while $\int P(\omega) d\omega$ scales with the concentration of moving RBCs alone, provided that the RBC concentration in tissue is low.

### 15.2.1 The Single Scattering Event

The starting point of this derivation is the interaction of a single photon and a scattering object, generally referred to as a *single scattering event* (Figure 15.1). If the incident wave $\mathbf{E}_i$ propagating in the direction $\mathbf{k}_i$ is denoted $\mathrm{E}_\mathrm{i} = E_{io} e^{i(\omega t + \mathrm{k}_i \mathrm{r})}$, where $E_{io}$ is the amplitude and $\omega$ the angular frequency, $t$ is time and $\mathbf{r}$ represents the spatial coordinates, the scattered wave $\mathbf{E}_s$ in the direction of $\mathbf{k}_s$ can be written in the form

$$\mathrm{E}_s = E_{so} e^{i((\omega - \mathrm{k}_i \mathrm{v})t - \mathrm{k}_i \mathrm{r}_o - \mathrm{k}_s(\mathrm{r}_1 - \mathrm{v}t))} = E_{so} e^{i\omega t} e^{-i(\mathrm{k}_i - \mathrm{k}_s)\mathrm{v}t} e^{-i(\mathrm{k}_i \mathrm{r}_0 + \mathrm{k}_s \mathrm{r}_1)}, \tag{15.1}$$

where $E_{so}$ is the amplitude of the scattered wave, $\mathbf{v}$ is the particle velocity and $e^{-i(\mathrm{k}_i \mathrm{r}_0 + \mathrm{k}_s \mathrm{r}_1)}$ is a time-invariant phase factor. Defining the scattering vector as $\mathbf{q} = \mathbf{k}_i - \mathbf{k}_i$, and neglecting the time-invariant phase factor, the scattered wave can be written as being composed of the product of the carrier frequency factor and the time variant phase factor according to

$$\mathrm{E}_\mathrm{s} = E_{so} e^{i\omega t} e^{-i\mathrm{q}\mathrm{v}t}. \tag{15.2}$$

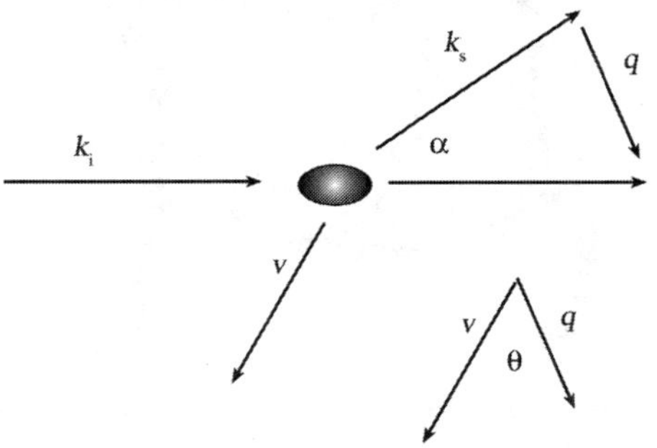


**FIGURE 15.1** The single scattering event.

If the scattering object is stationary ($\mathbf{v} = 0$), if the wave is not scattered by the object ($\mathbf{q} = 0$) or if the $\mathbf{v}$ and $\mathbf{q}$ vectors are perpendicular, the wave remains unshifted in frequency. Because $k_i \approx k_s$, the magnitude of the scattering vector equals $q = 2k\sin\left(\frac{\alpha}{2}\right) = \frac{4\pi}{\lambda_t}\sin\left(\frac{\alpha}{2}\right)$, where $\alpha$ is the scattering angle and $\lambda_t$ the wavelength in tissue. The frequency-determining argument of the time-variant phase factor $\mathbf{qv} = qv\cos(\theta)$ constitutes the angular Doppler frequency $\omega_D$, which amounts to

$$\omega_D = \frac{4\pi}{\lambda_t}\sin\left(\frac{\alpha}{2}\right)v\cos(\theta)\,, \tag{15.3}$$

where $\theta$ is the angle between $\mathbf{v}$ and $\mathbf{q}$.

### 15.2.2 Detection

When a laser beam impinges on the tissue surface, the individual photons migrate through the tissue matrix in a random pattern to a depth determined by the optical properties of the tissue. This migration process implies a multitude of single scattering events with or without associated frequency shifts (Figure 15.2). A portion of all photons injected return to the tissue surface, where they become available for detection. This backscattered light is Doppler-broadened by an amount dependent on the average individual RBC velocity and scattering vectors. Therefore, the expectation value of the time-variant phase factor $\langle e^{-iqv\tau}\rangle$, according to Equation 15.2, needs to be determined in order to calculate the tissue perfusion.

When the backscattered light impinges on the surface of a photodetector, a fluctuating speckle pattern is formed. The frequency and magnitude of intensity fluctuations in the individual coherence areas of this speckle pattern are related to the average speed and number of moving objects within the scattering volume. The extension of a single coherence area $\mathbf{A_c}$ in the speckle pattern is determined by the wavelength

**FIGURE 15.2** Migration of photons through the tissue matrix including numerous single scattering events where scattering in moving objects involves a frequency shift according to the Doppler principle. $\lambda_0$ constitutes the laser wavelength while $\lambda_1$, $\lambda_2$, $\lambda_3$, and $\lambda_4$ represent the wavelengths after single or multiple scattering in moving objects.

in vacuum ($\lambda$) and the solid angle ($\Omega$) under which the laser spot on the object is seen from the detector surface according to References 37 through 39:

$$A_c = \frac{\lambda^2}{\Omega}. \tag{15.4}$$

In the photodetector, the instantaneous light intensity is converted to a photocurrent. If the average and the fluctuating portions of the photocurrent produced by coherence area $j$ are denoted as $\langle i_c \rangle$ and $\Delta i_{cj}(t)$, respectively, the total photocurrent $i(t)$ proportional to the instantaneous light intensity can be expressed as[40]

$$i(t) = \sum_{j=1}^{K} \left( \langle i_c \rangle + \Delta i_{cj}(t) \right), \tag{15.5}$$

where $K$ is the total number of coherence areas on the detector surface and all noise factors are neglected. Since $i(t)$ is a stochastic process, its characteristics may be described by the autocorrelation function

$$\begin{aligned} \langle i(0) i^*(\tau) \rangle &= \sum_{j=1}^{K} \left( \langle i_c \rangle + \Delta i_{cj}(0) \right) \sum_{l=1}^{K} \left( \langle i_c^* \rangle + \Delta i_{cl}^*(\tau) \right) \\ &= \sum_{j=1}^{K} \sum_{l=1}^{K} \langle i_c \rangle \langle i_c^* \rangle + \sum_{j=1}^{K} \sum_{l=1}^{K} \Delta i_{cj}(0) \Delta i_{cl}^*(\tau) \\ &= K^2 \langle i_c \rangle^2 + K \langle \Delta i_c(0) \Delta i_c^*(\tau) \rangle, \end{aligned} \tag{15.6}$$

where $\Delta i_c(\tau)$ represents the average fluctuating portion of the photocurrent produced by a single coherence area. Equation 15.6 is valid under the assumption that the fluctuating portions of the photocurrents produced by the individual coherence areas are independent, i.e., all elements $\Delta i_{cj} \, \Delta i_{cl}^*(\tau)$ are nonzero only for $j = l$, and each fluctuating component has an average value equal to zero. From Equation 15.6 it can be concluded that the Doppler-related portion of the photocurrent autocorrelation function scales linearly with the number of coherence areas, while the stationary part scales with the square of the number of coherence areas. Furthermore, the performance of the entire system is uniquely described by the statistics of the photocurrent produced by a single coherence area.

Because the photodetector is a square-law detector, the photocurrent produced by a single coherence area is related to the electromagnetic field $\mathbf{E}(t)$ as $i_c(t) \propto \mathbf{E}(t)\, \mathbf{E}^*(t)$. Hence, the autocorrelation of the photocurrent produced by a single coherence area can be expressed as

$$\langle i_c(0) i^*_c(\tau) \rangle \propto \langle \mathbf{E}(0) \mathbf{E}^*(0) \mathbf{E}(\tau) \mathbf{E}^*(\tau) \rangle, \tag{15.7}$$

where the **E**-vector represents the sum of the electromagnetic field vectors impinging on the actual coherence area. This sum can be split into two parts, of which one is the sum of all field vectors representing photons that have undergone a Doppler shift, and the other is the sum of the field vectors representing photons with no frequency shift.

Evaluating Equation 15.7 thus results in 16 terms, 10 of which have an average value that equals zero because their fluctuating parts are independent and uncorrelated. Three terms include components that are time-invariant and thus represent the average photocurrent produced by the coherence area in question. Two terms represent *heterodyne* mixing of beams shifted in frequency with beams not shifted in frequency, while one term represents *homodyne* mixing of beams that have all undergone at least one frequency shift. The overall autocorrelation function of the photocurrent produced by a single coherence area can therefore be written in the form

$$\langle i_c(0)i^*_c(\tau)\rangle = \underbrace{2i_{Re}i_{Sc} + i^2_{Re} + i^2_{Sc}}_{\text{Stationary}} + \underbrace{i_{Re}i_{Sc}\left(\langle e^{iqv\tau}\rangle + \langle e^{-iqv\tau}\rangle\right)}_{\text{heterodyne -mixing}} + \underbrace{i^2_{Sc}\left\langle \frac{1}{S^2}\sum_{k=1,k\neq l}^{S}\sum_{l=1}^{S} e^{i(q_k v_k - q_l v_l)\tau}\right\rangle}_{\text{homodyne -mixing}}, \tag{15.8}$$

where $i_{Re}$ and $i_{Sc}$ represent the average currents produced, respectively, by photons not shifted in frequency and those shifted in frequency within a single coherence area, S denotes the total number of photons with Doppler shifts, and $\langle e^{iqv\tau}\rangle = \left\langle \frac{1}{S}\sum_{k=1}^{S} e^{i(q_k v_k \tau)}\right\rangle$. For a low and moderate RBC concentration ($C_{RBC}$), $i_{Sc}$ is proportional to the total photocurrent produced by a single coherence area and $i_{Sc} \approx C_{RBC} i_{Re}$ for $i_{Re} >> i_{Sc}$.

In most applications, including low and moderate tissue blood volumes, the heterodyne terms dominate over the homodyne term. Under those circumstances, the homodyne term may be disregarded.

## 15.2.3 Signal Processing

Correctly processing the detected fluctuations in the photocurrent and development of an algorithm producing an output signal that scales with the product of RBC speed and concentration requires derivation of the relationship between velocity distribution, the scattering vector distribution at an RBC single scattering event and the recorded autocorrelation function of the photocurrent. This derivation starts with calculation of $\langle e^{iqv\tau}\rangle_{\mathbf{v}}$ for the actual three-dimensional velocity distribution $N_0(\mathbf{v})$. Then the average value of this quantity $\langle\langle e^{iqv\tau}\rangle_{\mathbf{v}}\rangle_{\mathbf{q}}$ with respect to the scattering vector distribution $S_0(\mathbf{q}(\alpha))$ is calculated. According to the Wiener–Khintchine theorem,[41] the Fourier transform of $\langle\langle e^{iqv\tau}\rangle_{\mathbf{v}}\rangle_{\mathbf{q}}$ gives the photocurrent power spectral density. These successive calculations allow the recorded power spectral density to be linked to the RBC velocity distribution for the actual laser wavelength.

Certain assumptions need to be made regarding the velocity distribution of the RBCs. These assumptions include (1) a velocity distribution $N_0(\mathbf{v})$ that is independent of the spatial coordinates, (2) independent movements of the individual RBCs, and (3) an RBC concentration low enough to justify negligence of the homodyne term in the photocurrent autocorrelation function. These assumptions may not be entirely fulfilled in real tissues. Nevertheless, they are important for the derivation of a suitable and robust algorithm for laser Doppler flowmeters and imaging systems.

### 15.2.3.1 Derivation of $\langle e^{iqv\tau}\rangle_{\mathbf{v}}$

The expectation value of $\langle e^{iqv\tau}\rangle_{\mathbf{v}}$ with reference to **v** for a fixed value of **q** can be calculated according to

$$\langle e^{iqv\tau}\rangle_{\mathbf{v}} = \int_{\mathbf{v}} N_0(\mathbf{v}) e^{iqv\tau} d\mathbf{v}. \tag{15.9}$$

Assuming that the RBCs move randomly without any preferential direction, and making the transformation to spherical coordinates, Equation 15.9 can be written in the form

$$\langle e^{iqv\tau}\rangle_{\mathbf{v}} = \int_{v=0}^{\infty} N(v)\frac{\sin(qv\tau)}{qv\tau}\,dv, \tag{15.10}$$

where the one-dimensional velocity distribution $N(v) = 4\pi v^2 N_0(\mathbf{v})$.

#### 15.2.3.2 Derivation of $\langle\langle e^{iqv\tau}\rangle_v\rangle_q$

The expectation value of $\langle e^{iqv\tau}\rangle_v$ with reference to **q** can be calculated according to

$$\langle\langle e^{iqv\tau}\rangle_v\rangle_q = \int_{\alpha=0}^{\pi} \langle e^{iqv\tau}\rangle_v S_0(\mathbf{q}(\alpha))d\alpha\,. \qquad (15.11)$$

Assuming that $S_0(\mathbf{q}(\alpha))$ possesses circular symmetry with respect to $\mathbf{k}_i$, the one-dimensional scattering vector distribution $S(q(\alpha))$ may be written in the form

$$S(q(\alpha)) = \frac{2S_0(\mathbf{q}(\alpha))}{\sin(\alpha)}\,. \qquad (15.12)$$

Inserting Equations 15.10 and 15.12 into Equation 15.11 yields

$$\langle\langle e^{iqv\tau}\rangle_v\rangle_q = \int_{\alpha=0}^{\pi}\int_{v=0}^{\infty} N(v)\frac{\sin(qv\tau)}{2qv\tau}\, S(q(\alpha))\sin(\alpha)dvd\alpha\,. \qquad (15.13)$$

Equation 15.13 relates the RBC one-dimensional velocity $(N(v))$ and scattering vector $(S(q(\alpha))$ distributions to the measurable photocurrent autocorrelation function produced by the heterodyne mixing term in Equation 15.8.

#### 15.2.3.3 Power Spectral Density

The power spectral density $P_c(\omega)$ of the photocurrent produced by the heterodyne mixing term in a single coherence area can thus be calculated by applying the Wiener-Khintchine theorem to Equation 15.8, neglecting the stationary and the homodyne mixing terms

$$P_c(\omega) \propto C_{RBC} i_{Re}^{\,2} \int_{v=0}^{\infty} N(v) \int_{\alpha=0}^{\pi} \sin(\alpha)S(q(\alpha)) \int_{\tau=-\infty}^{\infty} \frac{\sin(qv\tau)}{qv\tau} e^{i\omega\tau} d\tau d\alpha dv\,. \qquad (15.14)$$

Using the fact that $q = \frac{4\pi}{\lambda_t}\sin\left(\frac{\alpha}{2}\right)$, Equation 15.14 can, by way of pure mathematical manipulations, be reduced to

$$P_c(\omega) \propto C_{RBC} i_{Re}^{\,2} \int_{v=\frac{\lambda_t\omega}{4\pi}}^{v_{max}} \frac{N(v)}{v} \int_{q=\frac{\omega}{v}}^{\frac{4\pi}{\lambda_t}} S(q)dqdv\,. \qquad (15.15)$$

After normalization with the total light intensity in order to render the output signal independent of the power of the laser, and taking into account that stationary components scale with $K^2$ while the Doppler-related portion of the signal scales with K alone, according to Equation 15.6, the total photocurrent power spectral density $P(\omega)$ corresponding to the heterodyne mixing term can be written in the form:

$$P(\omega) \propto \frac{C_{RBC}}{K} \int_{v=\frac{\lambda_t \omega}{4\pi}}^{v_{max}} \frac{N(v)}{v} \int_{q=\frac{\omega}{v}}^{\frac{4\pi}{\lambda_t}} S(q)dq dv . \qquad (15.16)$$

The integration limits of Equation 15.16 indicate that, in order for an RBC single scattering event to contribute to the photocurrent power spectral density at the angular frequency $\omega$, the associated scattering vector q needs to have a magnitude of at least $q = \frac{\omega}{v}$. In addition, the magnitude of the RBC velocity needs to be higher than $v = \frac{\lambda_t \omega}{4\pi}$.

As a final step we can now calculate $\int \omega^n P(\omega) d\omega$ and demonstrate that this quantity scales linearly with the RBC concentration for $n = 0$ and the product of the RBC speed and concentration for $n = 1$. An arbitrary velocity distribution can be written as the sum of its individual parts $N(v) = \sum_{v_0} N_{v_0} \delta(v - v_0)$.

The contribution to $\int \omega^n P(\omega) d\omega$ from $N(v) = N_{v_0} \delta(v - v_0)$ can then be written in the form

$$\begin{aligned}
\int \omega^n P(\omega) d\omega &\propto \frac{C_{RBC}}{K} \int_{\omega=0}^{\infty} \omega^n \int_{v=\frac{\lambda_t \omega}{4\pi}}^{v_{max}} \frac{N(v)}{v} \int_{q=\frac{\omega}{v}}^{\frac{4\pi}{\lambda_t}} S(q) dq dv d\omega \\
&= \frac{C_{RBC} N_{v_0}}{K} \int_{\omega=0}^{\infty} \omega^n \int_{v=\frac{\lambda_t \omega}{4\pi}}^{\infty} \frac{\delta(v - v_0)}{v} \int_{q=\frac{\omega}{v}}^{\frac{4\pi}{\lambda_t}} S(q) dq dv d\omega \\
&= \frac{C_{RBC} N_{v_0}}{K} \int_{x=0}^{\infty} (x v_0)^n \int_{v=\frac{\lambda_t x v_0}{4\pi}}^{\infty} \frac{\delta(v - v_0) v_0}{v} \int_{q=\frac{x v_0}{v}}^{\frac{4\pi}{\lambda_t}} S(q) dq dv dx \\
&= \frac{C_{RBC} N_{v_0} v_0^n}{K} \underbrace{\int_{x=0}^{\infty} x^n \int_{q=x}^{\frac{4\pi}{\lambda_t}} S(q) dq dx}_{constant}
\end{aligned} \qquad (15.17)$$

and the contribution from the complete velocity spectrum yields

$$\int \omega^n P(\omega) d\omega \propto \frac{C_{RBC}}{K} \underbrace{\int_{v=0}^{\infty} v^n N(v) \, dv}_{\langle v^n \rangle} \propto C_{RBC} \langle v^n \rangle . \qquad (15.18)$$

This processor algorithm scales with the RBC concentration ($n = 0$) and with the product of RBC concentration and speed ($n = 1$) for an arbitrary velocity distribution.

### 15.2.4 Sampling Volume

If the spatial distribution of photons available for detection within the sampling volume is described by a weighing function H(**r**) that is dependent on the tissue optical properties at location **r**, the output signal from the laser Doppler flowmeter is proposed to be[23]

$$\int_{r^3} H(\mathbf{r})v(\mathbf{r})C_{RBC}(\mathbf{r})d\mathbf{r}^3 \tag{15.19}$$

H(**r**) can be estimated for a specific combination of probe design and tissue optical properties by use of the Monte Carlo simulation, allowing a derivation of the median sampling depth.[42] In contrast to analytical models, Monte Carlo models have the added advantage of accounting for spatial coordinate-dependent distributions of velocity v(**r**) and concentration $C_{RBC}$ (**r**). The contribution to the output signal of blood flow in single vessels located at a specific depth[43] and the presence of multiple Doppler shifts for a given microvascular architecture can be readily predicted.[44,45]

## 15.3 Instrumentation

This section describes the instrumentation principles of LDPM and LDPI. In most LDPM devices the light is brought to and from the tissue by optical fibers. LDPM is well suited for continuous real-time monitoring of tissue blood perfusion at a single site. In LDPI, an airborne laser beam scans the tissue and successively records the tissue perfusion from a number of sites; from these data a two-dimensional map of the microvascular perfusion is generated. LDPI is primarily intended for studies of the spatial variability of tissue blood perfusion.

### 15.3.1 Laser Doppler Perfusion Monitoring

#### 15.3.1.1 First Experimental Setup

Following the demonstration of backscattered Doppler-broadened light from a perfused tissue,[18] the first prototype of a laser Doppler flowmeter was constructed.[46] This early experimental setup utilized as light source a 15-mW He-Ne laser, emitting coherent light at 632.8 nm. The large laser and a photomultiplier tube were positioned about 1 m above the tissue. In order to limit stray light, the returning light passed through a 2-mm aperture positioned at the surface of the tissue. A 0.5-mm pinhole was placed in front of the detector in such a way that only a single coherence area was selected. A flow parameter was obtained by use of an algorithm calculating the root mean square (rms) bandwidth of the un-normalized second moment of the Doppler signal power spectral density. With measurements on skin subjected to ultraviolet-induced erythema, a linear relationship with a first-order correlation coefficient of 88% between this flow parameter and values obtained by the $^{133}$Xe-clearance technique could be demonstrated.

#### 15.3.1.2 LDPM Devices

Although the early experimental prototypes worked well in the laboratory, they proved to be too bulky for use in the clinical environment. LDPM devices, in which optical fibers guided the light to and from the tissue and a semiconductor diode replaced the photomultiplier tube, were therefore developed.[19] In LDPM, step-index or graded-index optical fibers are used as light guides. In modern equipment these fibers have a core diameter ranging from 50 to 120 μm. The light is coupled via a microlens from the laser to the transmitting fiber that guides the light to the probe attached to the tissue by double adhesive tape. One or more fibers pick up a portion of the backscattered and Doppler-broadened light and bring it back to the photodetector unit in the instrument. In the probe tip, the fibers are typically positioned with a core center spacing of 250 to 500 μm. In the tissue the photons migrate in random pathways from the transmitting to the receiving fiber.

Widening the distance between the fiber tips in the probe tends to increase the average path length of the detected photons as well as the measurement depth.[42] Both of those quantities are dependent on the optical properties of the tissue. Consequently, perfusion recorded in different organs cannot be directly compared, and no universal absolute flow unit has so far been presented for LDF applications. Depending on the intended use of the instrument, the probes may be designed for attachment to the tissue surface or as small-diameter needle probes for recording deep muscle perfusion.[47,48] Because skin perfusion is especially sensitive to the ambient temperature, probe-holders incorporating a thermostatic element that allows the skin temperature to be kept at a preset value have been developed.[49]

Incorporating optical fibers as light guides between the tissue and the device, however, generally implies that the number of coherence areas on the photodetector surface increases and renders the system less coherent. According to Equation 15.6, this reduces the ratio of the photocurrent produced by the Doppler signal to the total light intensity by an amount proportional to the number of coherence areas. When fiber-optic-based devices were first tested, the superimposition of large spikes on the output signal at regular intervals significantly distorted the measurement. Closer analysis[50] identified optical mixing of longitudinal laser modes on the detector surface, giving rise to audio-range frequencies sweeping through the bandwidth of the device, as the origin of the problem. Since these signals are of common-mode origin, their adverse effect on the signal-to-noise ratio increases with the number of coherence areas. This problem was overcome by the introduction of a differential detector system[21] that effectively suppressed the common-mode signals, while not affecting the difference of the uncorrelated Doppler signals from the two detectors. Reducing the diameter of the fiber renders the detected signal more coherent and suppresses the influence of laser-mode artifacts[51] as well as the adverse effects of fiber motion artifacts.[49,52–54]

As LDPM technology matured, a diode laser replaced the He-Ne laser as the light source,[55,56] making a range of wavelengths available. The most commonly used wavelength today (780 nm), offers a deeper penetration as well as reduced dependence on skin color and is also close to the isobestic points, thus eliminating dependence on oxygen saturation.[23] In order to avoid beat notes in the Doppler signal caused by mode interference, the diode lasers need to be thermally stabilized.

In the signal processing unit, a high-pass filter with a cutoff frequency set to 20 Hz extracts the Doppler signal from the average photocurrent and pulse-synchronous light intensity fluctuations. This signal is amplified and normalized by the average photocurrent in order to make it independent of the light intensity. By using the algorithm derived in Equation 15.18 (with $n = 1$) an output signal that scales with perfusion can be implemented by calculating the instantaneous power after feeding the normalized Doppler signal through an $\omega$-weighing filter. The upper bandwidth limit is generally set to a value between 3 and 23 kHz,[23] depending on the expected highest RBC velocity. This output signal can be fed to a recorder or a display for continuous tracking of the perfusion, or to a computer for further postprocessing analysis.[57–59] In order to obtain an output signal that scales linearly with perfusion in tissues with an RBC concentration so high that the multiple generation of Doppler shifts cannot be neglected, a linearizer circuit may be utilized.[60] In this linearizer circuit, the output signal, implemented according to Equation 15.18 ($n = 1$) is multiplied by a compensation factor derived directly from the RBC concentration signal (Equation 15.18 with $n = 0$).

Double wavelength LDPM devices have been designed with the aim of penetrating different tissue depths and discriminating between superficial and global blood perfusion.[61] Green laser light (543 nm) with a short penetration depth has been tested especially with regard to its ability to record nutritive blood flow selectively, but with negative results.[62] Modulation of the average measurement depth can also be attained by widening the distance between the transmitting and receiving fiber,[63] but no clear indication has been obtained regarding whether this approach can discriminate between capillary and global tissue perfusion.[64] In order to suppress the adverse effect of the high spatial resolution of LDPM standard probes, integrating probes that record the average blood perfusion have been designed and tested with positive results.[65] The use of laser diodes integrated in the probe with photodetectors further eliminates the need for fiber optics and makes the device less sensitive to movement artifacts.[66]

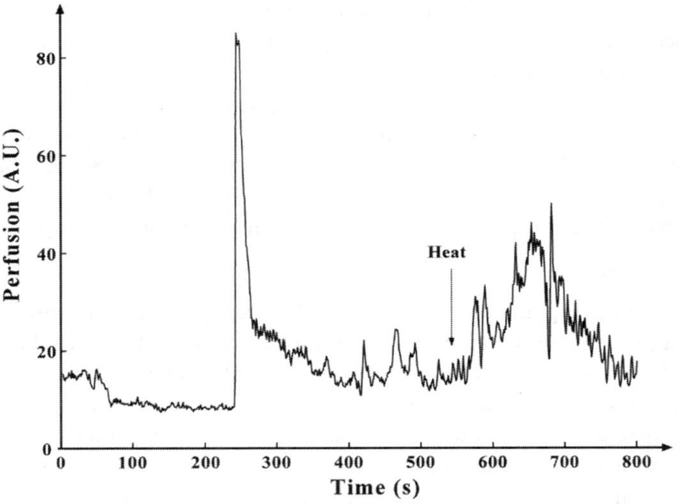


**FIGURE 15.3** Forearm skin blood perfusion recorded with LDPM following occlusion (at $t$ = 70 s), release of occlusion (at $t$ = 240 s), and hyperemia caused by heating (starting at $t$ = 550 s).

#### 15.3.1.3 Recording Tissue Perfusion

In their present design, LDPM devices are generally better suited for measurements where the probe is attached to a fixed tissue position than for measurements at multiple sites with intermittent movement of the probe.[49] The reason for this constraint is that moving the probe from one skin site to another may disturb the microcirculation under study. A typical registration demonstrating (1) the influence of a temporary occlusion of forearm skin perfusion, (2) the resulting hyperemic response following release of the occlusion, and (3) the slow increase in perfusion instituted by heating the skin may serve as an example (Figure 15.3).

The recording in Figure 15.3 was performed in accordance with the following recommended protocol.[34] Prior to recording, the test subject refrains from smoking and from the intake of food and beverages for 2 h and rests in a relaxed position for at least 20 min. The preparations start with positioning the probe in a probe holder attached to the skin and by placing a pressure cuff around the upper arm. Initially, the resting perfusion is recorded for about a minute. When the pressure cuff is inflated to above systolic pressure, the perfusion value is reduced to a stable level representing "biological zero" — which does not generally coincide with the zero output level of the instrument.[67] The predominant contribution to biological zero has been demonstrated to be Brownian motion arising from interstitial tissue.[68] When the pressure in the cuff is again released, the perfusion rapidly and intermittently increases to a peak value above the resting perfusion level. This phenomenon, generally referred to as "reactive hyperemia," represents a rapid flow of blood into the fully dilated microvascular bed. The perfusion value then returns to preocclusion levels. A successive increase in skin perfusion is then recorded as the skin temperature increases.

In the interpretation of the results of this recording, the biological zero has generally been taken as the zero perfusion level[69,70] from which the actual perfusion is calculated. Because the LDPM output signal is expressed in arbitrary units, it is generally better to present recorded alterations in blood perfusion as percentages of the resting value. In many situations, the latency time from the onset of stimuli to the appearance of the peak perfusion value may provide useful information.[71,72]

### 15.3.2 Laser Doppler Perfusion Imaging

#### 15.3.2.1 From Monitoring to Imaging

When repetitive recordings in adjacent sites on a tissue with seemingly homogenous perfusion are made, identical perfusion values are generally not recorded. By comparing biopsies with results obtained using

LDPM, Braverman and co-workers[4] demonstrated that the heterogeneous skin perfusion pattern recorded by LDPM at adjacent sites coincides with the underlying microvascular architecture. High, pulsatile perfusion values were found directly over ascending arterioles, whereas low values correlated with areas containing only capillaries and postcapillary venules. Topographic maps generated by successively moving the LDPM probe in 1-mm steps further demonstrated that high-perfusion spots correlated well with the presence of arterioles and low-perfusion areas with avascular zones.[73] To visualize this spatial variability, tissue perfusion needs to be investigated by use of a multipoint LDPM system[27] or presented as a two-dimensional flow map,[74] rather than by a curve trace displaying temporal variations at a single site alone.

#### 15.3.2.2 LDPI Devices

In LDPI two-dimensional mapping of tissue blood perfusion is performed by moving the laser beam in a raster scan over a predetermined area with no need for physical contact with the tissue (Figure 15.4). The light is guided from a low-power laser operating in the visible range to the tissue via a moving mirror system. A portion of the backscattered, Doppler-broadened light is brought back via the same mirror system to detectors placed in the proximity of the laser. This arrangement facilitates automatic focusing on and tracking of the laser spot as the beam moves over the tissue surface, thereby effectively suppressing the adverse effects of ambient stray light. Further rejection of ambient light is attained by utilizing a differential detector technology similar to that used in LDPM. In LDPI the laser and detector units are placed in a camera-like scanner head positioned 0.1 to 1 m above the tissue surface.

Two fundamentally different LDPI systems — one that utilizes a stepwise scanning beam,[28,30] and one that is based on a continuously moving beam[29] — have been presented in the literature. In stepwise scanning LDPI, the beam is stopped at each measurement site while the backscattered, Doppler-broadened light is sampled. This arrangement avoids the interference caused by a laser beam moving in relation to the tissue, which may generate Doppler components not related to tissue perfusion. The exact duration of the beam stop is determined by the trade-off between image integrity and maximum permissible time for capturing the image. If the beam is arrested for 50 ms at each measurement site, a full-format image, including 64 × 64 measurement sites, is generally captured in about 4 min (Figure 15.5).

In the LDPI system based on a continuously moving beam, the backscattered Doppler-broadened light is captured "on the fly." This method allows more points to be sampled per unit time, and a full-format image may typically include 256 × 256 true measurement sites. The relative motion between the laser beam and the tissue when the backscattered light is sampled may, however, result in the generation of nonperfusion-related Doppler components and other intensity fluctuations in the backscattered light.[29] Those Doppler components are generally of low frequency, and their adverse influence on the perfusion

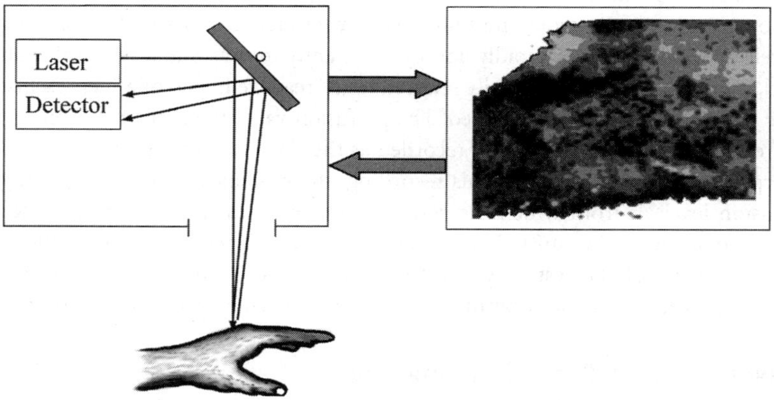


**FIGURE 15.4** Operating principle of LDPI. The laser beam moves stepwise or continuously in a raster scan over the tissue by moving the mirror. A portion of the backscattered light is brought back via the same mirror to impinge on the surface of a photodetector.

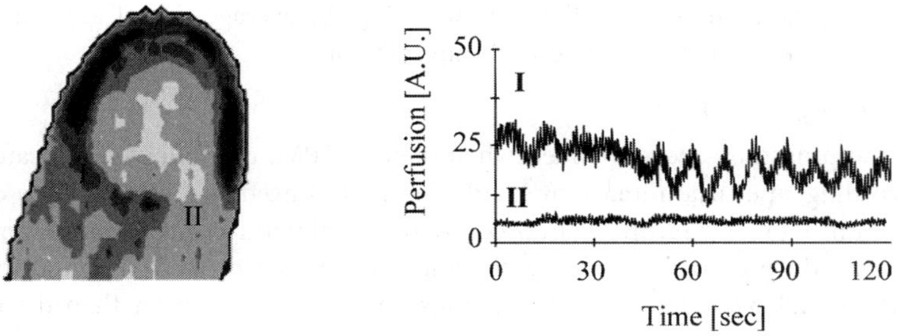


**FIGURE 15.5** Perfusion image of fingertip with continuous recordings made at two single skin areas. The continuous recordings each comprise the average perfusion value recorded in four adjacent sites. (From Wårdell, K., *Proc. 10th Nordic-Baltic Conf. Biomed. Eng.*, 271, June 9–13, 1996. With permission).

signal can be reduced by increasing the lower cutoff frequency of the instrument bandwidth or by reducing the scanning speed. Increasing the lower cutoff frequency, however, also tends to reject the low-frequency Doppler components of the blood perfusion–related signal;[75] increased scanning speed distorts the image[76] and reduced scanning speed prolongs the image capturing process.

In contrast to the situation in LDPM, with the use of LDPI devices the number of coherence areas and thus the inherent amplification factor change as the distance between the detector and object is altered. This difficulty has been overcome by introducing as a light source a slightly divergent laser beam maintaining a fixed solid angle under which the laser spot on the tissue is seen from the detector, thereby keeping the number of coherence areas constant, according to Equation 15.4.[30] It has been demonstrated empirically that the distance-dependent amplification factor of LDPI systems utilizing a collimated beam can also be reduced by using an algorithm that normalizes the perfusion signal with the light intensity raised to the power of $\beta$, where $\beta$ is close to unity (rather than 2, as suggested by Equation 15.15).

LDPI systems utilize a signal processor similar to that in LDPM and can generate full-format images of objects from the size of a fingertip to a complete torso. The intensity of the backscattered light can be used for discriminating the object from its background. Therefore, whenever possible the object should be placed on a light-absorbing background to facilitate this discrimination process.

In LDPI, all parameters controlling image size, resolution, scanning speed, background threshold, etc. are set in the system software. The image can generally be displayed in an individual color scale, in which the lowest and highest perfusion values constitute the endpoints of the scale. This option is probably the best choice for color representation if only the relative variation in perfusion within a single image is to be investigated. If several images are to be compared, they must be set to the same color scale before a comparison is made. Basic image-processing routines are available in the LDPI system software packages and the images can readily be exported to more sophisticated packages for further analysis as needed.

#### 15.3.2.3 Monitoring Mode

One important feature of LDPI is the capability of operating the system in a monitoring mode. In this mode a small image comprising (for example) $4 \times 4$ measurement sites is continuously updated, and the result can be displayed as a series of mini-images or as a time trace continuously displaying the average perfusion. This feature makes the system less dependent on the exact measurement site than single-point LDPM.[77]

#### 15.3.2.4 High-Resolution LDPI

Lateral resolution of standard LDPI is limited by the effective diameter of the laser beam (approximately 0.6 to 0.8 mm) and by the step-length in the stepwise scanning LDPI. By keeping the distance between the scanner head and the tissue constant, focusing the laser beam onto one spot on the tissue surface, and reducing the step-length, the resolution can be significantly increased and full-format images of tissue sites less than 1 cm$^2$ generated.[78] This arrangement also reduces the solid angle under which the laser spot is seen on the tissue surface, thereby increasing the average size of a coherence area on the detector surface and the inherent system amplification factor.

#### 15.3.2.5 Recording an Image

In addition to the procedure recommended when using LDPM, the following precautions should be noted when recording a perfusion image by LDPI. The subject needs to be placed in a comfortable and relaxing position during image recording in order to avoid unintentional influence on microcirculation. If the subject moves during image recording, the image will have isolated horizontal stripes, indicating a movement artifact. Likewise, temporal fluctuations on a timescale shorter than the image recording time will result in the appearance of horizontal perfusion stripes in the image with no correspondence in the microvascular architecture. A liquid film on the skin surface may cause direct reflections of the laser light to the detector at a single spot, mimicking a (falsely) reduced tissue perfusion.

### 15.3.3 Performance Check and Calibration

The algorithm for LDF devices based on ω-weighting filtering of the power spectral density of the unnormalized Doppler signal was derived under assumptions that are not necessarily fulfilled in *in vitro* and *in vivo* situations. Therefore, both flow simulators, with separate controls for the speed and concentration of RBCs or other scatterers, and *in vivo* models have been used in the evaluation of the processor algorithm.

Most flow simulator experiments demonstrate that a linear relationship between RBC speed and LDF output signal can be obtained, provided compatibility with the instrument bandwidth[21,54] is maintained. In the evaluation of the bandwidth requirements, however, it must be borne in mind that the size of the moving scatterer substantially influences the scattering angle distribution and thus the Doppler shift generated for a given speed according to Equations 15.3 and 15.16. For low and moderate RBC concentration, the probability for multiple Doppler shift generation and homodyne mixing is low. This implies that the LDF output signal also scales linearly with RBC concentration.[21] Recently, a novel flow simulator based on rotating disks with embedded scatterers has been constructed[79] and used to assess the influence of optical properties and separation of the transmitting and receiving fiber on measurement depth.[80]

*In vivo* calibration involves comparison of results obtained by LDF devices and other methods for assessment of the microvascular perfusion. Since no "gold standard" for recording tissue blood perfusion exists, the correlation of LDF data and reference method data is highly dependent on whether the two methods sample the same microvascular compartment and the degree of noninvasiveness of the reference method. Moving the LDPM probe from one position to an adjacent site while the global flow through an organ is used as the independent reference generally reveals a linear relationship with global flow at each site, albeit with a difference in scale factors caused by the heterogeneity of the tissue perfusion.[81] One of the most promising animal models for evaluation of LDPM performance is the feline intestine because of its uniform and easily controlled blood perfusion. In this model, a specific segment of the intestine is drained through a single vein, the blood flow of which can be accurately determined by a drop-counting technique.[82] A linear relationship between LDPM output signal and intestinal blood perfusion has been demonstrated within the range from zero to 300 ml $g^{-1}$.[83]

Because the processing algorithm of LDPI is identical to that of LDPM, it is expected and has been demonstrated[21] that results similar to those for LDPM are obtained when using LDPI. In addition to those performance tests, the uniformity in sensitivity over the entire image plane and the dependence on the distance between the scanner head and tissue also need to be controlled.

The day-to-day sensitivity of LDPM instruments, including the probe, is checked by placing the probe tip in a suspension of latex spheres in random motion.[49] The well-defined mobility pattern of such particles should result in a stable and reproducible output signal. By use of those "motility standards" two LDPM devices with identical specifications can be calibrated to the same sensitivity. A new calibration standard that includes latex spheres undergoing Brownian motion while diffusing through a porous polyethylene material has proven to give Doppler spectra closely resembling those obtained from human skin.[84] In LDPI the day-to-day calibration procedure can, in accordance with the procedure used in

LDPM, be made by way of a two-point calibration method based on a calibration box incorporating a transparent container with "motility standard" and a white surface "zero-reference," giving the high and zero calibration points, respectively.[35]

## 15.4 Applications

### 15.4.1 LDPM Applications

Because of the ease of use of LDPM, experimental investigations and clinical trials aiming at evaluating the method's practical usefulness have been undertaken on many organs and in a variety of disciplines. The most commonly investigated organs include the kidney,[85,86] the liver,[87,88] the intestines,[89,90] the brain,[91–93] and the skin.[19,94–99] By use of needle probes, the perfusion can furthermore be continuously monitored with minimal tissue trauma, for example, in skeletal muscle and tendons[47,48,100] and bone tissue.[101–107] Among the clinical applications, evaluation of the tissue perfusion in association with Raynaud's phenomenon,[108–110] diabetes microangiopathy,[111–115] plastic surgery and flap monitoring,[116–120] peripheral vascular disease,[71,72,121–124] thermal injury,[125–127] bowel ischemia and gastric blood perfusion,[128–132] pharmacological trials,[133,134] and numerous skin diseases[135–137] are most frequently represented in the literature. A thorough overview of many of those applications can be found in a review paper,[25] in two books on laser Doppler flowmetry, [24,26] and in the reference lists of the LDPM manufacturers.[138,139]

### 15.4.2 LDPI Applications

Following the introduction of LDPI in the early 1990s, this emerging technology was soon evaluated in numerous applications in which the spatial heterogeneity of tissue perfusion is significant.

In irritant[140,141] and allergy[142–144] patch testing, perfusion images visualize and quantify the erythema produced by a compound applied topically to the skin. The LDPI technology produces data that are well in accordance with visual scoring, in the assessment of skin irritation[145] and allergy.[142] It is therefore possible to grade irritation and type irritant reactions for individual compounds. In allergy patch testing, the optimal dose, influence of the vehicle, and the application and reaction time for different allergens can be readily detected and quantified.

Malignant skin tumors are highly vascularized compared to benign lesions; LDPI has been used to characterize the perfusion pattern of lesions such as basal cell carcinoma,[146] malignant melanomas, and nevi.[147] Furthermore, LDPI has been used in the assessment of basal cell carcinoma perfusion in conjunction with cryotherapy and photodynamic therapy.[148] This report suggests that LDPI can be used, not only to follow up on the healing process, but also to find recurrent or residual tumor growth.

LDPI has further been used to assess the depth and the healing process of burns[149,150] and a recommendation has been issued to include LDPI as a clinical tool in the evaluation of severely burned patients.[33] Characterization of perfusion in leg ulcers in combination with capillary microscopy has been performed by Gschwandtner and co-workers,[32,151,152] while perfusion in and around pressure sores was investigated by Mayrovitz.[153–156]

Other studies and applications of LDPI include assessment of diabetes neuropathy following contralateral cooling,[157] endothelial cell malfunction investigated by iontophoresis infusion of acethylcholine and nitroprusside in diabetes[158] and Alzheimer's disease,[159] angiogenesis and growth factor research,[160,161] psoriasis,[162,163] scleroderma,[164] and Raynaud's disease,[165] as well as flap surveillance,[166] microdialysis,[167]and phototesting.[168]

## 15.5 Conclusions

This chapter has covered the history and development of LDF, the theoretical background of this technology, its realization in versatile devices such as the laser Doppler perfusion monitor and imager and associated applications. Although it is a relatively recent medical technology, many reports on its use, especially in research applications, are already available in the scientific literature. The much slower introduction of the technology in clinical routines is probably because tissue perfusion is poorly predictable at any given tissue site and time. Furthermore, the natural variability of tissue perfusion in terms of temporal fluctuations and spatial heterogeneity makes it difficult to define any rules of thumb for normal conditions. Instead, stimuli–response experiments need to be applied in order to challenge the tissue and reveal possible malfunctioning of the minute vessels of the vascular tree. The main advantages and disadvantages of LDF are listed below.

Advantages:

- LDF makes possible noninvasive recording and imaging of tissue perfusion with minimal impact on microcirculation.
- LDF devices are easy to use.
- Continuous recordings over unlimited periods of time can be made with LDPM.
- Two-dimensional perfusion maps can be visualized by LDPI.
- The theoretical basis of LDF is well established.

Disadvantages:

- No absolute calibration is possible, and results obtained from different organs cannot be directly compared because of variations in photon path lengths due to the different optical properties of the tissue.
- Results obtained by recording at a single site using LDPM may not be representative for the entire tissue.
- LDF does not distinguish between nutritive (capillary) perfusion and global tissue perfusion.
- LDPI assumes steady-state conditions in perfusion during the image-capturing period, which may amount to 4 min or longer.

One of the main features of LDF in monitoring and imaging is that time traces and images can easily be recorded, while the interpretation of the results implies a thorough understanding of microcirculation and its regulation.

## References


1. Rhodin, J.A.G., Anatomy of the microcirculation, in *Microcirculation: Current Physiologic, Medical and Surgical Concepts.*, Effros, R.M., Schmid-Schoenbein, H., and Dietzel, J., Eds., Academic Press, New York, 1981, p. 11.
2. Tenland, T., Salerud, E.G., Nilsson, G.E., and Öberg, P.Å., Spatial and temporal variations in human skin blood flow, *Int. J. Microcirc. Clin. Exp.*, 2(2), 81, 1983.
3. Salerud, E.G., Tenland, T., Nilsson, G.E., and Öberg, P.Å., Rhythmical variations in human skin blood flow, Int. *J. Microcirc. Clin. Exp.*, 2(2), 91, 1983.
4. Braverman, I.M., Keh, K., and Goldminz, D., Correlation of laser Doppler wave patterns with underlying microvascular anatomy, *J. Invest. Dermatol.*, 95(3), 283, 1990.
5. Hertzman, A.B. and Spealman, C.R., Observation on the finger volume pulse recorded photoelectrically, *Am. J. Physiol.*, 119, 334, 1937.
6. Challoner, A.V.J., Photoelectric plethysmography for estimating cutaneous blood flow, in *Non-Invasive Physiological Measurements*, Vol. 1, Rolfe, P., Ed., Academic Press, London, 1979, p. 125.
7. Holti, G. and Mitchell, K.W., Estimation of the nutrient skin blood flow using a non-invasive segmented thermal probe, in *Non-Invasive Physiological Measurements*, Vol. 1, Rolfe, P., Ed., Academic Press, London, 1979, p. 113.

8. Kety, S.S., Measurement of regional circulation by the local clearance of radioactive sodium, *Am. Heart J.*, 38, 321, 1949.
9. Sejrsen, P., Measurement of cutaneous blood flow by freely diffusable radioactive isotopes. Methodological studies on the washout of krypton-85 and xenon-13 from the cutaneous tissue in man, *Dan. Med. Bull.*, 18(Suppl. 3), 9, 1971.
10. Groner, W., Winkelman, J.W., Harris, A.G., Ince, C., Bouma, G.J., Messmer, K., and Nadeau, R.G., Orthogonal polarization spectral imaging: a new method for study of the microcirculation, *Nat. Med.*, 5(10), 1209, 1999.
11. Nadeau, R.G. and Groner, W., The role of a new noninvasive imaging technology in the diagnosis of anemia, *J. Nutr.*, 131(5), 1610S, 2001.
12. Wayland, H. and Johnson, P.C., Erythrocyte velocity measurements in microvessels by a two-slit photometric method, *Am. J. Physiol.*, 22, 333, 1967.
13. Bollinger, A., Butti, P., Barras, J.P., Trachsler, H., and Siegenthaler, W., Red blood cell velocity in nailfold capillaries of man measured by a television microscopy technique, *Microvasc. Res.*, 7, 61, 1974.
14. Doppler, J.C., Über das farbige Licht der Doppelsterne und einiger anderer Gestirne des Himmels [On the colored light from double stars and some other heavenly bodies], *Abh. Konigl. Böhmischen Ges. Wiss.*, 5, 465, 1842 *(Treatises of the Royal Bohemian Society of Science).*
15. Riva, C., Ross, B. and Benedek, G.B., Laser Doppler measurements of blood flow in capillary tubes and retinal arteries, *Invest. Ophthalmol.*, 11(11), 936, 1972.
16. Mishina, H., Koyama, T., and Asakura, T., Velocity measurements of blood flow in the capillary and vein using the laser Doppler microscope,. *Appl. Opt.*, 14, 2326, 1974.
17. Einav, S., Berman, H.J., Fuhro, R.L., DiGiovanni, P.R., Fridman, J.D., and Fine, S., Measurement of blood flow *in vivo* by laser Doppler anemometry through a microscope, *Biorheology*, 12(3–4), 203, 1975.
18. Stern, M.D., *In vivo* evaluation of microcirculation by coherent light scattering, *Nature*, 254(5495), 56, 1975.
19. Holloway, G.A., Jr. and Watkins, D.W., Laser Doppler measurement of cutaneous blood flow, *J. Invest. Dermatol.*, 69(3), 306, 1977.
20. Nilsson, G.E., Tenland, T., and Öberg, P.Å., A new instrument for continuous measurement of tissue blood flow by light beating spectroscopy, *IEEE Trans. Biomed. Eng.*, 27(1), 12, 1980.
21. Nilsson, G.E., Tenland, T., and Öberg, P.Å., Evaluation of a laser Doppler flowmeter for measurement of tissue blood flow, *IEEE Trans. Biomed. Eng.*, 27(10), 597, 1980.
22. Bonner, R. and Nossal, R., Model for laser Doppler measurements of blood flow in tissue, *Appl. Opt.*, 20, 2097, 1981.
23. Leahy, M.J., de Mul, F.F., Nilsson, G.E., and Maniewski, R., Principles and practice of the laser-Doppler perfusion technique, *Technol. Health Care*, 7(2–3), 143, 1999.
24. Shepherd, A.P. and Öberg, P.Å., Eds., *Laser-Doppler Blood Flowmetry*, Kluwer Academic, Boston, 1990.
25. Öberg, P.Å., Laser Doppler flowmetry, *Crit. Rev. Biomed. Eng.*, 18(2), 125, 1990.
26. Belcaro, G., Hoffmann, U., Bollinger, A., and Nicolaides, A., Eds., *Laser Doppler*, Med-Orion, London, 1994.
27. Hill, S.A., Pigott, K.H., Saunders, M.I., Powell, M.E., Arnold, S., Obeid, A., Ward, G., Leahy, M., Hoskin, P.J., and Chaplin, D.J., Microregional blood flow in murine and human tumours assessed using laser Doppler microprobes, *Br. J. Cancer*, Suppl. 27, S260, 1996.
28. Nilsson, G.E., Jakobsson, A., and Wårdell, K., Imaging of tissue blood flow by coherent light scattering, in *IEEE 11th Annu. EMBS Conf.*, Seattle, WA, 1989.
29. Essex, T.J. and Byrne, P.O., A laser Doppler scanner for imaging blood flow in skin, *J. Biomed. Eng.*, 13(3), 189, 1991.
30. Wårdell, K., Jakobsson, A., and Nilsson, G.E., Laser Doppler perfusion imaging by dynamic light scattering, *IEEE Trans. Biomed. Eng.*, 40(4), 309, 1993.

31. Sommer, A., Laser Doppler imaging in dermatology — experimental and clinical applications — time to replace the dermatologist's eye? in *Dermatology*, University of Maastricht, Maastricht, the Netherlands, 2001.
32. Gschwandtner, M.E., Ambrozy, E., Schneider, B., Fasching, S., Willfort, A., and Ehringer, H., Laser Doppler imaging and capillary microscopy in ischemic ulcers, *Atherosclerosis*, 142(1), 225, 1999.
33. Pape, S.A., Skouras, C.A., and Byrne, P.O., An audit of the use of laser Doppler imaging (LDI) in the assessment of burns of intermediate depth, *Burns*, 27(3), 233, 2001.
34. Bircher, A., de Boer, E.M., Agner, T., Wahlberg, J.E., and Serup, J., Guidelines for measurement of cutaneous blood flow by laser Doppler flowmetry. A report from the Standardization Group of the European Society of Contact Dermatitis, *Contact Dermatitis*, 30(2), 65, 1994.
35. Fullerton, A., Stücker, M., Wilhelm, K., Wårdell, K., Anderson, C., Fischer, T., Nilsson, G.E., and Serup, J., Guidelines for visualisation of cutaneous blood flow by laser Doppler perfusion imaging, *Contact Dermatitis*, 46(3), 129, 2002.
36. Nilsson, G.E., Jakobsson, A., and Wårdell, K., Tissue perfusion monitoring and imaging by coherent light scattering, in *Biooptics: Optics in Biomedicine and Environmental Studies*, Soares, O.D.D. and Scheggi, A.M., Eds., *Proc. SPIE*, 1524, 90, 1992.
37. Steenbergen, W. and deMul, F.F.M., Role of speckles in laser Doppler flowmetry, in *Optical Diagnostics of Biological Fluids, Proc. SPIE,* San Jose, CA, 1998.
38. Siegman, A.E., The antenna properties of optical heterodyne receiver, *Appl. Opt.*, 5(10), 1588, 1966.
39. Serov, A., Steenbergen, W., and deMul, F.F.M., Prediction of the photodetector signal generated by Doppler-induced speckle fluctuations, theory and some validations, *J. Opt. Soc. Am. A*, 18(3), 622, 2001.
40. Berne, B.J. and Pecora, R., *Dynamic Light Scattering*, New York, Wiley, 1976.
41. Cummins, H.Z. and Swinney, H.L., Light beating spectroscopy, in *Progr. Opt.*, Wolf, E., Ed., North-Holland Publishing Co., Amsterdam, 1970, p. 133.
42. Jakobsson, A. and Nilsson, G.E., Prediction of sampling depth and photon pathlength in laser Doppler flowmetry, *Med. Biol. Eng. Comput.*, 31(3), 301, 1993.
43. Nilsson, H. and Nilsson, G., Monte Carlo simulations of the light interaction with blood vessels in human skin in the red wavelength region, in *Optical Diagnostics of Biological Fluids III*, BiOS'98, SPIE, San Jose, CA, 1998.
44. Jentink, H.W., Demul, F.F.M., Graaff, R., Suichies, H.E., Aarnoudse, J.G., and Greve, J., Laser Doppler flowmetry — measurements in a layered perfusion model and Monte-Carlo simulations of measurements, *Appl. Opt.*, 30(18), 2592, 1991.
45. Koelink, M.H., deMul, F.F.M., Greve, J., Graaf, R., Dassel, A.C.M., and Aarnoudse, J.G., Laser-Doppler blood flowmetry using two wavelengths: Monte Carlo simulations and measurements, *Appl. Opt.*, 33(16), 3549, 1994.
46. Stern, M.D., Lappe, D.L., Bowen, P.D., Chimosky, J.E., Holloway, G.A., Jr., Keiser, H.R., and Bowman, R.L., Continuous measurement of tissue blood flow by laser-Doppler spectroscopy, *Am. J. Physiol.*, 232, H441, 1977.
47. Salerud, E.G. and Öberg, P.Å., Single-fiber laser Doppler flowmetry, a method for deep tissue perfusion measurements, *Med. Biol. Eng. Comput.*, 25, 329, 1987.
48. Kvernebo, K., Staxrud, L.E., and Salerud, E.G., Assessment of human muscle blood perfusion with single-fiber laser Doppler flowmetry, *Microvasc. Res.*, 39, 376, 1990.
49. Nilsson, G.E., Perimed's LDV flowmeter, in *Laser-Doppler Flowmetry*, Shepherd, A.P. and Öberg, P.Å., Eds., 1990, Kluwer Academic Publisher: Boston, Dordrecht, London, 57.
50. Watkins, D. and Holloway, G.A., An instrument to measure cutaneous blood flow using the Doppler shift of laser light, *IEEE Trans. Biomed. Eng.*, 28, 1978.
51. Holloway, G.A., Medpacific's LDV blood flowmeter, in *Laser-Doppler Blood Flowmetry*, Shepherd, A.P. and Öberg, P.Å., Eds., Kluwer Academic, Dordrecht, 1990, p. 47.
52. Gush, R.J. and King, T.A., Investigation and improved performance of optical fibre probes in laser Doppler blood flow measurement, *Med. Biol. Eng. Comput.*, 25, 391, 1987.

53. Newson, T.P., Obeid, A., Wolton, R.S., Boggett, D., and Rolfe, P., Laser Doppler velocimetry: the problem of fibre movement artefact, *J. Biomed. Eng.*, 9, 169, 1987.
54. Obeid, A.N., Barnett, N.J., Dougherty, G., and Ward, G., A critical review of laser Doppler flowmetry, *J. Med. Eng. Technol.*, 14, 178, 1990.
55. Kolari, P., Optoelectronic Doppler velocimetry based on semiconductor laser diode for measurements of cutaneous blood flow, *Int. J. Microcirc. Clin. Exp.*, 3, 476, 1984.
56. deMul, F.F.M., van Spijker, J., van der Plas, D., Greve, J., Aarnoudse, J.G., and Smits, T.M., Mini laser-Doppler (blood) flow monitor with diode laser source and detection integrated in the probe, *Appl. Opt.*, 23, 2970, 1984.
57. Kvernmo, H.D., Stefanovska, A., Bracic, M., Kirkeboen, K.A., and Kvernebo, K., Spectral analysis of the laser Doppler perfusion signal in human skin before and after exercise, *Microvasc. Res.*, 56, 173, 1998.
58. Kano, T., Shimoda, O., Higashi, H., Sadanaga, M., and Sakamoto, S., Fundamental patterns and characteristics of the laser-Doppler skin blood flow waves recorded from the finger and toe, *J. Auton. Nerv. Syst.*, 45, 191, 1993.
59. Kano, T., Shimoda, O., Higashi, K., and Sadanaga, M., Effects of neural blockade and general anesthesia on the laser-Doppler skin blood flow waves recorded from the finger or toe, *J. Auton. Nerv. Syst.*, 48, 257, 1994.
60. Nilsson, G.E., Signal processor for laser Doppler tissue flowmeters, *Med. Biol. Eng. Comput.*, 22, 343, 1984.
61. Duteil, L., Bernengo, J., and Schalla, W., A double wavelength laser Doppler system to investigate skin microcirculation, *IEEE Trans. Biomed. Eng.*, 439, 1985.
62. Gush, R.J. and King, T.A., Discrimination of capillary and arterio-venular blood flow in skin by laser Doppler flowmetry, *Med. Biol. Eng. Comput.*, 29, 387, 1991.
63. Johansson, K., Jakobsson, A., Lindahl, K., Lindhagen, J., Lundgren, O., and Nilsson, G.E., Influence of fibre diameter and probe geometry on the measuring depth of laser Doppler flowmetry in the gastrointestinal application, *Int. J. Microcirc. Clin. Exp.*, 10, 219, 1991.
64. Gush, R.J., King, T.A., and Jayson, M.I., Aspects of laser light scattering from skin tissue with application to laser Doppler blood flow measurement, *Phys. Med. Biol.*, 29, 1463, 1984.
65. Salerud, E.G. and Nilsson, G.E., Integrating probe for tissue laser Doppler flowmeters. *Med. Biol. Eng. Comput.*, 24, 415, 1986.
66. deMul, F.F.M., Van Spijker, J., Van der Place, D., Greve, J., Aarndoudse, J.D., and Smits, T.M., Mini laser-Doppler (blood) flow monitor with diode laser source and detection integrated in the probe, *Appl. Opt.*, 23, 2970, 1984.
67. Wahlberg, E.P., Effects of local hyperemia and edema on the biological zero in laser Doppler fluxmetry (LD), *Int. J. Microcirc. Clin. Exp.*, 11, 157, 1984.
68. Kernick, D.P., Tooke, J.E., and Shore, A.C., The biological zero signal in laser Doppler fluximetry — origins and practical implications, *Pfluegers Arch. Eur. J. Physiol.*, 437, 624, 1999.
69. Colantuoni, A., Bertuglia, S., and Intaglietta, M., Biological zero of laser Doppler fluxmetry, microcirculatory correlates in the hamster cheek pouch during flow and no flow conditions, *Int. J. Microcirc. Clin. Exp.*, 13, 125, 1993.
70. Abbot, N.C. and Beck, J.S., Biological zero in laser Doppler measurements in normal, ischaemic and inflamed human skin, *Int. J. Microcirc. Clin. Exp.*, 12, 89, 1993.
71. Kvernebo, K., Slagsvold, C.E., and Gjolberg, T., Laser Doppler flux reappearance time (FRT) in patients with lower limb atherosclerosis and healthy controls, *Eur. J. Vasc. Surg.*, 2, 171, 1988.
72. Kvernebo, K., Laser Doppler flowmetry in evaluation of lower limb atherosclerosis, *J. Oslo City Hosp.*, 38, 127, 1988.
73. Braverman, I.M. and Schechner, J.S., Contour mapping of the cutaneous microvasculature by computerized laser Doppler velocimetry, *J. Invest. Dermatol.*, 97, 1013, 1991.
74. Wårdell, K., Braverman, I.M., Silverman, D.G., and Nilsson, G.E., Spatial heterogeneity in normal skin perfusion recorded with laser Doppler imaging and flowmetry, *Microvasc. Res.*, 48, 26, 1994.

75. Arildsson, M., Nilsson, G.E., and Wårdell, K., Critical design parameters in laser Doppler perfusion imaging, *SPIE Proc., Opt. Diagn. Living Cells Biofluids*, 2678, 401, 1996.
76. Mack, G.W., Assessment of cutaneous blood flow by using topographical perfusion mapping techniques, *J. Appl. Physiol.*, 85, 353, 1998.
77. Wårdell, K. and Nilsson, G.E., Duplex laser Doppler perfusion imaging, *Microvasc Res.*, 52, 171, 1996.
78. Lindén, M., Sirsjö, A., Lindbom, L., Nilsson, G., and Gidlof, A., Laser-Doppler perfusion imaging of microvascular blood flow in rabbit tenuissimus muscle, *Am. J. Physiol.*, 269(4 Pt 2), H1496, 1995.
79. Steenbergen, W. and deMul, F.F.M., New optical tissue phantom, and its use for studying laser Doppler blood flowmetry, in *Optical and Imaging Techniques for Biomonitoring*, Foth, H.-J., Marchesini, R., and Podbielska, H., Eds., *Proc. SPIE*, 1997.
80. Larsson, M., Steenbergen, W., and Strömberg, T., Influence of tissue phantom optical properties and emitting – receiving fiber distance on laser Doppler flowmetry, in *Optical Diagnostics of Biological Fluids*, Priezzhev, A.V. and Asahura, T., Eds., *Proc. SPIE*, 2000.
81. Shepherd, A.P., Riedel, G.L., Kiel, J.W., Haumschild, D.J., and Maxwell, L.C., Evaluation of an infrared laser-Doppler blood flowmeter, *Am. J. Physiol.*, 252(6 Part 1), G832, 1987.
82. Ahn, H., Lindhagen, J., Nilsson, G.E., Salerud, E.G., Jodal, M., and Lundgren, O., Evaluation of laser Doppler flowmetry in the assessment of intestinal blood flow in cat, *Gastroenterology*, 88, 951, 1985.
83. Ahn, H., Johansson, K., Lundgren, O., and Nilsson, G.E., *In vivo* evaluation of signal processors for laser Doppler tissue flowmeters, *Med. Biol. Eng. Comput.*, 25, 207, 1987.
84. Liebert, A., Leahy, M., and Maniewski, R., A calibration standard for laser-Doppler perfusion measurements, *Rev. Sci. Instrum.*, 66, 5169, 1995.
85. Stern, M.D., Bowen, P.D., Parma, R., Osgood, R.W., Bowman, R.L., and Stein, J.H., Measurement of renal cortical and medullary blood flow by laser-Doppler spectroscopy in the rat, *Am. J. Physiol.*, 236, F80, 1979.
86. Roman, R.J. and Smits, C., Laser-Doppler determination of papillary blood flow in young and adult rats, *Am. J. Physiol.*, 251(1 Pt 2), F115, 1986.
87. Arvidsson, D., Svensson, H., and Haglund, U., Laser-Doppler flowmetry for estimating liver blood flow, *Am. J. Physiol.*, 254(4 Pt 1), G471, 1988.
88. Pedrosa, M.E., Montero, E.F., and Nigro, A.J., Liver microcirculation after selective denervation, *Microsurgery*, 21, 163, 2001.
89. Feld, A.D., Fondacaro, J.D., Holloway, G.A., and Jacobson, E.D., Laser Doppler velocimetry, a new technique for the measurement of intestinal mucosal blood flow, *Gastrointestinal Endosc.*, 30, 225, 1984.
90. Ahn, H., Lindhagen, J., Nilsson, G.E., Öberg, P.Å., and Lundgren, O., Assessment of blood flow in the small intestine with laser Doppler flowmetry, *Scand. J. Gastroenterol.*, 21, 863, 1986.
91. Bogaert, L., O'Neill, M.J., Moonen, J., Sarre, S., Smolders, I., Ebinger, G., and Michotte, Y., The effects of LY393613, nimodipine and verapamil, in focal cerebral ischaemia, *Eur. J. Pharmacol.*, 411, 71, 2001.
92. Fabricius, M., Akgoren, N., Dirnagl, U., and Lauritzen, M., Laminar analysis of cerebral blood flow in cortex of rats by laser-Doppler flowmetry: a pilot study, *J. Cereb. Blood Flow Metab.*, 17, 1326, 1997.
93. Skarphedinsson, J.O., Sandberg, M., Hagberg, H., Carlsson, S., and Thoren, P.R., Relative cerebral ischemia in SHR due to hypotensive hemorrhage: cerebral function, blood flow and extracellular levels of lactate and purine catabolites, *J. Cereb. Blood Flow Metab.*, 9, 364, 1989.
94. Holloway, G.A., Cutaneous blood flow responses to injection trauma measured by laser Doppler velocimetry, *J. Invest. Dermatol.*, 74, 1, 1980.
95. Nilsson, G.E., Otto, U., and Wahlberg, J.E., Assessment of skin irritancy in man by laser Doppler flowmetry, *Contact Dermatitis*, 8, 401, 1982.

96. Engelhart, M. and Kristensen, J.K., Evaluation of cutaneous blood flow responses by 133Xenon washout and a laser-Doppler flowmeter, *J. Invest. Dermatol.*, 80, 12, 1983.
97. Wahlberg, J.E., Skin irritancy from alkaline solutions assessed by laser Doppler flowmetry, *Contact Dermatitis*, 10, 111, 1984.
98. Wahlberg, J.E. and Wahlberg, E., Patch test irritancy quantified by laser Doppler flowmetry, *Contact Dermatitis*, 11, 257, 1984.
99. Fischer, J.C., Parker, P.M., and Shaw, W.W., Laser Doppler flowmeter measurements of skin perfusion changes associated with arterial and venous compromise in the cutaneous island flap, *Microsurgery*, 6, 238, 1985.
100. Astrom, M. and Westlin, N., Blood flow in the human Achilles tendon assessed by laser Doppler flowmetry, *J. Orthopaed. Res.*, 12, 246, 1994.
101. Hellem, S., Jacobsson, L.S., and Nilsson, G.E., Microvascular response in cancellous bone to halothane-induced hypotension in pigs, *Int. J. Oral Surg.*, 12, 178, 1983.
102. Hellem, S., Jacobsson, L.S., Nilsson, G.E., and Lewis, D.H., Measurement of microvascular blood flow in cancellous bone using laser Doppler flowmetry and 133Xe-clearance, *Int. J. Oral Surg.*, 12, 165, 1983.
103. Swiontkowski, M.F., Tepic, S., Perren, S.M., Moor, R., Ganz, R., and Rahn, B.A., Laser Doppler flowmetry for bone blood flow measurement, correlation with microsphere estimates and evaluation of the effect of intracapsular pressure on femoral head blood flow, *J. Orthopaed. Res.*, 4, 362, 1986.
104. Swiontkowski, M.F., Ganz, R., Schlegel, U., and Perren, S.M., Laser Doppler flowmetry for clinical evaluation of femoral head osteonecrosis. Preliminary experience, *Clin. Orthopaed. Relat. Res.*, 218, 181, 1987.
105. Swiontkowski, M.F., Hagan, K., and Shack, R.B., Adjunctive use of laser Doppler flowmetry for debridement of osteomyelitis, *J. Orthopaed. Trauma*, 3, 1, 1989.
106. Swiontkowski, M.F., Surgical approaches in osteomyelitis. Use of laser Doppler flowmetry to determine nonviable bone, *Infect. Dis. Clin. N. Am.*, 4, 501, 1990.
107. Hobbs, C.M. and Watkins, P.E., Evaluation of the viability of bone fragments, *J. Bone Joint Surg.*, — Br. Vol., 83, 130, 2001.
108. Engelhart, M. and Kristensen, J.K., Raynaud's phenomenon, blood supply to fingers during indirect cooling, evaluated by laser Doppler flowmetry, *Clin. Physiol.*, 6, 481, 1986.
109. Gush, R.J., Taylor, L.J., and Jayson, M.I., Acute effects of sublingual nifedipine in patients with Raynaud's phenomenon, *J. Cardiovasc. Pharmacol.*, 9, 628, 1987.
110. Anderson, M.E., Campbell, F., Hollis, S., Moore, T., Jayson, M.I., and Herrick, A.L., Non-invasive assessment of digital vascular reactivity in patients with primary Raynaud's phenonenon and systemic sclerosis, *Clin. Exp. Rheumatol.*, 17, 49, 1999.
111. Tooke, J.E., Lins, P.E., Östergren, J., and Fagrell, B., Skin microvascular autoregulatory responses in type I diabetes: the influence of duration and control, *Int. J. Microcirc. Clin. Exp.*, 4, 249, 1985.
112. Rayman, G., Williams, S.A., Spencer, P.D., Smaje, L.H.,Wise, P.H., and Tooke, J.E., Impaired microvascular hyperaemic response to minor skin trauma in type I diabetes, *Br. Med. J. Clin. Res. Ed.*, 292, 1295, 1986.
113. Westerman, R.A., Widdop, R.E., Hogan, C., and Zimmet, P., Non-invasive tests of neurovascular function, reduced responses in diabetes mellitus, *Neurosci. Lett.*, 81, 177, 1987.
114. Caballero, A.E., Arora, S., Saouaf, R., Lim, S.C., Smakowski, P., Park, J.Y., King, G.L., LoGerfo, F.W., Horton, E.S., and Veves, A., Microvascular and macrovascular reactivity is reduced in subjects at risk for type 2 diabetes, *Diabetes*, 48, 1856, 1999.
115. Khan, F., Elhadd, T.A., Greene, S.A., and Belch, J.J., Impaired skin microvascular function in children, adolescents, and young adults with type 1 diabetes, *Diabetes Care*, 23, 215, 2000.
116. Larrabee, W.F., Jr., G.A. Holloway, Jr., R.E. Trachy and D. Sutton, Skin flap tension and wound slough, correlation with laser Doppler velocimetry, *Otolaryngol. — Head Neck Surg.*, 90, 185, 1982.

117. Svensson, H., Svedman, P., Holmberg, J., and Jacobsson, S., Detecting arterial and venous obstruction in flaps, *Ann. Plast. Surg.*, 14, 20, 1985.
118. Svensson, H., Holmberg, J., and Svedman, P., Interpreting laser Doppler recordings from free flaps, *Scand. J. Plast. Reconstr. Surg. Hand Surg.*, 27, 81, 1993.
119. Banic, A., Sigurdsson, G.H., and Wheatley, A.M., Continuous perioperative monitoring of microcirculatory blood flow in pectoralis musculocutaneous flaps, *Microsurgery*, 16, 469, 1995.
120. Cheng, M.H., H.C. Chen, F.C. Wei, S.W. Su, S.H. Lian and E. Brey, Devices for ischemic preconditioning of the pedicled groin fla, *J. Trauma-Injury Infect. Critical Care*, 48, 552, 2000.
121. Belcaro, G., Nicolaides, A.N., Laurora, G., Cesarone, M.R., De Sanctis, M.T., Incandela, L., and Labropoulos, N., Laser Doppler flux in normal and arteriosclerotic carotid artery wall, *Vasa*, 25, 221, 1996.
122. Kvernebo, K., Slasgsvold, C.E., and Stranden, E., Laser Doppler flowmetry in evaluation of skin post-ischaemic reactive hyperaemia. A study in healthy volunteers and atherosclerotic patients, *J. Cardiovasc. Surg.*, 30, 70, 1989.
123. Ghajar, A.W. and Miles, J.B., The differential effect of the level of spinal cord stimulation on patients with advanced peripheral vascular disease in the lower limbs, *Br. J. Neurosurg.*, 12, 402, 1998.
124. Duan, J., Murohara, T., Ikeda, H., Sasaki, K., Shintani, S., Akita, T., Shimada, T., and Imaizumi, T., Hyperhomocysteinemia impairs angiogenesis in response to hindlimb ischemia, *Arteriosclerosis, Thrombosis Vasc. Biol.*, 20, 2579, 2000.
125. Micheels, J., Alsbjorn, B., and Sorensen, B., Clinical use of laser Doppler flowmetry in a burns unit, *Scand. J. Plast. Reconstr. Surg.*, 18, 65, 1984.
126. Alsbjorn, B., Micheels, J., and Sorensen, B., Laser Doppler flowmetry measurements of superficial dermal, deep dermal and subdermal burns, *Scand. J. Plast. Reconstr. Surg.*, 18, 75, 1984.
127. Schiller, W.R., Garren, R.L., Bay, R.C., Ruddell, M.H., Holloway, G.A., Jr., Mohty, A., and Luekens, C.A., Laser Doppler evaluation of burned hands predicts need for surgical grafting, *J. Trauma-Injury Infect. Crit. Care*, 43, 35, 1997; discussion, 39.
128. Lunde, O.C., Kvernebo, K., and Larsen, S., Evaluation of endoscopic laser Doppler flowmetry for measurement of human gastric blood flow. Methodologic aspects, *Scand. J. Gastroenterol.*, 23, 1072, 1988.
129. Lunde, O.C., Kvernebo, K., and Larsen, S., Effect of pentagastrin and cimetidine on gastric blood flow measured by laser Doppler flowmetry, *Scand. J. Gastroenterol.*, 23, 151, 1988.
130. Lunde, O.C. and Kvernebo, K., Gastric blood flow in patients with gastric ulcer measured by endoscopic laser Doppler flowmetry, *Scand. J. Gastroenterol.*, 23, 546, 1988.
131. Lunde, O.C., Endoscopic laser Doppler flowmetry in evaluation of human gastric blood flow, *J. Oslo City Hosp.*, 38(11), 113, 1988.
132. Krogh-Sorensen, K. and Lunde, O.C., Perfusion of the human distal colon and rectum evaluated with endoscopic laser Doppler flowmetry. Methodologic aspects, *Scand. J. Gastroenterol.*, 28, 104, 1993.
133. Trimarco, B., Lembo, G., DeLuca, N., Ricciardelli, B., Rosiello, G., Volpe, M., Orofino, G., and Condorelli, M., Long-term reduction of peripheral resistance with celiprolol and effects on left ventricular mass, *J. Int. Med. Res.*, 16(Suppl. 1), 62A, 1988.
134. Iabichella, M.L., Dell'Omo, G., Melillo, E., and Pedrinelli, R., Calcium channel blockers blunt postural cutaneous vasoconstriction in hypertensive patients, *Hypertension*, 29, 751, 1997.
135. Mustakallio, K.K. and Kolari, P.J., Irritation and staining by dithranol (anthralin) and related compounds. IV. Visual estimation of erythema compared with contact thermometry and laser Doppler flowmetry, *Acta Dermato-Venereol.*, 63, 513, 1983.
136. Fullerton, A., Avnstorp, C., Agner, T., Dahl, J.C., Olsen, L.O., and Serup, J., Patch test study with calcipotriol ointment in different patient groups, including psoriatic patients with and without adverse dermatitis, *Acta Dermato-Venereol.*, 76, 194, 19962.
137. Hammarlund, A., Olsson, P., and Pipkorn, U., Blood flow in dermal allergen-induced immediate and late-phase reactions, *Clin. Exp. Allergy*, 19, 197, 19892.

138. Moor Instruments, L., reference list, http,//www.moor.co.uk, 2001, Moor Instruments Ltd.
139. Perimed, A., reference list, http,//www.perimed.se, 2001, Perimed AB.
140. Fullerton, A., Benfeldt, E., Petersen, J.R., Jensen, S.B., and Serup, J., The calcipotriol dose-irritation relationship, 48 hour occlusive testing in healthy volunteers using Finn Chambers, *Br. J. Dermatol.*, 138, 259, 1998.
141. Issachar, N., Gall, Y., Borrel, M.T., and Poelman, M.C., Correlation between percutaneous penetration of methyl nicotinate and sensitive skin, using laser Doppler imaging, *Contact Dermatitis*, 39, 182, 1998.
142. Bjarnason, B. and Fischer, T., Objective assessment of nickel sulfate patch test reactions with laser Doppler perfusion imaging, *Contact Dermatitis*, 39, 112, 1998.
143. Fischer, T., Dahlen, A., and Bjarnason, B., Influence of patch-test application tape on reactions to sodium dodecyl sulfate, *Contact Dermatitis*, 40, 32, 1999.
144. Bjarnason, B., Flosadottir, E., and Fischer, T., Assessment of budesonide patch tests, *Contact Dermatitis*, 41, 211, 1999.
145. Fullerton, A. and Serup, J., Laser doppler image scanning for assessment of skin irritation, in *Irritant Dermatitis. New Clinical and Experimental Aspects. Current Problems in Dermatology*, Burg, G., Ed., Zurich, 1995, p. 159.
146. Wang, I., Andersson-Engels, S., Nilsson, G.E., Wårdell, K., and Svanberg, K., Superficial blood flow following photodynamic therapy of malignant non-melanoma skin tumours measured by laser Doppler perfusion imaging, *Br. J. Dermatol.*, 136, 184, 1997.
147. Stücker, M., Horstmann, I., Nuchel, C., Rochling, A., Hoffmann, K., and Altmeyer, P., Blood flow compared in benign melanocytic naevi, malignant melanomas and basal cell carcinomas, *Clin. Exp. Dermatol.*, 24, 107, 1999.
148. Enejder, A.M., af Klinteberg, C., Wang, I., Andersson-Engels, S., Bendsoe, N., Svanberg, S., and Svanberg, K., Blood perfusion studies on basal cell carcinomas in conjunction with photodynamic therapy and cryotherapy employing laser-Doppler perfusion imaging, *Acta Derm. Venereol.*, 80, 19, 2000.
149. Kloppenberg, F.W., Beerthuizen, G.I., and ten Duis, H.J., Perfusion of burn wounds assessed by laser doppler imaging is related to burn depth and healing time, *Burns*, 27, 359, 2001.
150. Droog, E.J., Steenbergen, W., and Sjöberg, F., Measurement of depth of burns by laser Doppler perfusion imaging, *Burns*, 27, 561, 2001.
151. Gschwandtner, M.E., Ambrozy, E., Fasching, S., Willfort, A., Schneider, B., Bohler, K., Gaggl, U., and Ehringer, H., Microcirculation in venous ulcers and the surrounding skin, findings with capillary microscopy and a laser Doppler imager, *Eur. J. Clin. Invest.*, 29, 708, 1999.
152. Gschwandtner, M.E., Ambrozy, E., Maric, S., Willfort, A., Schneider, B., Böhler, K., Gaggl, U., and Ehringer, H., Microcirculation is similar in ischemic and venous ulcers, *Microvasc. Res.*, 62, 226, 20015.
153. Mayrovitz, H.N., Smith, J., and Delgado, M., Variability in skin microvascular vasodilatory responses assessed by laser-Doppler imaging, *Ostomy Wound Manage.*, 43, 66, 1997.
154. Mayrovitz, H.N., Smith, J., Delgado, M., and Regan, M.B., Heel blood perfusion responses to pressure loading and unloading in women, *Ostomy Wound Manage.*, 43, 16, 1997.
155. Mayrovitz, H.N. and Smith, J., Heel-skin microvascular blood perfusion responses to sustained pressure loading and unloading, *Microcirculation*, 5, 227, 1998.
156. Mayrovitz, H.N., Macdonald, J., and Smith, J.R., Blood perfusion hyperaemia in response to graded loading of human heels assessed by laser-Doppler imaging, *Clin. Physiol.*, 19, 351, 1999.
157. Bornmyr, S., Svensson, H., Lilja, B., and Sundkvist, G., Cutaneous vasomotor responses in young type I diabetic patients, *J. Diabetes Complications*, 11, 21, 1997.
158. Morris, S.J., Shore, A.C., and Tooke, J.E., Responses of the skin microcirculation to acetylcholine and sodium nitroprusside in patients with NIDDM, *Diabetologia*, 38, 1337, 1995.
159. Algotsson, A., Nordberg, A., Almkvist, O., and Winblad, B., Skin vessel reactivity is impaired in Alzheimer's disease, Neuro*biol. Aging*, 16, 577, 1995.

160. Couffinhal, T., Silver, M., Zheng, L., Kearney, M., Witzenbichler, B., and Isner, J.M., Mouse model of angiogenesis, *Am. J. Pathol.*, 152, 1667, 1998.
161. Rivard, A., Fabre, J.E., Silver, M., Chen, D., Murohara, T., Kearney, M., Magner, M., Asahara, T., and Isner, J.M., Age-dependent impairment of angiogenesis, *Circulation*, 99, 111, 1999.
162. Krogstad, A.L., Swanbeck, G., and Wallin, B.G., Axon-reflex-mediated vasodilatation in the psoriatic plaque? *J. Invest. Dermatol.*, 104, 872, 1995.
163. Krogstad, A.L., Lonnroth, P., Larson, G., and Wallin, B.G., Capsaicin treatment induces histamine release and perfusion changes in psoriatic skin, *Br. J. Dermatol.*, 141, 87, 1999.
164. Aghassi, D., Monoson, T., and Braverman, I., Reproducible measurements to quantify cutaneous involvement in scleroderma, *Arch. Dermatol.*, 131, 1160, 1995.
165. Picart, C., Carpentier, P.H., Brasseur, S., Galliard, H., and Piau, J.M., Systemic sclerosis, blood rheometry and laser Doppler imaging of digital cutaneous microcirculation during local cold exposure, *Clin. Hemorheol. Microcirc.*, 18, 47, 1998.
166. Eichhorn, W., Auer, T., Voy, E.D., and Hoffmann, K., Laser Doppler imaging of axial and random pattern flaps in the maxillo-facial area. A preliminary report, *J. Cranio-Maxillo-Facial Surg.*, 22, 301, 1994.
167. Anderson, C., Andersson, T., and Wårdell, K., Changes in skin circulation after insertion of a microdialysis probe visualized by laser Doppler perfusion imaging, *J. Invest. Dermatol.*, 102, 807, 1994.
168. Ilias, M.A., Wårdell, K., Falk, M. and Anderson, C., Phototesting based on a divergent beam — a study on normal subjects, *Photodermatol. Photoimmunol. Photomed.*, 17, 189, 2001.

# 16

# Light Scatter Spectroscopy and Imaging of Cellular and Subcellular Events

Nada N. Boustany
*Johns Hopkins University School of Medicine Baltimore, Maryland*

Nitish V. Thakor
*Johns Hopkins University School of Medicine Baltimore, Maryland*



## Overview

Thin *ex vivo* biological specimens, such as single cells and tissue slices, are widely used in biology and medicine. Optical microscopic imaging and spectroscopic techniques are ideal tools for probing the composition of these specimens nondestructively and with minimal sample preparation and handling. This chapter focuses on elastic scattering as one of these optically based analytical techniques. Unlike most turbid tissues analyzed *in vivo*, light scattering in these relatively thin *ex vivo* specimens is dominated by single scattering. Here, we describe the methods currently used to collect and interpret elastic scattering data from thin biological specimens, and biomedical applications in this field.

## 16.1 Introduction

The optical analysis of single cells or thin biological specimens, such as cells and tissue slices, plays an important role in many clinical and biological diagnostic studies, including the evaluation of diagnostic biopsies and the examination of live tissue metabolism in real time. Several optical techniques are available to study *ex vivo* biological specimens and serve as probes of human disease and biological function. These

0-8493-1116-0/03/$0.00+$1.50

include microscopic imaging techniques and spectroscopy. Diagnostic tests based on fluorescent labels of cellular metabolism as well as intrinsic fluorescence, elastic scattering, and Raman scattering from natural biomolecules have all been described. Typically, fluorescence and Raman scattering techniques are used to identify or localize specific biochemical entities. However, an important aspect of tissue diagnosis is based on assessing cellular and subcellular morphology. The morphological analysis of cells and subcellular organelles is the primary objective of biomedical optical techniques based on elastic light scattering.

Alterations in tissue morphology and composition will result in detectable changes in the way light is transmitted, refracted, diffracted, or reflected from a given tissue specimen. Experimental measurements, such as light scatter intensity and its angular dependence, can be used to infer changes in size and shape or refractive index of the specimen under study. Biological analysis methods based on elastic light scattering include microscopic imaging techniques, such as dark field, phase contrast, or differential interference contrast, as well as quantitative spectroscopic methods, in which the sample is not visualized. In phase contrast and differential interference contrast microscopy, variations in the refractive index of the tissue are utilized to optically manipulate the scattered wavefronts and produce a high-resolution image of the biological specimen under study. These types of microscopic images are widely used, for example, to visualize the morphology of the cells in culture or track cell movement. In these cases, morphological analysis of the tissue stems from direct observation by the user.

On the other hand, spectroscopy-based techniques find their way into applications that require automated quantification of cellular and subcellular morphology, without visualization of the specimen by a qualified technician. Such techniques are very useful in cell and tissue screening procedures in clinical and biological studies. Emerging techniques that combine imaging and spectroscopy are also increasingly utilized to localize the scatter information within a tissue slice or a monolayer of cells.

The constituent parts of cells and biological tissue, such as organelles or connective tissue fibers, are often at the limit of the resolution of optical microscopes. Alterations in organelle or subtissue morphology can be important indicators of underlying biochemical activity in living cells. While these changes could be quantified by electron microscopy in fixed tissue, greater insight about a biological process can be gained from minimally invasive techniques that require minimal sample preparation and are suited for live tissue monitoring. Quantitative light scattering techniques are an ideal tool to address this problem and complement the existing microscopic techniques. They are noninvasive and sensitive to changes in the dimension and optical properties of particles with size on the order of the wavelength.

Light scatter measurements have had a significant impact in medicine and biology. Applications in flow cytometry include cell identification and differential blood cell count[1–5] and human and bacterial cell response to various agents.[6–9] In static suspensions, light scattering has been used to monitor platelet aggregation,[10–12] the mitochondrial permeability transition,[13–16] and the optical properties of normal and tumor cells for future tissue diagnosis.[17,18] Optical analysis of thin biological specimens may involve monolayers of cells in culture and tissue slices. These *ex vivo* experimental models are used extensively to study important biological processes, and are crucial to advancing our understanding of biological processes at the cellular and molecular level in a controlled laboratory environment. As we will see, numerous applications of quantitative light scattering exist for the analysis of cells, organelles, and tissue slices. These biological specimens are usually thin enough to be dominated by single scattering, as opposed to studies of turbid whole tissue samples, where multiple scattering prevails.

The use of light scattering techniques to analyze thin biological specimens is the subject of this chapter. In particular, this chapter focuses on applications of quantitative light scattering, where a specific light scattering parameter is measured, such as intensity of the scattered light, or angular dependence of the scattered light. Methods for data acquisition and interpretation are discussed, as well as ongoing work in this field.

## 16.2 Brief Theoretical Overview

A very brief overview of the light scattering problem is given here. The general treatment of light scattering by a single particle can be found in Van de Hulst[19] and Bohren and Huffman.[20]

### 16.2.1 General Formulation of Scattering by a Single Particle

The biological samples considered in this chapter will contain many scattering particles; however, the samples studied will be sufficiently thin to safely assume single scattering. This condition may be satisfied if $e^{-\mu_s z} << 1$, where $\mu_s$ is the tissue scattering coefficient and $z$ is the sample thickness. $\mu_s$ is a function of wavelength, and $1/\mu_s$ represents the mean free path of the light before a scattering event occurs. Values of $\mu_s$ have been tabulated in the literature.[21] For example, $\mu_s$ is on the order of 100 cm$^{-1}$ at 780 nm for biological tissue.[22] When single particle scattering is considered, the following conditions ensue.

- Multiple scattering will be neglected.
- Each scatterer within the tissue will be exposed only to the radiation of the original incident beam.
- Light scattered from one particle will not be subjected to further scatter by another particle.

Thus, when only a single scattering event is considered for each particle making up the tissue, the total amount of scatter intensity by $N$ particles will be equal to the sum of the individual scatter intensities by each of the $N$ particles. If the particles in the sample are of varying size, the number density distribution of the particles may be taken into account. A case commonly considered is a distribution of spherical particles with different radii, $a$, in which

$$N = \int_0^\infty N(a)da.$$

The geometry of the scattering problem for an arbitrary scatterer is shown in Figure 16.1A. Given a particle of a given size and refractive index, $m$, relative to the surrounding medium, and given a plane wave of intensity $I_o$ and wavelength $\lambda$, incident on this particle, then the far-field scattered wave will be

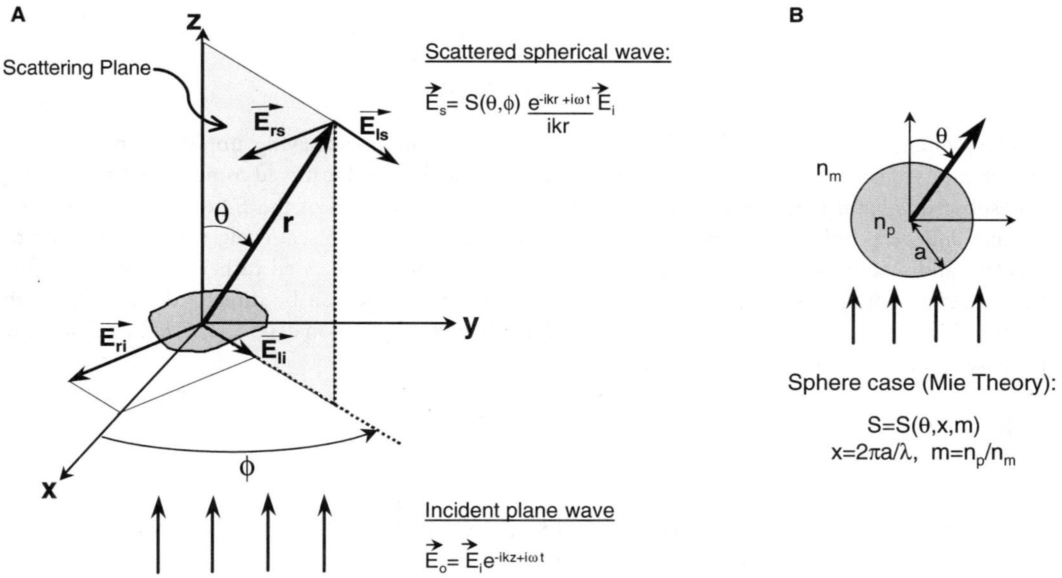


**FIGURE 16.1** Relationship between incident and scattered fields. (A) The scattered field by an arbitrary scatterer is related to the incident field by a complex amplitude function, $S$, which depends on scattering angles, $\theta$ and $\phi$, the geometry of the scatterer, and the refractive index of the scatterer compared to the surrounding medium. (B) For a sphere with symmetry around $\phi$, $S$ is a function of $\theta$, $x$, and $m$, where $x$ is a normalized size parameter equal to the ratio of the sphere circumference to the wavelength of the light, $x = 2\pi a/\lambda$, and $m$ is the ratio of the sphere's refractive index to the medium's refractive index, $m = n_p/n_m$.

a spherical wave originating at the particle. The intensity of the scattered wave, $I_s$, at any distance, $r$, in the far field can be written as $I_s = I_o F(\theta,\phi)/k^2r^2$, where the scattering direction is defined by the angles $\theta$ and $\phi$, and $k$ is the wave number with $k = 2\pi/\lambda$. The angle $\theta$ is the angle between the incident direction and the scattered direction and $\phi$ is the azimuthal angle of scatter. The scattering cross section of the particle is defined as

$$C_{sca} = \frac{1}{k^2}\int F(\theta,\phi)d\omega$$

where $d\omega$ is a solid angle differential element with $d\omega = \sin\theta d\theta d\phi$. The function $F/(C_{sca}k^2)$ is the nondimensional phase function, whose integral over solid angle is equal to 1.

Solving the scattering problem typically consists of solving for $F(\theta,\phi)$, which can therefore be used to calculate the scattered light intensity in all directions. Solving for $F(\theta,\phi)$ involves solving for the electromagnetic field everywhere in space. Maxwell's equations are solved to calculate the electromagnetic fields inside the particle and in the medium outside the particle. The field outside the particle will be a superposition of the incident field and the field scattered by the particle. The boundary conditions at the particle/medium interface require that the tangential components of the electric and magnetic field be continuous to satisfy conservation of energy at the interface. The problem is then reformulated in terms of electric fields, where the scattered field, $E_s$, is related to the incident field $E_i$ by a complex amplitude function, $S(\theta,\phi)$. The function $S$ may be represented as a matrix, such that

$$\begin{pmatrix} E_{ls} \\ E_{rs} \end{pmatrix} = \begin{pmatrix} S_2 & S_3 \\ S_4 & S_1 \end{pmatrix} \cdot \frac{e^{-ikr+ikz}}{ikr} \begin{pmatrix} E_{li} \\ E_{ri} \end{pmatrix} \tag{16.1}$$

where

$$S = \begin{pmatrix} S_2(\theta,\phi) & S_3(\theta,\phi) \\ S_4(\theta,\phi) & S_1(\theta,\phi) \end{pmatrix}.$$

The elements of the matrix S in Equation 16.1 are complex numbers having amplitude and phase and are functions of $\theta$ and $\phi$. The subscripts $i$ and $s$ denote the incident and scattered components, respectively. The subscripts $l$ and $r$ denote parallel and perpendicular polarization of the $E$ fields, respectively. The parallel and perpendicular directions are defined with respect to the scattering plane defined by the incident and scattering directions (Figure 16.1A). The function $F$ (discussed earlier), and which defines the *intensity* relationship between the incident and scattered light, can be directly deduced from the relationship between the incident and scattered fields given by the matrix $S$. For light of arbitrary polarization it is common to rewrite Equation 16.1 as

$$\begin{bmatrix} I_s \\ Q_s \\ U_s \\ V_s \end{bmatrix} = \frac{1}{k^2r^2} \begin{bmatrix} S_{11} & S_{12} & S_{13} & S_{14} \\ S_{21} & S_{22} & S_{23} & S_{24} \\ S_{31} & S_{32} & S_{33} & S_{34} \\ S_{41} & S_{42} & S_{43} & S_{44} \end{bmatrix} \begin{bmatrix} I_i \\ Q_i \\ U_i \\ V_i \end{bmatrix} \tag{16.2}$$

$I$, $Q$, $U$, and $V$ are the Stokes parameters, which can be given in terms of the electric field components as

$$\begin{aligned} I &= E_lE_l^* + E_rE_r^*, \\ Q &= E_lE_l^* - E_rE_r^*, \\ U &= E_lE_r^* + E_rE_l^*, \\ V &= i(E_lE_r^* - E_rE_l^*). \end{aligned}$$

The asterisk indicates the complex conjugate and the parameter $I$ represents the intensity of the light. In this notation, unpolarized (or natural light) is represented by the Stokes vector (1,0,0,0). The matrix in Equation 16.2 is known as the Mueller matrix. For an explanation of the explicit relationship between the elements of the Mueller scattering matrix and those of the matrix $S$ in Equation 16.1 see Reference 20 (p. 65). Also note that the 16 elements of the scattering Mueller matrix are not all independent. Only seven of them are independent, corresponding to the magnitudes of the $S_1$, $S_2$, $S_3$, and $S_4$ of Equation 16.1, and the three possible independent phase differences between these $S_j$.

## 16.2.2 Common Approximations to Solve for the Scattered Field of Biological Particles

The matrix $S$ in Equation 16.1 is a general expression that describes light scattering by a single scatterer. To predict the scattered field, the matrix elements of $S$ need to be determined. In general, these are complex numbers with magnitude and phase dependent on $\theta$ and $\phi$, as well as the dimensions of the particle, the wavelength of light, and the refractive index ratio $m = n_p/n_m$, between the particle (index $n_p$) and the surrounding medium (index $n_m$). The elements of $S$ are rarely solved analytically in the general case of a scatterer with arbitrary shape and refractive index. Depending on the biological system at hand, an approximation is usually made to simplify the problem. Commonly used approximations are discussed below.

### 16.2.2.1 Rayleigh-Gans Theory for Scattering Particles with Refractive Index Ratio Close to 1

Qualitatively, a scattering particle may be viewed as composed of different microscopic regions. An oscillating dipole moment is induced by the applied incident electric field in each of these microscopic regions. In turn, these driven dipoles scatter radiations in all directions. Thus the scattered wave originating from the particle is the sum of all the dipole radiations. The angular intensity dependence of the scattered wave will therefore depend on the phase relationships between the radiated waves and the separations between the particle dipoles relative to the incident wavelength. If the particle is very small compared to the wavelength, it may be approximated by a single dipole, and one can use Rayleigh's theory of scattering. In this case the elements of the matrix S in Equation 16.1 are $S_3 = S_4 = 0$, $S_1 = ik^3\alpha$, and $S_2 = ik^3\alpha\cos\theta$. $\alpha$ is the polarizability of the particle, and $k$ is the wavenumber $2\pi/\lambda$. For a sphere with radius $r$ and refractive index ratio, $m$, $\alpha = r^3(m^2 - 1)/(m^2 + 2)$, and for a homogeneous particle with refractive index ratio close to 1, $\alpha = (m^2 - 1)(V/4\pi)$, where $V$ is the particle's volume (see Chapter 6 of Reference 19).

To satisfy the Rayleigh approximation, $|m|ka << 1$, where $a$ is a lengthscale on the order of the size of the particle. A similar situation arises if the refractive index ratio, $m$, is close to one, such that $|m - 1| << 1$, and $2ka|m - 1| << 1$. These two latter conditions imply that the field inside the particle is close to the incident field, and that the particle may be assumed to be composed of volume elements dV that are subjected to the same incident field. Thus, instead of assuming that the particle is a single dipole, the particle is now composed of independently scattering dipoles corresponding to the different volume elements. These dipoles are therefore driven by the same applied field, and the scattered wave is the sum of the waves scattered by these dipoles. In this case, $S_3 = S_4 = 0$, as for Rayleigh scattering, and:

$$S_1 = \frac{ik^3(m-1)}{2\pi} VR(\theta,\phi)$$

$$S_2 = \frac{ik^3(m-1)}{2\pi} VR(\theta,\phi)\cos\theta \tag{16.3}$$

$$\text{with } R(\theta,\phi) = \frac{1}{V}\int e^{i\delta} dV$$

The phase $\delta$ in Equation 16.3 refers to the phase of the scattered waves with respect to a common origin in the reference coordinate system. After reformulating $\delta$ in terms of the problem's geometry, $R$ can then be integrated for a given particle shape. Calculations of $R$ for spheres, ellipsoids, and cylinders are discussed in Reference 19 (Chapter 7) and Reference 20 (Chapter 6). The treatment of scattering resulting in Equation 16.3 is referred to as the Rayleigh-Gans, or Rayleigh-Debye-Gans, theory.

Because the refractive index of biological organelles, such as mitochondria, is typically close to that of the surrounding medium,[23,24] various light scattering studies based on Rayleigh-Gans theory can be found in the literature. These include studies of scattering by bacteria,[24,25] macromolecules,[26,27] or nucleated lymphocytes.[28] Moreover, if the size and refractive index of the scattering particles satisfy the Rayleigh-Gans conditions, light scattering by a three-dimensional scattering object may be approximated as Fraunhofer diffraction by a two-dimensional aperture function.[29] This diffraction-based approach allows the Fourier optical treatment of diffraction and can be used to extract cellular geometric parameters from the diffraction pattern of biological cells.[30,31]

#### 16.2.2.2 Mie Theory for Spherical Particles of Arbitrary Size and Index

For a sphere with symmetry around $\phi$ (Figure 16.1B), $S$ may be re-expressed as a function of $\theta$, $x$, and $m$, where $\theta$ is the angle of scatter, $x$ is a dimensionless size parameter equal to the ratio of the sphere circumference to the wavelength, $x = 2\pi a/\lambda$, $a$ = particle radius, and $\lambda$ = wavelength; $m = n_p/n_m$ is the ratio of the particle's refractive index, $n_p$, to the surrounding medium's refractive index, $n_m$. The analytical solution for spheres was given by Mie in 1908, and can be found in Chapter 9 of Reference 19 or Chapter 4 of Reference 20. For a sphere, $S_3$ and $S_4$ are 0 in Equation 16.1, while $S_1$ and $S_2$ can be expressed as infinite sums that can be calculated on a computer. Fortran computing routines to solve for the angular scattering function for a homogeneous or coated sphere can be found in Reference 20. Graaff et al.[32] present a simple numerical approximation of Mie scattering for $5 < x < 50$, and $1 < m < 1.1$. Combined with a model of light propagation in a microscope with high numerical aperture, Mie theory was also used successfully to predict high-resolution images of spheres.[33]

The existence of an analytical solution for the case of a spherical scatterer has prompted many to approximate biological particles as spheres as a first-order approach to understanding light scatter from cells and tissues. Despite the complicated morphologies of biological particles, studies have been successful in utilizing Mie theory to model the angular scattering response of bulk biological tissue. For example, the angular scattering functions of brain and muscle were successfully predicted by use of Mie theory and assuming that the tissue is composed of spheres with sizes distributed according to a skewed logarithmic function.[34] Similarly, a model based on Mie theory was able to approximate reflectance spectra of colon tissue adequately.[35]

### 16.2.3 Solving the Scattering Problem for a Scatterer of Arbitrary Shape and Index

Although biological particles are close to spherical in some cases, such as when considering the nuclei of certain cells, in general this assumption is not necessarily warranted, especially when considering mitochondria, which appear rather filamentous *in situ*, or when considering neuronal dendritic structures or scattering collagen and elastin fibers in connective tissue. Thus when considering these tissue components individually, light scattering studies should take into account their potential nonspherical and sometime complicated geometries. Moreover, some tissue components, such as lipids, collagen, or melanin, have refractive index ratio $m > 1$.[36,37] Numerical approaches to solving the problem of scattering by particles of arbitrary shape and refractive index have been gradually emerging in applications of light scatter to biological systems. Two popular approaches for the study of nonspherical biological particles are the transition matrix (T-matrix) method originally developed by Waterman,[38,39] and the finite-difference time-domain (FDTD) technique originally proposed by Yee.[40] With recent advances in computer hardware, numerical computations of angular scatter intensities can now be achieved on a personal computer. Details on the T-matrix method with accompanying software can be found in Barber and

Hill.[41] The T-matrix and FDTD methods are also discussed in detail with relevant references in Mishchenko et al.[42]

In the T-matrix method, the incident and scattered field are expanded into vector spherical wave functions. Due to the linearity of Maxwell's equations, the incident and scattered field can be related by means of a transition matrix (the T-matrix), which depends solely on the particle geometry and refractive index. Solving for the elements of the T-matrix allows prediction of the scattered field. For example, the T-matrix method was used to study light scattering by red blood cells.[43] On the other hand, the FDTD technique is based on the discretization of Maxwell's curl equations in time and space. Numerical calculation of the electric field as a function of time near the scattering particle is computed after applying the appropriate boundary conditions at the edges of each grid element in space. The near-field values thus computed are then transformed to yield the scattered far field. At present, the FDTD technique has been used to predict scattering by inhomogeneous cell models composed of a spherical nucleus and of ellipsoidal organelles.[17,37] By using an incident time-limited pulse instead of an incident monochromatic plane wave, the frequency response of the scattered far-field response can be calculated. The pulsed FDTD approach was implemented to predict the angular scattering response of two-dimensional models of inhomogeneous cells as a function of wavelength.[44]

## 16.3 Scatter Data Interpretation

Typically, changes in light scatter result from changes in the scattering particle's size, shape, refractive index, and concentration. Changes in these parameters may accompany important biochemical events in organelles, cells, and tissues, and thus serve as diagnostic markers. If the particles that will change have already been identified, it may be feasible to predict the scattering behavior of these particles by utilizing the theoretical approaches described in Section 16.2. Combining the scatter measurements with the expected theoretical predictions will then serve to quantify the cellular or tissue events under study. In this case the light scattering methods can be utilized to optically track known variations in cell or tissue substructure. The light scatter technique can serve to sort the data, for example, or to automate a well understood diagnostic process.

Light scattering can also be used to probe tissue dynamics in which the scattering sources have not yet been identified. Interpreting the scatter data from such measurements can be particularly complicated however, because one must solve the "inverse problem" consisting of extracting hitherto unknown tissue properties from the available scatter data. Typically, biological systems are inhomogeneous, and contain particles of various geometries and optical properties, all of which could potentially result in the observed light scatter changes. As seen in the previous section, the dependence of scatter intensity on angle of scatter is complicated. Thus, in order to extract an absolute optical or morphological parameter, extensive angular scatter data may need to be gathered such that sufficient measurements are available to fully characterize the angular scatter and cross-section properties of the tissue under study.

Such extensive data are rarely available experimentally, and theoretical prediction about the scatterers is often limited by the set of measurements at hand. In most cases, to identify and characterize the possible sources of the tissue scatter, including organelles and other substructures, the biological system must be approximated by a model that can be easily plugged into the theoretical frameworks available. For example, the particles within the biological specimen are often assumed to be spherical or randomly oriented. Although such simplifying assumptions are necessary to begin analyzing a given scatter problem, such assumptions restrict data interpretation. Refining the initial simplifying assumptions will often prove necessary to ensure that the model is adequately taking into account the important variables in a given biological situation. In light of the difficulty of specifically identifying the sources of scatter change in a given tissue, light scattering techniques are best used in conjunction with other methods. For example, light scatter can be used to detect and localize possible morphological changes, while additional biochemical manipulation can be used to modulate scatter changes and help identify the molecules or tissue components leading to the observed change.

## 16.4 Methods and Applications of Light Scatter Measurements to the Study of Cells, Organelles, and Tissue Slices

As we saw briefly in Section 16.2, the amount of light scattered from a given particle typically depends on its size compared to the incident wavelength, and on the ratio of its refractive index compared to the surrounding medium. In general, the larger the particle and its refractive index ratio are, the larger the amount of scatter it will produce. Moreover, the shape and size of the particle will also affect the angular dependence of the scattered light. Thus, light scattered sideways as a fraction of the total light scattered will be larger for small spheres than for larger ones. Several techniques have been used to track such changes in the light scattering properties of biological samples. These techniques typically consist of measuring the angular scattering properties of a given sample. For example, changes in angular scatter as a function of sample composition or experimental condition can be monitored to deduce the size, shape or index of the scatterers in each case. In some studies, the dependence of light scatter on incident wavelength or polarization is also taken into account.

Scatter intensities at multiple angles could be studied by collecting angular scatter intensity with the aid of a motorized goniometer. On the other hand, the pattern of diffraction by the sample could be analyzed utilizing Fourier optics to infer angular scatter. Quantitative transmission and reflection microscopy have also been used to generate images in which the scatter intensity signals are measured locally in different parts of the specimen. Representative examples of these scatter measurement methods and their applications are discussed in this section.

### 16.4.1 Light Scattering Spectroscopy of Cells and Organelles in Suspensions

#### 16.4.1.1 Methods to Study Scattering by Particle Suspensions

Particle suspensions are usually probed spectroscopically without optically resolving the individual scatterers. Scatter intensities at multiple angles can be collected with a rotating goniometer and a single photodetector[17,18,45] (Figure 16.2A), or by many detectors positioned at different angles around the specimen[27] (Figure 16.2B). The cell or organelle suspensions are usually contained in a cuvette, or the particles may be flowed in single file through the optical analysis chamber of a flow cytometer (Figure 16.2C). Scatter intensities are plotted vs. angle around the specimen, and could then be compared to theoretical predictions. By varying the polarization of the incident light and analyzing the polarization of the scattered light (if polarizers are used in Figure 16.2A), additional elements of the scattering Mueller matrix (see Section 16.2.1) can be measured.[46] Bohren and Huffman list all the possible relationships between the intensity of the scattered light at the different polarizations and the intensity of the polarized incident beam in terms of the Mueller matrix elements, $S_{ij}$ (Table 13.1 in Reference 20).

#### 16.4.1.2 Applications

##### *16.4.1.2.1 Flow Cytometry*

Light scattering spectroscopy has extensively been employed in cell analysis, such as flow cytometry, to probe intracellular morphology. In a flow cytometer, scattering near the forward direction and side scatter near 90° are typically collected. Although data at many angles are necessary to characterize the scatterers in the suspensions fully, limited measurements, such as the amount of near-forward scattered light and side-scattered light, can also provide very useful information. These two measurements of forward- and side-scatter can be used individually, or combined into a ratio to yield discriminating data that allow cell identification.

In clinical flow cytometry, cells flowing in a single file through an optical analysis chamber can be analyzed, sorted, and counted based on the ratio of forward- to side-scatter intensity. These cells may originate from a patient's blood or from a tissue biopsy. By sorting and counting the cells with specific forward-to-side scatter signatures, the status of a patient may be evaluated, and disease may be ruled in or out. For example, utilizing the elastic scattering properties of the different blood cells, human leukocytes may be sorted and counted.[47] Differentiation of leukocyte can also be improved by measuring

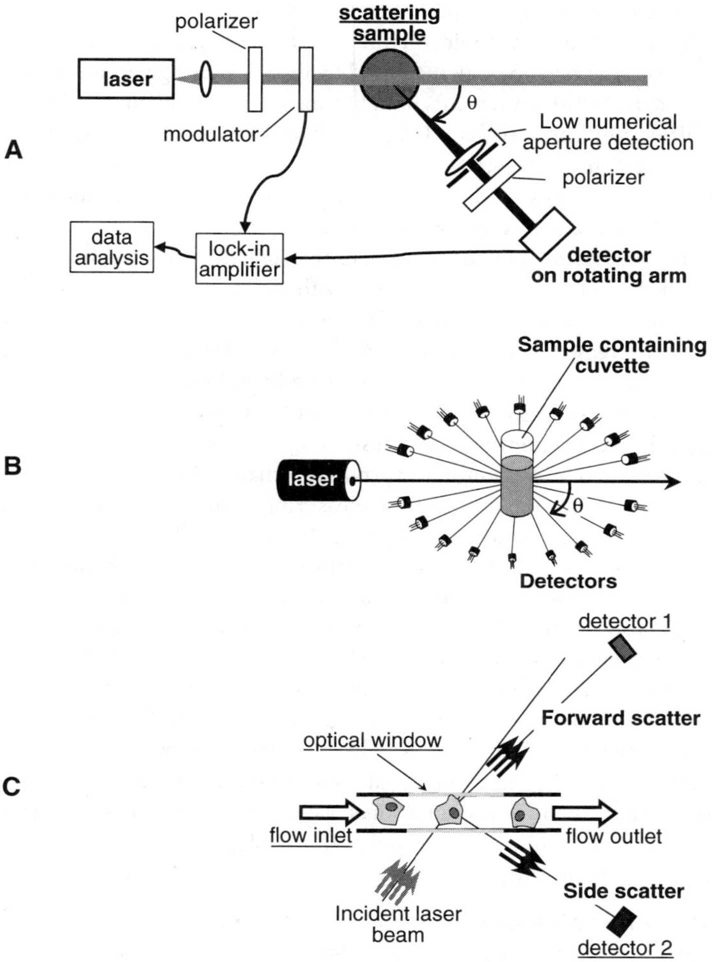


**FIGURE 16.2** Schematics illustrating the principle behind experimental setups for collecting angular scatter from particle suspensions. (A) Scanning goniometer. Light scattered by the sample at a given scattering angle, θ, is collected by a low numerical aperture setup. The collection-detection setup is mounted on the arm of a rotating goniometer. The polarizers and modulator are optional. (B) Multidetector setup. Light scattered by the sample is collected as a function of scattering angle, θ, by several detectors placed around the sample. In contrast to the setup in (A), here the angular measurements are made simultaneously. (Adapted from Wyatt, P.J., *Anal. Chim. Acta*, 272, 1, 1993.) (C) Flow cytometry. The diagram shows cells flowing in single file through a flow cytometer channel, while the scattered light is collected through an optical window. Near-forward and near-90° scatter are typically collected.

changes in the amount of depolarization of the side-scattered light.[48] A discussion of light scattering as it applies to flow cytometry can be found in Salzman et al.[2] (See also Chapter 25 of this handbook.)

#### *16.4.1.2.2 Angular Scatter Measurements of Isolated Mitochondria*

Light scattering spectroscopy has been extensively used to study mitochondrial swelling and changes in mitochondrial matrix conformations. Although change in mitochondrial morphology could be assessed by electron microscopy, dynamic studies of *viable* mitochondria typically utilize light scattering to study this organelle, whose size is close to the optical resolution of microscopes. Light scattering is a simple and convenient method that is sensitive to changes in the size and shape of particles with dimensions on the order of the wavelength. Moreover, as an optical method, light scattering permits rapid detection commensurate with the rates at which mitochondria are expected to change. The light scattering measurement may be carried out in a spectrophotometer with the mitochondria suspension contained in a regular cuvette or by flow cytometry. Studies on mitochondria isolated from tissue date back to the 1950s. Alterations in mitochondrial morphology measured by light scattering have been associated with mitochondrial metabolic state.[49–56] Measurements of light transmission or angular light scattering at 90° from a suspension of isolated mitochondria have long been correlated with the morphology of mitochondria in the orthodox and condensed states.[14,51,52,57] Since these early studies, light scattering has become the technique of choice to detect mitochondrial size change. Light scattering techniques have proved essential in studying the mitochondrial permeability transition.[14–16,58–62]

The first scattering studies of mitochondria were interpreted by correlating the absorbance or 90° scatter intensity from the mitochondrial suspension with electron micrographs of the tested mitochondria. In most cases, mitochondrial scatter at 90° decreased as the number of mitochondria in the "aggregated" configuration decreased.[14] This aggregated form was typically characterized by a shrunken, electron-dense matrix space with large intercristal space.[54] In addition, the absorbance and 90° scatter by mitochondrial suspensions were shown to decrease with increased mitochondrial swelling[63,64] (Figure 16.3). More recently, this relationship between swelling and mitochondrial absorbance and 90° scatter was utilized in the detection of mitochondrial morphology change during apoptosis. These recent studies were conducted utilizing flow cytometry[65] and by measuring changes either in 90° scatter or absorbance by a suspension of isolated mitochondria in a spectrophotometer cell.[66–69]

Nonetheless, one should interpret single angle scatter, or absorbance measurements with great care. Under certain conditions the early scattering studies of mitochondrial scatter have provided good correlation between light absorbance, or 90° scatter intensity, and mitochondrial morphology;[14,63,64] however, these methods could present some shortcomings. The general relationship between transmitted light, or light scattered at one single angle, and particle volume is not always monotonic.[70,71] Moreover, changes in refractive index also contribute to the change in light scatter in addition to morphology change, thus confounding data interpretation. A study by Knight et al.[72] shows how changes in light scattered at 90° may not necessarily correlate with mitochondrial volume change and points at the difficulty in interpreting single angle scatter data. Thus, additional validation by means of electron microscopy, for example, will prove necessary to infer the particles' morphologic configurations correctly from absorbance or single angle scatter measurements.

#### *16.4.1.2.3 Angular Scatter Measurements of Cellular Suspensions*

With the recent applications of diffuse light scattering techniques to the diagnosis of tissue *in vivo* (see accompanying chapter on elastic scattering and diffuse reflectance), interest in studying the scattering properties of cells, organelles, and subtissue structure has increased. Scattering parameters can be used to define the morphological organization of biological tissue and to better understand the different sources of scatter that contribute to the bulk tissue signal. Scattering from cell suspensions was used to show that cells have a broad distribution of scatterer sizes. Significant cell scatter was shown to originate from particles between 0.2 and 1 μm; the small tissue particles are expected to contribute to wide angle scatter, while larger particles will contribute mainly to forward directed light scatter.[18]

Moreover, the nuclei angular scatter spectrum most closely resembled that of the whole cells (Figure 16.4). In this study Mourant et al. assumed that the cells comprise spherical scatterers and used Mie theory to analyze the angular scatter distributions. Further studies by the same group have shown that cell suspensions do not depolarize light significantly.[73] Angular scatter measurements of cell suspensions were also utilized as experimental validation in the construction of a cell model based on the FDTD technique[17] (see Section 16.2.3). In contrast to the study by Mourant et al.,[18] the cell model in this case does not assume a distribution of spherical scatters. Instead, the model considers a 15-μm diameter spherical cell containing a nucleus with subnuclear refractive index variations, and ellipsoidal organelles. In particular, such optical cell models were used to explain the effect of adding acetic acid to cells, suggesting that acetic acid increases the frequency of fluctuations in nuclear refractive index; acetic acid was also found to increase the amplitude of these index variations.

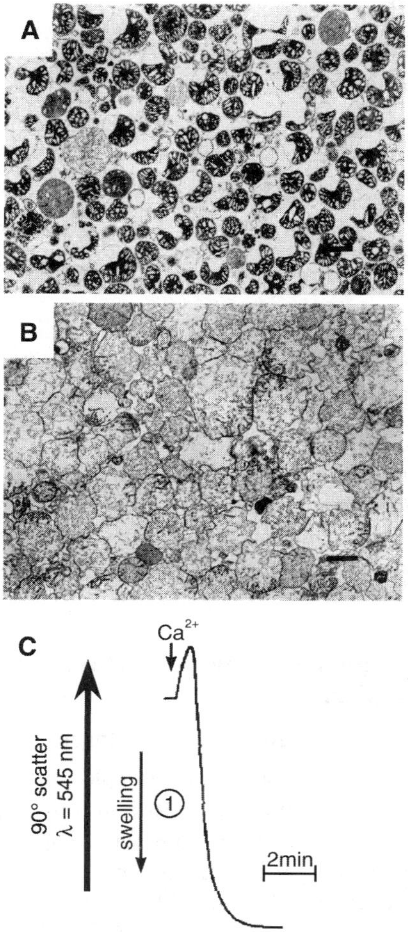


**FIGURE 16.3** Electron microcrographs and absorbance measurements from mitochondrial suspensions subjected to calcium overload. (A) Isolated liver mitochondria suspended in calcium-free incubation medium. (B) The medium was supplemented with 150 μ*M* $Ca^{2+}$. (C) The decrease in the measured 90° light scattering at 545 nm correlates with mitochondrial swelling upon addition of 150 μ*M* $Ca^{2+}$ to the control medium. Bar = 1 μm. (From Petronilli, V. et al., *J. Biol. Chem.*, 268, 1939, 1993. With permission.)

Acetic acid addition is very relevant to cancer diagnosis. Topical application of acetic acid to tissue is a very common method used by colposcopists to enhance contrast between normal and diseased regions of the cervical epithelium. Thus, by understanding how different conditions may change the optical scattering properties of the cells under study, cell modeling, together with scattering studies of cell suspensions, represents an important set of data, which will undoubtedly be helpful when optimizing and designing current and future optical diagnostic tools.

***16.4.1.2.4 Angular Scatter Measurements of Bacteria, Macromolecules, and Vesicles***

Scattering spectroscopy of cells and organelles has direct applications to understanding the scattering properties of biological tissues. It is important to note that the methods described here for collecting angular scatter data from particle suspensions may also have other biologically relevant applications. In particular, angular scatter has been used to identify bacteria.[25] The state of polarization of the light scattered by bacteria was also shown to be sensitive to very small changes in bacterial structure.[46] Angular

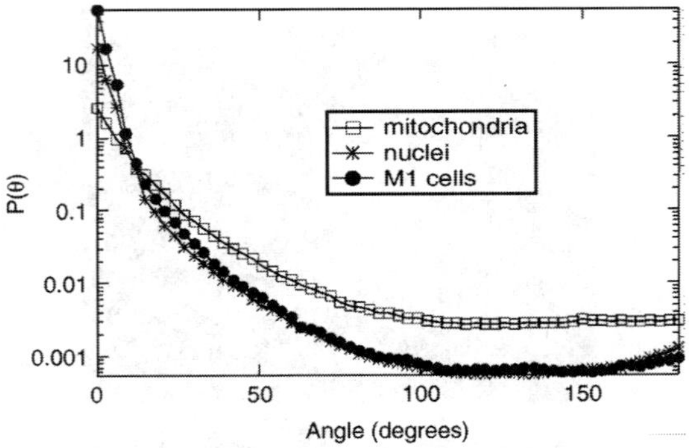


**FIGURE 16.4** Normalized angular scatter intensity, $P(\theta)$, of fibroblast cells (M1 cells), isolated fibroblast nuclei, and isolated fibroblast mitochondria. Values below 9 and 168° were extrapolated. (From Mourant, J.R. et al., *Appl. Opt.*, 30, 3586, 1998. With permission.)

scattering was used to characterize the size of macromolecules in suspension.[27] In addition, in a system where angular scatter distributions were measured as a function of time, the dynamics of time varying systems were characterized. Thus, time-dependent angular scatter measurement was used to track the polymerization of microtubules as well as dynamic changes in the size of chromaffin granules subjected to osmotic stress.[26]

## 16.4.2 Light Scattering Spectroscopy of Cellular Monolayers and Thin Tissue Slices

### 16.4.2.1 Methods for Collecting Angular Scatter Measurements by Diffraction

Another way to infer angular scatter is by analyzing the sample's diffraction pattern with Fourier optics. The principle behind this method is shown in Figure 16.5. In this setup, the sample is illuminated by a plane wave of light obtained by a collimated laser beam, for example. The light scattered by the sample is collected by a lens, whose numerical aperture will determine the highest angle of scatter that can be collected by the setup. As shown in Figure 16.5, the diffraction pattern of the sample is formed in the back focal plane, *F*, of the collection lens. Because the incident laser beam is collimated, the diffraction pattern is generated from light scattered by the sample. The angles of scatter are mapped onto the plane *F* in increasing order, moving radially away from the optical axis. The laser light that is transmitted without being scattered by the sample will be focused in the center of the plane *F*, and can be subtracted by a beam block at this point. Usually the diffraction pattern in *F* is reimaged by a second lens onto a photodetector array, such as a charge-coupled device (CCD) camera.[30,31,74] Changes in angular scattering by the sample can be studied by analyzing its diffraction pattern. The cell sample can be plated on a microscope slide in this diffraction-based setup, so this method is particularly useful for analyzing cells in a monolayer, as opposed to in suspension, as was discussed in the previous section.

As for the angular scattering measurements of particle suspensions, the diffraction technique is of spectroscopic nature: the diffraction pattern corresponds to scattering by the entire sample region illuminated by the laser beam. The size of this illuminated region can be a few hundreds of microns in width. To correlate the angular scatter pattern with a specific region of the sample, Valentine et al.[74] used a microscope condenser in the illumination path and were able to set the laser beam diameter to 70–100 μm, thus selectively analyzing small regions of a porcine skin specimen (Figure 16.6). In that system the optical microscope was also equipped with a beam splitter after the collection lens, such that an image of the sampled 70- to 100-μm region could be collected simultaneously with an image of its diffraction pattern on two separate cameras.

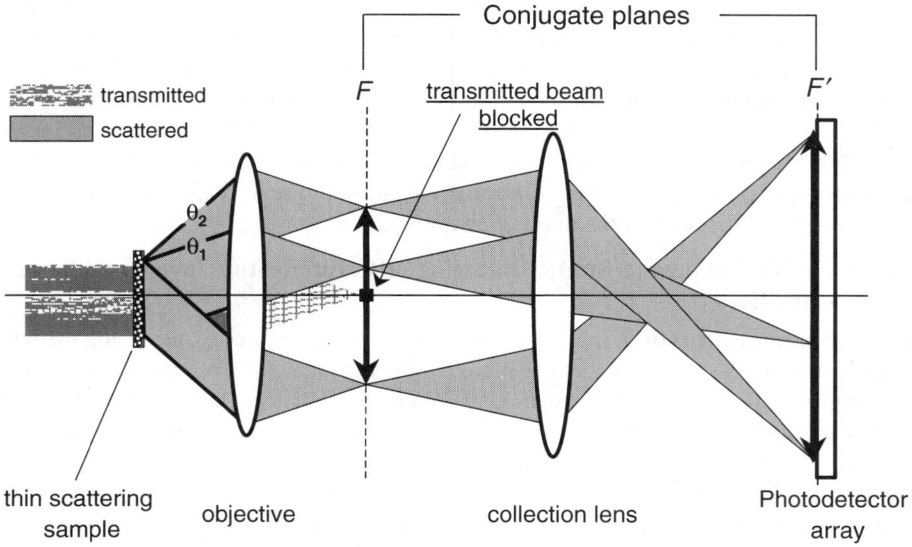


**FIGURE 16.5** Experimental setup for imaging the diffraction pattern of cells or tissue slices plated on a microscope slide. *F* = objective's back focal plane. *F* and *F′* are conjugate Fourier planes. The scattered light (gray beam), which forms the diffraction pattern of the sample, is reimaged onto a photodetector array, while the transmitted light (cross-hatched beam) is blocked at the center of the plane *F*.

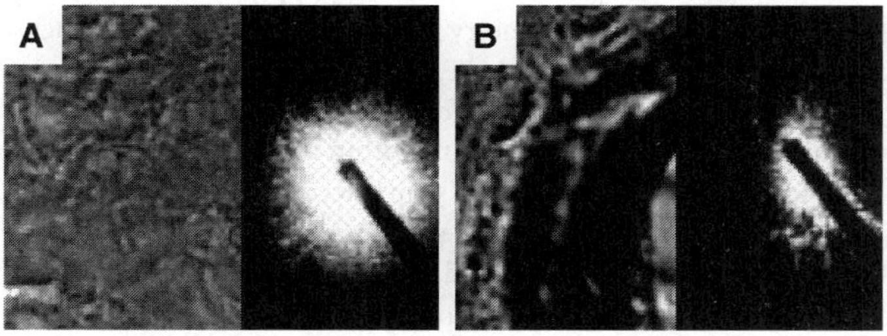


**FIGURE 16.6** Images and diffraction patterns of two porcine skin specimens, 20 μm thin. (A) The tissue region sampled is homogeneous and the diffraction pattern is isotropic. (B) A region in the vicinity of a hair results in an anisotropic scattering pattern. The black rod in the diffraction images corresponds to the transmitted light beam block. The scattering angle, θ, increases in the radial direction. Transmitted light and light scattered at 0° will be focused in the center of the diffraction pattern. (From Valentine, M.T. et al., *Opt. Lett.*, 26, 890, 2001. With permission.)

### 16.4.2.2 Applications of Diffraction to Cellular Analysis

Analysis of angular scattering by modeling the scatter as a Fraunhofer diffraction field from a two-dimensional flat object was used to measure the diameters of the nucleus and cytoplasm of stained cervical cells.[30] The nucleated cells were modeled as two circular concentric regions having different optical densities, and diameters $d_N$ and $d_C$ corresponding to the nucleus and cytoplasm, respectively. In some cases the effects of offsetting the nucleus from the center of the cell were also considered. The solution of the Fraunhofer diffraction model gives the radial dependence of the light intensity in the diffraction pattern of the sample. By analyzing only one radial scan of the diffraction pattern, nuclear and cytoplasmic diameters were calculated and compared with the actual dimensions of normal, dysplastic and cancerous cells. The results show that the correct nuclear diameter was inferred in more than 80% of the 378 cells tested.

A similar approach was taken by Burger et al.[31] to extract nuclear and cytoplasmic diameters from the Fraunhofer diffraction pattern of nucleated cell models and nucleated erythrocytes. The cellular dimensions inferred from the light scatter analysis in the diffraction pattern matched the microscopically determined dimensions of the cytoplasm and nucleus very well. An additional radial scan of the diffraction pattern taken at 90° to the first one also helped differentiate diameters of the major and minor axes these elliptical erythrocytes. More recently, modeling of light scattering as a Fraunhofer diffraction pattern was used to monitor the rounding of initially elongated cells in response to follicle-stimulating hormone[75] and the changes in cell diameter at the onset of apoptosis.[76]

#### 16.4.2.3 Other Techniques to Study Scattering of Cellular Monolayers and Thin Tissue Slices

Angular scatter measurements from thin tissue slices can be made by directly measuring the intensity of the forward scattered light from a tissue slice mounted on a cover slip. Direct measurements of angular light scattering were used to extract information about order and spacing between the collagen fibers in cartilage. In particular, the average scatter angle from 40-μm thin cartilage slices was shown to decrease as a function of the aggregate compressive modulus (measured separately from the bulk samples prior to slicing).[77] This relationship between angular scatter and modulus could be explained by a theoretical analysis relating average scatter angle, compressive modulus, and a short-range order parameter. From this analytical model, this short-range order parameter, which describes the spatial correlation length between the collagen fibers, was found to be 8.2 μm.

Other scattering spectroscopic techniques of monolayers of cells and thin tissue slices have also been described. By analyzing the refractive index fluctuations of mouse liver tissue obtained by phase-contrast microscopy, Schmitt and Kumar[78] showed that the power spectrum of refractive index variations follows an inverse power law as a function of spatial frequency suggesting a fractal packing of tissue mass.

Angle-dependent low coherence interferometry was used to measure the backscatter intensity as a function of angle from monolayers of cultured HT29 epithelial cells.[79] As for optical coherence tomography, this technique offers a depth resolution defined by the coherence length of the light source and allows scattering measurements from a specific point within the penetration depth of the sample. When probing a point close to the sample surface, or thin monolayers of cells, the angular scattering is dominated by single scattering and can be analyzed by Mie theory. Assuming spherical scatterers, the angle-dependent low coherence method was used to extract nuclear diameter and nuclear refractive index. Once the nuclear contribution to the angular scattering is subtracted, the remaining angular scatter spectrum can be analyzed to extract information about subcellular organelles smaller than the nucleus. From this residual angle distribution, a two-point spatial correlation of the optical field was obtained. This spatial correlation followed an inverse power law indicating fractal packing of subcellular structure as previously suggested by Schmitt and Kumar.[78]

### 16.4.3 Combining Spectrscopy and Imaging

#### 16.4.3.1 Transmission and Reflectance Images of Brain Slices

The techniques described in the two previous subsections probe the angular dependence of light scattering by the specimen under study, but without imaging the sample. The scatter intensities are typically collected from an ensemble of particles, and information about the location of the scatter sources within the specimen is not always saved. Cells often respond differentially to a given treatment; tracking the location of the scatter change within a monolayer of cells or within a tissue slice could provide a better understanding of the basic dynamics of a time dependent biological process. To this end, *imaging* methods based on scattering signals can be used to record relative changes in light scatter within the full field of view. Scattering information can be collected directly from bright-field and dark-field microscopic images collected at a specific wavelength.

The imaging wavelength is typically chosen in the red or near infrared, which penetrate tissue deeper than shorter wavelengths. The intensity distributions within transmission and reflectance microscopy images depend on the way the light is absorbed or scattered by the tissue under study. Changes in the biological composition of the tissue can affect absorbance and transmission through the specimen and can therefore be used to track subtissue dynamics. Transmission and reflectance imaging have been used to map neural activity within brain slices by differentiating the response of the different neuronal layers within the slice in response to a stimulus. Intrinsic optical signals such as transmittance or reflectance are often recorded in conjunction with fluorescent images using voltage-sensitive dyes or mitochondrial potential dyes, for example. These fluorescent images help correlate the optical scatter signal with biochemical and electrophysiological properties of the different tissue regions.

A short review of the application of intrinsic optical imaging techniques to brain slices can be found in Aitken et al.[80] Transmitted and reflected light signals are shown in the CA1 region of the hippocampus as a function of hypotonic stress,[80,81] neural stimulation,[80] or spreading depression.[80] Changes in light transmission were also compared between the CA1 and CA3 regions of the hippocampus as a function of *N*-methyl D-aspartate (NMDA) and kainate mediated excito-toxic injury.[82] Changes in transmitted light and mitochondrial depolarization measured with the fluorescent dye rhodamine 123 were used to track the spatio-temporal dynamics of hypoxia and spreading depression within hippocampal slices.[83] Using a fiber-optic excitation/collection bundle, reflected light was measured *in vivo* from the cat hippocampus and correlated with evoked potentials.[84]

Brain slices are advantageous compared to isolated neuronal cultures in that they preserve the physiological relationship between neurons and glia, as well as the connectivity between different neuronal regions. However, to preserve these relationships adequately and maintain an experimentally viable tissue, the slices are often several microns in thickness. As a result, multiple scattering may affect the optical signals detected, and the single particle scattering approaches presented in Section 16.2 may not be applicable. Moreover, because most studies do not record the full angular scatter response from the tissue, the optical signal interpretation becomes limited.

#### 16.4.3.2 Dual Angle Scatter Imaging of Brain Slices

To explore angular scatter by brain slices, Johnson et al. measured light scattered by hippocampal slices after illuminating the tissue at two different angles (Figure 16.7).[85] These brain slices were still 310 μm thick and multiple scatter may not be negligible. However, by comparing the individual images collected at each of the two illumination angles and their ratio, Johnson et al. were able to differentiate the hippocampal response to hypotonic stress from the response to NMDA-mediated injury (Figure 16.8).[85] When each illumination angle was considered individually, both hypotonic and NMDA treatments showed a relative decrease in the intensity of the light scattered by the CA1 region. However, when the ratio of the images collected at the two angles was considered, changes could only be measured after the NMDA treatment.

Although several scattering components, such as dendritic processes, axonal varicosities or cellular organelles, can contribute to transmission, reflection, and scattering measurements of brain slices,[86] the study by Johnson et al. still shows that intrinsic optical signals can be successfully used to differentiate functional neuronal responses. As such, despite the difficulty in fully interpreting the scattering response of relatively thick tissue slices, transmission and reflectance imaging of brain slices remain very valuable because they provide a simple method to record spatio-temporal dynamics that reflect morphological change and that can be measured simultaneously in the whole preparation, unlike focal electrode recordings.

#### 16.4.3.3 Optical Scatter Imaging of Cellular Monolayers

Recently, Boustany et al. demonstrated an optical scatter imaging (OSI) technique that produces images that directly encode a morphometric parameter within the full field of view of the microscope.[87] This OSI method combines Fourier filtering with central dark-field microscopy to detect alterations in the size of particles with wavelength-scale dimensions. A "scatter ratio" image is generated by taking the ratio of images collected at high and low numerical aperture in central dark-field microscopy. Such an image spatially encodes the ratio of wide to narrow angle scatter (or optical scatter image ratio, designated here as "OSIR") and hence provides a measure of local particle size.

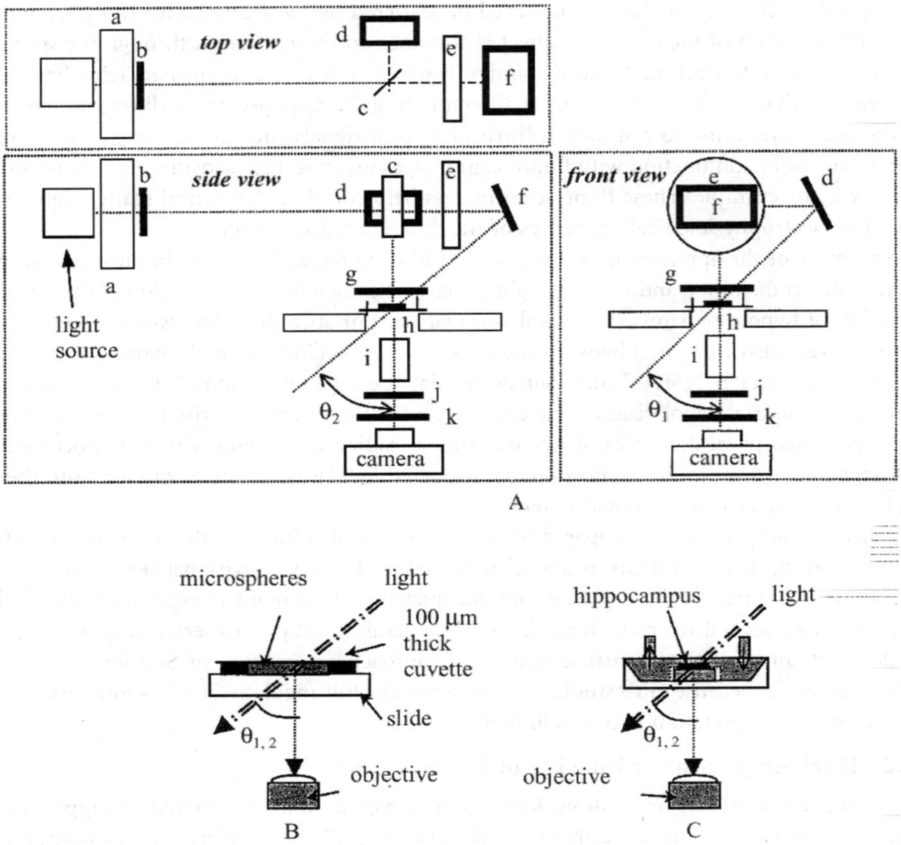


**FIGURE 16.7** (A) Setup for dual-angle scattering images of hippocampal slices. a = shutter; b = infrared filter; c = beam splitter; d = mirror; e = shutter; f = mirror; g = polarizer; h = specimen; i = low numerical aperture objective; j = polarizer; k = interference filter. (B) Stage design for experiments on microsphere suspensions. (C): Stage design for brain slice experiment. $\theta_{1,2}$ represent the two scattering angles, 31 and 34°, for which the images were acquired. (From Johnson, L.J., Hanley, D.F., and Thakor, N.V., *J. Neurosci. Methods*, 98, 21, 2000. With permission.)

Figure 16.9 shows the OSI microscopy setup. The specimens are mounted on the stage of an inverted microscope, which can also be fitted with an epi-fluorescence and differential interference contrast (DIC) imaging capabilities. The microscope condenser is adjusted to central Kohler illumination, with a condenser numerical aperture (NA) of 0.03 (condenser front aperture closed). For illumination, light from the microscope's Halogen lamp is filtered to yield an incident red beam, $\lambda = 630 \pm 5$ nm. The images were collected with a 60× oil immersion objective, NA = 1.4, and displayed on a CCD camera. In a Fourier plane conjugate to the back focal plane of the objective, a beam stop was placed in the center of an iris with variable diameter. As the inset in Figure 16.9 shows, the variable iris collects light scattered within a solid angle, bound by $2° < \theta < 10°$ for low NA, and $2° < \theta < 67°$ for high NA. Two sequential dark-field images are acquired at high and low NA by manually adjusting the diameter of the variable iris. The scatter ratio image is obtained by dividing the background subtracted high NA image by the background subtracted low NA image.

OSI was validated on sphere suspensions and live cells.[87] Figure 16.10A shows the mean pixel value and standard deviation of OSI images collected from polystyrene sphere suspensions and plotted against sphere diameter (open circles). The optical scatter images of the suspensions are displayed to the left of the graph. The solid line represents OSIR predictions as calculated from Mie theory for $m = 1.2$, and shows excellent agreement with experiment. For comparison, the dashed line shows the theoretical

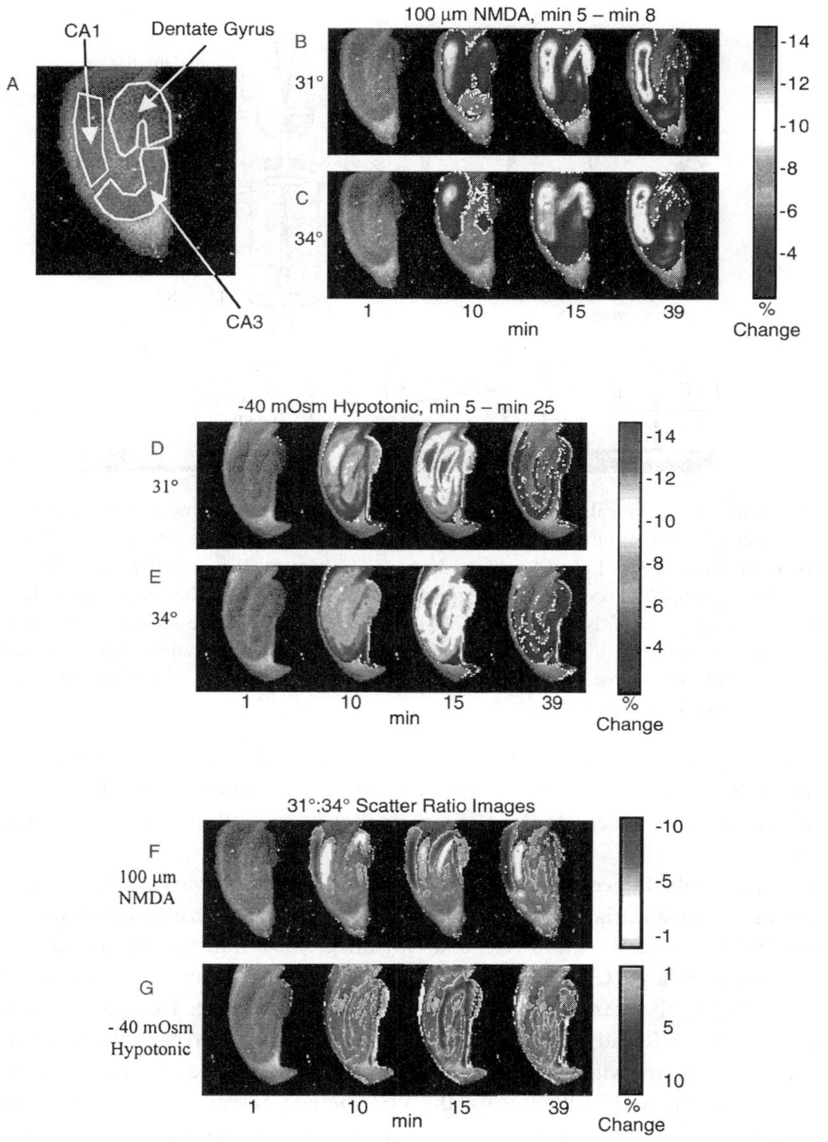


**FIGURE 16.8 (Color figure follows p. 28-30.)** Single-angle and dual-angle scatter images of hippocampal slices under osmotic stress and subjected to NMDA. In each case, the color bars indicate the percent change in image intensity. In regions of the slice where the change is less than ±2%, the gray scale images of the hippocampus are shown. (A) The scatter image of a hippocampal slice. The three main regions of the hippocampus are CA1, CA3, and the dentate gyrus. Also shown are single-angle scatter images at 31° (B) and 34° (C) of an NMDA-treated slice. In both (B) and (C) there is a large change in the CA1 region. There is also a significant change in the dentate gyrus that fades by minute 39 of the experiment. Single-angle scatter images at 31° (D) and 34° (E) are shown for hypotonic treatment. In both (D) and (E), a large change is indicative of cellular swelling in the CA1 region. There is also a significant change in the dendate gyrus. Dual-angle scatter ratio images are shown for NMDA (F) and hypotonic (G) treatments. The NMDA-treated slices in (F) undergo a relatively larger change in the dual-angle scatter ratio in CA1. In CA1 the scatter ratio change is negative, possibly indicating particle shrinkage. In (G), hypotonic treatment, the magnitude of the change in the scatter ratio is less in CA1. In addition, the overall location of the scatter change is more spread out than for NMDA treatment. In both (F) and (G) the white matter areas reveal a positive change in the scatter ratio. (From Johnson, L.J., Hanley, D.F., and Thakor, N.V., *J. Neurosci. Methods*, 98, 21, 2000. With permission.) Refer to color figure for accurate data rendering.

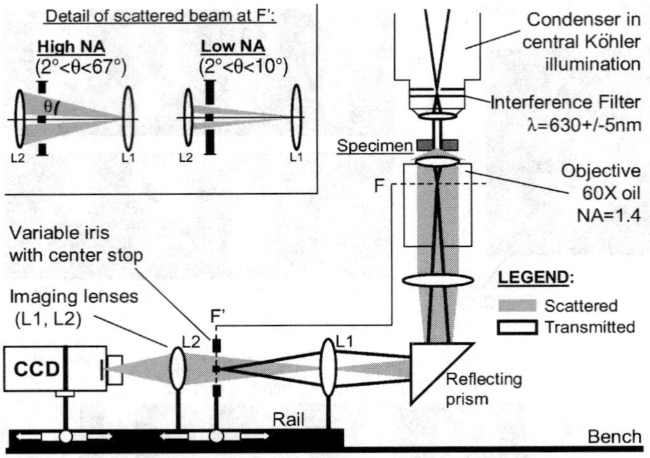


**FIGURE 16.9** Apparatus for optical scatter imaging (OSI) of cellular monolayers. A collimated beam, $\lambda = 630$ nm ($\pm 5$ nm), illuminates a sample mounted on the stage of an inverted microscope. The scattered light is collected by an oil immersion objective, NA = 1.4, and measured on a CCD camera. $F$ = objective's back focal plane. $F$ and $F'$ are conjugate Fourier planes. The scattered light (gray beam) is used to image the specimen on the CCD, while the transmitted light (black ray traces) is blocked at $F'$. In this setup, two images are acquired sequentially by manually varying the aperture diameter in the plane $F'$. The inset shows the scattered angles passed at the high and low numerical aperture (NA) settings of the variable diameter iris. (From Boustany, N.N., Kuo, S.C., and Thakor, N.V., *Opt. Lett.*, 26, 1063, 2001. With permission.)

prediction for $m = 1.06$. The OSIR parameter has the advantage of decreasing monotonically over a large range of sphere diameters from 0.2 to 1.5 μm. In addition, the OSIR is not significantly sensitive to refractive index changes for spheres in this size range and therefore reflects changes in morphology rather than composition.

The method was applied to cells that naturally contain scatterers of varying size, such as the nucleus, (4 to 15 μm) and mitochondria (0.5 to 2 μm). As expected, particle size variation was seen across the cell. Figure 16.10B (top panels) shows DIC and OSI images of a normal endothelial cell. The nucleus region (N) and the cytoplasm (C) containing the mitochondria are clearly differentiated in the OSI image despite a much larger pixel size than the accompanying DIC image. Due to the nonlinear inverse relationship between OSIR and particle diameter, in the OSI image, regions with small particles (C) appear brighter than regions with large particles (N). By monitoring the cells after treatment with 2 μm staurosporine (STS) (Figure 16.10B, $t > 0$), the OSI method revealed very early, apoptosis-induced subcellular changes, which were not apparent in conventional differential interference contrast images.[87] The OSI method was also used to quantify calcium-induced changes in mitochondrial shape resulting from the mitochondrial permeability transition.[88]

## 16.5 Summary and Conclusion

Light scattering techniques present simple and effective methods to detect subtle morphological changes, *in situ*, for particles with wavelength-scale dimensions, such as organelles or connective tissue fibers. The particles being probed need not be individually resolved and measured by traditional morphometric methods, thus avoiding a tedious process of image recognition, particle sizing and counting. Light scattering techniques complement other microscopic methods and, in contrast to electron microscopy, require no potentially damaging cell preparation procedures.

Unlike fluorescence labeling, which results in illuminating specific biochemical markers, methods based on light scattering do not require labeling. As a result, scattering is not biochemically specific,

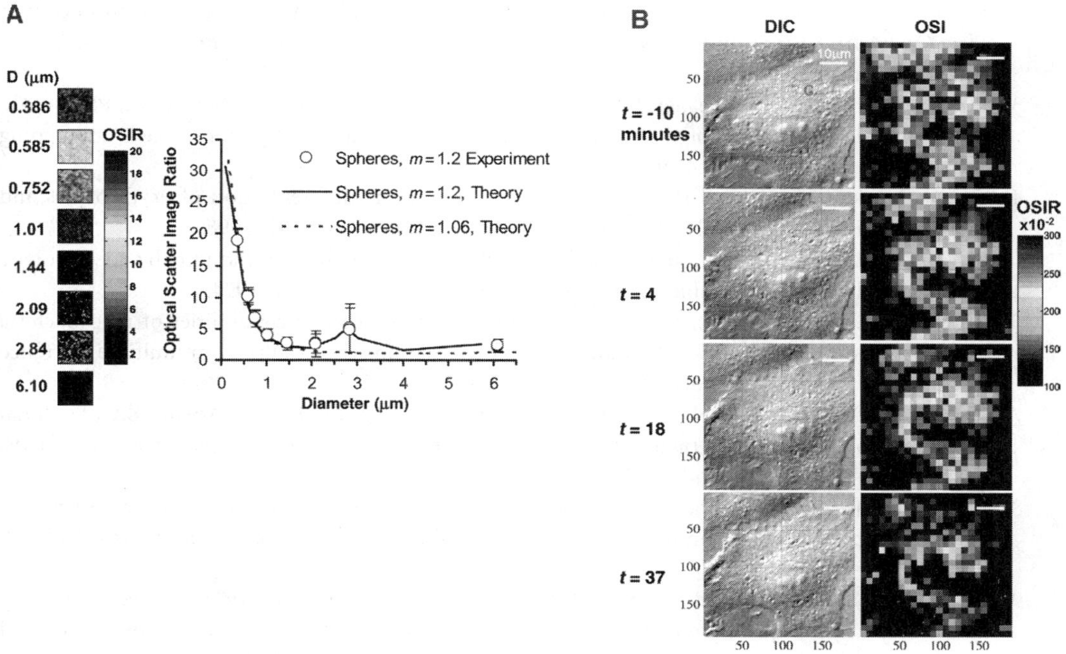


**FIGURE 16.10 (Color figure follows p. 28-30.)** (A) OSI images and measurement of the OSIR in aqueous suspensions of polystyrene spheres ($m$ = 1.2). The OSIR is a measure of wide to narrow angle scatter. Experimental data (open circles), and theoretical predictions (solid line, $m$ = 1.2; dashed line; $m$ = 1.06) are shown. The experimental data points show the mean pixel value and standard deviation in the scatter images displayed to the left of the graph. (From Boustany, N.N., Kuo, S.C., and Thakor, N.V., *Opt. Lett.*, 26, 1063, 2001. With permission.) (B) Representative cell undergoing apoptosis after treatment with 2 μ*M* staurosporine (STS). The cell was imaged in DIC (left panels) and OSI (right panels) at different time points. C = cytoplasm, N = nucleus. Images are displayed at times $t$ = –10, 4, 18, and 37 min from STS addition at $t$ = 0. The ratiometric scatter images show a decreasing scatter ratio within the cytoplasm (C). Refer to color figure for accurate data rendering.

and may originate from different tissue and cellular structures, often requiring further biochemical elucidation. Still, light scattering data can have important diagnostic value despite having limited specificity. One of the strengths of light scattering techniques is that they can reveal the presence of cellular and subcellular *morphological* dynamics noninvasively. Morphological information complements biochemical data, and could be particularly valuable if the biochemical events and time sequence underlying a given biological behavior are not yet known to allow specific biochemical manipulation. Moreover, under certain conditions in which only a few parameters are known to vary, scattering data can be used to automate cell differentiation and sorting efficiently during high-throughput cell analysis, as in flow cytometry.

Combining theory and experiment, various researchers have been striving to provide novel approaches to solve the "inverse problem" and infer, from a limited light scattering data set, the morphologic and optical properties of cells and organelles. These studies are invaluable in transforming hitherto phenomenological results into data that could be used to design optical instruments with a high impact on biology and medicine.

## References

1. Darynkiewicz, Z., Juan, G., Li, X., Goreczyca, W., Murakami, T., and Traganos, F., Cytometry in cell necrobiology: analysis of apoptosis and accidental cell death, *Cytometry*, 27, 1, 1997.

2. Salzman, G.C., Sigham, S.B., Johnston, R.G., and Bohren, C.F., Light scattering and cytometry, in *Flow Cytometry Sorting*, 2nd ed., Melamed, M.R., Lindmo, T., and Mendelsohn, M.L., Eds., Wiley-Liss, New York, 1990, p. 81.
3. Lizard, G., Fournel, S., Genestier, L., Dhedin, N., Chaput, C., Flacher, M., Mutin, M., Panaye, G., and Revillard, J.-P., Kinetics of plasma membrane and mitochondrial alterations in cells undergoing apoptosis, *Cytometry*, 21, 275, 1995.
4. Ost, V., Neukammer, J., and Rinneberg, H., Flow cytometric differentiation of erythrocytes and leukocytes in dilute whole blood by light scattering, *Cytometry*, 32, 191, 1998.
5. Weston, K.M., Alsalami, M., and Raison, R., Cell membrane changes induced by the cytolytic peptide, mellitin, are detectable by 90° laser scatter, *Cytometry*, 15, 141, 1994.
6. Conville, P.S., Witebsky, F.G., and MacLowry, J.D., Antimicrobial susceptibilities of mycobacteria as determined by differential light scattering and correlation with results from multiple reference laboratories, *J. Clin. Microbiol.*, 32, 1554, 1994.
7. Anderson, A.M., Angyal, G.N., Weaver, C.M., Felkner, I.C., Wolf, W.R., and Worthy, B.E., Potential application of laser/microbe bioassay technology for determining water-soluble vitamins in foods, *J. AOAC Int.*, 76, 682, 1993.
8. Lavergne-Mazeau, F., Maftah, A., Cenatiempo, Y., and Julien, R., Linear correlation between bacterial overexpression of recombinant peptides and cell light scatter, *Appl. Environ. Microbiol.*, 62, 3042, 1996.
9. Smeraldi, C., Berardi, E., and Porro, D., Monitoring of peroxisome induction and degradation by flow cytometric analysis of *hansenula polymorpha* cells grown in methanol and glucose media: cell volume refractive index and FITC retention, *Microbiology*, 140, 3161, 1994.
10. Hubbell, J.A., Pohl, P.I., and Wagner, W.R., The use of laser light scattering and controlled shear in platelet aggregometry, *Thrombosis Haemostasis*, 65, 601, 1991.
11. Tohgi, H., Takashi, H., Watanabe, K., and Hiroyuki, K., Development of large platelet aggregates from small aggregates as determined by laser light scattering: effect of aggregant concentration and antiplatelet medication, *Thrombosis Haemostasis*, 75, 838, 1996.
12. Ozaki, Y., Satoh, K., Yatomi, Y., Yamamoto, T., and Shirasawa, Y., Detection of platelet aggregates with particle counting method using light scattering, *Anal. Biochem.*, 218, 284, 1994.
13. Tedeshi, H. and Harris, D., Some observations on the photometric estimation of mitochondrial volume, *Biochim. Biophys. Acta*, 28, 392, 1958.
14. Hunter, D.R. and Haworth, R.A., The $Ca^{2+}$-induced membrane transition in mitochondria, *Arch. Biochem. Biophys.*, 195, 453, 1979.
15. Bernardi, P., Vassanelli, S., Veronese, P., Colonna, R., Szabo, I., and Zoratti, M., Modulation of the mitochondrial permeability transition pore, *J. Biol. Chem.*, 267, 2934, 1992.
16. Kristal, B.S. and Dubinsky, J.M., Mitochondrial permeability transition in the central nervous system: induction by calcium cycling-dependent and -independent pathways, *J. Neurochem.*, 69, 524, 1997.
17. Drezek, R., Dunn, A., and Richards-Kortum, R., Light scattering from cells: finite-difference time-domain simulations and goniometric measurements, *Appl. Opt.*, 38, 3651, 1999.
18. Mourant, J.R., Freyer, J.P., Hielscher, A.H., Eick, A.A., Chen, D., and Johnson, T.M., Mechanisms of light scattering from biological cells relevant to noninvasive optical-tissue diagnostics, *Appl. Opt.*, 37, 3586, 1998.
19. Van de Hulst, H.C., *Light Scattering by Small Particles*, Dover, New York, 1981.
20. Bohren, C.F. and Huffman, D.R., *Absorption and Scattering of Light by Small Particles*, John Wiley & Sons, New York, 1983.
21. Cheong, W.F., Prahl, S.A., and.Welch, A.J., A review of the optical properties of biological tissues, *IEEE J. Quantum Electron.*, 26, 2166, 1990.
22. Beauvoit, B., Evans, S.M., Jenkins, T.W., Miller, E.E., and Chance, B., Correlation between the light scattering and the mitochondrial content of normal tissues and transplantable rodent tumors, *Anal. Biochem.*, 226, 167, 1995.

23. Beuthan, J., Minet, O., Helfmann, J., Herrig, M., and Muller, G., The spatial variation of the refractive index in biological cells, *Phys. Med. Biol.*, 41, 369, 1996.
24. Koch, A.L., Some calculations on the turbidity of mitochondria and bacteria, *Biochim. Biophys. Acta*, 51, 429, 1961.
25. Wyatt, P.J., Differential light scattering: a physical method for identifying living bacterial cells, *Appl. Opt.*, 7, 1879, 1968.
26. Morris, S.J., Shultens, H.A., Hellweg, M.A., Striker, G., and Jovin, T.M., Dynamics of structural changes in biological particles from rapid light scattering measurements, *Appl. Opt.*, 18, 303, 1979.
27. Wyatt, P.J., Light scattering and the absolute characterization of macromolecules, *Anal. Chim. Acta*, 272, 1, 1993.
28. Sloot, P.M.A., Hoekstra, A.G., and Figdor, C.G., Osmotic response of lymphocytes measured by means of forward light scattering: theoretical considerations, *Cytometry*, 9, 636, 1988.
29. Evans, E., Comparison of the diffraction theory of image formation with the three-dimensional, first Born scattering approximation in lens systems, *Opt. Commun.*, 2, 317, 1970.
30. Turke, B., Seger, G., Achatz, M., and Seelen, W.v., Fourier optical approach to the extraction of morphological parameters from the diffraction pattern of biological cells, *Appl. Opt.*, 17, 2754, 1978.
31. Burger, D.E., Jett, J.H., and Mullaney, P.F., Extraction of morphological features from biological models and cells by Fourier analysis of static light scatter measurements, *Cytometry*, 2, 327, 1982.
32. Graaff, R., Aarnoudse, J.G., Zijp, J.R., Sloot, P.M.A., Mul, F.F.M.D., Grieve, J., and Kolink, M.H., Reduced light scattering properties for mixtures of spherical particles: a simple approximation derived form Mie calculation, *Appl. Opt.*, 31, 1370, 1992.
33. Ovryn, B. and Izen, S.H., Imaging of transparent spheres through a planar interface using a high-numerical-aperture optical microscope, *J. Opt. Soc. Am. A*, 17, 1202, 2000.
34. Schmitt, J.M. and Kumar, G., Optical scattering properties of soft tissue: a discrete particle model, *Appl. Opt.*, 37, 2788, 1998.
35. Zonios, G., Perelman, L.T., Backman, V., Manoharan, R., Fitzmaurice, M., Dam, J.V., and Feld, M.S., Diffuse reflectance spectroscopy of human adenomatous colon polyps *in vivo, Appl. Opt.*, 38, 6628, 1999.
36. Johnsen, S. and Widder, E.A., The physical basis of transparency in biological tissue: ultrastructure and the minimization of light scattering, *J. Theor. Biol.*, 199, 181, 1999.
37. Dunn, A. and Richards-Kortum, R., Three-dimensional computation of light scattering from cells, *IEEE J. Sel, Top. Quantum Electron.*, 2, 898, 1996.
38. Waterman, P.C., Matrix formulation of electro-magnetic scattering, *Proc. IEEE*, 53, 805, 1965.
39. Waterman, P.C., Symmetry, unitarity, and geometry in electro-magnetic scattering, *Phys. Rev. D*, 3, 825, 1971.
40. Yee, S.K., Numerical solution of initial boundary value problems involving Maxwell's equations in isotropic media, *IEEE Trans. Antennas Propag.*, 14, 302, 1966.
41. Barber, P.W. and Hill, S.C., *Light Scattering by Particles: Computational Methods*, World Scientific, Singapore, 1990.
42. Mishchenko, M.I., Travis, L.D., and Macke, A., T-matrix method and its applications, and Yang P. and Liou K.N., Finite difference time domain method for light scattering by nonspherical an inhomogeneous particles, in *Light Scattering by Nonspherical Particles: Theory Measurements and Applications*, Mishchenko, M.I., Hovenier, J.W., and Travis, L.D., Eds., Academic Press, San Diego, 2000, pp. 147–221.
43. Nilsson, A.M.K., Alsholm, P.L., Karlsson, A., and Andersson-Engles, S., T-matrix computations of light scattering by red blood cells, *Appl. Opt.*, 37, 2735, 1998.
44. Drezek, R., Dunn, A., and Richards-Kortum, R., A pulsed finite-difference time-domain (FDTD, method for calculating light scattering from biological cells over broad wavelength ranges, *Opt. Express*, 6, 148, 2000.

45. Bolt, R.A. and deMul, F.F.M., Goniometric instrument for light scattering measurment of biological tissues and phantoms, *Rev. Sci. Instrum.*, 73, 2211, 2002.
46. Bickel, W.S., Davidson, J.F., Huffman, D.R., and Kilkson, R., Application of polarization effects in light scattering: a new biophysical tool, *Proc. Nat. Acad. Sci. U.S.A.*, 73, 486, 1976.
47. Salzman, G.C., Crowell, J.M., Martin, J.C., Trujillo, T.T., Romero, A., Mullaney, P.F., and Labauve, P.M., Cell classification by laser light scattering: identification and separation of unstained leukocytes, *Acta Cytol. Praha*, 19, 374, 1975.
48. deGrooth, B.G., Terstappen, L.W.M.M., Puppels, G.J., and Greve, J., Light-scattering polarization measurements is a new parameter in flow cytometry, *Cytometry*, 8, 539, 1987.
49. Lehninger, A.L., Reversal of thyroxine-induced swelling of rat liver mitochondria by adenosine triphosphate, *J. Biol. Chem.*, 234, 2187, 1959.
50. Packer, L., Metabolic and structural states of mitochondria, *J. Biol. Chem.*, 235, 242, 1960.
51. Hackenbrock, C.R., Ultrastructural bases for metabolically linked mechanical acivity in mitochondria I, *J. Cell Biol.*, 30, 269, 1966.
52. Packer, L., Energy-linked low amplitude mitochondrial swelling, in *Methods in Enzymology*, Vol. 10, Estabrook, R.W. and Pullman, M.E., Eds., New York, Academic Press, 1967, p. 685.
53. Harris, R.A., Asbell, M.A., Asai, J., Jolly, W.W., and Green, D.E., The conformational basis of energy transduction in membrane systems.V. Measurement of configurational changes by light scattering, *Arch. Biochem. Biophys.*, 132, 545, 1969.
54. Hunter, D.R., Haworth, R.A., and Southward, J.H., Relationship between configuration, function, and permeability in calcium-treated mitochondria, *J. Biol. Chem.*, 251, 5069, 1976.
55. Halestrap, A.P., Regulation of mitochondrial metabolism through changes in matrix volume, *Biochem. Soc. Trans.*, 22, 522, 1994.
56. Territo, P.R., French, S.A., Dunleavy, M.C., Evans, F.J., and Balaban, R.S., Calcium activation of heart mitochondrial oxidative phosphorilation, *J. Biol. Chem.*, 276, 2586, 2001.
57. Hunter, F.E. Jr., and Smith, E.E., Measurement of mitochndrial swelling and shrinking — high amplitude, in *Methods in Enzymology*, Estabrook, R.W. and Pullman, M.E. Eds., Academic Press: New York, 1967, p. 689.
58. Bernardi, P., Modulation of the mitochondrial cyclosporin a-sensitive permeability transition pore by the proton electrochemical gradient, *J. Biol. Chem.*, 267, 8834, 1992.
59. Petronilli, V., Constantini, P., Scorrano, L., Colonna, R., Passamonti, S., and Bernardi, P., the voltage sensor of the mitochondrial permeability transition pore is tuned by the oxidation-reduction state of vicinal thiols, *J. Biol. Chem.*, 269, 16638, 1994.
60. Hoek, J.B., Farber, J.L., Thomas, A.P., and Wang, X., Calcium ion-dependent signaling and mitochondrial dysfunction: mitochondrial calcium uptake during hormonal stimulation in intact liver cells and its implication for the mitochondrial permeability transition, *Biochim. Biophys. Acta*, 1271, 93, 1995.
61. Constantini, P., Chernyak, B.V., Petronilli, V., and Bernardi, P., Modulation of the mitochondrial permeability transition pore by pyridine nucleotides and dithiol oxidation at two separate sites, *J. Biol. Chem.*, 271, 6746, 1996.
62. Scorrano, L., Petronilli, V., and Bernardi, P., On the voltage dependence of the mitochondrial permeability transition pore, *J. Biol. Chem.*, 272, 12295, 1997.
63. Pfeiffer, D.R., Kuo, T.H., and Chen, T.T., Some effect of $Ca^{2+}$, $Mg^{2+}$, and $Mn^{2+}$ on the ultrastructure, light scattering properties, and malic enzyme activity of adrenal cortex mitochondria, *Arch. Biochem. Biophys.*, 176, 556, 1976.
64. Petronilli, V., Cola, C., Massari, S., Colonna, R., and Bernardi, P., Physiological effectors modify voltage sensing by the cyclosporin a-sensitive permeability transition pore of mitochondria, *J. Biol. Chem.*, 268, 21939, 1993.
65. Vander-Heiden, M.G., Chandel, N.S., Williamson, E.K., Schumacker, P.T., and Thompson, C.B., Bcl-xL regulates the membrane potential and volume homeostasis of mitochondria, *Cell*, 91, 627, 1997.

66. Zamzami, N., Susin, S.A., Marchetti, P., Hirsh, T., Gomez-Monterrey, I., Castedo, M., and Kroemer, G., Mitochondrial control of apoptosis, *J. Exp. Med.*, 183, 1533, 1996.
67. Jurgensmeier, J.M., Xie, Z., Deveraux, Q., Ellerby, L., Bredesen, D., and Reed, J.C., Bax directly induces release of cytochrome *c* from isolated mitochondria, *Proc. Natl. Acad. Sci. U.S.A.*, 95, 4997, 1998.
68. Narita, M., Shimizu, S., Ito, T., Chittenden, T., Lutz, R.J., Matsuda, H., and Tsujimoto, Y., Bax interacts with the permeability transition pore to induce permeability transition and cytochrome c release in isolated mitochondria, *Proc. Natl. Acad. Sci. USA*, 95, 14681, 1998.
69. Finucane, D.M., Bossy-Wetzel, E., Waterhouse, N.J., Cotter, T.G., and Green, D.R., Bax-induced caspase activation and apoptosis via cytochrome *c* release from mitochondria is inhibitable by Bcl-xL, *J. Biol. Chem.*, 274, 2225, 1999.
70. Bryant, F.D., Latimer, P., and Seiber, B.A., Changes in total light scattering and absorption caused by changes in particle conformation — a test of theory, *Arch. Biochem. Biophys.*, 135, 109, 1969.
71. Latimer, P. and Pyle, B.E., Light scattering at various angles, theoretical predictions of the effects of particle volume change, *Biophys. J.*, 12, 764, 1972.
72. Knight, V.A., Wiggins, P.M., Harvey, J.D., and O'Brien, J.A., The relationship between the size of mitochondria and the intensity of light that they scatter in different states, *Biochim. Biophys. Acta*, 637, 146, 1981.
73. Mourant, J.R., Johnson, T.M., and Freyer, J.P., Characterizing mammalian cells and cell phantoms by polarized backscattering fiber-optic measurements, *Appl. Opt.*, 40, 5114, 2001.
74. Valentine, M.T., Popp, A.K., Weitz, D.A., and Kaplan, P.D., Microscope-based static light-scattering instrument, *Opt. Lett.*, 26, 890, 2001.
75. Schiffer, Z., Ashkenazy, Y., Tirosh, R., and Deutsch, M., Fourier analysis of light scattered by elongated scatterers, *Appl. Opt.*, 38, 3626, 1999.
76. Shiffer, Z., Zurgil, N., Shafran, Y., and Deutsch, M., Analysis of laser scattering pattern as an early measure of apoptosis, *Biochem. Biophys. Res. Commun.*, 289, 1320, 2001.
77. Kovach, I.S. and Athanasiou, K.A., Small-angle HeNe laser light scatter and the compressive modulus of articular cartilage, *J. Orthop. Res.*, 15, 437, 1997.
78. Schmitt, J.M. and Kumar, G., Turbulent nature of refractive index variations in biological tissue, *Opt. Lett.*, 21, 1310, 1996.
79. Wax, A., Yang, C., Backman, V., Badizadegan, K., Boone, C.W., Dasari, R.R., and Feld, M.S., Cellular organization and substructure measured using angle-resolved low-coherence interferometry, *Biophys. J.*, 82, 2256, 2002.
80. Aitken, P.G., Fayuk, D., Somjen, G.G., and Turner, D.A., Use of intrinsic optical signals to monitor physiological changes in brain tissue slices, *Methods: Companion Methods Enzymol.*, 18, 91, 1999.
81. Andrew, R.D., Lobinowich, M.E., and Osehobo, E.P., Evidence against volume regulation by cortical brain cells during acute osmotic stress, *Exp. Neurol.*, 143, 300, 1997.
82. Andrew, R.D., Adams, J.R., and Polischuk, T.M., Imaging NMDA- and kainate-induced intrinsic optical signals from the hippocampal slice, *J. Neurophys.*, 76, 2707, 1996.
83. Bahar, S., Fayuk, D., Somjen, G.G., Aitken, P.G., and Turner, D.A., Mitochondrial and intrinsic optical signals imaged during hypoxia and spreading depression in rat hippocampal slices, *J. Neurophys.*, 84, 311, 2000.
84. Rector, D.M., Poe, G.R., Kristensen, M.P., and Harper, R.M., Light scattering changes follow evoked potentials from hippocampal schaeffer collateral stimulation, *J. Neurophys.*, 78, 1707, 1997.
85. Johnson, L.J., Hanley, D.F., and Thakor, N.V., Optical light scatter imaging of cellular and subcellular morphology changes in stressed rat hippocampal slices, *J. Neurosci. Methods*, 98, 21, 2000.
86. Andrew, R.D., Jarvis, C.R., and Obeidat, A.S., Potential sources of intrinsic optical signals imaged in live brain slices, *Methods Companion Methods Enzymol.*, 18, 185, 1999.
87. Boustany, N.N., Kuo, S.C., and Thakor, N.V., Optical scatter imaging: subcellular morphometry *in situ* with Fourier filtering, *Opt. Lett.*, 26, 1063, 2001.
88. Boustany, N.N., Drezek, R., and Thakor, N.V., Calcium-induced alterations in mitochondrial morphology quantified *in situ* with optical scatter imaging, *Biophys. J.*, 88, 1691, 2002.

# 17

# Thermal Imaging for Biological and Medical Diagnostics

Jay P. Gore
*Purdue University*
*West Lafayette, Indiana*

Lisa X. Xu
*Purdue University*
*West Lafayette, Indiana*



## 17.1 Introduction

The change in local temperature of blood, tissue, and skin in biological systems is determined by the difference in the thermal energies received and lost by conduction transfer and advection, as well as by the thermal energy released by chemical reaction processes such as metabolism. The use of temperature (measured at a convenient location such as the ear, armpit, or mouth), as an indicator of health has been in routine use for a very long time. However, even in such use, it is widely recognized that different healthy individuals can have different temperatures. In fact, the same healthy individual can have different temperatures at different times, based on the local environmental conditions, blood perfusion rates as controlled by the nervous system, and metabolic rates as controlled by a variety of complex factors.

In addition, the single temperature measured at a chosen location does not represent the temperatures of the different parts of the body even on the skin surface. The temperature of a human being typically decreases from the core to the skin by 3 to 5°C to support the heat dissipation to the surroundings. The transport processes involving conduction and advection (blood perfusion) within the body and the thermal boundary conditions at the skin determine the local skin temperature. In addition, the sweat glands in the human skin respond to different stimuli and the surface temperature is affected by the evaporation of water at the surface.

Diagnostic techniques using single-point temperature measurements have long benefited patients. The theoretical possibility exists that temperature at multiple points, if properly interpreted, can provide information about conditions affecting blood circulation, local metabolism, sweat gland malfunction, inflammation and healing, and energy imbalances created by hyperactivity of glands such as the thyroid. Temperatures of internal surfaces such as the colon, lung, and arteries could also be indicators of the state of health. Measuring such internal surface temperatures using endoscopes is a possibility.

Thermal imaging involves measurements of surface temperature using an array of infrared sensors installed in an infrared camera. This imaging allows the simultaneous measurement of temperatures of

0-8493-1116-0/03/$0.00+$1.50

multiple points on the skin and is also a reference for surrounding temperature. The reference measurement could be subtracted from local measurements to provide an improved estimate of the temperature. In addition, measurement of differences in temperature between two points or in different regions of the image may be better indicators of the state of health. Motivated by this, thermal imaging using infrared cameras for medical diagnostics was initiated in the early 1970s. Over the subsequent three decades the technique has received mixed reviews and produced mixed results. Although the fundamentals of the thermal imaging process are sound, the relationship to a particular anomaly is not necessarily specific; therefore, the technique has greater promise as an adjunct diagnostic tool. The technique could also become valuable for monitoring the changes in skin temperature pattern over a specified period under highly controlled conditions.

Thermal imaging was considered for early diagnosis of breast cancer based on the theoretical conjecture that a rapidly growing tumor has high metabolism and blood perfusion rates leading to an increase in the local tumor temperature. If the tumor were to be located in a region close to the skin, then the local high temperature would also appear as a high temperature spot on the skin. Pattern recognition techniques, heuristic diagnosis, and left-breast-to-right-breast temperature pattern comparisons were used to interpret analog images visually. The results of the trials were highly inconclusive and controversial.

The controversy continues today with those opposed to the technique still basing their opinions on past failures. They have not changed their opinions even though the trials were in an era before the personal computer and advances in image processing techniques that can increase measurement accuracy and yield better computerized algorithms and objective tests for specific maladies. In addition, they also discount the possible use of thermal imaging as an adjunct functional diagnosis tool, for example, to determine if a positive mammography result could be combined with a high metabolic rate indication to separate malignancies, thereby reducing the number of false biopsies. Tests conducted for use as stand-alone diagnostics are generally not indicative of the value of a method for adjunct use.

Temperature patterns on the skin are the result of biological heat and mass transfer, metabolic processes in response to specific surrounding conditions, and mental, neurological and hormonal states of the subjects. Therefore, interpretation of single thermal images must not be based on use of arbitrary pattern recognition methods because such a diagnosis can be highly nonspecific and misleading. However, interpretation using proper accounting of the biological processes that determine the skin surface temperature, as well as multiple images taken during annual examinations and of dynamic thermal response to stimuli, can yield useful results, particularly when used as a function indicator technique in conjunction with a structure indication technique such as x-ray or ultrasound imaging. Therefore, we next review the fundamentals of infrared radiation and thermal imaging, followed by abstracts of some recent work in thermal imaging for biology and diagnostics. A table of available infrared cameras is also provided, followed by two examples of the use of the bioheat transfer equation for the interpretation of thermal images.

## 17.2 Infrared Radiation and Thermal Imaging

Thermal imaging relies on sensing the infrared radiation emitted by all objects above absolute zero temperature. All objects emit photons as a result of transitions from a high-energy to a low-energy state. In solids, such transitions lead to a continuous distribution of energy between different wavelengths according to the Planck distribution. Objects at a relatively high temperature, such as carbon particles in a flame or a hot iron rod, emit visible light, while the emission from objects closer to normal room temperature is at much longer wavelengths. At the normal temperature of the human body, the peak of the Planck function occurs in the mid-infrared region between 9- and 10-μm wavelengths. This radiation is not visible to the human eye but, in sufficient intensity, can be felt by the human skin, one function of which is a low-sensitivity infrared array detector.

The Planck function is exponentially nonlinear in temperature. This means that the lower-temperature objects emit orders of magnitude less energy than do higher-temperature objects. Therefore, detection of infrared energy accurately is a challenging task. In addition, lower-temperature objects and their surroundings

**TABLE 17.1** Infrared Camera Manufacturers, Web Sites, Camera Models, and Array Sizes

| Company | WWW Addresses | Camera | Array Size |
|---|---|---|---|
| FLIR | http://www.flir.com | ThermaCAM ® E1 | 160*120 |
| | | ThermaCAM ® SC300 | 160*120 |
| | | ThermaCAM ® SC3000 | 320*240 |
| Sierra Pacific | http://www.x20.org | PD300 IR | 320*244 |
| | | IR 747 | 320*240 |
| CMC Elec. | http://www.cinele.com/ | TVS 8500 | 256*256 |
| Indigo Sys. | http://www.indigosystems.com/index.html | Mirlin®-Mid | 320*256 |
| | | Mirlin®-Uncooled | 320*240 |
| | | Omega | 160*120 |
| | | Alpha NIR | 320*256 |
| | | TVS-620 | 320*240 |
| | | Phoenix-Mid | 640*512 |
| | | Phoenix-Long | 640*512 |
| Infrared Sys. | http://www.infraredsys.com/ | Raytheon IR500D | 320*240 |
| | | Raytheon IRPro4 | 320*240 |
| | | Mikron 5104 | 320*240 |
| | | Mikron 7102 | 320*240 |
| | | Monroe Scientific | 320*240 |
| Cantronic Sys. Inc. | http://www.cantronic.com | IR805 | 120*120 |
| | | IR860 | 320*240 |
| Ircon | http://www.ircon.com | DigiCam-IR | 120*120 |
| | 800–323–7660 | Stinger | 320*240 |
| Land Instruments International | http://www.landinst.com | Cyclops PPM+ | 120*120 |
| | | FTI 6 | |
| Mikron | http://www.mikroninst.com | 7200 | 320*240 |
| | | 7515 | 320*240 |
| | | 5104 | 255*223 |
| | | 5102 | 255*223 |
| | | 1100 | 207*344 |
| | | MHS500 | 255*223 |

emit comparable amounts of infrared energy, which leads to difficulties in measurements. Often the signal-to-noise ratios (SNRs) are of comparable magnitude, requiring specialized background correction instrumentation, mode-locked signal processing techniques, and careful analysis of the resulting data.

The technology of infrared array detectors, associated electronics, image processing, and noise reduction has significantly improved over the last 10 years. The accuracy with which temperature and temperature changes can be measured has reached $10^{-3}$ K. This is better by a factor of 100 compared to the techniques used in past studies — a fact missed by many readers of the thermal imaging literature.

The infrared or thermal imaging cameras available commercially are listed in Table 17.1. The most relevant attributes of these cameras are listed in Tables 17.1 through 17.4. The most important parameters for biological applications are the range and accuracy. The array size and the pixel size determine the spatial resolution possible for the measurements. The wavelength sensitivity is important, even in thermal imaging, because the object imaged must have a high enough emissivity in the spectral sensitivity range of the camera. The cameras with highest sensitivity near 9-μm wavelength are the most popular for biological applications because of the Planck function's peak near this wavelength. Most biological solids also have a significant emissivity in this region, yielding a high SNR.

The cameras must be calibrated routinely using a blackbody standard source; however, the sensitivity of the detectors has long surpassed many blackbody standards used in general purpose laboratories. In addition to the blackbody calibration standard, lenses, mirrors, filters, and prisms made from infrared transmitting materials form the core tool set of a thermal imaging laboratory. Many laboratories work in the visible and the infrared portion of the light spectrum; most visible optics do not transmit infrared but most infrared optics transmit visible. Therefore, care must be taken to separate and label the optical components properly.

**TABLE 17.2** IR Camera Models, Spatial Resolution, Spectral Range, and Detector Materials

| Camera | Spatial Resolution (mrad) | Spectral Range (μm) | Detector Material |
|---|---|---|---|
| ThermaCAM E1 | 2.6 | 7.5 to 13 | FPA uncooled microbolometer |
| ThermaCAM SC300 | 2.6 | 7.5 to 13 | FPA uncooled microbolometer |
| ThermaCAM SC3000 | 1.1 | 8 to 9 | GaAs |
| PD300 IR | 1.0 | 3.6 to 5 | PtSi |
| IR 747 | 1.3 | 7.5 to 13 | Uncooled microbolometer |
| TVS 8500 | 1.0 | 2.5 to 5.0 | InSb |
| Mirlin®-Mid | | 1 to 5.4 | InSb |
| Mirlin®-Uncooled | | 7.5 to 13.5 | Uncooled microbolometer |
| Omega | 1.7 | 7.5 to 13.5 | Uncooled microbolometer |
| Alpha NIR | 0.6 | 9 to 17 | InGaAs |
| TVS-620 | 0.7 | 8 to 14 | Uncooled microbolometer |
| Phoenix-Mid | 0.25 | 2 to 5 | InSb |
| Phoenix-Long | 0.25 | 8 to 9.2 | GaAs (QWIP) |
| Raytheon IR500D | | 7 to 14 | FPA uncooled BST |
| Raytheon IRPro4 | | 7 to 14 | Uncooled BST |
| Mikron 5104 | | 3 to 5.2 | Radiometric TE MCT |
| Mikron 7102 | | 8 to 14 | Radiometric uncooled microbolometer |
| Monroe Scientific | | 2 to 14 | Uncooled BST |
| IR805 | | 8 to 12 | Thermoelectric detector |
| IR860 | | 8 to 14 | Radiometric uncooled microbolometer |
| DigiCam-IR | 0.25 | 8 to 12 | Thermopile (TE) detector |
| Stinger | Not available | 8 to 14 | FPA uncooled |
| Cyclops PPM+ | 0.98 | 7 to 14 | FPA uncooled |
| FTI 6 | | | |
| 7200 | 1.58 | 8 to 14 | FPA uncooled microbolometer |
| 7515 | 1.58 | 8 to 14 | FPA uncooled microbolometer |
| 5104 | 2.00 | 3 to 5.2 | Mercury cadmium telluride |
| 5102 | 1.50 | 8 to 12 | Mercury cadmium telluride |
| 1100 | 1.50 | 8 to 13 | Mercury cadmium telluride |
| MHS500 | | | Mercury cadmium telluride |

## 17.3 Applications of Infrared Thermal Imaging

Many applications of infrared thermal imaging have been reported in the literature.[1-5] The basic measurement involves skin temperature distribution resulting from a variety of internal and external conditions affecting the circulation and metabolic processes. Therefore, there is an existing nonspecificity in the process, which must be recognized and addressed by improving the analytical tools and recognizing that, even with the most improved tools, a fundamentally nonspecific diagnostic technique such as thermal imaging can only be used as a powerful adjunct tool. Such recognition will avoid the controversy and confusion often surrounding this issue. More scientific work is needed to distinguish the differences between enhanced blood flow because of tumor and because of stress of a polygraph test. The present problem is that proponents of thermal imaging claim that the technique can be used as a diagnostic tool in both these scenarios, while opponents of the technique claim that a health parameter as basic as temperature and its distribution on different parts of the body has no diagnostic value. Both positions are extreme because the former ignores the nonspecificity of the technique while the later ignores the fact that, along with pulse and appearance, temperature is a health diagnostic modality proven over thousands of years.

Some of the recent literature involving use of infrared thermal imaging as an experimental tool in biology and health studies is summarized briefly in the next paragraph. This summary points to the broad applications as well as the nonspecificity inherent in this measurement modality.

**TABLE 17.3** Infrared Camera Models, Temperature Range, Sensitivity, Framing Rate, and Bits

| Camera | Temperature Range | Sensitivity | Framing Rate (Hz) | Bits |
|---|---|---|---|---|
| ThermaCAM E1 | –20 to 250°C | 120 mK | 60 | 14 |
| ThermaCAM SC300 | –40 to 500°C | 100 mK at 30°C | 50 to 60 | 14 |
| ThermaCAM SC3000 | –20 to 2000°C | 20 mK at 30°C | 50 to 60 | 14 |
| PD300 IR | –10 to 1500°C | 100 mK at 30°C | 60 | 12 |
| IR 747 | –15 to 45°C | 100 mK at 30°C | | |
| TVS 8500 | –40 to 2000°C | 20 mK at 30°C | 120 | 14 |
| Mirlin®-Mid | 0 to 2000°C | 25 mK | 60 | 12 |
| Mirlin®-Uncooled | 0 to 1000°C | 100 mK | 60 | 12 |
| Omega | –40 to 400°C | 40 mK | 30 | 14 |
| Alpha NIR | | | 30 | 12 |
| TVS-620 | –20 to 900°C | 150 mK at 30°C | 30 | 12 |
| Phoenix-Mid | | 25 mK | 30 | 14 |
| Phoenix-Long | | 30 mK | 30 | 14 |
| Raytheon IR500D | 0 to 300°C | 100 mK | 60 | 12 |
| Raytheon IRPro4 | 0 to 500°C | 100 mK | 30 | 12 |
| Mikron 5104 | –10 to 1500°C | 100 mK at 30°C | 60 | 12 |
| Mikron 7102 | –40 to 1500°C | 200 mK at 30°C | 60 | 14 |
| Monroe Scientific | –18 to 538°C | 100 mK at 25°C | 60 | 10 |
| IR805 | 0 to 350°C | 200 mK | 30 | 16 |
| IR860 | –10 to 1000°C | 100 mK at 30°C | 30 | Not available |
| DigiCam-IR | –10 to 1100°C | 350 mK at 30°C | 60 | 16 |
| Stinger | 0 to 500°C | 100 mK at 25°C | 30 | 12 |
| Cyclops PPM+ | –15 to 300°C | 100 mK at 30°C | 60 | 12 |
| FTI 6 | –20 to 2000°C | | | |
| 7200 | –40 to 120°C | 80 mK at 30°C | 60 | 14 |
| 7515 | –40 to 500°C | 80 mK at 30°C | 60 | 14 |
| 5104 | –10 to 800°C | 100 mK at 30°C | 22 | 12 |
| 5102 | –10 to 200°C | 30 mK at 30°C | 22 | 12 |
| 1100 | –50 to 2000°C | 100 mK at 30°C | 4 | 13 |
| MHS500 | –0 to 70°C | 100 mK | 2 | 14 |

Barnett et al.[6] used thermal imaging to conduct a preliminary investigation of the human gingiva and suggested that, with cooling and rewarming, it can be used as an indicator of inflammatory state. Chan et al.[7] state that thermal imaging is useful for identifying the health status of the thyroid gland. Whole-body thermal images are adequate indicators of ectodermal displasia according to Clark and co-workers,[8] and Dickey et al.[9] report that infrared thermography can be used to measure the depth of burns in burn victims. Thermal imaging systems in the management of pain were used by Hasegawa et al.[10] Stein and co-workers[11] demonstrated the use of thermal imaging in monitoring surgical tendon repair in veterinary medicine.

One of the oldest, most controversial, but possibly most promising applications of thermal imaging involves breast cancer detection.[12] The literature in this area is briefly reviewed in Section 17.3.1. Villringer and Dirnagl[13] used infrared thermal imaging to measure brain activity. Applications of thermal imaging to internal surfaces have begun appearing, with papers on the use of infrared fiber optic catheter for thermal imaging of atherosclerotic plaque by Naghavi-Morteza et al.[14]

Thermal imaging is being used in the two basic fields of biology, zoology and botany. In addition to human health studies, the applications are wide and therefore care is required in claims of specificity like those in health studies. Thermal imaging has been used in population studies of burrowing mammals by Hubbs et al.[15] Fuller and Wisnewski[16] use thermal imaging to study freezing in plants, and Jones[17] uses it to measure stomatal conductance of plants.

The most recent claims for thermal imaging involve its use in detecting anxiety by observing changes in facial temperature patterns resulting from neurological responses of the circulatory system.[18,19]

TABLE 17.4 Infrared Camera Models, Cooling Method, Cooling Temperature, and Space Required

| Camera | Cooling Method | Cooling Temperature (K) | Size (mm) |
|---|---|---|---|
| ThermaCAM E1 | Uncooled | — | 265*80*105 |
| ThermaCAM SC300 | Uncooled | — | 212*121*127 |
| ThermaCAM SC3000 | Stirling | 77 | 220*135*130 |
| PD300 IR | Stirling | 77 | 222*127*140 |
| IR 747 | Uncooled | — | 203*121*112 |
| TVS 8500 | Stirling | 77 | 200*250*120 |
| Mirlin®-Mid | Stirling | 77 | 140*127*250 |
| Mirlin®-Uncooled | Uncooled | — | 102*115*203 |
| Omega | Uncooled | — | 35*37*48 |
| Alpha NIR | Passive conduction and convection | — | 53*64*95 |
| TVS-620 | Uncooled | — | 115*217*142 |
| Phoenix-Mid | Stirling | 77 | 153*153*127 |
| Phoenix-Long | Stirling | 77 | 533*508*407 |
| Raytheon IR500D | Uncooled | — | 254*140*102 |
| Raytheon IRPro4 | Uncooled | — | 254*140*102 |
| Mikron 5104 | Thermoelectrically cooled | | 203*90*220 |
| Mikron 7102 | Uncooled | — | 95*110*170 |
| Monroe Scientific | Uncooled | — | 140*114*114 |
| IR805 | Uncooled | — | 240*100*130 |
| IR860 | Uncooled | — | 177*110*142 |
| DigiCam-IR | Uncooled | — | 240*100*130 |
| Stinger | Uncooled | — | 344*127*98 |
| Cyclops PPM+ | Uncooled | — | 150*245*275 |
| FTI 6 | | | |
| 7200 | Uncooled | — | 97*109*170 |
| 7515 | Uncooled | — | 97*109*170 |
| 5104 | | | 203*89*221 |
| 5102 | Stirling | | 198*98*235 |
| 1100 | Liquid nitrogen | 77 | 135*159*228 |
| MHS500 | Thermoelectrically cooled | | 221*147*270 |

Although the publicity value of such claims is high, care must be exercised in the interpretation of results based on the nonspecificity of the primary measure.

The following section describes an example of using temperature measurements in conjunction with a mathematical model. The mathematical model is then used to generate expected thermal images, which show their effectiveness in indicating high blood-perfusion and metabolic rates typically associated with a tumor. Very few studies of this nature aimed at evaluating the scientific basis for thermal imaging as a breast cancer diagnostic have been reported. Based on the results of this study, it is established that computational models of the bioheat transfer process can render thermal imaging a viable adjunct tool.

## 17.3.1 Calculations of Temperature Profiles in a Female Breast with and without a Tumor

### 17.3.1.1 Introduction

Breast cancer is one of the leading causes of death from cancer in women in the United States. One out of eight women suffers from breast cancer during her lifetime. The American Cancer Society estimated that in 2001, approximately 192,000 women in the United States would be diagnosed with breast cancer and approximately 40,600 would die as a result of this disease.[20] Early detection is considered to be the best defense against breast cancer.

Tumors grow by signaling for and receiving higher arterial blood supply, which supports the higher metabolic rates required for rapid cell division. Many diagnostic and therapeutic technologies based on these fundamentals are being considered. The differences in the energy consumption of normal and cancerous tissue lead to small but detectable local temperature changes.[21] For breast tumors near the skin surface the local temperature changes at the tumor location can be large enough to cause a local perturbation in temperature on the patient's skin.

Gautherie[22] examined 147 patients with malignant breast cancers and measured local temperatures and thermal conductivity using sterile fine-needle thermoelectric probes. He used a model based on electrical analogy to investigate correlations between thermal and geometrical parameters that appeared during heat interactions between the tumor and its surrounding tissue. The tumor metabolic heat generation rate was inferred from the model. The differences in effective thermal conductivity of healthy and cancerous tissues *in vivo* and *in vitro* as postoperative specimens were described as a function of local blood perfusion rate.

Osman et al.[23] developed a mathematical model of the three-dimensional temperature distribution in women's breasts using a finite element method. Based on their literature survey, the blood flow to the tumor was considered to be two- to threefold that of the normal tissue. They correlated the tumor metabolic heat generation rate with the doubling time of tumor volume as measured by Gautherie et al.[24] They modeled the tumor as a point heat source with the heat generation equal to the total metabolic heat production of the neoplastic tissue. Surface temperature distributions were plotted for the malignant breasts; however, the temperature distributions inside the breasts were not discussed and compared to the measurements of Gautherie.[22]

Infrared images of skin temperature can be used as an adjunct tool for detection combined with mammography in cases in which significant thermal expression of the tumor occurs on the skin surface. Some past studies have shown significant benefits of such a detection scheme, increasing the sensitivity rate from 85 to 95%.[25] However, many other studies were inconclusive, primarily because of incorrect measurements and qualitative interpretation. As a result, thermography is currently not used or recommended as an early detection technique. The accuracy of interpretation can be significantly enhanced if solutions of the bioheat equation are used to relate the infrared images obtained with modern high-accuracy cameras to the functional changes inside the breast. Such work is being conducted primarily outside the United States.[25–29]

#### 17.3.1.2 Bioheat Transfer Equation

The steady-state form of the bioheat transfer equation proposed by Pennes[21] is

$$\nabla \cdot k\nabla T + \rho_b \cdot c_b \cdot \omega_b \cdot (T_a - T) + \dot{q}_m = 0 \tag{17.1}$$

where $k$ denotes thermal conductivity of tissue, $\rho_b$, $c_b$ are the density and the specific heat of the blood; $\omega_b$ (ml/s/ml) is the blood perfusion rate, $\dot{q}_m$ is the metabolic heat generation rate (W/m$^3$), $T_a$ is the arterial blood temperature (K), and $T$ is the local temperature of the breast tissue. Equation 17.1 gives the quantitative relationship of the heat transfer characteristics in human tissue and includes the effects of blood perfusion ($\omega_b$) and metabolic heat generation ( $\dot{q}_m$ ). The temperature of the arterial blood is approximated to be the core temperature of the body.

In the bioheat equation proposed by Weinbaum et al.,[30] the expression derived for the tensor conductivity of the tissue is a function of the local vascular geometry and flow velocity in thermally significant small vessels. The solution of this equation gives rise to an effective thermal conductivity to account for the enhanced heat transfer due to incomplete countercurrent heat exchange between paired vessels. Structural properties of the breast in terms of size and distribution of blood vessels are necessary, but not available, for the solution of this equation; therefore, Equation 17.1 was used in the present simulations.

#### 17.3.1.3 Mathematical Model

A hemisphere of radius 9 cm was considered to match the patient in Gautherie's study.[22] The normal breast was approximated to be composed of a single layer with averaged properties. The breast with tumor included a sphere of radius 1.1 cm with its center located at 2.1 cm beneath the surface. The tumor was assigned a higher blood perfusion and metabolic heat generation rate compared to the

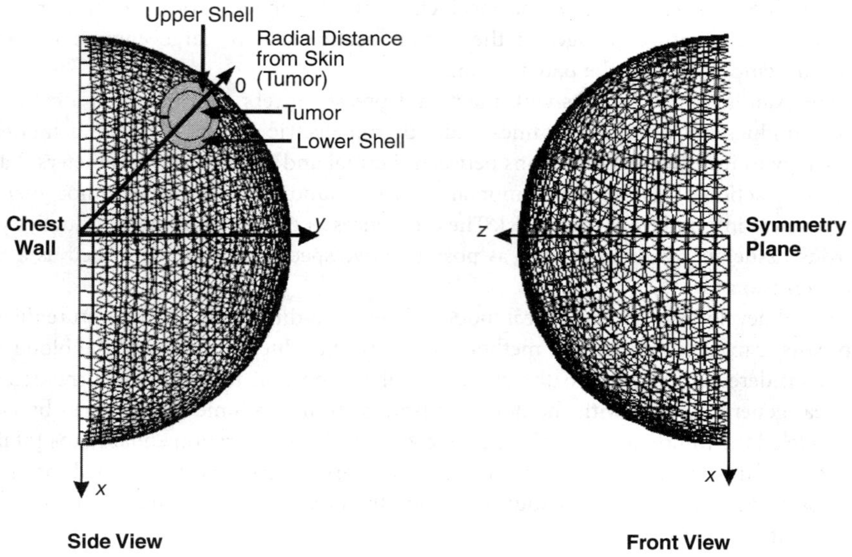


**FIGURE 17.1** Numerical grid for bioheat transfer calculations in a breast model for the generation of a computational thermal image.

corresponding quantities for the healthy tissue. Gautherie observed regions of increased metabolic activity around the tumor. Because metabolic heating depends on the supply of metabolites from the blood perfusion, the two quantities change in an interdependent monotonic manner. The region of abnormal metabolic rates around the tumor was modeled as 0.7-cm-thick lower and upper hemispherical shells of different properties around the tumor. The two different shells are used to represent the differences in the upper and lower parts of the breast that surround the tumor.

A convective boundary condition is specified at the skin surface. Natural convection for a sphere (Nusselt number ≅ 2) gives a heat transfer coefficient of approximately 5 W/m$^2$ K at the surface for ambient air at 21°C. The inner boundary of the breast is the pectorals muscle, which covers the ribs and is considered as adiabatic in the present simulations.

### 17.3.1.4 Solution

Figure 17.1 illustrates the computational grid created using the computer program GAMBIT 1.3. The volume is discretized into finite elements (tetrahedral and triangular). There are 96,410 cells and 238,324 faces in the grid and it satisfies a mesh independent solution. During meshing, a boundary layer was considered at the surface to account for the high temperature gradient. The grid is exported to the computer program FLUENT5 to solve for the temperature distributions. Blood perfusion and metabolic heat generation terms are added to the computations with a user-defined function written in C++. The function also increases the blood perfusion and metabolic heat generation at the tumor and the surrounding shells.

### 17.3.1.5 Optimum Results

The metabolic heat generation and the blood perfusion inside the normal breast were varied. Initially, the metabolic heat generation was assumed to be $\dot{q}_m$ = 420 W/m$^3$.[31] The temperatures were calculated at three different angles, namely, 0, 45, and 90° with respect to the *x*-axis. The maximum difference in temperature for these three angles was about 0.1°C.

Gautherie[22] determined the metabolic heat generation of the tumor to be 29,000 W/m$^3$ using the electrical analogy. Thus, for this metabolic heat generation inside the tumor, the blood perfusion rate varied. The thermal effect of higher blood perfusion is to reduce the highest temperature attained inside the tumor.

**TABLE 17.5** Blood Perfusion and Metabolic Heat Generation Values Used for an Optimum Fit with the Gautherie Data

| | $\dot{q}_m$ (W/m³) | $\omega_b$ (ml/s/ml of tissue) |
|---|---|---|
| Normal | 450 | 0.00018 |
| Tumor | 29,000 | 0.00900 |
| Lower Shell | 11,700 | 0.00360 |
| Upper Shell | 4,725 | 0.00144 |

Data from Gautherie, M., *Ann. N.Y. Acad. Sci.*, 335, 383, 1980.

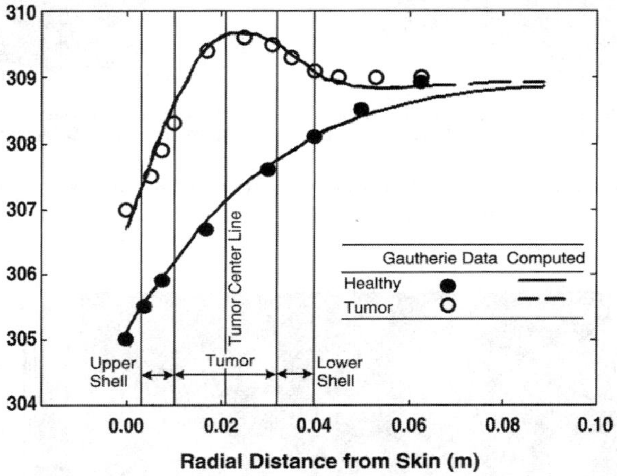


**FIGURE 17.2** Verification of the numberical model parameters used in the bioheat transfer simulations.

Table 17.5 gives the blood perfusion and metabolic heat generation rates used to give the best match with Gautherie's[22] temperature distributions using the least mean square error method. At the optimum, the root mean square error over the entire profile is 0.112°C for the normal breast and 0.293°C for the malignant breast. The blood perfusion rates for all domains are within the range specified by Vaupel.[32] The predicted radial temperature profile is plotted in Figure 17.2.

#### 17.3.1.6 Conclusion, Discussion, and Future Work

The calculated temperature distributions agree very well with the measurements based on the model structure and the parametric values used in the present simulation. This suggests that the tumor has approximately 1.5 to 2 orders of magnitude higher metabolic rate and close to 10 to 20 times blood perfusion rate. As long as the orders of magnitude of the metabolic rate and blood perfusion rate are correct, reasonable temperature profiles can be obtained. A more significant result is that the tumor introduces a local temperature rise on the breast surface that is accurately detectable by modern infrared cameras. Further inverse bioheat transfer calculations may provide a method of locating the tumor using the surface thermograms.

Figure 17.3 shows computational thermograms created using the results of the calculations. These data indicate that, for the tumor measured by Gautherie, the temperature differences apparent on the skin surface would be significant and detectable with thermal imaging. Furthermore, the right-hand panel shows that modern image processing techniques can enhance the thermal imaging signal significantly.

Presently, experimental verification of the combined use of thermal imaging and bioheat transfer computations is in progress in our laboratory. The initial thermal imaging results are shown in Figure 17.4. The

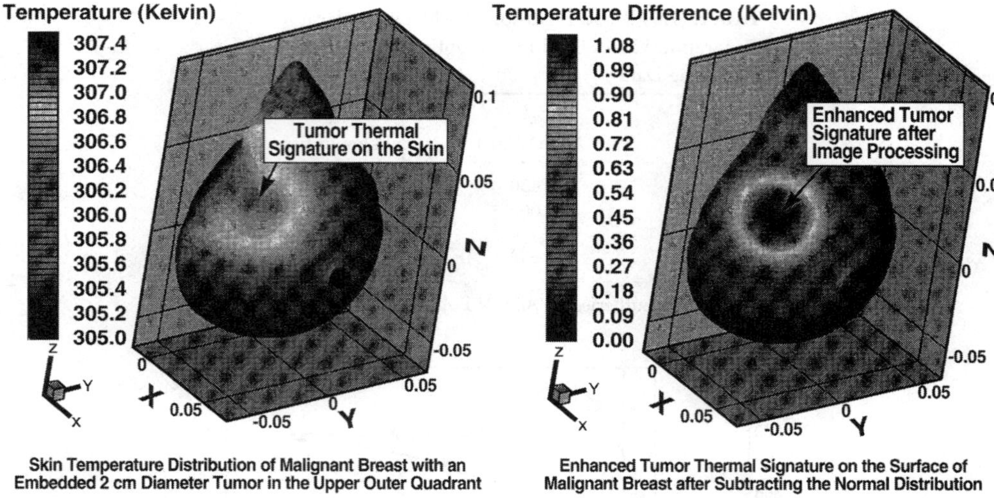


**FIGURE 17.3** Numerical thermal images demonstrating the fundamental ability of a thermal image to capture an abnormality and the power of image processing.

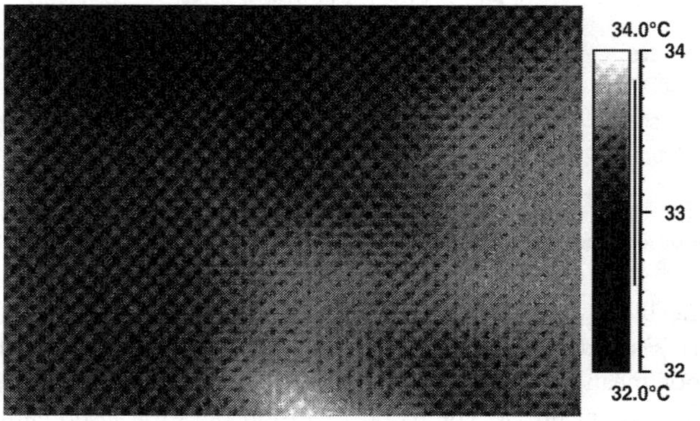


**FIGURE 17.4** A sample infrared image demonstrating the high temperature sensitivity of modern IR cameras.

data show that it is feasible to measure the type of temperature changes of interest in biological systems with very high accuracy and precision.

## 17.4 Summary and Conclusions

Infrared thermal imaging of temperature changes in a biological system is a measurement modality with a fundamental scientific basis. However, fundamental science also clearly points out the nonspecific nature of this measurement for monitoring any disease state. Therefore, this technique must be used as an adjunct to other diagnostic techniques and in conjunction with newly emerging analytical and numerical computational tools such as the one demonstrated previously.

The technique is clearly not ready for general purpose clinical applications without the establishment of a fundamental science and technology base aimed at addressing the nonspecificity issue by developing adjunct instrumentation and numerical tools. If such work is completed, thermal imaging can add a powerful measure of function in biological systems because of its fundamental basis in blood perfusion and metabolic activity.

## Acknowledgments

The Indiana 21st Century Research and Technology Funds (Dr. Karl Kohler, Program Director) supported this research. The contributions of Ph.D. students Ashish Gupta and Dawn Sabados of Purdue University are gratefully acknowledged.

### References

1. Incropera, F.P. and DeWitt, D.P., *Fundamentals of Heat and Mass Transfer*, John Wiley & Sons, New York, 1995.
2. Siegel, R. and Howell, J.R., *Thermal Radiation Heat Transfer*, Hemisphere, Washington, D.C., 1992.
3. Modest, M.F., *Radiative Heat Transfer*, McGraw-Hill, New York, 1993.
4. DeWitt, D.P. and Nutter, G.D., *Theory and Practice of Radiation Thermometry*, John Wiley & Sons, New York, 1988.
5. Welch, A.J. and Van Gemert, M.J.C., *Optical-Thermal Response of Laser Irradiated Tissue*, Plenum Press, New York, 1995.
6. Barnett, M.L., Gilman, R.M., Charles, C.H., and Bartels, L.L., Computer-based thermal imaging of human gingiva preliminary investigation, *J. Periodontol.*, 60, 628, 1989.
7. Chan, F.H.Y., So, A.T.P., Kung, A.W.C., Lam, F.K., and Yip, H.C.L., Thyroid diagnosis by thermogram sequence analysis, *Biomed. Mater. Eng.*, 5, 169, 1995.
8. Clark, R.P., Goff, M.R., and Macdermot, K.D., Identification of functioning sweat pores and visualization of skin temperature patterns in x-linked hypohidrotic ectodermal dysplasia by whole body thermography, *Hum. Genet.*, 86, 7, 1990.
9. Dickey, F.M., Holswade, S.C., and Yee, M.L., Burn depth estimation using thermal excitation and imaging, *Proc. SPIE*, 3595, 9, 1999.
10. Hasegawa, J. and Takaya, K., Development of microcomputer-aided pyroelectric thermal imaging system and application to pain management, *Med. Biol. Eng. Comput.*, 24, 275, 1986.
11. Stein, L.E., Pijanowski, G.J., Johnson, A.L., Maccoy, D.M., and Chato, J.C.A., Comparison of steady state and transient thermography techniques using a healing tendon model, *Vet. Surg.*, 17, 90, 1988.
12. Parisky, Y.R., Skinner, K.A., Cothren, R., DeWittey, R.L., Birbeck, J.S., Conti, P.S., Rich, J.K., and Dougherty, W.R., Computerized thermal breast imaging revisited: an adjunctive tool to mammography, *Proc. IEE Eng. Med. Biol. Soc.*, 20, 919, 1998.
13. Villringer, A. and Dirnagl, U., Optical imaging of brain function and metabolism 2: Physiological basis and comparison to other functional neuroimaging methods, *Adv. Exp. Med. Biol.*, 413, 15, 1997.
14. Naghavi, M., Melling, P., Gul, K., Madjid, M., Willerson, J.T., and Casscells, W., First prototype of a 4 French 180° side-viewing infrared fiber optic catheter for thermal imaging of atherosclerotic plaque, *J. Am. Coll. Cardiol.*, 37(Suppl. 2), 3A, 2001.
15. Hubbs, A.H., Karels, T., and Boonstra, R., Indices of population size for burrowing mammals, *J. Wildlife Manage.*, 64, 296, 2000.
16. Fuller, M.P. and Wisniewski, M., The use of infrared thermal imaging in the study of ice nucleation and freezing of plants, *J. Thermal Biol.*, 23, 81, 1998.
17. Jones, H.G., Use of thermography for quantitative studies of spatial and temporal variation of stomatal conductance over leaf surfaces, *Plant Cell Environ.*, 22, 1043, 1999.
18. Kobel, J., Holowacz, W., and Podbielska, H., Thermal imaging for face recognition in optical security systems. Optical sensing for public safety, health, and security, *Proc. SPIE*, 4535, 154, 2001.
19. Pavlidis, I., Levine, J., and Baukol, P., Thermal imaging for anxiety detection, *Proc. IEEE Workshop on Computer Vision beyond the Visible Spectrum: Methods and Applications*, 2000, p. 104.
20. http://www.nabco.org/.
21. Pennes, H.H., Analysis of tissue and arterial blood temperature in resting human forearms, *J. Appl. Physiol.*, 2, 93, 1948.

22. Gautherie, M., Thermopathology of breast cancer: measurements and analysis of *in vivo* temperature and blood flow, *Ann. N.Y. Acad. Sci.*, 335, 383, 1980.
23. Osman, M.M. and Afify, E.M., Thermal modeling of malignant women's breast, *ASME J. Biomech. Eng.*, 110, 269, 1988.
24. Gautherie, M., Quenneville, Y., and Gros, C., Metabolic heat production, growth rate and prognosis of early breast carcinomas, *Biomedicine*, 22, 328, 1975.
25. Keyserlingk, J.R., Ahlgren, P.D., Yu, E., and Belliveay, N., Infrared imaging of the breast: initial reappraisal using high-resolution digital technology in 100 successive cases of stage I and II breast cancer, *Breast J.*, 4, 245, 1998.
26. Ng, E.Y.K. and Sudharsan, N.M., An improved three-dimensional direct numerical modeling and thermal analysis of a female breast with tumor, *J. Eng. Med.*, 215, 25, 2001.
27. Ng, E.Y.K. and Sudharsan, N.M., Can numerical simulation adjunct to thermography be an early detection tool? *J. Thermol. Int.* [formerly *Eur. J. Thermol.*], 10, 119, 2000.
28. Sudharsan, N.M. and Ng, E.Y.K., Parametric optimization for tumor identification: bioheat equation using ANOVA and Taguchi method, *J. Eng. Med.*, 214, 505, 2000.
29. Sudharsan, N.M., Ng, E.Y.K., and Teh, S.L., Surface temperature distribution of a breast with/without tumor, *Comput. Methods Biomech. Biomed. Eng.*, 187, 1999.
30. Weinbaum, S. and Jiji, L.M., A new simplified bioheat equation for the effect of blood flow on local average tissue temperature, *ASME J. Biomech. Eng.*, 107, 131, 1985.
31. Liu, J. and Xu, L.X., Boundary information based diagnostics on the thermal states of biological bodies, *Int. J. Heat Mass Transfer*, 43, 2827, 2000.
32. Vaupel, P., *Tumor Blood Flow, Blood Perfusion and Microenvironment of Human Tumors: Implications for Clinical Radio Oncology*, Springer-Verlag, Berlin, 1998, p. 41.

# IV

# Biomedical Diagnostics I

# 18

# Glucose Diagnostics

Gerard L. Coté
*Texas A&M University*
*College Station, Texas*

Roger J. McNichols
*BioTex, Inc.*
*Houston, Texas*



## 18.1 Introduction

Glucose is one of the most important carbohydrate nutrient sources and is fundamental to almost all biological processes. Quantification of glucose concentration is important in monitoring and analysis of agricultural products, control and regulation of cell culture, and diagnosis and control of human diseases including diabetes. A wide range of parameters, including glucose concentration range, volume of glucose solution available, and required accuracy, exist across these applications. For instance, the sugar concentration in many agricultural products (e.g., fruit juices) is hundreds of grams per liter while the glucose concentration for an on-line cell culture system may be in the milligrams per deciliter range. Additionally, the volume or, more importantly for optical approaches, the available optical path length for on-line process control and agricultural applications can be tens of centimeters while *in vivo* these are typically millimeters to centimeters. Consequently, the required glucose sensitivity *in vivo* needs to be orders of magnitude better than in the agricultural industry. Finally, the environmental challenges associated with *in vivo* monitoring make this problem more difficult than typical industrial applications because of a range of potential confounders that cannot be controlled. These include temperature and pH variations, confounding chemical species, pressure changes, and correlated physiological changes.

For biotechnological processes such as cell culture, measurements of analytes such as glucose are currently obtained using off-line methods that require sample extraction from the process or bioreactor. The off-line measurements may cause cell contamination, can be time consuming, can be expensive, and may not reflect the real-time status of the cells. In order to overcome many of these limitations, nondestructive optical methods have been proposed as a solution. Three of the most common optical approaches for on-line glucose monitoring and process control are described in Section 18.2.

A significant role for physiological glucose monitoring is in the diagnosis and management of diabetes. Diabetes mellitus is a metabolic disease that presents as a complex group of syndromes related in large part to problems with insulin production and/or utilization. Insulin is the primary hormone responsible for glucose regulation in the body; in diabetics, blood glucose levels, which would normally be maintained between 90 and 120 mg/dl, may fluctuate between 40 and 400 mg/dl or more. Over time, elevated glucose levels may damage the kidneys, eyes, nerves, and heart, and severely low glucose levels may cause a patient

0-8493-1116-0/03/$0.00+$1.50

to go into shock or even die. Therefore, a goal of diabetes management is tight maintenance of blood glucose levels via insulin injection, modified diet, exercise, or a combination of these. In order to guide this therapy, regular measurement of blood glucose levels (up to five times per day) is required. Because current glucose sensing methods require a painful and inconvenient puncture of the skin to obtain a blood sample for analysis, efforts to develop a noninvasive or implantable glucose meter using optical methods have been strong.

A number of invasive and noninvasive techniques are being investigated for glucose monitoring, including direct transmission through blood vessels, measurement of glucose in interstitial fluid in the dermis, light transmission through blood containing body parts (including the earlobe, the finger, and the forearm), and optical interrogation of the aqueous humor of the eye. The invasive approaches have the advantage of direct contact with the media of interest and thereby enhanced specificity, but they often suffer from biocompatibility issues. The noninvasive approaches do not suffer from the biocompatibility issues but are less specific and must account for the fact that they simultaneously probe multiple volumes including interstitial fluid, blood, and intracellular fluid. The anterior chamber of the eye has been suggested as a noninvasive glucose monitoring site because it is uniquely transmissive to optical radiation. It has been shown that the aqueous humor (AH), the fluid just beneath the cornea, exhibits an age-dependent steady-state value of glucose approximately 70% of that found in blood and that the time lag between AH and blood glucose is on the order of minutes. The lack of significant scattering components and the relative simplicity of AH compared to blood or blood and tissue make it an attractive site for optical measurement techniques. Section 18.3 focuses on the optical approaches currently used and being developed for glucose monitoring.

## 18.2 On-Line Glucose Monitoring and Process Control

### 18.2.1 Near-Infrared Spectroscopy

The near-infrared (NIR) region of the optical spectrum extends from 700 to 2500 nm (0.7 to 2.5 μm) and can be used for quantitative measurement of organic functional groups, especially C–H, O–H, N–H, and C=O. Absorption bands in the NIR are composed primarily of overtone and combination bands of stretching and vibrational modes of organic molecules. These bands tend to be rather broad and overlap considerably, so the NIR region is not particularly well suited for qualitative analysis. However, the significantly reduced intensity of NIR absorption bands does allow for use of reasonable pathlengths in the millimeter to centimeter range; when used with multivariate statistics, suitable quantitative information involving multiple analytes may be obtained.

NIR transmission spectroscopy (NIRS) and diffuse reflectance spectroscopy have been the primary optical approaches investigated for on-line glucose monitoring and process control.[1–15] NIRS has been developed extensively in the agricultural and food industries for the past 30 years. An advancement that has enabled NIRS as an analytical tool has been the development of sophisticated multivariate data analysis or chemometric techniques. The use of NIRS for glucose monitoring in industrial and agricultural applications has been described in References 16 through 18. In this section we will focus on the application of NIRS to on-line measurement of glucose concentration in bioreactors and cell cultures. These applications represent the most challenging application of NIR glucose measurement because the media involved are more complex and the analyte concentrations involved are significantly lower than those found in current industrial applications.

Cell culture applications require careful maintenance of a number of factors, including pH, temperature and, importantly, glucose concentration. The automated control of glucose concentration via on-line continuous monitoring could result in improved culture productivity and efficiency. Optical measurement of glucose is complicated by the complex nature of cell culture media, which contain proteins, amino acids, salts, and other small carbohydrates and nutrients including glucose. A typical NIR spectrum of GTSF-2 cell culture medium in the 2- to 2.5-μm range with respect to an air background is shown in

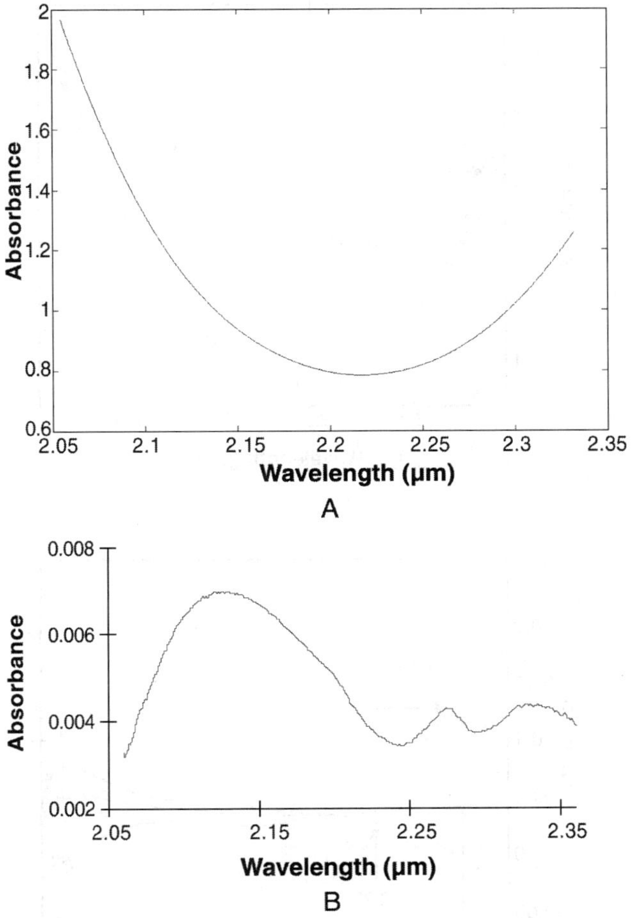


**FIGURE 18.1** (A) NIR absorbance spectrum of cell culture media with air as the background. (B) NIR absorbance spectrum for glucose in the 2- to 2.5-μm wavelength range after baseline removal of the other cell culture media components. Note that the glucose signal amplitude is orders of magnitude smaller than the raw absorption, which has strong water and protein bands.

Figure 18.1A. The NIR spectrum for aqueous glucose at a concentration of 500 mg/dl (27.5 m*M*), with the other cell culture medium background removed, is shown in Figure 18.1B. Of particular importance is the magnitude of the absorbance units on the y-axis. Note that the absorbance by water and proteins is at least three orders of magnitude greater than that of glucose. Thus, the stability and signal-to-noise ratio of an NIR instrument must be very good if it is to be used to distinguish changes in the glucose concentration. Further, the temperature of the sample must also be very tightly controlled because, with even a slight change in temperature, the absorbance of the background water spectrum will shift, severely impacting measurement of the glucose signal.

Figure 18.2 demonstrates the effect of varying temperature on the NIR spectrum of cell culture medium. Fortunately, techniques such as digital filtering combined with high-order multivariate statistics have been able to compensate for and model such variations. Additionally, cell culture applications typically exhibit well-controlled temperature and are amenable to inclusion of a fixed pathlength for optical measurements, thus making on-line NIR glucose monitoring in cell culture promising. Detailed discussion of specific NIR spectroscopic studies of glucose measurement in cell culture media may be found in References 6, 15, and 19.

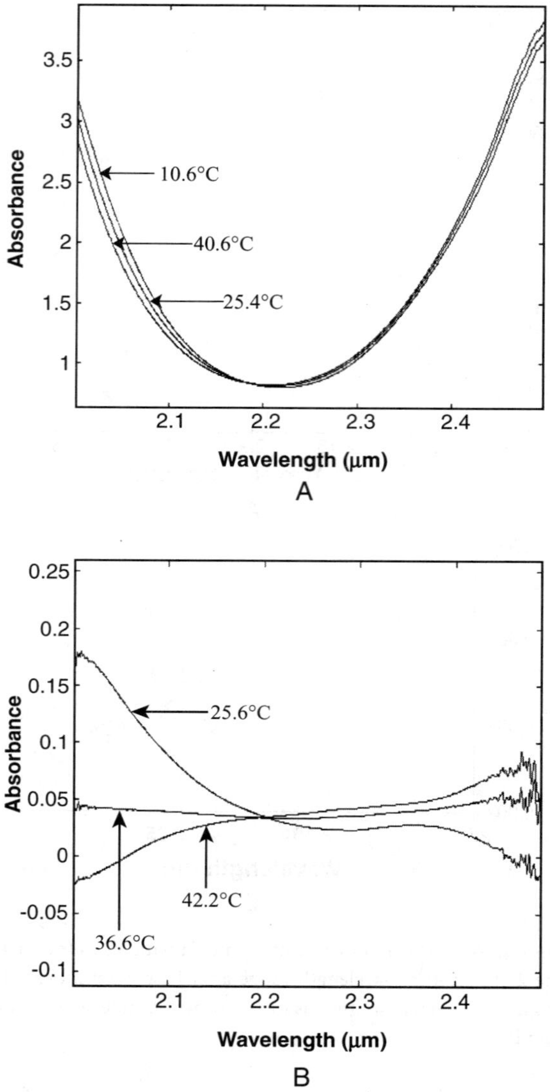


**FIGURE 18.2** (A) Water absorbance spectrum, referenced to air, showing the shift in the water peaks with temperature. (B) Glucose (500 mg/dl) absorbance spectrum taken at different temperatures, referenced to a water background taken at one temperature showing the baseline water peak shifts.

### 18.2.2 Raman Spectroscopy for Biological Glucose Analysis

In addition to NIR, the optical method known as Raman spectroscopy may provide for potentially rapid, precise, and accurate analysis of glucose concentration and biochemical composition.[20] Raman spectroscopy provides information about the inelastic scattering that occurs when vibrational or rotational energy is exchanged with incident probe radiation. When monochromatic radiation is incident upon Raman active media, most of the incident light of frequency $\nu_o$ is elastically scattered at the same frequency (Rayleigh scattering); however, a very small portion of the light undergoes Raman scattering and exhibits frequency shifts of $\pm\nu_m$ that are associated with transitions between the rotational and vibrational levels.[21,22] Energy decreases ($\nu_o - \nu_m$) are referred to as Stokes shifts and energy increases ($\nu_o + \nu_m$) are referred to as anti-Stokes shifts. Because Stokes shifts are much more prevalent, most studies utilize the

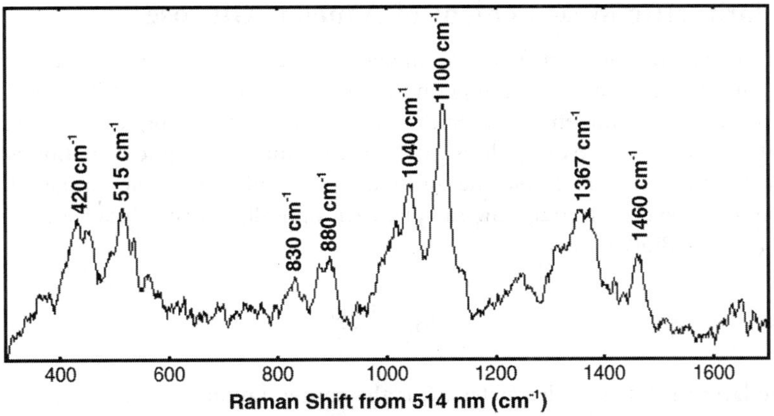


**FIGURE 18.3** Baseline corrected Raman spectra of glucose showing the narrow peaks that can be obtained when compared with the NIR spectra of Figure 18.1B.

Stokes type of scattering bands. Therefore, the Raman bands of interest are shifted to longer wavelengths relative to the excitation wavelength.

As with infrared spectroscopic techniques, Raman spectra can be utilized to identify molecules such as glucose because these spectra are characteristic of variations in the molecular polarizability and dipole moments. However, in contrast to infrared and NIRS, Raman spectroscopy has a spectral signature that is less influenced by water. In addition, as shown in Figure 18.3, Raman spectral bands are considerably narrower than those produced in NIR spectral experiments. Raman also has the ability to permit the simultaneous estimation of multiple analytes, requires minimum sample preparation, and would allow for direct sample analysis (i.e., would be on-line). Different bonds scatter different wavelengths of EM radiation, so this method gives quantitative information about the total biochemical composition of a sample, without its destruction.

The fundamental instrumentation required to produce a Raman spectrum includes a monochromatic light source such as a laser, a wavelength separation system such as a dispersive monochromater or nondispersive Fourier transform (FT) spectrometer, and a detection system such as a CCD array for the dispersive case or point detector for the FT Raman case. In either system, it is very important to eliminate the elastically scattered light, which is roughly $10^5$ times greater in intensity and occurs at the same wavelength. Normal optical filters are not adequate to remove enough of this elastically scattered light and therefore holographic notch filters are typically required.

One application of dispersive Raman spectroscopy is the noninvasive, on-line determination of the biotransformation by yeast of glucose to ethanol.[23] Using a diode-laser exciting at 780 nm and a CCD detector along with partial least squares multivariate analysis, glucose and ethanol could be measured to within 5%.

Although Raman scattering can be used in on-line monitoring applications of analytes at the millimolar level, it is a rather weak effect that typically requires a fairly powerful laser and cooled CCD array; it has substantial fluorescence interference from biological samples. However, the Raman effect can be greatly enhanced (by approximately $10^6$- to $10^8$-fold) if the molecules of interest in the monitoring process are attached to, or microscopically close to, a suitably undulating (roughened) surface, usually of the coinage metals Cu, Ag, or Au.[24] This method, called surface-enhanced Raman scattering (SERS),[25,26] has received much attention recently within biomedicine[26] and genomics.[27] Since the development of reproducible SERS on a silver-coated alumina substrate by Vo-Dinh and colleagues,[28] this technology has emerged as a very powerful biochemical detection method. It is noteworthy as well that the fluorescence background typically observed in many on-line samples using dispersive Raman may potentially be reduced by using SERS.

### 18.2.3 Polarimetric Measurement of Aqueous Glucose

Polarimetry is a sensitive, nondestructive technique for measuring the optical activity exhibited by inorganic and organic compounds. A compound is considered to be chiral if it has at least one center about which no structural symmetry exists. Such molecules are said to be optically active because linearly polarized light is rotated when passing through them. The amount of optical rotation is determined by the molecular structure of the molecule, the concentration of chiral molecules in the substance, and the pathlength the light traverses through the sample. Each optically active substance has its own specific rotation as defined by Biot's law:

$$[\alpha] = \frac{100\alpha}{LC} \tag{18.1}$$

where $L$ is the layer thickness in decimeters, $C$ is the concentration of solute in grams per 100 ml of solution, and $\alpha$ is the observed rotation in degrees. In the preceding equation the specific rotation $[\alpha]$ of a molecule is dependent on temperature, $T$, wavelength, $\lambda$, and the pH of the solvent.

The polarimetric method is employed in quality control, process control, and research in the pharmaceutical, chemical, essential oil, flavor, and food industries. It is so well established that the United States Pharmacopoeia and the Food and Drug Administration include polarimetric specifications for numerous substances.[29] Historically, one of the earliest applications of polarimetry has been the development of bench-top polarimeters, known as saccharimeters, designed solely for the estimation of starch and sugar in foods and beverage manufacturing and in the sugar industry.[30] For this agricultural industry application, in which the concentrations and path lengths are high, the commercial bench-top units are adequate. However, for the low glucose concentrations found in cell culture systems and *in vivo*, much more sophisticated polarimetry systems are required to monitor the glucose, as described next.

## 18.3 Diabetic Monitoring

### 18.3.1 Commercial Colorimetric Glucose Meters

Many of the home blood glucose monitors currently in use rely on the so-called colorimetric approach to glucose sensing. The colorimetric method consists of three basic steps: (1) the invasive withdrawal of a small blood sample, (2) the application of the blood sample to a specially formulated "test strip," and (3) the automated reading of the test strip results via optical means. In this section we focus on the second and third steps of this process.

Glucose monitor test strips are available from a number of manufacturers and in several varieties. However, all commercially available glucose monitor test strips (electroenzymatic or colorimetric) rely on the quantification of reaction of glucose with the naturally occurring enzyme glucose oxidase (GOX). While a number of red-ox reaction-based methods for determining the concentration of reducing sugars exist, the specificity afforded by GOX makes it the standardly practiced assay. GOX is a protein approximately 160,000 Da in size and is composed of two identical 80-kDa subunits linked by disulfide bonds. Each subunit contains 1 mol of Fe and 1 mol of flavin–adenine–dinucleotide (FAD). During the oxidation of glucose, the FAD groups become temporarily reduced; thus, GOX is one of the family of flavoenzymes. The reaction that is catalyzed by GOX is shown in Figure 18.4. In the presence of oxygen, reduced GOX further reacts to drive the reaction to the right. This reaction, which results in the production of hydrogen peroxide, is shown in Figure 18.5.

Commercially available colorimetric sensors employ a peroxidase enzyme (commonly horseradish peroxidase) and a red-ox-coupled dye pair to generate chromophore concentrations proportional to the amount of hydrogen peroxide produced. One such dye pair is the oxygen acceptor 3-methyl-2-benzothiazolinone hydrazone plus 3-(dimethylamino)benzoic acid (MBTH-DMAB), which has an absorption peak at 635 nm. Production of hydrogen peroxide by oxidation of glucose and reduction of oxygen drives

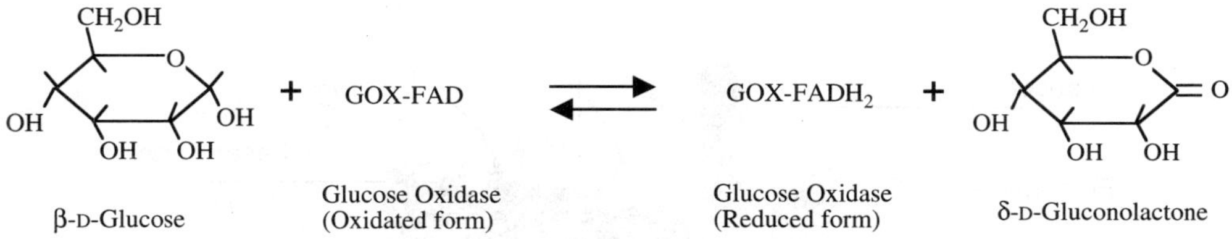


**FIGURE 18.4** Oxidation of glucose to gluconolactone by the flavoenzyme GOX.

$$\text{GOX-FADH}_2 + O_2 \longrightarrow \text{GOX-FAD} + H_2O_2$$

**FIGURE 18.5** Production of hydrogen peroxide by reduced GOX in the presence of oxygen.

the peroxidase catalyzed production of active MBTH-DMAB chormophore and thus results in a measurable increase in absorbance of light at 635 nm. Other oxygen acceptor dyes that can be used include O-dianisidine, benzidine, and 4-aminoantipyrene and chromotropic acid (AAP-CTA), among others.[31]

Optical measurement of the dye product (and hence glucose concentration) is based on the Beer's law increase in absorbance of the dye product. In practice, this increase in absorbance is usually quantified by measuring the attendant decrease in reflectance of the active site on a test strip. The relationship between absorbance and reflectance is described by the Kubelka–Munk equation (see, for example, Reference 16), $K/S = (1 - R)^2/2R$, where $K$ is the concentration-dependent absorbance, $S$ is a constant related to the scattering coefficient, and $R$ is the measured reflectance.

Practical colorimetric measurement systems typically employ a multi- (at least two) wavelength detection scheme to increase accuracy and account for variables like the background absorbance of blood, the oxygenation state of blood, and the hematocrit of the blood under test. For the case of the MBTH-DMAB dye pair, the absorbance peak at 635 nm is fairly sharp and absorbance by the dye at 700 nm is negligible. Therefore, a second reflectance measurement at 700 nm affords a correction factor, which is used to account for background and sample variations. Since the GOX reaction is a progressive one, the timing of measurements is also extremely important. Early measurement systems required that a timer be started when blood was placed on the strip and that a reading be taken after a specific period of time (typically 15 to 60 s). Modern systems overcome this inconvenience by taking continuous reflectance measurements as soon as a new test strip is inserted into the meter. When a drop of blood is placed on the meter, an immediate change in the reflectance signals the start of the reaction; multiple reflectance readings may be taken during the reaction to further increase accuracy or dynamic range of the sensor.

## 18.3.2 Laser Perforation and Poration Devices for Fluid Extraction

High-powered medical lasers have been used for surgical applications for decades, but over the last few years they have found their way into glucose monitoring. Currently, two laser techniques can penetrate skin tissue: one that perforates deep enough into the skin tissue to provide for the collection of a capillary blood sample, and a second poration technique that only removes the stratum corneum (the dead uppermost layer of the epidermis) of the skin to allow interstitial fluid to be collected. A standard glucose analysis method like the colorimetric one described earlier could be used on the blood sample, while a modified system would be needed for the interstitial fluid sample because it has different chemical confounding components and would have an overall lower glucose concentration level.

The laser perforation systems for collecting capillary blood typically use a pulsed laser such as an erbium yttrium aluminum garnet (Er:YAG) for ablation of the upper layers of skin.[32,33] This particular laser has a spectral peak of 2.94 μm, which is a primary absorption wavelength band for water. These systems differ from previous surgical laser ablation systems in that they are small and battery operated, provide for a limited cutting depth, and do not heat collateral tissue or cauterize the delicate capillary vessels. The typical opening it provides is on the order of 0.3 to 0.5 mm in diameter. Overall, the

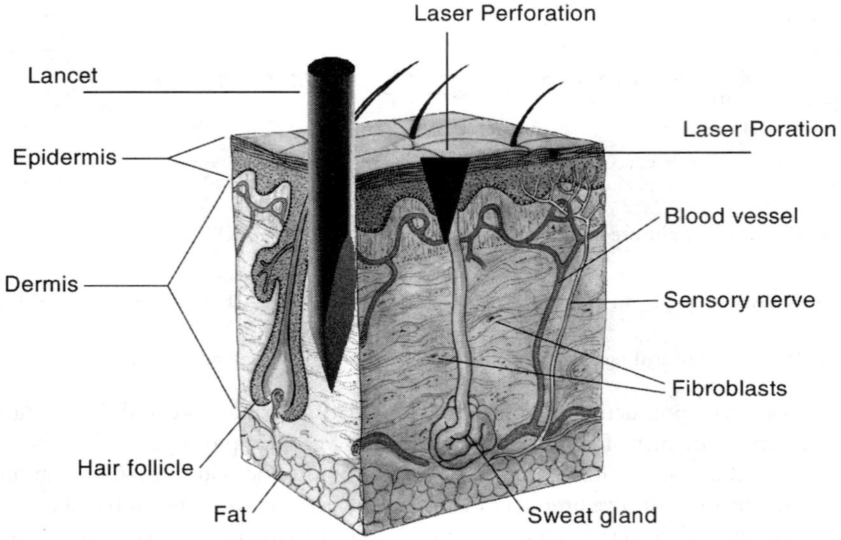


**FIGURE 18.6** **(Color figure follows p. 28-30.)** Cross section of skin comparing the depth of penetration incurred for a standard lancet to that of both the laser perforation and laser poration approaches.

advantages of using laser light, as opposed to a stainless steel lancet tip, are that it reduces the blunt tissue trauma, provides for less bruising and more rapid healing, and eliminates sharps as a potential source of cross contamination.

The laser poration approach has been described in the literature[34] and is being commercially pursued by SpectRx, Inc.[35] The individuals at SpectRx are developing the laser system as well as the sample handling system, which includes a suction system to remove the fluid and the sensor for measuring glucose. As mentioned, the biggest difference between this approach and the laser perforation system commercially available is that this approach removes only the stratum corneum and thus interstitial fluid (ISF) is obtained instead of capillary blood. This is illustrated in Figure 18.6, which shows a cross section of the skin comparing a standard lancet, laser perforation, and laser poration for drawing blood or ISF. Initially, a very short ultraviolet (UV) laser such as an argon-fluoride excimer was used, which operates at 193 nm. The strong absorption of the UV radiation by the protein of the stratum corneum allows as little as 0.25 μm of tissue to be removed per pulse.[34] Firing 80 pulses from the laser would therefore create a pore in the roughly 15-μm-thick stratum corneum. Er:YAG lasers similar to the type used in the laser perforation approach and NIR diode lasers have been investigated as alternatives to excimer because of their strong absorbance by tissue water.[34] After poration, an ISF harvesting device that potentially houses the glucose assay must be employed, followed by a vacuum pump to help in fluid withdrawal and an ISF glucose-sensing meter.

## 18.3.3 Spectroscopic Methods for *in Vivo* Glucose Diagnostics

### 18.3.3.1 Fluorescence Spectroscopy

#### *18.3.3.1.1 Glucose Oxidase and $O_2$-Based Fluorescent Sensors*

The specificity afforded by the catalytic enzyme GOX has been exploited to create a number of fiber-optic fluorescence-based sensors for glucose. Recall that, in the reaction of Figures 18.4 and 18.5, $O_2$ and glucose are consumed to produce gluconolactone and $H_2O_2$. A simple approach for creating a sensor based on this reaction is to incorporate an oxygen-sensitive dye such as a ruthenium complex. Ruthenium(II) complexes exhibit fluorescence quenching in the presence of oxygen; therefore, a decrease in local oxygen concentration can be detected as an increase in fluorescence of a ruthenium-based dye. Incorporating GOX and a ruthenium dye together results in a sensor whose fluorescence increases with increased glucose concentration since increased glucose concentration will lead to the consumption of more $O_2$. One example of such a sensor is described in References 36 and 37. A disadvantage of this simple approach is that a decrease in local oxygen content may not be distinguished from a rise in glucose concentration. Li et al. have solved this problem by creating a dual-channel sensor with a ruthenium–GOX channel and a second ruthenium-only channel.[38] Thus, the second $O_2$-sensitive channel can be used to correct for changes in local oxygen tension. Development of GOX-based fiber-optic sensors continues, and more elaborate sensors have been described.[39,40]

#### *18.3.3.1.2 Nonoxygen-Based Fluorescent Sensors*

One of the first nonoxygen-based fluorescence assays investigated for glucose monitoring *in vivo* used an indwelling fiber optic approach and a semipermeable membrane.[41,42] In this initial approach a fluorophore was linked to dextran, which was bound on the inner surface of the membrane to a lectin, concanavalin A (Con A) at the tip of the fiber, as shown in Figure 18.7A. Glucose has a higher affinity for the membrane-based lectin and thus displaces the dextran with the fluorophore causing the fluorescent light to be returned through the fiber. The use of concanavalin A (Con A) as the membrane-bound molecule affords the technique fairly good sensitivity because of Con A's strong affinity for glucose. Because the approach only used a single fluorophore, the probe was sensitive to alignment and had to be made to ensure the beam did not shine on the walls of the membrane.

A second approach, which was independent of the beam path into the fiber, was also developed by this group.[43,44] This glucose monitoring approach is based on fluorescent resonance energy transfer (FRET) between two fluorescent molecules — in this case fluorescein isothiocyanate (FITC) bound to dextran and tetramethylrhodamine isothiocyanate (TRITC) bound to Con A, as depicted in Figure 18.7B. When TRITC-Con A binds FITC-dextran, the fluorescent labels are brought to within the required molecular proximity (~50 Å) for FRET-based quenching to occur.[44] In this case, energy from excited FITC is donated to the TRITC acceptor, resulting in quenching of the observed FITC fluorescence peak. Using this approach, glucose concentrations can be measured by a rise in amplitude of the FITC fluorescence peak with increasing glucose concentration as depicted in Figure 18.8. Another potential means of monitoring the change with glucose is based on phase-modulation fluorimetry, a technique that measures fluorescent lifetimes.[45]

In either approach, the fluorescence techniques are very specific to glucose and sensitive to glucose concentration, without interference from other constituents frequently found in blood. Other advantages of the fiber-optic chemical sensors, specifically over electrochemical sensors, include miniaturization, geometric flexibility, and the lack of electrical contact between the sensor and sample. However, inherent problems with an indwelling optical fiber-based approach are similar to the problems associated with electrochemical approaches and include membrane fouling, encapsulation, increase in response time, and increased potential for infection. In addition, a fiber is small and thereby yields very low signal-to-noise-ratios.

As an alternative to fiber-optic-based assay sensing, it has been proposed that fully implantable fluorescence-based glucose assays, encapsulated using various membranes or polymers, could be used.[46,47] It has been suggested that small particles containing the FRET-based assay chemistry described previously could be implanted in the superficial dermis and interrogated by external means. In order for this type of implantable device to be useful, it must be biocompatible, must not exhibit acute reagent consumption or degradation, and must provide a means of communicating the sensor output to the physician or patient. Also, once injected, the implant will not be exposed to blood but will rather be exposed to interstitial fluid. The fluorescent assay has the potential problem of limited chemical and photochemical stability over the long term and the construct, such as the hydrogel or dialysis membrane that encases the assay, must be biocompatible.

### 18.3.3.2 Infrared and Near-Infrared Absorption Spectroscopy

Infrared and NIR absorption spectroscopy techniques have long been mainstays of nondestructive chemometric analysis and therefore hold great potential for the development of noninvasive blood glucose

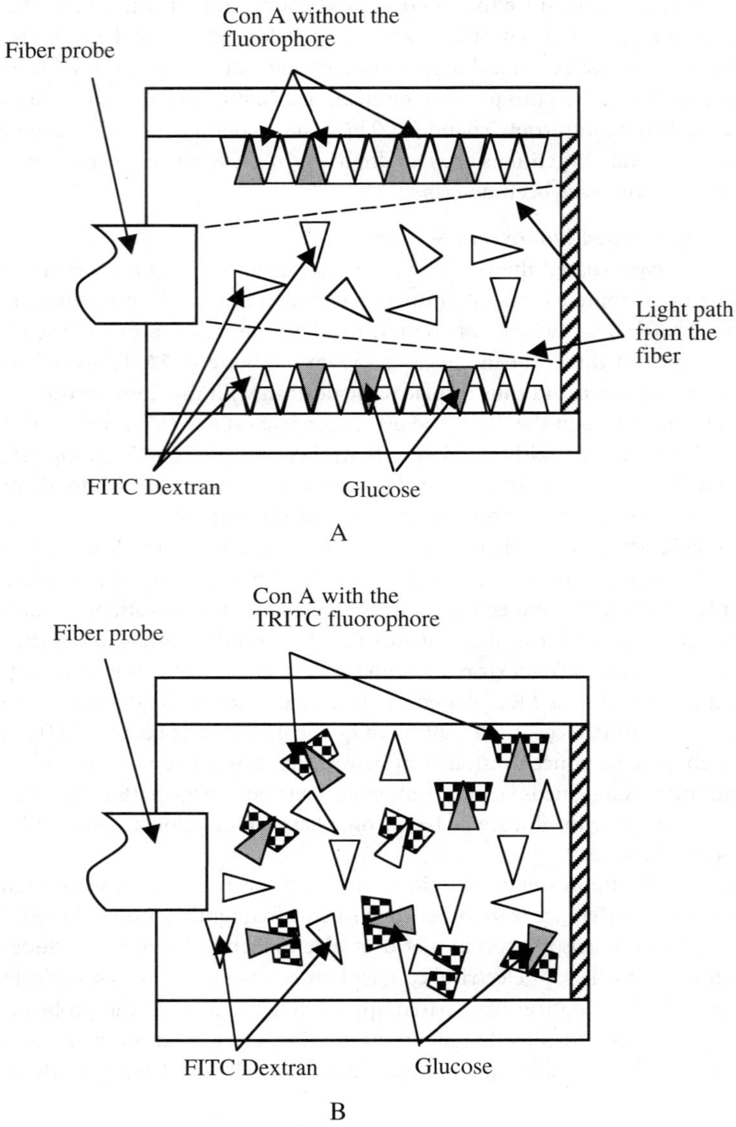


**FIGURE 18.7** Characterization of a fiber optic fluorescence assay using (A) a single fluorescent molecule bound to dextran. Note that the light path is critical in this configuration since it is necessary to measure fluorescence only from the free dextran and not from that bound to Con A on the walls of the sensor. (B) A FRET approach that uses two fluorescent molecules (FITC and TRITC), one bound to Con A and the other to dextran. Note that the FITC is quenched when placed in close proximity to the TRITC, as would be the case in the absence of glucose. The beam alignment is no longer critical.

measurement techniques. Optical absorption methods are based on the concentration-dependent absorption of specific wavelengths of light by glucose or other compounds of interest. The methods are attractive because, in theory, a beam of nonionizing radiation may be directed through a blood-containing portion of the body and the exiting light analyzed to determine the content of glucose or other molecules of interest. In the infrared region and, more specifically, the mid-infrared (MIR) region (from about 5 to 50 μm), the so-called "fingerprint" spectrum can be found. Absorption bands in this region are due to fundamental resonances of specific functional groups and bonds contained within the molecule, and

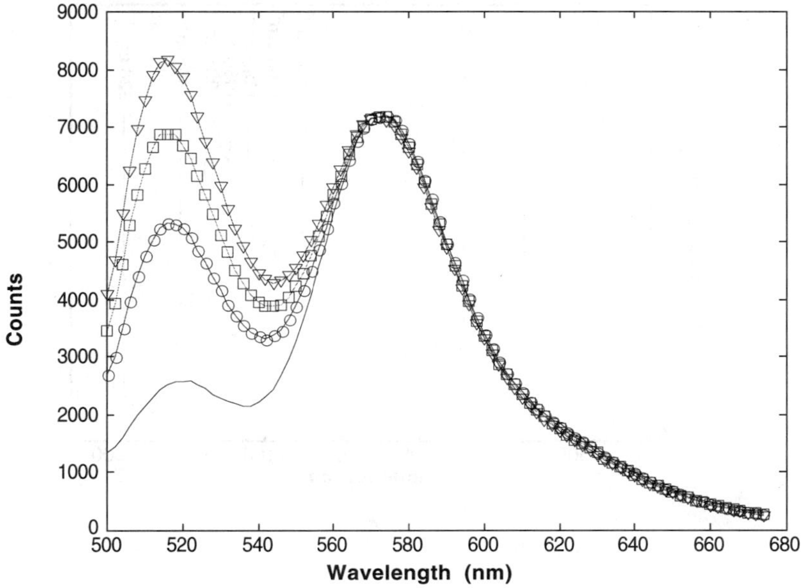


**FIGURE 18.8** Fluorescence spectra from a combined FITC-Dextran and TRITC-Con A FRET assay with glucose concentrations of 0 mg/dl (–), 200 mg/dl (○), 400 mg/dl (□), and 600 mg/dl (▽). An increase in the 520-nm FITC fluorescence relative to the 580-nm TRITC fluorescence is observed with increasing glucose concentration.

they tend to be rather sharp. The fingerprint spectrum, therefore, offers highly specific information about the molecular makeup of the compound or mixture under study. In contrast, absorption bands in the near infrared are composed of complex overtone or combination bands and tend to be broad, overlapping, and much less specific. Figures 18.9A and 18.9B show, respectively, portions of the MIR and NIR absorption spectra for glucose.

Despite the specificity offered by infrared absorption spectroscopy, its application to quantitative blood glucose measurement is limited. A strong background absorption by water and other components of blood and tissue severely limits the pathlength that may be used for transmission spectroscopy to roughly 100 μm or less. Further, the magnitude of the absorption peaks and the dynamic range required to record them make quantitation based on these sharp peaks difficult. Nonetheless, attempts have been made to quantify blood glucose using infrared absorption spectroscopy *in vitro* and *in vivo*.[48–52]

In contrast to the infrared and MIR spectrum, the NIR spectrum passes relatively easily through water and body tissues allowing moderate pathlengths to be used for measurements. Further, NIR instrumentation is readily available and relatively easy to use. Thus, a large amount of effort has been devoted to the development of NIR spectroscopy techniques for noninvasive measurement of blood glucose.

Though all optical glucose sensing methodologies require use of a prediction model relating optical measurements to glucose concentration, the broad overlapping peaks and complicated nature of multi-component, NIR spectra make single or dual wavelength models inadequate. NIR absorption bands may be significantly influenced by factors such as temperature, pH, and the degree of hydrogen bonding present; the unknown influence of background spectra further complicates the problem. For this reason, quantitative NIR spectroscopy has long relied on the development of very high-order multivariate prediction models and empirical calibration techniques. A brief analysis of the signal variation due to glucose in a typical sample should serve to illustrate the problem at hand.

The physiological range of glucose values seen in the normal human body may range from 80 to 120 mg/dl and should ideally remain around 100 mg/dl (5.5 m*M*). In diabetics, blood sugar may rise as high as 500 mg/dl. The generally accepted figure of merit for required accuracy of a useful glucose meter is 10 mg/dl (0.55 m*M*). For the most identifiable NIR glucose peak at approximately 2.27 μm, molar

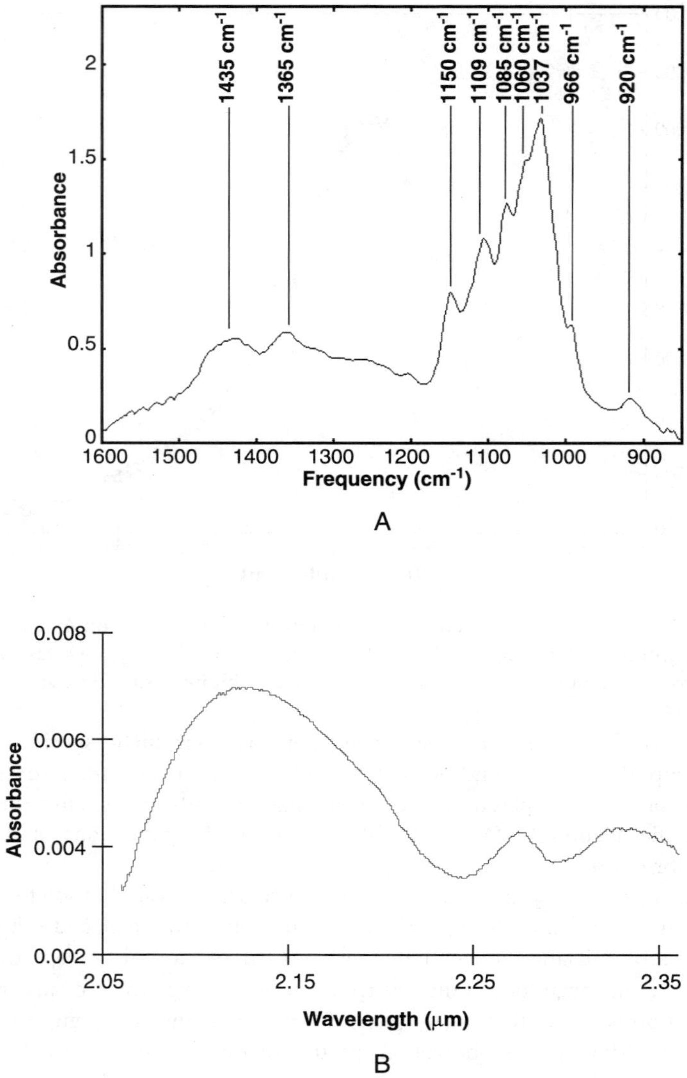


**FIGURE 18.9** Optical absorption spectra for glucose. (A) MIR region extending from 1600 to 900 $cm^{-1}$ or 6.25 to 11 μm and showing absorption peak assignments. (B) NIR region extending from 2.0 to 2.5 μm or 5000 to 4000 $cm^{-1}$. Note that the magnitude of the three absorbance peaks in the NIR region is much smaller.

absorptivity is roughly 0.23 $M^{-1}cm^{-1}$. The molar absorptivity of water at the same wavelength is 0.41 $M^{-1}cm^{-1}$ and the concentration of the water in typical body tissues is approximately 39 *M*. Consider a transmission measurement made through 1 mm of body tissue. The background absorbance due to water will be about 1.6 and that due to glucose will be about $1.26 \cdot 10^{-4}$. To meet the accuracy requirement, we must further be able to discern an absorbance change of about $1.26 \cdot 10^{-5}$ on a background of 1.6, and it is evident that even a change in tissue hydration of 1/1000 of a percent would result in a larger signal change than would a 10 mg/dl change in glucose concentration.

For this reason, high-order multivariate models that incorporate analysis of entire spectra must be used to extract NIR glucose information. A number of excellent and authoritative sources on such methods are available,[18,53,54] and these methods are not discussed here other than to introduce some of the terminology associated with their formulation, use, and practice.

Multivariate calibration techniques focus on finding a response matrix suitable to describe spectral observations obtained when analyzing a solution containing particular concentrations of a number of analytes. Once this response matrix has been found, it may then be inverted in order to relate analyte concentrations to observed spectral readings. The most direct method is the classical least squares method in which spectral observations are made of a number of samples with known analyte concentrations. The problem can be written mathematically as $\mathbf{A} = \mathbf{KC}$, where $\mathbf{A}$ is the matrix containing spectral observations for each sample, $\mathbf{K}$ is the response matrix, and $\mathbf{C}$ is the concentration matrix containing analyte concentrations for each known sample. In the classical least squares solution, $\mathbf{K}$ is computed from $\mathbf{A}$ and $\mathbf{C}$ as $\mathbf{K} = \mathbf{AC}^T(\mathbf{CC}^T)^{-1}$, and estimates of the concentration of analytes in subsequent samples may be determined from the spectral observations $\mathbf{a}$ as $\mathbf{c} = (\mathbf{K}^T\mathbf{K})^{-1}\mathbf{K}^T\mathbf{a}$.

This approach assumes that all measurement errors occur in the spectral observations and that the concentration matrix $\mathbf{C}$ is known exactly. This is often not the case, however, and alternative formulations for the least squares calibration model solution have been proposed that include inverse least squares and partial least squares (PLS). The latter has become overwhelmingly popular due to its robust ability to model complex data interactions and to formulate powerful prediction models. The PLS technique employs an iterative singular value decomposition of the $\mathbf{A}$ matrix to extricate spectral changes most strongly correlated to analyte concentration changes and thereby creates a number of model "factors," which may be linear combinations of many of the observations in $\mathbf{A}$.

In the practical application of PLS, data observations are typically split into two groups: one is used for formulation of the model (calibration) and the other is used for validation of the model (prediction). Two common statistics are the standard error of calibration (SEC), which is the RMS prediction error of the model compared to the calibration data set, and standard error of prediction (SEP), which is the RMS prediction error of the model when applied to the validation set. A problem that can arise in PLS modeling is "overfitting" of the data in which noise or spurious trends evident in the calibration data may actually be factored into the model if chance correlation to analyte concentrations exists. The separation of calibration and validation data sets is one technique used to detect potential overfitting.

For the purposes of discussion of the literature, it is useful to break the NIR region into the very NIR region (from 700 to 1300 nm) and the NIR region (from 2.0 to 5.0 μm). The attraction to the very NIR region is that optical detectors and sources in this region are particularly easy to come by, transmission through tissue is rather good, and transmissive fiber optics may be employed to facilitate probe design. However, glucose absorption bands are particularly weak in this region, and it may be difficult to acquire signals with substantial signal to noise to allow robust measurement. Examples of very NIR studies of glucose determination include References 55 through 58.

Longer in the NIR spectrum, a relative dip in the water absorbance spectrum opens a unique window in the 2.0 to 2.5 μm wavelength region. This window, saddled between two large water absorbance peaks, allows pathlengths or penetration depths on the order of millimeters and contains specific glucose peaks at 2.13, 2.27, and 2.34 μm. So far, this region has offered the most promising results for quantifiable glucose measurement using NIR spectroscopy. The instrumentation required for spectral measurements in this region is somewhat more expensive. Typically, FTIR spectrometers with InSb, HgCdTe, or extended InGaAs are required to obtain spectra with sufficient signal to noise. A further challenge is that fiber-optic materials appropriate for this spectral region are not as readily available. Chalcogenide fibers have typically high refactive indices leading to large Fresnel losses during coupling. They also exhibit sharp bands in transmission that may be bend dependent. Very low –OH glass fibers provide partial transmission out to about 2.3 μm, but fall off very rapidly in this region. Sapphire fibers and now fluoride are transmissive in this region and offer promise; however, they are quite expensive and typically more brittle and less flexible than glass fibers. Investigations of glucose determination using this spectral region include those recounted in References 59 through 62.

### 18.3.3.3 Raman Spectroscopy

Early attempts to employ Raman techniques to measure glucose concentration directly in aqueous solutions, serum, and plasma have met with some success *in vitro*.[63–67] However, efforts to utilize these techniques *in vivo* for transcutaneous measurement of whole blood glucose levels have met with considerable difficulty.[67,68] This is partly because whole blood and most tissues are highly absorptive; in addition, most tissues contain many fluorescent and Raman-active confounders.[68] As a surrogate to blood glucose measurement, several investigators have suggested using Raman spectroscopy to obtain the glucose signal from the aqueous humor of the eye.[68–70]

The glucose content of the aqueous humor reflects an age-dependent steady-state value of approximately 70% of that found in blood, and the time lag between aqueous humor and blood glucose has been shown to be on the order of minutes. For Raman sensing, the aqueous humor of the eye is relatively nonabsorptive and contains many fewer Raman-active molecules than whole blood. However, even though only four or five Raman-active constituents are present in significant concentrations in the aqueous humor, it is still necessary to use both linear and nonlinear multifactor analytical techniques to obtain accurate estimates of glucose concentrations from the total Raman spectrum.[71] In addition, as with other tissues, when excited in the near-infrared region (700 to 1300 nm) Raman spectra encounter less fluorescence background. However, although the background fluorescence falls off when excitation is moved to the NIR region, the Raman signal also falls off to the fourth power with wavelength. Thus it remains to be seen if the Raman approach could be measured at laser intensities that can be used safely in the eye.

#### 18.3.3.4 Polarimetric Glucose Sensing

As mentioned in the *in vitro* analysis section earlier, molecules such as glucose have the ability to rotate plane polarized light and this rotation is proportional to the concentration of the molecule. For polarimetry to be used as a noninvasive technique for blood glucose monitoring, the polarized light signal must be able to pass from the source, through the body, and to a detector without total depolarization of the beam. Since most tissues, including the skin, possess high scattering coefficients, maintaining polarization information in a beam passing through a thick piece of tissue (i.e., 1 cm), which includes skin, would not be feasible. Even highly scattering tissue thicknesses of less than 4 mm incur 95% depolarization of the light due to scattering from the tissue. In addition, in these highly scattering tissues a large amount of proteins and other chiral substances have a much stronger rotation than that due to glucose, making it very difficult to measure the millidegree rotations required for physiologic glucose concentrations. Similar to the Raman approach, the aqueous humor of the eye as a site for detection of *in vivo* glucose concentrations has been suggested as an alternative to transmitting light through the skin.[72–77] For the polarimetric approach, the cornea and the aqueous humor provide a low scattering window into the body and, because the aqueous humor is virtually devoid of proteins, the main optically chiral molecule is glucose.

Unlike on line processing systems that measure relatively high sugar concentrations (grams) and allow for longer pathlengths (10 cm), the *in vivo* glucose concentration is smaller and, if the eye is to be used as the sensing site, the pathlength is only 1 cm. Therefore, advances in polarimetric instrumentation have only recently made it possible to measure the small rotations due to glucose at physiological levels. For example, at a wavelength of 670 nm (red), glucose will rotate the linear polarization of a light beam approximately 0.4 millidegrees per 10 mg/dl for a 1-cm sample pathlength. In order to measure such small rotations, a very sensitive and stable polarimeter must be employed. Throughout the past decade many researchers have investigated the development of such sensitive units.[74–77] The primary approach being investigated today includes a light source (typically a laser), a means of modulating the polarization (photoelastic modulator, faraday rotator, etc.), and a means of detecting the signal, typically by using a lock-in amplifier or some means of heterodyne detection.

The eye as a sensing site for polarimetric glucose monitoring is not without its share of problems that need to be overcome. The cornea is birefringent, which means it can change the polarization vector of the light (i.e., like a quartz crystal); this change is variable in the presence of eye motion, which can confound the polarization signal due to glucose. For polarimetric monitoring, the light must travel across the anterior chamber of the eye but, due to its curvature and change in refractive index, the eye naturally tries to focus all the light impinging on it toward the retina. Therefore, the key issues to be addressed for this approach to be feasible for monitoring glucose include building an instrument that can accommodate for the large birefringence of the cornea, particularly in the presence of motion artifact, and to develop an instrument that can safely and simply couple the light source across the anterior chamber of the eye.

### 18.3.3.5 Other Optical Glucose Diagnostic Approaches

#### *18.3.3.5.1 Photoacoustic Spectroscopy*

Photoacoustic spectroscopy (PAS) can be used to acquire absorption spectra noninvasively from samples, including biological ones. The photoacoustic signal is obtained by probing the sample with a monochromatic radiation that is modulated or pulsed. Absorption of probe radiation by the sample results in localized short-duration heating. Thermal expansion then gives rise to a pressure wave, which can be detected with a suitable transducer. An absorption spectrum for the sample can be obtained by recording the amplitude of generated pressure waves as a function of probe beam wavelength. Because high signal-to-noise measurements require reasonable penetration of the sample by the probe radiation, the NIR region holds the same attraction as for more conventional NIR spectroscopy approaches. A purported advantage of PAS, however, is that the signal recorded is a direct result of absorption alone, and scatter or dispersion does not play a role in the acquired signal.

PAS techniques in the NIR region face many of the same challenges attendant to other NIR spectroscopic methods, namely, that complicated and empirical calibration models must be created in order to account for the broad overlapping spectral bands associated with the NIR spectrum. Further, the technique requires expensive instrumentation and the measurements are particularly sensitive to variables including temperature, transducer pressure, and sample morphology. Some examples of PAS techniques applied to glucose sensing include those found in References 78 and 79.

#### *18.3.3.5.2. Optical Property Measurements as Indicators of Glucose*

The refractive index (RI) of a material is routinely used to determine sugar concentrations in syrups, honey, molasses, tomato products, and jams for which the glucose concentration is very high. The RI of a solution of carbohydrate such as glucose increases with increasing concentration and so can be used to measure the amount of carbohydrate present. The RI of a liquid can easily be determined by optically measuring the angle of refraction at a boundary between the liquid and a solid of known RI. The RI is also temperature and wavelength dependent and so measurements are usually made at a specific temperature (20°C) and wavelength (589.3 nm). This method is quick and easy to carry out and can be performed with simple handheld instruments.

The change in refractive index is also directly related to a change in the elastic scatter of a molecule. Therefore, the measurement of light scatter in the NIR region has been investigated for potentially quantifying glucose noninvasively; both *in vitro* and *in vivo* results have been obtained.[80–82] This is much more challenging than for the agricultural measurement because *in vivo* blood glucose, as opposed to *in vitro* sugars such as that found in syrup, has very small concentrations. In addition, the specificity is potentially problematic because other physiologic effects, unrelated to glucose concentration, could produce similar variations of the reduced scattering coefficient with time. The measurement precision of the reduced scattering coefficient and separation of scatter and absorption changes is another potential problem with this approach. Tissue scattering is caused by a variety of substances and organelles (membranes, mitochondria, nucleus, etc.) and all have different RIs; thus, this approach also needs to take into account the different RIs of tissue. Lastly, there is a need to account for factors that might change the reduced scattering coefficient such as variations in temperature, red blood cell concentration, electrolyte levels, and movements of extracellular and intracellular water.

#### *18.3.3.5.3 Optical Coherence Tomography*

Optical coherence tomography (OCT) provides a particularly sensitive means for measuring the optical properties of turbid media including scatter and RI. OCT is an optical ranging technique in which a very short coherence length source is coupled to an interferometer with a sample in one arm and a reference reflector in the other. As the reference reflector is scanned, depth resolved interference fringes are produced with amplitude dependent on the amplitude of backscattered radiation. Because the scattering coefficient of tissue is dependent on the bulk index of refraction, an increase in RI and a decrease in scatter can be detected as a change in the slope of fall-off of the depth-resolved OCT amplitude. At least one group has explored this detection technique for measurement of glucose *in vivo*.[83]

## References


1. Schugerl, K., Progress in monitoring, modeling and control of bioprocesses during the last 20 years, *J. Biotechnol.*, 85, 149, 2001.
2. Fayolle, P., Picque, D., and Corrieu, G., On-line monitoring of fermentation processes by a new remote dispersive middle-infrared spectrometer, *Food Control*, 11, 291, 2000.
3. Ducommun, P., Bolzonella, I., Rhiel, M., Pugeaud, M., Stockar, U.V., and Marison, I.W., On-line determination of animal cell concentration, *Appl. Spectrosc.*, 72, 515, 2001.
4. Hagman, A. and Sivertsson, P., The use of NIR spectroscopy in monitoring and controlling bioprocesses, *Process Control Qual.*, 11, 125, 1998.
5. Heise, H.M., Bittner, A., and Marbach, R., Near-infrared reflectance spectroscopy for noninvasive monitoring of metabolites, *Clin. Chem. Lab. Med.*, 38, 137, 2000.
6. Lewis, C.B., McNichols, R.J., Gowda, A., and Coté, G.L., Investigation of near-infrared spectroscopy for periodic determination of glucose in cell culture media *in situ*, *Appl. Spectrosc.*, 54, 1453, 2000.
7. Hazen, K.H., Arnold, M.A., and Small, G.W., Measurement of glucose and other analytes in undiluted human serum with near-infrared transmission spectroscopy, *Appl. Spectrosc.*, 52, 1597, 1998.
8. Riley, M.R., Rhiel, M., Zhou, X., Arnold, M.A., and Murhammer, D.W., Simultaneous measurement of glucose and glutamine in insect cell culture media by near infrared spectroscopy, *Biotechnol. Bioeng.*, 55, 11, 1997.
9. Riley, M.R., Arnold, M.A., Murhammer, D.W., Walls, E.L., and Delacruz, N., Adaptive calibration scheme for quantification of nutrients and byproducts in insect cell bioreactors by near-infrared spectroscopy, *Biotechnol. Prog.*, 14, 527, 1998.
10. Yeung, K.S.Y., Hoare, M., Thornhill, N.F., Williams, T., and Vaghjiani, J.D., Near-infrared spectroscopy for bioprocess monitoring and control, *Biotechnol. Bioeng.*, 63, 684, 1999.
11. Fayolle, P., Picque, D., and Corrieu, G., Monitoring of fermentation processes producing lactic acid bacteria by mid-infrared spectroscopy, *Vibrator Spectrosc.*, 14, 247, 1997.
12. Li, Y., Brown, C.W., Sun, F.M., Mccrady, J.W., and Traxler, R.W., Non-invasive fermentation analysis using an artificial neural network algorithm for processing near infrared spectra, *J. Near Infrared Spectrosc.*, 7, 101, 1999.
13. Cavinato, A.G., Mayes, D.M., Ge, Z., and Callis, J.B., Noninvasive method for monitoring ethanol in fermentations processes using fiberoptic near-infrared spectroscopy, *Anal. Chem.*, 62, 1977, 1990.
14. Vaccari, G., Dosi, E., Campi, A.L., Gonzalezvaraa A., Matteuzzi, D., and Mantovani, G., A near-infrared spectroscopy technique for the control of fermentation processes — an application to lactic-acid fermentation, *Biotechnol. Bioeng.*, 43, 913, 1994.
15. McShane, M.J. and Coté, G.L., Near-infrared spectroscopy for determination of glucose lactate, and ammonia in cell culture media, *Appl. Spectrosc.*, 52, 1073, 1998.
16. Osbourne, B.G. and Fearn, T., *Near Infrared Spectroscopy in Food Analysis*, Longman Scientific and Technical, Harlow, Essex, U.K., 1986.
17. Burns, D.A. and Ciurczak, E.W., *Handbook of Near-Infrared Analysis*, Marcel Dekker, New York, 1992.
18. McClure, G.L., *Computerized Quantitative Infrared Analysis*, American Society for Testing and Materials, Philadelphia, PA, 1987.
19. Mattu, M.J., Small, G.W., and Arnold, M.A., Application of multivariate calibration techniques to quantitative analysis of bandpass-filtered Fourier transform infrared interferogram data, *Appl. Spectrosc.*, 51, 1369, 1997.

20. Goodacre, R., Monitoring microbial bioprocesses for metabolite concentrations using dispersive and surface enhanced Raman spectroscopies and machine learning, The Aberystwyth Quantitative Biology Group Institute of Biological Sciences, http://gepasi.dbs.aber.ac.uk/roy/advert/pg00.htm, site accessed November 2001.
21. Berger, A.J., Introduction to concepts in laser technology for glucose monitoring, *Diabetes Technol. Therapeutics*, 1, 121, 1999.
22. Long, D.A., *Raman Spectroscopy*, McGraw-Hill, New York, 1977, pp. 1–12.
23. Shaw A.D., Kaderbhai, N., Jones, A., Woodward, A.M., Goodacre, R., Rowland, J.J., and Kell, D.B., Non-invasive, on-line monitoring of the biotransformation by yeast of glucose to ethanol using dispersive raman spectroscopy and chemometrics, *Appl. Spectrosc.*, 53, 1419, 1999.
24. Moskovits, M., Surface enhanced spectroscopy, *Rev. Mod. Phys.*, 57, 783, 1985.
25. Chang, R.K. and Furtak, T.E., *Surface Enhanced Raman Scattering*, Plenum Press, New York, 1982.
26. Kneipp, K., Kneipp, H., Itzkan, I., Dasari, R.R., and Feld, M.S., Surface-enhanced Raman scattering: a new tool for biomedical spectroscopy, *Curr. Sci.*, 77, 915, 1999.
27. Vo-Dinh, T., Stokes, D.L., Griffin, G.D., Volkan, M., Kim.U.J., and Simon M.I., Surface-enhanced Raman scattering (SERS) method and instrumentation for genomics and biomedical analysis, *J. Raman Spectrosc.*, 30, 785, 1999.
28. Vo-Dinh, T., Surface enhanced Raman spectroscoppy using metallic nanostructures, *Trends Anal. Chem.*, 17, 557, 1998.
29. Rudolph Research Analytical, 354 Route 206 Flanders, NJ 07836, http://www.rudolphresearch.com/polarimetry.htm, site accessed November 2001.
30. Browne, C.A. and Zerban, F.W., *Physical and Chemical Methods of Sugar Analysis*, 3rd ed., John Wiley & Sons, New York, 1941, p. 263.
31. Phillips, R., McGarraugh, G., Jurik, F.A., and Underwood, R.D., U.S. patent 5,563,042, Whole blood glucose test strip, October 8, 1996.
32. Marcus, A., Medical lasers bid to supplant needles, *EEtimes.com*, April 4, 2001, http://www.eetimes.com/story/OEG20010404S0054, site accessed November 2001.
33. Lasette Plus, Cell Robotics International, Inc., 2715 Broadbent Parkway NE, Albuquerque, NM 87107 USA, http://www.cellrobotics.com/, site accessed November 2001.
34. Jacques, S., Laser poration of skin for transdermal drug delivery and interstitial fluid and blood gas collection, NewsEtc., from the Oregon Medical Laser Center, May, 1998, http://omlc.ogi.e.,du/news/may98/gallery_may98.html, site accessed November 2001.
35. SpectRx, Inc., Continuous Glucose Monitoring, http://www.spectrx.com/DevGlucose.html, site accessed November 2001.
36. Moreno-Bondi, M.C., Wolfbeis, O.S., Leiner, M.J., and Schaffar, B.P., Oxygen optrode for use in a fiber-optic glucose biosensor, *Anal. Chem.*, 62, 2377, 1990.
37. Rosenzweig, Z. and Kopelman, R., Analytical properties and sensor size effects of a micrometer-sized optical fiber glucose biosensor, *Anal. Chem.*, 68, 1408, 1996.
38. Li, L. and Walt, D.R., Dual-analyte fiber-optic sensor for the simultaneous and continuous measurement of glucose and oxygen, *Anal. Chem.*, 67, 3746, 1995.
39. Gunsingham, H., Tan, C.H., and Seow, J.K.L., Fiber-optic glucose sensor with electrochemical generation of indicator reagent, *Anal. Chem.*, 62, 755, 1990.
40. Abdel-Latif, M.S. and Guilbault, G.G., Fiber-optic sensor for the determination of glucose using micellar enhanced chemiluminescence of the peroxylate reaction, *Anal. Chem.*, 60, 2671, 1988.
41. Schultz, J.S., A miniature optical glucose sensor based affinity binding, *Biotechnology*, 2, 885, 1984.
42. Schultz, J.S., Mansouri, S., and Goldstein, I.J., Affinity sensor: a new technique for developing implantable sensors for glucose and other metabolites, *Diabetes Care*, 5, 245, 1982.
43. Meadows, D. and Schultz, J., Fiber-optic biosensors based on fluorescence energy-transfer, *Talanta*, 35, 150, 1988.
44. Meadows, D. and Schultz, J., Design, manufacture and characterization of an optical-fiber glucose affinity sensor-based on an homogeneous flouroscence energy-transfer assay system, *Anal. Chem. Acta*, 280, 21, 1993.

45. Lakowicz, J.R. and Maliwal, B., Optical sensing of glucose using phase-modulation fluorimetry, *Anal. Chim. Acta*, 271, 155, 1993.
46. Ballerstadt, R. and Schultz, J.S., A fluorescence affinity hollow fiber sensor for continuous transdermal glucose monitoring, *Anal. Chem.*, 72, 4185, 2000.
47. Russell, R., Gefrides, C., McShane, M., Coté, G.L., and Pishko, M., A fluorescence-based glucose biosensor based on Concanavalin A and Dextran encapsulated in a poly(ethylene glycol) hydrogel, *Anal. Chem.*, 71, 3126, 1999.
48. Bhandare, P., Mendelson, Y., Peura, R.A., Jantsch, G., Krüse-Jarres, J.D., Marbach, R., and Heise, H.M., Multivariate determination of glucose in whole blood using partial least-squares and artificial neural networks based on mid-infrared spectroscopy, *Appl. Spectrosc.*, 47, 1214, 1993.
49. Zeller, F., Novak, P., and Landgraf, R., Blood glucose measurement by infrared spectroscopy, *Int. J. Artif. Organs*, 12, 129, 1989.
50. Bhandare, P., Mendelson, Y., Stohr, E., and Peura, R., Glucose determination in simulated plasma solutions using infrared spectrophotometry, *Proc. IEEE-EMBS*, 14, 163, 1992.
51. Heise, H.M., Marbach, R., Janatsch, G., and Krüse-Jarres, J.D., Multivariate determination of glucose in whole blood by attenuated total reflection infrared spectroscopy, *Anal. Chem.*, 61, 2009, 1989.
52. Robinson, K., Blood analysis: noninvasive methods hover on horizon, *Biophotonics Int.*, 5, 48, 1998.
53. Martens, H. and Naes, T., *Multivariate Calibration*, John Wiley & Sons, New York, 1989.
54. Haaland, D.M. and Thomas, E.V., Partial least-squares methods for spectral analyses. 1. Relation to other quantitative calibration methods and the extraction of qualitative information, *Anal. Chem.*, 60, 1193, 1988.
55. Amato, I., Race quickens for non-stick blood monitoring technology, *Science*, 258, 892, 1992.
56. Robinson, M.R., Eaton, R.P., Haaland, D.M., Koepp, G.W., Thomas, E.V., Stallard, B.R., and Robinson, P.L., Noninvasive glucose monitoring in diabetic patients: a preliminary evaluation, *Clin. Chem.*, 38, 1618, 1992.
57. Rosenthal, R.D., Paynter, L.N., and Mackie, L.H., Non-invasive measurement of blood glucose, U.S. patent 5,028,787, 1991.
58. Danzer, K., Fischbacher, C., Jagemann, K.U., and Reichelt, K.J., Near-infrared diffuse reflection spectroscopy for non-invasive blood-glucose monitoring, *IEEE-LEOS Newslett.*, 12, 9, 1998.
59. Haaland, D.M., Robinson, M.R., Koepp, G.W., Thomas, E.V., and Eaton, R.P., Reagentless near-infrared determination of glucose in whole blood using multivariate calibration, *Appl. Spectrosc.*, 46, 1575, 1992.
60. Marbach, R., Koschinsky, T., Gries, F.A., and Heise, H.M., Noninvasive blood glucose assay by near-infrared diffuse reflectance spectroscopy of the human inner lip, *Appl. Spectrosc.*, 47, 875, 1993.
61. Shengtian, P., Chung, H., and Arnold, M.A., Near-infrared spectroscopic measurement of physiological glucose levels in variable matrices of protein and triglycerides, *Anal. Chem.*, 68, 1124, 1996.
62. Chung, H., Arnold, M.A., Rhiel, M., and Murhammer, D.W., Simultaneous measurements of glucose, glutamine, ammonia, lactate, and glutamate in aqueous solutions by near-infrared spectroscopy, *Appl. Spectrosc.*, 50, 270, 1996.
63. Wang, S.Y., Hasty, C.E., Watson, P.A., et al., Analysis of metabolites in aqueous solutions by using laser Raman spectroscopy, *Appl. Opt.*, 32, 925, 1993.
64. Goetz, M.J., Coté, G.L., Ercken, R., et al., Application of a multivariate technique to Raman spectra for quantification of body chemicals, *IEEE Trans. Biomed. Eng.*, 42, 728, 1995.
65. Dou, X.M., Yamaguchi, Y., Yamamoto, H., Uenoyama, H., and Ozaki, Y., Biological applications of anti-Stokes Raman spectroscopy-quantitative analysis of glucose in plasma and serum by a highly sensitive multichannel Raman spectrometer, *Appl. Spectrosc.*, 50, 1301, 1996.
66. Berger, A.J., Itzkan, I., and Feld, M.S., Feasibility of measuring blood glucose concentration by near-infrared Raman-spectroscopy, *Spectrochim. Acta A Mol. Biomol. Spectrosc.*, 53, 287, 1997.
67. Koo, T.-W., Berger, A.J., Itzkan, I., Horowitz, G., and Feld, M.S., Reagentless blood analysis by near-infrared Raman spectroscopy, *Diabetes Technol. Ther.*, 1, 153, 1999.

68. Lambert, J.L., Storrie-Lombardi, M.C., and Borchert, M.S., Measurement of physiologic glucose levels using Raman spectroscopy in a rabbit aqueous humor model, *LEOS Newslett.*, 12, 19, 1998.
69. Tarr, R.V. and Steffes, P.G., Non-invasive blood glucose measurement system and method using stimulated Raman spectroscopy, U.S. patent 5243983, Sept. 14, 1993.
70. Wicksted, J.P., Erckens, R.J., Motamedi, M., and March, W.F., Raman spectroscopy studies of metabolic concentrations in aqueous solutions and aqueous humor specimens, *Appl. Spectrosc.*, 49, 987, 1995.
71. Borchert, M.S., Storrie-Lombardi, M.C., and Lambert, J.L., A noninvasive glucose monitor: preliminary results in rabbits, *Diabetes Technol. Ther.*, 1, 145, 1999.
72. March, W.F., Rabinovitch, B., and Adams, R.L., Noninvasive glucose monitoring of the aqueous humor of the eye: part II. Animal studies and the scleral lens, *Diabetes Care,* 5, 259, 1982.
73. Rabinovitch, B., March, W.F., and Adams, R.L., Noninvasive glucose monitoring of the aqueous humor of the eye. Part I. Measurement of very small optical rotations, *Diabetes Care,* 5, 254, 1982.
74. Coté, G.L., Fox, M.D., and Northrop, R.B., Noninvasive optical polarimetric glucose sensing using a true phase measurement technique, *IEEE Trans. Biomed. Eng.*, 39, 752, 1992.
75. Cameron, B.D. and Coté, G.L., Noninvasive glucose sensing utilizing a digital closed-loop polarimetric approach, *IEEE Trans. Biomed. Eng.*, 44, 1221, 1997.
76. Chou, C., Han, C.Y., Kuo, W.C., Huang, Y.C., Feng, C.M., and Shyu, J.C., Noninvasive glucose monitoring *in vivo* with an optical heterodyne polarimeter, *Appl. Opt.*, 37, 3553, 1998.
77. Cameron, B.D., Baba, J.S., and Coté, G.L., Measurement of the glucose transport time-delay between blood and aqueous humor of the eye for the eventual development of a noninvasive glucose sensor, *Diabetes Technol. Ther.*, 3, 201, 2001.
78. Christison, G.B. and MacKenzie, H.A., Laser photoacoustic determination of physiological glucose concentration in human whole blood, *Med. Biol. Eng. Comput.*, 31, S17, 1993.
79. Quan, K.M., Christison, G.B., MacKenzie, H.A., and Hodgson, P., Glucose determination by a pulsed photoacoustic technique: an experimental study using a gelatin-based tissue phantom, *Phys. Med. Biol.*, 38, 1911, 1993.
80. Kohl, M. and Cope, M., Influence of glucose concentration on light scattering in tissue-simulating phantoms, *Opt. Lett.*, 19, 2170, 1994.
81. Maier, J.S., Walker, S.A., Fantini, S., Franceschini, M.A., and Gratton, E., Possible correlation between blood glucose concentration and the reduced scattering coefficient of tissues in the near infrared, *Opt. Lett.*, 19, 2062, 1994.
82. Bruulsema, J.T., Hayward, J.E., Farrell, J.T., Patterson, M.S., Heinemann, L., Berger, M., Koschinsky, T., Sandahl-Christiansen, J., Orskov, H., Essenpreis, M., Schmelzeisen-Redeker, G., and Böcker, D., Correlation between blood glucose concentration in diabetics and noninvasively measured tissue optical scattering coefficient, *Opt. Lett.*, 22, 190, 1997.
83. Larin, K.V., Larina, I.V., Motamedi, M., Gelikonov, V.M., Kuranov, R.V., and Esenaliev, R.O., Potential application of optical coherence tomography for noninvasive monitoring of glucose concentration, *Proc. SPIE,* 4263, 83, 2001.

# 19

# *In Vitro* Clinical Diagnostic Instrumentation

Gerald E. Cohn
*Cyber Tech Applied Science, Inc.*
*Evanston, Illinois*

Richard Domanik
*Xomix, Ltd.*
*Chicago, Illinois*



## 19.1 Introduction

Clinical diagnostic tests can be performed *in situ* or on physiological specimens that have been removed from the living organism. These latter tests are referred to as *in vitro* diagnostics. Specimens for *in vitro* diagnostic testing range from tissue and cell samples (or extracts from them) to respired or dissolved gasses and all types of physiological fluids and exudates. Analytes of interest span the gamut from inorganic ions and small organic molecules to cellular entities such as bacteria and viruses. Testing methods and their underlying technologies are almost as diverse as the samples evaluated. Furthermore, these tests are delivered on a variety of platforms ranging from dedicated-function, single-use, noninstrumented devices to high-throughput automated analyzers capable of performing a wide range of tests in a batch or a random access mode.

This chapter surveys optically based detection and transduction modalities frequently encountered in instrumented *in vitro* clinical diagnostic applications. Our approach is to summarize the characteristics of these system components with particular emphasis on the function, precision, and application of the different system components.

Automated *in vitro* diagnostic instruments capable of performing assays for multiple analytes on multiple samples may be dated to the introduction of the Technicon Auto Analyzer in 1957.[1] This innovation, and other early automated optical absorbance-based clinical analyzers, led the transformation of the laboratory measurement paradigm from individual samples analyzed manually to automated high-throughput instrumental analysis. Since the debut of the original Auto Analyzer, numerous signal generation, transduction, and systems have been introduced to expand and enhance the repertoire and performance of *in vitro* clinical laboratory instruments greatly.[2,3] Today, these instrument systems are being pushed to provide ever higher levels of analytical accuracy and precision — at progressively lower

0-8493-1116-0/03/$0.00+$1.50

limits of detection in a rapidly broadening array of assay types and in the face of significant legal and regulatory requirements, reduced laboratory staffing, and severe cost constraints.[4–6]

From a systems engineering perspective, a clinical diagnostic instrument can be viewed as a sequence of black boxes that transduce a parameter of clinical interest, such as the concentration of an analyte, into a form that can be used by the clinician. In most typical clinical diagnostic tests, the sequence of black boxes required to perform this function include sample acquisition and preparation, signal generation, signal transduction, data reduction, and results reporting. These black boxes, and the details of the operations that they perform, are tied together by the "intended use" of the test. The concept of intended use is particularly important to the developer of *in vitro* diagnostic tests because it often dictates the overall approach to be taken, the methods to be used, and the compromises to be made during the development process. For the purposes of this chapter, the focus is on the "black box" labeled signal transduction. It must, however, be recognized that the other "boxes" play critical roles in the overall system.

## 19.2 Assay Chemistry

By convention, the sample acquisition, sample preparation, and signal generation functions are usually lumped together under the rubric of "assay chemistry." The key design goals in the development of an assay chemistry are the delivery of an optical signal that is a tightly controlled function of the concentration of the analyte of interest and that exhibits a significant optical contrast between the analyte and its matrix. The actual signal delivered by the assay chemistry defines the type, spectral characteristics, magnitude, dynamic range, and signal-to-noise ratio (SNR) of the signal presented to the signal transduction element, while the intended use of the test defines the required measurement resolution. The assay chemistry also establishes the characteristics of the noise component of the input signal and operational parameters such as timing constraints.[7]

The assay chemistry can generate the necessary optical signal in any of a diverse variety of ways depending on the characteristics of the target analyte and specimen as well as on the intended use of the test. In the simplest cases, the analyte has some intrinsic characteristic that allows it to be directly differentiated from its matrix. In the majority of cases, however, the specimen must be treated with some extrinsic reagent to generate the necessary contrast. Chemical reagents that bind to (stain, label) or react with the analyte to generate an optical contrast are the basis of many traditional clinical chemistry, cytological, and histological tests. Tests based upon these types of reagents can be quite sensitive, but frequently are somewhat limited in specificity.

Increased specificity in signal generation is often obtained by incorporating some type of biological "recognition" moiety such as an enzyme, antibody, or nucleic acid probe into the reagent. Enzymatic reagent(s) are used to convert the target analyte into some other molecular form in a highly specific manner.[8] Depending on the specifics of the particular analyte–enzyme combination used, the contrast is generated by conversion of the analyte into some product or through ancillary chemical or enzymatic reactions stoicheometrically linked to the primary conversion.[7,9]

Antibodies, as well as certain other biological molecules such as nucleic acid probes, lectins, and certain nonantibody proteins, have the characteristic of being able to bind to an analyte with a high degree of specificity and tenacity.[10,11] In cases such as agglutination tests, this binding is used directly to cross-link multiple instances of the analyte, leading to the formation of a detectable aggregate or precipitate.[12,13] In other cases, the specific binding agent is used as a carrier for some indicator moiety such as a chromaphore or fluorophore.[12,13] Alternatively, the binding agent may serve as a carrier for an enzyme that, in turn, stoicheometrically causes a detectable transformation in some indicator species.

Generation of contrast is implemented in one of two ways: "end point" or "kinetic." In an end point assay, where the contrast-generating reaction is run to completion, the total amount of contrast produced is directly limited by the amount of analyte present. Kinetic assays, on the other hand, are performed under conditions where the rate of contrast formation is directly related to analyte concentration. End point assays tend to be relatively insensitive to process parameters such as reagent concentrations and reaction temperature, while kinetic assays are by definition sensitive to any and all factors that affect the rate of the reaction. Kinetic assays, therefore, impose stringent process control requirements that must be met in order to obtain valid results. However, accurate kinetic results can be obtained well before the assay chemistry has proceeded to completion. Thus, a kinetic assay is often the method of choice in an instrument-based system where sample throughput is a critical issue.

The methods that assay developers use to generate and optimize the necessary signal are many, varied, and, in any case, well beyond the scope of this chapter. However, it should be kept in mind that the assay chemistry can be and often is one of the larger contributors to the total system error budget. Some factors that contribute to this, such as accuracy of fluid transfers, rates and thoroughness of mixing, concentrations of reagents, and stability of temperature control, are operational issues. Chemical and biochemical factors, such as the purity and specificity of the assay reagent for the target analyte, can also be significant. The sample is yet another common contributor to the system error budget because it may contain substances that interfere with the detection reaction, signal transduction, or even physical operations such as fluid transfer. During the design process, accounting for and managing errors originating in the assay chemistry are as important as attending to those associated with the instrumentation.[7,14]

## 19.3 System Components

The components within the signal transduction black box usually consist of an optical system to deliver the signal produced by the assay to the transducer (detector), a detector that transduces the signal from the optical to the electronic domain, and supporting electronics that condition the detector output and convert it into a form suitable to be supplied to the data reduction black box. With few exceptions such as photomultiplier tubes (PMTs) and avalanche photodiodes used in the photon counting mode, almost all detectors commonly used in clinical diagnostic instruments intrinsically produce an analog output that is a continuous representation of the optical input to the detector. Many modern instruments, however, perform most, if not all, of the necessary signal conditioning and data reduction operations in the digital domain. The location in the signal processing chain at which analog-to-digital (A/D) conversion is performed and the characteristics of the circuitry utilized to perform this conversion are design decisions that can have significant impacts upon the error budget for the signal transduction process.[15–18]

In most types of clinical diagnostic instruments, an illumination subsystem is an essential component of the signal transduction optics. The major requirement placed on the sample illumination subsystem is that it illuminate the portion of the specimen to be interrogated in a uniform and stable manner, with light of suitable application specific intensity, spatial distribution, and spectral characteristics. Certain applications such as interferometric or polarization-based techniques require that additional controls be placed on the illuminating wavefront.

Numerous illuminator designs have been developed to meet the needs of different applications.[19,20] Incandescent and arc lamps remain the most popular illumination sources; however, with the advent of high power and "white" light-emitting diodes (LEDs) and relatively low cost diode lasers, solid-state illuminators are appearing in some applications. Most illuminators utilize free space optics, but optical waveguide–based light delivery systems can be useful in some applications.

Almost all clinical laboratory instruments utilize some form of optical arrangement to capture the light passing through or emitted by the sample and to deliver this light to the detector. In addition to light capture and delivery, most such optical systems process the captured light in some manner. Most commonly, this processing involves operations such as reformatting the size or shape of the captured beam of light to match the detector, wavelength selection, and aperturing or masking to control stray light. Again, numerous optical designs have been developed to meet the needs of specific applications.[19,20] Most of these designs are of the imaging or pseudo-imaging type, although true nonimaging optics are employed in some specialized applications.

Imaging optics are exemplified by photographic lenses and microscope objectives and are intended to deliver a spatially resolved image of the sample to the detector. The lenses or mirrors used in these applications are typically required to exhibit a uniform transfer function over the field of view and often must incorporate corrections for geometric and chromatic aberrations. The pseudo-imaging optics used in most photometer designs are intended to deliver an image of the sample to the detector without regard for maintaining spatial resolution. Efficient delivery of optical power from the sample to the detector is a major design consideration in this type of system, and transfer function uniformity and aberration correction are lesser concerns. Nonimaging optics, usually in the form of integrating spheres and compound parabolic or elliptical concentrators, are primarily used in chemiluminescent and similar photon-limited applications in which optical collection and throughput efficiency must be maximized while spatial resolution and the light intensity distribution at the detector are irrelevant.[21,22]

## 19.4 Detection Modalities

This section considers the transduction of several of the more widely used types of optical signals from the perspective of the physics of the measurement and the sources and effects of potential measurement errors. As implied earlier, the designs of the signal transduction subsystems of most clinical laboratory instruments are unique reflections of numerous application-specific design decisions and trade-offs made during the course of the instrument development process. Because a detailed mathematical analysis of a measurement system is sensitive to the specifics of the design and multiple designs need to be analyzed in order to represent current practices adequately, space constraints dictate that a number of the topics presented in this section be addressed in broad qualitative generalities rather than through formal derivations. For similar reasons, equations will be presented in this section without derivation or proof. Table 19.1 provides a summary of *in vitro* clinical diagnostic applications of various detection modalities.

### 19.4.1 Optical Absorbance

Optical absorbance is a contrast parameter widely utilized in instrumental and in visual clinical diagnostic tests. Traditionally, instrument-based absorbance assays are based upon the photometric measurement of optical transmission or reflectance of a sample at one or a small number of the wavelengths in the UV or visible spectral region. A few recently introduced tests, however, extend this spectral range into the near-infrared (NIR) region or utilize data captured at a large number of wavelengths.

Optical absorbance measures the reduction in the intensity of an incident beam by absorptive interaction with the sample. In this case, the governing equation for optical transmittance ($T$) under monochromatic illumination is

$$T = (I/I_0) = 10^{(-\varepsilon LC)} \tag{19.1}$$

where $I_0$ is the incident light intensity, $I$ is the measured light intensity after passage through the sample, $L$ is the optical path length through the sample in centimeters, $C$ is the analyte concentration in molar units, and $\varepsilon$ is the molar absorptivity (extinction coefficient) of the analyte in units of l/$M$-cm.

Equation 19.2 is the linearized form of Equation 19.1. This equation, which is known as Beer's law of absorbance, is used far more frequently in clinical diagnostic applications than is Equation 19.1 because absorbance ($A$) is a linear function of analyte concentration.

$$A = \log_{10}(I_0/I) = -\log_{10} T = \varepsilon LC \tag{19.2}$$

The measurements of $I_0$ and $I$ are subject to noncompensating errors from a number of sources. Classical single- and dual-beam optical arrangements, for example, carry the implicit assumption that the sample and reference paths differ only in concentration of the target analyte. Single-beam measurements further assume that the intensity and spectral distribution of the illumination source is stable over the entire period of time required to make the measurements. For these reasons, many clinical photometers use a single-beam arrangement in which $I_0$ is determined from a sample of the illuminating beam that is "picked off" before it enters the cuvette. These systems treat a "reagent blank" as a separate

**TABLE 19.1** *In Vitro* Clinical Diagnostic Applications of Detection Modalities

| Detection | Typical Assay Types | Routine Analytes | Features |
|---|---|---|---|
| Absorbance | | | |
| Single wavelength | Chemical, EIA | Clinical chemistries, drugs | |
| Dual wavelength | Chemical, EIA | Clinical chemistries, drugs | Partially self-compensating |
| Diffuse Reflectance | | | |
| | Lateral flow immunoassays | Drugs, small molecules | Simplified processing |
| | Transverse flow immunoassays | Drugs, small molecules, large molecules, bacteria, viruses | Simplified processing |
| Fluorescence | | | |
| Intensity | Direct staining, EIA, DNA probe | Drugs, small molecules, large molecules, bacteria, viruses | Sensitive |
| Polarization | Direct binding immunoassays | Drugs, small molecules | Assay mixture separtion not required |
| Chemiluminescence | Direct binding immunoassays | Drugs, small molecules, large molecules, bacteria, viruses | Very sensitive |
| Waveguide Sensors | | | |
| Absorbance | Competitive immunoassays | Drugs, small molecules | Simplified processing |
| Fluorescence | Competitive immunoassays | Drugs, small molecules | Simplified processing |
| Refractive index | Specific binding | Drugs, small molecules, large molecules, bacteria, viruses | No extrinsic reagents |
| Imaging Systems | | | |
| Classical | Image analysis | Cytology, histology, microbiology, agglutination | Combines chromatic and morphological |
| Multiplexed | Photometric | Drugs, small molecules, large molecules, bacteria, viruses | Improved throughput |
| Spectrographic | Hyperspectral | Cellular analysis | Detailed characterization and morphology |

sample, the apparent analyte concentration of which is subtracted from those of the corresponding experimental samples.[17,23]

Equation 19.2 can be used to estimate the effects of noise on the limits of detection obtainable in absorbance measurements. For example, at low absorbances, a noise level equivalent to $0.001 \times I_0$ corresponds to an uncertainty in $A$ of approximately $4.4 \times 10^{-4}$. If the detection limit is taken as three times the noise level, the corresponding lower limit of detection (LOD) is approximately 0.0013$A$. Applying the same calculation at high absorbances gives an upper LOD of about 3.00$A$. This noise also limits the resolution with which measurements can be made between the lower and upper LODs.[24]

Measurement uncertainties also play a role in the accuracy of determination of $T$ or $A$. At analyte concentrations that correspond to high values of $T$ (low $A$), $I$ and $I_0$ are large and of nearly equal magnitude. Under these conditions, measurement resolution limitations and optical and electronic noise in the system degrade the accuracy of the measurements and thus the accuracy of $T$ or $A$. At the opposite extreme, $I << I_0$, electronic noise and stray light in the system effectively introduce a "floor" under the measurement of $I$. Due to the large difference between $I$ and $I_0$, limitations on the dynamic range and resolution of the measurement device may also affect the accuracy with which $T$ and $A$ are determined.[24] The effects of these factors can be illustrated by differentiating Equation 19.2 to obtain

$$\delta A/A = 0.434\ dT/(T \log_{10} T) = \delta C/C = \text{RCE} \tag{19.3}$$

where RCE is the relative concentration error. Qualitatively,

At high *T*/low *A*, *T* approaches 1 and log *T* approaches 0, so the RCE increases.
At low *T*/high *A*, the term *T* log *T* approaches 0 and the RCE increases.
At some intermediate value of *A*, the RCE passes through a minimum with a range of low RCE centered about that minimum.[15,17]

Deriving Equation 19.3 in a form that includes a suitable noise model for the system allows determination of the value of *A* at which the RCE will be at a minimum. Assuming a shot-noise limited model in which uncertainties in $I_0$ are neglected, for example, yields $A = 0.869$. The range of A over which the RCE remains within preselected limits can also be determined. In this particular example, assuming that the RCE is less than twice its minimum value translates to the lowest measurement errors occurring in the range $0.201 < A < 2.326$.[25–27]

Both ε and *L* can be sources of analytical error when *T* or *A* is translated into analyte concentration by means of Equation 19.1 or 19.2, respectively. The value of ε is wavelength dependent and thus takes on a range of values over the spectral passband of the instrument. Similarly, the optical path length through a clinical laboratory sample is not necessarily well controlled over the entire region being interrogated, particularly when a disposable cuvette is used for each measurement. In cases such as microscopy specimens, the specimen can be the source of significant path length variability. In the slope of the *A* vs. *C* relationship, the assay/instrument combination in question can be experimentally determined and used as an empirical constant that subsumes ε and *L* as well as fixed errors in these parameters. This empirical constant can be used in place of the ε*L* term in Equations 19.1 and 19.2. To the extent that it cannot be controlled by design or process, variability must be included in the system error budget. The assay chemistry can also contribute to the measurement error and has been found to produce RCE ~ $1/A$.[28]

Bichromatic absorption photometry can be a useful technique in situations where the effects of interfering factors such as light scattering vary more slowly with wavelength than does the absorbance of the analyte of interest. If the absorbance of the sample is measured at two wavelengths, one at the peak absorbance of the analyte and the other at a (typically longer) wavelength where the analyte has no or minimal absorbance, an approximate correction for the interfering factors can be performed by subtracting the second measurement from the first.[29]

### 19.4.2 Reflectance

Lateral and transverse flow assays are performed and "read" on porous surfaces such as chromatographic and depth filtration media. Many of these assays utilize a color-generating indicator reaction and can be read out visually or by means of a photometric instrument. Because the substrates used in lateral and transverse flow assays tend to be highly scattering and opaque, the measurement instrument of choice is almost always a reflectance photometer. Light reflected from a surface consists of specularly reflected and diffusely reflected components. The intensity of the specular component is largely determined by the surface properties of the sample and is generally not of analytical utility. The intensity of the diffuse component, which includes contributions from the absorbance of light by the specimen and the scattering of light by the specimen and substrate, can be used to determine the concentration of the indicator species. In the ideal limiting case where all reflection is diffuse, light scattering is independent of the incident wavelength, the specimen is infinitely thick, and the indicator species is infinitely diluted in a nonabsorbing matrix, the relationship between diffuse reflectance, absorbance, and light scattering is given by the Kubelka–Munk equation[30]

$$f_{(R)} = (1 - R)^2/(2R) = K/S \tag{19.4}$$

where $R$ is the absolute reflectance, $K$ is the molar absorbance coefficient, and $S$ is the scattering coefficient of the specimen. By convention, quantitative diffuse reflectance measurements are presented in "log $1/R$" units that are directly analogous to the absorbance units (log $1/T$) described previously. In other words, the magnitude of the diffuse reflectance is linearly proportional to sample absorbance and, therefore, to indicator concentration.[31]

Although Equation 19.4 was derived for an ideal limiting case, it is generally applicable to real-world specimens so long as:

Light scattering and absorbance by the sample matrix are largely independent of wavelength over the wavelength band of interest.
The specimen is of sufficient thickness.
The species of interest is relatively dilute (typically <10% concentration).

The scattering constraint is generally satisfied if the sizes of the scattering centers are substantially greater than the wavelength of the incident light. The other constraints are addressed through the mechanics of the assay implementation and the design of the assay chemistry. Most of the remaining deviations from the ideal can be addressed through careful calibration of the system. It should be noted that the effective optical path length in a diffuse reflectance measurement can be exceptionally long due to the scattering and diffusion of photons in the specimen matrix. This means that the sensitivity of a reflectance measurement is often substantially greater than that of the corresponding measurement made in transmission.

As with the methods discussed earlier, the Kubelka–Monk model for light-sample interaction permits a minimum RCE analysis. Solving the dependence of the reflectance function on absorber concentration $C$ for $C$ and obtaining an expression for $\delta C/C$, which is then minimized, yields a minimum in RCE at $R = 0.414$. The RCE is within 10% of this minimum value for values of $R$ between 0.13 to 0.77.[30] This analysis can be combined with models for detector uncertainty, chemical noise, and other system uncertainties for a more complete understanding.[32]

### 19.4.3 Fluorescence

Fluorescence emission is the result of a complex series of events that occur at the molecular level. The net effect is that a fluorescent molecule absorbs light within a molecule-specific absorption band and consequently emits light isotropically at a longer wavelength. The wavelength shift accompanying fluorescent emission and the isotropic nature of these emissions facilitate in discriminating between exciting and emitted light in fluorescence measurements.

One form of the relationship between measured fluorescence intensity and analyte concentration is given in

$$I_F = I_0 KQ(1 - 10^{-\varepsilon LC}) \tag{19.5}$$

where $I_F$ is the measured fluorescence intensity, $K$ is a geometric instrument factor, $Q$ is the quantum efficiency, and the remaining terms have their previous meanings. This equation indicates that, for low analyte concentrations (typically under about $10^{-5}$ M), fluorescence intensity is linearly proportional to the analyte concentration. At very low analyte concentrations, the RCE for fluorescence measurements is similar to that for absorbance in that measurement noise and stray light predominate and result in values of $I_F$ that are systematically higher than expected. At high analyte concentrations, a variety of intermolecular interactions act to quench fluorescence emissions, thus leading to systematically low values of $I_F$.[33]

Many of the same factors that contribute to errors in absorbance measurements are of concern in fluorescence. Fluorescence emissions are isotropic, so geometry dictates that most practical optical systems can capture only a small fraction of the emitted light. Consequently, signal levels at the detector are low even at high analyte concentrations. Noise control is, therefore, a major consideration in the design of fluorescence-based instrumentation. A source of errors that is unique to fluorescence measurements is that, for Equation 19.4 to remain valid over any significant range of analyte concentrations, the incident photon flux must be sufficient to "saturate" the fluorophores at the highest analyte concentration of interest. This ensures that the amount of time that a fluorophore spends in its ground state is minimized and that $I_F$ will not be affected by fluctuations in illumination intensity.

The high illumination intensity implied by this consideration directly conflicts with the propensity of many fluorophores used in clinical assays to "photobleach" (i.e., be photochemically degraded) over time. Photobleaching, which leads to systematically low results, is generally addressed by limiting the amount of time that fluorescent samples are exposed to light and by limiting the illumination intensity during the measurement period to the minimum necessary to ensure saturation. In some cases, the assay chemists can select a fluorophore that is resistant to photobleaching, or can add stabilizer (usually a free radical scavenger or antioxidant) to the assay reagents. Many constituents of biological samples, as well as some assay reagents and some materials used in instrument construction, are intrinsically fluorescent. Unless considerable care is taken in the selection of the fluorophore, intrinsic system fluorescence may interfere with analyte quantitation.[24,34]

Fluorescent emission is time delayed from excitation by the fluorescence lifetime of the fluorophore. This time delay can, in some cases, be used to assist in discrimination of the fluorescence signal of interest from extraneous sources such as emissions from other fluorophores. Because fluorescence lifetimes are typically less than tens of nanoseconds, the instrumentation required to perform this time-based discrimination tends to be relatively sophisticated. Although direct time gating of the source and detector is routinely used in research laboratories, the few clinical laboratory instruments that perform "lifetime" measurements generally measure the phase shift between sinusoidally modulated exciting light and the correspondingly modulated fluorescence emissions.

Some molecules that are occasionally used in clinical assays have fluorescent lifetimes considerably in excess of 100-ns. These molecules are typically described as phosphorescent (typically considered as molecules having lifetimes in the 100-ns to 100-ms range) or luminescent (lifetimes over 100 ms). Fluorescent species are ubiquitous in biological materials and routinely introduce unwanted background fluorescence into clinical assays. Many materials commonly used in instrument construction also exhibit low levels of fluorescence. The use of a phosphorescent or luminescent indicator species in an assay permits the signal of interest to be discriminated from this background fluorescence by time gating, thus improving the SNR and lower LOD of the assay.[35,36]

### 19.4.4 Fluorescence Polarization (Fluorescence Anisotropy)

The excitation of a fluorophore can be modeled as excitation of a dipolar oscillator by an incident photon followed, after a time delay, by the emission of a longer wavelength photon by this same dipole. The net effect is that the fluorescent emission will be polarized in the plane defined by the emitting dipole. Furthermore, the fluorophore is not fixed in space, but rather is in a constant state of diffusional motion on several different timescales, with rotational diffusion typically being the fastest. Since the orientation of the emitting dipole is intrinsically linked to the molecular and electronic structure of the fluorophore, the orientation of the plane of polarization of the emitted photon will vary as a function of the excited state lifetime and the rotational rate of the fluorophore. The faster the rotational rate is relative to the excited state lifetime, the greater the change in the orientation of the polarization. This effect is employed in several different forms in clinical assays.[37]

The most common implementation is the measurement of steady-state fluorescence anisotropy. Typically, an external frame of reference is imposed by illuminating the sample with plane polarized light at the desired excitation wavelength, resulting in the dipole of each excited fluorophore being initially oriented in the same plane. If rotational diffusion rate of the excited state fluorophore is very slow with respect to the excited state lifetime, the emitted fluorescence retains this initial polarization. However, if the rotational rate is comparable to or faster than the excited state lifetime, the plane of polarization will change by some random amount. Taken as an ensemble over all of the fluorophores in a sample, this latter case results in at least partial depolarization of the fluorescence emissions.[38] The amount of this depolarization is usually determined by measuring the intensities of fluorescence emission having polarizations parallel ($I_{\|}$) and perpendicular ($I_{\perp}$) to that of the exciting light. The resulting polarization ($P$) is then defined as:

$$P = (I_{\|} - I_{\perp})/(I_{\|} + I_{\perp}) \tag{19.6}$$

Due to the requirement that the rotational rate and fluorescence lifetime of the fluorophore be roughly comparable, fluorescence polarization assays are largely restricted to small molecule analytes. These assays are most often of the competitive type in which the fluorescent reagent (tracer) is constructed by fluorescently labeling the target small molecule analyte or an analogue thereof. The other primary reagent in these assays is typically an unlabeled antibody directed against the target analyte. When the tracer and antibody are combined with the sample, the tracer competes with any analyte present in the sample for antibody binding sites.[39]

If several operational constraints are met, the amount of antibody-bound tracer present at the end of the reaction is inversely proportional to the amount of analyte that was initially present in the sample. Because the antibody–tracer complex is much larger and more massive than the free tracer, the polarization ($P$) of the fluorescent emissions decreases with increasing analyte concentration. A calibration curve is used to determine the analyte concentration from the value of $P$. For operational reasons, the assay chemistry is usually configured such that $I_{\|} > I_{\perp}$, which means that $\Delta P < 0.5$. Measurement resolution is usually in the range of 0.01 to 0.001 $P$.[40,41]

## 19.4.5 Chemiluminescence

All of the previously described modalities rely on the illumination of the sample in order to generate a detectable signal. Stray light and other adverse consequences of this illumination are frequently the controlling factors in determining the detection limit of the assay. Chemiluminescence, in which the detectable light signal is generated chemically, does not require an illuminator and can, therefore, often achieve significantly lower limits of detection than can the previously described methods.[42] The chemical light generating systems used in this modality typically consist of an acridinium ester or a cyclic hydrazide as the lumiphore and a peroxide as the trigger. Because only one photon is produced per lumiphore, photon counting techniques are generally required in order to quantitate light emission. Both photon emission rate and total integrated emission measurements have been used in clinical assays. Stray light and electronic noise in the detection circuitry are the usual nonchemical limits on detection sensitivity, although photon statistics also play a role.[43]

## 19.4.6 Guided Wave Optical Sensors

Guided wave sensors rely on total internal reflection[44] to confine incident light within a guiding medium (core) that is surrounded by a medium of lower refractive index (cladding).[45,46] The critical angle in such a device is given by

$$\sin \theta_c = \eta_{\text{clad}}/\eta_{\text{core}}. \tag{19.7}$$

Under conditions where the light propagating in the core is fully confined by total internal reflection, propagation of this energy is accompanied by an exponentially decaying field ("evanescent wave"), which extends into the cladding to a depth that is a function of the refractive indices of the core and cladding.[45,46] This evanescent wave provides a means of coupling energy between the core and molecules located external to the core, and is the basis of most types of guided wave sensors.[47–51] In order to maximize the efficiency of this bidirectional coupling process in sensing applications, the waveguide cladding generally consists of or incorporates some indicator species that produces an optical signal in response to the presence of an analyte.[52] The assay chemistries used in these applications are usually similar, if not identical, to those used in absorbance and fluorescence assays read out by conventional means.[53,54]

If a waveguide sensor is configured for an absorbance assay, energy is coupled from the core into the indicator species via the evanescent wave.[55] The measured property in this instance is the reduction in the optical energy propagating in the waveguide core. Fluorescence sensors operate in a similar manner except that the fluorescence signal radiated into the medium surrounding the waveguide or the fluorescence coupled from the fluorophore into the core via the evanescent wave can be measured.[56,57]

Another implementation of a waveguide-based sensor interposes a thin, (typically) discontinuous layer of metal (usually silver, gold, or aluminum) between the core and the fluorophore.[58–60] Surface plasmon modes excited at the metal surface by energy coupled from the core can couple to indicator species via the evanescent wave.[61,62] As the plasmon resonances act to amplify or concentrate the electric field at the surface of the metal, the amount of energy available for coupling to an indicator species is greatly increased over what is available in the absence of the metal.[63,64] This increases the sensitivity of the sensor when used in either of the two operating modes described previously.[65] In addition, surface plasmon sensors are sensitive to species that are neither absorbing nor fluorescent. Interaction of such a species with the surface plasmon is detectable through the resulting change in the energy distribution between the core and the metal.[66,67]

In all of these implementations, a major advantage of guided wave sensors over their conventional counterparts is that the "active" volume at the surface of the sensor is controlled and restricted by the penetration depth of the evanescent wave.[68] This means that guided wave sensors tend to be relatively insensitive to any species not in direct contact with the surface of the waveguide. Thus, because the assay chemistry is designed to concentrate the analyte at the surface of the waveguide, the contribution of interfering substances (which are not in close proximity to the surface) is greatly reduced.

### 19.4.7 Imaging Systems

For the purposes of this chapter, imaging systems are defined as those that capture specimen data in a spatially resolved manner. This definition includes classical *in vitro* diagnostic imaging applications such as microscopy or the reading of agglutination and microbial culture assays, as well as more recent applications in which imaging techniques are employed to capture data simultaneously from multiple independent samples.

Qualitative and quantitative imaging is used in *in vitro* diagnostic testing. Qualitative images are usually intended for visual evaluation, thus placing a premium on the appearance of the image and on the visual fidelity of the image to the specimen. In most cases, this translates to the operational requirements of using spectrally balanced, broadband illumination, employing a broadband detector, and maximizing spatial resolution at the specimen. Quantitative images are, in effect, spatially resolved arrays of photometric measurements that are mathematically processed in order to extract the information of interest. In this case, the emphasis is on the numerical quality and integrity of the data. In order to minimize the effects of stray light on this quality, illumination and detection are generally performed in one or more narrow spectral bands.

Images captured using an electronic imaging system comprise arrays of individual picture elements or pixels. (A pixel represents the smallest unit of spatial resolution in the captured image.) One approach to electronic imaging, exemplified by flying spot scanners and most forms of confocal microscopes, is to capture each individual pixel in the image sequentially using a single-element detector such as a photomultiplier tube or photodiode. The other common approach to electronic image capture is based on the use of a detector comprising an array of individual detection elements, each element corresponding to a pixel in the captured image.

Array detectors generally permit faster image capture than is possible using a single-point detector because all pixels in the image are captured simultaneously rather than sequentially. Another benefit of array detectors is that the spatial relationships between the individual pixels in the image are fixed during the manufacture of the sensor, rather than being dependent upon the characteristics of some scanning process. For various reasons related to device physics and construction, discrete detectors offer substantially greater useful sensitivity, dynamic range, and photometric resolution than do currently available array detectors. Array detectors, on the other hand, offer convenience and higher throughput, particularly in the case of high-pixel-count images.

It should be noted that the vast majority of the charge-coupled device (CCD) array detectors currently available for imaging applications have been explicitly designed and are intended for use in consumer applications in which low cost and the generation of a "pretty" picture are of paramount importance. As a consequence, certain operational and performance parameters, such as photometric resolution, that are of importance in quantitative imaging have been compromised. For example, one very widely used class of CCD array detectors is advertised as delivering "8-bit" or 256 level photometric resolution. When evaluated, however, many members of this class actually deliver only 5 to 6 bits (32 to 64 levels) of photometric resolution. The remaining 2 to 3 bits of the output signal are noise. This is generally not a significant issue in qualitative applications because the human eye has difficulty in discriminating between adjacent intensity levels in an image having 32 to 64 levels of photometric resolution, but it is unacceptable in many quantitative imaging applications.

Array detectors for quantitative applications must therefore be carefully selected to deliver the number of "true" bits of photometric resolution required by the application. The SNR of the detector is a useful indicator in this regard. A detector capable of delivering true 8-bit (256) photometric resolution will have an SNR in excess of approximately 50 dB. The SNR requirement increases by 3 dB for each additional bit of photometric resolution required. The poor spectral characteristics of the polymeric filters used in most "1-chip" color CCD sensors can also cause significant problems in quantitative imaging applications.

One of the primary challenges in clinical diagnostic imaging is matching the region of interest at the specimen to the field of view of the detector. In order to maximize the utilization of the detector array while minimizing the effects of optical aberrations, it is usually desirable to size the detector such that it occupies approximately the central 70% of the field of view (FOV) of the image-forming optics. The spatial resolution and depth of focus of the optical system are determined by the NA of the optics, while the spatial resolution of the detector in the specimen plane is determined by detector pixel size divided by the magnification of the optical system. It is usually desirable that system spatial resolution be limited by the optics in qualitative applications and by the detector in quantitative applications. Matching these parameters to each other, the requirements of the applications, and the characteristics of available components often requires design compromises.

A detailed discussion of the analysis of quantitative images is beyond the scope of this chapter. It is, however, generally desirable to perform a "shading correction" on the captured image to minimize the effects of variations in illumination, as well as optical and detector response nonuniformities, before analyzing the data. Chromatic corrections are usually also desirable when analyzing data captured using a "1-chip" array detector. Photometric calculations can be performed on the captured and corrected data using the equations given previously in this chapter and can be combined with morphological analysis in order to extract information as to the spatial distribution of the indicator species.

## 19.5 Conclusion

Successful clinical laboratory instrument system development is centered around two critical factors: clear and complete definition of what the instrument is intended to accomplish and the characteristics of the assay chemistry to be utilized. Given these prerequisites, selection of appropriate detection technology and design of the instrument can then proceed in a systematic manner. Adhering to the principles and "rules of thumb" presented in this chapter does not guarantee the successful development of a clinical laboratory instrument. However, the guidelines and processes described will usually increase the probability of success by a significant factor.

### References

1. Ekelman, K.B., Ed., *New Medical Devices: Invention, Development and Use*, National Academy Press, Washington, D.C., 1988, chap. 2.

2. Geddes, L.A. and Baker, L.E., *Principles of Applied Biomedical Instrumentation,* John Wiley & Sons, New York, 1975.
3. Schall, R.F. and Tenoso, H.J., Alternatives to radioimmunoassay: labels and methods, *Clin. Chem.*, 27, 1157, 1981.
4. Albertini, A. and Ekins, R., Eds., *Monoclonal Antibodies and Developments in Immunoassay,* Elsevier/North Holland Biomedical Press, Amsterdam, 1983.
5. Aloisi, R.M., *Principles of Immunodiagnostics,* C.V. Mosby, St. Louis, 1979.
6. Statland, B.E., *Clinical Decision Levels for Lab Tests,* 2nd ed., Medical Economics Books, Oradell, NJ, 1987.
7. Caraway, W.T., Photometry, in *Textbook of Clinical Chemistry,* Tietz, N.W., Ed., W.B. Saunders, Philadelphia, 1986, chap. 1.
8. Voller, A. and Bidwell, D.E., Enzyme immunoassays, in *Alternative Immunoassays,* Collins, W.P., Ed., John Wiley & Sons, New York, 1985.
9. Neumaier, M., Braun, A., and Wagener, C., Fundamentals of quality assessment of molecular amplification methods in clinical diagnostics, *Clin. Chem.*, 44, 12, 1998.
10. Kopetzki, E., Lehnert, K., and Buckel, P., Enzymes in diagnostics: achievements and possibilities of recombinant DNA technology, *Clin. Chem.*, 40, 688, 1994.
11. Painter, P., Automated nonisotopic immunoassay, *Diagn. Clin. Testing*, 28, 40, 1990.
12. Ling, N.R. and Catty, D., Haemagglutination and haemolysis assays, in *Antibodies Volume: A Practical Approach*, Catty, D., Ed., IRL Press, Oxford, 1988, p. 169.
13. Winzor, D.J. and Sawyer, W,H., *Quantitative Characterization of Ligand Binding*, Wiley-Liss, New York, 1995.
14. Wild, D., Improving immunoassay performance and convenience, *Clin. Chem.*, 42, 1137, 1996.
15. Ingle, J.D. and Crouch, S.R., *Spectrochemical Analysis,* Prentice-Hall, Englewood Cliffs, NJ, 1988.
16. Robinson, J.W., *Undergraduate Instrumental Analysis,* 4th ed., Marcel Dekker, New York, 1987.
17. Willard, H.H. et al., *Instrumental Methods of Analysis,* 7th ed., D. Van Nostrand Company, New York, 1988.
18. Skoog, D.A. and Leary, J.J., *Principles of Instrumental Analysis,* 5th ed., Saunders College Publishing, New York, 1998, chap. 7.
19. O'Shea, D.C., *Elements of Modern Optical Design*, John Wiley & Sons, New York, 1985, chap. 9.
20. Smith, W.J., *Modern Optical Engineering,* 2nd ed., McGraw-Hill, New York, 1990.
21. Sawyer, R.A., *Experimental Spectroscopy,* Dover Publications, New York, 1965.
22. James, J.F. and Sternberg, R.S., *The Design of Optical Spectrometers*, Chapman & Hall, London, 1969.
23. Ewing, G.W., *Instrumental Methods of Chemical Analysis*, McGraw-Hill, New York, 1985.
24. Khalil, O.S., Photophysics of heterogeneous immunoassays, in *Immunochemistry of Solid-Phase Immunoassays*, CRC Press, Boca Raton, FL, 1991, p. 67.
25. Cohn, G.E., Error budget considerations in diagnostic instrumentation, *Proc. SPIE*, 2985, 224, 1997.
26. Cohn, G.E., Interaction of instrumental and process uncertainties in the overall precision of clinical diagnostics, *Proc. SPIE*, 2386, 185, 1995.
27. Cohn, G.E., Instrument factors influencing the precision of measurement of clinical spectrometers, *Proc. SPIE*, 2680, 100, 1996.
28. Voigtman, E., Spectrophotometric precision: a case by case simulation study, *Anal. Instrum.*, 21, 43, 1993.
29. Chance, B., Principles of differential spectrophotometry with special reference to the dual wavelength method, in *Methods in Enzymology,* Vol. 24, *Photosynthesis and Nitrogen Fixation*, San Pietro, A., Ed., Academic Press, New York, 1972, p. 322.
30. Kubelka, P., New contributions to the optics of intensely light-scattering materials. Part I, *J. Opt. Soc. Am.*, 38, 448, 1948.
31. Frei, R.W. and MacNeil, J.D., *Diffuse Reflectance Spectroscopy in Environmental Problem-Solving,* CRC Press, Cleveland, 1973, chaps. 1–4.

32. Cohn, G.E., *In-vitro* diagnostic instrumentation: theory and practice (notes for short course taught at SPIE Bios symposium, San Jose, CA, 2002).
33. Guilbault, G.G., *Practical Fluorescence*, 2nd ed., Marcel Dekker, New York, 1990.
34. Fiore, M. et al., The Abbott IMx™ automated benchtop immunochemistry analyzer system, *Clin. Chem.*, 34, 1726, 1988.
35. Diamandis, E.P., Immunoassays with time resolved fluorescence spectroscopy, *Clin. Chem.*, 21, 139, 1988.
36. Jackson, T.M. and Ekins, R.P., Theoretical limitations on immunoassay sensitivity: current practice and potential advantages of fluorescent $Eu^{+}$ chelates as non-radioisotopic tracers, *J. Immunol. Methods*, 87, 13, 1986.
37. Lim, C.S. and Miller, J.N., The use of polarizers to improve detection limits in fluorimetric analysis, *Anal. Chim. Acta*, 100, 235, 1978.
38. Lakowicz, J.R., *Principles of Fluorescence Spectroscopy*, Vol. 1, Plenum Press, New York, 1986.
39. Popelka, S.R. et al., Fluorescence polarization immunoassay II: analyzer for the rapid and precise measurement of fluorescence polarization using disposable cuvettes, *Clin. Chem.*, 27, 1198, 1981.
40. Jolley, M.E., Fluorescence polarization immunoassay I: monitoring aminoglycoside antibiotics in serum and plasma, *Clin. Chem.*, 27, 1190, 1981.
41. Jolley, M.E. et al., Fluorescence polarization immunoassay. III. An automated system for therapeutic drug determination, *Clin. Chem.*, 27, 1575, 1981.
42. Ekins, R., Chu, F., and Micallef, J., High specific activity chemiluminescent and fluorescent markers, *J. Biolum. Chemilum.*, 4, 59, 1989.
43. Boland, J. et al., The Ciba Corning ACS: 180™ benchtop immunoassay analyzer, *Clin. Chem.*, 36, 1598, 1990.
44. Harrick, N.J., *Internal Reflection Spectroscopy*, John Wiley & Sons, New York, 1967, chaps. 1–2.
45. Kogelnik, H., Theory of dielectric waveguides, in *Topics in Applied Physics — Vol. 7, Integrated Optics*, 2nd ed., Tamir, T., Ed., Springer-Verlag, Berlin, 1982, chap. 2.
46. Hunsperger, R.G., Ed., *Integrated Optics: Theory and Technology*, 2nd ed., Springer-Verlag, Berlin, 1984, chaps. 1–7.
47. Coulet, P.R., What is a biosensor? in *Biosensor Principles and Applications*, Blum, L.J. and Coulet, P.R., Eds., Marcel Dekker, New York, 1991, chap. 1.
48. Robinson, G.A., Optical immunosensors: an overview, in *Biosensors*, Hall, E.A.H., Ed., Prentice-Hall, Englewood Cliffs, NJ, 1991, p. 229.
49. Seitz, W.R., Chemiluminescence and bioluminescence analysis: fundamentals and biomedical applications, *Crit. Rev. Anal. Chem.*, 1, 1, 1981.
50. Wadkins, R.M. and Ligler, F.S., Immunobiosensors based on evanescent wave excitation, in *Affinity Biosensors: Techniques and Protocols*, Rogers, K.R. and Mulchandani, A., Eds., Humana Press, Totowa, NJ, 1998.
51. Ctyroky, J. et al., Theory and modeling of optical waveguide sensors utilizing surface plasmon resonance, *Sensors Actuators*, B54, 66, 1999.
52. Lavers, C.R. et al., Planar optical waveguides for sensing applications, *Sensors Actuators*, B69, 85, 2000.
53. Sutherland, R. and Dähne, C., IRS devices for optical immunoassays, in *Biosensors: Fundamentals and Applications*, Turner, A.P.F, Karube, I., and Wilson, G.S., Eds., Oxford University Press, Oxford, 1987.
54. Sutherland, R. and Dähne, C., Interface immunoassays using the evanescent wave, in *Nonisotopic Immunoassay*, Ngo, T.T., Ed., Plenum Publishing, New York, 1988.
55. Reichert, W.M. et al., Evanescent detection of adsorbed protein concentration-distance profiles: fit of simple models to variable-angle total internal reflection fluorescence data, *Appl. Spectrosc.*, 41, 503, 1987.
56. Kronick, M. and Little, W.A., A new immunoassay based on fluorescence excitation by internal reflection spectroscopy, *J. Immunol. Methods*, 8, 235, 1975.

57. Herron, J.N. et al., Planar waveguide biosensors for nucleic acid hybridization reactions, *Proc. SPIE*, 3913, 177, 2000.
58. Flanagan, M.T. and Pantell, R.H., Surface plasmon resonance and immunosensors, *Electron. Lett.*, 20, 968, 1984.
59. Matsubara, K., Kawata, S., and Minami, S., Optical chemical sensor based on surface plasmon measurement, *Appl. Opt.*, 27, 1160, 1988.
60. Liedberg, B. and Johansen, K., Affinity biosensing based on surface plasmon resonance detection, in *Affinity Biosensors: Techniques and Protocols*, Rogers, K.R. and Mulchandani, A., Eds., Humana Press, Totowa, NJ, 1998, p. 31.
61. Ferrell, T.L., Callcott, T.A., and Warmack, R.J., Plasmons and surfaces, *Am. Sci.*, 73, 344, 1985.
62. Homola, J., Yee, S.S., and Gauglitz, G., Surface plasmon resonance sensors: review, *Sensors Actuators*, B54, 3, 1999.
63. Ford, G.W. and Weber, W.H., Electromagnetic interactions of molecules with metal surfaces, *Phys. Rep.*, 113, 195, 1984.
64. Parriaux, O. and Voirin, G., Plasmon wave versus dielectric waveguiding for surface wave sensing, *Sensors Actuators*, A21–A23, 1137, 1990.
65. Stemmler, I., Brecht, A., and Gauglitz, G., Compact surface plasmon resonance-transducers with spectral readout for biosensing applications, *Sensors Actuators*, B54, 98, 1999.
66. O'Shannesy, D., Brigham-Burke, M., and Peck, K., Immobilization chemistries suitable for use in the BIAcore™ surface plasmon resonance detector, *Anal. Biochem.*, 205, 132, 1992.
67. Nieba, L., Krebber, A., and Pluckthun, A., Competition BIAcore™ for measuring true affinities: large differences from values determined from binding kinetics, *Anal. Biochem.*, 234, 155, 1996.
68. Parriaux, O., Normalized analysis for the sensitivity optimization of integrated optical evanescent-wave sensors, *J. Lightwave Technol.*, 16, 573, 1998.

# 20

# Biosensors for Medical Applications




Tuan Vo-Dinh
*Oak Ridge National Laboratory*
*Oak Ridge, Tennessee*

Leonardo Allain
*Oak Ridge National Laboratory*
*Oak Ridge, Tennessee*


## 20.1 Introduction

Human beings, along with other mammals, consciously interact with the surrounding world by means of seven sensing mechanisms. In addition to the five senses, the abilities to detect temperature and variations in elevation are almost as important. One cannot help but be awed by the evolutionary process that brought about the development of such senses and by their integration into a brain capable of information processing and storage.

If a better awareness of our surroundings, food supplies, and predators was the main driver for this evolutionary process, a parallel might be drawn with our own human enterprise of creating sensors that help us understand the world with which we interact. A sensor can be viewed as the "primary element of a measurement chain, which converts the input variable into a signal suitable for measurement."[1] A sensing scheme is usually based on a transduction principle or mechanism; an input variable is transformed into an output variable through a transduction mechanism. Transduction principles are known physical or chemical effects that correlate observations in different domains. For example, the photoelectric effect is used to correlate number of photons with electric current. The piezoelectric effect does the same for stress and electricity, and Biot-Savart's law correlates magnetic field and electric current. In other words, the operating principle of a sensor involves transforming signals between different domains, from a domain we cannot directly access to one we can measure.

A biosensor is a special type of sensor often used in bioanalysis. Humankind has been performing bioanalysis since the dawn of time, using the sensory nerve cells of the nose to detect scents and those of the tongue to taste dissolved substances. As time has progressed, so has our level of understanding about the function of living organisms in detecting trace amounts of biochemicals in complex systems. The abilities of biological organisms to recognize foreign substances are unparalleled and have to some extent been mimicked by researchers in the development of biosensors. Using bioreceptors from biological organisms or receptors patterned after biological systems, scientists have developed a new means of chemical analysis that often has the high selectivity of biological recognition systems. These biorecognition elements, in combination with various transduction methods, have helped to create the rapidly expanding fields of bioanalysis and related technologies known as biosensors and biochemical sensors.

## 20.2 Biosensors: Definition and Classification

Two fundamental operating principles of a biosensor are"biological recognition" and "sensing." Therefore, a biosensor can be generally defined as a device that consists of two basic components connected in series: (1) a biological recognition system, often called a bioreceptor, and (2) a transducer. The basic principle of a biosensor is to detect this molecular recognition and to transform it into another type of signal using a transducer. The main purpose of the recognition system is to provide the sensor with a high degree of selectivity for the analyte to be measured. The interaction of the analyte with the bioreceptor is designed to produce an effect measured by the transducer, which converts the information into a measurable effect such as an electrical signal. Figure 20.1 illustrates the conceptual principle of the biosensing process.

Biosensors can be classified by bioreceptor or transducer type (Figure 20.2). A bioreceptor is a biological molecular species (e.g., an antibody, an enzyme, a protein, or a nucleic acid) or a living biological system (e.g., cells, tissue, or whole organisms) that utilizes a biochemical mechanism for recognition. The sampling component of a biosensor contains a biosensitive layer that can contain

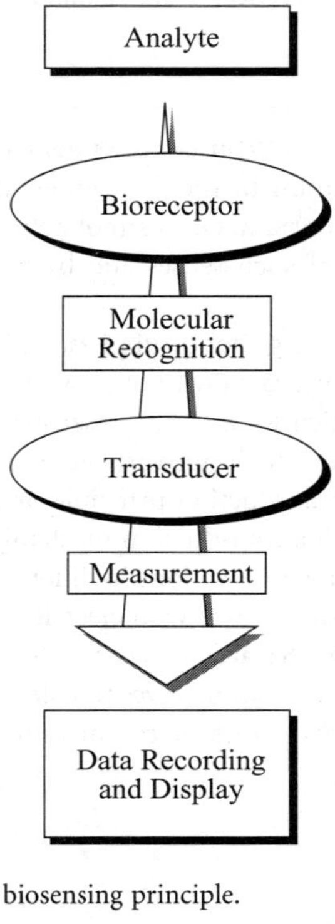


**FIGURE 20.1** Conceptual diagram of the biosensing principle.

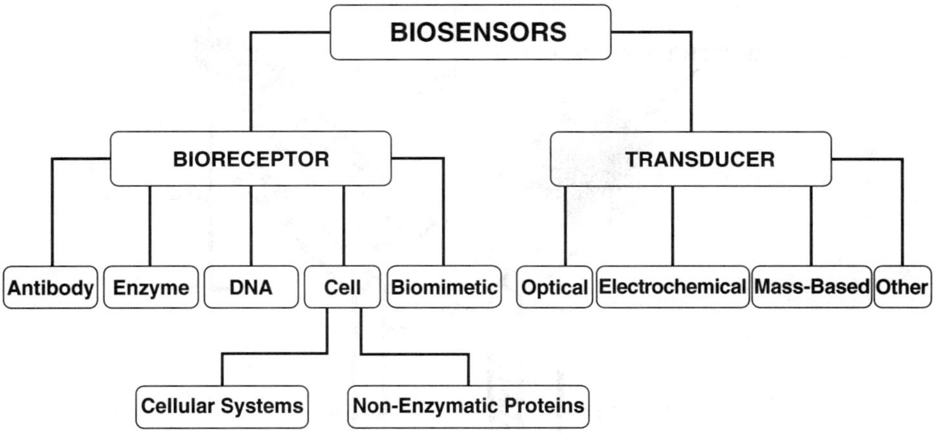


**FIGURE 20.2** Biosensor classification schemes.

bioreceptors or be made of bioreceptors covalently attached to the transducer. The most common forms of bioreceptors used in biosensing are based on (1) antibody/antigen interactions, (2) nucleic acid interactions, (3) enzymatic interactions, (4) cellular interactions (i.e., microorganisms, proteins), and (5) interactions using biomimetic materials (i.e., synthetic bioreceptors). For transducer classification, the previously mentioned techniques (optical, electrochemical, and mass-sensitive) are used.

Bioreceptors are the key to specificity for biosensor technologies. They are responsible for binding the analyte of interest to the sensor for the measurement. These bioreceptors can take many forms; the different bioreceptors that have been used are as numerous as the different analytes that have been monitored using biosensors. However, bioreceptors can generally be classified into five different major categories: (1) antibody/antigen, (2) enzymes, (3) nucleic acids/DNA, (4) cellular structures/cells, and (5) biomimetic. Figure 20.3 shows a schematic diagram of two types of bioreceptors: the structure of an immunoglobulin G (IgG) antibody molecule (Figure 20.3A), and DNA and the principle of base pairing in hybridization (Figure 20.3B).

Since the first biosensors were reported in the early 1960s,[2] growth of research activities in this area has accelerated.[3–28] Biosensors have seen a wide variety of applications, primarily in three major areas: biological monitoring, biomedical diagnostics and environmental sensing applications.

## 20.3 Transduction Systems

Biosensors can be classified based on the transduction methods they employ. Transduction can be accomplished through a large variety of methods. Most forms can be categorized in one of three main classes: (1) optical detection methods, (2) electrochemical detection methods, and (3) mass-based detection methods. Other detection methods include voltaic and magnetic methods, and new types of transducers are constantly being developed for use in biosensors. Each of these three main classes contains many different subclasses, creating a large number of possible transduction methods or combination of methods. This section provides a brief overview of the various detection methods used in biosensors. Special emphasis will be placed on the description of optical transducing principles, which is the focus of this chapter.

### 20.3.1 Optical Detection

Optical detection offers the largest number of possible subcategories of all three of the transducer classes. This is because optical biosensors can be used with many different types of spectroscopies (e.g., absorption, fluorescence, phosphorescence, Raman, surface-enhanced Raman scattering

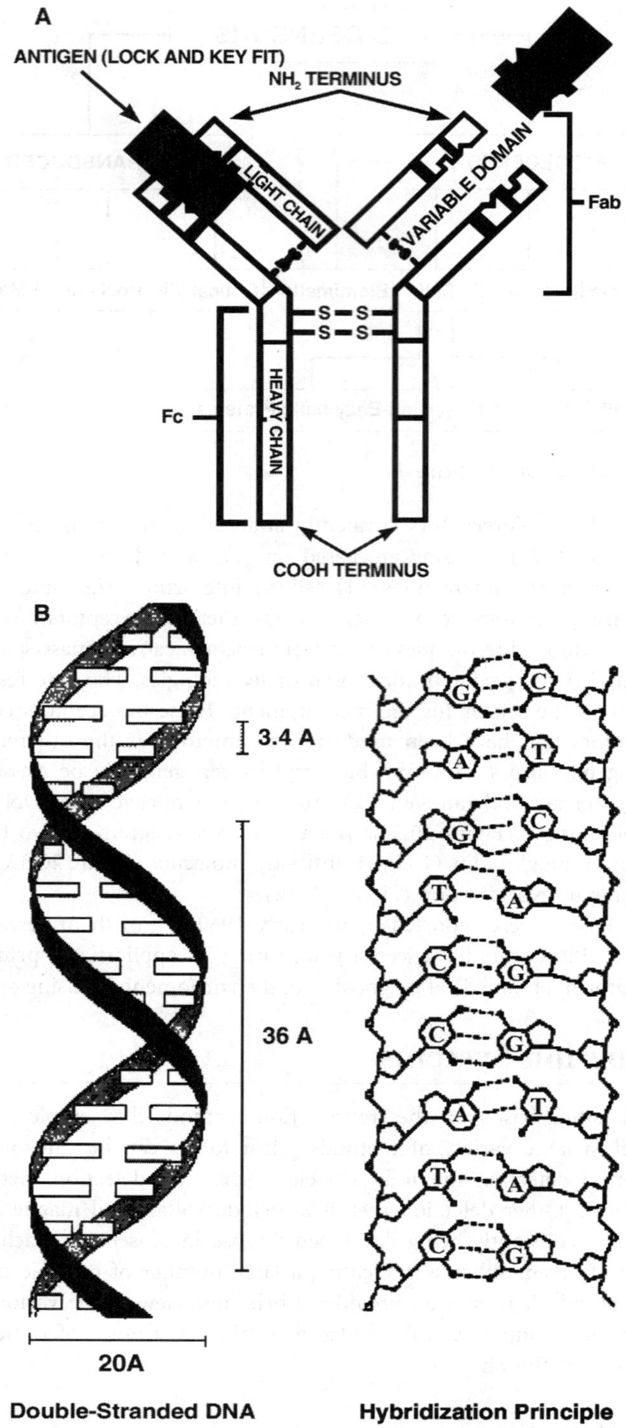


**FIGURE 20.3** Schematic diagrams of two types of bioreceptors: (A) IgG antibody; (B) DNA and the hybridization principle. (From Vo-Dinh, T. et al., *Fresenius J. Anal. Chem.*, 366, 540, 2000. With permission.)

(SERS), refraction, dispersion spectrometry) to measure different spectrochemical properties of target species. These properties include amplitude, energy, polarization, decay time, and phase. Amplitude is the most commonly measured parameter of the electromagnetic spectrum because it can generally be correlated with the concentration of the analyte of interest.

The energy of the electromagnetic radiation measured can often provide information about changes in the local environment surrounding the analyte, its molecular vibrations (i.e., Raman or infrared absorption spectroscopies), or the formation of new energy levels. Measurement of the interaction of a free molecule with a fixed surface can often be investigated with polarization measurements. Polarization of emitted light is usually random when emitted from a free molecule in solution; however, when a molecule becomes bound to a fixed surface, the emitted light often remains polarized. The decay time of a specific emission signal (i.e., fluorescence or phosphorescence) can also be used to gain information about molecular interactions because these decay times are highly dependent upon the excited state of the molecules and their local molecular environment. Another property that can be measured is the phase of the emitted radiation. When electromagnetic radiation interacts with a surface, the speed or phase of that radiation is altered, based on the refractive index of the medium (analyte). When the medium changes via binding of an analyte, the refractive index may change, thus changing the phase of the impinging radiation. This property of electromagnetic radiation has been successfully exploited in commercial applications using surface plasmon resonance sensors.

### 20.3.1.1 Fluorescence

Fluorescence is one of the most sensitive spectroscopic techniques, and its sensitivity makes it uniquely suited for the detection of very low concentrations of bioanalytes. Background information on the photophysical principles of the fluorescence emission process can be found in Chapter 28 of this handbook. When coupled with a high-power light source such as a laser, it can yield very high signal-to-noise (S/N) values. Single-molecule detection using laser-induced fluorescence has been reported in many studies. Because of its inherently high sensitivity, fluorescence has traditionally been the technique of choice for optical detection of trace-level analytes (at the femtomole level or lower). For high-quantum-yield fluorophores, the effective fluorescence cross sections can be as high as $10^{-16}$ cm$^2$/molecule.

A typical optical setup for a fluorescence biosensor using a laser as the light source is shown in Figure 20.4.[29] The instrument consists of an optical fiber with antibodies immobilized at the sensor tip. Excitation light from a laser is sent through a beam splitter onto the incidence end of the optical fiber. The laser beam is transmitted inside the fiber onto the sensor tip, where it excites the analyte molecules bound to the antibodies. The excited antigen fluorescence is collected and retransmitted to

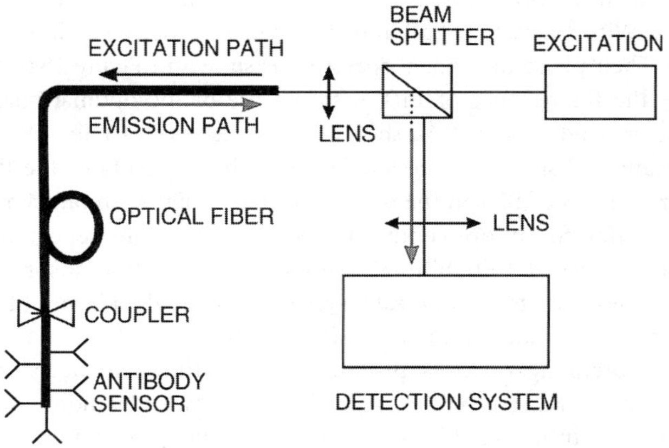


**FIGURE 20.4** Schematic diagram of an optical system for an antibody-based biosensor. (From Vo-Dinh, T. et al., *Appl. Spectrosc.*, 41, 735, 1987. With permission.)

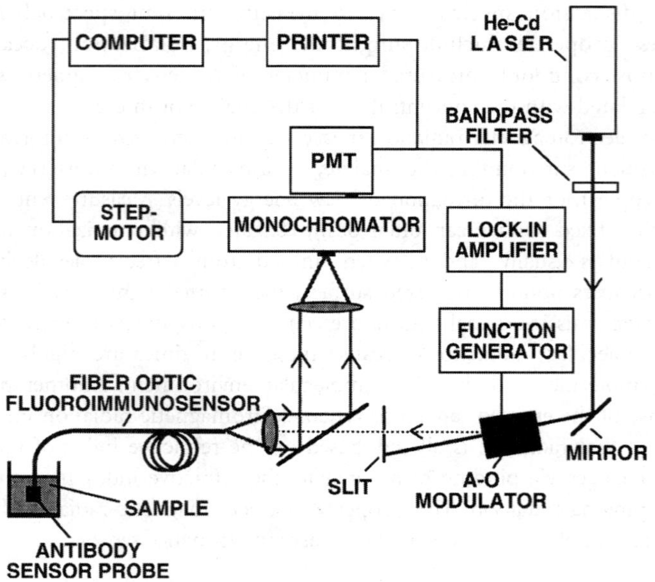


**FIGURE 20.5** Schematic diagram of an antibody-based biosensor with phase-reolution. (From Vo-Dinh, T. et al., *Appl. Spectrosc.*, 44, 128, 1990. With permission.)

the incidence end of the fiber, directed by the beam splitter onto the entrance slit of a monochromator, and recorded by a photomultiplier. This fluoroimmunosensor (FIS) was used to detect the carcinogen benzo[a]pyrene (BaP).[29]

Fluorescence detection is also suitable for time- or phase-resolved measurements, yielding additional information from the system of interest. Vo-Dinh and co-workers reported the development of a phase-resolved fiberoptic fluoroimmunosensor (PR-FIS) that can differentiate BaP and its metabolite benzopyrene tetrol (BPT) based on the difference in their fluorescence lifetimes.

A diagram for a phase-resolved optical setup is shown in Figure 20.5.[30] The excitation laser beam is modulated with an acousto-optic modulation system. A function generator provides the waveforms to drive the modulator. Laser light is delivered to the sample by an optical fiber and the fluorescence is collected by the same fiber. The fluorescence from the sensing probe is collimated by appropriate optics and focused onto the entrance slit of a monochromator equipped with a photomultiplier. A lock-in amplifier synchronized with the function generator is used to measure phase-resolved signals. With this setup, BaP and BPT could be detected simultaneously using phase-resolved fluorescence. Their phase-dependent spectrum is shown in Figure 20.6.[30]

Figure 20.6A shows the fluorescence of BPT with the use of the optimal phase shift for maximum BPT signal. On the other hand, Figure 20.6B shows the fluorescence of BaP with the use of the optimal phase shift for maximum BaP signal. The results illustrate the capability of the PR-FIS device to reveal the spectrum of 30 femtomoles of BPT in the presence of much higher amounts of interfering BaP.

Femtomolar sensitivities for fluorescently labeled proteins were reported by Herron and co-workers using a channel-etched thin film waveguide fluoroimmunosensor.[31] A silicon-oxynitride thin optical waveguide film was etched to create a channel for small volumes of analyte. Two different types of assays were performed and compared using this biosensor. The first was a direct assay of a fluorescently tagged protein ligand to a protein receptor that had been immobilized onto the waveguide. The second assay was an indirect sandwich-type assay of a nonfluorescent protein ligand, where the analyte (the protein ligand) binds to a protein bioreceptor immobilized on the waveguide; then a fluorescently tagged secondary receptor was used for measurement purposes. The fluorescent dye used to tag the proteins was Cy-5, a red emitting cyanine dye that reduced the chance of excitation of possible interferents.

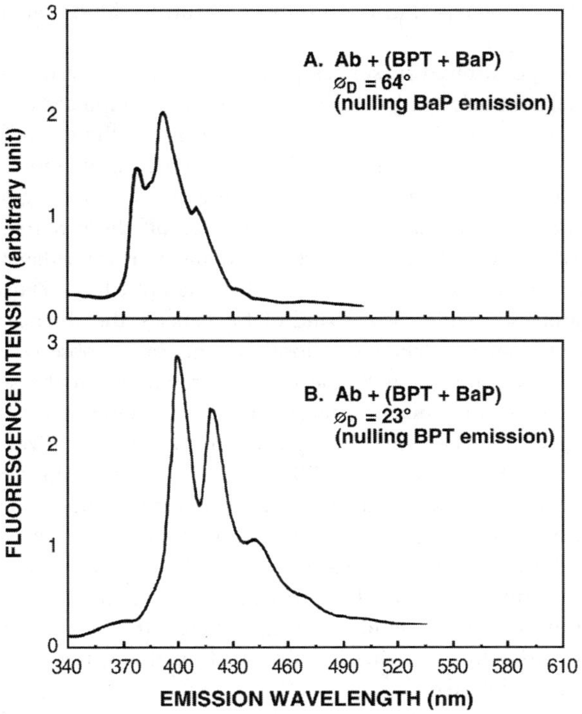


**FIGURE 20.6** Phase-resolved fluorescence spectra of BaP and BPT. (From Vo-Dinh, T. et al., *Appl. Spectrosc.*, 44, 128, 1990. With permission.)

An interesting application of fluorescence spectroscopy involves the detection of lipopolysaccharide endotoxin (LPS), which is the most powerful immune stimulant known and a causative agent in the clinical syndrome known as sepsis. Sepsis is responsible for more than 100,000 deaths annually, in large part due to the lack of a rapid, reliable, and sensitive diagnostic technique. LPS has been detected in *Escherichia coli* at concentrations as low as 10 ng/ml in 30 s, using an evanescent wave fiber-optic biosensor. Polymyxin B covalently immobilized onto the surface of the fiber-optic probe selectively bound fluorescently labeled LPS. The competitive assay format worked in buffer and in plasma with similar sensitivities. This method can be used with other LPS capture molecules such as antibodies, lectins, or antibiotics to detect LPS and determine the LPS serotype simultaneously. This LPS assay using the fiber-optic biosensor can be applied in clinical and environmental testing.[32]

### 20.3.1.2 Surface Plasmon Resonance

Since the first application of the surface plasmon resonance (SPR) phenomenon for sensing almost two decades ago, this method has made great strides in terms of instrumentation development and in applications.[33] SPR sensor technology has been commercially available and SPR biosensors have become a useful tool for characterizing and quantifying biomolecular interactions.

SPR makes it possible to monitor the binding process as a function of time by following the increase in refractive index that occurs when one of the interacting partners binds to its ligand immobilized on the surface of a SPR sensor substrate.[34] A technique that does not require the reactants to be labeled is a major advantage, simplifying the data collection process. Biosensor binding data are also useful for selecting peptides to be used in diagnostic solid-phase immunoassays. Very small changes in binding affinity can be measured with good precision, which is a prerequisite for analyzing the functional effect and thermodynamic implications of limited structural changes in interacting molecules. For example, the on-rate ($k_a$) and off-rate ($k_d$) kinetic constants of the interaction between a protein and an antibody can be readily measured and the equilibrium affinity constant, $K$, can be calculated from the ratio $k_a/k_d = K$.[34]

The transduction principle involved in surface plasmon resonance sensors is based on the arrangement of a dielectric/metal/dielectric sandwich so that, when light impinges on a metal surface, a wave is excited within the plasma formed by the conduction electrons of the metal.[35,36] A surface plasmon is a surface charge density wave occurring at a metal surface. When a plasmon resonance is induced in the surface of a metal conductor by the impact of light of a critical wavelength and angle, the effect is observed as a minimum in the intensity of the light reflected off the metal surface. The critical angle is naturally very sensitive to the dielectric constant of the medium immediately adjacent to the metal and therefore lends itself to exploitation for bioassay. For example, the metal can be deposited as, or on, a grating; upon illumination with a wide band of frequencies, the absence of reflected light at the frequencies at which the resonance matching conditions are met can be observed.

Because of the intrinsic dependence with the index of refraction at the surface, surface plasmon resonance can be used as a sensor transducer to indicate when alterations at the surface happen. The binding event involving antibody–antigen recognition or DNA hybridization at the SPR sensor surface is the most common SPR application. SPR is able to detect small variations of the index of refraction at the metal-coated interface caused by changes in a few monolayers above the surface.

In biosensor devices, the surface plasmon resonance is detected as a very sharp decrease of the light reflectance when the angle of incidence is varied. The resonance angle is very sensitive to variations in the refractive index of the medium just outside the metal film. Because the electric field probes the medium within only a few hundreds of nanometers from the metal surface, the condition for resonance is very sensitive to variations in thin films on this surface. Changes in the refractive index of about $10^{-5}$ are easily detected.

The surface plasmon wave penetrates in both directions normal to the interface; consequently, the incident angle or frequency at which resonance is observed is dependent on the refractive index of dielectric at the interface. Liedberg and co-workers[35] have shown that surface plasmon resonance can be used as the basis of a genuine reagentless immunosensor if large analytes are to be monitored. The antibody is immobilized on the metal. When a large antibody binds, it displaces solution (having a refractive index of approximately 1.34) with, for example, protein (having a refractive index of 1.5). The effective refractive index of the dielectric adjacent to the metal is thus changed in proportion to the amount of analyte bound, and the surface plasmon resonance (incident angle or resonance frequency) is shifted accordingly. Flanagan and Pantell[36] have shown that the amount of analyte bound can be directly related to the resonance shift even when the resonance curve is distorted by scattering caused by surface roughness, thus relieving one of the constraints of precise control of metalization, which would be unattractive in the mass production of inexpensive sensors.

SPR biosensors can provide qualitative information on macromolecular assembly processes under a variety of conditions. Quantitative information, in the form of affinity constants for complex formation, can be obtained in a manner similar to conventional solid phase assays. The major advantage of SPR biosensors is that the formation and breakdown of complexes can be monitored in real time. This offers the possibility of determining the mechanism and kinetic rate constants associated with a binding event. This information is essential for understanding how biological systems function at the molecular level.

However, accurate interpretation of biosensor data is not always straightforward.[37] A few software programs can interpolate SPR data and provide an estimate binding constants. The program CLAMP is software developed to interpret complex interactions recorded on biosensors.[37] It combines numerical integration and nonlinear global curve-fitting routines. The BIAcore™ system is one of the most used among the several commercially available optical biosensors.

For example, the interactions between adenylate kinase (AK) and a monoclonal antibody against AK (McAb3D3) were examined with an optical biosensor, and the sensograms were fitted to four models using numerical integration algorithms.[38] The interaction of AK in solution with immobilized McAb3D3 followed a single exponential function and the data fitted well to a pseudo first-order reaction model.

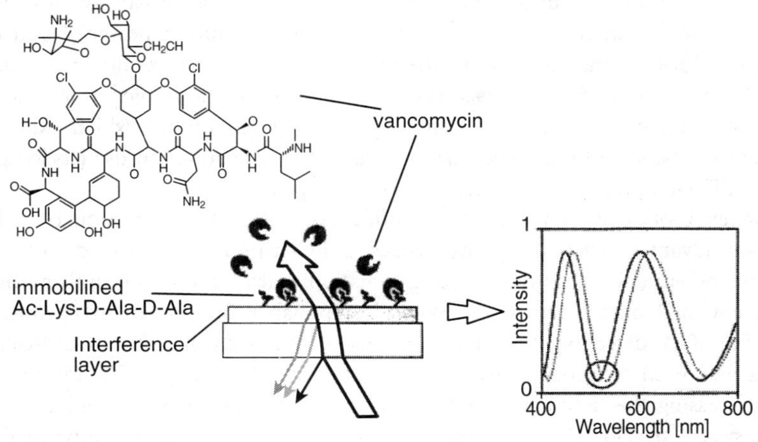


**FIGURE 20.7** Principle of reflectometric interference spectrometry. (From Tunnemann, R. et al., *Anal. Chem.*, 73, 4313, 2001. With permission.)

The application of surface plasmon resonance biosensors in life sciences and pharmaceutical research continues to increase. Several reviews providing a comprehensive analysis of the commercial SPR biosensor literature and highlights of emerging applications are available.[33,39,40] Some general guidelines to help increase confidence in the results reported from biosensor analyses (because of the variability in the quality of published biosensor data) have been compiled as well.[39]

### 20.3.1.3 Near-Infrared Absorption

Near-infrared (NIR) spectroscopy utilizes wavelengths above 800 nm to excite vibrational overtones and low-energy electronic levels of chemical species. Use of NIR usually profits from lower fluorescence background and higher specificity for appropriate dyes. Longer wavelengths also offer better penetration of translucent tissue, another major advantage of NIR for biomedical diagnostics.

A biomolecular probe utilizing NIR for the detection of biological molecules (immunochemical samples) with a semiconductor laser diode (780 nm) has been reported.[41] This probe consists of a modified fiber tip binding site, an NIR dye, and a photodiode detector. Preparation of the NIR biosensor involved the immobilization of anti-IgG antibody to the activated binding site of the fiber, followed by coating with IgG for a sandwich-type probe. The antibody was labeled with the commercially available NIR dye IR-144. The low background signal of the detector allowed the detection of 2.72 ng/ml of IgG in a probe coated with 10 ng/ml antibody at 820 nm.[41]

### 20.3.1.4 Reflectometric Interference

Gauglitz and co-workers have developed a unique technique based on reflectometric interference spectroscopy (RifS) for detection in biosensors.[42–44] RifS was used for the detection of biomolecular interactions and applied for small-molecule detection by chemical sensor surfaces. The principle of RifS, which does not require the use of labels, is illustrated in Figure 20.7.[42] A thin silica layer on a glass substrate is illuminated from the back side using white light. Light beams are reflected at the different layers and superimposed to form a characteristic interference pattern. Changes in the thickness of the transducer surface caused by biomolecular interactions lead to a shift of the interference pattern, which can be analyzed in real time.

### 20.3.1.5 Raman

The possibility of using Raman or surface-enhanced Raman scattering (SERS) labels as gene probes has been reported.[45–47] The SERS technique has been recently applied to the detection of DNA fragments of the human immunodeficiency virus (HIV)[20,45,46] and of the breast cancer gene.[48] Raman spectroscopy has also proved to be a very useful tool for chemical analysis because it can identify chemical groups. This technique, however, suffers from poor sensitivity, often requiring powerful and expensive laser excitation sources. However, the discovery of the SERS effect,[49–51] which results in increased sensitivities of up to $10^8$-fold for some compounds, has renewed interest in this technology for analytical purposes.[4] The feasibility of using surface-enhanced Raman gene (SERGen) probes, which exhibit an extremely narrow small spectral bandwidth, has been demonstrated.[45–47] The ability to use labeled primers extends the utility of the SERGen probes for medical diagnostic purposes.

Because SERGen probes rely on chemical identification rather than emission of radioactivity, they have a significant advantage over radioactive probes. SERGen probes are formed with stable chemicals that do not emit potentially dangerous ionizing radiation. Furthermore, the probes offer the excellent specificity inherent in Raman spectroscopy. While isotope labels are few, many chemicals can be used to label DNA for SERS detection. Potentially, dozens to hundreds of different SERGen probes could be constructed and used to probe several DNA sequences of interest (label multiplexing) simultaneously, thus decreasing the time and cost for gene diagnostics and DNA mapping. The multispectral imaging (MSI) system developed in this work, with its rapid wavelength switching of the acoustooptic tunable filter (AOTF) system could allow very rapid scanning and high-throughput data collection.[45]

For biomedical diagnostics, the SERGen probe could have a wide variety of applications in areas where nucleic acid identification is involved. The SERGen probes may also be used in polymerase chain reaction (PCR) applications for medical diagnostic applications, e.g., for HIV detection. In genomics applications as well as in high-throughput analysis, the SERS gene probe technology could lead to the development of detection methods that minimize the time, expense, and variability of preparing samples by combining the BAC mapping approach with SERS "label multiplex" detection. Large numbers of DNA samples can be simultaneously prepared by automated devices. With the SERGen technique, multiple samples can be separated and directly analyzed using multiple SERGen labels simultaneously (label multiplex scheme). The use of the SERS technique for biomedical application is further described in Chapter 64 of this handbook.

## 20.3.2 Electrochemical Detection

Electrochemical detection is another possible means of transduction that has been extensively used in biosensors.[52–56] This technique is complementary to optical detection methods such as fluorescence, the most sensitive of the optical techniques. Since many analytes of interest are not strongly fluorescent and tagging a molecule with a fluorescent label is often labor-intensive, electrochemical transduction can be very useful. By combining the sensitivity of electrochemical measurements with the selectivity provided by bioreception, detection limits comparable to those of fluorescence biosensors are often achievable.

Electrochemical detection is usually based on the chemical potential of a particular species in solution (the analyte), as measured by comparison to a reference electrode. Therefore, the electrochemical response is dependent on the activity of the analyte species, not their concentration. However, for dilute solutions of low ionic strength, the thermodynamic parameter activity approaches the physical parameter concentration (in molar terms). In comparison, the signal intensity associated with optical detection is usually directly proportional to the number of a specific chromophore within a certain pathlength and, therefore, directly dependent on the concentration of the chromophore. The linear relationship between signal intensity and concentration of a species is known as Beer–Lambert's law. The cases in which discrepancies from this linear relationship occur are usually caused by secondary effects such as self-absorbance, and equilibrium conditions.

Multiple examples of electrochemical sensors applied to biological systems are known. For example, electrochemical flow-through enzyme-based biosensors for the detection of glucose and lactate have been developed by Cammann and co-workers.[57] Glucose oxidase and lactate oxidase were immobilized in conducting polymers generated from pyrrole, *N*-methylpyrrole, aniline, and *o*-phenylenediamine on platinum surfaces. These various sensor matrices were compared on basis of amperometric measurements of glucose and lactate, and the *o*-phenylenediamine polymer was found to be the most sensitive.

This polymer matrix was also deposited on a piece of graphite felt and used as an enzyme reactor as well as a working electrode in an electrochemical detection system. Using this system, a linear dynamic range of 500 μ*M* to 10 m*M* glucose was determined with a detection limit of <500 μ*M*. For lactate, the linear dynamic range covered concentrations from 50 μ*M* to 1 m*M* with a detection limit of <50 μ*M*.

A biosensor for protein and amino acid estimation has been reported by Sarkar and Turner.[58] A screen-printed biosensor based on a rhodinized carbon-paste working electrode was used in the three-electrode configuration for a two-step detection method. Electrolysis of an acidic potassium bromide electrolyte at the working electrode produced bromine, which was consumed by the proteins and amino acids. The bromine production occurred at one potential while monitoring of the bromine consumption was performed using a lower potential. The method proved very sensitive to almost all of the amino acids, as well as some common proteins, and was even capable of measuring L- and D-proline, which gave no response to enzyme-based biosensors. This sensor has been tested by measuring proteins and amino acids in fruit juice, milk, and urine.

Scheller and co-workers have developed an electrochemical biosensor for the indirect detection of L-phenylalanine via NADH.[59] This sensor is based on a three-step multi-enzymatic/electrochemical reaction. Three enzymes — L-phenylalanine dehydrogenase, salicylate hydroxylase, and tyrosinase — are immobilized in a carbon paste electrode. The principle behind this reaction/detection scheme is as follows. First, the L-phenylalanine dehydrogenase, upon binding and reacting with L-phenylalanine, produces NADH. The second enzyme, salicylate hydroxylase, then converts salicylate to catechol in the presence of oxygen and NADH. The tyrosinase then oxidizes the catechol to *o*-quinone, which is detected electrochemically, and reduced back to catechol with an electrode potential of –50 mV vs. an Ag/AgCl reference electrode.

This reduction step results in an amplification of signal due to the recycling of catechol from o-quinone. Prior to the addition of the L-phenylalanine dehydrogenase to the electrode, it was tested for its sensitivity to NADH, its pH dependence, and its response to possible interferents, urea, and ascorbic acid. From these measurements, it was found that the sensor sensitivity for NADH increased 33-fold by introducing the recycling step over the salicylate hydroxylase system alone. When this sensor was tested for the detection of L-phenylalanine in human serum, the linear dynamic range was found to cover concentrations ranging from 20 to 150 μ*M* with a detection limit of 5 μ*M*, which is well within the clinical range of 78 to 206 μ*M*.

### 20.3.3 Mass-Sensitive Detection

Another form of transduction has also been used in biosensors to measure small changes in mass.[60–62] Mass-based detection is the newest of the three classes of transducers and has already been shown to be capable of sensitive measurements. Mass analysis relies on the use of piezoelectric crystals that can be made to vibrate at a specific frequency with the application of an electrical signal of a specific frequency. The frequency of oscillation is dependent on the electrical frequency applied to the crystal as well as the crystal's mass. Therefore, when the mass increases due to binding of chemicals, the oscillation frequency of the crystal changes and the resulting change can be measured electrically and used to determine the mass added to the crystal. In most cases the added mass consists of antibodies or DNA fragments bound to their biospecific counterparts that have been immobilized on the sensor surface.

Guilbault and co-workers developed a quartz crystal microbalance biosensor for the detection of *Listeria monocytogenes*.[63] Several approaches were tested for the immobilization of *Listeria* onto the quartz crystal through a gold film on the surface. Once bound, the microbalance was then placed in a liquid flow-cell, where the antibody and antigen were allowed to form a complex, and measurements were obtained. Calibration of the sensor was accomplished using a displacement assay and was found to have a response range from $2.5 \times 10^5$ to $2.5 \times 10^7$ cells/crystal. More recently, Guilbault and co-workers have also developed a method for covalently binding antibodies to the surface of piezoelectric crystals via sulfur-based self-assembled monolayers.[64] Prior to antibody binding, the monolayers are activated with 1-ethyl-3-[3-(dimethylamino)propyl] carbodiimide hydrochloride and *N*-hydroxysulfosuccinimide. Using this binding technique, a real-time capture assay based on mouse IgG was performed and results were reported.

The first use of a horizontally polarized surface acoustic wave biosensor has been reported by Hunklinger and co-workers.[65] This sensor has a dual path configuration, with one path acting as an analyte-sensitive path and the other acting as a reference path. A theoretical detection limit of 33 pg was calculated based on these experiments, and a sensitivity of 100 kHz/(ng/mm$^2$) is reported. In addition, a means of inductively coupling a surface acoustic wave biosensor to its radio-frequency generating circuitry has been reported recently.[66] This technique could greatly reduce problems associated with wire bonding for measurements made in liquids because the electrodes are coated with a layer of $SiO_2$.

A relatively new type of mass-based detection system uses microcantilevers. Constructed of silicon, these devices are generally shaped like a microsize diving board. Their advantages include their miniature size, high degree of sensitivity, simplicity, low power consumption, low manufacturing cost, and compatibility with array designs.[67] The extremely low mass of the device allows it to sense perturbing forces because of the adsorbed masses at the picogram level, the viscosity of a gas or liquid over several orders of magnitude, and the acoustic and seismic vibrations. Special coatings on the silicon will adapt the cantilever to sense relative humidity, temperature, mercury, lead, ultraviolet radiation, and infrared (IR) radiation. By using current micromachining technology, multiple arrays could be used to make multielement or multitarget sensor arrays involving hundreds of cantilevers without significantly increasing the size, complexity, or overall package costs.

## 20.4 Bioreceptors and Biosensor Systems

### 20.4.1 Antibody

#### *20.4.1.1 Antibody Bioreceptors*

The basis for the specificity of immunoassays is the antigen-antibody (Ag-Ab) binding reaction, which is a key mechanism by which the immune system detects and eliminates foreign matter.[68] The enormous range of potential applications of immunosensors is due, at least in part, to the astonishing diversity possible in one of their key components, antibody molecules. Antibodies are complex biomolecules made up of hundreds of individual amino acids arranged in a highly ordered sequence. The structure of an IgG antibody molecule is schematically illustrated in Figure 20.3A. The antibodies are actually produced by immune system cells (B cells) when such cells are exposed to substances or molecules called antigens. The antibodies called forth following antigen exposure have recognition/binding sites for specific molecular structures (or substructures) of the antigen.

The way in which an antigen and an antigen-specific antibody interact may perhaps be understood as analogous to a lock-and-key fit, in which the specific configurations of a unique key enable it to open a lock. In the same way, an antigen-specific antibody fits its unique antigen in a highly specific manner, so that hollows, protrusions, planes, and ridges on the antigen and the antibody molecules. (in a word, the total three-dimensional structure) are complementary. Further details of how such complementarity is achieved will be discussed later in this chapter. It is sufficient at this point simply to indicate that, due to this three-dimensional shape fitting, and the diversity inherent in individual antibody makeup, it is possible to find an antibody that can recognize and bind to any one of a huge variety of molecular shapes.

This unique property of antibodies is the key to their usefulness in immunosensors; the ability to recognize molecular structures allows one to develop antibodies that bind specifically to chemicals, biomolecules, microorganism components, etc. One can then use such antibodies as specific detectors to identify and find an analyte of interest that is present, even in extremely small amounts, in a myriad of other chemical substances.

The other antibody property of paramount importance to their analytical role in immunosensors is the strength or avidity/affinity of the antigen–antibody interaction. Because of the variety of interactions that can take place as the Ag–Ab surfaces lie in close proximity to each other, the overall strength of the interaction can be considerable, with correspondingly favorable association and equilibrium constants. What this means in practical terms is that the Ag–Ab interactions can take place very rapidly (for small antigen molecules, almost as rapidly as diffusion processes can bring antigen and antibody together) and that, once formed, the Ag–Ab complex has a reasonable lifetime. Figure 20.10 shows a schematic of how biosensor probes can be prepared from antibody production, isolation, and binding to the sensing surface.

For an immune response to be produced against a particular molecule, a certain molecular size and complexity are necessary: proteins with molecular weights greater then 5000 Da are generally immunogenic. Radioimmunoassay (RIA), which utilizes radioactive labels, has been one of the most widely used immunoassay methods. RIA has been applied to a number of fields, including pharmacology, clinical chemistry, forensic science, environmental monitoring, molecular epidemiology and agricultural science. The usefulness of RIA, however, is limited by several shortcomings, including the cost of instrumentation, the limited shelf life of radioisotopes, and the potential deleterious biological effects inherent to radioactive materials. For these reasons, extensive research efforts are aimed at developing simpler, more practical immunochemical techniques and instrumentation that offer comparable sensitivity and selectivity to RIA. In the 1980s, advances in spectrochemical instrumentation, laser miniaturization, biotechnology, and fiber-optic research provided opportunities for novel approaches to the development of sensors for the detection of chemicals and biological materials of environmental and biomedical interest.

Since the development of a remote fiber-optic immunosensor for *in situ* detection of the chemical carcinogen benzo[a]pyrene,[7,29] antibodies have become common bioreceptors used in biosensors today.[8–12,69] The schematics for the optical detection of a bioanalyte are shown in Figure 20.11. In the arrangement shown, a single optical fiber carries the excitation light source to the sample and the fluorescence signal back to a spectrometer.

Due to fiber-to-fiber differences in fiber-optic biosensors, there is often a great difficulty in normalizing the spectral signal obtained with one fiber to another fiber. Ligler and co-workers reported on a method for calibrating antibody-based biosensors using two different fluorescent dyes.[70] To accomplish this, they labeled the capture antibodies bound to the fiber with one fluorescent dye and the antigen with a different dye. Both dyes were excited at the same wavelength and their fluorescence was monitored. The resultant emission spectrum of the fluorescence signal from the capture antibodies was used to normalize the signal from the tagged antigen.

Another example of antibody-based biosensors for bioanalysis is the development of an electrochemical immunoassay for whole blood.[71] This work involved the development of a sandwich type of separationless amperometric immunoassy without any washing steps. The assay is performed on a conducting redox hydrogel on a carbon electrode on which avidin and choline oxidase have been co-immobilized. Biotinylated antibody is then bound to the gel. When the antigen binds to the sensor, another solution of complementary horseradish peroxidase-labeled antibody is bound to the antigen, thus creating an electrical contact between the redox hydrogel and the peroxidase. The hydrogel then acts as an electrocatalyst for the reduction of hydrogen peroxide water.

An important aspect of biosensor fabrication is the binding of the bioreceptor to the sensor solid support or to the transducer. Vogel and co-workers report on a method for the immobilization of histidine-tagged antibodies onto a gold surface for surface plasmon resonance measurements.[72] A synthetic thio-alkane chelator is self-assembled on a gold surface. Reversible binding of an anti-lysozyme Fab fragment with a hexahistidine-modified extension on the C terminal end is then performed. Infrared spectroscopy was used to determine that the secondary structure of the protein was unaffected by the immobilization process. Retention of antibody functionality upon immobilization was also demonstrated.

Due to the reversible binding of such a technique, this could prove a valuable method for regeneration of biosensors for various applications.[72] Enzyme immunoassays can further increase the sensitivity of detection of antigen–antibody interactions by the chemical amplification process, whereby one measures the accumulated products after the enzyme has been allowed to react with excess substrate for a period of time.[73]

With the use of nanotechnology, submicron fiber-optic antibody-based biosensors have been developed by Vo-Dinh and co-workers for the measurements of biochemicals inside a single cell.[74–76] Nanometer-scale fiberoptic biosensors were used for monitoring biomarkers related to human health effects associated with exposure to polycyclic aromatic hydrocarbons (PAHs). These sensors use a monoclonal antibody for benzo[a]pyrene tetrol (BPT), a metabolite of the carcinogen benzo[a]pyrene, as the bioreceptor. Excitation light is launched into the fiber and the resulting evanescent field at the tip of the fiber is used to excite any of the BPT molecules that have bound to the antibody. The fluorescent light is then collected via a microscope. Using these antibody-based nanosensors, absolute detection limits for BPT of approximately 300 zeptomol ($10^{-21}$ mol) have been reported.[74] These nanosensors allow probing of cellular and subcellular environments in single cells.[19,75,76] The development and applications of optical nanosensors are described in further detail in Chapter 60 of this handbook.

### 20.4.1.2 Immunoassay Formats

Biomolecular interactions can be classified into two categories according to the test format performed (direct or indirect). In a direct format the immobilized target molecule interacts with a ligand molecule or the immobilized ligand interacts with a target molecule directly. For immunosensors, the simplest situation involves *in situ* incubation followed by direct measurement of a naturally fluorescent analyte.[7] For nonfluorescent analyte systems, *in situ* incubation is followed by development of a fluorophor-labeled second antibody. The resulting antibody sandwich produces a fluorescence signal directly proportional to the amount of bound antigen. The sensitivity obtained when using these techniques increases with increasing amounts of immobilized receptor. The indirect format involves competition between fluorophor-labeled and unlabeled antigens.[69] In this case, the unlabeled analyte competes with the labeled analyte for a limited number of receptor binding sites. Assay sensitivity therefore increases with decreasing amounts of immobilized reagent. Figure 20.8 illustrates the principles of (1) a competitive assay, (2) a direct assay, and (3) a sandwich assay.

A sandwich immunoassay using fluorescently labeled tracer antibodies has been developed to detect cholera toxin (CT).[77] Using this fluorescence-based biosensor, researchers analyzed six samples simultaneously in 20 min. The biochemical assays utilized a ganglioside-capture format: ganglioside GM1, utilized for capture of the analyte, was immobilized in discrete locations on the surface of the optical waveguide. Binding of CT to the immobilized GM1 was demonstrated with direct assays (using fluorescently labeled CT). The limits of detection for CT were 200 ng/ml in direct assays and 40 ng/ml and 1 μg/ml in sandwich-type assays performed using rabbit and goat tracer antibodies. Binding of CT to other glycolipid capture reagents was also observed. While significant CT binding to loci patterned with GD1b, Gb3, and Gb4 was observed, CT did not bind significantly to immobilized GT1b at the concentrations tested.

A similar planar array, equipped with a charge-coupled device (CCD) as a detector, was used to detect three toxic analytes simultaneously.[16] Wells approximately 2 mm in diameter were formed on glass slides using a photoactivated optical adhesive. Antibodies against staphylococcal enterotoxin B (SEB), ricin, and *Yersinia pestis* were covalently attached to the bottoms of the circular wells to form the sensing surface. Rectangular wells containing chicken immunoglobulin were used as alignment markers and to generate control signals. After the optical adhesive was removed, the slides were mounted over a CCD operating at ambient temperature in inverted (multipin phasing) mode. Cy5-labeled antibodies were used to determine the identity and amount of toxin bound at each location, using quantitative image analysis. Concentrations as low as 25 ng/ml of ricin, 15 ng/ml of pestis F1 antigen, and 5 ng/ml of SEB could be routinely measured.

A similar assay for ricin, using an immobilized anti-ricin IgG on an optical fiber surface, was reported.[78] Two immobilization methods were tried: in the first, the antibody was directly coated onto the silanized fiber using a cross linker; in the second, avidin-coated fibers were incubated with biotinylated antiricin IgG to immobilize the antibody using an avidin–biotin bridge. The assay using the avidin–biotin-linked antibody demonstrated higher sensitivity and wider linear dynamic range than the assay using antibody directly conjugated to the surface. The limits of detection for ricin in buffer solution and river water are 100 pg/ml and 1 ng/ml, respectively.

The use of protein A, an immunoglobulin-binding protein, for antibody immobilization on the surface of these fiber probes has been investigated as an alternative immobilization method to the

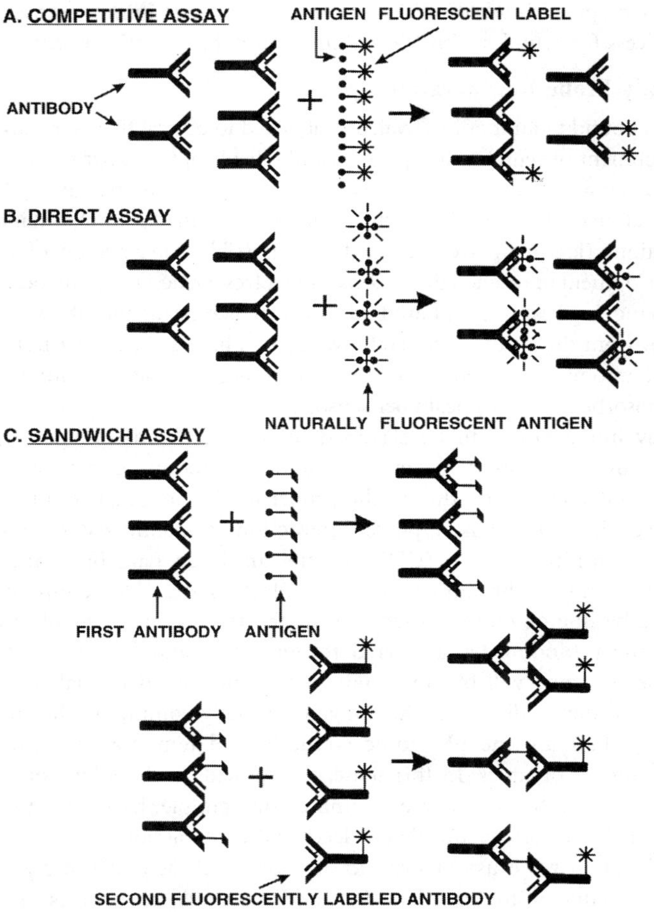


**FIGURE 20.8** Immunoassay formats: (A) competitive assay; (B) direct assay; (C) sandwich assay.

classical avidin–biotin and IgG–anti-IgG interactions.[14] No difference was observed in the binding of fluorescently labeled goat IgG by rabbit anti-goat IgG, regardless of whether the capture antibody was bound to the probe surface via protein A or covalently attached. However, in a sandwich immunoassay for the F1 antigen of *Yersinia pestis*, probes with rabbit anti-plague IgG bound to the surface via protein A generated twice the signal generated by the probes with the antibody covalently attached. Assay regeneration was also examined with protein A probes because Ab–Ag complexes have been successfully eluted from protein A under low pH conditions.

The regeneration of antibodies covalently immobilized to an optical fiber surface is also an important parameter that classifies the usefulness of a biosensor. Ideally, the antibody–antigen complex can be dissociated under mild conditions to regenerate the sensor. In a study by Liegler and co-workers,[79] three different restoring solutions were tested and compared: (1) 0.1 *M* glycine hydrochloride in 50% (v/v) ethylene glycol, pH = 1.75; (2) a basic solution 0.05 *M* tetraethylamine in 50% (v/v) ethylene glycol, pH = 11.0; and (3) 50% (v/v) ethanol in PBS. In this study, optical fibers coated with polyclonal rabbit anti-goat antibody against a large protein retained 70 and 65% of the original signal after five consecutive regenerations with acidic and basic solvent systems, respectively. The fibers coated with monoclonal mouse anti-trinitrobenzene antibody specific for a small organic molecule retained over 90% of the original signal when regenerated with basic and ethanol solutions.

Using a 635-nm laser diode light source, a biosensor based on labeled Cy5 antibodies was used to detect the F1 antigen of *Y. pestis* and the protective antigen of *Bacillus anthracis*. In a blind test containing F1 antigen spiked into 30 of 173 serum samples, this immunosensor was able to achieve 100% detection success for samples with 100 ng/ml or more of F1 antigen, with a specificity of 88%.[13]

#### 20.4.1.3 Antibody Probe Regeneration

Removal of antigens bound to antibodies covalently attached to optical fiber surfaces is one of the limiting factors in the development of reusable, inexpensive, and reliable optical fiber immunosensors for environmental and clinical analysis. Chemical reagents were supposed to cleave the binding between antibody and antigen, thus regenerating the biosensor[80] — a chemical procedure that is simple but ineffective after multiple regeneration operations (less than five cycles), due to possible denaturation of the antibody. Another approach in the development of regenerable biosensors involves the design of microcapillary systems capable of delivering and removing reagents and antibody-coated microbeads into the sensing chamber without removing the sensor from the sample.[81] Several investigators have searched for fast dissociation protocols able to regenerate immobilized antibodies while maintaining their stability for use in routine analysis, commercial immunosorbents,[82] or optical fiber sensors.[79]

Antigen–antibody interactions can be classified in three different groups:[83] hydrophobic interactions, electrostatic (or Coulombic) interactions, and interactions due to a combination of both forces. Hydrophobic interactions are due to the propensity of nonpolar groups and chains to aggregate when immersed in water. This type of interaction is maximized between the hydrophobic complementary determining regions (CDR) of the antibody parotopes and the predominantly hydrophilic groups found in the antigen epitope. Electrostatic interactions between antigen and antibody are caused by one or more ionized sites of the epitope and ions of opposite charge on the parotope. After primary binding has occurred through hydrophobic and electrostatic interactions, the epitope and the parotope will be close enough to allow van der Waals and hydrogen bonds to become operative. In order to dissociate the antigen–antibody complexes, the strength of these forces may be reduced by changing the pH, ionic strength, and temperature through the addition of dehydrating agents and/or organics. In this sense, strong acids such as HCl or $H_2SO_4$, mixtures such as glycine-HCl, or basic solutions of tetraethylamine, for example, have been used when the primary attractive forces in the bond can be considered electrostatic interactions.[84]

Lu et al. pointed out that the use of organic solutions such as ethylene glycol could improve the washing efficiency by reducing the van der Waals and the Coulombic forces maintaining the bond;[85] Wijesuriya et al. made a similar observation.[79] Nevertheless, every antigen–antibody pair may differ with respect to the nature of the forces implicated in the binding site. Haga et al. studied the effect of 28-kHz ultrasonic radiation with an intensity of 0.83 W/cm$^2$ on the dissociation of antigen–antibody complexes immobilized on CH-Sepharose gels.[86] They observed that the percent of dissociation increased with irradiation time and input wattage, obtaining a 22% dissociation after 20 min, whereas the immune reactivity decreased 8% without degradation of the dissolved antibody upon exposure to the ultrasound for times of up to 120 min.

Higher-frequency ultrasound (high-kHz to low-MHz range) is used to remove small (micrometer to submicrometer) particles from the surfaces of silicon wafers.[87,88] This so-called megasonic cleaning process does not damage the surfaces, suggesting that its mode of action does not depend on the strong effects of inertial cavitation that occur with lower frequency (e.g., 20 kHz) ultrasonic horns and baths. The mechanisms of megasonic cleaning likely involve nonlinear effects such as acoustic streaming and radiation pressure as well as the oscillatory linear forces.[88] Stable cavitation (a less intense process than inertial cavitation) may also play an important role, particularly in enhancing streaming effects. The success of the gentle but effective megasonic cleaning process suggests that MHz-range ultrasound may have some utility in regeneration of biosensor surfaces.

Vo-Dinh and co-workers described a novel procedure for regenerating antibodies immobilized on a fiberoptic surface with ultrasonic irradiation using a broadband imaging transducer operating near 5 MHz.[22] This type of ultrasound device is commonly used for the detection of flaws in the nondestructive evaluation of engineering materials and measurement of the mechanical properties of various media,[89] including biological tissues.[90] The use of ultrasound for the regeneration of optical fiber immunosensors could be an important advance in the application of these devices for *in vivo* and *in situ* measurements because it would no longer be necessary to supply a regeneration solution to the sensor system that could lead to the denaturation of the immobilized antibody.

## 20.4.2 Enzyme

Enzymes are often used as bioreceptors because of their specific binding capabilities as well as their catalytic activity. In biocatalytic recognition mechanisms, the detection is amplified by a catalytic reaction. This is the basis for the now commonplace enzyme-linked immunosorbent assay (ELISA) technique.

With the exception of a small group of catalytic ribonucleic acid molecules, all enzymes are proteins. Some enzymes require no chemical groups other than their amino acid residues for activity. Others require an additional component called a cofactor, which may be either one or more inorganic ions, such as $Fe^{2+}$, $Mg^{2+}$, $Mn^{2+}$, or $Zn^{2+}$, or a more complex organic or organometallic molecule called a coenzyme. The catalytic activity provided by enzymes allows for much lower limits of detection than would be obtained with common binding techniques. As expected, the catalytic activity of enzymes depends upon the integrity of their native protein conformation. If an enzyme is denatured, dissociated into its subunits, or broken down into its component amino acids, its catalytic activity is destroyed. Enzyme-coupled receptors can also be used to modify the recognition mechanisms. For instance, the activity of an enzyme can be modulated when a ligand binds at the receptor. This enzymatic activity is often greatly enhanced by an enzyme cascade, which leads to complex reactions in the cell.[91,92]

Gauglitz and co-workers have immobilized enzymes onto an array of optical fibers for use in the simultaneous detection of penicillin and ampicillin.[93,94] These biosensors provide an interferometric technique for measuring penicillin and ampicillin based on pH changes during their hydrolysis by penicillinase. Immobilized onto the fibers with the penicillinase is a pH indicator, phenol red. As the enzyme hydrolyzes the two substrates, shifts in the reflectance spectrum of the pH indicator are measured. Various types of data analysis of the spectral information were evaluated using a multivariate calibration method for the sensor array, which consisted of different biosensors

Rosenzweig and Kopelman described the development and use of a micrometer-sized fiber-optic biosensor for the detection of glucose.[95] These biosensors are 100 times smaller than existing glucose optodes and represent the beginning of a new trend in nanosensor technology.[96] They are based on the enzymatic reaction of glucose oxidase, which catalyses the oxidation of glucose and oxygen into gluconic acid and hydrogen peroxide. To monitor the reaction, an oxygen indicator, tris(1,10-phenanthroline)ruthenium chloride, is immobilized into an acrylamide polymer with the glucose oxidase, and this polymer is attached to the optical fiber via photopolymerization. A comparison of the response of glucose sensors created on different sizes of fibers found that the micrometer-size sensors have response times at least 25 times faster (2 s) than the larger fibers. In addition, these sensors are reported to have absolute detection limits of approximately 10 to 15 mol and an absolute sensitivity five to six orders of magnitude greater than current glucose optodes.[96]

A fiber-optic evanescent wave immunosensor for the detection of lactate dehydrogenase has been developed.[97] Two different assay methods, a one-step and a two-step process, using the sensor based on polyclonal antibody recognition were described. The response of this evanescent wave immunosensor was then compared to a commercially available SPR-based biosensor for lactate dehydrogenase detection using similar assay techniques, and similar results were obtained. It was also demonstrated that, although the same polyclonal antibody can be used for the one- and the two-step assay techniques, the two-step technique is significantly better when the antigen is large.

## 20.4.3 Nucleic Acid

### 20.4.3.1 Nucleic Acid Bioreceptors

Another biorecognition mechanism involves hybridization of DNA or RNA. In the last decade, interest in nucleic acids as bioreceptors for biosensor and biochip technologies has increased.[96,98–102] The complementarity of the pairing of the nucleotides adenine:thymine (A:T) and cytosine:guanine

(C:G) in a DNA ladder (Figure 20.3B) forms the basis for the specificity of biorecognition in DNA biosensors, often referred to as genosensors. If the sequence of bases composing a certain part of the DNA molecule is known, then the complementary sequence, often called a probe, can be synthesized and labeled with an optically detectable compound (e.g., a fluorescent label). By unwinding the double-stranded DNA into single strands, adding the probe, and then annealing the strands, the labeled probe can be made to hybridize to its complementary sequence on the target molecule.

Grabley and co-workers have reported the use of DNA biosensors for monitoring DNA–ligand interactions.[103] Surface plasmon resonance was the analytical method used to monitor real-time binding of low-molecular-weight ligands to DNA fragments that were irreversibly bound to the sensor surface via Coulombic interactions. The sensor was capable of detecting binding effects between 10 and 400 pg/mm$^2$. Binding rates and equilibrium coverages were determined for various ligands by changing the ligand concentration. In addition, affinity constants, association rates, and dissociation rates were also determined for these various ligands.

Sandwich-type biosensors based on liquid-crystalline dispersions formed from DNA-polycation complexes have been described by Yevdokimov and co-workers.[104] These sandwich biosensors have been shown to be useful for detection of compounds and physical factors that affect the ability of specific DNA cross linkers — polycationic molecules — to bind between adjacent DNA molecules. The specific case of dispersions from DNA/protamine complexes was investigated and it was demonstrated that, by using this type of sensor with this complex, the hydrolytic enzyme trypsin could be measured down to concentrations of approximately $10^{-14}$ *M*.

Karube and co-workers demonstrated another type of biosensor that uses a peptide nucleic acid as the biorecognition element.[99] The peptide nucleic acid is an artificial oligo amide capable of binding very strongly to complementary oligonucleotide sequences. By use of a surface plasmon resonance sensor, the direct detection of double stranded DNA that had been amplified by a polymerase chain reaction (PCR) has been demonstrated. This technique was capable of monitoring the target DNA over a concentration range of 40 to 160 n*M*, corresponding to an absolute detection limit of 7.5 pmol.

Using a unique analytical technique, Vo-Dinh and co-workers have developed a new type of DNA gene probe based on SERS detection.[20,45–47,105] The SERS probes do not require the use of radioactive labels and have great potential to provide sensitivity and selectivity via label multiplexing due to the intrinsically narrow bandwiths of Raman peaks. The effectiveness of the new detection scheme is demonstrated using the *gag* gene sequence of the human immunodefficiency (HIV) virus.[20] A SERS-based DNA assay for the breast cancer susceptibility gene (*BRCA1*) has also been developed. The assay is based on the immobilization of oligonucleotides on a thin silver surface and hybridization with Rhodamine-labeled probes.[21] The silver surface serves as the hybridization support and the means for Raman signal enhancement. The development of a biosensor for DNA diagnostics using visible and NIR dyes has been reported.[23] The system employed a two-dimensional charge-coupled device and was used to detect the cancer suppressor *p53* gene.

#### 20.4.3.2 DNA Biosensors

DNA sensors are usually based on hybridization assays and may incorporate simultaneous analytical capability to detect a large number of oligonucleotide fragments. A fiber-optic DNA biosensor microarray for the analysis of gene expression has been reported for the simultaneous analysis of multiple DNA sequences using fluorescent probes.[106]

Fluorescent intercalating and groove-binding dyes that can associate with double-stranded DNA (dsDNA) are used for detection of hybridization in some sensor and biochip designs.[107] It is possible that dye–dye interactions at concentrations relevant to biosensor use can lead to unexpected and undesired emission wavelength shifts and fluorescence quenching interactions. To maximize signal to noise, many biosensors utilize dye concentrations in large excess in comparison to the quantity of immobilized DNA. The linearity of fluorescence intensity response of dyes intercalated to dsDNA may vary with different dye:base pair ratios.[107]

A very common alternative to intercalating fluorescent dyes for the detection of dsDNA are covalently bound dyes. Dyes attached to the terminus of a strand of DNA through a short hydrocarbon chain (also known as a tether) are continuously available for hybridization and allow the biosensor to be fully reversible.[25]

Another important characteristic of the biosensor assay is the choice of the fluorescent dye that acts as the signal transducer. Traditionally, ethidium bromide (EB) has been used extensively to detect hybridization of DNA in applications such as electrophoresis, gene chips, and biosensors. A number of dyes with greater quantum efficiency than EB for detection of hybridization have been reported. Furthermore, other practical spectroscopic advantages can be gained in terms of improved signal-to-noise ratio (SNR) by use of dyes with excitation that is red-shifted relative to EB. Pyrilium iodide has been shown to be an intercalator of high quantum efficiency and long excitation wavelength.[25]

One type of DNA sensor utilized the direct synthesis of an ssDNA sequence directly onto optical fibers, using the well established solid-phase phosphoramidite methodology. The covalently immobilized oligomers were able to hybridize with available complementary ssDNA, which was introduced into the local environment to form dsDNA. This event was detected by the use of the fluorescent DNA stain ethidium bromide (EB). The sampling configuration utilized total internal reflection of optical radiation within the fiber, resulting in an intrinsic-mode optical sensor. The nonoptimized procedure used standard hybridization assay techniques to provide a detection limit of 86 ng $ml^{-1}$ cDNA, a sensitivity of 83% fluorescence intensity increase per 100 ng $ml^{-1}$ of cDNA initially present, with a hybridization analysis time of 46 min. The sensor has been observed to sustain activity after prolonged storage times (3 months) and harsh washing conditions (sonication).[26] A similar sensor fabricated using quartz optical fibers as the support provided very similar results.[108]

A very interesting biosensor capable of detecting triple-helical DNA formation was also based on the direct synthesis of oligonucleotides on the surface of fused silica optical fibers, using a DNA synthesizer. Two sets of oligonucleotides on different fibers were grown in the 3′ to 5′ and 5′ to 3′ directions, respectively. Fluorescence studies of hybridization showed unequivocal hybridization between oligomers immobilized on the fibers and complementary oligonucleotides from the solution phase, as detected by fluorescence from intercalated EB. The complementary origonucleotide, dT(10), which was expected to hybridize when the system was cooled below the duplex melting temperature, provided a fluorescence intensity with a negative temperature coefficient. Upon further cooling, to the point where the pyrimidine motif T*AT triple-helix formation occurred, a fluorescence intensity change with a positive temperature coefficient was observed.[27]

In another study, the same type of sensor, with directly immobilized ssDNA on optical fibers, was used to monitor variation in the melt temperature of dsDNA. Because of microenvironment conditions, the local ionic strength, the pH, and the dielectric constant at the surface can be substantially different from those in bulk electrolyte solution. The local conditions influence the thermodynamics of hybridization and can be studied by the melt temperature of dsDNA. Fiber-optic biosensors with dT(20) oligonucleotides attached to their surfaces were used to determine the Tm from the dissociation of duplexes of mixtures of fluorescein-labeled and unlabeled dA(20) and d(A(9)GA(10)). Each thermal denaturation of dsDNA at the surface of the optical fibers was accompanied by a two- to threefold reduction in standard enthalpy change relative to values determined for denaturation in bulk solution. The experimental results suggested that the thermodynamic stability of duplexes immobilized on a surface is dependent on the density of immobilized DNA.

Additionally, the deviation in melt temperature, arising as a result of the presence of a centrally located single base-pair mismatch, was significantly larger for thermal denaturation occurring at the surface of the optical fibers ($\Delta T_m$ = 6 to 10°C) relative to that observed in bulk solution ($\Delta T_m$ = 3.8 to 6.1°C). These results suggest that hybridization at an interface occurs in a significantly different physical environment from hybridization in bulk solution, and that surface density can be tuned to design analytical figures of merit.[109] Increased immobilization density resulted in significantly higher sensitivity but reduced dynamic range in all hybridization assays conducted. Sensitivity and selectivity were functions of temperature; however, the selectivity of hybridization assays done using the sensors could not be predicted by consideration of thermal denaturation temperatures alone.[28]

### 20.4.4 Cell-Based Systems

Cellular structures and cells comprise a broad category of bioreceptors that have been used in the development of biosensors and biochips.[110–120] These bioreceptors are based on biorecognition by an entire cell or microorganism or by a specific cellular component capable of specific binding to certain species. There are presently three major subclasses of this category: (1) cellular systems, (2) enzymes, and (3) nonenzymatic proteins. Due to the importance and large number of biosensors based on enzymes, these have been given their own classification and were previously discussed in Section 20.4.2. This section deals with cellular systems and nonenzymatic proteins.

#### 20.4.4.1 Cellular Bioreceptors

Microorganisms offer a form of bioreceptor that often allows a whole class of compounds to be monitored. Generally these microorganism biosensors rely on the uptake of certain chemicals into the microorganism for digestion. Often, a class of chemicals is ingested by a microorganism, therefore allowing a class-specific biosensor to be created. Microorganisms such as bacteria and fungi have been used as indicators of toxicity or for the measurement of specific substances. For example, cell metabolism (e.g., growth inhibition, cell viability, substrate uptake), cell respiration, and bacterial bioluminescence have been used to evaluate the effects of toxic heavy metals. Many cell organelles can be isolated and used as bioreceptors. Cell organelles are essentially closed systems, so they can be used over long periods of time. Whole mammalian tissue slices or *in vitro* cultured mammalian cells are used as biosensing elements in bioreceptors. Plant tissues are also used in plant-based biosensors; they are effective catalysts because of the enzymatic pathways they possess.[92]

Bilitewski and co-workers have developed a microbial biosensor for monitoring short-chain fatty acids in milk.[121] *Arthrobacter nicotianae* microorganisms were immobilized in a calcium-alginate gel on an electrode surface. To this gel was added 0.5 m*M* $CaCl_2$ to help stabilize it. Monitoring the oxygen consumption of the *A. nicotianae* electrochemically allowed its respiratory activity to be monitored, thereby providing an indirect means of monitoring fatty acid consumption. Detection of short-chain fatty acids in milk, ranging from 4 to 12 carbons in length, was accomplished with butyric acid as the major substrate. A linear dynamic range from 9.5 to 165.5 μ*M* was reported, with a response time of 3 min.

#### 20.4.4.2 Nonenzymatic Proteins

Many proteins found within cells often serve the purpose of bioreception for intracellular reactions that will take place later or in another part of the cell. These proteins could simply be used for transport of a chemical from one place to another, e.g., a carrier protein or channel protein on a cellular surface. In any case, these proteins provide a means of molecular recognition through one or another type of mechanism (i.e., active site or potential sensitive site). By attaching these proteins to various types of transducers, many researchers have constructed biosensors based on nonenzymatic protein biorecognition.

In one recent application, Cusanovich and co-workers developed micro- and nanobiosensors for nitric oxide that are free from most potential interferents.[96] These sensors are based on bioreception of nitric oxide by cytochrome *c*′. Two different techniques of immobilization of the cytochrome *c*′ to fibers were tested: polymerization in an acrylamide gel and reversible binding using a gold colloid-based attachment. The cytochrome used in this work was labeled with a fluorescent dye that is excited via an energy transfer from the hemoprotein. Response times of faster than 1 s were reported along with a detection limit of 20 μ*M*. Cytochrome *c*′ samples from three different species of bacteria were evaluated.

Vogel and co-workers have reported on the use of lipopeptides as bioreceptors for biosensors.[122] A lipopeptide containing an antigenic peptide segment of VP1, a capsid protein of the picornavirus that causes foot-and-mouth diseases in cattle, was evaluated as a technique for monitoring antigen–antibody interactions. The protein was characterized via circular dichroism and infrared spectroscopy to verify that upon self-assembly onto a solid surface it retained the same structure as in its free form. Based on surface plasmon resonance measurements, it was found that the protein was still fully accessible for antibody binding. This technique could provide an effective means of developing biomimetic ligands for binding to cell surfaces.

### 20.4.5 Biomimetic Receptors

A receptor fabricated and designed to mimic a bioreceptor is often termed a "biomimetic receptor." Several different methods have been developed over the years for the construction of biomimetic receptors.[123–128] These methods include genetically engineered molecules, artificial membrane fabrication, and molecular imprinting. The molecular imprinting technique, which has recently received great interest, consists of mixing analyte molecules with monomers and a large number of cross linkers. Following polymerization, the hard polymer is ground into a powder, and the analyte molecules are extracted with organic solvents to remove them from the polymer network. As a result, the polymer has molecular holes or binding sites that are complementary to the selected analyte.

Recombinant techniques, which allow for the synthesis or modification of a wide variety of binding sites using chemical means, have provided powerful tools for designing synthetic bioreceptors with desired properties. Hellinga and co-workers reported the development of a genetically engineered single-chain antibody fragment for monitoring phosphorylcholine.[129] In this work, protein engineering techniques are used to fuse a peptide sequence that mimics the binding properties of biotin to the carboxyterminus of the phosphorylcholine-binding fragment of IgA. This genetically engineered molecule can be attached to a streptavidin monolayer, and total internal reflection fluorescence was used to monitor the binding of a fluorescently labeled phosphorylcholine analog.

Artificial membranes have been developed for many different bioreception applications. Stevens and co-workers developed one by incorporating gangliosides into a matrix of diacetylenic lipids (5 to 10% of which were derivatized with sialic acid).[130] The lipids were allowed to self-assemble into Langmuir-Blodgett layers and were then photopolymerized via ultraviolet irradiation into polydiacetylene membranes. When cholera toxins bind to the membrane, its natural blue color changes to red; absorption measurements were used to monitor the toxin concentration. Using these polydiacetylenic lipid membranes coupled with absorption measurements, concentrations of cholera toxin as low as 20 μg/ml could be monitored.

Molecular imprinting has been used for the construction of a biosensor based on electrochemical detection of morphine.[131] A molecularly imprinted polymer for the detection of morphine was fabricated on a platinum wire using agarose and a cross-linking process. The resulting imprinted polymer was used to specifically bind morphine to the electrode. Following morphine binding, an electroinactive competitor, codeine, was used to wash the electrode and thus release some of the bound morphine. The freed morphine was then measured by oxidation at the electrode; concentrations ranging from 0.1 to 10 μg/ml were analyzed, with a reported detection limit of 0.05 μg/ml. One of the major advantages of the molecular imprinting technique is the rugged nature of a polymer relative to a biological sample. The molecularly imprinted polymer can withstand harsh environments such as those experienced in an autoclave or chemicals that would denature a protein.

## 20.5 Probe Development: Immobilization of Biomolecules

Many of the methods used in biosensor fabrication involve binding the recognition probe (oligonucleotide strand, antibody, etc.) to a sensor-sensitive surface or to an optically active tag (fluorescent dye, etc.). For enzyme-based sensors, immobilization of an enzyme is a critical step and can be accomplished via simple physical adsorption or through more elaborate covalent binding schemes.

Molecules may be physically immobilized in a solid support through hydrophobic or ionic interactions or covalently immobilized by attachment to activated surface groups.[132] Noncovalent immobilization is effective for many applications and usually requires easier and faster preparation steps.[133–135] In addition, the adsorbed molecules usually preserve their original properties (e.g., wavelength of absorption, excitation or emission, enzymatic activity) because they do not require the structural modification inherent in covalent immobilization to a solid support. However, continuous leaching of the adsorbate from the solid support may reduce the sensor's durability and even render it useless in the worst cases.

Covalent immobilization is often necessary for binding molecules that do not adsorb, adsorb very weakly, or adsorb with improper orientation and conformation to noncovalent surfaces. Covalent immobilization may provide greater stability, reduced nonspecific adsorption, and greater durability.[136,137]

Several synthetic techniques are available for the covalent immobilization of biomolecules or labeling of a sensor probe with a fluorescent dye.[138] Most of these techniques use free amine groups in a polypeptide (enzymes, antibodies, antigens, etc.) or in an amino-labeled DNA strand to react with a carboxylic acid moiety to form amide bonds. As a general rule, a more active intermediate (labile ester) is first formed with the carboxylic acid moiety and in a later stage reacted with the free amine, increasing the coupling yield. A few coupling procedures are described below.

*Carbodiimide coupling.* Surfaces modified with mercaptoalkyldiols can be activated with 1,1′-carbonyldiimidazole (CDI) to form a carbonylimidazole intermediate. A biomolecule with an available amine group displaces the imidazole to form a carbamate linkage to the alkylthiol tethered to the surface[139]

*N-hydroxysuccinimide (NHS) and its derivatives.* Using a succinimide ester intermediate in acylation reactions of 5′-amino-labeled DNA is also a very efficient protocol. The NHS-activated carboxyl group has a much longer lifetime than the reaction intermediates produced by carbodiimide coupling.[140] NHS can also be used to facilitate amide formation between a carboxylic acid moiety and free amine groups in a polypeptide (enzymes, antibodies, antigens, etc.). NHS reacts almost exclusively with primary amine groups, with the exception of mercaptans. This nucleophilic substitution reaction covalently immobilizes biomolecules via available amine moieties by forming stable amide bonds. Covalent immobilization can be achieved in as little as 30 min. This surface has been shown to immobilize 5′ amine-modified oligonucleotides, providing an ideal template for hybridization and amplification. Because the DNA is bound at one end rather than at numerous sites along the molecule, the result is high specificity and low background. Because $H_2O$ competes with $-NH_2$ in reactions involving these very labile esters, it is important to consider the hydrolysis kinetics of the esters used in this type of coupling. A derivative of NHS, *O*-(*N*-succinimidyl)-*N*,*N*,*N*′,*N*′-tetramethyluronium tetrafluoroborate, increases the coupling yield by utilizing a leaving group that is converted to urea during the carboxylic acid activation, hence favorably increasing the negative enthalpy of the reaction. The schematic in Figure 20.9 illustrates this approach.

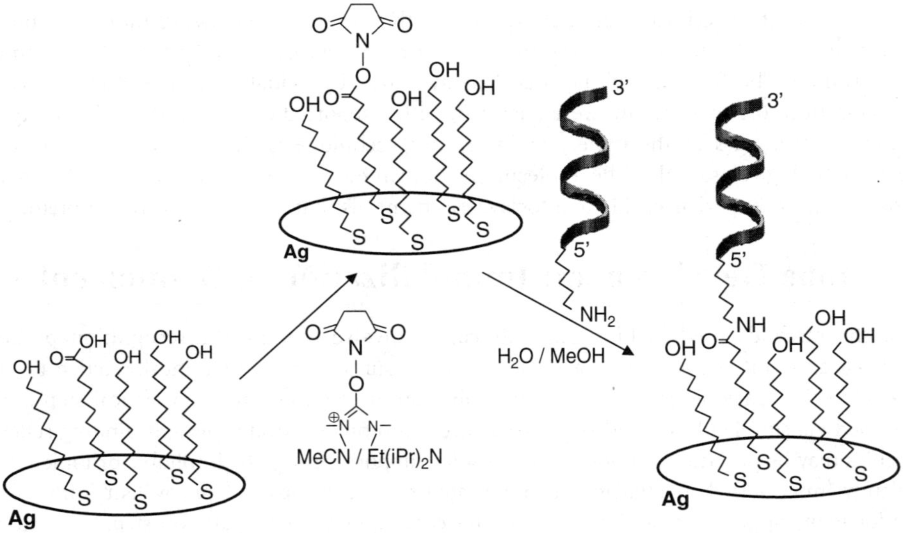


**FIGURE 20.9** Binding scheme for biomolecule immobilization.

*Maleimide.* Maleimide can be used to immobilize biomolecules through available –SH moieties. Coupling schemes with maleimide have been proven useful for the site-specific immobilization of antibodies, Fab fragments, peptides, and SH-modified DNA strands. Sample preparation for the maleimide coupling of a protein involves the simple reduction of disulfide bonds between two cysteine residues with a mild reducing agent, such as dithiothreitol, 2-mercaptoethanol, or tris(2-carboxyethyl)phosphine hydrochloride. However, disulfide reduction will usually lead to the protein losing its natural conformation and might impair enzymatic acitivity or antibody recognition. The modification of primary amine groups with 2-iminothiolane hydrochloride (Traut's reagent) to introduce sulfydryl groups is an alternative for biomolecules lacking them. Free sulfhydryls are immobilized to the maleimide surface by an additional reaction to unsaturated carbon–carbon bonds.

*Hydrazide.* Hydrazide is used for the covalent coupling of periodate-activated carbohydrates or glycosylated biomolecules. It can be used for the site-specific immobilization of antibodies, carbohydrates, glycolipids, glycoproteins, and many enzymes. Antibodies are immobilized to the hydrazide surface through the carbohydrate moieties on the Fc region, which allows the Fab regions to be properly oriented.

Non-covalent immobilization of biomolecules includes sol-gel or polymer entrapment of enzymes,[141–146] and adsorption of oligonuleotides onto poly-cationic membranes (poly-lysine, polymers with quaternarium ammonium groups, etc).[95,147,148] For sol-gel or polymeric immobilization, solubility of the biomolecules in the polymerization medium is a major constraint for the use of this technique. In the case of surface adsorption, conformation changes of proteins that might diminish antibody recognition or enzyme activity need to be considered.

A sensor for the nitrate ion based on the encapsulation of an enzyme in a sol-gel structure is reported to be effective even after a storage period of up to 6 months.[141] The enzyme of choice was the periplasmic nitrate reductase (Nap), extracted from the denitrifiying bacterium *Thiosphaera pantotropha*, which reacts specifically with the nitrate ($NO_3^-$) anion.

Edmiston and co-workers found evidence that the immobilization of proteins by entrapment in a porous silica matrix prepared by sol-gel techniques may significantly change the conformation of the proteins. These researchers examined two model proteins, bovine serum albumin (BSA) and horse heart myoglobin (Mb), entrapped in wet sol-gel glass bulks. They investigated the fluorescence behavior of dissolved and entrapped BSA in the presence of acid, a chemical denaturant, and a collisional quencher. The results show that a large fraction of the BSA added to the sol is entrapped within the gelled glass in a native conformation. However, the reversible conformational transitions that BSA undergoes in solution are sterically restricted in the gel. In contrast, the native properties of Mb are largely lost upon entrapment, as judged by the changes in the visible absorbance spectra of dissolved and entrapped Mb in acidic solutions. Fluorescence studies of dissolved and entrapped apomyoglobin supported this conclusion.[149]

## 20.6 Biomedical Applications

This section presents an overview of various medical applications of biosensors, especially in the diagnosis of diseases. A large body of work was accomplished using SPR commercial instruments and involves the study of protein or DNA interactions relevant for medical applications. Other widely used methods include fluorescence spectroscopy, NIR, and circular dichroism.

### 2.6.1 Cellular Processes

Staining cellular organelles is a classic laboratory method that utilizes visible or fluorescent dyes that have high affinity for specific organelles. After staining, visible or UV illumination is used for microscopy identification. Newer techniques explore the same principles to obtain more information on cellular processes as well.

For example, G protein-coupled receptors (GPCRs) represent one of the most important drug targets for medical therapy, and information from genome sequencing and genomic databases has substantially accelerated their discovery.[150] The lack of a systematic approach to identify the function of a new GPCR or to associate it with a cognate ligand has added to the growing number of orphan receptors. A novel approach to this problem using optical detection of a beta-arrestin2/green fluorescent protein conjugate (beta arr2-GFP) has been reported. Confocal microscopy demonstrates the translocation of beta arr2-GFP to more than 15 different ligand-activated GPCRs, providing a real-time and single-cell-based assay to monitor GPCR activation and GPCR-Gr protein-coupled receptor kinase or GPCR–arrestin interactions. The use of beta-arr2-GFP as a biosensor to recognize the activation of pharmacologically distinct GPCRs should accelerate the identification of orphan receptors and permit the optical study of their signal transduction biology, which is intractable to ordinary biochemical methods.[150]

Abscisic acid (ABA) is a plant hormone involved in many developmental and physiological processes, but no ABA receptor has been identified yet.[151] In an attempt to demonstrate that the monoclonal antibody JIM19 recognizes carbohydrate epitopes of cell surface glycoproteins, researchers have used flow cytometry of rice protoplasts and immunoblotting of purified plasma membranes (PMs).[151] Through use of SPR technology, specific binding of PMs to JIM19 was observed. The interaction was antagonized significantly by ABA but not by the biologically inactive ABA catabolite, phaseic acid. Pretreatment with JIM19 resulted in significant inhibition of ABA-inducible gene expression. Taken together, these data suggest that JIM19 interacts with a functional PM complex involved in ABA signaling.[151]

Another receptor binding study done with optical biosensor technology investigated the affinity and specificity of the putative proximal tubular scavenging receptor for protein reabsorption and the specificity of AGE-modified protein interactions with primary human mesangial cells.[152] An SPR biosensor with a carboxy-methyl dextran surface was used for binding competition analysis of five different proteins of the LLCPK cell line (ranging in size and charge). The biosensor data show evidence to support the existence of a single scavenging receptor for all the proteins tested. The proteins competed with each other, differing only in their relative binding affinity for the common receptor. This study has also showed that human mesangial cells can bind to AGE-modified human serum albumin (AGE-HSA) immobilized onto the carboxylate surface and that binding can be inhibited by using increasing concentrations of soluble AGE-HSA. However, increasing concentrations of soluble non-AGE-modified HSA can also inhibit binding to a similar extent, which implies relatively little AGE-receptor expression on cultured primary human mesangial cells. The SPR biosensor is a potential tool to explore cellular interactions with renal cells.[152]

### 20.6.2 Viral Agents

The use of biosensors to detect specific viruses in biological samples offers a great diagnostic tool for medical applications. To date, studies have targeted several viruses, including HIV (discussed separately), measles virus, herpes simplex virus, rhinoviruses, and foot-and-mouth disease prions. Some of the techniques used in their detection are outlined below.

Identifying viruses in clinical materials during the acute phase of infections could give necessary information for the treatment of infections by human immunoglobulin (hIg) or interferon (IF). However, because of lack of information, most virus infections are not treated. A real-time detection system for viruses in general has been developed using an optical biosensor and a model virus: herpes simplex virus Type 1 (HSV-1). The HSV-1 virus was found to propagate in Vero cells and, when diluted in minimum essential medium (MEM) with 10% fetal bovine serum (FBS), could be detected with an SPR sensor with high sensitivity and a detection limit of 10 infectious units (50% tissue culture infective dose [TCID50] units). When a crude homemade rabbit antiserum was used against measles virus with host cell debris as a ligand, the SPR sensor performed with lower sensitivity, detecting less than 500 infectious (TCID50) units of virus in a 100-μl solution. This real-time viral detection and titration system has sensitivity high enough for clinical purposes.[153]

The herpes virus was also the object of a study of epitope mapping using an SPR optical biosensor. The human herpes virus entry mediator C (HveC), also known as the poliovirus receptor-related protein 1 (PRR1) and as nectin-1, allows the entry of HSV-1 and HSV-2 into mammalian cells. The interaction of the virus envelope glycoprotein D (gD) with such a receptor is an essential step in the process leading to membrane fusion.[154] HveC is a member of the immunoglobulin (Ig) superfamily and contains three Ig-like domains in its extracellular portion. The gD binding site is located within the first Ig-like domain (V domain) of HveC. In a careful study using SPR, 11 monoclonal antibodies (MAbs) against the ectodomain of HveC were chosen to detect linear or conformational epitopes within the V domain. Besides the biosensor analysis, the HveC was detected by enzyme-linked immunosorbent assay, Western blotting, and directly on the surface of HeLa cells and human neuroblastoma cell lines, as well as simian Vero cells. A few of the 11 monoclonal antibodis blocked HSV entry. Competition assays on an optical biosensor showed that CK6 and CK8 (linear epitopes) inhibited the binding of CK41 and R1.302 (conformational epitopes) to HveC and vice versa.[154]

One of the reasons for the traditionally low success for the direct identification of viruses by simple immunological assays is the large variability of their surface epitopes. For example, more than 100 immunologically distinct serotypes of human rhinoviruses (HRV) have been discovered, making detection of surface-exposed capsid antigens impractical. However, the nonstructural protein 3C protease (3Cpro) is essential for viral replication and is relatively highly conserved among serotypes, making it a potential target for diagnostic testing of HRVs. An SPR biosensor with a modified silicon surface with broadly reactive serotype antibodies to 3Cpro has been developed.[155]

The *in vitro* sensitivity, specificity and multiserotype cross reactivity of the 3Cpro assay were tested using the SPR sensor in a 28-min, noninstrumented room-temperature test with a visual detection limit of 12 pM of 3Cpro (1000 TCID50 equivalents). Nasal washes from naturally infected individuals were used as test samples. The assay detected 87% (45 of 52) of the HRV serotypes tested but showed no cross reactivity to common respiratory viruses or bacteria. The SPR assay detected 3Cpro in expelled nasal secretions from a symptomatic individual on the first day of illness. In addition, 82% (9 of 11) concentrated nasal wash specimens from HRV-infected children were positive in the 3Cpro test. Thus the assay is suitable as a diagnostic test for a point-of-care setting, where rapid HRV diagnostic test results could contribute to clinical decisions regarding appropriate antibiotic or antiviral therapy.[155]

Another SPR system was applied to the quantitative analysis of the binding of HSV-1 to Vero cells. A commercially available sulfonated human immunoglobulin preparation was used as the neutralization antibody titer against this virus.[156]

Virus-like particles (VLPs) are multimeric proteins expressed by *Saccharomyces cerevisiae*. These particles are approximately 80 nm in diameter and are used as a framework for a range of biological products, including carriers of viral antigens. An SPR biosensor was developed for rapid monitoring of purified VLPs; this device can be used for real-time bioprocess monitoring of VLPs. Problems of mass transfer of the analyte were overcome through selection of a planar biosensor surface instead of the traditional polymer-coated surface. To prolong the surface activity for interaction analysis, a sandwich assay was developed that involved the use of a secondary capture species. It was shown that VLP concentration in pure solution could be determined within 10 min.[157]

An SPR biosensor has been used for screening synthetic peptides mimicking the immunodominant region of C-S8c1 foot-and-mouth disease virus. The main antigenic site (site A) of the foot-and-mouth disease virus (FMDV, strain C-S8c1) may be adequately reproduced by a 15-peptide with the amino acid sequence H-YTASARGDLAHLTTT-NH2 (A15), corresponding to the residues 136 to 150 of the viral protein VPI.[158] The SPR sensor surface was modified with monoclonal antibodies raised against antigenic site A. Although these antigenicities have previously been determined from ELISA methods, the SPR-based technique is superior in that it allows a fast and straightforward screening of antigens while simultaneously providing kinetic data for the Ag–Ab interaction.[158]

### 20.6.3 Human Immunodeficiency Virus (HIV)

The human immunodeficiency virus (HIV) has been the target of intense research in the past two decades. Some of the research efforts involving optical biosensors are outlined below.

The two main proteins on the HIV envelope are glycoproteins gp120 and gp41 (named for their approximate size in kilodaltons). Glycoproteins gp120 and gp41 are associated together noncovalently. Binding of

HIV-1 gp120 to T-cell receptor CD4 initiates conformational changes in the viral envelope that trigger viral entry into the host cells.[159]

SPR has been used in a number of HIV studies. This technique was applied to observe the conformational changes in gp120 upon binding to certain ligands, to compare the gp120-activation effects of CD4 mimetics, and to examine for CD4 competition and gp120 activation.[160–162] SPR optical biosensors provide a means of looking at the interaction between macromolecules as it occurs in real time, providing information about the kinetics of the interaction, in addition to estimating affinity constants.

SPR optical biosensor assays for the screening of low-molecular-weight compounds, using an immobilized protein target, have been developed. HIV-1 proteinase was immobilized on the sensor surface by direct amine coupling. A large number of inhibitors and noninteracting reference drugs were applied to the sensor surface in a continuous flow of buffer to estimate binding constants. The optimized assay could correctly distinguish HIV-1 inhibitors from other compounds in a randomized series, indicate differences in their interaction kinetics, and reveal artifacts due to nonspecific signals, incomplete regeneration, or carryover.[163]

SPR biosensors have also been used to study the interaction between HIV-1 protease and reversible inhibitors. The steady-state binding level and the time course of association and dissociation could be observed by measuring the binding of inhibitors injected in a continuous flow of buffer to the enzyme immobilized on the biosensor surface. Fourteen low-molecular-weight inhibitors (500? to 700 Da), including four clinically used HIV-1 protease inhibitors (indinavir, nelfinavir, ritonavir, and saquinavir) were analyzed. Inhibition constants (*Ki*) were determined by a separate enzyme inhibition assay. Indinavir had the highest affinity ($B_{50}$ = 11 n*M*) and the fastest dissociation ($t_{1/2}$ = 500 s) among the clinically used inhibitors, while saquinavir had a lower affinity ($B_{50}$ = 25 n*M*) and the slowest dissociation rate ($t_{1/2}$ = 6500 s). Because these two inhibitors have similar affinities, the differences in dissociation rates reveal important characteristics in the interaction that cannot be obtained by the inhibition studies alone.[164,165] Characterization of another set of HIV-1 protease inhibitors using binding kinetics data from an SPR biosensor-based screen has also been reported.[166]

Fluorescence polarization, circular dichroism, and SPR optical biosensor binding studies were used to investigate the novel virucidal protein cyanovirin-n (cv-n). Cv-n binds with equally high affinity to soluble forms of either h9 cell-produced or recombinant glycosylated HIV-1 gp120 (sgp120) or gp160 (sgp160). Studies showed that cv-n is also capable of binding to the glycosylated ectodomain of the HIV-envelope protein gp41 (sgp41), albeit with considerably lower affinity than the sgp120/cv-n interaction. These optical techniques shed light on the binding of cv-n with sgp120 and sgp41, providing direct evidence that conformational changes are a consequence of cv-n interactions with both HIV-1 envelope glycoproteins.[167]

### 20.6.4 Bacterial Pathogens

Several physicochemical instrumental techniques for direct and indirect identification of bacteria such as IR and fluorescence spectroscopy, flow cytometry, chromatography, and chemiluminescence have been reviewed as feasible biosensor technologies.[168]

*Staphylococcus aureus* is a pathogen that commonly causes human infections and intoxication. An evanescent-wave optical sensor was developed for the detection of protein A, a product secreted only by *S. aureus.* A 488-nm laser was used in conjunction with a plastic optical fiber with adsorbed antibodies for protein. A sandwich immunoassay with fluorescein isothiocyanate conjugated with anti-(protein A) IgG was used to monitor the Ag–Ab reaction. The detection limit was 1 ng/ml of protein A.[169]

In a different approach, an optical biosensor based on resonant mirrors was used in the detection of whole cells of *S. aureus* (Cowan-1).[170] The bacterium cells, which express protein-A at their surface, were detected through their binding to human IgG immobilized on an aminosilane-derivatized sensor surface at concentrations in the range $8 \times 10^6$ to $8 \times 10^7$ cells/ml. A control *S. aureus* strain (Wood-46) that does not express protein-A produced no significant response. The sensitivity of the technique was increased by three orders of magnitude when a human IgG-colloidal gold conjugate

(30 nm) was used in a sandwich assay format. *S. aureus* (Cowan-1) cells were detected in spiked milk samples at cell concentrations from $4 \times 10^3$ to $1.6 \times 10^6$ cells/ml using the sandwich assay.[170]

The same resonant mirror optical biosensor technology was used to characterize *Helicobacter pylori* strains according to their sialic acid binding. In that work, intact bacteria were used in real-time measurements of competition and displacement assays using different glycoconjugates. The authors found that that several, but not all, *H. pylori* strains express sialic acid-binding adhesin specific for alpha-2,3-sialyllactose. The adhesin, removable from the bacterial surface by water extraction, is not related to other reported *H. pylori* cell surface proteins with binding ability to sialylated compounds such as sialylglycoceramides.[171,172]

A fiber-optic evanescent-wave sensing system that features all-fiber optical design and red semiconductor-laser excitation has been developed and tested. A 650-nm laser was used because biological matrices demonstrate minimal fluorescent background in the red; this helps reduce the background signal of nonessential biomolecules. The fiber directs the fluorescent signal of a sandwich immunoassay to detect *Salmonella* back to a charge-coupled device (CCD) fiber spectrophotometer. The system could detect *Salmonella* with a concentration as low as $10^4$ colony-forming units (CFU) per milliliter.[173]

With a similar biodetection approach utilizing laser-induced fluorescence, an optic biosensor was used to detect the fraction 1 (F1) antigen from *Y. pestis*, the etiologic agent of plague.[15] An argon ion laser (514 nm) was used to launch light into a long-clad fiber, and the fluorescence produced by an immunofluorescent complex formed in the evanescent wave region was measured with a photodiode. Capture antibodies, which bind to F1 antigen, were immobilized on the core surface to form the basis of the sandwich fluoroimmunoassay. The evanescent wave has a limited penetration depth (<1 lambda), which restricts detection of the fluorescent complexes bound to the fiber's surface. The direct correlation between the concentration of the F1 antigen and the signal provided an effective method for sample quantitation; the method was able to detect F1 antigen concentrations from 50 to 400 ng/ml in phosphate-buffered saline, serum, plasma, and whole blood, with a 5-ng/ml detection limit.[15]

A very different detection approach for *Salmonella typhimurium* involved immunomagnetic separation and a subsequent enzyme-linked assay with alkaline phosphatase. The magnetic microbeads coated with anti-*Salmonella* were used to separate *Salmonella* from sample solutions at room temperature for 30 min. A sandwich complex with alkaline phosphatase and the *Salmonella* immobilized on the magnetic beads was formed, separated from the solution by a magnetic filtration, and incubated with a p-nitrophenyl phosphate substrate at 37°C for 30 min to produce p-nitrophenol by the enzymatic hydrolysis. *Salmonella* was detected by measuring the absorbance of p-nitrophenol at 404 nm, with a linear response between $2.2 \times 10^4$ and $2.2 \times 10^6$ CFU/ml.[174]

## 20.6.5 Cancer

Optical biosensors have been utilized as tools to aid in the direct diagnosis of carcinogenesis, the identification of genetic markers associated with it, and the quantification of known carcinogens.

Fluorescent detection has become a technique of choice for oligonucleotide hybridization detection. Pairs of high density oligonucleotide arrays (DNA chips) consisting of more than 96,000 oligonucleotides were designed to screen the entire 5.53-kb coding region of the hereditary breast and ovarian cancer BRCA1 gene for all possible sequence changes in the homozygous and heterozygous states. Fluorescent hybridization signals from targets, containing the 4 natural bases to more than 5592 different, fully complementary 25mer oligonucleotide probes on the chip, varied over two orders of magnitude. To examine the thermodynamic contribution of rU.dA and rA.dT target probe base pairs to this variability, modified uridine [5-methyluridine and 5-(1-propynyl)-uridine)] and modified adenosine (2,6- diaminopurine riboside) 5′-triphosphates were incorporated into *BRCA1* targets. Hybridization specificity was assessed based upon hybridization signals from >33,200 probes containing centrally localized single base pair mismatches relative to target sequence. Targets containing 5-methyluridine displayed promising localized enhancements in the hybridization signal, especially in pyrimidine-rich target tracts, while maintaining single nucleotide mismatch hybridization specificities comparable with those of unmodified targets.[175]

In another study, the breast cancer susceptibility gene *BRCA1* was also detected by a relatively new technique based on SERS. A single 24-mer sequence was used as the capture probe and was immobilized on a silver-coated microarray platform for hybridization.[21] Breast cancer was also the target of a biosensor design to measure the interaction of S100A4 and potential binding partners.[176] Elevated levels of S100A4 induce a metastatic phenotype in benign mammary tumor cells *in vivo*. In humans, the presence of S100A4 in breast cancer cells correlates strongly with reduced patient survival. There was significant interaction of S100A4 with nonmuscle myosin and p53 but not with actin, tropomyosin, or tubulin.[176]

A regenerable immunosensor utilizing an antibody against breast cancer antigen has been described. A 65% removal of the antigens bound to the Mab immobilized on the fiber surface is attained after ultrasound regeneration.[22] A multiarray biosensor utilizing DNA probes labeled with visible and NIR dyes has also been developed. The detection system uses a two-dimensional CCD to detect the *p53* cancer gene.[23]

Prostate cancer is the cause of death of many thousands of men worldwide. Screening men for elevated prostate-specific antigen (PSA) levels is believed to be an important tool for the diagnosis and management of the disease. An assay for measuring PSA in whole blood using the fluorescence capillary fill device has been developed for use in prostate cancer screening programs.[177]

A unique example of the technology being developed is optical nano-biosensors capable of interrogating the contents of a single isolated cell.[178] These submicrometer fiber-optic biosensors have been used to measure carcinogens within single cells. Optical fibers were pulled to a distal-end diameter of 40 nm and coated with antibodies to selectively bind benzo[a]pyrene tetrol (BPT), a metabolite of benzo[a]pyrene, an extremely potent carcinogen. Two different cell lines have been investigated: human mammary carcinoma cells and rat liver epithelial cells. The detection limit of these nanosensors has been determined to be $(0.64 \pm 0.17) \times 10^{-11}$ M for BPT.[75] The development and application of nanosensors are further described in the chapter on nanosensors for single-cell analysis in this handbook.

The carcinogen benzo[a]pyrene (BAP) was the target of an antibody-based fiberoptics biosensor.[7] In that biosensor, BAP was the analyte and fluorophore because it has a large fluorescence cross section. An antibody with high specificity for BAP was immobilized on the tip of an optical fiber. Upon exposure to contaminated samples, the optical fiber was irradiated with a laser and the resulting fluorescence correlated with the BAP concentration.

## 20.6.6 Parasites

Detection of antibodies specific for the parasite *L. donovani* in human serum samples has been reported. The method is based on an evanescent wave fluorescence collected by optical fibers that have the purified cell surface protein of *L. donovani* immobilized on their surface. The sensing fibers are incubated with the patient serum for 10 min and then incubated with goat anti-human IgG. Fluorescence was proportional to *L. donovani*–specific antibodies present in the test sera.[179]

## 20.6.7 Toxins

Ricin, a potently toxic protein, has been detected with an evanescent-wave fiber-optic biosensor with a detection limit of 100 pg/ml and 1 ng/ml for buffer solutions and river water, respectively. Athis detection was based on a sandwich immunoassay scheme, using an immobilized anti-ricin IgG on the surface of the optical fiber. Two coupling methods were used. In the first, the antibody was directly coated to the silanized fiber using a cross linker; the second method utilized avidin-coated fibers incubated with biotinylated antiricin IgG to immobilize the antibody using an avidin-biotin bridge. The assay using the avidin–biotin-linked antibody demonstrated higher sensitivity and a wider linear dynamic range than the assay using the antibody directly conjugated to the surface. The linear dynamic range of detection for ricin in buffer using the avidin–biotin chemistry is 100 pg/ml to 250 ng/ml.[78]

The lipopolysaccharide (LPS) endotoxin is the most powerful immune stimulant known and a causative agent in the clinical syndrome known as sepsis. Sepsis is responsible for more than 100,000 deaths annually, in large part due to the lack of a rapid, reliable, and sensitive diagnostic technique.[18] An evanescent wave fiberoptic biosensor was developed for the detection of LPS from *E. coli* at concentrations as low as 10 ng/ml in 30 s.[18] Polymyxin B covalently immobilized onto the surface of the fiber-optic probe was able to bind fluorescently labeled LPS selectively. Unlabeled LPS present in the biological samples was detected in a competitive assay format, by displacing the labeled LPS. The competitive assay format worked in buffer and in plasma with similar sensitivities. This method might also be used with other LPS capture molecules, such as antibodies, lectins, or antibiotics, to simultaneously detect LPS and determine the LPS serotype.[18]

An immunoaffinity fluorometric biosensor was developed for the detection of aflatoxins, a family of potent fungi-produced carcinogens commonly found in a variety of agriculture products. Developed into a fully automated instrument based on immunoassays with fluorescent tags, the detection system was able to detect aflatoxins from 0.1 to 50 ppb in 2 min with a 1-ml sample volume.[180]

Parathion was detected with a biosensor based on total internal reflection using a competitive-displacement immunoassay. This biosensor utilizes casein–parathion conjugates, immobilized by adsorption on quartz fibers, and selectively adsorbed antiparathion rabbit antibodies raised against bovine serum albumin(BSA)–parathion conjugates from polyclonal immune sera. The presence of free parathion inhibited the binding of the rabbit's anti-(BSA)-parathion. Fluorescein isothiocyanate (FITC) goat antirabbit IgG was used to generate the optical signal. It could detect 0.3 ppb of parathion and had a detection limit 100-fold higher for the detection of its oxygen analog, paraoxon.[181] Other biosensor approaches for the detection of parathion include another competition assay that inhibits the alkaline phosphatase generation of a chemiluminescent substance[182] and a direct detection method based on another enzymatic sensor.[183,184]

A very interesting approach for biosensing toxins involves the integration of multiple transducers with different affinities for a large range of biotoxins. The different transducers are based on membranes made of mixtures of biologically occurring lipids deposited on the sensing surface of an SPR optical biosensor. Eight surfaces were prepared, some of which contained various glycolipids as minor components, and one was supplemented with membrane proteins. The researchers analyzed the binding of six protein toxins (cholera toxin, cholera toxin B subunit, diphtheria toxin, ricin, ricin B subunit, and staphylococcal enterotoxin B) and of bovine serum albumin at pH 7.4 and pH 5.2 to each of the sensor surfaces.[185] Each of the seven proteins produced a distinct binding pattern to the multitransducer sensor. The same concept had been used earlier for the development of "artificial noses," which are sensor arrays able to detect a small collection of distinct analytes.

Cholera toxin (CT) has been also detected with a fluorescence-based biosensor using a waveguide platform. The biochemical assay utilized a ganglioside-capture format, where the ganglioside (GM1) that captures the analyte was immobilized in discrete locations on the surface of an optical waveguide. Binding of CT to immobilized GM1 was demonstrated with direct assays (using fluorescently labeled CT) and with "sandwich" immunoassays (using fluorescently labeled tracer antibodies). The detection limits for CT were 200 ng/mL in direct assays and 40 ng/ml and 1 μg/ml in sandwich-type assays performed using rabbit and goat tracer antibodies.[77] A slightly different biosensing approach was also used to detect cholera toxin. Instead of direct fluorescence detection, fluorescence quenching was used. The ganglioside GM1 was again used as the recognition unit for CT, and was covalently labeled with fluorophores and then incorporated into a biomimetic membrane surface.[186]

In a very nice application of surface plasmon resonance, SPR biosensors have been used to estimate the immunoreactivity of tetanus toxin and *Vipera aspis* venom against new pasteurized preparations of their horse F(ab′)(2) antidotes, in order to investigate immunoreactivity–immunoprotection efficacy relationships. The immunoreactivity data were compared with seroneutralization titres. The association–dissociation rate and affinity constants of the current and the new tetanus toxin-specific F(ab′)(2) preparations were similar, at about $10^4\ M^{-1}\ s^{-1}$, $10^{-4}\ s^{-1}$ and $10^8\ M^{-1}$, respectively.[187]

### 20.6.8 Blood Factors

SPR was used to determine absolute heparin concentration in human blood plasma. Protamine and polyethylene-imine (PEI) were used to modify the sensor surface and were evaluated for their affinity to heparin. Heparin adsorption onto protamine in blood plasma was specific with a lowest detection limit of 0.2 U/ml and a linear detection range of 0.2 to 2 U/ml. Although heparin adsorption onto PEI in buffer solution had indicated superior sensitivity to that on protamine, in blood plasma it was not specific for heparin and adsorbed plasma species to a steady-state equilibrium. By reducing the incubation time and diluting the plasma samples with buffer to 50%, the nonspecific adsorption of plasma could be controlled and a PEI pretreated with blood plasma could be used successfully for heparin determination. Heparin adsorption in 50% plasma was linear between 0.05 and 1 U/ml so that heparin plasma levels of 0.1 to 2 U/ml could be determined with a relative error of 11% and an accuracy of 0.05 U/ml.[188]

### 20.6.9 Congenital Diseases

SPR and biospecific interaction analysis (BIA) have been used to detect the Delta F508 mutation (F508del) of the cystic fibrosis transmembrane regulator (CFTR) gene in homozygous as well as heterozygous human subjects.[189] The detection method involved the immobilization on an SA5 sensor chip of two biotinylated oligonucleotide probes (one normal, N-508, and the other mutant, Delta F508) that are able to hybridize to the CFTR gene region involved in F508del mutation. A hybridization step between the oligonucleotide probes immobilized on the sensor chips and (1) wild-type or mutant oligonucleotides, as well as (2) ssDNA. These nucleic acid samples were obtained using asymmetric polymerase chain reaction (PCR), performed using genomic DNA from normal individuals and from F508del heterozygous and Delta 508del homozygous patients. The different stabilities of DNA/DNA molecular complexes generated after hybridization of normal and Delta F508 probes immobilized on the sensor chips were then evaluated. The results strongly suggest that the SPR technology enables a one-step, nonradioactive protocol for the molecular diagnosis of F508del mutation of the CFTR gene. This approach could be of interest in clinical genetics, because the hybridization step is often required to detect microdeletions present within PCR products.[189]

## 20.7 Conclusions

The past decade has witnessed the rapid development of a wide variety of biosensors for many different analyses and has even seen them begin to advance to clinical and, in some cases, commercially available technologies, such as SPR biosensors. The increasing interest in the field of optical biosensors has provided a great deal of information about the biochemistry of clinically important ailments and has provided drug discovery research with faster analytical tools to investigate drug–receptor interactions. Optical methods have also been extensively applied to DNA fingerprinting and genotyping, techniques that might find enormous applications in clinical diagnosis in the near future.

As a very positive sign, optical biosensors are coming of age as a bioanalytical tool. A larger number of researchers in different areas now have access to a more user-friendly technology, and are able to develop custom applications specific to their research needs, expanding the range of applications for the technology.

For practical medical diagnostic applications, there is a strong need for a truly integrated biosensor system that can be easily operated by relatively unskilled personnel. Some of the current commercially available technologies have dramatically simplified the data collection operation. Although these systems have demonstrated their usefulness in genomic detection, protein interaction analysis, carcinogen monitoring, etc., they are laboratory oriented and involve relatively expensive equipment and trained, supervised operation.

The outlook is promising for optical biosensor systems; the near future should bring a larger number of multichannel applications for the simultaneous detection of multiple biotargets; improvements in size and performance; and lower production costs due to a more integrated package. Highly integrated systems lead to reduction in noise and increase in signal due to the improved efficiency of sample collection and the reduction of interfaces. The capability of large-scale production using low-cost integrated circuit (IC) technology is an important advantage that cannot be overlooked. For medical applications, the development of low-cost, disposable biosensor surfaces that can be used for clinical diagnostics at the point of care can be a major driving force for the expansion of optical biosensor technologies.

## Acknowledgments

This work was sponsored by the U.S. Department of Energy (DOE) Office of Biological and Environmental Research and by the DOE NN-20 Chemical and Biological Program under contract DEAC05-00OR22725 with UT-Battelle, LLC.

### References

1. Nauta, J.M., vanLeengoed, H., Star, W.M., Roodenburg, J.L.N., Witjes, M.J.H., and Vermey, A., Photodynamic therapy of oral cancer — a review of basic mechanisms and clinical applications, *Eur. J. Oral Sci.*, 104, 69, 1996.
2. Clark, L.C., Jr. and Lions, C., Electrode systems for continuous monitoring in cardiovascular surgery, *Ann. N.Y. Acad. Sci.*, 102, 29, 1962.
3. Nice, E. and Catimel, B., Instrumental biosensors: new perspectives for the analysis of biomolecular interactions, *Bioessays*, 21, 339, 1999.
4. Braguglia, C.M., Biosensors: an outline of general principles and application, *Chem. Biochem. Eng. Q.*, 12, 183, 1998.
5. Weetall, H.H., Chemical sensors and biosensors, update, what, where, when and how, *Biosensors Bioelectron.*, 14, 237, 1999.
6. Tess, M.E. and Cox, J.A., Chemical and biochemical sensors based on advances in materials chemistry, *J. Pharm. Biomed. Anal.*, 19, 55, 1999.
7. Vo-Dinh, T., Tromberg, B.J., Griffin, G.D., Ambrose, K.R., Sepaniak, M.J., and Gardenhire, E.M., Antibody-based fiberoptics biosensor for the carcinogen benzo(a)pyrene, *Appl. Spectrosc.*, 41, 735, 1987.
8. Vo-Dinh, T., Alarie, J.P., Johnson, R.W., Sepaniak, M.J., and Santella, R.M., Evaluation of the fiberoptic antibody-based fluoroimmunosensor for DNA adducts in human placenta samples, *Clin. Chem.*, 37, 532, 1991.
9. Kienle, S., Lingler, S., Kraas, W., Offenhausser, A., Knoll, W., and Jung, G., Electropolymerization of a phenol-modified peptide for use in receptor-ligand interactions studied by surface plasmon resonance, *Biosensors Bioelectron.*, 12, 779, 1997.
10. Pathak, S.S. and Savelkoul, H.F.J., Biosensors in immunology: the story so far, *Immunol. Today*, 18, 464, 1997.
11. Regnault, V., Arvieux, J., Vallar, L., and Lecompte, T., Both kinetic data and epitope mapping provide clues for understanding the anti-coagulant effect of five murine monoclonal antibodies to human beta(2)-glycoprotein I, *Immunology*, 97, 400, 1999.
12. Huber, A., Demartis, S., and Neri, D., The use of biosensor technology for the engineering of antibodies and enzymes, *J. Mol. Recognition*, 12, 198, 1999.
13. Anderson, G.P., Breslin, K.A., and Ligler, F.S., Assay development for a portable fiberoptic biosensor, *Asaio J.*, 42, 942, 1996.
14. Anderson, G.P., Jacoby, M.A., Ligler, F.S., and King, K.D., Effectiveness of protein A for antibody immobilization for a fiber optic biosensor, *Biosensors Bioelectron.*, 12, 329, 1997.
15. Cao, L.K., Anderson, G.P., Ligler, F.S., and Ezzell, J., Detection of *Yersinia pestis* fraction-1 antigen with a fiber optic biosensor, *J. Clin. Microbiol.*, 33, 336, 1995.
16. Wadkins, R.M., Golden, J.P., Pritsiolas, L.M., and Ligler, F.S., Detection of multiple toxic agents using a planar array immunosensor, *Biosensors Bioelectron.*, 13, 407, 1998.

17. King, K., Anderson, G.P., Bullock, K.E., Regina, M.J., Saaski, E.W., and Ligler, F.S., Detecting staphylococcal enterotoxin B using an automated fiber optic biosensor, *Biosensors Bioelectron.*, 14, 163, 1999.
18. James, E.A., Schmeltzer, K., and Ligler, F.S., Detection of endotoxin using an evanescent wave fiber-optic biosensor, *Appl. Biochem. Biotechnol.*, 60, 189, 1996.
19. Cullum, B.M. and Vo-Dinh, T., The development of optical nanosensors for biological measurements, *Trends Biotechnol.*, 18, 388, 2000.
20. Isola, N.R., Stokes, D.L., and Vo-Dinh, T., Surface enhanced Raman gene probe for HIV detection, *Anal. Chem.*, 70, 1352, 1998.
21. Leonardo, R., Allain, L., and Vo-Dinh, T., Surface-enhanced Raman scattering detection of the breast cancer susceptibility gene *BRCA1* using a silver-coated microarray platform, *Anal. Chim. Acta*, 469, 149, 2002.
22. Moreno-Bondi, M.C., Mobley, J., Alarie, J.P., and Vo-Dinh, T., Antibody-based biosensor for breast cancer with ultrasonic regeneration, *J. Biomed. Opt.*, 5, 350, 2000.
23. Vo-Dinh, T., Isola, N., Alarie, J.P., Landis, D., Griffin, G.D., and Allison, S., Development of a multiarray biosensor for DNA diagnostics, *Instrum. Sci. Technol.*, 26, 503, 1998.
24. Dremel, B.A.A. and Schmid, R.D., Optical sensors for bioprocess control, *Chem.-Ing.-Tech.*, 64, 510, 1992.
25. Jakeway, S.C. and Krull, U.J., Consideration of end effects of DNA hybridization in selection of fluorescent dyes for development of optical biosensors, *Can. J. Chem.*, 77, 2083, 1999.
26. Piunno, P.A.E., Krull, U.J., Hudson, R.H.E., Damha, M.J., and Cohen, H., Fiber optic biosensor for fluorometric detection of DNA hybridization, *Anal. Chim. Acta*, 288, 205, 1994.
27. Uddin, A.H., Piunno, P.A.E., Hudson, R.H.E., Damha, M.J., and Krull, U.J., A fiber optic biosensor for fluorimetric detection of triple-helical DNA, *Nucleic Acids Res.*, 25, 4139, 1997.
28. Watterson, J.H., Piunno, P.A.E., Wust, C.C., and Krull, U.J., Controlling the density of nucleic acid oligomers on fiber optic sensors for enhancement of selectivity and sensitivity, *Sensors Actuators B Chem.*, 74 , 27, 2001.
29. Vo-Dinh, T., Tromberg, B.J., Griffin, G.D., Ambrose, K.R., Sepaniak, M.J., and Gardenhire, E.M., Antibody-based fiberoptics biosensor for the carcinogen benzo(a)pyrene, *Appl. Spectrosc.*, 41, 735, 1987.
30. Vo-Dinh, T., Nolan, T., Cheng, Y.F., Sepaniak, M.J., and Alarie, J.P., Phase-resolved fiberoptics fluoroimmunosensor, *Appl. Spectrosc.*, 44, 128, 1990.
31. Plowman, T.E., Reichert, W.M., Peters, C.R., Wang, H.K., Christensen, D.A., and Herron, J.N., Femtomolar sensitivity using a channel-etched thin film waveguide fluoroimmunosensor, *Biosensors Bioelectron.*, 11, 149, 1996.
32. Koncki, R. and Wolfbeis, O.S., Composite films of Prussian blue and *N*-substituted polypyrroles: covalent immobilization of enzymes and application to near infrared optical biosensing, *Biosensors Bioelectron.*, 14, 87, 1999.
33. Homola, J., Yee, S.S., and Gauglitz, G., Surface plasmon resonance sensors: review, *Sensors Actuators B Chem.*, 54, 3, 1999.
34. Van Regenmortel, M.H.V., Altschuh, D., and Chatellier, J., Uses of biosensors in the study of viral antigens, *Immunol. Invest.*, 26, 67, 1997.
35. Liedberg, B., Nylander, C., and Lundstrom, I., Surface plasmon resonance for gas detection and biosensing, *Sensors Actuators*, 4, 299, 1983.
36. Flanagan, M.T. and Pantell, R.H., Surface plasmon resonance and immunosensors, *Electron. Lett.*, 20, 968, 1984.
37. Morton, T.A. and Myszka, D.G., Kinetic analysis of macromolecular interactions using surface plasmon resonance biosensors, *Methods Enzymol.*, 295, 268, 1998.

38. Luo, J., Zhou, J.M., Zou, W., and Shen, P., Antibody–antigen interactions measured by surface plasmon resonance: global fitting of numerical integration algorithms. Production and molecular characterization of clinical phase I anti-melanoma mouse IgG3 monoclonal antibody R24, *J. Biochem.,* 130, 553, 2001.
39. Rich, R.L. and Myszka, D.G., Survey of the 1999 surface plasmon resonance biosensor literature, *J. Mol. Recognition,* 13, 388, 2000.
40. Wang, S., Boussaad, S., and Tao, N.J., Surface plasmon resonance enhanced optical absorption spectroscopy for studying molecular adsorbates, *Rev. Sci. Instrum.,* 72, 3055, 2001.
41. Casay, G.A., Daneshvar, M.I., and Patonay, G., Development of a fiber optic biomolecular probe instrument using near-infrared dyes and semiconductor-laser diodes, *Instrum. Sci. Technol.,* 22, 323, 1994.
42. Tunnemann, R., Mehlmann, M., Sussmuth, R.D., Buhler, B., Pelzer, S., Wohlleben, W., Fiedler, H.P., Wiesmuller, K.H., Gauglitz, G., and Jung, G., Optical biosensors. Monitoring studies of glycopeptide antibiotic fermentation using white light interference, *Anal. Chem.,* 73, 4313, 2001.
43. Birkert, O., Tunnernann, R., Jung, G., and Gauglitz, G., Label-free parallel screening of combinatorial triazine libraries using reflectometric interference spectroscopy, *Anal. Chem.,* 74, 834, 2002.
44. Birkert, O., Haake, H.M., Schutz, A., Mack, J., Brecht, A., Jung, G., and Gauglitz, G., A streptavidin surface on planar glass substrates for the detection of biomolecular interaction, *Anal. Biochem.,* 282, 200, 2000.
45. Vo-Dinh, T., Stokes, D.L., Griffin, G.D., Volkan, M., Kim, U.J., and Simon, M.I., Surface-enhanced Raman scattering (SERS) method and instrumentation for genomics and biomedical analysis, *J. Raman Spectrosc.,* 30, 785, 1999.
46. Vo-Dinh, T., Surface-enhanced Raman spectroscopy using metallic nanostructures, *Trends Anal. Chem.,* 17, 557, 1998.
47. Deckert, V., Zeisel, D., Zenobi, R., and Vo-Dinh, T., Near-field surface enhanced Raman imaging of dye-labeled DNA with 100-nm resolution, *Anal. Chem.,* 70, 2646, 1998.
48. Vo-Dinh, T., Allain, L.R., and Stokes, D.L., Cancer gene detection using SERS, *J. Raman Spectrosc.,* 33, 511, 2002.
49. Albrecht, M.G. and Creighton, J.A., Anomalously intense Raman spectra of pyridine at a silver electrode, *J. Am. Chem. Soc.,* 99, 5215, 1977.
50. Jeanmaire, D.J. and Van Dyune, R.P., Surface Raman spectroelectrochemistry teterocycli, aromatic, and aliphatic amines adsorbed on anodized silver electrode, *J. Electroanal. Chem.,* 84, 1, 1977.
51. Fleischmann, M., Hendra, P.J., and McQuillan, A.J., Raman spectra of pyridine adsorbed at a silver electrode, *Chem. Phys. Lett.,* 26, 63, 1974.
52. Gyurcsanyi, R.E., Vagfoldi, Z., Toth, K., and Nagy, G., Fast response potentiometric acetylcholine biosensor, *Electroanalysis,* 11, 712, 1999.
53. Dobay, R., Harsanyi, G., and Visy, C., Conducting polymer based electrochemical sensors on thick film substrate, *Electroanalysis,* 11, 804, 1999.
54. Coche-Guerente, L., Desprez, V., Diard, J.P., and Labbe, P., Amplification of amperometric biosensor responses by electrochemical substrate recycling Part I. Theoretical treatment of the catechol-polyphenol oxidase system, *J. Electroanal. Chem.,* 470, 53, 1999.
55. Dall'Orto, V.C., Danilowicz, C., Rezzano, I., Del Carlo, M., and Mascini, M., Comparison between three amperometric sensors for phenol determination in olive oil samples, *Anal. Lett.,* 32, 1981, 1999.
56. Karyakin, A.A., Vuki, M., Lukachova, L.V., Karyakina, E.E., Orlov, A.V., Karpachova, G.P., and Wang, J., Processible polyaniline as an advanced potentiometric pH transducer. Application to biosensors, *Anal. Chem.,* 71, 2534, 1999.
57. Rudel, U., Geschke, O., and Cammann, K., Entrapment of enzymes in electropolymers for biosensors and graphite felt based flow through enzyme reactors, *Electroanalysis,* 8, 1135, 1996.
58. Sarkar, P. and Turner, A,P.F., Application of dual-step potential on single screen-printed modified carbon paste electrodes for detection of amino acids and proteins, *Fresenius J. Anal. Chem.,* 364, 154, 1999.

59. Huang, T., Warsinke, A., Kuwana, T., and Scheller, F.W., Determination of l-phenylalanine based on a NADH-detecting biosensor, *Anal. Chem.*, 70, 991, 1998.
60. Mo, Z.H., Long, X.H., and Fu, W.L., A new sandwich-type assay of estrogen using piezoelectric biosensor immobilized with estrogen response element, *Anal. Commun.*, 36, 281, 1999.
61. Hengerer, A., Kosslinger, C., Decker, J., Hauck, S., Queitsch, I., Wolf, H., and Dubel, S., Determination of phage antibody affinities to antigen by a microbalance sensor system, *Biotechniques*, 26, 956, 1999.
62. Wessa, T., Rapp, M., and Ache, H.J., New immobilization method for SAW-biosensors: covalent attachment of antibodies via CNBr, *Biosensors Bioelectron.*, 14, 93, 1999.
63. Minunni, M., Mascini, M., Carter, R.M., Jacobs, M.B., Lubrano, G.J., and Guilbault, G.C., A quartz crystal microbalance displacement assay for *Listeria* monocytogenes, *Anal. Chim. Acta*, 325, 169, 1996.
64. Vaughan, R.D., Sullivan, C.K., and Guilbault, G.G., Sulfur based self-assembled monolayers (SAM's) on piezoelectric crystals for immunosensor development, *Fresenius J. Anal. Chem.*, 364, 54, 1999.
65. Welsch, W., Klein, C., von Schickfus, M., and Hunklinger, S., Development of a surface acoustic wave immunosensor, *Anal. Chem.*, 68, 2000, 1996.
66. Freudenberg, J., Schelle, S., Beck, K., von Schickfus, M., and Hunklinger, S., A contactless surface acoustic wave biosensor, *Biosensors Bioelectron.*, 14, 423, 1999.
67. Ji, H.F. and Thundat, T., *In situ* detection of calcium ions with chemically modified microcantilevers, *Biosensors Bioelectron.*, 17, 337, 2002.
68. Smith, D.S, Hassan, M., and Nargessi, R.D., in *Principles, Practice of Fluoroimmunoassay Procedures in Modern Fluorescence Spectroscopy*, Wehry, E.L., Ed., Plenum Press, New York.
69. Tromberg, B.J., Sepaniak, M.J., Vo-Dinh, T., and Griffin, G.D., Fiberoptic chemical sensors for competitive-binding fluoroimmunoassay, *Anal. Chem.*, 59, 1226, 1987.
70. Wadkins, R.M., Golden, J.P., and Ligler, F.S., Calibration of biosensor response using simultaneous evanescent-wave excitation of cyanine-labeled capture antibodies and antigens, *Anal. Biochem.*, 232, 73, 1995.
71. Campbell, C.N., de Lumley-Woodyear, T., and Heller, A., Towards immunoassay in whole blood: separationless sandwich-type electrochemical immunoassay based on *in-situ* generation of the substrate of the labeling enzyme, *Fresenius J. Anal. Chem.*, 364, 165, 1993.
72. Kroger, D., Liley, M., Schiweck, W., Skerra, A., and Vogel, H., Immobilization of histidine-tagged proteins on gold surfaces using chelator thioalkanes, *Biosensors Bioelectron.*, 14, 155, 1999.
73. Vo-Dinh, T., Griffin, G.D., and Ambrose, K.R., A portable fiberoptic monitor for fluorometric bioassays, *Appl. Spectrosc.*, 40, 696, 1986.
74. Alarie, J.P. and Vo-Dinh, T., Antibody-based submicron biosensor for benzo a pyrene DNA adduct, *Polycyclic Aromatic Compounds*, 8, 45, 1996.
75. Vo-Dinh, T., Rurie, J.R., Cullum, B., and Griffin, G.D., Antibody-based nanoprobe for measurement of a fluorescent analyte in a single cell, *Nat. Biotechnol.*, 18, 764, 2000.
76. Vo-Dinh, T., Alarie, J.P., Cullum, B.M., and Griffin, G.D., Antibody-based nanoprobe for measurements in a single cell, *Nat. Biotechnol.*, 18, 76, 2000.
77. Rowe-Taitt, C.A., Cras, J.J., Patterson, C.H., Golden, J.P., and Ligler, F.S., A ganglioside-based assay for cholera toxin using an array biosensor, *Anal. Biochem.*, 281, 123, 2000.
78. Narang, U., Anderson, G.P., Ligler, F.S., and Burans, J., Fiber optic-based biosensor for ricin, *Biosensors Bioelectron.*, 12, 937, 1997.
79. Wijesuriya, D., Breslin, K., Anderson, G., Shriverlake, L., and Ligler, F.S., Regeneration of immobilized antibodies on fiber optic probes, *Biosensors Bioelectron.*, 9, 585, 1994.
80. Vo-Dinh, T., Griffin, G.D., Ambrose, K.R., Thomson, R.M., Murchison, C.M., McManis, M., and St. Weclur, P.G.R., Production and characterization of antibodies to benzo(a)pyrene, in *Polynuclear Aromatic Hydrocarbons: A Decade of Progress*, Dennis, A.J. and Cooke, M., Eds., Battelle Press, Seattle, WA, 1988, p. 885.

81. Alarie, J.P., Bowyer, J.R., Sepaniak, M.J., Hoyt, A.M., and Vo-Dinh, T., Fluorescence monitoring of a benzo-a-pyrene metabolite using a regenerable immunochemical-based fiberoptic sensor, *Anal. Chim. Acta*, 236, 237, 1990.
82. Blanchard, G.C., Taylor, C.G., Busey, B.R., and Williamson, M.L., Regeneration of immunosorbent surfaces used in clinical, industrial and environmental biosensors — role of covalent and non-covalent interactions, *J. Immunol. Methods*, 130, 263, 1990.
83. Absolom, D.R. and Vanoss, C.J., The nature of the antigen-antibody bond and the factors affecting its association and dissociation, *CRC Crit. Rev. Immunol.*, 6, 1, 1986.
84. Shriverlake, L.C., Breslin, K.A., Charles, P.T., Conrad, D.W., Golden, J.P., and Ligler, F.S., Detection of TNT in water using an evanescent-wave fiberoptic biosensor, *Anal. Chem.*, 67, 2431, 1995.
85. Lu, B., Lu, C.L., and Wei, Y., A planar quartz wave-guide immunosensor based on TIRF principle, *Anal. Lett.*, 25, 1, 1992.
86. Haga, M., Shimura, T., Nakamura, T., Kato, Y., and Suzuki, Y., effect of ultrasonic irradiation on the dissociation of antigen-antibody complexes — application to homogeneous enzyme-immunoassay, *Chem. Pharm. Bull.*, 35, 3822, 1987.
87. Busnaina, A.A., Kashkoush, I.I., and Gale, G.W., An experimental-study of megasonic cleaning of silicon-wafers, *J. Electrochem. Soc.*, 142, 2812, 1995.
88. Qi, Q. and Brereton, G.J., Mechanisms of removal of micron-sized particles by high-frequency ultrasonic-waves, *IEEE Trans. Ultrason. Ferroelectr. Freq. Control*, 42, 619, 1995.
89. Mobley, J., Marsh, J.N., Hall, C.S., Hughes, M.S., Brandenburger, G.H., and Miller, J.G., Broadband measurements of phase velocity in Albunex (R) suspensions, *J. Acoust. Soc. Am.*, 103, 2145, 1998.
90. Mobley, J., Kasili, P., Norton, S., and Vo-Dinh, T., Application of ultrasonic techniques for brain injury diagnostics, in *Biomedical Diagnostics, Guidance and Surgical-Assist Systems*, Vo-Dinh, T., Grundfest, W.S., and Benaron, D., SPIE, Bellingham, WA, 1999.
91. Alberts, B., Bray, D., Lewis, J., Rolf, M., Roberts, K., and Watson, J.D., *Molecular Biology of the Cell*, Garland Publishing, New York, 1994.
92. Diamond, D., Ed., *Chemical and Biological Sensors*, Wiley, New York, 1998.
93. Polster, J., Prestel, G., Wollenweber, M., Kraus, G., and Gauglitz, G., Simultaneous determination of penicillin and ampicillin by spectral fibre-optical enzyme optodes and multivariate data analysis based on transient signals obtained by flow injection analysis, *Talanta*, 42, 2065, 1995.
94. Brecht, A. and Gauglitz, G., Recent developments in optical transducers for chemical and biochemical applications, *Sensors Actuators B*, 38, 1, 1997.
95. Rosenzweig, Z. and Kopelman, R., Analytical properties and sensor size effects of a micrometer-sized optical fiber glucose biosensor, *Anal. Chem.*, 68, 1408, 1996.
96. Barker, S.L.R., Kopelman, R., Meyer, T.E., and Cusanovich, M.A., Fiber-optic nitric oxide-selective biosensors and nanosensors, *Anal. Chem.*, 70, 971, 1998.
97. McCormack, T., O'Keeffe, G., MacCraith, B.D, and O'Kennedy, R., Optical imunosensing of lactate dehydrogenase (LDH), *Sensor Actuators B-Chem.*, 41, 89, 1997.
98. Vo-Dinh, T. and Cullum, B., Biosensors and biochips: advances in biological and medical diagnostics, *Fresenius J. Anal. Chem.*, 366, 540, 2000.
99. Sawata, S., Kai, E., Ikebukuro, K., Iida, T., Honda, T., and Karube, I., Application of peptide nucleic acid to the direct detection of deoxyribonucleic acid amplified by polymerase chain reaction, *Biosensors Bioelectron.*, 14, 397, 1999.
100. Niemeyer, C.M., Boldt, L., Ceyhan, B., and Blohm, D., DNA-directed immobilization: efficient, reversible, and site-selective surface binding of proteins by means of covalent DNA-streptavidin conjugates, *Anal. Biochem.*, 268, 54, 1999.
101. Marrazza, G., Chianella, I., and Mascini, M., Disposable DNA electrochemical sensor for hybridization detection, *Biosensors Bioelectron.*, 14, 43, 1999.
102. Wang, J., Rivas, G., Fernandes, J.R., Paz, J.L.L., Jiang, M., and Waymire, R., Indicator-free electrochemical DNA hybridization biosensor, *Anal. Chim. Acta*, 375, 197, 1998.

103. Piehler, J., Brecht, A., Gauglitz, G., Zerlin, M., Maul, C., Thiericke, R., and Grabley, S., Label-free monitoring of DNA-ligand interactions, *Anal. Biochem.*, 249, 94, 1997.
104. Skuridin, S.G., Yevdokimov, Y.M., Efimov, V.S., Hall, J.M., and Turner, A.P.F., A new approach for creating double-stranded DNA biosensors, *Biosensors Bioelectron.*, 11, 903, 1996.
105. Vo-Dinh, T., Houck, K., and Stokes, D.L., Surface-enhanced Raman gene probes, *Anal. Chem.*, 66, 33793, 1994.
106. Ferguson, J.A., Boles, T.C., Adams, C.P., and Walt, D.R., A fiber-optic DNA biosensor microarray for the analysis of gene expression, *Nat. Biotechnol.*, 14, 1681, 1996.
107. Hanafi-Bagby, D., Piunno, P.A. E., Wust, C.C., and Krull, U.J., Concentration dependence of a thiazole orange derivative that is used to determine nucleic acid hybridization by an optical biosensor, *Anal. Chim. Acta*, 411, 19, 2000.
108. Piunno, P.A.E., Krull, U.J., Hudson, R.H.E., Damha, M.J., and Cohen, H., Fiberoptic DNA sensor for fluorometric nuclei acid determination, *Anal. Chem.*, 67, 2635, 1995.
109. Watterson, J.H., Piunno, P.A.E., Wust, C.C., and Krull, U.J., Effects of oligonucleotide immobilization density on selectivity of quantitative transduction of hybridization of immobilized DNA, *Langmuir*, 16, 4984, 2000.
110. Franchina, J.G., Lackowski, W.M., Dermody, D.L., Crooks, R.M., Bergbreiter, D.E., Sirkar, K., Russell, R.J., and Pishko, M.V., Electrostatic immobilization of glucose oxidase in a weak acid, polyelectrolyte hyperbranched ultrathin film on gold: fabrication, characterization, and enzymatic activity, *Anal. Chem.*, 71, 3133, 1999.
111. Schuler, R., Wittkampf, M., and Chemnitius, G.C., Modified gas-permeable silicone rubber membranes for covalent immobilisation of enzymes and their use in biosensor development, *Analyst*, 124, 1181, 1999.
112. Houshmand, H., Froman, G., and Magnusson, G., Use of bacteriophage T7 displayed peptides for determination of monoclonal antibody specificity and biosensor analysis of the binding reaction, *Anal. Biochem.*, 268, 363, 1999.
113. Lebron, J.A. and Bjorkman, P.J., The transferrin receptor binding site on HFE, the Class I MHC-related protein mutated in hereditary hemochromatosis, *J. Mol. Biol.*, 289, 1109, 1999.
114. Kim, H.J., Hyun, M.S., Chang, I.S., and Kim, B.H., A microbial fuel cell type lactate biosensor using a metal-reducing bacterium, *Shewanella putrefaciens*, *J. Microbiol. Biotechnol.*, 9, 365, 1999.
115. Nelson, R.W., Jarvik, J.W., Taillon, B.E., and Tubbs, K.A., BIA MS of epitope-tagged peptides directly from *E. coli* lysate: multiplex detection and protein identification at low femtomole to subfemtomole levels, *Anal. Chem.*, 71, 2858, 1999.
116. Hara-Kuge, S., Ohkura, T., Seko, A., and Yamashita, K., Vesicular-integral membrane protein, VIP 36, recognizes high-mannose type glycans containing alpha 1 -> 2 mannosyl residues in MDCK cells, *Glycobiology*, 9, 833, 1999.
117. Pemberton, R.M., Hart, J.P., Stoddard, P., and Foulkes, J.A., A comparison of l-naphthyl phosphate and 4 aminophenyl phosphate as enzyme substrates for use with a screen-printed anperometric immunosensor for progesterone in cows' milk, *Biosensors Bioelectron.*, 14, 495, 1999.
118. Blake, R.C., Pavlov, A.R., and Blake, D.A., Automated kinetic exclusion assays to quantify protein binding interactions in homogeneous solution, *Anal. Biochem.*, 272, 123, 1999.
119. Patolsky, F., Zayats, M., Katz, E., and Willner, I., Precipitation of an insoluble product on enzyme monolayer electrodes for biosensor applications: characterization by faradaic impedance spectroscopy, cyclic voltammetry, and microgravimetric quartz crystal microbalance analyses, *Anal. Chem.*, 71, 3171, 1999.
120. Barker, S.L.R., Zhao, Y.D., Marletta, M.A., and Kopelman, R., Cellular applications of a sensitive and selective fiber optic nitric oxide biosensor based on a dye-labeled heme domain of soluble guanylate cyclase, *Anal. Chem.*, 71, 2071, 1999.
121. Schmidt, A., Standfuss Gabisch, C., and Bilitewski, U., Microbial biosensor for free fatty acids using an oxygen electrode based on thick film technology, *Biosensors Bioelectron.*, 11, 1139, 1996.

122. Boncheva, M., Duschl, C., Beck, W., Jung, G., and Vogel, H., Formation and characterization of lipopeptide layers at interfaces for the molecular recognition of antibodies, *Langmuir,* 12, 5636, 1996.
123. Cornell, B.A., Braach Maksvytis, V.L.B., King, L.G., Osman, P.D.J., Raguse, B., Wieczorek, L., and Pace, R.J., A biosensor that uses ion-channel switches, *Nature*, 387, 580, 1997.
124. Wollenberger, U., Neumann, B., and Scheller, F.W., Development of a biomometic alkane sensor, *Electrochim. Acta*, 43, 3581, 1998.
125. Ramsden, J.J., Biomimetic protein immobilization using lipid bilayers, *Biosensors Bioelectron.,* 13, 593, 1998.
126. Song, X.D. and Swanson, B.I., Direct, ultrasensitive, and selective optical detection of protein toxins using multivalent interactions, *Anal. Chem.*, 71, 2097, 1999.
127. Cotton, G.J., Ayers, B., Xu, R., and Muir, T.W., Insertion of a synthetic peptide into a recombinant protein framework: a protein biosensor, *J. Am. Chem. Soc.,* 121, 1100, 1999.
128. Zhang, W.T., Canziani, G., Plugariu, C., Wyatt, R., Sodroski, J., Sweet, R., Kwong, P., Hendrickson, W., and Chaiken, L., Conformational changes of gpl20 in epitopes near the CCR5 binding site are induced by CD4 and a CD4 miniprotein mimetic, *Biochemistry,* 38, 9405, 1999.
129. Piervincenzi, R.T., Reichert, W.M., and Hellinga, H.W., Genetic engineering of a single-chain antibody fragment for surface immobilization in an optical biosensor, *Biosensors Bioelectron.,* 13, 305, 1998.
130. Charych, D., Cheng, Q., Reichert, A., Kuziemko, G., Stroh, M., Nagy, J.O., Spevak, W., and Stevens, R.C., A "litmus test" for molecular recognition using artificial membranes, *Chem. Biol.,* 3, 113, 1996.
131. Kriz, D. and Mosbach, K., Competitive amperometric morphine sensor-based on an agarose immobilized molecularly imprinted polymer, *Anal. Chim. Acta,* 300, 71, 1995.
132. Jill, K. and Veilleux, L.W.D., Covalent immobilization of bio-molecules to preactivated surfaces, *IVD Technol. Mag.,* March, 26, 1996.
133. Allain, L.R., Sorasaenee, K., and Xue, Z.L., Doped thin-film sensors via a sol-gel process for high acidity determination, *Anal. Chem.,* 69, 3076, 1997.
134. Allain, L.R. and Xue, Z.L., Optical sensors for the determination of concentrated hydroxide, *Anal. Chem.,* 72, 1078, 2000.
135. Yost, T.L., Fagan, B.C., Allain, L.R., Barnes, C.E., Dai, S., Sepaniak, M.J., and Xue, Z.L., Crown ether-doped sol-gel materials for strontium(II) separation, *Anal. Chem.,* 72, 5516, 2000.
136. Rasmussen, S.R., Larsen, M.R., and Rasmussen, S.E., Covalent immobilization of DNA onto polystyrene microwells: the molecules are only bound at the 5′ end, *Anal. Biochem.*, 198, 138, 1991.
137. Larsson, P.H., Johansson, S.G.O., Hult, A., and Gothe, S., Covalent binding of proteins to grafted plastic surfaces suitable for immunoassays. I. Binding-capacity and characteristics of grafted polymers, *J. Immunol. Methods,* 98, 129, 1987.
138. Veilleux, J.K. and Duran, L.W., Covalent immobilization of biomolecules to preactivated surfaces, *IVD Technol. Mag.,* March, 26, 1996.
139. Potyrailo, R.A., Conrad, R.C., Ellington, A.D., and Hiefje, G.M., Adapting selected nucleic acid ligands (aptamers) to biosensors, *Anal. Chem.*, 70, 3419, 1998.
140. Yoo, S.K., Yoon, M., Park, U.J., Han, H.S., Kim, J.H., and Hwang, H., A radioimmunoassay method for detection of DNA based on chemical immobilization of anti-DNA antibody, *J. Exp. Mol. Med.*, 31, 122, 1999.
141. Aylott, J.W., Richardson, D.J., and Russell, D.A., Optical biosensing of nitrate ions using a sol-gel immobilized nitrate reductase, *Analyst,* 122, 77, 1997.
142. Blyth, D.J., Poynter, S.J., and Russell, D.A., Calcium biosensing with a sol-gel immobilized photoprotein, *Analyst,* 121, 1975, 1996.
143. Blyth, D.J., Aylott, J.W., Richardson, D.J., and Russell, D.A., Sol-gel encapsulation of metalloproteins for the development of optical biosensors for nitrogen-monoxide and carbon-monoxide, *Analyst,* 120, 2725, 1995.
144. Doong, R.A. and Tsai, H.C., Immobilization and characterization of sol-gel-encapsulated acetylcholinesterase fiber-optic biosensor, *Anal. Chim. Acta,* 434, 239, 2001.

145. Li, J., Wang, K.M., Yang, X.H., and Xiao, D., Sol-gel horseradish peroxidase biosensor for the chemiluminescent flow determination of hydrogen peroxide, *Anal. Commun.*, 36, 195, 1999.
146. Lin, J. and Brown, C.W., Sol-gel glass as a matrix for chemical and biochemical sensing, *Trends Anal. Chem.*, 16, 200, 1997.
147. Situmorang, M., Gooding, J.J., and Hibbert, D.B., Immobilisation of enzyme throughout a polytyramine matrix: a versatile procedure for fabricating biosensors, *Anal. Chim. Acta*, 394, 211, 1999.
148. Lee, Y.C. and Huh, M.H., Development of a biosensor with immobilized L-amino acid oxidase for determination of L-amino acids, *J. Food Biochem.*, 23, 173, 1999.
149. Edmiston, P.L., Wambolt, C.L., Smith, M.K., and Saavedra, S.S., Spectroscopic characterization of albumin and myoglobin entrapped in bulk sol-gel glasses, *J. Colloid Interface Sci.*, 163, 395, 1994.
150. Barak, L.S., Ferguson, S.S.G., Zhang, J., and Caron, M.G., A beta-arrestin green fluorescent protein biosensor for detecting G protein-coupled receptor activation, *J. Biol. Chem.*, 272, 27497, 1997.
151. Desikan, R., Hagenbeek, D., Neill, S.J., and Rock, C.D., Flow cytometry and surface plasmon resonance analyses demonstrate that the monoclonal antibody JIM19 interacts with a rice cell surface component involved in abscisic acid signalling in protoplasts, *FEBS Lett.*, 456, 257, 1999.
152. Newman, D.J., Thakkar, H., Lam-Po-Tang, M.K., and Kwan, J.T.C., The use of optical sensors to understand cellular interactions with renal cells, *Renal Failure*, 21, 349, 1999.
153. Inoue, K., Arai, T., and Aoyagi, M., Sensitivity of real time viral detection by an optical biosensor system using a crude home-made antiserum against measles virus as a ligand, *Biol. Pharm. Bull.*, 22, 210, 1999.
154. Krummenacher, C., Baribaud, I., de Leon, M.P., Whitbeck, J.C., Lou, H., Cohen, G.H., and Eisenberg, R.J., Localization of a binding site for herpes simplex virus glycoprotein D on herpesvirus entry mediator C by using antireceptor monoclonal antibodies, *J. Virol.*, 74, 10863, 2000.
155. Ostroff, R., Ettinger, A., La, H., Rihanek, M., Zalman, L., Meador, J., Patick, A.K., Worland, S., and Polisky, B., Rapid multiserotype detection of human rhinoviruses on optically coated silicon surfaces, *J. Clin. Virol.*, 21, 105, 2001.
156. Inoue, K., Arai, T., and Aoyagi, M., Real time observation of binding of herpes simplex virus type 1 (HSV-1) to Vero cells and neutralization of HSV-1 by sulfonated human immunoglobulin, *J. Biochem.*, 121, 633, 1997.
157. Tsoka, S., Gill, A., Brookman, J.L., and Hoare, M., Rapid monitoring of virus-like particles using an optical biosensor: a feasibility study, *J. Biotechnol.*, 63, 147, 1998.
158. Gomes, P., Giralt, E., and Andreu, D., Surface plasmon resonance screening of synthetic peptides mimicking the immunodominant region of C-S8c1 foot-and-mouth disease virus, *Vaccine*, 18, 362, 1999.
159. Li, C., Dowd, C.S., Zhang, W., and Chaiken, I.M., Phage randomization in a charybdotoxin scaffold leads to CD4-mimetic recognition motifs that bind HIV-1 envelope through non-aromatic sequences, *J. Peptide Res.*, 57, 507, 2001.
160. Hoffman, T.L., Canziani, G., Jia, L., Rucker, J., and Doms, R.W., A biosensor assay for studying ligand-membrane receptor interactions: binding of antibodies and HIV-1 Env to chemokine receptors, *Proc. Natl. Acad. Sci. U.S.A.*, 97, 11215, 2000.
161. Dowd, C.S., Zhang, W.T., Li, C.Z., and Chaiken, I.M., From receptor recognition mechanisms to bioinspired mimetic antagonists in HIV-1/cell docking, *J. Chromatogr. B*, 753, 327, 2001.
162. Zeng, X.X., Nakaaki, Y., Murata, T., and Usui, T., Chemoenzymatic synthesis of glycopolypeptides carrying alpha- Neu5Ac-(2 -> 3)-beta-D-Gal-(1 -> 3)-alpha-D-GalNAc, beta-D-Gal-(1 -> 3)-alpha-D-GalNAc, and related compounds and analysis of their specific interactions with lectins, *Arch. Biochem. Biophys.*, 383, 28, 2000.
163. Markgren, P.O., Hamalainen, M., and Danielson, U.H., Screening of compounds interacting with HIV-1 proteinase using optical biosensor technology, *Anal. Biochem.*, 265, 340, 1998.
164. Markgren, P.O., Hamalainen, M., and Danielson, U.H., Kinetic analysis of the interaction between HIV-1 protease and inhibitors using optical biosensor technology, *Anal. Biochem.*, 279, 71, 2000.

165. Markgren, P.O., Lindgren, M.T., Gertow, K., Karlsson, R., Hamalainen, M., and Danielson, U.H., Determination of interaction kinetic constants for HIV-1 protease inhibitors using optical biosensor technology, *Anal. Biochem.*, 291, 207, 2001.
166. Hamalainen, M.D., Markgren, P.O., Schaal, W., Karlen, A., Classon, B., Vrang, L., Samuelsson, B., Hallberg, A., and Danielson, U.H., Characterization of a set of HIV-1 protease inhibitors using binding kinetics data from a biosensor-based screen, *J. Biomol. Screening*, 5, 353, 2000.
167. O'Keefe, B.R., Shenoy, S.R., Xie, D., Zhang, W.T., Muschik, J.M., Currens, M.J., Chaiken, I., and Boyd, M.R., Analysis of the interaction between the HIV-Inactivating protein cyanovirin-N and soluble forms of the envelope glycoproteins gp120 and gp41, *Mol. Pharmacol.*, 58, 982, 2000.
168. Ivnitski, D., Abdel-Hamid, I., Atanasov, P., and Wilkins, E., Biosensors for detection of pathogenic bacteria, *Biosensors Bioelectron.*, 14, 599, 1999.
169. Chang, Y.H., Chang, T.C., Kao, E.F., and Chou, C., Detection of protein a produced by *Staphylococcus aureus* with a fiber-optic-based biosensor, *Biosci. Biotechnol. Biochem.*, 60, 1571, 1996.
170. Watts, H.J., Lowe, C.R., and Pollardknight, D.V., Optical biosensor for monitoring microbial-cells, *Anal. Chem.*, 66, 2465, 1994.
171. Hirmo, S., Artursson, E., Puu, G., Wadstrom, T., and Nilsson, B., Characterization of *Helicobacter pylori* interactions with sialylglycoconjugates using a resonant mirror biosensor, *Anal. Biochem.*, 257, 63, 1998.
172. Hirmo, S., Artursson, E., Puu, G., Wadstrom, T., and Nilsson, B., *Helicobacter pylori* interactions with human gastric mucin studied with a resonant mirror biosensor, *J. Microbiol. Methods*, 37, 177, 1999.
173. Zhou, C.H., Pivarnik, P., Auger, S., Rand, A., and Letcher, S., A compact fiber-optic immunosensor for *Salmonella* based on evanescent wave excitation, *Sensors Actuators B Chem.*, 42, 169, 1997.
174. Liu, Y.C., Che, Y.H., and Li, Y.B., Rapid detection of *Salmonella typhimurium* using immunomagnetic separation and immune-optical sensing method, *Sensors Actuators B Chem.*, 72, 214, 2001.
175. Hacia, J.G., Woski, S.A., Fidanza, J., Edgemon, K., Hunt, N., McGall, G., Fodor, S.P.A., and Collins, F.S., Enhanced high density oligonucleotide array-based sequence analysis using modified nucleoside triphosphates, *Nucleic Acids Res.*, 26, 4975, 1998.
176. Chen, H.L., Fernig, D.G., Rudland, P.S., Sparks, A., Wilkinson, M.C., and Barraclough, R., Binding to intracellular targets of the metastasis-inducing protein, S100A4 (p9Ka), *Biochem. Biophys. Res. Commun.*, 286, 1212, 2001.
177. O'Neill, P.M., Fletcher, J.E., Stafford, C.G., Daniels, P.B., and Bacaresehamilton, T., Use of an optical biosensor to measure prostate-specific antigen in whole-blood, *Sensors Actuators B Chem.*, 29, 79, 1995.
178. Vo-Dinh, T., Alarie, J.P., Cullum, B.M., and Griffin, G.D., Antibody-based nanoprobe for measurement of a fluorescent analyte in a single cell, *Nat. Biotechnol.*, 18, 764, 2000.
179. Nath, N., Jain, S.R., and Anand, S., Evanescent wave fibre optic sensor for detection of *L. donovani*–specific antibodies in sera of kala azar patients, *Biosensors Bioelectron.*, 12, 491, 1997.
180. Carlson, M.A., Bargeron, C.B., Benson, R.C., Fraser, A.B., Phillips, T.E., Velky, J.T., Groopman, J.D., Strickland, P.T., and Ko, H.W., An automated, handheld biosensor for aflatoxin, *Biosensors Bioelectron.*, 14, 841, 2000.
181. Anis, N.A., Wright, J., Rogers, K.R., Thompson, R.G., Valdes, J.J., and Eldefrawi, M.E., A fiberoptic immunosensor for detecting parathion, *Anal. Lett.*, 25, 627, 1992.
182. Ayyagari, M.S., Kamtekar, S., Pande, R., Marx, K.A., Kumar, J., Tripathy, S.K., Akkara, J., and Kaplan, D.L., Chemiluminescence-based inhibition-kinetics of alkaline-phosphatase in the development of a pesticide biosensor, *Biotechnol. Prog.*, 11, 699, 1995.
183. Mulchandani, A., Pan, S.T., and Chen, W., Fiber-optic enzyme biosensor for direct determination of organophosphate nerve agents, *Biotechnol. Prog.*, 15, 130, 1999.
184. Mulchandani, A., Kaneva, I., and Chen, W., Biosensor for direct determination of organophosphate nerve agents using recombinant *Escherichia coli* with surface-expressed organophosphorus hydrolase. 2. Fiber optic microbial biosensor, *Anal. Chem.*, 70, 5042, 1998.

185. Puu, G., An approach for analysis of protein toxins based on thin films of lipid mixtures in an optical biosensor, *Anal. Chem.,* 73, 72, 2001.
186. Song, X.D. and Swanson, B.I., Direct, ultrasensitive, and selective optical detection of protein toxins using multivalent interactions, *Anal. Chem.,* 71, 2097, 1999.
187. PepinCovatta, S., Lutsch, C., Grandgeorge, M., and Scherrmann, J.M., Immunoreactivity of a new generation of horse F(ab′)(2) preparations against European viper venoms and the tetanus toxin, *Toxicon,* 35, 411, 1997.
188. Gaus, K. and Hall, E.A.H., Surface plasmon resonance sensor for heparin measurements in blood plasma, *Biosensors Bioelectron.,* 13, 1307, 1998.
189. Feriotto, G., Lucci, M., Bianchi, N., Mischiati, C., and Gambari, R., Detection of the Delta F508 (F508del) mutation of the cystic fibrosis gene by surface plasmon resonance and biosensor technology, *Hum. Mutation,* 13, 390, 1999.

# 21

# Functional Imaging with Diffusing Light

Arjun G. Yodh
*University of Pennsylvania*
*Philadelphia, Pennsylvania*

David A. Boas
*Harvard Medical School*
*Massachusetts General Hospital*
*Athinoula A. Martinos Center for Biomedical Imaging*
*Charlestown, Massachusetts*



## 21.1 Introduction

Many materials are visually opaque because photons traveling within them are predominantly scattered rather than absorbed. Some common examples of these highly scattering media include white paint, foam, mayonnaise, and human tissue. Indeed, anyone who has held a flashlight up to his or her hand will notice some of this light is transmitted, albeit after experiencing many scattering events. Light travels through these materials in a process similar to heat diffusion.

What does it mean to say light transport is diffusive? Consider a simple experiment in which an optical fiber is used to inject light into a highly scattering material such as paint or tissue. Microscopically, the injected photons experience thousands of elastic scattering events in the media. A few of the photons will be absorbed by chromophores and will be lost. The remaining photons travel along pathways that resemble a random walk. These individual trajectories are composed of straight-line segments with sudden interruptions where the photon propagation direction is randomly changed. The average length of the straight-line segments is called the random walk steplength of the traveling photon. By summing all trajectories one can compute the photon concentration or photon fluence rate as a function of time and position within the media.

It is then straightforward to show that the collective migration of photon concentration is described by a diffusion equation. In practice one can carry out a variety of measurements to confirm the diffusive

0-8493-1116-0/03/$0.00+$1.50

nature of light transport. For example, if a short pulse of light is injected into the medium and a second optical fiber is used to detect transmitted photons, then, when the transport is diffusive, the most probable arrival times for the detected photons will scale with the square of the source-detector separation divided by the random walk steplength.

Diffuse light imaging and spectroscopy aims to investigate tissue physiology millimeters to centimeters below the tissue surface.[1–5] The cost of this goal is that we must abandon traditional optical spectroscopies and traditional microscopy because traditional methodologies require optically thin samples. In addition, light penetration must be large in order to reach tissue located centimeters below the surface. Fortunately, a spectral window exists within tissues in the near-infrared from 700 to 900 nm, wherein photon transport is dominated by scattering rather than absorption. The absorption of hemoglobin and water is small in the near-infrared, but elastic scattering from organelles and other microscopic interfaces is large. These are precisely the conditions required for application of the diffusion model. The recognition and widespread acceptance that light transport over long distances in tissues is well approximated as a diffusive process has propelled the field. Using this physical model it is possible to separate tissue scattering from tissue absorption quantitatively, and to incorporate the influence of boundaries, such as the air-tissue interface, into the transport theory accurately. The diffusion approximation also provides a tractable basis for tomographic approaches to image reconstruction using highly scattered light. Tomographic methods were not employed in early transillumination patient studies, and are crucial for recovery of information about tissue optical property heterogeneity.

Even though absorption in the near-infrared is relatively small, the spectra of major tissue chromophores, particularly oxy- and deoxyhemoglobin and water, differ significantly in the near-infrared. As a result, the diffuse optical methods are sensitive to blood dynamics, blood volume, blood oxygen saturation, and water and lipid content of interrogated tissues. In addition, one can induce optical contrast in tissues with exogenous contrast agents, for example, chemical species that occupy vascular and extravascular space and preferentially accumulate in diseased tissue. Together these sensitivities provide experimenters with access to a wide spectrum of biophysical problems. The greater blood supply and metabolism of tumors compared to surrounding tissues provides target heterogeneity for tissue maps based on absorption.[6–25] Similar maps can be applied for studies of brain bleeding[26–28] and cerebral oxygen dynamics associated with activation by mental and physical stimulation.[29–41] Other applications of the deep tissue methods include the study of mitochondrial diseases,[42–44] of muscle function and physiology,[45,46] and of photodynamic therapy.[47–51]

Biomedical applications for diffusing near-infrared light probes parallel the application of nuclear magnetic resonance to tissue study. Generally, the categories of measurement can be termed spectroscopy and imaging. Spectroscopy is useful for measurement of time-dependent variations in the absorption and scattering of large tissue volumes. For example, brain oximetry (hemoglobin spectroscopy) of the frontal, parietal, or occipital regions can reveal reduced brain perfusion caused by head injury. Imaging is important when a localized heterogeneity of tissue is involved, for example, an early breast or brain tumor, a small amount of bleeding in the brain, or an early aneurysm. Images enable one to identify the site of the trauma and differentiate it from background tissue. Imaging is also important because it improves the accuracy of a spectroscopic measurement. Typically, spectroscopic methods employ over-simplified assumptions about the scattering media. Imaging relaxes some of these assumptions, usually at the cost of a more complex experimental instrument and computation, and ultimately improves the fidelity of the gathered optical property information.

The purpose of this chapter is to discuss functional imaging with diffusing photons. Our emphasis will be on imaging rather than spectroscopy, but it will be necessary to briefly review the basics of diffuse optical spectroscopy. This chapter is intended as a tutorial about what can be done with diffuse optical imaging, how to do it, and how to understand it. We intend to give a tutorial snapshot of the field with selected examples, but not a comprehensive review of research in the field. The remainder of this tutorial consists of sections on theory, instrumentation, and imaging examples, and a discussion about limitations and compromises associated with the technique.

## 21.2 Theory

### 21.2.1 Diffusion Approximation

Many researchers (e.g., References 52 through 56 and others) have shown that the photon fluence rate, $\Phi(\mathrm{r}, t)$ (photons/[cm$^2$ · s]), obeys the following diffusion equation in highly scattering media:

$$\nabla \cdot D(\mathrm{r})\nabla\Phi(\mathrm{r},t) - v\mu_a(\mathrm{r})\Phi(\mathrm{r},t) + vS(\mathrm{r},t) = \frac{\partial\Phi(\mathrm{r},t)}{\partial t}. \tag{21.1}$$

$\Phi(\mathrm{r},t)$ is proportional to the photon number density $U(\mathrm{r}, t)$ (photons/cm$^3$), i.e., $\Phi(\mathrm{r}, t) = vU(\mathrm{r}, t)$. The turbid medium is characterized by a speed of light, $v$, an absorption coefficient $\mu_a$ (i.e., the multiplicative inverse of the photon absorption length), and a photon diffusion coefficient, $D = v/3(\mu_s' + \mu_a) \cong v/3(\mu_s')$; the dependence of $D$ on $\mu_a$ is a subject of recent debate,[53, 57–68] but the latter relation follows in most tissues wherein $\mu_s' >> \mu_a$. The medium's reduced scattering coefficient is defined as $\mu_s' = (1 - g)\,\mu_s$ and represents the multiplicative inverse of the photon random walk steplength, $l^*$. Here $\mu_s$ is the reciprocal of the photon scattering length, $l$, and $g = <\cos\theta>$ is the ensemble-averaged cosine of the scattering angle $\theta$ associated with a typical single scattering event in the sample; $g$ accounts for the fact that light is more typically scattered in the forward direction, so that many scattering events are required before the initial photon propagation direction is truly randomized. $S(r, t)$ is an isotropic source term that gives the number of photons emitted at position $r$ and time $t$ per unit volume per unit time.

The right-hand side of Equation 21.1 represents the rate of increase of photons within a sample volume element. This rate equals the number of photons scattered into the volume element per unit time from its surroundings, *minus* the number of photons absorbed per unit time within the volume element, *plus* the number of photons emitted per unit time from any sources in the volume element.

The diffusion equation is based upon the *P*1 approximation of the linear transport equation.[69,70] It is valid when the reduced albedo $\alpha' = \mu_s'/(\mu_a + \mu_s')$ is close to unity, i.e., the reduced scattering coefficient is much greater than the absorption coefficient ($\mu_s' >> \mu_a$). The near-infrared (NIR) spectral window (commonly called the "therapeutic" window) of biological tissue lies between the intense visible absorption bands of hemoglobin and the NIR absorption band of water. In this window the reduced scattering coefficient is often 10 to 1000 times greater than the absorption coefficient,[71] for example, $\mu_s' \approx 10$ cm$^{-1}$ and $\mu_a \approx 0.03$ cm$^{-1}$ at 800 nm in human breast tissues. Of course tissues are not homogeneous, but they can be accurately divided into domains of piecewise homogeneous turbid media, each obeying Equation 21.1. Measurements are accomplished using sources and detectors arranged on the surfaces of or embedded within the tissue. Strictly speaking, it is also important for the source-detector separation to be of order three photon random walk steps (i.e., $3l^*$) or larger; otherwise the photon scattering angles will not be sufficiently randomized at the point of detection for rigorous application of the diffusion approximation.[72,73]

### 21.2.2 Sources of Diffusing Photons

Three types of sources are commonly employed in diffusive light measurements (see Figure 21.1). The simplest and easiest method to use is the continuous-wave (CW) device. In this case the source amplitude is constant, and the transmitted amplitude is measured as a function of source-detector separation or wavelength. The second method is the pulsed-time or time-resolved technique. In this scheme a short, usually subnanosecond light pulse is launched into the medium, and the temporal point spread function of the transmitted pulse is measured. The third method is the intensity modulated or frequency-domain technique. In this case the amplitude of the input source is sinusoidally modulated, producing a diffusive wave within the medium. The amplitude and phase of the transmitted diffuse light wave are then measured. These methods are related; the time-resolved and frequency-domain approaches are Fourier

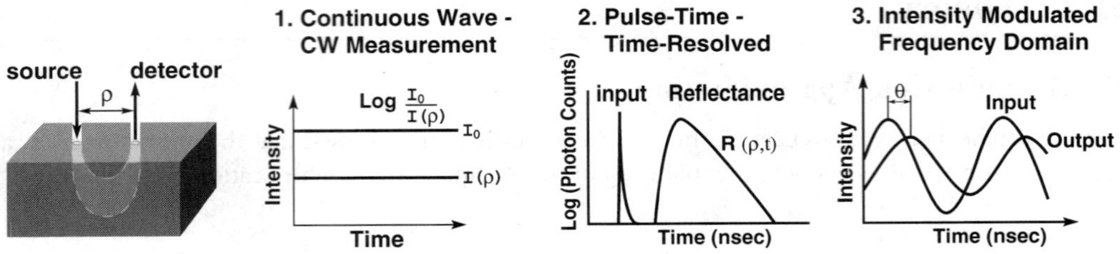


**FIGURE 21.1** Three source-detector schemes are generally employed in the photon migration field. On the far left we illustrate a typical remission geometry: (1) continuous-wave, called CW spectroscopy; (2) time-pulsed or time-resolved technique (often called TRS); (3) intensity amplitude modulation, i.e., often referred to as the frequency-domain method.

transformations of one another, and the CW approach is a special case of the frequency-domain approach wherein the modulation frequency is zero. Each of these approaches has strengths and weaknesses.

Briefly, the CW scheme is inexpensive and provides for rapid data collection. However, because it measures amplitude only, it lacks the capability for characterizing simultaneously the absorption and scattering of even a homogeneous medium from a measurement using only a single source-detector pair. The more expensive time-resolved scheme collects the full temporal point spread function, which is equivalent to a frequency domain measurement over a wide range of modulation frequencies. In this case, when the medium is homogeneous, $\mu_a$ and $\mu_s'$ can be obtained simultaneously from a single source-detector separation. The photon counting, however, can be slow and the technique is often limited by shot noise. The frequency domain technique is a compromise between CW and time-resolved techniques, with respect to cost and speed. It concentrates all the light energy into a single modulation frequency. It measures amplitude and phase, which ideally enable us to obtain $\mu_a$ and $\mu_s'$ for a homogeneous medium using a single source-detector separation. In practice all of these methods benefit significantly from use of many source-detector pairs and many optical wavelengths. In this chapter we focus on frequency domain sources, but the results can be applied to time-resolved and CW methods.

## 21.2.3 Diffuse Photon Density Waves in Homogeneous Turbid Media

Consider a light source at the origin with its intensity sinusoidally modulated at a modulation frequency $f$, e.g., the source term in Equation 21.1 is $S(r,t) = (M_{dc} + M_{ac}e^{-i\omega t})\,\delta(\mathbf{r})$, where $\omega = 2\pi f$ is the angular source modulation frequency, $M_{dc}$ and $M_{ac}$ are the source strengths of the DC and AC source components. The diffusion equation continues to be valid for light derived from these highly modulated sources as long as the modulation frequency $\omega$ is significantly smaller than the scattering frequency $v\mu'_s$; that is, photons must experience many scattering events during a single modulation period. Photons leaving the source and traveling along different random walk trajectories within the turbid medium will add incoherently to form a macroscopic scalar wave of photon concentration or fluence rate.

The total fluence rate consists of a DC and an AC component, i.e., $\Phi_{total}(\mathbf{r},t) = \Phi_{DC}(\mathbf{r}) + \Phi_{AC}(\mathbf{r},t)$. We focus on the AC component $\Phi_{AC}(\mathbf{r},t) = \Phi(\mathbf{r})e^{-i\omega t}$. The photon fluence will oscillate at the source of modulation frequency $\omega$. Plugging the AC source term into Equation 21.1 we obtain the following Helmholtz equation for the oscillating part of the photon fluence:

$$\left(\nabla^2 + k^2\right)\Phi(\mathbf{r}) = -\left(\frac{vM_{ac}}{D}\right)\delta(\mathbf{r})\,. \qquad (21.2)$$

We refer to this disturbance as a diffuse photon density wave (DPDW).[74,75] The DPDW has wavelike properties; for example, refractive,[76] diffractive,[77] and dispersive[78] behaviors of the DPDW have been demonstrated.

The photon density wave has a simple spherical wave solution for an infinite homogeneous highly scattering medium of the form:

$$\Phi_{AC}(\mathbf{r},t) = \left(\frac{vM_{ac}}{4\pi Dr}\right)\exp(ikr)\exp(-i\omega t). \tag{21.3}$$

The diffuse photon density wave wavenumber is complex, $k = k_r + ik_i$, and $k^2 = (-\nu\mu_a + i\omega)/D$. The real and imaginary parts of the wavenumber are:

$$\begin{aligned} k_r &= \left(\frac{\nu\mu_a}{2D}\right)^{1/2}\left(\left(1+\left(\frac{\omega^2}{\nu\mu_a}\right)\right)^{1/2} - 1\right)^{1/2} \\ k_i &= \left(\frac{\nu\mu_a}{2D}\right)^{1/2}\left(\left(1+\left(\frac{\omega^2}{\nu\mu_a}\right)\right)^{1/2} + 1\right)^{1/2}. \end{aligned} \tag{21.4}$$

In Figure 21.2 the measured wave is demonstrated within a tank of homogeneous highly scattering Intralipid. Constant-phase contours are shown in 20° intervals about the source at the origin. We see that the wave contours are circular and that their radii can be extrapolated back to the source. In the inset we exhibit the phase shift and a simple function of the wave amplitude plotted vs. the source-detector separation. From the slopes of these linear position-dependent measurements, one can deduce the wavelength of the disturbance, as well as the absorption and scattering factors of the homogeneous turbid medium via Equations 21.3 and 21.4.

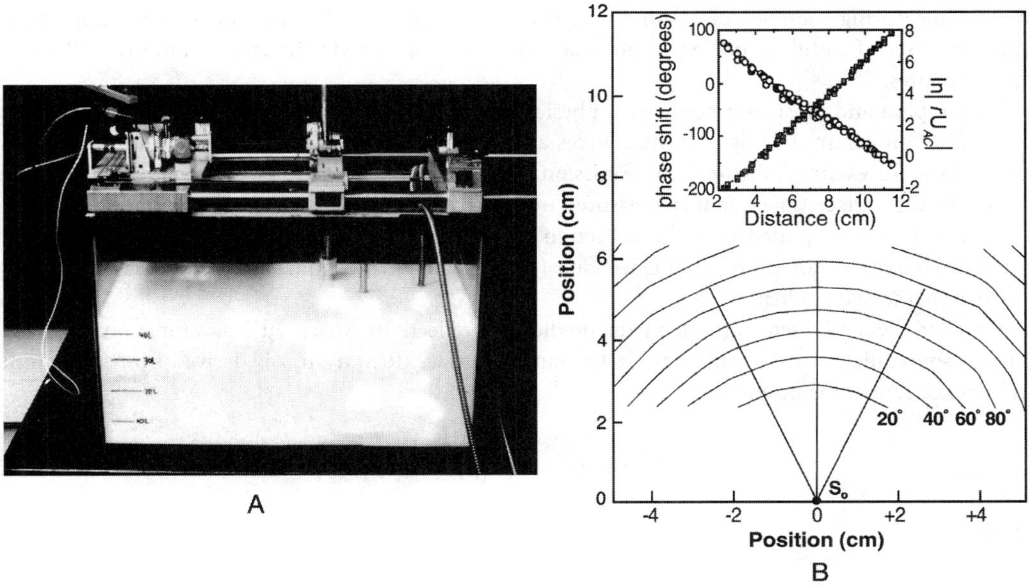


**FIGURE 21.2** (A) An aquarium used for model experiments. The aquarium is filled with Intralipid, a polydisperse emulsion whose absorption and scattering coefficients in the NIR region can be adjusted to approximate those of tissue. (B) Constant-phase contours of diffuse photon-density waves in the homogeneous sample of Intralipid. The source for this measurement is a 1-mW laser diode operating at 780 nm and modulated at 200 MHz. Inset: Measured phase-shift and a dimensionless (logarithmic) function of the amplitude as a function of source-detector separation.

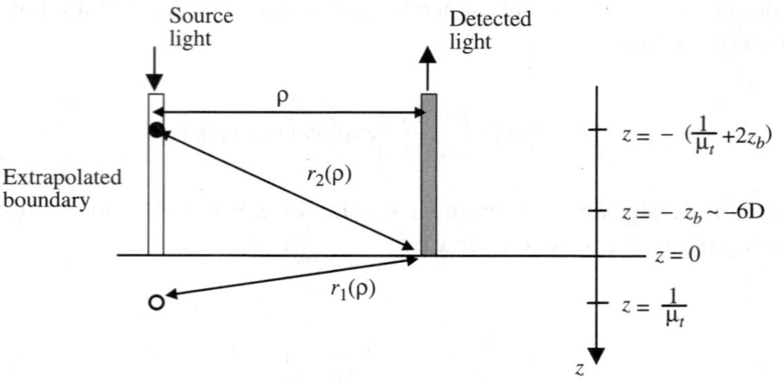


**FIGURE 21.3** Schematic of the experimental fiber configuration showing the relative positions of surface boundary ($z = 0$), extrapolated boundaries, $r_1(\rho)$ and $r_2(\rho)$ as defined in Equations 21.5, 21.6a, and 21.6b.

For homogeneous media in more complex geometries, one can still derive a set of phase and amplitude curves as a function of source-detector separation. The functional relationships may not be linear, but it is still readily possible to derive the average absorption and scattering factors of the underlying media by fitting to this data.

### 21.2.4 Spectroscopy of Homogeneous Turbid Media

The absorption factor, $\mu_a$, depends on the chromophore concentrations, and their extinction coefficients. The predominant endogenous absorbers in tissues are oxy- and deoxyhemoglobin, and water. The scattering factor, $\mu_s'$, depends on other tissue properties such as organelle (e.g., mitochondria) concentration and the index of refraction of the background fluids. If the medium is sufficiently homogenous, then by measuring the absorption and scattering coefficients as a function of light wavelength, one generates a set of simultaneous equations that can be solved to yield the concentrations of the tissue chromophores.

The simplest and most commonly used physical model for tissue spectroscopy treats the sample as a semi-infinite medium. In this case the sources and detectors are placed on the "air side" of the tissue surface (see, for example, Figure 21.3). Emission and detection take place through optical fibers placed flush with the surface. The quantity measured in practice at position **r**, time $t$, and along the direction $n$ is the radiance integrated over the collection solid angle. Within the diffusion approximation, the radiance consists of an isotropic fluence rate ($\Phi(\mathbf{r},t)$) and a directional photon flux ($J(\mathbf{r},t)$) that is proportional to the gradient of $\Phi$.

Diffusion theory for semi-infinite media predicts the reflectivity $R(\rho;\mu_a,\mu_s')$ as a function of ρ, where ρ is the source-detector separation along the sample surface. $R(\rho;\mu_a,\mu_s')$ is derived from photon flux and fluence rate at the boundary[79,80]

$$R(\rho;\mu_a,\mu_s') = C_1\Phi(\rho) + C_2 J_z(\rho) \tag{21.5}$$

where

$$\Phi(\rho) = \frac{1}{4\pi D}\left(\frac{\exp(-\mu_{eff} r_1(\rho))}{r_1(\rho)} - \frac{\exp(-\mu_{eff} r_2(\rho))}{r_2(\rho)}\right) \tag{21.6a}$$

and

$$J_z(\rho) = \frac{1}{4\pi\mu_t}\left[\left(\mu_{eff} + \frac{1}{r_1(\rho)}\right)\frac{\exp(-\mu_{eff} r_1(\rho))}{r_1^2(\rho)} + \left(\frac{1}{\mu_t} + 2z_b\right)\left(\mu_{eff} + \frac{1}{r_2(\rho)}\right)\frac{\exp(-\mu_{eff} r_2(\rho))}{r_2^2(\rho)}\right] . \tag{21.6b}$$

Here, $\mu_t = \mu_a + \mu_s'$, and $\mu_{eff} = [3\mu_a(\mu_a + \mu_s')]^{1/2}$. $C_1$ and $C_2$ are constants that depend on the relative refractive index mismatch between the tissue and the detector fiber, and the numerical aperture of the detection fibers. The parameters $r_i(\rho)$ are defined in Figure 21.2. Briefly, $r_1(\rho)$ is the distance from the point of contact of the detector fiber on the tissue surface to the effective source position in the tissue located $1/\mu_t'$ directly beneath the source fiber; $r_2(\rho)$ is the distance between the point of contact of the detector fiber and a point located $1/\mu_t + 2z_b$ directly above the source; $z_b$ is the extrapolated boundary length above the surface of the medium. Here the $z$-direction has been taken normal to the tissue surface (located at $z = 0$), so that $J_z$ is the directional flux normal to the surface.

The tissue optical properties at a fixed wavelength are derived from the measured reflectance by fitting with Equation 21.5. Many schemes have been developed to search for the optimal parameters;[80–82] their relative success depends on the measurement signal-to-noise ratio and the accuracy of the physical model. When everything works, one obtains a best estimate of the absorption factor and scattering factor at one or more optical wavelengths. We then decompose the absorption coefficient into contributions from different tissue chromophores, i.e.,

$$\mu_a(\lambda) = \sum_i \varepsilon_i(\lambda)\, c_i \, . \tag{21.7}$$

Here the sum is over the different tissue chromophores; $\varepsilon_i(\lambda)$ is the extinction coefficient as a function of wavelength for the $i$th chromophore and $c_i$ is the concentration of the $i$th chromophore. The $c_i$ are unknowns to be reconstructed from the wavelength-dependent absorption factors. Three unknowns require measurements at a minimum of three optical wavelengths (generally more, because tissue scattering is also an unknown).

Oxy- and deoxyhemoglobin concentrations (e.g., $c_{HbO_2}$, $c_{Hb}$, respectively) along with water concentration are the most significant tissue absorbers in the NIR. They can be combined to obtain blood volume (which is proportional to total hemoglobin concentration ($[c_{Hb} + c_{HbO_2}]$) and blood oxygen saturation (i.e., $[c_{HbO_2}/(c_{Hb} + c_{HbO_2})] \times 100$), which in turn provide useful physiological information. The same schemes are often extended to derive information about exogenous agents such as photodynamic therapy (PDT) drugs, indocyanine green (ICG), etc.; in such cases the effect of these other chromophores is accounted for by adding their contribution to the sum in Equation 21.7.

## 21.2.5 Imaging in Heterogeneous Media

### 21.2.5.1 Brief History

Tissue is often quite heterogeneous, so it is natural to contemplate making images with the diffusive waves. While high spatial resolution is desirable (e.g., a few millimeters), resolutions of about 1 cm are useful for many problems. A simple example of the utility of imaging is the early localization of a head injury that causes brain bleeding or hematomas. Tumors are another type of structural anomaly that one wants to detect, localize, and classify. The diffuse optical methods probe a variety of properties associated with tumor growth: larger blood volume resulting from a larger number density and volume fraction of blood vessels residing within the tumor; blood deoxygenation arising from relatively high metabolic activity within the tumor; increased concentration of the intracellular organelles necessary for the energy production associated with rapid growth; and the accumulation of highly scattering calcium precipitates.

Some of these properties may prove helpful in classifying tumors as benign or malignant. In the long term it should be possible to design contrast agents that respond to specific tumor properties. Other types of tissue of interest for functional imaging include the neonatal brain and a variety of animal models. For example, physiological studies of hemodynamics in relation to the oxygen demand probe important changes in the functional brain, especially during mental activity. In Section 21.4 we describe current research that investigates many of the clinical issues just outlined.

Optical characterization of the heterogeneous tissues has been attempted since 1929[83] when the term *diaphanography* was applied to shadowgraphs of breast tissue. This class of transillumination measurement was renewed in the early 1980s.[84–95] Even in the region of low tissue absorption, however, the high degree of tissue scattering distorted spectroscopic information and blurred optical images as a result of the large distribution of photon pathways through the tissue. Widebeam transillumination proved largely inadequate for clinical use because the two-dimensional "photographic" data were poorly suited for image reconstruction. The mathematical modeling of light transport in tissues was not developed sufficiently for optical tomography to be readily employed.

The diffusion approximation now provides a tractable basis for tomographic approaches to image reconstruction using highly scattered light. Tomographic methods crucial for recovery of information about breast heterogeneities were not employed in the early transillumination patient studies. Several approaches have been developed for diffuse optical tomography; these include: backprojection methods,[96,97] diffraction tomography in *k*-space,[98–101] perturbation approaches,[102–107] the Taylor series expansion approach,[108–113] gradient-based iterative techniques,[114] elliptic systems method (ESM),[115,116] and Bayesian conditioning.[117] Backprojection methods, borrowed from CT, produce images quickly and use few computational resources. However, they lack quantitative information and rely on simple geometries. Perturbation approaches based on Born or Rytov approximations can use analytic forms or iterative techniques based on numerical solutions. The analytic forms are relatively fast, but require the use of simple boundary conditions and geometries, and generally underestimate the properties of the perturbations. The numerical solutions are relatively slow and computationally intensive; however, in principle, realistic boundaries present no significant limitations for these methods.

### 21.2.5.2 Formulation of the Imaging Problem

In this section we formulate the imaging problem in the frequency domain. The starting point of this analysis is the time-independent form of the diffusion equation (Equation 21.1), where we have divided out all of the $e^{i\omega t}$ dependencies:

$$\nabla \cdot D(\mathbf{r})\nabla\Phi(\mathbf{r}) - \left(v\mu_a(\mathbf{r}) - i\omega\right)\Phi(\mathbf{r}) = -vS(\mathbf{r},\omega)\,. \tag{21.8}$$

The problem is difficult because the diffusion coefficient and the absorption coefficient vary with spatial position. We write $D(\mathbf{r}) = D_o + \delta D(\mathbf{r})$, and $\mu_a(\mathbf{r}) = \mu_{ao} + \delta\mu_a(\mathbf{r})$; here $D_o$ and $\mu_{ao}$ are constant, "background" optical properties. The source can have any form, but typically we assume point sources of the form $A\delta(\mathbf{r} - \mathrm{r}_s)$.

The goal of diffuse optical imaging is to derive $D(\mathbf{r})$ and $\mu_a(\mathbf{r})$ from measurements of $\Phi(\mathbf{r})$ on the sample surface. Two common forms are used for $\Phi(\mathbf{r})$ in the formulation of the inversion problem. The Born-type approach writes $\Phi(\mathbf{r}) = \Phi_o(\mathbf{r}) + \Phi_{sc}(\mathbf{r})$; traditionally one can view $\Phi_o(\mathbf{r})$ as the incident wave and $\Phi_{sc}(\mathbf{r})$ as the wave produced by the scattering of this incident wave off the absorptive and diffusive heterogeneities. The Rytov approach writes $\Phi(\mathbf{r}) = \Phi_o(\mathbf{r}) \exp[\Phi_{sc}(\mathbf{r})]$. We will focus on the Born approximation for our analysis, and indicate when possible the corresponding Rytov results.

We next substitute $D(\mathbf{r})$, $\mu_a(\mathbf{r})$, and $\Phi(\mathbf{r}) = \Phi_o(\mathbf{r}) + \Phi_{sc}(\mathbf{r})$ into Equation 21.8 to obtain a differential equation for $\Phi_{sc}(\mathbf{r})$ with general solution:

$$\Phi_{sc}(\mathbf{r}_d,\mathbf{r}_s) = \int\left(\frac{-v\delta\mu_a(\mathbf{r})}{D_o}\right)G(\mathbf{r}_d,\mathbf{r})\Phi(\mathbf{r},\mathbf{r}_s)d\mathbf{r} + \int\left(\frac{\delta D(\mathbf{r})}{D_o}\right)\nabla G(\mathbf{r}_d,\mathbf{r})\cdot\nabla\Phi(\mathbf{r},\mathbf{r}_s)d\mathbf{r}\,. \tag{21.9}$$

Here $\mathbf{r}_s$ is the source position, $\mathbf{r}_d$ is the detector position, $\mathbf{r}$ is a position within the sample. The integration is over the entire sample volume. $G(\mathbf{r},\mathbf{r}')$ is the Green's function associated with Equation 21.8. Examination of Equation 21.9 reveals some of the intrinsic challenges of the inverse problem. In a typical experiment one measures $\Phi$ on the sample surface and then extracts $\Phi_{sc}$ on the surface by subtracting $\Phi_o$ from $\Phi$. The problem of deriving $\delta\mu_a(\mathbf{r})$ and $\delta D(\mathbf{r})$ from $\Phi$ is intrinsically nonlinear because $\Phi$ and $G$ are nonlinear functions of $\delta\mu_a(\mathbf{r})$ and $\delta D(\mathbf{r})$.

***The Linearized Problem***

The simplest and most direct route to inverting Equation 21.9 starts by replacing $\Phi$ by $\Phi_o$ and $G$ by $G_o$. Here $\Phi_o$ and $G_o$ are solutions of the homogeneous version of Equation 21.8 with $D(\mathbf{r}) = D_o$, and $\mu_a(\mathbf{r}) = \mu_{ao}$. This approximation is good when $\Phi_o << \Phi$, and when the perturbations are very small compared to the background. It is also important that we have accurate estimates of $D_o$ and $\mu_{ao}$. In this case, Equation 21.9 is readily discretized in Cartesian coordinates and written in the following form:

$$\Phi_{sc}\left(\mathbf{r}_d,\mathbf{r}_s\right)=\sum_{j=1}^{NV}\left(W_{a,j}\delta\mu_a\left(\mathbf{r}_j\right)+W_{s,j}\delta D\left(\mathbf{r}_j\right)\right). \tag{21.10}$$

The sum is taken over $NV$ volume elements (i.e., voxels) within the sample; the absorption and scattering weights are, respectively, $W_{a,j} = G_o(\mathbf{r}_d,\mathbf{r}_j)\ \Phi_o(\mathbf{r}_j,\mathbf{r}s)\ (-v\Delta x\ \Delta y\ \Delta z/D_o)$, and $W_{s,j} = \nabla G_o(\mathbf{r}_d,\mathbf{r}_j)\cdot\nabla\Phi_o(\mathbf{r}_j,\mathbf{r}_s)\ (\Delta x\ \Delta y\ \Delta z/D_o)$. In any practical situation there will be $NS$ sources and $ND$ detectors, and so there will be up to $NM = NS \times ND$ measurements of $\Phi$ on the sample surface. For the multisource-detector problem one naturally transforms Equation 21.10 into a matrix equation, i.e.,

$$\left[\mathbf{W}_{a,jt},\mathbf{W}_{s,ij}\right]\left\{\delta\mu_a\left(\mathbf{r}_j\right),\delta D\left(\mathbf{r}_j\right)\right\}^T=\left\{\Phi_{sc}\left(\mathbf{r}_d,\mathbf{r}_s\right)_i\right\}. \tag{21.11}$$

Here, the index $i$ refers to source-detector pair, and the index $j$ refers to position within the sample. The perturbation vector $\{\delta\mu_a(\mathbf{r}_j),\ \delta D(\mathbf{r}_j)\}^T$ is $2NV$ in length, the measurement vector $\{\Phi_{sc}(\mathbf{r}_d,\mathbf{r}_s)_i\}$ is $NM$ in length, and the matrix [W] has dimensions $NM \times (2NV)$. In the Rytov scheme, the formulation in the weak perturbation limit is almost exactly the same, except that $W_{a,j} = G_o(\mathbf{r}_d,\mathbf{r}_j)\ \Phi_o(\mathbf{r}_j,\mathbf{r}_s)(-v\Delta x\ \Delta y\ \Delta z/\Phi_o(\mathbf{r}_d,\mathbf{r}_s)\ D_o)$, $W_{s,j} = \nabla G_o(\mathbf{r}_d,\mathbf{r}_j)\cdot\nabla\Phi_o(\mathbf{r}_j,\mathbf{r}_s)\ (\Delta x\ \Delta y\ \Delta z/\Phi_o(\mathbf{r}_d,\mathbf{r}_s)\ D_o)$, and the vector $\{\Phi_{sc}(\mathbf{r}_d,\mathbf{r}_s)_i\}$ is set equal to $\{\ln[\Phi(\mathbf{r}_d,\mathbf{r}_s)/\Phi_o(\mathbf{r}_d,\mathbf{r}_s)]_i\}$ rather than $\{[\Phi(\mathbf{r}_d,\mathbf{r}_s) - \Phi_o(\mathbf{r}_d,\mathbf{r}_s)]_i\}$. The Rytov scheme has some experimental advantages because it is intrinsically normalized (the Born scheme, however, can be modified so that it is normalized in essentially the same way); the major approximations of the Rytov scheme are associated with the gradients of $\Phi$, in particular that $(\nabla\Phi_{sc})^2$ is small relative to the perturbation terms in Equation 21.9. Thus both Born and Rytov approaches give rise to an inverse problem of the form $[W]\ \{x\} = \{b\}$; the unknown vector $\{x\}$ can be determined from this set of linear equations by a number of standard mathematical techniques. The numerical elements in [W] are often assigned in simple geometries using analytic forms of $G_o$ and $\Phi_o$ (e.g., Equation 21.3 and variants), or more generally by numerically solving Equation 21.8 and its Green's function analog for $\Phi_o$ and $G_o$.

***The Nonlinear Problem***

The linear formulation described above works well when perturbations are small and isolated, and when the background media are relatively uniform. However, Equation 21.9 is intrinsically nonlinear because $\Phi$ and $G$ are also functions of the variables we are trying to determine by inversion. The most broadly useful image reconstruction schemes are iterative. These approaches follow similar algorithms: (1) the optical properties ($\mu_a$ and $D$) are initialized; (2) the forward problem is solved; (3) a chi-squared is calculated and convergence is checked; (4) the inverse problem is set up; (5) the inverse problem is solved; (6) the optical properties are updated and a return to step 1 occurs.

The forward problem is defined as calculating the diffuse photon density, $\Phi_C(\mathbf{r},\mathbf{r}_s)$, for each source position $\mathbf{r}_s$ and is typically found using finite elements or finite difference methods using Equation 21.8. The boundary conditions are defined as:

$$\frac{\partial \Phi_C}{\partial \hat{n}} = -\alpha \Phi_C, \quad \alpha = \frac{\left(1-R_{eff}\right)}{\left(1+R_{eff}\right)} \frac{3\mu_s'}{2}. \tag{21.12}$$

$R_{eff}$ is the effective reflection coefficient and can be approximated by: $R_{eff} = -1.440n^{-2} + 0.710n^{-1} + 0.668 + 0.0636n$, $n = n_{in}/n_{out}$ the relative index of refraction.[55] The chi-squared ($\chi^2$) is generally defined as:

$$\chi^2 = \sum_{NM} \left( \frac{\Phi_M(r_d^i) - \Phi_C(r_d^i)}{\sigma^i} \right)^2. \tag{21.13}$$

Here $NM$ = number of measurements, $M$ = measured, $C$ = calculated, $\mathbf{r}_d^i$ is the $i$th detector position, and $\sigma^i$ is the $i$th measurement error. By comparing $\chi^2$ to some defined $\varepsilon$, a convergence criterion is defined and checked.

We then need a way of updating the optical properties from their previous values. A standard Taylor method expands $\Phi_C$ about its assumed optical property distribution, which is a perturbation away from another distribution presumed closer to the true value. In particular we set the measured photon density wave for each source-detector pair equal to the calculated photon density wave at the corresponding source-detector pair plus the first-order Taylor series perturbation expansion terms in $\mu_a$ and $D$, i.e., $\Phi_M = \Phi_C + (\partial\Phi_C/\partial\mu_a)\Delta\mu_a + (\partial\Phi_C/\partial D)\Delta D$.

The inverse problem is defined from this relationship:

$$[\mathrm{J}]\{\Delta\mu_a, \Delta D\}^T = -\{\Phi_M(\mathbf{r}_d) - \Phi_C(\mathbf{r}_d)\}. \tag{21.14}$$

Here $[\mathrm{J}] = [\partial\Phi_C/\partial\mu_a, \partial\Phi_C/\partial D]$ is called the Jacobian. The Jacobian matrix will have the following entries:

$$\left[\frac{\partial \Phi_C}{\partial \mu_a}\right]_{ij} = \frac{-v\Delta x \Delta y \Delta z}{D_o} G(\mathbf{r}_{di}, \mathbf{r}_j)\Phi_C(\mathbf{r}_j, \mathbf{r}_{si}) \tag{21.15a}$$

$$\left[\frac{\partial \Phi_C}{\partial D}\right]_{ij} = \frac{\Delta x \Delta y \Delta z}{D_o} \nabla G(\mathbf{r}_{di}, \mathbf{r}_j) \cdot \nabla \Phi_C(\mathbf{r}_j, \mathbf{r}_{si}). \tag{21.15b}$$

It is illuminating at this point to compare Equation 21.14 with Equation 21.11. The two expressions are essentially the same if we associate $\Phi_M$ with $\Phi$, $\Phi_C$ with $\Phi_o$, $\Delta\mu_a$ with $\delta\mu_a$, $\Delta D$ with $\delta D$, and if we use the true Green's function $G$ rather than $G_o$. The same set of substitutions in the Rytov formulation gives a Rytov version of the nonlinear inversion scheme. Thus the iterative formulation of the inverse problem is based on the same underlying integral relationship (Equation 21.9), and one readily sees that each step of the "nonlinear" iteration process is a linear inverse problem of the form [J] {x} = {b}.

#### 21.2.5.3 Methods for Solving the Inverse Problem

The inverse problem may be solved using a wide range of methods (an excellent review of these methods was given by Arridge[118]). The solution method chosen depends in part on the determination of the *implicit* or *explicit* Jacobian. For the explicit Jacobian two methods are commonly employed: the Newton-Raphson and the conjugate gradient techniques. It is also possible to combine these methods with Bayesian conditioning or regularization to improve reconstruction. For the implicit Jacobian, the methods of choice are the gradient-based iterative technique and ART (algebraic reconstruction technique).

There are essentially two ways to construct the Jacobian, [J] *explicitly*: direct and adjoint. The *direct* approach explicitly takes the derivative of the forward problem (Equation 21.8) with respect to the optical properties to determine the Jacobian. For example, suppose [A]{$\Phi_C$} = {S} is the forward problem; here [A] is the operator on the left side of Equation 21.8 and {S} is the source on the right side of Equation 21.8.

Then the equation $[A]\{\partial\Phi_C/\partial\mu_a\} = \{\partial S/\partial\mu_a\}\,[\partial A/\partial\mu_a]\{\Phi_C\}$ enables one to compute $\partial\Phi_C/\partial\mu_a$ (and a similar forward problem enables the computation of $\partial\Phi_C/\partial D$).[108] This approach is optimal with finite elements; because the numerical formulation lends itself easily to taking the derivative of [A]. $\Phi_C$, [A] and $[A]^{-1}$ are updated on each iteration. The *adjoint* approach solves the forward problem $[A]\{\Phi_C\} = \{S\}$ to determine $\Phi_C$, and an adjoint problem $[A^*]\{G\} = \{S'\}$ to determine Green's function $G(\mathbf{r}_d,\mathbf{r})$ due to a unit source at the detector position.[1,102–105,107,119,120] G and $\Phi_c$ then fix the elements of [J] according to Equation 21.15. Both $\Phi_c$ and $G$ are updated at each iteration. The impact of the two different approaches on convergence and accuracy is not well understood. However, from the computational perspective, for each iteration the direct approach requires 3NS solves per iteration, while the adjoint Born or Rytov approach requires *NS* plus *ND* solves per iteration.

The inverse problem, $[J]\,\{\Delta\mu_a, \Delta D\}^T = \{\Phi_M - \Phi_C\}$, which is also a matrix equation, is significantly more costly than these other subproblems because the Jacobian is a full nonsquare matrix $\{NM \times 2NV\}$. However, the inverse problem only needs to be solved once per iteration cycle. The Jacobian is ill-conditioned and is thus singular or close to singular, which makes it difficult to invert directly. Generally, the approach to address these issues is twofold. First, the matrix is made square by multiplying the inverse problem by the transpose of the Jacobian, i.e.,

$$[J]^T[J]\{\Delta\mu_a, \Delta D\}^T = [J]^T\{\Phi_M - \Phi_C\}. \tag{21.16}$$

Unfortunately, by squaring the Jacobian the equation becomes even more ill conditioned. This problem is solved by regularization[121,122] so that the equation becomes

$$\left([J]^T[J] + \lambda[C]^T[C]\right)\{\Delta\mu_a, \Delta D\}^T = [J]^T\{\Phi_M - \Phi_C\}. \tag{21.17}$$

Here $\lambda$ is called the regularization parameter and [C] is the regularizing operator, which is sometimes taken to be the identity matrix. The regularization parameter generally is related to the measurement signal-to-noise. It is a theoretical "knob" that can be adjusted and it will affect image quality by introducing a trade-off between spatial resolution and contrast.[122,123] Nevertheless, its use converts the inverse problem into a readily solvable problem. $\{\Delta\mu_a, \Delta D\}^T$ can now be determined using Equation 21.17 and any number of mathematical techniques that solve systems of linear equations. A particularly useful and common solution scheme is the conjugate gradient method.

A qualitatively different scheme due to Arridge involves the implicit determination of the Jacobian. Briefly, this technique utilizes the Born or Rytov approximation for the inverse problem, and has at least two known solution methods: gradient-based iterative technique and ART (algebraic reconstruction technique). The Jacobian is not calculated; instead an objective function (e.g., a chi-square function) is defined whose gradient, for example, can be used to derive subsequent search directions (see the original papers, References 114 and 118, for details.) Note that the implicit formulation is particularly attractive for experimental systems that rely on many detectors rather than many sources.

#### 21.2.5.4 Challenges for Implementation

The main barrier for full three-dimensional reconstruction is the significant memory and processing time it requires. There are three costly steps of the algorithms: (1) solving the forward problem for each source position, (2) determining the Jacobian, and (3) solving the inverse problem. The forward problem requires the solution of a matrix equation banded in finite difference or sparse in finite elements. The solvers in three-dimensional reconstruction are necessarily iterative because direct solvers require very large storage space for full matrices. Subsequently, multiple source positions demand multiple forward solves. Determination of the Jacobian requires additional matrix equation solutions. For explicit determination, the Green's function for each detector position is needed or the vectors $\partial\Phi_C/\partial\mu_a$ and $\partial\Phi_C/\partial D$ are needed.

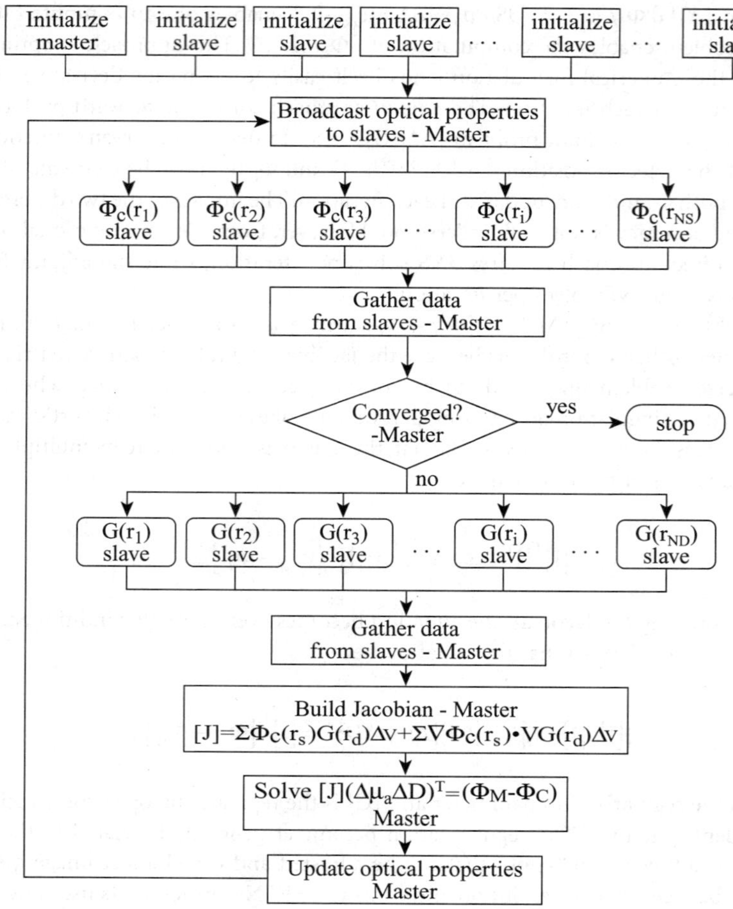


**FIGURE 21.4** Schematic of the University of Pennsylvania algorithm for image reconstruction using parallel computing. In the first step each node has its own set of arrays and variables that are initialized. The master must initialize the Jacobian array, the forward solution vector array, and the Green's function vector array. In the second step the master sends out the optical properties, which (in the first iteration) are just the background values. In the third step, the slaves calculate the solutions to the forward problem; *NS* slaves are used in this step. The solution vectors are then returned to the master and at this point the master checks for convergence. If convergence has not been achieved then the slaves calculate the solutions to the Green's function; *ND* slaves are used in this step. The solution vectors are returned to the master and the Jacobian is determined. Next the master solves the inverse problem, in this case by using a spatially variant regularized conjugate gradient optimization method. Finally, the optical properties are updated on the master and the algorithm repeats until convergence has been achieved.

Consider a system with one frequency — *NS* sources and *ND* detectors; assume further that all the detectors are used for each source. The number of measurements is $NM = NS \times ND$. For each iteration we require *NS* forward solves and, depending on the Jacobian determination, either *ND* Green's function or 2*NS* solves for the vectors $\partial\Phi_C/\partial\mu_a$ and $\partial\Phi_C/\partial D$ (implicit would require *NS* solves). The final costly step of the algorithm is the solution of the inverse problem. However, the inverse problem only needs to be solved once per iteration cycle. For many practical problems the calculations take a long time, and limited memory is available to store the Jacobian explicitly. An important solution to this large-scale computational problem is parallel computing — the execution of many computations at one time using many processors. It has been used successfully in areas of medical imaging, such as positron emission tomography (PET),[124–128] single photon emission computed tomography (SPECT),[129,130] computed tomography (CT),[131–133] electrical impedance tomography (EIT),[134] and optical sectioning microscopy

(OSM).[135] It is just beginning to catch on in the diffuse optical tomography (DOT) community.[136,137] In Figure 21.4, we illustrate how the problem is parallelized at the University of Pennsylvania.

## 21.2.6 Diffusion of Light Correlations: Blood Flow

Thus far our discussion has centered around the determination of tissue absorption and scattering properties. Among other things, these measurements provide access to concentrations of oxygenated and deoxygenated hemoglobin. Even more information, however, is impressed upon these diffusing light fields. Speckle fluctuations of the scattered light are sensitive to the motions of scatterers such as red blood cells.

The means for using light fluctuations and light frequency shifts to study motions have appeared with numerous names over the years.[138–158] In most of these experiments the quantity of interest is the electric field temporal autocorrelation function $G_1(\mathbf{r},\tau) = \langle E(\mathbf{r},t)\, E^*(\mathbf{r}, t + \tau)\rangle$ or its Fourier transform. Here the angle brackets $\langle\ \rangle$ denote ensemble averages or averages over time for most systems of practical interest. $\tau$ is called the correlation time. The field correlation function is explicitly related to the motions of scatterers within the samples that we study.

The study of these motions in deep tissues is possible because the electric field temporal autocorrelation function for light traveling in highly scattering media also obeys a diffusion equation.[152] In steady-state (i.e., $\omega = 0$) and in homogeneous media, this correlation diffusion equation is quite simple:

$$\left(D\nabla^2 - v\mu_a - \alpha k_o^2 \mu_s{}' \left\langle \Delta\mathbf{r}^2(\tau)\right\rangle / 3\right) G_1(\mathbf{r},\tau) = -vS(\mathbf{r}) \,. \tag{21.18}$$

Here $k_o$ is the wavevector of the photons in the medium and $\langle\Delta\mathbf{r}^2(\tau)\rangle$ is the mean-square displacement in time $\tau$ of the scattering particles (e.g., blood cells); it can have different forms depending on the nature of the particle motion and can also vary with position. $S(\mathbf{r})$ is the source light distribution, and $\alpha$ represents the fraction of photon scattering events in the tissue that occur from moving cells or particles.

Notice that, for $\tau \to 0$, $\langle\Delta\mathbf{r}^2(\tau)\rangle \to 0$, and Equation 21.18 reduces to the steady-state diffusion equation for diffuse photon fluence rate (i.e., Equation 21.8 with $\omega = 0$). Notice also that the homogenous version of Equation 21.18 can be recast as a Helmholtz-like equation for the temporal field autocorrelation function, $G_1(\mathbf{r},\tau)$, i.e.,

$$\left(\nabla^2 + K^2(\tau)\right) G_1(\mathbf{r},\tau) = -S(\mathbf{r})/D \,, \tag{21.19a}$$

$$K^2(\tau) = (-v/D)\left(\mu_a + \alpha k_o^2 \mu_s' \left\langle \Delta\mathbf{r}^2(\tau)\right\rangle / 3\right). \tag{21.19b}$$

For an infinite homogenous medium with a point source at the origin, this equation will also have the well known spherical-wave solution, i.e., $\sim[\exp\{-K(\tau)\mathbf{r}\}/D4\pi\mathbf{r}]$.

The mean-square displacement $\langle\Delta\mathbf{r}^2(\tau)\rangle = 6D_B\tau$ for organelles or cells undergoing Brownian motion with "particle" diffusion coefficient $D_B$. For the important case of random flow that can arise in the tissue vasculature, $\langle\Delta\mathbf{r}^2(\tau)\rangle = \langle V^2\rangle\tau^2$, where $\langle V^2\rangle$ is the second moment of the cell speed distribution. In the latter case the correlation function will decay exponentially as $\tau$.

Multidistance measurements of $G_1(\mathbf{r},\tau)$ provide dynamical information about the motions within the sample in exactly the same way that diffusive waves provide information about scattering and absorption properties. The layout of the sources and detectors is similar to the diffusive wave schemes, but the correlation measurements are a little more complex. For the measurements, one needs a special piece of equipment called an autocorrelator, which takes the detector output and uses the photon arrival times to compute $G_1(\mathbf{r},\tau)$ or (more precisely) its light *intensity* analog.

The entire set of formalisms for imaging outlined in Sections 21.2.5.2 and 21.2.5.3 is applicable to diffuse light temporal correlations. The technique is attractive because it enables us to measure the blood flow (i.e., $\langle \Delta \mathbf{r}^2(\tau) \rangle$) within deep tissues. The ability to measure relative changes concurrently in blood flow, hemoglobin concentration, and hemoglobin oxygenation within a single instrument makes possible a range of cerebral studies in animal models and in infants and neonates (see References 159 through 161).

## 21.2.7 Contrast Agents

The use of contrast agents in DOT and spectroscopy for disease diagnostics and for probing tissue functionality follows established clinical imaging modalities such as magnetic resonance imaging (MRI),[162–164] ultrasound,[164–166] and x-ray computed tomography (CT).[164,167,168] Contrast agent administration provides for accurate difference images of the same heterogenous tissue volume under nearly identical experimental conditions; this approach often yields superior diagnostic information. Although contrast agents most commonly induce changes in absorption, recently fluorescent/phosphorescent agents have also been considered as means to increase specificity and sensitivity for tumor detection and imaging.[77,169–184] In addition to concentration changes, fluorophore lifetime is sensitive to physiological environment, for example, through oxygen quenching or pH. Most contrast agent schemes rely on the fact that the exogenous macromolecular structures accumulate preferentially in abnormal tissues.

### 21.2.7.1 Fluorescent Contrast Agents

A chapter in this handbook already reviews contrast agents in photon migration, so we will be very brief. (Our discussion follows from References 185 through 187.) Suppose a heterogeneous turbid medium with fluorophore distribution $N(\mathbf{r})$ is excited by an excitation diffusive wave, $\Phi(\mathbf{r},\mathbf{r}_s)$, emitted from $\mathbf{r}_s$ and whose optical wavelength is in the absorption band of the fluorophore. A fluorescent diffuse photon density wave, $\Phi_{fl}(\mathbf{r},\mathbf{r}_s)$, is produced in the medium, and

$$\Phi_{fl}(\mathbf{r},\mathbf{r}_s) = \int G_{fl}(\mathbf{r},\mathbf{r}')T(\mathbf{r}')\Phi(\mathbf{r}',\mathbf{r}_s)d\mathbf{r} \qquad (21.20a)$$

where

$$T(\mathbf{r}') = \varepsilon \frac{\tau}{\tau_o} \frac{\eta N(r)}{1 - \omega\tau}. \qquad (21.20b)$$

The integration is over the sample volume. Here $T(\mathbf{r})$ is a fluorescence transfer function, which depends on the fluorophore radiative and total lifetimes, $\tau_o$ and $\tau$ respectively, the source modulation frequency $\omega$, the fluorophore extinction coefficient $\varepsilon$, the fluorescence quantum yield $\eta$, and the fluorophore distribution $N(\mathbf{r})$. In principle, $\tau_o$ and $\tau$ (and even $\varepsilon$ and $\eta$) can also depend on position. The Green's function $G_{fl}(\mathbf{r}, \mathbf{r}')$ is derived with a diffusion equation (Equation 21.8) for the fluorescent diffuse photon density wave at the emission wavelength, i.e., the absorption and scattering coefficients in this diffusion equation are defined at the fluorescent emission wavelength. Equation 21.20 is very similar to the equations we inverted in Sections 21.2.5.2 and 21.2.5.3. However, it is deceptively simple because heterogeneity information is embedded in $\Phi$ and $G_{fl}$, in addition to $N$ and $\tau$. Nevertheless, it can be inverted numerically using similar techniques.

### 21.2.7.2 Differential Absorption

In these measurements optical data are typically obtained before and after administration of the absorbing optical contrast agent (e.g., the intraveneous administration of ICG. In principle, DOT images taken before and after administration may be reconstructed and subtracted; however, in practice a more robust approach derives images of the differential changes due to the extrinsic perturbtion. In the latter case experimental measurements use the exact same geometry within a short time of one another, thereby minimizing positional and movement errors and instrumental drift. Furthermore, the use of differential measurements eliminates systematic errors associated with the different medium often required to calibrate operational parameters of the instrument or to provide a baseline measurement for independent reconstructions. Finally, the effect of surface absorbers such as hair- or skin-color variation is also minimized.

The main analytical difficulties of the differential approach arise because the media are inhomogeneous. Thus the total diffuse light field in the contrast agent perturbation problem does not separate into a homogeneous background field and a scattered field in a straightforward way. Furthermore, the average background optical properties, particularly the absorption, can change as a result of contrast agent administration.

In the typical experiment, relative absorption changes are much larger than scattering changes and one can ignore the relative changes in scattering. Under these circumstances the differential signal $\Delta\Phi$, can be related to the differential absorption $\Delta\mu_a$ via the integral relation:[188]

$$\Delta\Phi(\mathbf{r},\mathbf{r}_s) = \int W_a'(\mathbf{r},\mathbf{r}',\mathbf{r}_s)\Delta\mu_a(\mathbf{r}')d\mathbf{r} + C\,. \tag{21.21}$$

Here $W_a'$ is a weight function very similar to the functions discussed in Section 21.2.5.2. $C$ is a correction; it is an integral that depends on source-detector geometry and on the weight function before and after administration of the contrast agent. For many geometries (e.g., the transmission geometry) $C$ is small and can be ignored. In those cases we can directly invert Equation 21.21 according to the ideas described in Sections 21.2.5.2 and 21.2.5.3, and thus obtain the differential absorption changes directly from difference measurements.

## 21.3 Instrumentation

The basic imaging geometry for diffuse optical tomography consists of a set of distinguishable point-like sources and a set of photodetectors, each covering a small area of $<10$ mm$^2$ on the surface of the medium. In general some type of source encoding strategy must be used so that the origin of the detected signals can be traced to specific sources. In this way, measurements with differing spatial sensitivities are obtained, and an image can be reconstructed.

There are three common measurement geometries: (1) planar transillumination measurements, (2) cylindrical measurements, and (3) reflectance measurements. All three are used for breast imaging (transillumination,[10–13,189] cylindrical,[15,190–192] and reflectance.[119,137,193]). The cylindrical and reflectance geometries are used for imaging animals, human baby heads,[28,194–197] and limbs on the human body (e.g., arm or leg). The reflectance geometry is used for imaging human adult heads.[159,198–200]

In Section 21.2.2 we identified three diffuse light excitation schemes: (1) illumination by subnanosecond pulses of light, (2) CW illumination, and (3) radio-frequency (RF) amplitude-modulated illumination. Short-pulse systems[201–205] detect the temporal distribution of photons as they leave the tissue. The shape of the distribution provides information about tissue optical parameters. Although these systems provide the most information per source-detector pair for characterizing optical properties, their relatively poor signal-to-noise ratio (SNR) leads to longer image acquisition times (typically a few minutes in systems used today).

CW systems[195,206–208] emit light of constant intensity or amplitude. (Sometimes the emitted intensity is amplitude modulated at frequencies less than a few tens of kilohertz.) Detectors measure the amplitude of the light transmitted through the tissue. These systems are simple to build and provide fast image rates (presently up to 100 Hz), but their lack of temporal information makes quantitation of tissue absorption and scattering more difficult. In RF systems[209–211] the light source intensity is amplitude modulated at frequencies of tens to hundreds of megahertz. Information about the absorption and scattering properties of tissue is obtained by recording amplitude decay and phase shift (delay) of the detected signal with respect to the incident wave.[211,212] These systems offer the fast image acquisition rate of CW systems and contain information sufficient for quantitative characterization of absorption and

**TABLE 21.1** Comparison of Relative Advantages and Disadvantages of Time-Domain, Frequency-Domain, and Continuous-Wave Instrumentation

| Instrumentation | Advantages | Disadvantages |
|---|---|---|
| Time-Domain | Full temporal impulse response<br>Quantitative | Difficult to maintain<br>Expensive optoelectronics |
| Frequency-Domain | Diffusive wave phase and amplitude<br>Faster than time-domain<br>Lower cost than time-domain | Difficult RF electronics |
| Continuous-Wave | Lowest cost<br>Easy electronics<br>Fastest<br>Accurate for differential measurements of optical properties | Diffuse light amplitude only<br>Less accurate for extimates of absolute optical properties |

scattering optical properties. The advantages and disadvantages of the three systems are outlined in Table 21.1.

In DOT it is desirable to make a large number of measurements for image reconstruction in a short period of time so that the data are not confounded by physiological or movement artifacts. The balance between number of measurements and image acquisition time is dictated by application. CW systems are popular for imaging spatial variations of absorption changes on timescales of seconds to minutes, for example, imaging muscle[213,214] and brain activation.[195,196,198] CW systems usually have the best SNR and lend themselves well to several encoding strategies enabling massively parallel measurements. Furthermore, although CW systems are poor at quantifying static absorption and scattering properties uniquely, they excel at quantifying spectroscopic changes in absorption and scattering, particularly when *a priori* knowledge of the spectroscopic features is available. On the other hand, RF and time-domain systems are popular for imaging static optical properties when quantitative accuracy is required and when data acquisition times of one to several minutes are acceptable.

## 21.3.1 Source Encoding Strategies

If cost is not an issue, then increasing the number of parallel detectors or detection systems is a straightforward approach to increasing the number of measurements per unit time. Often, however, measurements with multiple source wavelengths and source positions are desired. In this case an encoding strategy enabling the separation of source wavelengths and positions must be employed. The encoding strategies currently used in CW systems and applicable to RF systems are switched-source time-division multiplexing (SS-TDM), phase-division multiplexing (PDM), pulse-modulated time-division multiplexing (PM-TDM), frequency-division multiplexing (FDM), and wavelength-division multiplexing (WDM). SS-TDM and WDM are applicable to time-domain imaging systems as well.

For SS-TDM, sources are modulated at the same frequency and cycled through consecutively; the detectors synchronously obtain the source signal through their own demodulators. This is the easiest system to design and build. Because at any given time only a single source illuminates the sample, interchannel crosstalk is low, and simple circuit construction techniques (point-to-point wiring, Protoboards, etc.) can be used successfully.

For PDM, two sources are modulated with a square-wave at the same frequency, but in phase quadrature (i.e., at a 90° phase difference). Each of the detectors synchronously detects each source through two separate demodulators and low-pass filters, each of which is "tuned" to the in-phase source. Source pairs can then be cycled through consecutively. This is an easy system to design and build, but component layout affects performance, particularly interchannel crosstalk.

For PM-TDM, M sources are cycled on and off in sequence, but at a rapid (~kHz) rate. For *N* detectors, each source is synchronously detected through individual demodulators, each of which is time-gated to one source. This approach has all the benefits of SS-TDM, but with no temporal skew. The system can be difficult to design and construct due to the complex interdigitation and fast switching speeds. Because only one source operates at any one time, the background level is very low.

For FDM, each of the sources is modulated at one of a number of anharmonically related frequencies (to minimize the effects of intermodulation distortion). For each of the detectors, the sources are demodulated coherently (synchronous detection) or incoherently (envelope detection). This is the most complex system to design and build due to the high potential for interchannel crosstalk. The high background flux arising because all sources are on simultaneously raises the shot noise floor and can saturate photo-detectors. However, because each source is "on" all the time, the scheme is more parallel than the sequential approaches described previously.

WDM simply uses bandpass optical filters in front of the photodetectors to distinguish different wavelength sources. This method can be used in combination with any of the encoding strategies described earier to distinguish light from different spatial locations.

### 21.3.1.1 Continuous-Wave Imaging System

As an example of the CW imaging system, we consider one used at the Massachusetts General Hospital-Nuclear Magnetic Resonance (MGH-NMR) Center. This system employs the FDM scheme to detect 32 lasers with 32 detectors (see Figure 21.5). At present, the 32 lasers are divided into 16 lasers at 690 nm and 16 at 830 nm. These 16 laser pairs are fiber coupled and deliver light to 16 positions on the medium to be imaged. The detectors are avalanche photodiodes (APDs, Hamamatsu C5460-01). A master clock generates the 32 distinct frequencies between 6.4 and 12.6 kHz in approximately 200-Hz steps. These frequencies are then used to drive the individual lasers with current stabilized square-wave modulation. Following each APD module is a bandpass filter, cut-on frequency of ~500 Hz to reduce 1/f noise and the 60-Hz room light signal, and a cut-off frequency of ~16 kHz to reduce the third harmonics of the square-wave signals. After the bandpass filter is a programmable gain stage to match the signal levels with the acquisition level on the analog-to-digital converter within the computer. Each detector is digitized at ~45 kHz and the individual source signals are then obtained by use of a digital bandpass filter — for example, a discrete Fourier transform or an infinite-impulse-response filter.

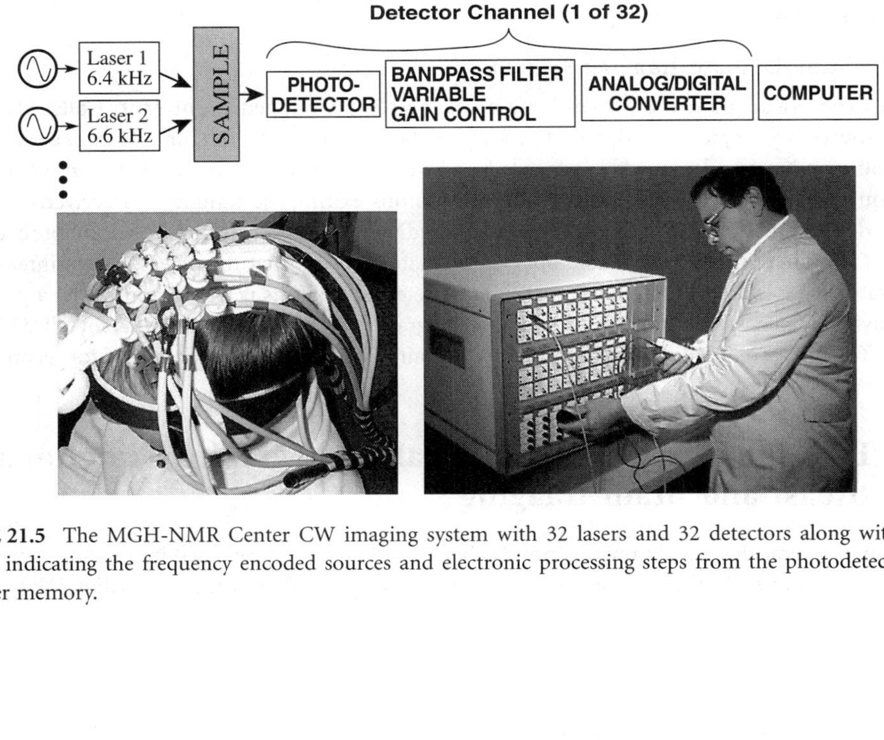


**FIGURE 21.5** The MGH-NMR Center CW imaging system with 32 lasers and 32 detectors along with a block diagram indicating the frequency encoded sources and electronic processing steps from the photodetector to the computer memory.

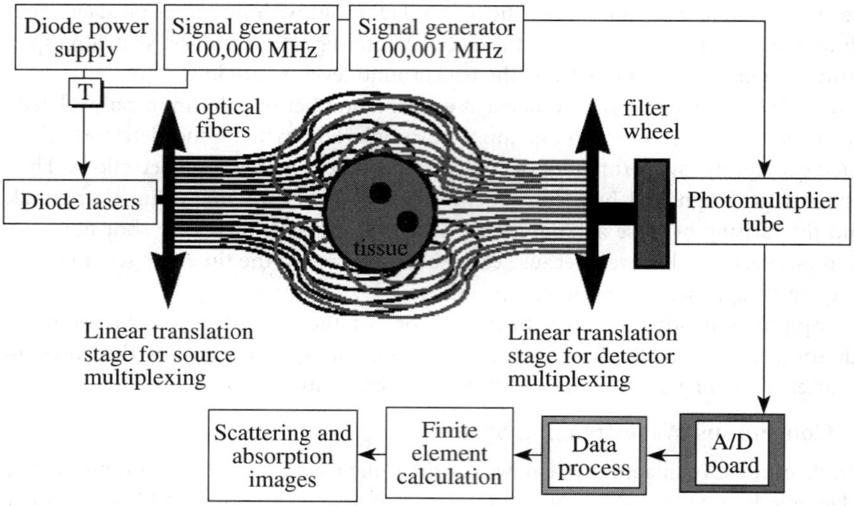


**FIGURE 21.6** Schematic of the frequency-domain imaging system developed at Dartmouth College. (From Pogue, B.W. et al., *Opt. Express*, 1, 391, 1997. With permission.)

### 21.3.1.2 Frequency-Domain Imaging System

A typical frequency-domain imaging system is illustrated in Figure 21.6. Developed at Dartmouth College,[215] this system modulates the intensity of a laser diode at 100.000 MHz and couples the light sequentially into different fiber optics (the SS-TDM encoding scheme). Diffusely remitted light is captured by detector fibers that are coupled into a single photomultiplier tube (PMT). The PMT is time-shared between the individual detector fibers, so the instrument employs a TDM encoding scheme. A filter wheel is positioned before the PMT to prevent saturation of the detector by stronger optical signals coming from closer to the source. The photo-electric signal within the PMT is modulated at 100.001 MHz to heterodyne the signal down to 1 kHz. The 1-kHz signal is acquired by an analog-to-digital converter, from which a computer calculates the relative amplitude and phase of the detected 100-MHz diffuse photon density wave.

### 21.3.1.3 Time-Domain Imaging System

Figure 21.7 shows a diagram of the time-domain image system developed at University College London.[216] The light source is a subpicosecond pulsed Ti:Sapphire laser with wavelength tunable from approximately 750 to 850 nm. Laser pulses are fiber coupled and fiber-optically multiplexed between 32 fibers that deliver source light to 32 independent positions. Photons exiting the sample are received by 32 large diameter detector fibers that relay the light to 32 individual time-correlated single photon counting channels. Computer controlled variable optical attenuators ensure that the detected light intensity does not saturate the photodetectors. The arrival time of detected photons is histogrammed by a picosecond time analyzer, which simultaneously produces histograms of the measured TPSF for all 32 detector channels. The full set of TPSFs for all source-detector pairs represents the raw data used for reconstructing images.

## 21.4 Experimental Diffuse Optical Tomography: Functional Breast and Brain Imaging

The imaging systems described previously and others like them have been used in phantom, animal, and human subject studies. In this section we provide a coherent snapshot of these activities. We begin with

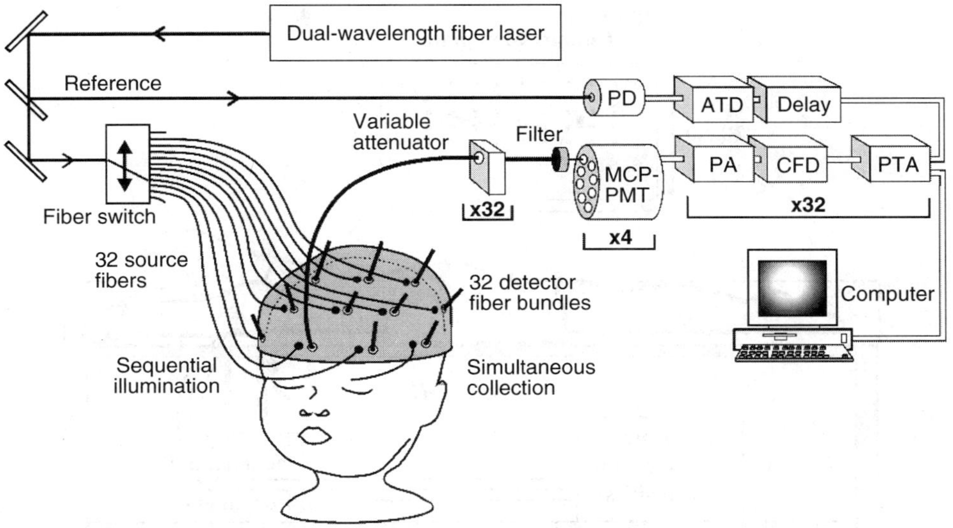


FIGURE 21.7 Schematic diagram of the time-domain imaging system. For clarity, only one detector fiber bundle is shown. PD = photo diode; PA = pre-amplifier; ATD = amplitude timing discriminator; CFD = constant fraction discriminator; PTA = picosecond time analyzer.

a tissue phantom experiment in order to illustrate image reconstruction on a controlled sample; the DOT reconstructions produce acceptable images by striking a balance between image noise, contrast, and resolution. Next we review the application to breast imaging, with recent clinical results. We describe measurements of normal breast optical properties, of tumors based on endogenous contrast, and of tumors using exogenous contrast agents to enhance tumor absorption characteristics preferentially.

Finally, we review the application to functional brain imaging. In this case we show images of blood flow, blood volume, and oxygen saturation changes in a rat stroke model. Then we show images that reveal localized variations in cerebral hemodynamics due to a sensory stimulation of a rat. Finally, we provide an example of functional brain imaging in adult humans with motor-sensory stimuli. The experiments presented are not exhaustive and do not represent the full range of results from the community. Nevertheless, the selected experiments indicate the promise of DOT for *in vivo* functional imaging.

### 21.4.1 Multiple Absorbers in a Slab Phantom

In this subsection we show how DOT is able to resolve multiple perturbations in the optical properties of a highly scattering medium. We also illustrate the trade-off between image resolution and the contrast-to-noise ratio. This type of experiment is important because it validates DOT techniques in well controlled samples.

The experimental data were collected using a hybrid CW and RF system developed at the University of Pennsylvania[217,218] (see Figure 21.8). Laser light is multiplexed to 45 positions (a 9 × 5 array) on one side of the slab phantom (i.e., parallel planes, see Figure 21.8). Nine detector fibers are interspersed within the 9 × 5 array to receive a 70 MHz frequency-domain signal, simultaneously, from which the amplitude and phase is determined by an IQ-homodyne demodulation.[219] This frequency-domain information was used to determine the background optical properties of the medium. A 16-bit CCD camera (Roper Scientific, NTE1340) was positioned to image the CW light transmitted through the phantom. The CCD camera had 800 × 1120 pixels, which were binned 24 × 24. The measurements were cropped to within a radius of 6 cm of the maximum transmitted signal producing approximately 250 independent measurements (i.e., a 21 × 13 detector array) per source position.

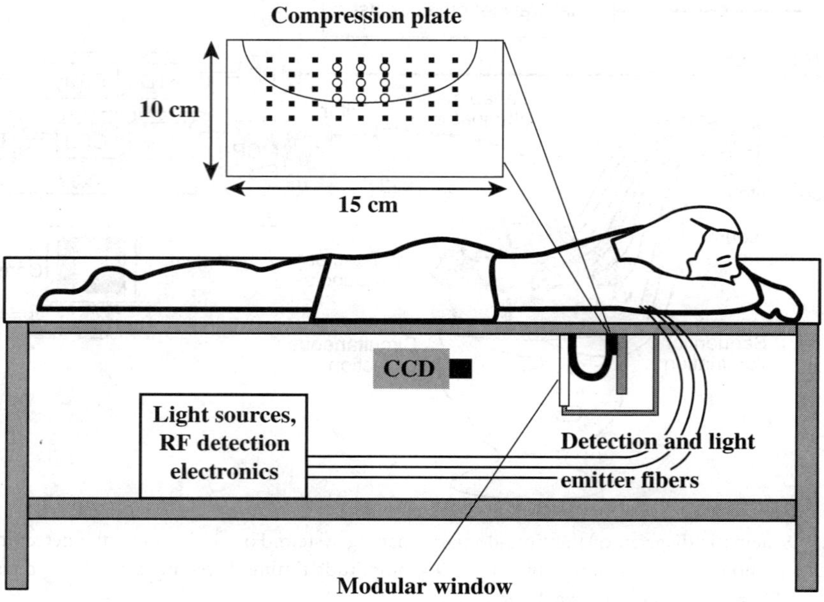


**FIGURE 21.8** Phantom geometry for the transillumination measurements. The array of 9 × 5 light sources is indicated in the compression plate by the black squares. The 3 × 3 array of frequency-domain photodetectors intermingled with the source fibers is indicated by the open circles. A CCD camera images the transmission of the CW light from each source position individually. A slightly modified version of this experimental system was used for the clinical measurements described in Section 21.4.1, with the CCD camera replaced by a scanning frequency-domain photodetector.

For the phantom experiment, the breast tank was filled with an Intralipid/ink solution (see Figure 21.8)[220,221] with $\mu_s' = 8$ cm$^{-1}$ and $\mu_a = 0.05$ cm$^{-1}$. In one experiment, two highly absorbing spheres ($\mu_a =$ 2 cm$^{-1}$ and $\mu_s' = 8$ cm$^{-1}$) were suspended in the Intralipid solution with a 5-cm separation; in another experiment many more spheres were suspended in the Intralipid solution. The image reconstruction algorithm was formulated using the Rytov approximation to the integral solution of the diffusion equation (see Section 21.2.5.2). A finite difference scheme was used to solve the forward problem in the rectangular geometry, and a conjugate gradient method was used to solve the inverse problem. Regularization was used for the inverse problem (see Section 21.2.5.3), and the entire scheme was iterative. Convergence was consistently obtained after 15 iterations.

An image of the two spheres was reconstructed with optimal regularization (see Figure 21.9). The three-dimensional image shows that two absorbers are easily resolved. The optimal regularization parameter was determined by examining the dependence of the regularization parameter on image norm, image variance, full-width at half-maximum (FWHM) of the imaged absorber, and measurement residual (see Figure 21.10). When the numerical value of the regularization parameter was increased, the image norm and image variance decreased, but at the expense of decreased image resolution (i.e., an increase in FWHM of the spheres) and increased measurement residual. A plot of the image contrast-to-noise ratio vs. the regularization parameter indicates that contrast-to-noise ratio is optimized with a regularization parameter that balances image noise and resolution (see arrows in Figure 21.10).[217] Similar trade-offs between image noise and resolution can be expected when imaging animals and human subjects. Culver et al.[217] also demonstrated images of many (i.e., >10) spheres in the same sample volume, thus indicating the potential of DOT for reconstructing multiple heterogeneities (as opposed to isolated heterogeneities in a homogenous background).

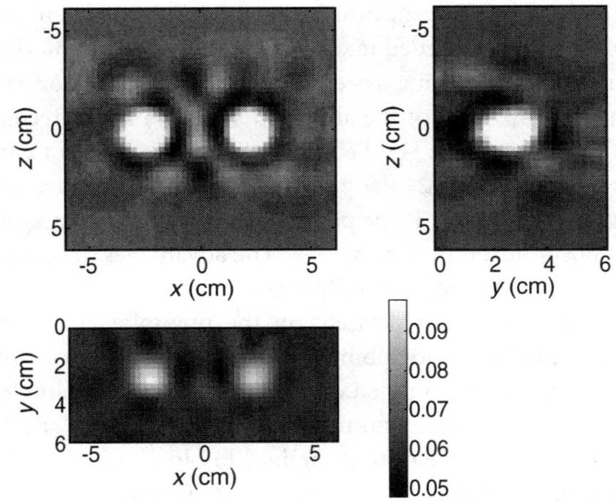


**FIGURE 21.9** Three slices of the three-dimensional absorption image reconstructed from the phantom data are shown. The *x-z* plane is parallel to the measurement planes of the phantom. The scale bar indicates the range of values for the absorption coefficient.

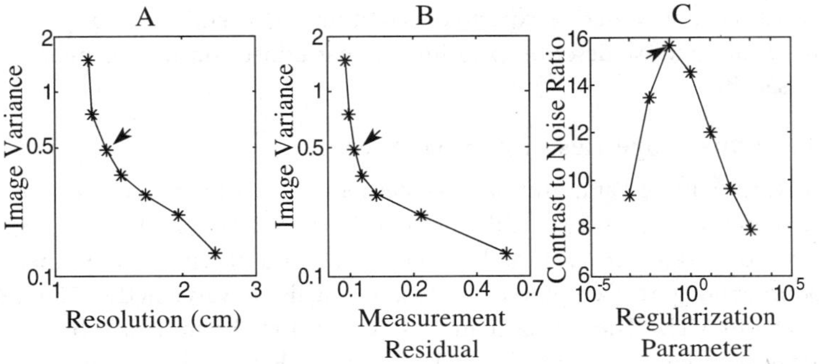


**FIGURE 21.10** The dependence of objective measures of image quality on the regularization parameter is shown. (A) Increasing the regularization parameter controls the trade-off between image noise and resolution. (B) Increasing the regularization parameter also increases the residual of the fit to the experimental data. (C) The optimal regularization parameter as determined by the maximum image contrast to noise ratio corresponds to the points in (A) and (B) that balance image noise and resolution/measurement residual (indicated by the arrows in each figure).

## 21.4.2 Breast Imaging

The use of light to detect tumors in the breast was first proposed by Cutler in 1929,[83] who hoped to be able to distinguish between solid tumors and cysts in the breast by illuminating the breast and shadowing the tumors. However, he found it difficult to produce the necessary light intensity without exposing the patient's skin to extreme heat, while the low resolution of the technique severely limited its clinical applications. With the advent of more modern optical sources and detectors, optical transillumination breast imaging was tried again in the 1970s and 1980s, but was abandoned because of lack of sensitivity and specificity.[93,222,223] The burst of activity in the 1990s can be traced to widespread acceptance of the diffusion approximation, which provides a tractable basis for tomographic approaches to image reconstruction using highly scattered light. Tomographic methods are crucial for recovery of information about heterogeneous breast tissues.

Functional, tomography-based diffuse optical breast imaging has been demonstrated, but further understanding and improvements are needed in order for it to become clinically useful. Current research is focusing on the construction of imaging systems with higher resolution and increased quantitative accuracy, as well as on the improvement of the algorithms used in image reconstruction. Thus far three general patient positions have been used: (1) the patient lies prone on a cot with a breast (or breasts) in a pendant position for imaging;[15,190–192] (2) the patient sits or stands with a breast held in compression similar to x-ray mammography;[10–13] and (3) the patient lies supine and is imaged with a hand-held probe that is moved to different positions on the breast.[119,193] The advantages of one approach over another for extraction of optical properties have not been fully explored.

Current research is also focusing more closely on the physiological information available to the technique. For example, quantitative hemoglobin images of the female breast showed localized increases in hemoglobin concentration that corresponded with biopsy-confirmed pathological abnormalities, suggesting that NIRS is capable of characterizing tumors as small as $0.8 \pm 0.1$ cm.[15] Other investigators have focused on the intrinsic sensitivity of the optical method to blood, water, and adipose — the principal components of the breast. Many of these researchers are studying the changes in breast optical properties associated with age, exogenous hormone levels, and menopausal status as well as, to a lesser extent, fluctuations in menstrual cycles.[14,224] Still other researchers are exploring the use of contrast agents[13,225,226] and the combination of DOT with other imaging techniques such as MRI and ultrasound.

In this section we briefly review some recent advances in the application of DOT to breast imaging. Our first example establishes the baseline optical properties of the normal breast; such information provides a useful benchmark about the requirements for detection of tumors based on endogenous (and exogenous) contrast. Then we describe experiments that image the endogeneous properties of the breast, revealing tumor signals, and we describe experiments that utilize contrast agents to create images with improved tumor sensitivity.

#### 21.4.2.1 Endogenous Properties of Normal Breast

Our group at the University of Pennsylvania has collected optical breast data with the subject in a prone position, using essentially the same system discussed in Reference 218 (see Figure 21.8). The goal of these measurements was to establish *in-vivo* the baseline optical and physiological properties of the normal breast. The experimental system employed diode lasers at three wavelengths (750, 786, and 830 nm), each modulated at 140 MHz. The source light was delivered to the moveable breast-stabilization plate through an optical fiber. For detection, the CCD camera in Figure 21.8 was not used; instead a fiber-coupled PMT was scanned to multiple positions on the breast with the source fiber fixed at the center of the tissue and scan region. Prior to collecting data, the tank surrounding the breast is filled with a scattering liquid whose optical properties closely approximated the breast tissue. The amplitude and phase of the transmitted DPDW was measured using a homodyne IQ demodulation scheme.[219] In 15 min, 153 measurements ($17 \times 9$) are obtained over an area of 10 cm $\times$ 7 cm $(x,y)$.

Measurements were collected on 52 healthy volunteers. The analysis had one new innovative feature: the data were fit simultaneously at all three wavelengths and "reconstructed" using spectral responses of oxy- and deoxyhemoglobin, and a Mie-based model for the wavelength dependence of the scattering coefficient. This spectroscopically self-consistent approach reduced crosstalk between scattering and absorption, overcoming some of the limitations of the homogeneous tissue model often used for characterization. The average absorption and scattering coefficients of the normal breast tissue were found to be $0.041 \pm 0.025$ cm$^{-1}$ and $8.5 \pm 2.1$ cm$^{-1}$, respectively, at 786 nm. The mean blood volume and blood oxygen saturation were found to be $34 \pm 9$ $\mu M$ and $68 \pm 8\%$, respectively.

A scatter plot of the blood volume and blood saturation results is shown in Figure 21.11. We see from this plot that the optical method will be sensitive to tumors with large blood volume and low blood saturation (outside the dashed lines); this condition might prevail in many tumors. A weak correlation of total hemoglobin concentration and scattering with body mass index (BMI) was found (i.e., decreasing with increasing BMI), but no statistically significant correlation with age was observed (20 to 65 years). This information provides insight into the types of intrinsic contrast available to optical breast imaging

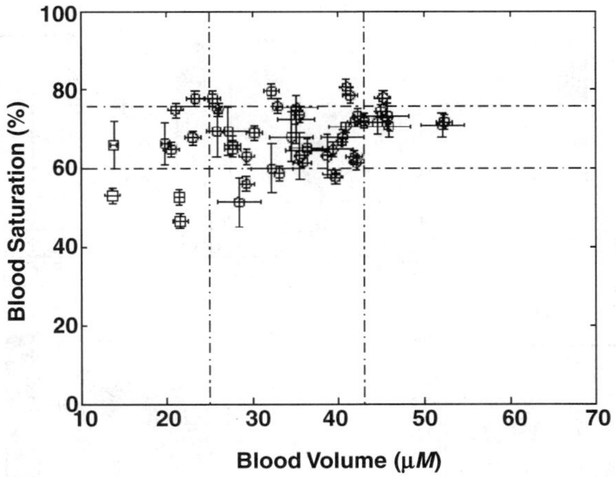


**FIGURE 21.11** Hemoglobin saturation vs. total hemoglobin concentration; the dashed lines indicate the ranges for normal tissue from the mean and standard deviation of the healthy breast tissue.

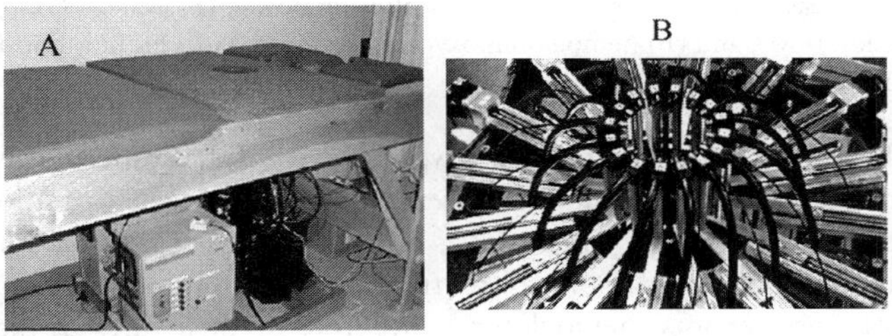


**FIGURE 21.12** (A) Patient bed for the optical breast imaging system developed at Dartmouth. (B) The fiber optic array for coupling light to and from the patient. The larger black cables are the detector fibers. The fibers are mounted on posts that translate radially for making contact with the tissue.

and, importantly, indicates the range of values that malignant tumors must have to be clearly distinguishable from the typical variation of normal breast tissue.

#### 21.4.2.2 Clinical Optical Images of Breast Lesions

The optical breast imaging system developed by the group at Dartmouth is shown in Figure 21.12.[15,191] It is a pendant system, similar in many respects to the one at the University of Pennsylvania, but it obtains a coronal image of one breast at a time with a ring of interspersed 16 source and 16 detector fibers. The fiber ring is vertically positioned by a translation stage to image different planes of interest. The laser diodes emit light at 761, 785, 808, and 826 nm, are modulated at 100 MHz, and are optically combined before being serially multiplexed to each of the 16 source fibers. The lasers are driven sequentially so that the signal at each of the 16 photodetectors distinguishes different wavelengths and positions by time-division multiplexing. At each of the 4 wavelengths, 256 independent measurements are obtained and then processed by a finite element program to reconstruct absorption and scattering images (as described in Pogue et al.[15] and McBride et al.[191]).

Figure 21.13 shows optical images for a 73-year-old female volunteer who had undergone routine mammography that demonstrated a 2.5-cm focal density with a larger ~6-cm-diameter area of associated architectural distortion. The imaged lesion corresponded to a palpable mass; needle biopsy subsequently diagnosed the mass as an invasive ductal carcinoma. The optical image was acquired 2 weeks after the biopsy and was aligned such that the coronal imaging plane was centered on the palpable lump. From

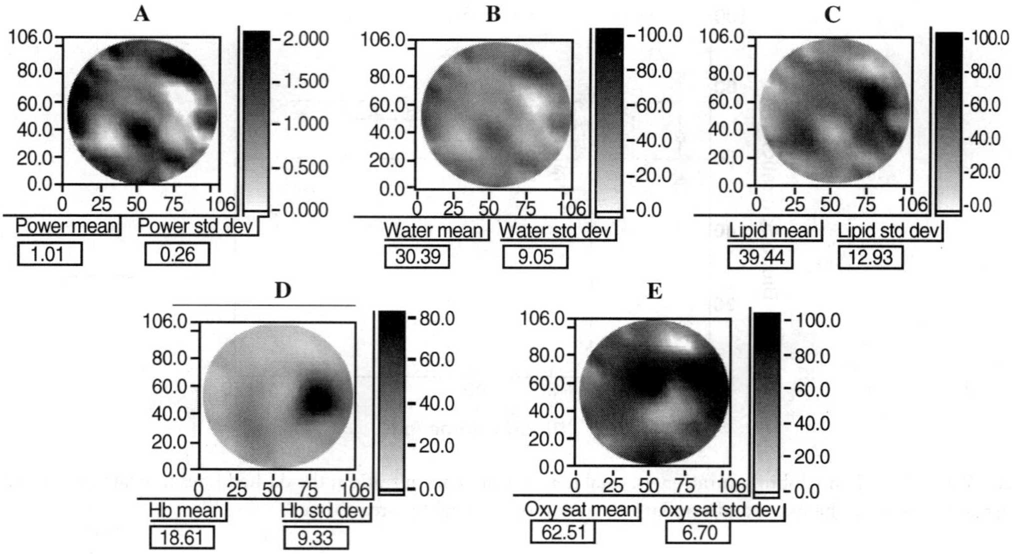


**FIGURE 21.13** Clinical breast images are shown of (A) scattering power, (B) water (%), (C) lipids (%), (D) total hemoglobin concentration ($\mu M$), and (E) hemoglobin oxygen saturation (%). (From McBride, T.O. et al., *J. Biomed. Opt.*, 7, 72, 2002. With permission.)

the scattering and absorption information, images were derived of the scatter power,[191,227] water concentration, lipid concentration, total hemoglobin concentration, and oxygen saturation. A lesion is clearly visible in the total hemoglobin image, which shows a threefold contrast-to-background ratio. In comparison to the variation of total hemoglobin concentration observed in normal tissue (Figure 21.11), the tumor has significantly greater total hemoglobin concentration. Structures in the other physiological images do not correspond well to the hemoglobin image; however, sensitivity to water and lipid percentages is expected to be weak because of the chosen wavelengths. The similarity between the scattering power, water, and lipid images suggests that water and lipid images are susceptible to crosstalk from the scattering power image.

### 21.4.2.3 Contrast Agents to Enhance Breast Lesion Detection

Another experiment at the University of Pennsylvania employed a time-domain imaging system in transmission mode to acquire *contrast enhanced* breast images while simultaneously obtaining magnetic resonance images. The time-domain optical imager has been described in detail in References 13 and 228. The instrument uses time-correlated single photon counting to measure the TPSF of photons diffusing through the breast tissue. The TPSF is then Fourier transformed to produce data at multiple modulation frequencies, which are then used to reconstruct differential absorption images[188] of the contrast agent ICG. The 830-nm pulsed laser source is coupled to 24 source fibers through a fiber optic switch. The detection collects light from eight positions simultaneously. Thus images are obtained using $24 \times 8$ measurements times the number of modulation frequencies. All source fibers are mounted on one plate forming a $3 \times 8$ array of fibers spaced by intervals of 1.25 cm. The detector fibers are mounted on the other plate to form a $2 \times 4$ array with a 2.5-cm separation. These two plates stabilize the breast and also contain the RF coils for the MRI. The MR studies were performed with a 1.5 T Signa, GE Medical Systems imager.

Figure 21.14 shows the results from a 70-year-old patient with an infiltrating ductal carcinoma about 1 cm in diameter. The pre-gadolinium (Gd) enhanced sagittal MR image slice passing through the carcinoma is shown in Figure 21.14A in gray scale, while the relative signal increase due to the Gd is

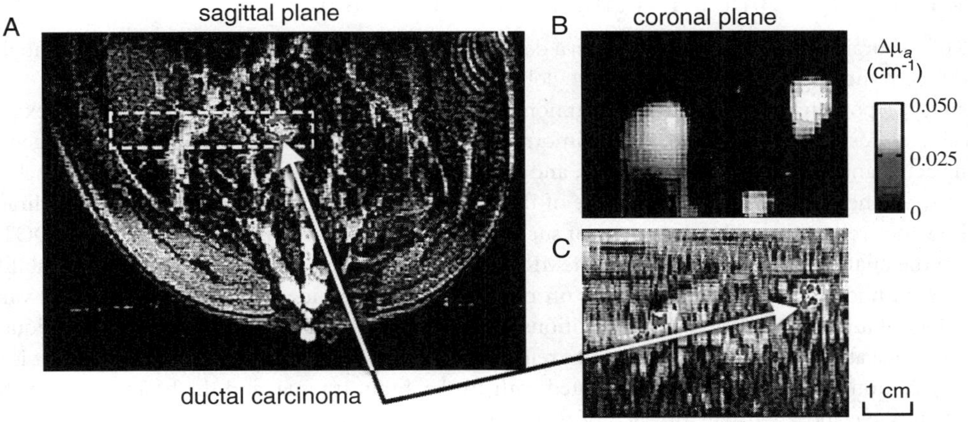


**FIGURE 21.14** **(Color figure follows p. 28-30.)** A dynamic MRI and contrast-enhanced DOT image of a ductal carcinoma. (A) A dynamic sagittal MR image after Gd contrast enhancement passing through the center of the malignant lesion. (B) The coronal DOT image, perpendicular to the plane of the MRI image in (A), but in the volume of interest indicated in (A) by the dashed-line box. (C) The dynamic MR coronal image resliced from the same volume of interest and same dimensions as (B).

shown in color. A rectangle surrounding the carcinoma indicates the sagittal cut of the region of interest that was reconstructed in the optical image (shown in Figure 21.14B) in the coronal plane, and the corresponding MR coronal slice shown in Figure 21.14C. A strong optical contrast-induced absorption increase is seen in the upper right of the optical image, congruent with the position of the carcinoma revealed in the coronal MR image. Another lesion is revealed in the left of the optical image, congruent with the Gd enhancement observed in the MR image, but with a different size and shape. The differences in size and shape can be explained by the low spatial resolution of this implementation of DOT and the chosen threshold level for displaying the images. Similar results were seen in other patients with malignant and benign lesions. A summary of tissue ICG absorption enhancement seen in a group of patients is shown in Figure 21.15. This summary suggests that malignant lesions enhance absorption more than benign lesions by a factor of two, and more than normal tissue by a factor of three to four.

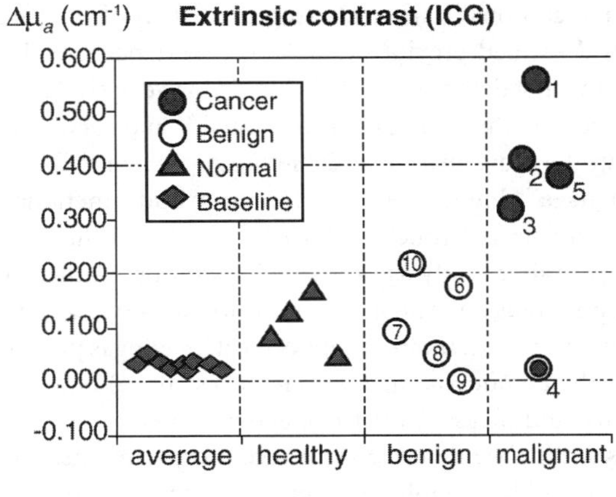


**FIGURE 21.15** The ICG-induced absorption enhancement in healthy tissue and benign and malignant lesions compared with the nonenhanced absorption coefficient of normal healthy tissue.

### 21.4.3 Diffuse Optical Imaging of Brain Function

DOT of the brain affords continuous, *in vivo* deep tissue[161,229] measurements of cerebral oxyhemoglobin (HbO), deoxyhemoglobin (Hb), total hemoglobin (HbT = HbO + Hb), and even blood flow.[197,230–232] It is therefore a potentially new and important noninvasive technique for bedside monitoring. For example, optical spectroscopy can provide crucial information about cerebral hemodynamics and oxygenation during acute and chronic brain conditions, and optical imaging enables the detection of brain ischemia, necrosis, and hemorrhage.[233,234] Knowledge of the most appropriate optical parameters to use clinically, and how they reflect the pathophysiology of such conditions, is still evolving. It is unlikely that DOT will achieve the anatomical resolution of CT or MRI, but its noninvasive nature, low cost, and capability to obtain continuous, real-time information on cerebral hemodynamics and oxygenation under various physiological and pathophysiological conditions constitute a major advantage over other techniques.

Knowledge about how optical parameters reflect tissue pathophysiology is still evolving, but it is clear that application areas for DOT are associated with stroke, brain trauma, and the basic science of brain activation. We elaborate briefly on each of these target areas next.

Diffuse optical tomography offers an attractive alternative for diagnosing and monitoring internal bleeding and ischemic stroke when CT and MRI are not available. There are several possible optical signal indications of stroke. Using a bolus of intravascular contrast agent, it is possible to monitor mean cortical transit times; transit time changes there can reveal flow variations associated with ischemic stroke. Quantitative differential spectroscopy can indicate alterations in hemoglobin saturation in affected tissue due to reduced perfusion or modulated metabolic demand. Bleeding stroke can also be detected spectroscopically, in this case via a large blood volume increase. Scattering contrasts are also possible. Edema and swelling may provide heterogeneities in images of tissue scattering.

The feasibility of near-infrared spectroscopy (NIRS) to detect brain hemorrhage in patients with head trauma has been demonstrated.[27] Studies that compare primitive diffuse optical measurements of brain hematomas with CT images clearly demonstrate a correlation between the optical signal and hematoma location.[26,27,235] These correlation studies have shown that subcortical hemorrhages as deep as 3 cm beneath the cortical surface and as small as 5 mm in diameter can be detected. Furthermore, the different classes of hemorrhages give characteristically different signals. These pilot NIRS results are promising, but they are based on simple relative NIRS measures. Quantitative images of hemoglobin saturation and fast hemodynamics will provide great improvement in available diagnostic information over the data involved in these pilot studies. A means to detect early hemorrhage in at-risk patients — i.e., those treated with thrombolysis or anticoagulants — would have tremendous value. In addition, swelling and edema may be detectable through alterations in the scattering coefficient of the tissue.

Beyond the clinical uses described previously, DOT is experiencing widespread application in functional brain imaging. Since the first demonstrations of noninvasive optical measures of brain function,[29–31] studies have been performed on adult humans using visual,[32] auditory,[33] and somatosensory stimuli.[34] In addition, a number of studies have investigated the motor system.[35–37] Other areas of scientific investigation have included language,[38] higher cognitive function, and functional studies of patient populations.[39–41] Following in the footsteps of functional MRI,[236,237] DOT is likely to play an important role in increasing our knowledge of brain activity associated with various stimulation paradigms, as well as our understanding of cerebral physiology, particularly the coupling between neuronal activity and the associated metabolic and vascular response. Interestingly, optical imaging is potentially the only neuroimaging modality that can measure the hemodynamic (see references above) and metabolic response[238–244] associated with neuronal activity, and measure neuronal activity directly.[245–247]

In the remainder of this section we describe three recent experiments. The first experiments measure flow, hemoglobin saturation, and hemoglobin concentration in a rat during stroke; this combination of parameters ultimately makes possible the assigment of oxygen metabolism changes to specific tissue volumes. The second and third sets of experiments image functional activation in the rat and in humans, with high spatial and temporal resolution.

### 21.4.3.1 Flow and Blood Oxygen Saturation Images of Rat Stroke

Experimenters at the University of Pennsylvania have used DOT to examine the spatial–temporal evolution of focal ischemia in a rat model.[248] Their measurements probe through the *intact* rat skull and combine "static" diffuse photon density wave measurements of Hb and HbO concentrations with "dynamic" diffuse correlation flowmetry (see Section 21.2.6). Their results are the first application of DOT correlation flowmetry to experimental stroke models.

Figure 21.16A shows the TTC stain of the infarct region of a focal ischemia induced in the rats by intraluminal suture occlusion of the middle cerebral artery. After 60 min of occlusion the suture was retracted for reperfusion. The animals recovered and, at 24 h after occlusion, were sacrificed for TTC staining. Differential images of the concentrations were reconstructed in a slice at a depth of 2 mm below the skull surface, extending 5 mm either side of midline and from 2 mm anterior of bregma to 8 mm posterior of bregma. The hemoglobin and flow images were obtained using the linear Rytov approach described in Sections 21.2.5.2 and 21.2.5.3.

Figure 21.16B shows image slices of total hemoglobin concentration ([HbT]), blood oxygen saturation $StO_2$, and relative changes in cerebral blood flow (rCBF) reconstructed with DOT. Note that the measurements cover the predominantly penumbral tissue and an equivalent tissue volume on the contralateral side. Image stacks were reconstructed from measurements averaged over five animals. Images are shown for time points representing baseline (–8 min), occlusion (+30 min), and reperfusion (+80 min). Regions of interest (ROIs) were defined for the contralateral and ipsilateral sides, consisting of $4 \times 8$ mm areas centered in the respective half of each image; the time traces for [Hb], $StO_2$, and (rCBF) in these regions are plotted in Figure 21.16C. Occlusion and saturation decreased by about 40% from baseline in the affected hemisphere. The numbers for flow are in reasonable agreement with near-surface laser Doppler measurements of penumbral tissues. The cerebral blood volume, on the other hand, showed much smaller percentage changes during occlusion.

These images demonstrate the feasibility of continuously imaging an integrated set of hemodynamic parameters through the time course of ischemia and reperfusion in experimental focal ischemia models. The combined measurements also offer the possibility to make quantitative maps of differential oxygen metabolism. A simplified model for oxygen metabolism relates two of the measurements made. If we assume that the product of the blood perfusion rate with the difference in oxyhemoglobin concentration between the artery perfusing the tissue and the vein draining the tissue equals the oxygen consumption rate, then the measured changes enable us to construct a map of local variations in cerebral oxygen metabolism in deep tissues.[197] This exciting prospect further enhances the attractiveness of the diffuse optical method.

### 21.4.3.2 Activation Imaging of Brain Function in a Rat Model

In this section we describe applications to functional activation of the rat brain. By using classical functional stimulation paradigms, these studies provide an opportunity to evaluate the imaging capabilities of diffuse optical tomography through comparisons with exposed cortex and fMRI studies.

These studies employed the CW system developed at the MGH-NMR center (see Section 21.3.1.1). The dual wavelength sources were positioned on a $3 \times 3$ grid interspersed within the $4 \times 4$ grid of detectors. In accordance with standard procedures used in exposed cortex studies of functional activation, the baseline hemoglobin concentrations were assumed — i.e., [Hb] = [HbO] = 50 $\mu M$.[249] Absorption coefficients were calculated using published spectra. The scattering coefficient was assumed to be equal to 10.0 $cm^{-1}$.

The experiments were performed on adult male Sprague-Dawley rats weighing 300 to 325 g. The rats were fasted overnight before measurements. The animals were anesthetized (Halothane 1 to 1.5%, $N_2O$ 70%, $O_2$ 30%) and catheters were placed into a femoral artery to monitor the arterial blood pressure and into a femoral vein for drug delivery. The animals were tracheotomized, mechanically ventilated, and fixed on a stereotaxic frame. The probe was then placed symmetrically about midline. It covered a region from 2 mm anterior to 6 mm posterior of the rhinal fissure. After the surgical procedures,

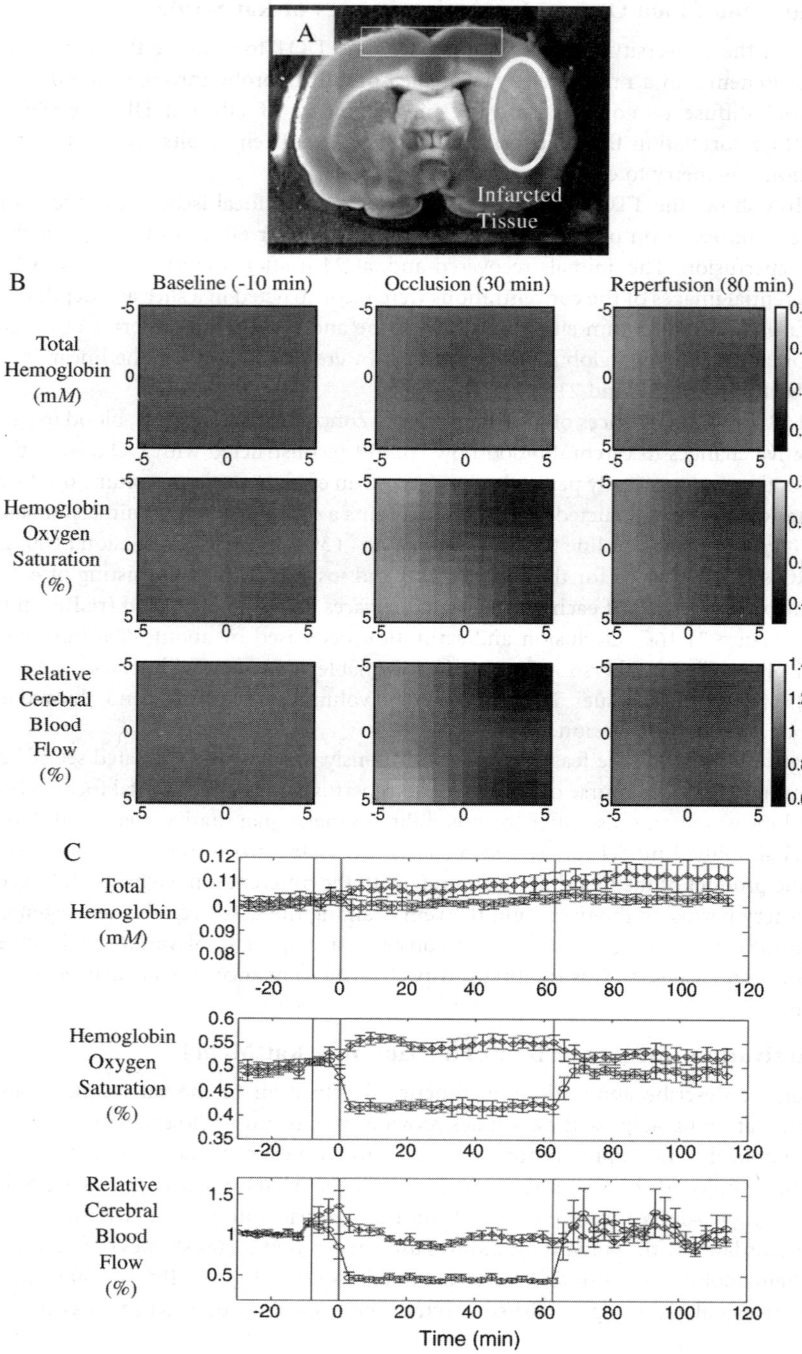


**FIGURE 21.16** (A) TTC staining of infarct area; rectangle indicates slice position of DOT image reconstruction. (B–C) Diffuse optical tomography images of total hemoglobin concentrations, tissue averaged hemoglobin saturation, and relative cerebral blood flow (rCBF). A middle cerebral artery occlusion was performed during the time from $t = -5$ min to $t = 0$. The suture was retracted at $t = 60$ min resulting in reperfusion. (B) Images at baseline, during occlusion (+30 min) and +80 min. The spatial dimensions are given in millimeters. The scale bars indicate the concentration of hemoglobin, the percent oxygen saturation, and the relative blood flow change where 1 is no change and 1.4 is a 40% change. (C) The time traces for ipsilateral and contralateral ROIs.

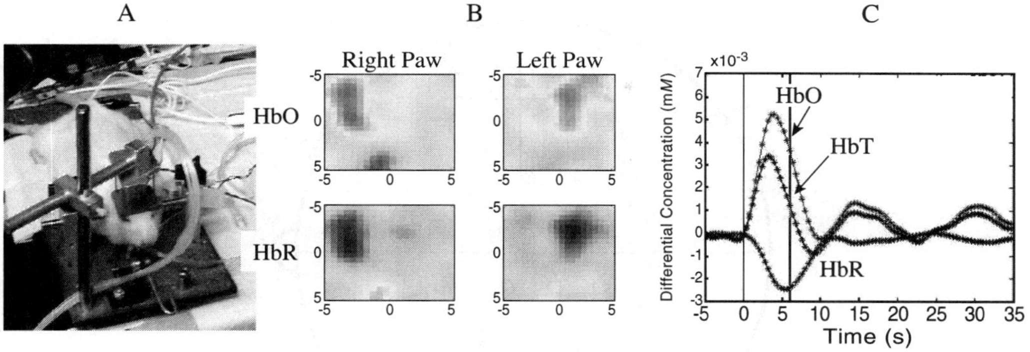


**FIGURE 21.17** (A) Photograph of the DOT fibers on the rat scalp. (B) DOT images of oxy- and deoxyhemoglobin concentrations during functional activation of the somatosensory cortex. The oxyhemoglobin (HbO) images exhibit a concentration increase. The deoxyhemoglobin (HbR) images exhibit a concentration decrease. Images are shown for left- and right-forepaw stimulation, the activation showing up on the contralateral side. (C) Time-course of the hemoglobin concentration changes in the region of interest defined by the significant focal concentration change seen in (B).

Halothane was discontinued, and anesthesia was maintained with a 50 mg/kg intravenous bolus of α-chloralose followed by continuous intravenous infusion at 40 mg/kg/hr.

Electrical forepaw stimulation was performed using two subdermal needle electrodes inserted into the dorsal forepaw. The stimulus pattern was relayed to an isolated, pulsed current supply to provide 300 $\mu_s$ constant current pulses at programmed times. The current was maintained at 1.0 mA. A 3 Hz, 6-s stimulus was provided with 54-s interstimulus interval. The measurements were then averaged over 42 stimulus intervals. The final averaged temporal data stack was reconstructed frame by frame using the methods described in Sections 21.2.5.2 and 21.2.5.3. Each slice was reconstructed at a depth of 2 mm from the skull surface extending 5 mm either side of midline, and from 2 mm anterior of bregma to 8 mm posterior of bregma. The absorption image stacks were then converted to oxy- and deoxyhemoglobin image stacks.

The series of DOT [Hb] and [HbO] images show focal activation contralateral to the stimulated forepaw. The images are frames taken every 5 s. The time traces were extracted for a 3 × 3 mm area centered at maximal activation (see Figure 21.17). The oscillations seen after stimulus are similar to vasomotion signals seen by optical studies in exposed cortex.[250]

### 21.4.3.3 Images of Brain Function in Humans

The same MGH CW imaging system described in Section 21.3.2.1 was used to study 15 subjects during finger tapping, finger tactile stimulation, and median nerve electrical stimulation. The finger tapping and finger tactile protocols consisted of series of 10 stimulation/rest sequences for each hand (i.e., 20-s stimulation and 20-s rest), wherein the stimulation occurs at a frequency of ~4 to 5 Hz (the stimulus frequency of each subject was adjusted so as to be anharmonic with the heart rate). The median nerve electrical stimulation protocol consisted of a series of 18 stimulation/rest sequences (i.e., 10-s stimulation and 20-s rest) with stimulus intensity slightly above the motor threshold (i.e., using rectangular electrical pulses, current peak: <10 mA, duration: 0.2 ms, repetition rate: 4 to 5 Hz). Motion sensors were used on the fingers of the subjects to synchronize the stimuli with the optical signals recorded on the head.

The optical data were band-pass filtered between 0.02 to 0.50 Hz to correct for slow drifts and to reduce the ~1-Hz arterial pulsation amplitude. Finally, the multiple stimulation sequences were block averaged to achieve better statistics. This resulted in a time series of the measured signal intensity for each source-detector pair. Source-detector pairs near a region of brain activation showed changes similar to those seen in a rat as shown in Figure 21.18, while the other source-detector pairs showed little signal variation. Absorption images at the different wavelengths were converted into images of changes in oxy- and deoxyhemoglobin concentrations.

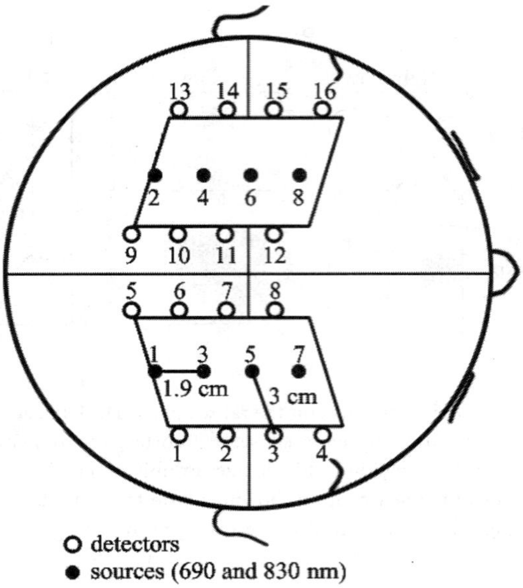


**FIGURE 21.18** Geometrical arrangement of source and detector fibers on the head.

The hemodynamic response to stimulation was always visible in the optical data. During the finger tapping experiment, the oxy- and deoxyhemoglobin concentration variations were not sharply localized. This is probably because of a systemic elevation in blood volume, presumably resulting from a corresponding heart rate increase. During finger tactile stimulation, oxyhemoglobin increases were two to three times smaller than during finger tapping; also, oxy- and deoxyhemoglobin were sharply localized. During electrical stimulation, the hemoglobin changes were sharply localized and had a magnitude comparable with finger tactile stimulation. In the latter experiments a peculiar feature of ipsilateral decrease in blood volume was observed during the stimulation period (see Figure 21.19). By contrast, magnetoencephalography (MEG) and functional magnetic resonance imaging (fMRI) studies during median nerve stimulation have shown an activation of only the contralateral primary sensory-motor cortex.[251,252] A deactivation of the ipsilateral primary sensory-motor cortex has been observed in previous fMRI experiments during finger tapping;[253] these optical results suggest that a similar decrease in blood flow occurs in the ipsilateral cortex during electrical somatosensory stimulation.

In summary, DOT offers an exciting new method for studying human brain function, and should play an important role in the brain sciences when other imaging modalities (e.g., fMRI, EEG, MEG) are not attractive because of cost, sensitivity to motion artifacts, or confinement of the research subject.

## 21.5 Fundamental and Practical Issues: Problems and Solutions

In this final section we touch on some of the fundamental and practical difficulties associated with DOT. In particular, we discuss the fundamental limits of detection, characterization, and resolution, we discuss practical problems of source-detector calibration, and we briefly outline some of the ways researchers are overcoming these barriers.

### 21.5.1 Detection, Localization, Characterization, and Resolution Limits

The detection limits of DOT are set by the smallest signal perturbation that can be detected above the noise level. The spatial localization limit is generally the same as the detection limit because the maximum signal perturbation often occurs when an object is located directly between a source and detector. Full characterization of object optical properties, shape, size, and position is more difficult than detection and

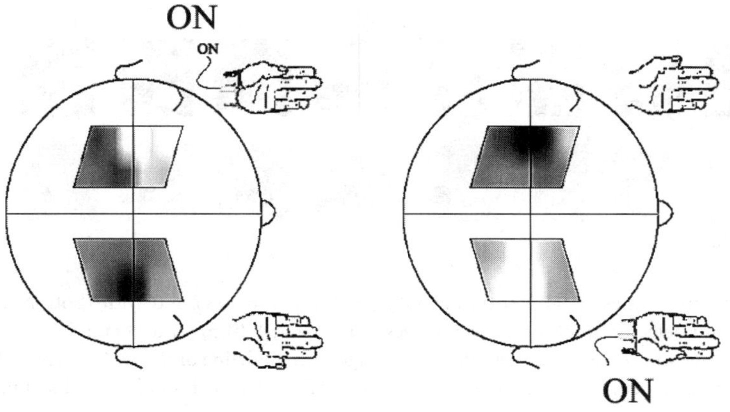


**FIGURE 21.19** Block average hemoglobin maps at the end of the electrical stimulation period. Top panels: deoxyhemoglobin changes; bottom panels: oxyhemoglobin changes. Left panels: left wrist median nerve stimulation; right panels: right wrist median nerve stimulation.

localization. For example, a fixed signal perturbation can be caused by a small, strongly absorbing object or a large, weakly absorbing object. The distinction between the large and small object must be derived from relative spatial differences in the perturbation of the measured fluence.

These limits have been extensively explored[254] for a transmission geometry applicable to breast imaging. In this work, a best-case scenario noise model assumed shot noise and small positional uncertainties gave rise to a random noise on the order of 0.1% of the signal intensity. Measurements were simulated in the transmission geometry with a single modulation frequency. With these assumptions about the noise, it was possible to detect 3-mm-diameter-absorbing objects that possessed a threefold greater absorption than the background. A similar analysis of the characterization limits indicated that simultaneous determination of object diameter and absorption coefficient required that object diameters were a minimum of 8 to 10 mm (for a 100% contrast); see Reference 254 for details. Another investigation,[255] focusing on resolution (defined as the FWHM of the fluence point-spread function caused by a localized perturbation,[255–257]) found that, if proper deconvolution of the measured data is performed, then resolutions of order 5 to 7 mm are possible with DOT.

There are potentially many ways to overcome some of these barriers, and many researchers are exploring the following possibilities, as well as other ideas. The brute force approach is to increase the number of measurements or decrease the measurement noise. This can be done, by increasing numbers of source-detector pairs, modulation frequencies, or optical wavelengths. The use of many optical wavelengths is beneficial because relatively few significant chromophores are in tissue and their spectra are well known. The use of prior spatial information, for example, the assignment of particular tissue types to specific volume elements of the image,[106,137,258] enables one to reduce the number of unknowns in the inverse problem and thus can effectively improve images. Optical techniques can also be combined with other imaging modalities (e.g., ultrasound[119,137,259] or MRI[13,260,261]) to constrain the DOT problem. Singular-value analysis of the tomographic weight matrix associated with specific data types, geometries, and optode arrangements has been developed recently[123] and should provide experimenters with quantitative tools to optimize for the spatial sampling interval, field-of-view parameters, resolution trade-offs, and, ultimately, physiology. Finally, it is now possible to reduce the systematic errors associated with source-detector amplitudes by directly incorporating these unknowns into the image reconstruction problem. This is the subject of the last section of this chapter.

### 21.5.2 Calibration of Source and Detector Amplitudes

The modeling of the DOT forward problem requires accurate knowledge of source and detector amplitudes and their positions. Systematic errors in the calibration of these parameters will result in absorption

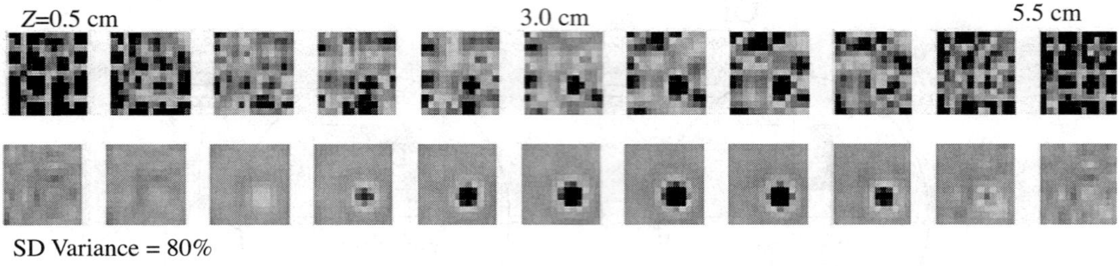


**FIGURE 21.20** A three-dimensional absorption image without (top row) and with (bottom row) modeling the unknown source and detector amplitudes. The images span $X$ and $Y$ from –3 to 3 cm; $Z$-slices are indicated from 0.5 to 5.5 cm. The absorbing object is observable in the image with modeling of the 80% source and detector variance; however, its contrast is obscured by the large amplitude variation near the boundaries. Modeling for the unknown variance provides an image with significantly improved image quality.

and scattering image artifacts, e.g., image spikes near the positions of the sources and detectors. This type of artifact has been observed by a number of groups.[262] Some schemes to minimize these artifacts include median filtering[263] and spatial regularization; the latter approach penalizes the reconstruction near sources and detectors with a weight that varies exponentially from the sample surfaces.[113,262,264] The result is an image with suppressed variations near the sources and detectors, but with improved image quality further from the boundaries. However, the scheme has the undesirable effect of biasing the reconstruction away from the boundaries of the medium.

A recent and relatively simple solution to this problem models the uncertainties of source and detector amplitude and includes them in the inverse problem.[265] Within the linear Rytov approximation, the unknown source and detector amplitudes can be solved simultaneously with the unknown optical properties of the medium. We briefly outline this procedure next, assuming for simplicity a sample with absorption heterogeneities, but without scattering heterogeneities.

Recall the linear approximation for calculating fluence perturbations from spatial variation in the absorption and scattering coefficients, discussed in Section 21.2.5.2. The problem was written in the form $[W]\{x\} = \{b\}$ where [W] represented the weight matrix, $\{x\}$ represented the unknown optical properties, and $\{b\}$ represented the difference between experiment and calculation (or the scattered wave). If we include the unknown source and detector amplitudes into the model, then the matrix equation can be reformulated as $[B]\{\xi\} = \{b\}$ where $[B] = [W' \; S \; D]$ and

$$\{\xi\} = \left\{ \frac{\delta\mu_{a,1}}{\mu_{ao}} \cdots \frac{\delta\mu_{a,NV}}{\mu_{ao}} \ln s_1 \;\; \cdots \;\; \ln s_{NS} \;\; \ln d_1 \;\; \cdots \;\; \ln d_{ND} \right\}. \tag{21.22}$$

Here $s_i$ and $d_i$ represent the amplitude of the $i$th source and detector, respectively, $NV$ is the number of voxels, $NS$ is the number of sources, and $ND$ is the number of detectors. Scaling $\delta\mu_{a,j}$ by $\mu_{ao}$ makes the elements dimensionless and of the same order as ln $s$ and ln $d$. $\mathbf{W}' = \mu_{ao}\mathbf{W}$ is the rescaling of the standard weight matrix. **S** and **D** are simple, well-defined matrices with block diagonal form that have 1 or 0 in the elements corresponding to particular sources and detectors.

Simulation results have demonstrated that incorporation of the unknown source and detector amplitudes into the inverse problem maintains image quality despite amplitude uncertainties greater than 50%. This is illustrated in Figure 21.20; a simulation of transmission through a 6-cm slab was considered with background optical properties of $\mu_{so}' = 10$ cm$^{-1}$ and $\mu_{ao} = 0.05$ cm$^{-1}$ and a 1.6-cm diameter absorbing sphere with $\mu_a = 0.15$ cm$^{-1}$, centered at $(x,y,z) = (1,1,3)$ cm. Measurements were made with 16 sources and 16 detectors equally spaced from –3 to 3 cm, and continuous-wave measurements were simulated. There was no additive measurement noise (i.e., shot or detector electronic noise) in the simulated data, only the multiplicative model error associated with the source and detector amplitudes.

## Acknowledgments

We have benefited from discussions with many colleagues and collaborators and many of their observations are found in this chapter. In particular, we thank Joe Culver, Maria Angela Franceschini, and Gary Boas for significant assistance in the preparation of this chapter. We also thank Turgut Durduran, Jeremy Hebden, Vasilis Ntziachristos, and Brian Pogue for discussion and for providing figures for the chapter. For other useful discussions over many years we thank Simon Arridge, Britton Chance, Regine Choe, Alper Corlu, Anders Dale, Joel Greenberg, Monica Holboke, Xingde Li, Eric Miller, Bruce Tromberg, Guoqiang Yu, and Tim Zhu.

D.A.B. acknowledges funding from Advanced Research Technologies, from National Institutes of Health grants R29-NS38842 and P41-RR14075, from the Center for Innovative Minimally Invasive Therapies, and from the U.S. Army under Cooperative Agreement DAMD17-99-2-9001. The material presented does not necessarily reflect the position or the policy of the government, and no official endorsement should be inferred. A.G.Y. acknowledges partial support from NIH grants 2-R01-CA-75124-04 and 2-RO1-HL-57835-04.

## References

1. Yodh, A. and Chance, B., Spectroscopy and imaging with diffusing light, *Phys. Today*, 48, 34, 1995.
2. Tromberg, B., Yodh, A., Sevick, E., and Pine, D., Diffusing photons in turbid media: introduction to the feature, *Appl. Opt.*, 36, 9, 1997.
3. Yodh, A., Tromberg, B., Sevick-Muraca, E., and Pine, D., Diffusing photons in turbid media, *J. Opt. Soc. Am. A*, 14, 136, 1997.
4. Boas, D.A., Brooks, D.H., Miller, E.L., DiMarzio, C.A., Kilmer, M., Gaudette, R.J., and Zhang, Q., Imaging the body with diffuse optical tomography, *IEEE Signal Process. Mag.*, 18, 57, 2001.
5. Miller, E., Focus issue: diffuse optical tomography — introduction, *Opt. Express*, 7, 461, 2000.
6. Chance, B., Near-infrared images using continuous, phase-modulated, and pulsed light with quantitation of blood and blood oxygenation, *Ann. N.Y. Acad. Sci.*, 838, 19, 1998.
7. Kang, K.A., Chance, B., Zhao, S., Srinivasan, S., Patterson, E., and Trouping, R., Breast tumor characterization using near-infrared spectroscopy, *Proc. SPIE*, 1888, 1993.
8. Suzuki, K., Yamashita, Y., Ohta, K., and Chance, B., Quantitative measurement of optical-parameters in the breast using time-resolved spectroscopy — phantom and preliminary *in-vivo* results, *Invest. Radiol.*, 29, 410, 1994.
9. Fishkin, J.B., Coquoz, O., Anderson, E.R., Brenner, M., and Tromberg, B., Frequency-domain photon migration measurements of normal and malignant tissue optical properties in a human subject, *Appl. Opt.*, 36, 10, 1997.
10. Grosenick, D., Wabnitz, H., Rinneberg, H.H., Moesta, K.T., and Schlag, P.M., Development of a time-domain optical mammograph and first *in vivo* applications, *Appl. Opt.*, 38, 2927, 1999.
11. Franceschini, M.A., Moesta, K.T., Fantini, S., Gaida, G., Gratton, E., Jess, H., Mantulin, W.W., Seeber, M., Schlag, P.M., and Kaschke, M., Frequency-domain techniques enhance optical mammography: initial clinical results, *Proc. Natl. Acad. Sci. U.S.A.*, 94, 6468, 1997.
12. Nioka, S., Miwa, M., Orel, S., Shnall, M., Haida, M., Zhao, S., and Chance, B., Optical imaging of human breast cancer, *Adv. Exp. Med. Biol.*, 361, 171, 1994.
13. Ntziachristos, V., Yodh, A.G., Schnall, M., and Chance, B., Concurrent MRI and diffuse optical tomography of breast after indocyanine green enhancement, *Proc. Natl. Acad. Sci. U.S.A.*, 97, 2767, 2000.
14. Cerussi, A.E., Berger, A.J., Bevilacqua, F., Shah, N., Jakubowski, D., Butler, J., Holcombe, R.F., and Tromberg, B.J., Sources of absorption and scattering contrast for near-infrared optical mammography, *Acad. Radiol.*, 8, 211, 2001.
15. Pogue, B.W., Poplack, S.P., McBride, T.O., Wells, W.A., Osterman, K.S., Osterberg, U.L., and Paulsen, K.D., Quantitative hemoglobin tomography with diffuse near-infrared spectroscopy: pilot results in the breast, *Radiology*, 218, 261, 2001.

16. McBride, T.O., Pogue, B.W., Gerety, E.D., Poplack, S.B., Osterberg, U.L., and Paulsen, K.D., Spectroscopic diffuse optical tomography for the quantitative assessment of hemoglobin concentration and oxygen saturation in breast tissue, *Appl. Opt.*, 38, 5480, 1999.
17. Delpy, D.T. and Cope, M., Quantification in tissue near-infrared spectroscopy, *Philos. Trans. R. Soc. London B*, 352, 649, 1997.
18. Painchaud, Y., Mailloux, A., Harvey, E., Verreault, S., Frechette, J., Gilbert, C., Vernon, M.L., and Beaudry, P., Multi-port time-domain laser mammography: results on solid phantom and volunteers, *Int. Symp. Biomed. Opt.*, 3597, 548, 1999.
19. Tromberg, B., Coquoz, O., Fishkin, J., Pham, T., Anderson, E.R., Butler, J., Cahn, M., Gross, J.D., Venugopalan, V., and Pham, D., Non-invasive measurements of breast tissue optical properties using frequency-domain photon migration, *Phil. Trans. R. Soc. London B*, 352, 661, 1997.
20. Sickles, E.A., Breast cancer detection with transillumination and mammography, *Am. J. Roentgenol.*, 142, 841, 1984.
21. Hoogenraad, J.H., van der Mark, M.B., Colak, S.B., Hooft, G.W., and van der Linden, E.S., First results from the Phillips Optical Mammoscope, in *Photon Propagation of Tissues III*, 31294, Benaron, D.A., Chance, B., and Ferrari, M., Eds., Plenum Press, New York, 1997, p. 184.
22. Ntziachristos, V., Ma, X.H., and Chance, B., Time-correlated single photon counting imager for simultaneous magnetic resonance and near-infrared mammography, *Rev. Sci. Instrum.*, 69, 4221, 1998.
23. Ntziachristos, V., Yodh, A.G., Schnall, M., and Chance, B., Comparison between intrinsic and extrinsic contrast for malignancy detection using NIR mammography, *Proc. SPIE*, 3597, 565, 1999.
24. Pogue, B.W., Poplack, S.D., McBride, T.O., Jiang, S., Osterberg, U.L., and Paulsen, K.D., Breast tissue and tumor hemoglobin and oxygen saturation imaging with multi-spectral near infrared computed tomography, in *Advances in Experimental Medicine and Biology Series*, Plenum Press, New York, 2001.
25. Pogue, B.W., Poplack, S.D., McBride, T.O., Jiang, S., Osterberg, U.L., and Paulsen, K.D., Near-infrared tomography: status of Dartmouth imaging studies and future directions, in progress.
26. Gopinath, S.P., Robertson, C.S., Grossman, R.G., and Chance, B., Near-infrared spectroscopic localization of intracranial hematomas, *J. Neurosurg.*, 79, 43, 1993.
27. Robertson, C.S., Gopinath, S.P., and Chance, B., A new application for near-infrared spectroscopy: detection of delayed intracranial hematomas after head injury, *J. Neurotrauma*, 12, 591, 1995.
28. Hintz, S.R., Cheong, W.F., Van Houten, J. P., Stevenson, D.K., and Benaron, D.A., Bedside imaging of intracranial hemorrhage in the neonate using light: comparison with ultrasound, computed tomography, and magnetic resonance imaging, *Pediatr. Res.*, 45, 54, 1999.
29. Hoshi, Y. and Tamura, M., Detection of dynamic changes in cerebral oxygenation coupled to neuronal function during mental work in man, *Neurosci. Lett.*, 150, 5, 1993.
30. Villringer, A., Planck, J., Hock, C., Schleinkofer, L., and Dirnagl, U., Near infrared spectroscopy (NIRS): a new tool to study hemodynamic changes during activation of brain function in human adults, *Neurosci. Lett.*, 154, 101, 1993.
31. Okada, F., Tokumitsu, Y., Hoshi, Y., and Tamura, M., Gender- and handedness-related differences of forebrain oxygenation and hemodynamics, *Brain Res.*, 601, 337, 1993.
32. Ruben, J., Wenzel, R., Obrig, H., Villringer, K., Bernarding, J., Hirth, C., Heekeren, H., Dirnagl, U., and Villringer, A., Haemoglobin oxygenation changes during visual stimulation in the occipital cortex, *Adv. Exp. Med. Biol.*, 428, 181, 1997.
33. Sakatani, K., Chen, S., Lichty, W., Zuo, H., and Wang, Y.P., Cerebral blood oxygenation changes induced by auditory stimulation in newborn infants measured by near infrared spectroscopy, *Early Hum. Dev.*, 55, 229, 1999.
34. Obrig, H., Wolf, T., Doge, C., Hulsing, J.J., Dirnagl, U., and Villringer, A., Cerebral oxygenation changes during motor and somatosensory stimulation in humans, as measured by near-infrared spectroscopy, *Adv. Exp. Med. Biol.*, 388, 219, 1996.

35. Colier, W.N., Quaresima, V., Oeseburg, B., and Ferrari, M., Human motor-cortex oxygenation changes induced by cyclic coupled movements of hand and foot, *Exp. Brain Res.*, 129, 457, 1999.
36. Kleinschmidt, A., Obrig, H., Requardt, M., Merboldt, K.D., Dirnagl, U., Villringer, A., and Frahm, J., Simultaneous recording of cerebral blood oxygenation changes during human brain activation by magnetic resonance imaging and near-infrared spectroscopy, *J. Cereb. Blood Flow Metab.*, 16, 817, 1996.
37. Hirth, C., Obrig, H., Villringer, K., Thiel, A., Bernarding, J., Muhlnickel, W., Flor, H., Dirnagl, U., and Villringer, A., Non-invasive functional mapping of the human motor cortex using near-infrared spectroscopy, *Neuroreport*, 7, 1977, 1996.
38. Sato, H., Takeuchi, T., and Sakai, K.L., Temporal cortex activation during speech recognition: an optical topography study, *Cognition*, 73, B55, 1999.
39. Hock, C., Villringer, K., Muller-Spahn, F., Wenzel, R., Heekeren, H., Schuh-Hofer, S., Hofmann, M., Minoshima, S., Schwaiger, M., Dirnagl, U., and Villringer, A., Decrease in parietal cerebral hemoglobin oxygenation during performance of a verbal fluency task in patients with Alzheimer's disease monitored by means of near-infrared spectroscopy (NIRS) — correlation with simultaneous rCBF-PET measurements, *Brain Res.*, 755, 293, 1997.
40. Hock, C., Villringer, K., Heekeren, H., Hofmann, M., Wenzel, R., Villringer, A., and Muller-Spahn, F., A role for near infrared spectroscopy in psychiatry? *Adv. Exp. Med. Biol.*, 413, 105, 1997.
41. Fallgatter, A.J. and Strik, W.K., Reduced frontal functional asymmetry in schizophrenia during a cued continuous performance test assessed with near-infrared spectroscopy, *Schizophr. Bull.*, 26, 913, 2000.
42. Bank, W. and Chance, B., Diagnosis of mitochondrial disease by NIRS, *Proc. SPIE,* 2383, 1995.
43. Bank, W. and Chance, B., An oxidative defect in metabolic myopathies: diagnosed by non-invasive tissue oximetry, *Ann. Neurol.*, 36, 830, 1994.
44. Chance, B. and Bank, W., Genetic disease of mitochondrial function evaluated by NMR and NIR spectroscopy of skeletal tissue, *Biochim. Biophys. Acta*, 1271, 7, 1995.
45. Nioka, S., Moser, D., Lech, G., Evengelisti, M., Verde, T., Chance, B., and Kuno, S., Muscle deoxygenation in aerobic and anaerobic exercise, *Adv. Exp. Med. Biol.*, 454, 63, 1998.
46. Nioka, S., Chance, B., and Nakayama, K., Possibility of monitoring mitochondrial activity in isometric exercise using NIRS, in *Oxygen Transport to Tissue*, Chance, B., Ed., Plenum Press, New York, 1998, p. 454.
47. Dougherty, T.J., Gomer, C.J., Henderson, B.W., Jori, G., Kessel, D., Korbelik, M., Moan, J., and Peng, Q., Photodynamic therapy, *J. Nat. Cancer Inst.*, 90, 889, 1998.
48. Patterson, M.S., Wilson, B.C., and Graff, R., *In vivo* tests of the concept of photodynamic threshold dose in normal rat liver photosensitized by aluminum chlorosulphonated phthalocyanine, *Photochem. Photobiol.*, 51, 343, 1990.
49. van Gemert, J.C., Berenbaum, M.C., and Gijsberg, G.H.M., Wavelength and light dose dependence in tumor phototherapy with haematoporphyrin derivative, *Br. J. Cancer*, 52, 43, 1985.
50. Farrell, T.J., Wilson, B.C., Patterson, M.S., and Olivio, M.C., Comparison of the *in vivo* photodynamic threshold dose for photofrin, mono- and tetrasulfonated aluminum phthalocyanine using a rat liver model, *Photochem. Photobiol.*, 68, 394, 1998.
51. Lilge, L., Olivo, M.C., Shatz, S.W., MaGuire, J.A., Patterson, M.S., and Wilson, B.C., The sensitivity of the normal brain and intracranially implanted VX2 tumour to intestinal photodynamic therapy, *Br. J. Cancer*, 73, 332, 1996.
52. Johnson, C.C., Optical diffusion in blood, *IEEE Trans. Biomed. Eng.*, 129, 1970.
53. Ishimaru, A., *Wave Propagation and Scattering in Random Media*, Academic Press, San Diego, 1978.
54. Furutsu, K., On the diffusion equation derived from the space-time transport equation, *J. Opt. Soc. Am. A*, 70, 360, 1980.
55. Groenhuis, R.A.J., Ferwerda, H.A., and Ten Bosch, J.J., Scattering and absorption of turbid materials determined from reflection measurements. I. Theory, *Appl. Opt.*, 22, 2456, 1983.

56. Patterson, M.S., Chance, B., and Wilson, B.C., Time resolved reflectance and transmittance for the noninvasive measurement of tissue optical properties, *Appl. Opt.*, 28, 2331, 1989.
57. Durduran, T., Chance, B., Yodh, A.G., and Boas, D.A., Does the photon diffusion coefficient depend on absorption? *J. Opt. Soc. Am. A*, 14, 3358, 1997.
58. Case, K.M. and Zweifel, P.F., *Linear Transport Theory*, Addison-Wesley, Boston, 1967.
59. Furutsu, K. and Yamada, Y., Diffusion approximation for a dissipative random medium and the applications, *Phys. Rev. E*, 50, 3634, 1994.
60. Furutsu, K., Pulse wave scattering by an absorber and integrated attenuation in the diffusion approximation, *J. Opt. Soc. Am.*, 14, 267, 1997.
61. Aronson, R. and Corngold, N., Photon diffusion coefficient in an absorbing medium, *J. Opt. Soc. Am. A*, 16, 1066, 1999.
62. Durian, D.J., The diffusion coefficient depends on absorption, *Opt. Lett.*, 23, 1502, 1998.
63. Glasstone, S. and Edlund, M.C., *The Elements of Nuclear Reactor Theory*, Van Nostrand, New York, 1952.
64. Star, W.M., Marijnissen, J.P.A., and van Gemert, M.J.C., Light dosimetry in optical phantoms and in tissues: I. Multiple flux and transport theory, *Phys. Med. Biol.*, 33, 437, 1988.
65. Davidson, B., *Neutron Transport Theory*, Clarendon, Oxford, 1957.
66. Duderstadt, J.J. and Martin, W.R., *Transport Theory*, Wiley, New York, 1979.
67. Graaff, R. and Ten Bosch, J.J., Diffusion coefficient in photon density theory, *Opt. Lett.*, 25, 43, 2000.
68. Martelli, F., Bassani, M., Alianelli, L., Zangheri, L., and Zaccanti, G., Accuracy of the diffusion equation to describe photon migration through an infinite medium: numerical and experimental investigation, *Phys. Med. Biol.*, 45, 2235, 2000.
69. Boas, D., Diffuse photon probes of structural and dynamical properties of turbid media: theory and biomedical applications, Ph.D. dissertation, University of Pennsylvania, Philadelphia, 1996.
70. Li, X., Fluorescence and diffusive wave diffraction tomographic probes in turbid media, Ph.D. dissertation, University of Pennsylvania, Philadelphia, 1998.
71. Wilson, B.C. and Patterson, M.S., The physics of photodynamic therapy, *Phys. Med. Biol.*, 31, 327, 1986.
72. Kaplan, P.D., Kao, M.H., Yodh, A.G., and Pine, D.J., Geometric constraints for the design of diffusing-wave spectroscopy experiments, *Appl. Opt.*, 32, 3828, 1993.
73. Kaplan, P.D., Optical studies of the structure and dynamics of opaque colloids, Ph.D. dissertation, University of Pennsylvania, Philadelphia, 1992.
74. Barbieri, B., Piccoli, F.D., van de Ven, M., and Gratton, E., What determines the uncertainty of phase and modulation measurements in frequency domain fluorometry? *SPIE Time Resolved Laser Spectrosc. Biochem. II*, 1204, 158, 1990.
75. Fishkin, J.B. and Gratton, E., Propagation of photon density waves in strongly scattering media containing an absorbing "semi-infinite" plane bounded by a straight edge, *J. Opt. Soc. Am. A*, 10, 127, 1993.
76. O'Leary, M.A., Boas, D.A., Chance, B., and Yodh, A.G., Refraction of diffuse photon density waves, *Phys. Rev. Lett.*, 69, 2658, 1992.
77. Boas, D.A., Oleary, M.A., Chance, B., and Yodh, A.G., Scattering and wavelength transduction of diffuse photon density waves, *Phys. Rev. E*, 47, R2999, 1993.
78. Tromberg, B.J., Svaasand, L.O., Tsay, T., and Haskell, R.C., Properties of photon density waves in multiple-scattering media, *Appl. Opt.*, 32, 607, 1993.
79. Farrell, T.J., Patterson, M.S., and Wilson, B., A diffusion theory model of spatially resolved, steady state diffuse reflectance for the noninvasive determination of tissue optical properties *in vivo*, *Med. Phys.*, 19, 879, 1992.
80. Hull, E.L., Nichols, M.G., and Foster, T.H., Quantitative broadband near-infrared spectroscopy of tissue-simulating phantoms containing erthrocytes, *Phys. Med. Biol.*, 43, 2281, 1998.
81. Solonenko, M., Cheung, R., Busch, T.M., Kachur, A., Griffin, G.M., Vulcan, T., Zhu, T.C., Wang, H.-W., Hahn, S.M., and Yodh, A.G., *In vivo* reflectance measurement of optical properties, blood oxygenation and motexafin lutetium uptake in canine large bowel, kidneys and prostates, *Phys. Med. Biol.*, 47, 1, 2002.

82. Fantini, S., Franceschini, M.A., Maier, J.S., Walker, S., and Gratton, E., Frequency domain multi-source optical spectrometer and oximeter, *Proc. SPIE*, 2326, 108, 1994.
83. Cutler, M., Transillumination of the breast, *Surg. Gynecol. Obstet.*, 48, 721, 1929.
84. Watmough, D.J., Transillumination of breast tissues: factors governing optimal imaging of lesions, *Radiology*, 147, 89, 1983.
85. Sickles, E.A., Periodic mammographic follow-up of probably benign lesions: results in 3184 consecutive cases, *Radiology*, 179, 463, 1991.
86. Wallberg, H., Alveryd, A., Bergvall, U., Nasiell, K., Sundelin, P., and Troel, S., Diaphanography in breast carcinoma, *Acta Radiol. Diagn.*, 26, 33, 1985.
87. Profio, A.E. and Navarro, G.A., Scientific basis of breast diaphanography, *Med. Phys.*, 16, 60, 1989.
88. Pera, A. and Freimanis, A.K., The choice of radiologic procedures in the diagnosis of breast disease, *Obstet. Gynecol. Clin. N. Am.*, 14, 635, 1987.
89. Homer, M.J., Breast imaging: pitfalls, controversies, and some practical thoughts, *Radiol. Clin. North Am.*, 14, 635, 1985.
90. Gisvold, J.J., Brown, L.R., Swee, R.G., Raygor, D.J., Dickerson, N., and Ranfranz, M.K., Comparison of mammography and transillumination light scanning in the detection of breast lesions, *Am. J. Roentgenol.*, 147, 191, 1986.
91. Bartrum, R.J.J. and Crow, H.C., Transillumination lightscanning to diagnose breast cancer: a feasibility study, *Am. J. Roentgenol.*, 142, 409, 1984.
92. Merrit, C.R.B., Sullivan, M.A., and Segaloff, A., Real time transillumination lightscanning of the breast, *Radiographics*, 4, 989, 1984.
93. Carlsen, E.N., Transillumination light scanning, *Diagn. Imaging*, 4, 28, 1982.
94. Marshall, V., Williams, D.C., and Smith, K.D., Diaphanography as a means of detecting breast cancer, *Radiology*, 150, 339, 1984.
95. Monsees, B., Destouet, J.M., and Totty, W.G., Light scanning versus mammography in breast cancer detection, *Radiology*, 163, 463, 1987.
96. Walker, S.A., Fantini, S., and Gratton, E., Image reconstruction by back-projection from frequency-domain optical measurements in highly scattering media, *Appl. Opt.*, 36, 170, 1997.
97. Colak, S.B., Papaioannou, D.G., 't Hooft, G.W., van der Mark, M.B., Schomberg, H., Paasschens, J.C.J., Melissen, J.B.M., and Van Asten, N.A.A.J., Tomographic image reconstruction from optical projections in light-diffusing media, *Appl. Opt.*, 36, 180, 1997.
98. Schotland, J.C., Continuous-wave diffusion imaging, *J. Opt. Soc. Am.*, 14, 275, 1997.
99. Li, X.D., Durduran, T., Yodh, A.G., Chance, B., and Pattanayak, D.N., Diffraction tomography for biochemical imaging with diffuse-photon density waves, *Opt. Lett.*, 22, 573, 1997.
100. Cheng, X. and Boas, D.A., Diffuse optical reflectance tomography with continuous-wave illumination, *Opt. Express*, 3, 118, 1998.
101. Matson, C.L. and Liu, H.L., Analysis of the forward problem with diffuse photon density waves in turbid media by use of a diffraction tomography model, *J. Opt. Soc. Am. A*, 16, 455, 1999.
102. Schotland, J.C., Haselgrove, J.C., and Leigh, J.S., Photon hitting density, *Appl. Opt.*, 32, 448, 1993.
103. O'Leary, M.A., Boas, D.A., Chance, B., and Yodh, A.G., Experimental images of heterogeneous turbid media by frequency-domain diffusing-photon tomography, *Opt. Lett.*, 20, 426, 1995.
104. Arridge, S.R., Photo-measurement density functions. Part 1: analytical forms, *Appl. Opt.*, 34, 7395, 1995.
105. Arridge, S.R. and Schweiger, M., Photon-measurement density-functions. 2. Finite-element-method calculations, *Appl. Opt.*, 34, 8026, 1995.
106. Barbour, R.L., Graber, H.L., Chang, J., Barbour, S.S., Koo, P.C., and Aronson, R., MRI-guided optical tomography: prospects and computation for a new imaging method, *IEEE Comput. Sci. Eng.*, 2, 63, 1995.
107. Yao, Y., Wang, Y., Pei, Y., Zhu, W., and Barbour, R.L., Frequency-domain optical imaging of absorption and scattering distributions using a Born iterative method, *J. Opt. Soc. Am. A*, 14, 325, 1997.

108. Paulsen, K.D. and Jiang, H., Spatially varying optical property reconstruction using a finite element diffusion equation approximation, *Med. Phys.*, 22, 691, 1995.
109. Jiang, H., Paulsen, K.D., Osterberg, U.L., Pogue, B.W., and Patterson, M.S., Optical image reconstruction using frequency-domain data: simulations and experiments, *J. Opt. Soc. Am. A*, 13, 253, 1996.
110. Paulsen, K.D. and Jiang, H., Enhanced frequency-domain optical image reconstruction in tissues through total variation minimization, *Appl. Opt.*, 35, 3447, 1996.
111. Jiang, H.B., Paulsen, K.D., Osterberg, U.L., and Patterson, M.S., Frequency-domain optical image reconstruction in turbid media: an experimental study of single-target detectability, *Appl. Opt.*, 36, 52, 1997.
112. Jiang, H., Paulsen, K.D., Osterberg, U.L., and Patterson, M.S., Improved continuous light diffusion imaging in single- and multi-target tissue-like phantoms, *Phys. Med. Biol.*, 43, 675, 1998.
113. Pogue, B.W., McBride, T.O., Prewitt, J., Osterberg, U.L., and Paulsen, K.D., Spatially variant regularization improves diffuse optical tomography, *Appl. Opt.*, 38, 2950, 1999.
114. Arridge, S.R. and Schweiger, M., A gradient-based optimisation scheme for optical tomography, *Opt. Express*, 2, 213, 1998.
115. Klibanov, M.V., Lucas, T.R., and Frank, R.M., A fast and accurate imaging algorithm in optical/diffusion tomography, *Inverse Probl.*, 13, 1341, 1997.
116. Gryazin, Y.A., Klibanov, M.V., and Lucas, T.R., Imaging the diffusion coefficient in a parabolic inverse problem in optical tomography, *Inverse Probl.*, 1, 373, 1999.
117. Eppstein, M.J., Dougherty, D.E., Troy, T.L., and Sevick-Muraca, E.M., Biomedical optical tomography using dynamic parameterization and Bayesian conditioning on photon migration measurements, *Appl. Opt.*, 38, 2138, 1999.
118. Arridge, S.R., Optical tomography in medical imaging, *Inverse Probl.*, 15, R41, 1999.
119. Zhu, Q., Durduran, T., Ntziachristos, V., Holboke, M., and Yodh, A.G., Imager that combines near-infrared diffusive light and ultrasound, *Opt. Lett.*, 24, 1050, 1999.
120. Yao, Y., Pei, Y., Wang, Y., and Barbour, R.L., A Born type iterative method for imaging of heterogeneous scattering media and its application to simulated breast tissue, *Proc. SPIE*, 2979, 232, 1997.
121. Lagendijk, A. and Biemond, J., *Iterative Identification and Restoration of Images*, Kluwer Academic, Dordrecht, the Netherlands, 1991.
122. Hansen, P.C., *Rank-Deficient and Discrete Ill-Posed Problems*, SIAM, Philadelphia, 1998.
123. Culver, J.P., Ntziachristos, V., Holboke, M.J., and Yodh, A.G., Optimization of optodearrangements for diffuse optical tomography: a singular-value analysis, *Opt. Lett.*, 26, 701, 2001.
124. Atkins, M.S., Murray, D., and Harrop, R., Use of transputers in a three-dimensional positron emission tomograph, *IEEE Trans. Med. Imaging*, 10, 276, 1991.
125. Rajan, K., Patnaik, L.M., and Ramakrishna, J., High-speed computation of the EM algorithm for PET image reconstruction, *IEEE Trans. Nucl. Sci.*, 41, 1721, 1994.
126. Chen, C.-M., An efficient four-connected parallel system for PET image reconstruction, *Parallel Comput.*, 24, 1499, 1998.
127. Gregor, J. and Huff, D.A., A computational study of the focus-of-attention EM-ML algorithm for PET reconstruction, *Parallel Comput.*, 24, 1998.
128. Zaidi, H., Labbe, C., and Morel, C., Implementation of an environment for Monte Carlo simulation of fully three-dimensional positron tomography on a high-performance parallel platform, *Parallel Comput.*, 24, 1523, 1998.
129. Miller, M.I. and Butler, C.S., Three-dimensional maximum a posteriori estimation for single photon emission computed tomography on massively-parallel computers, *IEEE Trans. Med. Imaging*, 12, 560, 1993.
130. Butler, C.S., Miller, M.I., Miller, T.R., and Wallis, J.W., Massively parallel computers for three-dimensional single-photon-emission computed tomography, *Phys. Med. Biol.*, 39, 575, 1994.
131. Chen, C.M., Lee, S.-Y., and Cho, Z.H., A parallel implementation of three-dimensional image reconstruction on hypercube multiprocessor, *IEEE Trans. Nucl. Sci.*, 37, 1333, 1990.

132. Rajan, K., Patnaik, L.M., and Ramakrishna, J., High-speed implementation of a modified PBR algorithm on DSP-based EH topology, *IEEE Trans. Nucl. Sci.*, 44, 1658, 1997.
133. Laurent, C., Peyrin, F., Chassery, J.-M., and Amiel, M., Parallel image reconstruction on MIMD computers for three-dimensional cone-beam tomography, *Parallel Comput.*, 24, 1461, 1998.
134. Woo, E.J., Hua, P., Webster, J.G., and Tompkins, W.J., A robust image reconstruction algorithm and its parallel implementation in electrical impedance tomography, *IEEE Trans. Med. Imaging*, 12, 137, 1993.
135. Joshi, S. and Miller, M., Maximum a posteriori estimation with Good's roughness for three-dimensional optical-sectioning microscopy, *J. Opt. Soc. Am.*, 10, 1078, 1993.
136. Schweiger, M., Zhukov, L., Arridge, S.R., and Johnson, C.R., Optical tomography using the SCIRun problem solving environment: preliminary results for three-dimensional geometries and parallel processing, *Opt. Express*, 4, 263, 1999.
137. Holboke, M.J., Tromberg, B.J., Li, X., Shah, N., Fishkin, J., Kidney, D., Butler, J., Chance, B., and Yodh, A.G., Three-dimensional diffuse optical mammography with ultrasound localization in a human subject, *J. Biomed. Opt.*, 5, 237, 2000.
138. Boas, D.A. and Yodh, A.G., Spatially varying dynamical properties of turbid media probed with diffusing temporal light correlation, *J. Opt. Soc. Am. A*, 14, 192, 1997.
139. Clark, N.A., Lunacek, J. H., and Benedek, G.B., A study of Brownian motion using light scattering, *Am. J. Phys.*, 38, 575, 1970.
140. Berne, P.J. and Pecora, R., *Dynamic Light Scattering*, Wiley, New York, 1976.
141. Fuller, G.G., Rallison, J.M., Schmidt, R.L., and Leal, L.G., The measurement of velocity gradients in laminar flow by homodyne light-scattering spectroscopy, *J. Fluid Mech.*, 100, 555, 1980.
142. Tong, P., Goldburg, W.I., Chan, C.K., and Sirivat, A., Turbulent transition by photon-correlation spectroscopy, *Phys. Rev. A*, 37, 2125, 1988.
143. Bertolotti, M., Crosignani, B., Di Porto, P., and Sette, D., Light scattering by particles suspended in a turbulent fluid, *J. Phys. A*, 2, 126, 1969.
144. Bourke, P.J., Butterworth, J., Drain, L.E., Egelstaff, P.A., Jakeman, E., and Pike, E.R., A study of the spatial structure of turbulent flow by intensity-fluctuation spectroscopy, *J. Phys. A*, 3, 216, 1970.
145. Tanaka, T., Riva, C., and Ben-Sira, I., Blood velocity measurements in human retinal vessels, *Science*, 186, 830, 1974.
146. Stern, M.D., *In vivo* evaluation of microcirculation by coherent light scattering, *Nature*, 254, 56, 1975.
147. Bonner, R. and Nossal, R., Model for laser Doppler measurements of blood flow in tissue, *Appl. Opt.*, 20, 2097, 1981.
148. Pine, D.J., Weitz, D.A., Chaikin, P.M., and Herbolzheimer, E., Diffusing-wave spectroscopy, *Phys. Rev. Lett.*, 60, 1134, 1988.
149. MacKintosh, F.C. and John, S., Diffusing-wave spectroscopy and multiple scattering of light in correlated random media, *Phys. Rev. B*, 40, 2382, 1989.
150. Maret, G. and Wolf, P.E., Multiple light scattering from disordered media. The effect of Brownian motion of scatterers, *Z. Phys. B*, 65, 409, 1987.
151. Pusey, P.N. and Vaughan, J.M., *Dielectric and Related Molecular Processes X XKW*, Vol. 2, Specialist Periodical Report, Davies, M., Ed., Chemical Society, London, 1975.
152. Boas, D.A., Campbell, L.E., and Yodh, A.G., Scattering and imaging with diffusing temporal field correlations, *Phys. Rev. Lett.*, 75, 1855, 1995.
153. Berne, B.J. and Pecora, R., *Dynamic Light Scattering with Applications to Chemistry, Biology, and Physics*, Krieger, Malabar, FL, 1990.
154. Brown, W., *Dynamic Light Scattering: The Method and Some Applications*, Clarendon, New York, 1993.
155. Cummings, H.Z. and Pike, E.R., *Photon Correlation and Light-Bearing Spectroscopy*, Plenum, New York, 1974.

156. Val'kov, A.Y. and Romanov, V.P., Characteristics of propagation and scattering of light in nematic liquid crystals, *Sov. Phys. JETP*, 63, 737, 1986.
157. Rice, S.O., Mathematical analysis of random noise, in *Noise and Stochastic Processes*, Wax, N., Ed., Dover, New York, 1954, p. 133.
158. Hackmeier, M., Skipetrov, S.E., Maret, G., and Maynard, R., Imaging of dynamic heterogeneities in mulitiple-scattering media, *J. Opt. Soc. Am. A*, 14, 185, 1997.
159. Benaron, D.A., Hintz, S.R., Villringer, A., Boas, D., Kleinschmidt, A., Frahm, J., Hirth, C., Obrig, H., van Houten, J.C., Kermit, E.L., Cheong, W.F., and Stevenson, D.K., Noninvasive functional imaging of human brain using light, *J. Cereb. Blood Flow Metab.*, 20, 469, 2000.
160. Danen, R.M., Wang, Y., Li, X.D., Thayer, W.S., and Yodh, A.G., Regional imager for low-resolution functional imaging of the brain with diffusing near-infrared light, *Photochem. Photobiol.*, 67, 33, 1998.
161. Villringer, A. and Chance, B., Non-invasive optical spectroscopy and imaging of human brain function, *Trends Neurosci.*, 20, 435, 1997.
162. Kelcz, F. and Santyr, G., Gadolinium-enhanced breast MRI, *Crit. Rev. Diagn. Imaging*, 36, 287, 1995.
163. Barkhof, F., Valk, J., Hommes, O.R., and Scheltens, P., Meningeal Gd-DTPA enhancement in multiple-sclerosis, *Am. J. Neuroradiol.*, 13, 397, 1992.
164. Tilcock, C., Delivery of contrast agents for magnetic resonance imaging, computed tomography, nuclear medicine and ultrasound, *Adv. Drug Delivery Rev.*, 37, 33, 1999.
165. Kedar, R.P., Cosgrove, D., McCready, V.R., Bamber, J.C., and Carter, E.R., Microbubble contrast agent for color Doppler US: effect on breast masses, *Radiology*, 198, 679, 1996.
166. Melany, M.L., Grant, E.G., Farooki, S., McElroy, D., and Kimme-Smith, C., Effect of US contrast agents on spectral velocities: *in vitro* evaluation, *Radiology*, 211, 427, 1999.
167. Thompson, S.E., Raptopoulos, V., Sheiman, R.L., McNicholas, M.M.J., and Prassopoulos, P., Abdominal helical CT: milk as a low-attenuation oral contrast agent, *Radiology*, 211, 870, 1999.
168. Jain, R., Sawhney, S., Sahni, P., Taneja, K., and Berry, M., CT portography by direct intrasplenic contrast injection: a new technique, *Abdom. Imaging*, 24, 272, 1999.
169. Knuttel, A., Schmitt, J.M., Barnes, R., and Knutson, J.R., Acousto-optic scanning and interfering photon density waves for precise localization of an absorbing (or fluorescent) body in a turbid medium, *Rev. Sci. Instrum.*, 64, 638, 1993.
170. O'Leary, M.A., Boas, D.A., Chance, B., and Yodh, A.G., Reradiation and imaging of diffuse photon density waves using fluorescent inhomogeneities, *J. Luminesc.*, 60, 789, 1994.
171. Li, X.D., Beauvoit, B., White, R., Nioka, S., Chance, B., and Yodh, A.G., Tumor localization using fluorescence of indocyanine green (ICG) in rat model, *SPIE Proc.*, 2389, 789, 1995.
172. Wu, J., Wang, Y., Perelman, L., Itzkan, I., Dasari, R.R., and Feld, M.S., Time-resolved multichannel imaging of fluorescent objects embedded in turbid media, *Opt. Lett.*, 20, 489, 1995.
173. Bambot, S.B., Lakowicz, J.R., Sipior, J., Carter, G., and Rao, G., Bioprocess and clinical monitoring using lifetime-based phase-modulation fluorometry, *Abstr. Papers Am. Chem Soc.*, 209, 11-BTEC Part 2, 1995.
174. Rumsey, W.L., Vanderkooi, J.M., and Wilson, D.F., Imaging of phosphorescence: a novel method for measuring oxygen distribution in perfused tissue, *Science*, 241, 1649, 1988.
175. Vinogradov, S.A., Lo, L.W., Jenkins, W.T., Evans, S.M., Koch, C., and Wilson, D.F., Noninvasive imaging of the distribution of oxygen in tissue *in vivo* using infrared phosphors, *Biophys. J.*, 70, 1609, 1996.
176. Lakowicz, J.R., *Principles of Fluorescence Spectroscopy*, Plenum Press, New York, 1983.
177. Mordon, S., Devoisselle, J.M., and Maunoury, *In-vivo* pH measurement and imaging of tumor-issue using a pH-sensitive fluorescent-probe (56-carboxyfluorescein) — instrumental and experimental studies, *Photochem. Photobiol.*, 60, 274, 1994.
178. Russell, D.A., Pottier, R.H., and Valenzeno, D.P., Continuous noninvasive measurement of *in-vivo* pH in conscious mice, *Photochem. Photobiol.*, 59, 309, 1994.
179. Sevick-Muraca, E.M. and Burch, C.L., The origin of phosphorescent and fluorescent signals in tissues, *Opt. Lett.*, 19, 1928, 1994.

180. Hutchinson, C.L., Lakowicz, J.R., and Sevick-Muraca, E.M., Fluorescence lifetime-based sensing in tissues — a computational study, *Biophys. J.*, 68, 1574, 1995.
181. Patterson, M.S. and Pogue, B.W., Mathematical-model for time-resolved and frequency-domain fluorescence spectroscopy in biological tissue, *Appl. Opt.*, 33, 1963, 1994.
182. Wu, J., Feld, M.S., and Rava, R.P., Analytical model for extracting intrinsic fluorescence in turbid media, *Appl. Opt.*, 32, 3585, 1993.
183. Hull, E.L., Nichols, M.G., and Foster, T.H., Localization of luminescent inhomogeneities in turbid media with spatially resolved measurements of CW diffuse luminescence emittance, *Appl. Opt.*, 37, 2755, 1998.
184. Feldmann, H.J., Molls, M., and Vaupel, P.W., Blood flow and oxygenation status of human tumors — clinical investigations, *Strahlenther. Onkol.*, 175, 1, 1999.
185. O'Leary, M.A., Boas, D.A., Li, X.D., Chance, B., and Yodh, A.G., Fluorescent lifetime imaging in turbid media, *Opt. Lett.*, 21, 158, 1996.
186. Li, X.D., O'Leary, M.A., Boas, D.A., Chance, B., and Yodh, A.G., Fluorescent diffuse photon density waves in homogenous and heterogeneous turbid media: analytic solutions and applications, *Appl. Opt.*, 35, 3746, 1996.
187. Li, X.D., Chance, B., and Yodh, A.G., Fluorescent heterogeneities in turbid media: limits for detection, characterization, and comparison with absorption, *Appl. Opt.*, 37, 6833, 1998.
188. Ntziachristos, V., Chance, B., and Yodh, A.G., Differential diffuse optical tomography, *Opt. Express*, 5, 230, 1999.
189. Chernomordik, V., Hattery, D.W., Grosenick, D., Wabnitz, H., Rinneberg, H., Moesta, K.T., Schlag, P.M., and Gandjbakhche, A., Quantification of optical properties of a breast tumor using random walk theory, *J. Biomed. Opt.*, 7, 80, 2002.
190. Colak, S.B., van der Mark, M.B., Hooft, G.W., Hoogenraad, J. H., van der Linden, E.S., and Kuijpers, F.A., Clinical optical tomography and NIR spectroscopy for breast cancer detection, *IEEE J. Sel. Top. Quantum Electron.*, 5, 1143, 1999.
191. McBride, T.O., Pogue, B.W., Poplack, S., Soho, S., Wells, W.A., Jiang, S., Osterberg, U.L., and Paulsen, K.D., Multispectral near-infrared tomography: a case study in compensating for water and lipid content in hemoglobin imaging of the breast, *J. Biomed. Opt.*, 7, 72, 2002.
192. Jiang, H., Iftimia, N.V., Xu, Y., Eggert, J.A., Fajardo, L.L., and Klove, K.L., Near-infrared optical imaging of the breast with model-based reconstruction, *Acad. Radiol.*, 9, 186, 2002.
193. Nioka, S., Yung, Y., Shnall, M., Zhao, S., Orel, S., Xie, C., Chance, B., and Solin, L., Optical imaging of breast tumor by means of continuous waves, *Adv. Exp. Med. Biol.*, 411, 227, 1997.
194. Pogue, B.W. and Paulsen, K.D., High-resolution near-infrared tomographic imaging simulations of the rat cranium by use of a priori magnetic resonance imaging structural information, *Opt. Lett.*, 23, 1716, 1998.
195. Siegel, A.M., Marota, J.J.A., and Boas, D.A., Design and evaluation of a continuous-wave diffuse optical tomography system, *Opt. Express*, 4, 287, 1999.
196. Hintz, S.R., Benaron, D.A., Siegel, A.M., Zourabian, A., Stevenson, D.K., and Boas, D.A., Bedside functional imaging of the premature infant brain during passive motor activation, *J. Perinat. Med.*, 29, 335, 2001.
197. Cheung, C., Culver, J.P., Takahashi, K., Greenberg, J.H., and Yodh, A.G., *In vivo* cerebrovascular measurement combining diffuse near-infrared absorption and correlation spectroscopies, *Phys. Med. Biol.*, 46, 2053, 2001.
198. Maki, A., Yamashita, Y., Ito, Y., Watanabe, E., Mayanagi, Y., and Koizumi, H., Spatial and temporal analysis of human motor activity using noninvasive NIR topography, *Med. Phys.*, 22, 1997, 1995.
199. Takahashi, K., Ogata, S., Atsumi, Y., Yamamoto, R., Shiotsuka, S., Maki, A., Yamashita, Y., Yamamoto, T., Koizumi, H., Hirasawa, H., and Igawa, M., Activation of the visual cortex imaged by 24-channel near-infrared spectroscopy, *J. Biomed. Opt.*, 5, 93, 2000.
200. Franceschini, M.A., Toronov, V., Filiaci, M., Gratton, E., and Fanini, S., On-line optical imaging of the human brain with 160-ms temporal resolution, *Opt. Express*, 6, 49, 2000.

201. Benaron, D.A. and Stevenson, D.K., Optical time-of-flight and absorbance imaging of biologic media, *Science*, 259, 1463, 1993.
202. Chance, B., Leigh, J. S., Miyake, H., Smith, D.S., Nioka, S., Greenfeld, R., Finander, M., Kaufmann, K., Levy, W., Young, M., et al., Comparison of time-resolved and unresolved measurements of deoxyhemoglobin in brain, *Proc. Natl. Acad. Sci. U.S.A*, 85, 4971, 1988.
203. Cubeddu, R., Pifferi, A., Taroni, P., Torricelli, A., and Valentini, G., Time-resolved imaging on a realistic tissue phantom: $\mu_s'$ and $\mu_a$ images versus time-integrated images, *Appl. Opt.*, 35, 4533, 1996.
204. Hebden, J. C., Arridge, S.R., and Delpy, D.T., Optical imaging in medicine: I. Experimental techniques, *Phys. Med. Biol.*, 42, 825, 1997.
205. Grosenick, D., Wabnitz, H., and Rinneberg, H., Time-resolved imaging of solid phantoms for optical mammography, *Appl. Opt.*, 36, 221, 1997.
206. Nioka, S., Luo, Q., and Chance, B., Human brain functional imaging with reflectance CWS, *Adv. Exp. Med. Biol.*, 428, 237, 1997.
207. Maki, A., Yamashita, Y., Watanabe, E., and Koizumi, H., Visualizing human motor activity by using non-invasive optical topography, *Front. Med. Biol. Eng.*, 7, 285, 1996.
208. Colier, W., van der Sluijs, M.C., Menssen, J., and Oeseburg, B., A new and highly sensitive optical brain imager with 50 Hz sample rate, *NeuroImage*, 11, 542, 2000.
209. Gratton, E., Fantini, S., Franceschini, M.A., Gratton, G., and Fabiani, M., Measurements of scattering and absorption changes in muscle and brain, *Philos. Trans. R. Soc. London B Biol. Sci.*, 352, 727, 1997.
210. Pogue, B.W., Patterson, M.S., Jiang, H., and Paulsen, K.D., Initial assessment of a simple system for frequency domain diffuse optical tomography, *Phys. Med. Biol.*, 40, 1709, 1995.
211. Chance, B., Cope, M., Gratton, E., Ramanujam, N., and Tromberg, B., Phase measurement of light absorption and scattering in human tissues, *Rev. Sci. Instrum.*, 689, 3457, 1998.
212. Fishkin, J.B., So, P.T. C., Cerussi, A.E., Fantini, S., Franceschini, M.A., and Gratton, E., A frequency-domain method for measuring spectral properties in multiply scattering media: methemoglobin absorption spectrum in a tissue-like phantom, *Appl. Opt.*, 34, 1143, 1995.
213. Quaresima, V., Colier, W.N., van der Sluijs, M., and Ferrari, M., Non-uniform quadriceps $O_2$ consumption revealed by near infrared multipoint measurements, *Biochem. Biophys. Res. Commun.*, 285, 1034, 2001.
214. Miura, H., McCully, K., Hong, L., Nioka, S., and Chance, B., Exercise-induced changes in oxygen status in calf muscle of ederly subjects with peripheral vascular disease using functional near infrared imaging machine, *Ther. Res.*, 2, 1585, 2000.
215. Pogue, B.W., Testorf, M., McBride, T., Osterberg, U., and Paulsen, K., Instrumentation and design of a frequency-domain diffuse optical tomography imager for breast cancer detection, *Opt. Express*, 1, 391, 1997.
216. Schmidt, F.E., Fry, M.E., Hillman, E.M.C., Hebden, J.C., and Delpy, D.T., A 32-channel time-resolved instrument for medical optical tomography, *Rev. Sci. Instrum.*, 71, 256, 2000.
217. Culver, J.P., Choe, R., Holboke, M.J., Zubkov, L., Durduran, T., Slemp, A., Ntziachristos, V., Chance, B., and Yodh, A.G., 3D diffuse optical tomography in the plane parallel transmission geometry: evaluation of a hybrid frequency domain/continuous wave clinical system for breast imaging, *Med. Phys.*, accepted.
218. Durduran, T., Choe, R., Culver, J. P., Zubkov, L., Holboke, M.J., Giammarco, J., Chance, B., and Yodh, A.G., Bulk optical properties of healthy female breast tissue, *Phys. Med. Biol.*, 47, 2847, 2002.
219. Yang, Y., Liu, H., Li, X., and Chance, B., Low-cost frequency-domain photon migration instrument for tissue spectroscopy, oximetry and imaging, *Opt. Eng.*, 36, 1562, 1997.
220. Flock, S.T., Jacques, S.L., Wilson, B.C., Star, W.M., and van Gemert, M.J.C., Optical properties of Intralipid: a phantom medium for light propagation studies, *Lasers Surg. Med.*, 12, 510, 1992.
221. van Staveren, H.J., Moes, C.J.M., van Marle, J., Prahl, S.A., and van Gemert, M.J.C., Light scattering in Intralipid — 10% in the wavelength range of 400 to 1100 nm, *Appl. Opt.*, 30, 4507, 1991.

222. Gros, C., Quenneville, Y., and Hummel, Y., Breast diaphanology, *J. Radiol. Electrol. Med. Nucl.*, 53, 297, 1972.
223. Alveryd, A., Andersson, I., Aspegren, K., Balldin, G., Bjurstam, N., Edstrom, G., Fagerberg, G., Glas, U., Jarlman, O., Larsson, S.A., et al., Lightscanning versus mammography for the detection of breast cancer in screening and clinical practice. A Swedish multicenter study, *Cancer*, 65, 1671, 1990.
224. Shah, N., Cerussi, A., Eker, C., Espinoza, J., Butler, J., Fishkin, J., Hornung, R., and Tromberg, B., Noninvasive functional optical spectroscopy of human breast tissue, *Proc. Natl. Acad. Sci. U.S.A.*, 98, 4420, 2001.
225. Sevick-Muraca, E.M., Reynolds, J. S., Troy, T.L., Lopez, G., and Paithankar, D.Y., Fluorescence lifetime spectroscopic imaging with measurements of photon migration, *Ann. N.Y. Acad. Sci.*, 838, 46, 1998.
226. Hawrysz, D.J. and Sevick-Muraca, E.M., Developments toward diagnostic breast cancer imaging using near-infrared optical measurements and fluorescent contrast agents, *Neoplasia*, 2, 388, 2000.
227. Tromberg, B.J., Shah, N., Lanning, R., Cerussi, A., Espinoza, J., Pham, T., Svaasand, L., and Butler, J., Non-invasive *in vivo* characterization of breast tumors using photon migration spectroscopy, *Neoplasia*, 2, 26, 2000.
228. Ntziachristos, V., Ma, X.H., Yodh, A.G., and Chance, B., Multichannel photon counting instrument for spatially resolved near infrared spectroscopy, *Rev. Sci. Instrum.*, 70, 193, 1999.
229. Benaron, D.A., Hintz, S.R., Villringer, A., Boas, D., Kleinschmidt, A., Frahm, J., Hirth, C., Obrig, H., van Houten, J.C., Kermit, E.L., Cheong, W.F., and Stevenson, D.K., Noninvasive functional imaging of human brain using light, *J. Cereb. Blood Flow Metab.*, 20, 469, 2000.
230. Boas, D.A., Meglinsky, I.V., Zemany, L., Campbell, L.E., Chance, B., and Yodh, A.G., Diffusion of temporal field correlation with selected applications, *Proc. SPIE*, 2732, 34, 1996.
231. Briers, J.D., Laser Doppler and time-varying speckle: a reconciliation, *J. Opt. Soc. Am. A*, 13, 345, 1996.
232. Dunn, A.K., Bolay, H., Moskowitz, M.A., and Boas, D.A., Dynamic imaging of cerebral blood flow using laser speckle, *J. Cereb. Blood Flow Metab.*, 21, 195, 2001.
233. Kuebler, W.M., Sckell, A., Habler, O., Kleen, M., Kuhnle, G.E.H., Welte, M., Messmer, K., and Goetz, A.E., Noninvasive measurement of regional cerebral blood flow by near-infrared spectroscopy and indocyanine green, *J. Cereb. Blood Flow Metab.*, 18, 445, 1998.
234. Patel, J., Marks, K., Roberts, I., Azzopardi, D., and Edwards, A.D., Measurement of cerebral blood flow in newborn infants using near infrared spectroscopy with indocyanine green, *Pediatr. Res.*, 43, 34, 1998.
235. Gopinath, S.P., Robertson, C.S., Contant, C.F., Narayan, R.K., Grossman, R.G., and Chance, B., Early detection of delayed traumatic intracranial hematomas using near-infrared spectroscopy, *J. Neurosurg.*, 83, 438, 1995.
236. Ogawa, S., Tank, D., Menon, R., Ellermann, J., Kim, S.-G., Merkel, H., and Ugurbil, K., Intrinsic signal changes accompanying sensory stimulation: functional brain mapping with magnetic resonance imaging, *Proc. Natl. Acad. Sci. U.S.A.*, 89, 5951, 1992.
237. Kwong, K.K., Belliveau, J.W., Chesler, D.A., Goldberg, I.E., Weisskoff, R.M., Poncelet, B.P., Kennedy, D.N., Hoppel, B.E., Cohen, M.S., Turner, R., Cheng, H.-M., Brady, T.J., and Rosen, B.R., Dynamic magnetic resonance imaging of human brain activity during primary sensory stimulation, *Proc. Natl. Acad. Sci. U.S.A.*, 89, 5675, 1992.
238. Jobsis, F.F., Nonivasive, infrared monitoring of cerebral and myocardial oxygen sufficiency and circulatory parameters, *Science*, 198, 1264, 1977.
239. Wyatt, J.S., Cope, M., Delpy, D.T., Wray, S., and Reynolds, E.O.R., Quantification of cerebral oxygenation and haemodynamics in sick newborn infants by near infrared spectrophotometry, *Lancet*, ii, 1063, 1986.
240. Chance, B. and Williams, G.R., The respiratory chain and oxidative phosphorylation, *Adv. Enzymol.*, 17, 65, 1956.

241. Lockwood, A.H., LaManna, J.C., Snyder, S., and Rosenthal, M., Effects of acetazolamide and electrical stimulation on cerebral oxidative metabolism as indicated by cytochrome oxidase redox state, *Brain Res.*, 308, 9, 1984.
242. Wong Riley, M.T., Cytochrome oxidase: an endogenous metabolic marker for neuronal activity, *Trends Neurosci.*, 12, 94, 1989.
243. Kohl, M., Nolte, C., Heekeren, H.R., Horst, S., Scholz, U., Obrig, H., and Villringer, A., Changes in cytochrome-oxidase oxidation in the occipital cortex during visual stimulation: improvement in sensitivity by the determination of the wavelength dependence of the differential pathlength factor, *Proc. SPIE*, 3194, 18, 1998.
244. Wobst, P., Wenzel, R., Kohl, M., Obrig, H., and Villringer, A., Linear aspects of changes in deoxygenated hemoglobin concentration and cytochrome oxidase oxidation during brain activation, *Neuroimage*, 13, 520, 2001.
245. Gratton, E., Fantini, S., Franceschini, M.A., Gratton, G., and Fabiani, M., Measurements of scattering and absorption changes in muscle and brain, *Philos. Trans. R. Soc. London B Biol. Sci.*, 352, 727, 1997.
246. Steinbrink, J., Kohl, M., Obrig, H., Curio, G., Syre, F., Thomas, F., Wabnitz, H., Rinneberg, H., and Villringer, A., Somatosensory evoked fast optical intensity changes detected non-invasively in the adult human head, *Neurosci. Lett.*, 291, 105, 2000.
247. Stepnoski, R.A., LaPorta, A., Raccuia-Behling, F., Blonder, G.E., Slusher, R.E., and Kleinfeld, D., Noninvasive detection of changes in membrane potential in cultured neurons by light scattering, *Proc. Natl. Acad. Sci. U.S.A.*, 88, 9382, 1991.
248. Cheung, C., Culver, J. P., Takahashi, K., Greenberg, J.H., and Yodh, A.G., *In vivo* cerebrovascular measurement combining diffuse near-infrared absorption and correlation spectroscopies, *Phys. Med. Biol.*, 46, 2053, 2001.
249. Mayhew, J., Johnston, D., Martindale, J., Jones, M., Berwick, J., and Zheng, Y., Increased oxygen consumption following activation of brain: theoretical footnotes using spectroscopic data from barrel cortex, *Neuroimage*, 13, 975, 2001.
250. Mayhew, J., Zheng, Y., Hou, Y., Vuksanovic, B., Berwick, J., Askew, S., and Coffey, P., Spectroscopic analysis of changes in remitted illumination: the response to increased neural activity in brain, *Neuroimage*, 10, 304, 1999.
251. Simoes, C.H.R., Relationship between responses to contra- and ipsilateral stimuli in the human second somatosensory cortex SII, *Neuroimage*, 10, 408, 1999.
252. Spiegel, J.T.J., Gawehn, J., Stoeter, P., and Treede, R.D., Functional MRI of human primary somatosensory and motor cortex during median nerve stimulation., *Clin. Neurophysiol.*, 110, 47, 1999.
253. Allison, J.D.M.K., Loring, D.W., Figueroa, R.E., and Wright, J.C., Functional MRI cerebral activation and deactivation during finger movement, *Neurology*, 54, 135, 2000.
254. Boas, D.A., O'Leary, M.A., Chance, B., and Yodh, A.G., Detection and characterization of optical inhomogeneities with diffuse photon density waves: a signal-to-noise analysis, *Appl. Opt.*, 36, 75, 1997.
255. Matson, C.L., Deconvolution-based spatial resolution in optical diffusion tomography, *Appl. Opt.*, 40, 5791, 2001.
256. Moon, J.A. and Reintjes, J., Image resolution by use of multiply scattered light, *Opt. Lett.*, 19, 521, 1994.
257. Moon, J.A., Mahon, R., Duncan, M.D., and Reintjes, J., Resolution limits for imaging through turbid media with diffuse light, *Opt. Lett.*, 1591, 1993.
258. Pogue, B.W. and Paulsen, K.D., High-resolution near-infrared tomographic imaging simulations of the rat cranium by use of a priori magnetic resonance imaging structural information, *Opt. Lett.*, 23, 1716, 1998.
259. Zhu, Q., Conant, E., and Chance, B., Optical imaging as an adjunct to sonograph in differentiating benign from malignant breast lesions, *J. Biomed. Opt.*, 5, 229, 2000.

260. Barbour, R.L., Graber, H.L., Chang, J., Barbour, S.S., Koo, P.C., and Aronson, R., MRI-guided optical tomography: prospects and computation for a new imaging method, *IEEE Comput. Sci. Eng.*, 2, 63, 1995.
261. Ntziachristos, V., Yodh, A.G., Schnall, M., and Chance, B., MRI-guided diffuse optical spectroscopy of malignant and benign breast lesions, *Neoplasia*, 4, 347, 2002.
262. Arridge, S.R., Schweiger, M., Hiraoka, M., and Delpy, D.T., Performance of an iterative reconstruction algorithm for near infrared absorption and scatter imaging, *Proc. SPIE*, 1888, 360, 1993.
263. Schweiger, M., Arridge, S.R., and Delpy, D.T., Application of the finite-element method for the forward and inverse models in optical tomography, *J. Math. Imaging Vision*, 3, 263, 1993.
264. Arridge, S.R. and Schweiger, M., Inverse methods for optical tomography, in *Information Processing in Medical Imaging (IPMI'93 Proceedings)*, Lecture Notes in Computer Science, 687, Springer-Verlag, Berlin, 1993, p. 259.
265. Boas, D.A., Gaudette, T.J., and Arridge, S.R., Simultaneous imaging and optode calibration with diffuse optical tomography, *Opt. Express*, 8, 263, 2001.

# 22

# Photon Migration Spectroscopy Frequency-Domain Techniques


Albert E. Cerussi
*Beckman Laser Institute*
*University of California, Irvine*
*Irvine, California*

Bruce J. Tromberg
*Beckman Laser Institute*
*University of California, Irvine*
*Irvine, California*




## 22.1 Photon Migration Spectroscopy

### 22.1.1 What Is Photon Migration Spectroscopy?

Photon migration spectroscopy (PMS) is a technique that combines experimental measurements and model-based data analysis to measure the bulk absorption and scattering properties of highly scattering media. PMS typically uses red and near-infrared (NIR) light, especially from 600 to 1000 nm, where light propagation in tissue is heavily dominated by scattering.[1] PMS measurements of tissue optical properties (i.e., absorption and scattering) are assumed to contain information about tissue structure and function. The term "photon migration" refers to photons that propagate diffusively throughout the tissue (i.e., in a random direction). PMS utilizes the photons that conventional optical techniques discard; instead of collimated or coherent photons, PMS measures incoherent, multiply scattered photons that are spread


0-8493-1116-0/03/$0.00+$1.50

out over space and time. Diffusive photons probe a large sample volume, providing macroscopically averaged absorption and scattering properties at depths up to a few centimeters in tissues.

In the red and NIR spectral regions, the dominant molecular absorbers in tissue are oxygenated ($Hb\text{-}O_2$) and reduced hemoglobin (Hb-R), water, and lipids.[2,3] Myoglobin (which is indistinguishable from hemoglobin in the NIR), cytochrome $aa_3$, and other hemoglobin states also absorb in the NIR but are found in small concentrations. (Note that the validity of this statement depends upon tissue site and physiological condition.)

The frequency-domain PMS techniques featured in this article measure absorption independently from scattering; not all NIR diffusive measurements do this. Such distinction is important because NIR photon attenuation is chiefly due to scattering. Traditional absorption spectroscopy assumes negligible scattering. With the help of PMS techniques, one can recover undistorted absorption spectra from within tissues and, ultimately, quantify absorber concentrations. Scattering spectra also can provide useful information about tissue structure and composition.

### 22.1.2 Historical Development

The dramatic growth of clinical laser applications during the 1990s has stimulated intense research into the fundamental nature of light–tissue interactions. Although studies in highly scattering biological materials were performed in the 1930s,[4] significant advances in the quantitative characterization of tissue optical properties *in vivo* were not made until the early 1990s. Bonner et al. introduced "photon migration" in tissue by proposing a random-walk theory to infer bulk tissue scattering, absorption, and photon pathlengths.[5] Delpy[6] and Chance[7] employed time-resolved spectroscopy to measure photon pathlengths and thereby determine hemoglobin concentration in tissue. Transport theory models had been developed earlier,[8] but did not become widely accepted for modeling in tissue optics until the late 1980s.[9,10] Patterson et al. measured absorption independently from scattering using analytic time- and frequency-domain models.[11,12]

Fishkin and Gratton first applied frequency-domain methods to turbid media and described the propagation of intensity-modulated light in terms of photon density waves (PDW).[13] Svaasand and Tromberg characterized the frequency-dependence of photon density waves and employed measurements of PDW dispersion to quantify absorption and scattering in turbid media.[14] Rapid and inexpensive means for measuring absorption and scattering in turbid media soon followed.[15] There are several advantages to the frequency-domain method in terms of information content, measurement speed, and cost.[16]

Tissue physiology can be inferred from reliable measurements of tissue absorption.[2] PMS systems typically feature two wavelengths, the bare minimum needed to recover [$Hb\text{-}O_2$] and [Hb-R]. (Brackets [ ] will be used to denote concentration.) Restricting systems to only two or three wavelengths can lead to significant errors, even for relative parameters such as the hemoglobin saturation, $S_tO_2$.[17] (Hemoglobin saturation is defined as $S_tO_2 = [Hb\text{-}O_2]/THC$, with $THC = [Hb\text{-}R]+[Hb\text{-}O_2]$.) Increased spectral bandwidth adds more information content and improves accuracy.[18]

Recent reviews of diffuse tissue optics suggest that photon migration methods can provide unique information and will play an important role in medicine.[19–21] As of this writing, several commercial devices are based upon the technology described in this chapter. Although many possibilities exist, the most important general application of PMS lies in quantifying thick tissue hemodynamics. Because the PMS signal represents a volume average of tissue, the diffusively-measured $S_tO_2$ is primarily representative of the tissue microvasculature.[22] This exquisite sensitivity to microvasculature suggests many broad applications in medicine (see Section 22.5 for examples).

## 22.2 Working in the Frequency Domain

### 22.2.1 The Basics of the Frequency-Domain Method

The time- and frequency-domain methods are general tools used to study light propagation in tissues. Time-domain spectroscopy measures the temporal broadening of an initial narrow pulse of light as it

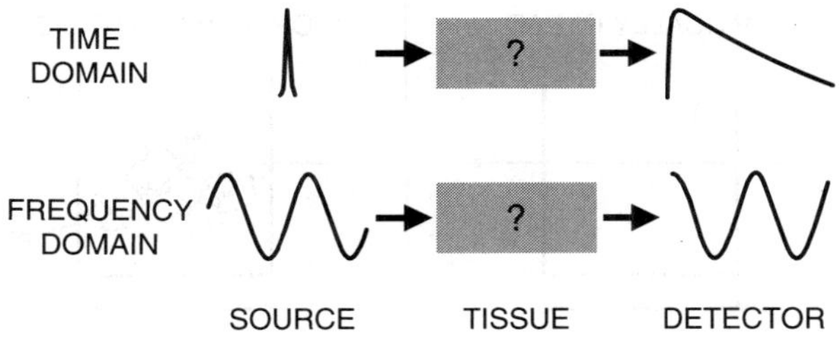


**FIGURE 22.1** Time-domain spectroscopy measures the temporal broadening in an incident pulse. Frequency-domain spectroscopy measures changes in wave characteristics relative to the incident wave.

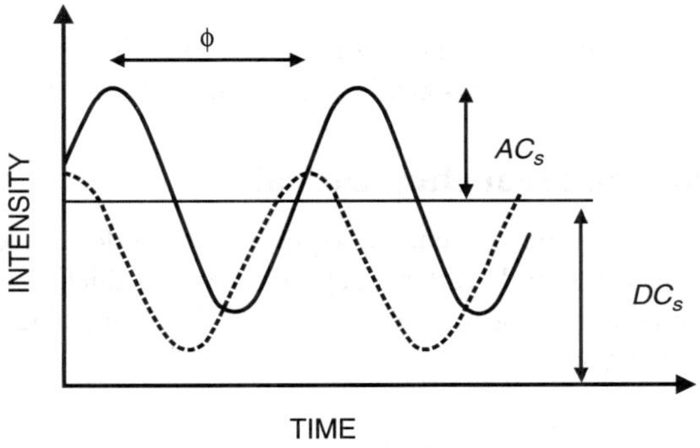


**FIGURE 22.2** The measured frequency-domain parameters are a phase shift ($\phi$) and a modulation decrease (i.e., ratio of AC to DC components) of the detected wave with respect to the source. Source and detected waves oscillate at the same frequency *f*.

interacts with a sample (Figure 22.1). Typical mechanisms behind these propagation delays include photon scattering and fluorescence lifetimes. The broadened pulse shape contains quantitative information about the decay mechanisms involved.

Frequency-domain spectroscopy measures the propagation of harmonic signals (Figure 22.2). We start with an intensity-modulated light source, characterized by an average intensity (*DC*), an oscillating intensity (*AC*), and a phase ($\phi$), which has the form:

$$I_s = DC_s + AC_s \cos(\phi - \omega t)\,. \tag{22.1}$$

In the preceding expression, *I* is the total intensity, $\omega$ is the angular frequency (such that $\omega = 2\pi f$, where *f* is the modulation frequency), and the subscript *s* denotes the source. The modulation frequency is the on–off rate of the source and is not related to the photon energy. The source need not be modulated at a single frequency; any waveform may be described as a sum of the terms of Equation 22.1. The detected wave and source wave repeat at the same frequency.

Figure 22.2 demonstrates the measured frequency-domain parameters. The detected response (dotted wave) is shifted in phase ($\phi$) and reduced in modulation (*M*) with respect to the source (solid wave). The term "modulation" or "modulation depth" refers to $AC_s/DC_s$ at *f* and ideally is equal to unity. In frequency-domain spectroscopy, there is no carrier wave; we do not modulate the frequency (frequency modulation, or FM), the AC amplitude (amplitude modulation, or AM), or the phase (phase modulation, or PM) of the optical wave.

An important point is that $\phi$ and *M* denote relative quantities that describe absolute values. Such relative measurements are insensitive to artifacts that can complicate absolute intensity measurements. Modulated signals can be discriminated against background signals at different frequencies (such as unwanted room light). Measurements in time and frequency domains are equivalent, being related by a

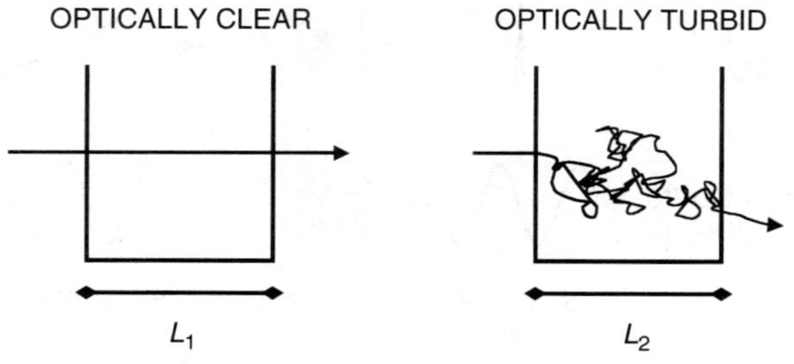


**FIGURE 22.3** Multiple scattering increases the photon path length. Although the sample thicknesses are the same, $L_2 >> L_1$ due to intense multiple scattering. The absorber concentration cannot be determined unless $L$ is known.

temporal Fourier transform. Because it is possible to obtain enough information for PMS using a single frequency, frequency-domain measurements can be very rapid and cost effective.[15]

## 22.2.2 The Need for the Frequency Domain

Richer frequency-domain information content (or equivalent time-domain content) is usually needed to perform quantitative absorption and scattering spectroscopy in thick tissues. The intensity of light transmitted through a purely absorbing medium (i.e., negligible scattering) is related to the incident intensity $I_0$ via the Beer-Lambert law:

$$AB = \log(I_0/I) = \varepsilon\ [C]L\ , \tag{22.2}$$

where $AB$ is the absorbance, $L$ is the photon path length (cm), $[C]$ is the absorber concentration, and $\varepsilon$ (based upon a base-10 logarithm scale) is the molar extinction coefficient (moles/liter cm$^{-1}$ or cm$^2$/mole). In terms of $\mu_a$, the Beer-Lambert law is $I = I_0 \exp(-\mu_a L)$, so that $\mu_a$ is proportional to the absorber concentration:

$$\mu_a \approx 2.303\varepsilon[C]. \tag{22.3}$$

In purely absorbing media, it is relatively simple to measure $\mu_a$ (or $[C]$ if $\varepsilon$ is known) because $L$ is equal to the width of the sample. However, in the case of multiple scattering, $L$ is not known *a priori.* Scattering increases the photon pathlength beyond the geometrical path of the light (Figure 22.3); $L$ must be known in order to determine $[C]$ accurately. Frequency-domain spectroscopy provides a method that effectively measures $L$ so that $[C]$ may be determined.

There is another way to view this dilemma. In the infinite medium geometry, diffusion theory predicts that DC intensity at a detector placed a distance $r$ from the source has the form:

$$DC_d \propto \frac{1}{r}\exp\left[-r\sqrt{\mu_a\mu_s'}\right], \tag{22.4}$$

where the subscript $d$ refers to the detected intensity. It is not possible to separate analytically the effects of $\mu_a$ from $\mu_s'$ using a single measurement because of the product between $\mu_a$ and $\mu_s'$. Under certain conditions, it is possible to separate $\mu_a$ from $\mu_s'$ while performing steady-state, multiple-distance measurements on a tissue surface (i.e., spatially resolved reflectance), although the method involves source-detector separations too small for characterizing tissue optical properties at depths beyond several millimeters.[17]

## 22.3 Frequency-Domain Solution to the Diffusion Equation

### 22.3.1 General Transport of Light in Turbid Media

Transport theory is a statistical bookkeeping scheme that treats photons as noninteracting point-particles undergoing elastic interactions. The quantity of interest is the photon density (photons $mm^{-3}$). Note that some texts use radiance as the fundamental quantity, which is the power per unit area per solid angle. The magnitude of this vector, averaged over the solid angle, divided by the photon energy, gives the photon density. The transport equation is easy to assemble, but difficult to solve analytically, even in simple cases.[23] PMS can use numerical solutions, but important physical insight has come from simple analytic models. The general $P_n$ approximation allows tractable analytic solutions. The $P_1$ approximation keeps only first-order spherical harmonic expansion terms of the transport equation.[23] Forcing the photon density to discard its true angular dependence introduces limitations that must be carefully understood. The $P_1$ approximation is sometimes referred to as the diffusion approximation, but there are some important differences between these approximations.[24] Excellent reviews exist on this topic.[25,26]

### 22.3.2 The $P_1$ Approximation: Infinite Medium Solution

#### 22.3.2.1 Formal Theory

Assuming an isotropic point source, keeping terms to first order reveals the $P_1$ equation, which expresses the photon density, $U(\mathbf{r},t)$, in terms of the optical properties of the tissue:[24,27]

$$vD\left[\frac{3}{v^2}\frac{\partial^2 U(\mathbf{r},t)}{\partial t^2} - \nabla^2 U(\mathbf{r},t)\right] + (1+3\mu_a D)\frac{\partial U(\mathbf{r},t)}{\partial t} + v\mu_a U(\mathbf{r},t) = Q(\mathbf{r},t) + \frac{3D}{v}\frac{\partial Q(\mathbf{r},t)}{\partial t}, \tag{22.5}$$

where $v$ is the speed of light inside the medium, $Q(\mathbf{r},t)$ is the photon density per unit time injected by the source, and $vD \equiv v(3\mu_s' + 3\mu_a)^{-1} \equiv v(3\mu_{tr})^{-1}$ is the optical diffusion coefficient ($mm^2\,s^{-1}$). (Note that $\mu_s'$ represents a scattering length where the direction of the photon has been randomized. Actual scattering events occur on a much smaller length scale.) The variables $|\mathbf{r}|$, and $t$ represent distance from the source and time, respectively. Three important assumptions make this expansion possible: (1) the medium is assumed to be macroscopically homogeneous, (2) we must be far from sources and boundaries (i.e., more than one transport mean-free-path, mfp), and (3) the medium must be strongly scattering (i.e., $\mu_s >> \mu_a$). These limitations stem from restricting the angular dependence of $U$; thus Equation 22.5 is only valid in regions where anisotropy is low.

The $P_1$ equation is simple to solve in the frequency domain. Noting that $i^2 = -1$, a frequency-domain point source has the form:

$$Q(\mathbf{r},\omega,t) = S(\omega)\exp\left[-i\left(\omega t - \phi_0(\omega)\right)\right]\delta(\mathbf{r}). \tag{22.6}$$

We have not broken the source into AC and DC components as in Equation 22.1 because the DC component results from setting $\omega = 0$. The terms $S(\omega)$ and $\phi_0(\omega)$ are frequency-dependent instrumental factors, called the source strength (photons $mm^{-3}\,s^{-1}$) and source phase (degrees), respectively. The delta function represents a point source centered at $\mathbf{r} = 0$. In the frequency domain, all time-dependent factors become $\exp(-i\omega t)$, as in Equation 22.6. If we substitute $U(\mathbf{r},\omega)\exp(-i\omega t)$ for $U(\mathbf{r},t)$ and so on, we arrive at the following expression for Equation 22.5:

$$\left[\nabla^2 - k^2(\omega)\right] U(\mathbf{r},\omega) = -\frac{S(\omega)}{vD}\exp\left[-i\phi_o(\omega)\right]\delta(\mathbf{r}) \tag{22.7}$$

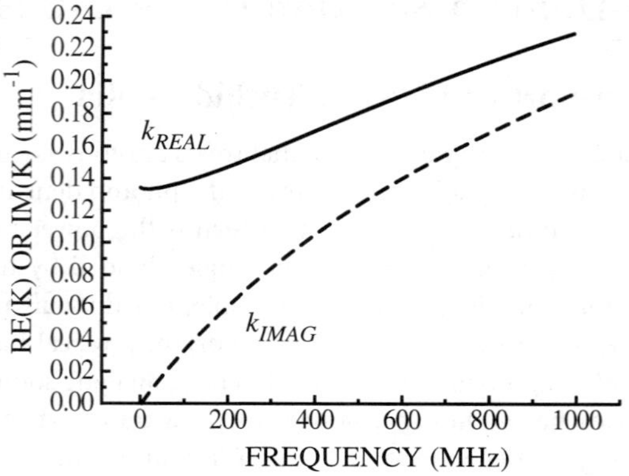


**FIGURE 22.4** PDW characteristics are determined by the real (solid line) and imaginary (dashed line) parts of the wave vector, $k$. The plot uses $\mu_a$ = 0.006 mm$^{-1}$ and $\mu_s'$ = 1.0 mm$^{-1}$. The sensitivity of $k$ to $\mu_a$ and $\mu_s'$ depends upon $f$ (i.e., wave dispersion).

$$k^2(\omega) \equiv \frac{\mu_a}{D}\left[1 - i\frac{\omega}{v\mu_a}(1+3\mu_a D)\right]. \tag{22.8}$$

Equation 22.7 is a Helmholtz equation, whose solutions are known to be waves. Within this framework, $ik$ is a wave vector (mm$^{-1}$). It is critical to note that $k$ depends upon $\mu_a$, $\mu_s'$, and $\omega$. Using infinite-medium boundary conditions (i.e., $U(r,\omega) \to 0$ as $r \to \infty$), we can write a one-dimensional solution to Equation 22.7:

$$U(r,\omega) = \frac{S(\omega)\exp[-i\phi_0(\omega)]}{4\pi v D}\frac{1}{r}\exp[-k(\omega)r]. \tag{22.9}$$

The frequency-domain solution emerges as a damped-spherical PDW. These PDWs exhibit classical wave phenomena such as diffraction, reflection, and refraction.[13,28] The phase and amplitude of the PDW described by Equation 22.7 are:

$$\Theta_{inf}(r,\omega) = k_{imag}(\omega)\, r + \phi_0 \qquad A_{inf}(r,\omega) = \frac{S(\omega)}{4\pi v D\, r}\exp[-k_{real}(\omega)\, r]. \tag{22.10}$$

The subscript *inf* reminds us these equations are valid in infinite media. The real and imaginary parts of $k$ are represented by $k_{real}$ and $k_{imag}$, respectively. We have written Equation 22.10 as functions of $r$ and $\omega$ because measurements of $\mu_a$ and $\mu_s'$ depend upon changing $r$ and/or $\omega$.

### 22.3.2.2 Frequency Dependence

Figure 22.4 demonstrates the dispersion of the components of $k$ for a medium with optical properties of $\mu_a$ = 0.006 mm$^{-1}$ and $\mu_s'$ = 1.0 mm$^{-1}$. Wave attenuation is representatived by $k_{real}$; higher frequency components have higher $k_{real}$ and higher attenuation. The sensitivity of $k$ to the optical properties changes with $f$. Intuitively, we may rewrite absorption and transport coefficients ($\mu_{tr}$) as relaxation frequencies: $f_a \equiv v\mu_a$ and $f_{tr} \equiv v\mu_{tr} = v(\mu_a + \mu_s')$, respectively, or as relaxation times: $\tau_a = 1/f_a$ and $\tau_{tr} = 1/f_{tr}$. In the DC case ($\omega$ = 0), the wave vector values become:

$$k_{real}(0) = \sqrt{3\mu_a\mu_s'} \quad \text{and} \quad k_{imag}(0) = 0. \tag{22.11}$$

In the DC regime, the product $\mu_a\mu_s'$ determines the attenuation and there is no phase information. $k_{real}$ is very sensitive to changes in $\mu_a$ and $\mu_s'$, but changes in $\mu_a$ cannot be distinguished from changes in $\mu_s'$. If we increase $f$ with a low-frequency expansion (i.e., $\omega << v\mu_a$) of Equation 22.8:

$$k_{real}(\omega) \approx \sqrt{3\mu_a\mu_s'}\left\{1+\frac{1}{8}(\omega\tau_a)^2\right\} \qquad k_{imag}(\omega) \approx \sqrt{3\mu_a\mu_s'}\,\frac{\omega\tau_a}{2} \tag{22.12}$$

where we have made the assumption $\mu_a << \mu_s'$ for diffusive media. The term "low-frequency" implies frequencies well below $f_a$, where most frequency-domain measurements have been performed. In most tissues $\mu_a$ ranges from 0.002 to 0.02 mm$^{-1}$, and thus $f_a$ ranges from 60 to 680 MHz. Within this low-frequency approximation, $\phi$ is linear in $f$ (see Equation 22.10).

As $f$ increases above $f_a$, the character of the $k$ changes considerably. A higher-frequency expansion of the real and imaginary parts of Equation 22.8 reveals ($\omega >> f_a$, but $\omega^2 << f_a f_{tr}$):[29]

$$k_{real}(\omega) = k_{imag}(\omega) = \sqrt{\frac{3}{2}\frac{\omega}{v}\mu_s'}\,. \tag{22.13}$$

In this limit, the wave vector loses dependence upon $\mu_a$. Eventually this behavior changes; in the limit where $\omega^2 >> f_a f_{tr}$, transport theory breaks down, leaving us with:

$$k_{real}(\omega \to \infty) = 0 \quad \text{and} \quad k_{imag}(\omega \to \infty) = 3\frac{\omega}{v}\,. \tag{22.14}$$

Photons described by this behavior are ballistic in nature and are not scattered. These expressions also describe the leading edge of the time-domain pulse.[24]

#### 22.3.2.3 Diffusion Wavelength

The diffusion wavelength, or $\lambda_{PDW} \equiv 2\pi/k_{imag}(\omega)$, has nothing to do with the photon energy. $\lambda_{PDW}$ depends upon the optical properties of the medium. Given a modulation frequency of 200 MHz, $\mu_a = 0.006$ mm$^{-1}$, and $\mu_s' = 1.0$ mm$^{-1}$ $\lambda_{PDW}$ is about 105 mm. The wave front speed is given by $v_{PDW} = \lambda_{PDW} f$, setting $v_{PDW} \sim 2.1 \times 10^{10}$ mm s$^{-1}$. The diffusive wave front speed proceeds at approximately one order of magnitude slower than the speed of an individual photon. $\lambda_{PDW}$ is typically much greater than our typical sampling distances of a few centimeters so that, in general, we record PDW phase and amplitude as "near-field" region. Higher modulation frequencies reduce $\lambda_{PDW}$, signifying greater attenuation and lower penetration.

### 22.3.3 The $P_1$ Approximation: Semi-Infinite Medium Solution

#### 22.3.3.1 Changes in the Theory

Two types of discontinuities impose important boundaries for light propagation in tissues: index of refraction and optical property. For example, the abrupt index of refraction mismatch at the air–tissue interface can lead to total internal reflection. Encounters with this boundary will alter the spatial distribution of the PDW. In order to quantify average tissue optical properties accurately, it is essential to modify the light transport model to account for this interface. In practical terms, this permits the use of measurement geometries that place the source and detector outside the tissue (i.e., in a nonscattering medium). Our approach uses the method of images to place an image source at the same point above the boundary as the actual source below the boundary. This configuration forces $U$ to vanish along a single plane perpendicular to the surface normal. Moving the condition of $U = 0$ to a plane external to the medium forces a sort of quasi-isotropic condition at the boundary by using the reflected flux at the boundary.[31] The location of the extrapolated boundary, $z_b$, is given by:

$$z_b = \frac{2}{3}\frac{1+R_{eff}}{1-R_{eff}} l_{tr}, \tag{22.15}$$

where $R_{eff}$ represents an effective Fresnel coefficient for the tissue–outside medium interface. Typical values for $z_b$ are about 2 mm beyond the physical boundary. Groenhuis et al. have provided an empirical expression for $R_{eff}$, whereas Haskell et al. have calculated this coefficient to be 0.493 and 0.431 when the air–tissue interfaces are $n = 1.4$ and 1.33, respectively.[30,31]

Using the subscript *si* to denote semi-infinite, solutions to the extrapolated boundary problem take the form of $U_{SI}(r,\omega,t) = U_{INF}(r_a,\omega,t) - U_{INF}(r_i,\omega,t)$. Solving this equation with a modified extrapolated boundary condition yields the phase shift and amplitude of a photon density wave in a semi-infinite medium,[31] where is the distance between source and detector on the medium surface:

$$\Theta_{si}(\rho,\omega) = k_{imag}(\omega) r_a - \arctan\left(\frac{\eta}{\xi}\right) + \phi_0(\omega) \tag{22.16}$$

$$A_{si}(\rho,\omega) = \frac{S(\omega)}{4\pi v D}\sqrt{\xi^2 + \eta^2} \tag{22.17}$$

where

$$\begin{aligned} \xi &= \frac{1}{r_a}\exp\left[-k_{real}(\omega) r_a\right] - \cos\left[k_{imag}(\omega)(r_i - r_a)\right]\frac{1}{r_i}\exp\left[-k_{real}(\omega) r_i\right] \\ \eta &= \sin\left[k_{imag}(\omega)(r_i - r_a)\right]\frac{1}{r_i}\exp\left[-k_{real}(\omega) r_i\right] \end{aligned} \tag{22.18}$$

and

$$r_a = \sqrt{\rho^2 + (l_{tr})^2} \quad r_i = \sqrt{\rho^2 + (2z_b l_{tr})^2}. \tag{22.19}$$

### 22.3.3.2 Sensitivity to the Optical Properties

Figures 22.5 and 22.6 present the photon density as functions of the experimental variables $r$ and $f$, respectively. For the sake of comparison each line style represents the same set of optical coefficients: solid line ($\mu_a$ = 0.01 mm$^{-1}$, $\mu_s'$ = 1 mm$^{-1}$), dashed line ($\mu_a$ = 0.01 mm$^{-1}$, $\mu_s'$ = 0.5 mm$^{-1}$), and dotted line ($\mu_a$ = 0.002 mm$^{-1}$, $\mu_s'$ = 1 mm$^{-1}$). In each graph, Equations 22.16 and 22.17 provided the AC intensity and the phase. In Figure 22.5, intensities and phases have been normalized to the $r$ = 10 mm value.

Changes in $\mu_s'$ and in $\mu_a$ have different effects upon $A$ and $\Theta$. For example, assume $\mu_s'$ increases while keeping $\mu_a$ constant (dashed line vs. solid line). Increasing $\mu_s'$ results in greater attenuation (i.e., lower $A$). This same increased $\mu_s'$allows longer pathlength photons to be detected (i.e., higher $\Theta$); these increase the overall phase of the detected signal. However, if we increase $\mu_a$ and keep $\mu_s'$ constant (dotted line vs. solid line), the attenuation increases as before, but the longer pathlength photons will be deleted and hence decrease the phase. Thus, $A$ and $\Theta$ have different sensitivities to $\mu_a$ and $\mu_s'$, allowing them to be measured independently from each other.

## 22.3.4 The Standard Diffusion Equation

In the standard diffusion equation (SDE), a few more approximations simplify the mathematics: (1) the scattering mfp must be much smaller than the absorption mfp, or $\mu_s' >> \mu_a$, and (2) $3\omega D v^{-1} << 1$. The physical meaning of the term $3\omega D v^{-1}$ is an isotropic collision period ($\tau_{coll} = 1/(v\mu_s')$). The source has its own period too: $\tau_{mod} = 1/f = 2\pi/\omega$. The condition $3\omega D v^{-1} << 1$ is another way of saying that there must

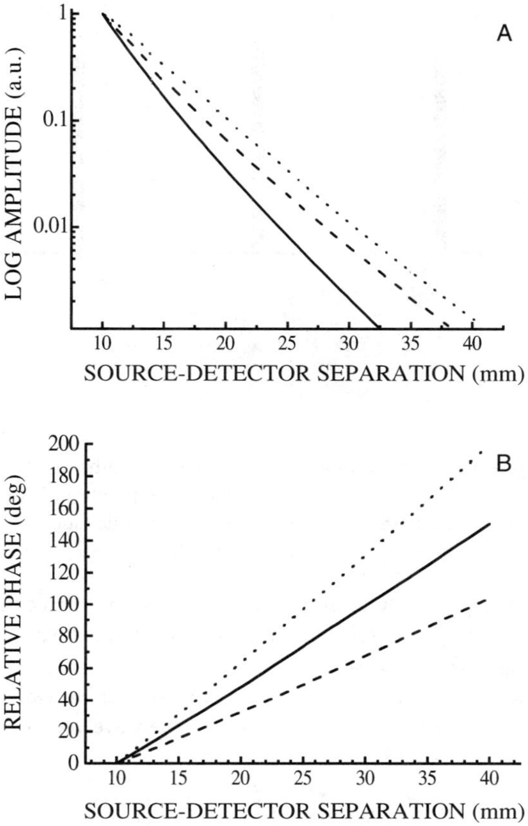


**FIGURE 22.5** Calculated changes in PDW amplitude (A) and phase (B) as a function of $r$ for three sets of optical properties: solid line ($\mu_a = 0.01\ \text{mm}^{-1}$, $\mu_s' = 1\ \text{mm}^{-1}$), dashed line ($\mu_a = 0.01\ \text{mm}^{-1}$, $\mu_s' = 0.5\ \text{mm}^{-1}$), and dotted line ($\mu_a = 0.002\ \text{mm}^{-1}$, $\mu_s' = 1\ \text{mm}^{-1}$). Intensities and phases have been normalized to the $r = 10$ mm value. Note the linearity of the phase and log (amplitude).

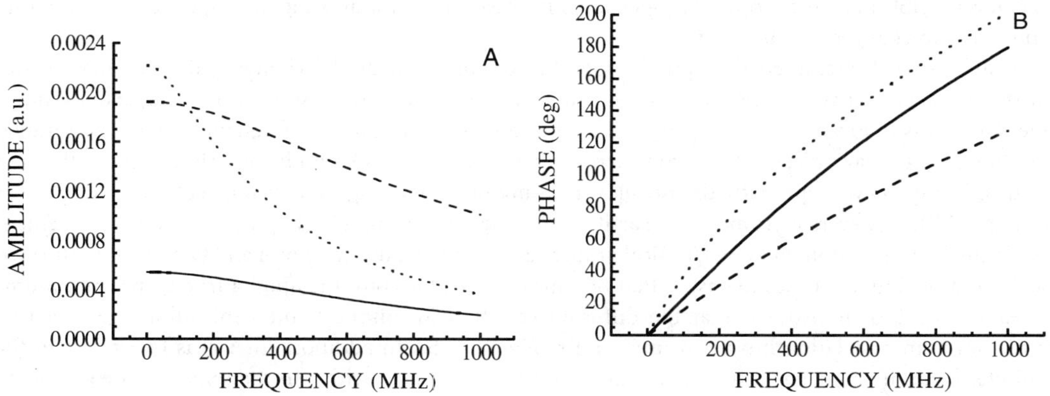


**FIGURE 22.6** Calculated changes in PDW amplitude (A) and phase (B) as a function of $f$. See Figure 22.5 for line definitions. Note that curvature changes vary with changes in $\mu_a$ and $\mu_s'$.

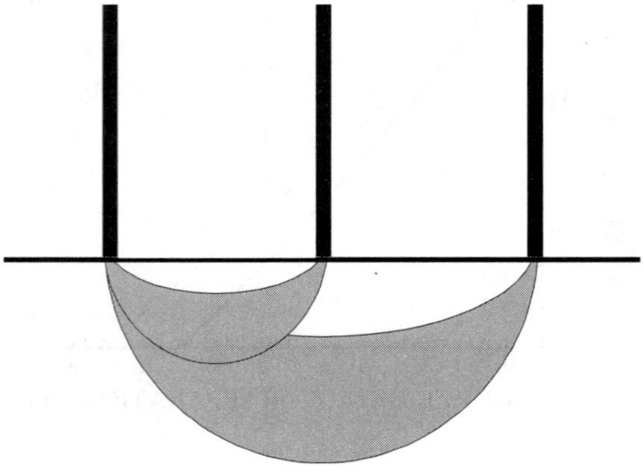


**FIGURE 22.7** Different source-detector separations sample different depths of tissue. The shaded areas represent the average volume probed by the light for a given source-detector separation. Differences in probed volume can cause problems in some layered (such as in the brain) or inhomogeneous (such as in a tumor) media. Note that changes in modulation frequency also change the probing depth, but to a lesser extent.

be many collisions within a single modulation period.[24] Tissues on average possess $\mu_s' \sim 1\ \text{mm}^{-1}$, so that $\tau_{\text{coll}}$ is typically about 5 ps, limiting $f < 1$ GHz. It has been shown conclusively that the diffusion approximation breaks down at higher frequencies.[24]

The preceding approximations simply change the layout of our previous Helmholtz equation. The solutions (infinite and semi-infinite) remain the same, but the wave vector simplifies to:

$$k^2(\omega) \equiv \frac{\mu_a}{D}\left(1 - i\frac{\omega}{v\mu_a}\right). \tag{22.20}$$

In many cases there is little difference between the $P_1$ and diffusion approximations. The slight edge in frequency response and allowed range of $\mu_s'$ make $P_1$ the more rigorous choice.

## 22.3.5 Measurements of PDW

Changing $r$ and $f$ has different effects upon the PDW intensity and phase (Figures 22.5 and 22.6). In each case it is possible to infer the optical properties of the medium by monitoring the response of the intensity and phase to changes in $r$ and/or $f$.

Tromberg et al. measured the optical properties of turbid media by changing the modulation frequency.[29] Sweeping $f$ assures the sampling of regions where $A$ and $\Theta$ are sensitive to $\mu_a$ and $\mu_s'$. For example, measurements of $\Theta$ vs. $f$ can recover $\mu_a$ and $\mu_s'$, but only when a sufficient frequency range that samples photon density wave dispersion is considered.[14,24] Fitting $A$ and $\Theta$ simultaneously produces the best results.[33] Frequency-swept methods also allow measurements at a single $r$, limiting depth-sampling errors (Figure 22.7). However, instrumentation artifacts $S$ and $\phi_0$ must be removed by a calibration measurement performed on a phantom of known optical properties. Simultaneous fitting of $A$ and $\Theta$ must be performed with multiple initial guesses to ensure that nonlinear global fits converge upon a true global minimum.

Fantini et al. demonstrated that the diffusion equation in infinite[15] and semi-infinite geometries[32] possesses a linear relationship between $\Theta$ and $r$ (Figure 22.5). In addition, $\ln(A\ r)$ is linear in $r$ in the infinite medium geometry; a function related to $\ln(A\ r^2)$ is linear with $r$ in the semi-infinite geometry. The optical properties are related to the slopes of the measured frequency-domain parameters vs. $r$. All source term information (i.e., $S$ and $\phi$) is constrained to the intercepts. Thus, simple linear fits of the intensity and phase slopes yield a direct calculation of $\mu_a$ and $\mu_s'$. Measurements can be performed at a

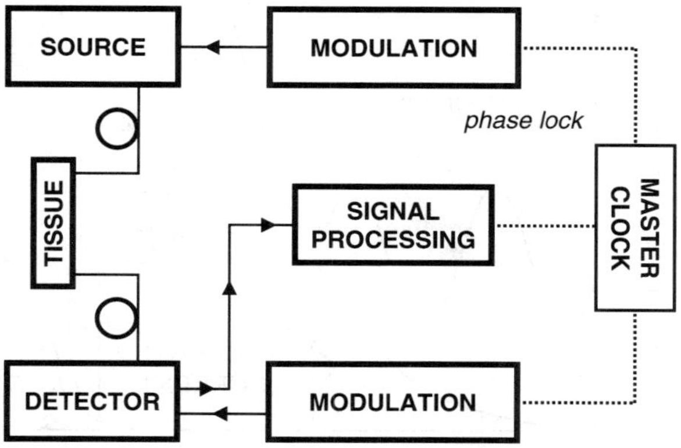


**FIGURE 22.8** Simple schematic for a frequency-domain instrument. The key element is that the modulation of source and detector must be synchronized (i.e., phase locked).

single modulation frequency, greatly simplifying the instrumentation. However, one must be careful not to use $f$ in a region of low sensitivity to $\mu_a$ or $\mu_s'$.[29] If multiple light sources are used for each $r$, then the intensities and phases of each must be calibrated, again on a phantom of known optical properties. Changing $r$ probes different tissue volumes (Figure 22.7), while changing $f$ over frequencies around $f_a$ changes the probed volume to a lesser extent.

# 22.4 Frequency-Domain Instrumentation

## 22.4.1 The Frequency-Domain Instrument

Figure 22.8 shows the essential elements of a frequency-domain instrument. There have been several detailed descriptions of frequency-domain photon migration (FDPM) devices.[16,33–35] An intensity-modulated light source (at frequency $f$) excites a sample. Diffusely reflected light is collected from the sample at frequency $f$ and guided to an optical detector. Typically, the detector is modulated at $f + \Delta f$ to process the high-frequency signal. The detected signal is then digitized and processed. Special considerations are required because some components must respond to $f > 100$ MHz. We will use the designation RF (radio frequencies) to describe the MHz regime.

In order to detect modulated optical signals, source and detector clocks must be phase-locked together. A phase shift between two unsynchronized waves is random, obscuring any physical meaning of the phase; locking together the clocks of these waves synchronizes them. Most frequency-domain systems use a 10-MHz master clock signal as a phase lock. One notable exception is the system developed by Tromberg, which uses a commercial network analyzer as RF source and RF detector, eliminating the need for external phase locking.[33] Typically, the light source is referenced to account for drifts in the source phase and intensity. When it is not possible or practical to employ this optical reference, an electronic reference may be used instead.

## 22.4.2 The Frequency-Domain Source

### 22.4.2.1 Internal Modulation

Light sources can be modulated internally by directly modulating their supply power. The modulation of the supply current of laser diodes (LD) and light-emitting diodes (LED) alters the optical output. Conventional lamps can also be modulated internally, but at frequencies far too low for use in PMS. Some lasers are modulated internally by mode locking, or coordinating the phases between different longitudinal modes within the laser cavity. A typical titanium-sapphire laser, such as the Coherent Mira system, is mode locked with 150-fs pulses firing at approximately 80 MHz.

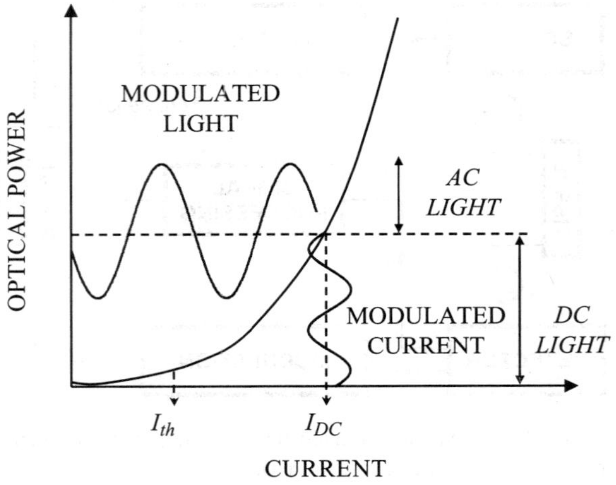


**FIGURE 22.9** Modulation of solid-state devices consists of exceeding the threshold current with a modulated current. For a given amount of RF power, the modulation depth decreases as the optical power increases; thus a balance must be made between modulation depth and total output power. Note that this same general diagram principle applies to modulating a solid-state detector such as an APD.

Semiconductor technology has substantially advanced biomedical optics by providing inexpensive and portable light sources. A DC bias current turns on the LD by marginally exceeding the threshold current $I_{th}$. A small RF current on top of the DC bias current drives the diode above and below $I_{th}$, switching the diode on and off (Figure 22.9). This condition yields the maximum modulation depth of the source (100%). One may trade modulation depth for increased optical power. Increasing the DC bias and RF currents results in more optical power, but to maintain the same modulation depth, more RF current is needed. Solid-state devices can only handle so much RF current, and will typically fail if the current drives negative. In general, NIR LEDs can be modulated in the 100-MHz range and LDs up to and beyond 1 GHz.

### 22.4.2.2 External Modulation

External modulation occurs outside the light source. Mechanical choppers can achieve modulation, but not at frequencies high enough to be useful in PMS. External devices, such as the acousto-optic modulator (AOM) and the Pockels cell, may be used on laser beams or lamps, although modulation efficiency is highly sensitive to input beam quality.

By inducing birefringence in a crystal from the application of an oscillating applied voltage, the Pockels cell is essentially a fast-switching polarizer.[36] Required voltages range from 0.1 to over 1 KV, depending upon the cell material and configuration. Pockels cells made from ammonium dihydrogen phosphase (ADP), for example, modulate light in the 0.3 to 1.2 μM spectral range up to about 500 MHz. Higher index materials such as lithium niobate can modulate light well above 1 GHz. The ability of the power source to produce voltages of adequate amplitude and frequency also limits the bandwidth of the Pockels cell. High-frequency modulation is achievable only with collimated light.

The AOM uses an applied voltage to induce ultrasonic standing waves inside a crystal to diffract light. The diffracted spots are modulated, whereas the center beam is unmodulated.[37] Two AOMs in series can be used to beat signals together, thereby increasing the frequency range.[38] Materials such as gallium phosphide can achieve bandwidth up to 1 GHz in the NIR, although a rather large supply voltage is required. Tellurium dioxide AOMs offer lower bandwidths of 300 MHz over a broader spectral region and with a considerable reduction in supply power. Compared to Pockels cells, AOMs have lower overall bandwidth and more modest voltage requirements.[36] The AOM working distance is large because the diffraction angles are small.

## 22.4.3 The Frequency-Domain Detector

### 22.4.3.1 Signal Detection

Accurately sampling a 100-MHz wave requires a clock running well into the GHz range. One method that allows accurate phase and amplitude measurements of 100-MHz signals is known as heterodyning, which is the same process used by AM radio receivers. Consider two signals, a source wave $A_1\sin(\omega t + \phi_1)$ and a detected wave $A_2\sin([\omega + \Delta\omega]t+\phi_2)$. In mixing these waves together (i.e., multiplying them), the product becomes:

$$\tfrac{1}{2}A_1A_2\left(\cos(\Delta\omega t+\Delta\phi)-\cos(2\omega t+\Delta\omega t+\phi_1+\phi_2)\right), \tag{22.21}$$

where $\Delta\phi \equiv \phi_2 - \phi_1$. Equation 22.21 expresses the sum and difference of the frequencies of two waves (i.e., a beat). Typically, $f$ is 100 MHz or greater, and $\Delta f$ is in the range of 1 KHz. Under these conditions, the first term of Equation 22.21 can be easily filtered from the second term. The usefulness of heterodyning is that the $\Delta\omega$ term is proportional to $A_2$ and depends upon $\Delta\phi$. Heterodyning translates a high frequency signal into a low frequency signal, while retaining the amplitude and phase information of the detected wave. This low-frequency signal may now be easily processed by standard electronic digitization methods.

Two additional configurations require comment. In the configuration devised by Tromberg, a network analyzer performs all heterodyning operations internally.[33] One can use any detector without concern for additional RF electronics. Secondly, lock-in detection can be used in principle, but in practice can be difficult because (1) both reference and detected signals must be carefully matched in power and (2) most lock-ins are not meant to work in the 100-MHz regime.

### 22.4.3.2 Photoemissive Detectors

Photo-multiplier tubes (PMT) have been the most commonly used detector in PMS. PMTs are highly inefficient in the NIR, resulting in quantum efficiencies well below 10% out to 800 nm. (See, for example, the PMT catalog of Hamamatsu, Japan). A commonly used PMT is the Hamamatsu R928, which has an effective NIR spectral range of 600 to 840 nm. Advantages of this PMT are its high gain of $10^7$ and low dark current (3 nA, with typical signal currents in the 30-μA range). The R928 also has a rise time of 2.2 ns, allowing in principle the detection of signals up to 400 MHz for well-focused light. Additional concerns for using any PMT are their high voltage considerations (the R928 requires about 1200 V).

Shutting off a dynode early in the chain and stopping the avalanche of electrons from reaching the anode modulates a PMT. Applying a sinusoidal voltage at frequency $f$ to the first dynode will turn the PMT on and off at the frequency $f$.[39] Microchannel plate (MCP) detectors have also been used in PMS.[24] The MCP offers immensely increased bandwidth, often exceeding several GHz. However, the dramatically lower signal currents require high-precision microwave electronics, making MCP use cumbersome.

### 22.4.3.3 Solid-State Detectors

The most useful solid-state detector for FDPM is the avalanche photodiode (APD). Photodiodes in general have excellent NIR sensitivity (Si and InGaAs), reaching quantum efficiencies near 100% at 800 nm. Conventional photodiodes have no gain mechanism, while the APD has internal gains that can reach $10^2$ to $10^3$ with a low noise figure. Photodiodes in general are very good frequency-domain detectors capable of processing optical signals in excess of 1 GHz. An APD can be modulated via its supply voltage, as with the PMT (see Figure 22.9). APDs used in PMS are usually reverse-biased, with bias voltages below 300 V. One major drawback of the APD is the small active area. An active area of 1 mm$^2$ translates into about a 600-MHz bandwidth for the Hamamatsu model S2383 APD; higher active areas generally decrease the frequency response. The APD has adequate sensitivity over the entire NIR range.

## 22.5 Current Clinical Examples

Next we provide a sample of frequency-domain PMS applied to medical problems. This list is not exhaustive, but only representative of frequency-domain PMS in tissues. Space does not permit mention of important works by other authors. Of course, frequency-domain PMS has not been the only method to investigate the examples we describe. The common thread in these emerging applications is that all measurements are quantitative and noninvasive, and reveal important aspects of tissue function not observed by other methods.

### 22.5.1 Breast Spectroscopy

#### 22.5.1.1 Past Efforts

Optical transillumination was applied to breast in 1929 as a tool to visualize the shadows cast by breast lesions. Attempts at spectroscopy started in the early 1980s by dividing images into red and NIR bands. Throughout the 1980s many clinical trials compared transillumination with mammography as a screening tool; a potpourri of conflicting clinical conclusions was the result. By 1990, transillumination sensitivity and specificity were found inferior to mammography.[18]

Classical transillumination (sometimes called diaphanography) made no attempt to distinguish between absorption and scattering and could not provide measures of tissue chromophore concentrations. Although direct visualization of lesion anatomy is difficult, if not impossible with visible and NIR optics, spectroscopic visualization of tissue function is an entirely different matter. Early studies have shown that red and NIR light absorption in breast is highly sensitive to hemoglobin. However, the presence of blood alone is not sufficient criteria for the diagnosis of cancer — a weakness that probably contributed to the high-false positive rate of transillumination. Neglecting scattering diminishes contrast even further. Additional information about water and lipid content may help improve sensitivity and specificity.

#### 22.5.1.2 New Contributions

Frequency-domain PMS measurements of breast tumors have shown increased absorption by hemoglobin and water, as well as lower $S_tO_2$ compared to normal tissue.[40] The sources of NIR contrast in healthy breast tissue have been quantified; studies have demonstrated that NIR-derived breast tissue composition and metabolism correlate with known histological changes that occur from long- and short-term hormonal use.[41,42] These measurements demonstrate clearly that PMS has the sensitivity to detect small functional changes in breast that may prove useful in detecting cancer and monitoring the effectiveness of cancer therapies. The use of endogenous contrast (i.e., hemoglobin) alleviates concerns for any unknown dynamics and undesirable toxicity of exogenous contrast agents.

Pogue et al. have constructed a frequency-domain imager that combines spectral information from 650 to 850 nm.[34] Pilot study results have shown that tumor hemoglobin concentrations correlate with tumor vascularity.[43] This finding indicates that PMS techniques could be used to monitor the progression of anti-angiogenesis drugs in the breast. It is important to note that this work uses model-based reconstructions of tissue optical properties and does not rely upon nonspectroscopic images of shadows. Wideband tissue spectroscopy coupled with this fast image capability may provide an important new approach for imaging tumor angiogenesis.

### 22.5.2 Functional Brain Monitoring

Rapid frequency-domain techniques have proved useful in probing human brain function. There have been numerous attempts to measure changes in brain hemoglobin concentrations using NIR spectroscopy (see Chapter 28 for more details about NIR functional imaging). Activation events, or the so-called "slow" brain signals, are hemodynamic in nature, occurring on a 2- to 10-s timescale. Relatively simple optical instruments can monitor these hemodynamic changes via absorption. However, it has been observed that changes in brain optical properties are also due to changes in scattering, the so-called "fast" events

(50 to 500 ms). Classic NIR methods that cannot separate absorption from scattering will not quantify these events.

E. Gratton, an early pioneer in the application of frequency-domain methods, used frequency-domain PMS to measure absorption and scattering changes in brain *in vivo*.[44] Gratton and colleagues have reported fast (i.e., <1 s) brain optical signal changes that are apparently the result of scattering fluctuations occurring during localized neural tissue activation. Using a commercial two-wavelength, multidistance device with a measurement time of 160 ms (ISS, Urbana, IL), Gratton et al. measured fast and slow changes in brain optical properties that correlate with brain activity in response to periodic stimuli. In addition, they have shown that these optical changes correlate with electrical changes in the brain. Further work by Cheung and co-workers has centered around measuring correlations in phase and amplitude as a means of quantifying brain optical property changes.[45]

## 22.5.3 Measurements of Tissue Physiology

### 22.5.3.1 Deep-Tissue Arterial and Venous Oximetry

Pulse oximetry provides the hemoglobin saturation in arterial blood ($S_aO_2$) by monitoring pulsitile changes in NIR attenuation (see Chapter 26 for greater detail). The technique is generally limited to thin sections of tissue, such as the finger tip or ear lobe. Frequency-domain PMS provides the $S_tO_2$, which is more representative of the tissue capillary bed (i.e., mixed arterial and venous). Franchescini and co-workers have used a multidistance, frequency-domain technique to measure arterial[46] and venous saturations ($S_vO_2$)[47] in thick tissues independently. The frequency-domain technique is required to provide the photon path length, which standard NIR techniques cannot do. The discrimination between $S_vO_2$ and $S_aO_2$ is accomplished by locking onto the harmonic content of changes in $\mu_a$ which, at the heartbeat frequency correspond to arterial signals; changes in $\mu_a$ at the respiratory frequency correspond to venous signals. This technique could be important in brain and in muscle physiology because no way to measure $S_vO_2$ noninvasively exists. This technique can be performed in real time.

### 22.5.3.2 Monitoring Photodynamic Therapy Response

Photodynamic therapy (PDT) drugs provide new treatment options in the management of cancer (see Chapters 40 and 41). A photosensitizing drug in the presence of oxygen will cause tissue necrosis after absorbing an appropriate amount of optical radiation. Without knowledge of the optical properties of the tissue, the exact amount of light dosage required for treatment remains an unknown variable, limiting the effectiveness of the treatment.

Pham et al. used a frequency-domain, multifrequency approach to measure changes in tumor physiology in response to a photosensitizer in a rat ovarian cancer model.[48] PMS revealed significant decreases in THC and $S_tO_2$ after the application and subsequent optical activation of the photosensitizer. Histologic inspection of the tumors revealed that the long-term efficacy of the treatment highly correlated with the measured changes in THC and $S_tO_2$. As PDT use increases in clinical practice, PMS could provide rapid feedback about the effectiveness of the delivered photosensitizer dose.

# Acknowledgments

We wish to thank many people who have worked with us in contributing to the basic development of PMS: Tuan Pham, Joshua Fishkin, Olivier Coqouz, Eric Anderson, Steen Madsen, and Lars Svaasand. The authors would also like to thank many organizations for their generous support during the development of the ideas in this chapter: National Institutes of Health (NIH) Laser Microbeam and Medical Program (RR-01192), NIH (R29-GM50958), Department of Energy (DOE DE-FG03–91ER61227), Air Force Office of Scientific Research (F49620-00-1-0371), and the Beckman, Whitaker, and Hewitt Foundations. A.E.C. cheerfully acknowledges support from the U.S. Army Medical Research and Material Command (DAMD17-98-1-8186).

## References


1. Wilson, B.C. et al., Tissue optical properties in relation to light propagation models and *in vivo* dosimetry, in *Photon Migration in Tissues*, Chance, B., Ed., Plenum Press, New York, 1988, p. 25.
2. Sevick, E.M. et al., Quantitation of time-resolved and frequency-resolved optical spectra for the determination of tissue oxygenation, *Anal. Biochem.*, 195, 330, 1991.
3. Ertefai, S. and Profio, A.E., Spectral transmittance and contrast in breast diaphanography, *Med. Phys.*, 12, 393, 1985.
4. Chance, B. et al., Photon migration in muscle and brain, in *Photon Migration in Tissues*, Chance, B., Ed., Plenum Press, New York, 1990, p. 121.
5. Bonner, R.F. et al., Model for photon migration in turbid biological media, *J. Opt. Soc. Am. A*, 4, 423, 1987.
6. Delpy, D.T. et al., Estimation of optical pathlength through tissue from direct time of flight measurement, *Phys. Med. Biol.*, 33, 1433, 1988.
7. Chance, B. et al., Comparison of time-resolved and time-unresolved measurements of deoxyhemoglobin in brain, *Proc. Natl. Acad. Sci. U.S.A.*, 85, 4971, 1988.
8. Ishimaru, A., *Wave Propagation and Scattering in Random Media,* Academic Press, New York, 1978, p. 572.
9. Star, W.M., Marijnissen, J.P.A., and van Gemert, M.J.C., Light dosimetry in optical phantoms and in tissues. I. Multiple flux and transport theory, *Phys. Med. Biol.*, 33, 435, 1988.
10. Profio, A.E. and Doiron, D.R., Transport of light in tissue in photodynamic therapy, *Photochem. Photobiol.*, 46, 591, 1987.
11. Patterson, M.S., Chance, B., and Wilson, B.C., Time resolved reflectance and transmittance for the non-invasive measurement of tissue optical properties, *Appl. Opt.*, 28, 2331, 1989.
12. Patterson, M.S. et al., Frequency-domain reflectance for the determination of the scattering and absorption properties of tissue, *Appl. Opt.*, 30, 4474, 1991.
13. Fishkin, J.B. and Gratton, E., Propagation of photon-density waves in strongly scattering media containing an absorbing semi-infinite plane bounded by a straight edge, *J. Opt. Soc. Am. A*, 10, 127, 1993.
14. Svaasand, L.O. and Tromberg, B.J., On the properties of optical waves in turbid media, in *Future Trends in Biomedical Applications of Lasers,* SPIE, Bellingham, WA, 1991, p. 41.
15. Fantini, S. et al., Quantitative determination of the absorption spectra of chromophores in strongly scattering media: a light-emitting-diode based technique, *Appl. Opt.*, 33, 5204, 1994.
16. Chance, B. et al., Phase measurement of light absorption and scatter in human tissue, *Rev. Sci. Instrum.*, 69, 3457, 1998.
17. Hull, E.L., Nichols, M.G., and Foster, T.H., Quantitative broadband near-infrared spectroscopy of tissue-simulating phantoms containing erythrocytes, *Phys. Med. Biol.*, 43, 3381, 1998.
18. Cerussi, A.E., Jakubowski, D., Shah, N., Bevilacqua, F., Lanning, R., Berger, A.J., Hsiang, D., Butler, J., Holcombe, R.F., and Tromberg, B.J., Spectroscopy enhances the information content of optical mammography, *J. Biomed. Opt.*, 7(1), 60, 2002.
19. See special issues of *J. Biomed. Opt.*, 5, 2000 and *Appl. Opt.*, 36, 1997.
20. Müller, G. et al., *Medical Optical Tomography: Functional Imaging and Monitoring,* SPIE, Bellingham, WA, 1993.
21. Chance, B., *Photon Migration in Tissues*, Plenum Press, New York, 1990.
22. Liu, H. et al., Influence of blood vessels on the measurement of hemoglobin oxygenation as determined by time-resolved reflectance spectroscopy, *Med. Phys.*, 22, 1209, 1995.
23. Duderstadt, J.J. and Hamilton, L.J., *Nuclear Reactor Analysis*, John Wiley & Sons, New York, 1976, p. 143.
24. Fishkin, J.B. et al., Gigahertz photon density waves in a turbid medium: theory and experiments, *Phys. Rev. E*, 53, 2307, 1996.

25. Starr, W.M., Diffusion theory of light transport, in *Optical-Thermal Response of Laser Irradiated Tissue*, van Gemert, M.J.C., Ed., Plenum Press, New York, 1995, p. 131.
26. Yodh, A. and Chance, B., Spectroscopy and imaging with diffusing light, *Phys. Today*, 48, 34, 1996.
27. Kaltenbach, J.M. and Kaschke, M., Frequency- and time-domain modeling of light transport in random media, in *Medical Optical Tomography: Functional Imaging and Monitoring*, Müller, G. et al., Eds., SPIE, Bellingham, WA, 1993, p. 65.
28. Boas, D.A. et al., Scattering and wavelength transduction of diffuse photon density waves, *Phys. Rev. E*, 47, R2999, 1993.
29. Tromberg, B.J. et al., Properties of photon density waves in multiple-scattering media, *Appl. Opt.*, 32, 607, 1993.
30. Groenhuis, R.A.J., Ferwerda, H.A., and Ten Bosch, J.J., Scattering and absorption of turbid materials determined from reflection measurements. I. Theory, *Appl. Opt.*, 22, 2456, 1983.
31. Haskell, R.C. et al., Boundary conditions for the diffusion equation in radiative transfer, *J. Opt. Soc. Am. A*, 11, 2727, 1994.
32. Fantini, S., Franceschini, M.A., and Gratton, E., Semi-infinite-geometry boundary problem for light migration in highly scattering media: a frequency-domain study in the diffusion approximation, *J. Opt. Soc. Am. B*, 11, 2128, 1994.
33. Pham, T. et al., A broad bandwidth frequency domain instrument for quantitative tissue optical spectroscopy, *Rev. Sci. Instrum.*, 71, 1, 2000.
34. McBride, T.O. et al., A parallel-detection frequency-domain near-infrared tomography system for hemoglobin imaging of the breast *in vivo*, *Rev. Sci. Instrum.*, 72, 1817, 2001.
35. Franceschini, M.A. et al., Optical study of the skeletal muscle during exercise with a second generation frequency-domain tissue oximeter, *Proc. SPIE*, 2979, 807, 1997.
36. Simcik, J., Electro-optic and acousto-optic devices, http://cord.org/cm/leot/course04_mod07/mod04_07.htm.
37. Saleh, B.E.A. and Teich, M.C., *Fundamentals of Photonics*, John Wiley & Sons, New York, 1991, p. 800.
38. Piston, D.W. et al., Wide-band acousto-optic light modulator for frequency domain fluorometry and phosphorimetry, *Rev. Sci. Instrum.*, 60, 2596, 1989.
39. Gratton, E. and Limkeman, M., A continuously variable frequency cross-correlation phase fluorometer with picosecond resolution, *Biophys. J.*, 44, 315, 1983.
40. Tromberg, B.J. et al., Non-invasive *in vivo* characterization of breast tumors using photon migration spectroscopy, *Neoplasia*, 2, 1, 2000.
41. Cerussi, A.E. et al., Sources of absorption and scattering contrast for non-invasive optical mammography, *Acad. Radiol.*, 8, 211, 2001.
42. Shah, N. et al., Non-invasive functional optical spectroscopy of human breast tissue, *Proc. Nat. Acad. Sci. U.S.A.*, 98(8), 4420, 2001.
43. Pogue, B.W. et al., Quantitative hemoglobin tomography with diffuse near-infrared spectroscopy: pilot results in the breast, *Radiology*, 218, 261, 2001.
44. Gratton, E. et al., Measurements of scattering and absorption changes in muscle and brain, *Philos. Trans. R. Soc. Lond. B*, 352, 727, 1997.
45. Cheung, C. et al., *In vivo* cerebrovascular measurement combining diffuse near-infrared absorption and correlation spectroscopies, *Phys. Med. Biol.*, 46, 2053, 2001.
46. Franceschini, M.A., Gratton, E., and Fantini, S., Noninvasive optical method of measuring tissue and arterial saturation: an application to absolute pulse oximetry of the brain, *Opt. Lett.*, 24, 829, 1999.
47. Franceschini, M.A., Boas, J.B., and Fantini, S., Near-infrared spiroximetry: non-invasive measurements of venous saturation in piglets and human subjects, *J. Appl. Physiol.*, 92(1), 372, 2002.
48. Pham, T.H. et al., Monitoring tumor response during photodynamic therapy using near-infrared photon-migration spectroscopy, *Photochem. Photobiol.*, 73, 669, 2001.

# 23

# Atomic Spectrometry in Biological and Clinical Analysis

Andrew Taylor
*University of Surrey*
*Guildford, Surrey, U.K.*



## 23.1 Atomic Spectrometry: Introduction

Within biological systems, elements may be classified as those essential to the well-being of the organism and those that have no known or demonstrable function and are therefore regarded as nonessential.[1,2] On the basis of the usual concentrations within tissues and body fluids they may also be classified as major or trace elements, trace elements being defined as those that individually contribute no more than 0.01% of the dry body mass.[3] All the major elements and a limited number of trace elements are essential (Table 23.1); all others that may be detected are nonessential. Essential elements are required for various biological functions,[1] e.g.,

- Enzyme structure and function
- Hormone structure and function
- Vitamin structure and function (e.g., $B_{12}$)
- Transport of oxygen
- Structure of macromolecules

When any are present in less than optimal concentrations, symptoms of morbidity will be evident — indeed, the severity of deficiency may be such that death is the eventual outcome (Figure 23.1). At the same time, all elements, whether essential or nonessential, are toxic if they accumulate in tissues to sufficiently large concentrations.[3]

Therefore, it is important that accurate measurements in biological and clinical specimens may be obtained for fundamental research involving mechanistic aspects of trace element biology and so that deficiencies or excess may be detected in various situations.[1,4] In addition to these concepts of essentiality

0-8493-1116-0/03/$0.00+$1.50

**TABLE 23.1** Major Elements and Essential Trace Elements in Clinical and Biological Samples

| | Minerals | Nonminerals |
|---|---|---|
| Major elements | Calcium, iron, magnesium, potassium, sodium | Carbon, chlorine, hydrogen, nitrogen, oxygen, phosphorus, sulfur |
| Essential elements (not all proven to have essential roles in man) | Chromium, cobalt, copper, iron, manganese, molybdenum, nickel, selenium, silicon, vanadium, zinc | Fluorine, iodine |

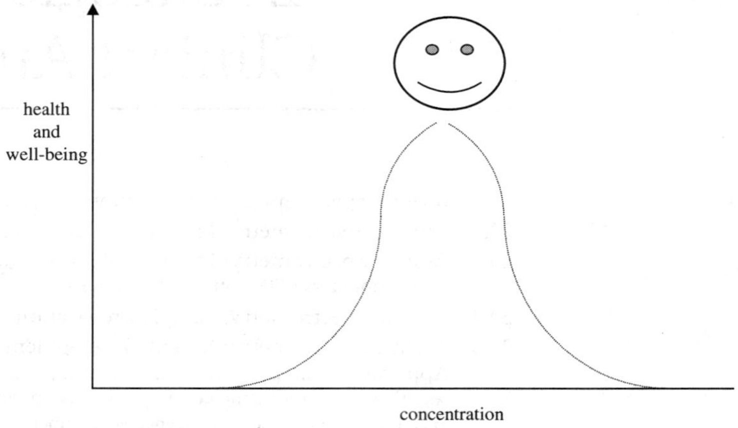


**FIGURE 23.1** Representation of the relationship between essential elements and well-being.

and nonessentiality and deficiency and toxicity, elements are also used therapeutically and concentrations in body fluids may be required to monitor the effectiveness and safety of the treatment.[3]

The analytical techniques used must afford the sensitivity necessary to measure concentrations below 1 ppm, often low ppb levels, in specimens of just a few microliters or milligrams with almost total specificity and relatively few interferences. These demands are met by the atomic spectrometry techniques described in this chapter.[5,6]

Put very simply, spectroscopy is concerned with the study of interactions between electromagnetic radiation and matter, while spectrometry is the exploitation of these interactions to gain analytical (quantitative or qualitative) information. As indicated by the terminology, the interactions studied in atomic spectrometry involve atoms (rather than molecules) and the purpose is to determine the concentration or, more rarely, simply the presence of an element within a sample. The appropriate energy required to interact with atoms is that derived from the UV-visible section, and by high-energy particles, of the electromagnetic spectrum. In practice, atomic spectrometry involves the emission, absorption, or fluorescence of such energy. Thus this chapter describes atomic emission spectrometry (AES), atomic absorption spectrometry (AAS), atomic fluorescence spectrometry (AFS), and x-ray fluorescence spectrometry (XRF). Although these procedures are used extensively, elements may also be determined by a number of other techniques; of particular relevance to biological and clinical specimens are those involving inorganic mass spectrometry, activation analysis, and anodic stripping voltametry. Recognizing the importance of these techniques, brief reference will be included, particularly where features overlap with the more conventional atomic spectrometry.

## 23.2 Atomic Spectrometry: Principles

Analytical AES, AAS, and AFS are quantitative techniques that exploit interactions between UV-visible light and the outer shell electrons of free, gaseous, uncharged atoms. In XRF and related techniques,

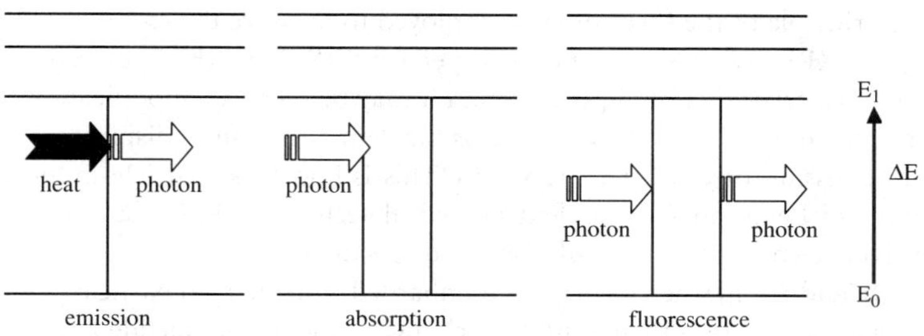


**FIGURE 23.2** Energy transitions associated with atomic emission, absorption, and fluorescence.

high-energy particles interact with inner shell electrons to initiate further electron transitions within the atom. Each element has a characteristic atomic structure with a positively charged nucleus surrounded by electrons in orbital shells to provide neutrality. These electrons occupy discrete energy levels, but it is possible for an electron to be moved from one level to another within the atom by the introduction of energy (Figure 23.2). This energy may be supplied by collisions with other atoms, i.e., heating (for AES), as photons of light (for AAS and AFS), or as high-energy particles (for XRF). Such transitions occur only if the available energy is equal to the difference between two levels ($\Delta E$). Uncharged atoms may exist at the lowest energy level or ground state ($E_0$), or at any one of a series of excited states ($E_n$) depending on how certain electrons have been moved to higher energy levels, although it is usual to consider just the first transition. Energy levels and the $\Delta E$s associated with electron transitions are unique for each element.

The $\Delta E$ for movements of outer shell electrons in most elements correspond to the energy equivalent to UV-visible radiation and these transitions are used for AES, AAS, and AFS. The energy of a photon (E) is characterized by

$$E = h\upsilon \tag{23.1}$$

where h = Planck's constant and $\upsilon$ = the frequency of the waveform corresponding to that photon. Furthermore, frequency and wavelength are related as

$$\upsilon = c/\lambda \tag{23.2}$$

where c = the velocity of light and $\lambda$ = the wavelength. Therefore,

$$E = hc/\lambda \tag{23.3}$$

and it follows that a specific transition, $\Delta E$, is associated with a unique wavelength.[6]

Under appropriate conditions, outer shell electrons of vaporized atoms may be excited by thermal energy (i.e., collisions with other atoms). As these electrons return to the more stable ground state, energy is lost. As Figure 23.2 shows, some of this energy will be in the form of emitted light, which can be measured with a detector; this is AES. When light (radiant energy) of a characteristic wavelength enters an analytical system, outer shell electrons of the corresponding atoms will be excited as energy is absorbed. Consequently, the amount of light transmitted from the system to the detector will be attenuated; this is understood as AAS. Finally, some of the radiant energy absorbed by ground state atoms can be emitted as light as the atom returns to the ground state, i.e., AFS.

When high-energy photons, electrons or protons strike a solid sample, an electron from the inner shells (K, L, or M) of a constituent atom may be displaced. The resulting orbital vacancy is filled by an outer shell electron, with an accompanying emission of an x-ray photon; its energy is equal to the difference between the energy levels involved. This emission is known as x-ray fluorescence (XRF). The energy of the emission, i.e., the wavelength, is characteristic of the atom (element) from which it originated, while the intensity of the emission is related to the concentration of the atoms in the sample.

Depending on the principle of the spectrometer employed to measure the emission, XRF is divided into wavelength dispersive XRF (WDXRF) or energy dispersive XRF (EDXRF). Total reflection XRF (TXRF) is usually described as a separate technique although it may be seen as a modification of EDXRF. High-energy hydrogen or helium ions may also be used as incident radiation to displace an electron from a K or L shell with an emission of characteristic x-rays. This is known as particle-induced x-ray emission (PIXE). In addition to being an independent analytical technique, PIXE can be used with electron microscopy to provide elemental analysis of visualized specimens.

Analogous to XRF and PIXE, when an atom is bombarded with charged particles a radioactive nuclide may be formed. Gamma radiation then emitted is characteristic of the nuclide and the intensity of the emission is proportional to the analyte concentration. Neutrons are most often used for excitation and the technique is then called neutron activation analysis. This multielement technique requires specialized facilities and is not widely available.

The high-temperature inductively coupled plasma (see below) is an effective ion source for a mass spectrometer; the technique of inductively coupled plasma-mass spectrometry (ICP-MS) is extensively used for measurements of trace elements in clinical and biological materials. It affords very sensitive multielement analysis and also provides for the determination of stable isotopes.[7,8]

Anodic stripping voltametry is an electrochemical procedure that offers exceptional sensitivity for some applications. It is ideally suited for large sample volumes such as water specimens, but it is widely applied to the measurement of lead in blood, particularly in North America.

It follows from Equations 23.1 to 23.3 that the wavelengths of the absorbed and emitted energies are unique to a given element. It is this that makes atomic spectrometric techniques specific, so that one element can be determined even in the presence of an enormous excess of a chemically similar element.[6]

## 23.3 Atomic Spectrometry: Instrumentation

Formation of the atomic vapor, i.e., atomization, is central to emission, absorption, and fluorescence by atoms. Atomizers and the devices for sample introduction are the heart of the instrumentation, with an associated spectrometer for wavelength separation and detection of light. Atomization involves the following steps: removal of solvent (drying), separation from anion or other components of the matrix, and reduction of ions to the ground state atom. Energy necessary to accomplish these steps is supplied as heat. The proportion of an atom population within the vapor, as the excited or the ground state atoms, is influenced by the temperature and the atomic structure of the element. At the temperatures of flame and electrothermal atomizers, around 2 to 3000 K, the ratio is at least $10^{-6}$:1 for most elements and AAS affords superior sensitivity to AES. With the much higher temperatures provided by an inductively coupled plasma atomizer, the proportion changes and AES may be favored.

### 23.3.1 Flame Atomizers

The flame provides for simple, rapid measurements with few interferences and is preferred wherever the analyte concentration is suitable. The typical pneumatic nebulizer for sample introduction is inefficient and although elements such as Na and K may be measured in biological specimens by flame AES, flame atomization is more usually suited to AAS and AFS. With AAS, measurements are possible with specimens where concentrations are around 1 μg/ml or more. Devices have been developed that overcome the limitations of the pneumatic nebulizer by by-passing the nebulizer so that 100% of the sample is atomized; they also introduce the sample as a single, rapid pulse rather than by continuous flow. These approaches are also features of electrothermal atomization and vapor generation procedures (see below). Lower detection limits are obtained with AFS but various constraints restrict this technique, with a few exceptions, to the hydride-forming elements (see below).[9]

### 23.3.2 Electrothermal Atomizers

Most systems use an electrically heated graphite tube, a technique often called graphite furnace atomization, although other materials are sometimes employed.[9,10] With a programmed temperature sequence, the test solution (10 to 50 µl) is dried, organic material destroyed, and the analyte ions dissociated from anions for reduction to ground state atoms. The temperatures achieved by this technique can be up to 3000 K so that refractory elements such as aluminium and chromium will form an atomic vapor. Because all of the sample is atomized and retained within the small volume of the furnace, a dense atom population is produced. The technique is, therefore, very sensitive and allows measurement of µg/l concentrations. Although the technique is widely used for AAS, electrothermal atomization is suitable for AES and for sample introduction into an inductively coupled plasma.[9,11]

### 23.3.3 Inductively Coupled Plasmas

As stated previously, high-temperature atomizers are required to provide useful numbers of excited atoms for AES. Historical sources include arcs and sparks but modern instruments use argon, or some other inert gas, in the form of a plasma. The plasma is formed when gas atoms are ionized, $Ar + e^- <=> Ar^+ + 2e^-$ — a process generated by seeding from a high-voltage spark — and is sustained with energy from an induction coil connected to a radio-frequency generator. This is known as an inductively coupled plasma (ICP). Plasmas exist at temperatures of up to 10,000 K and in the instrument have the appearance of a torch. Samples can be introduced via a nebulizer, by vapor generation procedures, by vaporization from a graphite atomizer or by laser ablation of solid specimens.

The main feature of AES is that it permits multielement analysis. Optical systems direct the emitted light via a monochromator to a single detector or to an array of monochromators and detectors positioned around the plasma. With the first arrangement, a sequential series of readings are made with the monochromator driven to give each of the wavelengths of interest in turn. Simultaneous readings can be made with the second arrangement. For most elements the analytical sensitivity for ICP-AES is similar to that obtained with flame AAS at the part per million level.

### 23.3.4 X-Ray Fluorescence

XRF requires that specimens be irradiated by high-energy photons. In most instruments the source is the polychromatic primary beam from x-ray tubes. Of interest to biological applications, however, is the use of radioactive isotopes as sources. Isotopes such as $^{244}Cm$, $^{241}Am$, $^{55}Fe$, and $^{109}Cd$ are used.[13,14] The latter is particularly important as the source in portable instruments developed for *in vivo* XRF (see below). A growing number of publications refer to the use of synchrotron radiation that acts as a high-resolution highly energetic source.[13]

Because sample matrix contributes considerably to signal intensity, calibration can be difficult, usually requiring the use of different reference materials and internal standardization. Fewer problems are encountered with samples prepared as very thin films and in TXRF. Together with the effect of the matrix, sensitivity is also influenced by wavelength, such that lighter elements present a difficult analytical challenge.

In WDXRF, high-intensity x-rays (e.g., from a 3-kW x-ray tube) are used to induce the fluorescence emission, which is dispersed into individual spectral lines by reflection at an analyzer crystal. The diffracted beams are collimated and directed onto a photon detector. As with ICP-AES, spectrometers may operate sequentially, with a number of interchangeable crystals, to permit the measurement of the full range of elements, or in a multichannel (simultaneous) mode usually preset for specific analytes. Detection limits for light elements are 10 to 100 times lower than with EDXRF. Resolution is good, although less so at shorter wavelengths. Sequential instruments require long analysis times to measure several elements compared with the more expensive simultaneous instruments or EDXRF technology.

For EDXRF, x-rays emitted from the sample are directed together into a crystal detector. A pulse of current is generated with a height proportional to the energy of the x-ray photon. The different energies

**TABLE 23.2** Approaches to Sample Preparation

| Procedure | Remarks |
|---|---|
| Dilution, protein precipitation | Using simple off-line arrangements or flow-injection manifold |
| Dry ashing | Using a muffle furnace or a low-temperature asher |
| Acid digestion | (1) In open vessels with convection or microwave heating |
| | (2) In sealed vessels to increase the reaction pressure |
| Base dissolution | Using quaternary ammonium hydroxides |
| Chelation and solvent extraction | For analyte enhancement and removal of interferences |
| Trapping onto solid phase media | For analyte enhancement and removal of interferences |

associated with the various atoms (elements) in the sample are sorted electronically. Lower energy sources (a low-power x-ray tube or an isotopic source) are used. The detector must be maintained at a very low temperature and in a clean vacuum. Analysis times are 10 to 30 times longer than with WDXRF but, as a truly multielement technique, the total time is not necessarily any greater.

When a collimated beam of x-rays is directed against an optically flat surface at an angle of around 5′, total reflection will occur. This is the principle of TXRF in which the sample is exposed to primary and total reflected beams and is excited to fluoresce. Emitted radiation is resolved and measured as an ED spectrum. There is effectively no absorption by the matrix, so measurement and calibration are much simpler and sensitivities are greater than with other x-ray techniques.

## 23.4 Atomic Spectrometry: Sample Preparation

The objectives for preparation of biomedical specimens are to remove interfering components from the matrix and to adjust the concentration of analyte to facilitate the actual measurement. These objectives may be realized by a number of approaches (Table 23.2).

Methods for destruction of the organic matrix by simple heating or by acid digestion have been used extensively and are thoroughly validated. Microwave heating is now well established for this purpose, with specifically constructed apparatus to avoid dangers of excessive pressure within reaction vessels. Although the number of specimens that can be processed is not large, microwave heating affords rapid digestion and low reagent blanks. More recent developments include continuous flow systems for automated digestion linked directly to the instrument for measurement of the analytes.

Preconcentration by liquid-liquid partitioning is a widely used procedure. Analyte atoms in a large volume of aqueous specimen are complexed with an appropriate agent and then extracted into a smaller volume of organic solvent. This leads to enhancement of concentration and also removes the analyte from potential or real interferences in the original matrix. It is used to measure lead in blood, metals in urine and for other applications. Although preconcentration by trapping onto solid phase media represents the area where much of the recent interest in FAAS has been focused, it is relevant to all sample preparation work.[14,15] The original work involved adsorption onto material such as charcoal or alumina but newer phases include ion-exchange resins and novel support systems to which functional groups are added to confer increased selectivity and capacity.[16] Trapping of analyte from dilute sample and elution into a small volume of release solution may be accomplished off-line; however, developments in flow-injection analysis provide for the assembly of simple on-line manifolds so that complete measurements may be carried through automatically.[14,17] Developments in these applications involving a wide range of biological and clinical sample types and elements are regularly published.

Vapor generation procedures were referred to in earlier sections. These permit the rapid introduction of 100% of the sample into the atomizer and are used for AAS, AFS, ICP-AES, and ICP-MS.[9,18] Certain elements such as arsenic, selenium and bismuth readily form gaseous hydrides, e.g., arsine ($AsH_3$), that are transferred by a flow of inert gas to a heated silica tube positioned in the light path. The tube is heated by the air-acetylene flame or by an electric current and the temperature is sufficient to cause dissociation of the hydride and atomization of the analyte. Thus, there is no loss of specimen, all the atoms enter the light path within a few seconds and they are trapped within the silica tube, which retards their dispersion. Hydride generation AAS allows the detection of a few nanograms of analyte from whatever sample volume is placed into the reaction flask. Mercury forms a vapor at ambient temperatures and this property is the basis for cold vapor generation. A reducing agent is added to the sample solution to convert $Hg^{2+}$ to the elemental mercury. Agitation or bubbling of gas through the solution causes rapid vaporization of the atomic mercury, which is then transferred to a flow-through cell placed in the light path. As with hydride generation, the detection limit is a few nanograms and common instrumentation to accomplish both procedures has been developed by some manufacturers.

Appreciation of the importance of determining not just the total concentration of an element in a specimen, but also something of its distribution, is now well established. This concept of speciation is applied to associations with different molecules such as proteins, to different organo-metallic compounds, and to different valence states.[19] A number of preparative procedures are available to separate or speciate the analyte,[20] but much innovation is directed to chromatographic and electrophoretic techniques that are coupled directly to the atomic spectrometric equipment to form an integrated analytical arrangement. Examples are presented in the following section.

## 23.5 Atomic Spectrometry: Recent Developments and Applications

Measurements of major and trace elements in biological and clinical specimens are required in many situations:

- Work to determine mechanisms of action within biological systems at cellular and biochemical levels, of essentiality and of toxicity, involves knowing concentrations within the experimental systems.
- Determining trace element and mineral physiology, routes of absorption, tissue distribution and concentrations in normal subjects and in patients with inborn errors involves these processes, e.g., copper and Wilson's disease.
- Nutritional studies investigate possible deficiencies of essential elements, e.g., in subjects with poor diets or patients receiving long-term total parenteral feeding (where protocols for regular monitoring are generally recommended).[1] In addition to measuring trace elements in blood and urine, such work may include analysis of foods and special investigations to assess intestinal absorption.
- Investigation of undue exposure to elements:[4] Increased exposure to minerals and trace elements can cause morbidity and some are carcinogenic. While the function of many organs may be perturbed by accumulation of metals, the kidney, liver, nervous, intestinal, and hemopoietec systems are more likely to be involved. Accidental, (or even deliberate suicidal or homicidal) exposures to trace elements feature in the differential diagnosis when considering signs and symptoms involving these sites. Increased exposure may be consequent on sources within the environment or in the home, associated with hobbies or from unusual cosmetics and remedies. In an occupational setting, biological monitoring is important in the implementation of health and safety regulations.[21]
- In other situations iatrogenic poisoning can occur. Profound toxicity has been observed for many elements including aluminium, bismuth, and manganese.[4]

### 23.5.1 Atomic Emission Spectrometry

Sodium, potassium and lithium are usually measured by flame AES or with ion-selective electrodes although they can also be determined by flame AAS. At the temperature of the flame there is no useful emission of other biologically important elements. However, with the greater energy of the ICP, much lower detection limits, typically around 1 μg/ml, are obtained and many elements may be determined simultaneously in solutions prepared from biological tissues. Furthermore, a few elements, such as boron, phosphorus, and sulfur, cannot be measured by AAS but are determined by ICP-AES.

Recent developments with the optical systems and array detectors have led to improvements in sensitivity and data collection. In consequence, elements such as copper, zinc, and aluminium may be measured in blood plasma,[22–24] while the expanded information caught by detectors is making it possible for powerful chemometric manipulation of individual signals to be undertaken. ICP-AES now has an established role for monitoring patients with possible compromised nutritional status[22] and those who receive hemodialysis to treat chronic renal failure and are at risk of developing aluminum toxicity.[23] With the multielement feature of the technique, ICP-AES is widely used for analysis of foodstuffs and tissues samples.[14] The convenience of this approach has, however, encouraged the dubious "diagnostic practice" of hair analysis among some laboratories.[25] Vapor generation techniques for sample introduction (see below) are possible.

### 23.5.2 Atomic Absorption Spectrometry

Measurement of calcium in serum was the first analysis to which AAS was applied and is an obvious example of how the technique is useful for biomedical analysis. Elements present in biological fluids at a sufficiently high concentration to be measured by flame AAS are lithium and gold, when used to treat depression and rheumatoid arthritis, respectively, and calcium, magnesium, iron, copper, and zinc. Flame AAS is used by the large majority of laboratories needing to measure these elements. It fulfills an important role in the investigation of patients with possible nutritional problems, genetic disorders, or other relevant clinical challenges. Other elements are present in fluids at too low a concentration to be measured by conventional FAAS with pneumatic nebulization. With more exotic fluids, e.g., seminal plasma or cerebrospinal fluid, analysis may just be possible for a very few elements.

The concentrations of many metals in plant, animal, or human tissues are usually much higher than in biological fluids and very often the weight of an available specimen is such that a relatively large mass of analyte is recovered into a small volume of solution, thus enhancing the concentration still further. For the analysis of tissues (including specimens such as hair and the cellular fractions of blood) following sample dissolution steps, FAAS is suitable for measurement of many of the biologically important elements.[14]

Atom traps, such as the slotted quartz tube, increase the sensitivity associated with FAAS for more volatile elements;[26] sporadic reports appear of their use in simplified methods for analysis of biological fluids. However, if concentrations are low it is more usual to take advantage of the lower limits of detection provided by electrothermal AAS.

Virtually all the trace elements of biological interest may be determined by electrothermal AAS. Although it is relatively slow, a single-element technique, and subject to various interferences, it is extensively used throughout the world for clinical investigations and for monitoring occupational exposures, as well as in the other settings mentioned earlier. The design and construction of furnaces are subject to continuous development to improve detection limits and to reduce interferences.[9,15] Devices to measure nonatomic absorption, e.g., Zeeman-effect background correction, are essential. Commercial furnaces are made from electrographite, electrographite with a pyrolytic coating or total pyrolytic graphite, although publications showing the advantages of other materials as coatings or for the furnace itself, regularly appear.[10] Typically these refer to analysis of elements that form extremely refractory carbides in graphite furnaces, e.g., molybdenum.

Design developments are introduced with the objective of separating the appearance of the atomic vapor from components that cause an interference with the atomization signal. These innovations include graphite platforms and probes but, as with automobiles, there are continuous refinements to the overall shape and dimensions to effect improvements in performance. An authoritative review of materials suitable for use in furnace construction and of recent developments in design has been prepared by Frech.[10] Several research groups have designed very novel atomizers with the purpose of separating the analyte from interfering species, to permit simple atomic absorption.[27–29] Although some appear to be effective, none are commercially available.

**TABLE 23.3** Chemical Modifiers Used in the Analysis of Biomedical Specimens by ETAAS

| Modifier | Purpose |
|---|---|
| Triton X-100 | To promote drying of protein-rich specimens, avoid a dried crust around a liquid core |
| Gaseous oxygen | To promote destruction of organic matrix, reduce smoke formation and particulates which give nonatomic absorption |
| Ni, Cu, Pd | To stabilize volatile elements, e.g., Se, As, during the dry and ash phases |
| Potassium dichromate | Stabilizes Hg up to a temperature of 200°C |
| $HNO_3$ or $NH_4NO_3$ | To stabilize analyte atoms by removal of halides as HCl or $NH_4Cl$ during the ash phase |
| $Mg(NO_3)_2$ | Becomes reduced to MgO, which traps the metals to reduce volatilization losses, delays atomization and separates the analyte signal from the background absorption |
| $NH_3H_2PO_4$ or $(NH_3)_2HPO_4$ | Usually used with $Mg(NO_3)_2$, reduces volatilization losses and delays atomization to separate the analyte signal from the background absorption |

It was shown some years ago that a 150-W tungsten filament from a light bulb could be used as an electrothermal atomizer. More recently, this concept has been used to develop very small portable instruments for on-site measurement of lead in blood.[30–32] Excellent results have been reported but a commercial model is still awaited.

In addition, effective analysis of most biological samples requires the addition of reagents that modify the behavior of the specimen during the heating program so as to reduce interferences.[14] The chemical modifiers most commonly employed with biological specimens are given in Table 23.3. Triton X-100 is used at a concentration of around 0.1% w/v and is included with the sample diluent. Gaseous oxygen or air is an effective ashing aid, but will cause rapid deterioration of the graphite furnace unless a desorption step is included before the temperature is increased for atomization. Other modifier solutions can be included with the sample diluent or separately added by the autosampler to the specimen inside the furnace. The choice of modifier often depends on the availability of a source material that is free from contamination.

### 23.5.3 Atomic Fluorescence Spectrometry

Recent innovations in AFS follow almost entirely from the development of commercial instruments specifically designed for use with hydride generation; the particular applications of interest will be considered in the next section. For various reasons earlier attempts to exploit the inherent sensitivity of AFS using flame systems were never fully realized. However, with improvements in light source technology and other instrumentation, this niche area has progressed rapidly in recent years.[9,14]

### 23.5.4 Vapor Generation Procedures

Depending on the nature of the exposure — inorganic salts, organomercury compounds, the metal or its vapor — it may be necessary to analyze specimens of urine, blood or tissues, and foods. Mercury is used extensively in industry and occupational monitoring continues to be relevant. There is now considerable interest in two particular environmental sources of exposure, i.e., from dental amalgam and from the diet, especially fish. Although no good evidence exists for mercury leaching from amalgam fillings in amounts that can cause toxicity, many members of the public believe that their health is affected and seek to have the mercury removed, even after analysis of blood or urine has failed to show increased concentrations of the metal.[33] Concern has been raised that undue exposure to methylmercury may occur from eating large amounts of seafoods. Those at risk are young children *in utero* and during early childhood. Long-term studies are in progress within communities where basic foods contain mercury and maternal hair concentrations of it are high. Mercury intakes and neurological development are being monitored within the target groups.[34]

Considerable interest in methods to measure arsenic and other hydride-forming elements has been evident in recent years. The basic procedure involves careful digestion of the specimens to convert all the different species to a single valency form, reduction with $BH_4^-$ and vaporization to the hydride, which is then transferred by a stream of inert gas to a quartz tube heated in an air-acetylene flame or with electrical thermal wire. Atomization is then achieved by the high temperature. Some work has been reported in which the hydride is transferred into a cold graphite furnace where it is trapped onto the surface. Atomization takes place as the furnace is rapidly heated and, as with conventional electrothermal AAS, improved sensitivity is observed due to the high atom density within the small volume of the furnace.

Trapping is more efficient when the graphite is coated with a metal salt, e.g., Ag, Pd, Ir.[35,36] The chemical hydride reaction is impaired by other hydride-forming elements and by transition metals so that careful calibration is essential. An emerging development involves an electrolytic process in which nascent hydrogen is produced as an alternative to chemical hydride generation. With this arrangement the interferences are much less and the reagents employed introduce less contamination.[37–39]

The biological and clinical importance of selenium receives much current interest; more attention is focused on measurement of this element than on any other. The stimuli to this flurry of activity are (1) the association between selenium status and cardiovascular disease coupled to the demonstration that dietary intakes are low in some regions and are declining in others,[40] and (2) epidemiological data suggesting a relationship between selenium status and the incidence of certain carcinomas.[41] Measurements of selenium in foods, biological fluids and tissues, using vapor generation techniques, electrothermal AAS, or ICP-MS, are integral to many large-scale studies now in progress or recently completed.

There is also much interest in the determination of arsenic. This element is important within the microelectronics industry and in other occupations, but extensive environmental exposure is also associated with naturally high concentrations in drinking water. Arsenic in drinking water is a problem in several areas of the world; however, the situation that now exists in Bangladesh and West Bengal, India, is extraordinary, with millions of people consuming highly toxic and carcinogenic water.[42] Measurement of total arsenic is not always entirely helpful. Fish contain large amounts of organoarsenic species that are absorbed and excreted without further metabolism and with no adverse health effects. These species will be included in a total arsenic determination and can mask attempts to measure toxic $As^{3+}$ and metabolites. Thus, methods to measure the individual species or related groups of compounds in urine or other samples provide more meaningful results. These methods include separation by chromatography or solvent extraction and pretreatment steps that transform only the species of interest into the reducible form.

As this speciation work has become more refined additional arsenic-containing compounds have been demonstrated and, very recently, some of these have been identified as methylated As species containing $As^{III}$. It is well known that methylated species with $As^{V}$ are found in blood and urine following exposure to inorganic arsenic and believed to represent steps in the detoxification pathway. Methylarsenic$^{III}$ species are potent enzyme inhibitors and cytotoxins and their formation may be involved in the mechanism of arsenic toxicity. In one recent study,[43] biliary and urinary arsenic species were determined in rats exposed to $As^{III}$ and As.$^{V}$ MonomethylAs$^{III}$arsonate (MMA$^{III}$) was present in bile but not in urine; the authors hypothesized that MMA$^{III}$ was subsequently oxidized to MMA$^{V}$ and excreted in urine. In a separate investigation, arsenic species were measured in water, urine and cultured cells and methylarsenic$^{III}$ species were identified in the urine of individuals who had consumed water contaminated with inorganic As.[44]

Other elements that form gaseous hydrides, such as antimony, bismuth, tellurium, etc., are also relevant to investigations in clinical and biological specimens.

### 23.5.5 X-Ray Fluorescence Spectrometry

XRF and other x-ray techniques offer no particular advantage over the many other procedures for simple quantitative measurement of minerals and trace elements in clinical and biochemical specimens. Alternative methods are widely available, well established, and relatively simple. Nevertheless, a few interesting applications have been reported with recent examples of gold and palladium in urine by TXRF, with detection limits around 2.5 ng/l, and the analysis of breast milk by PIXE.[45,46] In certain applications involving clinical and biochemical specimens, however, XRF does have a specific role. These include elemental mapping and *in vivo* analyses.

By virtue of the very narrow x-ray beam (100 μm or less) it is possible to take repeat measurements over a very small surface and develop a map of the distribution of elements within a structure. This approach is regularly applied to solid materials including hairs, teeth, and other calcareous materials. Results have also been reported in which single cells have been investigated, e.g., in a study to compare iron and other elements within neuromelanin aggregates in neuronal cells of patients with Parkinson's disease and their controls.[47]

Developments involving *in vivo* XRF have flourished in the last few years although the majority of the many publications originate from just a few centers. Most of the work is concerned with measurement of lead in bone by $^{109}$Cd-based XRF. Analytically, recent improvements refer to detection systems to reduce detection limits to low part per million levels, and to the methods for calibration. Phantoms (often plaster of Paris) containing known amounts of Pb are generally used. Recent alternative materials that are reported to behave more like bone are a synthetic apatite matrix[48] and a material with polyurethane and $CaCO_3$.[49] It has been shown that the measurement location, i.e., proximal-distal sites, influences the measured XRF intensity and its uncertainty. Considerable differences in mean bone Pb concentrations between the left and right legs of the same individual (0.8 and 2.0 μg/g bone mineral) have also been demonstrated.[50] Using this technique, cumulative exposures to lead at work have been assessed and results compared with other markers such as blood and urine lead concentrations.[51] Blood and bone lead concentrations were determined and related to the development of hypertension. A positive association was found between the baseline bone Pb level and the incidence of hypertension but no association was found with blood Pb level.[52] Bone lead has also been shown to be released into the circulation during pregnancy, with an implied risk to the fetus if the mother has a history of previous lead absorption.

There are reports of the determination of other elements by *in vivo* XRF. Skin iron concentrations correlated strongly with iron in the internal organs of rats injected with iron-dextran and it was concluded that this technique had great potential in the diagnosis and treatment of hereditary hemochromatosis and β-thalassemia.[53] A method for measuring platinum in kidneys of patients receiving Pt-based chemotherapy drugs has been developed.[54] These few examples illustrate that the potential for *in vivo* elemental analysis is hugely exciting.

## 23.6 Atomic Spectrometry: Quality Assurance

Apart from the actual analysis, adventitious contamination, which can occur during collection and storage of specimens, the preparative procedure, and the spectrometric measurement, has the greatest impact on the quality of results. Data from proficiency testing schemes indicate that specialist trace element centers tend to maintain the highest standards of analytical performance. This observation reflects the continuing application of practices that minimize contamination and the expertise and experience to ensure optimal functioning of equipment. Because of the stability of inorganic analytes and the purity with which standard materials can be prepared, there are reasonable numbers of reference materials available for use to validate methods and for internal and external quality control.[55] Proficiency testing schemes relating to occupational and environmental laboratory medicine are organized in most countries and all laboratories involved in this work should have access to an appropriate scheme to demonstrate the reliability of their analytical data.[56]

## References

1. Taylor, A., Detection and monitoring of disorders of essential trace elements, *Ann. Clin. Biochem.*, 33, 486, 1996.
2. Walker, A.W., Ed., *SAS Trace Element Laboratories. Clinical and Analytical Handbook*, 3rd ed., Royal Surrey County Hospital, Guildford, Surrey, U.K., 1997.

3. Taylor, A., Ed., *Trace Elements in Human Disease. Clinics in Endocrinology and Metabolism,* W.B. Saunders, Eastbourne, U.K., 1985.
4. Baldwin, D.R. and Marshall, W.J., Heavy metal poisoning and its laboratory investigation, *Ann. Clin. Biochem.,* 36, 267, 1999.
5. Lobinski, R. and Marczenko, Z., *Spectrochemical Trace Analysis for Metals and Metalloids,* Elsevier, Amsterdam, 1996.
6. Haswell, S., Ed., *Atomic Absorption Spectrometry. Theory, Design and Applications,* Elsevier, Amsterdam, 1991.
7. Holland, J.G. and Tanner, S.D., *Plasma Source Mass Spectrometry: The New Millenium,* Royal Society of Chemistry, Cambridge, U.K., 2001.
8. Giessman, U. and Greb, U., High resolution ICP-MS— a new concept for elemental mass spectrometry, *Fresenius J. Anal. Chem.*, 50, 186, 1994.
9. Evans, E.H., Dawson, J., Fisher, A., et al., Atomic spectrometry update — advances in atomic emission, absorption and fluorescence spectrometry, and related techniques, *J. Anal. At. Spectrom.,* 16, 672, 2001.
10. Frech, W., Recent developments in atomizers for electrothermal atomic absorption spectrometry, *Fresenius J. Anal. Chem.*, 355, 475, 1996.
11. Turner, J., Hill, S.J., Evans, E.H., et al., Accurate analysis of selenium in water and serum using ETV-ICP-MS with isotope dilution, *J. Anal. At. Spectrom.*, 15, 743, 2000.
12. Rahil-Khazen, R., Henriksen, H., Bolann, B., et al., Validation of ICP-AES for multi-element analysis of trace elements in human serum, *Scand. J. Clin. Lab. Invest.,* 60, 677, 2000.
13. Potts, P., Ellis, A., Holmes, M., et al., Atomic spectrometry update — X-ray fluorescence spectrometry, *J. Anal. At. Spectrom.*, 16, 1217, 2001.
14. Taylor, A., Branch, S., Halls, D.J., et al., Atomic spectrometry update — clinical and biological materials, foods and beverages, *J. Anal. At. Spectrom.,* 17, 414, 2002.
15. Taylor, A., Applications of recent developments for trace element analysis, *J. Trace Elem. Med. Biol.*, 11, 185, 1997.
16. Tsakalof, A. and Taylor, A., Pretreatment of biomedical samples for trace mental quantitation: the efficiency of solid phase extraction on iminidiacetate functionalised sorbents, *Int. Congr. Anal. Chem.*, Moscow, 1997.
17. Liu, X. and Fang, Z., Flame atomic absorption spectrometric determination of cobalt in biological materials using a flow injection system with on-line preconcentration by ion-pair adsorption, *Anal. Chim. Acta,* 316, 329, 1995.
18. Bacon, J., Crain, J., Van Vaeck, L., et al., Atomic spectrometry update — atomic mass spectrometry, *J. Anal. At. Spectrom.,* 16, 879, 2001.
19. Szpunar, J., Bio-inorganic speciation analysis by hyphenated techniques, *Analyst,* 125, 963, 2000.
20. Faure, H., Determination of the major zinc fractions in serum by ultrafiltration, *Biol. Trace. Elem. Res.*, 24, 25, 1990.
21. European Union, Council Directive 98/24/EC of 7 April 1998 on the protection of the health and safety of workers from the risks related to chemical agents at work (fourteenth individual Directive within the meaning of Article 16(1) of Directive 89/391/EEC), *Official Journal*, 131, 11, 1998.
22. Chappuis, P., Poupon, J., and Rousselet, F., A sequential and simple determination of zinc, copper and aluminium in blood samples by ICP-AES, *Clin. Chim. Acta,* 206, 155, 1992.
23. Lyon, T.D., Cunningham, C., Halls, D.J., et al., Determination of aluminium in serum, dialysate fluid and water by ICP-OES, *Ann. Clin. Biochem.,* 32, 160, 1995.
24. Norman, P.T., Joffe, P., Martinsen, I., et al., Quantification of gadodiamide as Gd in serum, peritoneal dialysate and faeces by inductively coupled plasma atomic emission spectroscopy and comparative analysis by high-performance liquid chromatography, *J. Pharm. Biomed. Anal.*, 22, 939, 2000.
25. Taylor, A., Usefulness of measurements of trace elements in hair, *Ann. Clin. Biochem.*, 23, 364, 1986.

26. Brown, A.A. and Taylor, A., Determination of copper and zinc in serum and urine by use of a slotted quartz tube and flame atomic absorption spectrometry, *Analyst,* 109, 1455, 1984.
27. Smith, C.M.M. and Harnley, J.M., Characterization of a modified two-step furnace for atomic absorption spectrometry for selective volatilization of iron species in hemin, *J. Anal. At. Spectrom.*, 11, 1055, 1996.
28. Kitagawa, K., Ohta, M., Kaneko, T., et al., Packed glassy carbon tube atomizer for direct determinations by atomic absorption spectrometry free of background absorption, *J. Anal. At. Spectr.,* 9, 1273, 1994.
29. Ohta, K., Koike, Y., and Mizuno, T., Determination of zinc in biological materials by sequential metal vapor elution analysis with atomic absorption detection, *Anal. Chim. Acta,* 329, 191, 1996.
30. Sanford, C.L., Thomas, S.E., and Jones, B.T., Portable, battery-powered, tungsten coil atomic absorption spectrometer for lead determinations, *Appl. Spectr.*, 50, 174, 1996.
31. Parsons, P.J., Qiao, H., Aldous, K.M., et al., A low-cost tungsten-filament atomizer for measuring lead in blood by atomic-absorption spectrometry, *Spectrochim Acta,* 50B, 1475, 1995.
32. Zhou, Y., Parsons, P.J., Aldous, K.M., et al., Atomization of lead from whole blood using novel tungsten filaments in electrothermal atomic absorption spectrometry, *J. Anal. At. Spectrom.*, 16, 82, 2001.
33. Bailer, J., Rist, F., Rudolf, A., et al., Adverse health effects related to mercury exposure from dental amalgam fillings: toxicological or psychological causes? *Psych. Med.,* 31, 255, 2001.
34. Murata. K., Weihe, P., Shunich, A., et al., Evoked potentials in Faroese children exposed to MeHg, *Neurotox. Teratol.,* 21, 471, 1999.
35. Zhe-Ming, Ni., Bin, H., and Heng-Bin, H., *In situ* concentration of selenium and tellurium hydrides in a silver-coated graphite atomizer, *J. Anal. At. Spectrom.*, 8, 995, 1993.
36. Tsalev, D L., D'ulivo, A., Lampugnani, L., et al., Thermally stabilized iridium on an integrated, carbide-coated platform as a permanent modifier for hydride-forming elements in electrothermal atomic absorption spectrometry, *J. Anal. At. Spectrom.*, 11, 979, 1996.
37. Schickling, C., Yang, J., and Broekaert, J., The optimization of electrochemical hydride generation coupled to microwave induced plasma atomic emission spectrometry for the determination of arsenic and its use for the analysis of biological tissue samples, *J. Anal. At. Spectrom.*, 11, 739, 1996.
38. Ding, W.W. and Sturgeon, R.E., Evaluation of electrochemical hydride generation for the determination of total antimony in natural waters by electrothermal atomic absorption with *in situ* concentration, *J. Anal. At. Spectrom.*, 11, 225, 1996.
39. Machado, L., Jacintho, A., and Gine, M., Electrochemical hydride generation for selenium determination in a flow injection system with air-GLP flame atomic absorption spectrometric detection, *Quim. Nova,* 23, 30, 2000.
40. Rayman, M., Dietary selenium: time to act, *Br. Med. J.,* 314, 387, 1997.
41. Clark, L.C., Coombs, G., Jr., Turnbull, B.W., et al., Effects of selenium supplementation for cancer prevention in patients with carcinoma of the skin, *J. Am. Med. Assoc.,* 276, 1957, 1996.
42. Chatterjee, A., Das, D., Mandal, B.K., et al., Arsenic in groundwater in six districts of West Bengal, India: the biggest arsenic calamity in the world, *Analyst,* 120, 643, 1995.
43. Gregus, Z., Gyurasics, A., and Csanaky, I., Monomethylarsonous acid as a major biliary metabolite in rats, *Toxicol. Sci.,* 56, 18, 2000.
44. Del Razo, L.M., Stybo, M., Cullen, W.R., et al., Determination of trivalent methylated arsenicals in biological matrices, *Toxicol. Appl. Pharmacol.,* 174, 282, 2001.
45. Messerschmidt, J., von Bohlen, A., Alt, F., et al., Separation and enrichment of palladium and gold in biological and environmental samples, adapted to the determination by total reflection X-ray fluorescence, *Analyst,* 125, 397, 2000.
46. Akanle, O.A., Balogun, F.A., Owa, J.A., et al., Study of the nutritional status of maternal breast milk in preterm infants in Nigeria, *J. Radioanal. Nucl. Chem.,* 244, 231, 2000.
47. Ektessabi, A., Yoshida, S., and Takada, K., Distribution of iron in a single neuron of patients with Parkinson's disease, *X-Ray Spectrom.*, 28, 456, 1999.

48. Todd, A.C., Coherent scattering and matrix correction in bone-lead measurements, *Phys. Med. Biol.*, 45, 1953, 2000.
49. Spitz, H., Jenkins, M., Lodwick, J., et al., A new anthropometric phantom for calibrating *in vivo* measurements of stable lead in the human leg using X-ray fluorescence, *Health Phys.*, 78, 159, 2000.
50. Hoppin, J.A., Aro, A., Hu, H., et al., Measurement variability associated with KXRF bone lead measurement in young adults, *Environ. Health Perspect.*, 108, 239, 2000.
51. Britto, J., McNeil, F., Chettle D., et al., Study of the relationships between bone lead levels and its variation with time and the cumulative blood lead index, in a repeated bone lead survey, *J. Environ. Monit.*, 2, 271, 2000.
52. Cheng, Y.W., Schwartz, D., Sparrow, A., et al., Bone lead and blood lead levels in relation to baseline blood pressure and the prospective development of hypertension — The Normative Aging Study, *Am. J. Epidemiol.*, 153, 164, 2001.
53. Farquharson, M.J., Bagshaw, A.P., Porter, J., et al., The use of skin Fe levels as a surrogate marker for organ Fe levels, to monitor treatment in cases of iron overload, *Phys. Med. Biol.*, 45, 1387, 2000.
54. Kadhim, R., Al-Hussany, A., Ali, P.A., et al., *In vivo* measurement of platinum in the kidneys using X-ray fluorescence, *In Vivo Body Comp. Stud.*, 90, 263, 2000.
55. Roelandts, I., Biological and environmental reference materials: update 1996, *Spectrochim. Acta*, 52B, 1073, 1997.
56. Morisi, G., Menditto, A., Patriarca, M., et al., European external quality assessment schemes in occupational and environmental medicine, *Ann. 1st. Super. Sanit.*, 32, 191, 1996.

# 24

# Capillary Electrophoresis Techniques in Biomedical Analysis

S. Douglass Gilman
*University of Tennessee*
*Knoxville, Tennessee*

Michael J. Sepaniak
*University of Tennessee*
*Knoxville, Tennessee*



## 24.1 Overview

Capillary electrophoresis (CE) is a microscale analytical separation technique that has matured rapidly over the past 20 years since the groundbreaking publications of James W. Jorgenson and co-workers.[1,2] Biomedical applications of CE are a leading factor driving the development of what has now evolved into a broad family of related separation techniques. Certainly the most prominent biomedical application of CE is the sequencing of the human genome.[3] The accelerated achievement of this goal depended on CE separations.[3–5] From 1981 through the writing of this text, more than 14,000 papers were published that included CE and related techniques. A survey of this literature over the past 12 months indicates that more than 70% of these reports include bioanalytical applications of CE.

Clearly, a comprehensive review of biomedical applications of CE is well beyond the scope of a single book chapter. This chapter first reviews basic CE separations; readers are directed to review articles for in-depth coverage of related CE separation methods. Next, the role of photonics in CE, primarily for detection purposes, is discussed. Finally, the chapter reviews specific examples of biomedical applications of CE that feature photonics, to provide the reader with a view of the possibilities of this approach for biomedical analysis.

Capillary electrophoresis typically is performed in 25- to 75-μm internal diameter (ID), fused-silica capillaries (15 to 100 cm in length) filled with aqueous, buffered solutions. Each end of the capillary is immersed in a buffer reservoir, and a potential of 10 to 30 kV is applied across the capillary using Pt electrodes in each reservoir. In its simplest form, capillary zone electrophoresis (CZE), different charged

0-8493-1116-0/03/$0.00+$1.50

compounds in an injected sample zone are separated based on their relative electrophoretic migration rates. For newcomers to this technique, an intuitive feel for this separation method can perhaps best be developed by comparing and contrasting CE with two more familiar and related separation techniques: high pressure liquid chromatography (HPLC) and slab gel electrophoresis.

Like HPLC, CE is a column-based separation technique. Sample plugs are injected into the CE column. Compounds separate as they pass through the column, and separated compound zones are detected as they elute past a fixed point. One obvious difference between the two separation techniques is scale. Typical internal diameters for CE are 25 to 75 μm compared to 1 to 5 mm for HPLC. A 50-μm ID CE column 1 m in length has a total column volume of only 2 μl. Typical CE injection volumes are only a few nanoliters; column flow rates are ordinarily a few hundred nanoliters per minute, although linear flow velocities are generally greater than in HPLC. One consequence of the scale of CE is that it is well suited for analysis of small-volume samples such as single cells.[6,7] In addition, insignificant volumes of solvent waste are generated compared to HPLC. Due to its small scale, CE is primarily an analytical technique, and is not as practical as HPLC for preparative work. Analyte detection is also more challenging at smaller scales, as discussed later in this chapter.

Capillary electrophoretic separations typically exhibit higher separation efficiencies (narrower peaks) and are rapid compared to HPLC. Typically, a CE separation will generate 10 to 100 times more theoretical plates than an HPLC separation for the same compounds and separation time. High separation efficiencies are primarily due to the plug-shaped electroosmotic flow profile and the electrophoretic separation mechanism of CZE. Peaks for HPLC are typically broader due to the parabolic flow profile of pressure-driven flow, effects of the particles used to pack stationary phases for HPLC, and resistance to mass transfer between the mobile phase and separation phase in HPLC. This last source of peak broadening for HPLC underscores the primary difference between the two techniques, which is separation mechanism. HPLC is a chromatographic technique where separation depends on the relative partitioning of different compounds between a mobile phase and a stationary phase. Separation in CZE is based on electrophoretic migration rates in free solution. Related CE techniques that include partitioning as part of the separation also have been developed and will be discussed later in this chapter.

The most common forms of slab gel electrophoresis, SDS-PAGE for proteins and PAGE for DNA analysis, separate molecules based on sieving behavior through a polymer gel while the separation mechanism for CZE is electrophoretic migration in free solution. Gel-free electrophoretic separations are possible with CE due to the small scale of the technique compared to slab gel electrophoresis. The reduced scale of CE results in lower electrophoretic currents and higher surface-to-volume ratios. The small electrophoretic current in CE generates relatively low joule heating, and the heat that is produced is dissipated more effectively. Capillary separations are much more rapid than slab gel separations, in part because much higher potential fields can be applied (hundreds of volts per centimeter) due to reduced electrophoretic currents and increased heat dissipation.

For slab gel electrophoresis, the gel is required to dissipate heat and reduce zone broadening due to convective flow in addition to providing a sieving medium. Like HPLC, slab gel electrophoresis is superior to CE for preparative work. Capillary electrophoresis, however, is better suited to volume-limited samples. In addition, slab gels are ideally suited for parallel separations and two-dimensional separations. Capillary array electrophoresis has been developed for parallel separations,[4,5] and CE has been used for two-dimensional separations,[8] but in both cases sophisticated instrumentation had to be developed to realize the advantages of these separation strategies with CE.

## 24.2 Capillary Electrophoresis Basics

### 24.2.1 Capillary Zone Electrophoresis

#### 24.2.1.1 Fundamentals

Capillary zone electrophoresis separations are based on differences in electrophoretic migration of ionic species in solution. For practical reasons, which will be explained later, a positive potential is normally

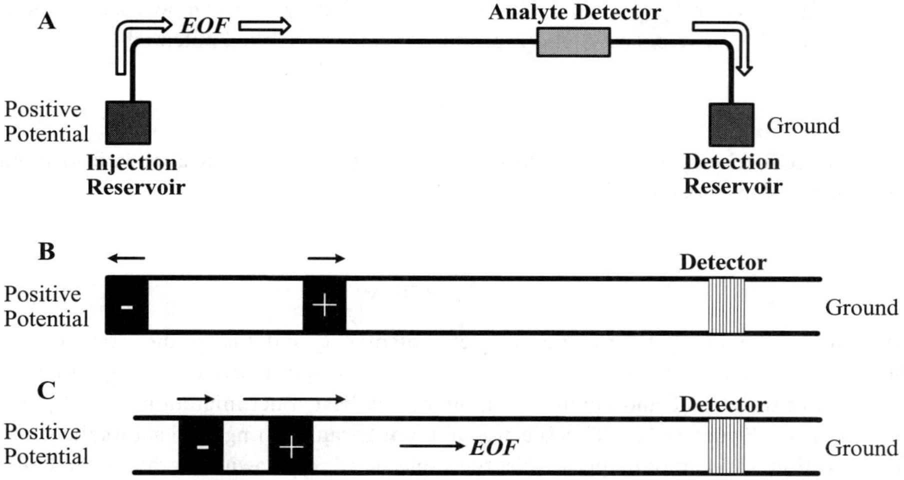


**FIGURE 24.1** (A) Capillary electrophoresis instrument. A potential is applied through Pt electrodes (not shown) in the injection and detection reservoirs. (B) Migration of anions and cations in the absence of EOF (fluid flow). (C) Migration of anions and cations in the presence of strong EOF.

applied at the injection end of the capillary, and the opposite end of the capillary is held at a relatively negative potential. Figure 24.1 illustrates a basic CZE instrument and separations with and without fluid flow.

Consider a sample zone at the injection end of a capillary column containing a mixture of cationic, anionic, and neutral compounds when a separation potential is applied (Figure 24.1B). In the absence of fluid flow, cationic compounds will migrate down the column toward the detector. Anionic compounds will immediately exit the column at the injection end. Neutral compounds (not shown) will be unaffected by the separation potential. The following relation describes electrophoretic migration of ionic species in free solution:

$$v_{ep} = u_{ep}E \tag{24.1}$$

where $v_{ep}$ is the electrophoretic velocity of a charged compound, $u_{ep}$ is the electrophoretic mobility, and $E$ is the applied field strength (V/cm). The electrophoretic mobility, $u_{ep}$ is defined as:

$$u_{ep} = \frac{q}{6\pi\eta r} \tag{24.2}$$

where $q$ is the charge of the compound, $\eta$ is the viscosity of the solution, and $r$ is the compound's hydrodynamic radius. In the absence of fluid flow, only cationic species can be separated with the experimental setup illustrated in Figure 24.1A. Anionic species could be separated by reversing the polarity of the applied potential.

In practice, CZE experiments typically include bulk fluid flow in the capillary due to electroosmosis.[9] Electroosmotic flow (EOF) is generated at the capillary surface in the presence of an applied potential by electrophoretic migration of solvated ions near the capillary surface. If the capillary surface includes charged functional groups (e.g., Si-$O^-$ for fused silica), a double-layer structure is formed in solution at the surface (typically nanometer in scale), which contains an excess of ions opposite in charge to the bound surface charge on the capillary inner wall. In the presence of an applied potential, these solvated counter ions in the diffuse part of the double layer will migrate toward one electrode, generating a bulk fluid flow. In effect, the entire capillary inner surface acts as a fluid pump for CE. For the experimental arrangement shown in Figure 24.1 with a fused-silica capillary filled with an aqueous solution near neutral pH, EOF is in the direction from the injection end of the capillary (positive polarity) to the detection end of the capillary (negative polarity) and is proportional to the applied potential:

$$v_{eof} = u_{eof}E \tag{24.3}$$

where $v_{eof}$ is the EOF velocity, $u_{eof}$ is the electroosmotic mobility, and $E$ is the applied field strength. The electroosmotic mobility is described by:

$$u_{eof} = \frac{\varepsilon\zeta}{4\pi\eta} \tag{24.4}$$

The structure of the double layer defines the zeta potential, $\zeta$, and $\varepsilon$ is the dielectric constant of the buffer filling the capillary. Electroosmosis is typically greater in magnitude than electrophoretic migration of charged species so cationic and anionic compounds will have a net migration toward the detection end of the capillary (Figure 24.1C). This is a critical point because strong EOF allows the simultaneous separation of cationic and anionic species. Neutral molecules (not shown) migrate at the EOF rate and are not separated from each other by CZE. Other CE techniques for separations of neutrals will be discussed later in this chapter. The net electrophoretic migration of ionic species in the presence of EOF is described by:

$$v_{net} = u_{net}E = (u_{ep} + u_{eof})E \tag{24.5}$$

where $v_{net}$ is the net velocity of a compound due to electrophoresis and EOF, and $u_{net}$ is the net electrophoretic mobility.

The resolving ability of CE is the primary reason for the technique's rapid development and widespread application. The resolution, $R$, for two compounds (1 and 2) by CE is:

$$R = \frac{0.177(u_{ep,1} - u_{ep,2})V^{1/2}}{\sqrt{(u_{avg} + u_{eof})D}} \tag{24.6}$$

Here $u_{avg}$ is the average electrophoretic mobility of the two compounds, $D$ is the diffusion coefficient of the compounds, and $V$ is the applied potential (volts). High separation efficiency is the primary reason that CE provides good resolution compared to other separation techniques. The number of theoretical plates for a CZE separation is given by:

$$N = \frac{(u_{ep} + u_{eof})V}{2D} \tag{24.7}$$

This equation shows that, ideally, CZE peaks are only broadened by axial diffusion. There is no stationary phase in a CZE separation, and, therefore, no mass transfer terms or particle packing terms are in Equation 24.7. Furthermore, EOF has a very flat, plug-shaped flow profile, so resistance to mass transfer in the mobile phase (running buffer) does not contribute to peak broadening. Separations with $10^5$ theoretical plates are common and plate counts over $10^6$ have been reported. Other factors can degrade separations for CE, however. Unwanted adsorption to the capillary surface is frequently encountered. Temperature changes due to excessive joule heating can cause peak broadening. Sample injection and detection can degrade separation efficiencies in some cases. Zone broadening for CE has been reviewed.[10,11]

### 24.2.1.2 Practical Considerations

Basic CE instruments are straightforward to construct in the laboratory, and much of the research in this field has relied on laboratory-constructed instrumentation. Commercial instrumentation for CE is widely available, however, and these instruments offer many advantages for new users of CE and experienced researchers who also use laboratory-constructed instruments. The most common instrument is a single-capillary device with a UV/VIS absorbance detector. Laser-induced fluorescence (LIF) detectors, electrochemical detectors, and instruments designed to interface with mass spectrometers are also available. Most complete CE systems include autoinjectors (pressure and electrokinetic) and capillary thermostats. Capillary array instruments are also available for applications requiring high throughput.

Most CE experiments are performed in bare fused-silica capillaries filled with buffered, aqueous solutions (running buffer) near neutral pH. Applied potentials typically range from 10 to 30 kV, and Pt electrodes are commonly used in the two buffer reservoirs for this purpose. One of the main advantages of CE is that selectivity can be changed substantially by simply altering the separation buffer. The simplest and most effective means to control selectivity is to change the running buffer pH to alter the net charge of analytes. The running buffer pH also influences EOF; as seen in Equation 24.6, this can have subtle to dramatic effects on resolution. Altering the buffer composition and ionic strength can also affect separation selectivity and EOF, but the effects of these changes are subtle and not as easy to predict as with a pH change. As described in following sections, compounds can be added to the buffer to alter the separation mechanism. Water-miscible organic solvents are often added to CE buffers, and nonaqueous CE has been studied extensively.[12]

Sample injections typically are performed by placing the injection end of the capillary into the sample container and applying pressure or an electrical potential. Both methods can be automated, and the precision for pressure injection can be as much as a factor of two better than for electrokinetic injection.[13,14] Electrokinetic injection suffers from injection bias for charged species due to differences in electrophoretic mobilities between sample components. Theoretical descriptions of these injection techniques, which include equations for calculating injection volumes, are available in the literature.[14] Sample stacking describes a family of related methods used to preconcentrate samples electrophoretically in the capillary column at the start of a CE separation. A number of variations of this technique have been developed to achieve sample preconcentration factors ranging from 10 to 1000 for charged analytes, and related methods have been developed to preconcentrate neutral analytes.[15,16] Preconcentration methods using solid phase supports also have been developed for CE.[16] Preconcentration is particularly important for CE methods due to the difficulty in detecting analytes at low concentrations with this separation method.

A typical CE capillary is 15 to 100 cm in length and has an ID from 25 to 75 μm. Capillaries made from materials other than fused silica are rarely used. One reason for this is the superior properties of fused-silica capillaries for on-column optical detection. The inner surface of the capillary is commonly modified, however. One goal of capillary surface modification is to reduce band broadening and sample loss due to adsorption of analytes to the capillary surface, which can be especially problematic for protein analysis. A second goal is to suppress EOF. Because EOF is sensitive to the solution composition and the surface chemistry at the capillary surface, it can be irreproducible or unstable, leading to poor analytical reproducibility. Coating the capillary surface can reduce these problems. Covalently bound coatings, adsorbed coatings, and separation buffer additives are used to modify the capillary surface.[17–20]

### 24.2.2 Biomedically Significant Variations on the Capillary Electrophoresis Theme

The facile addition of a wide variety of reagents to the CE running buffer to influence migration behavior and enhance separations is a major advantage of the technique. Among the additives that have been used are chelating agents such as crown ethers, surfactants above and below the critical micelle concentration (CMC), macrocyclic reagents such as cyclodextrins, antibiotics, and calixarenes, and sieving media such as soluble polymers.[17,21,22] In some cases, these reagents have served sufficiently unique and valuable purposes that specialized variants of the CE have been dubbed. We will cover the basic principles of a few of those variations on the CE theme.

A disadvantage of the conventional CZE technique is that neutral species migrate with EOF and co-elute at or very near the void time of the system. Given the preponderance of neutral compounds of biological and medical significance, this limitation is significant. Neutral compounds can be separated, however, if they acquire different effective electrophoretic mobilities due to differential association with charged running buffer additives. The most commonly used approach involves the addition of surfactant at concentrations above its CMC to form charged aggregates of surfactant molecules (micelles).

This technique, micellar electrokinetic capillary chromatography (MEKC), was first introduced by Terabe and co-workers.[23] Instrumental and operational aspects of MEKC are virtually indistinguishable from CZE; however, MEKC also shares many of the features of HPLC. The running buffer assumes the role of the mobile phase and is transported at the EOF rate, while the micellar phase constitutes a secondary (albeit not stationary) phase that migrates at a different velocity (usually slower than EOF as negatively charged micelles attempt to migrate in opposition to EOF). The technique differs from HPLC in that movement of the secondary phase creates a distinct elution window, bound by the void time of the capillary ($t_0$) and the effective migration time of a micelle ($t_{mic}$), within which neutral species elute. In a typical experiment the ratio of $t_0$ to $t_{mic}$ might take a value of 0.3. Efficiencies are generally not as high as in CZE because resistance to mass transfer between the running buffer and micellar phases is involved.[23,24] Efficiencies are much higher in MEKC than in HPLC, however. As with HPLC, resolution in MEKC depends on efficiency, selectivity, and system retention as seen in Equation 24.8:

$$R = \frac{\sqrt{N}}{4} \cdot \frac{\alpha - 1}{\alpha} \cdot \frac{k_2'}{1 + k_2'} \cdot \frac{1 - (t_0/t_{mic})}{1 + (t_0/t_{mic})k_1'} \tag{24.8}$$

Here, $k_1'$ and $k_2'$ are the capacity factors of the two analytes, and $\alpha$ is the selectivity factor defined as $k_2'/k_1'$.

A unique aspect of having micelles present in the running buffer is that biomedical samples containing charged and neutral species can be efficiently separated. Figure 24.2 is a separation of vitamin B6 (pyridoxine) and several of its metabolites using MEKC (sodium dodecyl sulfate as the surfactant) with LIF detection.[25] This group of metabolites contains both acidic and basic functionalities, and, as such, charged and neutral species are present at any pH.

There are two significant limitations of the MEKC technique. First, hydrophobic solutes associate very strongly with the micellar phase and tend to bunch up near $t_{mic}$. A ramification of the elution window is that $R$ (see Equation 24.8) tends to show an optimum at $k'$ values <5.[23] The use of mixed aqueous-organic running buffers expands the elution window and reduces the capacity factors of hydrophobic solutes.[26] Unfortunately, organic solvents at greater than about 25% v/v seriously inhibit micelle formation. A second limitation of MEKC is that control over selectivity is rather limited. A good number of surfactants are available, but many are not applicable to the MEKC technique, and when combining different types of surfactants, mixed micelles are formed with unpredictable selectivity effects.

These limitations have led to the use of other running buffer additives to address the separation of neutral solutes. The cyclodextrins (CD) are particularly useful reagents for this purpose. Native CDs are neutral, cylindrically shaped macrocyclic sugar molecules that possess a hydrophobic cavity and a hydrophilic exterior.[27] The size of the cavity depends on the number of sugar units in the structure ($\alpha = 6$, $\beta = 7$, and $\gamma = 8$ are most common). Inclusion complex formation between a solute and the cavity of the CD is very selective, and, given that native CDs can be derivatized with a wide variety of neutral and ionizable functional groups, the possibilities to tune selectivity are extensive.

CD-modified MEKC has been shown to be useful for separating hydrophobic compounds.[28] Alternately, systems comprising strictly CDs can be created with a technique referred to as cyclodextrin distribution capillary electrochromatography (CDCE).[29,30] With CDCE, running buffer-CD "cocktails" containing combinations of charged and neutral CDs are created that exhibit the correct selectivity for a given application. Figure 24.3 shows the process of a solute distributing between CD phases and a typical CDCE separation showing high efficiency and selectivity.

Cyclodextrins are also used as a secondary phase to carry out chiral CE separations. Because cyclodextrins are enantiomeric molecules, they can differentially bind pairs of small-molecule enantiomers such as amino acids or small molecules of pharmaceutical importance. This approach for chiral separations has been reviewed extensively.[21,31] A broad range of additional chiral molecules have been used to

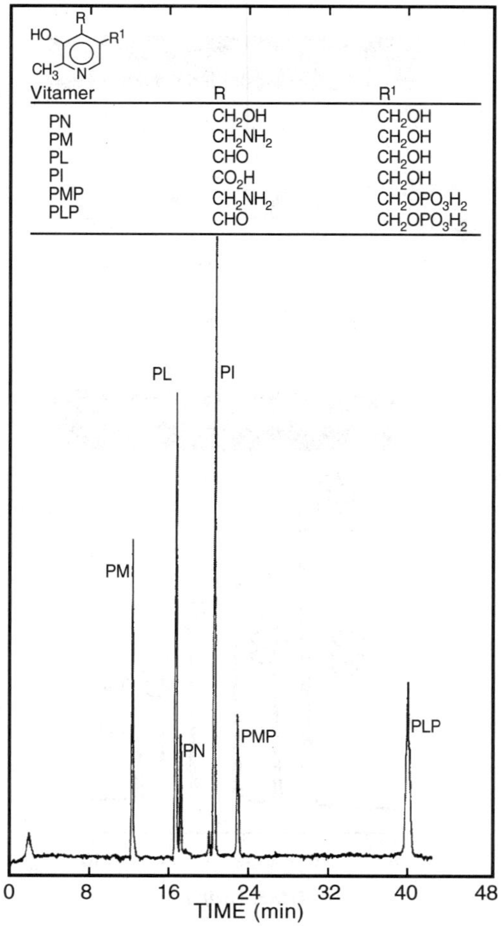


**FIGURE 24.2** Separation of vitamin B6 and its metabolites by MEKC using a running buffer containing 0.05 *M* sodium dodecyl sulfate.

form secondary phases for chiral CE separations. These include proteins, crown ethers, linear polysaccharides, macrocyclic antibiotics, and chiral surfactants.[21,31]

Analytical methodologies involving DNA and proteins are widely used in the life sciences, and a comparison of slab gel electrophoresis and CE approaches to DNA analysis was made earlier in this chapter. DNA fragments possess a charge and, hence, an electrophoretic mobility. Unfortunately, differences in migration rates for differently sized fragments are extremely small because they possess similar charge-to-mass ratios.[32,33] In order to separate DNA fragments, discrimination must be based on differences in size and is accomplished using sieving media.

One approach is to add a soluble polymer such as methylcellulose to the running buffer at concentrations above the entanglement threshold of the polymer. This creates a dynamic mesh in the capillary with a characteristic mesh size that can be varied by adjusting polymer concentration or molecular weight.[34] In practice, the size-selective capillary electrophoresis (SSCE) technique usually employs a surface-modified capillary that does not have appreciable EOF. Detection of the negatively charged DNA fragments, in order of increasing base pair number, is accomplished at the anodic side of the capillary. Because SSCE separations involve large biopolymers migrating through very dense media, axial diffusion is extremely slow and efficiencies numbering millions of theoretical plates are routinely achieved.[34] Figure 24.4A illustrates high efficiency in a temporal study of the digestion of a DNA sample,[35] and

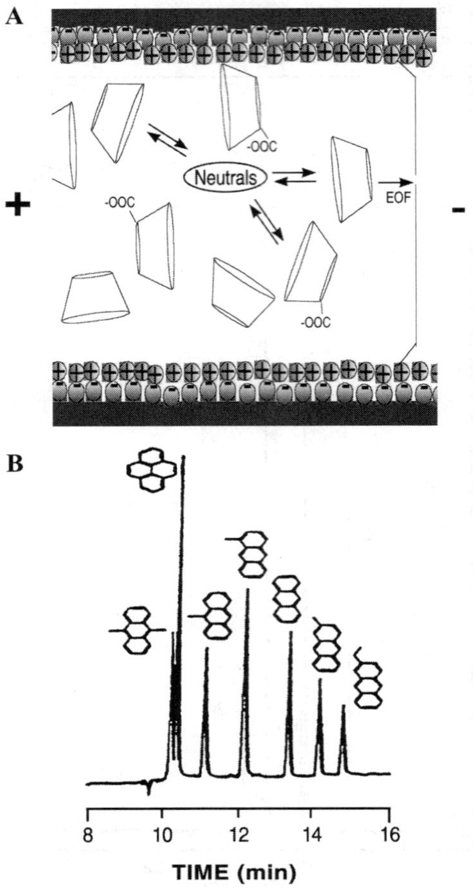


**FIGURE 24.3** (A) Depiction of the CDCE process. The neutral CDs migrate with EOF and the negative CDs migrate more slowly toward the cathode. Neutral compounds migrate at an intermediate rate that depends on their distribution between the CDs. (B) Separation of alkyl anthracene compounds using a running buffer with β-CD and carboxymethyl-β-CD.

Figure 24.4B illustrates the ability to study biomedically significant protein–DNA interactions using this technique.[36] In both instances, fluorescence labeling protocols and photonics-based detection played a critical role (see below). Moreover, the development of SSCE separations with polymer solutions paved the way for the practical development of large capillary array electrophoresis instruments and their application to sequencing the human genome.[3–5,22]

## 24.2.3 Additional Capillary Electrophoresis Separation Modes

During the active development of CE in the 1980s, most separations were based on CZE or on MEKC. Over the next decade, a number of additional separation modes were developed, and several of these have found widespread application (e.g., see preceding discussion). All of the CE-based separation methods currently being developed and applied cannot be discussed adequately in a single chapter; the reader is directed to detailed reviews of these CE-based separation methods. Isotachophoresis and isoelectric focusing have been used with CE technology.[37,38] Capillary electrochromatography is similar to reverse-phase HPLC, except that electrophoretic migration and EOF are used to propel analytes through a packed capillary chromatographic column.[39,40]

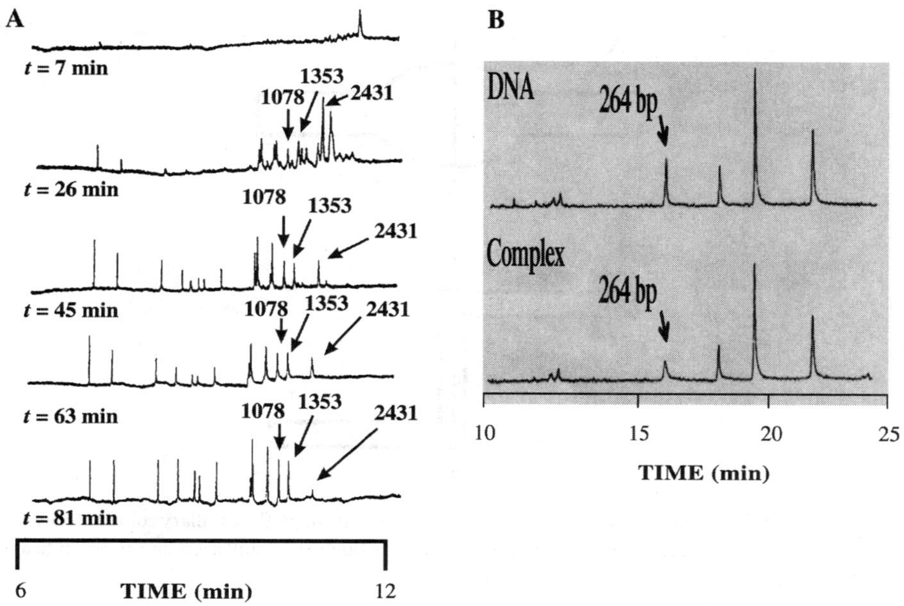


**FIGURE 24.4** (A) The use of SSCE for a temporal study of the digestion of ϕX-174 DNA with *HaeIII* restriction enzyme. The intermediate fragment at 2431 base pairs is digesting to the 1078 and 1353 fragments. On-column labeling with EB is employed. (B) The interaction of digested pBKH26 plasmid DNA with *trp* repressor protein. LIF detection is employed: top, prior to exposure of the digested plasmid to *trp* repressor protein and, bottom, after post-column EB labeling within a sheath flow cell.

## 24.3 Applications of Photonics to Capillary Electrophoresis

The most prominent role played by photonics in CE is detection of separated analytes. Many of the advantages offered by CE as a separation technique are a result of its small scale, which is a significant limitation for detection of analyte bands. The most common detection method used for liquid-phase separation methods is UV/VIS absorption, and it best illustrates the detection challenges for CE. Beer's law describes absorbance detection, $A = \varepsilon bc$. Here $A$ is the absorbance, $\varepsilon$ is the molar absorptivity, $b$ is the detection cell pathlength, and $c$ is the analyte concentration.

The pathlength for a standard cuvette used for absorption measurements is 1 cm. Optical detection for CE is most often carried out on-column using the fused-silica capillary as an optical cell. The pathlength for a 50-μm ID capillary is, therefore, only about 50 μm. An absorbance measurement of a compound in a capillary will produce an absorbance that is only $5 \times 10^{-3}$ that of the absorbance measured in a 1-cm cuvette. A 5-mm pathlength cell for an HPLC detection system will provide an absorbance reading 100× that for a typical CE capillary. Due to the short pathlength provided by a typical CE column and extremely small sample volumes, concentration detection limits are a critical issue when considering detection methods for CE.

### 24.3.1 Detection of Native Analytes

Ideally, all separated components in a CE separation would be detected sensitively and selectively in their native chemical form by a single detector. Of course, this ideal has not been realized in practice, and a wide range of detection strategies are used for CE separations. Although many of these detection methods require chemical modification of an analyte in order to detect it sensitively, a number of detection methods are able to detect compounds in their native state. Detection of an analyte in its native form is preferred if sufficient sensitivity and selectivity for the application at hand can be achieved. Native detection

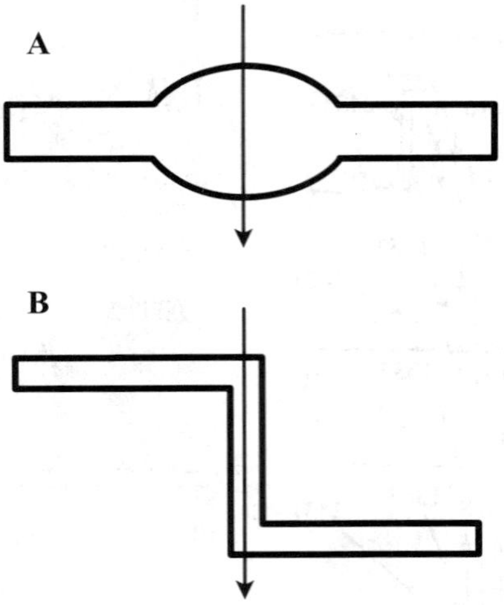

**FIGURE 24.5** (A) Bubble capillary to increase the optical pathlength through the capillary column (only the capillary bore is shown). (B) Z cell to increase the detection pathlength through the capillary. The arrows indicate the light path from the source.

simplifies the overall analytical method and typically results in better measurement precision and accuracy by reducing the total number of steps required for analysis.

#### 24.3.1.1 UV/VIS Absorbance

Despite the aforementioned sensitivity limitations due to short optical pathlengths, the most common detection method used for commercial and laboratory-constructed CE instruments is UV/VIS absorbance. The primary reasons for the popularity of UV/VIS absorbance detection are that it is a relatively general detection technique and can be used to detect a broad range of molecules in their native state with moderate sensitivity and selectivity. Single-wavelength UV/VIS absorbance detectors and diode-array based instruments capable of collecting spectra are commercially available. It is possible to use detection wavelengths near 200 nm when using aqueous solution with inorganic buffers, which do not absorb at these wavelengths. Peptides and proteins can be detected at these wavelengths regardless of sequence. Detection selectivity can be obtained by adjusting the detection wavelength using a variable, single- or dual-wavelength absorbance detector. Absorbance spectra can be used to help confirm peak identity or purity if a diode array-based detector is used.

A number of approaches have been used in an attempt to overcome the pathlength limitation of UV/VIS absorbance detection for CE. Most of these methods are based on increasing the optical pathlength by changing the capillary geometry in the region where on-column detection is performed.[41] Figure 24.5 illustrates two common approaches: the Z cell and the bubble cell. These methods have been used to reduce detection limits for CE; however, there is a limit to the usefulness of this approach. As the pathlength is increased using either of the approaches shown in Figure 24.5, eventually a point will be reached where separation efficiency will be compromised due to the increased volume in the detection zone. Other interesting approaches for reducing detection limits for UV/VIS absorbance detection have been reported, such as on-column signal averaging using diode array detectors[42] and thermal lens methods.[43] These approaches, however, are not commercially available and have not been widely used.

#### 24.3.1.2 Native Fluorescence

Laser-induced fluorescence is clearly the most sensitive detection method available for CE.[44] Single molecule detection has been demonstrated for CE.[45–47] Unfortunately, most molecules are not highly fluorescent in their native state. In cases where a molecule can be detected by native fluorescence and by UV/VIS absorbance, native fluorescence typically provides detection limits that are orders of magnitude

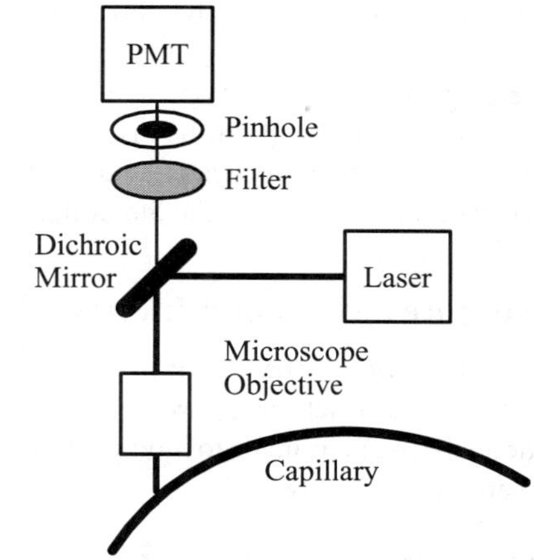


**FIGURE 24.6** Schematic of an LIF detector for CE. The same microscope objective is used to focus the laser beam on the capillary and to collect emitted fluorescence for detection at the PMT (photomultiplier tube).

lower.[48,49] Most proteins and some small peptides can be detected in their native state due to the fluorescence of tryptophan, tyrosine, and phenylalanine. Proteins have been detected after CE separation at low p*M* concentrations by native fluorescence.[50] Some nucleotides are also natively fluorescent, as well as catecholamine neurotransmitters.[51,52] NADH and NADPH are fluorescent; this property has been exploited to follow on-column enzyme reactions by CE.[53] These examples of native fluorescence all require excitation at UV wavelengths. Although NADH and NADPH can be excited effectively with a modestly priced He-Cd laser at 325 nm, native fluorescence of peptides, catecholamines, and nucleotides requires excitation at wavelengths below 300 nm. At these wavelengths, special UV optics are necessary and lasers are relatively expensive. Although these technological barriers will be overcome eventually, at present they still limit the widespread use of native fluorescence of peptides, proteins, and nucleotides with CE. Some compounds of biological interest, such as green fluorescent protein and porphyrins, are fluorescent at visible wavelengths, but this is relatively rare.[54,55]

Concentration detection limits for LIF are less affected by the short optical pathlength of CE because the fluorescence signal can be increased by increasing the laser power until substantial photobleaching begins to dominate.[56] With a small CE capillary, this typically requires a beam of only a few mW focused to the dimensions of the capillary bore. Figure 24.6 shows a common design for LIF detection with a 180° geometry. This optical design is essentially the same as a fluorescence microscope. Detection at 90° with separate objective lenses for focusing the laser beam and collecting fluorescence is also common. Although LIF detection is commercially available, it is less common than absorbance detection, and LIF detectors are substantially more expensive. In addition, it is often necessary to use different lasers for different applications. Most LIF detectors for CE are simple filter fluorometers. Laser-induced fluorescence detectors with CCD detection, which can rapidly collect emission spectra, have been developed, and have been applied to native fluorescence detection.[57,58] Recently, LIF detection with multiphoton excitation has been developed for CE and applied to study natively fluorescent species.[59]

#### 24.3.1.3 Additional Detection Methods

Refractive index (RI) detection is common for HPLC but has rarely been used with CE. This technique has the advantage of being universal. Unfortunately, RI detection is less selective and less sensitive than UV/VIS absorbance detection. Furthermore, miniaturizing RI detection for CE has proven to be challenging. Recently, Bornhop and co-workers have reported a simple but effective method for performing RI detection for CE.[60,61] This technique uses interference fringes from light backscattered through the capillary bore to detect RI changes. Instrumentally simple and inexpensive, it produces detection limits close to those obtained by UV/VIS absorbance for CE. One challenge is that it is necessary to thermostat the CE capillary carefully to obtain low detection limits because the RI detector response is extremely sensitive to temperature.

Electrochemical techniques are important for detection of native analytes separated by CE. These detection methods are not optical, however, and will not be discussed in detail in this chapter. Amperometry, voltammetry, potentiometry, and conductivity detection have all been used with CE, and detailed reviews about electrochemical detection are available in the literature.[62,63]

## 24.3.2 Detection Involving Reactions and Indirect Methods

In many instances the photonic responses of the analytes of interest are weak or nonexistent. However, such responses can be elicited via chemical reactions between the analyte and carefully tailored chromophoric reagents. Alternately, the analytes can be made to modulate a running buffer additive's photonic response in order to measure the analyte indirectly.

### 24.3.2.1 Derivatization (Fluorescence Labeling) in LIF

The low detection limits and selectivity that can be obtained with LIF provide strong motivation to extend the use of this detection method beyond the relatively few compounds that exhibit modest to high fluorescence quantum efficiencies (QE). This can be accomplished by derivatizing the analytes with suitable reagents. The derivatization can involve the formation of covalent bonds between the analyte and reagent or noncovalent association between these species. The reagent may be a natural fluorophore or it may fluoresce strongly only when associated with the analyte. The timing of the derivatization process relative to the CE separation is a critical factor; preseparation, on-column, or post-separation derivatization all have positive and negative characteristics.[64] These characteristics are much the same as in HPLC but are sometimes exacerbated by the small volumes and high efficiencies inherent to CE.

In some cases, preseparation derivatization does not involve specific reagents but, rather, a transformation of the analytes to fluorescent products. For example, the antibiotic ampicillin can be degraded preseparation to a fluorescent species.[65] In most cases, however, covalent interactions between functional groups on the analyte and on the derivatization reagent are involved. Examples of reagents commonly used for CE include *o*-phthalaldehyde (OPA), dansyl-Cl, and fluorescein isothiocyanate (FITC).[64] All of these reagents react to form fluorescent products with many amines, including those found in free amino acids, proteins, and many drugs. A recent review indicates that most covalent derivatization reagents used with CE react with amine groups, although a few reports include derivatization of thiols, carboxylic acids, carbonyls, hydroxyls, and the reducing end of saccharides.[64] A general review citing reagents that are specific for carboxylic acids has been published.[66] Guttman reported the use of 8-aminopyrene-3,6,8-trisulfonate to label sugar moieties and produce low femtomole quantification levels in CE–LIF.[67]

Relatively few compounds are natively fluorescent in the near-infrared (NIR) and, therefore, derivatization to produce molecules that fluoresce in the NIR offers high sensitivity with reduced interference from other compounds in a sample. McWhorter and Soper recently reviewed the use of NIR detection in CE–LIF.[68] Using a variety of newly developed derivatization dyes many measurements of biological samples have been performed. For example, subattomole limits of detection were achieved for amino acids derivatized preseparation with pyronin succinimidyl ester.[69] Another significant advantage of the NIR–LIF approach is that inexpensive diode lasers can be used for excitation.

Certain reagents form sufficiently strong noncovalent associations with analytes such that preseparation labeling can be performed and the separation and LIF detection can be conducted without dissociation of the label–analyte complex. For example, the cationic dye DTTCI was used to label proteins prior to CE separation[70] and certain cyanine intercalation dyes are used in this manner for DNA analysis.[71] Noncovalent labeling of proteins for CE–LIF has been reviewed by Colyer.[72]

The literature is replete with other examples of preseparation labeling.[64] Nevertheless, the preseparation derivatization technique is not always the approach of choice — it can be time consuming and often produces chemical by-products that complicate the sample. It is also possible to form multiple derivatization products when more than one functional group is available for reaction on the analyte. Swaile and Sepaniak compared native fluorescence and preseparation FITC labeling of proteins in terms of detection and of separation performance.[73] The 514-nm output of an Ar+ laser was used for excitation in the FITC case, and native fluorescence of the protein was excited via frequency doubling of the Ar+ laser output. In the case of conalbumin, which has over 100 side chain amine groups, the native fluorescence approach was less sensitive but produced a single high-efficiency peak, while the FITC labeling approach produced multiple unresolved peaks resulting from the creation of multiple products (differing labeling ratios) that spanned a range of electrophoretic mobilities.

Ultrahigh CE–LIF sensitivity is possible for analytes derivatized with strongly fluorescing reagents. For example, Dovichi and coworkers demonstrated single molecule detection for a highly fluorescent macromolecule, β-phycoerythin.[46] β-Phycoerythin has been used as a preseparation label.[74] It is important to note that derivatization is prohibitively slow in ultra-dilute solutions, and often the reported ultrahigh sensitivity involves dilution of samples that are derivatized at much higher concentrations. Nevertheless, it is still possible for derivatization at high concentration, when coupled with ultrahigh sensitivity, to provide unique analytical capabilities. For instance, Yeung and co-workers performed "single-molecule electrophoresis" to measure antibody bound and unbound β-phycoerythrin-digoxigenin.[74] The unique single-molecule electrophoresis technique is a direct product of advances in photonic detection. This technique truly takes advantage of the ability to detect single molecules for an analysis application. A relatively large number of migrating molecules are imaged individually and their mobilities measured simultaneously. Differences in the mobilities of the bound and unbound single molecules facilitated determination of the ratio of these species in the digoxigenin assay.[74]

On-column derivatization is a dynamic labeling process wherein a reagent is added to the running buffer that associates reversibly with analyte migrating in the capillary. Because the mobilities of the labeled and unlabeled analyte will usually differ, fast association–dissociation exchange is required to maintain efficiency. Thus, association constants generally are relatively modest in magnitude. It is essential that the spectral properties of the running buffer reagents be altered upon association. Swaile and Sepaniak investigated this mode of CE–LIF to measure metals in blood serum.[75] The reagent employed at low m*M* concentrations, 8-hydroxy-quinoline-5-sulfonic acid, is a nonfluorescent, bidentate ligand that forms fluorescent complexes with many metals.[76] Excitation of the complexes is conveniently supplied with the 325-nm output of an He-Cd laser. The same authors demonstrated the on-column dynamic labeling of proteins using hydrophobic probes such as 1-anilinonaohthalene-8-sulfonate.[73] This compound intercalates into the hydrophobic regions of proteins with association constants in the $10^{-4}$ to $10^{-6}$ range and experiences a considerable increase in fluorescence QE upon binding. Limits of detection using He-Cd (325 nm) excitation were not as low as with the preseparation labeling approach using FITC, partly because of an appreciable fluorescence background; however, the aforementioned multiple-labeling problem was not encountered.

The most significant use of on-column labeling involves the intercalation of dyes such as ethidium bromide (EB) or newer classes of monomeric and dimeric cyanine dyes into the structure of double-stranded DNA.[71,77–79] The combination of SSCE and on-column labeling provides a powerful tool for DNA analysis. The cyanine dyes generally outperform the more traditionally used EB in terms of sensitivity. However, EB is extremely reliable and always yields high separation efficiency. Upon intercalation, EB experiences roughly a 20-fold increase in fluorescence QE (less than is observed with the cyanine dyes) and also a bathochromic shift in the lowest energy absorption band convenient for excitation using the 543-nm output of an He-Ne "greenie" laser. Using low μ*M* concentrations of EB in the running buffer, it is relatively easy to achieve subattomole limits of detection for injections of modest-sized DNA fragments.[80] The separation shown in Figure 24.4A was obtained using SSCE–LIF with a running buffer containing 2.5 μ*M* EB.

Several types of reactors have been used for post-separation derivatization in CE–LIF; one of the most successful approaches involves sheath flow cells.[36,81,82] In this approach, a sheathing capillary or tube surrounds the end of the separation capillary and a coaxial flow is maintained. The sheathing flow serves to focus the effluent from the capillary hydrodynamically to minimize post-capillary band dispersion. It also has photonics-related advantages, such as reduced background due to Rayleigh scattering from the

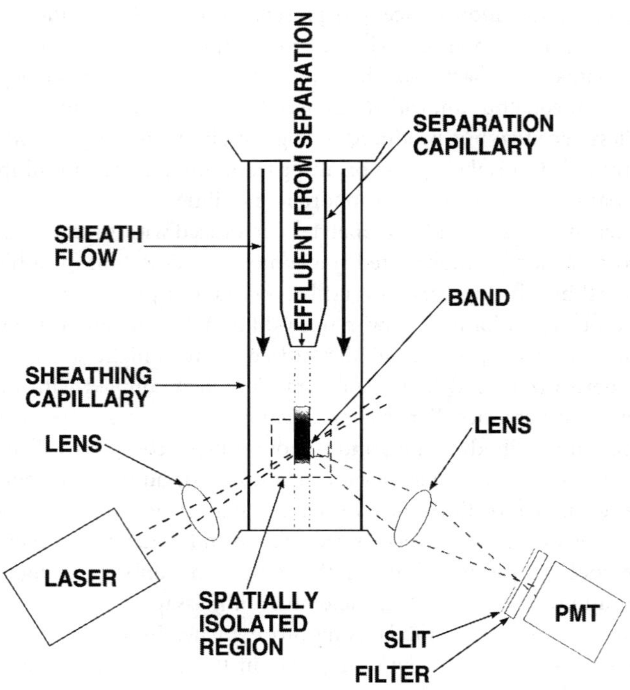


**FIGURE 24.7** Depiction of the sheath flow cell used for the experiment shown in Figure 24.4B. The sheathing fluid contained 2.5 μ*M* EB.

capillary (as compared to on-column detection). When used as a post-separation derivatization reactor, the reagents involved in the reaction are added to the sheathing fluid. To minimize post-separation band dispersion and maximize signal levels, it is critical that the derivatization reaction is rapid. The experiment shown in Figure 24.4B was performed using a sheath flow cell such as that depicted in Figure 24.7.[36]

Discerning the specifics of protein–DNA interactions is very important in genomic–proteomic studies. Unfortunately, preseparation labeling is problematic when dealing with biopolymers (see above). Similarly, on-column labeling with an intercalator can interfere with interactions between the biopolymers. *Try* repressor protein is responsible for the regulation of L-trytophan in *Escherichia coli*. When the amino acid concentration is low, *trp* operon is expressed. The separation in Figure 24.4B demonstrates that the correct (operand-containing) DNA fragment in a digested plasmid can be identified via SSCE–LIF. In this case EB in the sheathing fluid is used to label the separated DNA fragments as they exit the capillary. When EB is used in the normal manner (on-column labeling) in the running buffer, the protein–DNA interaction is suppressed (not shown).

### 24.3.2.2 Indirect Detection

Indirect detection involves the addition of a background electrolyte (BE) to the running buffer to create a relatively large optical background. The technique relies on the displacement of the BE from analyte bands by various mechanisms; the most efficient mechanism being charge displacement. A vacancy zone of BE within the analyte band results in a negative response peak as the band passes the point of detection. An obvious advantage of this technique is that detected analytes need not exhibit the measured optical property. A broad range of analytes have been detected using indirect absorbance and fluorescence, and many of these are of biological significance.[83] Typical BEs include creatine and chromate for indirect absorbance detection, and salicylate and quinine sulfate for indirect LIF.

Unfortunately, several problems are associated with indirect detection. The large BE background signal can be noisy, and baseline disturbances due to nonanalyte displacement of the BE can occur. The addition of a fairly large concentration of BE can also result in distorted (fronted or tailed) peaks due to field inhomogeneity in the analyte band.[84] This problem is exacerbated when the mobilities of the BE and analyte differ greatly. The linear dynamic range (dynamic reserve) for indirect detection is determined by the ratio of the background signal to the noise in that signal and can be small.

The inherent instability of laser sources translates into background fluctuations and results in far less sensitivity than with direct LIF. In early work in this area, Yeung and Kuhr achieved $10^{-7}$ M limits of detection for amino acids, but this required them to go to some length to reduce noise from the laser source.[85] More generally, the limits of detection for indirect LIF detection are in the $10^{-5}$ to $10^{-6}$ M range. This is about the same as for indirect absorbance detection where conventional light sources are employed. These sources exhibit better stability than laser sources.[83] Newer diode lasers have been reported to provide detection limits similar to those reported previously by Kuhr and Yeung due to their improved stability.[86] A technique that is somewhat similar to indirect fluorescence involves quenching the background emission of a phosphorescent running buffer system by analytes.[87] Limits of detection in the $10^{-7}$ to $10^{-8}$ M range were achieved with this technique.

#### 24.3.2.3 Chemiluminescence Detection

Chemiluminescence (CL) is an increasingly common detection method for CE and has been reviewed by several authors.[88–90] Chemiluminescence involves the excitation of an emitting species via a chemical reaction and can be classified as a direct or as a sensitized detection technique. In the direct case the reaction leaves the emitting analyte species in an excited state. In sensitized CL the reaction generates an excited state species that transfers its energy to a fluorescent analyte. An example of the latter is the peroxyoxalate system.[88–90] The obvious advantage of CL is that a light source is not needed, and the background is extremely low. However, a major challenge in implementing CL detection is executing the necessary reactions without degrading the separation. The efficiency of CL depends on the local chemical environment; hence, running buffer conditions such as pH are important.

A few basic configurations are used to implement CL detection in CE at the exit of the separation capillary. In many cases a coaxial design is used in which the separation capillary is slipped through a tee and also through a reaction capillary with an ID greater than the OD of the separation capillary.[82,88–90] The CL reagents flow in the reaction capillary and convectively mix with the effluent at the end of the CE capillary. Convectional mixing reduces problems with slow mixing kinetics but tends to degrade separation efficiency to values of $10^4$ theoretical plates or less. In one example, Ruberto and Grayeski employed a system wherein acridinium ester is oxidized by hydrogen peroxide to an excited state derivative of acridine.[91] This system is amenable to analytes tagged with an acridinium moiety or those that affect peroxide concentration.

A very simple approach employed by Zare and co-workers for CL detection that involves placing the reagents in a relatively large reservoir at the end of the CE capillary has been used with luminol CL and firefly luciferase bioluminescence systems.[92] Limits of detection were $2 \times 10^{-8}$ M for luminol, but separation efficiencies were still limited to less than $2 \times 10^4$ theoretical plates with this approach. This approach would be problematic with sensitized CL because the background level would increase over time as fluorescent species collect in the reservoir.

Dovichi and coworkers utilized a coaxial approach in a sheath flow cell to mix CL reagents contained in the sheathing fluid with CE effluent.[93] Since mixing is based on diffusion, sensitivity can be reduced due to slow mixing kinetics. The authors demonstrated that signals increased moving away from the end of the capillary up to a distance of about 4 mm and then diminished sharply. Absolute limits of detection in the subfemtomole level were quoted for the isoluminol thiocarbamyl derivative of valine. More importantly, the reported separation efficiencies were approximately $10^5$ theoretical plates.

### 24.3.3 Information-Rich Photonic Detection

Despite the qualitative analysis advantages afforded by the excellent resolving power of CE, keen interest in mating electrophoretic separations with information-rich (sometimes referred to as "hyphenated") detection techniques remains. Information-rich techniques provide unambiguous analyte identification, structural elucidation, and high selectivity for those cases where adequate resolution is not achieved. In addition, problems associated with poor EOF reproducibility that have inhibited widespread acceptance of CE by the research community are mitigated, to a degree, when one can track eluted analytes with certainty based on distinctive spectra. Nuclear magnetic resonance (NMR) spectrometry provides unmatched structural information and has been employed for detection in capillaries and in CE using microcoils directly surrounding the measurement capillary.[94] Unfortunately, the inherent insensitivity of the technique makes its implementation in CE detection very challenging.

Conversely, significant progress has been made in hyphenating mass spectrometry (MS) with CE.[95,96] Mass spectrometry has been interfaced with CE primarily using electrospray ionization (ESI). Interfacing CE to ESI–MS also poses some challenges. A compromise must be made between a buffer solution ideal for the CE separation and one optimal for ESI. The electrical interface between the CE capillary and the ESI capillary has proven to be a significant challenge. The coupling of CE to MS has been successful in that amol mass detection limits have been obtained for many analytes, and mass spectrometry can provide unambiguous identification of an analyte peak.

Although MS has been used with CE rather extensively, the expense and complications associated with CE–MS provide motivation to develop complementary photonic approaches to performing information-rich detection in CE. Fourier transform infrared (FTIR) spectrometry has been employed in conjunction with separations.[97] However, complications arise from short optical pathlengths and the absorption of IR radiation by running buffers and capillaries when FTIR is employed with CE. One fluorescence-based technique used in CE is fluorescence line narrowing spectrometry (FLNS). FLNS is a very low-temperature spectral technique that does not exhibit inhomogeneous broadening contributions to vibronic bands. Thus, distinctive, sharp line spectra are obtained. FLNS detection has been employed on-line in CE,[98] an approach that involves jacketing a portion of the CE capillary near its outlet to permit submersion in liquid helium. In operation, the CE flow is stopped and the capillary and its contents rapidly frozen prior to acquiring spectra. The obvious limitations of this approach are the stop flow aspect and the fact that it is limited to natural fluorophores.

The advantages of Raman spectroscopy, relative to FTIR, in accommodating glass sample cells and aqueous solvents are realized when the technique is applied to CE. Raman vibrational bands are distinctive and can be used to determine chemical structure and provide selectivity. Lasers operating in the visible to NIR spectral regions are generally used for excitation. Normal Raman spectroscopy suffers from poor sensitivity. Raman scattering cross sections are often ten orders of magnitude smaller than for more efficient photonic processes such as absorbance or fluorescence. Batchelder and co-workers recently reported an efficient hyperhemispherical detector configuration that provided mid- to low-millimolar detectability in CE.[99] Michael Morris and colleagues have made effective and impressive use of CE preconcentration (sample stacking) and zone-sharpening effects of capillary isotachophoresis to perform normal Raman detection.[100–102] In this approach, sub- to low-micromolar concentrations of analytes such as ribonucleotides, herbicides, and oxyanions are concentrated to detectable levels; distinctive spectra have been obtained, as shown in Figure 24.8.

The inherent inefficiency of normal Raman scattering has been overcome with remarkable success using resonance- and surface-enhanced techniques. Surface-enhanced Raman scattering (SERS) occurs when analytes are adsorbed onto or very near the nanofeatured surfaces of certain metals.[103–105] SERS active media range from simple silver colloidal solutions and silver island films on glass to sophisticated nanolithographically prepared planar substrates.[103–105] In some instances, extreme sensitivity reaching the single molecule level with enhancement factors greater than $10^{12}$ has been achieved by resonance-enhanced SERS.[106]

There have been a few reports of SERS detection in CE.[107–109] Sepaniak and co-workers have employed on-column and postseparation approaches. The on-column approach is simple and involves adding silver colloidal solution to the CE running buffer.[109] Spectra are acquired on the fly by positioning the microscope objective of a commercial Raman spectrometer directly above the CE capillary in a confocal optical arrangement. Figure 24.9 shows very good CE–SERS sensitivity with this

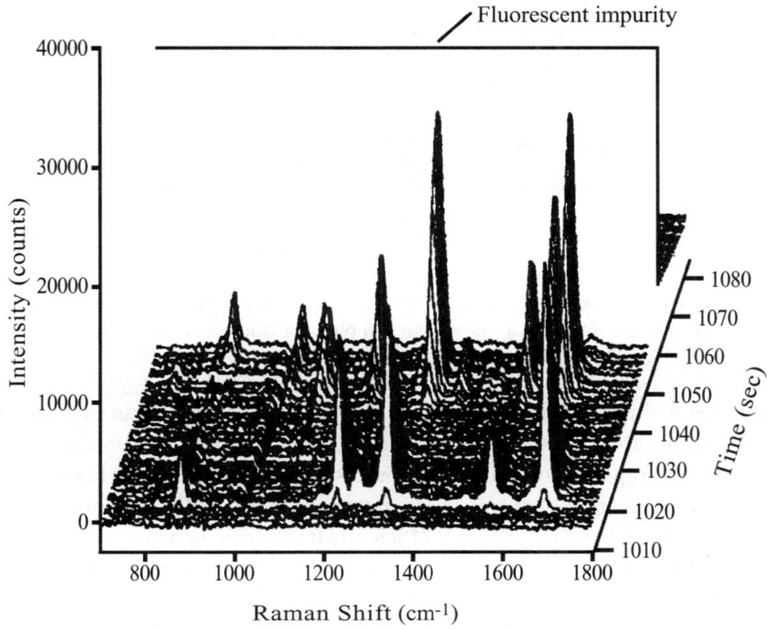


**FIGURE 24.8** On-column normal Raman spectra of $1.5 \times 10^{-5}$ *M* paraquat and diquat acquired in conjunction with a capillary isotachophoresis separation. (From Walker, P.A., III, Shaver, J.M., and Morris, M.D., *Appl. Spectrosc.*, 51, 1394, 1997. With permission.)

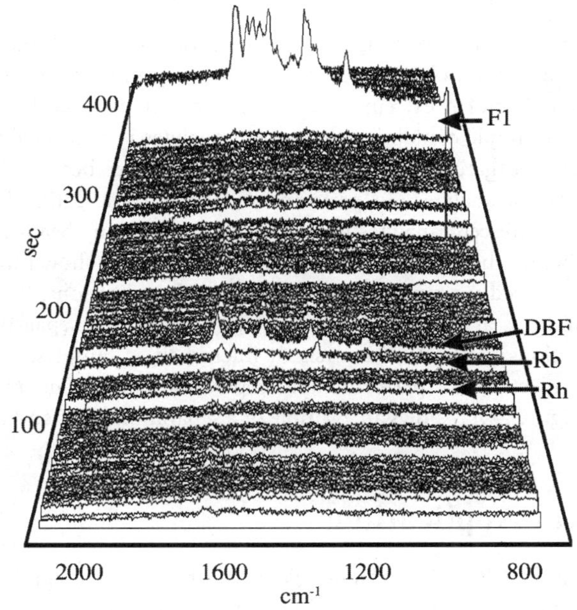


**FIGURE 24.9** SERS spectra obtained for the CE separation of rhodamine 6G (Rh), riboflavin (Rb), dibenzofluorescein (DBF) and fluorescein (Fl). The running buffer contained ~0.5% by weight Ag nanoparticles. Excitation was provided by an $Ar^+$ laser (515 nm, 5 mW). The weak Rh spectrum corresponds to the injection of $10^{-8}$ *M*.

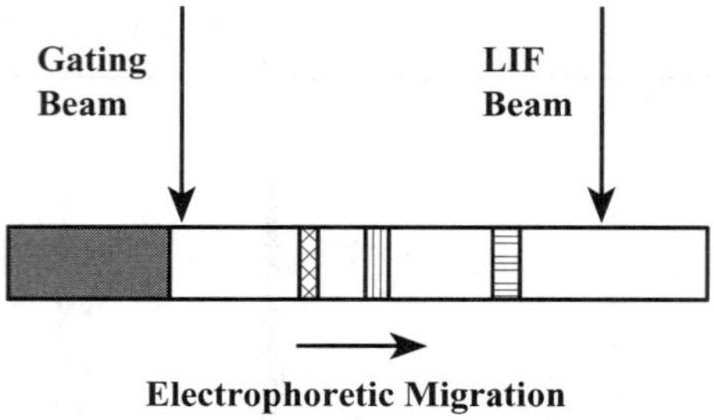


**FIGURE 24.10** Optically gated injection for CE. The gating beam bleaches the fluorescent sample migrating from the left. When the beam is briefly blocked, a fluorescent sample plug is injected. The schematic shows three analyte zones that have separated but not yet reached the LIF detector.

approach, although a strong-to-modest resonance enhancement is also occurring with these analytes. Unfortunately, in many cases the running buffer conditions needed for detection, e.g., silver colloid and electrolytes, degrade separation performance and shorten capillary life. However, the postseparation approach allows independent control over separation and detection conditions.

In one case, the effluent of the CE capillary is efficiently transferred to a moving, planar SERS substrate via a modification of the ESI sampling technique that is used extensively in MS.[108] In this case, the substrate was a frosted microscope slide onto which silver colloid had been deposited. The separation is deposited as a track of bands on the moving substrate, which acts as a semipermanent record of the separation and is available for performing manipulations (e.g., rinsing or derivatization) without the time constraints imposed by on-column detection. Spectra were demonstrated for low- to submicromolar concentrations of pharmaceutically significant analytes.[108] Moreover, the CE separation was transferred to the substrate with only a minor loss in separation efficiency. In the future, advances in SERS substrate design will probably improve this approach to performing information-rich photonic detection in CE.

### 24.3.4 Optically Gated Injection

Optically gated injection represents an unusual application of optical methods to CE, which has been applied to study research problems of biomedical interest. Monnig and Jorgenson first developed optically gated injection in order to perform extremely rapid CE separations (less than 2 s).[110] To perform an optically gated injection, a sample containing fluorescent analytes is continually introduced into a capillary by electrophoresis. The gating beam is a high-intensity laser beam, which almost completely photobleaches the fluorescent analytes. To perform an injection, the gating beam is briefly blocked, allowing a small plug of fluorescent analytes to pass through the gating beam area. The compounds in this plug are separated before they are detected by an LIF detector as shown in Figure 24.10. Optically gated injections have been used for a number of bioanalytical applications.[111] This technique has been used with fast CE separations as the second step of two-dimensional separation methods, as well as applied to perform injections for on-line analysis of microdialysis samples.[8,111] Moore and Jorgenson applied optically gated injections to perform rapid separations of *cis* and *trans* isomers of proline-containing peptides.[112] Figure 24.11 shows an example of a separation of *cis* and *trans* glycyl proline from this work.

## 24.4 Biomedical Applications

Much of the development of CE has been undertaken with biochemical applications in mind, and CE techniques have been applied to a wide range of biomedical problems. The chemical contents of single cells have been analyzed by CE, taking advantage of the technique's small-volume sampling capabilities.[6,7] Capillary electrophoresis has been applied extensively to the analysis of nucleic acids,[3–5,113] and applications to study protein conformational changes and protein folding have emerged in recent years.[114,115]

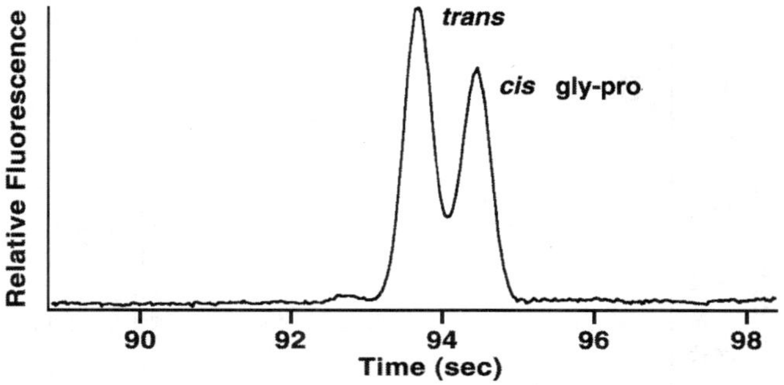


**FIGURE 24.11** Separation of fluorescein-labeled *trans* and *cis* gly-pro with optically gated injection (100 ms). The separation distance was 13.7 cm, and the applied field was 508 V/cm. (From Moore, A.W., Jr. and Jorgenson, J.W., *Anal. Chem.*, 67, 3464, 1995. With permission.)

Affinity CE has been developed to study the binding interactions of biological molecules with other chemical species by measuring shifts in electrophoretic mobilities when zones of interacting analytes electrophorese through each other.[116,117] Capillary electrophoresis methods have been applied to analyze and characterize enzymes and to quantify enzyme substrates.[118–120] Capillary electrophoresis is emerging as an important tool for clinical analysis.[121–123] Two specific examples of the application of CE to address biomedical problems are discussed in this section.

### 24.4.1 Analysis of Substance P Metabolites in Microdialysis Samples

Capillary electrophoresis was used by Lunte and co-workers to separate and detect metabolites of substance P from microdialysis samples collected from rat brains.[124] Substance P is an 11 amino acid neuropeptide, and it is metabolized in the brain by several cytosolic and membrane-bound peptidases. Microdialysis samples were collected from the rat striatum over 15-min intervals throughout the experiments. Substance P (100 μ*M*) was perfused through the microdialysis probe for the first 6 h of the experiments, and microdialysis samples were collected for an additional 5.5 h after substance P was removed from the perfusing solution. Figure 24.12B shows an electropherogram from a microdialysis sample collected at 120 min during the perfusion of substance P. The peak labels identify the corresponding fragment sequences. The peaks were identified based on co-elution with standard samples.

This application of CE takes advantage of the technique's strengths in the areas of separation and sampling and exposes its primary weakness — detection sensitivity. Resolution of the substance P fragments as shown in Figure 24.12B was accomplished using CD-modified MEKC. A very challenging separation was accomplished by simply adding a cyclodextrin (sulfobutyl ether (IV) β-cyclodextrin) and a surfactant (sodium cholate) to a simple aqueous buffer. The microdialysis samples were collected every 15 min at a flow rate of 0.2 μl/min, resulting in 3-μl sample volumes. Fast sampling rates and low probe flow rates are desirable for microdialysis, but they are limited by the sample volume requirements of the analytical methods used to analyze the samples. In principle, hundreds of nl-volume CE injections could be made from a 3-μl sample. In practice, the minimum sample volume required here was limited by the requirements for the derivatization reaction used to enhance analyte detection and by the volume required to immerse one end of the CE capillary in the sample for injection. Nonetheless, the low sample volume requirements of CE allow researchers to use lower flow rates and to collect samples over shorter time intervals compared to experiments using other analysis methods.[125,126] In addition, multiple injections can be made from a single sample to improve accuracy and reproducibility. Because CE typically uses buffered, aqueous solutions, microdialysates can be injected directly with minimal sample preparation.

Achieving adequate detection sensitivity for these experiments was challenging, however. Detection limits for the substance P fragments ranged from 2.5 to 26 n*M*, and several enhancement strategies were used to achieve these results. The substance P metabolites were detected by LIF, but these peptides are

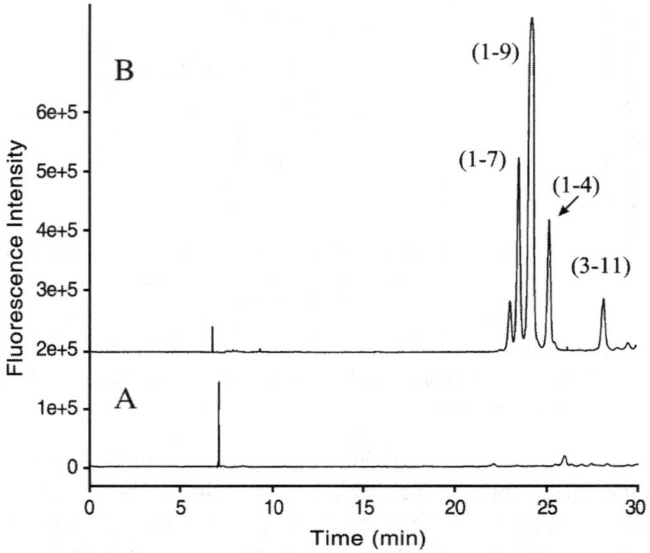


**FIGURE 24.12** (A) Blank microdialysis sample prior to introduction of substance P. (B) Brain microdialysis sample 120 min after substance P introduction. (From Freed, A.L. et al., *J. Neurosci. Methods*, 109, 23, 2001. With permission.)

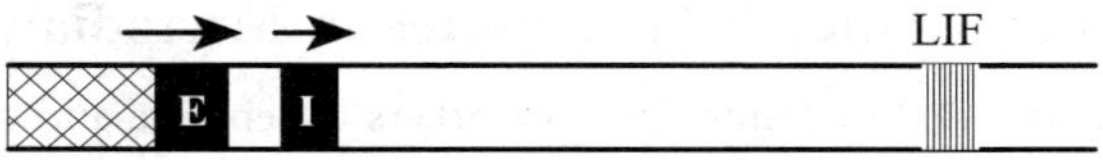


**FIGURE 24.13** CEEI assay. The enzyme-substrate zone (E) migrates faster than the inhibitor zone (I) and will overtake it in the capillary later in the experiment. The hashed region behind the enzyme–substrate zone is product formed by the enzyme-catalyzed reaction, which will be detected later by LIF.

either nonfluorescent or weakly fluorescent in their native state. The microdialysis samples were derivatized prior to CE analysis with naphthalene-2,3-dicarboxaldehyde and $CN^-$ to form highly fluorescent compounds ($\lambda_{ex}$ = 442 nm, $\lambda_{em}$ = 488 nm). A short reaction time was used (2 min) so that the reagent derivatized the amine groups on lysine side chains but did not react extensively with terminal amine groups. This resulted in less interference from other compounds in the dialysate, but it also limited the substance P metabolites detected to those fragments containing the lysine at position 3. In addition, sample stacking by injection of a plug of water preceding the analyte plug was used to lower detection limits. Finally, it was found that the cyclodextrin and sodium cholate added to the buffer enhanced the fluorescence of the labeled analytes by a factor of two to three.

### 24.4.2 Capillary Electrophoretic Enzyme Inhibition Assays

Whisnant and co-workers have developed an on-column CE method for studying enzyme inhibition.[127,128] Figure 24.13 illustrates the basic technique. The capillary is filled with a substrate for the enzyme under study. A zone of inhibitor is first injected into the capillary and an electrophoretic potential is applied for a few seconds. Then a zone of enzyme is injected into the capillary and an electrophoretic potential is applied again. A high concentration of substrate is used to ensure that the enzyme is saturated, and the enzyme zone is actually a zone of enzyme–substrate complex. Because the enzyme–substrate complex has a greater migration rate than the inhibitor zone, the two zones will merge in the capillary, and the enzyme-catalyzed reaction will be inhibited. The two zones will separate again as the experiment continues. The enzyme-catalyzed reaction rate is monitored throughout the experiment based on detection of the reaction product when it migrates past the CE detector. Saevels and co-workers have used a related

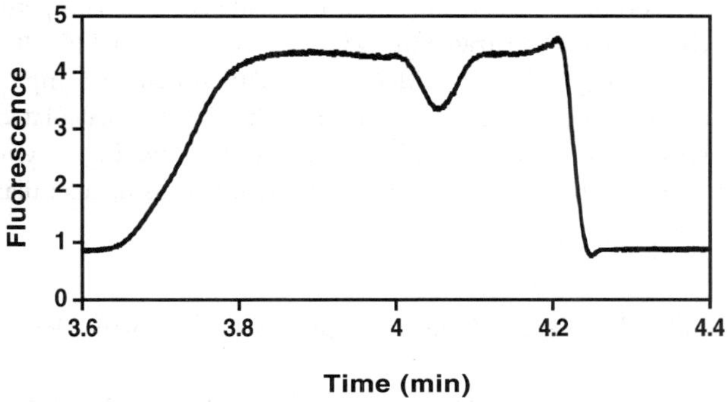


**FIGURE 24.14** CEEI assay showing the reversible inhibition of alkaline phosphatase (0.18 n*M*) by sodium arsenate (125 μ*M*). The sodium arsenate zone was injected first, and then the enzyme was injected after 60 s of electrophoresis. The fluorescence signal is due to product from the enzyme-catalyzed reaction of AttoPhos (0.1 m*M*).

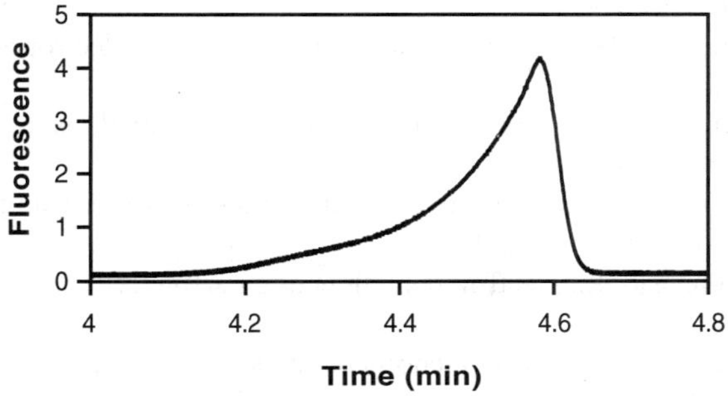


**FIGURE 24.15** CEEI assay showing the irreversible inhibition of alkaline phosphatase (0.18 n*M*) by EDTA (4 m*M*). The EDTA zone was injected first, and then the enzyme was injected after 180 s of electrophoresis.

approach to study the inhibition of adenosine deaminase by erythro-9-(2-hydroxyl-3-nonly)adenine.[129] In their work the capillary was filled with inhibitor and zones of enzyme and substrate were electrophoretically mixed in the capillary.

Figure 24.14 shows a capillary electrophoretic enzyme inhibition (CEEI) electropherogram for the inhibition of alkaline phosphatase by sodium arsenate.[128] Sodium arsenate is a reversible, competitive inhibitor for this enzyme. A commercial, fluorogenic substrate for alkaline phosphatase, AttoPhos, was used to determine alkaline phosphatase activity. The concentration of fluorescent reaction product detected by LIF indicates the reaction rate for the enzyme-catalyzed reaction as the enzyme traveled down the capillary. The negative peak at 4.05 min indicates inhibition when the zone of 125 μ*M* sodium arsenate overlaps with the zone of enzyme–substrate complex. When the inhibitor is not injected, the electropherogram appears as a flat plateau formed as the enzyme migrates from the injection end of the capillary to the LIF detector. A flat plateau indicates a constant reaction rate throughout the experiment. The enzyme concentration was 0.18 n*M*; only 1.9 amol of enzyme were injected.

The shape of the electropherogram in Figure 24.14 indicates that sodium arsenate is a reversible inhibitor. The enzyme returns to its original activity when the zones of enzyme–substrate complex and inhibitors separate again after mixing. Figure 24.15 shows a CEEI electropherogram for an irreversible inhibitor of alkaline phosphatase, EDTA, at 4 m*M*. In this case, the enzyme activity does not recover after the two zones mix, indicating irreversible inhibition. Note that the electropherogram is "reversed" compared to what one might expect. This is a result of the enzyme–substrate complex having a greater migration rate than the reaction product.[127,130] This CEEI technique also showed that EDTA unexpectedly activated alkaline phosphatase at concentrations from 20 to 400 μ*M*.[128]

Quantitative analysis of sodium arsenate and sodium vanadate, also a competitive reversible inhibitor of alkaline phosphatase, has been demonstrated using CEEI assays.[128] The CEEI data for a range of inhibitor concentrations was analyzed using a Michaelis–Menten-based treatment of competitive enzyme inhibition kinetics to construct a calibration curve for the inhibitors. The $K_i$ values calculated based on this analysis were consistent with values obtained using traditional methods for studying enzyme inhibition kinetics.[128] Theophylline, a noncompetitive, reversible inhibitor was quantified using a similar approach.[127,128]

## References


1. Jorgenson, J.W. and Lukacs, K.D., Zone electrophoresis in open-tubular capillaries, *Anal. Chem.*, 53, 1298, 1981.
2. Jorgenson, J.W. and Lukacs, K.D., Capillary zone electrophoresis, *Science,* 222, 266, 1983.
3. Venter, J.C. et al., The sequence of the human genome, *Science,* 291, 1304, 2001.
4. Kheterpal, I. and Mathies, R.A., Capillary array electrophoresis DNA sequencing, *Anal. Chem.*, 71, 31A, 1999.
5. Dovichi, N.J. and Zhang, J., How capillary electrophoresis sequenced the human genome, *Angew. Chem. Int. Ed.*, 39, 4463, 2000.
6. Shaner, L.M. and Brown, P.R., Single cell analysis using capillary electrophoresis, *J. Liq. Chromatogr. Relat. Technol.*, 23, 975, 2000.
7. Zabzdyr, J.L. and Lillard, S.J., New approaches to single-cell analysis by capillary electrophoresis, *Trends Anal. Chem.*, 20, 467, 2001.
8. Liu, Z. and Lee, M.L., Comprehensive two-dimensional separations using microcolumns, *J. Microcolumn Sep.*, 12, 241, 2000.
9. Rice, C.L. and Whitehead, R., Electrokinetic flow in a narrow cylindrical capillary, *J. Phys. Chem.*, 69, 4017, 1965.
10. Gas, B. and Kenndler, E., Dispersive phenomena in electromigration separation methods, *Electrophoresis,* 21, 3888, 2000.
11. Huang, X., Coleman, W.F., and Zare, R.N., Analysis of factors causing peak broadening in capillary zone electrophoresis, *J. Chromatogr.*, 480, 95, 1989.
12. Riekkola, M.-L., Jussila, M., Porras, S.P., and Valko, I.E., Non-aqueous capillary electrohoresis, *J. Chromatogr. A,* 892, 155, 2000.
13. Schaeper, J.P. and Sepaniak, M.J., Parameters affecting reproducibility in capillary electrophoresis, *Electrophoresis,* 21, 1421, 2000.
14. Rose, D.J., Jr. and Jorgenson, J.W., Characterization and automation of sample introduction methods for capillary zone electrophoresis, *Anal. Chem.*, 60, 642, 1988.
15. Quirino, J.P. and Terabe, S., Sample stacking of cationic and anionic analytes in capillary electrophoresis, *J. Chromatogr. A,* 902, 119, 2000.
16. Stroink, T., Paarlberg, E., Waterval, J.C.M., Bult, A., and Underberg, W.J.M., On-line sample preconcentration in capillary electrophoresis, focused on the determination of proteins and peptides, *Electrophoresis,* 22, 2374, 2001.
17. Corradini, D., Buffer additives other than the surfactant sodium dodecyl sulfate for protein separations by capillary electrophoresis, *J. Chromatogr. B,* 699, 221, 1997.
18. Horvath, J. and Dolnik, V., Polymer wall coatings for capillary electrophoresis, *Electrophoresis,* 22, 644, 2001.
19. Liu, C.-Y., Stationary phases for capillary electrophoresis and capillary electrochromatography, *Electrophoresis,* 22, 612, 2001.
20. Righetti, P.G., Gelfi, C., Verzola, B., and Castelletti, L., The state of the art of dynamic coatings, *Electrophoresis,* 22, 603, 2001.
21. Vespalec, R. and Bocek, P., Chiral separations in capillary electrophoresis, *Chem. Rev.*, 100, 3715, 2000.

22. Quesada, M.A. and Menchen, S., Replaceable polymers for DNA sequencing by capillary electrophoresis, in *Capillary Electrophoresis of Nucleic Acids,* Vol. I, Mitchelson, K.R. and Cheng, J., Eds., Humana Press, Totowa, NJ, 2001, p. 139.
23. Terabe, S., Otsuka, K., and Ando, T., Electrokinetic chromatography with micellar solution and open-tubular capillary, *Anal. Chem.,* 57, 834, 1985.
24. Sepaniak, M.J. and Cole, R.O., Column efficiency in micellar electrokinetic capillary chromatography, *Anal. Chem.,* 59, 472, 1987.
25. Burton, D.E., Sepaniak, M.J., and Maskarinec, M.P., Analysis of B6 vitamers by micellar electrokinetic capillary chromatography with laser-excited fluorescence detection, *J. Chromatogr. Sci.,* 24, 347, 1986.
26. Balchunas, A.T. and Sepaniak, M.J., Extension of elution range in micellar electrokinetic capillary chromatography, *Anal. Chem.,* 59, 1466, 1987.
27. Culha, M., Fox, S., and Sepaniak, M., Selectivity in capillary electrochromatography using native and single isomer anionic cyclodextrin reagents, *Anal. Chem.,* 72, 88, 2000.
28. Copper, C.L. and Sepaniak, M.J., Cyclodextrin-modified micellar electrokinetic capillary chromatography separations of benzopyrene isomers: correlation with computationally derived host-guest energies, *Anal. Chem.,* 66, 147, 1994.
29. Lurie, I.S., Klein, R.F.X., Dal Cason, T.A., LeBelle, M.J., Brenneisen, R., and Weinberger, R.E., Chiral resolution of cationic drugs of forensic interest by capillary electrophoresis with mixtures of neutral and anionic cyclodextrins, *Anal. Chem.,* 66, 4019, 1994.
30. Sepaniak, M.J., Copper, C.L., Whitaker, K.W., and Anigbogu, V.C., Evaluation of a dual-cyclodextrin phase variant of capillary electrokinetic chromatography for separations of nonionizable solutes, *Anal. Chem.,* 67, 2037, 1995.
31. Rizzi, A., Fundamental aspects of chiral separations by capillary electrophoresis, *Electrophoresis,* 22, 3079, 2001.
32. Dolnik, V. and Novotny, M., Capillary electrophoresis of DNA fragments in entangled polymer solutions: a study of separation variables, *J. Microcolumn Sep.,* 4, 515, 1993.
33. Grossman, P.D. and Soane, D.S., Experimental and theoretical studies of DNA separations by capillary electrophoresis in entangled polymer solutions, *Biopolymers,* 31, 1221, 1991.
34. Clark, B.K., Nickles, C.L., Morton, K.C., Kovac, J., and Sepaniak, M.J., Rapid separation of DNA restriction digests using size selective capillary electrophoresis with application to DNA fingerprinting, *J. Microcolumn Sep.,* 6, 503, 1994.
35. Stebbins, M.A., Schar, C.R., Peterson, C.B., and Sepaniak, M.J., Temporal analysis of DNA restriction digests by capillary electrophoresis, *J. Chromatogr. B,* 697, 181, 1997.
36. Nirode, W.F., Staller, T.D., Cole, R.O., and Sepaniak, M.J., Evaluation of a sheath flow cuvette for postcolumn fluorescence derivatization of DNA fragments separated by capillary electrophoresis, *Anal. Chem.,* 70, 182, 1998.
37. Rodriguez-Diaz, R., Wehr, T., and Zhu, M., Capillary isoelectric focusing, *Electrophoresis,* 18, 2134, 1997.
38. Gebauer, P. and Bocek, P., Recent progress in capillary isotachophoresis, *Electrophoresis,* 21, 3898, 2000.
39. Colon, L.A., Burgos, G., Maloney, T.D., Cintron, J.M., and Rodriguez, R.L., Recent progress in capillary electrochromatography, *Electrophoresis,* 21, 3965, 2000.
40. Unger, K.K., Huber, M., Walhagen, K., Hennessy, T.P., and Hearn, M.T.W., A critical appraisal of capillary electrochromatography, *Anal. Chem.,* 74, 200A, 2002.
41. Albin, M., Grossman, P.D., and Moring, S.E., Sensitivity enhancement for capillary electrophoresis, *Anal. Chem.,* 65, 489A, 1993.
42. Culbertson, C.T. and Jorgenson, J.W., Lowering the UV absorbance detection limit and increasing the sensitivity of capillary electrophoresis using a dual linear photodiode array detector and signal averaging, *J. Microcolumn Sep.,* 11, 652, 1999.

43. Seidel, B.S. and Faubel, W., Fiber optic modified thermal lens detector system for the determination of amino acids, *J. Chromatogr. A,* 817, 223, 1998.
44. Li, T. and Kennedy, R.T., Laser-induced fluorescence detection in microcolumn separations, *Trends Anal. Chem.,* 17, 484, 1998.
45. Haab, B.B. and Mathies, R.A., Single molecule fluorescence burst detection of DNA fragments separated by capillary electrophoresis, *Anal. Chem.,* 67, 3253, 1995.
46. Chen, D.Y. and Dovichi, N.J., Single-molecule detection in capillary electrophoresis: molecular shot noise as a fundamental limit to chemical analysis, *Anal. Chem.,* 68, 690, 1996.
47. Shortreed, M.R., Li, H., Huang, W.-H., and Yeung, E.S., High-throughput single-molecule DNA screening based on electrophoresis, *Anal. Chem.,* 72, 2879, 2000.
48. Yeung, E.S., Study of single cells by using capillary electrophoresis and native fluorescence detection, *J. Chromatogr. A,* 830, 243, 1999.
49. Gooijer, C., Kok, S.J., and Ariese, F., Capillary electrophoresis with laser-induced fluorescence detection for natively fluorescent analytes, *Analysis,* 28, 679, 2000.
50. Lee, T.T. and Yeung, E.S., High-sensitivity laser-induced fluorescence detection of native proteins in capillary electrophoresis, *J. Chromatogr.,* 595, 319, 1992.
51. Milofsky, R.E. and Yeung, E.S., Native fluorescence detection of nucleic acids and DNA restriction fragments in capillary electrophoresis, *Anal. Chem.,* 65, 153, 1993.
52. Chang, H.-T. and Yeung, E.S., Determination of catecholamines in single adrenal medullary cells by capillary electrophoresis and laser-induced native fluorescence, *Anal. Chem.,* 67, 1079, 1995.
53. Xue, Q. and Yeung, E.S., Variability of intracellular lactate dehydrogenase isoenzymes in single human erythrocytes, *Anal. Chem.,* 66, 1175, 1994.
54. Wu, N., Li, B., and Sweedler, J.V., Recent developments in porphyrin separations using capillary electrophoresis with native fluorescence detection, *J. Liq. Chromatogr.,* 17, 1917, 1994.
55. Korf, G.M., Landers, J.P., and O'Kane, D.J., Capillary electrophoresis with laser-induced fluorescence detection for the analysis of free and immune-complexed green fluorescent protein, *Anal. Biochem.,* 251, 210, 1997.
56. Mathies, R.A., Peck, K., and Stryer, L., Optimization of high-sensitivity fluorescence detection, *Anal. Chem.,* 62, 1786, 1990.
57. Timperman, A.T., Oldenburg, K.E., and Sweedler, J.V., Native fluorescence detection and spectral differentiation of peptides containing tryptophan and tyrosine in capillary electrophoresis, *Anal. Chem.,* 67, 3421, 1995.
58. Park, Y.H., Zhang, X., Rubakhin, S.S., and Sweedler, J.V., Independent optimization of capillary electrophoresis separation and native fluorescence detection conditions for indolamine and catecholamine measurements, *Anal. Chem.,* 71, 4997, 1999.
59. Shear, J.B., Multiphoton-excited fluorescence in bioanalytical chemistry, *Anal. Chem.,* 71, 598A, 1999.
60. Swinney, K., Markov, D., and Bornhop, D.J., Ultrasmall volume refractive index detection using microinterferometry, *Rev. Sci. Instrum.,* 71, 2684, 2000.
61. Swinney, K.A. and Bornhop, D.J., Universal detection for capillary electrophoresis using microinterferometric backscatter detection, *J. Microcolumn Sep.,* 11, 596, 1999.
62. Zemann, A.J., Conductivity detection in capillary electrophoresis, *Trends Anal. Chem.,* 20, 346, 2001.
63. Baldwin, R.P., Recent advances in electrochemical detection in capillary electrophoresis, *Electrophoresis,* 21, 4017, 2000.
64. Waterval, J.C.M., Lingeman, H., Bult, A., and Underberg, W.J.M., Derivatization trends in capillary electrophoresis, *Electrophoresis,* 21, 4029, 2000.
65. Miyazaki, K., Ohtani, K., Sunada, K., and Arita, T., Determination of ampicillin, amoxicillin, cephalexin, and cephradine in plasma by high-performance liquid chromatography using fluorometric detection, *J. Chromatogr.,* 276, 478, 1983.

66. Mukherjee, P.S. and Karnes, H.T., Ultraviolet and fluorescence derivatization reagents for carboxylic acids suitable for high performance liquid chromatography: a review, *Biomed. Chromatogr.*, 10, 193, 1996.
67. Guttman, A., Capillary gel electrophoresis of 8-aminopyrene-3,6,8-trisulfonate-labeled oligosaccharides, in *Techniques in Glycobiology*, Townsend, R.R. and Hotchkiss, A.T., Jr., Eds., Marcel Dekker, New York, 1997, p. 377.
68. McWhorter, S. and Soper, S.A., Near-infrared laser-induced fluorescence detection in capillary electrophoresis, *Electrophoresis*, 21, 1267, 2000.
69. Fuchigami, T., Imasaka, T., and Shiga, M., Subattomole detection of amino acids by capillary electrophoresis based on semiconductor laser fluorescence detection, *Anal. Chim. Acta*, 282, 209, 1993.
70. Legendre, B.L., Jr. and Soper, S.A., Binding properties of near-IR dyes to proteins and separation of the dye/protein complexes using capillary electrophoresis with laser-induced fluorescence detection, *Appl. Spectrosc.*, 50, 1196, 1996.
71. Gibson, T.J. and Sepaniak, M.J., Examination of cyanine intercalation dyes for rapid and sensitive detection of DNA fragments by capillary electrophoresis, *J. Cap. Elec.*, 5, 73, 1998.
72. Colyer, C., Noncovalent labeling of proteins in capillary electrophoresis with laser-induced fluorescence detection, *Cell Biochem. Biophys.*, 33, 323, 2000.
73. Swaile, D.F. and Sepaniak, M.J., Laser-based fluorometric detection schemes for the analysis of proteins by capillary zone electrophoresis, *J. Liq. Chromatogr.*, 14, 869, 1991.
74. Ma, Y., Shortreed, M.R., Li, H., Huang, W., and Yeung, E.S., Single-molecule immunoassay and DNA diagnosis, *Electrophoresis*, 22, 421, 2001.
75. Swaile, D.F. and Sepaniak, M.J., Determination of metal ions by capillary zone electrophoresis with on-column chelation using 8-hydroxyquinoline-5-sulfonic acid, *Anal. Chem.*, 63, 179, 1991.
76. Soroka, K., Vithanage, R.S., Phillips, D.A., Walker, B., and Dasgupta, P.K., Fluorescence properties of metal complexes of 8-hydroxyquinoline-5-sulfonic acid and chromatographic applications, *Anal. Chem.*, 59, 629, 1987.
77. Clark, S.M. and Mathies, R.A., Multiplex dsDNA fragment sizing using dimeric intercalation dyes and capillary array electrophoresis: ionic effects on the stability and electrophoretic mobility of DNA-dye complexes, *Anal. Chem.*, 69, 1355, 1997.
78. Glazer, A.N. and Rye, H.S., Stable dye-DNA intercalation complexes as reagents for high-sensitivity fluorescence detection, *Nature*, 359, 859, 1992.
79. Rye, H.S., Yue, S., Wemmer, D.E., Quesada, M.A., Haugland, R.P., Mathies, R.A., and Glazer, A.N., Stable fluorescent complexes of double-stranded DNA with bis-intercalating asymmetric cyanine dyes: properties and applications, *Nucl. Acids Res.*, 20, 2803, 1992.
80. Clark, B.K. and Sepaniak, M.J., Evaluation of on-column labeling with intercalating dyes for fluorescence detection of DNA fragments separated by capillary electrophoresis, *J. Microcolumn Sep.*, 5, 275, 1993.
81. Oldenburg, K.E., Xi, X., and Sweedler, J.V., Simple sheath flow reactor for post-column fluorescence derivatization in capillary electrophoresis, *Analyst*, 122, 1581, 1997.
82. Zhu, R. and Kok, W.T., Post-column derivatization for fluorescence and chemiluminescence detection in capillary electrophoresis, *J. Pharm. Biomed. Anal.*, 17, 985, 1998.
83. Doble, P. and Haddad, P.R., Indirect photometric detection of anions in capillary electrophoresis, *J. Chromatogr. A*, 834, 189, 1999.
84. Colburn, B.A., Starnes, S.D., Hinton, E.R., and Sepaniak, M.J., Quantitative aspects of rare earth metal determinations using capillary electrophoresis with indirect absorbance detection, *Sep. Sci. Technol.*, 30, 1511, 1995.
85. Yeung, E.S. and Kuhr, W.G., Indirect detection methods for capillary separations, *Anal. Chem.*, 63, 275A, 1991.
86. Melanson, J.E., Boulet, C.A., and Lucy, C.A., Indirect laser-induced fluorescence detection for capillary electrophoresis using a violet diode laser, *Anal. Chem.*, 73, 1809, 2001.

87. Kuijt, J., Brinkman, U.A.T., and Gooijer, C., Quenched phosphorescence, a new detection method in capillary electrophoresis, *Electrophoresis,* 21, 1305, 2000.
88. Staller, T.D. and Sepaniak, M.J., Chemiluminescence detection in capillary electrophoresis, *Electrophoresis,* 18, 2291, 1997.
89. Kuyper, C. and Milofsky, R., Recent developments in chemiluminescence and photochemical reaction detection for capillary electrophoresis, *Trends Anal. Chem.*, 20, 232, 2001.
90. Huang, X.-J. and Fang, Z.-L., Chemiluminescence detection in capillary electrophoresis, *Anal. Chim. Acta,* 414, 1, 2000.
91. Ruberto, M.A. and Grayeski, M.L. Investigation of acridinium labeling for chemiluminescence detection of peptides separated by capillary electrophoresis, *J. Microcolumn Sep.*, 6, 545, 1994.
92. Dadoo, R., Seto, A.G., Colon, L.A., and Zare, R.N., End-column chemiluminescence detector for capillary electrophoresis, *Anal. Chem.*, 66, 303, 1994.
93. Zhao, J.-Y., Labbe, J., and Dovichi, N.J., Use of a sheath flow cuvette for chemiluminescence detection of isoluminol thiocarbamyl-amino acids separated by capillary electrophoresis, *J. Microcolumn Sep.*, 5, 331, 1993.
94. Olson, D.L., Lacey, M.E., and Sweedler, J.V., The nanoliter niche, *Anal. Chem.*, 70, 257A, 1998.
95. von Brocke, A., Nicholson, G., and Bayer, E., Recent advances in capillary electrophoresis/electrospray-mass spectrometry, *Electrophoresis,* 22, 1251, 2001.
96. Naylor, S. and Tomlinson, A.J., Capillary electrophoresis–mass spectrometry of biologically active peptides and proteins, in *Clinical and Forensic Applications of Capillary Electrophoresis,* Petersen, J.R. and Mohammad, A.A., Eds., Humana Press, Totowa, NJ, 2001.
97. Somsen, G.W., Hooijschuur, E.W.J., Gooijer, C., Brinkman, U.A.Th., Velthorst, N.H., and Visser, T., Coupling of reversed-phase liquid column chromatography and Fourier transform infrared spectrometry using postcolumn on-line extraction and solvent elimination, *Anal. Chem.*, 68, 746, 1996.
98. Jankowiak, R., Roberts, K.P., and Small, G.J., Fluorescence line-narrowing detection in chromatography and electrophoresis, *Electrophoresis,* 21, 1251, 2000.
99. Ruddick, A., Batchelder, D.N., Bartle, K.D., Gilby, A.C., and Pitt, G.D., Development of a Raman detector for capillary electrophoresis, *Appl. Spectrosc.*, 54, 1857, 2000.
100. Kowalchyk, W.K., Walker, P.A., III, and Morris, M.D., Rapid normal Raman spectroscopy of sub-ppm oxy-anion solutions: the role of electrophoretic preconcentration, *Appl. Spectrosc.*, 49, 1183, 1995.
101. Walker, P.A., III, Shaver, J.M., and Morris, M.D., Identification of cationic herbicides in deionized water, municipal tap water, and river water by capillary isotachophoresis/on-line Raman spectroscopy, *Appl. Spectrosc.*, 51, 1394, 1997.
102. Walker, P.A., III and Morris, M.D., Capillary isotachophoresis with fiber-optic Raman spectroscopic detection. Performance and application to ribonucleotides, *J. Chromatogr. A,* 805, 269, 1998.
103. Vo-Dinh, T., Surface-enhanced Raman spectroscopy using metallic nanostructures, *Trends Anal. Chem.*, 17, 557, 1998.
104. Weaver, M.J., Zou, S., and Chan, H.Y.H., The new interfacial ubiquity of surface-enhanced Raman spectroscopy, *Anal. Chem.*, 72, 38A, 2000.
105. Campion, A. and Kambhampati, P., Surface-enhanced Raman scattering, *Chem. Soc. Rev.*, 27, 241, 1998.
106. Nie, S. and Emory, S.R., Probing single molecules and single nanoparticles by surface-enhanced Raman scattering, *Science,* 275, 1102, 1997.
107. He, L., Natan, M.J., and Keating, C.D., Surface-enhanced Raman scattering: a structure-specific detection method for capillary electrophoresis, *Anal. Chem.*, 72, 5348, 2000.
108. DeVault, G.L. and Sepaniak, M.J., Spatially focused deposition of capillary electrophoresis effluent onto surface-enhanced Raman-active substrates for off-column spectroscopy, *Electrophoresis,* 22, 2303, 2001.

109. Nirode, W.F., Devault, G.L., Sepaniak, M.J., and Cole, R.O., On-column surface-enhanced Raman spectroscopy detection in capillary electrophoresis using running buffers containing silver colloidal solutions, *Anal. Chem.,* 72, 1866, 2000.
110. Monnig, C.A. and Jorgenson, J.W., On-column sample gating for high-speed capillary zone electrophoresis, *Anal. Chem.,* 63, 802, 1991.
111. Kennedy, R.T., Bioanalytical applications of fast capillary electrophoresis, *Anal. Chim. Acta,* 400, 163, 1999.
112. Moore, A.W., Jr. and Jorgenson, J.W., Resolution of cis and trans isomers of peptides containing proline using capillary zone electrophoresis, *Anal. Chem.,* 67, 3464, 1995.
113. Mitchelson, K.R. and Cheng, J., Eds., *Capillary Electrophoresis of Nucleic Acids,* Vol. I and II, Humana Press, Totowa, NJ, 2001.
114. Righetti, P.G. and Verzola, B., Folding/unfolding/refolding of proteins: present methodologies in comparison with capillary zone electrophoresis, *Electrophoresis,* 22, 2359, 2001.
115. Rochu, D. and Masson, P., Mulitple advantages of capillary zone electrophoresis for exploring protein conformational stability, *Electrophoresis,* 23, 189, 2002.
116. Heegaard, N.H.H., Nissen, M.H., and Chen, D.D.Y., Applications of on-line weak affinity interactions in free solution capillary electrophoresis, *Electrophoresis,* 23, 815, 2002.
117. Duijn, R.M.G.-V., Frank, J., Dedem, G.W.K.V., and Baltussen, E., Recent advances in affinity capillary electrophoresis, *Electrophoresis,* 21, 3905, 2000.
118. Bao, J.J., Fujima, J.M., and Danielson, N.D., Determination of minute enzymatic activities by means of capillary electrophoretic techniques, *J. Chromatogr. B,* 699, 481, 1997.
119. Schultz, N.M., Tao, L., Rose, D.J., Jr., and Kennedy, R.T., Immunoassays and enzyme assays using capillary electrophoresis, in *Handbook of Capillary Electrophoresis,* 2nd ed., Landers, J.P., Ed., CRC Press, Boca Raton, FL, 1997, p. 611.
120. Harmon, B.J. and Regnier, F.E., Electrophoretically mediated microanalysis, *Chem. Anal.,* 146, 925, 1998.
121. Jenkins, M.A., Clinical applications of capillary electrophoresis: status at the new millennium, *Mol. Biotech.,* 15, 201, 2000.
122. Harvey, M.D., Paquette, D.M., and Banks, P.R., Clinical applications of CE, *J. Liq. Chromatogr. Relat. Technol.,* 24, 1871, 2001.
123. Petersen, J.R. and Mohammad, A.A., Eds., *Clinical and Forensic Applications of Capillary Electrophoresis,* Humana Press, Totowa, NJ, 2001.
124. Freed, A.L., Cooper, J.D., Davies, M.I., and Lunte, S.M., Investigation of the metabolism of substance P in rat striatum by microdialysis sampling and capillary electrophoresis with laser-induced fluorescence detection, *J. Neurosci. Methods,* 109, 23, 2001.
125. Dawson, L.A., Capillary electrophoresis and microdialysis: current technology and applications, *J. Chromatogr. B,* 697, 89, 1997.
126. Denoroy, L., Bert, L., Parrot, S., Robert, F., and Renaud, B., Assessment of pharmacodynamic and pharmacokinetic characteristics of drugs using microdialysis sampling and capillary electrophoresis, *Electrophoresis,* 19, 2841, 1998.
127. Whisnant, A.R., Johnston, S.E., and Gilman, S.D., Capillary electrophoretic analysis of alkaline phosphatase inhibition by theophylline, *Electrophoresis,* 21, 1341, 2000.
128. Whisnant, A.R. and Gilman, S.D., Studies of reversible inhibition, irreversible inhibition and activation of alkaline phosphatase by capillary electrophoresis, *Anal. Biochem.,* 307, 266, 2002.
129. Saevels, J., Van den Steen, K., Schepdael, A.V., and Hoogmartens, J., Study of competitive inhibition of adenosine deaminase by erythro-9-(2-hydroxy-3-nonyl)adenine using capillary zone electrophoresis, *J. Chromatogr. A,* 745, 293, 1996.
130. Bao, J. and Regnier, F.E., Ultramicro enzyme assays in a capillary electrophoretic system, *J. Chromatogr.,* 608, 217, 1992.

# 25

# Flow Cytometry




Francis Mandy
*Health Canada*
*Ottawa, Canada*

Rudi Varro
*BD Biosciences*
*San Jose, California*

Diether Recktenwald
*BD Biosciences*
*San Jose, California*


## 25.1 Introduction

Flow cytometry allows the analysis and sorting of particles of biological interest at rates of more than $10^4$ $s^{-1}$, based on the analysis of light scatter and fluorescence. It also permits the quantitative analysis of many cellular constituents based on fluorescence measurements. Measurements at the single-molecule level have been reported. Because of the versatility and richness of information of flow cytometry, it is used in biological and biomedical research, and for clinical data collection. This chapter provides an overview of flow cytometry and its biomedical and clinical applications. For readers interested in further details, we provide references to additional reviews of subtopics. Shapiro[1] provides a comprehensive account of flow cytometry, covering all aspects up to 1995.

## 25.2 Hardware

Flow cytometers measure multiple optical properties of particles, generally without spatial resolution, from about 20 μm to submicroscopic size at rates of several thousand per second. A typical flow cytometer consists of a fluidic system, the optical components, and analog or digital electronics for data processing, storage, and evaluation (Figure 25.1). Special cytometers also sort particles into different fractions based on optical particle properties. Several sorting mechanisms that have been used are described in this chapter.


0-8493-1116-0/03/$0.00+$1.50

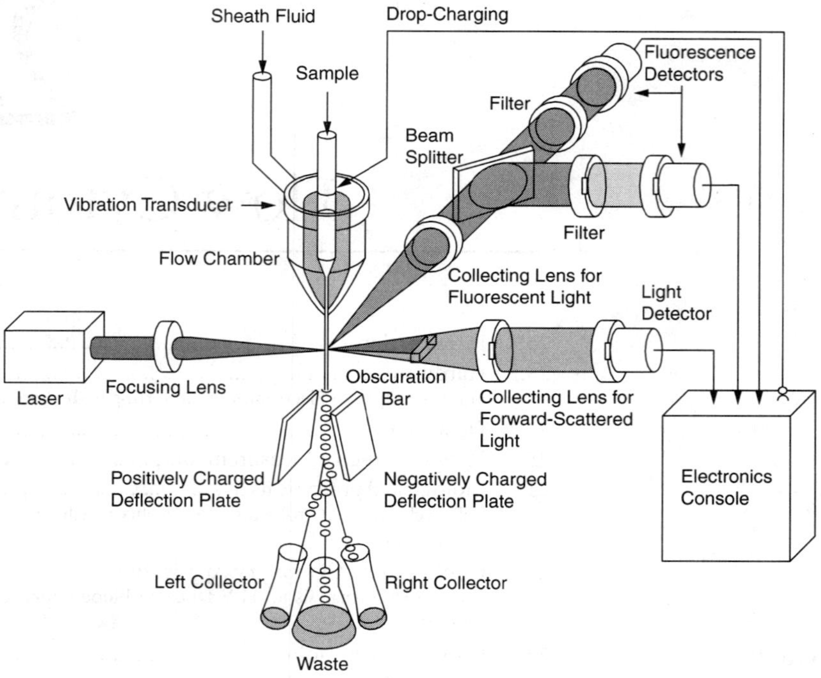


**FIGURE 25.1** Components of a typical flow cytometer with droplet cell sorting feature.

### 25.2.1 Fluidics

In the fluidics system of a cytometer a particle suspension from a tube or well of a microtiter plate is injected into a second fluid stream of aqueous sheath fluid — mostly saline — to create a very narrow, quasi one-dimensional file of particles for intersection with a light beam for optical measurements. Typical stream velocities are 10 m/s. The ratio of sheath fluid volume to sample volume with the size of the observation cuvette determines the diameter of the sample stream and influences the precision of the optical measurements. Typical total stream diameters are on the order of 100 μm, sample streams on the order of 10 μm. Coefficients of variation better than 2% are quite common for measurements of the DNA content of cell nuclei.

### 25.2.2 Optics

To perform optical measurements on the particle stream in a cytometer, a light beam, commonly from a laser, is focused on the center of the fluid stream in a cuvette or in a free-flowing stream in air for most cell sorters. Figure 25.2 shows a typical optics diagram for a three-laser sorter. Mercury arc lamps and LEDs are also used as excitation light sources. Cylindrical lenses are used frequently to achieve better uniformity of light intensity for the particle illumination, with a typical beam height of about 20 μm. Research flow cytometers are capable of measuring several light scatter and fluorescence emission intensities.

Many fluorescence measurements on biological systems require a very low limit of detection. Therefore high-numerical aperture lenses are used for the collection of emitted light. Dichroic filters separate scattered and fluorescent light into separate wavelength bands and photomultipliers (PMTs) measure the light intensities in the different bands. With an optimized instrument, photon-noise-limited measurements can be performed, and less than 200 molecules per particle (approximately $10^{-21}$ mol) of some fluorescent dyes can be detected. Some newer specialized systems use avalanche photodiodes in place of PMTs. Light scatter can be detected with photodiodes.

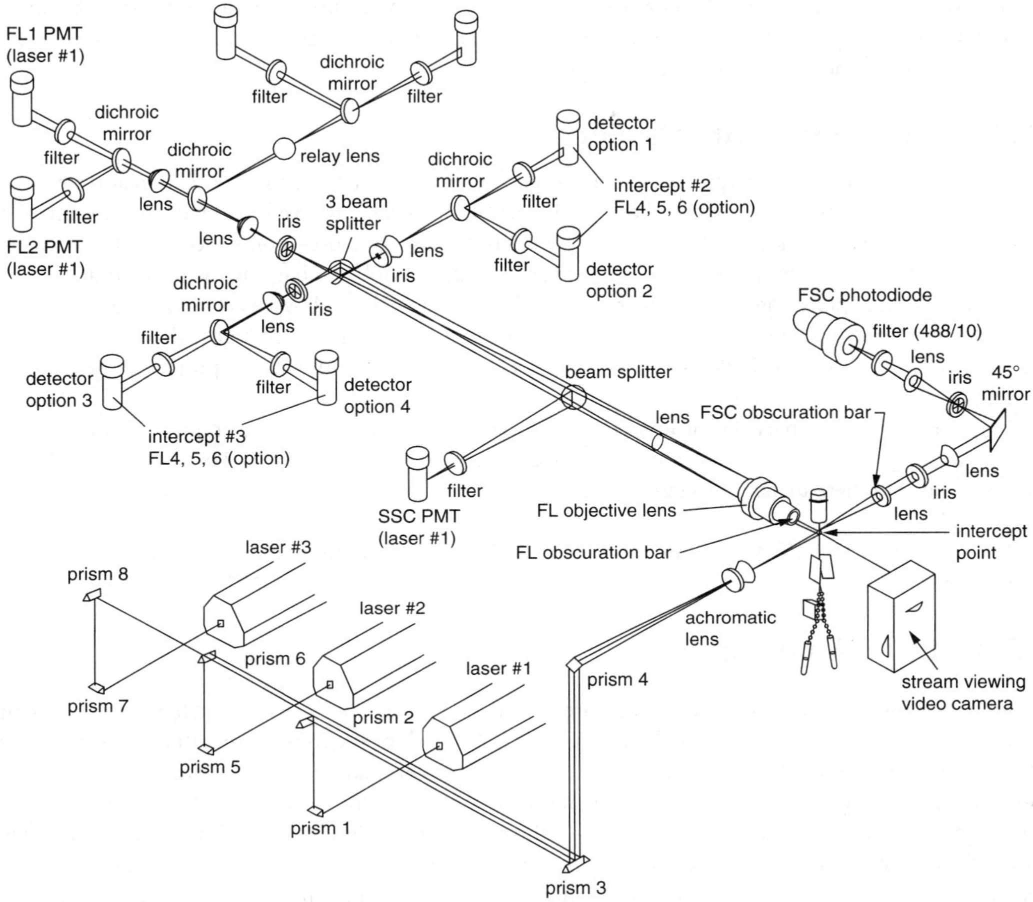


**FIGURE 25.2** Optical diagram of a three-laser flow cytometer with droplet sorting feature.

High-end research systems offer multiple excitation light sources, either in a colinear arrangement or for more flexibility for the resolution of multiple fluorophors with completely separate spatial optical paths and temporal synchronization.

## 25.2.3 Electronics

Signals from light detectors are amplified. To allow measurement of small and large signal intensity ranges, linear and logarithmic (typically four decades) ranges are provided on most instruments. Signal subtraction circuitry, linking adjacent spectral fluorescence emission bands, allows for the correction for spectral overlap between fluorescent dyes to express the intensity measurements in relative units of dye concentration rather than light intensities. After baseline subtraction, an analog pulse height or pulse area and a calculated pulse width are provided to an analog-to-digital converter (ADC). The ADC is triggered by a pulse height threshold, based on a Boolean combination of the measurement parameters; after digitization, all of the particle measurements, including those from separate light beams (after temporal synchronization), are stored as a record for the particle in the data matrix. Approaches for the extraction of population information from this data matrix are described in the data analysis section later.

Recently, digital electronics has been used to derive particle measurements from a continuous digitization of detector output through high-speed ADCs. All of the signal calculations on the pulses, including height, area, width, logarithms, and spectral overlap correction, are performed with high-speed digital signal processors. Calculations of signal parameters are performed with higher accuracy than by analog approximation, and the approach also provides more flexibility for future applications of modern signal signature analysis for flow cytometry.

### 25.2.4 Cell Sorting with Cloning

As mentioned earlier, several approaches have been used to sort particles in essentially real time, based on the measurements of a flow cytometer. Cell sorting has been reviewed in detail in a recent book chapter.[2] The most common method for flow cytometric cell sorting uses a piezoelectric actuator, which breaks the stream containing the particles under analysis into droplets. Charging the droplets at the right time with a charge pulse triggered by the results of the optical analysis allows electrostatic deflection into typically four positions, where a collection tube can be placed. Sorting rates of higher than $10^4$ per second can be achieved for the purification of millions of particles (cells) with specific properties in under an hour.

In another setup the particles of interest are deflected into a multiwell plate (typically 96 wells in an $8 \times 12$ well arrangement) or onto a plate with cell nutrient under xy control. For cloning, single cells can be deposited there to achieve a cultured cell population that can be traced back to an individual cell. (Figure 25.3).

A University of Cardiff Web site provides good additional information about the hardware of a flow cytometer.

## 25.3 Data Analysis

Data from flow cytometric measurements are stored in a data matrix, where each row contains the measurement for an individual particle with a column for each of the measurement parameters. One of the measurement parameters can be a time tag for kinetic data evaluation. For most analyses, information on at least 5000 particles is stored and, in many cases, substantially more. At least one byte of data is stored for each of the parameters; in many cases the resolution is higher, e.g., 10 bits per data item. This kind of data file is called a listmode file; the designation comes from the times when data storage was expensive, so, for memory-economy reasons, data were also stored as histogram data (see below). The listmode data matrix is the basis for all subsequent data analysis, but by itself cannot be evaluated easily without computer-based data transformation.

A simple analysis of the data consists of calculating a histogram of the number of particles at each parameter value for all of the parameters. In the simplest implementation, the digital parameter value for each of the particles in the matrix is used as an address for an array (1024 elements long for a 10-bit parameter resolution), and the content of the corresponding array element is incremented by 1 every time the respective value is observed. Data histograms (Figure 25.3A) provide a view of intensity distribution for the subpopulations in a particle ensemble; however, information about the correlation between parameters is lost.

A dotplot (Figure 25.3B) shows the intensity distribution with pairs of two measurement parameters. In a two-dimensional plot, dots are displayed for each particle at $x$-$y$ coordinates corresponding to two parameter values. This plot shows population locations and widths with some relative frequency information from the dot density. Quantitative frequency information is lost because overlapping dots show only as one.

Density and contour plots show quantitative population frequency information with two parameters. Both are based on two-dimensional histograms. A density plot uses a gray scale or colors to represent the $z$-axis (frequency information), whereas a contour plot uses different contour levels to show histogram height or population frequency.

All these data representations show only up to two parameters correlated simultaneously. The listmode data matrix may contain many parameters, for most measurements at least four. Therefore the problem remains to look at subpopulations of the particle ensemble in multidimensional space. A process called

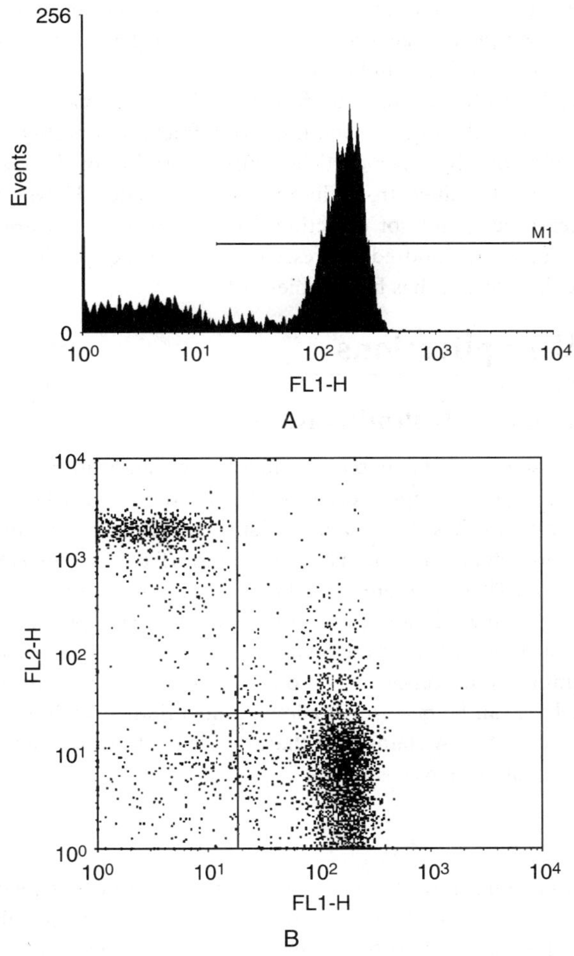


**FIGURE 25.3** Histogram (A) and dotplot (B) from a two-color immunofluorescence measurement of mammalian lymphocytes.

gating uses one- or two-dimensional displays to select parameter value sets to include or exclude from additional data displays. In this way properties of particle subpopulations can be determined for all measured parameters. A special case of novel multiparameter "gating" is detailed in Reference 3. Several software packages offer additional features for multiparameter data analysis. Paint-a-Gate™ (U.S. Patent 4,845,653)[4] uses color to highlight populations selected in one display in five more two-dimensional displays in real time. For automated multiparameter analysis, density-based or nearest-neighbor cluster algorithms have been used, among others. Further details on data analysis methods for flow cytometry have been discussed elsewhere.[5,6]

## 25.4 Flow Cytometry Measurements

A flow cytometric measurement characterizes one or more populations of a particle suspension with a count value and several optical parameter intensities and their distributions. From the partial count, particle concentrations can be calculated if the sample volume is known. Regardless of the capabilities of the instrumentation to record sample volumes, a known concentration of a reference particle with scatter and fluorescence properties different from the particles of the sample can be added to obtain concentrations of all the particle populations. However, many applications of flow cytometry only report the relative frequency of a sample particle subset as a fraction of the subset in a superset of the particle sample, e.g., T cells as a percentage of lymphocytes.

The optical parameter intensities are used to derive physical properties of the particles in a sample. Light scatter intensities are related to the size and index of refraction of particles; fluorescence intensities are related to the mass of fluorophors per particle. With calibration of the system, absolute masses of analyte per particle can be determined from fluorescence intensities. However, for most applications relative quantities are reported. Limits of detection for fluorescence measurements with commercial systems are on the order of a few hundred fluorescent molecules per particle;[7] with specialized instrumentation, single-molecule detection has been achieved.[8]

## 25.5 Biological Applications

### 25.5.1 Cell Cycle and Cell Proliferation

DNA content (including assessment of aneuploidy and S phase) can be measured with propidium iodide (PI), the Hoechst dyes, or other compounds. Propidium iodide intercalates in the DNA helix and fluoresces strongly orange-red. It has the advantage that it is excited by 488-nm light and can be used on most common flow cytometers. The Hoechst dyes bind to AT pairs in the DNA and enter viable cells without the need for fixation. These dyes are excited using UV light, so they cannot be used on benchtop flow cytometers; however, they may allow dye combinations that are not possible with propidium iodide.[9]

Bromodeoxyuridine (BrdU — a thymidine analog) incorporation for specific cell proliferation can also be measured by immunofluorescence. This is often performed along with DNA staining. Cell cycle and synchronization analysis can be performed allowing quantitative analysis of cell cycle phases. This can be used with or without antibody staining to characterize the transition between cell cycle phases in terms of expression of control proteins.[10]

### 25.5.2 Ca-Flux

$Ca^{2+}$ is the most common signal transduction element in cells, and it is involved in diverse cellular processes. Flow cytometry can be used to measure calcium flux[11] and the calcium flux assay can be combined with immunofluorescent stain to identify cell subpopulations by immunofluorescence and measure calcium flux in response to an activating signal.

### 25.5.3 Cellular Antigen Quantitation

Antigens per cell can be quantitated by standardized fluorescence measurements using flow cytometry.[12] Quantitative measurements can be used to characterize receptor expression and cellular activation. A number of bead-based methods are available to calibrate flow cytometers and provide quantitative results on density of antigen on cells.[13,14]

## 25.6 Clinical Flow Cytometry Applications

### 25.6.1 T-Cell Subset Analysis for HIV Disease

No disease has been impacted more by flow cytometry than the human immunodeficiency virus (HIV) infection that causes acquired immune deficiency syndrome (AIDS). Therefore a large part of this section covers applications related to monitoring this disease. Some of the principles discussed previously are further explained for a better understanding of the specific application of flow cytometry.

Contrary to earlier predictions, it was AIDS, not some high-volume oncological phenotype testing, that was responsible for the massive and rapid worldwide mobilization of flow cytometers into clinical immunology laboratories. In 1981, reports appeared from various parts of the United States about young men who had unusual immunosuppression coupled with opportunistic infections.[15,16] Soon it was discovered that the hallmark of this new disease, AIDS, was a decrease in numbers of CD4 T-cells (helper cells) in peripheral blood.[17] Next, the causative viral agent for AIDS was isolated: the human immunodeficiency virus (HIV).[18,19] It was established that the T-helper cell is the primary target for HIV and the CD4 receptor on the surface of these cells is the principal means for viral entry.[20] Current cocktails of antiretrovirals can suppress viral replications for years; hence, disease monitoring of patients may be required for decades because no current therapy can completely eliminate HIV from the host.

Throughout the course of disease, T-cell subsets are followed as the best surrogate marker for immune status.[21] Normal levels of T-helper cells are about 1000 cells/ml of blood (with a range from 600 to 1400); generally, levels below 500 cells/ml indicate that the virus has damaged the immune system.[22] The overall level of T cells remains constant until late stages of the disease.[23] For most of disease duration, the CD8 T-cell numbers increase as CD4 T cells diminish.[24] An AIDS definition was established based on a CD4 T-cell count of less than 200 cells/ml,[25] so accurate and reproducible measurement of T-cell subsets during therapy is an important part of managing treatment for HIV-infected patients.[26] T-helper cells often rise following combination antiretroviral therapy;[27] however, the rebound levels remain below normal values. Current therapies add years of quality life to patients living with HIV; nevertheless, the virus is never eliminated. Eventually, death from opportunistic infections results.

Fresh whole blood with anticoagulant is required for immunophenotyping T-cell subsets. In evaluating the effectiveness of therapies, viral load and T-cell subset assays are important clinical tools. Currently, clinical flow cytometers handle five or six distinct parameters: forward scatter (FS), side scatter (SS), and three or four fluorescence light (FL) signals: FL1, FL2, FL3, and FL4, respectively. The two light scatters are intrinsic parameters that define morphological features of leukocytes; they are size and granularity, respectively. The FL parameters measure extrinsic cell attributes, such as identification of surface antigens, via fluorescence scattering from fluorochrome-labeled monoclonal antibodies (MAbs).[28]

Fluorescein isothiocyanate (FITC) is the most universal fluorochrome for immunophenotyping; the second most commonly used dye is R-phycoerythrin (PE). Both dyes can be excited using 488-nm light. Natural and man-made tandem dyes are utilized as third and fourth dyes. Peridinin chlorophyll protein (PerCP) was the first directly conjugated commercially available natural tandem dye. Other dyes are energy-coupled dye (ECD), R-phycoerythrin-CyChrome 5 (PE-Cy5), allophycocyanin (APC), CyChrome 5 (Cy5) and allophycocyanin-CyChrome 7 (APC-Cy7). Currently, simultaneous four-color immunophenotyping is the advanced clinical method. This multicolor application is accomplished by adding a fourth PMT for the detection in the far-red region (Beckman Coulter Corporation, Miami, FL) or by using an additional laser as well as a fourth PMT (BD Biosciences, San Jose, CA). With a red diode laser that excites at 635 nm APC is used as the fourth fluorochrome (see Table 25.1).

For HIV immunophenotyping, single as well as dual laser instruments must produce similar results.[29,30] Samples are incubated with fluorochrome-labeled MAbs followed by red blood cell lysing.[31] Various types of leukocyte subsets can then be identified with combinations of intrinsic and extrinsic cell markers.[32] The resulting graphic presentation provides not only the relative location of lymphocytes in a dual-parameter dot plot, but also the clustering of monocyte and granulocyte populations. Thus, a three-part leukocyte differential is obtainable.

There is an inherent problem with displays based on intrinsic cell attributes alone. Although the three populations are easy to locate, the light scatter gating protocol does not resolve the interface between populations. For example, there is usually an interface zone between lymphocytes and monocytes where it is possible to find both types of cells (see Figure 25.4). The operator will often set a lymphocyte gate that includes all the lymphocytes but not only lymphocytes or includes only lymphocytes but not all of them. This type of gating strategy is predisposed to optimization of purity and or recovery of lymphocytes, but not both.

In the past decade, steady progress has been made toward the development of more intuitive software for data acquisition and manipulation. A T-gating protocol for simultaneous three-color application was developed in 1992.[33] With this gating method, T cells (CD3) are identified with a bivariate dotplot using heterogeneous parameters, SS (an intrinsic parameter), and CD3 fluorescence (an extrinsic parameter) (see Figure 25.1). From the T-gate, CD4/CD8 cells can be identified by conventional dual-color bivariate

**TABLE 25.1** Available MAB–Fluorochrome Conjugates for Single and Dual Laser Immunophenotyping

| Dye | Excitation with 488 1st laser | Excitation with 635 2nd laser | Emission (nm) | Extinction Coefficient ($cm^{-1}$ $M^{-1}$) | Quantum Yield | MW (Da) |
|---|---|---|---|---|---|---|
| FITC | x | | 519 | 67,000 | 0.71 | 389 |
| PE | x | | 578 | 1,960,000 | 0.68 | 240,000 |
| PerCP | x | | 675 | NA | 1 | 35,000 |
| ECD | x | | 613 | NA | NA | 250,000 |
| PE-Cy5 | x | | 675 | 1,960,000 | NA | 241,000 |
| APC | | x | 660 | 700,000 | 0.68 | 104,000 |
| Cy5 | | x | 670 | 250,000 | 0.28 | 792 |
| APC-Cy7 | | x | 767 | 700,000 | NA | 105,000 |

*Note:* This table includes most of the fluorochromes that are commercially available already conjugated to monoclonal antibodies for immunophenotyping T-cell subsets. Under the column of excitation, the first laser refers to the conventional argon ion source installed on most clinical instruments. The second laser is available as an option on some clinical flow cytometers. NA = not available.

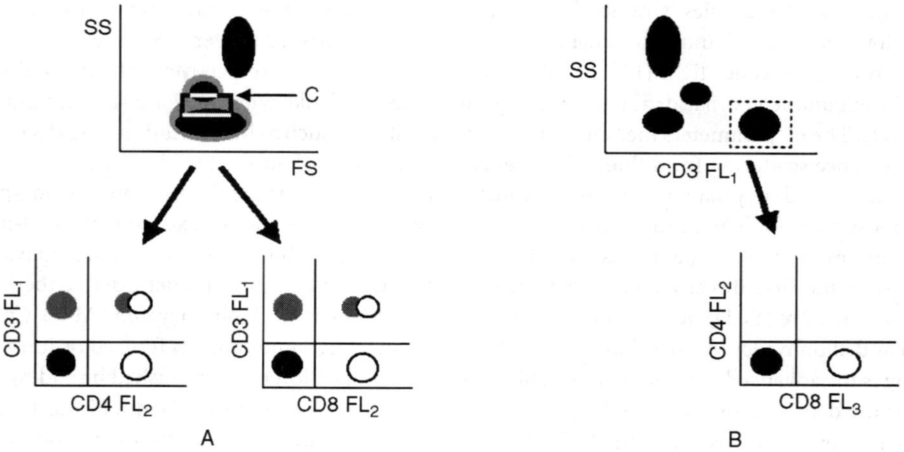


**FIGURE 25.4** Gating strategies based on intrinsic and extrinsic cell attributes. Illustrations of homogeneous (A) and heterogeneous (B) gating strategies as they are applied to simultaneous three-color immunophenotyping. (A) is an example of the traditional homogeneous approach to gating, where the two parameters are intrinsic attributes based on cellular morphology. The three regions in the upper part of (A) in descending order are granulocytes, monocytes, and lymphocytes, respectively. Arrow C points to an interface between lymphocytes and monocytes. It is focusing on a common dilemma: how to resolve where the monocyte gate ends and the lymphocyte gate begins. (B) illustrates the advantage of the heterogeneous gating strategy; an extrinsic attribute is harnessed (CD3 $FL_1$) in combination with an intrinsic one (SS).

quadrant analysis. Heterogeneous protocols can easily be adapted for absolute cell counting. A more recent heterogeneous parameter combination for simplifying CD4 T-cell determinations uses a single tube containing CD45, CD3, CD4, and CD8.[34] In this case, the heterogeneous gating strategy combines SS and bright CD45 fluorescence. CD45 is an epitope variably expressed on leukocyte populations in combination with SS; it allows the identification of lymphocytes. This innovative use of heterogeneous gating leads to improved diagnostic tools.[35]

Absolute cell count (cell concentration) can be obtained by using the single platform technology (SPT). Most clinical flow cytometers are designed to analyze distributions of positive cells within a gate; therefore, CD4/CD8 T cells have been reported as percentage of lymphocytes. However, the most recent guidelines

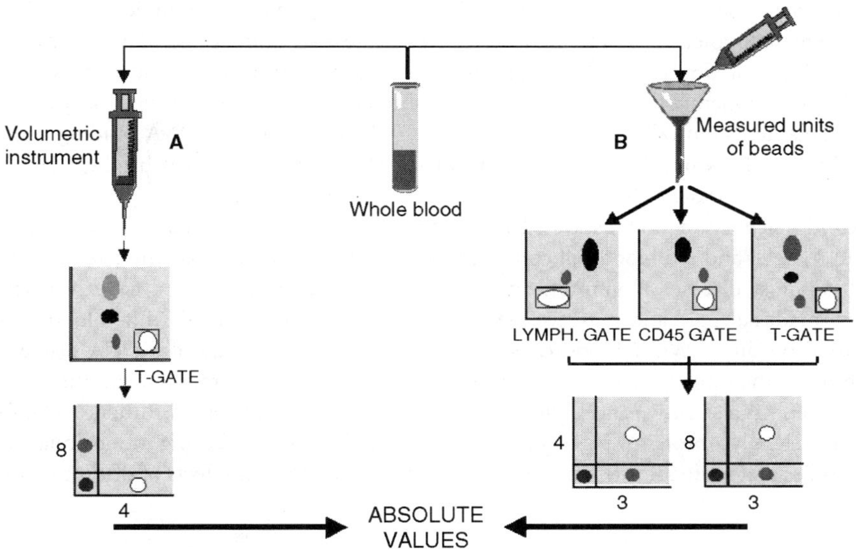


**FIGURE 25.5** Various single platform technologies for T-cell subset enumeration. Two SPT approaches are presented for T-cell subset enumeration in (A) and (B). (A) represents a dedicated volumetric instrument where a measured unit of blood is analyzed. (B) illustrates a situation where a known concentration of fluorospheres is added to a known volume of blood. The volume of the mixture is calculated. With the added fluorosphere unit, an accurate T-cell subset count per volume is possible. One homogeneous and two heterogeneous gating strategies are illustrated in the middle of (B).

for immunophenotyping recommend reporting absolute numbers for T-cell subsets.[36–38] When the absolute count is generated exclusively with a flow cytometer, the process is referred to as a single platform technology. Additional reagents, fluorospheres, are now available to produce absolute and percentage T-cell subsets without data from a hematology instrument.[38] SPT is becoming the default protocol for T-cell subset determination (see Figure 25.5). For most clinical instruments, a known number of fluorospheres are added to the whole blood specimen and calculations are made of the ratio between fluorospheres and the volume of sample analyzed with the cells of interest. Precision data have been reported for the single platform method.[39] The antigens most often monitored on CD8 T cells are CD57, CD28, and CD38.

Natural killer cells (CD16) decrease with HIV infection, particularly in the later phase of the disease.[40,41] Additional uses of flow cytometry in the detection of HIV infection include evaluation of cell functions such as activation, *in vitro* proliferation, cytotoxic T-lymphocyte responses, detection of various cytokine-producing cells, and measurement of *in situ* HIV antigens and apoptosis. Most of these additional applications are for clinical research use only. In evaluating the effectiveness of HIV therapies, T-cell subsets continue to be the essential clinical surrogate marker for immune status.

In the quest to eradicate the AIDS pandemic, perhaps the next generation of multicolor immunophenotyping instruments will accelerate significant discoveries in cellular immunology, which will in turn lead to long awaited breakthroughs in some overdue fundamental understanding of events in oncology.

## 25.6.2 Blood Banking

Flow cytometry interrogates the properties of every cell that moves by the laser beam. This feature is exploited in characterizing therapeutic blood products to count an enriched cell population or to validate the effectiveness of removal of unwanted cells.

An example for the first type of application is the enumeration of hematopoietic stem cells by using flow cytometry.[42] Transplantation of hematopoietic progenitor cells is used increasingly in the treatment of blood disorders, malignancies, and genetic abnormalities. A cell surface molecule, CD34, is present on immature hematopoietic cells and all hematopoietic colony-forming cells.[43] The relative frequency of the CD34+ cell population is 1 to 4% of the mononuclear cells in normal bone marrow, and less than 0.1% in normal peripheral blood.[44] An accurate measure of CD34 cell count is necessary for defining the dose of transplanted stem cells. Cell surface markers (CD34, CD45) and DNA binding dyes are useful to identify stem cells, while the absolute count of CD34 cells is determined by their proportion to a known number of fluorescent beads.[45]

The second category of flow assays identifies the presence of unwanted cells in a purified blood concentrate. Platelet and red blood cell concentrates are regularly used in patient care. Leukocytes could cause graft-vs.-host disease, alloimmunization, and other transfusion-related adverse reactions. Flow cytometry provides a rapid and sensitive method of enumerating "residual" white blood cells in leuco-reduced blood products. White blood cells are detected by their binding of a DNA dye; a fluorescent counting bead in the same tube is used as an internal standard for accurate cell counting.[46]

Flow cytometry is a sensitive and rapid research tool for study of platelet disorders.[47] Conformational changes in platelet glycoproteins, especially GPIIb-IIIa, can be measured using monoclonal antibodies.[48] These antibodies and multicolor testing strategies are useful to monitor effects of drugs to inhibit platelet activation.[49]

### 25.6.3 Cancer

Another clinical area in which flow cytometry has contributed to the understanding of disease is cancer diagnosis. Cells from leukemias differ in their molecular, genetic, and immunologic characteristics. Different leukemias have different prognoses and require different therapies. Multiparameter flow cytometry with fluorochrome-labeled monoclonal antibodies has substantially improved diagnosis and classification of the disease, as reviewed in detail by Weir et al.[50] and Dworzak.[51]

In leukemias and other cell proliferation diseases, the analysis of the cellular DNA content (DNA ploidy) has been used to differentiate between healthy and disease-associated cells. High-resolution DNA analysis with precision of close to 1% in DNA content provides information on abnormal DNA content, based on chromosome losses or additions and on the fraction of actively dividing cells (S-phase cells) — a measure for the growth rate of the tumor. More detail on S-phase and ploidy measurements in human cancer tissues can be found in the literature.[52]

## 25.7 Clinical Microbiology

Flow cytometry has been demonstrated in many clinical infectious disease applications. Nucleic acid dyes like ethidium bromide have been used to detect bacteria in blood after lysis of blood cells. Identification of bacteria can be performed by immunofluorescence followed by flow cytometric detection. Fungi and parasites can also be measured by flow cytometry. Detection and quantitation of viral antigens and nucleic acids and testing of antimicrobial agents — especially, susceptibility to antibiotics — are other microbiological applications. More detail on these and other applications in clinical microbiology can be found in a recent review with 347 references by Alvarez-Barrientos et al.[53] Another reference covers the measurement of microbial susceptibility measurements by flow cytometry.[54]

## 25.8 Biological and Medical Research

### 25.8.1 Antigen-Specific T Cells and Immune Function in Infectious Diseases

Quantitative and qualitative measurement of antigen-specific T cells is possible by flow cytometric methods. It offers effective monitoring of immune status during disease and in assessing vaccine efficacy. Single-cell flow assays of antigen-specific T cells include MHC-peptide tetramer staining[55] and intracellular cytokine assays.[56] Each of these assays can provide truly quantitative readouts because they enumerate antigen-specific cells.

Tetrameric complexes of HLA (human leucocyte antigen) molecules can be used to stain antigen-specific T cells in flow cytometric analysis. The enumeration and phenotypical analysis of antigen-specific cellular immune responses against viral, tumor, or transplantation antigens has applications in various experimental and clinical settings.

The intracellular cytokine assay utilizes a three-color combination of fluorochrome-labeled anti-cytokine, CD69, and CD4 antibodies. CD69 is an early activation antigen whose expression is induced during *in vitro* antigen stimulation. The T cells respond to stimulation by cytokine expression and the secretion of the cytokines is arrested by secretion inhibitors, such as Brefeldin A, inside the cells. The cells are fixed and permeabilized; the frequency of the T cells, which respond to specific stimulation, is determined by the ratio of the cytokine expressing T cells to the total T cell population. This technique allows the detection of functional populations of memory T cells that respond to specific soluble antigens in short-term restimulation assays.

Similar assays are described for detecting antigen specific CD8 T cells, using overlapping peptides as stimulants.[57] These functional assays assist in delineating the cellular immune response to natural infections or to vaccines.

### 25.8.2 Measurement of Soluble Analytes Using Multiplex Bead Assays

Flow cytometry can be used as readout for immunoassays. Bead-based, single-analyte flow immunoassays were described for $\alpha$-fetoprotein assay,[58] $\beta_2$-microglobulin,[59] and for immune complexes.[60] The single assay principle was expanded to the detection of multiple analytes from a single sample, using bead technology.

Beads of different sizes or colors are used for multiplexed immunoassays. McHugh[60] and Fulwyler et al.[61] used beads of different sizes as carriers for antigens or antibodies. The beads were differentiated by their scatter characteristics and the immunoassay signal was generated through the binding of fluorescent conjugates. Collins used a single bead to measure three different analytes.[62] Three antibodies were coated to the same bead and the corresponding antigens were detected through binding of monoclonal antibodies, each carrying a different fluorochrome. The three different fluorescent signals were simultaneously detected by using three-color flow cytometry.

Beads can also be identified by one type of fluorescence while the signal is generated by conjugates carrying a second type of fluorescent signal.[63] This concept is useful in creating low-complexity bead sets.[64]

Mixing two different fluorescent dyes for the identification of bead populations creates a larger set of distinguishable beads. The immunoassay signal is generated by conjugates coupled to a third type of fluorescent dye. Fulton described such a system, which could differentiate 64 unique bead populations by indexing their position in two-dimensional fluorescent space.[65]

Multiplex beads simplify panel assays. Only a single sample is needed to detect and quantify a number of analytes, an advantage when the sample volume is limited (e.g., pediatric samples, clinical studies, experiments in animal models). The multiple independent measurements within each bead population assure good precision; the wide detection range (at least three orders of magnitude) for fluorescent signal is another advantage. Bead assays are well suited to monitor antibody responses against a panel of antigens from infectious agents. McHugh et al. developed a prototype Hepatitis C virus antibody assay for potential use in the blood bank.[66]

Detection and quantitation of cytokines is a natural area for flow multiplex assays. Figure 25.6 shows a multiplex assay with six cytokines. Oliver, Camilla, and Chen and their co-workers have described multiplex methods to measure panels of secreted cytokines in serum.[63,64,67] Carson and Vignali demonstrated a multiplex bead assay for simultaneous quantitation of 15 cytokines, using the FlowMetrix™ system.[68]

Pei et al. developed a multiplex bead method to detect antibodies against HLA Class I and Class II antigens.[69] Pierangeli and co-workers described a flow multiplex assay for the simultaneous detection of

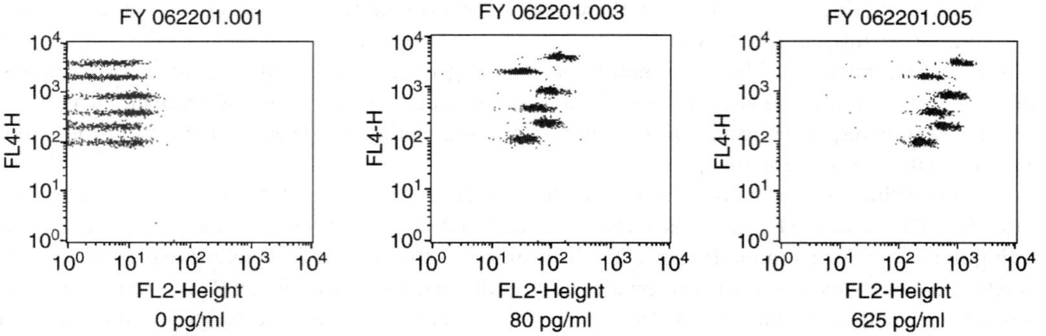


**FIGURE 25.6** Multiplex bead immunoassay of 6 human cytokines. Three levels (0, 80, and 625 pg/ml of each of the six cytokines) from a standard curve are shown. The order of the cytokines from top down: IL2, IL4, IL5, IL10, Interferon-$\gamma$ and tumor necrosis factor-$\alpha$. (C.-H. Chen and R. Varro, unpublished data.)

IgG and IgM antibodies to cardiolipin and phosphatidylserine.[70] Lund-Johansen and others[71] applied a bead multiplex assay to assess the level of phosphorylation of different proteins. Monoclonal antibodies were coupled to latex beads, which in turn bound protein kinase substrate proteins. Phosphorylation of the captured proteins was detected with PE-labeled (phycoerythrin) antiphosphotyrosine antibody.[7]

Stall and co-workers constructed a single-tube immunoassay for isotyping of monoclonal antibodies. Seven beads of different red fluorescence intensities were coated with mouse heavy-chain specific antibodies. A mixture of FITC-labeled anti-$\lambda$ and PE-labeled anti-$\kappa$ antibodies subsequently identified the heavy and light chain isotype of the captured monoclonal antibody.[72]

Molecular biology techniques are compatible with bead assays. Different amplification strategies were used to demonstrate flow methods to measure HIV viral load.[73,74] Defoort et al. and Fert developed a flow cytometric reverse transcriptase-polymerase chain reaction (RT-PCR) bead assay to quantitate HIV-1 mRNA.[75,76]

The multiplex bead method is also compatible with nucleic acid hybridization. Iannone and co-workers describe a multiplex bead method to detect single nucleotide polymorphisms (SNPs). The SNP sites were identified by oligonucleotide ligation (OLA) and subsequent bead capture. Nine SNP markers were simultaneously tested by this method.[77] Armstrong et al. used direct hybridization to determine the genotypes of eight polymorphic genes in a 32-plex flow assay.[78] Cai and collaborators developed a sensitive and rapid flow cytometry-based assay for the multiplexed analysis of SNPs based on polymerase-mediated primer extension, or minisequencing, using microspheres as solid supports.[79]

### 25.8.3 Other Cell Function Assays (Phagocytosis, Oxidative Burst, Basophils)

Flow cytometry is very useful for tracking apoptotic responses. Cellular viability (cytoplasmic membrane permeability) can be monitored using antitubulin antibody.[80] In the early phase of apoptosis the cells lose the asymmetry of their membrane phospholipids. Phosphatidylserine (PS), a negatively charged phospholipid located in the inner leaflet of the plasma membrane, becomes exposed at the cell surface. Annexin V, a calcium and phospholipid-binding protein, binds preferentially to PS, with high affinity. Apoptotic cells are stained by annexin V before the dying cell changes its morphology and hydrolyzes its DNA.[81] Other apoptosis assays include detection of Fas, FasL, caspase enzymes, and other biomolecules.[82] Whole blood flow assays are adaptable for cell function testing. The phagocytic function of granulocytes and monocytes may be measured by detection of ingested fluorescent, opsonized bacteria.[83] One can determine the overall percentage of monocytes and granulocytes showing phagocytosis in general (ingestion of one or more bacteria per cell) and the individual cellular phagocytic activity (number of bacteria per cell).

A variation of this method is used to evaluate the oxidative burst of leucocytes. Unlabeled opsonized bacteria are incubated with heparinized whole blood, and the generated reactive oxidants are detected

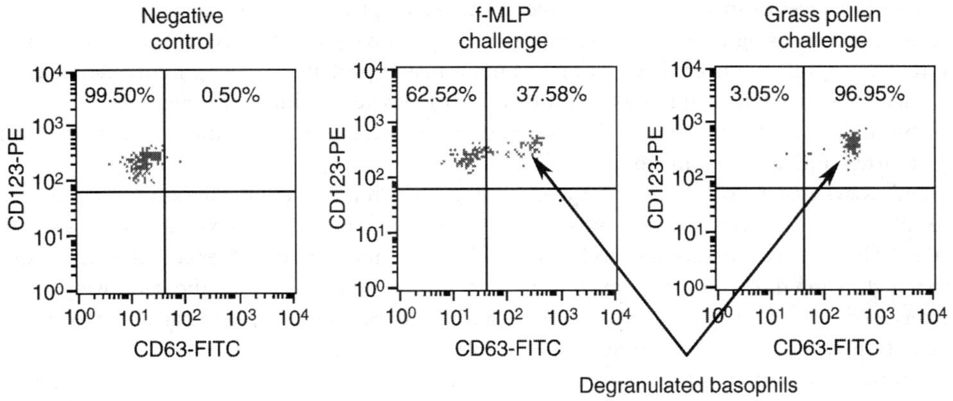


**FIGURE 25.7** Basophil degranulation is detected by the rapid appearance of a CD63+ population. (C.-H. Chen and R. Varro, unpublished data.)

by a fluorogenic substrate. Phagocytosis by polymorphonuclear neutrophils and monocytes constitutes an essential arm of host defense against bacterial or fungal infections; a reduction in the rate of oxidative burst may signal inborne defects.[84,85]

Flow cytometry is an effective method to detect expression levels of receptors on cell surfaces. An example of this type of application is the quantitative determination of low density lipoprotein (LDL) receptor expression on human monocytes. This method is suitable to identify people with the genetically inherited familial hypercholesterolemia (FH) disorder.[86] Individuals homozygous for FH carry mutant forms of the LDL receptor (frequency: 1 in 1,000,000), and have a high risk for atherosclerosis and coronary heart disease. Isolated peripheral blood mononuclear cells (lymphocytes and monocytes) are cultured and the amount of LDL receptors is determined by flow cytometry after staining the cells with a monoclonal antibody against the LDL receptor. This method is useful to identify the genetic defect.

Human basophils represent a very small portion of circulating leucocytes. They are hypersensitive effector cells of the immune system and play an important role in host defense mechanisms and allergic reactions. The binding of allergens to specific IgE fixed on the basophil membrane via the FceRI receptor triggers cellular activation, degranulation, and the release of potent mediators, including histamine and leukotriene $C_4$.[87,88] Multicolor whole blood flow cytometric assays were constructed to detect the degranulation of basophils upon allergen exposure. These assays utilize the appearance of CD63, a member of the tetraspanin superfamily, on activated basophils as a marker for basophil degranulation.[89] Such an assay is shown (Figure 25.7). In this case the basophil phenotype is defined as HLA-DR negative, CD123 positive cell population, and the degranulation is detected by the expression of CD63. A negative and a positive control are shown together with the effect of a specific allergen challenge.

The cytotoxic activity of human natural killer (NK) cells may be detected by flow cytometry. Fluorescently labeled K562 target cells are incubated with whole blood samples, and the killed target cells are identified by a DNA-stain, which penetrates the dead cells and specifically stains their nuclei. This method can detect altered NK function found in various disorders and can also be used to evaluate the effects of drugs on NK activity.[90–92]

## 25.8.4 Thermodynamic and Kinetic Analysis of Binding Phenomena

Flow cytometry is increasingly used for rapid analysis in drug screening.[93] When combined with autosampler pumps and microwell plates, the potential throughput can reach 100,000 samples or more per day.[94] Sample valve systems allow pressure-free introduction of very small samples, creating a "plug-flow" instrument for rapid sequential flow cytometric sample analysis.[95]

Flow cytometry is uniquely capable of making sensitive and quantitative measurements of molecular interactions. These measurements can be made in real time with subsecond kinetic resolution using purified biomolecules or living cells.[96] The role of ligand/receptor binding in signal transduction may be studied in real time using bead-bound receptors[97] or by studying cell surface receptor interactions and signaling pathways in engineered cells, which contain fusion proteins, carrying fluorescent signals. The use of variants of the green fluorescent protein (GFP) provides valuable *in vivo* methods to explore protein binding, transduction pathways, and binding reactions, exploiting the fluorescence resonance energy transfer (FRET) phenomenon.[98]

Live cell assays may be combined with methodology, which provide information on the longitudinal changes of assay parameters through subsequent generations of cells. Carboxyfluoroscein succinimidyl ester (CFSE) binds to cell membranes and then equally distributes into daughter cells during cell division. After a few days in culture, the amount of cell division may be measured by the decreasing amount of fluorescence of each progeny population. CFSE staining can be combined with calcium flux, mitochondrial activity, pH, and free radical production to measure live cell populations in real time.[99] Technologies are also available to enhance the detection of antigens expressed at low levels. A signal enhancement method, based on localized DNA amplification was demonstrated by Gusev[100] and substantially improved staining sensitivity of cellular antigens. Enzyme-based amplification of the fluorescent staining is another way to increase signal intensities.[101]

### 25.8.5 Molecular and Cellular Biology Research, Genomics, and Proteomics

High-speed sorting of chromosomes, especially at the U.S. National Laboratories in Livermore[102] and Los Alamos,[103] provided the early material for the human genome project. Chromosomes were prepared from cell division metaphase stages, when they are condensed and stable. After staining with fluorescent or fluorogenic dyes, different chromosomes were separated, based on fluorescence intensity. The nucleic acid base sequence of the chromosomes was determined mostly by restriction enzyme digestion and electrophoresis based sequencing. Based on the ability of specialized flow cytometers to detect single molecules,[8] it has been proposed to sequence long single nucleic acid strands by identifying bases released by an enzyme cleaving off terminal bases. Even though this proposal has not been implemented yet, a nucleic acid analyzer based on the analysis of single fragment molecules from a restriction enzyme digestion of RNAs has been described.[8]

Other applications of flow cytometry in genomics are the identification of specific nucleic acid sequences in single cells by fluorescence *in situ* hybridization and the fluorescence-based detection of the transfection of cells with genetic material in combination with a marker gene like an enzyme[103] or a fluorescent protein.[104,105] One of the earliest applications of flow cytometry is based on monoclonal antibody technology. In the past 20 years, thousands of antibodies have been developed against cellular structures. The reaction of fluorophor-labeled antibodies with cell suspensions, followed by a flow cytometric analysis, measures cell subsets carrying the structures identified by the antibodies. To date, 247 cell surface structures of white blood cells have been characterized with this method; they are tabulated by an international organization as numbered cluster designations.[106,107]

As mentioned in the previous section on medical applications, these surface markers have helped in our understanding of diseases like AIDS. The characterization of these molecules has also contributed substantially to the knowledge in cell biology. Combined with high-speed cell sorting, these markers provide the basis for large scale isolation of specific cell subsets, which can be further characterized by two-dimensional gel electrophoresis, and mass spectrometry for identifying the complete proteome of different cell types.

## 25.6 Industrial and Environmental Cytometry

Flow cytometry has found quality control and process monitoring applications in several industrial settings. Examples are the monitoring of industrial cell culture,[108] clone improvement by sorting and recloning highly productive single cells,[109] and the sorting of transfectants based on expression of co-expressed fluorescent proteins.[110]

In agriculture, flow cytometry is used for sperm counting[111] and sorting.[112] Methods for counting microbes in foods like milk have been described.[113] For environmental testing techniques have been proposed for water quality monitoring[114] and for measuring bioburden in air.[115] Marine biologists are using cytometry to count bacteria and viruses in seawater.[116]

## References

1. Shapiro, H.M., *Practical Flow Cytometry*, Wiley, New York, 1995.
2. Hoffman, R.A. and Houck, D.W., Cell separation using flow cytometric cell sorting, in *Cell Separation Methods and Applications*, Recktenwald, D. and Radbruch, A., Eds., Marcel Dekker, New York, 1998, chap. 11.
3. Bierre, P. and Thiel, D.E., Algorithmic engine for automated N-dimensional subset analysis, U.S. patent 5,739,000, April 14, 1998.
4. Conrad, M.P., Reichert, T.A., and Bezdek, J.C., Method of displaying multi-parameter data sets to aid in the analysis of data characteristics, U.S. patent 4,845,653, July 4, 1989.
5. Watson, J.V., *Flow Cytometry Data Analysis*, Cambridge University Press, Cambridge, U.K., 1992.
6. Boddy, L., Wilkins, M.F., and Morris, C.W., Pattern recognition in flow cytometry, *Cytometry*, 44, 195, 2001.
7. Coventry, B.J., Neoh, S.H., Mantzioris, B.X., Skinner, J.M., Zola, H., and Bradley, J., A comparison of the sensitivity of immunoperoxidase staining methods with high-sensitivity fluorescence flow cytometry-antibody quantitation on the cell surface, *J. Histochem. Cytochem.*, 42, 1143, 1994.
8. Harding, J.D. and Keller, R.A., Single-molecule detection as an approach to rapid DNA sequencing, *Trends Biotechnol.*, 10, 55, 1992.
9. van den Engh, G.J., Trask, B.J., and Gray, J.W., The binding kinetics and interaction of DNA fluorochromes used in the analysis of nuclei and chromosomes by flow cytometry, *Histochemistry*, 84, 501, 1986.
10. Jensen, P.O., Larsen, J.K., Christensen, I.J., and van Erp, P.E., Discrimination of bromodeoxyuridine labeled and unlabelled mitotic cells in flow cytometric bromodeoxyuridine/DNA analysis, *Cytometry*, 15, 154, 1994.
11. Vandenberghe, P.A. and Ceuppens, J.L., Flow cytometric measurement of cytoplasmic free calcium in human peripheral blood T lymphocytes with Fluo-3, a new fluorescent calcium indicator, *J. Immunol. Methods*, 127, 197, 1990.
12. Davis, K.A., Abrams, B., Iyer, S.B., Hoffman, R.A., and Bishop, J.E., Determination of CD4 antigen density on cells, role of antibody valency, avidity, clones, and conjugation, *Cytometry*, 33, 197, 1998.
13. Lenkei, R., Gratama, J.W., Rothe, G., Schmitz, G., D'hautcourt, J.L., Arekrans, A., Mandy, F., and Marti, G., Performance of calibration standards for antigen quantitation with flow cytometry, *Cytometry*, 33, 188, 1998.
14. Schwartz, A. and Fernandez-Repollet, E., Quantitative flow cytometry, *Clin. Lab. Med.*, 21, 743, 2001.
15. Gottlieb, M.S., Schroff, R., Schanker, H.M., Weisman, J. D., Fan, P.T., Wolf, R.A., and Saxon, A., *Pneumosystis carinii* pneumonia and mucosal candidiasis in previously healthy homosexual men, *N. Engl. J. Med.*, 305, 1426, 1981.
16. Siegal, F.P., Lopez, C., Hammer, G.S., Brown, A.E., Kornfeld, S.J., Gold, J., Hassett, J., Hirschman, S.Z., Cunningham-Rundles, C., Adelsberg, B.R., Parham, D.M., Siegal, M., Cunningham-Rundles, S., and Armstrong, D., Severe acquired immunodeficiency in male homosexuals, manifested by chronic perianal ulcerative herpes simplex lesions, *N. Engl. J. Med.*, 305, 1439, 1981.
17. Ammann, A.J., Abrams, D., Conant, M., Chudwin, D., Cowan, M., and Volberding, P., Acquired immune dysfunction in homosexual men: immunologic profiles, *Clin. Immunol. Immunopathol.*, 27, 315, 1983.

18. Barre-Sinoussi, R., Chermann, J.C., Rey, F., Nugeyre, M.T., Chamaret, S., Gruest, J., Dauguet, C., Axler-Blin, C., Vezinet-Brun, F., Rouzioux, C., Rozenbaum, W., and Montagnier, L., Isolation of a T-lymphotropic retrovirus from a patient at risk for acquired immune deficiency syndrome (AIDS), *Science*, 220, 868, 1983.
19. Gallo, R.C., Salahuddin, S.Z., Popovic, M., Shearer, G.M., Kaplan, M., Haynes, B.F., Palker, T.J., Redfield, R., Oleske, J., and Safai, B., Frequent detection and isolation of cytopathic retroviruses (HTLV-III) from patients with AIDS and at-risk for AIDS, *Science*, 224, 500, 1984.
20. McDougal, J.S., Kennedy, M.S., Sligh, J.M., Cort, S.P., Mawle, A., and Nicholson, J.K.A., Binding of HTLV-III/LAV to CD4+ T cells by a complex of the 100K viral protein and the T4 molecule, *Science*, 231, 382, 1986.
21. Stein, D.S., Korvick, J.A., and Vermund, S.H., CD4+ lymphocyte cell enumeration for prediction of clinical course of human immuno-deficiency virus disease: a review, *J. Infect. Dis.*, 165, 352, 1992.
22. Fahey, J.L., Taylor, J. M.G., Detels, R., Hofmann, B., Melmed, R., Nishanian, P., and Giorgi, J.V., The prognostic value of cellular and serologic markers in infection with human immunodeficiency virus type 1, *N. Engl. J. Med.*, 322, 166, 1990.
23. Nicholson, J.K.A. and Mandy, F.F., Immunophenotyping in HIV infection, in *Immunophenotyping*, Stewart, C.C. and Nicholson, J.K.A., Eds., Wiley-Liss, New York, 2000, chap. 11.
24. Giorgi, J. V., Fahey, J.L., Smith, D.C., Hultin, L.E., Cheng, H., and Mitsuyasu, R.T., Early effects of HIV on CD4 lymphocytes *in vivo*, *J. Immunol.*, 138, 3725, 1987.
25. Centers for Disease Control, Revised classification system for HIV infection and expanded surveillance case definition for AIDS among adolescents and adults, *Morbidity Mortality Weekly Rep.*, 41, 1, 1993.
26. Bergeron, M., Faucher, S., Minkus, T., Lacroix, F., Ding, T., Phaneuf, S., Somorjai, R., Summers, R., and Mandy, F., Impact of unified procedures as implemented in the Canadian Quality Assurance Program for T lymphocyte subset enumeration, Participating Flow Cytometry Laboratories of the Canadian Clinical Trials Network for HIV/AIDS Therapies, *Cytometry*, 33(2), 146, 1998.
27. Koot, M., Schellekens, P.T.A., Mulder, J.W., Lange, J. M.A., Roos, M.T.L., Coutinho, R.A., Tersmette, M., and Miedema, F., Viral phenotype and T cell reactivity in human immunodeficiency virus type 1-infected asymptomatic men treated with zidovudine, *J. Infect. Dis.*, 168, 733, 1993.
28. Shapiro, H.N., Parameters and probes, in *Practical Flow Cytometry*, Shapiro, A., Ed., Wiley-Liss, New York, 1994, chap. 7.
29. Mandy, F.F., Bergeron, M., and Minkus, T., Evolution of leukocyte immunophenotyping as influenced by the HIV/AIDS pandemic: a short history of the development of gating strategies for CD4 T-cell enumeration, *Cytometry*, 30, 157, 1997.
30. O'Gorman, M.R.G., and Nicholson, J. K.A., Adoption of single-platform technologies for enumeration of absolute T-lymphocyte subsets in peripheral blood, *Clin. Diag. Lab. Immunol.*, 7, 333, 2000.
31. Hoffman, R.A., Kung, P.C., Hansen, W.P., and Goldstein, G., Simple and rapid measurement of human T lymphocytes and their subclasses in peripheral blood, *Proc. Natl. Acad. Sci. U.S.A.*, 77, 4914, 1980.
32. Graedel, T.E. and McGill, R., Graphical presentation of results from scientific computer models, *Science*, 215, 1191, 1982.
33. Mandy, F.F., Bergeron, M., Recktenwald, D., and Izaguirre, C.A., A simultaneous three-color T cell subsets analysis with single laser flow cytometers using T cell gating protocol, *J. Immunol. Methods*, 156, 151, 1992.
34. Nicholson, J.K.A., Jones, B.M., and Hubbard, M., CD4 T-lymphocyte determinations on whole blood specimens using a single-tube three-color assay, *Cytometry*, 14, 685, 1993.
35. Mandy, F.F., Bergeron, M., and Minkus, T., Principles of flow cytometry, *Transf. Sci.*, 16, 303, 1995.
36. Calvelli, T., Denny, T.N., Paxton, H., Gelman, R., and Kagan, J., Guidelines for flow cytometric immunophenotyping: a report from the National Institute of Allergy and Infectious Diseases, Division of AIDS, *Cytometry*, 14, 702, 1993.

37. Centers for Disease Control and Prevention, 1997 revised guidelines for performing CD4+ T-cell determinations in persons infected with human immunodeficiency virus (HIV), *Morbidity Mortality Weekly Rep.*, 46, 1, 1997.
38. Mandy, F. and Brando, B., Enumeration of absolute cell counts using immunophenotyping techniques, *Curr. Protocols Cytometry*, 6.8.1, 2000.
39. Nicholson, J.K.A., Stein, D., Mui, T., Mack, R., Hubbard, M., and Denny, T., Evaluation of a method for counting absolute numbers of cells with a flow cytometer, *Clin. Diag. Lab. Immunol.*, 4, 309, 1997.
40. Hardy, N.M., Lymphocyte subset changes in persons infected with human immunodeficiency virus, *Ann. Clin. Lab. Sci.*, 22, 286, 1992.
41. Lucia, B., Jennings, C., Cauda, R., Ortona, L., and Landay, A.L., Evidence of a selective depletion of a CD16+ CD56+ CD8+ natural killer cell subset during HIV infection, *Cytometry*, 22, 10, 1995.
42. To, L.B., Haylock, D.N., Simmons, P.J., and Juttner, C.A., The biological and clinical uses of blood stem cells, *Blood*, 89, 2233, 1997.
43. Civin, C.I., Trischman, T.W., Fackler, M.J., et al., Summary of CD34 Cluster Workshop Section, in *Leucocyte Typing IV: White Cell Differentiation Antigens*, Knapp, W., Dorken, B., Gilks, W.R., et al., Eds., Oxford University Press, Oxford, U.K., 1989, p. 818.
44. Loken, M.R., Shah, V.O., Hollander, Z., and Civin, C.I., Flow cytometric analysis of normal B lymphoid development, *Pathol. Immunopathol. Res.*, 7, 357, 1988.
45. McNiece, I., Kern, B., Zilm, K., Brunaud, C., Dziem, G., and Briddell, R., Minimization of CD34+ cell enumeration variability using the ProCOUNT standardized methodology, *J. Hematother.*, 7, 499, 1998.
46. Barclay, R., Walker, B., Allan, R., Reid, C., Duffin, E., Kane, E., and Turner, M., Flow ceytometric determination of residual leucocytes in filter-depleted blood-products: an evaluation of Beckton Dickinson's LeucoCOUNT system, *Transfus. Sci.*, 19, 399, 1998.
47. Michelson, A.D., Flow cytometry: a clinical test of platelet function, *Blood*, 87, 4925, 1996.
48. Shattil, S.J., Cunningham, M., and Hoxie, J.A., Detection of activated platelets in whole blood using activation-dependent monoclonal antibodies and flow cytometry, *Blood*, 70, 307, 1987.
49. Coller, B.S., Anderson, K., and Weisman, H.F., New antiplatelet agents: platelets GpIIb/IIIa antagonists, *Thromb. Haemost.*, 74, 302, 1995.
50. Weir, E.G. and Borowitz, M.J., Flow cytometry in the diagnosis of acute leukemia, *Semin. Hematol.*, 38(2), 124, 2001.
51. Dworzak, M.N., Immunological detection of minimal residual disease in acute lymphoblastic leukemia, *Onkologie*, 24, 442, 2001.
52. Ross, J.S., DNA ploidy and cell cycle analysis in cancer diagnosis and prognosis, *Oncology*, 10, 867, 1996.
53. Alvarez-Barrientos, A., Arroyo, J., Canton, R., Nombela, C., and Sanchez-Perez, M., Applications of flow cytometry to clinical microbiology, *Clin. Microbiol. Rev.*, 13(2), 167, 2000.
54. Pore, R.S., Antibiotic susceptibility testing by flow cytometry, *J. Antimicrob. Chemother.*, 34, 613, 1994.
55. Savage, P.A., Boniface, J.J., and Davis, M.M., A kinetic basis for T cell receptor repertoire selection during an immune response, *Immunity*, 10, 485, 1999.
56. Suni, M.A,. Picker, L.J., and Maino, V.C., Detection of antigen-specific T cell cytokine expression in whole blood by flow cytometry, *J. Immunol. Methods*, 212, 89, 1998.
57. Maecker, H.T., Dunn, H.S., Suni, M.A., Khatamzas, E., Pitcher, C.J., Bunde, T., Persaud, N., Trigona, W., Fu, T.M., Sinclair, E., Bredt, B.M., McCune, J.M., Maino, V.C., Kern, F., and Picker, L.J., Use of overlapping peptide mixtures as antigens for cytokine flow cytometry, *J. Immunol. Methods*, 255, 27, 2001.
58. Frengen, J., Schmid, R., Kierulf, B., Nustad, K., Paus, E., Berge, A., and Lindmo, T., Homogeneous immunofluorometric assays for $\alpha$-fetoprotein with macroporous monosized particles and flow cytometry, *Clin. Chem.*, 39, 2174, 1993.

59. Bishop, J.E. and Davis, K.A., A flow cytometric immunoassay for $\beta_2$-microglobulin in whole blood, *J. Immunol. Methods*, 210, 79, 1997.
60. McHugh, T.M., Flow microsphere immunoassay for the quantitative and simultaneous detection of multiple soluble analytes, *Methods Cell Biol.*, 42, 575, 1994.
61. Fulwyler, M.J., McHugh, T.M., Schwadron, R., Scillian, J.J., Lau, D., Busch, M.P., Roy, S., and Vyas, G.N., Immunoreactive bead (IRB) assay for the quantitative and simultaneous flow cytometric detection of multiple soluble analytes, *Cytometry*, Suppl. 2, 19, 1988.
62. Collins, D.P., Luebering, B.J., and Shaut, D.M., T-lymphocyte functionality assessed by analysis of cytokine receptor expression, intracellular cytokine expression, and femtomolar detection of cytokine secretion by quantitative flow cytometry, *Cytometry*, 33, 249, 1998.
63. Camilla, C., Defoort, J.P., Delaage, M., Auer, R., Quintana, J., Lary, T., Hamelik, R., Prato, S., Casano, B., Martin, M., and Fert, V., A new flow cytometry-based multi-assay system. 1. Application to cytokine immunoassays, *Cytometry*, Suppl. 8, 132, 1998.
64. Chen, R., Lowe, L., Wilson, J.D., Crowther, E., Tzeggai, K., Bishop, J.E., and Varro, R., Simultaneous quantification of six human cytokines in a single sample using microparticle-based flow cytometric technology, *Clin. Chem.*, 9, 1693, 1999.
65. Fulton, R.J., McDade, R.L., Smith, P.L., Kienker, L.J., and Kettman, J.R., Advanced multiplexed analysis with the FlowMetrix system, *Clin. Chem.*, 43, 1749, 1997.
66. McHugh, T.M., Viele, M.K., Chase, E.S., and Recktenwald, D.J., The sensitive detection and quantitation of antibody to HCV using a microsphere-based immunoassay and flow cytometry, *Cytometry*, 29, 106, 1997.
67. Oliver, K.G., Kettman, J.R., and Fulton, R.J., Multiplexed analysis of human cytokines by use of the FlowMetrix system, *Clin. Chem.*, 44, 2057, 1998.
68. Carson, R.T. and Vignali, D.A., Simultaneous quantitation of fifteen cytokines using a multiplexed flow cytometric assay, *J. Immunol. Methods*, 227, 41, 1999.
69. Pei, R., Lee, J., Chen, T., Rojo, S., and Terasaki, P.I., Flow cytometric detection of HLA antibodies using a spectrum of microbeads, *Hum. Immunol.*, 60, 1293, 1999.
70. Pierangeli, S.S., Silva, L.K., and Harris, E.N., A flow cytometric assay for the detection of antiphospholipid antibodies, *Am. Clin. Lab.*, 18, 18, 1999.
71. Lund-Johansen, F., Davis, K., Bishop, J., and Malefyt, R. de W., Flow cytometric analysis of immunoprecipitates: high-throughput analysis of protein phosphorylation and protein-protein interactions, *Cytometry*, 39, 250, 2000.
72. Stall, A., Sun, Q., Varro, R., Lowe, L., Crowther, E., Abrams, B., Bishop, J., and Davis, K., A single tube flow cytometric multibead assay for isotyping mouse monoclonal antibodies, Abstract 1877, Experimental Biology Meeting, San Francisco, 1998.
73. Mehrpouyan, M., Bishop, J.E., Ostrerova, N., Van Cleve, M., and Lohman, K.L., A rapid and sensitive method for non-isotopic quantitation of HIV-1 RNA using thermophilic SDA and flow cytometry, *Mol. Cell Probes*, 11, 337, 1997.
74. Van Cleve, M., Ostrerova, N., Tietgen, K., Cao, W., Chang, C., Collins, M.L,. Kolberg, J., Urdea, M., and Lohman, K., Direct quantitation of HIV by flow cytometry using branched DNA signal amplification, *Mol. Cell Probes*, 12, 243, 1998.
75. Defoort, J.P., Camilla, C., Delaage, M., Auer, R., Quintana, J., Lary, T., Hamelik, R., Prato, S., Casano, B., Martin, M., and Fert, V., A new flow cytometry-based multi-assay system. 2. Application to HIV1 mRNA quantification, *Cytometry*, Suppl. 8, 132, 1998.
76. Fert, V., Simultaneous immuno-and nucleic acid probe assays on flow cytometer, critical parameters for assay design and applications as a screening tool for understanding immune disease, *Cytometry*, 38, 94, 1999.
77. Iannone, M.A., Taylor, J.D., Chen, J., Li, M.S., Rivers, P., Slentz-Kesler, K.A., and Weiner, M.P., Multiplexed single nucleotide polymorphism genotyping by oligonucleotide ligation and flow cytometry, *Cytometry*, 39, 131, 2000.

78. Armstrong, B., Stewart, M., and Mazumder, A., Suspension arrays for high throughput, multiplexed single nucleotide polymorphism genotyping, *Cytometry*, 40, 102, 2000.
79. Cai, H., White, P.S., Torney, D., Deshpande, A., Wang, Z., Keller, R.A., Marrone, B., and Nolan, J.P., Flow cytometry-based minisequencing: a new platform for high-throughput single-nucleotide polymorphism scoring, *Genomics*, 66, 135, 2000.
80. O'Brien, M.C. and Bolton, W.E., Comparison of cell viability probes compatible with fixation and permeabilization for combined surface and intracellular staining in flow cytometry, *Cytometry*, 19, 243, 1995.
81. Koopman, G., Reutelingsperger, C.P.M., Kuijten, G.A.M., Keehnen, R.M.J., Pals, S.T., and van Oers, M.H.J., Annexin V for flow cytometric detection of phosphatidylserine expression on B cells undergoing apoptosis, *Blood*, 84, 1415, 1994.
82. Labroille, G., Dumain, P., Lacombe, F., and Belloc, F., Flow cytometric evaluation of fas expression in relation to response and resistance to anthracyclines in leukemic cells, *Cytometry*, 39, 195, 2000.
83. Sawyer, D.W., Donowitz, G.R., and Mandell, G.L., Polymorphonuclear neutrophils: an effective antimicrobial force, *Rev. Infect. Dis.*, 11, S1532, 1989.
84. Donadebian, H.D., Congenital and acquired neutrophil abnormalities, in *Phagocytes and Disease*, Klempner, M.S., Styrt, B., and Ho, J., Eds., Kluwer, Dordrecht, 1989, p. 103.
85. Smith, R.M. and Curnutte, J.T., Molecular basis of chronic granulomatous disease, *Blood*, 77, 673, 1991.
86. Goldstein, J.L. and Brown, M.S., The LDL receptor locus and the genetics of familial hypercholesterolemia, *Ann. Rev. Genet.*, 13, 259, 1979.
87. Kurimoto, Y.A., de Week, A.L., and Dahinden, C.A., The effect of interleukin-3 upon IgE-dependent and IgE-independent basophil degranulation and leukotriene generation, *Eur. J. Immunol.*, 21, 361, 1991.
88. Dembo, M. and Goldstein, B., Theory of equilibrium bridging of symetric bivalent haptens to cell surface antibody. Application to histamine release from basophils, *J. Immunol.*, 121, 345, 1978.
89. Sainte-Laudy, J., Vallon, C., and Guerin, J.C., Diagnosis of latex allergy: comparison of histamine release and flow cytometric analysis of basophil activation, *Inflamm. Res.*, Suppl. 1, S35, 1996.
90. Rosenberg, Z.F. and Fauci, A.S., The immunopathogenesis of HIV infection, *Adv. Immunol.*, 47, 377, 1989.
91. Sibbitt, W.L. and Bankhurst, A.D., Natural killer cells in connective tissue disorders, *Clin. Rheum. Dis.*, 11, 507, 1985.
92. Henney, C.S., Kuribayashi, K., Kern, D.E., and Gillis, S., Interleukin-2 augments natural killer cell activity, *Nature*, 291, 335, 1981.
93. Nolan, J.P., Lauer, S., Prossnitz, E.R., and Sklar, L.A., Flow cytometry: a versatile tool for all phases of drug discovery, *Drug Discovery Today*, 4, 173, 1999.
94. Kuckuck, F.W., Edwards, B.S., and Sklar, L.A., High throughput flow cytometry, *Cytometry*, 44, 83, 2001.
95. Edwards, B.S., Kuckuck, F., and Sklar, L.A., Plug flow cytometry: an automated coupling device for rapid sequential flow cytometric sample analysis, *Cytometry*, 37, 156, 1999.
96. Nolan, J.P., and Sklar, L.A., The emergence of flow cytometry for sensitive, real-time measurements of molecular interactions, *Nat. Biotechnol.*, 16, 633, 1998.
97. Sklar, L.A., Vilven, J., Lynam, E., Neldon, D., Bennett, T.A., and Prossnitz, E., Solubilization and display of G protein-coupled receptors on beads for real-time fluorescence and flow cytometric analysis, *Biotechniques*, 28, 976, 2000.
98. Chan, F.K., Siegel, R.M., Zacharias, D., Swofford, R., Holmes, K.L., Tsien, R.Y., and Lenardo, M.J., Fluorescence resonance energy transfers analysis of cell surface receptor interactions and signaling using spectral variants of the green fluorescent protein, *Cytometry*, 44, 361, 2001.
99. Lyons, A.B. and Parish, C.R., Determination of lymphocyte division by flow cytometry, *J. Immunol. Methods*, 171, 131, 1994.

100. Gusev, Y., Sparkowski, J., Raghunathan, A., Ferguson, H., Jr,, Montano, J., Bogdan, N., Schweitzer, B., Wiltshire, S., Kingsmore, S.F., Maltzman, W., and Wheeler, V., Rolling circle amplification: a new approach to increase sensitivity for immunohistochemistry and flow cytometry, *Am. J. Pathol.*, 159, 63, 2001.
101. Kaplan, D. and Smith, D., Enzymatic amplification staining for flow cytometric analysis of cell surface molecules, *Cytometry*, 40, 81, 2000.
102. Gray, J.W., Dean, P.N., Fuscoe, J.C., Peters, D.C., Trask, B.J., van den Engh, G.J., and Van Dilla, M.A., High-speed chromosome sorting, *Science*, 238, 323, 1987.
103. Cram, L.S., Flow cytogenetics and chromosome sorting, *Hum. Cell*, 3, 99, 1990.
104. Fiering, S.N., Roederer, M., Nolan, G.P., Micklem, D.R., Parks, D.R., and Herzenberg, L.A., Improved FACS-Gal: flow cytometric analysis and sorting of viable eukaryotic cells expressing reporter gene constructs, *Cytometry*, 12, 291, 1991.
105. Mateus, C. and Avery, S.V., Destabilized green fluorescent protein for monitoring dynamic changes in yeast gene expression with flow cytometry, *Yeast*, 16, 1313, 2000.
106. Schlossman, S., Boumsell, L., Gilks, W., Harlan, J.M., Kishimoto, T., Morimoto, C., Ritz, J., Shaw, S., Silverstein, R., Springer, T., Tedder, T.F., and Todd, R.F., Eds., *Leucocyte Typing V, White Cell Differentiation Antigens,* Oxford University Press, Oxford, U.K., 1995, p. 980.
107. Goldsby, R.A., Kindt, T.J., and Osborne, B.A., Eds., *Kuby Immunology*, 4th ed., W.H. Freeman, San Francisco, 2000.
108. Zhao, R., Natarajan, A., and Srienc, F., A flow injection flow cytometry system for on-line monitoring of bioreactors, *Biotechnol. Bioeng.*, 62, 609, 1999.
109. Borth, N., Strutzenberger, K., Kunert, R., Steinfellner, W., and Katinger, H., Analysis of changes during subclone development and ageing of human antibody-producing heterohybridoma cells by northern blot and flow cytometry, *J. Biotechnol.*, 67, 57, 1999.
110. Van Tendeloo, V.F., Ponsaerts, P., Van Broeckhoven, C., Berneman, Z.N., and Van Bockstaele, D.R., Efficient generation of stably electrotransfected human hematopoietic cell lines without drug selection by consecutive FAC sorting, *Cytometry*, 41, 31, 2000.
111. Eustache, F., Jouannet, P., and Auger, J., Evaluation of flow cytometric methods to measure human sperm concentration, *J. Androl.*, 22, 558, 2001.
112. Johnson, L.A., Sexing mammalian sperm for production of offspring: the state-of-the-art, *Anim. Reprod. Sci.*, 60, 93, 2000.
113. Gunasekera, T.S., Attfield, P.V., and Veal, D.A., A flow cytometry method for rapid detection and enumeration of total bacteria in milk, *Appl. Environ. Microbiol.*, 66, 1228, 2000.
114. Vesey, G., Hutton, P., Champion, A., Ashbolt, N., Williams, K.L., Warton, A., and Veal, D., Application of flow cytometric methods for the routine detection of Cryptosporidium and Giardia in water, *Cytometry*, 16, 1, 1994.
115. Hairston, P.P., Ho, J., and Quant, F.R., Design of an instrument for real-time detection of bioaerosols using simultaneous measurement of particle aerodynamic size and intrinsic fluorescence, *J. Aerosol Sci.*, 28, 471, 1997.
116. Li, W.K. and Dickie, P.M., Monitoring phytoplankton, bacterioplankton, and virioplankton in a coastal inlet (Bedford Basin) by flow cytometry, *Cytometry*, 44, 236, 2001.

# 26

# X-Ray Diagnostic Techniques


Xizeng Wu
*University of Alabama at Birmingham*
*Birmingham, Alabama*

Abby E. Deans
*New York University*
*New York, New York*

Hong Liu
*University of Oklahoma*
*Norman, Oklahoma*




## Overview

X-ray diagnostic imaging is one of the most important tools of modern medicine. Approximately 120,000 x-ray rooms are in use in the United States; an estimated 240 million x-ray procedures are performed in the United States every year. X-ray diagnostic imaging is based on the tissue-differential contrast generated by x-ray and tissue interaction. In Section 26.1 we discuss x-ray–tissue interaction and tissue contrast mechanism. In addition to the traditional coverage of x-ray attenuation-based tissue contrast, we discuss x-ray phase-based tissue contrast, which arises from the very nature of x-ray as a wave. Such coverage is especially appropriate in view of the recent surge of research efforts in x-ray phase imaging. X-ray generation, spectra, and exposure control are discussed in Section 26.2. Section 26.3 discusses projection x-ray imaging (radiography and fluoroscopy), digital imaging detectors and x-ray image intensifiers, and image signal-to-noise ratio analysis of projection imaging. In Section 26.4 we discuss in-line phase-contrast imaging, an emerging x-ray imaging modality with great potential for biomedical applications.

## 26.1 Biological Tissue–X-Ray Interaction and Tissue Contrast

### 26.1.1 Attenuation-Based Tissue Contrast

X-rays are ionizing and invisible electromagnetic radiation that, compared to light, have much shorter wavelengths. For x-rays, a simple relation between the energy $E$ and wavelength $\lambda$ of an x-ray photon exists:

$$\lambda = \frac{12.4}{E(keV)} (\text{Å}) \tag{26.1}$$


0-8493-1116-0/03/$0.00+$1.50

For example, in medical diagnostic imaging, x-ray photon energy ranges from approximately 10 to 150 keV, so the wavelengths range from 1.2 to 0.083 Å. The amount of x-ray radiation exposure is often quantified by the amount of ionization generated by x-ray exposure, more specifically, by the electric charge generated per unit mass of air. The unit of x-ray exposure is defined as the Roentgen (R), and 1 R = 2.58 × $10^{-4}$ coulomb/kg.

Interacting with atomic electrons of biological tissue, x-ray photons can be transmitted, scattered (deflected), or absorbed by tissue. Due to the scatter and absorption the incident x-ray intensity $I_0$ is attenuated, and the transmitted x-ray intensity $I$ is given by

$$I = I_0 \exp\left(-\frac{\mu}{\rho}\rho t\right) = I_0 \exp(-\mu t) \tag{26.2}$$

where μ/ρ, ρ, and μ (product of μ/ρ and ρ) are the mass attenuation coefficient, mass density, and linear attenuation coefficient of the tissue, respectively, and $t$ is the thickness of the tissue. That is, x-ray attenuation depends on tissue's elemental composition (μ/ρ), mass density (ρ), and thickness ($t$). Apparently, x-ray attenuation increases with increasing μ/ρ, ρ, and $t$. A biological tissue's mass attenuation coefficient can be calculated as the weighted sum of the coefficients for its constituent elements

$$\left(\frac{\mu}{\rho}\right)_{tissue} = \sum_i w_i \left(\frac{\mu}{\rho}\right)_i \tag{26.3}$$

where $(\mu/\rho)_i$ is the mass attenuation coefficients for the $i$th element and $w_i$ is its weight fraction.

This mixture rule ignores any effect of changes in the atomic wave functions as a result of the molecular, chemical, and crystalline environment; however, the errors are less than a few percent. Different biological tissues have different elemental composition and mass densities and thus their linear attenuation coefficients are different. Along with the difference in tissue thickness, these differences in tissue linear attenuation coefficients result in the attenuation-based tissue contrast according to Equation 26.2.

Obviously bone has larger μ/ρ and ρ than that of soft tissue, and thus attenuates much more x-ray than soft tissue does for comparable thickness. That is why a high contrast exists between bone and tissue in x-ray images. However, μ/ρ and ρ of infiltrating invasive breast carcinoma are only slightly higher than those of glandular breast tissue; the small attenuation difference makes x-ray mammography the most technically challenging among all the radiological imaging tasks. A comprehensive compilation of attenuation can be found in the literature.[1] Table 26.1 provides the mass attenuation coefficients for water from 10 to 150 keV. The data are interpolated from the data of that compilation.

In x-ray diagnostic imaging, it is important to get good tissue contrast while keeping radiation dose as low as achievable. In order to accomplish this, it is important to understand that tissue attenuation-based contrast is x-ray photon energy-dependent because the mass attenuation coefficient μ/ρ of a given tissue depends on x-ray photon energy. In fact, three effects lead to the attenuation for photon energies up to 1 MeV (i.e., below the threshold for the pair production): coherent scattering, incoherent scattering, and photoelectric effects. Consequently, the mass attenuation coefficient μ/ρ has three contributions, one from each of the three:

$$\left(\frac{\mu}{\rho}\right)_{total} = \left(\frac{\mu}{\rho}\right)_{coh} + \left(\frac{\mu}{\rho}\right)_{incoh} + \left(\frac{\mu}{\rho}\right)_{photoel} \tag{26.4}$$

Coherent scattering is elastic scattering in which x-ray photon does not lose energy, but is deflected. Incoherent scattering is inelastic scattering; x-ray photon transfers energy to tissue while being deflected. With the photoelectric effect, x-ray photon is absorbed by an atom, and as a result, a bound electron is knocked out from the atom. As Table 26.2 shows, these three effects have different dependence on x-ray photon energy $E$, atomic number $Z$, and atomic weight $A$.[2] Radiological techniques utilize these relations

**TABLE 26.1** Water Mass Attenuation Coefficients

| X-Ray Photon $E$ (keV) | $\mu/\rho$ $m^2/kg$ |
|---|---|
| 10 | 0.5232 |
| 20 | 0.0795 |
| 30 | 0.0376 |
| 40 | 0.0268 |
| 50 | 0.0228 |
| 60 | 0.0207 |
| 70 | 0.0193 |
| 80 | 0.0183 |
| 90 | 0.0175 |
| 100 | 0.0169 |
| 110 | 0.0164 |
| 120 | 0.0159 |
| 130 | 0.0155 |
| 140 | 0.0152 |
| 150 | 0.0149 |

**TABLE 26.2** Dependence of Mass Attenuation Coefficients on X-Ray Photon Energy $E$, Atomic Number $Z$, and Atomic Weight $A$

| Mass-Attenuation Coefficients | Dependence on $E$ | Dependence on $Z$ and $A$ |
|---|---|---|
| $(\mu/\rho)_{coh}$ | $E^{-1.8}$ | $Z^{2.8}/A$ |
| $(\mu/\rho)_{incoh}$ | $E^{-0.2}$ | $Z/A$ |
| $(\mu/\rho)_{photoel}$ | $E^{-3.5}$ | $Z^{4.5}/A$ |

to determine optimal x-ray techniques. More specifically, tissue attenuation contrast in x-ray imaging is specified by the subject contrast ($SC$):

$$SC = \ln \frac{E_2}{E_1} \tag{26.5}$$

where $E_1$ and $E_2$ are the absorbed x-ray energy fluence in the detector for the two tissue projection areas. If $\mu_1$ and $\mu_2$ are the linear attenuation coefficients for the two tissues of thickness $t$, then the subject contrast between these two tissues for a monoenergetic incident x-ray can be found by using Equation 26.2:

$$SC = \ln \frac{\exp(-\mu_2 t)}{\exp(-\mu_1 t)} = (\mu_1 - \mu_2)t \tag{26.6}$$

Note that for polychromatic x-rays the difference ($\mu_1 - \mu_2$) in this formula should be weighted by the x-ray spectrum. Figure 26.1 shows the subject contrast between a 0.5-mm microcalcification and breast glandular tissue as a function of x-ray photon energy. It shows clearly that the tissue subject contrast decreases rapidly with increasing x-ray photon energy. This is because tissue $\mu/\rho$ decreases with increasing x-ray photon energy. Therefore, tissue contrast decreases with increasing x-ray photon energy.

Note that tissue contrast increases with differences in tissue atomic numbers. Therefore, different x-rays are employed for imaging different body parts. For breast imaging, where tissue contrast is low, photons of low energies from 15 to 25 keV are best to keep good tissue contrast and relatively low radiation dose to the breast.[3] For chest imaging, where tissue contrast among bone, soft tissue, and air is high, photons of much higher energies (up to 150 keV) are used to keep low dose and reasonable dynamic range.

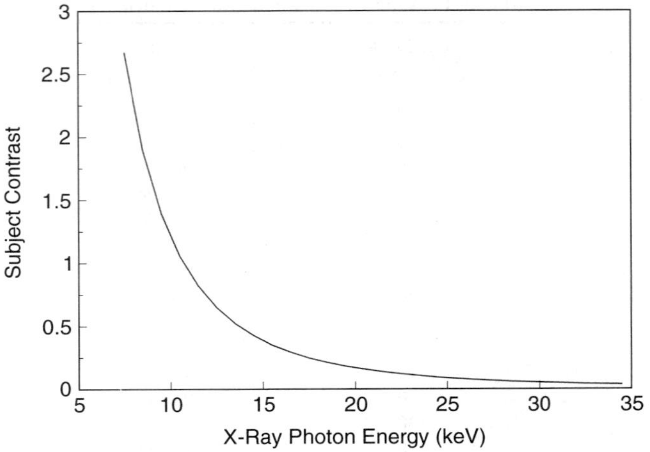


**FIGURE 26.1** The subject contrast between a 0.5-mm microcalcification and breast glandular tissue as a function of x-ray photon energy.

In general, the higher the photon energy, the smaller the $\mu/\rho$. However, when photon energy reaches the binding energy of atomic electrons of tissue, photon absorption jumps up because of increasing atom ionization; hence, the mass attenuation $\mu/\rho$ jumps up accordingly. The photon energies at which $\mu/\rho$ jumps up are called the absorption edges. The existence of absorption edges is of great significance for designing x-ray imaging detector and spectrum-shaping filters (see below.)

### 26.1.2 Phase-Based Tissue Contrast

In addition to the attenuation, tissue contrast can be realized from the x-ray phase change generated by tissues. Attenuation-based tissue contrast has been utilized in medical imaging for about 100 years, but phase-based tissue contrast was not recognized until quite recently.[4] X-ray phase change does not arise from new x-ray-tissue interaction; rather, it is a result of coherent x-ray scattering. As was mentioned previously, an x-ray, like light, is an electromagnetic wave. Like any wave field, a coherent x-ray wave field propagating along the $z$-axis can be described as

$$\psi(x,z) = \sqrt{I(x)}\exp(ikz) \tag{26.7}$$

where $x$ is the coordinate for the transverse direction, $k = 2\pi/\lambda$, $\lambda$ is the x-ray wavelength, and $I(x)$ is the x-ray intensity. In this case the phase of x-ray wave is $kz$. When an x-ray scatters from tissue, the phase of the x-ray wave field is changed. This phase change is the result of coherent scattering with small angles and contains diffraction and refraction effects from scattering. The amount of the phase change is determined by biological tissue dielectric susceptibility or, equivalently, by the refractive index of the tissue. The refractive index $n$ for an x-ray is complex and equal to

$$n = 1 - \delta - i\beta \tag{26.8}$$

where $\delta$, the refractive index decrement, is responsible for x-ray phase shift, and $\beta$ is responsible for x-ray absorption. The $\delta$ is given by Wilkins et al.:[4]

$$\delta = \left(\frac{r_e\lambda^2}{2\pi}\right)\sum_k N_k(Z_k + f_k^r) \tag{26.9}$$

where $r_e$, $N_k$, $Z_k$, and $f_k$ are the classical electron radius, atomic density, atomic number, and real part of the anomalous atomic scattering factor of the element $k$, respectively. If the x-ray is away from the absorption edge of tissue, this formula can be simplified to

$$\delta = \left(\frac{r_e\lambda^2}{2\pi}\right)\sum_k N_k(Z_k + f_k^r) \cong (4.49\times 10^{-16})\lambda^2 N_e \tag{26.10}$$

where $N_e$ is the electron density. Although we observe x-ray absorption by body parts everywhere in radiology, we nevertheless often do not realize that human tissue actually refracts x-rays. This is not because tissue $\delta$ is too small compared to $\beta$; on the contrary, tissue $\delta$ ($10^{-6}$ to $10^{-8}$) is about 1000 times greater than $\beta$ ($10^{-9}$ to $10^{-11}$) for x-rays in the 10 to 100 keV range. It is ironic that all x-ray clinical imaging techniques thus far have been designed to image tissue $\beta$ but not $\delta$. Using this formula for refractive index decrement $\delta$, the amount of x-ray phase change from biological tissue can be calculated as

$$\phi = \frac{2\pi}{\lambda}\int \delta(s)ds \tag{26.11}$$

where $s$ is the distance traveled by the x-ray beam along the vacuum propagation direction.

At this stage of development there are three major modes for phase imaging: x-ray interferometry, diffraction-enhanced imaging, and in-line phase-contrast imaging. The x-ray interferometry images the phase $\phi$ directly by using a monochromatic x-ray from synchrotron radiation and monochromator crystal.[5] The diffraction-enhanced imaging[6] measures the phase gradient $\nabla\phi$ directly by using monochromatic x-ray from synchrotron radiation as well. The in-line phase-contrast imaging measures the Laplacian of phase $\nabla^2\phi$ directly.[4] This mode can be implemented with polychromatic x-rays from an x-ray tube. Because x-ray tubes are compact and relatively readily available, we believe that this mode has great potential for clinical applications. The in-line phase-contrast mode is discussed in detail in Section 26.4.

## 26.2 X-Ray Spectra and Exposure Control

### 26.2.1 Bremstrahlung and Characteristic Radiation

X-ray attenuation and tissue contrast are x-ray photon energy-dependent, so it is very important to use x-ray photons of appropriate energies to image any given body parts. In order to achieve this, one must first learn how to control the spectrum of x-ray used in imaging. X-ray is usually generated by interaction of electromagnetic fields and charged particles such as electrons. Although an electromagnetic field surrounds a charged particle in a uniform motion of any velocity, it cannot radiate any electromagnetic waves. Only charged particles undergoing acceleration or deceleration can emit x-ray radiation. In medical imaging, the x-ray is usually generated by bombarding a metal target with energetic electrons.

When impacted with a metal target, the incident energetic electrons collide with atomic electrons and nuclei of the metal atoms. In metal the incident electrons slow down and are deflected. The swift deflections are caused primarily by the electrostatic attractive force between nuclei of metal atoms and incident electrons. With these swift deflections, incident electrons undergo huge momentum change, which means huge acceleration; hence, these deflected indent electrons emit x-rays. After repeatedly colliding with the metal atoms, the incident electrons eventually slow down to rest in metal and lose their ability to emit x-rays. X-ray radiation generated by the slowing down of energetic electrons in metal is called bremstrahlung. Bremstrahlung usually consists of x-ray photons of different wavelengths. In other words, bremstrahlung is polychromatic. According to Kramer's law,[2] bremstrahlung intensity $I_{br}(E)$ at photon energy $E$, is

$$I_{br}(E) = CZI_{br}(E_o - E), \quad \text{for } E < E_o, \qquad I_{br}(E) = 0, \quad \text{for } E > E_o \tag{26.12}$$

where $Z$ is the atomic number of the metal target, $I_{tube}$ is the current of incident electrons, and $E_o$ is the kinetic energy of incident electrons. In medical x-ray imaging equipment, incident electrons are accelerated by an applied voltage before impinging on a metal target. That is, for a bremstrahlung $E_o$ in Equation 26.12 equal to $eV_o$, where $V_o$ is the applied acceleration voltage, $e$ is the electron charge. Kramer's law predicts that below $E_o$ the spectral intensity increases with decreasing $E$ (Figure 26.2). However, due to x-ray attenuation by the metal target, one finds that the output bremstrahlung peaks somewhere around $E_o/2$ (Figure 26.2), depending on the target material and filtration (see below). The total approximate bremstrahlung intensity $I_{br\text{-}t}$ can be found by integrating Equation 26.12 over $E$:

$$I_{br-t} = CZI_{tube}V_o^2 = CZI_{tube}PV_o \tag{26.13}$$

where $P$ is the power loading of tube. Equation 26.13 shows that the total bremstrahlung intensity is approximately proportional to the square of applied acceleration voltage $V_o$, tube current $I_{tube}$, and the atomic number $Z$ of the metal target material. Therefore, in order to increase x-ray bremstrahlung intensity, one should apply high acceleration voltage, increase current, and adopt high-$Z$ target material. These three factors are among the most important considerations in the design of x-ray generating devices.

In addition to the bremstrahlung, there is another important mechanism of x-ray generation. When colliding with the atomic electrons of the target, the incident electrons can knock out atomic electrons from their atomic shell and cause the inner shell ionization of metal atoms. X-ray photons can be emitted when the ionized atoms return to their ground states. This kind of x-ray radiation is called the characteristic radiation because the energies of the emitted x-ray photons are discrete and equal to the binding energy differences of atomic shells. Characteristic radiation is named after the relevant inner atomic shells such as K-shell, L-shell, etc.

For example, the K-characteristic radiation results from knocking out of a K-shell electron by an incident electron. Let $eV_b$ be the binding energy of an atomic shell. Obviously, characteristic radiation from this shell can be generated only if $V > V_b$. The larger the $(V - V_b)$, the greater the characteristic radiation. The total

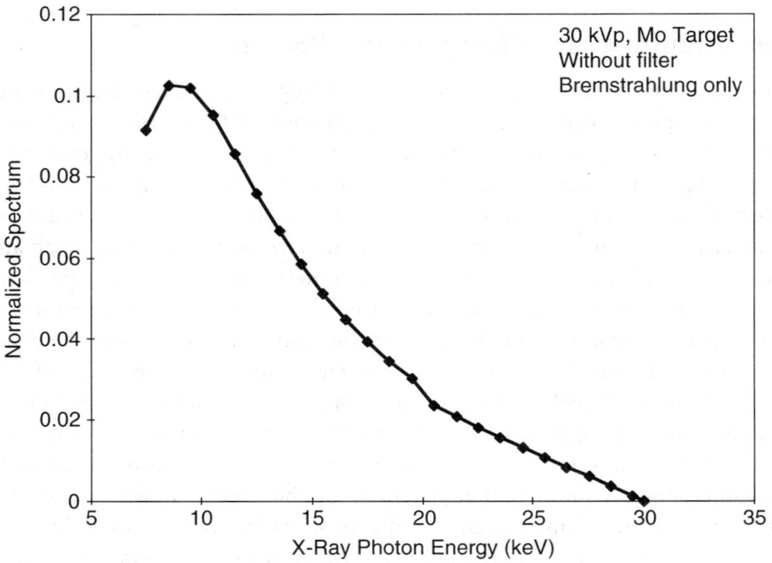


**FIGURE 26.2** X-ray bremstrahlung spectrum for an Mo-target at 30 kVp.

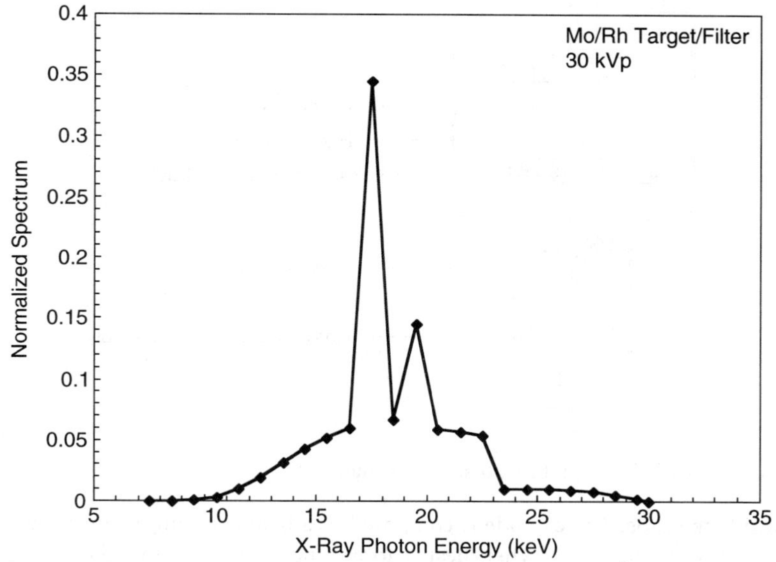


**FIGURE 26.3** X-ray spectrum for an Mo-target and Rh-filter at 30 kVp.

output x-ray consists of bremstrahlung and characteristic radiation, and the summed spectrum is shown in Figure 26.3. The spikes in this figure are superposition of these two kinds of x-ray radiation.

For tubes used for general radiography and fluoroscopy, the metal target layer is tungsten (W) alloyed with 5 to 10% of rhenium (Re). However, the target material for mammography is either molybdenum (Mo) or rhodium (Rh). Note that the atomic numbers for W, Mo, and Rh are 74, 42, and 45, respectively. Equation 26.13 shows that a W-target would generate much more bremstrahlung than would an Mo- or Rh-target. The reason for using an Mo- or Rh-target in mammography application is that the K-characteristic radiation photons generated from these targets are in the range of 17 to 23 keV, optimal for breast imaging.

In addition to the metal target material, one can use a filter, a metal sheet, to shape the output x-ray spectrum. During an exposure, low-energy photons are in the bremstrahlung as shown in Figure 26.2. These photons will be absorbed by the patient's body parts before reaching the imaging detector and hence contribute nothing to the imaging. An aluminum filter can reduce the number of low-energy photons, make the x-ray beam more penetrating, and reduce radiation doses to patients. This approach of shaping x-ray spectra is also called beam hardening by filtration. Usually an aluminum filter of 1 mm or more in thickness is used for this purpose. Sometimes, a copper filter of 0.1 mm or more in thickness can be used in fluoroscopy to reduce patient dose significantly.

For mammography, so-called absorption edge filters are used to shape the beam spectrum. As has been explained, the mass attenuation coefficient $\mu/\rho$ jumps up at the absorption edge. Using an absorption-edge filter one can selectively attenuate photons with energy higher than the absorption edge, hence shaping the output x-ray spectrum. For example, Mo has an absorption edge at 19.97 keV, and Rh at 23.19 keV. Using an Mo or Rh filter with x-ray tube with an Mo-target, one can generate a larger portion of x-ray photons with the optimal energy range (18 to 23 keV) to achieve better tissue contrast and reduction of radiation dose to breast glandular tissue.[7–9]

## 26.2.2 X-Ray Tubes

An x-ray tube is a device that generates x-rays by using the principle previously discussed; it consists of a cathode as the electron emitter, a rotating anode disk as the metal target, and a glass or metal envelope as a structural supporter (Figure 26.4). The glass or metal envelope is used to keep a vacuum of about

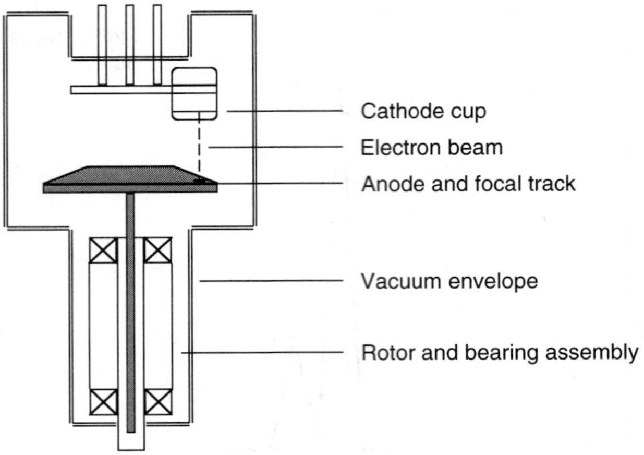


**FIGURE 26.4** Schematic of an x-ray tube with a rotating anode.

$5 \times 10^{-7}$ torr inside the tube. The cathode is composed of a filament (tungsten wire helix of 0.2 to 0.3 mm diameter) and a focusing cup. When heated to around 2400°C by the filament current, the hot filament emits electrons. Except for low tube potentials, the tube current $I_A$ that results from the emitted electrons is almost independent of tube potential and is a function of filament temperature:

$$I_A = CT^2 \exp(-W/kT)A_f \tag{26.14}$$

Here $T$ is the temperature in K, $W$ is the filament's work function (for tungsten $W = 4.5$ eV), $A_f$ is the area of the filament, $k$ is the Boltzmann constant, and $C$ is a material-dependent constant. Obviously, tube current $I_A$ increases with filament temperature. Usually, the filament current is around 5 A and the tube current ranges from 20 to 1000 mA for radiography. Note that Equation 26.14 is invalid for low tube potentials (<50 kV). When tube potential is low, not all the electrons emitted by filament are drawn to the anode, and the tube current $I_A$ is less than that predicted by Equation 26.14. $I_A$ becomes tube potential-dependent.

This behavior of $I_A$ is often called the space-charge effect. In breast imaging, tube potentials used are relatively low and vary from 25 to 35 kVp. In these cases the achievable $I_A$ is limited by the space-charge effect (especially with small focal spots) and the low $I_A$ imposes a technical challenge for mammography applications.

The tube focusing cup makes the electron beam focus so that the focal spot on the anode surface is kept small. A small focal spot is necessary for achieving a good spatial resolution of x-ray imaging. In projectional x-ray imaging, x-ray focal spot size determines the size of image blur:

$$B = (M-1)f \tag{26.15}$$

where $M$ is the geometric magnification factor, which is the ratio of the source-detector distance to the source-object distance, and $f$ is the focal spot size. So if $M = 2$ (as in magnification mammography), $B = f$. Any fine feature of sizes less than $f$ (at image plane) cannot be recognized. In general, a tube is equipped with two to three focal spot sizes for selection. For tubes used in general radiography, typical focal spot sizes are 1 to 1.2 mm for the large spot sizes and 0.5 to 0.6 mm for small spot sizes. For tubes used in mammography, typical focal spot sizes are 0.3 mm for the large spot and 0.1 mm for the small spot.

The tube anode is a rotating metal alloy disk. In addition to the metal target layer discussed in the previous subsection, the remainder of the anode disk is graphite or molybdenum (Mo). Because about 99% of the energy of incident electrons is transformed into heat at the anode, the anode disk materials should have high melting point and good thermal conductivity. During an exposure, the focal spot is hotter than the focal track, which is hotter than the anode body because the anode is rotating during an exposure. The anode as a rotor is driven by the stator through magnetic induction, running as an asynchronous motor. The stator driving current is provided by the high-speed starter of the x-ray generator (discussed in detail in the next subsection). The anode rotation speed usually ranges from 1200 to 3600 rpm, and can go as high as 10,800 rpm for the high-speed mode.

Operating "dry" in a vacuum at such a high temperature, the ball bearing must satisfy some exceptional specifications. To extend its life, the anode starts to rotate just before each exposure; after exposure it is brought to rest by a braking circuit of the generator. Note that electron impact time decreases with the velocity $\omega r$, where $\omega$ is the anode angular velocity and $r$ is the anode radius. Therefore, the focus spot temperature increases with the input power and x-ray exposure time and decreases with increasing focal spot size, rotation speed, and anode diameter. Because the practical limit of the focal spot temperature is 2500°C (although tungsten's melting point = 3370°C, the highest of all metals), each x-ray tube has its own maximum allowable power loading, which is called the tube power rating. During an exposure, the average power loading $P_{load}$ to an x-ray tube is determined by

$$P_{load} = wVI \tag{26.16}$$

where $w$ is the kV-wave-shape factor, $V$ is the tube kVp (the peak voltage in kV), and $I$ is the tube current in mA.

The kV-wave-shape factor $w$ depends on kV-wave form because the tube kVp is not equal to the average tube potential. For the 3-phase 12-pulse, or high-frequency generators, $w = 0.99$; for 3-phase, 6-pulse generators, $w = 0.95$; and for the single-phase generators, $w = 0.74$. For tube operation, the power loading of an exposure should be lower than the allowed tube power rating, which increases with increasing focal spot size, rotation speed, and anode diameter, but which decreases with exposure time. Power ratings for different exposure times can be determined from the manufacture-supplied tube rating chart or the so-called single-exposure rating chart. For the radiography and fluoroscopy applications, the tube power rating is 100 kW for a 0.1-s exposure with an anode rotation of 10,800 rpm and a focal spot of 1.2 mm. The tube power rating is reduced to 30 kW with a smaller focal spot of 0.6 mm. In addition to the power rating, x-ray tubes are also specified by anode heat capacity, the maximum anode heat load allowed. The heat load $H_L$ generated during an exposure, expressed in heat units (HU; 1 joule = 1.35 HU), is given by

$$H_L = 1.35 P_{load} T_{exp} \tag{26.17}$$

where $P_{load}$ is the power loading of the exposure and $T_{exp}$ is the exposure time (in seconds).

Use of large anode mass increases anode heat capacity. For radiography and fluoroscopy (R&F) applications, tube anode heat capacity usually is not less than 400 kHU. Although the power loading is not high for x-ray tubes in computed tomography (CT) applications, the heat loading is high due to long continuous scan time. The anode heat capacity for CT applications can be as high as 6.5 MHU or more. Tube heating during multiple exposures depends also on heat dissipation; in fact, the anode cools by radiating out heat. In order to increase heat dissipation, one can increase anode surface and use the emissive coating. In addition, tube housing is limited by its own maximal allowable heat load. Exceeding the housing maximum heat load can cause failure of the housing or the tube due to the thermal expansion of the insulation oil surrounding the tube. For some applications, like digital subtraction angiography, a tube assembly may be equipped with external oil-to-air or oil-to-water heat exchangers for fast cooling.

### 26.2.3 X-Ray Generators

In order to generate x-ray and control x-ray exposures one needs to apply controllable high voltage to an x-ray tube. The main function of an x-ray generator is to provide controllable high tube voltages and high tube currents for x-ray generation and exposure control. A generator is specified by the kVp rating

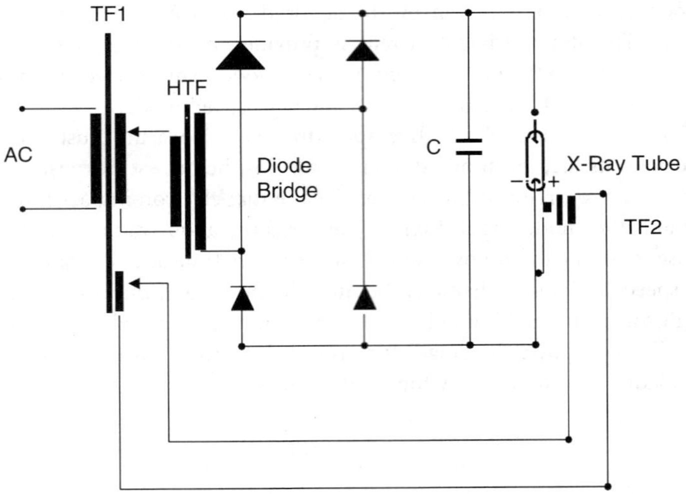


**FIGURE 26.5** Circuit schematic of a single-phase x-ray generator.

(the maximum kVp allowed), and the kV-ripple and power rating (the maximum power loading allowed for a 0.1-s exposure.) Here the kVp means the peak tube voltage in kV. The tube voltage is not a strict constant during an exposure, but with ripples around a mean value.

The amount of the kV-ripple is usually specified as a percentage of the peak voltage. As mentioned earlier, tube voltage determines the highest energy of the output x-ray photons. Therefore, in addition to tube kVp and tube current, the amount of kV-ripple affects x-ray radiation output and x-ray spectra. Small kV-ripple is desirable for x-ray exposure reproducibility, consistency of image quality, and radiation dose reduction. In general, the higher the kVp, the more penetrating the x-ray beam, but the lower the image contrast. Therefore, the tube kVp used varies depending on the body parts to image. Typically, one uses 125 to 150 kVp for chest x-rays, 110 to 130 kVp for computed tomography, 75 to 85 kVp for abdomen exams, 65 to 75 kVp for skull exams, and 24 to 35 kVp for mammography. Therefore, typical x-ray generator kVp ratings are 150 kVp for R&F generators and 45 kVp for mammography generators. The typical power ratings are 60 to 100 kW for digital subtraction angiography (DSA) generators, 50 to 60 kW for the R&F generators, and about 5 kW for mammography generators.

The heart of an x-ray generator is a high-power and high-voltage circuit. A basic single-phase generator is shown in Figure 26.5. The autotransformer TF1, which is fed by the line AC power, functions as the kVp control. The low-input AC voltage is transformed into high AC voltage by using the high-voltage transformer HTF. The diode bridge rectifies (in the full-wave) the high-voltage AC. The step-down transformer TF2 provides filament current control. The kV-ripple can go as high as 100% with a single-phase generator; however, if the line supply is replaced by a three-phase line supply, then the rectified sine waves from different phases can be interleaved; hence, kV-ripple can be reduced to 13 to 25% (for the 6-pulse interleaving) and 5 to 10% (for the 12-pulse interleaving.) Note that hat tube kVp generated by single- or three-phase generators depends on the input line regulation as well, and the kVp reproducibility is easily compromised by unstable line supply condition. The other drawback of single- or three-phase generators is the slow response for tube kVp control.

For medical imaging equipment, compactness is crucial. The high-voltage transformers of single- and three-phase generators are especially bulky in size and weight. However, note that the induced voltage, U, of a transformer is proportional to the product of $A$, $n$, and $f$, where $f$ is the transformer operating frequency, $A$ is the core cross section of transformer, and $n$ is the number of turns. It is clear that increase of operating frequency $f$ allows reduction in $A$ and $n$. Therefore, a generator operating at a high frequency can be much smaller in size and lighter in weight. This is why high-frequency generators operating at

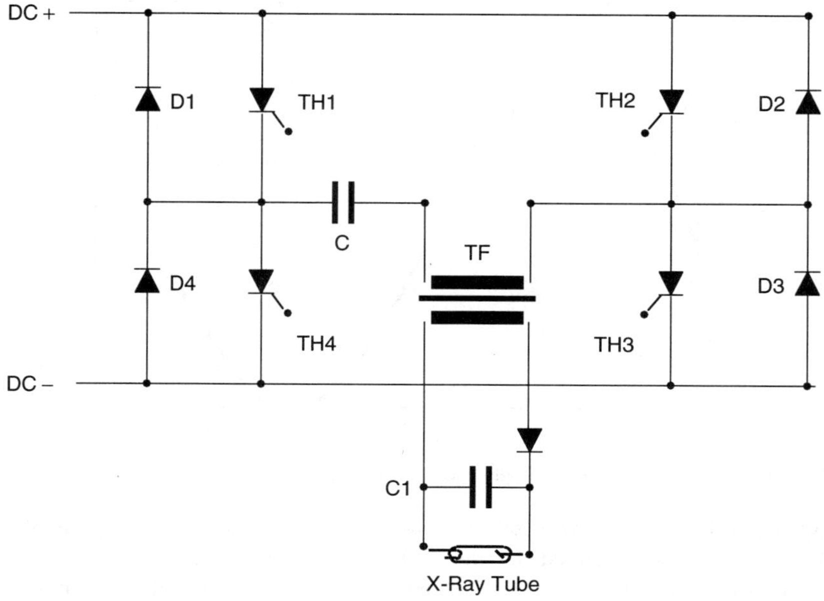


**FIGURE 26.6** Circuit schematic of a high frequency inverter.

a high frequency of a few to tens (or even up to about 100) kilohertz have proliferated in recent years. They reduce the size and weight by 50 to 80% compared to single- and three-phase generators, and they reduce manufacturing costs as well. Moreover, high-frequency generators provide fast response and greatly reduced kV ripples.

High-frequency generators are based on converter technology. The key is to convert the conventional AC power of 60 Hz and a few hundred volts to a high-frequency (HF) power supply of tens of kilohertz. This task is assumed by the HF inverter. The conventional AC power is first rectified and filtered to a DC power; the DC power is then inverted to HF pulses. Figure 26.6 shows a basic DC to HF inverter. Four thyristors, TH1 to TH4, work as controlled switches. A thyristor works like a diode with a control gate. The trigger pulses are applied to the gate of a thyristor control switching on and off the current flowing through the thyristor. The high-voltage transformer TF couples the x-ray tube as a load to the inverter and provides high voltage to the tube. The HF pulses are generated by a series RLC resonant circuit.

In this case, coupling capacitor *C*, inductance *L* of transformer TF, and the tube load consist of a series RLC resonant circuit controlled by "switches" TH1 to TH4. The RLC resonant circuit works as follows. When thyristors TH1 and TH3 are gated on, they work like two closed switches. The DC voltage is then applied to the RLC circuit. As is well known, capacitor *C* tends to oppose voltage changes and inductor *L* tends to resist current change because capacitor *C* stores electric energy and inductor *L* stores magnetic energy and energy transfer takes time. The interchange of the stored electric for magnetic energy and vice versa results in a current oscillation in the thyristors and RLC circuit. Due to energy dissipation on the load *R*, the oscillating current is really a dampened sinusoidal current. The frequency of the oscillating current is called the resonant frequency $f_{res}$, and is given by

$$f_{res} = \frac{1}{2\pi\sqrt{LC}} \tag{26.18}$$

During the negative half of the sinusoidal cycle, current flows through the antiparallel diodes D1 and D3, instead of thyristors TH1 and TH3, because thyristors do not allow a reverse current. Note that dampening the load makes the second peak smaller than the first in the cycle. However, when the full sinusoidal cycle is completed, the RLC resonator cannot start its second cycle because thyristors TH1

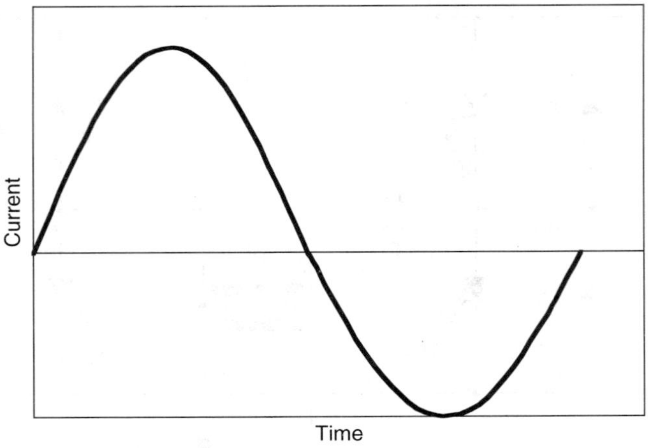


**FIGURE 26.7** Current pulse waveform in a high-frequency inverter.

and TH3 have already been switched off (self-communication) during the reversing half cycle. Therefore, only a single current pulse is generated (Figure 26.7) for each pair of trigger pulses to the gates of TH1 and TH3. To degauss the transformer core, TH2 and TH4 take turns generating the next current pulse in a similar way. The difference is that the polarity of the applied DC power is reversed this time; hence, the RLC resonance process leads to a "reversed" current pulse. The frequency at which the two trigger pulse pairs are applied is called the driving frequency $f_d$. The higher the driving frequency, the denser the current pulse train.

The secondary coil of the transformer TF couples a rectifying circuit with the RLC circuit. The rectifying circuit consists of a diode bridge and a capacitor $C_l$ for filtering. Because tube voltage is equal to the voltage across $C_l$, $C_l$ is called the load capacitor. The rectifying circuit converts the high-frequency current pulses into charge pockets stored in $C_l$. At the beginning of an exposure, the driving frequency $f_d$ is very high in order to charge $C_l$ rapidly and build up the desired tube voltage. The voltage of load capacitor $C_l$ can be measured by a precision resistor voltage divider. Once the desired tube voltage is established, a feedback signal will be sent to a voltage frequency converter V/F, the output of which controls the driving frequency $f_d$ of the gate trigger pulse circuit. The $f_d$ will then be reduced to deliver just enough charge pockets to $C_l$ for compensating the discharge consumed by the tube current. Therefore, although the resonant frequency $f_{res}$ is fixed by coupling capacitor $C$ and inductance $L$ through Equation 26.18, the driving frequency $f_d$ varies during exposures. However, the maximum driving frequency allowed is the resonance frequency $f_{res}$.

Using the closed-loop servo control described previously, tube voltage control is very fast and accurate, and the kV-ripple ranges from 2 to 15%, depending on the power rating. For a given resonant frequency $f_{res}$, the ripple decreases with increasing power. Therefore, in order to keep uniform kV-ripple, an x-ray generator may need to use different coupling capacitors when the tube focal spot is switched, because the power rating with the small focal spot is much lower than that with the large focal spot.

Another advantage of high-frequency generators is that they can use single-phase or three-phase line supply because the inverter circuits of the generators are powered by DC power. The inverter circuit described earlier can also provide power for the filament. In a filament inverter, the filament transformer is step down in voltage (step up in current) and the secondary coil of the transformer is coupled directly with the tube filament without rectifying circuitry. A precision resistor in series with the x-ray tube senses the tube current in real time, and the sensed signal is used as the feedback signal to control the driving frequency of the filament inverter. In this way tube current control is much faster and more accurate than that with single- or three-phase generators.

In addition to providing high voltage and power to the tube, an x-ray generator should also provide the means for x-ray exposure control. It is clear from Equation 26.13 that the exposure rate is proportional

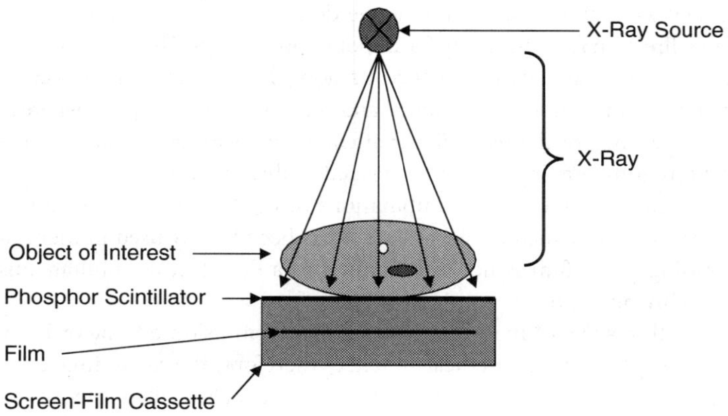


**FIGURE 26.8** Schematic of projection radiograph.

to tube current, and the total exposure is proportional to the integral of $I_{tube}$ over exposure time, which is called the total mAs (mA × seconds) of the exposure. All x-ray generators include an mAs switch to control the total mAs delivered during an exposure. A simple electronic timer (for a constant the mA during exposure) can be the mAs switch circuitry or a current integrator (for a falling $I_{tube}$ during exposure). However, modern x-ray generators use the automatic exposure control (AEC) circuitry as the mAs switch.

With AEC, exposures are turned off once a preset amount of exposure incident on the image receptor is reached. In this way AEC ensures that the imaging detector always receives almost the same amount of radiation exposure regardless of object attenuation. A typical AEC circuit consists of an x-ray sensor, a current integrator, and a comparator. The x-ray sensor generates a sensing current proportional to the exposure rate incident on image receptor. The current integrator measures the total exposure to the image receptor. Finally, the comparator sends out the exposure termination signal when the output of current integrator reaches the preset exposure value. The radiation sensor can be an ionization chamber that transforms x-ray to electric current. A fluorescent screen coupled with a photodiode can also function as the radiation sensor for AEC. In this case a fluorescent screen transforms the x-rays to light and a photodiode works as a light sensor.

## 26.3 Projection X-Ray Imaging

### 26.3.1 Conventional Radiography

Projection radiography is the most conventional technique used in radiological imaging. It uses an x-ray beam to generate a two-dimensional transmission image of patient anatomy. As shown in Figure 26.8, during x-ray exposure the x-ray beam passes through the body of the patient and is recorded by a two-dimensional detector. On the passage, the x-ray radiation is attenuated by absorption and scattering within the body; the x-ray gets diffracted as well. Different anatomical structure, as well as different body thickness, attenuates and diffracts the x-ray beam differently. However, in this section we discuss only attenuation-based projection radiography; phase-contrast imaging is discussed in Section 26.4.

The conventional projection radiograph is a spatially modulated attenuation pattern of the anatomical structure. In other words, a projection radiograph is a two-dimensional "x-ray shadow image" of the three-dimensional anatomy of the patient. More exactly, a projection radiograph is a two-dimensional map of $I(x, y)$:

$$I(x,y) = I_o \exp\left(-\int \mu(x,y;s)ds\right) \tag{26.19}$$

where $I_0$ is the incident x-ray intensity, $I(x, y, s)$ is the detected intensity at position $(x, y, s)$, and $s$ is the distance along the ray direction from the x-ray focal spot to point $(x, y)$. The integral in Equation 26.19 gives the projection sum of the linear attenuation coefficients along the ray path. This projection image is different from the images from the tomography (longitudinal and computed tomography), where an image of a tissue slice is a two-dimensional map of the tissue linear attenuation coefficients' map $\mu(x, y; s)$ of the slice at $s$. Due to space limitations, tomography will not be covered in this chapter.

The x-ray detectors can be screen–film combinations or digital imaging detectors. Film was the first radiographic detector used in medicine, and to date it has been widely used in medical imaging. In fact, the function of a radiographic film is not only limited to a detector; it simultaneously plays a role as display device and archiving medium.

Unexposed film is a thin sheet of inert plastic coated with emulsion on one or both sides. The film is insensitive to x-ray energies used in medical practice; therefore, the x-ray images are, in most cases, detected with a screen–film combination. Here the term screen, also called intensifying screen, refers to a luminescent screen that contains densely packed luminescent materials. The luminescent substances should be highly efficient for x-ray absorption and conversion of x-ray to light and should have light emission spectra matching the film's spectral sensitivity.

At present, most intensifying screens use gadolinium oxysulfide ($Gd_2O_2S$) as the luminescent material. The $Gd_2O_2S$-based intensifying screens convert each x-ray photon of 30 keV to about 1500 green light photons. A piece of film is placed in direct contact with a piece of intensifying screen in a cassette. During x-ray exposure, the x-ray photons exiting from the patient first interact with a $Gd_2O_2S$-based intensifying screen. The screen will convert the x-ray to green light, and the green light exposes the film. With a subsequent chemical development process, the original radiographic image of the patient is captured permanently on the film.

## 26.3.2 Digital Radiography

In recent years radiology has undergone a revolutionary change from screen–film-based radiology to digital radiology, which uses optoelectronic detectors to record images. A digital image detector detects transmitted x-ray pixel by pixel (the field of a single detector element). Thus, the digital image appears as an array of pixels. Figure 26.9 shows an example of how a simple one-dimensional image would appear in analog and digital format. To decrease the information loss from discontinuities in the image due to pixel averaging, the detector element size must be significantly smaller than the important features within the image. Although the digital image can be captured with a large dynamic range, interpretation of the radiograph requires that the image be depicted on screen or in hard-copy format. The information in these representations of the image is usually truncated to 8-bit data from 12- or 14-bit data. This is not necessarily a problem because features of interest are often within a smaller grayscale range than the entire dynamic range of the detector; however, high-fidelity electronic display and hard-copy printing devices are necessary to preserve diagnostic quality.

Various optoelectronic detectors have been developed and successfully applied in clinical practice. Three types of detectors are used in the majority of digital imaging systems: charge-coupled device (CCD)

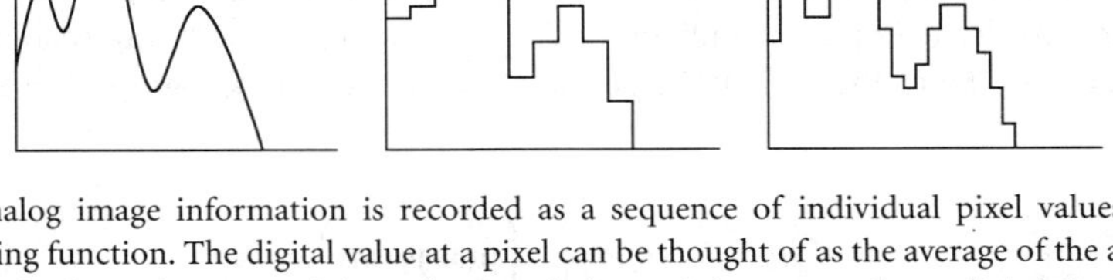

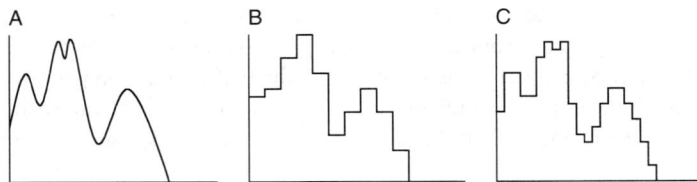


**FIGURE 26.9** Analog image information is recorded as a sequence of individual pixel values as opposed to a continually changing function. The digital value at a pixel can be thought of as the average of the analog value across the area of the pixel. The analog image (A) can lose resolution as it is converted to a digital signal (B). To improve resolution, the detector element size, i.e., pixel size, can be decreased (C).

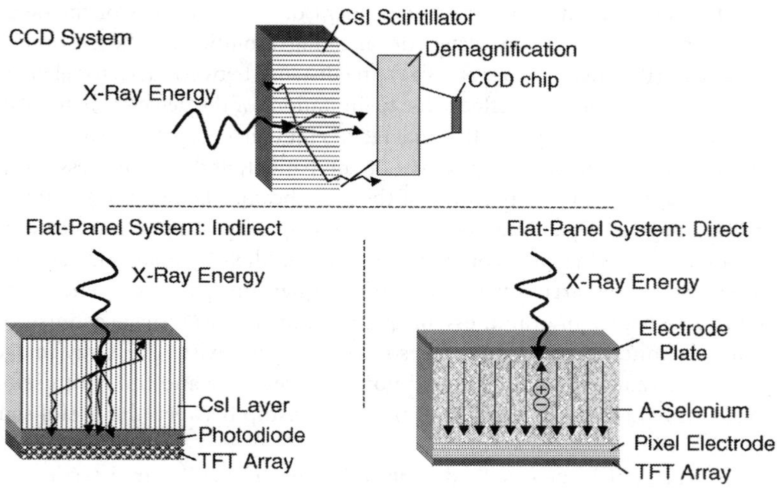


**FIGURE 26.10** Digital radiography detectors have two main types: the CCD system and the TFT flat-panel system. The CCD system converts the x-rays to visible light at a CsI scintillator screen before the signal is demagnified and detected at the CCD. One type of flat-panel system, the indirect system, also converts the x-ray signal to visible light before it is converted to charge by photodiodes at the detector. The direct flat-panel system converts the x-rays directly to charge in a layer of amorphous selenium. The charges are drawn to the detector by an applied electric field.

detector systems, flat-panel thin-film transistor (TFT) detector systems (Figure 26.10), and the photo stimulable phosphor imaging plate detector systems. Each detector type has its advantages and disadvantages; certainly they are most effective when applied to particular tasks. Before we describe and compare these detectors in detail, let us discuss the specification parameters for digital imaging detectors. These parameters include field coverage, detector sensitivity and quantum efficiency, spatial resolution, noise factors, and dynamic range.

The image size requirements are important factors in selecting a detector. Stereotactic breast biopsy imaging, for example, requires relatively small images to be acquired; thus, for this application, a single CCD detector including a demagnification step would be acceptable because minimal demagnification would be necessary.[10] For larger images, a higher degree of demagnification is necessary to map to the CCD, possibly decreasing image resolution to an unacceptable level. CCDs are combined into detector arrays for use in larger-size imaging applications. Another alternative, full-field TFT arrays, which are manufactured in standard film sizes, may have better performance potential for larger image fields.

Detector sensitivity, that is, how well the detector converts the x-ray signal into charge, is another factor. Sensitivity is directly related to the efficiency of this conversion, the so-called quantum efficiency. The x-ray quanta interact with the detector material primarily according to the photoelectric effect, wherein one or more photons are produced as the x-ray quanta loses energy to an atom (high Z) of the detector. The quantum efficiency, $\eta$, is related to the attenuation properties of the detector material as well as to detector thickness.[11] In other words, efficiency is dependent on an x-ray's effectiveness at giving up energy to a detector atom, as well as on the likelihood of the x-ray encountering a detector atom to donate this energy to, a quantity proportional to the thickness of the detector material.

The attenuation properties are determined by the composition of the detector material as well as by the energy of the incident x-rays. In general, higher-energy x-rays can penetrate more deeply into detector material (the attenuation coefficient is smaller for high energies), whereas low-energy x-rays penetrate more shallowly (the attenuation coefficient is larger for low energies). Another factor affecting quantum efficiency is the "fill factor" of the detector, i.e., the area of active detector material per total detector element area. As the fill factor approaches 100%, less undetected radiation must be used for an image of adequate intensity and thus the total radiation dose is minimized.

The spatial resolution of the detector plate is dependent on characteristics of the detector as well as on external factors. Significant detector characteristics affecting resolution are the size of the active portion of each detector element (the effective aperture size), the distance between detector elements (the spatial sampling interval), and lateral spreading effects due to limitations of the detector material or information transfer from the detector to the computer. External factors influencing spatial resolution include relative motion between the detector and the x-ray source or the patient, and unsharpness resulting from the magnification of the image due to the spreading of the x-ray beam.[11] Different resolutions are necessary for different types of imaging, depending upon the size and spacing of significant features.

Another factor affecting the sharpness of an image is noise. All x-ray images are statistical in nature. In other words, because x-rays arrive at the detector in quantum packets of energy, the image will inherently contain some degree of graininess, or quantum noise. Ideal image acquisition systems are called quantum noise limited because no other sources of noise exist. The x-ray image signal must be significantly stronger than the noise or it will not be discernable and thus not useful. The signal-to-noise ratio (SNR) is a quantity that indicates the relationship between the useful signal and other noise.

The detective quantum efficiency (DQE) describes the effectiveness of transferring the SNR of the incoming x-ray signal to the digital output, that is, adding as little extra noise in the process as possible. The mathematical formula for DQE and a detailed DQE analysis are the subject of the next subsection. Ideally, the DQE is equal to the quantum efficiency, $\eta$, but detectors add noise from different sources depending on their particular characteristics. Some detectors, for example, are sensitive to visible light as well as x-rays and therefore light leakage into the detector casing may increase noise. Other detectors are very sensitive to heat and must be operated at low temperatures to reduce so-called thermal noise. Background noise can often become insignificant if the incident image signal is intensified; however, a challenging balance exists between minimizing radiation dose and achieving reasonable SNR at the outcome.

The dynamic range of a detector describes the range of signal levels at which the detector can accurately record an image, considering intrinsic noise factors. The range must be adequate to measure the most and the least radio-opaque regions of the image and everywhere in between. The dynamic range is limited at low levels because the signal can become washed out by noise. At high levels, detector elements can become saturated and leak onto adjacent elements, blurring the final image. Although most digital detectors are capable of acquiring images with a significantly reduced x-ray dose compared to their screen–film counterparts, many sources of noise must be controlled to maintain adequate imaging ability at these low signal levels.

A common type of digital detector is the CCD (see Figure 26.10). Systems utilizing the CCD as the detector element have had a few variants in the decades since its advent. The CCD is manufactured as a 2- to 5-cm square chip tiled with 256 $\times$ 256 to 2048 $\times$ 2048 sensitive detectors that correspond to pixels in the digital image. CCD detectors utilize an indirect conversion system, incorporating a step to convert x-rays to light before the detection step. Incident x-rays are converted to visible light at a scintillation screen. Although so-called unstructured and structured scintillators are used, structured scintillators like crystalline cesium iodide have better resolution capabilities. Cesium iodide forms needle-like crystals that act like optical fibers to direct scattered light to the photodiode with improved spatial resolution.[12]

Although other scintillator materials are used, light produced by phosphor materials in unstructured screens is more likely to spread to adjacent pixels, reducing resolution properties. Because the CCD is smaller than what most imaging applications demand, image demagnification or various types of detector field expansion must take place. The detector field can be expanded by combining several CCDs into a detector array to obtain an image with a single exposure. The detector field can also be expanded by scanning a single CCD or a smaller CCD array throughout the image field to obtain piecewise images that are combined at the end of the imaging process.[10] Image demagnification can be accomplished with a system of lenses or with tapered optical fibers (Figure 26.11). The efficiency of lens coupling is limited by the sampled light signal that hits the collecting optics, and the image is susceptible to distortion due to lens defects.[11] Consequently, only a fraction of the signal emitted by the phosphor plate is transmitted

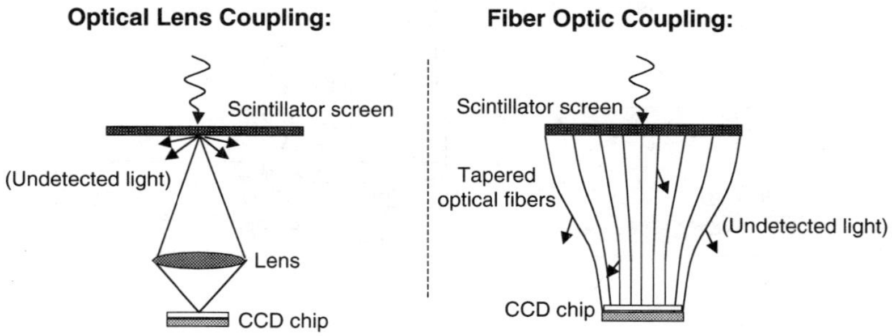


**FIGURE 26.11** The CCD detector receives the visible light image after demagnification by a lens system or tapered optical fiber system. The lens system transmits only a portion of the light from the CsI screen, leading to a reduced SNR at the detector (and thus a reduced DQE). The tapered optical fiber system is also inefficient, leaking some light as the reflection angle changes along the length of the fiber.

through the lens system to reach the detector. The SNR of this type of system is highest when little demagnification is necessary.

Optical fiber tapers are not completely efficient or distortion-free. Visible light output from the phosphor plate can leak from the optical fibers due to the changing reflection angle within the optical column, and bundling of multiple tapered fibers can produce image distortion (Figure 26.11). The inefficiency of the lens and fiber-optics systems causes the DQE of CCD detectors to be lower than systems not requiring demagnification. Systems that require very little demagnification have the highest DQE. The advantages to CCD detectors are that they have extremely low intrinsic noise factors, so a high degree of detector capability can be achieved despite inefficiencies due to demagnification. In addition, the CCD has a broad dynamic range and a highly linear response to the incident signal.[11]

The second broad system category is the flat-panel system, which utilizes a thin film transistor array as the detector (Figure 26.10). Within this category, there are two types of systems: those that convert x-rays directly to electric charge that can be read out by a computer as a digital signal (direct conversion), and those that include an intermediate conversion from x-ray photons to visible light photons (indirect conversion).

The indirect conversion detector has a few steps of signal modification before the analog image is converted to digital. The first modification occurs at a scintillation screen, where a phosphor, usually crystalline cesium iodide (CsI), absorbs the x-rays and re-emits energy in the visible spectrum, according to the photoelectric effect. High-energy x-ray quanta incident on the plate are each converted to several visible light quanta, resulting in an intensification of signal known as quantum gain.[11] The thicker the phosphor layer, the greater the likelihood that all of the x-rays will interact with phosphor to create photons and maximize quantum gain.

However, the light emitted from the phosphor is not necessarily emitted in the same direction in which the incident x-ray was traveling. Consequently, this conversion adds a scatter, or blurring, effect to the original signal that is directly related to the distance the light photons need to travel before detection, i.e., the thickness of the phosphor. Thus, the thickness of the layer must be determined by compromising these effects (Figure 26.12). The ideal thickness would maximize intensity while minimizing distortion of the signal at the output.[11] CsI crystals are actually grown on the surface of microelectronic detector plates to improve the detector's spatial resolution and efficiency. At the exit face of the scintillator screen, the visible light is detected by an array of amorphous silicon photodiodes that transmit a proportional quantity of charge to transistors, where it is stored until readout (Figure 26.13A).

The direct digital system is constructed as a layer of amorphous selenium sandwiched between a single electrode plate at the entry face and pixel-sized TFTs at the exit face. A voltage at the electrode plate during image acquisition creates an electric field within the selenium that maximizes image quality. The incident x-ray signal passes unaltered through the electrode plate and is absorbed in the selenium layer.

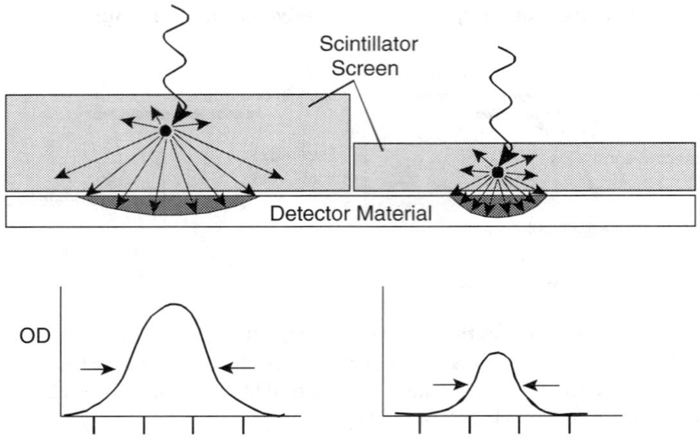


**FIGURE 26.12** The ideal scintillator screen would maximize intensity (increase thickness), while minimizing image distortion or wash-out due to scattering (decrease thickness).

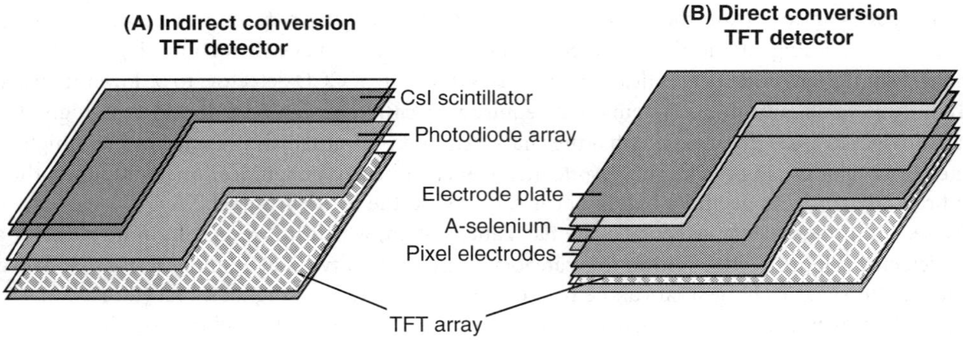


**FIGURE 26.13** The indirect conversion flat-panel detector (A) has three layers: a CsI scintillator plate, an array of pixel-sized photodiodes, and a layer of a-silicon thin film transistors. The direct conversion flat-panel detector (B) has four layers: an electrode plate, an amorphous selenium semiconductor layer, an array of pixel electrodes, and a layer of a-silicon thin film transistors.

The x-ray energy liberates electrons within the selenium layer which are drawn by the electric field straight to the pixel electrodes at the exit face. This design results in very little image scatter and a fill factor theoretically approaching 100% because the electric field funnels charge into the transistors and thus loses minimal signal from the area between active portions of detector elements (Figure 26.13B).

For direct or indirect conversion, information is retrieved in the same manner — through an active matrix readout method. The information in the TFTs is read row by row. An electric pulse is applied to the gate of the transistors one row at a time, allowing the charge stored at the drain to be measured at the source (Figure 26.14). The charge at each column location is amplified and multiplexed row by row until all of the transistors have been read out, at which point the information is compiled into a complete digital image file.[13] An indexed 12- or 14-bit numerical value — in base 10, a number from zero to several thousand — is stored in the file for each pixel in the image.

The third category system is computed radiography (CR), which utilizes the photostimulable phosphor imaging plate (also known as the storage phosphor) as the detector device. Photostimulable phosphor imaging plates are the most popular large-area electronically readable detectors currently available. CR (a misnomer) is based on the photostimulable phosphors. These phosphors are made up of barium fluorohalide family compounds (BaFX, where X is I, Br, or Cl) doped with bivalent europium ions that

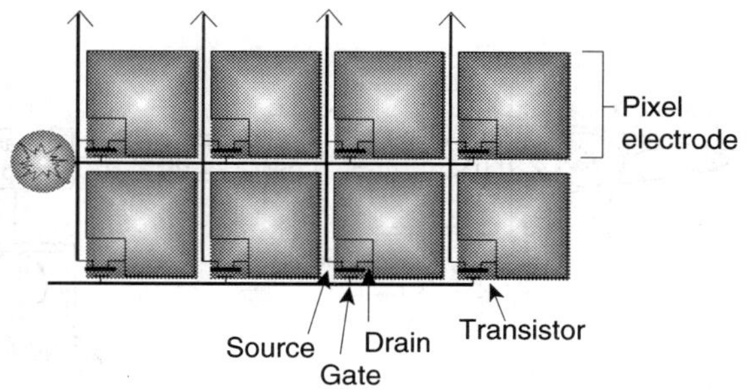


**FIGURE 26.14** Information is read from the detector through an active matrix readout method. An electric pulse is sent along a row (starburst), allowing the stored charge to be read at each pixel column location (arrows). The entire array is read out row by row and multiplexed into a digital image file.

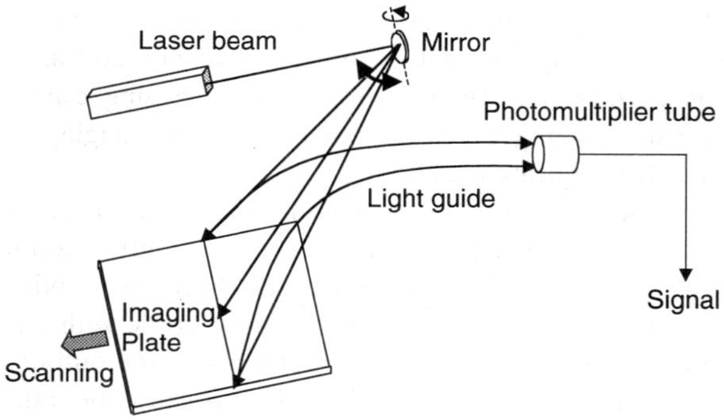


**FIGURE 26.15** Schematic of the laser scanner for reading exposed photostimulable phosphor plates.

provide the luminescence centers. Incident x-ray makes electron–hole pairs in the crystal. Some of the halogen ion vacancies trap electrons to create the metastable F-centers. These trapped electrons create a latent image.[14]

To retrieve the latent image, one places the exposed phosphor plate into a laser scanner. A red He-Ne laser beam raster scans the exposed phosphor and the trapped electrons are released from the traps. Undertaking atomic energy level transitions, these released electrons migrate to europium ions to release the stored energy as luminescence blue light. A photomultiplier tube collects the emitted blue light as signals to form the image (Figure 26.15). Note that the storage plate is reusable after erasure by intensive illumination. Obviously, the spatial resolutions depend on the spatial sampling frequency of the scan laser.

The major advantage of the storage plates is that they can be used exactly like conventional portable cassettes, but with latent images in traps rather than on film. The storage plates are available in different sizes matching the standard field of views (FOVs) and with large enough dynamic ranges for radiography. For these reasons, storage plates have been widely used in emergency and bedside radiography since their debut in the early 1990s. More than 10,000 CR units have been installed in the United States to date.

This technology also has several disadvantages. First, the laser light scattering in the phosphor reduces the spatial resolution. Spatial resolutions are approximately 1.5 lp/mm at 50% MTF; the limiting resolution is approximately 3.5 line pairs (lp)/mm. Second, the DQE($f$) is low due to the low efficiency of collecting the signal light and the granularity noise involved, especially at high spatial frequency. The DQE is about 15% for low spatial frequencies and decreases rapidly with increasing spatial frequency. This DQE is lower than that of screen–film-based radiography; therefore, the radiation doses to the patient are more than 50% higher than those with conventional radiography. Third, the work throughput of radiography with phosphors is lower than that of conventional radiography.

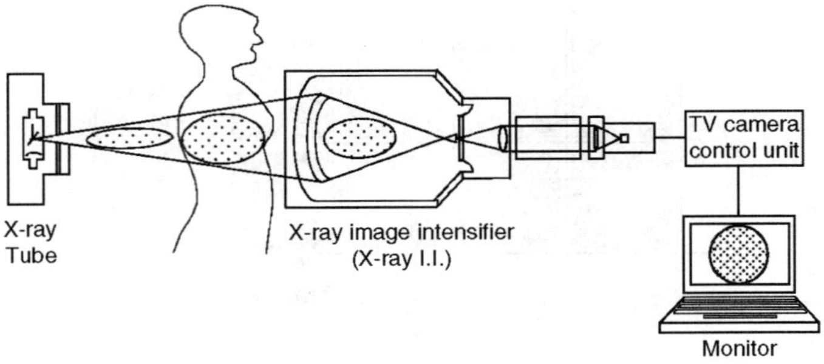


**FIGURE 26.16** The flow chart for fluoroscopic image formation.

### 26.3.3 Image Intensifier TV Chain and Fluoroscopy

Another important modality of projection x-ray imaging is fluoroscopy, in which exposures are continuous and acquired images are displayed on a TV monitor, usually on 30 frames per second. Therefore, fluoroscopy images body parts in real time. This real-time imaging capability makes fluoroscopy extremely useful for diagnostic imaging, therapeutic imaging (cardiac angioplasty, interventional procedures), and image guidance for surgical procedures.

The key device for fluoroscopy is the image intensifier-TV chain. The flow chart for fluoroscopic image formation is shown in Figure 26.16. As a projection imaging modality, a continuous x-ray exposure is projected on body parts of interest and the transmitted x-ray from these body parts is detected first by an image intensifier. An image intensifier comprises a vacuum bottle with an x-ray transparent input window and a glass or fiber-optic output window through which the output image is displayed. Figure 26.17 is a schematic drawing of an image intensifier (I.I.). Because the transmitted x-ray is incident upon a scintillation layer (commonly of CsI:Na) of the I.I., an x-ray photon is absorbed via the photoelectric effect (discussed in Section 26.1). The absorbed energy is eventually converted to low-energy secondary electrons and these electrons then recombine under emission of light. The wavelengths of the light depend on the scintillator and doping material used in the I.I.

For the commonly used CsI:Na scintillator the peak wavelength of light is about 420 nm. Each absorbed x-ray photon of 30 keV would produce about 1500 light photons. These photons then reach the semi-transparent photocathode of I.I., which is made up of antimony and various alkali metals such as cesium. For example, cesium antimonide ($Cs_3Sb$) is commonly used for photocathodes. In order to increase I.I. efficiency, the light sensitivity of the photocathode is matched to wavelength of the scintillation light. Again, through the photoelectric effect mentioned in Section 26.1, the light photons kick out electrons from the photocathode.

Thus far, the detected x-ray has completed conversion to photoelectrons. The spatial distribution of the photoelectrons is just a "contact print" of the light image in the input phosphor of I.I. Obviously,

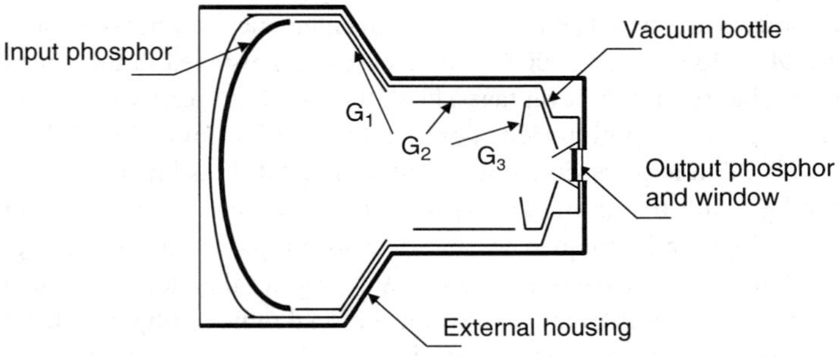


**FIGURE 26.17** Schematic of the image intensifier (I.I.).

the thicker the input phosphor (e.g., CsI scintillator layer), the more light is generated and, hence, more electrons are converted. However, an increase in thickness also leads to an increase in lateral diffusion of light inside the layer and consequently reduces its spatial resolution. Usually, the thickness of CsI layer varies from 300 to 450 μm; typically, each impinging x-ray photon of 30 keV generates about 200 electrons.[15] The x-ray attenuation-based image detected by the input phosphor converts to a photoelectronic image.

A photoelectronic image is not convenient for projection on a television camera for display, but is relatively easily "intensified." Therefore, one needs to amplify the photoelectronic image first and then convert the amplified photoelectronic image into a visible image for the TV camera. This amplification is done through electron acceleration by applied high voltage; an output phosphor (output screen) is employed to convert the amplified photoelectron image into a visible image, which is an intensified original x-ray attenuation-based image. For the amplification, a high voltage is applied between the photocathode ($V = 0$) and the output phosphor ($V = 25$ to 35 kV; see Figure 26.17). In this way each electron gains in energy by about 30 keV.

A common output phosphor is the P20 phosphor, which is a mixture of zinc cadmium sulphide doped with silver ($Zn_{0.6}Cd_{0.4}S{:}Ag$) and[15,16] emits green light of 520 and 540 nm. Impinging on the output phosphor, each of the accelerated electrons causes the emission of about 1000 green light photons. Because each absorbed x-ray photon of 30 keV generates about 200 electrons from photocathode as mentioned earlier, each absorbed x-ray photon of 30 keV will generate $2 \times 10^5$ green light photons at the output window. Comparing this figure to the intensifying screens ($Gd_2O_2S$-based screens) used in conventional radiography (Section 26.3), the light photon gain from an I.I. is about 100 to 200 times higher.

To further increase the light output luminance (flux intensity), demagnification technique is employed. The ratio of the input phosphor entrance size to the output phosphor size is called the demagnification ratio. The diameter of output phosphor is about 1.5 to 3.5 cm, much smaller than that of the I.I. entrance fields, which varies from 15 to 40 cm. The luminance of the output phosphors is the light flux density, so the luminance is proportional to the square of the demagnification ratio. Therefore, demagnification brings about 100-fold increase in output light luminance. Combining the I.I. intrinsic gain and demagnification gain, we can define the gain of an I.I. as the conversion factor. This factor specifies how much output light luminance is per unit exposure rate in units of mR/s. Modern I.I. have conversion factors more than 200 Nit/(mR/s)(for 112-in. I.I.), and when the conversion factor degrades to below 100, the image intensifier should be replaced; otherwise, poor image quality and high radiation doses to the patient and operator may result.

In addition to image intensification, an image intensifier should also provide images without much blurring and distortion. The function of electron optics of an I.I. is to ensure that the photoelectrons emitted by the photocathode are focused on the output phosphor with tolerable distortion. Three additional electrodes (G1, G2, and G3 in Figure 26.17) inserted between photocathode and output screen implement the electron optics. These five pieces form the so-called pentrode structure of electron optics. The G3 electrode allows one to zoom image by reducing the input screen portion projected onto the output phosphor. For example, by increasing G3 voltage, we can make a smaller portion of input phosphor project onto the output phosphor and the demagnification ratio reduces. Because the size of output phosphor is fixed during adjusting G3 voltage, reduction of the demagnification ratio magnifies the images.

Therefore image intensifiers of any given size always have magnification modes corresponding to smaller entrance sizes. For example, it is common that a 12-in. I.I. has magnification modes for 9-in. and 6-in. sizes. The G2 electrode is the focusing electrode. When fluoroscopy images show losing spatial resolution (say resolution <1 lp/mm for 12-in. I.I.), we usually recommend that G2 voltage (and the TV camera focusing) be checked. The G1 electrode adjusts the resolution uniformity.

Once the x-ray attenuation images have been converted to visible light images at output phosphor, a TV camera is used to generate the video signals for display on a TV monitor (Figure 26.16). The TV camera is optically coupled to the output screen of the I.I. by the so-called conjugate lens. Obviously the CCD device discussed in detail earlier is a good choice for a TV camera; however, the vacuum camera tube is commonly used instead. A camera tube is a vacuum tube of about 1 in. external diameter, consisting of a target plate made of light-sensitive photoconductive materials, e.g., antimony trisulphide ($Sb_2S_3$), an electron gun, electron scanning, and focusing coils. A green light image from I.I. output phosphor is projected by the lens onto the target of camera tube. The target is photoconductive, so one light photon creates a mobile electron–hole pair.

Hence, the local electric resistances of the target are changed according to the illumination intensities received locally. The higher the intensity, the lower the local resistance. In this way, the projected image forms a map of the resistance changes on the target. When an electron beam scans the target line by line (512 or 1024 lines), the electron beam picks up (senses) the resistance changes and converts them into electric current changes. The signal current (the displacement current) ranges from about 100 nA for an average image brightness to 2000 nA for a very bright image. The signal current changes with the illumination intensity as

$$I_s = cE^{\gamma} + I_d \tag{26.20}$$

where $I_s$ is the signal current, $I_d$ is the dark current, and $c$ is a constant. $E$ is the illumination intensity and $\gamma$ determines how fast the signal current changes with $E$. Therefore, $\gamma$ determines the signal transfer characteristics and has important effects on the image contrast scale and the dynamic range. While the beam scans the target line by line, the video signal changes accordingly. The camera tube functions as a signal current source and feeds the video amplifier. The video signal (voltage signal) is mixed with the TV synchronization signals to form the composite video signal and then is amplified. The amplified complete video signal is sent to a TV monitor for image display.

The most commonly used TV camera is the vidicon tube, which uses antimony trisulphide ($Sb_2S_3$) as the target photoconductive material. Vidicon tube has $\gamma$ of 0.7; hence, it has a good dynamic range. By the way, the video amplifier for fluoroscopy is linear, and the TV monitor has $\gamma$ of about 2. The overall system $\gamma$ is thus about 1.4. In addition to $\gamma$, other important parameters of a TV camera are the lag, sensitivity, noise, resolution, and burn resistance. The definition of lag requires explanation. Lag is the inertia that is the residual signal percentage after 20 scans. It is the most important parameter and essentially determines if the camera device can be used in an imaging task. Vidicon has a relatively large lag of 8 to 16%; it cannot be used for angiography and image-guided interventional procedures. However, it has the advantage of reduction of noise (by integration of image frames) for fluoroscopy. The target material mainly decides the lag of a camera tube. For example, plumbicon (PbO target) has a lag of 1% and saticon (SeAsTe target) has a lag of 6%.[16] The CCD camera has a very low lag as well. These camera tubes are used in angiography and interventional procedures. Note that all TV cameras (including CCD) other than vidicon have unit $\gamma$.

Another important aspect of fluoroscopy is automatic exposure control (AEC). Because different anatomical structures may have quite different attenuation, the x-ray factors (kVp and mA) should be adjusted accordingly in real time to keep comparable image brightness and reduce undue radiation exposure to the patient. To achieve this, one must have the means to sense the exposure rate of x-ray incident on I.I. entrance and to keep it constant during fluoroscopy by adjusting the x-ray factors (kV and mA) through a feedback control of the x-ray generator. The target I.I. entrance exposure should be set on a balance of good image quality and reduction of radiation doses to patients and medical staff. According to our experiences from clinical applications, target I.I. entrance exposure rates should be set so that the low contrast resolution (discussed in Section 26.3.4 in terms of SNR) is about 2.5 to 3.5% for features of about 1 cm in size. In our experience the I.I. entrance exposure rates should be set at approximately 1.2 to 2.4 mR/min for a 12-in. I.I., and scaled inversely with the square of the ratio of I.I. sizes. For example, the I.I. entrance exposure rates should be about 2.4 to 4.8 mR/min for a 9-in. I.I.

The photocathode current of I.I. and video signal from a TV camera can be used as the sensing signal AEC of fluoroscopy. With AEC, low x-ray factors (low kVp and mA) are used for imaging small and easily penetrated body parts; consequently, the x-ray exposure rate to the patient is relatively low as well. As the attenuation of imaged body parts increases, AEC automatically drives x-ray factors (kVp and mA) up in an amount just enough to maintain comparable image brightness. In this way constant average image brightness is achieved and inadequate radiation exposure to patients and medical staff is avoided. Of course, for the sake of radiation protection, the maximum exposure rate is limited. As increasing attenuation drives AEC to the maximal exposure rate, one must increase the video amplifier's gain to keep image brightness approximately constant. This video gain adjustment is implemented by the automatic gain control (AGC) circuitry. Note that the maximal exposure rate at patient skin entrance should not exceed 10 R/min according to U.S. federal regulations. If the x-ray system has a high-level control and special means of its activation, then the maximum exposure rate at the patient skin entrance should not exceed 20 R/min when high-level control is activated, also according to U.S. federal regulations.

## 26.3.4 Signal-to-Noise Ratio Analysis

In radiological imaging, x-ray quantum efficiency and corresponding imaging characteristics such as SNR, noise power spectrum (NPS), and DQE are crucial to design considerations and performance evaluations such as lesion detectability and patient radiation dose. The following modeling of SNR, NPS, and DQE provides a comprehensive summary of theoretical and of practical aspects of radiological imaging apparatus. Here we use the optically coupled CCD system described in the previous subsection as an example for analysis. This concept can be applied to many other radiological imaging systems.

In an optically coupled CCD x-ray imaging system (no image intensifier in the imaging chain), the transfer of light from an x-ray intensifying screen to a CCD imager can be accomplished by directed contact of these components. However, currently available CCDs require a demagnifying optical coupler (either a lens or fiber taper) between the intensifying screen and the detector to image an area larger than the size of the detector. In the quantum transfer of image data between an intensifying screen and imager, the optical coupling component represents a weak link in the cascaded imaging chain. Thus, in the 1960s and 1970s, image intensifiers were considered integral components for achieving quantum noise limitation in electronic imaging systems. To determine whether this assumption holds for newer electronic detectors, we derived an equation for analyzing the SNR for these systems.

The noise in an optically coupled CCD system is composed of quantum noise (x-ray quantum noise and secondary quantum noise generated through the cascaded process) and additive noise (noise related to the detector or to the detector electronics). Although each noise component can be analyzed separately using existing mathematical methods,[17] we have derived an SNR equation encompassing the combined effects of quantum and additive noise.[18] This equation was derived based on the principle of cascaded imaging analysis and other SNR models.[17] Although it can be applied to many imaging systems, it is particularly convenient for analyzing performance or optimizing design tradeoffs for optically coupled CCD x-ray imaging systems. The details and derivation of this equation can be found in Reference 18.

$$SNR = C(\eta N_i)^{1/2}\left[1+\frac{1}{g_1 g_2}+\frac{1}{g_1 g_2 g_3}+\frac{N_a^2}{(g_1 g_2 g_3)^2 \eta N_i}\right]^{-1/2} \tag{26.21}$$

where $\eta$ is the quantum efficiency of the intensifying screen or other scintillator used in the imaging chain.

It is assumed that all the statistical processes of absorption, photon transport within the screen, and photoemission can be described and characterized by one number: the quantum efficiency $\eta$. Also, $N_i$ is the x-ray photon flux, which is the number of x-ray photons per pixel at the entrance of the intensifying screen; $N_a$ is the total additive noise of the overall imaging chain; $g_1$ is the x-ray-to-light conversion ratio, or quantum gain of the intensifying screen; $g_2$ is the optical coupling efficiency of the lens, fiber taper, or other coupler; $g_3$ is the quantum efficiency of the CCD or other electronic detector; and $C$ is the subject contrast. The product of $g_1 g_2 g_3$ is the overall quantum efficiency, or total quantum gain of the system; its unit is the number of electrons produced in the CCD per x-ray photon absorbed in the scintillator (electrons per x-ray photon).

The quantum efficiency of a Min-R medium mammographic screen has been estimated at $\eta = 0.65$.[19] Assuming a typical mammographic screen entrance exposure of 12 mR and an average x-ray photon energy of 20 keV, the number of x-ray photons per pixel at the entrance of the intensifying screen for a detector having a 0.048 × 0.048 mm pixel size is $\eta N_i = 1600$ x-ray photons/pixel. Therefore, the number of absorbed x-ray photons per pixel is $\eta N_i = 1040$ x-ray photons/pixel. For this analysis, a contrast ($C$) of 1 was assumed. For a system using a cooled CCD, the total additive noise $N_a$ is usually less than 15 electrons; for TV tubes, $N_a$ is usually larger than 500 electrons. The x-ray-to-light conversion ratio ($g_1$) of a high-resolution mammography intensifying screen is reported at 400 to 600 in the forward direction.[20] Conservative estimates of $g_3$ are (1) 0.35 for a front-illuminated CCD; (2) 0.60 for a back-illuminated CCD such as the Tektronix (TK2048EB); and (3) 0.20 for a TV tube. From our previous work,[21] we have estimated the optical coupling efficiency of a lens to be:

$$g_2 = \frac{0.75}{4F^2(1+M)^2 + 1} \tag{26.22}$$

where $F$ is the F-number (the ratio of the focal length to the effective diameter) of the lens and $M$ is the demagnification ratio. For example, for an $F/0.8$ ($F = 0.8$) lens working at a demagnification factor of 2, $g_2 = 3.1\%$.

Equation 26.21 can be used to illustrate the dominant noise component in a cascaded electronic imaging system. For example, if the total quantum gain, $g_1g_2g_3$, of the imaging chain is small or the additive noise, $N_a$, is large, the value of the denominator is larger than 1. Then the SNR of the system is dominated by one of the cascaded stages or by additive noise. While $g_1g_2g_3$ is small, the additive noise, $N_a$, will affect the performance of the system significantly. As $g_1g_2g_3$ becomes large, the value of the denominator approaches 1 and the equation reduces to SNR = $C/\eta N_i$, representing a perfect x-ray quantum noise limited, additive noise-free imaging system.

To illustrate an example application of Equation 26.21, we use parameters typical for lens-coupled TV camera systems, which were extensively investigated in the 1960s and 1970s. Substituting $g_1g_2g_3$ = (500)(0.031)(0.2) = 3.1 and $N_a = 500$ into Equation 26.21, we have:

$$SNR = \frac{C(\eta N_i)^{1/2}}{(1.32+25)^{1/2}} = 0.2C(\eta N_i)^{1/2} \tag{26.23}$$

This example illustrates that the lens-coupled TV system is not x-ray quantum noise limited and that it is the additive noise ($N_a$), not the total quantum gain ($g_1g_2g_3$), that dominates the SNR.

The introduction of the I.I., as discussed earlier, provided a solution to the additive noise problem by boosting the image signal and increasing quantum gain. The value for $g_1g_2g_3$ may approach 400; therefore, the SNR of an I.I. TV system can be expressed as:

$$SNR = \frac{C(\eta N_i)^{1/2}}{(1+0.004)^{1/2}} = 0.99C(\eta N_i)^{1/2} \tag{26.24}$$

The I.I. TV system is absolutely x-ray quantum noise limited and both analog and digital I.I. TV systems are used currently in many clinical procedures. However, electro-optical devices such as image intensifiers have limitations that reduce spatial resolution and contrast sensitivity. Mammography, which requires high spatial resolution and high contrast sensitivity, cannot be performed using a conventional I.I. TV system. Examining Equations 26.21 through 26.24 more closely, it is theoretically intuitive that x-ray quantum noise limitation might be achieved **without** an image intensifier if the additive noise level were greatly reduced. Modern CCD receptors, with additive noise as low as 15 electrons, are suitable

**TABLE 26.3** Total Quantum Gain and Additive Noise and Their Impact on SNR

| | Lens-TV | I.I.-TV | Lens-CCD |
|---|---|---|---|
| Additive noise $N_a$ | 500 Electrons | 500 Electrons | 15 Electrons |
| Quantum gain $g_1 g_2 g_3$ | 3.1 Electrons/x-ray | 400 Electrons/x-ray | 9.3 Electrons/x-ray |
| SNR | $0.2C(\eta N_i)^{1/2}$ | $0.99C(\eta N_i)^{1/2}$ | $0.95C(\eta N_i)^{1/2}$ |

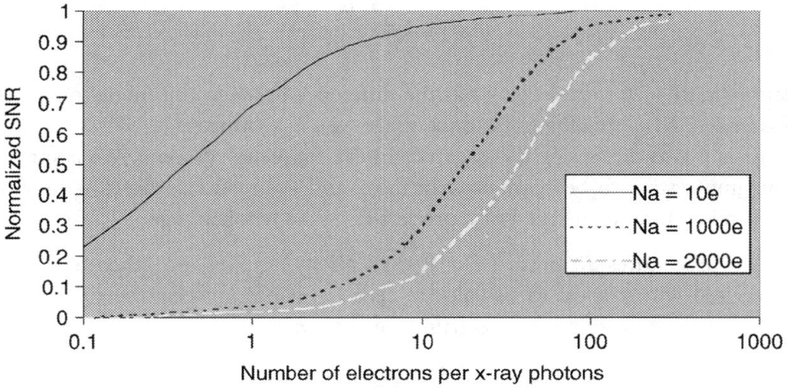


**FIGURE 26.18** Normalized SNR curves plotted as functions of total quantum gain (electrons/x-ray) and additive noise (electrons).

detector candidates. For current CCD devices, a typical total quantum gain can be expressed as $g_1 g_2 g_3$ = (500)(0.031)(0.6) = 9.3 electrons/x-ray photon. Using Equation 26.21, the SNR is calculated to be:

$$SNR = \frac{C(\eta N_i)^{1/2}}{(1.1 + 0.002)^{1/2}} = 0.95C(\eta N_i)^{1/2} \tag{26.25}$$

Therefore, the degradation in SNR caused by the cascaded stages and the additive noise is nearly negligible for a cooled CCD system. It illustrates that an optically coupled CCD imaging system can indeed be x-ray quantum noise limited.

Assuming a contrast ($C$) of 1 and a constant phosphor noise level for all three systems, the preceding analyses can be summarized as shown in Table 26.3. Note that, in this table, the value of $g_1 g_2 g_3$ for the lens TV was calculated assuming the quantum efficiency of the TV tube is 0.2. The value of $g_1 g_2 g_3$ for the lens CCD was calculated assuming the quantum efficiency of a back-illuminated CCD to be 0.6. Figure 26.18 plots the relationships of SNR to the total quantum gain (electrons/x-ray) and total additive noise (electrons) for three detector systems and illustrates that a total quantum gain of ten electrons per absorbed x-ray makes a CCD imaging system x-ray quantum noise limited at the low-frequency range. According to our calculations and experiments, it is feasible to design a lens-coupled or a fiber optically coupled CCD x-ray imaging system to meet these requirements under clinical conditions. In comparison, other electronic detectors have noise levels of a few hundreds or even a thousand electrons.[22] To achieve x-ray quantum noise limitation under these conditions, a total quantum gain of 100 to 200 electrons/absorbed x-ray photon is needed.

NPS analysis of image noise and noise transfer provides a complete description of noise. When characterizing the noise properties of real or proposed digital mammographic systems experimentally, we calculate the NPS directly from the noise fluctuations using the fast Fourier transform (FFT):

$$NPS(f_x, f_y) = FFT_{2D}[N(x, y)] \tag{26.26}$$

where $FFT_{2D}$ represents the fast Fourier transform and $N(x, y)$ is a two-dimensional data array of noise fluctuation or a so-called "noise-only" image. This image is acquired using the following procedure:

1. Acquire images $A(x, y)$ and $B_1(x, y)...B_{50}(x, y)$ under identical conditions simulating clinical mammography procedures. A 4.5-cm-thick lucite slab is used and the entire imaging field is uniformly irradiated.
2. Determine the "noiseless" image as an average of 50 images:

$$\langle B(x,y)\rangle = \frac{\sum B_n(x,y)}{50} \tag{26.27}$$

3. Calculate $N(x, y) = A(x, y) - B(x, y)$ as the difference between the single image $A(x, y)$ and the noiseless image; $N(x, y)$ is the noise-only image we are seeking.
4. $NPS(f_x, f_y)$ is a two-dimensional data array in the frequency domain. We assume the proposed CCD imaging system to be symmetric in the $x$ and $y$ directions; therefore, a slice of the data, $NPS(f_x)$, is taken to evaluate the noise properties of the whole image.

Repeating these procedures 20 times ($n = 20$), 20 $NPS(f_x)$ curves are generated. Furthermore, the $NPS(f_x)$ is normalized with respect to its value at zero frequency, making the value at zero frequency equal to 1; the normalized noise power spectrum is given by:

$$NPS'(f_x) = \frac{1}{20}\sum \frac{NPS(f_x)}{NPS(0)} \tag{26.28}$$

The averaging process in Equation 26.27 is usually necessary to avoid statistical uncertainty.

The preceding discussion on noise power spectrum provides a foundation for understanding the concept of noise propagation in a cascaded imaging chain, and its relation to the SNR and DQE. The measured NPS curves of various optically coupled CCD x-ray imaging systems can be found in the literature.[23,24]

The DQE expresses the SNR transfer characteristics of an imaging system as a function of the spatial frequency:[23,25]

$$DQE(f) = \frac{SNR_{out}^2(f)}{SNR_{in}^2(f)} \tag{26.29}$$

where $SNR_{out}(f)$ is the output SNR and $SNR_{in}(f)$ is the input SNR. The $DQE(f)$ provides a measure of how efficiently the imaging system transfers information in terms of spatial frequency. In order to determine $DQE(f)$, we conducted the following derivation:

$$SNR_{out}^2(f) = \frac{S_{out}^2(f)}{NPS(f)} = \frac{S_{out}^2(0)MTF^2(f)}{NPS(0)nps(f)} = \frac{SNR_{out}^2(0)MTF^2(f)}{nps(f)} \tag{26.30}$$

where $S_{out}(f)$ is the output signal. It can be expressed as the product of its zero frequency value, $S_{out}(0)$, and $MTF(f)$, the norm of the system transfer function.[26] Similarly, the $NPS(f)$ can be expressed by its normalized form, $nps(f)$, and the normalization factor $NPS(0)$. Assuming the input x-ray quanta obey Poisson statistics and have a flat noise spectrum within the range of the spatial frequency of practical interest, then

$$SNR_{in}^2(f) = SNR_{in}^2(f) = CN_i \tag{26.31}$$

where $N_i$ is the incident x-ray photon flux (number of x-ray photons per pixel at the entrance of the intensifying screen). Thus,

$$DQE(f) = \frac{DQE(0)MTF^2(f)}{nps(f)} \tag{26.32}$$

where the ratio $MTF^2(f)/nps(f)$ provides a spatial frequency modulation term for *DQE*.

*DQE*(0) can be determined using the following method. We have shown that, at zero frequency, repeating Equation 26.21 above:

$$SNR_{out}(f) = C(\eta N_i)^{1/2}\left[1 + \frac{1}{g_1 g_2} + \frac{1}{g_1 g_2 g_3} + \frac{N_a^2}{(g_1 g_2 g_3)^2 \eta N_i}\right]^{-1/2}, \tag{26.33}$$

$$SNR_{in}(f) \equiv CN_i^{1/2}$$

Therefore,

$$DQE(0) = \frac{SNR_{out}^2(0)}{SNR_{in}^2(0)} = \frac{\eta}{\left[1 + \frac{1}{g_1 g_2} + \frac{1}{g_1 g_2 g_3} + \frac{N_a^2}{(g_1 g_2 g_3)^2 \eta N_i}\right]} \tag{26.34}$$

Analyses of *DQE*(0) and of *SNR* at low frequency provide a useful tool for evaluating design tradeoffs for electronic x-ray imaging systems, such as choice of scintillator, optical coupling techniques, and electronic imagers.

Using Equation 26.34, the *DQE*(0) values for several electronic x-ray imaging systems were plotted as a function of total quantum gain ($g_1g_2g_3$) and the additive noise ($N_a$) in Figure 26.19. For a low additive noise system such as the optically coupled CCD, a total quantum gain of ten electrons per x-ray brings the DQE close to its maximum value. Further increasing the total gain will not improve the DQE significantly. The preceding analysis applies to the low-frequency range. To preserve high-frequency DQE and to ensure that the system is x-ray quantum limited at high frequencies, a total quantum gain of ten electrons per x-ray is again required.[23] These guidelines have been used in developing both lens- and fiber optically coupled CCD systems.

For a lens-coupled prototype, a large aperture lens was custom designed and a back-illuminated CCD used that had a quantum efficiency nearly two times greater than a front-illuminated CCD and met the required ten electrons/x-ray. A fiber optically coupled CCD could be designed with a total quantum gain

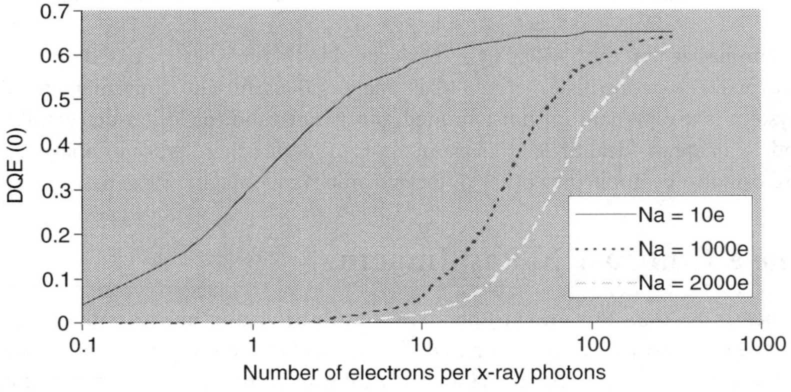


**FIGURE 26.19** Detective quantum efficiency curves plotted as functions of total quantum gain (electrons/x-ray) and additive noise (electrons). The quantum efficiency of the scintillating screen was assumed to be $\eta$ = 0.65. Note that the DQE values presented in this figure are the maximum achievable values.

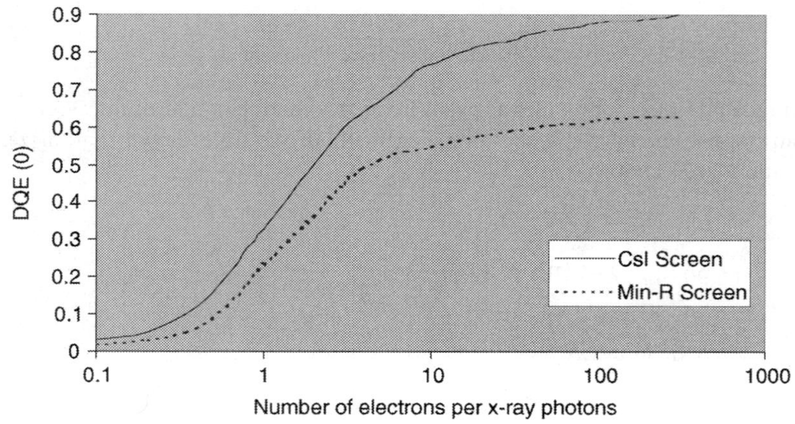


**FIGURE 26.20** The effect of the quantum efficiency of the scintillating screen ($\eta$) on the x-ray quantum efficiency for the overall system. In this figure, DQE(0) curves of two optically coupled CCD systems using different scintillator are plotted as a function of the quantum efficiency of the scintillator screen, ($\eta$), and total quantum gain (electronics/x-ray). Note that the DQE values presented in this figure represent the maximum achievable values.

much higher than ten electrons per x-ray due to the high efficiency of the optical fiber.[27] In a practical design tradeoff, however, we selected an optical fiber component with heavy extra-mural absorption material having only moderate efficiency in return for improved contrast and lower cost.[28] The curves presented in Figure 26.19 also show that 100 to 200 electrons per x-ray are needed to maximize the DQE for other electronic imaging systems with higher detector noise.

The quantum efficiency $\eta$ of the scintillating screen plays an important role in DQE and radiation efficiency. According to Equation 26.34, the maximum achievable DQE of an x-ray quantum noise limited system is limited primarily by $\eta$. In practice, a variety of scintillating screens are used in radiological imaging. In some of our early optically coupled CCD spot mammography systems,[28] Min-R screens with a quantum efficiency of 55 to 65% ($\eta = 0.55$ to 0.65) were used. Recently developed cesium iodide crystal offers a higher efficiency ($\eta = 0.90$ or higher), leading to a higher system DQE.[29] In Figure 26.20, the DQE(0) curves of two optically coupled CCD mammographic systems are plotted as a function of total quantum gain and the quantum efficiency of the scintillating screen. (Note that the DQE values presented in this figure represent the maximum achievable values.) This figure demonstrates that the scintillating screen used will significantly affect the DQE of the overall system. Based on the DQE analysis, we have developed a high-resolution digital imaging system; Figure 26.21 shows a bone image acquired by using this system.[29]

We wish to emphasize that the SNR analyses performed in this chapter are based on single pixel values (x-ray photons/pixel, electrons/pixel). Single pixel analysis is useful and convenient for determining if an imaging system is x-ray quantum noise limited, but is not valid for lesion detectability evaluations. The Rose model[30] or quasi-ideal SNR[31,32] is more appropriate for these types of analyses. The methods presented here can also be applied to other optoelectronic x-ray imaging systems.

## 26.4 Phase Contrast X-Ray Imaging

As we mentioned in Section 26.1, the in-line phase contrast imaging measures the Laplacian of phase $\nabla^2\phi$ directly, while the x-ray interferometry and diffraction-enhanced imaging directly measure the phase $\phi$ and phase gradient $\nabla\phi$, respectively. Using an x-ray tube as x-ray source, one can perform the in-line phase contrast imaging, so this mode holds great potential for clinical applications. Although this imaging modality is still in its infancy, we would like to discuss it in detail, considering its great potential in biomedical diagnostic imaging.

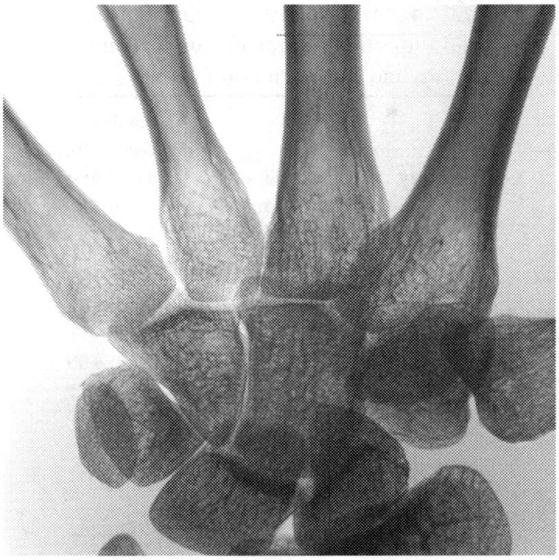

**FIGURE 26.21** Image of bones acquired by a prototype high-resolution digital x-ray imaging system.

The setting of in-line phase contrast imaging is similar to conventional radiography, but the object-detector distance is larger than that in CR. Consider a plane wave x-ray source. If the x-ray projection is along the $z$-axis direction, we can then model the phase shift and attenuation effects of a body part as a two-dimensional transmission function $q(x, y)$ in the $x$-$y$ plane

$$q(x,y) = \exp\left(i\phi(x,y) - \frac{\mu(x,y)}{2}\right) \quad (26.35)$$

where the phases $\phi(x,y)$ and $\mu(x,y)$ are the $z$-projection of the object phase and linear attenuation coefficient. More specifically, if the object has three-dimensional distributions of refraction index decrement $\delta(x, y, z)$ and linear attenuation coefficients $\mu(x, y, z)$), then the projected phase and attenuation are, respectively,

$$\phi(x,y) = \frac{2\pi}{\lambda}\int \delta(x,y,z)dz$$

and

$$\mu(x,y) = \int \mu(x,y,z)dz \quad (26.36)$$

Assume a plane x-ray wave $\exp(-ikz)$, where $k = 2\pi/\lambda$, the wave number, and $\lambda$ is the x-ray wavelength, illuminates the object at $z = 0$. Of course, in this scenario the object is assumed to be thin for x-ray so that the projection approximation holds. It can be shown that, as long as the size of the finest feature is larger than $\sqrt{\lambda}\sqrt{T}$, the object can be deemed thin. Here $T$ is the object thickness.

Table 26.4 lists the maximally allowed object thickness for valid projection approximation for different resolutions and x-ray photon energies. It is clear from this table that human body parts can be treated as thin objects for resolutions of 10 μm and photon energy of 10 to 150 keV. In general, the incident x-ray will be refracted and diffracted by the object. For diagnostic x-ray, the x-ray wavelength is much smaller to image than object features. The diffraction angles due to tissue structure are very small — in milliradians. Therefore, the diffracted x-ray wave field can be described by the small angle (i.e., paraxial) approximation of the Fresnel diffraction theory:[33]

**TABLE 26.4** Maximally Allowed Object Thickness for Valid Projection Approximation for Different Resolutions and X-Ray Photon Energies

| X-Ray Energy (keV) | Object's Finest Feature (mm) | Maximum Allowed Object Thickness (m) |
|---|---|---|
| 20 | 1.00E-02 | 8.06E-01 |
| 20 | 1.00E-03 | 8.06E-03 |
| 50 | 1.00E-02 | 2.02E+00 |
| 50 | 1.00E-03 | 2.02E-02 |
| 100 | 1.00E-02 | 4.03E+00 |
| 100 | 1.00E-03 | 4.03E-02 |

$$f(x,z)=\left(\frac{i}{\lambda z}\right)^{1/2}\exp(-ikz)\int q(X)\exp\left(-ik\frac{(x-X)^2}{2z}\right)dX \tag{26.37}$$

where $z$ is the distance in wave propagation from the object (body part) and $X$ is the coordinates in transverse plane. For the sake of concise notation, we omit the $y$-dimension without loss of generality. If we knows the object transmission function, we can then calculate the integral in Equation 26.37 and find the transmitted x-ray wave field $f(x, z)$ at $z$. The x-ray detector used in imaging is x-ray phase insensitive; therefore, what is detected is the intensity of transmitted x-ray at $z$:

$$I(x,z)=ff^{*}=|f(x,z)|^2 \tag{26.38}$$

Obviously, in conventional x-ray imaging the diffraction effects are negligible; then $f(x, z) = q(x)\exp(-ikz)$. Using Equation 26.35, we find

$$I(x,z)=|f(x,z)|^2=|q(x)|^2=\exp(-\mu(x)) \tag{26.39}$$

This $I(x, z)$ is exactly the conventional radiographic image of the object $q(x)$. Of course, the object does really diffract x-ray, and the diffracted wave $f(x, z)$ is different from $q(x)$ in general; thus, the detected $I(x, z)$ at $z$ can be considered an in-line hologram, and $I(x, z)$ is, in general, a complicated function of $q(x)$. This $I(x, z)$ is the phase contrast image of the object.

In order to explore the general features of phase contrast imaging, we can treat the Fresnel diffraction process as a linear filtering of the object transmission function.[33] The concept of linear filtering is not new to x-ray imaging. In fact (as mentioned in Section 26.4), a conventional x-ray image is a tissue attenuation map filtered by the imaging system's optical transfer function (e.g., the MTF analysis, etc.). Here a phase contrast image is a map of object transmission function filtered by a linear filter corresponding to the Fresnel diffraction process. In fact, the Fresnel integral for diffracted x-ray wave field $f(x, z)$ is of convolution type and the Fourier transform (FT) of $f(x, z)$ (Equation 26.37) with respect to $x$ is

$$F(u)=\exp(-ikz)Q(u)\exp(i\pi\lambda zu^2) \tag{26.40}$$

where $u$ is the spatial frequency in the object plane, and $F(u, z)$ and $Q(u)$ are the FTs of x-ray wave field $f(x, z)$ and object transmission function $q(x)$, respectively. From Equation 26.40 it is clear that the linear filter $\exp\{i\pi\lambda zu^2\}$ acting on the transmitted frequencies accounts for the Fresnel diffraction in the in-line phase contrast imaging. To see the effects of this filter, consider a pure phase object with

**TABLE 26.5** Phase Transfer Parameter as a Function of X-Ray Photon Energy, Object Detector Distance, and Targeted Spatial Frequency with a Plane Wave X-Ray Source

| X-Ray (keV) | X-Ray Wavelength $\lambda$ (nm) | Object-Detector Distance $Z$ (m) | Target freq · (1p/mm) | $\chi = \pi\lambda zu^2$ | At Target Frequency $\sin(\chi)$ |
|---|---|---|---|---|---|
| 10 | 1.24E-01 | 0.5 | 20.0 | 7.79E-02 | 7.78E-02 |
| 20 | 6.20E-02 | 0.5 | 20.0 | 3.90E-02 | 3.89E-02 |
| 50 | 2.48E-02 | 0.5 | 20.0 | 1.56E-02 | 1.56E-02 |
| 150 | 8.27E-03 | 0.5 | 20.0 | 5.19E-03 | 5.19E-03 |

small phase $\phi(x)$. In this case, the object transmission function can be approximated by using Equation 26.35:

$$q(x) \approx 1 + i\phi(x) \tag{26.41}$$

and the FT of $q(x)$ is

$$Q(u) \approx \delta(u) + i\Phi(u) \tag{26.42}$$

where $\delta(u)$ is the $\delta$-function and $\Phi(u)$ the FT of object phase $\phi(x)$. Substituting Equation 26.42 into Equation 26.40, one can find the FT of the diffracted x-ray wave field:

$$F(u) \approx \exp(-ikz)\{\delta(u) + i\Phi(u)\}\exp(i\pi\lambda zu^2) \tag{26.43}$$

and $F(u)$ can be further written as

$$F(u) \approx \exp(-ikz)\{\delta(u) - \Phi(u)\sin(\pi\lambda zu^2) + i\Phi(u)\cos(\pi\lambda zu^2)\} \tag{26.44}$$

In x-ray diagnostic imaging the energy of x-ray photons ranges from 10 to 150 keV, and the corresponding wavelength of x-rays varies from 0.124 to 0.0083 nm (from Equation 26.1 of Section 26.1.) For clinical applications the object detector distance $z$ cannot be larger than 1 m, and the maximum spatial resolution needed is about 20 lp/mm. We have calculated the phase transfer factor $\sin\chi$ as a function of x-ray photon energy, object detector distance and targeted spatial frequency with a plane wave x-ray source. The calculations are for cases relevant to clinical imaging and the results are shown in Table 26.5. This being so, $\sin(\pi\lambda zu^2) \approx \pi\lambda zu,^2$ $\cos(\pi\lambda zu^2) \approx 1$, and the FT of the diffracted x-ray wave becomes

$$F(u) \approx \exp(-ikz)\{\delta(u) - \pi\lambda zu^2\Phi(u) + i\Phi(u)\} \tag{26.45}$$

Using inverse FT, we find the diffracted x-ray wave field at $z$:

$$f(x,z) \approx \exp(-ikz)\left\{1 + \frac{\lambda z}{4\pi}\nabla^2\phi(x) + i\phi(x)\right\} \tag{26.46}$$

Using Equation 26.38, we find the phase image intensity for a pure phase object of small phase to first order in $\phi(x)$:

$$I(x,z) = |f(x,z)|^2 = 1 + \frac{\lambda z}{2\pi}\nabla^2\phi(x) \tag{26.47}$$

This result shows that the detected image contrast will be proportional to $\nabla^2\phi$, the Laplacian (second derivatives) of the object's projected phase $\phi$. Remember that $\phi$ is related to the tissue refractive index decrement $\delta$ by Equation 26.36, and is related to tissue composition by Equations 26.9 and 26.10 in Section 26.1. Away from the x-ray absorption edge of tissue, the object phase $\phi(x)$ is proportional to tissue electron density, and the phase contrast of image is proportional to the Laplacian of tissue electron density. Therefore, the boundary of areas with different tissue electron densities will be greatly enhanced, as if the images have been edge enhanced by using digital processing.

In addition, it is clear from the preceding equation that the phase contrast is proportional to the object detector distance. Apparently, if this distance $z = 0$, no phase contrast will be seen. Moreover, it is easy to see (Equation 26.47) that this technique works even with polychromatic x-ray sources such as x-ray tubes. In Section 26.2 we mentioned that x-ray from a x-ray tube consists of the bremstrahlung and characteristic radiation. The bremstrahlung is broadband x-ray; however, because the phase contrast here is not interferometry in nature, but rather proportional to the Laplacian of tissue electron densities, the generated phase contrast by photons of different wavelengths simply adds up, as long as these x-ray photons propagate along almost the same path.

It should be noted that, although temporal coherence (monochromacity) is not required, spatial coherence is critical for phase imaging. In practice a plane wave source emits a bundle of plane waves with a divergent angle $\alpha$. A plane wave x-ray beam's spatial coherence is characterized by the so-called lateral coherence width $2\lambda/\alpha$. The diffraction patterns generated by tissue from individual plane waves can blur each other. Obviously, the larger the lateral coherence width, the better phase image resolution. A quantitative analysis of the phase image resolution can found in Pogany et al.[33] Figure 26.22 is a phase contrast image of a mouse sample that clearly shows the edge enhancement of anatomic structures.

For clinical imaging applications, an x-ray tube is more likely to be used. For a point x-ray source the x-ray wave field from the source is spherical. Suppose the source object distance (SOD) is $R_1$, and the source detector distance (SID) is $R_1 + R_2$. The Fresnel diffraction wave field formula in this case is the same as that for plane wave (Equation 26.37) except to replace $z$ with $R_2/M$ and $x$ by $x/M$, where $M = (R_1+R_2)/R_1$ is the geometric magnification factor. For a point-like source such as an x-ray tube, the spatial coherence criteria for phase imaging require very small focal spot sizes of x-ray tubes in few tens of microns or smaller. As discussed in Section 26.2, conventional x-ray tubes have focal spot sizes ranging from 0.1 to 1.5 mm. An x-ray tube with a tiny focal spot is also limited in operating tube current and therefore limited in output exposure rates. These problems are the focus of current active research in phase contrast imaging.

On the theoretical analysis of phase contrast imaging, note that the phase image intensity formula of Equation 26.47 is derived under a weak phase assumption ($\phi(x) \ll 1$). One derives a similar formula to Equation 26.47 under the assumption that $\phi(x) \ll 1$ and $\mu(x) \ll 1$. We know that these assumptions will be grossly violated for body parts in possible clinical applications.

From the discussion of tissue attenuation in Section 26.1, even for a kVp as high as 120 kVp, average $\mu(x)$ for an anterior–posterior projection of an abdomen will be approximately 4 to 6. A phase change of $2\pi$ can result from a few tens of micron thickness of breast tissue for x-rays in mammography. Therefore, the phase image intensity formula of Equation 26.47 is invalid for these cases. With a goal for analyzing phase contrast in clinical imaging, we have recently derived a phase image intensity formula for a point-like x-ray source. This formula is valid for any strong or weak object phase $\phi(x)$ and attenuation $\mu(x)$.[34]

## Acknowledgments

The authors are grateful to Dr. Steve Wilkins for providing the phase contrast image of the mouse sample and to Dr. Yuhua Li for help with graphics.

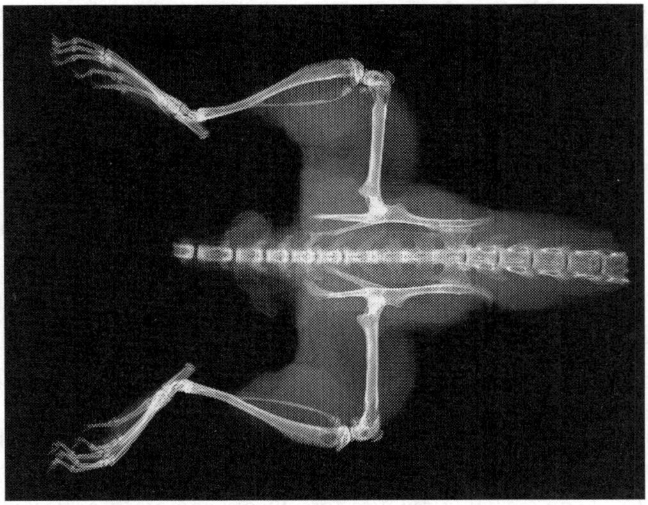

**FIGURE 26.22** Phase contrast image of a mouse sample. (Image courtesy of Steve Wilkins, Commonwealth Scientific & Industrial Research Organization, Sydney, Australia.)

## References


1. Hubbel, J.H., Photon mass attenuation and energy-absorption coefficients from 1keV to 20 MeV, *Int. J. Appl. Radiat. Isot.*, 33, 1269, 1982.
2. Dyson, N., *X-Rays in Atomic and Nuclear Physics*, Longman, London, 1973.
3. Piestrup, M.A., Wu, X., Kaplan, V.V., Uglov, S.R., Cremer, J.T., Rule, D.W., and Fiorito, D.B., A design of mammography units using a quasi-monochromatic x-ray source, *Rev. Sci. Instrum.*, 27, 2159, 2001.
4. Wilkins, S.W., Gureyev, T.E., Gao, D., Pogany, A., and Stevenson, A.W., Phase-contrast imaging using polychromatic hard x-rays, *Nature*, 384, 335, 1996.
5. Momose, A. and Fukuda, J., Phase-contrast radiograph of nonstained rat cerebellar specimen, *Med. Phys.*, 22, 375, 1995.
6. Chapman, D., Thomlinson, W., Johnson, R.E., Washburn, D., Pisano, E., Gmuer, N., Zhong, Z., Menk, R., Arfelli, F., and Sayers, D., Diffraction enhanced x-ray imaging, *Phys. Med. Biol.*, 42, 2015, 1997.
7. Wu, X., Barnes, G.T., and Tucker, D.M., Effect of filtration and kilovolt peak on image contrast and radiation dose in mammography, *Radiology*, 177, 244, 1990.
8. Wu, X., Gingold, E., Barnes, G.T., and Tucker, D.M., Normalized average glandular dose in molybdenum target-rhodium filter and rhodium target-rhodium filter mammography, *Radiology*, 193, 83, 1994.
9. Gingold, E.L., Wu, X., and Barnes, G.T., Contrast and dose in Mo/Mo, Mo/Rh and Rh/Rh target/filter mammography, *Radiology*, 195, 639, 1995.
10. Kimme-Smith, C., New digital mammography systems may require different x-ray spectra and, therefore, more general normalized glandular dose values, *Radiology*, 213, 7, 1999.
11. Yaffe, M.J. and Rowlands, J.A., X-Ray detectors for digital radiography, *Phys. Med. Biol.*, 42, 1, 1997.
12. Chotas, H.G., Dobbins, J.T., III, and Ravin, C.E., Principles of digital radiography with large-area, electronically readable detectors: a review of basics, *Radiology*, 210, 595, 1999.
13. Matsuura, N., Zhao, W., Huang, Z., and Rowlands, J.A., Digital radiography using active matrix readout: amplified pixel detector array for fluoroscopy, *Med. Phys.*, 26, 672, 1999.

14. Kato, H., Photostimulable phosphor radiography design considerations, in Seibert, J., Barnes, G., and Gould, R., Eds., *Specification, Acceptance Testing and Quality Control of Diagnostic X-Ray Imaging Equipment*, American Association of Physicists in Medicine, New York, 1994, p. 731.
15. De Groot, P., Image intensifier design and specifications, in Seibert, J., Barnes, G., and Gould, R., *Specification, Acceptance Testing and Quality Control of Diagnostic X-Ray Imaging Equipment*, American Association of Physicists in Medicine, New York, 1994, p. 429.
16. Krestel, E., *Imaging Systems for Medical Diagnostics*, Siemens, Berlin, 1990.
17. Macovski, A., *Medical Imaging Systems*, Prentice-Hall, Englewood Cliffs, NJ, 1983.
18. Liu, H., Digital fluoroscopy with an optically coupled charge-coupled device, Ph.D. dissertation, Worcester Polytechnic Institute, Worcester, MA, 1992.
19. Barnes, G.T. and Chakraborty, D.P., Radiographic mottle and patient exposure in mammography, *Radiology,* 145, 815, 1982.
20. Dick, C.E. and Motz, J.W., Utilization of monoenergetic x-ray beams to examine the properties of radiographic intensifying screen, *IEEE Trans Nucl. Sci.*, 28, 1554, 1981.
21. Liu, H., Karellas, A., Moore, S.C., Harris, L.J., and D'Orsi, C.J., Lesion detectability considerations for an optically coupled CCD x-ray imaging system, *IEEE Trans. Nucl. Sci.,* 41, 1506, 1994.
22. Gruner, S.M., CCD and vidicon x-ray detectors: theory and practice (invited), *Rev. Sci. Instrum.*, 60, 1545, 1989.
23. Maidment, A.D.A. and Yaffe, M.J., Analysis of the spatial-frequency-dependent DQE of the optically coupled digital mammography detectors, *Med. Phys.*, 21, 721, 1994.
24. Roehrig, H., Fajardo, L.L., Tong, Y., and Schempp, W.V., Signal, noise and detective quantum efficiency in CCD based x-ray imaging systems for use in mammography, *Proc. SPIE,* 2163, 320, 1994.
25. Nishikawa, R.M. and Yaffe, M.J., Model of the spatial-frequency-dependent detective quantum efficiency of phosphor screens, *Med. Phys.,* 17, 894, 1990.
26. Dainty, J.C. and Shaw, R., *Image Science*, Academic Press, Boston, 1974, p. 312.
27. Liu, H., Karellas, A., Harris, L., and D'Orsi, C., Optical properties of fiber tapers and their impact on the performance of a fiberoptically coupled CCD x-ray imaging system, *Proc. SPIE,* 1894, 136, 1993.
28. Liu, H., Fajardo, L.L., Buchanan, M., McAdoo, J., Halama, G., and Jalink, A., CCD-scanning techniques for full-size digital mammography, *Radiology,* 197, 291, 1995.
29. Liu, H., Fajardo, L.L., Barrett, J.R., Williams, M.B., and Baxter, R.A., Contrast-detail detectability analysis: comparison of a digital spot mammography system and an analog screen-film mammography system, *Acad. Radiol.,* 4, 197, 1996.
30. Rose, A., The sensitivity performance of the human eye on an absolute scale, *J. Opt. Soc. Am.,* 38, 196, 1948.
31. Wagner, R. and Brown, D., Unified SNR analysis of medical imaging systems, *Phys. Med. Biol.*, 30, 489, 1985.
32. Liu, H., Karellas, A., and Harris, L., Methods to calculate lens efficiency in optically coupled CCD x-ray imaging systems, *Med. Phys.*, 21, 1193, 1994.
33. Pogany, A., Gao, D., and Wilkins, S.W., Contrast and resolution in imaging with a microfocus x-ray source, *Rev. Sci. Instrum.*, 68, 2774, 1997.
34. Wu, X. and Liu, H., A general theoretical formalism for x-ray phase contrast imaging, *J. X-Ray Sci. Technol.*, accepted.

# 27

# Optical Pumping and MRI of Hyperpolarized Spins

Xizeng Wu
*University of Alabama at Birmingham*
*Birmingham, Alabama*

Thomas Nishino
*University of Texas Medical Branch*
*Galveston, Texas*

Hong Liu
*University of Oklahoma*
*Norman, Oklahoma*



## 27.1 Introduction

Nuclear spins and their interaction with the electromagnetic field form the basis of nuclear magnetic resonance (NMR). Based on NMR, magnetic resonance imaging (MRI) is one of the two most powerful clinical imaging modalities. To perform MRI, the targeted nuclear spins need to be placed in a strong magnetic field to become polarized. The polarization is proportional to the magnetic field strength in conventional MRI, as is the signal-to-noise-ratio (SNR). The ever-increasing demand on high SNR drives magnetic field strengths of MRI scanners ever higher. Presently, the standard high field strength is 1.5 T (1 tesla [T] = $10^4$ gauss [G]), and a new trend in this new century is redefining the standard to 3 T. Needless to say, the higher the field strength, the more expensive the MRI scanner. Currently, a 3-T MRI scanner costs at least $1 million more than a 1.5-T scanner. Moreover, high field strengths make scanners bulky and aggravate problems such as magnetic susceptibility artifacts and lengthening of the spin–lattice relaxation time.

Ironically, the nuclear spin polarizations at these high magnetic fields are really tiny because thermal spin polarizations are determined by the Boltzmann factor $e^{-h\mu/k_BT}$, where $T$ is the temperature and ν the

0-8493-1116-0/03/$0.00+$1.50

magnetic transition frequency. Here $h$ ($6.6266 \times 10^{-34}$ J/s) is Planck's constant and $k_B$ ($1.38 \times 10^{-23}$ J/K) is Boltzmann's constant. The magnetic transition frequencies are radio frequencies (RF) and range from $10^8$ to $10^9$ Hz in MRI. The resulting polarizations are approximately $10^{-6}$ to $10^{-5}$ only.

On the other hand, the optical photons are much more energetic than the RF quanta in MRI. The optical transition frequencies are much higher ($\nu \sim 10^{14}$ to $10^{15}$ Hz), and the thermal polarizations for optical transitions are of the order of unity. Optical pumping uses light photons to polarize the atomic electron spins first, and then the nuclear spins of noble gases are polarized through the spin exchange between atomic electron spins and nuclear spins. Using optical pumping, nuclear spins of noble gases such as $^3$He and $^{129}$Xe can be polarized to 50% or higher — 100,000 times higher than the proton polarization in conventional MRI. A powerful imaging technique developed in the mid-1990s,[1] MRI of hyperpolarized $^3$He and $^{129}$Xe shows great potential for clinical applications. Moreover, as we will show in this chapter, the SNR of MRI with hyperpolarized spins is almost independent of the magnetic field strength of the scanner. Therefore, relatively inexpensive low field scanners will benefit greatly from this SNR characteristic.

We discuss optical pumping from a biomedical photonics perspective in detail in Section 27.3. The special features of hyperpolarized spin MRI, such as the signal intensity estimates, the RF flip angle optimization, and diffusion-associated signal attenuation, are covered in Section 27.4. This relatively short chapter is not intended to present a comprehensive review of the field; rather, the emphasis is placed on working principles, analytical methods, and general techniques of optical pumping and MRI of hyperpolarized spins. A brief introduction to MRI is also presented in Section 27.2.

## 27.2 MRI Basics

### 27.2.1 Nuclear Magnetism

All atoms are composed of protons, neutrons, and electrons. The nucleus of an atom consists of protons and neutrons. Nuclei with an odd number of protons or neutrons possess a net nuclear spin-angular momentum. These nuclei have a magnetic moment $\mu$ that characterizes the magnetic field localized around the nucleus. This magnetic field is analogous to the magnetic field generated by a bar magnet.

In free space at room temperature, these magnetic moments (dipoles) are randomly oriented in space due to thermal fluctuations. However, in the presence of an external static magnetic field, these dipoles are inclined to align with the magnetic field. For hydrogen protons, the nuclear spin quantum number $I = 1/2$. The external magnetic field, denoted by $B_0$, creates two different energy states for hydrogen protons. Each energy state is identified by the magnetic quantum number $m_s = \pm 1/2$. The energy state $m_s = +1/2$ has a component parallel to the external magnetic field and has a lower energy than the energy state $m_s = -1/2$, which has a component antiparallel to $B_0$.

The phases of the magnetic moments are randomly distributed for both energy states with each ensemble of spins forming the surface of a cone. The net magnetization is the vector sum of all the individual magnetic moments. At room temperature, both energy states have approximately the same number of spins, which results in zero net longitudinal magnetization along the $z$-axis. Furthermore, because of the lack of phase coherence, no net transverse magnetization exists in the $x$-$y$ plane.

On the other hand, in the presence of a strong external magnetic field (e.g., 1.5 T = 15,000 G, which is about 20,000 times the Earth's magnetic field), the populations in the two energy states are no longer equal. In fact, the lower energy state will have an excess population of about 3 spins per million at 1.0 T. A more general formula for the thermal equilibrium polarization will be given in Equation 27.37. Thus, in 1 cc of water at 1.0 T there are roughly $10^{15}$ excess spins in the lower energy level. The number of excess spins in the lower energy state is directly related to the external magnetic field strength. This imbalance of spins in the two energy states gives rise to a net magnetization along the external magnetic field. Of course, thermal noise will cause spins to go from one energy state to the other, but as time passes, an equilibrium longitudinal magnetization, $M_0$ (along the $z$-axis), will be produced. Because the human body is composed primarily of water ($H_2O$), hydrogen (protons) is the preferred imaging isotope because of its relative abundance and sensitivity compared with other atomic nuclei found in the human body.

### 27.2.2 Magnetic Resonance

Resonance is a phenomenon through which energy can be transferred between objects or systems. In MRI, resonance refers to the induction of transitions between different energy states ($I = \pm 1/2$ for protons) by an RF wave with its magnetic field perpendicular to the static magnetic field $B_0$. The frequency of the RF wave must be such that the energy of the resonant RF quanta is equal to the difference in energy between the levels, which is given by:

$$\Delta E = \gamma \hbar B_0 \tag{27.1}$$

where $h = h/2\pi$ is the reduced Planck's constant. The resonance frequency, the so-called Larmor frequency $f_L$, is

$$f_L = \frac{\gamma}{2\pi} B_0 \tag{27.2}$$

where $\gamma$ is the gyromagnetic ratio specific to the nuclear isotope, for hydrogen, $(\gamma/2\pi) = 42.58$ MHz/T. This energy will cause the proton magnetic moments to "flip" from the lower energy state ($m = +1/2$, parallel to $B_0$) to their higher energy state ($m = -1/2$, antiparallel to $B_0$). By doing so, the net longitudinal magnetization can be diminished, reduced to zero, or even reversed entirely, depending on the amount of energy deposited to the protons.

The RF energy, in addition to causing spins to flip energy states, also forces the protons to precess in phase with each other at the resonant frequency. This phase coherence generates a transverse net magnetization, which is the only magnetization that we can physically measure, because the transverse magnetization precesses around the $B_0$-field and generates the inductive voltage in the receiver coil (Figure 27.1). The precession angular frequency $\omega_L$ is:

$$\omega_L = 2\pi f_L = \gamma B_0 \tag{27.3}$$

Because $f_L$ is the Larmor frequency, $\omega_L$ is called the angular Larmor frequency, or simply the Larmor frequency in the literature. Obviously, the Larmor frequency increases with the magnetic field $B_0$.

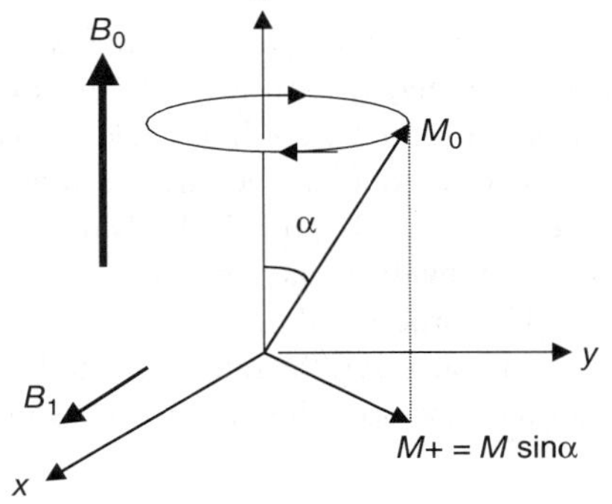


**FIGURE 27.1** The magnetization vector $M_0$ is processing about the static magnetic field $B_0$. The magnetic field $B_1$ of the RF pulse flips magnetization vector $M_0$ by an angle $\alpha$. The projection of magnetization vector $M_0$ onto $x$-$y$ plane is the transverse magnetization $M_+$.

The resonance process can also be viewed as exerting a torque generated by a resonance RF field (an RF wave with the resonant frequency) on the magnetization vector. This is like Earth's magnetic field exerting a torque on the magnetic needle of a compass. This is especially convenient to view the RF wave in the so-called rotating frame, a reference frame rotating about $B_0$ with Larmor frequency. In this frame the effect of a resonant RF wave is represented by a stationary magnetic field in the rotating frame. This field is also called the flip field, $B_1$, because it flips magnetization. This torque exerted by the flip field $B_1$ is equal to the vector product of the flip field and magnetization. As we know from mechanics, the effect of a torque is to change the angular momentum, and the speed of change of angular momentum is equal to the torque. Nuclear magnetization is proportional to nuclear spin angular momentum; hence,

$$\frac{d\bar{M}}{dt} = \gamma \bar{M} \times \bar{B}_1 \tag{27.4}$$

From this torque equation it is clear that the flip angle, $\alpha$, by which a magnetization vector is flipped away from $B_0$ field direction by the torque, is proportional to the resonant RF field strength $B_1$ and lasting time (the width) of the RF pulse:

$$\alpha = \gamma B_1 t_p \tag{27.5}$$

If magnetization $M_0$ is originally in the $z$-direction ($B_0$ field direction), then the longitudinal magnetization $M_z = M_0$. After the application of an RF pulse of $B_1$, and the transverse magnetization $M_+$ in the $x$-$y$ plane and longitudinal magnetization $M_z$ are (see Figure 27.1)

$$M_+ = M_0 \sin\alpha, \qquad M_z = M_0 \cos\alpha \tag{27.6}$$

From these equations it is obvious that a 90° RF pulse tips the equilibrium magnetization vector into the $x$-$y$ plane, yielding zero net longitudinal magnetization and maximum net transverse magnetization $M_+$.

When RF energy is no longer applied to the protons, the net magnetization continues to precess around the external magnetic field at the Larmor frequency. This precessing magnetization can be detected as a time-varying electrical signal across the leads of a coil of wire that can be represented as a complex number with real and imaginary parts. The net magnetization also decays exponentially, having a time constant $T_2$. Hence, because the induced voltage can be characterized by a decaying cosine function, it is often referred to as the free-induction decay, or FID.

### 27.2.3 Spin Relaxation, Tissue Characteristics, and Bloch Equation

The precessing protons interact with the surrounding tissue as well as with other protons that cause them to dephase. Each proton experiences a different local magnetic field depending on the orientation of other nearby protons. The fact that each proton feels a different local magnetic field leads to each proton having a different Larmor frequency. Consequently, because protons are now precessing at different frequencies, the phase coherence created by the RF pulse is slowly destroyed. Subsequently, the transverse magnetization decays because of the individual magnetic moments interacting with other magnetic moments causing them to dephase. This process is known as $T_2$ (or spin–spin) relaxation. The value of $T_2$ is different for various tissues and characterizes the rate at which the transverse magnetization decays to zero. The transverse magnetization decays according to the following relationship:

$$M_+(t) = M_+(0)e^{-t/T_2} \tag{27.7}$$

Note that this equation represents the decay of the transverse magnetization in a perfectly homogeneous static external magnetic field. Of course, it is impossible to create a perfectly uniform magnetic field. Pockets of tiny, but nonetheless significant, magnetic field inhomogeneities that affect the decay of

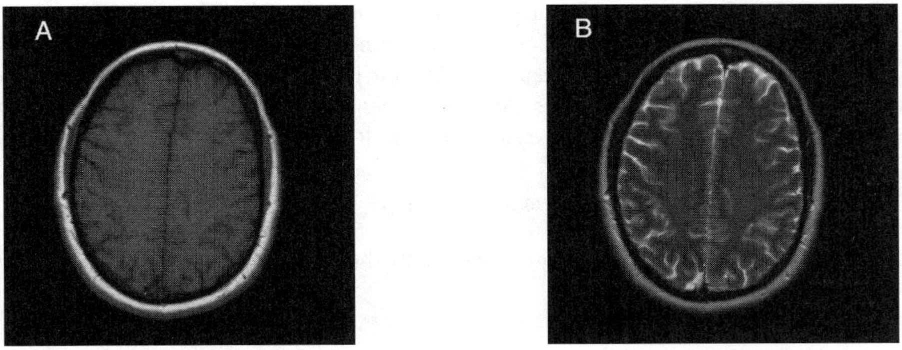


**FIGURE 27.2** (A) The $T_1$ weighted brain image of a patient. (B) The $T_2$ weighted brain image of the same patient.

the transverse magnetization always exist. When magnetic field inhomogeneities exist, the $T_2$ relaxation of the tissue speeds up, yielding $T_2^*$ relaxation.

When a short RF pulse is applied to the ensemble of protons with a flip angle of 90°, the longitudinal magnetization is tipped into the *x-y* plane, resulting in zero longitudinal magnetization. Once the RF pulse is removed from the system, the spins in the higher state will return to the lower state, in the process giving off energy to the surrounding tissue (lattice). As more spins return to the lower energy state, the longitudinal magnetization will grow back to its original value, $M_0$. The rate at which the longitudinal magnetization recovers is given by the $T_1$ (or spin–lattice) relaxation rate of the tissue and obeys

$$M_z(t) = M_{eq} + \left[M_z(t=0) - M_{eq}\right]\exp\left(\frac{-t}{T_1}\right) \tag{27.8}$$

where $M_{eq}$ is the thermal equilibrium magnetization in magnetic field $B_0$.

The value of $T_1$ depends on the dissipation of absorbed energy into the surrounding molecular lattice. Energy is most easily transferred if the proton precession frequency overlaps with the "vibrational frequencies" of the lattice. The more overlap, the easier it is to transfer energy, yielding shorter $T_1$ values. The less overlap, the harder it is to transfer energy, yielding longer $T_1$ values. It is important to note that the two processes, $T_1$ and $T_2$ relaxation, are independent of each other and often quite different from one another for the same tissue. Furthermore, $T_1$ is greatly dependent on external magnetic field strength, but $T_2$ is relatively independent of the field strength. Finally, in general, $T_1 > T_2 > T_2^*$.

Figure 27.2 shows two brain images of a patient at the same brain slice. The image shown in Figure 27.2A is $T_1$ weighted, that is, the image reflects the $T_1$ relaxation difference of different brain tissues. The longer the $T_1$, the lower the signal for that tissue. The image shown in Figure 27.2B is $T_2$ weighted, that is, the $T_2$ relaxation difference of different brain tissues is weighted in the image. In contrast to Figure 27.2A, in Figure 27.2B the longer the $T_2$, the higher the signal for that tissue.

Combining the torque equation with the spin relaxation effects as described previously, one can write the magnetization change rate due to the flip field and relaxation as the so-called Bloch equation

$$\frac{d\vec{M}}{dt} = \gamma\vec{M} \times \vec{B}_1 - \frac{1}{T_1}(M_z - M_{eq})\vec{k} - \frac{1}{T_2}(M_x\vec{i} + M_y\vec{j}) \tag{27.9}$$

where the vectors $\vec{i}$, $\vec{j}$, and $\vec{k}$ are the unit vectors of the *x*-, *y*-, and *z*-axes, respectively. The Bloch equation is the most fundamental equation of NMR and MRI.

## 27.2.4 Mapping Spatial Distribution of Spins

In MRI the spatial distribution of spins is mapped by applying magnetic field gradients and performing the Fourier transform. Consider a magnetic field gradient, $G_x$, in the *x*-direction superimposing on the

$B_0$ field of an MRI scanner's magnet. The magnetic field gradient specifies magnetic field change per unit length. For example, if $B_0$ = 1.5 T, and $G_x$ = 10 mT/m, then the resulting magnetic field is 1.5 T at $x$ = 0 and 1.501 T at $x$ = 10 cm. Therefore, spins at different position $x$ will experience slightly different magnetic fields. From Equation 27.3 it is clear that the spins at $x$ will have a precessing frequency $\omega(x)$ when the spin is flipped to the transverse plan by an RF pulse:

$$\omega(x) = \gamma \cdot B_0 + \gamma G_x x \tag{27.10}$$

Therefore the MRI signal intensity at time $t$ generated by these spins at $x$ will be

$$s(x,t) = CM_+(x)e^{(-i\omega(x)t)} = (Ce^{-i\omega_L t})M_+(x)e^{-i\gamma G_x xt} \tag{27.11}$$

where $C$ is a calibration constant for converting precessing $M_+$ to the signal. The receiver coil detects the total signal from spins all over the scan volume; hence, the total signal at time $t$ is

$$s(\gamma G_x t) = (Ce^{-i\omega_L t})\int M_+(x)e^{-i\gamma G_x xt}dx \tag{27.12}$$

Using Equation 27.6, the total signal can be related to the spin magnetizations as

$$s(\gamma G_x t) = (Ce^{-i\omega_L t})\sin\alpha\int M_z(x)e^{-i\gamma G_x xt}dx \tag{27.13}$$

Note that the integral

$$\int M_z(x)e^{-i\gamma G_x xt}dx \tag{27.14}$$

is a Fourier transform of the magnetization at $x$. Therefore, Equation 27.13 tells in words that the signal intensity received by the receiver coil essentially is a Fourier transform of spin's spatial distribution. This being so, collecting the coil signal as a function of time $t$ and performing an inverse Fourier transformation of the signal, we reconstruct the spin spatial distribution: the image of the spin along the $x$-axis. The field gradient $G_x$ is thus called the $x$-encoding field gradient. Using the magnetic field gradients $G_y$ and $G_z$ along the $y$- and $z$-axes and applying the same strategy to the other two dimensions ($y$ and $z$), we can reconstruct a three-dimensional spatial distribution of spins, i.e., a three-dimensional MRI image. In MRI, $G_x$ is also called the frequency encoding gradient, and $G_y$ the phase-encoding gradient. The encoding gradients used in the new generation MRI scanner can go as high as 40 to 50 mT/m.

From the MRI point of view, the spin spatial distribution $M_z(x)$ and the MRI signal $s(\gamma G_x\ t)$ form a Fourier pair; they are really two incarnations of the same identity. $M_z(x)$ represents an image characterized in the real space, and the signal $s(\gamma G_x\ t)$ represents the same image characterized in the $k$ space (Fourier space or the spatial frequency space) with $k_x = \gamma G_x\ t$. As we collect the signal data $s(\gamma G_x t)$ for different times such as $t_1$, $t_2$, ..., etc., one is really sampling the image at $k$ space positions (also called encoding steps) $k_{x1} = \gamma G_x\ t_1$, $k_{x2} = \gamma G_x\ t_2$, ..., etc. Simply stated, MRI is a technique to reconstruct $M_z$ by sampling it in the $k$ space through detecting NMR signals $s(k)$.

In order to sample an image in the $k$ space, one applies a sequence of RF pulses and field gradients. Figure 27.3 shows a typical gradient echo pulse sequence. Note that the label FEG denotes the frequency encoding gradient $G_x(t)$, PEG denotes the phase encoding gradient $G_y(t)$, PEG denotes the phase encoding gradient $G_y(t)$, and SEG denotes the slice encoding gradient $G_z(t)$. In this sequence an RF pulse of flip angle $\alpha$ establishes a transverse magnetization $M_+$ first, and then spins start to precess in the $x$-$y$ plane with the Larmor frequency. Meanwhile, a negative field $-G_{x1}$ gradient is applied for a short period $\delta t$, followed by a second positive field gradient $G_{x2}$, as shown in Figure 27.3. The data acquisition starts during the positive gradient period, which lasts for a sampling period of $T_s$ (from a few to ten millisecsonds). From Equation 27.14, the coil detects a time-varying signal proportional to

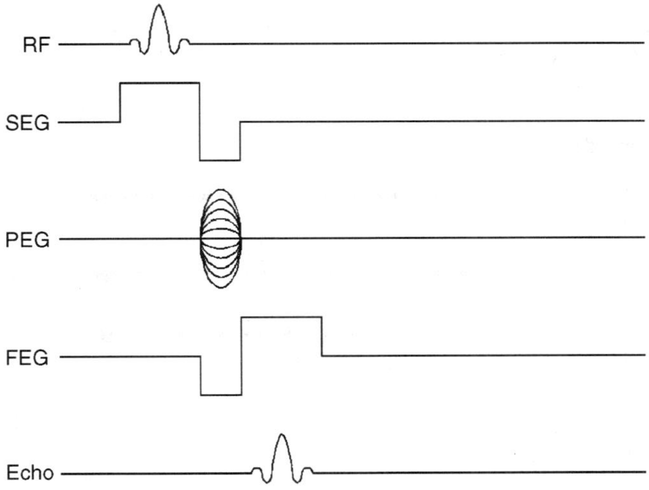


**FIGURE 27.3** A typical gradient echo pulse sequence.

$$\int M_z(x)e^{-i\gamma(-G_{x1}\delta t+G_{x2}t)x}dx, \qquad 0 \le t \le T_s \tag{27.15}$$

Again, as discussed previously, the signal detected during $0 < t < T_s$ samples magnetization $M_z(x)$ in the $k$ space from a negative value $-\gamma G_{x1}\delta t$ to a positive value $\gamma(G_{x2}T_s - G_{x1}\delta t)$. In this way a $k$ space line is completed. Note that, at $t = G_{x1}\delta t/G_{x2}$, the phase $(-G_{x1}\delta t + G_{x2}t)$ in the above equation becomes zero. This means all spins are precessing in the same phase at this moment; hence, the signal reaches the maximum, i.e., an echo is formed at this moment. That is why this type of pulse sequence is called the gradient echo pulse sequence. The gradients $G_y$ shown in Figure 27.3 encode the spin's $y$-position on the same principle. Each $G_y$ step encodes spins such that each $k_x$-line corresponds to different $k_y$. When the next spin-flipping RF pulse is applied, the same $k_x$ sampling repeats but with a different $G_y$ step. In this way, by repeating RF excitations (pulses) and using different $G_y$ steps after each pulse (RF excitation), one samples magnetization through the $k$ space line by line (Figure 27.4).

In MRI one often views a three-dimensional image by slices. The slice-selecting gradient $G_z$ in Figure 27.3 encodes the $z$-position of spins. For a detailed introduction to MRI, refer to Smith and Lange[2] and Chen and Hoult.[3] The higher the $G_z$, the thinner the resulting slice. It should be noted that the popular spin-echo pulse sequences of conventional MRI are not suitable for MRI with hyperpolarized spins (see Section 27.4.2).

## 27.3 Nuclear Spin Hyperpolarization by Optical Pumping

In the optical pumping and spin exchange approach for nuclear spin hyperpolarization, one shines a glass pumping cell (several hundred cubic centimeters in volume) containing alkali–metal vapors such as rubidium (Rb) vapor and noble gas ($^3$He or $^{129}$Xe) with a circularly polarized infrared light. The cell is heated to 80 to 150°C to maintain the desired Rb vapor density. $^3$He or $^{129}$Xe gas is kept at a pressure of several atmospheres in the cell, and approximately 100 torr of $N_2$ is also added to the cell for buffering. The circularly polarized infrared light polarizes the valence electrons of Rb atoms as explained later, and the polarized Rb atoms transfer their electronic spin polarization to nuclear spins of $^3$He or $^{129}$Xe by the spin-exchange collisions (Figure 27.5). In this way nuclear spin polarization of $^3$He or $^{129}$Xe can be as high as 50% or more.

It should be noted that $^3$He may also be polarized by the optical pumping and metastability exchange. Using 1.083 μm circularly polarized laser light, one can polarize the metastable $^3$He$^*$ (the $2^3S_1$ state) atoms in a $^3$He plasma at 1 mbar by optical pumping. Through the metastable exchange collisions between $^3$He$^*$ and $^3$He, the angular momentum is transferred from polarized $^3$He$^*$ to $^3$He nuclear spins of the ground

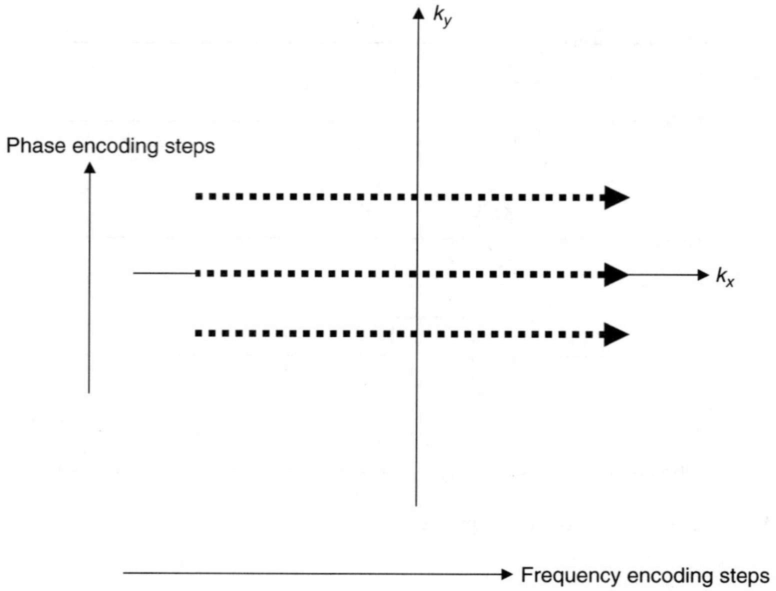


**FIGURE 27.4** The $k$-space sampling process for a gradient echo pulse sequence. Each dot line represents the $k_x$ sampling process during a signal read-out. Each dot represents a $k_x$ point for which the signal is acquired. After the next RF excitation (RF pulse) the $k_x$ sampling process repeats but with a new phase-encoding step (new $k_y$ step).

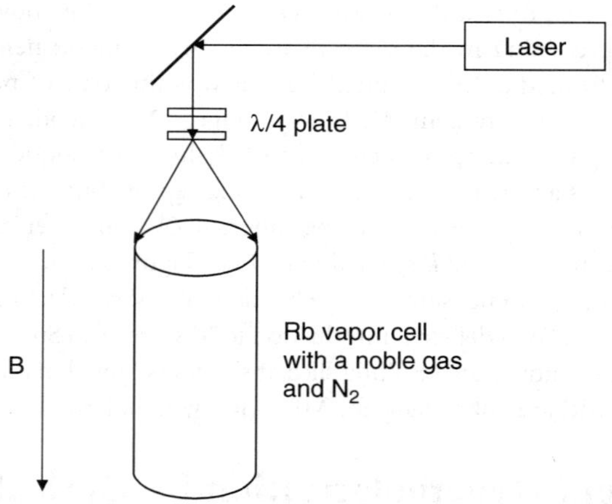


**FIGURE 27.5** Schematic of the optical pumping setup. The λ/4 plates are used to generate a circular polarization of the light.

state (the $1^1S_0$ state).[4] However, the metastability exchange polarization of $^{129}$Xe does not work so far. In this section we will discuss the optical pumping and spin exchange approach in detail.

## 27.3.1 Optical Depopulation Pumping of Alkali–Metal Atoms

In order to hyperpolarize nuclear spins, the first step is to use light to polarize the valence electronic spins of atoms of alkali–metal vapors such as rubidium (Rb), potassium (K), and cesium (Cs) vapors. This electron polarization process is based on the idea of depopulation pumping.[5] As is well known, an electron carries spin $s = 1/2$, just as the proton (hydrogen nucleus) carries spin $I = 1/2$. A spin-1/2 particle has two spin states, one up and one down, as discussed previously. Therefore, in ground state the valence

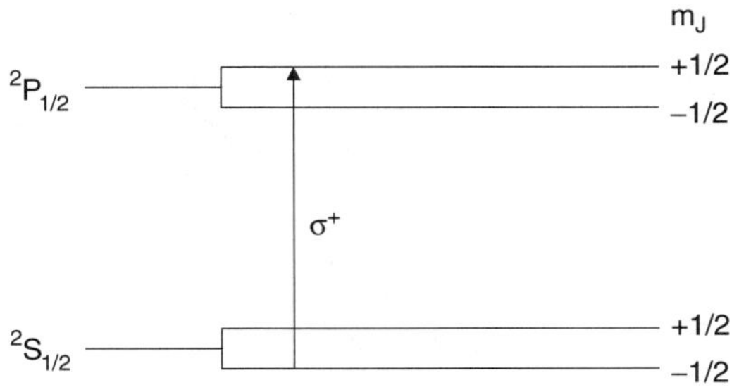


**FIGURE 27.6** Energy level diagram for Rb ground state and the lowest excited state. The energy differences between the sublevels are tiny because of the weakness of the magnetic holding field.

electron of an alkali metal atom has two different spin states, one up, one down. In a weak magnetic field (the holding field) of about 10 to 150 G, these two states have slightly different energy levels. In other words, the ground state has two sublevels (Figure 27.6).

Ordinarily these two spin states are almost equally populated; thus, the valence electron is not polarized. The idea of depopulation pumping is as follows. Shine light on the atoms. If one ground sublevel absorbs light much more strongly than the other, then atoms are pumped out more rapidly from the strongly absorbing sublevel. The populations of the two sublevels will not be equal any more as a result of the processes of the selective light absorption, atom excitation, and transition from the excited state to ground state via the spontaneous emission. Obviously the weakly absorbing sublevel will gain excess population and electron polarization results. In order to understand the factors affecting the yield of electron polarization, let us consider the optical pumping of rubidium atoms in detail.

## 27.3.2 Atomic States of Rubidium

For optical pumping, one places the rubidium vapor glass chamber in a weak magnetic field (e.g., 10 to 150 G), as shown in Figure 27.5. In order to understand the optical pumping of Rb atoms, one should know the atomic states and energy levels. Rubidium has two isotopes: $^{85}$Rb has a natural abundance of 72.17% and $^{87}$Rb a natural abundance of 27.83%. First, we assume that the nuclear spins of Rb ($I = 5/2$ for $^{85}$Rb and $I = 3/2$ for $^{87}$Rb) are not involved in the optical pumping process. This assumption will be justified later. With this assumption, the atomic states are labeled by conventional spectroscopic notation. A letter symbol indicates the total orbital angular momentum $L$ of the valence electrons. For example, symbols S, P, and D indicate $L = 0$, 1, and 2, respectively. A right subscript attached to the letter indicates the total angular momentum $J$ (the sum of orbital and spin angular momenta) of the state. A left superscript attached to the letter indicates the spin multiplicity $2s + 1$, where $s$ is the total spin of valence electron. Rubidium has only one valence electron; its ground state has zero orbital angular momentum, hence $L = 0$, $s = 1/2$, and $J = s = 1/2$, indicated as $^2S_{1/2}$ (more exactly, $5^2S_{1/2}$).

Note that the ground state has two spin states; one is the spin-up state of $m_J = m_s = 1/2$, and the other is the spin-down state of $m_J = m_s = -1/2$. Here the number $m_J$ denotes the projection of the total angular momentum along the applied holding magnetic field. Therefore the ground state has two subenergy levels. Although their energy levels are approximately the same due to the weakness of the holding field, they differ in electron spin orientations (Figure 27.6). The lowest excited state of Rb is $^2P_{1/2}$ (more exactly, $5^2P_{1/2}$). This state has two sublevels with $m_J = 1/2$ and $m_J = -1/2$ as well. (Note that, for the $^2P_{1/2}$ state, $m_J$ is a good quantum number and $m_s$ is not.) The energy difference between the ground state $^2S_{1/2}$ and the excited state $^2P_{1/2}$ is 1.56 eV, which corresponds to rubidium D1 absorption line at 794.8 nm.

## 27.3.3 Selective Absorption of Circularly Polarized Light

When a light with wavelength of approximately 794.8 nm shines on the glass cell of Rb vapor, the light propagates along the holding-field direction. The oscillating electric field of the light will induce the

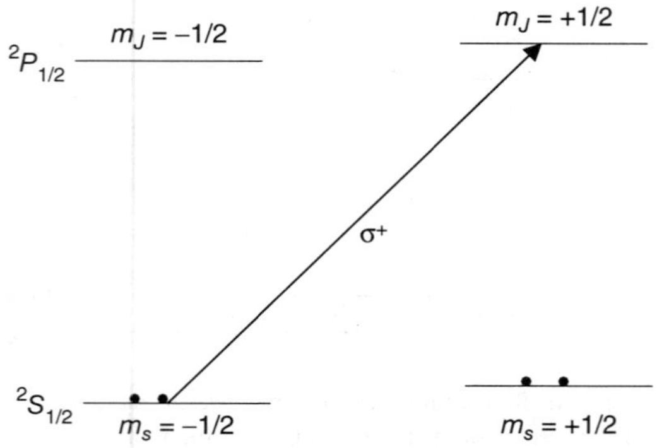


**FIGURE 27.7** Schematic of the ground state excitation by optical pumping using a $\sigma^+$-polarized light.

electric dipole and cause transitions between the ground state and the excited state. Remember that each has two sublevels with different $m_J$. At first glance it seems true that the valence electron from any ground state sublevel ($m_J$ = 1/2, −1/2) could be promoted to any sublevels of the state $^2P_{1/2}$ just by photon absorption; however, this is not true. In fact, the angular momentum conservation imposes the selection rules on what transitions are allowed.

Suppose that the input light is a left-hand circularly polarized light (i.e., a $\sigma^+$-polarized light). A circularly polarized light can be thought of as generated by two waves whose electric field components are orthogonal and 90° out of phase. This phase difference between these two waves causes an effective rotation of the total electric field vector about an axis in direction of propagation. For a left-hand circularly polarized light, the electric field vector is perpendicular to and rotating about the holding magnetic field direction in the left-hand sense.

As is well known, a photon carries a spin $s = 1$. A $\sigma^+$-polarized photon carries a spin pointing along the magnetic field and thus the spin projection $m_s = 1$. The angular momentum conservation dictates that a $\sigma^+$-polarized light can be absorbed by the $m_J = -1/2$ sublevel of ground state $^2S_{1/2}$ to induce a transition to the $m_J = 1/2$ sublevel of excited state $^2P_{1/2}$. This is because the total change of $m_J$ with the transition is $\Delta m_J = 1$, and this change can be compensated by absorption of a $\sigma$+-polarized photon. By the same reason of angular momentum conservation, a $\sigma^+$-polarized light cannot be absorbed by the electron in the $m_J = 1/2$ sublevel of the ground state. In this way the $\sigma$+-polarized light is absorbed only by the spin-down sublevel ($m_J = -1/2$) of ground state; the spin-up sublevel ($m_J = 1/2$) is "left in dark." Absorbing a light photon, the valence electron from the spin-down sublevel ($m_J = -1/2$) is promoted to the $m_J = 1/2$ sublevel of the excited state $^2P_{1/2}$ (Figure 27.7).

## 27.3.4 De-Excitation of Rb Atoms and Ground State Polarization

The atom cannot stay in the $m_J = 1/2$ sublevel of the excited state $^2P_{1/2}$ for long because of radiative decay via the spontaneous emission of light. When the excited state decays back to the ground state by spontaneous emission, it can decay to either of the two ground state sublevels. For an isolated atom, the probability of decay back to the $m_J = -1/2$ sublevel of ground state is twice that for returning to the $m_J = 1/2$ sublevel. From depopulation pumping point of view, this, of course, is not desirable, because it is against depopulation efforts. Fortunately, one can use the buffer gas to improve this problem.

In the Rb vapor cell there are usually noble gas atoms such as He and Xe gas for the spin exchange polarization (see below). If this gas has high enough density in the cell, the excited Rb atom will collide with these noble gas atoms many times during its lifetime. These collisions will redistribute the sublevel populations of the excited state. In other words, the sublevels of $m_J = 1/2$ and $m_J = -1/2$ are to be equally populated. In this way, when these excited state sublevels undergo radiative decay via spontaneous emission, the probabilities for returning to each of the ground state sublevels become the same (Figure 27.8).

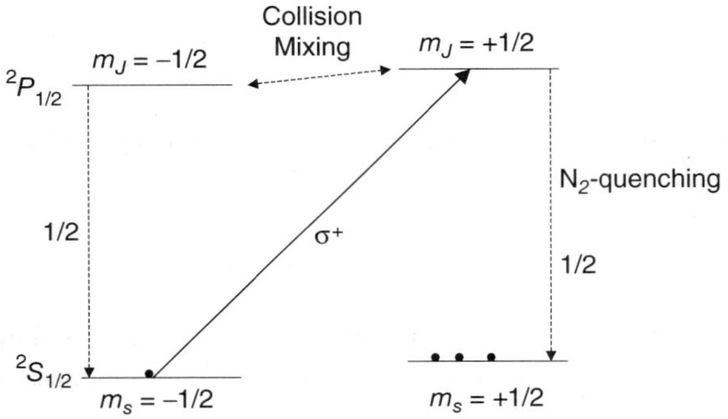


**FIGURE 27.8** Schematic of the state transitions via optical pumping, collision mixing, and $N_2$ gas quenching.

This process of collisional mixing of excited state sublevel populations ensures that the spin-up or spin-down sublevels will be repopulated at the same rate during the atomic de-excitation. In this way, because the spin-down sublevel of ground state is continuously pumped into the excited state by light absorption, eventually the spin-down sublevel is completely depopulated, and the valence electron is polarized to the spin-up state eventually.

In optical pumping practice, $N_2$ gas is added as the quenching gas into the Rb vapor cell to eliminate the radiation trapping problem (Figure 27.8). The remitted light (spontaneous decay) from the excited state is unpolarized and propagates in any direction. Once reabsorbed by Rb atoms, these remitted unpolarized photons can depolarize the atoms. Thus, if the mean free path of these photons is much less than the cell dimension, these unpolarized photons are trapped inside the cell and reduce the yield of electron polarization from optical pumping. The added $N_2$ gas allows the excited Rb atoms to decay nonradiatively (without emission of photons) from the excited state $^2P_{1/2}$ back to the ground state by collisions with the $N_2$.[6] Therefore the $N_2$ gas eliminates radiation trapping as a source of depolarization. The required $N_2$ densities are of 0.1 amagat or more.[5] Note that the density of an ideal gas is 1 amagat = $2.69 \times 10^{19}$ cm$^{-3}$. With adding $N_2$ gas to the Rb cell, the $N_2$ quenching becomes the main source of de-excitation.

## 27.3.5 Optical Pumping Dynamics and the Generalized Bloch Equation

The yield of rubidium ground state polarization depends obviously on the input light intensity, light spectral profile, and the lifetime of excited state. In order to understand the effects of these factors on polarization yield, one should find the equation of motion of ground state polarization. Similar to the longitudinal (along the magnetic field $B_0$) magnetization in NMR discussed earlier, the ground state polarization $S_z = [P(1/2) - P(-1/2)]/2$ specifies ground state polarization, where $P(/12)$ and $P(-1/2)$ are the occupation probability of the $m_J = 1/2$ sublevel and the $m_J = -1/2$ sublevel, respectively.

There may exist transverse (perpendicular to B) magnetizations $S_x$ and $S_y$ as well, since these transverse magnetizations may result from the coherence between ground state sublevel. Different from that in NMR, the pressing frequency of $S_x$ and $S_y$ about the magnetic field direction is faster than the Larmor frequency defined previously. The amount of difference, the so-called light shift, is proportional to the optical pumping rate and the light detuning (the amount of off-resonance of the light from the rubidium $D1$ line).[7] Fortunately, for typical cases where rubidium is in thermal equilibrium when the laser light is turned on, no transverse magnetization ($S_x$, $S_y$) will be generated by optical pumping. Therefore, for these cases optical pumping generates polarization $S_z$ only.

Similar to $M_z$ in NMR, the dynamics of $S_z$ can be described by a generalized Bloch equation. This is no surprise because the laser–atom interactions can be described very well by the optical Bloch equation.[8] Merging the optical Bloch equations and the magnetic resonance Bloch equation for spins (electron spins), we have a generalized Bloch equation of optical pumping:

$$\frac{dS_z}{dt} = -\frac{1}{T_{eff}}(S_z - S_{op-eq}) \tag{27.16}$$

where $S_{op-eq}$ is the equilibrium achieved using optical pumping and is given by

$$S_{op-eq} = \frac{1}{2}\frac{P_+(\Delta)}{\Gamma_{SD} + P_+(\Delta)} \tag{27.17}$$

where $\Gamma_{SD}$ is the $S_z$-destroying (spin-destruction) relaxation rate. The most important contribution to $\Gamma_{SD}$ is from collision between Rb atoms themselves. $P_+(\Delta)$ is called the optical pumping rate:

$$P_+(\Delta) = \frac{\Omega_1^2\Gamma_2^{\,2}}{4\Gamma_2(\Gamma_2^{\,2} + \Delta^2)} \tag{27.18}$$

Here $\Delta$ is the offset of the laser frequency from the resonance frequency ($D1$ line) $\omega_{D1}$, $\Delta = (\omega - \omega_{D1})$. $\Gamma_2$ is the optical coherence decay rate and equal to one half of the Rb $D1$ absorption linewidth broadened by collisions. Obviously, the optical pumping rate $P_+(\Delta)$ is a function of the frequency offset. $\Omega_1$ is the so-called optical Rabi frequency,

$$\hbar\Omega_1 = -\vec{E}\cdot\vec{D}_{ge} \tag{27.19}$$

The Rabi frequency is the scalar product of electric field $E$ and the induced atomic dipole $D_{ge}$ of the transition from the ground state to the excited state. Optical Rabi frequency characterizes the strength of the coupling between the incident light and Rb atoms. Because light is an electromagnetic wave, the light intensity is proportional to the square of the electric field. Therefore, squared Rabi frequency is proportional to the light intensity, and the optical pumping rate $P_+$ is proportional to light intensity as well.

Equations 27.16 through 27.18 determine the optical pumping dynamics. The spin polarization grows as

$$S_z(t) = S_{op-eq}\left(1 - e^{-\frac{t}{T_{eff}}}\right) \tag{27.20}$$

With optical pumping, the spin polarization increases with time exponentially according to this equation. Eventually spin polarization $S_z$ reaches the maximal value $S_{op-eq}$. The speed with which the maximal value is reached is determined by the effective relaxation rate $1/T_{eff}$:

$$\frac{1}{T_{eff}} = \Gamma_{SD} + P_+ \tag{27.21}$$

It is important to note that the maximal spin polarization $S_{op-eq}$ is not proportional to light intensity, as is shown by Equation 27.17. If we want to achieve 100% polarization ($S_{op-eq} = 1/2$), we should have $P_+ \gg \Gamma_{SD}$. For Rb vapor, $\Gamma_{SD} = k$ [Rb], where $k = 7.8 \times 10^{-13}$ cm$^3$/sec, and [Rb] is the Rb density. Obviously $\Gamma_{SD}$ is proportional to the Rb density. As shown later, in order to get more hyperpolarized He or Xe nuclear spins, the Rb densities used must be high enough. This requires even higher $P_+$, since $P_+ \gg \Gamma_{SD}$ must be kept for high Rb spin polarization. In order to see the effect of the frequency offset on the optical pumping rate of a narrow band laser light, we can rewrite Equation 27.18 as

$$P_+(\Delta) = \frac{\Gamma_2^{\,2}}{(\Gamma_2^{\,2} + \Delta^2)}P_+(0) = \frac{\Gamma_2^{\,2}}{(\Gamma_2^{\,2} + \Delta^2)}\sigma_0 I_0 \tag{27.22}$$

Here we use the fact that $P_+(0)$ is proportional to photon number intensity (photons per unit area per unit time), and $\sigma_0$ is the on-resonance photon scattering cross section that depends on the induced dipole of Rb atom and its *D*1 linewidth. Using the above equations, we can rewrite $\sigma_0$ in terms of the spontaneous emission rate and the *D*1 line wavelength:

$$\sigma_0 = \frac{\lambda_0^2 \Gamma_{nat}}{4\pi\Gamma_2} \tag{27.23}$$

where $\lambda_0$ is the *D*1 line wavelength, and $\Gamma_{nat}$ is the natural linewidth of Rb *D*1 line. Note that Rb has a small natural width $\Gamma_{nat}$ of 5.66 MHz for the D1 line because, for the excited state $5^2P_{1/2}$, $n = 5$ and the natural linewidth is $n^{-4.5}$ for alkali metals. Thus, $\sigma_0$ depends on total pressure in the pumping cell due to the D1 line pressure broadening (see below).

This equation shows that $P_+(\Delta)$ decreases with increasing offset $\Delta$ (or detuning) showing a Lorentz line shape. It is interesting to note that the NMR spectral line shapes for nuclear spins in liquid and many biological tissues are Lorentzian as well. This is no surprise; the similarity can be traced back to the underlying dynamics described by similar Bloch equations, i.e., by the optical Bloch equation for optical pumping and by the magnetization Bloch equation for NMR spectroscopy. Again, as in NMR, the full width of half-maximum (FWHM) of $P_+(\Delta)$ is given by the optical coherence decay rate; hence, FWHM of $P_+(\Delta)$ is equal to $2\Gamma_2$. Remember that $\Gamma_2$ is the optical coherence decay rate and equal to one half of the Rb *D*1 absorption linewidth. Note that Rb *D*1 absorption linewidth increases with pressure. The Rb *D*1 line is broadened by about 18 GHz per amagat of $^3$He.

### 27.3.6 Spin Exchange and Hyperpolarized Nuclear Spins

Once Rb atomic electrons are polarized, the Rb electronic polarization is transferred to nuclear spins of He and Xe by atomic collisions. The key process is the collision transfer. During collisions, the electronic spin *S* of a polarized Rb atom interacts with nuclear spin $I_{ng}$ of a noble gas atom through the hyperfine interaction

$$A(R)\vec{I}_{ng} \cdot \vec{S} \tag{27.24}$$

The hyperfine interaction arises from the magnetic field inside the nucleus of noble gas atom, and the magnetic field interacts with Rb electron spins. The factor $A(R)$ represents the coupling strength and strongly depends on the interatomic distance separation *R*. Because the hyperfine coupling is a scalar coupling, it is well known in NMR that a scalar coupling between spins causes mutual spin flipping. That is, during the collision the Rb electron spin and the nuclear spin of noble gas atoms are likely to flip in opposite directions. For example, an Rb electron spin flips from up to down during the collision, while the nuclear spin of the noble gas atom flips from down to up. Rb atoms have already been polarized via optical pumping; hence, certain polarization is transferred from Rb atoms to nuclear spins of noble gas during collision. That is, the polarization loss for Rb atoms is the polarization gain for nuclear spins of noble gas. Fortunately, the polarization of Rb can be constantly replenished by optical pumping. In this sense it is the angular momenta of circular polarized light that have been transferred to nuclear spins of noble gas.

The spin exchange rate equation is

$$\frac{dP_K}{dt} = \gamma_{SE}(2S_z - P_K) - \Gamma_K P_K \tag{27.25}$$

In the steady state, the noble gas nuclear spin polarization is

$$P_K = 2S_z \frac{\gamma_{SE}}{\gamma_{SE} + \Gamma_K} \tag{27.26}$$

where $P_K$ is the noble gas nuclear spin polarization, and $\Gamma_K$ is its relaxation rate.

The relaxation is often dominated by wall collision, although dipole–dipole coupling and impurities contribute as well. Obviously, the portion of Rb electron polarization transferred to noble gas nuclear spin is determined by $\gamma_{SE}/(\gamma_{SE} + \Gamma_K)$. If the light intensity is high enough and light reaches all parts of the pumping cell, this can reach the maximal value $S_{op-eq} \sim 1/2$, so $2S_z = 2S_{op-eq} \sim 1$. In order to achieve high polarization for noble gas nuclei, one must achieve a high Rb electron polarization and suppress $\Gamma_K$ such that $\gamma_{SE} \gg \Gamma_K$. Since $\gamma_{SE}$ is proportional to [Rb] (the Rb number density of the Rb vapor), one needs to increase the temperature of the cell (the oven temperature) to vaporize more rubidium. Then a higher light intensity is needed to maintain high Rb polarization such that $2S_{op-eq} \sim 1$. Therefore, the practical limitation is the light intensity achievable, i.e., the cost and complexity of the laser power employed.

### 27.3.7 Laser Source Considerations

From the previous discussion it is clear that a laser source with high output near the Rb *D*1 line is needed for optical pumping. However, we should also decide the bandwidth requirement for lasers. For example, we can use a narrow band laser source such as the Ti:Sapphire laser, or a broadband laser source such as the laser diode arrays (LDA) with power of a few tens of watts. A titanium-doped sapphire laser (Ti:Sapphire laser) is a narrow band laser; in fact, the lasing medium $Ti^{3+}:Al_2O_3$ has a very broad (about 300 nm) emission band with the peak emission at 790 nm.

However, laser source bandwidth will have important effects on the yield nuclear spin polarization of the noble gas. At first glance, it seems that a narrow band laser with output close to the Rb *D*1 line is better than a broadband laser, because the optical pumping rate $P_+(\Delta)$ decreases with increasing light detuning (offset). However, we must consider the light attenuation along propagation, since optical pumping needs to absorb those left circularly polarized photons. Let us denote the optical pumping rate at $z$ as $P_+(\Delta, z)$; then, similar to Equation 27.22, we have

$$P_+(\Delta, z) = \frac{\Gamma_2^{\,2}}{(\Gamma_2^{\,2} + \Delta^2)} \sigma_0 I(z) \tag{27.27}$$

$I(z)$ is the laser photon flux as $z$. Hence, the optical pumping rate $P_+(\Delta,z)$ is, in fact, decreasing with $I(z)$ due to light absorption. Traveling a very small distance $dz$, the laser photon flux decreases by

$$\frac{dI(z)}{dz} = -[Rb]P_+(\Delta, z)\frac{\Gamma_{SD}}{P_+(\Delta, z) + \Gamma_{SD}} \tag{27.28}$$

Note that the fraction at the right in this equation is the equilibrium occupation probability of the sublevel $m_s = -1/2$ of Rb ground state, since only that sublevel absorbs left-circularly polarized photons. It is clear from this equation that photon flux attenuation along propagation decreases with increasing light detuning. Therefore, we need to balance two things for optimal results of pumping. On the one hand, for increasing the polarization yield, we want to use high Rb density and have high optical pumping rate $P_+(\Delta, z)$. On the other hand, we want to have penetrating laser light to pump more volume fraction of Rb vapor in the cell. To find out if a narrow or broadband laser source is better, we must use the above equations to calculate optical pumping yield averaged for the whole cell. Of course, in the calculations the optical pumping rate should be summed over the line shape of a broadband laser source. That is, if the photon flux at a frequency $\nu$ and position $z$ is $\Phi(\nu, z)$, then the spectrum-summed optical pumping rate $P_+(z)$ at $z$ is

$$P_+(z) = \int \Phi(\upsilon, z) P_+(\Delta, z) dz \tag{27.29}$$

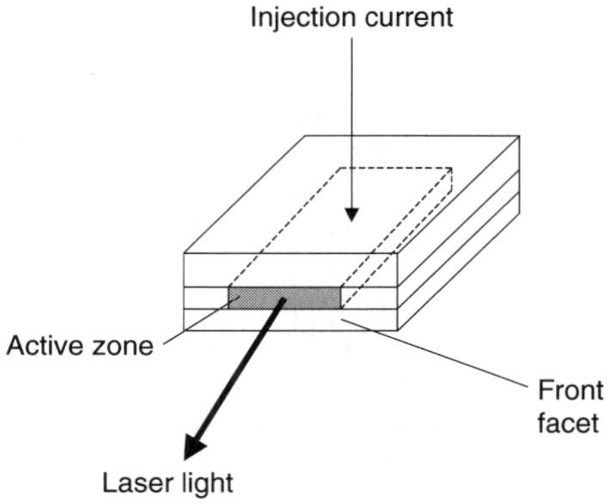


**FIGURE 27.9** Schematic of the laser diode.

With an 8-W Ti:Sapphire laser (with a linewidth of approximately 0.1 nm) the light penetration is approximately 6 cm for He gas of 5 amagat and the average Rb polarization is approximately 95%. For a 15-W laser diode array of 1.9 nm FWHM linewidth, the Rb polarization decreases from approximately 100% at $z = 0$ to approximately 25% at $z = 15$ cm.[9] These results show that, because the Rb $D1$ linewidth is very much broadened by the high-density He gas (45 to 75 GHz), the achieved Rb electron spin polarization averaged over the whole volume is about the same for the two laser sources. However, compared to the Ti:Sapphire laser, the laser diode array costs much less. A Ti:Sapphire laser pumped by an argon ion laser costs approximately $20K per watt; a laser diode array laser costs approximately $200 per watt.

The laser diode is a semiconductor device with the p–n junctions. The band structures of the semiconductor substrates largely determine the wavelength that a laser diode emits. Semiconductors have energy band structures consisting of the valence band and conducting band. GaAs, a III-V compound semiconductor, has a direct energy band structure such that the top of the valence band is directly below the bottom of the conducting band in the reciprocal space. Therefore, when an electron from the conducting band is recombined with a hole in the valence band, the energy released can be almost completely converted into that of the photon without energy loss to the crystal lattice. Doped with aluminum, gallium–aluminum arsenide [(GaAl) As] has an energy band structure such that a recombination of an electron–hole pair inside the p–n junction of (GaAl)As diode emits from 750 to 880 nm.

To cause an amplification of the light by the stimulated emission, the (GaAl)As is "pumped" up to the population inversion state by injecting sufficient minority carriers (electrons) into the p–n junction. The laser outputs are at wavelengths that satisfy two conditions. First, the wavelength must be within the bandwidth of the laser gain medium ([GaAl]As). Second, the wavelength must be in the passband of the laser resonator. The emitted light in laser diode is guided in the active zone (Figure 27.9). The common guiding provides a horizontal resonator structure. The guiding is realized by the built-in refractive index profiles (the index guiding) or by the concentration of the stimulating electric field.

A common type of resonator structure is the Fabry–Perot resonator. Using this resonator the emitted light will be selected and amplified by reflecting back and forth from the end-facet mirrors. For wavelengths for which the resonator length is an integral multiple of half the wavelength, all these light waves are transmitted in phase. As a result of in-phase addition, these waves are amplified greatly. Once the combination of amplification gain and end-facet reflectivity is larger than the loss, the device lases out light at these wavelengths. All these matched lasing wavelengths are called the longitudinal modes of the laser diode. The wavelengths satisfy

$$\lambda = 2l / k \tag{27.30}$$

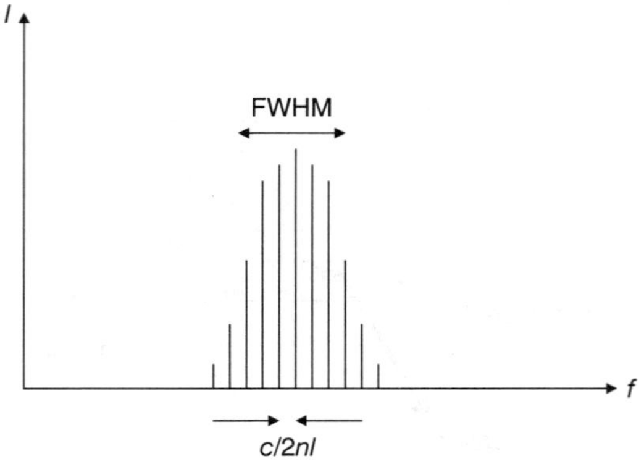


**FIGURE 27.10** Schematic of the multiple longitudinal modes of a laser diode array. The laser cavity length is $l$, and its refractive index is $n$. The mode spacing is $c/2nl$.

where $l$ is the laser resonator length, and $k$ an integer, so the spacing between these mode frequencies (Figure 27.10) is

$$\delta f = \frac{c}{2nl} \tag{27.31}$$

where $n$ is the refractive index of the resonator.

Obviously, the shorter the resonator, the larger the spacing of laser modes. For example, if a laser diode's cavity length is 1 mm, then the mode spacing is approximately 150 GHz. Combined with the 1 to 2 THz gain bandwidth, a laser diode's output spectrum consists of 150-GHz-spaced peaks modulated by the gain frequency curve (Figure 27.10). It should be noted that laser diodes are formed in arrays. A typical 40-W array of 795 nm consists of 19 diodes with emitting area 200 × 1 μm equally spaced along a 1-cm long stripe.[10] The arrays can also be coupled via the optical fiber for higher output power. External cavity effects from fiber coupling may perturb the output spectrum as well.

One important feature of LDAs is that the emitted wavelengths strongly depend on the operating temperature and injection current. Therefore, setting up the operating point in terms of the temperature and injection current is important. The temperature tuning is approximately 0.2 to 0.4 nm/K, and the current tuning coefficients obviously depend on the total number of diodes in LADs. For example, it was found that the current tuning is 0.18 to 0.42 nm/A for LDAs of 15-W power.[11] The total output power is approximately linear with the current; however, the wavelength tuning and the power saturate at large current.

As Equation 27.27 shows, the large spectral width of LADs (2 to 4 nm) limits the laser power absorbed by Rb vapor. Although one may increase the absorption by adding high-pressure buffer gas to increase the Rb $D1$ line broadening, mechanical constraints from cell strength and collision-induced polarization destruction limit the maximal buffer gas pressure to approximately 10 atm. At this pressure, the $D1$ line may be broadened to 0.4 nm only. Currently, some research efforts are directed to narrowing the LDA linewidth for increasing light absorption by Rb vapor. An approach that reduced the LDA linewidth by a factor of 2 with only 6% power loss by using the etalon reflection has been reported recently.[10]

## 27.4 MRI of Hyperpolarized $^3$He and $^{129}$Xe

In the last section we discussed how to hyperpolarize $^3$He and $^{129}$Xe nuclear spins. Because these nuclei have nuclear spin $I = 1/2$ — the same as a proton — the basic approaches of proton MRI can still be applied to these nuclear spins. However, there are many differences in imaging techniques as well. In this section we discuss special features of MRI with these hyperpolarized spins.

### 27.4.1 Signal Intensities

As mentioned in Section 27.2, in MRI the signal $S$ from the precessing transverse magnetization $M_+$ is proportional to

$$S = \gamma B_0 M_+ f(T_2, TE) \tag{27.32}$$

where $\gamma$ is the gyromagnetic ratio of the nuclei, $B_0$ is the field strength of the scanner magnet, and $M_+$ is the transverse magnetization. $f(T_2, TE)$ is a factor depending on scan parameters such as $TE$ and tissue relaxation times $T_2$.

Compared to the proton, $^3He$ and $^{129}Xe$ have smaller gyromagnetic ratios. While the proton's Larmor frequency is 63.9 MHz at 1.5 T, $^3He$'s Larmor frequency is 48.6 MHz (76% of the proton's) at the same field strength, $^{129}Xe$'s Larmor frequency is only 17.8 MHz (28% of the proton's). The signal is proportional to Larmor frequency $\gamma B_0$ because it is the Larmor frequency with which the transverse magnetization precesses about the $B_0$ direction (see Section 27.2). The different Larmor frequencies of $^3He$ and $^{129}Xe$ require broadband RF systems for excitation and signal reception. Many installed MRI systems are already equipped with broadband capabilities for P-31 Na-23, and C-13 MRI, and they can be used for $^3He$ and $^{129}Xe$ as well. The RF coils should be tunable to the appropriate Larmor frequencies. Usually these are manually tuned surface coils[12] or Helmholtz coils.[13]

To compare signal intensities, we rewrite the longitudinal magnetization $M_z$ before spin flipping by an RF pulse:

$$M_z = N\gamma\hbar\langle I_z\rangle \tag{27.33}$$

where $N$ is the nuclear spin density and $<I_z>$ is the average nuclear spin along the $B_0$-axis before flipping. Because $M_+$ is determined by the flip angle $\alpha$ (defined in Section 27.2) and $M_z$:

$$M_+ = M_z \sin\alpha = N\gamma\hbar\langle I_z\rangle \sin\alpha \tag{27.34}$$

Using these equations, we find that the signal intensity $S$ is

$$S = \gamma^2 B_0 N\hbar\langle I_z\rangle \sin\alpha f(T_2, TE) \tag{27.35}$$

We can then compare the signal intensities from MRI of protons and hyperpolarized spins to determine their signal intensity ratio, assuming the same magnetic field, the same flip angle, and the same $f$-factor:

$$\frac{S_{\mathrm{He}}}{S_{\mathrm{H}}} = \frac{\gamma_{\mathrm{He}}^2 N_{\mathrm{He}}\langle I_z\rangle_{\mathrm{He}}}{\gamma_{\mathrm{H}}^2 N_{\mathrm{H}}\langle I_z\rangle_{\mathrm{H}}} \tag{27.36}$$

For thermally polarized spins such as protons, the value of average nuclear spin $<I_z>$ depends on the particular pulse sequence used in the scan, but it never exceeds the thermal equilibrium value $<I_z>_{eq}$:

$$\langle I_z\rangle_{eq} = \frac{1}{4}\frac{\hbar\omega_L}{k_B T} \tag{27.37}$$

Using Equation 27.33, we obtain the equilibrium magnetization $M_0$

$$M_0 = \frac{N\gamma^2\hbar^2 B_0}{4k_B T} \tag{27.38}$$

Here $k_B = 1.38 \times 10^{-23}$, J/K is Boltzmann's constant, $T$ is the absolute temperature, and $\omega_L$ is the Larmor frequency of the nuclear spin. At $T = 310$ K and 1.5 T field strength, the proton's maximal $<I_z>_H = <I_z>_{eq} = 2.47 \times 10^{-6}$. Although the proton's spin polarization $<I_z>_{eq}$ is tiny, proton spin density $N_H$ is high in tissue water, $N_H = 6.69 \times 10^{22}/\text{cm}^3$. On the other hand, for hyperpolarized $^3$He, the polarization from the optical pumping and spin exchange can be as high as 50% or more. At 50% polarization, $<I_z>_{He} = 1/4$. Hence the optically pumped helium's spin polarization $<I_z>_{He}$ is $1.01 \times 10^5$ times higher than the proton spin polarization $<I_z>_H$. However, the spin density of $^3$He in the human body is much lower than the spin density of proton in human tissue water.

Suppose $^3$He is administered by inhalation in MRI, and at one bar pressure with dilution of 1/8 and body temperature ($T = 310$ K), the $^3$He spin density is $2.92 \times 10^{18}/\text{cm}^3$, assuming an ideal gas state equation. Using the signal ratio formula of Equation 27.36, we find the signal intensity ratio $S_{He}/S_H = 2.56$. It should also be noted that this calculation of $S_{He}/S_H$ assumes use of the same flip angle for both $^3$He and proton. We will see that the flip angle for $^3$He is usually smaller than that used for proton MRI. Use of a smaller flip angle tends to reduce the $^3$He signal. In any case, the small spin density of $^3$He is more than compensated for by its hyperpolarization to generate higher MRI signal intensity than that in proton MRI.

It should be noted that, in some organs such as the lung, proton MRI lung imaging encounters many difficulties. First of all, proton density is low in the lungs. Second, the differences of proton bulk magnetic susceptibilities between pulmonary tissue and air generate changes of local Larmor frequencies. This causes shot $T_2^*$ relaxation times and gross reduction of proton MRI signals by a factor $\exp(-TE/T_2^*)$, as pointed out in Section 27.2. Therefore, for lung imaging, the signal intensity turns more favorably to $^3$He MRI. It should also be noted that, in the above comparison, we have assumed proton spin polarization is the thermal equilibrium value. This is the maximum the spin polarization can reach for protons. However, in many cases, especially with short pulse sequence repeating time $TR$, the spin polarization is much lower than $<I_z>_{eq} = 2.47 \times 10^{-6}$ because, if $TR$ is short compared to the longitudinal relaxation time $T_1$, the spin polarization does not have enough time to be restored to the equilibrium value $<I_z>_{eq}$. As for $^{129}$Xe, the signal intensity ratio $S_{Xe}/S_H$ is reduced to 13.4% of $S_{Xe}/S_H$ if all other conditions are the same because of the small gyromagnetic ratio of $^{129}$Xe.

## 27.4.2 General Considerations for $^3$He and $^{129}$Xe as MRI Contrast Media

The above calculations are based on the inhalation administration of the noble gases. Note that helium is an inert and nontoxic gas. It is widely used in other cases such as in deep-sea diving, where helium is inhaled in quite a high amount and concentration (80% helium and 20% oxygen) without severe adverse effects.[14] As shown in Table 27.1, $^3$He has a tiny natural abundance (0.014%) produced from tritium radioactive decays and generated mainly in production of nuclear weapon material. Unpolarized $^3$He costs approximately $100/l and recovery after exhalation is necessary. Currently $^3$He can be polarized as high as 50% or more.

$^{129}$Xe has a natural abundance (26.4%). It is an inert and nontoxic gas used as an inhalation anesthetic (70% xenon and 30% oxygen). Xenon has a long history of use in medical imaging. As an inhalative contrast media, radioactive xenon isotope ($^{133}$Xe and $^{135}$Xe) imaging has been used for the γ-ray scintigraphy in nuclear medicine, and nonradioactive xenon ($^{131}$Xe) for CT brain perfusion studies using its high attenuation to x-ray.

One advantage of using xenon is its relatively high solubility in water and lipid; compared to helium, it is more than 10 times more soluble in water and 100 times more soluble in fatty tissue. This also makes

**TABLE 27.1** Physical Parameters for $^1$H, $^3$He, and $^{129}$Xe

| Isotope | $^1$H | $^3$He | $^{129}$Xe |
|---|---|---|---|
| Natural abundance | 99.9 % | 0.014 % | 26.4 % |
| Nuclear spin | 1/2 | 1/2 | 1/2 |
| Larmor freq. @ 1.5 T (MHz) | 63.9 | 48.6 | 17.8 |

xenon more suitable for preparation of $^{129}Xe$ carriers for injection. Moreover, among all NMR-sensitive nuclei, $^{129}Xe$ presents the largest chemical shift range; its chemical shift in tissue can be as high as a few hundred parts per million, making $^{129}Xe$ very sensitive to its chemical environment. Additionally, the chemical shift disperses from gas to blood and tissue, allowing selective imaging of xenon in blood, brain, and gas.[15] Xenon with natural abundance costs approximately a few dollars per liter. Compared to $^3He$, $^{129}Xe$ can only be polarized to approximately 20% because the rate of Rb electron spin depolarization in the spin–rotation collisions with $^{129}Xe$ is higher. (The spin–rotation interaction is the coupling between electron spin and the rotational angular momentum of an $Rb^{129}Xe$ pair.[16]) However, spin polarization up to 60% has been theoretically predicted, and the spin–exchange efficiency can be improved to increase polarization.[17] Compounded with its low gyromagnetic ratio, the signal intensity from $^{129}Xe$ will be much lower than that from hyperpolarized $^3He$.

### 27.4.3 Signal-to-Noise Ratio and Magnetic Field Strength

SNR is a key imaging performance measure. In MRI it is very important to find out the SNR's dependence on magnetic field strength. Looking at just the $B_0$-dependence of signal intensity, Equation 27.35 states that

$$S \propto B_0 \langle I_z \rangle \tag{27.39}$$

For thermally polarized spins like protons, its nuclear spin along the $B_0$-axis is, at most, $<I_z>_{eq}$. Using Equation 27.37, we find that

$$S_H \propto B_0^2 \tag{27.40}$$

since the equilibrium polarization is proportional to the field strength $B_0$ as well. But for hyperpolarized helium and xenon, their nuclear spin $<I_z>$ is achieved by optical pumping and spin exchange and is independent of field strength $B_0$; hence, the signal from hyperpolarized $^3He$ or $^{129}Xe$ is

$$S_{He/Xe} \propto B_0 \tag{27.41}$$

It differs from the quadratic dependence of the proton signal intensity on $B_0$. On the other hand, the Johnson noise voltage $V_{noise}$ in the RF receiver coil is equal to[3]

$$V_{noise} = \sqrt{4k_B T \Delta f (R_c + R_b)} \tag{27.42}$$

where $\Delta f$ is the signal frequency bandwidth, $R_c$ is the coil resistance, $R_b$ is the effective resistance caused by RF dissipation by the body parts, and $\Delta f$ is the frequency bandwidth of the MRI signals. Roughly speaking, all spins would generate signals in Larmor frequency. However, during the data acquisition period, spins at different locations actually experience slightly different Larmor frequencies due to the applied magnetic field gradient for position encoding. Therefore, the frequency bandwidth $\Delta f$ is given by

$$\Delta f = \frac{\gamma}{2\pi} G_x d_{FOV} \tag{27.43}$$

where $G_x$ is the magnetic field gradient in the $x$ direction, and $d_{FOV}$ is the dimension of image field of view. Hence, $\Delta f$ is independent of the field strength $B_0$.

However, the resistances $R_c$ and $R_b$ are field strength $B_0$ dependent. $R_c$ increases with Larmor frequency, due to the skin-depth effect, because the high-frequency current tends to stay in the skin layer of the coil's wire. This skin-depth effect effectively makes the coil's resistance increase. It turns out that $R_c$ is proportional to $B_0^{1/2}$. The RF dissipation-caused resistance $R_b$ increases with $B_0$ as $B_0^2$. For $B_0 > 0.2$ T, $R_b$ dominates over $R_c$. Therefore, for $B_0 > 0.2$ T, the noise varies with $B_0$ according to

$$V_{noise} \propto \sqrt{R_b} \propto B_0 \tag{27.44}$$

Using Equations 27.40 through 27.43, we find that the SNRs for proton and hyperpolarized He/Xe are

$$SNR_{\mathrm{H}} \propto \frac{B_0^2}{\sqrt{B_0^2}} \propto B_0, \qquad SNR_{\mathrm{He/Xe}} \propto \frac{B_0}{\sqrt{B_0^2}} \propto 1 \tag{27.45}$$

This result shows that, although the SNR of conventional MRI is proportional to $B_0$, the SNR of MRI with hyperpolarized spins is independent of $B_0$. As we mentioned at the beginning of this chapter, the ever-increasing demand on high SNR drives the magnetic field strengths employed in conventional MRI higher. The higher the field strength, however, the more expensive and bulky the MRI scanner. The high field strength aggravates problems such as magnetic susceptibility artifacts and spin–lattice relaxation time lengthening. Fortunately, the SNR for hyperpolarized spins is independent of $B_0$. A study of MRI using hyperpolarized $^3$He at 0.1 T has been reported in the literature.[18] Moreover, MRI images were obtained using MRI using hyperpolarized $^3$He at 21 G.[19] These efforts open doors for many potential applications such as portable MRI scanners for lung imaging.

Another advantage of low field imaging is the reduction of magnetic susceptibility artifacts. Consequently, the apparent spin–spin relaxation time $T_2^*$ can be greatly extended from a few to hundreds of milliseconds.

### 27.4.4 Pulse Sequence Considerations

Hyperpolarization is a spin state that is far away from the thermal equilibrium state, which has an important consequence for flip angle selection. Suppose one flips spin by 90° for maximal signal, since sin90° = 1. We also suppose the transverse magnetization becomes dephased after forming echo. After the flip, the longitudinal magnetization $M_z = 0$ and spins experience spin–lattice relaxation, regardless of whether the spins are thermally polarized, such as water protons, or hyperpolarized spins generated by optical pumping. As shown in Equation 27.8, the longitudinal magnetization $M_z$ evolves as a function of time $t$

$$M_z(t) = M_{eq} + (M_z(t=0) - M_{eq})\exp\left(\frac{-t}{T_1}\right) \tag{27.46}$$

where $T_1$ is the so-called spin–lattice relaxation time, as mentioned in Section 27.2, and $M_{eq}$ is the thermal equilibrium magnetization. More specifically, after a 90° flip, $M_z(t = 0) = 0$, $M_z(t)$ recovers according to

$$M_z(t) = M_{eq}\left(1 - e^{-\frac{t}{T_1}}\right) \tag{27.47}$$

Therefore, for proton, $M_z$ can be restored to its equilibrium $M_{eq}$ after a long enough time (approximately 95% of $M_{eq}$ at $3T_1$). This is true for hyperpolarized $^3$He and $^{129}$Xe as well. The difference is clear here: for proton, $M_{eq}$ is the maximum of $M_z$; hence the spin–lattice relaxation restores $M_z$. However, for hyperpolarized $^3$He and $^{129}$Xe, the equilibrium magnetization $M_{eq}$ for $^3$He and $^{129}$Xe are approximately $10^5$ times smaller than their initial $M_z$ before the flipping. Therefore, the RF pulses really destroy hyperpolarization irreversibly, and spin–lattice relaxation cannot restore it. With thermal equilibrium magnetizations, we cannot get strong enough signals because of the low concentration of noble gas spins. Due to this difference, we should avoid applying the conventional "pre-scan" calibration procedures to hyperpolarized $^3$He and $^{129}$Xe for preventing destroying hyperpolarization before the scan.

If we apply RF excitations of flip angle $\alpha$ to hyperpolarized $^3$He or $^{129}$Xe by $n$ times, and if all transverse magnetization after each RF excitation gets completely dephased, then the remaining longitudinal magnetization is

$$M_z(n) = M_z(0)\cos^n\alpha \tag{27.48}$$

Here we also assume that $T_1$ relaxation times for $^3$He and $^{129}$Xe are long compared to the scan duration. The signal-generating transverse magnetization $M_+$ is

$$M_+(n) = M_z(0)\cos^{n-1}\alpha\sin\alpha \tag{27.49}$$

Hence, the signal intensity gets weaker by a factor of $\cos\alpha$ for each RF pulse. Because RF excitations are applied one by one for acquiring scan data in phase-encoding direction or, equivalently, for sampling the data in the $k$ space line by line, then the later sampled lines get reduced signal due to the "consumption" of the initial hyperpolarization. Consequently, this results in nonuniform weighting of $k$-space scan data.

Depending on the $k$-space sampling order, this nonuniform weighting can cause some loss of image resolutions. As for selection of $\alpha$, Equation 27.49 suggests that we must balance two needs: we want a larger flip angle for a larger signal and we must ensure that enough $M_z$ is left for phase-encoding steps with large $n$. Image quality is determined largely by the data from the center of the $k$ space; therefore, we may want to ensure optimal signal intensity there. If sequential phase encoding is adopted, and $n_c$ is the phase-encoding step corresponding to the $k$-space center, then transverse magnetization $M_+$ at this phase-encoding step is

$$M_+(n_c) = M_z(0)\cos^{n_c-1}\alpha\sin\alpha \tag{27.50}$$

From the stationary point of $M_+(n_c)$ with respect to flip angle $\alpha$, we find the optimal $\alpha_0$ that maximizes the signal:

$$\alpha_0 = a\tan\left(\frac{1}{\sqrt{n_c - 1}}\right) \tag{27.51}$$

For example, if $n_c = 64$, the optical flip angle $\alpha_0 = 7.2°$. It is interesting to note that, with this flip angle, after the 127th RF pulse the remaining longitudinal magnetization is approximately 37% of the starting $M_+$.

To try to get constant signal as the RF excitation progresses, we must increase the flip angle to compensate the loss of hyperpolarization from previous excitation.[20] Note that if the flip angle varies as the RF-excitation progresses, then after $n$ pulses $M_z$ becomes

$$M_z(n) = M_z(0)\cos\alpha_1\cos\alpha_2\ldots\cos\alpha_{n-1}\cos\alpha_n \tag{27.52}$$

and the signal $M_+(n)$ after the $n$th RF pulse is

$$M_+(n) = M_z(0)\cos\alpha_1\cos\alpha_2\ldots\cos\alpha_{n-1}\sin\alpha_n$$

The signal $M_+(n + 1)$ after the $(n + 1)$th RF pulse is

$$M_+(n+1) = M_z(0)\cos\alpha_1\cos\alpha_2\ldots\cos\alpha_{n-1}\cos\alpha_n\sin\alpha_{n+1} \tag{27.53}$$

Comparing these two equations, we get the condition for a constant signal; on the other hand, in order to get a constant signal for all phase-encoding steps, we should require that

$$\tan\alpha_n = \sin\alpha_{n+1} \tag{27.54}$$

Let us suppose that we apply a 90° flip angle at the last phase-encoding step $N$. In this way we maximize the signal from what is left in $M_z$ and we need not save $M_z$ for later steps any more. From the above equation the flip angle for the $(N - 1)$th pulse should be

$$\tan\alpha_{N-1} = \sin\alpha_N = 1 \tag{27.55}$$

so the flip angle for the $(N-1)$th pulse should be 45°. Using Equation 27.54 we find the flip angle for the $n$th pulse is

$$\alpha_n = a\tan\left(\frac{1}{\sqrt{N-n}}\right) \tag{27.56}$$

So if $N = 128$, we start the sequence with a flip angle of 5.1° for the 1st pulse, and end the sequence with the 128th pulse of 90°. The variable flip angle approach can provide uniform $k$-space sampling and more efficient use of hyperpolarization.

Most studies performed with hyperpolarized spins, however, use a constant flip angle approach because it is difficult to implement the variable flip angles accurately, and the approach's performance is sensitive to variable flip angle errors from RF-$B_1$ field inhomogeneity and RF transmitter calibration errors. The variable flip angle technique can also be used in NMR spectroscopy. For example, to measure hyperpolarized xenon's $T_1$ relaxation times in blood, one usually applies multiple excitations to measure the relaxation-induced signal decay. With variable flip angle technique, the signal decay is truly contributed by the $T_1$ relaxation with the RF-induced loss of magnetization being compensated.[21] So if a variable flip angle sequence with eight pulses is used, then, according to Equation 27.56, we should start with a flip angle of 21° and end with a flip angle of 90°.

The self-diffusion coefficient of He gas is high: $D_{He}$ = 1.8 cm$^2$/sec at 1 atm and 20°C; self-diffusion coefficients of Xe gas are much lower: $D_{Xe}$ = 0.06 cm$^2$/sec at 1 atm and 20°C.[22] In the data acquisition period, the frequency-encoding gradient is applied. In a magnetic field gradient diffusion creates additional random and position-dependent phases to the transverse magnetizations of spins, since diffusion is a random translation. The coherence between transverse magnetizations of spins at different locations has been partially destroyed by the random phases. When these transverse magnetizations are summed over all the spins at different locations, the total dephasing effect of these random phases is to attenuate the total transverse magnetization by a factor:

$$\exp\left(-\frac{\gamma^2 G^2 D T_g^3}{12}\right) \tag{27.57}$$

where $G$ is the field gradient strength of a bipolar rectangular gradient waveform, and $T_g$ is the time period of the transverse magnetization under field gradient. For a gradient echo sequence, $G$ is the frequency-encoding gradient and $T_g = T_s$, the data sampling time. Because higher resolution corresponds to higher $G$, Equation 27.57 means that the diffusion-associated signal attenuation mainly affects the imaging of fine details. In fact, this equation can be equivalently written as

$$\exp\left(-\frac{\pi^2 D T_s}{3d^2}\right) \tag{27.58}$$

where $d$ is the size of detail (one line pair) resolved. Consequently, the signal reduction is resolution-dependent.

The signal decreases rapidly with decreasing $d$. For example, for He gas and with $T_s$ = 6 ms, 97% of the signal will be attenuated for $d$ = 1 mm, but only 80% of the signal will be attenuated for $d$ = 1.5 mm, and 33% of the signal will be attenuated for $d$ = 3 mm. Reducing sampling time improves resolution rapidly as well. In the previous example, if $T_s$ is reduced to 4 ms, 90.7% of the signal will be attenuated for $d$ = 1 mm, but only 65% of the signal will be attenuated for $d$ = 1.5 mm, and 23% of the signal attenuated will be for $d$ = 3 mm. Xe's self-diffusion coefficient is much smaller; hence the diffusion-associated signal intensity reductions are mild. For Xe gas and with $T_s$ = 6 ms, 11% of the signal will be attenuated for $d$ = 1 mm, but only 5% of the signal will be attenuated for $d = 1.5$ mm, and 1.2% of the signal will be attenuated for $d = 3$ mm. These results show the diffusion-associated signal attenuation as function of feature size. As for the minimal resolvable size, it depends on the feature's contrast as well. It should also be noted that the slower Xe diffusion allows one to use larger sampling time $T_s$. This means a smaller frequency bandwidth $\Delta f$ of the signals, and hence less noise (Equation 27.42). This helps improve the SNR for Xe imaging.

This analysis is based on pure He or Xe. Diluting He with heavier gases such as $N_2$ reduces helium's diffusion coefficient, thus reducing diffusion-associated signal attenuation. Furthermore, helium's apparent diffusion coefficient in lung parenchyma is further reduced due to the restricted diffusion imposed by the bronchial and alveolar walls. Some studies have reported that the apparent diffusion coefficient (ADC) of helium diluted with $N_2$ in lung parenchyma is reduced by an order of magnitude compared to self-diffusion coefficient.

On the flip side of the diffusion problem, one may very well use diffusion to help examine the lungs by performing diffusion-weighted MRI.[23] Using $^3$He-MRI it was determined that He gas diffusion in the lungs is anisotropic, because the diffusion restrictions are much less along the airway axis than perpenticular to it.[24] This technique is useful for emphysema studies. In emphysema the restrictions to diffusion are reduced because of expansion of the alveoli and airway and tissue destruction. It was found that the ADCs of lungs of patients with severe emphysema are increased by a factor of 2.7 compared to the normal lung, and all the transverse and longitudinal ADCs are elevated.[24]

## 27.4.5 Hyperpolarized Spin Relaxation

As noted earlier, the spin–lattice relaxation ($T_1$ relaxation) and spin–spin relaxation ($T_2$ relaxation) are the basis of tissue contrast in conventional MRI. This is true for hyperpolarized spins as well; however, spin–lattice relaxation becomes more important for hyperpolarized spins. This is because in proton MRI spin–lattice relaxation restores magnetization, but in $^3$He or$^{129}$Xe MRI spin–lattice relaxation destroys the magnetization of hyperpolarized spins. If the hyperpolarization is generated at $t = 0$, using Equation 27.8) for spin–lattice relaxation and the condition $M_z >> M_{eq}$, the magnetization of hyperpolarized spins relaxes as

$$M_z(t) \approx M_z(t=0)\exp\left(\frac{-t}{T_1}\right) \tag{27.59}$$

After $5T_1$ hyperpolarization is reduced to 0.6% of the initial hyperpolarization, and MRI signals from that much relaxed polarization will be diminished. Obviously, an MRI scan should be completed before the polarization consumed by relaxation. Relaxation time $T_1$ varies with the state and environment of hyperpolarized spins. Tables 27.2 and 27.3 show relaxation time $T_1$ for $^3$He and $^{129}$Xe in different conditions, respectively. Pure $^3$He and $^{129}$Xe would have long $T_1$ if there were no interaction with other impurities because the electrons of noble gases are in the filled orbitals. These electrons produce neither an electric field gradient nor a magnetic field at their nucleus. Also, noble gases are monoatomic, thus there is no intramolecular dipole–dipole coupling. That is why $^3$He has a $T_1$ of several days in a clean glass container as listed in Table 27.2. The listed $T_1$ in glass container is obtained with a holding magnetic field of 10 to 20 G to fend off the stray magnetic field gradients. The holding field can be provided by a permanent magnet or battery-powered coils. For $^{129}$Xe the distribution task is far more labored because $^{129}$Xe should

**TABLE 27.2** Spin–Lattice Relaxation Times for $^3$He

| $^3$He Condition | $T_1$ |
|---|---|
| Clean uncoated glass container | Several days |
| In a polyethylene bag | 4900 s |
| In the lungs | 28 s |

**TABLE 27.3** Spin–Lattice Relaxation Times for $^{129}Xe$

| $^{129}Xe$ Conditions | $T_1$ | Ref. |
|---|---|---|
| Stored at liquid $N_2$ temperature | 8500 s | 37 |
| Stored at liquid He temperature | 500 h | 37 |
| In a polyethylene bag | 120–350 s | |
| In the lungs | 31 s | |
| Artery blood, at 1.5 T, 37°C | 6.4 s | 21 |
| Venous blood, at 1.5 T, 37°C | 4 s | 21 |
| Saline: $\lambda_{solv}$ = 0.079 at 37°C | 66 s | 28 |
| Perfluorocarbon emulsion: $\lambda_{solv}$ = 0.6 at 37°C | 100 s | 28 |

be stored at low temperatures, as shown in Table 27.3. In fact, $^{129}Xe$'s relaxation time $T_1$ increases with higher $B/T$ for a holding field up to 1 kG.

The dominated relaxation mechanism for $^3He$ and $^{129}Xe$ in the human body is the relaxation by paramagnetic molecules and ions. Of clinical significance is the relaxation induced by paramagnetic molecular oxygen and by paramagnetic deoxyhemoglobin. Molecular oxygen has an atomic electronic spin $S = 1$. The intermolecular dipole–dipole coupling between the oxygen electronic spin and noble gas nuclear spins dominates the relaxation. $T_1$ of $^3He$ in the presence of oxygen has been measured:[25]

$$\frac{1}{T_1} = 0.45[O_2]\left(\frac{299}{T}\right)^{0.42} s^{-1} \Big/ \text{amagat} \tag{27.60}$$

where $T$ is the temperature in degrees Kelvin and $[O_2]$ is the oxygen concentration. This equation provides the quantitative basis for measuring $[O_2]$ using $^3He$ MRI. For human subject imaging, at the airway $T$ is approximately 37.5°C, so

$$\frac{1}{T_1} = \frac{1}{2.27}[O_2]s^{-1} \Big/ \text{amagat} \tag{27.61}$$

Another important aspect of noble gases MRI is the relaxation of noble gas hyperpolarization during vascular transport. The vascular transport is necessary for noble gases reaching other tissue such as brain tissues. Xe has much higher solubility in blood than He, so $^{129}Xe$ is more suitable than $^3He$ for brain imaging. Table 27.3 shows the spin–lattice relaxation times of $^{129}Xe$ in blood reported in Wolber et al.[21] At a temperature of 37°C and a field strength of 1.5 T, $T_1$ = 6.4 sec for arterial blood and $T_1$ = 4 for venous blood. This is because deoxyhemoglobin in blood is paramagnetic, while oxyhemoglobin is diamagnetic. The increased relaxation of $^{129}Xe$ for deoxygenated blood may originate from Xe interaction with paramagnetic Fe ions in the heme group of deoxygenated hemoglobin.[21] In fact, there are four iron atoms in each hemoglobin molecule, which can bind four oxygen atoms. However, this difference of relaxation times in oxygenated and deoxygenated blood could be used for functional MRI as well. The deoxygenated hemoglobin exposes their paramagnetic iron ions to Xe causing Xe spin relaxation. Therefore Xe's spin–lattice relaxation time $T_1$ increases with the blood oxygenation saturation $sO_2$ as measured by Wolber et al.[26] Note that the physiological oxygenation range from $sO_2$ = 0.6 for the venous blood to $sO_2$ = 0.98 for the arterial blood. These measurements of $T_1$ as a function of $sO_2$ also provide a quantitative basis for measuring blood and tissue oxygenation by using $^{129}Xe$ MRI.

To see the great potential of this capability of $^{129}Xe$ MRI, consider the functional proton MRI of the human brain, which has undergone explosive development in recent years. This functional proton MRI technique is based on the blood oxygenation level dependent (BOLD) effect. This BOLD effect is sensitive to changes in oxygenation and blood flow in response to neural stimulation. However, BOLD effect does not provide a direct measurement of blood oxygenation because blood flow changes are blended in as well. So $^{129}Xe$ MRI might be able to elucidate the relative contributions of perfusion and oxygenation to the BOLD effect. Finally, it should be noted that relative shift of intra- and extracellular $^{29}$Xe spectral peak is also sensitive to the blood oxygenation level.[27]

From this discussion it is also clear that delivery of hyperpolarized $^{129}$Xe from inhalation to other organs such as the brain is critically limited by loss of polarization during vascular transport. This calls for development of intravenous injection of hyperpolarized spins in biocompatible media. These media should allow much longer $T_1$ for $^{129}$Xe and high solubility. As listed in Table 27.3, $T_1$ for $^{129}$Xe has been increased to 66 s in saline and 100 s in perfluorocarbon emulsion.[28] However, perfluorocarbon emulsion has much higher Ostwald solubility $\lambda_{solv}$ than saline does, as shown in Table 27.3. (The Ostwald solubility is defined as the volume of gas dissolved per unit volume of solvent at equilibrium.) The search of biocompatible media for $^{129}$Xe and $^3$He is just beginning.[29]

For $^{129}$Xe dissolved in liquid, dipole–dipole coupling between xenon nuclear spins and dipole–dipole coupling between xenon nuclear spins and solution nuclear spins (e.g., protons) play an important role in determining xenon's $T_1$. Among these spins the spin polarization transfer from hyperpolarized xenon to solution nuclear spin is observed and called SPINOE, the spin–polarization-induced nuclear Overhauser effect.[30] The mechanism underlining the polarization transfer is the nuclear Overhauser effect generated by the cross-relaxation between Xe and solution nuclear spins. Therefore, through SPINOE, hyperpolarized Xe can "light up" solution nuclear spins by enhancing their polarization. In addition, differential SPINOE enhancements in solution should provide information of selective binding of xenon to biomolecules in solution.[31]

So far, all the spin–lattice relaxation is based on the assumption that hyperpolarized Xe relaxes in the same way as the thermally polarized Xe. In fact, some data of Xe $T_1$ in blood are measured by using thermally polarized Xe. We note that this assumption is not completely true.[32] First, the conventional relaxation laws are derived from the assumption of weak polarization near the thermal equilibrium. Second, we find that hyperpolarization and spin dipole–dipole coupling result in multipole spin orders for hyperpolarized spins. In another words, a hyperpolarized spin state may include not only the magnetization (the dipole spin order $\langle I_z \rangle$), but also other multipole order, such as the quadrupole spin order. These multipole spin orders cross-relax each other. The longitudinal magnetization relaxation of hyperpolarized spins is, in general, bi-exponential decay, although the ordinary $T_1$ component dominates in many but not all cases. These findings have potential applications in quantifying the paramagnetic source in lung and other tissue and in modeling hyperpolarized xenon spin auto-relaxation and cross relaxation in biocompatible media.[33]

### 27.4.6 MRI of Hyperpolarized $^3$He and $^{129}$Xe for Human Subjects

Hyperpolarized spin MRI has already been applied to human subjects in recent years. Here we discuss several examples in accordance with the scope of this chapter. Hyperpolarized $^3$He has been used to image human pulmonary morphology. This is done usually in a period of one breath after inhalation of $^3$He gas. The pulse sequences are usually the low-flip angle-spoiled gradient echo sequences. The spatial resolution of $^3$He MRI depends on the sampling time, $T_s$, used because, as shown by Equation 27.58, $T_s$ controls the adverse effect of the He diffusive movement in air space on the resolution. In any case, the image resolution is definitely better than that of ventilation scintigraphy.[34,35] Figure 27.11 shows $^3$He-MRI images of the lungs from a healthy volunteer and a smoker. The lung image of the smoker presented patchy and wedge-shaped signal defects due to reduced or absent entry of $^3$He.

In addition to morphological study, $^3$He gas can be used to measure the regional alveolar partial pressure of oxygen ($pO_2$). It is well known that, in the air space of the lungs, the partial pressure of oxygen varies regionally and temporally. These variations reflect the differences in ventilation-dependent oxygen delivery and perfusion-dependent oxygen uptake. Currently, there is no noninvasive means to measure $pO_2$ within occluded pulmonary segments, and the intrapulmonary $pO_2$ beyond the bronchi could not be measured even by invasive methods. However, as we pointed out earlier, the $T_1$ relaxation rate of He gas is proportional to the oxygen concentration $[O_2]$, as shown in Equation 27.61, because molecular oxygen is paramagnetic ($S = 1$).

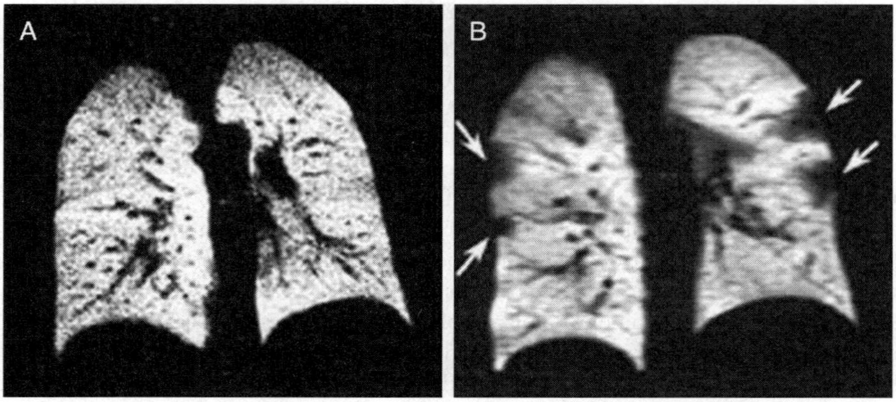


**FIGURE 27.11** (A) A $^3$He MRI image of the lungs of a healthy volunteer. (B) A $^3$He MRI image of the lungs of a smoker. (Images courtesy of Drs. W. Schreiber and R. Surkau, University of Mainz, Germany.)

Therefore, using $^3$He MRI one can derive the local alveolar $pO_2$ from the $^3$He signal decay curve generated from a series of images acquired at a series of times.[36] Briefly speaking, this can be done as follows. Suppose one acquires $N$ images, each is separated by a period $\Delta t$, and one uses a constant flip angle $\alpha$. Equation 27.49 gives the signal intensity with the RF depolarization effect, but without $T_1$ relaxation during the acquisitions. On the other hand, Equation 27.59 gives the relaxation law of the hyperpolarized magnetization of $^3$He. Compared to signal intensity of the $n$th image, the signal intensity of the $(n + 1)$th image should be reduced by the relaxation and the RF depolarization effects. Therefore, combining these two effects by using Equations 27.49 and 27.59, we can find the signal ratio of the two images:

$$q = \frac{S_{n+1}}{S_n} = \cos^{N_p - 1}\alpha \exp\left(-\frac{\Delta t}{T_1}\right) = \cos^{N_p - 1}\alpha \exp\left(-\frac{pO_2 \Delta t}{2.27}\right) \tag{27.62}$$

where $N_p$ is the total number of the phase-encoding steps per image, and the partial pressure of oxygen $pO_2$ is expressed in amagats. Once the ratio $q$ of a local region is found, then $pO_2$ can be calculated from the above equation. For further details of the technique, see Reference 36. This approach provides a quick and noninvasive way to assess ventilation–perfusion matching, since the local alveolar $pO_2$ manifests the local ventilation–perfusion ratio.

## Acknowledgments

We are grateful to Drs. W. Schreiber, R. Surkau, and J. Schmiedeskamp for providing $^3$He MRI images.

## References

1. Albert, M., Gates, G.D., Driehuys, B., Happer, W., Saam, B., Springer, C.S., and Wishinia, A., Biological magnetic resonance imaging using laser-polarized $^{129}$Xe, *Nature,* 370, 199, 1994.
2. Smith, R.C. and Lange, R.C., Eds., *Understanding Magnetic Resonance Imaging,* CRC Press, Boca Raton, FL, 1998.
3. Chen, C. and Hoult, D., *Biomedical Magnetic Resonance Technology,* Institute of Physics, London, 1989.
4. Surkau, R., Becker, J., Ebert, M., Grossman, T., Heil, W., Hofmann, D., Homblot, H., Leduc, M., Otten, E.W., Rohe, D., Siemensmeyer, K., Steiner, M., and Tasset, F., Realization of a broad band neutron spin filter with compressed, polarized $^3$He gas, *Nucl. Instr. Meth. Phys. Res.*, A 384, 444, 1997.
5. Happer, W., Optical pumping, *Rev. Mod. Phys.*, 44, 169, 1972.

6. Wagshul, M. and Chupp, T., Optical pumping of high density Rb with a broad-band dye-laser and GaAlAs diode-laser arrays: applications to $^3$He polarization, *Phys. Rev.*, A24: 4447, 1989.
7. Suter, D. and Mlynek, J., Laser excitation and detection of magnetic resonance, *Adv. Magn. Opt. Resonance,* 16, 1, 1991.
8. Cohen-Tannoudji, C., Dupont-Roc, J., and Grynberg, G., *Atom–Photon Interactions: Basic Processes and Applications,* John Wiley & Sons, New York, 1998.
9. Cummings, W., Haesser, O., Lorenzon, W., Swenson, D.R., and Larson, B., Optical pumping of Rb vapor using high-power GaAlAs diode laser arrays, *Phys. Rev.*, A51, 4842, 1995.
10. Romalis, M., Narrowing of high power diode laser arrays using reflection feedback from an etalon, *Appl. Phys. Lett.*, 77, 1080, 2000.
11. Phillips, D., Wong, G.P., Bear, D., Stoner, R.E., and Walsworth, R.L., Characterization and stabilization of fiber-coupled laser diode arrays, *Rev. Sci. Instrum.*, 70, 2905, 1999.
12. MacFall, J., Charles, H.C., Black, R.D., Middleton, H., Swartz, J.C., Saam, B., Driehuys, B., Erickson, C., Happer, W., Cates, G.D., Johnson, G.A., and Ravin, C.E., Human lung air spaces: potential for MR imaging with hyperpolarized $^3$He, *Radiology,* 200, 553, 1996.
13. Kauzor, H., Hofmann, D., Kreitner, K.F., Niljens, H., Surkay, R., Heil, W., Potthast, A., Knopp, M.V., Oten, E.W., and Thelen, M., Normal and abnormal pulmonary ventilation: visualization at hyperpolarized $^3$He MR imaging, *Radiology,* 201, 564, 1996.
14. Brauer, R., Hogan, P., Hugon, M., Macdonald, A., and Miller, K., Patterns of interaction of effects of light metabolically inert gases with those of hydrostatic pressure as such: a review, *Undersea Biomed. Res.*, 9, 353, 1982.
15. Swanson, S., Rosen, M.S., Coulter, K.P., Welsh, R.C., and Chupp, T.E., Distribution and dynamics of laser-polarized $^{129}$Xe magnetization *in vivo, Magn. Reson. Med.*, 42, 1137, 1999.
16. Walker, T. and Happer, W., Spin-exchange optical pumping of noble-gas nuclei, *Rev. Mod. Phys.*, 69, 629, 1997.
17. Driehuys, B., Cates, G.D., Miron, E., Sauer, K., Wlater, D.K., and Haper, W., High-volume production of laser polarized $^{129}$Xe, *Appl. Phys. Lett.*, 69, 1668, 1996.
18. Durand, E., Guillot, G., Darrasse, L., Tastevin, G., Nacher, P.J., Vignaud, A., Vattolo, D., and Bittoun, J., CPMG measurements and ultrafast imaging in human lungs with hyperpolarized $^3$He at low field (0.1 T), *Magn. Reson. Med.,* 47,75, 2002.
19. Tseng, C., Wong, G.P., Pomeroy, V.R., Mair, R.W., Hinton, D.P., Hoffmann, D., Stoner, R.E., Hersman, F.W., Cory, D.G., and Walsworth, R.L., Low-field MRI of laser polarized noble gas, *Phys. Rev. Lett.*, 81, 3785, 1998.
20. Zhao, L., Mulkern, R., Tseng, C., Williamson, D., Patz, S., Kraft, R., Walsworth, R., Jolez, F., and Albert, M., Pulse sequence considerations for biomedical imaging with hyperpolarized noble gas MRI, *J. Magn. Res.,* 113, 179, 1996.
21. Wolber, J., Cherubini, A., Dzik-Jurasz, A., Leach, M.O., and Bifone, A., Spin–lattice relaxation of laser-polarized xenon in human blood, *Proc. Natl. Acad. Sci. U.S.A.,* 96, 3664, 1999.
22. Patyal, B.R., Gao, J.-H., Williams, R.F., Roby, J., Saam, R., Rockwell, B.A., Thomas, R.J., Stolarski, D.J., and Fox, P.T., Longitudinal and diffusion measurements using magnetic resonance signals from laser-hyperpolarized $^{129}$Xe nuclei, *J. Mag. Res.,* 126, 58, 1997.
23. Brookeman, J., Mugler, J.R., III, Knight-Scott, J., Munger, T.M., de Lange, E.E., and Bogorad, P.L., Studies of $^3$He diffusion coefficients in the human lung: age-related distribution patterns, *Eur. Radiol.,* 9, B21, 1999.
24. Yablonskiy, D., Sukstanskii, A.L., Leawoods, J.C., Gierada, D.S., Bretthorst, L., Lefrak, S.S., Cooper, J.D., and Conradi, M.S., Quantitative *in vivo* assessment of lung microstructure at the alveolar level with hyperpolarized $^3$He diffusion MRI, *Proc. Natl. Acad. Sci. U.S.A.*, 99, 3111, 2002.
25. Saam, B., Happer, W., and Middleton, H., Nuclear relaxation of $^3$He in the presence of $O_2$, *Phys. Rev.*, A52, 862, 1995.

26. Wolber, J., Cherubini, A., Leach, M.O., and Bifone, A., On the oxygenation-dependent $^{129}Xe$ $T_1$ in blood, *NMR Biomed.,* 13, 234, 2000.
27. Wolber, J., Cherubini, A., Leach, M.O., and Bifone, A., Hyperpolarized $^{129}Xe$ NMR as a probe for blood oxygenation, *Magn. Res. Med.,* 43, 491, 2000.
28. Lavini, C., Payne, G.S., Leach, M.O., and Bifone, A., Intravenous delivery of hyperpolarized $^{129}Xe$: a compartment model, *NMR Biomed.*, 13, 238, 2000.
29. Goodson, B., Song, Y.Q., Taylor, R.E., Schepkin, V.D., Brennan, K.M., and Chingas, G.C., Budinger, T.F., Navon, G., and Pines, A., *In vivo* NMR and MRI using injection delivery of laser-polarized xenon, *Proc. Natl. Acad. Sci. U.S.A.*, 94, 14725, 1997.
30. Navon, G., Song, Y.Q., Room, T., Appelt, S., Taylor, R.E., and Pines, A., Enhancement of solution NMR and MRI with laser-polarized xenon, *Science,* 271,1848, 1996.
31. Luhmer, M., Godson, B., Song, Y.Q., Laws, D., Kaiser, L., Cyrier, M., and Pines, A., Study of xenon binding in cryptophane–a using laser-induced NMR polarization enhancement, *J. Am. Chem. Soc.,* 121, 3502, 1999.
32. Wu, X., Spin relaxation for laser-pumped hyperpolarized spins, *Proc. SPIE*, 3548, 67, 1998.
33. Wu, X., Autorelaxation and cross relaxation of hyperpolarized spins, *Radiology,* 221(P), 513, 2001.
34. Kauczor, H., Ebert, M., Kreitner, K.F., Nilgens, H., Surkau, R., Heil, W., Hofmann, D., Otten, E.W., and Thelen, M., Imaging of lung using $^{3}He$ MRI: preliminary clinical experience in 18 patients with and without lung disease, *J. Magn. Reson. Imag.*, 7, 538, 1997.
35. Donnelly, L., MacFall, J.R., McAdams, H.P., Majure, J.M., Smith, J., Frush, D.P., Bogonad, P., Charles, H.C., and Ravin, C.E., Cystic fibrosis: combined hyperpolarized $^{3}He$-enhanced and conventional proton MR imaging in the lung— preliminary observations, *Radiology,* 212, 885, 1999.
36. Deninger, A., Eberle, B., Ebert, M., Grossmann, T., Hanisch, G., Heil, W., Kauczor, H.U., Markstaller, K., Otten, E., Schreiber, W., Surkau, R., and Weiler, N., $^{3}He$-MRI-based measurements of intrapulmonary pO2 and its time course during apnea in healthy volunteers: first results, reproducibility, and technical limitations, *NMR Biomed.*, 13, 194, 2000.
37. Gatzke, M., Cates, G.D., Driehuys, B., Fox, D., Happer, W., and Saam, B., Extraordinary slow nuclear spin relaxation in frozen laser polarized $^{129}Xe$, *Phys. Rev. Lett.*, 70, 690, 1993.

# V

# Biomedical Diagnostics II: Optical Biopsy

# 28

# Fluorescence Spectroscopy for Biomedical Diagnostics

Tuan Vo-Dinh
*Oak Ridge National Laboratory*
*Oak Ridge, Tennessee*

Brian M. Cullum*
*Oak Ridge National Laboratory*
*Oak Ridge, Tennessee*



## 28.1 Introduction

In the past several decades, fluorescence spectroscopy has had a dramatic effect on many different fields of research. One field that has seen significant advancements is biomedical diagnostics. Within this field, fluorescence spectroscopy has been applied to the analysis of many different types of samples, ranging from individual biochemical species (e.g., NADH, tryptophan) to organs of living people. These studies have given rise to new methods for the early or noninvasive diagnosis of various medical conditions, including tooth decay, atherosclerosis, heart arrhythmia, cancer, and many others.

The medical condition that has seen the largest effort toward fluorescence-based analyses and truly demonstrates the potential of fluorescence-based diagnoses is cancer. Fluorescence spectroscopy and imaging have been investigated for the diagnosis of almost every type of cancer and early neoplastic difference found in humans. The detection of early neoplastic changes is important from an outcome viewpoint because, once invasive carcinoma and metastases have occurred, treatment is difficult. At present, excisional biopsy followed by histology is considered the "gold standard" for diagnosis of early neoplastic changes and carcinoma. In some cases, cytology rather than excisional biopsy is performed. These techniques are powerful diagnostic tools because they provide high-resolution spatial and morphological information of the cellular and subcellular structures of tissues. The use of staining and processing can enhance contrast and specificity of histopathology. However, both of these diagnostic procedures require physical removal of specimens, followed by tissue processing in a laboratory. These

*Current affiliation: University of Maryland Baltimore County, Baltimore, Maryland

procedures incur a relatively high cost because specimen handling is required; more importantly, diagnostic information is not available in real time. Moreover, in the context of detecting early neoplastic changes, excisional biopsy and cytology can have unacceptable false negative rates often arising from sampling errors.

Fluorescence techniques have the potential for performing *in vivo* diagnosis on tissue without the need for sample excision and processing. Another advantage of fluorescence-based diagnoses is that the resulting information can be available in real time. In addition, because removal of tissue is not required for optical diagnoses, a more complete examination of the organ of interest can be achieved than with excisional biopsy or cytology.

This chapter provides an overview of fluorescence spectroscopy and its basic principles as well as its biomedical applications. Because biomedical fluorescence spectroscopy represents an extremely large and growing field of research, we have attempted to classify the ongoing research and clinical studies into three different categories: (1) biochemical analyses of individual compounds, (2) *in vitro* analyses (cellular and tissue systems), and (3) *in vivo* analyses (animal and human studies). In addition to the overview of fluorescence spectroscopy in this chapter, special methods and instrumental aspects for fluorescence analyses are further described in other chapters in this handbook: Chapter 33 on near-infrared fluorescence imaging and spectroscopy in random media and tissues, Chapter 9 on lifetime-based imaging, microscopy, Chapter 10 on confocal miscroscopy, and Chapter 11 on two-photon excitation fluorescence microscopy.

## 28.2 Principles of Fluorescence Spectroscopy

### 28.2.1 Fluorescence Techniques

Luminescence is a branch of spectroscopy that deals primarily with the electronic states of an atom or molecule, as opposed to the vibrational, rotational, or nuclear energy states. Fluorescence and phosphorescence are the two sister techniques covered by the umbrella known as "luminescence spectroscopy." These techniques involve the optical detection and spectral analysis of light emitted by a substance undergoing a transition from an excited electronic state to a lower electronic state. Most luminescence-based medical diagnoses involve fluorescence spectroscopy because organic compounds fluoresce and do not phosphoresce at room temperature. In most phosphorescence studies, the measurements need to be performed using low-temperature frozen solvents to minimize collisional quenching. A unique technique called room temperature phosphorescence (RTP) uses various solid substrates to adsorb analytes into rigid matrices (e.g., cellulose, silica gel, and alumina) in order to allow phosphorescence measurements.[1] Because the vast majority of medical applications involve fluorescence analysis, this chapter focuses mainly on the theory of fluorescence. However, because phosphorescence is also involved in the field of photodynamic therapy (PDT), we will also describe the relationship between fluorescence and phosphorescence.

Fluorescence diagnostic methods can be grouped into two main categories:

1. Methods that detect endogenous fluorophores in tissues
2. Methods that detect or use exogenous fluorophores or fluorophore precursors [such as 5-aminolevulinic acid (ALA)]

Fluorescence that originates from native fluorescent chromophores already present in the tissue is often referred to as "autofluorescence." Fluorescence may also originate from administered exogenous chromophores that have been synthesized to target specific tissues (e.g., dysplastic vs. normal), or it may be activated by functional changes in the tissue. In a fluorescence analysis, excitation light at some specific wavelength (typically near-ultraviolet or visible) excites the tissue or exogenous fluorophore molecules and induces fluorescence emission. Then some measure of the fluorescent emission is obtained, typically an emission spectrum (fluorescence emission intensity vs. wavelength).

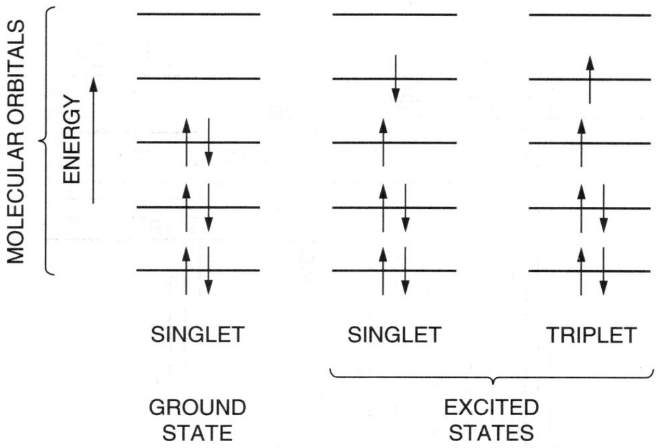


**FIGURE 28.1** Schematic diagram of typical spin arrangements in molecular orbitals for the ground, excited singlet, and triplet states.

## 28.2.2 Photophysical Basis of Luminescence

Most organic molecules contain an even number of electrons. In the ground state, the electrons fill the various atomic orbitals with the lowest energies in pairs. By the Pauli exclusion principle, two electrons in one given orbital must have spins in opposite directions, and the total spin $S$ must equal 0. For this reason, the ground state has no net electron spin; this state is called a "singlet" state.

Excitation of a molecule usually results in the promotion of one electron from the highest occupied orbital to a previously unoccupied orbital (also referred to as a "virtual" orbital). Whereas the ground state is generally a singlet, the excited state can be a singlet or a triplet, depending on the final spin state of the electron promoted to the higher orbital. For illustration, Figure 28.1 shows one ground state and two excited-state configurations. The spins of the two electrons in the unoccupied orbitals are no longer restricted by the Pauli exclusion principle. The singlet state has two antiparallel spins, whereas the triplet state has two parallel spins and a net spin, $S$, of 1. In the general case, the configuration is not so easily visualized by schematic diagrams.[2]

The nature of the orbitals involved in an electronic transition is an important factor in determining the luminescence characteristics of the molecule. Organic compounds that are known to be strongly luminescent are molecules with extensive $\pi$ electron systems. These compounds include aromatic molecules and a few unsaturated aliphatics. For molecules with no hetero-atom, electronic transitions generally involve promotion of an electron from a bonding $\pi$ orbital to an antibonding $\pi^*$ orbital. Such a transition is known as a $\pi - \pi^*$ transition, and the resulting electronic state is called the $\pi\pi^*$ excited state. For conjugated systems with hetero-atoms (N, O, or S) in the conjugated system, or with substituents containing the N, O, or S atoms, an electronic state may result from the promotion of an electron from a bonding $n$ orbital to an antibonding $\pi^*$ orbital; this electronic state is called an $n\pi^*$ excited state.

### 28.2.2.1 Molecular Electronic Energies

In the study of luminescence, we are first concerned with determination of the electronic states between which luminescence and other related processes such as intersystem crossing, vibrational relaxation, and internal conversion occur. The basic approach for studying luminescence behavior of molecules is to calculate the stationary-state electronic energies by solving the time-independent Schrödinger equation. A suitable wave function, $\Psi$, chosen to describe the properties of various molecular orbitals (MO) involved in the photophysical processes, is given by the Schrödinger equation

$$H(q,Q)\Psi(q,Q) = E\Psi(q,Q), \tag{28.1}$$

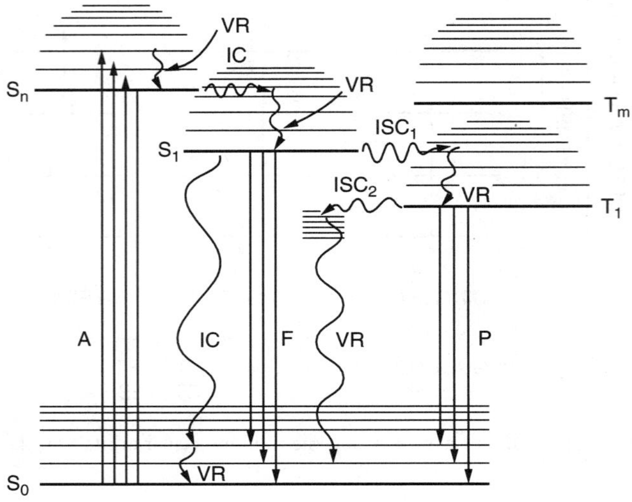


**FIGURE 28.2** Jablonskii diagram showing the different radiative and nonradiative transitions in a molecule upon excitation into a singlet state, $S_n$: $S_0$ = singlet ground state; $S_1$ = first excited singlet state; $T_1$ = first excited triplet state; $T_m$ = excited triplet state; ISC, $ISC_1$, and $ISC_2$ = intersystem crossings; A = absorption; F = fluorescence; IC = internal conversion; P = phosphorescence; and VR = vibrational relaxation.

where $H$ is the Hamiltonian operator of the molecule, $q$ is the electronic spatial coordinates, $Q$ is the nuclear spatial coordinates, and $E$ is the energy eigenvalue.

The exact solutions for the Schrödinger equation are impossible to calculate for any molecular system except $H^+$. Therefore, calculations of electronic energies for excited singlet and triplet states of larger systems usually involve some method of approximation. The most commonly used method is the MO method, which utilizes one-electron orbitals to form molecular orbitals. One of the simplest of these MO-based methods for calculating, or rather estimating, molecular electronic energies is the Hückel molecular orbital (HMO) method, which treats certain integrals as parameters and selects these parameters such that calculated values are in agreement with physical observables. This approach has been quite successful in describing bonding in aromatic and highly conjugated molecules.

Figure 28.2, known as the Jablonskii diagram, shows energy levels for an organic molecule. In this figure, $S_0 \ldots, S_n$ and $T_1 \ldots, T_m$ represent the discrete electronic energy levels of a molecule. The electronic state having the lowest energy, $S_0$ , is known as the ground state. $S_0 \ldots, S_n$ and $T_1 \ldots, T_m$ are the excited singlet and triplet states, respectively. Each electronic state has its own set of vibrational levels. Superimposed on each vibrational level is a series of closely spaced rotational levels not shown in Figure 28.2.

### 28.2.2.2 Population of the Excited Electronic States

The main photophysical processes involved in the population and deactivation of excited electronic states, both singlet and triplet, are shown in Figure 28.2. These processes — absorption (A), vibrational relaxation (VR), internal conversion (IC), fluorescence (F), intersystem crossing (ISC), and phosphorescence (P) — and related photophysical processes are discussed next.

***Absorption (A)***

At equilibrium, a group of molecules has a thermal distribution at the lowest vibrational and rotational levels of the ground state, $S_0$. When a molecule absorbs excitation energy, it is elevated from $S_0$ to some vibrational level of one of the excited singlet states, $S_n$, in the manifold $S_1, \ldots, S_n$. As postulated by the Franck–Condon principle, the time for an electronic transition (on the order of $10^{-15}$ s) is shorter than the time for nuclear rearrangement (on the order of $10^{-13}$ to $10^{-14}$ s). As a consequence, the molecular configuration following absorption is often a nonequilibrium nuclear configuration because the electronic absorption process is so rapid that the atomic nuclei do not have time to change positions or moments. Thus, the adiabatic (Born-Oppenheimer) approximation is often applied to solve the Schrödinger equation (Equation 28.1) and partially separates the eigenstates of the electronic ("fast") and nuclear ("slow") subsystems. The intensity of the absorption (i.e., fraction of ground-state molecules promoted to the electronic excited state) depends on the intensity of the excitation radiation (i.e., number of photons) and the probability of the transition with photons of the particular energy used. A term often used to characterize the intensity of an absorption (or induced emission) band is the oscillator strength, $f$, which may be defined from the integrated absorption spectrum by the relationship

$$f = 4.315 \times 10^{-9} \int \varepsilon_\nu d\nu, \tag{28.2}$$

where $\varepsilon_\nu$ is the molar extinction coefficient at the frequency $\nu$.

Oscillator strengths of unity or near unity correspond to strongly allowed transitions, whereas smaller values of $f$ indicate smaller transition dipole matrix elements of forbidden transitions.

### *Vibrational Relaxation (VR)*

In condensed media, the molecules in the $S_n$ state deactivate rapidly, within $10^{-13}$ to $10^{-11}$ s via *VR* processes, ensuring that they are in the lowest vibrational levels of $S_n$ possible as described by the thermal Boltzmann distribution. Because the VR process is faster than electronic transitions, any excess vibrational energy is rapidly lost as the molecules are deactivated to lower vibronic levels of the corresponding excited electronic state. This excess VR energy is released as thermal energy to the surrounding medium.

Only under special circumstances — for example, in the gas phase at very low pressures ($<10^{-6}$ torr) — may emission be observed from higher vibrational levels of the excited state. These emission mechanisms, known as single-vibronic-level (SVL) luminescence processes, do not generally occur in condensed media such as tissues.

### *Internal Conversion (IC)*

From the $S_n$ state, the molecule deactivates rapidly to the isoenergetic vibrational level of a lower electronic state such as $S_{n-1}$ via an IC process. IC processes are transitions between states of the same multiplicity. The molecule subsequently deactivates to the lowest vibronic levels of $S_{n-1}$ via a VR process. By a succession of IC processes immediately followed by VR processes, the molecule deactivates rapidly to $S_1$. As a result, a molecule may be excited to $S_1$ or higher excited states, $S_n$, depending on the excitation energy used, but the emission takes place only from the lowest excited electronic state, $S_1$. This is known as Kasha's rule. Fluorescence emission from upper electronic states (above $S_1$ or $T_1$) of polyatomic molecules, however, is a rare phenomenon in the condensed phase.

After vibrational relaxation or internal conversion, the thermal population at a given electronic state is determined by the Boltzmann distribution, which gives the relative population of species in a given electronic or vibronic state as compared with those in all other states. The Boltzmann distribution is expressed by

$$\frac{n_i}{\sum_j n_j} = \frac{g_i \exp(-E_i/kT)}{\sum_j g_j \exp(-E_i/kT)} \tag{28.3}$$

where $n_i$ is the number of molecules in electronic or vibronic state $I$, $g_i$ is the statistical weight of state of that state, $E_i$ is the energy of state $I$, $k$ is the Boltzmann constant, and $T$ is the temperature.

### *Fluorescence (F)*

Relaxation of the molecule from the $S_1$ state may occur by one of a several different of processes, including: (1) further deactivation to $S_0$ in a radiationless fashion via internal conversion (IC) and VR processes or (2) emission of a photon without a change in spin multiplicity. The latter $S_1 \rightarrow S_0$ transition is known as fluorescence. The energy of the resulting photon corresponds to the difference in energy between the lowest vibrational level of the excited state and the vibrational level of the ground state. Thus, the fluorescence spectrum primarily provides information about the vibrational structure of the ground state.

These radiative transitions are electric-dipole in nature. The probability of a radiative transition ($S_i \rightarrow S_j$) is proportional to the square of the matrix element, $|(\mathbf{M})_{ij}|^2$, where **M** is the electric dipole moment defined by

$$M\left(S_i \rightarrow S_j\right) = \left\langle \psi\left(S_i\right) \middle| \sum_q e\mathbf{r}_q \middle| \psi\left(S_j\right) \right\rangle \tag{28.4}$$

where $e$ is the electronic charge, $\mathbf{r}_q$ is the position vector of electron $q$, and $\psi$ $(S_i)$ is the wave function of electronic state $S_i$. For fluorescence ($i = 0, j = 1$) the electronic transition corresponds to $M_{1,0}{}^2$.

***Intersystem Crossing (ISC)***

Another process through which molecules in the $S_i$ state may relax is through a transition to some vibrational level of the triplet manifold via a mechanism known as an intersystem crossing (ISC) process ($ISC_1$ transition shown in Figure 28.2). The ISC process involves a change in spin multiplicity. The molecule then relaxes to the lowest vibrational level of $T_1$ via successive IC and VR processes.

From this $T_1$ state, the molecule may return to the ground state $S_0$ either by a radiationless deactivation path ($ISC_2$ process shown in Figure 28.2) or by the emission of a photon. This latter radiative transition is known as phosphorescence.

The radiationless deactivation that quite often competes with phosphorescence is the reverse ISC process ($ISC_2$). This process, which is often dominant, determines the observed phosphorescence lifetime. The nonradiative rate constant corresponding to the ISC process $T_1 \rightarrow S_0$ is strongly dependent on the magnitude of the energy gap $\Delta E(T_1 \rightarrow S_0)$. The rate constant decreases exponentially with increasing $\Delta E$; this feature is known as the energy gap law. The energy gap law not only is specific to ISC but also applies to other nonradiative transitions such as IC.

***Phosphorescence and the Triplet State***

The triplet state manifold consists of states $T_1$, $T_2$,... $T_n$, which have energy levels lower than the corresponding singlet states $S_1$, $S_2$,... $S_n$. This is a consequence of Hund's rule, which states that the energy associated with a state having parallel spins is always lower than that of the corresponding one with antiparallel spins. Qualitatively, Hund's rule may be explained by the tendency for electrons with parallel spins to be further separated in space (i.e., to lie in different molecular orbitals) and consequently experience less coulombic interaction. As a result of Hund's rule, phosphorescence occurs at longer wavelengths than does fluorescence.[1,2]

Properties of singlet and triplet states are significantly different. The singlet state is diamagnetic, whereas the triplet state is paramagnetic. The nomenclature "singlet" and "triplet" is derived from multiplicity considerations of the energy level splitting that occurs when the molecule is exposed to a magnetic field. Under the application of an external magnetic field, the triplet state with a spin magnetic momentum splits into three Zeeman levels. The singlet state, which has no magnetic momentum, is unchanged.

***Spin–Orbit Coupling***

Production of the triplet state in some compounds is important to photosensitization in certain biomedical applications, such as photodynamic therapy (PDT). A process that leads to the production of the triplet state is spin–orbit coupling. Transitions between different excited singlet states of the same multiplicity ($S_n \rightarrow S_m$ or $T_n \rightarrow T_m$) may occur easily, but transitions between pure spin states of different multiplicities ($S_n \rightarrow T_n$) are forbidden by quantum mechanics under the spin-selection rule ($\Delta S = 0$). Spin-forbidden transitions, however, do occur under certain conditions. If the spin and orbital motions of the electrons are independent of one another, triplet–singlet transitions do not give rise to electric dipole radiation because of the spin selection rule. However, the spin and orbital motions are actually not independent. Basically, the occurrence of spin-forbidden transitions — for example, singlet–triplet transitions — is made possible by coupling of the electron spin with the orbital angular momentum, a nonclassical phenomenon that produces a quantum mechanical mixing of states of different multiplicities. Qualitatively, this mechanism may be explained by the fact that the orbital motion of the electron induces a magnetic field that interacts with its spin magnetic moment. This interaction results in a change of the direction of the spin angular momentum of the electron.[2] This phenomenon is known as spin–orbit (S–O) coupling.

The S–O coupling mechanism actually mixes some singlet character into triplet states and vice versa. This process therefore removes the spin-forbidden nature of the transitions between pure-spin states. Because of the importance of S–O coupling in understanding the phosphorescence process, it is essential to consider briefly the theoretical treatment of S–O coupling.

***General Considerations for Nonradiative Properties***

The VR, IC, and ISC processes are known as "radiationless transitions." In the framework of the Born-Oppenheimer approximation, in which the nuclear and electronic motions are assumed to be independent of each other, the probability of $W_{mn}$ of a radiationless transition between two electronic states $n$ and $m$ can be treated by perturbation theory and is given by the following expression:[1,2]

$$W_{mn} = \frac{8\pi^2\tau}{h^2}\prod_{ij}\langle\theta_{mi}|\theta_{nj}\rangle^2\langle\Phi_m|H'|\Phi_n\rangle^2 \tag{28.5}$$

where $h$ is Planck's constant, $\theta_{mn}$ is the vibrational wave function $i$ of the electronic state $m$, $\theta_m$ is the electronic wave function of the electronic state $m$, $\tau$ is the relaxation time of the vibronic levels, $H'$ is the perturbing Hamiltonian and $\Pi_{ij}$ is the product over all vibronic states $i$ and $j$.

For radiationless transitions between states of similar multiplicity — for example, VR and IC processes — $H'$ may be the electron–electron repulsion term or the vibronic interaction term. For radiationless transitions between states of different multiplicity — for example, ISC processes — $H'$ is the spin–orbit coupling term.

A common expression that describes the probability $W_{mn}$ is the so-called Fermi Golden Rule for the statistical limit of a dense manifold for the final states. In this case $W_{mn}$ is given by

$$W_{mn} = \frac{2\pi}{h}V_{mn}\cdot\rho$$

where $V_{mn}$ is the matrix element of the perturbation between the initial state $m$ and final state $n$, $\rho$ is the density of the final state, and $h$ is Planck's constant.

***Delayed Fluorescence (DF)***

Fluorescence and phosphorescence are the two most common luminescence processes. A less common emission process is delayed fluorescence (DF). With some chemical systems, the molecule in the triplet state $T_1$ reverts back to the excited singlet state manifold. Because $T_1$ is always of lower energy than $S_1$, this transition requires some additional activation energy. Delayed fluorescence may occur from the $S_1$ state, exhibiting a spectrum identical to conventional fluorescence (often called "prompt fluorescence") but having a longer lifetime due to the additional stay in the triplet state. Two types of DF processes can be differentiated: (1) an eosin-type (E-type) DF produced by repopulation of $S_1$ from the triplet state by thermal activation, and (2) a pyrene-type (P-type) DF produced when pairs of triplet state molecules interact, providing an activation energy greater than or equal to the $S_1$ energy. At room temperature, E-type DF can be an important mechanism and has been observed for many organic compounds adsorbed on solid substrates.

## 28.3 Characterization of Luminescence

Several physical observables may be used to characterize fluorescence and phosphorescence. These include the following:

1. Emission, excitation, and synchronous spectra
2. Quantum yields
3. Lifetimes
4. Polarization

### 28.3.1 Emission, Excitation, and Synchronous Spectra

It is first important to determine the positions and nature of the energy levels of the electronic states involved in the luminescence excitation and emission processes because these energy levels are characteristic properties of the compounds of interest. In general, the simplest method of studying the energy levels of the excited state is absorption spectroscopy. An alternative method for investigating the absorption process for luminescent compounds is excitation spectroscopy.

The general relationship between the absorption and emission spectra and the vibrational levels of the electronic states is illustrated in Figure 28.3. Because a fluorescence emission spectrum is due to radiative decay from $S_i$ to different levels of $S_0$, the emission spectrum exhibits the vibrational frequencies that correspond to the vibrations of the molecule in its ground state, whereas the absorption spectrum exhibits frequencies of the excited state. The intensities of the vibrational bands are determined in quantum theory by the magnitude of the wave-function overlap (Franck–Condon coefficients) for the various vibrational levels in the ground and excited states (Figures 28.3B and 28.3C). The overlap coefficients are the Franck–Condon factors.

Figure 28.3A shows the potential energy curves of the excited and ground states and the transitions corresponding to an excitation (or absorption) and emission process. Figures 28.3B and 28.3C show two typical cases of electronic states with their potential minima at different relative positions on the internuclear distance coordinate $R$. In Figure 28.3B, the strongest transition is the 0–0 transition, at which the overlap is the largest. In Figure 28.3C, the most probable transition is the 0–4 transition involving the vibrational level $\nu' = 4$ of the ground state and the vibrational level $\nu' = 0$ of the excited state. The envelopes of the vibrational progressions illustrated in Figures 28.3B and 28.3C depend on the relative displacement $\delta R$ between the potential minima of the excited and ground states; for the simple case in which the ground vibrational level is thermally populated, the shape of the envelopes is determined by the Pekarian formula. For $\delta R = 0$, there is no vibrational progression; for large $\delta R$, the Pekarian distribution becomes a Gaussian distribution. Examination of the vibronic structure of the fluorescence, or phosphorescence, spectrum can therefore yield important information about the potential curves of the singlet and ground states or triplet and ground states, respectively.

With conventional emission (excitation) spectra, the excitation (emission) wavelength is fixed while the emission wavelength is scanned over the spectral region of interest. With synchronous spectra, excitation and emission wavelengths are scanned synchronously while maintaining a constant wavelength interval between them.[3,4]

### 28.3.2 Quantum Yields

Emission quantum efficiencies, also known as quantum yields, determine the sensitivity of the luminescence measurements. Quantum efficiencies are intrinsic molecular parameters of a compound under given conditions of temperature, solvent, and other environmental factors.

The fluorescence quantum efficiency $\Phi_F$ is defined as:

$$\Phi_F = \frac{\text{Number of luminescence photons}}{\text{Number of photons absorbed}} \tag{28.6}$$

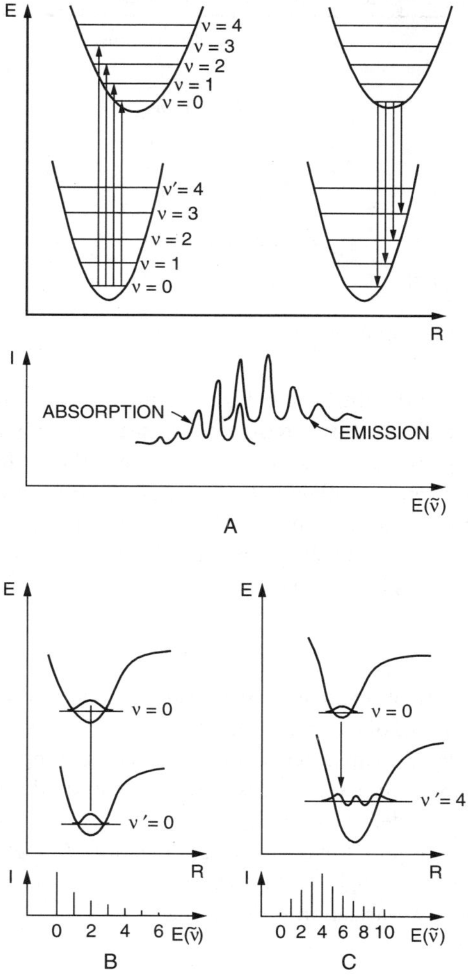


**FIGURE 28.3** (A) Diagrams of the potential energy curves of two electronic states showing the absorption and emission transitions between these two states and the typical intensity distribution of the corresponding spectra. (B) Potential energy curves with unchanged potential minima. The most intense transition is 0–0 (maximum overlap of the vibronic wave functions). (C) Potential energy curves with potential minima shifted with respect to each other. In this example, the most intense transition is 0–4.

The phosphorescence quantum efficiency $\Phi_P$ is defined in a similar manner. Quantum efficiencies can also be defined for processes other than emission and are used to define the fraction of molecules that undergoes a specific process.

## 28.3.3 Lifetimes

Another means for characterizing a luminescence emission process is the determination of its lifetime. The luminescence lifetime, $\tau$, is defined as the time required for the emission to decrease to 1/e of its original intensity following a $\delta$-pulse excitation.

If $I(t)$ is the luminescence intensity at time $t$ and $I_0$ is the intensity at $t = 0$, then $\tau$ is given by:

$$I_L(t) = I_0 \exp(-t/\tau) \tag{28.7}$$

**TABLE 28.1** Radiative Lifetimes and Oscillator Strengths in the Electronic Transitions for Polynuclear Aromatic Compounds

| Origin of Transitions | Radiative Lifetime (s) | Oscillator Strength |
|---|---|---|
| | ***$S_1 \rightarrow S_0$ Transitions*** | |
| $\pi\pi^*$ | $10^{-7}$–$10^{-9}$ | $10^{-3}$–1 |
| $\pi\sigma$ or $\sigma\pi^*$ | $> 10^{-6}$ | $< 10^{-2}$ |
| $n\pi^*$ | $\sim 10^{-6}$ | $\sim 10^{-6}$ |
| | ***$T_1 \rightarrow T_0$ Transitions*** | |
| $n\pi^*$ (or $\sigma\pi^*$ or $\pi\sigma^*$) | $10^{-2}$–$10^{-4}$ | $10^{-7}$–$10^{-5}$ |
| $\pi\pi^*$ in halo-aromatic | 1–$10^{-3}$ | $10^{-9}$–$10^{-6}$ |
| $\pi\pi^*$ in unsaturated hydrocarbons | $\sim 10^{-2}$ | $10^{-11}$ |

It is important to differentiate the intrinsic natural lifetime, $\tau^*$, from the observed lifetime, $\tau$, for a given radiative process. The intrinsic natural lifetime, $\tau^*$, is measured only when there is no radiationless deactivation process competing with the emission — that is, when $\Phi_L = 1$ for the resulting luminescence.

Experimentally, fluorescence can be differentiated from phosphorescence by its shorter lifetime. Fluorescence lifetimes of organic molecules are on the order of $10^{-9}$ to $10^{-7}$ s, while phosphorescence lifetimes range from $10^{-3}$ s to several seconds. Table 28.1 compares the oscillator strengths and radiative lifetimes of fluorescence ($S_1 \rightarrow S_0$) and phosphorescence ($T_1 \rightarrow S_0$) emission and absorption of different origins.

### 28.3.4 Polarization

Polarization is another physically observable characteristic of luminescence. Polarization is caused by unique symmetries and orientations of electric moment vectors and wavefunctions involved in electronic transistions. The electric dipole moment determines the direction along which charge is displaced in a molecule undergoing an electronic transition. It is possible to study the polarization of the transition using polarized light to excite and detect luminescence.

A common method for determining the degree of polarization, *P*, is photoselection. In this method, one excites the sample with polarized light and measures the luminescence intensity along two perpendicular directions. The degree of polarization is defined as

$$P = \frac{I_{EE} - I_{EB}}{I_{EE} + I_{EB}} \tag{28.8}$$

where $I_{EE}$ and $I_{EB}$ are the luminescence intensities measured along the direction parallel and perpendicular to the excitation electric vector, *E*, respectively. *P* is related to the angle, $\theta$, between the absorption vector and the emission vector by the relationship

$$P = \frac{3\cos^2\theta}{\cos^2\theta + 3} \tag{28.9}$$

If the absorption and emission vectors are parallel ($\theta = 0°$), $P = 0.5$. If these vectors are perpendicular ($\theta = 90°$), $P = -0.33$. In practice, *P* has a value between these two limiting cases.

## 28.4 Biomedical Applications

Biomedical fluorescence spectroscopy is an extremely large and growing field of research. In an attempt to classify the various types of research and clinical studies that are being performed, we have grouped this field into three categories:

1. Biochemical analyses of individual compounds
2. *In vitro* analyses
3. *In vivo* analyses

Biochemical analyses are the basis for the fundamental or general studies of individual biochemical compounds that form the basic constituents of biological samples (i.e., tryptophan, NAD, NADH, hemoglobin, etc.). From these studies the basic fluorescence properties of common fluorophores are generally determined and characterized. The second category, *in vitro* analyses, is further divided into two subcategories: cellular measurement and tissue measurement. These analyses typically provide additional information associated with complex systems such as cells and tissues. The third classification category, *in vivo* analyses, can also be subdivided into two more subcategories: animal studies and human clinical studies. These experiments are generally performed only for fluorescence analysis techniques meant to be used for optical diagnostic procedures. In addition to classification based upon the type of analysis performed, it is also possible to subcategorize based upon the type of analyte measured and whether it is an endogenous species or an exogenous fluorophore (e.g., used as a label or marker for the tissue area of interest).

## 28.4.1 Biochemical Analysis of Individual Species

Fluorescence analyses of a wide variety of biochemical species that have been applied to and extracted from biological samples have been performed for quite a long time. These analyses are generally classified based upon whether or not the fluorescent species is an endogenous (naturally occurring) or exogenous (externally administered) fluorophore.

### 28.4.1.1 Endogenous Fluorophores

A large number of endogenous fluorophores, autofluorescent species, exist within biological samples.[5–7] The majority of these species are typically associated with the structural matrix of tissues (e.g., collagen[8] and elastin[9]) or with various cellular metabolic pathways (e.g., NAD and NADH).[10] In the case of structural matrices, the most common fluorophores are collagen and elastin. The resulting fluorescence from these compounds is due primarily to the cross linking of various amino acids in their structure, which typically provides the conjugated systems that are generally found in most fluorescent molecules. In addition, because of the various degrees of cross linking that can occur in these species, the resulting fluorescence emission spectrum is typically very broad and featureless, ranging anywhere from approximately 325 to 600 nm.

Some of the most intense endogenous fluorophores that exist in humans and animals are involved in cellular metabolism. The predominant species in this category include the reduced form of nicotinamide adenine dinucleotide (NADH),[11] the various flavins (FAD, etc.),[12] and the strongly fluorescent lipopigments (e.g., lipofuscin, and ceroids).[13,14] In addition to these species, many other endogenous fluorophores exist, exhibiting emission of various strengths and covering various spectral ranges in the ultraviolet (UV) and visible regions of the electromagnetic spectrum. These fluorophores include aromatics, amino acids such as tryptophan, tyrosine, and phenylalanine,[6] various porphyrins (hemoglobin, myoglobin, etc.),[6] and in certain cases and locations, red porphyrin fluorescence due to bacteria.[10] Figure 28.4 shows the absorption and fluorescence spectra of various tissue fluorophores.

These biochemical species are generally structural or metabolic, so they can provide a significant amount of information about differences in tissues. One of the most common reasons for investigating these fluorescent species is the possibility for diagnosis of various diseases (e.g., cancer) without requiring exogenous fluorescent markers or tags. Because cells in various disease states often undergo different rates of metabolism, or have different structures, there are often distinct differences in their fluorescent emission spectra. These differences in fluorescence emission generally depend on at least one of the following parameters: fluorophore concentration or spatial distribution throughout the tissue, local microenvironment surrounding the fluorophores, the particular tissue architecture being interrogated, and a wavelength-dependent light attenuation due to differences in the amount of nonfluorescing chromophores.

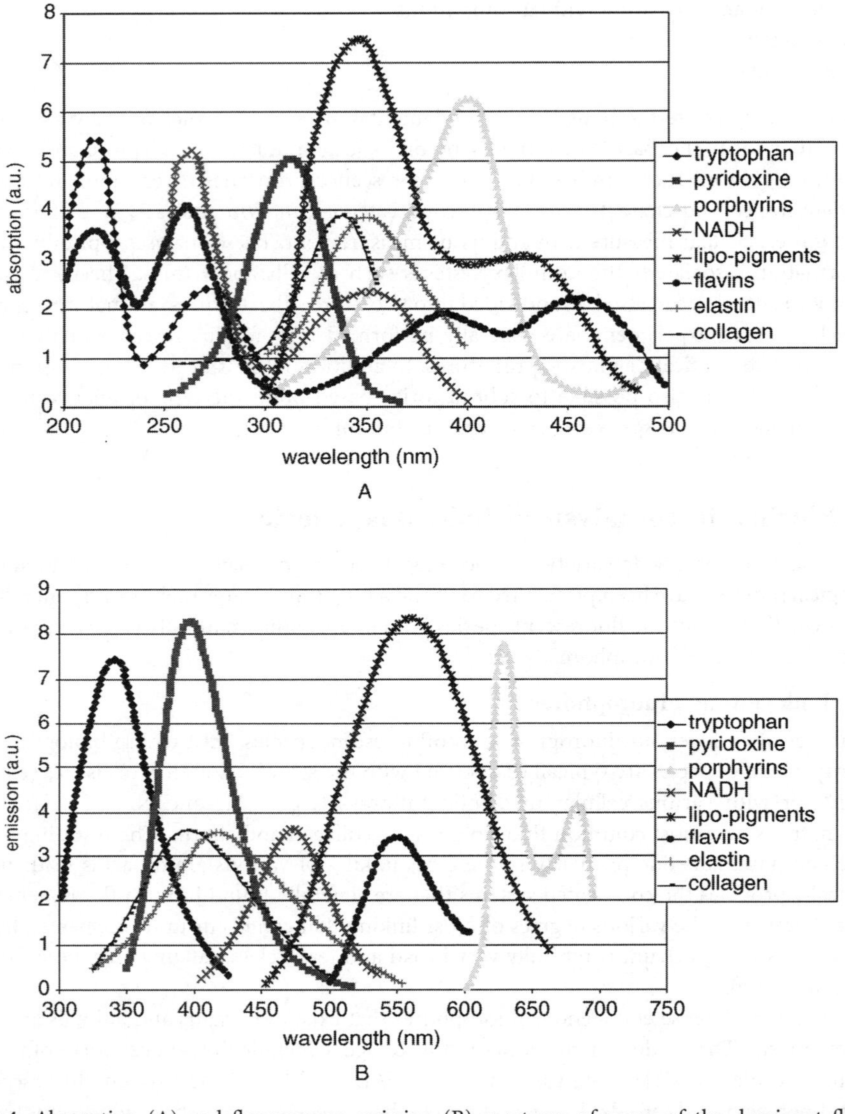


**FIGURE 28.4** Absorption (A) and fluorescence emission (B) spectrum of many of the dominant fluorophores present in tissue. (Adapted from Wagnières, G.A., Star, W.M., and Wilson, B.C., *Photochem. Photobiol.*, 68, 603, 1998.)

The first of these reasons, change in fluorophore concentration, can be attributed to many different events. In certain cases, various chemicals may not be produced in a diseased tissue. In other cases, the distribution or form of the fluorophore may vary throughout the tissue. For instance, NADH is an extremely fluorescent molecule in its reduced form but is nonfluorescent in the oxidized form. Therefore, depending on the specific phase of a particular metabolic reaction that may be occurring, the relative faction of NADH in the reduced and oxidized forms may vary greatly. Another reason for differences in the fluorescence emission from various tissue types is the difference in the local environment surrounding the fluorophore.

Change in the microstructure (local environment) surrounding a particular fluorophore can have significant effects on its fluorescence properties. Fluorescence properties that are often affected by such changes include the fluorescence quantum yield, the spectral position of its fluorescence emission maximum, its spectral line width, and its fluorescence lifetime. All these factors can have a significant impact on the overall fluorescence emission of the entire tissue sample. In addition to the local environment surrounding the fluorophore, the larger macrotissue structure or architecture also plays a significant role in the resulting fluorescence because of its specific optical properties (e.g., refractive index, scattering cross sections). This feature is described in more detail in Section I (Photons and Tissue Optics) chapters of this handbook.

An important factor that often plays a significant role in the differences associated with various disease states of tissue is the presence of nonfluorescing chromophores (hemoglobin, etc.). These chromophores, which can vary in concentration depending upon the state of the tissue, can absorb various wavelengths of fluorescent light emitted from the tissue or even absorb specific wavelengths of the excitation light used for the analysis. One example of this case is in the fluorescence diagnosis of tumor tissue, which typically has an increase in vascularization over normal tissues due to angiogenesis. Because of this increase in vascularization, the increased presence of hemoglobin often causes increased absorption of light in the visible region of the electromagnetic spectrum.

Due to this complex nature of biological tissues, extensive studies have been performed on individual species at the fundamental biochemical level. These fundamental studies of the absorption and fluorescence properties of various biochemical fluorophores give us a better understanding of the complex interactions that may be occurring in tissue samples.

In addition to conventional spectrally resolved fluorescence analyses, which are capable of providing a significant amount of information about the various fluorophores present, as well as some information about their local environment, some work in the field of time-resolved fluorescence analyses has also taken place. An in-depth analysis of this subject can be found in Chapter 9 of this handbook. Time-resolved spectroscopic analyses can provide information about molecular interactions and motions that occur in the picosecond–nanosecond time-scale range. Understanding these interactions can be especially useful in the analysis of biomolecular structure and dynamics. Recent advances in time-resolved fluorescence spectroscopy in the field of biological analyses have led to a better understanding of the origin of nonexponential fluorescence decay in proteins, the use of tryptophan analogs as unique spectroscopic probes of protein–protein interactions, the detailed characterization of protein-folding processes and intermediates, the development of new approaches to the study of DNA-protein interactions,[15] and the analysis of subdomains in proteins.[16,17]

### 28.4.1.2 Exogenous Fluorophores and Molecular Markers

In addition to the large number of endogenous fluorophores that have been used for biological diagnostics, many different exogenous fluorophores have been created and studied as well. These fluorophores have been created for many different purposes and applications in the field of biological monitoring, ranging from monitoring cellular function using fluorescent reporter dyes or molecules for various biochemical species ($Ca^{+2}$, $Mg^{+2}$, pH, nucleic acid sequences, etc.) to the demarcation of tumors with a relatively recent class of exogenous fluorophores. In the case of the monitoring of cellular function and chemical distribution, a large number of dyes with varying fluorescence properties for most of the more common cellular species can be obtained commercially from companies such as Molecular Probes, Inc. (Eugene, OR). These fluorescent reporter dyes are typically used to monitor the distribution of important chemical species throughout a cell by obtaining fluorescence microscopy images after injecting it with the dye. In addition to simply monitoring the distribution of certain chemicals, several dyes can be used to determine the viability of the cells or the permeability of their membranes. A more detailed description of these dyes and the types of analyses of which they are capable is presented in Section 28.4.2.

A second type of fluorescent reporter dye developed recently is known as a molecular beacon.[18–21] These molecular beacons are constructed by attaching a fluorescent dye (e.g., fluorescein) to one end of an oligonucleotide sequence, and a quencher molecule (e.g., DABSYL) to the other end. The particular oligonucleotide sequence chosen for the molecular beacon comprises the complementary sequence to the RNA or DNA sequence to be measured. For the molecular beacon to function properly, the last several nucleic acids on either end of the oligonucleotide strand must be complementary to each other. This forces the molecule to assume a hairpin form in its native state that brings the quencher molecule in close proximity to the fluorophore of the other end. The proximity of the quencher molecule to the fluorophore prevents the fluorophore from emitting any fluorescent light. However, when the molecular beacon is in the presence of its complementary sequence to the remainder of the oligonucleotide, the beacon is opened up, thus separating the quencher and fluorophore molecules and allowing the fluorescent dye to emit. This feature is described in more detail in Chapter 57 of this handbook.

In addition to fluorescent reporter dyes and molecular beacons, a third major area of research in biologically useful exogenous fluorophores has been in the area of tumor demarcation or sensitization. The most common of these tumor markers also act as sensitizers, or tumor-killing agents, for a unique scheme of cancer treatment known as photodynamic therapy (PDT). (A much more detailed description of PDT is offered in Chapter 36 of this handbook.) Some of the most commonly used and best characterized of these photosenstizers are hematoporphyrin derivatives (HpD),[22,23] phenophorbide-a,[24] mTHPC,[25] benzoporphyrin derivatives (BPD),[26,27] tin etiopurpurin ($SnET_2$),[28] hypericin,[29,30] and 5-aminolevulinic acid (ALA).[31–35] HpDs (i.e., Photofrin) are the best studied of these compounds.

HpDs were derived from the functionalization of hematophorphyrins (Hp) with hydrophobic substitutents that would allow these compounds to be taken up more readily in the tumor tissues than the unsubstituted Hps.[36] After these compounds had been developed, it was found that they could also serve as photosensitizers because they are capable of being excited and activated by the absorption of visible light. Although HpDs are still used quite extensively today for PDT treatments, many other compounds or classes of compounds with more appropriate characteristics for clinical drugs (such as those listed previously) have been developed.

The most desirable optical properties of PDT sensitizers are strong absorbtivity in the red or near-infrared (NIR) region of the electromagnetic spectrum and a high triplet-state quantum yield. The first of these two properties is important for several reasons. One reason is that the use of red or NIR wavelengths of light for excitation of the sensitizer causes much less damage to nearby healthy tissue. At red and NIR wavelengths, naturally occurring chromophores in the tissue absorb little light. In addition, because the absorption of these wavelengths by natural chromophores is minimal, the resulting background fluorescence from nearby healthy tissue is minimized, providing a much more accurate tumor marker as well as sensitizer. A third potential benefit gained by having photosensitizers with strong absorption spectra in the red and NIR is the ability to destroy tumors under thin layers of normal tissue.

When photosensitizers activated with visible or ultraviolet light are used, the absorption cross section of the healthy overlaying tissue is too great to allow a significant fraction of the light to penetrate deeper than approximately 100 μm. However, lower-energy photons, such as those in the red or NIR, penetrate significantly and could allow for the activation of sensitizers in tumors below healthy overlaying tissue. In addition to the desire to develop new sensitizers and markers with red-shifted absorption profiles, it is also important to obtain a high triplet-state conversion rate in these molecules. Triplet-state conversion is very important to formation of the reactive oxygen species responsible for the drug's activity.[10]

### 28.4.2 *In Vitro* Analyses and Diagnostics

*In vitro* analyses can typically be classified in two different categories: cellular analyses and tissue analyses. *In vitro* analyses are important for gaining a greater understanding of basic fundamental biological processes that are occurring and understanding any potential interferences or problems that may be present when investigating more complex biological systems (i.e., living animals).

In addition to gaining better fundamental biochemical understanding, *in vitro* fluorescence analyses are often performed to better understand the interactions or interferences that often take place in complex systems as opposed to individual chemicals. By investigating these effects on *in vitro* samples such as cells or tissues, investigators can determine potential diagnostic procedures that could be performed *in vivo* or identify problems with potential diagnostic procedures prior to involving live subjects. When performing analyses on neat biochemical systems, it is not possible to investigate the interactions of various biochemicals with others that are present in functional systems (i.e., cells, tissues, etc.). In addition, the effect of the local environment on the optical properties of the various fluorophores can be investigated much more accurately by using *in vitro* systems as compared to biochemical analyses alone.

#### 28.4.2.1 Cellular Analyses

Cells represent one of the most basic of biological systems capable of performing chemical reactions. Monitoring endogenous as well as exogenous fluorophores on the cellular level has provided a great deal of information about many different processes. Several different fluorescence techniques have been developed for the performance of such analyses. These techniques include probing with chemical or biochemical sensors as well as obtaining fluorescence images using any of several different forms of fluorescence microscopy. In this section, we provide a brief review of the various types of cellular processes that have been investigated using fluorescence spectroscopy.

Several other chapters in this handbook provide a more detailed description of the techniques and their applications. Chapter 54 discusses living cell analysis using optical techniques and provides a general review of various optical methods used for analyzing living cells. Chapter 20 on biosensors in medical applications, Chapter 59 on PEEBLE nanosensors for *in vitro* bioanalysis, and Chapter 60 on nanosensors for single-cell analysis provide detailed descriptions of chemical and biochemical sensors used for cellular analyses. Chapter 10 on confocal microscopy, Chapter 11 on two-photon excitation fluorescence microscopy, and Chapter 9 on lifetime-based imaging provide detailed reviews of the theory and application of fluorescence microscopy.

##### *28.4.2.1.1 Autofluorescence of Cells*

The vast majority of fluorescence microscopy is performed with the use of exogenous reporter dyes; however, a growing field of fluorescence microscopy relies on autofluorescent species inside the cells for diagnostic information. Such autofluorescence measurements provide an important tool for biomedical diagnostics. Much of the research in autofluorescence microscopy is performed as preliminary research prior to testing *in vivo* diagnostic procedures or as a medical diagnostic procedure itself. Some recent examples are described next.

*Cell Proliferation and Cell Analysis* — One of the most sought-after applications of fluorescence spectroscopy in biomedical optics is the ability to distinguish normal tissue from cancerous or even precancerous tissue in a real-time clinical setting, without administering any drugs. In order to determine the potential of this technique, several researchers have sought to answer the questions of whether or not unique fluorescence spectral patterns were associated with cell proliferation and whether differences between rapidly growing and slowly growing cells could be identified. Native cellular fluorescence was used to identify terminal squamous differentiation of normal oral epithelial cells in culture (Figure 28.5).[37]

In one such analysis, fluorescence excitation spectra were capable of distinguishing between slow- and rapidly growing cells in three different types of cells. To perform the analyses, excitation spectra were obtained by scanning the spectral range of 240 to 430 nm while monitoring the fluorescence emission at 450 nm. By taking the ratio of the intensity of the major broad-band peak at 320 to 350 nm to a point on the down slope of the curve at 370 nm, investigators found a statistical difference between slow- and rapidly growing cells. In addition, it was also possible to distinguish the slow-growing cells from the rapidly growing cells by obtaining fluorescence emission spectra of the various cells with an excitation wavelength at 340 nm and an emission wavelength range of 360 to 660 nm. These results demonstrate a great potential for discrimination of proliferating and nonproliferating cell populations *in vivo*.[38]

In a similar study, laser-induced fluorescence, with excitation by the 488-nm laser emission line of an argon ion laser, was used to differentiate between normal and tumor human urothelial cells. Experiments were performed using a confocal microspectrofluorimeter, allowing individual cells to be probed; the broadband autofluorescence emission between 550 and 560 nm, corresponding to oxidized flavoproteins, was monitored. An analysis of the data showed that the maximum autofluorescence intensity of normal urothelial cells was approximately ten times higher than that of any of the tumor cell types tested, thereby suggesting that the concentration of an oxidized flavoprotein in tumor urothelial cells is significantly less

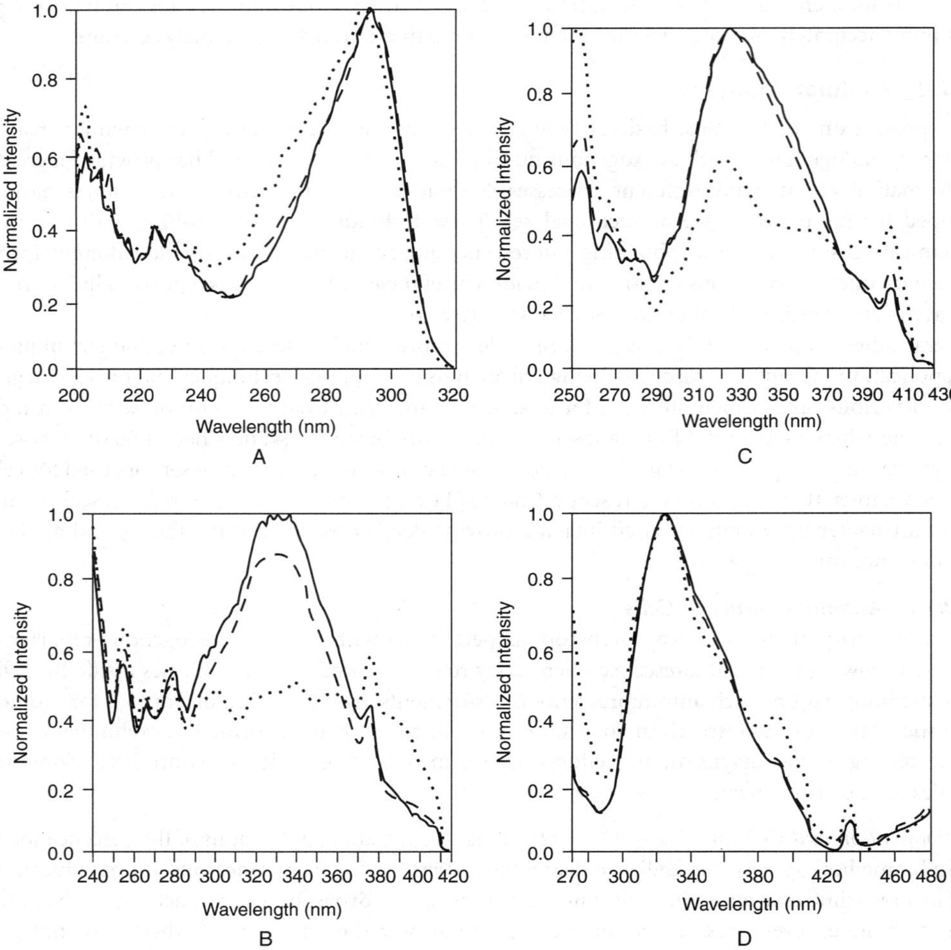


**FIGURE 28.5** Native cellular fluorescence of normal oral epithelial cells in culture. This figure shows several normalized fluorescence excitation spectra of normal oral epithelial cells that have been grown in several different media: KGM (dashed line), DMEM/F12/FCS (solid line), and DMEM/F12/NaCl (dotted line). The various culturing media were used to induce different stages of squamous differentiation in the cells, with KGM providing a medium for normal growth without differentiation, DMEM/F12/FCS providing a medium known to cause cell stratification within keratinocytes, and DMEM/F12/NaCl providing a medium capable of inducing the terminal stage of squamous cell differentiation in keratinocytes. The fluorescence excitation scans were taken under the following conditions: (A) $\lambda_{ex}$ 200 to 360 nm; $\lambda_{em}$ 380 nm; (B) $\lambda_{ex}$ 240 to 415 nm; $\lambda_{em}$ 450 nm; (C) $\lambda_{ex}$ 250 to 420 nm; $\lambda_{em}$ 480 nm; (D) $\lambda_{ex}$ 270 to 480 nm; $\lambda_{em}$ 520 nm. From such spectra, it is possible to differentiate between the various states of the cells *in vitro*. (Adapted from Sacks, P.G. et al., *Cancer Lett.*, 104, 171, 1996.)

than in a normal urothelial cell. Not only could this observation lead to the development of a clinical diagnostic procedure for the real-time differentiation between normal, cancerous, and potentially precancerous urothelial tissues, but it also provides a better understanding of the biochemical differences between normal and cancerous tissues.[39]

Synchronous luminescence (SL) is a unique approach that could improve the selectivity of fluorescence techniques for cancer diagnostics. As discussed previously, autofluorescence of neoplastic and normal tissues has been observed using fixed-wavelength laser excitation. However, the use of fixed excitation might not be sufficiently selective in some diagnostic applications due to strong overlap of the emission spectra from different fluorophores. An alternative approach is the SL method, which involves scanning excitation and emission wavelengths simultaneously while keeping a constant wavelength interval between them.[40,41] This method, also referred to as synchronous fluorescence (SF), has been developed for multicomponent analysis and used to obtain fingerprints of real-life samples, as well as for enhancing selectivity in the assay of complex systems. This SF procedure has been shown to simplify the emission spectrum and provides for greater selectivity when measuring fluorescence or phosphorescence from mixtures of compounds.[3,4,42,43]

Spectral differences in SF emission profiles are related to the specific macromolecules that differed between neoplastic and normal cells. The SF technique has been shown to improve spectral selectivity in detecting normal and cancer cells for potential use in biomedical diagnostics.[43] Whereas conventional fixed-excitation fluorescence could not show any differentiation between normal and cancerous cell lines, spectral difference between the fluorescence spectra of the normal rat liver epithelial (RLE) and neoplastic rat hepatoma McA cell lines were detected using SF. The results demonstrated the great potential of SF as an improved screening tool for cancer diagnosis in specific cases where conventional fixed-excitation methods are not sufficiently effective.

*Biological Fluids, Semen, Seminal Plasma, and Spermatozoa* — In addition to being used for the study of cancer, autofluorescence microscopy is also being investigated for its potential in many other forms of biomedical diagnostics. One potential application involves the determination of sperm viability and motility for real-time fertility diagnoses. In one study, fluorescence emission spectra of human semen, seminal plasma, and spermatozoa, excited at 488 nm, were obtained over the range of 500 to 700 nm.

Under these conditions, emission peaks from each component were observed at 622 nm. The intensity of the emission peaks from spermatozoa at 622 nm was strongly correlated with the concentration of spermatozoa ($r = 0.837$; $p = 0.0001$). In addition, sperm motility could also be correlated significantly, although to a poorer extent, with the intensity of the fluorescence emission peaks from spermatozoa ($r = 0.369$; $p = 0.019$) and semen plasma ($r = 0.356$; $p = 0.024$).[44] The SF method has been investigated as a rapid screening tool for monitoring DNA damage (DNA adduct metabolites) and in monitoring biological fluids in animal studies, thus providing a technique for early cancer prescreening at the DNA level.[42]

#### *28.4.1.1.2 Cellular Fluorescence Using Exogenous Dyes*

Fluorescence analyses of living cells by means of exogenous reporter molecules represent the vast majority of spectroscopic analyses performed on cells. Such analyses can generally be classified into one of two categories: optical sensor-based analyses or fluorescence imaging-based analyses. Both techniques have advantages and disadvantages for measurement of individual chemical species inside a single living cell. Optical sensors allow for accurate determination of chemical concentrations at individual points in the cell, while imaging-based analyses provide a means of monitoring the spatial distribution of the analyte throughout the cell.

Several types of optical sensors have been developed for minimally invasive fluorescence monitoring of individual chemical species inside a single cell.[45–52] These sensors are based on the same principles and concepts as those discussed in the chapter on biosensors in medical applications, but their dimensions are much smaller. Usually, an exogenous fluorescent dye whose properties (i.e., quantum yield, emission maximum, etc.) change in the presence of the analyte of interest is attached to the probing end of the sensor. Such sensors have been used to measure many analytes within living cells, thus providing information about many different biological processes. These analytes include calcium ions, magnesium ions, nitric oxide, glutamate, lactate, glucose, benzo[a]pyrene tetrol, and even oligonucleotides (i.e., DNA, RNA).[43–50] Because this relatively new field of sensors is described in greater detail in this handbook in Chapter 59 on PEEBLE nanosensors for *in vitro* analysis and Chapter 60 on nanosensors for single-cell analysis, this chapter will not go into any further detail on the subject.

Fluorescence microscopy is the most common technique for monitoring the spatial distribution of a particular analyte at many different locations simultaneously throughout a cell. In such analyses, one or several different exogenous fluorescent dyes are introduced into the cell and allowed to disperse. After dispersing throughout the cell, they begin to interact with the analyte of interest; this interaction, in turn, causes change in the fluorescence properties (intensity, spectral shift, etc.) of the dye. Therefore, by obtaining a fluorescence image of the cell at a specific wavelength corresponding to the maximum emission wavelength of the dye, one can determine the relative concentrations of the analyte at specific locations.

The great amount of information derived from such analyses has allowed researchers and companies (e.g., Molecular Probes) to develop many different fluorescent dyes for a wide variety of compounds. These dyes are available with many different excitation and emission maxima for most analytes, allowing the experimenter to choose the best dye for the particular application. The dyes most commonly used today for fluorescence microscopy have red or NIR excitation profiles, thus allowing the dye to be excited without providing the energy necessary to excite any of the autofluorescent species present inside the cell. In addition, the lower-energy light also causes less damage to the living cells than would visible or UV wavelengths, which are capable of cross linking DNA or proteins.

*Cell Proliferation* — Cell proliferation analyses are one of many types often performed using exogenous dyes. Learning about the part of the cell cycle that a particular cell or group of cells is in, or about the way in which it died (necrosis vs. apoptosis, for example) can provide a great deal of biomedically useful information. For instance, the determination of apoptotic cell death is of particular importance in the study of chemical carcinogenesis as well as in the development of novel PDT drugs. For this reason, many different types of fluorescence analyses have been developed to determine the extent of apoptosis in cells or tissues.

One such method employed a commercially available ApopTag kit (Serologicals Corporation, Norcross, GA) and automated fluorescence image analysis to quantify the distribution of apoptosis in formalin-fixed, paraffin-embedded liver tumor sections from rats whose tumors were induced by 2-acetylaminofluorene. In this work, specific treatments of tissue sections were developed for quenching the autofluorescence background from the cells of the tissue sample. Once the autofluorescence had been quenched, propidium iodide was used as a counterstain for the nuclei. Automated statistical evaluation of the percentage of nuclei stained positively for apoptosis was determined by using dual fluorescence detection and optical microscopy. The quantitative results indicated that the staining index for apoptosis in normal rat liver cells was $0.14 \pm 0.04\%$, whereas well- and poorly differentiated tumor cells showed increases of $3.48 \pm 0.59\%$ and $7.41 \pm 0.81\%$, respectively.[53]

*Cellular Response* — Another important application of fluorescence microscopy using exogenous fluorophores is in elucidating the role of particular chemicals in cellular biology and determining the associated kinetic constants for those processes. One chemical and cellular process that has been investigated using fluorescence microscopy and exogenous fluorophores is protein kinase C (PKC) and its role in signal transduction of many bioactive substances. In this study, phorbol-13-acetate-12-*N*-methyl-*N*-4-(*N,N'*-di(2-hydroxyethyl)amino)-7-nitrobenz-2-oxa-1,3diazole-aminododecanoate[*N*-C12-Ac(13)], the fluorescent derivative of 12-*o*-tetradecanoylphorbol-13-acetate (TPA), was synthesized to monitor the location of phorbol ester binding sites and evaluate its potential use as a probe for PKC in viable cells. The maximum excitation wavelength of *N*-C12-Ac(13) is close to 488 nm, making possible the use of an argon-ion laser for excitation of the dye molecule. When incubated with 100 n*M* *N*-C12-Ac(13), P3HR-1 Burkitt lymphoma cells accumulated the dye rapidly, reaching a maximum fluorescence (20-fold above the autofluorescence background) within 25 min. The subsequent addition of unlabeled TPA significantly decreased the fluorescence of *N*-C12-Ac(13) in the cells, in a dose-dependent manner, indicating specific displacement of the bound fluoroprobe.

Using this same competitive displacement technique, researchers displaced [3H]-phorbol-12,13dibutyrate ([3H]-PBu2) from rat brain cytosol with *N*-C12-Ac(13), which was found to exhibit an apparent dissociation constant and biological activity similar to that of TPA. In addition, like TPA, *N*-C12-Ac(13) also induced the expression of Epstein-Barr viral glycoprotein in P3HR-1 cells and differentiation of promyelocytic HL60 cells and caused predicted changes in the mitotic cycle of histiocytic DD cells. Microscopic fluorometric images of single cells for such analyses verified that the bound dye *N*-C12-Ac(13) showed bright fluorescence in the cytoplasm and dim fluorescence in the nuclear region, a phenomenon consistent with dye binding mainly to cytoplasmic structures or organelles.[54]

*Immunocytofluorometric Analyses* — In addition to using exogenous fluorescent dyes to monitor the concentration or distribution of various chemicals or ions within a cell, investigators have found useful applications for fluorescence spectroscopy in the identification of specific types of cells in immunocytofluorometric analyses. In such analyses, fluorescent molecules are attached to a monoclonal or polyclonal antibody that is specific to a surface protein or a class of surface proteins on a particular type of cell. The cells to be analyzed are then exposed to these fluorescently labeled antibodies. Once the antibodies have attached to the cells with the appropriate surface proteins, a fluorescence analysis system (e.g., a flow cytometer) can distinguish the cells of interest from the other cells present on the basis of a fluorescent property (e.g., spectral band) of the dye molecule attached to the antibody.

In a recent example, a time-resolved laser-induced fluorescence measurement system was used in conjunction with pyrene-derivatized molecules to perform immunocytofluorometric analyses. By using these pyrene-derivatized molecules [*N*-(1-pyrene)maleimide, 1-pyrenesulfonyl chloride and 1-pyreneisothiocyanate], fluorescent lifetimes for the labeled antimouse-IgG were in the range of 20 to 55 ns. Due to this long lifetime, it was possible to distinguish these cells easily from other nonlabeled cells whose fluorescence lifetimes were much shorter.[55]

### 28.4.2.2 Tissue Analyses and Diagnostics

Many studies have been performed on tissue samples[7,40,41,56–62] to help us understand the interaction between cells and to bring us closer to the much more complex area of *in vivo* analyses. Tissue studies provide a platform much closer to true *in vivo* analyses in terms of structural architecture on microscopic and macroscopic scales than do cellular analyses. *Ex vivo* tissues provide an excellent starting point for the development of many *in vivo* diagnostic procedures (e.g., optical biopsy). In addition, when compared with *in vivo* systems, *ex vivo* tissue samples are much easier to handle and manipulate.

Over the last two decades, many different types of fluorescent tissue analyses have been performed on a wide variety of samples. The types of tissues range from mouse and rat tumors[41] to human teeth.[60] Instrumentation has ranged from commercially available fluorescence microscopes to laboratory-built research instruments for specialized measurements. The development of fiber-optic instrumentation has made it possible to perform analyses in any area accessible to an endoscope.

#### *28.4.2.2.1 Autofluorescence of Tissues*

Many fluorescent species are present in biological tissues, and we want to develop diagnostic techniques that do not require chemical pretreatment of tissues. Therefore analyses based on autofluorescence represent a large portion of the fluorescence diagnostic procedures being developed. Some of the more promising autofluorescent diagnostic results as well as their state of development are described in the following sections.

*Cancer Diagnostics* — Recently, there has been a great deal of interest and research in the development of rapid, fluorescence-based clinical oncology.[6,10] Many different techniques have been developed for such analyses in different types of tissues and have shown a great deal of promise. These techniques and the instrumentation used for these measurements are described in greater detail in the following sections of this chapter. The effectiveness of the techniques is often determined by comparing their results with histological data, which are considered the "gold standard." Based upon the results of these comparisons, we can calculate three quantitative values to provide an objective comparison of various techniques.

Calculation of these three quantities (sensitivity, specificity, and accuracy) depends upon classifying a particular analysis as either a true positive (TP), which represents classification by both techniques as malignant; a false positive (FP), which corresponds to a classification as malignant by fluorescence spectroscopy and normal by histology; a false negative (FN), which corresponds to a normal classification by fluorescence and a malignant classification by histology; or a true negative (TN), which corresponds to classification by both techniques as normal. Once the appropriate classifications have been performed, sensitivity, specificity, and accuracy are calculated as follows:

Sensitivity = TP/(TP + FN)
Specificity = TN/(TN + FP)
Accuracy = (TP + TN)/(TP + TN + FP + FN)

An example of a laboratory-constructed instrument for tissue analyses is given by Alfano and co-workers, who developed a system for the differentiation of cancerous and normal human breast tissues and lung tissues.[61,62] In this work they developed a system that uses a pulsed or a continuous-wave (CW) excitation source for the differentiation of normal and cancerous tissues., The fluorescence emission spectrum for normal tissue caused by excitation with 488-nm light from an argon ion laser was found to be quite different from that for tumor tissue. The resulting spectra from tumor tissues had smooth emission curves with a maximum at approximately 530 nm. The resulting emission spectra from normal tissues appeared to have three peaks, at 530, 550, and 590 nm. In another study, Feld and co-workers attempted to determine the optimal excitation wavelength for distinguishing between normal and tumor tissues based upon their biochemical and histomorphological (architectural) components. They found that the most marked differences occurred when an excitation wavelength of 410 nm was used. Using this wavelength, they correctly diagnosed 20 of 22 samples studied.[63] This optimal wavelength, determined with *in vitro* studies, is consistent with results obtained by Vo-Dinh and co-workers from *in vivo* clinical studies of gastrointestinal cancer diagnostics using laser-induced fluorescence.[64,65]

*Cervical Cancer* — In the field of *in vitro* fluorescence diagnostics for the detection of cervical cancer, there is a relatively small amount of information. However, results that have been published forecast a promising future for this technique. In one such study, Richards-Kortum and co-workers investigated fluorescence excitation–emission matrices (EEMs) to analyze 18 cervical biopsies from 10 patients.[66] At all excitation–emission maxima, most prominently at 330-nm excitation and 385-nm emission, the average normalized fluorescence intensity of histologically normal tissue was significantly greater than that of histologically abnormal tissue. A diagnostic algorithm based upon this relative intensity difference was developed to differentiate between histologically normal and abnormal biopsies with a higher sensitivity but a lower positive predictive value and specificity than colposcopy. However, when comparisons of histologically normal and abnormal biopsies from the same patient were performed, sensitivity results of 75%, positive predictive value results of 86%, and specificity results of 88% were achieved for the spectroscopic identification of histological abnormality. These results compare favorably with colposcopy results. Based on these results, *in vivo* studies of cervical tissue fluorescence have also been conducted.[66]

In addition to research demonstrating the ability to distinguish between normal cervical tissues and invasive carcinomas in cervical tissue, research has also attempted to distinguish between dysplasia and invasive carcinomas. To this end, a technique developed recently for the *in situ* detection of melanomas has been applied for determining *in vitro* dysplasia and invasive carcinomas in the cervix uteri. Cervix uteri exhibit a fluorescence band with a peak at about 475 nm upon excitation with 365 nm light. At a fluorescence intensity of 475 nm, the fluorescence intensity increases concomitantly with the degree of dysplasia, ranging from 30 counts/100 ms (healthy) to approximately 200 counts/100 ms (carcinoma *in situ* 3). At the rim of a malignancy, the intensity is 250 counts/100 ms and higher. The excitation and emission spectra of the tissue suggest that the endogenous chromophore responsible for the observed fluorescence is NADH.[67]

*Colon Cancer* — Another major form of cancer that has seen a significant amount of research and advancement in terms of *in vitro* fluorescence diagnosis is colon cancer. In one study, unstained frozen sections of colon were studied by fluorescence microscopy to determine which structures fluoresce and to what extent in normal colon tissue and colonic adenomas. Tissues were excited by a 351- to 364-nm emission from an argon ion laser. The resulting fluorescence signals from various locations in the tissue samples were correlated to the tissue morphology at those locations via histological analyses with conventional stains (hematoxylin and eosin [H&E]), Movat pentachrome, mucicarmine, and oil red O dye).

In normal colon tissues, the measured fluorescence signals correlated morphologically with connective tissue fibers (mainly collagen) in all layers of the bowel wall and with cytoplasmic granules within eosinophils present between the crypts in the lamina propria of the mucosa. In addition, fluorescent signals from the crypts of normal cells were very faint. However, a significant fluorescence emission was observed in the cytoplasms of dysplastic epithelial cells in the crypts of colonic adenomas. In the lamina propria of colonic adenomas, a decrease in fluorescence intensity compared to that of normal colon tissue was found to exist. The decrease could be correlated to fewer fluorescent connective tissue fibers in the case of the adenomas. Finally, it was found that a larger number of fluorescent eosinophils exist in adenomatous colonic tissue than in normal colon tissue.[68] Similar results were also found in another study, in which a confocal fluorescence microscope was used.[69]

In an additional study, also aimed at determining the biochemical differences between fluorescence signals from normal and adenomatous colonic tissues, it was found that the fluorescent signals from plasma-soluble melanins derived from various sources were related to the overall autofluorescence signal of adenomatous tissue samples. It was also determined in this study that the fluorescence component of melanins derived from 3-hydroxyanthranilic acid, excited by 324-nm light and measured at 413 nm, was less than the fluorescence caused by melanins derived from other sources (i.e., dopa, catecholamines, catechol, and 3-hydroxykynurenine), with fluorescence excitation and emission maxima at 345 and 445 nm, respectively. [70]

In a quantitative determination of the efficacy of fluorescence spectroscopy to classify colonic tissues as adenoma, adenocarcinoma, or non-neoplastic, a series of analyses were performed on 83 biopsy specimens from several patients. In these tests, fluorescence emission spectra covering a spectral range of 450 to 800 nm were measured. From these measurements, it was found that the spectral properties were significantly different for the three different tissue classifications. In fact, the results of these analyses had a sensitivity of 80.6 and 88.2%, and a specificity of 90.5 and 95.2% in discriminating neoplastic from non-neoplastic mucosa and adenoma from non-neoplastic mucosa, respectively.[71]

With preliminary fluorescence analyses of colon tissues providing promise for a minimally invasive real-time classification procedure for colon cancer, a great deal of effort is being devoted toward the optimization of this procedure, both experimentally and through data analysis methods. In one such study, fluorescence EEMs were obtained to determine the optimal excitation regions for obtaining fluorescence emission spectra that can be used to differentiate normal and pathologic tissues. In the case of normal and adenomatous colon tissue, the optimal excitation wavelengths were found to be 330, 370, and 430 nm ± 10 nm. These excitation wavelengths were used with simple difference techniques as well as ratiometric analyses to determine the optimal emission wavelengths.

Based upon these results, the optimal excitation wavelength for discrimination between normal and adenomatous colon tissues *in vitro* was found to be 370 nm and the optimal emission wavelengths for analysis based upon this excitation were found to be 404, 480, and 680 nm.[72] In a second study, in which the optimization was based upon data analysis algorithms, 35 resected colonic tissue samples were excited by the 325-nm line of a helium cadmium laser and the resulting fluorescence spectra were measured over the spectral range of 350 to 600 nm. Scores were derived from these analyses by a multivariate linear regression analysis that accounted for six wavelengths in the emission spectra. Normal tissue samples, adenomatous tissue samples, and hyperplastic tissue samples were classified by using the resulting scores with accuracies of 100, 100, and 94, respectively. [73]

*Gastric Cancer* — Application of fluorescence spectroscopy to *in vitro* detection of gastric cancers has also been performed. In this work, fluorescence was measured following tissue excitation with the 325-nm light from a helium cadmium laser.[63] Fluorescence images of 72 surface areas of 21 resected tissue samples were recorded in six regions of the visible spectrum by a cooled charge-coupled device (CCD) camera. The fluorescence emission intensities measured at 440 and 395 nm, normalized to the intensity measured at 590 nm, differed significantly for malignant tissues, premaligant tissues, and normal gastric tissues. When these differences were used as a diagnostic parameter, classification of malignant tumor tissues with a sensitivity of 96% and a predictive value of 42% was possible. Additionally, the same approach applied to diagnosis of abnormal (but nontumor) stomach tissues gave values of 80 and 98%, respectively.[74]

*Oral Cancer* — Biopsy specimens from clinically suspicious lesions and normal-appearing oral mucosa were obtained from patients and fluorescence EEMs were measured.[56] From these EEMs, it was determined that the optimal excitation wavelength for differentiation between normal and dysplastic tissues was 410 nm. Based upon the fluorescence emission spectra produced by this wavelength for excitation, the 22 different resected samples, 12 histologically normal and 10 abnormal (dysplastic or malignant), were analyzed. From this analysis, 20 of the 22 samples were correctly classified as normal or abnormal tissues.[63]

*Atherosclerosis* — Laser-induced fluorescence has been used for the differentiation of calcified and non-calcified plaques in arterial tissues.[64] In this work, 248-nm laser light from a krypton-fluorine excimer laser was used to irradiate normal and severely atherosclerotic segments of human postmortem femoral arteries. In order to analyze deeper layers of the tissue while measuring the resulting fluorescence from each layer, 16-ns laser pulses were used, each having a fluence of 5 $J/cm^2$. Pulse powers of this magnitude were sufficient to ablate as well as excite the tissue. By synchronizing the detector with each of the laser pulses, it was possible to determine the characteristic composition of the ablated tissue layer. Normal tissue layers provided fluorescence spectra exhibiting a broad-continuum emission between 300 and 700 nm with peak fluorescence of equal intensity at wavelengths of 370 to 460 nm. Fluorescence maxima of atheromas without calcification occurred at the same wavelengths but with significantly reduced intensity at 460 nm.

In contrast to these broad-continuum fluorescence signals from normal and noncalcified atheromas, calcified plaques displayed multiple-line atomic fluorescence (atomic emission) with the most prominent peaks at wavelengths of 397, 442, 450, 461, 528, and 558 nm. These fluorescence/emission criteria identified the histologically classified target tissue precisely. Comparison of this technique to histological examination of the corresponding arterial layers indicated an extremely accurate characterization. These results demonstrated the potential for simultaneous real-time tissue identification and the feasibility of ablation by laser irradiation under strict laboratory conditions.[75]

To diagnose arterial tissue before ablation, many researchers have attempted to use less powerful and less energetic lasers for fluorescence excitation alone, and have experienced a great deal of success. In one such study, ultraviolet excited laser-induced fluorescence (LIF) using excitation wavelengths ranging from 306 to 310 nm was used to distinguish between normal and atherosclerotic arteries.[65] In this study, two distinct fluorescence emission bands were observed in the LIF spectra of normal and pathologic aorta: a short-wavelength band peaking at 340 nm (attributed to tryptophan) and a long-wavelength band peaking at 380 nm (attributed to a combination of collagen and elastin). In addition, the intensity of the short-wavelength band was found to be very sensitive to the choice of excitation wavelength while the long-wavelength band remained unchanged; this feature allowed for the relative contributions of each band to be controlled precisely by choice of excitation wavelength.

It was found that, by using 308-nm excitation to observe emissions from the short- and long-wavelength bands simultaneously, two ratios could be determined; from them it was possible to distinguish between normal and atherosclerotic aortic tissue. These two ratios characterized the relative tryptophan fluorescence content and the ratio of elastin to collagen. In addition, normal and atherosclerotic aorta were correctly distinguished in 56 of 60 total cases by a binary classification scheme in which these parameters were combined. Furthermore, atherosclerotic plaques, atheromatous plaques, and exposed calcifications could be classified individually with sensitivities and predictive values of 90 and 90%, 100 and 75%, and 82 and 82%, respectively.[76]

In another study, even less energetic light was used to excite autofluorescent species in blood vessels for the differentiation of normal tissue and atherosclerotic lesions. The use of longer wavelengths of light for excitation not only caused less photodamage to the irradiated tissue, but also allowed better transmission through optical fibers, which were necessary for *in vivo* light delivery. In this work, 325-nm light was used for excitation of the sample, and a ratio of the fluorescence intensity at 480 nm to the fluorescence intensity at 420 nm was determined for each of the different types of tissues. Based upon these ratios, a clear differentiation between normal and mild, as well as between normal and severe, atherosclerotic lesions was observed. In the case of normal tissue, there was an increased intensity in the range from

420 to 540 nm, whereas atherosclerotic lesions had no or only a small peak at 480 nm. In addition, the spectroscopic results showed no differences between samples taken from different types of vessels.[77]

Further studies have extended the wavelengths used for excitation of atherosclerotic tissue samples to 458 nm[78] and even 476 nm.[79] In both cases, even though the autofluorescent tissue components responsible for the fluorescence emission were different from those excited by ultraviolet light, good differentiation between normal arterial tissue, noncalcified arterial plaques, and calcified arterial plaques was demonstrated. When these longer wavelengths are used for excitation, differentiation is based upon the ratio of the fluorescence signals from structural proteins (elastin and collagen) and ceroid, which can be correlated to the concentration of these various components.

Several other studies have been performed to correlate LIF signals to the histochemical composition of human arteries and veins.[80,81] In one such study, unstained frozen sections of normal and atherosclerotic human aorta and coronary artery were examined by histochemical and fluorescence microscopy techniques to identify the species responsible for autofluorescence following excitation with light ranging from 351 to 364 nm. The species investigated included the structural proteins elastin and collagen in normal and atherosclerotic specimens, calcium deposits in calcified plaques, and granular or ring-shaped deposits histochemically identified as ceroid found in calcified and noncalcified plaques. The emission wavelength and intensity of ceroid autofluorescence differed greatly from those of elastin or collagen, with the ceroid emission red-shifted and showing a much greater resistance to photobleaching than the structural proteins, thus making it a prime candidate for investigation in the classification of arterial plaques.[80]

In another study aimed at determining the relationship between the histochemical and morphological characteristics of atherosclerotic tissue and their LIF spectra, unstained frozen sections of 47 normal and atherosclerotic human aortas and coronary arteries were excited by 476-nm laser light. The resulting fluorescence images were measured with a bright-field epifluorescence microscope. The samples were then stained with various dyes (including H&E, Movat pentachrome, and oil red O) for histological analysis. In the case of normal artery autofluorescence, the signals correlated with the structural proteins elastin and collagen in the intima, media, and adventitia. In atherosclerotic plaque, autofluorescence correlated morphologically with lipid or calcific deposits in the atheroma core. The autofluorescence of these deposits was different from that of elastin and collagen in distribution, intensity, and wavelength, and increased with the severity of the plaque.

Such excellent correlation qualitatively and quantitatively suggests that 476-nm excitation-based LIF analyses of arterial tissues would be useful.[81] However, it has recently been found that 476-nm-induced autofluorescence in arterial tissues suffers from changes in two prominent spectral characteristics of the emission spectrum. The changes related to alterations in the individual tissue chromophores being excited are a permanent decrease in the peak fluorescence intensity and a reversible change in the fluorescence emission profile (line shape). The permanent changes in absolute fluorescence intensity are due to irreversible photodamage to the tissue fluorophores; the reversible changes in fluorescence line shape are due to some type of alteration in tissue absorbers. An attempt has been made to minimize these effects; excitation intensity levels and exposure times have been investigated to establish the thresholds at which these alterations are minimized.[82,83]

To develop a fluorescence guidance system for the laser ablation of arterial plaques in femoral arteries, Gaffney and co-workers developed a technique that employs dual excitation wavelengths. Information about the fatty plaque content could be obtained from 325-nm excited fluorescence; information about structural proteins and calcific content could be obtained from fluorescence following 476-nm excitation. The fluorescence emission spectra obtained from each of the excitation wavelengths were used to perform ratiometric analyses. For 325-nm excited fluorescence spectra, 78 ratios were determined based upon 13 different wavelengths, and for 476-nm excited spectra, 55 ratios were determined based upon 11 different wavelengths. From these analyses, atherosclerotic lesions in human coronary arteries were characterized by an increase in normalized fluorescence intensity at longer wavelengths when excited with either ultraviolet or visible light. Calcific plaque content greater than 10% in lesions more than 1-mm thick were identified by increased normalized fluorescence intensity at 443 nm produced by excitation at 325 nm. Fatty plaque content was found to correlate with fluorescence intensity ratios produced by 325-nm excitation, whereas fibrous and calcific content correlated well with fluorescence ratios during 458-nm excitation.[78]

In addition to using multiple-excitation wavelengths to increase the accuracy and sensitivity of prediction of atherosclerotic tissue, there has also been a recent investigation into the use of time-resolved fluorescence detection rather than spectral-based detection for the differentiation of atherosclerotic tissue from normal tissue. Based upon the temporal differences in the fluorescence emission from atherosclerotic plaques and normal blood vessels, an enhanced differentiation between the two tissue types was reported.[84]

*Heart Arrhythmia* — Autofluorescence spectral properties of tissue have also been used in the diagnosis of heart-related illnesses. Heart arrythmias are often caused by formation of nodal conductive tissues; in severe cases that cannot be treated through conventional medical therapies, transcatheter ablation of the nodal conductive tissue can be performed. Therefore, prior to any tissue ablation, it is important to be able to classify the tissue of interest as nodal conductive tissue, atrial endocardium, or ventricular endocardium. To perform this diagnosis, Lucas and co-workers studied the fluorescence emission of *in vitro* heart tissue samples excited by the 308-nm laser line of an XeCl excimer laser (1.5 mJ/pulse, 10 Hz). Following excitation at 308 nm, the nodal tissue could clearly be distinguished from atrial endomyocardial tissue by a visible decrease in fluorescence emission intensity between 440 and 500 nm, peak area between 440 and 500 nm, and peak width. Nodal conduction tissue could also be distinguished from ventricular endocardium by its relative increase in fluorescence emission between 430 and 550 nm. Specificities of 73 and 88% and sensitivities of 73 and 60% were possible for sinus nodal and atrioventricular nodal conduction tissue identification, respectively.[85]

*Tooth Decay* — The field of dentistry has also seen a recent growth of research in the use of fluorescence spectroscopy. In work involving the use of a confocal scanning laser microscope (CLSM), autofluorescence spectra were taken of teeth with demineralized dentin on the root surfaces. The resulting spectra were then compared with spectra taken at other locations on the tooth with minimal to no demineralization in the dentin. When observed in CLSM images, demineralized dentin (excited at 488 nm) exhibited an increased fluorescence emission at 529 nm when compared with the spectra of healthy dentin. This difference in fluorescence intensity decreased deeper into the root, as the healthy dentin underneath the lesion was beginning to be excited.

In contrast, when fluorescence spectrophotometry was used with excitation around 460 and 488 nm, it yielded fluorescence emission intensity (about 520 nm) for demineralized dentin that was lower than for healthy dentin, but in a more pronounced peak. From excitation spectra obtained by a fixed emission wavelength of 520 nm, it could be seen that the contribution of excitation between 480 and 520 nm was more important in demineralized dentin than in healthy dentin. Because of the small sampling volume used in CLSM image acquisition, the recorded fluorescence was not affected by demineralization-induced changes in scattering and absorption properties. Thus the increased fluorescence for demineralized dentin implies an increased quantum yield. However, in fluorescence spectrophotometry, where the measurement volume is large relative to the lesions, changes in scattering and absorption properties do have an influence on the fluorescence signal. Therefore, increased absorption by nonfluorescing chromophores and increased reabsorption around the emission wavelength may compensate for the increase in quantum yield and absorption around the excitation wavelength by the fluorophores.[86]

*Eosinophils* — Eosinophils are rare granulocytes typically associated with allergic diseases or responses to various parasitic infections. Many types of human cancers, however, are also associated with extensive eosinophilia, either within the tumor, in the peripheral blood, or in both locations. Special techniques such as autofluorescence or immunohistochemistry are sometimes needed to detect the presence of intact and degranulating eosinophils within the tumors. With the help of these techniques, extensive amounts of eosinophilia have been found in hematologic tumors such as Hodgkin's disease and certain lymphomas. However, many other types of cancer, such as cancers of the colon, cervix, lung, breast, and ovary, also contain eosinophilia, and it can be identified if diligently sought. Although the presence or absence of eosinophilia within these tumors does not appear to have a major influence on the prognosis of the disease, eosinophils may play an important role in the host interaction with the tumor, perhaps by promoting angiogenesis and connective tissue formation adjacent to the cancer.

In addition, tumor-related eosinophilia provides some interesting clues into tumor biology, particularly with regard to production of cytokines by the tumor cells.[87] Characterization of the fluorescence properties of human eosinophils isolated from peripheral blood of normal donors has been performed by measuring EEMs over a wide range of wavelengths. Circulating eosinophils possess three fluorescence emission maxima: one at 330 nm following excitation at 280 nm, which can be attributed to tryptophan, a second peak at 440 nm following excitation at 360 nm, and the last at 415 nm following excitation with 380-nm light. Fluorescence microscopy studies also showed that the fluorescence of eosinophils appears to be site dependent. For instance, when observed following excitation by 365-nm light, circulating eosinophils fluoresce blue–violet, while tissue-dwelling eosinophils fluoresce amber–gold. Therefore, when fluorescence spectroscopy is used to develop optical biopsy techniques based upon eosinophils in human tissue, the differences in their local environments may have a significant impact on fluorescence spectra.[88]

***28.4.2.2.2 Tissue Analysis Using Exogenous Dyes***

Due to the significant differences in tissue uptake and storage of various exogenous fluorophores between *in vitro* and *in vivo* specimens, relatively few studies have been performed in this area in terms of developing diagnostic procedures. In most cases where exogenous fluorophores are used, the actual location and kinetics of tissue uptake are important; therefore, because *in vitro* tissues differ in these properties, such studies generally do not provide any useful information. However, properties such as tissue reactivity to a specific chemical can be studied by using *in vitro* systems to provide a better understanding of the mechanism by which specific reactions take place.

*Lipid Peroxidation* — One system studied by using fluorescence CLSM is the well-established experimental model of lipid peroxidation induced by haloalkane intoxication in liver tissues. In this study, the fluorescent reagent 3-hydroxy-2-naphtholic acid hydrazide was used to derivatize the carbonyl functional groups originating from the lipoperoxidative process in liver cyrostat sections from *in vivo* intoxicated rats, as well as isolated hepatocytes that were exposed *in vitro* to the haloalkanes. The resulting CLSM images were able to visualize the tissue areas and the subcellular sites first involved in oxidative stress and lipid peroxidation. The images also showed that haloalkane-induced lipid peroxidation in hepatocytes primarily involves the perinuclear endoplasmic reticulum, whereas the plasma membrane and the nuclear compartment are unaffected, and that lipid peroxidation also induces an increase of liver autofluorescence.[89]

## 28.4.3 *In Vivo* Analyses and Diagnostics

While *in vitro* studies can reduce the complexity of the biological system analyzed and provide useful information about basic biological functions or reactions, the need to develop *in vivo* analysis techniques for medical diagnosis of diseases is critical. The range of fluorescence-based *in vivo* diagnostic techniques spans from monitoring of atherosclerosis[90] to the detection of tooth decay.[60,91] One of the most common biomedical diagnostic procedures performed using fluorescence spectroscopy is known as "optical biopsy."[10,24,92–102] In this procedure, some form of optical spectroscopy, typically fluorescence spectroscopy, is used to identify differences between healthy, malignant, and premalignant tissues of various organs. Over the past two decades, a great deal of research has been performed in this field for the diagnosis of many different forms of cancer.[6,59,61,65,99,103–120] In order to ensure that the various diagnostic procedures will work in complex living systems, two types of analyses are generally performed: animal studies and clinical human trials.

### 28.4.3.1 Animal Studies

Animals typically provide an excellent test system for many different types of disease diagnoses prior to clinical studies. The key in choosing an animal for these studies is to ensure that the biochemical structure of the particular organ or location of the animal used is comparable to that of a human. In addition, the animal should be relatively easy to care for and should match any other criterion for the particular type of analysis to be performed. Mice and rats are often used as a suitable alternative for humans in many preliminary studies because of their 98% genetic compatibility with humans and because they are relatively easy to handle; however, many other animal models have also been investigated (e.g., hamster and pigs).

In the case of "optical biopsy" work or photodynamic therapy (PDT), transgenic mice with transplantable tumors are the most common model system. These mice have been genetically altered to prevent their providing an immune response to the tumors that are induced in them. By using these mice with transplantable tumors, valuable information useful to later clinical trials can be obtained. However, use of animal models could have artifacts as well. For instance, when using mice with transplantable tumors for tumor detection diagnostics, the implanted tumor tissue generally comes from a different type of tissue and therefore may not have the same biochemistry, architecture, and vasculature as the tissue into which it is being placed.

In addition, unlike spontaneous tumors in humans, transplanted tumors often remain mostly separated from the normal tissue. When exogenous fluorophores are used as contrast agents in fluorescence diagnostics, there is also an issue of species-dependent chemical distribution or pharmacokinetics. These problems also begin to become convoluted by the fact that drug pharmacokinetics are often very different between early stages and more advanced stages of cancer or transplanted tumors because vascularization differs.[121,122] One way to minimize these problems in tumor studies is to induce the tumor chemically or radiologically. This will allow the tumor to be integrated into the tissue as well as ensure that the tumor being investigated comes from the same type of tissue in which it is growing.

One well-established tumor model procedure is to use dimethyl benzanthracene (DMBA) to induce lesions in the cheek pouches of hamsters. This particular model is useful because it progresses through many different stages of cancer development: normal tissue to hyperplastic tissue to dysplastic tissue to carcinoma *in situ* (CIS), and finally to invasive carcinoma.[123] Endogenous and exogenous fluorophores have been used for *in vivo* diagnostics based on this model. In the case of autofluorescence diagnostics, it has been reported that 76% sensitivity and 86% specificity have been obtained in the detection of early neoplastic tissue. In addition, by intravenously administering the photosensitizer Photofrin™ 24 h prior to analysis, the sensitivity and the specificity could be increased to 100%.

#### *28.4.3.1.1 Animal Studies Using Autofluorescence*

With analyses on living animals, it is possible to gain a much more accurate picture in most cases than can be achieved *in vitro*. For instance, one of the strongest chromophores present in tissues is hemoglobin (oxy- and deoxy-); in *in vitro* studies, no blood flow is present, and the ratio of oxygenated hemoglobin to deoxygenated hemoglobin is very different from that in live animal studies. This factor alone could have a significant effect on the autofluorescence spectrum of the tissue by absorbing the fluorescent light or by changing the oxidative state of some of the autofluorescent chemicals present.

*Cancer Diagnostics — Esophageal Cancer* — The capacity to identify subclinical neoplastic diseases of the upper aerodigestive tract using tissue autofluorescence spectroscopy has significantly contributed to the field of fluorescence cancer screening. In 1993, the applicability of tissue autofluorescence for the early diagnosis of precancerous states in esophageal mucosa was studied through various model systems.[124] In an *N*-nitroso-*N*-methylbenzylamine (NMBA)-induced rat esophageal cancer model, alteration of the fluorescence emission pattern at 380 nm was found to correspond to disease progression from normal mucosa through dysplasia to invasive cancer. Although gross assessment of the tissue was indistinguishable from the saline-treated controls, histopathologic evaluation revealed NMBA-induced preneoplastic changes in the epithelium.[125] In addition, a multicellular tumor spheroid model, induced by transretinoic acid (RA), was also found to alter autofluorescence intensities at multiple wavelengths including 340, 450, and 520 nm. Such RA-induced alterations corresponded to changes in the state of spheroid differentiation.[124]

*Brain Cancer* — Brain tumors are one of the more difficult types of cancer to treat. Unlike many other forms, where tumor margining can be less important and healthy tissue surrounding the tumor can be removed to ensure that no malignant cells are left behind, the removal of excess brain tissue during tumor removal can have dramatic adverse effects. Because of this important requirement, a technique capable of providing an accurate demarcation between malignant cells and normal brain cells is extremely important. One method that is being developed is autofluorescence-based optical biopsy.

In initial animal studies, EEMs of rat gliomas revealed three distinct regions of decreased autofluorescence emission with respect to the normal rat brain tissue. These differences in the fluorescence emission spectra of the two different types of tissue samples occurred at 470, 520, and 630 nm, with corresponding excitation wavelengths of 360, 440, and 490 nm. The fluorescence emission at 470 nm corresponds to NAD(P)H, while the emissions of 520 nm and 630 nm correspond to various flavins and porphyrins, respectively. Due to the nature of the chemical differences between the normal tissue and the glioma tissue, this finding suggests a relationship between brain tissue autofluorescence and metabolic activity. This is in contrast to *in vitro* brain tissue studies, which also found that NAD(P)H fluorescence was lower in all measured human brain tumors, but, depending on their nature, flavin and porphyrin autofluorescence in neoplastic tissues was not always lower than in normal tissue.[126]

*Atherosclerosis* — Due to the high occurrence rate of atheriosclerosis in adults and the preliminary success of LIF in diagnosing the disease in its various stages *in vitro*, animal studies have been performed to ensure that the same fluorescence phenomena occur in living systems. In one such study, an XeCl excimer laser operating at 308 nm was used to excite proteins of human aortas containing early lipid-rich noncollagenous lesions. The emitted fluorescence exhibited significant red shifts and spectral broadening compared with spectra from nonatherosclerotic human aortas. Similar red-shifted and spectrally broadened autofluorescence profiles were observed from oxidatively modified low-density lipoproteins excited by 308-nm illumination. However, in the case of native low-density lipoproteins, neither the red shift nor the spectral broadening was found to exist. In order to compare these tissue results to the autofluorescence emission from live animals, LIF studies were performed on hypercholesterolemic rabbits with early foam cell lesions. The resulting autofluorescence spectra were similar to those of oxidized beta low-density lipoprotein, the major lipoprotein accumulating in arteries of rabbits fed cholesterol and an early indicator of atherosclerosis.[127]

#### *28.4.3.1.2 Animal Studies Using Exogenous Dyes*

The study of exogenous fluorophores in living animals can provide a great deal of information. In addition to being a model system that more closely resembles human when testing fluorescence-based diagnostic systems prior to clinical trials, they also play a large part in the field of pharmacokinetics. Using living animals with functioning circulatory systems, it is possible to watch the real-time distribution of various drugs throughout the body as well as study their uptake into various tissues and, finally, their excretion from the body. This is an extremely important field of research because it ensures timely deliverey of drugs to the target tissue areas of interest.

*Cancer Diagnostics* — Some of the more commonly studied exogenous fluorophores in biomedical analyses are used for cancer diagnosis or treatment, or both. These fluorophores, known as photolabels in the case of the former and photosensitizers in the case of the latter, represent an extensive class of compounds whose pharmacokinetics properties as well as their tumor demarcation abilities need to be studied prior to use in human trials. These studies are generally performed on animal subjects; several of the more common ones are described in the following sections.

*Pharmacokinetics of Photosensitizers* — Photosensitizers represent a unique class of compounds that, upon being activated by the absorption of a specific wavelength of light, are capable of killing the surrounding tissues. A detailed description of the mechanism of action of these photosensitizers can be found in Chapter 36 of this handbook. Therefore, only salient features are briefly described here. The concept behind these photosensitizers is to develop drugs that are preferentially taken up or localized in malignant tissues and can then be photoactivated. After the drug is administered and a specific amount of time has passed for optimal uptake, light is shown on the tissue area of interest and absorbed by the sensitizer. The sensitizer then kills the surrounding tumor tissue, leaving the healthy tissue undamaged.

Although the number of available photosensitizers approved for clinical use is relatively small, a large number of new potential candidates are continually being developed. These new agents are then compared with the better characterized sensitizers based upon many parameters, including tissue localization ability, effectiveness in promoting cell death, and toxicity, as well as several other parameters. In a study to determine the most effective photosensitizer of five commonly used drugs, their tissue localization properties were studied *in vivo* using "sandwich" observation chambers and tumors that were growing in thigh muscle. Several common dyes were studied, including hematoporphyrin derivative, PhotoFrin II, aluminum phthalocyanine tetrasulphonate, uroporphyrin I, and acridine red. Of the photodynamically active dyes (the first three), aluminum phthalocyanine tetrasulphonate was found to exhibit the best *in vivo* tumor localization properties as determined by fluorescence spectroscopy.[128]

LIF studies have also been used to investigate the pharmacokinetic properties of 5-aminolevulinic acid (ALA)-induced protoporphyrin IX (PpIX). These analyses were performed in normal and tumor tissues of rats following intravenous (i.v.) injection of ALA. The aim of the study was to investigate ALA-induced (PpIX) formation and its accumulation in different types of rat tissues after the systemic administration of ALA. Tissue types investigated included a malignant rat tumor and normal tissue from 13 different organs in 8 rats. The various rats were injected with two different doses of ALA (30 and 90 mg/kg body weight), and fluorescence analyses were performed 10, 30, and 240 min after injection.

Fluorescence analyses were performed by exciting the sample with 405-nm light and monitoring the resulting fluorescence with a fiber-optic probe over the spectral region of 400 to 750 nm. The fluorescence signal consisted of a broadband autofluorescence background with a maximum at approximately 500 nm and a characteristic dual-peak emission from the PpIX at 635 and 705 nm. From this work, it was found that the maximum tumor buildup of PpIX was achieved in less than 1 h after ALA injection, that the fluorescence demarcation between tumor and surrounding tissue was between 7:1 and 8:1 after 30 min, and that it decreased with longer retention times. Of the 13 different organs investigated in this study, PpIX buildup was found to be particularly high in the stomach and the intestine.[129]

Another class of exogenous dyes that have been used for tumor demarcation are hematoporphyrins and hemotoporphyrin derivatives. In a pharmacokinetic study of hematoporphyrin, rats were administered i.v. injections of the dye, followed by immediate and continuous analysis by LIF. Excitation of the hematoporphyrins was provided by the 337-nm output of a nitrogen laser; the signal was collected by an optical multichannel analyzer (OMA), which allowed for acquisition of entire fluorescence spectrum for each laser shot. Upon analysis of data from several of the rat's organs, it was found that the fluorescence emission from the hematoporphyrin at 630 nm exhibited an initial peak intensity as well as a delayed peak intensity. The delayed fluorescence peak was described as due to the chemical components of intracellularly transformed hematoporphyrin derivatives. In addition, it was also discovered that, by dividing the background-free 630-nm signal by the autofluorescence intensity at shorter (blue) wavelengths, a ratio exhibiting a larger contrast between tumor and surrounding tissue could be obtained that would greatly aid in tumor demarcation.[130]

A third and relatively recent photosensitizer used for the demarcation of tumor tissues is benzoporphyrin derivatized-monoacid (BPD-MA). In LIF studies designed to provide pharmacokinetic information about this compound, 337-nm light from a nitrogen laser was used to excite the BPD-MA in rats that had been administered the drug via i.v. The fluorescence emission, over the spectral range of 380 to 750 nm, was then monitored with a spectrometer equipped with a diode array detector. Three hours after the injection period, the fluorescence signals were measured from many different types of rat tissue, including malignant tumors that had been experimentally induced. These results were then compared with results from several other common sensitizers, such as hematoporphyrin (HP), polyhematoporphyrin ester (PHE), tetrasulfonated phthalocyanine (TSPc), and the commercially available Photofrin™. After three hours, the demarcation potential between tumor and surrounding tissue in terms of fluorescence signal for the tumor model used was 2:1 for BPD-MA. When compared with other drugs such as HP, it shows about the same demarcation potential, whereas Photofrin and PHE exhibit about 3 times better and TSPc about 1.5 times better demarcation. Employing the endogenous tissue fluorescence signature enhanced the contrast by a factor of about two for each of the five drugs.[131]

In work by Nilsson et al., the biodistribution of two recently developed tumor markers, trimethylated (CP[Me]3) and trimethoxylated (CP[OMe]3) carotenoporphyrin, was investigated by means of LIF. In this study, 38 tumor-bearing (MS-2 fibrosarcoma) female BALB/c mice were administered the drugs through i.v. injection. At 3, 24, 48, or 96 h after administration, the carotenoporphyrin fluorescence was measured in tumoral and peritumoral tissue, as well as in the abdominal, thoracic, and cranial cavities. Excitation of the exogenous dye was performed by using the 425-nm emission of a nitrogen-pumped dye laser and was measured by a spectrometer equipped with an intensified charge-coupled device (ICCD) at 490, 655, and 720 nm. The emission bands at 655 and 720 nm corresponded to carotenoporphyrin (CP), whereas the band at 490 nm represented a location near the maximum of the autofluorescence emission of the tissue. The tissues that showed the greatest extent of CP-related fluorescence were the tumors and the liver tissues, whereas the cerebral cortex and muscle tissues consistently exhibited weak CP-related fluorescence. Additionally, although the fluorescence intensity of most tissue types decreased over time, it was found that the fluorescence intensity in the liver remained constant the full 96 h over which it was investigated.[132]

*Oral Cancer* — Apart from pharmacokinetic analyses of various sensitizers, animal studies with these compounds have also been performed to verify and quantify the ability of these drugs for tumor demarcation in different types of tissues. One such study investigated the effectiveness of porfimer sodium-derived drugs for the detection of early neoplastic changes in the oral cavity of hamsters. Neoplasia was induced in the hampsters' cheek pouches by the application of 9,10-dimethyl-1,2-benzanthracene. Following formation of neoplastic tissue in the hamster cheek pouch, autofluorescence analyses as well as fluorescence analyses using porfimer sodium as an exogenous tumor demarcation drug were performed and compared with histologic results of the same tissue sites. When the two fluorescence diagnostic techniques were compared with the histological results, it was found that the autofluorescence analysis was capable of 76% sensitivity and 83% specificity, whereas the porfimer sodium-based fluorescence technique provided 100% sensitivity and specificity for the samples investigated.[133]

*Pancreatic Cancer* — Pheophorbide-a (Ph-a) represents another of the many different types of tumor demarcation drugs that have been used for optical differentiation of normal tissues and tumor tissues. To test its viability for the detection pancreatic tumors, LIF analyses of Ph-a were used to image six intrapancreatic tumors and six healthy pancreases *in vivo* in rats. Ph-a was intravenously administered to the rats at a concentration of 9 mg/kg body weight, and fluorescence images were acquired up to 48 h after injection of the drug. Excitation of the blue tissue autofluorescence was performed by using 355-nm light from a frequency-tripled Nd:YAG laser and excitation of the dye was performed by the 610-nm output from an Nd:YAG pumped dye laser. Fluorescence images were obtained at three different wavelengths; bandpass filters were used for wavelength selection. Images at 470 nm and 640 nm were used to monitor the autofluorescence of the tissue, while images at 680 nm were taken to obtain composite images of the dye and autofluorescence emissions together. In order to achieve a good contrast between the normal pancreatic tissue and the tumor tissue, the autofluorescence intensity of the 640-nm image was normalized to the background fluorescence intensity in the 680-nm image prior to being subtracted from the 680-nm image. Following this subtraction, the differential image was divided by the autofluorescence image at 470 nm. The resulting ratiometric images allowed for safe diagnoses to be made by providing well-contrasted tumor images.[24]

*Brain Tumors* — Animal models have also been used in studies in which hematoporphyrin derivative (HPD) was used for the diagnosis of brain tumors. In this study, adult Wistar rats had C6 glioma cells implanted into their brains to act as a cerebral glioma model. After the tumors had developed to a size of 7 to 12 mm in diameter, they were injected intravenously with 5 mg/kg body weight of HPD, and continuous fluorescence analyses were performed for 24 h using a fiber optic-based fluorimeter. The fluorescence intensity of the normal brain tissue reached a plateau 60 min after injection, while the glioma region reached a plateau 80 min after injection. Fluorescence analyses of glioma, brain tissue adjacent to the tumor (BTAT), and normal brain tissue 24 h after injection revealed that the fluorescence intensity of the glioma was 6.1 times greater than that of normal brain tissue; the BTAT also was 3.9 times greater than the normal tissue.[134]

*Liver Cancer* — Testing efficacy of PDT using ALA-induced PpIX sensitization for the treatment of heptic tumors was performed with rat models. Liver tumors were induced in the rats by local innoculation of tumors cells or by administration of the tumor cells through the portal vein. Following tumor formation, 60 mg/kg body weight of ALA was administered intravenously. After 60 min, the PpIX in the heptic tumor tissue was excited, initiating the PDT treatment as well as allowing the PpIX to be excited. The resulting fluorescence emission was monitored at 635 nm. Fluorescence analyses revealed that large accumulations of the PpIX occurred in the heptic tumor, as well as in normal liver tissue, but not in the abdominal wall muscles.

In addition to fluorescence analyses, laser Doppler imaging was used to determine changes in the superficial blood flow in connection with PDT. Histopathological examinations were also performed to evaluate the photodynamic therapy effects on the tumor and the surrounding liver tissue, including pathological features in the microvascular system. Laser Doppler imaging results indicated an effect on the vascular system in the tumor as well as on the surrounding tissue following PDT treatment, as determined by decreased blood flow in the treated area. In addition, the tumor growth rate decreased significantly when evaluated 3 and 6 days after the treatment, showing that ALA-induced PDT holds promise for treatment of heptic tumors.[135]

### 28.4.3.2 Human Studies and Clinical Diagnostics

By far the most common type of fluorescence-based biomedical diagnostic procedure used in clinical studies is the optical detection of malignant or premalignant tissues in various organs. Among the more common types of tumors that have been investigated with this technique are skin, urinary bladder, bronchus, gastrointestinal, head and neck, gynocological, breast, and brain cancers. Currently, two very different approaches to these fluorescence-based optical biopsy techniques are employed. The first of these techniques relies on subtle differences in the tissue composition and morphology between normal, dysplastic, and malignant tissues and their effect on the autofluorescence properties for differentiation. Examples of this type of diagnoses are discussed in the next section of this chapter. The second type of fluorescence-based diagnostic technique relies on the presence of exogenous fluorophores, such as PDT photosensitizers, for the tissue differentiation.

#### *28.4.3.2.1 Clinical Studies and Diagnostics Using Autofluorescence*

Autofluorescence-based optical biopsy techniques represent the ideal form of fluorescence-based diagnostic procedure. In these analyses, laser light is used to excite the naturally occurring fluorophores in the tissue; differences in chemical composition between the various types of tissue will allow for real-time diagnoses without removal of a tissue sample or treatment with a contrast-enhancing drug. Because of this and the great deal of promise that initial studies have shown, research in this field is currently experiencing tremendous growth.

*Cancer Diagnostics (Optical Biopsy)* — *Cervical Cancer* — Cervical cancer is the second most common malignancy in women worldwide and remains a significant health problem. Despite the widely used Papanicolaou smear (Pap smear) screening procedure and costly treatments, the overall survival rate remains 40%. For these reasons, efforts to strengthen screening and prevention are needed.[136] Autofluorescence optical biopsy techniques offer just such a screening technique with automated diagnosis in real time and comparable sensitivity and specificity to colposcopy.[137] Feld and co-workers have demonstrated the ability to distinguish between various types of tissues, *in vivo*, based upon multicomponent analysis (Figure 28.6).[63]

**COLOR FIGURE 1.1** Louis Pasteur used the microscope, which provided the central observation tool for a new style of research and out of which emerged the germ theory of disease. Albert Edelfelt's *Louis Pasteur* (1865), Musée d'Orsay, Paris, France. (©Réunion des Musées Nationaux/Art Resource, New York. Reproduced with permission of Artists Rights Society, New York.)

**COLOR FIGURE 1.4** Quantum theory also had a profound effect on many fields beyond science, such as art. It was no coincidence that, during the quantum revolution in science, Cubist and Surrealist art abolished realistic shapes referenced in fixed space and fixed time. Pablo Picasso's renderings of the human face with its multifaceted perspectives often reflected the dual nature of reality. *Portrait of Dora Maar* (1937), Musée Picasso, Paris, France. (©Giraudon/Art Resource, New York. Reproduced with permission of Artists Rights Society, New York.)

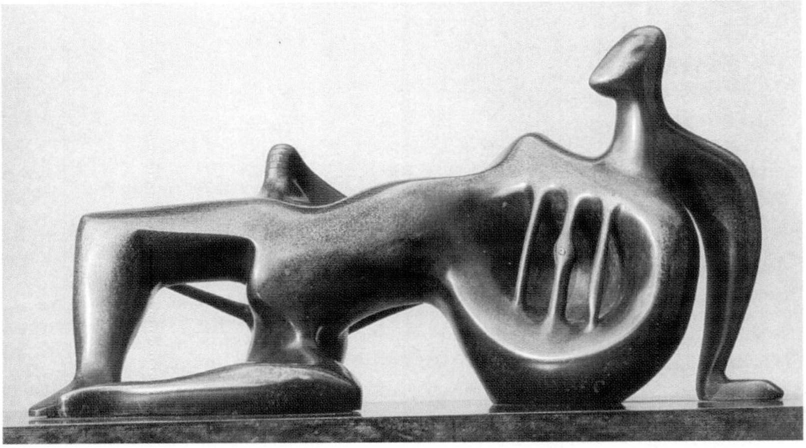

**COLOR FIGURE 1.5** It is noteworthy that modern art, with its seemingly strange distortions of visual reality, also appeared in the epoch of quantum theory. Henry Moore's sculptures reshaping the physical form of the human body marked a departure from geometric forms. *Reclining Figure* (1931), Dallas Museum of Art, Dallas, Texas. (From Kosinski, D., *Henry Moore: Sculpting the 20th Century*, Yale University Press, New Haven, CT, 2001, p. 91. Reproduced with permission from The Henry Moore Foundation.)

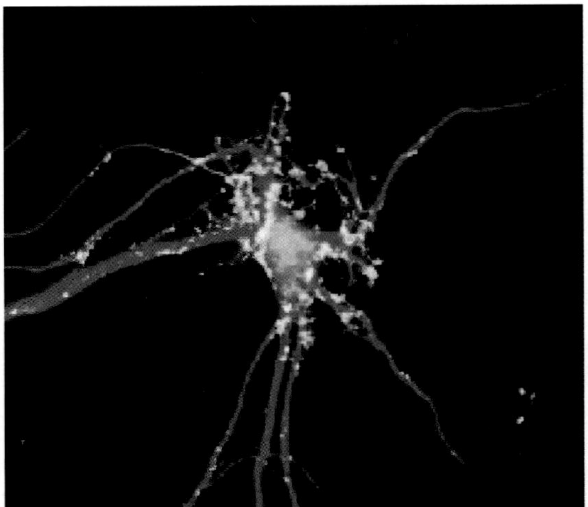

**COLOR FIGURE 1.8** Biochemical processes in nerve cells can be tracked using green fluorescent protein. Tracking biochemical processes can now be performed with the use of fluorescent probes. For example, there is great interest in understanding the origin and movement of neurotrophic factors such as brain-derived neurotrophic factor (BDNF) between nerve cells. This figure illustrates the use of BDNF tagged with green fluorescent protein to follow synaptic transport from axons to neurons to postsynaptic cells. (From Berman, D.E. and Dudai, Y.K., *Science*, 291, 2417, 2002. With permission.)

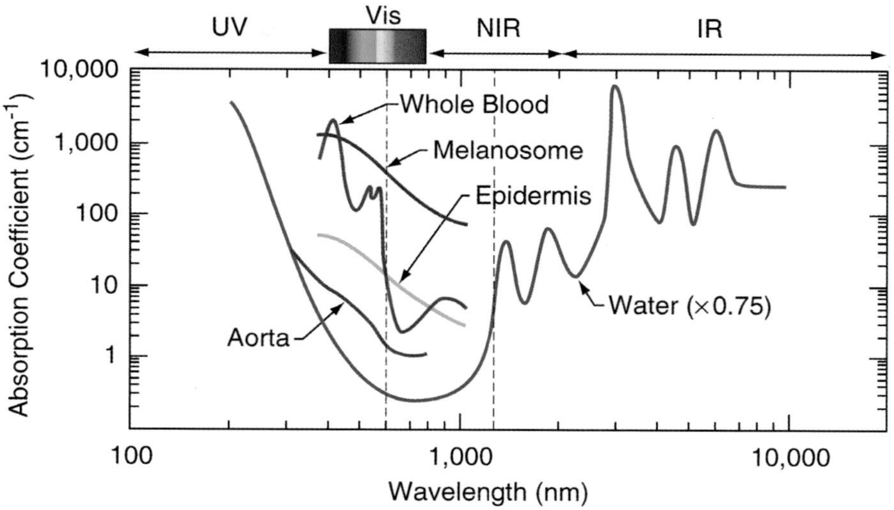


**COLOR FIGURE 2.16** The "therapeutic window" (region between the dashed lines) in tissue, which runs from 600 to 1400 nm, and the spectra of several important absorbers in this range.

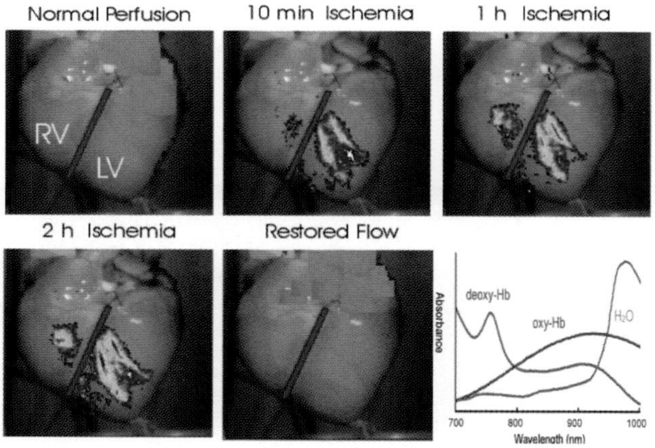


**COLOR FIGURE 8.7** Near-infrared spectroscopy can detect variations in tissue oxygen levels by means of hemoglobin absorption peaks (lower right). Wavelength control was achieved using a liquid crystal tunable filter in front of a CCD. Pseudocolor highlights left ventricle deoxygenation during occlusion of the artery that normally supplies it. (Images courtesy of Henry Mantsch, National Research Council, Winnipeg, Canada.)

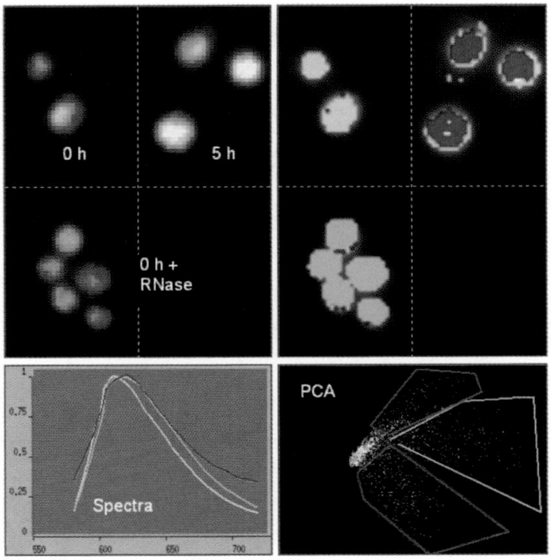


**COLOR FIGURE 8.8** Spectral imaging and image segmentation of yeast cells stained with propidium iodide (PI). Top left: Composite showing three panels of PI-stained yeast cells imaged under three conditions: immediately after transfer of yeast to new medium, after 5 h of culture, and after 5 h of culture plus the addition of RNase. Bottom left: Spectra from yeast in each culture condition obtained by imaging using a liquid crystal tunable filter. Small differences in peak position and shoulder configuration are visible. Bottom right: Scatterplot of the spectral image after principal components analysis (PCA). Three clusters are circled and the pixels contained in each cluster pseudocolored. Top right: Result of mapping pseudocolored PCA clusters back to the original image, resulting in robust segmentation. (Analysis: R. Levenson, CRI, Inc.)

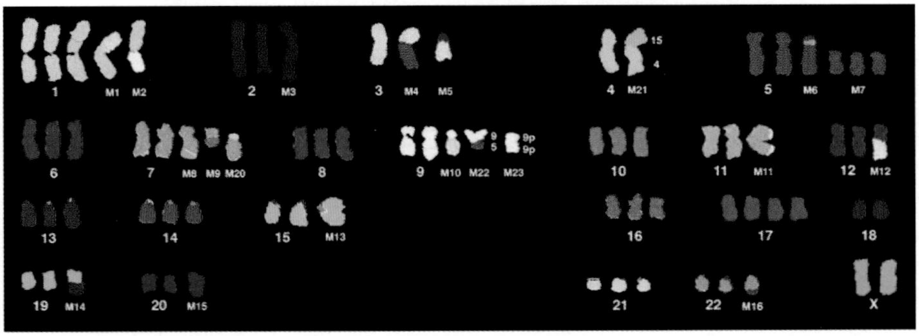


**COLOR FIGURE 8.9** Comprehensive cytogenetic analysis of a metaphase spread from a child (CK1) with dysmorphic features and developmental delay resembling an 18q-syndrome. Spectral karyotyping (SKY) was performed on a metaphase spread. The multicolor hybridization clearly reveals an aberrant chromosome (arrow) that contains chromosomes 18 (red) and X (dark green) material. The G-banding interpretation of a normal male. (From Schröck, E. et al., *Hum. Genet.*, 101, 255, 1997. With permission.)

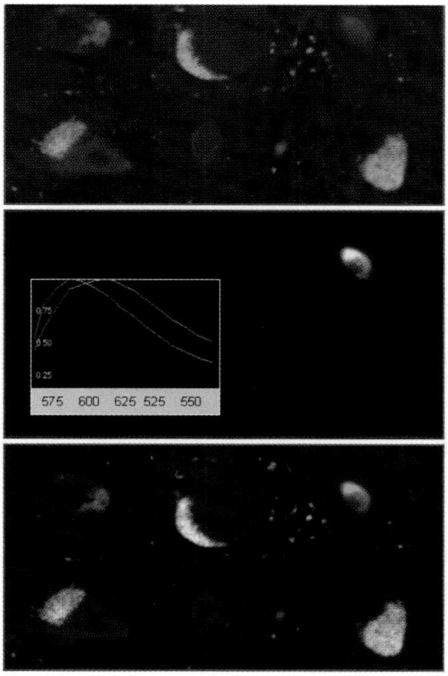


**COLOR FIGURE 8.10** Removal of autofluorescence using pixel unmixing. Top panel: Composite of six neurons in formalin-fixed, paraffin-embedded human brain stained with anti-GDNF labeled with Cy2. The bulk of the fluorescent signal is auto-fluorescence. Center panel: The single positively stained neuron is separated from the abundant autofluorescence using pixel unmixing (unmixing spectra shown in the insert). Bottom panel: Cy2 signal (in green) overlain on top of the autofluorescence signal (in gray). (Samples courtesy of Neelima Chauhan and George Siegel; analysis: R. Levenson, CRI, Inc.)

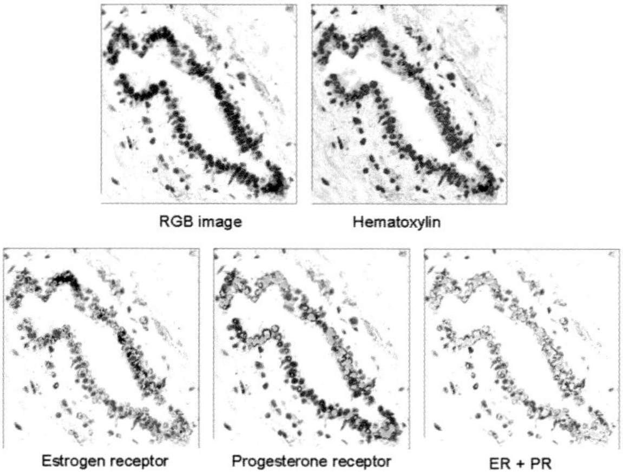


**COLOR FIGURE 8.11** Multicolor immunohistochemistry and spectral unmixing. Top left: RGB image of a cluster of breast cancer cells stained for the presence of estrogen receptor (ER, red), progesterone receptor (PR, brown); nuclei are counterstained with hematoxylin (blue). It is difficult to determine how much of each antigen is present in the cancer cell nuclei. After collecting a spectral stack, the signals corresponding to nucleus, ER, and PR are spectrally unmixed and shown in separate images. Bottom right shows where ER and PR are co-expressed (yellow signal). (Sample courtesy of DAKO, Inc.; analysis: R. Levenson, CRI, Inc.)

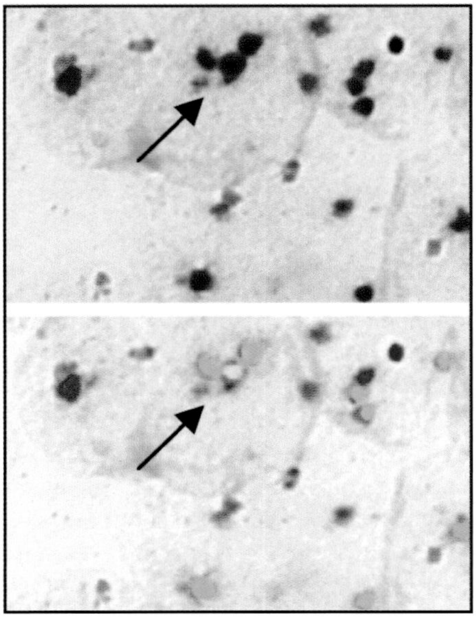

**COLOR FIGURE 8.12** Spectral unmixing of transmission *in situ* hybridization (TRISH). Nuclei of cytospun bladder carcinoma cells probed for three chromosome centromeres. Detection was performed using DAB, New Fuchsin, and TMB as chromogens. Lower panel: Spectral unmixing reveals overlap of brown and red signals (pseudocolored as green and red, respectively). The arrow points to a yellow spot representing the overlap. (Sample courtesy of Anton Hopman, University of Maastricht, Maastricht, the Netherlands; analysis: R. Levenson, CRI, Inc.)

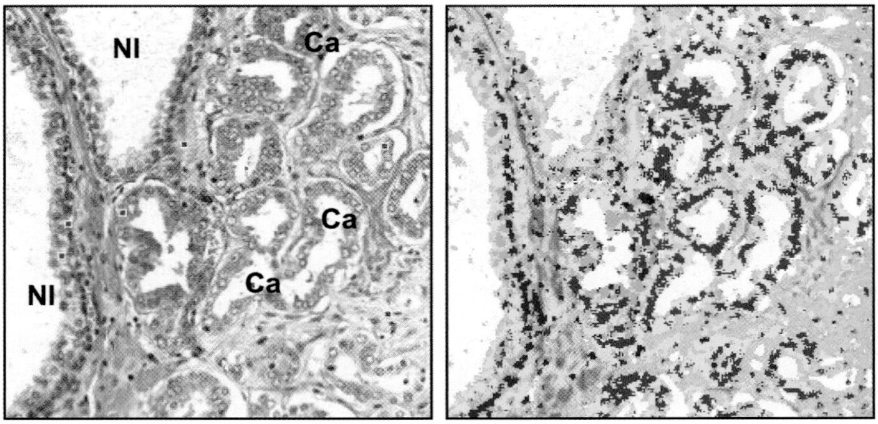


**COLOR FIGURE 8.13** Spectral segmentation of hematoxylin- and eosin-stained prostate cancer specimen. Left: RGB image of prostate cancer (Ca) and normal (Nl) glands. The normal glands are lined with a double cell layer consisting of epithelial and basal cells; the cancerous glands have a single cell layer. Right: Result of spectral segmentation using spectra chosen manually from representative pixels in the image and a minimum square error classification algorithm. The two cell layers (pseudocolored green and blue) are spectrally distinct from one another and from the cancer cells (pseudocolored red). (Analysis: R. Levenson, CRI, Inc.)

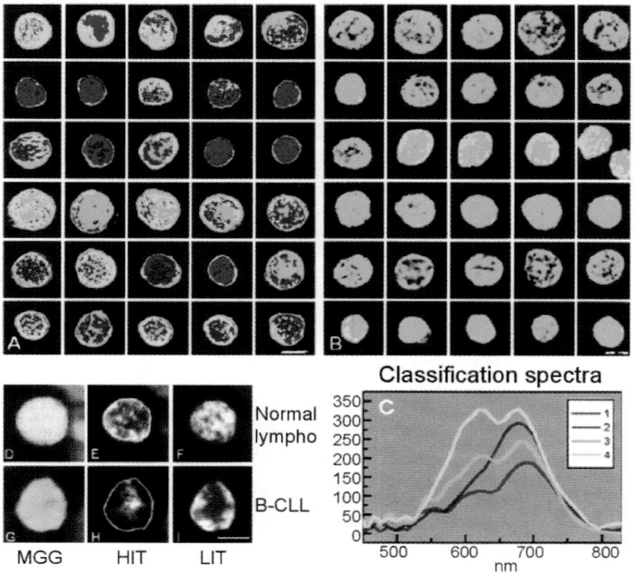


**COLOR FIGURE 8.14** Spectral classification and morphological analysis. Normal lymphocytes are compared to small B-cell lymphocytic leukemia cells, both stained with Giemsa. By eye these are virtually indistinguishable. Spectral classification, using the spectra shown in the lower right panel, reveals spectral differences in content and spatial distribution of the spectral features (top right and left panels). Lower left: Spectral similarity mapping algorithms indicate more clearly the differences in distribution of spectral features in normal vs. lymphocytic leukemia cells. (From Malik, Z. et al., *J. Histochem. Cytochem.*, 46, 1113, 1998. With permission.)

**COLOR FIGURE 11.6** Three-dimensional reconstructed two-photon images of dermal structures in a mouse ear tissue specimen. The three images show distinct structural layers: (A) epidermal keratinocytes; (B) basal cells; and (C) collagen/elastin fibers.

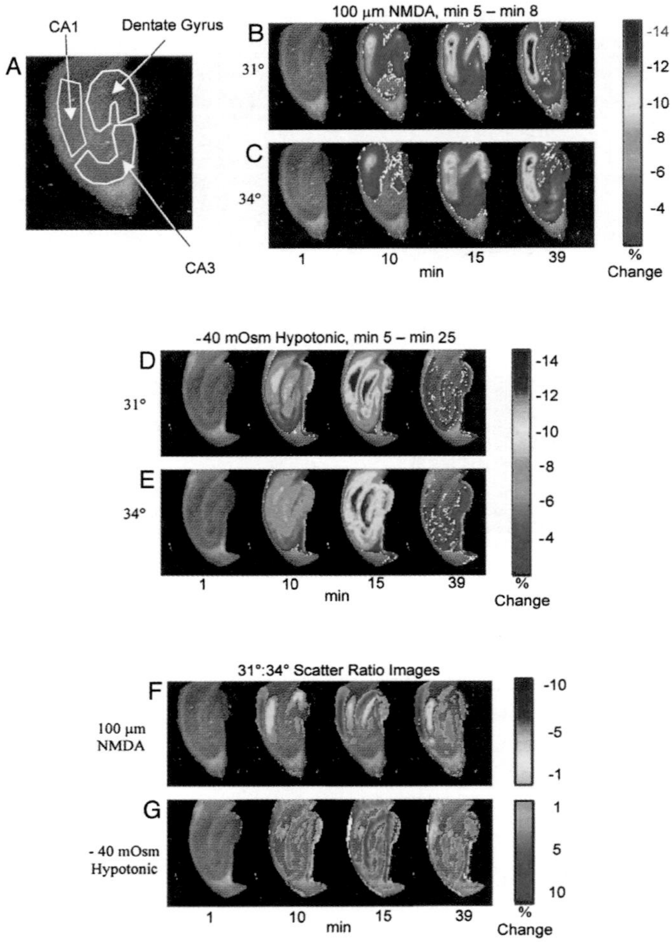


**COLOR FIGURE 16.8** Single-angle and dual-angle scatter images of hippocampal slices under osmotic stress and subjected to NMDA. In each case, the color bars indicate the percent change in image intensity. In regions of the slice where the change is less than ±2%, the gray scale images of the hippocampus are shown. (A) The scatter image of a hippocampal. The three main regions of the hippocampus are CA1, CA3, and the dentate gyrus. Also shown are single-scatter images at 31° (B) and 34° (C) of an NMDA-treated slice. In both (B) and (C) there is a large change in the CA1 region. There is also a significant change in the dentate gyrus that fades by minute 39 of the experiment. Single-angle scatter images at 31° (D) and 34° (E) are shown for hypotonic treatment. In both (D) and (E), a large change is indicative of cellular swelling in the CA1 region. There is also a significant change in the dentate gyrus. Dual-angle scatter ratio images are shown for NMDA (F) and hypotonic (G) treatments. The NMDA-treated slices in (F) undergo a relatively larger change in the dual-angle scatter ratio in CA1. In CA1 the scatter ratio change is negative, possibly indicating particle shrinkage. In (G), hypotonic treatment, the magnitude of the change in the scatter ratio is less in CA1. In addition, the overall location of the scatter change is more spread out than for NMDA treatment. In both (F) and (G) the white matter areas reveal a positive change in the scatter ratio. (From Johnson, L.J., Hanley, D.F., and Thakor, N.V., *J. Neurosci. Methods*, 98, 21, 2000. With permission.)

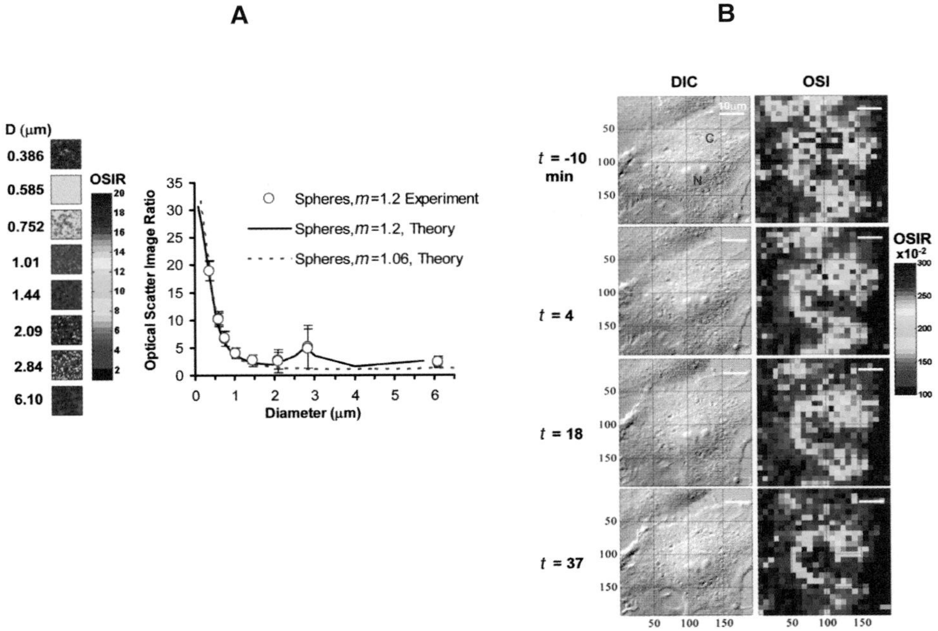


**COLOR FIGURE 16.10** (A) OSI images and measurement of the OSIR in aqueous suspensions of polystyrene spheres ($m = 1.2$). The OSIR is a measure of wide to narrow angle scatter. Experimental data (open circles) and theoretical predictions (solid line, $m = 1.2$; dashed line, $m = 1.06$) are shown. The experimental data points show the mean pixel value and standard deviation in the scatter images displayed to the left of the graph. (From Boustany, N.N., Juo, S.C., and Thakor, N.V., *Opt. Lett.*, 26, 1063, 2001. With permission.) (B) Representative cell undergoing apoptosis after treatment with 2 μ*M* staurosporine (STS). The cell was imaged in DIC (left panels) and OSI (right panels) at different time points. C = cytoplasm, N = nucleus. Images are displayed at times $t = -10$, 4, 18, and 37 min from STS addition at $t = 0$. The ratiometric scatter images show a decreasing scatter ratio within the cytoplasm (C).

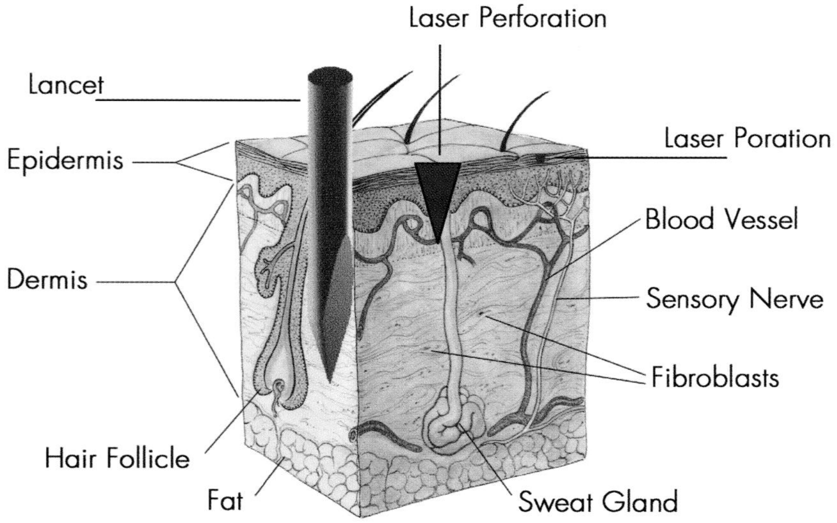


**COLOR FIGURE 18.6** Cross section of skin comparing the depth of penetration incurred for a standard lancet to that of both the laser performation and laser poration approaches.

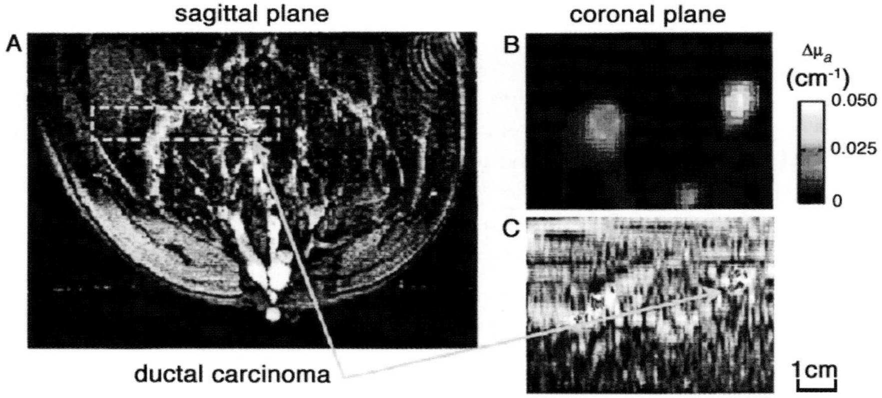


**COLOR FIGURE 21.14** A dynamic MRI and contrast-enhanced DOT image of a ductal carcinoma. (A) A dynamic sagittal MR image after Gd contrast enhancement passing through the center of the malignant lesion. (B) The coronal DOT image, perpendicular to the plane of the MRI image in (A), but in the volume of interest indicated in (A) by the dashed-line box. (C) The dynamic MR coronal image resliced from the same volume of interest and same dimensions as (B).

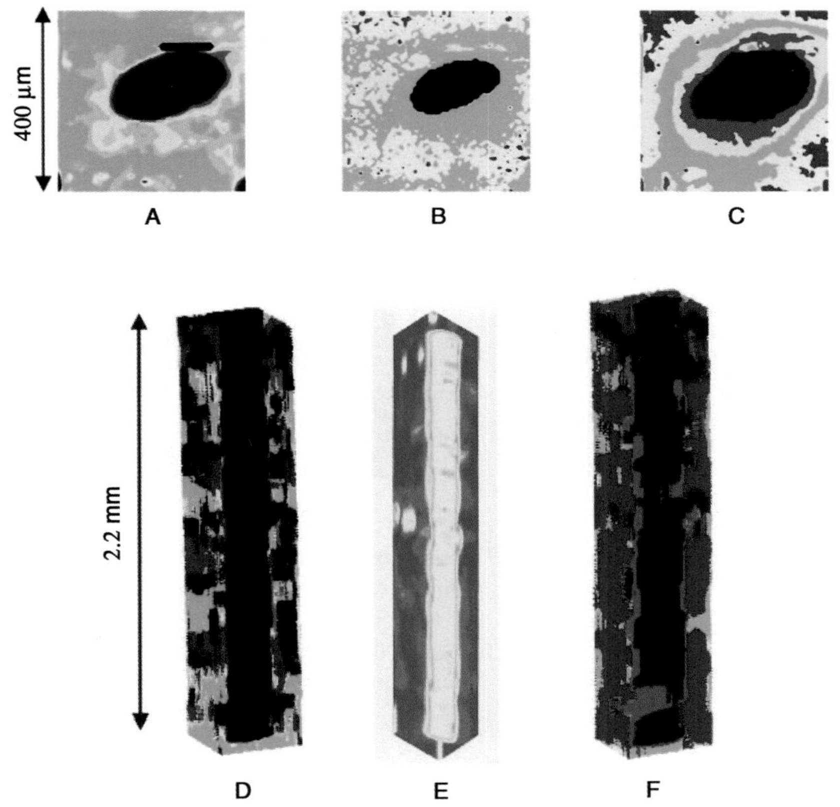


**COLOR FIGURE 32.5** (A–C) 2D images of three IR parameters for the cortical bone section (human tibia) shown as follows: (A) nonreducible/reducible collagen crosslinks, measured by the 1660/1690 peak height ratio in the amide I spectral region; (B) hydroxyapatite crystallinity index, measured by shape changes in $PO_4^{3-}$ $\nu_1$, $\nu_3$ mode and quantitated by the 1030/1020 peak height ratio; (C) mineral-to-matrix ratio, measured by the phosphate $\nu_1$, $\nu_3$ /amide I area ratio. Black and blue indicate low numerical values while green, yellow, and red indicate progressively increasing numerical values of the particular parameter. The dimensions of each image are 400 × 400 μm. (D–F) 3D reconstructed views of three IR parameters for the cortical bone section (human tibia), as follows: (D) view of the collagen crosslink parameter measured as the 1660/1690 height ratio in the amide I spectral region; (E) PMMA distribution in a cortical bone section. The relative concentration of PMMA is determined from the integrated intensity of the C = O stretching mode; (F) 3D reconstructed oblique view of the relative amounts of mineral-to-matrix as measured by the phosphate $\nu_1$, $\nu_3$ /amide I area ratio. The dimensions represented in each image are 400 × 400 μm × 2.2 mm. (From Ou-Yang, H. et al., *Appl. Spectrosc.*, 56, 419, 2002. With permission.)

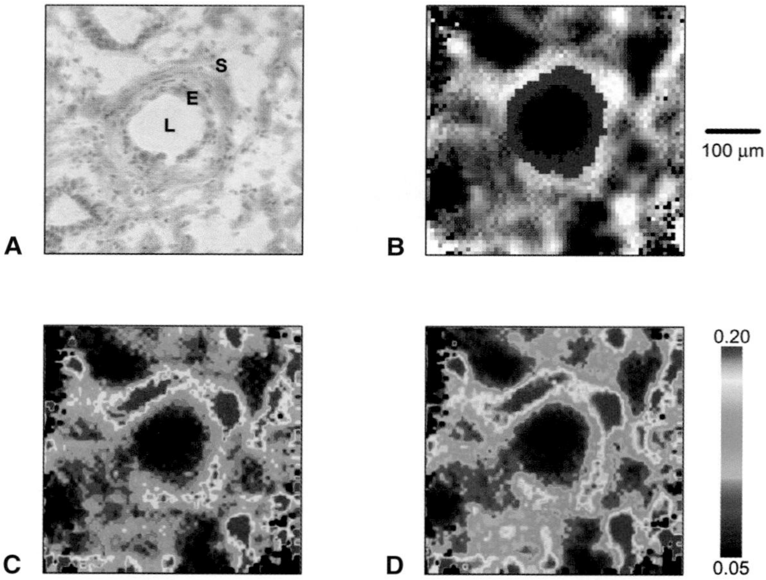


**COLOR FIGURE 32.6** (A) H&E stained section corresponding to the unstained section used to obtain spectroscopic images. The circular lumen (L) of prostatic gland in the center is lined by a double layer of epithelial cells (E) and supported by a smooth muscular and fibrocollagenous stroma (S). (B) Unstained section with the epithelial regions highlighted in red. Images of the distribution of absorbance obtained by applying (C) no gain ranging at a gain setting of one and (D) gain ranging for gains one and ten with a gain ranging radius of ten points. The images represent a sample of 500 μm × 500 μm as indicated by the size bar. The color-coded images show the distribution of the absorbance at 1245 $cm^{-1}$ between the indicated limits. (From Bhargava, R. et al., *Appl. Spectrosc.*, 55, 1580, 2001. With permission.)

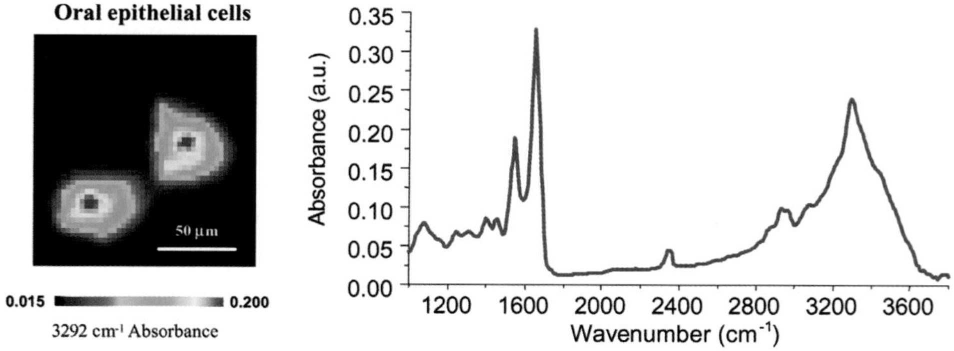


**COLOR FIGURE 32.7** IR spectroscopic image of single oral epithelial cells (left) and the corresponding spectrum from a part of the cell. The high signal-to-noise ratio of the data allows for detailed analysis of subtle chemical changes.

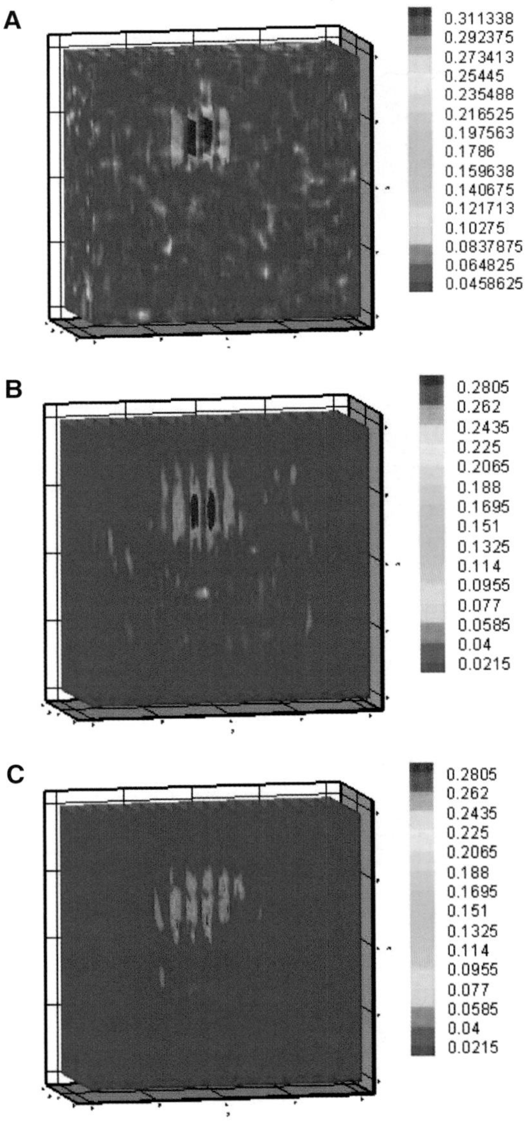


**COLOR FIGURE 33.21** Three-dimensional reconstruction from simply bound truncated Newton's method. (A) Actual distribution of fluorophore-absorption coefficient of background tissue variability of endogenous (50%) and exogenous (500%) properties. (B) Reconstructed fluorophore-absorption coefficient of background tissue variability of endogenous (50%) and exogenous (500%) properties using relative measurement of the emission fluence with respect to the excitation fluence at the same detector point, $\varepsilon = 0.0001$. (C) Reconstructed fluorophore absorption coefficient of background tissue variability of endogenous (50%) and exogenous (500%) properties using relative measurement of the emission fluence with respect to the excitation fluence at the same detector point. (From Roy, R., Godavarty, A., and Sevick-Muraca, E.M., *IEEE Trans Med. Imaging*, in press. With permission.)

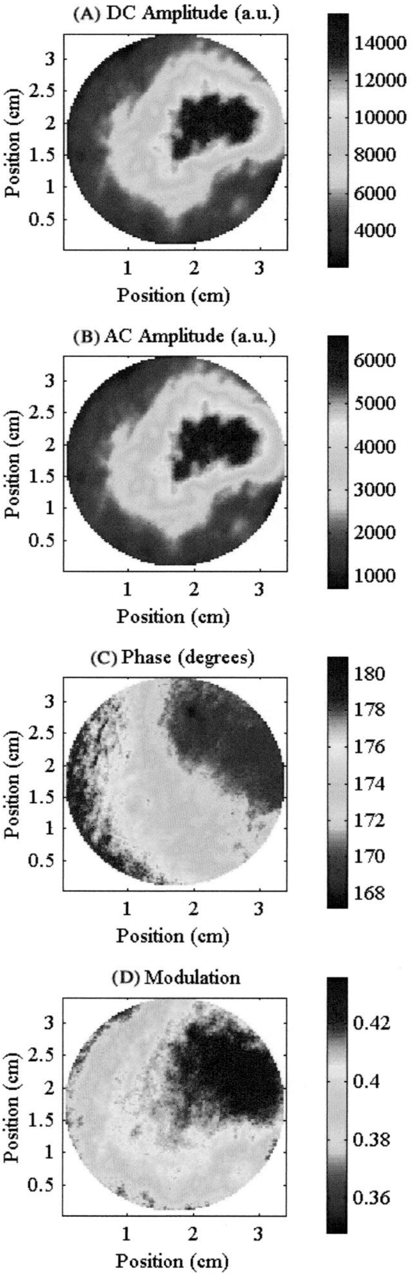


**COLOR FIGURE 33.23** The 128 × 128 pixel-based imaging of 830-nm fluorescence of (A) CW $I_{DC}$, (B) amplitude $I_{AC}$, (C) phase delay, and (D) modulation ratio of the detected fluorescence generated from the area cranial of the left fourth mammary gland of a canine. Illumination was accomplished with an expanded 780-nm laser diode. Modulation frequency was 100 MHz. (From Reynolds, J.S. et al., *Photochem. Photobiol.*, 70, 87, 1999. With permission.)

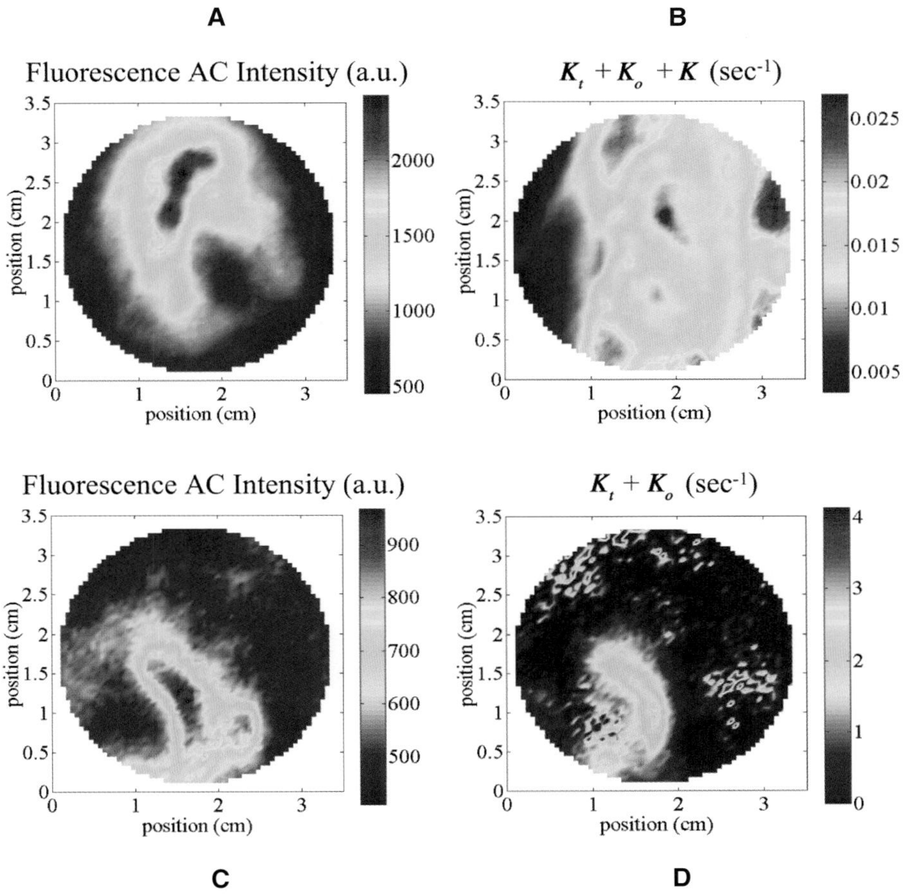


**COLOR FIGURE 33.25** (A) Fluorescence $I_{AC}$ intensity map from ICG delineating diseased tissue and (B) map of pharmacokinetic uptake parameters obtained from fitting the time sequences of fluorescence-intensity images showing no specific uptake of ICG in diseased tissue. (C) Fluorescence AC intensity map from HPPH-car delineating diseased tissue and (D) map of pharmacokinetic uptake parameters obtained from fitting the time sequences of fluorescence-intensity images showing specific uptake of HPPH-car in diseased tissue. (From Gurfinkel, M. et al., *Photochem. Photobiol.*, 72, 94, 2000. With permission.)

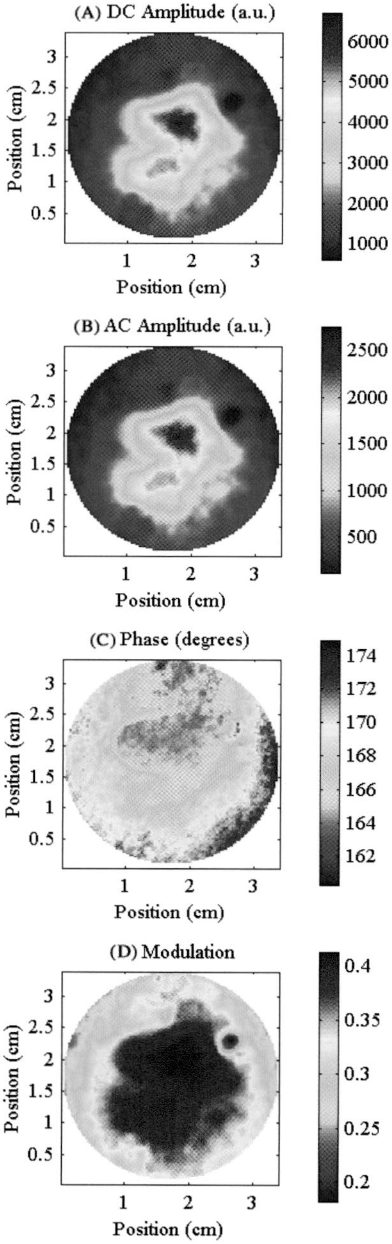


**COLOR FIGURE 33.26** The 128 × 128 pixel-based imaging of 830-nm fluorescence of (A) CW $I_{DC}$ (B) amplitude $I_{AC}$ (C) phase delay, and (D) modulation ratio of the detected fluorescence generated from the area cranial of the left fifth mammary gland of a canine. Illumination was accomplished with an expanded 780-nm laser diode. Modulation frequency was 100 MHz. (From Reynolds, J.S. et al., *Photochem. Photobiol.*, 70, 87, 1999. With permission.)

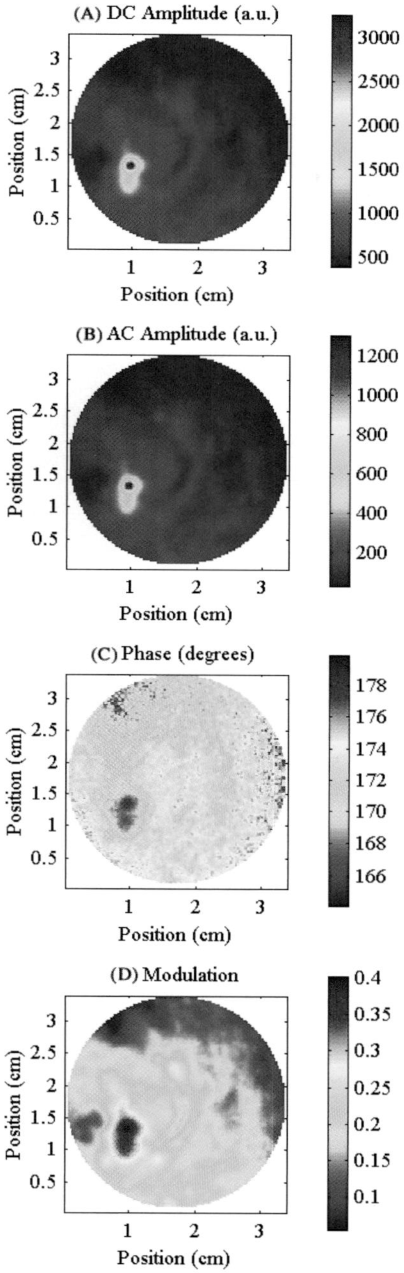


**COLOR FIGURE 33.27** The 128 × 128 pixel-based imaging of 830-nm fluorescence of (A) CW $I_{DC}$ (B) amplitude $I_{AC}$ (C) phase delay, and (D) modulation ratio of the detected fluorescence generated from a lymph node in the area of the right fifth mammary gland of a canine. Illumination was accomplished with an expanded 780-nm laser diode. Modulation frequency was 100 MHz. (From Reynolds, J.S. et al., *Photochem. Photobiol.*, 70, 87, 1999. With permission.)

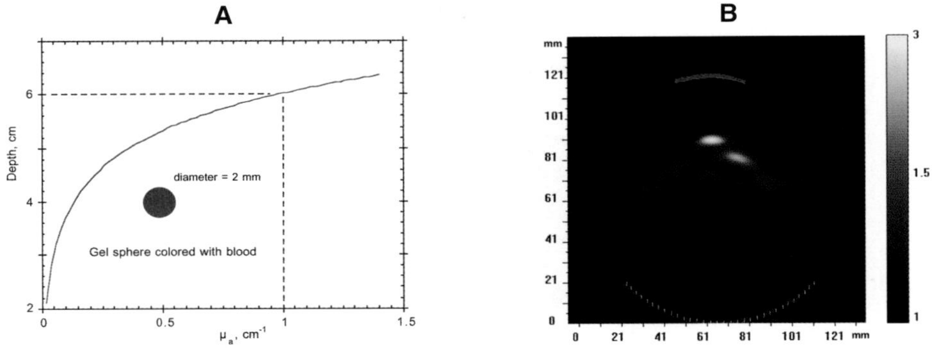


**COLOR FIGURE 34.10** (A) Minimally detectable tumor radius as a function of tumor depth in the bulk of normal tissue. Optical absorption coefficient of 1 $cm^{-1}$ corresponds to relative blood content of 12% in the tumor. (B) Optoacoustic image of two small spheres colored with hemoglobin in gelatin phantom resembling optical properties of breast tissue. An arc above the spheres indicates positions of optical fiber for delivery of laser pulses.

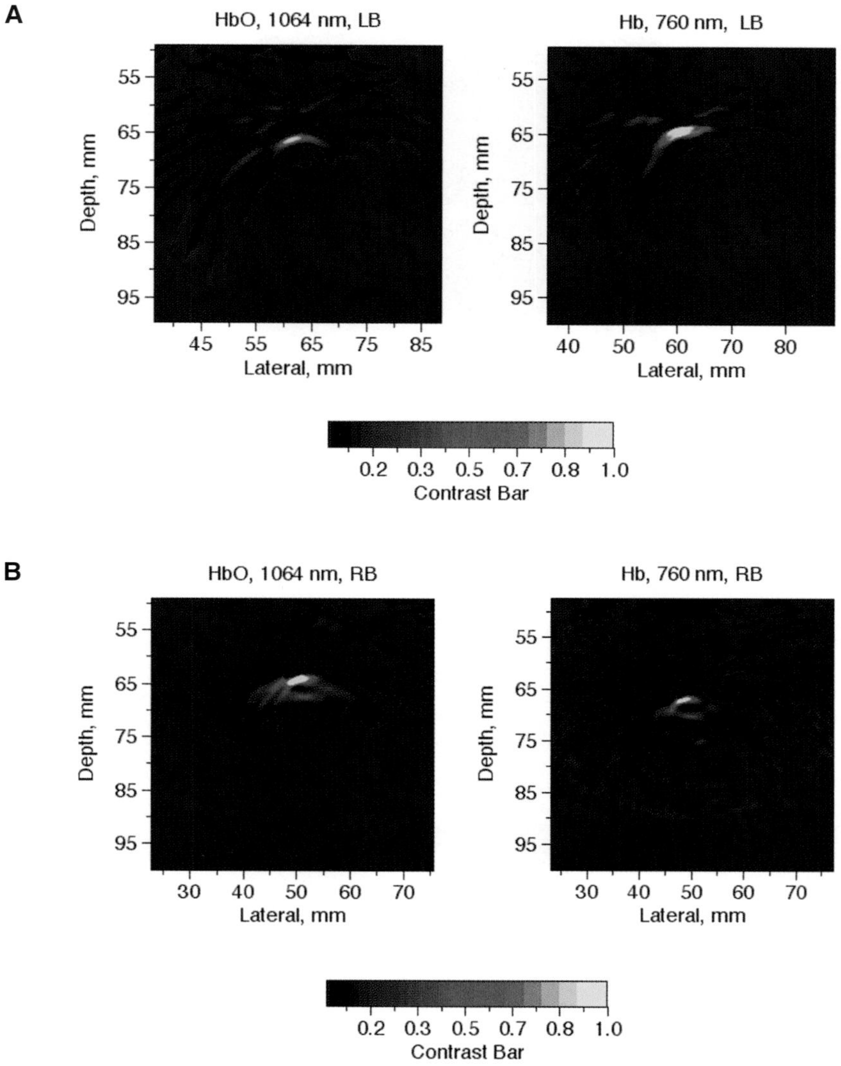


**COLOR FIGURE 34.13** (A) Two-dimensional optoacoustic *in vivo* images of a ductal-lobular carcinoma in the left breast (LB). Laser pulses were delivered to a single point on the skin surface above the tumor. Location of the piezoelectric detector array was at the bottom of the imaging field presented. Contrast between a tumor and surrounding normal tissue strongly depends on the laser wavelength. At the wavelength of 1064 nm the image brightness is significantly lower than the brightness at the wavelength of 757 nm, indicating tumor with reduced level of blood oxygenation. (B) Two-dimensional optoacoustic *in vivo* images of a benign fibroadenoma in the right breast (RB). Laser pulses were delivered to a single point on the skin surface above the tumor. Location of the piezoelectric detector array was at the bottom of the imaging field presented. Contrast between a tumor and surrounding normal tissue strongly depends on the laser wavelength. At the wavelength of 1064 nm the image brightness is significantly higher than the brightness at the wavelength of 757 nm, indicating tumor with normal level of blood oxygenation.

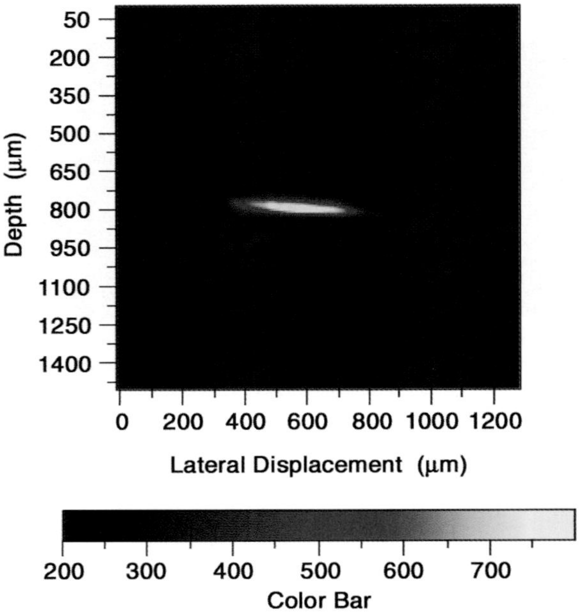


**COLOR FIGURE 34.16** Optoacoustic image of a human hair.

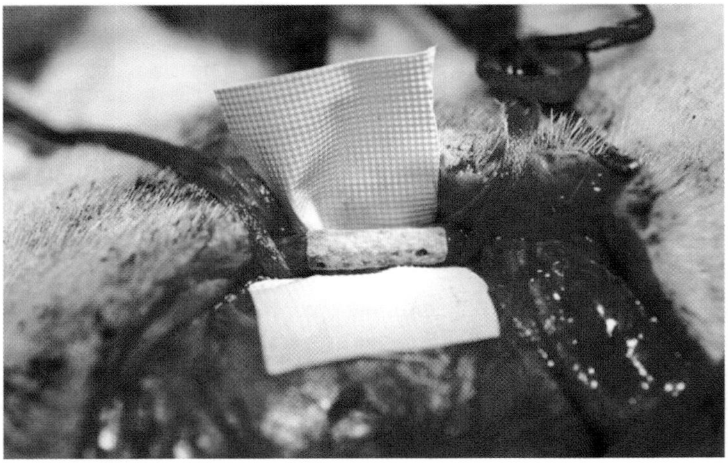

**COLOR FIGURE 39.9** Intraoperative photograph showing the right carotid artery of a canine repaired using a solder-doped polymer membrane in conjunction with a diode laser ($\lambda = 805$ nm, spot diameter = 2 mm, $E_0 = 15.9$ W/cm$^2$, $t = 100$ sec). The polymer membrane was fabricated using PLGA with an 85:15 copolymer ratio and 70% NaCl with particle sizes of 106 to 150 µm and doped with protein solder composed of 50% BSA, 0.5 mg/ml ICG, and deionized water. (From McNally-Heintzelman, K.M. et al., *J. Biomed. Opt.*, 6(1), 68, 2001. With permission.)

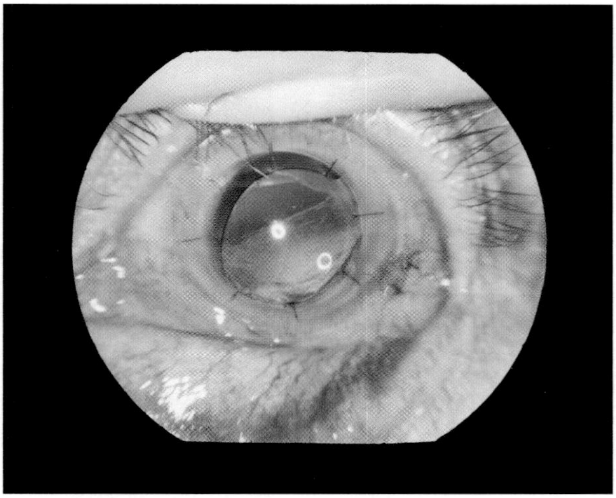

**COLOR FIGURE 39.15** Aspect of a human cornea at 3 months after diode-laser-assisted penetrating keratoplasty (transplantation of the cornea), where laser welding ($\lambda = 805$ nm, P ~ 70 mW, $t$ ~ 200 sec, chromophore: saturated ICG solution in saline) was used despite the conventional continuous suture (eight interrupted stitches had been applied during the surgery, as in the conventional PK procedure, to hold the corneal button). (Image courtesy of Roberto Pini, Istituto di Elettronica Quantistica–C.N.R., Florence, Italy.)

A

B

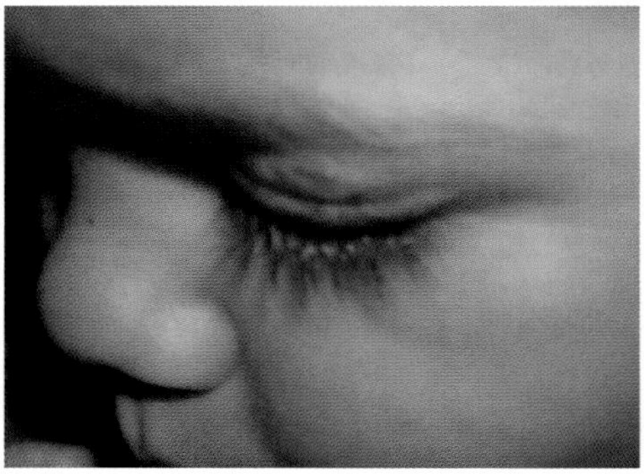

**COLOR FIGURE 40.2** (A) A hemangioma of the left upper eyelid prior to treatment. (B) After a series of eight treatments with the 595-nm pulsed-dye laser at 10 to 15 J/cm$^2$ and 7-mm spot size, the hemangioma cleared significantly.

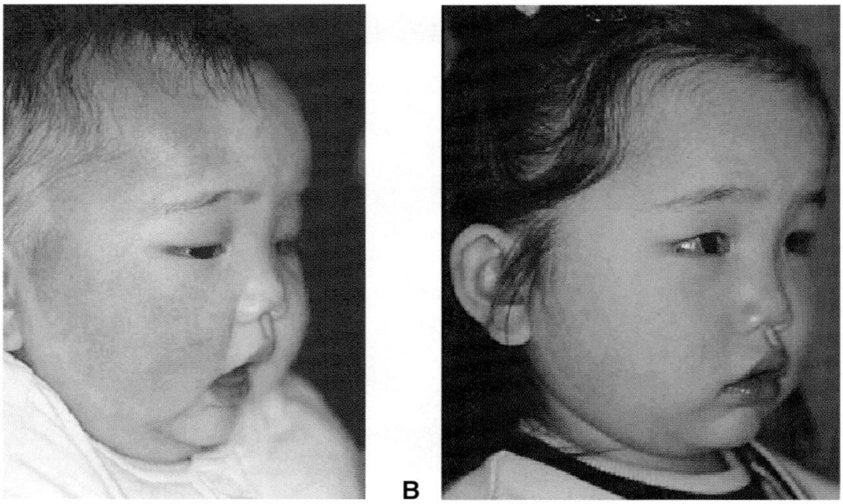

A B

**COLOR FIGURE 40.6** (A) A 4-month-old girl with nevus of Ota on her right face prior to treatment. (B) Almost complete clearing after five treatments with Q-switched alexandrite laser. (Images courtesy of Harue Suzuki, M.D.)

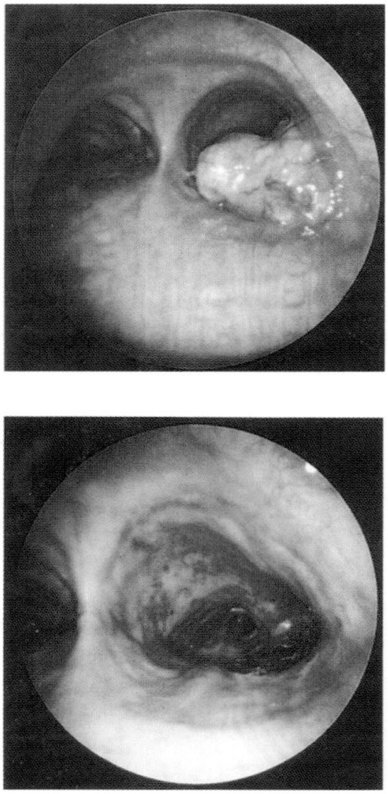

A

B

**COLOR FIGURE 41.1** Photographs illustrating the use of the Nd:YAG laser and mechanical resection to remove an endobronchial lesion from the lung. (A) Endobronchial lesion due to non-small-cell lung cancer seen in the right mainstem bronchius arising from the right upper lobe. (B) Endobronchial lesion has been removed using the Nd:YAG laser and mechanical resection.

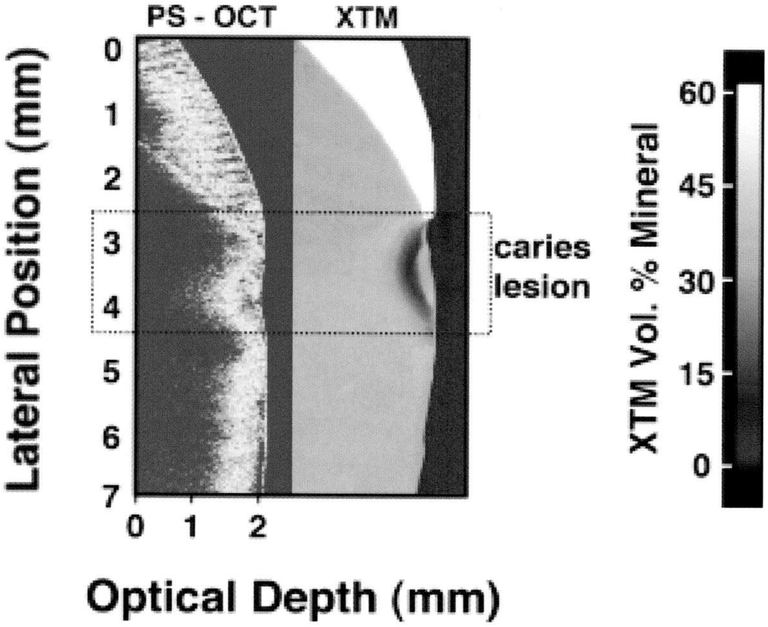


**FIGURE 50.6** Comparison of the PS-OCT fast-axis image (orthogonal polarization state) with the corresponding x-ray tomogram (XTM) of the mineral density taken from the same region of the tooth (right). A small root caries lesion is present just below the cementum–enamel junction shown in the box. The intensity of the OCT images ranges from 12 dB to –45 dB; areas with regions of intensity greater than –5db are shown in red, and those areas of less than –35 dB are shown in blue. In the XTM image on the right, normal dentin is yellow, enamel is white, the water outside the tooth is indicated in red, and the demineralized area of the lesion is blue (color bar on right). (From Fried, D. et al., *J. Biomed. Opt.*, 7(4), 618, 2002. With permission.)

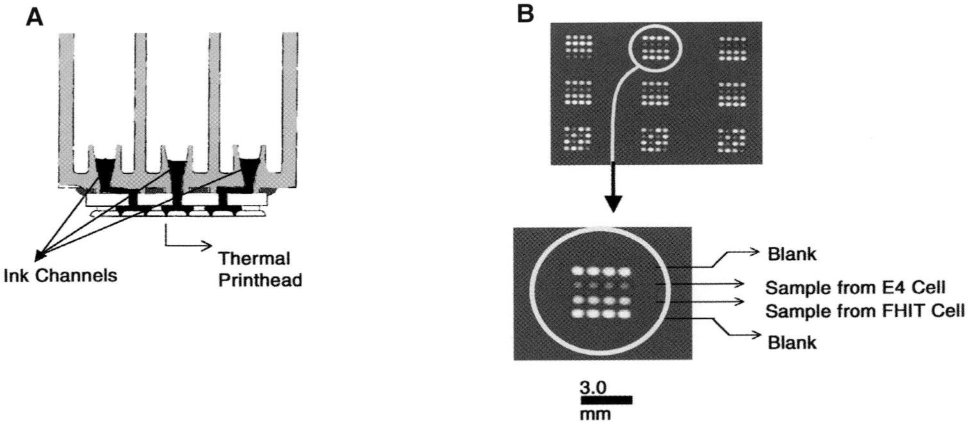


**COLOR FIGURE 51.5** (A) Schematic diagram of a generic bubble-jet cartridge illustrating the connection of ink channels to the printhead. (B) Membrane printed with biological materials using bubble-jet technology. For purposes of better visualization of the spotting, different fluorescent dyes were added for the preparation of this sample. (Adapted from Allain, L.R. et al., *Fresenius J. Anal. Chem.*, 371(2), 146, 2001.)

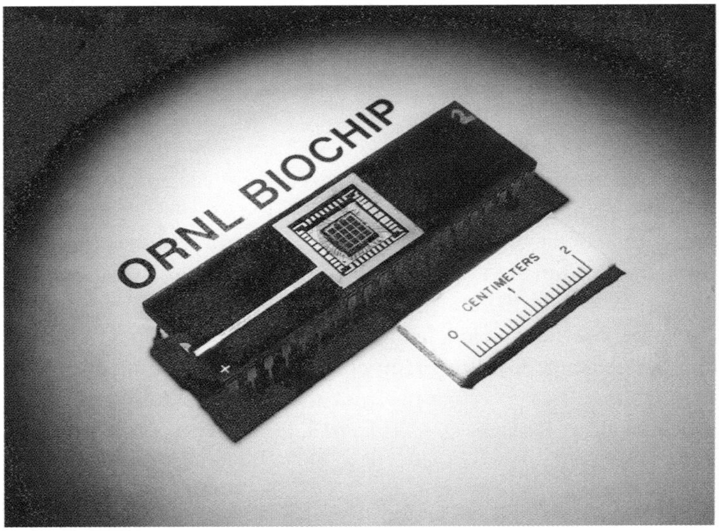


**COLOR FIGURE 51.6** Integrated circuit microchip for a biochip with $4 \times 4$ sensor array. (From Vo-Dinh, T., *Sensors Actuators*, B51, 52, 1998. With permission.)

A

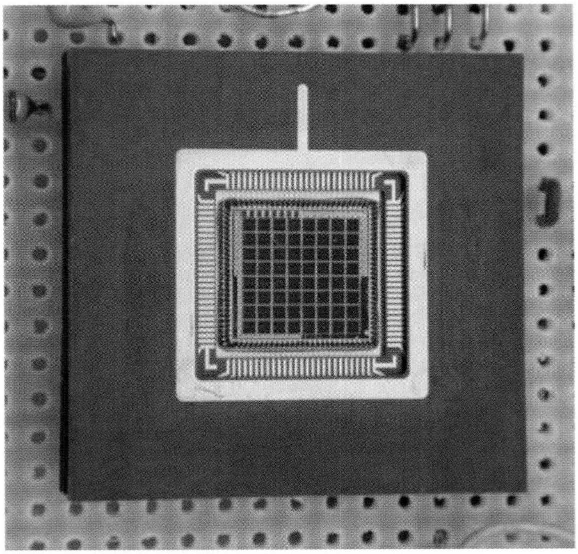

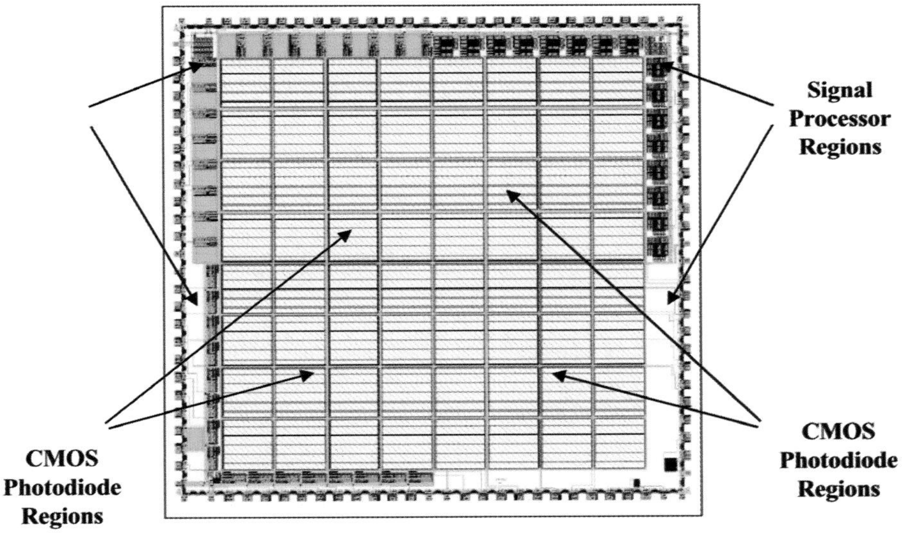


B

**COLOR FIGURE 51.7** (A) Photograph of the 8 × 8 integrated circuit microchip. (B) Schematic of the electronic design of the 8 × 8 microchip with CMOS photodiode regions and signal processor regions.

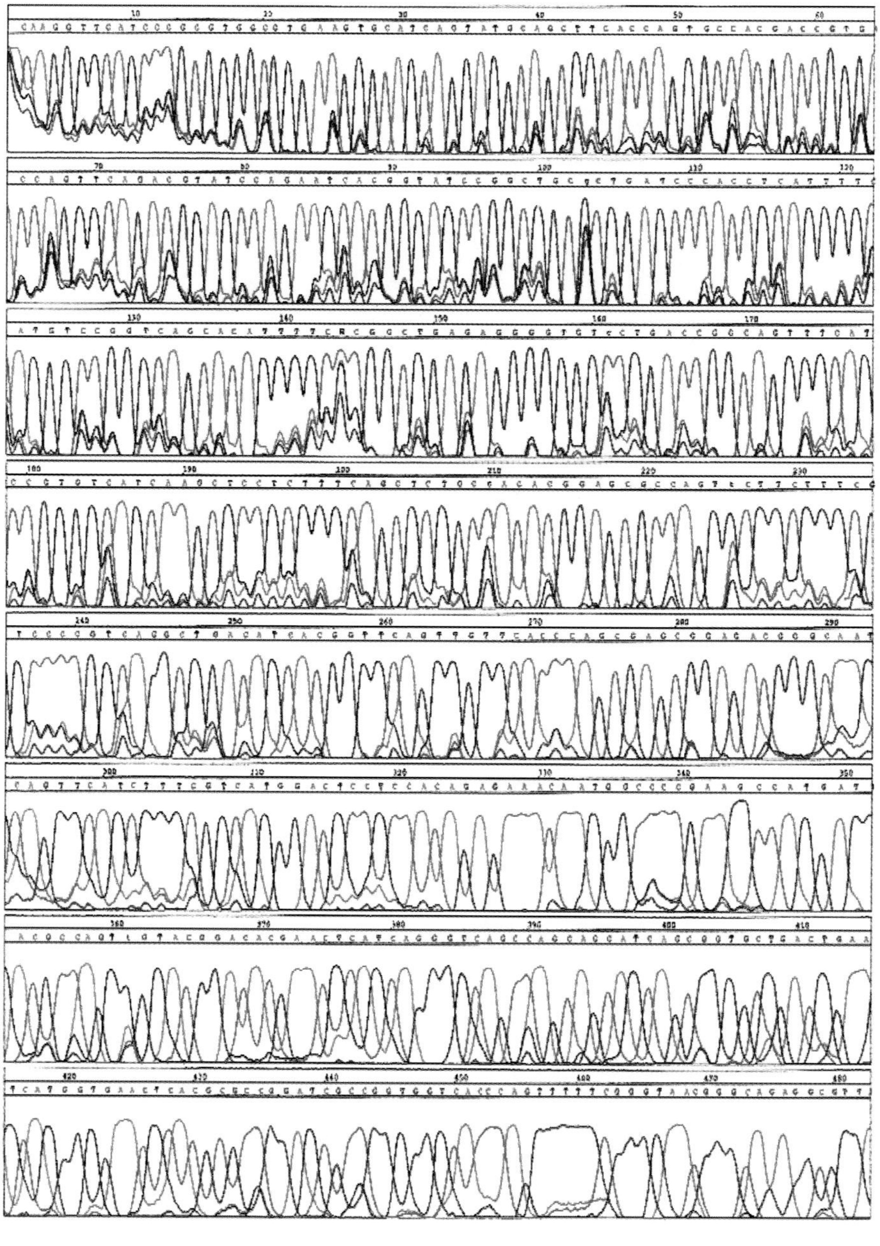

**FIGURE 53.18** Single-color/four-lane sequencing trace (called bases 1 to 480) of a PCR-amplified λ-bacteriophage template. The sequencing was performed using an IRD-800-labeled primer (21mer) and SGE instrument (Li-COR 4000). The gel consisted of 8% polyacrylamide with 7 *M* urea. The dimensions of the gel were 25 cm (width) by 41 cm (length). The traces from the four lanes were overlaid to reconstruct the sequence of the template.

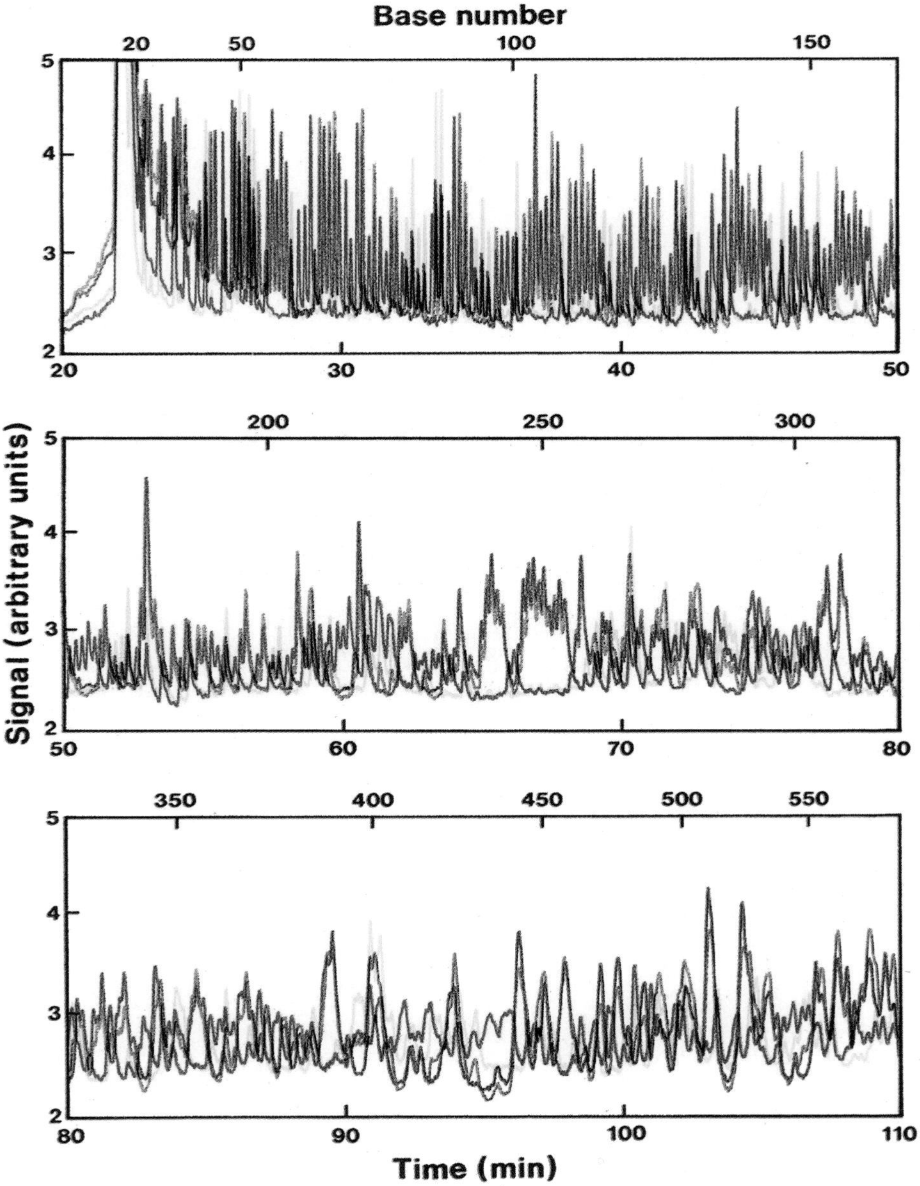


**COLOR FIGURE 53.21** Four-color/single-lane sequencing of an M13mp18 template with a histidine tRNA insert. The numbers along the top of the electropherogram represent cumulative bases from the primer-annealing site. The electrophoresis was performed in a capillary column with a length of 41 cm and an i.d. of 50 μm. The sieving gel consisted of a cross-linked polyacrylamide (6% T/5% C) and was run at a field strength of 150 V/cm. The dyes used for the labeling of the sequencing primers were FAM, JOE, TAMRA, and ROX. Each color represents fluorescence from a different wavelength region (blue = 540 nm, green = 560 nm, yellow = 580 nm, red = 610 nm). (Adapted from Swerdlow, H. et al., *Anal. Chem.*, 63, 2835, 1991).

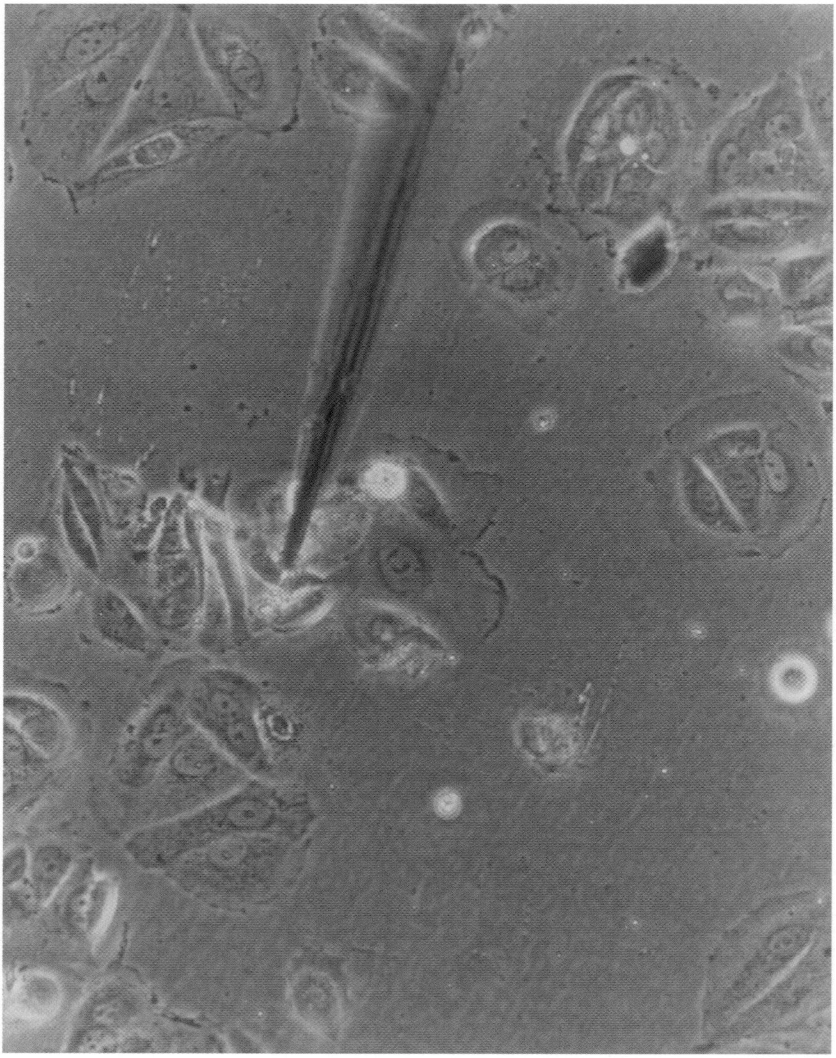

**COLOR FIGURE 54.6** Photograph of an antibody-based nanoprobe used to measure the presence of benzopyrene tetrol in a single cell. The small size of the fiber-optics probe allows manipulation of the nanoprobe at specific locations within the cell. (From Vo-Dinh, T. et al., *Nat. Biotechnol.*, 18, 76, 2000. With permission.)

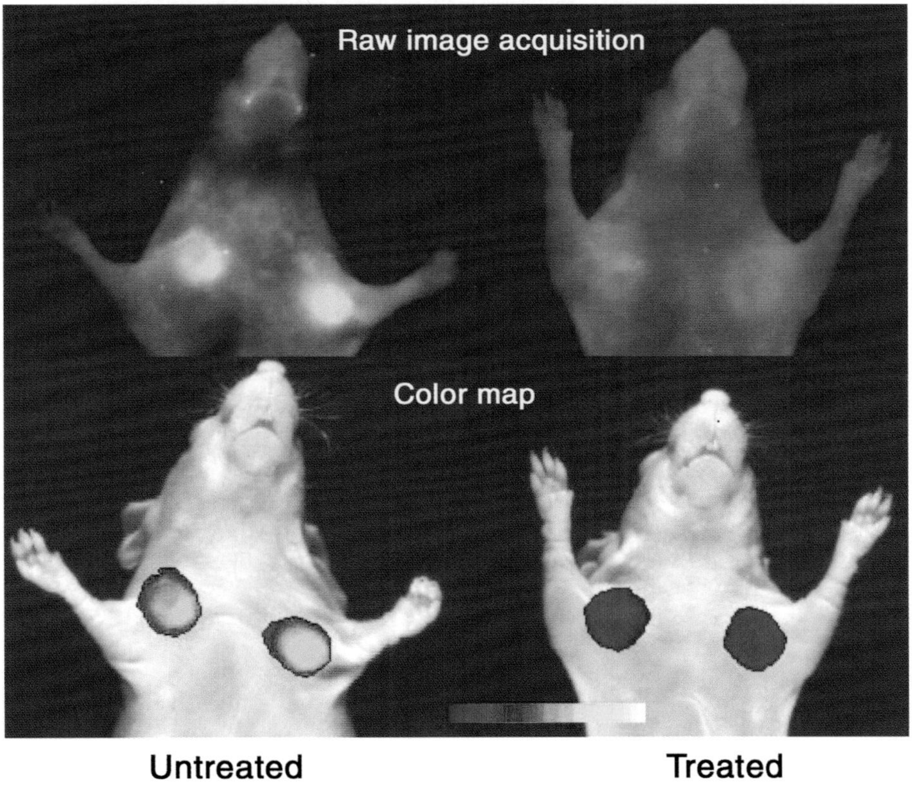


**COLOR FIGURE 56.8** Fluorescence imaging of HT1080 tumor-bearing mice 48 h after injection of Cy5.5-labeled poly(ethylene glycol) poly-(L-lysine) graft polymer (see Figure 56.6) into animals treated with the MMP-inhibitor primomastat vs. untreated animals. The top row shows raw fluorescence images, and the bottom row shows color-coded intensity profiles superimposed onto white-light images. The results show a significantly less fluorescence signal in treated animals relative to the untreated group. (From Bremer, C. et al., *Nat. Med.*, 7(6), 743, 2001. With permission.)

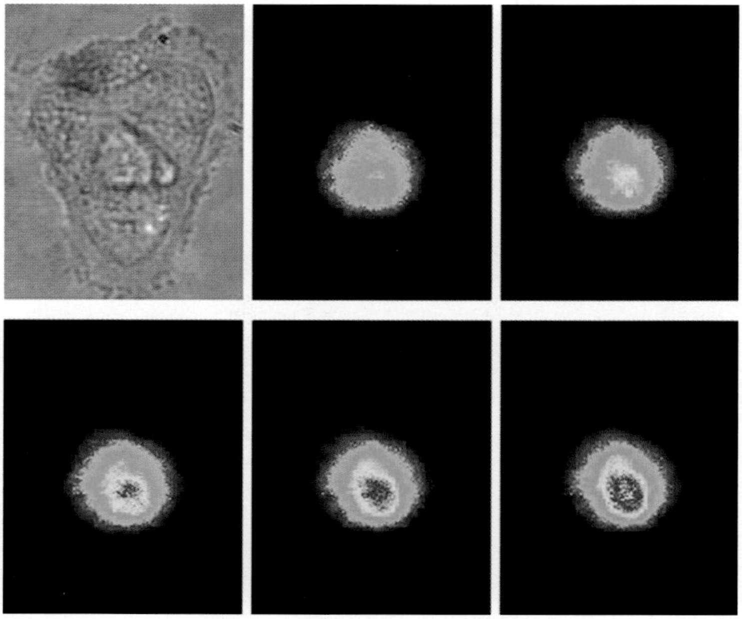

**COLOR FIGURE 57.6** Hybridization of the MB to mRNA in living cells. Clockwise, from top left: An optical image of the kangaroo rat kidney cell and the fluorescence images of the cell at 3, 6, 9, 12, and 15 min after injection of MB to the PtK2 cell.

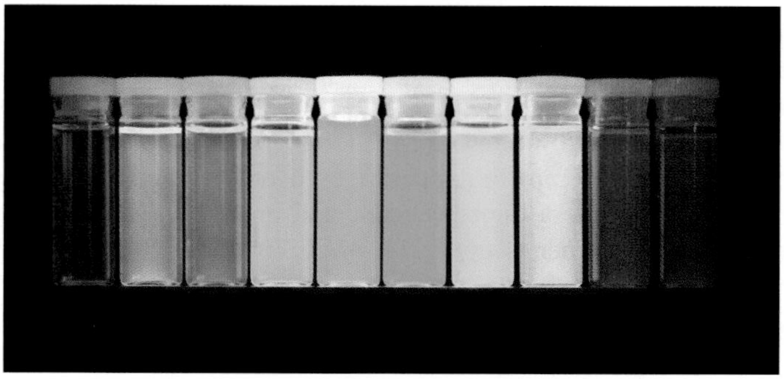

**COLOR FIGURE 58.3** Ten distinguishable emission colors of ZnS-capped CdSe quantum dots excited with a near-UV lamp. From left to right (blue to red), the emission maxima are located at 443, 473, 481, 500, 518, 543, 565, 587, 610, and 655 nm. (Adapted from Han, M.Y. et al., *Nat. Biotechnol.*, 19, 631, 2001. With permission.)

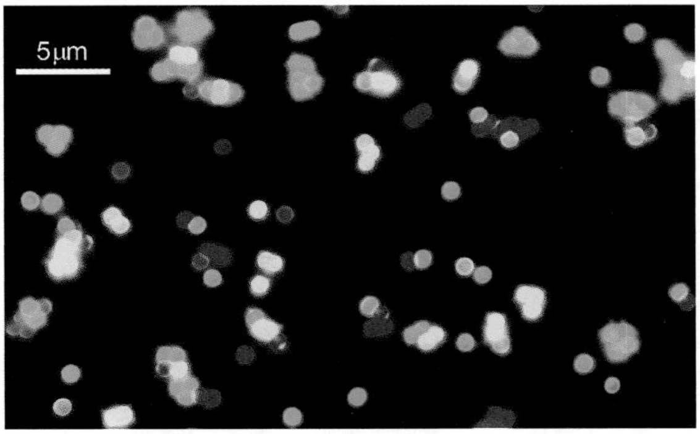


**COLOR FIGURE 58.8** Fluorescence micrograph of a mixture of CdSe/ZnS QD-tagged beads emitting single-color signals at 484, 508, 547, 575, and 611 nm. The beads were spread and immobilized on a polylysine-coated glass slide, which caused a slight clustering effect. (From Han, M.Y. et al., *Nat. Biotechnol.* 19, 631, 2001. With permission.)

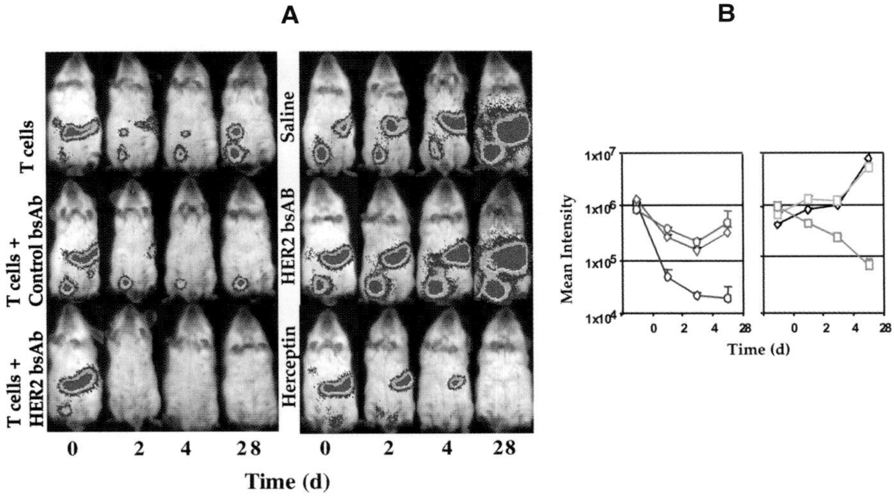


**COLOR FIGURE 62.1** Advancing complex therapeutic strategies for malignancy through imaging. Evaluation of cell-based therapies complexed with molecular therapies may be complicated by having "too many moving parts," and imaging approaches have been used to optimize these therapies by rapidly providing efficacy data without sacrificing the study subjects. In a study by Scheffold et al., the ability to redirect a tumoricidal NK T cell population to a tumor target using a bispecific antibody was revealed using BLI. The control groups in this study consisted of the NK T cells alone, the NK T cells with an irrelevant bispecific antibody, normal saline, the bispecific antibody without additional cells, and herceptin (as a positive control). Temporal analyses for representative animals in each treatment group are shown (A) and data from all animals are plotted (B). Color-coded NK T cells only = light blue, NK T cells and control bispecific antibody = red, NK T cells and bispecific antibody = dark blue, saline = aqua, control bispecific antibody = black, and herceptin = dark green. BLI offers the ability to rapidly study multiple animals in each of six treatment groups and provide accurate whole-body data that are quantitative.

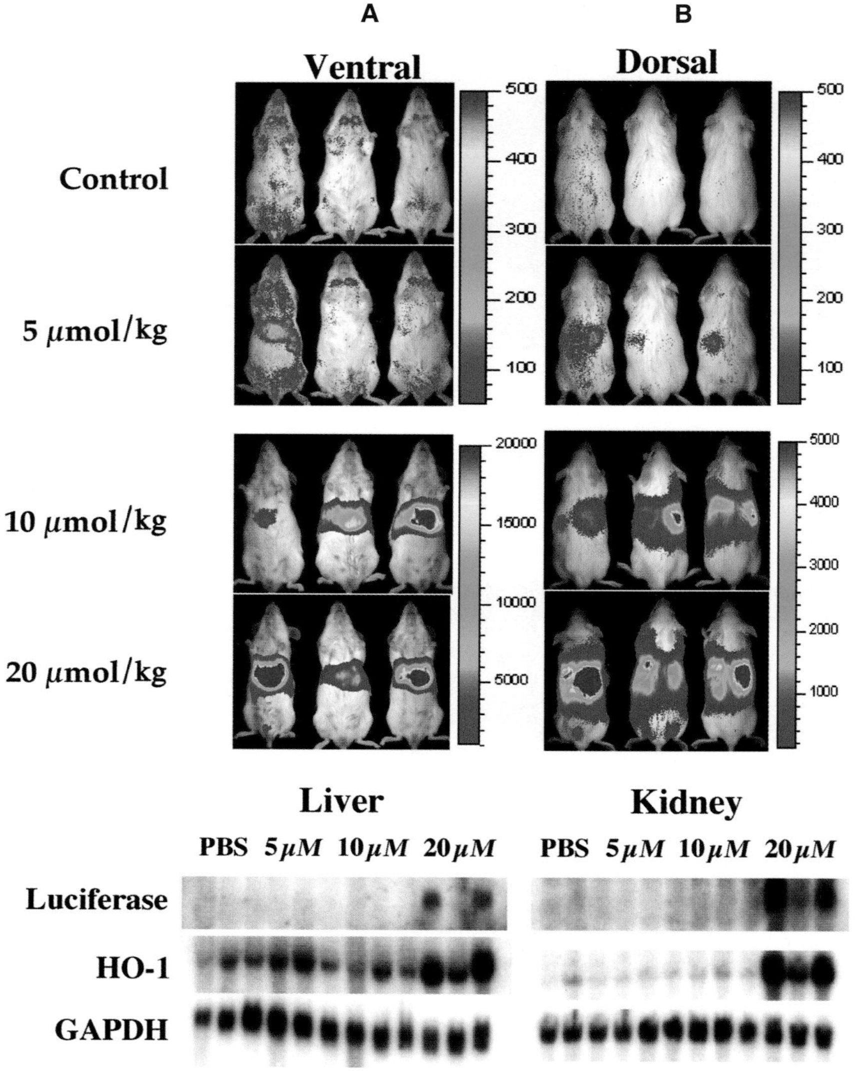


**COLOR FIGURE 62.2** Screening heavy-metal toxicity in a transgenic mouse model. Using a Tg model where the transgene consisted of the heme oxygenase promoter fused to the firefly luciferase coding sequence; dose-dependent increases in luciferase transcription in the liver and kidney following systemic treatment with PBS or three doses of $CdCl_2$ (5, 10, and 20 μmol/kg) were revealed by BLI (A). After imaging the animals were sacrificed aand the tissues removed, and total RNA was isolated from the liver and kidneys and analyzed by Northern blot hybridization (B). Levels of mRNA for HO-1, luciferase, and GAPDH were determined for both tissues from three mice at each concentration. Luc signals increased with increasing concentrations of $CdCl_2$ as measured by imaging, and levels of Luc and HO-1 mRNA were elevated in the treatment group that received the highest concentrations of $CdCl_2$.

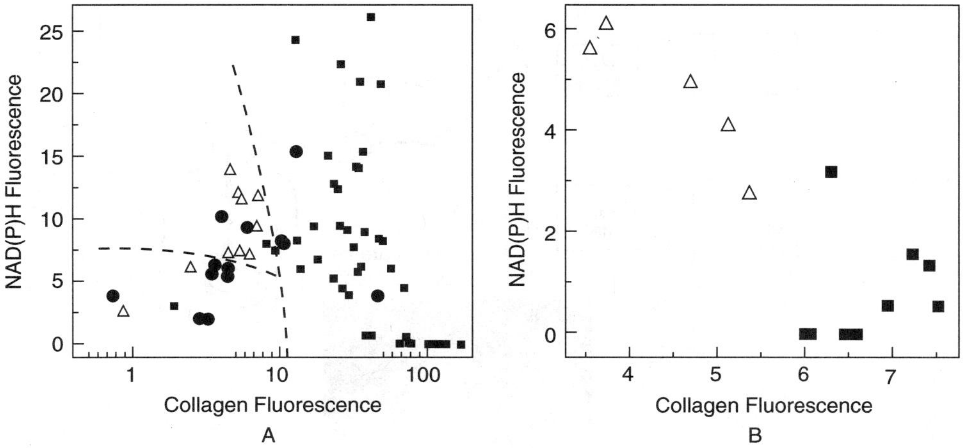


**FIGURE 28.6** Study of *in vivo* quantitative fluorescent biomarkers of epithelial precancerous changes. This figure demonstrates the ability to distinguish between various types of tissues, *in vivo*, based upon a multicomponent fluorescence analysis. (A) Demonstrates the ability to differentiate between normal ectocervical tissue (■), squamous metaplasia sites (●), and high-grade squamous intraepithelial lesions (△), based upon this two-dimensional analysis procedure. The dashed lines represent the boundary between the different tissues, based upon logistic regression analysis. (B) Demonstrates the ability to differentiate between nondysplatic tissues (■) and high-grade dysplastic tissues sites (△). (Adapted from Georgakoudi, I. et al., *Cancer Res.*, 62, 682, 2002.)

Richards-Kortum and co-workers have used laser-induced fluorescence, employing 337-nm excitation to differentiate *in vivo* cervical intraepithelial neoplasia (CIN), non-neoplastic abnormal, and normal cervical tissues from one another.[138] In this work, a colposcope was used to identify normal and abnormal sites on the cervix. These sites were then interrogated via fluorescence spectroscopy with an optical fiber probe. Based upon results from the fluorescence analyses, two algorithms were developed for the diagnosis of CIN. The first of these algorithms allowed for differentiation of histologically abnormal tissues from colposcopically normal tissues with a sensitivity, specificity, and positive predictive value of 92, 90, and 88%, respectively. The second algorithm then allowed for the differentiation of preneoplastic and neoplastic tissues from non-neoplastic abnormal tissues with a sensitivity, specificity, and positive predictive value of 87, 73, and 74%, respectively. These results found that as the tissue progresses from a normal state to an abnormal state in the same patient, it is accompanied by a decrease in the absolute fluorescence contribution of collagen, an increase in the absolute attenuation by oxyhemoglobin, and an increase in the relative contribution from reduced nicotinamide dinucleotide phosphate [NAD(P)H]. Differentiation of the various tissues is then determined by the extent of each of these factors. Such results provide a great deal of hope to the future of *in vivo* fluorescence spectroscopy for the diagnosis of CIN at colposcopy.[138,139]

*Esophageal Cancer* — Esophageal cancers, like most other cancers, generally progress through a series of stages. Barrett's esophagus, named after the physician who first identified the condition,[140] is an abnormality resulting from long-term acid reflux, and is associated with an increased occurrence of mucosal dysplasia and adenocarcinoma in the specialized glandular mucosa.[141–143] In addition, individuals with Barrett's esophagus have a tendency to develop cancer of the esophagus, with a 30- to 52-fold increase over individuals without this condition.[144,145] In esophageal cancer, the mucosa progresses from "normal" through low-grade dysplasia (LGD) to high-grade dysplasia (HGD), and, finally, cancer. If detected early, endoscopic treatment of dysplasia with photodynamic therapy or thermal ablation can be effective, thereby eliminating the need for surgical esophagectomy.

Vo-Dinh and co-workers have developed a laser-induced fluorescence diagnostic procedure for *in vivo* detection of GI cancer that uses 410-nm laser light from a nitrogen-pumped dye laser passed through a fiber-optic probe to excite the tissue.[104,105,119,120,146] After the tissue is excited, the resulting fluorescence

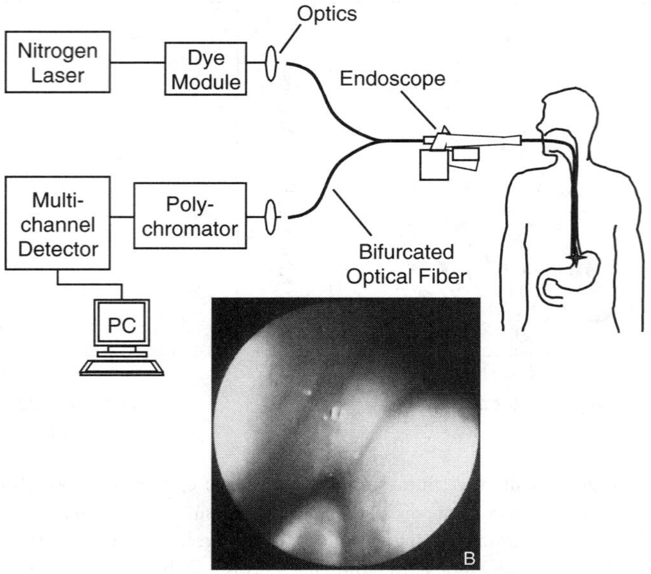


**FIGURE 28.7** Laser-induced fluorescence instrument for *in vivo* gastrointestinal cancer diagnostics. This figure shows a typical fluorescence-based optical biopsy system in which a fiber optic probe is transmitted down a biopsy channel of an endoscope for point-by-point types of analyses. Insert: photograph of the optical fiber inside the GI tract. (Modified from Vo-Dinh, T. et al., *Appl. Spectrosc.,* 51, 58, 1997.)

emission is collected by the same fiber-optic probe and is recorded on an OMA in 600 ms. Based upon the resulting fluorescence spectra, a diagnostic technique known as differential normalized fluorescence was employed to enhance the slight spectral differences between normal and malignant tissues.[64,65] This technique greatly improves the accuracy of diagnosis, compared to direct intensity measurements, because each spectrum is normalized with respect to its total integrated intensity and therefore becomes independent of the intensity factor. This in turn enhances small spectral features in weak fluorescence signals, making classification much easier. The sensitivity of the DNF method in classifying normal tissue and malignant tumors is 98%.[65,119,120,146]

The LIF methodology was also employed in clinical studies of over 100 patients, in which Barrett's mucosa without dysplasia was diagnosed with a specificity of 96% (208 of 216 data points) and high-grade dysplasia was diagnosed with a sensitivity of 90% (9 of 10 data points).[104,105] Figure 28.7 shows the instrument setup for GI cancer diagnostics.[64] A photograph of the optical fiber inside the GI tract is shown in the insert of this figure. Figure 28.8 shows clinical data demonstrating the capability of LIF to differentiate normal from malignant tissues.[64]

*Bladder Cancer* — Bladder cancer has also shown a significant amount of promise for clinical diagnosis by autofluorescence analyses. In one such study, a quartz fiber-optic probe, placed in the working channel of a cytoscope, was used to deliver 337-nm laser light from a nitrogen laser to the tissue area of interest. Following excitation, the resulting fluorescent light was collected by a second fiber and then spectrally dispersed before being detected by an OMA. Resulting fluorescence emission spectra were then analyzed by taking a ratio of the intensity of the fluorescent light at 385 nm to the fluorescence intensity at 455 nm. This ratiometric analysis was used to analyze 114 lesions, and it was possible to differentiate clearly between malignant and nonmalignant bladder tissues with a sensitivity of 98%.[147]

In another study, the ability of laser-induced autofluorescence spectroscopy to distinguish between neoplastic urothelial bladder lesions and normal or nonspecific inflammatory mucosa was investigated. In this study, three different pulsed-laser wavelengths were used successively for excitation: 308 nm (XeCl laser), 337 nm ($N_2$ laser), and 480 nm (coumarin dye laser). The excitation beam was delivered through

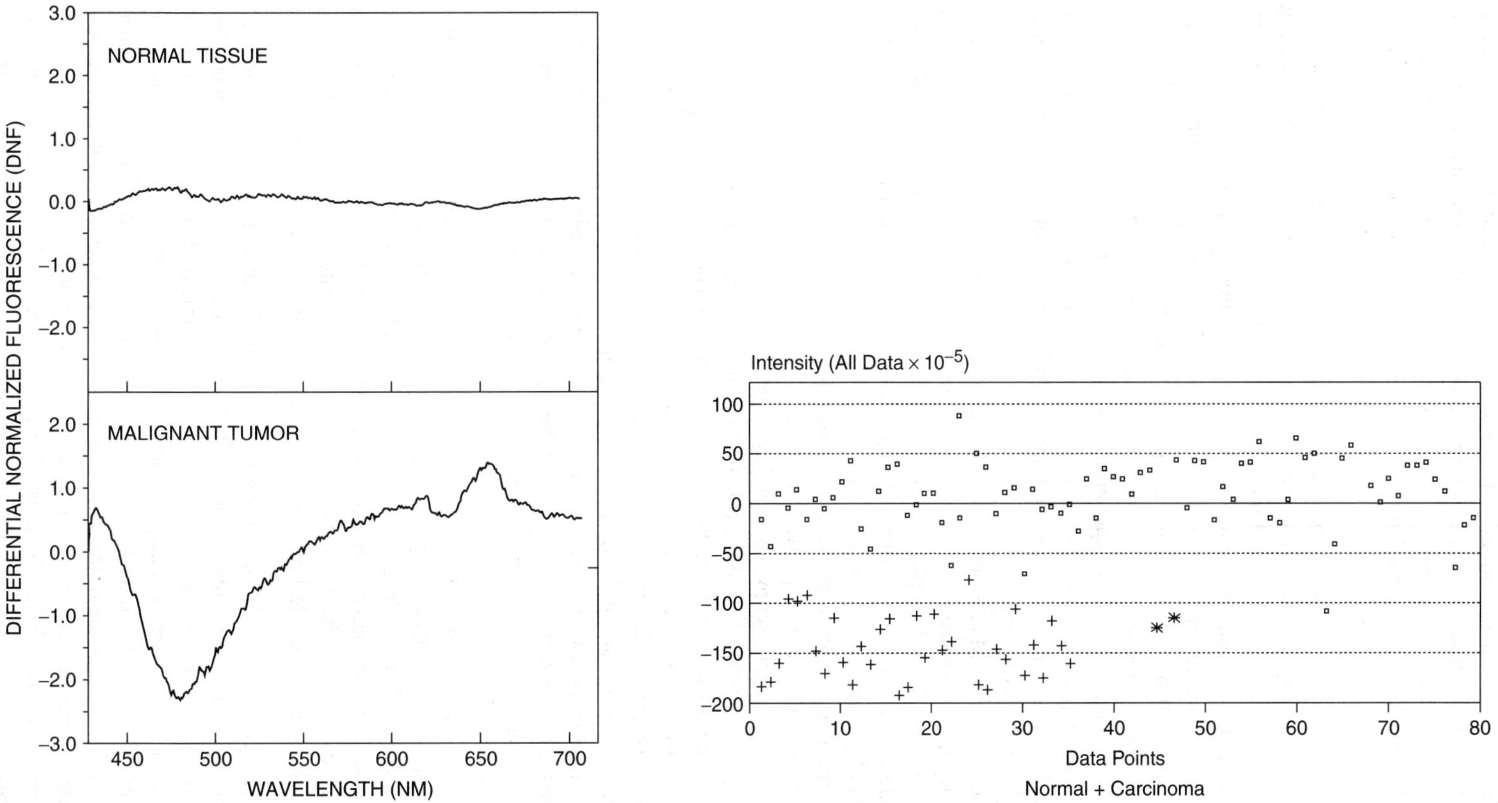


**FIGURE 28.8** Clinical data demonstrating the capability of LIF to differentiate normal from malignant tissues. Left: Differential normalized fluorescence (DNF) emission spectra of normal tissue (top) and malignant tissue (bottom). These spectra were obtained by normalizing the fluorescence intensity at every point in the spectrum to the overall area under the curve, thus emphasizing small differences between the two types of tissues. By subtracting the DNF spectrum of a particular sample from an average DNF normal tissue spectrum, a single index value at each wavelength can be determined. (From Vo-Dinh, T. et al., *Appl. Spectrosc.*, 51, 58, 1997.) Right: Plot of DNF indices at 480 nm for different tissue samples using *in vivo* LIF measurements (taken prior to biopsy). These tissue samples were then characterized via histology as normal (•) or as carcinoma (+). As expected, the normal tissue results are scattered about the index value of zero while the carcinoma values are significantly lower due to a decrease in fluorescence intensity at this wavelength. The samples denoted by (*) represent points initially classified by histology as normal but later reclassified as malignant. The sensitivity of the DNF method in classifying normal tissue and malignant tumors for GI cancer is 98%. (From Vo-Dinh, T., Panjehpour, M., and Overholt, B.F., in *Advances in Optical Biopsy and Optical Mammography*, New York Academy of Sciences, New York, 1998, p. 116; Vo-Dinh, T., et al., *Lasers Surg. Med.*, 16, 41, 1995.)

optical fibers placed in the working channel of a standard cystoscope, and the resulting fluorescence was detected using an OMA. When 337- and 480-nm excitation wavelengths were used, the overall fluorescence intensity of bladder tumors was clearly decreased compared with normal urothelial mucosa regardless of tumor stage and grade. However, when 308-nm excitation was employed, the shape of the tumor spectra, including CIS, was markedly different from that of normal or nonspecific inflammatory mucosa.[148]

*Colon Cancer* — Autofluorescence diagnostic procedures are used extensively to investigate colon cancers. In one such study, 337-nm light from a nitrogen laser was launched into a fiber-optic probe placed in the working channel of a colonoscope and was used for excitation of *in vivo* and *in vitro* colon tissues. The fluorescence spectra were then measured and analyzed. In all cases, the spectra exhibited peaks at 390 and 460 nm, which are believed to arise from collagen and NADH, as well as a minimum at 425 nm, consistent with absorption attributable to hemoglobin. Despite the presence of these bands in the *in vivo* and *in vitro* samples, the relative intensity of each was quite different, especially for the NADH component, whose intensity was found to decay exponentially with time after resection. Differentiation of normal colonic tissue from hyperplastic or adenomatous tissues of the same type (i.e., *in vivo* vs. *in vitro*) could be accomplished based upon a decrease in the collagen component of the autofluorescence and an increase in hemoglobin reabsorption.

When multivariate linear regression (MVLR) analysis based upon these differences was used, neoplastic tissues could be distinguished from non-neoplastic tissues with a sensitivity, specificity, positive predictive value, and negative predictive value toward neoplastic tissue of 80, 92, 82, and 91%, respectively. However, when this same MVLR technique was used to distinguish neoplastic polyps from non-neoplastic polyps, values of 86, 77, 86, and 77%, respectively, were obtained. This suggested that the LIF measurements sense changes in polyp morphology rather than changes in fluorophores specific to polyps, and that this change in morphology leads indirectly to polyp discrimination.[149]

Laser-induced autofluorescence spectroscopy has been investigated to detect colonic dysplasia *in vivo* using an excitation wavelength of 370 nm.[141] In this work, fluorescence emission data were used to devise an algorithm to classify colonic tissue as normal, hyperplastic, or adenomatous, based on probability distributions of the fluorescence intensity at 460 nm and the ratio of the intensity at 680 nm to that at 600 nm. Fluorescence spectra were then collected from normal mucosa and colonic polyps during routine colonoscopy exams; the predictive abilities of the diagnostic algorithm were tested in a blinded fashion, with histology results the standard against which they were measured. Results revealed that the algorithm correctly determined the tissue type in 88% of cases, equal to the agreement of independent pathologists. In addition, the sensitivity, specificity, and positive predictive value for the detection of dysplasia were 90, 95, and 90%, respectively.[150]

In additional work by Feld and co-workers, LIF studies were performed on 20 different patients, providing analyses on 31 colonic adenomas, 4 hyperplastic polyps, and 32 samples of normal mucosa. From these studies, it was found that classification of the tissue as adenomatous or normal colonic mucosa/hyperplastic polyp was correct, based upon an automated probabilistic algorithm, 97% of the time. The sensitivity, specificity, and positive predictive value of the technique were 100, 97, and 94%, respectively — once again showing a great deal of promise for this technique in the future.[151]

*Lung Cancer* — Lung cancer accounts for 25% of all cancer deaths and is currently the most common cause of cancer death among both men and women in the United States. Currently, patients are diagnosed only when the cancer is already in a severe state because no rapid, practical, and effective screening test is available for lung cancer. In lung cancer, premalignant epithelium usually progresses to a malignant tumor through the development of dysplasia, starting from LGD and progressing to HGD, which is then typically followed by carcinoma.[152,153] Although this well-known stepwise progression could provide the basis to survey premalignant epithelium by random biopsies, histology analyses from expert to expert typically have poor agreement, particularly for the most severe and highest-risk samples such as HGD or CIS.[154] Based on such results, it appears that optical biopsies could eventually be well suited for early diagnosis of lung cancer and could greatly increase the survival rate of this disease.

In one study determining the potential of fluorescence spectroscopy for the diagnosis of dysplastic tissues and CIS, fluorescence bronchoscopy was performed on 82 volunteers, 25 nonsmokers, 40 ex-smokers, and 17 current smokers with mean ages of 52, 55, and 49 years, respectively. Excitation of the tissue was performed using the 325-nm emission of an HeCd laser and the ratiometric analyses of the resulting fluorescence emission wavelengths were determined. Upon analysis of the results, it was found that the sensitivity of the autofluorescence bronchoscopy was 86%, which is 50% better than conventional white-light bronchoscopy for the detection of dysplasia and CIS.[93,114,]

*Upper Aerodigestive Tract Cancers* — In order to determine the potential of laser-induced autofluorescence for the diagnosis of dysplastic tissues in the upper aerodigestive tract, many different research groups have developed many types of analyses. In one study, various fluorescence excitation and emission scans were employed for differentiation of normal mucosa from neoplastic tissues. One spectral difference was the disappearance of an excitation peak at 330 nm in tumor tissue that existed in normal tissue, when an emission wavelength of 380 nm was used for fluorescence monitoring. Therefore, by ratioing the excitation scan intensity at 290 nm to that of 330 nm, the resulting value could be directly correlated to increases epithelial thickness resulting from carcinogenesis.

In addition to changes in the excitation spectrum between tumor and normal tissue, it was also found that the fluorescence emission intensity at 390 nm, upon excitation with 340-nm light, was also decreased in tumor tissue relative to normal tissue. A ratio of the fluorescence emission intensity at 390 and 450 nm correlated negatively with the mean epithelial thickness, again providing an indication of the stage of dysplasia.[155] In a similar study, the same autofluorescence ratios were used to differentiate between normal and neoplastic tissues in the oral cavity and pharynx of patients with previously untreated mucosal neoplasias. Using the same ratios, researchers found that significant differences existed between the neoplastic tissue and the contralateral normal sites, leading to the thought that this technique could represent a noninvasive screening method for head and neck squamous cell cancers.[156]

In another study, autofluorescence spectral characteristics of untreated oral and oropharyngeal lesions in patients were studied with excitation wavelengths of 370 and 410 nm generated by a nitrogen-pumped dye laser. Upon examining the resulting fluorescence emission of normal and neoplastic tissues, differences in the spectral profile were noted in two regions of the spectrum for both excitation sources. However, it was found that these differences were more significant when 410-nm light is used for excitation. By ratioing the fluorescence maximum intensity of dysplastic tissue to that of normal mucosa, a quantitative means of differentiation has been developed for the differentiation of the two types of tissue.[157]

In addition to being able to determine the presence of a cancerous or precancerous lesion in a patient, it is also important to be able to determine the point or margin where the tumor ends and the healthy tissue begins. Bohle and co-workers have used autofluorescence spectroscopy of tumor connective tissue to determine accurately where the tumor ends and the healthy skin begins. In this work, 365-nm light was used for the excitation of the tissue, and the resulting fluorescence, either from the elastic fibers of a tumor or the keratinization of precancerous lesions, was used for delineation of the tissue. Contrary to previous results in the literature, no homogeneous fluorescence gradient could be proved between darker marginal epithelium and the brighter tumor connective fibers. The greatest differentiation between tumor and nontumor tissues was exhibited in tissues from the same patient; however, comparisons between different patients also showed promise for the margining of lesions using autofluorescence.[158]

*Laryngeal and Oral Cancer* — Like other cancers involving mucosal membranes, oral and laryngeal carcinomas have also been studied by autofluorescence. Fluorescence spectroscopy has been used to differentiate normal tissue from dysplastic or cancerous tissue with a sensitivity of 90% and a specificity of 88% in a training set and a sensitivity of 100% and a specificity of 98% in a validation set (Figure 28.9).[159] A study of laryngeal cancer employed an HeCd laser operating at 325 nm for excitation of the sample and an ICCD for fluorescence image collection. Images were obtained at various wavelengths with optical filters for wavelength discrimination and were analyzed to provide diagnostic fluorescence images of the area of interest. This technique was used to evaluate 30 patients, of whom 18 had suspect malignancies that were confirmed via histopathological findings.[160]

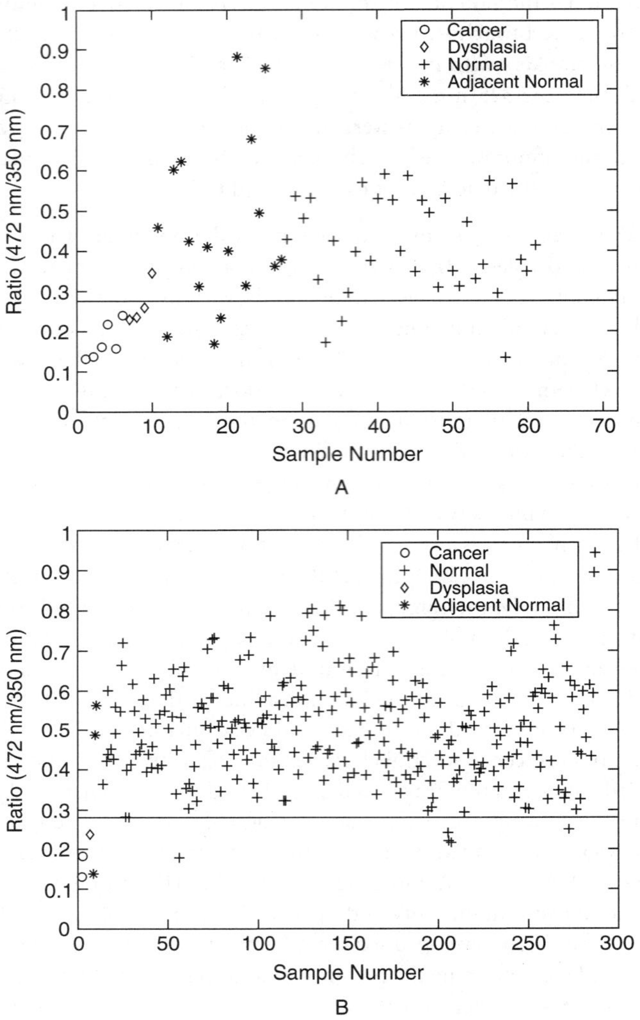


**FIGURE 28.9** *In vivo* detection of oral neoplasia using fluorescence spectroscopy The results demonstrate the ability of automated data analysis algorithms to classify tissues as normal (+), dysplastic (◊), cancerous (○), or normal tissue adjacent to cancerous (∗), based upon ratios of fluorescence emission data at multiple wavelengths following excitation of the oral tissue at 350 nm. Data in (A) represent the training set for automated analyses, while (B) represents the validation set. The solid line at 0.28 represents the threshold for separation of normal and dysplastic or cancerous tissue. Based upon these results, differentiation of normal tissue from dysplastic or cancerous tissue can be achieved with a sensitivity of 90% and a specificity of 88% in the training set, and a sensitivity of 100% and a specificity of 98% in the validation data set. (Adapted from Heintzelman, D.L. et al., *Photochem. Photobiol.*, 72, 103, 2000.)

*Skin Diagnostics* — To better understand the role of various tissue components in the autofluorescence of skin, Gratton and co-workers[161] have employed multiphoton fluorescence microscopy for the analysis of skin tissue *in vivo*. In this work, multiphoton fluorescence microscopy using excitation wavelengths of 730 and 960 nm was used to image *in vivo* human skin cells from the surface to a depth of approximately 200 μm. (Details on multiphoton fluorescence techniques are described in Chapter 11 of this handbook.) Fluorescence emission spectra and fluorescence lifetime images were obtained at selected locations near the surface (0 to 50 μm) and at deeper depths (100 to 150 μm) for both excitation wavelengths.

The resulting spectroscopic data suggest that reduced pyridine nucleotides and NADH are the primary source of skin autofluorescence when using 730-nm excitation. With 960-nm excitation, a two-photon fluorescence emission at 520 nm indicates the presence of a variable, position-dependent intensity component caused by flavoproteins. In addition, a second fluorescence emission component that starts at 425 nm is observed when using 960-nm excitation. Such fluorescence emission at wavelengths less than half the excitation wavelength suggests an excitation process involving three or more photons. This is further confirmed by observation of a super-quadratic dependence of the fluorescence intensity on the excitation power. Further work is still required to identify these emitting species spectroscopically; however, this study demonstrates the use of multiphoton fluorescence microscopy for functional imaging of the metabolic states of *in vivo* human skin cells.[161]

In addition to microscopic analyses, macroscopic autofluorescence-based analyses have also provided useful information about the tissue state of skin. One property about which autofluorescence analyses of skin can provide information is the extent of photoaging that an individual's skin has undergone. Two of the more common fluorescent components in the dermis of skin are the structural proteins elastin and collagen, which are altered by age and photoexposure. Based upon *in vivo* fluorescence analyses of skin from 28 volunteers, it was found that fluorimetry could provide a marker for the extent of photoaging that has occurred in a person. In this study, by monitoring the fluorescence emission from ultraviolet irradiated skin, it was possible to determine the extent of photoaging in a way that was independent of the age, pigmentation, and skin thickness of the individual. Such a marker could lead to development of a technique capable of determining an individual's risk of cancer due to ultraviolet exposure from the sun.[162]

#### *28.4.3.2.2 Clinical Studies and Diagnostics Using Exogenous Dyes*

Exogenous fluorophores are used for many reasons in clinical applications. The most common reason is to provide a contrasting agent that will make medical diagnoses easier, much like radionucleotides are sometimes used as contrast agents in circulation studies. However, due to the limited penetration of optical wavelengths into biological tissues, the most common type of fluorescence analyses performed *in vivo* are cancer diagnoses of optically accessible tissues. For this reason the majority of clinical fluorescence diagnoses using exogenous fluorophores is in the area of cancer visualization. The most common exogenous fluorophores used for these studies are photosensitizers that have been developed for PDT treatments.

These drugs generally exhibit strong fluorescence properties, and preferentially locate in malignant tissues. Figure 28.10 illustrates the use of a fluorescence method for monitoring the production of PDT products produced *in vivo* during treatment.[163] The development and applications of PDT drugs are described in this handbook in Chapter 36 on mechanistic principles of photodynamic therapy, Chapter 37 on synthetic strategies in designing prophyrin-based photosensitizers, and Chapter 38 on photodynamic therapy (PDT) and clinical applications.

*Photosensitizers* — Determining the optimal photosensitizer to use for various types of tumors is a continuous field of investigation. In one such study, fluorescence analyses of the HPD-type photosensitizer Photogem™ were tested in 22 patients with tumors of the lungs, larynx, and skin; gastric and esophageal carcinoma; and cancer of the gynecological organs. Retention of the drug in the various types of tumors and tissues was monitored by fluorescence spectroscopy after excitation of the drug with 510-nm light. The results demonstrated that tumor detection by fluorescence spectroscopy when using Photogem as a contrast agent was possible even in low selectivity of drug accumulation, which appeared to be dependent on the stage and type of the disease and the organ involved.[164]

*Cancer Diagnostics — Liver Cancer* — Monitoring and comparison of the results of ALA-induced PpIX tumor demarcation in chemically induced adenocarcinoma in the liver of rats and in an aggressive basal cell carcinoma in a patient were studied using LIF. In this study, *in vivo* point monitoring and fluorescence microscopy incorporating a CCD camera were used to study the fluorescence distribution of ALA-induced PpIX in tumors. Fluorescence analyses were performed after i.v. injection of 30 mg/kg body weight of ALA. These analyses revealed a slightly larger concentration of PpIX in the tumor than in surrounding healthy liver tissue or abdominal muscles in the rat. In the aggressive basal cell carcinoma of the human, a much greater concentration of PpIX was found in the visible regions of the basal cell carcinoma than

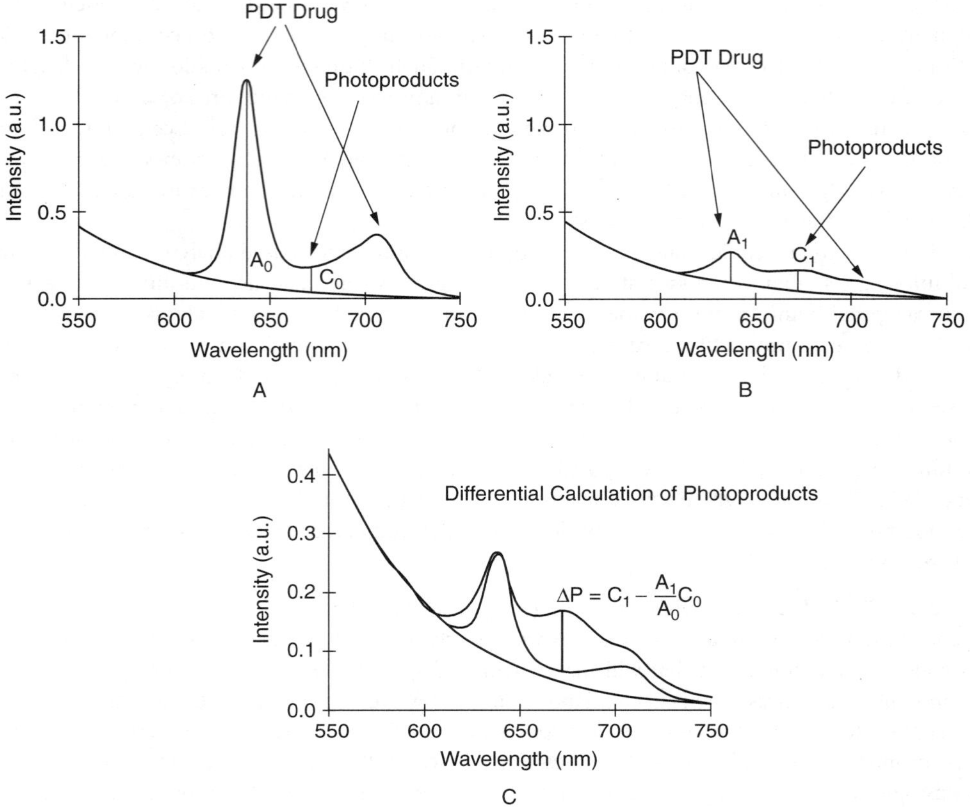


**FIGURE 28.10** Fluorescence method for monitoring the production of PDT photoproducts produced *in vivo* during treatment. Immediately prior to activation of the PDT drug, a fluorescence emission spectrum is measured (A). Then immediately following treatment a second fluorescence spectrum is obtained (B). The fluorescent signal due to the PDT drug is designated $A_0$ for measurements prior to irradiation, and $A_1$ for measurements following irradiation, while $C_0$ represents the fluorescence intensity of the photoproducts prior to irradiation and $C_1$ the fluorescence intensity of the photoproducts after irradiation. In order to determine the change in fluorescence intensity due to the formation of photoproducts during PDT treatment, at the appropriate wavelength, the fluorescence intensity of the irradiated tissue is subtracted from the fluorescence intensity of the nonirradiated tissue, as shown in (C). (Adapted from Klinteberg, C. et al., *J. Photochem. Photobiol. B*, 49, 120, 1999.)

in the necrotic regions and the surrounding normal skin, thereby demonstrating selective uptake and retention of PpIX in the carcinoma.[165]

*Oral Cancer* — Diagnosis of neoplastic lesions in the oral mucosa of 11 patients was investigated using ALA-induced PpIX as a contrast agent. In this work, semiquantitative fluorescence measurements were performed at regular intervals of 3 h, following 15 min of continuous topical application of 0.4% ALA solution to the lesions. Excitation of the PpIX was performed using violet light from a xenon lamp (375 to 440 nm). Fluorescence images in the red region of the visible spectrum were recorded with a CCD while quantitative spectral analyses were performed with an OMA. From these analyses, it was found that PpIX accumulated earlier in the necrotic tissue than in surrounding normal tissue and that a contrast ratio of 10:1 was found for tumor tissue relative to normal tissue 1 to 2 h after application of ALA.[166]

*Esophageal Cancer* — Investigation of Photofrin™-enhanced LIF differentiation of Barrett's metaplastic epithelium and esophageal adenocarcinoma in five patients was performed following low-dose intravenous injections (0.35 mg/kg body weight). In this work, LIF measurements were performed on tissue specimens of normal mucosa, Barrett's epithelium, and tumor tissue treated with Photofrin. Based upon fluorescence measurement ratios of the Photofrin intensity divided by the autofluorescence intensity, quantitative values demonstrating the differentiation ability for the various degrees of dysplasia were determined. The mean ratio ± standard deviation for each type of tissue was found to be 0.10 ± 0.058 for normal esophageal mucosa, 0.16 ± 0.073 for normal gastric mucosa, 0.205 ± 0.17 for Barrett's epithelium with moderate dysplasia, 0.79 ± 0.54 for severe dysplasia, and 0.78 ± 0.56 for adenocarcinoma.[167]

*Bladder Cancer* — Diagnosis of bladder cancer based upon the LIF of exogenous fluorophores has also been performed. In this study, a point-monitoring fluorescence diagnostic system based on a low-energy pulsed laser, fiber transmission optics, and an OMA was used for the diagnosis of 24 patients with bladder malignancies. Malignancies ranged from bladder carcinoma, CIS, and dysplasia. In order to provide a better contrast than can be achieved using autofluorescence, the hematoporphyrin derivative, Photofrin, was injected into the tissue at a concentration of 0.35 or of 0.5 mg/kg body weight. Fluorescence measurements were taken 48 h after injection and a ratio of the red photosensitizer fluorescence to the blue autofluorescence of the tissue was calculated. Based upon this ratio, excellent demarcation between papillary tumors and normal bladder wall tissue was achieved. In addition, these ratios allowed for objective differentiation of certain cases of dysplasia from normal mucosa and benign exophytic lesions such as malakoplakia from malignant tumors.[168]

*Skin Cancer* — Due to the great success in using fluorescence spectroscopy for diagnosis of many other types of malignancies, LIF has also been suggested for the noninvasive diagnosis of malignant melanomas. In one study, LIF was used for real-time monitoring of the PpIX distribution in tumor tissues and in the normal surrounding skin, before and after treatment in all patients. Based upon comparison of fluorescence intensities from normal and tumor tissue of the same pigmentation from the same individual, PpIX distribution demonstrated excellent demarcation between tumor and normal skin of approximately 15:1 for basal cell carcinoma BCC and Bowen's disease, and 5:1 for T-cell lymphomas.[169] However, the reliability of such measurements can be seriously compromised by spatial variations in the optical properties of the tissue that are not related to malignancy (e.g., pigmentation).

One approach to fluorescence-based analyses that minimizes this problem employs a double ratiometric analysis procedure and the use of the photosensitizer ALA. In this technique, two types of fluorescence ratios were calculated. The first of these ratios was the ratio of the skin's fluorescence emission between 660 and 750 nm to the emission between 550 and 600 nm following 405-nm excitation. The second ratio was calculated by dividing the fluorescence emission between 660 and 750 nm by the emission between 550 and 600 nm following excitation with 435 nm. These two excitation wavelengths were chosen to be close to the fluorescence emission wavelength of ALA-induced PpIX, but some distance from the Soret excitation band of the porphyrins. Analysis of either of the two single ratios showed a significant correlation to the presence of a lesion and also to the color of the skin tissue. However, when the 405-nm excited ratio is divided by the 435-nm excited ratio, a value independent of skin coloration can be determined and can be based primarily on the presence of the ALA-induced photosensitizer. Such a technique will enable *in vivo* studies of the pharmacokinetics of tumor-localizing agents in pigmented lesions and may significantly contribute to development of a noninvasive diagnostic tool for malignant melanoma.[170]

*Colon Cancer* — Another class of compounds beginning to be tested for *in vivo* tumor demarcation is fluorescently labeled antibodies. Based on previous experiments performed using nude mice, a feasibility study was performed to determine whether or not LIF of fluoresceinated monoclonal antibodies against carcinoembryonic antigens localized specifically in human carcinoma xenografts could be used for clinical colon cancer diagnoses. In this study, six patients with known primary colorectal carcinoma were given an i.v. injection of 4.5 or 9 mg of mouse–human chimeric anticarcinoembryonic antigen monoclonal antibody labeled with 0.10 to 0.28 mg of fluorescein. In addition, the monoclonal antibody was labeled with 0.2 to 0.4 mCi of $^{125}I$. Photodetection of the tumor was performed *ex vivo* on surgically resected tissues from all six patients and *in vivo* by fluorescence rectosigmoidoscopy for one of the six patients.

Fluorescence analyses revealed that the dye-labeled antibody localized preferentially in the tumor tissue at concentrations up to 0.059% of the injected dose per gram of tumor. This was ten times greater than in the normal tissue, which exhibited a concentration of 0.006% of the injected dose per gram of normal mucosa. Such immunophotodiagnoses may prove very useful in the clinical setting for rapid tumor diagnoses in the colon and, potentially, other organs.[171]

## 28.5 Conclusion

This chapter illustrates the usefulness of fluorescence spectroscopy in a wide variety of medical applications spanning cellular screening to tissue analysis and *in vivo* diagnostics. The major advantage of fluorescence (as well as of other techniques such as Raman scattering, diffuse scattering, elastic scattering, and NIR absorption) is that biochemical and morphological information about the native tissue state can be obtained without physically removing tissues. In addition, because tissue removal is not needed for diagnostics, a more comprehensive examination of the organ can be achieved than is possible with excised tissue. Fluorescence techniques can also be combined with imaging so that images of diseased tissue sites vs. normal sites can be constructed. Thus, these diagnostic techniques inherently provide superior coverage and do not suffer from the sampling errors that often occur with biopsy or cytology. Real-time diagnostic information is a significant added benefit.

At the early stage of development, fluorescence diagnostic techniques need to be verified by comparison with pathology, the current "gold standard," which also varies in reliability. Thus, the sensitivity and specificity of spectroscopic diagnostics depends upon the reliability of pathology in each specific desired application. If "optical biopsy" is to have a role in clinical diagnosis, the problem becomes which gold standard to use for calibrating optical measurements. Using pathology as the gold standard for intraepithelial neoplasia (preinvasive cancer) has its own difficulties because pathologists often disagree on their interpretations of these lesions, with the level of consistency varying for different organs. Therefore, building decision-making boundaries using optical measurements based on conventional pathology is problematic. Because under- and over-diagnosis are undesirable, it is imperative to establish objective and standardized pathological classification systems for grading preinvasive lesions for each separate organ area where these new optical technologies may be applied.

Another important application for optical diagnostic technologies such as fluorescence is the possible use of these techniques for guidance of surgical intervention and treatment. In such surgical-assist applications, the ability of fluorescence diagnostic technologies to provide real-time information would be critically useful. Fluorescence techniques may also be used to provide real-time assessment of tissue response to therapy — as in the assessment of tissue viability and necrosis in thermal, laser interstitial therapy, or PDT.

Fluorescence technologies could also be used for noninvasive measurement of concentrations of various drugs and biological species in tissues. This capability would provide a variety of benefits in medical research. One example of such a benefit is the use of chemotherapy drugs for treatment of various cancers. The therapeutic benefit is determined by the concentration of the drug in the tissues of the targeted organ or site; currently, the only minimally invasive check available to the oncologist is to track the blood serum concentration and to assume a relationship between the amount of drugs at the target organs and the tissue concentration. More generally, the ability to track concentrations of compounds in tissue noninvasively would be a tremendous advantage. Optical methods such as fluorescence could bypass many tedious and time-consuming trials that attempt to relate dosage to metabolic rates and target organ concentrations. Because of all of these desirable features, it is evident that fluorescence spectroscopy can be a powerful tool for the medical researcher to obtain a better understanding of the disease process and for the physician to perform real-time *in vivo* diagnoses and, ultimately, provide treatment at the point of care.

## Acknowledgments

This work was sponsored by the National Institutes of Health (RO1 CA88787-01) and by the U.S. Department of Energy (DOE) Office of Biological and Environmental Research, under contract DEAC05-00OR22725 with UT-Battelle, LLC. T.V.D. acknowledges contributions from Bergein F. Overholt and Masoud Panjehpour at the Thomson Cancer Survival Center.

### References


1. Vo-Dinh, T., *Room Temperature Phosphorimetry for Chemical Analysis,* John Wiley & Sons, New York, 1984.
2. McGlynn, S.P., Azumi, T., and Kinoshita, M., *Molecular Spectroscopy of the Triplet State,* Prentice-Hall, Engelwood Cliffs, NJ, 1969.
3. Vo-Dinh, T., Multicomponent analysis by synchronous luminescence spectrometry, *Anal. Chem.,* 50, 396, 1978.
4. Vo-Dinh, T., Synchronous luminescence spectroscopy — methodology and applicability, *Appl. Spectrosc.,* 36, 576, 1982.
5. Wolfbeis, O.S., Fluorescence or organic natural products, in *Molecular Luminescence Spectroscopy,* Schulman, S.G.J., Ed., John Wiley & Sons, New York, 1993, p. 167.
6. Richards-Kortum, R. and Sevick-Muraca, E., Quantitative optical spectroscopy for tissue diagnosis, *Annu. Rev. Phys. Chem.,* 47, 555, 1996.
7. Bottiroli, G.C., Locatelli, A.C.., Marchesini, R., Pignoli, E., Tomatis, S., Cuzzoni, C., Dipalma, S., Dalfante, M., and Spinelli, P, Natural fluorescence of normal and neoplastic human colon: a comprehensive *ex vivo* study, *Lasers Surg. Med.,* 16, 48, 1995.
8. Fujimoto, D., Akiba, K.Y., and Nakamura, N., Isolation and characterization of a fluorescent material in bovine Achilles-tendon collagen, *Biochem. Biophys. Res. Commun.,* 76, 1124, 1977.
9. Wilson, B., Laser reflectance spectroscopy of tissue, in *Optronic Techniques in Diagnostic and Therapeutic Medicine,* Pratesi, R., Ed., Plenum Press, New York, 1991.
10. Wagnières, G.A., Star, W.M., and Wilson, B.C., *In vivo* fluorescence spectroscopy and imaging for oncological applications, *Photochem. Photobiol.,* 68, 603, 1998.
11. Lakowicz, J.R., *Principles of Fluorescence Spectroscopy,* Plenum Press, New York, 1985.
12. Masters, B.R. and Chance, B., in *Fluorescent and Luminescent Probes for Biological Activity,* Maon, W.T., Ed., Academic Press, London, 1993, p. 44.
13. Tsuchida, M., Miura, T., and Aibara, K., Lipofuscin and lipofuscin-like substances, *Chem. Phys. Lipids,* 44, 297, 1987.
14. Eldred, G.E., Miller, G.V., Stark, W.S., and Feeney-Burns, L., Lipofuscin — resolution of discrepant fluorescence data, *Science,* 216, 757, 1982.
15. Millar, D.P., Time-resolved fluorescence spectroscopy, *Curr. Opin. Struct. Biol.,* 6, 637, 1996.
16. Viallet, P.M., Vo-Dinh, V., Bunde, T., Ribou, A.C., Vigo, J., and Salmon, J.M., Fluorescent molecular reporter for the 3-D conformation of protein subdomains: the Mag-Indo system, *J. Fluorescence,* 9, 153, 1999.
17. Viallet, P.M., Vo-Dinh, T., Ribou, A.C., Vigo, J., and Salmon, J.M., Native fluorescence and Mag-Indo-1-protein interaction as tools for probing unfolding and refolding sequences of the bovine serum albumin subdomain in the presence of guanidine hydrochloride, *J. Protein Chem.,* 19, 431, 2000.
18. Tyagi, S. and Kramer, F.R., Molecular beacons: probes that fluoresce upon hybridization, *Nat. Biotechnol.,* 14, 303, 1996.
19. Gao, W.W., Tyagi, S., Kramer, F.R., and Goldman, E., Messenger RNA release from ribosomes during 5′-translational blockage by consecutive low-usage arginine but not leucine codons in *Escherichia coli, Mol. Microbiol.,* 25, 707, 1997.

20. Tyagi, S., Bratu, D.P., and Kramer, F.R., Multicolor molecular beacons for allele discrimination, *Nat. Biotechnol.,* 16, 49, 1998.
21. Tyagi, S., Marras, S.A.E., and Kramer, F.R., Wavelength-shifting molecular beacons, *Nat. Biotechnol.,* 18, 1191, 2000.
22. Protio, A.E. and Sarniak, J., Fluorescence of HpD for tumor detection and dosimetery in photoradiation therapy, in *Porphyrin Localization and Treatment of Tumors*, Doiron, D.R. and Gomer, C.J., Eds., Alan R. Liss, New York, 1984, p. 163.
23. Lam, S.H.J. and Palcic, B., Detection of lung cancer by ratio fluorimetry with and without photofrin II, *Proc. SPIE,* 1201, 561, 1990.
24. Tassetti, V.H., Sowinska, M., Evrard, S., Heisel, F., Cheng, L.Q., Miehe, J.A., Marescaux, J., and Aprahamian, M., *In vivo* laser-induced fluorescence imaging of a rat pancreatic cancer with pheophorbide-a, *Photochem. Photobiol.,* 65, 997, 1997.
25. Kim, R.Y.H., Flotte, T.J., Gragoudas, E.S., and Young, L.H.Y., Digital angiography of experimental choroidal melanomas using benzoporphyrin derivative, *Am. J. Ophthalmol.,* 123, 810 , 1997.
26. Kollias, N.L.H., Wimberly, J., and Anderson, R.R., Monitoring of benzoporphyrin derivative monoacid ring a (bpd-ma) in skin tumors by fluorescence during photodynamic therapy, *Proc. SPIE,* 1881, 41, 1993.
27. Peyman, G.A., Moshfeghi, D.M., Moshfeghi, A., Khoobehi, B., Doiron, D.R., Primbs, G.B., and Crean, D.H., Photodynamic therapy for choriocapillaris using tin ethyl etiopurprin ($SnET_2$), *Ophthalmic Surg. Lasers,* 28, 409, 1997.
28. Koren, H., Schenk, G.M., Jindra, R.H., Alth, G., Ebermann, R., Kubin, A., Koderhold, G., and Kreitner, M., Hypericin in phototherapy, *J. Photochem. Photobiol.,* B36, 113, 1996.
29. Diwu, Z., Novel therapeutic and diagnostic applications of hypocrellins and hypericins, *Photochem. Photobiol.,* 61, 529, 1995.
30. Dets, S.M, Buryi, A.N., Melnik, I.S., Joffe, A.Y., and Rusina, T.V., Laser-induced fluorescence detection of stomach cancer using hypericin, *Proc. SPIE,* 2926, 51, 1996.
31. Peng, Q., Berg, K., Moan, J., Kongshaug, M., and Nesland, J.M., 5-Aminolevulinic acid-based photodynamic therapy — principles and experimental research, *Photochem. Photobiol.,* 65, 235, 1997.
32. Battle, A.M.D.C., Phorophyrins, phorphyrias, cancer and photodynamic therapy: a model of carcinogenesis, *J. Photochem. Photobiol.,* B20, 5, 1993.
33. Cox, G.S., Bobillier, C., and Whitten, D.G., Photooxidation and singlet oxygen sensitization by photophyrin ix and its photoxidiation products, *Photochem. Photobiol.,* 36, 401, 1982.
34. Peng, Q., Warloe, T., Berg, K., Moan, J., Kongshaug, M., Giercksky, K.E., and Nesland, J.M., 5-Aminolevunic acid based photodynamic therapy: clinical research and future challenges, *Cancer,* 79, 2282, 1997.
35. Marcus, S.L., Solbel, R.S., Golub, A.L., Carroll, R.L., Lundahl, S., and Shulman, D.G., Photodynamic therapy (PDT) and photodiagnosis (PD) using endogenous photosensitization induced by 5-aminolevulinic acid (ALA): current clinical and developmental status, *Laser Med. Surg.,* 14, 59, 1996.
36. Schwartz, S., Historical perspectives, in *Photodynamic Therapy: Basic Priniciples and Clinical Applications*, Henderson, B.W. and Dougherty, T.J., Eds., Marcel Dekker, New York, 1992, p. 1.
37. Sacks, P.G., Savage, H.E., Levine, J., Kolli, V.R., Alfano, R.R., and Schantz, S.P., Native cellular fluorescence identifies terminal squamous differentiation of normal oral epithelial cells in culture: a potential chemoprevention biomarker, *Cancer Lett.,* 104, 171, 1996.
38. Zhang, J.C., Savage, H.E., Sacks, P.G., Delohery, T., Alfano, R.R., Katz, A., and Schantz, S.P., Innate cellular fluorescence reflects alterations in cellular proliferation, *Lasers Surg. Med.,* 20, 319, 1997.
39. Anidjar, M., Cussenot, O., Blais, J., Bourdon, O., Avrillier, S., Ettori, D., Villette, J.M., Fiet, J., Teillac, P., and LeDuc, A., Argon laser induced autofluorescence may distinguish between normal and tumor human urothelial cells: a microspectrofluorimetric study., *J. Urol.,* 155, 1771, 1996.

40. Harth, R., Gerlach, M., Riederer, P., and Gotz, M.E., A highly sensitive method for the determination of protein bound 3,4-dihydroxyphenylalanine as a marker for post-translational protein hydroxylation in human tissues *ex vivo*, *Free Radical Res.*, 35, 167, 2001.
41. Alfano, R.R., Tata, D.B., Cordero, J., Tomashefsky, P., Longo, F.W., and Alfano, M.A., Laser-induced fluorescence spectroscopy from native cancerous and normal tissue, *IEEE J. Quantum Electron.*, 20, 1507, 1984.
42. Uziel, M., Ward, R.J., and Vo-Dinh, T., Synchronous fluorescence measurement of bap metabolites in human and animal urine, *Anal. Lett.*, 20, 761, 1987.
43. Watts, W.E., Isola, N.R., Frazier, D., and Vo-Dinh, T., Differentiation of normal and neoplastic cells by synchronous fluorescence: rat liver epithelial and rat hepatoma cell models, *Anal. Lett.*, 32, 2583, 1999.
44. Amano, T., Kunimi, K., and Ohkawa, M., Fluorescence spectra from human semen and their relationship with sperm parameters, *Arch. Androl.*, 36, 9, 1996.
45. Xu, H., Aylott, J., and Kopelman, R., Sol-gel PEBBLE sensors for biochemical analysis inside living cells, *Abstr. Papers Am. Chem. Soc.*, 219, 97-ANYL, 2000.
46. Tan, W.H., Thorsrud, B.A., Harris, C., and Kopelman, R., Real time pH measurements in the intact rat conceptus using ultramicrofiber-optic sensors, in *Polymers in Sensors*, American Chemical Society, Washington, D.C., 1998, p. 266.
47. Clark, H.A., Hoyer, M., Philbert, M.A., and Kopelman, R., Optical nanosensors for chemical analysis inside single living cells. 1. Fabrication, characterization, and methods for intracellular delivery of PEBBLE sensors, *Anal. Chem.*, 71, 4831, 1999.
48. Clark, H.A., Kopelman, R., Tjalkens, R., and Philbert, M.A., Optical nanosensors for chemical analysis inside single living cells. 2. Sensors for pH and calcium and the intracellular application of PEBBLE sensors, *Anal. Chem.*, 71, 4837, 1999.
49. Tan, W.H., Kopelman, R., Barker, S.L.R., and Miller, M.T., Ultrasmall optical sensors for cellular measurements, *Anal. Chem.*, 71, 606A, 1999.
50. Cullum, B.M., Griffin, G.D., Miller, G.H., and Vo-Dinh, T., Intracellular measurements in mammary carcinoma cells using fiber-optic nanosensors, *Anal. Biochem.*, 277, 25, 2000.
51. Cullum, B.M. and Vo-Dinh, T., The development of optical nanosensors for biological measurements, *Trends Biotechnol.*, 18, 388, 2000.
52. Vo-Dinh, T., Alarie, J.P., Cullum, B.M., and Griffin, G.D., Antibody-based nanoprobe for measurement of a fluorescent analyte in a single cell, *Nat. Biotechnol.*, 18, 764, 2000.
53. Kong, J.A.R.,D.P., Quantitative *in situ* image analysis of apoptosis in well and poorly differentiated tumors from rat liver, *Am. J. Pathol.*, 147, 1626, 1995.
54. Balazs, M.S.J., Lee, W.C., Haugland, R.P., Guzikowski, A.P., Fulwyler, M.J., Damjanovich, S., Feurstein, B.G., and Pershadsingh, H.A., Fluorescent tetradecanoylphorbol acetate: a novel probe of phorbol ester binding domains, *J. Cell Biochem.*, 46, 266, 1991.
55. Andeoni, A., Bottiroli, G., Colasanti, A., Giangare, M.C., Riccio, P., Roberti, G. and Vaghi, P., Fluorochromes with long-lived fluorescence as potential labels for pulsed laser immunocytofluorometry: photophysical characterization of pyrene derivatives, *J. Biochem. Biophys. Methods,* 29, 157, 1994.
56. Fritsch, C., Batz, J., Bolsen, K., Schulte, K.W., Zumdick, M., Ruzicka, T., and Goerz, G., *Ex vivo* application of delta-aminolevulinic acid induces high and specific porphyrin levels in human skin tumors: possible basis for selective photodynamic therapy, *Photochem. Photobiol.*, 66, 114, 1997.
57. DaCosta, R.S.L., Andersson, H.M.S., Cirocco, M., Kandel, G., Kortan, P., Haber, G., Marcon, N.E., and Wilson, B.C., Correlation of autofluorescence (AF) and ultrastructures of *ex vivo* colorectal tissues and isolated living epithelial cells from primary cell cultures (PCC) of normal colon and hyperplastic and dysplastic polyps: implications for early diagnosis, altered cell metabolism and cytopathology, *Gastroenterology,* 116, G1721, 1999.

58. Han, I., Saito, H., Fukatsu, K., Inoue, T., Yasuhara, H., Furukawa, S., Matsuda, T., Lin, M.T., and Ikeda, S., *Ex vivo* fluorescence microscopy provides simple and accurate assessment of neutrophil-endothelial adhesion in the rat lung, *Shock,* 16, 143, 2001.
59. Sculean, A., Auschill, T.M., Donos, N., Brecx, M., and Arweiler, N.B., Effect of an enamel matrix protein (Emdogain (R)) on *ex vivo* dental plaque vitality, *J. Clin. Periodontol.,* 28, 1074, 2001.
60. Alfano, R.R., Lam, W., Zarrabi, H.J., Alfano, M.A., Cordero, J., Tata, D.B., and Swenberg, C.E., Human teeth with and without caries studied by laser scattering, fluorescence, and absorption-spectroscopy, *IEEE J. Quantum Electron.,* 20, 1512, 1984.
61. Alfano, R.R., Tang, G.C., Pradhan, A., Lam, W., Choy, D.S.J., and Opher, E., Fluorescence spectra from cancerous and normal human-breast and lung tissues, *IEEE J. Quantum Electron.,* 23, 1806, 1987.
62. Tang, G.C., Pradhan, A., Sha, W., Chen, J., Liu, C.H., Wahl, S.J., and Alfano, R.R., Pulsed and cw laser fluorescence spectra from cancerous, normal, and chemically treated normal human-breast and lung tissues, *Appl. Opt.,* 28, 2337, 1989.
63. Ingrams, D.R., Dhingra, J.K., Roy, K., Perrault, D.F., Bottrill, I.D., Kabani, S., Rebeiz, E.E., Pankratov, M.M., Shapshay, S.M., Manoharan, R., Itzkan, I., and Feld, M.S., Autofluorescence characteristics of oral mucosa, *Head Neck J. Sci. Specialties Head Neck,* 19, 27, 1997.
64. Vo-Dinh, T., Panjehpour, M., Overholt, B.F., and Buckley, P., Laser-induced differential fluorescence for cancer diagnosis without biopsy, *Appl. Spectrosc.,* 51, 58, 1997.
65. Vo-Dinh, T., Panjehpour, M., and Overholt, B.F., Laser-induced fluorescence for esophageal cancer and dysplasia diagnosis, in *Advances in Optical Biopsy and Optical Mammography,* New York Academy of Sciences, New York, 1998, p. 116.
66. Richards-Kortum, R., Mitchell, M.F., Ramanujam, N., Mahadevan, A., and Thomsen, S., *In vivo* fluorescence spectroscopy: potential for non-invasive, automated diagnosis of cervical intraepithelial neoplasia and use as a surrogate endpoint biomarker, *J. Cell Biochem.,* Suppl. 19, 111, 1994.
67. Lohmann, W., Mussman, J., Lohmann, C., and Kunzel, W., Native fluorescence of cervix uteri as a marker for dysplasia and invasive carcinoma, *Eur. J. Obstet. Gynecol. Reprod. Biol.,* 31, 249, 1989.
68. Romer, T.J., Fitzmaurice, M., Cothren, R.M., Richardskortum, R., Petras, R., Sivak, M.V., and Kramer, J.R., Laser-induced fluorescence microscopy of normal colon and dysplasia in colonic adenomas — implications for spectroscopic diagnosis, *Am. J. Gastroenterol.,* 90, 81, 1995.
69. Fiarman, G.S., Nathanson, M.H., West, A.B., Deckelbaum, L.I., Kelly, L., and Kapadia, C.R., Differences in laser-induced autofluorescence between adenomatous and hyperplastic polyps and normal colonic mucosa by confocal microscopy, *Dig. Dis. Sci.,* 40, 1261, 1995.
70. Hegedus, Z.L. and Nayak, U., Relative fluorescence intensities of human plasma soluble melanins in normal adults, *Arch. Int. Physiol. Biochim. Biophys.,* 102, 311, 1994.
71. Marchesini, R., Brambilla, M., Pignoli, E., Bottiroli, G., Croce, A.C., Dal Fante, M., Spinelli, P., and diPalma, S., Light-induced fluorescence spectroscopy of adenomas, adenocarcinomas and non-neoplastic mucosa in human colon I. *In vitro* measurements, *J. Photochem. Photobiol. B.,* 14, 219, 1992.
72. Richards-Kortum, R., Rava, R.P., Petras, R.E., Fitzmaurice, M., Sivak, M., and Feld, M.S., Spectroscopic diagnosis of colonic dysplasia, *Photochem. Photobiol.,* 53, 777, 1991.
73. Kapadia, C.R., Cutruzzola, F.W., O'Brien, K.M., Stetz, M.L., Enriquez, R., and Deckelbaum, L.I., Laser-induced fluoresence spectroscopy of human colonic mucosa, *Gastroenterology,* 99, 150, 1990.
74. Chwirot, B.W., Chwirot, S., Jedrzejczyk, W., Jackowski, M., Raczynska, A.M., Winczakiewicz, J., and Dobber, J., Ultraviolet laser-induced fluorescence of human stomach tissues: detection of cancer tissues by imaging techniques, *Lasers Surg. Med.,* 21, 149, 1997.
75. Laufer, G.W.G., Hohla, K., Horvat, R., Henke, K.H., Buchelt, M., Wutzl, G., and Wolner, E., Excimer laser-induced simultaneous ablation and spectral identification of normal and atherosclerotic arterial tissue layers, *Circulation,* 78, 1031, 1988.

76. Baraga, J.J., Rava, R.P., Taroni, P., Kittrell, C., Fitzmaurice, M., and Feld, M.S., Laser induced fluorescence spectroscopy of normal and atherosclerotic human aorta using 306 to 310 nm excitation, *Lasers Surg. Med.*, 10, 245, 1990.
77. Bosshart, F.U., Hess, O.M., Wyser, J., Mueller, A., Schneider, J., Niederer, P., Anliker, M., and Krayenbuehl, H.P., Fluorescence spectroscopy for identification of atherosclerotic tissue, *Cardiovasc. Res.*, 26, 620, 1992.
78. Lucas, A., Radosavljevic, M.J., Lu, E., and Gaffney, E.J., Characterization of human coronary artery atherosclerotic plaque fluorescence emission, *Can. J. Cardiol.*, 6, 219, 1990.
79. Richards-Kortum, R., Rava, R.P., Fitzmaurice, M., Kramer, J.R., and Feld, M.S., 476 nm excited laser-induced fluorescence spectroscopy of human coronary arteries applications in cardiology, *Am. Heart J.*, 122, 1141, 1991.
80. Verbunt, R., Fitzmaurice, M.A., Kramer, J.R., Ratliff, N.B., Kittrell, C., Taroni, P., Cothren, R.M., Baraga, J., and Feld, M., Characterization of ultraviolet laser-induced autofluorescence of ceroid deposits and other structures in atherosclerotic plaques as a potential diagnostic for laser angiosurgery, *Am. Heart J.*, 123, 208, 1992.
81. Fitzmaurice, M., Bordagary, J.O., Engelmann, G.L., Richards-Kortum, R., Kolubayev, T., Feld, M.S., Ratliff, N.B., and Kramer, J.R., Argon ion laser-excited autofluorescence in normal and atherosclerotic aorta and coronary arteries: morphologic studies, *Am. Heart J.*, 118, 1028, 1989.
82. Chaudhry, H.W., Richards-Kortum, R., Kolubayev, T., Kittrell, C., Partovi, F., Kramer, J.R., and Feld, M.S., Alteration of spectral characteristics of human artery wall caused by 476-nm laser irradiation, *Lasers Surg. Med.*, 9, 572, 1989.
83. Andersson-Engels, S.J.J., Svanberg, K., and Svanberg, S., Fluorescence imaging and point measurements of tissue: applications to the demarcation of malignant tumors and atherosclerotic lesions from normal tissue, *Photochem. Photobiol.*, 53, 807, 1991.
84. Andersson-Engels, S.J.J., Stenram, U., Svanberg, K., and Svanberg, S., Time-resolved laser-induced fluorescence spectroscopy for enhanced demarction of human atherosclerotic plaques, *J. Photochem. Photobiol. B.*, 4, 363, 1990.
85. Perk, M., Flynn, G.J., Gulamhusein, S., Wen, Y., Smith, C., Bathgate, B., Tulip, J., Parfrey, N.A., and Lucas, A., Laser induced fluorescence identification of sinoatrial and atrioventricular nodal conduction tissue, *Pacing Clin. Electrophysiol.*, 16, 1701, 1993.
86. van der Veen, M.H. and ten Bosch, J.J., Autofluorescence of bulk sound and *in vitro* demineralized human root dentin, *Eur. J. Oral Sci.*, 103, 375, 1995.
87. Samoszuk, M., Eosinophils and human cancer, *Histol. Histopathol.*, 12, 807, 1997.
88. Barnes, D.A.S., Thomsen, S., Fitzmaurice, M., and Richards-Kortum, R., A characterization of the fluorescent properties of circulating human eosinophils, *Photochem. Photobiol.*, 58, 297, 1993.
89. Pompella, A. and Comporti, M., Imaging of oxidative stress at subcellular level by confocal laser scanning microscopy after fluorescent derivitization of cellular carbonyls, *Am. J. Pathol.*, 142, 1353, 1993.
90. Richards-Kortum, R., Mehta, A., Hayes, G., Cothren, R., Kolubayev, T., Kittrell, C., Ratliff, N.B., Kramer, J.R., and Feld, M.S., Spectral diagnosis of atherosclerosis using an optical fiber laser catheter, *Am. Heart J.*, 118, 381, 1989.
91. Alfano, R.R. and Yao, S.S., Human teeth with and without dental caries studied by visible luminescent spectroscopy, *J. Dent. Res.*, 60, 120, 1981.
92. Zellweger, M., Goujon, D., Conde, R., Forrer, M., van den Bergh, H., and Wagnières, G., Absolute autofluorescence spectra of human healthy, metaplastic, and early cancerous bronchial tissue *in vivo*, *Appl. Opt.*, 40, 3784, 2001.
93. Zellweger, M., Grosjean, P., Goujon, D., Monnier, P., van den Bergh, H., and Wagnières, G., *In vivo* autofluorescence spectroscopy of human bronchial tissue to optimize the detection and imaging of early cancers, *J. Biomed. Opt.*, 6, 41, 2001.

94. Wang, T.D., Crawford, J.M., Feld, M.S., Wang, Y., Itzkan, I., and Van Dam, J., *In vivo* identification of colonic dysplasia using fluorescence endoscopic imaging, *Gastrointest. Endosc.*, 49, 447, 1999.
95. Wennberg, A.M., Gudmundson, F., Stenquist, B., Ternesten, A., Molne, L., Rosen, A., and Larko, O., *In vivo* detection of basal cell carcinoma using imaging spectroscopy, *Acta Dermato-Venereolog.*, 79, 54, 1999.
96. Schantz, S.P., Kolli, V., Savage, H.E., Yu, G.P., Shah, J.P., Harris, D.E., Katz, A., Alfano, R.R., and Huvos, A.G., *In vivo* native cellular fluorescence and histological characteristics of head and neck cancer, *Clin. Cancer Res.*, 4, 1177, 1998.
97. Major, A.L., Rose, G.S., Chapman, C.F., Hiserodt, J.C., Tromberg, B.J., Krasieva, T.B., Tadir, Y., Haller, U., DiSaia, P.J., and Berns, M.W., *In vivo* fluorescence detection of ovarian cancer in the NuTu-19 epithelial ovarian cancer animal model using 5-aminolevulinic acid (ALA), *Gynecol. Oncol.*, 66, 122, 1997.
98. Andersson-Engels, S., Afklinteberg, C., Svanberg, K., and Svanberg, S., *In vivo* fluorescence imaging for tissue diagnostics, *Phys. Med. Biol.*, 42, 815, 1997.
99. Ramanujam, N., Chen, J.X., Gossage, K., Richards-Kortum, R., and Chance, B., Fast and noninvasive fluorescence imaging of biological tissues *in vivo* using a flying-spot scanner, *IEEE Trans. Biomed. Eng.*, 48, 10341, 2001.
100. Brancaleon, L., Durkin, A.J., Tu, J.H., Menaker, G., Fallon, J.D., and Kollias, N., *In vivo* fluorescence spectroscopy of nonmelanoma skin cancer, *Photochem. Photobiol.*, 73, 178, 2001.
101. Koenig, F., Knittel, J., and Stepp, H., Diagnosing cancer *in vivo*, *Science*, 292, 1401, 2001.
102. Brewer, M., Utzinger, U., Silva, E., Gershenson, D., Bast, R.C., Follen, M., and Richards-Kortum, R., Fluorescence spectroscopy for *in vivo* characterization of ovarian tissue, *Lasers Surg. Med.*, 29, 128, 2001.
103. Panjehpour, M., Overholt, B.F., Vo-Dinh, T., Farris, C., and Sneed, R., Fluorescence spectroscopy for detection of malignant-tissue in the esophagus, *Gastroenterology*, 104, A439, 1993.
104. Panjehpour, M., Overholt, B.F., Schmidhammer, J.L., Farris, C., Buckley, P.F., and Vo-Dinh, T., Spectroscopic diagnosis of esophageal cancer — new classification model, improved measurement system, *Gastrointest. Endosc.*, 41, 577, 1995.
105. Panjehpour, M., Overholt, B.F., Vo-Dinh, T., Haggitt, R.C., Edwards, D.H., and Buckley, F.P., Endoscopic fluorescence detection of high-grade dysplasia in Barrett's esophagus, *Gastroenterology*, 111, 93, 1996.
106. Panjehpour, M., Overholt, B.F., Vo-Dinh, T., Haggitt, R.C., Edwards, D.H., Buckley, F.P., and Decosta, J.F., Fluorescence spectroscopy for detection of dysplasia in Barrett's esophagus, *Gastroenterology*, 110, A574, 1996.
107. Overholt, B.F., Panjehpour, M., Vo-Dinh, T., Farris, C., Schmidhammer, J.L., Sneed, R.E., and Buckley, P.F., Spectroscopic diagnosis of esophageal cancer — improved technique, *Gastroenterology*, 106, A425, 1994.
108. Richards-Kortum, R. and Sevick-Muraca, E., Quantitative optical spectroscopy for tissue diagnosis, *Annu. Rev. Phys. Chem.*, 47, 555, 1996.
109. Ramanujam, N., Fluorescence spectroscopy of neoplastic and non-neoplastic tissues, *Neoplasia*, 2, 89, 2000.
110. Ramanujam, N., Mitchell, M.F., Mahadevan-Jansen, A., Thomsen, S.L., Staerkel, G., Malpica, A., Wright, T., Atkinson, N., and Richards-Kortum, R., Cervical precancer detection using a multivariate statistical algorithm based on laser-induced fluorescence spectra at multiple excitation wavelengths, *Photochem. Photobiol.*, 64, 720, 1996.
111. Lam, S. and Shibuya, H., Early diagnosis of lung cancer, *Clin. Chest Med.*, 20, 53, 1999.
112. Lam, S. and Palcic, B., Autofluorescence bronchoscopy in the detection of squamous metaplasia and dysplasia in current and former smokers, *J. Natl. Cancer Inst.*, 91, 561, 1999.
113. Lam, S., Macaulay, C., Hung, J., Leriche, J., Profio, A.E., and Palcic, B., Detection of dysplasia and carcinoma i*n situ* with a lung imaging fluorescence endoscope device, *J. Thorac. Cardiovasc. Surg.*, 105, 1035, 1993.

114. Lam, S., Hung, J.Y.C., Kennedy, S.M., Leriche, J.C., Vedal, S., Nelems, B., Macaulay, C.E., and Palcic, B., Detection of dysplasia and carcinoma *in situ* by ratio fluorometry, *Am. Rev. Respir. Dis.*, 146, 1458, 1992.
115. Alfano, R.R. and Alfano, M.A., Medical diagnostics — a new optical frontier, *Photonics Spectra*, 19, 55, 1985.
116. Alfano, R.R., Tang, G.C., Pradhan, A., and Wenling, S., Investigation of optical spectroscopy of cancerous and normal human tissues, *J. Electrochem. Soc.*, 135, C387, 1988.
117. Alfano, R.R., Tomaselli, V.P., Beuthan, J., Feld, M.S., Flotte, T.J., Fujimoto, J.G., and Thomsen, S., Advances in optical biopsy and optical mammography — panel discussion — review and summary of presentations, in *Advances in Optical Biopsy and Optical Mammography*, New York Academy of Sciences, New York, 1998, p. 194.
118. Alfano, R.R., Advances in optical biopsy and optical mammography — closing remarks, in *Advances in Optical Biopsy and Optical Mammography*, New York Academy of Sciences, New York, 1998, p. 197.
119. Vo-Dinh, T., Panjehpour, M., Overholt, B.F., Farris, C., Buckley, F.P., and Sneed, R., *In-vivo* cancer-diagnosis of the esophagus using differential normalized fluorescence (DNF) indexes, *Lasers Surg. Med.*, 16, 41, 1995.
120. Vo-Dinh, T., Panjehpour, M., Overholt, B.F., and Buckley, P., Laser-induced differential fluorescence for cancer diagnosis without biopsy, *Appl. Spectrosc.*, 51, 58, 1997.
121. Braichotte, D., Savary, J.F., Glanzmann, T., Westermann, P., Folli, S., Wagnieres, G., Monnier, P., and Vandenbergh, H., Clinical pharmacokinetic studies of tetra(meta-hydroxyphenyl)chlorin in squamous-cell carcinoma by fluorescence spectroscopy at 2 wavelengths, *Int. J. Cancer*, 63, 198, 1995.
122. Braichotte, D.R., Wagnières, G.A., Bays, R., Monnier, P., and Vandenbergh, H.E., Clinical pharmacokinetic studies of photofrin by fluorescence spectroscopy in the oral cavity, the esophagus, and the bronchi, *Cancer*, 75, 2768, 1995.
123. Andrejevic, S., Savary, J.-F., Fontolliet, C., Monnier, P., and van den Bergh, H.E., 7,12-Dimethylbenz(a)anthracene-induced early squamous-cell carcinoma in the golden Syrian hamster: evaluation of an animal model and comparison with early forms of human squamous-cell carcinoma in the upper aerodigestive tract., *Int. J. Exp. Pathol.*, 77, 7, 1996.
124. Schantz, S.P. and Alfano, R.R., Tissue autofluorescence as an intermediate end-point in cancer chemoprevention trials, *J. Cell. Biochem.*, 64, 199, 1993.
125. Glasgold, R., Glasgold, M., Savage, H., Pinto, J., Alfano, R., and Schantz, S., Tissue autofluorescence as an intermediate end-point in NMBA-induced esophageal carcinogenesis, *Cancer Lett.*, 82, 33, 1994.
126. Chung, Y.G., Schwartz, J.A., Gardner, C.M., Sawaya, R.E. and Jacques, S.L., Diagnostic potential of laser-induced autofluorescence emission in brain tissue, *J. Korean Med. Sci.*, 12, 135, 1997.
127. Oraevsky, A.A., Jacques, S.L., Pettit, G.H., Sauerbrey, R.A., Tittel, F.K., Nguy, J.H., and Henry, P.D., XeCl laser-induced fluorescence of atherosclerotic arteries. Spectral similarities between lipid-rich lesions and peroxidized lipoproteins, *Circ. Res.*, 72, 84, 1993.
128. van Leengoed, E.V.J., van der Veen, N., van der Berg-Blok, A., Marijnissen, H., and Star, W., Tissue-localizing properties of some photosensitizers studied by *in vivo* fluorescence imaging, *J. Photochem. Photobiol., B.* 6, 111, 1990.
129. Johansson, J., Berg, R., Svanberg, K., and Svanberg, S., Laser-induced fluorescence studies of normal and malignant tumour tissue of rat following intravenous injection of delta-amino levulinic acid, *Lasers Surg. Med.*, 20, 272, 1997.
130. Svanberg, K., Kjellen, E., Ankerst, J., Montan, S., Sjoholm, E., and Svanberg, S., Fluorescence studies of hematoporphyrin derivative in normal and malignant rat-tissue, *Cancer Res.*, 46, 3803, 1986.
131. Andersson-Engels, S., Ankerst, J., Johansson, J., Svanberg, K., and Svanberg, S., Laser-induced fluorescence in malignant and normal tissue of rats injected with benzoporphyrin derivative, *Photochem. Photobiol.*, 57, 978, 1993.

132. Nilsson, H.J.J., Svanberg, K., Svanberg, S., Jori, G., Reddi, E., Segalla, A., Gust, D., Moore, A.L., and Moore, T.A., Laser-induced fluorescence studies of the biodistribution of carotenoporphyrins in mice, *Br. J. Cancer,* 76, 355, 1997.
133. Pathak, I., Davis, N.L., Hsiang, Y.N., Quenville, N.F., and Palcic, B., Detection of squamous neoplasia by fluorescence imaging comparing porfimer sodium fluorescence to tissue autofluorescence in the hamster cheek pouch model, *Am. J. Surg.,* 170, 423, 1995.
134. Tsai, J.C., Kao, M.C., and Hsiao, Y.Y., Fluorospectral study of the rat brain and glioma *in vivo, Lasers Surg. Med.,* 13, 321, 1993.
135. Svanberg, K., Liu, D.L., Wang, I., Andersson-Engels, S., Stenram, U., and Svanberg, S., Photodynamic therapy using intravenous delta-aminolaevulinic acid-induced protoporphyrin IX sensitisation in experimental hepatic tumours in rats, *Br. J. Cancer,* 74, 1526, 1996.
136. Mitchell, M.F.-L.G., Wright, T., Sarkar, A., Richards-Kortum, R., Hong, W.K., and Schottenfeld, D., Cervical human papillomavirus infection and intraepithelial neoplasia: a review, *J. Natl. Cancer Inst. Monogr.,* 21, 17, 1996.
137. Mitchell, M.F., Hittelman, W.K., Lotan, R., Nishioka, K., Tortolero-Luna, G., Richards-Kortum, R., and Hong, W.K., Chemoprevention trials in the cervix: design, feasibility, and recruitment, *J. Cell. Biochem.,* Suppl. 23, 104, 1995.
138. Ramanujam, N., Mitchell, M.F., Mahadevan, A., Warren, S., Thomsen, S., Silva, E., and Richards-Kortum, R., *In vivo* diagnosis of cervical intra epithelial neoplasia using 337-nm-excited laser-induced fluorescence, *Proc. Natl. Acad. Sci. U.S.A.,* 91, 10193, 1994.
139. Ramanujam, N., Mitchell, M.F., Mahadevan, A., Thomsen, S., Silva, E., and Richards-Kortum, R., Fluorescence spectroscopy: a diagnostic tool for cervical intra epithelial neoplasia (CIN), *Gynecol. Oncol.,* 52, 31, 1994.
140. Barrett, N., Chronic peptic ulcer of the oesophagus, *Br. J. Surg.,* 38, 175, 1950.
141. Spechler, S.J. and Goyal, R.K., Barrett's esophagus, *New Engl. J. Med.,* 315, 362, 1987.
142. Burbige, E.J. and Radigan, J.J., Characteristics of the columnar-lined (Barrett's) esophagus, *Gastrointest. Endosc.,* 24, 133, 1979.
143. Cameron, A.J., Sinmeister, A.R., and Ballard, D.J., Prevalence of columnar-lined (Barrett's) esophagus: comparison of population-based and autopsy findings, *Gastroenterology,* 99, 918, 1990.
144. Spechler, S.J., Robbins, R.H., Rubins, H.B., Vincent, M.E., Heeren, T., Doos, W.G., Colton, T., and Schimmel, E.M., Adenocarcinoma and Barrett's esophagus, an overrated risk, *Gastroenterology,* 87, 927, 1984.
145. Polopalle, S.C. and McCallum, R.W., Barrett's esophagus: current assessment and future perspectives, *Gastroenterol. Clin. North Am.,* 19, 733, 1990.
146. Vo-Dinh, T. and Mathur, P., Optical diagnostics and therapeutic technologies in pulmonary medicine, in *Interventional Bronchoscopy,* Bollinger, D.T. and Mathur, P., Eds., Basel-Karger Publishers, Basel, 2000, p. 267.
147. Koenig, F., McGovern, F.J., Althausen, A.F., Deutsch, T.F., and Schomacker, K.T., Laser induced autofluorescence diagnosis of bladder cancer, *J. Urol.,* 156, 1597, 1996.
148. Anidjar, M., Ettori, D., Cussenot, O., Meria, P., Des Grandchamps, F., Cortesse, A., Teillac, P., LeDuc, A., and Avrillier, S., Laser induced autofluorescence diagnosis of bladder tumors: dependence on the excitation wavelength, *J. Urol.,* 156, 1590, 1996.
149. Schomacker, K.T., Frisoli, J.K., Compton, C.C., Flotte, T.J., Richter, J.M., Nishioka, N.S., and Deutsch, T.F., Ultraviolet laser-induced fluorescence of colonic tissue: basic biology and diagnostic potential, *Lasers Surg. Med.,* 12, 63, 1992.
150. Cothren, R.M., Sivak, M.V., VanDam, J., Petras, R.E., Fitzmaurice, M., Crawford, J.M., Wu, J., Brennan, J.F., Rava, R.P., Manoharan, R., and Feld, M.S., Detection of dysplasia at colonoscopy using laser-induced fluorescence: a blinded study, *Gastrointest. Endosc.,* 44, 168, 1996.

151. Cothren, R.M., Richards-Kortum, R., Sivak, M.V., Fitzmaurice, M., Rava, R.P., Boyce, G.A., Doxtader, M., Blackman, R., Ivanc, T.B., Hayes, G.B., Feld, M.S., and Petras, R.E., Gastrointestinal tissue diagnosis by laser-induced fluorescence spectroscopy at endoscopy, *Gastrointest. Endosc.*, 36, 105, 1990.
152. Lashner, B.A. and Brzezinski, A., Cancer mortality-rates in ulcerative-colitis surveillance programs, *Gastroenterology*, 106, 278, 1994.
153. Lashner, B.A., Provencher, K.S., Bozdech, J.M., and Brzezinski, A., Worsening risk for the development of dysplasia or cancer in patients with chronic ulcerative-colitis, *Gastroenterology*, 106, A718, 1994.
154. Jensen, P., Krogsgaard, M.R., Christiansen, J., Braendstrup, O., Johansen, A., and Olsen, J., Observer variability in the assessment of type and dysplasia of colorectal adenomas, analyzed using kappa-statistics, *Dis. Colon Rectum*, 38, 195, 1995.
155. Kolli, V.R., Shaha, A.R., Savage, H.E., Sacks, P.G., Casale, M.A., and Schantz, S.P., Native cellular fluorescence can identify changes in epithelial thickness *in vivo* in the upper aerodigestive tract, *Am. J. Surg.*, 170, 495, 1995.
156. Kolli, V.R., Savage, H.E., Yao, T.J., and Schantz, S.P., Native cellular fluorescence of neoplastic upper aerodigestive mucosa, *Arch. Otolaryngol.-Head Neck Surg.*, 121, 1287, 1995.
157. Dhingra, J.K., Perrault, D.F., McMillan, K., Rebeiz, E.E., Kabani, S., Manoharan, R., Itzkan, I., Feld, M.S., and Shapshay, S.M., Early diagnosis of upper aerodigestive tract cancer by autofluorescence, *Arch. Otolaryngol.-Head Neck Surg.*, 122, 1181, 1996.
158. Fryen, A., Glanz, H., lohmann, W., Dreyer, T., and Bohle, R.M., Significance of autofluorescence for the optical demarcation of field cancerisation in the upper aerodigestive tract, *Acta Otolaryngol, (Stockholm)*, 117, 316, 1997.
159. Heintzelman, D.L., Utzinger, U., Fuchs, H., Zuluaga, A., Gossage, K., Gillenwater, A.M., Jacob, R., Kemp, B., and Richards-Kortum, R.R., Optimal excitation wavelengths for in vivo detection of oral neoplasia using fluorescence spectroscopy, *Photochem. Photobiol.*, 72, 103, 2000.
160. Zargi, M.S.L., Fajdiga, I., Bubnic, B., Lenarcic, J., and Oblak, P., Detection and localization of early laryngeal cancer with laser-induced fluorescence: preliminary report, *Eur. Arch. Otorhinolaryngol.*, Suppl. 1, S113, 1997.
161. Masters, B.R., So, P.T., and Gratton, E., Multiphoton excitation fluorescence microscopy and spectroscopy of *in vivo* human skin, *Biophys. J.*, 72, 2405, 1997.
162. Leffell, D.J., Stetz, M.L., Milstone, L.M., and Deckelbaum, L.I., *In vivo* fluorescence of human skin. A potential marker of photoaging, *Arch. Dermatol.*, 124, 1514, 1988.
163. Klinteberg, C., Enejder, A.M.K., Wang, I., Andersson, E., Svanberg, S., and Svanberg, K., Kinetic fluorescence studies of 5-aminolaevulinic acid-induced protoporphyrin IX accumulation in basal cell carcinomas, *J. Photochem. Photobiol., B* 49, 120, 1999.
164. Chissov, V.I., Sokolov, V.V., Filonenko, E.V., Menenkov, V.D., Zharkova, N.N., Kozlov, D.N., Polivanov, I.N., Prokhorov, A.M., Pyhov, R.L., and Smirnov, V.V., Clinical fluorescent diagnosis of tumors using photosensitizer photogem., *Khirurgiia (Moskva)*, 5, 37, 1995.
165. Heyerdahl, H., Wang, I., Liu, D.L., Berg, R., Andersson-Engels, S., Peng, Q., Moan, J., Svanberg, S., and Svanberg, K., Pharmacokinetic studies on 5-aminolevulinic acid-induced protoporphyrin IX accumulation in tumours and normal tissues, *Cancer Lett.*, 112, 225, 1997.
166. Leunig, A., Rick, K., Stepp, H., Goetz, A., Baumgartner, R., and Feyh, J., Fluorescence photodetection of neoplastic lesions in the oral cavity following topical application of 5-aminolevulinic acid, *Laryngo-Rhino-Otologie*, 75, 459, 1996.
167. von Holstein, C.S., Nilsson, A.M., Andersson-Engels, S., Willen, R., Walther, B., and Svanberg, K., Detection of adenocarcinoma in Barrett's oesophagus by means of laser induced fluorescence, *Gut*, 39, 711, 1996.

168. Baert, L., Berg, R., Vandamme, B., Dhallewin, M.A., Johansson, J., Svanberg, K., and Svanberg, S., Clinical fluorescence diagnosis of human bladder-carcinoma following low-dose photofrin injection, *Urology,* 41, 322, 1993.
169. Svanberg, K., Andersson, T., Killander, D., Wang, I., Stenram, U., Andersson-Engels, S., Berg, R., Johansson, J., and Svanberg, S., Photodynamic therapy of nonmelanoma malignant tumors of the skin using topical delta-amino levulinic acid sensitization and laser irradiation, *Br. J. Dermatol.,* 130, 743, 1994.
170. Sterenborg, H., Saarnak, A.E., Frank, R., and Motamedi, M., Evaluation of spectral correction techniques for fluorescence measurements on pigmented lesions *in vivo, J. Photochem. Photobiol., B-Biol.,* 35, 159, 1996.
171. Folli, S., Wagnières, G., Pelegrin, A., Calmes, J.M., Braichotte, D., Buchegger, F., Chalandon, Y., Hardman, N., Heusser, C., Givel, J.C., Chapuis, G., Chatelain, A., Vandenbergh, H., and Mach, J.P., Immunophotodiagnosis of colon carcinomas in patients injected with fluoresceinated chimeric antibodies against carcinoembryonic antigen, *Proc. Natl. Acad. Sci. U.S.A.,* 89, 7973, 1992.

# 29

# Elastic-Scattering Spectroscopy and Diffuse Reflectance




Judith R. Mourant
*Los Alamos National Laboratory*
*Los Alamos, New Mexico*

Irving J. Bigio
*Boston University*
*Boston, Massachusetts*


## 29.1 Basic Concepts

Wavelength-dependent spectral measurements of elastically scattered light from tissue, performed in a manner that is sensitive to scattering and absorption properties, may be used to detect and diagnose tissue pathologies. Many tissue pathologies, including a majority of cancers, exhibit significant architectural changes at the cellular and subcellular level. In making a diagnosis, pathologists determine some of these architectural changes by examining surgically removed samples called biopsies. Microscopic assessment, often referred to as histopathology, is performed on the biopsy samples to determine cell and tissue architecture, including the sizes and shapes of cells, the ratio of nuclear to cellular volume, the form of the bilipid membrane, cell clustering patterns, etc. The properties of light elastically scattered in tissue also depend on architectural features. For example, the size of the structures in tissue responsible for the scattering of light determines how much more strongly a short wavelength, e.g., blue light, is scattered than a long wavelength, e.g., red light.

This concept is illustrated in Figure 29.1, where the likelihood of scattering is plotted as a function of wavelength for suspensions of two different sizes of spheres. If the scattering structures are much smaller than the measurement light wavelengths, the scattering probability decreases as $1/\lambda^4$, where $\lambda$ is wavelength. If the particles are of a size near that of the measurement light wavelengths or larger, scattering will not decrease nearly as rapidly with wavelength. Light is scattered by structures with a variety of shapes and sizes in tissue. Properties of light that has scattered inside tissue, such as the wavelength-dependent intensity, can provide information on cell and tissue structure.

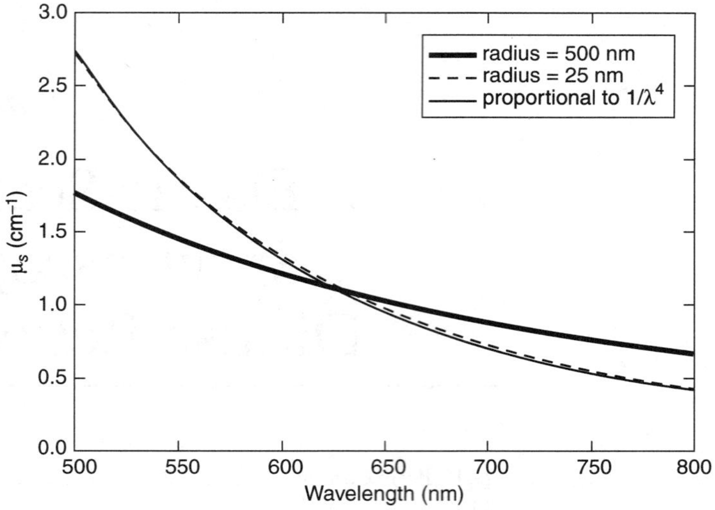


**FIGURE 29.1** The wavelength dependence of the scattering coefficient, $\mu_s$, for two sizes of spheres. The scattering coefficient, typically denoted by $\mu_s$, is the inverse of the mean free path between scattering events. The spheres had an index of refraction of 1.39 and were immersed in a medium of index 1.35. The concentration of spheres with a radius of 25 nm was 300 times greater than the concentration of spheres with a radius of 500 nm. All calculations were performed using a solution to Maxwell's equations for a plane wave incident on an object, in this case a sphere called Mie theory. (For a discussion of Mie theory, see Bohren, C.F. and Huffman, D.R., *Absorption and Scattering of Light by Small Particles*, Wiley-Interscience, New York, 1983.)

The full range of information available and its relationship to traditional histopathology are still under investigation. Most likely, noninvasive light scattering methods will provide a subset of traditional histopathologic characterization as well as some information not traditionally available. The propagation of light through tissue depends on absorption as well as scattering properties. The three primary light absorbing compounds in healthy, nonskin tissue are oxygenated hemoglobin, deoxygenated hemoglobin, and water. Absorption spectra of these chromophores are shown in Figure 29.2. Concentrations of these compounds can serve as important diagnostic criteria. The importance of hemoglobin oxygenation and the ability of optical systems to provide quantitative information have already been demonstrated by the pulsed oximeter, which is in widespread use.[1]

The basic geometry of the measurements described in this chapter is shown in Figure 29.3. Broad bandwidth light is incident on the tissue, typically through a flexible optical fiber with a diameter of 0.2 to 0.4 mm. This light travels into the tissue and can be absorbed and scattered. The absorption depends strongly on wavelength; it is greatest at visible wavelengths for which hemoglobin absorbs and much lower at near-infrared wavelengths between the hemoglobin and water absorption bands. Tissue is highly scattering, and the average distance between scattering events is on the order of 100 μm (0.1 mm). A small fraction of the light that is not absorbed will be scattered such that it can be collected by a probe on the surface near the original delivery location. Thus, the light collected and transmitted to the analyzing spectrometer has typically undergone several scattering events through a small volume of the tissue. No light is collected from surface reflection. The measurements typically take a fraction of a second and cause no damage to the tissue.

Systems that use light levels meeting FDA guidelines have been designed. The components of the optical system in Figure 29.3 have an intrinsic wavelength dependence; for example, silicon CCD detectors are more efficient at some wavelengths than at others. Consequently, the spectral dependence of the measurement system must be determined. Commonly, this is accomplished using a control sample that reflects equally at all wavelengths. An example of the elastic-scatter spectrum of tissue (from the esophagus), corrected for the wavelength dependence of the measurement system, is shown in Figure 29.4. The

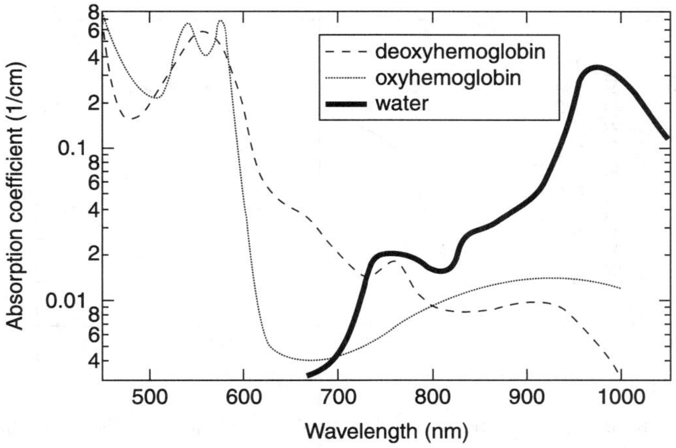


**FIGURE 29.2** Absorption spectra of oxyhemoglobin, deoxyhemoglobin, and water. The concentration of heme groups is taken to be 20 m*M*. The hemoglobin spectra are a combination of data taken by the authors and from van Assendelft, O.W., *Spectrophotometry of Haemoglobin Derivatives*, Royal Vangorcum Ltd., Assen, the Netherlands, 1970 and Wray, S., Cope, M., Delpy, D.T., Wyatt, J.S., and Reynolds, E.O.R., *Biochim. Biophys. Acta*, 933, 184, 1988. The absorption of water has been scaled by a factor of 0.7 to be consistent with the idea that tissue is about 70% water. The water spectra data are from Kou, L., Labrie, D., and Chylek, P., *Appl. Opt.*, 32, 3531, 1993.

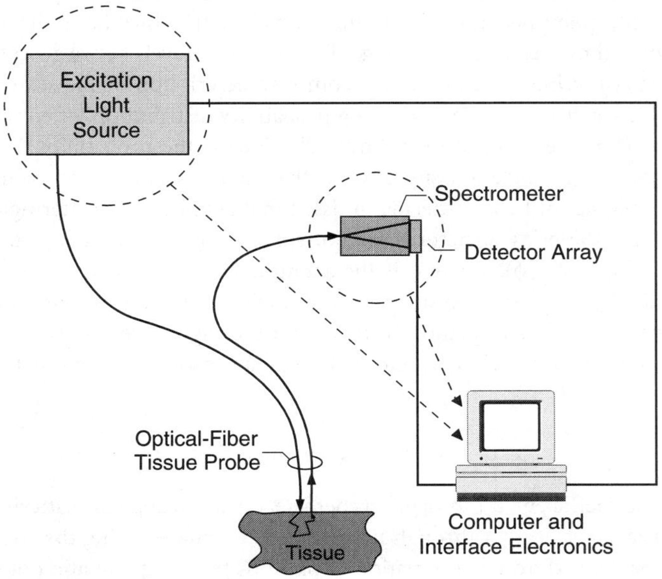


**FIGURE 29.3** A schematic of a system for measuring elastic scattering and diffuse reflectance. The excitation light source is typically a tungsten bulb or a mercury arc-lamp. The optical fibers are multimode, often with diameters of 200 μm, although larger diameter fibers are sometimes used. The spectrometer is commonly a CCD (charge-coupled device), although diode arrays have also been used.

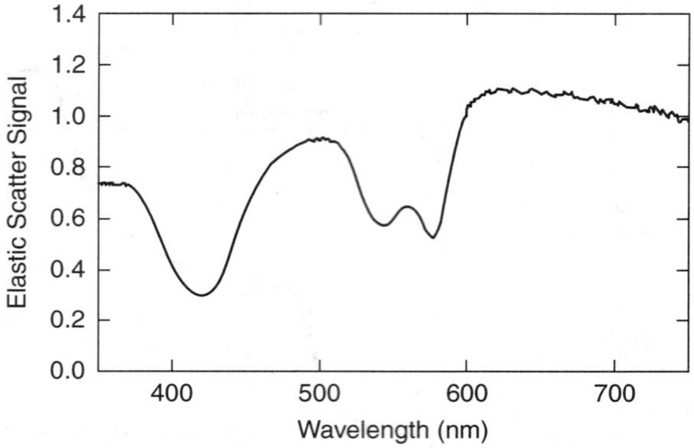


**FIGURE 29.4** *In vivo* elastic scatter spectra of Barrett's esophagus. Light was delivered to the tissue with a 400-μm diameter optical fiber and collected with an adjacent 200-μm diameter optical fiber.

dips at 419, 543, and 577 nm are due to absorption by hemoglobin. Past 650 nm and before water absorption becomes significant, the decreasing intensity is due to the decreased scattering probability with increasing wavelength. The distance the collected photons travel in the tissue is several times greater than the separation of the light delivery and detection points. Consequently, in addition to the scattering spectral sensitivity to microscopic tissue morphology, this type of system can have good sensitivity to the optical absorption bands of the tissue components.

Sensitivity to absorption increases as the source and detector are separated. Quantitative measurements of the compounds in Figure 29.2 as well as possibly other absorbing compounds, such as fat and melanin, may be made at the same time tissue structural properties are determined. The intensity and wavelength dependence of the elastic-scatter spectrum depend on several light transport properties. The likelihood of scattering is generally quantified as the scattering coefficient, $\mu_s$, which has units of inverse length. The scattering coefficient is the inverse of the average distance light travels between scattering events.

In addition to the scattering coefficient, $\mu_s$, a complete description of the scattering properties of a tissue requires knowledge of the angular scattering probability distribution, often written as $P(\theta)$, where the deflection angle, $\theta$, ranges from 0 to $\pi$. Physically, $P(\theta)$ is the probability that when a photon is scattered it is deflected by an angle $\theta$. Using $P(\theta)$ as the angular scattering probability distribution is a simplification of the physics of light scattering in tissue that assumes the scattering centers are spherical in shape. In tissue, the scattering structures are not spherically symmetric and the angular scattering probability is, in principle $P(\theta, \phi)$, where $\phi$ is the azimuthal angle, ranging from 0 to $2\pi$. To complete the description of the light transport properties of a medium, the absorption coefficient, $\mu_a$, must be known. The absorption coefficient quantifies the decrease in light intensity when light is absorbed. If light with an initial intensity $I_0$ travels a distance $L$ through an absorbing medium, the resultant intensity, $I$, is given by

$$I = I_0 \exp(-\mu_a L) \tag{29.1}$$

When the scattering coefficient, $\mu_s$, absorption coefficient, $\mu_a$, and angular scattering probability, $P(\theta,\phi)$ are all known, the transport of light through tissue can be calculated using the "transport equation."[2–4] (This equation can be derived from conservation of photons traveling into and out of a small volume.[2]) However, it can only be solved analytically for special geometries and consequently approximations are often used.

The most common approximation is the diffusion equation, which holds when the light collection is well separated from light delivery — typically a few centimeters.[5,6] (This approximation assumes that $P(\theta,\phi)$ and the light source term can be expanded in Legendre polynomials and only terms with $\ell \leq 1$ kept.) In this approximation, the light transport properties can be described by $\mu_a$ and the reduced scattering coefficient, $\mu_s' = \mu_s(1 - g)$, where $g = \langle\cos\theta\rangle$. The anisotropy parameter, $g$, is 0 when scattering

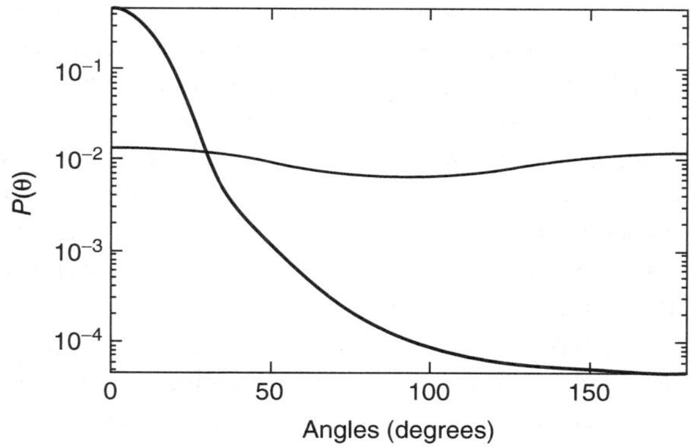


**FIGURE 29.5** The probability of scattering **s** to **s′**, where **s** is the incoming direction of propagation, **s′** is the direction of propagation after scattering, and $\cos\theta = \mathbf{s} \cdot \mathbf{s}'$. The thin line was calculated for spheres of 25-nm diameter with an index of 1.39 immersed in a medium of index 1.35 at 600 nm, $g = 0.025$. The thick line was calculated for spheres of 500-nm diameters with all other parameters the same, $g = 0.948$.

from an object is equally likely in the forward and backward directions and is near 1 when scattering is primarily forward directed. The reduced scattering coefficient, $\mu_s'$, is quite intuitive; it is greater for a substance that appears more scattering, such as whole milk compared to skim milk. In contrast, two scattering suspensions with the same scattering coefficients may appear quite different. A suspension in which scattering is forward directed will not appear to scatter light as strongly as a scattering medium in which scattering is equally likely in the forward and backward directions. In tissue, most light scattering is forward directed; in other words the deflection angle, θ, is on average quite small.

The identity of the morphological features that scatter light is not known with certainty and is an active area of research. Several structures have been proposed, including the cell membrane, the nucleus, mitochondria, and features within these and other organelles. The size, shape, and refractive index of tissue and cellular structural features determine how light is scattered. Figure 29.1 demonstrated how size affects the wavelength dependence of the scattering coefficient, $\mu_s$. The size shape and refractive index also strongly influence the angular dependence of scattering at each wavelength. For example, as shown in Figure 29.5, particles much smaller than the measurement wavelength scatter light equally in the forward and backward directions, while light scattering from particles with dimensions on the order of the measurement wavelength is primarily in the forward direction.

The intensity of the collected light depends strongly on $P(\theta)$ and consequently on details of tissue structure. When the light source and collection areas are adjacent[7] and especially if they overlap,[8] the intensity of collected light correlates with the probability of high angle scattering events. For these measurement geometries, the details of $P(\theta)$ are quite important. As the separation between the small area of light delivery and the small area of light collection is increased, the dependencies of the collected light on the probability of large angle scattering events and on the details of $P(\theta)$ decreases. In fact, at larger separations (centimeters), only $\mu_s'$ and $\mu_a$ are needed to predict light collection and the diffusion approximation becomes valid. In this chapter, we are only interested in measurements with small source–detector separations because that geometry provides the greatest sensitivity to details of light-scattering properties. A further advantage of small distances between light delivery and detection is that the measurements can be made using fiber optics through the working channel of an endoscope. This feature enables access to many of the epithelial layers of tissue where most adult cancers arise.

## 29.2 Clinical Studies

Several small-scale, *in vivo* clinical trials have demonstrated the potential of elastically scattered light to diagnose tissue pathologies. Results of these trials are commonly presented in terms of the sensitivity and specificity for detecting a specified abnormality. Sensitivity is defined as the percentage of abnormal samples that the test found to be abnormal, whereas specificity is the percentage of normal samples that the test found to be normal. These two metrics can sometimes provide a quantitative method for comparing studies; however, the numbers should be interpreted with care. In many of the early studies, separate training and testing data sets were not used. Consequently, the reported sensitivity and specificity may be higher than they would be if determined with an independent data set. Furthermore, the gold standard used for these studies, histopathology, is not perfect; therefore, a sensitivity of 100% means that the test agrees with the pathologists, not that 100% of the abnormal samples were detected.

The measurement methods used in the following studies varied significantly. Methods for optimizing sensitivity to tissue pathology based on a detailed understanding of the physics and biology of light scattering in tissue are discussed in subsequent sections. The clinical studies described here will hopefully motivate that discussion by demonstrating the potential of light transport to provide clinically relevant information.

Skin is the most accessible organ for testing light-scattering methods for cancer diagnosis. Consequently, the earliest work examined lesions of the skin. Research in this area is still motivated by a need for noninvasive methods to distinguish melanoma from benign pigmented lesions (nevi). The earliest work involved only a small number of *in vivo* measurements[9] and concentrated on the absorption properties of skin due to melanin and hemoglobin.[10,11] Attempts to understand these spectra have led to a detailed understanding of the optical properties of skin.[12–14] In studies with a greater number of patients, statistical methods were applied to determine the significance of the results. Measurements of 31 primary melanomas and 31 benign nevi were made using a modified integrating sphere with a standard ultraviolet–visible spectrophotometer over the spectral range of 420 to 780 nm. The data were used to develop discriminant functions. A sensitivity of 90.3% and a specificity of 77.4% were obtained for distinguishing the melanomas from nevi with leave-one-out cross validation.[15]

More recently, wavelength-dependent reflectance images of skin have been obtained.[16] Light that entered the tissue and was scattered back to the surface as well as reflection off of the tissue surface was measured, making interpretation of the multispectral images difficult. It was noted that images of benign lesions fade faster with increasing wavelength than do images of cutaneous melanoma. The analysis of the data, however, did not make use of the wavelength dependence of any features, but rather used properties of individual images, such as lesion size and border irregularity. A sensitivity of 80% and a specificity of only 51% were obtained on the training data set.

A recent study by V.P. Wallace et al. measured the diffuse-reflectance spectra of skin over the spectral range of 300 to 1100 nm without any artifacts due to surface reflectance.[17] These data were analyzed using multivariate discriminant analysis and artificial neural networks.[18] The best results were obtained with the artificial neural network and yielded a sensitivity of 83.6% and specificity of 85.3% for diagnosing melanoma compared to compound nevi for the training set. When the artificial neural network was applied to new cases, the sensitivity and specificity changed to 90.9 and 58.8%, respectively.

As with most cancers, the survival rate for bladder cancer increases with early detection. While most frank tumors can be detected by a clinician, some flat (early) cancerous lesions are indistinguishable from inflammation. One of the earliest endoscopic clinical studies of elastic-scattering spectroscopy involved ten patients undergoing examination for bladder cancer. Light was delivered using fiber optics with a center-to-center separation of about 350 μm at the distal end of the probe, which was placed in contact with the tissue. With knowledge of the pathology for each measured spectra, the slope from 330 to 370 nm found to be a diagnostic criteria with a sensitivity of 100% and a specificity of 97%.[19] Koenig et al.[20] have also tested scattering spectroscopy for sensing bladder cancer, but with distributed illumination of the tissue. They sensed the changes in hemoglobin absorption due to increased perfusion in neoplastic areas, but did not sense the spectral differences associated with structural changes. Consequently, the sensitivity was good (91%) but the specificity was poor (60%), because simple inflammation also causes increased perfusion. As with the previous study, the method has not been tested on a separate data set.

Several diseases of the gastrointestinal tract have been correlated with increased risk for the development of adenocarcinoma, including Barrett's esophagus, Crohn's disease (in the colon), and chronic ulcerative colitis of the colon. Consequently, clinical tests for diagnosis of cancer in the gastrointestinal tract have been performed. In an early study of 60 sites in the colon and rectum of 16 patients, a spectral metric was developed based on the regions of the hemoglobin absorption bands (400 to 440 nm and 540 to 580 nm). Eight sites diagnosed under histopathology as being dysplasia (a potentially premalignant condition), adenoma (premalignant growth), or adenocarcinoma were differentiated from sites of normal mucosa or more benign conditions such as hyperplastic polyps or quiescent colitis. The sensitivity of this metric on the set of data used to generate the metric was 100% with a specificity of 98%.[21]

A very clinically relevant study of the colon with both training and testing data sets has been published by Ge et al.[22] in which adenomatous polyps, which must be removed, and hyperplastic polyps, which do not need to be surgically resected, were distinguished. Three methods for spectral classification were tried: multiple linear regression, linear discriminant analysis, and neural network pattern recognition. The best predictive sensitivity and specificity were 89 and 75%, respectively, obtained using multiple linear regression. In a third study of the colon, diffuse reflectance spectra of 13 patients were measured and analyzed to obtain information on specific morphological and biochemical features.[23] These data were analyzed by applying the diffusion approximation to the transport equation in order to separate the effects of scattering and absorption. While the use of the diffusion approximation is possibly inappropriate because of the close proximity of the light detection and delivery fibers and may lead to systematic errors, this approach does have the advantage of yielding quantitative results. Hemoglobin concentration was significantly increased in the adenomatous polyps as compared to the normal mucosa, while hemoglobin oxygenation was essentially unchanged. Effective scatter density was decreased and effective scatterer size increased in the adenomatous polyps.

A few studies of light scattering for the diagnosis of dysplasia in patients with Barrett's esophagus have also been performed. M.B. Wallace et al. measured spectra at 76 sites in 13 patients.[24] These data were analyzed by a model assuming two components to the scattered light. One component is light that has been diffusely reflected and has an intensity that decreases monotonically with wavelength; the other is assumed to be an oscillatory component due to scattering from epithelial nuclei. After using eight samples to define the criteria for classifying the samples, the spectra were prospectively analyzed resulting in a sensitivity of 90% and a specificity of 90% for detecting dysplasia.

Bigio et al. measured spectra at 67 sites in 39 patients with Barrett's esophagus.[25] These data were analyzed using artificial neural networks and hierarchical cluster analysis. Sensitivity and specificity were calculated using 80% of the data for training, 20% for testing, and repeating this method five times until the sensitivity and specificity were determined for all of the data. The best results were obtained with hierarchical cluster analysis and resulted in an average sensitivity of 82% and an average specificity of 80%.

Light-scattering methods have also been combined with fluorescence to increase the accuracy of diagnosis of dysplasia in patients with Barrett's esophagus. In a study of 15 patients, the combined methods were able to separate dysplasia from nondysplastic Barrett's with a sensitivity and a specificity of 100% using leave-one-out cross validation.[26]

The advantages of early detection for cervical cancer have been demonstrated by the Papanicolaou (Pap) smear. However, the Pap smear has limitations such as sampling errors and a low sensitivity.[27] Nordstrom et al. have studied the ability of reflectance spectra to identify the stages of cervical intraepithelial neoplasia (CIN) in 41 patients.[28] Their measurements used flood illumination in a geometry that did not block the surface reflectance and consequently caused reduced sensitivity to the spectral differences associated with structural changes. They obtained a predictive sensitivity and specificity of 77 and 76%, respectively, for distinguishing CIN II/III from metaplasia.

Georgakoudi et al. demonstrated that when light-scattering methods are combined with fluorescence, the sensitivity and specificity for detecting squamous intraepithelial lesions increases substantially. Using pathology as the gold standard, and leave-one-out cross validation, their method had a sensitivity of 92% and a specificity of 71% for detecting squamous intraepithelial lesions vs. mature squamous epithelium or squamous metaplasia.[29] Richards-Kortum's group has investigated the application of reflectance spectroscopy to additional gynecological cancers. In an exploratory study of ovarian cancer with 18 patients, Utzinger et al. were able to separate ovarian cancers retrospectively from normal ovary and benign neoplasms with a sensitivity of 86% and specificity of 80%.[30]

Clinical studies have also addressed applications to assist in the diagnosis and management of breast cancer. If a reliable method of diagnosing lesions through a needle could be developed, thousands of lumpectomies could be avoided each year in the United States alone. In the surgical arena, elastic-scattering spectroscopy could potentially assist the surgeon in determining tumor margins. An *in vivo* study of 72 biopsy sites in 24 patients has been performed; 80% of the data was used for training and 20% for testing of an artificial neural network. This procedure was repeated five times with disjoint testing sets and the average sensitivity and specificity for separating cancerous breast tissue from noncancerous breast were calculated to be 69 and 85%, respectively.[31] Another important diagnostic criterion in the evaluation of breast cancer is the involvement of the sentinel node, which is the first lymph node that drains the part of the breast containing the tumor. In the same breast cancer study cited above, hierarchical cluster analysis was used to separate involved from noninvolved sentinel nodes with a sensitivity of 91% and a specificity of 77%.

Another potential clinical benefit of optical light-scattering measurements using small probes is identification of tissues during brain surgery. Johns et al.[32] investigated the use of scattering spectroscopy to distinguish gray matter from white matter *in vivo*. This identification is important for surgeries such as pallidotomy, a treatment for Parkinson's patients. For this treatment the boundary of the globus pallidus, a structure composed of gray matter surrounded by white matter, must be accurately determined. The ability of reflectance spectroscopy in combination with fluorescence spectroscopy to identify brain tumors and infiltrating margins *in vivo* has been investigated by Mahadevan-Jansen and co-workers.[33] Using a two-step empirical discrimination algorithm, they obtained a retrospective sensitivity of 100% and a specificity of 76%.

## 29.3 Increasing Sensitivity to Structures of Interest

The details of the measurement geometries employed in the studies described above varied. This variation is likely to have influenced the reported sensitivities and specificities. An active area of research has been, and continues to be, the development of measurement methods to enhance the ability of elastic light scattering to detect cancerous and precancerous changes. Most cancers originate in the epithelium, which is typically 100 to 300 μm thick. Consequently, methods that only probe this top layer of tissue are desired. The separation of the source and detector is one controlling factor for sampling the top layer. As the source and detector are moved closer together a more superficial region is probed. If the illumination and detection areas overlap, the sensitivity will be further restricted to the epithelium.

Some of the probes designed by the biomedical spectroscopy group at MIT have this quality. By placing a quartz wedge on the probe surface, they were able to have overlapping light delivery and collection without interference from surface reflectance.[23] The use of a single fiber with a beveled tip for light delivery and detection can also increase sensitivity to the top surface. Measurements with linearly polarized light can be used to reject some of the multiply scattered light from deeper than the tissue layer of interest. An example of linearly polarized light is light propagating perpendicular to and toward the tissue surface while the electric field is oscillating in a direction parallel to the tissue surface (such as parallel to a line connecting the source and detector fibers). If polarized light incident on tissue returns to the surface after only a few scattering events, the polarization is likely to be very similar to the incident light polarization. If light penetrates more deeply into the tissue, the polarization will be randomized.

Measurement systems can be set up for which the light incident on the tissue is linearly polarized and the intensities of light returning to the surface with the same polarization or a perpendicular polarization are separately collected. Subtraction of these two light intensities is a measure of scattering near the tissue surface. Jacques et al. have demonstrated that when imaging with polarized light, they could obtain images that only pertained to approximately the top 265 μm of skin.[34] Backman et al. used broad-wavelength, polarized light scattering to isolate light scattered from near the tissue surface. Measurements

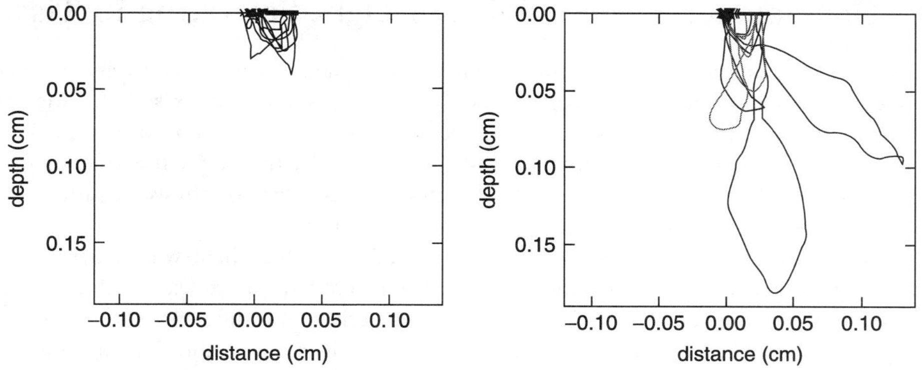


**FIGURE 29.6** Trajectories for light transport through tissue calculated using Monte Carlo simulations. For both simulations, the scatterers were 500-nm radii spheres at a concentration such that the scattering coefficient was 245 $cm^{-1}$ leading to a reduced scattering coefficient of 12.4 $cm^{-1}$ at 600 nm. For the graph on the left, there was no absorption; for the graph on the right absorption was 1 $cm^{-1}$. The entry points for all trajectories are marked with crosses that lie between –0.01 and 0.01 cm; a black bar marks the position of the collection fiber, 0.01 to 0.03 cm. The trajectories have been projected onto the plane containing the delivery and detection fibers. Many more photons were run in the Monte Carlo simulation. The average trajectory length for the case of no absorption was found to be 0.30 cm; when absorption was present, the average trajectory length was 0.17 cm.

were made individually of the light intensities scattered with polarizations parallel to or perpendicular to the incident light polarization, and the two intensities were subtracted.[35] Sokolov et al. have also made measurements of the diffuse reflectance of polarized light that are particularly sensitive to scattering near the surface.[36] In addition to localizing the measurement volume to the epithelium, these techniques are reported to provide quantitative characteristics of nuclei and are discussed in more detail in Section 29.6.

A third way to focus measurements primarily on the epithelium and not the underlying tissue is to make measurements of tissue morphology using wavelengths of light that do not penetrate deeply into tissue. When diffuse reflectance measurements are made at wavelengths that strongly absorb light, a shallower region of tissue is probed and the distance traveled in tissue is reduced. These results are illustrated in Figure 29.6, which shows typical trajectories for collected photons, as calculated by a Monte Carlo simulation. (This technique is described later in the chapter.) Representative trajectories for light transport with absorption are plotted on the left, while trajectories with no absorption are plotted on the right. For the case of absorption, the trajectories are quite short. In the case of no absorption, there are very few short trajectories; most are medium to long in length. By keeping the statistics on a large number of collected photons, the average pathlength for the case of no absorption was found to be 0.30 cm, and when absorption was present the pathlength was 0.17 cm. Similarly the median depth of scattering interactions changed from 0.062 to 0.037 cm. Light cannot penetrate deeply into tissue at wavelengths for which hemoglobin absorbs strongly. Therefore, if methods can be developed for measuring tissue morphology and architecture with only a single wavelength, then measurements that probe different tissue depths can be potentially achieved by using different wavelengths.

Techniques that use linearly polarized light, in a geometry different from that described above, are being developed to quantify the effective size and concentration of scattering centers in cells. This work will be described in more detail in Section 29.6. The basic physics of polarized light propagation in scattering media is still under active investigation. Yao and Wang have presented time-resolved movies of polarized light propagation in turbid media.[37] The preservation of polarization in lipid, myocardium, and polystyrene spheres has been studied in detail,[38,39] and methods for examining light scattered from tissues in the exact backscattering direction have also been developed.[40]

## 29.4 Understanding the Origins of Light Scattering in Tissue

The development of measurement systems to sense small changes in tissue and cellular architecture can be facilitated by an understanding of what structures in tissue are responsible for the scattering of light. In particular, the development of measurement systems may be improved with knowledge of the properties of light scattered from specific structures. Furthermore, the analysis of diffuse reflectance measurements, in terms of particular tissue or cellular structures, requires an understanding of scattering properties for specific architectural features.

Light scattering occurs at refractive index boundaries. As examples, light will scatter off an interface between two media of different indices, or from a sphere with an index of refraction different from its surrounding medium. Tissue is composed of materials with different refractive indices. For example, membranes have a higher index than water and will scatter light. Similarly high concentrations of DNA or protein can also scatter light. A wide variety of physical structures contain high concentrations of lipid, protein, or DNA within mammalian cells. Nuclei are on the order of 5 to 10 μm in diameter;[41] mitochondria, lysosomes, and peroxisomes have dimensions on the order of 0.2 to 2 μm.[42] Ribosomes are on the order of 20 nm in diameter[43] and structures within various organelles can have dimensions up to a few hundred nanometers. Refractive indices for some of these organelles have been compiled.[44] In liver tissue, mitochondria have been demonstrated to be a significant source of scattering.[45,46] Also for liver tissue, Schmitt and Kumar demonstrated that the spectrum of index variations exhibits a power-law behavior for a wide range of spatial frequencies.[47]

Models of tissue having a broad range of discrete particle sizes have been proposed; calculated scattering properties compared well with those available in the literature.[48] Further evidence for a broad size range of architectural features contributing to tissue light scattering comes from experimental measurements of light scattering from epithelial cells. Because the epithelium is primarily a cellular structure without much intracellular material, studying scattering from only the cells is quite relevant. The size distribution of refractive-index structure variations in epithelial cells was found to include particles with effective radii from smaller than a few hundred nanometers to ~2 μm. There are orders of magnitude more structures the size of macromolecules than there are particles the size of organelles scattering light.[49] Consistent with results that most scatterers are small, cell shape and cell–cell contact were found to have only very minor effects on light scattering.[50] Furthermore, the result that much of the light scattering occurs from structures a few hundred nanometers or less in radius implies that light scattering is likely to be sensitive to some structures smaller than those commonly investigated by standard pathology methods.

As discussed earlier, the importance of high angle scattering events for determining the intensity of collected light is increased when the separation between the illumination and light collection locations is reduced. Therefore, changing the source–detector separation may change the contribution of different structures to the elastic-scatter signal, if the contribution of various structure sizes to light scattering varies with scattering angle. There is evidence that different physical structures within cells are responsible for scattering at different angles. Cell size can be determined by measurements of scattering in the angle range of $\theta = 0.5$ to $1.5°$.[51] Scattering at angles between about 2 and 25° has been attributed to scattering from the nuclei. Measurements of Chinese hamster ovary (CHO) cells showed a fine structure between $\theta = 2.5$ and 25° that could be modeled as a coated sphere, i.e., a nucleus surrounded by cytoplasm, although for HeLa cells no fine structure was observed.[52]

Both experimental and modeling work indicate that small internal structures have a strong influence on the scattering pattern at angles greater than 40°.[44,53] Consequently, a decreased source–detector separation may increase the relative contribution of small structures to the measured diffuse reflectance.The contribution of the nucleus to light scattering from biological cells has been a topic of investigation by several groups because of the known changes in the nucleus accompanying carcinogenesis, including an increased nuclear-to-cytoplasmic ratio and changes in nuclear structure, one of which is apparent in histopathology as hyperchromaticity (dark irregular staining of DNA).

For fibroblast cells, less than 40% of the scattered light at any angle was determined to have been scattered from the nucleus.[53] Finite difference time-domain methods, which can calculate scattering from objects with very inhomogeneous index of refraction structures, have been used to study potential contributions of nuclei to light scattering.[44,54] Consistent with the idea that the nuclei contribute significantly to scattering at angles less than 25°, increasing the size of the nucleus increased the forward scattering.

If the nucleus is heterogeneous, high angle scattering also increases with nuclear size.[44] There is experimental evidence for nuclear heterogeneity. A determination of the size distribution of scattering centers within nuclei isolated from epithelial cells found that the size of scattering structures ranged from less than a few hundred nanometers to 2 μm.[49] Not only is the physical composition of biological cells very heterogeneous, but so also is the composition of the nuclei within them. A correlation between DNA content and light-scattering properties has been observed. When cells are growing exponentially, their DNA content is increased compared to cells in the plateau phase of growth. Light scattering measurements of cells harvested in the exponential phase of growth showed stronger backscattering and a steeper decrease in scattering as a function of wavelength than did cells harvested in the plateau phase of growth.[53] Both these scattering changes are indicative of an increased concentration of small scattering centers.

Nearly all of the research described here focused on understanding scattering from cells. Scattering from tissue layers underlying the epithelium has not been as well studied. Mechanical support for most epithelial layers is provided by the lamina propria, a thin layer of connective tissue with a large collagen content. Muscle tissue commonly underlies the lamina propria. Studies of skin provide a small amount of information relevant to the lamina propria because skin also contains a large amount of collagen, although the organization of collagen fibers in the lamina propria varies with the organ studied and, therefore, may be different from the organization of collagen fibers in skin. The large collagen fibers from the dermis of the skin can be one of the primary sources of scattering from skin; microscopic changes in collagen fibers have been correlated with macroscopically observed light-scattering changes.[55] Collagen fibers can also cause an anisotropy in the macroscopically measured light scattering. The measured reduced scattering coefficient of the skin can vary by a factor of two depending on the measurement orientation.[56] Muscle also demonstrates large scale anisotropy, particularly with polarized light.[57]

Several microscopic methods for examining the spatial variations of the angular scattering probability, $P(\theta)$, are under development. A microscope-based, light-scattering instrument that can measure the angular dependent scattering of an 80-μm spot from 0 to 60° has been developed.[58] The ratio of wide angle to narrow angle scatter can be examined at microscopic resolution;[59] an interferometric method for measuring of scattering at angles near 180° has been demonstrated[60] and could potentially be integrated into a microscopic imaging system.

## 29.5 Monte Carlo Methods

An important tool for developing and understanding elastic-scattering spectroscopy has been Monte Carlo simulation of light transport.[61,62] In a Monte Carlo simulation, a photon is injected into a scattering medium and then propagated through the medium based on knowledge of $\mu_s$, $\mu_a$, and $P(\theta, \phi)$ for the scattering medium. Photon propagation is then simulated, typically, for millions of photons. A common assumption is that the scattering centers are spherical. In this case, all $\phi$ are equally likely and $P(\theta)$ can be calculated using a solution to Maxwell's equations for a plane wave incident on a sphere, called Mie theory.[63] Henyey–Greenstein phase functions, which resemble phase functions computed by Mie theory, are sometimes used.[64] The advantage of Henyey–Greenstein phase functions is their simple analytical form; however, they cannot be used in simulations of polarized light scattering or to simulate scattering from nonspherical particles. In principle, the scattering centers could be ellipsoids of revolution or other particles because computational codes for calculating scattering parameters from these shapes are now available.[65] In fact, the evidence is that, when polarized light propagation in tissue is simulated, a model using spherical particles will not be adequate. Ellipsoids of revolution or other nonspherical particles must be used because the physical structures in epithelial cells do not scatter light like spherical scatterers.[49]

Monte Carlo simulations are particularly useful for determining features of light transport that are not easily measured, such as the depth to which light penetrates into tissue before being collected, the length of the path the light travels in tissue, and the polarization of light within the tissue. The degree of polarization in a scattering medium as a function of depth and lateral distance from the delivery point can be seen in simulations by Yao and Wang.[37] The depth probed as a function of the incident and detected light polarization for fixed source-detector separations has also been examined using Monte Carlo simulations.[66] The depth probed depends on the collected polarization relative to the incident polarization. As expected from the discussion in Section 29.3, a deeper depth can be probed when the collected light is polarized perpendicular to the incident light polarization. A less intuitive result is that the depth probed when the incident and collected polarizations are the same depends on the measurement geometry.

Consider two collection fibers, each 550 μm from the light delivery fiber, such that a line connecting the first collection fiber to the delivery fiber and a line connecting the second collection fiber to the delivery fiber form a right angle. If the polarization for delivery and collection is oriented along one of the lines, then the depth probed by the fiber connected by that line will be greater. For example, for $\mu_s' = 15$ cm$^{-1}$, $\mu_a = 2$ cm$^{-1}$, and $g = 0.61$, the median depth probed is 230 μm for one of the collection fibers and ~ 300 μm for the other. Regardless of polarization, the depth probed decreases with fiber separation. The depth probed when using a single fiber-optic for delivery and for detection of unpolarized light is 170 μm for a similar highly scattering medium with $\mu_s' = 15$ cm$^{-1}$, $\mu_a = 2$ cm$^{-1}$, and $g = 0.83$.[8] (There is an error in Table 2 of Reference 8: cm$^{-1}$ should be mm$^{-1}$.)

Monte Carlo simulations indicate that, with a fiber optic probe separation of 250 μm, primarily the epidermal layer and papillary dermis of skin are probed.[67] Monte Carlo simulations also make clear that the distance photons travel in tissue is much greater than the depth they probe. When the mean path length within myocardium for diffusely reflected light is 1.3 mm, the depth examined is only ~365 μm.[68] Measurements of the average distance traveled by light in tissue between the delivery and collection fiber have been calculated for a few different sets of light transport parameters.[69] For example, when the source and detector are separated by 1.75 mm and the absorption is 0.01 mm$^{-1}$, the distance traveled by light in the tissue is about 10 mm. As discussed in Section 29.7, at this fiber separation, the distance light travels between the source and detector does not depend strongly on the scattering properties. Finally, the distribution of scattering angles of the scattering events undergone by light that reaches the collection fiber can be calculated using Monte Carlo simulations. When the source and detector are in close proximity, e.g., 200 or 250 μm apart, the collected photons undergo significantly more high angle scattering events than do photons that entered the tissue, but were not collected.[7,70] Monte Carlo methods are also used for the determination of quantitative optical properties from experimental measurements as described in Section 29.6.

## 29.6 Quantification of Morphological and Biochemical Properties

An ultimate goal of several research groups is to provide quantitative biochemical and morphological information from noninvasive optical measurements of tissue. Georgakoudi et al. provide a review of much of the work in this area using a wide variety of techniques (see Chapter 31 of this handbook). Here, we discuss methods aimed at obtaining quantitative information from wavelength-dependent, light-scattering measurements.

In order to develop methods for quantifying optical properties or specific morphological parameters, it is necessary to have well-characterized media for which the optical and morphological properties are known. Tissue is much too complicated to use and requires special handling procedures. Consequently, tissue phantoms are frequently used. Two common phantoms are intralipid and suspensions of polystyrene spheres. Intralipid is a fat emulsion consisting of soybean oil, glycerin, lecithin, and water.[71,72] Small vesicles formed primarily of soybean oil scatter light; polystyrene spheres can be used to make tissue phantoms with precisely known properties. Measurements of these phantoms can be compared directly to theory. The disadvantages of phantoms made with polystyrene spheres are that the relative index of polystyrene in water is very different from that of biological materials in water and the broad distribution of structure sizes found in tissue is hard to reproduce with polystyrene spheres. Absorbing materials, such as oxy- or deoxyhemoglobin can be added to intralipid of polystyrene sphere suspensions and both types of phantoms have proven very useful in the development of quantitative methods.

When the source and detector are sufficiently separated, the diffusion approximation to the transport equation can be applied and the absorption coefficient, $\mu_a$, and reduced scattering coefficient, $\mu_s'$, calculated. Initially, this method was demonstrated to measure the absorption of methemoglobin[73] using a frequency-domain system — a more intricate measurement system that oscillates the incident light.[74,75] The addition of broad-wavelength, steady-state techniques to frequency-domain measurements performed only at a few wavelengths can improve the accuracy of absorption measurements.[76] Steady-state illumination alone can be used[77–81] and the concentration of cytochrome and hemoglobin quantified in scattering suspensions containing mitochondria and red blood cells.[81] A probe with an oblique angle of light incidence has also been developed for separately measuring absorption and the reduced scattering coefficient using a steady-state source.[82,83]

When the source–detector separation, $\rho$, and the reduced scattering coefficient, $\mu_s'$, have values such that their product, $\rho\mu_s'$ is less than about 2, the diffusion approximation is no longer valid.[7] This limit of applicability of the diffusion solution depends on several additional factors such as absorption, numerical aperture of the fiber, and details of $P(\theta)$.[84] Consequently, different researchers have found different limits for the validity of the diffusion approximation. The value of 2 stated above for the source-detector separation is smaller than values obtained and discussed by Bevilacqua and Depeursinge.[85]

Modifications to diffusion equation solutions have been made in order to increase their accuracy.[86] Venugoplan et al. have presented an extension of the standard diffusion approximation that shows improved agreement with experimental results at small fiber separations and high absorption.[87] A different improvement on the diffusion approximation is the $P_3$ approximation. For the diffusion approximation, the radiance and source term of the Boltzmann transport equation are expanded in spherical harmonics and terms up to $\ell = 1$ are kept. The $P_3$ solution keeps terms up to $\ell = 3$. This solution demonstrates significantly improved results for the determination of optical properties, particularly for the determination of large absorption coefficients (>0.25 mm$^{-1}$) when light transport is measured with relatively small (<1 cm) source–detector separations.[88]

Multiple polynomial regression methods[89] and an iterative Monte Carlo method[90] have been proposed as methods to more accurately determine tissue optical properties using source–detector separations less than 1 cm. An iterative Monte Carlo method has been applied to hyperspectral data. When reflectance data measured over the range of 0.5 to 2.5 mm from the source location were used, the accuracy of calculated $\mu_a$ and $\mu_s'$ were ±12 and ±4%, respectively.[91] A fast perturbative Monte Carlo method can accurately quantify changes in $\mu_a$ or $\mu_s'$. In test cases using layered media with a top layer thickness of only 0.5 mm and detectors up to 3 mm from the source, excellent results were obtained for $\mu_a$ and $\mu_s'$ perturbations in each layer.[92]

Many of the papers cited in the previous paragraph used Henyey–Greenstein phase functions. When $\rho\mu_s'$ is small enough that the diffusion approximation does not hold, the intensity of collected light will depend on details of the phase function. For example, with $\rho\mu_s' = 2.4$, Monte Carlo simulations of photon transport demonstrate that the intensity of light collected can be 60% greater when a Mie phase function is used rather than a Henyey–Greenstein phase function, assuming the other scattering parameters, $g$, $\mu_s$, and $\mu_a$, are held constant.[70] Bevilacqua and Depeursinge[85] have determined the effect of the moments of $P(\theta)$ on light transport. The first moment is $g$; the second moment is given by

$$g_2 = \int_0^{\pi} \tfrac{1}{2}(3\cos^2\theta - 1)P(\theta)\sin\theta d\theta \qquad (29.2)$$

An important result from their work is that, from $\rho\mu_s' > 0.5$ up to values for which the diffusion equation holds, light transport can be described by three parameters: $\mu_s'$, $\mu_a$, and $\gamma = (1 - g_2)/(1 - g_1)$.[85] A procedure to determine $\mu_s'$, $\mu_a$, and $\gamma$ from diffuse reflectance data, taken at several fiber separations from 0.3 to 1.4 mm, resulted in values of $\mu_s'$ with errors <10% and values of $\mu_a$ accurate to approximately 0.01 mm$^{-1}$.[93]

So far the quantification of morphological and biochemical properties has been stated as the quantification of light transport properties. As an alternative to quantifying $\mu_s'$ or $\mu_a$ and the properties of the angular dependent scattering function, it may be useful to determine the effective size of scattering structures and their concentrations directly. Methods using polarized light are being developed to quantify morphological properties directly. At least two separate approaches are being taken. One method is based on the observation of interesting optical patterns generated by the propagation of polarized light many years ago. A crosslike pattern was observed over the macular area of the eye when the macula was photographed using crossed polarizers as early as 1978.[94] The physical origin of pattern features is well understood[95–99] and methods for obtaining effective scatterer size and concentration from polarized backscattering images of polystyrene sphere suspensions have been described.[100] Recently a fiber-optic method for measuring polarized light scattering that can determine average scatterer size and density of polystyrene spheres was demonstrated.[66] Polarized fiber-optic and imaging measurements of cell suspensions also provide information about scattering structures.[66,101,102] For example, the apparent size of scattering centers in tumorigenic MR1 rat fibroblast cells is greater than the apparent size of scattering centers in nontumorigenic M1 rat fibroblast cells as determined by polarization images obtained with circularly polarized light.[102]

The other method, mentioned in Section 29.3, subtracts measurements of light returning to the same location on the surface with polarizations parallel and perpendicular to the incident light polarization. When the subtraction is performed as a function of wavelength, the resulting spectrum can be fit to a theoretical expression for scattering from spheres. Backman et al. have assumed that this spectrum is due to backscattering from nuclei, and they report that the distribution of cell-nucleus sizes obtained agrees with microscopic measurements of nuclear size.[35] This technique has also been implemented in an imaging geometry, for which the analyses to determine nuclear size and refractive index are performed for each 25 μm × 25 μm pixel.[103]

Sokolov et al.[36] have also made measurements of the diffuse reflectance of polarized light. Specifically, they measured the depolarization ratio,

$$[I_{||}(\lambda) - I\perp(\lambda)]/(I_{||}[\lambda) + I\perp(\lambda)],$$

for epithelial cells in phosphate buffered saline above a highly scattering medium. They assumed a cell in phosphate buffer consisted of two independent scatterers: a spherical nucleus in an environment with the refractive index of cytoplasm and a spherical cytoplasm in an environment with the refractive index of water. Additionally, their model assumes that the wavelength-dependent scattering is a linear sum of forward and backward scattering from these spheres. With this model they were able to obtain wavelength-dependent fits to the data using sphere sizes similar to those obtained for the cells and nuclei by phase-contrast microscopy.

## 29.7 Other Applications

In addition to the diagnosis and detection of cancer, elastic-scattering spectroscopy may be used for diagnosing other tissue pathologies, for tissue identification, and for monitoring changes to tissue. Freezing of tissue is used in cancer treatment (cryosurgery) and in tissue preservation and transplantation. For these applications, it is important to know the location of the ice-front boundaries. Benaron and co-workers found that there is a large overall intensity change in the transmitted light upon freezing.[104] The opposite of cryosurgery, thermotherapy, also affects the optical properties of tissue. There is an increase in absorption, probably due to accumulation of erythrocytes and a decrease in the size of scattering centers.[105]

Applications of reflectance spectrometry have also been developed for measurement of bilirubin concentration[106,107] and a small hand-held device is currently on the market. One of the difficulties with measuring absorption in tissue is that the distance traveled by light between the source and detector is not known and depends on the scattering properties, which are generally also not known. In order to mitigate this problem, an illumination–detection separation can be chosen for which the effect of scattering variations on the pathlength is weak. Mourant et al. have demonstrated that, for a source–detection separation of ~1.7 mm, the variation in pathlength for a large range of scattering properties found in tissue (5 to 20 $cm^{-1}$) is less than about 20%.[69] This result is being further developed to measure the pharmacokinetics of investigational drugs. The basis of the technique is a simple manipulation of Equation 29.1, given by

$$\log[I(t)/I_0] = \mu_a L \tag{29.3}$$

where $I(t)$ is the intensity of light measured at time $t$ after a drug was administered, $I_0$ is the intensity of light measured before the drug was administered, $\mu_a$ is the absorption coefficient, and $L$ is the distance light travels through the tissue. By measuring with a geometry (shown in Figure 29.3) using a distance of ~1.7 mm between the illuminating and collecting fibers, the dependence of pathlength on variations in scattering coefficient can be substantially ignored. However, as mentioned in Section 29.3, the pathlength will still depend on absorption. The dependence of $L$ on $\mu_a$ can be parameterized as $L = x_0 + x_1\exp(-x_2\mu_a)$, where $x_0$, $x_1$, and $x_2$ are parameters that depend on details of the measurement probe, such as the exact fiber separation and the numerical aperture of the fibers.

The measurement of drug concentrations in tissue as a function of time using the principles described previously has been named "optical pharmacokinetics" and a proof of principle demonstration of optical pharmacokinetics has been published.[108] The concentration of mitoxantrone (a chemotherapy drug) was followed as a function of time in subcutaneous human cancer tumors grown in nude (immune-suppressed) mice. The conventional method used for pharmacokinetics studies during drug development is animal sacrifice followed by tissue assay. This method is time- and labor-intensive as well as quite costly in terms of laboratory animals required — statistically significant studies require hundreds of animals (multiple animals for each time-point for each dose tested). An alternative method of measuring pharmacokinetics sometimes used in hospitals, microdialysis, is also invasive and utilizes a needle with a semipermeable membrane that may irritate surrounding tissue. Another drawback of microdialysis is that intracellular concentrations cannot be determined. Intracellular concentrations are more significant for understanding the therapeutic dose of a drug than the intercellular concentration.

Single photon emission computed tomography and positron emission tomography have been used to quantify *in vivo* imaging of radiolabeled drugs.[109–111] These techniques do not depend on the optical properties of tissue; however, they do require handling radioactive compounds and using expensive equipment. Also, the spatial resolution of the imaging is limited and subsecond imaging of fast drug dynamics is likely to be difficult. In contrast to animal sacrifice/tissue assay or microdialysis, optical pharmacokinetics is noninvasive and can provide real-time, site-specific measurements of absolute drug concentrations *in vivo*. Measurements over time can determine the entire pharmacokinetic time history (for a given dosage) of a drug at specific tissue sites with just one laboratory animal.

Optical measurements can also be used to assess properties of tumor neovasculature. A fast sequence of measurements made immediately following intravenous administration of a short bolus of optical contrast agent can be used to determine important parameters of angiogenesis. Because tumor capillaries (particularly those that are still immature) in the angiogenesis process are leaky,[112] macromolecules that do not leak from normal capillaries will often pass through the permeable walls of angiogenic microvessels and can be used to quantify the leakiness of these capillaries and, therefore, the angiogenesis process.[113] Very rapid sequences of optical pharmacokinetics measurements (>2 measurements/second) would permit tracking the fast dynamics of the first-pass kinetics of compounds in neovascularized tissue, and the leakage into the extracellular fluid space, following administration of a short bolus of an optical contrast agent with a known leakage rate.

Thus, possible clinical applications of optical pharmacokinetics include monitoring photodynamic therapy drug concentrations so that the drug can be activated by light at the optimum time, measuring blood pooling, and using first-pass kinetics to monitor neovasculature and assess the response to treatment by antiangiogenic drugs.

## 29.8 Summary

Elastic-scattering spectroscopy and diffuse reflectance are two terms for noninvasive methods of measuring the propagation of light through tissue to obtain biochemical and morphological information that can be used in diagnosing tissue pathologies. Light is incident on a small area, typically a spot 200 to 400 μm diameter, and enters the tissue where it is scattered or absorbed. Some of the light will return to the surface where properties such as intensity vs. wavelength and polarization can be measured. The properties of the light returning to the surface will depend on some of the structural and biochemical properties of the tissue. Clinical trials using elastic scattering and diffuse reflectance have demonstrated the potential of light scattering to diagnose cancerous and precancerous lesions. For many of these trials, the same set of data was used to calculate the sensitivity and specificity as was used to develop the distinguishing metrics. More prospective clinical studies that have independent testing data sets are now needed to prove the diagnostic accuracy.

The full power of elastic scattering and diffuse reflectance, however, has not yet been realized, partly because the relationship between the biology of the tissue and the physics of light propagation is not completely understood. Such an understanding will facilitate optimization of measurement methods and analysis to provide the most accurate and clinically relevant information possible. Significant progress has been made towards understanding which structural features in tissue are responsible for light scattering and the contribution of light scattering from these structural features to the properties of the detected light intensity.

New methods for quantifying physical scattering properties and biomedically important properties, such as nuclear characteristics or drug metabolism, are being developed. At present few of the methods have been implemented by more than one or two research groups and there are some apparent contradictions in results. For example, the idea that nuclear size can be determined from backscattering measurements may appear to be at odds with evidence that backscattering occurs primarily from structures much smaller than the nucleus and that the nucleus contributes primarily to forward scattering. It is important that published work include the details of measurements and analysis so that other research groups can validate, reconcile, and expand on results. The development of new analysis and measurement methods, along with new technical developments and ideas, will enable diffuse reflectance and elastic-scattering spectroscopy to reach their potentials as tools for the noninvasive characterization of tissue. Information about tissue presently only available by biopsy and histopathology, as well as new information not available via traditional methods, may be provided by light-scattering methods. We hope that in the coming years elastic scattering and diffuse reflectance will be developed into accurate, real-time, noninvasive, and, possibly, hand-held tools for measuring diagnostically relevant characteristics of tissue.

## Acknowledgments

Support for researching and writing this chapter was provided by NIH CA17898 and CA82104. The authors thank James P. Freyer, Paul W. Fenimore, and Toru Aida for comments on the manuscript.

## References

1. Sinex, J.E., Pulse oximetry: principles and limitations, *Am. J. Emergency Med.*, 17, 59, 1999.
2. Case, K.M. and Zweifel, P.F., *Linear Transport Theory*, Addison-Wesley, Reading, MA, 1967, chap. 1.
3. Ishimaru, A., *Wave Propagation and Scattering in Random Media*, Oxford University Press, Oxford, U.K., 1997, chap. 2.
4. Groenhuis, R.A.J., Ferwerda, H.A., and Ten Bosch, J.J., Scattering and absorption of turbid materials determined from reflection measurements. 1. Theory, *Appl. Opt.*, 22, 2456, 1983.
5. Farrell, T.J. and Patterson, M.S., A diffusion theory model of spatially resolved, steady-state diffuse reflectance for the noninvasive determination of tissue optical properties, *Med. Phys.*, 19, 8798, 1992.
6. Haskell, R.C., Svaasand, L.O., Tsay, T.-T., Feng, T.-C., McAdams, M.S., and Tromberg, B.J., Boundary condition for the diffusion equation in radiative transfer, *J. Opt. Soc. Am.*, 2727, 1994.
7. Canpolat, M. and Mourant, J.R., High-angle scattering events strongly affect light collection in clinically relavent measurement geometries for light transport through tissue, *Phys. Med. Biol.*, 45, 1127, 2000.
8. Canpolat, M. and Mourant, J.R., Particle size analysis of turbid media with a single optical fiber in contact with the medium to deliver and detect white light, *Appl. Opt.*, 40, 3792, 2001.
9. Kollias, N. and Baqer, A.H., Quantitative assessment of UV-induced pigmentation and erythema, *Photodermatology*, 5, 53, 1988.
10. Dawson, J.B., Barker, D.J., Ellis, D.J., Grassam, E., Cotteril, J.A., Fisher, G.W., and Feather, J.W., A theoretical and experimental study of light absorption and scattering by *in vivo* skin, *Phys. Med. Biol.*, 25, 6969, 1980.
11. Feather, J.W., Hajizadeh-Saffar, M., Leslie, G., and Dawson, J.B., A portable scanning reflectance spectrophotometer using visible wavelengths for the rapid measurement of skin pigments, *Phys. Med. Biol.*, 34, 807, 1989.
12. Anderson, R.R. and Parrish, J.A., The optics of human skin, *J. Invest. Derm.*, 77, 13, 1981.
13. Hajizadeh-Saffar, M., Feather, J.W., and Dawson, J.B., An investigation of factors affecting the accuracy of measurements of skin pigments by reflectance spectrophotometry, *Phys. Med. Biol.*, 35, 1301, 1990.
14. Saidi, I.S., Jacques, S.L., and Tittel, F.K., Mie and Rayleigh modeling of visible-light scattering in neonatal skin, *Appl. Opt.*, 34, 7410, 1995.
15. Marchesini, R., Cascinelli, N., Brambilla, M., Clemente, C., Mascheroni, L., Pignoli, E., Testori, A., and Ventroli, D.R., *In vivo* spectrophotometric evaluation of neoplastic and non-neoplastic skin pigmented lesions. II. Discriminant analysis between nevus and melanoma, *Photochem. Photobiol.*, 55, 515, 1992.
16. Farina, B., Bartoli, C., Bono, A., Colombo, A., Lualdi, M., Tragni, G., and Marchesini, R., Multispectral imaging approach in the diagnosis of cutaneous melanoma: potentiality and limits, *Phys. Med. Biol.*, 45, 1243, 2000.
17. Wallace, V.P., Crawford, D.C., Mortimer, P.S., Ott, R.J., and Bamber, J.C., Spectrophotometric assessment of pigmented skin lesions: methods and feature selection for evaluation of diagnostic performance, *Phys. Med. Biol.*, 45, 735, 2000.
18. Wallace V.P., Banber, J.C., Crawford, D.C., Ott, R.J., and Mortimer, P.S., Classification of reflectance spectra from pigmented skin lesions, a comparison of multivariate discriminant analysis and artificial neural networks, *Phys. Med. Biol.*, 45, 2859, 2000.
19. Mourant, J.R., Bigio, I.J., Boyer, J., Conn, R.L. Johnson, T., and Shimada T., Spectroscopic diagnosis of bladder cancer with elastic light scattering, *Lasers Surg. Med.*, 17, 350, 1995.
20. Koenig, F., Larne, R., Enquist, H., McGovern, F.J., Schomacker, K.T., Kollias, N., and Deutsch, T.F., Spectroscopic measurement of diffuse reflectance for enhanced detection of bladder carcinoma, *Urology*, 51, 342, 1998.

21. Mourant, J.R., Bigio, I.J., Boyer, J., Johnson, T.M., and Lacey, J., Elastic scattering spectroscopy as a diagnostic for differentiating pathologies in the gastrointestinal tract: preliminary testing, *J. Biomed. Opt.*, 1, 1, 1996.
22. Ge, Z., Schomacker, K.T., and Nishioka, N.S., Identification of colonic dysplasia and neoplasia by diffuse reflectance spectroscopy and pattern recognition techniques, *Appl. Spec.*, 52, 833, 1998.
23. Zonios, G., Perelman, L.T., Backman, V., Manahoran, R., Fitzmaurice, M., Van Dam, J., and Feld, M., Diffuse reflectance spectroscopy of human adenomatous colon polyps *in vivo, Appl. Opt.*, 38, 6628, 1999.
24. Wallace, M.B., Perelman, L.T., Backman, V., Crawford, J.M., Fitzmaurice, M., Seiler, M., Badizadigan, K., Shields, S.J., Itzkan, I., Dasari, R.R., Van Dam, J., and Feld, M.S., Endoscopic detection of dysplasia in patients with Barrett's esophagus using light-scattering spectroscopy, *Gastroenterology,* 119, 677, 2000.
25. Bigio, I.J., Bown, S.G., Kelly, C., Lovat, L., Pickarde, D., and Ripley, P.M., Developments in endoscopic technology for oesophageal cancer, *J. R. Coll. Surg. Edinburgh*, 25, 267, 2000.
26. Georgakoudi I., Jacobson B.C., van Dam, J., Backman, V., Wallace, M.B., Muller, M.G., Zhanv, Q., Dadizadegan, K., Sun, D., Thomas, G.A., Perelman, L.T., and Feld, M.S., Fluorescence, reflectance, and light-scattering spectroscopy for evaluating dysplasia in patients with Barrett's esophagus, *Gastroenterology,* 120, 1620, 2001.
27. Wall, J.M.E., Cervical cancer: developments in screening and evaluation of the abnormal smear, *West. J. Med.*, 169, 304, 1998.
28. Nordstrom, R.J., Burke, L., Niloff, J.M., and Myrtle, J.F., Identification of cervical intraepithelial neoplasia (CIN) using UV-excited fluorescence and diffuse-reflectance spectroscopy, *Lasers Surg. Med.*, 29, 118, 2001.
29. Georgakoudi, I., Sheets, E.E., Muller, M.G., Backman, V., Crum, C.P., Badizadegan, K., Dasarim, R.R., and Feld, M.S., Tri-modal spectroscopy for the detection and characterization of cervical precancers *in vivo, Am. J. Obstet.*, 186, 374, 2001.
30. Utzinger, U., Brewer, M., Silvio, E., Gershenson, D., Blast, R.C., Follen, M., and Richards-Kortum, R., Reflectance spectroscopy for *in vivo* characterization of ovarian tissue, *Lasers Surg. Med.,* 28, 56, 2001.
31. Bigio I.J., Bown, S.G., Briggs, G., Kelley, C., Lakhani, S., Pickard, D., Ripley, P.M., Rose, I.G., and Saunders, C., Diagnosis of breast cancer using elastic-scattering spectroscopy: preliminary clinical results, *J. Biomed. Opt.*, 5, 221, 2000.
32. Johns, M., Giller, C., and Liu, H., Computational and *in-vivo* investigation of optical reflectance from human brain to assist neurosurgery, *J. Biomed. Opt.*, 3, 437, 1998.
33. Lin, W.-C., Toms, S.A., Jonson, M., Jansen, E.D., and Mahadaven-Jansen, A., *In vivo* brain tumor demarcation using optical spectroscopy, *Photochem Photobiol.*, 73, 396, 2001.
34. Jacques, S.L., Roman, J.R., and Lee, K., Imaging superficial tissue with polarized light, *Lasers Surg. Med.*, 26, 119, 2000.
35. Backman, V., Gurjar, R., Badizadegan, K., Itzkan, I., Dasari R.R., Perelman, T., and Feld, M.S., Polarized light scattering spectroscopy for quantitative measurement of epithelial structures *in situ, IEEE J. Sel. Top. Quantum Electron.*, 5, 1019, 1999.
36. Sokolov, K., Drezek, R., Gossage, K., and Richards-Kortum, R., Reflectance spectroscopy with polarized light: is it sensitive to cellular and nuclear morphology, *Opt. Express,* 5, 302, 1999.
37. Yao, G. and Wang, L.V., Propagation of polarized light in turbid media: simulated animation sequences, *Opt. Express,* 7, 1983, 2000.
38. Vanitha, S., Everett, M.J., Maitland, D.J., and Walsh, J.T., Comparison of polarized-light propagation in biological tissue and phantoms, *Opt. Lett.*, 24, 1044, 1999.
39. Sankaran, V., Walsh, J.T., and Maitland, C.J. Polarized light propagation through tissue phantoms containing densely packed scatterers, *Opt. Lett.*, 25, 239, 2000.
40. Studinski, R.C.N. and Vitkin, I.A. Methodology for examining polarized light interactions with tissues and tissuelike media in the exact backscattering direction, *J. Biomed. Opt.*, 5, 330, 2000.

41. Junqueiram, L.C., Carneiro, J., and Kelley, R.O., *Basic Histology*, Appleton and Lange, Norwalk, CT, 1992.
42. Lodish, H., Baltimore, D., Berk, A., Zipursky, S.L., Matsudaira, P., and Darnell, J., *Molecular Cell Biology*, 3rd ed., Scientific American Books, New York, 1995, pp. 173 and 847.
43. Stryer, L., *Biochemistry*, 3rd ed., W.H. Freeman, New York, 1988, p. 760.
44. Drezek, R., Dunn, A., Richards-Kortum, R., Light scattering from cells: finite-difference time-domain simulations and goniometric measurements, *Appl. Opt.*, 38, 3651, 1999.
45. Beauvoit B., Kitai, T., and Chance, B., Contribution of the mitochondrial compartment to the optical propertires of the rat liver: a theoretical and practical approach, *Biophys. J.*, 67, 2501, 1994.
46. Beauvoit, B. and Chance, B., Time-resolved spectroscopy of mitochondira, cells and tissues under normal and pathological conditions, *Mol. Cell. Biochem.*, 184, 445, 1998.
47. Schmitt, J.M. and Kumar, G., Turbulent nature of refractive-index variations in biological tissue, *Opt. Lett.*, 21, 1310, 1996.
48. Schmitt, J.M. and Kumar, G., Optical scattering properties of soft tissue: a discrete particle model, *Appl. Opt.*, 37, 2788, 1998.
49. Mourant, J.R., Johnson T.M., Carpenter, S., Guerra, A., and Freyer, J.P., Polarized angular-dependent spectroscopy of epithelial cells and epithelial cell nuclei to determine the size scale of scattering structures, *J. Biomed. Opt.*, 7, 378, 2001.
50. Mourant, J.R., Johnson, T.M., and Freyer, J.P., Angular dependent light scattering from multicellular spheroids, *J. Biomed. Opt.*, 7, 93, 2002.
51. Watson, J.V., *Introduction to Flow Cytometry*, Cambridge University Press, Cambridge, U.K., 1991, chap. 10.
52. Brunsting, A. and Mullaney, P.F., Differential light scattering from spherical mammalian cells, *Biophys. J.*, 14, 439, 1974.
53. Mourant, J.R., Canpolat, M., Brocker, C., Esponda-Ramos, O., Johnson, T., Matanock, A., Stetter, K., and Freyer, J.P., Light scattering from cells: the contribution of the nucleus and the effects of proliferative status, *J. Biomed. Opt.*, 5, 131, 2000.
54. Drezek, R., Dunn, A., and Richards-Kortum, R., A pulsed finite-difference time-domain (FDTD) method for calculating light scattering from biological cells over broad wavelength ranges, *Opt. Express*, 6, 147, 2000.
55. Saidi, I.S., Jacques, S.L., and Tittel, F.K., Mie and Rayleigh modeling of visible-light scattering in neonatal skin, *Appl. Opt.*, 34, 7410, 1995.
56. Nickell, S., Hermann, M., Essenpreis, M., Farrell, T.J., Kramer, U., and Patterson, M.S., Anisotropy of light propagation in human skin, *Phys. Med. Biol.*, 45, 2873, 2000.
57. Jarry, G., Henry, F., and Kaiser, R., Anisotropy and multiple scattering in thick mammalian tissues, *J. Opt. Soc. Am. A*, 17, 149, 2000.
58. Valentine, M.T., Popp, A.K., Weitz, D.A., and Kaplan, P.D., Microscope-based light-scattering instrument, *Opt. Lett.*, 26, 890, 2001.
59. Boustany, N.N., Kuo, S.C., and Thakor, N.V., Optical scatter imaging, subcellular morphometry *in situ* with Fourier filtering, *Opt. Lett.*, 1063, 2001.
60. Wax, A., Yang, C., Dasari, R.R., and Feld, M.S., Measurement of angular distributions by use of low-coherence interferometry for light-scattering spectroscopy, *Opt. Lett.*, 26, 322, 2001.
61. Jacques, S.L. and Wang, L., Monte Carlo modeling of light transport in tissues, in *Optical-Thermal Response of Laser Irradiated Tissue*, Welch, A.J. and van Gemert, M.J.C., Eds., Plenum Press, New York, 1995, p. 73.
62. Wang, L., Jacques, S.L., and Zheng, L., MCML—Monte Carlo modeling of light transport in multi-layered tissues, *Comput. Methods Programs Biomed.*, 47, 131, 1995.
63. Bohren, C.F. and Huffman, D.R., *Absorption and Scattering of Light by Small Particles*, Wiley-Interscience, New York, 1983.
64. Henyey, L.G. and Greenstein, J.L., Diffuse radiation in the galaxy, *Astrophys. J.*, 93, 70, 1941.

65. Mishchenko, M.I., Hoveneir, J.W., and Travis, L.D., *Light Scattering by Nonspherical Particles. Theory, Measurements and Applications*, Academic Press, San Diego, 2000.
66. Mourant, J.R., Johnson, T.M., and Freyer, J.P., Characterizing mammalian cells and cell phantoms by polarized backscattering fiber-optic measurements, *Appl. Opt.*, 40, 5114, 2001.
67. Meglinsky, I.V. and Matcher, S.J., Modeling the sampling volume for skin blood oxygenation measruements, *Med. Biol. Eng. Comput.*, 39, 4, 2001.
68. Gandjbakhche, A.H., Bonner, R.F., Arai, A.E., and Balaban, R.S., Visible-light photon migration through myocardium *in vivo, Am. J. Physiol. Heart Circ. Physiol.*, 46, H698, 1999.
69. Mourant J.R., Bigio, I.J., Jack, D.A., Johnson, T.M., and Miller, H.D., Measuring absorption coefficients in small volumes of highly scattering media: source-detector separations for which path lengths do not depend on scattering properties, *Appl. Opt.*, 36, 5655, 1997.
70. Mourant, J.R., Boyer, J., Hielscher, A.H., and Bigio, I.J., Influence of the scattering phase function on light transport measurements in turbid media performed with small source-detector separations, *Opt. Lett.*, 21, 546, 1996.
71. van Staveran, H.J., Moes, C.J.M., van Marle, J., Prahl, S., and van Gemert, M.J.C., Light scattering in Intralipid 10% in the wavelength range of 400 to 1100 nm, *Appl. Opt.*, 30, 4507, 1991.
72. Flock, S.T., Jacques, S.L., Wilson, B.C., Star, W.M., and van Gemert, M.J.C. Optical properties of intralipids: a phantom medium for light propagation studies, *Lasers Surg. Med.*, 12, 510, 1992.
73. Fishkin, J.B., So, P.T.C., Cerussi, A.E., Fantini, S., Franceschini, M.A., and Gratton, E., Frequency-domain method for measuring spectral properties in multiple-scattering media: methemoglobin absorption spectrum in a tissue-like phantom, *Appl. Opt.*, 34, 1143, 1995.
74. Tuchin, V., *Tissue Optics Light Scattering Methods and Intruments for Medical Diagnosis*, SPIE Press, Bellingham, WA, 2000.
75. Chapter 22 of this book.
76. Bevilacqua, F., Berger, A.J., Cerussi, A.E., Jakubowski, D., and Tromberg, J., Broadband absorption spectroscopy in turbid media by combined frequency-domain and steady state methods, *Appl. Opt.*, 39, 6498, 2000.
77. Nichols, M.G., Hull, E.L., and Foster, T.H., Design and testing of a white-light, steady-state diffuse reflectance spectrometer for determination of optical properties of highly scattering systems, *Appl. Opt.*, 36, 93, 1997.
78. Mourant, J.R., Fuselier, T., Boyer, J., Johnson, T.M., and Bigio, I., Predictions and measurements of scattering and absorption over broad wavelength ranges in tissue phantoms, *Appl. Opt.*, 36, 949, 1997.
79. Hull E.L., Nichols, M.G., and Foster, T.H., Quantitative broadband near-infrared spectroscopy of tissue-simulating phantoms containing erythrocytes, *Phys. Med. Biol.*, 43, 3381, 1998.
80. Ghosh, N., Mohanty, S.K., Majumder, S.K., and Gupta, P.K., Measurement of optical transport properties of normal and malignant human breast tissue, *Appl. Opt.*, 40, 176, 2001.
81. Hull, E.L. and Foster, T.H., Cytochrome spectroscopy in scattering suspensions containing mitochondria and red blood cells, *Appl. Opt.*, 55, 149, 2001.
82. Wang, L. and Jacques, S.L., Use of laser beam with an oblique angle of incidence to measure the reduced scattering coefficient of a turbid media, *Appl. Opt.*, 34, 2362, 1995.
83. Lin, S.-P., Wang, L., Jacques, S.L., and Tittel, F.K., Measurements of tissue optical properties by use of oblique-incidence optical fiber reflectometry, *Appl. Opt.*, 136, 1997.
84. Keinle, A., Forster, F.K., and Hibst, R., Influence of the phase function on determination of the optical properties of biological tissue by spatially resolved reflectance, *Opt. Lett.*, 26, 1571, 2001.
85. Bevilacqua, F. and Depeursinge, C., Monte Carlo study of diffuse reflectance at source-detector separations close to one transport mean free path, *J. Opt. Soc. Am. A*, 16, 2935, 1999.
86. Keinle, A. and Patterson, M.S., Improved solutions of the steady-state and the time-resolved diffusion equations for reflectance from a semi-infinite turbid medium, *J. Opt. Soc. Am. A*, 14, 246, 1997.

87. Venugoplan, V., You, J.S., and Tromberg, B.J., Relative transport in the diffusion approximation: an extension for highly absorbing media and small source-detector separations, *Phys. Rev. E*, 58, 2395, 1998.
88. Hull, E.L. and Foster, T.H., Steady-state reflectance spectroscopy in the $P_3$ approximation, *J. Opt. Soc. Am. A*, 18, 584, 2001.
89. Dam, J.S., Pederson, C.B., Dalgaard, T., Fabricius, P.E., Aruna, P., and Andersson-Engels, S., Fiber-optics probe for noninvasive real-time determination of tissue optical properties at multiple wavelengths, *Appl. Opt.*, 40, 1155, 2001.
90. Pifferi, A., Taroni, P., Valentini, G., and Andersson-Engels, S., Real-time method for fitting time-resolved reflectance and transmittance measurements of a Monte Carlo model, *Appl. Opt.*, 37, 2774, 1998.
91. Pham, T.H., Bevilacqua, F., Spott, T., Dam, J.S., and Tromberg, B.J., Quantifying the absorption and reduced scattering coefficients of tissue-like turbid media over a broad spectral range with noncontact Fourier-transform hyperspectral imaging, *Appl. Opt.*, 39, 6487, 2000.
92. Hayakawa, C.K., Spanier, J., Bevilacqua, F., Dunn, A.K., You, J.S., Tromberg, B.J., and Venugoplan, V., Perturbation Monte Carlo methods to solve inverse photon migration problems in heterogeneous tissues, *Opt. Lett.*, 26, 1335, 2001.
93. Bevilacqua, F., Piguet, D., Marquet, P., Gross, J.D., Tromberg, B.J., and Depeusinge, C., *In vivo* local determination of tissue optical properties: applications to human brain, *Appl. Opt.*, 38, 4939, 1999.
94. Hochheimer, B.F., Polarized light retinal photography of a monkey eye, *Vision Res.*, 18, 19, 1978.
95. Johnson, T.M. and Mourant, J.R., Polarized wavelength-dependent measurements of turbid media, *Opt. Express*, 4, 200, 1997.
96. Pal, S.R. and Carswell, A.I., Polarization anisotropy in lidar multiple scattering from atmospheric clouds, *Appl. Opt.*, 24, 3464, 1985.
97. Dogariu, A., Dogariu, M., Richardson, K., Jacobs, S.D., and Boreman, G.D., Polarization asymmetry in waves backscattering from highly absorbance random media, *Appl. Opt.*, 36, 8159, 1997.
98. Dogariu, M. and Asakura, A., Polarization-dependent backscattering patterns from weakly scattering media, *J. Opt. (Paris)*, 24, 271, 1993.
99. Rakovic, M.J. and Kattawar, G.W., Theoretical analysis of polarization patterns from incoherent backscattering of light, *Appl. Opt.*, 37, 3333, 1998.
100. Hielscher, A.H., Mourant, J.R., and Bigio, I.J., Influence of particle size and concentration on the diffuse backscattering of polarized light from tissue phantoms and biological cell suspensions, *Appl. Opt.*, 36, 125, 1997.
101. Hielscher, A.H., Eick, A.A., Mourant, J.R., Shen, D., Freyer, J.P., and Bigio, I.J., Diffuse backscattering Mueller matrices of highly scattering medium, *Opt. Express*, 1, 441, 1997.
102. Mourant, J.R., Hielscher, A.H., Eick, A.A., Johnson, T.M., and Freyer, J.P., Evidence of intrinsic differences in the light scattering properties of tumorigenic and nontumorigenic cells, *Cancer Cytopathol.*, 84, 366, 1998.
103. Gurjar, R.S., Backman, V., Perelman, L.T., Georgakoudi, I., Badizadegan, K., Itzkan, I., Dasari, R.R., and Feld, M.S., Imaging human epithelial properties with polarized light-scattering spectroscopy, *Nat. Med.*, 7, 1245, 2001.
104. Ottne, D.M., Rubinsky, B., Cheong, W.-F., and Benaron, D.A., Ice-front propagation monitoring in tissue by the use of visible-light spectroscopy, *Appl. Opt.*, 37, 6006, 1998.
105. Nilsson, A.M.K., Sturesson, C., Liu, D.K., and Andersson-Engels, S., Changes in spectral shape of tissue optical properties in conjunction with laser-induced thermotherapy, *Appl. Opt.*, 37, 1256, 1998.
106. Optical transcutaneous bilirubin detector, U.S. patent 5,259,382, 1991.
107. Saidi, I.S., Transcutaneous optical measurements of hyperbilirubinemia in neonates, Ph.D. thesis, Rice University, Houston, TX, 1992.

108. Mourant, J.R., Johnson, T.M., Los, G., and Bigio, I.J., Non-invasive measurement of chemotherapy drug concentrations in tissue: preliminary demonstrations of *in vivo* measurements, *Phys. Med. Biol.,* 44, 1397, 1999.
109. Firnau, G., Maass, D., Wilson, B.C., and Jeeves, W.P., $^{64}Cu$ labeling of hematoporphyrin derivative for non-invasive *in-vivo* measurements of tumour uptake, *Prog. Clin. Biol. Res.,* 170, 629, 1984.
110. Jeeves, W.P., Wilson, B.C., Firnau, G., and Brown, K., *Methods in Porphyrin Photosensitization*, Plenum Press, New York, 1985, p. 51.
111. Scasnar, V. and van Lier, J.E., Biological activities of phthalocyanines. XV. Radiolabeling of the differently sulfonated $^{67}Ga$-phthalocyanines for photodynamic therapy and tumor imaging, *Nucl. Med. Biol.,* 20, 257, 1993.
112. Schwickert, H.C., Stiskal, M., Roberts, T.P., van Dijke, C.F., Mann, J., Muhler, A., Shames, D.M., Demsar, F., Disston, A., and Brasch, R.C., Contrast-enhanced MR imaging assessment of tumor capillary permeability: effect of irradiation on delivery of chemotherapy, *Radiology,* 198, 893, 1996.
113. Turetschek, K., Huber, S., Floyd, E., Helbich, T., Roberts, T.P., Shames, D.M., Tarlo, K.S., Wendland, M.F., and Brasch, R.C., MR imaging characterization of microvessels in experimental breast tumors by using a particulate contrast agent with histopathologic correlation, *Radiology,* 1218, 562, 2000.

# 30

# Raman Spectroscopy: From Benchtop to Bedside




Anita Mahadevan-Jansen
*Vanderbilt University*
*Nashville, Tennessee*


## 30.1 Introduction

Light can react with tissue in different ways and its response can yield information about the physiology and pathology of tissue. The light used to probe tissue does so in a nonintrusive manner; in contrast to therapeutic applications, very low levels of light are typically used. The use of fiber optics allows light to probe tissue in a minimally invasive manner. Because tissue response is virtually instantaneous, the results are obtained in real time, and data processing techniques and computers allow for automated detection of disease. These factors have resulted in a variety of applications that employ light and optical modalities for tissue diagnosis.

When a photon is incident on a molecule, it may be transmitted, absorbed, or scattered. The various techniques that are based on these different light–tissue interactions include:

- Absorption spectroscopy
- Reflectance spectroscopy
- Fluorescence spectroscopy
- Raman spectroscopy

Each of these four techniques has been studied for the purpose of tissue diagnosis with varying degrees of success. Of these techniques, fluorescence spectroscopy is perhaps the most researched one for tissue diagnosis.


0-8493-1116-0/03/$0.00+$1.50

Considerable evidence indicates that fluorescence spectroscopy of exogenous (see, for example, Reference 1) and endogenous chromophores can be used to identify neoplastic transformations in the breast,[2] lung,[3] bronchus,[4] oral cavity,[5] cervix[6] and gastrointestinal (GI) tract.[7,8] Despite the success of this technique, results indicate that fluorescence spectra of precancerous tissues of the cervix and colon and benign abnormalities such as inflammation and metaplasia are similar in many patients.[6–8] This suggests that the use of fluorescence diagnosis in a screening setting, where the incidence of precancer is expected to be low, may result in an unacceptably high false-positive rate. Many groups consider vibrational spectroscopy to be the solution to these problems because its molecular specificity enhances the specificity of spectroscopic diagnostics. The focus of this chapter is the application of Raman spectroscopy in biomedicine, especially as it pertains to disease detection.

Raman spectroscopy has been used for many years to probe the biochemistry of various biological molecules.[9,10] In recent years, there has been interest in using this technique in diagnostics.[11–14] Most of the diagnostic applications are still in the early stages of development. They range from the detection of (pre)malignant lesions in various organ sites to quantitative detection of blood analytes such as glucose, from measurement of skin hydration to the study of aging. The focus of all of these applications is on quantitative, *in vivo*, nonintrusive, automated detection in a real-time manner.

This chapter describes the concept, instrumentation, and application of Raman spectroscopy as it applies to biomedicine. We begin with a description of the basic concepts underlying Raman spectroscopy and Raman signatures of materials. We then review the past, present, and future directions of instrumentation that facilitates the use of this technique for detection of disease. Finally, we examine sample applications of Raman spectroscopy for disease diagnosis, with reference to work at the molecular and microscopic levels. Although many of the reported articles are based on *in vitro* work, this chapter focuses on the potential of the application of Raman spectroscopy to clinical diagnosis.

## 30.2 Principles of Raman Spectroscopy

Raman spectroscopy is based on the Raman effect that results from energy exchange between incident photons and the scattering molecules. When the energy of the incident photon is unaltered after collision with a molecule, the scattered photon has the same frequency as the incident photon. This is Rayleigh or elastic scattering.[15,16] During the collision, when energy is transferred from the molecule to the photon or vice versa, the scattered photon has less or more energy than that of the incident photon. This is inelastic or Raman scattering (Figure 30.1).

Quantum mechanically, Raman scattering is a result of an energy transition of the scattering molecule to a virtual excited state and its return to a higher (or lower) vibrational state with the emission of an altered incident photon (Figure 30.1). Because energy can be transferred from the photon to the molecule or from the molecule to the photon, the scattered photon can have less or more energy compared to the incident photon. When the scattered photon has less energy than the incident photons, the process is referred to as Raman Stokes scattering. When the scattered photon has more energy than the incident photons, the process is referred to as Raman anti-Stokes scattering.

A Raman spectrum is a plot of scattered intensity as a function of the frequency shift (which is proportional to the energy difference) between the incident and scattered photons (Figure 30.2). The frequency shift is characteristic of the molecule with which the photon collided, and the resulting spectrum is characterized by a series of Raman bands that correspond to the various vibrational modes of that molecule. The locations of the Raman bands are typically presented in terms of relative wavenumbers (or Raman shift) defined as shifts in wavenumbers (inverse of wavelength in $cm^{-1}$) from the incident frequency, which is set to zero. This is because the location of Raman bands for a given molecule in relative wavenumbers remains constant, regardless of the incident frequency (within the physical constraints of the light–tissue interaction).

Raman signals are usually weak and require powerful sources and sensitive detectors. Typically, Raman peaks are spectrally narrow (a few wavenumbers) and, in many cases, can be associated with the vibration of a particular chemical bond (or normal mode dominated by the vibration of a single functional group)

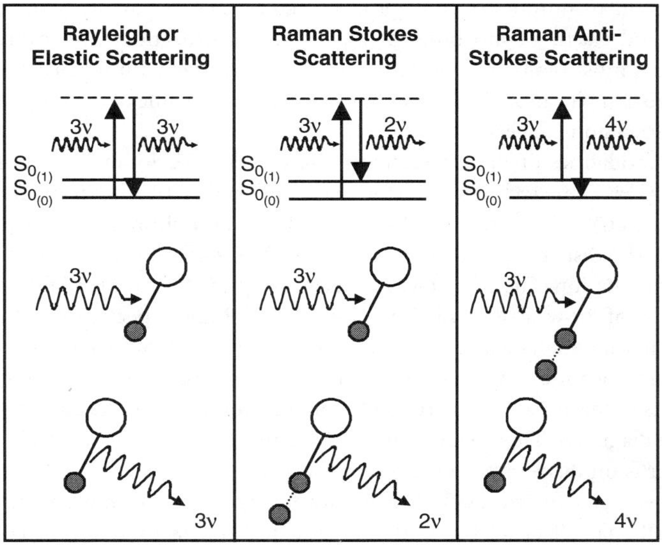


**FIGURE 30.1** Schematic illustration of classical and quantum mechanical explanation of scattering.

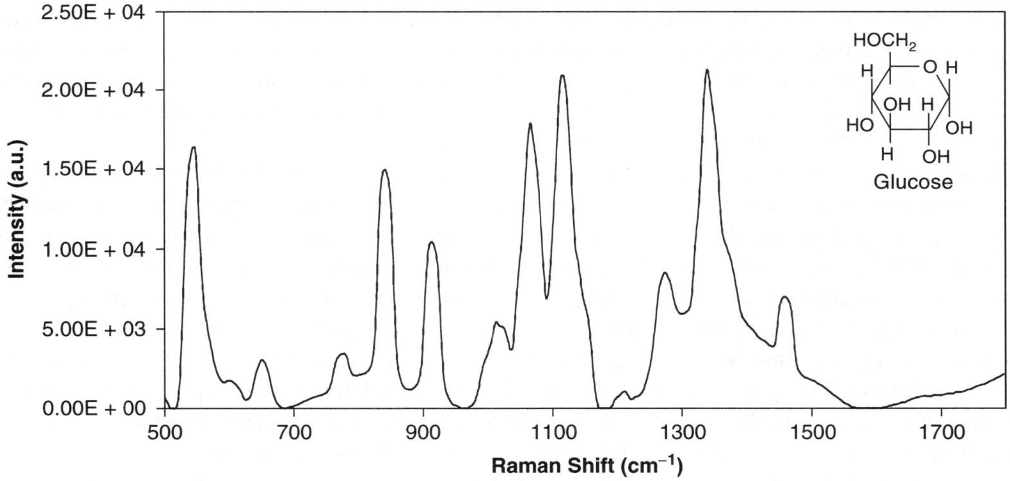


**FIGURE 30.2** Raman spectrum of glucose in powdered form obtained with dispersive Raman spectroscopy at 785-nm excitation.

in a molecule. Consider a complex molecule such as glucose (Figure 30.2). Each of the stretching and bending modes of vibration of this molecule has a characteristic frequency. When a photon is incident on this molecule, each mode of vibration results in a characteristic shift in frequency that is indicated in the Raman spectrum. Because each type of bond has characteristic modes of vibration, each type of molecule has its own spectral "fingerprint." Thus, in a complex mixture of molecules such as in tissue, the presence of the unique bands of glucose can be traced; this results in the quantitative evaluation of the sample's chemical composition, which can then be related to the tissue pathology for diagnosis.

Raman spectroscopy probes characteristics of materials other than fluorescence.[15] The energy transitions of molecules are solely between the vibrational levels. When a photon is incident on a molecule, it may be transmitted, absorbed, or scattered. Fluorescence results from the emission of absorbed energy; Raman scattering results from perturbations of the molecule that induce vibrational or rotational transitions. Only a limited number of biological molecules such as flavins, porphyrins, and structural proteins (collagen and elastin) contribute to tissue fluorescence, most with overlapping, broadband emission. In contrast, most biological molecules are Raman active with fingerprint spectral characteristics. Indications are that vibrational spectroscopy may overcome some of the limitations of fluorescence diagnosis of precancers and cancers.

Several different modalities of Raman scattering have been used to analyze the structure of various biological molecules.[17,18] These techniques include near-infrared (NIR) dispersive Raman, Fourier transform Raman (FT-Raman), surface-enhanced Raman (SERS), and ultraviolet resonance Raman (UVRR) spectroscopy.[15,16] In NIR dispersive Raman spectroscopy, NIR radiation typically in the range of 780 to 1100 nm is used for excitation. The advantage of this technique is that minimal fluorescence is produced, thus making detection of the weak Raman signal easier. In FT-Raman spectroscopy, the Fourier transform of the signal is detected and then inversely transformed to give the actual Raman signature. This technique yields improved signal-to-noise ratio (SNR) of hard-to-detect events, but requires high collection times.[13] Recent developments in spectroscopic instrumentation have resulted in reduced application of FT-Raman techniques in tissue diagnosis. Thus, while some early results from FT-Raman studies are referenced, the focus of this chapter is on dispersive Raman spectroscopy.

SERS is used to investigate the vibrational properties of single adsorbed molecules.[15,16] It was discovered that the rather weak Raman effect can be greatly strengthened (by a factor of up to 14 orders of magnitude) if single molecules are attached to nanometer-sized metal structures. Metal surfaces must be of high reflectivity and of a suitable roughness. In this way, many groups have detected single molecules attached to colloidal silver particles adhered to a glass slide or even in an aqueous solution. The advantages of this method are that it is fast, it can supply some structural information about the molecules, and it does not bleach the molecules. Single-molecule detection is of great practical interest in chemistry, biology, medicine, and pollution monitoring; examples include DNA sequencing and the tracing of interesting molecules such as those used in bioterrorism. Again, because the focus of this chapter is on clinical *in vivo* diagnosis, SERS will not be discussed here.

When the excitation frequency approaches or enters the region of electronic absorption of a molecule, the resonance Raman spectrum of that molecule is obtained.[16] By choosing appropriate excitation frequencies, a sample is selectively excited at the maximum absorption frequency of a characteristic molecule to detect that feature above all else. Resonance excitation increases the scattering intensity by several orders of magnitude. Typical absorption frequencies of biological molecules such as proteins and nucleic acids are in the ultraviolet portion of the spectrum, where the Raman and fluorescence signatures may be spectrally resolved. However, the high intensities may cause photolysis of the sample and destroy it over time. Besides, this signal may be attenuated by simultaneous intensified absorption and fluorescence emission.[16] In addition to these factors, the mutagenicity of UV radiation makes this technique inviable for clinical *in vivo* use.[19] Therefore, UVRR studies are excluded from this chapter.

Many groups have recognized the potential of Raman spectroscopy in the study and diagnosis of disease because of its successful application in biology. Early attempts to measure Raman spectra of cells and tissues were hindered by two factors: (1) the highly fluorescent nature of these samples and (2) instrument limitations, which necessitated long integration times and high-power densities to achieve spectra with good S/N ratios. Improvements in instrumentation in the last decade, particularly in the near-infrared region of the spectrum where fluorescence is reduced, have resulted in a dramatic increase in biomedical applications of Raman spectroscopy. Recent reviews of this field[11,20–22] illustrate the diversity of potential applications, ranging from monitoring of cataract formation *in vivo*[23] to the precise molecular diagnosis of atherosclerotic lesions in coronary arteries.[24,25] Reports have begun to appear validating the potential of Raman spectroscopy to diagnose disease in human tissues *in vivo*.[36,27]

## 30.3 Instrumentation Considerations

Despite the wealth of information provided by Raman spectroscopy about the structure of biological molecules, early attempts to measure Raman spectra of tissues were limited by two factors: (1) the highly fluorescent nature of these samples and (2) instrument limitations, which necessitated long integration times and high-power densities to achieve spectra with good S/N ratios. The initial Raman spectra of tissue were measured with visible laser excitation, using primarily the argon laser lines (see, for example, Yu and East[28] and Clarke et al.[29]). With the development of interferometers, FT-Raman spectroscopy was used to measure tissue Raman spectra, typically using 1064 nm (Nd:YAG) for excitation with germanium detectors.[10]

The development of diode lasers and cooled silicon CCD cameras sensitive in the near-IR has resulted in a wide range of applications beyond the laboratory. Diode lasers can provide excitation in the region of 750 to 850 nm, which allows the use of silicon detectors (sensitive only to 1100 nm). There are two advantages of this technique: (1) fluorescence emission is reduced and (2) spectra with acceptable SNRs ratios can be achieved with relatively short integration times on the order of a few seconds. Thus, Raman spectroscopy is now commonly used in environmental monitoring and manufacturing. Advances in the Raman instrument by the commercial sector that supplies this industry have positively impacted the application of Raman spectroscopy in biomedicine, making clinical use viable.

### 30.3.1 Basic Instrumentation

The basic instrument capable of measuring Raman spectra is similar to any spectroscopic system. It consists of a light source, which is typically a laser, light delivery and collection, and a detection system. Figure 30.3 shows a typical clinical Raman system used today.

The weak nature of the Raman phenomenon makes the details of this basic system challenging. These three basic components interact in terms of overall system performance, but as a first approximation they can be viewed as individual components contributing to the system. Significant technological advances in each of these areas have resulted in Raman instruments superior to similar systems from the previous decade. We consider each of the three components of this basic system individually; however each is dependent on the intended use of the system. For example, a system capable of confocality requires a single-mode laser and appropriate optics. A clinical fiber-based Raman system requires a multimode diode laser with stable frequency output and rugged design that facilitates portability.

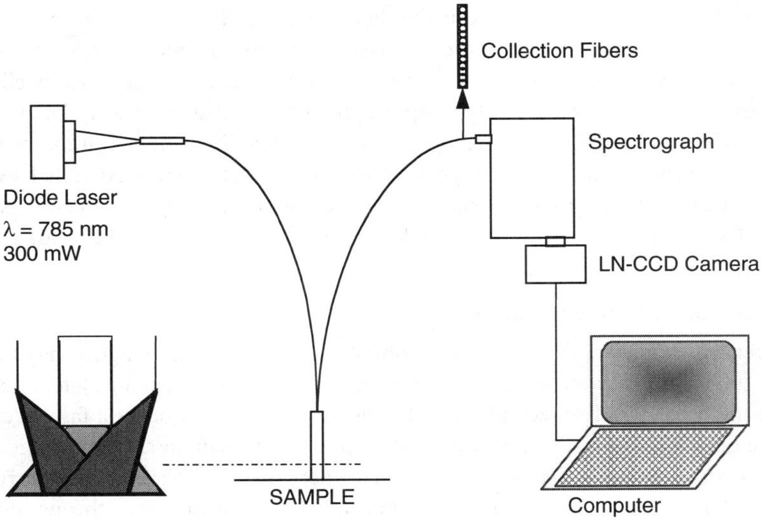


**FIGURE 30.3** Schematic of a typical near-infrared dispersive Raman spectrometer used for tissue diagnosis today. This system is shown with an Enviva Raman probe.

#### 30.3.1.1 Light Sources

Traditional Raman systems used the argon ion ($Ar^+$) laser for visible excitation, the ND:YAG laser for FT-Raman applications and the Titanium:Sapphire (Ti:Sapph) laser for NIR excitation due to their high output powers, single spatial and longitudinal mode operation, and Gaussian beam profile that allows for near-diffraction-limited optical performance.[30–33] However, the size and electronic and cooling requirements of these lasers make them impractical in a portable clinical system. Some current Raman instruments, especially those with confocal capabilities, still use the Ti:Sapph laser. However, advances in diode laser technology have completely changed the fingerprint of a typical Raman system.

Diode lasers utilize electro-optical components (diodes), which emit light as a function of applied current and operating temperature. The laser diodes are small (<1 $mm^3$) and require highly accurate controlling electronics to obtain stable output. Laser diodes require highly stabilized temperature controllers to minimize thermoelastic effects on the laser cavity length (and thus output frequency), and highly stabilized current sources to minimize output power fluctuations. However, even with highly accurate controlling electronics, laser diodes are still susceptible to output variations, so users of these lasers must be aware of their limitations. In addition to the frequency and power considerations described, laser diodes are also characterized by their elliptical beam output (due to the rectangular shape of the output facet) and astigmatism (due to unequal beam divergence from each dimension of the rectangular facet). These issues also need to be accounted for when using a diode laser. Most commercial diode lasers, however, are available with a pig-tail option where a fiber is coupled directly to the laser diode, thus minimizing losses due to astigmatism and elliptical nature of the beam.

A diode laser with tunability in its wavelength is available; such a system tends to be more susceptible to changes in the laser cavity length, resulting in so-called "mode-hops" in which the primary resonant mode of the laser jumps from one frequency to another. Because Raman scattering is very sensitive to the frequency of the incident photon, a source stable in frequency with high line widths is essential to obtain reliable spectra. Diode lasers that are optimized for highly stable frequency output and powers of the order of 300 mW specifically designed for Raman spectroscopy are now commercially available (such as that offered by Process Instruments Inc., Salt Lake City, UT).

Most published reports of confocal Raman microspectroscopy utilize bulky $Ar^+$ and Ti:Sapph lasers for excitation sources.[34] However, the size of these lasers, and their electronic and cooling requirement, make use in a portable clinical system impractical. Thus, in order to develop a truly portable confocal Raman microscope, an alternative excitation source is needed. By extending the length of the resonant cavity of a laser diode, the distance between the diode's longitudinal modes can be extended, thereby minimizing the effect of small thermoelastic changes on the output frequency. Such a source is called external cavity diode laser (ECDL). Compared to a bare laser diode, the ECDL nearly eliminates mode-hops, minimizes spectral bandwidth of the output light, allows substantial wave length tunability, and markedly decreases frequency dependence on temperature stabilization. A commercial ECDL single-mode system, previously available at a high price, is no longer on the market. An ECDL excitation source can be built in-house using off-the-shelf optics, hardware, and electronics at a price (~$6,000) far less than the cost of commercially available $Ar^+$ and Ti:Sapph laser systems (= $20,000) typically used in confocal Raman systems.

#### 30.3.1.2 Detectors and Spectrometers

A typical dispersive Raman detection system consists of a short focal length imaging spectrograph attached to a cooled CCD camera. Clinical implementation of the Raman system requires spectral acquisition on the order of a few seconds. This fast acquisition in turn needs a fast spectrograph and a highly sensitive detector, especially given the weak nature of the Raman signal. A typical CCD camera used in spectroscopy consists of a rectangular chip wherein the horizontal axis corresponds to the wavelength axis. The vertical axis is used to stack multiple fibers for increased throughput and is subsequently binned for improved SNR. Technological advances have led to CCD chips with quantum efficiencies on the order of 90% in the near-infrared. While various types of chips are commercially available for different applications, a back-illuminated, deep depletion CCD is highly recommended for NIR

Raman spectroscopy. These chips are known to be susceptible to the so-called "etaloning effect" wherein the thin silicon chip acts as an etalon, resulting in the introduction of sharp peaks in the sample signal that are hard to resolve from the narrow Raman signal. However, commercial cameras are now available (such as that offered by Roper Scientific Inc., Princeton, NJ) that incorporate technology that effectively eliminates this etaloning effect.

Most laboratory-grade systems currently utilize a liquid nitrogen–cooled CCD that allows the detector to be cooled to at least –80°C, resulting in extremely low dark noise as well as read-out noise. While this level of noise is important in a laboratory grade system where integration times and other system parameters are still being evaluated, a clinical system with a liquid nitrogen–cooled detector is cumbersome and impractical. Most recently, thermoelectrically cooled detectors are now available with multistage Peltier cooling systems that allow the camera to be cooled down to –80°C, thus solving this dilemma. It should be noted that shot noise due to tissue fluorescence continues to be a limiting factor.

Silicon-based detectors are relatively inexpensive and have high quantum efficiencies in the NIR, making them ideal detectors down to about 1100 nm. For longer wavelengths, other types of detectors such as indium gallium arsenide (InGaAs) and germanium (Ge) detectors need to be used. While the longer wavelengths do result in lower fluorescence interference, the quantum efficiency and noise exhibited by these detectors are less than silicon detectors. Research needs to be conducted in this area to verify that Raman spectra with adequate S/N can be acquired from tissue using 1064 nm excitation. New technological advances have yielded low-cost InGaAs and Ge detectors with compact Nd:YAG lasers resulting in extremely compact 1064 nm Raman systems (TRI, Inc., Laramie, WY).

A Raman-sensitive detection system requires an appropriate imaging spectrograph that couples to the light delivery device (such as a fiber probe) on one end and the CCD of choice at the other end. In order to resolve details of the Raman bands, the Raman system should have a spectral resolution of at least 8 $cm^{-1}$. Commercial spectrographs optimized for Raman use are now available that provide *f*-number matching with standard fiber optics and high throughput for rapid acquisition using holographic gratings. These spectrographs (such as that offered by Kaiser Optical Systems, Inc., Ann Arbor, MI) are compact by design and rugged for portable use. Additional components of the detection system include laser line rejection filters that remove any laser light as well as the elastically scattered light from the detected signal. Holographic notch filters can block the laser wavelength with an optical density of six with steep edges and provide 90% transmission elsewhere with a relatively flat curve.

Most Raman systems used in medical applications today are state-of-the-art, high-cost, high-performance Raman systems.[27,35] This system at its best consists of a Kaiser imaging spectrograph, a Roper Scientific liquid nitrogen–cooled CCD detector, a 300 mW Process Instruments diode laser, and a custom fiber probe. Typical integration times are on the order of 5 to 10 sec for this type of system.

#### 30.3.1.3 Light Delivery and Collection (Fiber Optics)

Raman scattering is a weak phenomenon, but most materials are Raman active, which allows for molecular-specific study of samples. On the other hand, because most molecules are Raman active, the materials used in the Raman system themselves interfere with the detection of sample signal. The development and availability of diode lasers, imaging spectrographs, and cooled CCD cameras has made it possible to build compact NIR Raman systems that acquire spectra with short integration times. However, the limiting factor remains the signal generated in the delivery systems (luminescence and Raman) used for remote sensing.[36] Light is typically delivered using optical fibers made of silica in a remote sensing spectroscopic system. However, silica has a strong Raman signal, which overrides sample signal. The signal generated is proportional to the fiber length and limits the detection capability of the technique.[37]

Figure 30.4 shows the Raman spectrum of fused silica fibers used to design a probe. Raman signal was observed to be generated from the core as well as the cladding and buffer of the fiber.[20,37] This signal can have magnitudes equal to and sometimes greater than those of the sample under study and thus demands careful consideration. Fiber signal is generated in the delivery fiber by the excitation light. In addition, background signal is generated in the collection fibers by the elastically scattered excitation light returning to the collection fibers. Mathematical techniques cannot be used to remove this unwanted fiber signal.

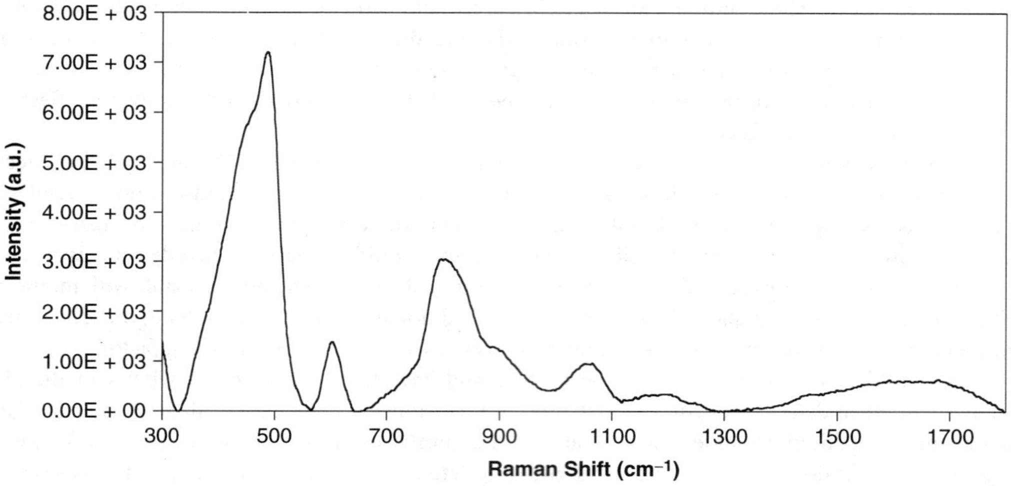


**FIGURE 30.4** Raman spectrum at 785-nm excitation of fused silica fiber. (Modified from Berger, A.J. et al., *Appl. Opt.*, 38, 2916, 1999.)

A feasible probe design must prevent unwanted signal generated in the delivery fiber from illuminating the sample as well as prevent elastically scattered excitation light from entering the collection fibers and generating unwanted signal.

Several different designs have been proposed for potential clinical acquisition of Raman spectra using fiber optic probes. Early on, Myrick and Angel developed different dual fiber probes that could be used under different conditions with maximum collection efficiency but minimum fiber interference.[36] Most fiber designs since have been based on similar concepts with modifications. In general, Raman probe designs utilize a bandpass filter placed after the excitation fiber lens, thus allowing only the transmission of the excitation light from the delivery fiber, and a longpass or notch filter placed in front of the collection fibers that blocks the transmission of the Fresnel reflected excitation light and also prevents the elastically scattered light from entering the collection fibers. These filters are placed at the tip of the probe for maximum effectiveness and they must be sized on the order of a few millimeters or smaller. There is a demand in biomedicine for high-quality optical coatings and micro-optical components to simplify the design of much-needed compact fiber optic probes for Raman spectroscopy.

Multiple fiber bundles have been utilized for fluorescence measurements of tissue by several groups.[6–8] Such a bundle typically consists of a central excitation fiber surrounded by many collection fibers linearly aligned in front of the spectrometer. McCreery et al. used this design and tested it on different samples.[38] Although spectra with good SNR ratio could be obtained from transparent samples, fiber background was still a serious problem in samples with high elastic scattering such as tissue. Feld et al. have adapted an old design (used in solar energy collection) for improved signal collection to allow spectral acquisition in a few seconds.[39] A hollow compound parabolic concentrator (CPC) was used at the distal tip of the probe to yield signals with seven times greater collection than a fiber probe without the CPC. Fiber background was reduced by using a dichroic mirror and separate excitation and collection fiber geometries. Excitation light was reflected by the mirror (placed at 45° with respect to the sample normal) through the CPC onto the sample. Raman spectra were collected by the CPC, transmitted through the mirror into a collection fiber bundle (with about 100 fibers) to a detector. This probe design was used to acquire Raman spectra for transcutaneous blood glucose measurements.

*In vivo* applications thus far have been mostly confined to exposed tissue areas where fiber background could be circumvented using a macroscopic arrangement — e.g., noncontact probe (DLT, Laramie, WY) used for breast tissues by McCreery et al.[32] and CPC probe for transcutaneous measurements of blood analytes by Feld et al.[42] However, other applications such as in the colon, cervix, and oral cavity, require

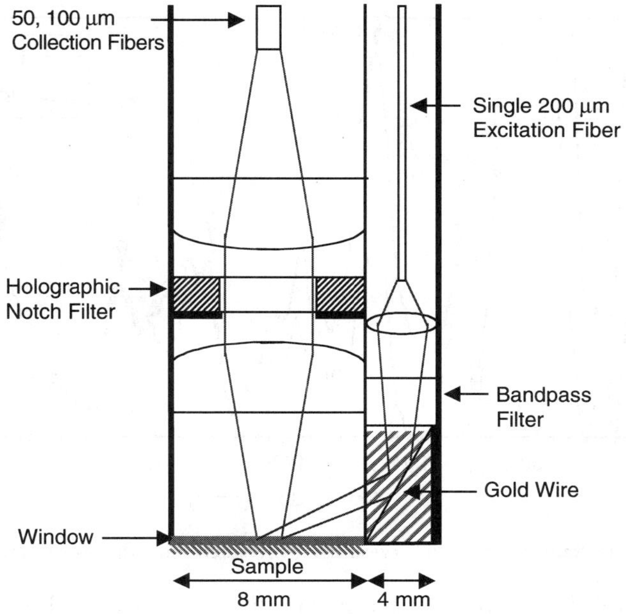


**FIGURE 30.5** Transverse section of the clinical Raman probe implemented in the cervix.

a more compact configuration and probe design. Thus, while many reports have been published assessing the feasibility of Raman spectroscopy for disease detection, few reports have applied this technique *in vivo*. The main obstacle continues to be fiber probe design.

One of the first designs of a compact fiber probe used clinically in the cervix was reported by the author.[40] As stated previously, a feasible probe design must prevent unwanted signal generated in the delivery fiber from illuminating the sample, as well as prevent elastically scattered excitation light from entering the collection fibers and generating unwanted signal. Experimental results show that significant proportions of silica signal are generated in the excitation *and* collection fibers, indicating the need for filters in the excitation and collection legs of the probe. Figure 30.5 shows a transverse section of the design that was implemented as a Raman probe for the cervix.[40] Rather than placing the excitation leg at an angle, a mirror surface is placed in the excitation path to deflect the beam onto the sample.

This probe was designed using the smallest available physical dimensions of the bandpass and the holographic notch filters at that time. An angularly polished gold wire was glued in place such that the deflected excitation and normal collection spots overlap. One of the collection fibers was used to provide an aiming beam during placement of the probe on the sample. A quartz shield was used at the tip of the common end of the probe, forming a barrier between the probe optics and the sample. Quartz was selected as the material of choice because its fluorescence and Raman signal were known and any additional background signal from the probe could be identified. The inner surfaces of the metal tubings used to house the probe optics were anodized to reduce the incidence of multiple reflections of light.

The probe was tested in a pilot clinical study approved by the Institutional Review Board (IRB). NIR Raman spectra of the cervix were successfully measured *in vivo*, as shown in Figure 30.6, using integration times in the order of 90 sec.[43] Since this initial report was produced, several designs for a Raman probe have been reported. One design in widespread use is the beam-steered Enviva Raman probe designed by Visionex Inc. (now Cirrex Corp., Atlanta, GA), which incorporates in-line filtering of the laser light and the scattered light at the tip of the probe.[41] The beam steering allows improved overlap of the excitation and collection volumes and thus improves signal collection efficiency (see Figure 30.3).

Another unique feature of this probe was its dimension in the order of 1 to 3 mm, which was not seen in other probes available commercially. However, the company no longer manufactures these probes and

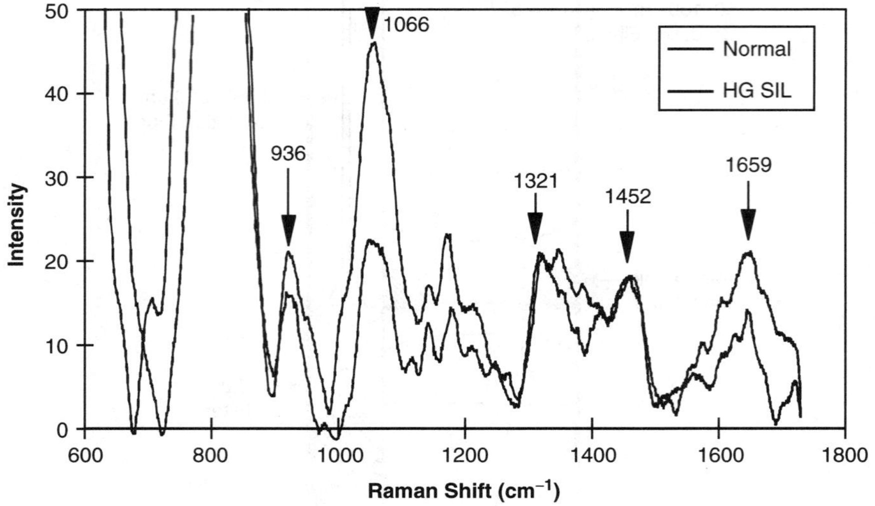


**FIGURE 30.6** *In vivo* NIR Raman spectra of cervical tissue measured using the probe shown in Figure 30.5.

thus a need for such slickly designed Raman probes still exists. Attempts have been made by other companies such as InPhotonics Inc. to fill the gap. In a survey of currently available probes for biomedical Raman use, it was found that all commercially available probes were bulky (up to several millimeters in diameter) and expensive (~$3500 or more). Proponents of Raman spectroscopy for tissue diagnosis must rely on custom-designed fiber probes built using commercial vendors.

Other new designs have been recently reported. Most recently, the Feld group has proposed a newer design for breast cancer detection, based on micro-optics and optical coatings.[42] Results of spectra acquired with this probe have not yet been reported *in vivo*. The recent developments in the field of microfabrication (MEMs) open a new avenue for compact probe designs using reflective optics. Others (including the author) continue to develop new ideas for feasible probe designs that should be tested and reported in the near future.

## 30.4 Effect of Sample Conditions

The design and construction of a Raman system is clearly a critical part of the process. An often forgotten aspect in the use of a new technique deals with the application itself. In developing and applying a new technique, the technology must first be tested in a model system before it can be applied *in vivo*. Subsequently, when it is ready for *in vivo* use, several factors unique to clinical use must also be considered. These issues are addressed below.

### 30.4.1 Tissue Model Selection

In the field of spectroscopy, there are two philosophies on the approach to use in testing the feasibility of a given technique. Some researchers prefer to use an animal model that resembles human tissues in function and structure to simulate the behavior of human tissues, while others prefer *in vitro* human tissue studies before tackling *in vivo* studies. Each of these approaches needs to be carefully considered in the context of the human *in vivo* model.

Although an animal model allows *in vivo* testing, Raman spectra of animal tissues can in some cases differ from those of human tissues, resulting in conclusions not valid for humans. One example of this is the use of mouse eye lenses for the study of lens aging.[43] In early studies, Yu suggested that the intensity of Raman bands at 2580 cm$^{-1}$ due to sulfhydryl groups (–SH) and 508 cm$^{-1}$ due to disulfide groups (–S–S) can be used to calculate their relative concentrations. These can then be related to aging and cataract formation. In an extensive study of mouse lenses, a fall in –SH concentration and a corresponding increase in –S–S concentration was observed along the visual axis of mouse lens nucleus from age 1 to 6 months. A tandem increase in protein development with a decrease in sulfhydryl was concluded to accompany the normal aging process. However, on repeating these studies on guinea pig and human lenses, a different phenomenon was observed. Although guinea pig lenses also showed a decrease in sulfhydryl intensity, no corresponding increase in disulfide band was observed. Human eye lenses behaved similarly to guinea pig lenses. This study clearly indicates the pitfalls in improper selection of an animal model and that independent verification of the model is needed to ensure that results will be applicable to human conditions.

A primary concern in devising a clinical diagnostic system based on Raman spectroscopy is whether the spectra of excised tissue resemble those of tissue spectra acquired *in vivo* and whether the information obtained from these *in vitro* studies can be applied to a clinical setting. It is well known that the spectra acquired from *in vitro* tissues (especially those that have undergone the freeze–thaw process) differ from those acquired *in vivo*.[44] Spectra from intact human stratum corneum were compared to those from excised human stratum corneum;[45] significant differences in the Raman spectral features were observed. Increased intensity of the C–C stretching vibrational bands at 1030 and 1130 $cm^{-1}$ was observed *in vivo*. An additional spectral band at 3230 $cm^{-1}$ of unknown origin is observed only *in vivo*. Thus, it is important to be aware of the potential differences that may occur when moving from *in vitro* studies to *in vivo* conditions. Most researchers who follow this approach use the *in vitro* studies as proof of concept before moving directly into *in vivo* studies.

## 30.4.2 *In Vivo* Considerations

*In vivo* clinical application of the technology is the goal of all researchers in this field. However, various issues specific to working *in vivo* must be considered in the planning of these clinical studies. Logistic issues such as IRB approval and consent forms have become challenging processes. When building a clinical system, some issues need to be considered. Because the probe must be sterilized or disinfected, depending on the application, it must be designed to withstand the physical and chemical conditions of these processes. The power consumption of the Raman system must not overload hospital and clinic systems. Such requirements can be verified by the biomedical engineering department of the hospital. Acquisition times must be on the order of a few seconds for the sake of both the patient and clinician participation. The system needs to be self-contained and nonobtrusive. Some of these issues may appear trivial, but they all contribute to smooth conduct of clinical studies.

Any technique that uses lasers must comply with certain safety standards to avoid potential hazards. Safety standards for laser exposure of skin and eye are established by the American National Standards Institute (ANSI). The power densities used for successful Raman spectral acquisitions can be quite high and thus warrant consideration as safety hazards. The maximum permissible exposures as set by ANSI are wavelength dependent and can be calculated for a given exposure time following the directions specified in the ANSI laser safety manual.[46]

Sources of variability that need to be evaluated experimentally include ambient effects, physician variation, pressure effects, and procedural interference. Regardless of the short duration of the integration times used for Raman acquisition, ambient effects can be significant for spectral integrity. Although room lights may be turned off and windows may be light-tight in a laboratory setting, these are more critical issues in a clinical setting and require careful planning. Variability from physician to physician in terms of single-site placement, as well as site-to-site placement for a given pathology, must be evaluated. The effect of varying pressures applied to the probe during acquisition should also be considered. Another critical parameter is the measurement-site to biopsy-site variability. Because histology continues to be the gold standard against which we must compare this technology, it is critical that the biopsy be obtained from the precise site of spectral measurement using such techniques as marking inks, etc. placed appropriately. It is essential to account for sources of variability when acquiring Raman spectra from human patients *in vivo*.

## 30.5 Processing

Once a Raman spectrum is acquired using a feasible Raman system, it must be processed for different reasons, including calibration, noise smoothing, and fluorescence elimination and binning. Other processing methods may be applied depending on the requirements of the analysis tools and the variability in the data.

### 30.5.1 Calibration

Because different researchers have different approaches in the development of a given application for Raman spectroscopy, a standard must be set to allow transferability of data across systems and methods. Instrument and data calibration are key components of data processing immediately following data acquisition. Some calibration must occur at the time of acquisition and other calibration may be implemented later. Most biomedical Raman systems currently in use are based on dispersive CCD systems where the spectra are recorded as a function of pixel numbers along the horizontal axis. Thus, the spectrum of a known calibrated source such as a neon lamp is typically taken to calibrate the horizontal axis. Raman spectrum of a strong Raman scatterer such as naphthalene, a strong Raman scatterer and fluorophore such as rhodamine 6G, and a weak Raman scatterer such as methylene blue are used as intensity standards to account for day-to-day variations in the spectral intensity. A spectrum of the laser source is taken to verify the location (and thus the wavelength) of the laser line. This spectrum is also used with the neon spectrum to convert the horizontal axis in terms of relative wavenumbers.

Finally, there exists a wavelength-dependent efficiency of all the various components of the detection system such as the grating, the filters, optics, quantum efficiency of the CCD chip, etc.; therefore, the wavelength dependency of the system response must be calibrated. This is typically done using an NIST-calibrated source such as a tungsten lamp to generate correction factors for the instrument variation.[67] As various researchers acquire calibrated Raman spectra, a library of tissue Raman spectra is created that can be used to study cross links in the chemical compositions and correlation in tissue pathologies as well as to strengthen and standardize the use of Raman spectroscopy for tissue diagnosis.

### 30.5.2 Fluorescence Elimination

In recent years we have witnessed an explosion in the use of Raman spectroscopy for biological purposes such as tissue diagnosis, blood analyte detection, and cellular examination. The greatest benefits of this technique are its high sensitivity to subtle molecular (biochemical) change and its capacity for nonintrusive application. Because biological applications of Raman spectroscopy involve turbid, chemically complex, and widely varying target sites, much of the challenge in using Raman spectroscopy for biological purposes is not only the acquisition of viable Raman signatures but also the suppression of inherent noise sources present in the target media. Perhaps the greatest contributor of "noise" to biological Raman spectra is the intrinsic fluorescence of many organic molecules in biological materials. This fluorescence is often several orders of magnitude more intense than the weak chemical transitions probed by Raman spectroscopy and, if left untreated, can dominate the Raman spectra and make analysis of tissue biochemistry as probed by the technique impractical. Therefore, in order to extract Raman signal from the raw spectra acquired, it is necessary to process the spectrum to remove this fluorescence.

A number of techniques, implemented in both hardware and software, have been proposed for fluorescence subtraction from raw Raman signals. Hardware methods such as wavelength shifting and time-gating have been shown to effectively minimize fluorescence interference in Raman spectra, but require modification of the spectroscopic system to achieve their results.[48,49] Mathematical methods implemented in software require no such system modifications, and have thus become the norm for fluorescence reduction. These methods include first- and second-order differentiation,[50,51] frequency-domain filtering,[42] wavelet transformation[52,53] and manual polynomial fitting.[20,54] Although each of these methods has been shown to be useful in certain situations, these methods are not without limitations.

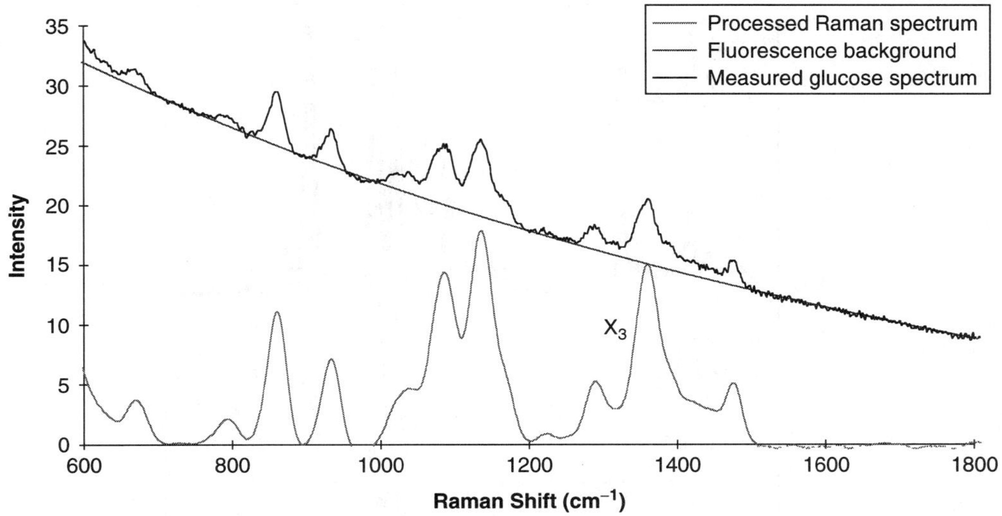


**FIGURE 30.7** Processing Raman spectra of rhodamine 6G for fluorescence removal using a manual fifth order polynomial fit.

Differentiation can be implemented in various ways. One way is to measure the spectra at two slightly shifted excitation wavelengths and take their difference.[55] The fluorescence remains unchanged at both excitation wavelengths whereas the Raman peaks are shifted. The difference of the two spectra is comparable to the first derivative of the Raman spectrum; integrating the difference spectrum yields the original Raman signal. A similar result can be obtained by measuring the spectrum at a single excitation wavelength and taking the first derivative of the spectrum. The Raman spectrum can be obtained by integrating the noise-smoothed derivative spectrum following baseline correction.[49,50] The derivative method is entirely unbiased and very efficient in fluorescence subtraction, yet it severely distorts Raman line shapes and relies on complex mathematical fitting algorithms to reproduce a traditional spectral form.[49]

Frequency filtering can be achieved by using the fast Fourier transform (FFT). In this technique, the measured spectrum is Fourier transformed to the frequency domain by taking the FFT of the signal. The FFT signal can then be multiplied with a linear digital filter to eliminate the fluorescence. The inverse FFT then yields the Raman spectrum free of fluorescence.[52] This method can cause artifacts to be generated in the processed spectra if the frequency elements of the Raman and noise features are not well separated. Wavelet transformation, a more recently utilized method, is highly dependent on the decomposition method used and the shape of the fluorescence background.

A simple, accurate, and more elegant method to subtract fluorescence is to fit the measured spectrum containing both Raman and fluorescence information to a polynomial of high enough order to describe the fluorescence line shape but not the higher-frequency Raman line shape (Figure 30.7). A fourth- or fifth-degree polynomial has been found to be optimal.[20] Polynomial curve-fitting has a distinct advantage over other fluorescence reduction techniques because of its ability to retain the spectral contours and intensities of the input Raman spectra. However, simply fitting a polynomial curve to the raw Raman spectrum in a least-squares manner will not efficiently reproduce the fluorescence background because the fit is based on minimizing differences between the fit and the measured, which includes the fluorescence background and the Raman peaks. Subsequent subtraction of this fit polynomial results in a spectrum that varies about the zero baseline.

This technique traditionally relies on user-selected spectral locations on which to base the fit from regions that do not include the more intense Raman bands. Unfortunately, this intervention has several drawbacks. It is time consuming because the user must process each spectrum individually to identify

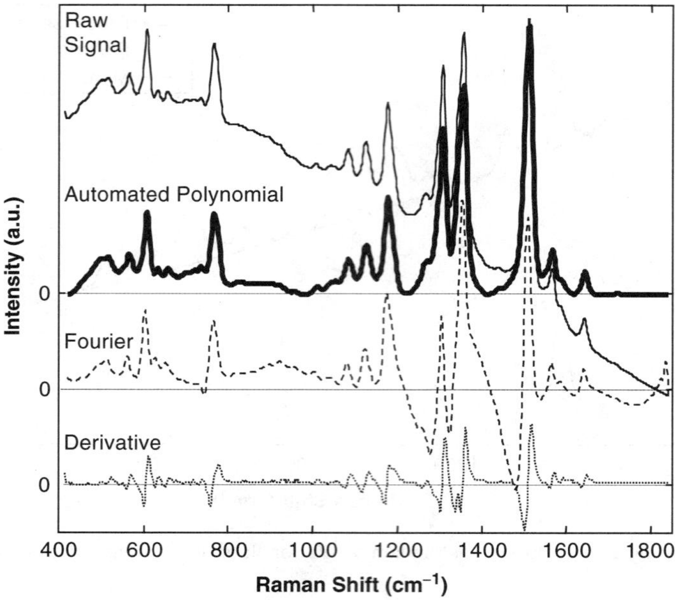


**FIGURE 30.8** Raman spectra of rhodamine 6G processed using the derivative, manual curve-fitting, and modified-mean methods.

non-Raman-active spectral regions to be used in the fit. In addition, identification of non-Raman-active frequencies is not always trivial because biological Raman spectra sometimes contain several adjacent peaks or peaks that are not immediately obvious. The end effect is a method that is highly subjective and prone to variability. In order to address the limitations of existing methods of fluorescence subtraction, various methods to automate polynomial curve-fitting have been developed that retain the benefits of manual curve fitting, without the need for user intervention.[56] One implementation of this automation is the modified-mean method.[56]

In one implementation of this automation, the basis is a simple sliding-window mean filter, in which the center pixel intensity in the window is set equal to the mean of all the intensities within the window. The modification involved in this method is that any filtered pixels with an intensity value higher than the original are automatically reassigned to the original intensity; therefore, the smoothing only eliminates high-frequency Raman peaks while retaining the underlying shape of the baseline fluorescence. This smoothing is repeated over the entire spectrum until a specified coefficient of determination ($R^2$) of the processed spectrum with an $n$th order polynomial (typically fifth order) is obtained. The smoothed spectrum is then subtracted from the raw spectrum to yield the Raman bands on a near-null baseline. This algorithm can be implemented easily in MATLAB, thus automating the entire procedure.

Figure 30.8 shows the Raman spectra of rhodamine 6G, as well as the processed Raman spectrum using derivative, FFT, and automated polynomial fluorescence subtraction techniques for direct comparison of the methods. The figure shows the effectiveness of the automated polynomial method for removing the slow varying fluorescence baseline while retaining the Raman features of rhodamine, especially in comparison to the other methods.

Each of the different techniques has advantages and disadvantages and the method should be selected based on the specific application and measurement technique used. Mosier-Boss et al. tested the use of the shifted excitation, first derivative, and FFT techniques for fluorescence subtraction; a preference for using the FFT technique was indicated due to its ability to filter out random noise from the spectrum.[49] In an analysis of the different techniques, the use of a polynomial fit was found to be the simplest technique from both experimental and computational points of view.

### 30.5.3 Other Preprocessing Methods

Because Raman scattering is such a weak phenomenon, the SNRs of most measured Raman spectra are such that noise-smoothing filters must be included in the processing procedure. Various types of filters that have been effectively used include the Gaussian filter, whose full-width-half-maximum is typically set equal to the spectral resolution of the system, and the Savitsky-Golay filter of various orders.[27] When using any method of noise smoothing, care should be taken to retain the integrity of the spectral line shape, especially in the case of multiple peaks.

Other preprocessing methods include binning of the spectral dataset for computational ease because of a large number of variables. Depending on the variability in the acquired data and the needs of the analysis methods used, other methods include normalization to intensity standards, normalization to its own maximum intensity, mean-scaling the spectra acquired from a given patient, etc.

## 30.6 Analysis

One advantage of spectroscopic diagnosis is automation, which allows objective and real-time diagnosis of pathologies. Differences in spectral features can be incorporated into diagnostic algorithms; several techniques have been identified and applied to enhance the differentiation and classification of tissues for potential automated clinical diagnosis. Because Raman spectroscopy is a molecular-specific technique, the contribution of the various participating chromophores can also be extracted from the measured spectra. These can, in turn, be used in diagnostic algorithms and in the understanding of the spectral signature as it pertains to the disease process. In addition, concentration of specific components such as glucose can be obtained for diagnostic use.

### 30.6.1 Automated Diagnosis

Early studies analyzed biological Raman spectra for differences in intensity, shape, and location of the various Raman bands between normal and non-normal materials such as tissues. Based on consistent differences observed among the various tissue categories, diagnostic algorithms are developed using empirical methods. These algorithms may be based on changes in intensity or ratios of intensities or number and location of peaks. For example, it has been observed that the intensity ratio of the $CH_2$ bending vibrational mode at 1450 $cm^{-1}$ to the amide I vibrational mode at 1655 $cm^{-1}$ varies with disease in several applications, including breast, gynecologic, and precancers.[57] These empirical algorithms indicate the specific changes that occur in the spectra acquired and provide information about the biochemical processes that result in these spectral differences.

This empirical analysis, however, has two important limitations. First, clinically useful diagnostic information is typically contained in more than just the few wavenumbers surrounding peaks or valleys observed in tissue; a method of analysis and classification that includes all the available spectral information can potentially improve the accuracy of detection. Second, empirical algorithms are optimized for the spectra within the study. Hence, the estimates of algorithm performances will be biased toward that tissue population. An unbiased estimate of the performance of the algorithms is required for an accurate evaluation of the performance of Raman spectroscopy for tissue diagnosis. To address these limitations, multivariate statistical techniques have become the practice to develop and evaluate algorithms that differentiate between normal and non-normal tissues.

In recent years, the potential for using multivariate techniques for spectroscopic data analysis in disease detection has been exploited with great success.[20] Discrimination techniques such as linear regression, as well as classification techniques such as neural networks, have been used.[20,22] Data compression tools such as principal component analysis (PCA) are used to account for variability in the data.[58] Subsequently, methods such as hierarchical cluster analysis (HCA),[11] linear discriminant analysis (LDA)[22] and others have been used to yield classification algorithms for disease differentiation. Partial least squares, a regression-based technique, and hybrid linear analysis have been used to extract accurate concentrations of analytes such as glucose using NIR Raman spectra for transcutaneous blood analysis.[59]

The consensus of the scientific community is that, while any of these methods may form the basis of performance estimates, the true measure of success of Raman spectroscopy for tissue diagnosis can be assessed by extensive unbiased (and independent) validation alone. This may best be accomplished by random distribution of the subject population into two equal datasets: one for calibration or testing and the other for validation. When this is not feasible, one can then rely (in a limited manner) on the leave-one-out or leave-five-out methods of cross validation.

### 30.6.2 Component Extraction

Analyzing Raman spectra for the purpose of disease classification may be sufficient when the goal of the study is to achieve tissue diagnosis alone; however, most researchers want to know if and why they can diagnose disease. In the case of Raman spectroscopy, the wealth of information provided in these spectra, especially with respect to biochemical composition, allows us to study the question of "why" in detail that can subsequently be used for the purpose of detection as well.

While most researchers study the biochemical basis of the measured Raman spectra from biological tissues, some groups have been particularly active in this area. For example, Puppels et al. have utilized skin Raman spectra to obtain quantitative information about skin hydration;[13] Feld et al. have developed comprehensive models for biochemical component extraction that have subsequently been used for disease classification.[60] Each of these methods is based on the acquisition of Raman spectra from individually identified chromophores as morphological tissue components or as extracted biochemical constituents.

Pixelated Raman microspectroscopy is typically used (with or without confocality) to measure Raman spectra from individual morphologic tissue components using tissue sections. A Raman spectrometer is coupled to a microscope and is scanned across the tissue section to obtain Raman images that can then be correlated with serial hematoxylin- and eosin-stained sections to identify relevant morphologic components and their Raman signature. Alternatively, tissue chromophores can be extracted from biochemical assays and Raman spectra can be acquired from each of these extracted chromophores.[60] By using mathematical models such as those developed by the Feld group, contributions of the various extracted or morphologic components to intact tissue spectra can then be extracted.[60]

## 30.7 Clinical Application of Raman Spectroscopy

*In vivo* studies are necessary to fully evaluate the potential of Raman spectroscopy for clinical diagnosis. Several groups have initiated this process with varying degrees of success.[26,27,62] This progression has been made possible by the development of sensitive instrumentation, use of fiber optics, and development of automated algorithms. However, despite these significant developments, the majority of reported studies on the application of Raman spectroscopy for tissue diagnosis continue to be from *in vitro* studies. Only a limited number of reports exist on the application of Raman spectroscopy *in vivo* in humans in a clinical setting; most of them are pilot clinical studies. Nevertheless, these critical studies create the road map for the future of Raman spectroscopy in biomedicine.

Several biological molecules such as nucleic acids, proteins, and lipids have distinctive Raman features that yield structural and environmental information. These molecules have been studied in solutions and in their natural microscopic environments.[20] The molecular and cellular changes that occur with disease result in distinct Raman spectra that can be used for diagnosis. The transitional changes in precancerous tissues as well as in benign abnormalities such as inflammation can also yield characteristic Raman features that allow their differentiation.

Several groups have indicated the potential of vibrational spectroscopy for disease diagnosis in various organ sites. These groups have shown that features of the vibrational spectrum can be related to molecular and structural changes associated with disease. Raman spectroscopy has been studied extensively for tissue diagnosis in four main organ sites: breast,[60] esophagus (and the GI tract),[63] cervix (and other gynecological tissues),[27] and skin.[64] Other organ sites include the brain, the eye, and biological fluids.[11,20,21] Some examples of the application of Raman spectroscopy for disease diagnosis are detailed in the following sections.

### 30.7.1 Breast

Until recently, perhaps the most widespread application of Raman spectroscopy in cancer research has been for breast cancer detection. Breast cancer is the most common cancer in women, accounting for 18% of all cancer deaths in women.[65] The breast consists of mammary glands arranged in lobes separated by fibrous connective tissue and a considerable amount of fatty tissue.[66] Fibrocystic changes are benign proliferative processes that vary from the innocuous to those associated with increased risk of carcinomas. These benign changes take three forms: (1) cyst formation and fibrosis, (2) hyperplasia, and (3) adenosis. In most afflicted breasts, all forms occur simultaneously in fibroadipose stroma.

Breast tumors may arise from the ductal and lobular epithelium or connective tissue. These tumors vary from benign adenomas and malignant adenocarcinomas to benign fibromas and malignant fibrosarcomas. Infiltrating ductal carcinoma (IDC), the most frequent invasive form of breast cancer, has been studied using Raman spectroscopy. IDCs typically show an increase in dense fibrous stromal tissue and malignant proliferation of ductal epithelium.[67] Although routine screening using mammography can aid in early detection of malignancy, lesions identified with this method must be biopsied and evaluated histopathologically to determine the presence of malignancy. In recent years, spectroscopic techniques have been used for breast cancer diagnosis.[31,68] Although fluorescence spectroscopy has shown some promise as a diagnostic tool, Raman spectroscopy may provide more definitive characteristics that allow differentiation of benign and malignant tumors.

Several groups have studied the potential of Raman spectroscopy for pathologies of the breast, from detection of breast cancers to study of capsules from breast implants.[69] Using an FT-Raman system, Alfano et al. obtained the first Raman spectra from excised normal human breast tissues and benign and malignant breast tumors and discussed the feasibility of using FT-Raman spectroscopy for differentiating normal and malignant breast tissues.[32] The vibrational spectra of benign breast tissues (which include normal tissues) showed four characteristic bands at 1078, 1300, 1445, and 1651 $cm^{-1}$. The spectra of benign tumors showed bands at 1240, 1445, and 1659 $cm^{-1}$, and malignant tumors displayed only two Raman bands at 1445 and 1651 $cm^{-1}$. The intensity ratio of 1445 to 1651 $cm^{-1}$ was found to be larger in benign tissues compared to benign tumors and the same ratio was found to be even lower in malignant tumors.

Redd et al.[30, 31] employed Raman spectroscopy using visible excitation to study excised human breast tissues. Spectra were also obtained from pure compounds, and the features observed in tissue spectra were determined to be primarily due to carotenoids, myoglobin, and lipids. The Raman spectra of normal breast tissues showed peaks assigned to carotenoids at 1005, 1157, and 1523 $cm^{-1}$ (not observed with IR excitation); peaks assigned to lipids, primarily oleic acid derivatives at 1302, 1442, and 1653 $cm^{-1}$; a peak at 1370 $cm^{-1}$ due to myoglobin; and several other smaller unassigned peaks.

A subsequent study by McCreery et al. assessed the feasibility of using NIR Raman spectroscopy for breast cancer detection. Chromophore contributions differed as excitation was shifted from the visible (yielding carotenoid and lipid bands) to the NIR (yielding only lipid bands) in normal tissues. NIR Raman spectroscopy yielded signals with lower fluorescence interference and higher SNR.

NIR Raman spectra were measured from excised normal human breast tissues, tissues with fibrocystic change, and infiltrating ductal carcinoma (IDC) at 784 nm excitation (Figure 30.9).[31] The ratio of the areas under the peaks at 1654 $cm^{-1}$ and 1439 $cm^{-1}$ were found to increase in malignant breast tissues as compared to normal breast tissues. This increase is consistent with the changes reported by Alfano et al.[32] The intensity of the 1654 $cm^{-1}$ C = C stretching band varies with the degree of fatty acid unsaturation; the $CH_2$ scissoring band at 1439 $cm^{-1}$ depends on the lipid-to-protein ratio. The spectra from IDC tissues showed an overall decrease in intensity with respect to normal tissue.

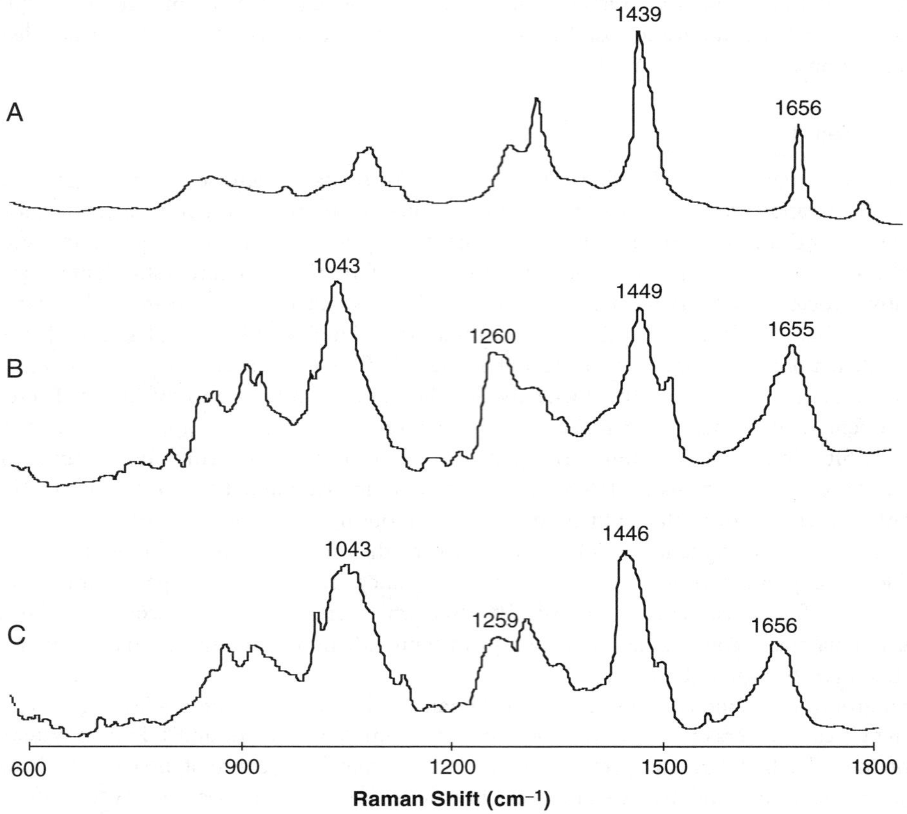


**FIGURE 30.9** Raman spectra of (A) normal, (B) malignant and (C) benign human breast tissue at 784-nm excitation. (Modified from Frank, C.J., McCreery, R.L., and Redd, D.C., *Anal. Chem.*, 67, 777, 1995.)

Several differences were also observed when IDC tissues were compared with benign abnormal tissue. In benign tissue, the intensities of the bands at 1656 $cm^{-1}$ and 1259 $cm^{-1}$ were smaller than the band at 1449 $cm^{-1}$ and this band is further shifted to 1446 $cm^{-1}$. The region of 850 to 950 $cm^{-1}$ showed only two bands in benign tissue as compared to four in IDC samples. The peaks observed in normal tissues were primarily attributed to oleic acid methyl ester, a lipid, and the peaks observed in IDC and benign tissues were primarily attributed to collagen I. This is consistent with histopathology where IDC and benign tissues show an increase in interstitial tissues microscopically. These spectral differences were found to be significant; subsequently, this group reported the development of a clinical probe and system for *in vivo* testing. Preliminary testing of the probe was reported on *in vitro* samples but no other reports have since been published.

Similar results were also obtained by Manoharan et al. using a comparable system.[68] The Raman spectra acquired from this study were analyzed using multivariate statistical techniques and results yielded no false negatives. This group has subsequently reported the development of a clinical Raman system and *in vivo* testing will be reported in the future.[42]

## 30.7.2 Cervix

The various gynecologic tissues differ both structurally and functionally[67,70] from one another and they have been studied using Raman spectroscopy. Because of its well-characterized disease process and its significance, the cervix has been studied extensively in the development of spectroscopic detection. The cervix is the most inferior portion of the uterus, which typically measures 2.5 to 3.0 cm in diameter in the human adult female. The cervix is covered by two types of epithelia. The multilayered squamous epithelium covers most of the ectocervix and is separated from the stroma by the basal layer; the columnar epithelium consists of a single layer of columnar cells and covers the surface of the endocervical canal. The interface of the two epithelia is called the squamo–columnar junction. Over time, the columnar epithelium is replaced by squamous epithelium, which causes the squamo–columnar junction to move toward the os. This transitional epithelium is termed squamous metaplasia.[70]

Virtually all squamous cervical neoplasias (new growth) begin at the functional squamo–columnar junction and the extent and limit of their precursors coincide with the distribution of the transformation zone.[70] Cervical intraepithelial neoplasia (CIN) refers to the precancerous stages of cervical carcinoma and is often also referred to as cervical dysplasia. Here we generally refer to them as "precancers." Other pathologies that affect the cervix include cervicitis or inflammation, which is usually the response of tissue to injury,[70] and the human papilloma viral (HPV) infection. Similarities observed in the morphological changes of the epithelial cells between those induced by HPV and precancer have led to the suggestion that certain strains of HPV may be involved in the incipient stages of cervical precancer and other strains may aid in the progression of the disease.[71] Thus, HPV is typically placed in the same category as mild precancers and is clinically treated as such. Endocervical cancers are typically adenocarcinomas, arising within the endocervix as opposed to cervical epithelial lesions, which arise in the squamo–columnar junction.[70]

Cervical cancer is the second most common malignancy found in women worldwide. It was estimated that in 2001, 4100 deaths occurred in the United States from this disease and 13,000 new cases of invasive cervical cancer were diagnosed.[65] Although early detection of cervical precancer has played a central role in reducing the mortality associated with this disease over the last 50 years, the incidence of preinvasive squamous carcinoma of the cervix has risen dramatically, especially among women under the age of 50. The primary screening tool for cervical precancer is the Papanicolaou (Pap) smear, where scrapings from the walls of the ecto- and endocervix, which contain a variable number of cells, are examined and diagnosed.[70] Although the widespread application of the Pap smear as a screening tool has greatly decreased the incidence of cervical cancer,[72] sampling and reading errors lead to high false-positive and -negative rates. Treatment ultimately relies on directed biopsies and subsequent pathological findings.

Alfano et al. were the first to report on the feasibility of using FT-Raman spectroscopy for detecting cancers from various gynecologic tissues.[57] Characteristic Raman eatures of normal tissues and malignant tumors from the cervix, uterus, endometrium, and ovary were described. Three significant peaks were noted to differ in the Raman spectra of normal and benign cervix compared to cancerous lesions. In cancerous tissues, the intensity of the amide I stretching vibration band at 1657 $cm^{-1}$ is less than the intensity of the C–H bending vibrational band at 1445 $cm^{-1}$. The amide III band at 1262 $cm^{-1}$ is broadened in cancerous lesions. An additional unidentified peak at 934 $cm^{-1}$ is observed only in normal and benign cervical samples. A possible diagnostic algorithm could be based on the relative intensities of the two peaks where $I_{1657} > I_{1445}$ in normal and benign tissues and $I_{1657} < I_{1445}$ in cancerous samples. Alfano et al. attributed these peaks primarily to collagen and elastin.

The author has reported *in vitro* as well as *in vivo* studies on the application of Raman spectroscopy for cervical *precancer* detection.[27,58] In her early work, Raman spectra of cervical tissues were measured to characterize the spectral signatures of the different cervical tissue types and to assess the feasibility of using Raman spectroscopy for cervical precancer diagnosis.[58] Primary tissue Raman peaks were observed at 626, 818, 978, 1070, 1246, 1330, 1454, and 1656 $cm^{-1}$ ($\pm 10$ $cm^{-1}$), present in all samples. The peaks at 626, 818, and 1070 were attributed as primarily due to silica from the optics of the system. Both empirical and multivariate techniques were used to explore the diagnostic capability of NIR Raman spectra from cervical tissues. A multivariate discrimination algorithm developed using the entire Raman spectrum could differentiate cervical precancers from nonprecancers with a sensitivity and specificity of 91 and 90%, respectively.[58] A discrimination algorithm developed using intensities at just 8 Raman bands gave an unbiased estimate of the sensitivity and specificity as 82 and 92%, respectively.[58] Thus, the success of the *in vitro* Raman studies for precancerous cervical tissue recognition warranted the development of a clinical system.

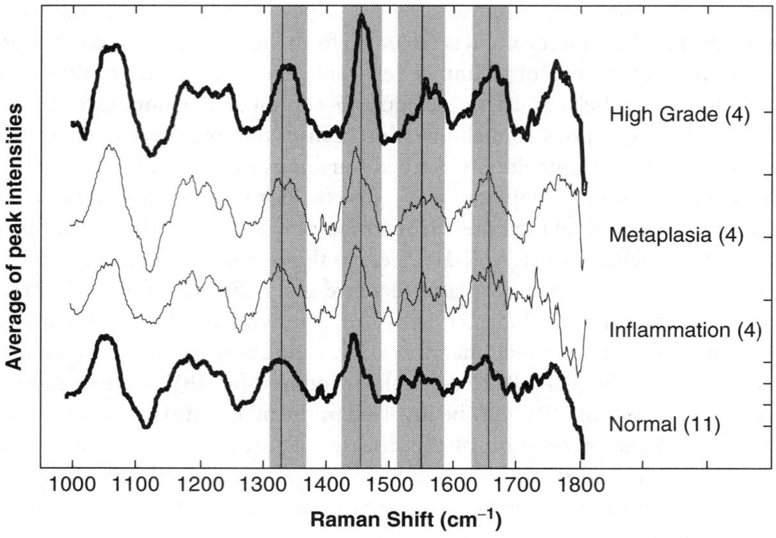


**FIGURE 30.10** *In vivo* NIR Raman spectra for different types of cervical tissues. (Modified from Utzinger, U. et al., *Appl. Spectrosc.*, 55, 955, 2001.)

In a pilot clinical study approved by the IRB, NIR Raman spectra of the cervix were successfully measured *in vivo* from 24 sites (11 normal, 4 inflammation, 4 metaplasia, 1 low-grade, and 4 high-grade lesions) in 13 patients.[27] Raman spectra were measured from normal and abnormal areas of the cervix using the fiber probe described in Figure 30.5 and a Raman system similar to the one shown in Figure 30.3. Figure 30.10 shows typical Raman spectra of the different types of cervical tissues acquired *in vivo*.[27]

Cervical tissue Raman spectra show peaks in the vicinity of 1070, 1180, 1210, 1245, 1270, 1330, 1400, 1454, 1580, and 1656 cm$^{-1}$. The ratio of intensities at 1454 to 1656 cm$^{-1}$ is greater for precancerous tissues than for all other tissue types, while the ratio of intensities at 1330 to 1454 cm$^{-1}$ is lower for samples with cervical precancer than for all other tissue types. A simple algorithm based on these two intensity ratios separates high-grade lesions from all others with a sensitivity of 100% and a specificity of 95%.[27] Preliminary results thus indicate that it is possible to measure Raman spectra *in vivo* and extract potentially diagnostically useful information.

*In vivo* Raman spectra, in general, appeared similar to *in vitro* Raman spectra previously obtained from cervical biopsies. Due to the small number of patients included, the number of samples within a particular category was small; only one of the investigated samples was histologically identified to be low grade. Thus, while promising results were obtained in discriminating high-grade lesions from all other tissue types, significant research still remains to be done to characterize the spectra of low-grade lesions and other tissue pathologies of the cervix before Raman spectroscopy can be validated as a viable method for the diagnosis and potential screening of cervical precancers.[73]

## 30.7.3 Skin

The skin is the largest organ of the body and has many different functions. It is divided into two main regions: the epidermis and the dermis. The epidermis primarily consists of keratinocyctes and varies in thickness throughout the body. Melanocytes, the pigment-producing cells of the skin, are found throughout the epidermis.[66] The dermis consists mostly of fibroblasts, which are responsible for secreting the collagen and elastin that give support and elasticity to the skin. There are two major groups of skin cancers: malignant melanoma and nonmelanoma skin cancers. Cancers that develop from melanocytes are called melanoma. Nonmelanoma skin cancers are the most common cancers of the skin. Two of the most common nonmelanoma types are basal cell carcinoma (BCC) and squamous cell carcinoma (SCC).

About 75% of all skin cancers are BCCs, which have a high likelihood of recurrence after treatment, either at the same site or elsewhere.[65] SCCs account for about 20% of all skin cancers; these carcinomas are more likely to invade tissues beneath the skin and distant parts of the body than are BCCs.

Although most people believe they are not at risk for skin cancer, cancers of the skin (including melanoma and nonmelanoma skin cancers) are the most common of all cancers and account for about half of all cases in the United States.[65] The American Cancer Society estimated approximately 9800 skin cancer deaths for 2002, 7800 from melanoma and 2000 from other skin cancers. Melanomas account for about 4% of skin cancer cases but cause about 77% of skin cancer deaths. The number of new cases of melanoma found in the United States is on the rise. The American Cancer Society predicted 53,600 new cases of melanoma in the United States in 2002. For most skin cancer patients (including melanoma and nonmelanoma skin cancer), early diagnosis and thorough treatment (i.e., complete resection) are critical for a favorable prognosis.

Current diagnostic methods for skin cancers rely on physical examination of the lesion in conjunction with skin biopsy. Suspicious areas are selected upon visual inspection by the clinician, after which those lesions are partially or wholly biopsied for complete histological evaluation.[65] The biopsy is then sectioned and stained for pathological investigation and diagnosis. This protocol for skin lesion diagnosis is accepted as the gold standard, but it is subjective, invasive, and time consuming; hence, there is considerable interest in developing a noninvasive diagnostic tool that can accurately detect skin lesions noninvasively in real time, especially in the early stages.

Previous studies on skin cancer detection using optical spectroscopy have been primarily limited to fluorescence and diffuse reflectance spectroscopy and have produced limited success.[74,75] The interference of skin pigment and external agents such as creams and soaps has been the major limitation to the success of these techniques. Despite the relative ease of studying the skin, the first published reports of skin Raman spectra appeared only in the early 1990s. These early studies, however, focused on characterizing skin components, and included research on skin hydration, skin aging, and the effect of UV radiation.[11]

Williams et al. utilized FT-Raman spectroscopy to examine a number of skin features and correlate the Raman spectra with the biochemical agents responsible.[76,77] Caspers et al. used confocal Raman microspectroscopy to ascertain the *in vivo* Raman signal characteristics emanating from each layer of normal human skin. They showed that each layer of the skin contained Raman features that could be highly correlated with the protein and chemical content of each respective layer.[62] A recent study has also shown that it is possible to extract carotenoid Raman spectra *in vivo* from various skin tissues.[78] Gniadecka et al. utilized FT-Raman spectroscopy to successfully differentiate BCC from normal, healthy skin *in vitro* in 16 patients.[79] Although promising results were reported, FT-Raman spectroscopy is not a feasible technique for clinical diagnosis.

A recent study exploring the use of confocal Raman microspectroscopy for skin cancer detection was also reported.[80] Results from two pilot studies performed on a novel bench top confocal Raman microspectrometer were encouraging. In the first study, Raman spectra were acquired *in vitro* from normal and various types of malignant skin tissues obtained from human patients undergoing excision.[80] Figure 30.11 shows the Raman spectra of normal skin, BCC, SCC, and melanoma tissues at a depth where optimal differences between the various tissue types were observed. Key differences in the spectra are observed at several Raman bands including those seen at 860, 940, 1120, 1220 to 1340, and 1550 $cm^{-1}$.

Another study assessed the capability of the confocal Raman system to acquire Raman spectra *in vivo* and the variability in the Raman spectra from various types of normal skin. Raman spectra were acquired from the dorsal area of the hand from volunteers of various ethnicity and skin color (white, African-American, east Asian, and south Asian) at various depths of tissue. Subsequent statistical analysis of the Raman spectra at various depths revealed significant variations of Raman signal in the upper strata of the skin, presumably due to differences in melanin content. However, Raman spectra from locations deeper in the skin (~60 μm) showed much more similarity visually and statistically. These results indicate that confocal Raman spectroscopy can limit the measurement volume to specified depths of the skin,

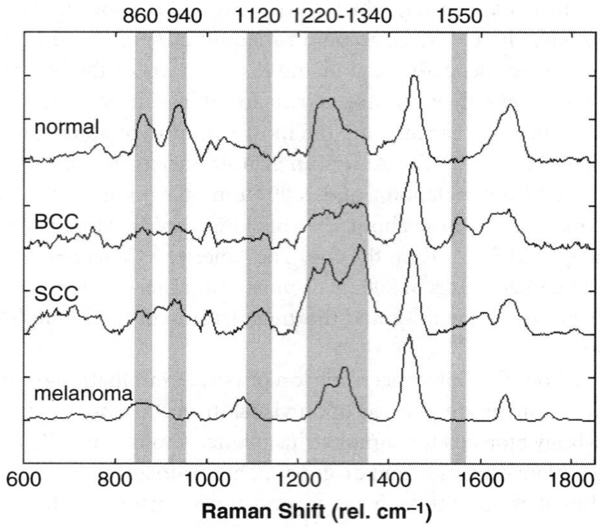


**FIGURE 30.11** NIR Raman spectra of normal and cancerous skin lesions acquired *in vitro* using a confocal Raman microspectrometer.

which may circumvent the problems in skin color differences traditionally associated with optical measurement.

### 30.7.4 Other Applications

The preceding examples show the various stages of development in the implementation of Raman spectroscopy for disease detection *in vivo*. Additional applications of this versatile technique not specifically reported here include the detection of various pathologies in the brain[81] and the GI tract, especially the esophagus,[63] ovary,[56] and colon.[82] Raman spectroscopy has also been used to determine various analytes such as those in the blood. For example, studies have been reported on the determination of glucose concentration in diabetes patients.[59] A recent study reported the transdermal measurement of blood glucose using tissue modulation for signal extraction.[83] Researchers are using Raman spectroscopy in clinical settings and results appear promising. Raman spectroscopy is poised to move to the next level of implementation.

## 30.8 Future Perspectives

Recent technological developments have made clinical implementation of Raman spectroscopy feasible. Advances in diode laser technology and CCD detector technology have contributed tremendously to this process. However, the need still exists for compact Raman probes and integrated systems that would make this technique even more versatile. Currently, several integrated systems are available for the application of Raman spectroscopy in environmental and manufacturing processes. A similar advancement is required in the field of biomedicine. The introduction of micro-optics and microfabrication can further aid in the development of compact small-diameter probes.

This chapter presents a review of concepts, instrumentation, and sample applications of nonresonance Raman spectroscopy for disease detection *in vivo*. The success of the technique has led to the development of feasible clinical systems that can measure Raman signals from tissue with short collection times. Many of these systems have already undergone preliminary testing and several others are currently in the process of being tested. These studies clearly indicate that clinical application of Raman spectroscopy is imminent and may be expected to change the face of disease detection in the near future. While more extensive studies are still needed to form a true assessment of the capability of Raman spectroscopy for disease detection, preliminary results are extremely encouraging. What makes this technique so invaluable is not only its use for disease classification, but also the possibility of evaluating the acquired spectra for biochemical composition. This technique is a viable clinical tool and an important research tool for furthering both the technology of patient care and the understanding of the disease process.

## References


1. Stummer, W., Novotny, A., Stepp, H., Goetz, C., Bise, K., and Reulen, H.J., Fluorescence-guided resection of glioblastoma multiforme by using 5-aminolevulinic acid-induced porphyrins: a prospective study in 52 consecutive patients, *J. Neurosurg.*, 93, 1003, 2000.
2. Alfano, R.R., Pradhan, A., Tang, C.G., and Wahl, S.J., Optical spectroscopic diagnosis of cancer in normal and breast tissues, *J. Opt. Soc. Am. B*, 6, 1015, 1989.
3. Kennedy, T.C., Lam, S. and Hirsch, F.R., Review of recent advances in fluorescence bronchoscopy in early localization of central airway lung cancer, *Oncologist*, 6, 257, 2001.
4. Hung, J., Lam, S., LeRiche, J.C., and Palcic, B., Autofluorescence of normal and malignant bronchial tissue, *Lasers Surg. Med.*, 11, 99, 1991.
5. Gillenwater, A., Jacob, R., Ganeshappa, R., Kemp, B., El-Naggar, A.K., Palmer, J.L., Clayman, G., Mitchell, M.F., and Richards-Kortum, R., Noninvasive diagnosis of oral neoplasia based on fluorescence spectroscopy and native tissue autofluorescence, *Arch. Otolaryngol. Head Neck Surg.*, 124, 1251, 1998.
6. Ramanujam, N., Mitchell, M.F., Mahadevan, A., Thomsen, S., Malpica, A., Wright, T., Atkinson, N., and Richards-Kortum, R., Spectroscopic diagnosis of cervical intraepithelial neoplasia (CIN) *in vivo* using laser-induced fluorescence spectra at multiple excitation wavelengths, *Lasers Surg. Med.*, 19, 63, 1996.
7. Cothren, R.M., Richards-Kortum, R.R., Sivak, M.V., Fitzmaurice, M., Rava, R.P., Boyce, G.A., Hayes, G.B., Doxtader, M., Blackman, R., Ivanc, T., Feld, M.S., and Petras, R.E.., Gastrointestinal tissue diagnosis by laser induced fluorescence spectroscopy at endoscopy, *Gastrointest. Endosc.*, 36, 105, 1990.
8. Schomacker, K.T., Frisoli, J.K., Compton, C.C., Flotte, T.J., Richter, J.M., Nishioka, N.S., and Deutsch, T.F., Ultraviolet laser-induced fluorescence of colonic tissue: basic biology and diagnostic potential, *Lasers Surg. Med.*, 12, 63, 1992.
9. Johnson, C.R., Ludwig, M., O'Donnell, S., and Asher, S.A., UV resonance Raman spectroscopy of the aromatic amino acids and myoglobin, *J. Am. Chem. Soc.*, 106, 5008, 1984.
10. Nie, S., Bergbauer, K.J., Ho, J.J., Kuck, J.F.R., Jr., and Yu, N.T., Applications of near-infrared Fourier transform Raman spectroscopy in biology and medicine, *Spectroscopy*, 5, 24, 1990.
11. Choo-Smith, L.-P., Edwards, H.G.M., Endtz, H.P., Kros, J.M., Heule, F., Barr, H., Robinson, J.S., Jr., Bruining, H.A., and Puppels, G.J., Medical applications of Raman spectroscopy: from proof of principle to clinical implementation, *Biopolymers*, 67, 1, 2002.
12. Pilotto, S., Pacheco, M.T., Silveira, L., Jr., Villaverde, A.B., and Zangaro, R.A., Analysis of near-infrared Raman spectroscopy as a new technique for a transcutaneous non-invasive diagnosis of blood components, *Lasers Med. Sci.*, 16, 2, 2001.
13. Caspers, P.J., Lucassen, G.W., Bruining, H.A., and Puppels, G.J., Automated depth-scanning confocal Raman microspectrometer for rapid *in vivo* determination of water concentration profiles in human skin, *J. Raman Spectrosc.*, 31, 813, 2000.
14. Gellermann, W., Ermakov, I.V., Ermakova, M.R., McClane, R.W., Zhao, D.Y., and Bernstein, P.S., *In vivo* resonant Raman measurement of macular carotenoid pigments in the young and the aging human retina, *J. Opt. Soc. Am. A Opt. Image Sci. Vis.*, 19, 1172, 2002.
15. Colthup, N.B., Daly, L.H., and Wiberley, S.E., *Introduction to Infrared and Raman Spectroscopy*, Academic Press, Boston, 1990.
16. Ferraro, J.R. and Nakamoto, K., *Introductory Raman Spectroscopy*, Academic Press, Boston, 1994.

17. Carey, P.R., *Biochemical Applications of Raman and Rresonance Raman Spectroscopies,* Academic Press, New York, 1982.
18. Twardowski, J. and Anzenbacher, P., *Raman and IR Spectroscopy in Biology and Biochemistry,* Ellis Horwood, New York, 1994.
19. Feld, M.S. and Kramer, J.R., Mutagenicity and the XeCl excimer laser: a relationship of consequence? *Am. Heart J.,* 122, 1803, 1991.
20. Mahadevan-Jansen, A. and Richards-Kortum, R., Raman spectroscopy for the detection of cancers and precancers, *J. Biomed. Opt.,* 1, 31, 1996.
21. Hanlon, E.B., Manoharan, R., Koo, T.W., Shafer, K.E., Motz, J.T., Fitzmaurice, M., Kramer, J.R., Itzkan, I., Dasari, R.R., and Feld, M.S., Prospects for *in vivo* Raman spectroscopy, *Phys. Med. Biol.,* 45, R1, 2000.
22. Petrich, W., Mid-infrared and Raman spectroscopy for medical diagnostics, *Appl. Spectrosc. Rev.,* 36, 181, 2001.
23. Mizuno, A. and Ozaki, Y., Aging and cataractous process of the lens detected by laser Raman spectroscopy, *Lens Eye Toxicol. Res.,* 8, 177, 1991.
24. Van de Poll, S.W., Romer, T.J., Puppels, G.J., and Van der Laarse, A., Raman spectroscopy of atherosclerosis, *J. Cardiovasc. Risk,* 9, 255, 2002.
25. Buschman, H.P., Deinum, G., Motz, J.T., Fitzmaurice, M., Kramer, J.R., van der Laarse, A., Bruschke, A.V., and Feld, M.S., Raman microspectroscopy of human coronary atherosclerosis: biochemical assessment of cellular and extracellular morphologic structures *in situ, Cardiovasc. Pathol.,* 10, 69, 2001.
26. Shim, M.G., Song, L.M., Marcon, N.E., and Wilson, B.C., *In vivo* near-infrared Raman spectroscopy: demonstration of feasibility during clinical gastrointestinal endoscopy, *Photochem. Photobiol.,* 72, 146, 2000.
27. Utzinger, U., Heintzelman, D.L., Mahadevan-Jansen, A., Malpica, A., Follen, M., and Richards-Kortum, R., Near infrared Raman spectroscopy for *in vivo* detection of cervical precancers, *Appl. Spectrosc.,* 55, 955, 2001.
28. Yu, N.T. and East, E.J., Laser Raman spectroscopic studies of ocular lens and its isolated protein fractions, *J. Biol. Chem.,* 250, 2196, 1975.
29. Clarke, R.H., Hanlon, E.B., Isner, J.M., and Brody, H., Laser Raman spectroscopy of calcified atherosclerotic lesions in cardiovascular tissue, *Appl. Opt.,* 26, 3175, 1987.
30. Redd, D.C.B., Feng, Z.C., Yue, K.T., and Gansler, T.S., Raman spectroscopic characterization of human breast tissues: implications for breast cancer diagnosis, *Appl. Spectrosc.,* 47, 787, 1993.
31. Frank, C.J., McCreery, R.L., and Redd, D.C., Raman spectroscopy of normal and diseased human breast tissues, *Anal. Chem.,* 67, 777, 1995.
32. Alfano, R.R., Lui, C.H., Sha, W.L., Zhu, H.R., Akins, D.L., Cleary, J., Prudente, R., and Cellmer, E., Human breast tissues studied by IR Fourier transform Raman spectroscopy, *Lasers Life Sci.,* 4, 23, 1991.
33. Puppels, G.J., de Mul, F.F., Otto, C., Greve, J., Robert-Nicoud, M., Arndt-Jovin, D.J., and Jovin, T.M., Studying single living cells and chromosomes by confocal Raman microspectroscopy, *Nature,* 347, 301, 1990.
34. Caspers, P.J., Lucassen, G.W., Wolthuis, R., Bruining, H.A., and Puppels, G.J., *In vitro* and *in vivo* Raman spectroscopy of human skin, *Biospectroscopy,* 4, S31, 1998.
35. Stone, N., Stavroulaki, P., Kendall, C., Birchall, M., and Barr, H., Raman spectroscopy for early detection of laryngeal malignancy: preliminary results, *Laryngoscope,* 110, 1756, 2000.
36. Myrick, M.L. and Angel, S.M., Elimination of background in fiber-optic Raman measurements, *Appl. Spectrosc.,* 44, 565, 1990.
37. Myrick, M.L. and Angel, S.M., Comparison of some fiber optic configurations for measurement of luminescence and Raman scattering, *Appl. Opt.,* 29, 1333, 1990.
38. Schwab, S.D. and McCreery, R.L., Versatile, efficient Raman sampling with fiber optics, *Anal. Chem.,* 56, 2199, 1984.

39. Tanaka, K., Pacheco, M.T.T., Brennan, J.F., Itzkan, I., Berger, A.J., Dasari, R.R., and Feld, M.S., Compound parabolic concentrator probe for efficient light collection in spectroscopy of biological tissue, *Appl. Opt.*, 35, 758, 1996.
40. Mahadevan-Jansen, A., Mitchell, M.F., Ramanujam, N., Utzinger, U., and Richards-Kortum, R., Development of a fiber optic probe to measure NIR Raman spectra of cervical tissue *in vivo, Photochem. Photobiol.*, 68, 427, 1998.
41. Shim, M., Wilson, B., Marple, E., and Wach, M., Study of fiber-optic probes for *in vivo* medical Raman spectroscopy, *Appl. Spectrosc.*, 53, 619, 1999.
42. Motz, J.T., Hunter, M., Galindo, L., Kramer, J.R., Dasari, R.R., and Feld, M.S., Development of optical fiber probes for biological Raman spectroscopy, in *Biomedical Optical Spectroscopy and Diagnostics, OSA Biomedical Topical Meetings Technical Digest*, 336, Optical Society of America, Washington, D.C., 2002.
43. Yu, N.-T., DeNagel, D.C., Pruett, P.L., and Kuck, J.F.R., Jr., Disulfide bond formation in the eye lens, *Proc. Natl. Acad. Sci. U.S.A.*, 82, 7965, 1985.
44. Palmer, G.M., Marshek, C.L., Vrotsos, K.M., and Ramanujam, N., Optimal methods for fluorescence and diffuse reflectance measurements of tissue biopsy samples, *Lasers Surg. Med.*, 30, 191, 2002.
45. Williams, A.C., Barry, B.W., Edwards, H.G., and Farwell, D.W., A critical comparison of some Raman spectroscopic techniques for studies of human stratum corneum, *Pharm. Res.*, 10, 1642, 1993.
46. ANSI, American National Standards for Safe Use of Lasers, American National Safety Institute, Orlando, FL, 2000.
47. Lakowicz, J.R., *Principles of Fluorescence Spectroscopy*, Plenum Press, New York, 1983.
48. Van Duyne, R.P., Jeanmaire, D.L., and Shriver, D.F., Mode-locked laser Raman spectroscopy: a new technique for the rejection of interfering background luminescence signals. *Anal. Chem.*, 46, 213, 1974.
49. Mosier-Boss, P.A., Lieberman, S.H., and Newberry, R., Fluorescence rejection in Raman spectroscopy by shifted-spectra, edge detection, and FFT filtering techniques, *Appl. Spectrosc.*, 49, 630, 1995.
50. Zhang, D.M. and Ben-Amotz, D., Enhanced chemical classification of Raman images in the presence of strong fluorescence interference, *Appl. Spectrosc.*, 54, 1379, 2000.
51. O'Grady, A., Dennis, A.C., Denvir, D., McGarvey, J.J., and Bell, S.E.J., Quantitative Raman spectroscopy of highly fluorescent samples using pseudosecond derivatives and multivariate analysis, *Anal. Chem.*, 73, 2058, 2001.
52. Barclay, V.J., Bonner, R.F., and Hamilton, I.P., Application of wavelet transforms to experimental spectra: smoothing, denoising, and data set compression, *Anal. Chem.*, 69, 78, 1997.
53. Cai, T.T., Zhang, D.M., and Ben-Amotz, D., Enhanced chemical classification of Raman images using multiresolution wavelet transformation, *Appl. Spectrosc.*, 55, 1124, 2001.
54. Vickers, T.J., Wambles, R.E., and Mann, C.K., Curve fitting and linearity: data processing in Raman spectroscopy, *Appl. Spectrosc.*, 55, 389, 2001.
55. Baraga, J.J., Feld, M.S., and Rava, R.P., Rapid near-infrared Raman spectroscopy of human tissue with a spectrograph and CCD detector, *Appl. Spectrosc.*, 46, 187, 1992.
56. Lieber, C.A., Molpus, K., Brader, K., and Mahadevan-Jansen, A., Diagnostic tool for early detection of ovarian cancers using Raman spectroscopy, in *Biomedical Spectroscopy: Vibrational Spectroscopy and Other Novel Techniques*, Mahadevan-Jansen, A. and Puppels, G.J., Eds., 3918, SPIE, Bellingham, WA, 2000.
57. Liu, C.H., Das, B.B., Sha Glassman, W.L., Tang, G.C., Yoo, K.M., Zhu, H.R., Akins, D.L., Lubicz, S.S., Cleary, J., Prudente, R., Cellmer, E., Caron, A., and Alfano, R.R., Raman, fluorescence and time-resolved light scattering as optical diagnostic techniques to separate diseased and normal biomedical media, *J. Photochem. Photobiol. B*, 16, 187, 1992.

58. Mahadevan-Jansen, A., Mitchell, M.F., Ramanujam, N., Malpica, A., Thomsen, S., Utzinger, U., and Richards-Kortum, R., Near-infrared Raman spectroscopy for *in vitro* detection of cervical precancers, *Photochem. Photobiol.*, 68, 123, 1998.
59. Berger, A.J., Koo, T.W., Itzkan, I., Horowitz, G., and Feld, M.S., Multicomponent blood analysis by near-infrared Raman spectroscopy, *Appl. Opt.*, 38, 2916, 1999.
60. Shafer-Peltier, K.E., Haka, A.S., Fitzmaurice, M., Crowe, J., Myles, J., Dasari, R.R., and Feld, M.S., Raman microspectroscopic model of human breast tissue: implications for breast cancer diagnosis *in vivo, J. Raman Spectrosc.*, 33, 552, 2002.
61. Koljenovic, S., Choo-Smith, L.P., Bakker Schut, T.C., Kros, J.M., van den Berge, H.J., and Puppels, G.J., Discriminating vital tumor from necrotic tissue in human glioblastoma tissue samples by Raman spectroscopy, *Lab Invest.*, 82, 1265, 2002.
62. Caspers, P.J., Lucassen, G.W., Carter, E.A., Bruining, H.A., and Puppels, G.J., *In vivo* confocal Raman microspectroscopy of the skin: noninvasive determination of molecular concentration profiles, *J. Invest. Dermatol.*, 116, 434, 2001.
63. Stone, N., Kendall, C., Shepherd, N., Crow, P., and Barr, H., Near-infrared Raman spectroscopy for the classification of epithelial pre-cancers and cancers, *J. Raman Spectrosc.*, 33, 564, 2002.
64. Nijssen, A., Bakker Schut, T.C., Heule, F., Caspers, P.J., Hayes, D.P., Neumann, M.H., and Puppels, G.J., Discriminating basal cell carcinoma from its surrounding tissue by Raman spectroscopy, *J. Invest. Dermatol.*, 119, 64, 2002.
65. Cancer Facts and Figures 2002, American Cancer Society, New York, 2002.
66. Pearce, E., *Anatomy and Physiology for Nurses,* Oxford University Press, New York, 1989.
67. Robbins, S.L., Cotran, R.S., and Kumar, V., *Pathologic Basis of Disease,* W.B. Saunders, Philadelphia, 1994.
68. Manoharan, R., Shafer, K., Perelman, L., Wu, J., Chen, K., Deinum, G., Fitzmaurice, M., Myles, J., Crowe, J., Dasari, R.R., and Feld, M.S., Raman spectroscopy and fluorescence photon migration for breast cancer diagnosis and imaging, *Photochem. Photobiol.*, 67, 15, 1998.
69. Centeno, J.A., Mullick, F.G., Panos, R.G., Miller, F.W., and Valenzuela-Espinoza, A., Laser-Raman microprobe identification of inclusions in capsules associated with silicone gel breast implants, *Mod. Pathol.*, 12, 714, 1999.
70. Ferenczy, A. and Winkler, B., Cervical intraepithelial neoplasia, in *Pathology of the Female Genital Tract,* Blaustein, A., Ed., Springer-Verlag, New York, 1982, p. 156.
71. Bosch, F.X., Munoz, N., and de Sanjose, S., Human papillomavirus and other risk factors for cervical cancer, *Biomed. Pharmacother.*, 51, 268, 1997.
72. Myers, E.R., McCrory, D.C., Subramanian, S., McCall, N., Nanda, K., Datta, S., and Matchar, D.B., Setting the target for a better cervical screening test: characteristics of a cost-effective test for cervical neoplasia screening, *Obstet. Gynecol.*, 96, 645, 2000.
73. Robichaux, A., Shappell, H., Huff, B., Jones, H., and Mahadevan-Jansen, A., *In vivo* detection of cervical dysplasia using near infrared Raman spectroscopy, *Lasers Surg. Med.*, Suppl. 14, 5, 2002.
74. Sterenborg, N.J., Thomsen, S., Jacques, S.L., Duvic, M., Motamedi, M., and Wagner, R.F., Jr., *In vivo* fluorescence spectroscopy and imaging of human skin tumors, *Dermatol. Surg.*, 21, 821, 1995.
75. Panjehpour, M., Julius, C.E., Phan, M.N., Vo-Dinh, T., and Overholt, S., Laser-induced fluorescence spectroscopy for *in vivo* diagnosis of non-melanoma skin cancers, *Lasers Surg. Med.*, 31, 367, 2002.
76. Williams, A.C., Edwards, H.G.M., and Barry, B.W., Fourier-transform Raman spectroscopy — a novel application for examining human stratum corneum, *Intl. J. Pharm.*, 81, R11, 1992.
77. Williams, A.C., Edwards, H.G.M., and Barry, B.W., Raman spectra of human keratotic biopolymers: skin, callus, hair and nail, *J. Raman Spectrosc.*, 25, 95, 1994.
78. Hata, T.R., Scholz, T.A., Ermakov, I.V., McClane, R.W., Khachik, F., Gellermann, W., and Pershing, L.K., Non-invasive Raman spectroscopic detection of carotenoids in human skin, *J. Invest. Dermatol.*, 115, 441, 2000.

79. Gniadecka, M., Wulf, H.C., Nielsen, O.F., Christensen, D.H., and Hercogova, J., Distinctive molecular abnormalities in benign and malignant skin lesions: studies by Raman spectroscopy, *Photochem. Photobiol.*, 66, 418, 1997.
80. Lieber, C., Robichaux, A., Ellis, D., and Mahadevan-Jansen, A., Detection of skin abnormalities using near infrared confocal Raman spectroscopy, *Lasers Surg. Med.*, Suppl. 14, 4, 2002.
81. Wolthuis, R., van Aken, M., Fountas, K., Robinson, J.S., Jr., Bruining, H.A., and Puppels, G.J., Determination of water concentration in brain tissue by Raman spectroscopy, *Anal. Chem.*, 73, 3915, 2001.
82. Boustany, N.N., Manoharan, R., and Dasari, R.R., Ultraviolet resonance Raman spectroscopy of bulk and microscopic human colon tissue, *Appl. Spectrosc.*, 54, 24, 2000.
83. Peterson, C.M., Peterson, K.P., Chaiken, J., Finney, W., Knudson, P.E., and Weinstock, R.S., Noninvasive monitoring of blood glucose using Raman spectroscopy, *Diabetes*, 49, 490, 2000.

# 31

# Quantitative Characterization of Biological Tissue Using Optical Spectroscopy


Irene Georgakoudi
*Massachusetts Institute of Technology*
*Cambridge, Massachusetts*

Jason T. Motz
*Massachusetts Institute of Technology*
*Cambridge, Massachusetts*

Vadim Backman
*Northwestern University*
*Evanston, Illinois*

George Angheloiu
*The Cleveland Clinic Foundation*
*Cleveland, Ohio*

Abigail S. Haka
*Massachusetts Institute of Technology*
*Cambridge, Massachusetts*

Markus Müller
*Massachusetts Institute of Technology*
*Cambridge, Massachusetts*

Ramachandra Dasari
*Massachusetts Institute of Technology*
*Cambridge, Massachusetts*

Michael S. Feld
*Massachusetts Institute of Technology*
*Cambridge, Massachusetts*




## 31.1 Introduction

Spectroscopic techniques examine different types of light–tissue interactions and provide biochemical and morphological information at the molecular, cell, and tissue levels in a noninvasive way. Because light delivery and collection are compatible with optical fibers and data analysis can be achieved in real time, spectroscopic information can serve as a powerful tool for assessing the state of tissue *in vivo* thus, guiding surgery or biopsy or monitoring the effects of treatment. Ultimately, optical techniques could eliminate the need for biopsy, at least in some cases, and allow for a single triage visit during which detection and treatment of a lesion could be combined. However, to optimize the use of optical methods in disease detection and treatment, it is important to understand and quantify the information provided by the detected optical signals. Spectroscopic techniques are ideal for this purpose.

Three different methods have been used to establish the presence of distinct spectral features for normal and diseased tissues *in vivo*. Empirical techniques use intensity information at specific wavelengths or


0-8493-1116-0/03/$0.00+$1.50

wavelengh ranges. Such techniques are easy to implement, but they do not use the characteristics of the full spectrum and lack quantitation. Statistical techniques, such as principal component analysis, are based on the analysis of the entire spectrum. However, this approach does not provide insights into the origins of the detected changes. In addition, statistical analysis methods usually assume a linear relationship between the components contributing to a given spectrum. Thus, extraction of quantitative information about the source of the spectroscopic signals is often difficult because of the turbid tissue nature or interference between different types of spectroscopic features (for example, fluorescence and Raman, or fluorescence, scattering, and absorption). Thus, it is necessary to develop models that take into account the optical, morphological, and biochemical properties of tissue in a quantitative way. Parameters extracted from model-based techniques can be used to classify tissue (e.g., normal vs. diseased) or to quantify tissue components (e.g., blood analyte concentrations).

In this chapter, we review how four spectroscopic techniques (fluorescence, diffuse reflectance, light scattering and Raman spectroscopy) can be used to extract quantitative information about tissue biochemistry, organization, morphology, and molecular composition and the corresponding changes that take place during the development of disease. In addition, we discuss the complementary character of these techniques and provide examples demonstrating the synergy and enhancement that can be achieved by combining spectroscopic information.

## 31.2 Characterization of Tissue Biochemistry Using Fluorescence Spectroscopy *in Vivo*

Fluorescence is one of the most widely used spectroscopic techniques at the clinical and the basic science levels.[1] It has been used as a tool for the detection of endogenous and exogenous chromophores as a means of localizing lesions and tailoring dosimetry for treatments such as photodynamic therapy.[2] In this section, we will focus on endogenous tissue fluorophores. Promising results have been reported on the use of steady-state and time-resolved fluorescence spectroscopy as a diagnostic modality in a number of tissues, including the gastrointestinal tract,[3–5] the uterine cervix,[6–8] the skin,[6–9] the bladder,[10] the oral cavity,[10–12] and the lung.[10–13]

For most epithelial tissues, a small number of endogenous fluorophores are present that can be excited in the 300- to 600-nm range. Among those, tryptophan, collagen, elastin, NAD(P)H, FAD, and porphyrins are the most prominent. Only two or three of these fluorophores are excited simultaneously for a specific excitation wavelength. Despite this fact, extraction of quantitative information about the fluorescence contributions of a specific chromophore is not trivial. Most of the difficulties are associated with the fact that measured tissue fluorescence can be highly distorted by tissue scattering and absorption.

In most cases, statistical or empirical algorithms are used to assess the sensitivity and specificity with which lesions can be distinguished from normal tissues based on their fluorescence characteristics. Empirical algorithms typically use the value or the ratio of the fluorescence intensity at specific excitation–emission wavelengths or wavelength ranges.[3] This is the approach adopted by most fluorescence imaging diagnostic systems used clinically.[13,14]

Principal component analysis is also a very useful statistical tool that is often used to decompose the spectra within a given data set into a linear combination of orthogonal basis spectra called principal components (PCs).[15] The first PC accounts for the spectral features that vary the most, and subsequent PCs represent features with progressively smaller variance. To describe a specific spectrum, a linear combination of PCs is used, with each PC weighed by the appropriate PC score. Typically, the values of these PC scores are used to determine the spectral features that are different between normal and diseased tissues and to develop diagnostic algorithms. For example, the scores of selected PCs describing measured fluorescence spectra at 337-, 380-, and 460-nm excitation were used to separate cervical squamous intraepithelial lesions (SILs) from non-SILs with a sensitivity of 82% and a specificity of 68%.[6] Statistical algorithms based on neural network nonlinear methods have also been developed and have shown, in some cases, superior diagnostic performance over multivariate statistical analysis-based algorithms.[16]

Such techniques are often useful in identifying spectral regions within which diagnostically useful fluorescence changes exist, and they can provide qualitative insights into the origins of these changes. However, in some cases subtle fluorescence differences can be masked by interferences introduced by tissue scattering and absorption. Thus, to acquire reliable quantitative information about the biochemical changes that take place during the development of disease and are represented in the tissue fluorescence, it is necessary to remove these distortions. Empirical and more theoretically rigorous models have been developed to achieve this.

An empirical model developed by Richards-Kortum et al.[17] expressed the measured fluorescence as the product of two factors: (1) a linear combination of all the fluorophore contributions representing intrinsic fluorescence and (2) an attenuation factor representing broadband attenuation due to scattering, and oxyhemoglobin attenuation due to blood absorption. This model has been used to describe fluorescence excited at 476 nm from normal and diseased human arterial *ex vivo* tissues in terms of the intrinsic fluorescence from structural proteins (collagen and elastin) and ceroid, and the attenuation due to hemoglobin and structural proteins.[17] Using the extracted fluorescence contributions for ceroid and structural proteins, diagnostic algorithms were developed that separated normal from diseased tissue with 91% sensitivity and 85% specificity.

This model has also been employed to describe measured fluorescence spectra excited at 337 nm and acquired *in vivo* from colposcopically normal and abnormal cervical tissues.[18] In this case, the measured fluorescence was described in terms of the intrinsic fluorescence from collagen, elastin, NAD(P)H, and FAD, and attenuation due to oxyhemoglobin and scattering. The latter was represented by a constant. After normalizing the collagen contribution for each site of each patient to the average collagen contribution to the fluorescence of the normal sites of the same patient, a decrease was observed in the collagen fluorescence of precancerous tissues compared to colposcopically normal ones.

A trend toward elevated levels of the normalized relative NAD(P)H contributions was also reported for precancerous tissues when compared to normal tissues in the same patient. To observe this trend, spectra were normalized to their peak intensity prior to model analysis. The relative NAD(P)H fluorescence contribution of each site was then divided by the average NAD(P)H contribution to the normal sites in the same patient. This type of analysis provided useful information for understanding the origins of some of the observed measured fluorescence changes. However, because of the required normalizations, it was not very useful diagnostically in a clinical setting.

It has been recognized for some time that the diffuse reflectance spectrum, measured simultaneously with fluorescence, can also provide a means to remove absorption and scattering distortions in the measured tissue fluorescence spectrum, and thus to extract the intrinsic fluorescence. This was implemented in early fluorescence studies designed to monitor NAD(P)H in metabolically active tissues, such as the brain, heart and liver. The basic rationale behind this approach is that fluorescence and reflectance photons are distorted similarly by scattering and absorption. A comprehensive review of relevant work has been presented by Ince et al.[19]

Initial expressions involved linear[20–24] or nonlinear[25–27] combinations of the fluorescence and reflectance spectral features at specific wavelengths. Unfortunately, these models are applicable only for the particular wavelength regimes for which they were developed and tested, and are not sufficient to recover the entire intrinsic tissue fluorescence spectrum line shape and intensity. To achieve that, it is necessary to take into account the wavelength-dependent optical properties of tissue in a more rigorous manner. To this effect, an analytic model based on a photon migration picture developed from Monte Carlo simulations was introduced by Wu et al.[28] This model combines fluorescence and reflectance measurements acquired over the same wavelength range using identical light delivery and collection geometries. An empirical modification of this model that takes into account changes in the optical properties at the excitation wavelength has been reported recently by Finlay et al.[29] This expression was used to follow quantitatively the kinetics of protoporphyrin IX photobleaching and photoproduct formation in the normal rat skin *in vivo*. Gardner et al.[30] used Monte Carlo simulations to relate the propagation of laser excitation light and fluorescence to the diffuse reflectance. With these empirically obtained expressions, an expression for the intrinsic fluorescence was derived. Durkin et al.[31] used Kubelka-Munk absorption and scattering coefficients obtained from diffuse reflectance and transmittance experiments to predict the intrinsic fluorescence spectrum. Although such a model could be very useful for *ex vivo* studies, the acquisition of parameters from transmittance measurements renders this approach impractical for *in vivo* tissue measurements. Zhadin et al.[32] expressed reflectance and fluorescence in terms of the medium's "darkness," a variable defined as the ratio of the absorption coefficient to the reduced scattering coefficient. The darkness parameter was extracted by inverting the reflectance and was used to extract the intrinsic fluorescence line shape from the measured fluorescence.

Most of these studies have been limited to wavelength ranges within which hemoglobin and water absorption are not very strong;[28–30,32,33] however, a large number of fluorescence studies have been conducted in the 400-nm emission range,[1,2] where important tissue chromophores, such as collagen and NAD(P)H, fluoresce strongly. Unfortunately, the Soret band of hemoglobin absorption is also in this spectral region, with a peak in tissue absorption at approximately 420 nm. This absorption can give rise to a dip in the measured bulk fluorescence that can be misinterpreted because the spectra are no longer simply a linear sum of spectral contributions from endogenous fluorophores such as collagen, NAD(P)H, elastin, etc. Such interference can not only lead to misinterpretation of the biochemical information conveyed by the measured fluorescence, but they can also mask small biochemical changes that take place in diseased tissue. Therefore, it is important to extract the tissue intrinsic fluorescence over a wide emission range that includes this important diagnostic region.

Recently, the photon-migration-based model developed by Wu et al.[28] was modified to extend its validity in regimes of significant absorption.[34] The ability of this model to recover the intrinsic fluorescence line shape *and* intensity has been validated using tissue phantoms with a wide range of physiologically relevant scattering and absorption properties.[35] This model has been used clinically to recover the intrinsic fluorescence of several types of tissue, including Barrett's esophagus,[36] the uterine cervix,[8] and the oral cavity.[12]

For these studies, fluorescence and reflectance spectra were acquired using a FastEEM instrument, depicted schematically in Figure 31.1.[37] This instrument allows collection of 11 fluorescence emission spectra excited between 337 and 610 nm and a white light (350 to 750 nm) reflectance spectrum in less than 1 s. Light is delivered to the tissue via an optical fiber probe,[38] which consists of a central light delivery fiber surrounded by six light collection fibers (all fibers have a 200-μm core diameter and a numerical aperture (NA) of 0.22). At the tip of the probe, all fibers are fused together, creating an optical shield approximately 1.5 mm in diameter. The shield is beveled at a 17° angle to eliminate detection of specular reflections. In addition, the shield provides a fixed geometry for light delivery and collection, which are identical for the measured fluorescence and reflectance spectra.

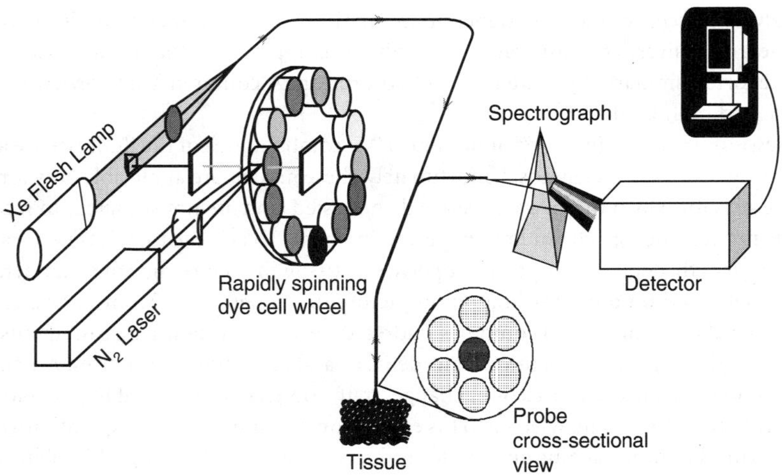


**FIGURE 31.1** Schematic representation of FastEEM instrument used to acquire fluorescence and reflectance tissue spectra *in vivo*. (From Georgakoudi, I. et al., *Gastroenterology*, 120, 1620, 2001. With permission.)

Extraction of intrinsic fluorescence spectra for *in vivo* tissues has allowed minimization of the variations in line shape and intensity in measured fluorescence spectra that result from the presence of physiological blood content variations.[12] In some cases, it is found that even significantly different measured fluorescence spectra correspond to similar intrinsic fluorescence spectra, mainly as a result of the removal of absorption distortions (Figure 31.2). Significant differences have been detected in the intrinsic fluorescence spectra of normal and diseased tissues in Barrett's esophagus, the uterine cervix, the oral cavity, and coronary arteries. Mean intrinsic fluorescence spectra for normal and diseased tissues based on the analysis of data from multiple patients are shown in Figure 31.3.

In contrast to the measured fluorescence spectra, which consist of nonlinear contributions from tissue fluorescence, scattering and absorption, intrinsic fluorescence spectra are composed of a linear combination of the fluorescence spectral features of the chromophores excited at a particular wavelength. Thus, once the component spectral features are identified, a simple linear decomposition can be performed to extract quantitative tissue biochemical information.

Biochemical decomposition of measured tissue fluorescence spectra has been performed previously using the spectral features of commercially available chromophores diluted in saline.[5,18] However, because fluorescence is sensitive to the local environment of the chromophore, it does not necessarily follow that such spectra are an accurate representation of the chromophores' spectral features *in vivo*. To extract the fluorescence signatures of two of the major tissue chromophores, namely collagen and NAD(P)H, in an *in vivo* environment, fluorescence EEMs and reflectance spectra were acquired during asphyxiation (i.e., elimination of blood flow) of human esophageal tissue *in vivo*.[39] The changes in the tissue redox state that take place during loss of oxygen were evident in the measured tissue reflectance spectra, and they were accompanied by spectral changes in the intrinsic tissue fluorescence.

These changes were expected because the levels of NAD(P)H should increase as the levels of tissue oxygen decrease. The observed changes in intrinsic fluorescence during tissue deoxygenation could be described accurately by two spectral components extracted by analysis of the intrinsic fluorescence using a multivariate curve resolution (MCR) algorithm.[39] The fluorescence spectral features of these two components are consistent with those of collagen (commercially available collagen Type I) and NAD(P)H (from isolated cervical tissue epithelial cells).[39] Thus, it could be concluded that the MCR extracted fluorescence EEMs (Figure 31.4) represent the spectral signatures of collagen and NAD(P)H *in vivo*.

A linear combination of the *in vivo* NAD(P)H and collagen EEMs extracted from the tissue asphyxiation measurements was fit to intrinsic fluorescence excitation–emission matrices of normal and diseased tissues. This decomposition provided quantitative information about the biochemical make-up of tissue and the changes that take place during the development of disease. The results of this decomposition for Barrett's esophagus, cervical, oral, and coronary artery tissue sites are included in Figure 31.5. Note that no normalizations were performed prior to biochemical decomposition of these data sets.

Barrett's esophagus is defined as the replacement of normal squamous esophageal tissue by metaplastic columnar epithelium. Patients with Barrett's eosphagus are at higher risk for developing adenocarcinoma of the esophagus. As a result, they undergo regular endoscopic surveillance procedures in an attempt to detect changes at the precancerous or dysplastic stage when treatment can be effective. Unfortunately, dysplastic changes are endoscopically invisible. Thus, random biopsies are acquired, typically, one at each quadrant of the esophagus for every 2 cm of the Barrett's segment. A technique that would allow the physician to detect precancerous lesions or to serve as a guide to biopsy could improve significantly the clinical management of these patients.

When a linear combination of NAD(P)H and collagen fluorescence spectra was fit to *in vivo* fluorescence spectra from seven patients with Barrett's esophagus, it was found that high-grade dysplastic tissues have lower levels of collagen and increased levels of NAD(P)H when compared to nondysplastic tissues. This information is diagnostically useful, and it provides important insights about the biochemical changes that occur in tissue during the development of premalignancies. For example, the increase in NAD(P)H for high-grade tissues is consistent with cellular hyper-proliferation or increased metabolic activity.[23] The decrease in collagen fluorescence could be the result of degradation in the collagenous network of the connective tissue. This in turn could be the result of an increase in the activity of

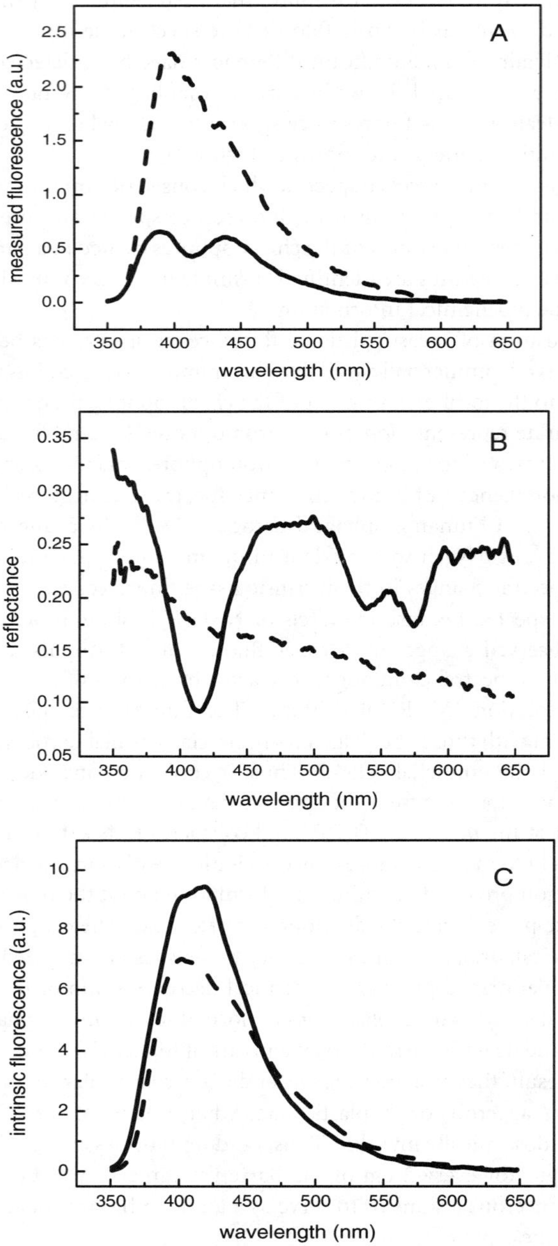


**FIGURE 31.2** (A) Measured fluorescence spectra from two normal oral epithelial tissue sites, 337-nm excitation. (B) Corresponding measured reflectance spectra. (C) Corresponding intrinsic fluorescence spectra. Notice that the intrinsic fluorescence spectra exhibit much more similar spectral features than the corresponding measured spectra.

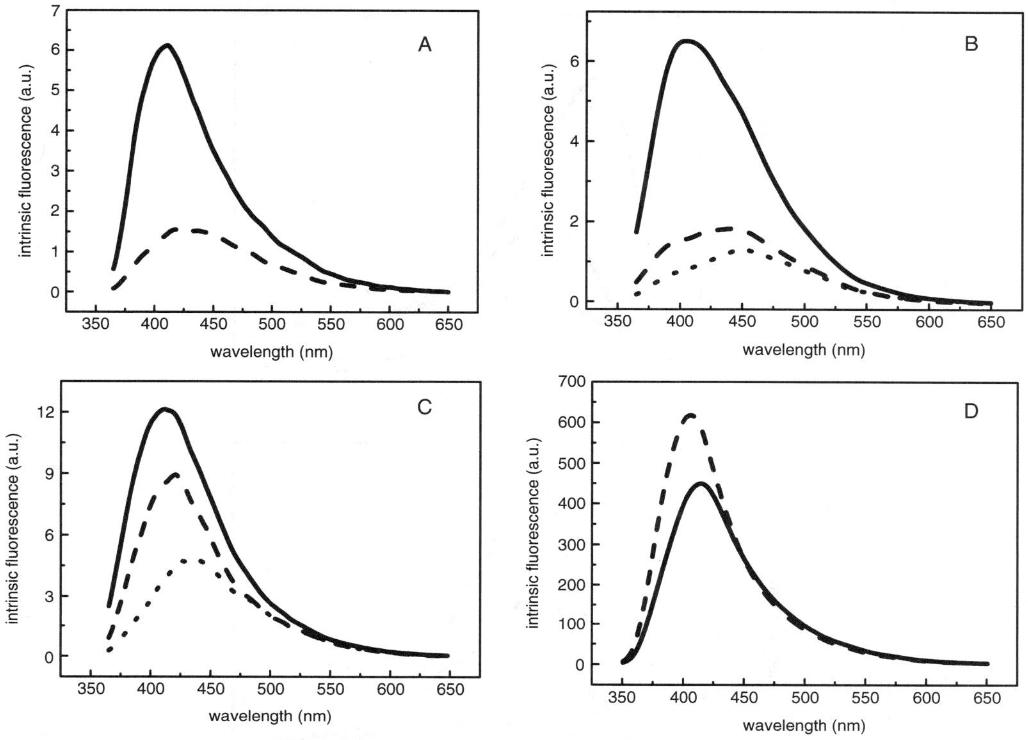


**FIGURE 31.3** Mean intrinsic fluorescence spectra from (A) ten nondysplastic (solid line) and five high-grade dysplastic lesions (dashed line) in seven Barrett's esophagus patients. (B) 43 normal squamous (solid line), 12 biopsied squamous metaplastic (dashed line), and 10 biopsied high-grade squamous intraepithelial lesions (dotted line) of the cervix from 35 patients. (From Georgakoudi, I. et al., *Am. J. Obstet. Gynecol.*, 186, 374, 2002. With permission). (C) 28 normal (solid line), 8 dysplastic (dashed line), and 9 cancerous (dotted line) oral epithelial tissue sites from 15 patients. (D) 22 normal/intimal fibroplasia (solid line) and 88 atherosclerotic and atheromatous (dashed line) coronary artery tissue sites. All spectra were acquired *in vivo* with the exception of the coronary artery spectra. Excitation wavelength = 337 nm.

collagenases, enzymes that cleave collagen. Indeed, an increase in the level of serine and cysteine proteases has been found in gastric and colorectal cancerous and precancerous lesions.[40] Such proteases are known to be activators of matrix metalloproteinases, a prominent class of tissue collagenases.[41] The decrease in collagen fluorescence could also be at least partially attributed to an increase in epithelial thickness, which would limit the amount of light that reaches the underlying stroma.

Similar differences are found in the NAD(P)H and collagen fluorescence levels of normal and precancerous uterine cervical tissue sites.[8,39] During the normal reproductive life of a woman, squamous epithelium, the tissue type lining the ectocervix (i.e., the vaginal portion of the cervix), starts gradually to replace the columnar epithelium of the endocervix (i.e., the cervical canal that leads to the uterus). This replacement is known as squamous metaplasia and it takes place within the transformation zone. Most precancerous and cancerous cervical changes occur at the transformation zone. During colposcopy, the cervix is visualized under 6× or 15× magnification and colposcopically abnormal tissues are biopsied and examined histopathologically to determine the presence or absence of disease.

Although colposcopy and biopsy are highly sensitive techniques for detecting cervical cancerous and precancerous lesions known as SILs, they have very low specificity. That means that a significant number of tissue sites that appeared colposcopically abnormal are histopathologically normal and did not have to be biopsied. A technique that would improve upon the specificity of colposcopy would enhance its efficiency and have a significant impact on the time and resources dedicated to these procedures.

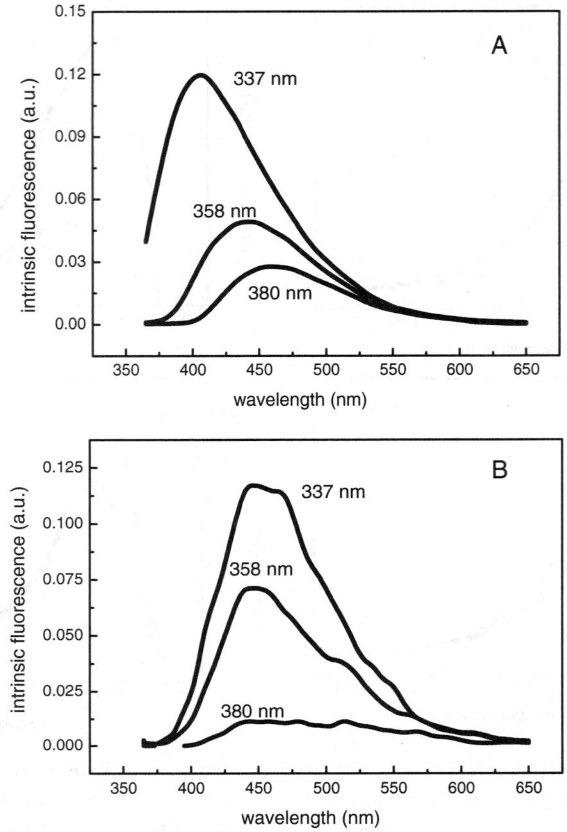


**FIGURE 31.4** (A) Intrinsic fluorescence spectra of tissue collagen extracted for 337-, 358-, and 380-nm excitation from measured fluorescence and reflectance spectra acquired during asphyxiation of human esophageal tissue *in vivo*. (B) Corresponding tissue NAD(P)H intrinsic fluorescence spectra.

Especially in the case of the cervix, a tool that could provide a diagnosis in real time would potentially allow the combination of detection and treatment of a lesion during a single patient visit, providing a further positive economic and psychological impact.

Results of the analysis of intrinsic fluorescence spectra acquired *in vivo* from colposcopically normal and colposcopically abnormal cervical tissue sites from 35 patients are shown in Figure 31.5B. Colposcopically abnormal tissue sites were biopsied and classified either as squamous metaplastic (i.e., benign) or SIL (i.e., precancerous). A significant decrease in collagen fluorescence was present between colposcopically normal and abnormal tissues.[8,39] As in the case of Barrett's esophagus, this decrease could be the result of collagenases. Indeed, it has been reported that differences can be found in the levels or patterns of expression of metalloproteinases in normal, squamous metaplastic and SIL cervical sites.[42]

As mentioned previously, most of the colposcopically abnormal tissue sites are found within the transformation zone of the cervix, an area of constant dynamic change. Higher levels of collagenase expression have been reported for tissues undergoing architectural changes as in tissue regeneration and wound healing.[43] In addition, we find that among the colposcopically abnormal tissues, SILs tend to have higher NAD(P)H fluorescence than squamous metaplastic sites. This increase could be attributed to increases in epithelial thickness or metabolic activity, as in the case of Barrett's esophagus. As indicated by the diagnostic threshold lines drawn based on logistic regression analysis, the colposcopically abnormal but histopathologically benign squamous metaplastic tissue sites can be separated fairly well from the

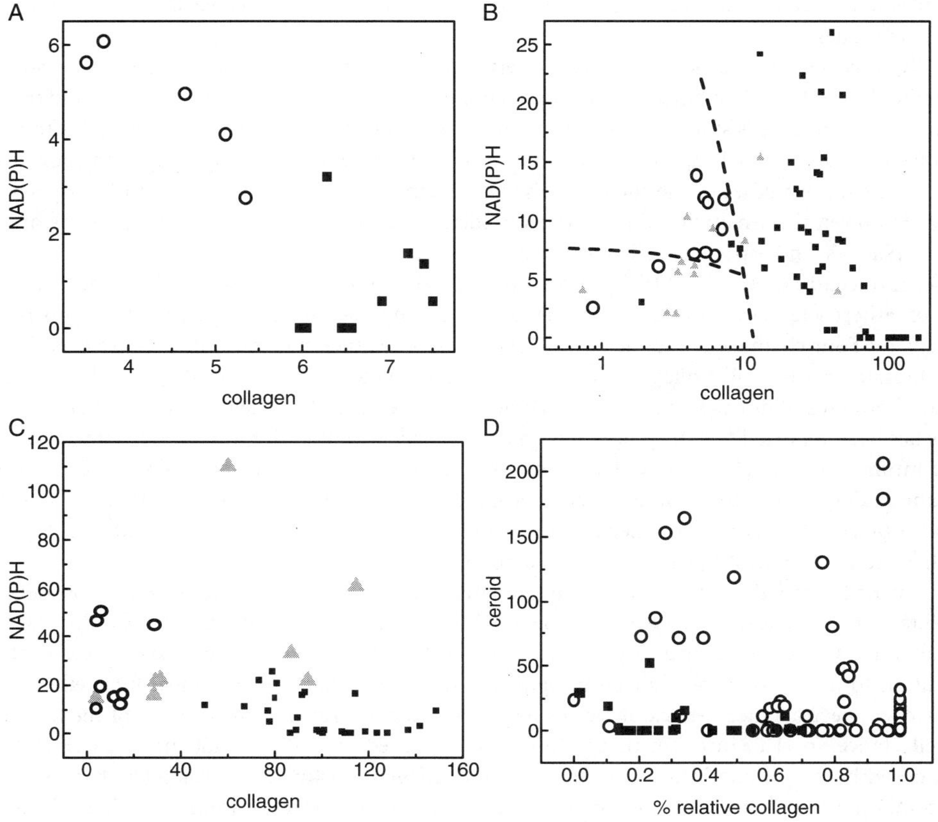


**FIGURE 31.5** NAD(P)H and collagen fluorescence contributions to intrinsic tissue fluorescence excitation–emission matrices (excitation = 337 to 425 nm; emission = 375 to 700 nm) of the patient populations described in Figure 31.3. (A) Nondysplastic (squares) and high-grade dysplastic (circles) Barrett's esophagus. (B) Normal squamous epithelium (squares), colposcopically abnormal squamous metaplastic, i.e., benign (triangles) and high-grade SILs (circles) of the uterine cervix. (C) Normal squamous (squares), dysplastic (triangles), and cancerous (circles) nonkeratinizing oral epithelial tissues. (D) Ceroid fluorescence extracted from analysis of intrinsic fluorescence spectra excited at 480 nm shown as a function of the percentage of the relative collagen contribution with respect to the sum of the collagen and elastin fluorescence contributions to intrinsic fluorescence spectra excited at 337 nm for normal/intimal fibroplasia (squares) and atheromatous/atherosclerotic (circles) coronary artery. The decision lines in (B) were drawn based on logistic regression analysis. ([A] and [B] from Georgakoudi, I. et al., *Cancer Res.*, 62, 682, 2002. With permission.)

histopathologically abnormal tissues, illustrating how quantitative biochemical information can be valuable in helping the physician decide which colposcopically abnormal tissue sites to biopsy.

A third tissue type for which intrinsic fluorescence spectra have been extracted and biochemically analyzed is oral squamous epithelium. Squamous cell carcinoma of the upper aerodigestive tract continues to be a major public health problem worldwide. In the U.S. alone, 40,000 new cases and 11,000 deaths were expected for the year 2000.[44] Despite advances in radiotherapy, chemotherapy, and surgery, patients' survival with oral cancer has not improved significantly. The entire oral cavity is lined with squamous epithelium; however, the detailed morphology and architecture of the tissue vary depending on its function. Initial studies indicate that it is diagnostically important to consider separately the intrinsic fluorescence properties of keratinized and nonkeratinized epithelia.[12] The NAD(P)H and collagen contributions to the intrinsic fluorescence of normal, dysplastic and cancerous nonkeratinized epithelial sites are shown in Figure 31.5C. A decrease in collagen and an increase in NAD(P)H fluorescence are observed in dysplastic and cancerous oral tissues, consistent with the changes observed in cervical and Barrett's esophagus tissues.

Finally, intrinsic fluorescence of coronary artery, a nonepithelial tissue, has been extracted and decomposed into fluorescent biochemical constituents. Despite major improvements in diagnosis and treatment, coronary artery disease is still the most significant cause of death in the United States, and is responsible for more than 3 million hospital admissions every year.[45] Coronary angiography, and more recently intravascular ultrasound, are the most widely used methods for invasive diagnosis of coronary artery disease. However, the sensitivity of these tools in identifying the morphological components of atherosclerotic plaques, and especially of necrotic core, is poor.

Laser-induced fluorescence (LIF) spectroscopy has been tested in the last decade in the *in vitro* diagnosis of atherosclerosis, with the purpose of improving the morphology-based diagnosis of the plaque.[17,46,47] Four chemical components with specific fluorescence spectral features have been identified in the human arterial wall: collagen, elastin, ceroid, and tryptophan.[17,46–48] Collagen and elastin are the structural proteins of normal and diseased arteries. An increase in the collagen content of atherosclerotic tissues has been reported.[49–51] Tryptophan is an amino acid present in the skeleton of numerous proteins in the intimal extracellular matrix that is increased in atherosclerotic plaques. Ceroid is an insoluble conglomerate with marked autofluorescence properties, present in the atherosclerotic macrophage cells and necrotic core.[52] Oxidized low-density lipoprotein is its main chemical component, which gives rise to a characteristic spectral fingerprint of the atherosclerotic lesion.[46,52,53]

After being harvested from cardiac transplant patients and postmortem autopsies, 110 coronary artery segments were investigated *ex vivo* with the FastEEM. While extracting the intrinsic fluorescence of normal and diseased coronary artery tissues, it was found that beta carotene, a second tissue absorber in addition to hemoglobin, was affecting significantly the line shapes of the measured reflectance and fluorescence spectra. Once beta carotene absorption was incorporated in the analysis of the fluorescence and reflectance spectra, intrinsic tissue fluorescence excited at 340 and 480 nm was extracted and decomposed biochemically into contributions from collagen, elastin, and ceroid. The basis spectra for these components were derived based on the fluorescence characteristics of commercially available elastin and collagen type I and MCR. A significant increase was found in the ceroid and collagen fluorescence of atheromatous and atherosclerotic specimens when compared to that of normal and intrimal fibroplasia tissues, consistent with the histopathological analysis of these specimens (Figure 31.5D).

In conclusion, in this section we summarized methodologies that can be used to extract quantitative tissue biochemical information based on fluorescence spectroscopic measurements. Removal of distortions introduced in measured tissue fluorescence by scattering and absorption is essential in achieving this goal. The extracted quantitative information can be diagnostically useful, providing important insights into the biochemical changes that take place *in vivo* during disease development. In addition, such information can be used to optimize the design of instruments to detect or monitor lesions based on fluorescence.

## 31.3 Characterization of Bulk Tissue Optical Properties Using Diffuse Reflectance Spectroscopy *in Vivo*

A diffuse reflectance spectrum arises from light that has been scattered multiple times within the sample of interest. The spectral features of light diffusely reflected from tissue depend on its scattering and absorption properties. Diffuse reflectance spectroscopy studies the changes in these optical properties associated with disease or therapy. Reflectance measurements can be performed with very short light pulses (time-domain), an intensity-modulated source (frequency-domain), or a steady-state broadband or monochromatic source. In this section, we will focus on the use of visible steady-state *in vivo* diffuse reflectance measurements to characterize tissue and provide clinically useful diagnostic information. A broad review of elastic light-scattering spectroscopy and diffuse reflectance is provided in Chapter 29 of this handbook.

A number of clinical studies performed with tissues such as colon,[54,55] bladder,[56] breast,[57] ovary,[58] and skin[59–63] demonstrate that diffuse reflectance spectra contain diagnostically useful information. Typically, an optical fiber probe is used to deliver white light from a high-power lamp onto the tissue and to collect the diffuse reflectance. Several algorithms have been developed based on statistical or on empirical approaches.

For example, the ratio of the area under the normalized reflectance curve between 540 and 580 nm to the area under the reflectance between 400 and 420 nm has been used to differentiate neoplastic from nonneoplastic colon tissues.[64] The slope of the reflectance in the 330- to 370-nm range was employed to distinguish between malignant and nonmalignant bladder tissues.[56] A combination of parameters related to the area under the reflectance curve, the reflectance slope and the mean intensity at specific wavelength bands were used by Wallace et al.[65] to distinguish between malignant melanoma and benign pigmented skin lesions. Such skin lesions were also separated with 80% sensitivity and 51% specificity using an imaging approach that combined reflectance-related parameters with morphological features of the lesion, such as dimension and roundness.[62] Neural network-based pattern recognition algorithms have been also used to identify characteristic skin,[63] breast,[57] and colonic[54] neoplastic reflectance features.

A semiempirical model developed to describe skin optical properties based on specific features of the reflectance was reported by Dawson et al.[60] In this model, skin was treated as a four-layer system with absorption due to fibrous protein, melanin, and hemoglobin in the first three layers, respectively, and scattering due to collagen and fat in the fourth layer. By assuming that the amount of reflected light from the first three layers is minimal, they found that the logarithm of the inverse skin reflectance (LIR) is equal to the sum of the absorbances of the first three layers minus the log of the reflectance of the fourth layer. The effects of scattering were empirically accounted for in the LIR spectrum by subtracting a line connecting the LIR intensities at 510 and 610 nm. Using the LIR values at 510, 543, 560, 576, and 610 nm, an erythema index was defined that was proportional to the hemoblobin content of the skin sample. An approximate correction for melanin scattering had to be implemented to ensure that the erythema index was not correlated with the pigmentation or melanin content of the skin. The melanin index was characterized empirically by using the slope of the LIR spectrum between 650 and 700 nm. However, this erythema index was dependent on the oxygen saturation levels of hemoglobin.

Feather et al.[66] showed that the hemoglobin index, which depended on the gradients of the LIR values at four isosbestic points between 527.5 and 573 nm, was independent of oxygen saturation and, thus, a more suitable parameter for quantifying the amount of cutaneous hemoglobin. Skin oxygen saturation levels could be estimated by combining the hemoglobin index with the LIR intensity at 558.5 nm. These empirical expressions for hemoglobin content and oxygen saturation were validated with *in vitro* measurements of red cells in plasma. To perform similar measurements *in vivo*, it was necessary to remove significant contributions from specular reflections at the skin surface. This was achieved by using crossed linear polarizers for light delivery and detection.[61] Further empirical modifications were also implemented to correct for hemoglobin absorption effects in the case of the melanin index, and for melanin absorption and scattering in the case of the hemoglobin index. The usefulness of these models for characterizing the optical properties of skin was illustrated by analyzing *in vivo* measurements performed with different skin colors and on a human forearm raised to different heights relative to the heart. However, there are no reports for this latter model being used clinically as an aid in detecting or treating skin lesions.

Another empirical model based on analysis of the LIR spectrum has been reported by Koenig et al.[67] and used to describe reflectance spectra collected from the bladder[67] and the colon.[54] First, measured reflectance spectra were converted into absorbance units using the expression, $A = -\log(R/Ro)$, where R and Ro were reflectance from tissue and a white reflectance standard, respectively. Then, a line was fit to the 640- to 820-nm range of the spectrum. This line was assumed to represent contributions from tissue scattering to the absorbance spectrum and was subtracted after it was extrapolated to cover the entire absorbance spectrum. Hemoglobin oxygen saturation was estimated based on the absorbance intensities at 555 and 577 nm, which are peak absorption wavelengths for deoxy- and oxyhemoglobin, respectively. The total amount of blood was estimated in the case of the bladder study based on these intensities and was found to be the most diagnostically useful parameter for discriminating neoplastic from nonneoplastic bladder tissue with high sensitivity (91%), but low specificity (60%).[67] The high blood content of inflammed tissues was thought to be the reason for the low specificity results. This model was also used to interpret the success of statistically based algorithms designed to differentiate between neoplastic from neoplastic colon tissues, as well as between adenomatous and hyperplastic colon polyps, which is the more clinically relevant problem in the colon.[54] However, the extracted parameters were not used as the basis of a diagnostic algorithm.

Zonios et al.[55] have developed a more theoretically rigorous model describing the measured tissue reflectance as a function of the absorption and reduced scattering coefficients for a light delivery/collection geometry consistent with that of an optical fiber probe. This model is based on an expression developed by Farrell et al.,[68] which describes the diffuse reflectance from a narrow pencil beam of light incident on the surface of a semi-infinite turbid medium in terms of the reduced scattering ($\mu_s'$) and absorption ($\mu_a$) coefficients and the source-collection fiber separation distance. The latter expression was derived using the diffusion approximation to the light transport equation. To acquire an analytical expression, Zonios et al. assumed point delivery of light and collection over a circular spot with an effective radius, extracted from the model, approximately equal to the radius of the optical fiber probe. This model is particularly appropriate for tissues that can be modeled by a single layer.

Some *a priori* knowledge with regard to the spectral features of the tissue's scattering and absorption is also required before using this model to describe tissue reflectance spectra. This is needed because the measurements employ a probe with a single source-collection fiber separation; hence, only a single piece of information is at each wavelength, insufficient to extract the diagnostic parameters, $\mu_s'$ and $\mu_a$, independently. In particular, the identity of the absorbing tissue chromophores and their corresponding tissue spectra must be known. In most cases, this does not present a serious limitation. For example, in the case of most epithelial tissues, oxy- and deoxyhemoglobin, whose extinction coefficients have been studied widely, are the only significant absorbers in the visible region of the spectrum. In the case of other tissues such as breast and artery, additional absorbers, such as beta carotene, need to be included. Melanin absorption also needs to be included for tisssues such as skin. The level of agreement between the observed reflectance and the model fit can be used to ascertain whether all the absorbers have been accounted for.

This model has been used to describe, with good agreement, reflectance spectra acquired *in vivo* from tissues such as the colon,[55] Barrett's esophagus[36] (Figure 31.6A), the cervix,[8] the oral cavity,[12] the artery (Figure 31.6B), and the breast. Notice the differences in the reflectance spectral features of Barrett's mucosa and coronary artery tissue, attributable mainly to the presence of beta carotene absorption in coronary artery. Based on this model analysis, quantitative information was extracted with regard to the bulk tissue scattering and absorption properties. For example, the extracted total hemoglobin concentration was significantly higher for adenomatous polyps than for normal colon mucosal tissues, while no significant changes in tissue oxygenation were detected.[55] Coronary artery tissue classified histopathologically as normal or as intimal fibroplasia exhibits consistently low levels of beta carotene absorption, in contrast to calcified and noncalcified atheromatous and atherosclerotic coronary arteries, which have varying amounts of beta carotene (Figure 31.6D).

In addition, the reduced scattering coefficient of colon polyps was generally lower and varied less as a function of wavelength than that of normal colon mucosa. This is a diagnostically useful trend that has also been detected between dysplastic and nondysplastic tissues such as Barrett's esophagus,[36] the cervix,[8] and the oral cavity.[12] To quantify these changes, the slope and intercept of a line fit to the wavelength-dependent $\mu_s'$ was used. The gradual decrease in the slope and intensity of $\mu_s'$ as Barrett's esophagus tissue is transformed from nondysplastic to low-grade and high-grade dysplasia is shown in Figure 31.6C. This decrease in the reduced scattering coefficient could be due to an increase in epithelial thickness, which in turn leads to less light reaching the highly scattering collagen fibers of the stroma; thus, less light is reflected within the collection angle of the fiber probe. In addition, changes in the density of the collagen fibers resulting from an increase in the activity of collagenases, as discussed in the fluorescence section, could also contribute to the observed decrease in the reduced scattering coefficient.

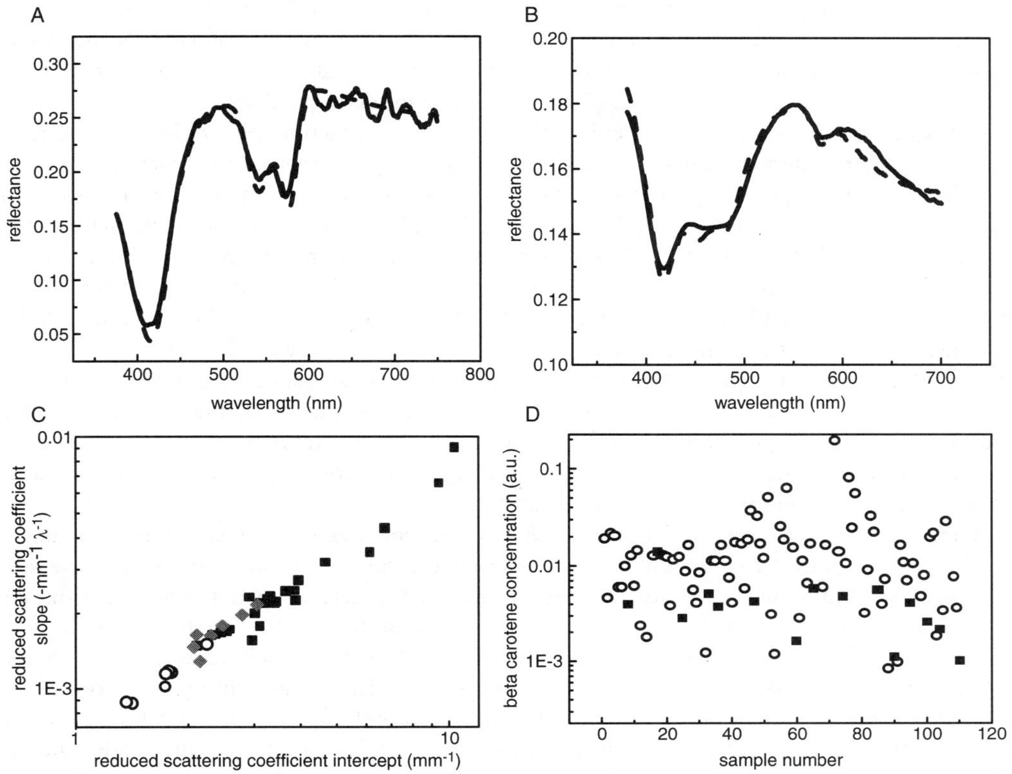


**FIGURE 31.6** Measured reflectance spectra (solid lines) from a nondysplastic Barrett's esophagus tissue site (A) and an atheromatous coronary artery tissue sample (B) with corresponding fits (dashed lines). (C) Slope and intercept of line fit to the wavelength-dependent reduced scattering coefficient for nondysplastic (squares), low-grade (diamonds), and high-grade (circles) dysplastic Barrett's esophagus tissues. (D) Beta-carotene concentration extracted from the absorption coefficient of normal/intimal fibroplasia (squares) and atheromatous/atherosclerotic (circles) coronary artery. ([A] and [C] from Georgakoudi, I. et al., *Gastroenterology*, 120, 1620, 2001. With permission.)

These results demonstrate that diffuse reflectance spectroscopy is a promising tool that can be used clinically to detect and quantify changes in tissue biochemistry and morphology. Further improvements in the models used to describe measured tissue diffuse reflectance spectra, that would account more rigorously for the layered tissue architecture[69,70] and small source-detector separations,[71,72] could enhance the sensitivity of these measurements and allow the detection of more subtle changes with higher accuracy.

## 31.4 Characterization of Cellular and Subcellular Morphology Using Light-Scattering Spectroscopy

Light-scattering spectroscopy (LSS) is a novel optical technology for imaging and characterizing the morphology of subcellular organelles within the epithelial linings of the body.[73,74] LSS is based on the fact that the spectral and angular distributions of light scattered by a particle depend on the size, shape, and internal structure of this particle. Thus, by analyzing light that has undergone a single backscattering event in tissue, we can extract quantitative morphological information about the scattering particle. As discussed next, the backscattering angle at which light is collected can be adjusted to gain some selectivity in terms of the size of the scattering particle that is characterized spectroscopically. For example, scattering in almost the exact backward direction provides detailed information about the cell nucleus.[8,36,73–75] Changes in the morphology of the nucleus, such as enlargement, pleomorphism (variation in size and shape), crowding (increase in the number of nuclei per unit area), and hyperchromatism (increase in chromatin content or nuclear material) comprise important histopathologic hallmarks for the diagnosis of precancerous and cancerous lesions; thus, this technique bears great promise as a noninvasive tool for characterizing important tissue morphological features and for detecting disease.

Light scattering measurements are performed routinely for extracting cell size in flow cytometry.[76] Goniometric light scattering studies, i.e., studies that examine light scattering as a function of scattering angle, have also been performed in the 2- to 170°-range with suspensions of cells and subcellular organelles to characterize their scattering properties.[77–80] Unfortunately, single scattering events are masked in biological tissue, because only a small portion of the light incident on tissue is returned after a single scattering.[81] The rest enters the tissue and undergoes multiple scattering from a variety of tissue constituents. As a result, it becomes randomized in direction, producing a large background of diffusely scattered light. (This is the diffuse reflectance component discussed in the previous section.) This diffuse component is influenced significantly by hemoglobin absorption and scattering by cells, as well as by noncellular structures, such as collagen. As discussed earlier, important information can be obtained about the overall tissue optical properties from analysis of the diffuse reflectance spectra. In this section, we focus on the light representing single backscattering events. To study this, the diffusely scattered component must be removed from the overall reflected light.

Several approaches may be used to accomplish this. One employs a mathematical model to describe the diffusive background. Examples of such models were described in the previous section. The diffuse reflectance model fit is subtracted from the overall measured reflectance spectrum to obtain the single scattering component. Polarized light can be used to accomplish this as well.[77,82] This approach uses the fact that incident light singly scattered in the backward direction retains its polarization, whereas multiply scattered light becomes depolarized. Thus, by using linearly polarized incident light, the contribution due to single scattering can be obtained as the difference between the components of light scattered from the tissue polarized parallel and perpendicular to the direction of polarization of the incident light.

Recent measurements on tissue phantoms consisting of water and polystyrene beads using a single 200-μm core diameter fiber for light delivery and collection indicate that such a probe geometry could also be sensitive to the size distribution and refractive index of the scattering particles near the surface of the sample. In addition, this study suggests that determination of the size and refractive index in this case could be achieved simply by studying the features of the intensity oscillations of the collected light (i.e., background removal would not be required).[83]

The spectrum of the single scattering component can be analyzed to obtain information about the properties of the scatterers.[77,82,83] The origin of LSS signals depends on the geometry of collection. For example, particles that are large compared to the wavelength have strong scattering components in the near-exact forward and backward directions (even though backscattering is much weaker than forward scattering).[84] Cells and nuclei have the right size to exhibit this type of scattering behavior. Early light scattering measurements performed with suspensions of HeLa cells, Chinese hamster oocytes, and white blood cells indicated that the detected scattering patterns were in good agreement with Mie scattering from a dense sphere embedded within a larger softer sphere.[85] However, in tissue, because cells are adjacent to one another, there is index matching among the cell membranes of neighboring cells; thus, the nucleus becomes the major large size scatterer.

In addition, the intensity of the backscattered light from large particles oscillates in intensity as a function of wavenumber with a frequency characteristic of the particle size and relative refractive index.[84] Particles that are small compared to the wavelength, such as the tubules of the endoplasmic reticulum, scatter light in an approximately isotropic fashion over all angles. The light-scattering intensity as a function of wavelength also exhibits very broad features.[84] The angular and wavelength-dependent distributions of the light intensity scattered by particles whose size is comparable to the wavelength, such as mitochondria and lysosomes, exhibit significantly broader features than the corresponding spectra of large particles — but not as broad as those of small particles.[84] For cells with a high volume fraction of mitochondria, such as hepatocytes, the scattering properties are dominated by those of mitochondria, even in the exact backward direction.[86] When particles of several sizes are present, the resulting signal is a superposition of these variations. Thus, the size distribution and refractive index of the scatterers can be determined from analysis of the spectrum of light backscattered by these particles. Once the size distribution and refractive index are known, quantitative measures characterizing alterations of morphology of the epithelial cells can be obtained, and corresponding diagnostic algorithms can be developed.

Preliminary *in vivo* studies have been performed to assess the potential of this technique as a tool for detecting precancerous and early cancerous changes in five different organs with three different types of epithelia: columnar epithelia of the colon and Barrett's esophagus,[36,73,75] transitional epithelium of the urinary bladder,[73] and stratified squamous epithelium of the oral cavity[35,73] and the uterine cervix.[8] Reflectance spectra were collected using the FastEEM instrument described in Section 31.2. The diffuse background was subtracted from measured reflectance spectra using modeling.[55] The light scattering spectra were analyzed based on the van de Hulst approximation for light scattering by particles that are large compared to the wavelength.[74] The extracted size distributions were then used to obtain quantitative measures characterizing the degree of nuclear enlargement and crowding. Figure 31.7 displays these LSS parameters in binary plots to show the degree of correlation with histological diagnoses. In all five organs there is a clear distinction between dysplastic and nondysplastic epithelium. Both dysplasia and CIS exhibit a higher population density of nuclei than normal tissues and either a higher percentage of enlarged nuclei (Figure 31.7A through D) or a larger variation in the distribution of nuclear size. These features can be used as the basis for spectroscopic tissue diagnosis.

Studies of backscattering of linearly polarized light from cells,[77,82] *ex vivo*[77,82,87–90] and *in vivo*[77] tissues have also shown great promise for characterizing cell and tissue morphology. Experiments have been performed during which the backscattered light collected from a small tissue area (approximately 3 mm$^2$) was detected along the parallel and the perpendicular polarization relative to the propagation of the incident light.[82] The light-scattering spectrum extracted by subtraction of the perpendicular from the parallel polarized light was analyzed to provide information about the size distribution of cell nuclei from cell monolayers and *ex vivo* tissues. Significant differences were detected in the nuclear size distributions from normal and tumor colon tissues, consistent with the results of histopathology from corresponding thin tissue sections.[82] Further measurements with tissue phantoms and *ex vivo* and *in vivo* tissues have been performed with a probe with a linear pollarizer at its tip, which consisted of a central light delivery fiber and four light collection fibers positioned symmetrically around the light collection fiber.[77] These studies concluded that the spectral features of backscattered light from a biological sample with the same polarization as the incident light are dependent on the internal morphology of the cell.

LSS measurements have also been performed in an imaging modality using polarization for background removal. Typically, the sample is illuminated with linearly polarized light, and the backscattered light is collected along the parallel and perpendicular polarizations. Jacques et al.[88] showed that, by combining the information in these two images, one can acquire an image sensitive only to the morphology of the upper few hundred micrometers of skin, where most skin cancers start to develop.

In a recent study by Gurjar et al.,[90] it was shown that, by imposing limitations on the angle of the backscattered light collected, quantitative information can be acquired about the size distribution and refractive index of epithelial cell nuclei (Figure 31.8). A 1.25 × 1.25-cm area was illuminated with narrowband collimated linearly polarized light. Light backscattered within one degree from the exact backward direction along the parallel and perpendicular polarizations with respect to the incident light was imaged onto a charge-coupled device (CCD) camera. By subtracting the perpendicular image from the parallel-polarized image pixel by pixel, the singly backscattered light intensity was extracted. A series of such images was acquired for 13 narrow wavelength bands, such that a light-scattering, wavelength-dependent spectrum could be constructed for each CCD pixel.

These spectra were analyzed using Mie scattering theory to determine the corresponding distributions of nuclear size and refractive index. While the spatial resolution of these images is limited by the CCD pixel size, the size distribution of the nuclei within that pixel can be determined with an accuracy of 0.1 μm, and the corresponding refractive index can be estimated with three-digit accuracy.[90] LSS images representing distinct regions of nuclear enlargement within the physical outline of an adenomatous colon polyp are shown in Figure 31.9. Similar maps have been obtained to represent an increase in the relative

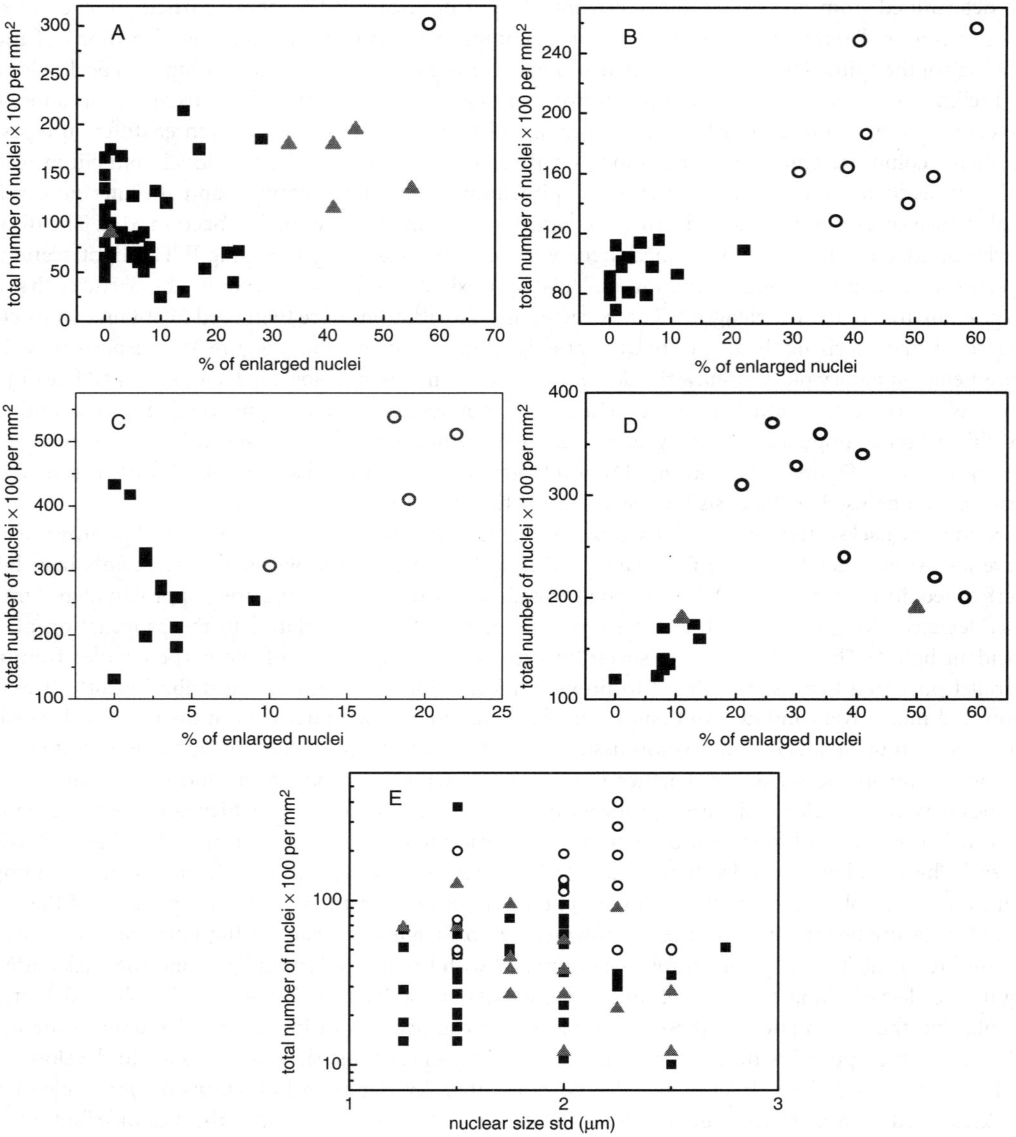


**FIGURE 31.7** The population density of nuclei is plotted against the percentage of enlarged nuclei (i.e., nuclei larger than 10 μm) for (A) Barrett's esophagus epithelium (squares, nondysplastic Barrett's mucosa; triangles, low-grade dysplasia; circles, high-grade dysplasia). (B) Colon epithelium (squares, normal colon mucosa; circles, adenomatous polyp). (C) Urinary bladder epithelium (squares, normal cells; circles, transitional cell carcinoma *in situ*). (D) Oral cavity epithelium (squares, normal cells; triangles, low-grade; circles, squamous cell carcinoma *in situ*). (E) Uterine cervical epithelium (squares, normal and squamous metaplastic epithelium; circles, SILs). Note that the standard deviation of the nuclear size distribution is plotted in the $x$ axis as opposed to the % enlarged nuclei in (E). ([A] through [D] from Backman, V. et al., *Nature*, 406, 35, 2000. [E} from Georgakoudi, I. et al., *Am. J. Obstet. Gynecol.*, 186, 374, 2002. With permission.)

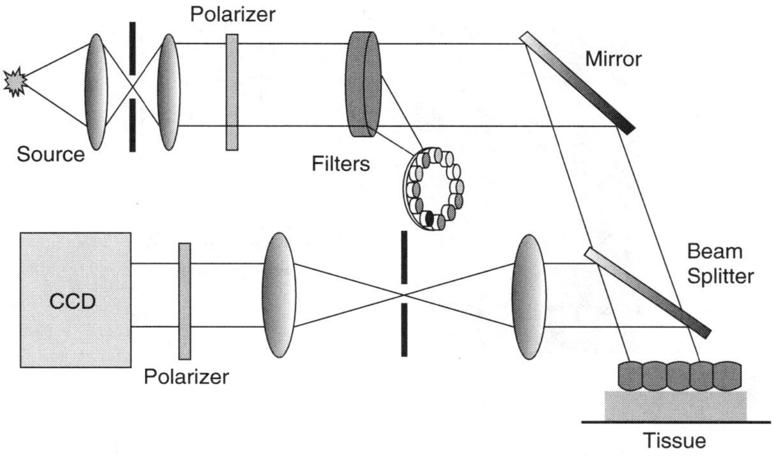


**FIGURE 31.8** Schematic diagram of instrument used for acquiring images with polarized LSS. (From Gurjar, R. et al., *Nat. Med.*, 7, 1245, 2001. With permission.)

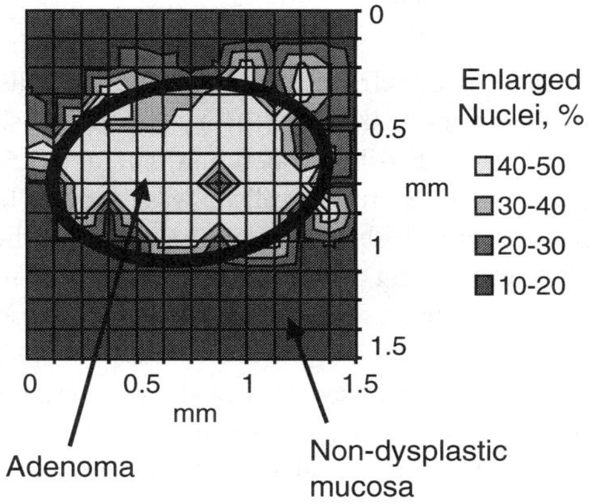


**FIGURE 31.9** LSS image showing the spatial distribution of the percentage of enlarged nuclei, i.e., nuclei larger than 10 μm, for an adenomatous colon polyp. (From Gurjar, R. et al., *Nat. Med.*, 7, 1245, 2001. With permission.)

refractive index of cell nuclei (i.e., hyperchromatism) within the adenomatous polyp.[90] Results of nuclear morphometry acquired in this fashion are consistent with hematoxylin and eosin stained thin sections prepared from the same polyp following acquisition of the LSS images.

Wavelength-dependent, light-scattering spectra have also been collected as a function of angle (angular LSS) from monolayer cultures of benign mesothelial intestinal cells and T84 tumor colonic cells.[91] Analysis of these measurements using Mie theory shows that the cell nuclei scatter predominantly within a narrow cone (1 to 2°) near the exact backward direction, in agreement with previous results as well as theoretical calculations (Figure 31.10). Thus, the signal measured within a sufficiently narrow solid angle provides information about the size and refractive index of the cell nuclei. The signal scattered into larger scattering angles (>3°), particularly with 45° azimuth with respect to the polarization of the incident light, provides information about subcellular structure at the submicron scale (see Figure 31.10).

In this angular regime, the LSS signals could not be fitted under the assumption that the particles are normally distributed for any given mean size from 0.3 to 15 μm. Moreover, other types of size distributions, such as top-hat, exponential, or skewed-normal could not fit the data. The spectra were found to be best described by an inverse power-law distribution for the number of particles, where the concentration $N(d)$ of particles with a diameter $d$, was given by $N(d) \propto d^{-\beta}$ with the exponent $\beta$ not dependent

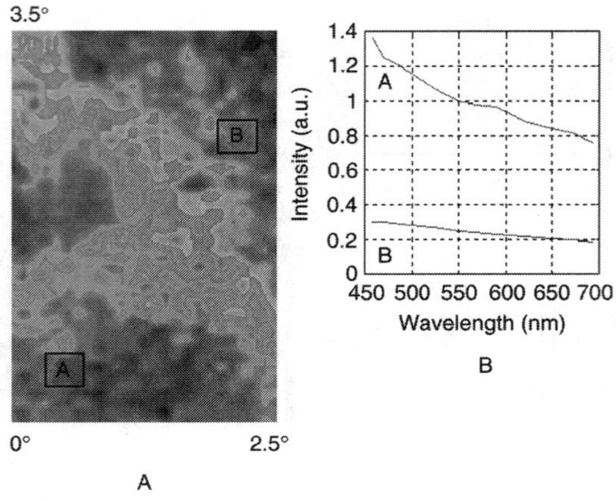


**FIGURE 31.10** Experiments with two-layer tissue models with the upper layer formed by a monolayer of T84 colon tumor cells. (A) A/LSS contour map at $\lambda = 532$ nm. (B) Spectra of light scattered by cells in regions A and B marked on map in (A). The analysis of these spectra enables extraction of quantitative characteristics of the size distributions of scattering particles responsible for the respective scattering spectra. (From Backman, V. et al., *IEEE J. Sel. Top. Quantum Electron.*, 7, 887, 2001. With permission.)

on *d* and characterizing the variations of refractive index inside the cells and, thus, the variations of intracellular density. To obtain the values of $\beta$, a Mie theory-based inversion procedure was used to fit the data. The sizes were assumed to be distributed between 0.01 and 1.5 μm according to an inverse power law, with fitting parameter $\beta$. For T84 tumor cells, $\beta$ was found to be approximately 2.2, whereas for normal mesothelial cells the best fit was obtained for $\beta = 2.7$. Thus, the size distribution of the T84 colon tumor cells is shifted with respect to that of the normal cells, with an increase in the relative number of large intracellular and intranuclear structures. We note that an inverse power-law distribution is a hallmark of fractal behavior.[92]

Inverse power-law distributions have been used previously to describe refractive index variations in frozen tissue sections[93,94] and have been associated with the fractal dimension of tissue. The smaller values of the size exponent for malignant cells can be correlated with certain alterations of normal cell structure associated with cancerous changes. For example, a smaller value of $\beta$ in the size distribution of the intranuclear structure correlates with the visual perception of clumped and rough chromatin when a stained tissue sample is microscopically evaluated. Pathologists evaluating hematoxylin and eosin–stained samples of precancerous cells frequently observe such changes. Smaller values of $\beta$ for malignant cells may also indicate higher structural entropy and, therefore, higher disorganization of the cell structure. Finally, a difference in $\beta$ may be related to change in the fractal properties of the malignant cells. Thus, angular LSS studies can provide information about the organization and packing of subcellular and subnuclear structures at scales significantly more sensitive than conventional microscopic imaging.

These initial studies demonstrate that LSS can be used not only to detect changes in nuclear morphology that are well established hallmarks of precancerous and cancerous lesions but also to characterize subtle changes in the morphology and organization and packing of cells and tissues that have not been explored previously.

## 31.5 Characterization of Tissue Molecular Composition Using Near-Infrared Raman Spectroscopy

Near-infrared (NIR) Raman spectroscopy is well suited for probing the detailed biochemical and morphological composition of human tissue.[95–97] Unlike fluorescence, reflectance, and NIR absorption spectroscopies, Raman spectroscopy provides narrow spectral bands with high information content that can be assigned to specific molecular vibrations. Whereas relatively few endogenous biological fluorophores exist, many of whose spectra significantly overlap, there exist a multitude of Raman active chemicals in tissue that have unique spectra and can be specifically related to healthy and diseased conditions.

Several features of NIR Raman spectroscopy are especially important to the study of human tissue. NIR excitation wavelengths have relatively small extinction coefficients and large penetration depths in human tissue (approximately 1 mm), providing the opportunity to observe subsurface structures. The small absorption coefficient also precludes photolytic sample decomposition. Additionally, in contrast to vibrational spectra obtained via mid-IR absorption, water is a relatively weak absorber in the NIR, and water interference is not a problem in Raman experiments. Furthermore, the strong fluorescence interference from biological tissue samples encountered with visible excitation wavelengths is significantly reduced in the NIR region, rendering Raman features more easily accessible.

Most importantly, the sharp bands in the IR fingerprint region can be conveniently studied as Raman shifts and allow identification and quantification of the chemical species involved. Quantitative analysis is achieved by examining the intensities of molecular vibrational spectral bands because they provide a direct reflection of molecular concentrations. Such an analysis is possible because the intensity of the Raman bands for a given moiety is linearly related to its concentration.

Several groups have investigated methods to employ the technique and accurately extract chemical concentrations.[98–104] The ability to quantitatively probe biochemical features that characterize normal tissue and accompany disease progression is of substantial importance in the utility of Raman spectroscopy as a diagnostic aid. Furthermore, in several diseases, prognosis is directly related to the concentration of key chemical constituents in the tissue. Although it is often difficult to determine the spectra of all the components of a Raman spectrum from a biological sample, due to its complexity, spectral analysis techniques can still be used to extract quantitative information. These techniques are known as multivariate analysis methods, or chemometrics, and can be implemented to extract chemical concentrations, even without complete knowledge of the chemical's Raman spectrum or cross section.[105,106] Although chemometric techniques are powerful algorithms in spectral analysis, the accuracy of a model built depends on the quality of the spectral data set and the accuracy of the reference measurements.

Several geometries can be employed for quantitative Raman spectroscopy. These include direct backscattering, excitation at various angles relative to the collection optics, microscopic illumination; they can include free-space optics, fiber optics, or a combination of the two. The particular geometry invoked depends on the nature of the sample and the conditions required for the particular application. The common overlying factor involved in Raman spectroscopy instrumentation is that the system must have a high throughput and efficiency because of the inherently weak nature of the Raman effect. Two specific instruments used in the development of quantitative spectroscopic models are presented here as examples. They have spectral resolutions of ~8 $cm^{-1}$, which is sufficient to resolve all relevant Raman bands.

The laboratory system used for studying biopsied biological tissue, shown in Figure 31.11A, is capable of collecting Raman and fluorescence spectra from macroscopic or from microscopic samples.[107] NIR Raman excitation at 830 nm is provided by an argon ion laser-pumped Ti:sapphire laser. The excitation laser beam traverses a bandpass filter and can be focused onto a bulk sample (~1-mm diameter spot) or coupled into the confocal microscope (~2-μm diameter spot) via a prism that can be moved in and out of position. If bulk tissue is studied, the Raman-scattered light is collimated, filtered, and coupled into an imaging spectrograph attached to a CCD detector.

For microscopic measurements, an epi-illuminated microscope is used. The microscope objective focuses the excitation light and collects the Raman scattered light in a backscattering geometry. A dichroic beamsplitter and mirror combination redirect the Raman-scattered light to the spectrograph system along the same optical path used for the bulk tissue system, after passing through a confocal pinhole to increase axial resolution. A CCD camera atop the microscope allows for registration of the focused laser spot with a white-light transilluminated image. Typically, a 63× infinity-corrected water immersion objective (0.9 NA) is used. The detector and the microscope translation stages are computer controlled; therefore, spectral maps of the tissue can be created by raster-scanning the translation stage under the microscope objective.

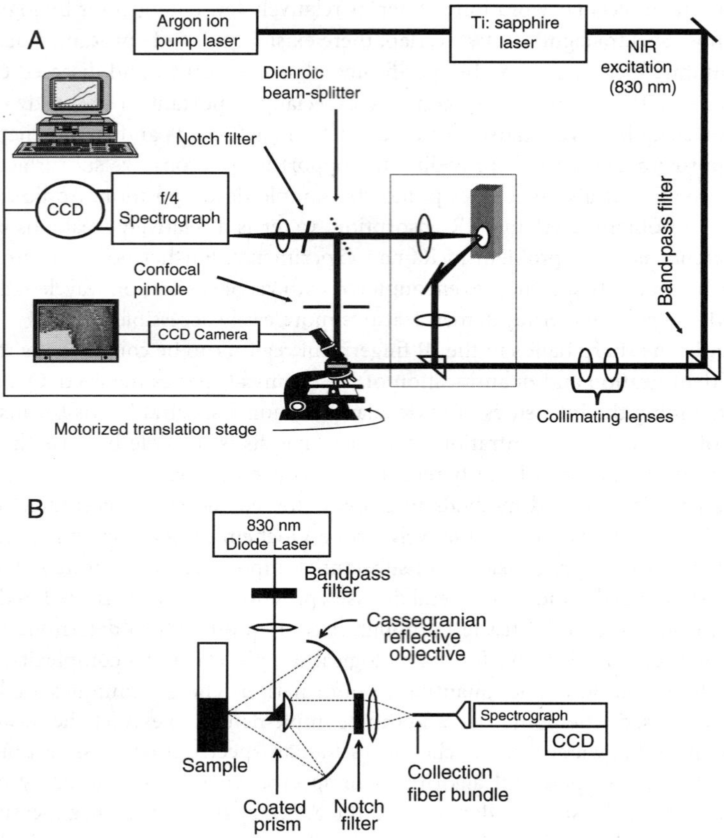


**FIGURE 31.11** (A) Schematic diagram of instrument designed to acquire fluorescence and Raman spectroscopy maps of thin samples and thick tissue specimens. (B) Instrument designed to acquired Raman spectra from blood samples. (From Berger, A.J. et al., *Appl. Opt.*, 38, 2916, 1999. With permission.)

Simply by changing the excitation source, the filters/beamsplitters, and spectrograph grating, the system can be adapted for collecting fluorescence. Three additional wavelengths, 352, 477, and 647 nm, are delivered to the system via optical fibers. The switch between Raman and fluorescence can be made in less than 2 min, thereby allowing collection of complementary data of point measurements or spectral maps.

The experimental setup designed for the measurement of aqueous biological analytes using NIR Raman spectroscopy is shown in Figure 31.11B.[103] Excitation light, provided by an 830 nm diode laser, is bandpass filtered and delivered to the sample via a prism mounted in a Cassegrain microscope objective. The Cassegrain provides coaxial excitation with wide-angle collection and the backscattered Raman light from the sample is notch filtered and reimaged onto an optical fiber bundle. The circular collection fiber bundle is converted to a linear array and coupled directly into an imaging spectrograph. The light is dispersed onto a CCD detector and the fiber bundle image is binned to produce a spectrum.

Raman spectra are often analyzed by looking at markers such as the ratio of the intensities of two Raman bands. Such approaches have been useful in developing diagnostic algorithms for diseases such as breast cancer and cervical cancer.[108–110] However, because the amount of information contained in Raman spectra is large and relevant to disease diagnosis and to the study of disease progression, it is prudent to develop spectroscopic models that take full advantage of this content. Methods that use information about the entire spectrum can be statistical or based on physical models.

Statistical methods are used when the spectra of the individual tissue components are not accurately known. There are two distinct types of problems in Raman spectral diagnosis of biological tissue:

classification analysis ("Is this specimen in category A or B?") and quantitative analysis ("How much of C is present?"); statistical methods can be applied to both. In the case of classification, statistical analysis is useful for proof-of-principle studies to evaluate the diagnostic capabilities of Raman spectroscopy when a new disease is being approached, or when there is an incomplete understanding of the basic constituents comprising the Raman spectra of the tissue so that a more physical model cannot be constructed. In this case, principal components are used to characterize the Raman spectra of a "calibration set" of samples containing A and B, as determined by pathology. As mentioned earlier, the principal components constitute an orthogonal set of spectra, a linear combination of which can accurately characterize each of the spectra in the sample set. By correlating the fit coefficients ("scores") of the spectra in the data set with their known classifications, a diagnostic algorithm can be established. This approach has been shown to be effective in several tissue types, including atherosclerosis, breast cancer, and gastrointestinal cancers.[108,111,112] However, its drawback is that the principal components are mathematical constructs, without direct physical meaning.

In the case of quantitative analysis of biological samples the procedure is similar, except that the calibration data set is composed of samples with various values of C, as determined by a standard reference technique. A variety of techniques to correlate the fit coefficients with the reference values for this type of problem include principal component analysis and partial least squares (PLS) analysis. An example of the use of PLS in determining the concentration of blood analytes is discussed next.

Methods of analyzing tissue "statistically," without full knowledge of the constituent spectra, are called implicit methods. On the other hand, if the spectra of the tissue constituents are accurately known, physically based "explicit" models can be developed. These can be based on the known chemical moieties in the tissue, its morphological constituents, or a combination of both.[107,113,114] Even if this approach cannot provide a complete characterization of all the spectral features from a given tissue type, it can usually account for the majority of them and subsequently lead to identification of more subtle, and yet important, constituents.

The use of physical models is based on the linear nature of the Raman effect; this allows the least-squares minimization reconstruction of the complex Raman spectrum of a heterogeneous macroscopic (~1 mm$^3$) sample from a set of "basis spectra" obtained separately from the individual chemical or morphological constituents. Such a model is mathematically described as follows:

$$R(\nu) = K\sum_i \rho_i r_i(\nu), \tag{31.1}$$

where K is a constant that depends on the experimental system and collection geometry, $R(\nu)$ is the Raman spectrum of the macroscopic sample, $r_i(\nu)$ is the Raman spectrum of a given basis spectrum, and $\rho_i$ is the concentration of the corresponding chemical or morphological component.

If the basis spectra are of individual chemicals, they can be obtained microscopically or macroscopically, and the intensity of various spectral features can be directly related to concentrations by controlling the experimental conditions when the model is developed. If the basis spectra are of individual morphological features, a confocal microscope of the type described previously can be employed to collect spectra of various structures. The latter approach has the advantage that the spectra are observed in their native environment. Although the morphological approach can be quantitative, in practice quantification is difficult because concentrations cannot be directly measured. Several studies have used the morphological approach to date;[107,113,114] however, they are only semiquantitative.

Different methods are appropriate for different applications. For example, in the case of measuring blood analytes, when chemical concentrations are derived, implicit methods are most appropriate. In other cases, such as breast or artery tissue analysis, when morphologic structures are characterized, explicit methods are the most appropriate. Below, we give examples of both. Combination methods that introduce physical information into statistical techniques, such as hybrid linear analysis,[115] are also possible.

Statistical methods can be used to build an accurate quantitative spectroscopic model. One such example is the measurement of concentration of blood analytes, with an ultimate goal of obtaining

**TABLE 31.1** Prediction Accuracy of Blood Analytes in Human Serum Using a PLS-Generated Algorithm

| Analyte | Reference Error | RMSEP | $r^2$ | Integration Time (s) |
|---|---|---|---|---|
| Glucose | 3 mg/dl | 26 mg/dl | 0.83 | 60 |
| Cholesterol | 4 mg/dl | 12 mg/dl | 0.83 | 60 |
| Triglyceride | 3 mg/dl | 29 mg/dl | 0.88 | 60 |
| BUN (urea) | 0.9 mg/dl | 3.8 mg/dl | 0.74 | 60 |
| Total protein | 0.1 g/dl | 0.19 g/dl | 0.77 | 10 |
| Albumin | 0.09 g/dl | 0.12 g/dl | 0.86 | 10 |

accurate transcutaneous measurements so that, for example, diabetic patients would not need to draw blood several times per day to monitor their blood glucose levels. The first step in such a study is to analyze human serum, thereby circumventing any complicating factors such as scattering from cells and absorption from hemoglobin.

Berger et al. have reported such a study where six analytes in serum could be predicted with clinical or near-clinical accuracy in less than 1 min.[103] The analytes studied include glucose, cholesterol, urea, total protein, albumin, and triglycerides. They also completed a preliminary study in whole blood that was able to predict hematocrit accurately as well. These studies used the implicit method of PLS analysis.

Berger et al. used the experimental setup shown in Figure 31.11B to collect Raman spectra of human serum. The experiment used 250 mW of excitation at 830 nm focused down to a 50-μm diameter spot; data were collected for 5 min at 10-s intervals. A total of 69 samples was analyzed over a 7-week period, and reference concentrations were provided by standard hospital chemical analysis methods. Data were analyzed with PLS in a leave-one-week-out cross validation manner. Accurate predictions were maintained for all analytes with 60-s integration times, and some predictions could be accurately made in 10 s. Accuracy for the spectroscopic analysis is reported as root mean square error of prediction (RMSEP) compared to the reference technique. The prediction accuracies, along with the reference errors, are shown in Table 31.1. As can be seen, there is good correlation for all six analytes; in fact the accuracy for some (total protein and albumin) is limited by the reference technique.

As an example of the use of explicit methods, we consider the analysis of coronary artery spectra. In such a case, one collects the Raman spectra of all of the chemical or morphological components individually, keeping track of the intensity of each Raman spectrum relative to its unit concentration. The composite tissue spectrum can then simply be reconstructed by summing the basis spectra with the appropriate weighting, thereby determining the abundance of each component in the tissue (Equation 31.1).

Using this approach, Brennan et al. have developed a quantitative model to characterize the relative amounts of chemical coronary artery tissue components.[116] This model consisted of basis spectra from commercially available beta carotene, cholesterol and cholesterol linoleate, chemically extracted lipids from arterial adventitia (i.e., noncholesterol lipids), isolated mineralizations dissected from highly diseased tissue (i.e., calcium hydroxyapatite), and proteins, characterized by two different spectra collected from delipidized normal and diseased arterial tissues. Appropriate calibration measurements were performed to estimate the weightings of the individual basis spectra.

To study the accuracy of the model, coronary arteries were obtained from explanted hearts, and therefore a wide range of disease states were represented. Samples were frozen and ground into minces, which were combined in various mixtures to vary the distribution of chemical concentrations further. It was necessary to use homogenized minces so that the Raman spectra, which were obtained from a small volume (~1 mm$^3$), were representative of the entire sample that was subjected to chemical analysis. Raman spectra were collected from 10 different sites from each mince using the laboratory system described above with 350 mW of excitation at 830 nm over a 100-μm diameter spot. The samples were then analyzed chemically to determine the amount of triglycerides, free cholesterol, cholesterol esters, phospholipids, and inorganic phosphorus (to characterize the calcium salts). Figure 31.12 shows examples

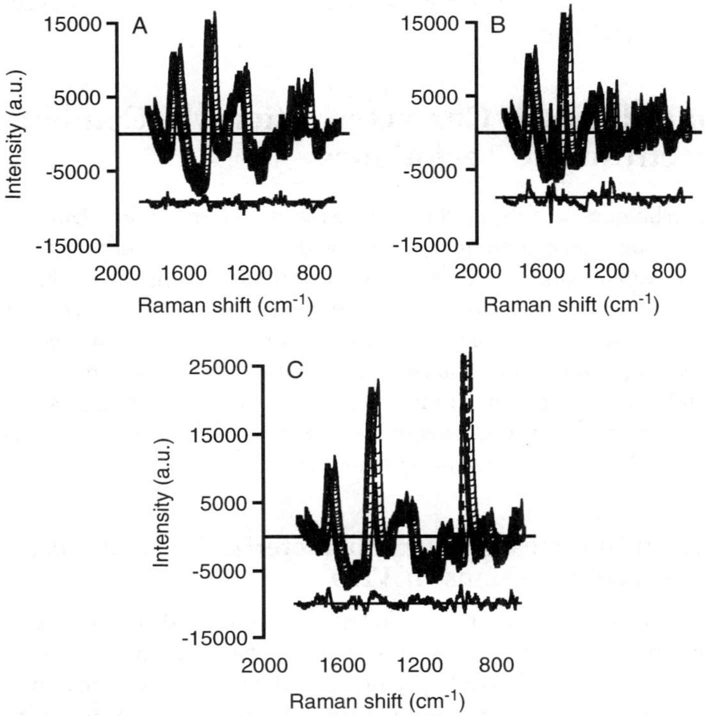


**FIGURE 31.12** Comparison between the model (line) and spectra (dots) of coronary artery that exhibited (A) intimal fibroplasia, (B) noncalcified atheromatous plaque, and (C) calcified plaque. The differences between a spectrum and its model fit are displayed below each comparison (same scale). (From Brennan, J.F. et al., *Circulation*, 96, 99, 1997. With permission.)

of the model fits (line) to the data from macroscopic samples (dots) for three classes of artery. The small residuals, whose size is on the order of the noise, show that the model is relatively complete.

Despite all of the assumptions used in generating the basis spectra and calibrating the model, macroscopic spectra can be accurately reconstructed, and quantitative spectroscopic analysis correlates well with the chemical assay. Table 31.2 shows correlation coefficients between the chemical assay and the Raman analysis, the bias (mean difference between the two methods), and the standard deviation ($\sigma$) of the difference between the measurements. All correlation coefficients are >0.9, with the exception of free cholesterol, which is slightly lower.

The two examples discussed here show the potential for Raman spectroscopy not only for identifying the chemicals present in a sample, but also as a means for providing an optical quantitative assay. In addition to providing chemical information, it is also possible to take this information and develop diagnostic algorithms to determine disease states.[117] Also, utilizing this chemical information opens up

**TABLE 31.2** Prediction Accuracy of Raman Artery Chemical Model Compared to Reference Chemical Assay

| Component | $r$ | Bias, % | $\sigma$, % |
|---|---|---|---|
| Total proteins | 0.998 | −0.7 | 2.6 |
| Fats (noncholesterol) | 0.999 | 1.6 | 2.1 |
| Total cholesterol | 0.951 | −2.3 | 1.5 |
| Free cholesterol | 0.848 | −2.2 | 1.4 |
| Cholesterol esters | 0.900 | −0.0 | 1.4 |
| Calcium salts | 0.994 | 1.5 | 1.3 |

the potential for studying disease progression, regression, and etiology *in vivo*. Such an approach may lead to advances in future therapies.

## 31.6 Enhanced Tissue Characterization via Combined Use of Spectroscopic Techniques

The spectroscopic techniques described in the previous sections provide complementary information on different aspects of tissue composition, morphology, and biochemistry. Diagnostic algorithms that separate normal from diseased tissues based on specific quantitative changes that take place during disease development can be established based on each technique individually. However, medical decisions and diagnoses are generally not made on the basis of one measure. Experience has shown that the combined use of several measures provides more robust and reliable results. In the same way, it is reasonable to expect that by combining information from different spectroscopic techniques we can enhance our capability to classify tissue and our understanding of the processes that lead to the creation of a lesion. Indeed, this has been achieved in the following examples.

### 31.6.1 Tri-Modal Spectroscopy for Characterization and Detection of Precancerous Lesions *in Vivo*

Tri-modal spectroscopy, or TMS, refers to the combined use of intrinsic fluorescence, diffuse reflectance and light scattering spectroscopy as a quantitative tool for characterizing tissue morphology and biochemistry. The fluorescence and reflectance spectra required for the implementation of each one of these techniques are acquired simultaneously with the FastEEM instrument shown in Figure 31.1. Thus, the information extracted from IFS, DRS, and LSS characterizes the same tissue site. In this small data set, this information can be combined simply by assigning a classification to a particular tissue site consistent with the diagnostic algorithms developed for at least two of the three techniques. As a result, a significant enhancement is observed in the sensitivity and specificity with which we can separate dysplastic (low- and high-grade) lesions from nondysplastic Barrett's esophagus or high-grade from nonhigh-grade (low-grade and nondysplastic) Barrett's esophagus. Results are presented for spectra acquired from 16 patients, 26 nondysplastic, 7 low-grade, and 7 high-grade dysplastic tissue sites (Table 31.3).[36]

A similar enhancement was observed when spectroscopic information from all three techniques was combined to separate colposcopically abnormal but histopathologically benign cervical tissues from histopathologically abnormal cervical lesions. The reported sensitivities and specificities were based on logistic regression and cross validation performed on a set of spectra from 44 patients including 50 colposcopically normal, 21 colposcopically abnormal but histopathologically benign and 13 colposcopically abnormal sites classified as low-grade (2) or high-grade (11) SILs (Table 31.4).[8] A similar enhancement is also observed for detection of oral cavity lesions.[35] Ultimately, it is expected that dedicated TMS algorithms will be developed based on the weighted contributions of the individual pieces of information extracted from each technique.

**TABLE 31.3** Accuracy of Spectroscopic Classification of Nondysplastic (NDB), Low-Grade (LGD) and High-Grade Dysplastic (HGD) Tissue in Barrett's Esophagus

| | HGD vs. (LGD and NDB) | | (LGD and HGD) vs. NDB | |
|---|---|---|---|---|
| | Sensitivity | Specificity | Sensitivity | Specificity |
| Intrinsic fluorescence (IF) | 100% | 97% | 79% | 88% |
| Diffuse reflectance (DR) | 86% | 100% | 79% | 88% |
| Light scattering (LS) | 100% | 91% | 93% | 96% |
| Combination of IF, DR, and LS | 100% | 100% | 93% | 100% |

**TABLE 31.4** Performance of Different Spectroscopic Techniques for Separating SILs from Non-SILs

| | Biopsied Non-SILs[a] vs. SILs | | Non-SILs[b] vs. SILs | |
|---|---|---|---|---|
| | Sensitivity | Specificity | Sensitivity | Specificity |
| IFS | 62% | 67% | 62% | 92% |
| DRS | 69% | 57% | 62% | 82% |
| LSS | 77% | 71% | 77% | 83% |
| TMS | 92% | 71% | 92% | 90% |

[a] Biopsied non-SILs include 21 colposcopically abnormal biopsied sites classified as mature squamous epithelium (5/21) or squamous metaplasia (16/21).
[b] Non-SILs in this case include 50 colposcopically normal sites and 21 biopsied sites classified as squamous metaplasia or mature squamous epithelium.

**TABLE 31.5** Sensitivity, Specificity, and Positive Predictive Value of Three Different Diagnostic Algorithms for Coronary Atherosclerosis

| | Sensitivity | Specificity | PPV* |
|---|---|---|---|
| IFS | 90% | 82% | 95% |
| DRS | 72% | 68% | 90% |
| BMS* | 95% | 91% | 98% |

* PPV = positive predictive value; BMS = Bi-modal spectroscopy.

## 31.6.2 Bi-Modal Spectroscopy for Characterizing Atherosclerotic *ex Vivo* Lesions

Analysis of intrinsic fluorescence and diffuse reflectance spectra acquired using a FastEEM instrument from normal and diseased coronary artery *ex vivo* tissues indicates that significant changes can be detected in tissue composition during the development of atherosclerotic lesions. For example, analysis of intrinsic fluorescence spectra at several excitation wavelengths yields the fluorescence contributions of collagen/elastin and ceroid. On the other hand, diffuse reflectance spectra convey important information about the levels of beta-carotene absorption. The combination of information concerning these important tissue constituents results in a significant enhancement in our ability to classify tissue spectroscopically in accordance with pathology (Table 31.5).

## 31.6.3 Fluorescence and Raman Spectroscopy for Characterizing Microscopic Ceroid Deposits in Coronary Artery Samples

Ceroid is postulated to be a complex of protein associated with oxidized lipids and may be responsible for causing irreversibility of certain atherosclerotic plaques.[118,119] In small uncomplicated plaques, ceroid initially appears within the cytoplasm of superficial macrophage-like foam cells. It has therefore been suggested that macrophages might be responsible for lipid oxidation in the plaque, with potentially damaging consequences, such as cell injury and necrosis and the release of insoluble material into the extracellular space, ultimately causing the plaque to become irreversible. In advanced plaques, most of the ceroid is extracellular.

Ceroid is histochemically identified by insolubility in a variety of lipid solvents and the uptake of lipid dyes. In addition, ceroid is characterized by the emission of intense yellow autofluorescence when it is excited with ultraviolet (UV) light.[53,120] However, the exact chemical composition of ceroid deposits in atherosclerotic plaques is not well defined, and may vary in different samples or tissues. Insight into the

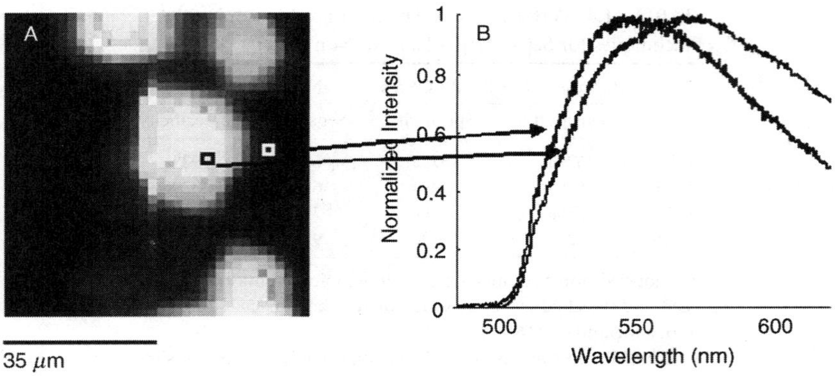


**FIGURE 31.13** (A) Fluorescence map showing the distribution of ceroid at 565-nm excitation. (B) Fluorescence spectra of collagen/elastin and ceroid acquired at 476-nm excitation.

chemical composition of ceroid may provide a better understanding of the mechanism of its formation and suggest avenues to induce plaque regression with medical therapy.

By spatially monitoring tissue autofluorescence, the location of ceroid within an unstained thin tissue section can be identified spectroscopically; subsequently, Raman measurements can be acquired to ascertain more detailed information on the chemical composition of the deposit. Figure 31.13A shows a fluorescence map, 477-nm excitation, that pinpoints the location of several ceroid deposits in a sample of coronary artery. Although collagen and elastin also exhibit autofluoresce in the UV, it is typically less intense than the emission from ceroid and peaks around 530 nm. This is blue shifted compared to the ceroid fluorescence maximum that occurs between 550 to 580 nm. The normalized fluorescence spectra of ceroid and collagen are displayed in Figure 31.13B. The fluorescence map in Figure 31.13A was generated by plotting the fluorescence peak intensity at 565 nm for each pixel and thus bright regions are indicative of the presence of ceroid.

Following identification of ceroid deposits through their characteristic fluorescence emission, Raman spectroscopy was performed on the same region of tissue in order to elicit the chemical composition of the deposits. Due to its narrow and distinct spectral features, Raman spectroscopy is ideally suited to the study of complex chemical mixtures such as those occurring in the core of unstable atherosclerotic plaques. Figure 31.14A depicts a Raman map showing the distribution of cholesterol in the same ceroid deposit displayed in Figure 31.13A. In addition to cholesterol, several other chemical moieties, such as cholesterol esters, apoproteins, and triolene, have been identified in the deposits. After acquisition of the Raman data, the identical tissue section was lipid extracted and subsequently stained with oil red O as further confirmation of the identity of the deposit. A white light image of this stained tissue section is presented in Figure 31.14B. In this image, the deposit appears red, indicating that the substance is histochemcially identified as ceroid.

By capitalizing on the well known fluorescence spectral profile of ceroid and the detailed chemical information provided by Raman spectroscopy, we were able to gain insight into the chemical composition of ceroid deposits. The combination of Raman and fluorescence spectroscopies provides a means to identify ceroid deposits and to probe their chemical composition. This might shed light on the process of lipoprotein degradation and deposition in atherosclerotic plaques.

## 31.7 Conclusion

In conclusion, the preceding examples illustrate that spectroscopy reveals biochemical and morphological information about the native tissue state, without the need for tissue removal. The extracted morphological information can be similar in nature to that obtained by histopathology, as in the characterization of cell nuclear morphology using LSS. In addition, spectroscopic techniques (such as intrinsic fluorescence

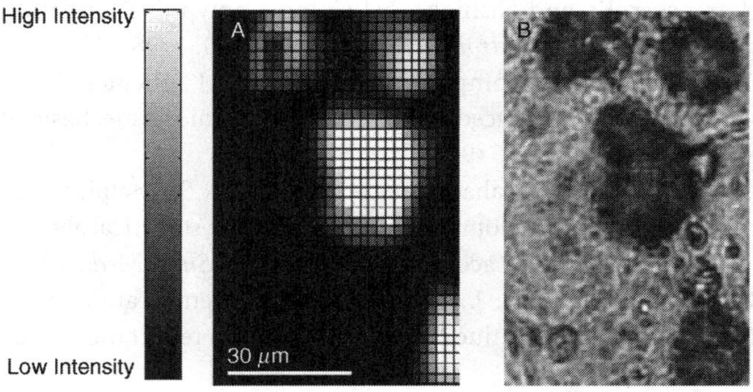


**FIGURE 31.14** (A) Raman map and (B) phase contrast image of a ceroid deposit in a sample of human coronary artery.

and Raman) also provide information about tissue biochemistry which cannot be directly obtained otherwise, because tissue removal and processing necessarily alter its biochemistry. Methods such as angular LSS can be sensitive to morphological structures and structural changes that cannot be resolved using even invasive conventional microscopic techniques. Furthermore, analytical models can be used to describe measured tissue spectra and extract information in a quantitative manner. Based on all of these desirable features, it is evident that spectroscopy can be a powerful tool for understanding better some of the fundamental processes that take place very early in the development of disease.

The combination of information extracted from complementary spectroscopic techniques offers the opportunity to acquire a more accurate picture of the tissue state, which in turn enables more robust diagnosis of disease. The capability to perform measurements over small and wide tissue areas enhances the clinical applicability of these techniques. Based on the work already performed in this field, it is clear that spectroscopy is on the way to serving as an important adjunct to clinical and histopathological evaluation of disease.

## Acknowledgments

The authors acknowledge contributions from Andrew Berger (now at the University of Rochester), Martin Hunter, Irving Itzkan, Maxim Kalashnikov, Tae-Woong Koo, Lev Perelman (now at Harvard Medical School), and Qingguo Zhang from the MIT Spectroscopy Laboratory; Jacques Van Dam (now at the Stanford University School of Medicine), Michael Wallace (now at the Medical University of South Carolina), Brian Jacobson, David Carr-Locke, Ellen Sheets, and Christopher Crum from the Brigham and Women's Hospital; Stanley Shapsay, Tulio Valdez, Cesar Fuentes, and Sandro Kabani from the Boston University Medical Center; John Kramer, Joseph Crowe, and Joseph Arendt from the Cleveland Clinic Foundation; Maryann Fitzmaurice from the Cleveland Clinic Foundation; and Gary Horowitz from Beth Israel/Deaconness Medical Center. The work presented in this chapter that involved collaborative projects with the MIT Spectroscopy Laboratory was supported by the National Institutes of Health grants P41RR02594, CA53717, and CA72517, and by generous support from the Bayer Corporation.

### References

1. Ramanujam, N., Fluorescence spectroscopy of neoplastic and non-neoplastic tissues, *Neoplasia*, 2, 89, 2000.
2. Wagnières, G.M., Star, W.M. and Wilson, B.C., *In vivo* fluorescence spectroscopy and imaging for oncological applications, *Photochem. Photobiol.*, 68, 603, 1998.
3. Panjehpour, M., Overholt, B., Vo-Dinh, T., Haggit, R., Edwards, D., and Buckley, F., Endoscopic fluorescence detection of high-grade dysplasia in Barrett's esophagus, *Gastroenterology*, 111, 93, 1996.

4. Mycek, M., Schomacker, K., and Nishioka, N., Colonic polyp differentiation using time-resolved autofluorescence spectroscopy, *Gastrointest. Endosc.*, 48, 390, 1998.
5. Schomacker, K.T., Frisoli, J.K., Compton, C.C., Flotte, T.J., Richter, J.M., Nishioka, N.S., and Deutsch, T.F., Ultraviolet laser-induced fluorescence of colonic tissue: basic biology and diagnostic potential, *Lasers Surg. Med.*, 12, 63, 1992.
6. Ramanujam, N., Mitchell, M.F., Mahadeevan, A., Thomsen, S., Malpica, A., Wright, T., Atkinson, N., and Richards-Kortum, R., Development of a multivariate statistical algorithm to analyze human cervical tissue fluorescence spectra acquired *in vivo*, *Lasers Surg. Med.*, 19, 46, 1996.
7. Nordstrom, R.J., Burke, L., Niloff, J., and Myrtle, J.F., Identification of cervical intraepithelial neoplasia (CIN) using UV-excited fluorescence and diffuse reflectance tissue spectroscopy, *Lasers Surg. Med.*, 29, 118, 2001.
8. Georgakoudi, I., Sheets, E.E., Müller, M.G., Backman, V., Crum, C.P., Badizadegan, K., Dasari, R.R., and Feld, M.S., Tri-modal spectroscopy for the detection and characterization of cervical precancers *in vivo*, *Am. J. Obstet. Gynecol.*, 186, 374, 2002.
9. Brancaleon, L., Durkin, A.J., Tu, J.H., Menaker, G., Fallon, J.D., and Kollias, N., *In vivo* fluorescence spectroscopy of nonmelanoma skin cancer, *Photochem. Photobiol.*, 73, 178, 2001.
10. Anidjar, M., Cussenot, O., Avrillier, S., Ettori, D., Teillac, P., and LeDuc, P., The role of laser-induced autofluorescence spectroscopy in bladder tumor detection, *Ann. N.Y. Acad. Sci.*, 838, 130, 1998.
11. Gillenwater, A., Jacob, R., Baneshappa, R., Kemp, B., El-Naggar, A., Palmer, J., Clayman, G., Mitchell, M.F., and Richards-Kortum, R., Noninvasive diagnosis of oral neoplasia based on fluorescence spectroscopy and native tissue autofluorescence, *Arch. Otolaryngol. Head Neck Surg.*, 124, 1251, 1998.
12. Müller, M.G., Valdez, T., Georgakoudi, I., Backman, V., Fuentes, C., Kabani, S., Laver, N., Boone, C., Dasari, R., Shapsay, S., and Feld, M.S., Tri-modal spectroscopy: a new technique for detecting and evaluating early human oral cancer, in review.
13. Lam, S., Kennedy, T.,Unger, M., Miller, Y., Gelmont, D., Rusch, V., Gipe, B., Howard, D., LeRiche, J., Coldman, A., and Gazdar, A., Localization of bronchial intraepithelial neoplastic lesions by fluorescence bronchoscopy, *Chest*, 113, 696, 1998.
14. Goujon, D., Glanzmann, T., Gabrecht, T., Zellweger, M., Radu, A., van den Bergh, H., Monnier, P., and Wagnières, G.A., Detection of early bronchial carcinoma by imaging of the tissue autofluorescence, in *Diagnostic Optical Spectroscopy in Biomedicine*, *Proc. SPIE*, Munich, Germany, 2001.
15. Jackson, J.E., *A User's Guide to Principal Components*, John Wiley & Sons, New York, 1991.
16. Tumer, K., Ramanujam, N., Ghosh, J., Richards-Kortum, R., Ensembles of radial basis function networks for spectroscopic detection of cervical precancer, *IEEE Trans. Biomed. Eng.*, 45, 953, 1998.
17. Richards-Kortum, R., Rava, R.P., Fitzmaurice, M., Tong, L., Ratliff, N.B., Kramer, J., and Feld, M.S., A one-layer model of laser induced fluorescence for diagnosis of disease in human tissue: applications to atherosclerosis, *IEEE Trans. Biomed. Eng.*, 36, 1222, 1989.
18. Ramanujam, N., Mitchell, M., Mahadevan, A., Warren, S., Thomsen, S., Silva, E., and Richards-Kortum, R., *In vivo* diagnosis of cervical intraepithelial neoplasia using 337-nm-excited laser-induced fluorescence, *Proc. Natl. Acad. Sci. U.S.A.*, 91, 10193, 1994.
19. Ince, C., Coremans, J.M.C.C., and Bruining, H.A., *In vivo* NADH fluorescence, *Adv. Exp. Med. Biol.*, 317, 277, 1992.
20. Jöbsis, F.F., O'Connor, M., Vitale, A., and Vreman, H., Intracellular redox changes in functioning cerebral cortex I. Metabolic effects of epileptiform activity, *J. Neurophysiol.*, 34, 735, 1971.
21. Mayevsky, A. and Chance, B., Repetitive patterns of metabolic changes during cortical spreading depression of the awake rat, *Brain Res.*, 65, 529, 1974.
22. Ji, S., Chance, B., Stuart, B.H., and Nathan, R., Two-dimensional analysis of the redox state of the rat cerebral cortex *in vivo* by NADH fluorescence photography, *Brain Res.*, 119, 357, 1977.
23. Mayevsky, A. and Chance, B., Intracellular oxidation-reduction state measured *in situ* by a multichannel fiber-optic surface fluorometer, *Science*, 217, 537, 1982.

24. Dóra, E., Gyulai, L., and Kovách, A.G.B., Determinants of brain activation-induced cortical NAD/NADH responses *in vivo*, *Brain Res.*, 299, 61, 1984.
25. Ji, S., Chance, B., Nishiki, K., Smith, T., and Rich, T., Micro-light guides: a new method for measuring tissue fluorescence and reflectance, *Am. J. Physiol.*, 236, C144, 1979.
26. Kobayashi, S., Kaede, K., and Ogata, E., Optical consequences of blood substitution on tissue oxidation-reduction state of the rat kidney *in situ*, *J. Appl. Physiol.*, 31, 93, 1971.
27. Renault, G., Raynal, E., Sinet, M., Muffat-Joly, M., Berthier, J.-P., Cornillault, J., Godard, B., and Pocidalo, J.-J., *In situ* double-beam NADH laser fluorimetry: a choice of a reference wavelength, *Am. J. Physiol.*, 246, H491, 1984.
28. Wu, J., Feld, M.S., and Rava, R.P., Analytical model for extracting intrinsic fluorescence in turbid media, *Appl. Opt.*, 32, 3585, 1993.
29. Finlay, J.C., Conover, D.L., Hull, E.L., and Foster, T.H., Porphyrin bleaching and PDT-induced spectral changes are irradiance dependent in ALA-sensitized normal rat skin *in vivo*, *Photochem. Photobiol.*, 73, 54, 2001.
30. Gardner, C.M., Jacques, S.L., and Welch, A.J., Fluorescence spectroscopy of tissue: recovery of intrinsic fluorescence from measured fluorescence, *Appl. Opt.*, 35, 1780, 1996.
31. Durkin, A.J., Jaikumar, S., Ramanujam, N., and Richards-Kortum, R., Relation between fluorescence-spectra of dilute and turbid samples, *Appl. Opt.*, 33, 414, 1994.
32. Zhadin, N.N. and Alfano, R.R., Correction of the internal absorption effect in fluorescence emission and excitation spectra from absorbing and highly scattering media: theory and experiment, *J. Biomed. Opt.*, 3, 171, 1998.
33. Patterson, M.S. and Pogue, B.W., Mathematical model for time-resolved and frequency-domain fluorescence spectroscopy in biological tissues, *Appl. Opt.*, 33, 1963, 1994.
34. Zhang, Q., Müller, M.G., Wu, J., and Feld, M.S., Turbidity-free fluorescence spectroscopy of biological tissue, *Opt. Lett.*, 25, 1451, 2000.
35. Müller, M.G., Georgakoudi, I., Zhang, Q., Wu, J., and Feld, M.S., Intrinsic fluorescence spectroscopy in turbid media: disentangling effects of scattering and absorption, *Appl. Opt.*, 40, 4633, 2001.
36. Georgakoudi, I., Jacobson, B., Van Dam, J., Backman, V., Wallace, M.B., Müller, M.G., Zhang, Q., Badizadegan, K., Sun, D., Thomas, G.A., Perelman, L.T., and Feld, M.S., Fluorescence, reflectance and light scattering spectroscopy for evaluating dysplasia in patients with Barrett's esophagus, *Gastroenterology*, 120, 1620, 2001.
37. Zângaro, R.A., Silveira, L., Manoharan, R., Zonios, G., Itzkan, I., Dasar, R.R., Van Dam, J., and Feld, M.S., Rapid multiexcitation fluorescence spectroscopy system for *in vivo* tissue diagnosis, *Appl. Opt.*, 35, 5211, 1996.
38. Cothren, R.M., Hayes, G.B., Kramer, J.R., Sacks, B.A., Kittrell, C., and Feld, M.S., A multifiber catheter with an optical shield for laser angiosurgery, *Lasers Life Sci.*, 1, 1, 1986.
39. Georgakoudi, I., Jacobson, B.C., Müller, M.G., Badizadegan, K., Carr-Locke, D.L., Sheets, E.E., Crum, C.P., Dasari, R.R., Van Dam, J., and Feld, M.S., NADPH and collagen as *in vivo* quantitative biomarkers of epithelial pre-cancerous changes, *Cancer Res.*, 62, 682, 2002.
40. Herszenyi, L., Plebani, M., Carraro, P., De Paoli, M., Roveroni, G., Cardin, R., Foschia, F., Tulassay, Z., Naccarato, R., and Farinati, F., Proteases in gastrointestinal neoplastic diseases, *Clin. Chim. Acta*, 291, 171, 2000.
41. Johansson, N., Ahonen, M., and Kähäri, V.-M., Matrix metalloproteinases in tumor invasion, *Cell. Mol. Life Sci.*, 57, 5, 2000.
42. Talvensaari, A., Apaja-Sarkkiner, M., Hoyhtya, M., Westerlund, A., Puistola, U., and Turpeernniemi-Hujanen, T., Matrix metalloproteinase 2 immunoreactive protein appears early in cervical epithelium dedifferentiation, *Gynecol. Oncol.*, 72, 306, 1999.
43. Ellis, D.L. and Yannas, I.V., Recent advances in tissue synthesis *in vivo* by use of collagen-glycosaminoglycan copolymers, *Biomaterials*, 17, 291, 1996.
44. American Cancer Society, *Cancer Facts and Figures*, American Cancer Society, New York, 2000, p. 4.
45. American Heart Association, *2000 Heart and Stroke Statistical Update*, Dallas, TX, 2000.

46. Richards-Kortum, R., Rava, R.P., Fitzmaurice, M., Kramer, J., and Feld, M.S., 476 nm excited laser-induced fluorescence spectroscopy of human coronary arteries: applications in cardiology, *Am. Heart J.*, 122, 1141, 1991.
47. Laifer, L.I., O'Brien, K.M., Stetz, M.L., Gindi, G.R., Garrand, T.J., and Deckelbaum, L.I., Biochemical basis for the difference between normal and atherosclerotic arterial fluorescence, *Circulation*, 80, 1893, 1989.
48. Baraga, J.J., Rava, R.P., Taroni, P., Kittrell, C., Fitzmaurice, M., and Feld, M.S., Laser induced fluorescence spectroscopy of normal and atherosclerotic human aorta using 306,0 nm excitation, *Lasers Surg. Med.*, 10, 245, 1990.
49. Rekhter, M.D., Zhang, K., Narayanan, A.S., Phan, S., Schork, M., and Gordon, D., Type I collagen gene expression in human atherosclerosis. Localization to specific plaque regions., *Am. J. Pathol.*, 143, 1634, 1993.
50. Tammi, M., Seppala, P.O., Lehtonen, A., and Mottonen, M., Connective tissue components in normal and atherosclerotic human coronary arteries, *Atherosclerosis*, 29, 191, 1978.
51. Levene, C.I. and Poole, J.C.F., The collagen content of the normal and atherosclerotic human aortic intima, *Br. J. Exp. Pathol.*, 43, 469, 1962.
52. Hoff, H.F. and Hoppe, G., Structure of cholesterol-containing particles accumulating in atherosclerotic lesions and the mechanisms of their derivation, *Curr. Opin. Lipidol.*, 6, 317, 1995.
53. Fitzmaurice, M., Bordagaray, J.O., Engelmann, G.L., Richards-Kortum, R., Kolubayev, T., Feld, M.S., Ratliff, N.B., and Kramer, J.R., Argon ion laser-excited autofluorescence in normal and atherosclerotic aorta and coronary arteries: morphologic studies, *Am. Heart J.*, 118, 1028, 1989.
54. Ge, Z., Schomacker, K.T., and Nishioka, N.S., Identification of colonic dysplasia and neoplasia by diffuse reflectance spectroscopy and pattern recognition techniques, *Appl. Spectrosc.*, 52, 833, 1998.
55. Zonios, G., Perelman, L.T., Backman, V., Manoharan, R., Fitzmaurice, M., Van Dam, J., and Feld, M.S., Diffuse reflectance spectroscopy of human ademonatous colon polyps *in vivo*, *Appl. Opt.*, 38, 6628, 1999.
56. Mourant, J.R., Bigio, I., Boyer, J., Conn, R., Johnson, T.M., and Shimada, T., Spectroscopic diagnosis of bladder cancer with elastic scattering spectroscopy, *Lasers Surg. Med.*, 17, 350, 1995.
57. Bigio, I., Bown, S., Briggs, G., Kelley, C., Lakhani, S., Rickard, D., Ripley, P., Rose, I., and Saunders, C., Diagnosis of breast cancer using elastic-scattering spectroscopy: preliminary clinical results, *J. Biomed. Opt.*, 5, 221, 2000.
58. Utzinger, U., Brewer, M., Silva, E., Gershenson, D., Blast, R.C., Follen, M., and Richards-Kortum, R., Reflectance spectroscopy for *in vivo* characterization of ovarian tissue, *Lasers Surg. Med.*, 28, 56, 2001.
59. Anderson, R.R. and Parrish, J.A., The optics of human skin, *J. Invest. Dermatol.*, 77, 13, 1981.
60. Dawson, J.B., Barker, D.J., Ellis, D.J., Grassam, E., Cotterill, J.A., Fisher, G.W., and Feather, J.W., A theoretical and experimental study of light absorption and scattering by *in vivo* skin, *Phys. Med. Biol.*, 25, 695, 1980.
61. Hajizadeh-Saffar, M., Feather, J.W., and Dawson, J.B. An investigation of factors affecting the accuracy of *in vivo* measurements of skin pigments by reflectance spectrophotometry, *Phys. Med. Biol.*, 35, 1301, 1990.
62. Farina, B., Bartoli, C., Bono, A., Colombo, A., Lualdi, M., Tragni, G., and Marchesini, R., Multispectral imaging approach in the diagnosis of cutaneous melanoma: potentiality and limits, *Phys. Med. Biol.*, 45, 1243, 2000.
63. Wallace, V., Bamber, J., Crawford, D., Ott, R., and Mortimer, P., Classification of reflectance spectra from pigmented skin lesions, a comparison of multivariate discriminant analysis and artificial neural networks, *Phys. Med. Biol.*, 45, 2859, 2000.
64. Mourant, J.R., Bigio, I., Boyer, J., Johnson, T.M., and Lacey, J., Elastic scattering spectroscopy as a diagnostic for differentiating pathologies in the gastrointestinal tract: preliminary testing, *J. Biomed. Opt.*, 1, 1, 1996.

65. Wallace, V.P., Crawford, D., Mortimer, P., Ott, R., and Bamber, J., Spectrophotometric assessment of pigmented skin lesions: methods and feature selection for evaluation of diagnostic performance, *Phys. Med. Biol.*, 45, 735, 2000.
66. Feather, J.W., Hajizadeh-Saffar, M., Leslie, G., and Dawson, J.B., A portable scanning reflectance spectrophotometer using visible wavelengths for the rapid measurement of skin pigments, *Phys. Med. Biol.*, 34, 807, 1989.
67. Koenig, F., Larne, R., Enquist, H., McGovern, F., Schomacker, K.T., Kollias, N., and Deutsch, T.F., Spectroscopic measurement of diffuse reflectance for enhanced detection of bladder carcinoma, *Urology*, 51, 342, 1998.
68. Farrell, T.J., Patterson, M.S., and Wilson, B., A diffusion theory model of spatially resolved, steady-state diffuse reflectance for the noninvasive determination of tissue optical properties, *Med. Phys.*, 19, 879, 1992.
69. Farrell, T.J., Patterson, M.S., and Essenpreis, M., Influence of layered tissue architecture on estimates of tissue optical properties obtained from spatially resolved diffuse reflectometry, *Appl. Opt.*, 37, 1958, 1998.
70. Alexandrakis, G., Farrell, T.J., and Patterson, M.S., Accuracy of the diffusion approximation in determining the optical properties of a two-layer turbid medium, *Appl. Opt.*, 37, 7401, 1998.
71. Venugopalan, V., You, J.S., and Tromberg, B.J., Radiative transport in the diffusion approximation: an extension for highly absorbing media and small source-detector separations, *Phys. Rev. E*, 58, 2395, 1998.
72. Bevilacqua, F. and Depeursinge, C., Monte Carlo study of diffuse reflectance at source-detector separations close to one transport mean free path, *J. Opt. Soc. Am. A*, 16, 2935, 1999.
73. Backman, V., Wallace, M., Perelman, L., Arendt, J., Gurjar, R., Muller, M., Zhang, Q., Van Dam, J., Badizadegan, K., Crawford, J., Fitzmaurice, M., Kabani, S., Levin, H., Seiler, M., Dasari, R., and Feld, M., Detection of preinvasive cancer cells, *Nature*, 406, 35, 2000.
74. Perelman, L., Backman, V., Wallace, M., Zonios, G., Manoharan, R., Nusrat, A., Shields, S., Seiler, M., Lima, C., Hamano, T., Itzkan, I., Van Dam, J., Crawford, J., and Feld, M., Observation of periodic fine structure in reflectance from biological tissue: a new technique for measuring nuclear size distribution, *Phys. Rev. Lett.*, 80, 627, 1998.
75. Wallace, M., Perelman, L.T., Backman, V., Crawford, J., Fitzmaurice, M., Seiler, M., Badizadegan, K., Shields, S., Itzkan, I., Dasari, R.R., Van Dam, J., and Feld, M.S., Endoscopic detection of dysplasia in patients with Barrett's esophagus using light-scattering spectroscopy, *Gastroenterology*, 119, 677, 2000.
76. Watson, J.V., *Introduction to Flow Cytometry*, Cambridge University Press, Cambridge, U.K., 1991.
77. Sokolov, K., Drezek, R., Gossage, K., and Richards-Kortum, R., Reflectance spectroscopy with polarized light: is it sensitive to cellular and nuclear morphology? *Opt. Express*, 5, 302, 1999.
78. Mourant, J.R., Canpolat, M., Brocker, C., Esponda-Ramos, O., Johnson, T.M., Matanock, A., Stetter, K., and Freyer, J.P., Light scattering from cells: the contribution of the nucleus and the effects of proliferative status, *J. Biomed. Opt.*, 5, 131, 2000.
79. Mourant, J.R., Freyer, J.P., Hielscher, A.H., Eick, A.A., Shen, D., and Johnson, T.M., Mechanisms of light scattering from biological cells relevant to noninvasive optical-tissue diagnostics, *Appl. Opt.*, 37, 3586, 1998.
80. Drezek, R., Dunn, A., and Richards-Kortum, R., Light scattering from cells: finite-difference time-domain simulations and goniometric measurements, *Appl. Opt.*, 38, 3651, 1999.
81. Yodh, A. and Chance, B., Spectroscopy and imaging with diffusing light, *Phys. Today*, 48, 34, 1995.
82. Backman, V., Gurjar, R., Badizadegan, K., Itzkan, I., Dasari, R., Perelman, L.T., and Feld, M.S., Polarized light scattering spectroscopy for quantitative measurement of epithelial cellular structures, *IEEE J. Sel. Top. Quantum Electron.*, 5, 1019, 1999.
83. Canpolat, M. and Mourant, J.R., Particle size analysis of turbid media with a single optical fiber in contact with the medium to deliver and detect white light, *Appl. Opt.*, 40, 3792, 2001.
84. Newton, R.G., *Scattering Theory of Waves and Particles*, McGraw-Hill, New York, 1969.

85. Brunsting, A. and Mullaney, F., Differential light scattering from spherical mammalian cells, *Biophys. J.*, 14, 439, 1974.
86. Beauvoit, B., Kitai, T., and Chance, B., Contribution of the mitochondrial compartment to the optical properties of rat liver: a theoretical and practical approach, *Biophys. J.*, 67, 2501, 1994.
87. Demos, S.G. and Alfano, R.R., Optical polarization imaging, *Appl. Opt.*, 36, 150, 1997.
88. Jacques, S.L., Roman, J.S., and Lee, K., Imaging superficial tissues with polarized light, *Lasers Surg. Med.*, 26, 119, 2000.
89. Johnson, T.M. and Mourant, J.R., Polarized wavelength-dependent measurements of turbid media, *Opt. Express*, 4, 200, 1999.
90. Gurjar, R.S., Backman, V., Perelman, L.T., Georgakoudi, I., Badizadegan, K., Itzkan, I., Dasari, R.R., and Feld, M.S., Imaging human epithelial properties with polarized light-scattering spectroscopy, *Nat. Med.*, 7, 1245, 2001.
91. Backman, V., Gopal, V., Kalashnikov, M., Badizadegan, K., Gurjar, R., Wax, A., Georgakoudi, I., Mueller, M., Boone, C.W., Dasari, R.R., and Feld, M.S., Measuring cellular structure at submicron scale with light scattering spectroscopy, *IEEE J. Sel. Top. Quantum Electron.*, 7, 887, 2001.
92. Bunde, A. and Havlin, S., *Fractals and Disordered Systems*, Springer-Verlag, New York, 1991.
93. Schmitt, J. and Kumar, G., Turbulent nature of refractive-index variations in biological tissue, *Opt. Lett.*, 21, 1210, 1996.
94. Schmitt, J. and Kumar, G., Optical scattering properties of soft tissue: a discrete particle model, *Appl. Opt.*, 37, 2788, 1998.
95. Hanlon, E.B., Manoharan, R., Koo, T.-W., Shafer, K.E., Motz, J.T., Fitzmaurice, M., Kramer, J.R., Itzkan, I., Dasari, R.R., and Feld, M.S., Prospects for *in vivo* Raman spectroscopy, *Phys. Med. Biol.*, 45, R1, 2000.
96. Mahadevan-Jansen, A. and Richards-Kortum, R., Raman spectroscopy for the detection of cancers and precancers, *J. Biomed. Opt.*, 1, 31, 1996.
97. Manoharan, R., Wang, Y., and Feld, M.S., Review: histochemical analysis of biological tissues using Raman spectroscopy, *Spectrochim. Acta A*, 52, 215, 1996.
98. O'Grady, A., Dennis, A.C., Denvir, D., McGarvey, J.J., and Bell, S.E.J., Quantitative Raman spectroscopy of highly fluorescent samples using pseudosecond derivatives and multivariate analysis, *Anal. Chem.*, 73, 2058, 2001.
99. Romer, T., Brennan, J., Bakker Schut, T., Woltuis, R., can den Hoogen, R., Emeis, J., van der Laarse, A., Bruschke, A., and Puppels, G., Raman spectroscopy for quantifying cholesterol in intact coronary artery wall, *Atherosclerosis*, 141, 117, 1998.
100. Romer, T.J., Brennan, J.F.B., Puppels, G.J., Zwinderman, A.H., van Duinen, S.G., van der Laarse, A., van der Steen, A.F.W., Born, N.A., and Bruschke, A.V.G., Intravascular ultrasound combined with Raman spectroscopy to localize and quantify cholesterol calcium salts in atherosclerotic cornary arteries, *Atheroscler. Thromb. Vasc. Biol.*, 20, 478, 2000.
101. Qu, J.N.Y., Wilson, B.C., and Suria, D., Concentration measurements of multiple analytes in human sera by near-infrared laser Raman spectroscopy, *Appl. Opt.*, 38, 5491, 1999.
102. Cacheux, P.L., Menard, G., Nguyen Quang, H., Weinmann, P., Jouan, M., and Mguyen Quy, D., Quantiative analysis of cholesterol and cholesterol ester mixtures using near-infrared Fourier transform Raman spectroscopy, *Appl. Spectrosc.*, 50, 1253, 1996.
103. Berger, A.J., Koo, T.-W., Itzkan, I., Horowitz, G., and Feld, M.S., Multicomponent blood analysis by near-infrared Raman spectroscopy, *Appl. Opt.*, 38, 2916, 1999.
104. Brennan, J.F., Wang, Y., Dasari, R.R., and Feld, M.S., Near-infrared Raman spectrometer systems for human tissue studies, *Appl. Spectrosc.*, 51, 201, 1997.
105. Kramer, R., *Chemometric Techniques for Quantitative Analysis*, Marcel Dekker, New York, 1998.
106. Sharma, S., *Applied Multivariate Techniques*, John Wiley & Sons, New York, 1996.

107. Buschman, H.P.J., Deinum, G., Motz, J.T., Fitzmaurice, M., Kramer, J.R., van der Laarse, A., Bruschke, A.V.G., and Feld, M.S., Raman microspectroscopy of human coronary atherosclerosis: biochemical assessment of cellular and extracellular morphologic structures *in situ*, *Cardiovasc. Pathol.*, 10, 69, 2001.
108. Manoharan, R., Shafer, K., Perelman, L., Wu, J., Chen, K., Deinum, G., Fitzmaurice, M., Myles, J., Crowe, J., Dasari, R., and Feld, M., Raman spectroscopy and fluorescence photon migration for breast cancer diagnosis and imaging, *Photochem. Photobiol.*, 67, 15, 1998.
109. Mahadevan-Jansen, A., Mitchell, M., Ramanujam, N., Utzinger, U., and Richards-Kortum, R., Development of a fiber optic probe to measure NIR Raman spectra of cervical tissue *in vivo*, *Photochem. Photobiol.*, 68, 427, 1998.
110. Utzinger, U., Heintzelman, D.L., Mahadevan-Jansen, A., Malpica, A., Follen, M., and Richards-Kortum, R., Near-infrared Raman spectroscopy for *in vivo* detection of cervical precancers, *Appl. Spectrosc.*, 55, 955, 2001.
111. Deinum, G., Rodrigues, D., Romer, T.J., Fitzmaurice, M., Kramer, J.R., and Feld, M.S., Histological classification of Raman spectra of human coronary artery atherosclerosis using principal component analysis, *Appl. Spectrosc.*, 53, 938, 1999.
112. Shim, M.G., Song, L.-M.W.K., Marcon, N.E., and Wilson, B.C., *In vivo* near-infrared Raman spectroscopy: demonstration of feasibility during clinical gastrointestinal endoscopy, *Photochem. Photobiol.*, 72, 146, 2000.
113. Buschman, H.P., Motz, J.T., Deinum, G., Romer, T.J., Fitzmaurice, M., Kramer, J.R., van der Laarse, A., Bruschke, A.V., and Feld, M.S., Diagnosis of human coronary atherosclerosis by morphology-based Raman spectroscopy, *Cardiovasc. Pathol.*, 10, 59, 2001.
114. Shafer-Peltier, K.E., Haka, A.S., Fitzmaurice, M., Crowe, J., Myles, J., Dasari, R.R., and Feld, M.S., Raman microspectroscopic model of human breast tissue: implications for breast cancer diagnosis *in vivo*, *J. Raman Spectrosc.*, in press.
115. Berger, A.J., Koo, T.-W., Itzkan, I., and Feld, M.S., An enhanced algorithm for linear multivariate calibration, *Anal. Chem.*, 70, 623, 1998.
116. Brennan, J.F., Romer, T.J., Lees, R.S., Tercyak, A.M., Kramer, J.R., and Feld, M.S., Determination of human coronary artery composition by Raman spectroscopy, *Circulation*, 96, 99, 1997.
117. Romer, T.J., Brennan, J.F.B., Fitzmaurice, M., Feldstein, M.L., Deinum, G., Myles, J.L., Kramer, J.R., Lees, R.S., and Feld, M.S., Histopathology of human coronary atherosclerosis by quantifying its chemical composition with Raman spectroscopy, *Circulation*, 97, 878, 1998.
118. Mitchinson, M.J., Insoluble lipids in human atherosclerotic plaques, *Atherosclerosis*, 45, 1982.
119. Mitchinson, M.J., Hothersall, D.C., Brooks, P.N., and DeBurbure, C.Y., The distribution of ceroid in human atherosclerosis, *J. Pathol.*, 145, 1985.
120. Verbunt, R.J.A.M., Fitzmaurice, M.A., Kramer, J.R., Ratliff, N.B., Kittrell, C., Taroni, P., Cothren, R.M., Baraga, J., and Feld, M.S., Characterization of ultraviolet laser-induced autofluorescence of ceroid deposits and other structures in atherosclerotic plaques as a potential diagnostic for laser angiosurgery, *Am. Heart J.*, 123, 208, 1992.

# 32

# Recent Developments in Fourier Transform Infrared (FTIR) Microspectroscopic Methods for Biomedical Analyses: From Single-Point Detection to Two-Dimensional Imaging

Rohit Bhargava
*Laboratory of Chemical Physics*
*National Institute of Diabetes and Digestive and Kidney Diseases*
*National Institutes of Health*
*Bethesda, Maryland*

Ira W. Levin
*Laboratory of Chemical Physics*
*National Institute of Diabetes and Digestive and Kidney Diseases*
*National Institutes of Health*
*Bethesda, Maryland*



## 32.1 Introduction

Since both biological structures and systems of interest within the materials research community are chemically heterogeneous at the microscopic level, the spatial distribution of the underlying chemical species defining the molecular complex often determines the function and properties of the macroscopic assembly. As a corollary, changes in the microscopic molecular heterogeneity may alter significantly the chemical and physical attributes of a substance. While optical microscopy techniques may be employed to survey molecular heterogeneity, they provide little information on, for example, the chemical composition, molecular structure, or local packing characteristics. In contrast, microspectroscopic techniques, such as those involving vibrational spectroscopy, afford potentially powerful complementary approaches for addressing these questions. Thus, a combination of microscopy methodologies and infrared spectroscopy, in particular, retains the spectroscopic advantages of experimental versatility and the availability of extensive databases, while introducing the spatial selectivity inherent in microscopic techniques. In particular, the mid-infrared spectral region (4000 to 400 $cm^{-1}$), which has been the focus of microspectroscopic examinations for many decades, has proved to be an invaluable source of compositional and structural information for a wide variety of materials. In this brief survey, we first review the instrumentation required for conducting mid-infrared microspectroscopy and then describe several examples illustrating recent developments in this area.

Infrared microspectroscopy involves the measurement of a vibrational spectral response from a prescribed, small region of an appropriately derived sample. Historically, attempts to harness infrared spectroscopy for microscopic measurements have been successful in various forms for more than 50 years.[1,2] In the last 20 years, a number of approaches have become commonly accepted, and numerous commercial instruments are now either available or being developed. Simultaneously, the growth in the number and variety of studies involving infrared microspectroscopy has been explosive. Although spectral specificity for this technique may be achieved in various ways, the most widely employed current methods involve infrared interferometry. In general, the three basic instrumental approaches for obtaining infrared microspectroscopic measurements are to (1) provide incident infrared radiation to a small area of the sample, (2) restrict detection of the signal from the sample to only a small spatial area, and (3) detect wide field signal radiation through the use of multichannel detectors. Based on integration of a single element or multichannel detector to an interferometer and microscope assembly, we divide our discussion into two parts: (1) single-element microscopy and (2) imaging using multichannel detectors.

## 32.2 Single-Element Detector FTIR Microspectroscopy

Fourier transform infrared (FTIR) spectroscopy became commercially viable approximately three decades ago, and within 10 years attempts were made to use interferometers for microscopic measurements. Specifically, the advantages of radiation throughput, spectral reproducibility, and time averaging afforded by FTIR spectroscopy were particularly conducive to examining small spatial regions. In addition, the development of stable, sensitive, fast-response cryogenic detectors allowed high-fidelity measurements of spectral intensities.[3] Infrared interferometers coupled to infrared microscopes incorporating these sensitive detectors were introduced in the 1980s.[4] Today, these systems and their advanced versions are used in microscopic infrared analyses in thousands of laboratory locations.

### 32.2.1 Instrumentation

In single-element microspectroscopic instrumentation,[5] modulated radiation from an interferometer is diverted to a set of optics that condense light to a prescribed spatial area, allowing spectral information to be obtained from small samples. Using opaque apertures of controlled size, smaller and better-defined spatial areas can be imaged by restricting the region illuminated by the infrared beam. Subsequently, the radiation is collected by another set of optical components and focused onto a detector. In this manner, infrared spectra may be obtained from highly specified microscopic locations. To uniquely identify the sample area to be spectroscopically examined, a corresponding white light optical image is also acquired. Clearly, focusing the infrared beam for maximal throughput and minimal dispersion in the sample plane requires the optical and infrared paths to be parfocal and collinear. Hence, an optical microscope is integrated into the infrared microscope assembly. A schematic drawing of the system is shown in Figure 32.1.

Important differences exist between a microscope used for infrared spectroscopy and one used for optical microscopy. While optical microscopy is performed by employing high-quality refractive optics made of glass, infrared microscopes usually consist of all-reflective optics. Glass cannot be employed because it does not transmit radiation wavelengths in the mid-infrared spectrum longer than ~5 μm.

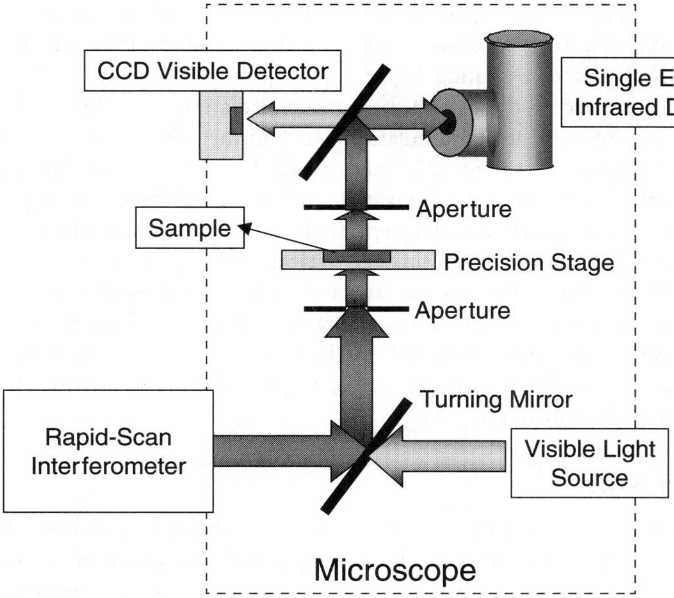


**FIGURE 32.1** Schematic diagram of a mapping spectrometer incorporating a single element detector

Further, spectral fidelity is of paramount concern in the infrared spectral domain, where optical aberrations resulting in spectral nonlinearities must be minimized. The source of modulated radiation in the infrared spectral region is derived from an attached rapid-scan interferometer. In principle, any type of modulator may be employed (for example, step-scan interferometers); rapid-scan interferometers are generally used because they offer cost advantages. Radiation corresponding to small spatial areas is governed by carbon-black-coated metal apertures. In some later designs, specialty infrared absorbing glasses were used, allowing visible imaging while providing restricted infrared imaging.

The instrumentation described above permits the examination of either microscopic samples or narrowly defined regions within larger sample areas and is capable of routinely examining samples in the microgram range. This approach is appropriate particularly for studying small impurities or defects in samples but is of limited utility in obtaining quantitative information from large, heterogeneous materials for which information on the presence and distribution of specific chemical species is desired. By moving the sample, which is easier than moving the optics, different sample locations may be spectroscopically examined. Thus, the sequential movement of the sample in a predetermined manner allows a large sample area to be mapped point by point.[6] Hence, this single-element infrared microspectroscopy using apertures is also referred to as "point mapping" or "point scanning" infrared spectroscopy. Clearly, a sample holder capable of precise microscopic movements must be employed, and for every spatial point a registration of the sample movement to the spectrum corresponding to each spatial index must be maintained. This is accomplished by an automated, programmable microscope stage and an attached computer, which can also be used to control the instrument and to synchronize the various events that are required to map large sample areas.

## 32.2.2 Capabilities and Limitations

While apertures permit infrared microspectroscopy, they also result in loss of light due to diffraction effects when the aperture is small. In practice, radiation transmitted through an aperture results in the formation of a diffraction pattern and can lead to the detector sampling light from outside the apertured region due to the secondary lobes of the diffraction pattern. The effects of this stray light may compromise the spectral content as far as 40 μm away from the points of interest in the sample.[7] This problem may be circumvented by employing a second aperture in tandem to reject radiation even further. However, this also results in loss of signal from the desired sample area under study, thus degrading the spectral signal-to-noise ratio (SNR) and necessitating larger data-acquisition times to improve data quality. Alternately, large apertures are required for permitting greater radiation throughput, which then incurs a tradeoff between the time required for accumulating data and the achieved spatial resolution. Thus, spatial fidelity and spectral quality are intimately linked in single-element systems incorporating apertures; a balance between the two is required to maximize the efficacy of the sampling protocol.

The design of infrared microscopes that employ apertures for mapping considers the largest aperture that may be employed during the course of experiments. Typically, the largest apertures are approximately on the order of 100 × 100 μm.[2] Thus, the spot size at the sample plane allowed by the optics is fixed at a larger spot size (hundreds of micrometers in diameter) to account for all aperture sizes that a user might utilize. The effective area available for light throughput is determined, however, by the area of the aperture opening. Thus, the flexibility to employ large aperture sizes implies that the efficiency of light utilization is anywhere from 0.25 to 100%.

### 32.2.3 Data Processing

Once the spectral response from a sample area is obtained, the absorbance magnitude of a specific vibrational mode is plotted to obtain a map of the distribution of that chemical species. Maps of the relative abundance of chemical species are variously termed "chemical maps" or "functional group maps." Clearly, both the large amount of time required to record data and the optical configuration of a single element detector are not suited either for high spatial resolution or for high spectral resolution mapping over large sample areas. Hence, the acquired data sets usually contain less than 1000 spectral resolution elements and hundreds of spatial resolution elements. Sophisticated spectral data processing may, however, be carried out readily in short time periods. Spatial data processing is of little use because small spatial features are usually below the resolution limit of the technique, and the small number of recorded spatial elements renders the results of many spatial processing routines meaningless. Thus, most data-processing strategies are usually extensions of those developed for a single spectrum acquired by conventional infrared spectrometers.

The usual baseline corrections, normalizations for thickness, and other such steps taken prior to plotting the data for a chemical map are routinely applied. Complex multivariate analyses may be implemented, and library matching of the entire data set is usually feasible. In general, even the most complex data-processing protocols are rapid due to the small data sizes and can be accomplished in a matter of minutes. Computation speeds and power requirements are moderate because less than 1000 spectra are usually processed. One common strategy for use in the spatial domain is to apply deconvolution methods after the point data are acquired, with the points of observation being separated by a distance smaller than the aperture dimension along the direction of separation. Since the collected spectra are often few in number, it is usually more effective to collect data for a longer time than to apply sophisticated techniques for improving the SNR ratio in order to extract molecular information.

## 32.3 FTIR Raster Imaging Using Multichannel Detectors

While single-element microspectroscopy provides the capability for obtaining spectra from small spatial regions, poor SNR characteristics, diffraction effects, and stray light issues resulting from the use of apertures limits the applicability of these methods. Further, the point-by-point mapping approach results in large collection times and poor utilization of the hundreds of focal plane spot sizes whose diameters measure in the micrometers. An approach to resolve some of these issues has recently been implemented[8] in which, as an improvement over single element detectors, a linear array detector is employed to image an area corresponding to a rectangular spatial area on the sample. Unfortunately, the imaged area may not be sufficiently representative of the distribution of chemical species in the sample, and, hence, a technique to image larger areas is required. To compensate for this the linear array is moved precisely to sequentially image a selected, relatively large spatial region of the sample. This is referred to as "push-broom" mapping or "raster scanning." The process is conceptually similar to point mapping but takes advantage of the multiple channels of detection. Imaging a large sample area is faster by a factor of *m* for a linear array containing *m* detector elements.

While detectors for point-by-point mapping are typically 100 to 250 μm in size, modern array detector pixels are sized in the tens of micrometers. Employment of a linear array thus eliminates the need for apertures because small array detectors directly image different spatial regions. The spatial resolution is determined by the optics; in one commercial implementation,[8] the detectors are 25 μm in size, and the instrument can be operated at either a 1:1 magnification or 4:1 magnification to provide a 25-μm or 6.25-μm spatial resolution. Such magnification ratios can be readily achieved by employing available infrared optics that are expected to be relatively aberration-free. However, the optical and infrared paths must be collinear, and the visible image must be referenced to acquire infrared data. Further, a precision stage that reproducibly steps in small spatial increments is required for any mapping larger than that achieved by one row of detectors. Since a precision stage is an integral part of the instrument and a record of the visible image is required, the instrument does not require any visual observation accessories. Once the sample is manually positioned a visual image can be constructed by using the visible-light camera and by moving the sample stage. The sample area from which infrared spectroscopic data are to be determined is therefore delineated, and acquisition may then be initiated.

## 32.3.1 Instrumentation

The schematic diagram of the instrumentation, shown in Figure 32.2, is similar to that required for single-element infrared microscopy but incorporates several important changes. First, no apertures are required; consequently, the unavoidable deficiency in utilizing radiation in point mapping systems is mitigated. Second, the requirements for high-quality optics are simpler and not as stringent as those for the single-element microscope. Third, the spot size can be smaller and matched to the size of the detector array. Since the infrared radiation is more effectively utilized, a reduced intensity source can be employed. Further, the small number of detectors allows fast readout times, and, with the interferometer being operated in a continuous scan mode, instrumental costs decrease. By combining a small multichannel

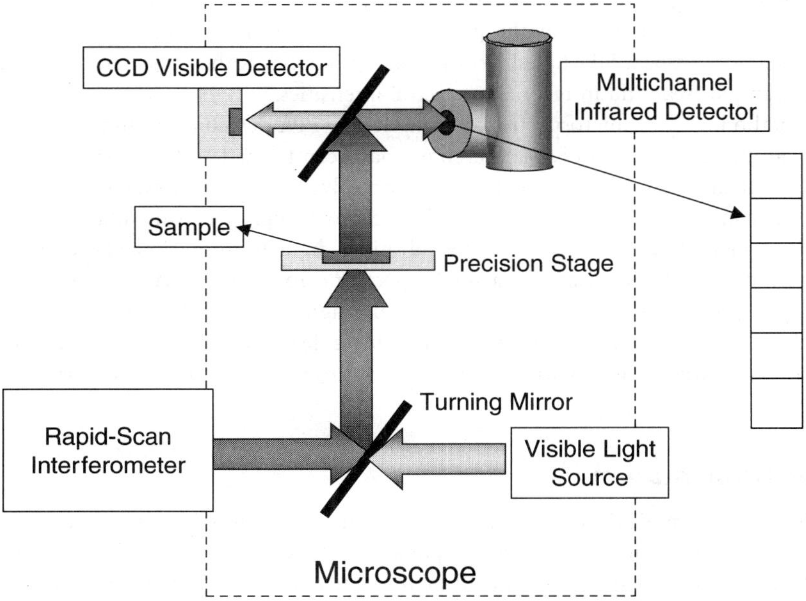


**FIGURE 32.2** Schematic diagram of raster-scanning instrumentation for FTIR microspectroscopy.

detector with rapid-scan spectroscopy to allow mirror scanning in a favorable Fourier frequency regime, and by utilizing frequency domain filtering, higher SNR data can be obtained. Compared to single element detector systems, the removal of apertures allows for higher radiation throughput and provides additional increases in the SNR of data obtained. Hence, with respect to point mapping procedures, the performance of raster-scanning systems is enhanced for both the spatial and spectral domains.

### 32.3.2 Data and Data Processing

A data set from a raster-scanning system can be quite large compared to a single element detector system. For the same data-acquisition time the data set increases by at least the number of channels present in the detector. Further, as discussed above, the SNR per measurement time is higher; hence, in a given time, larger data sets are obtained. The spatial resolution of raster-scanning systems is essentially diffraction-limited, resulting in chemical concentration plots over the field of view being comparable to the visualization afforded by optical microscopy methods. Many data-processing techniques for enhancing visible images can be effectively applied to the infrared spectroscopic data. In addition, since large numbers of spectra are accumulated in each data set, multivariate methods allow statistically rigorous conclusions to be determined.

## 32.4 Global FTIR Spectroscopic Imaging

The state of the art in FTIR microspectroscopic instrumentation is the combination of a focal plane array (FPA) detector and an interferometer.[9,10] FPA detectors consist of thousands of individual detectors placed in a two-dimensional grid pattern. Compared to a linear array, the increase in the number of detectors increases the multichannel advantage; for example, a $p \times p$ pixel focal plane array detector provides a $p^2$ time savings compared to a single-element detector and a $p^2/m$ time saving compared to a linear array detector, where $m$ is the number of detector elements in the linear array detector. For a $256 \times 256$ element detector compared to the single-element case, the advantage is a factor of 65,536; in comparison to the 16-element detector, the multichannel advantage is a factor of 4,096. Further, the two-dimensional detectors are capable of imaging large spatial areas. An FPA matched to the characteristics of the optical system is capable of imaging the entire field of view afforded by the optics and of utilizing a large fraction of the infrared radiation spot size at the plane of the sample. Among the available various infrared microspectroscopic methods, FPA-based global imaging represents the most versatile approach toward viewing large sample areas and in recording spectral intensities.

Since the methodology of microspectroscopy employing FPA detectors is similar to the acquisition of images from an optical microscope, the technique has been termed "infrared spectroscopic imaging." FTIR coupled with an FPA detector provides spatially resolved images across a wide field of view in data-collection times comparable to the time required to record a single spectrum at a single spatial location using a single element detector. This enormous advantage in acquisition time allows both spatially and spectrally specific analyses. For example, a single spectrum from a small sample region, comparable to a single FPA pixel, may be examined or a spectroscopic signal from a spectral feature representative of the entire field of view can be imaged. Due to the considerable reduction in experimental recording times both the imaging of large areas of static samples[11] and the examination of dynamic processes are feasible and rapid.

### 32.4.1 Instrumentation

The schematic of a typical FTIR microimaging system shown in Figure 32.3 interfaces an interferometer to an infrared microscope, yielding a specified spot size at the sample plane. This spot, which is typically hundreds of micrometers in size, is magnified by the optical train and imaged onto a focal plane array that is tens of millimeters in diameter. The spatial resolution is thus determined both by the optics that determine magnification and the size of an individual detector element on the focal plane array. Even

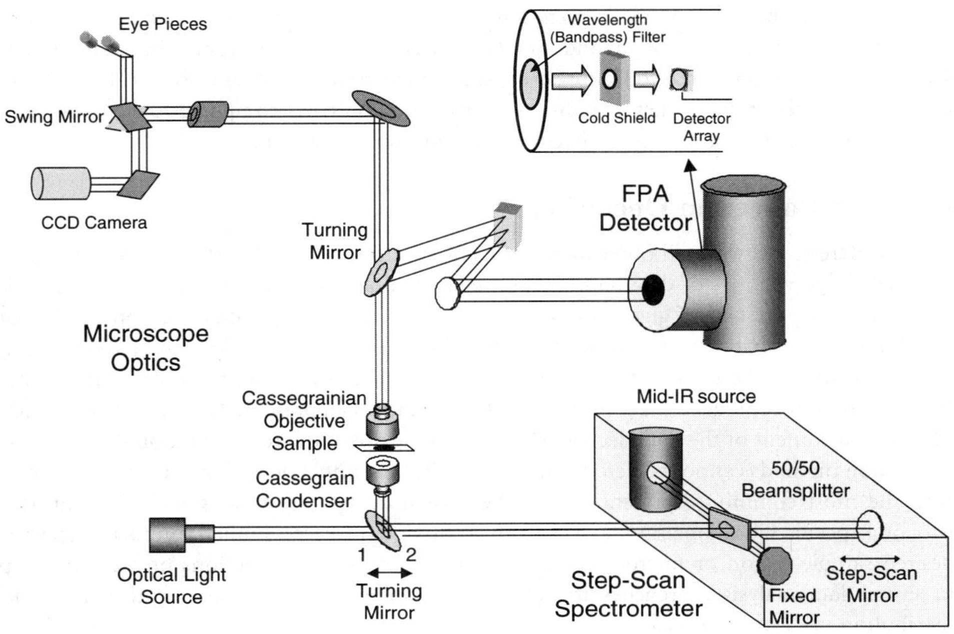


**FIGURE 32.3** Schematic diagram of an FTIR imaging spectrometer incorporating a focal plane array detector.

though the nominal spatial resolution is determined by the optics and the detector, the actual resolution limit usually depends upon the diffraction limit of the wavelengths of interest incident upon the sample.

The first and, to date, most popular approach to FTIR microimaging spectrometers incorporates a step-scan interferometer, which provides a means for maintaining constant optical retardations for desired (often large) time periods. A constant retardation over extended time periods is employed both for signal averaging purposes and for computer storage of the acquired data. An initial small time delay prior to data acquisition allows for mirror stabilization at each interferometer step. The data-acquisition and readout formats may be spatially sequential (rolling mode) or simultaneous (snapshot) across the array. In either case, the signal is integrated for only a fraction of the time required for the collection of each frame. The integration time, number of frames coadded, and number of interferometer retardation steps, which is determined by the desired spectral resolution, determine the time required for the experiment. Since signal integration governs the data quality, efforts are made to increase the ratio of the integration time to the total data-acquisition time.

Imaging configurations utilizing a continuous scan[12] or rapid-scan[13] interferometer have been proposed for small detector arrays. Since a large number of detector elements generally precludes regular rapid scanning velocities due to slow readout rates, many instrumental configurations employ a continuously-scanning interferometer at slow speeds. Some manufacturers employ fast step-scan modes where the mirror is partially stabilized to achieve the same retardation error as in continuous-scan spectroscopy; this approach is termed “slow-scan.” While slow-scan approaches appear to be similar to rapid-scan methods, they do not have the advantage afforded by the coupled Fourier frequencies as in rapid scanning and merely delay FPA nonlinearity corrections until after the completion of the interferometer scan. We believe that the utility of these slow-scan methods is limited and of little consequence in acquiring high SNR data. A generalized data-acquisition scheme that permits true rapid-scan data acquisition for large arrays or higher mirror speed acquisitions from small arrays has been proposed.[13] The integration time for individual frames collected by the FPA detector is small enough to be considered negligible on the time scale of interferogram collection. For most FPA detectors available today, however, the motion of the moving mirror in a continuous-scan mode does not allow the coaddition of frames per interferometer retardation step. Compared to step-scan approaches, rapid-scan data collection allows fast interferogram acquisition because there is no requirement for mirror stabilization; however, the image stored per resolution element is noisier. Although random FPA noise is dominant, the error arising from the deviation in mirror position during frame collection is considered the next largest contributor. The advantage of continuous-scan interferometry lies in reduced instrumentation costs compared to step-scan units and more efficient data collection because mirror-stabilization times are eliminated.

### 32.4.2 Post-Collection Operations

The major difference between all other forms of FTIR spectroscopy and FTIR spectroscopic imaging is the extremely large volumes of data handled in the latter mode of operation. For a 256 × 256 element FPA, a user may acquire tens of gigabytes of data per hour. Further, the human comprehension of such a large volume of data in its entirety is not only impossible but also undesirable. Hence, the most common operation is to reduce the data in some manner to allow compact visualization of desired features. A number of analytical approaches have been suggested with the data presented in a variety of forms. Since the information content of the data decreases as the complexity of the representation decreases, the best data-extraction methods cannot be determined *a priori* but are implemented on a case-by-case basis. For example, the most common representations either visualize the two-dimensional distribution of the intensity of a specific vibrational mode over the field of view, project a limited number of spectra from any desired sample region or microdomain, or present statistical distributions of a chemical species. Often, several data-analysis approaches are employed in tandem to provide an effective visualization of the distribution of chemical species.

Another facet of data processing not commonly encountered in conventional infrared spectroscopy is the need to reduce noise in the data set. The low SNR in FTIR images arises both from the characteristics of the FPA and from data-acquisition schemes. While efforts to improve FPAs for spectroscopic imaging have been suggested and advances are constantly being made, the modification of data-acquisition methods is an area of considerable attention. A common data-acquisition approach has been to coadd as many frames as possible during the available experimental time. While this approach results in significant gains, as expected, the SNR, as a function of coadded frames, achieves a plateau.[14] Thus, after the coaddition of a limited number of frames, the benefits are relatively minor compared to the large increase in experimental time expenditures. The coaddition of complete image data sets was shown to be effective in increasing the fidelity of the data, with an optimum sampling strategy being derived by combining both frame and image data-set coaddition. While image coaddition results in improved SNR characteristics, the process involves a $2n^2$ increase in collection time for an SNR improvement factor of $n$. This rules out image coaddition as a technique to increase the SNR for either real-time imaging of rapidly changing systems or for the expeditious examination of large numbers of samples.

A low-noise, single-beam background ratioed to a single-sample, single-beam image was shown to exhibit lower noise characteristics, allowing for imaging measurements of dynamic systems.[15] This method, termed "pseudo coaddition," is also limited in achievable benefits because noise from the sample's single beam tends to dominate the resultant spectra. Spectra from sample areas with the same true absorbances can be coadded in a similar statistical manner to yield low-noise-average spectra.[15] Mathematical noise reduction, using, for example, the minimum noise fraction (MNF) transform, is an alternate pathway toward obtaining higher-fidelity images after data acquisition.[16] The transformation and inverse transformation after eliminating components due to noise is computationally intensive but does not result in loss of image content or affect image-collection times. The gain in the SNR depends on the SNR characteristics of the original data. Noise is reduced by a factor greater than five if the noise in the initial data is sufficiently low.

## 32.5 Applications

A recent survey of FTIR techniques, including microspectroscopy, extended to biological materials has been compiled.[17] Although examples of global FTIR imaging applied to biological systems have been discussed elsewhere,[18] we include here several specific studies to illustrate the range of applications and to outline potential future directions. Since FTIR imaging is a noninvasive, nondestructive technique, analyses of intact cells or biopsied tissues are not only a relatively straightforward process but are also of enormous contemporary interest. Conventional cell and tissue examinations are conducted using visible microscopy at various magnification levels. To assist in discrimination between different constituent morphological features of the specimen, samples are stained with reagents responsive to specific chemical components. Although staining methods are generally useful for visualizing tissue structure, they are relatively nonspecific for obtaining detailed chemical information. Moreover, the choice of a specific staining protocol or application of other immunohistochemical techniques often depends on a preliminary biochemical knowledge of the system or disease under scrutiny. FTIR imaging presents, however, an opportunity to directly visualize detailed chemical and morphological characteristics without concern for sample history or disease etiology. As inherent chemical properties of the molecular composition dictate contrast between various components, no sample staining is required. In contrast to conventional IR microscopic methods, the large numbers of spectral data that are obtained through global imaging allow quantitative analyses based on statistical techniques. From the simple demonstration of imaging a lipid [C(16)-lysophosphatidylcholine] distribution in a KBr disk,[19] FTIR imaging applications have grown to include, for example, various tissue sections, single cells, pharmaceutical devices, and food grains.

## 32.5.1 Human and Animal Tissue Sections

### 32.5.1.1 Brain

Concentration distributions of various components in thin monkey brain sections[20] revealed specific lipid/protein distributions corresponding to morphological features. Concentration abundance images constructed from spectral data indicated that a greater proportion of lipids occurred in white-matter regions relative to the lipid proportion in gray-matter areas. Since visualization of the entire data set is difficult, a number of statistical methodologies, including scatter plots, histograms, and profile matching, were suggested for data analysis to extract the desired information. These techniques, based on a statistically significant number of observations (spectra), can then be employed to characterize different regions of the tissue. Once the characteristic spectral features of specific regions of a tissue and their inherent variations are known, significant deviations from these may be used to characterize biochemical imbalances, even in cases where morphological changes may be small or nonexistent. This is the fundamental philosophy of employing imaging for biomedical analyses.

Neurotoxic effects of an antineoplastic drug (cytarabine [Ara-C]) on Purkinje cells, which are cells in rat cerebella that have been shown to strongly influence motor coordination and memory processes, were visualized using FTIR imaging.[21,22] Based on the total absorbance inherent in the IR brightfield images, it was observed that the IR images contained more contrast compared to the visible microscopy images. The packing properties of the tissue influenced the refractive index in the infrared spectral region, resulting in better contrast. An analysis of the spectroscopic image from rat cerebella revealed a chemical basis for contrast between different parts of the tissue. This contrast was characterized by a ratio of the lipid-to-protein concentration. The relative lipid concentration exhibited significant differences among sample regions and could be employed as an index to characterize subtle structural changes. The higher protein-packing density of the Purkinje layer was shown to possess increased fractions of disordered lipid acyl chains. Hence, spatially resolved structural changes are determinable based on the molecular information contained in the vibrational spectra. Changes in these measures can be employed to detect and to quantify effects that result from diseases and the effects of drugs or foreign agents. The infrared spectroscopic images reflecting various tissue regions are shown in Figure 32.4.

Neuropathologic effects of a genetic lipid storage disease, Niemann–Pick type C, were investigated by examining sections of control and affected mice cerebella in an effort to understand the chemical basis of the observed morphological changes and to relate these changes to disease pathology.[23] By employing infrared imaging, various cellular layers were readily identified, and the diseased and control samples were distinguished without external, histological staining on the basis of spectral features

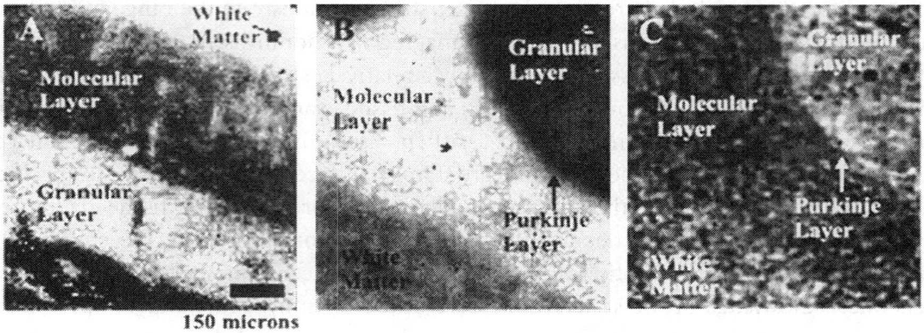


**FIGURE 32.4** Infrared spectroscopic images of thin cerebellar sections from a control and Ara-C-treated animal. Spectroscopic image showing the spatial distribution of lipid/protein ratios in a rat treated with (A) saline and (B) cytarabine rat. (C) Spectroscopic image depicting the distribution of phosphatidylcholine in a cerebellar section from a cytarabine treated animal. The image is derived from the intensity of a vibrational absorption band centered at approximately 3060 $cm^{-1}$, assigned to the methyl ($CH_3$) stretching vibration of the lipid choline headgroup (From Lester, D.S. et al., *Cell. Mol. Biol.*, 44, 29, 1998. With permission.)

characteristic of each layer. Lipid depletion was found in diseased samples in comparison to control tissue sections; other molecular differences were discerned in the granular-layer region of the tissue. The observed chemical changes were consistent with demyelination within the cerebellum of the diseased animal. In this manner, morphology and underlying chemical changes are readily correlated. Of particular interest is that separate analyses for each constituent chemical are not required. Instead, a single spectroscopic experiment yields information on the many types of chemical species that are present and their relative concentrations.

#### 32.5.1.2 Bone

As opposed to grinding bone for average spectroscopic analyses, which destroys local structure and may introduce artificial changes, spatially resolved measurements have been carried out on bone-related systems since the 1980s using point mapping. Spatially resolved FTIR imaging[24] has recently allowed the examination of bone in a nondestructive manner, where the advent of imaging allows the rapid characterization of different structures.[25] The spatial variations of chemical components representing the main structural ingredients in bone formation can be easily monitored. For example, hydroxyapatite and protein distributions, as shown in Figure 32.5, illustrate the morphological distributions of these chemicals. Not only can the two-dimensional structure of a thin section be observed, but FTIR imaging also allows the three-dimensional structure of bone, as shown in Figure 32.5, to be visualized in terms of the distribution of its chemical constituents.[26]

The nondestructive nature of the two-dimensional infrared imaging technique also allows the determination of molecular orientation in addition to simple concentration analyses.[24] Chemical-composition data from FTIR imaging and morphology were correlated for human iliac crest biopsies.[27] The known developmental processes in bone were related to observed gradients in mineral levels by examining spatial profiles from the middle of the osteon to the periphery. The observations were found to be consistent with the models for structure development in the Haversian model of bone development. Protein content, which could be readily monitored using the Amide I contour (1620 to 1680 $cm^{-1}$), revealed nonmineralized, high-protein-content regions, namely, the osteoid. A measure of crystallinity and bone maturity was determined using the phosphate $\nu_1$, $\nu_3$ contour and was found to increase away from the osteonal center. Due to the design of experiments, sample-selection limitations, and the myriad approaches for characterizing extracted chemicals, the available databases are often contradictory and often lead researchers to accept the historical understanding of osteoporosis as a mass deficiency of otherwise healthy bone materials. Using imaging techniques investigators have shown from their examination of spectral data from known, localized sites of mineralized tissue that the conventional view of the disease is not entirely correct. When imaging data from normal and osteoporotic human iliac crest biopsies were compared,[28]

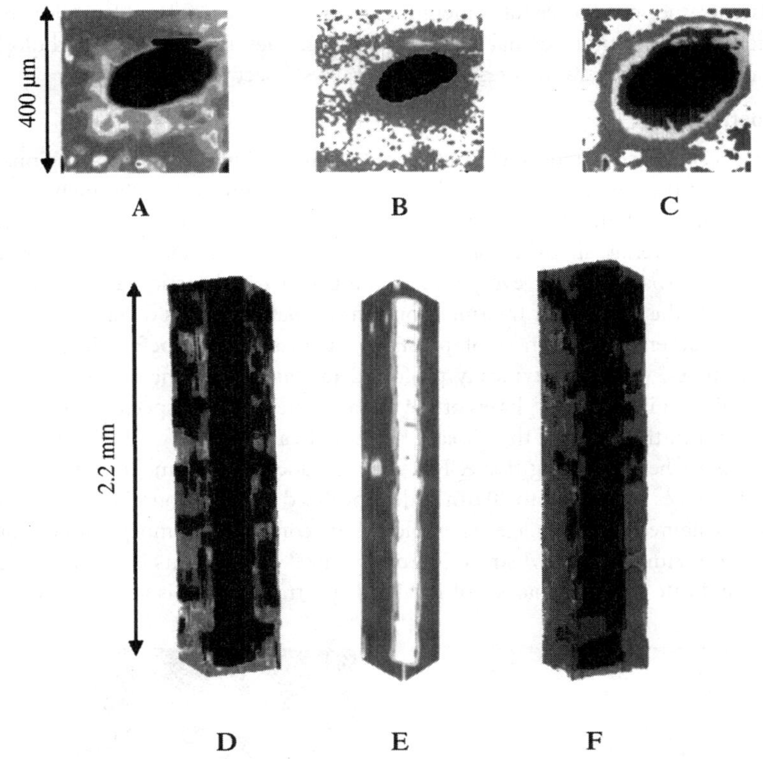


**FIGURE 32.5 (Color figure follows p. 28-30.)** (A–C) 2D images of three IR parameters for the cortical bone section (human tibia) shown as follows: (A) nonreducible/reducible collagen crosslinks, measured by the 1660/1690 peak height ratio in the amide I spectral region; (B) hydroxyapatite crystallinity index, measured by shape changes in $PO_4^{3-}$ $\nu_1$, $\nu_3$ mode and quantitated by the 1030/1020 peak height ratio; (C) mineral-to-matrix ratio, measured by the phosphate $\nu_1$, $\nu_3$ /amide I area ratio. Black and blue indicate low numerical values while green, yellow, and red indicate progressively increasing numerical values of the particular parameter. The dimensions of each image are 400 × 400 µm. (D–F) 3D reconstructed views of three IR parameters for the cortical bone section (human tibia), as follows: (D) view of the collagen crosslink parameter measured as the 1660/1690 height ratio in the amide I spectral region; (E ) PMMA distribution in a cortical bone section. The relative concentration of PMMA is determined from the integrated intensity of the C = O stretching mode; (F) 3D reconstructed oblique view of the relative amounts of mineral-to-matrix as measured by the phosphate $\nu_1$, $\nu_3$ /amide I area ratio. The dimensions represented in each image are 400 × 400 µm × 2.2 mm. (From Ou-Yang, H. et al., *Appl. Spectrosc.*, 56, 419, 2002. With permission.)

two spectral parameters were found useful in characterizing the biochemical changes. The first, which is indicative of the extent of mineral (hydroxyapatite) formation in the tissue, was obtained using a ratio of the integrated areas of the phosphate contour and the amide I peak. This parameter also served to normalize any thickness variations arising from the sample preparation process. The second parameter, indicative of the size and perfection of the crystals, was obtained by taking a ratio of absorbance at 1030 $cm^{-1}$ to the absorbance at 1020 $cm^{-1}$. It was shown that the average mineral levels in osteoporotic samples are considerably reduced compared to normal bone and that the crystal size and crystal perfection were substantially degraded.

The analysis protocol was applied to examine the effects of estrogen therapy on fracture healing in rat femurs. The mineral content per unit matrix in the estrogen-treated samples was found to be higher, demonstrating the effectiveness of the treatment. Similarly, the crystal sizes and perfection ratios were also found to be elevated in the treated samples. When healing bone was monitored, it was found that several sites along the fracture contained a disproportionately large protein content, indicating that these were sites for the formation of new mineral. In addition, cellular activity was detected at the fracture sites by signals corresponding to cellular-membrane fatty acids. It is clear that FTIR imaging may be employed to detect and monitor normal structure, structure development, and pathological changes in bone through the chemical alterations reflected by localized spectral changes.[24–28]

#### 32.5.1.3 Prostate

Prostate cancer, specifically prostatic adenocarcinoma originating in the epithelial regions of the prostate gland, is prevalent in the western world. In the United States, almost 180,000 men are diagnosed with prostate cancer each year. Similarly, cancer of the prostate is the most common male cancer affecting British men, with the average lifetime risk of development approximately 1 in 13, with nearly 21,000 men diagnosed with prostate cancer every year in the U.K. Early detection and correct diagnoses may increase considerably the number of treatment options available for this disease. We have attempted to discern chemical markers of the onset of prostatic adenocarcinoma before it becomes apparent in morphological changes, which are typically profound for late-stage cancers but extremely subtle and often difficult to discern in the early phases of the disease. The chemical specificity of FTIR spectroscopy is particularly useful in the study of this disease because it can be combined with the spatial selectivity afforded by imaging when analyzing tissue. For prostatic adenocarcinoma, it is particularly important to assess spectral changes in small (5 to 20 μm wide), localized regions around prostatic ducts, or lumens. A comparison of a stained visible image, as employed for conventional morphological analyses, and an infrared spectroscopic image of the distribution of chemical components is shown in Figure 32.6. Preliminary studies indicate that the analysis of localized spectral changes is an attractive means for char-

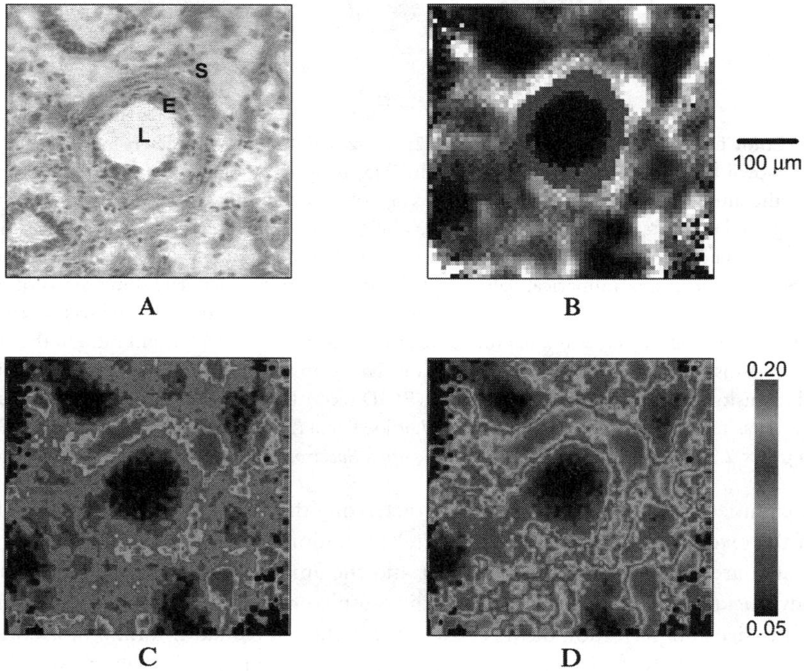


**FIGURE 32.6 (Color figure follows p. 28-30.)** (A) H&E stained section corresponding to the unstained section used to obtain spectroscopic images. The circular lumen (L) of prostatic gland in the center is lined by a double layer of epithelial cells (E) and supported by a smooth muscular and fibrocollagenous stroma (S). (B) Unstained section with the epithelial regions highlighted in red. Images of the distribution of absorbance obtained by applying (C) no gain ranging at a gain setting of one and (D) gain ranging for gains one and ten with a gain ranging radius of ten points. The images represent a sample area of 500 μm × 500 μm as indicated by the size bar. The color-coded images show the distribution of the absorbance at 1245 $cm^{-1}$ between the indicated limits. (From Bhargava, R. et al. *Appl. Spectrosc.*, 55, 1580, 2001. With permission.)

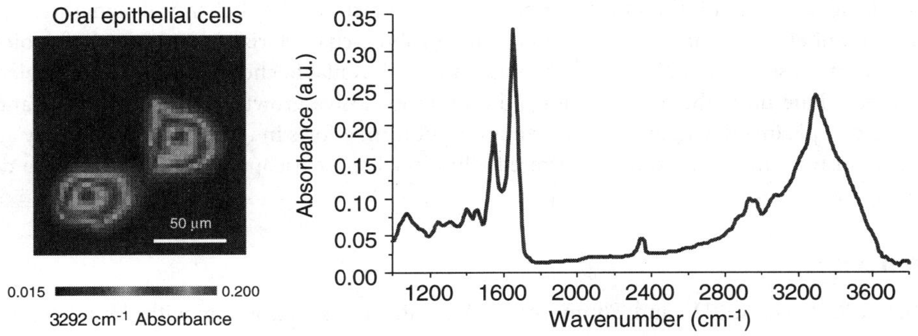


**FIGURE 32.7 (Color figure follows p. 28-30.)** IR spectroscopic image of single oral epithelial cells (left) and the corresponding spectrum from a part of the cell. The high signal-to-noise ratio of the data allows for detailed analysis of subtle chemical changes.

acterizing spectral changes for disease diagnosis and for monitoring its progression. Since the spectral changes involved are subtle, large numbers of samples must be examined, and FTIR imaging techniques that provide superior data quality, as indicated in Figure 32.6,[29] must be employed.

#### 32.5.1.4 Breast

Silicone breast implants, consisting of an elastomeric shell filled with silicone gel material, are used extensively for breast enhancement procedures. The silicone gel may leak from the implant due either to material failure as a consequence of aging or to a rupture in the casing. Among the complications arising from this leakage are capsular contracture, calcification, and connective tissue disorders. To reliably assess any histopathological changes it is first necessary to confirm the presence of implant material in the breast tissue. Since silicone gel differs in chemical composition from surrounding tissue, it is easily detected in tissue sections using FTIR spectroscopic imaging.[30] The presence of an inclusion was revealed by monitoring the Si-$CH_3$ characteristic stretching vibrations, which are normally absent in human tissue. Silicone gel contaminations in regions as small as ~10 μm could be observed even for inclusions where conventional optical microscopy contrast was poor and chemical specificity was lacking. Dacron, a commercial name for poly(ethylene terepthalate), threads due to fixative patches were also distinguished. In particular, the rapid analysis and minimal sample preparation allowed the detection of contamination within minutes of sectioning the tissue.[30] These advantages imply that FTIR imaging may potentially be useful in settings requiring real-time analysis, for example, in surgical facilities.

### 32.5.2 Single Cells

The high spatial resolution of infrared imaging, which has allowed the examination of small morphological features in tissues, may also be employed to image single biological cells.[31] An example of single-cell imaging and the associated spectral data that are collected are shown in Figure 32.7. It is expected that in the future the spatial-resolution limits and spectral SNR that can be obtained will allow relatively complete chemical characterizations of single cells and their intracellular components. In particular, the monitoring of biochemical changes within single cellular assemblies that have either been modified in a specific manner or harvested from specific types of tissue will make more detailed approaches to biomedical analyses achievable.

## 32.6 Summary

FTIR spectroscopic imaging employing an infrared interferometer and infrared-sensitive multichannel detectors have considerably enhanced the ability to rapidly acquire high-fidelity, spatially resolved data at unprecedented levels of spatial resolution and chemical detectivity. These approaches are particularly suited to detailed biochemical and biological studies in which localized chemical signatures combined with large numbers of recorded data points lead to robust characterization protocols for biomedical analyses. An increase in the number and sophistication of available chemometric tools coupled to this powerful technique undoubtedly will contribute to an explosive growth in the acceptance and use of FTIR spectroscopic imaging in biomedical venues. Significant efforts in the authors' laboratory are under way to demonstrate fully the utility of infrared vibrational spectroscopic imaging for disease detection and diagnosis.

## References


1. Barer, R., Cole, A.R.H., and Thompson, H.W., Infra-red spectroscopy with the reflecting microscope in physics, chemistry and biology, *Nature,* 163, 198, 1949.
2. Gore, R.C., Infrared spectrometry of small samples with the reflecting microscope, *Science*, 110, 710, 1949.
3. Reffner, J.A., Instrumental factors in infrared microscopy, *Cell. Mol. Biol.*, 44, 1, 1998.
4. Kwiatkoski, J.M. and Reffner, J.A., FT-IR microscopy advances, *Nature,* 328, 837, 1987.
5. Messerschmidt, R.G., *Practical Guide to Infrared Microspectroscopy,* Humecki, H.J., Ed., Marcel Dekker, New York, 1995, pp. 3–39.
6. Reffner, J.A., Molecular microspectral mapping with the FT-IR microscope, *Inst. Phys. Conf. Ser.,* 98, 559, 1989.
7. Sommer, A.J. and Katon, J.E., Diffraction-induced stray light in infraredmicrospectroscopy and its effect on spatial resolution, *Appl. Spectrosc.*, 45, 1663, 1991.
8. Spotlight FT-IR Imaging system by Perkin-Elmer Company.
9. Colarusso, P., Kidder, L.H., Levin, I.W., Fraser, J.C., Arens, J.F., and Lewis, E.N., Intrared spectroscopic imaging: from planetary to cellular systems, *Appl. Spectrosc.,* 52, 106A, 1998.
10. Lewis, E.N., Treado, P.J., Reeder, R.C., Story, G.M., Dowrey, A.E., Marcott, C., and Levin, I.W., Fourier-transform spectroscopic imaging using an ingrared focal-plane array detector, *Anal. Chem.,* 67, 3377, 1995.
11. Koenig, J.L. and Snively, C.M., Fast FT-IT imaging: theory and applications, *Spectroscopy,* 13, 22, 1998.
12. Snively, C.M., Katzenberger, S., Oskarsdottir, G., and Lauterbach, J., Fourier-transform infrared imaging using a rapid-scan spectrometer, *Appl. Opt.,* 24, 1841, 1999.
13. Huffman, S.W., Bhargava, R., and Levin, I.W., A generalized implementation of rapid-scan Fourier transform infrared spectroscopic imaging, *Appl. Spectrosc.,* 56, 965, 2002.
14. Snively, C.M. and Koenig, J.L., Characterizing the performance of a fast FT-IR imaging spectrometer, *Appl. Spectrosc.,* 53, 17, 1999.
15. Bhargava, R., Ribar, T., and Koenig, J.L., Towards faster FT-IR imaging by reducing noise, *Appl. Spectrosc.,* 53, 1313, 1999.
16. Bhargava, R., Wang, S.-Q., and Koenig, J.L., Processing FT-IR imaging data for morphology visualization, *Appl. Spectrosc.,* 54, 1690, 2000.
17. Gremlich, H.-U. and Yan, B. *Infrared and Raman Spectroscopy of Biological Materials,* Marcel Dekker, New York, 2000.
18. Bhargava, R. and Levin, I.W., Fourier transform infrared imaging: a new spectroscopic tool for microscropic analysis of biological tissue, *Trends Appl. Spectrosc.,* 3, 57, 2001.
19. Lewis, E.N. and Levin, I.W., Advances in vibrational imaging microscopy, *J. Microsc. Soc. Am.*, 1, 35, 1995.
20. Lewis, E.N., Gorbach, A.M., Marcott, C., and Levin, I.W., High-fidelity Fourier transform infrared spectroscopic imaging of primate brain tissue, microscopy in neurotoxicity, *Appl. Spectrosc.,* 50, 263, 1996.

21. Lewis, E.N., Kidder, L.H., Levin, I.W., Kalasinsky, V.F., Hanig, J.P., and Lester, D.S., High-fidelity Fourier transform infrared imaging microscopy in neurotoxicity, *Ann. N.Y. Acad. Sci.,* 820, 234, 1997.
22. Lester, D.S., Kidder, L.H., Levin, I.W., and Lewis, E.N., Infrared microspectroscopic imaging of the cerebellum of normal and cytarabine treated rats, *Cell. Mol. Biol.,* 44, 29, 1998.
23. Kidder, L.H., Colarusso, P., Stewart, S.A., Levin, I.W., Appel, N.M., Lester, D.S., Pentchev, P.G., and Lewis, E.N., Infrared spectroscopic imaging of the biochemical modifications induced in the cerebellum of the Niemann–Pick type C mouse, *J. Biomed. Opt.,* 4, 7, 1999.
24. Paschalis, E.P., DiCarlo, E., Betts, F., Sherman, P., Mendelsohn, R., and Boskey, A.L., FTIR microspectroscopic analysis of human osteonal bone, *Calcif. Tissue Int.*, 59, 480, 1996.
25. Marcott, C., Reeder, R.C., Paschalis, E.P., Boskey, A.L., and Mendelsohn, R., Infrared microspectroscopic imaging of biomineralized tissues using a mercury-cadmium-telluride focal-plane array detector, *Phosphorus Sulfur,* 146, 417, 1999.
26. Ou-Yang, H., Paschalis, E.P., Boskey, A.L., and Mendelsohn, R., Chemical structure-based three-dimensional reconstruction of human cortical bone from two-dimensional infrared images, *Appl. Spectrosc.,* 56, 419, 2002.
27. Mendelsohn, R., Paschalis, E.P., and Boskey, A.L., Infrared spectroscopy, microscopy, and microscopic imaging of mineralizing tissues: spectra-structure correlations from human iliac crest biopsies, *J. Biomed. Opt.,* 4, 14, 1999.
28. Mendelsohn, R., Paschalis, E.P., Sherman, P.J., and Boskey, A.L., IR microscopic imaging of pathological states and fracture healing of bone, *Appl. Spectrosc.,* 54, 1183, 2000.
29. Bhargava R., Schaeberle, M.D., Fernandez, D.C., and Levin, I.W., Novel route to faster Fourier transform infrared spectroscopic imaging, *Appl. Spectrosc.,* 55, 1580, 2001.
30. Kidder, L.H., Kalasinsky, V.F., Luke, J.L., Levin, I.W., and Lewis, E.N., Visualization of silicone gel in human breast tissue using new infrared imaging spectroscopy, *Nat. Med.,* 3, 235, 1997.
31. Bhargava, R., Fernandez, D.C., and Levin, I.W., FTIR Imaging of Biological Tissue for Histopathological Analyses. Presentation at Pittcon 02, New Orleans, LA, 2002.

# 33

# Near-Infrared Fluorescence Imaging and Spectroscopy in Random Media and Tissues

Eva M. Sevick-Muraca
*Texas A&M University*
*College Station, Texas*

Eddy Kuwana
*Texas A&M University*
*College Station, Texas*

Anuradha Godavarty
*Texas A&M University*
*College Station, Texas*

Jessica P. Houston
*Texas A&M University*
*College Station, Texas*

Alan B. Thompson
*Texas A&M University*
*College Station, Texas*

Ranadhir Roy
*Texas A&M University*
*College Station, Texas*



0-8493-1116-0/03/$0.00+$1.50

## 33.1 Introduction

Most radiation-based spectroscopic and imaging techniques typically depend on evaluation of a non-scattered or singly scattered signal for retrieval of quantitative information. For example, absorption spectroscopy depends on the survival of unscattered light across a known pathlength, *L*; dynamic-light scattering or photon-correlation spectroscopy, because of its scatter by Brownian motion out of the optical path, requires the fluctuation of light intensity; x-ray and computed x-ray tomography depend on the straight-line path of nonabsorbed x-rays; and so forth. Yet most systems of interest multiply scatter radiation of low energy and require diluted suspensions or nonscattering media when dealing with optical interrogation. Hence, optical techniques developed for imaging and spectroscopy are usually plagued by scatter influence. To expand the quantitative applicability of optical techniques to these real systems new techniques have been developed that focus on coherence properties, temporal and spatial correlation, and other properties; these allow for the extraction of nonscattered or singly scattered light from a multiply scattered signal. Yet these approaches neglect the scattered signal, which is the largest portion of the signal, in favor of that portion that possesses the smallest signal-to-noise ratio (SNR).

In this chapter, we first review continuous-wave and time-resolved techniques along with the associated diffusion equation for quantitative absorption, scattering, and fluorescence spectroscopy using multiply scattered light. In addition, since in the wavelength window of 600 to 900 nm light is multiply scattered by most biological tissues, we then focus on the development of these optical techniques for biomedical spectroscopy and imaging, i.e., optical tomography. Because of the limitations imposed by endogenous chromophores on tissues in this wavelength regime, we provide a comprehensive review of fluorescence-enhanced optical spectroscopy and imaging methods including measurement methods, solutions to forward- and inverse-imaging problems, and attributes of clinical and sensing fluorophore development.

## 33.2 Background: Probing Random Media with Multiply Scattered Light

Before presenting the measurement methods and analysis for probing random media with multiply scattered light, we will first consider traditional light-spectroscopy techniques that depend on monitoring light transmitted across a known pathlength, *L*.

### 33.2.1 Beer–Lambert Relation for Absorption and Turbidity Spectroscopy

Absorption and turbidity measurements consist of monitoring the attenuation of light intensity I(λ) at wavelength λ, given incident-light intensity $I_o$ (λ), to determine the absorption or scattering coefficients ($\mu_a$ [cm$^{-1}$] or $\mu_s$ [cm$^{-1}$], respectively):

$$\log\frac{I(\lambda)}{I_o(\lambda)} = -\mu_a(\lambda)L \quad \text{or} \quad -\mu_s(\lambda)L \tag{33.1}$$

where the absorption coefficient is provided by the product of the concentration of light-absorbing species, $[C_i]$ [mM], and its extinction coefficient at wavelength λ, $\varepsilon_i^\lambda$ [cm$^{-1}$ mM$^{-1}$]:

$$\mu_a(\lambda) = 2.303\sum_{i=1}^{N}\varepsilon_i^\lambda[C_i] \tag{33.2}$$

and the scattering coefficient can be predicted from:

$$\mu_s(\lambda) = \int_0^\pi \frac{12\phi}{k^2}\int_0^\infty \frac{f(x_i)}{x_i^3}\int_0^\infty \frac{f(x_j)}{x_j^3} F_{i,j}(n, x_i, x_j, \lambda, \theta)\cdot S_{i,j}(x_i, x_j, q, \phi)\sin\theta\, dx_j dx_i d\theta \tag{33.3}$$

where $F_{i,j}$ is the binary form factor between the particles with different sizes $x_i$ and $x_j$; $S_{i,j}$ is the corresponding partial structure factor, which describes the correction factor of coherent scattering due to particle interactions of particles $i$ and $j$; $n$ is the relative refractive index of the particles to the medium; $\lambda$ is wavelength; $\theta$ is the scattering angle; $\phi$ is the volume fraction of particles in the suspension; $f(x)$ is the particle-size distribution; and $q$ is the magnitude of the wave vector, $q = 2k\sin(\theta/2)$, where k is given by $2\pi m/\lambda$ and m is the refractive index of the medium. The structure factor is a direct measure of the local ordering of colloidal particles, and the values of $S_{i,j}$ are equal to unity in the absence of particle interactions (e.g., in a dilute suspension).

Determination of absorption and scattering coefficients through the Beer–Lambert relationship in Equation 33.1 assumes that light is absorbed or scattered *out of the path* and that no light scattered is back *into the path*. Absorption and scattering mechanisms can be considered simultaneously in dilute suspensions as an effective absorption cross section:

$$\mu_{eff} = \mu_a + \mu_s \tag{33.4}$$

## 33.2.2 Fluorescence Spectroscopy and Fluorescence-Lifetime Spectroscopy

Fluorescence spectroscopy, whether measured using time-resolved or continuous-wave (CW) techniques, is based on the absorption of excitation light at $\lambda_x$ across path length $L$ by fluorophores of concentration $[C_i]$. Quantum efficiency, $\alpha$, describes the fraction number of emission photons at fluorescence wave length $\lambda_m$ emitted for each excitation photon absorbed by the fluorophore; it is typically described as the rate of radiative decay, $\Gamma$, relative to the sum of radiative and nonradiative decay rates ($\Gamma + k_{nr}$). In other words, $\alpha = \Gamma/(\Gamma + k_{nr})$. The intensity of detected fluorescence light, $I_m$, in response to a constant intensity of incident excitation light, $I_o^{\lambda_x}$, can be expressed as:[1]

$$I_m \propto I_o^{\lambda_x}\alpha\left[\varepsilon_i^{\lambda_x}[C_i]\right]\cdot\int_0^\infty g(t)\,dt \tag{33.5}$$

Here $g(t)$ represents the time-dependent fluorescence decay that describes the process of radiative and nonradiative relaxation of the activated fluorophore elevated to an excited state by absorption of excitation light. For most laser dyes the relaxation is a first-order process described by a mean lifetime, $\tau$, of the activated state. Consequently, the time-invariant emission intensity predicted by Equation 33.5 can be rewritten as:

$$I_m \propto I_o^{\lambda_x}\alpha\left[\varepsilon_i^{\lambda_x}[C_i]\right]\cdot\int_0^\infty \exp\left[-\frac{t}{\tau}\right]dt \propto I_o^{\lambda_x}\alpha\left[\varepsilon_i^{\lambda_x}[C_i]\right]\tau \tag{33.6}$$

where the fluorescence lifetime, $\tau$, is influenced by the relative rates of radiative and nonradiative decay [i.e., $\tau = 1/(\Gamma + k_{nr})$].

Ratiometric fluorescent probes, in which re-emission is monitored across two or more wavelengths (such as *bis*-carboxyethyl carboxyfluorescein [BCECF] or seminaphthofluorescein [SNAFL]), also provide a means to monitor changes in decay kinetics using CW methods. The ratio of the emission intensities at $\lambda_{m1}$ and $\lambda_{m2}$ following excitation at a single excitation wavelength is independent of the concentration of fluorophore available and depends only on the decay kinetics probed at the two emission wavelengths:

$$\frac{I_m(\lambda_{m1})}{I_m(\lambda_{m2})} = \frac{I_o^{\lambda_x}\alpha^{\lambda_{m1}}\left[\varepsilon_i^{\lambda_x}[C_i]\right]\tau^{\lambda_{m1}}}{I_o^{\lambda_x}\alpha^{\lambda_{m2}}\left[\varepsilon_i^{\lambda_x}[C_i]\right]\tau^{\lambda_{m2}}} = \frac{\alpha^{\lambda_{m1}}}{\alpha^{\lambda_{m2}}}\cdot\frac{\tau^{\lambda_{m1}}}{\tau^{\lambda_{m2}}} \tag{33.7}$$

In time-domain measurements, where an incident impulse of excitation light is used to excite the sample, the resulting time-dependent emission intensity can be predicted by:

$$I_m(t) \propto I_o^{\lambda_x}(\delta)\alpha\left[\varepsilon_i^{\lambda_x}[C_i]\right]\cdot\int_0^t \exp\left[-\frac{t'}{\tau}\right]dt' \tag{33.8}$$

Thus, when a dilute fluorescence sample is excited with an impulse of excitation light and the emission intensity is monitored as a function of time, the lifetime or decay kinetics that govern the relaxation of the activated state to the ground state can be quantitated independently of the concentration of fluorophore present. The measurement of the time-dependent emission light following activation in a diluted, nonscattering suspension with an incident impulse of excitation light also serves as the basis of the time-domain measurements described below for random media.

Above, CW and time-domain analyses were presented for fluorophores exhibiting first-order decay kinetics in which the form of the decay kinetics, $g(t)$, is given by:

$$g(t) = \exp(-\frac{t}{\tau}) \tag{33.9}$$

However, most analyte fluorophores exhibit more complex decay kinetics such as multiexponential decays:

$$g(t) = \sum_{j=1}^{N} a_j \exp\left[\frac{-t}{\tau_j}\right] \tag{33.10}$$

or stretched-exponential decay kinetics, which indicates collisional quenching among species $j$:

$$g(t) = \sum_{j=1}^{N} a_j \exp\left[-\alpha_j\cdot t - \beta_j\sqrt{t}\right] \tag{33.11}$$

By monitoring the time dependence of the emitted fluorescence light as a function of time following excitation, the decay kinetics can best be ascertained and correlated with the local environment that impacts the relaxation process. For example, the Stern–Volmer equation relates the quencher concentration, $[Q]$, and fluorescence-intensity measurements made in the absence and presence of the quenchers $I_m^o$ and $I_m$, respectively:

$$\frac{I_m^o}{I_m} = 1 + K[Q] = 1 + \left(k_q\tau_o\right)[Q] \tag{33.12}$$

where $K$ is the Stern–Volmer constant, and $k_q$ and $\tau_o$ are the bimolecular quenching constant and the lifetime of the fluorophore in the absence of quencher.

The decay kinetics of many analyte-sensing fluorophores can be used to assess concentrations of analytes such as $H^+$ and $Ca^{2+}$, which may have no appreciable absorption cross section at the emission and excitation wavelengths used. Consequently, fluorescence-lifetime spectroscopy broadens the applicability of absorption spectroscopy, provided a fluorophore with analyte-sensitive decay kinetics can be identified.

While time-dependent techniques are the best way to assess fluorescence-decay kinetics, their need for Dirac pulses of excitation light complicates instrumentation and limits quantitation. Frequency-domain approaches provide an alternative approach to the impulse function by exciting with an intensity-modulated excitation light modulated at MHz–GHz modulation frequencies, $\omega$. Activation of the fluorophore creates isotropic, intensity-modulated fluorescent light that is both phase delayed and amplitude attenuated relative to the incident light owing to the kinetics of the relaxation process. For a simple, first-order system the decay kinetics, the phase delay, $\theta(\omega)$, and the modulation ratio, $M(\omega)$, at modulation frequency $\omega$ can be predicted from:

$$M(\omega)\exp(-i\theta(\omega)) = \int_0^\infty g(t)\exp[-i\omega t]dt = \int_0^\infty \exp\left[-\frac{t}{\tau}\right]\exp[-i\omega t]dt$$
$$M(\omega) = \frac{I_{AC}(\omega)}{I_{DC}(\omega)} = \frac{\mu_a\alpha}{\sqrt{1+(\omega\tau)^2}}; \quad \theta(\omega) = \tan^{-1}[\omega\tau] \tag{33.13}$$

where the amplitude and average of the modulated emission light is given by $I_{AC}(\omega)$ and $I_{DC}(\omega)$. In dilute, nonscattering media, the fluorescent emission is collected at right angles to the excitation illumination to avoid inadvertently collecting excitation light.

It has been proposed that absorption and scattering spectroscopy employing the Beer–Lambert relationship as well as the CW, time-domain, and frequency-domain fluorescence-spectroscopy approaches for quantitative spectroscopy when scattered back into the optical path do not corrupt attenuation or intensity measurements. However, most systems are comprised of random media, i.e., those that absorb, multiply scatter, and fluoresce. The techniques of CW, time and frequency domain, and associated approaches to performing quantitative spectroscopy and imaging in random media are outlined in the next section.

## 33.2.3 Measurement Approaches for Quantitative Spectroscopy and Imaging in Random Media

Here we restrict our discussion to quantitative spectroscopy and imaging in random media in which the diffusion approximation to the radiative transport equation holds. The conditions are (1) the source of incident (excitation) light is isotropic; (2) the scattering capacity of the tissue exceeds that of its absorption capacity, i.e., $\mu_a << (1 - g)\mu_s$, where $g$ is the mean cosine of angular scatter of the medium; and (3) the light that is collected has been multiply scattered. When referring to measurements of multiply scattered light, i.e., light that has traveled a distribution of pathlengths or "times of flight," we term them photon-migration measurements.

### 33.2.3.1 CW and Time-Resolved-Measurement Approaches

CW measurements employ a light source whose intensity nominally does not vary with time. The constant-power, isotropic source illuminates the random medium with light whose intensity becomes exponentially attenuated with increasing distance from the tissue surface. Increased absorption or scattering properties of the medium result in increased light attenuation as the light propagates deeper into the random medium. In CW measurements, the time-invariant intensity is measured as a function of distance from the incident source and is primarily a function of the product $\mu_a\mu_s'$ or $\mu_a\mu_s(1-g)$. The amount of generated fluorescent light at any position $\bar{r}$ is proportional to the product of the concentration of fluorophore, $[C_i]$, and the local excitation fluence, $\Phi_x(r)$, which is the concentration of excitation photons times the speed of light within the medium. Thus, the origin of emission light predominates from the region in which the excitation fluence, $\Phi_x$, is greatest. For time-invariant CW measurements, the region with the greatest excitation fluence always remains close to the point of incident-excitation illumination. Consequently, the origin of fluorescence is mainly from the surface or subsurface regions. For determination of fluorescent optical properties in a uniform medium with CW techniques, fluorescence spectroscopy may not be impacted by the confinement of the origin of emission light if the random medium is indeed homogeneous. However, in imaging scenarios where concentration of fluorophore is nonuniform, CW techniques will undoubtedly emphasize surface and subsurface regions. In imaging cases where the fluorescent dye acts as a contrast agent and has "perfect uptake," (i.e., partitioning of the dye occurs exclusively in the tissue of interest without any residual

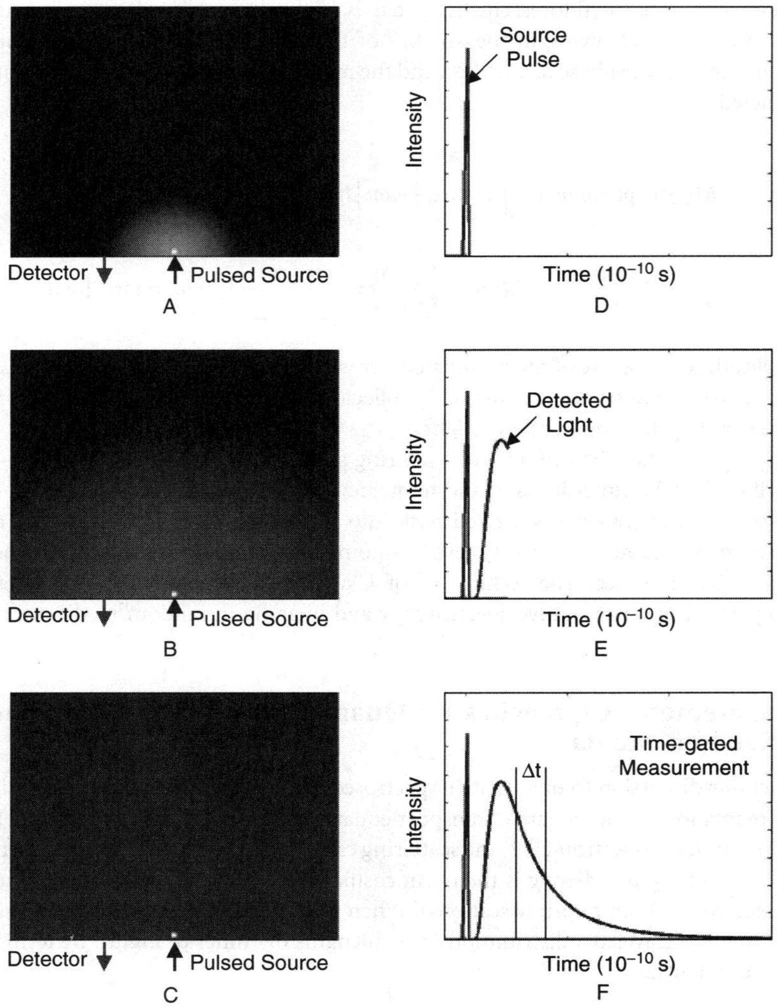


**FIGURE 33.1** Schematic of time-domain photon-migration (TDPM) measurement approach used in NIR optical spectroscopy and tomography. TDPM imaging approaches use an incident impulse of light that results in the propagation of the pulse, which attenuates as a function of distance from the source and time following its delivery. The detected pulse is measured as intensity vs. time, which represents the photon "time of flight." Panel (A) illustrates the light distribution in tissue from a pulse point source after $1 \times 10^{-10}$ s, (B) $25 \times 10^{-10}$ s, and (C) $150 \times 10^{-10}$ s following the incident impulse. The corresponding recorded data during the time intervals at the detector are illustrated in panels (D) through (F). A time-gated illumination measurement is shown in panel (F) in which the integrated intensity measured within a specified window is measured. (From Hawrysz, D.J. and Sevick-Muraca, E.M., *Neoplasia*, 2(5), 388, 2000. With permission.)

dye in the intervening tissues between the target and the surface), then CW techniques may be appropriate. However, the elusive "holy grail" of contrast-based imaging for all medical-imaging modalities is to develop agents that maximize their partitioning in the target region of interest. Near-infrared (NIR) techniques involving fluorescent contrast agents for clinical imaging will likely not involve CW measurement despite the simplicity of its instrumentation.

Time-domain photon-migration (TDPM) measurements employ a light source that delivers a pulse of excitation light that broadens and attenuates as it propagates through the random medium, as shown in Figure 33.1. TDPM techniques employ single-photon counting (sometimes called time-correlated counting) or gated integration measurements to acquire the emitted pulse broadened by as much as several nanoseconds of photon time of flight. As the absorption properties of the random media increase, the broadening of the excitation pulse lessens and greater attenuation occurs. In the case of increased scattering properties, the excitation pulse increasingly broadens and attenuates during its propagation away from the incident point source. Clearly the impact of absorption and scattering has differing effects on the photon time of flight: increased absorption decreases the path and travel time of migrating photons, while increased scattering enhances it. When fluorophores are excited within the random medium, a propagating excitation pulse generates a propagating emission pulse that is further broadened owing to the decay kinetics of the dye.

Since the region of highest excitation fluence is not stationary and propagates away from its incidence with time following delivery to the surface, the origin of fluorescent signals activated by the propagating excitation pulse may not be restricted to surface or subsurface tissues as in the case of CW measurements. If the time constant of the dye's decay kinetics is less than or comparable to the detected excitation photon time of flight, then the fluorescence measured at a distance away from the incident excitation source may originate deeply within the random medium. On the other hand, if the time constant of the dye's decay kinetics is greater than the detected excitation photon times of flight, then the fluorescence will originate from shallow locations within the random medium.[2]

The phenomenon can be explained by the fact that at the medium surface, the position of maximum excitation fluence travels from its point of incidence to deep within the medium and exponentially attenuates as it penetrates. Consider a fluorophore that has an instantaneous rate of radiative relaxation. The maximum emission fluence will likewise follow the trend of the propagating excitation pulse: at time t = 0 when the excitation pulse is launched, the emission fluence will be greatest at the point of incidence, and as time progresses the point of greatest emission fluence will propagate into the medium, attenuating as it does so. The emission light that reaches the surface will initially originate from regions closest to the incident source and then from deeper within the random medium with increased time after the initial impulse of excitation light. However, in the case of a phosphorescent agent with a slow radiative relaxation rate and an effective lifetime on the order of milliseconds (much larger that the measured photon-migration times of flight), the greatest concentration of activated fluorophore will reside close to the incident point of excitation illumination; while the pulse of excitation fluence transits deeper within the random medium, the slow decay of the activated fluorophore closest to the point of incident of excitation will result in a pulse of emission fluence that does not propagate spatially with time into the random medium and away from the point of the incident-excitation illumination. Consequently, for imaging and spectroscopic-imaging applications where information from within the random medium is desired, long-lived fluorophores cannot be employed.

The qualitative CW and TDPM measurements for fluorescence spectroscopy and imaging are described above. Quantitative prediction is also possible with the radiative transport equation, Monte Carlo, and diffusion equations, provided the proper model for fluorescent decay is incorporated. Here we restrict our analysis to media in which the diffusion approximation to the radiative-transfer equation applies. The photon-diffusion equation may be written to predict CW and TDPM measurements of excitation (subscript $x$) and emission (subscript $m$) fluence, $\Phi_{x,m}(\vec{r})$ and $\Phi_{x,m}(\vec{r},t)$, respectively:[3,4]

$$\vec{\nabla}\cdot\left(D_{x,m}\vec{\nabla}\Phi_{x,m}(\vec{r},t)\right)-\mu_{ax,m}\Phi_{x,m}(\vec{r},t)=\frac{1}{c}\frac{\partial\Phi_{x,m}(\vec{r},t)}{\partial t}-S_{x,m}(\vec{r},t) \tag{33.14}$$

where $D_{x,m}$ is the optical-diffusion coefficient at the excitation or emission wavelength in centimeters [cm] given by:

$$D_{x,m}=\frac{1}{3[\mu_{a_{x,m}}+\mu'_{s_{x,m}}]} \tag{33.15}$$

and $\mu'_{s_{x,m}}$ is the isotroptic scattering coefficient given by $(1-g)\mu_{s_{x,m}}$. The excitation or emission fluence, $\Phi_{x,m}(\vec{r},t)$ [W/m$^2$], is the angle-integrated scalar flux of photons and is defined as the power incident on an infinitesimally small sphere divided by its area. Again, it can also be thought of as the local concentration of photons times the speed of light, c, at a given position $\vec{r}$ and (for time-dependent cases) at time $t$. For continuous-wave spectroscopy or imaging there is no time dependence, and the source term, $S_x(\vec{r},t)$, becomes time invariant. Assuming the source is isotropic, this term is equivalent to the power deposited over its area. For TDPM measurements the source, $S_x(\vec{r},0)$ is assumed to be a Dirac delta function, assuming a finite value at time zero, but zero at all other times. For both CW and TDPM measurements Equation 33.14 can be solved to predict the excitation fluence, $\Phi_x(\vec{r})$ or $\Phi_x(\vec{r},t)$, in response to the known spatial distribution of absorption and scattering properties, $\mu_{ax}(\vec{r})$ and $\mu'_{s_x}(\vec{r})$, of the media volume at the excitation wavelength. Here the absorption coefficient at the excitation wavelength is comprised of contributions from the endogenous chromophores ($\mu_{a_{x\to}}(\vec{r})$) as well as the exogenous fluorophores ($\mu_{a_{x\to m}}(\vec{r})$).

Since a closed-form solution for $\Phi_x(\vec{r},t)$ exists only for simple geometries such as infinite and semi-infinite media of uniform optical properties, the solutions are otherwise developed numerically using finite-difference or finite-element methods.

Of the three boundary conditions commonly used, the partial-current condition is the most rigorous.[5,6] It states that a photon leaving a tissue never returns and uses a reflectance parameter to account for Fresnel reflection at the tissue–air surface. If the boundary condition is perfectly transmitting (thereby exhibiting no Fresnel reflection), then the fluence evaluated at the boundary must fall to zero, creating a discontinuity that violates the diffusion approximation that assumes isotropic radiance. When the Fresnel reflection is considered at the boundary, this violation is eased by modeling a portion of the photons to be reflected back into the medium. The partial-current-boundary condition can be expressed in terms of the fluence and its gradient normal to the boundary:

$$\Phi_{x,m}(\vec{r},t) = 2\cdot\frac{1+R_{eff_{x,m}}}{1-R_{eff_{x,m}}}\cdot D_{x,m}\vec{\nabla}_n\Phi_{x,m}(\vec{r},t) \tag{33.16}$$

where $R_{eff}$ is the effective reflection coefficient whose quantity predicts the amount of light reflection and degree of anisotropy at the boundary.

A slightly simpler condition, the extrapolated-boundary condition, is an approximation of the partial-current condition and yields solutions similar to those of the diffusion equation.[7,8] In this case, the fluence is set to zero at an extrapolated boundary located at a specified distance outside the medium to account for Fresnel reflection at the surface.

The third boundary condition, the zero condition, merely sets the fluence to zero on the boundary and is used for its simplicity. In a homogeneous scattering medium, the zero-boundary condition results in an analytical solution to the diffusion equation in terms of the absorption and scattering coefficients.[8,9]

The measured flux or photon current in CW or TDPM measurements, $J_{x,m}(\vec{r})$ or $J_{x,m}(\vec{r},t)$, is then determined by the gradient of fluence, $\vec{\nabla}\Phi_{x,m}(\vec{r})$ and $\vec{\nabla}\Phi_{x,m}(\vec{r},t)$, at the surface (also known as Fick's law):

$$J_{x,m}(\vec{r},t) = -D_{x,m}\vec{\nabla}\Phi_{x,m}(\vec{r},t) \tag{33.17}$$

As a result of combining Fick's law and the partial-current- or extrapolated-boundary conditions, the measured flux or photon density is simply proportional to the fluence at the surface.

Since red light multiply scatters as it transits tissues, it can excite exogenous fluorescent agents that, in turn, act as uniformly distributed sources of fluorescent light. The fluence at the emission wavelength, $\Phi_m(\vec{r})$ or $\Phi_m(\vec{r},t)$, is generated and propagates within the multiply scattering medium; it can also be described by Equation 33.14, provided that its source-emission kinetics are properly modeled in the source term, $S_m(\vec{r})$ or $S_m(\vec{r},t)$. The isotropic scattering coefficient at the emission wavelength, $\mu'_{s_m}$, may be considered to be different than that at the excitation wavelength. The absorption coefficient at the emission wavelength, $\mu_{a_m}(\vec{r})$, is due to both endogenous chromophores and, if reabsorption of the emission light from the fluorophore occurs, the reabsorption of fluorescent light by the exogenous agent. While including secondary reabsorption and photobleaching effects is relatively straightforward, when the excitation and emission spectra are well separated and the fluorophore is in dilute concentrations, we neglect this contribution to absorption at the emission wavelength. The general form for the emission source, $S_m(\vec{r},t)$, is:

$$S_m(\vec{r},t) = \mu_{a_{x \to m}}(\vec{r})\alpha\Xi_m \int_0^t \Phi_x(\vec{r},t')\cdot g(t')dt' \tag{33.18}$$

where $\Phi_x(\vec{r},t)$, $\alpha$, and $\Xi_m$ are, respectively, excitation photon density, quantum efficiency of the fluorophore, and the detection efficiency factor of the system at the emission wavelength (which contains the system-spectral response and the fluorophore spectral-emission efficiency[10]). The time-invariant source of CW measurements, $S_m(\vec{r})$, is described by Equation 33.18, with the upper limit of the time integral equal to infinity. The source of emission light from a mixture of fluorophores undergoing various decay kinetics is simply a combination of the above expressions.

If the solution for the excitation fluence, $\Phi_x(\vec{r})$ or $\Phi_x(\vec{r},t)$, is first obtained, and the decay kinetics and optical properties at the emission wavelength are known, the emission fluence, $\Phi_m(\vec{r})$ or $\Phi_m(\vec{r},t)$, can then be solved using one of the three commonly used boundary conditions described above. The measured flux or photon current in CW or TDPM measurements is then determined from the gradient of the emission fluence.

While CW measurements can be limited in information content regarding decay kinetics and spatial discrimination, TDPM measurements are tedious in that they require an incident Dirac pulse or convolution/deconvolution of the pulse, suffer from low SNR, and mathematically require the solution of an integral-differential equation for spectroscopy and imaging applications. Indeed, the large dynamic range of SNR over the entire distribution of photon times of flight in TDPM approaches can require significant data-acquisition times to resolve or reduce uncertainty in the resulting images. However, some developers prefer to employ TDPM measurements to construct optical property maps since their information content is the richest.[11] Frequency-domain approaches sidestep these issues with the use of an easily achievable sinusoidally modulated light source, measurements possessing high SNR, and, as shown in the next section, a more tractable set of equations for solution of the spectroscopy and imaging-inverse problems.

### 33.2.3.2 Frequency-Domain Measurement Approaches

Frequency-domain measurements in random media are similar to those described in the previous section. An intensity-modulated point source of excitation light is launched into a scattering medium, and the propagating "photon-density wave" is attenuated and phase delayed relative to the incident source as it propagates through the random medium, as shown in Figure 33.2. The detected phase delay and amplitude attenuation measured at the excitation wavelength can be used to determine the optical properties of the random medium, whether they are uniform (for solution of the inverse-spectroscopy problem) or nonuniform (for solution of the inverse-imaging problem). The diffusion equation for solution of the forward problem of predicting measurements also applies, with the difference that the equation is cast in the frequency domain rather than in the time domain:

$$\vec{\nabla}\cdot\left(D_{x,m}\vec{\nabla}\Phi_{x,m}(\vec{r},\omega)\right) - \left[\mu_{a x \to m} + \frac{i\omega}{c}\right]\Phi_{x,m}(\vec{r},\omega) + S_{x,m}(\vec{r},\omega) = 0 \tag{33.19}$$

Here the fluence, $\Phi_{x,m}(\vec{r},\omega)$, is now a complex number describing the characteristics of the photon-density wave at position $\vec{r}$ and modulated at angular frequency $\omega$. Moreover, the fluence is comprised of alternating, $\Phi_{AC_{x,m}}(\vec{r},\omega)$, and nonalternating, $\Phi_{DC_{x,m}}(\vec{r},0)$, components, of which the former provides an accurate description of the phase delay, $\theta_{x,m}$, and amplitude, $I_{AC_{x,m}}$, of the wave at position $\vec{r}$:

$$\begin{aligned}\Phi_{x,m}(\vec{r},\omega) &= \Phi_{ACx,m}(\vec{r},\omega) + \Phi_{DCx,m}(\vec{r},0)\\ &= I_{ACx,m}\exp(i\theta_{x,m}) + I_{DCx,m}(\vec{r},0)\end{aligned} \tag{33.20}$$

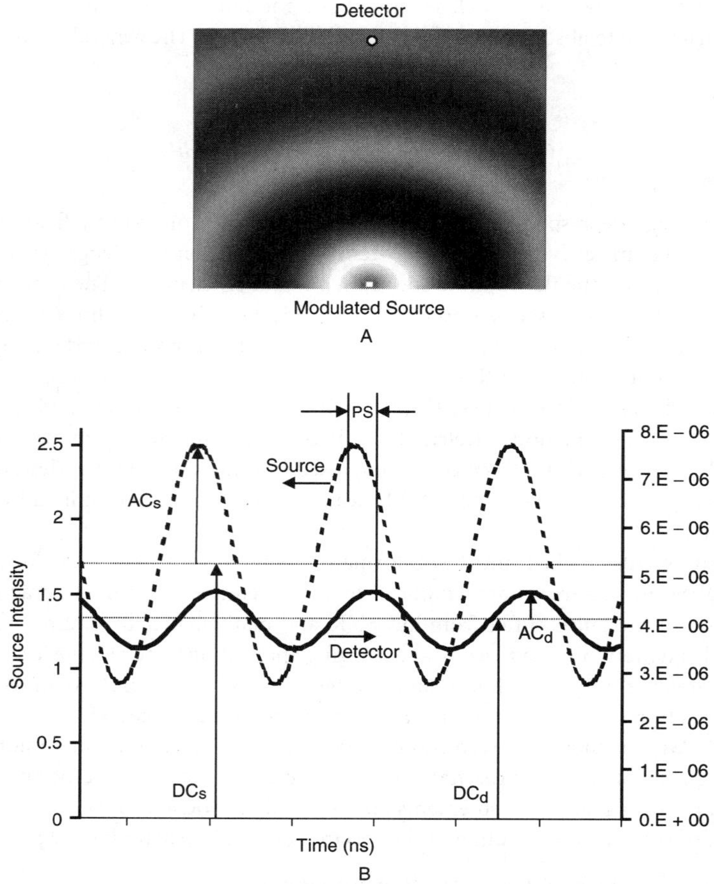


**FIGURE 33.2** Schematic of the frequency-domain photon-migration (FDPM) measurements used in NIR optical spectroscopy and tomography. FDPM traditionally consists of an incident, intensity-modulated light source that creates a "photon-density wave" that spherically propagates continuously throughout the tissue. Panel (A) shows light distribution in tissue due to a modulated source (exaggerated for purposes of illustration), and panel (B) illustrates the detected signal (solid line) in response to the source illumination (dotted line). The typical frequency domain data, where the measurable quantities are the phase shift θ, the amplitude of each wave $I_{AC}$, and the average value $I_{DC}$ of intensity. As shown in (B), the intensity wave that is detected some distance away from the source is amplitude-attenuated and phase-delayed relative to the source. (From Hawrysz, D.J. and Sevick-Muraca, E.M., *Neoplasia*, 2(5), 388, 2000. With permission.)

The nonalternating component of the fluence, $\Phi_{DC_{x,m}}(\vec{r},0)$, is simply the fluence that is measured when using a CW source ($\omega = 0$). The pre-exponential factor, $I_{AC_{x,m}}$, is the amplitude of the photon-density wave, and $\theta_{x,m}$, is the phase delay of the wave relative to the incident source. At larger modulation frequencies the photon-density wave attenuates more rapidly during its propagation and experiences greater phase lag. Consequently, the amplitude decreases with increasing modulation frequency, while the phase delay increases with modulation frequency. Often the amplitude or modulation ratio is reported as a measurement. The modulation ratio is simply the amplitude of the wave normalized by $I_{DC_{x,m}}(\vec{r},0)$.

For the solution of the excitation fluence via Equation 33.19 the source function is either a point source (as shown in Figure 33.2) or a plane source of modulated light:

$$S_x(\vec{r},\omega) = S(\vec{r}_s,\omega)\exp(i\theta_s(\vec{r}_s,\omega)) \tag{33.21}$$

where the strength (or amplitude) of the excitation source at its position of incidence, $\vec{r}_s$, is $S(\vec{r}_s,\omega)$, and its absolute phase is $\theta_s(\vec{r}_s,\omega)$. Typically, frequency-domain photon-migration (FDPM) measurements are conducted between a point source of illumination, and the amplitude and the phase of the detected light (also collected at a point) is determined relative to the source. Consequently, in most cases in the literature, the source strength is designated as unity, and the phase of the incident source is taken as zero. As described in the next section, emission fluence in response to incident-planar-wave excitation can be employed and predicted by the diffusion equation, provided that the spatial phase, $\theta_s(\vec{r}_s,\omega)$, and amplitude, $S(\vec{r}_s,\omega)$, of the source are properly accounted for.[12]

The boundary conditions for frequency-domain measurements in random media are identical to those described above for CW and TDPM techniques, and the partial-current-boundary condition is similarly written:

$$\Phi_{x,m}(\vec{r},\omega) = 2\cdot\frac{1+R_{eff_{x,m}}}{1-R_{eff_{x,m}}}\cdot D_{x,m}\vec{\nabla}_n\Phi_{x,m}(\vec{r},\omega) \tag{33.22}$$

The measured flux or photon current in frequency-domain measurements, $J_{x,m}(\vec{r},\omega)$, is then determined by Fick's law:

$$J_{x,m}(\vec{r},\omega) = -D_{x,m}\vec{\nabla}\Phi_{x,m}(\vec{r},\omega) \tag{33.23}$$

The result of combining Fick's law and the partial-current- or extrapolated-boundary conditions is that the measured flux or photon density of the wave (now a complex number) is simply proportional to the fluence at the surface. Consequently, the measured phase, $\theta_{x,m}$, and amplitude, $I_{ACx,m}$, are predicted from the fluence, $\Phi_{ACx,m} = I_{ACx,m}\exp(i\theta_{x,m})$.

As with CW and TDPM methods, the radiative relaxation of the activated fluorophore serves as a distributed source of emission light within the random medium. The emission source, $S_m(\vec{r},\omega)$, for single-exponential-decay kinetics is:

$$S_m(\vec{r},\omega) = \mu_{a_{x\to m}}\left(\frac{1}{1-i\omega\tau}\right)\Phi_x(\vec{r},\omega)\alpha\Xi_m \tag{33.24}$$

and for any arbitrary-decay kinetics expressed by $g(t)$, the source term can be generally derived from:

$$S_m(\vec{r},\omega) = \mu_{a_{x\to m}}\Phi_x(\vec{r},\omega)\,\alpha\,\Xi_m\int_0^{\infty}\left(g(t)e^{(i\omega t)}dt\right) \tag{33.25}$$

The solution of Equation 33.19 describes the propagation of excitation light; from this the excitation fluence, $\Phi_x(\vec{r},\omega)$, can be directly obtained and used as input for the source term to solve Equation 33.19 for the emission fluence, $\Phi_m(\vec{r},\omega)$.[13] Thus, the solution of the coupled equations with the specified boundary conditions, the phase and amplitude of the detected emission wave relative to the incident excitation source, can be directly determined.

As with TDPM measurements, the ability to use fluorescence to interrogate random media is afforded by FDPM measurements when the lifetime of the fluorescent agent is small when compared to the photon time-of-flights. As with TDPM approaches, effective contrasts for FDPM approaches are limited to fluorescence rather than phosphorescent or long-lived compounds. This was demonstrated in the Photon-Migration Laboratory (PML) by comparing FDPM contrast offered by *tris* (2,2′-bipyridyl) dichloro-ruthenium (II) $Ru(bpy)_3^{2+}$ with a lifetime of 600 ns and ICG with a lifetime of 0.56 ns. In this case, the FDPM contrast was defined as the change in the phase and amplitude of the emission light as the position of the target changed relative to the position of the point of excitation illumination and emission detection. Using a single target with 100-fold greater concentration than the background in a phantom (see Figure 33.3 for measurement geometry of the phantom), the phase and amplitude modulation contrast at each of the

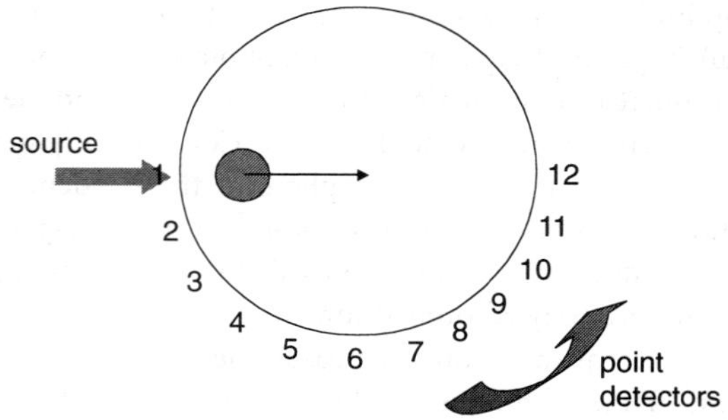


**FIGURE 33.3** Schematic of the phantom tests showing the change in emission-phase measurements as a fluorescent and phosphorescently tagged, 10-mm-diameter target was moved from the periphery toward the center of a 100-mm-diameter cylindrical vessel.

detectors could be seen when ICG was used (Figure 33.4A); however, no contrast was measured when ruthenium dye, $Ru(bpy)_3^{2+}$, was used as the contrast agent (Figure 33.4B). These results confirm computational predictions that effective contrast agents must possess shorter lifetimes than the time of flight of photon propagation.[14]

Consequently, in order to develop fluorescence-lifetime spectroscopy and imaging techniques, the time dependence of the photon-migration process must be accounted for in order to obtain lifetime information. Whether imaging or spectroscopy is used, the inverse problem becomes one of separating photon migration from fluorescence-decay kinetics. Some investigators have sought to avoid the problem by employing phosphorescent dyes wherein the photon migration times of flight of picoseconds to nanoseconds are insignificant compared to lifetimes on the order of micro- to milliseconds. Nonetheless, time-dependent emission measurements will be unable to interrogate beyond the surfaces or subsurfaces when long-lived dyes are employed. When used as contrast agents for imaging, these long-lived dyes do have utility, but only if their partitioning within the target is perfect and there is no residual dye in the background.

Finally, since the amplitude of the detected fluorescence, $I_{AC_m}$, is insensitive to the intensity due to the ambient light, the frequency-domain approach has clear advantages for application in environments that are not light tight. In addition, since frequency-domain approaches offer steady-state measurement of a time-dependent light-propagation process, they have comparatively high SNRs with respect to time-domain approaches and retain lifetime-dependent signals, which is otherwise missing in CW measurements. Due to the ease of instrumentation of frequency-domain over time-domain approaches, and due to the superior information of time-dependent techniques over CW measurements, the remainder of this chapter will focus on FDPM measurements, but studies conducted using CW and TDPM techniques will be cited.

## 33.3 Frequency-Domain Photon Migration (FDPM) Measurement Approaches

Two approaches have been employed in spectroscopy and imaging of random media: (1) point detection and point illumination and (2) area detection and area illumination. Point-detection schemes typically employ heterodyne or I and Q mixing techniques, which employ signal mixing at the photodetector following point detection to extract signals of phase and amplitude modulation at a single point. To conduct FDPM measurements among a number of sources and detectors, either scanning of the source/detector or transmitter/receiver pair or replication of the receiver/transmitter circuitry is required. Consequently, this restricts FDPM imaging and spectroscopy to sparse data sets for solving the inverse-spectroscopy and imaging problems. While sufficient for solving the problem of inverse spectroscopy (as discussed in Section 33.4), point illumination and point detection provide sparse sets for optical tomography or solution of the inverse-imaging problem (as discussed in Section 33.5). The use of an incident point of excitation light delivered by a fiber optic requires a number of measurements as its position is scanned or replicated along the surface for imaging purposes (Figure 33.5).[15] Since excitation fluence

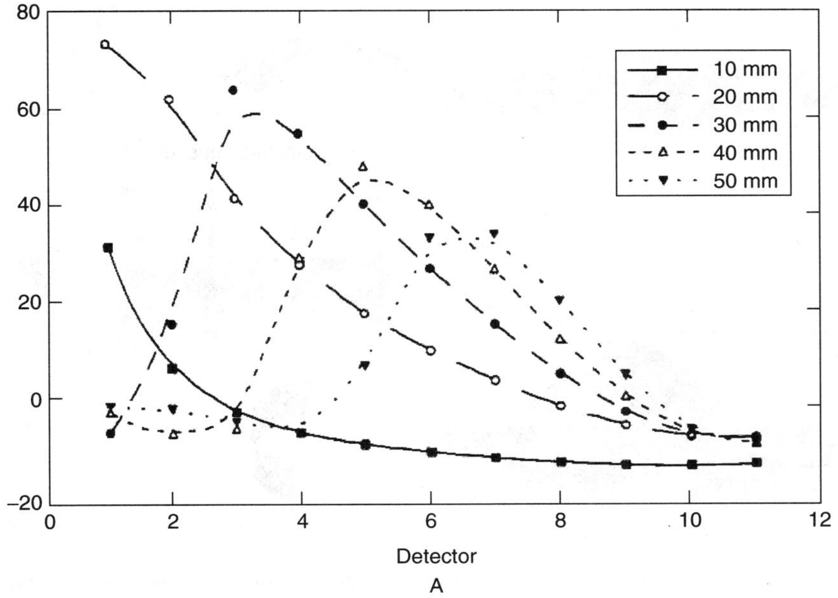


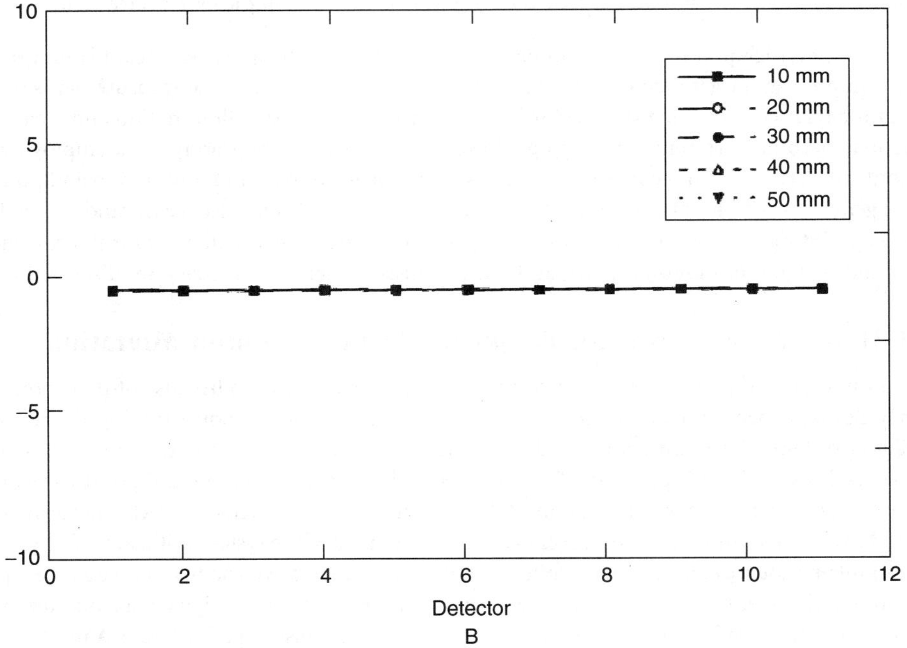


**FIGURE 33.4** The phase contrast in degrees (determined from the phase measured in the presence and absence of a target) measured at the emission light for a 100:1 target-to-background ratio for (A) phosphorescent dye with lifetime of 1 ms and (B) fluorescent dye with lifetime of 1 ns. The phase contrast is predicted from simulation of the target moving from the perimeter (10 mm) toward the center (50 mm) of a 10-cm-diameter cylinder under conditions of maximum phase contrast, i.e., $\omega\tau = 1$ and uniform lifetime. The detectors are located around half of the perimeter of the cylinder, as described in Figure 33.3.

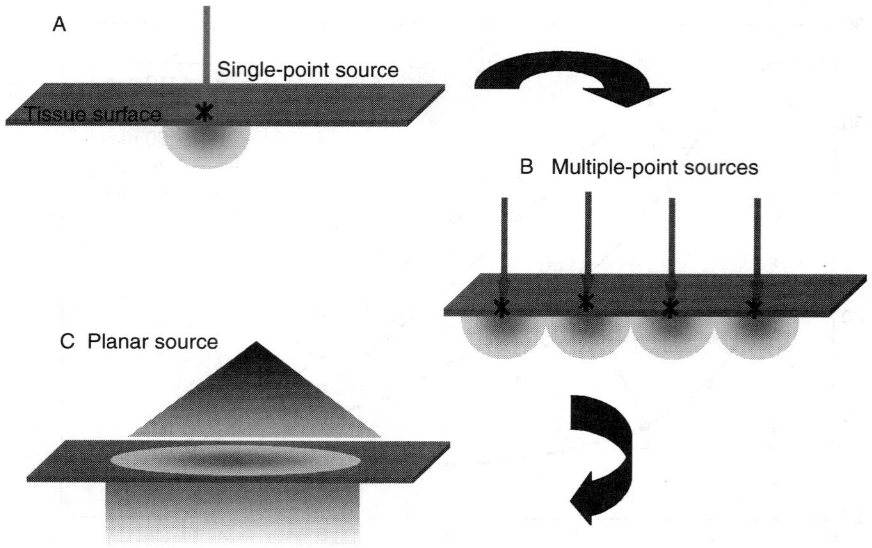


FIGURE 33.5 Illustration of different geometries for illumination of deep tissues. (A) Single-point source of excitation-light delivery. (B) Multiple-point sources, which in the extreme of high density of simultaneous sources is representative of (C). (C) Planar source of excitation light with illumination spread over an area of the tissue surface.

attenuates rapidly, each point-source illumination will not necessarily probe significant volumes and may "miss" the fluorescent target region of interest. Consequently, a high density of measurements is typically required for a relatively confined volume for imaging purposes, and area-illumination and area-detection schemes may become pertinent for imaging. However, for optical-tomography work employing endogenous contrast, measurement geometries are necessarily restricted to point illumination and point detection. The general principles of frequency-domain measurements of fluorescence in random media using heterodyne point measurements and homodyning area measurements, challenges for fluorescence spectroscopy and imaging in random media, and, finally, measurements geometries are discussed below.

## 33.3.1 Heterodyne Mixing for Frequency-Domain Photon Migration

Point-illumination and point-detection measurements are most common because of their prevalence in frequency-domain spectrometers for measurement of decay kinetics in nonscattering diluted samples. For imaging systems the point-source and point-detection schemes and tomographic approaches are developed exclusively for this geometry. The heterodyned point-illumination and point-detection measurements consist of three parts: (1) the modulated source; (2) the detector, which may also act as the mixer; and (3) the electronics to accomplish mixing. A schematic of the system is illustrated in Figure 33.6.

The modulated source can be either a coherent light source that is externally modulated via an electro-optic modulator, a laser diode modulated by use of a radio frequency (RF) signal via a bias tee, or, which is more complicated, a pulsed source with a constant and known pulse-repetition rate. A master oscillator drives the source, which is focused to illuminate a point on the surface of the random medium or coupled to the surface through the use of fiber optics. Modulation frequencies are typically on the order of 30 to 500 MHz. A fast detector — i.e., silicon photodiode, avalanche photodiode, or fast photomultiplier — is required to detect the amplitude and phase delay of the detected photon-density wave at excitation and emission wavelengths. To acquire the signal for standard data acquisition, the signal, $L$, of frequency $\omega$ is "mixed down" to a more manageable frequency, $\Delta\omega$, by mixing with another signal of frequency, $\omega + \Delta\omega$. For example, the mixing can be accomplished through direct-gain modulation of the photomultiplier tube or at a mixer, which receives the photomultiplier signal. (Figure 33.7). Consider the signal

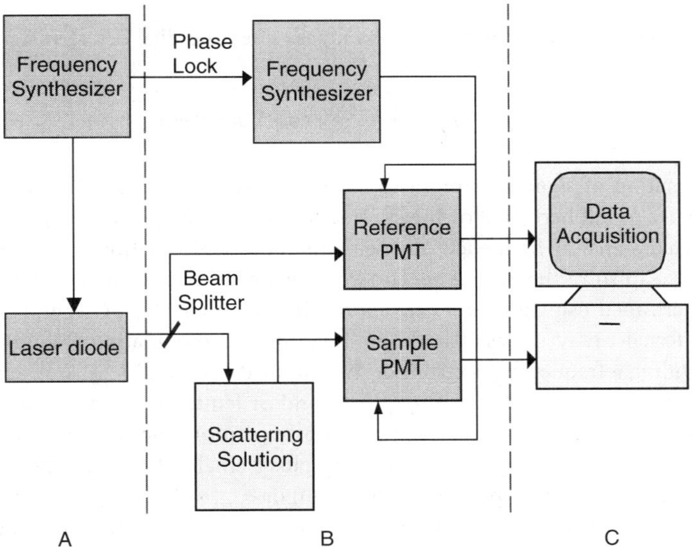


**FIGURE 33.6** Schematic of heterodyned FDPM system that consists of three parts: (A) the modulated source (shown as laser diode), (B) the detector (PMT), which also acts as the mixer, and (C) the data-acquisition hardware and software to accomplish mixing.

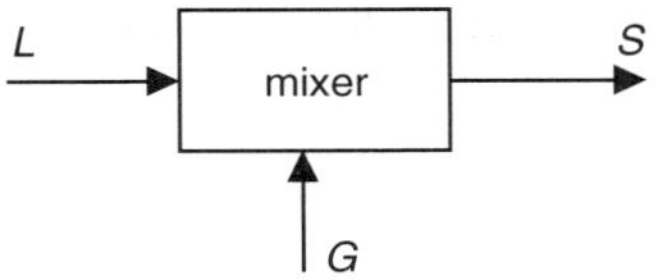


**FIGURE 33.7** Schematic of mixer for heterodyne and homodyne detection of $I_{AC_{x,m}}$ and $\theta_{x,m}$ for FDPM.

representing the detected light, $L$, which has propagated to a position and experienced phase delay $\theta$ and has amplitude $L_{AC}$ and average signal $L_{DC}$.

$$L = L_{DC} + L_{AC} \cdot \cos[\omega t + \theta] \tag{33.26}$$

Assume the signal $G$ generated by a slave oscillator is in phase with the master oscillator (i.e., there is no phase delay relative to the incident light) and has amplitude $G_{AC}$ and average signal level $G_{DC}$.

$$G = G_{DC} + G_{AC} \cdot \cos[(\omega + \Delta\omega)t + \theta_{inst}] \tag{33.27}$$

where $\theta_{inst}$ represents the phase delay introduced into the signal due to instrumentation rather than light propagation. Upon mixing the signals, one obtains $S$, which consists of high- ($2\omega + \Delta\omega$, $\omega + \Delta\omega$, $\omega$) and low-frequency ($\Delta\omega$) components:

$$\begin{aligned} S &= L \times G \\ S &= L_{DC} \cdot G_{DC} + L_{DC} \cdot G_{AC} \cdot \cos[(\omega + \Delta\omega)t + \theta_{inst}] \\ &\quad + G_{DC} \cdot L_{AC} \cdot \cos[\omega t + \theta] \\ &\quad + \frac{L_{AC} \cdot G_{AC}}{2} \cdot \cos[\Delta\omega t - \theta + \theta_{inst}] \\ &\quad + \frac{L_{AC} \cdot G_{AC}}{2} \cdot \cos[(2\omega + \Delta\omega)t + \theta + \theta_{inst}] \end{aligned} \tag{33.28}$$

When the mixed signal is filtered with a low-bandpass filter, the final detected signal is:

$$S = L_{DC} \cdot G_{DC} + \frac{L_{AC} \cdot G_{AC}}{2} \cos\left[\Delta\omega t - \theta + \theta_{inst}\right] \tag{33.29}$$

whereby the information of signal $L$ is preserved in the low-frequency signal, where $\Delta\omega$ is typically in the hundreds of hertz or kilohertz. Following subtraction of the average of the signal, $L_{DC} \cdot G_{DC}$, Fourier analysis of the mixed signal at frequency $\Delta\omega$ yields the phase information, $[-\theta + \theta_{inst}]$, as well as the product of $L_{AC} \cdot G_{AC}$. To solve the inverse spectroscopy or imaging problem, accurate assessment of $L_{AC}$ and $\theta$ must be determined using a referencing approach to eliminate $\theta_{inst}$, $G_{AC}$, and $G_{DC}$.

When using external-cavity or laser-diode modulation, the modulation frequencies can be swept continuously, providing a frequency spectrum of $\Phi_{x,m}(\omega)$ or $\theta_{x,m}(\omega)$ and $I_{ACx,m}(\omega)$. However, in the case of a pulsed light source, such as a Ti:sapphire picosecond or femtosecond laser, the master oscillator is set by the length of the laser cavity, which sets the laser repetition rate, $1/\omega$, and the signal $G$ is swept across the harmonics of the laser-repetition frequency plus a small offset, $n\omega + \Delta\omega$, $n = 1, 2, \ldots$. Thus, pulsed sources provide a frequency spectrum of $\Phi_{x,m}(n\omega)$ or $\theta_{x,m}(n\omega)$ and $I_{ACx,m}(n\omega)$ where $n = 1, 2, \ldots$. See Reference 16 for more information about frequency-domain measurements with pulsed laser sources.

## 33.3.2 Homodyne Mixing for Frequency-Domain Photon Migration

The homodyne approach is similar to the heterodyne approach described above except that the signal, $L$, of frequency $\omega$ is mixed down to DC signal through mixing with another signal of identical frequency. Consider the signal $G$ generated by the master oscillator with an introduced phase delay, $\eta$, with amplitude $G_{AC}$ and average signal level $G_{DC}$:

$$G = G_{DC} + G_{AC} \cdot \cos\left[(\omega)t + \theta_{inst} + \eta\right] \tag{33.30}$$

where $\theta_{inst}$ represents the phase delay introduced into the signal due to instrumentation rather than light propagation.

The mixing of signals $L$ and $G$ produces $S$, which consists of a high-frequency component ($2\omega$, $\omega$) and a DC component:

$$\begin{aligned} S &= L \times G \\ S &= L_{DC} \cdot G_{DC} + L_{DC} \cdot G_{AC} \cdot \cos\left[(\omega)t + \theta_{inst} + \eta\right] \\ &\quad + G_{DC} \cdot L_{AC} \cdot \cos\left[\omega t + \theta\right] \\ &\quad + \frac{L_{AC} \cdot G_{AC}}{2} \cdot \cos\left[-\theta + \theta_{inst} + \eta\right] \\ &\quad + \frac{L_{AC} \cdot G_{AC}}{2} \cdot \cos\left[(2\omega)t + \theta + \theta_{inst} + \eta\right] \end{aligned} \tag{33.31}$$

When the mixed signal is filtered with a low-bandpass filter, the final detected DC signal is:

$$S = L_{DC} \cdot G_{DC} + \frac{L_{AC} \cdot G_{AC}}{2} \cos\left[-\theta + \theta_{inst} + \eta\right] \tag{33.32}$$

When the phase delay $\eta$ is changed by known values, it is possible to evaluate $L_{DC} \cdot G_{DC}$, $\frac{L_{AC} \cdot G_{AC}}{2}$, $\left[-\theta + \theta_{inst}\right]$ and, by proper referencing, determine those quantities that define signal $L$ and the light propagation within the random medium.

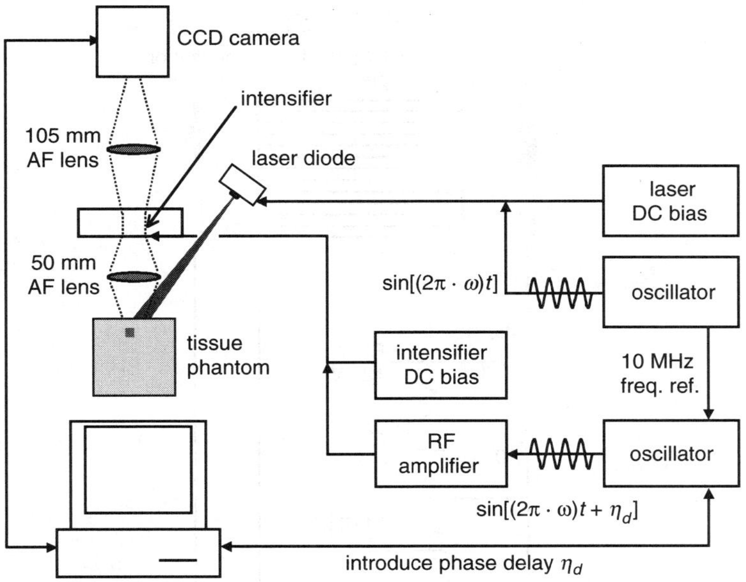


**FIGURE 33.8** Schematic of ICCD homodyne FDPM system in the Photon-Migration Laboratory (PML).

Because image intensifiers used in area-detection schemes have slow response times, the homodyne approach is typically used. Figure 33.8 illustrates the image-intensified charge-coupled device (ICCD) homodyne FDPM system consisting of three major components: (1) a CCD camera, which houses a multipixel array of photosensitive detectors; (2) a gain-modulated image intensifier, which acts as the mixer (see below); and (3) oscillators that sinusoidally modulate the laser-diode light source and the image intensifier's photocathode gain at the same frequency, $\omega$. A 10-MHz reference signal between the oscillators ensures that they operate at the same frequency with a constant phase difference. Emitted light from the tissue or phantom surface is imaged via a lens onto the photocathode of the image intensifier. As before, the light ($L$) that reaches the photocathode of the image intensifier has a phase delay, $\theta(r)$, average intensity, $L_{DC}(r)$, and amplitude intensity, $L_{AC}(r)$, that may vary as a function of position on the sample and, consequently, across the photocathode face.

The gain of the image intensifier has an average, $G_{DC}$, a possible phase delay due to the instrument response time, $\theta_{inst}$, and an amplitude, $G_{AC}$, at the modulation frequency. The modulated gain is accomplished by modulating the potential between the photocathode, which converts the NIR photons into electrons, and the multichannel plate (MCP), which multiplies the electrons before they are focused onto the phosphor screen (Figure 33.9). The resulting signal at the phosphor screen is a mixed homodyne signal ($S$) containing all the amplitude, DC, and phase information of the optical signal collected by the detector. Yet since the phosphor screen has response times on the order of submilliseconds, it acts as a low-pass filter so that the image transferred to the CCD camera is simply the homodyne signal represented in Equation 33.32. The time-invariant but phase-sensitive image on the phosphor screen is then imaged onto the CCD using a lens or fiber coupling.[12,17–19]

Rapid multipixel FDPM data acquisition proceeds as follows. The phase of the photocathode modulation is stepped, or delayed, at regular intervals between 0 and 360° relative to the phase of the laser-diode modulation. At each phase delay $\eta_d$, the CCD camera acquires a phase-sensitive image for a given exposure time (see Figure 33.10), which is on the order of milliseconds. A computer program then arranges the phase-sensitive images in the order acquired and performs a fast Fourier transform (FFT) to calculate modulation amplitude, $I_{AC}$, and phase, $\theta$, at each CCD pixel $(i,j)$ using the following relationships:

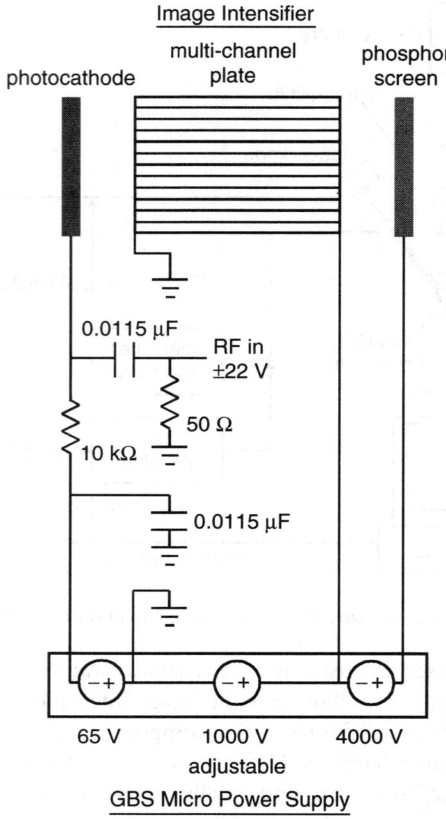


**FIGURE 33.9** Schematic of the image-intensifier circuit and system used in the homodyne ICCD system. (From Thompson, A.B. and Sevick-Muraca, E.M., *J. Biomed. Opt.*, 8, 111, 2003. With permission.)

$$I_{AC}(i,j) = \frac{[\{IMAG[I(f_{\max})_{ij}]\}^2 + \{REAL[I(f_{\max})_{ij}]\}^2]^{1/2}}{N/2} \tag{33.33}$$

$$\theta(i,j) = \arctan\left(\frac{IMAG[I(f_{\max})_{ij}]}{REAL[I(f_{\max})_{ij}]}\right) \tag{33.34}$$

$I(f)$ is the Fourier transform of the phase-sensitive intensity data and $I(\eta_d)$. $IMAG[I(f_{max})]$ and $REAL[I(f_{max})]$ symbolize, respectively, the imaginary and real components in the digital-frequency spectrum that best describe the sinusoidal data. $N$ relates the number of phase delays between the gain modulation of the image intensifier and the incident-light source.

Area illumination is accomplished simply by expanding a modulated laser-diode beam onto the surface of the phantom or tissue to be imaged. To date there has been no attempt to use area illumination and area detection for tomographic reconstructions because all formulations are based on the propagation of light from a point-excitation source to a point on the medium's surface. Yet despite its current lack of acceptance by the tomographic community (see Section 33.5), planar-wave illumination is by far the most common means of illuminating photodynamic therapy (PDT) agents for assessing therapeutic drug distribution and providing excitation for assessing diagnostic fluorochrome distribution in s.c. tumor-bearing rodents. Typically, CW light from a xenon or tungsten lamp, laser, or laser diode is expanded to

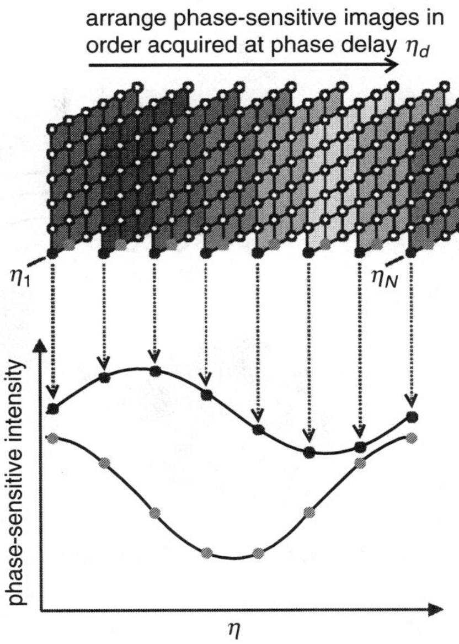


**FIGURE 33.10** The process in which the phase delay between the image intensifier and laser-diode modulation is adjusted between 0 and 360° yielding phase-sensitive yet constant-intensity images at the phosphor screen. When the intensities are compiled at each pixel, the sine wave is reconstructed and the phase and amplitude attenutation are obtained from simple FFT. (From Thompson, A.B. and Sevick-Muraca, E.M., *J. Biomed. Opt.*, 8, 111, 2003. With permission.)

illuminate an entire animal or portion of the animal. Incident powers range from μW/cm² to mW/cm², and area detection can be accomplished using a CCD with or without an image-intensifer coupling and with or without a spectrograph for spectral discrimination. Typically, CW measurements are conducted in mice and rats of small tissue volumes, while the frequency-domain measurements of fluorescence-enhanced contrast have been performed in canines (see Section 33.5).

In addition to the area measurements, the ICCD can be employed to rapidly conduct single-pixel measurements for tomographic reconstructions by simply using the ICCD to simultaneously measure the phase and amplitude of light collected by a number of fibers whose ends are affixed onto an interfacing plate focused on the photocathode of the image intensifier via a lens (Figure 33.11).

### 33.3.3 Homodyne I and Q

Another homodyning method for frequency-domain measurements that does not depend on conducting successive measurements at varying phase delays, $\eta$, imposed on signal $G$ employs $I$ and $Q$ demodulation.[20] The technique depends on mixing signal $L$ with two signals, $G_1$ and $G_2$, at the same frequency $\omega$, but phase shifting it by 90° (Figure 33.12):

$$G_1 = G_{AC}\cos(\omega t);\quad G_2 = G_{AC}\sin(\omega t) \tag{33.35}$$

When $L$ is mixed with $G_1$, the signal output, $S_1$, is given by:

$$S_1 = L_{DC}\cdot G_{AC}\cdot\cos(\omega t) + \frac{L_{AC}\cdot G_{AC}}{2}\cdot\left[\cos(2\omega t+\theta+\theta_{inst}+\frac{\pi}{2}) + \cos(\theta+\theta_{inst}+\frac{\pi}{2})\right] \tag{33.36}$$

When $L$ is mixed with $G_2$, the signal output, $S_2$, is given by:

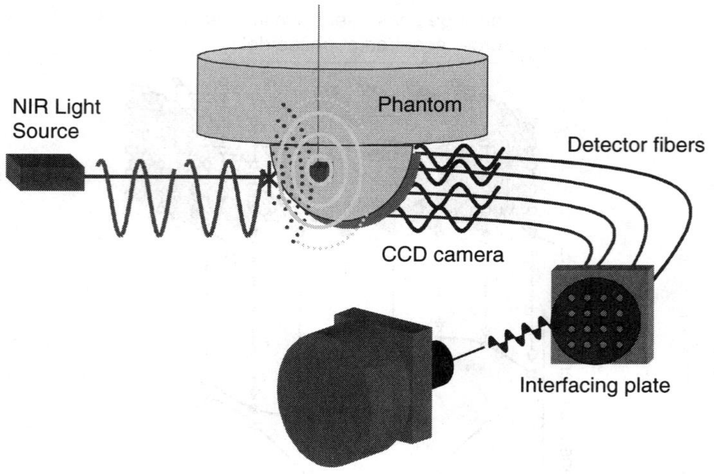


**FIGURE 33.11** The adaptation of a number of single fibers collecting detected light for imaging by the ICCD system, which is depicted in Figure 33.9.

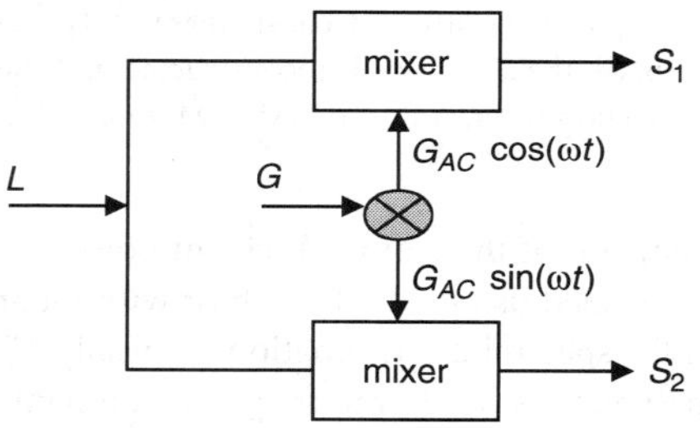


**FIGURE 33.12** Illustration of the homodyne FDPM detection employing the mixing of $L$, $G_1$, and $G_2$, where $G_2$ is phase shifted by 90° relative to $G_1$.

$$S_2 = L_{DC} \cdot G_{AC} \cdot \sin(\omega t) + \frac{L_{AC} \cdot G_{AC}}{2} \cdot \left[ \sin(2\omega t + \theta + \theta_{inst} + \frac{\pi}{2}) - \sin(\theta + \theta_{inst} + \frac{\pi}{2}) \right] \tag{33.37}$$

Upon passing through a low-pass filter, the high-frequency components at ω and 2ω can be eliminated, leaving two DC signals:

$$S_1 = \frac{L_{AC} \cdot G_{AC}}{2} \cdot \left[ \cos(\theta + \theta_{inst} + \frac{\pi}{2}) \right]$$

$$S_2 = \frac{L_{AC} \cdot G_{AC}}{2} \cdot \left[ -\sin(\theta + \theta_{inst} + \frac{\pi}{2}) \right] \tag{33.38}$$

When the two signals are combined, the quantities of $\frac{L_{AC} \cdot G_{AC}}{2}, \left[ -\theta + \theta_{inst} \right]$ can be determined, and by proper referencing the AC and phase delay associated with signal $L$ and with the light propagation within the random medium can be determined.

### 33.3.4 Excitation-Light Rejection Considerations

Regardless of the measurement method (i.e., CW, time-domain, or frequency-domain), and regardless of the measurement geometry (point or area illumination and detection), one of the greatest and largely unrecognized challenges in fluorescence spectroscopy and imaging in random media is the importance of excitation-light rejection. In fluorescence spectroscopy of dilute, nonscattering samples, the isotropic emission light is collected at right angles to the incident excitation light to avoid corruption of excitation light in the fluorescence measurements (see Section 33.4). Yet in random media, the excitation light is propagated isotropically, potentially corrupting measurements in random media. Generally, the Stoke's shift associated with many fluorophores is small; in the case of indocyanine green (ICG), an FDA-approved agent, it is 50 nm. The wavelength sensitivity at the photocathode does not discriminate between excitation and emission wavelengths requiring a mechanism for excitation-light rejection and passage of emission light only for accurate spectroscopy and imaging data. The excitation light reaching a detector at a location on the surface is the predominant signal and can be as little as $10^3$ times greater than the emission fluence when the fluorophore concentration is high and significantly greater as the fluorophore concentration is reduced to nanomolar and femtomolar levels, as might be expected in fluorescent contrast-agent identification of small cancer metastases. For planar-wave illumination, specularly reflected excitation light would create an even greater portion of the signal, further compounding the discrimination of the weak fluorescent signal emitted from the tissue surface (Figure 33.5). Generally, investigators employ interference filters with a rejection capability of OD 3 for excitation light, which generally sets the noise floor of the fluorescent measurement and limits the smallest amount of detectable fluorophore. When interference and bandpass filters for emission-light passage and holographic filters for excitation-light rejection are stacked, the noise floor can be reduced as much as nine orders of magnitude.[21] Clearly, the sensitivity of fluorescence spectroscopy and imaging in random media hinges on the success of excitation-light rejection.

The forward-spectroscopy and imaging problems consist of using the diffusion model to predict light propagation, fluorescence generation, and the resulting measurements at the medium–air interface given the spatial distribution of optical properties within the entire volume. To solve the inverse-spectroscopy problem we assume a uniform distribution of optical properties with the unknowns being (1) the optical properties at the excitation and emission wavelengths, (i.e., $\mu_{a_{x,m}}$, $\mu'_{s_{x,m}}$) and (2) the optical properties associated with the fluorescent agent (i.e., $\alpha\mu_{a_{x\to m}}$, $\tau$), which experiences first-order decay kinetics. Section 33.4 further develops the inverse solution to the imaging problem for tomographic imaging.

## 33.4 Fluorescence Spectroscopy in Random Media

Fluorescence-lifetime spectroscopy is advantageous for quantitative spectroscopy of analytes and metabolites because the measurement of fluorescence-decay kinetics (rather than fluorescence intensity) obviates the need to know the concentration of the sensing fluorophore.[1] As described in Section 33.1, frequency-domain techniques provide measurement of fluorescence lifetime ($\tau$) using simple relationships of the phase delay ($\theta$) and modulation ratio ($M$) of the re-emitted fluorescence as a function of the modulation frequency relative to the incident intensity-modulated excitation light. However, the development of fluorescence-lifetime spectroscopy for NIR biomedical tissue diagnostics for sensing through the use of systematically administered dyes[22,23] or implantable devices[24,25] requires deconvolving the influence of multiple scatter upon the measured emission phase delay and amplitude attenuation. As shown below, the addition of scatter increases the sensitivity of fluorescence-lifetime spectroscopy over traditional methods that focus on isotropic emission-light generation across a fixed path length, $L$.

Approaches to suitable modeling of multiply scattering of NIR excitation and fluorescence photons and to the use of diffusion models for quantitative spectroscopy have been previously demonstrated[10,26,27] for dyes exhibiting single-exponential-decay kinetics. Failure to properly account for multiply scattered excitation and emission-light propagation in random media can result in incorrect decay kinetics. For example, when conducting phase-modulation measurements on a solution of Intralipid® containing the NIR-excitable fluorophore ICG, Lakowicz et al.[28] did not incorporate the propagation of light into their calculations, yet they attribute multiexponential decay kinetics to this dye, which typically exhibits single-exponential decay kinetics.

## 33.4.1 Single-Exponential-Decay Spectroscopy

There are generally six unknowns to be solved for a uniform medium of unknown optical properties containing a fluorophore exhibiting a first-order radiative relaxation process: $\mu_{a_{\lambda_x}}, \mu_{a_{\lambda_m}}, \mu'_{s_{\lambda_x}}, \mu'_{s_{\lambda_m}}, \alpha\mu_{a_{x\to m}}, \tau$ with the subscripts altered slightly from the past nomenclature to emphasize the optical properties at two separate wavelengths, $\lambda_x$ and $\lambda_m$.

### 33.4.1.1 Optical Property Determination

For a uniform random medium, the optical properties can be accurately determined from multidistance frequency-domain measurements.[29,30] The analytical solution to the diffusion equation with point-source illumination at wavelength $\lambda$,

$$\vec{\nabla}\cdot\left(D_\lambda\vec{\nabla}\Phi_\lambda(\vec{r},\omega)\right)-\left[\mu_{a\lambda}+\frac{i\omega}{c}\right]\Phi_\lambda(\vec{r},\omega)+S_\lambda(\vec{r},\omega)=0 \tag{33.39}$$

in an infinite medium provides three equations for $I_{AC_\lambda}$, $\theta_\lambda$, and $I_{DC_\lambda}$ as a function of modulation frequency, $\omega$, distance away from the source, $r$, in terms of the optical properties $\mu_{a_\lambda}, \mu'_{s_\lambda}$.

When FDPM measurements are conducted as a function of modulation frequency, nonlinear regression can be performed to arrive at the optical properties of the medium. Conversely, when referencing the measurements of $I_{AC_\lambda}$, $\theta_\lambda$, and $I_{DC_\lambda}$ at position $r$ to a "reference" position $r_0$, linear regression of the following equations:

$$DC_{rel}\equiv\frac{DC(r)}{DC(r_0)}=\frac{r_0}{r}\exp[-(r-r_0)(\frac{\mu_a}{D})^{1/2}] \tag{33.40}$$

$$AC_{rel}\equiv\frac{AC(r)}{AC(r_0)}=\frac{r_0}{r}\exp\{-(r-r_0)(\frac{c^2\mu_a{}^2+\omega^2}{c^2D^2})^{1/4}\cos[\frac{1}{2}\tan^{-1}(\frac{\omega}{c\mu_a})]\} \tag{33.41}$$

$$\theta_{rel}\equiv\theta(r)-\theta(r_0)=(r-r_0)(\frac{c^2\mu_a{}^2+\omega^2}{c^2D^2})^{1/4}\sin[\frac{1}{2}\tan^{-1}(\frac{\omega}{c\mu_a})] \tag{33.42}$$

enables accurate estimation of the optical properties at a single modulation frequency using a single detector. Frequency-domain measurements must be referenced to eliminate contributions of $G_{AC}$, $G_{DC}$, and $\theta_{inst}$, as denoted in Equation 33.29, before data are regressed to Equations 33.40 through 33.42.

An alternative referencing method was devised by Mayer et al.[27] that involved measurement of modulation and phase delay at two positions, $r_1$ and $r_2$, using two unmatched detectors. Unlike traditional fluorescence-spectroscopy measurement across a known pathlength, $L$, the instrument-response function for frequency-domain measurements in scattering media can be corrected without the use of reference dyes. The correction can be obtained by multiplexing two unmatched detectors at two different positions in the sample.[27] The two fibers leading to the unmatched detectors shown in Figure 33.13 are of the same length, which ensures equal optical pathlengths of the two received signals, $L_1, L_2$.

The measured relative phase shift between detector 1 and detector 2, $\theta_{rel_{1,2}}$, and measured modulation ratio, $M_{rel_{1,2}}$, between the two detectors reflect light propagation (and fluorescence generation, in the case of fluorescence measurements) in the sample between the incident source and the two detectors, (i.e., $\theta_1, \theta_2, \left(L_{AC_1}/L_{DC_1}\right), \left(L_{AC_2}/L_{DC_2}\right)$) as well as the instrument function ($\theta_{inst_1}, \theta_{inst_2}, \left(G_{AC_1}/G_{DC_1}\right), \left(G_{AC_2}/G_{DC_2}\right)$):

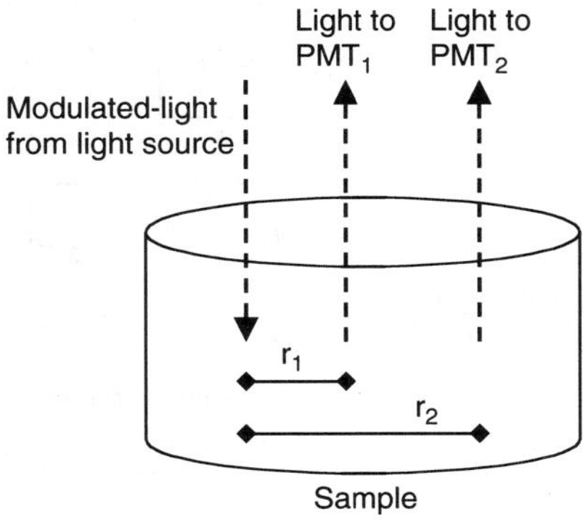


**FIGURE 33.13** Schematic of fiber optically coupled source and detector placement for fluorescence measurements in scattering media. The two detector fibers are of the same length to ensure equal optical pathlengths.

$$\theta_{rel_{1,2}} = \left(-\theta_1 + \theta_{inst_1}\right) - \left(-\theta_2 + \theta_{inst_2}\right) \tag{33.43}$$

$$M_{rel_{1,2}} = \frac{L_{AC_1} G_{AC_1}}{L_{AC_2} G_{AC_2}} \cdot \frac{L_{DC_2} G_{DC_2}}{L_{DC_1} G_{DC_1}} \tag{33.44}$$

After multiplexing, the measured relative phase shift between detector 2 and detector 1, $\theta_{rel_{2,1}}$, and measured modulation ratio, $M_{rel_{2,1}}$, between the two detectors continue to reflect light propagation (and fluorescence generation, in the case of fluorescence measurements) in the sample $\left(\theta_1, \theta_2, \left(L_{AC_1}/L_{DC_1}\right), \left(L_{AC_2}/L_{DC_2}\right)\right)$ as well as the instrument function $\left(\theta_{inst_1}, \theta_{inst_2}, \left(G_{AC_1}/G_{DC_1}\right), \left(G_{AC_2}/G_{DC_2}\right)\right)$:

$$\theta_{rel_{2,1}} = -\left(-\theta_1 + \theta_{inst_1}\right) + \left(-\theta_2 + \theta_{inst_2}\right) \tag{33.45}$$

$$M_{rel_{2,1}} = \frac{L_{AC_2} G_{AC_2}}{L_{AC_1} G_{AC_1}} \cdot \frac{L_{DC_1} G_{DC_1}}{L_{DC_2} G_{DC_2}} \tag{33.46}$$

where subscript 12 denotes the relative value of the signal detected at detector 1 to the signal detected at detector 2, and subscript 21 denotes the converse.

Combining Equations 33.43 and 33.45 and Equations 33.44 and 33.46 gives phase-shift and modulation ratios that are devoid of instrument function:

$$\Delta\theta = \theta_2 - \theta_1 = \frac{1}{2}\left(\theta_{rel_{1,2}} - \theta_{rel_{2,1}}\right) \tag{33.47}$$

$$M = \left(\frac{M_{rel_{1,2}}}{M_{rel_{2,1}}}\right)^{1/2} = \left(\frac{\dfrac{L_{AC_1} G_{AC_1}}{L_{AC_2} G_{AC_2}} \cdot \dfrac{L_{DC_2} G_{DC_2}}{L_{DC_1} G_{DC_1}}}{\dfrac{L_{AC_2} G_{AC_2}}{L_{AC_1} G_{AC_1}} \cdot \dfrac{L_{DC_1} G_{DC_1}}{L_{DC_2} G_{DC_2}}}\right)^{1/2} = \left(\frac{L_{AC_1}/L_{AC_2}}{L_{DC_2}/L_{DC_1}} \frac{G_{AC_1}/G_{AC_2}}{G_{DC_2}/G_{DC_1}}\right) \tag{33.48}$$

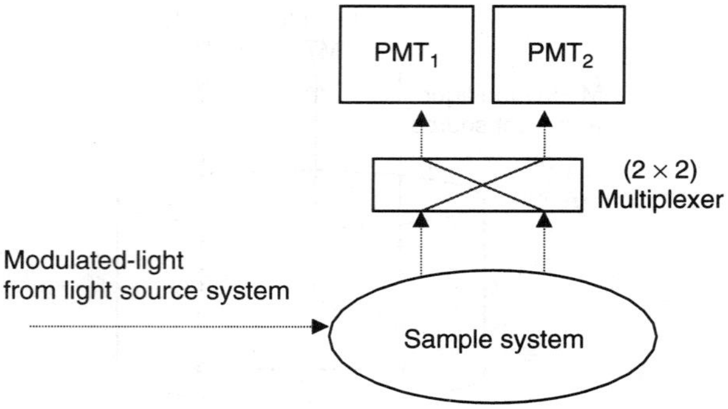


**FIGURE 33.14** Illustration of multiplexing system for the two-channel frequency-domain detection apparatus.

For matched detectors, $\left(\dfrac{G_{AC_1}/G_{AC_2}}{G_{DC_2}/G_{DC_1}}\right)$ is unity. For unmatched detectors, the ratio must be experimentally determined.[31]

Figure 33.14 illustrates the multiplexing system for the conventional two-channel frequency-domain system developed for fluorescence-lifetime spectroscopy.

The multiplexing method described above can be used for measurements performed with sources at the excitation and emission wavelengths, and hence single-distance (multifrequency) nonlinear regression[29] can be performed to obtain optical properties of the medium without corruption from the instrument functions. In a recent study, we showed that measurements made at multiple distances enable linear regression of parameters[29] and result in the most precise optical property estimation.[30] Regardless of whether from referenced or multiplexed frequency-domain measurements conducted with point-source illumination at wavelengths $\lambda_x$ and $\lambda_m$, the optical properties of $\mu_{a_{\lambda_x}}, \mu_{a_{\lambda_m}}, \mu'_{s_{\lambda_x}}, \mu'_{s_{\lambda_m}}$ can be accurately obtained.

### 33.4.1.2 Determination of Single-Exponential-Decay Lifetime

After the optical properties are estimated from frequency-domain measurements employing the two wavelength sources, the emission fluence is measured in response to point-source illumination at the excitation wavelength using excitation-light-rejection filters to reduce the noise floor. From the referenced measurement used to arrive at $\Phi_m(\vec{r},\omega) = I_{AC_m}\exp(i\theta_m)$, the fluorescent properties of $\alpha\mu_{a_{x\to m}}$ and $\tau$ are determined from the solution to Equations 33.19 and 33.24 for an infinite medium:

$$\Phi_m(\vec{r},\omega)_{ac} = \frac{\alpha\mu_{a_{x\to m}}\Xi_m(SA)}{4\pi c D_x D_m r}\frac{1}{\left[1+(\omega\tau)^2\right]}\{[\psi-\kappa\omega\tau]+i[\kappa+\psi\omega\tau]\} \tag{33.49}$$

where $r$ is the distance to the excitation point source, $SA$ is the complex fluence of the source describing its modulation depth and phase, and the terms $\psi$ and $\kappa$ are functions of optical properties ($\mu_a$ and $\mu'_s$), c, and $\omega$.[10,27]

$$\psi(\vec{r},\omega) = \frac{\delta(\vec{r},\omega)\xi+\zeta(\vec{r},\omega)\rho(\omega)}{\xi^2+\rho(\omega)^2} \tag{33.50}$$

$$\kappa(\vec{r},\omega) = \frac{\zeta(\vec{r},\omega)\xi-\delta(\vec{r},\omega)\rho(\omega)}{\xi^2+\rho(\omega)^2} \tag{33.51}$$

$$\delta(\vec{r},\omega) = \exp\left[-\beta_x(\omega)r)\right]\cos\left[\gamma_x(\omega)r\right]-\exp\left[-\beta_m(\omega)r)\right]\cos\left[\gamma_m(\omega)r\right] \tag{33.52}$$

$$\zeta(\vec{r},\omega) = \exp[-\beta_x(\omega)r)]\sin[\gamma_x(\omega)r] - \exp[-\beta_m(\omega)r)]\sin[\gamma_m(\omega)r] \quad (33.53)$$

$$\xi = \frac{\mu_{a_m}}{D_m} - \frac{\mu_{a_x}}{D_x} \quad (33.54)$$

$$\rho(\omega) = \frac{\omega}{c}\left[\frac{1}{D_x} - \frac{1}{D_m}\right] \quad (33.55)$$

Frequency-domain photon-migration measurements at both the excitation and emission wavelengths have demonstrated experimentally the ability to measure the single exponential lifetimes of ICG and 3,3′-diethylthiatricarbocyanine iodide (DTTCI),[27] rhodamine B,[10] and mixtures of ICG and DTTCI[13] in tissue-like scattering media of Intralipid.

Yet most analyte-sensing fluorphores exhibit multiexponential decays or stretched-exponential-decay kinetics, increasing the number of unknowns from two $(\alpha\mu_{a_{x\to m}}, \tau)$ to $2j+1$ for a fluorophore experiencing j-activated states (Equation 33.10): $(\mu_{a_{x\to m}}, \tau_j, a_j)$ and to $3j+1$ for a fluorophore undergoing collisional quenching (Equation 33.11): $(\mu_{a_{x\to m}}, \tau_j, a_j, \alpha_j, \beta_j)$.

## 33.4.2 Multiexponential-Decay Kinetics

In general, the solution for the emission fluence in an infinite medium of uniform optical properties is given by:

$$\Phi_m(\vec{r},\omega) = \frac{\alpha\mu_{a_{x\to m}}\Xi_m(SA)}{4\pi c D_x D_m r}\left\{(\psi - i\kappa)\int_0^\infty \left(g(t)e^{(i\omega t)}dt\right)\right\} \quad (33.56)$$

Generally, frequency-domain measurements are insensitive to the form of the decay kinetics used to describe the relaxation process. As an example, Figure 33.15 illustrates the phase and modulation ratio measured in a dilute, nonscattering sample using traditional fluorescence-lifetime-spectroscopy techniques. The emitted fluorescence results from a combination of two dyes, ICG and DTTCI, that individually exhibit first-order relaxation kinetics. The frequency-domain data are equally well fit using a single-exponential-decay (which represents the average of the decay times), a two-exponential-decay, and a stretched-exponential-decay model, indicating that the data at these modulation frequencies are insufficient to discern the relaxation processes. Typically, differences in decay kinetics are manifested at higher modulation frequencies, which unfortunately suffer from a small-instrument response function and low SNR. However, when predicting the phase-delay and modulation ratio from the solution to the diffusion equation for infinite media employing the various kinetic models for radiative relaxation, significant differences in frequency-domain data taken at modulation frequencies below 150 MHz (Figure 33.16) become apparent. Figure 33.16 shows model predictions for the two dyes within a scattering medium experiencing different relaxation mechanisms along with the data, indicating the potential for enhanced sensitivity of fluorescence-lifetime spectroscopy in random or multiply scattering media. Note that the phase-delay and modulation ratios in Figures 33.15 and 33.16 span a larger range in scattering media than in nonscattering media.

In summary, challenges remain in connection with solving the inverse-fluorescence-spectroscopy problem when it comes to extracting parameters that accurately predict changes in both decay kinetics and, therefore, analyte of metabolite concentrations. While a few studies in the literature indicate that the inverse-spectroscopy problem for fluorescence lifetime in random media can be solved, to date the inverse problem of multiexponential-decay functions that exist for analyte-sensing fluorophores has not been solved. However, as long as the number of unknown parameters remains smaller than the number of measurements, the solution to the inverse-spectroscopy problem entails a straightforward least-squares-minimization problem. In contrast to the inverse-spectroscopy problem, the inverse-imaging

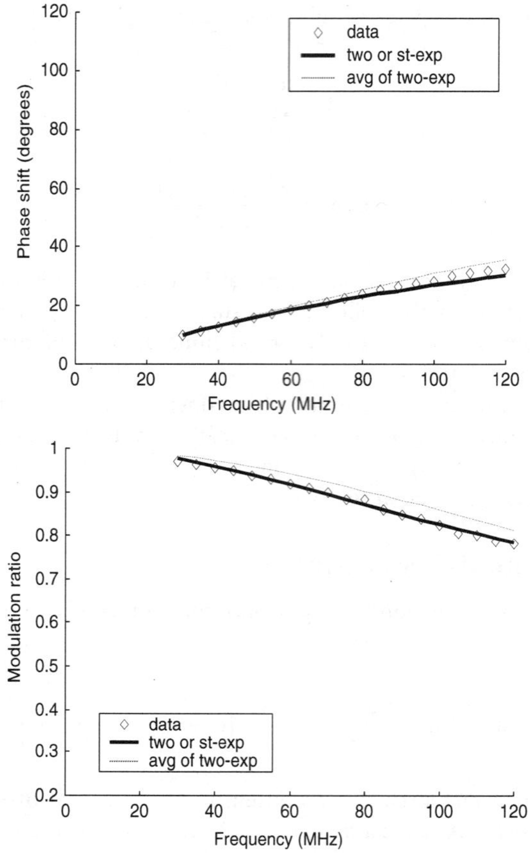


**FIGURE 33.15** The plot of fluorescence phase shift (top) and modulation ratio (bottom) as a function of modulation frequency on ICG–DTTCI mixture (ICG:DTTCI = 0.15 m*M*: 0.5 m*M*, in a dilute nonscattering medium) for corrected experimental measurements ( ◇ ), and that predicted by incorporating two-exponential-decay kinetics or stretched-exponential-decay kinetics (bold line) and average of two-exponential-decay kinetics (thin gray line).

problem entails a smaller number of measurements than unknown parameters, necessitating optimization approaches.

## 33.5 Fluorescence FDPM for Optical Tomography[15]

The solution to the inverse-imaging problem, known as optical tomography, has been motivated over the past decade by optical mammography, i.e., the use of deeply penetrating NIR light for detecting breast cancer on the basis of endogenous optical property contrast between normal and diseased tissues. It has been proposed that the optical property contrast in scattering and absorption is due to the increased size and density of neoplastic cells in a tumor region and the increased vascular blood supply (as a result of angiogenesis) that locally increases hemoglobin, a primary chromophore in tissues. While optical mammography has seen many advances in recent years, most notably in the application of FDPM,[32,33] TDPM,[11] and continuous-wave[34] measurements (see also Reference 22), the need for angiogenesis-mediated absorption contrast for diagnostic optical mammography limits the potential applications of optical mammography. Since the endogenous contrast from angiogenesis can be expected to be low in small lesions and nonspecific to cancer, NIR detection of nonpalpable disease in dense breast tissue is limited without the addition of contrast. Hence, moderately resolved, biochemical molecular imaging within tissues using

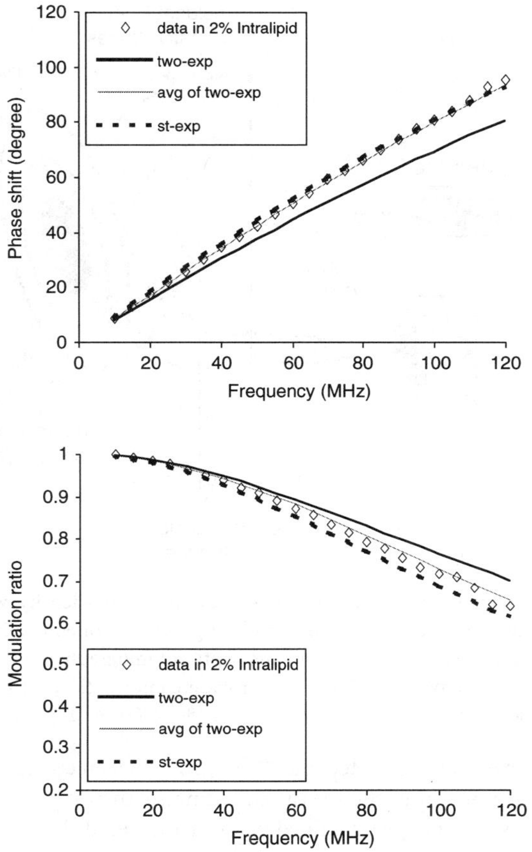


**FIGURE 33.16** The plot of fluorescence phase shift (top) and modulation ratio as (bottom) a function of modulation frequency on ICG–DTTCI mixture (ICG:DTTCI = 0.15 m*M*: 0.5 m*M*, in 2% Intralipid® solution) for corrected experimental measurements ( ◇ ) and that predicted by the propagation model incorporating two-exponential-decay kinetics (bold line), average of two-exponential-decay kinetics (thin gray line), and the stretched-exponential-decay kinetics (dashed line).

unassisted NIR optical techniques is somewhat limited in scope and can be expanded through the use of contrast-enhancing agents.

Fluorescent contrast agents have been proposed and independently confirmed as the most efficient means for inducing optical contrast when time-dependent measurements (i.e., TDPM or FDPM) are conducted.[35,36] Furthermore, in the near future these agents may offer a host of opportunities for molecular imaging that are limited only by synthetic design. The basic principles behind fluorescence-enhanced NIR optical tomography stem first from the kinetics of fluorescence generation. It is the kinetics of the fluorescence-decay process that imparts the superior contrast of fluorescence over absorbance when TDPM or FDPM measurements are made.

Figure 33.17 provides a simple schematic describing in physical terms why contrast by fluorescence is greater than contrast by absorption for FDPM imaging. Consider a tissue volume illuminated by an intensity-modulated light source at source position $r_s$. The propagating wave is denoted by dotted lines. As the propagating excitation wave transits through the tissue, it is attenuated and phase delayed due to the tissue optical properties. If the wave encounters a light-absorbing heterogeneity, such as a highly vascularized tumor, a portion of the intensity wave is reflected. The strength of the "reflected wave" (or dotted line) depends on the absorption contrast and the size and depth of the heterogeneity. This reflected wave makes a small contribution to the wave that ultimately is detected at detector position $r_d$. It is this

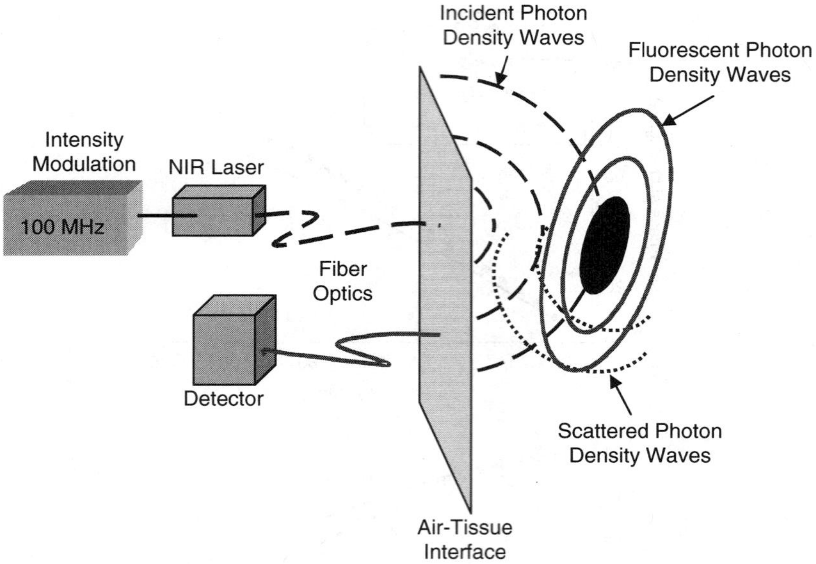


**FIGURE 33.17** Schematic detailing the propagation of excitation photon-density waves (solid lines) and their perturbation by absorbing heterogeneities (dotted lines) and the generation of emission photon-density waves (solid gray lines) within the tissues. Fluorescence contrast-enhance optical tomography provides greater localization capability because the detected emission waves act as "beacons" providing information regarding the tagged heterogeneity. (From Hawrysz, D.J. and Sevick-Muraca, E.M., *Neoplasia,* 2(5), 388, 2000. With permission.)

small, added contribution that is used to detect the heterogeneity when endogenous contrast is employed. If the added contribution is not within the measurement noise, then it provides information for image recovery in one of the inversion strategies outlined below. However, if the heterogeneity is contrasted by fluorescence, then upon reaching the heterogeneity the excitation wave generates an emission wave (solid lines). The emission wave then acts as a beacon, and, when appropriate filters are used to reject the excitation light, it can be measured to directly locate the heterogeneity. In small tissue volumes, such as the mouse or rat, inversion algorithms may not be required to detect the fluorescent heterogeneity, as shown in the literature results summarized in Section 33.6. However, in larger tissue volumes, the forward- and inverse-imaging problems require solution for effective image recovery. Most importantly, since a phase delay and amplitude attenuation occur between the activating excitation wave and the re-emitted fluorescent wave associated with the fluorescence-decay kinetics, the fluorescence decay increases the contrast or phase-delay and amplitude-attenuation change associated with the detected signal. Finally, since a fluorescent probe can be "tuned" to exhibit differing fluorescence-decay characteristics that depend on the probe's local environment, the re-emitted signal can contain diagnostic information about the tissue of interest.

Few studies have successfully inverted NIR tissue optical measurements to render images of exogenously contrasted tissues or tissue-mimicking phantoms. Table 33.1 outlines several investigations that have employed "synthetic" data sets and phantoms.[37–57] The approaches are basically similar to those used in NIR optical-tomography work, with the following three exceptions: (1) because of the low quantum yield of fluorescent dyes, the SNR for CW, TDPM, and FDPM measurements is inarguably lower, potentially making it more difficult to successfully reconstruct images; (2) due to the fluorescence-lifetime delay, both TDPM and FDPM approaches have additional contrast in the time-dependent photon-migration characteristics; and, finally, (3) because the fluorescence-kinetic parameters of fluorescence lifetime and quantum efficiency can be directly inverted, the technique can be used to perform quantitative imaging via dyes that report cancer.

**TABLE 33.1** Fluorescence-Enhanced Contrast Imaging: Literature of Image Reconstructions

| Ref. | Inversion Formulation | Data Type | Noise | Two- or Three-Dimensional | Forward Method | Measurement Method | Contrast Agent | Uptake Ratio |
|---|---|---|---|---|---|---|---|---|
| O'Leary, M.A. et al., 1994[37] | Localization | Experimental phantom | Yes | Two-dimensional | None | TDPM | ICG | Perfect uptake |
| Wu, J. et al., 1995[38] | Localization | Experimental phantom | Yes | Three-dimensional | None | TDPM | Diethylthiatric arbocyanine | Perfect uptake |
| Chang, J. et al., 1995[39] | POCS, CGD, SART | Experimental phantom | Yes | Two-dimensional | NS | CW | Rhodamine 6G dye | Perfect uptake and 1000:1 |
| O'Leary, M.A. et al., 1996[40] | Integral(SIRT) | Synthetic data | 0.1° in phase, 1% in amplitude | Two-dimensional | Analytical | FDPM | ICG | Perfect uptake, 20:1 |
| Paithankar, D.Y. et al., 1997[41] | Differential (Newton-Raphson) | Synthetic data | 0.1°–1° in phase; 0.01 in log AC (Gaussian) | Two-dimensional | MFD | FDPM | ICG | 20:1 |
| Wu, J. et al., 1997[42] | Laplace transform | Experimental phantom | Yes | Two-dimensional | NS | TDPM | HITCI iodide dye | Perfect uptake |
| Chang, J. et al., 1998[43] | Differential (conjugate gradient) | Synthetic data | 1–10% white noise | Two-dimensional or three-dimensional | NS | CW/FDPM | N/A | 100:1 10:1 |
| Jiang, H., 1998[44] | Differential (Newton's iterative method) | Synthetic data | 0–5% | Two-dimensional | FEM | FDPM | N/A | 2:1 contrast in $\phi$, $\tau$ |
| Hull, E.L. et al., 1998[45] | Localization | Experimental phantom | Yes | Two-dimensional | Monte Carlo | CW | Nile Blue A | Perfect uptake |
| Eppstein, M.J. et al., 1999[46,47] | Differential | Synthetic data | 0.1°in phase; 1% in log AC (Gaussian) | Two-dimensional, three-dimensional | MFD | FDPM | N/A | Perfect uptake and 100:1 10:1 |
| Chernomordik, V. et al., 1999[48] | Random walk theory | Experimental phantom | Yes | Two-dimensional and three-dimensional | NS | CCD | Rhodamine | Perfect uptake |
| Roy, R. and Sevick-Muraca, E.M., 1999, 2000[49,50] | Differential (gradient-based and truncated Newton method) | Synthetic data | 0.1°in phase; 1% in log AC (Gaussian) | Two-dimensional | FEM | FDPM | N/A | 2.5:1 5:1 10:1 |
| Yang, Y. et al., 2000[51] | Marquardt and Tikhonov regularization | *In vivo* (rats) | Yes | Two-dimensional images | FEM | FDPM | ICG and DTTCI | Perfect and imperfect uptake |
| Eppstein, M.J. et al., 2001;[52] Hawrysz, D.J. et al., 2001[53] | Differential | Experimental phantom | Yes | Three-dimensional | MFD | FDPM | ICG | 50:1 100:1 |
| Roy, R. and Sevick-Muraca, E.M., 2001[54,55] | Differential (gradient-based optimization and truncated Newton's method) | Synthetic data | 55 dB in excitation; 35 dB in emission | Two-dimensional, three-dimensional | FEM | FDPM | N/A | 10:1 |

**TABLE 33.1** Fluorescence-Enhanced Contrast Imaging: Literature of Image Reconstructions (continued)

| Ref. | Inversion Formulation | Data Type | Noise | Two- or Three-Dimensional | Forward Method | Measurement Method | Contrast Agent | Uptake Ratio |
|---|---|---|---|---|---|---|---|---|
| Ntziachristos, V. et al., 2001[56] | Integral (normalized Born expansion) | Experimental phantom | 2% amplitude noise in source | Two-dimensional along *z*-planes | FD | CW | ICG in background; Cy5.5 as contrast tumor | Perfect uptake |
| Lee, J. and Sevick-Muraca, E.M., 2002[57] | Integral (distorted BIM) | Experimental phantom | Yes | Three-dimensional | MFD | FDPM | ICG | 100:1 |

CCD = charge-coupled device; CGD = conjugate gradient descent; CW = continuous-wave imaging; FD = finite difference; FDPM = frequency-domain photon migration; FEM = finite-element method; MFD = multigrid finite difference; N/A = not applicable; NS = not specified; POCS = projection onto convex sets; SART/SIRT = simultaneous algebraic reconstruction techniques; TDPM = time-domain photon migration; $\phi$, $\tau$ = quantum efficiency and lifetime of contrast agent.

The latter characteristic of fluorescence-lifetime imaging reveals its similarity to magnetic resonance imaging (MRI). In MRI, imaging is accomplished by monitoring the RF signal arising from the relaxation of a magnetic dipole perturbed from its aligned state using a pulsed magnetic field. In fluorescence-contrast-enhanced optical tomography, imaging is accomplished by monitoring the emission signal arising from the electronic relaxation from an optically activated state to its ground state. Unfortunately, the emission light is multiply scattered; hence, the resolution afforded by MRI is unlikely to be matched by contrast-enhanced optical tomography. Unlike contrast agents for conventional imaging modalities, optical contrast, due to fluorescent agents, may be achieved in two ways: (1) through increased target:background concentration ratios and (2) through alteration in the fluorescence-decay kinetics upon partitioning within tissue regions of interest.[58] Section 33.6 summarizes the literature on fluorescent contrast agents and their development over the past decade. The approaches for solving the inverse-imaging problem are presented in the next section.

## 33.5.1 Approaches to the Inverse-Imaging Problem

Attempts to solve the fluorescence-enhanced optical-imaging problem have been made both by solving a formal inverse problem and by taking a less rigorous model-based approach. For example, localization of a fluorescent target has been demonstrated using localization techniques in which the strength of reradiating target is used to ascertain its central position within a background containing no fluorophore. Through the use of FDPM measurements between a point source and detector as the pair was scanned over the phantom surface, the center of a single fluorescent target could be accurately identified.[37] In yet another approach, an analytical solution to the spherical propagation of emission light in uniform scattering media was used to determine the $x,y,z$ position of the point source of fluorescence in an otherwise nonfluorescent background probed by CW measurements.[48] Using time-domain measurements, Wu et al.[38,42] developed a system for assessing the position of a fluorescent target in turbid media by evaluating early-arriving photons to determine the origin of the fluorescence generation. Hull et al. similarly used spatially resolved CW measurements to determine the location of the fluorescent target in scattering media.[45] Unlike the studies described above, when there is presence of background signal or if the goal of reporting fluorescence-decay kinetics prevents a localization approach, then a solution to the full inverse-imaging problem is required.

The solution of a formal inverse problem requires the use of the appropriate mathematical models described above in Section 33.2. Specifically, a guess of the interior optical or fluorescence properties is iteratively updated until the predicted measurements given by the solution of the forward problem match the actual measurements. Since the number of unknowns (or optical and fluorescent contrast-agent properties) is greater than the number of measurements, the problem is underdetermined and especially difficult. This inverse problem is unavoidably "ill-posed," which generally means that the solutions are nonunique and unstable in the presence of measurement error. In addition, the optical-tomography problem is generally highly nonlinear, and attempts to linearize it result in solution instabilities and often intractably long computational times if the update step is to remain within the range of accuracy of the linearization. The solution of the inverse problem is an intensive area of research in itself that is motivated by several different research areas including biomedical NIR optical tomography based on endogenous contrast.

To assess the performance of an inverse-problem algorithm, the achieved solution must be compared to the known distribution of optical properties. As a consequence, studies investigating the inverse optical-tomography problem are performed using either (1) "synthetic" measurements, i.e., measurements that are predicted by the forward problem to which artificial random "noise" is added to simulate measurement error; or (2) phantom studies in which tissue-mimicking scattering media of known optical properties are used to collect experimental CW, time, or frequency-domain measurements that are used for the inverse solution. In the following sections, we briefly review the methods used to solve fluorescence-enhanced optical-tomography problems, broadly classified into categories of integral and differential approaches and employing a number of parameter-updating schemes. Finally, sample reconstructions from actual experimental data are presented.

### 33.5.2 Integral Formulation of the Inverse Problem

One of the more common methods of formulating the inverse problem is by integral treatment. Since the emission diffusion equation is in the form of an nonhomogeneous differential equation, Green's function is used to obtain the analytical solution of the emission fluence $\Phi_m(\vec{r},\omega)$. Equation 33.19 can be rewritten to account for variation in the endogenous and exogenous optical properties including the absorption due to fluorophore, $\mu_{a_{x\to m}}(\vec{r})$, and the optical-diffusion coefficient, $D_m(\vec{r})$:

$$\vec{\nabla}^2\Phi_m(\vec{r},\omega)+k_m^2(\vec{r})\Phi_m(\vec{r},\omega)=-\frac{S_m(\vec{r})}{D_m(\vec{r})}-\frac{\vec{\nabla}D_m(\vec{r})\cdot\vec{\nabla}\Phi_m(\vec{r},\omega)}{D_m(\vec{r})} \tag{33.57}$$

where the complex diffusion wave number can be expressed as:

$$k_m^2(\vec{r})=\frac{1}{D_m(\vec{r})}\left(-\mu_{a_m}(\vec{r})+i\frac{\omega}{c}\right) \tag{33.58}$$

The term $\left[\vec{\nabla}D_m(\vec{r})/D_m(\vec{r})\right]\cdot\vec{\nabla}\Phi_m(\vec{r})$ in Equation 33.57 accounts for the discontinuity in $D_m(\vec{r})$. However, since there is little or no variation in the isotropic scattering coefficient (the component that constitutes the overwhelming contribution to $D_m(\vec{r})$ at the emission wavelength, $\lambda_m$) the above term is insignificant. The Green's function corresponding to Equation 33.57 consequently satisfies

$$\vec{\nabla}^2G_f(\vec{r},\vec{r}\,')+k_m^2(\vec{r})G_f(\vec{r},\vec{r}\,')=-\delta(\vec{r}-\vec{r}\,') \tag{33.59}$$

by manipulating Equations 33.57 and 33.19 (at the excitation wavelength), and with the use of Green's theorem, $\int_v\left(U\vec{\nabla}^2G-G\vec{\nabla}^2U\right)d^3r=\oint_S\left(U\vec{\nabla}G-G\vec{\nabla}U\right)\cdot dS$, one can obtain the expression of the emission fluence measured at $r_d$ following point excitation at $r_s$, $\Phi_m(r_d, r_s)$, which arises due to $S_m(r', r_s)$, the emission generation at point $r'$ following the incident illumination of excitation light at $r_s$, and due to the propagation of the emission light from position $r'$ to detector point $r_d$ as predicted from the Green's solution, $G_f(r_d, r_s)$:

$$\tilde{\Phi}_m(\vec{r}_d,\vec{r}_s)=\int_\Omega G_f(\vec{r}_d,\vec{r}\,')S_m(\vec{r}\,',\vec{r}_s)\,d\Omega=\int_\Omega G_f(\vec{r}_d,\vec{r}\,')\frac{\alpha\mu_{a_{x\to m}}(\vec{r}\,')}{D_m(\vec{r})(1+i\omega\tau)}\Phi_x(\vec{r}\,',\vec{r}_s)\,d\Omega \tag{33.60}$$

where $\Omega$ is the volume of integration, $\vec{r}_d$ is the point-detector location, and $\vec{r}_s$ is the point-source location.

To reconstruct the spatial map of $\mu_{a_{x\to m}}(\vec{r})$ detailing the heterogeneity, Equation 33.60 is discretized into a series of equations in terms of $G_f$, $\Phi_x(\vec{r}\,',\vec{r}_s)$, and the vector of measurements of $\tilde{\Phi}_m(\vec{r}_d,\vec{r}_s)$. We consider the excitation source to be amplitude-modulated by a frequency of $\omega$. Measurements of phase shift, $\theta_m$, and amplitude of AC component $I_{AC}$ (or $\Phi_m(\vec{r}_d,\vec{r}_s)=I_{AC_m}e^{-i\theta_m}$) are obtained at detector positions $\vec{r}_d$ in response to the excitation source at $\vec{r}_s$. [*Note: It is assumed that the phase shift and the AC component are predicted relative to the incident excitation light at* $\vec{r}_s$. *In addition, it is assumed that absolute measurements of* $I_{AC_m}$ *and* $\theta_m$ *are used as inputs.*] Discretizing Equation 33.60 yields:

$$\tilde{\Phi}_m(\vec{r}_d,\vec{r}_s)=\sum_{j=1}^{N}G_f(\vec{r}_j,\vec{r}_d)\Phi_x(\vec{r}_j,\vec{r}_s)\frac{\alpha\mu_{a_{x\to m}}(\vec{r}_j)\Delta}{D_m(\vec{r})(1+i\omega\tau)} \tag{33.61}$$

where $N$ is the total number of cells in the domain, and $\Delta$ is the area or volume of the pixel or voxel. If there are $K$ sources and $L$ detectors, then $\tilde{\Phi}_m = FX$ can be denoted as:

$$\begin{pmatrix} \tilde{\Phi}_m(\vec{r}_d,\vec{r}_s)_1 \\ \tilde{\Phi}_m(\vec{r}_d,\vec{r}_s)_2 \\ \vdots \\ \vdots \\ \tilde{\Phi}_m(\vec{r}_d,\vec{r}_s)_M \end{pmatrix} = \begin{pmatrix} F_{11} & \cdots & F_{1N} \\ F_{21} & \cdots & F_{2N} \\ \vdots & \ddots & \vdots \\ \vdots & \ddots & \vdots \\ F_{M1} & \cdots & F_{MN} \end{pmatrix} \begin{pmatrix} X(\vec{r}_1) \\ X(\vec{r}_2) \\ \vdots \\ \vdots \\ X(\vec{r}_N) \end{pmatrix} \tag{33.62}$$

$$F_{ij} = \frac{G_f(\vec{r}_{di},\vec{r}_j)\Phi_x(\vec{r}_{si},\vec{r}_j)\alpha\Delta}{D_m(\vec{r})(1+i\omega\tau)} \tag{33.63}$$

$$X(\vec{r}_j) = \mu_{a_{x\to m}}(\vec{r}_j) \tag{33.64}$$

where $F \in C^{M\times N}$, $X \in \Re^N$, $\tilde{\Phi}_m \in C^M$, and $M = K^*L$.

Equation 33.64 is also appropriate if imaging is performed on the basis of a fluorophore-absorption cross section with a constant lifetime $\tau$. The problem formulated in Equations 33.62 through 33.64 is nonlinear in $\mu_{a_{x\to m}}$ since the excitation fluence, $\Phi_x(\vec{r})$, is also a function of absorption.

However, if tomographic reconstruction on lifetime were pursued, the problem would become linear, and Equations 33.63 and 33.64 would be rewritten as:

$$F_{i,j} = \frac{G_f(\vec{r}_{di},\vec{r}_j)\Phi_x(\vec{r}_{si},\vec{r}_j)\Delta}{D_m(\vec{r})} \tag{33.65}$$

$$X(\vec{r}_j) = \frac{\alpha\mu_{a_{x\to m}}(\vec{r}_j)}{(1+i\omega\tau)} \tag{33.66}$$

where the vector X represents the sources of fluorescence within the random medium.

The formulation of the inverse-lifetime problem is called fluorescence-lifetime imaging[40,41,59] and has been the subject of tomographic reconstructions from synthetic measurements. Fluorescence-lifetime imaging is not possible for CW measurements ($\omega = 0$). To date fluorescence-lifetime imaging has been accomplished using actual experimental measurements only in limited works,[51] probably because few contrast agents exist that are designed with "tuneable" lifetimes.

Nonetheless, the opportunity for lifetime imaging is clearly evidenced by ICCD measurements using the instrumentation depicted in Figure 33.8. Two fluorescent targets were positioned 0.5 cm from the imaged surface of a tissue-simulating phantom and embedded in $10 \times 10 \times 10$ cm$^3$ of 0.5% Intralipid. The first target encapsulated a 1-μM solution of ICG, and the second encapsulated a 1.42-μM solution of 3,3′-DTTCI. These solution concentrations were chosen to equilibrate the number of fluorescent photons emitted from each target because an equivalent number of excitation photons encounter each target. Figures 33.18A and 33.18B confirm this equilibration, as it is difficult to differentiate the targets from the DC and AC measurements. However, Figures 33.18C and 33.18D plot the phase delay and the modulation ratio ($I_{AC}/I_{DC}$) and provide differentiation of the two volumes. The differentiation, which results from a disparity in lifetime (0.62 ns) between the two fluorescent agents, demonstrates the potential of fluorescence-lifetime imaging. While these images present raw data, quantitative tomographic recovery of fluorescence lifetime is possible with solution to the inverse-imaging problem.

The inverse-imaging problem can be approximated as a linear problem and iteratively solved using the Gauss–Newton method:

$$\Phi_m^{meas}(\vec{r}_d,\vec{r}_s) - \Phi_m^{comp}(\vec{r}_d,\vec{r}_s) = \mathbf{F}\cdot\Delta\mathbf{X}(\vec{r}) \tag{33.67}$$

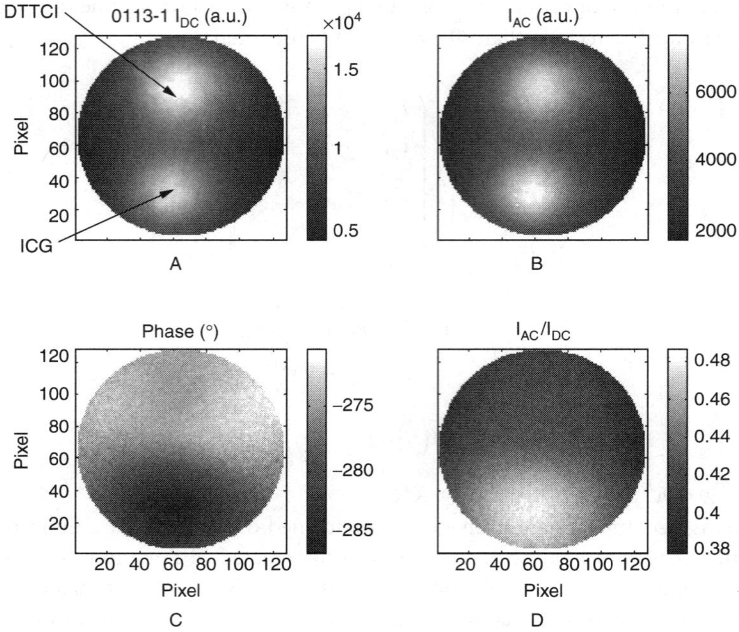


**FIGURE 33.18** The (A) $I_{DC}$, (B) $I_{AC}$, (C) phase, and (D) $I_{AC}/I_{DC}$ modulation ratio of two submerged targets of differing fluorescent lifetimes showing that $I_{DC}$ cannot distinguish the difference in fluorescence-decay kinetics.

Regardless of whether absorption or fluorescence-lifetime imaging is to be performed, parameter updating can be accomplished by noting that **F** is simply the Jacobian, **J**, and the difference between the measurement and the model-predicted fluence, $\Phi^{meas} - \Phi^{comp} = \Delta\Phi$, can be used to update the optical property map, $\Delta X(\vec{r})$, by the relationship specified above, or alternatively by:

$$\Delta\Phi = \mathbf{J} \cdot \Delta \mathrm{X} \tag{33.68}$$

While a number of investigators report reconstructions based on the integral approaches using frequency-domain or CW ($\omega = 0$) approaches, all assume that the measurements can be accurately measured relative to the incident light. Yet in practice, it is nearly impossible to perform such a measurement. Typically, a portion of the incident light is split for simultaneous measurement at a reference detector that cannot report the source strength or the phase delay of light that is incident on the medium surface. Calibration of the source via a "reference" phantom may also be conducted if a three-dimensional (3D) model can be used to predict the measurements and if the source strength can be accurately recovered from the measurements and model. Yet such a procedure is cumbersome and susceptible to errors and increases the challenges for incorporation into medical imaging in clinical situations. A method to eliminate "calibration" against an external standard would eliminate the number of measurements as well as improve measurement accuracy. The following sections present three general approaches to reference measurements for the recovery of optical properties in light of image-reconstruction formalisms: (1) emission measurements at detector position $r_d$ referenced to the background emission wave, $\Phi_m^b(\vec{r}_s,\vec{r}_d)$; (2) emission measurements at detector position $r_d$ referenced to the detected emission wave at a single reference position $r_r$, $\Phi_m(\vec{r}_s,\vec{r}_r)$; and (3) emission measurements at detector position $r_d$ referenced to the detected excitation wave at a single reference position $r_r$, $\Phi_x(\vec{r}_s,\vec{r}_r)$.

#### 33.5.2.1 Measurement Referenced to the Background Emission Wave $\Phi_m^b(\vec{r}_s,\vec{r}_r)$

In this approach, calibration of the detection system is achieved by normalizing the measurement in the presence of the heterogeneity with that measured or predicted in its absence. Typically, the "absence" case consists of a uniform medium with constant and known optical properties. In this case, the main problem associated with matching experimental data to the solution of Equation 33.57 has been the unknown source strength, the unknown phase delay from the timing characteristics of various components of the detection system, and the amplitude gain or loss. The advantage of using background or a well-defined reference phantom is that it allows for the elimination of the most systematic errors common to both background and actual measurement data sets. This approach has been widely employed in both frequency[60,61] and time-domain[62,63] photon-migration measurements. However, this is an unrealistic approach in a clinical sense because, unlike phantom studies, the absence case and "background" fluence measurements are impossible in clinical applications. For contrast-enhanced imaging, "absence" may be achieved prior to the administration of a contrast agent or activation-induced contrast such as that resulting from blood flow. Yet when the agent is fluorescent, there is no emission signal to measure the absence scenario; hence, the "absence" scenario is unrealistic in fluorescence imaging.

For heterodyne FDPM measurements the referenced measurements enable elimination of instrument responses. For example, an emission measurement referenced to the background case in which the target,

$$\frac{\Phi_{AC_m}(r_d,r_s)}{\Phi^b_{AC_m}(r_d,r_s)} = \frac{L_{AC_m} G_{AC}\exp(i(-\theta_m+\theta_{instr}))}{L^b_{AC_m} G_{AC}\exp(i(-\theta^b_m+\theta_{instr}))} = \frac{L_{ACm}}{L^b_{AC_m}}\exp\left[i(\theta^b_m-\theta_m)\right]$$

successfully eliminates the instrument response, and the background optical properties are known.

The mechanics of inverting data referenced to the background absence case may be illustrated by the following equation, which shows the relationship between the measurement and the inversion algorithm for a source–detector pair when a Born-type inversion scheme is used for fluorescent measurements, where the referenced measurement is represented by $\left.\frac{\Phi_{AC_m}(\vec{r}_s,\vec{r}_d)}{\Phi^b_{AC_m}(\vec{r}_s,\vec{r}_d)}\right]_{exp}$:

$$\left.\frac{\Phi_{AC_m}(\vec{r}_s,\vec{r}_d)}{\Phi^b_{AC_m}(\vec{r}_s,\vec{r}_d)}\right]_{exp} = \frac{1}{\left[\Phi^b_{AC_m}(\vec{r}_s,\vec{r}_d)\right]_{a\ priori}}\int_{\Omega}\frac{\alpha}{D_m(1+i\omega\tau)}G_f(\vec{r}_d,\vec{r})\Phi_x(\vec{r}_s,\vec{r})\mu_{a_{x\to m}}(\vec{r})\,d\Omega \tag{33.69}$$

Here, $\left[\Phi^b_{AC_m}(\vec{r}_s,\vec{r}_d)\right]_{a\ priori}$ is the background emission wave detected at the same detector position, $\vec{r}_d$, as in the presence of the heterogeneity case in response to excitation from the source position, $\vec{r}_s$. $\Phi^b_{AC_m}(\vec{r}_s,\vec{r}_r)$ is computed from the known optical properties of the assumed uniform background case or measured in the case of an absent heterogeneity.

In contrast to the absolute-measurement case, where the matrix, **F**, is itself the Jacobian matrix (Equation 33.63), relative-measurement schemes have the added complication of calculating the Jacobian matrix by using the integral equation. However, in this case, the $\Phi^b_{AC_m}(\vec{r}_s,\vec{r}_d)$ stays constant throughout the iteration due to its homogeneous nature, the inversion problem remains linear, and the calculation of the Jacobian matrix is straightforward. As a result, the reconstruction remains stable during the iterative process, while the source strength dependency and other calibration problems are eliminated. Even though the reference to the background-emission-fluence approach is relatively easy to implement in phantom studies and is a good benchmark to test the inversion algorithms, it can be difficult in actual application to clinical trials where a true absence condition does not occur or where the spatially distributed optical properties are not known.

#### 33.5.2.2 Measurement Referenced to the Emission Wave $\Phi_m(\vec{r}_s,\vec{r}_r)$

Because of the unrealistic nature of referencing to the background emission wave, the inversion scheme uses the emission fluence at the reference point on the tissue surface. This approach involves the measurements of the emission fluence relative to one another, where the referenced measurement is represented by $\left.\frac{\Phi_m(\vec{r}_s,\vec{r}_d)}{\Phi_m(\vec{r}_s,\vec{r}_r)}\right]_{exp}$ and

$$\frac{\Phi_{AC_m}(r_d,r_s)}{\Phi_{AC_m}(r_d,r_r)}=\frac{L_{AC_m}(r_d)\,G_{AC}\exp(i(-\theta_m+\theta_{instr}))}{L_{AC_m}(r_r)\,G_{AC}\exp(i(-\theta_m(r_r)+\theta_{instr}))}=\frac{L_{ACm}(r_d)}{L_{AC_m}(r_r)}\exp\left[i(\theta_m(r_r)-\theta_m(r_d))\right]$$

and the Fredholm's equation to be solved is written:

$$\frac{\Phi_m\left(\vec{r}_s,\vec{r}_d\right)}{\Phi_m\left(\vec{r}_s,\vec{r}_r\right)}=\frac{1}{\Phi_m(\vec{r}_s,\vec{r}_r)}\int_\Omega \frac{\alpha}{D_m\left(1+i\omega\tau\right)}G_f\left(\vec{r}_d,\vec{r}\right)\Phi_x\left(\vec{r}_s,\vec{r}\right)\mu_{ax\to m}(\vec{r})\,d\Omega \tag{33.70}$$

Here $\Phi_m(\vec{r}_s,\vec{r}_r)$ is the emission wave detected at the reference position, $\vec{r}_r$, in response to excitation from the source position, $\vec{r}_s$. This referencing approach is the most reasonable method for matching the actual measurement data with simulation data because it does not require the separate measurement of a homogeneous background wave. Also, this referencing approach has merit in that, unlike the method described below, it uses the measurements at the same wavelength.

While referencing to the emission wave, $\Phi_m(\vec{r}_s,\vec{r}_r)$, is more practical than referencing to the background emission wave, $\Phi^b_{AC_m}(\vec{r}_s,\vec{r}_d)$, normalization by $\Phi_m(\vec{r}_s,\vec{r}_r)$ renders the inversion algorithm using the Born-type integral approach highly nonlinear. This nonlinearity is absent when referencing to the background emission wave. Consequently, the calculation of the Jacobian matrix is not as straightforward as with the background emission wave (for details see Reference 57).

Hence, the formulated inverse-imaging problem is inherently unstable due to its nonlinear nature. Using the Marquardt-Levenberg regularization, Lee and Sevick-Muraca[57] were unable to recover optical property maps in two-dimensional (2D) or 3D synthetic data. In contrast, through the use of the approximate extended Kalman filter for nonlinear systems,[64–66] 3D reconstructions from sparse experimental measurements were possible in a large 256-cm$^3$ volume.[53]

#### 33.5.2.3 Measurement Referenced to the Excitation Wave $\Phi_x(\vec{r}_s,\vec{r}_r)$

Another practical referencing scheme utilizes measurements of excitation fluence at fixed reference positions to be used as the normalization factors; these would, in turn, eliminate the source strength dependency. This approach is more realistic than measurements referenced to the background-emission case because it does not require separate and impractical measurements with and without heterogeneity.

The relationship between the experimental measurements and the simulation for a given source and detector pair is represented by:

$$\frac{\Phi_{AC_m}(r_d,r_s)}{\Phi^b_{AC_x}(r_d,r_r)}=\frac{L_{AC_m}(r_d)\,G_{AC_m}\exp(i(-\theta_m(r_d)+\theta^m{}_{instr}))}{L_{AC_x}(r_r)\,G_{ACx}\exp(i(-\theta_x(r_r)+\theta^m{}_{instr}))}\approx\frac{L_{ACm}(r_d)}{L_{ACx}(r_r)}\exp\left[i(\theta_x(r_r)-\theta_m(r_d)\right]$$

and described by the following integral equation:

$$\left.\frac{\Phi_{AC_m}\left(\vec{r}_s,\vec{r}_d\right)}{\Phi_{AC_x}\left(\vec{r}_s,\vec{r}_r\right)}\right]_{exp}=\frac{1}{\Phi_{AC_x}(\vec{r}_s,\vec{r}_r)}\int_\Omega \frac{\alpha}{D_m\left(1+i\omega\tau\right)}G_f\left(\vec{r}_d,\vec{r}\right)\Phi_x\left(\vec{r}_s,\vec{r}\right)\mu_{a_{x\to m}}(\vec{r})\,d\Omega \tag{33.71}$$

where the referenced measurement is represented by $\left.\dfrac{\Phi_{AC_m}\left(\vec{r}_s,\vec{r}_d\right)}{\Phi_{AC_x}\left(\vec{r}_s,\vec{r}_r\right)}\right]_{exp}$

Here $\Phi_{AC_x}(\vec{r}_s,\vec{r}_r)$, represents the excitation fluence detected at a fixed reference position, $\vec{r}_r$, in response to excitation from the source position, $\vec{r}_s$.

Even though the excitation fluence, $\Phi_{AC_x}(\vec{r}_s,\vec{r}_r)$, is updated during the iteration, the Jacobian matrix can be directly calculated from Equation 33.71, and the change in $\Phi_x(\vec{r}_s,\vec{r}_r)$ is small compared to the change in $\Phi_m(\vec{r}_s,\vec{r}_r)$. As the source term of the emission-diffusion equation is modified after each iteration

(Equation 33.57), changes in the emission fluence are greater than those in the excitation fluence. The same consequence can be inferred from the integral equation (Equation 33.71). Moreover, the phase of the emission fluence is greater than that of the excitation fluence, and the normalized fluence, $\Phi_m/\Phi_x$, maintains a high phase contrast.

Due to noise and the ill condition of the Jacobian matrix for inverting systems of equations, updating can be accomplished using Newton's method[67] with Marquardt–Levenberg parameters λ:

$$(\mathbf{J}^T\mathbf{J}+\lambda\mathbf{I})\Delta X = \mathbf{J}^T\left[\left[\frac{\Phi_{AC_m}(\vec{r}_s,\vec{r}_d)}{\Phi_{AC_x}(\vec{r}_s,\vec{r}_r)}\right]^{meas} - \left[\frac{\Phi_{AC_m}(\vec{r}_s,\vec{r}_d)}{\Phi_{AC_x}(\vec{r}_s,\vec{r}_r)}\right]^{comp}\right] \tag{33.72}$$

Using excitation referencing at a single reference point, Lee and Sevick-Muraca[57] reconstructed an 8 × 4 × 8 cm³ phantom containing a 1 × 1 × 1 cm³ target with 100-fold greater ICG concentration by using 8 excitation sources, 24 detection fibers for collecting excitation light, and 2 reference detection fibers (one on either side of the reflectance and transillumination measurements) for collecting excitation light. Figure 33.19A is the original map containing 2D slices that demark the heterogeneity placement, while Figure 33.19B is the 3D reconstructed image.

While the results in Figure 33.19 represent reconstructions based on emission FDPM measurements relative to excitation FDPM measurements at a fixed reference position, Ntziachristos and Weissleder[56] successfully reconstructed two fluorescent targets in a 2.5-cm-diameter, 2.5-cm-long cylindrical vessel containing ICG and Cy5.5 dyes. In addition, they used CW emission measurements referenced to excitation measurements at each of the 36 detector fibers as a result of point excitation at 24 source fibers. The high density of measurements for reconstruction of the small simulated tissue volume is troublesome for validity of the diffusion equation used in the forward solver but is similar to that demonstrated by Yang et al.,[51] who reconstructed ICG and DTTCI in similarly sized phantoms and mice, presumably from absolute FDPM measurements at the emission wavelength alone.

The studies of the reconstruction presented above assumed that the absorption and scattering properties were known *a priori*. However, using a differential approach coupled with Bayesian reconstruction approaches (see below), Eppstein et al.[68] were able to demonstrate the insensitivity of reconstructions to changes in endogenous optical properties. Using a synthetic 256-cm³ volume containing 0.125-cm³ targets with 10:1 contrast in absorption due to fluorophore and surrounded on four sides by 68 sources and 408 detection fibers, Eppstein was able to show that when the absorption cross section at the excitation wavelength, $\mu_{a_{xi}}$, varied as much as 90% and was unmodeled, while the scattering coefficient, $\mu_{s_{xi}}$, varied 10% or less and was also unmodeled, the impact on the reconstruction was minimal or negligible. Recently, Roy et al.[69] produced similar results when they demonstrated unmodeled variations in all endogenous optical properties by as much as 50%, which did not impact reconstructions when emission FDPM measurements were individually self-referenced to excitation FDPM measurements, as was done with the CW measurements of Ntziachristos and Weissleder.[56] While it appears promising that fluorescence-enhanced optical tomography can be accomplished without much *a priori* information about the endogenous optical properties, these results nonetheless pertain to synthetic studies and must be conducted on actual tissues of substantive and clinically relevant volumes for validation.

### 33.5.3 Differential Formulation of the Inverse Problem

A second approach to the full-inverse-imaging problem may be the differential formulation, but this time it is rewritten for measurement $Z(\vec{r}_d,\vec{r}_s)$, whether absolute, relative to a reference measurement at the emission or excitation wavelength, or self-referenced relative to the excitation wavelength at each detector position, $\vec{r}_d$. We term this approach the differential formulation because a small change in the predicted measurements is directly expressed in terms of a small change in the optical properties, $\Delta X$, using a Jacobian matrix, $\mathbf{J}$, $\partial(\Delta Z_i)/\partial X_j$. Consider a number of detectors, $M$; the error function is then

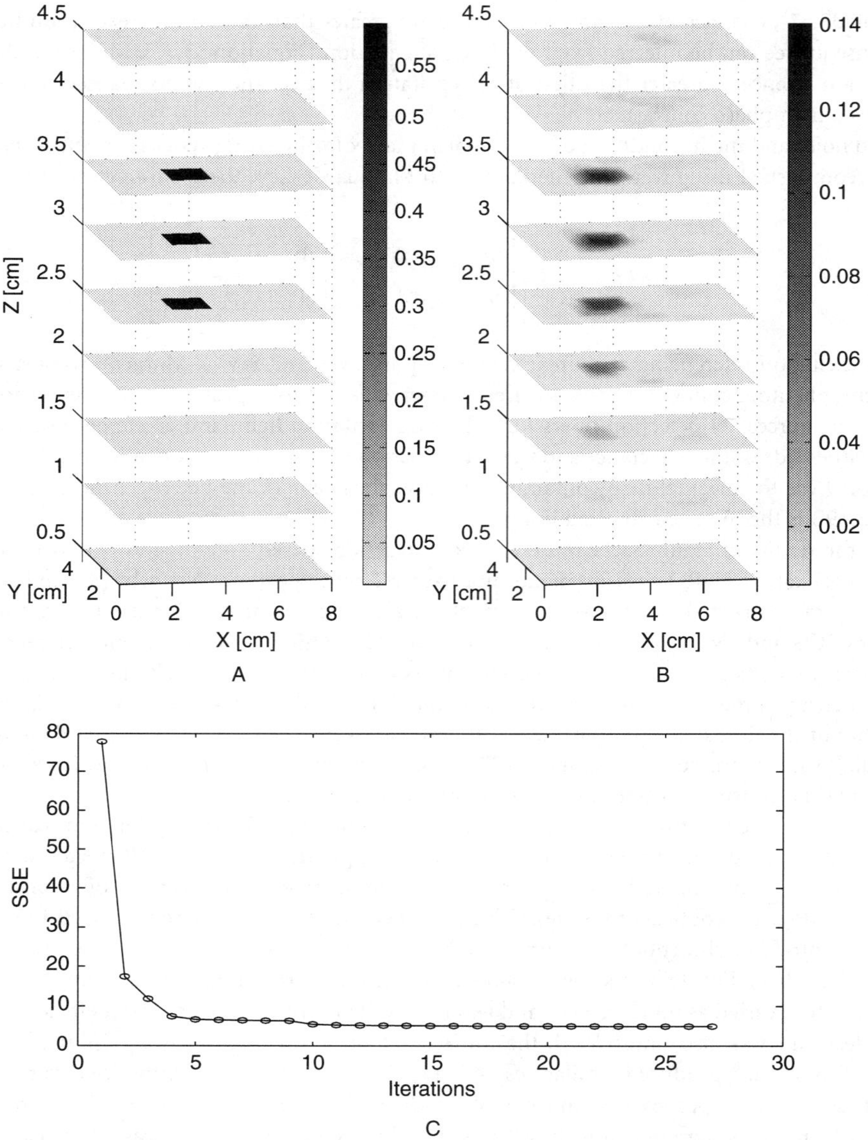


**FIGURE 33.19** The reconstruction of $\mu_{a\rightarrow m}$ using the excitation wave as a reference using the integral approach and Marquardt–Levenberg reconstruction. The image was required after 27 iterations with regularization parameters for $I_{AC}$ ratio (ACR), $\lambda_{AC} = 1.0$, and for relative phase shift (RPS), $\lambda_{\theta} = 0.02$. (A) Optical property maps of true $\mu_{a\rightarrow m}$ distribution and (B) reconstructed $\mu_{a\rightarrow m}$ distribution. Peak values of $\mu_{a\rightarrow m}$ reached 0.1205 cm$^{-1}$. (C) Iteration vs. SSE.

defined as the sum of the square of errors between the measured and calculated values at detector $i = 1\ldots M$:

$$F(\mathbf{X}) = \sum_{i=1}^{M}\left[(Z_i)^{(m)} - (Z_i)^{(c)}\right]^2 = \sum_{i=1}^{M}[f_i(X)]^2 \tag{33.73}$$

We refer to each $f_i$ as a residual and the gradients of the error function with respect to the property, $X$:

$$\nabla F\ (\mathbf{X}) = 2\mathbf{J}^T f(X) \tag{33.74}$$

$$\nabla^2 F(\mathbf{X}) = 2\left[\mathbf{J}^T\mathbf{J} + \sum_{i=1}^{M} f_i(X)\nabla^2 f_i(X)\right] \tag{33.75}$$

Consider the Taylor's expansion of function $F$ around a small perturbation of optical properties, $\Delta X$:

$$F(\mathbf{X}+\Delta\mathbf{X}) = F(\mathbf{X}) + \nabla F(\mathbf{X})\cdot\Delta\mathbf{X} + \frac{1}{2}\Delta\mathbf{X}^T\cdot\nabla^2 F(\mathbf{X})\cdot\Delta(\mathbf{X}) \tag{33.76}$$

which can be expressed as:

$$F(\mathbf{X}+\Delta\mathbf{X}) = F(\mathbf{X}) + 2\mathbf{J}^T f(\mathbf{X})\cdot\Delta\mathbf{X} + 2\cdot\Delta\mathbf{X}^T\left[\mathbf{J}^T\mathbf{J} + \sum_{i=1}^{M} f_i(X)\nabla^2 f_i(X)\right]\cdot\Delta\mathbf{X} \tag{33.77}$$

and the function to be minimized, $\Phi(\Delta X)$, can be explicitly written:

$$\Phi(\Delta\mathbf{X}) = F(\mathbf{X}+\Delta\mathbf{X}) - F(\mathbf{X}) = 2\mathbf{J}^T f(\mathbf{X})\cdot\Delta\mathbf{X} + 2\cdot\Delta\mathbf{X}^T\left[\mathbf{J}^T\mathbf{J} + \sum_{i=1}^{M} f_i(X)\nabla^2 f_i(X)\right]\cdot\Delta\mathbf{X} \tag{33.78}$$

For first-order Newton's methods, the term $2\cdot\Delta\mathbf{X}^T\left[\sum_{i=1}^{M} f_i(X)\nabla^2 f_i(X)\right]\cdot\Delta\mathbf{X}$ is neglected, and the Gauss–Newton method becomes one of minimizing:

$$\nabla\Phi(\Delta\mathbf{X}) \Rightarrow 0 = \mathbf{J}^T\mathbf{J}\cdot\Delta\mathbf{X} + \mathbf{J}^T f(\mathbf{X}) \tag{33.79}$$

$$\mathbf{J}^T\mathbf{J}\cdot\Delta\mathbf{X} = -\mathbf{J}^T f(\mathbf{X}) \tag{33.80}$$

The Levenberg–Marquardt method of optimization becomes[49]

$$\left[\mathbf{J}^T\mathbf{J} + \lambda\mathbf{I}\right]\cdot\Delta\mathbf{X} = -\mathbf{J}^T f(\mathbf{X}) \tag{33.81}$$

The gradient-based truncated Newton's method is based on retaining the second-order terms such that Equation 33.78 becomes[49]

$$\nabla\Phi(\Delta\mathbf{X}) \Rightarrow 0 = \mathbf{J}^T f(\mathbf{X}) + \left[\mathbf{J}^T\mathbf{J} + \sum_{i=1}^{M} f_i(X)\nabla^2 f_i(X)\right]\cdot\Delta\mathbf{X} \tag{33.82}$$

or alternatively,

$$\Phi'(\Delta\mathbf{X}) \Rightarrow 0 = \nabla F(\mathbf{X}) + \nabla^2 F(\mathbf{X})\cdot\Delta\mathbf{X} \tag{33.83}$$

Typically, the first-order Newton's methods were employed, with the exception of the work by Roy.[70] In the Newton's methods, it is assumed that $\Delta\Phi = \mathbf{J}\cdot\Delta\mathbf{X}$, and the solution is found using one of the several optimization approaches. The Jacobian matrix can be computed either directly from the stiffness matrices of the finite element formulation or simply, but more computationally time-consuming, from backward, forward, or central differencing approaches that compute the differences in the values of $Z(\vec{r}_d, \vec{r}_s)$, with small differences in the parameter to be updated, $X(\vec{r}_j)$. The Gauss–Newton and the Levenberg–Marquardt algorithms performed poorly in a large residual problem. Since the inverse is highly nonlinear and ill conditioned due to the error in measurement data, the residual at the solution will be large. It seems reasonable, therefore, to consider the truncated Newton's method.

For the truncated Newton's method, the additional computational cost of computing the Hessian [associated with $\nabla^2$ F(**X**)] is assisted by reverse automatic differentiation.[49,70] Using synthetic data, Roy has shown the feasibility of using the technique for 3D reconstruction of lifetime, $\tau$, and absorption-coefficient $\mu_{a_{x\to m}}$ changes in frustrum and slab geometries from synthetic data containing noise that mimics experimental data.[54]

### 33.5.4 Regularization and Other Approaches to Parameter Updating

In both the integral and differential formulations of the inverse problem, the tissue to be imaged must be mathematically discretized into a series of nodes or volume elements (voxels) in order to solve these inverse problems. The unknowns of the inverse problems are then comprised of the optical properties at each node or voxel. The final image resolution is naturally related to the density nodes or voxels. However, the dimensionality of the imaging problem is directly related to the number of nodes and can easily exceed 10,000 unknowns for a 3D image. In a problem of this scale, the calculation of Jacobian matrices and matrix inversions involved in updating the optical property map are computationally intensive and contribute to the long computing times required to reconstruct the image. The instability arises because the measurement noise in the data or errors associated with the validity of the diffusion approximation can result in large errors in the reconstructed image.

One of the greatest challenges associated with fluorescence-enhanced tomography is the propagation of error. In comparison with absorption imaging based on measurements of excitation light, fluorescence measurements have a reduced signal level and SNR. Lee and Sevick-Muraca[71] measured the SNR for single-pixel excitation and the emission-frequency domain at 100 MHz and found them to be 55 and 35 dB, respectively. In addition to the reduced signal, the noise floor of emission measurements can be expected to be elevated when excitation-light leakage constitutes an increased proportion of the detected signal. Consequently, for emission-tomography measurements, excitation-light leakage is crucial, and interference filters that attenuate excitation light four orders of magnitude (i.e., filters of OD 4) may be clearly insufficient. Excitation-light leakage will be a significant problem when emission measurements are conducted in tissue regions where the target is absent and fluorescent contrast agents are not activated. Unfortunately, this type of error is not present in synthetic studies and is undoubtedly underestimated in the vast proportion of tomography investigations to date.

#### 33.5.4.1 Regularization

Regularization is a mathematical tool used to stabilize the solution of the Newton's inverse problem and to make it more tolerant to measurement error. Regularization approaches will play an important role in the development of suitable algorithms for actual clinical screening. For example, when discretized, the differential and integral general formulations result in a set of linear Newton's equations generally denoted by $AY = Z$, where $Y$ is the unknown optical properties and $Z$ is the measurements. This system is commonly solved in the least-squares sense where the object function $Q = \|AY - Z\|^2 + \lambda\|Y\|^2$ is minimized and $\lambda$ is called the regularization parameter. The minimization of this function results in $Y = (A^TA + \lambda I)^{-1}A^TZ$. The regularization parameter is generally chosen either arbitrarily or by a Levenberg–Marquardt algorithm so that the object function is minimized.[72] Thus, the choice of regularization parameter is through *a priori* information and adds another degree of freedom to the inverse-problem solution. Finally, in a recent work, Pogue et al.[60] present a physically based rationale for empirically choosing a spatially varying regularization parameter to improve image reconstruction.

#### 33.5.4.2 Bayesian Regularization

Eppstein et al.[46,47,52,68] used actual measurement-error statistics to govern the choice of varying regularization parameters in their Kalman filter implementation in optical tomography. In their work, they developed a novel Bayesian reconstruction technique, called APPRIZE (automatic progressive parameter-reducing inverse zonation and estimation), specifically for groundwater problems and adapted them to fluorescence-enhanced optical tomography.[64–66] Unique components of the APPRIZE method are an approximate extended Kalman filter (AEKF), which employs measurement error and parameter uncertainty to

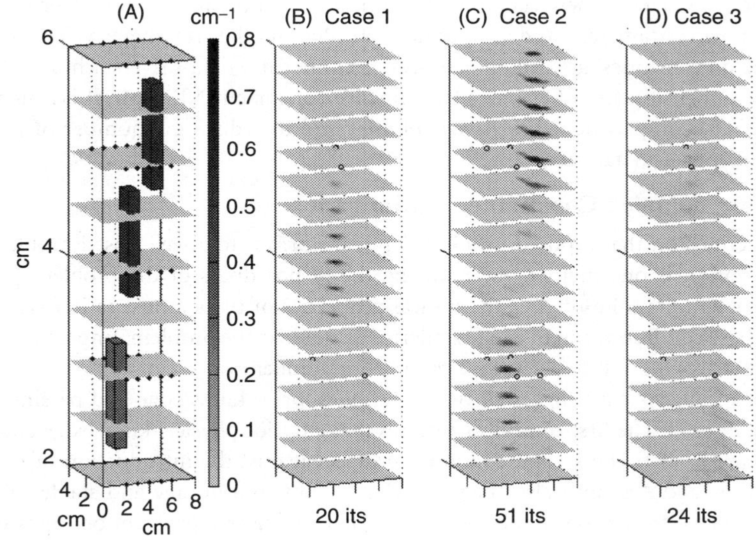


**FIGURE 33.20** Image reconstruction with APPRIZE. (A) The initial homogeneous estimate discretized onto the 9 × 17 × 17 grid used for the initial inversion iteration and shown with the true locations of the 3 heterogeneities and 50 detectors (small dots). (B) Case 1: the reconstructed absorption due to the middle fluorescing heterogeneity, interpolated onto the 17 × 33 × 33 grid used for prediction and shown with the locations of the four sources used (open circles). (C) Case 2: the reconstructed absorption due to the top and bottom fluorescing heterogeneities shown with the locations of the eight sources used (open circles). (D) Case 3: the reconstructed absorption of a homogeneous phantom shown with the locations of the four sources used (open circles). Although the phantoms and reconstructions were actually 8 cm in the vertical dimension, only the center four vertical centimeters are shown here. (From Eppstein, M.J. et al., *Proc. Natl. Acad. Sci. U.S.A.*, 99, 9619, 2002. With permission.)

regularize the inversion and compensate for spatial variability in SNR, and a unique approach to stabilizing and accelerating convergence called data-driven zonation (DDZ). Using the notation (ΔX, *f*(X)) as described in Section 33.5.3, the Newton's solution is formulated here as:[52]

$$\Delta \mathbf{X} = \left[ \left[ \mathbf{J}^T (\mathbf{Q}+\mathbf{R})^{-1} \mathbf{J} + \mathbf{P}_{xx}^{-1} \right]^{-1} \cdot \mathbf{J}^T (\mathbf{Q}+\mathbf{R})^{-1} \right] \cdot f(\mathbf{X}) \tag{33.84}$$

where Q is the system-noise covariance, which describes the inherent model mismatch between the forward model (the diffusion equation) and the actual physics of the problem; R is the covariance of the measurement error that is actually acquired in the measurement set; and $P_{xx}$ is the recursively updated error covariance of the parameters, X, and is estimated from the measurement error, *f*(X). The use of this spatially and dynamically variant covariance matrix results in the minimization of the variance of the estimated parameters, taking into account the measurement and system error.

The novel Bayesian minimum-variance reconstruction algorithm compensates for the spatial variability in SNR that must be expected to occur in actual NIR contrast-enhanced diagnostic medical imaging. Figure 33.20 illustrates the image reconstruction of 256-$cm^3$ tissue-mimicking phantoms containing none, one, or two 1-$cm^3$ heterogeneities with 50- to 100-fold greater concentration of ICG dye over background levels. The spatial-parameter estimate of absorption from the dye was reconstructed from only 160 to 296 surface reference measurements of emission light at 830 nm in response to incident 785-nm excitation light modulated at 100 MHz. Measurement error of acquired fluence at fluorescent emission wavelengths is shown to be highly variable.

Another important feature of the Bayesian APPRIZE algorithm is the use of DZZ. With DDZ, spatially adjacent voxels with similarly updated estimates are identified through cluster analysis and merged into larger stochastic parameter "zones" via random field union.[73] Thus, as the iterative process proceeds, the number of unknown parameters, *X*, decreases dramatically, and the size, shape, value, and covariance of the different "parameter zones" are simultaneously determined in a data-driven fashion. Other approaches to reducing the dimensionality of the problems involve concurrent NIR optical imaging with MRI[74–76] and ultrasound[77] to compartmentalize tissue volumes and to reduce the number of parameters to be recovered in the optical-image reconstruction.

#### 33.5.4.3 Simply Bounded Constrained Optimization

Imposing restrictions on the ill-posed problem can transform it to a well-posed problem, as discussed above. Regularization is one method for reducing the ill-posedness of the problem.[78] In the optical-tomography problem, its solution, i.e., the optical properties of tissue, must satisfy certain constraints, and imposing these conditions can in itself regularize or stabilize the problem. Imposing these constraints also explicitly restricts the solution sets and can restore uniqueness.

Provencher and Vogel[78] have suggested two techniques, prior knowledge and parsimony, for making the problem well posed. The first condition requires that all prior physical knowledge about the solution be included in the model. The second condition protects against the introduction of nonphysical phenomena. Tikhonov and Arsenin[79] also suggested that, to obtain a unique and stable solution from the data, supplementary information should be used so that the inverse problem becomes well posed. The basic principle of using *a priori* knowledge of the properties of the inverse problem is to restrict the space of possible solutions so that the data uniquely determine a stable solution.

Roy and Sevick-Muraca showed that the constrained optimization technique, which places simple bounds on a physical parameter to be estimated, might be more appropriate for solving the fluorescence-enhanced optical-tomography problem.[50] That is, a range of fluorescent optical properties is physically defined for the problem, and the recovered parameter, *X*, must always be positive. Specifically, Roy demonstrated the use of the bounding parameter, $\varepsilon$, both as a means of regularizing and accelerating convergence and as a means of setting the level of optical property contrast to be reconstructed using referenced emission measurements.[69] Here, the possible values of parameter estimates are stated to lie between an upper and lower bound. In the first pass of the iterative solution, the optical property map is recovered, and parameter estimates that lie within the upper and lower bounds plus and minus a small bounding parameter, $\varepsilon$, are recovered and held constant for the next iteration. Thus, the number of unknowns decreases with each iteration. Indeed, the value of the bounding parameter can be used to set the resolution and the performance of the tomographic image. For example, if the bounding parameter is large, then the tomographic image will "filter out" artifacts not associated with the target; but if the bounding parameter is small, the tomographic image may sensitively capture artifacts and heterogeneity that are not necessarily associated with the target. Figure 33.21 illustrates the reconstruction using the simply bounded truncated Newton's method, which shows that as the bounding parameter is increased, the recovered image becomes less sensitive to the background "noise." This approach may have significant application for increasing the target to background signals, an issue related to nuclear imaging that impairs tomographic reconstructions.

## 33.6 Fluorescent Contrast Agents for Optical Tomography[15]

Table 33.2 provides a chronological listing of studies reported in the literature over the past decade that involve a number of different fluorescent-contrast agents.[23,51,80–116] While the studies have progressed from using photodynamic agents; freely-diffusable agents, such as ICG; fluorochromes conjugated to monoclonal antibodies (MAb) and their fragments; small-peptide targeting agents similar to those employed in nuclear imaging; and, finally, activatable and "reporting agents." Unfortunately, the translation of these agents to human clinical studies has been limited. Furthermore, investigations have been largely confined to superficial or subcutaneous tumors, where the true advantages of NIR fluorescent agents, that is, deep-tissue penetration and optical tomography, cannot be aptly demonstrated. Nonetheless, the strategies for

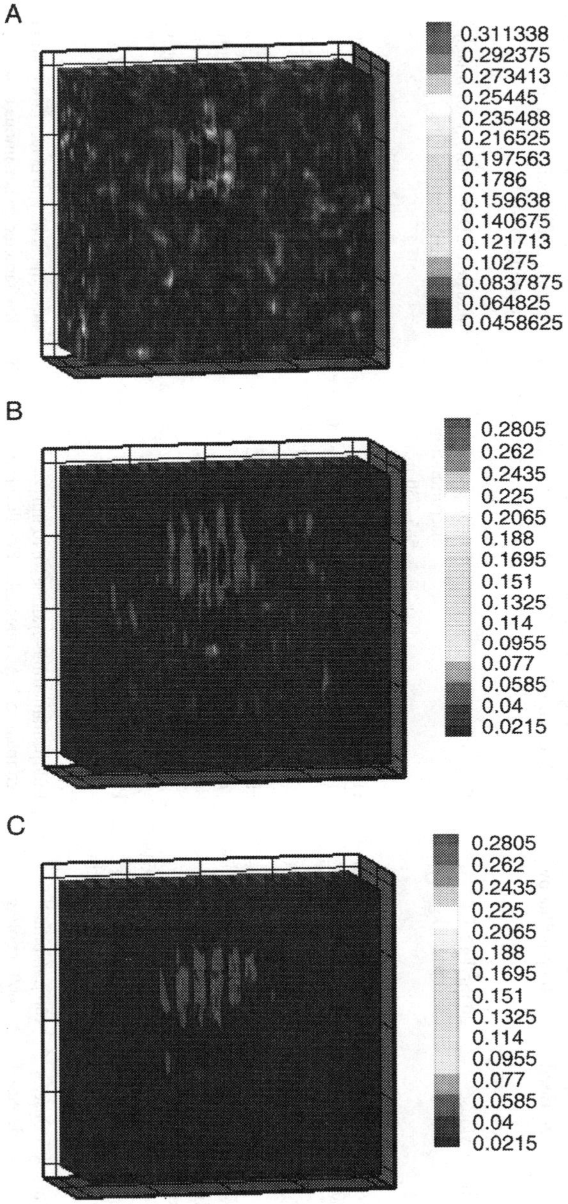


**FIGURE 33.21 (Color figure follows p. 28-30.)** Three-dimensional reconstruction from simply bound truncated Newton's method. (A) Actual distribution of fluorophore-absorption coefficient of background tissue variability of endogenous (50%) and exogenous (500%) properties. (B) Reconstructed fluorophore-absorption coefficient of background tissue variability of endogenous (50%) and exogenous (500%) properties using relative measurement of the emission fluence with respect to the excitation fluence at the same detector point, $\varepsilon = 0.0001$. (C) Reconstructed fluorophore absorption coefficient of background tissue variability of endogenous (50%) and exogenous (500%) properties using relative measurement of the emission fluence with respect to the excitation fluence at the same detector point. (From Roy, R., Godavarty, A., and Sevick-Muraca, E.M., *IEEE Trans. Med. Imaging*, in press. With permission.)

NIR fluorescent-contrast agents have been impressive and have included using simple blood-pooling agents to highlight hypervascularity, employing contrast provided by pharmacokinetic model-parameter estimates of uptake, and designing agents that specifically target membrane receptors of cells lining

**TABLE 33.2** Fluorescence-Enhanced Contrast Imaging: Literature of Agent Studies

| Ref. | Imaging System (incident fluence) | Animal Model | Dose | Contrast Agent | λ Excitation/Emission | Comments |
|---|---|---|---|---|---|---|
| Biolo et al., 1991[80] | Spectrograph for point detection of fluorescence following surface illumination | Mouse | 0.12mg/kg bw ~0.24 µmol/kg bw | ZnPc (phthalocyanine) in liposomes, spectral detection of fluorescence at λ | 600 nm | Provide measurements for assessing pharmacokinetics of PDT agent |
| Pelegrin et al., 1991[81] | Area illumination with area (133 mW/cm²) detection using photography | Mouse, deceased | 100 µg/animal ~600 pmol/kg bw | Fluorescein isothiocyanate (FITC) coupled to MAb | 488 nm Kodak Wratten filter #12 for excitation light rejection | Study the localization of dye targeted to human colon carcinoma in mice after coupling to MAb |
| Straight et al., 1991[82] | Interstitial illumination (20 mW) with area detection using CCD camera | Mouse | 20 mg/kg bw ~70 µmol/kg bw | Photofrin II | 514.5/spectral discrimination 585–730 nm | Validate CCD technology for imaging drug distribution in tumors |
| Folli et al., 1992[83] | Fiber through endoscope illumination (10 mW/cm²) with area detection using photography | Human | 0.1–0.28 mg/patient ~0.1 nmol to 0.28 nmol/patient | Fluorescein isothiocyanate (FITC) coupled to MAb | 488 nm Kodak Wratten filter #12 for excitation light rejection | Study immunophotodiagnosis in colon carcinoma patients |
| Cubeddu et al., 1993, 1997[84–85] | Area illumination (75 µW/cm²) with pulsed dye laser with gated CCD video camera | Mouse | 5–25 mg/kg bw ~17–87 µmol/kg bw, 1993<br>0.1 mg/kg ~0.35 µmol/kg bw, 1997 | HpD (hematopor-phyrin derivative) | 405 nm/>560 nm | Demonstrate the use of time-dependent measurements to identify HpD distinct from native fluorescence based upon long fluorescent lifetimes |
| Kohl et al., 1993[86] | Intensified CCD | Mouse | 0.2–1 mg/kg bw ~0.7–3.5 µmol/kg bw | Porphyrin-based photosensitizers | | Demonstrate the ability to image s.c. tumor |
| Folli et al., 1994[87] | Area illumination with area (13 mW/cm²) detection using photography | Mouse, deceased | 100 µg/animal ~600 pmol/kg bw[a] | Indopentamethine-cyanine coupled to MAb directed against squamous cell carcinoma | 640 nm Kodak Wratten filter #70 for excitation light rejection | Show the ability to detect indocyanine dye targeted to squamous cell carcinoma in the upper respiratory tract through MAb E48 without removing skin, as was necessary when fluorescein was employed |
| Haglund et al., 1994[88] | Area illumination (100-W tungsten-halogen bulb) and area detection with CCD camera | Rat | 1.0 mg/kg bw 1.3 µmol/kg bw | ICG | 780/830 | Distinguish rat gliomas from normal brain tissue through free-agent fluorophore imaging with CCD camera |

| | | | | | | |
|---|---|---|---|---|---|---|
| Mordon et al., 1994[89] | Area illumination (150W Xenon lamp, 2.5 mW/cm$^2$) with area detection using intensified CCD | Mouse | 5 mg/kg bw (~13 μmol) | 5,6-CF carboxyfluorescein (BCECF) | 465 nm/490 and 515 nm | Show the use of dual-wavelength measurements of a ratiometric dye to provide a 2D pH image of tumor tissues |
| Ballou et al., 1995[90] | Area illumination with area detection using intensified video camera or cooled CCD | Mouse | 10–100 μg/animal ~40 pmol to 6 nmol/animal | Cyanine fluorochromes coupled to MAb | 550–674 nm/565–694 nm | Demonstrate the use of tumor-targeting antibodies using Cy3.18, Cy5.18, and Cy5.5.18 cyanine fluorochromes |
| Devoisselle et al., 1995[91] | Area (50 mm$^2$) illumination (Xe lamp) with point detection using fiber optics and spectrograph | Mouse | 7.5 mg/kg bw ~10 μmol/kg bw | ICG emulsion | 720 nm/spectra discrimination of fluorescence | Demonstrate the use of emulsion preparation to alter the pharmacokinetics of ICG |
| Rokahr et al., 1995[92] | $N_2$ laser pulsed through fiber and detected with fiber to spectrometer | Human undergoing urinary-bladder cytoscopy | 50 mg/patient (ALA induces fluorescence) | Protoporphyrin IX | 337 and 405/380 through 685 spectral discrimination | Discriminate malignant and normal bladder tissue with ALA-induced protoporphyrin imaging |
| Haglund et al., 1996[93] | Area illumination with photography lights and area detection with CCD camera | Human (open brain) | 1 mg/kg bw ~1.3 mmol/kg bw | ICG | 790/805 nm | Study detection of human glioma with ICG imaging |
| Sakatani et al., 1997[94] | Cooled CCD camera, 100-mW laser diode | Rat | 554 pmol/rat | ICG-lipoprotein | 790 nm/840 nm | Conduct cerebrospinal imaging with ICG bound to lipoprotein, injected intracranially |
| Neri, 1997[95] | Method similar to Folli et al. (1994); 100-W tungsten lamp for area illumination with detection via an 8-bit CCD in a light-tight box | Mouse | 100 μl/mouse of a concentrated antibody solution of 1 mg/ml with dye:MAb ratio of 1:1 | Fragments of human antibodies directed against oncofetal fibronectin (B-FN) and labeled with Cy7 | 673–748 nm/765–855 nm | Demonstrate the use of B-FN targeting for providing diagnostic imaging and therapy of cancer targeting angiogenic vessels |
| Ballou et al., 1998[96] | Area illumination and area detection via CCD | Mouse | 50 μg/animal MAb:dye (1:2) ~600pmol/animal | Cy3, Cy5, Cy5.5, and Cy7 labeled antibodies against human nucleolin and stage-specific embryonic antigen-1 | | Demonstrate the ability to penetrate more deeply with Cy7 dye |

**TABLE 33.2** Fluorescence-Enhanced Contrast Imaging: Literature of Agent Studies (continued)

| Ref. | Imaging System (incident fluence) | Animal Model | Dose | Contrast Agent | λ Excitation/Emission | Comments |
|---|---|---|---|---|---|---|
| Eker et al., 1999[97] | Fiber for excitation and collection in a colonoscope for detection via a spectrometer | Human | 5mg/kg bw ~30 nmol/kg bw | Protoporphyrin IX, as a metabolized product of ALA (photosensitizer) | 337, 405, 436 nm excitation | Demonstrate the use of ALA as a contrast agent for detecting adenomatous polyps of the colon and showed promise for distinguishing adenomatous from hyperplastic polyps |
| Reynolds et al., 1999[98] | Area illumination with laser diode (1 mW/cm$^2$) and area detection using intensified FDPM CCD system | Canine | 1.0 mg/kg bw ~1.3 μmol/kg bw | ICG | 780 nm/830 nm | Demonstrate the ability to detect spontaneous disease of the canine mammary chain as well as reactive lymph nodes |
| Becker et al., 1999[99] | Area illumination and area detection with CCD and MRI | Mouse | 2 μmol/kg bw (1:2.4 or 2 for transferrin or HSA) | Indotricarbocyanine and ultrasmall superparamagnetic iron oxide particles coupled to transferrin or human serum albumin (HSA) | | Demonstrate targeting of tumors that express the transferring receptor using an optical agent as well as an MRI agent |
| Weissleder et al., 1999;[23] Mahmood et al., 1999[100] | Area illumination and area detection using CCD in a light-tight chamber; illumination with 150-W halogen lamp with interference filters, 10–100 μW/cm$^2$ | Mouse | 10 μmol/animal (92 MEG, 11 dye molecules); 250 pmol/animal | Cy5.5. loaded onto a polylysine and methoxypolyethylene glycol polymer backbone with cathepsin B and H-cleavage sites | 610–650 nm/>700 nm | Demonstrate that tumor proteases can be used as molecular targets SNR for 30-sec exposure 173 for 200 pmol in phantom |
| Becker et al., 2000[101] | Area illumination and area detection with CCD | Mouse | 2 μmol/kg bw | Transferrin and human serum albumin coupled with indotricarbocyanine dye | 740 nm/780–900 nm | Demonstrate targeting to the tumors expressing transferring receptor |

| | | | | | | |
|---|---|---|---|---|---|---|
| Gurfinkel et al., 2000[102] | Area illumination (1.98 and 5.5 mW/cm$^2$) and area detection using intensified FDPM CCD system | Canine | 1.1 and 1.0 mg/kg bw ~1.3 μmol/kg bw | ICG and carotene-modified PDT agent (HPPH) conjugated with carotene moiety for reduction of phototoxicity | 780 nm/830 nm (ICG) 660 nm/710 nm (HPPH-car) | Demonstrate the use of temporal AC measurements to image pharmacokinetic parameters to discern diseased tissues |
| Licha et al., 2000[103] | Single-point detection and point illumination (5 mW) using FDPM | Rat | 0.5 μmol/kg bw | Derivatives of ICG (unclear whether fluorescence was detected *in vivo*) | 750 and 786 nm excitation | Provide measurements of absorption at the excitation wavelength as a function of time to provide pharmacokinetic evaluation of ICG and its hydrophilic derivatives |
| Yang et al., 2000;[104] Hoffman et. al., 2001[105] | Area illumination and detection using CCD camera | Mouse | | GFP expressed *in vivo* | | Demonstrate visualization of tumors and tumor metastasis by whole-body fluorescence imaging |
| Ntziachristos et al., 2000[74] | Fiber bundle to PMT using time-domain photon migration, point illumination and point detection | Breast | 0.25 mg/kg bw ~0.32 μmol/kg bw | ICG | | Fluorescence was not used, but absorption provided contrast that was validated by simultaneous MRI images obtained with gadolinium contrast |
| Bugaj et al., 2001;[106] S. Achilefu et al., 2000, 2001[107–109] | Area illumination (40 mW) and area detection using CCD camera | Rat | 5.2–6.0 mg/kg bw ~6.7–7.7 μmol/kg bw | ICG, ICG small-peptide conjugates cytate and cybesin | 780 nm/830 nm | Targeting to rat tumor lines expressing the somatostatin and bombesin receptors |
| Becker et al., 2001[110] | Area illumination and detection using CCD camera | Mouse | 0.02 μmol/kg bw | Peptide-cyanine dye conjugate, indodicarbocyanine (IDCC), and indotricarbocyanine (ITCC) conjugated to octreotate, an analog of somatostatin | 740 nm/780–900 nm | Targeting to mouse tumor lines expressing the somatostatin receptors |

**TABLE 33.2** Fluorescence-Enhanced Contrast Imaging: Literature of Agent Studies (continued)

| Ref. | Imaging System (incident fluence) | Animal Model | Dose | Contrast Agent | λ Excitation/Emission | Comments |
|---|---|---|---|---|---|---|
| Bremer et al., 2001[111] | Area illumination and area detection using CCD in a light-tight chamber; illumination with 150-W halogen lamp with interference filters, 10–100 μW/cm² | Mouse | 167 pmol/animal, i.v. | Polylysine polymer coupled with mMP-2 peptide substrates holding Cy5.5 | 610–650 nm/>700 nm | Measure matrix metalloproteinase (MMP) activity *in vivo* for directing the therapeutic use of proteinase inhibitors |
| Ebert et al., 2001[112] | Area illumination with pulsed laser and detection using CCD camera (ambient light rejection) | Rat | 2 μmol/kg bw, i.v. | SIDAG (hydrophilic derivative of cyanine dye), 1–1′-*bis*-(4-sulfbutyl) indotricarbocyanine 5,5′-dicarboxylic acid diglucamide monosodium salt; Nd:YAG | 740 nm, 3 ns FWHM, 50 Hz/750–800 nm | Demonstrated localization of tumor and presented phantom data using FDPM with contrast ratios of 6:1 |
| Finlay et al., 2001[113] | Point illumination with fiber probe, point detection with fibers directed to a spectrograph and CCD | Rat | 200 mg/kg bw (ALA injected) | ALA-induced porphyrin | 514 nm/676 nm emission | Photobleaching kinetics of ALA-induced protoporphyrin measured |
| Rice et al., 2001[114] | Area detection of light-emitting probes with CCD | Mouse | Bioluminescence of fluorescent proteins | Firefly luciferase | >600 nm emission | Imaging light-emitting probes |

| | | | | | | |
|---|---|---|---|---|---|---|
| Soukos et al., 2001[115] | Area illumination using pumped dye laser (15 mW/cm$^2$) and detection using room temperature CCD camera | DMBA-induced tumor in hamster cheek pouch | 670 μg/animal ~3.3nmol/kg bw[a] | Anti-EGFR MAb (C225) coupled to Cy5.5 (1:2.1) IgG-Cy5.5 (1:2.3) | 670 nm/>700 nm | Demonstrate that the targeted MAb–dye complex could be used to provide immunophotodiagnostic information, thereby guiding therapeutic intervention |
| Yang et al., 2001[51] | Single-pixel FDPM using point source and point detector | Rat | 1.5 mg/kg bw ~2 μmol/kg bw | ICG, DTTCI | | Work toward demonstration of fluorescence imaging *in vivo* |
| Zaheer et al., 2001[116] | Area illumination (18 mW/cm$^2$) | Mice (hairless) | 100 nmol/kg (i.v) | (Indocyanine) IR Dye 78 conjugated to pamidronate with hydroxyapatite binding properties | 771 nm/796 nm | Assess osteoblastic activity for skeletal development, osteoblastic metastasis, and coronary atherosclerosis |

[a] Molecular weight of proteins are estimated on the order of $10^6$ g/mol.

neovasculatures as well as the neoplastic cells that the vasculature feeds. Strategies that focus on lysosomal activity and enzyme cleavage for fluorochrome activation and mediation of fluorescence decay as well as for fluorochrome accumulation specific to cancer cells have also been demonstrated. Table 33.1 outlines these fluorochromes used in *in vivo* studies. When available, the chronological listing also notes the excitation and emission wavelengths used, the incident illumination, the measurement geometry, and the type as well as number of fluorochrome molecules used to detect a signal. The fluorophores are broadly classified into PDT agents, nontargeting blood-pooling agents, agents that "report" or sense, targeting agents based on immunodiagnostics and small-peptide conjugation, and activatable agents.

## 33.6.1 Photodynamic Therapy Agents

Starting with the area of photodynamic imaging, the early studies relating to fluorescence-enhanced imaging date back to 1991 and focused on evaluating the spatial distribution as well as the pharmacokinetics of PDT agents for dosimetry purposes. Measurements were typically conducted in tumor-bearing mice with total agent administration between 0.1 and 90 μM/kg bw using area illumination and CCD camera detection. PDT agents are likely candidates for fluorescent-enhanced imaging, more than likely due to their existing Food and Drug Administration investigational-new-drug (IND) applications for therapeutic use. The ability to obtain an IND for a diagnostic agent previously approved for therapeutic use enhances the opportunity for fluorescence-enhanced optical imaging. Yet despite the attractiveness for their current and pending INDs, photodynamic agents do not possess the excitation and emission spectra favorable for fluorescent contrast agents for imaging deep tissues. First, to maximize penetration depth into tissues, excitation must be between 750 and 800 nm, and the Stokes shift must be significant (~50 nm) to enable discrimination of the small fluorescent component from the overwhelming large component of excitation light. An insufficient Stokes shift of 10 nm or less complicates the process of rejecting multiply scattered light as the efficiency of filters cannot be guaranteed. Next, the fluorochrome for systemic administration should not experience a net lifetime or long-lived decay kinetics that exceed the photon time of flight, as described in Section 33.2.3. For these reasons, the usefulness of PDT agents for contrast (and for therapy) is largely limited to epithelial linings of accessible tissues and not for deep tissues. However, one notable use of a PDT agent for contrast-enhanced surface imaging involves the use of its metabolic by product. Using exogenous δ aminolevulinic acid (ALA) for natural production of protoporphryin IX, Ecker et al.[97] showed the ability to detect adenomatous polyps of the colon in humans and provided evidence to suggest the ability to discriminate between hyperplastic and adenomatous polyps. While excitation at 337, 405, and 436 nm does not classify this agent as an NIR probe for deep-tissue penetration, this study is nonetheless significant in that it employs a natural "reporting" mechanism in which the nonfluorescent ALA is hypermetabolized in diseased tissue to the fluorescent porphyrin form.

## 33.6.2 Nontargeting Blood-Pooling Agents

ICG, with its 778-/830-nm excitation/emission maxima was an early contrast agent choice used as a blood-pooling agent for assessing hypervascularity and "leaky" angiogenic vessels of high permeability. While many advances in dye development have accelerated within the past 2 years, the majority of studies investigating NIR fluorescent-contrast agents have been limited to ICG, a compound with FDA approval for systemic administration for investigating hepatic function[117] and retinal angiography.[118] ICG is excited at 780 nm and emits at 830 nm. It has an extinction coefficient of 130,000 $M^{-1}cm^{-1}$, a fluorescent lifetime of 0.56 ns, and a quantum efficiency of 0.016 for the 780-/830-nm excitation/emission wavelengths in water.[35] These values are not necessarily what will be observed *in vivo*.

When dissolved in blood, ICG binds to proteins such as albumin and lipoproteins. The absorption maximum shifts up to 805 nm, but the wavelength of maximum fluorescence is stable near 830 nm, and the fluorescent intensity depends on its concentration.[119,120] ICG is a nonspecific agent and is cleared rapidly from the blood, but it tends to collect in regions of dense vascularity through extravasation.

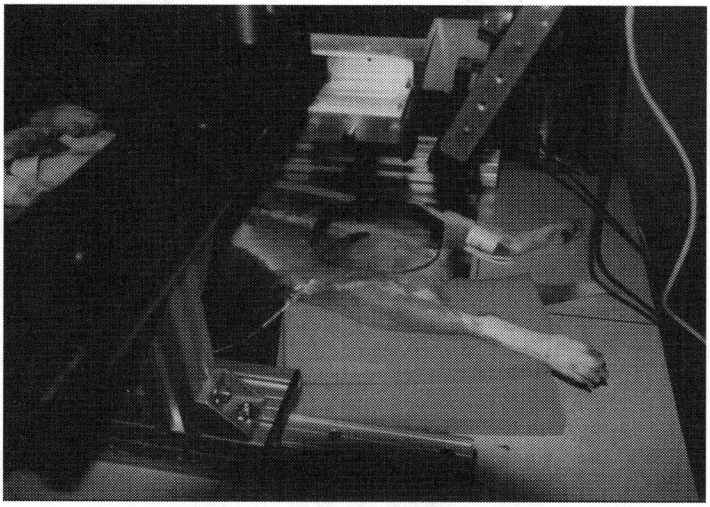

**FIGURE 33.22** Use of an incident-expanded beam on the mammary chain of the canine to excite systemically administered fluorophore and to collect the emission of generated light from the tissue surface. (From Hawrysz, D.J. and Sevick-Muraca, E.M., *Neoplasia*, 2(5), 388, 2000. With permission.)

Devoisselle et al.[91] demonstrated the use of ICG in an emulsion preparation at an administration of 10 μmol/kg bw in a tumor-bearing rat to measure its prolonged pharmacokinetics, while Reynolds et al.[98] used free ICG (1.3 μmol/kg bw) as a fluorescent agent in canines to image spontaneous mammary disease on a veterinary-outpatient basis.

In a study by Reynolds et al.,[98] frequency-domain approaches were employed whereby the canine mammary chain area was illuminated with intensity-modulated light, and the resulting amplitude of the generated fluorescent light that propagated to the surface was imaged by a gain-modulated image-intensified camera. The approach enabled rejection of room light and provided the first demonstration of fluorescence-enhanced imaging in a spontaneous tumor as in a large animal (Figure 33.22). Since the canine is the only species other than the human to naturally encounter mammary and prostate cancer,[121] this is an excellent animal model in which to assess the potential of detecting diseased tissue via a contrast agent. However, penetration depths are nonetheless limited to 0.5 to 2 cm, still not meeting the deep imaging potential of fluorescence-enhanced optical imaging.

In measurements of the canine mammary chain, also by Reynolds et al., a homodyned, gain-modulated image intensifier was used as described in Section 33.3.2. Excitation was accomplished by illuminating the tissue surface with a 4-cm-diameter expanded beam of a 20-mW, 780-nm laser, which was modulated at 100 MHz. Figure 33.23 shows the DC, amplitude, phase, and modulation ratio ($I_{AC}/I_{DC}$) of an 830-nm wavelength emitted from the left fourth mammary gland with a palpable 1.2-cm (longitudinal) by 0.5-cm (axial) papillary adenoma located approximately 1 cm deep within the mammary tissue. The image was acquired 23 min following i.v. injection of l mg/kg ICG. The diseased region is clearly shown in the raw, unprocessed DC, amplitude, phase, and modulation images.

The use of ICG for optical tomography has already been identified by the Chance group at the University of Pennsylvania. In a combined time-domain and MRI-imaging study of 11 patients, Ntziachristos et al.[74] administered 0.2 mg/kg ICG i.v. and conducted measurements in response to pulsed excitation at 780 nm. Their time-domain system involved pulsed laser diodes at 780 and 830 nm multiplexed into 24 source fibers and collected at 8 detection points.[122] Using the MRI images to validate their integral inversion results, they were able to reconstruct images of an infiltrating ductal carcinoma due to the enhanced signature from the vascular blood pooling of ICG. Unfortunately, fluorescence signals were not acquired, possibly due to the low SNR available with TDPM measurements, and the images were reconstructed from signals at the incident wavelength.

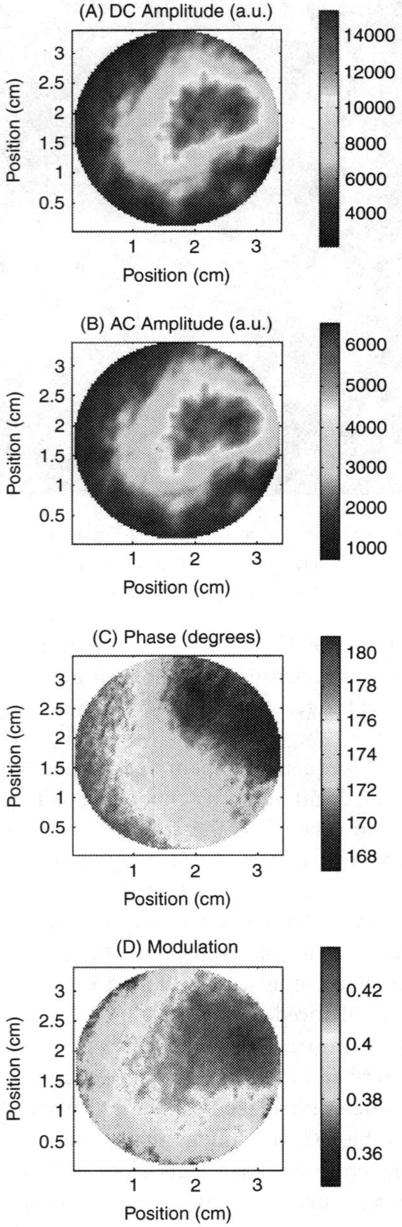


**FIGURE 33.23 (Color figure follows p. 28-30.)** The 128 × 128 pixel-based imaging of 830-nm fluorescence of (A) CW $I_{DC}$, (B) amplitude $I_{AC}$, (C) phase delay, and (D) modulation ratio of the detected fluorescence generated from the area cranial of the left fourth mammary gland of a canine. Illumination was accomplished with an expanded 780-nm laser diode. Modulation frequency was 100 MHz. (From Reynolds, J.S. et al., *Photochem. Photobiol.*, 70, 87, 1999. With permission.)

Later, using the modulated ICCD system Gurfinkel et al.[102] used FDPM measurements with ICG as a blood-pooling agent as well as with a photodynamic agent, carotene-conjugated 2-devinyl-2-(1-hexyloxyethyl) pyropheophorbide (HPPH-car), to provide the difference in pharmacokinetics. The time course of images was clearly able to discriminate between the nonselective uptake of the ICG blood-pooling agent, whose contrast was due mainly to the density of the microvasculature associated with the disease,

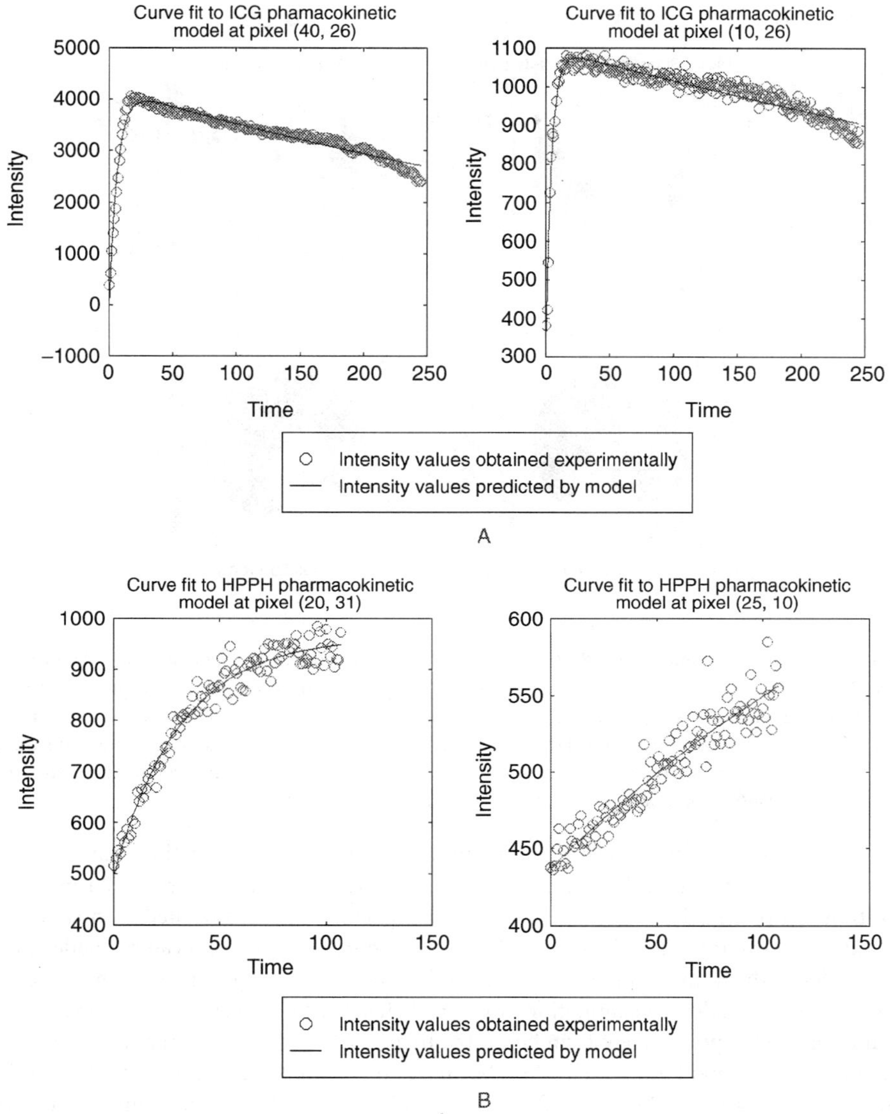


**FIGURE 33.24** AC fluorescent intensity as a function of time illustrating typical curve fits using (A) the ICG pharmacokinetic model and (B) the HPPH-car pharmacokinetic model. The symbols denote actual measurements, while the solid curve denotes the model fit. (From Gurfinkel, M. et al., *Photochem. Photobiol.*, 72, 94, 2000. With permission.)

and the specific uptake of the HPPH agent, whose uptake is hypothesized to be mediated with the enhanced overexpression of low-density lipoprotein (LDL) receptors on the surface of cancer cells and the association of HPPH and LDL in the blood compartment. Figure 33.24 represents the values of the AC intensity as a function of time at a single point in the area detection corresponding to the s.c. tumor following i.v. injection of ICG as well as HPPH-car. Upon fitting the time course of AC-intensity measurements with pharmacokinetic models, a map of uptake parameters shown in Figure 33.25 demonstrates the ability to enhance optical contrast based on pharmacokinetics, as is currently done in MRI. In an effort to tune the pharmacokinetics of cyanine dyes by changing the level of hydrophobicity/philicity, Licha et al.[103] and Ebert et al.[112] showed that a hydrophilic derivative of cyanine dyes could enhance

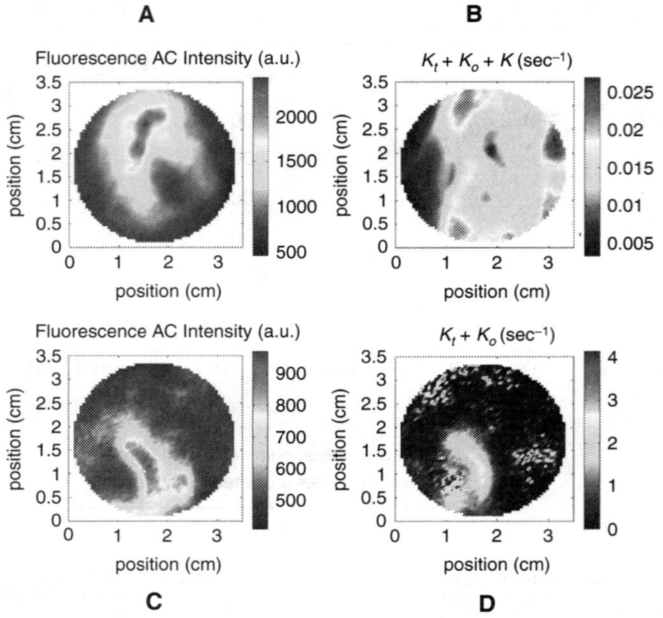


**FIGURE 33.25 (Color figure follows p. 28-30.)** (A) Fluorescence $I_{AC}$ intensity map from ICG delineating diseased tissue and (B) map of pharmacokinetic uptake parameters obtained from fitting the time sequences of fluorescence-intensity images showing no specific uptake of ICG in diseased tissue. (C) Fluorescence AC intensity map from HPPH-car delineating diseased tissue and (D) map of pharmacokinetic uptake parameters obtained from fitting the time sequences of fluorescence-intensity images showing specific uptake of HPPH-car in diseased tissue. (From Gurfinkel, M. et al., *Photochem. Photobiol.*, 72, 94, 2000. With permission.)

uptake and be detected using single-point illumination and detection at the excitation wavelength (0.5 µmol/kg bw) and area illumination and detection at the emission wavelength (2 µmol/kg bw).

In addition to using ICG as a means for assessing hypervascularity associated with cancer, ICG may also be used to assess lymph flow. Figure 33.26 is an *in vivo* ICCD image of a canine made cranial to the nipple of the left fifth gland 30 min after ICG injection. While the imaged area was not associated with a palpable nodule, pathologic examination confirmed that the fluorescence was attributed to a blood vessel that bifurcated approximately 1 cm below the tissue surface in an area cranial to a regional lymph node. Figure 33.27 represents the *in vivo* FDPM images of the fluorescence generated from the area of the right fifth mammary gland 43 min after injection of the ICG. Pathologic examination showed that the fluorescent source in this image corresponded to the regional lymph node.

The ability to detect fluorescence signals originating from regional lymph nodes suggests that FDPM fluorescence imaging coupled with improved fluorescent dyes could provide a valuable diagnostic method for assessing regional lymph-node status in breast cancer patients. Lymph-node status in breast cancer patients can be a powerful predictor of recurrence and survival, and the number of lymph nodes with metastases provides crucial prognostic information regarding the choice of adjuvant therapy.[123] Currently, lymph-node involvement is assessed by dissection and subsequent pathologic examination, but researchers are investigating the use of other diagnostic modalities including MRI, x-ray-computed tomography, and sonography.[123] More recently, nuclear imaging of a technetium-99 sulfur colloid injected into the tissue area of a known breast tumor has been used to identify the sentinel lymph nodes. With the simultaneous or sequential injection of a blue dye to visually aid in its location, the sentinel lymph node can then be surgically removed.[123,124] Moreover, with NIR fluorescent agents, sentinel-lymph-node mapping could possibly be achieved without the use of the radionucleotide or with the introduction of a second dye to aid in surgical incision. Furthermore, the development of peptide-, protein-, or antibody-conjugated fluorescent dye (see below) makes possible nonsurgical, optical diagnosis of nodal involvement.

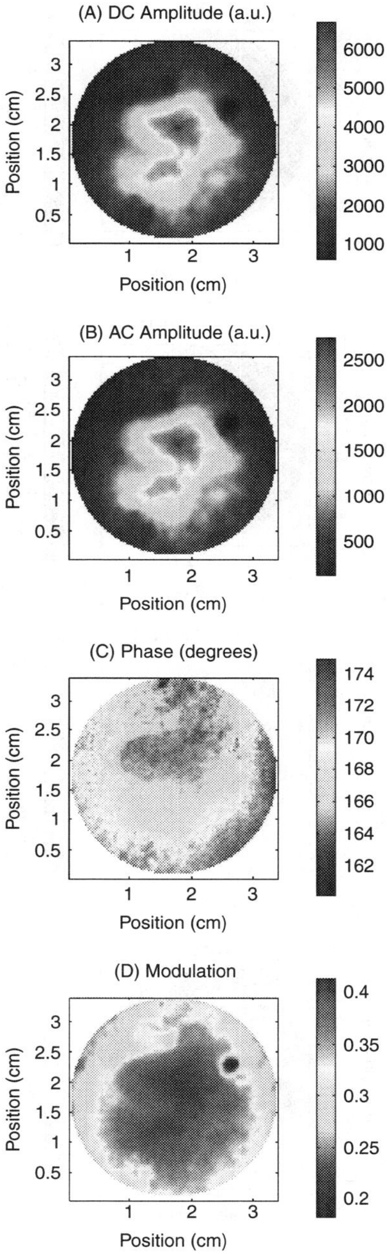


**FIGURE 33.26 (Color figure follows p. 28-30.)** The 128 × 128 pixel-based imaging of 830-nm fluorescence of (A) CW $I_{DC}$, (B) amplitude $I_{AC}$, (C) phase delay, and (D) modulatioOn ratio of the detected fluorescence generated from the area cranial of the left fifth mammary gland of a canine. Illumination was accomplished with an expanded 780-nm laser diode. Modulation frequency was 100 MHz. (From Reynolds, J.S. et al., *Photochem. Photobiol.*, 70, 87, 1999. With permission.)

Using intracranial injection of ICG bound to lipoprotein, Sakatani et al.[94] also showed the use of ICG in mapping the cerebrospinal fluid pathways in rats. Area illumination using a 100-mW laser diode and area detection with a cooled CCD camera was sufficient to detect meaningful images of fluid pathways following injection of 554 pmol ICG.

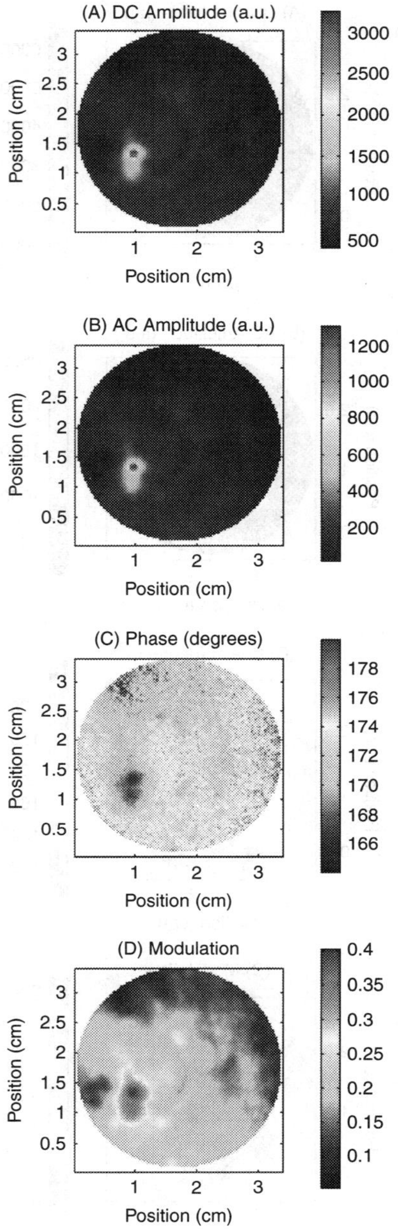


**FIGURE 33.27 (Color figure follows p. 28-30.)** The 128 × 128 pixel-based imaging of 830-nm fluorescence of (A) CW $I_{DC}$, (B) amplitude $I_{AC}$, (C) phase delay, and (D) modulation ratio of the detected fluorescence generated from a lymph node in the area of the right fifth mammary gland of a canine. Illumination was accomplished with an expanded 780-nm laser diode. Modulation frequency was 100 MHz. (From Reynolds, J.S. et al., *Photochem. Photobiol.*, 70, 87, 1999. With permission.)

### 33.6.3 Nontargeting Contrast Agents That "Report" or Sense Environment

While not employing the favorable excitation/emission characteristics, the studies of Mordon et al.[89] are especially intriguing because they are the first studies to employ a "reporting" dye, or a dye whose emission characteristics varied with tissue milieu. Specifically, they employed the pH-sensitive dye of 5,6 CF carboxyfluorescence (BCECF), which is a ratiometric dye sensitive to pH. Using CW, spectrally resolved measurements of fluorescence resulting from area illumination with excitation light at 465 and 490 nm, they employed area detection using an intensified CCD camera to determine changes in fluorescence with changes in wavelength of the excitation light. Using an *in vitro* calibration of the ratiomeric dye, Mordon et al. were able to correlate the 2D fluorescent images to provide 2D images of tumor pH in a s.c. mouse model. While CW measurements are not time-dependent methods, this study did not directly measure changes in the fluorescence-decay kinetics but rather used spectral ratiometric changes to demark the change in the radiative relaxation rates that arose owing to acidotic tissue conditions.

### 33.6.4 Targeting, Contrast Agents: Immunophotodiagnosis

Folli et al.[87] were the first to demonstrate the use of NIR targeting agents by coupling an indopentamethinecyanine dye coupled to a MAb E48 for targeting squamous cell carcinoma in the upper respiratory tract in mice. Approximately 600 pmol conjugated dye molecules were injected into mice and imaged postmortem using simple planar illumination and photography. This was the first demonstration of NIR immunophotodiagnosis that followed prior work to employ non-NIR, fluorescein-labeled antibodies targeted against carcinomembryonic antigen (CEA) in mouse[81] and clinical studies.[83] The study was repeated by Neri,[95] who used generated antibodies targeted to oncofetal fibronectin (B-FN), which is present in the angiogenic vessels of neoplasms but not in mature vessels. Further emphasizing the use of NIR fluorochromes in deep-tissue penetration for photoimmunodiagnosis, Ballou et al.[90,96] conjugated the cyanine dye class of Cy3, Cy5, Cy5.5, and Cy7 fluorochromes and not unsurprisingly found they could probe more deeply with the Cy7-conjugated MAbs in living tumor-bearing mice. They were able to image targeted delivery of fluorochromes of 40 pmol to 6 nmol/kg bw, and 600 pmol/animal using area illumination and an intensified or cooled CCD camera. More recently, Soukos et al.[115] used a 7,12-dimethylbenz(a)anthracene (DMBA)-induced tumor in the hamster-cheek-pouch model and showed the ability to target cyanine dye to express endothelial growth factor receptor using the antiendothelial growth factor receptor (EGFR) MAb (C225) in surface illumination and detection. Approximately 3.3 nmol fluorochrome was used in each animal.

### 33.6.5 Targeting Contrast Agents: Small-Peptide Conjugations

A significant advancement in the design of optical contrast agents mimicked those used in other medical imaging modalities. Achilefu et al.[107–109] and Bugaj et al.[106] used area illumination and area CCD detection in tumor-bearing rats in order to detect ~6 to 7 μmol/kg bw of cyanine dye conjugated to small peptides for targeting somatostatin and bombesin receptors. One commercial nuclear diagnostic agent, Octreoscan®, is based on targeting the somatostatin receptor, which is overexpressed in neuroendocrine tumors. Bugaj et al. showed that the optical imaging using cytate, the derivatized and peptide-conjugated ICG, is similar to the radiolabeled peptide analog for somatostatin. Similar results were reported from using the peptide-conjugated derivative of ICG, cybesin, which is similar to the radiolabeled peptide analog for bombesin. Becker et al.[110] reported similar work in tumor-bearing mice with detection limits reduced to 0.02 μmol/kg bw using a similar targeting construct. In their work, they conjugated the indodicarbocyanine dyes and the indotricarbocyanine (ITCC) dyes with analogs of somatostatin, somatostatin-14, and octreotate. In another approach, Becker et al.[99,101] followed the targeting approach previously used for methotrexate and PDT therapies, MRI gadolinium and magnetic-particle contrast, with human serum albumin (HSA) and transferrin (Tf) coupled to indotricarbocyanine dyes. While Tf binds to specific cell-surface receptors, HSA binds nonspecifically. Their studies show the contrast enhancement of targeting specificity using the Tf-ITCC.

In another study involving peptide conjugation, Zaheer et al.[116] conjugated a bisphosphonate derivative, pamidronate, which exhibits specific binding to hydroxyapatite to an indocyanine dye to image bone structure in hairless mouse. The system may be capable of NIR detection of osteoblastic activity, enabling NIR imaging of skeletal development, coronary atherosclerosis, and other diseases.

### 33.6.6 Reporting or Sensing Contrast Agents

A novel "reporting" optical-contrast design was reported by Weissleder et al.,[23] who employed fluorophore Cy5.5 loaded onto a polylysine backbone with methoxypolyethylene glycol polymer. When conjugated to the polymer backbone in high concentration, the fluorochrome tends to quench itself. However, when the polymer backbone is cleaved by cathespin B or H, lysosomal proteases whose activity may be enhanced in cancer cells, the fluorochromes become free and radiatively relax to produce fluorescence. In contrast to the small-peptide-conjugated dyes, this system requires fluorochrome internalization. The pioneering work enabled detection of 10 μmol of agent or 250 pmol of fluorochrome administered per tumor-bearing animal and represented the first time an optical-contrast agent based on an internalization construct had been demonstrated. Along the same lines, another agent that reported on the basis of protease activity was developed using the same design principles. Bremer et al.[111] coupled matrix metalloproteinase-2 (MMP-2) peptide substrates onto a poly-L-lysine polymer backbone and onto the peptides further conjugated with Cy5.5. The fluorochromes were sufficiently packed to be quenched upon activation. Upon action of the proteinase on the peptide, the Cy5.5 was freed and able to radiatively relax, reporting proteinase activity. MMPs are overexpressed in cancers, and MMP-2 in particular has been identified as the cause of collagen IV degradation. (Collagen IV is the major component of basement membranes.) The MMP-2 activity is thought to be responsible for the pathogenesis of cancer, including spread, metastasis, and angiogenesis. Using area illumination and detection, as little as 167 pmol per animal resulted in detected fluorescence to measure MMP activity *in vivo* for directing the therapeutic use of proteinase activity.

### 33.6.7 Combined Targeting and Reporting Dyes

Finally, Licha et al.[125] sought to combine fluorochrome targeting using membrane receptors, such as transferrin, or the somatostatin and bombesin receptors with acid-cleavable constructs that would enable internalization of the fluorochromes in the lysosomal compartments and recycling of the receptors. Such constructs to augment the accumulation and, therefore, concentrate the signal from the targeting fluorochrome would be enhanced only if the contrast agent had a long half-life in the circulation. Coupling these cyanine dyes to different acid-cleavable hydrazone links that were bound to peptides, proteins, and antibodies, Licha et al. furthermore sought to develop a pH-sensitive contrast agent whose fluorescence is mediated by tumor acidosis.

In another innovative development, Huber et al.[126] synthesized bifunctional contrast agents containing a metal chelator for binding of a paramagnetic ion such as gadolinium and a conjugated fluorescent dye such as tetramethylrhodamine to combine optical imaging and MRI of experimental animals. While rhodamine excites within the visible with maximum absorbance at 547 nm and emission at 572 nm, the approach was successful for imaging of *Xenopus laevis* embryos. With the conjugation of an NIR-excitable dye, the potential to develop bifunctional contrast agents for deep-tissue medical imaging could also be realized. Again the reader is urged to be cautious in assessing contrast-agent studies conducted in mice and rats. These tissue volumes are not comparable to those in humans, and it is unlikely that emission signals at these wavelengths can be detected with sufficient SNR for image reconstruction in large volumes.

### 33.6.8 Summary

In summary, the approaches for fluorescent contrast-agent development have to date focused on:

- Blood-pooling agents specific to increased microvessel density in neovascularized tumors
- Targeting agents based on:
  - Immunophotodiagnosis
  - Small-peptide conjugations targeting overexpressed receptors whose location of action is directed to:
    - Membrane receptors of endothelial cells that line angiogenic vessels of tumors
    - Membrane receptors of cancer cells

- Reporting agents that change their fluorescence-decay kinetics either through self-quenching or through environmental changes associated with:
  - Interstitial pH
  - Membrane associated proteases and receptors
  - Lysosomal, enzymatic degradation

In addition, the summary of the literature reports presented in Table 33.1 leads to the following conclusions:

- Fluorescence-enhanced optical imaging has not been demonstrated on large tissue volumes that are scalable to the clinic.
- The minimum number of fluorochrome molecules reported detected in a mouse or rat is 167 pmol.
- Favorable excitation and emission spectra are currently achievable using only the cyanine dye family.

Table 33.2 also contains limited references to endogenous fluorescence-enhanced contrast owing to green-fluorescence-protein-expressing tumors. In these systems, the green fluorescent protein has been transduced into cancer-cell lines to "report" tumors and metastases as well as their response to therapy. While the approach uses similar detection technology, it is not subject to the excitation-light-rejection issues that can plague fluorescence-enhanced contrast imaging. We present literature in this area for completeness.

In the following section, we summarize the theory and mathematics that enable us to predict the success of optical imaging using CW and time- and frequency-domain approaches as well as to develop the tomographic algorithms for fluorescence-enhanced optical tomography.

## 33.7 Challenges for NIR Fluorescence-Enhanced Imaging and Tomography

The preceding discussion presented an overview of the status of fluorescence-enhanced optical imaging. The opportunity to develop an emission-based tomographic imaging modality similar to that provided by nuclear imaging but without the use of radionucleotides is offered by NIR fluorescent agents. Yet the added challenge for NIR fluorescence-enhanced imaging over nuclear imaging is that, unlike nuclear techniques, an activating or excitation signal must first be delivered to the contrast agent before registration of the emission signal from the tissue. Preliminary data from animals (Table 33.2) and phantoms (not presented here) suggests that penetration depth and sensitivity may be comparable to nuclear techniques. A side-by-side comparison of NIR fluorescence-enhanced imaging with nuclear imaging is needed before the comparative performance can be ascertained.

Another opportunity for optical imaging is the ability for tomographic reconstruction and additional diagnostic information based on the fluorescence-decay kinetics of smartly designed probes. Tomography of large tissue-simulating volume has been demonstrated from experimental data as well as synthetic data (Table 33.1), albeit with the rather inconvenient point-source and point-detector geometries. The single-point-source and detector geometry is a throwback to NIR optical tomography from endogenous contrast studies and may not be the appropriate geometry for fluorescence-enhanced optical imaging, especially when transillumination through large tissues is required. Nonetheless, the tomographic algorithms as reviewed in Section 33.5 are already established for these systems. The challenge for the future is to develop tomographic algorithms for illumination and detection that are clinically feasible and adaptable for hybrid nuclear imaging.

While the field of NIR fluorescence-enhanced optical imaging is less than a decade old, the coming decade holds great promise for exciting new developments and, it is hoped, will result in an adjuvant tomographic imaging modality for nuclear imaging.

## Acknowledgments

This review was supported in part by the National Institutes of Health grants R01CA67176 and R01CA88082 and the State of Texas Advanced Research/Advanced Technology Program.

## References

1. Lakowicz, J.R., *Principles of Fluorescence Spectroscopy,* Plenum Press, New York, 1983.
2. Sevick-Muraca, E.M. and Burch, C.L., Origin of phosphorescence re-emitted from tissues, *Opt. Lett.*, 19, 1928, 1994.
3. Patterson, M.S. and Pogue, B.W., Mathematical model for time-resolved and frequency-domain fluorescence spectroscopy in biological tissue, *Appl. Opt.*, 33, 1963, 1994.
4. Sevick, E.M., Chance, B., Leigh, J., Maris, M., and Nioka, S., Quantitation of time-resolved and frequency-resolved optical-spectra for the determination of tissue oxygenation, *Anal. Biochem.*, 195(2), 330, 1991.
5. Haskell, R.C., Svassand, L.O., Tsay, T.-T., Feng, T.-C., McAdams, M.S., and Tromberg, B.J., Boundary conditions for the diffusion equation in radiative transfer, *J. Opt. Soc. Am. A*, 11, 2727, 1994.
6. Keijzer, M., Star, W.M., and Storchi, P.R.M., Optical diffusion in layered media, *Appl. Opt.*, 27, 1820, 1988.
7. Patterson, M.S., Chance, B., and Wilson, B., Time resolved reflectance and transmittance for the non-invasive measurement of tissue optical properties, *Appl. Opt.*, 28, 2331, 1989.
8. Hielscher, A.H., Jacques, S.L., Wang, L., and Tittel, F.K., The influence of boundary conditions on the accuracy of diffusion theory in time-resolved reflectance spectroscopy of biological tissues, *Phys. Med. Biol.*, 40, 1957, 1995.
9. Farrell, T.J., Patterson, M.S., and Wilson, B. A., Diffusion theory model of spatially resolved, steady-state diffuse reflectance for the noninvasive determination of tissue optical properties *in vivo, Med. Phys.*, 9, 879, 1992.
10. Cerussi, A.E., Maier, J.S., Fantini, S., Franceschini, M.A., Mantulin, W.W., and Gratton, E., Experimental verification of a theory for time-resolved fluorescence spectroscopy of thick tissues, *Appl. Opt.*, 36, 116, 1997.
11. Grosenick, D., Wabnitz, H., Rinnebert, H.H., Moesta, K.T., and Schlag, P.M., Development of a time-domain optical mammography and first *in vivo* applications, *Appl. Opt.*, 38, 2927, 1999.
12. Thompson, A.B. and Sevick-Muraca, E.M., NIR fluorescence contrast enhanced imaging with ICCD homodyne detection: measurement precision and accuracy, *J. Biomed. Opt.*, 8, 111, 2003.
13. Kuwana, E. and Sevick-Muraca, E.M., Fluorescence lifetime spectroscopy in multiply scattering media with dyes exhibiting multi-exponential decay kinetics, *Biophys. J.*, 83, 1165, 2002.
14. Chen, A., Effects of fluorescence and phosphorescence lifetime on frequency domain optical contrast for biomedical optical imaging, M.S. thesis, Purdue University, West Lafayette, IN, 1997.
15. Sevick-Muraca, E.M., Godavarty, A., Houston, J.P., Thompson, A.B., and Roy, R., Near-infrared imaging with fluorescent contrast agents, in *Fluorescence in Biomedicine,* Pogue, B. and Mycek, M., Eds., Marcel Dekker, New York, in press.
16. Alcala, J.R., Gratton, E., and Jameson, D.M., A multifrequency phase fluorometer using the harmonic content of a mode-locked laser, *Anal. Instrum.*, 14(3), 225, 1985.
17. Reynolds, J.S., Troy, T.L., and Sevick-Muraca, E.M., Multipixel techniques for frequency-domain photon migration imaging, *Biotechnol. Prog.*, 13(5), 669, 1997.
18. Lakowicz, J.R. and Berndt, K., Lifetime-sensitive fluorescence imaging using an rf phase-camera, *Rev. Sci. Instrum.*, 62, 1727, 1991.
19. Sevick, E.M., Lakowicz, J.R., Szmacinski, H., Nowaczyk, K., and Johnson, M., Frequency-domain imaging of obscure absorbers: principles and applications, *J. Photochem. Photobiol.*, 16, 169, 1992.
20. Yang, Y., Liu, H., Li, X., and Chance, B., Low-cost frequency-domain photon migration instrument for tissue spectroscopy, oximetry, and imaging, *Opt. Eng.*, 36(5), 1562, 1997.

21. Houston, J.P., Near-infrared fluorescence enhanced optical imaging: an analysis of penetration depth, M.S. thesis, Texas A&M University, College Station, TX, 2002.
22. Hawrysz, D.J. and Sevick-Muraca, E.M., Developments toward diagnostic breast cancer imaging using near-infrared optical measurements and fluorescent contrast agents, *Neoplasia,* 2(5), 388, 2000.
23. Weissleder, R., Tung, C.H., Mahmood, U., and Bogdanov, A., Jr., *In vivo* imaging of tumors with protease-activated near-infrared fluorescent probes, *Nat. Biotechnol.*, 17, 375, 1999.
24. Qing, C., Lakowicz, J.R., Murtaza, Z., and Rao, G., A fluorescence lifetime-based solid sensor for water, *Anal. Chim. Acta*, 350(1–2), 97, 1997.
25. Russell, R.J., Cote, G.L., Gefrides, C.C., McShane, M.J., and Pishko, M.V., A fluorescence-based glucose biosensor using concanavalin A and dextran encapsulated in a poly(ethylene glycol) hydrogel, *Anal. Chem.*, 71, 3126, 1999.
26. Hutchinson, C.L., Lakowicz, J.R., and Sevick-Muraca, E.M., Fluorescence lifetime-based sensing in tissues: a computational study, *Biophys. J.*, 68, 1574, 1995.
27. Mayer, R.H., Reynolds, J.S., and Sevick-Muraca, E.M., Measurement of fluorescence lifetime in scattering media using frequency-domain photon migration, *Appl. Opt.*, 38, 4930, 1999.
28. Lakowicz, J.R. and Abugo, O.O., Modulation sensing of fluorophores in tissue — a new approach to drug compliance monitoring, *J. Biomed. Opt.*, 4(4), 429, 1999.
29. Fishkin, J.B., Cerussi, A.E., Fantini, S., Franceschini, M.A., Gratton, E., and So, P.T.C., Frequency-domain method for measuring spectral properties in multiple scattering media — methemoglobin absorption spectrum in a tissue-like phantom, *Appl. Opt.*, 34, 1143, 1995.
30. Zhigang, S., Yingqing, H., and Sevick-Muraca, E.M., Precise analysis of frequency domain photon migration measurement for characterization of concentrated colloidal suspensions, *Rev. Sci. Instrum.*, 73(2), 383, 2002.
31. Lee, J., Fluorescence-enhanced biomedical optical imaging using frequency-domain photon migration, Ph.D. thesis, Purdue University, West Lafayette, IN, 2001.
32. Franceschini, M.A., Moesta, K.T., Fantini, S., Gaida, G., Gratton, E., Jess, H., Mantulin, W.W., Seeber, M., Schlag, P.M., and Kaschke, M., Frequency-domain techniques enhance optical mammography: initial clinical results, *Proc. Natl. Acad. Sci. U.S.A.*, 94, 6468, 1997.
33. McBride, T.O., Pogue, B., Gerety, E.D., Poplack, S.B., Osterberg, U.L., and Paulsen, K.D., Spectroscopic diffuse optical tomography for the quantitative assessment of hemoglobin and oxygen saturation in breast tissue, *Appl. Opt.*, 38, 5480, 1999.
34. Colak, S.B., van der Mark, M.B., Hooft, G.W., Hoogenraad, J.H., van der Linden, E.S., and Kuijpers, F.A., Clinical optical tomoraphy and NIR spectroscopy for breast cancer detection, *IEEE J. Sel. Top. Quantum Electron.*, 5, 1143, 1999.
35. Sevick-Muraca, E.M., Lopez, G., Troy, T.L., Reynolds, J.S., and Hutchinson, C.L., Fluorescence and absorption contrast mechanisms for biomedical optical imaging using frequency-domain techniques, *Photochem. Photobiol.*, 66, 55, 1997.
36. Li, X., Chance, B., and Yodh, A.G., Fluorescence heterogeneities in turbid media limits for detection, characterization, and comparison with absorption, *Appl. Opt.*, 37, 6833, 1998.
37. O'Leary, M.A., Boas, D.A., Chance, B., and Yodh, A.G., Reradiation and imaging of diffuse photon density waves using fluorescent inhomogeneities, *J. Luminescence*, 60(61), 281, 1994.
38. Wu, J., Wang, Y., Perleman, L., Itzkan, I., Dasari, R.R., and Feld, M.S., Time-resolved multichannel imaging of fluorescent objects embedded in turbid media, *Opt. Lett.*, 20, 489, 1995.
39. Chang, J., Barbour, R.L., Graber, H., and Aronson, R., Fluorescence optical tomography, *Proc. SPIE*, 2570, 59, 1995.
40. O'Leary, M.A., Boas, D.A., Li, X.D., Chance, B., and Yodh, A.G., Fluorescence lifetime imaging in turbid media, *Opt. Lett.*, 21(2), 158, 1996.
41. Paithankar, D.Y., Chen, A.U., Pogue, B.W., Patterson, M.S., and Sevick-Muraca, E.M., Imaging of fluorescent yield and lifetime from multiply scattered light reemitted from random media, *Appl. Opt.*, 36, 2260, 1997.

42. Wu, J., Perelman, L., Dasari, R.R., and Feld, M.S., Fluorescence tomographic imaging in turbid media using early-arriving photons and Laplace transforms, *Proc. Natl. Acad. Sci. U.S.A.*, 94, 8783, 1997.
43. Chang, J., Graber, H.L., and Barbour, R.L., Improved reconstruction algorithm for luminescence when background luminophore is present, *Appl. Opt.*, 37, 3547, 1998.
44. Jiang, H., Frequency-domain fluorescent diffusion tomography: a finite-element-based algorithm and simulations, *Appl. Opt.*, 37(22), 5337, 1998.
45. Hull, E.L., Nichols, M.G., and Foster, T.H., Localization of luminescent inhomogeneities in turbid media with spatially resolved measurements of cw diffuse luminescence emittance, *Appl. Opt.*, 37, 2755, 1998.
46. Eppstein, M.J., Dougherty, D.E., Hawrysz, D.J., and Sevick-Muraca, E.M., Three-dimensional optical tomography, *Proc. SPIE,* 3497, 97, 1999.
47. Eppstein, M.J., Dougherty, D.E., Troy, T.L., and Sevick-Muraca, E.M., Biomedical optical tomography using dynamic parameterization and Bayesian conditioning on photon migration measurements, *Appl. Opt.*, 38, 2138, 1999.
48. Chenomordik, V., Hattery, D., Gannot, I., and Gandjbakhche, A.H., Inverse method three-dimensional reconstruction of localized *in vivo* fluorescence — application to Sjogren syndrome, *IEEE J. Sel. Top. Quantum Electron.*, 54, 930, 1999.
49. Roy, R. and Sevick-Muraca, E.M., Truncated Newton's optimization scheme for absorption and fluorescence optical tomography. Part II. Reconstruction from synthetic measurements, *Opt. Express*, 4, 372, 1999.
50. Roy, R. and Sevick-Muraca, E.M., Active constrained truncated Newton method for simple-bound optical tomography, *J. Opt. Soc. Am. A,* 17(9), 1627, 2000.
51. Yang, Y., Iftimia, N., Xu, Y., and Jiang, H., Frequency-domain fluorescent diffusion tomography of turbid media and *in vivo* tissues, *Proc. SPIE,* 4250, 537, 2001.
52. Eppstein, M.J., Hawrysz, D.J., Godavarty, A., and Sevick-Muraca, E.M., Three-dimensional, near-infrared fluorescence tomography with Bayesian methodologies for image reconstruction from sparse and noisy data sets, *Proc. Natl. Acad. Sci. U.S.A.*, 99, 9619, 2002.
53. Hawrysz, D.J., Eppstein, M.J., Lee, J., and Sevick-Muraca, E.M., Error consideration in cotrast-enhanced three-dimensional optical tomography, *Opt. Lett.*, 26(10), 704, 2001.
54. Roy, R. and Sevick-Muraca, E.M., Three-dimensional unconstrained and constrained image-reconstruction techniques applied to fluorescence, frequency-domain photon migration, *Appl. Opt.*, 40(13), 2206, 2001.
55. Roy, R. and Sevick-Muraca, E.M., A numerical study of gradient-based nonlinear optimization methods for contrast-enhanced optical tomography, *Opt. Express*, 9(1), 49, 2001.
56. Ntziachristos, V. and Weissleder, R., Experimental three-dimensional fluorescence reconstruction of diffuse media by use of a normalized Born approximation, *Opt. Lett.*, 26(12), 893, 2001.
57. Lee, J. and Sevick-Muraca, E.M., Three-dimensional fluorescence enhanced optical tomography using references frequency-domain photon migration measurements at emission and excitation measurements, *J. Opt. Soc. Am. A*, 19, 759, 2002.
58. Sevick-Muraca, E.M. and Paithankar, D.Y., Fluorescence imaging system and measurement, U.S. patent 5,865,754, February 2, 1999.
59. Paithankar, D.Y., Chen, A. Sevick-Muraca, E.M., Fluorescence yield and lifetime imaging in tissues and other scattering media, *Proc. SPIE,* 2679, 162, 1996.
60. Pogue, B., McBride, T., Prewitt, J., Osterberg, U., and Paulsen, K., Spatially varying regularization improves diffuse optical tomography, *Appl. Opt.*, 38, 2950, 1999.
61. Holboke, M.J. and Yodh, A.G., Parallel three-dimensional diffuse optical tomography, *Biomed. Topical Meetings*, OSA, Miami Beach, FL, 2000, p. 177.
62. Xu, M., Lax, M., and Alfano, R.R., Time-resolved Fourier diffuse optical tomography, *Biomed. Topical Meetings*, OSA, Miami Beach, FL, 2000, p. 345.

63. Arridge, S.R., Hebden, J.C., Schweiger, M., Schmidt, F.E.W., Fry, M.E., Hillman, E.M.C., Dehghani, H., and Delpy, D.T., A method for three-dimensional time-resolved optical tomography, *Int. J. Imaging Syst. Technol.*, 11, 2, 2000.
64. Eppstein, M.J. and Dougherty, D.E., Optimal three-dimensional traveltime tomography, *Geophysics*, 63, 1053, 1998.
65. Eppstein, M.J. and Dougherty, D.E., Efficient three-dimensional data inversion: soil characterization and moisture monitoring from cross-well ground penetrating radar at a Vermont test site, *Water Resour. Res.*, 34, 1889, 1998.
66. Eppstein, M.J. and Dougherty, D.E., Three-dimensional stochastic tomography with upscaling, U.S. patent application 09/110,506, July 9, 1998.
67. Yorkey, T.J., Webster, J.G., and Tompkins, W.J., Comparing reconstruction algorithms for electrical impedance tomography, *IEEE Trans. Biomed. Eng. BME*, 34, 843, 1987.
68. Eppstein, M.J., Dougherty, D.E., Hawrysz, D.J., and Sevick-Muraca, E.M., Three-dimensional Bayesian optical image reconstruction with domain decomposition, *IEEE Trans. Med. Imaging*, 3, 147, 2000.
69. Roy, R., Godavarty, A., and Sevick-Muraca, E.M., Fluorescence-enhanced, optical tomography using referenced measurements of heterogeneous media, *IEEE Trans. Med. Imaging*, in press.
70. Roy, R. and Sevick-Muraca, E.M., Truncated Newton's optimization scheme for absorption and fluorescence optical tomography. I. Theory and formulation, *Opt. Express*, 4, 353, 1999.
71. Lee, J. and Sevick-Muraca, E.M., Fluorescence-enhanced absorption imaging using frequency-domain photon migration: tolerance to measurement error, *J. Biomed. Opt.*, 6(1), 58, 2000.
72. Arridge, S.R., Optical tomography in medical imaging, *Inverse Problems*, 15, R41, 1999.
73. Eppstein, M.J. and Dougherty, D.E., Simultaneous estimation of transmittivity values and zonation, *Water Resour. Res.*, 32, 3321, 1996.
74. Ntziachristos, V., Yodh, A.G., Schnall, M., and Chance, B., Concurrent MRI and diffuse optical tomography of the breast after indocyanine green enhancement, *Proc. Natl. Acad. Sci. U.S.A.*, 97, 2767, 2000.
75. Pei, Y., Lin, F-B., and Barbour, R.L., Modeling of sensitivity and resolution to an included object in homogeneous scattering media and in MRI-derived breast maps, *J. Biomed. Opt.*, 5, 302, 1999.
76. Pogue, B.W. and Paulsen, K.D., High-resolution near-infrared tomographic imaging simulations of the rat cranium by use of *a priori* magnetic resonance imaging structural information, *Opt. Lett.*, 23, 1716, 1998.
77. Holboke, M.J., Tromberg, B.J., Li, X., Shah, N., Fishkin, J., Kidney, D., Butler, J., Chance, B., and Yodh, A.G., Three-dimensional diffuse optical mammography with ultrasound localization in a human subject, *J. Biomed. Opt.*, 5, 237, 2000.
78. Provencher, S.W. and Vogel, R.H., Regularization techniques for inverse problems in molecular biology, in *Numerical Treatment of Inverse Problems in Differential and Integral Equations*, Deuflhard, P. and Hairer, E., Eds., Birkhauser Press, Boston, 1983.
79. Tikhonov, A.N. and Arsenin, V.Y., *Solution of Ill-Posed Problems*, V.H. Winston & Sons, Washington, D.C., 1977.
80. Biolo, R., Jori, G., Kennedy, J.C., Nadeau, P., Potteir, R., Reddi, E., and Weagle, G., A comparison of fluorescence methods used in the pharmacokinetic studies of Zn(II) phthalocyanine in mice, *Photochem. Photobiol.*, 53, 113, 1991.
81. Pelegrin, A., Folli, S., Buchegger, F., Mach, J.-P., Wagnières, G., and van den Bergh, H., Antibody-fluorescein conjugates for photoimmunodiagnosis of human colon carcinoma in nude mice, *Cancer*, 67, 2529, 1991.
82. Straight, R.C., Benner, R.E., McClane, R.W., Go, P.M.N.Y., Yoon, G., and Dixon, J.A., Application of charge-coupled device technology for measurement of laser light and fluorescence distribution in tumors for photodynamic therapy, *Photochem. Photobiol.*, 53, 787, 1991.

83. Folli, S., Wagnières, G., Pelegrin, A., Calmes, J.M., Braichotte, D., Buchegger, F., Chalandon, Y., Hardman, N., Heusser, D.G., Givel, J.C., Chapuis, G., Chatelain, A., van Den Bergh, H., and Mach, J.P., Immunophotodiagnosis of colon carcinomas in patients injected with fluoresceinated chimeric antibodies against carcinoembryonic antigen, *Proc. Natl. Acad. Sci. U.S.A.*, 89, 7973, 1992.
84. Cubeddu, R., Canti, G., Taroni, P., and Valentini, G., Time-gated fluorescence imaging for the diagnosis of tumors in a murine model, *Photochem. Photobiol.*, 57, 480, 1993.
85. Cubeddu, R., Canti, G., Pifferi, A., Taroni, P., and Valentini, G., Fluorescence lifetime imaging of experimental tumors in hematoporphyrin derivative-sensitized mice, *Photochem. Photobiol.*, 66, 229, 1997.
86. Kohl, M., Sukowski, U., Ebert, B., Neukammer, J., and Rinneberg, H.H., Imaging of superficially growing tumors by delayed observation of laser-induced fluorescence, *Proc. SPIE*, 1881, 206, 1993.
87. Folli, S., Westermann, P., Braichotte, D., Pelegrin, A., Wagnières, G., van den Bergh, H., and Mach, J.P., Antibody-indocyanin conjugates for immunophotodetection of human squamous cell carcinoma in nude mice, *Cancer Res.*, 54, 2643, 1994.
88. Haglund, M.M., Hochman, D.W., Spence, A.M., and Berger, M.S., Enhanced optical imaging of rat gliomas and tumor margins, *Neurosurgery*, 35, 930, 1994.
89. Mordon, S., Devoisselle, J.M., and Maunoury, V., *In vivo* pH measurement and imaging of tumor tissue using a pH-sensitive fluorescent probe (5,6-carboxyfluorescein): instrumental and experimental studies, *Photochem. Photobiol.*, 60, 274, 1994.
90. Ballou, B., Fisher, G.W., Waggoner, A.S., Farkas, D.L., Reiland, J.M., Jaffe, R., Mujumdar, R.B., Mujumdar, S.R., and Hakala, T.R., Tumor labeling *in vivo* using cyanine-conjugated monoclonal antibodies, *Cancer Immunol. Immunother.*, 41, 257, 1995.
91. Devoisselle, J.M., Soulie, S., Mordon, S.R., Mestres, G., Desmettre, T.M.D., and Maillols, H., Effect of indocyanine green formulation on blood clearance and *in vivo* fluorescence kinetic profile of skin, *Proc. SPIE*, 2627, 100, 1995.
92. Rokahr, I., Andersson-Engels, S., Svanberg, S., D'Hallewin, M.-A., Baert, L., Wang, I., and Svanberg, K., Optical detection of human urinary bladder carcinoma utilising tissue autofluorescence and protoporphyrin IX-induced fluorescence following low-dose ALA instillation, *Proc. SPIE*, 2627, 2, 1995.
93. Haglund, M.M., Berger, M.S., and Hochman, D.W., Enhanced optical imaging of human gliomas and tumor margins, *Neurosurgery*, 38, 308, 1996.
94. Sakatani, K., Kashiwasake-Jibu, M., Taka, Y., Wang, S., Zuo, H., Yamamoto, K., and Shimizu, K., Noninvasive optical imaging of the subarachnoid space and cerebrospinal fluid pathways based on near-infrared fluorescence, *J. Neurosurg.*, 87, 738, 1997.
95. Neri, D., Targeting by affinity-matured recombinant antibody fragments of an angiogenesis-associated fibronectin isoform, *Nat. Biotechnol.*, 15, 1271, 1997.
96. Ballou, B., Fisher, G.W., Deng, J.-S., Hakala, T.R., Srivastava, M., and Farkas, D.L., Fluorochrome-labeled antibodies *in vivo*: assessment of tumor imaging using Cy3, Cy5, Cy5.5, and Cy7, *Cancer Detection Prev.*, 22(3), 251, 1998.
97. Eker, C., Montan, S., Jaramillo, E., Koizumi, K., Rubio, C., Andersson-Engels, S., Svanberg, K., Svanberg, S., and Slezak, P., Clinical spectral characterization of colonic mucosal lesions using autofluorescence and D aminolevulinic acid sensitizations, *Gut*, 44, 511, 1999.
98. Reynolds, J.S., Troy, T.L., Mayer, R.H., Thompson, A.B., Waters, D.J., Cornell, K.K., Snyder, P.W., and Sevick-Muraca, E.M., Imaging of spontaneous canine mammary tumors using fluorescent contrast agents, *Photochem. Photobiol.*, 70, 87, 1999.
99. Becker, A., Licha, K., Kresse, M., Riefke, B., Sukowski, U., Ebert, B., Rinneberg, H., and Semmler, W., Transferrin-mediated tumor delivery of contrast media for optical imaging and magnetic resonance imaging, *Proc. SPIE*, 3600, 142, 1999.
100. Mahmood, U., Tung, C.-H., Bogdanov, A., and Weissleder, R., Near-infrared optical imaging of protease activity for tumor detection, *Radiology*, 21(3), 866, 1999.

101. Becker, A., Riefke, B., Ebert, B., Sukowski, U., Rinneberg, H., Semmler, W., and Licha, K., Macromolecular contrast agents for optical imaging of tumors: comparison of indotricarboyanine-labeled human serum albumin and transferrin, *Photochem. Photobiol.*, 72, 234, 2000.
102. Gurfinkel, M., Thompson A.B., Ralston, W., Troy, T.L., Moore, A.L., Moore, T.A., Gust, J.D., Tatman, D., Reynolds, J.S., Muggenburg, B., Nikula, K., Pandey, R., Mayer, R.H., Hawrysz, D.J., and Sevick-Muraca, E.M., Pharmacokinetics of ICG and HPPH-car for the detection of normal and tumor tissue using fluorescence, near-infrared reflectance imaging: a case study, *Photochem. Photobiol.*, 72, 94, 2000.
103. Licha, K., Riefke, B., Ntziachristos, V., Becker, A., Chance, B., and Semmler, W., Hydrophilic cyanine dyes as contrast agents for near-infrared tumor imaging: synthesis, photophysical properties and spectroscopic *in vivo* characterization, *Photochem. Photobiol.*, 72, 392, 2000.
104. Yang, M., Baranov, E., Sun, F., Li, X., Li, L., Hasegawa, S., Bouvet, M., Al-Tuwaijri, M., Chishima, T., Shimada, H., Moossa, A., Penman, S., and Hoffman, R., Whole-body optical imaging of green fluorescent protein-expressing tumors and metastases, *Proc. Natl. Acad. Sci. U.S.A.*, 97, 1206, 2000.
105. Hoffman, R.M., Visualization of GFP-expressing tumors and metastasis *in vivo*, *Biotechniques*, 30, 1016, 2001.
106. Bugaj, J.E., Achilefu, S., Dorshow, R.B., and Rajagopalan, R., Novel fluorescent contrast agents for optical imaging of *in vivo* tumors based on a receptor-targeted dye-peptide conjugate platform, *J. Biomed. Opt.*, 6, 122, 2001.
107. Achilefu, S., Dorshow, R.B., Bugaj, J.E., and Rajagopalan, R., Tumor specific fluorescent contrast agents, *Proc. SPIE*, 3917, 80, 2000.
108. Achilefu, S., Bugaj, J.E., Dorshow, R.B., Jimenez, H.N., and Rajagopalan, R., New approach to optical imaging of tumors, *Proc. SPIE*, 4259, 110, 2001.
109. Achilefu, S., Bugaj, J.E., Dorshow, R.B., Jimenez, H.N., Rajagopalan, R., Wilhelm, R.R., Webb, E.G., and Erion, J.L., Site-specific tumor targeted fluorescent contrast agents, *Proc. SPIE*, 4156, 69, 2001.
110. Becker, A., Hessenius, C., Licha, K., Ebert, B., Sukowski, U., Semmler, W., Wiedenmann, B., and Grotzinger, C., Receptor-targeted optical imaging of tumors with near-infrared fluorescent ligands, *Nat. Biotechnol.*, 19, 327, 2001.
111. Bremer, C., Tung, C., and Weissleder, R., *In vivo* molecular target assessment of matrix metalloproteinase inhibition, *Nat. Med.*, 7, 743, 2001.
112. Ebert, B., Sukowski, U., Grosenick, D., Wabnitz, H., Moesta, K.T., Licha, K., Becker, A., Semmler, W., Schlag, P.M., and Rinneberg, H., Near-infrared fluorescent dyes for enhanced contrast in optical mammography: phantom experiments, *J. Biomed. Opt.*, 6(2), 134, 2001.
113. Finlay, J.C., Conover, D.L., Hull, E.L., and Foster, T.H., Porphyrin bleaching and PDT-induced spectral changes are irradiance dependent in ALA-sensitized normal rat skin *in vivo*, *Photochem. Photobiol.*, 73, 54, 2001.
114. Rice, B.W., Cable, M.D., and Nelson, M.B., *In vivo* imaging of light-emitting probes, *J. Biomed. Opt.*, 6(4), 432, 2001.
115. Soukos, N.S., Hamblin, M.R., Keelm, S., Fabian, R.L., Deutsch, T.F., and Hasan, T., Epidermal growth factor receptor-targeted immunophotodiagnosis and photoimmunotherapy of oral precancer *in vivo*, *Cancer Res.*, 61, 4490, 2001.
116. Zaheer, A., Lenkinski, R.E., Mahmood, A., Jones, A.G., Cantley, L.C., and Frangioni, J.V., *In vivo* near-infrared fluorescence imaging of osteoblastic activity, *Nat. Biotechnol.*, 19(12), 1148, 2001.
117. Leevy, C.M., Smith, F., and Longueville, J., Indocyanine green clearance as a test for hepatic function. Evaluation by dichromatic ear densitometry, *J. Am. Med. Assoc.*, 200, 236, 1967.
118. Kogure, K., David, N.J., Yamanouchi, U., and Choromokos, E., Infrared absorption angiography of the fundus circulation, *Arch. Ophthalmol.*, 83(2), 209, 1970.
119. Landsman, M.L., Kwant, G., Mook, G., and Zijlstra, W.G., Light-absorbing properties, stability, and spectral stabilization of indocyanine green, *J. Appl. Physiol.*, 40, 575, 1976.
120. Mordon, S., Devoisselle, J.M., Soulie, S.-Begu, and Desmettre, T., Indocyanine green: physiochemical factors affecting its fluorescence *in vivo*, *Microvasc. Res.*, 55, 146, 1998.

121. Schafer, K.A., Kelly, G., Schrader, R., Griffith, W.C., Muggenburg, B.A., Tierney, L.A., Lechner, J.F., Janovitz, E.B., and Hahn, F.F., A canine model of familial mammary gland neoplasia, *Vet. Pathol.*, 35, 168, 1998.
122. Ntziachristos, V., Ma, X., and Chance, B., Time-correlated single photon counting imager for simultaneous magnetic resonance and near-infrared mammography, *Rev. Sci. Instrum.*, 69, 4221, 1998.
123. McMaster, K.M., Giuliano, A.E., Ross, M.I., Reintgen, D.S., Hunt, K.K., Klimberg, V.S., Whitworth, P.W., Tafra, L.C., and Edwards, M.J., Sentinel lymph node biopsy for breast cancer – not yet the standard of care, *New Engl. J. Med.*, 339, 990, 1998.
124. Krag, D., Weaver, D., Ashikaga, T., Moffat, F., Klimberg, S., Shriver, C., Feldman, S., Kusminsky, R., Gadd, M., Kuhn, J., Harlow, S., Beitsch, P., Whitworth, P., Foster, R., Jr., and Dowlatshahi, K., The sentinel node in breast cancer — a multicenter validation study, *New Engl. J. Med.*, 339, 941, 1998.
125. Licha, K., Becker, A., Kratz, F., and Semmler, W., New contrast agents for optical imaging: acid-cleavable conjugates of cyanine dyes with biomolecules, *Proc. SPIE*, 3600, 29, 1999.
126. Huber, M.M., Staubili, A.B., Kustedjo, K., Gray, M.H.B., Shih, J., Fraser, S., Jacobs, R.E., and Meade, T.J., Fluorescently detectable magnetic resonance imaging agents, *Bioconjug. Chem.*, 9, 242, 1998.

# 34

# Optoacoustic Tomography




Alexander A. Oraevsky*
*University of Texas Medical Branch*
*Galveston, Texas*

Alexander A. Karabutov**
*University of Texas Medical Branch*
*Galveston, Texas*


## 34.1 Definition of Optoacoustic Tomography

Optoacoustic tomography (OAT) is a method of image reconstruction based on time-resolved detection of acoustic-pressure profiles induced in tissue through absorption of optical pulses under irradiation conditions of temporal pressure confinement during optical-energy deposition.[1,2] The phrase "irradiation conditions of temporal pressure confinement" means that optical energy (or other heat-generating energy) must be delivered to tissue faster than the resulting acoustic wave can propagate the distance in tissue equal to the desirable spatial resolution. For example, to achieve a resolution of optoacoustic images


Current affiliation: * Fairway Medical Technologies, Houston, Texas
** Moscow State University, Moscow, Russia



0-8493-1116-0/03/$0.00+$1.50

of 15 μm with a speed of sound propagation in tissue of 1.5 mm per microsecond, optical pulses shorter than 0.01 μsec are required. Thus, the use of short (nanosecond) optical pulses represents a necessary (but not sufficient) condition for achieving the desired spatial (axial) resolution of OAT.[3,4] Achievement of the desired spatial resolution requires acoustic-wave detectors that have a temporal response function not slower than the optical-pulse duration.

Irradiation conditions of temporal pressure confinement must also be satisfied if optoacoustic signals are to accurately resemble profiles of absorbed optical energy in tissue.[5] The distribution of absorbed optical energy can be used to visualize and quantitatively characterize various tissue structures and their physiological functions based on variations in tissue optical properties. To relate tissue structure to optoacoustic images the acoustic detectors must be capable of resolving rapid changes in optoacoustic signals associated with sharp edges and boundaries in tissues and reproducing slow changes associated with smooth variation in optical properties within one type of tissue. In other words, acoustic detectors must detect both high and low ultrasonic frequencies of acoustic pressure at once. These types of acoustic detectors are called ultrawide-band acoustic transducers.[6,7] These transducers have relatively equal detection sensitivity over the entire ultrasonic range from 20 kHz to 20 MHz (and in some cases even higher up to 100 MHz).[8] The ultrasonic detection bandwidth of acoustic transducers defines the limits of depth resolution. The lateral resolution of OAT, on the other hand, is defined by the dimensions of each acoustic transducer and the dimensions and geometry of the acoustic transducers in an array.[9,10] Only an array of transducers provides lateral resolution of optoacoustic images. The array can be simulated by scanning a single transducer along tissue surface.[11]

In this chapter we will elaborate on every aspect given in the definition of OAT including an explanation of optoacoustic profiles, ultrawide-band acoustic transducers, algorithms of optoacoustic image reconstruction, and medical applications of OAT. The chapter provides introductory and background information for students, scientists, engineers, and medical professionals interested in imaging methods that use light and ultrasound. Space limitations do not allow either a comprehensive theoretical treatment of optoacoustic phenomena in biological tissue or a detailed description of optoacoustic imaging technology.

## 34.2 History of Development

Theoretical and experimental optoacoustic studies from the 1970s and early 1980s conducted with nonbiological media created a theoretical and experimental background for the development of biomedical optoacoustics.[12] The initial idea driving advances in laser optoacoustics was the well-known exceptional sensitivity of piezoelectric detection.[13] Optoacoustic-signal profiles were first studied with temporal resolution after successful development of acoustic transducers with fast (nanosecond) response times.[14] The first proposed optoacoustic applications aimed at sensitive detection of lesions in biological tissue based on measurements of arrival time of acoustic signals with no regard for information contained in the profile of optoacoustic signals.[15] Later it was realized that the optoacoustic method with temporal resolution could be applied to monitoring laser-energy deposition in optically absorbing media during the ablation process.[16–18] The lack of a theoretical basis, however, prevented researchers from achieving quantitative results at that time. Developing an optoacoustic method suitable for applications in biological tissue was challenging due to its complex heterogeneous structure and optical properties, which were determined by three parameters: absorption coefficient, scattering coefficient, and scattering anisotropy. Rapid progress in the area of biomedical optoacoustics came with the understanding that under irradiation conditions of temporal pressure confinement the profile of laser-induced pressure transients accurately replicates the distribution of absorbed optical energy in the irradiated volume of tissue.[3,4] In other words, under experimental conditions, when laser energy is deposited faster than the propagation of acoustic waves along the volume of tissue with characteristic dimensions of desirable spatial resolution, quantitative information on tissue optical properties can be obtained from time-resolved profiles of laser-induced pressure. It was then realized that a time-resolved optoacoustic profile would allow visualization of the distribution of absorbed optical energy in two and three dimensions inside opaque highly scattering media such as biological tissue.[19–22] The imaging technology is now known as OAT.[23,24]

Thermal pulses of microwave radiation can also be used instead of optical energy for generating acoustic transients in tissue. This type of imaging technology, called thermoacoustic computed tomography, provides deeper penetration of excitation energy in tissue but at the expense of lower tissue contrast.[25,26] Initial experiments on optoacoustic imaging of gelatin phantoms resembling tissue optical properties were performed less than a decade ago.[23,24,27–37] Those first experiments were followed by studies on human tissues *in vitro*.[38–40] Recently OAT was established as a new method of medical imaging after being successfully used for two-dimensional (2D) OAT of tissues *in vivo*, mainly for noninvasive detection of breast cancer with submillimeter resolution[41–43] and early detection of microscopic superficial lesions of oral cancer.[44–46]

Advancement of the time-resolved stress-detection (TRSD) method was made possible after the development of sensitive piezoelectric detectors with an ultrawide ultrasonic detection band of up to 100 MHz.[6–8,11] Piezoelectric transducers are the ultrasonic detectors of choice, especially when low thermal noise and sensitive detection is required within a very wide frequency band.[8] Ultrawide-band piezoelectric detectors can be miniaturized for use in various medical applications including image-guided biopsy of prostate cancer and optoacoustic endoscopy. Optical oblique incidence reflectance and interferometric techniques can also be applied to detect wide-band acoustic signals.[47–49] Optical detection has two main advantages over other methods; it (1) has the potential for remote (noncontact) optoacoustic detection[50,51] and (2) provides rapid monitoring of large areas.[52] The disadvantage of this method relative to piezoelectric detection is a lower sensitivity and high noise level in the range of acoustic frequencies greater than 1 MHz.[8]

## 34.3 Laser Generation of Optoacoustic Transients

The generation of acoustic waves by the consecutive transformation of optical energy into heat and then to mechanical stress is known as the thermooptical mechanism of stress generation.[12] Short laser pulses allow the most efficient induction of the thermooptical stress. Absorption of laser radiation in a medium followed by a fast nonradiative relaxation of the excited states converts laser energy into heat. Subsequently, thermal expansion of the tissue heated under irradiation conditions of temporal pressure confinement causes a pressure rise, $\Delta P$, in the irradiated volume:[5,12]

$$\Delta P = \frac{1}{\gamma}\frac{\Delta V}{V} = \frac{1}{\gamma}\beta\Delta T = \frac{1}{\gamma}\frac{\beta E_a(z)}{\rho C_V} = \frac{\beta c_0^2}{C_p}\mu_a F(z) = \Gamma\mu_a F(z) \tag{34.1}$$

where $\gamma$ [bar$^{-1}$] is the thermodynamic coefficient of isothermal compressibility ($4.59 \times 10^{-5}$ bar$^{-1}$ for water):

$$\gamma = \frac{1}{\rho c_0^2}\frac{C_p}{C_V} \tag{34.2}$$

where $c_0$ [m/sec] is the sound velocity in tissue, $\Delta V$ [cm$^3$] is the volume increase caused by the thermal expansion, $V$ is the laser-irradiated volume initially at room temperature, $\rho$ [g/cm$^3$] is the density of a medium, $C_p$ [J/gK] is the heat capacity at constant pressure, and $C_v$ is the heat capacity at constant volume. The pressure increase is proportional to the thermal coefficient of volume expansion, $\beta$[K$^{-1}$], of the given medium and the absorbed energy density, $E_a$ [J/cm$^3$], which in turn equals the product of the laser fluence, $F(z)$ [J/cm$^2$], and the absorption coefficient of tissue, $\mu_a$ [cm$^{-1}$]. Pressure measured in [J/cm$^3$] can be equally expressed in [MPa] or decabar (see SI and CGS units). The expression $(\beta c_s^2/C_p)$ in Equation 34.1 represents thermoacoustic efficiency, often called Grüneisen parameter, $\Gamma$, which equals 0.1 for water and aqueous solutions at room temperature, T = 20°C. In comparison, $\Gamma$ = 12.7 for fats, lipids, and oils, the most efficient biological media for generation of optoacoustic pressure (data from References 53 through 55). The Grüneisen parameter is a dimensionless, temperature-dependent factor proportional to the fraction of thermal energy converted into mechanical stress. For liquid water within a temperature range of 4 to 100°C, $\Gamma$ can be expressed by the following empirical formula:[5,56]

$$\beta(T) = -0.033 + 0.007T - 0.0000236T^2 \tag{34.3}$$

where temperature, $T$, is measured in degrees Celsius.

Equation 34.1 is valid only when the heating process is much faster than the medium expansion and conditions of temporal pressure confinement are satisfied. When the laser pulse does not satisfy the conditions of temporal pressure confinement, optoacoustic imaging will lose both sensitivity and resolution.

# 34.4 Optoacoustic Profiles and Detection Geometry

## 34.4.1 Two Modes of Optoacoustic Detection

Unlike the majority of optical imaging technologies, where image information is obtained through measurements of the signal amplitude, OAT is based on measurements and analysis of optoacoustic-signal profiles. Optoacoustic profiles generated under irradiation conditions of temporal pressure confinement replicate axial profiles of the absorbed optical energy (where the axis crosses the centers of the acoustic transducer and irradiated volume). Irradiation of tissue in OAT is generally performed through an optically and acoustically transparent medium (optoacoustic buffer) designed to provide coupling of both the incident light from the laser source into the tissue and the resulting ultrasonic waves from the tissue into the acoustic detector.

The two modes of optoacoustic detection schematically depicted in Figure 34.1A and B are known as optoacoustic "forward" and "backward" modes.[4,56] Pulsed laser heating of the absorbing tissue under planar wave geometry results in generation of two ultrasonic transients being launched in opposite directions — forward into the volume of irradiated medium and backward into the transparent medium. The temporal optoacoustic profile depends on the position of the acoustic transducer relative to the irradiated tissue and on the ratio of acoustic impedances of the absorbing and transparent media.[5,12] In the forward detection mode, the optoacoustic signal propagated forward along the laser beam into absorbing medium is detected at the rear surface of the irradiated medium (Figure 34.1A). In the backward mode, the optoacoustic transient, propagated backward in the opposite direction of the incident laser beam, is detected at the site of laser irradiation. The two detection modes enable measurements of optical absorption and scattering over a wide range of tissue optical properties and ultrasonic absorption and scattering over a wide ultrasonic frequency range.

Each detection mode has its advantages and disadvantages. Both detection modes allow a certain amount of flexibility for *in vivo* and *in vitro* measurements with OAT. The forward detection mode provides signals that are easy to interpret. Possible medical applications of optoacoustic detection in the forward mode are, however, limited by applications in human organs with free access to the two opposite surfaces (such as breast). For various medical applications with only one tissue-surface access (such as skin or internal hollow organs), the backward detection mode is often the only choice. At the same time,

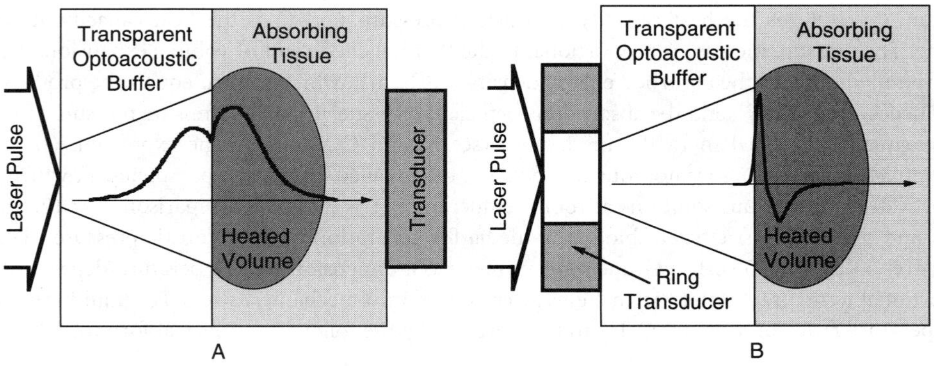


**FIGURE 34.1** Forward (A) and backward (B) modes of the optoacoustic signal detection.

the backward detection mode, which uses solid optically and acoustically transparent material (such as fused silica, see Figure 34.1B) placed on the irradiated surface of tissue for optoacoustic coupling, provides the tissue-surface flatness that is required for generation (and detection) of an ultrawide ultrasonic frequency band in tissue, which in turn provides high spatial resolution of optoacoustic images. On the other hand, the detection of laser-induced transient pressure at the front-irradiated surface makes the design of the piezoelectric detector more complex. Different designs of optoacoustic transducers operating in the backward mode permit tissue irradiation from either the top or the side of the transparent medium coupled to the absorbing tissue.[6,7,56] An ultrawide-band acoustic transducer coupled to the top surface of the optoacoustic transparent medium can detect optoacoustic signals with high temporal resolution sufficient for medical purposes (up to 10 μm). That is why the backward mode enables measurements of optical absorption distribution for a variety of tissues and over the entire optical spectral range.

## 34.4.2 One-Dimensional Optoacoustic Profiles

### 34.4.2.1 Depth Profiles in Plane-Wave Geometry

The optoacoustic profile generated in absorbing tissue, propagated through this tissue to the acoustic detector and detected with temporal resolution, is well established and has been quantitatively investigated theoretically and experimentally.[56] Figure 34.2A and B shows typical optoacoustic profiles detected in forward- (A) and backward- (B) mode geometries. The temporal profiles of the detected optoacoustic (pressure) signals can be presented as a solution:

$$p^{tr}\left(\tau^{tr}=t+\frac{z}{c_0^{tr}}\right)=T_{\alpha c}\frac{c_0^2\beta}{2c_p}\int_0^{+\infty}\left(\tau^{tr}-\theta\right)Q\left(c_0\theta\right)d\theta \tag{34.4}$$

$$p'\left(\tau=t-\frac{z}{c_0}\right)=T_{\alpha c}\frac{c_0^2\beta}{2c_p}\int_{\infty}^{+\infty}\left(\tau^{tr}-\theta\right)Q^{tr}\left(-c_0\theta\right)d\theta \tag{34.5}$$

of the nonstationary thermal-wave equation:[12]

$$\frac{\partial^2\rho'}{\partial t^2}-c_0^2\,\Delta p'=\rho c_0^2\beta\frac{\partial^2 T'}{\partial t^2} \tag{34.6}$$

Here we used the following terms: $c_0$ , $c_0^{tr}$ is the speed of sound in tissue and in the transparent optoacoustic buffer, $\beta$ is the thermal-expansion coefficient for tissue, and $Q(z)$ and $Q^{tr}(z)$ represent distributions of absorbed optical (thermal) energy in tissue (the zone of positive $z$) and in optoacoustic buffer (the zone of negative $z$) that also accounts for reflection of acoustic waves from the boundary.

$$Q=\mu_a F(z),\quad Q^{tr}(z)=\begin{cases}Q(z), & z>0;\\ R_{ac}Q(-z), & z<0;\end{cases} \tag{34.7}$$

Coefficients of acoustic transmission and reflection from the boundary are defined by the acoustic impedances of the transparent buffer, $z_b=\rho^b c_0^b$, and the absorbing tissue, $z_t=\rho^t c_0^t$, as follows:

$$T_{ac}=\frac{2z_b}{z_b+z_t},\quad R_{ac}=\frac{z_b-z_t}{z_b+z_t}, \tag{34.8}$$

where $\rho^b,\rho^t$ [g/m$^3$] are the densities, and $c_0^b$, $c_0^t$ [m/sec] are the velocities of ultrasound waves in the buffer and in tissue.

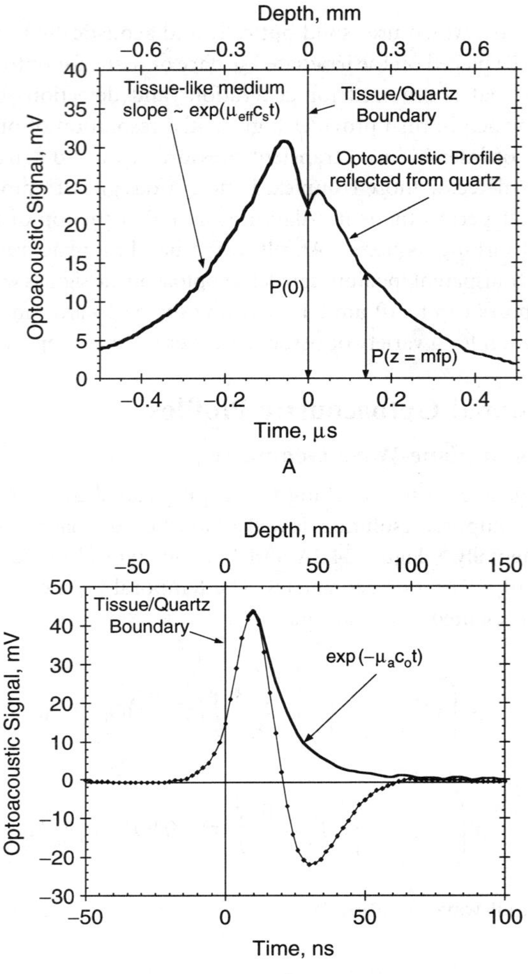


**FIGURE 34.2** Optoacoustic profiles generated in optically absorbing and scattering tissue-like medium covered by an optically transparent window with significantly higher acoustic impedance (quartz). (A) Optoacoustic profile detected in forward mode. (B) Optoacoustic profile detected in backward mode and the same profile corrected for acoustic diffraction. The detection bandwidth of acoustic transducer was approximately 80 MHz.

Therefore, the reflected profile can either be in phase with the original profile ($z_b > z_t$, as at the quartz–tissue boundary) or in contraphase with the original profile ($z_b < z_t$, as at the air–tissue boundary). The optoacoustic profile in the forward mode is described by convolution of the spatial distribution of laser-induced acoustic sources and the temporal shape of the laser pulse.[12,56] Therefore, for sufficiently short laser pulses, $\mu_{eff} c_0 \tau_L << 1$ (where $\mu_{eff}$ is the effective optical attenuation in tissue and $\tau_L = \int_{-\infty}^{+\infty} f(t)dt$ is the laser-pulse duration), the optoacoustic profiles replicate the distribution profile of the absorbed laser energy:

$$p^{tr}\left(\tau^{tr} = t + \frac{z}{c_0^{tr}}\right) = \begin{cases} 0, & \tau^{tr} < 0; \\ T_{ac}\dfrac{c_0^2 \beta \tau_L}{2c_p} Q\left(c_0 \tau^{tr}\right), & \tau^{tr} > 0; \end{cases} \tag{34.9}$$

$$p\left(\tau=t-\frac{z}{c_0}\right)=\begin{cases}\frac{c_0^2\beta\tau_L}{2c_p}Q(-c_0\tau), & \tau<0;\\ R_{ac}\frac{c_0^2\beta\tau_L}{2c_p}Q(c_0\tau), & \tau>0;\end{cases} \tag{34.10}$$

The temporal profile of optoacoustic transients generated under irradiation conditions of temporal pressure confinement and detected in the forward mode without signal distortion replicates the profile of the spatial distribution of absorbed optical energy (Figure 34.2A). The signals from acoustic sources generated by laser pulses deep in the tissue are the first to arrive at the transducer (detector) placed at the rear surface of tissue. These signals from deep inside the tissue are followed by signals from sources located closer to the irradiated surface. A plane-wave optoacoustic profile generated in the tissue by a wide laser beam propagates in two opposite directions. Therefore, the temporal profile consists of two components moving with the speed of sound: the rising (original) spatial profile followed by a falling profile that corresponds to the reflection from the tissue–quartz boundary with acoustic reflection coefficient, $R_{ac}$. Depending on the sign of $R_{ac}$, the falling slope could be either a compressive ($z_b > z_t$, as in the case of rigid quartz–tissue) or tensile ($z_b < z_t$, as in the case of free air–tissue) boundary. The transition zone at the boundary has a duration equal to that of the laser pulse.

With the backward detection mode (see Figure 34.2B) the profile consists of a sharp rising front at $\tau_{tr} = 0$ with the duration of the integrated laser pulse followed by an exponential trailing slope. The trailing slope resembles the profile of absorbed optical energy in tissue, which is determined by the optical attenuation coefficient and the speed of sound. This optoacoustic profile can be represented either by a compression transient wave (when $\beta > 0$, as in tissue at room temperature) or by a tensile transient wave (when $\beta < 0$, as in water in the temperature range of 0 to 4°C). However, due to acoustic diffraction in the quartz optoacoustic buffer, the signal detected in the backward mode is described by the bipolar derivative of the original laser-induced optoacoustic profile, as shown in Figure 34.2B.

The finite duration of laser pulses leads to a widening of all components of the optoacoustic profile. Initially the pulse duration extends the central transition zone of the optoacoustic transients detected in the forward mode and the front of the optoacoustic profile detected in the backward mode. Further increase in the pulse duration results in an optoacoustic profile that does not resemble the distribution of laser-induced acoustic sources and, instead, in most cases replicates the shape of the laser pulse. The optoacoustic profile detected in the forward mode from tissue having a free irradiated surface (free acoustic boundary produced between tissue and air) represents the exception, when the optoacoustic profile follows the derivative of the laser pulse.[12,56]

### 34.4.2.2 Optoacoustic Profiles from an Absorbing Sphere

Spherical objects in tissue irradiated with near-infrared (NIR) light may be considered optically semi-transparent, which means that the effective optical penetration depth in tissue is greater than the linear dimensions (radius, $a$) of the absorbing objects (such as tumors). Furthermore, when $\mu_{eff} << 1/a$, distribution of optical energy inside spherical objects may be considered homogeneous.

For the sake of simplicity, let us consider a spherical object in tissue with optical properties dominated by optical absorption (scattering is apparent but may be neglected). In the spherical coordinates whose center coincides with the center of the object, the radial distribution of thermal energy in and around this object could be approximated by the following function:

$$Q(r)=\begin{cases}\mu_\alpha F_o\, ch(\mu_\alpha r)/ch(\mu_\alpha r), & 0\le r\le a;\\ 0, & a<r.\end{cases} \tag{34.11}$$

where $F_o$ is the laser fluence at the surface of the absorbing sphere, which depends on the effective optical attenuation by absorption and scattering in tissue:

$$F_o = F_0 \exp(-\mu_{eff} z) \tag{34.12}$$

The transient acoustic wave resulting from laser heating of the spherical object will have spherical symmetry and can be given by the following equation:

$$\frac{\partial^2(rp')}{\partial t^2} - c_0^2 \frac{\partial^2(rp')}{\partial r^2} = \frac{c_0^2 \beta}{c_p} rQ(r)\frac{df}{dt} \tag{34.13}$$

with the boundary condition associated with the finite pressure in the center of the sphere:

$$(rp')\big|_{r=0} = 0. \tag{34.14}$$

The reflection of the acoustic wave from the center of the sphere occurs in the contraphase. The system of Equations 34.13 and 39.14 is similar to the planar case with a free (air–tissue) boundary, Equation 34.6. Therefore, Equation 34.5, with a reflection coefficient $R_{ac} = -1$, can be used here:

$$p'\left(\tau = t - \frac{r}{c_0}\right) = \frac{\mu_\alpha F_o c_0^2 \beta}{2rC_p} \int_{-\alpha/c_0}^{+\alpha/c_0} f(\tau - \theta)(-c_0\theta)\frac{ch(\mu_\alpha c_0 \theta)}{ch(\mu_\alpha a)} d\theta. \tag{34.15}$$

As with planar acoustic waves, the optoacoustic profile of a spherical acoustic wave can be expressed as the convolution of the temporal profile of the laser pulse with the depth profile of the absorbed optical energy modified for the contraphase reflection of the acoustic wave from the center of the sphere. Under irradiation conditions of temporal pressure confinement (short laser pulses), so that $\mu_\alpha c_0 \tau_L << 1$ and for relatively large dimensions of the absorbing sphere, $c_0 \tau_L << a$, the optoacoustic profile replicates that of the absorbed (thermal) energy distribution:

$$p'\left(\tau = t - \frac{r}{c_0}\right) = \frac{\mu_\alpha F_o c_0^2 \beta}{2rc_p} \begin{cases} (-c_0\tau)ch(\mu_\alpha c_0 \tau)/ch(\mu_\alpha c_0 \tau), & |\tau| < a/c_0; \\ 0, & |\tau| > a/c_0. \end{cases} \tag{34.16}$$

The optoacoustic profile for relatively weakly absorbing objects, $\mu_\alpha a < 1$, possesses a so-called N-shape, with duration $\tau_N = 2a/c_0$ associated with sound propagation through the object. The optoacoustic profile detected from a strongly absorbing object, $\mu_\alpha a >> 1$, can be described by two short pulses, where a tensile pulse follows the leading compression pulse with a delay equal to the time of sound propagation through the object.

Figure 34.3 shows a typical optoacoustic profile detected in forward mode from absorbing spheres inside a bulk tissue-like phantom irradiated with NIR laser pulses. A wide laser beam was used for illumination of this phantom, which resulted in a planar acoustic wave with a profile resembling depth (axial) distribution of the absorbed optical energy. On the background of the main exponential trend, which is due to the effective optical attenuation, this optoacoustic profile contains two N-shaped wavelets resulting from spherical acoustic waves from the two absorbing spheres. Since both spheres were positioned on one axis with the irradiation-detection system, the pressure profile in Figure 34.3 includes signals simultaneously recorded from both spheres. These phantom studies demonstrated that the accuracy of localization and the accuracy of "tumor" dimension measurements could be greater than 0.5 mm.

In the case of microscopic absorbing objects in tissue ($c_0 \tau_L >> a$, Equation 34.15 can be transformed into the following expression for the optoacoustic profile that replicates the derivative of the laser pulse:

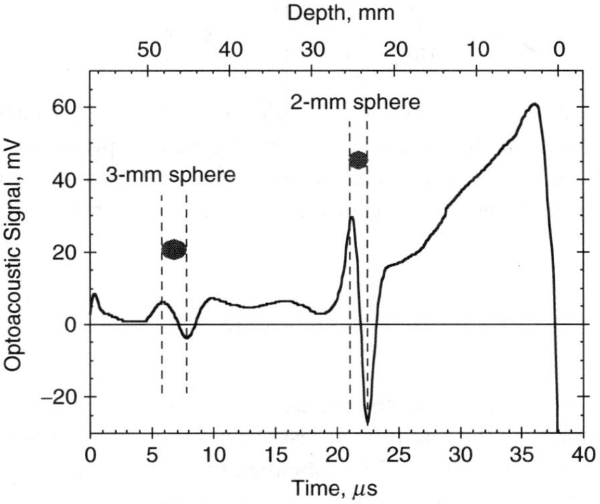


**FIGURE 34.3** Laser-induced optoacoustic profile measured with a wide-band acoustic transducer along the axis with two gel spheres 2 and 3 mm in diameter colored with hemoglobin of blood. Two N-shaped signals are shown. The signal at $t = 0$ representing the transducer response to direct illumination was used for calibration of the time scale. Optical attenuation of the bulk phantom was $\mu_{eff} = 1.4\ cm^{-1}$.

$$p'\left(\tau = t - \frac{r}{c_0}\right) = \frac{\beta F_0 4\pi a^2}{4\pi r c_p}\left[th(\mu_\alpha a) - \frac{2(\mu_\alpha a - th(\mu_\alpha a))}{(\mu_\alpha a)^2}\right]\frac{df}{d\tau}. \tag{34.17}$$

If small (microscopic) structures in tissue also possess a relatively small optical absorbance, i.e., $\mu_a a << 1$, then the optoacoustic amplitude linearly increases with the absorption coefficient:

$$p'\left(\tau = t - \frac{r}{c_0}\right) = \frac{\beta I_s 4\pi a^2}{4\pi r c_p}\frac{\mu_\alpha a}{3}\frac{df}{d\tau}. \tag{34.18}$$

If those microspheres possess strong absorption, then the optoacoustic amplitude saturates:

$$p'\left(\tau = t - \frac{r}{c_0}\right) = \frac{\beta I_s 4\pi a^2}{4\pi r c_p}\frac{df}{d\tau}. \tag{34.19}$$

This type of behavior is explained by the fact that the optoacoustic signal amplitude is determined by the total absorbed optical energy.

### 34.4.3 Acoustic Diffraction and Other Factors of Signal Distortion

The optoacoustic profile can be altered in the course of propagation to the transducer through tissue and a transparent optoacoustic buffer. The main phenomena that contribute to the distortion of original optoacoustic profile are acoustic diffraction, acoustic attenuation, and the detector response dependent on the ultrasonic frequency of the detected signals.[5,12,36] Additional distortion can occur due to roughness of the irradiated tissue surface and oblique incidence of the acoustic wave to the transducer aperture.[10]

The equations for planar (34.3 and 34.4) and spherical (34.15) optoacoustic profiles did not include acoustic diffraction. Generally, acoustic diffraction must be included, so that the optoacoustic profile is described by the following equation based on Equation 34.18:

$$p'(t,\mathbf{r}) = \frac{\beta}{4\pi c_p} \frac{\partial}{\partial t} \int \frac{Q(\mathbf{r}')}{|\mathbf{r}-\mathbf{r}'|} f\left(t - c_0^{-1}|\mathbf{r}-\mathbf{r}'|\right) dV' \tag{34.20}$$

which considers the divergent acoustic wave only outside the volume occupied with laser-induced acoustic sources, $Q(r)$. The solution (Equation 34.20) represents the main expression for OAT, whose goal is to invert Equation 34.20 to image the spatial distribution of the acoustic sources, $Q(r)$.

The acoustic diffraction of the optoacoustic profiles can be described by the following parabolic equation:[12,57]

$$\frac{\partial^2 p'}{\partial \tau \partial z} = \frac{c_0}{2} \Delta_\perp p', \tag{34.21}$$

where $\Delta_\perp$ is the Laplacian in the transversal coordinates $\{x,y\}$, and $z$ is the depth coordinate. The boundary conditions for Equation 34.21 in the case of wide laser beams (commonly used in OAT) can be taken in the form of solutions (Equations 34.4 and 39.5) obtained upon consideration of planar acoustic waves.

Since optical illumination usually has a Gaussian cross-sectional profile, the first approximation for the distribution of laser-induced acoustic sources in the plane perpendicular to the axis of optical illumination may also be taken as Gaussian:

$$p'(z=0,\tau,\mathbf{r}_\perp) = p_0(\tau)\exp\left(-\mathbf{r}_\perp^2/a_0^2\right), \tag{34.22}$$

where $p_0(\tau)$ is the temporal profile of the optoacoustic signal at the irradiated tissue surface.

A specific feature of optoacoustic signals is their ultrawide range of ultrasonic frequencies. The effective length, $L_D$, of the acoustic diffraction depends on the ultrasonic frequency (acoustic wavelength). The acoustic-beam cross section, $\pi a_0^2$, doubles after propagation of a distance equal to the effective diffraction length:

$$L_D = \pi a_0^2/\lambda_{ac} = a_0^2 \omega_{ac}/c_0 \tag{34.23}$$

where $\lambda_{ac}$ is the wavelength of the acoustic wave with ultrasonic frequency, $\omega_{ac}$.[56] Therefore, different spectral components of the optoacoustic profiles undergo different degrees of acoustic diffraction; the lower ultrasonic frequencies contribute the most pronounced changes in the optoacoustic profile. The higher frequencies undergo minimal acoustic diffraction, thereby propagating mainly along the axis perpendicular to the acoustic front. Considering the importance of the effective diffraction length as a function of ultrasound frequency, the solution for the system of Equations 34.21 and 34.22 can be found in the spectral domain:

$$p'\left(z,\tau=t-\frac{z}{c_0},\mathbf{r}_\perp\right) = \int_{-\infty}^{\infty} \exp\left(-i\omega(\tau-t) - \frac{\mathbf{r}_\perp^2}{a_0^2}\frac{\omega}{\omega+i\frac{2c_0 z}{a_0^2}}\right)\left(\omega+i\frac{2c_0 z}{a_0^2}\right)^{-1} \omega d\omega. \tag{34.24}$$

Since high ultrasonic frequencies concentrate near the axis of acoustic beam, the highest resolution in OAT can be obtained by detecting optoacoustic profiles propagating along the axis of the laser beam. The optoacoustic profile at axis $\mathbf{r}_\perp = 0$ can be obtained from Equation 34.24 and expressed as:

$$p'(z,\tau,\mathbf{r}_\perp=0) = p_0(\tau) - \int_{-\infty}^{\tau} \omega_D \exp\left(-\omega_D(\tau-t)\right) p_0(t)dt, \tag{34.25}$$

where $\omega_D = 2c_0 z / a_0^2$ is the characteristic frequency of acoustic diffraction defined as the frequency at which the characteristic diffraction length is equal to the length of propagation, $z$. The first item in equation 34.24 replicates the optoacoustic profile at the irradiated tissue surface, and the second item describes the influence of the limited radial dimensions of the acoustic wave. In the far diffraction zone ($z \to \infty$), the solution (Equation 34.24) can be simplified as:

$$p'(z,\tau,\mathbf{r}_\perp = 0) = \frac{a_0^2}{2c_0 z}\frac{dp_0}{d\tau}. \tag{34.26}$$

The effect of the acoustic diffraction is to produce a derivative of the original optoacoustic profile in the far zone. The task of OAT is to use optoacoustic profiles at the irradiated surface for image reconstruction. Thus, for the purposes of OAT Equation 34.25 must be inverted, which yields the following result:

$$p'(\tau,z=0,\mathbf{r}_\perp = 0) = p'(\tau,z,\mathbf{r}_\perp = 0) + \omega_D \int_{-\infty}^{\tau} p'(\vartheta,z,\mathbf{r}_\perp = 0)\,d\vartheta. \tag{34.27}$$

Acoustic diffraction is most prominent in the backward detection mode, where a quartz acoustic buffer (with high acoustic impedance) is placed between the tissue and the transducer. The influence of acoustic diffraction can be compensated for, and the original laser-induced profile can be reconstructed with a procedure that convolves the detected optoacoustic profile and the reference profile detected from a highly absorbing medium (with $\mu_{eff} > \Delta f_{at}/c_0$, where $\Delta f_{at}$ is the ultrasonic detection bandwidth of the acoustic transducer).[36] The detected reference signal replicates the profile of the laser pulse, with the Fourier spectrum altered by the signal-transmission path and the detection system, which include all distortion factors including acoustic diffraction, acoustic attenuation, and the sensitivity response of the acoustic transducer.

## 34.5 Two-Dimensional Imaging with Array vs. Scanning Mode

Two-dimensional optoacoustic images can be acquired either by using a stationary array of acoustic transducers or by scanning a single transducer along the tissue surface. Figure 34.4 depicts schematic diagrams of the two image-acquisition modes. One mode may have better lateral resolution for a specific clinical application. Stationary arrays of acoustic transducers allow real-time imaging deep inside tissue within a large field of view. On the other hand, all arrays have a limited zone where all transducers perform equally. Therefore, transducer arrays yield much better lateral resolution when imaging large tissue volumes in the far zone of the detector, especially in human organs that can be surrounded by the array (such as breast).

Figure 34.4A shows an array of acoustic transducers employed for breast imaging in the forward mode. Optical pulses can be delivered to the tissue surface with one or more optical fibers, as shown in the drawing. As an alternative, a wide laser beam can be used to illuminate the entire volume of the breast. Breast tumors having stronger optical absorption (primarily due to enhanced blood content) and greater optical scattering (primarily due to enlarged cell nuclei) produce a larger amplitude of optoacoustic signals in response to laser heating. The resulting acoustic waves propagate in tissue in all directions and can be detected with an array of acoustic transducers. The product of the temporal width of the optoacoustic signal produced in a tumor and the speed of sound yields tumor dimensions in the direction of the axis between the tumor and the detector. The product of the speed of sound in the tissue and the delay time between the laser pulse and the arrival of the transient pressure wave at the transducer is equal to the distance between the tumor and the transducer. A linear or arc-shaped array of acoustic transducers permits reconstruction of 2D images of optically heterogeneous structures located in the breast.[42]

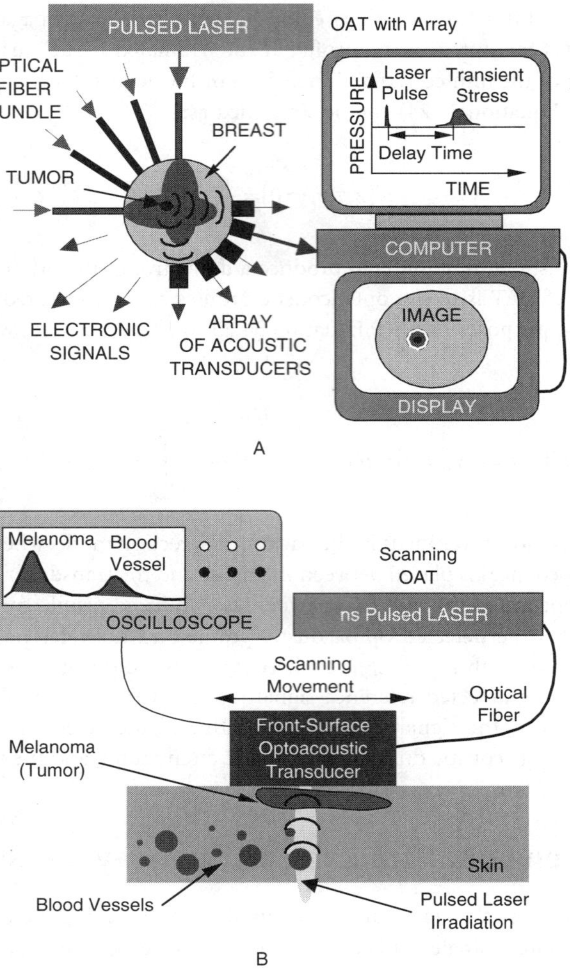


**FIGURE 34.4** Schematic diagrams of optoacoustic tomography. (A) Forward mode imaging with an array of optoacoustic transducers. (B) Backward mode imaging by scanning a single transducer.

Figure 34.4B shows scanning of a single acoustic transducer along the skin surface for acquisition of 2D images of subsurface layers. Laser pulses are delivered to the tissue surface with a single optical fiber. Scanning a single transducer is most appropriate for the backward mode of acoustic detection at the site of optical irradiation. Translation of a single transducer along straight lines can produce 2D images with high lateral resolution defined by the resolution of a single transducer. The scanning mode is the only way to produce 2D imaging in the near zone of subsurface tissue volume because only one transducer with finite dimensions can image tissue located underneath that transducer. Acoustic transducers for the scanning mode must be designed with a very narrow angle of acceptance to provide lateral resolution. The best lateral resolution can be obtained with confocal optoacoustic transducers.

## 34.6 The Laser Optoacoustic Imaging System

The laser optoacoustic imaging system (LOIS) was developed for OAT of tissue structures located at a depth of up to 6 to 7 cm. LOIS combines the most compelling features of optical and ultrasound imaging systems, pronounced optical contrast, and high resolution of ultrawide-band ultrasound detection, yielding a more sensitive imaging modality with substantially improved accuracy of the localization (<1 mm). The hybrid of optical and acoustic technologies overcomes the problems of pure optical imaging associated with the loss of resolution and sensitivity due to strong light scattering as well as problems of pure ultrasound imaging associated with poor tissue contrast and low signal-to-noise ratio (SNR).

## 34.6.1 Arc-Array of Ultrawide-Band Piezoelectric Transducers

The heart of LOIS is an array of ultrawide-band acoustic transducers. The greater the distance between the end elements in the array and the smaller the individual elements, the better spatial resolution can be achieved.[57,58] The first LOIS had only 12 relatively wide piezoelectric elements in a relatively short linear array, which provided very limited spatial resolution.[39] Advances in OAT over last 3 years have resulted in the design and development of longer transducer arrays that have a larger number of smaller-dimension piezoelectric transducers and are shaped like an arc to improve spatial (lateral) resolution. The array employed in experiments described in this chapter had 32 rectangular piezoelectric transducers measuring 1 mm × 12.5 mm and a distance of 3.85 mm between the transducers (Figure 34.5).[40] The 110-μm-thick piezoelectric polymer PVDF was used to make the transducer. Low acoustic impedance and an ability to operate in a wide ultrasonic-frequency band are the advantages of PVDF for detecting optoacoustic profiles.[8] Alternatively, piezoceramic materials, such as PZT-5H or composite polymer-ceramic materials, can be used.[8]

The transducers were located on the arc surface with a radius of 60 mm. Such geometry provided optimal resolution in the entire 60-mm × 60-mm field of view within the image plane. The shape and length of individual piezoelectric elements determine spatial resolution of this array in the plane perpendicular to the image plane. Flat elements provide a resolution equal to the linear size of the element. Backing polymer material with acoustic impedance similar to that of PVDF was used for mechanical matching of the piezoelectric material and damping reverberations after detection. OAT requires that the piezoelectric element not vibrate at its thickness-related resonance frequency after detecting an optoacoustic signal, and the frequency band of ultrasonic detection is widened in the range defined by the application. The front surface of the transducer array was covered by an aluminum film to reduce external electric noise. The thickness of the aluminum layer was such that the balance between ultrasound penetration and electric-noise reduction was optimized. The transducer-array housing was also made of aluminum. Acoustic insulating material was placed between the piezoelectric elements and the housing to prevent ringing of the housing that follows the vibration of the piezoelectric elements themselves.

## 34.6.2 Sensitivity of Detection

The theoretical absolute sensitivity of transducers, $S$ = 19.8 μV/Pa, can be calculated on the basis of piezoelectric properties of the PVDF material used:

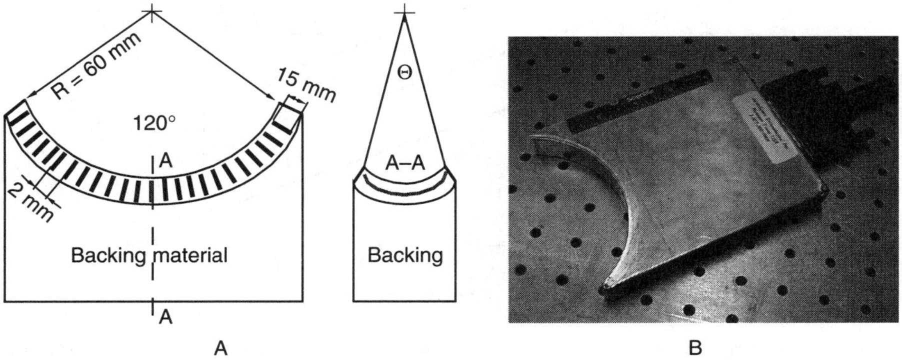


**FIGURE 34.5** Schematic diagram (A) and photograph (B) of the arc-shaped transducer array.

$$S = g_{33}d\,[\text{V/Pa}]. \tag{34.28}$$

where $g_{33}$ = 0.18 V m/N is the piezoelectric pressure modulus of the transducer material, and $d$ = 110 μm is the thickness of the piezoelectric element. Some portion of acoustic energy, however, can be lost due to reflection at the boundary of the shielding layer and the piezoelectric film. The defects of bonding of these two layers cause additional losses.

Detection of the optoacoustic signal obliquely incident on a piezoelectric transducer with finite dimensions can alter its profile.[10,59] Therefore, an ideal piezoelectric element would have dimensions significantly smaller than the shortest detected acoustic wavelength, $\lambda_{ac}$. However, small dimensions yield a small value of electric capacitance of the transducer piezoelement, $C_{pe}$, which can be derived from the following formula:

$$C_{pe} = \frac{\varepsilon\varepsilon_0 A}{d} \tag{34.29}$$

where $A$ is the surface area of the piezoelement, and $\varepsilon\varepsilon_0$ is the dielectric permittivity of the transducer material. To avoid the shunting effect of the electronic circuitry, the capacitance of each transducer element should not be lowered to a level comparable to the stray capacitance of the connecting wires and input capacitance of the electronic preamplifier. The shunting effect can potentially be the source of significant losses of sensitivity in the detection system.

The ratio of the absolute sensitivity and the minimal detectable pressure dependent on thermal noise in the piezoelectric element defines the detection capability of the acoustic transducer. The minimally detectable pressure of the ultrawide-band ultrasonic transducer can be derived from the following expression:[8]

$$p_{min} = \frac{\text{M}\, f_{max}}{g_{33}c_l}\sqrt{\frac{kT_0}{C_{pe}}} \tag{34.30}$$

where $f_{max}$ is the upper ultrasonic frequency of the transducer detection band, $c_l$ is the longitudinal speed of sound in the piezoelectric element, and $kT_0$ is the product of the Boltzmann constant ($1.38 \cdot 10^{-23}$ J/K) and ambient temperature (300 K). The numerical coefficient, $M$, in the numerator depends on whether the transducer operates in an open ($M$ = 4.6) or short ($M$ = 6.5) electrical circuit.[8]

Minimally detectable pressure is presented in Figure 34.6 as a function of the transducer capacitance, the main variable parameter for the PVDF material of the piezoelectric element operating in the open

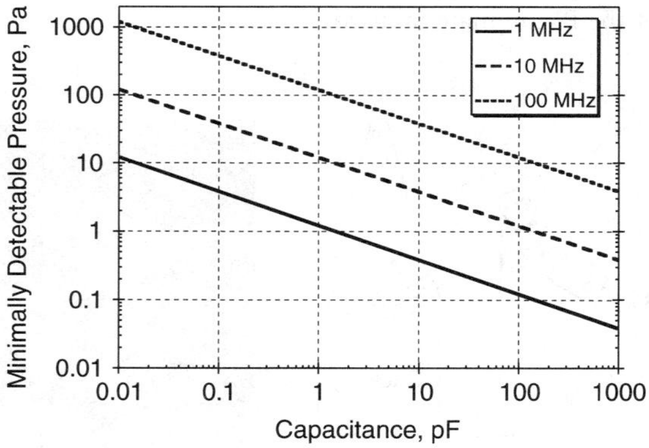


**FIGURE 34.6** Acoustic pressure generated by thermal noise as a function of capacitance of acoustic transducer made of PVDF film and operating in the open circuit mode. Three curves for three values of upper ultrasonic frequency (1, 10, and 100 MHz) of the transducer detection band are shown.

circuit. This plot demonstrates that pressures in the range of 1 Pa can be detected with LOIS. It is difficult to calculate all possible losses of acoustic energy. Therefore, direct sensitivity measurement for all transducers in an array is very important. The experimental variability of the sensitivity between array elements is usually about 10 to 15% and must be taken into account in LOIS calibrations.

### 34.6.3 Spatial Resolution

The axial (depth) resolution of a pulsed optoacoustic system operating under irradiation conditions of temporal pressure confinement is directly defined by the detection bandwidth of ultrasonic transducers. The temporal response time of an ultrawide-band acoustic transducer is defined as $1.5/f_{max}$, where $f_{max}$ is the upper frequency limit of the piezoelectric element. The coefficient 1.5 accounts for the fact that for resolved detection of a sphere with diameter $2a_{min}$ at least three data samples must be recorded. According to Sparrow's criteria, two coherently emitting identical spherical acoustic sources with radius $a_{min}$ will be detected as separate objects if the space between their centers is equal to or greater than $\delta_z = 2a_{min}$.[58] The use of this transducer for imaging in tissue where the speed of sound equals $c_0$ makes it possible to obtain spatial resolution of:

$$\delta_z = 2a_{min} = 3\frac{c_0}{f_{max}}. \tag{34.31}$$

The upper frequency limit of the ultrasonic detection band, $f_{max} = c_1/2d$, is defined by the transducer thickness and the frequency bandwidth of the preamplifier used. In LOIS design using a 110-μm-thick PVDF film, the upper frequency limit was 4 MHz, which provided 1.1-mm axial (depth) resolution.

The arc geometry of the transducer array provides enhanced lateral resolution of the area near the center of the arc curvature. This array geometry is the most beneficial for imaging within a circular field of view. In this area, the lateral resolution can be estimated as the width of the synthetic directivity pattern produced by the entire aperture of the array (see diagram in Figure 34.7):

$$\delta_l = 1.22\frac{\lambda_{ac}}{\arcsin(r/R)} \approx c_0\tau_a / 2\sin\theta, \tag{34.32}$$

where $\theta = \arcsin(r/R)$ is the angular aperture of the array, $R$ is the radius of the arc curvature, $r$ is the aperture radius, and $\tau_a = 2a_{min}/c_0$ is the duration of the acoustic pulse from the minimally detectable sphere. If optoacoustic transients are detected, the spatial pulse duration $c_0\tau_a$ should be used as the equivalent of the shortest detectable acoustic wavelength, $\lambda_{ac}$. Equation 34.32 is equivalent to the equation for the waist size of a focused transducer with the same configuration obtained for harmonic-wave radiation.[10] This formula works well as long as all transducers in the array make equal contributions to the formation of the synthetic aperture. Diffraction can reduce the sensitivity of some transducers when

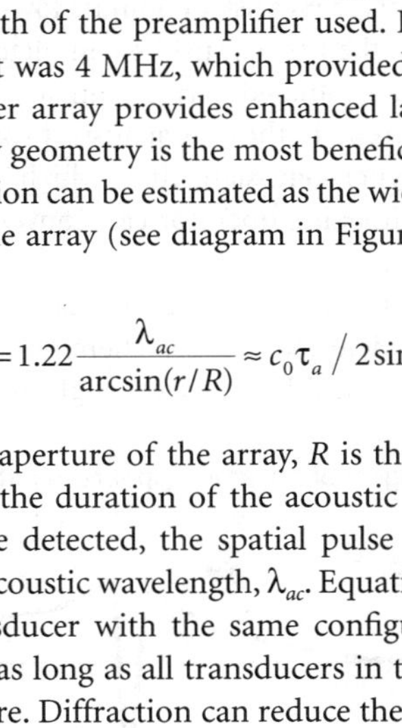

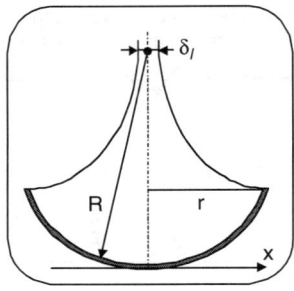


**FIGURE 34.7** Schematic diagram of directivity of acoustic detection for an arc-shaped array of piezoelectric transducers.

they receive acoustic transients from points outside the central zone. The acoustic diffraction can, therefore, reduce the lateral resolution, which depends on the position of the acoustic source relative to the array.

With the 32-element arc-shaped array depicted in Figure 34.5, lateral resolution of 1 mm can be obtained. In the entire field of view, the lateral resolution is in the range of 1 mm < $\delta_l$ < 2 mm. For the arc-shaped transducer array described above, the ratio $r/R$ equals approximately 0.87; therefore, Equation 34.32 can be simplified, yielding $\delta_l \approx 1.22\lambda_{ac}$. This estimation is valid for monofrequency acoustic waves. With ultrawide-band detection, the detection-frequency spectrum must be accounted for and the correct value of the effective $\lambda_{ac}$ selected.

## 34.6.4 Signal Processing in Optoacoustic Tomography

The signal-processing algorithm realized in LOIS is shown schematically in Figure 34.8. Each signal detected by acoustic transducers in the array consists of three components: (1) a sharp peak produced at the very surface of the medium under investigation, (2) a low-frequency, smoothly ascending exponential slope produced by the attenuation of optical intensity inside the tissue, and (3) a short N-shaped pulse generated by a small absorbing object (tumor). We assumed here that the tumor had a spherical shape with radius $a$. The duration of the N-shaped pulse is defined by the time of sound propagation through the sphere diameter. The delay in time of signal arrival depends on tumor location relative to the transducer. Therefore, the position of the tumor and its dimensions can be determined from the optoacoustic signal.

Before the signal can be used for image reconstruction, its low-frequency component associated with the effective optical attenuation inside the tissue must be completely eliminated because the high amplitude of the optoacoustic signal at the irradiated surface can significantly decrease image contrast. The removal of the smooth (exponential) trend in the optoacoustic profile can be achieved with either wavelet filtering[60] or a high-pass numeric hyper-Gaussian filter, which performs better than a simple set of resistor-capacitor (RC) cells.[40,42] The imaging operator can conveniently vary the cut-off frequency and the slope of the filter transfer function.

The signal generated by laser pulses in the subsurface layer (very close to the surface) possesses maximum amplitude over the entire detected course. Therefore, the signal gradient is maximal at the tissue surface, which yields high ultrasound frequencies. High ultrasound frequencies that fall near the

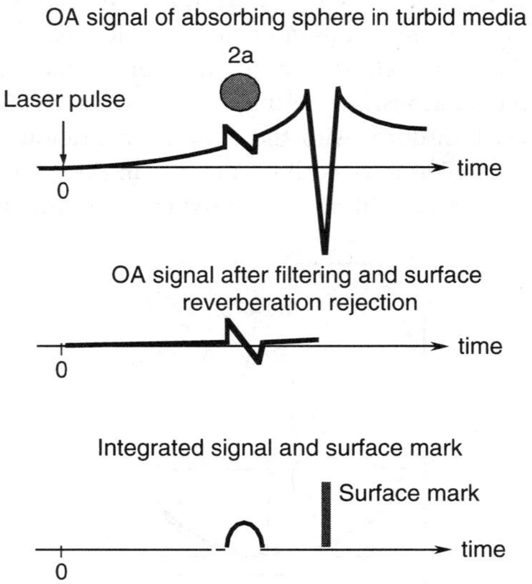


**FIGURE 34.8** Optoacoustic signal processing. Top: Detected optoacoustic profile. Center: Exponential trend caused by effective optical attenuation is removed. Bottom: Original signal from an absorbing sphere restored.

resonance frequency, $f_{max}$, of piezoelements can induce significant reverberation in the transducer. The transducers used in LOIS are designed to dampen resonance frequencies and widen the range of detectable frequencies. Nevertheless, a strong reverberation takes place after the arrival of the signal from the tissue surface. These reverberations, however, do not affect the useful part of the optoacoustic profile. Besides, it is convenient to use a sharp high-gradient (derivative) peak for automatic determination of the tissue surface. The position of the irradiated tissue surface was considered to be the position of absolute minimum in the signal. The controlling computer generated the numerical signal at the surface position determined for each transducer and stored these data in a separate file. All data samples were cut off from the signal after arrival of the signal generated at the tissue surface. Integrated signals were used as input data for the image-reconstruction code. The computer-generated surface marks were used for surface visualization, so that the location of internal tissue structures could be related to the irradiated surface of tissue.

## 34.6.5 Algorithm of Image Reconstruction

A radial back-projection algorithm, presented schematically in Figure 34.9, was developed for optoacoustic imaging with an array of ultrawide-band transducers.[40,42,61] For the reconstruction of 2D optoacoustic images we used signal integrals projected back onto the 2D grid, taking into account the directivity pattern (angle of acceptance) of each transducer in the array. The image represents the distribution of a product of thermoacoustic efficiency, optical absorption coefficient, and effective laser fluence in tissue.

Consider a pulsed laser excitation of acoustic waves in a homogeneous medium with optical-absorption coefficient $\mu_a$. The temporal integral of acoustic pressure detected by the transducer located at point $r$ can be expressed in the following form:[58]

$$u(\vec{r},t) = \int_{-\infty}^{t} p(\vec{r},t')dt' = \frac{\beta}{4\pi C_p} \int_V \frac{\mu_a(\vec{r}')I(\vec{r}')L\left(t - \frac{|\vec{r}-\vec{r}'|}{c_s}\right)}{|\vec{r}-\vec{r}'|} d\vec{r}'. \tag{34.33}$$

where $\beta$ is the thermal coefficient of volume expansion, $C_p$ is the specific heat capacity of the medium, and $L$ is the laser pulse temporal profile. All other notations used in Equation 34.33 are shown in Figure 34.9.

The integral in Equation 34.33 is calculated over the entire space. This means that the acoustic pressure $p(\vec{r},t)$ at the time moment $t$ and at point $r$ is determined by the acoustic sources located in the spherical shell with radius $|\vec{r}-\vec{r}'|$ and thickness $d\vec{r}'$. Acoustic waves arrive at measurement point $\vec{r}$, with time delay $|\vec{r}-\vec{r}'|/c_s$. If the laser pulses are sufficiently short, then generation of thermal sources may be considered instantaneous, and the laser waveform function can be expressed as $L(t) = \tau_L\delta(t)$, where $\tau_L$ is the laser-pulse duration, and $\delta(t)$ is the delta function. Equation 34.33 for short laser pulses can be written as follows:

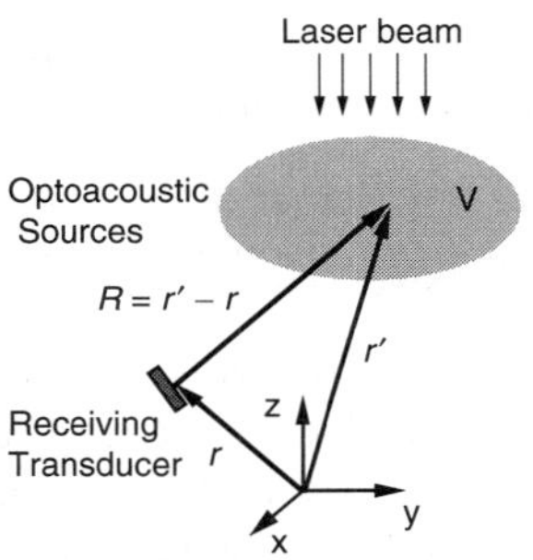


**FIGURE 34.9** Geometry of the optoacoustic detection.

$$u(\vec{r},t) = \frac{\beta}{4\pi C_p} \int_V \frac{\mu_a(\vec{r}')F(\vec{r}')\delta(t - \frac{|\vec{r}-\vec{r}'|}{c_s})}{|\vec{r}-\vec{r}'|} d\vec{r}' \quad (34.34)$$

where $F$ is the laser fluence. Let surface $\Sigma$ be the spherical surface of radius, $R$, with its origin at the position of the receiving transducer. The radius of the sphere is defined by the time of sound propagation from the surface to the transducer. Vector $\vec{R}$ can be presented as $\vec{R} = c_s t\vec{n}$, where $\vec{n}$ is the unit vector of the surface element $d\Sigma$. Thus, the temporal integral of acoustic pressure can be expressed as:

$$u(\vec{r},t) = \frac{\beta}{4\pi C_p c_s t} \int_\Sigma Q(\vec{r} + c_s t\vec{n}) d\Sigma \quad (34.35)$$

where function $Q(r) = \mu_a(r)F(r)$ describes the spatial distribution of absorbed optical energy.

Equation 34.35 can be considered fundamental for OAT. The temporal integral of the pressure profile detected by transducer at time $t$ is the superposition of the laser-induced acoustic sources located on the sphere of radius $R = c_s t$. The product of the optical absorption coefficient and the effective optical fluence at this spherical surface determines the optoacoustic amplitude. The back-projected images acquired with a limited number of detectors display artifacts associated with image reconstruction using an incomplete data set. The contrast of these images can be improved through the image-filtration procedure applied to the entire 2D image.[40] However, unprocessed optoacoustic images contain quantitative information about the distribution of absorbed optical energy in tissue.[42] This quantitative information may be used for functional imaging and measurements of tissue optical properties.

### 34.6.6 Image Processing

OAT adopted the radial back-projection algorithm, earlier developed for ultrasound imaging.[62] For the reconstruction of 2D optoacoustic images LOIS calculates integrals of the optoacoustic signals and projects them back onto the 2D grid, taking into account the directivity pattern (angle of acceptance) of each acoustic transducer. The field of view that results from the radial back-projection algorithm can be expressed in the following form:

$$\xi(\vec{r}) = \sum_{n=1}^{m} u_n(|\vec{r} - \vec{r}_n|/c_s)|\vec{r} - \vec{r}_n| \quad (34.36)$$

where $u_n$ is the integrated optoacoustic signal of the $n$th transducer, $\vec{r}_n$ is the radius vector of the $n$th transducer, and $m$ is the number of transducers in the array. The image represents the distribution of a product of thermoacoustic efficiency, optical absorption, and laser fluence. The back-projected images of very small or low-contrast objects might be noisy and their brightness sometimes may be insufficient for object recognition. Therefore, an image-filtration procedure was employed:

$$\xi_2(\vec{r}) = \frac{1}{4\pi^2} \int F_\xi(\vec{\omega}) H(\vec{\omega})\, e^{i\vec{\omega}\vec{r}} d\vec{\omega} \quad (34.37)$$

where $F_\xi(\vec{\omega}) = \int \xi(\vec{r}) e^{-i\vec{\omega}\vec{r}} d\vec{\omega}$ is the 2D Fourier spectrum of the back-projected image, $H(\vec{\omega}) = |\vec{\omega}| e^{-(|\vec{\omega}|/\sigma)^2}$ is the filter-transfer function, and $\vec{\omega} = \{\omega_x, \omega_y\}$ is the vector of spatial frequencies. Parameter $\sigma$ permits variation of the filter bandwidth and the steepness of the filter characteristics. Optimization of $\sigma$ can be achieved manually by operator or automatically. In the automatic procedure, parameter $\sigma$ can be chosen from the analysis of image quality based on the contrast–dimension relationship. The filter parameter, $\sigma$, can be further optimized by employing contrast loss in the resulting image.[63]

An image-reconstruction algorithm based on the Radon transform with a limited data set was proposed for OAT.[61] If used directly, this method does not yield satisfactory resolution, even when a significant number of transducers is used. However, use of the Radon transform is possible if the integrals along the arcs are substituted by the integrals along the straight lines tangential to the corresponding arcs in limited areas containing suspicious objects inside the total field of view and using an iterative approach for the solution of the inverse problem.[63]

# 34.7 Application of LOIS to Detection of Breast Cancer

At present, detection of breast cancer is the main biomedical application for OAT. LOIS combines advantages of pronounced optical contrast and sensitive, high-resolution ultrasonic detection, resulting in a more advanced imaging system for breast cancer.[1,19,21,42] As discovered by Folkman, rapidly growing cancer cells need additional blood supply and gradually develop a dense microvascular network inside and especially around tumors required for tumor growth.[64,65] Solid tumors require enhanced energy and oxygen supply through the microvascular network. Tumor-associated neovascularization allows tumor cells to express their critical growth advantage. Rapid tumor growth, bleeding, and metastasis usually follow this new vascular phase of growth.

Malignant breast tumors (carcinoma) have enhanced blood content and contain noticeably hypoxic blood.[66,67] In contrast, benign tumors have a relatively normal level of blood oxygenation. Spectroscopic imaging, i.e., optical imaging using several optical wavelengths, allows for noninvasive *in vivo* imaging and quantification of oxygenated and deoxygenated hemoglobin, which reflects the physiologic functions of tumors.[68,69]

Strong absorption of pulsed laser radiation leads to preferential energy deposition in tumors compared with normal tissue. The absorbed laser energy yields a temperature rise and a simultaneous pressure jump in the irradiated volume, thereby converting heated tumors into sources of ultrasonic waves. Short-pulse (nanosecond) energy deposition results in high thermoacoustic efficiency of pressure generation (up to 0.8) in live tissues and enables a superior spatial resolution of 0.5 mm and a field of view extending to a depth of up to 60 mm.[42] The optically generated pressure profile resembles the distribution of absorbed energy in the irradiated tissue. The signal generated by the optoacoustic imaging system is initially a spatial replicate of the tumor in the direction of detection. This signal mimics the spatial distribution of optical absorption, creating an ultrasonic wave that mimics the shape of the object. The propagation of this signal to the surface follows the rules of ultrasound transmission, but the signal itself is distinct from typical ultrasound. Tumors measuring 1 to 10 mm and irradiated with laser pulses represent sources of pressure waves with an ultrawide ultrasonic frequency range of 150 kHz to 1.5 MHz. Such ultrasonic waves can propagate in tissues with insignificant attenuation and deliver spatially resolved information to the surface of tissue.[6,21]

In contrast to pure optical tomography, where diagnostic information about tissue structure is integrated over the entire optical path, optoacoustic imaging permits direct reconstruction of the absorbed energy distribution from the profile of laser-induced pressure.[1,5,36,43] The sensitive time-resolved piezoelectric detection and analysis of the laser-induced ultrasonic profiles offer a unique opportunity to visualize objects at substantial depth in optically turbid and opaque tissues and to provide quantitative information about tissue structure with high-resolution ultrasound. Unlike pure ultrasonic imaging, where ultrasonic waves are delivered to tissue from outside and only a fraction of the total ultrasonic energy is reflected back from tissue boundaries, optoacoustic imaging converts tumors into bright acoustic sources standing out on a dark (low absorbing) background of normal tissues and detects 100% of the ultrasonic energy emitted by tumors. Thus, LOIS, which uses sensitive detection of laser-induced ultrasonic waves instead of detecting scattered photons, is free of the limitations associated with pure optical and pure ultrasound technologies.

## 34.7.1 Sensitivity of LOIS for Small-Tumor Detection

One of the important parameters of LOIS is its ability to detect small (early) tumors in the breast. Consider a tumor as a sphere of radius $a$ with optical absorption coefficient $\mu_a$. Let this tumor be located at such a depth where the energy of laser pulses exponentially attenuated in the normal breast tissue can generate optoacoustic signals with a SNR of 1. The maximum depth of the tumor detection can be evaluated as:[40,42]

$$z_{max} = -\frac{1}{\mu_{eff}} \ln\left(\frac{2p_{noise}r}{\Gamma\mu_a aF_0}\right) \tag{34.38}$$

where $\mu_{eff}$ is the effective optical-attenuation coefficient in breast tissue, and $r$ is the distance between the tumor and the transducer.

Fluence, $F_0$, of laser pulses at the breast surface must be less than 40 mJ/cm$^2$ due to safety requirements for medical laser procedures with repetitive pulses.[70] The increase in effective optical fluence in scattering tissues relative to incident fluence makes it possible to use several-fold-lower pulsed energy to achieve the projected sensitivity. Parameter $\Gamma$ characterizes the efficiency of thermoacoustic excitation in the media. For live biological tissues at 37°C this parameter varies from 0.40 to 0.85 and is maximal in fatty tissues. The effective noise pressure, $p_{N,}$ is calculated for a SNR equal to 1, according to Equation 34.30. The root mean square of noise voltage $U_{RMS}$ produced by the transducer is defined by the thermal noise of the transducer capacitance and the input noise of the preamplifiers. For LOIS, the anticipated value of $U_{RMS}$ will be about 10 μV. The noise voltage is reduced $\sqrt{N}$ times if the procedure of averaging by $N$ pulses is used. The optimal number of acquisitions for optoacoustic signal averaging is $N$ = 16, which should yield an effective noise pressure of less than 0.5 Pa.

Using the experimental data obtained in the course of preliminary studies we were able to calculate the minimum diameter of spherical tumors with various blood content detectable inside the breast by the current system.[26,39–42] The radius of the minimum detectable spherical tumor is plotted in Figure 34.10A as a function of depth from the irradiated surface for optical absorption coefficient $\mu_a$ = 1 cm$^{-1}$ (at λ = 757 nm) corresponding to the pathologically average blood content in tumors of 12%. Figure 34.10A shows that a 2-mm tumor can be detected under the current system at a depth of 6 cm. Substituting the noise pressure $p_N$ characteristic of LOIS in Equation 34.38 yields a maximum depth of tumor detection of about 6 cm, assuming the tumor is a sphere 2 mm in diameter with $\mu_a$ = 0.6 cm$^{-1}$, or a greater depth for tumors with an absorption of 1 cm$^{-1}$.

Figure 34.10B depicts an optoacoustic image of two 7-mm spheres colored with hemoglobin (with an effective optical absorption of 12% of blood) and embedded in a gel phantom with optical attenuation of $\mu_{eff}$ = 1.2 cm$^{-1}$. The anticipated blood content in actively growing small tumors is about 8 to 20%.[65–67] The area above the curve in Figure 34.10A represents the range beyond the limit of detection for these transducers, and all tumors with a combination of absorption–diameter below the curve can be detected by ultrawide-band acoustic transducers made of piezoelectric polymer materials (sensitivity >1 V/bar) designed and fabricated in collaboration with LaserSonix Technologies, Inc. (Houston, TX) and used in the experiments described in this chapter. The detection limits of LOIS may be further improved with advances in the piezoelectric transducer arrays supported by advanced electronics and image-processing algorithms.

The 2D image of blood-colored spheres differs from the correct circular shape, presenting them as ellipses. This fact has a simple explanation. When the angle of acceptance of these objects by the transducer array was about 120°, only one third of the data needed for complete image reconstruction was collected and used in image reconstruction. This factor limited lateral resolution in the far zone of the images presented in Figure 34.10B. An increase in the acceptance angle for the absorbing volumes located farther from the irradiation site but closer to the detector array results in an enhanced lateral resolution in general and better reproduction of spherical objects. The angle of acceptance in the case of the near zone may be over 180°, and, as a consequence, the shape of imaged objects would highly correlate with their real shapes (see Figure 34.11).

As depicted in the optoacoustic image in Figure 34.11, an artificial blood vessel (polyethylene tubing shaped as a loop) was placed at the substantial depth of 60 mm in the phantom with optical absorption

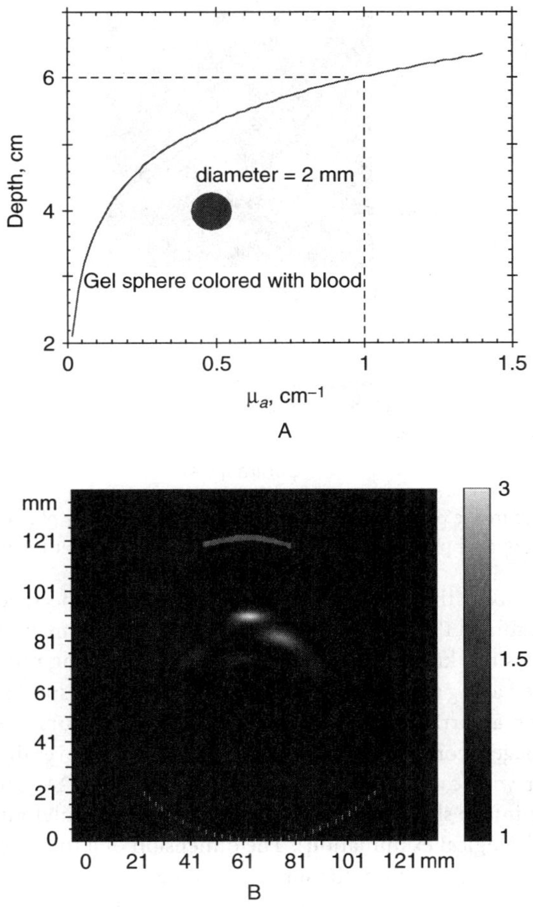


**FIGURE 34.10 (Color figure follows p. 28-30.)** (A) Minimally detectable tumor radius as a function of tumor depth in the bulk of normal tissue. Optical absorption coefficient of 1 $cm^{-1}$ corresponds to relative blood content of 12% in the tumor. (B) Optoacoustic image of two small spheres colored with hemoglobin in gelatin phantom resembling optical properties of breast tissue. An arc above the spheres indicates positions of optical fiber for delivery of laser pulses.

and scattering properties mimicking the breast tissues. The transducer array was at the bottom of the phantom, so that the field of view of the array overlapped with the blood-vessel loop. A single fiber was used for irradiation at only one location on the phantom surface, as depicted in the top portion of Figure 34.11. The phantom surface position is visible as a white dash with the conic divergent optical beam spreading into the depth. This optoacoustic image has a relatively dark background with some lighter structures due to the heterogeneity of the phantom. The optoacoustic contrast (~500%) on the image in Figure 34.11 corresponds to the difference in both background optical absorption and blood at $\lambda = 1064$ nm. An experiment with a blood vessel embedded inside a gel phantom demonstrated that LOIS could image small-diameter vessels with sufficient resolution. In addition, this image gives evidence for the utility of optoacoustic imaging in the quantitative monitoring of blood in tumors.

## 34.7.2 *In Vitro* Imaging of Breast Cancer (Mastectomy Specimens)

Before optoacoustic imaging was applied in patients, extensive studies were conducted *in vitro* in mastectomy specimens.[39,40] Breast specimens were obtained after radical mastectomy and examined within the first 10 min after surgical excision. It provided specimens that underwent minimal alterations of

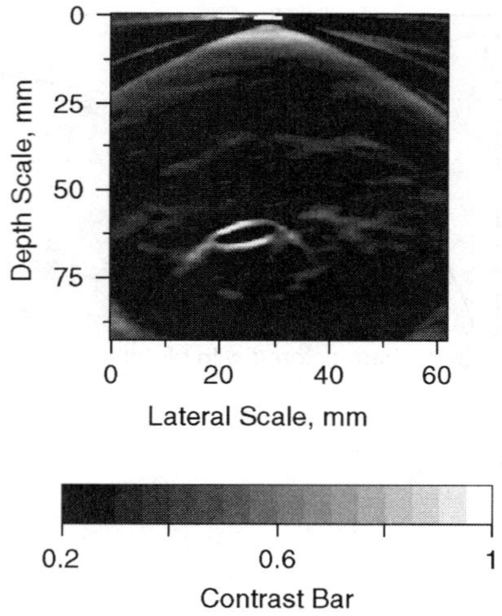


**FIGURE 34.11** Optoacoustic image of artificial "blood vessel" loop in a milky phantom. Optical properties of this phantom were similar to the optical properties of the breast with real arteries at the wavelength of 1064 nm.

tissue optical properties, except that some blood was drained from the specimens. After optoacoustic imaging, a pathologist examined the mastectomy specimens to determine tumor location and geometry. The location of tumors was not known before the optoacoustic imaging procedure was performed.

Optical fiber was scanned along the surface of the radical mastectomy specimen with a 3.5-mm interval. Optoacoustic signals (from an array of 32 ultrawide-band piezodetectors) were employed for data collection and subsequent image reconstruction. One scan passed directly over the tumor. A 2D optoacoustic image of part of a breast with carcinoma is presented in Figure 34.12A. The field of view was about 36 $cm^2$. The optoacoustic image shows clearly the tumor core, horizontal tail, and small sprouts (confirmed by subsequent pathological examination). The dimension of a tumor core can be estimated as 10 $\times$ 7 mm. A tumor was detected 11 mm from the surface, and skin was removed from most of the specimen; thus, the irradiated surface is depicted in Figure 34.12A as a rough line. Optoacoustic contrast between the tumor and surrounding normal tissues in this image exceeds 3.5-fold, making surrounding normal tissue look dark on the image. The x-ray mammography image of the same tumor, in a wider field of view, is also presented in Figure 34.12B. The core of the tumor and its long tail are also visible in this image. However, small sprouts are not visible due to limited radiographic contrast. Note that the x-ray image reveals two suspicious areas in the breast, only one of which was proven cancerous. The optoacoustic imaging system visualized only one bright object in the breast — a malignant tumor. The results presented in Figure 34.12 indicate that LOIS has a potential for early detection of cancerous tumors. A greater challenge for LOIS is to provide noninvasive differentiation of malignant and benign tumors. Therefore, the next series of studies were devoted to imaging blood with two colors of NIR light to compare blood-oxygenation levels in benign and malignant tumors.

## 67.7.3 Diagnostic Imaging in a Breast Cancer Patient

An eligible patient was chosen from a group of patients with breast tumors detected by x-ray mammography in both breasts, and one or more breast tumors were suspected of being cancerous. Two-dimensional optoacoustic images were acquired in several locations on each breast. Two images at two laser wavelengths of 1064 and 757 nm were acquired in succession. Each image required irradiation with 16 100-nsec pulses performed over the course of 0.8 sec, and collection of the imaging information was done on a 32-element array of acoustic transducers. Image reconstruction took about 1 sec on a 700-MHz PC. Laser pulses were delivered at one site of the breast surface, and the optoacoustic transducer array was placed on the opposite side of the breast approximately beneath the point of irradiation. The

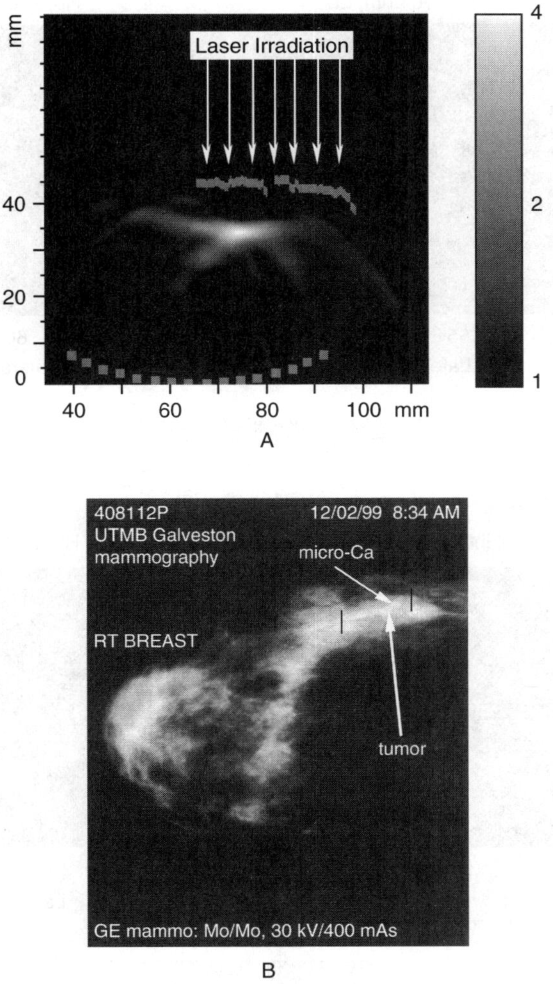


**FIGURE 34.12** Optoacoustic image (A) and x-ray mammography image (B) of a malignant tumor in the breast of UTMB Patient #40811. Note the two major suspicious areas in x-ray mammography image, only one of which proved cancerous. Note also the improved contrast in optoacoustic image. The rough line in (A) represents points on the breast surface illuminated with laser pulses. In order to display detailed structure of the breast tumor, 12 points of the breast surface were irradiated with laser pulses at 1064 nm.

irradiation point was placed approximately in the center of the imaging field suspected of having a tumor. The exact location of tumors was not known prior to optoacoustic imaging; however, the approximate location could be determined from the mammogram. One tumor (in the left breast) was not palpable; however, at least one tumor in the right breast was palpable. Tumors in both breasts were located several centimeters deep within the breast.

Optoacoustic images of breast ductal-lobular carcinoma and benign fibroadenoma obtained prior to biopsy are presented in Figure 34.13A and B, respectively. Optoacoustic images depict the dimensions, shape, location, and structure of the tumor angiogenesis network of microscopic blood vessels with enhanced optical absorption. However, the optical scattering and thermoelastic properties of the tumors might also have contributed to the optoacoustic contrast. The location of the tumors inside the breast and the tumor dimensions could be accurately determined from the optoacoustic images. However, direct comparison with x-ray mammography or ultrasound images based on tissue density was not appropriate.

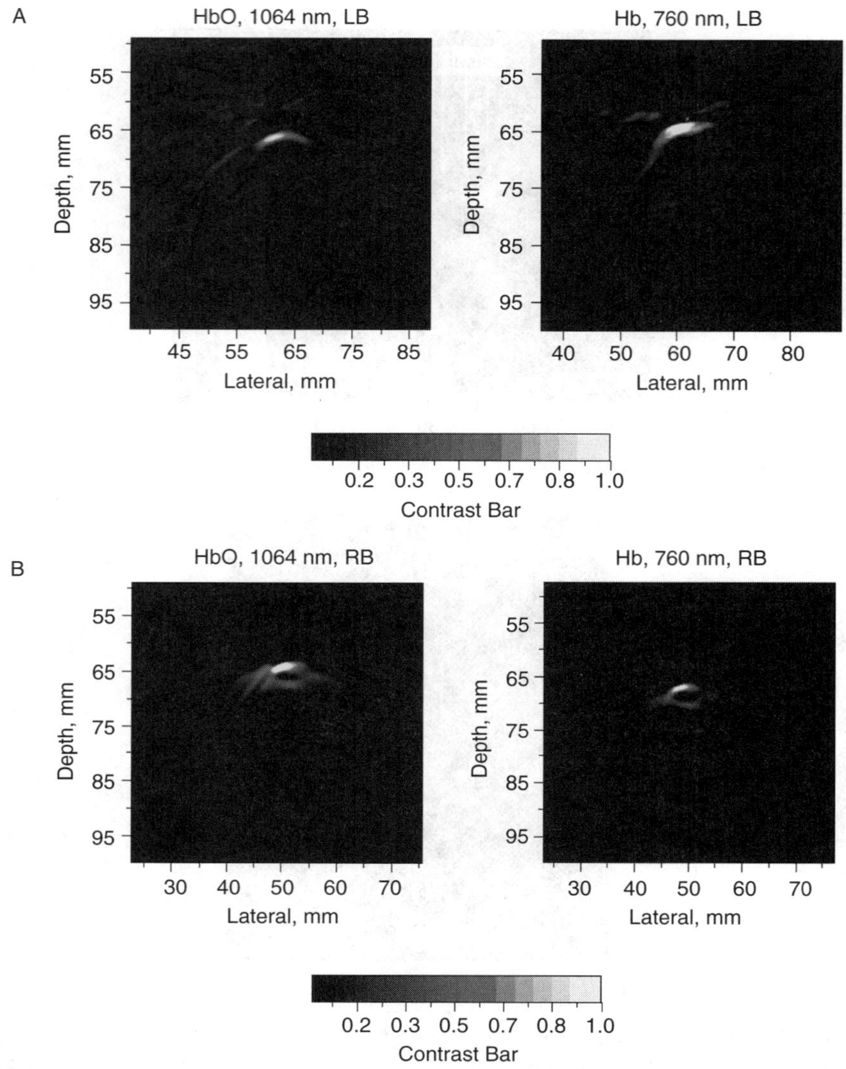


**FIGURE 34.13 (Color figure follows p. 28-30.)** (A) Two-dimensional optoacoustic *in vivo* images of a ductal-lobular carcinoma in the left breast (LB). Laser pulses were delivered to a single point on the skin surface above the tumor. Location of the piezoelectric detector array was at the bottom of the imaging field presented. Contrast between a tumor and surrounding normal tissue strongly depends on the laser wavelength. At the wavelength of 1064 nm the image brightness is significantly lower than the brightness at the wavelength of 757 nm, indicating tumor with reduced level of blood oxygenation. (B) Two-dimensional optoacoustic *in vivo* images of a benign fibroadenoma in the right breast (RB). Laser pulses were delivered to a single point on the skin surface above the tumor. Location of the piezoelectric detector array was at the bottom of the imaging field presented. Contrast between a tumor and surrounding normal tissue strongly depends on the laser wavelength. At the wavelength of 1064 nm the image brightness is significantly higher than the brightness at the wavelength of 757 nm, indicating tumor with normal level of blood oxygenation.

Note that LOIS provided only a small amount of light through a single optical fiber, which limited significantly the field of view on the optoacoustic images. Resolution of the image visualization with LOIS (~1 mm along the depth axis and 1.5 mm in lateral dimension) was comparable to that of x-ray mammography and ultrasound.

Two main conclusions may be drawn from clinical experiments on breast cancer patients. First, OAT substantially enhances the contrast between normal tissues and tumors. Second, the optoacoustic contrast correlates to the level of oxygen saturation of hemoglobin in the tumor microcirculation. The tissue contrast was naturally variable; however, no tumors examined had a contrast of less than twofold relative to the background in normal glandular or fatty tissue. Since optoacoustic contrast in breast tumors is angiogenesis dependent; it may potentially serve as a means for noninvasive diagnosis of malignancy. A correlation between microvessel density in tumor angiogenesis (i.e., total concentration of hemoglobin) and brightness of tumors on optoacoustic images is currently being studied with the help of immunohistology.

### 34.7.4 Advantages and Limitations of LOIS

Five major characteristics of laser OAT that make it attractive for diagnostic imaging of breast cancer are: (1) high detection sensitivity, which results from efficient generation of acoustic sources in tumors by short laser pulses (thermoacoustic efficiency of ~0.7 to 0.8), and sensitive detection of acoustic waves with piezoelectric transducers; (2) high in-depth resolution of ~0.5 to 1.0 mm, which results from short-pulse laser excitation and fast time-resolved detection (ultrawide-band ultrasound detection); (3) high (submillimeter) lateral resolution that results from the geometry of the detection array; (4) high optical contrast (up to sixfold) between tumors and normal tissues, which most likely results from enhanced concentration of blood in malignant tissues due to angiogenesis and a laser wavelength correctly selected for its ability to penetrate tissue and target tumors; and (5) substantial depth of imaging without noticeable loss of resolution (up to 7 cm), which results from an optimal laser-excitation wavelength and efficient conversion of laser energy into ultrasonic waves propagating in soft tissues with insignificant attenuation and minimal distortion.

The two main limitations of the current design of LOIS are (1) limited field of view dictated by the limited number of acoustic transducers in an array that, in turn, was limited by the number (32) of electronic channels of data acquisition and processing and (2) inability to perform imaging in very large breasts.

### 34.7.5 Three-Dimensional Optoacoustic Imaging

One of the merits of OAT is its real-time performance. Three-dimensional (3D) optoacoustic images may be beneficial for some clinical applications. However, for real-time 3D imaging LOIS seems unreasonably expensive due to the large number (over $10^3$) of electronic channels that must operate in parallel to acquire data from a large organ such as breast. The real-time imaging capability of LOIS is very important for enabling functional imaging with high spatial resolution. The laser-pulse-repetition rate cannot be increased much over 100 Hz due to thermal superheating of tissue (especially blood vessels). Therefore, the only realistic approach to 3D imaging is the one currently being developed by LaserSonix Technologies, Inc. The system shown in Figure 34.14 has a bifocal array of 64 piezoelectric transducers

**FIGURE 34.14** Bifocal array of 64 acoustic transducers.

and 64 data-acquisition channels. With a laser-pulse-repetition rate of 32 Hz, only half a second is required to produce a 2D image with an average of 16 signals from each transducer. Data-transmission time using a high-speed interface and 2D image reconstruction takes another half second with a 2-GHz pentium computer. This corresponds to close-to-real-time image acquisition and processing in LOIS. An optimal curvature and length of arc-shaped elements provide a resolution of about 3 mm in the plane perpendicular to the image plane, so that thin frontal slices of the breast can be visualized. Full-field-of-view 3D images can be reconstructed by fusing 60 2D slices.

# 34.8 High-Resolution Subsurface Imaging in Scanning Mode

In a continuous effort to develop an imaging modality with better spatial resolution applicable to sensitive detection of a variety of cancerous lesions, we have designed the optoacoustic front-surface transducer (OAFST), which operates in the backward mode, i.e., detects acoustic profiles at the site of laser irradiation.[6,11,30,36] A feasibility study characterizing tissue structure performed with an OAFST confirmed its utility for *in vivo* imaging of various skin lesions and cancer staging in oral cavity.[38,44–46] An ultrawide band of ultrasonic detection realized in an OAFST yields an in-depth resolution of about 15 μm. A focused OAFST was developed to achieve lateral resolution using an OAFST comparable to an in-depth resolution.[46] The new transducer design is known as the confocal optoacoustic transducer (COAT). The performance of the COAT was initially tested in optically homogeneous and heterogeneous phantoms.[43] Experiments *in vivo* were also performed demonstrating the developed transducer's ability to image and distinguish early stages of squamous-cell carcinoma in oral mucous of golden hamsters.[43–46]

## 34.8.1 Confocal Optoacoustic Transducer

The COAT provides sharp focusing of the incident optical beam and a narrow, long waist of acoustic focus in the ultrasonic detection system. The distribution of focused light and the caustic of the ultrasonic detection form the volume of monitoring for the COAT. A schematic diagram of the COAT is presented in Figure 34.15. Pulsed laser radiation was delivered to the COAT via optical fiber focused by a lens onto the tissue surface through an optoacoustic lens. Ultrasonic waves induced in tissue by laser pulses and propagating backward through the optoacoustic lens were focused with the optoacoustic lens on the ring-shaped piezoelectric detector. The combination of a flat piezoelectric transducer and an optoacoustic lens is aimed at obtaining a focused receiving ultrasonic transducer with a narrow, long waist of the focal zone directly under the irradiated tissue surface. The shape of the lens curvature allows for the transformation of a spherical acoustic wave from tissue into a planar acoustic wave without phase distortion. A charge preamplifier was used to transform acoustic transducer signals into amplified electronic signals for further data acquisition and processing.

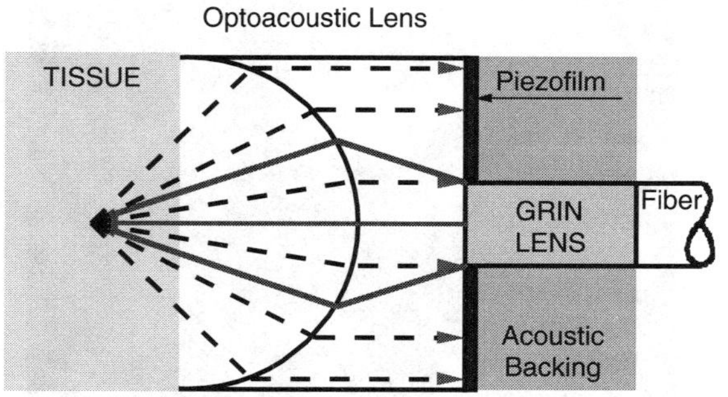


**FIGURE 34.15** Principle schematic diagram of confocal optoacoustic transducer.

Laser pulses from a Q-switched Nd:YAG laser operating at the fundamental ($\lambda = 1.064$ μm) and second ($\lambda = 0.532$ μm) harmonics were used in studies of light-absorption distribution in phantoms and tissues. The temporal shape of laser pulses was close to Gaussian with full width at the max/e level $2\tau_L = 12$ ns for $\lambda = 532$ nm. The energy of each laser pulse was measured with a calibrated Joulemeter (ED-200, GenTec, Canada). The sensitivity of the COAT permitted detection of ultrasonic waves with a SNR of 1 in tissues with an absorption of 20 $cm^{-1}$ while using incident laser fluence of $F_0 < 1$ $mJ/cm^2$, suitable for safe application in medical imaging.[45]

## 34.8.2 Test Experiments in Tissue Phantoms

Optoacoustic imaging of human hair was carried out to demonstrate the spatial resolution of the COAT. Human hair with a diameter of 80 μm was placed in water and irradiated with laser pulses at 532 nm. Scanning across the hair axis was performed with a lateral step of 25.4 μm. The signals obtained were processed using absorption reconstruction, and the resulting image was created. The image is presented in Figure 34.16. The image shows the hair as a bright strip with a thickness corresponding to the thickness of the hair. The in-depth resolution was high, yielding correct reproduction of the hair thickness on the optoacoustic image. The width of the hair as defined by the optoacoustic image is several times greater than its thickness due to the relatively low angular aperture of the transducer. One may conclude from this that in confocal laser optoacoustic microscopy, the size of the image in the lateral direction will be greater than the in-depth dimension according to the angular aperture of the optoacoustic lens.

## 34.8.3 Noninvasive Detection of Oral Cancer *in Vivo*

A well-established animal model of 9,10-dimethyl-1,2-benzanthracene (DMBA)-induced squamous-cell carcinoma in oral cavity was provided for this study by Dr. M. Motamedi.[71] Syrian golden hamsters weighing ~200 g were employed. The motivation for using hamster models was the development of a hamster buccal pouch carcinoma comparable to human oral cancer, as shown in References 72 and 73. DMBA solution

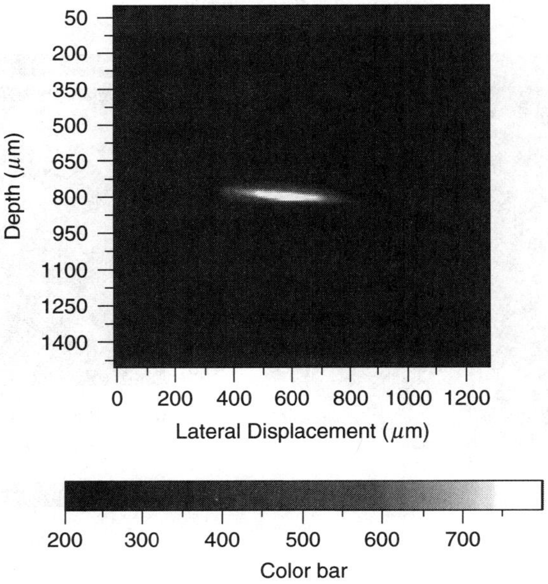


**FIGURE 34.16 (Color figure follows p. 28-30.)** Optoacoustic image of a human hair.

(0.5% of 9,10-dimethyl-1,2-benzanthracene dissolved in mineral oil) was applied twice a week onto the buccal mucosa of the left cheek pouch using a small brush until the first lesion was visible to the naked eye. Hamsters were euthanized and their pouches cut, stretched, and fixed with pins on a flat cork and then embedded in paraffin. Tissue sections were stained with hematoxylin-eosin (H&E). Histological sections were used as a gold standard to compare with optoacoustic images of different types of lesions and stages of squamous-cell carcinoma (from dysplasia to carcinoma *in situ* to invasive malignant lesion).

The optoacoustic-imaging procedure consisted of translating the transducer along the wet surface of pouch mucosa. Slight pressure was applied for better acoustic contact; this resulted in flattening of the irradiated tissue surface. A COAT was applied to the pouch surface with a thin water layer to provide acoustic coupling between the transducer and the tissue.

The main goal of this study was to test the feasibility of the COAT in the detection and staging of cancer in oral mucosa.[43–46] Previous studies performed with well-defined models of animal oral carcinoma consistently reported the following sequence of histological changes that occur in the affected pouch mucosa: (1) hyperplasia, (2) dysplasia, (3) carcinoma *in situ*, and (4) invasive carcinoma. Images presented in Figures 34.17 through 34.19 show a thin (10- to 30-μm) water layer as the top layer, so that the first mucosal surface is located at a depth of about 10 to 30 μm. The recorded signals were processed using the procedure described in Section 34.7, which converted optoacoustic signals into profiles of the absorbed optical energy and plotted on the 2D plane as 25-μm-wide columns. Thus, reconstruction of 2D images was made in hamsters in various stages of cancer development. Optoacoustic images were compared with histological microscopic sections stained with H&E.

Histological H&E sections of normal hamster cheek pouch have been described previously in the literature.[72,73] A normal cheek pouch has several distinct layers. Figure 34.17 shows a microscopic histology section (A) and the corresponding optoacoustic image of a normal hamster cheek pouch (B). Both the histology and the OAT display distinct layers that correspond to keratinized stratified squamous epithelium (E), a very thin basal membrane separating the epithelium from mucosa of lamina propria (L), muscle-fiber mucosa (M), and submucosa connective tissue (S).

The process of cancer development leads to the gradual destruction in a layered structure of mucosa. Precancer in mucosa is characterized by hyperplasia and irregularly increased thickness of epithelium

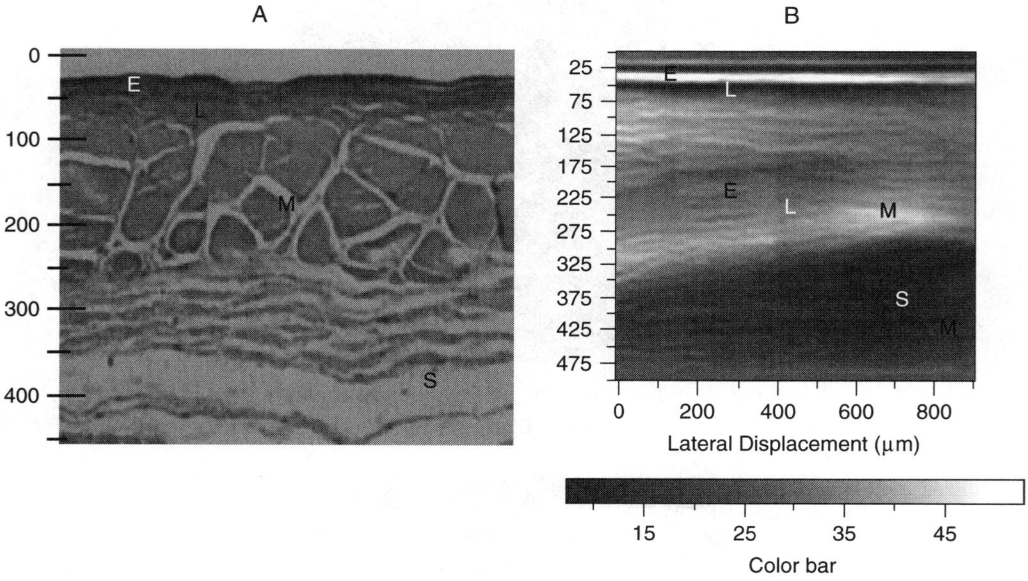


**FIGURE 34.17** Histological section (A) and optoacoustic image (B) of hamster normal cheek pouch. Normal buccal mucosa pouch is characterized layered structure with distinct layers: keratin on top of epithelium (E), lamina propria (L) muscle fiber mucosa (M), and submucosa connective tissue (S).

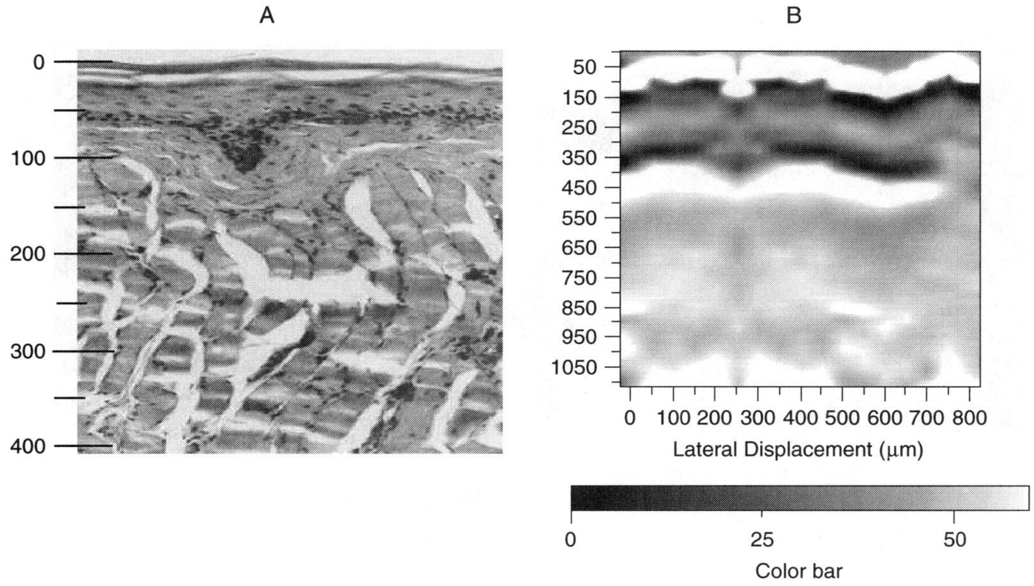


**FIGURE 34.18** Histology section (A) and corresponding optoacoustic image (B) of microscopic tumor dysplasia in a hamster cheek pouch.

without, however, the layered structure being compromised. An early stage of cancer, dysplasia, is characterized by a more prominent thickening of the epithelium layer with rare and irregular breaks in the layered structure, formed by the local accumulation of epithelial cells with larger nuclei and a higher concentration of nuclei. These abnormal epithelial cells start proliferating into deeper mucosal layers. A histological section of dysplasia is depicted in Figure 34.18A. The optoacoustic image presented in Figure 34.18B shows a focal point of dysplasia. The epithelium layer looks normal around the microscopic region with dysplasia.

Figure 34.19A presents a histological section and optoacoustic image of a hamster pouch treated with DMBA for 6 weeks. At this time microscopic malignant lesions proliferating into deeper layers as fingers could be found throughout the treated area. The optoacoustic image in Figure 34.19B shows partial loss of the layered tissue structure with local invasions of epithelial cells into deeper mucosa and irregularly shaped areas in mucosa of lamina propria. Further progression of carcinoma development was observed in all animals by the presence of invasive carcinoma, which was detectable by the naked eye.

### 34.8.4 Advantages and Limitations of the Confocal Optical Transducer

The COAT represents a new technology for 2D imaging of layered tissue structure *in vivo*. This transducer combines a focused optical system for laser irradiation of tissue and a focused piezoelectric detection system operating in the ultrasonic band of 1 MHz to 100 MHz. The axial (depth) resolution of the transducer is 18 μm for biological tissue. Lateral resolution is not worse than 60 μm at a depth of up to 600 μm. The major advantage of the COAT over the OAFST is improved lateral resolution over the range of depths in tissue from 0 to 1.5 mm. A deconvolution procedure for reconstruction of absorbed-energy distribution was developed and tested yielding a satisfactory accuracy of quantitative analysis.

Confocal optoacoustic imaging is sufficiently sensitive to depict structural changes in mucosa tissues much earlier than any changes could be detected on the tissue surface with the naked eye. Furthermore, optoacoustic images yield information on subsurface tissue structure that closely resembles structures provided by the optical microscopy of histology sections. Microscopic dysplasia can be differentiated from normal mucosa, and carcinoma *in situ* can be differentiated from dysplasia. Changes associated

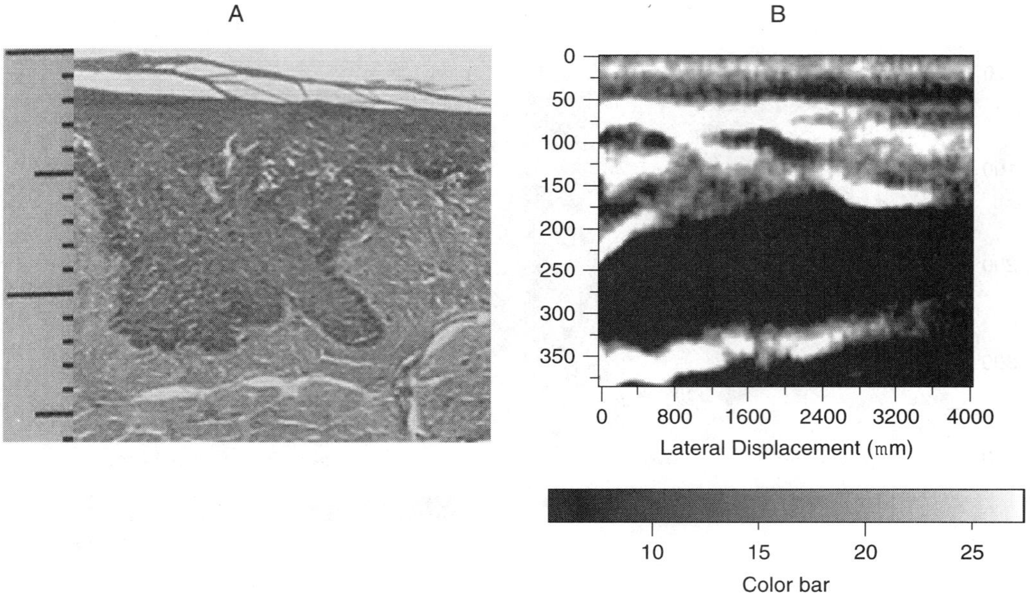


**FIGURE 34.19** Histology section (A) and an optoacoustic image (B) of carcinoma *in situ* of a hamster cheek pouch. Embedding in parafin resulted in some mismatch of H&E image relative to COAT image.

with gradual cancer development could be characterized as a transformation of a tissue-layered structure into a heterogeneous structure. The nature of the optoacoustic contrast in mucosa and carcinoma must be established to enhance further the imaging sensitivity. We speculate that one factor contributing to the differentiation of cancer is the tumor angiogenesis network of microscopic blood vessels. The results from the *in vivo* study suggest that COAT may have applications for imaging, monitoring, and guiding biopsy.

## 34.9 Summary and Conclusions

About a decade ago our group began basic research and development of laser optoacoustic imaging systems for cancer detection in various organs. Presently, LOIS is being evaluated in patients with breast cancer, and COAT is being tested in an animal model of oral cancer. Future advancements of LOIS are envisioned in the direction of developing diagnostic utility and improved imaging parameters (sensitivity and resolution), accommodating the needs of continuous monitoring for ambulatory subjects, including those with implants, and in producing displays that intelligently combine structural, chemical, and functional information. Emphasis continues to be on maximizing accuracy of quantitative information, minimizing image-acquisition and processing time and costs, eliminating patient discomfort, and improving ease of image interpretation. The estimated manufacturing cost of the laser optoacoustic imaging system will not exceed the cost of an ultrasound imaging machine. The affordable price of LOIS and its portability will permit utilization of optoacoustic technology in diagnostic clinics. The application of low-level nonionizing NIR radiation will allow for continuous, safe, and frequent screening. The noninvasive diagnostics is painless and much less expensive than biopsy. In addition, LOIS can be used for image-guided biopsy and image-guided therapy, especially for monitoring the effect of chemotherapy with antiangiogenesis drugs. This hybrid technology employs the most useful features of optical spectroscopy and ultrasonic imaging: nonionizing NIR radiation (to achieve maximum tissue contrast) and low-noise piezoelectric transducers for ultrawide-band ultrasonic detection (to visualize both masses of abnormal tissue and fine details of tumor microcirculation). The diagnostic capability of LOIS to noninvasively differentiate breast carcinoma, fibroadenoma, and cyst will soon be tested in a statistically significant number of patients. LOIS promises to become the first real-time, high-resolution imaging system that uses tissue contrast based on tumor angiogenesis. Two-color OAT, being less expensive than existing methods of invasive biopsy, may provide noninvasive diagnosis. Recent breakthroughs in all-solid-state laser technology and miniaturization and integration of multifunction electronic components allow for a portable and compact design of future commercial laser optoacoustic imaging systems.

## Acknowledgments

The authors would like to express their deep gratitude to Dr. Valeriy Andreev, Dr. Sergey Solomatin, Dr. Vladimir Solomatin, Ms. Elena Savateeva, and other researchers at the Optoacoustic Imaging and Spectroscopy Laboratory; Dr. Massoud Motamedi and Dr. Rinat Esenaliev from the Center for Biomedical Engineering at the University of Texas Medical Branch at Galveston; and Dr. Declan Fleming, Dr. Zoran Gatalica, Dr. Tuenchit Khamapirad, and other UTMB clinicians who contributed to the development of OAT. The authors acknowledge support from the Texas Advanced Technology Program, the National Cancer Institute, the U.S. Army Medical Research and Materiel Command, and LaserSonix Technologies, Inc., Houston, Texas.

## References

1. Oraevsky, A.A., Jacques, S.L., and Esenaliev, R.O., Laser optoacoustic imaging for medical diagnostics, USPTO serial no. 05,840,023, January 31, 1996.
2. Oraevsky, A.A., Jacques, S.L., Esenaliev, R.O., and Tittel, F.K., Direct measurement of laser fluence distribution and optoacoustic imaging in heterogeneous tissues, *Proc. SPIE,* 2323, 250, 1994.
3. Oraevsky, A.A., Jacques, S.L., and Tittel, F.K., Determination of tissue optical properties by time-resolved detection of laser-induced stress waves, *Proc. SPIE,* 1882, 86, 1993.
4. Oraevsky, A.A., Esenaliev, R.O., Jacques, S.L., and Tittel, F.K., Laser optoacoustic tomography for medical diagnostics principles, *Proc. SPIE,* 2676, 22, 1996.
5. Oraevsky, A.A., Jacques, S.L., and Tittel, F.K., Measurement of tissue optical properties by time-resolved detection of laser-induced transient stress, *Appl. Opt.,* 36, 402, 1997.
6. Oraevsky, A.A., A nanosecond acoustic transducer with applications in laser medicine, *LEOS Newslett.,* 8(1), 6, 1994.
7. Karabutov, A.A., Podymova, N.B., and Letokhov, V.S., Time-resolved optoacoustic detection of absorbing particles in scattering media, *J. Modern Opt.,* 42, 7, 1995.
8. Karabutov, A.A. and Oraevsky, A.A., Ultimate sensitivity of wide-band detection for laser-induced ultrasonic transients, *Proc. SPIE,* 3916, 228, 2000.
9. Oraevsky, A.A., Andreev, V.G., Karabutov, A.A., and Esenaliev, R.O., Two-dimensional optoacoustic tomography, transducer array and image reconstruction algorithm, *Proc. SPIE,* 3601, 256, 1999.
10. Andreev, V.G., Karabutov, A.A., and Oraevsky, A.A., Wide-band pulse detection in optoacoustic tomography system for breast cancer imaging, *IEEE Trans. UFFC,* in press.
11. Oraevsky, A.A., Esenaliev, R.O., Jacques, S.L., Thomsen, S.L., and Tittel, F.K., Lateral and z-axial resolution in laser optoacoustic imaging with ultrasonic transducers, *Proc. SPIE,* 2389, 198, 1995.
12. Gusev, V.E. and Karabutov, A.A., *Laser Optoacoustics,* American Institute of Physics, New York, 1993.
13. Tam, A.C., Applications of photoacoustic sensing techniques, *Rev. Modern Phys.,* 58(2), 381, 1986.
14. Sigrist, M.W., Laser generated acoustic waves in liquids and solids, *J. Appl. Phys.,* 60, R83, 1986.
15. Wolbarsht, M.L., A proposal to localize an intraocular melanoma by photoacoustic spectroscopy, *Sov. J. Quantum Electron.,* 11(12), 1623, 1981.
16. Cross, F.W., Al-Dhahir, R.K., Dyer, P.E., and MacRobert, A.J., Time-resolved photoacoustic studies of vascular tissue ablation at three laser wavelengths, *Appl. Phys. Lett.,* 50(15), 1019, 1987.
17. Cross, F.W., Al-Dhahir, R.K., and Dyer, P.E., Ablative and acoustic response of pulsed UV laser-irradiated vascular tissue in a liquid environment, *J. Appl. Phys.,* 64, 2194, 1988.

18. Oraevsky, A.A., Esenaliev, R.O., and Letokhov, V.S., Temporal characteristics and mechanism of atherosclerotic tissue ablation by picosecond and nanosecond laser pulses, *Lasers Life Sci.*, 5(1–2), 75, 1992.
19. Esenaliev, R.O., Karabutov, A.A., and Oraevsky, A.A., Sensitivity of laser optoacoustic imaging in detection of small deeply embedded tumors, *IEEE J. Sel. Top. Quantum Electron.*, 5(4), 981, 1999.
20. Kruger, R.A., Pingyu, L., Fang, Y., and Appledorn, C.R., Photoacoustic ultrasoud — reconstruction tomography, *Med. Phys.*, 22(10), 1605, 1995.
21. Oraevsky, A.A., Esenaliev, R.O., Jacques, S.L., Tittel, F.K., and Medina, D., Breast cancer diagnostics by laser optoacoustic tomography, in *Trends in Optics and Photonics*, Vol. 2, Alfano, R.R. and Fujimoto, J.G., Eds., OSA Publishing, Washington, D.C., 1996, pp. 316–321.
22. Oraevsky, A.A., Laser optoacoustic imaging for cancer diagnosis, *LEOS NewsLett.*, 10(6), 17, 1996.
23. Karabutov, A.A., Podymova, N.B., and Letokhov, V.S., Time-resolved laser optoacoustic tomography of inhomogeneous media, *Appl. Phys. B*, B61, 545, 1996.
24. Esenaliev, R.O., Tittel, F.K., Thomsen, S.L., Fornage, B., Stelling, C., Karabutov, A.A., and Oraevsky, A.A., Laser optoacoustic imaging for breast cancer diagnostics, limit of detection and comparison with X-ray and ultrasound imaging, *Proc. SPIE*, 2979, 71, 1997.
25. Kruger, R.A., Kopecky, K.K., Aisen, A.M., Reinecke, D.R., Kruger, G.A., and Kiser, W.L., Jr., Thermoacoustic CT with radio waves, a medical imaging paradigm, *Radiology*, 211(1), 275, 1999.
26. Ku, G. and Wang, L.V., Scanning thermoacoustic tomography in biological tissue, *Med. Phys.*, 27(5), 1195, 2000.
27. Kruger, R.A. and Liu, P., Photoacoustic ultrasound, pulse production and detection in 0.5% Liposyn, *Med. Phys.*, 21(7), 1179, 1994.
28. Karabutov, A.A., Podymova, N.B., and Letokhov, V.S., Time-resolved optoacoustic measurement of absorption of light by inhomogeneous media, *Appl. Opt.*, 34, 1484, 1995.
29. Hoelen, C.G.A., de Mul, F.F.M., Pongers, R., and Dekker, A., Three dimensional imaging of blood vessels in tissue, *Opt. Lett.*, 28(3), 648, 1998.
30. Oraevsky, A.A., Esenaliev, R.O., Tittel, F.K., Ostermeyer, M., Wang, L., and Jacques, S.L., Laser opto-acoustic imaging of turbid media, determination of optical properties by comparison with diffusion theory and Monte Carlo simulation, *Proc. SPIE*, 2681, 277, 1996.
31. Paltauf, G., Schmidt-Kloiber, H., and Guss, H., Optical detection of laser-induced stress waves for measurement of the light distribution in living tissue, *Proc. SPIE*, 2923, 127, 1996.
32. Beard, P.C. and Mills, T.N., Characterization of post mortem arterial tissue using time-resolved photoacoustic spectroscopy at 436, 461 and 532 nm, *Phys. Med. Biol.*, 42(1), 177, 1997.
33. Hoelen, C.G.A., Pongers, R., Hamhuis, G., de Mul, F.F.M., and Greve, J., Photo-acoustic blood cell detection and imaging of blood vessels in phantom tissue, *Proc. SPIE*, 3196, 142, 1998.
34. Kopp, C. and Niessner, R., Depth-resolved analysis of aqueous samples by optoacoustic spectroscopy, *Appl. Phys. B*, 68, 719, 1999.
35. Köstli, K.P., Frenz, M., Weber, H.P., Paltauf, G., and Schmidt-Kloiber, H., Pulsed optoacoustic tomography of soft tissue with piezoelectric ring sensor, *Proc. SPIE*, 3916, 67, 2000.
36. Karabutov, A.A., Savateeva, E.V., Podymova, N.B., and Oraevsky, A.A., Backward mode detection of laser-induced wide-band ultrasonic transients with optoacoustic transducer, *J. Appl. Phys.*, 87(4), 2003, 2000.
37. Paltauf, G., Schmidt-Kloiber, H., and Guss, H., Light distribution measurements in absorbing materials by optical detection of laser-induced stress waves, *Appl. Phys. Lett.*, 69, 1526, 1996.
38. Karabutov, A.A., Savateeva, E.V., and Oraevsky, A.A., Imaging vascular and layered structure of skin with optoacoustic (front surface) transducer, *Proc. SPIE*, 3601, 284, 1999.
39. Oraevsky, A.A., Andreev, V.G., Karabutov, A.A., Fleming, R.Y.D., Gatalica, Z., Sindh, H., and Esenaliev, R.O., Laser optoacoustic imaging of the breast: detection of cancer angiogenesis, *Proc. SPIE*, 3601, 352, 1999.

40. Andreev, V.G., Karabutov, A.A., Solomatin, V.S., Savateeva, E.V., Aleynikov, V.A., Julina, Yu. V., Fleming, R.Y.D., and Oraevsky, A.A., Optoacoustic tomography of breast cancer with arc-array-transducer, *Proc. SPIE*, 3916, 36, 2000.
41. Oraevsky, A.A., Karabutov, A.A., Andreev, V.G., Singh, H., Gatalica, Z., and Fleming, R.Y.D., Optoacoustic imaging of breast cancer, technology and tissue contrast, *Radiology*, 213, 234, 2002.
42. Oraevsky, A.A., Karabutov, A.A., Solomatin, V.S., Savateeva, E.V., Andreev, V.G., Gatalica, Z., Singh, H., and Fleming, R.Y.D., Laser optoacoustic imaging of breast cancer *in vivo*, *Proc. SPIE*, 4256, 12, 2001.
43. Karabutov, A.A., Savateeva, E.V., Andreev, V.G., Solomatin, S.V., Fleming, R.D.Y., Gatalica, Z., Singh, H., Henrichs, P.M., and Oraevsky, A.A., Optoacoustic images of early cancer in forward and backward modes, *Proc. SPIE*, 4443, 21, 2001.
44. Oraevsky, A.A., Karabutov, A.A., Savateeva, E.V., Bell, B., Motamedi, M., Thomsen, S.L., and Pasricha, J.P., Optoacoustic detection of oral cancer, feasibility studies in hamster model of squamous cell carcinoma, *Proc. SPIE*, 3597, 385, 1999.
45. Savateeva, E.V., Karabutov, A.A., Bell, B., Johnigan, R., Motamedi, M., and Oraevsky, A.A., Noninvasive detection and staging of oral cancer *in vivo* with confocal optoacoustic tomography, *Proc. SPIE*, 3916, 55, 2000.
46. Oraevsky A.A., Savateeva, E.V., Karabutov, A.A., Bell, B., Johnigan, R., Pasricha, J.P., and Motamedi, M., Application of confocal opto-acoustic tomography in detection of squamous epithelial carcinoma at early stages, in *In Vivo Optical Imaging*, Workshop at National Institutes of Health, September 15, 1999, Gandjbakhche, A., Ed., Optical Society of America Press, Washington, D.C., 2000, p. 153.
47. Paltauf, G. and Schmidt-Kloiber, H., Measurement of laser-induced acoustic waves with a calibrated optical transducer, *J. Appl. Phys.*, 82(4), 1525, 1997.
48. Kostli, K.P., Frenz, M., Weber, H.P., Paltauf, G., and Schmidt-Kloiber, H., Optoacoustic tomography, time-gated measurement of pressure distributions and image reconstruction, *Appl. Opt.*, 40(22), 3800, 2001.
49. Beard, P.C. and Mills, T.N., An optical detection system for biomedical photoacoustic imaging, *Proc. SPIE*, 3916, 100, 2000
50. Lyamshev, L.M., Optoacoustic probing of heterogeneous condensed media, *Sov. Phys. Dokl.*, 24(6), 463, 1979.
51. Egerev, S.V. and Pashin, A.A., Optoacoustic diagnostics of micro inhomogeneous liquid media, *Acoust. Phys.*, 39(1), 43, 1993.
52. Kostli, K.P., Frenz, M., Bebie, H., and Weber, H.P., Temporal backward projection of optoacoustic pressure transients using fourier transform methods, *Phys. Med. Biol.*, 46(7), 1863, 2001.
53. Kikoin, I.K., *Tables of Physical Parameters,* Atomizdat, Moscow, 1976, and *Handbook of Chemistry and Physics*, 54th ed., CRC Press, Boca Raton, FL, 1974.
54. Goss, S.A., Johnston, R.L., and Dunn, F., Comprehensive compilation of empirical ultrasonic properties of mammalian tissues, *J. Acoust. Soc. Am.*, 64(2), 423, 1978.
55. Duck, F.A., *Physical Properties of Tissue*, Academic Press, London, 1990.
56. Oraevsky, A.A. and Karabutov, A.A., Time-resolved detection of optoacoustic profiles for measurement of optical energy distribution in tissues, in *Handbook of Optical Biomedical Diagnostics*, Tuchin, V.V., Ed., SPIE Press, Bellingham, WA, 2002, pp. 585–646.
57. Morse, P.M. and Ingard, K.U., *Theoretical Acoustics*, McGraw-Hill, New York, 1968.
58. Kino, G.S., *Acoustic Waves, Devices, Imaging and Analog Signal Processing*, Prentice-Hall, Englewood Cliffs, NJ, 1987.
59. Andreev, V.G., Ponomarev, A.E., Karabutov, A.A., and Oraevsky, A.A., Detection of optoacoustic transients with rectangular transducer of finite dimensions, *Proc. SPIE*, 4619, 153, 2002.
60. Oraevsky, A.A., Esenaliev, R.O., and Karabutov, A.A., Laser optoacoustic tomography of layered tissues: signal processing, *Proc. SPIE*, 2979, 59, 1997.

61. Liu, P., Image reconstruction from photoacoustic pressure signals, *Proc. SPIE*, 2681, 285, 1999.
62. Ternovoy, K.S. and Sinkov, M.V., *Introduction to Modern Tomography*, Naukova Dumka, Kyiv, Ukraine, 1996.
63. Andreev, V.G., Popov, D.A., Sushko, D.V., Karabutov, A.A., and Oraevsky, A.A., Inverse radon transform for optoacoustic imaging, *Proc. SPIE*, 4256, 119, 2001.
64. Folkman, J., Clinical applications of research on angiogenesis, *New Engl. J. Med.*, 333, 1757, 1995.
65. Weidner, N., Semple, J.P., Welch, W.R., and Folkman, J., Tumor angiogenesis and metastasis — correlation in invasive breast carcinoma, *New Engl. J. Med.*, 324, 1, 1991.
66. Marks, F.A., Optical spectroscopy of breast biopsies and human breast cancer xenografts in nude mice, *Frontiers Biosci.*, 2, a1, 1998.
67. Van der Mark, M.B., Hooft, G.W., Wachters, A.J.H., de Vries, U.H., Jansen, J.P., and Wasser, M.N.J.M., Clinical study of the female breast using spectroscopic diffuse optical tomography, in OSA Technical Digest, Biomedical Topic Meetings, April 2000, Optical Society of America Press, Washington, D.C., pp. PD71-PD75.
68. Ntziachristos, V., Chance, B., probing physiology and molecular function using optical imaging: applications to breast cancer, *Breast Cancer Res.*, 3(1), 41, 2001.
69. Oraevsky, A.A., Savateeva, E.V., Solomatin, S.V., Karabutov, A.A., Andreev, V.G., Gatalica, Z., and Khamapirad, T., Optoacoustic imaging of blood for visualization and diagnostics of breast cancer, *Proc. SPIE*, 4619, 81, 2002.
70. American National Standard for Safe Use of Lasers, ANSI Z136.1–2000, American Laser Institute, New York, 2000.
71. Slaga, T.J. and Gimenez-Conti, I.B., An animal model for oral cancer, *J. Natl. Cancer Inst. Rev.*, 13, 55, 1992.
72. Santis, H., Shlar, G., and Chauncey, H.H., Histochemistry of experimentally induced leukoplakia and carcinoma of the hamster buccal pouch, *Oral Surg.*, 17, 307, 1964.
73. Schribner, J.D. and Suss, R., Tumor initiation and promotion, *Int. Rev. Exp. Pathol.*, 8, 137, 1978.

Juliette Selb
*Ecole Supérieure de Physique et Chimie Industrielles Paris, France*

Sandrine Lévêque-Fort
*Université Paris Sud Orsay Cedex, France*

Arnaud Dubois
*Ecole Supérieure de Physique et Chimie Industrielles Paris, France*

Benoît C. Forget
*Ecole Supérieure de Physique et Chimie Industrielles Paris, France*

Lionel Pottier
*Ecole Supérieure de Physique et Chimie Industrielles Paris, France*

François Ramaz
*Ecole Supérieure de Physique et Chimie Industrielles Paris, France*

Claude Boccara
*Ecole Supérieure de Physique et Chimie Industrielles Paris, France*

# 35 Ultrasonically Modulated Optical Imaging



## 35.1 Introduction

Medical imaging technology has experienced many major advances over the last 20 years. A number of powerful tools are now available to clinically investigate large structures in the human body. The most widespread techniques are x-ray radiography, magnetic-resonance imaging, computed tomography, and echography.

In recent years, light as a possible technique for biological tissue imaging has received much attention. In histological observations, abnormal tissues can be easily distinguished from normal ones because of differences in optical properties (optical absorption or reflection, scattering, texture). Optical imaging could therefore reveal these optical contrasts, providing additional information for medical diagnosis. In addition, light has the advantages over x-rays of being noninvasive and nonionizing. Moreover, the cost of optical-imaging techniques would likely be less than that of the most common techniques currently in use.

Biological tissues are relatively transparent in the near infrared (NIR). However, they are highly scattering media; thus, conventional optical methods are unable to provide good-quality images. To overcome this difficulty a number of techniques have been proposed: time-resolved optical imaging,[1,2] frequency-domain optical imaging,[3] and optical coherence tomography.[4–8] These techniques have proved successful for the determination of scattering and optical-absorption distributions in biological tissues. Nevertheless, resolution is no higher than 1 cm through samples several centimeters thick. Besides these purely optical methods, hybrid techniques that combine light and ultrasounds have been proposed. They include photoacoustic imaging[9–11] and ultrasonically modulated optical tomography.[12–26] The basic idea is to use ultrasonic waves, which scatter much less than light waves, to provide better localization

0-8493-1116-0/03/$0.00+$1.50

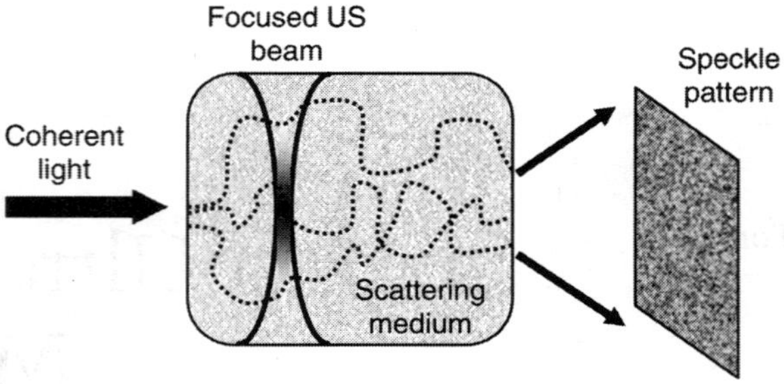


**FIGURE 35.1** Schematic of AO imaging principle.

information for imaging than purely optical techniques. In ultrasonically modulated optical imaging, part of the light is modulated by an ultrasonic beam focused inside the biological tissue. Such tagged photons can be discriminated from background-unmodulated photons, and their origin can be directly derived from the position of the focused ultrasonic beam. Three-dimensional (3D) images can be built up by moving the focused ultrasonic beam. Several systems, using a single detector, have been developed.

Our laboratory has recently proposed a scheme for parallel detection using a charge-coupled device (CCD) camera as a detector array, leading to a dramatic increase in the signal-to-noise ratio (SNR).[17,19,21] We present here recent improvements of our ultrasonically modulated optical imaging setup. The principle of the technique is detailed and the performances are reported. In addition, 3D acousto-optic (AO) images of real biological tissues are shown. Finally, future possible developments and improvements are discussed.

## 35.2 The Principle of Acousto-Optic Imaging

The principle of AO imaging is schematically represented in Figure 35.1. A biological sample is illuminated by coherent laser light. Due to the highly scattering nature of the sample, a speckle field is generated everywhere around it. An ultrasound (US) beam focused on the sample induces periodic displacements of the scatterers (amplitude of a few nanometers),[14,15] and a modulation of the refractive index,[26] mainly in the focal zone. As a result, the optical paths of light passing through the US focal zone are also modulated. Finally, the intensity of the speckle pattern is modulated at the US frequency (a few megahertz). The amplitude of the speckle modulation is directly related to the optical properties of the medium inside the US focal zone. In particular, if the optical absorption in the US focal zone is high, the probability that light will escape from this zone is low; consequently, the modulation of the speckle is low. Local differences in the optical absorption of the sample can thus be revealed. The influence of other optical properties, such as the light-scattering coefficient, is currently being studied.

A simple picture to help understand the principle of AO imaging is that of the "virtual modulated source." The tissue region where both light and ultrasound are present becomes a "virtual source," i.e., a source of modulated light that can be scanned through the sample. When it is scanned over an absorbing region, its intensity decreases. The distribution of light in the medium is roughly uniform; thus, the virtual source coincides with the US beam. The spatial resolution of this imaging technique is given by the size of the US focal zone (typically a few millimeters in diameter and several centimeters in length).

## 35.3 Setup

The experimental setup is shown in Figure 35.2. The sample is placed in a water tank (20 × 25 × 25 cm) to ensure good acoustic impedance matching with the transducer. Two empty tubes closed at one end by a glass window hold the sample. In this configuration, the sample, slightly pressed between the two glass windows, presents a uniform thickness over the entire observed region, and the light paths avoid

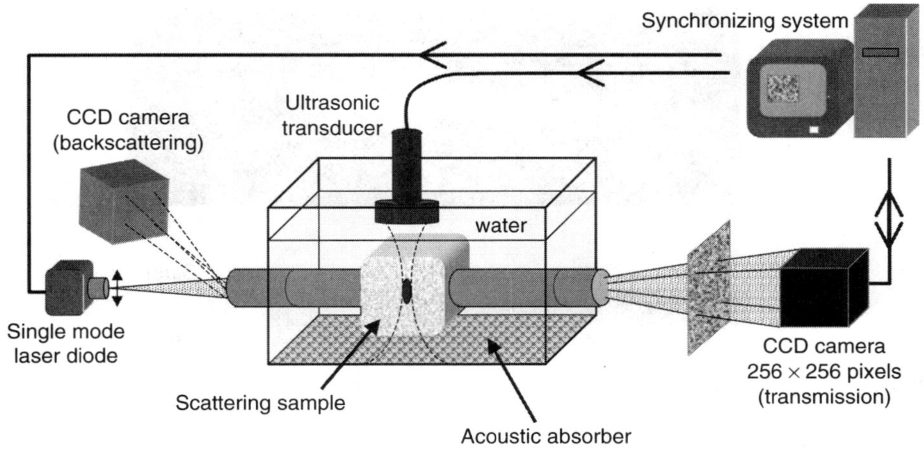


**FIGURE 35.2** Experimental setup.

the water. Indeed, an experiment performed with light traveling in water had revealed that the speckle decorrelated very quickly due to water turbulence, inducing a very poor SNR. The sample is illuminated with a NIR (840-nm) single-mode laser diode (100 mW). This source was chosen for its high coherence length (several meters) and its ability to be modulated at high frequency, which is essential for our detection system (see Section 35.4 below). A diverging lens expands the light beam so that its diameter is ~2.5 cm upon entering the sample. The speckle pattern formed by the emerging scattered light is recorded on a CCD camera array of 256 × 256 pixels. The camera can be placed either opposite to the laser diode (transmission configuration) or close to it on the same side (backscattering configuration). The results in both configurations are compared in Section 35.5. The average diameter of a speckle grain on the CCD is $\phi_s \approx \lambda\, d/r$, where $\lambda$ is the laser wavelength (840 nm), $d$ is the distance between the exit face of the sample and the camera, and $r$ is the effective radius of the aperture through which light exits the sample (1.4 cm). We match the speckle grain size to the pixel size (16 μm × 16 μm) to exploit the modulation of each speckle grain separately by adjusting the distance between the camera and the sample ($d$ = 30 cm; we varied this distance to experimentally check the computed optimum value). However, with very thick samples we reduce the sample-to-camera spacing to collect light more efficiently.

The focused ultrasonic beam is emitted perpendicular to the incident-light beam. We have used two different acoustic transducers. The first one is a simple transducer of fixed focal length (69 mm) driven by a sinusoidal signal. The second one consists of eight independent concentric rings, each driven by its own electronics board. The operating frequency is 3 MHz. The focal length is varied electronically by suitable phase shifts of the otherwise identical driving waveforms. This transducer is mounted on a motorized angular rotation. This more sophisticated transducer can emit in both continuous wave (CW) and pulsed modes. The CW mode enables us to make AO images (as with the first transducer). In pulsed mode, the repetition rate is ~1 kHz. Each pulse typically contains two US periods (duration ~0.7 μsec at 3 MHz). The transducer also acts as a receiver for the echoes of the pulse. When its angular motor is at rest, the transducer records the echoes of the pulse along the focused US beam. By scanning the emitter angularly step by step and repeating the same process at each step, we can reconstruct a 2D image of the acoustic properties of the medium. In this mode (the "B mode" of echography), where the amplitude of the echo is converted into brightness of the spot on a gray scale, we obtain an echographic 2D section of the medium (≈ 4 × 4 cm) in about 0.1 sec. Suitable software allows discrete scanning of the US focus over an array of preselected locations as well as the acquisition of CCD camera frames at each of these locations.

By scanning a calibrated hydrophone in front of the second transducer, we mapped the pressure field emitted in water (Figure 35.3). The full width at half maximum (FWHM) of the focal zone is measured to be 1.4 mm transversely and 12.5 mm along the axis. Moreover, the US beam outside the focal zone

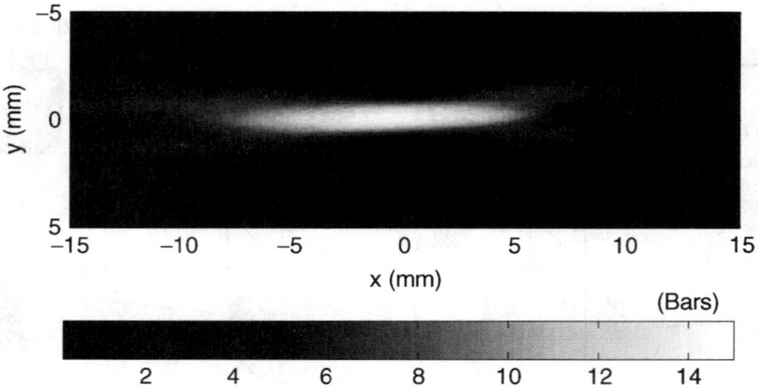


**FIGURE 35.3** Map of pressure field emitted by ultrasonic transducer (right).

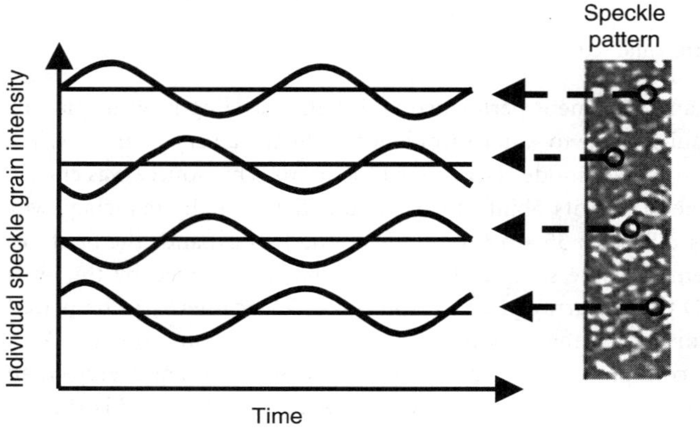


**FIGURE 35.4** Evolution in the intensity of individual speckle grains.

also contributes somewhat to the speckle modulation. Consequently, while the resolution is very sharp in any direction transverse to the US beam, it is somewhat lower in the axial direction.

## 35.4 Parallel Multichannel Lock-In Detection

As explained above in Section 35.2, our aim is to detect the AO modulation of the speckle by ultrasound. As mentioned years ago by other authors,[15] "perhaps the most striking feature of AO imaging is the fact that the signal essentially resides in a single coherence area." Consequently, detecting simultaneously yet independently the modulations in a large number $N$ of coherence areas (or "speckle grains") improves the SNR by a factor $\sqrt{N}$. We have chosen the 256 × 256 pixels of a CCD array camera as the $N$ independent detectors. We set the sample-to-camera spacing to match the coherence area of the speckle to the size of a pixel. On each pixel we detect the amplitude of the modulation at the US frequency. We define our signal as the sum of $N$ amplitudes. (*Note:* Since the relative phases of their modulations are random, if several coherence areas were matched to one single pixel, the sum would cancel out on the average [see Figure 35.4]).

In experiments where a single grain of speckle is detected by a single detector, the signal is usually analyzed with a lock-in amplifier. This extraction of the amplitude (and eventually of the phase*) of the modulated part of the signal can also be achieved by sampling and applying the discrete Fourier transform. Because we are dealing with 256 × 256 individual detectors, we will, of course, use the second approach.

*The phase of each pixel is of little interest here. It is, however, convenient for testing the system.

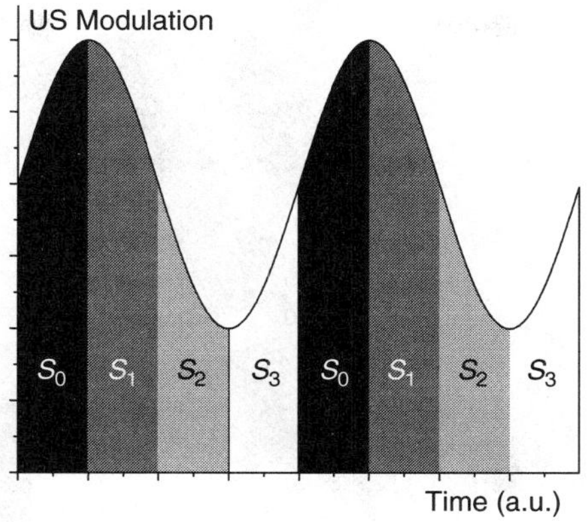


**FIGURE 35.5** Principle of light chopping.

Using a CCD camera as a parallel detector raises two points of concern. First, it is an integrating detector; second, it has a limited frame rate (thus a low sampling frequency). Integration by the camera can be treated as a filtering process, and the computation of the amplitude from the integrated samples is readily carried out.

Dealing with the low frame rate requires careful mixing of the acoustic modulation and the optical-detection signals. The idea is to perform homodyne detection by modulating the laser source at the same frequency $f$ as the acoustic transducer.[19,21,27] Our signal is proportional to the product of the acoustic displacement by the laser intensity, and the signal detected by the CCD camera is thus the product of the two waveforms: the sum of a direct current (DC) component and a harmonic component at frequency $2f$.

Comparing this method with conventional lock-in amplifier detection, our reference is the driving voltage of the ultrasonic emitter, our mixing is performed by the AO effect,* and the CCD camera is both the (multichannel) detector and the output filtering stage. The last component we need to emulate is the phase shifter. In our system, the phase shifter is implemented in the electronic chopping of the laser beam. This slightly tricky chopping is illustrated in Figure 35.5.

The electronic system (which may be called a sequencer) relies on a master clock that actually oscillates at frequency $4f$. The idea is to switch the laser on during one period of this master clock (one quarter period of the signal) and off during the three remaining quarters. This goes on for a preselected number $K$ of signal periods. The frame is then read, yielding the sum of the signals integrated over quarter periods, denoted by $S_1$ in Figure 35.5. For the next $K$ periods the laser is on during the next quarter $S_2$, then a frame is read again. We proceed similarly for the third and then fourth quarter, $S_3$ and $S_4$, respectively.

Let $S_1$, $S_2$, $S_3$, and $S_4$ denote the signals acquired on a given pixel during these four consecutive sequences of $K$ periods each. It is not difficult to show that the amplitude $A$ of the modulation seen by the considered pixel is (except for some unimportant constant factor) given by:

$$A \propto \sqrt{\left(S_1 - S_3\right)^2 + \left(S_2 - S_4\right)^2}\,. \tag{35.1}$$

A final point of concern in our implementation of this technique is the limited dynamic range of the CCD camera. The 8-bit resolution of the analog-to-digital converter is insufficient because the rounding-off noise (1 over 255 for 8 bits) is unacceptably high. However, it is well known that, provided some other noise of the order of the rounding-off noise is present, accumulating $K$ periods in one frame reduces

*Consequently, in this method the mixing is performed not by an electronic instrument but by the studied physical system. In addition, there is only one "mixer," and the multichannel character appears only at the level of the "output-filtering and detection" stage, i.e., the 64K pixels of the CCD camera.

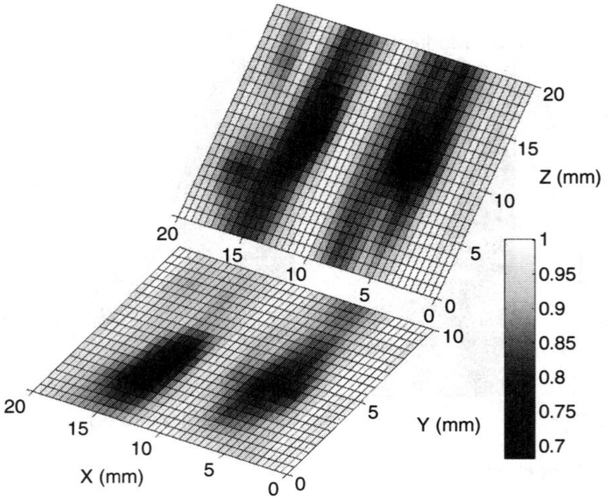


**FIGURE 35.6** AO 2D images of two absorbing spheres embedded in a turkey breast sample. The upper section was recorded along the direction of US propagation. The lower section was recorded in a plane perpendicular to the US beam. Note that the axes are different.

the rounding-off noise by a factor $\sqrt{K}$ . In other words, summing over $K$ periods increases the effective dynamic range by a factor of $\sqrt{K}$ .

At first glance, it seems that $K$ should be as large as possible, but in practice, there is a trade-off between this effective dynamic range and other limitations. First, in our sequencer scheme, the sum over $K$ periods is performed by the CCD camera itself through the integration of the signal. The number of periods is thus limited by the saturation of the camera. Second, the total measurement time ($4K$ times the frame-acquisition time) must remain short enough to limit drifts in the signal.

## 35.5 Results

### 35.5.1 Samples

We worked mainly with two kinds of samples: turkey breast tissue and artificial phantoms ranging in thickness from 1 to 4.5 cm. The latter consist of a gelatin, in which agar grains are embedded to create small ultrasonic reflectors, and a latex, i.e., a suspension of polystyrene microspheres (diameter 220 nm), to scatter light. By varying the concentration of the latex we can change the scattering coefficient of the medium and adjust the acoustic properties of the medium by changing the gelatin and the agar concentrations. The advantages of these artificial phantoms are their homogeneity and their well-controlled properties. However, we believe that biological tissues should also be studied to tackle the concrete problems they will raise. This is why most of our experiments were performed on turkey breast samples, whose optical properties have already been studied.[28] Despite the absence of blood circulation in such samples, we believe they provide more realistic conditions than artificial phantoms.

### 35.5.2 Validation of the Principle on Simple Structures

Our first experiments were performed on turkey breast tissues in which we embedded artificial absorbing objects (dark ink or soft modeling clay). Figure 35.6 shows two 2D AO sections of a 35-mm-thick turkey breast sample in which were embedded two small absorbing spheres of soft modeling clay (3-mm diameter). One section was recorded scanning the US transducer along the direction of the US beam, and the second one perpendicular to it. The two absorbing spheres are clearly revealed on both images. However, the spatial resolution is very different. This is due to the shape of the US beam, which is much longer than it is wide, as discussed in Section 35.3.

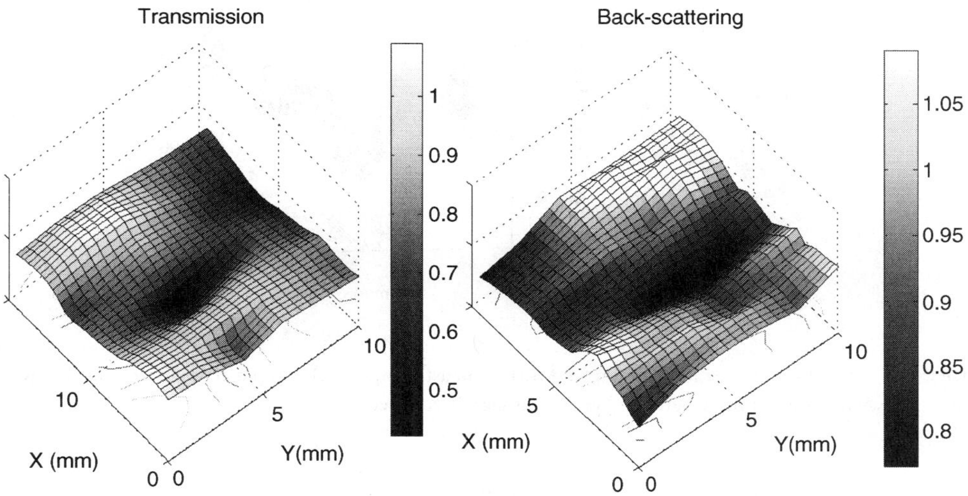


**FIGURE 35.7** AO 2D images of an absorbing sphere embedded in a turkey breast sample recorded in transmission (left) and backscattering (right) configurations.

## 35.5.3 Validation of the Backscattering Configuration

In its current state, our setup is obviously not appropriate for making a real medical apparatus. In particular, the transmission configuration might be difficult to implement on thick organs. That is why we wanted to validate the backscattering configuration, where the light source and the detector were on the same side of the sample.[22,24] Figure 35.7 compares results recorded in transmission (left) and backscattering (right) configurations. The studied object was an absorbing sphere (∅ 3 mm) buried 10 mm deep inside a 35-mm-thick turkey breast sample. On both images the buried structure is clearly revealed at its actual position. However, the reconstruction of the image is more difficult in the backscattering configuration because we must take into account the decrease of the modulated light with increasing depth. Moreover, in the backscattering configuration, the sphere appears larger in the depth direction than its actual size. We believe this is a shadow effect: the photons have a low probability of reaching the region just behind the sphere and traveling back to the detector without being absorbed.

## 35.5.4 Intrinsic Structures in Biological Tissues

We also performed some experiments on *ex vivo* animal or human tissues presenting intrinsic contrasts. We revealed an unexpected naturally present white structure (probably a ligament) in a 2-cm-thick turkey breast sample. Figure 35.8A shows a picture of the sample after dissection, and Figure 35.8B shows the AO 2D image. The size of the imaged ligament agrees with its actual size measured after dissection (~3 mm).

We studied an *ex vivo* human womb presenting tumorous tissues. Figure 35.9 shows the image of the organ and an AO section taken along the black line. On the picture, the normal tissues appear darker than the abnormal ones. On the AO image, the normal, darker tissues yield a lower AO signal because they are more absorbent.

## 35.5.5 Comparison of the Acoustic and Acousto-Optic Contrasts

The second, more sophisticated, transducer (see Section 35.3) enabled us to study the origin of the AO contrast on various samples. In fact, AO contrast can originate in optical contrast (as explained in Section 35.2) or in mere acoustic contrast. If an US-absorbing structure is present in front of the acoustic beam, this will decrease the AO signal and can be misinterpreted as a light-absorbing structure. Our transducer

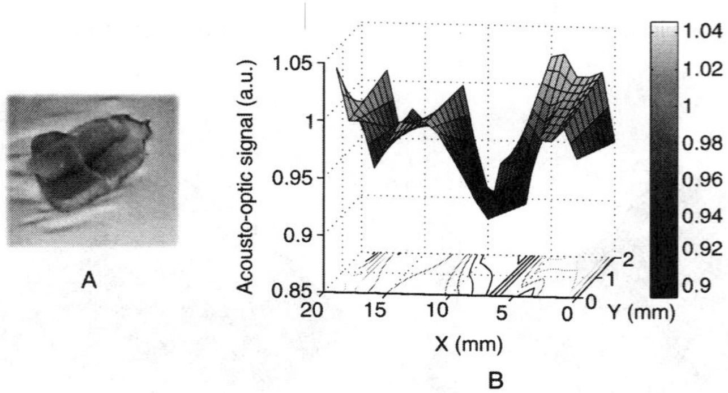


**FIGURE 35.8** Detection of a white filament in a turkey breast sample. (A) Sample after dissection. (B) AO image of the sample before dissection, where the ligament is clearly revealed.

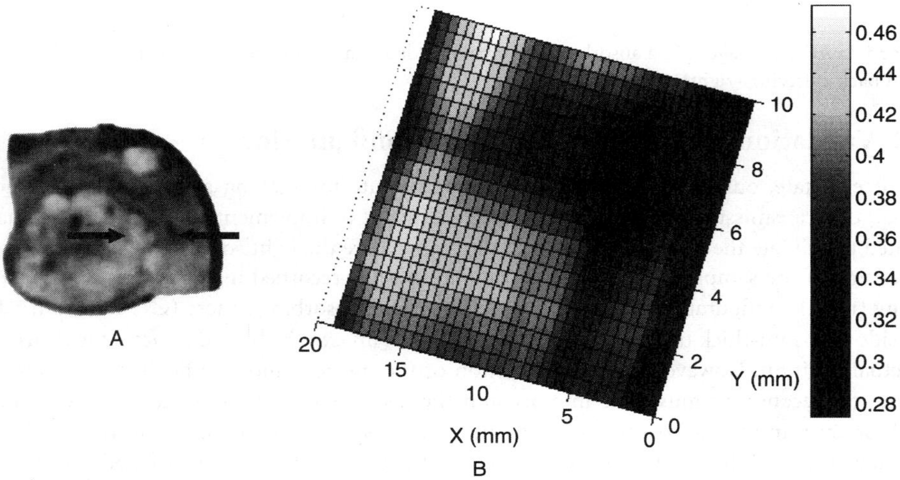


**FIGURE 35.9** AO contrast of tumorous tissues on a human female womb. (A) Picture of the organ where the abnormal tissues appear darker than the normal ones. (B) The AO image shown was recorded in a plane perpendicular to (A) between the two arrays.

enables us to distinguish between these two possibilities by looking at the echographic, i.e., purely acoustic, image of the studied samples.

We performed a simple experiment on turkey breast samples in which we embedded a small cylinder (∅ 7 mm, length 30 mm) absorbing either light or US. The light-absorbent structure consisted in a turkey core that was colored with black ink and inserted back inside the sample. The US-absorbent cylinder was made of tough modeling clay. Figure 35.10 presents the comparison between the echographic 2D images and the AO profiles taken along the dotted lines for both objects. In the first case (modeling clay cylinder), the structure is clearly revealed by the echography. There the AO contrast is either a purely acoustic contrast or a combination of acoustic and optical contrast. In the second case (blackened turkey cylinder), the echographic image does not reveal any contrast. On the contrary, the cylinder is revealed by the AO scan. Here we can be sure that the AO contrast is purely optical. However, the contrast in the second case is poor.

### 35.5.6 Visualization of the "Virtual Modulated Source"

We can directly visualize the "virtual modulated source" (defined in Section 35.2) using a slight variant of the setup described above. We simply insert an objective in front of the CCD camera to image the exit

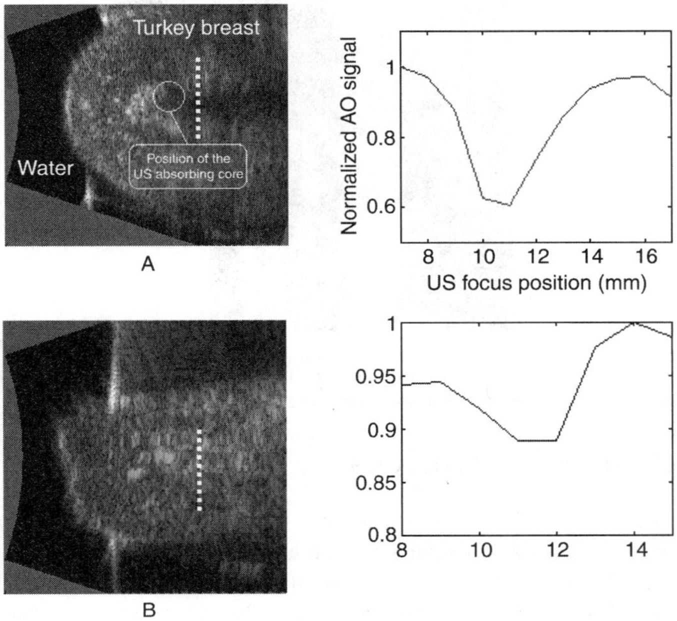


**FIGURE 35.10** Comparison of echographic 2D sections (left) and AO profiles (right), for a US-absorbent cylinder (A) and a light-absorbent cylinder (B) embedded in a turkey breast sample. The AO profiles were recorded by scanning the US beam along the dotted white line.

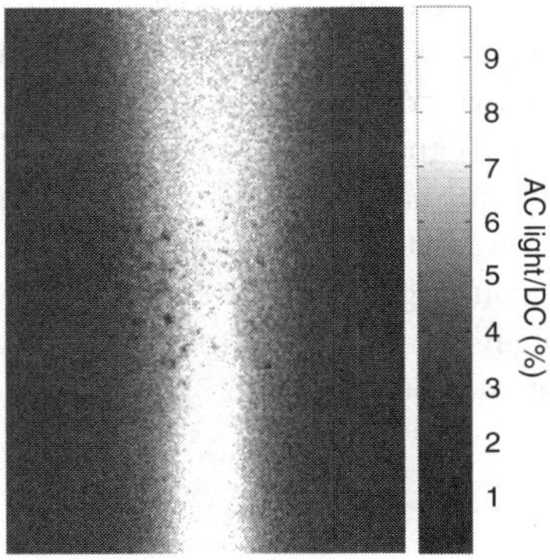


**FIGURE 35.11** Distribution of the modulated light on the exit face of a scattering sample, normalized by the distribution of the DC light. This representation shows the "virtual modulated source" brought close to the exit face.

face of the sample on the CCD array. Moreover, now we look at the spatial distribution of this modulation on the exit face. We typically divide the CCD array into 40 × 40 squares of 6 × 6 pixels each, and we extract the modulation amplitude in each square separately (i.e., for each square, we sum the amplitudes of the 36 pixels). We thus obtain a map of the modulated light emerging from the sample. If we bring the US beam close to the exit face, the light coming from the modulation zone barely scatters before exiting the sample; thus, we obtain directly an image of the modulation zone. Figure 35.11 shows the distribution of the modulated light normalized by the distribution of the unmodulated light (the normalization avoids effects due to the nonuniform distribution of light in the medium). A shape very similar to that of the acoustic pressure field is apparent.

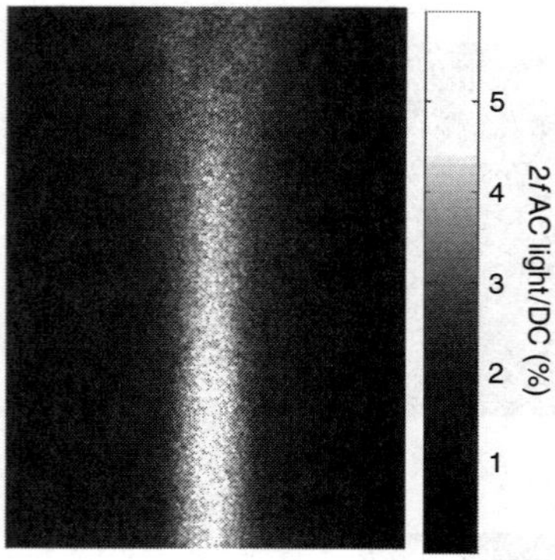


**FIGURE 35.12** Visualization of the virtual modulated source at frequency $2f$ = 6 MHz. Comparison with Figure 35.11 shows that the size of the virtual source is noticeably reduced.

# 35.6 Prospects and Conclusions

Due to the shape of the US beam, the spatial resolution in the axial direction is lower than in the transverse direction. Furthermore, the contrast of structures that are small with respect to the US volume is low, as can be seen in Figure 35.10. To improve these two aspects, the ideal solution would be to use pulsed ultrasounds whose length (usually about two US wavelengths) determines the axial resolution. To our knowledge, this approach has not yet been used for thick light-scattering media or biological tissues. One alternative method, developed by Wang et al.,[18,23] is to use a frequency-swept ultrasonic wave. Nevertheless, the contrast and the SNR must be improved to make this approach suitable for thick samples.

We have begun to explore another approach using the nonlinear response of the system.[29] We expect AO nonlinear signal generation to be restricted to a region of sufficiently high acoustic pressure, i.e., to a region smaller than that of the linear signal. A second harmonic AO signal ($f$ = 6 MHz) varying as the square of the ultrasonic amplitude has been found. The virtual source at this second harmonic frequency is smaller than that at the fundamental frequency, as shown in Figure 35.12.

The next step we foresee is to use two crossed beams and detect the AO signal at the beat frequency. When that happens, we expect the AO signal to come mainly from the small crossed zone.

To conclude, an AO imaging technique has been used to detect artificial and intrinsic optical contrasts through biological tissues several centimeters thick with a millimetric resolution. The first results of experiments on human *ex vivo* tumorous organs are very encouraging with regard to medical applications. Moreover, the preliminary experiments using the AO second harmonic signal open the door to a new way of improving both the contrast and the resolution of this technique.

## Acknowledgments

The authors thank the personnel of Laboratory of Waves and Acoustic, ESPCI, for their collaboration, especially in characterizing the pressure field. They also thank D.A. Boas from Massachusetts General Hospital for fruitful discussions on the concept of the virtual source. They are grateful to PRL Corélec[30] for the design of a specific transducer. Financial support from the French Ministry of National Education, Research and Technology (grants 99 B 0510/0512/0513) is gratefully acknowledged.

## References

1. Hebden, J. C. and Delpy, D.T., Enhanced time-resolved imaging with a diffusion model of photon transport, *Opt. Lett.*, 9, 311, 1994.
2. Hebden, J.C., Schmidt, F.E.W., Fry, M.E., Schweiger, M., Hillman, E.M.C., Delpy, D.T., and Arridge, S.R., Simultaneous reconstruction of absorption and scattering images using multichannel measurement of purely temporal data, *Opt. Lett.*, 24, 534, 1999.

3. O'Leary, M.A., Boas, D.A., Chance, B., and Yodh, A.G., Experimental images of heterogeneous turbid media by frequency-domain diffusing-photon tomography, *Opt. Lett.*, 20, 426, 1995.
4. Huang, D., Swanson, E.A., Lin, C.P., Schuman, J.S., Stinson, W.G., Chang, W., Hee, M.R., Flotte, T., Gregory, K., Puliafito, C.A., and Fujimoto, J.G., Optical coherence tomography, *Science*, 254, 1178, 1991.
5. Fujimoto, J.G., Brezinski, M.E., Tearney, G.J., Boppart, S.A., Bouma, B.E., Hee, M.R., Southern, J.F., and Swanson, E.A., Optical biopsy and imaging using optical coherence tomography, *Nat. Med.*, 1, 970, 1995.
6. Fercher, A.F., Optical coherence tomography, *J. Biomed. Opt.*, 1, 157, 1996.
7. Tearney, G.J., Brezinski, M.E., Bouma, B.E., Bopart, S.A., Pitris, C., Southern, J.F., and Fujimoto, J.G., *In vivo* endoscopic optical biopsy with optical coherence tomography, *Science*, 276, 2037, 1997.
8. Izatt, J.A., Hee, M.R., Owen, G.M., Swanson, E.A., and Fujimoto, J.G., Optical coherence microscopy in scattering media, *Opt. Lett.*, 19, 590, 1994.
9. Kruger, R.A. and Liu, P., Photoacoustic ultrasound: theory, *Proc. SPIE*, 2134A, 114, 1994.
10. Oraevsky, A.A., Esenaliev, R.O., Jacques, S.L., and Tittel, F.K., Laser optoacoustic tomography for medical diagnostics: principles, in *Biomedical Sensing, Imaging, and Tracking Technologies I*, Lieberman, R.A., Podbielska, H., and Vo-Dinh, T., Eds., *Proc. SPIE*, 2676, 22, 1996.
11. Hoelen, C.G.A., de Mul, F.F.M., Pongers, R., and Dekker, A., Three-dimensional photoacoustic imaging of blood vessels in tissue, *Opt. Lett.*, 23, 648, 1998.
12. Marks, F.A., Tomlinson, H.W., and Brooksby, G.W., A comprehensive approach to breast cancer detection using light: photon localisation by ultrasound modulation and tissue characterization by spectral discrimination, in *Photon Migration and Imaging in Random Media and Tissues*, Chance, B. and Alfa, R.R., Eds., *Proc. SPIE*, 1888, 500, 1993.
13. Wang, L., Jacques, S.L., and Zhao, X., Continuous-wave ultrasonic modulation of scattered laser light to image objects in turbid media, *Opt. Lett.*, 20, 629, 1995.
14. Leutz, W. and Maret, G., Ultrasonic modulation of multiply scattered light, *Physica B*, 204, 14, 1995.
15. Kempe, M., Larionov, M., Zaslavsky, D., and Genack, A.Z., Acousto-optic tomography with multiply scattered light, *J. Opt. Soc. Am. A*, 14, 1151, 1997.
16. Wang, L. and Zhao, X., Ultrasound-modulated optical tomography of absorbing objects buried in dense tissue-simulating turbid media, *Appl. Opt.*, 36, 7277, 1997.
17. Lévêque, S., Boccara, A.C., Lebec, M., and Saint-Jalmes, H., A multidetector approach to ultrasonic speckle modulation imaging, in *Advances in Optical Imaging and Photon Migration*, Vol. 23, Fujimoto, J.G. and Patterson, M.S., OSA Trends in Optics and Photonics Series, Washington, D.C., 1998.
18. Wang, L. and Ku, G., Frequency-swept ultrasound-modulated optical tomography of scattering media, *Opt. Lett.*, 23, 975, 1998.
19. Lévêque, S., Boccara, A.C., Lebec, M., and Saint-Jalmes, H., Ultrasonic tagging of photon paths in scattering media: parallel speckle modulation processing, *Opt. Lett.*, 24, 181, 1999.
20. Yao, G. and Wang, L.V., Theoretical and experimental studies of ultrasound-modulated optical tomography in biological tissue, *Appl. Opt.*, 39, 659, 2000.
21. Lévêque-Fort, S., Three-dimensional acousto-optic imaging in biological tissues with parallel signal processing, *Appl. Opt.*, 40, 1029, 2000.
22. Lev, A., Kotler, Z., and Sfev, B.G., Ultrasound tagged light imaging in turbid media in a reflectance geometry, *Opt. Lett.*, 25, 378, 2000.
23. Yao, G., Jiao, S., and Wang, L.V., Frequency-swept ultrasound-modulated optical tomography in biological tissue by use of parallel detection, *Opt. Lett.*, 25, 734, 2000.
24. Lévêque-Fort, S., Selb, J., Pottier, L., and Boccara, A.C., *In situ* local tissue characterization and imaging by backscattering acousto-optic imaging, *Opt. Commun.*, 196, 127, 2001.
25. Granot, E., Lev, A., Kotler, Z., Sfev, B.G., and Taitelbaum, H., Detection of inhomogeneities with ultrasound tagging of light, *J. Opt. Soc. Am. A*, 18, 1962, 2001.
26. Wang, L.V., Mechanisms of ultrasonic modulation of multiply scattered coherent light: an analytic model, *Phys. Rev. Lett.*, 87, 1, 2001.

27. Gleyzes, P., Guernet, F., and Boccara, A.C., Profilométrie picométrique. II. L'approche multidétecteur et la détection synchrone multiplexée, *J. Opt.*, 26, 251, 1995.
28. Marquez, G., Wang, L.H., Lin, S.-P., Schwartz, J.A., and Thomsen, S.L., Anisotropy in the absorption and scattering spectra of chicken breast tissue, *Appl. Opt.*, 37, 798, 1998.
29. Selb, J., Pottier, L., and Boccara, A.C., Non-linear effects in acousto-optic imaging, *Opt. Lett.*, 27, 918, 2002.
30. Société PRL-Corélec, 19 Route Nationale, F-77580 Crécy-la-Chapelle, France; fax: +33 1 64 63 69 75.

# VI

# Intervention and Treatment Techniques

# 36

# Mechanistic Principles of Photodynamic Therapy




Barbara W. Henderson
*Roswell Park Cancer Institute*
*Buffalo, New York*

Sandra O. Gollnick
*Roswell Park Cancer Institute*
*Buffalo, New York*


## 36.1 Introduction

Photodynamic therapy (PDT) exploits the biological consequences of localized oxidative damage inflicted by photodynamic processes. A schematic outline of the major steps that lead to tumor destruction by PDT is given in Figure 36.1. Three critical elements are required for the initial photodynamic processes to occur: a drug that can be activated by light (a photosensitizer), light, and oxygen.[1] Interaction of light at the appropriate wavelength with a photosensitizer produces an excited triplet-state photosensitizer that can interact with ground-state oxygen via two different pathways, designated as Type I and Type II. The individual steps of these pathways are shown in Figure 36.2. The Type II reaction that gives rise to singlet oxygen ($^1O_2$) is believed to be the dominant pathway because elimination of oxygen or scavenging of $^1O_2$ from the system essentially eliminates the cytocidal effects of PDT.[2,3] Type I reactions, however, may become important under hypoxic conditions or where photosensitizers are highly concentrated.[4] The highly reactive $^1O_2$ has a short lifetime (<0.04 msec) in the biological milieu and therefore a short radius of action (<0.02 μm).[5] Consequently, $^1O_2$-mediated oxidative damage will occur in the immediate vicinity of the subcellular site of photosensitizer localization. Depending on photosensitizer pharmacokinetics,


0-8493-1116-0/03/$0.00+$1.50

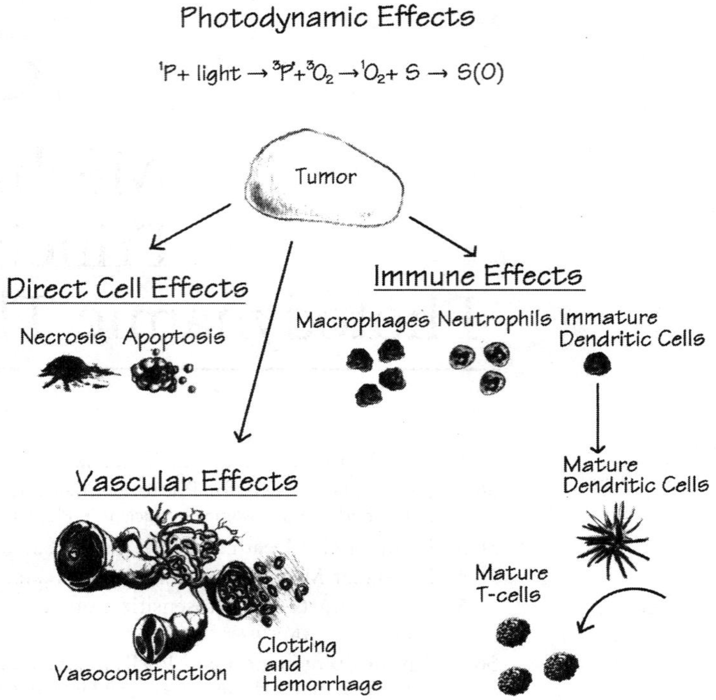


**FIGURE 36.1** Illustration of the three major tissue targets affected by the photodynamic effect. Tumor cells can be damaged or killed directly by the effects of singlet oxygen generated within them, succumb to oxygen and nutrient deprivation due to vascular damage inflicted by PDT, or be attacked by the inflammatory/immune system activated by PDT.

these sites can be varied and numerous, resulting in a large and complex array of cellular effects. Similarly, on a tissue level, tumor cells as well as various normal cells can take up the photosensitizer, which upon activation by light can lead to effects on such targets as the tumor cells proper, the tumor and normal microvasculature, and the inflammatory and immune host system. PDT effects on all these targets may influence each other, producing a plethora of responses; the relative importance of each has yet to be fully defined. It seems clear, however, that the combination of all these components is required for long-term tumor control.

## 36.2 Subcellular Targets for Photosensitization

The potential cellular targets of PDT are shown schematically in Figure 36.3. They depend on the specific pharmacokinetic characteristics, such as lipophilicity, amphiphilicity, aggregation, and serum protein interactions (see Reference 1 for a review) and, therefore, the localization, of the photosensitizer but appear to be largely independent of cell type (Table 36.1). Localization studies have generally been carried out in *in vitro* cell systems where exposure conditions to the photosensitizer can be easily controlled or varied. Such studies have revealed that cellular photosensitizer distribution can be a dynamic process, influenced by such parameters as length of exposure and drug concentration.[6–8] Photosensitizers may even relocalize after photodamage to an initial site of accumulation such as from lysosomes to other, possibly more sensitive, cellular locations, where they will then be available for activation.[9,10] That subcellular localization of a photosensitizer, and consequently the target of PDT, may influence *in vivo* treatment outcome was demonstrated in a series of studies that related structure and activity of a

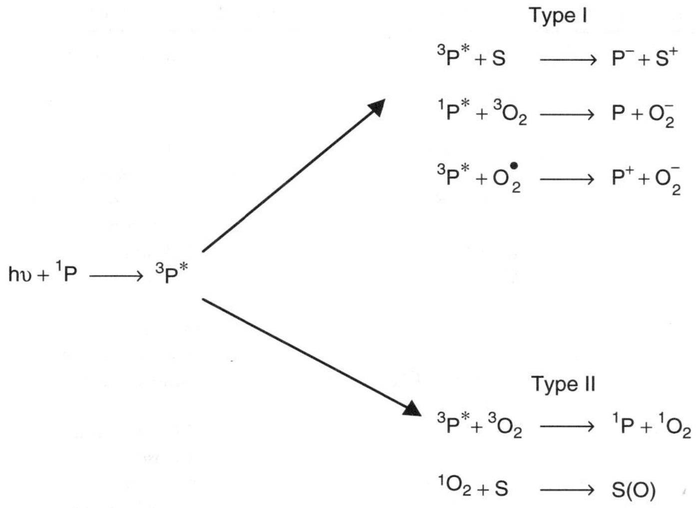


**FIGURE 36.2** Photoreaction pathways emanating from the interaction of a photosensitizer with light where $^1P$ is a photosensitizer in a singlet ground state, $^3P^*$ is a photosensitizer in a triplet excited state, S is a substrate molecule, $P^-$ is a reduced photosensitizer molecule, $S^+$ is an oxidized substrate molecule, $^3O_2$ is molecular oxygen (triplet ground state), $O_2^-$ is the superoxide anion, $O_2^{\bullet}$ is the superoxide radical, $P^+$ is the oxidized photosensitizer, $^1O_2$ is oxygen in a singlet excited state, and S(O) is an oxygen adduct of a substrate. (From MacDonald, I.J. and Dougherty, T.J., *J. Porphyrins Phthalocyanines*, 5, 105, 2001. With permission.)

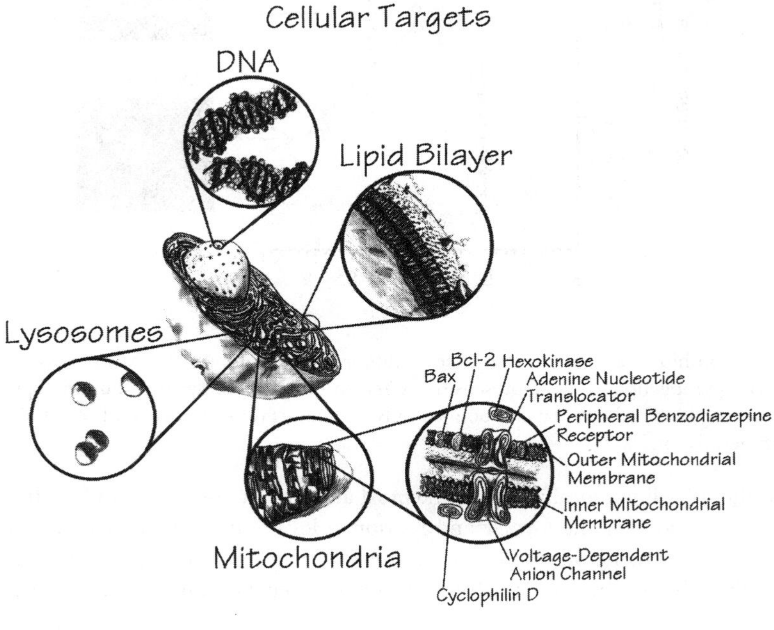


**FIGURE 36.3** Illustration of the major cellular targets affected by PDT.

congeneric series of photosensitizing compounds to their subcellular localization (Figure 36.4).[8,11] Pyropheophorbide-a ether derivatives, designed to possess progressively increasing degrees of lipophilicity, exhibited drug uptake in tumor cells that increased linearly with lipophilicity, while PDT activity tested in an *in vivo* murine tumor system showed a parabolic quantitative structure activity relationship (QSAR).

**TABLE 36.1** Intracellular Localization Sites of Major Photosensitizers in Cells *in Vitro*

| Photosensitizer | Cell Line | Localization Site | Ref. |
|---|---|---|---|
| BPD-MA<br>Verteporfin | NHIK3025 | ER/Golgi | 31 |
| Phthalocyanines | NHIK3025, RIF, LOX, V79 | Lysosomes<br>Cytoplasm | 9, 225–227 |
| Hematoporphyrin/Photofrin | V79, L1210 | Plasma membrane<br>Mitochondria | 7, 228 |
| Lutex | EMT6 | Lysosomes | 32, 229 |
| mTHPC | V79 | Cytoplasm, diffuse | 230 |
| $Npe_6$ and analogs | CHO, L1210 | Lysosomes<br>Mitochondria<br>Plasma membrane | 18, 19, 32, 231 |
| PpIX from ALA | WiDr, NHIK3025, V79<br>FaDu, RIF | Mitochondria | 16, 17, 21 |
| Pyropheophorbide derivatives | FaDu, RIF | Mitochondria<br>Lysosomes | 8 |
| TPPS analogs | NHIK3025 | Lysosomes<br>Mitochondria<br>Perinuclear | 47, 232–234 |
| Purpurin analogs<br>$SnEt_2$ | P388, OVCAR5 | Lysosomes | 52, 235 |

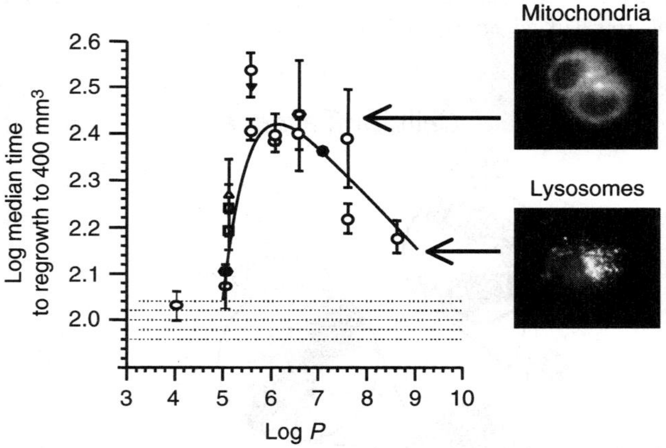


**FIGURE 36.4** Relationship of the log median time of tumor regrowth to 400-mm³ tumor volume to log *P* for a series of congeneric pyropheophorbide photosensitizers and to their subcellular distribution. Optimal compounds accumulate in mitochondria, suboptimal compounds in lysosomes. (From Henderson, B.W. et al., *Cancer Res.*, 57, 4000, 1997. With permission.)

Comparison of the subcellular localization of the most and least active compounds in the series revealed mitochondrial localization for the former and lysosomal localization for the latter. Nevertheless, such less active compounds can be very effective photosensitizers because the lack of sensitivity can be compensated with higher dosing, provided that treatment selectivity can be maintained.

## 36.2.1 Mitochondria

Although the first-generation photosensitizer Photofrin® and its forerunners are chemically heterogeneous and therefore exhibit heterogeneous subcellular distributions,[7] the major target also appears to be the mitochondria.[12–15] The development of chemically pure and well-defined photosensitizers and the use of 5-aminolevulinic acid (ALA) for the intramitochondrial generation of photosensitizing protoporphyrin IX (PpIX)[16,17] have allowed the targeting of specific subcellular sites and the analysis of the relative sensitivity to photodamage of these sites[18–21] (also see Table 36.1). Again, the mitochondria have emerged as especially sensitive targets. Recent interest has focused on the mitochondrial permeability transition pore, a protein complex consisting of hexokinase, the peripheral benzodiazepine receptor (PBR), the voltage-dependent anion channel (VDAC), creatinin kinase, the adenine nucleotide translocator (ANT), and cyclophyllin D. Several of its components have been implicated as PDT targets.[22–29]

### 36.2.2 Lysosomes

Hydrophilic or highly aggregated sensitizers may be taken up by pinocytosis or endocytosis and therefore accumulate in cytoplasmic vesicles.[8,9,18,30] While PDT-induced rupture of lysosomes and release of damaging hydrolytic enzymes into the cytoplasm have been described,[30,31] it appears now that lysosomes per se are not very effective targets for PDT.[8,18] In a recent study, selective lysosomal photodamage that was associated with release of cathepsin B was followed by a gradual loss of mitochondrial membrane potential, release of cytochrome *c* into the cytosol, caspase 3 activation, and a limited apoptotic response.[32] Thus, lysosomal damage may secondarily mediate mitochondrial damage. It has also been suggested that lysosomes might serve as reservoirs from which photosensitizers may be released after vesicle rupture and migrate to more sensitive sites.[9] All these studies assume that initial sensitizer accumulation is restricted to the lysosomes, but it is still possible that high lysosomal concentration may mask small but highly effective sensitizer accumulations at other sites.[8]

### 36.2.3 Nucleus

Photosensitizers used in PDT generally do not accumulate in the cell nucleus, and therefore frank DNA damage, such as strand breaks, sister-chromatid exchanges, and chromatid aberrations, are much less frequently observed in PDT than in ionizing or UV irradiation.[33–35] Consistent with these findings, the mutation potential of PDT was found to be significantly less than that of ionizing radiation or UV.[36]

### 36.2.4 Plasma Membrane

The plasma membrane may become a target of PDT, especially if the sensitizer is highly lipophilic and the incubation time of cells *in vitro* is brief. Membrane damage manifests itself rapidly through blebbing[20,37] as well as leakage of cytosolic enzymes[38,39] and chromium.[40] Leakage of lactate dehydrogenase (LDH) showed the same kinetics as generation of prostaglandin $E_2$ from tumor cells *in vitro*, directly relating membrane disruption to production of eicosanoids.[39] The biosynthesis of eicosanoids, which play an important role in mediating the PDT tissue response (discussed below), is set in motion when phospholipases catalyze the liberation of free fatty acids from membrane phospholipids. Unsaturated phospholipids and cholesterol are important membrane targets of photodamage, and lipid peroxidation by PDT has been extensively studied and linked with cell lethality.[41,42] Oxidation of intrinsic membrane proteins[43] and protein crosslinking[44] have also been observed. PDT can inhibit the plasma-membrane enzymes $Na^+K^+$-adenosine triphosphate (ATPase) and $Mg^{2+}$-ATPase.[45] $Ca^{2+}$ flux may be affected, and the plasma membrane may become depolarized.[46] Certain membrane-localized photosensitizers, such as mesotetraphenylporphines (TPPSs), can destroy microtubules in interphase cells and lead to arrest of cells in mitosis.[31,47]

## 36.3 Mode of Cell Death

PDT can induce cell death via either the apoptotic or necrotic pathways.[48] Necrosis is metabolism-independent and a result of a massive insult. In contrast, the apoptotic pathway is an intrinsic physiological process that is dependent on active cellular metabolism and characterized by chromatin condensation, DNA fragmentation, and the formation of apoptotic bodies.[49] The dominant mode of cell death depends on the photosensitizer used, the localization of the photosensitizer, and the treatment protocol.[1] Dellinger[50] demonstrated that increasing the drug incubation time prior to administration of light altered the mode of cell death induced by Photofrin-PDT; longer incubation times (1 d) resulted in apoptosis, while shorter incubation times (1 h) induced necrosis. The cellular localization of Photofrin changes with incubation time, moving from the cell membrane to the nucleus as the incubation time increases.[8] Similar results have recently been published for zinc II phthalocyanine (Pc).[51] Studies by Kessel and Luo[19] demonstrated that photosensitizers that localize to the mitochondria (porphycenes PcM and PcD) resulted in apoptotic cell death, while those that localize to the lysosome (chlorin LCP) or cell membrane (cationic porphyrin, MCP) did not. Kessel et al.[52] also demonstrated that photodamage of the cell membrane may inhibit the induction of apoptosis by photodamage to the mitochondria. These results were confirmed by Luo and Kessel,[53] and this study also demonstrated that the mode of cell death shifts in response to the photodynamic dose such that high doses result in a necrotic cell death, while lower doses result in an apoptotic mode of death.

The kinetics and extent of PDT-induced apoptosis also vary with photosensitizer and cell type treated.[52–57] Photodynamic treatment of L5178Y murine lymphoma cells with ClAlPc resulted in apoptosis within 30 min of treatment.[57] Treatment of the lymphoma line P388 with ClAlPc or tin etiopurpurin ($SnEt_2$), which localize to the lysosome and the mitochondria, induced apoptotic nuclei within 1 h.[52,53] However, treatment of the same cell line with tin octaethylpurpurin amidine, which localizes to the mitochondria, the lysosome, and the cell membranes, did not result in apoptotic nuclei until 24 h after treatment.

The rapid induction of apoptosis suggests that PDT triggers late-stage apoptotic processes directly. Release of cytochrome *c* from the mitochondria triggers caspase 3 activation and results in initiation of apoptosis at a late stage in the pathway.[58] Cytochrome *c* release is associated with a loss of mitochondrial potential. PDT results in the loss of mitochondrial potential and the rapid release of cytochrome *c*.[59–64] Several studies have demonstrated that PDT also induces rapid activation of caspases 3, 6, 7, and 8 and cleavage of poly(ADP-ribose) polymerase (PARP).[56,59] Kessel and Luo have shown that PDT-induced release of cytochrome *c* is sufficient to directly initiate a caspase-dependent apoptotic cell death.[65] However, a recent study by Xue et al.[66] indicates that while caspase 3 is required for the late stages of apoptosis, it is not the critical lethal event in PDT-induced apoptosis.

Bcl-2 has been shown to block the release of cytochrome *c* from the mitochondria and thus prevent apoptosis.[58] Several groups have altered the expression of bcl-2, through overexpression or antisense technology, and demonstrated that increasing the levels of bcl-2 enhances cellular resistance to PDT-induced apoptotic cell death.[67–70] Xue et al.[71] have demonstrated that Pc 4-PDT results in photochemical destruction of bcl-2. Similar results have been reported for AlPc-PDT.[72] Interestingly, a study by Kim et al.[72] reported that overexpression of bcl-2 enhanced the apoptotic response. In these studies, overexpression of bcl-2 was accompanied by an increase in bax, a proapoptotic member of the bcl-2 family. PDT resulted in the selective destruction of bcl-2 but had no effect on bax. The greater apoptotic response in the cells overexpressing bcl-2 was attributed to a higher bax:bcl-2 ratio after PDT. A PDT-induced shift in the bax:bcl-2 ratio toward apoptosis has also been reported by others.[67,73]

PDT also affects other proteins implicated in apoptosis. Phospholipases C and $A_2$, which are involved in the transient increases in intracellular calcium levels and DNA fragmentation, are activated by Photofrin-PDT.[74] The apoptotic-associated protooncogenes c-jun and c-fos are also activated by PII-PDT.[75] Ceramide, which has been linked to apoptosis in several malignant cell lines, accumulates following PDT and has been associated with PDT-induced apoptosis and cytotoxicity.[76] Gupta et al.[77] have demonstrated that Pc 4-PDT induces expression of nitric oxide and suggested that it may be involved in PDT-induced apoptosis. Finally, the tumor-suppressor gene p53, which can force damaged cells into the apoptotic pathway, is associated with Photofrin-PDT resistance,[78,79] although there have been dichotomous reports on the role of p53 in PDT sensitivity.[80]

# 36.4 Molecular Mechanisms

Alterations in gene expression are generally due to changes in the molecular mediators present in the cell. PDT has been shown to alter the expression of several molecular mediators including the redox-regulated transcription factor, AP-1.[81,82] AP-1 is induced by changes in the redox potential[83–85] and hypoxia.[86] It is composed of homodimers of the products of the c-jun gene family or heterodimeric combinations of c-jun and c-fos family members. PDT induces prolonged expression of both c-fos and c-jun as a result of oxidative stress.[75,87] Therefore, modulation of AP-1 activity by PDT might be mediated by changes in oxidative potential as a result of the generation of singlet oxygen or PDT-induced hypoxia.

NF-κB plays a critical role in the expression of immunomodulatory and proinflammatory genes.[88] Activated NF-κB is a heterodimeric protein most commonly comprised of the p50 and p65 (Rel A) species. It is also activated in response to cellular oxidative stress.[89] NF-κB is sequestered in the cytoplasm by IκB proteins; phosphorylation of IκB results in its proteosomal degradation and release of NF-κB.[90,91] NF-κB binding activity was found to be induced by PDT[92–94] and has been proposed to play a role in determining the cellular response to PDT.[95] Granville et al.[92] demonstrated that cellular levels of IκB were transiently depressed following Verteporfrin®-PDT. Pyropheophorbide-a methyl ester-PDT also leads to IκB degradation.[94] In contrast to these studies, other groups have failed to demonstrate NF-κB activation following Photofrin-PDT.[82,81] Thus, NF-κB activation may be photosensitizer specific. It has also been suggested that activation is related to PDT dose because higher doses do not result in decreases in IκB.[92]

In addition to AP-1 and NF-κB activation, PDT has been shown to stimulate stress-kinase-signaling pathways (SAPK/JNK, p38/HOG1)[96,97] and HS1 phosphorylation,[98] as well as stress-response transcriptional activators. HIF-1 (hypoxia inducible factor-1) is a key regulator of the cellular response to hypoxia and induced in response to PDT.[99,100] PDT also enhances the expression of the HIF-1 target gene, vascular endothelial growth factor (VEGF),[100] and it has been proposed that HIF-1 expression may act as a predictor of PDT responsiveness.[99] Finally, PDT also activates the heat-shock-protein (HSP) family promoters[101] and induces the expression of HSPs and the related glucose-regulated proteins[102–106] as well as heme oxygenase (HSP 32).[107]

# 36.5 Tissue Targets of Photosensitization

## 36.5.1 Tumor Cells

The ability of PDT to eliminate tumor cells through direct photodamage has been most effectively studied by *in vivo–in vitro* tumor-explant methodology.[108] Such in-depth analysis has revealed that the full potential of direct photodynamic tumor-cell kill, provided by the gross tumor-photosensitizer concentration and absorbed-light dose, is generally not realized by *in vivo* PDT treatment.[109] Clonogenic assays carried out with a number of different photosensitizers and tumor systems immediately after potentially curative PDT exposures *in vivo* have revealed that at most 1 to 2 logs of direct tumor-cell kill have been achieved,[110–115] far less than the 7 to 8 logs required for tumor cure. Clearly, limitations to direct tumor-cell kill exist *in vivo*, the most important of which may be (1) inhomogeneous photosensitizer distribution within the tumor including a gradual decrease of photosensitizer concentration with distance from blood vessels,[116] (2) insufficient light penetration through the tissue (light is attenuated exponentially with depth of tissue penetration[117,118]), and (3) insufficient oxygen availability (discussed below under separate heading). That intrinsic tumor-cell sensitivity contributes to the overall PDT response was suggested by studies using tumor cells selected to be resistant to PDT.[119] Tumors of resistant phenotype were less responsive to PDT than sensitive tumors while exhibiting equal vascular responses. The studies did not exclude the possibility, however, that tumor immunogenicity might have differed in the two tumor lines, thereby affecting the response. On the other hand, tumors expressing a multidrug-resistance phenotype that prevented the uptake of a cationic photosensitizer were nevertheless found responsive to PDT carried out with that agent, the antitumor effect being attributed to vascular disruption.[120] Studies like these illustrate PDT's ability to exert its antitumor action through several different tissue targets.

**TABLE 36.2** Tumor Perfusion and Oxygenation Following PDT

| Photosensitizer | Tumor Model | Technique | Ref. |
|---|---|---|---|
| Hematoporphyrin derivative | RMA rat mammary tumor | Window chamber observation | 236 |
| Photofrin | Cremaster muscle, rat | Window chamber observation | 139 |
| Phthalocyanine analogs | Chondrosarcoma, rat | Window chamber observation | 141 |
| Npe6 | Chondrosarcoma, rat | Window chamber observation | 140 |
| Photofrin | RIF-1 mouse fibrosarcoma | Radiobiological assay | 237 |
| Hematoporphyrin derivative | AY-27 rat urothelial tumor | $^{103}$Ru, $^{141}$Ca | 238 |
| Photofrin | RIF-1 mouse fibrosarcoma | $^{86}$Rb extraction Hoechst 33342 | 239 |
| Polyhematoporphyrin ALA | $LSBD_1$ rat fibrosarcoma | Microspheres, $^{86}$Rb extraction | 240 |
| Hematoporphyrin derivative | T50 80 mouse mammary tumor | NMR imaging | 241 |
| Photofrin | R3230A rat mammary tumor | NMR imaging | 242 |
| Bacteriochlorophyll-serine | M2R melanotic melanoma xenograft | Contrast-enhanced MRI | 243 |
| Photofrin | Chondrosarcoma, rat | Oxygen microelectrode | 244 |
| Photofrin | VX-2 rabbit skin carcinoma | Transcutaneous oxygen electrodes | 157 |
| Photofrin | Mammary carcinoma, mouse | Oxygen microelectrode | 158 |
| Photofrin | RIF-1 mouse fibrosarcoma | Oxygen microelectrode, $^{86}$Rb extraction, fluorescein | 134 |
| Photofrin | A673 sarcoma, human xenograft | Laser Doppler | 245 |
| Photofrin | RIF-1 mouse fibrosarcoma | Hypoxia marker | 166 |

Initial studies trying to establish a pattern of selectivity of photosensitizer uptake in malignant vs. normal cells *in vitro* were largely unsuccessful.[121] *In vivo*, moderately favorable tumor-to-normal-tissue ratios can be found for almost all photosensitizers, with establishment of these ratios depending on the specific pharmacokinetics of the compound as well as the pathophysiology of the tumor (for detailed reviews see References 1, 122, and 123). Mechanisms invoked for this "selectivity" range from leaky vasculature and impaired lymphatic drainage in tumors to low tumor pH and an increase in low-density lipoprotein and other membrane receptors on tumor cells.[124,125] Carrier systems, such as antibody conjugates, have been designed to direct the photosensitizer directly to the tumor cells. Such immunoconjugates have been directed against epitopes on ovarian[126] and colon-cancer cells[127] as well as against the epidermal growth factor receptor (EGFR) that is overexpressed in oral precancer.[128]

## 36.5.2 Microvasculature

The microvasculature was one of the earliest tissue targets identified because vascular PDT effects are rapid and dramatic, especially with the use of the sensitizer Photofrin and its forerunners. Reduction or cessation of tumor microcirculation following *in vivo* PDT exposure employing a variety of photosensitizers has been demonstrated in preclinical models through numerous different techniques, summarized in Table 36.2. Since the kinetics of vascular shutdown and tumor-cell death have been found to coincide[111,129] and inhibition of shutdown retards tumor response,[130] it has been argued that disruption of the tumor microcirculation is a major factor contributing to tumor control by PDT. Differences in response between tumor and normal microvasculature are subtle and difficult to discern.[131] Careful dose-ranging studies have revealed that the tumor vasculature is slightly more susceptible to shutdown than its normal counterpart.[132,133] It has also been demonstrated that Photofrin-PDT at high fluence rate can protect the normal skin microvasculature, but the same treatment fails to protect tumor vessels.[134,135] Occlusion of the tumor-surrounding vasculature can contribute to tumor control, at least in preclinical models, presumably by adding to the nutrient deprivation and retardation of vascular resupply of the tumor.[132,136] Another approach to retardation of vessel regrowth after PDT is the use of antiangiogenic agents in combination with PDT, which has been shown to enhance long-term tumor control.[100,137]

The acute manifestations of PDT-induced vascular damage greatly depend on the type and dose of photosensitizer used. On the microscopic level, vascular changes in a rat cremaster muscle preparation after Photofrin-PDT included vessel constriction, occlusive platelet thrombi in arteries, arterioles, veins

**TABLE 36.3** Vascular Response Mediators Generated or Affected by PDT

| Photosensitizer | Cell/Tumor Type | Mediator | Ref. |
|---|---|---|---|
| Hematoporphyrin derivative | Mast cells, rat | Histamine | 246 |
| Protoporphyrin | Mast cells, rat | $PGD_2$, $PGE_2$, $PGI_2$ | 247 |
| Hematoporphyrin | Platelets, human | Serotonin | 248 |
| Phthalocyanine (ClAl-S) | Endothelial cells, bovine | Clotting factors | 249 |
| Photofrin | Endothelial cells, human | Von Willebrand factor | 250 |
| Photofrin | EMT6 mammary tumor, mouse; RIF fibrosarcoma, mouse | $PGE_2$ | 39 |
| Photofrin | Macrophages, mouse | $PGE_2$ | 39 |
| Photofrin phthalocyanine (ZnS) | Endothelial cells, bovine; human | $PGI_2$, AA, $F_{2\alpha}$, HETES | 143 |
| Photofrin | Mouse serum | Thromboxane $B_2$ | 144 |

and venules, edema, and neutrophil margination and migration.[138,139] Ultrastructural features included damage to numerous endothelial-cell organelles as well as perivascular changes such as degranulation of perivascular mast cells and damage to myocytes.[138] Npe6 (mono-L-aspartyl chlorin e6)-PDT in a rat chondrosarcoma model resulted in obstructive platelet thrombi but no vasoconstriction.[140] PDT with variously substituted zinc Pcs, tested in the same tumor model, showed a range of effects, including vessel constriction and leakage, with one compound (disulfonated zinc Pc) exhibiting no apparent effects.[141]

Numerous vascular-response mediators seem to be involved in these processes, summarized in Table 36.3. Most studied and prominent among them are the eicosanoids. They are products of the release from the cell membrane of arachidonic acid that is subsequently metabolized by cyclooxygenase to generate prostaglandins and thromboxanes and by lipoxygenase to form leukotrienes and hydroxy acids. The generation of a wide spectrum of arachidonic-acid metabolites by PDT has been described in mast cells, macrophages, endothelial cells, platelets, and tumor cells, as well as in animals. Interestingly, *in vitro* PDT exposure of platelets blocks their capacity to aggregate,[142,143] and *in vitro* exposure of endothelial cells is dominated by the release of prostacyclin ($PGI_2$), which inhibits platelet aggregation.[143] These *in vitro* findings contradict the well-established proaggregating mechanisms observed *in vivo* and demonstrate the difficulty of translating *in vitro* data to the *in vivo* situation. The latter represents a much more complex interplay of mechanistic components that likely involves platelets, endothelial cells, leukocytes, macrophages, and other stromal cells. Thromboxane appears to play a major role in mediating the observed vascular effects, at least in rat models.[144] Antiinflammatory drugs, such as indomethacine or aspirin, can block the release of eicosanoids *in vitro*[39,143] and *in vivo*.[141,144–146] Rats that are made thrombocytopenic, and thus deprived of a source for thromboxane generation, also show a diminished vascular response.[147] Eicosanoids, in addition to serotonin, also appear to be involved in changes of tumor interstitial pressure observed after Photofrin-PDT, a consequence of fluid leakage from the vasculature and possibly contributing to occlusion of the tumor microvasculature.[148]

## 36.6 Tumor Oxygenation and PDT

Any restriction of tissue oxygen supply *during* PDT light delivery will reduce $^1O_2$ production and therefore have negative consequences for treatment outcome. Such restrictions can arise from numerous sources including preexisting tumor hypoxia, acute vascular damage, and photochemical oxygen depletion. All of these can interact, and the dynamics of these interactions may determine treatment success.

### 36.6.1 Preexisting Tumor Hypoxia

Preexisting tumor hypoxia, a therapeutic problem still grappled with by radiation oncologists because of the oxygen dependence of sparsely ionizing radiation treatment, may limit the oxygen supply for PDT as well. Solid tumors are prone to develop hypoxic tumor regions due to deteriorating diffusion geometry, structural abnormalities of tumor microvessels, and disturbed microcirculation.[149] Hypoxic, but viable, tumor cells located in these areas may be protected from PDT-induced photodamage. In a preclinical study, Fingar et al.[150] manipulated tumors pharmacologically and physically to induce a wide range of hypoxic tumor fractions (~2% to ~40%) and followed this with aggressive Photofrin-PDT. It was found that hypoxic fractions below 5% did not adversely affect tumor control, those of ~10% slightly diminished tumor control, and fractions of ~40% totally blocked tumor control. Vascular shutdown and nutrient deprivation following PDT were believed to be responsible for the elimination of small numbers of initially surviving hypoxic tumor cells. Early preclinical attempts to raise tumor oxygenation prior to PDT through the administration of a perfluorochemical emulsion and carbogen breathing were highly successful in increasing tumor oxygen levels (~tenfold) up to 1 h after PDT. The intervention did not alter long-term tumor control,[151] probably because the vascular damage induced by the aggressive PDT regime overwhelmed any subtle improvement in treatment outcome that might have been attributed to increased tumor oxygenation. More recent studies, both preclinical and clinical, have demonstrated significantly improved PDT efficiency with adjuvant administration of hyperbaric oxygen or carbogen.[69,152,153]

However, the factor that probably influences preexisting tumor hypoxia most profoundly is one that accompanies most PDT treatments, namely, changes in tumor temperature. Temperature increases due to PDT light delivery have been analyzed by Svaasand et al.[154] and have been recorded in preclinical models[155] and in patients' tumors.[156] Measurements of baseline intratumoral temperature and $pO_2$ in nodular basal-cell carcinomas have demonstrated a linear relationship between increasing tumor $pO_2$ and lesion temperature in ranges between 0 and 20 mm Hg and 30 to 35°C.[156] Upon laser illumination during Photofrin-PDT, the temperature in these lesions increased further in a fluence-rate-dependent manner. Surprisingly, even low-fluence-rate light produced significant temperature rises (150 mW/cm$^2$: median temperature change +1.9°C [range 1.0 to 6.2], 30 mW/cm$^2$: median temperature change +1.5°C [range 1.3 to 3.6]), and these correlated with increased tumor $pO_2$. Whether these relationships hold true for tumors other than skin tumors remains to be seen.

### 36.6.2 Oxygen Limitation through Vascular Damage

With photosensitizers that can acutely constrict and occlude vessels, blood-flow obstruction can be marked, very rapidly limiting the oxygen supply to the tumor. Photofrin-PDT in mouse models, for example, rendered up to 10% of tumor cells hypoxic within a very moderate light exposure (45 J/cm,$^2$ 10 min).[109] This hypoxia was persistent and progressive with time, 50% of tumor cells being hypoxic within 1 h of such light exposure. Similar observations were reported for a rabbit-skin-tumor model, where a series of brief light exposures resulted in induction of irreversible tumor hypoxia that was cumulative with the number of exposures, i.e., fluence.[157] Hypoxia induction by PDT depends greatly on the vascular supply of a given tumor and even on the site of tumor implantation in rodent models.[158] Transient reoxygenation may occur depending on PDT dose.[158] With certain second-generation sensitizers, many of which exert less severe acute effects on the vasculature than Photofrin,[141] and a tendency toward the use of lower drug doses, such acute vascular effects are less likely to occur. As discussed above, the extent and timing of vascular damage, and therefore induction of tumor hypoxia, are very important for treatment outcome. Occurrence of vascular occlusion can be detrimental *during* treatment but beneficial *after* completion of the PDT tumor treatment.

### 36.6.3 Oxygen Limitation through Photochemical Oxygen Depletion

Photochemical oxygen depletion is roughly characterized by instantaneous or near-instantaneous development of tumor hypoxia upon light exposure of a photosensitized tumor/tissue and equally rapid reoxygenation upon cessation of light. The theoretical basis for this phenomenon has been provided through mathematical modeling of the dynamic changes to be expected in tissue when oxygen is consumed in the process of $^1O_2$ generation.[159,160] Photochemical oxygen depletion will occur in tissue if the rate of photodynamic-oxygen consumption is faster than the rate of oxygen resupply from the vasculature. The major parameters that determine whether or not photochemical oxygen depletion will occur are (1) the absorption coefficient of the photosensitizer, (2) the tissue concentration of the photosensitizer, (3) the fluence rate of light, and (4) the vascular supply of the tissue.[159,161,162] If the first three parameters are high, $^1O_2$ production will be rapid and oxygen depletion favored; if the vascular supply of the tissue is poor, oxygen depletion will also be favored. The mathematical predictions have been validated in tightly defined *in vitro* systems,[163,164] in tumor models,[134,157,165,166] and in humans.[156] Light fluence rate is the most easily controlled parameter, one that can be readily modulated during light delivery. Therefore, much attention has been paid to the effects of fluence rate on PDT oxygen consumption. It is clear that lowering of the fluence rate can diminish or eliminate photochemical oxygen consumption through lowering the $^1O_2$-generation rate. However, the optimal fluence rate will depend on the other parameters listed above and therefore vary from situation to situation. In human tumors, the variability is great, both among patients and among lesions in the same patient.[156]

## 36.6.4 The Role of Photobleaching in Photochemical Oxygen Depletion

Photobleaching is the destruction of the photosensitizer by light-mediated processes.[167–170] Since photosensitizer concentration is one of the major determinants for photochemical oxygen depletion, it stands to reason that the destruction of photosensitizer through photobleaching during PDT will reduce the likelihood that oxygen depletion will occur. The most detailed studies of photobleaching have been carried out in an *in vitro* multicell tumor spheroid model.[171] Sustained illumination of Photofrin-photosensitized spheroids was shown to lead to a progressive decrease of photochemical oxygen depletion, implying reduction of photosensitizer levels and consistent with a theoretical model in which bleaching occurs via a self-sensitized singlet-oxygen reaction with the photosensitizer ground state. Similarly, PpIX was degraded by $^1O_2$-mediated mechanisms, while another photosensitizer (Nile blue selenium) was degraded by $^1O_2$-independent mechanisms.[172] Oxygen measurements in rodent tumor models and human basal-cell carcinomas also implicated photobleaching in influencing photochemical oxygen depletion.[134,156] In these studies, significant oxygen depletion was observed during the early time periods of illumination at high fluence rates of Photofrin-photosensitized lesions, but less or no oxygen depletion was detected toward the end of illumination, implying that the Photofrin concentration had been reduced through photobleaching below the threshold needed for oxygen depletion.

## 36.6.5 Enhancement of PDT Efficiency through Modified Light-Delivery Schemes

It is evident that any means by which the well-oxygenated tumor volume can be increased during light exposure should have beneficial effects on treatment outcome. Downward adjustment of treatment fluence rate is one such means; fractionation of light delivery is another.[134,173–176] While the former allows for continuous maintenance of oxygen levels sufficient for $^1O_2$ production, the latter facilitates reoxygenation of the tissue between light exposures. Significant enhancements of tumor response have been observed with either of these alternatives for PDT with Photofrin, ALA/PpIX, and mTHPC. In part, this may be due to a significant but moderate increase in direct photodamage to tumor cells. Direct tumor-cell death increased with low fluence rate PDT by ~1/2–1 log in the RIF mouse model.[115] However, the microvasculature, especially the normal tumor-surrounding microvasculature, can also be affected by modification of fluence rate.[135] One practical drawback of low-fluence-rate as compared with high-fluence-rate treatment is the increase in exposure time required to deliver a given fluence. Due to the higher treatment efficiency this can be somewhat, but not entirely, compensated for by a reduction of the total fluence delivered. In fact, a reduction of the total fluence may be necessary with low-fluence-rate treatment because the PDT efficiency for causing vascular and normal tissue effects will also increase, thereby decreasing treatment selectivity.[135,176]

Enhanced responses to light-dose fractionation in ALA-PDT of murine tumors and normal rat colon may involve relocalization or resupply of PpIX during dark periods in addition to reoxygenation.[177,178]

# 36.7 Immune Effects of PDT

## 36.7.1 Immune Suppression

The effect of PDT on the host immune system is dichotomous, resulting in either immune suppression or immune activation. Although the mechanism leading to either immune activation or suppression is unclear, recent studies have indicated that the nature and extent of the tissue treated and the dose play a major role.[179,180] Cutaneous PDT suppresses allograft rejection[181–183] and contact hypersensitivity (CHS) reactions.[179,184–187] PDT suppression of CHS in murine models has been demonstrated with a number of different photosensitizers[184–186,188–191] and is antigen-specific.[184,185] The kinetics of CHS suppression is photosensitizer-dependent;[185] $TPPS_4$ and HpD-PDT induce immediate CHS suppression, while *m*-THPC and Photofrin-PDT suppression requires 72 h to develop. Kinetic differences have been linked to photosensitizer localization.[185] The mechanism of PDT-induced suppression appears to be associated with the induction of immunosuppressive cytokines. PDT induces tumor necrosis factor (TNF)-α,[187,192] which is involved in some aspects of UV-mediated immunosuppression.[193] However, it is not responsible for PDT-induced CHS suppression.[187] PDT also induces interleukin (IL)-10,[179,189,194] which has been shown to inhibit cell-mediated immune responses including CHS.[189,195,196] Gollnick et al.[179] have shown that *in vitro* PDT induces IL-10 expression from keratinocytes as a result of activation of the IL-10 gene promoter by enhanced expression of AP-1 and prolonged IL-10 mRNA half-life. IL-10 is thought to mediate its suppressive activity through inhibition of the ability of antigen-presenting cells (APCs) to stimulate $CD4^+$ T helper (Th1) cells.[195] Th1 cells, which secrete IL-2 and interferon (IFN)-γ, are necessary for induction of cell-mediated immune responses. The duration and strength of CHS reactions are increased in the presence of Th1 cells.[189] Th1 cells depend on the presence of IL-12, secreted from APCs such as macrophages and dendritic cells, for their development.[197,198] Cytotoxic T (Tc1) cells are the major effector cells in the CHS response[196] and also depend on IL-12 for their development. IL-12 secretion from APCs is inhibited by IL-10.[195,196] Interestingly, Lynch et al.[186] demonstrated that adoptive transfer of macrophages from mice treated with peritoneal Photofrin-PDT, which involves treatment of the entire peritoneal cavity, suppressed the CHS response to di-nitrofluorobenzene (DNFB) in the adoptively transferred animals, suggesting that this PDT regime altered the ability of the macrophages to induce Th1 and Tc1 cells.

Direct mechanistic studies into the role of IL-10 in PDT-induced suppression of CHS have yielded contradictory results that can be explained, at least in part, by the treatment regime and dose of PDT. Treatment with cutaneous PDT, using Photofrin and blue light centered at 430 nm, results in irradiation that is mostly limited to the skin. In contrast, transdermal PDT, with BPD-MA and illumination with 690-nm light, involves whole-body irradiation. Also, due to the greater depth of light penetration of this wavelength some irradiation of internal sites may occur. Both transdermal and cutaneous PDT result in CHS suppression and induce IL-10 expression in the skin.[179,189,194] Simkin et al.[189] implicated IL-10 as the active mediator in transdermal PDT-induced suppression of CHS by demonstrating that IL-10 knockout (KO) mice did not undergo transdermal PDT-induced CHS suppression. In contrast, cutaneous PDT treatment of IL-10 KO mice did induce CHS suppression.[179] The lack of involvement of IL-10 in the suppression of CHS by cutaneous PDT was further confirmed when studies using neutralizing anti-IL-10 antibodies failed to inhibit cutaneous PDT-induced CHS suppression.[179,191] Thus, it appears that the mechanism of PDT-induced suppression of CHS depends on the treatment regime. Regimes that result in large treatment fields and internal exposure, i.e., transdermal and potentially peritoneal PDT, mediate CHS suppression via IL-10. PDT regimes that employ lower doses and superficial cutaneous exposure suppress CHS reactions via an IL-10-independent mechanism.

Interestingly, CHS suppression by both transdermal and cutaneous PDT was reversed by administration of exogenous IL-12,[179,189] suggesting that these processes share a common regulatory point, perhaps in the development of Th1 or Tc1 cells. A second potential suppressor of Th1 and Tc1 development, and thus of CHS, is IL-4. IL-4 has been shown to directly inhibit Th1-cell proliferation[199] and to induce anergy in Tc1 cells.[200] Th2 cells, which are critical for the development of humoral immunity[199] and secrete IL-4, have been shown to negatively regulate CHS responses.[79] Induction of systemic IL-4 by cutaneous PDT may result in direct inhibition of Th1/Tc1 cells and suppression of CHS responses. Preliminary data from our group suggest that cutaneous PDT results in the induction of systemic IL-4, and a recent study demonstrated that cutaneous PDT suppression could be transferred by a $CD4^+$ T-cell population.[180]

In addition to the effects of cytokines on PDT-induced suppression of immune responses, the effect of PDT on the expression of immune molecules critical to immune-system activation should be considered. Transdermal PDT has been shown to inhibit the ability of Langerhans cells (LC) to stimulate alloreactive T cells,[181] and LC treated *ex vivo* expressed lower levels of major histocompatibility (MHC) antigens and CD80 and CD86 costimulatory molecules, which are needed for T-cell activation. Additionally, *in vitro* PDT-treated murine dendritic cells (DC) had a reduced ability to stimulate alloreactive T cells and exhibited lower levels of MHC molecules, costimulatory molecules, and adhesion molecules.[201] Thus, in addition to altering the function of APCs, PDT has the ability to disrupt the APC–T-cell cognate, which is needed for T-cell activation.

## 36.7.2 Immune Potentiation

In contrast to the immunosuppressive effects of transdermal or cutaneous PDT, tumor-directed PDT has been shown by a number of preclinical studies to trigger an inflammatory response and enhance a specific antitumor response.[123,202] PDT is characterized by the onset of an inflammatory reaction that is dominated by an influx of neutrophils.[194,203] The strength of the inflammatory response varies with photosensitizer; Photofrin-PDT induces a strong inflammatory response and a rapid influx of neutrophils, which appear to be critical to long-term tumor control.[204–207] HPPH-PDT induces a mild inflammatory response that is associated with reduced neutrophil infiltration,[208] which is not critical to long-term tumor control following HPPH-PDT.[208] The influx of neutrophils into the treatment site is preceded by an induction of chemokines and adhesion molecules critical to neutrophil migrations.[208] Neutrophil infiltration is followed by an influx of mast cells and macrophages.[194,203] PDT enhances macrophage tumoricidal activity[209–211] and stimulates macrophage release of TNF-$\alpha$ and tumor-cell recognition.[75] It has been suggested that nonspecific killing of tumor cells by inflammatory cells, potentially through the release of reactive oxygen species, contributes to the overall tumor kill by PDT.[206] This hypothesis is supported by studies showing that if the PDT-induced inflammatory response is further stimulated by the addition of adjuvants or macrophage-activating factors, the overall tumor response to PDT is greater.[212–214] The role of the innate immune response in overall tumor kill by PDT was also shown to involve NK (natural killer) cells. Depletion of NK cells reduced the long-term tumor control by PDT,[202] and augmentation of NK activity enhanced PDT tumor control.[215]

PDT has also been shown to enhance the adaptive or specific host antitumor response.[123,206] Tumor-draining lymph-node cells isolated from PDT-treated mice are able to suppress subsequent tumor challenges when transferred to a naïve host.[216,217] Canti et al.[217] have shown that PDT-treated mice that remain tumor free for 100 d post-PDT are resistant to subsequent tumor challenges, suggesting the presence of immune memory cells. The establishment of immune memory by PDT was recently confirmed by studies of Korbelik and Dougherty[210] showing that immune cells isolated at protracted times post-PDT could transfer tumor immunity. The importance of the immune response in PDT was definitively shown by a series of experiments in *scid* (severe combined immunodeficient) and nude mice.[204,210] PDT treatment, at a dose that was curative in immunocompetent BALB/c mice, provided only short-term cures of EMT6 tumors in *scid* and nude mice. The ability to provide long-term cures was restored when immunodeficient animals were reconstituted with bone-marrow cells from BALB/c mice. Depletion studies showed that the critical cells involved in the PDT-induced immune response were $CD8^+$ cells, although $CD4^+$ cells played a supportive role.[210] Interestingly, a recent clinical study has demonstrated an infiltration of $CD8^+$ cells into PDT-treated tumor tissue.[218] These studies suggest that a functional immune system is necessary for long-term tumor control by PDT; however, the mechanism behind the potentiation of the immune response by PDT is unknown.

Several factors have been postulated to contribute to the ability of PDT to enhance the host antitumor immune response, including release of tumor antigens and stimulation of various proinflammatory cytokines and other immunologically important genes, via activation of stress-response factors like AP-1 and NF-κB.[194,210] PDT has been shown to enhance the expression of a number of cytokines including IL-6, TNF-$\alpha$, IL-1$\beta$, IL-2, and granulocyte-macrophage colony-stimulating factor (GM-CSF).[187,192,194,219,220] PDT modulation of IL-6 is the result of increased AP-1 activity.[81] Preliminary mechanistic studies have suggested that IL-6 plays a critical role in long-term tumor control by PDT;[208] however, its role in induction of the antitumor immune response is unknown. The release of tumor antigens by PDT is also likely to play a role in the initiation of the antitumor immune response. The mechanism by which tumor cells are killed and the environment into which tumor antigens are released have been shown to dictate the ensuing immune response.[221,222] Release of tumor antigens into an inflammatory milieu is thought to promote an effective antitumor immune response,[221] and the presence of apoptotic and necrotic tissue acts to stimulate DCs,[222] which are necessary for initiation of the immune response.[223] Thus PDT, by stimulating inflammation and generating apoptotic and necrotic tissue, provides an ideal environment for initiation of an antitumor immune response. Evidence from our laboratory demonstrating that PDT-generated tumor-cell lysates are able to activate DC in the absence of cofactors[224] supports this hypothesis. An understanding of the mechanisms leading to PDT enhancement of the host antitumor immune response may allow for amplification of the response, thereby amplifying the effectiveness of PDT and potentially providing for protection against metastases outside the treatment field.

## 36.8 Conclusion

The scientific effort that supports this new cancer therapy has led to numerous significant advances. The development of new photosensitizers is essentially eliminating the problem of prolonged cutaneous photosensitivity and is extending treatment depth, an issue dealt with in detail elsewhere in this volume. The complex dynamics of tumor oxygenation in response to PDT are now largely understood. The oxidative stress effects of PDT on redox-sensitive transcription factors and the genes they control are being uncovered. The complex interplay of biological mechanisms governing the PDT tumor response has been realized. Given these accomplishments, it remains for them to be translated into actual patient benefit. Noninvasive probes need to be perfected that will allow the monitoring of photosensitizer levels, oxygen status, or, ideally, directly of singlet oxygen, the cytotoxic agent. New light-delivery regimes need to be devised for clinical use that will minimize oxygen limitations. The ways in which such regimes might influence redox-sensitive gene regulation and how these genes might affect treatment outcome need to be explored. Finally, our expanding understanding of the complex effects of PDT on host immunity needs to be exploited to modulate the PDT response.

### Acknowledgment

The authors thank Dr. Ian MacDonald, Photodynamic Therapy Center, Roswell Park Cancer Institute, for the creation and preparation of figures presented in this chapter.

### References

1. MacDonald, I.J. and Dougherty, T.J., Basic principles of photodynamic therapy, *J. Porphyrins Phthalocyanines*, 5, 105, 2001.
2. Moan, J. and Sommer, S., Oxygen dependence of the photosensitizing effect of hematoporphyrin derivative in NHIK 3025 cells, *Cancer Res.*, 45, 1608, 1985.
3. Henderson, B.W. and Miller, A.C., Effects of scavengers of reactive oxygen and radical species on cell survival following photodynamic treatment *in vitro*: comparison to ionizing radiation, *Radiat. Res.*, 108, 196, 1986.
4. Foote, C.S., Mechanisms of photooxygenation, *Prog. Clin. Biol. Res.*, 170, 3, 1984.
5. Moan, J. and Berg, K., The photodegradation of porphyrins in cells can be used to estimate the lifetime of singlet oxygen, *Photochem. Photobiol.*, 53, 549, 1991.

6. Kessel, D., Chemical and biochemical determinants of porphyrin localization, *Prog. Clin. Biol. Res.*, 170, 405, 1984.
7. Kessel, D., Sites of photosensitization by derivatives of hematoporphyrin, *Photochem. Photobiol.*, 44, 489, 1986.
8. MacDonald, I.J., Morgan, J., Bellnier, D.A., Paszkiewicz, G., Whitaker, J.E., Litchfield, D.J., and Dougherty, T.J., Subcellular localization patterns and their relationship to photodynamic activity of pyropheophorbide-α derivatives, *Photochem. Photobiol.*, 70, 789, 1999.
9. Moan, J., Berg, K., Anholt, A., and Madslien, K., Sulfonated aluminum phthalocyanines as sensitizers for photochemotherapy. Effects of small doses on localization, dye fluorescence and photosensitivity in V79 cells, *Int. J. Cancer*, 58, 865, 1994.
10. Berg, K., Madslien, K., Bommer, J.C., Oftebro, R., Winkelman, J.W., and Moan, J., Light induced relocalization of sulfonated meso-tetraphenylporphines in NHIK 3025 cells and effects of dose fractionation, *Photochem. Photobiol.*, 53, 203, 1991.
11. Henderson, B.W., Bellnier, D.A., Greco, W.R., Sharma, A., Pandey, R.K., Vaughan, L.A., Weishaupt, K.R., and Dougherty, T.J., An *in vivo* quantitative structure-activity relationship for a congeneric series of pyropheophorbide derivatives as photosensitizers for photodynamic therapy, *Cancer Res.*, 57, 4000, 1997.
12. Murant, R.S., Gibson, S.L., and Hilf, R., Photosensitizing effects of Photofrin II on the site-selected mitochondrial enzymes adenylate kinase and monoamine oxidase, *Cancer Res.*, 47, 4323, 1987.
13. Gibson, S.L. and Hilf, R., Photosensitization of mitochondrial cytochrome *c* oxidase by hematoporphyrin derivative and related porphyrins *in vitro* and *in vivo, Cancer Res.*, 43, 4191, 1983.
14. Salet, C. and Moreno, G., Photosensitization of mitochondria. Molecular and cellular aspects, *J. Photochem. Photobiol. B*, 5, 133, 1990.
15. Salet, C., Hematoporphyrin and hematoporphyrin-derivative photosensitization of mitochondria, *Biochimie*, 68, 865, 1986.
16. Kennedy, J.C. and Pottier, R.H., Endogenous protoporphyrin IX, a clinically useful photosensitizer for photodynamic therapy, *J. Photochem. Photobiol. B Biol.*, 14, 275, 1992.
17. Peng, Q., Berg, K., Moan, J., Kongshaug, M., and Nesland, J.M., 5-aminolevulinic acid-based photodynamic therapy: principles and experimental research, *Photochem. Photobiol.*, 65, 235, 1997.
18. Kessel, D., Woodburn, K., Gomer, C.J., Jagerovic, N., and Smith, K.M., Photosensitization with derivatives of chlorin p6, *J. Photochem. Photobiol. B*, 28, 13, 1995.
19. Kessel, D. and Luo, Y., Mitochondrial photodamage and PDT-induced apoptosis, *J. Photochem. Photobiol. B*, 42, 89, 1998.
20. Kessel, D., Woodburn, K., Henderson, B.W., and Chang, C.K., Sites of photodamage *in vivo* and *in vitro* by a cationic porphyrin, *Photochem. Photobiol.*, 62(5), 875, 1995.
21. Morgan, J., Potter, W.R., and Oseroff, A.R., Comparison of photodynamic targets in a carcinoma cell line and its mitochondrial DNA-deficient derivative, *Photochem. Photobiol.*, 71, 747, 2000.
22. Verma, A., Focchina, S.L., Hirsch, S., Dillahey, L., Williams, J., and Snyder, S.H., Photodynamic tumor therapy: mitochondrial benzodiazepine receptors as a therapeutic target, *Mol. Med.*, 4, 40, 1998.
23. Tsuchida, T., Zheng, G., Pandey, R.K., Potter, W.R., Bellnier, D.A., Henderson, B.W., Kato, H., and Dougherty, T.J., Correlation between Site II-specific human serum albumin (HSA) binding affinity and murine *in vivo* photosensitizing efficacy of some Photofrin® components, *Photochem. Photobiol.*, 66, 224, 1997.
24. Dubbelman, T.M.A.R., Prinsze, C., Penning, L.C., and van Steveninck, J., Photodynamic therapy: membrane and enzyme photobiology, in *Photodynamic Therapy*, Henderson, B.W. and Dougherty, T.J., Eds., Marcel Dekker, New York, 1992, pp. 37–46.
25. Miccoli, L., Beurdeley-Thomas, A., DePinieux G., Sureau, F., Oudard, S., Dutrillaux, B., and Poupon, M.-F., Light induced photoactivation of hypericin affects the energy metabolism of human glioma cells by inhibiting hexokinase bound to mitochondria, *Cancer Res.*, 58, 5777, 1998.
26. Atlante, A., Moore, G., Passarella, S., and Salet, C., Hematoporphyrin derivative (Photofrin II): impairment of anion translocation, *Biochem. Biophys. Res. Commun.*, 141, 584, 1986.

27. Perlin, D.S., Murant, R.S., Gibson, S.L., and Hilf, R., Effects of photosensitization by hematoporphyrin derivative on mitochondrial adenosine triphosphatase-mediated proton transport and membrane integrity of R3230AC mammary adenocarcinoma, *Cancer Res.*, 45, 653, 1995.
28. Ratcliffe, S.L. and Matthews, E.K., Modification of the photodynamic action of $\alpha$-aminolaevulinic acid (ALA) on rat pancreatoma cells by mitochondrial benzodiazepine receptor ligands, *Br. J. Cancer*, 71, 300, 1995.
29. Morgan, J. and Oseroff, A.R., Mitochondria-based photodynamic anti-cancer therapy, *Advanced Drug Delivery Rev.*, 49, 71, 2001.
30. Geze, M., Morliere, P., Maziere, J.C., Smith, K.M., and Santus, R., Lysomes, a key target of hydrophobic photosensitizers proposed for chemotherapeutic applications, *J. Photochem. Photobiol. B*, 20, 23, 1993.
31. Berg, K. and Moan, J., Lysosomes and microtubules as targets for photochemotherapy of cancer, *Photochem. Photobiol.*, 65, 403, 1997.
32. Kessel, D., Luo, Y., Mathieu, P., and Reiners, J.J., Jr., Determinants of the apoptotic response to lysosomal photodamage, *Photochem. Photobiol.*, 71, 196, 2000.
33. Gomer, C.J., Rucker, N., Banerjee, A., and Benedict, W.F., Comparison of mutagenicity and induction of sister chromatid exchange in Chinese hamster cells exposed to hematoporphyrin derivative photoradiation, ionizing radiation, or ultraviolet radiation, *Cancer Res.*, 43, 2622, 1983.
34. Evensen, J.F. and Moan, J., Photodynamic action and chromosomal damage: a comparison of haematoporphyrin derivative (HpD) and light with X-irradiation, *Br. J. Cancer*, 45, 456, 1982.
35. McNair, F.L., Marples, B., West, C.M.L., and Moore, J.V., A comet assay of DNA damage and repair in K562 cells after photodynamic therapy using hematoporphyrin derivative, methylene blue and *meso*-tetrahydroxyphenylchlorin, *Br. J. Cancer*, 75, 1721, 1997.
36. Evans, H.H., Horng, M.-F., Ricanati, M., Deahl, J.T., and Oleinick, N.L., Mutagenicity of photodynamic therapy as compared to UVC and ionizing radiation in human and murine lymphoblast cell lines, *Photochem. Photobiol.*, 66, 690, 1997.
37. Moan, J., Pettersen, E.O., and Christensen, T., The mechanism of photodynamic inactivation of human cells *in vitro* in the presence of haematoporphyrin, *Br. J. Cancer*, 39, 398, 1979.
38. Christensen, T., Volden, G., Moan, J., and Sandquist, T., Release of lysosomal enzymes and lactate dehydrogenase due to hematoporphyrin derivative and light irradiation of NHIK 3025 cells *in vitro*, *Ann. Clin. Res.*, 14, 46, 1982.
39. Henderson, B.W. and Donovan, J.M., Release of prostaglandin E2 from cells by photodynamic treatment *in vitro*, *Cancer Res.*, 49, 6896, 1989.
40. Bellnier, D.A. and Dougherty, T.J., Membrane lysis in Chinese hamster ovary cells treated with hematoporphyrin derivative plus light, *Photochem. Photobiol.*, 36, 43, 1982.
41. Thomas, J.P. and Girotti, A.W., Role of lipid peroxidation in hematoporphyrin derivative-sensitized photokilling of tumor cells: protective effects of glutathione peroxidase, *Cancer Res.*, 49, 1682, 1989.
42. Girotti, A.W., Photodynamic lipid peroxidation in biological systems, *Photochem. Photobiol.*, 51, 497, 1990.
43. Deuticke, B., Henseleit, U., Haest, C.W., Heller, K.B., and Dubbelman, T.M., Enhancement of transbilayer mobility of a membrane lipid probe accompanies formation of membrane leaks during photodynamic treatment of erythrocytes, *Biochim. Biophys. Acta*, 982, 53, 1989.
44. Moan, J. and Vistnes, A.I., Porphyrin photosensitization of proteins in cell membranes as studied by spin-labelling and by quantification of DTNB-reactive SH-groups, *Photochem. Photobiol.*, 44, 15, 1986.
45. Gibson, S.L., Murant, R.S., and Hilf, R., Photosensitizing effects of hematoporphyrin derivative and photofrin II on the plasma membrane enzymes 5′-nucleotidase, $Na^+K^+$-ATPase, and $Mg^{2+}$-ATPase in R3230AC mammary adenocarcinomas, *Cancer Res.*, 48, 3360, 1988.
46. Specht, K.G. and Rodgers, M.A.J., Plasma membrane depolarization and calcium influx during cell injury by photodynamic action, *Biochim. Biophys. Acta*, 1070, 60, 1991.

47. Berg, K., Steen, H.B., Winkelman, J.W., and Moan, J., Synergistic effects of photoactivated tetra(4-sulfonatophenyl)porphine and nocodazole on microtubule assembly, accumulation of cells in mitosis and cell survival, *J. Photochem. Photobiol. B*, 13, 59, 1992.
48. Oleinick, N.L. and Evans, H.H., The photobiology of photodynamic therapy: cellular targets and mechanisms, *Radiat. Res.*, 150(5 Suppl.), S146, 1998.
49. Henkel, T., Apoptosis: corralling the corpses, *Cell*, 104, 325, 2001.
50. Dellinger, M., Apoptosis or necrosis following photofrin photosensitization: influence of the incubation protocol, *Photochem. Photobiol.*, 64, 182, 1996.
51. Fabris, C., Valduga, G., Miotto, G., Borsetto, L., Jori, G., Garbisa, S., and Reddi, E., Photosensitization with zinc (II) phthalocyanine as a switch in the decision between apoptosis and necrosis, *Cancer Res.*, 61, 7495, 2001.
52. Kessel, D., Luo, Y., Deng, Y., and Chang, C.K., The role of subcellular localization in the initiation of apoptosis by photodynamic therapy, *Photochem. Photobiol.*, 65, 422, 1997.
53. Luo, Y. and Kessel, D., Initiation of apoptosis versus necrosis by photodynamic therapy with chloroaluminum phthalocyanine, *Photochem. Photobiol.*, 66, 479, 1997.
54. He, X.-Y., Sikes, R.A., Thomsen, S., Chung, L.W.K., and Jaques, S.L., Photodynamic therapy with Photofrin II induces programmed cell death in carcinoma cell lines, *Photochem. Photobiol.*, 59, 468, 1994.
55. Luo, Y., Chang, C.K., and Kessel, D., Rapid initiation of apoptosis by photodynamic therapy, *Photochem. Photobiol.*, 63, 528, 1996.
56. Granville, D.J., Levy, J.G., and Hunt, D.W.C., Photodynamic therapy induces caspase-3 activation in HL-60 cells, *Cell Death Differ.*, 4, 623, 1997.
57. Agarwal, M.L., Clay, M.E., Harvey, E.J., Evans, H.H., Antunez, A.R., and Oleinick, N.L., Photodynamic therapy induces rapid cell death by apoptosis in L5178Y mouse lymphoma cells, *Cancer Res.*, 51, 5993, 1991.
58. Yang, J., Bhalla, K., Kim, C.N., Ibrado, A.M., Peng, T.I., Jones, D.P., and Wang, X., Prevention of apoptosis by Bcl-2: release of cytochrome *c* from mitochondria blocked, *Science*, 275, 1129, 1997.
59. Granville, D.J., Carthy, C.M., Jiang, H., Shore, G.C., McManus, B.M., and Hunt, D.W.C., Rapid cytochrome *c* release, activation of caspases 3, 6, 7 and 8 followed by Bap31 cleavage in HeLa cells treated with photodynamic therapy, *FEBS Lett.*, 437, 5, 1998.
60. Chiu, S.M., Evans, H.H., Lam, M., Nieminen, A, and Oleinick, N.L., Phthalocyanine 4 photodynamic therapy-induced apoptosis of mouse L5178Y-R cells results from a delayed but extensive release of cytochrome *c* from mitochondria, *Cancer Lett.*, 165, 51, 2001.
61. Granville, D.J., Cassidy, B.A., Ruehlmann, D.O., Choy, J.C., Brenner, C., Kroemer, G., van Breemen, C., Margaron, P., Hunt, D.W.C., and McManus, B.M., Mitochondrial release of apoptosis-inducing factor and cytochrome *c* during smooth muscle cell apoptosis, *Am. J. Pathol.*, 159, 305, 2001.
62. Chiu, S.M. and Oleinick, N.L., Dissociation of mitochondrial depolarization from cytochrome *c* release during apoptosis induced by photodynamic therapy, *Br. J. Cancer*, 84, 1099, 2001.
63. Varnes, M.E., Chiu, S.M., Xue, L.Y., and Oleinick, N.L., Photodynamic therapy-induced apoptosis in lymphoma cells: translocation of cytochrome *c* causes inhibition of respiration as well as caspase activation, *Biochem. Biophys. Res. Commun.*, 255, 679, 1999.
64. Vantieghem, A., Xu, Y., Declercq, W., Vandenabeele, P., Denecker, G., Vandenheede, J.R., Merlevede, W., de Witte, P.A., and Agostinis, P., Different pathways mediate cytochrome *c* release after photodynamic therapy with hypericin, *Photochem. Photobiol.*, 74, 133, 2001.
65. Kessel, D. and Luo, Y., Photodynamic therapy: a mitochondrial inducer of apoptosis, *Cell Death Differ.*, 6, 28, 1999.
66. Xue, L.Y., Chiu, S.M., and Oleinick, N.L., Photodynamic therapy-induced death of MCF-7 human breast cancer cells: a role for caspase-3 in the late steps of apoptosis but not for the critical lethal event, *Exp. Cell Res.*, 263, 145, 2001.
67. Srivastava, M., Ahmad, N., Gupta, S., and Mukhtar, H., Involvement of bcl-2 and bax in photodynamic therapy-mediated apoptosis, *J. Biol. Chem.*, 276, 15481, 2001.

68. Granville, D.J., Jiang, H., An, M.T., Levy, J.G., McManus, B.M., and Hunt, D.W.C., Bcl-2 overexpression blocks caspase activation and downstream apoptotic events instigated by photodynamic therapy, *Br. J. Cancer*, 79, 95, 1999.
69. Schouwink, H., Ruevekamp, M., Oppelaar H., van Veen, R., Bass, P., and Stewart, F.A., Photodynamic therapy for malignant mesothelioma: preclinical studies for optimization of treatment protocols, *Photochem. Photobiol.*, 73, 410, 2001.
70. He, J., Agarwal, M.L., Larkin, H.E., Friedman, L.R., Xue, L.Y., and Oleinick, N.L., The induction of partial resistance to photodynamic therapy by the protooncogene bcl-2, *Photochem. Photobiol.*, 64, 845, 1996.
71. Xue, L.Y., Chiu, S.M., and Oleinick, N.L., Photochemical destruction of the bcl-2 oncoprotein during photodynamic therapy with the phthalocyanine photosensitizer Pc 4, *Oncogene*, 20, 3420, 2001.
72. Kim, H., Luo, Y., Li, G., and Kessel, D., Enhanced apoptotic response to photodynamic therapy after bcl-2 transfection, *Cancer Res.*, 59, 3429, 1999.
73. Usuda, J., Okunaka, T., Furukawa, K., Tsuchida, T., Kuroiwa, Y., Ohe, Y., Saijo, N., Nishio, K., Konaka, C., and Kato, H., Increased cytotoxic effects of photodynamic therapy in IL-6 gene transfected cells via enhanced apoptosis, *Int. J. Cancer*, 93, 475, 2001.
74. Agarwal, M.L., Larkin, H.E., Zaidi, S.I.A., Mukhtar, H., and Oleinick, N.L., Phospholipase activation triggers apoptosis in photosensitized mouse lymphoma cells, *Cancer,* 53, 5897, 1993.
75. Kick, G., Messer, G., Plewig, G., Kind, P., and Goetz, A.E., Strong and prolonged injuction of c-*jun* and c-*fos* proto-oncogenes by photodynamic therapy, *Br. J. Cancer*, 74, 30, 1996.
76. Separovic, D., Mann, K.J., and Oleinick, N.L., Association of ceramide accumulation with photodynamic treatment-induced cell death, *Photochem. Photobiol.*, 68, 101, 1998.
77. Gupta, S., Ahmad, N., and Mukhtar, H., Involvement of nitric oxide during phthalocyanine (Pc4) photodynamic therapy-mediated apoptosis, *Cancer Res.*, 58, 1785, 1998.
78. Fisher, A.M., Danenberg, K., Banerjee, D., Bertino, J.R., Danenberg, P., and Gomer, C.J., Increased photosensitivity in HL60 cells expressing wild-type p53, *Photochem. Photobiol.*, 66, 265, 1997.
79. Tong, Z., Singh, G., and Rainbow, A.J., The role of the p53 tumor suppressor in the response of human cells to Photofrin-mediated photodynamic therapy, *Photochem. Photobiol.*, 71, 201, 2000.
80. Fisher, A.M., Ferrario, A., Rucker, N., Zhang, S., and Gomer, C.J., Photodynamic therapy sensitivity is not altered in human tumor cells after abrogation of p53 function, *Cancer Res.*, 331, 1999.
81. Kick, G., Messer, G., Goetz, A., Plewig, G., and Kind, P., Photodynamic therapy induces expression of interleukin 6 by activation of AP-1 but not NF-kB DNA binding, *Cancer Res.*, 55, 2373, 1995.
82. Gollnick, S.O., Lee, B.Y., Vaughan, L., Owczarczak, B., and Henderson, B.W., Activation of the IL-10 gene promoter following photodynamic therapy of murine keratinocytes, *Photochem. Photobiol.*, 73, 170, 2001.
83. Abate, C., Patel, L., Rauscher, F.J., and Curran, T., Redox regulation of fos and jun DNA-binding activity *in vitro*, *Science*, 249, 1157, 1990.
84. Xanthoudakis, S., Miao, G., Wang, F., Pan, Y.-C.E., and Curran, T., Redox activation of Fos-Jun DNA binding activity is mediated by a DNA repair enzyme, *EMBO J.*, 11(9), 3323, 1992.
85. Meyer, M., Schreck, R., and Baeuerle, P.A., $H_2O_2$ and antioxidants have opposite effects on activation of NF-κB and AP-1 in intact cells: AP-1 as secondary antioxidant-responsive factor, *EMBO J.*, 12, 2005, 1993.
86. Yao, K.-S., Zanthoudakis, S., Curran, T., and O'Dwyer, P.J., Activation of AP-1 and of nuclear redox factor, Ref1, in the response of HT29 colon cancer cells to hypoxia, *Mol. Cell. Biol.*, 14, 5997, 1994.
87. Luna, M.C., Wong, S., and Gomer, C.J., Photodynamic therapy mediated induction of early response genes, *Cancer Res.*, 54, 1374, 1994.
88. Baeuerle, P.A. and Henkel, T., Function and activation of NF-B in the immune system, *Ann. Rev. Immunol.*, 12, 141, 1994.
89. Schreck, R., Albermann, K., and Baeuerle, P.A., Nuclear factor κB: an oxidative stress-responsive transcription factor of eukaryotic cells, *Radiat. Res. Commun.*, 17, 221, 1992.

90. DiDonato, J.A., Mercurio, R., Rosette, C., Wu-Li, J., Suyang, H., Ghosh, S., and Karin, M., Mapping of the inducible IκB phosphorylation sites that signal its ubiquination and degradation, *Mol. Cell. Biol.*, 16, 1295, 1996.
91. DiDonato, J.A., Hayakawa, M., Rothwarf, D.M., Zandi, E., and Karin, M., A cytokine-responsive IκB kinase tht activates the transcription factor NFκB, *Nature*, 388, 548, 1997.
92. Granville, D.J., Carthy, C.M., Jiang, H., Levy, J.G., McManus, B.M., Matroule, J.Y., Piette, J., and Hunt, D.W.C., Nuclear factor-κB activation by the photochemotherapeutic agent verteporfin, *Blood*, 95, 256, 2000.
93. Ryter, S.W. and Gomer, C.J., Nuclear factor κB binding activity in mouse L1210 cells following Photofrin II-mediated photosensitization, *Photochem. Photobiol.*, 58, 753, 1993.
94. Matroule, J.Y., Bonizzi, G., Morliere, P., Paillous, N., Santus, R., Bours, V., and Piette, J., Pyropheophorbide-a methyl ester-mediated photosensitization activates transcription factor NF-κB through the interleukin-1 receptor-dependent signaling pathway, *J. Biol. Chem.*, 270, 2899, 1999.
95. Legrand-Poels, S., Schoonbroodts, S., Matroule, J.Y., and Piette, J., NF-κB: an important transcription factor in photobiology, *J. Photochem. Photobiol. B*, 45, 1, 1998.
96. Tao, J.-S., Sanghera, S., Pelech, S.L., Wong, G., and Levy, J.G., Stimulation of stress-activated protein kinase and p38 HOG1 kinase in murine keratinocytes following photodynamic therapy with benzoporphyrin derivative, *J. Biol. Chem.*, 271, 27107, 1996.
97. Oleinick, N.L., He, J., Xue, L.Y., and Separovic, D., Stress-activated signalling responses leading to apoptosis following photodynamic therapy, *Proc. SPIE Optical Methods Tumor Treat. Detections Mechanisms Tech. Photodyn. Ther. VII*, Vol. 3247, 1998, p. 82.
98. Xue, L.Y., He, J., and Oleinick, N.L., Rapid tyrosine phosphorylation of HS1 in the response of mouse lymphoma L5178Y-R cells to photodynamic treatment sensitized by the phthalocyanine Pc4, *Photochem. Photobiol.*, 66, 105, 1997.
99. Koukourakis, M.I., Giatromanolaki, A., Skarlatos, J., Corti, L., Blandamura, S., Piazza, M., Gatter, K.C., and Harris, A.L., Hypoxia inducible factor (HIF-1a and HIF-2a) expression in early esophageal cancer and response to photodynamic therapy and radiotherapy, *Cancer Res.*, 611, 1830, 2001.
100. Ferrario, A., von Tiehl, K.F., Rucker, N., Schwarz, M.A., Gill, P.S., and Gomer, C.J., Antiangiogenic treatment enhances photodynamic therapy responsiveness in a mouse mammary carcinoma, *Cancer Res.*, 60, 4066, 2000.
101. Luna, M.C., Ferrario, A., Wong, S., Fisher, A.M.R., and Gomer, C.J., Photodynamic therapy-mediated oxidative stress as a molecular switch for the temporal expression of genes ligated to the human heat shock promoter, *Cancer Res.*, 60, 1637, 2000.
102. Gomer, C.J., Ryter, S.W., Ferrario, A., Rucker, N., Wong, S., and Fisher, A.M.R., Photodynamic therapy-mediated oxidative stress can induce expression of heat shock proteins, *Cancer Res.*, 56, 2355, 1996.
103. Curry, P.M. and Levy, J.G., Stress protein expression in murine tumor cells following photodynamic therapy with benzoporphyrin derivative, *Photochem. Photobiol.*, 58, 374, 1993.
104. Gomer, C.J., Ferrario, A., Rucker, N., Wong, S., and Lee, A.S., Glucose regulated protein induction and cellular resistance to oxidative stress mediated by porphyrin photosensitization, *Cancer Res.*, 51, 6574, 1991.
105. Xue, L.Y., Agarwal, M.L., and Varnes, M.E., Elevation of GRP-78 and loss of HSP-70 following photodynamic treatment of V79 cells: sensitization by nigericin, *Photochem. Photobiol.*, 62, 135, 1995.
106. Morgan, J., Whitaker, J.E., and Oseroff, A.R., GRP-78 induction by calcium ionophore potentiates photodynamic therapy using the mitochondrial targeting dye Victoria Blue BO, *Photochem. Photobiol.*, 67, 155, 1998.
107. Gomer, C.J., Luna, M., Ferrario, A., and Rucker, N., Increased transcription and translation of heme oxygenase in Chinese hamster fibroblasts following photodynamic stress or Photofrin II incubation, *Photochem. Photobiol.*, 53, 275, 1991.

108. Henderson, B.W., Probing the effects of photodynamic therapy through *in vivo-in vitro* methods, in *Photodynamic Therapy of Neoplastic Disease*, Vol. I, Kessel, D., Ed., CRC Press, Boca Raton, FL, 1990, p. 169.
109. Henderson, B.W. and Fingar, V.H., Oxygen limitation of direct tumor cell kill during photodynamic treatment of a murine tumor model, *Photochem. Photobiol.*, 49, 299, 1989.
110. Henderson, B.W., Waldow, S.M., Mang, T.S., Potter, W.R., Malone, P.B., and Dougherty, T.J., Tumor destruction and kinetics of tumor cell death in two experimental mouse tumors following photodynamic therapy, *Cancer Res.*, 45, 572, 1985.
111. Henderson, B.W., Sumlin, A.B., Owczarczak, B.L., and Dougherty, T.J., Bacteriochlorophyll-a as photosensitizer for photodynamic treatment of transplantable murine tumors, *J. Photochem. Photobiol. B*, 10, 303, 1991.
112. Henderson, B.W., Vaughan, L., Bellnier, D.A., vanLeengoed, H., Johnson, P.G., and Oseroff, A.R., Photosensitization of murine tumor, vasculature and skin by 5-aminolevulinic acid-induced porphyrin, *Photochem. Photobiol.*, 62, 780, 1995.
113. Cincotta, L., Foley, J.W., MacEachern, T., Lampros, E., and Cincotta, A.H., Novel photodynamic effects of a benzophenothiazine on two different murine sarcomas, *Cancer Res.*, 54, 1249, 1994.
114. Chan, W.-S., Brasseur, N., La Madeleine, C., and van Lier, J.E., Evidence for different mechanisms of EMT-6 tumor necrosis by photodynamic therapy with disulfonated aluminum phthalocyanine or Photofrin: tumor cell survival and blood flow, *Anticancer Res.*, 16, 1887, 1996.
115. Sitnik, T.M. and Henderson, B.W., Effects of fluence rate on cytotoxicity during photodynamic therapy, in *Proc. SPIE Opt. Methods Tumor Treat. Detection Mechanisms Tech. Photodyn. Ther. VI*, Dougherty, T.J., Ed., SPIE Press, Bellingham, WA, 1997, p. 95.
116. Korbelik, M. and Krosl, G., Cellular levels of photosensitisers in tumours: the role of proximity to the blood supply, *Br. J. Cancer*, 70, 604, 1994.
117. Svaasand, L.O., Optical dosimetry for direct and interstitial photoradiation therapy of malignant tumors, *Prog. Clin. Biol. Res.*, 170, 91, 1984.
118. Wilson, B.C., Jeeves, W.P., Lowe, D.M., and Adam, G., Light propagation in animal tissues in the wavelength range 375–825 nanometers, *Prog. Clin. Biol. Res.*, 170, 115, 1984.
119. Adams, K., Rainbow, A.J., Wilson, B.C., and Singh, G., *In vivo* resistance to Photofrin-mediated photodynamic therapy in radiation-induced fibrosarcoma cells resistant to *in vitro* Photofrin-mediated photodynamic therapy, *J. Photochem. Photobiol. B Biol.*, 49, 136, 1999.
120. Kessel, D., Hampton, J., Fingar, V., and Morgan, A., Tumor versus vascular photodamage in a rat tumor model, *J. Photochem. Photobiol. B Biol.*, 45, 25, 1998.
121. Pass, H.I., Evans, S., Matthews, W.A., Perry, R., Venzon, D., Roth, J.A., and Smith, P., Photodynamic therapy of oncogene-transformed cells, *J. Thorac. Cardiovasc. Surg.*, 101, 795, 1991.
122. Henderson, B.W. and Dougherty, T.J., How does photodynamic therapy work? *Photochem. Photobiol.*, 55, 145, 1992.
123. Dougherty, T.J., Gomer, C.J., Henderson, B.W., Jori, G., Kessel, D., Korbelik, M., Moan, J., and Peng, Q., Photodynamic therapy, *J. Natl. Cancer Inst.*, 90, 889, 1998.
124. Hamblin, M.R. and Newman, E.L., Photosensitizer targeting in photodynamic therapy II. Conjugates of hematoporphyrin with serum lipoproteins, *J. Photochem. Photobiol. B Biol.*, 26, 147, 1994.
125. Hamblin, M.R. and Newman, E.L., Photosensitizer targeting in photodynamic therapy I. Conjugates of hematoporphyrin with albumin and transferrin, *J. Photochem. Photobiol. B Biol.*, 26, 45, 1994.
126. Molpus, K.L., Hamblin, M.R., Rizvi, I., and Hasan, T., Intraperitoneal photoimmunotherapy of ovarian carcinoma xenografts in nude mice using charged photoimmunoconjugates, *Gynecol. Oncol.*, 76, 397, 2000.
127. DelGovernatore, M., Hamblin, M.R., Shea, C.R., Rizvi, I., and Hasan, T., Experimental photoimmunotherapy of hepatic metastases of colorectal cancer with a 17.1A chlorin (e6) immunoconjugate, *Cancer Res.*, 60, 4200, 2000.

128. Soukos, N.S., Hamblin, M.R., Keel, S., Fabian, R.L., Deutsch, T.F., and Hasan, T., Epidermal growth factor receptor-targeted immunophotodiagnosis and photoimmunotherapy of oral precancer *in vivo*, *Cancer Res.*, 61, 4490, 2001.
129. Henderson, B.W., Dougherty, T.J., and Malone, P.B., Studies on the mechanism of tumor destruction by photoradiation therapy, *Prog. Clin. Biol. Res.*, 170, 601, 1984.
130. Fingar, V.H., Siegel, K.A., Wieman, T.J., and Doak, K.W., The effects of thromboxane inhibitors on the microvascular and tumor response to photodynamic therapy, *Photochem. Photobiol.*, 58(3), 393, 1993.
131. Reed, M.W.R., Wieman, T.J., Schuschke, D.A., Tseng, M.T., and Miller, F.N., A comparison of the effects of photodynamic therapy on normal and tumor blood vessels in the rat microcirculation, *Radiat. Res.*, 119, 542, 1989.
132. Fingar, V.H. and Henderson, B.W., Drug and light dose dependence of photodynamic therapy: a study of tumor and normal tissue response, *Photochem. Photobiol.*, 46, 837, 1987.
133. Fingar, V.H., Kik, P.K., Haydon, P.S., Cerrito, P.B., Tseng, M., Abang, E., and Wieman, T.J., Analysis of acute vascular damage after photodynamic therapy using benzoporphyrin derivative (BPD), *Br. J. Cancer*, 79, 1702, 1999.
134. Sitnik, T.M., Hampton, J.A., and Henderson, B.W., Reduction of tumor oxygenation during and after photodynamic therapy *in vivo*: effects of fluence rate, *Br. J. Cancer*, 77, 1386, 1998.
135. Sitnik, T. and Henderson, B.W., The effect of fluence rate on tumor and normal tissue responses to photodynamic therapy, *Photochem. Photobiol.*, 67, 462, 1998.
136. Henderson, B.W., Sitnik-Busch, T.M., and Vaughan, L.A., Potentiation of PDT anti-tumor activity in mice by nitric oxide synthase inhibition is fluence rate dependent, *Photochem. Photobiol.*, 70, 64, 1999.
137. Dimitroff, C.J, Klohs, W., Sharma, A., Pera, P., Driscoll, D., Veith, J., Steinkampf, R., Schroeder, M., Klutchko, K., Sumlin, A., Henderson, B., Dougherty, T.J., and Bernacki, R.J., Anti-angiogenic activity of selected receptor tyrosine kinase inhibitors, PD166285 and PD173074: implications for combination treatment with photodynamic therapy, *Invest. New Drugs*, 17, 121, 1999.
138. Tseng, M.T., Reed, M.W., Ackermann, D.M., Schuschke, D.A., Wieman, T.J., and Miller, F.N., Photodynamic therapy induced ultrastructural alterations in microvasculature of the rat cremaster muscle, *Photochem. Photobiol.*, 48, 675, 1988.
139. Fingar, V.H., Wieman, J., Wiehle, S.A., and Cerrito, P.B., The role of microvascular damage in photodynamic therapy: the effect of treatment on vessel constriction, permeability, and leukocyte adhesion, *Cancer Res.*, 53, 4914, 1992.
140. McMahon, K.S., Wieman, T.J., Moore, P.H., and Fingar, V.H., Effects of photodynamic therapy using mono-L-aspartyl chlorin $e_6$ on vessel constriction, vessel leakage, and tumor response, *Cancer Res.*, 54, 5374, 1994.
141. Fingar, V.H., Wieman, T.J., Karavolos, P.S., Doak, K.W., Ouellet, R., and van Lier, J.E., The effects of photodynamic therapy using differently substituted zinc phthalocyanines on vessel constriction, vessel leakage and tumor response, *Photochem. Photobiol.*, 58, 251, 1993.
142. Zieve, P.D., Solomon, H.M., and Krevans, J.R., The effect of hematoporphyrin and light on human platelets. I. Morphologic, functional, and biochemical changes, *J. Cell Physiol.*, 67, 271, 1966.
143. Henderson, B.W., Owczarczak, B., Sweeney, J., and Gessner, T., Effects of photodynamic treatment of platelets or endothelial cells *in vitro* on platelet aggregation, *Photochem. Photobiol.*, 56, 513, 1992.
144. Fingar, V.H., Wieman, T.J., and Doak, K.W., Role of thromboxane and prostacyclin release on photodynamic therapy-induced tumor destruction, *Cancer Res.*, 50, 2599, 1990.
145. Reed, M.W., Schuschke, D.A., and Miller, F.N., Prostanoid antagonists inhibit the response of the microcirculation to "early" photodynamic therapy, *Radiat. Res.*, 127, 292, 1991.
146. Taber, S.W., Wieman, T.J., and Fingar, V.H., The effects of aspirin on microvasculature after photodynamic therapy, *Photochem. Photobiol.*, 57, 856, 1993.

147. Fingar, V.H., Wieman, T.J., and Haydon, P.S., The effects of thrombocytopenia on vessel stasis and macromolecular leakage after photodynamic therapy using Photofrin, *Photochem. Photobiol.*, 66, 513, 1997.
148. Fingar, V.H., Wieman, T.J., and Doak, K.W., Changes in tumor interstitial pressure induced by photodynamic therapy, *Photochem. Photobiol.*, 53, 763, 1991.
149. Höckel, M. and Vaupel, P., Tumor hypoxia: definitions and current clinical, biologic, and molecular aspects, *J. Natl. Cancer Inst.*, 93, 266, 2001.
150. Fingar, V.H., Wieman, T.J., Park, Y.J., and Henderson, B.W., Implications of a pre-existing tumor hypoxic fraction on photodynamic therapy, *J. Surg. Res.*, 53, 524, 1992.
151. Fingar, V.H., Mang, T.S., and Henderson, B.W., Modification of photodynamic therapy-induced hypoxia by fluosol-DA (20%) and carbogen breathing in mice, *Cancer Res.*, 48, 3350, 1988.
152. Jirsa, M.J., Pouckova, P., Dolezal, J., Pospisil, J., and Jirsa, M., Hyperbaric oxygen and photodynamic therapy in tumour-bearing nude mice [letter], *Eur. J. Cancer*, 27, 109, 1991.
153. Maier, A., Tomaselli, F., Anegg, U., Rehak, P., Fell, B., Luznik, S., Pinter, H., and Smolle-Juttner, F.M., Combined photodynamic therapy and hyperbaric oxygenation in carcinoma of the esophagus and the esophago-gastric junction, *Eur. J. Cardio-Thorac. Surg.*, 18, 649, 2001.
154. Svaasand, L.O., Doiron, D.R., and Dougherty, T.J., Temperature rise during photoradiation therapy of malignant tumors, *Med. Phys.*, 10, 10, 1983.
155. Mattiello, J., Hetzel, F., and Vandenheede, L., Intratumor temperature measurements during photodynamic therapy, *Photochem. Photobiol.*, 46, 873, 1987.
156. Henderson, B.W., Busch, T.M., Vaughan, L.A., Frawley, N.P, Babich, D., Sosa, T.A., Zolo, J.D., Dee, A.S., Cooper, M.T., Bellnier, D.A., Greco, W.R., and Oseroff, A.R., Photofrin photodynamic therapy can significantly deplete or preserve oxygenation in human basal cell carcinomas during treatment, depending on fluence rate, *Cancer Res.*, 60, 525, 2000.
157. Tromberg, B.J., Orenstein, A., Kimel, S., Barker, S.J., Hyatt, J., Nelson, J.S., and Berns, M.W., *In vivo* tumor oxygen tension measurements for the evaluation of the efficiency of photodynamic therapy, *Photochem. Photobiol.*, 52, 375, 1990.
158. Chen, Q., Chen, H., and Hetzel, F.W., Tumor oxygenation changes post-photodynamic therapy, *Photochem. Photobiol.*, 63, 128, 1996.
159. Foster, T.H., Murant, R.S., Bryant, R.G., Knox, R.S., Gibson, S.L., and Hilf, R., Oxygen consumption and diffusion effects in photodynamic therapy, *Radiat. Res.*, 126, 296, 1991.
160. Henning, J.P., Fournier, R.L., and Hampton, J.A., A transient mathematical model of oxygen depletion during photodynamic therapy, *Radiat. Res.*, 142, 221, 1995.
161. Foster, T.H. and Gao, L., Dosimetry in photodynamic therapy: oxygen and the critical importance of capillary density, *Radiat. Res.*, 130, 379, 1992.
162. Pogue, B.W. and Hasan, T., A theoretical study of light fractionation and dose-rate effects in photodynamic therapy, *Radiat. Res.*, 147, 551, 1997.
163. Foster, T.H., Hartley, D.F., Nichols, M.G., and Hilf, R., Fluence rate effects in photodynamic therapy of multicell tumor spheroids, *Cancer Res.*, 53, 1249, 1993.
164. Mitra, S., Finlay, J.C., McNeill, D., Conover, D.L., and Foster, T.H., Photochemical oxygen consumption, oxygen evolution and spectral changes during UVA irradiation of EMT6 spheroids, *Photochem. Photobiol.*, 73, 703, 2001.
165. Zilberstein, J., Bromberg, A., Frantz, A., Rosenbach-Belkin, V., Kritzmann, A., Pfefermann, R., Salomon, Y., and Scherz, A., Light-dependent oxygen consumption in bacteriochlorophyll-serine-treated melanoma tumors: on-line determination using a tissue-inserted oxygen microsensor, *Photochem. Photobiol.*, 65, 1012, 1997.
166. Busch, T.M., Hahn, S.M., Evans, S.M., and Koch, C.J., Depletion of tumor oxygenation during photodynamic therapy: detection by the hypoxia marker EF3, *Cancer Res.*, 60, 2636, 2000.
167. Mang, T.S., Dougherty, T.J., Potter, W.R., Boyle, D.G., Somer, S., and Moan, J., Photobleaching of porphyrins used in photodynamic therapy and implications for therapy, *Photochem. Photobiol.*, 45, 501, 1987.

168. Spikes, J.D., Quantum yields and kinetics of the photobleaching of hematoporphyrin, Photofrin II, tetra(4-sulfonatophenyl)porphine and uroporphyrin, *Photochem. Photobiol.*, 55(6), 797, 1992.
169. Coutier, S., Mitra, S., Bezdetnaya, L.N., Parache, R.M., Georgakoudi, I., Foster, T.H., and Guillemin, F., Effects of fluence rate on cell survival and photobleaching in meta-tetra-(hydroxyphenyl)chlorin-photosensitized Colo 26 multicell tumor spheroids, *Photochem. Photobiol.*, 73, 297, 2001.
170. Finlay, J.C., Conover, D.L., Hull, E.L., and Foster, T.H., Porphyrin bleaching and PDT-induced spectral changes are irradiance dependent in ALA-sensitized normal rat skin *in vivo*, *Photochem. Photobiol.*, 73, 54, 2001.
171. Georgakoudi, I., Nichols, M.G., and Foster, T.H., The mechanism of Photofrin photobleaching and its consequences for photodynamic dosimetry, *Photochem. Photobiol.*, 65, 135, 1997.
172. Georgakoudi, I. and Foster, T.H., Singlet oxygen- versus nonsinglet oxygen-mediated mechanisms of sensitizer photobleaching and their effects on photodynamic dosimetry, *Photochem. Photobiol.*, 67, 612, 1998.
173. Gibson, S.L., van der Meid, K.R., Murant, R.S., Raubertas, R.F., and Hilf, R., Effects of various photoradiation regimens on the antitumor efficacy of photodynamic therapy for R3230AC mammary carcinomas, *Cancer Res.*, 50, 7236, 1990.
174. Iinuma, S., Schomacker, K.T., Wagnières, G., Rajadhyaksha, M., Bamberg, M., Momma, T., and Hasan, T., *In vivo* fluence rate and fractionation effects on tumor response and photobleaching: photodynamic therapy with two photosensitizers in an orthotopic rat tumor model, *Cancer Res.*, 59, 6164, 1999.
175. Van Geel, I.P.J., Oppelaar, H., Marijnissen, J.P.A., and Stewart, F.A., Influence of fractionation and fluence rate in photodynamic therapy with Photofrin or mTHPC, *Radiat. Res.*, 145, 602, 1996.
176. Blant, S.A., Woodtli, A., Wagnières, G., Fontolliet, C., van den Bergh, H., and Monnier, P., *In vivo* fluence rate effect in photodynamic therapy of early cancers with tetra(m-hydroxyphenyl)chlorin, *Photochem. Photobiol.*, 64, 963, 1996.
177. Curnow, A., McIlroy, B.W., Postle-Hacon, M.J., MacRobert, A.J., and Bown, S.G., Light dose fractionation to enhance photodynamic therapy using 5-aminolevulinic acid in the normal rat colon, *Photochem. Photobiol.*, 69, 71, 1999.
178. DeBruijn, H.S., van der Veen, N., Robinson, D.J., and Star, W.M., Improvement of systemic 5-aminolevulinic acid-based photodynamic therapy *in vivo* using light fractionation with a 75-minute interval, *Cancer Res.*, 59, 901, 1999.
179. Gollnick, S.O., Musser, D.A., Oseroff, A.R., Vaughan, L.A., Owczarczak, B., and Henderson, B.W., IL-10 does not play a role in cutaneous Photofrin® photodynamic therapy-induced suppression of the contact hypersensitivity response, *Photochem. Photobiol.*, 74, 811, 2001.
180. Musser, D. and Oseroff, A.R., Characteristics of the immunosuppression induced by cutaneous photodynamic therapy: persistence, antigen specificity and cell type involved, *Photochem. Photobiol.*, 73, 518, 2001.
181. Obochi, M.O.K., Ratkay, L.G., and Levy, J.G., Prolonged skin allograft survival after photodynamic therapy associated with modification of donor skin antigenicity, *Transplantation*, 63, 810, 1997.
182. Gruner, S., Meffert, H., Volk, H.D., Grunow, R., and Jahn, S., The influence of haematoporphyrin derivative and visible light on murine skin graft survival, epidermal Langerhans cells and stimulation of the allogeneic mixed leucocyte reaction, *Scand. J. Immunol.*, 21, 267, 1985.
183. Qin, B., Selman, S.H., Payne, K.M., Keck, R.W., and Metzger, D.W., Enhanced allograft survival after photodynamic therapy: association with lymphocyte inactivation and macrophage stimulation, *Transplantation*, 56, 1481, 1993.
184. Elmets, C.A. and Bowen, K.D., Immunological suppression in mice treated with hematoporphyrin derivative photoradiation, *Cancer Res.*, 46, 1608, 1986.
185. Musser, D.A. and Fiel, R.J., Cutaneous photosensitizing and immunosuppressive effects of a series of tumor localizing porphyrins, *Photochem. Photobiol.*, 53, 119, 1991.
186. Lynch, D.H., Haddad, S., King, V.J., Ott, M.J., Straight, R.C., and Jolles, C.J., Systemic immunosuppression induced by photodynamic therapy (PDT) is adoptively transferred by macrophages, *Photochem. Photobiol.*, 49, 453, 1989.

187. Anderson, C., Hrabovsky, S., McKinley, Y., Tubesing, K., Tang, H.-P., Dunbar, R., Mukhtar, H., and Elmets, C.A., Phthalocyanine photodynamic therapy: disparate effects of pharmacologic inhibitors on cutaneous photosensitivity and on tumor regression, *Photochem. Photobiol.*, 65, 895, 1997.
188. Simkin, G., Obochi, M., Hunt, D.W. C., Chan, A.H., and Levy, J.G., Effect of photodynamic therapy using benzoporphyrin derivative on the cutaneous immune response, in *Proc. SPIE Opt. Methods Tumor Treat. Detection Mechanisms Tech. Photodyn. Ther. IV*, Dougherty, T.J., Ed., SPIE Press, Bellingham, WA, 1995, p. 23.
189. Simkin, G., Tao, J.-S., Levy, J.G., and Hunt, D.W.C., IL-10 contributes to the inhibition of contact hypersensitivity in mice treated with photodynamic therapy, *J. Immunol.*, 164, 2457, 2000.
190. Musser, D.A., Gollnick, S.O., Oseroff, A.R., and Henderson, B.W., Photodynamic therapy (PDT) induces long term suppression of CHS which is correlated with systemic and localized expression of IL-10, *Photochem. Photobiol.*, 67S, 102S, 1998.
191. Reddan, J.C., Anderson, C., Xu, H., Hrabovsky, S., Freye, K., Fairchild, R., Tubesing, K.A., and Elmets, C.A., Immunosuppressive effects of silicon phthalocyanine photodynamic therapy, *Photochem. Photobiol.*, 70, 72, 1999.
192. Ziolkowski, P., Symonowicz, K., Milach, J., and Szkudlarek, T., *In vivo* tumor necrosis factor-alpha induction following chlorin $e_6$-photodynamic therapy in Buffalo rats, *Neoplasma*, 44, 192, 1996.
193. Rivas, J.M. and Ullrich, S.E., The role of IL-4, IL-10, and TNFα in the immune supression induced by ultraviolet radiation, *J. Leukoc. Biol.*, 56, 769, 1994.
194. Gollnick, S.O., Liu, X., Owczarczak, B., Musser, D.A., and Henderson, B.W., Altered expression of interleukin 6 and interleukin 10 as a result of photodynamic therapy *in vivo*, *Cancer Res.*, 57, 3904, 1997.
195. Moore, K.W., de Waal Malefyt, R., Coffman, R.L., and O'Garra, A., Interleukin-10 and the interleukin-10 receptor, *Ann. Rev. Immunol.*, 19, 683, 2001.
196. Xu, H., DiIulio, N.A., and Fairchild, R.L., T cell populations primed by hapten sensitization in contact sensitivity are distinguished by polarized patterns of cytokine production: interferon γ-producing (Tc1) effector $CD8^+$ T cells and interleukin (Il)4/Il-10-producing (Th2) negative regulatory $CD4^+$ T cells, *J. Exp. Med.*, 183, 1012, 1996.
197. Macatonia, S.E., Hsieh, C.-S., Murphy, K.M., and O'Garra, A., Dendritic cells and macrophages are required for Th1 development of $CD4^+$ T cells from αβ TCR trangenic mice: IL-12 substitution for macrophages to stimulate IFN-γ production is IFN-γ-dependent, *Int. J. Immunol.*, 5, 1119, 1993.
198. Macatonia, S.E., Hosken, N.A., Litton, M., Vieira, P., Hsieh, C.-S., Culpepper, J.A., Wysocka, M., Trinchieri, G., Murphy, K.M., and O'Garra, A., Dendritic cells produce IL-12 and direct the development of Th1 cells from naive $CD4^+$ T cells, *J. Immunol.*, 154, 5071, 1995.
199. O'Garra, A., Cytokines induce the development of functionally heterogeneous T helper cell subsets, *Immunity*, 8, 275, 1998.
200. Sad, S. and Mosmann, T.R., Interleukin (IL)-4, in the absence of antigen stimulation, induces an anergy-like state in differentiated CD8+ Tc1 cells: loss of IL-2 synthesis and autonomous proliferation but retention of cytotoxicity and synthesis of other cytokines, *J. Exp. Med.*, 182, 1505, 1995.
201. King, D.E., Jiang, H., Simkin, G., Obochi, M., Levy, J.G., and Hunt, D.W.C., Photodynamic alteration of the surface receptor expression pattern of murine splenic dendritic cells, *Scand. J. Immunol.*, 49, 184, 1999.
202. Korbelik, M. and Dougherty, G.J., Photodynamic therapy-mediated immune response against subcutaneous mouse tumors, *Cancer Res.*, 59, 1941, 1999.
203. Krosl, G., Korbelik, M., and Dougherty, G.J., Induction of immune cell infiltration into murine SCCVII tumour by Photofrin-based photodynamic therapy, *Br. J. Cancer*, 71, 549, 1995.
204. Korbelik, M., Krosl, G., Krosl, J., and Dougherty, G.J., The role of host lymphoid populations in the response of mouse EMT6 tumor to photodynamic therapy, *Cancer Res.*, 56, 5647, 1996.
205. DeVree, W.J.A., Essers, M.C., DeBruijn, H.S., Star, W.M., Koster, J.F., and Sluiter, W., Evidence for an important role of neutrophils in the efficacy of photodynamic therapy *in vivo*, *Cancer Res.*, 56, 2908, 1996.

206. Korbelik, M., Induction of tumor immunity by photodynamic therapy, *J. Clin. Laser Med. Surg.*, 14, 329, 1996.
207. Korbelik, M. and Cecic, I., Contribution of myeloid and lymphoid host cells to the curative outcome of mouse sarcoma treatment by photodynamic therapy, *Cancer Lett.*, 137, 91, 1999.
208. Gollnick, S.O., Evans, S.S., Baumann, H., Owczarczak, B., Maier, P., Vaughan, L., Wang, W.C., Unger, E., and Henderson, B.W., The role of cytokines in photodynamic therapy (PDT) induced local and systemic inflammation, *Br. J. Cancer*, in press.
209. Yamamoto, N., Hoober, J.K., and Yamamoto, S., Tumoricidal capacities of macrophages photodynamically activated with hematoporphyrin derivative, *Photochem. Photobiol.*, 56, 245, 1992.
210. Korbelik, M. and Krosl, G., Enhanced macrophage cytotoxicity against tumor cells treated with photodynamic therapy, *Photochem. Photobiol.*, 60, 497, 1994.
211. Rousset, N., Vonarx, V., Eléouet, E., Carré, J., Kerninon, E., Lajat, Y., and Patrice, T., Effects of photodynamic therapy on adhesion molecules and metastasis, *J. Photochem. Photobiol. B*, 52, 65, 1999.
212. Korbelik, M., Sun, J., and Posakony, J.J., Interaction between photodynamic therapy and BCG immunotherapy responsible for the reduced recurrence of treated mouse tumors, *Photochem. Photobiol.*, 73, 403, 2001.
213. Korbelik, M., Naraparaju, V.R., and Yamamoto, N., Macrophage-directed immunotherapy as adjuvant therapy, *Br. J. Cancer*, 75, 202, 1997.
214. Korbelik, M. and Cecic, I., Enhancement of tumour response to photodynamic therapy by adjuvant mycobacterium cell-wall treatment, *J. Photochem. Photobiol. B*, 44, 151, 1998.
215. Korbelik, M. and Sun, J., Cancer treatment by photodynamic therapy combined with adoptive immunotherapy using genetically altered natural killer cell line, *Int. J. Cancer*, 93, 269, 2001.
216. Curry, P.M. and Levy, J.G., Tumor inhibitory lymphocytes derived from the lymph nodes of mice treated with photodynamic therapy, *Photochem. Photobiol.*, 61S, 72S,1995.
217. Canti, G., Lattuada, D., Nicolin, A., Taroni, P., Valentini, G., and Cubeddu, R., Immunopharmacology studies on photosensitizers used in photodynamic therapy (PDT), *Proc. SPIE Photodyn. Ther. Cancer*, Vol. 2078, Jori, G., Moan, J., and Star, W.M., Eds., SPIE Press, Bellingham, WA, 1994, p. 268.
218. Abdel-Hady, E.S., Martin-Hirsch, P., Duggan-Keen, M., Stern, P.L., Moore, J.V., Corbitt, G., Kitchener, H.C., and Hampson, I.N., Immunological and viral factors associated with the response of vulval intraepithelial neoplasia to photodynamic therapy, *Cancer Res.*, 61, 192, 2001.
219. Nseyo, U.O., Whalen, R.K., Duncan, M.R., Berman, B., and Lundahl, S.L., Urinary cytokines following photodynamic therapy for bladder cancer: a preliminary report, *Urology*, 36, 167, 1990.
220. DeVree, W.J.A., Essers, M.C., Koster, J.F., and Sluiter, W., Role of interleukin 1 and granulocyte colony-stimulating factor in Photofrin-based photodynamic therapy of rat rhabdomyosarcoma tumors, *Cancer Res.*, 57, 2555, 1997.
221. Matzinger, P., Tolerance, danger, and the extended family, *Ann. Rev. Immunol.*, 12, 991, 1994.
222. Sauter, B., Albert, M.L., Francisco, L.M., Larsson, M., Somersan, S., and Bhardwaj, N., Consequences of cell death: exposure to necrotic tumor cells, but not primary tissue cells or apoptotic cells, induces the maturation of immunostimulatory dendritic cells, *J. Exp. Med.*, 191, 423, 2000.
223. Lanzavecchi, A. and Sallusto, F., Regulation of T cell immunity by dendritic cells, *Cell*, 106, 263, 2001.
224. Gollnick, S.O., Vaughan, L.A., and Henderson, B.W., Generation of effective antitumor vaccines using photodynamic therapy, *Cancer Res.*, 62, 1604, 2002.
225. Peng, Q., Farrants, G.W., Madslien, K., Bommer, J.C., Moan, J., Danielsen, H.E., and Nesland, J.M., Subcellular localization, redistribution and photobleaching of sulfonated aluminum phthalocyanines in a human melanoma cell line, *Int. J. Cancer*, 49, 290, 1991.
226. Peng, Q., Moan, J., Farrants, G.W., Danielsen, H.E., and Rimington, C., Location of P-II and AlPCS4 in human tumor LOX *in vitro* and *in vivo* by means of computer-enhanced video fluorescence microscopy, *Cancer Lett.*, 58, 37, 1991.

227. Wood, S.R., Holroyd, J.A., and Brown, S.B., The subcellular localization of Zn(II) phthalocyanines and their redistribution on exposure to light, *Photochem. Photobiol.*, 65, 397, 1997.
228. Singh, G., Jeeves, W.P., Wilson, B.C., and Jang, D., Mitochondrial photosensitization by Photofrin II, *Photochem. Photobiol.*, 46, 645, 1987.
229. Woodburn, K.W., Fan, Q., Miles, D.R., Kessel, D., Luo, Y., and Young, S.W., Localization and efficacy ananlysis of the phototherapeutic lutetium texaphyrin (PCI-0123) in the murine EMT6 sarcoma model, *Photochem. Photobiol.*, 65, 410, 1997.
230. Ma, L., Moan, J., and Berg, K., Evaluation of a new photosensitizer, meso-tetra-hydroxyphenyl-chlorin, for use in photodynamic therapy: a comparison of its photobiological properties with those of two other photosensitizers, *Int. J. Cancer*, 87, 883, 1994.
231. Kessel, D., Determinants of photosensitization by mono-L-aspartyl chlorin e6 [published erratum appears in *Photochem. Photobiol.*, 50, 1, 1989], *Photochem. Photobiol.*, 49, 447, 1989.
232. Kessel, D., Thompson, P., Saatio, K., and Nantwi, K.D., Tumor localization and photosensitization by sulfonated derivatives of tetraphenylporphine, *Photochem. Photobiol.*, 45, 787, 1987.
233. Berg, K., Moan, J., Bommer, J.C., and Winkelman, J.W., Cellular inhibition of microtubule assembly by photoactivated sulphonated meso-tetraphenylporphines, *Int. J. Radiat. Biol.*, 58, 475, 1990.
234. Berg, K. and Moan, J., Lysosomes as photochemical targets, *Int. J. Biochem.*, 59, 814, 1994.
235. Pogue, B.W., Ortel, B., Chen, N., Redmond, R.W., and Hasan, T., A photobiological and photo-physical-based study of phototoxicity of two chlorins, *Cancer Res.*, 61, 717, 2001.
236. Star, W.M., Marijnissen, H.P., van den Berg Blok, A.E., Versteeg, J.A., Franken, K.A., and Reinhold, H.S., Destruction of rat mammary tumor and normal tissue microcirculation by hematoporphyrin derivative photoradiation observed *in vivo* in sandwich observation chambers, *Cancer Res.*, 46, 2532, 1986.
237. Henderson, B.W. and Fingar, V.H., Relationship of tumor hypoxia and response to photodynamic treatment in an experimental mouse tumor, *Cancer Res.*, 47, 3110, 1987.
238. Selman, S.H., Kreimer Birnbaum, M., Klaunig, J.E., Goldblatt, P.J., Keck, R.W., and Britton, S.L., Blood flow in transplantable bladder tumors treated with hematoporphyrin derivative and light, *Cancer Res.*, 44, 1924, 1984.
239. Van Geel, I.P.J., Oppelaar, H., Rijken, P.F.J.W., Bernsen, H.J.J.A., Hagemeier, N.E.M., van der Kogel, A.J., Hodgkiss, R.J., and Stewart, F.A., Vascular perfusion and hypoxic areas in RIF-1 tumours after photodynamic therapy, *Br. J. Cancer*, 73, 288, 1996.
240. Roberts, D.J.H., Cairnduff, F., Driver, I., Dixon, B., and Brown, S.B., Tumour vascular shutdown following photodynamic therapy based on polyhaematoporphyrin or 5-aminolaevulinic acid, *Int. J. Oncol.*, 5, 763, 1994.
241. Dodd, N.F.J., Moore, J.V., Poppitt, D.G., and Wood, B., *In vivo* magnetic resonance imaging of the effects of photodynamic therapy, *Br. J. Cancer*, 60, 164, 1989.
242. Chapman, J.D., McPhee, M.S., Walz, N., Chetner, M.P., Stobbe, C.C., Soderlind, K., Arnfield, M., Meeker, B.E., Trimble, L., and Allen, P.S., Nuclear magnetic resonance spectroscopy and sensitizer-adduct measurements of photodynamic therapy-induced ischemia in solid tumors, *J. Natl. Cancer Inst.*, 83, 1650, 1991.
243. Zilbertstein, J., Schreiber, S., Bloemers, M.C., Bendel, P., Neeman, M., Schechtman, E., Kohen, F., Scherz, A., and Salomon, Y., Antivascular treatment of solid melanoma tumors with bacteriochlo-rophyll-serine-based photodynamic therapy, *Photochem. Photobiol.*, 73, 257, 2001.
244. Reed, M.W., Mullins, A.P., Anderson, G.L., Miller, F.N., and Wieman, T.J., The effect of photody-namic therapy on tumor oxygenation, *Surgery*, 106, 94, 1989.
245. Engbrecht, B.W., Menon, C., Kachur, A.V., Hahn, S.M., and Fraker, D.L., Photofrin-mediated photodynamic therapy induces vascular occlusion and apoptosis in a human sarcoma xenograft model, *Cancer Res.*, 59, 4334, 2001.
246. Kerdel, F.A., Soter, N.A., and Lim, H.W., *In vivo* mediator release and degranulation of mast cells in hematoporphyrin derivative-induced phototoxicity in mice, *J. Invest. Derm.*, 88, 277, 1987.

247. Lim, H.W., Effects of porphyrins on skin, in *Photosensitizing Compounds: Their Chemistry, Biology and Clinical Use*, Ciba Foundation Symposium 146, John Wiley & Sons, Chichester, U.K., 1989, p. 148.
248. Zieve, P.D. and Solomon, H.M., The effect of hematoporphyrin and light on human platelets. 3. Release of potassium and acid phosphatase, *J. Cell Physiol.*, 68, 109, 1966.
249. Ben-Hur, E., Heldman, E., Crane, S.W., and Rosenthal, I., Release of clotting factors from photosensitized endothelial cells: a possible trigger for blood vessel occlusion by photodynamic therapy, *FEBS Lett.*, 236, 105, 1988.
250. Foster, T.H., Primavera, M.C., Marder, V.J., Hilf, R., and Sporn, L.A., Photosensitized release of von Willebrand factor from cultured human endothelial cells, *Cancer Res.*, 51, 3261, 1991.

# 37

# Synthetic Strategies in Designing Porphyrin-Based Photosensitizers for Photodynamic Therapy

Ravindra K. Pandey
*Roswell Park Cancer Institute*
*Buffalo, New York*



## 37.1 Introduction

Photodynamic therapy (PDT) is a promising cancer treatment that involves the combination of visible light and a photosensitizer.[1] Each factor is harmless by itself, but when combined with oxygen they can produce lethal cytotoxic agents, initially singlet oxygen, that inactivate the tumor cells.[2] This enables greater selectivity toward diseased tissue, as only those cells that are simultaneously exposed to the photosensitizer, light, and oxygen are exposed to the cytotoxic effect. The dual selectivity of PDT is produced by both a preferential uptake of the photosensitizer by the diseased tissue and the ability to confine activation of the photosensitizer to this diseased tissue by restricting the illumination to the specific site.

0-8493-1116-0/03/$0.00+$1.50

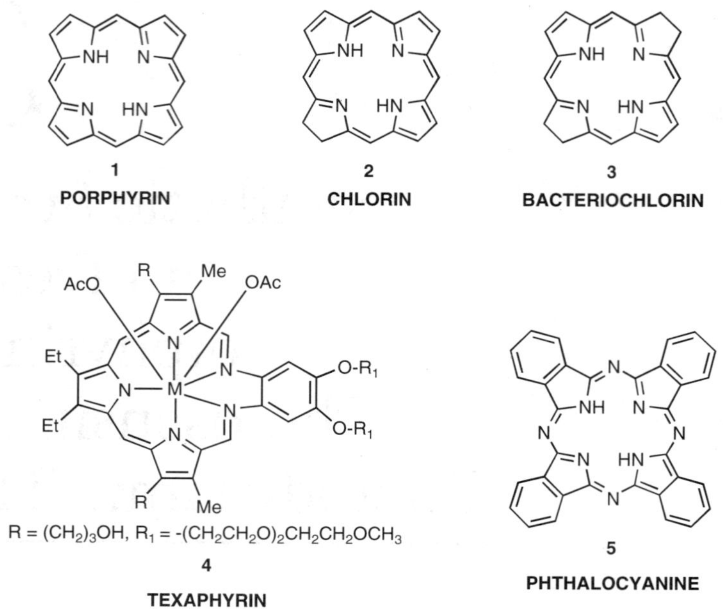


**FIGURE 37.1** Porphyrins and related tetrapyrrolic systems.

The porphyrins and related tetrapyrrolic systems are among the most widely studied compounds for use as photosensitizers in PDT.[3] Porphyrins are 18$\pi$-electron aromatic macrocycles that exhibit characteristic optical spectra with a strong $\pi$–$\pi^*$ transition around 400 nm (Soret band) and usually four Q bands in the visible region. As shown in Figure 37.1, two of the peripheral double bonds in opposite pyrrolic rings are cross-conjugated and are not required to maintain aromaticity. Thus, reduction of one or both of these cross-conjugated double bonds (to give chlorins and bacteriochlorins, respectively) maintains much of the aromaticity, but the change in symmetry results in bathochromically shifted Q bands with high extinction coefficients.[4] Nature uses these optical properties of the reduced porphyrins to harvest solar energy for photosynthesis with chlorophylls and bacteriochlorophylls as both antenna and reaction-center pigments.[5] The long-wavelength absorption of these natural chromophores led to explorations of their use as photosensitizers in PDT.

As indicated above, PDT is based on the interaction of a photosensitizer retained in tumors with photons of visible light, resulting in the formation of singlet oxygen ($^1O_2$), the putative lethal agent.[6] To achieve an effective destruction of tumor cells, a high quantum yield of singlet oxygen is required. Even in the absence of heavy-atom substitutions and coordination of transition-metal ions, porphyrin systems generally satisfy these criteria, which is why most sensitizers currently under clinical evaluation for PDT are porphyrins or porphyrin-based molecules.

At present, Photofrin®, a hematoporphyrin derivative,[7–11] is the only photosensitizer approved worldwide for the treatment of various types of cancer by PDT. It fits some of the criteria for ideal photosensitizers, but it also has several drawbacks. First, it is a complex mixture of various monomeric, dimeric, and oligomeric forms.[7–11] Second, its long-wavelength absorption falls at 630 nm, which lies well below the wavelength necessary for maximum tissue penetration. Finally, it induces prolonged cutaneous phototoxicity, a major adverse effect associated with most of the porphyrin-based photosensitizers.

It is well established that both absorption and scattering of light by tissue increases as the wavelength decreases and that most efficient sensitizers are those that have strong absorption bands from 700 to 800 nm.[12] Light transmission by tissues drops rapidly below 550 nm; however, it doubles from 550 to 630 nm and doubles again from 630 to 700 nm. This is followed by an additional 10% increase in tissue penetration as the wavelength increases toward 800 nm.[3] Another reason to set the ideal wavelength for

PDT at 700 to 800 nm is due to the availability of easy-to-use diode lasers. Although diode lasers are now available at 630 nm (where clinically approved Photofrin absorbs), photosensitizers with absorptions between 700 and 800 nm in conjunction with diode lasers are still desirable for treating deeply seated tumors. Therefore, in recent years a variety of photosensitizers related to chlorins, bacteriochlorins, porphycenes, phthalocyanines, naphthalocyanines, and expanded porphyrins have been synthesized and evaluated for PDT efficacy. However, for designing improved photosensitizers for PDT it becomes necessary to consider several other factors, such as overall lipophilicity (i.e., a proper balance between hydrophobicity and hydrophilicity), pH, lymphatic drainage, and lipoprotein binding, that could influence the biodistribution and localization of sensitizers in tissue and tumors.[13]

The main focus of this chapter is to summarize the various synthetic strategies used by several research groups in designing long-wavelength-absorbing photosensitizers related to chlorins, bacteriochlorins, expanded porphyrins, and Pcs. An ongoing interest in developing target-specific photosensitizers will also be briefly reviewed. The majority of chlorins and bacteriochlorins have been generated through three different approaches. The first approach involves the modification of a preformed porphyrin. The second approach utilizes the use of chlorophyll *a* as the starting material for the synthesis of other chlorins and bacteriochlorins. The third approach utilizes the unstable bacteriochlorophyll *a* as a substrate for the synthesis of stable bacteriochlorins. Each procedure has been used successfully for the preparation of sensitizers that show promise in PDT, and each is discussed in terms of synthetic methodology and biological significance.

## 37.2 Chlorins and Bacteriochlorins from Porphyrins

### 37.2.1 Chlorins and Bacteriochlorins by Diimide Reduction

Almost 30 years ago, Whitlock et al.[14] developed an efficient diimide-reduction method for the synthesis of bacteriochlorins and isobacteriochlorins from porphyrins. Diimide reduction of metal-free tetraphenyl chlorin afforded tetraphenyl bacteriochlorin, while reduction of the corresponding zinc analog produced the related tetraphenylisobacteriochlorin. It is now accepted that the reduced double bond in chlorins induces a pathway for the delocalized $\pi$ electrons that "isolates" the diagonal crossing-conjugated pyrrolic double bond such that reduction of this double bond is favored due to minimal loss of $\pi$ energy over the double bond present in the adjacent ring. The presence of a metal changes the delocalization of the $\pi$ electrons, which makes the adjacent pyrrolic ring more reactive, and diimide reduction produces mainly the corresponding isobacteriochlorin. To avoid the formation of an isomeric mixture this approach is useful only for the reduction of symmetrical porphyrins. This diimide-reduction approach was later employed by Bonnett et al.[15] for preparing the *meso*-tetra (*m*-hydroxyphenyl)-chlorin (*m*-THPC) **7** (650 nm) and the bacteriochlorin (*m*-THPBC) **8**. The formation of these components was found to depend on the amount of the reductant used.

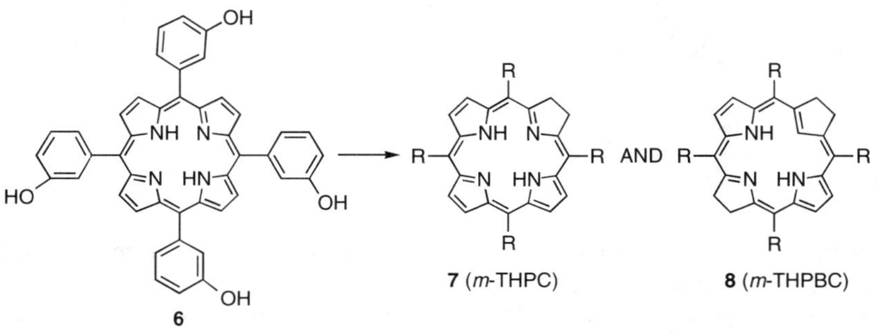


**SCHEME 37.1**

Although the formation of a bacteriochlorin resulted in further red shift in the electronic absorption spectrum with long-wavelength absorption near 750 nm, these molecules were generally found to be air-sensitive. Among various chlorin analogs, *m*-THPC (Foscan®) **7** (Scheme 37.1) appears to be quite effective and is currently in phase III human clinical trials (see Chapter 38).

## 37.2.2 Chlorins and Bacteriochlorins by Diels–Alder Reaction

Cycloaddition reactions are among the most powerful reactions available to the organic chemists.[16] The ability to simultaneously form and break several bonds, with a wide variety of atomic substitution patterns, and often with a high degree of stereocontrol, has made cycloaddition reactions the subject of intense study. In porphyrin chemistry, the [4 + 2] Diels–Alder reactions have been used by various investigators for converting porphyrins into chlorin systems. Callot et al. were the first to show that protoporphyrin IX (PpIX) dimethyl ester **9** could undergo cycloaddition reactions with various dienophiles[17] (Scheme 37.2). A few years later Dolphin and co-workers discovered the utility of one such

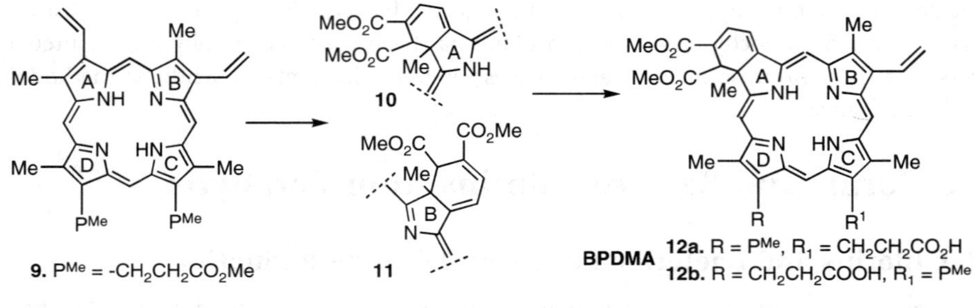


SCHEME 37.2

analog named as benzoporphyrin derivative monocarboxylic acid (BPDMA) **12a** and **12b** for treating age-related macular degeneration (AMD) when activated with light at 690 nm.[18] This treatment has already received approval worldwide. BPDMA has also been used for the treatment of cancer by PDT (see Chapter 38). However, due to its rapid clearance, it was found to be effective only if the tumors were treated with light at 3 h post injection of the drug. Pandey et al.[15] developed another approach for preparing these analogs by using 8-acetyl-3-vinyl deuteroporphyrin IX dimethyl ester as a starting material (Scheme 37.3). The vinyl group was then replaced with various alkyl ether functionalities. Among

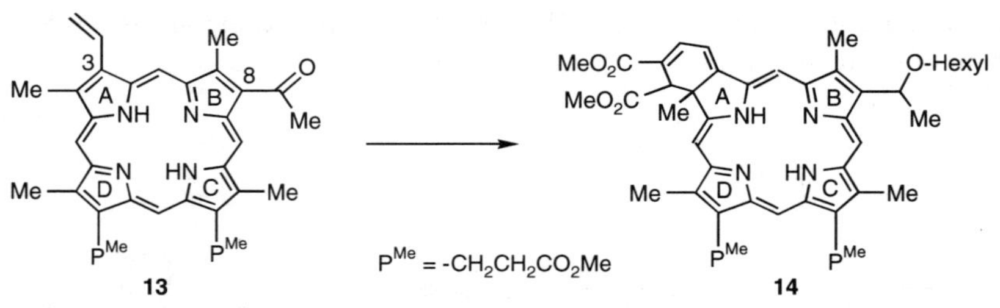


SCHEME 37.3

these analogs the related 8-(1-hexyloxy ethyl)-derivative **14** was found to be more effective than BPDMA in eradicating tumors in mice bearing SMT-F tumors.[19–21] This methodology was further extended independently by Pandey et al.[22] and You-Hin et al.[23] for the synthesis of novel bacteriochlorins that involved a double Diels–Alder reaction on divinylporphyrins **15** and **17** (Scheme 37.4). These bacteriochlorins **16** and **18** exhibit long-wavelength-absorption maxima near 800 nm PDT efficacy.

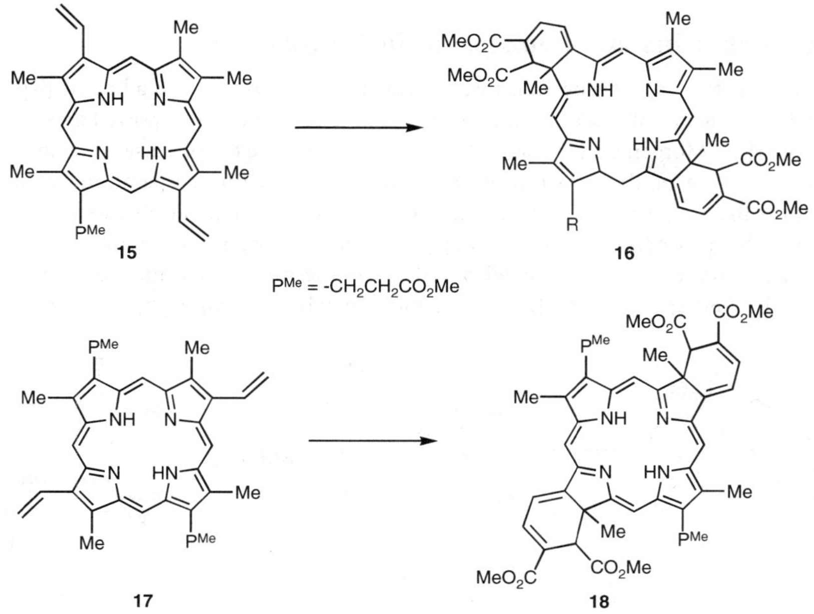


SCHEME 37.4

Morgan et al.[24] have shown that bacteriochlorin-like macrocycles can also be generated by cyclization of either 5,10- or 5,15-*bis*[(ethoxycarbonyl)vinyl]-porphyrins. However, the resulting products rapidly decompose upon exposure to air, thus precluding their use as photosensitizers for PDT. For developing a general synthesis of stable bacteriochlorins, the same authors (A.R. Morgan, unpublished results) followed the pinacol-pinacolone approach in preparing ketochlorins **22** and **23**. In brief, dehydration of **19** produced a mixture of **20** and **21,** which on reaction with dimethylacetylenedicarboxylate (DMAD) produced the corresponding bacteriochlorins **22** and **23** as an isomeric mixture. This isomeric mixture showed some photodynamic activity in a mouse tumor model; 75% of the mice treated at a dose of 1 mg/kg were found to be free from palpable tumor 12 d after the light treatment. However, in this class of compounds, the spectroscopic properties of **22** and **23** resemble those of porphyrinones (long-wavelength absorption near 700 nm) rather than bacteriochlorins (Scheme 37.5).

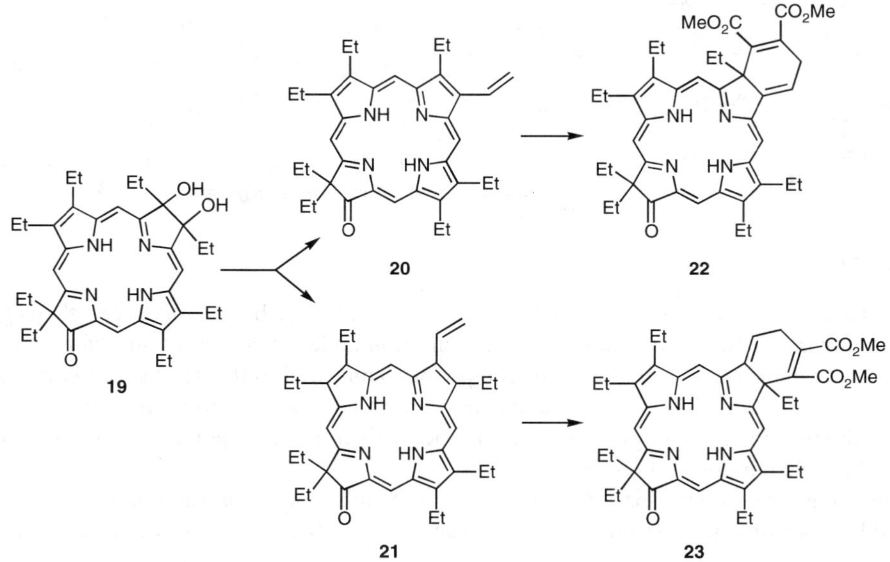


SCHEME 37.5

### 37.2.3 Benzochlorins and Benzobacteriochlorins

Benzochlorin consists of a benzene ring fused between the *meso*- and the adjacent β-position of the pyrrole ring. In a sequence of reactions, this class of compounds was first reported by Arnold et al. from octaethylporphyrin.[25] Morgan et al.[26,27] were the first to demonstrate the photosensitizing efficacy of these analogs (e.g., **28**). One of the major problems associated with this preparation is the difficulty in metalation at the final step of the synthesis, and it is also difficult to chemically modify these benzochlorins. Therefore, this procedure has limited application in preparing a series of analogs with variable lipophilicity. This problem can be avoided by following the method recently reported by Li et al.[28,29] (Scheme 37.6). In their approach, Ni(II)meso-(2-formylvinyl)octaethylporphyrin **26** was reacted with a

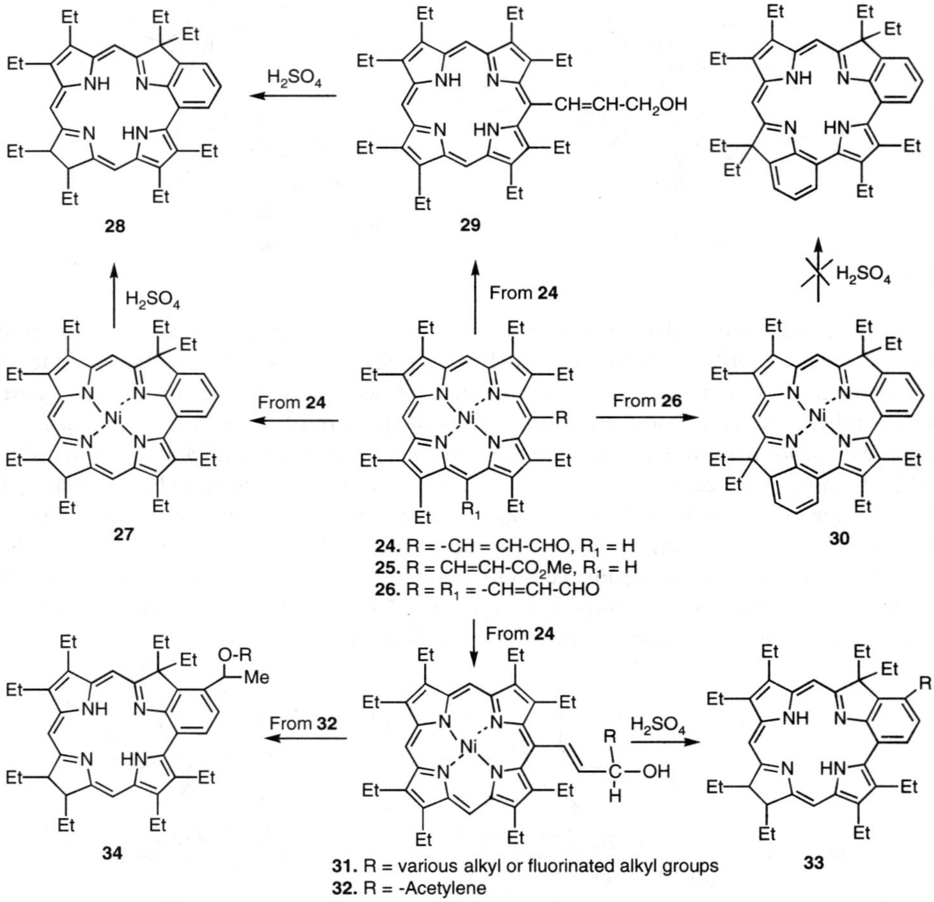


SCHEME 37.6

Grignard's reagent of various fluorinated or nonfluorinated alkyl halides or Ruppert's reagent. The corresponding intermediates via intramolecular cyclization under acidic conditions afforded the related free-base benzochlorins **34**. In this series of compounds, compared to the free-base analogs, the related Zn(II) complexes (671 to 677 nm) were found to be more effective both *in vitro* and *in vivo*. In preliminary screening, the fluorinated analogs showed better efficacy than the corresponding nonfluorinated derivatives (G. Li et al., unpublished results).

This methodology was later extended by Vicente and Smith[30] for the preparation of octaethylporphyrin-based benzochlorin **30** by intramolecular cyclization, of Ni(II)5,10 *bis*-(2-formylvinyl) porphyrin **21**

(Scheme 37.7). Unfortunately, attempts to remove the Ni(II) metal were unsuccessful. When Ni(II)5-(2-formyl-vinyl)-10-(2-ethoxycarbonyl-vinyl)octaethyl porphyrin was used as a substrate, the formation of

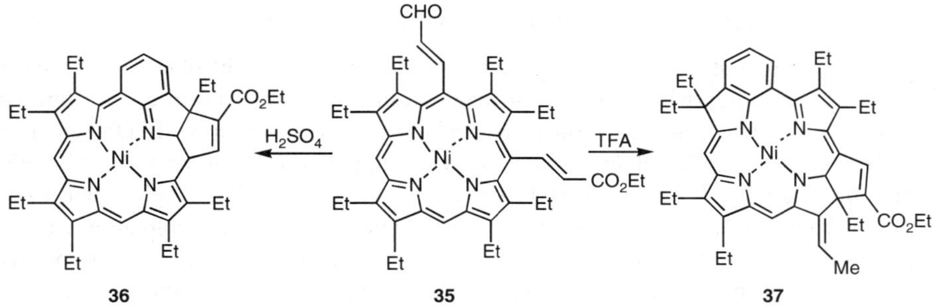

**SCHEME 37.7**

the reaction product was found to depend on the strength of the acid used.[31] For example, reaction of **35** with sulfuric acid produced a chlorin containing both six- and five-member rings fused at the same pyrrole unit **36**. Replacing sulfuric acid with trifluoroactic acid (TFA) produced **37**, the Ni (II) complex of bacteriochlorin containing an ethylidine group at the peripheral position ($\lambda_{max}$ 895 nm). Attempts to prepare the desired free-base analogs for investigating their application as photosensitizers were unsuccessful.

### 37.2.4 Purpurins (Tin Etiopurpurin Dichloride)

Purpurins have been known for quite some time as degradation products of chlorophyll. The first synthesis of this class of compounds was reported by Woodward[32] during the synthesis of chlorophyll *a* by intramolecular cyclization of a *meso*-acrylate functionality to a β-pyrrolic position. This methodology was later followed by Morgan et al.[33] and others[34,35] to synthesize a series of octaetylporphyrin, etioporphyrin, 5,10-diphenyl- and 5,10-dipyridylporphyrin-based purpurin analogs. Among all the purpurins evaluated for PDT efficacy, the Sn etiopurpurin ($SnEt_2$) **40** (Scheme 37.8) is considered the most effective *in vivo*.[33] Its long-wavelength absorption falls at 650 nm and produces a high singlet-oxygen quantum

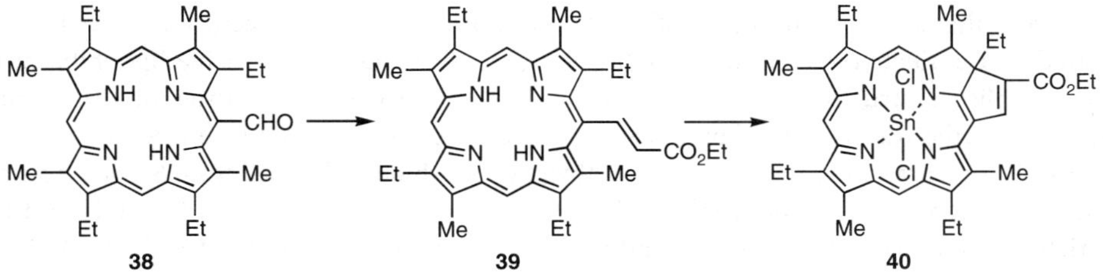

**SCHEME 37.8**

yield. This product is currently in phase III clinical trials for the treatment of age-related macular degeneration (AMD), a major cause of blindness among people over 50 years of age.

## 37.3 Chlorins and Bacteriochlorins from Chlorophyll

Chlorophyll *a*, the green photosynthetic pigment, is one of the prototypes of the chlorin class of natural product. Because of its ready availability, a large amount of work has been done by several investigators to modify and synthesize other chlorin-like chromophores. The photosensitizers derived from chlorophyll *a* can be divided into three categories in which either the five-member isocyclic ring was cleaved or kept intact or was replaced with other ring systems. Some of the photosensitizers in these series have attracted enormous attention; a description of them follows.

## 37.3.1 Aspartic Acid Derivative of Chlorin $e_6$ (Np$e_6$)

Chlorophyllin, a water-soluble degradation product of chlorophyll *a*, can be obtained by the cleavage of the isocyclic ring of chlorophyll.[36] Removal of magnesium resulted in chlorin $e_6$ with limited *in vivo* photosensitizing efficacy. It has been shown that replacing the vinyl group with alkyl ether groups of variable carbon units generally enhances the photosensitizing efficacy.[30] However, in this series, better results were obtained with the monoaspartyl derivative known as Np$e_6$.[31] This photosensitizer appears to clear rapidly from skin, and good tumor response was obtained only after irradiation within 3 to 4 h of sensitizer administration. Np$e_6$ is in human clinical trials in Japan for treatment of endobronchial lung cancer. The recent extensive NMR studies of Np$e_6$ confirmed that in Np$e_6$, the aspartic-acid functionality is linked with an amide bond at position-15 (**41**) of chlorin $e_6$,[39] instead of at position-17 (**42**) as reported in several publications[40,41] (Scheme 37.9).

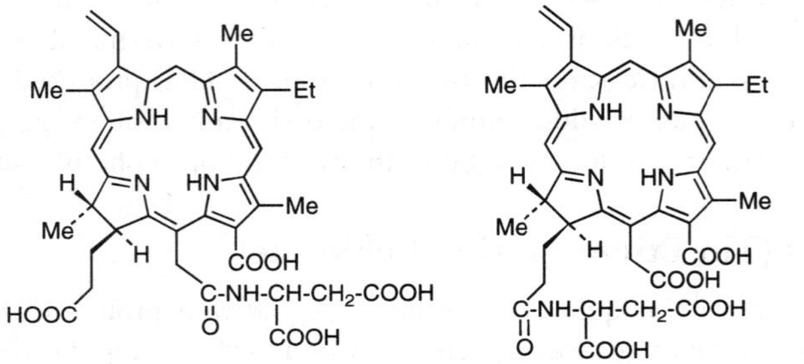


(CORRECT STRUCTURE) **41**

(INCORRECT STRUCTURE) **42**

SCHEME 37.9

## 37.3.2 Alkyl Ether Derivatives of Pyropheophorbide *a*

To understand the effect of various substituents on photosensitizing efficacy, the Roswell Park group synthesized and evaluated a series of pyropheophorbide *a* analogs with variable lipophilicity. In their effort to establish a structure-activity relationship (QSAR), a congeneric series of the primary and secondary alkyl ether derivatives of pyropheophorbide *a* were synthesized (the isocyclic ring was kept intact). For the preparation of these analogs, methylpheophorbide *a* obtained from *Spirulina pacifica* was converted into more stable pyropheophorbide *a* **43**, which on reacting with HBr/AcOH and then the appropriate alcohol produced the corresponding ether analogs in excellent yield[42] (Figure 37.10). At the

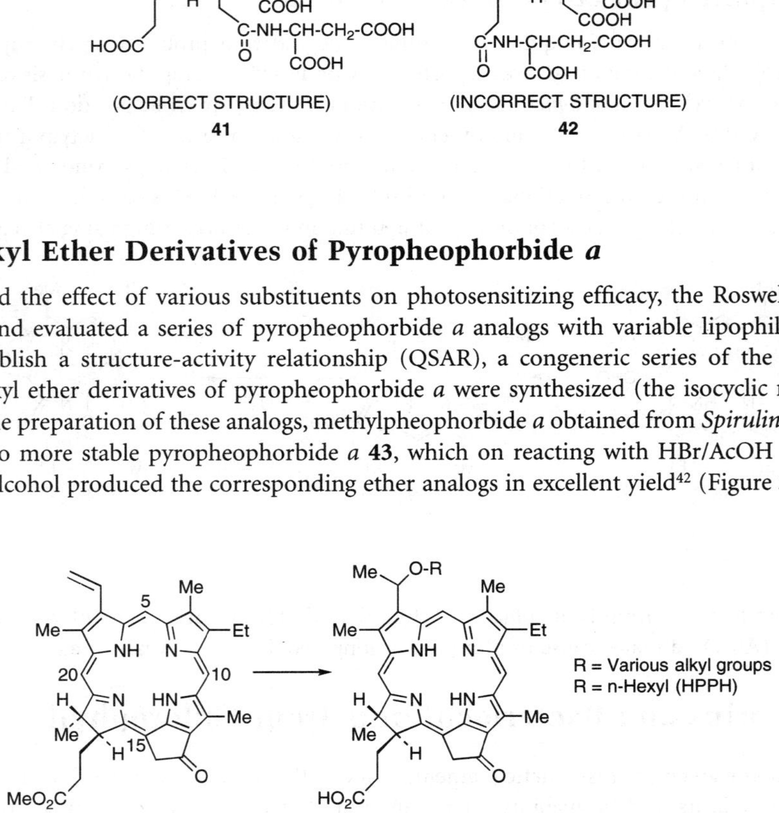


**43** **44**

SCHEME 37.10

final step, the methyl ester functionality was hydrolyzed into the corresponding carboxylic acid **44** (HPPH: R = *n*-hexyl). These analogs exhibit long-wavelength absorption near 665 nm (*in vivo*) and showed excellent singlet-oxygen-producing efficiency (45%). The results obtained from the *in vivo* studies in mice demonstrated that the photodynamic efficacy of these photosensitizers increased by when the length of the carbon chain was increased, reaching a maximum in compounds with *n*-hexyl and *n*-heptyl chains at position-3. Interestingly, the PDT efficacy decreased when the length of alkylether carbon-units was further increased. When compensated for differences in tumor photosensitizer concentration, the *n*-hexyl derivative (HPPH) [optimal lipophilicity] was fivefold more potent than the n-dodecyl derivative (more lipophilic) and threefold more potent than the *n*-pentyl analog (less lipophilic). Interestingly, the introduction of the hexyl ether side chain at other positions of the macrocycle (position-8 or position-20) significantly reduced the *in vivo* efficacy.[43] These data suggest that, besides the lipophilicity, the presence and position of the substituent may play an important role in drug efficacy. HPPH is currently at phase I/II human clinical trials for the treatment of a variety of cancers. Among the patients treated so far no long-term skin phototoxicity has been observed.[44]

### 37.3.3 Alkyl Ether Analogs of Purpurinimides

Having developed a QSAR for the alkyl ether analogs of the pyropheophorbide series, the Roswell Park group extended its approach to photosensitizers with longer-wavelength absorption. For this study, the purpurin-18 methyl ester obtained from methylpheophorbide *a*[45] was converted into purpurin-18-*N*-alkyl imides **46**.[38] The vinyl group at position-3 was then replaced with a variety of alkyl ether analogs **47** (Figure 37.11) with variable carbon units with log *P* values ranging from 5.32 to 16.44 and exhibiting long-wavelength absorption near 700 nm.[39]

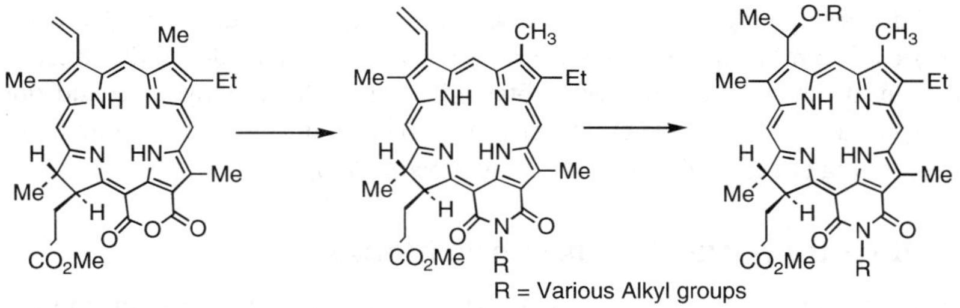


**45** **46** **47**

**SCHEME 37.11**

In animal studies, this class of compounds was found to be quite effective *in vivo*. The results obtained from a set of photosensitizers with similar lipophilicity (log *P* = 10.68 to 10.88) indicate that, like the pyropheophorbide series, in addition to the overall lipophilicity, the presence and position of the alkyl groups (*O*-alkyl vs. *N*-alkyl) in a molecule also play an important role in tumor uptake, tumor selectivity, and *in vivo* PDT efficacy.[47–49]

### 37.3.4 Benzoporphyrin Derivatives Derived from Pyropheophorbide *a* and Purpurinimides

One of the main synthetic problems associated with PpIX-based benzoporphyrin derivatives is to isolate the most effective analog (ring-A reduced, monocarboxylic acid) from the complex reaction mixture. To solve this problem the Roswell Park and Vancouver groups[50,51] have reported the preparation of various BPD analogs (e.g.. **49**) from phylloerythrin and methyl 9-deoxypyropheophorbide (Scheme 37.12). Among these compounds, the benzoporphyrin derivative (*cis*- isomer) obtained from rhodoporphyrin XV di-*tert*-butyl aspartate was found to have PDT efficacy similar to BPDMA. This methodology was

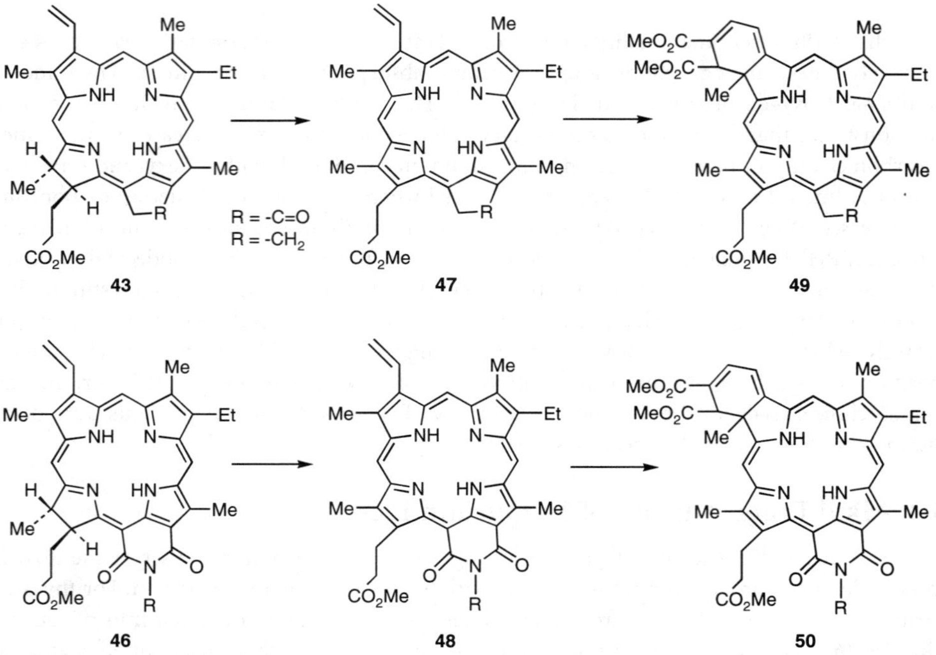


SCHEME 37.12

also extended in the purpurinimde series, and the lipophilicity was altered by introducing *N*-alkyl groups with variable carbon units at the imide ring system 50.[52] In preliminary *in vivo* testing, the corresponding *N*-hexyl and *N*-dodecyl analogs were found to be quite effective at a dose of 0.5 μ*M*/kg when treated with light (135 J/cm,[2] 75 mW/cm[2]) at 728 nm and 24 h post injection. Under similar treatment conditions (treated with light at 690 nm) the BPDMA obtained from PpIX dimethyl ester did not produce any photosensitizing efficacy.[33] Thus, the Diels–Alder approach in the purpurinimide system provides a simple approach to generating effective photosensitizers with variable lipophilicity.

## 37.3.5 *vic*-Dihydroxy- and Ketobacteriochlorins

Osmium tetroxide has very frequently been used for the conversion of porphyrins to the corresponding *vic*-dihydroxy chlorins and tetrahydroxy bacteriochlorins as a mixture of isomers.[45] The overall lipophilicity of these analogs can be altered by subjecting them to pinacolpinacolone recation conditions. The formation of the corresponding keto- analog is not straightforward and depends not only on the intrinsic

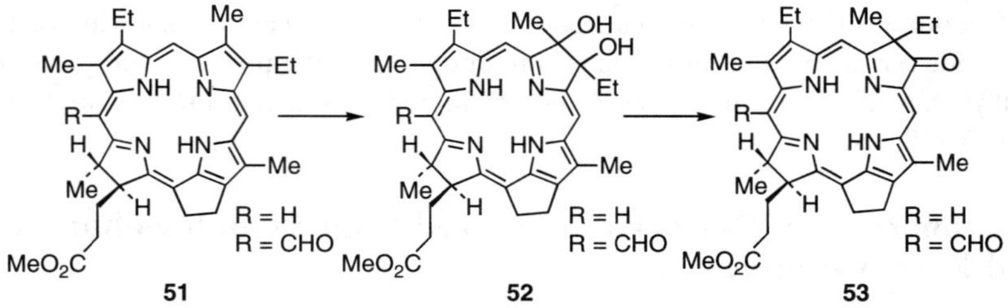


SCHEME 37.13

nature of the migratory group but also on the electronic and steric factors elsewhere on the porphyrin nucleus.[55] Therefore the concept of designing chlorin and bacteriochlorin analogs from porphyrins by following this approach was unsuccessful. A few years ago Chang et al.[56] showed that chlorins under certain conditions could be converted into *vic*-dihydroxybacteriochlorins upon reaction with osmium tetroxide. The Roswell Park group extended this methodology to the pheophorbide chlorin $e_6$, and a series of *vic*-hydroxy- and ketobacteriochlorins was synthesized. The stable ketobacteriochlorins had strong absorptions in the 710- to 760-nm region but showed no significant photosensitizing activity in mice (DBA/2) transplanted with SMT-F tumors.[57] However, the ketobacteriochlorins obtained from 9-deoxypyropheophorbide *a* **51** and the related *meso*-formyl derivative with long-wavelength absorption showed long-wavelength absorptions at 734 and 758 nm, respectively (Scheme 37.13). Among these bacteriochlorins the triplet states were quenched by ground-state molecular oxygen in a relatively similar manner, yielding comparative singlet-oxygen quantum yields. In preliminary *in vivo* screening, the ketochlorins **53** (R = CHO) were found to be more photodynamically active than the related *vic*-dihydroxy analogs. Replacement of the methyl ester functionalities with di-*tert*-butylaspartic acids enhanced the *in vivo* efficacy.[58] It appears to be cleared rapidly from skin, and good tumor responses can be obtained only after irradiation within 3 to 4 h of the sensitizer administration.

## 37.3.6 Bacteriochlorins Derived from 8-Vinyl Chlorins

The Roswell Park group combined the use of osmium tetroxide and the Diels–Alder approach for the construction of stable bacteriochlorins. In this approach, mesopurpurin-18 methyl ester **54** obtained from methylpheophorbide *a* was reacted with osmium tetroxide. The resulting *vic*-dihydroxy bacteriochlorin on reacting with *p*-toluenesulfonic acid in refluxing benzene produced the 8-vinyl derivative **55,** which on reacting with dimethylacetylene dicarboxylate (DMAD) under Diels–Alder reaction conditions produced bacteriochlorin **56** with long-wavelength absorption near 800 nm.[55] Unfortunately, the utility of this compound for use in PDT was diminished due to the unstable nature of the six-member anhydride ring system (Scheme 37.14). In a recent approach, the anhydride ring was replaced with an *N*-hexyl-imide ring system, and these compounds were found to be quite stable *in vivo* (G. Li et al., unpublished results). The preliminary *in vitro* photosensitizing results obtained from this compound are quite promising. This system also presents a unique opportunity to prepare a series of *N*-alkyl ether derivatives with variable carbon units and to establish the structure/activity relationship in a particular series of compounds.

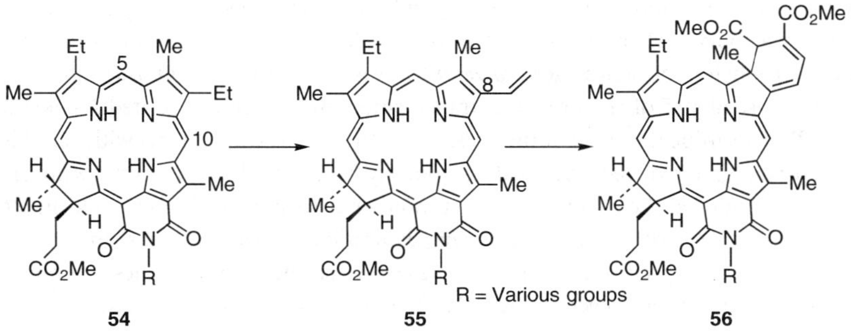


54 55 56

SCHEME 37.14

## 37.3.7 Bacteriochlorins from Bacteriochlorophyll

Most of the naturally occurring bacteriochlorins have absorptions between 760 and 780 nm and have been studied by various investigators for their use as photosensitizers for PDT.[60] They were found to be extremely sensitive to oxidation, resulting in a rapid transformation into the chlorin state, which generally has an absorption maxima at or below 660 nm.[61] Furthermore, if a laser is used to excite the bacteriochlorin *in vivo*, oxidation may result in the formation of a new chromophore absorbing outside the laser window, reducing the photodynamic efficacy. Due to the desirable photophysical properties and promising *in vitro/in vivo* photosensitizing efficacy of bacteriochlorins, there has been increasing interest in the synthesis of stable bacteriochlorins either from bacteriochlorophyll *a* or from the other related tetrapyrrolic systems.

In general, for designing improved photosensitizers, overall lipophilicity has proven to be one of the important factors. For example, among porphyrin-based photosensitizers, the hydrophobic porphyrins are preferentially accumulated and partitioned into corresponding hydrophobic loci *in vivo*. Moan et al.[62] have shown that among diether derivatives of hematoporphyrin, retention in cells increases with decreasing polarity. The Roswell Park group has studied the uptake of a series of alkyl ether derivatives of pyropheophorbide *a* and found that a strong correlation existed between uptake and hydrophobicity, although each correlation cannot be extended to the *in vivo* PDT efficacy. On the other hand, photosensitizers with high partition-coefficient values (increased hydrophobicity) induced sensitizer insolubility, thereby preventing drugs from entering circulation. Therefore, a proper balance between hydrophobicity and hydrophilicity is probably the most important factor influencing tumor localization of sensitizers.

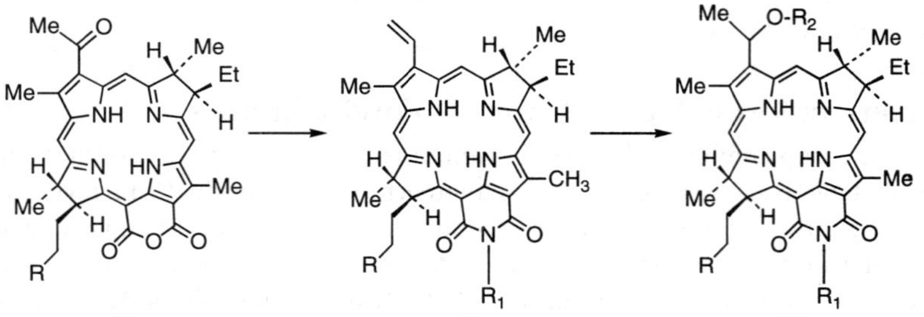


SCHEME 37.15

A simple approach used by the Roswell Park group was to vary the overall lipophilicity of various types of photosensitizers such as pyropheophorbide *a*, benzoporphyrin derivatives, benzochlorins, and purpurinimides by altering the length of carbon units in alkyl ether substituents; this approach has been quite successful. It was demonstrated that replacing an anhydride ring system in purpurin-18 (a chlorophyll *a* analog) with a six-member imide ring substantially enhanced its *in vivo* stability and retained effective *in vivo* photodynamic activity.[63] Therefore, to investigate the effect of such substitutions in the bacteriochlorin series, bacteriochlorophyll *a*, present in *Rb. sphaeroides*, was first converted (*in situ*) into bacteriopurpurin-18 (**57**),[64] which in a sequence of reactions was transformed into a series of related *N*-alkyl derivatives **58**[65] (Scheme 37.15). To determine the effect of the presence of these alkyl substituents with variable carbon units, the acetyl- group was first reduced with sodium borohydride, which on reacting with HBr gas and an appropriate alcohol produced the corresponding alkyl ether derivatives of bacteriochlorin **59** in high yield. These compounds are stable both *in vitro and in vivo*, exhibit long-wavelength absorption near 790 nm, and show high tumor uptake. In preliminary *in vitro* and *in vivo* studies, some of these compounds have been found to be quite effective at low injected doses (Y. Chen et al., unpublished results).[58]

## 37.4 Expanded Porphyrins

### 37.4.1 Texaphyrin

The texaphyrins are aromatic tripyrrolic, penta-aza, Schiff-base macrocycles that bear a strong but "expanded" resemblance to the porphyrins and other naturally occurring tetrapyrrolic prosthetic groups.[66] Like porphyrins, the texaphyrins are fully aromatic and colored compounds (Scheme 37.16). However they are 22 $\pi$-electron electron systems rather than 18 $\pi$-electron ones. This class of compounds exhibits long-wavelength absorption at >700 nm depending on the nature of substituents present at the peripheral position. Also, unlike porphyrins, the texaphyrins are monoanionic ligands that contain five, rather than four, coordinating nitrogen atoms within the central core, which is roughly 20% larger than that of the porphyrins. High-yield production of long-lived triplet states and their remarkable singlet-oxygen-producing efficiency are important

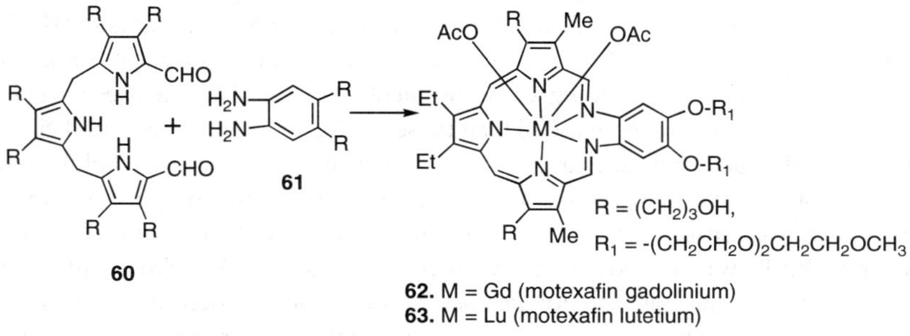


SCHEME 37.16

features of this class of photosensitizers. Currently, two different water-solubilized lanthanide(III)texaphyrin complexes — i.e., the gadolinium(III) (**62**)[67] and lutetium(III) (**63**)[68] derivatives (Gd-Tex and Lu-Tex, respectively) — are being tested clinically. The first of these, XCYTRIN™, is in a pivotal phase III clinical trial as a potential enhancer of radiation therapy for patients with metastatic cancers of the brain receiving whole-brain radiation therapy. The second, in various formulations, is being tested as a photosensitizer for use in the treatment of recurrent breast cancer (LUTRIN); is in phase II clinical trials for treatment of photoangioplastic reduction of atherosclerosis involving peripheral arteries (ANTRIN); and is in phase I clinical trials for light-based treatment of AMD (OPRTIN).[65]

## 37.5 Phthalocyanines and Naphthalocyanines

Phthalocyanines (Pcs) and naphthalocyanines (Ncs) can be regarded as azaporphyrins containing four isoindoles linked by nitrogen atoms[69] (Scheme 37.17). Compared to porphyrins, Pcs and Ncs offer high molar-extinction coefficients and red-shift maximums at 680 nm for Pcs and 780 nm for Ncs resulting from the benzene or naphthalene rings condensed at the periphery of the porphyrin-like macrocycle. They possess high singlet-oxygen-producing efficiency, and interestingly chelation of the metal ions such as zinc or aluminum increases the singlet-oxygen yield to nearly 100%. Therefore, metal complexes of Pcs and Ncs have attracted attention for their use as photosensitizers in PDT. In recent years, a large number of metalated or nonmetalated Pc-based photosensitizers have been synthesized by introducing a variety of substituents at the peripheral positions. If the valency of the central metal is higher than two, it binds various axial ligands. All of these chemical changes in the Pc and Nc skeleton alter their PDT efficacy. Aggregated Pc and Nc are inactive photochemically because of a greately enhanced rate of excited singlet-state deactivation by internal conversion of the ground state. In the Pc series, Olenick et al. in collaboration with Kenney[70] synthesized and evaluated four silicon analogs with variable lipophilicity to learn more about the structural features that

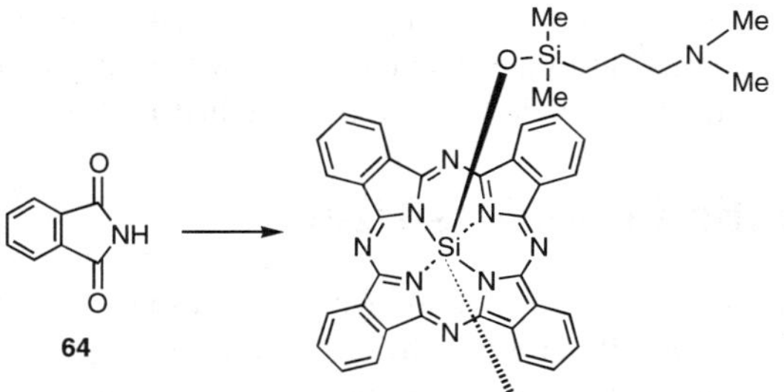


SCHEME 37.17

silicon Pc must have to be a good PDT photosensitizer. All these analogs produced similar photophysical properties; however, the photosensitizer denoted as Pc4, bearing a long-chain amino axial ligand **65,** has shown promising results both *in vitro* and *in vivo* and is presently entering clinical trials.[71] Further, it was concluded that the structural features leading to improvement in the association between the photosensitizers and important cellular targets were more useful than those leading to improvements in their already acceptable photophysical and photochemical characteristics. A series of benzyl-substituted phthalonitriles, substituted at the 3,4- and 4,5-positions, was converted into a series of Zn(II) hydroxyphthalocyanines (Pc phenol analogs). Their efficacy as sensitizers for PDT was evaluated on the EMT-B mammary tumor cell line. *In vitro,* the 2-hydroxy Zn Pc was the most active, followed by 2,3- and 2,9-dihydroxy ZnPc, with the 2,9,16-trihydroxy ZnPc exhibiting the least activity. *In vivo,* the monohydroxy derivative and the 2,3-dihydroxy analog were both efficient in inducing tumor necrosis, but complete tumor regression was poor even at high doses. In contrast, the 2,9-dihydroxy isomer at 2 μ*M*/Kg induced tumor necrosis in all animals treated, with 75% complete regression. These results underline the importance of the position of the substituents on the Pc macrocycle to optimize tumor response and confirm the PDT potential of the unsymmetrical Pcs bearing functional groups on adjacent benzene rings.[72]

Rodgers et al. in collaboration with Kenney and collaborators[73] developed a new route to silicon-substituted Pcs and Pc-like compounds that are robust and flexible. One of the siloxysilicon compounds with the ligand 5,9,12,16,23,28,32-octabutoxy-33H, 35H[b,g]dinaphtho[2,3–1:2′,3′-q]porphyrazine **69** has a q band at a wavelength of 804 nm and an extinction coefficient of $1.9 \times 10^5\ M^{-1}\ cm^{-1}$ (Scheme 37.18).

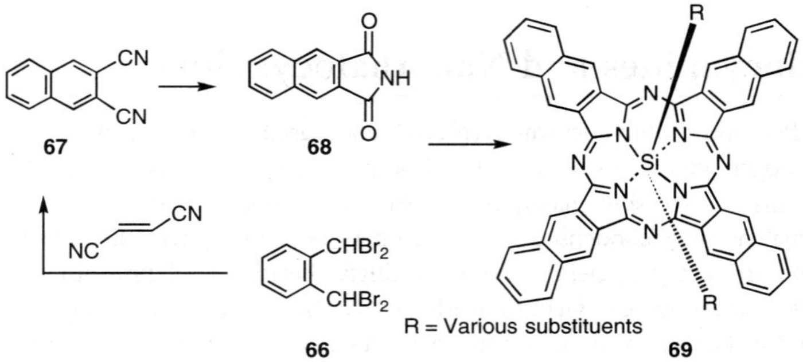


**SCHEME 37.18**

These compounds showed promising *in vitro* photosensitizing efficacy *in vitro;* however, in animal studies, compared to Pc (Pc 4), they were found to be less effective. A first photophysical study of a member of the family of subnaphthalocyanines (SubNcs) has been described.[74] The cone-shaped unsubstituted SubNcs, synthesized in 35% yield, showed distinctive photophysical properties that are better than those of the related planar Pcs and Ncs. SubNc absorbs in the red part of the spectrum and has substantial fluorescense. Triplet- and singlet-oxygen quantum yield are substantially higher than those of the related Pcs and Ncs. These results, together with their synthetic availability, high solubility, and low tendency to aggregate, make this class of sensitizers worthy of further study with a view to investigating their PDT applications.

## 37.6 Target-Specific Photosensitizers

Since the introduction of the first PDT drug Photofrin there has not been much success in improving the photosensitizer's tumor selectivity and specificity because tumor cells in general have nonspecific affinity to porphyrins.[75,76] Although the mechanism of porphyrin retention by tumors is not well understood, the balance between lipophilicity and hydrophilicity is recognized as an important factor.[77] Some attempts have been made to direct photosensitizers to known cellular targets by creating a photosensitizer conjugate, where the other molecule is a ligand specific to the target. For example, to improve localization to cell-membrane cholesterol[78] and antibody-conjugates have also been prepared to direct photosensitizers to specific tumor antigens.[79–81] Certain chemotherapeutic agents have also been attached to porphyrin chromophores to increase the lethality of the PDT treatment.[82] Certain protein and microsphere conjugates were made to improve the pharmacology of the compounds.[83] These strategies seldom work well because the pharmacological properties of both compounds are drastically altered.[84]

Recently, the bovine serum albumin conjugate of a sulfonated Pc (BSA-AlPc$S_4$) prepared by Brasseur et al. has been shown to target the scavenger receptor of macrophages.[85] Relative photocytotoxicities were reported using a receptor-positive and a receptor-negative cell line, where their lethal effects correlated with its receptor affinity. Also, ceratin adenoviral proteins have been employed to target lung cancer cells rich in the appropriate class integrin receptor.[86] Using the EMT-6 murine model, *in vivo* results were encouraging with Pc–adenoviral protein conjugates. Further, both AlPc and CoPc have been covalently labeled with epidermal growth factor (EGF), and this conjugate produced a fivefold increase in photocytotoxicity as compared with nonconjugated Pc.[87]

Since oligosaccharides play essential roles in molecular recognition,[88] porphyrins with sugar moieties should have both good aqueous solubility and possible specific membrane interaction. In addition to providing the molecule with polar hydroxy groups, the sugar moiety may lead the conjugate to a cell-surface target through specific binding to its receptor. Oligosaccharides play essential roles in various cellular activities as antigens, growth signals, targets of bacterial and viral infection, and glues in cell adhesion and metastasis, where the saccharide–receptor interactions are usually specific and multivalent.[89] This specificity suggests a potential utility of synthetic-saccharide derivatives as carriers in directed drug delivery. Therefore, in recent years, a number of carbohydrate derivatives of various photosensitizers have been synthesized. Hombrecher et al.[90] reported the preparation of a galactopyranosyl-cholesteryloxy-substituted porphyrin by following

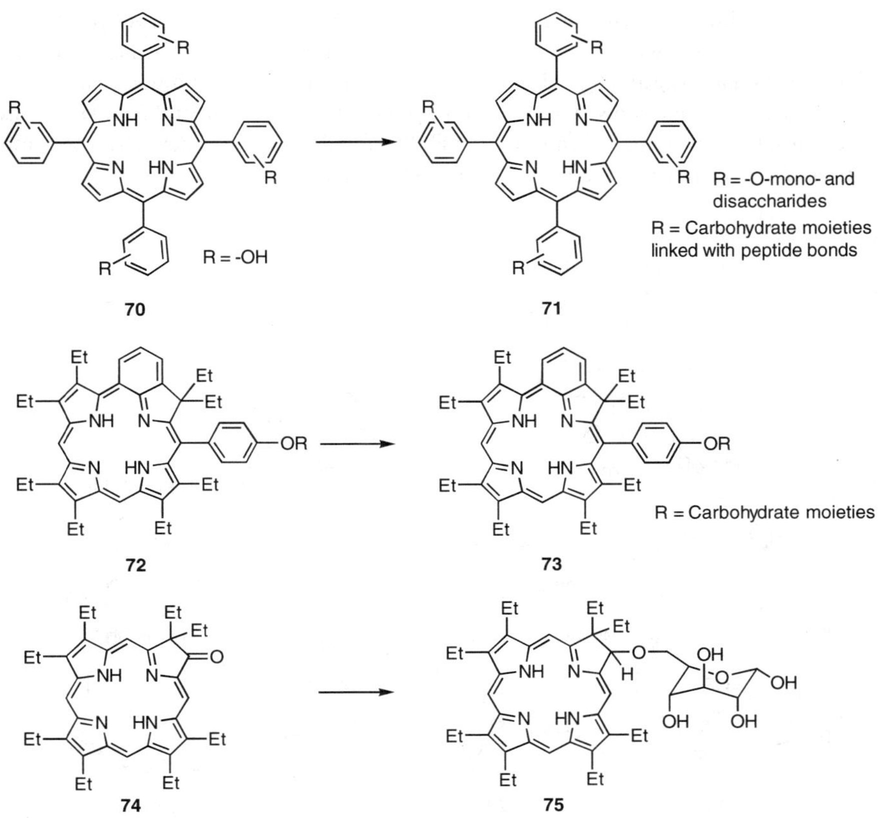


**SCHEME 37.19**

the McDonald approach. Several tetra- and octaglycoconjugated tetraphenylporphyrins were also reported by Yano et al.,[91] Cornia et al.,[92] Momenteau et al.,[93] and Ono et al.[94] A glycosylated peptide porphyrin synthesized by Krausz et al.[95,96] showed promising *in vitro* PDT efficacy. Maillard et al.[97] reported the synthesis of a glycoconjugated *meso*-monoarylbenzochlorin (related to **71** and **73**; see Scheme 37.19). This compound displayed good *in vitro* PDT efficacy in tumor cell lines. A few years ago, Bonnett et al.[98] showed that, compared to β-hydroxy-octaethylchlorin **75**, the related glycosylated analog was more effective as a photosensitizing agent *in vitro* as well as *in vivo*. Montforts et al.[99] further explored this approach by attaching hydrophilic carbohydrate structural units to certain chlorins. Such conjugation increased the water solubility of the parent chlorins by introducing an estradiol with a diethyl spacer. The chlorin–estrogen conjugate was then prepared in the hope that it would bind to an estrogen receptor that could induce destruction of a mammalian carcinoma. In a recent report, Aoyama et al.[100] reported the synthesis of certain tetraphenylporphyrin-based saccharide-functionalized porphyrins and demonstrated the importance of hydrophobicity masking for the saccharide-directed cell recognition. Since the saccharide–receptor interactions are ubiquitous, well-defined/well-designed synthetic saccharide clusters may serve as a new tool in glycoscience and glycotechnology.

It is known that galectins are involved in the modulation of cell adhesion, cell growth, immune response, and angiogenesis; therefore, it is likely that their expression plays a critical role in tumor progression.[101] Gal-1 mRNA levels increase 20-fold in low tumorigenic and up to 100-fold in highly tumorigenic cells.[102] The Roswell Park group has recently reported the synthesis and biological significance of certain β-galactose-conjugated purpurinimides (a class of chlorins containing a six-member fused imide ring system) **76** to **80** (Scheme 37.20) as a Gal-1-recognized photosensitizer via enyne metathesis.[103] Molecular-modeling analysis using model photosensitizers and the available crystal structures of galectin–carbohydrate moiety indicated that when the β-galactose moiety is placed at the appropriate position the photosensitizer does not interfere with the galectin–carbohydrate recognition. Intracellular studies with known cell-surface counterstains confirmed the cell-surface recognition of the conjugates. Under similar drug and light doses, compared to the free purpurinimide analog, the galactose- and lactose-conjugated analog showed a considerable increase both *in vitro* and *in vivo* photosensitizing efficacy. These results, therefore, indicate the possibility for developing a new class of tumor-specific photosensitizers for PDT based on recognition of a cellular receptor.[103]

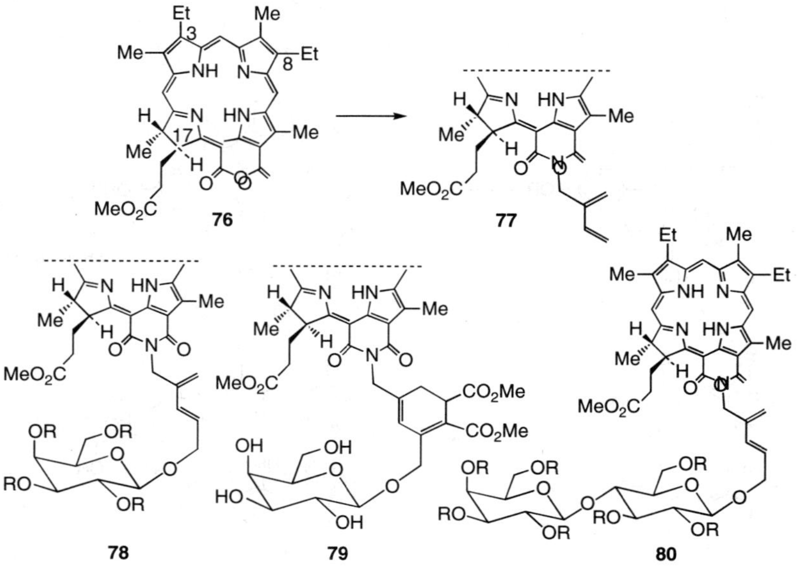


**SCHEME 37.20**

## 37.7 Summary

This chapter has focused largely on the chemical rather than the biological aspects of various porphyrin-based photosensitizers for use in PDT. It is important to note that porphyrin-based compounds, in addition to their use in cancer, have also shown potential for application in other areas: treatment of AMD, tumor imaging by MRI (magnetic resonance imaging), psoriasis, bone-marrow purging, and purification of blood infected with various viruses including HIV. Since the discovery of Photofrin, enormous progress has been made in the development of various porphyrin-based compounds with improved photophysical characteristics. In recent years, a number of pharmaceutical companies — such as QLT Pharmaceuticals, Vancouver, B.C., Canada; Miravant, Santa Barbara, CA; Scotia Pharmaceuticals, Stirling, U.K.; Ciba Vision, Atlanta, GA; Pharmacia, Stockholm, Sweden; DUSA, Wilmington, MA; Pharmacyclics, Palo Alto, CA; and Light Sciences Corp., Seattle, WA — have shown major interest in PDT. However, given the possible implications for the use of porphyrin-based compounds (porphyrins, chlorins, bacteriochlorins, expanded porphyrins, Pcs) for the treatment of cancer by PDT, future design strategies for new agents should be directed toward tumor-specific drug molecules. Such compounds might show greater tumor selectivity with reduced skin phototoxicity, a major problem associated with most of the porphyrin-based compounds.

## References

1. Dougherty, T.J., Gomer, C., Henderson, B.W., Jori, G., Kessel, D., Korbelik, M., Moan, J., and Peng, Q., Photodynamic therapy, *J. Natl. Cancer Inst.*, 90, 889, 1998.
2. Weishaupt, K.R., Gomer, C.J., and Dougherty, T.J., Identification of singlet oxygen as the cytotoxic agent in photoinactivation of murine tumor, *Cancer Res.*, 36, 2326, 1976.
3. Pandey, R.K. and Zheng, G., Porphyrins as photosensitizers in photodynamic therapy, in *The Porphyrin Handbook*, Vol. 6, Kadish, K.M., Smith, K.M., and Guilard, R., Eds., Academic Press, San Diego, 2000, chap. 43.
4. Chang, C.K., Cation radicals of ferrous and free base isobacteriochlorins: models for siroheme and sirohydrochlorin, *Proc. Natl. Acad. Sci. U.S.A.*, 78, 2653, 1981.
5. Barkigia, K.M. and Fajer, J., in *The Photosynthetic Reaction Center*, Vol. 2, Deisenhofer, H. and Norris, J.R., Eds., Academic Press, San Diego, 1993, p. 514.
6. Sherman, W.M., Allen, C.M., and van Lier, J.E., Role of activated oxygen species in photodynamic therapy, *Meth. Enzymol.*, 319, 376, 2000.
7. Dougherty, T.J., Kaufman, J.H., Goldfrab, A., Weishaupt, K.R., Boyle, D., and Mittleman, A., Photoradiation therapy for the treatment of malignant tumors, *Cancer Res.*, 38, 2628, 1978.
8. Pandey, R.K., Marshall, M.S., Tsao, R., McReynolds, J.H., and Dougherty, T.J., Fast atom bombardment mass spectral analyses of Photofrin II and its synthetic analogs, *Biomed. Environ. Mass Spectrosc.*, 19, 405, 1990 and references therein.
9. Pandey, R.K., Smith, K.M., and Dougherty, T.J., Porphyrin dimers as photosensitizers in photodynamic therapy, *J. Med. Chem.*, 33, 2032, 1990.
10. Pandey, R.K. and Dougherty, T.J., Syntheses and photosensitizing activity of porphyrins joined with ester linkages, *Cancer Res.*, 49, 2042, 1989.
11. Pandey, R.K., Shiau, F.-Y., Dougherty, T.J., and Smith, K.M., Regioselective syntheses of ether linked porphyrin dimers and trimers related to Photofrin II, *Tetrahedron*, 47, 9571, 1991 and references therein.
12. Dolphin, D., Photomedicine and photodynamic therapy, *Can. J. Chem.*, 72, 1005, 1994 and references therein.
13. MacDonald, I. and Dougherty, T.J., Basic principles of photodynamic therapy, *J. Porphyrins Phthalocyanines*, 5, 105, 2001.
14. Whitlock, H.W., Hanaurer, R., Oester, M.Y., and Bower, B.K., Diimide reduction of porphyrins, *J. Am. Chem. Soc.*, 91, 7485, 1969.

15. Bonnett, R., Photosensitizers of the porphyrin and phthalocyanine series for photodynamic therapy, *Chem. Soc. Rev.*, 24, 19, 1995.
16. Hamata, M., *Advances in Cycloaddition*, Vol. 4, Hamata, M., Ed., Jai Press, London, 1998.
17. Callot, H.L., Johnson, A.W., and Sweeney, A., Addition to porphyrins involving the formation of new carbon-carbon bonds, *J. Chem. Soc. Perkin Trans. 1*, 1424, 1973.
18. Morgan, A.R., Pangka, V.S., and Dolphin, D., Ready syntheses of benzoporphyrins via Diels-Alder reactions with protoporphyrin IX, *Chem. Commun.*, 1047, 1984.
19. Meunier, I., Pandey, R.K., Walker, M.M., Senge, M.O., Dougherty, T.J., and Smith, K.M., New syntheses of benzoporphyrin derivatives and analogs for use in photodynamic therapy, *Bioorg. Med. Chem. Lett.*, 2, 1575, 1992.
20. Meunier, I., Pandey, R.K., Senge, M.O., Dougherty, T.J., and Smith, K.M., Benzoporphyrin derivatives: synthesis, structure and preliminary biological activity, *J. Chem. Soc. Perkin Trans. 1*, 961, 1994.
21. Pandey, R.K., Potter, W.R., Meunier, I., Sumlin, A.B., and Smith, K.M., Structure–activity relationships among benzoporphyrin derivatives, *Photochem. Photobiol.*, 62, 764, 1995.
22. Pandey, R.K., Shiau, F.Y., Ramachandran, K., Dougherty, T.J., and Smith, K.M., Long wavelength photosensitizers related to chlorins and bacteriochlorins for the use in photodynamic therapy, *J. Chem. Soc. Perkin Trans. 1*, 1377, 1992.
23. Yon-Hin, P., Wijesekera, T.P., and Dolphin, D., A convenient synthetic route for the bacteriochlorin chromophore, *Tetrahedron Lett.*, 32, 2875, 1991.
24. Morgan, A.R., Skalkos, D., Garbo, G.M., Keck, R.W., and Selman, S.H., Synthesis and *in vivo* photodynamic activity of some bacteriochlorin derivatives against bladder tumors in rodents, *J. Med. Chem.*, 34, 2126, 1991.
25. Arnold, D.P., Johnson, A.W., and Williams, G.A., Wittig condensation products from nickel meso-formyl-octaethylporphyrin and aetioporphyrin-I and some cyclization reactions, *J. Chem. Soc. Perkin Trans. 1*, 1660, 1978.
26. Morgan, A.R., Garbo, G.M., Ranapersaud, A., Shalkos, D., Keck, R.W., and Selman, S.H., Photodynamic action of benzochlorins, *Proc. SPIE*, 146, 1065, 1989.
27. Morgan, A.R., Skalkos, D., Maguire, G., Ranpersaud, A., Garbo, G., Keck, K., and Selman, S.H., Observations of the synthesis and *in vivo* photodynamic activity of some benzochlorins, *Photochem. Photobiol.*, 55, 133, 1992.
28. Li, G., Graham, A., Potter, W.R., Grossman, Z.D., Oseroff, A., Dougherty, T.J., and Pandey, R.K., A simple and efficient approach for the synthesis of fluorinated and non-fluorinated octaethylporphyrin-based benzochlorins with variable lipophilicity: their *in vivo* tumor uptake, and the preliminary *in vitro* photosensitizing efficacy, *J. Org. Chem.*, 66, 1316, 2001.
29. Li, G., Chen, Y., Missert, J.R. Rungta, A., Dougherty, T.J., and Pandey, R.K., Application of Ruppert's reagent in preparing novel perfluorinated porphyrins, chlorins and bacteriochlorins, *J. Chem. Soc. Perkin Trans. 1*, 1785, 1999.
30. Vicente, M.G.H., Rezanno, I.N., and Smith, K.M., Efficient new syntheses of benzochlorins, isobacteriochlorins and bacteriochlorins, *Tetrahedron Lett.*, 31, 1365, 1990.
31. Morgan, A.R. and Gupta, S., Synthesis of benzopurpurins, isobacteriobenzopurpurins and bacteriobenzopurpurins, *Tetrahedron Lett.*, 35, 4291, 1994.
32. Woodward, R.B. et al., The total synthesis of chlorophyll *a*, *J. Am. Chem. Soc.*, 82, 3800, 1960, and *Tetrahedron*, 46, 7599, 1990.
33. Morgan, A.R., Garbo, G.M., Keck, R.W., and Selman, S.H., New photosensitizers for photodynamic therapy: combined effect of metallopurpurin derivatives and light in transplantable bladder, *Cancer Res.*, 48, 194, 1988.
34. Gunter, M.J. and Robinson, B.C., Purpurins bearing functionality at the 6,16 meso-positions: synthesis from 5,15-disubstituted meso-[β-(methoxycarbonyl)vinyl] porphyrins, *Aust. J. Chem.*, 43, 1839, 1990.

35. Forsyth, T.P., Nurco, D.J., Pandey R.K., and Smith, K.M., Syntheses and structure of 5,15-*bis*-(4-pyridyl)purpurins, *Tetrahedron Lett.*, 36, 9093, 1995.
36. Smith, K.M., Ed., *Porphyrins and Metalloporphyrins*, Elsevier Science Publishers, Amsterdam, 1975.
37. Pandey, R.K., Bellnier, D.A., Smith, K.M., and Dougherty, T.J., Chlorin and porphyrin derivatives as potential photosensitizers in photodynamic therapy, *Photochem. Photobiol.*, 53, 65, 1991.
38. Bommer, J.C. and Burnham, B.F., Tetrapyrrole compounds, Europe patent 169831, 1986.
39. Gomi, S., Nishizuka, T., Ushiroda, O., Uchida, N., Takahashi, H., and Sumi, S., The structures of mono-L-aspartyl chlorin $e_6$ and its related compounds, *Heterocycles*, 48, 2231, 1998.
40. Nelson, J.S., Roberts, W.G., and Berns, M.W., *In vivo* studies on the utilization of mono-L-aspartyl chlorin ($NPe_6$) in photodynamic therapy, *Cancer Res.*, 47, 5681, 1987.
41. Roberts, W.G., Shiau, F.Y., Nelson, J.S., Smith, K.M., and Berns, M.W., *In vitro* characterization of monoaspatyl chlorin $e_6$ and diaspartyl chlorin $e_6$ for photodynamic therapy, *J. Natl. Cancer Inst.*, 80, 330, 1988.
42. Pandey, R.K., Sumlin, A.B., Constantine, S., Potter, W.R., Bellnier, D.A., Henderson, B.W., Rodgers, M.A., Smith, K.M., and Dougherty, T.J., Alkyl ether analogs of chlorophyll-*a* derivatives. Synthesis, photophysical properties and photodynamic efficacy of chlorophyll-*a* derivatives, *Photochem. Photobiol.*, 64, 194, 1996.
43. Henderson, B.W., Bellnier, D.A., Greco, W.R., Sharma, A., Pandey, R.K., Vaughan, L.A., Weishaupt, K.R., and Dougherty, T.J., An *in-vivo* quantitative structure–activity relationship for a congeneric series of pyropheophorbide derivatives as photosensitizers for photodynamic therapy, *Cancer Res.*, 57, 4000, 1997.
44. Dougherty, T.J., Pandey, R.K., Nava, H.R., Smith, J.A., Douglass, H.O., Edge, S.B., Bellnier, D.A., and Cooper, M., Preliminary clinical data on a new photodynamic therapy photosensitizer-HPPH, *Proc. SPIE*, 3909, 25, 2000.
45. Lee, S.H., Jagerovic, N., and Smith, K.M., Use of chlorophyll derivative purpurin-18 for syntheses of sensitizers for use in photodynamic therapy, *J. Chem. Soc. Perkin Trans. 1*, 2369, 1993.
46. Kozyrev, A.N., Zheng, G., Lazarou, E., Dougherty, T.J., Smith, K.M., and Pandey, R.K., Synthesis of emeraldin and purpurin-18 analogs as target-specific photosensitizers for photodynamic therapy, *Tetrahedron Lett.*, 38, 3335, 1997.
47. Zheng, G., Potter, W.R., Camacho, S.H., Missert, J.R., Wang, G., Bellnier, D.A., Henderson, B.W., Rodgers, M.A.J., Dougherty, T.J., and Pandey, R.K., Synthesis, photophysical properties, tumor uptake and preliminary *in vivo* photosensitizing efficacy of a homologous series of 3-(1-alkoloxy)ethyl-3-divinyl-purpurin-18-*N*-alkylimides with variable lipophilicity, *J. Med. Chem.*, 44, 1540, 2001.
48. Zheng, G., Potter, W.R., Sumlin, A.B., Dougherty, T.J., and Pandey, R.K., Photosensitizers related to purpurin-18-*N*-alkylimides: a comparative *in vivo* tumoricidal ability of ester versus amide functionalities, *Bioorg. Med. Chem. Lett.*, 10, 123, 2000.
49. Rungta, A., Zheng, G., Missert, J.R., Potter, W.R., Dougherty, T.J., and Pandey, R.K., Purpurinimides as photosensitizers: effects of the position and presence of the substituents in the *in vivo* photodynamic efficacy, *Bioorg. Med. Chem. Lett.*, 10, 1463, 2000.
50. Pandey, R.K., Jagerovic, N., Ryan, J.M., Dougherty, T.J., and Smith, K.M., Syntheses and preliminary *in vivo* photodynamic efficacy of benzoporphyrin derivatives from phylloerythrin and rhodoporphyrin XV methyl ester and aspartyl amides, *Tetrahedron*, 52, 5349, 1996.
51. Ma, L. and Dolphin, D., Chemical modification of chlorophyll-*a*: synthesis of new regiochemically pure benzoporphyrin and dibenzoporphyrin derivatives, *Can. J. Chem.*, 75, 262, 1997.
52. Mettath, S., Dougherty, T.J., and Pandey, R.K., Cycloaddition reaction of 2-vinylemeraldines: formation of unexpected porphyrins with seven member exocyclic ring systems, *Tetrahedron Lett.*, 40, 6171, 1999.
53. Mettath, S., Synthesis of new photosensitizers and their biological significance, Ph.D. thesis, Roswell Park Graduate Division, State University of New York at Buffalo, 2000.

54. Chang, C.K. and Sotiriou, C.J., Migratory aptitudes in pinacol-pinacolone rearrangement of *vic*-dihydroxychlorins, *Heterocyclic Chem.*, 22, 1739, 1985 and references therein.
55. Pandey, R.K., Issac, M., MacDonald, I., Medforth, C.J., Senge, M.O., Dougherty, T.J., and Smith, K.M., Pinacol-Pinacolone rearrangements in *vic*-dihydroxychlorins and bacteriochlorins: effect of substituents at the peripheral positions, *J. Org. Chem.*, 62, 1463, 1997.
56. Chang, C.K., Sotiriou, C., and Weishih, W., Differentiation of bacteriochlorin and isobacteriochlorin formation by metallation: high yield synthesis of porphyrindiones via $OsO_4$ oxidation, *Chem. Commun.*, 1213, 1986.
57. Kessel, D., Smith, K.M., Pandey, R.K., Shiau, F.Y., and Henderson, B.W., Photosensitization with bacteriochlorins, *Photochem. Photobiol,*, 58, 200, 1993.
58. Pandey, R.K., Tsuchida, T., Constantine, S., Zheng, G., Medforth, C., Kozyrev, A., Mohammad, A., Rodgers, M.A.J., Smith, K.M., and Dougherty, T.J., Synthesis, photophysical properties and *in vivo* photosensitizing activity of some novel bacteriochlorins, *J. Med. Chem.*, 40, 3770, 1997.
59. Zheng, G., Kozyrev, A.N., Dougherty, T.J., Smith, K.M., and Pandey, R.K., Synthesis of novel bacteriopurpurinimides by Diels-Alder cycloaddition, *Chem. Lett.*, 1119, 1996.
60. Beems, E.M., Dubbelman, T.M.A.R., Lugtenberg, J., Best, J.A.V., Smeets, M.F.M.A., and Boefheim, J.P.J., Photosensitizing properties of bacteriochlorophyll-*a* and bacteriochlorin-*a*, *Photochem. Photobiol.*, 46, 639, 1987.
61. Henderson, B.W., Sumlin, A.B., Owczarczak, B.L., and Dougherty, T.J., Bacteriochlorophyll-*a* as photosensitizer for photodynamic treatment of transplantable murine tumors, *J. Photochem. Photobiol.*, 10, 303, 1991.
62. Moan, J., Peng, Q., Evenson, J.F., Berg, K., Western, A., and Rimington, C., Photosensitizing efficacy, tumor and cellular uptake of different photosensitizing drugs for photodynamic therapy, *Photochem. Photobiol.*, 46, 713, 1987.
63. Pandey, R.K. and Herman, C., Shedding some light on tumors, *Chemistry Industry* (*London*), 739, 1998.
64. Kozyrev, A.N., Zheng, G., Dougherty, T.J., Smith, K.M., and Pandey, R.K., Syntheses of stable bacteriochlorophyll-*a* derivatives as potential photosensitizers for PDT, *Tetrahedron Lett.*, 37, 6431, 1996.
65. Chen, Y., Graham, A., Potter, W.R., Morgan, J., Vaughan, L., Bellnier, D.A., Henderson, B.W., Oseroff, A., Dougherty, T.J., and Pandey, R.K., Bacteriopurpurinimides: highly stable and potent photosensitizer for photodynamic therapy, *J. Med. Chem.*, 45, 255, 2002.
66. Mody, T.R. and Sessler, J.L., Texaphyrins: a new approach to drug development, *J. Porphyrins Phthalocyanines*, 5, 892, 1996.
67. Young, S.W., Woodburn, K.W., Wright, M., Modt, T.D., Fan, Q., Sessler, J.L., Dow, W.C., and Miller, R.A., Lutetium texaphyrin (PC-0123): a near-infrared water-soluble photosensitizer, *Photochem. Photobiol.*, 63, 692, 1996.
68. Young, S.W., Quing, F., Harriman, A., Sessler, J.L., Dow, W.C., Mody, T.D., Hemmi, G.W., Hao, Y., and Miller, A., Gadolinium(III) texaphyrin: a tumor selective radiation sensitizer that is detectable by MRI, *Proc. Natl. Acad. Sci. U.S.A.*, 93, 6610, 1996.
69. Allen, C.M., Sharman, W.M., and van Lier, J.E., Current status of phthalocyanines in the photodynamic therapy of cancer, *J. Porphyrins Phthalocyanines*, 5, 161, 2001.
70. Olenick, N.L., Antunez, A.R., Clay, M.E., Rihter, B.D., and Kenney, M.E., New phthalocyanine photosensitizers for photodynamic therapy, *Photochem. Photobiol.*, 57, 242, 1993.
71. Sherman, W.M., Allen, C.M., and van Lier, J.E., Photodynamic therapeutics: basic principles and clinical applications, in *Current Trends*, Vol. 4, Elsevier Scientific, London, 1999, p. 507.
72. Hu, M., Brasseur, N., Yildiz, S.Z., van Lier, J.E., and Leznoff, C.C., Hydroxyphthalocyanines as potential photodynamic agents for cancer therapy, *J. Med. Chem.*, 41, 1789, 1998.
73. Aoudia, M., Cheng, G.Z., Kennedy, V.O., Kenney, M.E., and Rodgers, M.A.J., Synthesis of a series of octabutoxy- and octabutoxybenzonaphthalocyanines and photophysical properties of two members of the series, *J. Am. Chem. Soc.*, 119, 6029, 1997.

74. He, J., Larkin, J.E., Li, Y.S., Rihter, B.D., Zaidi, S.I.A., Rodgers, M.A.J., Mukhtar, H., Kenney, M.E., and Oleinick, N.L., The synthesis, photophysical and photobiological properties and *in vitro* structure–activity relationships of a set of silicon-phthalocyanine PDT photosensitizers, *Photochem. Photobiol.*, 65, 581, 1997.
75. Schmidt-Erfurth, U., Diddens, H., Birngruber, R., and Hasan, T., Photodynamic targeting of human retinoblastoma cells using covalent low density lipoprotein conjugates, *Br. J. Cancer*, 75, 54, 1997.
76. Finger, V.H., Guo, H.H., Lu, Z.H., and Peiper, S.C., Expression of chemokine receptors by endothelial cells: detection by intravital microscopy using chemokine-located fluorescent microspheres, *Meth. Enzymol.*, 28, 148, 1997.
77. Pandey, R.K., Photosensitizers related to chlorins and bacteriochlorins: effect of lipophilicity in PDT efficacy, *First Int. Conf. Porphyrins Phthalocyanines*, Dijon, France, Abstract SYM 149, June 25–30, 2001.
78. Hombrecher, H.K., Schell, C., and Thiem, J., Synthesis and investigation of galactopyranosyl-cholesteryloxy substituted porphyrins, *Bioorg. Med. Chem. Lett.*, 6, 1199, 1999.
79. Donald, P.J., Cardiff, R.D., He, D., and Kendell, K., Monoclonal antibody-porphyrin conjugate for head and neck cancer, the possible magic bullet, *Head Neck Surg.*, 105, 781, 1991.
80. Hemming, A.W., Davis, N.L., Dubois, B., et al., Photodynamic therapy of squamous cell carcinoma. An evaluation of a new photosensitization agent, benzoporphyrin derivatives and new photoimmunoconjugate, *Surg. Oncol.*, 2, 187, 1993.
81. Vrouenraets, M.B., Visser, G.W.M., Stewart, F.A., et al., Development of *meta*-tetrahydroxyphenyl-chlorin-monoclonal antibody conjugate for photoimmunotherapy, *Cancer Res.*, 59, 1505, 1999.
82. Karagianis, G., Reiss, J.A., Marchesini, R., et al., Biophysical and biological evaluation of porphyrin-bisacridine conjugates, *Anti-Cancer Drug Design*, 11, 205, 1996.
83. Bachor, B.S., Shea, C.R., Gillies, R., and Hasan, T., Photosensitized destruction of human bladder carcinoma cells treated with chlorine e6-conjugated microspheres, *Proc. Natl. Acd. Sci. U.S.A.*, 88, 1580, 1991.
84. Ali, H. and van Lier, J.E., Metal complexes as photo- and radiosensitizers, *Chem. Rev.*, 99, 2379, 1999.
85. Brasseur, N., Langlois, R., La Madeleine, C., Ouellet, R., and van Lier, J.E., Receptor-mediated targeting of phthalocyanones to macrophages via covalent coupling to native or maleylated bovine serum albumin, *Photochem. Photobiol.*, 69, 345, 1999.
86. Allen, C.M., Sharman, W.M., La Madeleine, C., Weber, J.M., Langlois, R., Guellet, R., and van Lier, J.E., Photodynamic therapy: tumor targeting with adenoviral proteins, *Photochem. Photobiol.*, 70, 512, 1999.
87. Lutsenko, S.V., Feldman, N.B., Finakova, G.V., Posypanova, G.A., Severin, S.E., Skryabin, K.G., Kirpichnikov, M.P., Lukyanets, E.A., and Vorozhtsov, G.N., Targeting phthalocyanines to tumor cells using epidermal growth factor conjugates, *Tumor Biol.*, 20, 218, 1999.
88. Sears, P. and Wong, C.H., Carbohydrate mimetics: a new strategy for tackling the problem of carbohydrate-mediated biological recognition, *Angew. Chem. Intl. Ed. Eng.*, 38, 2301, 1999.
89. Lee, Y.C. and Lee, R.T., Carbohydrate-protein interactions: basis of glycobiology, *Acc. Chem. Res.*, 28, 321, 1995.
90. Schell, C. and Hombrecher, H.K., Synthesis and investigation of glycosylated mono- and diarylporphyrins for photodynamic therapy, *Bioorg. Med. Chem. Lett.*, 6, 1199, 1996.
91. Mikata, Y., Onchi, Y., Tabata, K., Ogure, S., Okura, I., Ono, H., and Yano, S., Sugar-dependent photocytotoxic property of tetra- and octa-glycoconjugated tetraphenyl porphyrins, *Tetrahedron Lett.*, 19, 4505, 1998.
92. Cornia, M., Valenti, C., Capacchi, S., and Cozzini, P., Synthesis, characterization and conformational studies of lipophilic, amphiphilic and water-soluble c-glyco-conjugated porphyrins, *Tetrahedron*, 54, 8091, 1998.

93. Millard, P., Guerquin-Kern, J.L., Huel, C., and Momenteau, M., Glycoconjugated porphyrins: Synthesis of sterically constructed polyglycosylated compounds derived from tetraphenylporphyrins, *J. Org. Chem.*, 58, 2774, 1993.
94. Ono, N., Bougauchi, M., and Marutama, K., Water-soluble porphyrins with four sugar molecules, *Tetrahedron Lett.*, 33, 1629, 1992.
95. Sol, V., Blais, J.C., Bolbach, G., Carre, V., Granet, R., Guillonton, M., Spiro, M., and Krausz, P., Toward glycosylated peptide porphyrins: a new strategy for PDT, *Tetrahedron Lett.*, 38, 6391, 1997.
96. Sol, V., Blais, J.C., Bolbach, G., Carre, V., Granet, R., Guillonton, M., Spiro, M., and Krausz, P., Synthesis, spectroscopy, and phototoxicity of glycosylated amino acid porphyrin derivatives as promising molecules for cancer therapy, *J. Org. Chem.*, 64, 4431, 1999.
97. Millard, P., Hery, C., and Momenteau, M., Synthesis, characterization and phototoxicity of a glycoconjugated *meso*-monoarylbenzochlorin, *Tetrahedron Lett.*, 38, 3731, 1997.
98. Adams, K.R., Berenbaum, M.C., Bonnett, R., Nizhnik, A.N., Salgado, A., and Valles, M.A., Second generation tumour photosensitizers: the syntheses and biological activity of octaethyl chlorines and bacteriochlorins with graded amphiphilic character, *J. Chem Soc. Perkin Trans. 1*, 1465, 1992.
99. Montforts, F.P., Gerlach, B., Haake, G., Hoper, F., Kusch, D., Meier, A., Scheurich, G., Brauer, H.D., Schiwon, K., and Schermann, G., Selective synthesis and photophysical properties of tailor-made chlorins for photodynamic therapy, *Proc. SPIE*, 29, 2325, 1994.
100. Fujimoto, K., Miyata, T., and Aoyama, Y., Saccharide-directed cell recognition using monocyclic saccharide clusters: masking of hydrophobicity to enhance specificity, *J. Am. Chem. Soc.*, 122, 3558, 2000.
101. Chiariotti, L., Berlingieri, M.T., De Rosa, P., Battaglia, C., Berger, N., Bruni, C.B., and Fusco, A., Increased expression of the negative growth factor, galactoside-binding gene in transformed thyroid cells and in human thyroid carcinoma, *Oncogene*, 7, 2507, 1992.
102. Chiariotti, L., Salvatore, P., Benvenuto, G., and Bruni, C.B., Control of galectin gene expression, *Biochimie*, 81, 381, 1999.
103. Zheng, G., Graham, A., Shibata, M., Missert, J.R., Oseroff, A.R., Dougherty, T.J., and Pandey, R.K., Synthesis of β-galactose conjugated chlorines by enyne metathesis as galectin-specific photosensitizers for photodynamic therapy, *J. Org. Chem.*, 66, 8709, 2001.

# 38

# Photodynamic Therapy (PDT) and Clinical Applications




Thomas J. Dougherty
*Roswell Park Cancer Institute*
*Buffalo, New York*

Julia G. Levy
*QLT Inc.*
*Vancouver, Canada*


## 38.1 Introduction

Photodynamic therapy (PDT) refers to photodynamic action *in vivo* to destroy or modify tissue. Originally developed for treatment of various solid cancers, its applications have been expanded to include treatment of precancerous conditions, e.g., actinic keratoses, high-grade dysplasia in Barrett's esophagus, and noncancerous conditions, e.g., various eye diseases and age-related macular degeneration (AMD). PDT currently is approved for commercialization worldwide both for various cancers (lung, esophagus) and for AMD.

PDT requires a drug (photosensitizer), light corresponding to an absorption band of the photosensitizer, and endogenous oxygen. The absence of any of these agents precludes the photodynamic effect.


0-8493-1116-0/03/$0.00+$1.50

The putative cytotoxic agent is singlet oxygen,[1] an electronically excited state of ground-state triplet oxygen formed according to the Type II photochemical process, as follows:

| | |
|---|---|
| $P + h\nu \rightarrow P^*(S)$ | Absorption |
| $P^*(S) \rightarrow P^*(T)$ | Intersystem crossing |
| $P^*(T) + O_2 \rightarrow {}^1O^*_2 + P$ | Energy transfer |

where P = photosensitizer, $P^*(S)$ = excited singlet state, $P^*(T)$ = excited triplet state, and ${}^1O^*_2$ = singlet excited state of oxygen.

## 38.2 The Photosensitizer

Only photosensitizers that undergo efficient intersystem crossing to the excited triplet state and whose triplet state is relatively long lived (to allow time for collision with oxygen) and have few other competing pathways will produce high yields of singlet oxygen. Most photosensitizers in clinical use have triplet-quantum yields in the range of 40 to 60%, with the singlet oxygen yield being slightly lower. Competing processes include loss of energy by deactivation to ground state by fluorescence or internal conversion (loss of energy to the environment). However, while a high yield of singlet oxygen is desirable, it is by no means sufficient for a photosensitizer to be clinically useful. Pharmacokinetics, pharmacodynamics, stability *in vivo*, and acceptable toxicity play critical roles as well. For example, it is desirable to have relatively selective uptake in the tumor or other tissue being treated relative to the normal tissue that necessarily will be exposed to the exciting light as well. Pharmacodynamic issues, such as the subcellular localization of the photosensitizer, may be important as certain organelles appear to be more sensitive to PDT damage than others (e.g., the mitochondria). Toxicity can become an issue if high doses of photosensitizer are necessary to obtain a complete response to treatment. For example, although PDT is rarely carried out on the liver, dark toxicity to the liver is often the main limiting toxicity seen in animal toxicity studies. Along the same lines, lack of *in vivo* stability would be a further issue relating to toxicity, as the toxicity profile of the breakdown products may need to be evaluated as well. In spite of all these impediments, there is a large number of photosensitizers potentially useful in PDT, several of which are currently in various stages of clinical trials for U.S. Food and Drug Administration (FDA) approval. Additional information on photosensitizers and mechanisms involved in PDT are provided in Chapter 36 of this handbook.

### 38.2.1 Light

A limiting factor in PDT is the penetration of the activating light into tissue. Visible light in the red region of the spectrum (the most penetrating) has a general penetration depth ($l/\varepsilon$) in the range of 2 to 6 mm, depending on the wavelength and the tissue. However, the depth of the biological effect of PDT (necrosis, apoptosis, vascular collapse) is generally found to occur at approximately twice this depth, i.e., 4 to 12 mm, meaning that approximately 10% of the incident light, i.e., the intensity drop-off at two penetration depths, is sufficient to elicit the photodynamic effect in tissue. As would be expected, the liver is one of the most opaque of tissues (due to pigmentation) and the muscle one of the most transparent. Most solid cancers originate in epithelial tissue (carcinomas) and have optical penetration depths of 3 to 5 mm (at ~600 nm) to approximately 6 to 10 mm (at 800 nm, where an optical window exists between absorption by blood and that of water).

Light sources generally are gas-vapor lasers, e.g., argon-pumped, wavelength-tunable dye lasers or solid-state lasers such as frequency-doubled neodymium:yttrium-aluminum-garnet (Nd:YAG) pumped-dye lasers or diode lasers, which are currently being used more frequently. However, appropriate lamps can be used as well, as was the case in the early days of PDT and is currently the case for treatment of actinic keratoses. Light from the laser is generally coupled into quartz fibers of various configurations. For lesions involving the skin and in certain other instances, a lens fiber with a microlens on the tip is used to produce a homogeneous spot; for treatment of esophageal and lung tumors as well as for interstitial placement a fiber with a diffuser on the end is used, allowing for scattering of the light laterally from the fiber over a distance of 1 to 5 cm generally; for bladder treatment a bulb-type tip is used to produce isotropic distribution over the entire bladder wall. Details of the light dosimetry for particular types of treatments will be described below.

## 38.3 Methodology of Clinical Photodynamic Therapy

The following discussion refers to the photosensitizer Photofrin® as used in PDT where the most information is available, but with some differences the basic discussion applies to the use of most photosensitizers.

### 38.3.1 Photosensitizer Injection

Patients receive an intravenous bolus injection of Photofrin in doses ranging from 1 to 2 mg/kg body weight. With Photofrin, as well as certain other photosensitizers that are taken up by the skin, this immediately renders the patient susceptible to severe burns from bright light exposure (sunlight being the most likely source), a condition that generally lasts 4 to 6 weeks but may be longer. Therefore, the patient, family, and others are educated prior to injection to take appropriate precautions, e.g., by wearing clothing to cover the body completely, a broad-brimmed hat, etc. However, patients should not remain in darkened rooms during the day as photobleaching by low-level light enhances clearance of the drug from the skin. Patients are asked to expose a small area on the back of their hand to bright sunlight for 10 min beginning about 4 weeks after injection before slowly beginning to go into the sun. Should a reaction occur, patients are advised to wait an additional week and retest their sun sensitivity. While cutaneous photosensitivity is a bothersome side effect of certain photosensitizers, it is our experience that patients rarely refuse PDT treatment for this reason.

### 38.3.2 Injection-Light Treatment Time Interval

The amount of time between injection and light exposure varies considerably between photosensitizers, ranging from 1 to 2 d for Photofrin to 15 min for AMD treatment. In general, for solid tumor treatment it is best (but not always necessary) to allow for photosensitizer clearance from normal tissue that may also be exposed to the activating light during treatment to minimize damage as much as possible. However, some photosensitizers demonstrate little retention in tumors and therefore are generally activated shortly after injection. While this risks the normal tissue, in some cases the dosimetry can be adjusted to minimize the damage to normal tissue. Selectivity can also be increased by using a minimum amount of photosensitizer and increasing the light dose to compensate because photobleaching of the lower amount of photosensitizer generally found in the normal tissue can lead to protection. Interstitial placement of the fibers into the tumor also tends to protect surrounding normal tissue but is not always feasible or desirable, e.g., attempts to insert fibers into obstructing esophageal tumors can result in perforations; in this case, they are generally placed within the lumen adjacent to the tumor, while in the case of endobronchial tumors insertion into the tumor can often be done without risking perforation.

### 38.3.3 Light Dosimetry

Light dosimetry is probably the most difficult variable to define in PDT because the optimum light dose in one organ may be ineffective or toxic in another. Time-consuming light-dose ranging protocols must be carried out to define this parameter for each indication. (Recently it has been recognized that the light-dose rate may also be important; see Chapter 36 of this handbook.) Table 38.1 summarizes the general range of PDT variables used for various photosensitizers, showing the large range of time interval, drug doses, and light fluence. However, even for a single photosensitizer like Photofrin, light doses range from 15 $J/cm^2$ in bladder to over 200 $J/cm^2$ for skin lesions.

**TABLE 38.1** Photosensitizers in Clinical Trials

| Photosensitizer[a] | Photosensitizer Dose (mg/kg) | Interval (h) | Energy Dose[b] J/cm$^2$ or J/cm | Wavelength (nm) | Indications |
|---|---|---|---|---|---|
| Photofrin | 1–2 | 24–48 | 15–215 (1)<br>100–300 (2) | 630 | Esophageal, lung, skin, bladder, head and neck, brain, and certain other sites |
| ALA (Levulan) | Topical ALA (20% in vehicle) | 14–18 | 10 (1) | Blue (peak at 417 nm) | Actinic keratosis |
| mTHPc (Foscan) | 0.15 | 72–96 | 10–20 (1) | 650 | Oral (approved in Europe) |
| $SnET_2$ (Purlytin) | 1.2 | 4 | 200 (1) | 660 | Breast metastases |
| Lutetium Texaphyrin | 1.5–3 | 3 | 100–150 | 732 | Prostate, photoangioplasty, AMD |
| BPD-MA (Visudyne) | 6 mg/m$^2$ (~0.15 mg/kg) | 0.25 | 50 (1) | 699 | AMD (also in Phase III trial for skin cancer) |
| HPPH (Photochlor) | 3–6 mg/m$^2$ (~0.08–0.15 mg/kg) | 24–48 | 50–150 (1)<br>150 (2) | 665 | Skin, lung, esophagus |
| PC4 | — | — | — | — | Cutaneous cancers (no further information available) |

*Note:* Photofrin = polyhematoporphyrin ethers; ALA = aminolevulinic acid; mTHPc = meta-tetetrahydroxyphenylchlorin; SnET2 = tin etiopurpurin; Lutetium Texaphyrin = a pentadentate aromatic metalloporphyrin; BPD-MA = benzoporphyrin derivative mono acid; HPPH = 2-[1-hexyloxyethyl]-devinylpyropheophorbide-a; PC4 = a silicon phthalocyanine; AMD = age-related macular degeneration.

[a] Chemical structures can be found in Chapter 37.

[b] J = Joules; J/cm$^2$ refers to surface dose; J/cm refers to Joules delivered per cm from a diffusing fiber.

### 38.3.4 Photodynamic Therapy with Photofrin

Photofrin (polyhematoporphyrin ethers), the first and, thus far, only photosensitizer approved by health agencies worldwide for treatment of cancer, remains the most widely used in routine clinical practice for treatment of esophageal and lung cancers (approved), as well as in a host of off-label indications (e.g., skin cancer, bladder cancer, brain tumors, head and neck cancer, etc.).

The method of treatment and outcomes for the various indications beginning with those approved by various health agencies worldwide are described in the following sections.

## 38.4 Health Agency–Approved Indications for Photofrin-Photodynamic Therapy

### 38.4.1 Obstructive Esophageal Cancer

This indication is for palliation of partially or totally obstructing tumors in the esophagus. Photofrin, obtained as lyophilized powder, is dissolved in 5% dextrose for injection shortly before a single bolus injection of 2 mg/kg body weight. Because this photosensitizer is accumulated and retained in the skin (as well as in most other tissues and tumor), patients are immediately rendered photosensitive and must take precautions against exposure to bright lights (e.g., in a dental office) and especially sunlight. While the supplier of Photofrin suggests precautions continue for 90 d, it is our experience that clearance occurs within 30 to 60 d, although there have been some reported cases of up to 90 d. We ask our patients to expose a small patch of skin on the hand to bright sunlight for 10 min at 30 d post injection and, if redness and swelling occur, to continue taking precautions for another week and then until reaction no longer occurs, at which time they can slowly increase exposure. Patients should not remain in the dark, as low-level light aids clearance by photobleaching the drug. In addition, since Photofrin is activated by visible light, sunscreens and window glass are not protective. As a further precaution, we supply sunglasses

**TABLE 38.2** Objective Tumor Response — PDT vs. Nd:YAG Laser in Partially Obstructing Esophageal Cancer

| | Week 1 | | Month 1 | |
|---|---|---|---|---|
| | PDT | Nd:YAG | PDT | Nd:YAG |
| Complete + partial response | 45 | 40 | 32[a] | 20 |
| Stable disease | 24 | 22 | 15 | 14 |
| Progressive disease | 3 | 1 | 11 | 14 |

[a] Statistically different, $p < 0.05$; 118 patients in each arm.

to wear when outside. This cutaneous photosensitivity is an annoyance, but we have rarely had a patient refuse treatment because of it.

Approximately 48 h post injection the photosensitizer localized to the tumor is activated by light at 630 nm (generally derived from a laser) that is directed via a single-quartz fiber optic and delivered to the tumor through the biopsy channel of an endoscope. The distal end of the fiber is fitted with a diffusing surface, generally 2.5 to 5.0 cm in length, which allows the light to scatter laterally to the tumor on the wall of the esophagus. In some instances, the fiber may be inserted directly into the tumor, although care must be taken to avoid perforation of the esophageal wall. This can be problematical because obstructing tumors may cause distortion of the esophagus, making insertion of the fiber somewhat blind. In most instances, the fiber is placed in the remaining lumen adjacent to the tumor. Some physicians place patients under general anesthesia, but this is generally unnecessary, and conscious sedation is used most often. The light dose is 300 J/cm of diffuser length with a power density of 400 mW/cm diffuser length, requiring 12.5 min to deliver the prescribed dose. Patients will often experience mild to severe chest pain (due to treatment-induced edema) controlled with analgesics. Other adverse reactions include fever, abdominal pain, constipation, and substernal pain. Patients are examined endoscopically 2 d following treatment to judge initial response, to obtain biopsies, and to remove necrotic debris (not generally necessary). Also, if necessary, a second light treatment can be administered at this time (an advantage of long tumor retention of Photofrin). The treated tumor becomes blanched, and noninvolved areas in the light field generally are found to be edematous and erythemic. Table 38.2 summarizes the outcome of the randomized clinical trial of PDT vs. Nd:YAG laser thermal ablation carried out for the U.S. FDA and other health agency approvals.[2]

Only at 1 month was the advantage of PDT evident; this outcome is similar to that seen for PDT treatment of obstructive endobronchial tumors vs. Nd:YAG ablation (see the following section). This may reflect the ability of the light to reach more of the tumor than is possible with the Nd:YAG laser, which must be aimed at the tumor; care must be taken to avoid perforation. (In the clinical trial noted above there were seven perforations in the Nd:YAG group vs. one in the PDT group.)

## 38.4.2 Obstructing Endobronchial Tumors (Nonsmall Cell Lung Cancer)

Also approved by the U.S. FDA and other health agencies is the palliative treatment of obstructive endobronchial tumors. The treatment approach is similar to that for the obstructing esophageal tumors (2.0 mg/kg body weight Photofrin, 630 nm light treatment approximately 48 h later) with certain notable differences. The light dose is 200 J/cm diffuser (8.3 min), generally delivered interstitially when feasible with 1.0- or 2.5-cm diffuser fibers. Also, 2 d following treatment patients are re-endoscoped, and all necrotic tumor debris and exudate must be removed because they can further obstruct the airway. This step is especially critical when the treatment is near the carina and debris and exudate can extend into both mainstem bronchi, presenting a potentially life-threatening outcome.[3]

As in the esophageal study, patients in this multicenter phase III trial were randomized to PDT or thermal ablation with the Nd:YAG laser. Also, as was seen in the esophageal trial, there was no statistical difference in outcome until after 1 week (see Table 38.3).[4]

**TABLE 38.3** Objective Tumor Response of PDT vs. Nd:YAG Laser for Obstructing Endobronchial Cancer (211 patients)

| | Week 1 | | Month 1 | |
|---|---|---|---|---|
| | PDT | Nd:YAG | PDT | Nd:YAG |
| Complete and partial response | 59 | 58 | 60 | 41 |

**TABLE 38.4** Outcome of Patients with Microinvasive Lung Cancer

| | 11 Patients | 62 Patients |
|---|---|---|
| CR (3 months) (biopsy proven) | 3/11 | 31/62 |
| Recurrence | 1 | 11 |
| Median time to recurrence | | >2.7 years |
| Median survival | | 2.9 years |
| Median disease-specific survival | | 4.1 years |

For example, at 1 month 40% of PDT-treated patients demonstrated improvement in symptoms (dyspnea, cough, hemoptysis) vs. 27% of Nd:YAG patients. Adverse reactions included photosensitivity reaction (20% in PDT arm only), hemoptysis, cough, dyspnea, chest pain, and fever, which occurred with similar frequency in both arms.

### 38.4.3 Early Stage Endobronchial Tumors

Treatment of microinvasive, nonsmall cell endobronchial tumors using Photofrin-PDT is the only indication thus far approved by the U.S. FDA and other health agencies where the intent is not palliation but cure. Also, unlike the palliative approval previously described, this study was approved based on several studies not specifically designed and carried out by the sponsoring company but, rather, independently in three noncomparative studies. Patients in this study included those with CIS (carcinoma *in situ*) or microinvasive tumors. Also included were patients not considered candidates for surgery (or radiation therapy in some cases).

Patients received 2 mg/kg Photofrin and 2 d later were treated endoscopically using a diffuser fiber (usually 1 to 2.5 cm) delivering 200 J/cm of diffuser length at 630 nm. Because these lesions are very thin, the fiber was held in the lumen adjacent to the lesion. Patients were rescoped 2 d following treatment, and the treated area was debrided.

Results with 102 patients indicated biopsy-proven complete response in 79% of patients (initial followup), with more than half continuing to be complete responders often more than 2 years post treatment. The median survival was 3.5 years and disease-specific survival 5.7 years. In a smaller segment of this trial population with microinvasive lung cancer (the actual approved indication), 62 inoperable patients were treated, 11 of whom were considered not to be candidates for surgery or radiation therapy. Table 38.4 indicates the outcome in this group.[5,6]

### 38.4.4 Early Stage Esophageal Cancer

Although not yet approved by the U.S. FDA, a Phase III randomized, controlled, multicenter trial of Photofrin-PDT in patients with high-grade dysplasia (HGD) in Barrett's esophagus was recently completed for filing with the U.S. FDA. This study randomized patients to receive an acid suppressor, Omeprazole (Prilosec), or PDT plus Omeprazole. Endoscopic PDT was applied using a specially designed balloon light applicator with windows of 3, 5, or 7 cm in length produced from diffuser fibers of 5, 7, or 9 cm in length. The last centimeter on each end is blocked off because the power at the ends of the diffuser is less than that emitted in the center of these fibers. The balloon was inserted deflated and then inflated in place to an appropriate pressure to allow "unfolding" of the esophageal wall without shutting down the blood flow. The Photofrin dose was 2.0 mg/kg injected approximately 2 d prior to light delivery of 130 J/cm of diffuser length. A preliminary report[7] at 6-month follow-up indicated that HGD was eliminated in 72% of patients in the PDT group but only in 31% of patients in the control group (Omeprazole alone). Moreover, at 2 years 70% of responders in the PDT group continued to be complete responders, while none of the patients in the Omeprazole-alone group were without HGD. Significantly, only 10% of patients in the PDT group progressed from HGD to esophageal cancer vs. 19% of patients receiving only Omeprazole.

Overholt et al. have reported on 100 patients treated for this indication with Photofrin and the balloon light emitter. Of 73 patients with HGD, 56 were complete responders, and 32 had complete elimination of the Barrett's.[8] The current practice is to subject patients with HGD to an esophagectomy, a surgical procedure associated with high mortality (5 to 10%) and morbidity.

Barrett's esophagus is also eliminated by PDT treatment by 70 to 100% in most patients, being replaced with normal squamous mucosa. While not a specific endpoint in the trial, PDT appears to be one of the few methods currently available to remove this acid reflux–induced inflammatory process, a risk factor for developing esophageal cancer (from 0.5 to 1.9% of patients with Barrett's esophagus will progress to HGD, and up to 50% of patients with HGD will go on to esophageal cancer over a 3-year period[9]).

## 38.5 Other Indications for Photofrin-Photodynamic Therapy

In addition to the approved or soon-to-be-approved indications of PDT using Photofrin, numerous other types of cancers have been treated including those of the head and neck, especially early stage,[10] skin,[11,12] brain,[13,14] bladder,[15] intrathoracic cavity (especially as an adjunct to surgery for mesothelioma),[16,17] and intraperitoneal tumors, such as ovarian studding[18] and cholangiocarcinoma.[19] Several of these are discussed in the following sections.

### 38.5.1 Cholangiocarcinoma

A potentially very useful application of PDT is in the treatment of cholangiocarcinoma cancer. These tumors are generally diagnosed late when surgery is rarely curative. In nonresectable cases, drainage or biliary bypass is used for palliation. Radiation therapy and chemotherapy appear to be of little benefit. In a series of 21 patients reported by Ortner et al.,[19] nonresectable patients were treated with Photofrin (2 mg/kg) 2 d prior to endoscopic fiber placement in the lumen (630 nm, diffuser fibers, 2.5 or 4.0 cm of diffuser length delivering 180 J/cm at a constant total power output of 800 mW). There was a dramatic drop in bilirubin in the PDT group (69 μmol/l) vs. stenting alone (20 μmol/L), and the Karnofsky score improved from 49 to 73 in the PDT arm. Eight patients were alive at 82 to 739 d post treatment, and median survival was about 1 year. A small randomized trial has been reported[20] with 35 patients, 20 to PDT + endoprostheses insertion (EP), and 15 to EP alone. An additional 34 patients who refused randomization received compassionate PDT + EP and were analyzed separately. The PDT treatment was as above. The median survival for the PDT + ES was 493 d vs. EP-alone group of 98 d. The compassionate PDT + EP group had median survival of 406 d. All patients in the control arm (EP alone) are dead; six are alive in the PDT + EP compassionate group and two in the PDT + EP group. Significant improvement in bilirubin levels was seen in both groups receiving PDT + ED, but none was seen in the EP-alone group. Karnofsky scale improved only in the PDT + EP compassionate group. It was concluded that addition of PDT to EP was a valuable improvement in palliation treatment because of significantly increased survival and improvement in quality of life. As a result of this outcome the randomized trial was stopped for ethical considerations.

### 38.5.2 Head and Neck Cancers

Application of PDT to treatment of head and neck cancer was initially limited to treatment of patients with advanced disease who had failed or recurred on standard therapy. These patients rarely benefited from PDT. However, more recently, Biel[10] has shown a high success rate in PDT treatment of early stage disease of the oral cavity and larynx. Twenty-five patients who had failed radiation therapy for early laryngeal cancers (CIS or T1) received Photofrin (2 mg/kg) followed 2 days later by 630 nm light at an energy dose of 80 J/cm². All patients received a complete response and remain free of disease with a mean follow-up of 27 months. An additional 23 patients with early carcinomas (CIS T1, T2) of the oral cavity, nasal cavity, and nasopharynx were treated similarly, and all were initially complete responders, with a mean follow-up of 26 months. One patient with recurrent leukoplakia of the tongue recurred at 12 months, one with recurrent cancer of the tongue and floor of the mouth recurred at 8 months, one patient developed regional metastases at 2 months, and another patient with nasal cancer recurred at 3 months post-PDT.

## 38.6 Photodynamic Therapy as Adjuvant Treatment

### 38.6.1 Head and Neck Cancers

PDT has been used frequently as an adjuvant to surgery in an attempt to "clean up" the remaining cancer cells in the operative bed. Biel has reported on five patients with recurrent infiltrating squamous cell carcinoma all of whom had failed previous surgery, radiation therapy, and chemotherapy. Patients received 2 mg/kg Photofrin followed 2 d later by surgery and PDT at 50 J/cm². He reported no postoperative complications. With a follow-up of 18 to 30 months only one patient had recurred in an area outside the surgical and PDT field.

### 38.6.2 Brain Tumors

Muller has treated over 100 patients with surgery and PDT.[13] Most had glioblastoma or astrocytoma, newly diagnosed or recurrent. These tumors are rarely curable by surgery, radiation therapy, or chemotherapy. The adjuvant studies were undertaken to determine the safety and feasibility of adding PDT to the surgical resection. Photofrin (2 mg/kg) was injected 12 to 36 h prior to light treatment using a specially designed expandable balloon applicator that fit within the surgical cavity and was filled with a light-scattering material (1:1000 dilution of lipid). The delivered light at 630 nm was measured using a "pick-up" fiber placed adjacent to the balloon. Total energy doses ranged from 440 to 4500 J, with an energy density of 8 to 110 J/cm². The mortality rate was 3%, with a combined serious mortality and morbidity rate of 8%. Eight patients with recurrent gliomas who received more than 60 J/cm² had a median survival of 58 weeks, while 24 who received less than 60 J/cm² had a median survival of 29 weeks.

Twenty patients with all types of brain tumors who received PDT at first operation had median survival of 44 weeks, with a 1-year survival of 40% and 2-year survival of 15%. Again, a light-dose effect was found, with those patients who received less than a total light dose of 1200 J having a median survival of 39 weeks and those with light doses greater than 1200 J having a median survival of 52 weeks. Patient characteristics in the two groups were similar. Patients with high Karnofsky score (>70) and who received light doses over 1260 J had a median survival of 92 weeks, with 1- and 2-year survival of 83% and 33%, respectively. The authors conclude that PDT with Photofrin is safe and appears to increase survival, especially at the higher light doses. A multicenter, randomized (high or low light dose) trial is currently ongoing to verify this conclusion.

### 38.6.3 Mesothelioma

This asbestos-induced disease of the mesothelium can be very advanced before becoming symptomatic and, like the brain tumors discussed above, is rarely curable by surgery, radiation therapy, or chemotherapy. Moskal et al. have reported on a series of 40 patients treated by a combination of surgery and PDT.[17] Patients received 2 mg/kg Photofrin, and 2 d later as much tumor was removed as possible or debulked to 5 mm or less. PDT was then carried out by delivering 630 nm light to the entire thoracic cavity via four to six diffuser-type delivery fibers (2.5 to 5.0 cm in diffuser length) placed so as to deliver 20 to 30 $J/cm^2$ as evenly as possible. This was done in two placements of the fibers, posteriorly and anteriorly. Because of the large area to be treated (estimated from the CT scans) total light-treatment time can range up to 4 h at a power density of approximately 10 $J/cm^2$. Postoperative staging indicated 12 patients with Stage I disease, 10 being completely resected; 11 patients with Stage II disease with complete resection; 25 patients were Stage III, only 5 of whom could be completely resected (3 died); and 2 patients with Stage IV disease. Morbidity and mortality rates for patients with Stage I or II were 38% and 0%, and for Stage III or IV, 48% and 11%, respectively. Patients requiring extrapleural pneumonectomy did not benefit from the combined treatment, as 43% of these patients had a bronchopleural fistula, much higher than the 1 to 4% reported in the literature. However, Stage I and II patients not requiring pneumonectomy appear to have increased survival (41% at 2 years) compared to those receiving surgery alone.

## 38.7 Other Photosensitizers

Because of prolonged cutaneous photosensitivity induced by Photofrin (30 to 60 d) and less-than-optimum absorption in the red region of the spectrum, a host of newer photosensitizers are undergoing clinical trials. (See Table 38.1 and Chapter 41 of this handbook.)

### 38.7.1 ALA

δ-Aminolevulinic acid (ALA) is involved in the heme pathway in which the final metal-insertion step can be obviated with a large excess of ALA, thereby causing the build-up of protoporphyrin IX (PpIX), a porphyrin photosensitizer. Injectable PpIX is not a useful photosensitizer because it is not well retained by tumors or other tissues and clears the system very rapidly. However, the intracellular accumulation of PpIX through excess ALA can be effective. In fact, topically applied ALA is approved by the U.S. FDA for the treatment of actinic keratosis, a precancerous lesion that can lead to squamous cell carcinoma. The sponsor has also developed a special blue-light source that can be fitted on the patient's head to allow treatment of the entire face for patients with numerous lesions. In the Phase III trials,[21] patients were treated topically with 20% ALA in a proprietary vehicle (Levulan®) and the next day received the blue-light treatment (approximately 15-min exposure). In one study of 117 patients, 80% of lesions had a complete response that increased to 94% with a second treatment, while in the placebo group (placebo + light) only 32% of lesions responded completely after two treatments. Adverse reactions included a stinging, burning sensation, which was well tolerated.

Topical ALA has also been applied to the treatment of superficial skin cancers — with excellent results. It is less effective for nodular lesions.[12] In this treatment, the ALA (10 to 40%) was applied topically in leucerin cream for various periods (usually 4 or 24 h) and then activated at 635 nm derived from a dye laser delivering light doses up to 200 $J/cm^2$. In this treatment, the pain is sufficient to require local injection of an anesthetic prior to treatment. Most patients tolerate this well.

ALA is currently being investigated as a fluorescent localizing agent in bladder cancer (see Section 38.8) and for treatment of low-grade dysplasia in Barrett's esophagus.[22]

### 38.7.2 mTHPc (Foscan)

This chlorin-type photosensitizer has been approved in Europe for treatment of oral cancers. Fan et al. have reported preliminary clinical results.[23] Nineteen patients with various stages of disease (8 with field change disease) were injected with mTHPc (0.15 mg/kg) and 72 to 96 h later treated with varying light doses at 652 nm. Three patients with single T3 lesions received a complete response, as did three of six T4 tumors. Field change disease responded less well, with 9 of 14 T1 and T2 lesions responding completely. Adverse reactions included one patient with tongue tethering and necrosis of normal tissue in another. However, while mTHPc is approved in Europe for this indication, the U.S. FDA chose not to approve it in the United States, although as of this writing this decision is under review.

### 38.7.3 SnET2 (Purlytin, Tin Etiopurpurin)

While trials have been completed with this photosensitizer in cutaneous breast cancer metastasis, the sponsoring company chose not to file for approval for the indication. However, they have completed two placebo-controlled clinical trials with 933 patients in 59 centers for AMD.[24] These trials concluded in December 2001 (a 2-year follow-up period is required).

### 38.7.4 Lutetium Texaphrin (Lu-Tex)

This is an example of a so-called expanded metalloporphyrin system resulting in a strong shift in the red absorption band from around 620 to 732 nm, where nearly optimum tissue penetration occurs. Current clinical trials are in locally recurrent prostate cancer and noncancerous indications, AMD and photoangioplasty.[25] Details are not available. Moderate to severe pain during treatment has hampered studies of treatment of various cutaneous malignancies.

### 38.7.5 HPPH (Photochlor)

Three Phase I/II clinical trials of PDT with this new chlorin-type photosensitizer are under way at Roswell Park Cancer Institute: partially obstructing esophageal cancer, early stage lung cancer, and basal cell carcinoma.

In obstructing esophageal cancer eight patients have been treated at 6 mg/m$^2$ (0.15 to 0.19 mg/kg body weight) and 150 J/cm of 665-nm light delivered at approximately 48 h post infusion of HPPH. All patients have achieved partial or complete opening of the esophageal obstruction, and most have had improvement in dysphagia. In these trials, patients are tested daily for cutaneous photosensitivity for 3 to 4 d using a solar simulator (1-cm spot on the underside of the wrist) at doses up to 133 J/cm$^2$. Only two patients have had a mild erythemic reaction. Although the in-house solar simulator indicates very little cutaneous photosensitivity, the patients still are advised to test their reactions to sunlight by exposing a small area on the back of their hands for 10 to 30 min daily until no HPPH-induced reaction is found before ceasing precautions.

### 38.7.6 PC4

This silicon phthalocyanine, with two axial ligands on silicon, a hydroxy group, and a dimethylaminopropyl siloxy group, has just recently entered Phase I dose-escalation clinical studies with cutaneous lesions. No information is available as of this writing.

### 38.7.7 BPD-MA (Verteporfin®)

#### 38.7.7.1 AMD

The macula constitutes a small, highly specialized region of the human retina. The central part is rich in photoreceptor cells and is the area responsible for high visual acuity in humans. The end result of macular degeneration is the ultimate destruction of photoreceptor cells in the macula as well as the underlying cells of the retinal pigmented epithelium (RPE). Macular degeneration is the leading cause of blindness in the developed world and is most prevalent in people over the age of 60. There are two principal forms of AMD: dry or nonexudative AMD and wet or exudative AMD. While wet AMD constitutes only about 10% of all cases of AMD, it is responsible for most of the vision loss associated with the disease. Approximately 500,000 new cases of wet AMD are reported every year worldwide. Because this is a disease predominantly of caucasians, the main areas of disease incidence are North America and Europe. Wet AMD occurs when rapidly growing abnormal blood vessels develop under the macula in a process described as subretinal neovascularization. The newly formed vessels do not mature properly and leak into the subretinal space, effecting detachment of the RPE and the ultimate degeneration and destruction of the photoreceptor cells in the region of the lesion.

The underlying causes of wet or dry AMD are not well understood, but there is evidence for genetic tendencies since there are familial as well as racial propensities in disease occurrence. While there is strong circumstantial evidence for a genetic base for the disease, the genes relating to susceptibility are clearly complex and are not well understood at this time. Familial studies on patients with Stargardt's disease have clearly identified a specific gene mutation associated with this severe macular disorder. The gene identified is a photoreceptor-cell-specific ATP-binding cassette transporter gene (ABCA4), and patients with Stargardt's disease, as well as a number of other disorders associated with the retina, express mutations in this gene, which is polymorphic.[26] A recent study has shown that patients who express particular mutations in at least one allele of this same gene are at either a three- or fivefold greater risk of developing wet AMD.[27] These studies support the genetic basis for susceptibility to AMD but also demonstrate that the genetic basis is complex and no doubt involves additional gene loci. It has been suggested that the ABCA4 gene plays a functional role in the eye. Most ABC transporters act as "flippases," i.e., as transporters of the protonated complex of all-trans retinal and phosphatidylethanolamine (*N*-retinylidene-PE). In knockout mice lacking the ABCA4 gene, phosphatidylethanolamine accumulates with resulting deposition of a major lipofuscin fluorophore (A2-E or a pyridinium bis-retinoid, which is derived from two molecules of vitamin A aldehyde and one molecule of ethanolamine) in the retina. Accumulation of lipofuscin in the macular region of aging eyes is characteristic of AMD.[28]

The exudative or wet form of AMD is diagnosed using standard fluorescein angiography. Serial fluorescent photography of the macular area can readily identify where vessel leakage is occurring because fluorescein will pool in the area of neovasculature. Exudative AMD is subdivided into two categories based on the speed of visible leakage on angiographs. The "classic" form of AMD is considered the most aggressive form of the disease, and leakage occurs rapidly during the angiography. This form of AMD is characterized by rapid progression and concurrent vision loss. Lesions that leak slowly and show up on angiograms late are defined as "occult" wet AMD. These lesions lead to vision loss at a slower rate than the classic form.

Although the overwhelming majority of research and clinical testing with PDT has addressed the application of this technology in the treatment of solid tumors, during the past decade and a half possible applications in other disease areas have been investigated. Because many photosensitizers accumulate or are retained somewhat selectively in both rapidly dividing cells as well as neovascular endothelial cells, the PDT effect, at least on solid tumors, has been elegantly demonstrated using Photofrin to effect both the tumor cells themselves and the neovascular bed underlying rapidly growing tumors.[29,30]

Ocular conditions involving either neoplasias or hypervascularity lend themselves to PDT because of the accessibility to light of all parts of the eye. Early investigator-sponsored studies using Photofrin for the treatment of ocular tumors demonstrated the effectiveness of this therapy in destroying local tumors as well as the neovascular bed surrounding them. The Verteporfin molecule (also termed BPD-MA or Visudyne) is a hydrophobic photosensitizer with pharmacokinetic properties that make it a desirable drug to use in indications such as AMD. It is formulated as a liposome in order to render it soluble in water. However, once in association with plasma the drug demonstrates an affinity for lipoproteins much greater than its affinity for the liposomal constituents. Consequently, it is almost instantaneously released from the liposome and transferred to LDL and HDL in blood. This somewhat unusual characteristic results in very rapid uptake of the drug by neovascular endothelia and a treatment window for selective accumulation at between 15 and 30 min following intravenous administration. Therefore, unlike most photosensitizers, which rely on selective retention by target cells, Verteporfin achieves selectivity by differential uptake by target tissue.[31] Verteporfin has the additional advantage of having a very short half-life (4 to 5 h in humans). Thus, treatment at an early time interval permits use of relatively low drug doses, and with the rapid half-life patients are photosensitive for only 24 h.

Research was undertaken in 1992 to investigate the feasibility of using Verteporfin for the treatment of AMD. Initially two models in the rabbit were used — induced corneal neovasculature and implanted ocular tumors. Results from these studies were encouraging and showed in both instances that the expected thrombotic closure of neovasculature was seen in PDT-treated eyes. These studies also suggested a window of selectivity in which normal vasculature might be preserved.[32] Final proof-of-concept work involved the primate model for AMD in which thermal laser damage to the retina induces a neovascular lesion on the choroid. Although self-limiting, this model is considered the most reliable one for testing treatments for AMD. Results from these studies were encouraging and continued to demonstrate effective closure of neovascular vessels at time intervals from 10 to 60 min following intravenous administration.[33]

Phase I/II clinical work with Verteporfin was initiated in 1996 in which patients with AMD received a single treatment with Verteporfin and were followed for 12 weeks by angiographic examination and visual-acuity evaluation. A variety of regimens were tested involving two different drug doses, five different light doses, and two different times of treatment. These studies revealed that at 7 d post-PDT all lesions treated showed complete closure of neovasculature with no accompanying vision loss or closure of normal retinal blood vessels, with the exception of two patients who received the highest light dose (150 J/cm$^2$ at 690 nm and 600 mW/cm). In these patients, closure of normal retinal vessels was observed, thereby demonstrating the level at which selectivity was lost. Interestingly, while patients on average did not lose significant visual acuity at either 4 weeks or 12 weeks following treatment, a majority of them did show the presence of vascular leakage at the 12-week time point, indicating that repeat treatment would likely be necessary if a sustained benefit of PDT were to be maintained.[34]

These preliminary clinical trials, in which approximately 150 patients had participated, led to development of a protocol for two pivotal Phase III trials for treatment of patients with AMD with lesions that contained a classic component. The study design involved randomization of patients in a 2:1 ratio of Verteporfin treatment vs. placebo. A total of 609 patients were accrued in the two trials; approximately 400 received treatment, while 200 received placebo. Patients were tested for baseline visual acuity and lesion assessment by angiography. They received either Verteporfin intravenously at a dose of 6 mg/m$^2$ or placebo. Fifteen minutes after drug administration, light treatment of 50 J/cm$^2$ at 600 mW/cm was administered to the lesion using a diode laser emitting at 690 nm. Patients returned every 3 months, at which time they were tested by angiography for evidence of vessel leakage. If leakage was present, they received another treatment. Visual acuity was tested at each visit. The primary endpoint was designated as showing a significant difference between treatment vs. placebo in a loss of three lines of vision (approximately 15 letters on a standard eye chart) at 12 months. Secondary endpoints included serious vision loss (six lines), lesion size and progression, contrast sensitivity (a measure of reading ability), and actual visual acuity. Although this was a 2-year study, agreement with regulatory bodies permitted filing for approval if the 12-month primary endpoint was achieved. All primary endpoints achieved significance at the 12-month time point.[35] These patients have now been followed for nearly 4 years. After 2 years all patients, including placebo patients, have been receiving treatment if needed. Visual acuity in treated patients stabilized within the first year, and no loss of vision has been observed in years two and three.[36]

Visudyne (Verteporfin) therapy has now been approved for wet predominantly classic AMD in more than 50 countries, and over 50,000 patients have been treated.

Further clinical trials have been carried out to test the ability of PDT to arrest progression of the occult form of AMD. A placebo-controlled study of similar design to that used for the classic form containing lesions was carried out with similar endpoints. In this study, the significance of the primary endpoint was not reached at 12 months, but patients on treatment did stabilize during the second year, and all endpoints were significant at 2 years. Because occult disease is slower progressing than the classic form, it is not surprising that the difference between placebo and treatment arms took longer to show a significant difference.[37]

In addition to AMD, other ocular conditions have been treated with Visudyne therapy. These diseases include conditions that cause choroidal neovasculature similar to that seen in AMD. Among these are pathologic myopia, a complication of extreme nearsightedness, and acular histoplasmosis syndrome, a complication of a parasitic infection. Regulatory approval for these conditions has also been granted in both North America and Europe.[38]

These studies have established PDT as first-line therapy for AMD and other conditions. The safety of repeated treatments has been established, with very few side effects noted. In the occult trial, about 4% of patients experienced severe vision loss within the first week after treatment. A majority of these patients recovered from this event but are not recommended for further treatment.

Two other photosensitizers have been tested clinically for their effect on choroidal neovasculature. These are tin ethyl etiopurpurin (Purlytin) and texaphrin. Phase I/II studies with Purlytin showed vessel closure similar to that seen with Visudyne.[39] A Phase III placebo-controlled trial has almost completed 2 years of follow-up, and results should be made available early in 2003. Since this also is a hydrophobic photosensitizer, one would assume that vessel closure similar to that seen with Visudyne should be achieved. Texaphrin has been tested in Phase I/II clinical trials in approximately 70 patients. These results, reported at various ophthalmology meetings, suggested that vessel closure was difficult to achieve, even at high drug doses. It is possible that the relative water solubility of this drug vs. Visudyne and Purlytin to some extent does not lend itself to rapid cell uptake.

PDT is now established as part of the ophthalmologist's armamentarium for the treatment of abnormal vascular conditions in the eye. The therapy has proven itself to be both safe and efficacious. In addition to the diseases mentioned above, there are a number of other ocular conditions in which PDT could play a role including diabetic macular edema, diabetic retinopathy, corneal neovasculature, and ocular tumors. While alternative therapies for these conditions exist, unlike AMD, for which there was no treatment, it is possible that PDT may offer some advantages over existing therapies.

## 38.8 Photodetection

Photosensitizer fluorescence as a marker of early stage cancers has been studied in several indications. PpIX fluorescence derived *in situ* from ALA has proven capable of detecting lesions in the bladder not detectable by ordinary cystoscopy.[40] Riedl et al. have recently demonstrated the clinical significance of this increased detection rate.[41] In a prospective, randomized study of 102 bladder cancer patients with tumors ranging from TaG1 to T1G3, 51 received transurethral resection under white-light cystoscopy and 51 received ALA-induced fluorescence guided transurethral resection. Fluorescence detection was carried out by instilling into the bladder 1.5 g ALA in 50 cc sodium bicarbonate (to achieve a pH of 6.5) and left in place 1 to 4 h (the time did not appear to affect the outcome). A white-light source coupled with a band-pass filter to produce blue light was used to excite the red fluorescence of PpIX. Six weeks later all patients underwent ALA-assisted transurethral resection. Tumor was detected in 39% (20 of 51) of patients who initially received the standard method, and 16% (8 of 51) of the ALA group experienced a reduction in tumor recurrence of 59% ($p = 0.005$). The authors conclude that ALA-assisted transurethral resection is an inexpensive and effective method for significantly increased tumor detection in the bladder.

### 38.8.1 Detection of Early Stage Lung Cancer

As is well known, early stage lung cancer is difficult to detect, and earlier screening studies of high-risk patients using chest x-ray and sputum cytology have shown no survival benefit. Recently, however, spiral CT and fluorescence bronchoscopy have been investigated for detection of early lung lesions — the former to detect peripheral lesions and the latter mainly for centrally located lesions. At this institution we have initiated a screening program of high-risk patients (previous history of lung cancer or head and neck cancer or exposure to asbestos) using an FDA-approved fluorescence-endoscopic method called LIFE (Light Induced Fluorescence Excitation) from Xillix Corporation for examination of the central airways. This system has been shown to be nearly three times as likely to detect an early lesion when used with white-light bronchoscopy as white-light bronchoscopy alone. It is based on the principle that even very early dysplasia or CIS tends to accumulate endogenous chromophores (e.g., porphyrins), thereby presenting a reddish-brown fluorescence compared to normal bronchial mucosa, which is seen as greenish in color.[42] The suspicious areas are confirmed by biopsy. In addition, each patient is screened using spiral CT for detection of peripheral lesions. Appropriate patients are treated by PDT. While this is a relatively modest trial of 500 patients, the intention is not only to investigate these techniques for efficacy in detection of early lesions but also to determine effect on survival after appropriate treatment (and not, incidentally, to assist in convincing third-party payers to reimburse for screening).

## 38.9 Future Directions for Photodynamic Therapy

While PDT has been mainly a cancer therapy, it is clear that its applications are extending to many noncancerous conditions, as is evident from the previous discussion. Other areas currently being investigated are prevention of restenosis after balloon angioplasty for cardiac artery disease, treatment of rheumatoid arthritis (inflammatory cells are particularly prone to photosensitizer uptake), and certain early-detection indications.

### References

1. Weishaupt, K.R., Gomer, C.J., and Dougherty, T.J., Identification of singlet oxygen as the cytotoxic agent in photoinactivation of a murine tumor, *Cancer Res.*, 36, 2326, 1976.
2. Lightdale, C.J., Heier, S.K., Marcon, N.E., McCaughan, J.S., Jr., Gerdes, H., Overholt, B.F., Sivak, M.V., Jr., Stiegmann, G.V., and Nava, H.R., Photodynamic therapy with porfimer sodium versus thermal ablation therapy with Nd:YAG laser for palliation of esophageal cancer: a multicenter randomized trial, *Gastrointest. Endosc.*, 42, 507, 1995.
3. Vincent, R.G., Dougherty, T.J., Rao, U., Boyle, D.G., and Potter, W.R., Photoradiation therapy in advanced carcinoma of the trachea and bronchus, *Chest*, 85, 29, 1984.
4. Package insert: Photofrin® (porfimer sodium) for injection, Axcanpharma, Inc., 1999.
5. Lam, S., Haussinger, K., Leroy, M., Sutedja, T., and Huber, R., Photodynamic therapy (PDT) with Photofrin®, a treatment with curative potential for early stage superficial lung cancer, *Proc. 34th Annu. Meeting Am. Soc. Clin. Oncol.*, 1998, abstract 1781.
6. Sutedja, T., Lam, S., LeRiche, J., and Postmus, P., Response and pattern of failure after photodynamic therapy for intraluminal Stage I lung cancer, *J. Bronchol.*, 295, 1994.
7. Axcan website: www.axcan.com.
8. Overholt, B.F., Panjehpour, M., and Haydek, J.M., Photodynamic therapy for Barrett's esophagus: follow-up in 100 patients, *Gastrointest. Endosc.*, 49, 1, 1999.
9. Buttar, N., Wang, K., Sebo, T., Riehle, D., Krishnadath, K., Lutzke, L., Anderson, M., Petterson, T., and Burgart, L., Extent of high grade dysplasia in Barrett's esophagus correlates with risk of adenocarcinoma, *Gastroenterology,* 120, 1630, 2001.
10. Biel, M., Photodynamic therapy and the treatment of head and neck neoplasia, *Laryngoscope*, 108, 1259, 1998.
11. Schweitzer, V., Photofrin-mediated photodynamic therapy for treatment of aggressive head and neck non-melanomatous skin tumors in elderly patients, *Laryngoscope*, 111, 1091, 2001.
12. Oseroff, A., Shieh, S., Frawley, N., Blumenson, L., Parsons, J., Potter, W., Graham, A., Henderson, B., and Dougherty, T., PDT for skin cancer: what do we know about how to go, *29th Annu. Meeting Am. Soc. Photobiol.*, 2001, abstract 146.
13. Muller, P. and Wilson, B., Photodynamic therapy for recurrent supratentorial gliomas, *Semin. Surg. Oncol.*, 11, 346, 1995.
14. Popovic, E., Kaye, A., and Hill, J., Photodynamic therapy for brain tumors, *J. Clin. Laser Med. Surg.*, 14, 251, 1996.
15. Nseyo, U., Shumaker, B., Klein, E., and Sutherland, K., Photodynamic therapy using porfimer sodium as an alternative to cystectomy in patients with refractory transitional cell carcinoma-in-situ of the bladder, *J. Urol.*, 160, 39, 1998.
16. Pass, H.I. and Donington, J.S., Use of photodynamic therapy for the management of pleural malignancies, *Semin. Surg. Oncol.*, 11, 360, 1995.
17. Moskal, T.L., Dougherty, T.J., Urschel, J.D., Antkowiak, J.G., Regal, A.-M., Driscoll, D.A., and Takita, H., Operation and photodynamic therapy for pleural mesothelioma: six year follow-up., *Ann. Thorac. Surg.*, 66, 1128, 1998.
18. Delaney, T.F., Sindelar, W.F., Tochner, Z.S., Friauf, W.S., Thomas, G., Dachowski, L., Cole, J.W., Steinberg, S.M., and Glatstein, E., Phase I study of debulking surgery and photodynamic therapy for disseminated intraperitoneal tumors, *Int. J. Radiat. Oncol. Biol. Phys.*, 25, 445, 1993.

19. Ortner, M., Photodynamic therapy of cholangiocarcinoma, *Cancer GI Endosc. Clin. North Am.*, 10, 481, 2000.
20. Ortner, M., Photodynamic therapy for cholangiocarcinoma, *J. Hepato-Biliary-Pancreatic Surg.*, 8, 137, 2001.
21. Jeffes, E., McCullogh, J., Weinstein, G., Kaplan, R., Glazer, S., and Taylor, R., Photodynamic therapy of actinic keratoses with topical aminolevulinic acid hydrochloride and fluorescent blue light, *J. Am. Acad. Dermatol.*, 45, 96, 2001.
22. DUSA website: www.dusapharma.com.
23. Fan, K., Hopper, C., Speight, P., Buonaccorsi, G., and Bown, S., Photodynamic therapy using mTHPc for malignant disease in the oral cavity, *Int. J. Cancer*, 73, 25, 1997.
24. Miravant website: www.miravant.com.
25. Sessler, J.L. and Miller, R.A., Texaphyrins: new drugs with diverse clinical applications in radiation and photodynamic therapy, *Biochem. Pharmacol.*, 59, 733, 2000.
26. Maugeri, A., Klevering, B.J., Rohrschneider, K., Blankenagel, A., Brunner, H.G., Deutman, A.F., Hoyng, C.B., and Cremers, F.P.M., Mutations in the ABCA4 (ABCR) gene are the major cause of autosomal recessive cone-rod dystrophy, *Am. J. Hum. Genet.*, 67, 960, 2000.
27. Allikmets, R. and International ABCR Screening Consortium, Further evidence for an association of ABCR alleles with age-related macular degeneration, *Am. J. Hum. Genet.*, 67, 487, 2000.
28. Allikmets, R., Simple and complex ABCR: genetic predisposition to retinal disease, *Am. J. Hum. Genet.*, 67, 791, 2000.
29. Fingar, V.H., Vascular effects of photodynamic therapy, *J. Clin. Laser Med. Surg.*, 14, 323, 1996.
30. Henderson, B.W. and Dougherty, T.J., How does photodynamic therapy work? *Photochem. Photobiol.*, 55, 145, 1992.
31. Richter, A.M., Waterfield, E., Jain, A.K., Canaan, A.J., Allison, B.A., and Levy, J.G., Liposomal delivery of a photosensitizer, benzoporphyrin derivative monoacid ring A (BPD) to tumor tissue in a mouse tumor model, *Photochem. Photobiol.*, 57, 1000, 1993.
32. Schmidt-Erfurth, U. and Hasan, T., Mechanisms of actions of photodynamic therapy with verteporfin for the treamtent of age-related macular degeneration, *Surv. Ophthalmol.*, 45, 195, 2000.
33. Kramer, M., Miller, J.M., Michaud, N., Moulton, R.S., Hasan, T., Flotte, T.J., and Gragoudas, E.S., Liposomal benzoporphyrin derivative verteporfin photodynamic therapy. Selective treatment of choroidal neovascularization in monkeys, *Ophthalmology*, 103, 427, 1996.
34. Miller, J.W., Schmidt-Erfurth, U., and Sickenburg, M., Photodynamic therapy with verteporfin for choroidal neovascularization caused by age related macular degeneration: results of a single treatment in a Phase I and 2 study, *Arch. Ophthalmol.*, 117, 1161, 1999.
35. TAP Study Group, Treatment of age related macular degeneration with photodynamic therapy (TAP), photodynamic therapy of subfoveal choroidal neovascularization in age related macular degeneration with verteporfin: one year results of two randomized clinical trials — TAP report 1, *Arch. Ophthalmol.*, 117, 1329, 1999.
36. TAP Study Group, Treatment of age-related macular degeneration with photodynamic therapy (TAP). Photodynamic therapy of subfoveal choroidal neovascularization in age related macular degeneration with verteporfin: two year results of two randomized clinical trials — TAP report 2, *Arch. Ophthalmol.*, 119, 198, 2001.
37. Verteporfin in Photodynamic Study Group, Verteporfin therapy of subfoveal choroidal neovascularization in age-related macular degeneration: two year results of a randomized trial including lesions with occult with no classic choroidal neovascularization. Verteporfin in photodynamic therapy report 2, *Am. J. Ophthalmol.*, 131, 541, 2001.
38. Verteporfin in Photodynamic Study Group, Verteporfin therapy of subfoveal choroidal neovascularization in pathologic myopia: twelve month results of a randomized trial, *Ophthalmology*, 108, 841, 2001.
39. El, T., Li, X.Y., and Paciotti N.M., Fluorescein angiographic characteristics of subfoveal choroidal neovascular lesions secondary to age related maculopathy in pivotal Phase III trial for SnET2 PDT, Abstract, Association for Research in Vision and Ophthalmology, 2352, 2001.

40. Zaak, D., Kriegmair, M., Stepp, H., Baumgartner, R., Oberneder, R., Schneede, P., Corvin, S., Frimberger, D., Knuchel, R., and Hofstetter, A., Endoscopic detection of transitional cell carcinoma with 5-aminolevulinic acid: results of 1012 fluorescence endoscopies, *Urology*, 57, 690, 2001.
41. Riedl, C., Danilchenko, D., Koenig, S., Simak, R., Loenig, S., and Pflueger, H., Fluorescence endscopy with 5-aminolevulinic acid reduces early recurrent rate in superficial bladder cancer, *J. Urol.*, 165, 1121, 2001.
42. Lam, S., Kennedy, T., Unger, M., Miller-York, E., Gelmont, D., Rusch, V., Gipe, B., Howard, D., LeRiche, J.C., Coldman, A., and Gazdar, A.F., Localization of bronchial intraepithelial neoplastic lesions by fluorescence bronchoscopy, *Chest*, 113, 696, 1998.

# 39

# Laser Tissue Welding




Karen M. McNally-Heintzelman
*Rose-Hulman Institute of Technology*
*Terre Haute, Indiana*

Ashley J. Welch
*University of Texas at Austin*
*Austin, Texas*


## 39.1 Introduction

The joining of tissue by the application of heat through the use of hot-loop forceps was first described in the 1960s. In 1964, a neodymium laser was used to join small blood vessels — the first reported use of a laser for thermal welding of tissue. Since then, numerous experimental studies have been conducted using a variety of lasers for welding of soft tissues including blood vessels, the genitourinary tract, the gastrointestinal tract, liver, spleen, nerves, dura mater, skin, sclera, trachea, and cartilage. As we enter the 21st century, laser tissue welding has reached the threshold at which it is moving from the laboratory bench to clinical application, making this an exciting time for all involved. This chapter reviews the principles, theory, and application of laser tissue welding. An insight is also provided into several important developments that have been made involving the use of light-activated surgical adhesives to


0-8493-1116-0/03/$0.00+$1.50

assist in the welding procedure, infrared (IR) temperature feedback control of the laser device, and computer modeling of the welding process. In addition, this chapter provides a comprehensive review of current and future clinical applications of laser tissue welding including developments in the fields of endoscopic and laparoscopic surgeries.

## 39.2 Background Principles and Theory

Lasers have been demonstrated to be extremely useful in many medical and surgical applications as both diagnostic and therapeutic tools. The use of laser energy to induce thermal changes in connective-tissue proteins is of particular interest for the joining of severed tissue, where proteins within the target tissue are coagulated to form a bond between the two adjoining edges. As the degree of success of all new surgical techniques must ultimately be judged in comparison with existing techniques, the following sections review the most common closure methods currently in use.

### 39.2.1 Conventional Closure Methods

The conventional methods for tissue repair use sutures, staples, or clips.[1] Sutures are favored because they are cost-effective, reliable, and, more importantly, suitable for almost any type of tissue. However, due to their mechanical intrusion, the use of any of these conventional fasteners causes tissue injury. By their very nature they result in a foreign body being left in the tissue. Tissue injury and foreign-body reaction can give rise to inflammation (nonspecific immune response of tissue to injury resulting in redness and swelling), granuloma formation (development of a grain-like tissue lesion), scarring (permanent marking of tissue), and stenosis (abnormal narrowing or constriction of tissue lumens). Sutures become difficult or tedious to execute in microsurgical or minimally invasive endoscopic applications, where staples or clips are better suited. Staples and clips are not easily adapted to different tissue dimensions, however, and maintaining precision of alignment of the tissue is difficult because of the relatively large force required to deploy them. Finally, none of these fasteners produces a watertight seal over the repair.

### 39.2.2 Other Closure Methods

Besides sutures, staples, and clips, other closure techniques, including both biological and nonbiological adhesives as well as various coupling devices, have been tested. Fibrin glue has long been used as a biological surgical adhesive.[2–4] Fibrin glue imitates the final step of blood coagulation and thus has been used effectively for hemostasis. However, repairs formed with fibrin glue alone are typically weak in comparison with suture repairs.[5] Consequently, the glue is almost always used in conjunction with stay sutures.[4] Among the various synthetic adhesives available, cyanoacrylates are the most popular.[6–8] However, they are toxic to the tissues, not absorbed in the normal wound healing process, and cause foreign-body granulomas and allergic reactions. Two examples of anastomotic coupling devices are the UNILINK (3M, St. Paul, MN) anastomotic system[9,10] and the 3M Precise Microanastomotic System (3M).[11,12] While these devices and several others like them have produced superior repairs with regard to anastomotic time and patency rate when compared to suture, they suffer from size inflexibility and often involve the sacrifice of some length of the structure being anastomosed.

## 39.3 The Recognition and Development of Laser Tissue Welding

The feasibility of laser tissue welding was first shown by Yahr and Strully in 1964, when they used a neodymium laser to join small blood vessels.[13] In 1979, Jain and Gorisch reported the first reproducible experimental use of laser tissue welding.[14] In that report, a neodymium:yttrium-aluminum-garnet (Nd:YAG) laser was used for microvascular anastomosis of rat carotid and femoral arteries. The work was followed up with a report on a small series of clinical studies.[15] Since then a wide range of lasers

**TABLE 39.1** Absorption ($\mu_a$), Reduced Scattering ($\mu_s'$), and Optical Penetration Depth (OPD) of Water and Tissue for Wavelengths of Commonly Available Medical Lasers

| Laser | $\mu_a$ – Water (cm$^{-1}$) | $\mu_a$ – Tissue (cm$^{-1}$) | $\mu_s'$ – Water (cm$^{-1}$) | $\mu_s'$ – Tissue (cm$^{-1}$) | OPD – Water (μm) | OPD – Tissue (μm) |
|---|---|---|---|---|---|---|
| $CO_2$ | | | | | | |
| (10.6 μm) | 950[32,a] | 600[32] | | | 11[32] | 17[32] |
| THC:YAG | | | | | | |
| (2.15 μm) | 34[86] | | | | 290 | |
| Ho:YAG | | | | | | |
| (2.09 μm) | 37[318] | 35[32] | | | 270 | 286[32] |
| Tm:YAG | | | | | | |
| (2.01 μm) | 75[318] | 67 | | | 133 | 150[319] |
| Diode | | | | | | |
| (1.9 μm) | 100 | 80 | | | 100[63] | 125[180] |
| Nd:YAG | | | | | | |
| (1.320 μm) | 1.2[32] | 4.3[317] | | 18[317] | 8,333 | 600 |
| (1.064 μm) | 0.1[32] | 0.7[317] | | 22[317] | 100,000 | 1,400 |
| GaAlAs diode | | | | | | |
| (805 nm) | 0.01[32] | 3.8[316] | | 29[316] | 1,000,000 | 520 |
| KTP | | | | | | |
| (532 nm) | | 13[316] | | 60[316] | | 190 |
| Argon | | | | | | |
| (514.5 nm) | | 12[316] | | 63[316] | | 190 |
| (488.0 nm) | | 16[316] | | 68[316] | | 160 |

[a]Superscripts indicate reference numbers in end-of-chapter list.

have been used for laser tissue welding. Infrared sources include carbon dioxide ($CO_2$), thulium-holmium-chromium, holmium, thulium, and neodymium rare-earth doped garnets (THC:YAG, Ho:YAG, Tm:YAG, and Nd:YAG, respectively), and gallium aluminum arsenide diode (GaAlAs) lasers. Visible sources include potassium-titanyl-phosphate (KTP) frequency-doubled Nd:YAG, and argon lasers. The laser energy is absorbed by water at the infrared wavelengths and by hemoglobin and melanin at the visible wavelengths, thereby heating proteins within the target tissue. As temperatures rise and heating times are prolonged, cellular and tissue structural proteins undergo denaturation and conformational changes, a process defined as coagulation.[16] Table 39.1 lists the absorption coefficients, reduced scattering coefficients, and optical penetration depths, in water and tissue, of commonly available medical lasers.

## 39.3.1 Benefits of Laser Tissue Welding

While laser tissue welding is unlikely to replace sutures in all applications, it has been shown to achieve functionality comparable to that of conventional suturing techniques,[17–19] with the added advantage of moderately reduced operation times,[20,21] reduced skill requirements,[22,23] reduced suture and needle trauma,[24] reduced foreign-body reaction,[25,26] and reduced bleeding.[27] Repairs formed using laser tissue welding tend to heal faster,[24,28,29] have the ability to grow,[30] and exhibit better cosmetic appearances.[29,31] Welding also has the potential to form complete closures, enabling an immediate watertight anastomosis intraoperatively in the case of vascular,[32] genitourinary-tract,[33] and gastrointestinal repairs.[34] A watertight closure is also advantageous for neural repairs as it discourages the exit of regenerating axons and the entry of fibroblasts.[35,36]

## 39.3.2 Limitations of Laser Tissue Welding

Although some success has been achieved in experimental applications, two disadvantages of the laser-assisted procedure are foreseen for the clinical application of the technique: the first is the low strength of the resulting anastomosis, especially in the acute-healing phase up to 5 d postoperative; the second is thermal damage of tissue by direct laser heating and heat transfer. Technical difficulties with the ambiguity of the endpoint for the procedure,[32,37] tissue apposition,[38] and poor reproducibility[39,40] are also concerns.

Low-strength anastomoses can lead to an anastomotic aneurysm (swelling) or even rupture of the repair. Dehiscence rates of 8 to 90% have been reported for laser repair of tissue in comparison to 0 to 12% for suture-repaired tissue.[41–44] Most researchers have used one to four temporary or permanent stay sutures to improve the success rate of their laser anastomoses.[23,45,46] This complicates the operative procedure and does not eliminate the sutures or their consequent problems. Another approach has been to apply an extra layer of tissue, such as venous grafts in vascular repair[47,48] or the epineurium in nerve repair,[43] to the anastomotic site. This approach also avoids foreign-body reactions, but suitable tissue must be prepared, and the operation is technically challenging.

Thermal damage to tissue is a particular concern in medical laser applications.[49–52] Tissue welding is achieved by laser-induced heating of tissue to levels that can cause immediate irreversible damage of the structural proteins of the tissue (refer to Section 39.5 for a discussion on the mechanisms of laser tissue welding). The resulting denaturation of protein is not specific to the target tissue, and all irradiated tissues are heated. The general objective of laser tissue-repair techniques is to obtain coagulation of a desired volume of tissue with minimal effects in the surrounding healthy tissue. The $CO_2$ laser has been favored for laser repair of thin tissues due to its short penetration depth (17 μm) in tissue.[19,36,46,53] However, in thick tissues, it does not permit complete sealing of the anastomotic site. In such tissues, welding is achieved by using either high irradiances or long exposure times.[41,54–57] High irradiances present the risk of tissue charring, which may cause secondary disruption of the repair or aneurysm formation. Long exposure times result in the conduction of heat from the initial absorption zone. This is a particular problem when heat is conducted in a horizontal direction from the anastomotic site to the surrounding healthy tissue. The argon laser has a much deeper penetration depth in tissue (500 to 714 μm) and is thus better suited to thicker tissues.[58] However, difficulty has been experienced in accurately determining the dosimetry of the laser energy due to variations in the concentration of hemoglobin and melanin present in the tissue, local tissue thickness, and state of hydration.[42,59,60] A narrow margin exists between a successful and an unsuccessful weld. Too low an irradiance results in a weak bond, whereas excessive laser irradiances result in excessive thermal damage to the tissue with drying and shrinkage of the tissue proteins, thereby reducing the flexibility and strength of the weld.

Surgeons generally look for visible signs of the tissue being heated, such as tissue coaptation, discoloration, drying, and contraction, as markers of the endpoint for tissue fusion.[61–65] These rather subjective indicators are not always reliable. For example, disparities in tissue properties can cause differences in the absorption characteristics. In addition, the irradiance at the tissue surface can vary significantly for identical procedures if the laser-spot size or energy-delivery rate is varied. These differences will cause the tissue temperature to fluctuate during the laser procedure. Since the thermal damage to the irradiated tissue strongly depends on the tissue temperature achieved during the laser procedure,[49–52] the resulting surgical outcome can be highly variable.

The use of cryogen spray to cool the anastomotic site has been investigated as a means of reducing thermal damage and hence increasing weld strength, and it has met with good results.[66] The possibility of irradiating the tissue through water or saline has also been investigated.[56,67] As the thermal conductivity of water is approximately twice that of air, the tissue cools more efficiently by releasing heat into the water. However, due to the short penetration depth in water of many of the infrared laser wavelengths commonly used for laser tissue welding (Table 39.1), insufficient energy may reach the tissue and prevent the desired coagulation effect from being achieved.

Good apposition of the tissue edges is also essential for tissue welding, where even the smallest gaps can act as a lead point for failure of the entire repair. To solve this problem suture devices (Figure 39.1) and various mechanical apposition devices have been developed to hold the tissue edges together during the laser-welding procedure.[23,68,69] Building on this concept, the resulting tensile strength of welded tissue has been shown to increase significantly with apposition pressure.[69] Another concern associated with laser anastomosis of blood vessels, the genitourinary tract, and the gastrointestinal tract includes the

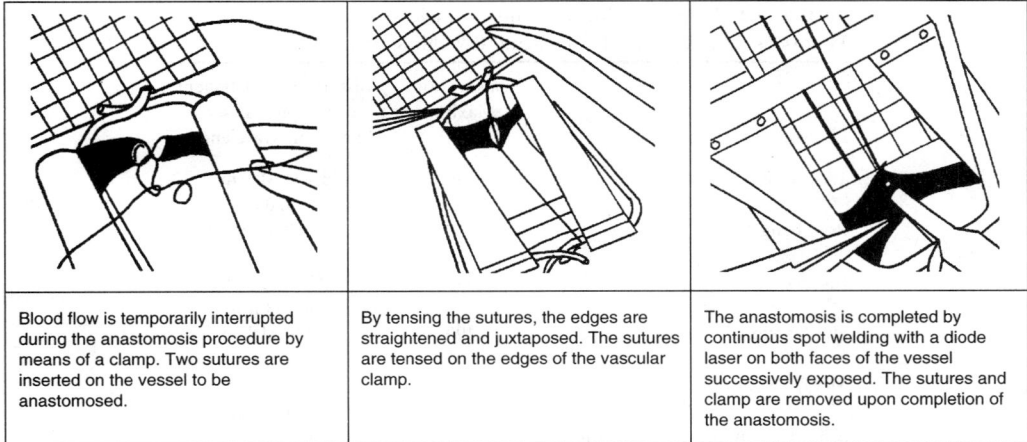

| | | |
|---|---|---|
| Blood flow is temporarily interrupted during the anastomosis procedure by means of a clamp. Two sutures are inserted on the vessel to be anastomosed. | By tensing the sutures, the edges are straightened and juxtaposed. The sutures are tensed on the edges of the vascular clamp. | The anastomosis is completed by continuous spot welding with a diode laser on both faces of the vessel successively exposed. The sutures and clamp are removed upon completion of the anastomosis. |

A

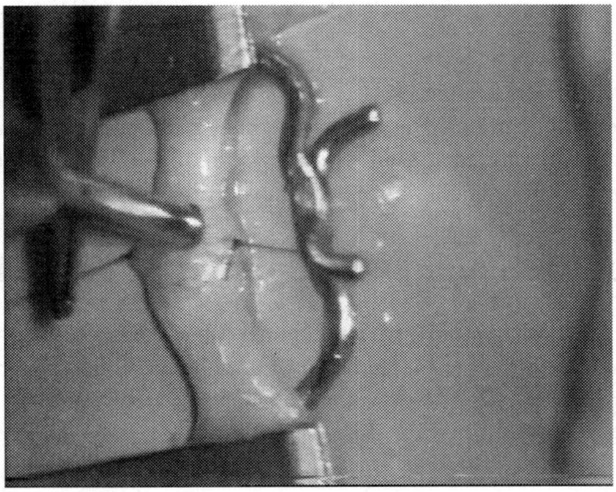

B

**FIGURE 39.1** (A) Different steps of laser-assisted microvascular anastomosis technique using the diode laser ($\lambda$ = 1.9 μm, spot diameter = 500 μm, $P$ = 125 mW, $t$ = 1.2 sec: 10 to 15 spots on each face) for digital replantation in humans; (B) completion of the anastomosis with a diode laser. (Courtesy of Serge Mordon, INSERM, Lille University Hospital, Lille, France.)

possibility of stenosis at the irradiated sites.[70] Dissolvable stents have been used to successfully separate the anterior and posterior walls of such tissues, allowing for an easier and faster welding procedure.[19,71,72]

### 39.3.3 Photochemical Welding

Photochemical welding of tissue has been investigated as an alternative method for tissue repair without the use of heat and its associated tissue damage.[73–78] The technique utilizes chemical cross-linking agents that, when light-activated, produce covalent cross-links between the collagen fibers contained within the tissue.[74] In theory, this technique should produce stronger bonds than the noncovalent bonds produced by photothermal welding. Agents used for photochemical welding include 1,8-naphthalimide,[75,76] rose bengal (RB),[78] riboflavin-5-phosphate (R-5-P),[77,78] fluorescein (Fl),[78] methylene blue (MB),[78] and *N*-hydroxypyridine-2-(1H)-thione (N-HPT).[78] Table 39.2 lists the absorption maxima and absorption coefficients at typical laser-excitation wavelengths for each of these photosensitizers. Of particular interest is the fact that photoactive

**TABLE 39.2** Absorption Maxima and Coefficients for Common Photosensitizers Used for Photochemical Welding of Tissue

| Photosensitizer | Absorption Maxima (nm) | Absorption Coefficient at Typical Laser Excitation Wavelengths |
|---|---|---|
| 1,8-Naphthalimide | 440 | 13,100 $M^{-1}cm^{-1}$ at 420 nm |
| Rose bengal | 550 | 33,000 $M^{-1}cm^{-1}$ at 514 nm |
| Riboflavin-5-phosphate | 445 | 4330 $M^{-1}cm^{-1}$ at 488 nm |
| Fluorescein | 490 | 88,300 $M^{-1}cm^{-1}$ at 488 nm |
| Methylene blue | 664 | 15,600 $M^{-1}cm^{-1}$ at 661 nm |
| *N*-Hydroxypyridine-2-(1H)-thione | 314 | 2110 $M^{-1}cm^{-1}$ at 351 nm |

*Source*: From Mulroy, L. et al., *Invest. Ophthalmol. Vis. Sci.*, 41(11), 3335, 2000. With permission.

tissue welding does not rely on the healing properties of blood vessels. The ability of the technique to repair avascular tissue could prove to be quite advantageous to many orthopedic applications.[75] Further studies to determine optimal treatment parameters in terms of cross-linking ability, strength, and long-term healing response are necessary before clinical acceptance of this procedure will be gained.

## 39.4 The Recognition and Development of Laser Tissue Soldering

Two advances have been useful in addressing the issues of low repair strength and thermal damage associated with laser tissue welding: (1) the addition of endogenous and exogenous materials to be used as solders and (2) the application of laser-wavelength-specific chromophores.

The addition of endogenous and exogenous materials helps to maintain edge alignment and to strengthen the wound, particularly during the acute postoperative healing phase, while shielding the underlying tissue from excessive thermal damage caused by direct absorption of the laser light. Useful materials include blood,[79–81] cryoprecipitate (typically plasma),[82–84] fibrinogen,[62,85,86] and albumin.[87–91]

The application of wavelength-specific chromophores provides for differential absorption between the dyed region and the surrounding tissue. An advantage of this technique is that the area may be bathed by the laser radiation while energy is absorbed selectively only by the target. Hence the requirement for precise focusing and aiming of the laser beam may be removed. Furthermore, due to the increased absorption characteristics of the dyed tissue or solder, lower laser irradiances may be used to achieve the required effect, increasing the safety of the technique. Examples of dyes that have been used to assist laser tissue-welding procedures include carbon black and Fen 6 for use with Nd:YAG lasers,[92–94] indocyanine green (ICG) for use with ~800-nm diode lasers,[61,88–91,95] iron oxide and fluorescein for use with KTP frequency-doubled Nd:YAG lasers,[96,97] and basic fuchsin, methyl violet, crystal violet, chlorin(e6), and fluorescein isothiocyanate for use with argon lasers.[60,92,98,99]

The combination of serum albumin and ICG dye with an ~800-nm diode laser, first described by Poppas et al., has gained the most widespread popularity (Figure 39.2).[100] The light-sensitive ingredient in the protein solder is the ICG dye, which has a maximum absorption coefficient at 805 nm of $2 \times 10^5$ $M^{-1}$ $cm^{-1}$.[101] The dye is water soluble and preferentially binds to serum protein, thereby ensuring that the absorbed heat is efficiently transferred to the protein solder. Since direct tissue absorption at this laser wavelength is low, the dye-containing solder absorbs most of the incident-light energy. Consequently, the amount of thermal damage to the surrounding tissue caused by direct laser-light absorption is minimized. The increased temperature at the repair site during laser treatment denatures the albumin solder, enabling it to bond with the tissue. The resulting solder coagulum is nonimmunogenic[102] and is

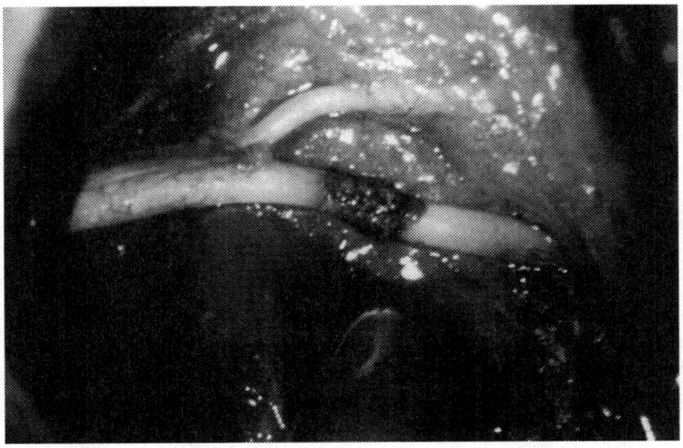

**FIGURE 39.2** Intraoperative photograph showing the tibial nerve of a rat repaired using four solid-protein solder strips (60% BSA, 2.5 mg/ml ICG, and deionized water) in conjunction with a diode laser ($\lambda$ = 805 nm, spot diameter = 1 mm, $E_0$ = 12.7 W/cm,[2] $t$ = 40 sec). (Image courtesy of Karen McNally and Judith Dawes, Macquarie University, Sydney, Australia.)

gradually absorbed in the normal wound-healing process.[103] Moreover, the presence of the solder coagulum increases the initial laser repair strength.[32]

The efficacy of diode-laser tissue soldering using ICG-doped albumin-protein solders has been demonstrated, with good results, in a wide range of tissues including blood vessels,[104–108] the genitourinary tract,[91,109–112] the gastrointestinal tract,[34] liver,[113] nerves,[114,115] dura mater,[116] skin,[117–119] trachea,[120,121] and cartilage.[122] In addition to these successes, much has been learned about the optimal solder specifications.

Initial studies with protein solders employed liquid formulations containing low protein concentrations (e.g., 25% serum albumin). Application of these fluid solders to the repair site is often difficult due to problems associated with "runaway" of the low-viscosity material, which leads to large variations in the surface area and thickness of the applied solder. This variation makes determination of the optimal laser dosage difficult. As a result, the strength of repairs formed using liquid-protein solders is often inferior to that of sutured repairs.[123] Lauto et al.[103] and Lauto[124] have shown that the strength of solder–tissue bonds depends on protein concentration. Bovine serum albumin (BSA) concentrations of 25, 50, and 66% (2.5 mg/ml ICG) formed bonds having acute tensile strengths of 10.4, 37.5, and 69.2%, respectively, of that of sutured nerves.[103] An attempt to improve on liquid solders was made by further concentrating the albumin solder to produce solid-protein solder strips.[89,115,125] Repairs formed using the solid-protein solder have been found to be comparable in strength to native tissue.[89] In addition, solid-protein solder strips are easier to handle and apply to the tissue surface than their liquid counterparts, and slight rehydration of the solid solder surface during application improves adhesion so as to hold the tissue edges in close approximation during the repair procedure. Finally, the uniform dimensions of the solid-protein solder strips allow preselection of optimal irradiation parameters, reducing any ambiguity regarding the endpoint for the procedure.

Chan et al. investigated the effects of hydration on tissue repairs formed with liquid-protein solder using scanning electron microscopy analysis.[126] The solder was observed to detach from the tissue substrate once the specimens were submerged in a hydrated environment of phosphate-buffered aqueous saline. This was traced to inadequate heat transmission through the solder, which produced unstable protein globules, rather than a homogeneous solder coagulum. Subsequent studies showed that the tensile strength of repairs could be improved by an order of magnitude (257 N/cm$^2$ vs. 5.2 N/cm$^2$,[32] 31 N/cm$^2$,[127] and 2.8 N/cm$^2$ — see Reference 102) if a series of small (0.2-μl) droplets of solder (25% BSA and 2.5 mg/ml ICG) were applied to the tissue and each droplet were coagulated individually.[128,129]

The laser solder-repair technique has been reported to be much faster to perform than conventional suture techniques while providing an immediate leak-free closure with improved histological behavior, improved cosmetic appearance (Figure 39.3), and fewer intraoperative complications including disruptions.

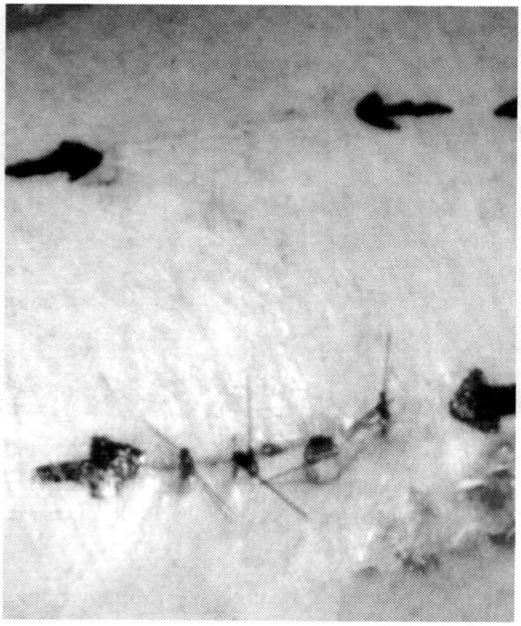

**FIGURE 39.3** Suturing with 4–0 nylon thread (bottom) vs. soldering with 50% albumin solder and a $CO_2$ laser (top) on rabbit skin at 10 d postoperative. Infrared temperature feedback control was used to maintain a constant surface temperature of 60°C during the laser soldering procedure. (Image courtesy of Avi Ravid and Abraham Katzir, Tel Aviv University, Tel Aviv, Israel.)

More recent advancements in light-activated surgical adhesives for tissue soldering are discussed in Section 39.6.

# 39.5 Mechanisms of Laser Tissue Welding and Laser Tissue Soldering

The precise mechanism of laser tissue welding and soldering is still unknown. An understanding of both the tissue and the solder proteins involved in the welding and soldering processes as well as the heat-induced interactions that occur within them is fundamental to the establishment of their physical bonding mechanisms.

Proteins are composed of many amino acids polymerized by condensation between their amino and carboxyl groups. The amino-acid sequence of the protein's polypeptide chain is its primary structure. The important distinguishing features of proteins are found in their secondary structure comprising α-helices and β-sheets.[130] These are discussed below in relation to the structural properties of both collagen and the most common solder protein, serum albumin.

## 39.5.1 Structural Properties of Collagen

Collagen proteins are a major constituent of all tissue extracellular matrices. They are involved in both structural maintenance and tissue growth and are thought to be the principal agents involved in thermal repair procedures.[32] There are more than 20 types of collagen present in the body. Although the basic structure of all types is the same, there are variations from type to type, depending on the function of the collagen. Collagen constitutes about 20% of the dry weight of the large elastic arteries such as the aorta (up to 50% in smaller vessels), about 80% of that of skin,[131] and about 50% of that of peripheral nerves. Table 39.3 lists the types of collagen found in the various layers of blood vessels, skin, and peripheral nerves.

**TABLE 39.3** Types of Collagen Found in Blood Vessels, Skin, and Nerves

| Tissue | Collagen Type |
|---|---|
| Blood Vessel[a] | |
| Tunica adventitia | I |
| Tunica media | I, III |
| Tunica intima | I, III, VI |
| Skin | |
| Epidermis | — |
| Dermal/epidermal junction | VII |
| Dermis | I, III, V |
| Peripheral Nerve | |
| Epineurium | I |
| Perineurium | I, III |
| Endoneurium | III |

[a]Collagen types IV, V, and VI appear in small amounts in the wall of blood vessels.

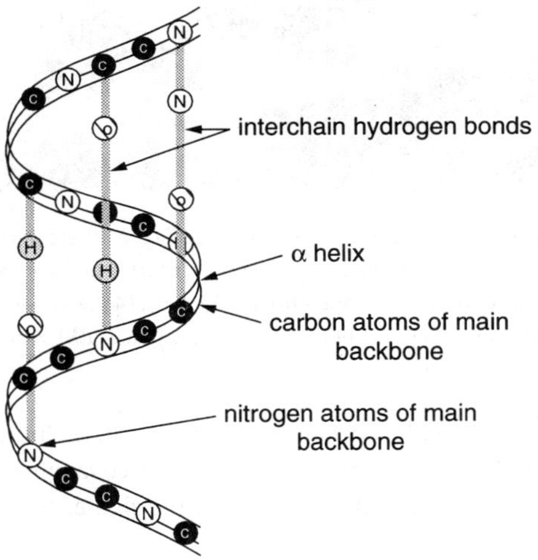


**FIGURE 39.4** Schematic showing hydrogen bonds occurring between the adjacent α-helix turns in protein secondary structures. (Adapted from Yudkin, M. and Offord, R., *Comprehensive Biochemistry*, Longman, London, 1973.)

The α-chains of collagen consist of a fiber (called the D fiber) that, along with the amino acids, cause the primary left-handed helical conformation of the α-chain and the superhelical conformation of the overall molecule. Each α-chain coils into a left-handed helix and intertwines with two other such helices to form a triple-stranded superhelix known as a tropocollagen.[132] A tropocollagen molecule is a rod-like particle of about 3000 Å in length and 15 Å in thickness.[131] The long molecules typically run parallel in a fibril and overlap each other by about one quarter of their length. This causes cross striations in the fibril bands at intervals of 640 Å.[133] The three helical chains forming the tropocollagen are then wrapped around one another in a higher-order rope-like fashion to produce the tight, triple-helical structure of the molecule.[134]

The stability of the collagen superhelix is due to many factors. Interchain hydrogen bonds strengthen the helix, as shown in Figure 39.4. The hydrogen donors are the peptide NH groups, while the hydrogen acceptors are the peptide CO groups on the other chains. The hydrogen bonds are directed perpendicular to the long axis α-chains, giving the protein elasticity.[130] Weaker van der Waals bonds between residues on different strands further strengthen the helix. Steric repulsion between pyrolidine rings in the proline

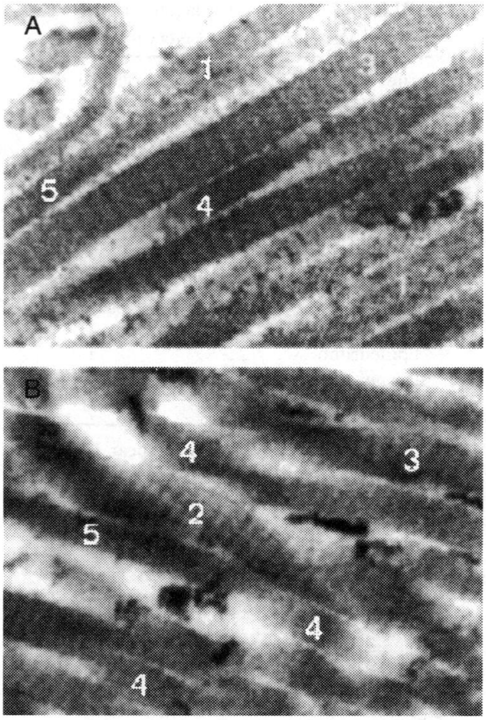


**FIGURE 39.5** (A and B) Electron microscopic view showing collagen fibers' alterations after diode laser welding. (1) Fused collagen fibers with coalescence of their periodicity. (2) Intertwined and rope-like collagen fiber. (3) Swollen collagen fibers. Note the normal periodicity of swollen fibers (55 nm). (4) Dissolved collagen fibers with disappearing of their periodicity. (5) Note normal collagen fibers in the same place. EM × 60,000. (From Tang, J. et al., *Lasers Surg. Med.*, 21(5), 438, 1997. With permission.)

residues also help to stabilize the helix.[135] Some types of collagen, including types I and III collagen, are stabilized even further by intermolecular and intramolecular covalent cross-links. The degree of each type of cross-linking depends on the age and function of the tissue.[131]

## 39.5.2 Mechanism of Laser Tissue Welding

### 39.5.2.1 Photothermal Welding

Many mechanisms of laser tissue welding have been proposed including collagen-to-collagen fusion,[38,136,137] the covalent cross-linking of tissue proteins,[38,56,59,138] denaturation of structural tissue proteins,[139,140] acceleration of natural fibrinogen polymerization,[141] formation of noncovalent bonding between collagen molecules,[142] and interdigitation and rope-like twisting of collagen fibers (Figure 39.5).[143–148] Menovsky et al. proposed that the mechanism of bonding achieved with the $CO_2$ laser is collagen-to-collagen bonding.[137] After laser irradiation, collagen fibrils of both dura mater (the connective tissue that covers the spinal cord and brain) and peripheral nerves appeared to be swollen, densely packed, and fused together when viewed with scanning electron microscopy. White et al. investigated the mechanism of tissue fusion in argon laser-welded vein and arterial anastomoses using electron-microscopy examination and measuring the formation of [3H] hydroxyproline as an index of collagen synthesis.[38] Follow-up studies concluded that tissue fusion is governed by a biochemical mechanism involving covalent cross-linking between collagen bundles in the extracellular matrix.[56,59,138] Based on scanning electron microscopy analysis of carotid arteries, the mechanism of Nd:YAG-induced laser welding has been suggested to be the interdigitation of collagen fibers, which appear to be increased in caliber and swollen and to lose their periodicity immediately after irradiation.[143,145] Two theories incorporating these observations and others found in the literature have been postulated to describe the mechanism of laser tissue welding.

The first theory is based on triple-helix reformation within the collagen fibers. An increase in the local temperature of collagen fibers modifies the triple-helix structure of tropocollagen, inducing alterations of viscosity and periodicity.[133,149] Heat-denatured collagen unwinds its helix to expose sites for intermolecular interactions, especially in transversely cut fibers.[143] As the collagen superhelix is mostly stabilized by cooperative interactions from a large number of weak, reinforcing bonds, it has been hypothesized that when one bond is lost, adjacent bonds are quickly broken, resulting in a domino effect.[149] When cooled over a few minutes, the α-strands then conform themselves into their lowest energy states.[144] The glycine residue from each α-chain is folded into the helix center for extensive hydrogen bonding, and hydroxyproline molecules further stabilize the whole structure.[143] Using this theory, strong welds are created by transincisional fiber fusion but not side-on-side fiber fusion.

The second theory is based on the random noncovalent bonding of opposed collagen layers caused by heat denaturation and, consequently, coagulation. Lemole et al. observed transected rat tail tendons to bind together when heated in a constant temperature water bath at 62°C for 4 min.[144] Rapid quenching of the tendon repairs in a cooler bath immediately after treatment produced stronger bonds than slow cooling. Based on these observations the authors suggested that some form of hydrogen, ionic, or van der Waals bonding between the transected collagen fibers occurred during heat treatment of tissue. Covalent bonding was disputed by Bass et al. based on identical gel electrophoretic mobility patterns and circular dichroism measurements obtained for both laser-treated and control rat-tail tendons.[142,150] In these studies, an 808-nm diode laser and ICG were used to weld rat-tail tendon and blood vessels. Laser treatment was shown to induce strong noncovalent interactions among the collagen strands that sealed the vessel edges together, with the denatured collagen exhibiting random triple-helix-coil transitions and aggregations of randomly oriented fibrils.

The results of Tang et al. support the theory of noncovalent bonding.[147] Collagen fibers in rat aortas were observed to lose a proportion of their birefringence after diode laser welding at 830 nm, suggesting that the protein helices had been disrupted. Under electron-microscopic examination, collagen fibers were visualized as being either fused, "roped," swollen or dissolved, and surrounded by normal collagen fibers situated in the same zone. This was thought to be due to the presence of several types of collagen in the vessel wall, where each type of collagen has a different stability to temperature resulting from differences in chemical composition. Tang et al. suggested that the main type of adhesion in laser fusion was the "roping" of two parallel collagen fibers. Cross-linking was observed to occur at the cut ends of the collagen fibers and between their parallel faces.

The view shared by both of these theories is that laser irradiation induces thermal restructuring within the extracellular matrix. In particular, the thermal restructuring is thought to take place within the tissue collagen, where new bonds and interactions with adjacent proteins are formed and then stabilized upon cooling.

#### *39.5.2.1.1 Temperatures Required for Laser Tissue Welding*

While it appears that the thermal effect is fundamental to the mechanisms behind laser tissue welding, the optimum temperature range for tissue welding remains undetermined. Temperatures at which laser welding has been reported to be successful range from 45 to 140°C, depending on the laser source.

Using a $CO_2$ laser for microvascular anastomosis, tissue fusion has been observed to begin at approximately 70°C with coagulation of denatured collagen in the media and adventitia and fibrin polymerization.[56,151] Histological examination of the repairs showed the optimum temperature for $CO_2$-induced welding to be around 80 to 100°C, as inferred from thermographic images. In contrast, weld-strength analysis and silver-halide fiber-optic radiometry have been correlated to show the optimal temperature for $CO_2$ laser-assisted welding to be 55 ± 5°C in the urinary bladder,[52,152] skin,[153] and corneal[154] models.

Successful welds with the argon laser have been achieved at reported temperatures of 45 to 50°C for arterial vein anastomoses under concomitant saline irrigation using thermography and acute histological examination.[56] Contrary to these results, Martinot et al.[155] and Klioze et al.[49] demonstrated that the temperature range suitable for argon laser fusion was really 60 to 77°C, also using thermography and histological examination. Cilesiz et al. have reported even higher optimal temperatures for tissue welding using an argon laser.[156] Based on bursting-strength analysis, optimal welds of canine intestine were achieved between 90 and 95°C using the argon laser and controlled-temperature feedback with an infrared radiometer.

The optimum temperature for welding of porcine skin using an Nd:YAG laser was found to be 65 to 75°C using infrared thermometry when tensile-strength measurements were made.[37] Much higher temperatures of 90 to 140°C, measured by 3-mm-diameter thermistors, were determined to be optimum when welding human plaque–arterial-wall separations *in vitro* based on shear-strength analysis.[157] Finally, the optimum temperature for welding of vascular repairs using a 1.9-μm diode laser was found to be 80°C in a rat model.[158]

In summary, the effective welding temperature depends on wavelength, solder and tissue type, irradiation parameters, and boundary conditions such as the presence of convective cooling by air or water and blood perfusion. It appears that required temperatures for laser tissue welding, with collagen as the major bonding component, are in the range of 60 to 80°C.[32] The determination of more definitive temperatures is more difficult because thermal coagulation of proteins is a rate process driven by energy and not by temperature thresholds.[16,159] The temperatures outside this range observed in some investigations may relate to differences in the initial tissue conditions (e.g., tissue temperature and level of hydration), distribution and duration of heating, accuracy of the thermal sensor used, and the portion of the tissue that is monitored. Of particular note are the differences observed between measurements made on *in vitro* and *in vivo* tissue. Temperatures measured from *in vitro* tissue specimens are typically higher than those measured for *in vivo* tissue due primarily to conductive heat loss secondary to perfusion.[160] However, variations in tissue hydration and in tissue optical and thermal properties also contribute to this discrepancy. In addition, discrepancies will be observed when different criteria, such as tensile strength and histological examination, are used to judge the "success" of the resultant repairs.

#### 39.5.2.2 Photochemical Welding

The feasibility of photochemical tissue welding was first demonstrated by Judy et al. in a study performed using a newly designed and synthesized class of photochemical dyes to weld dura mater.[73] In this study, photochemical investigation strongly suggested that the dye functioned as a photoalkylation agent following activation to an intermediate state by visible light. The activated species was found to react readily with nucleophilic amino-acid residues such as tryptophan, tyrosine, cysteine, and methionine. The production of covalent cross-links in type I collagen was determined to be the underlying mechanism involved in photochemical tissue welding using brominated 1,8-naphthalimide dyes activated by 420-nm light. This finding was consistent with previous studies where light and photosensitizers were reported to produce intermolecular covalent cross-links in proteins.[74,161–166] More recent studies utilizing 1,8-naphthalimide for photochemical welding of meniscus and articular cartilage[167] and rose bengal for welding of corneal tissue[78] support these findings.

### 39.5.3 Structural Properties of Serum Albumin

Serum albumin is commonly used as a constituent of solder for laser tissue-repair procedures (refer to Section 39.4). Albumin has a molecular mass of about 66,400 Da[168] and is highly soluble in water. It is a globular protein composed of three triple-helical domains shaped as an ellipsoid with a diameter of 38 Å and a length of 150 Å. The triple-helical domains consist of both α-chains and β-strands. The β-strands are aligned adjacent to each other such that hydrogen bonds can form between both the NH-groups and the CO-groups of the adjacent β-strands.[130] The β-sheets formed from these β-strands are pleated with carbon atoms, as shown in Figure 39.6. A series of cysteine disulfide bridges, which constitute

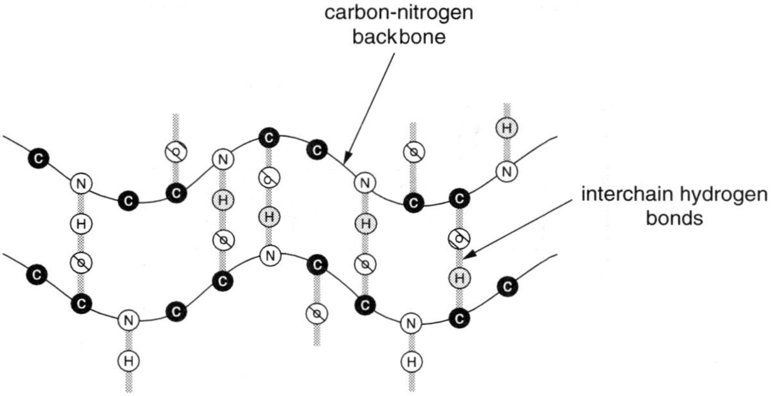


**FIGURE 39.6** Schematic showing the β pleated sheet in protein secondary structures. (Adapted from Yudkin, M. and Offord, R., *Comprehensive Biochemistry*, Longman, London, 1973.)

covalent bonds, stabilizes the albumin structure. Due to the strong covalent bonds in the β-sheet, albumin has much less elasticity than collagen.

## 39.5.4 Mechanism of Laser Tissue Soldering

Several studies investigating the thermal denaturation process of albumin have been conducted.[169–172] With increased temperature, intramolecular motion has been shown to increase and irreversible structural alterations to occur.[172] Loss of α-helix and gain of β-form have been shown to occur between 62 and 75°C, according to infrared laser and Raman studies.[171] The denaturation point of human serum albumin has been found to be 63°C using differential scanning calorimetry.[173] In another study using differential scanning calorimetry, the peak denaturation temperatures for bovine, porcine, canine, and human serum albumin containing fatty acid were found to be 65.31 ± 0.50°C, 65.32 ± 0.20°C, 66.97 ± 0.69°C, and 67.77 ± 0.88°C, respectively.[174] Fatty-acid-free albumins were found to denature at significantly lower temperatures. Nuclear magnetic resonance studies suggest comparable results with denaturation occurring at 72°C.[172]

Marx showed that the much smaller albumin monomers (80 to 150 Å) were able to diffuse among the collagen fibers (500-nm diameter) as they underwent thermally driven conformational transitions due to laser irradiation.[175] The albumin monomers then polymerized by hydrophobic effects to form an interdigitated matrix among the collagen fibers. Using scanning electron microscopy to investigate laser tissue repair of dura mater and peripheral nerves, Menovsky et al. reported that coagulated albumin forms a solid bridge between the tissue edges, which is melted on and between the collagen fibrils.[137] This finding was supported in a later study by Park and Min using albumin solder in conjunction with a $CO_2$ laser for ossicular reconstruction.[176] Using a fluid protein solder, Kirsch et al. showed that a scaffold-like internal lattice was formed when the water content of the solder vaporized, causing the albumin component to bubble and desiccate.[177] This scaffolding was incorporated into the healing tissue layers. After 5 to 7 d, cellular ingrowth into the solder was observed with the deposition of collagen extracellular matrix proteins. These results were similar to those of Tang et al., where a large number of new collagen fibers were observed to have proliferated in the vessel walls by the tenth day of regeneration after treatment with an 830-nm diode laser.[147,148] Finally, in a study by McNally et al., scanning electron microscopy suggested albumin intertwining within the tissue collagen matrix and subsequent fusion with the collagen as the mechanism for laser tissue soldering.[178]

### 39.5.4.1 Temperatures Required for Laser Tissue Soldering

If thermal restructuring is the main mechanism of laser tissue soldering, then the strength of the solder–tissue bond depends on appropriate temperatures being reached for a suitable duration (set by the choice of laser and solder parameters) to obtain protein restructuring and on the contact area between

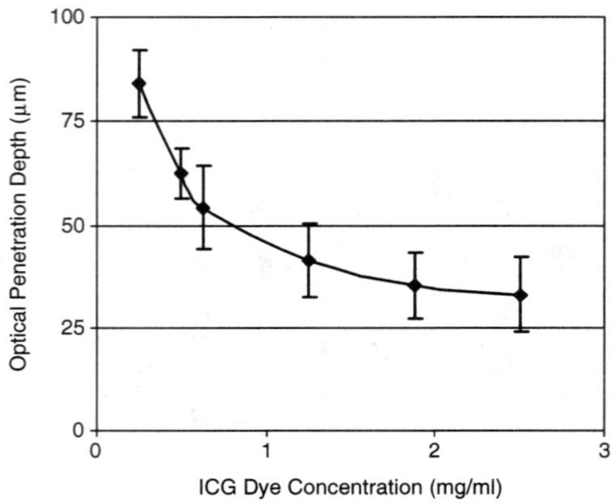


**FIGURE 39.7** Optical penetration depth of 805-nm light in solid-protein solder (60% BSA) at a range of ICG dye concentrations.

the solder and the tissue, which determines the number of available bond sites. The proteins at the solder–tissue interface must denature simultaneously to allow the solder proteins (albumin) to bind to the adjacent tissue proteins (collagen). The energy applied to the solder and the tissue must be high enough to initiate molecular transformation, i.e., hydrogen-bond breaking, to permit polypeptide rearrangement. This energy represents the activation energy required for rotation around the C–N bond that links the amino acids.[179] Subsequent to hydrogen-bond breaking, polypeptide uncoiling in the tissue collagen would require additional energy.[180] An activation energy of 480 J/g (energy content of hydrogen bond in tissue with two hydrogen bonds per peptide) with an additional energy requirement in excess of 800 J/g was suggested in a study by Kung et al.[180]

## 39.6 Recent Advances in Light-Activated Surgical Adhesives

The results of the various studies discussed above are indeed promising for the future of laser tissue repair; however, several problems associated with the technique still must be overcome. In particular, four disadvantages of laser tissue soldering have been identified. (1) Solid-protein solders are initially malleable, but they quickly dry and become very brittle and inflexible. As such they are not easily adapted to different tissue geometries. Liquid solders, on the other hand, tend to "run away" from the application site before they can be denatured with the laser. (2) Protein solders are soluble in physiological solutions prior to denaturation, which can be problematic during application. In addition, as these solders are often subjected to blood dilution during operation, the solder may undergo mechanical alteration that can weaken the solder–tissue bond.[181] (3) Whether water or an absorbing dye is used as the chromophore, more energy is generally absorbed near the upper portion of the solder, closer to the laser source. Irradiation of the solder produces a temperature gradient over the depth of the solder. The gradient is a function of the irradiance and the absorption coefficient, $\mu_a$, of the solder/dye combination. The 1/e penetration depth of the laser light is defined as $1/\mu_a$, neglecting scattering of the naturally clear solder. Figure 39.7 shows a plot of the optical penetration depth of laser light at 805 nm in albumin-protein solders containing different concentrations of ICG dye. The optical penetration depth is observed to decrease with increased ICG-dye concentration. Depending on the temperature gradient and the laser exposure, the upper portion of the solder can become overcoagulated while the most critical region — the solder–tissue interface — remains uncoagulated. Nonuniform denaturation across the solder thickness can result in the formation of unstable solder–tissue bonds.[89,126] (4) The achievement of good apposition between tissue edges during tubal anastomoses is particularly difficult. In addition, there is a high risk of fusing the anterior and posterior walls together during the laser-welding or laser-soldering procedure, which can result in stenosis at the irradiated site.[70] Stay sutures can be used to combat this problem; however, they only tend to complicate the operative procedure and introduce the additional risk that a foreign-body reaction will occur.

Researchers continue to search for a new range of light-activated surgical adhesives that can combat the four identified problems associated with current laser tissue-soldering techniques. Recent advancements have included the development of (1) semi-solid solders and polymer-reinforced solders as a means of improving the flexibility of the adhesive, (2) predenaturation of the solders as a means of enhancing the stability of the adhesive in physiological fluids prior to laser denaturation, (3) layered solders containing different chromophore concentrations as a means to improve control of the heat-source gradient through the adhesive during laser irradiation, and (4) stents as a means to provide good tissue apposition during the laser procedure.

## 39.6.1 Semi-Solid and Polymer-Reinforced Solders

Improvements on liquid- and solid-protein solders have been made with the introduction of semi-solid solder made of hydroxypropylmethylcellulose (HPMC), albumin, and water.[174] HPMC is a synthetic water-soluble, nonstarch polysaccharide that has been safely used in humans and animals to lower cholesterol.[182,183] Bleustein et al. have shown that repairs formed using semi-solid solder in conjunction with an Nd:YAG laser in canine small-bowel and porcine-skin models are much stronger than both 50% liquid solder and 60% solid solder and that they remain stronger than liquid solder repairs at 7 d.[174] The semi-solid solder is observed to have better handling characteristics.

Another improvement on traditional liquid- and solid-protein solders involves the use of synthetic polymers, which when doped with serum albumin provide better flexibility along with improved repair strength over previous published results using albumin-protein solders alone.[184] Poly(lactic-co-glycolic acid) (PLGA) is one of several types of poly(alpha ester)s, a class of synthetic biodegradable polymers and copolymers, under investigation. PLGA, a copolymer of glycolic acid and lactic acid, is degraded *in vivo* by hydrolysis of the ester bonds. The products are eliminated through normal metabolic pathways.[185] In addition to current usage in biodegradable sutures, these materials have been researched as drug-delivery systems. Thin membranes of PLGA of various thickness are easily fabricated in the laboratory using a solvent-casting technique.[186,187] By adding salt to the polymer mixture during the casting stage, and later leaching it out, one can produce membranes of a desired level of porosity (Figure 39.8). The resulting porous membranes quickly absorb the traditional protein-solder mix of serum albumin and an absorbing chromophore. A range of dopants, including hemostatic and thrombogenic agents, antibiotics, anesthetics, and various growth factors, can be added to the solder-doped polymer membranes to enhance the rate and quality of wound healing.[32] Finally, the *in vivo* degradation rate, and consequently the drug-delivery rate, can be tailored by altering the composition of the polymer.[185,188] McNally et al. have shown that the bursting pressure of repairs performed on both heparinized and unheparinized blood vessels using the reinforced solder (Figure 39.9) is comparable to that measured in native vessels.[108,189] In addition, repairs formed with the adhesive technique were achieved more rapidly than suturing, and acute leakage was observed less frequently. Reinforcement of traditional liquid-solder coagulums with thin strips of PLGA membranes has also been investigated with good success rates.[190,191]

## 39.6.2 Predenatured Solder

Prior to laser denaturation, protein solders tend to dissolve within a few minutes of application in a hydrated environment.[192] This can be quite problematic for most surgical procedures using the protein solder, where ultimately there must be some time lag between placement of the solder and laser treatment. Predenaturation of albumin-protein solders using hot-water baths set at temperatures between 65 and 80°C or low concentrations of tissue fixatives are currently under investigation as a means for reducing the solubility of the materials in physiological solutions prior to laser irradiation.[90,184] McNally et al. have shown predenaturation of solid protein solder at 75°C to be a feasible option for reducing the solubility of the solder and ensuring that minimal mechanical alteration occurs during the surgical procedure that

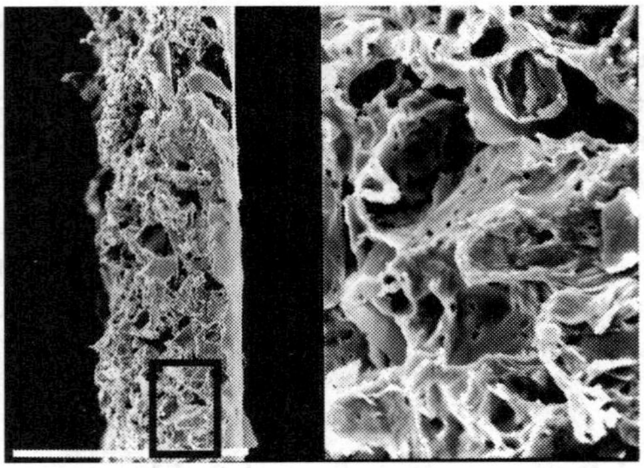

A

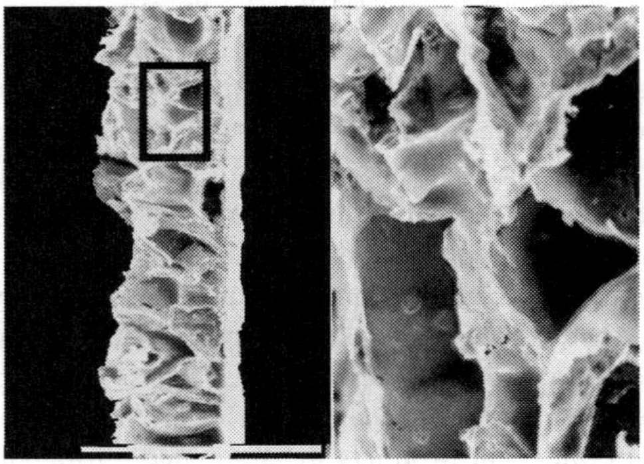

B

**FIGURE 39.8** Scanning electron micrographs of PLGA scaffolds prior to doping with protein solder. By using salt particles of different sizes, membranes were prepared with the same porosity but with different pore diameters. (A) Scaffold cast using salt particle sizes ≤ 106 μm in diameter. (B) Scaffold cast using salt particle sizes of 106 to 150 μm in diameter. The right half of each micrograph shows a higher magnification view (526 times) of the section of the micrograph highlighted by a rectangular frame on the left half of the micrograph (88 times). The scaffolds exhibit a uniform pore morphology, which is interconnected, thus yielding an open-cell polymer. (From McNally, K.M. et al., *Lasers Surg. Med.*, 27(2), 147, 2000. With permission.)

could weaken the solder-tissue bond.[184] Predenaturation of the solid-protein solder improves the flexibility of the solder, a feature that has proven beneficial for the formation of flexible stents (Figure 39.10) now being used for sutureless end-to-end ureteral anastomoses.[90,105] Tanning with glutaraldehyde had previously been applied to human umbilical vein grafts,[193] collagen grafts prepared from bovine carotid arteries,[194] and albumin-coated knitted polyester grafts[195] to improve the tensile strength of the grafts. The strength of repairs formed using solder predenatured with 0.5 to 1.0% glutaraldehyde, however, were not as strong as their heat-denatured counterparts.[184] This was attributed to the fact that glutaraldehyde is a known fixative, and some degree of fixation would have occurred in the albumin-protein solder.

## 39.6.3 Chromophore Concentration Gradient

Wavelength-specific chromophores such as ICG are typically added to protein solders to provide for selective absorption of the laser light as a means of minimizing the degree of collateral thermal damage caused to the surrounding tissue during the laser procedure (see Section 39.4). The optical penetration depth of 805-nm light in solid-albumin-protein solders containing 2.5 mg/ml ICG, the most common ICG dye concentration used by researchers, is approximately 35 μm. Thus, approximately 63% of the laser light is absorbed in the anterior surface of the solder, and thermal conduction is relied upon to transfer the heat to the posterior

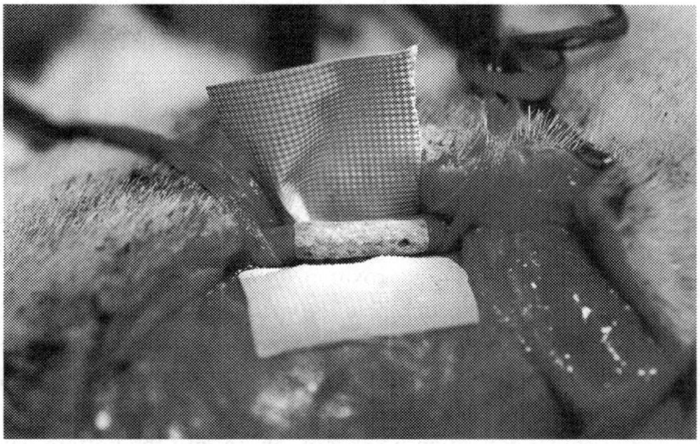

**FIGURE 39.9 (Color figure follows p. 28-30.)** Intraoperative photograph showing the right carotid artery of a canine repaired using a solder-doped polymer membrane in conjunction with a diode laser ($\lambda$ = 805 nm, spot diameter = 2 mm, $E_0$ = 15.9 W/cm$^2$, $t$ = 100 sec). The polymer membrane was fabricated using PLGA with an 85:15 copolymer ratio and 70% NaCl with particle sizes of 106 to 150 μm and doped with protein solder composed of 50% BSA, 0.5 mg/ml ICG, and deionized water. (From McNally-Heintzelman, K.M. et al., *J. Biomed. Opt.*, 6(1), 68, 2001. With permission.)

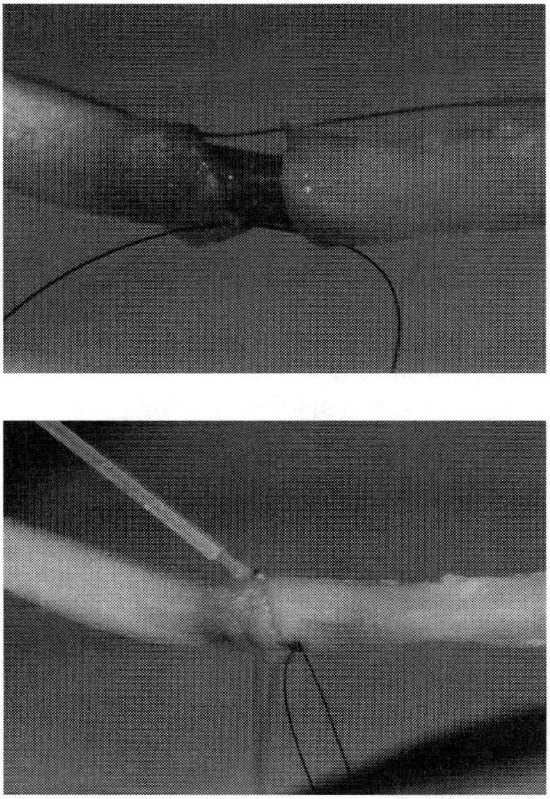

**FIGURE 39.10** Use of an albumin stent (50% albumin, 10 m*M* ICG) for laser-assisted ($\lambda$ = 810 nm, spot diameter = 1 to 2 mm, $P$ = 1 W, pulse width = 0.1 sec, repetition rate = 5 Hz) uretal anastomosis. The addition of dye to the stent maximizes heat conduction at the solder/tissue interface. (Image courtesy of Hua Xie, Oregon Medical Laser Center, Portland, Oregon.)

surface of the solder, where it is needed to provide adequate solder–tissue fusion. As solders often range in thickness between 100 and 250 μm, this assumption is not always valid. In a study by McNally et. al., two 75-m-thick solid-protein solder strips containing different ICG dye concentrations were pressed together using a vice to form a single solid solder strip 150 μm thick.[196] The solder strips were pressed together immediately after preparation, prior to drying, so as to create a chromophore-concentration gradient across the thickness of the solder, rather than two separate layers. Repairs made with solders containing a chromophore-concentration gradient formed by placing a protein solder strip containing 0.25 mg/ml ICG on top of a strip containing 2.5 mg/ml ICG were found to be approximately 58% stronger than those formed using solder containing 2.5 mg/ml ICG and approximately 33% stronger than those formed using solder containing 0.25 mg/ml ICG. The improved strength of repairs formed using solders containing this concentration gradient was attributed to more even heating across the thickness of the protein solder, eliminating the problems of "overcooking" the anterior portion of the solder while having insufficient coagulation at the vital solder–tissue interface. Lauto et al. applied a similar principle when designing their two-layered solders.[197] In this study, the top layer of solder contained no chromophore, and the bottom layer of solder contained 0.39 ± 0.08% carbon black. Each layer was approximately 55 to 75 μm thick. Heat diffusion from the black midplane of the two-layered solder was found to decrease the difference in temperature recorded on the solder surface and at the solder–tissue interface.

### 39.6.4 Biocompatible Stents

Biocompatible stents have been used for several years to assist surgical procedures during conventional microsuturing of blood vessels.[198] The function of the stent is to separate the anterior and posterior walls of the vessel, allowing for an easier and faster repair procedure with minimal tissue injury. He et al.[19] and He and Tang[199] have extended this idea to obtain a hybrid technique that they refer to as the "stented welding" technique. Stents fabricated using a mixture of glyceride, diglyceride, and triglyceride — elements found naturally in the blood circulatory system — and a melting temperature of 39°C have been used for small-vessel (0.4 to 0.6 mm) anastomosis by $CO_2$ laser welding in a rat model.[19] This melting temperature was selected to match the blood temperature of rats, ensuring that the stent would become completely molten once blood flow was restored after surgery. All repaired arteries remained patent at 1 h post surgery, with no detectable stenosis or dilation at the repaired sites. The stent was dispersed in the bloodstream with no evidence of adverse effects. When compared with conventional microsuturing techniques, the stented welding procedure offers the benefits of having a high patency rate, being less time-consuming· and involving less hand manipulation.[199] As such this new technique holds great promise for extension to laser-soldering techniques as well.

More recently, researchers at New York Hospital–Cornell University Medical Center have designed a biocompatible stent with a self-expandable mechanism to facilitate laser-assisted tubal anastomoses. Thin films, prepared from deacetylated chitosan dissolved in acetic-acid solution, are pulled and wound around a cylindrical rod in a helical fashion to manufacture the stents (Figure 39.11). The elastic nature of the chitosan· combined with the helical shape, makes the stent self-expand when inserted in small ducts. Upon stent insertion, tubal anastomoses can be achieved using standard laser-welding and soldering techniques. Lauto et al. have used these new stents to perform laser-soldered vasal anastomoses in wistar rats, where 12 of 13 anastomoses were observed to have remained open inside the vas deferens at 8 weeks postoperative despite a high incidence of sperm granuloma formation.[200]

## 39.7 Beyond the Laser

### 39.7.1 Automated Dosimetry

Although several successful experimental studies have been conducted using laser tissue welding and soldering techniques, they have not yet gained widespread clinical acceptance because reliable results have been difficult to obtain. As discussed in Section 39.2.2 above, a major limitation of this approach

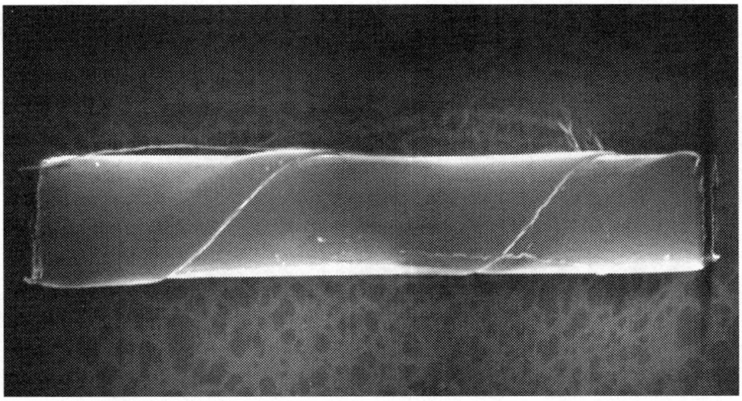

**FIGURE 39.11** Chitosan stent prepared from deacetylated chitosan film pulled and wound around a cylindrical rod in a helical fashion. (Image courtesy of Antonio Lauto, New York Hospital–Cornell University Medical Center, New York.)

seems to be the lack of a satisfactory objective criterion for the optimal laser exposure required to produce a strong and durable repair. Several studies have been conducted to develop a system that could monitor the condition of the tissue undergoing the welding/soldering process and regulate the laser exposure accordingly. The successful implementation of such a system could remove much of the uncertainty from laser tissue welding and soldering techniques, rendering them both simpler and more effective.

Protein denaturation is a rate process governed by the local temperature–time response.[201,202] It can be approximated by a simple Arrhenius model:[156,158,159]

$$\Omega(z,t) = -\ln\left[\frac{C(t)}{C(t_0)}\right] = \int_0^t A\exp\left(-\frac{E_a}{RT}\right)dt \tag{39.1}$$

where $\Omega$ is the damage integral [no units], $C(t_0)$ represents the amount of native (undamaged) tissue originally present at time $t_0$ [M], $C(t)$ is the amount of native tissue at time $t$ [M], $A$ is the frequency factor or preexponential factor [sec$^{-1}$], $E_a$ is the activation energy [J/mol], $R$ represents the universal gas constant [8.314 J/mol K], and $T$ is the absolute temperature [K].

This equation is used to describe kinetic processes in which a population of reactants, $C(t_0)$, is converted to reaction products, $C(t)$, at a rate that depends on temperature. Coagulation (observable damage) is defined as $\Omega = 1.0$, which corresponds to a reduction in reactant concentration by 1/e.[203] The total damage, $\Omega$, is the summation of all damage increments over time.

The values of $A$ and $Ea$ depend on the tissue or solder irradiated and are determined by plotting experimental data in the form of $\ln(t_0)$ vs. $1/T$, with $\Omega = 1.0$. Agah et al. determined values of $A$ and $E_a$ for human aorta to be $5.6 \times 10^{63}$ sec$^{-1}$ and $4.30 \times 10^5$ J/mol, respectively.[204] The rate of damage formation

$$\frac{d\Omega(z,t)}{dt} = A\exp\left(-\frac{E_a}{RT}\right) \tag{39.2}$$

for these coefficients is illustrated in Figure 39.12. The figure graphically illustrates the importance of the temperature–time relationship in the occurrence of thermal damage. Rate coefficients for various tissues have been published by Pearce and Thomsen.[16]

The extent of protein denaturation is linearly proportional to time and an exponential function of temperature. As a result of this exponential dependence, temperature is the dominant parameter for governing the extent of the reaction. Thus, an obvious variable to monitor is the tissue temperature.[205] Several research groups have attempted temperature-controlled tissue bonding for laser tissue welding[50,52,158,206] and laser tissue soldering[37,65,207,208] techniques.

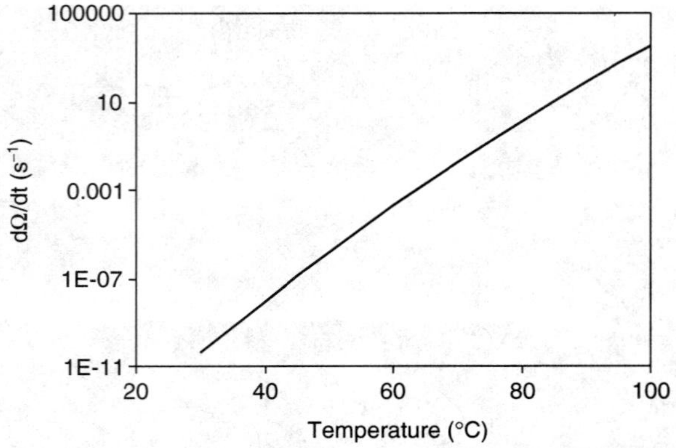


**FIGURE 39.12** Rates of accumulation of damage in human aorta. Rate coefficients ($A = 5.6 \times 10^{63}$ sec$^{-1}$ and $E_a = 4.30 \times 10^5$ J/mol). (From Agah, R. et al., *Lasers Surg. Med.*, 15(2), 176, 1994. With permission.)

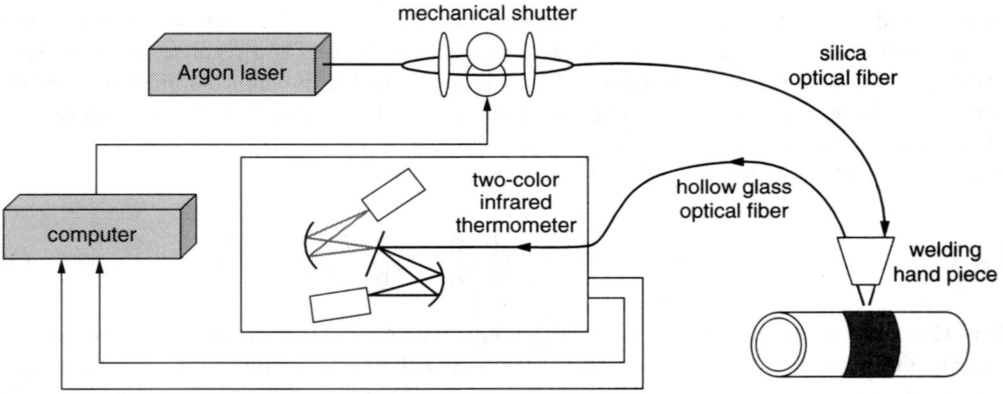


**FIGURE 39.13** Schematic of the LLNL temperature feedback system. A two-color infrared temperature sensor is used to control the laser used to heat the weld site.

The difficulty of radiometric temperature measurement is control of the distance between the detector and the surface of the tissue. To measure temperature with a hand-held device IR-remitted power must be measured at two wavelengths. The ratio of these measurements provides a unique measurement of temperature, based on Planck's radiation law, which is independent of the distance between the detector and tissue.[209] Systems that incorporate two-wavelength detection for tissue welding have been developed at Tel Aviv University and Lawrence Livermore National Laboratories (LLNL) (Livermore, CA).

Researchers at Tel Aviv University have used the concept of infrared radiometry to develop a noncontact technique to measure the surface temperature of biological materials during laser tissue welding/soldering. Their system is based on a $CO_2$ laser and on infrared-transmitting AgClBr fibers.[52,152,205] The laser beam is focused on the tissue–solder surface, and a remote infrared thermometer is used to monitor the black-body radiation emitted from the tissue–solder surface during the laser heating procedure. A feedback loop is then used to control the surface temperature during the laser procedure. Using this system, temperature control to ± 0.1°C has been achieved with a response time of 0.1 to 0.2 sec.[52] This fiber-optic system could be adapted for endoscopic laser welding.

A schematic of the sensor developed at LLNL is shown in Figure 39.13. The sensor has a single 700-μm core and a hollow-glass optical fiber mounted on one prong of a two-pronged welding handpiece. The other prong is the silica optical fiber that transmits the laser beam that welds the tissue. While the laser beam is heating the tissue surface, the sensor's hollow fiber collects and transmits the infrared radiation to the detectors. Using a single optical fiber for both wavelength bands guarantees that the radiation transmitted in each path originates from the same geometric region on the target, providing true values for the temperature and emissivity. The hollow fiber optic has a small numerical aperture that can accurately monitor the surface temperature of a 0.8-mm spot at a distance of 1 cm. The variable working distance means that the surgeon can more easily maneuver around the surgical site while still obtaining accurate temperature measurements.

Another technique for monitoring the temperature of laser-irradiated tissue has involved the use of temperature-sensitive liposome-encapsulated dyes.[210] Heating of the liposomes by means of a laser causes the lysis of the liposomal membrane.[211,212] The fluorescence emission is then measured with a fluorescent imaging system. The amount of the drug release has been shown to increase linearly with increasing temperature, allowing measurement of the temperature of the laser-irradiated tissue.[210] While the response time of this technique depends on the release latency time of the liposomes, using different liposome formulations for different temperature ranges, e.g., di-stearoyl-phosphatidyl-choline (DSPC) for temperatures between 53 and 62°C and di-stearoyl-phosphatidyl-ethanolamine (DSPE) for temperatures between 65 and 75°C, can reduce this limitation.

Finally, Tang et al. have used native fluorescence spectroscopy to monitor molecular changes in laser-treated tissue and thus evaluate laser-induced thermal tissue damage.[147,148,213,214] Examples of natural-tissue fluorophores include collagen and elastin. Most recently, they have shown that fluorescence emission from collagen at 380 nm with 340-nm excitation and from elastin at 450 nm with 380-nm excitation decreases with thermal damage in a predictable fashion following laser skin welding.[213,214]

## 39.7.2 Optical-Thermal Models of Laser Tissue Repair

Computational modeling of the laser-tissue-interaction process can be used to explore and reduce the experimental parameter space including laser irradiance, exposure time, protein concentration, and chromophore concentration. The use of numerical models, such as the radiative transport equation,[215] the diffusion approximation,[216–218] Beer's law,[219] or the Monte Carlo technique[220] for optical modeling and the finite difference or finite element techniques[221] for thermal modeling, have helped in the understanding of the basic optical and thermal mechanisms of laser–tissue interactions that are pertinent to laser tissue welding and soldering. Using the bioheat-transfer equation

$$\frac{1}{\alpha}\frac{dT}{dt} = \nabla^2 T + \frac{Q}{k} + \rho c_p \omega (T_f - T) \tag{39.3}$$

where $T$ is temperature [K], $t$ is time [sec], $\alpha$ is thermal diffusivity [m$^2$/sec] ($\alpha = k/\rho c$, where $\rho$ [kg/m$^3$] is the tissue density, and $c_p$ [J/kg/K] is the tissue heat capacity), $k$ is the thermal conductivity of the tissue [W/m/K], $Q$ is the heat-source term [W/m$^3$], $\omega$ is tissue perfusion [kg of perfusate/kg of tissue/sec], and $Tf$ is the fluid temperature [K], as well as the Arrhenius rate process relation (Equation 39.2), thermal destruction of tissue has also been simulated by a rate process that is a linear function of time and an exponential function of temperature.[16,222] These optical and thermal models are now being combined together to account for local irreversible changes in the optical and thermal coefficients of the tissue due to thermal damage.[223–226]

Researchers at LLNL have developed a two-dimensional (2D) model, LATIS (LAser-TISsue interaction), which uses coupled Monte Carlo, thermal-transport, and mass-transport models to simulate tissue welding of multilayered tissue.[225,227–230] In this model, light transport is calculated with the Monte Carlo technique. The absorbed energy serves as the source term to a modified version of the heat-conduction equation, which takes account of the tissue hydrodynamic response including evaporation and water diffusion. The dynamic optical properties of each layer are modeled as a kinetic function of tissue temperature rise and thermal damage as governed by the Arrhenius relation. Thus, full coupling of the optical and thermal response of the tissue is achieved. In an *in vitro* study of laser tissue welding mediated with a dye-enhanced protein solder for repair of arteriotomies approximately 4 mm in length, temperature histories of the surface of the weld site were obtained using a fiberoptic-based infrared thermometer.[230] Comparison of the experimental and simulated thermal results showed that the inclusion of water transport and evaporative losses in the model, a concept first reported by Torres et al.,[231] was necessary to determine the thermal distributions and hydration state of the tissue. Such volumetric information, used in conjunction with weld strength assessment, provides an interesting tool for investigating the mechanisms of laser–tissue fusion.

McNally et al. have developed a transient 2D optical-thermal model that accounts for dynamic changes in optical and thermal properties with temperature to investigate the mechanisms leading to thermal damage during laser tissue soldering.[196] The model was implemented using the electrical circuit simulator SPICE (Simulation Program with Integrated Circuit Emphasis).[232] Electrical analogies for the optical and thermal behavior of the solder and tissue were established. With these analogies, light was propagated using a flux representation of the light, and the electrical simulator was used to calculate heat transfer with an algorithm based on the finite-difference technique. Thermal damage was calculated using the Arrhenius rate process relation. Temperature-dependent absorption and scattering coefficients, thermal conductivity, and thermal diffusivity[226] were incorporated in the SPice Optical-Thermal Simulation (SPOTS), as well as the time-domain behavior of a scanning laser source.

Experimental results from an *in vitro* study performed using an 808-nm diode laser in conjunction with ICG-doped albumin-protein solders to repair bovine aorta specimens compared favorably with numerical results obtained from SPOTS using dynamic optical and thermal properties.[196] In these studies, infrared radiometry and histological analyses were used to determine the solder surface temperature and the extent of thermal damage in the tissue substrate, respectively, as a result of varying laser and solder parameters. The maximum surface temperature was overestimated by almost 10% when dynamic properties were not taken into account. This difference corresponds to over two orders of magnitude difference in terms of the Arrhenius tissue-damage integral. Thus, the incorporation of dynamic changes in optical and thermal properties of tissue during laser-induced heating does appear to represent a significant advance in computer modeling of laser–tissue interactions.

## 39.8 Clinical Applications

### 39.8.1 Cardiothoracic and Vascular Surgery

In cardiothoracic and vascular surgery, it is often necessary to perform anastomoses on vessels with diameters in the range of 0.5 to 5 mm. Suture repair of such vessels can be technically challenging.[27] This fact has prompted several researchers to commence the search for an alternative technique that can produce strong, reliable, and reproducible repairs. Laser welding for achieving vascular anastomoses was first proposed in the early 1960s.[13] Since then, extensive effort has been expended in performing vascular anastomoses using a wide range of lasers including $CO_2$,[19,53,83,233,234] THC:YAG,[71] 1.9-μm diode,[21,23,107,180] Nd:YAG,[21,107,235] ~800-nm diode,[104,108] and argon[27,58,236,237] lasers.

Laser welding has been shown to produce excellent patency with faster anastomotic times and reduced problems of foreign-body reaction in comparison with suturing.[30,58,236,238] In addition, histological studies have shown that laser welding can lead to improved healing, with an increased ability for expansion and growth at the anastomotic site[58] and reduced neointimal hyperplasia compared with suture repairs.[27,41,58] One major complication associated with the laser welding is the risk of thermal injury to the vessel wall, which can result in thrombogenesis.[180,234] If thermal injury is limited to the superficial layers, thrombogenesis may be prevented.[53,56,60]

The first clinical application of laser tissue welding was performed in 1985 on a 44-year-old female patient with chronic renal failure for whom an arteriovenous fistula was created using a low power $CO_2$ laser and a few stay sutures.[233,239] Based on the excellent results of this study, laser-assisted arteriovenous and arterio-arterial anastomoses of the peripheral vessels in 35 patients were subsequently attempted. Pressure-tolerance tests, tensile-strength tests, and microscopic examinations at the sites of the laser anastomoses were compared with the conventional suture method. No significant differences between laser and suture methods were observed.[233]

In 1990, White et al. investigated the feasibility of forming vascular anastomoses using an argon laser and four evenly spaced biodegradable stay sutures. In initial animal studies, femoral arteriovenous fistulas were created bilaterally in each of ten dogs and studied histologically over a period of 2 to 24 weeks.[236] All anastomoses remained patent without hematomas, aneurysms, or luminal narrowing during this evaluation period. Control suture specimens at the same intervals exhibited an organized fibrous-tissue response to the suture material. Clinical adaptability of the laser-welding technique was subsequently evaluated in five patients at 10 to 27 months by physical examination and duplex scanning. No evidence of abnormal healing was observed.

One significant advance in the area of plastic and reconstructive surgery has been the microsurgical transfer of free flaps. During these procedures, microsurgical-vessel anastomoses are necessary to restore blood circulation. These anastomoses are performed on vessels whose diameters are less than 5 mm. Conventional anastomoses use very fine sutures, typically around 22 μm in diameter, and as such patency of these anastomoses is highly dependent on the surgeon's skill.[240] Laser tissue welding and soldering offer a potential solution to this problem.[18,23,71] In 1999, Maitz et al. demonstrated the use of a fully biodegradable, diode-laser-activated protein tube for sutureless microvascular anastomoses.[105] The protein tube is used to achieve good apposition of the tissue edges during end-to-end anastomoses while separating the anterior and posterior walls of the vessel to prevent vascular occlusion. The combination of these biodegradable protein tubes with a diode laser provides a quick and reliable technique to successfully anastomose small-diameter arteries, avoiding vessel-wall fibrosis by eliminating any permanent implanted devices.

Diagnostic catheterization and interventional cardiac procedures including coronary angioplasty require hemostasis at the puncture site immediately following the procedure. Several techniques have been investigated as a means for achieving rapid hemostasis including manual compression as well as the application of either collagen vascular hemostatic devices or suture-mediated devices.[241–243] While these conventional techniques are effective in achieving adequate hemostasis in most patients, they are not without complications. Manual compression is uncomfortable and often painful for the patient. Risks associated with the collagen vascular hemostasis devices include arterial occlusive complications, local infection,[244] and allergies to the collagen.[245] Although suture-mediated devices can achieve closure of the site just as well as open surgical methods, they are limited by the complexity of the devices and a steep learning curve associated with the procedures.[244,246] One possible alternative to the conventional techniques for achieving rapid hemostasis at the puncture site is laser tissue soldering. In a 2000 study, McNally et al. demonstrated the feasibility for using protein-solder-doped polymer membranes in conjunction with a diode laser for achieving rapid vascular hemostasis following cardiac procedures in both canine and porcine models.[108,189] In the porcine study, the patency rate of both unheparinized and heparinized vessels was 100% at 3 d postoperative with evidence of intraluminal thrombosis seen in only 1 of the 24 repaired vessels.[108] In addition, repairs formed with the adhesive technique were achieved more rapidly than suturing, and acute leakage was observed less frequently.

Sealing of prosthetic vascular grafts to native vessels has also been attempted.[247–251] Limited success has been achieved using tissue-welding techniques; however, laser soldering, used in conjunction with a few stay sutures to provide good apposition between the tissue edges and the graft material, has been particularly useful in providing a watertight seal over the repair.[248,251]

Lastly, laser tissue welding and soldering techniques have been used to repair air leaks after lung biopsy or wedge resection[252] and bronchial stump leaks.[104] In these studies, laser soldering was found to be superior to both conventional suturing and fibrin-glue techniques. Attempts have also been made to weld an expanded polytetrafluoroethylene (e-PTFE) patch to the trachea to promote healing.[253]

## 39.8.2 Dermatology

Laser welding and soldering techniques offer many potential advantages over standard suture and staple closure of cutaneous tissue including minimal tissue handling, quicker closure times, minimal foreign-body

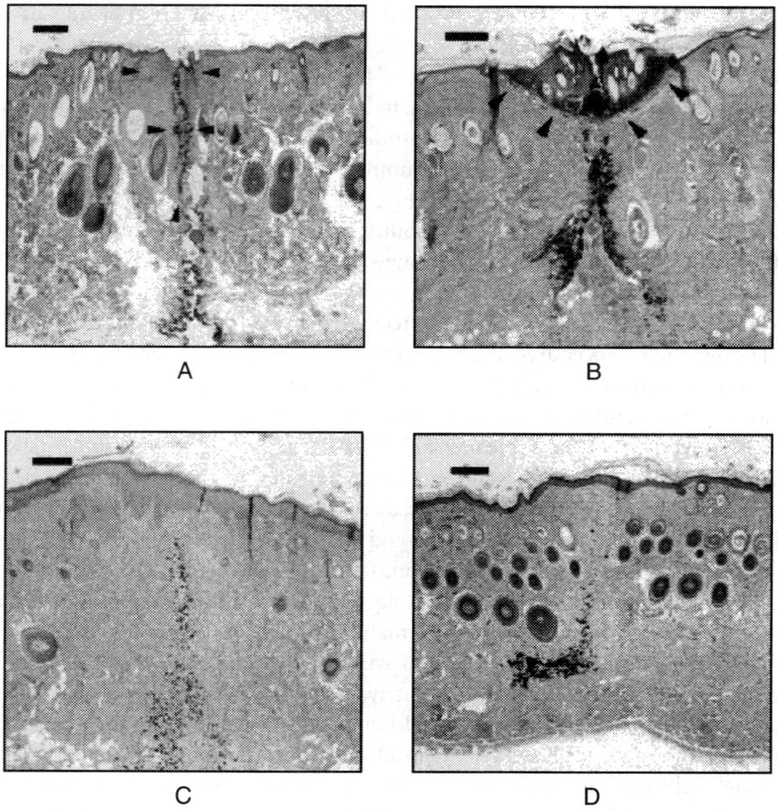

**FIGURE 39.14** Photomicrographs showing wounds closed by Nd:YAG laser ($\lambda$ = 1.06 μm, spot diameter = 4 mm, $P$ = 10 W, pulse width ~80 msec, pulse interval = 8 sec): (A) day 0, (B) day 6, (C) day 14, and (D) day 28. Arrows indicate border of the thermal damage zone. The chromophore (India ink) is visible along the borders of the incision. At day 0 a "funnel-shaped" thermal denaturation zone with an average width of 200 μm near the epidermis and 50 μm in the mid-dermis is seen. At day 6, the epidermis had hypertrophied to 300 μm, and a complete epithelial bridge had formed under the necrotic tissue contained within the thermal damage zone. Wound healing progressed between days 14 and 28, with granulation tissue appearing in the papillary dermis, thus filling the gap created after sloughing of thermally damaged tissue (around day 10) from the wound site. Scale bars = 200 μm. (From Fried, N.M. et al., *Lasers Surg. Med.*, 27(1), 55, 2000. With permission.)

and inflammatory response, maximal tissue alignment, early reepithelialization, watertight closure, and, most important, improved cosmesis.[254] Laser welding has been explored as an alternative method of skin closure by many researchers using a variety of sources including $CO_2$,[55,254,255] Nd:YAG, [254,256,257] diode,[31] and argon[254] lasers.

As mentioned above, a superior cosmetic result is claimed as the principal advantage of laser skin closure. The principal disadvantage is dehiscence. It is nearly impossible to obtain full-thickness welds while limiting thermal denaturation lateral to the weld site in the epidermis and papillary dermis. While thermal denaturation of tissue is necessary to produce a strong weld, excessive thermal damage may result in initially high-weld strengths, which then decrease because of sloughing of the necrotic tissue during wound remodeling, eventually leading to unnecessary scarring and dehiscence.[66,94] Dye enhancement of the laser-welding procedure has been investigated to minimize unwanted collateral thermal injury by ensuring selective energy deposition, thereby allowing the use of low-power lasers (Figure 39.14). Combinations include ICG with a diode laser[95,258] and India ink with an Nd:YAG laser.[66,94,259] In 2000, Fried and Walsh used India ink as an absorber to create welds with immediate tensile strengths higher than sutured wounds.[94] Application of a variety of soldering agents has also been shown to greatly improve the strength of cutaneous repairs.[29,51,118,119,177]

More recent studies have focused on optimizing the welding technique to minimize the effects of collateral thermal damage. In a 2000 study, Fried and Walsh showed that the typical thermal-denaturation profile demonstrated significant denaturation near the tissue surface. The extent of denaturation then decreased with depth as both absorption and scattering attenuated the incident radiation.[66] Surface cooling with either a water or cryogen spray can effectively invert the denaturation gradient to minimize the problem of scar formation in the epidermis and papillary dermis that can affect cosmesis.[66,260] Additionally, weld strength may be improved when denaturation is confined to the mid- and deep dermis resulting in a larger cross-sectional area for collagen bonding across the weld site. In the study by Fried and Walsh, dynamic cryogen cooling of the epidermis and papillary dermis was proven effective in reducing the amount of collateral thermal tissue damage.[66]

### 39.8.3 General Surgery

Laser tissue welding has been investigated as an alternative technique for repair of most tissue types in the gastrointestinal tract, where, compared with suture closure, it has been found to be a much easier technique with greater speed, a better healing response, and no stone formation. The first published study about laser bowel welding was reported in 1986 by Sauer et al.[261] A $CO_2$ laser was used to repair longitudinal transmural incisions in otherwise intact rabbit ileum, producing strong hermetic closures. Visual observation of a slight desiccation and shrinkage of the tissue indicated the endpoint of the procedure. In 1989, Sauer et al. proposed the use of an Nd:YAG laser to create end-to-end small bowel anastomoses in a rabbit model. India ink, applied to the tissue edges prior to laser irradiation, was used as an absorbing chromophore. The anastomoses were performed using a biocompatible, water-soluble, intraluminal stent that held both intestinal edges in contact without any sutures or staples.[262] In 1989, Oz et al. investigated the feasibility of using a pulsed THC:YAG (2.15 $\mu$m) and a CW argon laser (488 to 514 nm) for welding of biliary tissue.[61] Visual indicators similar to those used by Sauer et al. for the endpoint of the welding procedure were used. In this study, bowel anastomoses formed by argon-laser welding withstood the highest pressures prior to breaking. The tissue surfaces of bowel anastomoses formed with the THC:YAG were frequently charred, possibly explaining the lower breaking pressures. In 1994, Rabau et al. described the healing process of $CO_2$ laser intestinal welding in a rat model.[263] Laser intestinal anastomoses were found to be more prone to dehiscence than their suture alternatives during the first 4 d postoperative to only improve with time, becoming at least as effective as sutured repairs by 7 to 10 d postoperative. In 1997, Cilesiz et al. investigated the use of argon[50] and Ho:YAG[264] lasers for temperature-controlled laser intestinal welding with good results. The healing response of repairs formed with the argon and Ho:YAG lasers compared well with control sutured anastomoses, although wound abscesses were more prevalent in the Ho:YAG group, leading to delay in mucosal healing.

In a 2001 study, Wadia et al. evaluated laser soldering using ICG-doped liquid-albumin solders in conjunction with a diode laser for repair of liver injuries.[113] Major liver trauma has a high mortality rate because of immediate exsanguination and a delayed morbidity from septicemia, peritonitis, biliary fistulae, and delayed secondary hemorrhage.[265,266] Common surgical techniques for repair of liver lacerations include mass ligation with absorbable sutures, packing with reexploration,[265] mesh hepatorrhaphy,[266] and fibrin sealant.[267] These techniques are notoriously difficult to apply — a particular concern in such a surgical situation where time is of the essence. Thus, an alternative technique for achieving rapid hemostasis following liver injuries would be greatly beneficial. The laser-soldering technique was shown to effectively seal the liver surface and join lacerations with minimal thermal injury while working independently of the patient's coagulation status.

### 39.8.4 Gynecology and Obstetrics

Laser welding techniques hold great potential for many surgical procedures on the fallopian tube and ovaries, where reconstruction is required to preserve fertility.[268] Laser welding was first proposed by von

Klilizing et al. in 1978 for fallopian-tube reconstructive surgery.[269] In this study, a $CO_2$ laser was used to perform end-to-end-anastomoses of cornua uteri with a 70% pregnancy success rate. Since then, experimental studies on tubal sterilization-reversal surgeries using lasers have had mixed results. With zero success in the rate of pregnancy, Fayez et al. presented less-than-satisfactory results for $CO_2$ laser reanastomosis of rabbit uterine horns in a 1983 study.[270] In a 1984 study by Choe et al., $CO_2$ laser microsurgery was shown to significantly reduce postoperative adhesions in microsurgical uterine-tube reanastomosis compared with conventional microsurgery using a scalpel; however, they were unable to weld with the $CO_2$ laser.[271] In these studies, the main problem with laser repairs appeared to be nonuniform heating of the tissue. More successful results were seen in a 1991 study by Vilos et al., where a $CO_2$ laser was used to perform a series of intraluminal-isthmic fallopian-tube anastomoses with a patency rate of 100%.[272]

The use of a low-power diode laser in conjunction with protein solder could avoid some of the earlier problems associated with uneven heating of the tissue. In 1995, Kao and Giles compared laser-welding and soldering techniques with conventional suture techniques for tubal reconstruction in a rabbit model.[273] Operative time was significantly reduced for both laser techniques compared with the suture technique, and all techniques resulted in excellent healing. With the protection of serum albumin, the laser-assisted anastomoses did not cause any thermal damage, and the anastomotic sites could tolerate the distension pressure of pregnancy and parturition without problems. Laser welding without protection of serum albumin, however, showed potential for thermal damage and dehiscence. Finally, the implantation and pregnancy rates were comparable for all three types of procedures.[273]

Laparoscopic tubal procedures using conventional suture techniques have been extremely tedious and time-consuming.[274] With further improvement in laser welding and soldering techniques, expeditious laparoscopic reanastomosis of fallopian tubes may be possible. In addition, the use of absorbable stents should make tubal welding and soldering techniques more feasible (see Section 39.6.4).

## 39.8.5 Laparoscopic and Endoscopic Surgery

Laparoscopic and endoscopic surgeries provide an ideal environment for laser-tissue-welding and soldering techniques, where conventional suture techniques remain time-consuming and tedious, and stapling devices are not usable in all applications.[275,276] Although laser welding and soldering techniques have not yet eliminated the need for suturing laparoscopically, particularly for tubal anastomoses, their potential in achieving this end is promising.[3,96,277–280] In a 1993 study, Poppas et al. used a 532-nm KTP laser with fluorescein as a chromophore for laser anastomosis of ureters in a canine model.[96] Fluorescein emits a visible fluorescence when exposed to laser energy, thereby allowing the laparoscopic surgeon to see when the laser light is hitting the target. This helps reduce the incident of collateral thermal damage as a result of a wandering laser beam. This system was subsequently tested laparoscopically in a porcine model, and the results compared favorably with those of conventional suturing.[3] Closures formed with the laser-welding technique were achieved more rapidly than suturing techniques with superior histological results, and acute leakage was observed less frequently. Laparoscopic surgery should become more appealing to clinicians as soon as an approximation technique is discovered that is easy to use, safe, and able to reliably hold tissue together in the laparoscopic environment.[280]

## 39.8.6 Neurosurgery

Techniques for nerve closure using sutures are well established. In a human patient, suture repair of a severed nerve may involve the individual joining of 40 fascicles, with three or four microsutures per fascicle. Since meticulous care is required, such an operation tends to be prolonged.[43] In addition, problems involving poor coaptation of the nerve fascicles, with loss of regenerating axons and scar formation, and mechanical damage to nerve tissue and intraneural circulation are also concerns.[32] Early attempts at laser welding of nerves demonstrated the technique to be significantly faster than conventional microsuture techniques; however, the hoped-for superiority of nerve regeneration after laser repair was not seen, with dehiscence rates of 12 to 60%.[43,281,282] Experimental studies by Maragh et al. suggest that these high dehiscence rates stem from the fact that the high power densities required to coapt the nerve edges induce thermal damage in the epineurium, while low power densities do not result in sufficient tensile strength.[20]

More recent studies with $CO_2$ lasers[36,81,283,284] and diode lasers combined with a dye-enhanced protein solder[81,114,115,123,284–286] hold greater promise for neurosurgical application of laser repair techniques. In 2000, Hajek et al. used a $CO_2$ laser for nerve coaptation in rat sciatic nerves. A larger degree of scarring and constriction was observed in nerves anastomosed by classical microsuture than that of the laser technique, with no significant difference in the quality of the action potentials of both groups.[283] Likewise, in a similar study by Happak et al. published in 2000, better results were obtained during functional evaluation of the sciatic-function index and the toe-spread index from the laser nerve coaptations than conventional suture coaptations.[36] In addition, the morphologic evaluations of the fiber density and area fraction were better in the laser group.

Laser tissue soldering reduces the effects of thermal damage while reinforcing wound strength, providing a promising technique for many neurosurgical applications.[81,114,285] In a 1995 study by Trickett et al., laser repair was noted to be a significantly faster procedure, inducing no inflammation and no foreign-body reaction.[123] Promising results in terms of axonal regeneration and functional results have also been reported.[103,115,123,284]

Laser-soldering techniques were also shown to possess several advantages over microsuture techniques for repair of the inferior alveolar nerve in two 1998 studies by Curtis et al.[287,288] In these studies, histochemical analysis showed comparable mean neuron counts and mean tracer uptake by neurons for the microsuture and laser-weld groups. Giant cell reactions were identified in two of the primary suture cases and axon deflection in three cases, demonstrating possible advantages of the laser-weld technique that showed no adverse reactions by axons or epineurium to the coagulative repair with the solder.

Finally, in a 1996 study by Foyt et al., laser soldering was shown to significantly decrease the incidence of cerebrospinal-fluid leak after dural closure, with the additional benefit of making dural closure possible where space constraints make traditional suture closure difficult.[116]

## 39.8.7 Ophthalmology

Conventional ophthalmic incision closure using corneal sutures is very demanding and time-consuming, even in the hands of a skilled surgeon.[289] Postoperative complications include wound leakage, foreign-body reaction to stitches, corneal astigmatism as a result of tight sutures, sutures loosening as a result of wound contraction providing a portal of entry for microorganisms or epithelial ingrowth, and irritation of the cornea and conjunctiva with giant papillary conjunctivitis from exposed knots.[290] Sutureless closure would overcome these problems.

The first attempts to weld corneal tissue with a $CO_2$ laser were unsuccessful because the heat created by the absorbed laser light induced corneal contraction, which widened rather than closed the incisions.[291,292] Corneal-tissue closure with a 1.9-μm diode laser, on the other hand, has been used to achieve leak-proof, full-thickness closure of corneal incisions.[293] In a 1995 study by Williams et al., laser welding at 1.9 μm was shown to stimulate homogenous wound healing by activating leukocytic and fibroblastic infiltration uniformly across the wound, thereby reducing the risk of postoperative astigmatism.[293] Success with a hydrogen fluoride laser ($\lambda$ = 2.3 to 2.6 μm), which has a penetration depth similar to that of the 1.9-μm diode laser, suggests that treatment depth is important when selecting a laser for welding of corneas.[294,295] Despite these relative successes, the $CO_2$, diode, and HF wavelengths are all strongly absorbed in tissue, and as such there is a relatively high risk of thermal damage associated with the laser-welding procedure, which could result in irreversible damage to the transparent structures of the eye.[294,296,297]

The ideal welding technique for opthalmic applications would utilize low temperatures to minimize thermal damage, thereby reducing the resulting contraction of the ocular tissue, which tends to result in tissue separation rather than fusion. Two attempts have been made to reduce the risk of thermal damage associated with ophthalmic laser-welding techniques: (1) the use of temperature-controlled welding

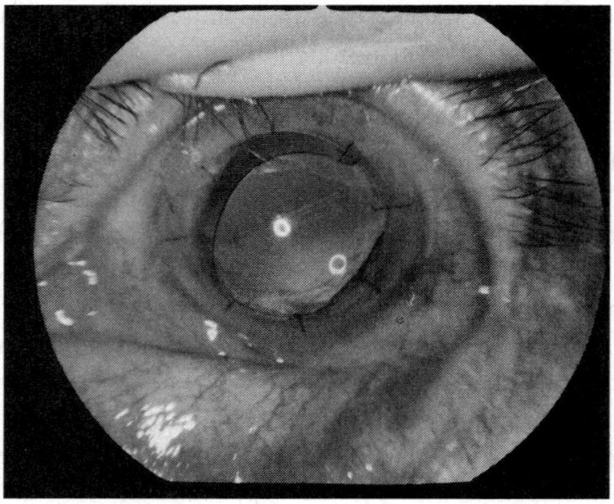

**FIGURE 39.15 (Color figure follows p. 28-30.)** Aspect of a human cornea at 3 months after diode-laser-assisted penetrating keratoplasty (transplantation of the cornea), where laser welding ($\lambda$ = 805 nm, $P$ ~70 mW, $t$ ~200 sec, chromophore: saturated ICG solution in saline) was used despite the conventional continuous suture (eight interrupted stitches had been applied during the surgery, as in the conventional PK procedure, to hold the corneal button). (Image courtesy of Roberto Pini, Istituto di Elettronica Quantistica–C.N.R., Florence, Italy.)

systems[154,208] and (2) the application of a chromophore to the weld site to make the welding technique more selective and localized.[298,299] In a 2001 study by Pini et al., the first application of laser welding to clinical transplantation of the cornea, the welding effect was observed to be localized in the incision with no thermal side effects or inflammatory response when low-power irradiation was used.[299] In addition, the healing process appeared to occur earlier and be more effective using the laser welding technique than suture techniques (Figure 39.15). Laser-solder closure has also been investigated as a means of improving the initial strength of the repairs.[208,300]

## 39.8.8 Orthopedic Surgery

One potential use of laser welding in orthopedics is repair of tendons. Currently, microsurgical suture techniques are used to treat ruptured or lacerated tendons. Suture material produces a foreign-body reaction, which may inhibit natural healing and promote adhesions.[301,302] The laser welding technique avoids this problem, and it could prove to be a useful tool in many orthopedic reconstructive procedures. The feasibility of laser welding of tendinous tissue was investigated in a 1996 study by Kilkelly et al. using $CO_2$ and Nd:YAG lasers in conjunction with a 25% human-albumin solder, and an argon laser in conjunction with a fluorescein-dye-doped albumin solder.[303] Morphologic analysis showed greater tissue injury and inflammatory response in laser-treated tendons than in those treated with suture. In addition, they were unable to produce a weld sufficient to withstand the significant tensile loads to which tendons are subjected immediately after repair. After 14 d, however, there was no significant difference between the strength of the laser and suture repairs.

Laser welding of meniscus tears were reported in a 1995 study by Forman et al. using an argon laser in conjunction with a fibrinogen solder.[304] Meniscal laser repairs also suffered from low initial strength; however, the solder was observed to provide a biological scaffold that was compatible with reparative cell migration and proliferation, resulting in cartilage formation with the biomechanical properties of native meniscus. The fibrocartilage produced after conventional repair is typically not as compatible.

Photothermal welding techniques are less successful in the avascular tissue often encountered in orthopedic applications because they rely on the presence of blood vessels to mediate a healing response in the form of scar tissue. Tendons, for example, are quite unique in their structure, as they are almost exclusively (>90%) composed of densely packed, highly organized type I collagen bundles. For this reason, photochemical-welding techniques, which utilize light-activated chemical cross-linking agents to produce covalent cross-links between collagen fibers, might be better suited for orthopedic applications as they do not rely on the healing properties of blood vessels (see Section 39.2.2.3).

## 39.8.9 Otolaryngology

One possible application of laser tissue welding in the field of otolaryngology is the formation of tracheobronchial anastomoses.[120,305,306] A major and often fatal complication of lung and combined heart–lung transplantation is dehiscence of the tracheobronchial anastomosis.[307] Conventional suture techniques suffer from suture-line leaks, microabscesses, and perianastomotic infection, eventually leading to granulation formation, stenosis, or dehiscence. In 1990, Auteri et al. investigated the possibility of performing laser-welded tracheobronchial anastomoses using an 808-nm diode laser and ICG-dye-enhance fibrinogen and a single suture.[120] Laser-welded and sutured anastomoses exhibited similar bursting pressures and histology immediately following the repairs; however, at 21 and 28 d postoperative, the laser-welded anastomoses exhibited less peritracheal inflammatory reach and showed visibly smoother luminal surfaces.

Other possible applications of laser welding and soldering in the field of otolaryngology include the fixation of flaps to vocal cords or other mucosal flaps,[121] the repair of fascial grafts in tympanoplasty,[308] and the sealing of pharyngocutaneous fistulas.[309] In addition, the vascular and neural techniques discussed in Sections 39.2.7.1 and 39.2.7.5, respectively, would be applicable.

## 39.8.10 Urology

Laser tissue welding has been tested on almost every organ and tissue in the genitourinary tract.[33,310] Urologic applications demand watertight, nonlithogenic closures due to the continuous flow of urine. Urine lacks the clotting ability of blood and thus requires a tighter anastomotic closure than is required in vascular surgical applications. Consequently, suture techniques in the urinary tract tend to be relatively technically demanding and time-consuming. In addition, the suture material itself can act as a nidus for stone formation. Laser welding can provide an immediate watertight, nonlithogenic anastomosis with tensile strength that is superior to that of traditional closure techniques.[310] Applications of laser tissue welding to open urologic surgery include vasovasostomy,[110,311–313] urethral reconstruction (hypospadias repair),[91,117,314,315] ureterotomy repair,[3,97] and pyeloplasty.[5]

The majority of clinical successes using laser tissue-welding and soldering techniques have been made in the field of urology. In 1988, Rosemberg reported the results of 14 patients who had a vasovasostomy by $CO_2$ laser.[311] He demonstrated postoperative sperm counts of more than 20 million/ml in 86% of his patients in comparison with 43% for those patients receiving standard microsurgical repair. A significant reduction in total operative time was also noted. In a 1989 study, Gilbert and Beckert also performed vasovasostomies on four males using an Nd:YAG laser and achieved normal sperm counts in two of these patients.[312] Of the remaining two patients, one was reported to have cryptospermia and the other oligoasthenospermia. In 1996, Kirsch et al. examined the use of laser tissue soldering using ICG-doped albumin solder in conjunction with an 808-nm diode laser for hypospadias repair in 30 children.[315] The laser solder-repair technique performed faster than conventional suture techniques, providing an immediate leak-free closure (intraoperative leak pressures for laser solder anastomoses were 94.2 ± 24.2 mm Hg as opposed to 20.0 ± 2.9 mm Hg for suture controls). In addition, no intraoperative complications were noted, and the laser-soldering technique had an overall complication rate of 19 vs. 24% of those with suture repair. In a 2001 study, Kirsch et al. compared the results of laser tissue soldering with ICG-doped albumin-protein solder ($n$ = 54) and conventional suturing ($n$ = 84) for hypospadias repair on 138 boys between the ages of 4 months and 8 years (mean age 15 months).[91] While stay sutures were used for tissue alignment in the laser solder-repair group, the total number of sutures was reduced by a factor of three compared to the controls. In addition, the mean operative time was reduced by a factor of five compared to the controls. The complication rate was 4.7% in the laser solder-repair group and 10.7% in the control group, with fistula in 2 of the 54 laser-solder cases and fistula and meatal stenosis in 7 and 2 of the 84 control cases, respectively. The ease of the laser technique and the lower complication rate in the laser group indicate that laser tissue soldering is an acceptable means of tissue closure in hypospadias repair. In addition to these clinical results, a tremendous amount of laboratory work has shown that laser tissue welding and soldering techniques can be applied to closure of genitourinary tissue using several different wavelengths.[3,5,97,100,110]

## 39.9 Summary

Laser tissue welding is a technological advance for tissue closure that offers many beneficial features over conventional closure techniques including reduced suture and needle trauma, reduced foreign-body reaction, better cosmetic appearance, reduced bleeding, the potential to form an immediate watertight anastomosis intraoperatively, and shorter operating times. Major advances have addressed the problems that have prevented the widespread clinical use of laser welding. The incorporation of polymer-membrane scaffolds has reduced the runoff associated with liquid solder alone. The development of biodegradable stents has improved tissue approximation. The incorporation of temperature-monitoring systems for feedback control has provided the basis for objectifying the endpoint of laser irradiation, thus eliminating reduced weld strength secondary to under- or overheating. Future advancements in the use of light-activated surgical adhesives and the development of laser-delivery systems will certainly open up more applications for this growth.

### References


1. Werker, P.M.N. and Kon, M.K., Review of facilitated approaches to vascular anastomosis surgery, *Ann. Thor. Surg.*, 63, S122, 1997.
2. Houston, K.A. and Rotstein, O.D., Fibrin sealant in high-risk colonic anastomoses, *Arch. Surg.*, 123, 230, 1988.
3. Wolf, J.S., Jr., Soble, J.J., Nakada, S.Y., Rayala, H.J., Humphrey, P.A., Clayman, R.V., and Poppas, D.P., Comparison of fibrin glue, laser weld, and mechanical suturing device for the laparoscopic closure of ureterotomy in a porcine model, *J. Urol.*, 157(4), 1487, 1997.
4. Detweiler, M.B., Detweiler, J.G., and Fenton, J., Sutureless and reduced suture anastomosis of hollow vessels with fibrin glue: a review, *J. Invest. Surg.*, 12(5), 245, 1999.
5. Barrieras, D., Reddy, P.P., McLorie, G.A., Bagli, D., Khoury, A.E., Farhat, W., Lilge, L., and Merguerian, P.A., Lessons learned from laser tissue soldering and fibrin glue pyeloplasty in an *in vivo* porcine model, *J. Urol.*, 164, 3, 1106, 2000.
6. Linden, C.L. and Shalaby, S.W., Performance of modified cyanoacrylate composition as tissue adhesives for soft and hard tissues, *J. Biomed. Mater. Res.*, 38(4), 348, 1997.
7. Wang, M.Y., Levy, M.L., Mittler, M.A., Liu, C.Y., Johnston, S., and McComb, J.G., A prospective analysis of the use of octylcyanoacrylate tissue adhesive for wound closure in pediatric neurosurgery, *Pediatr. Neurosurg.*, 30(4), 186, 1999.
8. Larrazabal, R., Pelz, D. and Findlay, J. M., Endovascular treatment of a lenticulostriate artery aneurysm with *N*-butyl cyanoacrylate, *Can. J. Neurol. Sci.*, 28(3), 256, 2001.
9. Ragnarsson, R., Berggren, A., and Ostrup, L.T., Long term evaluation of the unilink anastomotic system. A study with light and scanning electron microscopy, *Scand. J. Plast. Reconstr. Surg. Hand Surg.*, 26(2), 167, 1992.
10. Zhang, L., Kolker, A.R., Choe, E.I., Bakshandeh, N., Josephson, G., Wu, F.C., Siebert, J.W., and Kasabian, A.K., Venous microanastomosis with the Unilink system, sleeve, and suture techniques: a comparative study in the rat, *J. Reconstr. Microsurg.*, 13(4), 257, 1997; discussion, 261–262.
11. Lanzetta, M. and Owen, E.R., Use of the 3M precise microvascular anastomotic system in grafting 1-mm diameter arteries with polytetrafluoroethylene prostheses: a long-term study, *J. Reconstr. Microsurg.*, 9(3), 173, 1993.

12. Oguchi, H. and van der Lei, B., The 3M precise microvascular anastomotic system for implanting PTFE microvenous prostheses into the rat femoral vein, *Plast. Reconstr. Surg.*, 97(3), 662, 1996.
13. Yahr, W.Z. and Strully, K.J., Non-occlusive small arterial anastomosis with neodymium laser, *Surg. Forum*, 15, 224, 1964.
14. Jain, K.K. and Gorisch, W., Repair of small blood vessels with the neodymium-YAG laser: a preliminary report, *Surgery*, 85, 684, 1979.
15. Jain, K.K., Sutureless extra-intracranial anastomosis by laser, *Lancet*, 8406, 816, 1984.
16. Pearce, J. and Thomsen, S., Rate process analysis of thermal damage, in *Optical-Thermal Response of Laser-Irradiated Tissue*, Welch, A.J. and van Gemert, M.J.C., Eds., Plenum Press, New York, 1995, chap. 17.
17. Dubuisson, A.S. and Kline, D.G., Is laser repair effective for secondary repair of a focal lesion in continuity? *Microsurgery*, 14, 398, 1993.
18. Tang, J., Godlewski, G., Rouy, S., Dauzat, M., Juan, J., Chambettaz, F., and Salathe, R., Microarterial anastomosis using a noncontact diode laser versus a control study, *Lasers Surg. Med.*, 14(3), 229, 1994.
19. He, F.C., Wei, L.P., Lanzetta, M., and Owen, E.R., Assessment of tissue blood flow following small artery welding with an intraluminal dissolvable stent, *Microsurgery*, 19(3), 148, 1999.
20. Maragh, H., Hawn, R., Gould, J., and Terzis, J., Is laser nerve repair comparable to microsuture coaptation? *J. Reconstr. Microsurg.*, 4, 189, 1988.
21. Phillips, A.B., Ginsburg, B.Y., Shin, S.J., Soslow, R., Ko, W., and Poppas, D.P., Laser welding for vascular anastomosis using albumin solder: an approach for MID-CAB, *Lasers Surg. Med.*, 24(4), 264, 1999.
22. Burger, R.A., Gerharz, C.D., Rothe, H., Engelmann, U.H., and Hohenfellner, R., $CO_2$ and Nd:YAG laser systems in microsurgical venous anastomoses, *Urol. Res.*, 19, 253, 1991.
23. Mordon, S.R., Schoofs, M., Martinot, V.L., Capon, A., Buys, B., Patenotre, P., and Pellerin, P.N., 1.9 μm diode-laser-assisted anastomoses (LAMA) in reconstructive microsurgery: results of the preliminary clinical study, *Proc. SPIE*, 4244, 272, 2001.
24. Godlewski, G., Rouy, S., Tang, J., Dauzat, M., Chambettaz, F., and Salathe, R.P., Scanning electron microscopy of microarterial anastomoses with a diode laser: comparison with conventional manual suture, *J. Reconstr. Microsurg.*, 11(1), 37, 1995.
25. Dalsing, M.C., Packer, C.S., Kueppers, P., Griffith, S.L., and Davis, T.E., Laser and suture anastomosis: passive compliance and active force production, *Lasers Surg. Med.*, 12(20), 190, 1992.
26. Hasegawa, M., Sakurai, T., Matsushita, M., Nishikimi, N., Nimura, Y., and Kobayashi, M., Comparison of argon-laser welded and sutured repair of inferior vena cava in a canine model, *Lasers Surg. Med.*, 29(1), 62, 2001.
27. Chikamatsu, E., Sakurai, T., Nishikimi, N., Yano, T., and Nimura, Y., Comparison of laser vascular welding, interrupted sutures, and continuous sutures in growing vascular anastomoses, *Lasers Surg. Med.*, 16(1), 34, 1995.
28. Vale, B.H., Frenkel, A., Trenka-Benthin, S., and Matlaga, B.F., Microsurgical anastomoses of rat carotid arteries with the $CO_2$ laser, *Plast. Reconstr. Surg.*, 77, 759, 1986.
29. Wider, T.M., Libutti, S.K., Greenwald, D.P., Oz, M.C., Yager, J.S., Treat, M.R., and Hugo, N.E., Skin closure with dye-enhanced laser welding and fibrinogen, *Plast. Reconstr. Surg.*, 88(6), 1018, 1991.
30. Frazier, O.H., Painvin, G.A., Morris, J.R., Thomsen, S., and Neblett, C.R., Laser-assisted microvascular anastomoses: angiographic and anatomopathologic studies on growing microvascular anastomoses: preliminary report, *Surgery*, 97, 585, 1985.
31. Capon, A., Souil, E., Gauthier, B., Sumian, C., Bachelet, M., Buys, B., Polla, B.S., and Mordon S., Laser assisted skin closure (LASC) by using a 815-nm diode-laser system accelerates and improves wound healing, *Lasers Surg. Med.*, 28(2), 168, 2001.
32. Bass, L. and Treat, M., Laser tissue welding: a comprehensive review of current and future clinical applications, *Lasers Surg. Med.*, 17, 315, 1995.

33. Scherr, D.S. and Poppas, D.P., Laser tissue welding, *Urol. Clin. North Am.*, 25(1), 123, 1998.
34. Xie, H., Buckley, L.A., Prahl, S.A., Shaffer, B.S., and Gregory, K.W., Thermal damage control of dye-assisted laser tissue welding: effect of dye concentration, *Proc. SPIE*, 4244, 189, 2001.
35. Eppley, B.L., Kalenderian, E., Winkelmann, T., and Delfino, J.J., Facial nerve graft repair: suture versus laser-assisted anastomosis, *J. Oral Maxillofac. Surg.*, 18(1), 50, 1989.
36. Happak, W., Neumayer, C., Holak, G., Kuzbari, R., Burggasser, G., and Gruber, H., Morphometric and functional results after $CO_2$ laser welding of nerve coaptations, *Lasers Surg. Med.*, 27(1), 66, 2000.
37. Poppas, D.P., Stewart, R.B., Massicotte, J.M., Wolga, A.E., Kung, R.T.V., Retik, A.B., and Freeman, M.R., Temperature-controlled laser photocoagulation of soft tissue: *in vivo* evaluation using a tissue welding model, *Lasers Surg. Med.*, 18(4), 335, 1996.
38. White, R.A., Kopchok, G.E., Donayre, C.E., Peng, S.K., Fujitani, R.M., White, G.H., and Uitto, J., Mechanism of tissue fusion in argon laser-welded vein-artery anastomoses, *Lasers Surg. Med.*, 8(1), 83, 1988.
39. Thomsen, S., Chan, E., Stubig, I., Menovsky, T., and Welch, A.J., Importance of wound stabilization in early wound healing of laser skin welds, *Proc. SPIE*, 2395, 490, 1995.
40. Hui, K.C., Zhang, F., Shaw, W.W., Kryger, Z., Piccolo, N.S., Harper, A., and Lineaweaver, W.C., Learning curve of microvascular venous anastomosis: a never ending struggle? *Microsurgery*, 20(1), 22, 2000.
41. Quigley, M., Bailes, J., Kwann, H., Cerullo, L., and Brown, J.T., Aneurysm formation after low power $CO_2$ laser-assisted vascular anastomosis, *Neurosurgery*, 18, 292, 1986.
42. Lawrence, P.F., Merrell S.W., and Goodman, G.R., A comparison of absorbable suture and argon laser welding for lateral repair of arteries, *J. Vasc. Surg.*, 14, 183, 1991.
43. Korff, M., Bent, S.W., Havig, M.T., Schwaber, M.K., Ossoff, R.H., and Zealer, D.L., An investigation of the potential for laser nerve welding, *Otolaryngol. Head Neck Surg.*, 106(4), 345, 1992.
44. Zelt, D.T., LaMuraglia, G.M., L'Italien, G.J., Megerman, J., Kung, R.T.V., Stewart, R.B., and Abbott, W.M., Arterial laser welding with a 1.9 micrometer Raman-shifted laser, *J. Vasc. Surg.*, 15, 1025, 1992.
45. Lobel, B., Eyal, O., Belotserkovsky, E., Shenfeld, O., Kariv, N., Goldwasser, B., and Katzir, A., *In vivo* $CO_2$ laser rat urinary bladder welding with silver halide fiberoptic radiometer temperature control, *Proc. SPIE*, 2395, 517, 1994.
46. Menovsky, T., van den Bergh Weerman, M., and Beek, J.F., Effect of $CO_2$ milliwatt laser on peripheral nerves. I. A dose-response study, *Microsurgery*, 17(10), 562, 1996.
47. Abramson, D.L., Shaw, W.W., Kamat, B.R., Harper, A., and Rosenberg, C.R., Laser-assisted venous anastomosis: a comparison study, *J. Reconstruct. Microsurg.*, 7(3), 199, 1991.
48. Burger, R.A., Gerharz, C.D., Draws, J., Engelmann, U.H., and Hohenfellner, R., Sutureless laser-welded anastomosis of the femoral artery and vein in rats using $CO_2$ and Nd:YAG lasers, *J. Reconstr. Microsurg.*, 9(3), 213, 1993.
49. Klioze, S.D., Poppas, D.P., Rooke, C.T., Choma, T.J., and Schlossberg, S.M., Development and initial application of a real time thermal control system for laser tissue welding, *J. Urol.*, 152(2), 744, 1994.
50. Cilesiz, I., Thomsen, S., Welch, A.J., and Chan, E.K., Controlled temperature tissue fusion: Ho:YAG laser welding of rat intestine *in vivo*. II., *Lasers Surg. Med.*, 21(3), 278, 1997.
51. Fung, L.C., Mingin, G.C., Massicotte, M., Felsen, D., and Poppas, D.P., Effects of temperature on tissue thermal injury and wound strength after photothermal wound closure, *Lasers Surg. Med.*, 25(4), 285, 1999.
52. Lobel, B., Eyal, O., Kariv, N., and Katzir, A., Temperature controlled $CO_2$ laser welding of soft tissues: urinary bladder welding in different animal models (rats, rabbits, and cats), *Lasers Surg. Med.*, 26(1), 4, 2000.
53. Gennaro, M., Ascer, E., Mohan, C., and Wang, S., A comparison of $CO_2$ laser-assisted venous anastomoses and conventional suture techniques: patency, aneurysm formation, and histologic differences, *J. Vasc. Surg.*, 14(5), 605, 1991.

54. Thomsen, S., Mueller, J., and Serure, A., Pathology of rat femoral arteries exposed to a low-energy $CO_2$ laser beam, *Lab. Invest.*, 52, 68A, 1985.
55. Robinson, J.K., Garden, J.M., Taute, P.M., Leibovich, S.J., Lautenschlager, E.P., and Hartz, R.S., Wound healing in porcine skin following low-output carbon dioxide laser irradiation of the incision, *Ann. Plast. Surg.*, 18, 499, 1987.
56. Kopchok, G.E., White, R.A., White, G.H., Fujitani, R., Vlasak, J., Dykhovsky, L., and Grundfest, W.S., $CO_2$ and argon laser vascular welding: acute histological and thermodynamic comparison, *Lasers Surg. Med.*, 8(6), 584, 1988.
57. Chow, J.W.N. and Flemming, A.F.S., Laser assisted microvascular anastomosis: a histological study, *Laser Med. Sci.*, 5, 281, 1990.
58. White, R.A., Kopchok, G., Donayre, C., Lyons, R., White, G., Klein, S.R., Pizzurro, D., Abergel, R.P., Dwyer, R.M., and Uitto, J., Large vessel sealing with the argon laser, *Lasers Surg. Med.*, 7(3), 229, 1987.
59. Murray, L.W., Su, L., Kopchok, G.E., and White, R.A., Crosslinking of extracellular matrix proteins: a preliminary report on a possible mechanism of argon laser-welding, *Lasers Surg. Med.*, 9(5), 490, 1989.
60. Self, A.B., Coe, D.A., and Seeger, J.M., Limited thrombogenicity of low temperature, laser-welded vascular anastomoses, *Lasers Surg. Med.*, 18(3), 241, 1996.
61. Oz, M.C., Bass, L.S., Popp, H.W., Chuck, R.S., Johnson, J.P., Trokel, S.L., and Treat, M.R., *In vitro* comparison of Thulium-Holmium-Chromium:YAG and argon ion lasers for welding biliary tissue, *Lasers Surg. Med.*, 9(3), 248, 1989.
62. Menovsky, T., Beek, J.F., and van Gemert, M.J., $CO_2$ laser nerve welding: optimal laser parameters and the use of solders *in vitro*, *Microsurgery*, 15(1), 44, 1994.
63. Mordon, S., Martinot, V., and Mitchell, V., End-to-end microvascular anastomoses with a 1.9 µm diode laser, *J. Clin. Laser Med. Surg.*, 13(6), 357, 1995.
64. Trickett, R.I., Dawes, J.M., Knowles, D.S., Lanzetta, M., and Owen, E.R., *In vitro* laser nerve repair: protein solder strip irradiation or irradiation alone, *Int. Surg.*, 82(1), 38, 1997.
65. Pohl, D., Bass, L.S., Stewart, R., and Chiu, D.T., Effect of optical temperature feedback control on patency in laser-soldered microvascular anastomosis, *J. Reconstr. Microsurg.*, 14(1), 23, 1998; discussion, 29–30),
66. Fried, N.M. and Walsh, J.T., Jr., Cryogen spray cooling during laser tissue welding, *Phys. Med. Biol.*, 45(3), 753, 2000.
67. Shalhav, A., Wallach-Kapon, R., Akselrod, S., and Katzir, A., Laser irradiation of biological tissue through water as a means of reducing thermal damage, *Lasers Surg. Med.*, 19(4), 407, 1996.
68. Sauer, J.S., McGuire, K.P., and Hinshaw, J.R., Exposure update: automated laser welding of circumferential tissue anastomoses, *Proc. SPIE*, 1066, 53, 1989.
69. Wu, P.J. and Walsh, J.T., Relationship between apposition pressure during welding and tensile strength of the acute weld, *Proc. SPIE*, 4244, 175, 2001.
70. Cooley, B.C., Heat-induced tissue fusion for microvascular anastomosis, *Microsurgery*, 17, 198, 1996.
71. Bass, L.S., Treat, M.R., Dzakonski, C., and Trokel, S.L., Sutureless microvascular anastomosis using the THC:YAG laser: a preliminary report, *Microsurgery*, 10(3), 189, 1989.
72. Nijima, K.H., Yonekawa, Y., Handa, H., and Taki, W., Nonsuture microvascular anastomosis using an Nd-YAG laser and a water-soluble polyvinyl alcohol splint, *J. Neurosurg.*, 67, 579, 1987.
73. Judy, M.M., Matthews, J.L., Boriack, R.L., Burlacu, A., Lewis, D.E., and Utecht, R.E., Heat-free photochemical tissue welding with 1,8-naphthalimide dyes using visible (420nm) light, *Proc. SPIE*, 1876, 175, 1993.
74. Judy, M.M., Fuh, L., Matthews, J.L., Lewis, D.E., and Utecht, R.E., Gel electrophoretic studies of photochemical cross-linking of type I collagen with brominated 1,8-naphthalimide dyes and visible light, *Proc. SPIE*, 2128, 506, 1994.

75. Judy, M.M., Chen, L., Fuh, L., Nosir, H., Jackson, R.W., Matthews, J.L., Lewis, D.E., Utecht, R.E., and Yuan, D., Photochemical cross-linking of type I collagen with hydrophobic and hydrophilic 1,8-naphthalimide dyes, *Proc. SPIE*, 2681, 53, 1996.
76. Judy, M.M., Nosir, H.R., Jackson, R.W., Matthews, J.L., Utecht, R.E., Lewis, D.E., and Yuan, D., Photochemical bonding of skin with 1,8-naphthalimide dyes, *Proc. SPIE*, 3195, 21, 1998.
77. Merguerian, P.A., Pugach, J.L., and Lilge, L.D., Nonthermal ureteral tissue bonding: comparison of photochemical collagen crosslinking with thermal laser bonding, *Proc. SPIE*, 3590, 194, 1999.
78. Mulroy, L., Kim, J., Wu, I., Scharper, P., Melki, S.A., Azar, D.T., Redmond, R.W., and Kochevar, I.E., Photochemical keratodesmos for repair of lamellar corneal incisions, *Invest. Ophthalmol. Vis. Sci.*, 41(11), 3335, 2000.
79. Krueger, R.R. and Almquist, E.E., Argon laser coagulation of blood for anastomosis of small vessels, *Lasers Surg. Med.*, 5(1), 55, 1985.
80. Wang, S., Grubbs, P.E., Basu, S., Robertazzi, R.R., Thomsen, S., Rose, D.M., Jacobwitz, I.J., and Cunningham, J.N., Jr., Effect of blood bonding on bursting strength of laser-assisted microvascular anastomosis, *Microsurgery*, 9, 10, 1988.
81. Menovsky T., $CO_2$ and Nd:YAG laser-assisted nerve repair: a study of bonding strength and thermal damage, *Acta Chir. Plast.*, 42(1), 16, 2000.
82. Grubbs, P.E., Wang, S., Corrado, M., Basu, S., Rose, D.M., and Cunningham, J.N., Jr., Enhancement of $CO_2$ laser microvascular anastomosis by fibrin glue, *J. Surg. Res.*, 45, 112, 1988.
83. Cikrit, D.F., Dalsing, M.C., Weinstein, T.S., Palmer, K., Lalka, S.G., and Unthank, J.L., $CO_2$ welded venous anastomosis: enhancement of weld strength with heterologous fibrin glue, *Lasers Surg. Med.*, 10(6), 584, 1990.
84. Mendoza, G.A., Acuna, E., Allen, M., Arroyo, J., and Quintero, R.A., *In vitro* laser welding of amniotic membranes, *Lasers Surg. Med.*, 24(5), 315, 1999.
85. Oz, M.C., Treat, M.R., Libutti, S.K., Popp, H.W., Bass, L.S., and Popilskis, S., Preliminary report: laser welding and fibrinogen soldering are superior to sutured cholecystostomy closure in canine model, in *Laser Surgery: Advanced Characterization, Therapeutics, and Systems*, Vol. 2, Jaffe, S.N. and Atsumi, K., Eds., SPIE, Bellingham, WA, 1990, pp. 55–60.
86. Bass, L.S., Libutti, S.K., Oz, M.C., Rosen, J., Williams, M.R., Nowygrod, R., and Treat, M.R., Canine choledochotomy closure with diode laser-activated fibrinogen solder, *Surgery*, 115(3), 398, 1994.
87. Bass, L.S., Libutti, S.K., and Eaton, A.M., New solders for laser welding and sealing, *Lasers Surg. Med.*, 5, 63A, 1993.
88. Poppas, D.P., Wright, E.J., Guthrie, P.D., Shlahet, L.T., and Retik, A.B., Human albumin solders for clinical application during laser tissue welding, *Lasers Surg. Med.*, 19(1), 2, 1996.
89. McNally, K.M., Sorg, B.S., Chan, E.K., Welch, A.J., Dawes, J.M., and Owen, E.R., Optimal parameters for laser tissue soldering. I. Tensile strength and scanning electron microscopy analysis, *Lasers Surg. Med.*, 24(5), 319, 1999.
90. Xie, H., Shaffer, B.S., Prahl, S.A., and Gregory, K.W., Laser welding with an albumin stent: experimental ureteral end-to-end anastomosis, *Proc. SPIE*, 3907, 215, 2000.
91. Kirsch, A.J., Cooper, C.S., Gatti, J., Scherz, H.C., Canning, D.A., Zderic, S.A., and Snyder, H.M., Laser tissue soldering for hypospadias repair: results of a controlled prospective clinical trial, *J. Urol.*, 165(2), 574, 2001.
92. Brooks, S.G., Ashley, S., Fisher, J., Davies, G.A., Griffiths, J., Kester, R.C., and Rees, M.R., Exogenous chromophores for the argon and Nd:YAG lasers: a potential application to laser-tissue interactions, *Lasers Surg. Med.*, 12, 294, 1992.
93. Kokosa, J.M., Przjazny, A., Bartels, K.E., Motamedi, M.E., Hayes, D.J., Wallace, D.B., and Frederickson, C.J., Laser-initiated decomposition products of indocyanine green (ICG) and carbon black sensitized biological tissues, *Proc. SPIE*, 2974, 205, 1997.
94. Fried, N.M. and Walsh, J.T., Jr., Laser skin welding: *in vivo* tensile strength and wound healing results, *Lasers Surg. Med.*, 27(1), 55, 2000.

95. DeCoste, S.D., Farinelli, W., Flotte, T., and Anderson, R.R., Dye-enhanced laser welding for skin closure, *Lasers Surg. Med.*, 12, 25, 1992.
96. Poppas, D., Sutaria, P., Sosa, R.E., Mininberg, D., and Scholssberg, S., Chromophore enhanced laser welding of canine ureters *in vitro* using a human protein solder: a preliminary step for laparoscopic tissue welding, *J. Urol.*, 150, 1052, 1993.
97. Wright, E.J. and Poppas, D.P., Effect of laser wavelength and protein solder concentration on acute tissue repair using laser welding: initial results in a canine ureter model, *Tech. Urol.*, 3(3), 176, 1997.
98. Khadem, J., Veloso, A.A., Jr, Tolentino, F., Hasan, T., and Hamblin, M.R., Photodynamic tissue adhesion with chlorin(e6) protein conjugates, *Invest. Ophthalmol. Vis. Sci.*, 40(13), 3132, 1999.
99. Birch, J.F., Mandley, D.J., Williams, S.L., Worrall, D.R., Trotter, P.J., Wilkinson, F., and Bell, P.R., Methylene blue based protein solder for vascular anastomoses: an *in vitro* burst pressure study, *Lasers Surg. Med.*, 26(3), 323, 2000.
100. Poppas, D.P., Schlossberg, S.M., Richmond, I.L., Gilbert, D.A., and Devine, C.J., Laser welding in urethral surgery: improved results with a protein solder, *J. Urol.*, 139, 415, 1988.
101. Sauda, K., Imasaka, T., and Ishibashi, N., Determination of protein in human serum by high performance liquid chromatography, *Anal. Chem.*, 58, 2649, 1986.
102. Poppas, D.P., Massicotte, J.M., Stewart, R.B., Roberts, A.B., Atala, A., Retik, A.B., and Freeman, M.R., Human albumin solder supplemented with TGF-β1 accelerates healing following laser welded wound closure, *Lasers Surg. Med.*, 19(3), 360, 1996.
103. Lauto, A., Trickett, R., Malik, R., Dawes, J., and Owen, E., Laser activated solid protein bands for peripheral nerve repair: an *in vivo* study, *Lasers Surg. Med.*, 21, 134, 1997.
104. Oz, M.C., Libutti, S.K., Ashton, R.C., Lontz, J.F., Lemole, G.M., and Nowygrod, R., Comparison of laser-assisted fibrinogen-bonded and sutured canine arteriovenous anastomoses, *Surgery*, 112, 76, 1992.
105. Maitz, P.K., Trickett, R.I., Dekker, P., Tos, P., Dawes, J.M., Piper, J.A., Lanzetta, M., and Owen, E.R., Sutureless microvascular anastomoses by a biodegradable laser-activated solid protein solder, *Plast. Reconstr. Surg.*, 104(6), 1726, 1999.
106. Ott, B., Zuger, B.J., Erni, D., Banic, A., Schaffner, T., Weber, H.P., and Frenz, M., Laser balloon vascular welding using a dye-enhanced albumin solder, *Proc. SPIE*, 4244, 183, 2001.
107. Lauto, A., Hamawy, A.H., Phillips, A.B., Petratos, P.B., Raman, J., Felsen, D., Ko, W., and Poppas, D.P., Carotid artery anastomosis with albumin solder and near infrared lasers: a comparative study, *Lasers Surg. Med.*, 28(1), 50, 2001.
108. McNally-Heintzelman, K.M., Riley, J.N., Dickson, T.J., Hou, D.M., Rogers, P., and March, K.L., *In vivo* tissue repair using light-activated surgical adhesive in a porcine model, *Proc. SPIE*, 4244, 226, 2001.
109. Kirsch, A.J., Canning, D.A., Zderic, S.A., Hensle, T.W., and Duckett, J.W., Laser soldering technique for sutureless urethral surgery, *Tech. Urol.*, 3(2), 108, 1997.
110. Trickett, R.I., Wang, D., Maitz, P., Lanzetta, M., and Owen, E.R., Laser welding of vas deferens in rodents: initial experience with fluid solders, *Microsurgery*, 18(7), 414, 1998.
111. Kirsch, A.J., Cooper, C., Camming, D., Snyder, H., and Zderic, S., Hypospadias repair using laser tissue soldering (LTS): preliminary results of a prospective randomized study, *Proc. SPIE*, 3245, 309, 1998.
112. Barrieras, D., Reddy, P.P., McLorie, G.A., Bagli, D., Khoury, A.E., Farhat, W., Lilge, L., and Merguerian, P.A., Lessons learned from laser tissue soldering and fibrin glue pyeloplasty in an *in vivo* porcine model, *J. Urol.*, 164 (3 Part 2), 1106, 2000.
113. Wadia, Y., Xie, H., and Kajitani, M., Sutureless liver repair and hemorrhage control using laser-mediated fusion of human albumin as a solder, *J. Trauma*, 51(1), 51, 2001.
114. Bass, L.S., Moazami, N., Avellino, A., Trosaborg, W., and Treat, M.R., Feasibility studies for laser solder neurorrhaphy, *Proc. SPIE*, 2128, 472, 1994.
115. Lauto, A., Dawes, J.M., Piper, J.A., and Owen, E.R., Laser nerve repair by solid protein band technique. II: assessment of long-term nerve regeneration, *Microsurgery*, 18(1), 60, 1998.

116. Foyt, D., Johnson, J.P., Kirsch, A.J., Bruce, J.N., and Wazen, J.J., Dural closure with laser tissue welding, *Otolaryngol. Head Neck Surg.*, 115(6), 513, 1996.
117. Kirsch, A.J., Chang, D.T., Kayton, M.L., Libutti, S.K., Treat, M.R., and Hensle, T.W., Laser welding with albumin-based solder: experimental full-tubed skin graft urethroplasty, *Lasers Surg. Med.*, 18, 225, 1996.
118. Suh, D.D., Schwartz, I.P., Canning, D.A., Snyder, H.M., Zderic, S.A., and Kirsch, A.J., Comparison of dermal and epithelial approaches to laser tissue soldering for skin flap closure, *Lasers Surg. Med.*, 22(5), 268, 1998.
119. Cooper, C.S., Schwartz, I.P., Suh, D., and Kirsch, A.J., Optimal solder and power density for diode laser tissue soldering (LTS), *Lasers Surg. Med.*, 29(1), 53, 2001.
120. Auteri, J.A., Oz, M.C., Jeevanandem, V., Treat, M.R., and Smith, C.A., Tracheal anastomosis using tissue welding with diode laser energy, *J. Invest. Surg.*, 3, 301, 1990.
121. Wang, Z., Pankratov, M.M., Gleich, L.L., Rebeiz, E.E., and Shapshay, S.M., New technique for laryngotracheal mucosa transplantation. 'Stamp' welding using indocyanine green dye and albumin interaction with diode laser, *Arch. Otolaryngol. Head Neck Surg.*, 121(7), 773, 1995.
122. Zuger, B.J., Ott, B., Mainil-Varlet, P., Schaffner, T., Clemence, J.F., Weber, H.P., and Frenz, M., Laser solder welding of articular cartilage: tensile strength and chondrocyte viability, *Lasers Surg. Med.*, 28(5), 427, 2001.
123. Trickett, R., Lauto, A., Dawes, J., and Owen, E., Laser activated solder for peripheral nerve repair, *Proc. SPIE*, 2395, 542, 1995.
124. Lauto, A., Repair strength dependence on solder protein concentration: a study in laser tissue welding, *Lasers Surg. Med.*, 22(2), 120, 1998.
125. McNally, K.M., Sorg, B.S., Chan, E.K., Welch, A.J., Dawes, J.M., and Owen, E.R., Optimal parameters for laser tissue soldering. II. Premixed versus separate dye/solder methods, *Lasers Surg. Med.*, 26(4), 346, 2000.
126. Chan, E., Kovach, I., Brown, D., and Welch, A.J., Effects of hydration on laser soldering, *Proc. SPIE*, 2970, 244, 1997.
127. Kirsch, A.J., Chang, D.T., Kayton, M.L., Newhouse, J., Libutti, S.K., Treat, M.R., Connor, J.P., and Hensle, T.W., Sutureless rabbit bladder mucosa patch graft urethroplasty using diode laser and solder, *J. Urol.*, 153, 1303, 1995.
128. Chan, E.K., Lu, Q., Bell, B., Motamedi, M., Frederickson, C., Brown, D.T., Kovach, I.S., and Welch, A.J., Laser assisted soldering: microdroplet accumulation with a microjet device, *Lasers Surg. Med.*, 23(4), 213, 1998.
129. Chan, E.K., Welch, A.J., Shay, E.L., Springer, T., Frederickson, C.J., and Motamedi, M., Accumulative small droplet soldering, *Proc. SPIE*, 3245, 268, 1998.
130. Yudkin, M. and Offord, R., *Comprehensive Biochemistry*, Longman, London, 1973.
131. Nimni, M.E., *Collagen, Vol. 1: Biochemistry*, CRC Press, Boca Raton, FL, 1988.
132. Ramachandran, G.N. and Ramakrishnan, C.F., Molecular structure, in *Biochemistry of Collagen*, Ramachandran, G.N. and Reddi, A.H., Eds., Plenum Press, New York, 1976, pp. 45–84.
133. Gustavson, K.H., *The Chemistry and Reactivity of Collagen*, Academic Press, New York, 1956.
134. Traub, W. and Piez, K.A. The chemistry and structure of collagen, *Adv. Protein Chem.*, 25, 243, 1971.
135. Crewther, W.G., *Symposium on Fibrous Proteins*, Butterworths, Sydney, Australia, 1968.
136. Godlewski, G., Rouy, S., and Dauzat, M., Ultrastructural study of arterial wall repair after argon laser micro-anastomosis, *Lasers Surg. Med.*, 7(3), 258, 1987.
137. Menovsky, T., Beek, J.F., and van Gemert, M.J., Laser tissue welding of dura mater and peripheral nerves: a scanning electron microscopy study, *Lasers Surg. Med.*, 19(2), 152, 1996.
138. White, R.A., White, G.H., Fujitani, J.M., Vlassak, J.W., Donayre, C.E., Kopchok, G.E., and Peng, S.K., Initial human evaluation of argon laser-assisted vascular anastomoses, *J. Vasc. Surg.*, 9(4), 542, 1989.
139. Kada, O., Shimizu, K., Ikuta, H., Horii, H., and Nakamura, K., An alternative method of vascular anastomosis by laser: experimental and clinical study, *Lasers Surg. Med.*, 7(3), 240, 1987.

140. Dew, D.K., Supik, L., Darrow, C., and Price, G.F., Tissue repair using lasers: a review, *Orthopedics*, 16, 581, 1993.
141. Vale, B.H., Frenkel, A., Trenka-Benthin, S., and Matlaga, B.F., Microsurgical anastomoses of rat carotid arteries with the $CO_2$ laser, *Plast. Reconstr. Surg.*, 77, 759, 1986.
142. Bass, L.S., Moazami, N., Pocsidio, J., Oz, M.C., LoGerfo, P., and Treat, M.R., Changes in type I collagen following laser welding, *Lasers Surg. Med.*, 12(5), 500, 1992.
143. Schober, R., Ulrich, F., Sander, T., Durselen, H., and Hessel, S., Laser induced alteration of collagen substructure allows microsurgical tissue welding, *Science*, 232, 1421, 1986.
144. Lemole, G.M., Anderson, R.R., and DeCoste, S., Preliminary evaluation of collagen as a component in the thermally induced "weld," *Proc. SPIE*, 1422, 116, 1991.
145. Back, M.R., Kopchok, G.E., White, R.A., Cavaye, D.M., Donayre, C.E., and Peng, S.K., Nd:YAG laser-welded canine arteriovenous anastomoses, *Lasers Surg. Med.*, 14(2), 111, 1994.
146. Solhpour, S., Weldon, E., Foster, T.E., and Anderson, R.R., Mechanism of thermal tissue "welding," *Lasers Surg. Med.*, 6A, 56, 1994.
147. Tang, J., Godlewski, G., Rouy, S., and Delacretaz, G., Morphologic changes in collagen fibers after 830 nm diode laser welding, *Lasers Surg. Med.*, 21(5), 438, 1997.
148. Tang, J., O'Callaghan, D., Rouy, S., and Godlewski, G., Quantitative changes in collagen levels following 830-nm diode laser welding, *Lasers Surg. Med.*, 22(4), 207, 1998.
149. Stryer, L., *Biochemistry*, Freeman and Company, San Francisco, 1975.
150. Bass, L.S., Moazami, N., Pocsidio, J., Oz, M.C., LoGerfo, P., and Treat, M.R., Electrophoretic mobility patterns of collagen following laser welding, *Proc. SPIE*, 1422, 123, 1991.
151. Badeau, A.F., Lee, C.E., Morris, J.R., Thomsen, S., Malk, E.G., and Welch, A.J., Temperature response during microvascular anastomoses using milliwatt $CO_2$ laser, *Lasers Surg. Med.*, 6, 179, 1986.
152. Shenfeld, O., Ophir, E., Goldwasser, B., and Katzir, A., Silver halide fibre optic radiometric temperature measurement and control of $CO_2$ laser-irradiated tissues and application to tissue welding, *Lasers Surg. Med.*, 14(4), 323, 1994.
153. Simhon, D., Ravid, A., Halpern, M., Levanon, D., Brosh, T., Kariv, N., and Katzir, A., Laser soldering of rabbit skin using IR fiber optic temperature control system, *Proc. SPIE*, 4244, 242, 2001.
154. Barak, A., Eyal, O., Rosner, M., Belotzerkousky, E., Solomon, A., Belkin, M., and Katzir, A., Temperature controlled $CO_2$ laser tissue welding of ocular tissue, *Surv. Opthalmol.*, 42(Suppl.), S77, 1997.
155. Martinot, V.L., Mordon, S.R., Mitchell, V.A., Pellerin, P.N., and Brunetaud, J.M., Determination of efficient parameters for argon laser-assisted anastomoses in rats: macroscopic, thermal and histological evaluation, *Lasers Surg. Med.*, 15, 168, 1994.
156. Cilesiz, I., Springer, T., Thomsen, S., and Welch, A.J., Controlled temperature tissue fusion: argon laser welding of canine intestine *in vitro*, *Lasers Surg. Med.*, 18(4), 325, 1996.
157. Jenkins, R.D., Sinclair, I.N., Anand, R., Kalil, A.G., Schoen, F.J., and Spears, J.R., Laser balloon angioplasty: effect of tissue temperature on weld strength of human postmortem intima-media separations, *Lasers Surg. Med.*, 8(1), 30, 1988.
158. Stewart, R.B, Benbrahim, A., LaMuraglia, G.M., Rosenberg, M., L'Italien, G.L., Abbot, W.M., and Kung, R.T, Laser assisted vascular welding with real time temperature control, *Lasers Surg. Med.*, 19, 9, 1996.
159. Welch, A.J., The thermal response of laser irradiated tissue, *IEEE J. Quantum Electron.*, 20(12), 1471, 1984.
160. Welch, A.J., Wissler, E.H., and Priebe, L.A., Significance of blood flow in calculations of temperature in laser irradiated tissue, *IEEE Trans. Biomed. Eng.*, 27(3), 164, 1980.
161. Shen, H., Spikes, J.D., Kopeckova, P., and Kopecek, J., Photodynamic crosslinking of proteins, II: photocrosslinking of a model protein-ribonuclease, *J. Photochem. Photobiol.*, 35(3), 213, 1996.
162. Girotti, A.W., Photosensitized crosslinking of erythrocyte membrane proteins: evidence against participation of amino groups in the reaction, *Biochim. Biophys. Acta*, 602, 45, 1980.

163. Dubbelman, T.M., de Goeij, A.F., and Stevenick, J.V., Photodynamic effects of protoporphyrin on human erythrocytes: nature of the cross-linking of membrane proteins, *Biochim. Biophys. Acta*, 511(2), 141, 1978.
164. Verweij, H., Dubbelman, T.M.A.R., and Steveninck, J.V., Photodynamic protein crosslinking, *Photochem. Photobiol.*, 28, 87, 1981.
165. Ramshaw, J.A.M., Stephens, L.J., and Tulloch, P.A., Methylene blue sensitized photo-oxidation of collagen fibrils, *Biochim. Biophys. Acta*, 1206, 225, 1994.
166. Spoerl, E., Huhle, M., and Seiler, T., Induction of cross-links in corneal tissue, *Exp. Eye Res.*, 66, 97, 1998.
167. Judy, M.M., Jackson, R.W., Nosir, H.R., Matthews, J.L., Loyd, J.D., Lewis, D.E., Utecht, R.E., and Yuan, D., Healing results in meniscus and articular cartilage photochemically welded with 1,8-naphthalimide dyes, *Proc. SPIE*, 2970, 257, 1997.
168. Peters, T., *All About Albumin*, Academic Press, New York, 1996.
169. Aoki, K., Sato, K., Nagaoka, S., Kamada, M., and Hiramatsu, K., Heat denaturation of bovine serum albumin in alkaline pH region, *Biochim. Biophys. Acta*, 328, 323, 1973.
170. Wetzel, R., Becker, M., Behlke, J., Billwitz, H., Bohm, S., Ebert, B., Hamann, H., Krumbiegel, J., and Lassmann, G., Temperature behaviour of human serum albumin, *Eur. J. Biochem.*, 104, 469, 1980.
171. Clark, D.C., Smith, L.J., and Wilson, D.R., A spectroscopic study of the conformational properties of foamed bovine serum albumin, *J. Colloid. Interface Sci.*, 121, 136, 1981.
172. Gallier, J., Rivet, P., and de Certaines, J., 1H- and 2H-NMR study of bovine serum albumin solutions, *Biochim. Biophys. Acta*, 915, 1, 1987.
173. Pico, G., Thermodynamic aspects of the thermal stability of human serum albumin, *Biochem. Mol. Biol. Int.*, 36(5), 1017, 1995.
174. Bleustein, C.B., Sennett, M., Kung, R.T., Felsen, D., Poppas, D.P., and Stewart, R.B., Differential scanning calorimetry of albumin solders: interspecies differences and fatty acid binding effects on protein denaturation, *Lasers Surg. Med.*, 27(5), 465, 2000.
175. Marx, G., Mechanisms of photothermal wound closing with "soldering grade" albumin, *Am. Soc. Laser Med. Surg.*, A, 48, 1997.
176. Park, M.S. and Min, H.K., Laser soldering and welding for ossicular reconstruction: an *in vitro* test, *Otolaryngol. Head Neck Surg.*, 122(6), 803, 2000.
177. Kirsch, A.J., Duckett, J.W., Snyder, H.M., Canning, D.A., Harshaw, D.W., Howard, P., Macarak, E.J., and Zderic, S.A., Skin flap closure by dermal laser soldering: a wound healing model for sutureless hypospadius repair, *Urology*, 50(2), 263, 1997.
178. McNally, K.M., Sorg, B.S., Welch, A.J., Dawes, J.M., and Owen, E.R., Photothermal effects of laser tissue soldering, *Phys. Med. Biol.*, 44, 983, 1999.
179. Godlewski, G., Prudhomme, M., and Tang, J., Applications and mechanisms of laser tissue welding in 1995: review, *Proc. SPIE*, 2623, 334, 1996.
180. Kung, R.T.V., Stewart, R.B., Zelt, D.T., Litalien, G.J., and LaMuraglia, G.M., Absorption characteristics at 1.9 μm: effect on vascular welding, *Lasers Surg. Med.*, 13, 12, 1993.
181. Lauto, A., Poppas, D.P. and Murrell, G.A.C., Solubility study of albumin solders for laser tissue-welding, *Lasers Surg. Med.*, 23(5), 258, 1998.
182. Reppas, C., Adair, C.H., Barnett, J.L., Berardi, R.R., Du Ross, D., Sidan, S.Z., Thill, P.F., Tobey, S.W., and Dressman, J.B., High viscosity hydroxypropylmethylcellulose reduces postprandial blood glucose concentrations in NIDDM patients, *Diabetes Res. Clin. Pract.*, 22, 61, 1993.
183. Topping, D., Hydroxypropylmethylcellulose, viscosity, and plasma cholesterol control, *Nutr. Rev.*, 52, 176, 1994.
184. McNally, K.M., Sorg, B.S., and Welch, A.J., Novel solid protein solder designs for laser-assisted tissue repair, *Lasers Surg. Med.*, 27(2), 147, 2000.

185. Holland, S.J., Tighe, B.J., and Gould, P.L., Polymers for biodegradable medical devices. 1. The potential of polyesters as controlled macromolecular release systems, *J. Controlled Release*, 4, 155, 1986.
186. Mikos, A.G., Thorsen, A.J., Czerwonka, L.A., Bao, Y., Winslow, D.N., and Vacanti, J.P., Preparation and characterization of poly(L-lactic acid) foams, *J. Biomed. Mater. Res.*, 27(2), 183, 1993.
187. Wake, M.C., Gupta, P.K., and Mikos, A.G., Fabrication of pliable biodegradable polymers foams to engineer soft tissues, *Cell Transplant.*, 5(4), 465, 1996.
188. Reed, A.M. and Gilding, D.K., Biodegradable polymers for use in surgery-poly(glycolic)/poly(lactic acid) homo and copolymers. II. *In vitro* degradation, *Polymer*, 22(4), 499, 1981.
189. McNally, K.M., Sorg, B.S., Hammer, D.X., Heintzelman, D.L., Hodges, D.E., and Welch, A.J., Improved vascular tissue fusion using new light-activated surgical adhesive on a canine model, *J. Biomed. Opt.*, 6(1), 68, 2001.
190. Sorg, B.S., McNally-Heintzelman, K.M., and Welch, A.J., Biodegradable polymer thin film for enhancement of laser-assisted incision closure with an indocyanine-green-doped liquid albumin solder, *Proc. SPIE*, 3907, 50, 2000.
191. Sorg, B.S. and Welch, A.J., Tissue soldering with biodegradable polymer films: *in-vitro* investigation of hydration effects on weld strength, *Proc. SPIE*, 4244, 180, 2001.
192. Lauto, A., Stewart, R., Ohebshalom, M., Nikkoi, N.D., Felsen, D., and Poppas, D.P., Impact of solubility on laser tissue-welding with albumin solid solders, *Lasers Surg. Med.*, 28(1), 44, 2001.
193. How, T.V., Mechanical properties of arteries and arterial grafts, in *Cardiovascular Biomaterials*, Hastings, G.W., Ed., Springer-Verlag, London, 1992, pp. 1–35.
194. Rosenberg, N., Gaughran, E.R.L., Henderson, J., Lord, G.H., and Douglas, J.F., The use of segmental arterial implants prepared by enzymatic modification of heterologous blood vessels, *Surg. Forum*, 6, 242, 1982.
195. Kottke-Marchant, K., Anderson, J.M., Umemura, Y., and Marchant, R.E., Effect of albumin coating on the *in vitro* blood compatibility of dacro arterial prostheses, *Biomaterials*, 10, 147, 1989.
196. McNally, K.M., Parker, A.E., Heintzelman, D.L., Sorg, B.S., Dawes, J.M., Pfefer, T.J., and Welch, A.J., Dynamic optical-thermal modeling of laser tissue soldering with a scanning source, *IEEE J. Sel. Top. Quantum Electron.*, 5(4), 1, 1999.
197. Lauto, A., Kerman, I., Ohebshalon, M., Felsen, D., and Poppas, D.P., Two-layer film as a laser soldering biomaterial, *Lasers Surg. Med.*, 25, 3, 250, 1999.
198. Zhang, C., Tang, N.X., Zheng, C.F., Xu, Y.W., and Wang, T.D., Experimental study on microvascular anastomosis using a dissolvable stent support in the lumen, *Microsurgery*, 12, 67, 1991.
199. He, F.C. and Tang, N.X., Stent assisted laser welding in a rabbit model, *Med. J. PLA*, 19, 383, 1994.
200. Lauto, A., Ohebshalom, M., Esposito, M., Mingin, J., Li, P.S., Felsen, D., Goldstein, M., and Poppas, D.P., Self-expandable chitosan stent: design and preparation, *Biomaterials*, 22(13), 1869, 2001.
201. Moritz, A.R. and Henriques, F.C., Studies of thermal injury II: the relative importance of time and surface temperature in the causation of cutaneous burns, *Am. J. Pathol.*, 23, 695, 1947.
202. Pearce, J. and Thomsen, S., Kinetic models of laser tissue fusion processes, *Biomed. Sci. Instrum.*, 29, 355, 1993.
203. Henriques, F.C., Studies of thermal injury, *Arch. Pathol.*, 43, 489, 1947.
204. Agah, R., Pearce, J.A., Welch, A.J., and Motamedi, M., Rate process model for arterial tissue thermal damage: implications on vessel photocoagulation, *Lasers Surg. Med.*, 15(2), 176, 1994.
205. Eyal, O., Zur, A., Shenfeld, O., Gilo, M., and Katzir, A., Infrared radiometry using silver halide fibers and a cooled photonic detector, *Opt. Eng.*, 33(2), 502, 1994.
206. Springer, T.A. and Welch, A.J., Temperature control during laser vessel welding, *Appl. Opt.*, 32(4), 517, 1993.
207. Simhon, D., Ravid, A., Halpern, M., Levanon, D., Brosh, T., Kariv, N., and Katzir, A., Laser soldering of rabbit skin using IR fiber optic temperature control system, *Proc. SPIE*, 4244, 242, 2001.

208. Strassman, E., Loya, N., Gaton, D.D., Ravid, A., Kariv, N., Weinberger, D., and Katzir, A., Laser soldering of the cornea in a rabbit model using a controlled-temperature $CO_2$ laser system, *Proc. SPIE*, 4244, 253, 2001.
209. Valvano, J.W. and Pearce, J., Temperature measurements, in *Optical-Thermal Response of Laser-Irradiated Tissue*, Welch, A.J. and van Gemert, M.J.C., Eds., Plenum Press, New York, 1995, chap. 15.
210. Mordon, S., Desmettre, T., and Devoisselle, J.M., Laser-induced release of liposome-encapsulated dye to monitor tissue temperature: a preliminary *in vivo* study, *Lasers Surg. Med.*, 16(3), 246, 1995.
211. Khoobehi, B., Char, C.A., and Peyman, G.A., Assessment of laser-induced release of drugs from liposomes: an *in vitro* study, *Lasers Surg. Med.*, 10(1), 60, 1990.
212. Khoobehi, B., Char, C.A., Peyman, G.A., and Schuele, K.M., Study of the mechanisms of laser-induced release of liposome-encapsulated dye, *Lasers Surg. Med.*, 10(3), 303, 1990.
213. Tang, J., Zeng, F., Savage, H., Ho, P.P., and Alfano, R.R., Fluorescence spectroscopic imaging to detect changes in collagen and elastin following laser tissue welding, *J. Clin. Laser Med. Surg.*, 18(1), 3, 2000.
214. Tang, J., Zeng, F., Savage, H., Ho, P.P., and Alfano, R.R., Laser irradiative tissue probed *in situ* by collagen 380-nm fluorescence imaging, *Lasers Surg. Med.*, 27(2), 158, 2000.
215. Chandrasekhar, S., *Radiative Transfer*, Dover, New York, 1960.
216. Ishimaru, A., *Wave Propagation and Scattering in Random Media*, Vol. 1, Academic Press, New York, 1978.
217. Anvari, B., Rastegar, S., and Motamedi, M., Modeling of intraluminal heating of biological tissue: implications for treatment of benign prostatic hyperplasia, *IEEE Trans. Biomed. Eng.*, 41(9), 854, 1994.
218. Kim, B.M., Jacques, S.L., Rastegar, S., Thomsen, S., and Motamedi, M., The role of dynamic changes in blood perfusion and optical properties in thermal coagulation of the prostate, *Proc. SPIE*, 2391, 443, 1995.
219. Sagi, A., Shitzer, A., Katzir, A., and Akselrod, S., Heating of biological tissue by laser irradiation: theoretical model, *Opt. Eng.*, 31(7), 1417, 1992.
220. Jacques, S.L. and Wang, L., Monte Carlo modeling of light transport in tissues, in *Optical-Thermal Response of Laser-Irradiated Tissue*, Welch, A.J. and van Gemert, M.J.C., Eds., Plenum Press, New York, 1995, chap. 4.
221. Roider, J. and Birngruber, B., Solution of the heat conduction equation, in *Optical-Thermal Response of Laser-Irradiated Tissue*, Welch, A.J. and van Gemert, M.J.C., Eds., Plenum Press, New York, 1995, chap. 12.
222. Pfefer, T.J., Barton, J.K., Chan, E.K., Ducros, M.G., Sorg, B.S., Milner, T.E., Nelson, J.S., and Welch, A.J., A three dimensional modular adaptable grid numerical model for light propagation during laser irradiation of skin tissue, *IEEE J. Sel. Top. Quantum Electron.*, 2(4), 934, 1996.
223. Rastegar, S., Kim, B., and Jacques, S.L., Role of temperature dependence of optical properties in laser irradiation of biological tissue, *Proc. SPIE*, 1646, 228, 1992.
224. Pfefer, T.J., Barton, J.K., Smithies, D.J., Milner, T.E., Nelson, J.S., van Gemert, M.J.C., and Welch, A.J., Laser treatment of port wine stains: three-dimensional simulation using a biopsy-defined geometry in an optical-thermal model, *Proc. SPIE*, 3245, 322, 1998.
225. Small, W., Heredia, N.J., Maitland, D.J., Eder, D.C., Celliers, P.M., Da Silva, L.B., London, R.A., and Matthews, D.L., Experimental and computational laser tissue welding using a protein patch, *J. Biomed. Opt.*, 3(1), 96, 1998.
226. McNally, K.M., Sorg, B.S., Bhavaraju, N.C., Ducros, M.J., Welch, A.J., and Dawes, J.M., Optical-thermal characterization of albumin protein solders, *Appl. Opt.*, 38(31), 6661, 1999.
227. Glinsky, M.E., London, R.A., Zimmerman, G.B., and Jacques, S.L., Modeling of endovascular patch welding using the computer program LATIS, *Proc. SPIE*, 2623, 349, 1995.
228. Maitland, D.J., Eder, D.C., London, R.A., Glinsky, M.E., and Soltz, B.A., Dynamic simulations of tissue welding, *Proc. SPIE*, 2671, 234, 1996.

229. London, R.A., Glinsky, M.E., Zimmerman, G.B., Bailey, D.S., Eder, D.C., and Jacques, S.L., Laser-tissue interaction modeling with LATIS, *Appl. Opt.*, 36(34), 9068, 1997.
230. Small, W., Maitland, D.J., Heredia, N.J., Eder, D.C., Celliers, P.M., Da Dilva, L.B., London, R.A., and Matthews, D.L., Investigation of laser tissue welding dynamics via experiment and modeling, *J. Clin. Laser Med. Surg.*, 15(1), 3, 1997.
231. Torres, J.H., Motamedi, M., Pearce, J.A., and Welch, A.J., Experimental evaluation of mathematical models for predicting the thermal response of tissue to laser irradiation, *Appl. Opt.*, 32(4), 597, 1993.
232. Quarles, T.L., SPICE3 version 3C1 Users Guide, Memorandum No. UCB/ERL M89/46, Electronics Research Laboratory, College of Engineering, University of California, Berkeley, April 24, 1989.
233. Okada, M., Shimizu, K., Ikuta, H., Horii, H., and Nakamura, K., An alternative method of vascular anastomosis by laser: experimental and clinical study, *Lasers Surg. Med.*, 7(3), 240, 1987.
234. Thomsen, S., Morris, J.R., Neblett, C.R., and Mueller, J., Tissue welding using a low energy microsurgical $CO_2$ laser, *Med. Instrum.*, 21(4), 231, 1987.
235. Ulrich, F., Durselen, R., and Schober, R., Long-term investigations of laser-assisted microvascular anastomoses with the 1.318-micron Nd:YAG laser, *Lasers Surg. Med.*, 8(2), 104, 1988.
236. White, R.A., Kopchok, G.E., Vlasak, J., Hsiang, Y., Fujitani, R.M., White, G.H., and Peng, S.K., Experimental and early clinical evaluation of vascular anastomoses with argon laser fusion and the use of absorbable guy sutures: a preliminary report, *J. Vasc. Surg.*, 12(4), 401, 1990; discussion, 406–408.
237. Kuroyanagi, Y., Taguchi, M., Yano, T., Jones, D.N., and Shionoya, S., Argon laser-assisted anastomoses in medium-size vessels: one-year follow-up, *Lasers Surg. Med.*, 11(3), 223, 1991.
238. Godlewski, G., Pradal, P., Rouy, S., Charras, A., Dauzat, M., Lan, O., and Lopez, F.M., Microvascular carotid end-to-end anastomosis with the argon laser, *World J. Surg.*, 10(5), 829, 1986.
239. Okada, M., Ikuta, H., Shimizu, K., Horii, H., Tsuji, Y., Yoshida, M., and Nakamura, K., Experimental and clinical studies on the laser application in the cardiovascular surgery: analysis of clinical experience of 112 patients, *Nippon Geka Gakkai Zasshi*, 90(9), 1589, 1989.
240. Serure, A., Withers, E.H., Thomsen, S., and Morris, J., Comparison of carbon dioxide laser-assisted microvascular anastomosis and conventional microvascular sutured anastomosis, *Surg. Forum*, 34, 634, 1983.
241. Brachmann, J., Anasa, M., Kosinski, E.J., and Schuler, G.C., Improved clinical effectiveness with a collagen vascular hemostasis device for shortened immobilization time following diagnostic angiography and percuataneous transluminal coronary angioplasty, *Am. J. Cardiol.*, 81(12), 1502, 1998.
242. Chamberlin, J.R., Lardi A.B., McKeever, L.S., Wang, M.H., Ramadurai, G., Grunenwald, P., Towne, W.P., Grassman, E.D., Leya, F.S., Lewis, B.E., and Stein, L.H., Use of vascular sealing devices (VasoSeal and Perclose) versus assisted manual compression (Femostop) in transcatheter coronary interventions requiring Abciximab (ReoPro), *Catheter Cardiovasc. Interv.*, 47, 143, 1999.
243. Eidt, J.F., Habibipour, S., Saucedo, J.F., McKee, J., Southern, F., Barone, G.W., Talley, J.D., and Moursi, M., Surgical complications from hemostatic puncture closure devices, *Am. J. Surg.*, 178, 551, 1999.
244. Lumsden, A.B., Miller, J.M., Kosinski, A.S., Allen, R.C., Dodson, T.F., Salam, A.A., and Smith, R.B., A prospective evaluation of surgically treated groin complications following percutaneous cardiac procedures, *Am. J. Surg.*, 60, 132, 1994.
245. Popma, J.J., Satler, L.F., Pichard, D., Kent, K.M., Campbell, A., Chuang, Y.C., Clark, C., Merritt, A.J., Bucher, T.A., and Leon, M.B., Vascular complications after balloon and new device angioplasty, *Circulation*, 88, 1569, 1993.
246. Camenzind, E., Grossholz, M., Urban, P., Dorsaz, P.A., Didier, D., and Meier, B., Collagen application versus manul compression: a prospective randomized trial for arterial puncture site closure after coronary angioplasty, *J. Am. Coll. Cardiol.*, 24(3), 655, 1994.

247. Ninomiya, J., Shoji, T., Tanaka, S., Tamura, K., and Noishiki, Y., Laser vascular welding in biologic grafts, *Am. Soc. Artif. Intern. Organs Trans.*, 35(3), 208, 1989.
248. Auteri, J.S., Libutti, S.K., Oz, M.C., Bass, L.S., and Treat, M.R., Reduced blood loss in canine PTFE-femoral anastomoses using a dye enhanced laser activated protein solder, *Symp. Surg. Tissue Adhesives* (Abstract), Atlanta, October 1993.
249. Bentz, M.L., Parva, B., Dickson, C.S., Futrell, J.W., and Johnson, P.C., Laser-assisted microvascular anastomosis of human adult and placental arteries with expanded polytetrafluoroethylene microconduit, *Plast. Reconstr. Surg.*, 91(6), 1124, 1993; discussion, 1132–1133.
250. Mueller, M.P., Kopchok, G.E., Tabbara, M.R., Cavaye, D.M., and White, R.A., Argon laser-welded bovine heterograft anastomoses, *J. Clin. Laser Med. Surg.*, 11(1), 1, 1993.
251. Tabbara, M., White, R.A., Kopchok, G., Mirsch, W., Cormier, F., and Cavaye, D., Laser-fused biologic vascular graft anastomoses, *J. Invest. Surg.*, 6(3), 289, 1993.
252. Lo Cicero, J., III, Frederickson, J.W., Hartz, R.S., Kaufman, M.W., and Michaelis, L.L., Experimental air leaks in lung sealed by low-energy carbon dioxide laser irradiation, *Chest*, 87(6), 820, 1985.
253. Arai, T., Tanaka, S., and Kikuchi, M., Investigation of possible mechanism of laser welding between artificial film and bronchial tissue, *Proc. SPIE*, 2128, 510, 1994.
254. Abergel, R.P., Lyons, R., Dwyer, R., White, R.R., and Uitto, J., Use of lasers for closure of cutaneous wounds: experience with Nd:YAG, argon, and $CO_2$ lasers, *J. Dermatol. Surg. Oncol.*, 12, 1181, 1986.
255. Garden, J.M., Robinson, J.K., Taute, P.M., Lautenschlager, E.P., Leibovich, S.J., and Hartz, R.S., The low-output carbon dioxide laser for cutaneous wound closure of scalpel incisions: comparative tensile strength studies of the laser to the suture and staple for wound closure, *Lasers Surg. Med.*, 6(1), 67, 1986.
256. Dew, D.K., Hsu, T.M., Halpern, S.J., and Michaels, C.E., Laser assisted skin closure at 1.32 microns: the use of a software driven medical laser system, *Proc. SPIE*, 1422, 111, 1991.
257. Romanos, G.E., Pelekanos, S., and Strub, J.R., A comparative histological study of wound healing following Nd:YAG laser with different energy parameters and conventional surgical incision in rat skin, *J. Clin. Laser Med. Surg.*, 13(1), 11, 1995.
258. Reali, U.M., Borgognoni, L., Martini, L., Chiarugi, C., Gori, F., Pini, R., Toncelli, F., and Vanni, U., Preliminary experiences on diode laser welding of skin, *Proc. SPIE*, 2327, 211, 1994.
259. Fried, N.M. and Walsh, J.T., Dye-assisted laser skin closure with pulsed radiation: an *in vitro* study of weld strength and thermal damage, *J. Biomed. Opt.*, 3(4), 401, 1998.
260. Majaron, B., Kimel, S., Verkruysse, W., Aguilar, G., Pope, K., Svaasand, L.O., Lavernia, E.J., and Nelson, J.S., Cryogen spray cooling in laser dermatology: effects of ambient humidity and frost formation, *Lasers Surg. Med.*, 28(5), 469, 2001.
261. Sauer, J.S., Rogers, D.W., and Hinshaw, J.R., Bursting pressures of $CO_2$ laser-welded rabbit ileum, *Lasers Surg. Med.*, 6(2), 106, 1986.
262. Sauer, J.S., Hinshaw, J.R., and McGuire, K.P., The first sutureless, laser-welded, end-to-end bowel anastomosis, *Lasers Surg. Med.*, 9(1), 70, 1989.
263. Rabau, M.Y., Wasserman, I., and Shoshan, S., Healing process of laser-welded intestinal anastomosis, *Lasers Surg. Med.*, 14(1), 13, 1994.
264. Cilesiz, I., Thomsen, S., Welch, A.J., and Chan, E.K., Controlled temperature tissue fusion: Ho:YAG laser welding of rat intestine *in vivo*. Part two, *Lasers Surg. Med.*, 21(3), 278, 1997.
265. Cue, J.I., Cryer, H.G., and Miller, F.B., Packing and planned reexploration for hepatic and retroperitoneal hemorrhage: critical refinements of a useful technique, *J. Trauma*, 30, 1007, 1990.
266. Reed, R.L., Merrell, R.C., Meyers, W.C., and Fischer, R.P., Continuing evolution in the approach to severe liver trauma, *Ann. Surg.*, 216, 524, 1992.
267. Holcomb, J.B., Pusateri, A.E., Harris, R.A., Reid, T.J., Beall, L.D., Hess, J.R., and MacPhee, M.J., Dry fibrin sealant dressing reduce blood loss, resuscitation volume, and improve survival in hypothermic coagulopathic swine with grade V liver injuries, *J. Trauma*, 47, 233, 1999.
268. Bellina, J.H., Fick, A.C., and Jackson, J.D., Application of the $CO_2$ laser to infertility surgery, *Surg. Clin. North Am.*, 64(5), 899, 1984.

269. Von Klilizing, L., Grosspietzsch, R., Klink, F., Endell, W., Husstedt, W., and Oberheuser, F., Surgical refertilization by means of a laser technique: animal experimental study, *Fortschr. Med.*, 96, 357, 1978.
270. Fayez, J.A., McComb, J.S., and Harper, M.A., Comparison of tubal surgery with the $CO_2$ laser and the unipolar microelectrode, *Fertil. Steril.*, 40(4), 476, 1983.
271. Choe, J.K., Dawood, M.Y., Bardawil, W.A., and Andrews, A.H., Clinical and histologic evaluation of laser reanastomosis of the uterine tube, *Fertil. Steril.*, 41(5), 754, 1984.
272. Vilos, G.A., Intramural-isthmic fallopian tube anastomosis facilitated by the carbon dioxide laser, *Fertil. Steril.*, 56, 571, 1991.
273. Kao, L.W. and Giles, H.R., Laser-assisted tubal anastomosis, *J. Reprod. Med.*, 40(8), 585, 1995.
274. Dawood, M.Y., Laparoscopic surgery of the fallopian tubes and ovaries, *Semin. Laparosc. Surg.*, 6(2), 58, 1999.
275. Lezoche, E., Paganini, A., Feliciotti, F., and Chan, R., Laparoscopic suture technique after common bile duct exploration, *Surg. Laparosc. Endosc.*, 3(3), 209, 1993.
276. Swanstrom, L.L., Common bile duct exploration, in *Minimally Invasive Surgery*, Hunter, J.F. and Sackier, J.M., Eds., McGraw-Hill, New York, 1993, pp. 231–244.
277. Bass, L.S., Oz, M.C., Auteri, J.S., Williams, M.R., Rosen, J., Libutti, S.K., Eaton, A.M., Lontz, J., Nowygrod, R., and Treat, M.R., Laparoscopic application of laser-activated tissue glues, *Proc. SPIE*, 1421, 164, 1991.
278. Kirsch, A.J., Dean, G.E., Oz, M.C., Libutti, S.K., Treat, M.R., Nowygrod, R., and Hensle, T.W., Preliminary results of laser tissue welding in extravesical reimplantation of the ureters, *J. Urol.*, 151, 514, 1994.
279. Eden, C.G. and Coptcoat, M.J., Assessment of alternative tissue approximation techniques for laparoscopy, *Br. J. Urol.*, 78(2), 234, 1996.
280. Kumar, U. and Albala, D.M., Newer techniques in intracorporeal tissue approximation: suturing, tissue adhesives, and microclips, *Urol. Clin. North Am.*, 28(1), 15, 2001.
281. Bailes, J.E., Quigley, M.R., Cerullo, L.J., Kline, D.G., and Sahga, V., Sutureless $CO_2$ laser nerve anastomosis. histological and electrophysiological analysis, *Lasers Surg. Med.*, 6, 248, 1986.
282. Huang, T.C., Blanks, R.H., Berns, M.W., and Crumley, R.L., Laser vs. suture nerve anastomosis, *Otolaryngol. Head Neck Surg.*, 107, 14, 1992.
283. Hajek, P., Malec, R., Danek, T., Broz, T., and Zinek, K., Repair of traumatic lesions in peripheral nerves — initial experience with laser anastomosis, *Rozhl. Chir.*, 79(9), 403, 2000.
284. Menovsky, T., van den Bergh Weerman, M., and Beek, J.F., Effect of $CO_2$-milliwatt laser on peripheral nerves. II. A histological and functional study, *Microsurgery*, 20(3), 150, 2000.
285. Menovsky, T., Beek, J.F., and Thomsen, S., Laser (-assisted) nerve repair: a review, *Neurosurgery*, 18(4), 225, 1995.
286. McNally, K.M., Dawes, J.M., Lauto, A., Parker, A.E., Piper, J.A., and Owen, E.R., Laser-solder repair technique for nerve anastomosis, *Laser Med. Sci.*, 14, 228, 1999.
287. Curtis, N.J., Lauto, A., Trickett, R., Owen, E., and Walker, D.M., Preliminary study of microsurgical repairs of the inferior alveolar nerve in rats using primary suturing and laser weld techniques, *Int. J. Oral Maxillofac. Surg.*, 27(6), 476, 1998.
288. Curtis, N.J., Trickett, R.I., Owen, E., and Lanzetta, M., Intraosseous repair of the inferior alveolar nerve in rats: an experimental model, *J. Reconstr. Microsurg.*, 14(6), 391, 1998.
289. Nordon, L.T., Maxwell, W.A., and Davison, J.A., Eds., *The Surgical Rehabilitation of Vision: An Integrated Approach to Anterior Segment Surgery*, Gower Medical Publishing, New York, 1992.
290. Wilhelmus, K.R., Huang, A.J.W., Hwang, D.G., Parish, C.M., Sutphin, J.E., Jr., and Whitsett, J.C., External disease and cornea. Clinical approach to corneal transplantation, *Am. J. Opthalmol.*, 8, 420, 1999.
291. Keates, R.H., Fried, S., Levy, S.N., and Morris, J.R., Carbon dioxide laser use in wound sealing and epikeratophakia, *J. Cataract Refract. Surg.*, 13(3), 290, 1987.
292. Gailitis, R.P., Thompson, K.P., Ren, Q., Morris, J., and Waring, G.O., Laser welding of synthetic epikeratoplasty lenticules to the cornea, *Refract. Corneal Surg.*, 6(6), 430, 1990.

293. Williams, J.M., Burstein, N.L., Nowicki, M.J., Zietkiewicz, C.J., and Jeffers, W.Q., Infrared laser welding of the rabbit cornea in vivo, *Proc. SPIE*, 2393, 131, 1995.
294. Burnstein, N.L., Williams, J.M., Nowicki, M.J., Johnson, D.E., and Jeffers, W.Q., Corneal welding using hydrogen fluoride lasers, *Arch. Opthalmol.*, 110(1), 12, 1992.
295. Williams, J.M., Burnstein, N.L., Nowicki, M.J., and Jeffers, W.Q., Corneal welding with the hydrogen fluoride laser, *Proc. SPIE*, 2126, 193, 1994.
296. Desmettre, T.J., Mordon, S.R., and Mitchell, V., Tissue welding of corneal wound suture with CW 1.9 μm diode laser: an *in vivo* preliminary study, *Proc. SPIE*, 2623, 372, 1996.
297. Trabucchi, G., Gobbi, P.G., Brancato, R., Carones, F., Resti, A.G., Martina, E., Jensen, A., and Pini, R., Laser welding of corneal tissue: preliminary experiences using 810 nm and 1950 nm lasers, *Proc. SPIE*, 2623, 380, 1996.
298. Menabuoni, L., Mincione, F., Mincione, G.P., and Pini, R., Laser welding to assist penetrating keratoplasty: *in vivo* studies, *Proc. SPIE*, 3195, 25, 1998.
299. Pini, R., Menabuoni, L., and Starnotti, L., First application of laser welding in clinical transplantation of the cornea, *Proc. SPIE*, 4244, 266, 2001.
300. Eaton, A.M., Bass, L.S., Libutti, S.K., Schubert, H.D., and Treat, M., Sutureless cataract incision closure using laser activated tissue glues, *Proc. SPIE*, 1423, 52, 1991.
301. Popovic, N.A., Johnstone, F.L., Kilkelly, F.X., McKinney, L., van de Merve, W.P., and Smith, A.C., Comparison of $CO_2$ laser welding with suture technique for repair of tendons, *Proc. SPIE*, 2395, 502, 1995.
302. Zobitz, M.E., Zhao, C., and Amadio, P.C., Comparison of mechanical properties of various suture repair techniques in a partially lacerated tendon, *J. Biomech. Eng.*, 122(6), 604, 2000.
303. Kilkelly, F.X., Choma, T.J., Popovic, N., Miller, D.W., and Sweet, D.E., Tendon repair by laser welding: a histologic and biomechanical comparison and suture repair with $CO_2$ and argon lasers, *Lasers Surg. Med.*, 19(4), 487, 1996.
304. Forman, S.K., Oz, M.C., Lontz, J.F., Treat, M.R., Forman, T.A., and Kiernan, H.A., Laser-assisted fibrin clot soldering of human menisci, *Clin. Orthop.*, 310, 37, 1995.
305. Pan, Y.M., Shen, Y.Z., Wu, K.L., and Hu, Z.Y., An experimental study of trachea anastomosis in rabbits using carbon dioxide laser, *J. Tongji Med. Univ.*, 12(4), 243, 1992.
306. Gleich, L.L., Wang, Z., Pankratov, M.M., Aretz, H.T., and Shapshay, S.M., Tracheal anastomosis with the diode laser and fibrin tissue adhesive: an *in vitro* and *in vivo* investigation, *Laryngoscope*, 105(5 Part 1), 494, 1995.
307. Cooper, J.D., Pearson, F.G., Patterson, G.A., Todd, T.R., Ginsberg, R.J., Goldberg, M., and DeMajo, W.A., Technique of successful lung transplantation in humans, *J. Thorac. Cardiovasc. Surg.*, 93, 173, 1987.
308. Hanna, E., Eliachar, I., Cothren, R., Ivanc, T., and Hughes, G., Laser welding of fascial grafts and its potential application in tympanoplasty: an animal model, *Otolaryngol. Head Neck Surg.*, 108(4), 356, 1993.
309. Shohet, J.A., Reinisch, L., and Ossoff, R.H., Prevention of pharyngocutaneous fistulas by means of laser-weld techniques, *Laryngoscope*, 105(7 Part 1), 717, 1995.
310. Poppas, D.P. and Scherr, D.S., Laser tissue welding: a urological surgeon's perspective, *Haemophilia*, 4(4), 456, 1998.
311. Rosemberg, S.K., Further clinical experience with a $CO_2$ laser in microsurgical vasovasostomy, *Urology*, 32, 225, 1988.
312. Gilbert, P.T. and Beckert, R., Laser-assisted vasovasostomy, *Lasers Surg. Med.*, 9(1), 42, 1989.
313. Seaman, E.K., Kim, E.D., Kirsch, A.J., Pan, Y.C., Lewitton, S., and Lipshultz, L.I., Results of laser tissue soldering in vasovasotomy and epididymovasostomy: experience in the rat animal model, *J. Urol.*, 158(2), 642, 1997.
314. Kirsch, A.J., Miller, M.I., Hensle, T.W., Chang, D.T., Shabsigh, R., Olsson, C.A., and Connor, J.P., Laser tissue soldering in urinary tract reconstruction: first human experience, *Urology*, 46(2), 261, 1995.

315. Kirsch, A.J., De Vries, G.M., Chang, D.T., Olsson, C.A., Connor, J.P., and Hensle, T.W., Hypospadias repair by laser tissue soldering: intraoperative results and follow-up in 30 children, *Urology*, 48, 616, 1996.
316. Keijzer, M., Jacques, S.L., Prahl, S.A., and Welch, A.J., Light distributions in artery tissue: Monte Carlo simulations for finite-diameter laser beams, *Lasers Surg. Med.*, 9(2), 148, 1989.
317. Cheong, W.F., Summary of optical properties, in *Optical-Thermal Response of Laser-Irradiated Tissue*, Welch, A.J. and van Gemert, M.J.C., Eds., Plenum Press, New York, 1995, appendix to chap. 8.
318. Jansen, E.D., van Leeuwen, T.G., Motamedi, M., Borst, C., and Welch, A.J., Temperature dependence of the absorption coefficient of water for midinfrared laser radiation, *Lasers Surg. Med.*, 14(3), 258, 1994.
319. Welch, A.J., van Gemert, M.J.C., Star, W.M., and Wilson, B.C., Overview of tissue optics, in *Optical-Thermal Response of Laser-Irradiated Tissue*, Welch, A.J. and van Gemert, M.J.C., Eds., Plenum Press, New York, 1995, chap. 2.

# 40

# Lasers in Dermatology




Kittisak Suthamjariya
*Wellman Laboratories of Photomedicine*
*Harvard Medical School*
*Boston, Massachusetts*

R. Rox Anderson
*Wellman Laboratories of Photomedicine*
*Harvard Medical School*
*Boston, Massachusetts*


## 40.1 Historical Overview

In 1960, long after Einstein proposed the theoretical concept of stimulated photon emission, Maiman built the first laser.[1] Leon Goldman then pioneered the use of lasers in dermatology.[2,3] Treatment of basal cell carcinoma with ruby and neodymium lasers and tattoo removal with Q-switched ruby laser were reported in 1964 and 1965, respectively.[4,5] Ruby and other pulsed lasers were essentially abandoned when


0-8493-1116-0/03/$0.00+$1.50

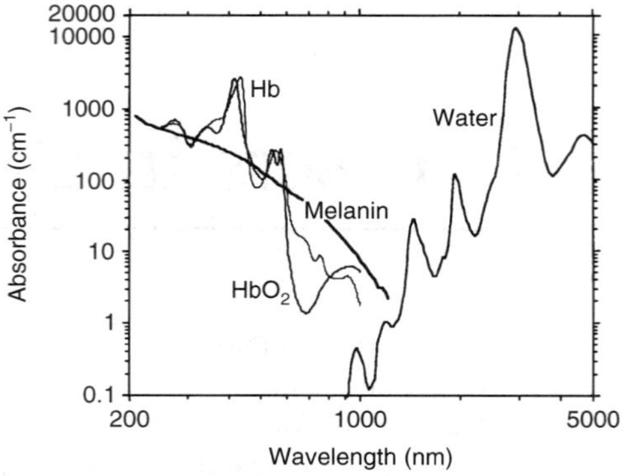


**FIGURE 40.1** Absorption spectra of major skin chromophores.

argon-ion and carbon dioxide ($CO_2$) lasers were used for coagulation and cutting during the 1970s. A new era began in the early 1980s when Anderson and Parrish proposed a theory for target-selective injury with pulsed lasers called selective photothermolysis.[6] A 450-μsec yellow-dye laser was initially developed for treating port-wine lesions, the first laser-cavity design intrinsically motivated by a medical need.[7] At present, selective photothermolysis is the basis for low-risk treatment of microvascular lesions, tattoos, and benign pigmented lesions.

## 40.2 Skin Optics

Optical penetration into skin is determined by a combination of absorption and scattering. The principal chromophores (absorbing molecules) for optical radiation in skin are water, melanin, and hemoglobin. When absorption occurs, photon energy is transferred to the chromophore. The excited chromophore can dissipate this energy in various ways including by photochemical reaction, heat, or re-emission of light. Absorption spectra of the major chromophores in skin throughout the ultraviolet (UV), visible, and infrared regions are shown in Figure 40.1. The absorption coefficient ($\mu_a$) is the probability per unit path length that a photon at a particular wavelength will be absorbed. Absorption depends on the concentration of chromophores present. Scattering occurs when the photon changes its direction of propagation. About 4 to 7% of light is reflected upon striking the skin surface because of the sudden change in refractive index between air and stratum corneum.[8] The remaining light penetrates into the skin and can be either absorbed or scattered by molecules, particles, and structures in the tissue.

Absorption is the primary process limiting penetration of UV and visible light through the epidermis. Proteins, melanin, urocanic acid, and DNA absorb UV wavelengths shorter than about 320 nm. For wavelengths from 320 to 1200 nm, especially at the shorter wavelengths, melanin is the primary absorber of light in the epidermis. Water is the dominant skin chromophore for mid- and far-infrared wavelengths. In the dermis, scattering by collagen fibers is very important for determining optical penetration of light.

In general, scattering is an inverse function of wavelength, which accounts in part for a large increase in optical penetration into the dermis, with increasing wavelength from UV through visible and near-infrared (NIR) wavelengths. The absorption coefficient of the dermis is low throughout the visible and NIR spectrum, except in blood vessels. Blood has strong absorption in the blue, green, and yellow wavelengths and a weak but significant absorption band in the 800- to 1000-nm range. An optical "window" exists between 600 and 1300 nm, which are the most penetrating wavelengths. The least-penetrating wavelengths are in the far-UV and far-infrared due to protein and water absorption, respectively. For visible and NIR wavelengths, the exposure spot size also affects the effective penetration depth of a laser beam into skin due to scattering.[9]

## 40.3 Laser–Skin Interactions

The Grotthus-Draper law states that no effect of light occurs without absorption. In other words, laser light can impose a tissue effect only when it is absorbed by chromophores. There are three basic effects: photothermal, photochemical, and photomechanical effects. In practice, these modes of laser interaction frequently coexist, but one or two usually dominate. Photothermal effects are the mainstay of current clinical laser use. The majority of current laser applications in dermatology use photothermal effects.

## 40.4 Photothermal Interactions

Unlike photochemistry, which requires a given photon quantum energy, heating can occur after absorption at any wavelength. Thermal effects on tissue are both time- and temperature-dependent. The Arrhenius model states that the rate of thermal denaturation of a given molecule is exponentially related to temperature. Therefore, the accumulation of denatured material rises exponentially with temperature and proportionally to time. Temperature is an expression of average molecular kinetic motion. At high temperatures, the specially configured macromolecules necessary for life are shaken open, resulting in loss of function. Most proteins are denatured above 60°C and DNA above 70°C. Type I collagen, the major protein in dermis, has a sharp melting transition for the fibrillar form near 70°C, causing irreversible coagulation and shrinkage of the connective-tissue matrix. Above 100°C, tissue water is vaporized. If desiccated tissue is heated further, carbonization occurs.

One popular use of lasers in dermatology is for precise tissue ablation, called laser resurfacing. Two main factors influence laser vaporization or ablation of tissue. The first is the penetration depth for the laser radiation, which depends on wavelength. The second is the rate of tissue vaporization, which depends on the amount and rate of laser energy absorption; this is discussed in the next section.

## 40.5 Selective Photothermolysis

The term "selective photothermolysis" was coined to describe site-specific, thermally mediated injury of pigmented tissue targets by pulses of radiation. This technique relies on selective absorption of a brief optical pulse to generate and confine heat at certain pigmented targets. Selective photothermolysis has three basic requirements. First, the wavelength used must be preferentially absorbed by the targeted structure. This can be achieved by choosing a wavelength within absorption bands for a chromophore associated with the target, e.g., melanin in a pigmented hair follicle or hemoglobins in a blood vessel. Second, a sufficient fluence (fluence = energy/area) is required to achieve a damaging temperature in the targets. Third, the pulse duration or exposure time must be about equal to or less than the time needed for the targets to cool. As soon as heat is created in the chromophores, it begins to dissipate by conduction. Thus, competition between active heating and passive cooling determines how hot the targets become. One useful concept is that of the "thermal relaxation time," defined as the time needed for substantial cooling of a target structure. When the laser exposure is less than the thermal relaxation time for a given light-absorbing target, maximal thermal confinement occurs. Thermal relaxation time is proportional to the square of the target's size and to shape. Thus, small objects cool much faster than large ones. For a given diameter, spheres cool faster than cylinders, which cool faster than planes. For most tissue targets the simple rule of thumb is that the thermal relaxation time in seconds is about equal to the square of the target dimension in millimeters.

Many of the cutaneous target structures treated by selective photothermolysis are heterogeneous such that the light-absorbing chromophore is not identical to the actual target. For example, blood-vessel walls, rather than the light-absorbing blood contained within these vessels, are the actual targets when treating a vascular birthmark. Because heat must flow from blood to the surrounding vessel wall, ideally one should use a pulse duration approximately equal to the thermal relaxation time of the entire vessel structure. A similar example is laser hair removal, in which heat generated by light absorption in the pigmented hair shaft must flow into surrounding structures of the hair follicle.

## 40.6 Photomechanical Interactions

Photomechanical interactions, sometimes called photoacoustic interactions, are characterized by light-induced stress and strain, leading to mechanical disruption of tissue structures, organelles, membranes, and cells. Rapid thermal expansion, high-pressure stress waves, low-pressure tensile waves, cavitation, and local vaporization can occur when laser pulses are absorbed by tissue. Rapid heating with thermal expansion and vaporization is often involved. For example, hemorrhage occurred when a 1-μsec 577-nm pulsed dye laser (PDL) was tested for the treatment of vascular lesions, prior to the present generation of longer PDLs.[10] In contrast, mechanical effects may be the primary therapeutic mechanism for the treatment of pigmented lesions and tattoo removal with Q-switched lasers.

## 40.7 Photochemical Interactions

Classic photobiology of skin is based on photochemical reactions. The major application for laser-induced photochemistry at present is photodynamic therapy (PDT). PDT involves the therapeutic use of photochemical reactions mediated through the interaction of photosensitizing agents, light, and oxygen for the treatment of malignant or benign diseases. PDT is a two-step procedure. In the first step, the photosensitizer is administered to the patient and is allowed to be absorbed by the target tissues. The second step involves activation of the photosensitizer, in the presence of oxygen, with a specific wavelength and fluence of light delivered to the target tissue. In general, PDT drugs transfer the excitation energy to oxygen, creating a reactive intermediate, singlet oxygen. Selectivity for tissue damage depends on the distributions of the three factors necessary for PDT effect: drug, light, and oxygen.

## 40.8 Laser Safety

The potential hazards from laser use are injury to the skin and eyes, aerosolization of hazardous biologic materials, ignition of fires, and electrocution. Eye injuries result from absorption of laser light by the structures in the eye including the retina, lens, sclera, and cornea. Lasers operating between 400 and 1400 nm are particularly dangerous to the retina. Wavelengths between 295 and 320 nm and between 1 and 2 μm may injure the lens. Injury to the cornea is possible from the wavelengths in the UV and most of the infrared at wavelengths beyond 1400 nm.[11] Appropriate safety goggles that filter specific wavelengths of laser light must be worn at all times while the laser is in operation. Each pair of goggles should be marked with the appropriate wavelength of protection and optical density (OD) for the specific lasers in use. Prescription glasses or clear plastic wrap-around goggles made of polycarbonate are generally satisfactory when using far-infrared sources such as $CO_2$ and erbium:yttrium-aluminum-garnet (Er:YAG) lasers.

Skin burns can be caused directly by laser-beam exposures, or by laser-ignited fires. Despite the shallow depth of photon penetration, surgical $CO_2$ laser-beam burns can occur rapidly and deeply. NIR wavelengths penetrate deeply into skin; continuous or pulsed 1064-nm neodymium:yttrium-aluminum-garnet (Nd:YAG) lasers can cause deep, severe burns. Continuous or quasicontinuous visible-light lasers such as potassium-titanyl-phosphate (KTP) (532 nm) or argon lasers tend to produce an intermediate burn depth.[11] Surgical instruments used in the operating room should have black anodized or sandblasted, roughened surfaces to reduce the potential eye or skin hazard from the reflected laser beam.

The smoke plume generated by laser vaporization of tissue may contain bacteria, viral DNA, or viable cells that may have infective potential. Intact DNA from papilloma virus has been documented in $CO_2$ laser plume.[12] Although the p24 HIV gag antigen has been detected in $CO_2$ laser–vaporized plume, infectivity is low.[13] Apart from infectious risk, the vaporized material can be an irritant to respiratory mucosa.[14] The Q-switched and dye lasers used for tattoo and pigmented lesions can eject large tissue fragments from the skin due to subsurface vaporization. Protective cylinders and shields that attach to the end of the laser handpiece should be used to contain fumes, vaporized particles, and splattered tissue. The surgeon and laser-room staff must also wear specifically designed laser-filter facemasks during the procedure, and a good local exhaust ventilation must be used.

Fires ignited by laser exposure are a major hazard, especially in the presence of oxygen and flammable items such as drapes, gauze, clothing, and flammable solvents like alcohol. The use of plastic endotracheal tubes with greater than 40% oxygen is particularly hazardous when $CO_2$ lasers are used near the tube. Endotrachial tube combustion is rapid and releases toxic fumes that can permanently destroy the patient's airway. Drapes should be wet with sterile water or saline solution and be moistened repeatedly throughout the surgery. Ignition of dry hair is a common hazard, with or without oxygen in the surgical field. Hair should be wetted.

Lasers can pose an electrical hazard if accidental discharge occurs. Electrical risks can persist within the laser itself even after power has been disconnected when the energy-storage capacitor may still hold a substantial charge. A high-energy xenon flashlamp is also a potential source of electrical hazard. Appropriate grounding and other electrical safety procedures are essential.

## 40.9 Lasers for Vascular Skin Lesions

### 40.9.1 Mechanisms of Action

Laser wavelengths absorbed by hemoglobin are used to treat cutaneous vascular lesions such as port-wine stains (PWSs), hemangioma, spider angioma, cherry angioma, venous lake, and telangiectasia of face and legs. The basic objective in treating vascular lesions is to irreversibly damage abnormal blood vessels but spare normal skin tissue. Argon and other continuous and quasicontinuous visible-light lasers were first used to treat PWSs. The risk of scarring in the pink PWS of children was high.[15] The PDL was then developed based on the concept of selective photothermolysis.[6] In the first theoretical formulation of selective photothermolysis,[7] a 1-msec, 577-nm pulse of at least 2 J/cm$^2$ was predicted as theoretically ideal for most pediatric PWSs. Initially, however, such long pulses were not available at this wavelength. A 1-μsec PDL was tested at 577 nm in normal human and animal skin and was noted to cause histologically selective injury with extensive hemorrhage to dermal vessels.[10,16] In an attempt to minimize vaporization injury and maximize thermal coagulation of vessels, the pulse duration was lengthened. A 400-μsec, 577-nm PDL was constructed, tested clinically on PWSs, and noted to work well with a low risk of scarring, even in children.[17] Subsequently, a longer wavelength (585 nm), approximately twice as penetrating, was shown to offer similar vascular selectivity with greater depth of effect.[18] A flashlamp-pumped, tunable dye laser near 585 nm (0.5-msec pulse duration) is now a standard method for treating PWS because of the very low risk of scarring when used in single pulses at 6 to 8 J/cm$^2$ fluence. With its short pulse duration, this laser produces transient purpura because of hemorrhage and a delayed vasculitis. Dierickx et al. demonstrated experimentally that the thermal relaxation time for vessels roughly 60 μm in diameter was between 1 and 10 msec.[19] PDLs with a pulsewidth of 1.5 msec have subsequently shown better clearing of PWSs. Pulses longer than about 10 msec tend to produce less immediate purpura (hemorrhage).

Purpura occurring immediately after laser treatment is generally due to blood vaporization, vessel ruptures, and hemorrhages. When the pulse duration exceeds about 20 msec, immediate purpura can generally be avoided. Recently, a number of tunable pulse-width sources, including pulsed dye (585 nm), frequency-doubled Nd:YAG (532 nm), alexandrite (755 nm), Nd:YAG (1064 nm), diode (800 nm), and flashlamps (500 to 1200 nm), have become available. Another potential advantage of long pulses is that skin cooling can more effectively protect the epidermis than can shorter pulses. In general, good results can be achieved with pulse durations that nearly match the thermal relaxation time of the target vessels, using sufficient fluence in combination with active skin-cooling methods, which are described below.

Wavelength affects the depth of vascular-lesion treatment and uniformity of heating luminal blood. Generally, the longer the laser wavelength up to 1200 nm the deeper the penetration depth[6,17,20] due to lower scattering and absorption. At wavelengths weakly absorbed by hemoglobins, such as 700 to 1100 nm, high treatment fluences are needed. Spot size also plays a significant role in determining the depth of effective treatment. The intensity of a narrow beam (small spot size) tends to suffer greater loss with depth, as light is scattered outside the beam diameter. Larger exposure spots are therefore desirable when target vessels are deep in skin.[21] Increasing the spot size has been shown to increase clearing with the 585-nm PDL in PWSs and hemangiomas.[22,23] Another significant issue is the number of pulses delivered to a single skin site. The Arrhenius model suggests that thermal injury is cumulative over time; therefore, in theory, multiple lower-fluence pulses that do not cause hemorrhage might be used to accumulate selective, gentler, and more complete damage to microvessels. This has been reported in animal and human studies.[24,25]

Fast events occurring during exposures of blood vessels to 532-nm and 1064-nm laser pulses (1- to 50-msec pulse duration) have recently been observed using fast imaging microscopy. In order of increasing fluence, the fast events were blood coagulation, vasoconstriction, thread-like appearance of the treated vascular segment, vessel disappearance, intravascular cavitation, extravascular cavitation, vessel-wall rupture and hemorrhage, and shrinkage of perivascular collagen. The apparent mechanism for vessel disappearance, a desired endpoint for treating vascular lesions, is contraction of intravascular blood and perivascular-collagen shrinkage after thermal denaturation. This occurs most reliably when the pulse duration is approximately equal to the thermal relaxation time of the exposed blood vessels.

## 40.10 Laser Devices

In addition to pulsed lasers, continuous and quasicontinuous visible-light lasers have been used to treat vascular lesions. These include argon (488 to 514 nm), krypton (520 to 530, 568 nm), argon-pumped tunable dye (488 to 638 nm), and copper-vapor and copper-bromide (578 nm) lasers. A variety of new lasers have been developed over the past decade. These include standard and long-pulsed PDL, 532-nm pulsed lasers with pulse duration of 1 msec to 100 msec, pulsed alexandrite laser at 755 nm with pulse duration up to 20 msec, pulsed 800-nm diode lasers with pulse durations from 5 to 1000 msec, pulsed 1064-nm Nd:YAG laser with pulse duration up to 100 msec, and a filtered broad-spectrum flash-lamp intense pulsed light source (IPLS).

PDLs contain a rhodamine dye, which is excited by a xenon flashlamp to produce yellow light. The traditional PDL emits 585-nm-wavelength light and 450-μsec pulse duration that penetrates the dermis to a depth of 1.2 mm.[17,18] Newer versions emit tunable 585- to 600-nm wavelengths at a pulse duration of 1.5 msec, or tunable pulse durations from 0.4 to 20 msec in a stuttered-pulse mode. One manufacturer offers an integrated cryogen-spray skin-cooling apparatus, which is less painful and in pigmented skin allows treatment at somewhat higher fluences without epidermal damage. Copper-vapor and copper-bromide lasers emit light at 511 nm and 578 nm in a "pseudocontinuous" emission consisting of 20-nsec pulses at a frequency of 15,000 Hz. The KTP laser emits 532-nm light that has a similar hemoglobin-absorption coefficient as 585-nm light. Melanin absorption is somewhat higher at 532 nm, however. These devices produce millisecond-domain pulses, either as a true millisecond pulse or as a millisecond "macropulse" of rapid delivery of shorter "micropulses." In practice, the experience and skill of the laser operator is a large variable affecting clinical outcome, regardless of the type of laser used.

## 40.11 Port-Wine Stain

The PWS is a congenital malformation of dermal microvasculature present in 0.3 to 0.5% of newborns and persists throughout life. Histologically, PWSs reveal an increased number of venules in the papillary and upper reticular dermis.[26] PWSs darken progressively with age, from pink to red to deep violet, and are frequently associated with progressive hypertrophy, a raised and nodular surface.[27] The darkening correlates with vessel dilation, not an increase in the number of vessels.

Treatments of PWSs have included skin grafting, ionizing radiation, cryosurgery, tattooing, dermabrasion, and laser treatments. At present, lasers clearly offer the treatment of choice. The current standard treatment of PWSs is with PDL at a wavelength ranging from 585 to 600 nm, 0.4 to 1.5 msec pulsewidth, and fluence of 3 to 10 J/cm$^2$ with 3- to 10-mm exposure spots delivered with minimal overlap. The somewhat shorter pulses are ideal for pediatric PWS in which the vessel diameters are relatively small, while long-pulsed PDLs may be more suitable for resistant or adult PWS.[28,29] Approximately 50 to 75% lightening of PWSs is achieved within two to three treatments with PDL therapy. Complete resolution of the lesions can be achieved with repetitive laser treatments in about half of patients.[30] Immediate purpura in the treated areas is the treatment endpoint and takes 10 or more days to subside.

Several prognostic criteria have been used in predicting the outcome of PWS treatments. These include age of patient, site, size, color, and microvascular pattern of the lesion. In one study, it was shown that the major determinants of treatment response in order of decreasing importance were PWS location and size and patient's age.[31] The site of a PWS clearly has prognostic significance. PWSs on the face and neck tend to respond better than elsewhere on the body, and the worst results are seen on the distal limb.[32–34] For PWS of the head and neck, centrofacial lesions (medial aspect of the cheek, upper cutaneous lip, and nose) and lesions involving dermatome V2 respond less favorably than lesions located elsewhere on the head and neck (periorbital, forehead/temple, lateral aspect of the cheek, neck, and chin).[35] Central forehead PWSs respond most favorably.[31] Another prognostic factor for the response of PWS to laser treatment is the initial size of the lesion being treated. Small PWSs have shown a greater response to laser treatment than larger lesions.[31,36]

There are contradictory reports concerning the influence of patient age and color of PWSs on the outcome of laser treatment. Although young children are generally considered to respond better to PDL treatment,[17,28,32] some authors found no age dependence.[37–39] However, there are strong arguments in favor of treating PWS during infancy. These include rapid expansion of the PWS area as the child grows, ease of treatment under topical anesthesia only, completion of therapy prior to the child's formation of physical self-image, and avoidance of the psychological trauma associated with disfigurement during socialization and early school years. With regard to color of lesions, some authors report best results in pink lesion,[32] while others report better results in red lesions.[40] In one study, color of PWSs was found to be of no prognostic value.[33] Fiskerstrand et al. demonstrated the correlation of the treatment outcome and the size and color of PWSs.[41,42] The vessels of the good responders were located significantly more superficially than the vessels of the moderate and poor responders. The poor responders had significantly smaller vessels than the moderate and good responders. The moderate responders had deeper but larger vessels than the poor responders. The vessel diameter was correlated to the color, with pink lesions having the smallest diameter vessels and purple lesions the largest. Vessel depth was partly reflected in the lesional color, as the pink and purple lesions had significantly deeper vessels than the red ones. These results indicate that pink lesions predict poor blanching due to deeply located small vessels, while red lesions predict a good therapeutic result because of more superficially located vessels.

The two major microvascular patterns examined by a videomicroscope are ectasia localized to the capillary loops (type 1) and dilated ectatic vessels in the superficial horizontal plexus in a ring pattern (type 2). It has been shown that type 1 lesions are more likely to respond to laser therapy than are type 2 or mixed lesions.[43] Eubanks et al. subsequently demonstrated that PWSs in areas that typically respond well to laser treatment (V3, neck, and trunk) are more likely to have a type 1 pattern, whereas PWSs in areas that have a poorer response to therapy (V2, distal extremities) are more likely to have a deeper type 2 pattern.[44] The videomicroscope study of PWSs may in the future provide a better prognostic guide for treating PWS.

Although PDLs at present have been established as the treatment of choice for PWS, the results are variable and unpredictable.[33] Treatment is somewhat more effective using higher fluences, e.g., PDL with cryogen spray cooling[45–49] or long-pulsed KTP laser.[50] A controversy is the apparent "recurrence" of PWS after treatment with lasers. The natural course of a PWS is to darken and dilate over years, and residual

PWS vessels follow this course. One study reported that almost 50% of PWSs darken 3 to 4 years following discontinuation of treatment.[39] Michel et al. reported 16.3% recurrence rate in 147 PWS patients treated with PDL at least 1 year after the last treatment. Since children under 10 years of age did not show any PWS recurrence, the authors pointed out that age at the beginning of treatment may have an influence on the recurrence rate.[51] In theory, a high-energy pulsed 1064-nm Nd:YAG laser could offer a combination of vascular selectivity and deep penetration that may be superior to PDL treatment. However, controlled studies of appropriate pulsed Nd:YAG lasers for PWS treatment have not yet been reported.

Topical anesthesia is usually satisfactory for the treatment of PWS lesions in adults, but local or regional anesthesia may be required. General anesthesia should be considered in children under 12 years of age to minimize the risk of eye injury and pain associated with the multiple treatments.[52]

Side effects from PDL treatment of PWSs occur infrequently. In a study of 701 PWS patients with 3877 PDL treatments, hyperpigmentation was the most frequent side effect observed (9.1%) followed by blistering, atrophic scar (4.3%), hypopigmentation, crusting, and hypertrophic scar (0.7%).[53]

## 40.12 Hemangioma

Hemangiomas are common "benign" proliferations of blood vessels occurring in up to 10% of all children by the age of 1 year. They typically undergo a rapid growth phase during the first year of life followed by complete or partial involution by age 12 years.[54] Histologically, there are dense collections of thin-walled vessels interspersed with sheets of vascular endothelial cells. Because of the natural course with spontaneous involution of hemangioma, conservative management has been advocated for the majority of these lesions. Indications for early interventions include alarming hemangiomas, hemangiomas with associated complications such as recurrent bleeding, infection, and ulceration, hemangiomas in certain locations (such as the nose, lip, and ear) that are more likely to result in permanent anatomic distortion and scarring, and extreme familial or patient distress.[55–57] The alarming hemangiomas are those lesions that cause, or threaten to cause, ocular compromise, respiratory distress, congestive heart failure, gastrointestinal bleeding, and extensive skin ulceration.[55] Hemangiomas have been treated with varying success using compression therapy, ionizing radiation, intralesional or systemic corticosteroids, interferon $\alpha$, cryosurgery, embolization, and various laser therapies. The treatment methods used depend on the age of patient, site and size of lesions, and stage of the disease. Early surgery is an option in some cases of pedunculated hemangiomas. Superficial hemangiomas are generally treated with local corticosteroids (potent topical or intralesional), PDL, or both. Larger lesions with a deeper component and alarming hemangiomas usually require the administration of systemic corticosteroids.[57]

Laser treatment of infantile hemangiomas is very controversial. One report evaluating 617 cases of hemangiomas treated with PDL noted that laser treatment halted progression or induced regression in the majority of patients. Total regression was achieved in nearly half of the small superficial hemangiomas; however, a tightly controlled study of these self-regressing lesions is still lacking.[58] In clinical practice, treatment with PDL appears to be effective and may be the treatment of choice for small, superficial cutaneous hemangiomas (Figure 40.2). Hemangiomas with a deep component do not benefit from PDL treatment due to the limitation of the laser-penetration depth. Early therapeutic intervention with PDL may or may not prevent proliferative growth of the deeper or subcutaneous component of the hemangioma.[59–61] Even in carefully selected patients with relatively superficial lesions, PDL treatment is variably effective, presumably because of continued growth of a deeper component not reached by the laser.[62,63] In large disfiguring hemangiomas, PDL has been used as an adjunct to glucocorticosteroid or interferon-$\alpha$ treatments.[64] Systemic interferon treatment in infants is associated with development of spastic diplegia, a major complication limiting this form of therapy. Laser treatment is generally accepted for control of ulcerated hemangiomas, leading to a decrease in pain, promotion of healing, and acceleration of involution.[65–67] One to three laser treatments have been reported to heal most ulcerated lesions.[65,66] However, laser treatment has also been reported to induce ulceration.[68] The continuous 1064-nm Nd:YAG laser is sometimes used to debulk and photocoagulate hemangiomas, often by interstitial fiber-optic delivery. In

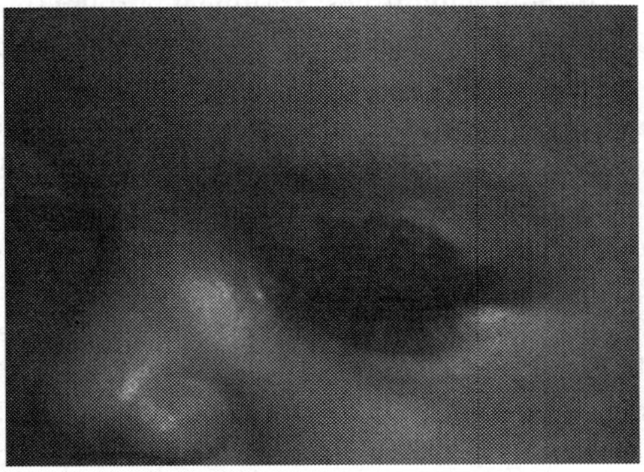

A

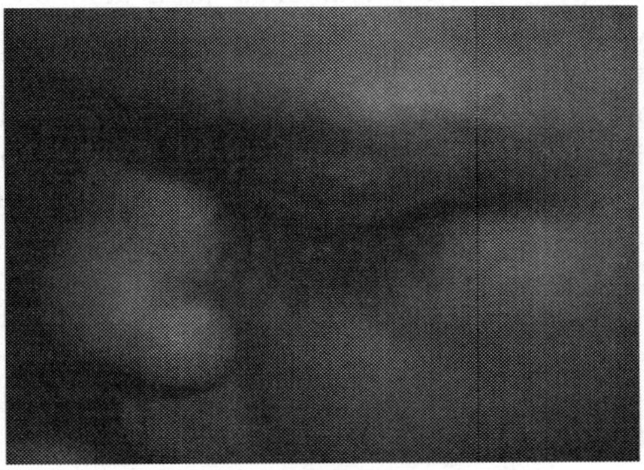

B

**FIGURE 40.2 (Color figure follows p. 28-30.)** (A) A hemangioma of the left upper eyelid prior to treatment. (B) After a series of eight treatments with the 595-nm pulsed-dye laser at 10 to 15 J/cm$^2$ and 7-mm spot size, the hemangioma cleared significantly.

a study of 25 children with bulky lesions, 72% of cases showed good to excellent response to Nd:YAG laser photocoagulation.[69]

## 40.13 Telangiectases and Dilated Leg Veins

Telangiectases are common dilated, superficial blood vessels associated with aging, chronic sun damage, scars, malformations, acne rosacea, and other conditions. Dilated veins are also common in both sexes due to poor venous return from incompetent vein valves, pregnancy, prolonged standing or sitting, aging, and heredity. Sclerotherapy, which consists of local injection of substances that kill endothelial cells lining blood vessels, remains the gold standard of treatment for telangiectasias of the lower extremities. Facial telangiectases are less responsive to sclerotherapy than those located on the leg and more prone to complications.[70] Therefore, laser treatment in general has a greater role for treating facial telangiectases.

Red linear and arborizing telangiectases frequently occur on the face, particularly on the nose, cheeks, and chin. They measure 0.1 to 1.0 mm in diameter and represent a dilated venule, capillary, or arteriole.[71] With a 532-nm KTP laser and pulse duration in the millisecond range, good to excellent results can be obtained on facial telangiectasia with minimal side effects.[72] Immediately after appropriate laser treatment with green, yellow, or infrared laser pulses between about 10 and 50 msec, the abnormal vessels appear subtly darker and coagulated or simply disappear from view. In a study comparing the copper-vapor laser and PDL in the treatment of facial telangiectases, both lasers provided satisfactory and comparable results at 2 and 6 weeks, but PDL induced more post-treatment purpura (bruising) and other side effects. Postoperative swelling, healing time, and incidence of postinflammatory hyperpigmentation were greater with the PDL compared to the copper-vapor laser.[73] The results of KTP laser and PDL in the treatment of facial telangiectasia are also comparable. A comparison of long-pulsed dye (590 nm) and KTP lasers for facial telangiectasia showed a somewhat better result with long-pulsed dye laser according to blinded grading of response.[74] However, patients strongly preferred the KTP laser due to its lower incidence of post-treatment purpura and other side effects. This stimulated development of so-called "stuttered-pulse" dye lasers (SPDLs), which emit a train of several 0.4-msec pulses during about 20 msec. The SPDL produces less purpura than conventional PDLs but still more than a simple KTP-laser pulse of similar duration.

Laser treatment of leg telangiectasia has improved greatly over the past 5 years, to the point that it is almost comparable in efficacy to sclerotherapy. Before this, argon, argon-dye, copper-vapor, KTP:YAG, pulsed-dye, and krypton lasers were used with disappointing results. Traditional PDLs (585 nm, 0.4 msec, 6 to 8.5 J/cm,$^2$ 5-mm spot size) are ineffective for vessels greater than 0.2 mm in diameter.[75] Based on theoretical considerations, longer wavelengths and longer pulse durations would be more appropriate for treatment of most leg veins. Longer wavelengths are needed to penetrate deeper into skin. Longer pulse width produces less mechanical injury, with more uniform thermal damage of large vessels, and no purpura. Attempts to modify yellow PDLs for leg-vein therapy have met with limited success. At a wavelength of 595 nm or 600 nm and a pulse duration of 1.5 msec, with $2 \times 7$ mm elliptical spot at fluences of 15 and 20 J/cm$^2$, there was only 50 to 75% clearance of leg veins less than 1.5 mm in diameter after one to three treatment sessions.[76–78] A comparison of 595-nm PDLs at 1.5- and 4-msec pulse duration found neither to be effective for vessels greater than 1 mm in diameter.[79]

Vessel size and depth largely determine the appropriate choice of laser for treatment. Recent studies suggest that NIR, millisecond-domain alexandrite, diode, and Nd:YAG lasers are considerably superior to visible lasers for treating vessels larger than about 1 mm in diameter but less effective for small vessels. One study evaluated 28 patients with leg telangiectasia treated with a 755-nm long-pulsed alexandrite laser. Each patient received three laser treatments given at 4-week intervals with a fluence of 20 J/cm$^2$, 5-msec pulse duration, 10-mm spot size, and single or double pulse at 1 Hz. The greatest response was seen in medium-diameter (0.4- to 1.0-mm) leg vessels. Reduced response was seen in large-diameter (1.0- to 3.0-mm) vessels, while small-diameter (less than 0.4-mm) vessels did not respond at all.[80] In another study, each of the 20 leg telangiectatic patients was treated once using one to three passes of a 755-nm, 3-msec alexandrite laser at fluences of 60 to 80 J/cm$^2$. By 12 weeks, 65% of 51 treatment sites showed greater than 75% clearance, and an additional 22% showed greater than 50% clearance.[81] One study evaluated an 800-nm diode-array laser on leg veins using fluences of 15 to 40 J/cm$^2$ with 5- to 30-msec double or triple pulses. Three treatment sessions were given every 4 weeks. By 2 months after the last treatment, complete clearing occurred in 22%, 75% clearing in 42%, and 50% clearing in 32% of area treated.[82] A diode laser with scanner was also evaluated for leg-vein treatment. After 2 treatment sessions, 2 to 4 weeks apart, 18% had 75 to 90% clearance, 21% had 50 to 75% clearance, 18% had 25 to 50% clearance, and 36% had less than 25% clearance.[83]

The NIR, 1064-nm laser provides good absorption by hemoglobin and deeply penetrates into the skin to the depth of a few millimeters. This allows selective and deep vascular injury of larger leg veins. Recent studies confirm the effectiveness of millisecond-domain 1064-nm Nd:YAG laser treatment of venulectasia and reticular veins up to 4 mm in diameter[84–86] (Figures 40.3 and 40.4). In one study, 25 women were treated up to three times with a 1064-nm Nd:YAG laser at 6-week intervals. A double pulse of 7 msec at 120 J/cm$^2$ was used for vessels 0.2 to 2 mm in diameter, and a single pulse of 14 msec, fluence 130 J/cm$^2$, with 6-mm spot size was used for vessels 2 to 4 mm diameter. A total 64% of patients achieved 75% or greater clearing of vessels after a maximum of three treatment sessions.[85] The results of leg-vein treatment with the IPLS are variable. Although some studies showed encouraging results,[87–89] contradictory results have been reported,[90] and the treatment technique requires significant experience to achieve good results.[91]

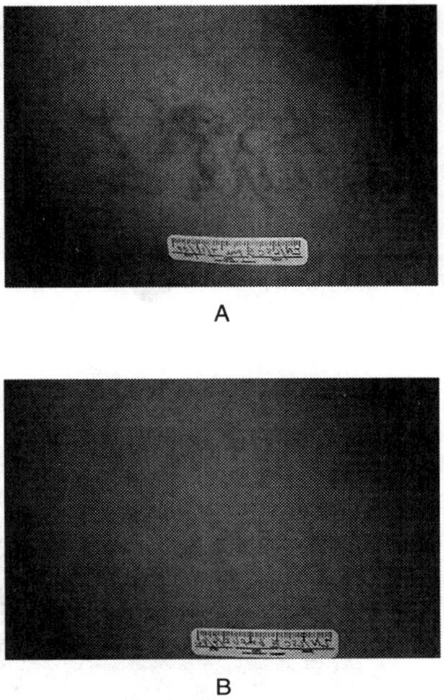

FIGURE 40.3 (A) Leg veins in the thigh of a 54-year-old woman before treatment. (B) 6 months after one treatment with long-pulsed Nd:YAG laser (at 350 J/cm$^2$, 50 msec, and 3-mm spot size).

## 40.14 Other Vascular Lesions

Other cutaneous vascular lesions that have been reported to respond to laser treatment include venous lakes, angiofibromas, angiokeratomas, poikiloderma of Civatte, cherry angiomas, spider angioma, acne rosacea, adenoma sebaceum, angiolymphoid hyperplasia with eosinophilia, pyogenic granulomas, and granuloma faciale.[92–94]

## 40.15 Other Applications

### 40.15.1 Warts

PDL-treated warts had reported clearance rates of 45 to 99%.[95–99] Kauvar et al. treated 142 patients with 703 recalcitrant and 25 previously untreated warts with PDL at fluences of 7.25 to 9.5 J/cm$^2$. The overall response rates for warts on all sites were 74% after one treatment and 93% after an average of 2.5 treatments.[96] Subsequent studies were performed to analyze the treatment technique using an infrared thermal camera to compare differences in heat accumulation at sites treated with single- vs. multiple-laser pulsing with varying pulse-repetition rate. The results suggest that multiple pulses at a fast repetition rate (at least 1 Hz) contribute to wart response by accumulating heat within the lesions.[100] Another study prospectively evaluated the efficacy of PDL therapy (9.0 to 9.5 J/cm$^2$, 5-mm spot size) vs. conventional liquid-nitrogen cryotherapy or cantharidin in the treatment of warts. A total of 194 warts were evaluated by the conclusion of the study. Complete response was noted in an average of 70% of the warts treated with conventional therapy and in 66% of those in the PDL group.[99] Thus, laser therapy is one among many alternative treatments for warts. KTP lasers have also been reported to improve recalcitrant warts in 80% of patients.[101] The combination of laser with other therapy has not been well studied.

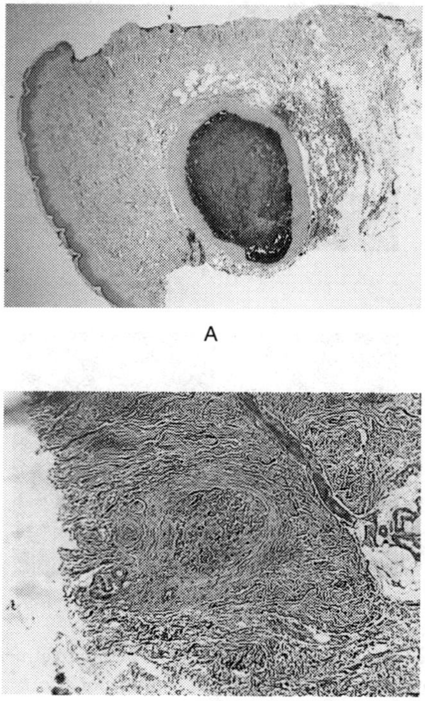

**FIGURE 40.4** Histology of leg veins. (A) 1 day after treatment with long-pulsed Nd:YAG laser. Thrombosis and necrosis of the whole vascular wall are demonstrated. (B) 1 month after treatment. Fibrosis of the entire vascular structure is shown.

## 40.15.2 Hypertrophic Scars, Keloids, and Striae

Many studies address PDL treatment of hypertrophic scars, keloids, and striae distesae ("stretch marks"), but these indications remain controversial.[102–107] Striae may respond better to lower fluences (3.0 J/cm$^2$) in one or two treatments.[106] A recent prospective, single-blind, randomized, controlled study assessed the efficacy of PDL and silicone-gel sheeting in hypertrophic scar treatment. There was no significant difference between either PDL or gel sheeting and the untreated control areas.[108]

## 40.15.3 Psoriasis

Several studies indicate that PDL offers an effective treatment for stable plaque-type psoriasis, and the device has been cleared by the U.S. Food and Drug Administration (FDA) for this purpose.[109–111] In one study, 0.4-msec and 1.5-msec PDLs were compared, and no significant effect by pulse duration was found. Patients responding to treatment with the PDL remained in remission for up to 13 months.[111]

## 40.15.4 Laser Treatment of Pigmented Lesions (Melanin)

Melanocytes at the base of the epidermis synthesize melanin in the form of 0.5-μm intracellular organelles called melanosomes. The thermal relaxation time of a melanosome is just under 1 μsec. Selective photothermolysis of melanosomes was first described in 1983[6] using 351-nm, submicrosecond excimer laser pulses at less than 1 J/cm$^2$ fluence. Melanin absorbs across the UV, visible, and NIR spectrum, with decreasing absorption at longer wavelengths.[7] Across this wide spectral range, any laser with sufficient power or pulse energy can cause thermal injury due to absorption by melanin. Shorter wavelengths damage pigmented cells with relatively lower fluences, while longer wavelengths penetrate deeper into the skin but need more energy to induce melanosome disruption. In clinical terms, short wavelengths affect only superficial pigmented lesions, leaving deeper structures intact, while longer wavelengths can affect pigmented targets in dermis such as hair follicles.

Target specificity of laser pulses depends on wavelength and pulse width. The appropriate pulse width depends primarily on the size of the target and its thermal relaxation time. The calculated thermal relaxation time for melanosomes is about 250 to 1000 nsec. Submicrosecond or even femtosecond laser pulses cause individual pigmented cell death by violent cavitation, a microscopic steam bubble erupting around each melanosome after the laser pulse, but longer pulse widths, in the hundreds-of-microseconds domain, do not appear to cause specific melanosome damage.[112] The immediate effect of submicrosecond near-UV, visible, or NIR laser pulses in pigmented skin is immediate whitening, which fades away after several minutes. Immediate skin whitening correlates well with melanosome rupture seen by electron microscopy[113] and is presumably due to residual gas bubbles after cavitation of melanosomes. Nearly identical but deeper whitening occurs with Q-switched laser exposure of tattoos. This offers a clinically useful immediate endpoint. Larger pigmented targets, such as hair follicles and "nests" of congenital nevus cells, are more effectively treated with much longer pulses, e.g., in the millisecond domain.

The clinical importance of matching laser pulse width with the intended pigmented targets' thermal relaxation time is well illustrated by comparing skin responses to Q-switched (nsec) vs. long-pulsed (msec) ruby lasers, both at a wavelength of 694 nm. At typical fluences, the Q-switched ruby laser is excellent for treating nevus of Ota, in which individual pigmented cells are the target, but it is incapable of permanent hair removal. In contrast, the long-pulsed ruby laser is capable of permanent hair removal but has little effect on nevus of Ota.

## 40.15.5 Epidermal Pigmented Lesions

Epidermal pigmented lesions are typically responsive to many different treatments that damage the epidermis but spare the dermis including lasers, cryotherapy, and electrosurgery. These common lesions include freckles, benign lentigo, and café-au-lait macules (CALMs). Pigment-selective submicrosecond pulsed lasers include green-light lasers (510-nm pulsed-dye, 532-nm frequency-double Nd:YAG), red-light lasers (694-nm ruby and 755-nm alexandrite), and NIR (1064-nm Nd:YAG) lasers. Lasers with little or no selective absorption, such as $CO_2$, erbium:YAG (Er:YAG), and some low-power continuous or quasicontinuous wave lasers, e.g., argon, krypton, or copper-vapor lasers, are also effective for treating pigmented epidermal lesions. Superficial pigmented lesions, such as ephelides (freckles), solar and labial lentigines, and seborrheic keratoses, can be effectively treated with any of the pigment-specific and nonspecific lasers.[114–117] Successful treatment of freckles in Asian skin with Q-switched alexandrite laser has been reported. Substantial improvement was achieved with an average of 1.5 laser-treatment sessions at 7.0 J/cm$^2$ without long-term side effects.[118] Todd et al. compared three lasers (Q-switched frequency-doubled Nd:YAG laser, krypton laser, 532-nm diode-pumped laser) with liquid nitrogen for treatment of solar lentigines. Of the 27 patients treated, 25 preferred laser therapy to cryotherapy, with the Q-switched frequency-doubled Nd:YAG laser being the most preferred[119] (Figure 40.5). Hyperpigmentation and hypopigmentation are common post-treatment side effects. In general, postinflammatory hyperpigmentation is most frequent in darker skin types and suntanned individuals. The risk of hypopigmentation may be somewhat higher after Q-switched ruby laser than after Q-switched alexandrite or Nd:YAG lasers.

Treatment of CALMs yields variable results, whether pigment-selective or nonselective lasers are used.[114,120,121] There is a controversial study reporting complete lightening of all 34 large CALMs treated with a 510-nm pigmented-dye laser with no evidence of recurrence at 1-year follow-up. Each lesion received an average of 8.4 laser treatments with energy densities ranging from 2.0 to 4.0 J/cm$^2$ at intervals of 6 to 8 weeks.[122] In another study, 16 CALM patients were treated with 511-nm copper-vapor laser and automated scanner. Patients received one to four treatment sessions using a mean fluence of 8.9 J/cm$^2$ (range 7 to 22 J/cm$^2$). The responses were graded as good to excellent in 15 cases and poor in one case. Transient pigmentary changes have been observed in some cases.[123] A recent study reports a "Q-switched

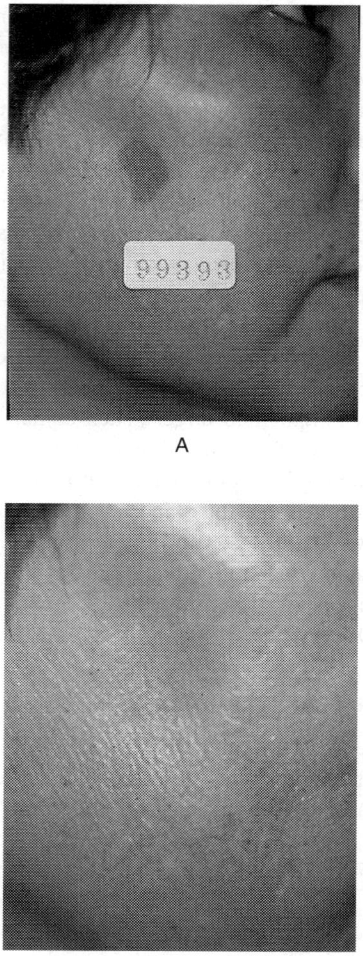

A

B

**FIGURE 40.5** (A) Senile lentigo at the right cheek of a 50-year-old woman prior to treatment. (B) 6 months after one single treatment with 532-nm Q-switched Nd:YAG laser, the lesion faded almost completely. (Images courtesy of Hirotaka Akita, M.D.)

laser-resistant" CALM being successfully removed with Er:YAG laser resurfacing.[124] In general, CALMs require repeated treatment over a period of months or even years to achieve maximal clearing.

The term "atypical" is used to describe pigmented lesions containing cellular dysplasia or precursor lesions for melanoma. Treatment of these lesions must be approached with extreme caution. For atypical melanocytic lesions, laser treatment may not be adequate for eradication of lesions. In one study, two patients with atypical-appearing solar lentigines were treated with Q-switched ruby laser. In both cases, the initial results were excellent, but they recurred several months later and biopsy showed lentigo maligna, a form of melanoma *in situ*.[125]

## 40.15.6 Dermal and Mixed Epidermal/Dermal Pigmented Lesions

Laser treatment of melasma is difficult, the results are variable and unpredictable, and recurrence is the rule. Taylor and Anderson assessed the efficacy of Q-switched ruby-laser treatment of melasma and found that, regardless of fluence used, no permanent improvement was observed. Transient darkening was

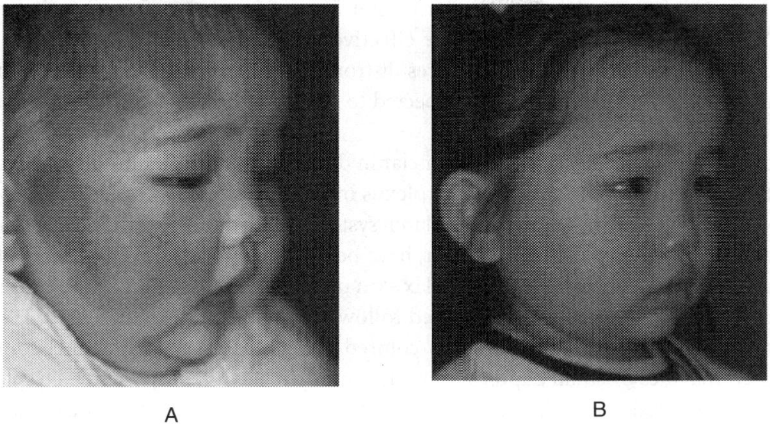

FIGURE 40.6 (Color figure follows p. 28-30.) (A) A 4-month-old girl with nevus of Ota on her right face prior to treatment. (B) Almost complete clearing after five treatments with Q-switched alexandrite laser. (Images courtesy of Harue Suzuki, M.D.)

noted after laser treatment in some patients. No textural changes were seen after healing, except for a slight depression at high fluences on black patients.[126] Recently, resurfacing lasers (pulsed/scanned carbon dioxide or Er:YAG) have been used in the treatment of refractory melasma, with satisfactory results. In one study, full face-skin resurfacing using an Er:YAG laser was performed using 5.1 to 7.6 J/cm$^2$ energy to ten refractory melasma patients. Marked improvement of melasma was noted immediately after laser surgery; however, all patients exhibited postinflammatory hyperpigmentation between 3 and 6 weeks postoperatively, requiring biweekly glycolic-acid peels. At the end of 6 months, significant improvement in the melasma was demonstrated using clinical scoring and melanin reflectance spectroscopy assessment.[127] In another study, complete resolution of dermal-type melasma was achieved with a combination of pulsed $CO_2$-laser resurfacing followed by Q-switched alexandrite-laser treatment. Abnormal melanocytes are first destroyed with pulsed $CO_2$ laser, and then dermal melanins are selectively eliminated with the alexandrite laser.[128]

Dermal melanocytosis includes nevus of Ota, nevus of Ito, and mongolian spot. The Q-switched ruby, Q-switched alexandrite, and Q-switched Nd:YAG lasers are all highly effective and are the treatment of choice for these pigmented lesions (Figure 40.6). Significant or complete clearing of lesions can be achieved after an average of four to six treatment sessions with fluences ranging from 5 to 12 J/cm$^2$.[129–132] Ueda et al. classified 151 Japanese patients with nevus of Ota according to lesion color; the patients were treated with Q-switched ruby laser at 5 J/cm$^2$ and 6.5-mm spot size. In general, the degree of lightening of lesion was directly proportional to the number of treatments performed. To achieve good to excellent results in brown, brown-violet, violet-blue, and blue-green nevi at least three, four, five, and six treatment sessions were required, respectively.[133]

In a clinicopathologic study of nevi of Ota treated with Q-switched alexandrite laser, lesions less than 1 mm deep achieved the best results.[134] Chan et al. retrospectively reviewed the complications following treatment of nevi of Ota with Q-switched alexandrite, Q-switched Nd:YAG, or a combination of both lasers. Hypopigmentation was the most common complication encountered in all three groups with the incidences of 38, 10.5, and 7.6% in the combined, Q-switched alexandrite, and Q-switched Nd:YAG laser-treated groups, respectively.[135] The same author also proposed a new classification of nevi of Ota and noted that periorbital underresponse (panda's sign) may be due to overly cautious treatment in that area.[136] Once the nevus of Ota has fully cleared after laser treatment, the results are usually permanent. However, repigmentation has been observed in one patient during pregnancy.[137]

Acquired bilateral nevus-of-Ota-like macules (ABNLM), also called nevus fuscoceruleus zygomaticus or nevus of Hori, is a dermal melanocytic lesion characterized histopathologically by the presence of elongated melanocytes in upper dermis similar to that found in nevus of Ota. Q-switched ruby and Q-switched Nd:YAG lasers have been shown to be effective treatments for these conditions.[138–140]

Postinflammatory hyperpigmentation may result from any inflammatory insult causing dermal melanin deposition. Pigment-specific lasers are expected to improve these lesions, but the results so far have been disappointing.[126]

Dark infraorbital circles result from dermal melanin deposition, postinflammatory pigmentation from atopic or contact allergic dermatitis, superficial plexus of blood vessels, and shadowing from lax skin and infraorbital swelling. Several pigment-specific laser systems, such as Q-switched ruby laser, Q-switched alexandrite laser, and Q-switched Nd:YAG laser, have been shown to improve infraorbital pigmentation related to dermal melanin deposition.[141] When lax skin or infraorbital swelling is the cause of infraorbital darkening, significant lightening can be achieved following $CO_2$ laser resurfacing.[142]

Nevus spilus presents as a circumscribed tan-colored macule (CALM component) containing smaller darkly pigmented macules or small papules (nevo-melanocytic nevus component). Dysplastic nevus and malignant melanoma arising within these lesions have been described;[143–145] thus, caution should be exercised in treating nevus spilus with lasers. These pigmented lesions have a variable response to laser treatment. The CALM component may be more resistant to treatment, requires more treatment sessions, and tends to recur. In one study, two lesions of nevus spilus were treated with the Q-switched ruby laser. Complete clearances of lesions occurred after two to seven treatments, but the CALM component recurred within 6 months after the last treatment.[146] However, in another study, six patients with nevus spilus were treated with the Q-switched ruby laser and the Q-switched Nd:YAG laser at 532 or 1064 nm. All lesions showed >80% or complete response to Q-switched ruby laser at 5.5 to 10 J/cm$^2$. The Q-switched ruby laser produced significantly better results than both Q-switched Nd:YAG lasers.[147]

Becker's nevus is an acquired patch of hyperpigmentation plus hypertrichosis in pubescent boys. These lesions are typically located on the shoulder, chest, and proximal arm. Satisfactory results are not easily obtained with the Q-switched lasers. Although hyperpigmentation has been treated successfully, recurrence rates have been high, and the terminal hairs within the nevi may decrease temporarily.[115,148] With the new long-pulsed red or infrared lasers for hair removal more effective treatment results can be achieved. Nanni and Alster reported successful results treating a case of Becker's nevus with long-pulsed (3-msec) ruby laser. Ninety percent reduction of hair and pigmentation was achieved after three laser-treatment sessions at fluences of 18 to 22 J/cm$^2$, 10-mm spot size, and at 6- to 8-week intervals. The clinical reduction in hair and pigment continued to be evident 10 months after the final treatment without scarring or skin textural change.[149]

### 40.15.7 Nevomelanocytic Nevus

Nevomelanocytic nevi are usually pigmented "moles" consisting of melanocyte-related cells present at the base of the epidermis (junctional nevi), within the dermis (dermal and congenital nevi), or both (compound nevi). The lesions may be acquired or congenital and exhibit cellular atypia ranging from benign to malignant (melanoma arising within a nevus). The use of pigment-specific lasers to remove congenital and acquired melanocytic nevi remains controversial. The standard method of nevus removal is surgical excision with complete histologic evaluation. For the time being, laser treatment should be considered only in selected cases in which surgical excision cannot be performed; otherwise, it may result in unsightly scars. Continuous-wave (CW) lasers, such as the argon and $CO_2$ lasers, have been used in the past to remove nevomelanocytic lesions, including congenital nevi, but they resulted in significant scarring. Recurrences were not uncommon, and occasional pseudomelanoma (melanoma-like lesions occurring in the context of wound healing) could be seen.

Despite controversy, the Q-switched ruby laser has been used for the treatment of small and medium congenital nevi, with some good clinical improvement after several sessions.[150–152] Vibhagool et al. reported that Q-switched ruby laser at 8 J/cm$^2$ and 5-mm spot size is moderately effective at treating small nevomelanocytic nevi. Two thirds of the lesions showed complete response, and one third showed partial response. In cases of partial response, residual nevomelanocytic nests were identified at depths of

0.1 to 0.4 mm from the surface.[152] Grevelink et al. also reported that Q-switched ruby and Q-switched Nd:YAG lasers were capable of removing the superficial portions of congenital nevi to depths of 0.16 mm and 0.44 mm, respectively.[151] Similar results were noted in other studies including a comparative study of benign melanocytic nevi treated with the Q-switched alexandrite and the Q-switched Nd:YAG lasers.[153] However, in these reports, congenital melanocytic nevi (CMN) were either incompletely removed or recurred within 1 year of laser treatment. Recently, Nelson and Kelly reported treating an infant with medium-sized congenital CMN with Q-switched ruby laser. Five years after only one treatment, the lesion remained clinically and histologically clear.[154]

Pulse duration appears to be important for laser treatment of acquired and congenital nevi. In a study of ruby-laser pulse duration and fluence for treatment of extensive congenital nevi, complete clearing was reported after four treatments at 15 J/cm$^2$ at 1-msec pulse duration, but Q-switched ruby laser (nanosecond-domain pulses) was ineffective for clearing the lesions.[155] Subsequent long-term histological study revealed the presence of residual nevus cells in the dermis at approximately 1 mm below the skin surface covered by a thin zone of fibroplasia that masked the residual nevomelanocytes. No clinical or histologic evidence of malignant degeneration was observed in the treated areas up to 8 years after laser treatment.[156] In another study of benign acquired and dysplasic melanocytic nevi treated once with normal-mode vs. Q-switched ruby lasers, 52% of the nevi showed significant decrease in pigment, but no lesion had complete histologic response. The authors suggested that currently lasers should not be used in the treatment of atypical nevi.[157]

The effects of pigment-specific laser pulses on melanocytes with the potential for malignant transformation are not well understood. Removal of the superficial portion of nevi might be beneficial by decreasing the number of melanocytic cells capable of malignant degeneration.[150,153] On the other hand, healing after laser treatment may stimulate or select for persistent nevus cells with neoplastic change. To date, there have been no reports of melanoma arising in benign pigmented lesions treated with lasers. However, patients who have had a nevomelanocytic lesion removed by laser must continue to be followed, and any signs of recurrence should be approached cautiously.

## 40.15.8 Tattoos

Tattoos consist of phagocytosed submicrometer ink particles trapped in the lysosomes of phagocytic dermal cells, mostly fibroblasts, macrophages, and mast cells. Most tattoos can effectively be removed or substantially lightened using Q-switched lasers, with a low risk of scarring. The mechanisms of tattoo-ink removal involve fracture of ink granules followed by release from cells into the dermis. Some of the ink is eliminated by lymphatic and transepidermal elimination, but most of it is rephagocytosed by somatic dermal cells within a few days.[158] In general, different laser wavelengths are needed to remove different tattoo-ink colors. Black tattoo ink, which is typically carbon or black iron oxide, has a broad absorption spectrum and is well treated by either Q-switched ruby, Q-switched alexandrite, or Q-switched Nd:YAG lasers. Red and orange tattoo inks can be removed with green pulses (510-nm pulsed-dye or 532-nm Q-switched frequency-doubled Nd:YAG lasers). For green colors, Q-switched ruby and to some extent Q-switched alexandrite lasers can usually achieve successful clearing, but the Q-switched Nd:YAG laser is generally ineffective.[159–161] (Figure 40.7) Green and yellow ink are usually the most difficult to remove. Resistant tattoos can be ablated with $CO_2$ or Er:YAG lasers, but this produces scarring. In general, the responses of tattoos to laser treatment are fluence-dependent, and amateur tattoos require fewer treatments than professional ones.[162,163] Traumatic tattoos usually contain less pigment than others and require even fewer treatments. Q-switched Nd:YAG laser is preferred in darkly pigmented patients with black tattoos because of fewer epidermal side effects due to melanin absorption. Pulse duration also affects the treatment result. For example, picosecond laser pulses were more efficient than nanosecond laser pulses for tattoo removal.[164]

Cosmetic tattoos containing red iron oxide ($Fe_2O_3$) and titanium oxide ($TiO_2$) can turn black on exposure by Q-switched lasers.[165] Ink darkening is probably a combination of chemical reduction and changes in particle size. A similar reaction has been reported as localized chrysiasis after Q-switched ruby

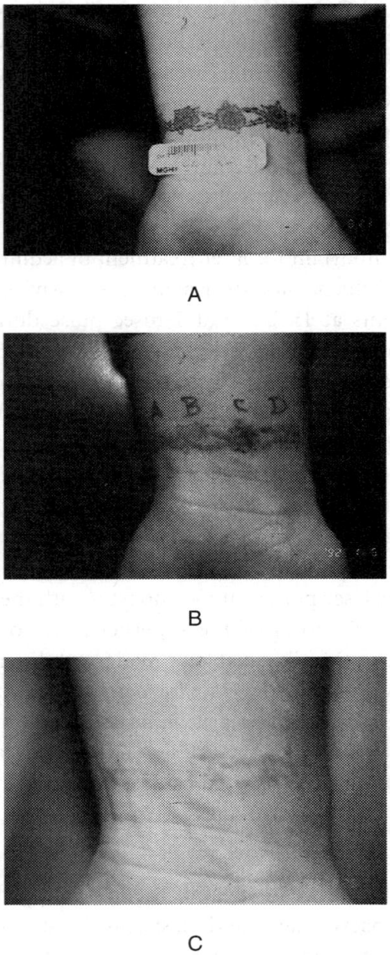


**FIGURE 40.7** (A) Multicolored tattoo prior to treatment. (B) Site "C" after three treatments with Q-switched ruby (694-nm) laser shows loss of green and black inks but not red. Other sites treated with Q-switched Nd:YAG (1064, 532-nm) laser show loss of red and black inks, but not green. (C) After six further treatments using both lasers on the entire tattoo, all ink colors have substantially faded.

laser treatment in patients receiving parenteral gold therapy.[166] Recently, Ross et al. reported a significant association between the presence of $TiO_2$ in tattoos and poor response to laser treatment.[167]

## 40.16 Other Exogenous Pigments

Drug-induced pigmentations are typically due to a complex of drug and melanin deposits in dermal macrophages and hence are similar in some ways to tattoos. Exogenous ochronosis from hydroquinone treatment of melasma has been successfully treated with Q-switched ruby laser.[168] Minocycline-induced cutaneous pigmentation has also been successfully treated with Q-switched alexandrite,[169] Q-switched KTP,[170] and Q-switched ruby lasers.[171–173] Similarly, amiodarone-induced blue-gray hyperpigmentation can resolve with Q-switched ruby laser,[174] and imipramine-induced hyperpigmentation was effectively treated with Q-switched alexandrite and Q-switched ruby lasers.[175]

Dermal hemosiderin in Kaposi's sarcoma and dermal melanin pigments after antimalarial therapy have also been reported to be effectively treated with Q-switched ruby laser.[176] Circumscribed siderosis after intramuscular or intravenous iron injection has been effectively treated with Q-switched ruby and Nd:YAG lasers after 3 to 16 treatment sessions.[177]

### 40.16.1 Laser Hair Removal

There are many medical and surgical settings in which excessive hair growth is a crucial problem. Hirsutism often is associated with ovarian, adrenal, or pituitary disorders. Impressive hypertrichosis may be part of a congenital syndrome or drug-induced.[178] Hair-removal methods include shaving, plucking, chemical depilatories, waxing, electrolysis (DC current), electrothermolysis (RF current), and laser hair removal. Only the last three methods are known to produce permanent hair loss.

There are two potentially important targets for permanent hair removal: the "bulb" and "bulge" areas of hair follicles. The bulb is the neurovascular papilla at the base of anagen (active growth phase) hair follicles, responsible for hair-shaft growth, and typically about 3 to 7 mm below the skin surface. The bulge or stem-cell area is a region near the insertion of arrector pili muscle, about 1.5 mm below the skin surface, and is the source of new matrix cells with each hair cycle.[179,180] These stem cells appear to be necessary for cycling of the hair follicles into the anagen phase. Perifollicular vessel damage may also play a role in the laser hair-removal process.[181]

Various lasers and light sources have been developed for hair removal. These include ruby, alexandrite, diode, and Nd:YAG lasers and IPLSs. These devices target melanin, the endogenous pigment. Therefore, laser hair removal is more effective for darker hair colors.

Photomechanical destruction of hair has been attempted with low-fluence, Q-switched, 1064-nm Nd:YAG lasers in combination with a topically applied carbon suspension, which was supposed to act as an exogenous chromophore.[182] This was the first approved laser hair-removal treatment; it did not produce permanent hair loss but was effective for temporary control of hair growth, with few side effects. In theory, the carbon was intended to penetrate into hair follicles and provide a target for Q-switched laser pulses to generate photomechanical injury. However, there is no evidence that the carbon particles actually penetrate the follicle or that they offer an advantage over the laser alone, and complete hair regrowth was reported within 6 months after treatment.[183] If greater uptake of carbon could be achieved and a longer, more energetic Nd:YAG laser pulse were used, the approach might be more effective. Alternatively, a dye other than carbon could be used that stains the viable follicular epithelium and absorbs red or NIR wavelengths.

Photochemical destruction of hair follicles by PDT has been reported by Grossman et al. A 20% solution of aminolevulinic acid (ALA) was applied topically under occlusion for 3 h, followed by 100 to 200 J/cm$^2$ fluences of 630-nm light delivered over about 30 min from an argon-pumped dye laser. Three months after treatment there was significant loss of hair in sites treated with high ALA and high fluence, without scarring.[184]

Photothermal destruction of hair follicle relies on melanin in the hair shaft and the matrix to provide local absorption of a pulse of light, which is timed to produce heating and destruction of the follicles. The wavelength region capable of reaching follicular targets, with sufficiently strong and selective absorption by melanin, lies between about 600 and 1100 nm. Pulsed optical sources that operate in these wavelength regions include 694-nm ruby laser, 755-nm alexandrite laser, 800-nm diode laser, 1064-nm Nd:YAG laser, and xenon flashlamp, which is an incoherent source spectrally filtered to emit in this spectrum. These optical hair-removal devices have pulse durations in the millisecond domain. Laser hair removal requires destructive heating of hair follicles due to melanin absorption, without significant injury to the epidermis, which also contains melanin. For patients with fair skin and darkly pigmented hair there is a wide range of safe and effective laser fluences using any of these devices. For darkly pigmented or tanned skin, however, various combinations of longer wavelength, longer pulse duration, and active skin cooling are used to minimize the risk of unwanted epidermal injury. Dynamic (epidermal) and bulk (epidermal and dermal) cooling devices have been developed that are most useful with short and long pulses, respectively. Temporary hair removal for a few months is easily achieved even at low fluences and appears to be due to induction of the catagen and then the telogen (resting) phase. Transition to catagen

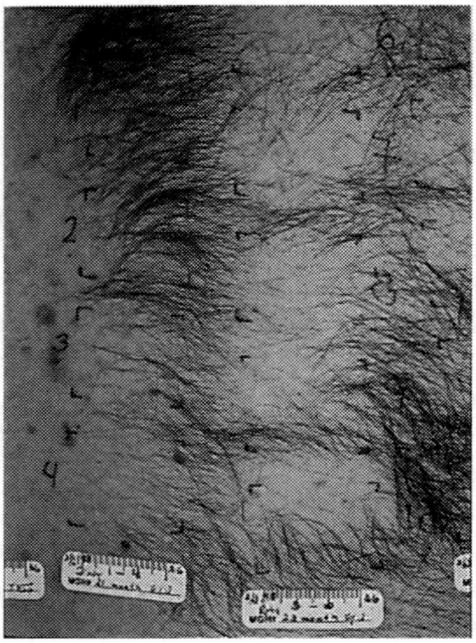

**FIGURE 40.8** Hair removal more than 20 months after treatment with 800-nm diode laser (left) and 694-nm ruby laser (right), in test sites on the back.

and telogen is a common follicular response to injury, for example, from stress, drugs, hypoxia, surgery, and after laser treatment, from photothermal injury. Permanent hair reduction requires higher fluences, which are capable of necrosing follicular target regions, and has been defined as "a significant decrease in terminal hairs after a given treatment, which is sustained for a period of time longer than the complete growth cycle of hair follicles at the given body site" (Figure 40.8).

A ruby laser was first reported by Grossman et al.[185] to remove hair from normal human skin. In a fluence-response study of fair-skinned volunteers with dark hair, test sites on the back or thigh were either wax-epilated or shaved before laser exposures with a range of 0 to 60 J/cm$^2$ at 694 nm, 0.3 msec, and 6-mm diameter in contiguous exposure spots delivered through an ice-cold sapphire window pressed against the skin. Significant temporary hair loss lasting 1 to 3 months was induced in all laser-treated sites in all subjects, while the hair loss at 6 months was fluence-dependent. Wax epilation protected hair follicles from laser-induced hair loss at 6 months, suggesting that absorption of light by the pigmented hair shaft is involved in the mechanisms for long-term hair removal. Subsequently, laser-induced hair loss up to 2 years after a single ruby-laser treatment with high fluence (>30 J/cm$^2$) was reported in 4 of 13 subjects.[186] No scarring occurred, and about one third of the subjects had transient pigmentary changes. Selective thermal injury of superficial and deep follicular epithelium was observed histologically. Ruby lasers were the first devices marketed for permanent laser hair removal and remain useful, especially in patients with light-colored hair and relatively fair skin.

Alexandrite lasers, with a slightly longer wavelength than ruby lasers, behave very similarly to ruby lasers for hair removal. Alexandrite lasers provide slightly greater depth of penetration but lower absorption by pheomelanin, the dominant form of melanin in "red" hair. Several studies have shown that alexandrite lasers are effective in removing unwanted hair.[187–189] In one study, 11 patients received five treatments at 3-week intervals in the groin area using alexandrite laser at 20 J/cm$^2$, 20 msec, and 10-mm spot size. At 1-year follow-up, an average of 78% clearance of hair was noted, with no evidence of scarring and pigmentary changes.[190] Another study compared alexandrite laser and electrolysis for hair removal. The right axilla was treated by four electrolysis sessions with an intensity of 4 to 8 mA at 3-week intervals, and the left axilla was treated three times at 4-week intervals with 3-msec alexandrite laser at 30 to 50 J/cm$^2$. The average clearance rate of the hairs was 74% by laser and 35% by electrolysis 6 months after the initial treatment.[191] Recently, equivalent clinical and histologic responses with a long-pulsed alexandrite (2-msec pulse, 10-mm spot) and a long-pulsed diode laser (12.5 msec or 25 msec, 9-mm spot) at either 25 J/cm$^2$ or 40 J/cm$^2$ for hair removal were reported.[192]

A widely used and versatile laser for hair removal is a pulsed 800-nm diode laser, with variable pulse width of 5 to 30 msec, spot size of either 9 × 9 or 12 × 12 mm, delivered through a cold sapphire window. The 9 × 9-mm version was used to remove hair from 38 subjects with average fluences of 33.4 J/cm$^2$, in one to four treatment sessions (mean, 2.7). More than 60% of the subjects showed more than 75% hair reduction after a mean interval of 8 months after the last treatment.[193] In another study, 50 volunteers with Fitzpatrick skin types II and III and dark brown or black hair were treated using the same laser system. After one treatment, hair regrowth ranged from 22 to 31% at 1 month, then remained stable at 65 to 75% at 3 months and thereafter for 20-month average follow-up. There were relatively longer growth delays after two treatments, with hair regrowth beginning at 6 months after treatment and ranging from 47 to 66% for the remainder of the follow-up evaluations. Side effects were limited to transient pigmentary changes.[194] Baugh et al. treated 36 subjects with varying shades of brown or black hair using an 800-nm diode laser guided by a scanner. Patients received one to four treatments during the course of the study. Significant fluence-dependent hair reduction was demonstrated between treatment and control groups. At 48 J/cm$^2$, the highest dose, a mean hair reduction of 43% was achieved 30 d after the final treatment, and 34% was achieved 90 d after the final treatment. Darker hairs were more effectively treated than lighter hairs, which is typically the case for all lasers and flashlamps used for hair removal.[195]

Millisecond-domain, long-pulsed Nd:YAG (1064-nm) lasers also are capable of inducing long-term hair loss; however, higher fluence is needed to compensate for the lower melanin absorption. These lasers, when delivered in combination with skin cooling, are excellent for controlling hair growth in dark-skinned individuals, e.g., for pseudofolliculitis barbae in men of African descent. In one study, 15 patients were treated with a 30-msec pulsed Q-switched Nd:YAG laser at fluences between 125 and 150 J/cm$^2$. The average hair reduction was 36% at 7 d, 52% at 30 d, and 59% at 90 d.[196] In another study, 208 subjects were treated using a long-pulsed Nd:YAG laser with a fluence of 23 to 56 J/cm$^2$ at 1-month intervals until desirable results were obtained. Every session resulted in a 20 to 40% hair loss, depending on the color of hair. Complete epilation was obtained in four to six sessions. Only white hair did not respond, and no patients had blistering or hypo- or hyperpigmentation.[197] In another study, 20 dark-skinned women with skin phototypes IV through VI and dark brown to black terminal hair on the face, axillae, or legs were treated with a long-pulse (50-msec) 1064-nm Nd:YAG laser. A series of three treatments at fluences ranging from 40 to 50 J/cm$^2$ were delivered to the identified treatment areas on a monthly basis. Substantial hair reduction was seen after each of the three treatment sessions. Prolonged hair loss was observed 12 months after the final laser treatment (70 to 90% hair reduction).[198]

Filtered xenon flashlamps, also called intense pulsed light (IPL), emit a wide-wavelength band of noncoherent light pulses. A variety of these sources have been shown to produce permanent hair removal. Gold et al. performed a study using IPL to remove hair in 37 subjects. Patients received a single treatment with energies of 34 to 55 J/cm$^2$. Approximately 60% hair removal was noted at 12-week follow-up.[199] In a multicenter study, 40 women with hirsutism on the upper lip and chin were treated with an IPL, with an average fluence of 38.7 J/cm$^2$ and a mean wavelength of 585 nm. Within six treatments, 76.7% of the hair was removed.[200] In another study, 48 patients with Fitzpatrick skin types I through V randomly enrolled for two treatments 1 month apart with an IPL system. Parameters used were a 2.8- to 3.2-msec pulse duration typically for three pulses with thermal relaxation intervals of 20 to 30 msec with a total fluence of 40 to 42 J/cm$^2$. Hair reductions of 42 and 33% were noted at 8-week and 6-month follow-ups, respectively.[201] A prospective study compared long-term results of single vs. multiple treatments with IPL in 34 patients. Parameters used were a cut-on wavelength of 615 to 695 nm, pulse duration of 2.6 to 3.3 msec, fluence of 34 to 42 J/cm$^2$, and 10 × 45-mm exposure field. The mean hair-removal efficiency achieved was 76% after a mean of 3.7 treatments and a mean 21.1-month follow-up; more than 94% of the treated sites reached >50% hair loss.[202]

Before initiating laser hair-removal procedures, factors to consider include skin color, hair color, hair diameter, and density of hair. These factors affect laser parameters to be used and in turn affect the efficacy and side effects of treatment. For any given laser type, efficacy is laser-fluence-, hair-color-, and hair-diameter-dependent, but the side effects are laser-fluence-, skin-type-, and hair-density-dependent. Perifollicular erythema and edema are expected as an endpoint for laser hair removal. In general, laser hair removal is effective and safe in the majority of patients with dark hair and relatively fair skin. With a combination of longer wavelength, longer pulse duration, and parallel cooling, dark skin can be safely and effectively treated.[203]

## 40.17 Lasers Targeting Water as Chromophore

Laser skin resurfacing (LSR) uses water in tissue as a target chromophore to rapidly deposit energy in a thin layer of tissue, which vaporizes in a controlled way. Due to strong water absorption in the far infrared, either $CO_2$ lasers (10,600 nm) or Er:YAG lasers (2940 nm) are used. The epidermis and upper dermis are precisely ablated, with a predictable zone of residual thermal damage followed by reepithelialization from the residual hair follicles and remodeling of dermal collagen during wound healing. The depth of residual thermal injury must be limited, which is accomplished by vaporizing a thin layer at the surface in a time short enough to limit the amount of heat conducted into the dermis.

### 40.17.1 $CO_2$ Lasers

With pulsed $CO_2$ lasers at fluences of 4 to 19 J/cm$^2$ per pulse, precise tissue ablation of tissue (20 to 40 μm thick) with less than 100-μm residual thermal damage (RTD) can be produced.[204,205] The minimum ablation depth per "pass" of laser exposure is consistent with the optical penetration depth, which is about 20 μm for the $CO_2$ laser (and is equal to $1/\mu_a$, where $\mu_a$ is the tissue absorption coefficient of about 500 cm$^{-1}$) at this laser wavelength.[204] Selective tissue ablation with minimal RTD occurs when the laser energy is deposited faster than the thermal relaxation time of the heated tissue layer, which for $CO_2$ lasers is around 1 msec (Figure 40.9). To cleanly vaporize the superficial layer of skin with minimal RTD, a $CO_2$ laser delivering energy in less than 1 msec must produce a fluence greater than 5 J/cm$^2$ (the vaporization threshold).[204–207]

For illustration, a popular pulsed laser and a scanned-CW laser will be compared, i.e., the UltraPulse 5000 laser (Coherent, Palo Alto, CA) and the SilkTouch laser (Sharplan, Allendale, NJ), respectively. The UltraPulse laser delivers up to 500 mJ in single pulses, with a duration of about 1 msec. This energy is delivered in a 3-mm collimated beam, generally guided by a computer-pattern generator (CPG) at an average power of up to about 80 W. Rapid precise placement of these high-energy pulses can be done manually, but the CPG provides very rapid and precise arrays. A typical mode uses a 2.25-mm exposure spot spacing, 300 mJ per pulse, and a power of 60 W, resulting in placement of pulses in one of six chosen patterns at a rate of 220 Hz. The amount of overlap between adjacent pulses is programmable via the CPG, from zero to 60% overlap. In contrast, the SilkTouch laser system uses a standard continuous $CO_2$ beam focused to a 100- to 200-μm-diameter spot that is manipulated by a pair of rotating mirrors in the delivery handpiece to scan various patterns at the tissue rapidly, resulting in a tissue dwell time of about 1 msec. Round or rectangular scanning patterns typically ranging from 4 to 9 mm may be chosen.

The $CO_2$ laser ablation threshold is about 5 J/cm$^2$ for submillisecond pulses and may be somewhat higher for rapidly scanned beams (around 15 J/cm$^2$).[208,209] In general, treatment is performed well above the ablation threshold, whether a pulsed or scanned-CW laser is being used. Thermal damage and dermal temperatures have also been shown to be about 15% to 20% higher for scanned-CW vs. pulsed lasers.[209] Despite these differences, the clinical efficacy and side effects have been shown to be similar in side-by-side comparisons treating perioral and periorbital wrinkles.[210] In typical clinical use, two passes with the SilkTouch laser appear to be approximately equal to three passes with the UltraPulse laser in terms of the amount of tissue removed and the depth of residual thermal necrosis.[211,212] It must be emphasized that clinical judgment and careful assessment of ablation depth at the time of treatment is absolutely necessary as a guide to laser resurfacing. Experienced physicians obtain essentially equivalent clinical results with any of the commercially available lasers for skin resurfacing. Treating patients in a cookbook fashion is, in general, hazardous. One very helpful sign of tissue depth during laser resurfacing is immediate shrinkage. This obvious reaction, due to thermal denaturation of type I collagen at about

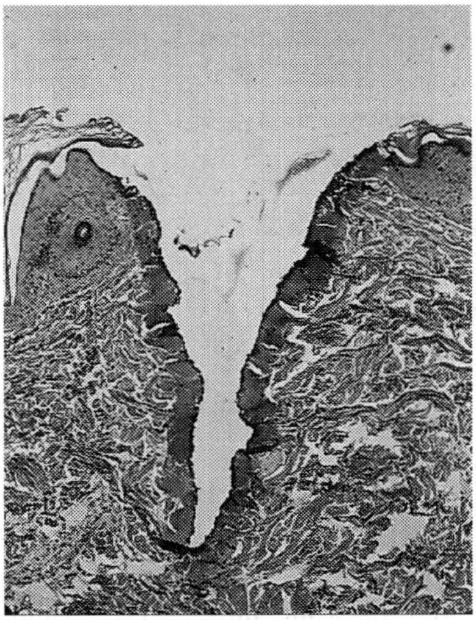

A

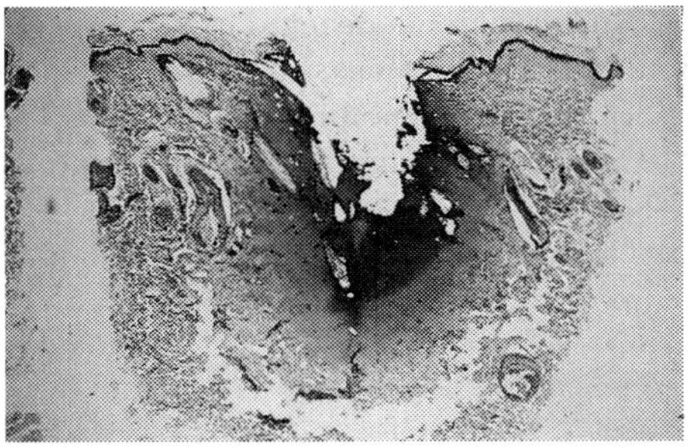

B

**FIGURE 40.9** Photomicrograph of skin biopsy specimens comparing the amount of residual thermal damage (RTD) after $CO_2$ laser ablation. (A) After submillisecond $CO_2$ laser. (B) After continuous $CO_2$ laser exposure. Note minimal amount of RTD in submillisecond exposure.

70°C, occurs only when the dermis is initially subjected to a layer of residual thermal damage. Clinically, it is well established that some dermal injury is necessary for improvement of moderate wrinkles, regardless of the type of laser being used. In contrast, epidermal pigmented lesions, such as freckles, lentigines, and CALMs, can be removed without dermal injury. Typically, the first pass with pulsed lasers, either $CO_2$ or Er:YAG, causes little or no immediate shrinkage.

## 40.17.2 Er:YAG Lasers

The Er:YAG laser, operating at 2940 nm, is even more strongly absorbed by water than the $CO_2$ laser. The $\mu_a$ of pure water at this wavelength is approximately 13,000 cm$^{-1}$. In tissue, which is typically 70% water, the $\mu_a$ is approximately 9000 cm$^{-1}$, and the optical penetration depth, $1/\mu_a$, is approximately 1 μm.[213,214] However, thermally induced changes in $\mu_a$ during Er:YAG laser pulses tend to increase the optical penetration depth. The estimated thermal relaxation time of the layer in which the laser energy is absorbed is approximately 1 to 10 μsec. The calculated threshold fluence for tissue ablation is 0.28 J/cm$^2$, which is close to that measured for submicrosecond Er:YAG laser pulses. However, longer (e.g., 100 to 1000 μsec) Er:YAG laser pulses are used for LSR to provide better hemostasis. For the Er:YAG lasers used for skin resurfacing, the tissue-ablation threshold is typically 1.5 J/cm$^2$.[213,215] With submicrosecond Er:YAG laser pulses, the RTD zone is only 5 to 10 μm.[214] Short-pulse Er:YAG lasers are therefore capable of ablating only one or two cell layers at a time with minimal residual injury. This is excellent for extremely fine ablation but a poor choice when hemostasis is needed.[216,217] When operated to produce longer pulses, hemostasis can be achieved (and thermal injury increased intentionally) by allowing more thermal conduction to occur. Thus, Er:YAG lasers can be made to perform more like a $CO_2$ laser, and vice versa, by manipulating fluence and pulse duration.[218,219] When the same absorbed power density (W/cm$^3$ deposited at the surface of the tissue) is used for both systems, one generally sees similar effects. Some Er:YAG-laser manufacturers have made this laser behave like the $CO_2$ laser (increasing RTD and improving hemostasis) by either adding a simultaneous $CO_2$ laser or manipulating the pulse duration and fluence of the Er:YAG pulses.

## 40.18 Mechanisms of Action

Tissue removal (ablation) was initially thought to be the most important mechanism in LSR, as if one were literally sculpting the skin. While sculpting may play some role, it is also well established that other, nonselective agents, such as dermabrasion, acids, and solvents (chemical "peels"), can achieve equivalent clinical results. Furthermore, when ablation-plus-RTD depth into the dermis is less than the wrinkle depths, marked cosmetic improvement is often observed. It appears that a combination of collagen shrinkage and dermal extracellular matrix remodeling during wound healing contributes to the final cosmetic improvement. When the $CO_2$ laser interacts with tissue, there are three distinct zones of tissue alteration correlating with the degree of tissue heating: (1) a zone of vaporization; (2) underlying this, a zone of irreversible thermal damage and denaturation, resulting in tissue necrosis; (3) and below this, a zone of reversible, sublethal thermal damage.

Immediate collagen shrinkage occurs in the zone of irreversible injury and, to some extent, the zone of reversible injury. This accounts for the visible tissue tightening observable as the $CO_2$ laser interacts with the dermis, but there is controversy about the role of immediate collagen shrinkage in achieving the final outcome. Using two to five passes with a pulsed $CO_2$ laser, shrinkage increases somewhat with the number of passes, while the zone of dermal collagen denaturation stays constant (~115 μm).[209] In pigs, pulsed $CO_2$ laser resurfacing produces immediate tissue contraction (maximum of approximately 38%) and residual thermal damage that is saturable for multiple passes and high fluences.[220] A subsequent study in pigs showed that $CO_2$ laser resurfacing produces short- and long-term wound contraction — greater than that induced by purely ablative methods (mechanical or Er:YAG laser) for the same total depth of injury. Wound contraction over time was dependent on both the depth of RTD and depth of ablation. Wounds with more than 70 μm RTD layer healed with greater fibrosis (microscopic scarring), as evidenced by greater compaction and horizontal orientation of collagen fibers in the superficial dermis after healing. These data suggest that initial collagen contraction and thermal damage modulate wound healing.[221] Also supporting this concept is a greater and more sustained elevation in collagen content after $CO_2$ laser injury vs. 50% trichloroacetic acid "peel" in mouse skin.[222] New collagen deposition and remodeling in the dermis appear to be responsible for the long-term clinical improvement (up to 1 year) after LSR, although immediate tissue shrinkage is present for only a short period.[212,223–225] Another study found increased elements other than collagen, e.g., factor 13, vimentin, and actin expression, after pulsed $CO_2$ laser injury in a pig.[226] Contraction of these filaments may provide a scaffold for deposition and organization of collagen during wound healing, which may contribute to skin tightening and decreased surface irregularities seen after $CO_2$ laser resurfacing.

It is interesting to contrast mechanisms of LSR with those of a "face-lift," a common surgical procedure for wrinkles and skin laxity in which facial skin is stretched. Two nearby points on facial skin typically become closer together after laser resurfacing, whereas after a face-lift the same two points typically grow further apart. The combination of LSR and plastic surgery is often more effective than either modality alone.

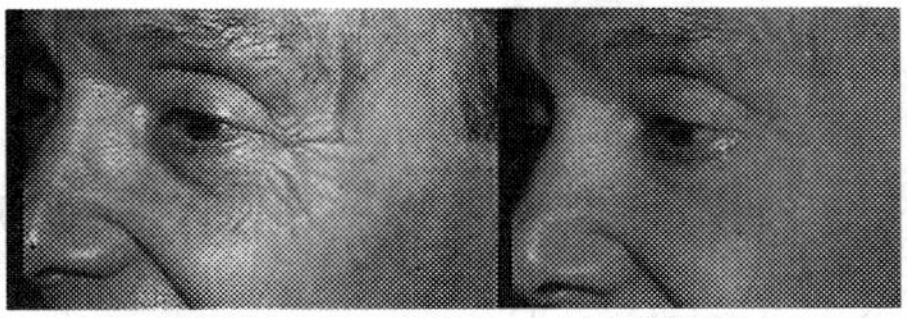

**FIGURE 40.10** Skin resurfacing. (A) Before treatment. (B) 6 months after three passes of pulsed $CO_2$ laser resurfacing (fluences 5 and 3.5 J/cm$^2$).

# 40.19 Indications for Laser Skin Resurfacing

## 40.19.1 Acne Scars

LSR is moderately beneficial in treating atrophic partial-thickness nondistensible scars,[227] although results vary.[228] Kye reported 40% improvement of acne scar at 3-month follow-up after Er:YAG-laser treatment.[229] With acne scar treatment, a longer postoperative interval (12 to 18 months) prior to assessment for retreatment was advocated because continued clinical improvement was observed as long as 18 months after $CO_2$ laser resurfacing, with an 11% increase in improvement observed between 6 and 18 months postoperatively.[230] Kwon and Kye treated 12 patients with hypertrophic scars, 20 patients with depressed scars, and 4 patients with burn scars using Er:YAG laser, 2-mm handpiece, 500 to 1200 mJ/pulse at 3.5 to 9 W. In this series, 9 of 12 hypertrophic scars, 17 of 20 depressed scars, and 2 of 4 burn scars were substantially improved.[231]

## 40.19.2 Photoaging

Chronic photoaging, comprised of rhytides, dyschromias, solar lentigines, ephelides, and actinic keratoses, has been shown to respond very favorably to $CO_2$ LSR (Figure 40.10). Ross et al. showed that scanning laser with a short dwell time performed just as well as a pulsed laser in a side-by-side comparison trial.[210] Er:YAG lasers have also been used successfully in this context.[232–235] It is now apparent that clinical improvement following equivalent-depth dermal laser wounding with the Er:YAG laser is less than that observed following $CO_2$ LSR, although the time of re-epithelialization and postoperative erythema are shorter.[221,225,236] Side-by-side comparison studies using $CO_2$ laser alone and combined $CO_2$/Er:YAG laser resurfacing have demonstrated that the overall healing times and erythema can be reduced without compromising clinical efficacy in the combined $CO_2$ and Er:YAG laser treatment group.[237–239] Kauvar et al. reported a reduction of residual thermal damage from 120 to 150 μm to 80 to 100 μm, re-epithelialization times by an average of 3 d, and duration of erythema by approximately 1 month, in the combined laser treatment group, but there were no differences in the relative depths of granulation tissue, fibroplasia formation, and the degree of clinical improvement. It appears that excellent clinical outcomes probably require residual thermal damage zones in the 60- to 70-μm range or higher.[238] The Er:YAG laser may prove safe and effective for photoaged skin of the neck, arms and hands, and chest.[233,240]

In clinical practice, both lasers are useful for deep photodamage. The $CO_2$ laser is used to efficiently remove the epidermis in a single pass, followed by a second pass to produce collagen tightening. A third pass is used in selective areas of deeper photodamage. This may also be followed by one to two passes with the Er:YAG laser. For superficial therapy, when the patient desires rapid healing and is willing to accept improvement only in texture and pigmentation, a single pass of the $CO_2$ laser can be performed without wiping the epidermal debris away. For single-pass wounds, not wiping decreased the level of wounding. In contrast, not wiping in multiple-pass wounds significantly increased the depth and variability of residual thermal damage and necrosis, resulting in prolonged healing.[245] Deeper wrinkles caused by facial expression, for example, the glabella and nasolabial folds, do not respond as well as finer wrinkles, especially perioral and periorbital rhytides.[212,223,227,242–245] These are best managed with other techniques, e.g., injections of botulinum toxin.

## 40.20 Other Indications

Almost any skin lesion treatable with electrodesiccation or currettage or exhibiting irregular surface topography can be considered for LSR. Epidermal processes, such as actinic and seborrheic dermatoses, diffuse solar lentigines, actinic cheilitis, superficial basal cell carcinoma, and squamous cell carcinoma *in situ,* often respond nicely to resurfacing. Dermal processes, including nonmovement-associated rhytides and papular elastosis, xanthelasma, sebaceous hyperplasia, rhinophyma, and angiofibroma, may also be responsive to pulsed $CO_2$ or Er:YAG laser therapy.[246] Laser-assisted hair transplantation[247] and various unrelated conditions, including xanthelasmas, viral warts, Hailey-Hailey disease, and lymphangioma circumscriptum, have been reported to respond to $CO_2$ laser treatment.[248,249] Recalcitrant psoriatic plaques have been reported to be effectively treated with Er:YAG laser resurfacing.[250]

## 40.21 Contraindications

Patients with a history of keloid generally should not undergo LSR. Any patient with severe systemic disease or diseases complicated by immunosuppression should not be treated, as these may alter wound healing. Isotretinoin has been reported to produce atypical scarring with resurfacing methods. It is recommended that patients discontinue isotretinoin for at least 1 year before laser treatment.[251] Any condition in which the adnexal structures are compromised (e.g., radiation therapy, collagen vascular disease or are nonabundant (the skin on the neck and extremities) may retard re-epithelialization.[251,252]

## 40.22 Side Effects and Complications

The most common short-term side effects include postoperative erythema, edema, pain, pruritus, burning discomfort, a sensation of tightness, and textural changes. Postoperative erythema occurs in all patients, and its duration is mainly a reflection of the depth of thermal damage.[246] The risk of complications is largely determined by the depth and extent of the resurfacing. Before re-epithelialization occurs, the major complications are bleeding, particularly following Er:YAG laser resurfacing, and infection. *Staphylococcus aureus* and *Pseudomonas aeruginosa* appear to be the most common infective agents, followed by *S. epidermidis* and *Candida albicans.*[253] Antiherpetic prophylaxis should be used in patients having full facial or lower facial laser peels, irrespective of a history of HSV. The role of prophylactic antibiotics is controversial. A retrospective study of 130 patients undergoing laser resurfacing revealed a significantly higher rate of infection in patients receiving combination intraoperative and postoperative antibiotic prophylaxis.[254] Late complications include acneiform folliculitis and milia formation, contact dermatitis, pigmentary change, scarring, and ectopion.[255]

Generally, pigmentary changes are related to the level of injury. Injury just to the papillary dermis is more likely to cause hyperpigmentation and is dependent on skin type. With deeper passes, hypopigmentation is possible or even common. Normally, hyperpigmentation is seen 2 to 4 weeks after injury, but hypopigmentation is typically more delayed, sometimes occurring several months after surgery. Two types of hypopigmentation can be recognized. One is a relative hypopigmentation compared to the mottled coloration of background photodamaged untreated skin that can be prevented by feathering the perimeter of the area or performing a full-face procedure. Another type is the true hypopigmentation, which usually occurs 6 to 12 months after LSR. This common delayed, and apparently permanent, side effect may be caused by loss of epidermal melanin or by underlying fibrosis.[256] Permanent hypopigmentation is a significant concern limiting LSR in moderately pigmented skin.

## 40.23 Nonablative Laser "Rejuvenation"

LSR has excellent efficacy but also requires weeks of recovery and occasionally causes unwanted side effects and complications. A novel nonablative approach for treating wrinkles and scars was recently introduced and is variously called nonablative laser resurfacing, nonablative rejuvenation, and subsurfacing. As discussed, clinical improvement following LSR is partly due to sculpting by ablation, partly due to remodeling during healing after controlled thermal injury, and perhaps partly due to shrinkage of dermal collagen, immediately tightening the dermis. Nonablative rejuvenation attempts to achieve all of these effects, except sculpting, by creating an "upside-down" thermal injury in which the epidermis is spared while the dermis is heated to temperatures that denature collagen or stimulate a healing response. This is easily accomplished by cooling the epidermis while propagating optical energy through it that is absorbed in the dermis. Apparently, the optical energy may be absorbed by any dermal structure or chromophore.

The first devices for nonablative rejuvenation used mid-infrared lasers, which are weakly or moderately absorbed by water, but it has also been observed that selective photothermolysis using visible light pulses absorbed by blood vessels produces beneficial effects on photoaged skin. The present popularity of this approach is such that many lasers and light sources are being used for this purpose including Q-switched 1064-nm Nd:YAG lasers, 585-nm pulsed-dye lasers (0.3 to 20 msec, at fluences below the threshold for immediate purpura), 980-nm diode laser, 1320-nm long-pulsed Nd:YAG and 1540-nm Er:glass lasers in combination with skin cooling devices, 1440-nm diode lasers, and IPL sources. All of these seem to offer real but subtle and minimal improvement in wrinkles compared with skin resurfacing. New dermal collagen formation and remodeling has been shown histologically.[257–263] At present, only a 0.3-msec, 585-nm PDLs used at fluences less than about 3 J/cm$^2$ have been cleared by the U.S. FDA specifically for this indication. There was a significant (about 10% of cases) risk of scarring associated with the 1320-nm Nd:YAG laser system, which apparently has been successfully minimized by monitoring peak skin temperature as a technique for calibrating individual treatment fluence. Promising results have also been reported of photorejuvenation using IPL.[264,265] Among 97 patients treated with IPL, "good" or "excellent" results were subjectively rated for improvement in pigmentation in 90% of the patients, in 83% for telangiectasia, and in 65% for skin texture, without major side effects.[265] However, there may be nothing unique about optical energy for stimulating an upside-down minor burn of the skin. Radiofrequency current, microwaves, and ultrasound are other radiant-energy sources capable of dermal heating.

## 40.24 Ultraviolet Laser Phototherapy

Excimer laser phototherapy has recently emerged in dermatology. Psoriasis is a genetic autoimmune disease affecting about 3% of the population worldwide and leads to annoying and occasionally life-threatening skin lesions, or arthritis. Psoriasis has traditionally been treated with drugs and with UV-light therapy. Initially, sunlight or broadband UV sources were used, followed by artificial UV-B (290 to 320 nm) sources, and then by narrow-band UV-B lamps (maximum 311 nm). The action spectrum for clearing psoriasis shows a sharp maximum of efficiency between 300 and 310 nm.[266] The 308-nm XeCl excimer laser has recently been shown to clear psoriasis faster than conventional phototherapy.[267,268] The excimer laser beam is selectively directed toward lesional skin, thereby sparing the surrounding normal skin from unnecessary radiation exposure and allowing a higher fluence to be used on the skin lesions. Bonis et al. first demonstrated that the cumulative dose needed for clearing of lesions was much less (by a factor of more than six) with XeCl excimer laser than with narrow-band UV-B therapy.[267] In animal models and human epidemiological studies, cumulative UV exposure is associated with increased risk of nonmelanoma skin cancer, which is one of the side effects of UV phototherapy for psoriasis and other skin diseases. Presumably, the excimer laser poses a much lower risk of UV-induced skin cancer compared with conventional phototherapy because of lower cumulative dose and treatment of a much smaller fraction of the total body surface area.

The commonly used lasers and light sources with their applications in dermatology are summarized in Table 40.1.

**TABLE 40.1** Lasers and Light Sources Used in Dermatology

| Laser | Wavelength (nm) | Mode | Pulse Duration (sec) | Typical Uses |
|---|---|---|---|---|
| Excimer | 308 | Pulsed | $10^{-7}$ | Psoriasis, vitiligo, lichen planus |
| Argon | 488, 514 | CW | | Vascular |
| Copper vapor/bromide | 512, 578 | QCW | $10^{-8}$ | Vascular, epidermal pigment |
| Pulsed dye (green) | 510 | Pulsed | $3 \times 10^{-4}$ | Epidermal pigment |
| KTP/df Q-switched Nd:YAG | 532 | QCW/Q-switched | $10^{-8}$ | Vascular/epidermal pigment |
| Pulsed dye (yellow) | 585-600 | Pulsed | $4 \times 10^{-4}$ | Vascular, warts, striae, hypertrophic scar, and keloid |
| Argon-dye | 630 | CW | | PDT |
| Ruby-Q-switched | 694 | Pulsed/Q-switched | $10^{-3}/10^{-8}$ | Hair removal/epidermal and dermal pigment, tatoo |
| Alexandrite/ Q-switched alexandrite | 755 | Pulsed/Q-switched | $10^{-3}/10^{-7}$ | Hair removal, leg vein/epidermal and dermal pigment, tattoo |
| Diode | 800 | Pulsed | $10^{-3}$ | Hair removal, vascular (telangiectases, leg vein) |
| Nd:YAG/Q-switched Nd:YAG | 1064 | Pulsed/Q-switched | $10^{-3}/10^{-8}$ | Hair removal, vascular/epidermal and dermal pigment, tattoo |
| Nd:YAG | 1320 | Pulsed | $10^{-3}$ | Nonablative dermal remodeling |
| Diode | 1450 | Pulsed | $10^{-3}$ | Nonablative dermal remodeling |
| Erbium:glass | 1540 | Pulsed | $10^{-3}$ | Nonablative dermal remodeling |
| Erbium:YAG | 2940 | Pulsed | $10^{-3}$ | Skin resurfacing, ablation of epidermal lesions |
| $CO_2$ | 10,600 | CW/pulsed | $10^{-4}$ | Skin resurfacing, ablation of epidermal lesions |
| Xenon flashlamp | 500–1200 | Pulsed | $10^{-3}$ | Hair removal, telangietases, epidermal pigment, and non-ablative dermal remodeling |

*Note:* CW = continuous wave; QCW = quasi-continuous wave (rapid, low energy pulses); pulsed (high-energy pulses).

## References


1. Maiman, T.H., Stimulated optical radiation in ruby, *Nature,* 187, 493, 1960.
2. Goldman, L., Blaney, D.J., Kindel, D.J., and Franke, E.K., Effect of the laser beam on the skin: preliminary report, *J. Invest. Dermatol.,* 40, 121, 1963.
3. Goldman, L., Blaney, D.J., Kindel, D.J., Richfield, D., and Franke, E.K., Pathology of the effect of the laser beam on the skin, *Nature,* 197, 912, 1963.
4. Goldman, L. and Wilson, R.G., Treatment of basal cell epithelioma by laser radiation, *J. Am. Med. Soc.,* 189(10), 773, 1964.
5. Goldman, L., Wilson, R., and Hornby, P., Radiation from a Q-switched ruby laser: effect of repeated impacts of power output of 10 megawatts on a tatoo of man, *J. Invest. Dermatol.*, 44, 69, 1965.
6. Anderson, R.R. and Parrish, J.A., Selective photothermolysis: precise microsurgery by selective absorption of pulsed radiation, *Science*, 220(4596), 524, 1983.
7. Anderson, R.R. and Parrish, J.A., Microvasculature can be selectively damaged using dye lasers: a basic theory and experimental evidence in human skin, *Lasers Surg. Med.*, 1(3), 263, 1981.
8. Anderson, R.R. and Parrish, J.A., The optics of human skin, *J. Invest. Dermatol.*, 77(1), 13, 1981.
9. Anderson, R.R. and Ross, E.V., Laser tissue interactions, in *Cosmetic Laser Surgery,* Fitzpatrick, R.E. and Goldman, M.P., Eds., Mosby, St. Louis, 2000.
10. Garden, J.M., Tan, O.T., Kerschmann, R., Boll, J., Furumoto, H., and Anderson, R.R., Effect of dye laser pulse duration on selective cutaneous vascular injury, *J. Invest. Dermatol.*, 87(5), 653, 1986.
11. Sliney, D.H., Laser safety, *Lasers Surg. Med.*, 16(3), 215, 1995.

12. Garden, J.M., O'Banion, M.K., Shelnitz, L.S., Pinski, K.S., Bakus, A.D., and Reichmann, M.E., Papillomavirus in the vapor of carbon dioxide laser-treated verrucae, *J. Am. Med. Soc.*, 259(8), 1199, 1988.
13. Baggish, M.S., Poiesz, B.J., Joret, D., Williamson, P., and Refai, A., Presence of human immunodeficiency virus DNA in laser smoke, *Lasers Surg. Med.*, 11(3), 197, 1991.
14. Baggish, M.S. and Elbakry, M., The effects of laser smoke on the lungs of rats, *Am. J. Obstet. Gynecol.*, 156(5), 1260, 1987.
15. Silver, L., Argon laser photocoagulation of port wine stain hemangiomas, *Lasers Surg. Med.*, 6(1), 24, 1986.
16. Paul, B.S., Anderson, R.R., Jarve, J., and Parrish, J.A., The effect of temperature and other factors on selective microvascular damage caused by pulsed dye laser, *J. Invest. Dermatol.*, 81(4), 333, 1983.
17. Tan, O.T., Sherwood, K., and Gilchrest, B.A., Treatment of children with port-wine stains using the flashlamp-pulsed tunable dye laser, *New Engl. J. Med.*, 320(7), 416, 1989.
18. Tan, O.T., Morrison, P., and Kurban, A.K., 585 nm for the treatment of port-wine stains, *Plast. Reconstr. Surg.*, 86(6), 1112, 1990.
19. Dierickx, C.C., Casparian, J.M., Venugopalan, V., Farinelli, W.A., and Anderson, R.R., Thermal relaxation of port-wine stain vessels probed *in vivo*: the need for 1–10-millisecond laser pulse treatment, *J. Invest. Dermatol.*, 105(5), 709, 1995.
20. Van Gemert, M.J., Welch, A.J., and Amin, A.P., Is there an optimal laser treatment for port wine stains? *Lasers Surg. Med.*, 6(1), 76, 1986.
21. Jacques, S.L., Simple optical theory for light dosimetry during PDT, *Proc. SPIE*, 1645, 155, 1992.
22. McMeekin, T. and Goodwin, D., A comparison of spot size: 7 mm versus 5 mm of pulsed dye laser treatment of benign cutaneous vascular lesions, *Lasers Surg. Med.*, Suppl. 7, 55, 1995.
23. Kauvar, A., Waldorf, H., and Geronemus, R., Effect of 7 mm versus 5 mm spot size on pulsed dye treatment of port wine stains and hemangiomas, *Lasers Surg. Med.*, Suppl. 7, 56, 1995.
24. Roider, J., Traccoli, J., Michaud, N., Flotte, T., Anderson, R., and Birngruber, R., Selective vascular occlusion by repetitive short laser pulse, *Ophthalmologe*, 91(3), 274, 1994.
25. Dierickx, C., Farinelli, W., and Anderson, R., Multiple pulse photocoagulation of port wine stain blood vessels with a 585 nm pulsed dye laser, *Lasers Surg. Med.*, Suppl. 7, 56, 1995.
26. Barsky, S.H., Rosen, S., Geer, D.E., and Noe, J.M., The nature and evolution of port wine stains: a computer-assisted study, *J. Invest. Dermatol.*, 74(3), 154, 1984.
27. Noe, J.M., Barsky, S.H., Geer, D.E., and Rosen, S., Port-wine stains and the response to argon laser therapy: successful treatment and predictive role of color, age and biopsy, *Plast. Reconstr. Surg.*, 65, 130, 1980.
28. Ashinoff, R. and Geronemus, R.G., Flashlamp-pumped pulsed dye laser for port-wine stains in infancy: earlier versus later treatment, *J. Am. Acad. Dermatol.*, 24(3), 467, 1991.
29. Garden, J.M. and Bakus, A.D., Laser treatment of port-wine stains and hemangiomas, *Dermatol. Clin.*, 15(3), 373, 1997.
30. Kauvar, A.N. and Geronemus, R.G., Repetitive pulsed dye laser treatments improve persistent port-wine stains. *Dermatol. Surg.*, 21(6), 515, 1995.
31. Nguyen, C.M., Yohn, J.J., Huff, C., Weston, W.L., and Morelli, J.G., Facial port wine stains in childhood: prediction of the rate of improvement as a function of the age of the patient, size and location of the port wine stain and the number of treatments with the pulsed dye (585 nm) laser, *Br. J. Dermatol.*, 138(5), 821, 1998.
32. Fitzpatrick, R.E., Lowe, N.J., Goldman, M.P., Borden, H., Behr, K.L., and Ruiz-Esparza, J., Flashlamp-pumped pulsed dye laser treatment of port-wine stains, *J. Dermatol. Surg. Oncol.*, 20(11), 743, 1994.
33. Katugampola, G.A. and Lanigan, S.W., Laser treatment of port-wine stains: therapeutic outcome in relation to morphological parameters, *Br. J. Dermatol.*, 136(3), 467, 1997.
34. Lanigan, S.W., Port wine stains on the lower limb: response to pulsed dye laser therapy, *Clin. Exp. Dermatol.*, 21(2), 88, 1996.

35. Renfro, L. and Geronemus, R.G., Anatomical differences of port-wine stains in response to treatment with the pulsed dye laser, *Arch. Dermatol.*, 129(2), 182, 1993.
36. Yohn, J.J., Huff, J.C., Aeling, J.L., Walsh, P., and Morelli, J.G., Lesion size is a factor for determining the rate of port-wine stain clearing following pulsed dye laser treatment in adults, *Cutis*, 59(5), 267, 1997.
37. Van der Horst, C.M., Koster, P.H., de Borgie, C.A., Bossuyt, P.M., and van Gemert, M.J., Effect of the timing of treatment of port-wine stains with the flash-lamp-pumped pulsed-dye laser, *New Engl. J. Med.*, 338(15), 1028, 1998.
38. Alster, T.S. and Wilson, F., Treatment of port-wine stains with the flashlamp-pumped pulsed dye laser: extended clinical experience in children and adults, *Ann. Plast. Surg.*, 32(5), 478, 1994.
39. Orten, S.S., Waner, M., Flock, S., Roberson, P.K., and Kincannon, J., Port-wine stains. An assessment of 5 years of treatment, *Arch. Otolaryngol. Head Neck Surg.*, 122(11), 1174, 1996.
40. Taieb, A., Touati, L., Cony, M., Leaute-Labreze, C., Mortureux, P., Renaud, P., et al., Treatment of port-wine stains with the 585-nm flashlamp-pulsed tunable dye laser: a study of 74 patients, *Dermatology*, 188(4), 276, 1994.
41. Fiskerstrand, E.J., Svaasand, L.O., Kopstad, G., Ryggen, K., and Aase, S., Photothermally induced vessel-wall necrosis after pulsed dye laser treatment: lack of response in port-wine stains with small sized or deeply located vessels, *J. Invest. Dermatol.*, 107(5), 671, 1996.
42. Fiskerstrand, E.J., Svaasand, L.O., Kopstad, G., Dalaker, M., Norvang, L.T., and Volden, G., Laser treatment of port wine stains: therapeutic outcome in relation to morphological parameters, *Br. J. Dermatol.*, 134(6), 1039, 1996.
43. Motley, R.J., Lanigan, S.W., and Katugampola, G.A., Videomicroscopy predicts outcome in treatment of port-wine stains, *Arch. Dermatol.*, 133(7), 921, 1997.
44. Eubanks, L.E. and McBurney, E.I., Videomicroscopy of port-wine stains: correlation of location and depth of lesion, *J. Am. Acad. Dermatol.*, 44(6), 948, 2001.
45. Chang, C.J. and Nelson, J.S., Cryogen spray cooling and higher fluence pulsed dye laser treatment improve port-wine stain clearance while minimizing epidermal damage, *Dermatol. Surg.*, 25(10), 767, 1999.
46. Bencini, P.L., The multilayer technique: a new and fast approach for flashlamp-pumped pulsed (FLPP) dye laser treatment of port-wine stains (preliminary reports), *Dermatol. Surg.*, 25(10,) 786, 1999.
47. Geronemus, R.G., Quintana, A.T., Lou, W.W., and Kauvar, A.N., High-fluence modified pulsed dye laser photocoagulation with dynamic cooling of port-wine stains in infancy, *Arch. Dermatol.*, 136(7), 942, 2000.
48. Bernstein, E.F., Treatment of a resistant port-wine stain with the 1.5-msec pulse duration, tunable, pulsed dye laser, *Dermatol. Surg.*, 26(11), 1007, 2000.
49. Scherer, K., Lorenz, S., Wimmershoff, M., Landthaler, M., and Hohenleutner, U., Both the flashlamp-pumped dye laser and the long-pulsed tunable dye laser can improve results in port-wine stain therapy, *Br. J. Dermatol.*, 145(1), 79, 2001.
50. Chowdhury, M.M., Harris, S., and Lanigan, S.W., Potassium titanyl phosphate laser treatment of resistant port-wine stains, *Br. J. Dermatol.*, 144(4), 814, 2001.
51. Michel, S., Landthaler, M., and Hohenleutner, U., Recurrence of port-wine stains after treatment with the flashlamp-pumped pulsed dye laser, *Br. J. Dermatol.*, 143(6), 1230, 2000.
52. Grevelink, J.M., White, V.R., Bonoan, R., and Denman, W.T., Pulsed laser treatment in children and the use of anesthesia, *J. Am. Acad. Dermatol.*, 37(1), 75, 1997.
53. Seukeran, D.C., Collins, P., and Sheehan-Dare, R.A., Adverse reactions following pulsed tunable dye laser treatment of port wine stains in 701 patients, *Br. J. Dermatol.*, 136(5), 725, 1997.
54. Simpson, J.R., Natural history of cavernous hemangioma, *Lancet*, 2, 1057, 1959.
55. Enjolras, O., Riche, M.C., Merland, J.J., and Escande, J.P., Management of alarming hemangiomas in infancy: a review of 25 cases, *Pediatrics*, 85(4), 491, 1990.

56. Enjolras, O. and Mulliken, J.B., The current management of vascular birthmarks, *Pediatr. Dermatol.*, 10(4), 311, 1993.
57. Frieden, I.J., Which hemangiomas to treat — and how? *Arch. Dermatol.*, 133(12), 1593, 1997.
58. Hohenleutner, S., Badur-Ganter, E., Landthaler, M., and Hohenleutner, U., Long-term results in the treatment of childhood hemangioma with the flashlamp-pumped pulsed dye laser: an evaluation of 617 cases, *Lasers Surg. Med.*, 28(3), 273, 2001.
59. Landthaler, M., Hohenleutner, U., and el-Raheem, T.A., Laser therapy of childhood haemangiomas, *Br. J. Dermatol.*, 133(2), 275, 1995.
60. Hohenleutner, U., Baumler, W., Karrer, S., Michel, S., and Landthaler, M., Treatment of pediatric hemangiomas with the flashlamp-pumped pulsed dye laser, *Hautarzt*, 47(3), 183, 1996.
61. Poetke, M., Philipp, C., and Berlien, H.P., Flashlamp-pumped pulsed dye laser for hemangiomas in infancy: treatment of superficial vs mixed hemangiomas, *Arch. Dermatol.*, 136(5), 628, 2000.
62. Scheepers, J.H. and Quaba, A.A., Does the pulsed tunable dye laser have a role in the management of infantile hemangiomas? Observations based on 3 years' experience, *Plast. Reconstr. Surg.*, 95(2), 305, 1995.
63. Ashinoff, R. and Geronemus, R.G., Failure of the flashlamp-pumped pulsed dye laser to prevent progression to deep hemangioma, *Pediatr. Dermatol.*, 10(1), 77, 1993.
64. Low, D.W., Hemangiomas and vascular malformations, *Semin. Pediatr. Surg.*, 3(2), 40, 1994.
65. Morelli, J.G., Tan, O.T., and Weston, W.L., Treatment of ulcerated hemangiomas with the pulsed tunable dye laser, *Am. J. Dis. Child.*, 145(9), 1062, 1991.
66. Morelli, J.G., Tan, O.T., Yohn, J.J., and Weston, W.L., Treatment of ulcerated hemangiomas infancy, *Arch. Pediatr. Adolesc. Med.*, 148(10), 1104, 1994.
67. Barlow, R.J., Walker, N.P., and Markey, A.C., Treatment of proliferative haemangiomas with the 585 nm pulsed dye laser, *Br. J. Dermatol.*, 134(4), 700, 1996.
68. Goldberg, G.N., Special symposium: management of hemangiomas, *Pediatr. Dermatol.*, 14, 71, 1997.
69. Achauer, B.M. and Van der Kam, V.M., Capillary hemangioma (strawberry mark) of infancy: comparison of argon and Nd:YAG laser treatment, *Plast. Reconstr. Surg.*, 84(1), 60, 1989; discussion, 70.
70. Goldman, M.P., Weiss, R.A., Brody, H.J., Coleman, W.P., III, and Fitzpatrick, R.E., Treatment of facial telangiectasia with sclerotherapy, laser surgery, and/or electrodesiccation: a review, *J. Dermatol. Surg. Oncol.*, 19(10), 899, 1993; quiz, 9–10.
71. Goldman, M.P. and Bennett, R.G., Treatment of telangiectasia: a review, *J. Am. Acad. Dermatol.*, 17(2 Part 1), 167, 1987.
72. Goldberg, D.J. and Meine, J.G., A comparison of four frequency-doubled Nd:YAG (532 nm) laser systems for treatment of facial telangiectases, *Dermatol. Surg.*, 25(6), 463, 1999.
73. Waner, M., Dinehart, S.M., Wilson, M.B., and Flock, S.T., A comparison of copper vapor and flashlamp pumped dye lasers in the treatment of facial telangiectasia, *J. Dermatol. Surg. Oncol.*, 19(11), 992, 1993.
74. West, T.B. and Alster, T.S., Comparison of the long-pulse dye (590–595 nm) and KTP (532 nm) lasers in the treatment of facial and leg telangiectasias, *Dermatol. Surg.*, 24(2), 221, 1998.
75. Goldman, M.P. and Fitzpatrick, R.E., Pulsed-dye laser treatment of leg telangiectasia: with and without simultaneous sclerotherapy, *J. Dermatol. Surg. Oncol.*, 16(4), 338, 1990.
76. Hsia, J., Lowery, J.A., and Zelickson, B., Treatment of leg telangiectasia using a long-pulse dye laser at 595 nm, *Lasers Surg. Med.*, 20(1), 1, 1997.
77. Grossman, M.C., Bernstein, L.J., Kauvar, A.N.B., Laughlin, S., Dudley, D., and Geronemus, R.G., Treatment of leg veins with a long pulse tunable dye laser, *Lasers Surg. Med.*, Suppl. 8, 35, 1996.
78. Bernstein, E.F., Lee, J., Lowery, J., Brown, D.B., Geronemus, R., Lask, G., and Hsia, J., Treatment of spider veins with the 595 nm pulsed-dye laser, *J. Am. Acad. Dermatol.*, 39(5 Part 1), 746, 1998.
79. Alora, M.B., Stern, R.S., Arndt, K.A., and Dover, J.S., Comparison of the 595 nm long-pulse (1.5 msec) and ultralong-pulse (4 msec) lasers in the treatment of leg veins, *Dermatol. Surg.*, 25(6), 445, 1999.

80. McDaniel, D.H., Ash, K., Lord, J., Newman, J., Adrian, R.M., and Zukowski, M., Laser therapy of spider leg veins: clinical evaluation of a new long pulsed alexandrite laser, *Dermatol. Surg.*, 25(1), 52, 1999.
81. Kauvar, A.N. and Lou, W.W., Pulsed alexandrite laser for the treatment of leg telangiectasia and reticular veins, *Arch. Dermatol.*, 136(11), 1371, 2000.
82. Dierickx, C.C., Duque, V., and Anderson, R.R., Treatment of leg telangiectasia by a pulsed, infrared laser system, *Lasers Surg. Med.*, Suppl. 10, 40, 1998.
83. Garden, J.M., Bakus, A.D., and Miller, I.D., Diode laser treatment of leg veins, *Lasers Surg. Med.*, Suppl. 10, 38, 1998.
84. Weiss, R.A. and Weiss, M.A., Early clinical results with a multiple synchronized pulse 1064 NM laser for leg telangiectasias and reticular veins, *Dermatol. Surg.*, 25(5), 399, 1999.
85. Sadick, N.S., Long-term results with a multiple synchronized-pulse 1064 nm Nd:YAG laser for the treatment of leg venulectasias and reticular veins, *Dermatol. Surg.*, 27(4), 365, 2001.
86. Sadick, N.S., Prieto, V.G., Shea, C.R., Nicholson. J., and McCaffrey, T., Clinical and pathophysiologic correlates of 1064-nm Nd:Yag laser treatment of reticular veins and venulectasias, *Arch. Dermatol.*, 137(5), 613, 2001.
87. Goldman, M.P. and Eckhouse, S., Photothermal sclerosis of leg veins. ESC Medical Systems, LTD Photoderm VL Cooperative Study Group, *Dermatol. Surg.*, 22(4), 323, 1996.
88. Schroeter, C.A. and Neumann, H.A., An intense light source. The photoderm VL-flashlamp as a new treatment possibility for vascular skin lesions, *Dermatol. Surg.*, 24(7), 743, 1998.
89. Weiss, R.A. and Sadick, N.S., Epidermal cooling crystal collar device for improved results and reduced side effects on leg telangiectasias using intense pulsed light, *Dermatol. Surg.*, 26(11), 1015, 2000.
90. Green, D., Photothermal removal of telangiectases of the lower extremities with the PhotodermVL, *J. Am. Acad. Dermatol.*, 38(1), 61, 1998.
91. Dover, J.S., Sadick, N.S., and Goldman, M.P., The role of lasers and light sources in the treatment of leg veins, *Dermatol. Surg.*, 25(4), 328, 1999; discussion, 35–36.
92. Ross, B.S., Levine, V.J., and Ashinoff, R., Laser treatment of acquired vascular lesions, *Dermatol. Clin.*, 15(3), 385, 1997.
93. Fosko, S.W., Glaser, D.A., and Rogers, C.J., Eradication of angiolymphoid hyperplasia with eosinophilia by copper vapor laser, *Arch. Dermatol.*, 137(7), 863, 2001.
94. Ammirati, C.T. and Hruza, G.J., Treatment of granuloma faciale with the 585-nm pulsed dye laser, *Arch. Dermatol.*, 135(8), 903, 1999.
95. Tan, O.T., Hurwitz, R.M., and Stafford, T.J., Pulsed dye laser treatment of recalcitrant verrucae: a preliminary report, *Lasers Surg. Med.*, 13(1), 127, 1993.
96. Kauvar, A.N., McDaniel, D.H., and Geronemus, R.G., Pulsed dye laser treatment of warts, *Arch. Fam. Med.*, 4(12), 1035, 1995.
97. Jain, A. and Storwick, G.S., Effectiveness of the 585nm flashlamp-pulsed tunable dye laser (PTDL) for treatment of plantar verrucae, *Lasers Surg. Med.*, 21(5), 500, 1997.
98. Ross, B.S., Levine, V.J., Nehal, K., Tse, Y., and Ashinoff, R., Pulsed dye laser treatment of warts: an update. *Dermatol. Surg.*, 25(5), 377, 1999.
99. Robson, K.J., Cunningham, N.M., Kruzan, K.L., Patel, D.S., Kreiter, C.D., O'Donnell, M.J., and Arpey, C.J., Pulsed-dye laser versus conventional therapy in the treatment of warts: a prospective randomized trial, *J. Am. Acad. Dermatol.*, 43(2 Part 1), 275, 2000.
100. Ross, E.V., McDaniel, D.H., Anderson, R.R., Kauvar, A.N., and Geronemus, R.G., Pulsed dye (585 nm) treatment of warts: a comparison of single versus multiple pulse techniques examining clinical response, fast infrared thermal camera measurements, and light electron microscopy, *Lasers Surg. Med.*, Suppl. 7, 59, 1995.
101. Gooptu, C. and James, M.P., Recalcitrant viral warts: results of treatment with the KTP laser, *Clin. Exp. Dermatol.*, 24(2), 60, 1999.
102. Alster, T.S. and Williams, C.M., Treatment of keloid sternotomy scars with 585 nm flashlamp-pumped pulsed-dye laser, *Lancet*, 345(8959), 1198, 1995.

103. Alster, T.S. and McMeekin, T.O., Improvement of facial acne scars by the 585 nm flashlamp-pumped pulsed dye laser, *J. Am. Acad. Dermatol.*, 35(1), 79, 1996.
104. Dierickx, C., Goldman, M.P., and Fitzpatrick, R.E., Laser treatment of erythematous/hypertrophic and pigmented scars in 26 patients, *Plast. Reconstr. Surg.*, 95(1), 84, 1995; discussion, 91–92.
105. Manuskiatti, W., Fitzpatrick, R.E., and Goldman, M.P., Energy density and numbers of treatment affect response of keloidal and hypertrophic sternotomy scars to the 585-nm flashlamp-pumped pulsed-dye laser, *J. Am. Acad. Dermatol.*, 45(4), 557, 2001.
106. Alster, T.S., Laser treatment of hypertrophic scars, keloids, and striae, *Dermatol. Clin.*, 15(3), 419, 1997.
107. McDaniel, D.H., Ash, K., and Zukowski, M., Treatment of stretch marks with the 585-nm flashlamp-pumped pulsed dye laser, *Dermatol. Surg.*, 22(4), 332, 1996.
108. Wittenberg, G.P., Fabian, B.G., Bogomilsky, J.L., Schultz, L.R., Rudner, E.J., Chaffins, M.L., Saed, G.M., Burns, R.L., and Fivenson, D.P., Prospective, single-blind, randomized, controlled study to assess the efficacy of the 585-nm flashlamp-pumped pulsed-dye laser and silicone gel sheeting in hypertrophic scar treatment, *Arch. Dermatol.*, 135(9), 1049, 1999.
109. Katugampola, G.A., Rees, A.M., and Lanigan, S.W., Laser treatment of psoriasis, *Br. J. Dermatol.*, 133(6), 909, 1995.
110. Ros, A.M., Garden, J.M., Bakus, A.D., and Hedblad, M.A., Psoriasis response to the pulsed dye laser, *Lasers Surg. Med.*, 19(3), 331, 1996.
111. Zelickson, B.D., Mehregan, D.A., Wendelschfer-Crabb, G., Ruppman, D., Cook, A., O'Connell, P., and Kennedy, W.R., Clinical and histologic evaluation of psoriatic plaques treated with a flashlamp pulsed dye laser, *J. Am. Acad. Dermatol.*, 35(1), 64, 1996.
112. Watanabe, S., Flotte, T., Margolis, R.J., Dover, J.S., Hruza, G.J., and Polla, L., and Anderson, R., The effect of pulse duration on selective pigmented cell injury by dye lasers, *J. Invest. Dermatol.*, 88(4), 523, 1987.
113. Tong, A.K., Tan, O.T., Boll, J., Parrish, J.A., and Murphy, G.F., Ultrastructure: effects of melanin pigment on target specificity using a pulsed dye laser (577 nm), *J. Invest. Dermatol.*, 88(6), 747, 1987.
114. Fitzpatrick, R.E., Goldman, M.P., and Ruiz-Esparza, J., Laser treatment of benign pigmented epidermal lesions using a 300 nsecond pulse and 510 nm wavelength, *J. Dermatol. Surg. Oncol.*, 19(4), 341, 1993.
115. Goldberg, D.J., Benign pigmented lesions of the skin. Treatment with the Q-switched ruby laser, *J. Dermatol. Surg. Oncol.*, 19(4), 376, 1993.
116. Kilmer, S.L., Wheeland, R.G., Goldberg, D.J., and Anderson, R.R., Treatment of epidermal pigmented lesions with the frequency-doubled Q-switched Nd:YAG laser: a controlled, single-impact, dose-response, multicenter trial, *Arch. Dermatol.*, 130(12), 1515, 1994.
117. Fitzpatrick, R.E., Goldman, M.P., and Ruiz-Esparza, J., Clinical advantage of the CO2 laser superpulsed mode. Treatment of verruca vulgaris, seborrheic keratoses, lentigines, and actinic cheilitis, *J. Dermatol. Surg. Oncol.*, 20(7), 449, 1994.
118. Jang, K.A., Chung, E.C., Choi, J.H., Sung, K.J., Moon, K.C., and Koh, J.K., Successful removal of freckles in Asian skin with a Q-switched alexandrite laser, *Dermatol. Surg.*, 26(3), 231, 2000.
119. Todd, M.M., Rallis, T.M., Gerwels, J.W., and Hata, T.R., A comparison of 3 lasers and liquid nitrogen in the treatment of solar lentigines: a randomized, controlled, comparative trial, *Arch. Dermatol.*, 136(7), 841, 2000.
120. Scheepers, J.H. and Quaba, A.A., Clinical experience with the PLDL-1 (Pigmented Lesion Dye Laser) in the treatment of pigmented birthmarks: a preliminary report, *Br. J. Plast. Surg.*, 46(3), 247, 1993.
121. Grossman, M.C., Anderson, R.R., Farinelli, W., Flotte, T.J., and Grevelink, J.M., Treatment of cafe au lait macules with lasers: a clinicopathologic correlation, *Arch. Dermatol.*, 131(12), 1416, 1995.
122. Alster, T.S., Complete elimination of large cafe-au-lait birthmarks by the 510-nm pulsed dye laser, *Plast. Reconstr. Surg.*, 96(7), 1660, 1995.

123. Somyos, K., Boonchu, K., Somsak, K., Panadda, L., and Leopairut, J., Copper vapour laser treatment of cafe-au-lait macules, *Br. J. Dermatol.*, 135(6), 964, 1996.
124. Alora, M.B. and Arndt, K.A., Treatment of a cafe-au-lait macule with the erbium:YAG laser, *J. Am. Acad. Dermatol.*, 45(4), 566, 2001.
125. Lee, P.K., Rosenberg, C.N., Tsao, H., and Sober, A.J., Failure of Q-switched ruby laser to eradicate atypical-appearing solar lentigo: report of two cases, *J. Am. Acad. Dermatol.*, 38(2 Part 2), 314, 1998.
126. Taylor, C.R. and Anderson, R.R., Ineffective treatment of refractory melasma and postinflammatory hyperpigmentation by Q-switched ruby laser, *J. Dermatol. Surg. Oncol.*, 20(9), 592, 1994.
127. Manaloto, R.M. and Alster, T., Erbium:YAG laser resurfacing for refractory melasma, *Dermatol. Surg.*, 25(2), 121, 1999.
128. Nouri, K., Bowes, L., Chartier, T., Romagosa, R., and Spencer, J., Combination treatment of melasma with pulsed $CO_2$ laser followed by Q-switched alexandrite laser: a pilot study, *Dermatol. Surg.*, 25(6), 494, 1999.
129. Geronemus, R.G., Q-switched ruby laser therapy of nevus of Ota, *Arch. Dermatol.*, 128(12), 1618, 1992.
130. Watanabe, S. and Takahashi, H., Treatment of nevus of Ota with the Q-switched ruby laser, *New Engl. J. Med.*, 331(26), 1745, 1994.
131. Apfelberg, D.B., Argon and Q-switched yttrium-aluminum-garnet laser treatment of nevus of Ota, *Ann. Plast. Surg.*, 35(2), 150, 1995.
132. Alster, T.S. and Williams, C.M., Treatment of nevus of Ota by the Q-switched alexandrite laser, *Dermatol. Surg.*, 21(7), 592, 1995.
133. Ueda, S., Isoda, M., and Imayama, S., Response of naevus of Ota to Q-switched ruby laser treatment according to lesion colour, *Br. J. Dermatol.*, 142(1), 77, 2000.
134. Kang, W., Lee, E., and Choi, G.S., Treatment of Ota's nevus by Q-switched alexandrite laser: therapeutic outcome in relation to clinical and histopathological findings, *Eur. J. Dermatol.*, 9(8), 639, 1999.
135. Chan, H.H., Leung, R.S., Ying, S.Y., Lai, C.F., Kono, T., Chua, J.K., and Ho, W.S., A retrospective analysis of complications in the treatment of nevus of Ota with the Q-switched alexandrite and Q-switched Nd:YAG lasers, *Dermatol. Surg.*, 26(11), 1000, 2000.
136. Chan, H.H., Lam, L.K., Wong, D.S., Leung, R.S., Ying, S.Y., Lai, C.F., Ho, W.S., and Chua, J.K.H., Nevus of Ota: a new classification based on the response to laser treatment, *Lasers Surg. Med.*, 28(3), 267, 2001.
137. Kono, T., Nozaki, M., Chan, H.H., and Mikashima, Y., A retrospective study looking at the long-term complications of Q-switched ruby laser in the treatment of nevus of Ota, *Lasers Surg. Med.*, 29(2), 156, 2001.
138. Kunachak, S., Leelaudomlipi, P., and Sirikulchayanonta, V., Q-switched ruby laser therapy of acquired bilateral nevus of Ota-like macules, *Dermatol. Surg.*, 25(12), 938, 1999.
139. Polnikorn, N., Tanrattanakorn, S., and Goldberg, D.J., Treatment of Hori's nevus with the Q-switched Nd:YAG laser, *Dermatol. Surg.*, 26(5), 477, 2000.
140. Kunachak, S. and Leelaudomlipi, P., Q-switched Nd:YAG laser treatment for acquired bilateral nevus of Ota-like maculae: a long-term follow-up, *Lasers Surg. Med.*, 26(4), 376, 2000.
141. Lowe, N.J., Wieder, J.M., Shorr, N., Boxrud, C., Saucer, D., and Chalet, M., Infraorbital pigmented skin. Preliminary observations of laser therapy, *Dermatol. Surg.*, 21(9), 767, 1995.
142. West, T.B. and Alster, T.S., Improvement of infraorbital hyperpigmentation following carbon dioxide laser resurfacing, *Dermatol. Surg.*, 24(6), 615, 1998.
143. Rhodes, A.R. and Mihm, M.C., Jr., Origin of cutaneous melanoma in a congenital dysplastic nevus spilus, *Arch. Dermatol.*, 126(4), 500, 1990.
144. Wagner, R.F., Jr. and Cottel, W.I., *In situ* malignant melanoma arising in a speckled lentiginous nevus, *J. Am. Acad. Dermatol.*, 20(1), 125, 1989.
145. Bolognia, J.L., Fatal melanoma arising in a zosteriform speckled lentiginous nevus, *Arch. Dermatol.*, 127(8), 1240, 1991.

146. Taylor, C.R. and Anderson, R.R., Treatment of benign pigmented epidermal lesions by Q-switched ruby laser, *Int. J. Dermatol.*, 32(12), 908, 1993.
147. Grevelink, J.M., Gonzalez, S., Bonoan, R., Vibhagool, C., and Gonzalez, E., Treatment of nevus spilus with the Q-switched ruby laser, *Dermatol. Surg.*, 23(5), 365, 1997; discussion, 69–70.
148. Nelson, J.S. and Applebaum, J., Treatment of superficial cutaneous pigmented lesions by melanin-specific selective photothermolysis using the Q-switched ruby laser, *Ann. Plast. Surg.*, 29(3), 231, 1992.
149. Nanni, C.A. and Alster, T.S., Treatment of a Becker's nevus using a 694-nm long-pulsed ruby laser, *Dermatol. Surg.*, 24(9), 1032, 1998.
150. Waldorf, H.A., Kauvar, A.N., and Geronemus, R.G., Treatment of small and medium congenital nevi with the Q-switched ruby laser, *Arch. Dermatol.*, 132(3), 301, 1996.
151. Grevelink, J.M., van Leeuwen, R.L., Anderson, R.R., and Byers, H.R., Clinical and histological responses of congenital melanocytic nevi after single treatment with Q-switched lasers, *Arch. Dermatol.*, 133(3), 349, 1997.
152. Vibhagool, C., Byers, H.R., and Grevelink, J.M., Treatment of small nevomelanocytic nevi with a Q-switched ruby laser, *J. Am. Acad. Dermatol.*, 36(5 Part 1), 738, 1997.
153. Rosenbach, A., Williams, C.M., and Alster, T.S., Comparison of the Q-switched alexandrite (755 nm) and Q-switched Nd:YAG (1064 nm) lasers in the treatment of benign melanocytic nevi, *Dermatol. Surg.*, 23(4), 239, 1997; discussion, 44–45.
154. Nelson, J.S. and Kelly, K.M., Q-switched ruby laser treatment of a congenital melanocytic nevus, *Dermatol. Surg.*, 25(4), 274, 1999.
155. Ueda, S. and Imayama, S., Normal-mode ruby laser for treating congenital nevi, *Arch. Dermatol.*, 133(3), 355, 1997.
156. Imayama, S. and Ueda, S., Long- and short-term histological observations of congenital nevi treated with the normal-mode ruby laser, *Arch. Dermatol.*, 135(10), 1211, 1999.
157. Duke, D., Byers, H.R., Sober, A.J., Anderson, R.R., and Grevelink, J.M., Treatment of benign and atypical nevi with the normal-mode ruby laser and the Q-switched ruby laser: clinical improvement but failure to completely eliminate nevomelanocytes, *Arch. Dermatol.*, 135(3), 290, 1999.
158. Taylor, C.R., Anderson, R.R., Gange, R.W., Michaud, N.A., and Flotte, T.J., Light and electron microscopic analysis of tattoos treated by Q-switched ruby laser, *J. Invest. Dermatol.*, 97(1), 131, 1991.
159. Kilmer, S.L., Lee, M.S., Grevelink, J.M., Flotte, T.J., and Anderson, R.R., The Q-switched Nd:YAG laser effectively treats tattoos: a controlled, dose-response study, *Arch. Dermatol.*, 129(8), 971, 1993.
160. Stafford, T.J., Lizek, R., and Tan, O.T., Role of the alexandrite laser for removal of tattoos, *Lasers Surg. Med.*, 17(1), 32, 1995.
161. Goyal, S., Arndt, K.A., Stern, R.S., O'Hare, D., and Dover, J.S., Laser treatment of tattoos: a prospective, paired, comparison study of the Q-switched Nd:YAG (1064 nm), frequency-doubled Q-switched Nd:YAG (532 nm), and Q-switched ruby lasers, *J. Am. Acad. Dermatol.*, 36(1), 122, 1997.
162. Taylor, C.R., Gange, R.W., Dover, J.S., Flotte, T.J., Gonzalez, E., Michaud, N., and Anderson, R.R., Treatment of tattoos by Q-switched ruby laser: a dose-response study, *Arch. Dermatol.*, 126(7), 893, 1990.
163. Kilmer, S.L. and Anderson, R.R., Clinical use of the Q-switched ruby and the Q-switched Nd:YAG (1064 nm and 532 nm) lasers for treatment of tattoos, *J. Dermatol. Surg. Oncol.*, 19(4), 330, 1993.
164. Ross, V., Naseef, G., Lin, G., Kelly, M., Michaud, N., Flotte, T.J., Raythen, J., and Anderson, R.R., Comparison of responses of tattoos to picosecond and nanosecond Q-switched neodymium: YAG lasers, *Arch. Dermatol.*, 134(2), 167, 1998.
165. Anderson, R.R., Geronemus, R., Kilmer, S.L., Farinelli, W., and Fitzpatrick, R.E., Cosmetic tattoo ink darkening. A complication of Q-switched and pulsed-laser treatment, *Arch. Dermatol.*, 129(8), 1010, 1993.
166. Trotter, M.J., Tron, V.A., Hollingdale, J., and Rivers, J.K., Localized chrysiasis induced by laser therapy, *Arch. Dermatol.*, 131(12), 1411, 1995.

167. Ross, E.V., Yashar, S., Michaud, N., Fitzpatrick, R., Geronemus, R., Tope, W.D., and Anderson, R.R., Tattoo darkening and nonresponse after laser treatment: a possible role for titanium dioxide, *Arch. Dermatol.*, 137(1), 33, 2001.
168. Kramer, K.E., Lopez, A., Stefanato, C.M., and Phillips, T.J., Exogenous ochronosis, *J. Am. Acad. Dermatol.*, 42(5 Part 2), 869, 2000.
169. Green, D. and Friedman, K.J., Treatment of minocycline-induced cutaneous pigmentation with the Q- switched Alexandrite laser and a review of the literature, *J. Am. Acad. Dermatol.*, 44(Suppl. 2), 342, 2001.
170. Wilde, J.L., English, J.C., III, and Finley, E.M., Minocycline-induced hyperpigmentation: treatment with the neodymium:YAG laser, *Arch. Dermatol.*, 133(11), 1344, 1997.
171. Tsao, H., Busam, K., Barnhill, R.L., and Dover, J.S., Treatment of minocycline-induced hyperpigmentation with the Q-switched ruby laser, *Arch. Dermatol.*, 132(10), 1250, 1996.
172. Knoell, K.A., Milgraum, S.S., and Kutenplon, M., Q-switched ruby laser treatment of minocycline-induced cutaneous hyperpigmentation, *Arch. Dermatol.*, 132(10), 1251, 1996.
173. Collins, P. and Cotterill, J.A., Minocycline-induced pigmentation resolves after treatment with the Q-switched ruby laser, *Br. J. Dermatol.*, 135(2), 317, 1996.
174. Karrer, S., Hohenleutner, U., Szeimies, R.M., and Landthaler, M., Amiodarone-induced pigmentation resolves after treatment with the Q-switched ruby laser, *Arch. Dermatol.*, 135(3), 251, 1999.
175. Atkin, D.H. and Fitzpatrick, R.E., Laser treatment of imipramine-induced hyperpigmentation, *J. Am. Acad. Dermatol.*, 43(1 Part 1), 77, 2000.
176. Becker-Wegerich, P.M., Kuhn, A., Malek, L., Lehmann, P., Megahed, M., and Ruzicka, T., Treatment of nonmelanotic hyperpigmentation with the Q-switched ruby laser, *J. Am. Acad. Dermatol.*, 43(2 Part 1), 272, 2000.
177. Raulin, C., Werner, S., and Greve, B., Circumscripted pigmentations after iron injections-treatment with Q-switched laser systems, *Lasers Surg. Med.*, 28(5), 456, 2001.
178. Olsen, E.A., Methods of hair removal, *J. Am. Acad. Dermatol.*, 40(2 Part 1), 143, 1999; quiz, 56–57.
179. Akiyama, M., Dale, B.A., Sun, T.T., and Holbrook, K.A., Characterization of hair follicle bulge in human fetal skin: the human fetal bulge is a pool of undifferentiated keratinocytes, *J. Invest. Dermatol.*, 105(6), 844, 1995.
180. Akiyama, M., Smith, L.T., and Holbrook, K.A., Growth factor and growth factor receptor localization in the hair follicle bulge and associated tissue in human fetus, *J. Invest. Dermatol.*, 106(3), 391, 1996.
181. Adrian, R.M., Vascular mechanisms in laser hair removal, *J. Cutan. Laser Ther.*, 2(1), 49, 2000.
182. Goldberg, D.J., Littler, C.M., and Wheeland, R.G., Topical suspension-assisted Q-switched Nd:YAG laser hair removal, *Dermatol. Surg.*, 23(9), 741, 1997.
183. Nanni, C.A. and Alster, T.S., Optimizing treatment parameters for hair removal using a topical carbon-based solution and 1064-nm Q-switched neodymium:YAG laser energy, *Arch. Dermatol.*, 133(12), 1546, 1997.
184. Grossman, M., Wimberly, J., Dwyer, P., Flotte, T., and Anderson, R.R., PDT for hirsutism, *Lasers Surg. Med.*, Suppl. 7, 205, 1995.
185. Grossman, M.C., Dierickx, C., Farinelli, W., Flotte, T., and Anderson, R.R., Damage to hair follicles by normal-mode ruby laser pulses, *J. Am. Acad. Dermatol.*, 35(6), 889, 1996.
186. Dierickx, C.C., Grossman, M.C., Farinelli, W.A., and Anderson, R.R., Permanent hair removal by normal-mode ruby laser, *Arch. Dermatol.*, 134(7), 837, 1998.
187. Finkel, B., Eliezri, Y.D., Waldman, A., and Slatkine, M., Pulsed alexandrite laser technology for noninvasive hair removal, *J. Clin. Laser Med. Surg.*, 15(5), 225, 1997.
188. Goldberg, D.J. and Ahkami, R., Evaluation comparing multiple treatments with a 2-msec and 10-msec alexandrite laser for hair removal, *Lasers Surg. Med.*, 25(3), 223, 1999.
189. Garcia, C., Alamoudi, H., Nakib, M., and Zimmo, S., Alexandrite laser hair removal is safe for Fitzpatrick skin types IV–VI, *Dermatol. Surg.*, 26(2), 130, 2000.

190. Lloyd, J.R. and Mirkov, M., Long-term evaluation of the long-pulsed alexandrite laser for the removal of bikini hair at shortened treatment intervals, *Dermatol. Surg.*, 26(7), 633, 2000.
191. Gorgu, M., Aslan, G., Akoz, T., and Erdogan, B., Comparison of alexandrite laser and electrolysis for hair removal, *Dermatol. Surg.*, 26(1), 37, 2000.
192. Handrick, C. and Alster, T.S., Comparison of long-pulsed diode and long-pulsed alexandrite lasers for hair removal: a long-term clinical and histologic study, *Dermatol. Surg.*, 27(7), 622, 2001.
193. Campos, V.B., Dierickx, C.C., Farinelli, W.A., Lin, T.Y., Manuskiatti, W., and Anderson, R.R., Hair removal with an 800-nm pulsed diode laser, *J. Am. Acad. Dermatol.*, 43(3), 442, 2000.
194. Lou, W.W., Quintana, A.T., Geronemus, R.G., and Grossman, M.C., Prospective study of hair reduction by diode laser (800 nm) with long- term follow-up, *Dermatol. Surg.*, 26(5), 428, 2000.
195. Baugh, W.P., Trafeli, J.P., Barnette, D.J., Jr., and Ross, E.V., Hair reduction using a scanning 800 nm diode laser, *Dermatol. Surg.*, 27(4), 358, 2001.
196. Goldberg, D.J. and Samady, J.A., Evaluation of a long-pulse Q-switched Nd:YAG laser for hair removal, *Dermatol. Surg.*, 26(2), 109, 2000.
197. Bencini, P.L., Luci, A., Galimberti, M., and Ferranti, G., Long-term epilation with long-pulsed neodymium:YAG laser, *Dermatol. Surg.*, 2(3), 175, 1999.
198. Alster, T.S., Bryan, H., and Williams, C.M., Long-pulsed Nd:YAG laser-assisted hair removal in pigmented skin: a clinical and histological evaluation, *Arch. Dermatol.*, 137(7), 885, 2001.
199. Gold, M.H., Bell, M.W., Foster, T.D., and Street, S., Long-term epilation using the EpiLight broad band, intense pulsed light hair removal system, *Dermatol. Surg.*, 23(10), 909, 1997.
200. Schroeter, C.A., Raulin, C., Thurlimann, W., Reineke, T., De Potter, C., and Neumann, H.A., Hair removal in 40 hirsute women with an intense laser-like light source, *Eur. J. Dermatol.*, 9(5), 374, 1999.
201. Weiss, R.A., Weiss, M.A., Marwaha, S., and Harrington, A.C., Hair removal with a non-coherent filtered flashlamp intense pulsed light source, *Lasers Surg. Med.*, 24(2), 128, 1999.
202. Sadick, N.S., Weiss, R.A., Shea, C.R., Nagel, H., Nicholson, J., and Prieto, V.G., Long-term photo-epilation using a broad-spectrum intense pulsed light source, *Arch. Dermatol.*, 136(11), 1336, 2000.
203. Battle, E.F., Suthamjariya, K., Alora, M.B., Palli, K., and Anderson, R.R., Very long-pulsed (20–200 ms) diode laser for hair removal on all skin types, *Lasers Surg. Med.*, Suppl. 12, 21, 2000.
204. Walsh, J.T., Jr., Flotte, T.J., Anderson, R.R., and Deutsch, T.F., Pulsed $CO_2$ laser tissue ablation: effect of tissue type and pulse duration on thermal damage, *Lasers Surg. Med.*, 8(2), 108, 1988.
205. Green, H.A., Domankevitz, Y., and Nishioka, N.S., Pulsed carbon dioxide laser ablation of burned skin: *in vitro* and *in vivo* analysis, *Lasers Surg. Med.*, 10(5), 476, 1990.
206. Green, H.A., Burd, E., Nishioka, N.S., Bruggemann, U., and Compton, C.C., Middermal wound healing. A comparison between dermatomal excision and pulsed carbon dioxide laser ablation, *Arch. Dermatol.*, 128(5), 639, 1992.
207. Yang, C.C. and Chai, C.Y., Animal study of skin resurfacing using the ultrapulse carbon dioxide laser, *Ann. Plast. Surg.*, 35(2), 154, 1995.
208. Ross, E.V., Grossman, M.C., Anderson, R.R., and Grevelink, J.M., Treatment of facial rhytides: comparing a pulsed $CO_2$ laser with a collimated beam to a $CO_2$ laser enahnced by a flashscanner, *Lasers Surg. Med.*, Suppl. 16, 50, 1995.
209. Gardner, E.S., Reinisch, L., Stricklin, G.P., and Ellis, D.L., *In vitro* changes in non-facial human skin following $CO_2$ laser resurfacing: a comparison study, *Lasers Surg. Med.*, 19(4), 379, 1996.
210. Ross, E.V., Grossman, M.C., Duke, D., and Grevelink, J.M., Long-term results after $CO_2$ laser skin resurfacing: a comparison of scanned and pulsed systems, *J. Am. Acad. Dermatol.*, 37(5 Part 1), 709, 1997.
211. Alster, T.S., Kauvar, A.N., and Geronemus, R.G., Histology of high-energy pulsed $CO_2$ laser resurfacing, *Semin. Cutan. Med. Surg.*, 15(3), 189, 1996.
212. Fitzpatrick, R.E., Goldman, M.P., Satur, N.M., and Tope, W.D., Pulsed carbon dioxide laser resurfacing of photo-aged facial skin, *Arch. Dermatol.*, 132(4), 395, 1996.

213. Walsh, J.T., Jr. and Deutsch, T.F., Er:YAG laser ablation of tissue: measurement of ablation rates, *Lasers Surg. Med.*, 9(4), 327, 1989.
214. Walsh, J.T., Jr., Flotte, T.J., and Deutsch, T.F., Er:YAG laser ablation of tissue: effect of pulse duration and tissue type on thermal damage, *Lasers Surg. Med.*, 9(4), 314, 1989.
215. Hohenleutner, U., Hohenleutner, S., Baumler, W., and Landthaler, M., Fast and effective skin ablation with an Er:YAG laser: determination of ablation rates and thermal damage zones, *Lasers Surg. Med.*, 20(3), 242, 1997.
216. Kaufmann, R. and Hibst, R., Pulsed erbium:YAG laser ablation in cutaneous surgery, *Lasers Surg. Med.*, 19(3), 324, 1996.
217. Kaufmann, R. and Hibst, R., Pulsed Er:YAG- and 308 nm UV-excimer laser: an *in vitro* and *in vivo* study of skin-ablative effects, *Lasers Surg. Med.*, 9(2), 132, 1989.
218. Majaron, B., Lukac, M., Drnovsek-Olup, B., Vedlin, B., and Rotter, A., Heat diffusion and ablation front dynamics in Er:YAG laser skin resurfacing, *Proc. SPIE*, 2970, 350, 1997.
219. Majaron, B., Plestenjak, P., and Lukac, M., Quantitative investigation of thermal damage in Er:YAG laser skin resurfacing, *Proc. SPIE*, 3245, 366, 1998.
220. Ross, E.V., Yashar, S.S., Naseef, G.S., Barnette, D.J., Skrobal, M., Grevelink, J., and Anderson, R.R., A pilot study of *in vivo* immediate tissue contraction with $CO_2$ skin laser resurfacing in a live farm pig, *Dermatol. Surg.*, 25(11), 851, 1999.
221. Ross, E.V., Naseef, G.S., McKinlay, J.R., Barnette, D.J., Skrobal, M., Grevelink, J., and Anderson, R.R., Comparison of carbon dioxide laser, erbium:YAG laser, dermabrasion, and dermatome: a study of thermal damage, wound contraction, and wound healing in a live pig model: implications for skin resurfacing. *J. Am. Acad. Dermatol.*, 42(1 Part 1), 92, 2000.
222. Margolis, S., Butler, P., and Randolph, M., et al., Quantitative comparison of $CO_2$ laser and TCA peel in skin resurfacing, *Proc. New Engl. Soc. Plast. Surg. Annu. Meeting*, Boston, 1996.
223. Lowe, N.J., Lask, G., Griffin, M.E., Maxwell, A., Lowe, P., and Quilada, F., Skin resurfacing with the Ultrapulse carbon dioxide laser: observations on 100 patients, *Dermatol. Surg.*, 21(12), 1025, 1995.
224. Shim, E., Tse, Y., Velazquez, E., Kamino, H., Levine, V., and Ashinoff, R., Short-pulse carbon dioxide laser resurfacing in the treatment of rhytides and scars: a clinical and histopathological study, *Dermatol. Surg.*, 24(1), 113, 1998.
225. Khatri, K.A., Ross, V., Grevelink, J.M., Magro, C.M., and Anderson, R.R., Comparison of erbium:YAG and carbon dioxide lasers in resurfacing of facial rhytides, *Arch. Dermatol.*, 135(4), 391, 1999.
226. Smith, K.J., Skelton, H.G., Graham, J.S., Hamilton, T.A., Hackley, B.E., Jr., and Hurst, C.G., Depth of morphologic skin damage and viability after one, two, and three passes of a high-energy, short-pulse $CO_2$ laser (Tru-Pulse) in pig skin, *J. Am. Acad. Dermatol.*, 37(2 Part 1), 204, 1997.
227. Alster, T.S. and West, T.B., Resurfacing of atrophic facial acne scars with a high-energy, pulsed carbon dioxide laser, *Dermatol. Surg.*, 22(2), 151, 1996; discussion, 54–55.
228. West, T.B., Laser resurfacing of atrophic scars, *Dermatol. Clin.*, 15(3), 449, 1997.
229. Kye, Y.C., Resurfacing of pitted facial scars with a pulsed Er:YAG laser, *Dermatol. Surg.*, 23(10), 880, 1997.
230. Walia, S. and Alster, T.S., Prolonged clinical and histologic effects from $CO_2$ laser resurfacing of atrophic acne scars, *Dermatol. Surg.*, 25(12), 926, 1999.
231. Kwon, S.D. and Kye, Y.C., Treatment of scars with a pulsed Er:YAG laser, *J. Cutan. Laser Ther.*, 2(1), 27, 2000.
232. Teikemeier, G. and Goldberg, D.J., Skin resurfacing with the erbium:YAG laser, *Dermatol. Surg.*, 23(8), 685, 1997.
233. Weinstein, C., Computerized scanning erbium:YAG laser for skin resurfacing, *Dermatol. Surg.*, 24(1), 83, 1998.
234. Weiss, R.A., Harrington, A.C., Pfau, R.C., Weiss, M.A., and Marwaha, S., Periorbital skin resurfacing using high energy erbium:YAG laser: results in 50 patients, *Lasers Surg. Med.*, 24(2), 81, 1999.

235. Alster, T.S., Clinical and histologic evaluation of six erbium:YAG lasers for cutaneous resurfacing, *Lasers Surg. Med.*, 24(2), 87, 1999.
236. Ross, E.V., McKinlay, J.R., and Anderson, R.R., Why does carbon dioxide resurfacing work? A review, *Arch. Dermatol.*, 135(4), 444, 1999.
237. Goldman, M.P. and Manuskiatti, W., Combined laser resurfacing with the 950-microsec pulsed $CO_2$ + Er:YAG lasers, *Dermatol. Surg.*, 25(3), 160, 1999.
238. Kauvar, A.N.B., Lou, W.W., and Quintana, A.T., Equivalent depth resurfacing: a clinical and histological evaluation of erbium:YAG, carbon dioxide and combined laser procedures, *Lasers Surg. Med.*, Suppl. 11, 63, 1999.
239. McDaniel, D.H., Lord, J., Ash, K., and Newman, J., Combined $CO_2$/erbium:YAG laser resurfacing of peri-oral rhytides and side-by-side comparison with carbon dioxide laser alone, *Dermatol. Surg.*, 25(4), 285, 1999.
240. Goldman, M.P., Fitzpatrick, R.E., and Manuskiatti, W., Laser resurfacing of the neck with the Erbium: YAG laser, *Dermatol. Surg.*, 25(3), 164, 1999; discussion, 67–68.
241. Ross, E.V., Mowlavi, A., Barnette, D., Glatter, R.D., and Grevelink, J.M., The effect of wiping on skin resurfacing in a pig model using a high energy pulsed $CO_2$ laser system, *Dermatol. Surg.*, 25(2), 81, 1999.
242. Fitzpatrick, R.E., Williams, B., and Goldman, M.P., Preoperative anesthesia and postoperative considerations in laser resurfacing, *Semin. Cutan. Med. Surg.*, 15(3), 170, 1996.
243. Ho, C., Nguyen, Q., Lowe, N.J., Griffin, M.E., and Lask, G., Laser resurfacing in pigmented skin, *Dermatol. Surg.*, 21(12), 1035, 1995.
244. Lask, G., Keller, G., Lowe, N., and Gormley, D., Laser skin resurfacing with the SilkTouch flash-scanner for facial rhytides, *Dermatol. Surg.*, 21(12), 1021, 1995.
245. Waldorf, H.A., Kauvar, A.N., and Geronemus, R.G., Skin resurfacing of fine to deep rhytides using a char-free carbon dioxide laser in 47 patients, *Dermatol. Surg.*, 21(11), 940, 1995.
246. Ratner, D., Tse, Y., Marchell, N., Goldman, M.P., Fitzpatrick, R.E., and Fader, D.J., Cutaneous laser resurfacing, *J. Am. Acad. Dermatol.*, 41(3 Part 1), 365, 1999; quiz, 90–92.
247. Grevelink, J.M., Laser hair transplantation, *Dermatol. Clin.*, 15(3), 479, 1997.
248. Alster, T.S. and Lewis, A.B., Dermatologic laser surgery: a review, *Dermatol. Surg.*, 22(9), 797, 1996.
249. Hruza, G.J., Laser treatment of warts and other epidermal and dermal lesions, *Dermatol. Clin.*, 15(3), 487, 1997.
250. Boehncke, W.H., Ochsendorf, F., Wolter, M., and Kaufmann, R., Ablative techniques in psoriasis vulgaris resistant to conventional therapies, *Dermatol. Surg.*, 25(8), 618, 1999.
251. Weinstein, C., Carbon dioxide laser resurfacing: long-term follow-up in 2123 patients, *Clin. Plast. Surg.*, 25(1), 109, 1998.
252. Apfelberg, D.B., Varga, J., and Greenbaum, S.S., Carbon dioxide laser resurfacing of peri-oral rhytides in scleroderma patients, *Dermatol. Surg.*, 24(5), 517, 1998.
253. Sriprachya-Anunt, S., Fitzpatrick, R.E., Goldman, M.P., and Smith, S.R., Infections complicating pulsed carbon dioxide laser resurfacing for photoaged facial skin, *Dermatol. Surg.*, 23(7), 527, 1997; discussion, 35–36.
254. Walia, S. and Alster, T.S., Laser resurfacing infection rate with and without prophylactic antibiotics, *Lasers Surg. Med.*, Suppl. 24, 64, 1999.
255. Bernstein, L.J., Kauvar, A.N., Grossman, M.C., and Geronemus, R.G., The short- and long-term side effects of carbon dioxide laser resurfacing, *Dermatol. Surg.*, 23(7), 519, 1997.
256. Laws, R.A., Finley, E.M., McCollough, M.L., and Grabski, W.J., Alabaster skin after carbon dioxide laser resurfacing with histologic correlation, *Dermatol. Surg.*, 24(6), 633, 1998.
257. Muccini, J.A., Jr., O'Donnell, F.E., Jr., Fuller, T., and Reinisch, L., Laser treatment of solar elastosis with epithelial preservation, *Lasers Surg. Med.*, 23(3), 121, 1998.
258. Menaker, G.M., Wrone, D.A., Williams, R.M., and Moy, R.L., Treatment of facial rhytids with a nonablative laser: a clinical and histologic study, *Dermatol. Surg.*, 25(6), 440, 1999.

259. Kelly, K.M., Nelson, J.S., Lask, G.P., Geronemus, R.G., and Bernstein, L.J., Cryogen spray cooling in combination with nonablative laser treatment of facial rhytides, *Arch. Dermatol.*, 135(6), 691, 1999.
260. Ross, E.V., Sajben, F.P., Hsia, J., Barnette, D., Miller, C.H., and McKinlay, J.R., Nonablative skin remodeling: selective dermal heating with a mid-infrared laser and contact cooling combination, *Lasers Surg. Med.*, 26(2), 186, 2000.
261. Goldberg, D.J., Full-face nonablative dermal remodeling with a 1320 nm Nd:YAG laser, *Dermatol. Surg.*, 26(10), 915, 2000.
262. Trelles, M.A., Allones, I., and Luna, R., Facial rejuvenation with a nonablative 1320 nm Nd:YAG laser: a preliminary clinical and histologic evaluation, *Dermatol. Surg.*, 27(2), 111, 2001.
263. Fournier, N., Dahan, S., Barneon, G., Diridollou, S., Lagarde, J.M., and Gall, Y., and Mordon, S., Nonablative remodeling: clinical, histologic, ultrasound imaging, and profilometric evaluation of a 1540 nm Er:glass laser, *Dermatol. Surg.*, 27(9), 799, 2001.
264. Bitter, P.H., Noninvasive rejuvenation of photodamaged skin using serial, full-face intense pulsed light treatments, *Dermatol. Surg.*, 26(9), 835, 2000; discussion, 43.
265. Negishi, K., Tezuka, Y., Kushikata, N., and Wakamatsu, S., Photorejuvenation for Asian skin by intense pulsed light, *Dermatol. Surg.*, 27(7), 627, 2001; discussion, 32.
266. Parrish, J.A. and Jaenicke, K.F., Action spectrum for phototherapy of psoriasis, *J. Invest. Dermatol.*, 76(5), 359, 1981.
267. Bonis, B., Kemeny, L., Dobozy, A., Bor, Z., Szabo, G., and Ignacz, F., 308 nm UVB excimer laser for psoriasis, *Lancet*, 350(9090), 1522, 1997.
268. Asawanonda, P., Anderson, R.R., Chang, Y., and Taylor, C.R., 308-nm excimer laser for the treatment of psoriasis: a dose-response study, *Arch. Dermatol.*, 136(5), 619, 2000.
269. Anderson, R.R., Lasers in dermatology — a critical update, *J. Dermatol.*, 27(11), 700, 2000.

# 41

# Lasers in Interventional Pulmonology




Anubhav N. Mathur
*Indiana University Medical School*
*Indianapolis, Indiana*

Praveen N. Mathur
*Indiana University Medical School*
*Indianapolis, Indiana*


## 41.1 Introduction

The lung cancer epidemic has resulted in many patients with malignant obstruction of the central airways. Conventional therapy, including radiation therapy, chemotherapy, or dilation techniques with rigid bronchoscopy, has proven to be inadequate in relieving symptomatic airway obstruction. Thus, the application of laser technology for the treatment of various tracheobronchial obstructions, using standard rigid and flexible bronchoscopes, has proven to be a major advance in the treatment of these patients and was the original stimulus for the specialty of interventional pulmonology. The successes of laser bronchoscopy have stimulated interest in other endobronchial therapies such as cryotherapy, electrocautery, and brachytherapy.

Laser bronchoscopists must be experts in diagnostic bronchoscopy techniques, become knowledgeable in laser physics and laser–tissue interaction, and be comfortable dealing with patients with respiratory insufficiency, acute hypoxemia, and airway bleeding. This chapter reviews laser physics, laser–tissue interaction, and laser safety as it relates to bronchoscopic applications. Indications, contraindications, and techniques of laser bronchoscopy are also addressed.

## 41.2 Laser Light

A light amplification by stimulated emission of radiation (laser) beam is an intense form of electromagnetic energy channeled into parallel, synchronized rays of light of the same wavelength that are used in many industrial, research, and medical applications. Traditionally, the portion of the electromagnetic spectrum that has had the greatest diagnostic and therapeutic application in medicine ranges from 0.01 to 200,000 Å. Included in this range are gamma rays, x-rays, and ultraviolet (UV), visible, and infrared


0-8493-1116-0/03/$0.00+$1.50

light. The human eye perceives visible light as different colors, with the shorter wavelengths of 400 nm seen as violet blue and the longer wavelengths of 700 nm seen as red light.

All light is produced by the spontaneous or stimulated emission of energy by atoms, the smallest components of a molecule. The Rutherford-Bohr model of the atom, although not wholly accurate, provides a good model for understanding the structure of the atom where the nucleus contains positively charged particles called protons and neutral particles called neutrons. Orbiting the nucleus at a distance are negatively charged electrons that balance the positively charged protons. Each electron orbit corresponds to a particular quantized energy level where the energy of each orbital is directly proportional to its distance from the nucleus and increases in incremental or quantum steps.

In the spontaneous emission of light, an input of energy is required to "excite" electrons from a low- to a high-energy orbit. A photon, which is a discrete particle of electromagnetic energy, is released when excited electrons surrounding the nucleus spontaneously drop from their high-energy orbit back to a low-energy orbit, which is detected as "light." The energy of the photon released is equal to the energy input that was required to excite the electron and is calculated by multiplying the frequency of the photon (measured in hertz) by Planck's constant ($6.623 \times 10^{-34} \times 10^{-34}$ joule seconds). This implies the energy of an electromagnetic wave is directly proportional to its frequency and inversely proportional to its wavelength. As an example, a high-frequency, short-wavelength photon, such as a photon of blue light, is more energetic than a low-frequency, long-wavelength photon of red light.

Lasers differ in that photons that contribute to laser light are, in large part, released via induced or stimulated emission as opposed to spontaneous emission. In stimulated emission, an atom is already in an excited or high-energy state. A passing photon from an external source and with the correct energy is absorbed by the excited atom and induces the release of two photons that travel in the same direction and have the same energy and whose electromagnetic waves are perfectly in phase. The energy requirement of the passing photon (that is responsible for the initiation of the stimulated photon emission) must equal the energy difference between the high- and low-energy orbits of the excited atom. Although the excited atom would have released one photon via spontaneous emission, without the aid of the passing external photon, the transition from an excited state to ground state happens sooner after being induced by the passing, external photon.[1] This release is termed stimulated emission of radiation and is the basis for laser-light energy.

When the process of stimulated emission is amplified to an intense level by an external energy source, a specialized form of light energy is created, which we term laser light. Lasers create their form of light energy by using an optical cylinder containing a medium consisting of solids, gases, or dyes. The cylinder contains a fully reflective mirror at one end and a partially reflective mirror at the other end. Atoms of the laser medium become excited when an external source adds thermal, electrical, or optical energy to the system. A chain reaction of photon release is initiated by spontaneous emission of energetic photons as electrons switch from high- to low-energy orbits. A spontaneously emitted photon with the right energy can induce an excited atom to release two photons, each of which can induce an excited atom to release two more photons. This process continues, doubling the number of photons released at each step; all the photons are moving in the same direction and with the same energy, and their electromagnetic waves are perfectly in phase. This intense laser-light energy exits through the partially transmissive mirror at a rate directly proportional to the incoming external energy source. The energy is emitted as laser light and has special properties of stimulated light emission.[2,3]

Three unique properties of laser light make lasers particularly useful.[1,3] First, all waves of laser light are in phase with each other and have identical frequencies, wavelengths, and velocities as they travel through space. This property is known as coherence, and it results in a concentration of laser power at the target point. Second, the electromagnetic waves emitted by a laser are monochromatic, meaning that all waves have nearly the same frequency and thus the same wavelength, color, and energy. This characteristic permits all of the radiant energy to be centered in a narrow electromagnetic spectral band. Finally, all waves of laser light are essentially parallel and are thus considered collimated. This property permits lasers to travel great distances with minimal loss of intensity of light energy. Ordinary light energy (i.e., incandescent bulbs) will scatter in multiple directions at varied wavelengths, making it good for illumination but inadequate for more precise medical applications. Laser light, on the other hand, is emitted in a narrow, parallel bundle with minimal scatter into space,[3,4] making it particularly useful as a surgical tool. Compared with most other forms of light, lasers also have a high luminosity (brightness), which makes them useful for light therapy.

Two other important characteristics of laser light determine a laser's impact on a target. The transverse electromagnetic mode (TEM) is one measure of a laser's energy distribution per unit of cross-sectional area of the beam. Although lasers are collimated and coherent, the impact configuration on a target is not perfectly uniform, as illustrated in the ideal mode. Instead, the actual cross-sectional-impact configuration of most lasers is symmetric in a bell-shaped or Gaussian distribution, with more beam intensity being concentrated in the center of the target. This impact configuration is termed basic mode.

In addition to the TEM, its power density also determines a laser beam's intensity, which is the concentration of power (energy per unit time) per unit of cross-sectional area of the beam. Optical lenses focus the laser beam into a small focal spot, creating a high power density in that area. For each laser, the beam focus is fixed to prevent divergence and loss of power density. The power density is a function of the size of the focal spot. The smaller the focal spot, the higher the power density. The power density can also be intensified by increasing the incoming power of the beam, which is usually in the range of 20 to 100 W for most medical lasers.[1] Therefore, power density is directly proportional to incoming power and inversely proportional to the area of the focal spot. This intensity of a concentrated beam makes it useful for cutting, coagulating, and vaporizing tissue.

A laser's TEM and power density are important determinants of the laser beam's effect on the target tissue. Because the TEM determines the profile of the beam's effective cross-sectional area, it also determines the pattern of tissue interaction. The tissue cavity created by laser destruction has the same configuration as the intensity profile (basic mode) of the laser beam. The extent of tissue destruction is a function of the amount of energy absorption, the length of exposure, and the power density of the laser beam. When laser light is applied to tissue, some of the energy penetrates into the tissue and is absorbed, producing heat; some energy is dissipated, and the rest of the energy is reflected back. Power density at the surface of the tissue will be higher than the power density farther inside because of the partial reflection of energy.

Absorption attenuates the original power density incrementally as it passes deeper into the tissues. The distance below the surface of the tissue at which the power density has decreased to a percentage of the original level (usually 1%) is termed the attenuation depth.[3] It is differentiated from penetration depth, which is the distance at which the laser energy no longer has significant effects on the tissue. Laser beams can still have a considerable effect on tissues even after 99% attenuation if the exposure is prolonged or if the power density of the original beam is very high. For instance, if the laser beam has an original power density at the surface of 100 W/cm$^2$, only 1 W/cm$^2$ will be present at the 99% attenuation depth. However, if the original power density at the surface is 10,000 W/cm$^2$, 100 W/cm$^2$ will be present at the 99% attenuation depth. The attenuation depth is a function of the tissue type and the wavelength of the laser light, whereas the penetration depth depends on the tissue type, wavelength, and power density.

The amount of attenuation that occurs secondary to absorption is termed the attenuation coefficient.[3] Each laser has a different scattering characteristic when the beam penetrates living tissues, but the amount of scattering within tissues generally depends on the tissue type and wavelength of the laser beam. Lasers that have high scatter coefficients function well as coagulators, whereas lasers that have high absorption coefficients are more useful for precision cutting. For instance, a carbon dioxide ($CO_2$) laser has a high absorption coefficient that produces rapid tissue vaporization and makes it an ideal cutting instrument. Its low scattering coefficient, however, limits its ability to photocoagulate bleeding tissues. In contrast, a neodymium:yttrium-aluminum-garnet (Nd:YAG) laser functions well as a coagulator because of its high scattering coefficient, but it is a relatively poor cutting instrument. Nevertheless, the Nd:YAG laser is capable of serving both functions because at higher power settings or within dark tissues more absorption and vaporization occur. The main determinant of a laser's effect on a particular tissue is the proportion of the scattering and absorption coefficients at the wavelength of the individual laser.

Lasers can interact with living tissues in different ways.[2] Light energy is converted into heat energy, enabling the thermal effects of laser radiation to be used for cutting, vaporization, and coagulation of tissues. Because of their high water content, living tissues readily absorb laser energy. This laser energy is rapidly converted into heat, which results in cell destruction. Lasers can also mechanically disrupt tissues either by pressure waves or by inducing vaporization of intracellular or extracellular water, resulting in separation of tissue layers. This interaction is useful in the removal of cataracts or intravascular atheromatous plaques. Laser-light energy can also stimulate various biochemical reactions by interacting with photosensitizing chemicals absorbed by abnormal tissues. This use of laser light is termed photodynamic therapy, which is used in the diagnosis and treatment of patients with tracheobronchial malignancies.[5–8]

The substance used as the laser medium determines the performance characteristics of individual lasers. Each substance produces varying wavelengths of emitted light. The first laser used a synthetic ruby crystal as a laser medium; now, commonly used materials include argon, krypton, fluorescent organic dyes, gold vapor gas, helium-neon, potassium titanyl phosphate (KTP), $CO_2$, and a combination of erbium (Er) or Nd:YAG.

The argon gas laser has a wavelength of 514 nm in the visible blue–green spectrum and can be transmitted through optical fibers. Because of its blue color, however, much of its energy is absorbed in hemoglobin, which greatly limits its use in photoresection of endobronchial malignancies.[9] The gold vapor and liquid-dye lasers are used to excite the photosensitizer dihematoporphyrin ether, which is selectively retained by tumor cells.[7,8] By exposing the tissue to appropriate laser light, clusters of tumor cells can easily be identified and treated. The helium-neon gas laser has no therapeutic role in laser bronchoscopy, but it produces light in the visible spectrum, making it useful as an aiming guide for other lasers, such as the $CO_2$ and Nd:YAG laser. The KTP laser uses a solid medium similar to the Nd:YAG laser and has been used in photocoagulation and resection of tracheobronchial stenosis.[10]

The $CO_2$ laser has commonly been used for lesions of the aero digestive tract.[11] The $CO_2$ laser has a long wavelength of 10,600 nm, resulting in prompt absorption by most living tissues. This invisible infrared electromagnetic energy is rapidly converted to thermal energy, resulting in vaporization of the water in the tissue. If the tissue has high water content, absorption is further enhanced, and tissue destruction is more precise. The $CO_2$ laser's high absorption and low scatter coefficients give it a predictable depth of penetration and make it ideal for precise surgical applications.[3,12] Tonsillectomy, adenoidectomy, and uvulopalatopharyngoplasty (UPPP) have been performed with the use of the $CO_2$ laser.[13] Although well accepted for the treatment of benign and malignant lesions of the head and neck,[14] its use in tracheobronchial malignancies has been limited. The $CO_2$ laser can seal small vessels and lymphatics up to 0.5 mm in diameter,[15,16] but it is an ineffective photocoagulator and cannot reliably control bleeding from larger vessels. Another important drawback of the $CO_2$ laser is its cumbersome articulated mirror delivery system, which requires a micromanipulator and surgical microscope for operation. Use of this system for lower airway tumors requires precise alignment of the laser beam down the entire length of a rigid bronchoscope. This complex setup makes resection difficult and often precludes the use of certain suction devices.

The Nd:YAG laser has been used extensively for the treatment of patients with tracheobronchial malignancies because the laser beam is delivered through a quartz optical fiber, which is easily adaptable for use with either the flexible or rigid bronchoscope.[17] The Nd:YAG laser operates in the invisible near-infrared (NIR) range of the electromagnetic spectrum at a wavelength of 1064 nm. At this wavelength, scattering exceeds absorption for most lighter-colored tissues, which makes the Nd:YAG laser effective at coagulating large amounts of tissue. If the tissue is darker pigmented or if a higher power setting is used, absorption and vaporization can be increased. The wide power range (5 to 100 W) of the Nd:YAG laser contributes to its versatility by permitting photocoagulation at low settings and vaporization at higher settings. The Nd:YAG laser penetrates to about 6 mm and will coagulate vessels of up to 2 mm in diameter. A new type of Nd:YAG laser has been developed with a wavelength of 1320 nm, which may have advantages over the current system because of slightly less tissue scatter.[9] The Nd:YAG laser fiber can be fitted with a sapphire tip to permit a contact operational mode, but because these tips limit viewing

**TABLE 41.1** Indications for Laser Bronchoscopy

| Author | Year | Patients | Bronchogenic Carcinoma | Metastatic Carcinoma | Tracheal Stenosis | Benign Tumor | Uncertain Prognosis | Miscellaneous |
|---|---|---|---|---|---|---|---|---|
| Cavaliere et al.[19] | 1994 | 1884 | 1419 (75%) | 105 (5.6%) | Not included | 156 (8%) | 143 (8%) | 61 (3.2%) |
| Cavaliere et al.[22] | 1988 | 1000 | 612 (61%) | 37 (3.7%) | 139 (13.9%) | 59 (5.9%) | 64 (6.4%) | 89(8.9%) |
| Brutinel et al.[23] | 1987 | 116 | 71 (61%) | 17 (14.7) | Not included | 9 (8%) | 13 (11%) | 6 (5%) |
| Beamis et al.[20] | 1991 | 269 | 200 (74%) | 36 (18%) | 27 (10%) | 6 (2%) | Not included | Not included |
| Personne et al.[21] | 1986 | 1310 | 643 (49%) | | 389 (27%) | 75 (5.7%) | 57 (4%) | 146 (11%) |
| Dumon[18] | 1985 | 544 | 278 (51%) | | 132 (24%) | 31 (6%) | 20 (4%) | 83 (15%)[a] |

[a]Suture removal, 29; hemorrhage, 20; foreign body, 4.

contact probes are usually not used in tracheobronchial endoscopy.[4] Because the noncontact Nd:YAG laser is the most commonly used laser for tracheobronchial disease, the rest of this chapter will concentrate on the use of this laser for bronchoscopic applications.

## 41.3 Indications

The Nd:YAG laser has been used to reestablish airway patency in patients with primary bronchogenic and metastatic malignancies, benign tumors, tracheal stenosis, and tumors of uncertain prognosis.[18] Malignant tracheobronchial obstruction from primary bronchogenic carcinoma is the most common indication for Nd:YAG laser bronchoscopy and accounts for 49 to 75% of the cases (Table 41.1).[18–23] Tracheobronchial obstruction as a result of endobronchial metastasis from renal, thyroid, colon, breast, and esophageal carcinomas account for between 6 and 18% of laser bronchoscopies and are the next most common indication for the procedure. Although they are responsible for less than 10% of the cases of tracheobronchial obstruction, benign tumors, such as hamartomas, fibromas, papillomas, lipomas, and amyloidomas, are an excellent indication for use of the Nd:YAG laser.[21,24] Endobronchial amyloid and tracheopathia osteoplastica are usually not curable by laser resection, but these lesions are relatively slow growing and often do not require multiple treatments.[22]

At many institutions, tracheal stenosis represents a relatively common indication for laser bronchoscopy, accounting for 10 to 27% of the cases. Tracheal stenosis usually occurs after translaryngeal intubation or tracheotomy, but it can occur without a known cause or in response to granulomatous or diphtheria infections. Personne et al.[21] have identified three main types of iatrogenic stenosis. The diaphragm type of stenosis (type I) consists of a concentric fibrotic intraluminal stricture without involvement of the tracheal wall and is ideal for laser resection. The second type of stenosis (type II) involves extrinsic, "bottle-neck narrowing" of the trachea secondary to collapse of the tracheal wall. Because no intraluminal component is present, use of the Nd:YAG laser for this type of narrowing is contraindicated. The last type of stenosis (type III) is a combination of the first two types and may be acceptable for laser therapy of the intraluminal fibrotic ring in certain situations. If stenosis is suitable for endoscopic therapy, the laser is used to make radial incisions in the tracheal scar before gentle rigid bronchoscopic dilation of the tracheal or subglottic stenosis.[25,26] Some patients with severe obstruction of the lower airway may require a permanent tracheotomy or implantation of a silicone stent or a Montgomery tracheal T tube to maintain patency of the airway.[25–29]

A less common indication for endobronchial laser therapy is treatment of tumors of uncertain prognosis, which include carcinoid, adenoid cystic, spindle cell, mucoepidermoid, and other mixed cell types. These lesions should be resected surgically if possible, but in many instances, surgery is impossible because the lesions are widespread or the patient is a poor operative candidate. Other infrequent uses for the Nd:YAG laser include treatment of tracheobronchial granulomas,[22] removal of suture threads, control of hemorrhage from biopsy sites or necrotic mucosa,[18] and even vaporization of granulation tissue surrounding an aspirated foreign body.[30]

**TABLE 41.2** Characteristics of Lesions for Laser Bronchoscopy

| Favorable Lesions for Laser Resection | Unfavorable Lesions for Laser Resection |
|---|---|
| Tracheal and main stem lesions | Extrinsic lesions |
| Polypoid lesions | Diffuse lesions with extensive submucosal involvement |
| Short lesion length | Upper lobe and segmental lesions |
| Large endobronchial involvement | Long tapering lesions |
| Visible distal bronchial lumen | Total bronchial obstruction |
| Functional lung distal to the obstruction | |

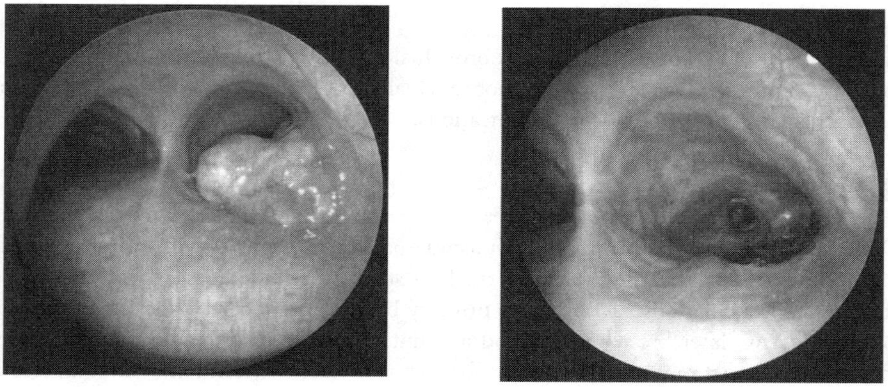


**FIGURE 41.1 (Color figure follows p. 28-30.)** Photographs illustrating the use of the Nd:YAG laser and mechanical resection to remove an endobronchial lesion from the lung. (A) Endobronchial lesion due to nonsmall-cell lung cancer seen in the right mainstem bronchius arising from the right upper lobe. (B) Endobronchial lesion has been removed using the Nd:YAG laser and mechanical resection.

Nd:YAG laser bronchoscopy is indicated for any centrally obstructing lesion within the airway lumen of the tracheobronchial tree, but some lesions are more favorable for treatment than others (Table 41.2). Although location of the lesion is a primary concern, the appearance of a lesion is also important. For instance, mechanical resection of short, polypoid lesions is technically easier than resection of lesions that involve a longer portion of the airway. Because much of laser bronchoscopy involves the use of the rigid bronchoscope to physically resect the tumor, long tapering lesions with extensive submucosal infiltration make the procedure technically more difficult and often reduce the chances for a successful outcome. Complete obstruction of the airway lumen also complicates the procedure because it makes it difficult to know the course of the bronchus beyond the area of obstruction. Most patients do not have complete airway obstruction, and a passage beyond the lesion can often be found by gently probing the area with a suction catheter or a flexible bronchoscope. If the course of the bronchus can be determined, mechanical and laser resection can proceed. Otherwise, complete obstruction of the airway lumen is a contraindication to laser bronchoscopy, and it represents the most common reason for not performing laser resection.[31] Figure 41.1 illustrates the use of the Nd:YAG laser and mechanical resection to remove an endobronchial lesion from the lung.

## 41.4 Contraindications

Because the procedure is often used as palliative therapy for individuals with severe malignant tracheobronchial stenosis, there are actually few contraindications to laser bronchoscopy. The only absolute contraindication to laser bronchoscopy is lack of an intraluminal lesion. Because lesions extrinsic to the

**TABLE 41.3** Contraindications to Laser Bronchoscopy

| |
|---|
| Absolute contraindications |
| Extraluminal disease |
| Relative contraindications |
| Cardiovascular |
| Recent myocardial infarction |
| Ventricular arrhythmias |
| Conduction abnormalities |
| Hypotension |
| Decompensated congestive heart failure |
| Respiratory |
| Severe obstructive lung disease |
| Extensive tumor involvement |
| Involvement of pulmonary artery |
| Unsalvageable lung distal to obstruction |
| Chronic collapse |
| Small cell carcinoma cell type |
| Posterior tracheal wall location or tracheoesophageal fistula |
| Extensive preprocedure radiation therapy |
| Electrolyte abnormalities |
| Bleeding diathesis |
| Sepsis |
| Tracheobronchial malacia |

airway are not amenable for treatment, it is important to confirm that the airway narrowing is not the result of compression from mediastinal tumors, lymphadenopathy, or lobar collapse or the result of type II tracheal stenosis. Failure to recognize extrinsic airway compression as a cause of airway narrowing can result in inadvertent perforation of the tracheobronchial wall. Except for extraluminal disease, there are few other reasons not to consider laser resection even in patients who are at a higher surgical risk. Personne et al.[21] performed laser resection on 267 patients *in extremis* with severe hypoxia or coma and recorded only a 2.7% mortality rate. A low rate of fatal complications of 0.3 to 3% has been confirmed by three other large series.[18,20,22] Although there are few absolute contraindications to laser bronchoscopy, a number of relative contraindications need to be considered (Table 41.3). A preoperative evaluation is often used to identify cardiovascular and respiratory problems that might place the patient at an unacceptably high risk for complications. The palliative benefits of laser bronchoscopy frequently outweigh the small risk of postprocedural complications. In many instances, respiratory insufficiency is both the primary indication for the procedure and the reason why the patient is a higher surgical risk. For some patients, bronchoscopic laser resection represents the only chance for rapid relief of dyspnea.

It is important to know whether there is functional lung distal to the obstruction. Lung parenchyma that is severely damaged by radiation treatments or recurrent postobstructive pneumonia is unable to participate in gas exchange even when the obstruction is removed.[32] In addition, significant tumor involvement of the pulmonary artery can result in irreversible ventilation–perfusion inequalities, negating any effects of laser resection. Computed tomography of the chest can be helpful in evaluating lung parenchyma in the region of the obstruction.

The type and location of the lesion may be additional relative contraindications to laser bronchoscopy because they influence the likelihood of procedural complications. Tumors involving the posterior tracheal wall can extend into the esophagus, increasing the risk of perforation or fistula formation during laser applications. The risk of airway erosion and perforation is also increased by extensive preprocedure radiation therapy, which can distort and soften the tracheal wall.[18] Particular caution should be taken with upper-lobe lesions because laser treatment can result in excessive bleeding owing to the proximity of major vessels.[39] Coagulopathies, electrolyte abnormalities, hypotension, and sepsis represent other obvious contraindications to laser endoscopy and should be corrected before the procedure.

# 41.5 Anesthesia

The method of anesthesia delivery for Nd:YAG laser bronchoscopy depends on the type of bronchoscopic technique.[34] Some bronchoscopists prefer to operate through the open end of the rigid bronchoscope, and thus a different type of ventilation is required. Several authors prefer spontaneous assisted ventilation,[35] but others[20,36,38,39] use a manual Venturi jet-ventilation technique. This method of ventilation enables the surgeon to operate through the open end of the rigid bronchoscope while ventilation is maintained by manually injecting an air–oxygen mixture at 20 to 60 breaths per minute.[37,38,40] A high-frequency jet-ventilation technique has also been described wherein a machine delivers approximately 300 breaths per minute.[41] The effectiveness of manual Venturi jet ventilation has been well validated,[42,43] and the use of continuous $CO_2$ monitoring is not required. Standard $O_2$ pulse oximetry is recommended to avoid inadvertent desaturation events during the procedure. After rigid bronchoscopy, most patients are easily extubated in the operating room after clearance of the intravenous anesthesia.

# 41.6 Techniques for Laser Bronchoscopy

## 41.6.1 Rigid vs. Flexible Bronchoscopy

The Nd:YAG laser can be used with either a flexible or rigid bronchoscope, and which instrument is better has been discussed widely.[44,45] Ironically, the original papers in the early 1980s by Toty[46] and Dumon[47] described the use of the Nd:YAG laser with both the flexible and rigid bronchoscopes. For the next few years, the flexible bronchoscope was used alone or in conjunction with an endotracheal tube or rigid scope in the majority of laser resections.[23,31,48,49] Subsequently, reports of major complications, including deaths resulting from massive hemorrhage[23,34,50] and airway fire,[51] began to surface, and by the late 1980s many authors[23,52] advocated using the rigid bronchoscope for most Nd:YAG laser resections.

## 41.6.2 Flexible Bronchoscopy

The major advantage of the flexible bronchoscope is its versatility and familiarity to the pulmonologist. In the *Bronchoscopy Survey* by the American College of Chest Physicians,[53] the flexible scope was used by more than 80% of responding physicians at some point in their practice. The popularity of the flexible scope for laser procedures may be partly related to the limited number of physicians who are trained in rigid bronchoscopy. In this survey, only 8.4% of the responders used the rigid scope in their practice. This trend away from the use of rigid bronchoscopy limits the choice of methods to apply the Nd:YAG laser, but in some instances, flexible bronchoscopy may actually be the preferred technique. The flexible scope is small and maneuverable, which permits easier access to distal airway lesions and improves visualization of the bronchial tree. Aiming the laser beam is also easier with the flexible scope because the operator has more control over the tip of the laser fiber.[54] The versatility and familiarity of the flexible scope make it well adapted for the safe and effective laser resection of specific lesions within the tracheobronchial tree.[54]

Another theoretical advantage of the flexible bronchoscope is the potential to avoid the inherent risks of general anesthesia including myocardial infarction, arrhythmia, hemodynamic instability, $CO_2$ retention, barotrauma, and prolonged neuromuscular weakness.[34,36,55] Unger[45] and Toty et al.[46] advocate the use of flexible bronchoscopy for most tracheobronchial laser procedures and reports no significant anesthesia-related hemodynamic or ventilatory complications in more than 290 procedures performed under topical anesthesia. Cardiovascular complications, such as myocardial infarction, however, can still result from flexible bronchoscopy,[54] especially if severe hemorrhage or airway compromise occurs. In fact, the only deaths in some studies using flexible bronchoscopy for laser procedures have not been the result of anesthetic complications but rather due to uncontrollable hemorrhage.[23,34,54] Although complications related to general anesthesia for rigid bronchoscopy have been well documented,[20–22,35] the overall rate of occurrence is exceedingly low. Cavaliere et al.[19] reported respiratory failure in only 0.5%, cardiac arrest in 0.4%, and myocardial infarction in 0.2% of 2253 cases in which more than 90% were performed under general anesthesia using a rigid bronchoscope. In a separate prospective study[35] of 124 interventional rigid bronchoscopies, no cardiac complications occurred, and transient reversible hypoxemia was seen in only 15% of cases.

### 41.6.3 Rigid Bronchoscopy

Although there may be advantages to flexible bronchoscopy for certain situations, the rigid bronchoscope is now generally preferred for most large obstructing tracheobronchial lesions.[20–22,52] Rigid laser bronchoscopy has a number of important advantages over flexible bronchoscopy including efficiency, effectiveness, and, most important, safety. It is usually more efficient because larger pieces of tumor material can be removed through the rigid scope, significantly reducing the time of the procedure and the total number of procedures required.[44,54] Other advantages of rigid bronchoscopy include the ability to perform stent implantation or removal at the time of the laser intervention. Both silicone stents and self-expandable metal stents can be placed through a rigid bronchoscope.

Although improved efficiency is desirable, the most important advantage of rigid bronchoscopy is safety. Three safety priorities for laser bronchoscopy[56–59] are maintenance of ventilation, effective suction, and a clear visual field. Thus, the rigid bronchoscope is a safer instrument for laser application in patients at higher risk for procedural complications.

Bronchoscopists performing laser photoresection of tracheobronchial lesions should be proficient in both techniques to assess the risks and benefits of each technique as it applies to a particular patient.

## 41.7 Safety

Laser safety requires knowledge of the properties of individual lasers and the potential hazards that can arise with improper use. Each laser has a different capacity for producing injury, and the American National Standards Institute (ANSI) has classified lasers (I to IV) based on the degree of their potential hazard.[60] This classification system uses an accessible emission limit (AEL) and a maximum permissible exposure (MPE) limit for various organs.[1] The AEL is the amount of laser radiation that can be collected by a specific detector per unit of area. The MPE corresponds to the total energy absorbed by the tissue and is a function of the wavelength of the laser and the duration of exposure. A class I rating is given to lasers that do not emit hazardous levels of radiation and represent the lowest risk for injury. Class II lasers are also very safe if exposure duration is limited to less than 0.25 sec in the visible band of light because the natural aversion reflex of the eye is rapid enough to avoid injury by this type of laser. Their low power gives an AEL of below 1 mW, which is the arbitrary power level that separates class II lasers from the next laser class. Class III lasers have a wide power range between 1 and 500 mW and can be potentially dangerous even if viewed momentarily. Most surgical lasers are classified as class IV because they have a power output above 500 mW. The main safety risks for surgical lasers are thermal and ocular injuries related to accidental reflection of laser light.

Inadvertent injury to the patient and operating-room personnel is an important safety concern with the use of class IV surgical lasers. Misdirected laser-light energy can reflect off instruments and shiny surfaces located near the operating field, causing corneal, retinal, and thermal injury to exposed areas. Surgical instruments can be roughened or blackened to help diffuse and absorb these potentially dangerous reflections of laser light.[61] Even if a laser beam is reflected, it retains much of its focus and intensity because of the collimation and coherence properties of laser light.[62] The dangers of reflected laser light are often not readily apparent because Nd:YAG and $CO_2$ lasers have wavelengths in the invisible range of the electromagnetic spectrum. The bronchoscopist may actually be at the highest risk for injury because Nd:YAG laser light can scatter and reflect back along the endoscope. This is usually not a concern with the $CO_2$ laser as the optics of the system protects the operator from injury. To avoid inadvertent injury from reflected laser light, safety glasses or eye shields are mandatory for all operating room personnel as well as for the patient. In addition, the windows in the procedure room should be covered to protect anyone outside the room from becoming accidentally exposed. Signs indicating the use of laser light should also be placed on the doors entering the room. Certain built-in precautions, such as foot pedals for laser activation and a standby mode, help to avoid accidental deployment of laser energy.

One of the most recognized hazards of class IV surgical lasers is retinal or corneal injury. Although the eye has a natural lid-aversion reflex that limits exposure to bright light, the power density of certain lasers is great enough to produce significant damage. Because of their various wavelengths, lasers affect the eye in different ways. Laser beams with short wavelengths in the UV region (100 to 400 nm) or long wavelengths in the far-infrared region (1400 to 10,000 nm) of the electromagnetic spectrum are absorbed by the cornea and lens of the eye. Lasers with wavelengths in the visible and NIR region (400 to 1400 nm), such as the Nd:YAG laser, are more dangerous to the retina. This area of the electromagnetic spectrum has been termed the retinal hazard region because laser light of this wavelength can be focused by the lens onto the retina with a magnitude of up to 100,000 times.[62] If the beam becomes focused onto a small point on the retina, significant damage can occur from thermal effects. If the damage is in an area such as the fovea of the macula, irreversible loss of fine vision can occur. Unfortunately, the loss of vision is usually permanent because neural tissue has limited ability to regenerate itself. For this reason, all lasers in classes II to IV should be considered potentially dangerous to the eye if adequate precautions are not taken. The type of eye protection is important because safety glasses are designed for use with lasers of a specific wavelength. Eye protection manufactured for use with one type of laser may not be protective against lasers with a different wavelength. For instance, clear plastic safety glasses are adequate for brief exposures to $CO_2$ laser energy because of the limited penetration of $CO_2$ laser light, but an Nd:YAG laser can easily penetrate plastic or glass. For protection against the Nd:YAG laser, special eyewear with an optical density of 7.0 at 1064 nm is required.

Although ocular damage is the most feared safety risk, the skin and airway mucosa are also susceptible to injury from the laser. The likelihood of injury to the skin may actually be greater than to the eye because of the larger area of exposure. Lasers in the visible and NIR region (400 to 1400 nm), such as the Nd:YAG laser, can produce thermal injury severe enough to create third-degree burns.[62] The $CO_2$ laser, however, has a longer wavelength (10,600 nm), which limits skin penetration and reduces the risk of a severe burn. Nevertheless, thermal injury can still occur if exposure is prolonged or if plastic, paper, or cloth items near the surgical field are accidentally ignited by the laser. The airway mucosa is also vulnerable to thermal injury because airway fires can occur if the flexible bronchoscope, endotracheal tube, or plastic suction catheter is accidentally ignited by the laser.[51] A polyvinyl chloride (PVC) endotracheal tube should not be used with a $CO_2$ laser because it is flammable, and toxic gases can be released from the burned plastic. The use of a red rubber tube with aluminum wrapping can reduce the risk of airway fires caused by the $CO_2$ laser.[63] This special tube should not be used with an Nd:YAG lasers because the darker rubber tube may actually absorb more laser energy. The risk of airway fires can also be reduced by limiting the power of the Nd:YAG laser exposure time and the inspired oxygen $FiO_2$.[64] The use of a metallic rigid bronchoscope will also substantially reduce the risk of airway fires, although with proper precautions a flexible bronchoscope or endotracheal tube can be used safely during laser applications. Although airway fires and burns are rare events, they are important safety considerations when using most types of surgical lasers.

The composition of the smoke plume generated by the laser has also recently come under investigation.[65] In 1988, a questionnaire study by Lobraico et al.[66] demonstrated an association between $CO_2$ laser treatment of papillomas and the subsequent development of these lesions in the treating physicians. Furthermore, Garden et al.[67] demonstrated the presence of intact human papilloma virus (HPV) DNA in the vapors of laser-treated verrucae and cautioned practitioners about the potential transmissibility of infection. Abramson et al.,[68] however, were unable to demonstrate HPV DNA present in the smoke plume from vaporization of laryngeal papillomas unless direct suction contact was made with the tissue during surgery. The presence of human immunodeficiency virus (HIV) DNA in laser smoke has also been documented, but sustained growth of the cultured cells did not occur, possibly because of damage produced by the laser energy.[69] The risk of clinical disease in surgical personnel from viral transmission has not yet been established, but special safety precautions are recommended until further studies can be performed.

Certain guidelines need to be followed for the safe operation of the Nd:YAG laser during bronchoscopy. For flexible bronchoscopy these recommendations concentrate on the prevention of hemorrhage and airway fires. To help avoid these complications and achieve successful results, the location of the tip of the laser fiber should be known at all times. When firing the laser, the tip of the fiber should be at least 4 cm from the distal end of an endotracheal tube and at least 4 mm from the end of the bronchoscope to prevent ignition of the plastic. The concentration of inspired oxygen should also be 40% or less to decrease the risk of airway fires. The tip of the laser fiber is best positioned 0.4 cm to 1 cm from the lesion and should be kept clean at all times. When appropriately positioned, the laser should also be fired parallel to the airway wall in pulses of 0.4 to 1 sec in duration with a setting of 20 to 40 W. The goal of the Nd:YAG laser application should not be to vaporize the tumor completely but rather to photocoagulate the lesion to produce hemostasis and necrosis. Residual tumor can be removed mechanically with biopsy forceps, a balloon catheter, or vigorous suctioning, eliminating the need for vaporization. Aggressive use of laser energy should also be avoided because penetration of the Nd:YAG laser is often deeper than is apparent on the surface, and the full extent of tissue necrosis is not complete until 24 to 48 h after the procedure.[45] Instead, repeat bronchoscopy is recommended in a few days to assess the extent of the remaining tumor and need for further treatment. For smaller lesions subsequent bronchoscopy may not be necessary, but for large obstructing lesions a multiple-stage procedure is often the safest approach. If these guidelines are followed, Nd:YAG laser bronchoscopy can safely and effectively be performed through the flexible bronchoscope with successful palliation in more than 80% of patients.[56]

For rigid bronchoscopy, the same Nd:YAG laser fiber is passed down the barrel of the scope along with a suction catheter and telescope. The laser fiber is aimed at the lesion by manipulating the rigid scope itself. The laser power settings remain at 20 to 40 W, with a pulse duration of 0.4 to 1 sec; this results in the application of about 22 J laser energy per pulse to the lesion. The rigid metal bronchoscope is nonflammable, and thus the laser fiber does not have to extend beyond the end of the scope. It is important, however, to keep the flexible suction catheter out of the laser field to avoid any chance of fire. The laser fiber should be in close proximity (<1 cm) to the lesion because laser energy disperses as it exits the tip of the fiber, and the distance to the lesion determines the effective power density. As with flexible laser bronchoscopy, the initial goal should be photocoagulation of the tumor and not vaporization. Debridement of the remaining tissue can be accomplished by alligator forceps or a suction catheter. If needed, the beveled end of the rigid bronchoscope can also be used to shear off parts of the tumor once photocoagulation has devitalized the tissue. Care must be taken not to perforate the airway wall with the rigid scope. It is always important to know the path of the distal airway before aggressively advancing the scope beyond the tumor. The flexible scope can be helpful for blunt dissection of the tumor and in evaluating the airways distal to the main obstruction. After most of the tumor has been removed by these means, hemostasis can be achieved through further laser photocoagulation.

## 41.8 Outcome

Although the use of laser bronchoscopy has become more common for benign and malignant tracheobronchial obstruction,[18–21] no randomized controlled trials exist proving a survival benefit in treated patients. Numerous studies, however, have shown a significant palliative benefit with regard to relief of dyspnea, control of hemoptysis, and facilitation of weaning from mechanical ventilation. In 1419 patients with bronchogenic carcinoma, Cavaliere et al.[19] achieved satisfactory immediate results corresponding to a normal airway lumen gauge or significant improvement in ventilation in 93% of patients. Brutinel et al.[23] found that airway caliber was improved in 79.7% of 182 sites and that even completely obstructed airways could be opened 57.7% of the time. In a previous study at the Lahey Clinic,[20] subjective symptomatic relief was demonstrated after 313 of 400 procedures (78.3%) in patients with benign and malignant disease.

Although many of the larger studies relied on subjective measurements of improvement, other studies have documented objective benefits in pulmonary function tests, performance scores, and even ventilation-perfusion scans after laser bronchoscopy. Spirometry and flow-volume loops have shown immediate functional improvement, with increases in peak expiratory flow rates ranging from 26 to 512%.[70,71] Waller et al.[72] showed a mean overall improvement in $FEV_1$ of 27% and symptomatic relief in 103 of 116 patients whose main complaint was dyspnea. Gelb and Epstein[49] reported significant improvement in Karnofsky performance scores, dyspnea index, in 23 of 27 patients with incomplete malignant obstruction of the tracheobronchial tree. Jam et al.[73] also used Karnofsky scores in patients treated by Nd:YAG laser photoresection and reported an improved performance status in 13 of 15 patients lasting between 2 and 13 months. Other performance scores, such as the simple 0 to 4 symptom score used by Clarke et al.,[74] confirmed good symptomatic relief in 73% of their first 200 patients consecutively treated with laser bronchoscopy.

Ventilation-perfusion relationships have also been studied in patients undergoing laser bronchoscopy. George et al.[75] found that both ventilation and perfusion improved after laser treatments in 23 of 28 patients (82%) and that spirometric values, 6-mm walking distance, Karnofsky performance index, and breathlessness scores also improved significantly. In addition, it appears that improvements are not simply transient. Although some patients with benign and malignant tracheobronchial lesions may require multiple procedures, Emslander et al.[76] reported complete or partial removal of tumor in 74% of 224 laser applications and showed that in 72% of the successful results, the stenoses were still open after 4 to 6 months.

Nd:YAG laser photocoagulation is also effective at controlling hemoptysis[9,73,74] from malignant endobronchial lesions, with a response rate estimated to be 60%.[9] Although photocoagulation for hemoptysis has shown initial success, long-term benefits were not as easily maintained as with treatment for relief of dyspnea. Clarke et al.[74] reported that patients who were treated principally for hemoptysis tended to have a recurrence of their symptoms within 30 d of treatment, whereas those who were treated for dyspnea or cough tended to experience recurrence much later.

Laser bronchoscopy has also been shown to be beneficial for patients with impending respiratory failure as well as for those who require mechanical ventilation. Radiation therapy can relieve cough, chest pain, and dyspnea in 60 to 80% of patients, but it only relieves atelectasis in 25% of patients, and improvement is usually delayed.[77] External-beam-radiation therapy is a common palliative intervention for patients with bronchogenic carcinoma, but only laser bronchoscopy provides the means for rapid resolution of symptoms. In one series,[70] laser bronchoscopy successfully relieved breathlessness and avoided mechanical ventilation in 11 of 14 patients who had respiratory distress on presentation. The authors[70] concluded that laser treatment provided an excellent method of resuscitating patients with life-threatening tracheal obstruction, enabling subsequent management such as radiation therapy, chemotherapy, or tracheobronchial stenting. Laser bronchoscopy may also facilitate weaning from mechanical ventilation in patients with respiratory failure caused by endoluminal obstruction. Stanopoulos et al.[78] reported that laser bronchoscopy improved the clinical status of 9 of 17 mechanically ventilated patients by permitting successful extubation. This subgroup of patients had appreciably shorter requirements for mechanical ventilation and a longer survival. The initial use of Nd:YAG laser bronchoscopy for treatment of an endobronchial lesion can be debated, but the palliative use of laser therapy in an inoperable, symptomatic patient has been shown to be beneficial for relief of dyspnea, control of hemoptysis, and weaning from mechanical ventilation.

Laser bronchoscopy is clearly a valuable tool in providing immediate symptomatic relief, but its survival benefit is more difficult to demonstrate. Randomized studies[79] evaluating this palliative treatment in terminal patients have been considered unethical. Although the issue of survival is a difficult one, it appears that there may be a survival advantage for patients undergoing laser resection of malignant tracheobronchial obstruction. In a preliminary study of 19 patients treated with laser photoresection, Eichenhorn et al.[80] found a median survival of 340 d, which was greater than the 198 to 266 d found in historical control subjects using irradiation alone.[81,82] In another study using historical controls, Brutinel et al.[23] compared 71 patients with bronchogenic carcinoma treated with Nd:YAG laser photoresection with 25 patients who would have received laser treatment had it been available. They found that 76% of the control subjects were dead within 4 months, and all were dead within 7 months. By contrast, 60% of patients in the laser-treated group were alive at 7 months and 28% were alive at 1 year. Desai et al.[79]

also found a significant increase in survival in a subgroup of patients treated with emergent laser resection compared with historical control subjects who had emergent radiation therapy alone. In mechanically ventilated patients, Stanopoulos et al.[78] showed an increased survival (98 d vs. 85 d) in patients who were weaned after emergent laser resection of an obstructing lesion. The addition of brachytherapy to laser resection may also contribute to increased survival. Shea et al.[83] found an improved mean survival time (41 weeks vs. 16 weeks) in patients treated with both endobronchial brachytherapy and Nd:YAG laser photoresection.

Although these data suggested a survival advantage, the retrospective design of the studies makes statistical validation difficult. Furthermore, other authors[84] failed to show any overall survival benefit for patients undergoing laser treatment. Although Desai et al.[79] found improved survival in a subgroup of patients treated emergently, they did not find an overall increase in survival in patients who underwent elective laser photoresection. Clarke et al.[74] also gave no evidence that laser treatment prolongs life and suggested that patients die of progression of their primary disease rather than obstruction of the bronchi. Nevertheless, patients on mechanical ventilation or with impending respiratory failure likely represent a subgroup of patients who may have a survival benefit from laser photoresection. A survival advantage may also be seen if endobronchial brachytherapy is combined with laser resection.

Absolute survival, however, may not be the best way to judge the importance of a palliative intervention in symptomatic patients. The rapid relief of dyspnea and improved performance provide justification for laser bronchoscopy in most patients. Laser bronchoscopy has been shown to be an effective method for immediate relief of respiratory symptoms related to both benign and malignant tracheobronchial obstruction. It is complementary to other modes of therapy, such as radiation, chemotherapy, stenting, or surgery, and often provides the time necessary to initiate these treatments.

## References


1. Wright, C.V. and Riopelle, M.A., *Surgical $CO_2$ Laser Fundamentals*, Biomedical Communications, Houston, TX, 1988.
2. Beamis, J.F. and Shapshay, S.M., Nd:YAG laser therapy for tracheobronchial disorders, *Postgrad. Med.*, 75, 173, 1984.
3. Polanyi, T.G., Laser physics, *Otolaryngol. Clin. North Am.*, 16, 753, 1983.
4. Dumon, J.F. and Corsini, A., Cicatricial tracheostenosis, in *Bronchoscopic Laser Resection Manual,* Biomedical Communications, Houston, TX, 1989.
5. Pass, H.I. and Pogrebniak, H., Photodynamic therapy for thoracic malignancies, *Semin. Surg. Oncol.*, 8, 217, 1992.
6. Hayata, Y., Kato, H., Konaka, C., Ono, J., and Takizawa, N., Hematoporphyrin derivative and laser photoradiation in the treatment of lung cancer, *Chest,* 81, 269, 1982.
7. Edell, E.S. and Cortese, D.A., Bronchoscopic phototherapy with hematoporphyrin derivative for treatment of localized bronchogenic carcinoma: a 5-year experience, *Mayo Clin. Proc.*, 62, 8, 1987.
8. Kato, H. and Cortese, D.A., Early detection of lung cancer by means of hematoporphyrin derivative fluorescence and laser photoradiation, *Clin. Chest Med.*, 6, 237, 1985.
9. Hetzel, M.R. and Smith, S.G., Endoscopic palliation of tracheobronchial malignancies, *Thorax*, 46, 325, 1991.
10. Rimell, F.L., Shapiro, A.M., Mitskavich, M.T., Modreck, P., Post, J.C., and Maisel, R.H., Pediatric fiberoptic laser rigid bronchoscopy, *Otolaryngol. Head Neck Surg.*, 114, 413, 1996.
11. Strong, M.S., Jako, G.J., Polanyi, T.G., and Wallace, R.A., Laser surgery in the aerodigestive tract, *Am. J. Surg.*, 126, 529, 1973.
12. Polanyi, T.G., Physics of surgery with lasers, *Clin. Chest Med.*, 6, 179, 1985.
13. Ossoff, R.H., Coleman, J.A., Courey, M.S., Duncavage, J.A., Werkhaven, J.A., and Reinisch, L., Clinical applications of lasers in otolaryngology — head and neck surgery, *Lasers Surg. Med.*, 15, 217, 1994.
14. Coleman, J.A., Jr., van Duyne, M.J., and Ossoff, R.H., Laser treatment of lower airway stenosis, *Otolaryngol. Clin. North Am.*, 28, 771, 1995.

15. Shapshay, S.M. and Simpson, G.T., Lasers in bronchology, *Otolaryngol. Clin. North Am.*, 16, 879, 1983.
16. Gillis, T.M. and Strong, M.S., Surgical lasers and soft tissue interactions, *Otolaryngol. Clin. North Am.*, 16, 775, 1983.
17. Mehta, A.C., Laser applications in respiratory care, In *Current Respiratory Care,* Kacmarek, R.M. and Stoller, J.K., Eds., BC Decker, Toronto, 1988, p. 100.
18. Dumon, I.F., *YAG Laser Bronchoscopy,* Praeger Publishers, New York, 1985.
19. Cavaliere, S., Foccoli, P., Toninelli, C., and Feijo, S., Nd:YAG laser therapy in lung cancer: an 11-year experience with 2,253 applications in 1,585 patients, *J. Bronchol.*, 1, 105, 1994.
20. Beamis, J.F., Jr., Vergos, K., Rebeiz, E.E., and Shapshay, S.M., Endoscopic laser therapy for obstructing tracheobronchial lesions, *Ann. Otol. Rhino. Laryngol.*, 100, 413, 1991.
21. Personne, C., Colchen, A., Leroy, M., Vourc'h, G., and Toty, L., Indications and technique for endoscopic laser resections in bronchology: a critical analysis based upon 2,284 resections, *J. Thorac. Cardiovasc. Surg.*, 91, 710, 1986.
22. Cavaliere, S., Foccoli, P., and Farina, P.L., Nd:YAG laser bronchoscopy: a five-year experience with 1,396 applications in 1,000 patients, *Chest*, 94, 15, 1988.
23. Brutinel, W.M., Cortese, D.A., McDougall, J.C., Gillio, R.G., and Bergstralh, El., A two-year experience with neodymium-YAG laser in endobronchial obstruction, *Chest*, 91, 159, 1987.
24. Shah, H., Garbe, L., Nusshauni, E., Dumon, J.F., Chiodera, P.L., and Cayaliere, S., Benign tumors of the tracheobronchial tree: endoscopic characteristics and role of laser resection, *Chest,* 107, 1744, 1995.
25. Shapshay, S.M., Beamis, J.F., Jr., and Dumon, J.P., Total cervical tracheal stenosis: treatment by laser, dilation, and stenting, *Ann. Otol. Rhino. Laryngol.*, 98, 890, 1989.
26. Mehta, A.C., Lee, F.Y., Cordasco, E.M., Kirby, T., Eliachar, I., and DeBoer, G., Concentric tracheal and subglottic stenosis: Management using the Nd:YAG laser for mucosal sparing followed by gentle dilatation, *Chest*, 104, 673, 1993.
27. Wanamaker, J.R. and Eliachar, I., An overview of the treatment options for lower airway obstruction, *Otolaryngol. Clin. North Am.*, 28, 751, 1995
28. Montgomery, W.W. and Montgomery, S.K., Manual for use of Montgomery laryngeal, tracheal, and esophageal prostheses: update 1990, *Ann. Otol. Rhino. Laryngol. Suppl.*, 150, 2, 1990.
29. Gelb, A.F., Tashkin, D.P., Epstein, J.D., and Zamel, N., Nd:YAG laser surgery for severe tracheal stenosis physiologically and clinically masked by severe diffuse obstructive pulmonary disease, *Chest*, 91, 166, 1987.
30. Hayashi, A.H., Gillis, D.A., Bethune, D., Hughes, D., and O'Neil, M., Management of foreign-body bronchial obstruction using endoscopic laser therapy, *J. Pediatr. Surg.*, 25, 1174, 1990.
31. Kvale, P.A., Eichenhorn, M.S., Radke, J.R., and Miks, V., YAG laser photoresection of lesions obstructing the central airways, *Chest*, 87, 283, 1985.
32. Dierkesmann, R., Indication and results of endobronchial laser therapy, *Lung*, 168(Suppl.), 1095, 1990.
33. Dierkesmann, R. and Huzly, A., The significance of the pulmonary artery in endobronchial laser treatment (abstr.), *Lasers Surg. Med.*, 3, 197, 1983.
34. Warner, M.E., Warner, M.A., and Leonard, P.F., Anesthesia for neodymium-YAG (Nd:YAG) laser resection of major airway obstructing tumors, *Anesthesiology*, 60, 230, 1984.
35. Perrin, C., Colt, H.G., Martin, C., Mak, M.A., Dumon, J.F., and Gouin, F., Safety of interventional rigid bronchoscopy using intravenous anesthesia and spontaneous assisted ventilation: a prospective study, *Chest*, 102(5), 1526, 1992.
36. Duckett, J.E., McDonnell, T.J., Unger, M., and Parr, G.V., General anaesthesia for Nd:YAG laser resection of obstructing endobronchial tumours using the rigid bronchoscope, *Can. Anaesthesiol. Soc. J.*, 32, 67, 1985.

37. Vourc'h, G., Fischler, M.F., Michon, F., Meichior, J.C., and Seigneur, F., High frequency jet ventilation manual jet ventilation during bronchoscopy in patients with tracheo-bronchial stenosis, *Br. J. Anaesthesiol.*, 55, 969, 1983.
38. Vourc'h, G., Fischler, M.F., Michon, 13 , Meichior, J.C., and Seigneur, F., Manual jet ventilation v. high frequency jet ventilation during laser resection of tracheo-bronchial stenosis, *Br. J. Anaesthesiol.*, 55, 973, 1983.
39. Sanders, R.D., Two ventilating attachments for bronchoscopes, *Del. Med. J.*, 39, 170, 1967.
40. Ramser, E.R. and Beamis, J.F., Jr., Laser bronchoscopy, *Clin. Chest Med.*, 6, 415, 1995.
41. Schlenkhoff, D., Droste, H., Scieszka, S., and Vogt, H., The use of high-frequency jet-ventilation in operative bronchoscopy, *Endoscopy*, 18, 192, 1986.
42. Lennon, R.L., Hosking, M.P., Warner, M.A., Cortese, D.A., McDougall, J.C., Brutinel, W.M., and Leonard, P.F., Monitoring and analysis of oxygenation and ventilation during rigid bronchoscopic neodymium-YAG laser resection of airway tumors, *Mayo Clin. Proc.*, 62, 584, 1987.
43. Goidhill, D.R., Hill, A.J., Whithurn, R.H., Feneck, R.O., George, P.J., and Keeling, P., Carboxyhaemoglobin concentrations, pulse oximetry and arterial blood-gas tensions during jet ventilation for Nd:YAG laser bronchoscopy. *Br. J. Anaesthesiol.*, 65, 749, 1990.
44. Chan, A.L., Tharratt, R.S., Siefkin, A.D., Albertson, T.E., Volz, W.G., and Allen, R.P., Nd:YAG laser bronchoscopy: rigid or fiberoptic mode? *Chest*, 98, 271, 1990.
45. Unger, M., Rigid vs. flexible bronchoscope in laser bronchoscopy, *J. Bronchol.*, 3, 69, 1994.
46. Toty, L., Personne, C., Colchen, A., and Vourc'h, G., Bronchoscopic management of tracheal lesions using the neodymium yttrium aluminum garnet laser, *Thorax*, 36, 175, 1981.
47. Durnon, J.F., Reboud, E., Garbe, L., Aucomte, F., and Meric, B., Treatment of tracheobronchial lesions by laser photoresection, *Chest*, 81, 278, 1982.
48. Hetzel, M.R., Nixon, C., Edmondstone, W.M., Mitchell, D.M., Millard, F.J., Nanson, E.M., Woodcock, A.A., Bridges, C.E., and Humberstone, A.M., Laser therapy in 100 tracheohronchial tumours, *Thorax*, 40, 341, 1985.
49. Gelb, A.F. and Epstein, J.D., Laser in treatment of lung cancer, *Chest*, 86, 662, 1984.
50. Arabian, A. and Spagnolo, S.V., Laser therapy in patients with primary lung cancer, *Chest*, 86, 519, 1984.
51. Casey, K.R., Fairfax, W.R., Smith, S.J., and Nixon, J.A., Intratracheal fire ignited by the Nd:YAG laser during treatment of tracheal stenosis, *Chest*, 84, 295, 1983.
52. Dumon, J.F., Shapshay, S., Bourcereau, J., Cavaliere, S., Meric, B., Garbi, N., and Beamis, J., Principles for safety in application of neodymium:YAG laser in bronchology, *Chest*, 86, 163, 1984.
53. Prakash, U.B., Offord, K.P., and Stubbs, S.E., Bronchoscopy in North America: the ACCP survey, *Chest*, 100, 1668, 1991.
54. George, P.J., Garrett, C.P., Nixon, C., Hetzel, M.R., Nanson, E.M., and Millard, F.J., Laser treatment for tracheobronchial tumours: local or general anesthesia? *Thorax*, 42, 656, 1987.
55. Hanowell, L.H., Martin, W.R., Savelle, J.E., and Foppiano, L.E., Complications of general anesthesia for Nd:YAG laser resection of endobronchial tumors, *Chest*, 99, 72, 1991.
56. Mehta, A.C., Golish, J.A., Ahmad, M., Zurick, A., Padua, N.S., and O'Donnell, J., Palliative treatment of malignant airway obstruction by Nd:YAG laser, *Cleveland Clin. Q.*, 52, 513, 1985.
57. Unger, M., Neodymium:YAG laser therapy for malignant and benign endobronchial obstructions, *Clin. Chest Med.*, 6, 227, 1985.
58. Personne, C., Coichen, A., Bonnette, P., Leroy, M., and Bisson, A., Laser in bronchology: methods of application. *Lung*, 168(Suppl.), 1085, 1990.
59. Unger, M., Parr, G.V., and Lugano, E., Various applications of Nd:YAG laser in tracheobronchial pathology, in *Nd:YAG Laser in Medicine and Surgery: Fundamental and Clinical Aspects,* Ogura, Y., Ed., Professional Postgraduate Services, Tokyo, 1986, p. 261.
60. American National Standards for the Safe Use of Lasers in Health Care Facilities, American National Standards Institute, The Laser Institute of America, Orlando, FL, 1996.

61. Wood, R.L., Jr., Sliney, D.H., and Basye, R.A., Laser reflections from surgical instruments, *Lasers Surg. Med.*, 12, 675, 1992.
62. Sliney, D.H., Laser safety, *Lasers Surg. Med.*, 16, 215, 1995.
63. Schramm, V.L., Jr., Mattox. D.E., and Stool, S.E., Acute management of laser-ignited intratracheal explosion, *Laryngoscope*, 91, 1417, 1981.
64. Shapshay, S.M. and Beamis, J.P., Jr., Safety precautions for bronchoscopic Nd:YAG laser surgery, *Otolaryngol. Head Neck Surg.*, 94, 175, 1986.
65. Wenig, B.L., Stenson, K.M., Wenig, B.M., and Tracey, D., Effects of plume produced by the Nd:YAG laser and electrocautery on the respiratory system, *Lasers Surg. Med.*, 13, 242, 1993.
66. Lobraico, R.V., Schifano, M.J., and Brader, K.R., A retrospective study on the hazards of the carbon dioxide laser plume, *J. Laser Appl.*, 1, 6, 1988.
67. Garden, J.M., O'Banion, M.K., Shelnitz, L.S., Pinski, K.S., Bakus, A.D., Reichmann, M.E., and Sundberg, J.P., Papillomavirus in the vapor of carbon dioxide laser-treated verrucae, *J. Am. Med. Assoc.*, 259, 1199, 1988.
68. Abramson, A.L., DiLorenzo, T.P., and Steinberg, B.M., Is papillomavirus detectable in the plume of laser-treated laryngeal papilloma? *Arch. Otolaryngol. Head Neck Surg.*, 116, 604, 1990.
69. Baggish, M.S., Poiesz, B.J., Joret, D., Williamson, P., and Refai, A., Presence of human immunodeficiency virus DNA in laser smoke, *Lasers Surg. Med.*, 11, 197, 1991.
70. George, P.J., Garrett, C.P., and Hetzel, M.R., Role of the neodymium YAG laser in the management of tracheal tumours, *Thorax*, 42, 440, 1987.
71. Mohsenifar, Z., Jasper, A.C., and Koerner, S.K., Physiologic assessment of lung function in patients undergoing laser photoresection of tracheobronchial tumors, *Chest*, 93, 65, 1988.
72. Waller, D.A., Gower, A., Kashyap, A.P., Conacher, I.D., and Morritt, G.N., Carbon dioxide laser bronchoscopy — a review of its use in the treatment of malignant tracheobronchial tumours in 142 patients, *Respir. Med.*, 88, 737, 1994.
73. Jam, P.R., Dedhia, H.V., Lapp, N.L., Thompson, A.B., and Frich, J.C., Jr., Nd:YAG laser followed by radiation for treatment of malignant airway lesions, *Lasers Surg. Med.*, 5, 47, 1985.
74. Clarke, C.P., Ball, D.L., and Sephton, R., Follow-up of patients having Nd:YAG laser resection of bronchostenotic lesions, *J. Bronchol.*, 1, 19, 1994.
75. George, P.J., Clarke, G., Tolfree, S., Garrett, C.P., and Hetzel, M.R., Changes in regional ventilation and perfusion of the lung after endoscopic laser treatment, *Thorax*, 45, 248, 1990.
76. Emslander, H.P., Munteanu, J., Prauer, H.J., Heinl, K.W., Hinke, K.W., Sebenning, H., and Daum, S., Palliative endobronchial tumor reduction by laser therapy: procedure — immediate results — long-term results, *Respiration*, 51, 73, 1987.
77. Slawson, R.G. and Scott, R.M., Radiation therapy in bronchogenic carcinoma, *Radiology*, 132, 175, 1979.
78. Stanopoulos, I.T., Beamis, J.F., Jr., Martinez, F.J., Vergos, K., and Shapshay, S.M., Laser bronchoscopy in respiratory failure from malignant airway obstruction, *Crit. Care Med.*, 21, 286, 1993.
79. Desai, S.J., Mehta, A.C., vander Brug Medendorp, S., Golish, J.A., and Ahmad, M., Survival experience following Nd:YAG laser photo-resection for primary bronchogenic carcinoma, *Chest*, 94, 939, 1988.
80. Eichenhorn, M.S., Kvale, P.A., Miks, V.M., Seydel, F.I.G., Horowitz, B., and Radke, J.R., Initial combination therapy with YAG laser photoresection and irradiation for inoperable non-small cell carcinoma of the lung: a preliminary report, *Chest*, 89, 782, 1986.
81. Petrovich, Z., Stanley, K., Cox, J.D., and Paig, C., Radiotherapy in the management of locally advanced lung cancer of all cell types: final report of randomized trial, *Cancer*, 48, 1335, 1981.
82. Roswit, B., Patno, M.E., Rapp, R., Veinbergs, A., Feder, B., Stuhlbarg, J., and Reid, C.B., The survival of patients with inoperable lung cancer: a large-scale randomized study of radiation therapy versus placebo, *Radiology*, 90, 688, 1968.

83. Shea, J.M., Allen, R.P., Tharratt, R.S., Chan, A.L., and Siefkin, A.D., Survival of patients undergoing Nd:YAG laser therapy compared with Nd:YAG laser therapy and brachytherapy for malignant airway disease, *Chest*, 103, 1028, 1993.
84. Quinj, A., Letsou, G.V., Tanoue, L.T., Matthay, R.A., Higgins, R.S., and Baldwin, J.C., Use of neodymium yttrium aluminum garnet laser in long-term palliation of airway lesions, *Conn. Med.*, 59, 407, 1995.

# 42

# Lasers in Neurosurgery

Devin K. Binder
*University of California, San Francisco*
*San Francisco, California*

Meic H. Schmidt
*University of California, San Francisco*
*San Francisco, California*

Rose Du
*University of California, San Francisco*
*San Francisco, California*

Mitchel S. Berger
*University of California, San Francisco*
*San Francisco, California*



## 42.1 Lasers and Neurosurgery

The physical basis for most laser (*L*ight *A*mplification by *S*timulated *E*mission of *R*adiation) applications in neurosurgery is the conversion of laser light into heat. Laser light is monochromatic, coherent, and collimated. Laser photon absorption by tissue stimulates oscillation and rotation of tissue molecules, and subsequent inelastic collisions release kinetic energy to produce heat. Depending on the duration of irradiation, and thus the degree of heat, several biologic effects can be seen including hyperthermia ($T$ = 45°C), coagulation (involving protein denaturation) ($T$ = 65°C), vaporization ($T$ = 100°C), and carbonization ($T$ > 100°C).[1] Both the laser-emission parameters (wavelength, fluence, and mode) as well as the exposed tissue's optical and thermal properties (water and hemoglobin content, thermal conductivity, and specific heat) are critical in determining the tissue effect (hemostasis, cutting, and vaporization). The main potential disadvantage is diffusion of heat energy resulting in thermal damage to surrounding normal tissue. Furthermore, tissue factors such as degree of vascularity affect kinetics of cooling following hyperthermia from a laser pulse. In addition, laser irradiation may kill tumor cells but of course does not remove them, which must still be done surgically.

0-8493-1116-0/03/$0.00+$1.50

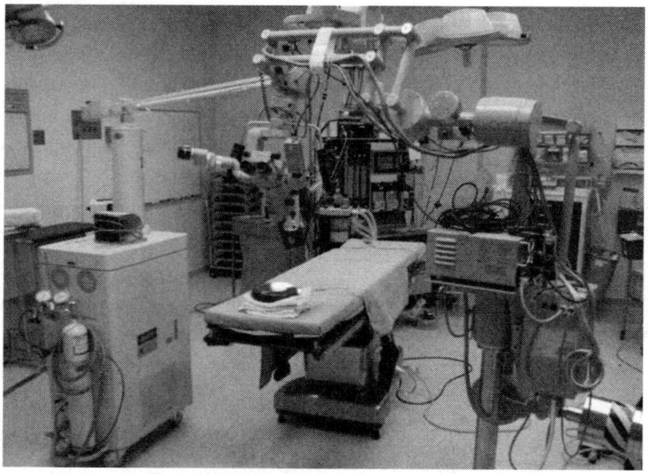

**FIGURE 42.1** Cooper model 500Z $CO_2$ surgical laser system setup. The laser wavelength is 10.6 μm. The laser power can be varied from 1 to 50 W at the treatment site in increments of 1 W. The irradiation time can be adjusted from 10 to 990 msec either continuously or in increments of 1.0 msec. The aiming beam consists of a helium–neon laser projected to the same position and spot size as the $CO_2$ laser. The spot size can be varied from 0.8 to 3.5 mm.

## 42.2 Development of the $CO_2$ Laser

The world's first laser, a pulsed ruby laser, was developed in 1960.[2] Rosomoff and Carroll in 1966[3] were the first to focus pulsed-ruby-laser beams on brain tumors and found some evidence of tumor necrosis. However, they did not attempt to use the laser to remove tumor tissue. The ruby laser did not possess desirable surgical characteristics,[4] and its thermal effects were difficult to control. The continuous-wave $CO_2$ laser was introduced in 1967 and immediately replaced the ruby laser for surgical applications (Figure 42.1). The $CO_2$ laser was superior at vaporization, cutting, hemostasis, and sterilization, and its tissue-damage characteristics were more predictable. Stellar et al. found that precise incisions could be made with the $CO_2$ laser in the cat brain and spinal cord with little collateral tissue damage.[5] They were the first to use a $CO_2$ laser to subtotally vaporize a glioblastoma.[6] In the mid- to late 1970s, other surgeons began to report case series of patients who had a variety of procedures performed with the $CO_2$ laser.[7–11] From 1976 to 1979, Ascher and Heppner performed more than 250 procedures using the $CO_2$ laser, including 200 brain tumors and 13 spinal cord tumors.[7–9] In 1980, Takizawa et al. reported on treatment of meningiomas with the $CO_2$ laser.[11]

During the late 1970s and early 1980s, two principal technical changes dramatically increased the ease of applicability of the laser in neurosurgery. First, a helium laser was added as a pilot to target the invisible $CO_2$ laser; the pilot beam served as a marker for the treatment beam. Second was the coupling of the laser to the operating microscope. Initially, most of these procedures were performed with a freehand laser not attached to an operating microscope,[9,12] which made precise targeting quite difficult.

After some experience with the $CO_2$ laser, investigators found it particularly useful for the excision of extraaxial tumors such as meningiomas, vestibular schwannomas, and craniopharyngiomas. Poorly accessible deep brain tumors and recurrent gliomas were other common uses. The growing interest in lasers in neurosurgery culminated in the First American Congress of Laser Neurosurgery, held in Chicago in October 1981.

## 42.3 Development of the Nd:YAG Laser

Another laser, the neodymium:yttrium-aluminum-garnet (Nd:YAG) laser, was found to have unique biologic properties, suggesting applications within neurosurgery distinct from the $CO_2$ laser.[13] In particular, its selective absorption by blood and blood vessels suggested its use in neurovascular surgery. Early experimental studies had shown that it was possible to perform blood-vessel microanastomosis and repair with the Nd:YAG laser.[14–16] Experimental comparison of $CO_2$ and Nd:YAG lasers applied to rabbit brains demonstrated that the Nd:YAG laser penetrated more deeply and also had a stronger effect on vascular tissue.[17] This led to the clinical use of the Nd:YAG laser by Beck et al.[10] and Takeuchi et al.,[18] especially in the excision of vascular meningiomas and arteriovenous malformations. It was also used to perforate sphenoid bones in transsphenoidal surgery by Takeuchi et al.[18] The primary advantage noted was that blood loss was lower than with conventional techniques or with the $CO_2$ laser due to the superior coagulating ability of the Nd:YAG laser.[19]

Subsequently, Sundt and colleagues carried out experimental and clinical studies of the Nd:YAG laser.[20,21] In studies on cat brains, they found a blood:brain absorption ratio of 100:1, indicating selective absorption by hemoglobin and concurrent selective heating of blood compared with brain tissue.[20] This clarified the role and technique for using the Nd:YAG laser in hemostasis.

One of the only studies specifically reporting complications of the use of the Nd:YAG laser was published by Jain in 1985.[22] In a review of 32 patients who underwent brain-tumor procedures with the Nd:YAG laser, five complications were presented. Two mortalities, one due to intratumor explosion and the other to heat necrosis of the brainstem following laser irradiation of a fourth ventricular ependymoma, were described. Protection of normal tissue from thermal spread-induced damage during use of the laser was emphasized, with adequate venting of hot gases created by vaporization of tissue and covering surrounding normal tissue with moist cottonoids.

## 42.4 Development of the Argon Ion Laser

The initial use of the argon ion laser was in ophthalmology for the treatment of retinal detachment.[13] Its water transmissibility, good hemostatic ability, and small size made it ideal for ophthalmologic procedures. Subsequently, it was introduced into otolaryngology.[23] Glasscock et al.[24] were the first to report use of the argon laser in removal of vestibular schwannomas.

In preclinical studies, Boggan et al.[25] directly compared the response of brain tissue to argon vs. $CO_2$ lasers. Pulses were given at identical power densities to 32 rat brains. Although there was an initial larger increase in blood-barrier defect caused by the $CO_2$ laser, by 1 month after injury there was no difference in the size of the core of coagulation necrosis and surrounding gliosis.

Powers et al.[26] described the use of the argon laser in 68 procedures including tumors, spinal-cord fenestration for syringomyelia, and production of dorsal root entry zone (DREZ) lesions for pain control. They emphasized that the useful characteristics of the argon laser as a microneurosurgical instrument included a small laser spot size (0.15 to 1.5 mm), a single laser-aiming and treatment beam, the transmission of the argon laser light through aqueous media, and excellent hemostasis. At the same time, however, the same group[27] compared standard microsurgical techniques, the $CO_2$ laser, and the argon laser in a rat gliosarcoma model and found identical gross total resection and rate of tumor recurrence by all three methods. Therefore, in the case of malignant glial tumors, it was not clear that the option of laser therapy had any particular advantage over standard microsurgical techniques.

## 42.5 Development of the KTP Laser

The potassium-titanyl-phosphate (KTP) laser was initially developed as a second-generation laser to the argon laser but has been little used. Gamache and Morgello[28] compared the histopathologic effects of the KTP laser vs. the $CO_2$ laser on canine brain and spinal cord. Each laser led to a central zone of tissue vaporization surrounded by coagulation necrosis; the only difference appeared to be slightly more histologic evidence of hemorrhage in $CO_2$-laser-induced lesions. Clinical experience with this laser has been limited.[29]

## 42.6 Meningiomas

In 1980, Takizawa et al.[11] reported on the treatment of meningiomas with the $CO_2$ laser. Beck[10] described 51 cases of meningiomas treated with the Nd:YAG laser, but outcome was not reported. Later, Roux et al.[30] described treatment of 17 meningiomas with a combined $CO_2$ and Nd:YAG laser. They noted that the combined laser penetrated more deeply than the $CO_2$ laser alone with the same power and the hemostasis was contributed by the Nd:YAG component. Outcome was not reported. Lombard et al.[31] studied 220 meningiomas operated on with a laser or with conventional techniques and examined postoperative morbidity and mortality. Both patients in the conventional-surgery and laser groups improved after surgery, but there was a significantly better outcome for patients with meningiomas located in functional areas treated with a laser compared to conventional surgery. Desgeorges et al.[32] reported on 164 meningiomas treated with various lasers over 6 years. Complete tumor removal was documented in 83% of cases, and mortality was 3%. They commented particularly on the ability to remove deep meningiomas that might be difficult with conventional microsurgery. Specific advantages included reduced brain retraction, ability to operate with smaller exposures, reduced mechanical manipulation, and decreased intraoperative blood loss. Waidhauser et al.[33] reported on 43 patients with frontobasal meningiomas. In this series, they used the Nd:YAG laser for contact-free shrinkage of the tumor, lysis of the dural and bony attachments, and coagulation (in addition to standard bipolar cautery). In their experience, the Nd:YAG laser facilitated microsurgical dissection and reduced blood loss.

## 42.7 Vestibular Schwannomas

Several authors have described laser-assisted excision of vestibular schwannomas.[26,34,35] In this procedure, preservation of cranial nerve function is critical. Irrigation with suction of the laser plume has been used to minimize thermal injury to structures around the tumor. In 1993, Eiras et al.[35] compared a group of 12 cases of giant vestibular schwannomas operated with the $CO_2$ laser to a group of 12 similar cases undergoing conventional microsurgery. In this small study, the main finding was that while duration of surgery was slightly longer with the use of the $CO_2$ laser, other parameters studied, such as facial nerve preservation, were slightly better in the laser-treated group.

## 42.8 Pituitary Tumors

Takeuchi et al.[18] were the first to use the Nd:YAG laser to penetrate the sellar floor during transsphenoidal pituitary operations. Subsequently, Oekler et al.[36] reported 15 cases of pituitary adenomas in which the sellar floor was irradiated with the Nd:YAG laser to minimize bleeding from the capsule. Powers et al.[26] reported resection of a fibrous prolactinoma with the aid of the argon laser. In this case, the laser was used to open the sellar floor, cut through the dura, and vaporize the tumor. However, the application of lasers in pituitary tumors is likely to be limited in most cases to brief focal photocoagulation since thermal spread to adjacent structures (optic nerves, optic chiasm, cavernous sinus, internal carotid artery) could be devastating.

## 42.9 Intraaxial Brain Tumors

While there was initial enthusiasm over the use of lasers in the treatment of intraaxial brain tumors, conventional microsurgery, especially with modern image-guidance techniques, seems more appropriate in most cases. For example, there is no clear advantage to performing the corticotomy required to expose brain tumors with a laser. However, Kelly et al. integrated a $CO_2$ laser into a computer-assisted volumetric stereotactic system for the resection of intraaxial neoplasms.[37–40] Other investigators have demonstrated the ability to remove intraaxial brain neoplasms with lasers.[26,41,42] Devaux et al. recently reported on the new high-power semiconductor diode laser in the treatment of intraaxial brain tumors.[43] There remains no clearly demonstrated advantage to lasers for intraaxial tumors that can be reached with conventional techniques.

# 42.10 Spinal Lesions

## 42.10.1 Intramedullary Tumors

$CO_2$ and argon lasers have been used for the excision of intramedullary spinal-cord tumors.[9,26,34,41,44] Usually, the myelotomy is made with the laser followed by laser vaporization of the spinal tumor. While theoretically attractive, it is not clear that outcome following laser treatment of intramedullary tumors is better than with conventional techniques, from the point of view of either reduced damage to surrounding tissue or increased efficacy of tumor removal.

## 42.10.2 Lipomas

Lasers have more commonly been used in the treatment of spinal lipomas and lipomyelomeningoceles. With these lesions, the goal is to untether the spinal cord and nerve roots from the lipoma. However, the fibroadipose tissue of the lipoma may have a large and poorly defined border with the conus medullaris and nerve roots. In this regard, the $CO_2$ laser has proven to be quite effective at vaporization of fat. McLone and Naidich[45] reported a series of 50 cases of pediatric spinal lipoma treated with $CO_2$ laser. They found that use of the laser reduced the length of the operation, intraoperative blood loss, and degree of manipulation of spinal cord and nerve roots. In addition, many patients' neurologic function improved postoperatively including recovery of urinary continence and improved motor function. Likewise, Maira et al.[46] reported using the $CO_2$ laser to remove large spinal lipomas in two adult patients.

## 42.10.3 Laser Discectomy

The surgical treatment of herniated intervertebral discs has progressively improved with the development of less invasive techniques.[47] While nearly all operations are performed open, endoscopic methods of discectomy have been introduced.[48,49] Distinct procedures have been developed to perform percutaneous discectomy with the use of lasers.[50–53] Choy et al. first reported percutaneous laser discectomy in 12 patients using the Nd:YAG laser introduced through an 18-gauge needle.[52,53] Unfortunately, 5 of the initial 12 patients ultimately required open operation. Mayer et al.[50] used a rigid endocope with a flexible Nd:YAG laser fiber to perform discectomy in six patients. No study has yet been done comparing percutaneous laser discectomy with open microdiscectomy in an appropriately selected patient population. The current clinical role of percutaneous discectomy is negligible compared with open discectomy.

# 42.11 Arteriovenous Malformations

As described above, the development of the Nd:YAG laser was found to have superior properties to the $CO_2$ laser for coagulation of blood vessels and hemostasis. It was first used as an adjunct in treatment of arteriovenous malformations (AVMs) by Beck et al. in 1980.[10] In this case, it was used selectively to coagulate feeding vessels. Subsequently, Fasano et al.[54] reported on the combined use of the $CO_2$ laser and Nd:YAG laser in the treatment of three vascular malformations. In a larger follow-up study, Fasano et al.[55] used either the Nd:YAG or the argon laser to photocoagulate six AVMs. Sundt et al. demonstrated the use of the Nd:YAG laser in the resection of ten arteriovenous malformations (AVMs).[21] In these cases, the laser was found helpful for defining the plane between the AVM and brain and for achieving hemostasis of the bed following AVM resection but could not arrest high-flow bleeding from the thin-walled vessels of the AVM and did not appreciably reduce operating time. Similarly, Powers et al.[26] found the argon laser helpful in removal of a low-flow AVM but not a large, high-flow AVM. Zuccarello et al.[56] resected ten AVMs with the Nd:YAG laser, studied the histologic effect of the laser on AVMs, and found shrinkage of collagen in the AVM vessels leading to vessel narrowing.

Strugar and Chyatte reported five cases of spinal dural AV fistulas treated by microsurgical exposure of the dural nidus followed by nidus treatment with the Nd:YAG laser. Satisfactory obliteration was confirmed with postoperative angiography, and there were no complications.[57]

## 42.12 Interstitial Laser Thermotherapy

Hyperthermia has been a much-studied adjunctive therapy in the treatment of malignant brain tumors.[58–60] It has been produced by ultrasound, microwaves, radiosurgery, and indwelling probes. Interstitial laser thermotherapy using the Nd:YAG laser as a thermal source has been studied both in animal models and in humans.[61–65] Conical sapphire-tipped optical fibers delivering a point source of Nd:YAG laser energy can be positioned at any point within a tumor.[61] As with typical laser tissue effects, the size of the lesion depends on the energy and exposure time.[66] However, despite some encouraging results in the treatment of glioblastoma,[59] the current role of hyperthermia in clinical neurosurgery is limited.

## 42.13 Laser-Assisted Vascular Anastomosis

In the mid-1960s, Yahr et al.[14] and Yahr and Strully[15] were the first to demonstrate laser-assisted vascular anastomoses.[14,15] This was initially done with microvessels (0.7- to 2.0-mm diameter) and subsequently performed on larger vessels (3- to 8-mm diameter). Exactly how laser treatment welds blood vessels together is unclear, but apparently denaturation of collagen in the blood vessel walls is partially responsible.[67,68] In 1979, Jain and Gorisch[16] reported repair of small blood vessels (0.3 to 1.0 mm) in the rat with the Nd:YAG laser. In 1985, Frazier et al.[67] compared end-to-end anastomoses of femoral arteries (1.6-mm diameter) in miniature swine done with $CO_2$ laser vs. conventional microvascular suture anastomosis and found that the laser anastomosis was shorter (20 vs. 30 min) and patency rates were superior for laser anastomosis. Quigley et al.[69] showed that intimal hyperplasia was reduced in laser-assisted compared with suture anastomosis. However, a very important finding was the association of laser treatment with the development of aneurysms in the laser-joined vessels.[70] Shapiro et al.[71] demonstrated successful end-to-side anastomosis of rat carotid arteries using the Nd:YAG laser. While anastomotic patency was 86%, aneurysm formation occurred in 23% of anastomoses. This high rate of laser-induced aneurysm formation and the efficacy of conventional microsurgical anastomosis undermined enthusiasm for any clinical role for laser-assisted vascular anastomosis in neurosurgery. However, Tulleken and colleagues have since developed a novel high-flow bypass procedure using an excimer laser to revascularize the ischemic brain and shown excellent patency without aneurysm formation.[72–74] Another application under study is the potential use of lasers in reversing cerebral vasospasm.[75–77]

## 42.14 Laser-Assisted Nerve Repair

Despite refinements in microsurgical nerve repair, results can be unsatisfactory and may include fascicle mismatching, scar formation, foreign-body response to suture material, and neuroma formation.[78] Therefore, investigators have been interested in the potential for using the laser in nerve repair (laser neurorrhaphy).[79–88] Almquist et al. first reported the use of the argon laser to repair peripheral nerves in rats and monkeys in 1984.[79] Fischer et al.[80] found similar results using the $CO_2$ laser in rat sciatic-nerve repair. Maragh et al.[81] compared standard microsurgical vs. laser nerve repair in a rat-sciatic-nerve model and found better nerve conduction velocities in the microsuture group. Bailes et al.[82] compared $CO_2$ laser with 9–0 nylon perineurial neurorrhaphy in a primate model of sural to peroneal nerve graft and found comparable results on nerve conduction velocity and histology at 5, 8, 10, and 12 months. Seifert and Stolke[83] demonstrated feasibility of microsurgical $CO_2$ laser–assisted repair of the oculomotor nerve in the cat. Campion et al.[84] claimed to demonstrate improved neuromuscular function of argon laser-assisted nerve repair compared to conventional microsurgical epineurial-suture repair in a rabbit model. Huang et al.[85] compared microsuture vs. $CO_2$ laser repair of transected sciatic nerves in rats and found similar results on EMG and nerve conduction velocities. Korff et al.[86] found a similar amount of regeneration 2 months after repair of severed rat sciatic nerves repaired with suture or laser. More recently, Menovsky et al.[88] have compared $CO_2$ laser-assisted nerve repair with fibrin glue or absorbable sutures in the rat-sciatic-nerve model. There were no significant differences in motor function at 16 weeks and no significant histologic differences.

Taken together, the above preclinical studies demonstrate both advantages and disadvantages of laser nerve repair.[78] Advantages include reduced neuroma and scar formation and shorter repair time. The main disadvantages are inferior tensile strength and the cost associated with laser treatment. While these studies demonstrate that laser-assisted nerve repair is feasible, application of the technique is likely to be limited if the outcome is not shown to be superior to conventional microsurgical neurorrhaphy.[89]

## 42.15 Fluorescence Imaging and Photodynamic Therapy

Surgical resection of malignant astrocytomas improves quality of life and survival. However, complete resection of these tumors is frequently not possible because tumor margins are not readily distinguished from normal brain tissue. Laser-induced fluorescence (LIF) spectroscopy and imaging have been used to distinguish neoplastic from normal tissue.[90] Normal brain tissue has a characteristic pattern of fluorescence when exposed to laser light.[91] This is due to endogenous fluorophores present in normal brain tissue (autofluorescence). By combining autofluorescence with diffuse-reflectance spectroscopy one can readily distinguish tumor from infiltrated brain tissue. Lin et al. recently published their results with this method in 26 brain-tumor patients.[90] The sensitivity and specificity for differentiating normal brain from tumor and tumor-infiltrated brain was 100 and 76%, respectively. In order to enhance the demarcation further, exogenous fluorophores can be administered. Phthalocyanine tetrasulfonate is an exogenous fluorophore that accumulates preferentially in brain tumors. In a rat glioma model, surgical resection was guided by LIF.[92] Thirty percent of animals in the LIF resection-guided group were tumor-free at 2 weeks compared to none in the visual-resection group. A high contrast ratio (40:1) between normal brain and tumor was achieved. A similar technique was used by Stummer et al. in 52 malignant-glioma patients.[93] They utilized 5-aminolevulinic acid (5-ALA) to induce fluorescence of tumor cells in 52 glioblastoma patients and demonstrated that fluorescence-guided resection increased completeness of resection on postoperative MRI. Sensitivity and specificity of fluorescence-guided tumor identification of tumor were 85 and 100%, respectively.[94]

Despite the ability of LIF to demonstrate tumor margins, resection is sometimes not possible because the tumor has infiltrated functional (eloquent) brain tissue. Many of the fluorescent agents used for LIF diagnosis and resection guidance can also be used for photodynamic therapy (PDT). In PDT, laser light is used to activate a photosensitizing molecule to produce cytotoxic singlet oxygen that can result in selective destruction of microscopic tumor cells. The most common photosensitizer used for brain-tumor therapy has been hematoporphyrin derivative (HPD).[95] Studies indicate that when HPD is activated by laser light at 630 nm, patients with brain tumors can show significant tumor responses.[96–98] Two large clinical studies for patients with recurrent gliomas demonstrated that PDT with laser light at 630 nm can increase survival.[96,97] Higher laser-light doses appeared to be associated with increased survival.[97] The most common cause of treatment failure was local recurrence, which is most likely secondary to the limited tissue penetration of laser light at 630 nm. Muller and Wilson recorded light penetration of 630 nm laser light in human brain and gliomas intraoperatively.[99] The mean penetration depths for tumor,

tumor-infiltrated brain, and normal brain were 2.9 ± 1.5 mm, 2.4 ± 1.2 mm, and 1.5 ± 0.4 mm, respectively.[99] In general, increasing the wavelength of the activating laser light increases the penetration in brain tissue. However, the laser-light absorption of HPD over 630 nm is limited. Newer photosensitizers that can absorb laser light at longer wavelengths may allow for deeper tissue penetration and might be able to improve local recurrence rates. Benzoporphyrin derivative (BPD) is a second-generation photosensitizer that can effectively absorb laser light at 680 nm.[100] This could result in improved local tumor control but may also result in increased toxicity to normal brain tissue. In a canine glioma model, normal-brain-tissue toxicity occurred at high photosensitizer doses with Photofrin, a purified HPD.[95] This study illustrated that reducing the photosensitizer dose limited normal-brain-tissue toxicity without abrogating therapeutic efficacy. Animal studies with BPD indicate that the increased tissue penetration does not result in increased toxicity.[100]

Encouraging results have been obtained with the application of lasers in LIF and PDT of brain tumors. Other light sources are being developed but are limited to broad-spectrum light output and low-energy light.[101] The development of cheaper diode lasers and photosensitizers that absorb light at longer wavelengths for fluorescence-guided resection and adjuvant PDT after maximal resection could potentially improve the results of prior studies.

## 42.16 Lasers in Stereotactic and Functional Neurosurgery

As mentioned above, Kelly and colleagues integrated a $CO_2$ laser into a computer-assisted volumetric stereotactic system for the resection of intraaxial neoplasms.[37–40,102,103] In 1986, they described this system[38] and its results in 41 patients. It involved initial reconstruction of tumor volume from computed tomography (CT) and magnetic resonance imaging (MRI) imaging data followed by a computer-monitored, stereotactically directed $CO_2$ laser to vaporize the intracranial tumor. A computer terminal was used to monitor the position of the laser in relation to reformatted planar slices through the tumor. In a separate paper, Kelly et al.[39] reported on 83 computer-assisted stereotactic laser procedures for tumor in 78 patients. No direct comparison to standard freehand microsurgery was performed, but the authors comment that the stereotactic laser technique was particularly useful in aggressive resection of deep-seated tumors from eloquent brain areas with acceptable postoperative results.[104] Kelly et al. also reported excellent results in excision of 12 colloid cysts of the third ventricle using this stereotactic microsurgical laser technique.[105]

Neuroablative properties of lasers have also been used in functional neurosurgery. DREZ lesions and commissural myelotomies have been made by laser.[106–109] DREZ lesions, originally described by Nashold,[110] are used primarily for intractable pain following spinal-cord injury or brachial plexus avulsion, postherpetic neuralgia, and phantom limb pain. While the DREZ lesion is most often accomplished with radiofrequency ablation, lasers were used in several cases. For example, Powers et al.[109] used argon and $CO_2$ lasers to produce DREZ lesions in 21 patients with denervation pain syndromes.

Lasers have been used in adjunctive management of epileptic syndromes. Kelly et al.[111] treated 18 patients with medically intractable epilepsy with stereotactic amygdalohippocampectomy using a $CO_2$ laser. In addition, the same group reported on stereotactic laser treatment of epileptogenic lesions associated with tuberous sclerosis.[112] While effective, it is not clear that the laser technique has any advantage over conventional microsurgery in the treatment of epileptic syndromes.

## 42.17 Lasers in Neuroendoscopy

Endoscopic techniques in neurosurgery most commonly involve procedures to treat obstructive hydrocephalus. This includes endoscopic third ventriculocisternostomy and procedures to provide communication between loculated ventricular cavities. Lasers have been used to assist with these procedures. Powers[113] used the argon laser to perform endoscopic fenestration in two infants with hydrocephalus and compartmentalization of the lateral ventricles due to ventriculitis. He later reported using the argon laser and a flexible steerable endoscope to accomplish endoscopic intraventricular cyst fenestration in seven patients.[114] Bucholz and Pittman[115] used an Nd:YAG laser to endoscopically coagulate the choroid plexus in an infant with hydrocephalus. Vandertop et al.[116] performed endoscopic laser third ventriculocisternostomy with an Nd:YAG laser and a new-generation diode laser in 33 patients without morbidity. Recently, Buki et al.[117] used a newly developed combined pulsed holmium-YAG (Ho:YAG) and Nd:YAG laser to perform a variety of endoscopic procedures successfully.

Other investigators have coupled lasers with endoscopic systems to treat other lesions as well. Zamorano et al.[118,119] integrated image-guided stereotaxis with rigid–flexible endoscopy and the Nd:YAG laser (endoscopic laser stereotaxis) and found this system useful in cystic and intraventricular lesions. Similarly, Otsuki et al.[120,121] used a stereotactic guiding tube and endoscopes to deliver laser irradiation to deep-seated brain tumors.

## 42.18 Summary

Initial experiments with ruby, $CO_2$, Nd:YAG, and argon lasers led to many preclinical and a few clinical studies of laser application in various neurosurgical procedures.[122,123] The major advantages described include improved precision, reduction of surgically related mechanical trauma, reduction of blood loss, and decreased operative time.[10,124] Disadvantages include collateral thermal damage, potential danger to operating-room personnel, cost, and complexity. While effective cutting, vaporization, and hemostasis have been demonstrated, there are few instances where lasers are routinely used in modern clinical neurosurgery. In isolated instances, lasers are a useful adjunct therapy in tumor, vascular, and functional procedures. Lasers may be particularly useful in the treatment of lesions situated in critical areas (e.g., brainstem), or lesions that are difficult to reach surgically (such as intraventricular meningiomas or extraaxial skull-base tumors). However, the efficacy of bipolar cautery, ultrasonic aspiration, radiofrequency lesion making, and radiosurgery have left little room for application of lasers in current neurosurgical practice. Future applications of lasers in neurosurgery may involve evaluation of more precise thermal and nonthermal interactions with brain tissue. New high-power diode lasers may offer superior tissue-cutting and hemostatic properties as well as reduced cost.[43] Combined laser and ultrasound aspiration units are in development that could combine tissue disruption by ultrasound with coagulation by laser.[125] Furthermore, a new technique of "cold" laser ablation is being studied that involves ultrashort laser pulses to cause plasma-induced ablation of tumor tissue with minimal thermal spread.[123,126]

### References

1. Goldman, I. and Rockwell, R.J., Laser action at cellular level, *J. Am. Med. Assoc.*, 198, 173, 1966.
2. Maiman, T.H., Stimulated optic radiation in ruby, *Nature*, 187, 493, 1960.
3. Rosomoff, H.L. and Carroll, F., Reaction of neoplasm and brain to laser, *Arch. Neurol.*, 14, 143, 1966.
4. Stellar, S. and Polanyi, T.G., Lasers in neurosurgery: a historical overview, *J. Clin. Laser Med. Surg.*, 10, 399, 1992.
5. Stellar, S., Polanyi, T.G., and Bredemeier, H.C., Experimental studies with the carbon dioxide laser as a neurosurgical instrument, *Med. Biol. Eng.*, 8, 549, 1970.
6. Stellar, S., Polayni, R.A., and Bredemeier, H.C., Lasers in surgery, in *Laser Application in Biology and Medicine*, Wolbarsht, M.L., Ed., Plenum, New York, 1970, pp. 241–293.
7. Heppner, F., The laser scalpel on the nervous system, in *Laser Surgery II*, Kaplan, I., Ed., Jerusalem Academic Press, Jerusalem, 1978, pp. 79–80.
8. Ascher, P.W., Neurosurgery, in *Microscopic and Endoscopic Surgery with the $CO_2$ Laser*, Kaplan, I., Ed., John Wright-PSG, Boston, 1982, pp. 298–314.
9. Ascher, P.W. and Heppner, F., $CO_2$ laser in neurosurgery, *Neurosurg. Rev.*, 7, 123, 1984.
10. Beck, O.J., The use of the Nd-YAG and the $CO_2$ laser in neurosurgery, *Neurosurg. Rev.*, 3, 261, 1980.
11. Takizawa, T., Yamazaki, T., Miura, N., Matsumoto, M., Tanaka, Y., Takeuchi, K., Nakata, Y., Togashi, O., Nagai, M., Ariga, T., Nishimura, T., Mizutani, H., and Sano, K., Laser surgery of basal, orbital and ventricular meningiomas which are difficult to extirpate by conventional methods, *Neurol. Med. Chir.*, 20, 729, 1980.

12. Robertson, J.H. and Clark, W.C., Carbon dioxide laser in neurosurgery, *Contemp. Neurosurg.*, 5, 1, 1984.
13. Jain, K.K., Lasers in neurosurgery: a review, *Lasers Surg. Med.*, 2, 217, 1983.
14. Yahr, W.Z., Strully, K.J., and Hurwitt, E.S., Non-occlusive small vessel arterial anastomosis with a neodymium laser, *Surg. Forum*, 15, 224, 1964.
15. Yahr, W.Z. and Strully, K.J., Blood vessel anastomosis by laser and other biomedical applications, *J. Assoc. Adv. Med. Instrum.*, 1, 28, 1966.
16. Jain, K.K. and Gorisch, W., Repair of small blood vessels with the neodymium-YAG laser: a preliminary report, *Surgery*, 85, 684, 1979.
17. Beck, O.J., Wilske, J., Schonberger, J.L., and Gorisch, W., Tissue changes following application of laser to the rabbit brain, *Neurosurg. Rev.*, 1, 31, 1979.
18. Takeuchi, J., Handa, H., Taki, W., and Yamagami, T., The Nd:YAG laser in neurological surgery, *Surg. Neurol.*, 18, 140, 1982.
19. Beck, O.J., Use of the Nd-YAG laser in neurosurgery, *Neurosurg. Rev.*, 7, 151, 1984.
20. Wharen, R.E., Jr., Anderson, R.E., Scheithauer, B., and Sundt, T.M., Jr., The Nd:YAG laser in neurosurgery. Part 1. Laboratory investigations: dose-related biological response of neural tissue, *J. Neurosurg.*, 60, 531, 1984.
21. Wharen, R.E., Jr., Anderson, R.E., and Sundt, T.M., Jr., The Nd:YAG laser in neurosurgery. Part 2. Clinical studies: an adjunctive measure for hemostasis in resection of arteriovenous malformations, *J. Neurosurg.*, 60, 540, 1984.
22. Jain, K.K., Complications of use of the neodymium:yttrium-aluminum-garnet laser in neurosurgery, *Neurosurgery*, 16, 759, 1985.
23. Hobeika, C.P., and Rockwell, R.J., Jr., Laser microsurgery in experimental otolaryngology, *Trans. Am. Acad. Ophthalmol. Otolaryngol.*, 76, 325, 1972.
24. Glasscock, M.E., Jackson, C.G., and Whitaker, S.R., The argon laser in acoustic tumor surgery, *Laryngoscope*, 41, 1405, 1981.
25. Boggan, J.E., Edwards, M.S., Davis, R.L., Bolger, C.A., and Martin, N., Comparison of the brain tissue response in rats to injury by argon and carbon dioxide lasers, *Neurosurgery*, 11, 609, 1982.
26. Powers, S.K., Edwards, M.S., Boggan, J.E., Pitts, L.H., Gutin, P.H., Hosobuchi, Y., Adams, J.E., and Wilson, C.B., Use of the argon surgical laser in neurosurgery, *J. Neurosurg.*, 60, 523, 1984.
27. Edwards, M.S., Boggan, J.E., Bolger, C.A., and Davis, R.L., Effect of microsurgical and carbon dioxide and argon laser resection on recurrence of the intracerebral 9L rat gliosarcoma, *Neurosurgery*, 14, 52, 1984.
28. Gamache, F.W., Jr. and Morgello, S., The histopathological effects of the $CO_2$ versus the KTP laser on the brain and spinal cord: a canine model, *Neurosurgery*, 32, 100, 1993.
29. Gamache, F.W., Jr. and Patterson, R.H., Jr., The use of the potassium titanyl phosphate (KTP) laser in neurosurgery, *Neurosurgery*, 26, 1010, 1990.
30. Roux, F.X., Leriche, B., Cioloca, C., Devaux, B., Turak, B., and Nohra, G., Combined $CO_2$ and Nd-YAG laser in neurosurgical practice: a 1st experience apropos of 40 intracranial procedures, *Neurochirurgie*, 38, 235, 1992.
31. Lombard, G.F., Luparello, V., and Peretta, P., Statistical comparison of surgical results with or without laser in neurosurgery, *Neurochirurgie*, 38, 226, 1992.
32. Desgeorges, M., Sterkers, O., Ducolombier, A., Pernot, P., Hor, F., Rosseau, G., Yedeas, M., Elabbadi, N., and Le Bars, M., Laser microsurgery of meningioma: an analysis of a consecutive series of 164 cases treated surgically by using different lasers, *Neurochirurgie*, 38, 217, 1992.
33. Waidhauser, E., Beck, O.J., and Oeckler, R.C., Nd:YAG-laser in the microsurgery of frontobasal meningiomas, *Lasers Surg. Med.*, 10, 544, 1990.
34. Cerullo, L.J. and Burke, L.P., Use of the laser in neurosurgery, *Surg. Clin. North Am.*, 64, 995, 1984.
35. Eiras, J., Alberdi, J., and Gomez, J., $CO_2$ laser in the surgery of acoustic neuroma, *Neurochirurgie*, 39, 16, 1993.

36. Oekler, R.T.C., Beck, H.C., and Frank, F., Surgery of the sellar region with the Nd:YAG laser, *Fortschr. Med.*, 9, 218, 1984.
37. Kelly, P.J., Alker, G.J., Jr., and Goerss, S., Computer-assisted stereotactic microsurgery for the treatment of intracranial neoplasms, *Neurosurgery*, 10, 324, 1982.
38. Kelly, P.J., Kall, B.A., Goerss, S., and Cascino, T.L., Results of computer-assisted stereotactic laser resection of deep-seated intracranial lesions, *Mayo Clin. Proc.*, 61, 20, 1986.
39. Kelly, P.J., Kall, B.A., Goerss, S., and Earnest, F., Computer-assisted stereotaxic laser resection of intra-axial brain neoplasms, *J. Neurosurg.*, 64, 427, 1986.
40. Camacho, A. and Kelly, P.J., Volumetric stereotactic resection of superficial and deep seated intraaxial brain lesions, *Acta Neurochir. Suppl.*, 54, 83, 1992.
41. Edwards, M.S. and Boggan, J.E., Argon laser surgery of pediatric neural neoplasms, *Child's Brain* 11, 171, 1984.
42. Tobler, W.D., Sawaya, R., and Tew, J.M., Successful laser-assisted excision of a metastatic midbrain tumor. *Neurosurgery*, 18, 795, 1986.
43. Devaux, B.C., Roux, F.X., Nataf, F., Turak, B., and Cioloca, C., High-power diode laser in neurosurgery: clinical experience in 30 cases, *Surg. Neurol.*, 50, 33, 1998.
44. Edwards, M.S., Boggan, J.E., and Fuller, T.A., The laser in neurological surgery, *J. Neurosurg.*, 59, 555, 1983.
45. McLone, D.G. and Naidich, T.P., Laser resection of fifty spinal lipomas, *Neurosurgery*, 18, 611, 1986.
46. Maira, G., Fernandez, E., Pallini, R., and Puca, A., Total excision of spinal lipomas using $CO_2$ laser at low power: experimental and clinical observations, *Neurol. Res.*, 8, 225, 1986.
47. Caspar, W., Campbell, B., Barbier, D.D., Kretschmmer, R., and Gotfried, Y., The Caspar microsurgical discectomy and comparison with a conventional standard lumbar disc procedure, *Neurosurgery*, 28, 78, 1991.
48. Mayer, H.M. and Brock, M., Percutaneous endoscopic discectomy: surgical technique and preliminary results compared to microsurgical discectomy, *J. Neurosurg.*, 78, 216, 1993.
49. Dickman, C.A., Detweiler, P.W., and Porter, R.W., Endoscopic spine surgery, *Clin. Neurosurg.*, 46, 526, 2000.
50. Mayer, H.M., Brock, M., Berlien, H.P., and Weber, B., Percutaneous endoscopic laser discectomy (PELD): a new surgical technique for non-sequestrated lumbar discs, *Acta Neurochir. Suppl.*, 54, 53, 1992.
51. Boult, M., Fraser, R.D., Jones, N., Osti, O., Dohrmann, P., Donnelly, P., Liddell, J., and Maddern, G.J., Percutaneous endoscopic laser discectomy, *Aust. N.Z. J. Surg.*, 70, 475, 2000.
52. Choy, D.S., Case, R.B., Fielding, W., Hughes, J., Liebler, W., Ascher, P., Percutaneous laser nucleolysis of lumbar disks, *New Engl. J. Med.*, 317, 771, 1987.
53. Choy, D.S., Ascher, P.W., Ranu, H.S., Saddekni, S., Alkaitis, D., Liebler, W., Hughes, J., Diwan, S., and Altman, P., Percutaneous laser disc decompression: a new therapeutic modality, *Spine*, 17, 949, 1992.
54. Fasano, V.A., The treatment of vascular malformation of the brain with laser source, *Lasers Surg. Med.*, 1, 347, 1981.
55. Fasano, V.A., Urciuoli, R., and Ponzio, R.M., Photocoagulation of cerebral arteriovenous malformations and arterial aneurysms with the neodymium:yttrium-aluminum-garnet or argon laser: preliminary results in twelve patients, *Neurosurgery*, 11, 754, 1982.
56. Zuccarello, M., Mandybur, T.I., Tew, J.M., Jr., and Tobler, W.D., Acute effect of the Nd:YAG laser on the cerebral arteriovenous malformation: a histological study, *Neurosurgery*, 24, 328, 1989.
57. Strugar, J. and Chyatte, D., *In situ* photocoagulation of spinal dural arteriovenous malformations using the Nd:YAG laser, *J. Neurosurg.*, 77, 571, 1992.
58. Salcman, M. and Samaras, G.M., Hyperthermia for brain tumors: biophysical rationale, *Neurosurgery*, 9, 327, 1981.

59. Sneed, P.K., Stauffer, P.R., McDermott, M.W., Diederich, C.J., Lamborn, K.R., Prados, M.D., Chang, S., Weaver, K.A., Spry, L., Malec, M.K., Lamb, S.A., Voss, B., Davis, R.L., Wara, W.M., Larson, D.A., Phillips, T.L., and Gutin, P.H., Survival benefit of hyperthermia in a prospective randomized trial of brachytherapy boost +/– hyperthermia for glioblastoma multiforme, *Int. J. Radiat. Oncol. Biol. Phys.*, 40, 287, 1998.
60. Sneed, P.K., Larson, D.A., and Gutin, P.H., Brachytherapy and hyperthermia for malignant astrocytomas, *Semin. Oncol.*, 21, 186, 1994.
61. Sugiyama, K., Sakai, T., Fujishima, I., Ryu, H., Uemura, K., and Yokoyama, T., Stereotactic interstitial laser-hyperthermia using Nd-YAG laser, *Stereotactic Funct. Neurosurg.*, 54–55, 501, 1990.
62. Terzis, A.J., Nowak, G., Mueller, E., Rentzsch, O., and Arnold, H., Induced hyperthermia in brain tissue *in vivo, Acta Neurochir. Suppl.*, 60, 406, 1994.
63. Nowak, G., Rentzsch, O., Terzis, A.J., and Arnold, H., Induced hyperthermia in brain tissue: comparison between contact Nd:YAG laser system and automatically controlled high frequency current, *Acta Neurochir.*, 102, 76, 1990.
64. Menovsky, T., Beek, J.F., Roux, F.X., and Bown, S.G., Interstitial laser thermotherapy: developments in the treatment of small deep-seated brain tumors, *Surg. Neurol.*, 46, 568, 1996.
65. Menovsky, T., Beek, J.F., van Gemert, M.J., Roux, F.X., and Bown, S.G., Interstitial laser thermotherapy in neurosurgery: a review, *Acta Neurochir.*, 138, 1019, 1996.
66. Bettag, M., Ulrich, F., Schober, R., Furst, G., Langen, K.J., Sabel, M., and Kiwit, J.C., Stereotactic laser therapy in cerebral gliomas, *Acta Neurochir. Suppl.*, 52, 81, 1991.
67. Frazier, O.H., Painvin, G.A., Morris, J.R., Thomsen, S., and Neblett, C.R., Laser-assisted microvascular anastomoses: angiographic and anatomopathologic studies on growing microvascular anastomoses: preliminary report, *Surgery*, 97, 585, 1985.
68. White, R.A., Kopchok, G.E., Donayre, C.E., Peng, S.K., Fujitani, R.M., White, G.H., and Uitto, J., Mechanism of tissue fusion in argon laser-welded vein-artery anastomoses, *Lasers Surg. Med.*, 8, 83, 1988.
69. Quigley, M.R., Bailes, J.E., Kwaan, H.C., Cerullo, L.J., and Block, S., Comparison of myointimal hyperplasia in laser-assisted and suture anastomosed arteries: a preliminary report, *J. Vasc. Surg.*, 4, 217, 1986.
70. Quigley, M.R., Bailes, J.E., Kwaan, H.C., Cerullo, L.J., and Brown, J.T., Aneurysm formation after low power carbon dioxide laser-assisted vascular anastomosis, *Neurosurgery*, 18, 292, 1986.
71. Shapiro, S., Sartorius, C., Sanders, S., and Clark, S., Microvascular end-to-side arterial anastomosis using the Nd:YAG laser, *Neurosurgery*, 25, 584, 1989.
72. Tulleken, C.A., van der Zwan, A., van Rooij, W.J., and Ramos, L.M., High-flow bypass using nonocclusive excimer laser-assisted end-to-side anastomosis of the external carotid artery to the P1 segment of the posterior cerebral artery via the sylvian route: technical note, *J. Neurosurg.*, 88, 925, 1998.
73. Tulleken, C.A., Verdaasdonk, R.M., Berendsen, W., and Mali, W.P., Use of the excimer laser in high-flow bypass surgery of the brain, *J. Neurosurg.*, 78, 477, 1993.
74. Tulleken, C.A., van Dieren, A., Verdaasdonk, R.M., and Berendsen, W., End-to-side anastomosis of small vessels using an Nd:YAG laser with a hemispherical contact probe: technical note, *J. Neurosurg.*, 76, 546, 1992.
75. Teramura, A., Macfarlane, R., Owen, C.J., de la Torre, R., Gregory, K.W., Birngruber, R., Parrish, J.A., Peterson, J.W., and Zervas, N.T., Application of the 1-microsecond pulsed-dye laser to the treatment of experimental cerebral vasospasm, *J. Neurosurg.*, 75, 271, 1991.
76. Macfarlane, R., Teramura, A., Owen, C.J., Chase, S., de la Torre, R., Gregory, K.W., Peterson, J.W., Birngruber, R., Parrish, J.A., and Zervas, N.T., Treatment of vasospasm with a 480-nm pulsed-dye laser, *J. Neurosurg.*, 75, 613, 1991.
77. Kaoutzanis, M.C., Peterson, J.W., Anderson, R.R., McAuliffe, D.J., Sibilia, R.F., and Zervas, N.T., Basic mechanism of *in vitro* pulsed-dye laser-induced vasodilation, *J. Neurosurg.*, 82, 256, 1995.

78. Menovsky, T., Beek, J.F., and Thomsen, S.L., Laser-assisted nerve repair: a review, *Neurosurg. Rev.*, 18, 225, 1995.
79. Almquist, E.E., Nachemson, A., Auth, D., Almquist, B., and Hall, S., Evaluation of the use of the argon laser in repairing rat and primate nerves, *J. Hand Surg. Am.*, 9, 792, 1984.
80. Fischer, D.W., Beggs, J.W., Kenshalo, D.J., and Shetter, A.G., Comparative study of microepineurial anastomoses with the use of $CO_2$ laser and suture techniques in rat sciatic nerves. Part 1. Surgical technique, nerve action potentials, and morphological studies, *Neurosurgery*, 17, 300, 1985.
81. Maragh, H., Hawn, R.S., Gould, J.D., and Terzis, J.K., Is laser nerve repair comparable to microsuture coaptation? *J. Reconstr. Microsurg.*, 4, 189, 1988.
82. Bailes, J.E., Cozzens, J.W., Hudson, A.R., Kline, D.G., Ciric, I., Gianaris, P., Bernstein, L.P., and Hunter, D., Laser-assisted nerve repair in primates, *J. Neurosurg.*, 71, 266, 1989.
83. Seifert, V. and Stolke, D., Laser-assisted reconstruction of the oculomotor nerve: experimental study on the feasibility of cranial nerve repair, *Neurosurgery*, 25, 579, 1989.
84. Campion, E.R., Bynum, D.K., and Powers, S.K., Repair of peripheral nerves with the argon laser: a functional and histological evaluation, *J. Bone Joint Surg. Am.*, 72, 715, 1990.
85. Huang, T.C., Blanks, R.H., Berns, M.W., and Crumley, R.L., Laser vs. suture nerve anastomosis. *Otolaryngol. Head Neck Surg.*, 107, 14, 1992.
86. Korff, M., Bent, S.W., Havig, M.T., Schwaber, M.K., Ossoff, R.H., and Zealear, D.L., An investigation of the potential for laser nerve welding, *Otolaryngol. Head Neck Surg.*, 106, 345, 1992.
87. Menovsky, T., $CO_2$ and Nd:YAG laser-assisted nerve repair: a study of bonding strength and thermal damage, *Acta Chir. Plast.*, 42, 16, 2000.
88. Menovsky, T. and Beek, J.F., Laser, fibrin glue, or suture repair of peripheral nerves: a comparative functional, histological, and morphometric study in the rat sciatic nerve, *J. Neurosurg.*, 95, 694, 2001.
89. Terris, D.J. and Fee, W.E., Jr., Current issues in nerve repair, *Arch. Otolaryngol. Head Neck Surg.*, 119, 725, 1993.
90. Lin, W.C., Toms, S.A., Johnson, M., Jansen, E.D., and Mahadevan-Jansen, A., *In vivo* brain tumor demarcation using optical spectroscopy, *Photochem. Photobiol.*, 73, 396, 2001.
91. Eggert, H.R. and Blazek, V., Optical properties of normal human intracranial tissues in the spectral range of 400 to 2500 nm, *Adv. Exp. Med. Biol.*, 333, 47, 1993.
92. Poon, W.S., Schomacker, K.T., Deutsch, T.F., and Martuza, R.L., Laser-induced fluorescence: experimental intraoperative delineation of tumor resection margins, *J. Neurosurg.*, 76, 679, 1992.
93. Stummer, W., Novotny, A., Med, C.N., Stepp, H., Goetz, C., Bise, K., and Reulen, H.J., Fluorescence-guided resection of glioblastoma multiforme by using 5-aminolevulinic acid-induced porphyrins: a prospective study in 52 consecutive patients, *J. Neurosurg.*, 93, 1003, 2000.
94. Stummer, W., Stocker, S., Wagner, S., Stepp, H., Fritsch, C., Goetz, C., Goetz, A., Kiefmann, R., and Reulen, H.J., Intraoperative detection of malignant gliomas by 5-aminolevulinic acid-induced porphyrin fluorescence, *Neurosurgery*, 42, 518, 1998.
95. Whelan, H.T., Schmidt, M.H., Segura, A.D., McAuliffe, T.L., Bajic, D.M., Murray, K.J., Moulder, J.E., Strother, D.R., Thomas, J.P., and Meyer, G.A., The role of photodynamic therapy in posterior fossa brain tumors: a preclinical study in a canine glioma model, *J. Neurosurg.*, 79, 562, 1993.
96. Muller, P.J. and Wilson, B.C., Photodynamic therapy for recurrent supratentorial gliomas, *Semin. Surg. Oncol.*, 11, 346, 1995.
97. Popovic, E.A., Kaye, A.H., and Hill, J.S., Photodynamic therapy of brain tumors, *Semin. Surg. Oncol.*, 11, 335, 1995.
98. Powers, S.K., Cush, S.S., Walstad, D.L., and Kwock, L., Stereotactic intratumoral photodynamic therapy for recurrent malignant brain tumors, *Neurosurgery*, 29, 688, 1991.
99. Muller, P.J. and Wilson, B.C., An update on the penetration depth of 630 nm light in normal and malignant human brain tissue *in vivo*, *Phys. Med. Biol.*, 31, 1295, 1986.

100. Schmidt, M.H., Reichert, K.W., Ozker, K., Meyer, G.A., Donohoe, D.L., Bajic, D.M., Whelan, N.T., and Whelan, H.T., Preclinical evaluation of benzoporphyrin derivative combined with a light-emitting diode array for photodynamic therapy of brain tumors, *Pediatr. Neurosurg.*, 30, 225, 1999.
101. Schmidt, M.H., Bajic, D.M., Reichert, K.W., Martin, T.S., Meyer, G.A., and Whelan, H.T., Light-emitting diodes as a light source for intraoperative photodynamic therapy, *Neurosurgery*, 38, 552, 1996.
102. Alker, G., Kelly, P.J., Kall, B., and Goerss, S., Stereotaxic laser ablation of intracranial lesions, *Am. J. Neuroradiol.*, 4, 727, 1983.
103. Kall, B.A., Kelly, P.J., and Goerss, S.J., Interactive stereotactic surgical system for the removal of intracranial tumors utilizing the $CO_2$ laser and CT-derived database, *IEEE Trans. Biomed. Eng.*, 32, 112, 1985.
104. Kelly, P.J., Computer assisted volumetric stereotactic resection of superficial and deep seated intra-axial brain mass lesions, *Acta Neurochir. Suppl.*, 52, 26, 1991.
105. Abernathey, C.D., Davis, D.H., and Kelly, P.J., Treatment of colloid cysts of the third ventricle by stereotaxic microsurgical laser craniotomy, *J. Neurosurg.*, 70, 525, 1989.
106. Levy, W.J., Nutkiewicz, A., Ditmore, Q.M., and Watts, C., Laser-induced dorsal root entry zone lesions for pain control: report of three cases, *J. Neurosurg.*, 59, 884, 1983.
107. Fink, R.A., Neurosurgical treatment of nonmalignant intractable rectal pain: microsurgical commissural myelotomy with the carbon dioxide laser, *Neurosurgery*, 14, 64, 1984.
108. Powers, S.K., Laser-induced dorsal root entry zone lesions for pain control, *J. Neurosurg.*, 60, 871, 1984.
109. Powers, S.K., Adams, J.E., Edwards, M.S., Boggan, J.E., and Hosobuchi, Y., Pain relief from dorsal root entry zone lesions made with argon and carbon dioxide microsurgical lasers, *J. Neurosurg.*, 61, 841, 1984.
110. Nashold, B.S., Jr. and Ostdahl, R.H., Dorsal root entry zone lesions for pain relief, *J. Neurosurg.*, 51, 59, 1979.
111. Kelly, P.J., Sharbrough, F.W., Kall, B.A., and Goerss, S.J., Magnetic resonance imaging-based computer-assisted stereotactic resection of the hippocampus and amygdala in patients with temporal lobe epilepsy, *Mayo Clin. Proc.*, 62, 103, 1987.
112. Bebin, E.M., Kelly, P.J., and Gomez, M.R., Surgical treatment for epilepsy in cerebral tuberous sclerosis, *Epilepsia*, 34, 651, 1993.
113. Powers, S.K., Fenestration of intraventricular cysts using a flexible, steerable endoscope and the argon laser, *Neurosurgery*, 18, 637, 1986.
114. Powers, S.K., Fenestration of intraventricular cysts using a flexible, steerable endoscope, *Acta Neurochir. Suppl.*, 54, 42, 1992.
115. Bucholz, R.D. and Pittman, T., Endoscopic coagulation of the choroid plexus using the Nd:YAG laser: initial experience and proposal for management, *Neurosurgery*, 28, 421, 1991.
116. Vandertop, W.P., Verdaasdonk, R.M., and van Swol, C.F., Laser-assisted neuroendoscopy using a neodymium-yttrium aluminum garnet or diode contact laser with pretreated fiber tips, *J. Neurosurg.*, 88, 82, 1998.
117. Buki, A., Doczi, T., Veto, F., Horvath, Z., and Gallyas, F., Initial clinical experience with a combined pulsed holmium-neodymium-YAG laser in minimally invasive neurosurgery, *Minim. Invas. Neurosurg.*, 42, 35, 1999.
118. Zamorano, L., Dujovny, M., Chavantes, C., Malik, G., and Ausman, J., Image-guided stereotactic centered craniotomy and laser resection of solid intracranial lesions, *Stereotactic Funct. Neurosurg.*, 54–55, 398, 1990.
119. Zamorano, L., Chavantes, C., Dujovny, M., Malik, G., and Ausman, J., Stereotactic endoscopic interventions in cystic and intraventricular brain lesions, *Acta Neurochir. Suppl.*, 54, 69, 1992.
120. Otsuki, T., Jokura, H., and Yoshimoto, T., Stereotactic guiding tube for open-system endoscopy: a new approach for the stereotactic endoscopic resection of intra-axial brain tumors, *Neurosurgery*, 27, 326, 1990.

121. Otsuki, T., Yoshimoto, T., Jokura, H., and Katakura, R., Stereotactic laser surgery for deep-seated brain tumors by open-system endoscopy, *Stereotactic Funct. Neurosurg.*, 54–55, 404, 1990.
122. Krishnamurthy, S. and Powers, S.K., Lasers in neurosurgery, *Lasers Surg. Med.*, 15, 126, 1994.
123. Goetz, M.H., Fischer, S.K., Velten, A., Bille, J.F., and Sturm, V., Computer-guided laser probe for ablation of brain tumours with ultrashort laser pulses, *Phys. Med. Biol.*, 44, N119, 1999.
124. Saunders, M.L., Young, H.F., Becker, D.P., Greenberg, R.P., Newlon, P.G., Corales, R.L., Ham, W.T., and Povlishock, J.T., The use of the laser in neurological surgery, *Surg. Neurol.*, 14, 1, 1980.
125. Desinger, K., Liebold, K., Helfmann, J., Stein, T., and Muller, G., A new system for a combined laser and ultrasound application in neurosurgery, *Neurol. Res.*, 21, 84, 1999.
126. Suhm, N., Gotz, M.H., Fischer, J.P., Loesel, F., Schlegel, W., Sturm, V., Bille, J., and Schroder, R., Ablation of neural tissue by short-pulsed lasers — a technical report, *Acta Neurochir.*, 138, 346, 1996.

# 43

# Lasers in Ophthalmology

Ezra Maguen
*Ophthalmology Research Laboratories, Cedars-Sinai Medical Center*
*The Jules Stein Eye Institute, UCLA School of Medicine*
*Los Angeles, California*

Thomas G. Chu
*The Retina Vitreous Associates Medical Group*
*Doheny Eye Institute, Keck-USC School of Medicine*
*Los Angeles, California*

David Boyer
*Ophthalmology Research Laboratories, Cedars-Sinai Medical Center*
*The Retina Vitreous Associates Medical Group*
*Doheny Eye Institute, Keck-USC School of Medicine*
*Los Angeles, California*



## 43.1 Laser Surgery of the Anterior Segment of the Eye

### 43.1.1 Introduction

The eye is a transparent organ, and for the past 200 years its structural elements could be viewed by optical devices such as ophthalmoscopes and slit-lamp biomicroscopes. With the emergence of coherent-light sources, early efforts were made to use such sources in order to affect disease processes in the eye. Early on, other properties of laser were found to be helpful in the treatment of eye disease. Small spot sizes made it possible to avoid exposing unwanted structures. Coherence of light was found to be synonymous with a significant decrease of collateral damage to surrounding tissues. The ability to reflect visible laser light off mirrors made it possible to treat ocular structures that would otherwise be unreachable. The ability to drive coherent light through flexible optical fibers also made it possible to use lasers in the operating room, both with "open-sky" and endoscopic techniques.

At the same time that range lasers were developing, intraocular surgery techniques, which had been performed with magnifying eyeglasses, started being performed with operating microscopes. These microscopes, along with slit lamps and ophthalmoscopes, became the obvious delivery systems for coherent light into the eye. They were modified to allow coaxial placement of the light source along with the optical system and an aiming beam, mostly HeNe lasers of power small enough not to damage eye structures.

This chapter focuses on how different structures in the eye can be treated with different lasers; brief descriptions of the most common diseases involved are provided.

0-8493-1116-0/03/$0.00+$1.50

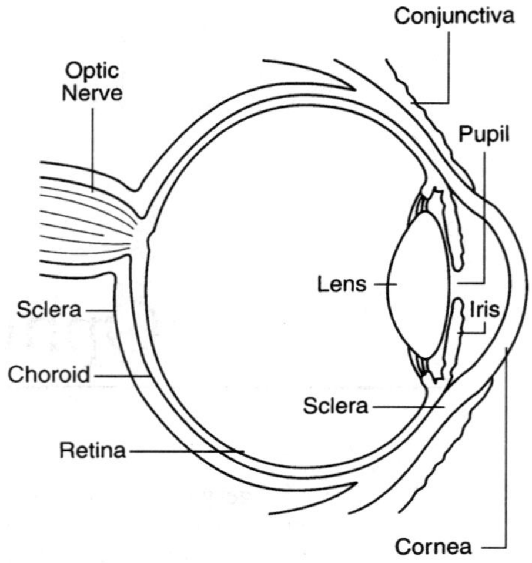


**FIGURE 43.1** Cross section of the eye.

### 43.1.2 Functional Anatomy of the Eye

Figure 43.1 shows a cross section of the eye where the cornea is anterior and the optic nerve is posterior. The eye is arbitrarily divided into an anterior segment and a posterior segment whereby the dividing line between the two is a vertical line tangent to the posterior end of the ciliary body.

The cornea is a transparent organ responsible for most of the refracting power of the eye. There are no blood vessels within the cornea. Transparency is achieved by a structure of collagen-fiber packs that are perfectly parallel to one another. The bundles appear to be at various angles to one another in a two-dimensional view. However, in a three-dimensional (3D) view, they are parallel to each other. The cornea is continued by the sclera, which is made of collagen arranged in intersecting fibers. It is therefore nontransparent and serves as a wall protecting the inner structures of the eye. It has several holes within it to allow passage of several nerves and blood vessels, the largest of which is the optic nerve.

Behind the cornea is a space called the anterior chamber (AC). It is filled with fluid leaked from vessels in the ciliary body. These fluids travel in the space between the lens and iris, access the AC through the pupil, and are drained out of the eye via a drainage system located at the anterior chamber angle (ACA). A cross-section view shows that the ACA is located at the junction between the backside of the cornea and the periphery of the iris. A decreased rate of fluid drainage out of the eye does not reduce leakage; this produces a buildup in intraocular pressure called glaucoma. If excessive pressure is allowed to remain, blindness can result. Decrease in drainage may be induced by a disturbance in the structures of the ACA, hence the name open-angle glaucoma. If an eye has an inherent narrow access to the ACA and pressure builds up, it is called narrow-angle glaucoma. With a narrow angle, due to the buildup in pressure, the iris may be pushed forward and obstruct fluid access to the ACA. This condition is called angle-closure glaucoma. This condition is urgent and requires prompt treatment; otherwise, the eye could lose vision. A similar acute condition known as pupillary block can be induced by blockage of the passage between the iris, lens, and pupil.

The iris consists of two muscles designed to vary the diameter of the central aperture, the pupil. The sphincter decreases pupil size, while the dilator increases it. Behind the iris is a system consisting of the ciliary body and the lens. Both are connected by a network of thin fibers called zonules. The base of the ciliary body is anchored solidly to the sclera. The lens is made of a "bag" or capsule containing proteins of different densities. With respect to the cornea, the 3D molecular arrangement of both capsule and contents makes the lens transparent. The contraction and relaxation of the ciliary body makes the lens stretch and relax. This movement induces a change of the curvature of the front surface of the lens, making it possible to vary the focal distance of the refracting system of the eye (cornea and lens). This capability decreases with age because of the decreased flexibility of the lens (presbyopia), hence the need for near correction with glasses. The lens can also become more opaque (cataract). When the opacity induces a significant decrease in vision, the lens is removed and replaced with a prosthetic lens. Behind the plane formed by the ciliary body and the back of the lens is a space filled by the vitreous. The posterior segment of the eye is comprised of clear vitreous gel and the back wall of the globe. The anatomic features of the back wall are the retina, choroids, and sclera. The vitreous cavity occupies four fifths of the volume of the globe. The vitreous gel is transparent, consisting of 99% water, mucopolysaccharide, and hyaluronic acid. It absorbs laser energy negligibly within the visible-light spectrum (Table 43.1).

The retina processes light into electrical impulses, which are transmitted by a network of nerves and fibers to the back of the brain (the occipital lobe), where these data are processed into images. The retina can be divided into the central retina, which includes the optic-nerve exit, and the macula. The macula has a higher density of retinal cells than the rest of the retina. As a result, it is responsible for better visual acuity with better resolution and color vision. The peripheral retina provides peripheral vision with less resolution and night vision. The retina is comprised of two layers: (1) an inner transparent neurosensory retina, consisting of neural (photoreceptors), glial, and vascular elements, and (2) an outer pigmented layer called retinal pigment epithelium (RPE), consisting of hexagonal-shaped cells important in the metabolic maintenance of the neurosensory retina. Of note, the cytoplasm of RPE cells contains multiple round and ovoid pigment granules or melanosomes. The retina has no physical attachment to the next layer, the choroid. It is maintained in place by capillarity. Therefore, if the retina breaks and fluid is allowed to invade the space between the retina and the choroid, the retina will detach (retinal detachment).

The choroid — the posterior portion of the uveal tract — nourishes the outer portion of the retina. It is a highly vascularized structure consisting of multiple layers of blood vessels and is separated from the RPE by Bruch's membrane, which is not a true membrane but rather a layer of PAS positive material. Bruch's membrane presents a barrier to large molecules and blood vessels but is permeable to small molecules.

The sclera is an essentially avascular structure consisting of dense connective tissue, fibroblasts, collagen, and ground substance.

## 43.1.3 Laser Surgery of the Cornea

The goal of laser surgery of the cornea is either to make incisions (linear,[1] curved, or circular) or, most commonly, alter the curvature of the front surface of the cornea. Table 43.1 details the relevant physical properties of the cornea. These numbers make clear that the anterior surface of the cornea has the greatest influence on the refractive power of the eye. This, combined with easy access, makes the anterior surface of the cornea the obvious target organ for surgery designed to alter the refraction of the eye.

### 43.1.3.1 Ultraviolet Lasers

Srinivasan[2] and Trokel supplied the basic ideas and tools to apply excimer lasers to refractive corneal surgery. Their experiments made it clear that the 193-nm ArF excimer laser provides a smooth ablation pattern, minimal thermal damage, and exquisite control of the depth of ablation. Based on the simplified Munnerlyn[3] formula, an ablation depth ($t_0$) of 12 μm per diopter ($D = 1$) is carried out for an ablation diameter, or optical zone ($Z$) of 6 mm that reflects the tight tolerances of such a system in order to be accurate.

**TABLE 43.1** Optical Properties of the Cornea and Lens Based on Gullstrand's "Schematic Eye"

| | |
|---|---|
| Diameter | 12 mm |
| Thickness | 0.55 mm |
| Refracting power, anterior surface | 48.83 $D$ |
| Refractive power, posterior surface | –5.88 $D$ |
| Total refracting power of eye | 58.64 $D$ |

$$t_0 = \frac{Z^2 \cdot D}{3} \tag{43.1}$$

The first clinical models of excimer lasers for refractive surgery were built by Meditec (Oberkochen, Germany),[1] VISX (Santa Clara, CA),[4] and Summit Technologies (Greenwood Village, CO).[5] They included a delivery system similar to an operating microscope and a system of quartz lenses to guide the beam from the laser chamber to the eye. The laser beam was delivered via a scanning mechanism (Meditec) and as a broad beam (VISX and Summit). Beam-shape control was achieved with the use of physical masks (Meditec) and a shutter mechanism designed to vary the diameter of the beam (VISX and Summit). When these were used, a lens-like profile could be obtained, thereby flattening the central cornea and correcting spherical myopia.

An additional challenge in refractive surgery with the excimer laser was the correction of astigmatism. This involved an ablation pattern that would transform a toroidal surface into a spherical surface. This problem was solved by combining a variable slit opening to the shutter mechanism (VISX) and by interposing a proprietary "erodible mask" designed to vary the ablation profile in a customized fashion (Summit).

Additional strategies needed to be devised to treat high refractive errors. In the case of high myopia, an ablation pattern varying the optical zone has been devised. It allows for the significant reduction of the ablation depth by decreasing the size of the zone as the ablation becomes more central. Indeed, an ablation pattern of 6 mm initially reduced to 5 mm in mid-depth and to 4 mm at the deepest could save significant amounts of corneal tissue. Barraquer[6] recognized the limit of ablation depth beyond which a significant number of complications would occur. The thickness of the residual cornea underlying the laser treatment area cannot be thinner than 250 μm.

Additional refinements in the delivery of the beam to the target organ included a scanning wide beam (Nidek EC5000) and a combination of the same with an Axicon-type prism (VISX)[7] designed to increase the ablation diameter to 9 mm to treat hyperopia (far sightedness). Indeed, the ablation profile for hyperopia involves steepening the original corneal profile. To accomplish this most of the ablation must be performed in the periphery of the ablation zone, thereby allowing a minimum of a 4- to 5-mm optical zone.

Further improvements in the delivery system included the technology of randomly applying a laser spot with a diameter of 1 mm. This method can theoretically create any possible ablation pattern over any size optical zone. The use of larger optical zones can in many cases reduce night glare if the border of the zone exceeds the diameter of the dilated pupil as it occurs in the dark. This delivery method carries with it the need for precise spot application and centration of the ablation pattern, hence the need for a tracking device of the eye as ablation proceeds.[8] Alcon Autonomous Technologies (Fort Worth, TX) was the first to develop such a system and have it approved by the U.S. Food and Drug Administration (FDA) for all refractive surgical applications. Other manufacturers of lasers with similar technology are Bausch & Lomb (Rochester, NY) (Technolas 217) and LaserSight (Green Bay, WI).

The most recent advance in laser vision correction involves the use of wavefront technology in an attempt to correct higher-order optical aberrations. Most laser manufacturers have included this capability into their systems. In the United States, Alcon Autonomous Technologies and VISX are proceeding with FDA-guided clinical trials. The procedure includes obtaining data pertaining to higher-order aberrations as produced by a Tchernig Aberrometer or a Hartman Schack device. These data are used to modify the ablation pattern accordingly. Once the prospective trials are complete, an assessment could be made as to whether this technology improves visual acuity or quality.

There are two procedures for performing laser vision correction: photorefractive keratectomy (PRK) and laser-assisted *in situ* keratomileusis (LASIK). A third procedure, laser-assisted subepithelial keratomileusis (LASEK), has recently become popular.

PRK consists of removal of the superficial layer of the cornea (epithelium), either manually or by laser ablation. The refractive portion of the laser ablation is performed on the bulk of the cornea (stroma). A contact lens is inserted, and under the contact lens the epithelium regrows over the ablated area.

LASIK consists of using a mechanical device called a microkeratome to perform a partial-thickness round cut of the cornea. The cut is intentionally incomplete so that a hinge of tissue is left securing the flap of tissue thus formed in place. The flap is then lifted and the ablation performed on the exposed corneal stroma. The flap is replaced, and the anterior surface flattens or steepens as the flap conforms to the new profile made by the laser ablation. LASEK requires carefully displacing the corneal epithelium from the stroma after loosening the attachment between the two layers with diluted alcohol. A flap of epithelium secured by a hinge is formed and reflected. Ablation then proceeds on the underlying stroma. The epithelium is then replaced and smoothened.

Both procedures have pros and cons. PRK is simpler, and the surgery is safer because it does not involve the use of a microkeratome. After surgery, significant pain is experienced for 24 to 48 h. The healing is slow and may take several weeks. Superficial hazing of the cornea is relatively common, especially with higher corrections. LASIK is relatively less safe as it involves the use of a microkeratome with the complications specific to that surgical procedure. This is outweighed by very little pain after surgery and a fast recovery of optimal vision. With LASEK, no microkeratome is used. There is less pain than with PRK, and visual recovery time is better than with PRK but slower than with LASIK.

#### 43.1.3.2 Infrared and Other Lasers

The use of ultraviolet lasers is complicated by the fact that they are gas lasers. Solid-state lasers could be simpler to build and maintain. Such systems could be miniaturized with relative ease. In addition, solid-state lasers within certain ranges of infrared could be used for other ophthalmic procedures on the iris, ACA, and lens.

The Novatec laser[8] is a proprietary solid-state laser with a wavelength of 0.2 μm, fluence of 100 mJ/cm$^2$, a variable spot size between 10 and 500 μm, and optical zone of up to 10 mm. The laser showed promise in initial clinical trials but ultimately was not produced. Other infrared solid-state lasers were built for PRK and LASIK, but they were not commercially produced.

Other strategies of using infrared lasers for vision correction included shortening the pulse width of the beam. Earlier models showed that in addition to performing corneal surgery on the surface of the end organ, the beam could be focused inside the cornea and tissue altered with relatively little collateral tissue damage. Picosecond[9] lasers were used with some success but never came to market. One of the most recent attempts at using this strategy is the emergence of a femtosecond laser (Intralase, Irvine, CA). The system can create a corneal flap without using a mechanical device and remove tissue within the deeper corneal stroma to create a change in curvature. Clinical trials are ongoing to assess the safety and efficacy of the system.

The Hyperion solid-state infrared laser produced by Sunrise Technologies (Winston-Salem, NC)[9] uses a different strategy and is FDA approved to correct hyperopia. It is a solid-state holmium:YAG (Ho:YAG) laser operating at a wavelength of 2.13 μm and pulse duration of 250 μsec. The spot size is 500 to 600 μm. The treatment strategy (laser thermokeratoplasty, or LTK) consists of applying 8 to 16 simultaneous applications on the cornea in a ring pattern at a given distance from the corneal center, thereby inducing central steepening and correcting hyperopia by inducing collagen shrinkage around the laser applications. It is simple to perform and carries relatively low intraoperative risk. At the same time, the amount of correction is not as controlled as with excimer-laser procedures. The range of correction is limited, and in many cases regression of the correction occurs over time.

## 43.1.4 Laser Surgery of the Anterior Chamber Angle

Laser procedures here are designed to treat open-angle glaucoma. They target a collagen meshwork called the trabecular meshwork. This is the initial part of the system that drains fluid out of the eye. The surgical technique consists of driving a laser beam through a slit-lamp delivery system into a reflecting mirror. From there the beam travels at an angle into the target organ. Of the 360° of the ACA, 180° are treated at one time with an average of 50 applications per treatment. The mechanism of action is unclear. The most plausible explanation is that collagen shrinks at the treatment area, stretching wider open the meshwork adjacent to the laser spots. The success rate of the procedure is reportedly up to 90% in the short term and decreases to 50% at 5 years after surgery.

The argon laser has been exclusively used for the past 20 years for this type of treatment, hence the name argon laser trabeculoplasty (ALT).[10] The treatment parameters range as follows: energy = 800 to 1000 mW, spot size = 50 μm, and exposure time = 0.1 sec. Recently, a new treatment modality of the ACA was introduced using a 532-nm, frequency-doubled, Q-switched neodymium:yttrium-aluminum-garnet (Nd:YAG) laser (Selecta 7000 model, Lumenis, Inc., Santa Clara, CA). The treatment strategy is similar to that of ALT, using 54 to 55 spots per treatment. Treatment parameters are as follows: spot size = 400 μm, pulse duration = 3 nsec, and energy = 0.6 to 1.2 mJ. The mechanism of action is purported to produce a biological effect by targeting pigmented cells within the trabecular meshwork, hence the name selective laser trabeculoplasty (SLT).[11] During FDA-guided trials, a 70% success rate was reported.

## 43.1.5 Laser Surgery of the Iris

Several laser surgical procedures are performed on the iris. There are three main goals with this type of surgery: (1) create a bypass between the space behind the iris and the AC to reduce eye pressure, (2) enlarge the approach to the ACA, and (3) enlarge pupil size. The argon laser described above and Nd:YAG lasers can be used. The Nd:YAG lasers (wavelength = 1064 nm) are mostly of the fundamental Q-switched type, paired with a HeNe laser for aiming. They induce photodisruption, which in turn releases a shockwave, thereby creating disruption of the iris fibers. The argon lasers induce mostly thermal damage of the tissue. Iris color is directly related to the efficacy of the two lasers. Argon lasers are more efficient in more pigmented (darker color) irides, whereas in lighter irides, the Nd:YAG laser is preferred. A collimating lens placed on the eye can reduce laser energy.

### 43.1.5.1 Laser Iridotomy

This procedure is designed to create a bypassing hole in the iris to allow fluid trapped behind it to flow to the AC. Fluid blockage occurs due to closure of the ACA or to pupillary block. Both argon[12] and Nd:YAG[13] lasers can be used — separately or in combination. Laser parameters for the argon laser are energy = 800 to 1000 mW, diameter = 50 μm, and exposure time = 0.1 sec. A total of 40 to 80 applications are used. For the Nd:YAG laser, energy = 3 to 7 mJ, and four to ten applications are used.

### 43.1.5.2 Laser Iridoplasty (Gonioplasty)

This procedure[14] is designed to enlarge the approach to the ACA by thinning the peripheral iris using the thermal damage to tissue induced by the argon laser. Thermal damage is followed by atrophy of the tissue and therefore is thinned. Several applications are made in a 360° circular pattern. One or more rows can be applied. Laser parameters are energy = 400 to 600 mW, diameter = 400 μm, and exposure time = 0.2 to 0.4 sec.

### 43.1.5.3 Laser Photomydriasis

This procedure is designed to increase the size of the pupil by using a technique similar to laser iridoplasty. Laser applications are guided as close as possible to the margin of the pupil. The same type of laser–tissue interaction eventually induces atrophy of the iris muscle, which closes the pupil (iris sphincter). A small pupil can be induced by drugs or by previous disease, causing adhesion of the iris to the lens. Its enlargement is sought mainly to better view the inside of the eye.

## 43.1.6 Laser Surgery of the Ciliary Body

The ciliary body is the structure least accessible to viewing and laser treatment because of its location, which is immediately behind the peripheral iris. The goal of laser treatment of the ciliary body is to decrease intraocular pressure. Indeed, vessels in this structure leak fluid into the eye, and a decrease in fluid outflow could regulate better eye pressure. There are three approaches to laser treatment of the ciliary body.

#### 43.1.6.1 Laser Cyclophotocoagulation

Direct treatment of the ciliary body with the argon laser is possible only if a sector of the iris was previously removed by surgery. Typical laser settings are mean energy = 400 mW, diameter = 100 μm, and exposure time = 0.1 to 0.2 sec.

#### 43.1.6.2 Endolaser Cyclophotocoagulation

In this procedure,[15] the ciliary body is visualized and treated via a fiber optic with the argon laser. This involves a major surgical procedure, as the surrounding vitreous must be removed. The treatment cannot be applied if the lens has not been removed.

#### 43.1.6.3 Transscleral Cyclophotocoagulation

The third approach to the ciliary body involves using a laser beam to irradiate the overlying sclera. Laser energy is transmitted via the sclera to the ciliary body. Laser–tissue interaction reflects mostly transformation of laser energy to thermal energy. Two methodologies can be applied: noncontact treatment can be provided with an Nd:YAG laser. A typical treatment consists of 32 applications of 8 J each to the 360° circumference of the ciliary body. The LASAG Microruptor® laser can be used for the treatment.[16] Contact transscleral treatment (the probe is applied to the sclera overlying the ciliary body) can be performed with a gallium-aluminum-arsenide diode laser emitting at 810 nm.[17]

### 43.1.7 Laser Surgery of the Lens

The lens may become opaque over time, leading to a condition called cataract. Modern cataract surgery involves opening the lens capsule (bag), liquefying and aspirating the protein contents, and inserting a prosthetic lens implant into the remaining bag. After surgery the back of the bag (posterior capsule) may get opaque, and an opening must be made to recover good vision. Lasers can be used for both liquefying and aspirating (phakoemulsification) and for perforating the posterior capsule (capsulotomy), if needed later.

#### 43.1.7.1 Laser Phakoemulsification

Erbium YAG (Er:YAG)[18] and Nd:YAG[19] laser systems are available for lens emulsification. The Nd:YAG system is manufactured by A.R.C. Laser AG (Jona, Switzerland) and is named The Dodick Photolysis System® after its inventor. The energy is delivered through a handheld probe containing a quartz fiber optic and is focused on a titanium target within the probe tip. Each pulse releases 12 mJ over 14 nsec. The pulse impacts the titanium target, leading to plasma formation. The ensuing shockwave fragments the lens proteins, which are aspirated through the same hand piece. The Er:YAG system is made by several manufacturers. It includes a hand piece containing a zirconium fluoride fiber optic to drive laser energy and an irrigation and aspiration system. Because of the water content of the lens, this laser can effectively emulsify this tissue. Data published for the Aesculap-Meditec MCL 29 model show that application frequency was between 20 to 60 Hz. For a typical phakoemulsification, the mean number of pulses was 1740, mean total energy was 38.5 J, and mean treatment time was 3 min.

#### 43.1.7.2 Laser Posterior Capsulotomy

This procedure[20] was the first application of the Nd:YAG laser in ophthalmology and is now a well-established routine procedure. The system consists of an Nd:YAG laser delivered via a slit lamp and guided with a HeNe beam. A collimating lens can be applied to the eye to decrease the energy necessary to induce a central rupture of the posterior capsule. Typical settings are energy = 2 to 3 mJ and number of applications = 5 to 20. The laser–tissue interaction includes plasma formation followed by tissue disruption, mostly by the shockwave induced.

# 43.2 Laser Surgery of the Posterior Segment of the Eye

## 43.2.1 Pathophysiologic Considerations

Diseases within the posterior segment of the eye can be classified as conditions that (1) overlie the neurosensory retina, (2) lie within the neurosensory retina, (3) lie completely through the neurosensory retina, and (4) underlie the neurosensory retina. Various laser therapies have been developed to address all these conditions.

Abnormal tissue growth above the neurosensory retina can be vascular tissue or fibrous tissue. Such disease conditions include epiretinal membranes and abnormal neovascularization (known as neovascular proliferation) associated with retinal ischemia (i.e., diabetic retinopathy, sickle-cell retinopathy, and retinopathy of prematurity). The various laser treatments discussed below have been designed to remove such abnormal tissue.

Disease processes that occur within the substance of the neurosensory retina are typically associated with abnormal leakage of serum or incompetence from retinal vascular. Leakage into the neurosensory retina causes damage to the normal retinal architecture, thereby destroying vision. Many vascular diseases, such as diabetic macular edema, retinal-vein occlusions, retinal-vessel telangiectasia, and retinal-arterial macroaneurysms, can cause this type of direct damage to the neurosensory retina.

Disease processes that involve the entire neurosensory retina are typically retinal holes or retinal tears. These conditions are significant because they can develop into retinal detachments, which are vision-threatening conditions.

Lastly, abnormal conditions can develop below or under the neurosensory retina, causing retinal disease and vision loss. Abnormal vascular tissue (choroidal neovascular membrane) can grow under the retina, causing damage to the normal retinal architecture and resulting in vision loss. Examples of these conditions include exudative macular degeneration, myopic degeneration, and ocular histoplasmosis. In addition, abnormal tumor cells and other tissues (hamartomas), such as choroidal melanoma, metastatic carcinomas, and choroidal hemagioma, can grow under the neurosensory retina.

### 43.2.1.1 Light Absorption

Melanin, hemoglobin, and macular xanthophyll represent the three most significant light absorbers in the retina and choroid. Melanin is the most significant chorioretinal light absorber. Its light absorption gradually decreases with increasing wavelength, thereby allowing longer laser wavelengths, such as krypton, red light, and diode infrared light, to produce deeper chorioretinal lesions than occurs with shorter-wavelength argon green laser light (Figure 43.2).

Hemoglobin is the next most effective light absorber. Hemoglobin absorption also decreases with increasing wavelength; however, there are two peaks in the absorption spectrum of oxyhemoglobin (542 nm, green; 577 nm, yellow) and one absorption peak for reduced hemoglobin (555 nm, yellow). Hemoglobin absorbs blue, green, and yellow light well but has poor red-light absorption.

Xanthophyll in the macular region of the retina is the least effective light absorber. Macular xanthophylls absorb blue-wavelength light well, green light minimally, and yellow to red light poorly.

### 43.2.1.2 Temperature Changes

Various forms of laser photocoagulation cause different temperature rises within the retina. Melanin within the RPE and choroid acts as the primary light absorber in retinal photocoagulation. Light absorption converts laser radiation into heat energy, increasing the temperature of light-absorbing tissues. Temperature rise is proportional to retinal irradiance (laser power/area) for a particular wavelength, spot size, and exposure duration.[21] Heat conduction spreads temperature increases from the light-absorbing RPE and choroid to contiguous tissues. For very long exposures, heat convection, because of choroidal blood flow, moderates chorioretinal temperature rise.

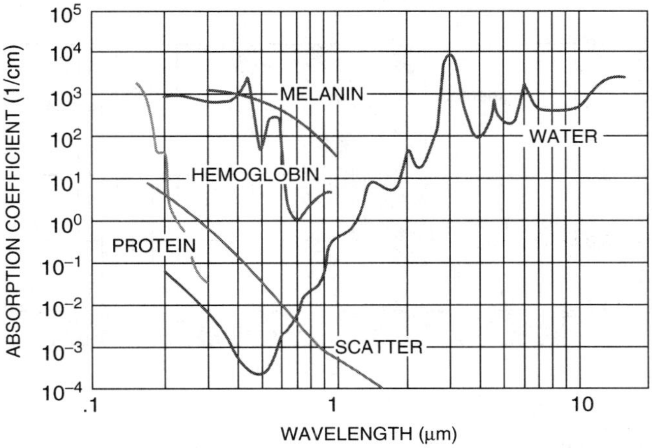


**FIGURE 43.2** Laser absorption and tissue penetration.

## 43.2.2 Tissue Photocoagulation

Conventional short-pulse photocoagulation is a highly suprathreshold procedure, associated with temperature increases of 40 to 60°C above normal body temperature of 37°C.[21] Most ophthalmic lasers in use today exemplify this concept of short-pulse tissue photocoagulation. Various laser wavelengths are used to accomplish this: (1) argon blue (488 nm), (2) argon blue/green (514.5 nm), (3) krypton red (647 nm), (4) dye (577 to 630 nm), (5) diode (810 nm), and (6) frequency-doubled Nd:YAG (532 nm).

This concept of short-pulse tissue photocoagulation has been successfully applied to treat a wide variety of ocular conditions. In essence, they can be divided into five broad categories:

1. Focal treatment of leaking microvascular or macrovascular abnormalities. Focal-tissue obliteration of leaks and grid-laser treatment to damage leaking capillary beds are used. Examples are diabetic macular edema, branch-vein occlusion, radiation retinopathy, and idiopathic perifoveal telangiectasias.[22,23]
2. Panretinal photocoagulation for treatment of proliferative retinopathies or neovascularization of the iris. Examples include proliferative diabetic retinopathy, sickle cell retinopathy, central retinal-vein occlusion with rubeosis, proliferative disease following radiation retinopathy, and proliferation associated with branch-vein occlusion.[24–26]
3. Treatment or prevention of retinal detachment. Examples include treatment of retinal breaks, demarcation of subclinical retinal detachment, and treatment of lattice degeneration.[27]
4. Treatment of ocular tumors and neoplasms. Examples include choroidal melanoma, retinoblastoma, choroidal cavernous hemangioma, and angioma (Von Hippel disease).[28,29]
5. Treatment of choroidal neovascular membrane or focal choroidal lesions. Examples include central serous retinopathy, age-related macular degeneration (AMD), presumed ocular histoplasmosis, myopia degeneration, and idiopathic polypoidal choroidal vasculopathy.[30–33]

Multiple different wavelengths have been used to treat the above conditions. Several generalities exist regarding the use of various wavelengths. Argon green appears to have advantages over blue light only if there is less damage when treatment is placed near the fovea due to less absorption by the retinal xanthophyll and there is more light scatter. Krypton red may have advantages in patients with hazy ocular media. The deeper the penetration of laser energy into the choroid (krypton red and diode), the more pain is associated with treatment.

Diode lasers emit laser energy at a wavelength of 810 nm. This longer wavelength can be absorbed by all three ocular pigments — hemoglobin, melanin, and xanthophyll — albeit inefficiently. Because of the

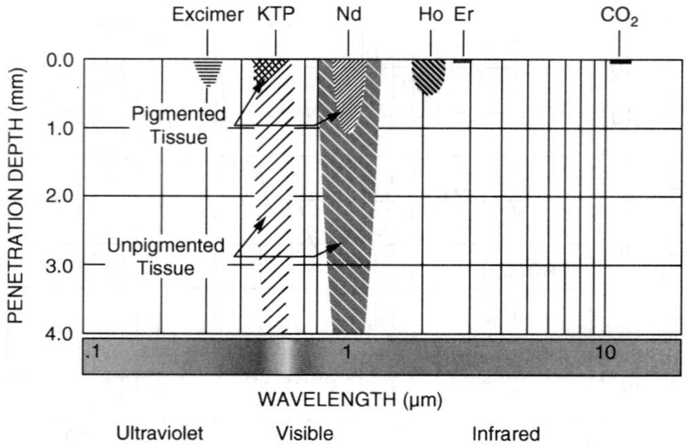


**FIGURE 43.3** Tissue penetration characteristics.

inefficient absorption of this wavelength, diode-laser energy can penetrate deeper into the choroidal layer of the eye (Figure 43.3). It therefore has theoretical advantages over other wavelengths in treating retinal pathology located deeper within the eye. This theoretical advantage, however, has not been noted in clinical studies. Diode-laser photocoagulation has been used to treat all the conditions effectively treated by argon green and yellow and krypton red laser photocoagulation, with similar, but not better, results. It has proven particularly effective in the treatment of retinopathy of prematurity, a blinding condition in neonatal infants, and is the laser wavelength of choice.[34] Unfortunately, the deeper penetration of the diode laser energy in the eye also causes more pain stimulation than with shorter wavelengths.

## 43.2.3 Thermotherapy

Thermotherapy, the treatment of tissue disorders by causing a rise in tissue temperature, is another novel medical approach. Retinal temperature rise in laser therapy is proportional to retinal irradiance (laser power/area) for a particular spot size, exposure duration, and wavelength.[1] Transpupillary thermotherapy (TTT) is a low-irradiance, large-spot-size, and prolonged-exposure (long-pulse) laser treatment using infrared diode-laser energy. TTT has been successfully used to treat intraocular tumors such as choroidal melanoma and choroidal hemangioma.[35–37] Recently, long-pulse TTT has been advocated in the treatment of choroidal neovascularization in AMD.[38,39]

Maximal chorioretinal temperature rise in short- or long-pulse intraocular or transpupillary photocoagulation occurs at the inner surface of the RPE and in the center of the laser spot. In long-pulse TTT for choroidal neovascularization in macular degeneration, an 810-nm infrared diode laser is used with a 3-mm-diameter retinal spot size and a 60-sec exposure. For these parameters, maximal temperature rise is roughly 10°C for a typical 800-mW clinical power setting used in treating CNVM in a lightly pigmented fundus.[38] In comparison, even a 50-mW 0.1-sec, 200-µm argon laser exposure produces a 42°C RPE temperature rise.[40] In photodynamic therapy (PDT), discussed later in this chapter, a 3-mm-diameter, 0.6-W/cm$^2$, 83-sec, 689-nm exposure causes a maximal RPE temperature rise of less than 2°C.[21] Actual temperature increases are somewhat lower than maximal increases calculated for TTT and PDT because of the cooling effect of choroidal circulation.

## 43.2.4 Photodynamic Therapy

PDT requires administration of a photosensitizing dye and activation of the dye by light irradiation at the target tissue. This usually causes activation of the dye to a higher state, releasing singlet oxygen and causing tissue-cell injury. This process accomplishes destruction of abnormal blood vessels.

Many photosensitizing dyes are available including rose bengal, hematoporphyrin, Photofrin®, benzoporphyrin, tin etiopurpura, chlorinae 6, bacterochlorinae, and chaloalimium sulfonated phthalocyanine.

The laser light used to activate the photosensitized dye coincides with the absorption maximum of the photosensitizer. A better dye would localize preferentially to the target, minimizing collateral damage. An additional desirable feature would be a relatively short half-life, thereby minimizing the potential general side effects of environmental light exposure.

In ophthalmology, PDT lends itself to treatment of CNVMs and tumors. The photosensitizing dye can be allowed to accumulate in these pathologic tissues without accumulating in normal tissues and structures. In CNVM and tumor treatment, the damage is done by singlet-oxygen damage of the endothelial wall, which leads to vessel closure. PDT is therefore a nonthermal treatment. Today, only Vertiporphyrin (Visudyne™, QLT Pharmaceuticals, Vancouver, Canada) has been approved for clinical use for the treatment of classic CNVM in AMD.[41–44]

## 43.2.5 Selective Retinal Pigment Epithelial Photocoagulation

Micropulse laser treatment, with pulse durations from nanoseconds to microseconds, has been shown to cause selective damage to RPE tissue, sparing the neurosensory retina altogether. Again this has theoretical advantages over shorter-wavelength treatment, especially in conditions originating in the RPE and choroidal layers. This treatment is limited by the lack of a visible change in the retinal appearance with laser application, making it difficult to assess adequate treatment. Several laser systems have been tested to create this selective type of tissue damage including Q-switched Nd:YAG, micropulsed diode, micropulsed argon, and neodymium:yttrium-lithium-fluoride (Nd:YLF). It has been suggested that treatment above the ablation threshold may reduce untoward thermal and mechanical effects.[45,46]

## 43.2.6 Photodisruption

Various laser systems have been used to create tissue photodisruption in the posterior segment of the eye. Practical considerations for all these systems include their method of delivery and ease of use. Fiber optics wear out or are not easily bendable without breaking their tips and can become coated, losing their effectiveness. Also, cavitation can occur, making visualization during surgery difficult.

### 43.2.6.1 Carbon Dioxide Laser

Wavelength of 10.6 um (infrared spectrum) has been used intraocularly to phototransect and simultaneously coagulate vitreoretinal membranes. Because of the absorption of energy by water, all of the effect is at the point of impact at the target tissue. An advantage of this wavelength is its ability to treat on top of critical structures. The $CO_2$ laser does not require pigment to accomplish cutting or coagulation. The major disadvantage of the $CO_2$ laser is that it requires direct contact with the treating tissue. With too much pressure, a tear in the retina may occur. It cannot be delivered easily by fiber optic, an additional disadvantage.[47,48]

### 43.2.6.2 Ho:YAG Laser

Wavelength of 2.12 μm can be transmitted by flexible optical fibers with little power attenuation. Due to transmission, the beam must be shielded to prevent underlying retinal burns or hemorrhage. It has been used successfully to cut experimental vitreous membranes in rabbits.[49]

### 43.2.6.3 Sapphire-Tipped Nd:YAG Laser

Sapphire crystal has been used at the tip of various laser probes to deliver energy to the target tissue. The sapphire crystals do not alter the laser energy or transmission. The crystal has a high meeting point and low thermal conductivity and is mechanically strong. This tip can deliver Nd:YAG and argon energy. The Nd:YAG has been used clinically to coagulate retina and internally to transect vitreoretinal membranes.[50]

#### 43.2.5.4 Er:YAG Laser (2.94 μm)

The Er:YAG laser offers almost complete absorption of laser energy in water with little transmission. It has been used through intraocular fiber-optic delivery systems to photovaporize membranes in rabbits. In addition, it has been used in humans in two clinical studies, treating elevated membranes and vascularized diabetic preretinal membranes, creating retinotomies, and ablating epiretinal membranes. Recent trials showed that ablation was time-consuming and coagulation poor. Complications noted included hemorrhage, retinal breaks, and intraocular lens damage.[51,52]

#### 43.2.5.5 Picosecond Nd:YLF Laser

A power level of 1053 nm (operating in picoseconds) may have high power, short pulse, and high repotion and may reduce breakdown of ocular tissues adjacent to target tissue. A pulse rate of 1000 per second for up to 10 sec allows segmentation of tissue near the retinal surface and has been used clinically in diabetic patients.[53]

#### 43.2.5.6 Nd:YAG Phototransection (1.06 μm)

Vitreoretinal membranes may be cut in vitreous or AC. Because of explosive cutting, this should not be used within 3 mm of the retina. A contact lens is used to focus, and burst mode is used to cut. A 100-psec Nd:YAG laser pulsing at 50 to 200 Hz achieved optical breakdown and vitreous cutting using only 70 uJ of energy.[54,55]

## 43.3 Summary

Coherent-light sources with novel laser-energy delivery systems are well suited in the treatment of many ophthalmologic conditions. Various laser-energy wavelengths have been used effectively in ophthalmic medicine and continue to be the mainstay of many treatment therapies. Novel uses of new laser-energy sources and wavelengths will undoubtedly become possible in the future.

### Acknowledgment

The research for this chapter was supported by the Discovery Fund for Eye Research, Los Angeles, CA.

### References

1. Tenner, A., Neuhann, T., and Schroeder, E., Excimer laser radial keratotomy in the living human eye: a preliminary report, *J. Refract. Surg.*, 4, 5, 1988.
2. Srinivasan, R. and Sutcliffe, E., Dynamics of the ultraviolet laser ablation of corneal tissue, *Am. J. Ophthalmol.*, 103, 470, 1987.
3. Munnerlyn, C.R., Koons, S.J., and Marshall, L., Photorefractive keratectomy: a technique for laser refractive surgery, *J. Cataract Refract. Surg.*, 14, 46, 1988.
4. McDonald, M.B., Frantz, J.M., Klyce, S.D., et al., Central photorefractive keratectomy for myopia: the blind eye study, *Arch. Ophthalmol.*, 108, 799, 1990.
5. Seiler, T., Kahle, G., and Kriegerowski, M., Excimer laser (193 nm) myopic keratomileusis in sighted and blind human eyes, *Refract. Corneal Surg.*, 6, 165, 1990.
6. Barraquer, J.I., *Queratomileusis y Queratofaquia,* Instituto Barraquer de America, p. 115.
7. Jackson, W.B., Casson, E., and Hodge, W.G., Mintsioulis, G., and Agapitos, P.J., Laser vision correction for low hyperopia. An 18-month assessment of safety and efficacy, *Ophthalmology*, 105, 1727, 1998.
8. Swinger, C.A. and Lai, S.T., Solid state photoablative decomposition, the Novatec laser, in *Corneal Laser Surgery*, Salz, J.J., Ed., C.V. Mosby, St. Louis, 1995, p. 261.
9. Rowsey, J.J., Koch, D.D., et al., Alternative lasers and strategies for corneal modification, in *Corneal Laser Surgery*, Salz, J.J., Ed., C.V. Mosby, St. Louis, 1995, p. 269.

10. Wise, J.B. and Witter, S.L., Argon laser therapy for open-angle glaucoma, *Arch. Ophthalmol.*, 97, 319, 1979.
11. Latina, M.A., Sibayan, S.A., Shin, D.H., Noecker, R.J., and Marcellino, O., Q-switched 532 nm Nd:YAG laser trabeculoplasty (selective laser trabeculoplasty): a multicenter pilot clinical study, *Ophthalmology*, 105, 2082, 1998.
12. Abraham, R.K. and Miller, G.E., Outpatient argon laser iridectomy for angle closure glaucoma: a 2-year study, *Trans. Am. Acad. Ophthalmol. Otolaryngol.*, 79, 529, 1975.
13. Tomey, K.F., Traverso, C.E., and Shammas, I.V., Neodymium-YAG laser iridotomy in the treatment and prevention of angle closure glaucoma: a review of 373 eyes, *Arch. Ophthalmol.*, 105, 476, 1987.
14. Shin, D., Argon laser iris photocoagulation to relieve acute angle-closure glaucoma, *Am. J. Ophthalmol.*, 93, 348, 1982.
15. Chen, J., Cohn, R.A., and Lin, S.C., Endoscopic photocoagulation of the ciliary body for treatment of refractory glaucomas, *Am. J. Ophthalmol.*, 124, 787, 1997.
16. van der Zypen, E., Kwasniewska, S., Roe, P., and England, C., Transscleral cyclophotocoagulation using a neodymium:YAG laser, *Ophthalmic Surg.*, 17, 94, 1986.
17. Bloom, P.A., Tsai, J.C., Sharma, K., et al., "Cyclodiode": trans-scleral diode laser cyclophotocoagulation in the treatment of advanced refractory glaucoma, *Ophthalmology*, 104, 1508, 1997.
18. Stevens, G., Jr., Long, B., Hamman, J.M., and Allen, R.C., Erbium:YAG laser-assisted cataract surgery,. *Ophthal. Surg. Lasers*, 29(3), 185, 1998.
19. Kanellopoulos, A.J., Dodick, J.M., Brauweiler, P., and Alzner, E., Dodick photolysis for cataract surgery. Early experience with the Q-switched neodymium:YAG laser in 100 consecutive patients, *Ophthalmology*, 106, 2197, 1999.
20. Francois, J.H. and Aladlouni, T., Treatment of opacification of the posterior crystalline capsule after extracapsular extraction: surgery or laser, *Bull. Soc. Ophtalmol. Fr.*, 89(11), 1297, 1989.
21. Mainster, M.A., White, T.J., Tips, J.H., and Wilson, P.W., Retinal-temperature increases produced by intense light sources, *J. Opt. Soc. Am.*, 60, 264, 1970.
22. Early Treatment Diabetic Retinopathy Study Research Group, Photocoagulation for diabetic macular edema, Early Treatment Diabetic Retinopathy Study report number 1, *Arch. Ophthalmol.*, 103, 1796, 1985.
23. Branch Vein Occlusion Study Group, Argon laser photocoagulation for macular edema in branch vein occlusion, *Am. J. Ophthalmol.*, 98, 271, 1984.
24. Photocoagulation treatment of proliferative diabetic retinopathy: the second report of diabetic retinopathy study findings, *Ophthalmology*, 85, 82, 1978.
25. Kimmel, A.S., Magargal, L.E., and Tasman, W.S., Proliferative sickle retinopathy and neovascularization of the disc: regression following treatment with peripheral retinal scatter laser photocoagulation,. *Ophthalmic Surg.*, 17, 20, 1986.
26. Central Vein Occlusion Study Group, Natural history and clinical management of central retinal vein occlusion, *Arch. Ophthalmol.*, 115, 486, 1997.
27. Pollak, A. and Oliver, M., Argon laser photocoagulation of symptomatic flap tears and retinal breaks of fellow eyes, *Br. J. Ophthalmol.*, 65, 469, 1981.
28. Shields, C.L., Shields, J.A., Kiratli, H., and De Potter, P.V., Treatment of retinoblastoma with indirect ophthalmoscope laser photocoagulation, *J. Pediatr. Ophthalmol. Strabismus*, 32, 317, 1995.
29. Shields, J.A., The expanding role of laser photocoagulation for intraocular tumors: the 1993 H. Christian Zweng Memorial Lecture, *Retina*, 14, 310, 1994.
30. Argon laser photocoagulation for senile macular degeneration: results of a randomized clinical trial, *Arch. Ophthalmol.*, 100, 912, 1982.
31. Macular Photocoagulation Study Group, Krypton laser photocoagulation for idiopathic neovascular lesions: results of a randomized clinical trial, *Arch. Ophthalmol.*, 108, 832, 1990.
32. Macular Photocoagulation Study Group, Laser photocoagulation for juxtafoveal choroidal neovascularization: five-year results from randomized clinical trials, *Arch. Ophthalmol.*, 12, 500, 1994.

33. Macular Photocoagulation Study (MPS) Group, Evaluation of argon green vs. krypton red laser for photocoagulation of subfoveal choroidal neovascularization in the macular photocoagulation study, *Arch. Ophthalmol.*, 112, 1176, 1994.
34. Hunter, D.G. and Repka, M.X., Diode laser photocoagulation for threshold retinopathy of prematurity: a randomized study, *Ophthalmology*, 100, 238, 1993.
35. Journee-de Korver, J.G., Oosterhuis, J.A., Kakebeeke-Kemme, H.M., and de Wolff-Rouendaal, D., Transpupillary thermotherapy (TTT) by infrared irradiation of choroidal melanoma, *Doc. Ophthalmol.*, 82, 185, 1992.
36. Rapizzi, E., Grizzard, W.S., and Capone, A., Jr., Transpupillary thermotherapy in the management of circumscribed choroidal hemangioma, *Am. J. Ophthalmol.*, 127, 481, 1999.
37. Shields, C.L., Shields, J.A., DePotter, P., and Kheterpal, S., Transpupillary thermotherapy in the management of choroidal melanoma, *Ophthalmology*, 103, 1642, 1996.
38. Reichel, E., Berrocal, A.M., Ip, M., Kroll, A.J., Desai, V., Duker, J.S., and Puliafito, C.A., Transpupillary thermotherapy of occult subfoveal choroidal neovascularization in patients with age-related macular degeneration, *Ophthalmology*, 106, 1908, 1999.
39. Newsom, R.S., McAlister, J.C., Saeed, M., and McHugh, J.D., Transpupillary thermotherapy (TTT) for the treatment of choroidal neovascularisation, *Br. J. Ophthalmol.*, 85, 173, 2001.
40. Mainster, M.A., White, T.J., and Allen, R.G., Spectral dependence of retinal damage produced by intense light sources, *J. Opt. Soc. Am.*, 60, 848, 1970.
41. Photodynamic therapy of subfoveal choroidal neovascularization in age-related macular degeneration with verteporfin: one-year results of 2 randomized clinical trials — TAP report, Treatment of age-related macular degeneration with photodynamic therapy (TAP) Study Group, *Arch. Ophthalmol.*, 117, 1329, 1999.
42. Verteporfin therapy of subfoveal choroidal neovascularization in age-related macular degeneration: two-year results of a randomized clinical trial including lesions with occult with no classic choroidal neovascularization — verteporfin in photodynamic therapy report 2, *Am. J. Ophthalmol.*, 131, 541, 2001.
43. Photodynamic therapy of subfoveal choroidal neovascularization in pathologic myopia with verteporfin: 1-year results of a randomized clinical trial — VIP report no. 1, *Ophthalmology*, 108, 841, 2001.
44. American Academy of Ophthalmology, Photodynamic therapy with verteporfin for age-related macular degeneration, *Ophthalmology*, 107, 2314, 2000.
45. Roider, J., Hillenkamp, F., Flotte, T., and Birngruber, R., Microphotocoagulation: selective effects of repetitive short laser pulses, *Proc. Natl. Acad. Sci. U.S.A.*, 90, 8643, 1993.
46. Brinkmann, R., Huttmann, G., Rogener, J., Roider, J., Birngruber, R., and Lin, C.P., Origin of retinal pigment epithelium cell damage by pulsed laser irradiance in the nanosecond to microsecond time regimen, *Lasers Surg. Med.*, 27, 451, 2000.
47. Karlin, D., Jakobiec, F., Harrison, W., et al., Endophotocoagulation in vitrectomy with a carbon dioxide laser, *Am. J. Ophthalmol.*, 101, 445, 1986.
48. Meyers, S.M., Bonner, R.F., Rodrigues, M.M., and Ballintine, E.J., Phototransection of vitreal membranes with the carbon dioxide laser in rabbits, *Ophthalmology*, 90, 563, 1983.
49. Borirakchanyavat, S., Puliafito, C.A., Kliman, G.H., Margolis, T.I., and Galler, E.L., Holmium-YAG laser surgery on experimental vitreous membranes, *Arch. Ophthalmol.*, 109, 1605, 1991.
50. Peyman, G.A., Katoh, N., Tawakol, M., Khoobehi, B., and Desai, A., Contact application of Nd:YAG laser through a fiberoptic and a sapphire tip, *Int. Ophthalmol.*, 11, 3, 1987.
51. Brazitikos, P.D., D'Amico, D.J., Bernal, M.T., and Walsh, A.W., Erbium:YAG laser surgery of the vitreous and retina, *Ophthalmology*, 102, 278, 1995.
52. D'Amico, D.J., Blumenkranz, M.S., Lavin, M.J., Quiroz-Mercado, H., Pallikaris, I.O., Marcellino, G.R., and Brooks, G.E., Multicenter clinical experience using an erbium:YAG laser for vitreoretinal surgery, *Ophthalmology*, 103, 1575, 1996.

53. Cohen, B.Z., Wald, K.J., and Toyama, K., Neodymium:YLF picosecond laser segmentation for retinal traction associated with proliferative diabetic retinopathy, *Am. J. Ophthalmol.*, 123, 515, 1997.
54. Peyman, G.A., Contact lenses for Nd:YAG application in the vitreous, *Retina*, 4, 129, 1984.
55. Lin, C.P., Weaver, Y.K., Birngruber, R., Fujimoto, J.G., and Puliafito, C.A., Intraocular microsurgery with a picosecond Nd:YAG laser, *Lasers Surg. Med.*, 15, 44, 1994.

# 44

# Lasers in Otolaryngology




Lou Reinisch*
*Vanderbilt University Medical Center*
*Nashville, Tennessee*


## 44.1 Introduction

The laser and its role in otolaryngology begin just after Theodore Maiman made the first ruby laser on May 16, 1960.[1] In the early 1960s, as ophthalmologists[2] and dermatologists[3,4] experimented with lasers, otolaryngologists first considered different methods of using pulsed-laser systems in the middle ear and labyrinth.[5,6] During this time, Geza Jako began studying the effects of laser energy on human vocal folds.[7] His first attempts at tissue ablation were made with the neodymium-glass laser, with a wavelength of 1.06 μm. The absorption characteristics of the tissue were not suitable for precise excision with this wavelength of light. In 1965, Strully and Yahr tried to enhance the absorption of the tissue by painting the tissue with a copper sulfate solution.[8] The results were still unsatisfactory. They found at least three problems associated with using the laser: (1) they needed higher intensity levels; (2) they could produce only small lesions; and (3) they were left with significant thermal destruction of the tissue surrounding the laser-ablation site.

Starting in 1967, Polanyi experimented with the $CO_2$ laser in a human cadaver larynx and was encouraged by its ability to produce discrete wounds.[9] The 10.6-μm wavelength of the $CO_2$ laser is strongly absorbed by water ($\alpha = 250$ cm$^{-1}$).[10] Therefore, biological tissue, which is high in water content, absorbs laser energy well. The energy is concentrated at the point of laser impact, and comparatively minimal spread through surrounding tissue occurs. In addition, the longer wavelength at 10.6 μm shows minimal scattering of the laser light in tissue. Polanyi's work was spurred by the development of an endoscopic delivery system that made it possible to test the laser *in vivo*.[11,12] In 1972, Jako[7] and Strong and Jako[13] reported the initial use of this new equipment in a canine model.

The most common application of the medical laser in otolaryngology is for tissue ablation. It is therefore not surprising that the most commonly used laser in otolaryngology is the $CO_2$ laser. The 10.6-μm light from the $CO_2$ laser can create intense localized heating sufficient to vaporize both extra- and intracellular water, producing a coagulative necrosis.[14–16]

*Current affiliation: Department of Physics and Astronomy, University of Canterbury, Christchurch, New Zealand


0-8493-1116-0/03/$0.00+$1.50

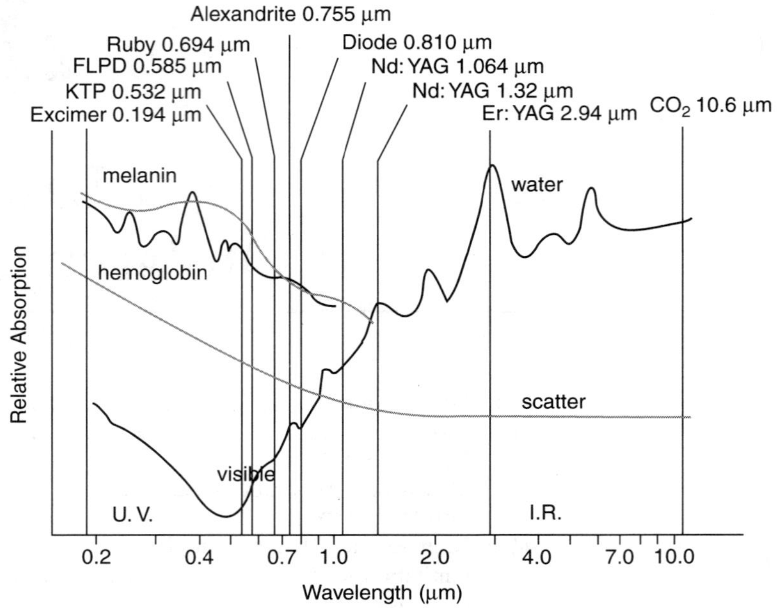


**FIGURE 44.1** The electromagnetic spectrum from the ultraviolet, through the visible, to the mid-infrared. The absorption spectra of major tissue chromophores, hemoglobin, melanin, and water are shown. The limit of penetration depth due to scatter is shown. The wavelengths of several important medical lasers have been added.

## 44.2 Laser Use

The tissue effects produced by the laser vary with the wavelength and pulse structure of the laser (Figure 44.1). The interaction of laser energy with living tissue can produce at least three distinct reactions. First, the laser energy can be absorbed by chromophores within the tissue.[17,18] For example, water is the chromophore for the $CO_2$ laser, and hemoglobin is often the chromophore for the potassium-titanyl-phosphate (KTP or KTP/532) laser. The absorption of the light energy by the tissue is then converted into heat. This is the thermal effect used today with most conventional surgical laser systems.

The second reaction, the radiant energy of a laser, can stimulate or react with molecules within a cell.[19–22] These molecules, after the absorption of the light energy, effect a biomolecular chemical change within the cell. This change is termed photochemical. An example of a photochemical process is the reaction that occurs with the injection of a photosensitizing drug into tissue and the subsequent biochemical effect that is produced when the drug is activated by the laser energy.

Third, the use of short pulses of high-intensity laser light can disrupt cellular architecture because of the production of stress-transient waves or photoacoustic shockwaves. The short burst of light causes rapid heating and thermal expansion of a small volume of the tissue. The expansion causes an acoustic wave to propagate from the source. This mechanical disruption of tissue is an example of a nonthermal tissue effect.[23]

Several studies comparing the histologic and tensile properties of wounds after laser and scalpel-produced incisions on experimental animals have been performed (Figure 44.2). As early as 1971, Hall demonstrated that the tensile strength in a $CO_2$ laser–induced incision was less than the scalpel incision up to the 20th day of healing and became the same by the 40th day.[14] Later, in 1981, Norris and Mullarky studied the healing properties of $CO_2$ laser incisions using a porcine model histologically.[24] They showed that scalpel-induced incisions exhibited better tissue reconstruction than laser-induced incisions up to the 30th day, after which time both incisions exhibited similar results. In a similar study presented in

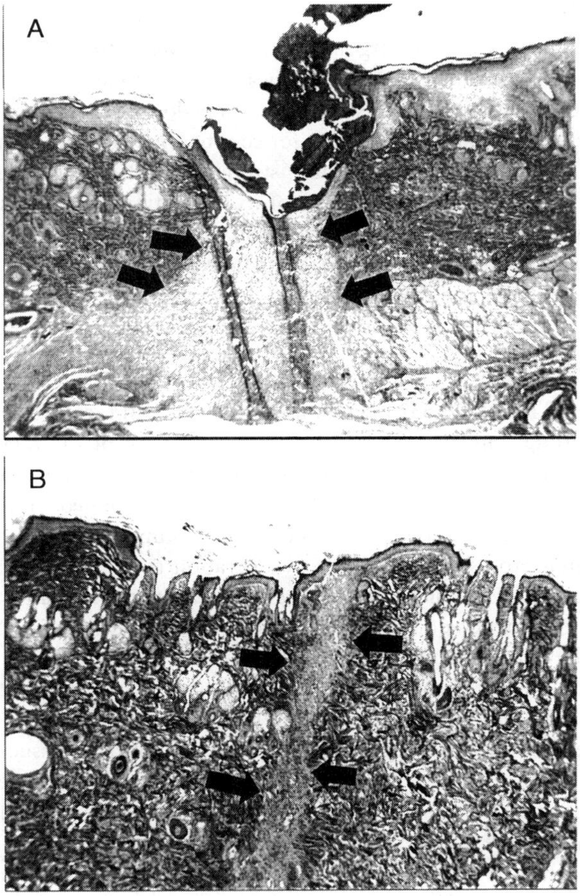


FIGURE 44.2 Histological sections of incisions made to rat skin after (A) $CO_2$ laser incision made with 5-W, 0.2-sec repeat-pulse mode and focused to 125 μm and (B) scalpel incision. The wounds were harvested 7 d after the incisions were made. A wide band of denatured collagen (marked with arrows) is shown in (A). The tissue has reepithelialized, and a large amount of coagulum is seen on the surface. A narrow band of new collagen (marked with arrows) is shown in (B). The laser incision (A) shows the most pronounced delay in wound healing at its early (7-d) time point.

1983, Buell and Schuller created $CO_2$ laser incisions in pigs and compared them to scalpel incisions.[25] They found the laser wound to be weaker in tensile strength than the scalpel wounds for the first 3 weeks.

For all the advantages of the laser, including hands-off operation and hemostasis, the delay in wound healing caused by the lateral thermal damage remains a problem. When the otolaryngologist decides to use the laser, all of the advantages must be weighed against the disadvantages and the potential for increased morbidity. Thus, the surgeon must be informed and aware of how the laser interacts with the tissue.

### 44.2.1 Types of Lasers

Various types of lasers are currently used in otolaryngology head and neck surgery. Each laser will be explained and the range of applications given. Of course, this list is constantly changing. In general, any laser is first tried in many different applications, and use becomes better defined and more specific as the surgeon gains experience with the laser.

#### 44.2.1.1 The Argon Ion Laser

The argon ion laser (frequently called the argon laser), with wavelengths of 514 nm (green in color) and 488 nm (blue in color), is operated in the continuous-wave (CW) mode, can be delivered through optical fibers, and is used in cutaneous applications as well as stapedotomies. This laser normally has special electrical-power requirements and needs tap water for cooling the laser. The argon laser energy is poorly absorbed by clear liquids. Hemoglobin and melanin strongly absorb the laser light. The argon laser can be used to treat vascular cutaneous lesions because of its absorption by melanin and hemoglobin.[26]

Focusing the argon laser beam to a small spot size results in high-power densities sufficient to vaporize tissue. Otologists have used the argon laser to perform stapedectomy procedures because of its ability to be focused to the small spot size and the optical-fiber delivery system.[27,28] Another otologic application of this laser is the lysis of middle ear adhesions.[29]

#### 44.2.1.2 The KTP/532 Laser

Like the argon laser, the KTP or KTP/532 laser emits light at 532 nm (green color) in a quasi-CW mode and can be delivered through optical fibers. The single wavelength of the KTP laser is centered on a hemoglobin-absorption band. The laser normally does not have any special power requirements and does not require external water to cool the laser. The lasing source is a neodymium:yttrium-aluminum-garnet (Nd:YAG) laser. The Nd:YAG laser rod is continuously pumped with a krypton arc lamp and Q-switched. The Q-switching process changes the CW operation to quasi-CW. The light is emitted in a series of short pulses that repeat so quickly that the light appears to the human eye to be CW. The pulsed 1.06-μm light traverses a frequency-doubling KTP crystal, yielding the quasi-CW 532-nm green light.

Like the argon laser, the radiant energy from the KTP laser is readily transmitted through clear aqueous tissues because it has a low water-absorption coefficient. Certain tissue pigments, such as melanin and hemoglobin, absorb the KTP-laser light effectively. When low levels of green laser light interact with highly pigmented tissues, a localized coagulation takes place within these tissues. The KTP laser can be selected for procedures requiring precise surgical excision with minimal damage to surrounding tissue, vaporization, or photocoagulation. The power density chosen for a given application determines the tissue interaction achieved at the operative site.

The KTP laser is transmitted through a flexible fiber-optic delivery system, which can be used in association with a micromanipulator attached to an operating microscope or freehand in association with various hand-held delivery probes having several different tip angles. These hand-held probes facilitate use of the KTP laser for functional endoscopic sinus surgery and other intranasal,[30] otologic,[31] and microlaryngeal applications.[32] The optical-fiber delivery of the 532-nm laser light can be manipulated through a rigid pediatric bronchoscope as small as 3.0 mm, facilitating lower tracheal and endobronchial lesion treatment in infants and neonates.[33] Examples of handheld KTP laser applications include tonsillectomy,[34–38] stapedectomy,[39,40] excision of acoustic neuroma,[41] and excision of benign and malignant laryngeal lesions.[42]

#### 44.2.1.3 The Nd:YAG Laser

The Nd:YAG laser at 1.06 μm (near-infrared [NIR] region of the spectrum) has the deepest penetration depth of any of the common surgical lasers. This laser can produce a zone of thermal coagulation and necrosis that extends up to 4 mm from the impact site. The CW light can also be delivered through optical fibers. The laser normally does not have any special power requirements and does not need external cooling water.

The primary applications for the Nd:YAG laser in otolaryngology head and neck surgery include palliation of obstructing tracheobronchial lesions,[43–47] palliation of obstructing esophageal lesions,[48] photocoagulation of vascular lesions of the head and neck,[49,50] and photocoagulation of lymphatic malformations.[51] The Nd:YAG laser has several distinct advantages in the management of obstructing lesions of the tracheobronchial tree. Hemorrhage is a frequent and dangerous complication associated with laser bronchoscopy, and its control is extremely important. Control of hemorrhage is more secure with this laser because of its deep penetration in tissue. Nd:YAG-laser application through an open, rigid bronchoscope allows for multiple distal suction capabilities simultaneous with laser application and rapid removal of tumor fragments and debris to prevent hypoxemia.

The major disadvantage of the Nd:YAG laser is its less predictable depth of tissue penetration. This laser is used primarily to photocoagulate tumor masses rapidly at power settings in the upper and lower aerodigestive tract of 40 to 50 W and 0.5- to 1.0-sec exposures. Whenever possible, the laser beam is applied parallel to the wall of the tracheobronchial tree. The rigid tip of the bronchoscope is used mechanically to separate the devascularized tumor mass from the wall of the tracheobronchial tree.

#### 44.2.1.4 Other Lasers

Other lasers include the flashlamp excited-dye laser (FEDL). This laser operates at 585 nm with pulses of light approximately 0.4 msec long. The parameters have been optimized for the selective treatment of vascular lesions with minimal damage to the dermis. The light is delivered through an optical fiber. The laser normally requires 220 V and cooling water. Another laser developed for cutaneous applications is the argon pumped-dye laser.[52,53] Here the wavelength can be varied from 488 nm through the red region of the spectrum (800 nm). The CW light is delivered through an optical fiber. This laser has special electrical requirements and requires a significant flow of tap water for cooling.

Recent investigations have considered the diode laser. These small lasers operated in the red to NIR region of the spectrum. They are small and require no water for cooling. The lasers operate in CW mode, and the light is delivered through an optical fiber. The intensities are relatively low. They have been investigated for photodynamic therapy and tissue welding.[54–56] The diode laser has also been investigated to shrink the tonsillar tissue without damaging the overlying mucosal layer.[57] This is the first step toward tonsillectomies that are relatively painless and with no loss of blood.

#### 44.2.1.5 The $CO_2$ Laser

The workhorse of surgery, the $CO_2$ laser operates at 10.6 μm. The invisible infrared $CO_2$ laser beam has a coaxial helium-neon laser or diode laser beam to act as a pointing beam. This laser does not have special electrical requirements and is normally air-cooled. The wavelength of the $CO_2$ laser is strongly absorbed by water.[9] Therefore, the light energy of this laser is well absorbed by all tissues of high water content. The mid-infrared light at 10.6 μm cannot propagate through glass or sapphire optical fibers and is normally delivered through an articulated arm. In addition, silver halide optical fibers for the $CO_2$ laser as well as waveguides have recently been introduced to the market.[58] The laser energy can be delivered to tissue either through a hand piece for macroscopic surgery or adapted to an operating microscope for microscopic surgery. The universal endoscopic coupler also allows for delivery of the laser energy through a rigid bronchoscope.[59,60]

## 44.3 Microscopic Applications

Laser technology is being coupled with additional new instrumentation for the laser surgeon, especially newer devices that have recently been introduced to deliver the laser beam. Often hospitals will update their equipment on a regular basis. It is mandatory for the surgeon to be completely familiar with the laser unit and the delivery system before any patient application is attempted. Power output, spot size, and, therefore, power density can vary and cannot always be extrapolated from one unit to another.

The lasers and the wavelengths presented above may all be adapted for use with the binocular operating microscope. The argon, KTP, diode, and Nd:YAG lasers may be used with optical-fiber delivery probes that are threaded through a suction catheter. The optical fibers are used in either a contact or noncontact mode. The $CO_2$ and KTP lasers have the added feature of being delivered by an optical system with a micromanipulator. This allows precise, noncontact delivery at a predetermined focal length.

Microscopic delivery systems for the $CO_2$ laser have evolved since the initial endoscopic couplers. Several problems, including parallax error and a large spot size, have been overcome in the newest generation of micromanipulators. Perhaps the most significant change in the micromanipulator has been the addition of the partially reflective dichroic mirror. This mirror allows reflection of the $CO_2$ laser while

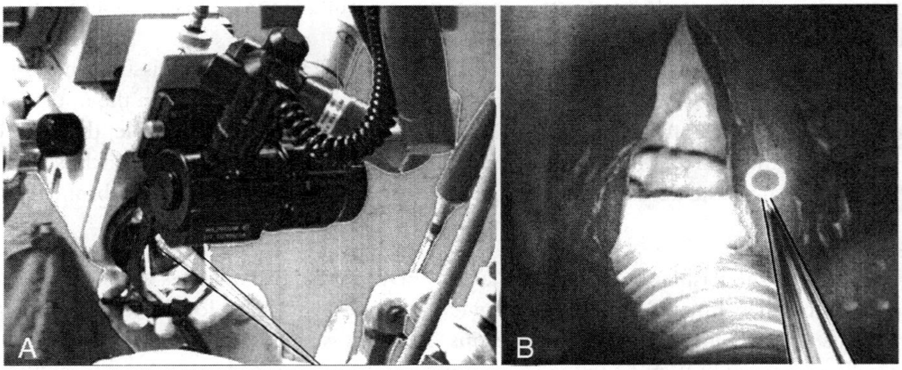


**FIGURE 44.3** (A) A scanning device is coupled with a microscope micromanipulator for laryngeal surgery. This scanner scans the beam in a spiral pattern. (B) The area that will be ablated is shown by the circle drawn by the aiming beam. This area ablation is effective for removing papilloma. The beam is aiming at the vocal fold. A laser-safe endotracheal tube is used, and the cuff is protected by saline-soaked cottonoids.

permitting transmission of almost 75% of visible light.[61] The larger beam diameter possible with the dichroic mirror also allows the spot size to be reduced from 800 μm to 200 to 250 μm.[62] As the thermal effects of laser applications have become better understood the desirability of using a smaller spot size for tissue interaction has become evident.

Two devices that work with the micromanipulator to scan the beam in a straight line or over a circular pattern are now available (Figure 44.3). These devices allow the lasers to operate in the superpulse mode and significantly reduce lateral thermal damage. They also permit an even fluence of laser energy to impact the tissue. Although it seems that these devices should offer many advantages, few clinical reports are available at present.

## 44.3.1 Laryngeal Surgery

Over the past 10 years, the role and limitations of lasers in microlaryngoscopic applications have become better understood. The $CO_2$ laser remains the laser of choice for many microlaryngoscopic applications. Certain benign laryngeal diseases, such as the removal of recurrent respiratory papillomatosis, retain the laser as the instrument of choice. The surgeon is quick to realize the advantages of microscopic control and decreased postoperative edema. Although there was early disappointment when the laser failed to cure the papillomatosis, the disappointment has been overcome by the laser's ability to precisely remove papilloma and spare normal laryngeal tissue. Other applications appropriate for the $CO_2$ laser include subglottic stenosis, webs, granuloma, and capillary hemangiomas. Surgery for other benign laryngeal disease processes, such as polyps, nodules, leukoplakia, and cysts, may also be performed with the $CO_2$ laser; however, cold-knife excision has been shown to produce equal, if not improved, postoperative results. Surgery in the pediatric group for webs, subglottic stenosis, and capillary meningiomas has all been significantly improved by the precision, preservation of normal tissue, and decreased postoperative edema associated with the $CO_2$ laser.

The Nd:YAG laser has much more limited application with regard to intralaryngeal use. However, there are applications where the Nd:YAG is appropriate and often superior to the $CO_2$ laser. Specifically, increased tissue absorption with an increased depth of penetration make the Nd:YAG laser ideal for the treatment of vascular lesions, such as cavernous hemangiomas, where hemostasis and vessel coagulation are the treatment goals.

### 44.3.1.1 Stenosis

Stenotic lesions appropriate for endoscopic management have certain features in common as determined by retrospective analysis.[63] First, all lesions treated with endoscopic techniques must have intact external

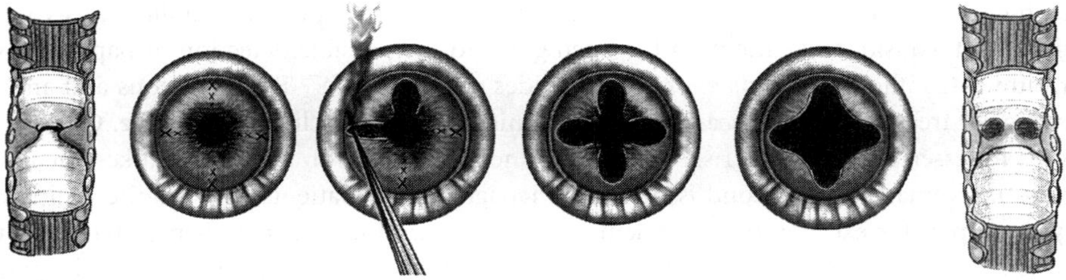

**FIGURE 44.4** On the far left, a subglottic stenosis is shown in the trachea with an exposed view. The images progressing from left to right represent the stenosis as viewed by the otolaryngologist. The radial incisions are made with the laser and then dilated. The resulting airway is shown at the far right with a cutaway view of the trachea.

cartilaginous support. Attempted endoscopic incision or excision of areas of tracheomalacia can have disastrous results if surrounding structures are perforated. Second, lesions appropriate for endoscopic management are usually less than 1 cm in vertical length. Favorable results, however, have been reported for lesions up to 3 cm in length when endoscopic incision is combined with prolonged stenting.[64,65] Finally, total cervical tracheal or subglottic stenosis does not usually respond well to endoscopic management. Again, however, successful case reports exist for endoscopic management when it is combined with prolonged stenting of the stenotic area.[64]

Still, the management of laryngotracheal stenosis is a difficult problem for the otolaryngologist. The first decision — whether open management is necessary or if endoscopic techniques alone are adequate — is probably the most demanding question. All patients with laryngotracheal stenosis need staging direct laryngoscopy and bronchoscopy to determine the extent and degree of stenosis. The $CO_2$ laser should always be available during the staging laryngoscopy. With the laser on standby, any scarring can be easily removed or incised with the $CO_2$ laser. Supplemental dilation with the bronchoscope or stent placement may then be beneficial in further management of the stenotic area.[64]

Endoscopic management of laryngotracheal stenosis relies on mucosal preservation. The two techniques advocated for this task are radial incision with bronchoscopic dilation[64] and the microtrapdoor flap.[66–69] In bronchoscopic dilation, the laser is used to make radial incision, like the spokes of a wheel, in the stenotic area (Figure 44.4). Bronchoscopes are then sequentially passed through the stenosis to dilate it. The microtrapdoor flap is more complicated but uses a flap of mucosa to drape over the excised tissue.

### 44.3.1.2 Recurrent Respiratory Papillomatosis

The $CO_2$ laser has become the standard treatment modality for patients with recurrent respiratory papillomatosis. Some surgeons have reported successful treatment results with the KTP laser for this disorder; however, the less predictable depth of penetration and, therefore, increased potential for thermal injury in surrounding normal tissue make it less than ideal.[70] As stated above, the $CO_2$ laser cannot cure the disease. The greater absorption and decreased scatter potential in laryngeal tissue of the 10.6-μm wavelength of the $CO_2$ laser account for its greater effectiveness in preserving normal laryngeal structures and maintaining the airway.

$CO_2$ laser treatment should be directed at removing as much papilloma as possible from one vocal cord and then as much as possible from the other. Although some papilloma may remain, it is important to preserve a 2- to 3-mm strut of covering over the anterior part of one vocal cord. This decreases the possibility of creating an anterior web. The papilloma overlying the true vocal cord can be vaporized to the vocal ligament. Following the initial removal of papilloma by $CO_2$ laser, a planned repeat operation should be performed in approximately 6 weeks. In a series reported by Ossoff et al., 22 patients underwent 105 $CO_2$ laser excisions. The intraoperative soft-tissue complication rate was zero. The delayed soft-tissue complication rate, consisting of two patients with slight true vocal-fold scarring and one patient with a small posterior web, was 13.6%.[71] This compares favorably with other published complication rates of 28.7 and 45%.[72,73]

Two further considerations regarding the treatment of recurrent respiratory papillomatosis are worthy of mention. First, considerable attention has been given to the possible detection of papilloma virus in the laser plume. Conflicting reports exist on both sides of the issue.[74,75] Both surgeons and anesthesiologists have been treated for the disease that was manifest only after clinical exposure. Current recommendations to lessen the potential risk of exposure include the use of adequate smoke evacuation and high-filtration facemasks. The second issue is the management of patients who require multiple repeat laser laryngoscopies for excision on a frequent basis. Treatment intervals can then be based on the rate of reexpression. There are a number of adult and pediatric patients, however, who cannot be placed into clinical remission with this regimen. Therapeutic trials with interferon alpha n-3 have shown no long-term benefit of administration. An investigation with interferon alpha n-1, however, seems to show more promise in producing remission after a 3- to 6-month drug trial.[76]

## 44.4 Laser-Assisted Uvulo-Palatoplasty

Laser-assisted uvulo-palatoplasty (LAUP) is a technique that was developed by Dr. Yves-Victor Kamami in Paris, France in the late 1980s.[77] The procedure is designed to correct snoring caused by airway obstruction and soft-tissue vibration at the level of the soft palate by reducing the amount of tissue in the velum and uvula. The procedure can be performed in an ambulatory setting under local anesthesia and is performed over several stages.

The patient should first be evaluated completely to determine the degree of severity of airway obstruction during sleep. Apnea cannot be ruled out on the basis of history and physical appearance alone.[78,79] If the patient is found to have apnea, then treatment of this condition is mandatory.

One can consider this the equivalent of a dental procedure. If the patient needs subacute bacterial endocarditis prophylaxis, it will need to be administered in conjunction with this procedure. The procedure is performed with the patient sitting in an exam chair in the upright position. A local anesthetic is administered. The procedure is performed using a $CO_2$ laser. The laser is delivered with a hand piece equipped with a backstop to protect the posterior pharyngeal wall from stray laser energy. The $CO_2$ laser was selected because of its high efficiency for making incisions. Additionally, the high absorption of the infrared light creates very fine incisions, and the tissue interactions are favorable for this type of procedure. Although the $CO_2$ laser is not the best coagulating laser available, it has adequate coagulation for the diameter of most vessels encountered in the soft palate. The $CO_2$ laser is used at a power setting of between 15 and 20 W in continuous mode. The beam is used focused for cutting and defocused for ablation and vaporization.

Once the adequate level of anesthesia has been achieved in the soft palate, the $CO_2$ laser is then used to make bilateral vertical incisions through the palate at the base of the uvula (Figure 44.5). The uvula can be reduced by approximately 50% of its length and is reshaped in a curved fashion. One can retract the tip of the uvula anteriorly and vaporize the central portion of the uvula muscle while leaving the mucosa intact. Maintaining the mucosa results in less postoperative pain and less delay in wound healing.

After the initial visit, the patient is scheduled for a subsequent treatment 4 weeks later. When the patient returns to the clinic, she is requested to report on the amount of pain experienced, any changes in voice or swallowing, and bleeding and infection. Subjective improvements in the patient's sleep are also elicited, as are any objective changes in snoring. The subsequent procedures are basically the same as the initial procedure. During the treatment course, the otolaryngologist should always examine the palate and determine if attention is needed in other areas. As the retraction occurs, the palate will advance anteriorly and superiorly, and this may tether the posterior tonsillar pillars, causing them to medialize. Should this occur and the pillars themselves begin to obstruct the airway, they can be released by making a horizontal incision at the superior aspect of the pillar, through and through the pillar. If the tonsils are still present, they can be superficially vaporized after this surface is first anesthetized. A crescent-shaped incision can be made vaporizing lateral to the incision; this may also prevent some of the side-to-side healing that would tend to take place in the palate with just simple through and through incisions. When

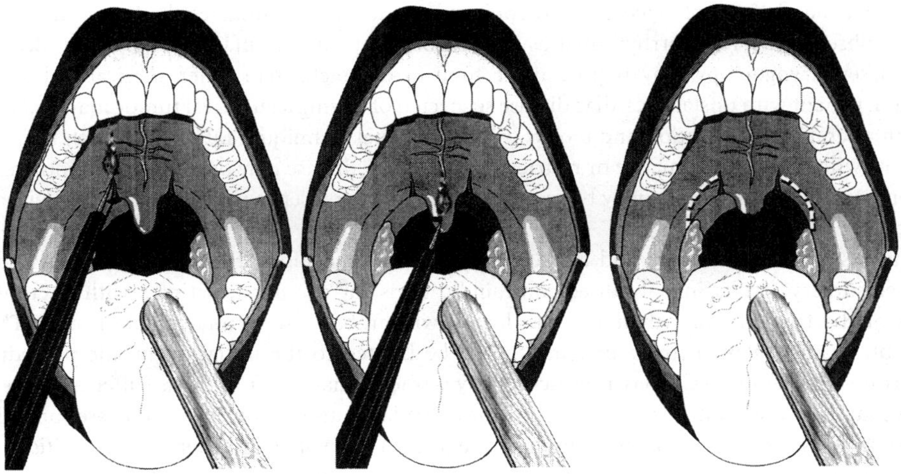

**FIGURE 44.5** The steps of the laser-assisted uvulo-palatoplasty. From left to right, first the vertical incisions are made using a laser hand piece with a backstop. Next, the uvula is trimmed. If necessary, incisions can be made along the dashed line if tissue contraction pulls the airway closed.

side-to-side healing is reduced, more retracting of the palate occurs; over the course of the treatment sessions, this will reduce the number of sessions necessary to achieve the end result.

The endpoint for this procedure is determined by any one or a combination of the following factors: (1) if the snoring has stopped; (2) if the patient is satisfied — unfortunately, many patients accept a small improvement; or (3) if the patient is unable to make a snorting sound. The patient might continue to have noisy breathing but be unable to make a snorting sound. At this point the procedure should stop. Further resection of the soft palate after the patient is no longer able to snort runs the risk of excising too much tissue and creating velopharyngeal incompetence. The patient may experience noisy breathing from the vibration of other tissues, such as in the region of the base of the tongue, and not necessarily at the soft palate.

The complication rate for this procedure has been extremely low. There has been some bleeding in the immediate postoperative period that has responded to silver-nitrate cauterization. Permanent voice changes, velopharyngeal insufficiency, temporary or permanent nasal reflux, and dehydration from the inability to take food or liquid orally have not been observed with this procedure.

Although there is pain associated with the procedure, it is not of an incapacitating nature, and patients are able to attend classes or work in their normal routine. Most patients report some degree of weight loss, at most between 10 and 15 lb over the course of the treatment. In most of our patients, this has been a desirable side effect.

## 44.5 Limitations of Laser Use

Prior to any surgical procedure, the surgeon must consider whether the laser is the best method for treating a particular laryngeal disorder. As noted earlier, such entities as vocal nodules may require no therapy whatsoever. Although the laser is a highly precise instrument, thermal injury and carbonaceous debris may stimulate submucosal fibrosis that may be unacceptable in certain types of patients such as the professional voice user. If the laser is felt to be the optimal method of treatment, limitations may be imposed by the patient's particular anatomic configuration. Mandible size and position as well as spinal-column flexibility may limit the type of laryngoscope that can be used and thereby affect visualization of the larynx. Patients with cervical arthritis, retrognathia, prominent teeth, or hypertrophy of the base of the tongue may be difficult candidates for endoscopy. Patients with ischemic cardiovascular disease may not withstand the prolonged laryngoscopic suspension that stimulates the vagus nerve and may produce subsequent cardiac arrhythmias as well as silent myocardial infarctions. Even in ideal circumstances, laser laryngoscopy may require more time to complete than other surgical techniques. For example, a patient with unstable cardiac disease suspected of having a laryngeal tumor may well be treated best with operative visualization and biopsy by conventional techniques rather than with the laser.[80]

Patients with chronic obstructive or restrictive pulmonary disease may be difficult to ventilate. In such situations, an endotracheal tube may be mandatory and by its presence may limit laryngeal visualization and access.

The precise tissue interaction and the relatively minimal morbidity rate associated with endoscopic $CO_2$ laser surgery results in only three contraindications for use. First, patients with circumferential stenosis of the trachea wider than 1 cm in length will probably have unfavorable results. Therefore, endoscopic management of these patients should be limited to the initial diagnostic procedure. The second contraindication to $CO_2$ laser bronchoscopy is tracheomalacia. In patients suffering from tracheal stenosis caused by tracheomalacia or loss of tracheal cartilaginous support, use of the laser bronchoscope can be dangerous. It may result in perforation of the trachea wall with subsequent mediastinitis or rupture of a great vessel. This complication obviously can be devastating. The third and final contraindication to $CO_2$ laser bronchoscopy is in patients with airway obstruction caused by extrinsic compression of the tracheobronchial tree. Extrinsic compression is not amenable to any form of endoscopic treatment.

## 44.6 Conclusions

The laser continues to be an important tool for the otolaryngologist. In surgery, the importance of the laser results from the ability to direct the beam and make precise incisions. The more recent additions of computer-controlled microscanners will add to the usefulness of the laser. The microscanners permit precise control and the ability to ablate lines or patterns with minimal loss of material. They can also reduce lateral thermal damage and thus reduce the delay in wound healing often associated with the laser.

The user must be appropriately trained before using any laser and fully understand all the advantages and possible complications. Appropriate safety measures are also necessary to prevent unnecessary morbidity and mortality.

Finally, the laser–tissue interactions must be understood. The otolaryngologist must always consider whether or not the laser is necessary. The benefits of laser use must always outweigh any potential problems or risks. Used appropriately the laser is both an important and a critical tool for otolaryngology.

### References

1. Maiman, T.H., Stimulated optical radiation in ruby, *Nature*, 187, 493, 1960.
2. Koester, C.J., Snitzer, E., Campbell, C.J., and Rittler, M.C., Experimental laser retina photocoagulation, *J. Opt. Soc. Am.*, 52, 607, 1962.
3. Goldman, L., Blaney, D., Kindel, D., and Franke, E.K., Effect of the laser beam on the skin, *J. Invest. Dermatol.*, 40, 121, 1963.
4. Goldman, L., Blaney, D., Kindel, D.J., Jr., Richfield, D., and Franke, E.K., Pathology of the effect of the laser beam on the skin, *Nature*, 197, 912, 1963.
5. Sataloff, J., Experimental use of laser in otosclerotic stapes, *Arch. Otolaryngol.*, 85, 614, 1967.
6. Hogberg, L., Stahle, J., and Vogel, K., The transmission of high-powered ruby laser beam through bone, *Acta Soc. Medicorum Upsaliensis*, 72, 223, 1967.
7. Jako, G.J., Laser surgery of the vocal cords: an experimental study with carbon dioxide laser on dogs, *Laryngoscope*, 82, 2204, 1972.
8. Strully, K.J. and Yahr, W., Biological effects of laser radiation enhancements by selective stains, *Fed. Proc.*, 24, S-81, 1965.
9. Polanyi, T.G., Bredemeier, H.C., and Davis, T.W., Jr., Lasers for surgical research, *Med. Biol. Eng.*, 8, 541, 1970.

10. Hale, G.M. and Querry, M.R., Optical constants of water in the 200-nm to 200-μm wavelength region, *Appl. Opt.*, 12, 555, 1973.
11. Bredemeier, H.C.,1969, Laser accessory for surgical applications, U.S. patent 3,659,613, issued 1972.
12. Bredemeier, H.C., 1973, Stereo laser endoscope, U.S. patent 3,796,220, issued 1974.
13. Strong, M.S. and Jako, G.J., Laser surgery in the larynx: early clinical experience with continuous $CO_2$ laser, *Ann. Otol. Rhinol. Laryngol.*, 81, 791, 1972.
14. Hall, R.R., The healing of tissues incised by carbon-dioxide laser, *Br. J. Surg.*, 58, 222, 1971.
15. Fleischer, D., Lasers and gastroenterology, a review, *Am. J. Gastroenterol.*, 79, 406, 1984.
16. Cochrane, J.P.S, Beacon, J.P., Creasey, G.H., and Russel, C.G., Wound healing after laser surgery: an experimental study, *Br. J. Surg.*, 67, 740, 1980.
17. LeCarpentier, G.L., Motamedi, M., McMath, L.P., Rastegar, S., and Welch, A.J., Continuous wave laser ablation of tissue: analysis of thermal and mechanical events, *IEEE Trans. Biomed. Eng.*, 40, 188, 1993.
18. Verdaasdonk, R.M., Borst, C., and van Gemert, M.J., Explosive onset of continuous wave laser tissue ablation, *Phys. Med. Biol.*, 35, 1129, 1990.
19. Wenig, B.L., Kurtzman, D.M., Grossweiner, L.I., Mafee, M.F., Harris, D.M., Lobraico, R.V., Prycz, R.A., and Appelbaum, E.L., Photodynamic therapy in the treatment of squamous cell carcinoma of the head and neck, *Arch. Otolaryngol. Head Neck Surg.*, 116, 1267, 1990.
20. Rausch, P.C., Rolfs, F., Winkler, M.R., Kottysch, A., Schauer, A., and Steiner, W., Pulsed versus continuous wave excitation mechanisms in photodynamic therapy of differently graded squamous cell carcinomas in tumor-implanted nude mice, *Eur. Arch. Oto-Rhino-Laryngol.*, 250, 82, 1993.
21. Okunaka, T., Kato, H., Conaka, C., Yamamoto, H., Bonaminio, A., and Eckhauser, M.L., Photodynamic therapy of esophageal carcinoma, *Surg. Endosc.*, 4, 150, 1990.
22. Shikowitz, M.J., Comparison of pulsed and continuous wave light in photodynamic therapy of papillomas: an experimental study, *Laryngoscope*, 102, 300, 1992.
23. Doukas, A.G., McAuliffe, D.J., and Flotte, T.J., Biological effects of laser-induced shock waves: structural and functional cell damage *in vitro*, *Ultrasound Med. Biol.*, 19, 137, 1993.
24. Norris, C.W. and Mullarky, M.B., Experimental skin incision made with the carbon dioxide laser, *Laryngoscope*, 92, 416, 1982.
25. Buell, B.R. and Schuller, D.E., Comparison of tensile strength in $CO_2$ laser and scalpel skin incisions, *Arch. Otolaryngol.*, 109, 465, 1983.
26. Apfelberg, D.B., Maser, M.R., Lash, H., and Rivers, J., The argon laser for cutaneous lesions, *J. Am. Med. Assoc.*, 245, 2073, 1981.
27. Perkins, R.C., Laser stapedotomy for otosclerosis, *Laryngoscope*, 90, 228, 1980.
28. Strunk, C.L., Jr. and Quinn, F.B., Jr., Stapedectomy surgery in residency: KTP-532 laser versus argon laser, *Am. J. Otol.*, 14, 113, 1993.
29. DiBartolomeo, J.R. and Ellis, M., The argon laser in otology, *Laryngoscope*, 90, 1786, 1980.
30. Levine, H.L., Endoscopy and the KTP/532 laser for nasal sinus disease, *Ann. Otol. Rhinol. Laryngol.*, 98, 46, 1989.
31. Thedinger, B.S., Applications of the KTP laser in chronic ear surgery, *Am. J. Otol.*, 11, 79, 1990.
32. Atiyah, R.A., Friedman, C.D., and Sisson, G.A., The KTP/532 laser in glossal surgery: KTP/532 clinical update, Laserscope, San Jose, CA, Report No. 22, 1988.
33. Ward, R.F., Treatment of tracheal and endobronchial lesions with the potassium titanyl phosphate laser, *Ann. Otol. Rhinol. Laryngol.*, 101, 205, 1992.
34. Joseph, M., Reardon, E., and Goodman, M., Lingual tonsillectomy: a treatment for inflammatory lesions of the lingual tonsil, *Laryngoscope*, 94, 179, 1984.
35. Krespi, Y.P., Har-El,G., Levine, T.M., Ossoff, R.H., Wurster, C.F., and Paulsen, J.W., Laser laryngeal tonsillectomy, *Laryngoscope*, 99, 131, 1989.
36. Kuhn, F., The KTP/532 laser in tonsillectomy: KTP/532 clinical update, Laserscope, San Jose, CA, Report No. 06, 1988.

37. Strunk, C.L. and Nichols, M.L., A comparison of the KTP/532-laser tonsillectomy vs. traditional dissection/snare tonsillectomy, *Otolaryngol. Head Neck Surg.*, 103, 966, 1990.
38. Linden, B.E., Gross, C.W., Long, T.E., and Lazar, R.H., Morbidity in pediatric tonsillectomy, *Laryngoscope*, 100, 120, 1990.
39. Bartels, L.J., KTP laser stapedotomy: is it safe? *Otolaryngol. Head Neck Surg.*, 103, 685, 1990.
40. Strunk, C.L., Quinn, F.B., Jr., and Bailey, B.J., Stapedectomy techniques in residency training, *Laryngoscope*, 102, 121, 1992.
41. McGee, T.M., The KTP/532 laser in otology: KTP/532 clinical update, Laserscope, San Jose, CA, Report No. 08, 1988.
42. Atiyah, R.A., The KTP/532 laser in laryngeal surgery: KTP/532 clinical update, Laserscope, San Jose, CA, Report No. 21, 1988.
43. McDougall, J.C. and Cortese, D.A., Neodymium-YAG laser therapy of malignant airway obstruction, *Mayo Clin. Proc.*, 58, 35, 1983.
44. Shapshay, S.M. and Simpson, G.T., Lasers in bronchology, *Otolaryngol. Clin. North Am.*, 16, 879, 1983.
45. Toty, A., Personne, C., Colchen, A., and Vourc'h, G., Bronchoscopic management of tracheal lesions using the Nd:YAG laser, *Thorax*, 36, 175, 1981.
46. Dumon, J.F., Reboud, E., Garbe, L., Aucomte, F., and Meric, B., Treatment of tracheobronchial lesions by laser photoresection, *Chest*, 81, 278, 1982.
47. Bemis, J.F., Jr., Vergos, K., Rebeiz, E.E., and Shapshay, S.M., Endoscopic laser therapy for obstructing tracheobronchial lesions, *Ann. Otol. Rhinol. Laryngol.*, 100, 413, 1991.
48. Fleischer, D., Endoscopic laser therapy for gastrointestinal neoplasms, *Otolaryngol. Clin. North Am.*, 64, 947, 1984.
49. Shapshay, S.M. and Oliver, P., Treatment of hereditary hemorrhagic telangiectasia by Nd-YAG laser photocoagulation, *Laryngoscope*, 94, 1554, 1984.
50. Rebeiz, E., April, M.M., Bohigian, R.K., and Shapshay, S.M., Nd-YAG laser treatment of venous malformations of the head and neck: an update, *Otolaryngol. Head Neck Surg.*, 105, 655, 1991.
51. April, M.M., Rebeiz, E.E., Friedman, E.M., Healy, G.B., and Shapshay, S.M., Laser therapy for lymphatic malformations of the upper aerodigestive tract: an evolving experience, *Arch. Otolaryngol. Head Neck Surg.*, 118, 205, 1992.
52. Cosman, B., Experience in the argon laser therapy for port-wine stains, *Plast. Reconstr. Surg.*, 65, 119, 1980.
53. Parkin, J.L. and Dixon, J.A., Argon laser treatment of head and neck vascular lesions, *Otolaryngol. Head Neck Surg.*, 93, 211, 1985.
54. Spitzer, M. and Krumholz, B.A., Photodynamic therapy in gynecology, *Obstet. Gynecol. Clin. North Am.*, 18, 649, 1991.
55. Wang, Z., Pankratov, M.M., Gleich, L.L., Rebeiz, E.E., and Shapshay, S.M., New technique for laryngotracheal mucosa transplantation: "stamp" welding using indocyanine green dye and albumin interaction with diode laser, *Arch. Otolaryngol. Head Neck Surg.*, 121, 773, 1995.
56. Wang, Z., Pankratov, M.M., Rebeiz, E.E., Perrault, D.F., Jr., and Shapshay, S.M., Endoscopic diode laser welding of mucosal grafts on the larynx: a new technique, *Laryngoscope*, 105, 49, 1995.
57. Volk, M.S., Wang, Z., Pankratov, M.M., Perrault, D.F., Jr., Ingrams, D.R., and Shapshay, S.M., Mucosal intact laser tonsillar ablation, *Arch. Otolaryngol. Head Neck Surg.*, 122, 1355, 1996.
58. Kao, M.C., Video endoscopic sympathectomy using a fiberoptic $CO_2$ laser to treat palmar hyperhidrosis, *Neurosurgery*, 30, 131, 1992.
59. Ossoff, R.H., Duncavage, J.A., Gluckman, J.L., Adkins, J.P., Karlan, M.S., Toohill, R.J., Keane, W.M., Norris, C.W., and Tucker, J.A., The universal endoscopic coupler bronchoscopic carbon dioxide laser surgery: a multi-institutional clinical trial, *Otolaryngol. Head Neck Surg.*, 93, 824, 1985.
60. Ossoff, R.H., Sisson, G.A., and Shapshay, S.M., Endoscopic management of selected early vocal cord carcinoma, *Ann. Otol. Rhinol. Laryngol.*, 94, 560, 1985.

61. Ossoff, R.H., Werkhaven, J.A., Raif, J., and Abraham, M., Advanced microspot microslad for the $CO_2$ laser, *Otolaryngol. Head Neck Surg.*, 105, 411, 1991.
62. Shapshay, S.M., Wallace, R.A., Kveton, J.F., Hybels, R.L., and Setzer, S.E., New microspot micromanipulator for $CO_2$ laser application in otolaryngology — head and neck surgery, *Otolaryngol. Head Neck Surg.*, 98, 179, 1988.
63. Simpson, G.T. and Polanyi, T.G., History of the carbon dioxide laser in otolaryngologic surgery, *Otolaryngol. Clin. North Am.*, 16, 739, 1983.
64. Shapshay, S.M., Beamis, J.F., Jr., and Dumon, J.F., Total cervical tracheal stenosis: treatment by laser, dilation, and stenting, *Ann. Otol. Rhinol. Laryngol.*, 98, 890, 1989.
65. Whitehead, E. and Salam, M.A., Use of the carbon dioxide laser with the Montgomery T-tube in the management of extensive subglottic stenosis, *J. Laryngol. Otol.*, 106, 829, 1992.
66. Beste, D.J. and Toohill, R.J., Microtrapdoor flap repair of laryngeal and tracheal stenosis, *Ann. Otol. Rhinol. Laryngol.*, 100, 420, 1991.
67. Dedo, H.H. and Sooy, C.D., Endoscopic laser repair of posterior glottic, subglottic and tracheal stenosis by division or micro-trapdoor flap, *Laryngoscope*, 94, 445, 1984.
68. Duncavage, J.A., Ossoff, R.H., and Toohill, R.J., Laryngotracheal reconstruction with composite nasal septal cartilage grafts, *Ann. Otol. Rhinol. Laryngol.*, 98, 565, 1989.
69. Duncavage, J.A., Piazza, L.S., Ossoff, R.H., and Toohill, R.J., Microtrapdoor technique for the management of laryngeal stenosis, *Laryngoscope*, 97, 825, 1987.
70. Strong, M.S., Vaughan, C.W., Cooperband, S.R., Healy, G.B., and Clemente, M.A., Recurrent respiratory papillomatosis management with the $CO_2$ laser, *Ann. Otol. Rhinol. Laryngol.*, 85, 508, 1976.
71. Ossoff, R.H., Werkhaven, J.A., and Dere, H., Soft-tissue complications of laser surgery for recurrent respiratory papillomatosis, *Laryngoscope*, 101, 1162, 1991.
72. Crockett, D.M., McCabe, B.F., and Shive, C.J., Complications of laser surgery for recurrent respiratory papillomatosis, *Ann. Otol. Rhinol. Laryngol.*, 96, 639, 1987.
73. Wetmore, S.J., Key, J.M., and Suen, J.Y., Complications of laser surgery for laryngeal papillomatosis, *Laryngoscope*, 95, 798, 1985.
74. Abramson, A.L., DiLorenzo, T.P., and Steinberg, B.M., Is papillomavirus detectable in the plume of laser-treated laryngeal papilloma? *Arch. Otol. Head Neck Surg.*, 116, 604, 1990.
75. Garden, J.M., O'Banion, M.K., Shelnitz, L.S., Pinski, K.S., Bakus, A.D., Reichmann, M.E., and Sundberg, J.P., Papillomavirus in the vapor of carbon dioxide laser-treated verrucae, *J. Am. Med. Assoc.*, 259, 1199, 1988.
76. Leventhal, B.G., Kashima, H.K., Mounts, P., Thurmond, L., Chapman, S., Buckley, S., and Wold, D., Long-term response of recurrent respiratory papillomatosis to treatment with lymphoblastoid interferon alfa-N1. Papilloma Study Group, *New Engl. J. Med.*, 325, 613, 1991.
77. Kamami, Y.-V., Laser $CO_2$ for snoring: preliminary results, *Acta Oto-Rhino-Laryngol. Belg.*, 44, 451, 1990.
78. Croaker, B.D., Allison, G.L., Saunders, N.A., Henley, M.J., McKeon, J.L., Allen, K.M., and Gyulay, S.G., Estimation of the probability of disturbed breathing during sleep before a sleep study, *Am. Rev. Respir. Dis.*, 142, 14, 1990.
79. Young, T., Palta, M., Dempsey, J., Skatrud, J., Weber, S., and Badr, S., The occurrence of sleep disordered breathing among adults, *New Engl. J. Med.*, 328, 1230, 1993.
80. Fried, M.P., Kelly, J.H., and Strome, M., *Complications of Laser Surgery of the Head and Neck*, Year Book Medical Publishers, Chicago, 1986.

# 45

# Lasers in Urology

John T. Leyland II
*Medical College of Ohio*
*Toledo, Ohio*

Detlef F. Albrecht
*Alza Corporation*
*Mountain View, California*

Steven H. Selman
*Medical College of Ohio*
*Toledo, Ohio*



## 45.1 Introduction

The modern era of urology was inaugurated when incandescent light was cystoscopically introduced into the urinary bladder. Lasers represent one aspect of the continued development and application of light technology in urology. The photothermal, photochemical, and photomechanical properties of lasers are used for both the diagnosis and treatment of urologic disease.

## 45.2 Nonmalignant Urologic Disease

### 45.2.1 Lasers for the Treatment of Urinary Lithiasis

Although early investigators demonstrated the feasibility of the use of laser energy for stone fragmentation, the first effective clinical application of laser lithotripsy was reported in 1987. A pulsed-dye laser (504 nm, 1 μsec) was coupled to a 250-μm quartz fiber for endoscopic stone fragmentation. Excellent fragmentation of all but the hardest cystine and calcium oxalate monohydrate stones was achieved.[1] The mechanism of stone fragmentation was determined to be a concentrically propagating cavitation bubble created at the laser–stone interface, collapsing and producing a photoacoustic shockwave and causing the calculus to fragment.[2,3] There are no direct photothermal effects from the pulsed-dye laser, and ureteral trauma from the acoustic shockwave is minimal.[4] The pulsed-dye laser continues to be a "workhorse" for urologic laser lithotripsy at many institutions with overall success rates reported between 80 and 95%.[5]

Subsequent investigations with the Q-switched neodymium:yttrium-aluminum-garnet (Nd:YAG) and alexandrite lasers demonstrated that these lasers were effective in fragmenting calculi.[6,7] Success rates

Portions of this chapter were published in Grafstein, L.A. and Selman, S.H., *J. Clin. Laser Med. Surg.*, 16(1), 49, 1998. Reprinted with permission.

0-8493-1116-0/03/$0.00+$1.50

were reported in excess of 90%. However, both were limited in their clinical application by high material costs and the lack of success in the treatment of stones composed of cystine and calcium oxalate monohydrate.

Recently the holmium:YAG (Ho:YAG) laser has moved to the forefront of laser lithotripsy technology. In contrast to the pulsed-dye laser whose therapeutic efficacy is based on photoacoustic properties, the Ho:YAG laser's effectiveness as a lithotripter is related to its thermal mechanics. The energy of the Ho:YAG laser also forms a plasma vapor bubble at the fiber tip, but this bubble is pear shaped rather than concentric.[8] Laser energy is transmitted efficiently through the bubble of water vapor (the so-called "Moses effect") directly to the stone itself.[9] Water trapped in the stone interstices is then superheated, causing fracture and disintegration. The photoacoustic properties of the cavitation bubble in this case are negligible.[10] Sayer et al. in 1993 reported effective fragmentation of all varieties of stone with the Ho:YAG laser including cystine and calcium oxalate monohydrate (brushite) calculi.[11] Results reported by Densted et al.,[12] Grasso et al.,[13] Devarajan et al.,[14] and others[15–17] describe stone-free success rates in excess of 90%. Equally impressive is the success this laser has had when used in combination with miniaturized actively deflectable ureteroscopes to reach the proximal ureter and even intrarenal calculi.[18,19] With a 200-μm fiber and an actively deflectable ureteroscope, the entire intrarenal collecting system is accessible for retrograde approach. Grasso has reported an 88.5% success rate in these situations — clearing 23 of 26 patients of their stone burden in one session.[20] Other reports cite a single-session success rate of 60 to 79% for these upper-tract calculi.[21,22] Such an approach may be prohibitively time-consuming or more technically difficult than the percutaneous approach.

The Ho:YAG laser has been shown to produce smaller stone fragments than other modes of lithotripsy.[23] In an *in vitro* study by Teichman et al., the size of stone fragments produced by the Ho:YAG was compared to stone fragments collected following treatment with the electrohydraulic lithotripter, mechanical lithoclast, and pulsed-dye laser. The Ho:YAG was the only instrument that yielded no fragments >4 mm and had a greater percentage of fragments <1 mm. Reducing the stone to "dust" or "sand" rather than fragments considerably improves the success of treatment as measured by "stone-free" status.[23] The laser itself causes very little retrograde propulsion of the stone to more proximal locations. Where proximal migration has been described in association with this laser, it is more often related to irrigation solution than the laser effect itself.[10,14]

Concerns for the safety and integrity of surrounding mucosal lining are minimized by careful and accurate placement of the fiber tip in direct visual contact with the stone. Care must be taken to avoid blindly drilling straight through a stone into adjacent tissue, where injury may occur. If the laser is fired with the fiber tip in contact with mucosal tissue, thermal injury and perforation will result. The Ho:YAG is a "thermal" laser and does possess tissue coagulative and ablative properties. While microsecond pulsatile delivery and the high degree of energy absorption by the water medium significantly increase the margin of safety, direct-contact mucosal irradiation will vaporize tissue up to a depth of 0.5 mm.[10,24] For this reason it is essential that this laser be used only under direct vision.[25] Side firing tips have proven useful in this regard, particularly when approaching stones in the bladder or in the renal pelvis or calyces. During percutaneous nephrolithotomy, where it may be difficult to position the laser-fiber tip perpendicular to the stone surface or where maneuverability for access is limited, a side-firing fiber will allow the endoscopist to maintain the proper 0° incident angle for efficient energy delivery.[26]

The most frequently reported complication following Ho:YAG laser lithotripsy has been ureteral stricture formation. In the majority of cases, this has been attributed to the semirigid ureteroscopy rather than to the laser itself.[14,16,20,24] Previous stone impaction and unsuccessful attempts at stone manipulation are also noted as contributing risk factors for subsequent stricture formation observed following endoscopic laser lithotripsy.[24] Other potential complications involve metal endoscopic stone baskets and guidewires that may be cut inadvertently with the Ho:YAG laser. Resulting sharp wire barbs may be traumatic to tissue, and therefore extreme care must be taken when using these instruments in conjunction with this laser.[27] The development of preformed soft-wire baskets have reduced the risk of injury if the wire is transected.

One interesting theoretical concern has been raised involving the thermochemical interactions of the Ho:YAG laser with uric-acid stones. As reported by Teichman et al., when uric-acid stones are heated by irradiation with Ho:YAG energy, hydrogen cyanide is released as a chemical-breakdown product.[28] In their series of 18 patients, there was no evidence of cyanide toxicity; however, actual blood levels were not measured. The minimum lethal oral dose of cyanide is 50 mg. *In vitro* studies, also by Teichman et al., used a 940-μm fiber at 1.0-J pulse to deliver 1 kJ of Ho:YAG irradiation to a 97% pure uric-acid stone.[29] As a result, 400 mg of cyanide was produced. In a clinical series, several patients with uric-acid bladder calculi received upwards of 300 kJ of energy. No apparent ill effects were observed. The absorption capacity of urothelium is not known, and irrigation fluid used during lithotripsy may play a significant role in minimizing the absorption potential. The authors conclude that there is a high margin of safety based on these findings but encourage caution in situations where increased absorption might be observed due to bleeding or high-pressure irrigation.[29]

## 45.2.2 Lasers for the Treatment of Benign Prostatic Hyperplasia

Urinary frequency, incomplete bladder emptying, and nocturia can be caused by enlargement of the prostate. The surgical "gold standard" for treatment has been the transurethral electrosurgical resection of obstructing prostatic tissue. Although transurethral resection of the prostate is a minimally invasive technique, it is not without significant morbidity principally due to blood loss. Laser technology is evolving as a viable treatment option for this common problem. The Nd:YAG laser was first used for the treatment of prostatic enlargement practice in the early 1990s. Two technologies were developed, both dependent on the thermal coagulative effects of the Nd:YAG laser. The transurethral ultrasound-guided laser-induced prostatectomy (TULIP) incorporated a side-firing Nd:YAG laser that used real-time ultrasound monitoring during treatment.[30] Ultrasound was incorporated for optimal guidance of laser energy. The expense and complexity of the procedure led to its abandonment, although its success in treatment was reported in a multicenter trial.

Visual laser ablation of the prostate (VLAP) as introduced was more appealing to urologists in that it incorporated endoscopic rather than ultrasonic techniques for the direct application of laser energy (Nd:YAG) to the obstructing prostatic tissue. Pioneered by Costello et al., the technique incorporated a gold alloy tip at the end of the laser-delivery fiber, which allowed right-angle delivery of laser energy directly into the obstructing prostate tissue.[31] Coagulation of treated prostatic tissue created the desired therapeutic effect. A number of reports demonstrated the utility of this approach, as evidenced by improvement in symptom scores and urinary-flow rates. However, some patients required prolonged catheterization in the postoperative period, while others developed severe irritative voiding symptoms that only slowly resolved as treated tissue gradually sloughed or resorbed. Because of these problems this technique is no longer widely practiced. As an alternative to the free-beam technology, contact Nd:YAG-laser techniques were developed for the removal of obstructing tissue.[32] However, the slow pace of this technique has led to its lack of widespread acceptance.

Transurethral interstitial laser-energy application, designed to cause coagulation necrosis of obstructing prostatic tissue while preserving the urethral mucosa, is another laser-based approach to the treatment of benign prostatic hyperplasia.[33] A 600-um PTFE-coated fiber with a 1-cm diffusing tip is placed transurethrally through the urethral wall into the targeted prostatic tissue. An 830-um diode laser provides the energy for treatment. Treatment is directed at heating the targeted tissue to 100°C, and temperature in the treatment area is controlled by a feedback system. Preliminary clinical results have demonstrated an improvement in symptom scores and maximum flow rates after treatment.[34]

A technique for enucleation of the prostate using the Ho:YAG laser has been pioneered by Gilling et al.[35] Sections of obstructing prostate are removed by a combination of the cutting and ablative effects of the laser. The procedure has been associated with little blood loss and creates a prostatic cavity similar to that produced by transurethral electrosurgical resection. The large pieces of enucleated material fall into the bladder and are subsequently morcellated by a mechanical device. Extremely large obstructing prostate adenomas have been removed using this approach.[36]

The potassium titanyl phosphate (KTP) laser has been used clinically to vaporize obstructing prostatic tissue. There are reports of its use in combination with the Nd:YAG. The KTP vaporizes the bladder-neck tissue, and the Nd:YAG is used on the remainder of the obstructing tissue.[37] This combination of lasers has been reported to be associated with fewer irritative voiding symptoms than when the Nd:YAG is used alone. A high-power (60-W) KTP laser has been developed as the sole laser for prostate vaporization. Initial reports on this laser have been encouraging, with all men catheter free in 24 h and none requiring blood transfusion.[38]

### 45.2.3 Lasers for the Treatment of Urethral Stricture Disease

Strictures of the male urethra arise from inflammatory, traumatic, or iatrogenic causes. Cicatrix formation compromises the urethra lumen, which restricts normal urinary flow. Techniques for management included periodic dilatation (bouginage), cold-knife incision, or open surgical repair with or without tissue transfer.

Bulow et al. were among the early investigators using the Nd:YAG for the treatment of urethra stricture disease.[39] Five patients were treated without complications; however, long-term follow-up was not reported. In 1980, Rothauge reported on the use of the argon laser in 40 patients with stricture disease. Six patients required a second treatment.[40] Again, long-term follow-up was not provided. Smith and Dixon reported on the use of the Nd:YAG in 17 patients. Of these patients, 11 (64%) suffered recurrent strictures within 6 months of treatment, leading the investigators to conclude that Nd:YAG treatment offered no advantage over conventional therapy. However, within this series all patients had undergone multiple urethral dilatations, and 13 had been treated prior to laser with cold-knife urethrotomy.[41] Subsequent studies using the Nd:YAG laser have offered more optimistic results. Perkash has described the use of a contact chisel tip to achieve a 93% success rate at 28.2 months for 42 spinal-cord-injury patients.[42] Another recent investigation has employed an 805-nm aluminum-gallium-arsenide diode laser. Kamal has reported on 22 patients treated in this manner with 23.2-month mean follow-up. A 78.6% success rate is described for those patients not previously treated. Of 8 patients with more complex strictures previously treated with cold-knife urethrotomy, 7 had further recurrence. However, the mean time to recurrence in these patients did increase from 8 to 16 months.[43]

Bladder-neck contractures after prostatectomy represent stricture formation at the vesicourethral junction. Contact-laser ablation of the narrowed area has been reported with the use of the Nd:YAG laser. Silber and Servadio reported on 21 patients with excellent results.[44]

### 45.2.4 Miscellaneous Use of Lasers

Sporadic use of lasers for a variety of urologic problems have been reported. Among these are posterior urethral valves,[45,46] bladder hemangiomas,[47] and ureteral strictures.[48–50] Each of these has been studied in small series, and further long-term follow-up is needed, though the initial results have been favorable.

## 45.3 Urologic Oncology

### 45.3.1 Thermal Lasers in Urologic Oncology*

Thermal lasers refer to those lasers developing heat in targeted tissue to create their therapeutic effects. The production of heat in laser-targeted tissues depends on the emitting laser wavelength and operating mode (pulsed vs. continuous) in combination with the absorbing and scattering characteristics of the

*Parts of the following appeared in Grafstein, L.A. and Selman, S.H., *J. Clin. Laser Med. Surg.*, 16(1), 49, 1998.

tissues. Thermal effects in tissues will also vary with laser-beam power density, time of irradiation, and the presence of heat sinks (blood flow). Within tissue, water content as well as the presence of absorbing pigments (chromophores) will largely determine the extent of tissue penetration and scattering of laser energy.

Tissue effects produced by thermal lasers will differ depending on the above parameters. Once laser energy is deposited in the targeted tissue the effects will vary with type of laser delivery (free beam, contact tip) as well as tissue characteristics. When temperatures within the targeted tissue rise above 60°C, protein denaturation occurs with tissue "coagulation." When temperatures rise above 100°C, intracellular water begins to boil and evaporate. This evaporation of tissue is referred to as ablation, since tissue is removed from the target as a laser plume of debris. Further heating results in tissue carbonization, the process by which carbon remains after vaporization. In a sense, the coagulative and evaporative effects are competitive in their effects on the target tissue. High power delivered over a short interval creates the ablative effects, while lower power delivered over a longer time interval creates a coagulative effect.

In some clinical situations, precise localization of thermal energy is required. Localization of thermal effects can be achieved through the use of contact tips placed on the end of quartz laser fibers. These tips convert light energy to thermal energy within the tip. This energy then serves to cut or coagulate tissue.

#### 45.3.1.1 Carcinoma of the Penis

The challenge in the treatment of squamous-cell carcinoma of the penis is preservation of cosmetic and functional integrity while eradicating the neoplastic process. Standard surgical treatment of distal penile lesions includes circumcision for tumors of the prepuce or distal penectomy for tumors of the glans penis, while total penectomy may be necessary for penile-shaft lesions. Carcinoma *in situ* (CIS) (Erythroplasia of Queyrat) can be effectively treated with topical 5-fluorouracil cream, but in some patients severe normal-tissue reaction limits its use. It is toward the organ-preserving treatment of superficial penile cancers that thermal lasers have shown greatest promise and utility.

Parsons et al., in one of the earliest genitourinary applications of thermal laser technology, used the ruby laser for the treatment of carcinoma of the penis. The patient was treated with pulsed laser energy in two sessions. In the first, black shoe polish was applied to the lesion to enhance absorption. Although the treatment proved less than completely successful, the author clearly showed the potential for laser energy to be used for the treatment of the neoplastic process of the penis.[52] Rosemberg reported the first case of $CO_2$-laser tumor vaporization for a patient with Erythroplasia of Queyrat.[53] Rosemberg subsequently published his results on the treatment of three cases of penile cancer in which both the $CO_2$ and Nd:YAG lasers were used. In two of the cases, the $CO_2$ laser was used as a scalpel to perform partial penectomy, while in the other laser energy was used to vaporize and ablate the neoplastic lesions.[54] Bandieramonte et al. used laser microsurgery for the treatment of superficial lesions of the penis, some of which were nonneoplastic or premalignant. The $CO_2$ laser was used in conjunction with the operating microscope for resecting lesions. The aim of the procedure was to remove tissues while preserving the architecture for pathologic evaluation. The investigators felt that this approach served both aims well in addition to maintaining cosmesis.[55]

Between 1977 and 1981, Hofstetter utilized the Nd:YAG laser in 17 patients with T1 and T2 penile carcinomas; 7 of these patients had no local relapse or metastases in 4 years of follow-up.[56] Seeking to preserve penile integrity, Rothenberger utilized the Nd:YAG laser as an adjunct to local surgical resection in 19 patients (13, T1:6, T2). Immediately after surgical resection of all gross tumor, the tumor bed and a 0.5-cm margin were treated with the laser using 45 W for 3 sec. A total of 17 of the 19 patients evaluated at 1 year were free of disease.[57] Malloy et al. treated 16 men (5 TIS, 9 T1, and 2 T2) with squamous-cell carcinoma of the penis with the Nd:YAG laser. Laser irradiation was applied with a focusing hand piece, and iced saline was applied to prevent surface carbonization. Within the follow-up period ranging from 18 to 36 months none of the TIS patients had a recurrence, while 6 of the 9 T1 patients were free of disease. Neither of the T2 patients was free of cancer, and one of these had developed a urethrocutaneous fistula at the site of therapy.[58] Malek reported the Mayo Clinic experience on the use of the $CO_2$, Nd:YAG, and KTP lasers for the ablation of premalignant and malignant penile lesions. A total of 25 of 30 patients had either premalignant carcinoma or CIS. Although the number of T1 and T2 lesions was small, the author concluded that laser treatment was clinically equal and functionally superior to standard surgical therapy.[59]

These studies suggest that thermal lasers are capable of eradicating superficial neoplasms of the penis with excellent cosmetic and functional results. One caveat is that patients must be informed of the prolonged healing time, which may be up to 3 to 4 months when the Nd:YAG laser is used.

#### 45.3.1.2 Prostate Cancer

Carcinoma of the prostate is the most common neoplasm in men over the age of 50. Although both radiation therapy and radical prostatectomy have proved to be excellent forms of therapy, alternative methods of treatment are being pursued. To date, thermal lasers have not been widely applied for the treatment of carcinoma of the prostate.

A series of articles by Beisland et al. explored the use of thermal lasers (Nd:YAG) for residual prostate ablation subsequent to "radical" transurethral resection of the prostate (TURP). In the initial study, 5 patients (3 T1; 2 T2) were treated with transurethral Nd:YAG laser energy (50 to 55 W, 4 sec) 4 to 6 weeks after TURP. The laser tip, introduced through a standard cytoscope, was rotated between 12 and 6 o'clock and gradually drawn back from the bladder neck toward the prostatic apex. No bleeding, perforations, or infection was encountered.[60] In a subsequent report, 16 patients were treated. A technique modification had been added in which the laser endoscopic was introduced through a suprapubic cystostomy. This was done to facilitate treatment of the prostatic apex.[61] In a third publication, the number of patients increased to 63 (36 T1, 27 T2). Rectal temperatures were monitored with a thermistor. In addition, transrectal ultrasound was used to determine whether an ultrasound signature of thermally treated prostate could be ascertained. The investigators found none. With a mean follow-up of 22 months, 56 patients were "free of disease." Unfortunately, prostatic-specific antigen (PSA) was not measured, and clinical follow-up was short.[62] Although this approach to the treatment of prostate cancer is appealing, the inability to adequately monitor temperatures in the critical areas, such as the prostatic capsule, bladder neck, and prostatic apex, is a drawback. However, adjunctive technologies, such as transrectal MRI or interstitial thermistors, may make this approach feasible.

The standard treatment of metastatic carcinoma of the prostate is hormonal manipulation with the aim of lowering serum testosterone to anorchid levels. The use of luteinizing hormone-releasing hormone (LHRH) agonists has supplanted oral estrogen therapy. Some men with metastatic prostate cancer, however, are treated with bilateral orchiectomy. Wishnow and Johnson reported on the use of the $CO_2$ laser for subcapsular orchiectomy. This surgical approach to castration has the putative advantage of fewer complications such as bleeding and pain. The authors used a $CO_2$ laser as a surgical knife to incise the scrotum, dartos fascia, and tunica albuginea of the testicle. The testicular parenchyma was then removed by dissection with a moist sponge. The visceral surface of the tunica albuginea was ablated with the laser to assure complete removal of testicular tissue and assure hemostasis. Of the 13 patients so treated, one developed a scrotal hematoma. Postoperative serum testosterone levels were in the anorchid range.[63] Bolton and Costello used the same approach on 17 patients. Their mean operating time was 32 min. None of the patients developed complications. Beta human chorionic gonadotrophin stimulation (5,000 U) administered to 10 patients confirmed the adequacy of orchiectomy.[64]

#### 45.3.1.3 Bladder Cancer

Thermal lasers have been widely used in the treatment of bladder tumors. The Nd:YAG, argon, and Ho:YAG lasers have all been used in the treatment of bladder tumors. Because there is no fiber system for transmission of $CO_2$ laser energy, the $CO_2$ laser has minimal usefulness for the treatment of bladder tumors. The development of a new cystoscope that enables the transmission of a $CO_2$ laser beam in a $CO_2$-gas-filled bladder may render this laser useful in the future.[65]

Smith and Dixon reported on the treatment of 11 patients with superficial transitional-cell carcinoma (TCC) of the bladder with the argon laser. Of these 11 patients, 9 had low-grade papillary lesions, while

2 patients had CIS. Follow-up revealed no local tumor recurrence in the nine patients with papillary lesions. However, three patients had remote tumor recurrence outside the treatment area. One patient with CIS had persistent lesions and positive cytology on follow-up. The authors concluded that the argon laser should be limited to patients with known low-grade, low-stage TCC in whom histologic confirmation is not necessary. Lesions treated in this manner should be either less than 1.0 cm in size or CIS.[66]

The Nd:YAG laser is the most widely used laser for the treatment of bladder neoplasms. This laser can be used without general anesthesia and is associated with minimal blood loss and obturator-nerve stimulation. Catheter drainage is usually not necessary, and tumor ablation can often be done on an outpatient basis.[67–70]

It has been postulated that the "no touch " technique of the Nd:YAG bladder-tumor ablation decreases exfoliation of tumor cells and therefore decreases tumor recurrences. Several series have compared the number of tumor recurrences after standard transurethral electrosurgical resection (TUR) to those seen after the use of the Nd:YAG laser. Hofstetter found a significantly decreased ($p < 0.001$) rate of recurrence in 66 patients (Ta-T2) treated with the Nd:YAG compared to standard laser. He also observed a decreased local recurrence rate of 1% with the Nd:YAG vs. 40% with TUR.[71] Beisland and Seland in a prospective randomized study of 122 patients with stage Ta-T2 disease showed a decreased local recurrence rate of 4.8% with the Nd:YAG compared to 31.6% with TUR. However, the appearance of new foci of TCC outside the area of previous resection or ablation was similar with the Nd:YAG and TUR (20 vs. 22%) despite the "no touch" approach of the former.[72] Beer et al. found similar results, with a 7% recurrence rate for the "no touch" Nd:YAG ablation and 25% recurrence after TUR. Neither technique offered a clear benefit to preventing the appearance of new lesions outside the original margin of resection.[73]

Johnson reported on the use of Ho:YAG in the treatment of 15 patients with superficial bladder cancer. Follow-up revealed 27% of patients had no recurrent disease; 53% had out-of-field recurrences, and 20% had in-field recurrences.[74] The local recurrence rate is higher than the 4.8% reported for the Nd:YAG.[72] Further clinical experience should clarify the efficacy of the Ho:YAG laser in the treatment of bladder cancer.

Muscle-invasive bladder cancer is usually treated by radical cystectomy; however, in selected cases the Nd:YAG laser, because of its deep coagulation effects, can be used for treatment. The Nd:YAG laser has the potential to treat these lesions. Two problems exist limiting the effectiveness of the Nd:YAG for the general treatment of muscle-invasive disease — (1) errors in staging and (2) inability of the Nd:YAG to reliably and consistently achieve full-thickness penetration of the bladder.[75] Smith reported on 21 patients treated with the Nd:YAG for muscle-invasive bladder cancer. Patients had clinical stages B1 to D disease. Patients were preselected because they were either poor surgical risks or they refused cystectomy. On follow-up he obtained negative biopsies in 80% of B1 lesions, 50% of B2 lesions, 25% of C lesions, and 0% of D lesions.[76] McPhee's results were not as encouraging. He had recurrences or residual disease in 41.6% of his stage T2(B1) lesions, 57% of T3a(B2) lesions, and 100% of his T3b© lesions.[75] Beisland and Sander treated 15 patients with the Nd:YAG laser for stage T2 disease. Of these 15 patients, 4 failed treatment.[78] These authors concluded that laser treatment for invasive TCC should be reserved for selected patients who are either poor surgical candidates for cystectomy or who refuse cystectomy.

Clinical data have not proved the laser more efficacious in reducing overall tumor recurrence rates as compared to TUR, although there is evidence of decreased local tumor recurrence in the laser-treated area. The laser is effective in eradicating small superficial lesions and is a viable option when the pathologic stage is apparent and no tissue is needed for staging. Radical cystectomy remains the standard treatment for muscle-invasive bladder cancer; however, laser treatment may be an alternative treatment option for high-risk surgical candidates or for those refusing cystectomy.

#### 45.3.1.4 Ureteral and Renal Pelvis Carcinoma

The traditional approach to the treatment of transitional-cell carcinoma of the ureter and renal pelvis has been nephroureterectomy. However, it has been recognized for some time that conservative (parenchymal sparing) procedures (e.g., partial ureterectomy for ureteral tumors) have their place in the management of patients with upper-tract urothelial cancers.[79] This is especially true in patients with compromised renal function, bilateral cancer, or solitary kidney. Survival after ureteral resection for low-grade lesions is comparable to those achieved with nephroureterectomy.[80] The widespread availability of small urologic endoscopic instruments, both flexible and rigid, facilitating access to the entire collecting system has naturally led to the use of thermal lasers for endoscopic upper-urinary-tract-tumor ablation. Both percutaneous and transureteral approaches have been used.

Schilling et al. treated 13 ureteral tumors in ten patients transureteroscopically using the Nd:YAG laser (40 W, 1 to 3 sec). They reported excellent patient tolerance with no complications. Complete healing of the lasered site was documented in one patient who underwent nephroureterectomy 3 weeks following laser therapy. The authors noted that the small working space within the ureter limited the approach to the tumor to a tangential one creating a shadow area within the targeted tissue.[81] Smith et al. reported on percutaneous treatment of renal pelvic tumors in nine selected patients. The investigators first gained access to the renal pelvis using standard percutaneous techniques and then undertook tumor resection using electrocautery. Subsequently (2 to 28 d later), Nd:YAG laser treatment (15 to 20 W, 2 sec) of the previously resected site was undertaken. Within the short mean follow-up (9.5 months), five patients remained free of tumors. However, adjunctive intraluminal chemotherapy/immunotherapy was also used in some patients. No conclusions could be reached regarding the additional benefit of laser ablation.[82]

Schmeller and Hofstetter found 36 ureteral tumors between 1984 and 1987 of which endoscopic treatment was attempted in 20 using the laser alone or in combination with electrocautery. A 600-μm fiber was used to deliver between 20 and 30 W of power for no longer than 3 sec. The authors noted that when tumors were greater than 5 mm, complete laser coagulation was difficult. In these cases, the superficial coagulated neoplastic tissue was removed by "scratching" it away with the tip of the fiber. Stents were placed after treatment and removed 4 to 6 weeks later, at which time ureteroscopy was repeated with biopsy and, if necessary, retreatment. One patient treated for a circumferential tumor developed narrowing at the treatment site. In the 8 patients treated with laser therapy only, 2 developed recurrent tumors within the median follow-up of 22 months.[83]

Of 12 patients with upper-urinary-tract urothelial tumors felt to be candidates for endoscopic ablation, Grossman et al. were able to treat 8. Of the 4 not treated, 2 had tumors found to be too large, while 2 had tumors within the renal pelvis not endoscopically accessible with the rigid instruments available to the investigators. Tumors were treated with an Nd:YAG laser (30 W) using a 600-μm fiber. When necessary, an internal ureteral stent was left after treatment and removed several days later. No serious complications or strictures developed. As would be expected, recurrences were common (5 of 7 not lost to follow-up). The authors concluded that transureteroscopic laser therapy was appropriate for low-grade superficial ureteral tumors. However, renal pelvis tumors were not felt to be amenable to transureteral rigid endoscopic laser therapy because of limited accessibility.[84]

Kaufman and Carson selected 9 patients with low-grade, low-stage ureteral tumors to treat endoscopically. Using cold-cup forceps, tumors were first avulsed and debulked. The base of the removed tumor was then treated with 30 W of Nd:YAG laser energy through a 400-μm fiber. Ureteral stents were placed in all patients. Using this approach, 8 of 9 patients had no recurrences with a mean follow-up of 28 months. Two "minor" strictures developed that were subsequently treated with balloon dilatation. The investigators conclude that the combination of tumor debulking and Nd:YAG laser ablation is an effective treatment in selected patients with ureteral tumors.[85]

At the Mayo Clinic 44 patients with transitional carcinoma of the upper urinary tract were treated either with electrofulguration or Nd:YAG laser energy. Of the 44, 21 had tumors of the renal pelvis and 23 of the ureter. Patients with tumors greater than 2 cm were excluded from this approach. Both the percutaneous (24%) and the transureteroscopic (76%) approach utilizing rigid and flexible instruments were used for tumors of the renal pelvis. Ureteral tumors were treated without prior avulsion by cold-cup forceps. Adjunctive intraluminal chemo/immunotherapy was used in some patients. Three patients treated transureteroscopically with the laser developed strictures. Tumor-recurrence rates — 33% using Nd:YAG laser and 44% using electrocautery — were not statistically different. The authors conclude that in selected patients the endourological approach is an acceptable treatment of upper-tract neoplasms.[86]

The use of the Ho:YAG laser for the treatment of ureteral tumors has been explored by Bagley and Erhard. The very limited tissue penetration of energy of this wavelength led the investigators to use the laser to resect small tumors.[87] In a subsequent article, this laser was used alone and in combination with the Nd:YAG for endoscopic tumor ablation and coagulation.[88]

As evidenced by the above studies, the use of the laser for endoscopic removal of ureteral and renal pelvic tumors is gaining acceptance in the urologic community. It appears to be associated with minimal morbidity and excellent tumor control.

#### 45.3.1.5 Renal Tumors

In the last decade, there has been increasing interest in partial (renal sparing) nephrectomy in the treatment of carcinomas arising in the renal parenchyma. To this end, there has been interest in the use of the laser both as a tool for incising the renal parenchyma and as an adjunct to resection for thermal ablation of the surgical bed.

Malloy et al. reported on the use of the Nd:YAG laser for partial nephrectomies in six patients. In three patients with invasive TCCs arising in either the upper- or lower-pole calyx, a small pyelotomy was first made in the exposed renal pelvis, and a flexible nephroscope and laser were then introduced and used to ablate/coagulate visible tumor. Polar nephrectomies were then performed using the laser to aid in hemostasis of the cut surface of the kidney. Three patients with renal-cell carcinomas underwent partial nephrectomies. In these patients, the perimeter of the tumor was first "treated" with laser energy with the aim of sealing draining lymphatics. Subsequently, the laser was used as an aid in tumor enucleation or partial nephrectomy. In none of the six patients was blood transfusion necessary.[89]

Contact-Nd:YAG-laser incision, in combination with regional hypothermia and renal artery occlusion, was used for partial nephrectomy in six patients with renal-cell carcinoma by Korhonen et al. These investigators felt that clamping the renal artery limited the extent of laser-energy damage in the surrounding normal parenchyma. With a mean follow-up of 15.3 months, no recurrences or metastasis occurred.[90]

In a report by Merguerian and Seremetis, KTP and Nd:YAG lasers were used in combination for partial nephrectomies in patients with bilateral Wilms' tumors. The KTP (40 W, 0.5-mm spot size) laser was used to incise, while the Nd:YAG (50 W, 0.5-mm spot size) laser was used to coagulate blood vessels. Blood loss was minimal in both cases. The authors point out that the small size of the blood vessels in the pediatric kidney is ideally suited for laser coagulation.[91]

### 45.3.2 Photodynamic Therapy

PDT is a two-step procedure based on the localization of an administered drug (photosensitizer) within targeted tissue and the subsequent excitation of the photosensitizer with light of the appropriate wavelength to generate toxic photoproducts.[92] To achieve the desired effect, sufficient amounts of the photosensitizer, oxygen, and light must be present at the optimal time (determined by the pharmacokinetics of the sensitizer) for light delivery.

#### 45.3.2.1 Photosensitizers

Hematoporphyrin derivative (HPD), the first synthetic photosensitizer, has been the predominant photosensitizer for PDT of malignant tissues. HPD is a mixture of porphyrins, hematoporphyrin, and its dehydration products. The putative active component of HPD has been identified as dihematoporphyrin ether (DHE, porfimer sodium) and is currently marketed as Photofrin®.[93] HPD and DHE are excited from their ground-state energy levels to higher energy levels by red light of 630-nm wavelength. Return to the ground-state energy level results in energy transfer to molecular oxygen and subsequent tissue destruction. The depth of the photodynamic effect, and thus tissue necrosis, is limited by the tissue penetration depth of red light at this wavelength.[94] The skin, which can retain the drug for 6 weeks or longer, is especially troublesome because patients must limit sun exposure for significant periods of time.

Second-generation photosensitizers with absorption peaks at higher wavelengths, higher chemical purity, and the potential for less prolonged photosensitivity have been developed and tested in preclinical and clinical studies. Newer sensitizers currently in clinical development include benzoporphyrin derivative (BPD), tin ethyl etiopurpurin (SnET2), 5-animolevulinic acid (ALA), meso-tetra (hydroxyphenyl) chlorin (mTHPC), and lutetium texaphyrin (Lutex) (Figure 45.1). Of these, ALA is unique in that it requires metabolization to its active form, protoporphyrin IX (PpIX), before becoming a photosensitizer.

#### 45.3.2.2 Light Sources and Dosimetry

Tunable-dye lasers, principally the argon pumped-dye laser, producing light of 630 nm to match an absorption peak of HPD have been the most widely used lasers. Other dye-laser systems that have been used include frequency-doubled KTP/dye, dye/Nd:YAG lasers, copper-vapor/dye lasers, excimer/dye lasers, and gold-vapor lasers. Most dye lasers are expensive and bulky, require special cooling, and are difficult to maintain.[95] Smaller and less expensive diode lasers that are easier to handle and maintain may be able to overcome these problems. In preclinical and clinical studies, diode lasers have proved to be just as effective as dye lasers in inducing tissue destruction after sensitization with the photosensitizers BPD, Lutex, CASPc, and SnET2.[95–97]

Although shorter wavelengths (e.g., 514 nm) of light have proved effective in the destruction of cultured human bladder cells, the minimal depth of penetration at these wavelengths has prevented their clinical use.[98] On average, red light at 630 nm used for clinical PDT with DHE or ALA penetrates between 1 and 3 mm.[99] Light wavelengths from 700 to 850 nm penetrate nearly twice this depth, making longer visible wavelengths more attractive for photodestruction of tumors, particularly solid tumors. The total light dose is usually delivered in one session using continuous-light application. The optional properties of targeted tissue must be considered to optimize treatment plans in PDT. The bladder acts as a light-integrating sphere and must be considered as such when planning treatment dosimetry.

#### 45.3.2.3 Clinical Studies

##### *45.3.2.3.1 Patient Selection*

The indications for the use of PDT in treatment of bladder cancer have yet to be clearly defined. Most studies have been performed in patients with recurrent bladder cancer confined to the mucosa (Ta), submucosa (T1), or carcinoma *in situ* (Tis). The majority of treatments reported in the literature were performed in "worst-case" patients who had failed other conventional bladder-cancer therapies such as instillation therapy (BCG or mitomycin) or transurethral resection.[100,101] In general, long-term results have documented that patients with tumor stages greater than T1 are not good candidates for PDT. The best results have been obtained in patients with Tis.[102]

##### *45.3.2.3.2 PDT Technique for Bladder Cancer Treatment*

A number of photosensitizers have been proposed for the PDT of bladder cancers, but HPD and DHE have been the most widely investigated. Table 45.1 summarizes the results of 11 recent studies performed with 2.5 to 5 mg/kg HPD or 1.5 to 2 mg/kg DHE.[103–113] The interval between drug administration and light treatment was usually 24 to 72 h. The homogenous application of light to the bladder wall to achieve a uniform and consistent illumination of each part of the bladder has been problematic.[114] For whole-bladder irradiation (integral PDT) several methods have been tested to achieve a more homogenous light distribution. Laser-light scattering optics,[115] light-diffusing medium, e.g., intralipid,[116,117] or specific catheters for light application in the bladder have been developed.[114]

For focal therapy of lesions a flat-cut fiber (front-end light emission) is suitable, whereas for whole-bladder irradiation spherical bulb-type diffusers produce a homogenous light distribution.[100] Scattering of light during whole-bladder treatment increases the actual light dose by a factor of five to six in a bladder model, and there is considerable interpatient variability in light scattering.[118,119]

The treatment surface area, treatment time, and power density must be calculated individually to achieve the optimal treatment.[102,120] Intravesical light treatment was usually performed under general or spinal anesthesia.[102,120] For focal tumor treatment a light dose of approximately 150 $J/cm^2$ has been reported as sufficient.[100] Commonly only 15 to 25/30 $J/cm^2$ light are necessary for whole-bladder

**TABLE 45.1** Clinical Experiences with PDT for Bladder Cancer

| Author/ Ref. No. | Indication | N | Drug and Dose | Light Energy | Results at 3 Months | Long-Term Results | Adverse Effects |
|---|---|---|---|---|---|---|---|
| Nseyo[103] | Recurrent Ta to T3 and CIS | 23 | DHE 2 mg/kg | 5–60 J/cm$^2$ | CR 7 (30%)<br>PR 10 (43%) | Not reported | All patients with urgency, frequency, nocturia, bladder spasms; 4 patients with bladder shrinkage |
| Prout[104] | Recurrent and newly diagnosed Ta to T1 | 19 | DHE 2 mg/kg | Focal:<br>100–200 J/cm$^2$<br>Whole bladder:<br>5.5 J/cm$^2$ | CR 9 (47%)<br>PR 9 (47%) | Not reported | 18 (95%) with adverse events |
| Heng Li[105] | Bladder | 50 | HPD 5 mg/kg | Focal:<br>100–300 J/cm$^2$ | CR 40 (80%)<br>PR 4 (8%) | Not reported | Not reported |
| Jocham[106] | Recurrent multifocal Ta/T1 | 20 | DHE<br>1.5–2 mg/kg<br>HPD 3 mg/kg | Whole bladder:<br>15–70 J/cm$^2$ | CR 83% of tumors | 6 NED* up to 5 years | All patients with pollakisuria, dysuria, and temporary bladder shrinkage |
| Naito[107] | Recurrent multifocal Ta/T1 and CIS | 35 | HPD 3 mg/kg<br>DHE 2 mg/kg | Whole bladder:<br>10–30 J/cm$^2$ | CR 24 (69%)<br>PR 5 (14%) | 10 NED average of 20.9 months (5 months to 5 years) | All patients with bladder irrigation including hematuria, frequency, and burning urination |
| Windahl[108] | Recurrent Ta to T3 and CIS | 11 | HPD 2.5 mg/kg<br>DHE 2 mg/kg | Whole bladder:<br>for HPD:<br>30–53 J/cm$^2$<br>for DHE: 15 J/cm$^2$ | CR 8 (73%)<br>PR 1 (9%) | 4 NED for 12–30 months | 9 (82%) patients with bladder symptoms: frequency, urgency and dysuria, hematuria |
| D'Hallewin[109] | Multifocal Ta and CIS | 15 | DHE 2 mg/kg | Whole bladder:<br>75–100 J/cm$^2$ | CR 15 (100%) | 9 NED at 36 months | All patients with irritative symptoms: bladder shrinkage (2 patients required cystectomy for permanent fibrotic shrinkage) |
| Kriegmair[110] | Recurrent Ta to T1 and CIS | 21 | DHE 2 mg/kg | Whole bladder:<br>15–30 J/cm$^2$ | CR 12 (57%) | 5 NED for 15–42 months | 19 (90%) patients with dysuria and urgency; 1 patient with bladder shrinkage |
| Uchibayashi[111] | Recurrent Ta to T1 and CIS | 34 | HPD 3 mg/kg<br>DHE 2 mg/kg | Whole bladder:<br>dose not reported | CR 25 (74%) | 4 NED at 24 months | 34 (100%) hematuria, burning, frequency; temporary bladder shrinkage |
| Walther[112] | Recurrent Ta to T1 and CIS | 20 | DHE<br>1.5–2 mg/kg | Whole bladder and focal:<br>5.1–25.6 J/cm$^2$ | CR 9 (45%) | 4 NED average of 36.3 months (23–56 months) | 20 (100%) urgency, frequency, and suprapubic pain |
| Nseyo[113] | Recurrent Ta to T1 and CIS | 58 | DHE<br>1.5–2 mg/kg | Whole bladder:<br>10–60 J/cm$^2$ | CR 75–90% | 3 NED average of 50 months (9–110 months) | Not reported |

*NED = No evidence of disease.

PDT.[100,114,121] Usually a power density of less than 50 $mW/cm^2$ and bladder-filling pressure lower than 30 cm $H_2O$ are adequate.[121]

#### 45.3.2.3.3 Results

As noted in Table 45.1, recent studies demonstrate the excellent short-term responses of PDT in high-risk populations with complete tumor-remission rates between 30 and 100% at 3 months post treatment.[103–113] Patients with no evidence of disease for up to 10 years after PDT have been reported.[113] However, the majority of patients with initial complete responses experienced recurrence of the cancer. Recurrences occurred more frequently in patients with large tumors, higher tumor stages, and higher tumor grades and less frequently in patients with carcinoma *in situ*.

#### 45.3.2.3.4 Adverse Reactions

Increased and prolonged skin photosensitivity is the most frequent treatment-related adverse event in PDT.[122–124] Photosensitivity or phototoxic reactions resemble sunburns or erythemas of different severity grades from reddening to the occurrence of edema or bulla formation.[125] Unfortunately, the documentation of these events is inconsistent in the literature, which results in reported frequencies of skin photosensitivity reactions in 10 to 100% of patients treated with photodynamic therapy.[103,104,112,122–124,126–128] Skin-photosensitivity reactions are reported to occur within days to months after photosensitization.[124,129–131] Depending on the pharmacological properties of the photosensitizer, patients should avoid direct sunlight and bright indoor light and wear protective clothing for 6 to 8 weeks following PDT with DHE, or 1 to 2 weeks following BPD-PDT.[124,131]

The most frequent local side effect of PDT in the bladder is irritability of various degrees starting immediately after the treatment and usually lasting for up to 4 weeks, in rare cases for several months.[103,106,107,109,112] Almost all patients experience urinary frequency, urgency, and mild hematuria.[107,120] These acute inflammatory-type reactions seem to be light-dose-dependent and are more severe if the bladder has extensive disease or has undergone previous radiation therapy.[103,108,109] One immediate consequence of the inflammatory response is a temporary decrease in bladder capacity and compliance. Bladder capacity can be reduced by 20 to 80% within the first few days after PDT but usually returns to pretreatment values. The loss of capacity seems to correlate with the light dose/drug doses applied and the resulting amount of tissue damage.[106,107,109] Some authors have recommended excluding patients with bladder capacities less than 200 ml from PDT.[107] While most of these changes are reversible, cases of permanent fibrotic changes of the bladder wall requiring cystectomy have been reported.[108,132,133] Patients who have undergone radiation therapy for bladder cancer are at a higher risk for irreversible fibrosis of the bladder.[108] However, the problem of permanent bladder fibrosis may reflect overtreatment with light, drug, or both.[120] Choosing the correct treatment parameters should reduce damage to the bladder muscle layers.[120,134] Other rare local adverse effects reported include bladder calcifications presumed to occur because of tissue injury and rectovesical fistula.[108,112] Almost all side effects are reversible except fibrotic bladder shrinkage.[100]

#### 45.3.2.3.5 PDT after Intravesical ALA Instillation

More recently, PDT after intravesical instillation of ALA has been advocated as an effective treatment for superficial bladder cancer.[135,136] This technique may reduce the problem of long-term detrusor-muscle damage and skin photosensitivity. Patients retained a 30-ml sodium bicarbonate solution containing 5 g ALA for 4 to 8 h in their bladders. Thereafter bladders were illuminated with green (514 nm) and red light (635 nm) with the patients under general or spinal anesthesia.[136] At 3 months after therapy four of ten patients had a complete and two of ten had a partial response. No skin phototoxicity or bladder-capacity reduction was observed. All patients reported dysuria and pollakiuria and six had a macrohematuria. Long-term results have yet to be reported.

#### 45.3.2.3.6 Photodynamic Detection (PDD) of Bladder Cancer

The fluorescent properties of several photosensitizers have been studied for their use in the detection of neoplastic diseases. Fluorescence is the emission of light from an excited molecule and has been used to determine the uptake of a photosensitizer by malignant tissues.[99,137–139] Fluorescence detection after tumor sensitization with a photosensitizer has been investigated for its potential to increase the diagnostic accuracy of cystoscopy for bladder cancer.[140–143] Preclinical and clinical studies have confirmed increased fluorescence of bladder tumor tissue. A sensitivity of 97% in detection of bladder tumors has been reported for fluorescence cystoscopy as compared to standard white-light cystoscopy (73% sensitivity), while the specificity did not differ significantly.[144] However, the value of this technology is controversial, and it remains to be integrated as a standard tool into the diagnostic armamentarium for bladder cancer detection.[144]

## Acknowledgment

The research for this chapter was supported in part by the Stranahan Chair for Oncologic Research, Medical College of Ohio Foundation.

## References

1. Watson, G.M., Murray, S., Dretler, S.P., and Parrish, J.A., The pulsed dye laser for fragmenting urinary calculi, *J. Urol.*, 138, 195, 1987.
2. Rudhart, M. and Hirth, A., Use of an absorbent in laser lithotripsy with dye lasers: *in vitro* study of fragmentation efficacy and jet formation, *J. Urol.*, 152, 1005, 1994.
3. Larizogitin, I. and Pons, J.M.V., A systematic review of the clinical efficacy and effectiveness of the Holmium:YAG laser in urology, *Br. J. Urol. Int.*, 84, 1, 1999.
4. Zerbib, M., Flam, T., Belas, M., Debre, B., and Steg, A., Clinical experience with a new pulsed dye laser for ureteral stone lithotripsy, *J. Urol.*, 143, 483, 1990.
5. Dretler, S.P., An evaluation of ureteral laser lithotripsy: 225 consecutive patients, *J. Urol.*, 142, 267, 1990.
6. Hofmann, R., Hartung, R., Schmidt-Kloiber, H., and Reichal, F., First clinical experience with a Q-switched neodynium:YAG laser for urinary calculi, *J. Urol.*, 141, 275, 1989.
7. Thomas, S., Pensel, J., Engelhardt, R., Meyer, W., and Hofstetter, A.G., The pulsed dye laser versus the Q-switched Nd:YAG laser in laser induced shock wave lithotripsy, *Lasers Surg. Med.*, 8, 363, 1988.
8. Vassar, G.J., Chan, K.F., Teichman, J.M.H., et al., Holmium: YAG lithotripsy: photothermal mechanism, *J. Endourol.*, 13, 181, 1999.
9. Zhong, P., Tong, H.-L., Cocks, F.H., Pearle, M.S., and Preminger, G.M., Transient cavitation and acoustic emission produced by different laser lithotripters, *J. Endourol.*, 12, 371, 1998.
10. Teichman, J.M.H., The use of Holmium:YAG laser in urology, *AUA Update Series*, 19, 20, 2001.
11. Sayer, J., Johnson, D.E., Price, R.E., and Cromeens, D.M., Ureteral lithotripsy with Holmium: YAG laser, *J. Clin. Laser Med. Surg.*, 11, 61, 1993.
12. Denstedt, J.D., Razvi, H.A., Sales, J.L., and Eberwein, P.M., Preliminary experience with Holmium:YAG laser lithotripsy, *J. Endourol.*, 9, 255, 1995.
13. Grasso, M. and Chalik, Y., Principles and Applications of laser lithotripsy: experience with the holmium laser lithotrite, *J. Clin. Laser Med. Surg.*, 16, 1, 3, 1998.
14. Devarajan, R., Ashraf, M., Beck, R.O., Lemberger, R.J., and Taylor, M.C., Holmium:YAG lasertripsy for ureteric calculi: an experience of 300 procedures, *Br. J. Urol. Int.*, 82, 342, 1998.
15. Matsuoka, K., Iida, S., Inoue, M., Yoshe, S., Arai, K., Tomiyasu, K., and Noda, S., Endoscopic lithotripsy with the Holmium: YAG laser, *Lasers Surg. Med.*, 25, 389, 1999.
16. Shroff, S., Watson, G.M., Parikh, A., Thomas, R., Soonawalla, P.F., and Pope, A., The Holmium:YAG laser for ureteric stones, *Br. J. Urol.* 78, 836, 1996.
17. Yiu, M.K., Liu, P.L., Yiu, T.F., and Chau, A.Y.T., Clinical experience with Holmium:YAG laser lithotripsy of ureteral calculi, *Lasers Surg. Med.*, 19, 103, 1996.
18. Razvi, H.A., Denstedt, J.D., Chun, S.S., and Sales, J.L., Intracorporeal lithotripsy with the Holmium:YAG laser, *J. Urol.*, 156, 912, 1996.
19. Das, A., Erhard, J., and Bagley, D.H., Intrarenal use of the Holmium laser, *J. Urol.*, 160, 630, 1998.
20. Grasso, M., Experience with the holmium laser as an endoscopic lithotrite, *Urology*, 48, 199, 1996.
21. Grasso, M., Conlin, M., and Bagley, D.H., Retrograde uretero-peloscopic treatment of 2 cm or greater upper urinary tract and minor staghorn calculi, *J. Urol.*, 160, 346, 1998.

22. Grasso, M. and Bagley, D.H., Small diameter, actively deflectable, flexible ureteropyeloscopy, *J. Urol.*, 160, 1648, 1998.
23. Teichman, J.M.H., Vassar, G.J., Bishoff, J.T., and Bellman, G.C., Holmium:YAG efficiency yields smaller fragments than lithoclast, pulse dye or electrohydraulic lithotripsy, *J. Urol.*, 159, 17, 1998.
24. Wollin, T.A. and Denstedt, J.D., The Holmium laser in urology, *J. Clin. Laser Med. Surg.*, 16, 13, 1998.
25. Beaghler, M., Poon, M., Ruckle, H., Stewart, S., and Weil, D., Complications employing the Holmium:YAG laser, *J. Endourol.*, 12, 533, 1998.
26. Teichman, J.M.H., Rao, R.D., Glickman, R.D., and Harris, J.M., Holmium: the laser incident angle matters, *J. Urol.*, 159, 690, 1998.
27. Freiha, G.S., Glickman, R.D., and Teichman, J.M.H., Holmium:YAG laser induced damage to guidewires: an experimental study, *J. Endourol.*, 11, 331, 1997.
28. Teichman, J.M.H., Vassar, G.J., Glickman, R.D., Beserra, C.M., Cina, S.J., and Thompson, I.M., Holmium:YAG lithotripsy: photothermal mechanism converts uric acid calculi to cyanide, *J. Urol.*, 160, 320, 1998.
29. Teichman, J.M.H., Champion, P.C., Wollin, T.A., and Denstedt, J.D., Holmium:YAG lithotripsy of uric acid calculi, *J. Urol.*, 160, 2130, 1998.
30. McCullough, D.L., Roth, R.A., Babayan, R.K., Gordon, J.O., Reese, J.H., Crawford, E.D., Fuselier, H.A., Smith, J. A., Murchison, R.J., and Kaye, K.W., Transurethral ultrasound-guided laser-induced prostatectomy: national human cooperative study results, *J. Urol.*, 150, 1607, 1993.
31. Costello, A.J., Bowsher, W.G., Bolton, D.M., Braslis, K.G., and Burt, J., Laser ablation of the prostate in patients with benign prostatic hypertrophy, *Br. J. Urol.*, 69, 603, 1992.
32. Keoghane, S.R., Lawrence, K.C., Gray, A.M., Hancock, A., and Cranston, D.W., The Oxford Laser Prostate Trial: economic issues surrounding contact laser prostatectomy, *Br. J. Urol.*, 77, 386, 1996.
33. Muschter, R., de la Rosette, J.J., Whitfield, H., Pellerin, J.-P., Madersbacher, S., and Gillatt, D., Initial human clinical experience with diode laser interstitial treatment of benign prostatic hyperplasia, *Urology*, 48, 223, 1996.
34. Virdi, J.S., Chandrasekar, P., and Kapasi, F., Interstitial laser ablation (Indigo™) of the prostate — a randomised prospective study, three year followup, *J. Urol.*, 165(5 Suppl.), 368, 2001.
35. Gilling, P.J., Cass, C.B., Cresswell, M.D., and Fraundorfer, M.R., Holmium laser resection of the prostate: preliminary results of a new method for the treatment of benign prostatic hyperplasia, *Urology*, 47(1), 48, 1996.
36. Moody, J.A. and Lingeman, J.E., Holmium laser enucleation for prostate adenoma greater than 100 gm: comparison to open prostatectomy, *J. Urol.*, 165(2), 459, 2001.
37. Carter, A., Sells, H., Speakman, M., Ewings, P., MacDonagh, R., and O'Boyle, P., A prospective randomized controlled trial of hybrid laser treatment or transurethral resection of the prostate, with a 1-year follow-up, *B.J.U. Int.*, 83(3), 254, 1999.
38. Malik, R.S., Kuntzman, R.S., and Barrett, D.M., KTP laser prostatectomy: longterm experience, *J. Urol.*, 165(5 Suppl.), 369, 2001.
39. Bulow, H., Bulow, U., and Frohmuller, H.G., Transurethral laser urethrotomy in man: preliminary report, *J. Urol.*, 121, 286, 1979.
40. Rothauge, C.F., Urethroscopic recanalization of urethral stenosis using argon laser, *Urology*, 17(2), 158, 1980.
41. Smith, J.A. and Dixon, J.A., Neodymium:YAG laser treatment of benign urethral strictures, *J. Urol.*, 131, 1080, 1984.
42. Perkash, I., Ablation of urethral strictures using contact chisel crystal firing Neodymium:YAG laser, *J. Urol.*, 157, 809, 1996.
43. Kamal, B.A., The use of the diode laser for treating urethral strictures, *Br. J. Urol. Int.*, 87, 831, 2001.
44. Silber, N. and Servadio, C., Neodymium:YAG laser treatment of bladder neck contracture following prostatectomy, *Lasers Surg. Med.*, 12, 370, 1992.

45. Biewald, W. and Schier, F., Laser treatment of posterior urethral valves in neonates, *Br. J. Urol.*, 69, 425, 1992.
46. Bhatnagar, V., Agarwala, S., Lal, R., and Mitra, D.K., Fulguration of posterior urethral valves using the Nd:YAG, *Pediatr. Surg. Int.*, 16(1–2), 69, 2000.
47. Smith, J. J., Laser treatment of bladder haemangioma, *J. Urol.*, 143, 282, 1990.
48. Singal, R.K., Denstedt, J. D., Razvi, H.A., and Chun, S.S., Holmium:YAG laser endoureterotomy for treatment of ureteral stricture, *Urology*, 50(6), 875, 1997.
49. Hibi, H., Mitsui, K., Taki, T., Mizumoto, H., Yamamda, Y., Honda, N., and Fukatsu, H., Holmium laser incision technique for ureteral stricture using a small-caliber ureteroscope, *J. Soc. Laparoendosc. Surg.*, 4(3), 215, 2000.
50. Giddens, J.L. and Grasso, M., Retrograde ureteroscopic endopyelotomy using the Holmium:YAG laser, *J. Urol.*, 164, 1509, 2000.
51. Grafstein, L.A. and Selman, S.H., Thermal lasers in urologic oncology, *J. Clin. Laser Med. Surg.*, 16(1), 49, 1998.
52. Parsons, R.L., Campbell, J.L., and Thomley, M.W., Carcinoma of the penis treated by the ruby laser, *J. Urol.*, 100, 38, 1968.
53. Rosemberg, S.K. and Fuller, T.A., Carbon dioxide rapid superpulsed laser treatment of erythroplasia of Queyrat, *Urology*, 19, 539, 1982.
54. Rosemberg, S.K., Lasers and squamous cell carcinoma of external genitalia, *Urology*, 27, 430, 1986.
55. Bandieramonte, G., Lepera, P., Marchesini, R., and Pizzocaro, G., Laser microsurgery for superficial lesions of the penis, *J. Urol.*, 138, 315, 1987.
56. Hofstetter, A.G., Laser in urology, *Lasers Surg. Med.*, 5, 412, 1986.
57. Rothenberger, K.H., Value of the neodymium-yag laser in the therapy of penile carcinoma, *Eur. Urol.*, 12(Suppl. 1), 34, 1986.
58. Malloy, T.R., Wein, A.J., and Carpiniello, V.L., Carcinoma of the penis treated with neodymium yag laser, *Urology*, 31(1), 26, 1988.
59. Malek, R.S., Laser treatment of premalignant and malignant squamous cell lesions of the penis, *Lasers Surg. Med.*, 12, 246, 1992.
60. Sander, S., Beisland, H.O., and Fossberg, E., Neodymium:YAG laser in the treatment of prostatic cancer, *Urol. Res.*, 10, 85, 1982.
61. Sander, S. and Beisland, H.O., Laser in the treatment of localized prostatic carcinoma, *J. Urol.*, 132, 280, 1984.
62. Beisland, H.O., Neodymium-YAG laser in the treatment of urinary bladder carcinoma and localized prostatic carcinoma, *J. Oslo City Hospitals*, 36, 63, 1986.
63. Wishnow, K.I. and Johnson, D.E., Subcapsular orchiectomy using the $CO_2$ laser: a new technique, *Lasers Surg. Med.*, 8, 604, 1988.
64. Bolton, D.M. and Costello, A.J., CO2 laser subcapsular orchidectomy in the treatment of metastatic prostate cancer, *Lasers Surg. Med.*, 14, 88, 1994.
65. Fournier, G.R.J., Kung, A.H., and Trost, D., High powered $CO_2$ laser cystoscope for endoscopic bladder surgery, *Lasers Surg. Med.*, 16, 390, 1995.
66. Smith, J.A. and Dixon, J.A., Argon laser phototherapy of superficial transitional cell carcinoma of the bladder, *J. Urol.*, 131, 655, 1984.
67. Stein, B.S. and Kendall, R.A., Lasers in urology. II. Laser therapy, *Urology*, 5, 411, 1984.
68. Orihuela, E. and Smith, D., Laser treatment of transitional cell cancer of the bladder and upper urinary tract, *Cancer Treat. Res.*, 46, 123, 1989.
69. Smith, J.D., Laser Surgery for transitional cell carcinoma: technique, advantages, and limitations, *Urol. Clin. North Am.*, 19, 473, 1992.
70. Malloy, T.R. and Wein, A.J., Laser treatment of bladder carcinoma and genital condylomata, *Urol. Clin. North Am.*, 14, 121, 1987.
71. Hofstetter, A.G., Treatment of urological tumors by Neodymium-YAG laser, *Eur. Urol.*, 12(Suppl. 1), 21, 1986.

72. Beisland, H.O. and Seland, P.A., Prospective randomized study on Neodymium-YAG laser irradiation versus TUR in the treatment of urinary bladder cancer, *Scand. J. Urol. Nephrol.*, 20, 209, 1986.
73. Beer, M., Jocham, D., Beer, A., and Staehler, G., Adjuvant laser treatment of bladder cancer: 8 years' experience with the Nd:YAG laser 1064 nm, *Br. J. Urol.*, 63, 476, 1989.
74. Johnson, D.E., Use of the Holmium: YAG (Ho:YAG) laser for treatment of superficial bladder carcinoma, *Lasers Surg. Med.*, 14, 213, 1994.
75. McPhee, M.S., Arnfield, M.R., Tulip, J., and Lakey, W.H., Neodymium:YAG laser therapy for infiltrating bladder cancer, *J. Urol.*, 140, 44, 1988.
76. Smith, J. A., Treatment of invasive bladder cancer with a Neodymium:YAG laser, *J. Urol.*, 135, 55, 1986.
77. Smith, J.A.J. and Landau, S., Neodymium:YAG laser specifications for safe intravesicle therapy, *J. Urol.*, 141, 1238, 1989.
78. Beisland, H.O. and Sander, S., Neodymium-YAG laser irradiation of stage T2 muscle-invasive bladder cancer: long term results, *Br. J. Urol.*, 65, 24, 1990.
79. Petkovic, S.D., Conservation of the kidney in operations for tumours of the renal pelvis and calyces: a report of 26 cases, *B.J.U.* 44, 1, 1972.
80. Zincke, H. and Neves, R.J., Feasibility of conservative surgery for transitional cell cancer of the upper urinary tract, *Urol. Clin. North Am.*, 11, 717, 1984.
81. Schilling, A., Bowering, R., and Keiditsch, E., Use of the Neodymium-YAG laser in the treatment of ureteral tumors and urethral condylomata acuminata, *Eur. Urol.*, 12(Suppl. 1), 1986.
82. Smith, A.D., Orihuela, E., and Crowley, A.R., Percutaneous management of renal pelvic tumors: a treatment option in selected cases, *J. Urol.*, 137, 852, 1987.
83. Schmeller, N.T. and Hofstetter, A.G., Laser treatment of ureteral tumors, *J. Urol.*, 141, 840, 1989.
84. Grossman, H.B., Schwartz, S.L., and Konnak, J.W., Ureteroscopic treatment of urothelial carcinoma of the ureter and renal pelvis, *J. Urol.*, 148, 275, 1992.
85. Kaufman, R.P. and Carson, C.C., Ureteroscopic management of transitional cell carcinoma of the ureter using the Neodymium:YAG laser, *Lasers Surg. Med.*, 13, 625, 1993.
86. Elliot, D.S., Blute, M.L., Patterson, D.E., Bergstralh, E.J., and Segura, J.W., Long-term follow-up of endoscopically treated upper urinary tract transitional cell carcinoma, *Urology*, 47(6), 819, 1996.
87. Bagley, D.H. and Erhard, M., Use of the Holmium laser in the upper urinary tract, *Tech. Urol.*, 1(1), 25, 1995.
88. Keeley, F.X., Bibbo, M., and Bagley, D.H., Ureterscopic treatment and surveillance of upper tract transitional cell carcinoma, *J. Urol.*, 157, 1560, 1997.
89. Malloy, T.R., Schultz, R.E., Wein, A.J., and Carpiniello, V.L., Renal preservation utilizing Neodymium:YAG laser, *Urology*, 27(2), 99, 1986.
90. Korhonen, A.-K., Talja, M., Karlsson, H., and Tuhkanen, K., Contact Nd:YAG laser and regional renal hypothermia in partial nephrectomy, *Ann. Chir. Gynaecol.*, 82, 59, 1993.
91. Merguerian, P.A. and Serementis, G., Laser-assisted partial nephrectomy in children, *J. Pediatr. Surg.*, 29(7), 934, 1994.
92. Moan, J. and Berg, K., Photochemotherapy of cancer: experimental research, *Photochem. Photobiol.*, 55, 931, 1992.
93. Photofrin product information, in *Physician's Desk Reference,* Medical Economics, Montvale, NJ, 1999, p. 2795.
94. Wilson, B.C. and Jeeves, W.P., Photodynamic therapy of cancer, in *Photomedicine*, Ben-Hur, E. and Rosenthal, I., Eds., CRC Press, Boca Raton, FL, 1987.
95. Hammer-Wilson, M.J., Sun, C.-H., Ghahramanlou, M., and Berns, M.W., *In-vitro* and *in-vivo* comparison of argon-pumped and diode lasers for photodynamic therapy using second-generation photosensitizers, *Lasers Surg. Med.*, 23, 274, 1998.
96. Lytle, A.C., Doiron, D.R., and Selman, S.H., Red diode laser for photodynamic therapy: a small animal efficacy study, *Proc. SPIE*, 2133, 178, 1994.

97. Kaplan, M., Wieman, T., Glaspy, J., Mang, T., Rifkin, R., Panella, T., and Albrecht, D., Phase II/III controlled multicenter study of single-dose tin ethyl etiopurpurin (SnET2) photodynamic therapy (PDT) in cutaneous metastatic breast cancer, *Proc. Am. Soc. Clin. Oncol.*, May 1999.
98. Bellnier, D.A., Prout, G.R., and Lin, C., Effect of 514.5-nm argon ion laser radiation on hematoporphyrin-derivative-treated bladder tumor cells *in vitro* and vivo, *J. Nat. Cancer Inst.*, 74(3), 617, 1985.
99. Manyak, M.J., Photodynamic therapy, in *Lasers in Urologic Surgery*, Smith, J.A., Stein, B.S., and Benson, R.C., Eds., Mosby, St. Louis, 1994, pp. 230–258.
100. Benson, R.C., Phototherapy of bladder cancer, in *Phototherapy of Cancer*, Morstyn, G. and Kaye, A., Eds., Harwood Academic, New York, 1990, pp. 1–22.
101. Jocham, D., Photodynamische Verfahren in der Urologie, *Urologe [A]*, 33, 547, 1994.
102. Manyak, M.J., Practical aspects of photodynamic therapy for superficial bladder carcinoma, *Tech. Urol.*, 1(2), 84, 1995.
103. Nseyo, U.O., Dougherty, T.J., and Sullivan, L., Photodynamic therapy in the management of resistant lower urinary tract carcinoma, *Cancer*, 60, 3113, 1987.
104. Prout, G.R., Lin, C.-W., Benson, R.C., Nseyo, U.O., Daly, J. J., Griffin, P.P., Kinsey, J., Tian, M.-E., Lao, Y.-H., Mian, Y.-Z., Chen, X., Ren, F.-M., and Qiao, S.-J., Photodynamic therapy with hematoporphyrin derivative in the treatment of superficial transitional-cell carcinoma of the bladder, *New Engl. J. Med.*, 317(20), 1251, 1987.
105. Li, J.-H., Guo, Z.-H., Jin, M.-L., Zhao, F.-Y., Cai, W.-M., Gao, M.-L., Shu, M.-Y., and Zou, J., Photodynamic therapy in the treatment of malignant tumors: an analysis of 540 cases, *Photochem. Photobiol.*, 6, 149, 1990.
106. Jocham, D., Baumgartner, R., Stepp, H., and Unsold, E., Clinical experience with the integral photodynmic therapy of bladder carcinoma, *Photochem. Photobiol.*, 6, 183, 1990.
107. Naito, K., Hisazumi, H., Uchibayashi, T., Amano, T., Hirata, A., Komatsu, K., Ishida, T., and Miyoshi, N., Integral laser photodynamic treatment of refractory multifocal bladder tumors, *J. Urol.*, 146, 1541, 1991.
108. Windahl, T. and Lofgren, L.A., Two years' experience with photodynamic therapy of bladder carcinoma, *Br. J. Urol.*, 71, 187, 1993.
109. D'Hallewin, M.-A. and Baert, L., Long-term results of whole bladder wall photodynamic therapy for carcinoma *in situ* of the bladder, *Urology*, 45(5), 763, 1995.
110. Kriegmair, M., Waidelich, R., Lumper, W., Ehsan, A., Baumgartner, R., and Hofstetter, A.G., Integral photodynamic treatment of refractory superficial bladder cancer, *J. Urol.*, 154, 1339, 1995.
111. Uchibayashi, T., Koshida, K., Kunimi, K., and Hisazumi, H., Whole bladder wall photodynamic therapy for refractory carcinoma *in situ* of the bladder, *Br. J. Cancer*, 71, 625, 1995.
112. Walther, M.M., Delaney, T.F., Smith, P.D., Friauf, W.S., Thomas, G.F., Shawker, T.H., Vargas, M.P., Choyke, P.L., Linehan, W.M., Abraham, E.H., Okunieff, P.G., and Glatstein, E., Phase I trial of photodynamic therapy in the treatment of recurrent superficial transitional-cell carcinoma of the bladder, *Urology*, 50, 199, 1997.
113. Nseyo, U.O., DeHaven, J., Dougherty, T.J., Potter, M.R., Merrill, D.L., Lundahl, S.L., and Lamm, D.L., Photodynamic therapy (PDT) in the treatment of patients with resistant superficial bladder cancer: a long-term experience, *J. Clin. Laser Med. Surg.*, 16(1), 61, 1998.
114. Beyer, W., Systems for light application and dosimetry in photodynamic therapy, *Photochem. Photobiol., B* 36, 153, 1996.
115. Hisazumi, H., Miyoshi, N., and Misaki, T., A trial manufacture of a motor-driven laser light scattering optic for whole bladder wall irradiation, *Prog. Clin. Biol. Res.*, 17, 239, 1984.
116. Jocham, D., Unsold, E., and Staehler, G., Use of a light dispersion medium for an integrated dye laser therapy of bladder tumors after photosensitization with hematoporphyrin derivative (HpD), in *Porphyrin Photosensitization*, Doiron, D.R. and Gomer, C.J., Eds., Alan R. Liss, New York, 1984, pp. 249–256.

117. Baghdassarian, R., Wright, M.W., Vaughn, S.A., Berns, M.W., Martin, D.C., and Wile, A.G., The use of lipid emulsion as an intravesical medium to dispense light in the potential treatment of bladder tumors, *J. Urol.*, 133, 120, 1985.
118. Star, W.M., Marijnissen, J.P.A., Jansen, H., Keijzer, M., and van Gemert, M.J.C., Light dosimetry for photodynamic therapy by whole bladder wall irradiation, *Photochem. Photobiol.*, 46(5), 619, 1987.
119. Marijnissen, J.P.A., Star, W.M., Zandt, H.J.A., D'Hallewin, M.-A., and Baert, L., *In-situ* light dosimetry during whole bladder wall photodynamic therapy: clinical results and experiemental verification, *Phys. Med. Biol.*, 38, 567, 1993.
120. Nseyo, U.O., Photodynamic therapy, *Urol. Clin. North Am.*, 19, 3, 1992.
121. Nseyo, U.O., Dougherty, T.J., Boyle, D.G., and Potter, M.R., Study of factors mediating effect of photodynamic therapy on bladder in canine bladder model, *Urology*, 32(1), 41, 1988.
122. Dougherty, T.J., Photodynamic therapy (PDT) of malignant tumors, *Crit. Rev. Oncol. Hematol.*, 2, 83, 1984.
123. Manyak, M.J., Russo, A., Smith, P.D., and Glatstein, E., Photodynamic therapy, *J. Clin. Oncol.*, 6, 830, 1988.
124. Mullooly, V.M., Abramson, A.L., and Shikowitz, M.J., Dihematoporphyrin ether-induced photosensitivity in laryngeal papilloma patients, *Lasers Surg. Med.*, 10, 349, 1990.
125. Wilkinson, J.D. and Rucroft, R.J.G., Photosensitization reactions, in *Textbook of Dermatology*, Rooke, A. et al., Eds., Blackwell Scientific, Oxford, U.K., 1986, pp. 512–515.
126. Dougherty, T.J., Cooper, M.T., and Mang, T.S., Cutaneous phototoxic occurences on patients receiving photorin, *Lasers Surg. Med.*, 10, 485, 1990.
127. Calzavara-Pinton, P.G., Szeimies, R.M., and Ortel, B., and Zane, C., Photodynamic therapy with systemic administration of photosensitizers in dermatology, *Photochem. Photobiol. B Biol.*, 36, 225, 1996.
128. Fingar, V.H. and Henderson, B.W., Drug and light dose dependence of photodynamic therapy: a study of tumor and normal tissue response, *Photochem. Photobiol.*, 46, 837, 1987.
129. Schuh, M., Nseyo, U.O., and Potter, M.R., Dao, T.L., and Dougherty, T.J., Photodynamic therapy for palliation of locally recurrent breast carcinoma, *J. Clin. Oncol.*, 5(11), 1766, 1987.
130. Wilson, B.C., Mang, T., Stoll, H., Jones, C., Cooper, M., and Dougherty, T.J., Photodynamic therapy for the treatment of basal cell carcinoma, *Arch. Dermatol.*, 128, 1597, 1992.
131. Lui, H., Photodynamic therapy in dermatology with porfimer sodium and benzoporphyrin derivative: an update, *Semin. Oncol.*, 21(6 Suppl. 15), 11, 1994.
132. Harty, J.I., Amin, M., Wieman, T.J., Tseng, M.T., Ackerman, D., and Broghamer, W., Complications of whole bladder dihematoporphyrin ether photodynamic therapy, *J. Urol.*, 141, 1341, 1989.
133. D'Hallewin, M.-A., Baert, L., Marijnissen, J.P.A., and Star, W.M., Whole bladder wall photodynamic therapy with *in situ* light dosimetry for carcinoma *in situ* of the bladder, *J. Urol.*, 148, 1152, 1992.
134. Pope, A.J. and Bown, S.G., The morphological and functional changes in rat bladder following photodynamic therapy with phthalocyanine photosensitization, *J. Urol.*, 145, 1064, 1991.
135. Bachor, R., Reich, E., Ruck, A., and Hautmann, R., Aminolevulinic acid for photodynamic therapy of bladder carcinoma cells, *Urol. Res.*, 24, 285, 1996.
136. Kriegmair, M., Baumgartner, R., Lumper, W., Waidelich, R., and Hofstetter, A.G., Early clinical experience with 5-aminolevulinic acid for the photodynamic therapy of superficial bladder cancer, *Br. J. Urol.*, 77, 667, 1996.
137. Auler, H. and Banzer, G., Untersuchungen über die Rolle der Porphyrine bei geschwulstkranken Menschen und Tieren, *Z. Krebsforsch.*, 53, 65, 1942.
138. Figge, F.H.J., Weiland, G.S., and Manganiello, L.O.J., Cancer detection and therapy. Affinity of neoplastic, embryonic, and traumatized tissues for porphyrins and metalloporphyrins, *Proc. Soc. Exp. Biol. Med.*, 68, 640, 1948.
139. Gregorie, H.B., Horger, E.O., Ward, J.L., Green, J.F., Richards, T., Robertson, H.C., Jr., and Stevenson, T.B., Hematoporphyrin derivative fluorescence in malignant neoplasms, *Ann. Surg.*, 176, 820, 1968.

140. Benson, R.C., Farrow, G.M., Kinsey, J., Cortese, D.A., Zincke, H., and Utz., D.C., Detection and localization of *in situ* carcinoma of the bladder with hematoporphyrin derivative, *Mayo Clin. Proc.*, 548, 1982.
141. Tsuchiya, A., Obara, N., Miwa, M., Ohi, T., Kato, H., and Hayata, Y., Hematoporphyrin derivative and laser photoradiation in the diagnosis and treatment of bladder cancer, *J. Urol.*, 130, 79, 1983.
142. Baumgartner, R., Fuchs, N., Jocham, D., Stepp, H., and Unsold, E., Pharmacokinetics of fluorescent polyporphyrin photofrin II in normal rat tissue and rat bladder tumor, *Photochem. Photobiol.*, 55, 569, 1992.
143. Kriegmair, M., Stepp, H., Steinbach, P., Lumper, W., Ehsan, A., Stepp, H.G., Rick, K., Knuchel, R., Baumgartner, R., and Hofstetter, A.G., Fluorescence cystoscopy following intravesical instillation of 5-aminolevulinic acid: a new procedure with high sensitivity for detection of hardly visible urothelial neoplasias, *Urol. Int.*, 55, 190, 1995.
144. Kriegmair, M., Baumgartner, R., Knuechel, R., Stepp, H., Hofstaedter, F., and Hofstetter, A.G., Detection of early bladder cancer by 5-aminolevulinic acid induced porphyrin fluorescence, *J. Urol.*, 155, 105, 1996.

# 46

# Therapeutic Applications of Lasers in Gastroenterology

Masoud Panjehpour
*Thompson Cancer Survival Center*
*Knoxville, Tennessee*

Bergein F. Overholt
*Thompson Cancer Survival Center*
*Knoxville, Tennessee*



## 46.1 Introduction

The use of lasers in the field of gastroenterology has evolved mainly because of the availability of flexible endoscopes that allow easy access to the lower and upper gastrointestinal (GI) tract. Laser energy may be easily delivered using a fiber optic passed through the working channel of the endoscope. In general, lasers have been used for either therapeutic or diagnostic applications. This review focuses on the therapeutic application of lasers in gastroenterology.

The major therapeutic applications of lasers in gastroenterology have been for ablation of early and advanced cancers, stopping of hemorrhage from ulcers, treatment of vascular malformations, and lithotripsy.

## 46.2 Lasers for Destruction of Tumors

In general, these lasers can be categorized as either photodynamic therapy (PDT) lasers or thermal lasers. PDT lasers are used to activate a specific photosensitizer in the tumor. Thermal lasers deliver concentrated laser energy to the tissue, resulting in coagulation or vaporization of tumors.

### 46.2.1 Lasers for PDT

PDT is a new class of cancer treatment that uses a combination of photosensitizer and laser light to destroy malignant tissues. First, a photosensitizer is administered, typically intravenously. After the photosensitizer accumulates in the tumor, light from a laser is delivered to the tumor, resulting in a reaction between the photosensitizer and the oxygen in the tissue. The resulting singlet oxygen, or free radicals, is highly cytotoxic, causing necrosis of the tissue. PDT for palliation of dysphagia in patients with partially or completely obstructing esophageal cancer has been reported.[1–4] Photofrin® (porfimer

0-8493-1116-0/03/$0.00+$1.50

sodium) from Axcan Pharma, Inc. (Mont-Saint-Hilaire, Quebec, Canada) is currently the only photosensitizer approved by the U.S. Food and Drug Administration for treatment of esophageal cancer and endobronchial lung cancer. There have also been reports of PDT in the GI tract for patients with early esophageal cancer and dysplasia in Barrett's esophagus.[5,6] The phase III multicenter clinical study for treatment of patients with high-grade dysplasia in Barrett's esophagus was recently completed.

An important component of PDT is the delivery of light to the tumor. This chapter discusses different lasers and light-delivery devices used for PDT of premalignant and malignant conditions in the GI tract. The choice of laser is determined by the photosensitizer in addition to the desired depth of necrosis. For example, porfimer sodium is activated at a wavelength of 630 nm. Meta-tetrahydroxyphenylchlorin (mTHPC), a second-generation photosensitizer, is typically activated at 652 nm. These wavelengths are in the red region of the spectrum and penetrate the tissue effectively. On the other hand, a specific laser may be employed to limit the depth of necrosis in the treated tissue. An argon laser at a single wavelength of 514 nm may be used to produce minimal depth-of-tissue necrosis, thereby minimizing the risk of perforation in treatment of superficial lesions in the esophagus.

In gastroenterology, laser light must be delivered to the tissue using optical fibers. Different fiber configurations and light-delivery devices have been developed for treatment of different conditions. Following the discussion of PDT lasers, light-delivery devices for treatment of advanced esophageal cancer, dysplasia, and early cancer in Barrett's esophagus, gastric cancer, and colorectal tumors will be reviewed.

#### 46.2.1.1 Argon-Pumped Dye Laser (Argon/Dye Laser)

The most widely used laser for PDT has been the continuous-wave (CW) argon-pumped dye laser (argon/dye laser). Argon/dye lasers can generate powerful monochromatic light that can effectively be coupled into an optical fiber for delivery to the tissue. In addition, the wavelength of the dye laser can be adjusted to match the optimum absorption of the photosensitizer. For example, porfimer sodium and its older versions hematoporphyrin derivative (HPD) and dihematoporphyrin ether (DHE) are activated at 630 nm generated from an argon/dye laser.[3–5,7–13] Aminolevulinic acid (ALA)-induced protoporphyrin IX (PpIX) is activated at 635 nm,[14] which can easily be obtained from the same argon/dye laser with minor adjustment to the birefringent filter in the dye laser. An argon/dye laser tuned at a wavelength of 652 nm can be used to activate mTHPC.[15,16]

The amount of power from a dye laser is mostly a function of the size of the pumping argon laser. High-power argon lasers, in the range of 20 W, are typically required to obtain sufficient power conversion from the dye laser.[3] Different laser manufacturers such as Spectra Physics (Mountain View, CA) make scientific argon/dye laser that have been used extensively.[7,15] Scientific lasers require frequent minor optical adjustments for optimal performance.

Lumenis, Inc. (Santa Clara, CA) formerly known as Coherent, manufactures a clinical argon/dye laser, Lambda plus PDT laser, specifically designed for PDT with Photofrin.[4,17] This laser can generate 2.7 W of 630-nm light. All optical adjustments are done automatically during the warm-up period, which takes about 2 min.

#### 46.2.1.2 KTP-Pumped Dye Laser (KTP/Dye Laser)

KTP/dye laser is a clinical PDT laser system manufactured by Laserscope (San Jose, CA). It is a quasi-CW (25 kHz) pulsed laser system (consisting of a KTP laser and a dye module) that several studies have shown to be equivalent to a CW light source such as argon/dye laser.[18,19] The dye module is designed to operate at various wavelengths. This laser has been used at 630 nm for activation of Photofrin,[5] at 652 nm for activation of mTHPC,[20] and at 635 nm for activation of ALA-induced PpIX.[21,22]

The Laserscope system employs a commonly used surgical KTP laser (532 nm). The 532-nm light from the KTP laser is used to optically pump a specially designed dye laser (600 Series dye module), which generates the 630 nm. Two models are available. The standard model has a maximum power of 3.2 W when pumped by a 700 Series KTP laser. The high-power XP 600 Series dye module has a maximum power of 7 W when pumped by 30 W of 532 nm from an 800 Series KTP laser. A high-power system is useful when treating a long segment of Barrett's esophagus using a 7-cm balloon.[5] An advantage of this system is the application of a KTP to pump a dye module. The laser energy from the KTP laser is delivered to the dye module using a fiber that allows independent movement of the dye module and the KTP laser. A PDT dye module can be purchased as an accessory to the KTP laser.

#### 46.2.1.3 Gold-Vapor Laser

Gold-vapor lasers generate a monochromatic 627.8-nm pulsed light with a typical repetition-rate range of 5 to 15 kHz.[23–26] Gold-vapor lasers have been used for activation of ALA-induced PpIX and Photofrin (and HPD and DHE). Nakamura et al.[23] used a gold-vapor laser to treat early gastric cancer using HPD PDT. Mlkvy et al.[27,28] used a gold-vapor laser to activate Photofrin and ALA-induced PpIX for treatment of a variety of GI tumors. The use of this laser eliminated the need for a dye laser. However, since the beam diameter of a gold-vapor laser is large, a 600-μm-diameter fiber is typically required for efficient coupling of the laser beam to the fiber.[29] Since gold-vapor lasers are pulsed, several studies have shown their equivalence to the CW argon/dye laser in inducing PDT response.[24–26,29]

#### 46.2.1.4 Copper-Vapor Pumped Dye Laser

While gold-vapor lasers can produce high powers at a single wavelength of 627.8 nm for activation of Photofrin, their use for activation of other photosensitizers is limited. However, gold-vapor lasers can be modified into copper-vapor lasers that in turn can be used to pump a dye laser to generate suitable wavelengths for PDT.[29] Barr et al.[30] showed that a copper-vapor pumped dye laser produced the same PDT effect in the normal rat colon as that produced with a CW argon/dye laser.

#### 46.2.1.5 Diode Lasers

Diode lasers are semiconductor light sources that are compact, user-friendly, and less expensive than conventional lasers. Diode lasers have been available at longer wavelengths such as 664 nm for activation of tin ethyl etiopurpurin, or SnET2 (Miravant Medical Technologies, Santa Barbara, CA) for treatment of metastatic breast cancer and Kaposi's sarcoma. Diode lasers with sufficient power at 630 nm (for activation of Photofrin) have been more difficult to manufacture. Diomed (Andover, MA) manufactures a diode laser with a wavelength of 630 nm ± 3 nm with a maximum calibrated power of 2.0 W from the fiber-optic delivery system. An internal power meter allows measurement of power from the fiber delivery system. Diomed also manufactures diode lasers at other wavelengths such as 635, 652, and 730 nm. All diode lasers operate on standard electrical power supply and require no water cooling. They are mobile and rugged, require less maintenance, and are cheaper than other PDT lasers.

#### 46.2.1.6 Argon Laser

The choice of laser is sometimes dictated by the desired depth of tissue necrosis. Most photosensitizers can be effectively activated at both short and long wavelengths. Therefore, a suitable short wavelength that is strongly absorbed by tissue may be used to limit depth of necrosis in the treated area. Green light (514 nm) from an argon laser has a limited penetration depth in tissue due to its strong absorption by hemoglobin. Using mTHPC, 514-nm green light from an argon laser has been recommended to eliminate the possibility of through-the-wall necrosis in the esophagus.[9,15,16] In contrast, esophageal perforations after mTHPC PDT were reported when using 652-nm red light.[9,15,16]

### 46.2.2 Light-Delivery Devices for PDT

Light-delivery devices for GI PDT are fiber optics with special configurations for treatment of various tissue sites. Historical data on the evolution of devices will be presented with emphasis on the most recent developments.

#### 46.2.2.1 Esophageal Cancer

Before the development of cylindrical diffusers many investigators used a straight-tip bare fiber that delivered the laser light from the tip of the fiber, illuminating the surface of the lesion.[3,4,10,12] These fibers were passed through the scope, and the tip of the fiber was held at a specific distance from the tissue during the treatment. Often the fiber tip was inserted into the tumor for interstitial treatment of the tumor.[4,10,12]

Cylindrical diffusers were developed for endoscopic intraluminal treatment of partial or completely obstructing esophageal cancers.[3,10,31] Cylindrical diffusers circumferentially illuminate a specific length of the inner surface of a tubular organ such as the esophagus. The distal end of the fiber is modified such that the light is emitted uniformly in a circumferential manner throughout the length of the diffuser section. These fibers are manufactured at different lengths of 1, 1.5, 2.0, 2.5, and 5 cm for treatment of esophageal cancer and endobronchial cancers (Optiguide, Fibersdirect.com, Kirkland, WA). The choice of diffuser length is dictated by the length of the tumor. These fibers are semirigid but can easily be passed through a standard GI scope and may be implanted into a tumor such as in completely obstructing esophageal cancer. Typical light dose for palliation of dysphagia in advanced esophageal cancer is 300 J/cm. These diffusers are designed for use at 630 nm.

Flexible cylindrical diffusers have been developed for intraluminal illumination of tissue (CardioFocus, Inc., West Yarmouth, MA). Flexible diffusers are available in several lengths (1 to 5 cm). These fibers are easily passed through the scope regardless of scope angulation. They are especially useful in endobronchial PDT, where the bronchoscope channel is much smaller.

Van den Bergh[32] described a reusable light-delivery device called esophageal light distributor based on the standard Savary-Gilliard dilator with a diameter of 15 mm. The distal end of the light distributor is shaped like a Savary-Gilliard dilator. The illuminating section is rigid and allows either 180 or 240° noncircumferential illumination of the esophageal lumen. This design was recommended over a circumferential illuminator to reduce stenosis after PDT.[9,15,32] This esophageal light distributor is typically used under general anesthesia.[32]

### 46.2.2.2 Barrett's Esophagus and Early Esophageal Cancer

Several techniques have been used to treat superficial cancer of the esophagus and dysplasia in Barrett's esophagus. Before development of cylindrical diffusers and balloons, a straight-tip fiber was used to treat early esophageal cancer.[11,12] Later, cylindrical diffusers were developed to treat a specific length of the esophagus. Overholt et al.[7] used a 2-cm cylindrical diffuser to treat two patients with early invasive cancer in Barrett's esophagus. Laukka and Wang[8] used a 2-cm cylindrical diffuser for treatment of patients with dysplastic Barrett's mucosa. Long segments were treated sequentially in 2-cm intervals by repositioning of the diffuser. Barr et al.[21] used a 3-cm cylindrical diffuser in a specially designed 10- to 14-mm perspex dilator to provide even light distribution. Long segments were treated by withdrawing the light source and repeating the treatment with some overlapping of the fields. Gossner et al.[22] used a 2-cm cylindrical diffuser for treatment of short segments of Barrett's and early cancer. Long segments were treated sequentially by repositioning of the diffuser.

Grosjean et al.[15] and Savary et al.[9] used a previously described cylindrical light distributor with 180 and 240° windows and a diameter of 15 mm to treat patients with early squamous cell carcinoma of the esophagus. They indicated that by using a noncircumferential light-delivery device the incidence of stenosis was reduced by limiting tissue damage to part of the lumen. The design of the cylindrical light distributor has been described.[32]

A balloon light-delivery device was specifically developed for treatment of Barrett's esophagus.[33] The balloon was constructed from a polyurethane membrane and had a diameter of 25 mm to effectively flatten the esophageal folds to allow uniform illumination of specific length of the esophagus. Increasing the balloon diameter reduced the blood flow in tissue, resulting in no PDT injury even at higher light doses.[34]

A balloon with a 2- or 3-cm window was first used clinically to treat four patients.[17] A 2-cm cylindrical diffuser was used in the 2-cm windowed balloon. A 2.5-cm cylindrical diffuser was used in the 3-cm windowed balloon (3-cm diffusers were not available). Long segments were treated sequentially by repositioning of the balloon. This balloon was used to determine the proper light dosimetry and for developing the methodology.[35]

PDT of long segments of Barrett's using a 2- or 3-cm windowed balloon required sequential treatments of the esophagus during each session. Later, 5- and 7-cm-long windowed balloons were developed to allow treatment of a longer segment of the Barrett's esophagus without repositioning of the balloon.[5,36] Flexible 5- or 7-cm cylindrical diffusers were used in, respectively, the 5- and 7-cm windowed balloons. In addition, the balloon design was modified for guidewire positioning of the balloon in the esophagus. The light dose of 175 to 200 J/cm using these balloons is typically required for destruction of high-grade dysplasia in Barrett's esophagus.

These balloons were modified with a 180° window to illuminate half the circumference of the lumen. While the concept of semicircumferential treatment was demonstrated in the canine esophagus,[37] it was never used clinically because suitable cases were not found. We believe that the entire circumference of the esophagus should be treated to eliminate all Barrett's mucosa. While this increases the likelihood of potential complications from strictures, it improves results in the ultimate goal of eliminating all Barrett's mucosa.

The above balloon was modified for the phase III multicenter study for treatment of high-grade dysplasia in Barrett's esophagus. The interior surface of the distal and proximal capped portions of the balloon was coated with a reflective material (Wilson-Cook Medical, Inc., Winston-Salem, NC). A cylindrical diffuser 2 cm longer that the window length was used inside the central channel, extending 1 cm proximally and 1 cm distally beyond the window margins. Using this diffuser/balloon configuration, the uniformity of light emitted from the balloon window was improved (unpublished data from QLT). In addition, laboratory testing and canine studies showed that the light intensity emitted from the balloon window was 1.5 times higher using the reflective-balloon design vs. that from the initial balloon design (M. Panjehpour et al., unpublished data). Using this balloon, a light dose of 130 J/cm was used for treatment of high-grade dysplasia in Barrett's esophagus. The reflective balloons were available for the study in window lengths of 3, 5, and 7 cm. The balloon material was changed to nonstretchable polyethylene terephthalate (PET). It is critical to inflate the balloon to about 20 mm Hg to reduce the pressure on the esophageal wall, minimizing the reduction of oxygen supply to the tissue.[5]

#### 46.2.2.3 Gastric Cancer

Straight-tip bare fibers have been used for treatment of gastric cancer.[4,11,12] These were simple fiber optics with cleaved and polished ends that allowed surface illumination of the tumor by positioning the fiber tip at a specific distance from the tissue. Often, the fiber tip was inserted into the tumor for intralesional (interstitial) illumination of tumors.[5,12] In most recent works, microlens fibers have been developed to deliver light to the surface of gastric tumors.[14,20] Microlens fibers generate a circular beam of light with excellent edge-to-edge uniformity and a well-defined illumination field. Microlens fibers are routinely used for cutaneous PDT. It should be noted that the beam diameter (and ultimately the energy density) is strongly dependent on the distance between the tissue and the tip of fiber when using a microlens fiber (or bare straight-tip fiber). Therefore, endoscopic illumination of a lesion should be performed taking into account the relative movements of fiber and tissue inherent in any endoscopy procedure.

#### 46.2.2.4 Lower-GI-Tract Cancers

The use of PDT has been reported for treatment of a variety of lower-GI malignancies. Much of the treatments were delivered using a bare straight-tip fiber. Loh et al.[13] used a bare fiber by removing the cladding from the distal end. They inserted the tip of the fiber into villous adenomas and treated the polyps intralesionally. Mlkvy et al.[27] used a bare fiber and inserted the tip 1 to 2 mm deep into adenomatous polyps. Patrice et al.[4] used bare fibers to either superficially or intralesionally treat rectosigmoid adenocarcinoma in inoperable patients. In the case of superficial treatment, the tip of the fiber was positioned 2 to 2.5 cm from the surface of tumor during the treatment. For intralesional treatment, the tip of the fiber was implanted into the tumor. Mlkvy et al.[28] treated colorectal and duodenal tumors using a bare fiber placed interstitially or intraluminally using a 1-cm cylindrical diffuser.

## 46.2.3 Lasers for Thermal Ablation of Tumors

### 46.2.3.1 Neodymium:Yttrium-Aluminum-Garnet (Nd:YAG) Laser and Argon Laser

The initial application of laser in gastroenterology was for thermal ablation. Laser energy is always delivered through the working channel of an endoscope using an optical fiber. The tissue effect depends on several parameters including the wavelength, power density, and energy density of the laser delivered to the tissue. Vaporization of tissue is achieved when the power density is sufficiently high. However, tissue ablation is also possible at lower power densities where coagulation is achieved, resulting in tissue necrosis. The necrotic tissue is either sloughed or can be debrided endoscopically at a later session. While several lasers have been applied, the Nd:YAG laser has been the most widely used system. Nd:YAG lasers are solid-state lasers that can generate high powers in the range of 60 to 120 W at a wavelength of 1064 nm. The laser energy is easily transmitted through a 600-μm quartz fiber, which is passed through the biopsy channel of the endoscope.

Fleischer et al.[38] reported the first palliative application of Nd:YAG for esophageal carcinoma. Significant clinical, endoscopic, and radiographical improvements were noted in all patients. Jensen et al.[39] compared the Nd:YAG laser and electrocautery technique for palliation of esophageal cancer. They concluded that electrocautery and the Nd:YAG laser were similar for treatment of circumferential tumors, while the laser was more effective for treatment of noncircumferential tumors.

The Nd:YAG laser has typically been delivered as a noncontact technique using a quartz fiber where the tip of the fiber is held about 1.5 cm from the tissue during the laser application. Later, contact probes were introduced by Joffe[40] to allow more precise application of laser energy to the lesions by touching the probe to the tissue. The contact probe is attached to the distal end of a quartz fiber. The contact probes are constructed from synthetic sapphire crystal that has proven superior to conventional noncontact quartz fiber for delivery of Nd:YAG. The advantages of using the contact technique include greater precision in delivering the laser, protection of the quartz-fiber tip from damage, and lower power requirement from the laser. In addition, there is less smoke generated using the contact technique. Contact probes are geometrically designed for each specific application and desired effect (coagulation, vaporization, cutting). The rounded contact probe is typically used for coagulative endoscopic applications using a laser power of 12 to 15 W vs. the noncontact technique requiring 80 to 100 W of laser power.[41] Overholt[42] has provided a clinical review of both contact and noncontact laser ablation for esophageal cancer. Regardless of technique, multiple sessions of laser therapy are required to establish an open lumen in esophageal cancer patients. While laser therapy is not intended to increase survival, Karlin et al.[43] reported significantly longer survival in esophageal cancer patients who were treated with Nd:YAG laser.

Another technique was described by Sander and Poesl[44] for delivering Nd:YAG laser energy for coagulation of tissue. In this technique, a water jet was used to guide the laser energy to the tissue. The advantages of this technique include ease of use, absence of smoke and carbonization, reduction in organ distention, and deeper coagulation in tissue. They suggested the use of a water-jet Nd:YAG laser for treatment of tumors in the GI tract.

While surgery remains the standard treatment for patients with colorectal cancers, lasers have been used for palliation of symptoms in inoperable or high-risk patients. The majority of treated tumors are in the rectum or rectosigmoid due to better access and less patient preparation. The typical laser power is about 100 W using a noncontact laser.[45] Mathus-Vliegen[46] has thoroughly reviewed the application of lasers for colorectal cancer. In general, lasers were effective for those with bleeding tumors. There was 75% efficacy in palliation of obstructing colorectal cancers. While the majority of laser applications have been for palliation of symptoms in advanced lower GI cancers, Lambert et al.[47] reported successful treatment of early colorectal cancer using laser.

Thermal lasers have also been used for treatment of benign lower-GI lesions. Brunetaud et al.[48] reported using argon laser and Nd:YAG laser for rectal and rectosigmoid villous adenomas. Total tumor ablation was achieved in 92% of patients.

Yasuda et al.[49] reported on the use of Nd:YAG laser for treatment of early gastric cancer. They indicated that the Nd:YAG laser was curative in 96% of 28 cases of early gastric cancer. They recommended using laser therapy for those cases that were confirmed as mucosal lesions by endoscopic ultrasound.

#### 46.2.3.2 Mid-Infrared Lasers

Mid-infrared lasers such as the erbium:YAG laser (2.94 μm) and thulium-holmium-chromium:YAG (THC:YAG) laser (2.15 μm)[50] may also be applied endoscopically because the laser energy can be transmitted through quartz fibers. In addition, as with $CO_2$ lasers a large absorption peak of water in the 2- to 3-μm range makes these lasers attractive for precision cutting. Treat et al.[50] reported preclinical results using the above lasers in human colon (*in vitro*) and in rabbit stomach (*in vivo*). They indicated that the depth of penetration was controllable with minimal spreading of injury 24 h after treatment. Bass et al.[51] compared pulsed THC:YAG laser with a clinical Nd:YAG laser in canine colonic mucosa and concluded that the THC:YAG laser created significantly less collateral thermal damage as compared with the Nd:YAG laser. They recommended this laser to reduce the risk of perforation when removing sessile polyps. Similarly, Nishioka et al.[52] tested a flashlamp-excited pulsed holmium-yttrium-scandium-gallium-garnet laser (2.1-μm, 250-μsec pulses) in rabbit liver, stomach, and colon. They indicated that this laser produced less thermal necrosis than the Nd:YAG laser and the ablation rate could be controlled in a more precise manner; they recommended it as an alternative method for endoscopic ablation of tissues.

## 46.3 Lasers for Endoscopic Control of Hemorrhage in Ulcers

Both the argon and Nd:YAG lasers have been used for photocoagulation of hemorrhage from peptic ulcers.[53–55] However, the Nd:YAG has proven to be superior to the argon laser and has been used most commonly. Typically, the Nd:YAG laser is delivered to the hemorrhage via a quartz fiber at a high power of 70 to 90 W using the noncontact method where several applications are delivered around the ulcer. The efficacy of laser therapy for hemorrhaging ulcers was reported by Swain et al.[55] and Rutgeerts et al.[56] where the need for emergency surgery and mortality was effectively reduced. The exact role of laser for this application is not clear because other nonlaser techniques, such as bipolar electrocoagulation, have been shown to be as effective at a lower cost.[57]

## 46.4 Lasers for Treatment of Vascular Malformations

Angiodysplasia and vascular malformations associated with hereditary hemorrhage telangiectasia (HHT) are the major causes of hemorrhage in the GI tract. Surgical treatment of these types of hemorrhage has been associated with considerable morbidity and mortality and is not an option for patients with multiple lesions throughout the GI tract. The argon and Nd:YAG lasers have been used for endoscopic treatment of such vascular lesions.

The argon laser has the advantage that its blue-green light is strongly absorbed by hemoglobin, resulting in relatively shallow treatment depth. This may reduce the risk of perforation given a careful control of delivered energy. However, deeper lesions within the submucosa may not be treated effectively. Waitman et al.[58] used an argon laser for treating hemorrhage secondary to telangiectasia, indicating that two thirds of patients had no recurrence of bleeding. Jensen et al.[59] used an argon laser for treatment of GI angioma. They reported a significant reduction in bleeding and transfusions after laser therapy.

The Nd:YAG laser has better penetration in tissue than the argon laser. Therefore, the Nd:YAG laser can coagulate vessels that are deeper within the submucosa. Rutgeerts et al.[60] used an Nd:YAG laser to treat vascular lesions in the upper and lower GI tract. The laser therapy resulted in significant reduction in bleedings as well as the need for transfusions. The treatment was most effective for patients with angiodysplasia. The treatment was not effective in patients with HHT. Bown et al.[61] used both the argon laser and the Nd:YAG laser for treatment of vascular lesions (HHT, single and multiple angiodysplasias).

The majority of patients required minimal transfusions after one or more laser therapies. The Nd:YAG laser appeared to achieve better long-term results due to its greater depth of penetration for destroying submucosal vessels.

Lasers have also been used for treating hemorrhage from other vascular lesions such as watermelon stomach and radiation-induced vascular lesions. Gostout et al.[62] reported using an Nd:YAG laser for treatment of watermelon stomach. Reduction in bleeding was achieved in 92% of patients. Bjorkman and Buchi[63] used an argon laser to effectively treat watermelon stomach within the mucosal layer. Viggiano et al.[64] used an Nd:YAG laser for treating radiation-induced proctopathy. After laser therapy, the number of patients with daily bleeding was reduced from 85 to 5%.

## 46.5 Lasers for Lithotripsy

High-energy pulsed lasers may be used to deliver a large amount of optical energy, resulting in the shattering of the biliary stones. The laser energy is delivered to the stone using a small optical fiber. The laser fiber should be in direct contact with the stone to improve the efficiency of fragmentation. The efficacy of stone fragmentation also depends on other laser parameters such as pulse duration.[65] Longer pulses have a reported tendency to melt renal stones[66] rather than fragment them. However, pulse durations ranging from 20 nsec to 2 msec have been used effectively.

Several pulsed lasers have been used for fragmenting stones in the common bile duct or intrahepatic ducts. Ell et al.[67] used a pulsed Nd:YAG laser to treat patients with large common bile-duct stones. The laser energy was transmitted through a 200-μm quartz fiber under direct visualization through a choledochoscope. Two thirds of patients were completely cleared of their stones. In another study by Cotton et al.,[68] a pulsed tunable dye laser at a wavelength of 504 nm was used for treating patients who were unresponsive to standard treatments. The pulse duration was 1 μsec. Some degree of fragmentation was detected in 92% of patients. A total of 8% of patients had clearance of their stones. The authors concluded that laser lithotripsy was a safe but challenging alternative to surgery in patients with large bile-duct stones. Dawson et al.[69] also used a pulsed tunable dye laser at a wavelength of 504 nm to fragment stones in the hepatic ducts or common bile duct. The pulse energy was 60 mJ. They concluded that laser lithotripsy was a safe nonsurgical alternative for patients who had failed standard treatments. Ell et al.[70] described a smart system that could detect whether the delivery fiber was in contact with a stone or tissue during treatment. If the tip of the probe came into contact with tissue, the laser treatment was terminated to minimize tissue injury. Using a flashlamp-pumped tunable dye laser to fragment biliary stone *in vitro*, Nishioka et al.[71] showed that the energy threshold was increased when the wavelength was increased from 450 nm to 700 nm. However, changing the pulse duration from 0.8 to 360 μsec did not affect the efficacy of fragmentation. The efficiency of lithotripsy was enhanced if the procedure was performed in an aqueous medium.[71] Due to the complexity of the procedure and the expense of the system, laser lithotripsy is used in a small percentage of patients with common bile-duct stones that have not responded to standard treatments.

While not a lithotripsy application, an interesting use of a high-power pulsed laser is worth noting here. Lam et al.[72] reported use of a holmium:YAG laser for fragmenting a denture that was impacted in the esophagus of a patient. After disimpacting the denture into the stomach, the laser was used to successfully fracture the denture into three pieces. The pieces were then successfully removed from the stomach.

## 46.6 Closing Remarks

This chapter reviewed the use of therapeutic lasers in gastroenterology. PDT is the newest development in the use of lasers in gastroenterology; hence, emphasis on the use of lasers and techniques for PDT in the esophagus was noted. This chapter is not intended as a thorough review of clinical literature but rather as an introduction to various applications of lasers in gastroenterology.

## References

1. Marcon, N.E., Photodynamic therapy and cancer of the esophagus, *Semin. Oncol.*, 21, 20, 1994.
2. Lightdale, C.J., Heier, S.K., Marcon, N.E., McCaughan, J.S., Jr., Gerdes, H. Overholt, B.F., Sivak, M.V., Jr., Stiegmann, G.V., and Nava, H.R., Photodynamic therapy with porfimer sodium versus thermal ablation therapy with Nd:YAG laser for palliation of esophageal cancer: a multicenter randomized trial, *Gastrointest. Endosc.*, 42, 507, 1995.
3. McCaughan, J.S., Jr., Hicks, W., Laufman, L., May, E., and Roach, R., Palliation of esophageal malignancy with photoradiation therapy, *Cancer*, 54, 2905, 1984.
4. Patrice, T., Foultier, M.T., Yactayo, S., Adam, F., Galmiche, J.P., Douet, M.C., and Le Bodic, L., Endoscopic photodynamic therapy with hematoporphyrin derivative for primary treatment of gastrointestinal neoplasms in inoperable patients, *Dig. Dis. Sci.*, 35, 545, 1990.
5. Overholt, B.F., Panjehpour, M., and Haydek, J.M., Photodynamic therapy for Barrett's esophagus: follow-up in 100 patients, *Gastrointest. Endosc.*, 49, 1, 1999.
6. Wang, K.K., Current status of photodynamic therapy of Barrett's esophagus, *Gastrointest. Endosc.*, 49, S20, 1999.
7. Overholt, B., Panjehpour, M., Teffteller, E., and Rose, M., Photodynamic therapy for treatment of early adenocarcinoma in Barrett's esophagus, *Gastrointest. Endosc.*, 39, 73, 1993.
8. Laukka, M.A. and Wang, K.K., Initial results using low-dose photodynamic therapy in the treatment of Barrett's esophagus, *Gastrointest. Endosc.*, 42, 59, 1995.
9. Savary, J.F., Grosjean, P., Monnier, P., Fontolliet, C., Wagnières, G., Braichotte, D., and Van den Bergh, H., Photodynamic therapy of early squamous cell carcinoma of the esophagus: a review of 31 cases, *Endoscopy*, 30, 258, 1998.
10. Okunaka, T., Kato, H., Conaka, C., Yamamoto, H., Bonaminio, A., and Eckhauser, M.L., Photodynamic therapy of esophageal carcinoma, *Surg. Endosc.*, 4, 150, 1990.
11. Tajiri, H., Daikuzono, N., Joffe, S.N., and Oguro, Y., Photoradiation therapy in early gastrointestinal cancer, *Gastrointest. Endosc.*, 33, 88, 1987.
12. Hayata, Y., Kato, H., Okitsu, H., Kawaguchi, M., and Konaka, C., Photodynamic therapy with hematoporphyrin derivative in cancer of the upper gastrointestinal tract, *Semin. Surg. Oncol.*, 1, 1, 1985.
13. Loh, C.S., Bliss, P., Bown, S.G., and Krasner, N., Photodynamic therapy for villous adenomas of the colon and rectum, *Endoscopy*, 26, 243, 1994.
14. Gossner, L., Sroka, R., Hahn, E.G., and Ell, C., Photodynamic therapy: successful destruction of gastrointestinal cancer after oral administration of aminolevulinic acid, *Gastrointest. Endosc.*, 41, 55, 1995.
15. Grosjean, P., Savary, J.F., Mizeret, J., Wagnières, G., Woodtli, A., Theumann, J.F., Fontolliet, C., Van den Bergh, H., and Monnier, P., Photodynamic therapy for cancer of the upper aerodigestive tract using tetra(m-hydroxyphenyl) chlorin, *J. Clin. Laser Med. Surg.*, 14, 281, 1996.
16. Savary, J.F., Monnier, P., Fontolliet, C., Mizeret, J., Wagnières, G., Braichotte, D., and Van den Bergh, H., Photodynamic therapy for early squamous cell carcinoma of the esophagus, bronchi, and mouth with m-tetra (hydroxyphenyl) chlorin, *Arch. Otolaryngol. Head Neck Surg.*, 123, 162, 1997.
17. Overholt, B.F. and Panjehpour, M., Barrett's esophagus: photodynamic therapy for ablation of dysplasia, reduction of specialized mucosa, and treatment of superficial esophageal cancer, *Gastrointest. Endosc.*, 42, 64, 1995.
18. Ferrario, A., Rucker, N., Ryter, S.W., Doiron, D.R., and Gomer, C.J., Direct comparison of *in vitro* and *in vivo* Photofrin-II mediated photosensitization using a pulsed KTP pumped dye laser and a continuous wave argon ion pumped dye laser, *Lasers Surg. Med.*, 11, 404, 1991.
19. Panjehpour, M., Overholt, B.F., DeNovo, R.C., Petersen, M.G., and Sneed, R.E., Comparative study between pulsed and continuous wave lasers for Photofrin photodynamic therapy, *Lasers Surg. Med.*, 13, 296, 1993.

20. Ell, C., Gossner, L., May, A., Schneider, H.T., Hahn, E.G., Stolte, M., and Sroka, R., Photodynamic ablation of early cancers of the stomach by means of mTHPC and laser irradiation: preliminary clinical experience, *Gut*, 43, 345, 1998.
21. Barr, H., Shepherd, N.A., Dix, A., Roberts, D.J., Tan, W.C., and Krasner, N., Eradication of high-grade dysplasia in columnar-line (Barrett's) oesophagus by photodynamic therapy with endogenously generated protoporphyrin IX, *Lancet*, 348, 584, 1996.
22. Gossner, L., Stolte, M., Sroka, R., Rick, K., May, A., Hahn, E.G., and Ell, C., Photodynamic ablation of high-grade dysplasia and early cancer in Barrett's esophagus by means of 5-aminolevulinic acid, *Gastroenterology*, 114, 448, 1998.
23. Nakamura, T., Ejiri, M., Fujisawa, T., Akiyama, H., Ejiri, K., Ishida, M., Fujimori, T., Maeda, S., Saeki, S., and Baba, S., Photodynamic therapy for early gastric cancer using a pulsed gold vapor laser, *J. Clin. Laser Med. Surg.*, 8, 63, 1990.
24. Cowled, P.A., Grace, J.R., and Forbes, I.J., Comparison of the efficacy of pulsed and continuous-wave red laser light in induction of photocytotoxicity by haematoporphyrin derivative, *Photochem. Photobiol.*, 39, 115, 1984.
25. McCaughan, J.S., Jr., Barabash, R.D., Hatch, D.R., and McMahon, D.C., Gold vapor laser versus tunable argon-dye laser for endobronchial photodynamic therapy, *Lasers Surg. Med.*, 19, 347, 1996.
26. LaPlant, M., Parker, J., Stewart, B., Waner, M., and Straight, R.C., Comparison of the optical transmission properties of pulsed and continuous wave light in biological tissue, *Lasers Surg. Med.*, 7, 336, 1987.
27. Mlkvy, P., Messmann, H., Debinski, H., Regula, J., Conio, M., MacRobert, A., Spigelman, A., Phillips R., and Bown, S.G., Photodynamic therapy for polyps in familial adenomatous polyposis: a pilot study, *Eur. J. Cancer*, 31A, 1160, 1995.
28. Mlkvy, P., Messmann, H., Regula, J., Conio, M., Pauer, M., Millson, C.E., MacRobert, A.J., and Bown, S.G., Photodynamic therapy for gastrointestinal tumors using three photosensitizers — ALA induced PPIX, Photofrin and MTHPC: a pilot study, *Neoplasma*, 45, 157, 1998.
29. Mckenzie, A.L. and Carruth, J.A.S., A comparison of gold-vapour and dye lasers for photodynamic therapy, *Lasers Med. Sci.*, 1,117, 1986.
30. Barr, H., Boulos, P.B., and MacRobert, A.J., Comparison of lasers for photodynamic therapy with a phthalocyanine photosensitizer, *Lasers Med. Sci.*, 4, 7, 1989.
31. Mimura, S., Ito, Y., Nagayo, T., Ichii, M., Kato, H., Sakai, H., Goto, K., Noguchi, Y., Tanimura, H., Nagai, Y., Suzuki, S., Hiki, Y., and Hayata, Y., Cooperative clinical trial of photodynamic therapy with Photofrin II and excimer dye laser for early gastric cancer, *Lasers Surg. Med.*, 19, 168, 1996.
32. Van den Bergh, H., On the evolution of some endoscopic light delivery systems for photodynamic therapy, *Endoscopy*, 30, 392, 1998.
33. Panjehpour, M., Overholt, B.F., DeNovo, R.C., Sneed, R.E., and Petersen, M.G., Centering balloon to improve esophageal photodynamic therapy, *Lasers Surg. Med.*, 12, 631, 1992.
34. Overholt, B.F., Panjehpour, M., DeNovo, R.C., Petersen, M.G., and Jenkins, C., Balloon photodynamic therapy of esophageal cancer: effect of increasing balloon size, *Lasers Surg. Med.*, 18, 248, 1996.
35. Overholt, B.F. and Panjehpour, M., Photodynamic therapy in Barrett's esophagus, *J. Clin. Laser Med. Surg.*, 14, 245, 1996.
36. Overholt, B.F. and Panjehpour, M., Photodynamic therapy for Barrett's esophagus, *Gastrointest. Endosc. Clin. North Am.*, 7, 207, 1997.
37. Overholt, B.F., Panjehpour, M., DeNovo, R.C., and Petersen, M.G., Photodynamic therapy for esophageal cancer using a 180 degree windowed esophageal balloon, *Lasers Surg. Med.*, 14, 27, 1994.
38. Fleischer, D., Kessler, F., and Haye, O., Endoscopic Nd:YAG laser therapy for carcinoma of the esophagus: a new palliative approach, *Am. J. Surg.*, 143, 280, 1982.
39. Jensen, D.M., Machicado, G., Randall, G., Tung, L.A., and English-Zych, S., Comparison of low power YAG laser and BICAP tumor probe for palliation of esophageal cancer strictures, *Gastroenterology*, 94, 1263, 1988.

40. Joffe, S.N., Contact neodymium:YAG laser surgery in gastroenterology. An updated report, *Surg. Endosc.*, 1, 25, 1987.
41. Sander, R.R. and Poesl, H., Cancer of the oesophagus — palliation — laser treatment and combined procedures, *Endoscopy*, 25(Suppl.), 679, 1993.
42. Overholt, B.F., Photodynamic therapy and thermal treatment of esophageal cancer, *Gastrointest. Endosc. Clin. North Am.*, 2, 433, 1992.
43. Karlin, D.A, Fisher, R.S., and Krevsky, B., Prolonged survival and effective palliation in patients with squamous cell carcinoma of the esophagus following endoscopic laser therapy, *Cancer*, 59, 1969, 1987.
44. Sander, R. and Poesl, H., Water jet guided Nd:YAG laser coagulation — its application in the field of gastroenterology, *Endosc. Surg. Allied Technol.*, 1, 233, 1993.
45. Escourrou, J., Delvaux, M., deBellison, F., et al., Laser for curative treatment of rectal cancer: indications and follow-up [abstr.], *Gastrointest. Endosc.*, 34, 195, 1988
46. Mathus-Vliegen, E.M.H., Treatment modalities in colorectal cancer, in *Lasers in Gastroenterology*, Krasner, N., Ed., Wiley-Liss, New York, 1991, p. 151.
47. Lambert, R., Sabben, G., Guyot, P., Chavaillon, A., and Descos, F., Cancer of the rectum: results of laser treatment, *Lasers Surg. Med.*, 3, 342, 1984.
48. Brunetaud, J.M., Maunoury, V., and Ducrott, E., Palliative treatment of rectosigmoid carcinoma by laser endoscopic photoablation, *Gastroenterology*, 92, 663, 1987.
49. Yasuda, K., Nakajima, M., and Kawai, K., Endoscopic diagnosis and treatment of early gastric cancer, *Gastrointest. Endosc. Clin. North Am.*, 3, 495, 1992.
50. Treat, M.R., Trokel, S.L., DeFilippi, V.J., Andrew, J., and Cohen, M.G., Mid-infrared lasers for endoscopic surgery: a new class of surgical lasers, *Am. Surgeon*, 55, 81, 1989.
51. Bass, L.S., Oz, M.C., Trokel, S.L., and Treat, M.R., Alternative lasers for endoscopic surgery: comparison of pulsed thulium-holmium-chromium:YAG with continuous-wave neodymium:YAG laser for ablation of colonic mucosa, *Lasers Surg. Med.*, 11, 545, 1991.
52. Nishioka, N.S., Demankevitz, Y., Flotte, T.J., and Anderson, R.R., Ablation of rabbit liver, stomach, and colon with a pulsed holmium laser, *Gastroenterology*, 96, 831, 1989.
53. Fruhmorgan, P., Bodem, F., Reidenback, H.D., et al., The first endoscopic laser coagulation in the human GI tract, *Endoscopy*, 7, 156, 1975.
54. Swain, C.P., Bown, S.G., Storey, D.W., Kirkham, J.S., Northfield, T.C., and Salmon, P.R., Controlled trial of argon laser photocoagulation in bleeding peptic ulcers, *Lancet*, 2, 1313, 1981.
55. Swain, C.P., Kirkham, J.S., Salmon, P.R., Bown, S.G., and Northfield, T.C., Controlled trial of Nd:YAG laser photocoagulation in bleeding peptic ulcers, *Lancet*, 1, 1113, 1986.
56. Rutgeerts, P., Vantrappen, G., Broeckaert, L., Coremans, G., Janssen, J., and Geboes, K., A new and effective technique of YAG laser photocoagulation for severe upper gastrointestinal bleeding, *Endoscopy*, 16, 115, 1984.
57. Rutgeerts, P., Vantrappen, G., Van Hootegem, P., Broeckaert, L., Janssen, J., Coremans, G., and Geboes, K., Nd:YAG laser photocoagulation versus multipolar electrocoagulation for the treatment of severely bleeding ulcers: a randomized comparison, *Gastrointest. Endosc.*, 33, 199, 1987.
58. Waitman, A.M., Graut, D.Z., and Chateau, F., Argon laser photocoagulation treatment of patients with acute and chronic bleeding secondary to telangiectasia, *Gastrointest. Endosc.*, 28, 153, 1982.
59. Jensen, D.M., Machicado, G.M., and Silpa, M.L., Treatment of GI angioma with argon laser, heater probe or bipolar electrocoagulation, *Gastrointest. Endosc.*, 30, 134, 1984.
60. Rutgeerts, P., Van Gompel, F., Geboes, K., Vantrappen, G., Broeckaert, L., and Coremans, G., Long term result of treatment of vascular malformations of the gastrointestinal tract by neodymium-YAG laser photocoagulation, *Gut*, 26, 586, 1985.
61. Bown, S.G., Swain, C.P., Storey, D.W., Collins, C., Matthewson, K., Salmon P.R., and Clark, C.G., Endoscopic laser treatment of vascular anomalies of the upper gastrointestinal tract, *Gut*, 26, 1338, 1985.

62. Gostout, C.J., Ahlquis, D.A., Radford, C.M., Viggiano, T.R., Bowyer, B.A., and Balm, R.K., Endoscopic laser therapy for watermelon stomach, *Gastroenterology,* 96, 1462, 1989.
63. Bjorkman, D.J. and Buchi, K.N., Endoscopic laser therapy of watermelon stomach, *Lasers Surg. Med.*, 12, 478, 1992.
64. Viggiano, T.R., Zighelboim, J., Ahlquist, D.A., Gostout, C.J., Wang, K.K., and Larson, M.V., Endoscopic Nd:YAG laser coagulation of bleeding from radiation proctopathy, *Gastrointest. Endosc.*, 39, 513, 1993.
65. Nishioka, N.S., Laser lithotripsy of biliary calculi, *Semin. Interventional Radiol.*, 5, 202, 1988.
66. Watson, G.M., Wickham, J.E.A., Mills, A.T.N., Bown, S.G., Swain, P., and Salmon, P.R., Laser fragmentation of renal calculi, *Br. J. Urol.*, 55, 613, 1983.
67. Ell, C., Lux, G., Hochberger, J., Muller, D., and Demling, L., Laser lithotripsy of common bile duct stones, *Gut*, 29, 746, 1988.
68. Cotton, P.B., Kozarek, R.A., Schapiro, R.H., Nishioka, N.S., Kelsey, P.B., Ball, T.J., Putnam, W.S., and Weinerth, J., Endoscopic laser lithotripsy of large bile duct stones, *Gastroenterology,* 99, 1128, 1990.
69. Dawson, S.L., Mueller, P.R., Lee, M.J., Saini, S., Kelsey, P., and Nishioka, N.S., Treatment of bile duct stones by laser lithotripsy: results in 12 patients, *Am. J. Roentgenol.*, 158, 1007, 1992.
70. Ell, C., Hochberger, J., May, A., Fleig, W.E., Bauer, R., Mendez, L., and Hahn, E.G., Laser lithotripsy of difficult bile duct stones by means of a rhodamine-6G laser and an integrated automatic stone-tissue detection system, *Gastrointest. Endosc.*, 39, 755, 1993.
71. Nishioka, N.S., Levins, P.C., Murray, S.C., Parrish, J.A., and Anderson, R.R., Fragmentation of biliary calculi with tunable dye laser, *Gastroenterology,* 93, 250, 1987.
72. Lam, Y.H., Ng, E.K., Chung, S.C., and Li, A.K., Laser-assisted removal of a foreign body impacted in the esophagus, *Lasers Surg. Med.*, 20, 480, 1997.

# 47

# Laser Treatment of Breast Tumors

Gavin M. Briggs
*National Medical Laser Centre*
*Royal Free and University College Medical School*
*London, U.K.*

Andrew C. Lee
*National Medical Laser Centre*
*Royal Free and University College Medical School*
*London, U.K.*

Stephen G. Bown
*National Medical Laser Centre*
*Royal Free and University College Medical School*
*London, U.K.*



## 47.1 Introduction

Approximately 570,000 new cases of breast cancer occur in the world each year, and breast cancer is the second leading cause of cancer death in women.[1] Breast cancer survival rates are extremely variable, with a current 5-year survival in the U.K. of 68%. This compares poorly with the rest of Europe (average 73%)[2] and the United States (average 84%).[3] No other cancer produces such an emotive response among the general public, due mainly to increased public education and high-profile coverage in the popular press. With ever increasing activity of patient-advocacy groups, attention has been refocused on all aspects of breast cancer care, causing renewed efforts to improve prevention, detection, and treatment. Despite many recent advances, however, a universal cure for breast cancer is still some way off. Research is currently under way to determine if laser treatment has a role to play in the treatment of breast cancer, but at present it can only be regarded as a research technique. Benign conditions of the breast attract much less media attention, but laser therapy appears particularly promising for the treatment of fibroadenomas of the breast, a common condition that causes hard lumps to appear in the breasts of young women.

## 47.2 Breast Cancer

### 47.2.1 Surgery

Little has changed substantively over the past century in the surgical treatment of breast cancer in terms of technology. Most surgeons still perform the operation with a scalpel, some technique like cautery to control bleeding and sutures. What has changed is a move toward more conservative, breast-preserving surgery. This shift has essentially taken place only in the last 20 years. This has followed the publication of several large studies that evaluated the long-term survival after localized surgery in conjunction with radiotherapy vs. traditional mastectomy alone.[4–6] Breast-conservation surgery generally means the excision of the cancer with a surrounding rim of normal tissue. Most would accept that a 1-cm margin

0-8493-1116-0/03/$0.00+$1.50

around the tumor is adequate, although there is still much debate about this. Although local recurrence rates are higher with lumpectomy as compared with wide local excision, overall survival remains the same.[4] There is a trade-off between removing enough tissue to keep local recurrence to a minimum against removing too much, with a resultant poor cosmetic outcome.

Lesion size is the main predictor of whether breast-conservation surgery is possible. An arbitrary cut-off figure of less than 4 cm is often quoted, but it is highly dependent on the relative size of the breast and the position of the lesion, i.e., excision of a moderate-sized tumor in a small breast may have a worse outcome than a larger lesion in a large breast. Also, excision of tumors in the lower aspect of the breast often has a poorer cosmetic outcome than those in the upper part due to a pulling down of the nipple through removal of supporting tissue and contraction of the scar.

What is evident from these trials is that surgery to remove the primary breast tumor serves only to achieve local control of disease and may have little effect on the overall outcome. Any technique that can destroy tumor tissue in the breast equally well without surgical excision has obvious attractions.

## 47.2.2 Thermal Ablation

The use of heat to treat breast cancer has been documented as far back as 3000 to 2500 B.C. with mention of "cauterization with a fire stick."[7] As heat has the obvious ability to destroy tissue, early physicians saw its potential to destroy cancerous lesions, with the added benefit of achieving cautery to blood vessels at the same time. Thermal destruction of tissue can occur in two ways. Traditionally, any thermal ablative technique has relied on explosive vaporization (true ablation) by using a high-power energy source to cause intracellular water to reach 100°C and therefore vaporize, causing rupture of the cell membrane. The second method is by protein denaturing (thermal coagulation), which utilizes a lower power source to raise the temperature above 50°C for a set period of time. It has been shown that temperatures as low as 41 to 42°C will cause cell death if exposure is prolonged,[8] but as the temperature is increased the time to achieve this decreases.[9] As a rule of thumb, the time required to kill tissue halves for each 1°C increase in temperature. The mechanism of tissue destruction is complex but most probably is related to reduced pH, hypoglycemia, vasoconstriction, increased capillary permeability, and edema.[10] Temperature changes in the opposite direction are also effective. Extreme cold has been used for many years as a method for destroying tissue.[11] Indeed "frostbite" can be just as damaging to tissue, if not more so, than a burn.

As the precision and sophistication of energy-delivery devices has evolved, interest in interstitial thermal therapy has increased. This is the direct delivery of heat into lesions in the center of solid organs, minimizing the effects in the overlying normal tissues. The rate of heat delivery is much slower than with techniques for immediate vaporization of tissue to avoid explosive effects deep within a solid organ, but the precision is usually much better. This is the "gentle-casserole" rather than the "flame-grilled" effect. Such methods are gaining interest for the treatment of a variety of benign and malignant tumors where evaporative ablation is undesirable. All interstitial therapy is critically dependent on imaging to define the limits of the lesion to be treated, to insert the energy-delivering device, and to monitor the efficacy of treatment. Appropriate imaging techniques are evolving rapidly[12] and are discussed later in this chapter.

### 47.2.2.1 Thermal Ablative Techniques

There are five main methods of tissue ablation by changing tissue temperature, all of which can be adapted for minimally invasive use. Three of these — laser photocoagulation, radio-frequency heating, and microwaves — use electromagnetic radiation. The other two are ultrasound (pressure waves) and cryotherapy (direct damage by freezing). For interstitial use laser and radio-frequency heating are the most convenient because the probes can be inserted into tissue through thin needles. To some extent this is also true for microwaves, but cryotherapy requires a wide enough probe to permit circulation of a coolant such as liquid nitrogen. Therapeutic ultrasound is usually delivered via phased arrays of transducers, and thus it is not necessary to insert anything into the tissue, but it is difficult to produce effects accurately more than a few centimeters deep in tissue or if there are extra tissue planes through which the beam must pass before reaching the target lesion. The discussion in this chapter will be limited to interstitial laser photocoagulation (ILP).

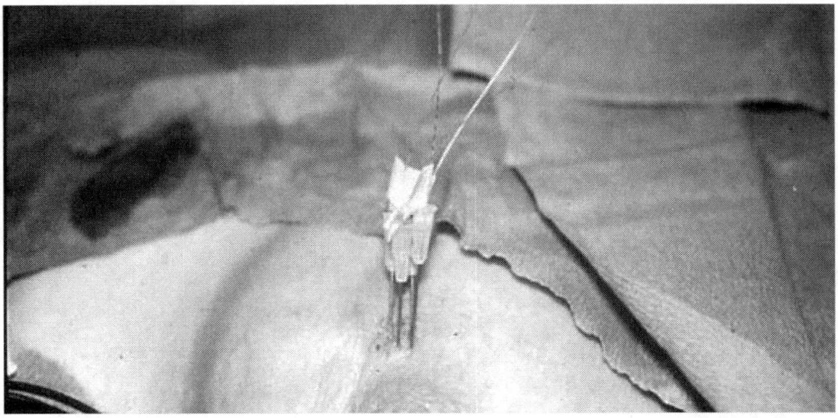

**FIGURE 47.1** Needles inserted into breast with fibers passed through needles into the tumor to be treated.

#### 47.2.2.2 Interstitial Laser Photocoagulation

This procedure was first described in 1983.[13] Simply put, the concept involves a laser fiber (typically 0.2- to 0.4-mm core diameter) being inserted directly into the target tissue (usually through a needle of about 18 gauge) so that light is delivered to the tissue from the end of the fiber as a point source (bare-tip fiber) or emitted from the end section of the fiber (diffuser fiber, where the diffuser section can be up to several centimeters long). The energy is absorbed within the surrounding tissues, causing local thermal necrosis. One or more fibres can be used.[14] Various lasers have been tried, but the most convenient are either a semiconductor diode or a neodymium:yttrium-aluminum-garnet (Nd:YAG) laser, at wavelengths between 805 and 1064 nm. The technique is sometimes referred to as laser interstitial thermal therapy (LITT) or interstitial laser thermal therapy (ILTT).

For treatment of breast tumors, the procedure can be carried out under local anesthetic and mild sedation. Up to four needles can be placed directly into the tumor either under ultrasound or MR guidance (Figure 47.1). Optical fibers are then inserted through each needle. Once the fibers are in place the needle is withdrawn slightly so that the bare ends of the fibers lie within the tumor. Previous work has demonstrated that the size of the laser-induced necrosis is more predictable when the bare end of the fiber is precharred before insertion.[15,16] This leaves a deposit of carbon particles on the tip of the fiber that absorb the laser energy, making the fiber tip a point source of heat. Without this precharring, some patients have shown no discernible treatment effect, probably because the available energy is absorbed in a larger volume of tissue so there is not enough energy per unit volume of tissue to produce a biological effect. The volume of tissue treated from each fiber position can be increased by using diffuser fibers instead of bare-tip fibers, although these are usually larger in diameter than the bare-tip ones and thus require larger needles for insertion. This causes more trauma to the overlying tissue through which the needle must pass to reach the target lesion. Diffuser fibers are also relatively difficult to manufacture in such a way that they tolerate the high temperatures generated during light delivery. Once the fibers are safely secured in position in the target tissue the laser is activated, typically for 10 min at 2 to 3 W per fiber. This gives a characteristic zone of tissue destruction (necrosis) with central charring (sometimes with liquefaction), a surrounding area of heat-fixed tissue consistent with coagulative necrosis, and an outer hemorrhagic rim. These changes are seen in surgical specimens removed a few days after ILP.

Microscopically, the central area contains carbonized and necrotic debris. Within the surrounding heat-fixed zone, the features are of morphologically normal cells but with featureless, heavily stained nuclei consistent with coagulative necrosis (Figure 47.2). Within the hemorrhagic peripheral zone the cells are less severely disrupted, with the only indication of cell damage being some increased staining in the nuclei. Also within this zone are proliferating fibroblasts (cells that lay down scar tissue) and damaged blood vessels with extravasation of red blood cells. It has been shown using enzyme histochemical

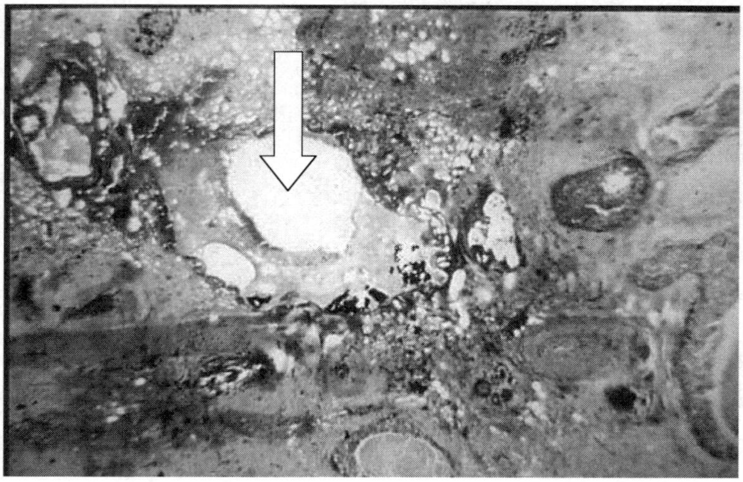

**FIGURE 47.2** Photomicrograph of a breast cancer treated with ILP and removed by surgery a few days later. A small amount of charring is visible around the needle track (arrow) with coagulation of the cancer glands outside this zone.

staining (NADH-diaphorase) that all these areas contain dead tissue.[17] These findings have been confirmed recently by a group at Stanford University that is evaluating radio-frequency ablation of breast tumors.[18]

### 47.2.2.3 Clinical Results

The earliest reported case of a laser-treated breast cancer was by Steger et al. in 1989.[19] Since then several studies investigating the feasibility of treating breast cancers in this way have been published.[15,17,20,21] In all cases, the laser treatment was either purely a research procedure (in which case ILP was followed by surgical removal of the treated cancer for microscopic examination of the treated tissue) or a palliative procedure undertaken when other treatments such as surgery or radiotherapy had failed or the patient had refused alternative treatments.

Harries et al.[15] treated 44 patients with breast cancer prior to routine surgical excision. Predictable areas of tumor necrosis (median diameter 14 mm) were achieved using a precharred fiber. Only one minor complication occurred — a small bleed from the puncture site. Mumtaz et al.[17] reported the outcomes of a further 20 patients with similar results but with more sophisticated imaging, as discussed below.

More recently, a group in Ukraine[20] presented its results in the treatment of 35 breast-cancer patients. The researchers used an Nd:YAG laser at slightly higher power outputs (2.5 to 6 W). A total of 28 patients underwent laser therapy prior to surgery, and a further 7 had ILP as their only invasive treatment. All tumors again showed laser-induced necrosis, with increasing diameter of destruction with increasing power and energy levels. There were slightly more complications, however, with four minor skin burns and one gaseous rupture. These were probably related to the increased power employed. As the first two studies have shown, 2.5 W is adequate to produce cell death without evaporative ablation. Of the seven patients who did not undergo surgery, three were postmenopausal with severe coexisting medical disease. Following ILP they all showed a good response to treatment with tumor shrinkage A further three patients with metastases at the time of diagnosis underwent palliative ILP. Two of the three showed a reasonable local response to treatment. The third, however, showed no change in tumor size. The final patient was a young woman who refused surgery. Unfortunately, she progressed quickly with aggressive metastatic disease. Overall, however, their results were again encouraging with few complications.

The most recent study by Dowlatshahi et al.[21] reported the results of 36 patients with tumors less than 2 cm in diameter treated with an 805-nm diode laser. They employed a single water-cooled diffusing fiber that allowed light to spread from the tip equally in all directions and also maintained the temperature below 100°C to avoid charring, although this type of fiber is considerably larger in diameter than those used in the earlier studies. All patients underwent surgery 8 weeks later. Examination of the excised tissue showed that there was complete tumor ablation in 66% of cases. Only two minor skin burns were encountered with no serious complications.

The only other study of thermal ablation of human breast tumors *in vivo* was by Jeffrey et al. who described the effects of radio-frequency ablation prior to routine surgery in five patients.[18] Using a single 15-gauge electrode they demonstrated partial tumor ablation in all patients with no complications. The aim was to assess the feasibility of destroying breast tumors by this method, and they intentionally left part of the tumor untreated to evaluate the margin between ablated and nonablated tissue. The size of necrosis was similar to laser therapy, although the treatment times were much longer — 30 min vs. an average of 10 min for ILP.

These studies have demonstrated the feasibility of producing local destruction of breast cancers using thermal ablative techniques. None of the studies set out to completely destroy breast cancers, although this was achieved in some cases.

Having shown that interstitial therapy can ablate cancerous tissue in the breast, clinicians now face the next major challenges — to produce a noninvasive means of assessing *in vivo* whether the cancer ablation is complete or whether viable malignant tissue has been left behind and to be sure that laser-treated areas heal safely. These aspects are addressed in the following sections of this chapter.

#### 47.2.2.4 Imaging for ILP of Breast Cancer

The main goal of any thermal treatment is to achieve complete local destruction of tumors with minimal damage to surrounding normal tissues and fewer complications than after conventional surgery. One of the key factors is the ability to accurately assess the extent of tumor destruction *in vivo*, either at the time of treatment or shortly afterwards. The gold standard for assessing completeness of tumor clearance is microscopic examination of surgically excised tissue to show that there is no residual cancer at the resection margins. The challenge for those advocating interstitial therapy is to find an equally reliable way of detecting residual viable cancer after treatment without having to remove the treated tissue from the patient. An ideal imaging modality would provide real-time assessment of the extent of tissue destruction. This would allow adjustment of treatment parameters (laser power and treatment time) or repositioning of probes/fibers during treatment to ensure complete ablation of the tumor with a "safe" surrounding rim of normal tissue. Alternatively, it may be easier to look some time after treatment to detect tumor persisting or recurring in the treated area.

Four imaging modalities are being evaluated for assessing the immediate and longer-term efficacy of ILP: ultrasound scanning (US), computerized tomography scanning (CT), magnetic resonance imaging (MRI), and positron emission tomography (PET). US is the simplest option and was tried in the early work with ILP on the liver and breast.[15,19,22] Steger found US was useful for placing the needles and fibers prior to treatment; he also found that some changes could be detected during energy delivery. However, US images did not give an accurate indication of the extent of thermal damage either at the time of treatment or subsequently. The studies by Amin[22] and Harries[15] reported similar findings. In the more recent study by Dowlatshahi, it was suggested that by using an intravenous contrast agent (Optison), US could detect the loss of blood flow in coagulated tissue,[21] although there is some doubt about how accurately the boundaries of the necrosed tissue could be defined as the US results were not correlated with histological findings. The same study presented data on the use of PET scanning pre- and postlaser treatment. They found reasonable correlation between isotope uptake and residual tumor, but this was only done on four patients because it is an expensive and time-consuming option.

Before the widespread availability of MRI, dynamic enhanced CT was investigated for determining the extent of tissue destruction and residual tumor after ILP.[15,22] Characteristically, the necrotic tissue appeared as avascular areas (no uptake of contrast material injected intravenously). This was confirmed by core-cut biopsies, which did not find any residual viable tumor in these areas. The paper by Harries reported six patients who had CT with subsequent histopathological correlation on the resected specimen. This showed reasonable accuracy for detecting the volume of treated tissue. However, there have been no larger trials evaluating CT in the breast. CT is not a conventional imaging modality for breast cancer, and it exposes patients to ionizing radiation, which ultrasound and MRI do not, so it is unlikely to become the investigation method of choice for this indication. CT has proved much more effective for assessing the results of ILP on liver tumors.[22]

It now appears that MRI has the best chance of achieving the main goals required for monitoring breast-cancer ILP. This applies both to real-time monitoring to see the extent of tissue damage as the laser is firing and for follow-up imaging to look for persistent or recurrent cancer. Thermally destroyed tumors show signal change on both T1- and T2-weighted MRI sequences.[23,24] For breast MRI other than at the actual time of light delivery, an intravascular paramagnetic "enhancement" agent (usually containing gadolinium) is used. This concentrates preferentially within breast tumors (most likely in regions of angiogenesis, the new blood vessels that develop in the regions where cancers are spreading into adjacent normal tissues) and therefore appears as a bright area on the image (increased signal intensity). Characteristically, thermally coagulated tissue loses this enhancement, whereas in residual viable tumor the increased signal intensity is maintained. The use of dynamic fast-sequence scanning (taking repeated scans at intervals of a minute or two after injection of the contrast agent) allows the enhancement characteristics of specified areas to be studied. There has been much debate about the ability of variations in signal intensity with time to differentiate between benign and malignant enhancing lesions, but for follow-up after ILP the most important aspect is to see if any of the tumor enhancement that was present prior to ILP remains after treatment. Care must be taken when interpreting the images in the first month or so after treatment to see if there is an inflammatory reaction in the surrounding tissue at the edges of the ILP-treated zone, which may itself give rise to a zone of increased uptake of contrast agent (Figure 47.3). There are not yet enough data available to allow for judgment on whether dynamic-enhancement characteristics will have any role to play in determining the nature of residual enhancement after ILP.[17]

There have been a multitude of papers in the recent literature exploring the possibilities of using MRI to guide, monitor, and follow up thermally ablated tumors, although only one has specifically assessed its accuracy following ILP in breast cancer.[17] Pre- and post-treatment scans were performed and compared to the pathological specimen following routine surgical excision a few days after ILP. Imaging immediately after ILP did not show good correlation with the final histology, whereas delayed imaging (more than 24 h after ILP) was much more accurate (Figure 47.4).

Contrast-enhanced imaging gives an accurate picture of viable cancer in the breast, but it cannot be used to follow real-time changes during light delivery because the time taken for the contrast agent to clear after one injection is comparable to the time required for laser-light delivery, so it is only possible to get an image of the situation at one point in time. Further, it has been well documented that contrast-enhanced scans taken within a few hours of ILP do not give a true picture of the extent of tumor destruction; this takes a day or two to evolve. However, MRI can also be used without administration of a contrast agent for real-time temperature mapping during treatment. The MRI parameters that vary with temperature are signal intensity on T1-weighted images,[25] signal intensity change due to diffusion,[26] and signal phase change due to proton-frequency shift.[27] Most investigators have concentrated on T1 signal-intensity change and proton resonance-frequency shift (PRFS). Both have their relative merits but also certain disadvantages.

With increasing temperature, the T1 relaxation time lengthens, and signal intensity decreases. This occurs in a relatively linear fashion until the temperature reaches around 60°C, when a plateau is reached.[12] Acquisition times are fast (around 3 sec per slice), and thermal color-coded images can be obtained (Thermo-TurboFLASH, Siemens). The main disadvantage with this method is the antagonistic decrease in T1 relaxation time within coagulated tissue. This fact has led some investigators to concentrate on frequency-shift imaging because this is dependent on temperature change only.[28] PRFS has also been shown to be more accurate than signal-intensity change for temperature mapping.[29] The main drawbacks are slow acquisition times (up to 1 min per slice) and marked sensitivity to motion artefacts. These problems can be partly overcome with faster imaging (spoiled-gradient echo — 20 sec per slice) and immobilization of the tissue in question. The need for sub-20-sec imaging is also questionable as most coagulation methods have treatment times extending over several minutes. Of greater concern, however,

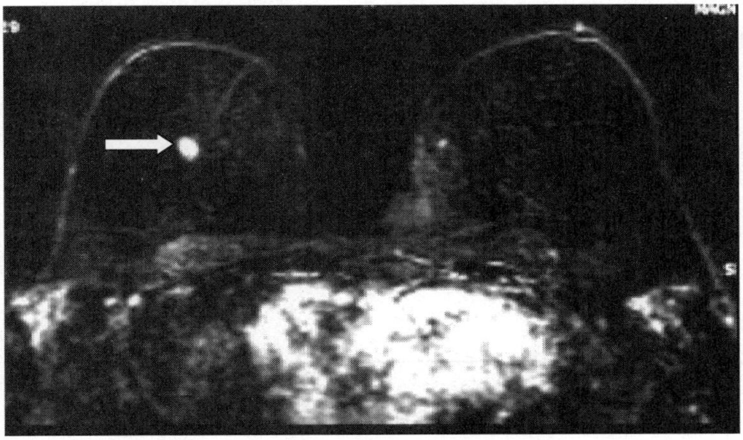

A

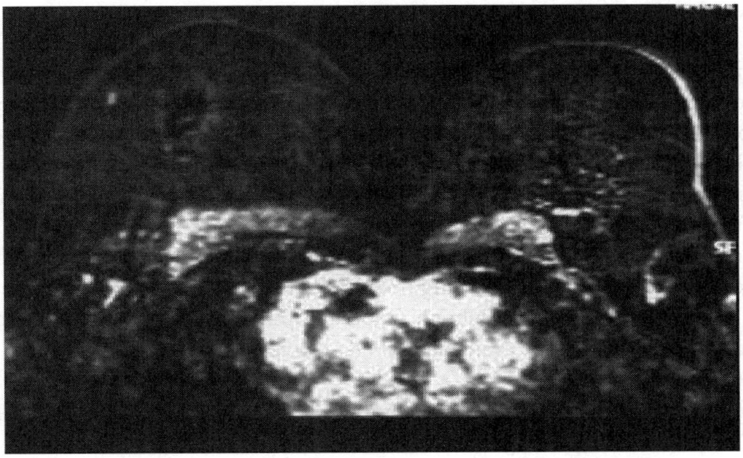

B

**FIGURE 47.3** Pre- and post-laser contrast-enhanced MRI scans of a small breast cancer. (A) Pretreatment scan showing marked enhancement at the site of the cancer (arrow). (B) Scan 2 d after ILP showing loss of enhancement in the cancer with a new, fainter ring of enhancement around the treated area indicating inflammatory changes in the surrounding normal tissue as a result of the treatment.

is the difficulty of using this technique in tissues (like the breast) that contain fat as this causes temperature-induced susceptibility changes.[27] This makes signal-intensity changes more suitable for monitoring thermal ablative methods within the breast.

Careful studies in normal rat liver have shown good correlation between real-time MR images and the extent of ILP-induced thermal necrosis.[30] Figure 47.5, from this study, shows the MRI changes at the end of a 500-sec treatment and an image taken 360 sec after the laser was turned off, when the tissue temperature had returned to normal. A subsequent clinical paper from the same group reported real-time MR images of ILP being applied to liver metastases and showed that the extent of laser-induced necrosis correlated well with contrast-enhanced CT scans taken a few days later.[31]

Real-time dynamic imaging (looking at the temperature changes) is the ideal option for monitoring ILP in the breast as the images can be compared with contrast-enhanced images taken prior to ILP. However, much research remains to be done to establish what changes on the scans really mean in terms of whether or not the tissue being imaged is viable. The MRI demonstrates changes due to the increase

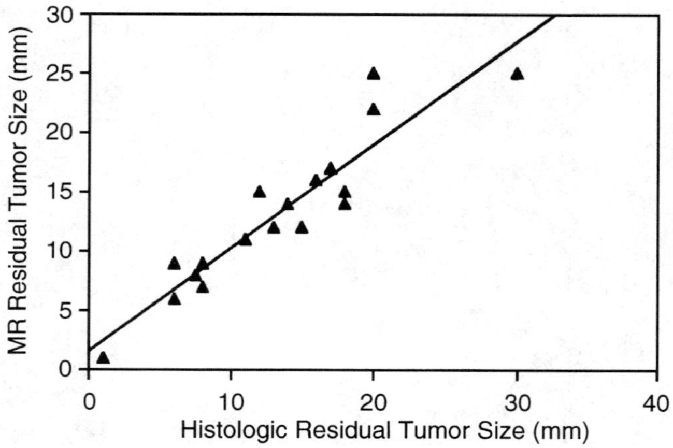


**FIGURE 47.4** Graph correlating the maximum diameter of remaining viable cancer after ILP as measured by contrast-enhanced MRI and by conventional histology on the tumor removed surgically a few days after ILP. (From Mumtaz, H. et al., *Radiology*, 200, 651, 1996. With permission.)

in tissue temperature and the consequent biological changes, but it is difficult to be sure how severe these changes need to be to indicate a loss of tissue viability, which depends on both the temperature reached and the time for which that temperature increase is maintained. The ultimate aim is to continue energy delivery until the extent of thermal tissue destruction covers the entire tumor volume, together with a surrounding rim of normal tissue (in an attempt to ablate nests of cancer cells that may have spread beyond the cancer volume detected on the scans). This is still some way in the future and requires further developments in MRI scanners. Adjusting the position of laser fibers during ILP requires an open-magnet scanner (to give easy access to the tissue being treated), but current open magnets have magnetic fields only up to 0.5 T, which is not really high enough. Good dynamic imaging of temperature changes requires around 1.5 T, which is currently only available in larger closed scanners, although this is likely to change in the relatively near future.

Thus, contrast-enhanced MRI is the most accurate technique for documenting the extent of thermal necrosis in the tumor, but it cannot give answers at the time of treatment. If residual cancer is found some time after ILP, treatment would have to be repeated. ILP does have the major attraction that, unlike radiotherapy, it can be repeated at the same site, several times if necessary.

## 47.3 Treatment of Benign Fibroadenomas and Healing of Breast Tissue after Interstitial Laser Photocoagulation

Until MRI or some other form of noninvasive imaging makes it possible to ascertain with confidence whether or not laser ablation of a cancer in the breast is complete, it would be unethical not to follow laser treatment by surgical excision. This makes it impossible with the current state of knowledge to follow how laser-treated breast cancers heal. An ethical way around this dilemma is to treat benign lesions in the breast and follow how the zone of thermal necrosis heals as the body's response to any dead tissue in the breast, cancer or otherwise, is likely to be the same. By serendipity, this approach has led to the realization that ILP is an attractive option for treating fibroadenomas.

Benign fibroadenomas have probably caused more unnecessary anguish to young women than any other breast problem. They are by far the most common cause of a discrete breast lump in young women and reach a peak incidence between the ages of 20 and 30,[32] although they are not an uncommon finding in older women. They are often only discovered through routine screening mammography, where they appear as well-circumscribed lesions, often with characteristic "popcorn calcification." Originally classified as a true benign tumor, they are now regarded as an aberration in the normal changes that occur

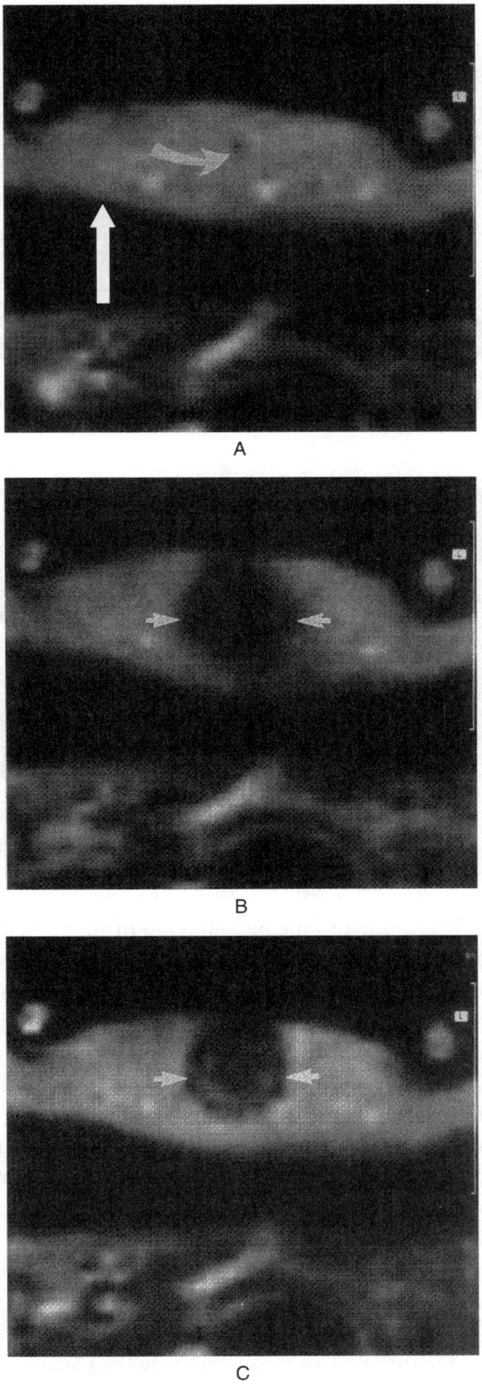


**FIGURE 47.5** Representative images from dynamic FLASH MRT monitoring of ILP to a normal rat liver. Under general anesthesia, the abdomen of the rat was opened and a lobe of liver (straight arrow) gently eased out of the abdominal cavity onto the skin. (A) Baseline showing site of insertion of fiber (curved arrow). (B) 500 sec after turning on laser to deliver 1 W of laser light at 805 nm through a bare-tip fiber. (C) 360 sec after laser had been turned off. The limits of the laser-induced effects in (B) and (C) are shown with arrows and corresponded closely to subsequent microscopic examination of the treated tissue. (From Roberts, H.R.S. et al., *Minimally Invasive Ther. Allied Technol.*, 6, 41, 1997. With permission.)

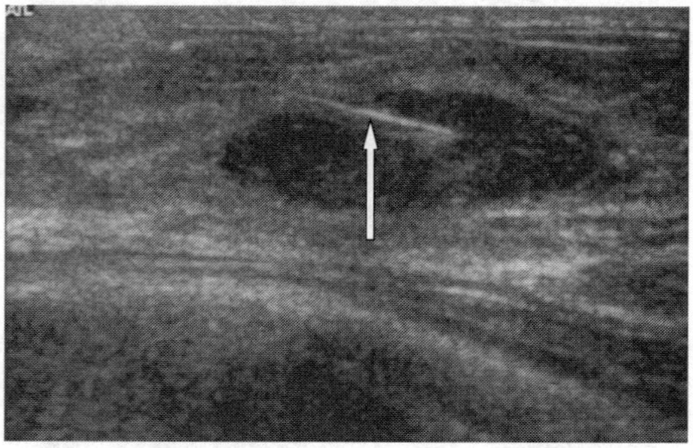

**FIGURE 47.6** Ultrasound scan showing a laser fiber (arrow) positioned in a fibroadenoma.

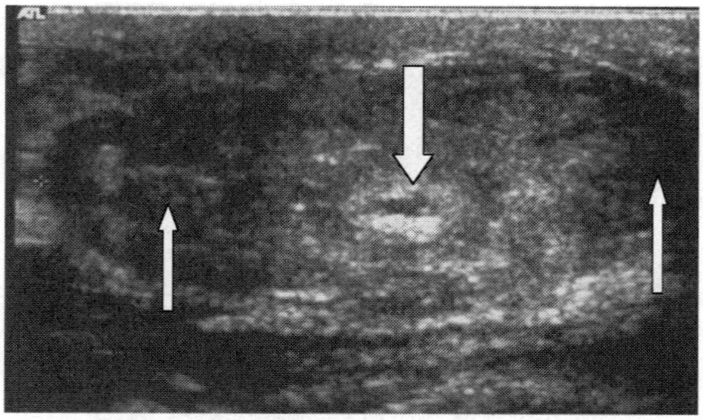

**FIGURE 47.7** Ultrasound scan to show a partly treated fibroadenoma. The central area (thick arrow) has been treated and has a bright echo pattern. The sections on either side (thin arrows) have not been treated and have very few echoes.

in the breast throughout life. This has led to the introduction of the term ANDI (aberrations of normal development and involution). Much debate has arisen in recent years about the natural history of fibroadenomas, raising the question of which ones should be treated. In simple terms, around a third will get larger, a third smaller, and a third will remain the same size.[33–35] There is no evidence that they ever turn into cancer, although the two diseases can coexist.

It can be argued that all but the most rapidly growing fibroadenomas should receive a trial of conservative management (i.e., no active treatment), and this is becoming normal practice in the U.K. There is, however, a subgroup of women who request early excision due to fear of leaving a lump *in situ* or of cancer being missed. Others request surgical treatment at some point down the line should their lump fail to resolve spontaneously. Few women would admit to being totally happy about leaving a fibroadenoma indefinitely, especially if it is particularly prominent.

As these lesions are benign, it is not critical that 100% of the lesion be treated, as residual fibroadenoma tissue does not pose a danger to the patient and can be treated again later if required. Further, the lesions have well-defined margins, unlike cancers, from which nests of cancer cells may infiltrate the tissue surrounding the main bulk of the tumor. This makes imaging simpler, so ultrasound scanning, which is cheaper and more readily available than MRI, can be used for guiding ILP and for follow-up assessment (Figure 47.6). For larger fibroadenomas multiple fibers can be used (typically up to a maximum of four). These are spaced as equally as possible within the tumor. It is important to check the position of each needle in two planes prior to light delivery because although a needle can appear central on one view, it can in fact be near the edge of the target lesion in another plane. For very large fibroadenomas, it is sometimes not feasible to coagulate it all at one sitting. Areas of coagulated fibroadenoma can, however, be distinguished from nontreated areas on ultrasound scans (Figure 47.7).

The original aim of treating fibroadenomas was to see how zones of thermally coagulated tissue in the breast healed, and this has been assessed by serial US scans. These have shown the change in size of the lesions and the characteristics of laser-coagulated tissue. Typically, ILP-treated areas of fibroadenoma become hyperechoic (bright echo pattern) compared to the normal hypoechoic (dark echo) pattern (Figure 47.7). After ILP some of the smaller fibroadenomas can become quite indistinct and difficult to distinguish from the normal surrounding breast tissue, presumably because the area of coagulation extended into the surrounding normal tissue. Nevertheless, the remarkable finding has been that the coagulated tissue is gradually absorbed by the body's normal healing processes over a period of months. As long as the lump has been completely treated, in due course it disappears entirely, although this may take up to 1 year. Patients comment on some soreness in the treated area for a few days after treatment, and the lump may initially swell (due to edema) before shrinking, but the patients do not feel generally unwell in any way.

There have only been two published studies using ILP for fibroadenomas. Lai et al.[36] treated 29 fibroadenomas with a mean diameter of 25 mm (range 14 to 35 mm). These patients were scheduled for conventional surgical excision but were offered ILP as an alternative, on the understanding that they could still have surgery at a later date if they so wished. Of the 29 lesions treated, 28 decreased in size on serial US. Only six patients continued with the original plan of surgery, and most of these chose surgery soon after ILP. It later became apparent that if they had waited longer, the fibroadenomas would probably have shrunk as a result of the ILP. The remaining patients were happy with the gradual reduction in size following ILP and declined further intervention. The average decreases in size documented were 38, 60, and 92% at 3, 6, and 12 months, respectively. Fourteen patients were followed up for 1 year, by which time none had a palpable lump and only one lesion was still detectable on US. There were no serious complications. Three patients had minor skin burns around the puncture site, either because the fiber slipped during treatment or the fiber had been positioned too superficially in the breast. This problem has been solved in more recent treatments by cooling the skin with iced water during light delivery.

Basu et al.[37] used a similar technique, the only real difference being the use of an Nd:YAG laser operating at 1064 nm rather than the semiconductor diode laser (805 nm) used by Lai et al.[36] A total of 27 patients were treated for lesions averaging 16 mm in diameter, smaller than those in Lai's study. Follow-up was shorter (8 weeks), but there was a larger average decrease in size at this time point (60%). Ten patients with residual lumps underwent excision biopsy. Again no serious complications were observed, although eight patients had minor skin burns around the puncture sites and two had a sterile, self-limiting discharge.

These papers have demonstrated the safety and efficacy of ILP for treating fibroadenomas. The smaller number of skin burns in the first study is probably attributable to the use of an oblique course for needle placement when the lumps were superficial, rather than going directly through the skin overlying the lesion. Also, cooling of the breast with cold saline may play a positive role in preventing this complication. Both studies demonstrated an increase in size over the first 2 weeks after ILP, presumably due to swelling from an inflammatory response, before a gradual decrease in size over the following months. It is important to warn patients undergoing ILP that this is likely to happen.

## 47.4 Interstitial Laser Photocoagulation in Other Organs

Much of the early experimental work on ILP was done on normal animal livers, as the liver is a forgiving organ and heals well after a localized thermal insult. In consequence, the first clinical studies were for the treatment of isolated metastatic tumors in the liver, most often in patients who had previously had cancers of the colon removed surgically.[19] Amin et al. described their experience in the treatment of 55 liver tumors in 21 patients. A total of 82% of patients fulfilled the UICC (Union Internationale contre le Cancer) criteria for at least a partial response (>50% reduction in tumor volume).[22] Recent results have concentrated on the long-term outcome in over 500 treated patients with a median survival of 27 months and a 5-year survival rate of 26%. This is comparable to patient outcomes following operative treatment for liver metastases at the same institution of 33-month median survival and 30% 5-year survival, respectively.[38] Another group reported a mean survival of 35 months[39] after ILP for liver metastases.

Some work has been published on the treatment of benign prostatic hyperplasia. Here, the aim of treatment is to reduce the volume of the prostate gland to improve urinary flow,[40] although there is some doubt about whether ILP is really any better than the conventional surgical approach of shaving out the part of the prostate causing obstruction. Another promising clinical use is in the treatment of moderate-sized uterine fibroids. These are benign tumors of the womb that can cause heavy and painful periods or infertility. Two groups have shown that ILP can shrink these lesions. Law et al. demonstrated a mean reduction in tumor volume following ILP of 37.5% at 3 months in 12 female volunteers, although they gave no indication as to clinical benefit.[41] Visvanathan et al. treated 30 fibroids in 24 symptomatic patients and documented shrinkage in 23 lesions.[42] A total of 13 patients reported an improvement in symptoms of abdominal discomfort, urinary frequency, or dysmenorrhea (painful periods), and one delivered a healthy child, having been unable to conceive prior to ILP. Yet another indication is for treating osteoid osteomas. These are benign but painful tumors of bone that often occur in sites that are very difficult to get to by conventional surgery, but where it is straightforward to insert one or more needles under CT guidance for ILP. ILP gives an almost immediate cure.[43] It has even been used for small but inoperable lung tumors.[44]

## 47.5 Conclusion

The idea of using ILP as an alternative to surgery for the destruction of small breast cancers is attractive, but it is important to keep a clear picture of the role it might play in the overall management of these patients. Local destruction of the cancer is only one part of the management of patients with this unpleasant disease, and most will require additional treatment such as radiotherapy, hormones, or chemotherapy, whether they have ILP or surgery. Further, it must be recognized that for surgeons not to follow ILP by a conventional surgical excision they will have to reach the point where they have enough confidence that MRI (real-time or delayed) can detect any viable cancer persisting after ILP. This stage remains in the distant future. However, if it is reached, the benefits will be the general simplicity and safety of the technique without the need for a general anesthetic and good cosmetic outcome (no scarring and minimal change in the shape and size of the breast). In an age where breast-conserving surgery is standard practice, it could provide an alternative for some patients with breast cancer. There are particular attractions for patients whose general condition is poor or who simply refuse surgery. For benign fibroadenomas, where precise imaging is not so vital, the evidence of efficacy and safety is now strong enough to offer ILP to all patients as an alternative to surgery.

### References

1. Cancer research campaign. http://www.cancerresearchuk.org/aboutcancer/statistics/9517?version=1, 1996.
2. Anon., *Survival of Cancer Patients in Europe: the EUROCARE-2 Study*, IARC Science, Lyon, France, 1999, 572 pp.
3. Office for National Statistics, Ref. type: Internet communication, 1999.
4. Veronesi, U., Luini, A., Galimberti, V., and Zurrida, S., Conservation approaches for the management of stage I/II carcinoma of the breast: Milan Cancer Institute trials, *World J. Surg.*, 18, 70, 1994.
5. Jacobson, J.A., Danforth, D.N., Cowan, K.H., d'Angelo, T., Steinberg, S.M., Pierce, L., Lippman, M.E., Lichter, A.S., Glatstein, E., and Okunieff, P., Ten-year results of a comparison of conservation with mastectomy in the treatment of stage I and II breast cancer, *New Engl. J. Med.*, 332, 907, 1995.
6. Fisher, B., Anderson, S., Redmond, C.K., Wolmark, N., Wickerham, D.L., and Cronin, W.M., Reanalysis and results after 12 years of follow-up in a randomized clinical trial comparing total mastectomy with lumpectomy with or without irradiation in the treatment of breast cancer, *New Engl. J. Med.*, 333, 1456, 1995.

7. Bearsted, J.H., *The Edwin Smith Surgical Papyrus*, University of Chicago Press, Chicago, 1930.
8. Overgaard, J. and Suit, H.D., Time-temperature relationship in the hyperthermic treatment of malignant and normal tissue *in vivo*, *Cancer Res.*, 39, 3248, 1979.
9. Borrelli, M.J., Thompson, L.L., Cain, C.A., and Dewey, W.C., Time-temperature analysis of cell killing of BHK cells heated at temperatures in the range of 43.5 degrees C to 57.0 degrees C, *Int. J. Radiat. Oncol. Biol. Phys.*, 19, 389, 1990.
10. Vaupel, P., Kallinowski, F., and Kluge, M., Pathophysiology of tumours in hyperthermia, *Recent Results Cancer Res.*, 107, 65, 1988.
11. Gage, A.A., History of cryosurgery, *Semin. Surg. Oncol.*, 14, 99, 1998.
12. Lamb, G.M. and Gedroyc, W.M., Interventional magnetic resonance imaging, *Br. J. Radiol.*, 70, 81, 1997.
13. Bown, S.G., Phototherapy of tumours, *World J. Surg.*, 7, 700, 1983.
14. Steger, A.C., Lees, W.R., Shorvon, P., Walmsley, K., and Bown, S.G., Multiple-fibre low-power interstitial laser hyperthermia: studies in the normal liver, *Br. J. Surg.*, 79, 139, 1992.
15. Harries, S.A., Amin, Z., Smith, M.E.F., and Bown, S.G., Interstitial laser photocoagulation as a treatment for breast cancer, *Br. J. Surg.*, 81, 1617, 1994.
16. Amin, Z., Buonaccorsi, G., Mills, T.N., Harries, S.A., Lees, W.R., and Bown, S.G., Interstitial laser photocoagulation: evaluation of a 1320 nm nd:yag and an 805 nm diode laser, the significance of charring and the value of pre-charring the fibre-tip, *Laser Med. Sci.*, 8, 113, 1993.
17. Mumtaz, H., Hall-Craggs, M.A., Wotherspoon, A., Paley, M., Buonaccorsi, G., Amin, Z., Wilkinson, I., Kissin, M.W., Davidson, T.I., Taylor, I., and Bown, S.G., Laser therapy for breast cancer: MR imaging and histopathologic correlation, *Radiology*, 200, 651, 1996.
18. Jeffrey, S.S., Birdwell, R.L., Ikeda, D.M., Daniel, B.L., Nowels, K.W., Dirbas, F.M., and Griffey, S.M., Radio-frequency ablation of breast cancer, *Arch. Surg.*, 134, 1064, 1999.
19. Steger, A.C., Lees, W.R., Walmsley, K., and Bown, S.G., Interstitial laser hyperthermia: a new approach to local destruction of tumours, *Br. Med. J.*, 299, 362, 1989.
20. Akimov, A.B., Seregin, V.E., Rusanov, K.V., Tyurina, E.G., Glushko, T.A., Nevzorov, V.P., Nevzorova, O.F., and Akimora, E.V., Nd: YAG interstitial laser thermotherapy in the treatment of breast cancer, *Lasers Surg. Med.*, 22, 257, 1998.
21. Dowlatshahi, K., Fan, M., Gould, V.E., Bloom, K.J., and Ali, A., Stereotactically guided laser therapy of occult breast tumors: work-in-progress report, *Arch. Surg.*, 135, 1345, 2000.
22. Amin, Z., Donald, J.J., Masters, A., Kant, R., Steger, A.C., Bown, S.G., and Lees, W.R., Hepatic metastases: interstitial laser photocoagulation with real-time US monitoring and dynamic CT evaluation of treatment, *Radiology*, 187, 339, 1993.
23. Solbiati, L., Goldberg, S.N., Ierace, T., Livraghi, T., Meloni, F., Dellanoce, M., Sironi, S., and Gazelle, G.S., Hepatic metastases: percutaneaous radiofrequency ablation with cooled-tip electrodes, *Radiology*, 205, 367, 1997.
24. Klotz, H.P., Flury, R., Erhart, P., Steiner, P., Debatin, J., Uhlschmid, G., and Largiader, F., Magnetic resonance-guided laparoscopic interstitial laser therapy of the liver, *Am. J. Surg.*, 174, 1997.
25. Cline, H.E., Hynynen, K., Hardy, C.J., Watkins, R.D., Schenck, J.F., and Jolesz, F.A., MR temperature mapping of focused ultrasound surgery, *Magn. Resonance Med.*, 31, 628, 1994.
26. Bleier, A.R., Jolesz, F.A., Cohen, M.S., Weisskoff, R.M., Dalcanton, J.J., Higuchi, N., Feinberg, D.A., Rosen, B.R., and McKinstry, R.C., Real-time magnetic resonance imaging of laser heat deposition in tissue, *Magn. Resonance Med.*, 21, 132, 1991.
27. De Poorter, J., de Wagter, C., de Deene, Y., Thomsen, C., Stahlberg, F., and Achten, E., Noninvasive MRI thermotherapy with the proton resonance frequency (PRF) method: *in vivo* results in human tissue, *Magn. Resonance Med.*, 33, 74, 1995.
28. Graham, S.J., Chen, L., Leitch, M., Peters, R.D., Bronskill, M.J., Foster, F.S., Henkelman, R.M., and Plewes, D.B., Quantifying tissue damage due to focused ultrasound heating observed by MRI, *Magn. Resonance Med.*, 41, 321, 1999.

29. Moriarty, J.A., Chen, J.C., Purcell, C.M., Ang, L.C., Hinks, R.S., Peters, R.D., Henkelman, R.M., Plewes, D.B., Bronskill, M.J., and Kucharczyk, W., MRI monitoring of interstitial microwave-induced heating and thermal lesions in rabbit brain *in-vivo*, *J. Magn. Resonance Imaging*, 8, 128, 1998.
30. Roberts, H.R.S., Paley, M., Sams, V.R., Wilkinson, I.D., Lees, W.R., Hall-Craggs, M.A., and Bown, S.G., Magnetic resonance imaging of interstitial laser photocoagulation of normal rat liver: imaging-histopathological correlation, *Minimally Invasive Ther. Allied Technol.*, 6, 41, 1997.
31. Roberts, H.R.S., Paley, M., Sams, V.R., Wilkinson, I.D., Lees, W.R., Hall-Craggs, M.A., and Bown, S.G., Magnetic resonance imaging control of laser destruction of hepatic metastases: correlation with post-operative dynamic helical CT, *Minimally Invasive Ther. Allied Technol.*, 6, 64, 2002.
32. Dixon, J.M., Dobie, V., Lamb, J., Walsh, J.S., and Chetty, U., Assessment of the acceptability of conservative management of fibroadenoma of the breast, *Br. J. Surg.*, 83, 264, 1996.
33. Carty, N.J., Carter, C., Rubin, C., Ravichandran, D., Royle, G.T., and Taylor, I., Management of fibroadenoma of the breast, *Ann. R. Coll. Surgeons Engl.*, 77, 127, 1995.
34. Dent, D.M. and Cant, P.J., Fibroadenoma, *World J. Surg.*, 13, 706, 1989.
35. Wilkinson, S., Anderson, T.J., Rifkind, E., Chetty, U., and Forrest, A.P., Fibroadenoma of the breast: a follow-up of conservative management, *Br. J. Surg.*, 76, 390, 1989.
36. Lai, L.M., Hall-Craggs, M.A., Mumtaz, H., Ripley, P.M., Davidson, T.I., Kissin, M.W., et al., Interstitial laser photocoagulation for fibroadenomas of the breast, *The Breast*, 8, 89, 1999.
37. Basu, S., Ravi, B., Kant, R., Saunders, C., Taylor, I., and Bown, S.G., Interstitial laser hyperthermia, a new method in the management of fibroadenoma of the breast: a pilot study, *Lasers Surg. Med.*, 25, 148, 1999.
38. Dodd, G.D., Soulen, M.C., Kane, R.A., Livraghi, T., Lees, W.R., Yamashita, Y., Gillams, A.R., Karahan, O.I., and Rhian, H., Minimally invasive treatment of malignant hepatic tumors: at the threshold of a major breakthrough, *Radiographics*, 20, 9, 2000.
39. Vogl, T.J., Mack, M.G., Straub, R., Roggan, A., and Felix, R., Magnetic resonance imaging-guided abdominal interventional radiology: laser-induced thermotherapy of liver metastases, *Endoscopy*, 29, 577, 1997.
40. De la Rosette, J.J., Muschter, R., Lopez, M.A., and Gillatt, D., Interstitial laser coagulation in the treatment of benign prostatic hyperplasia using a diode-laser system with temperature feedback, *Br. J. Urol.*, 80, 433, 1997.
41. Law, P., Gedroyc, W.M., and Regan, L., Magnetic-resonance-guided percutaneous laser ablation of uterine fibroids [letter], *Lancet*, 354, 2049, 1999.
42. Visvanathan, D., Connell, R., Hall-Craggs, M.A., Cutner, A.S., and Bown, S.G., Interstitial laser photocoagulation for uterine myomas, *Am. J. Obstet. Gynaecol.*, 187(2), 382, 2002.
43. Witt, J.D., Hall-Craggs, M.A., Ripley, P., Cobb, J.P., and Bown, S.G., Interstitial laser photocoagulation for the treatment of osteoid osteoma, *J. Bone Joint Surg. Br.*, 82, 1125, 2000.
44. Brookes, J.A., Lees, W.R., and Bown, S.G., Interstitial laser photocoagulation for the treatment of lung cancer, *Am. J. Roentgenol.*, 168, 357, 1997.

# 48

# Low-Power Laser Therapy




Tiina I. Karu

*Institute of Laser and Information Technologies*
*Russian Academy of Sciences*
*Troitsk, Moscow Region, Russian Federation*


## 48.1 Introduction

The first publications about low-power laser therapy (then called laser biostimulation) appeared more than 30 years ago. Since then, approximately 2000 studies have been published on this still controversial topic.[1] In the 1960s and 1970s, doctors in Eastern Europe, and especially in the Soviet Union and Hungary, actively developed laser biostimulation. However, scientists around the world harbored an open skepticism about the credibility of studies stating that low-intensity visible-laser radiation acts directly on an organism at the molecular level. The coherence of laser radiation for achieving stimulative effects on biological objects was more than suspect. Supporters in Western countries, such as Italy, France, and Spain, as well as in Japan and China also adopted and developed this method, but the method was — and still remains — outside mainstream medicine. The controversial points of laser biostimulation,[2–4] which were topics of great interest at that time, were analyzed in reviews that appeared in the late 1980s.

Since then, medical treatment with coherent-light sources (lasers) or noncoherent light (light-emitting diodes, LEDs) has passed through its childhood and adolescence. Most of the controversial points from "childhood" are no longer topical. Currently, low-power laser therapy — or low-level laser therapy (LLLT) or photobiomodulation — is considered part of light therapy as well as part of physiotherapy. In fact, light therapy is one of the oldest therapeutic methods used by humans (historically as sun therapy, later as color light therapy and UV therapy). A short history of experimental work with colored light on


0-8493-1116-0/03/$0.00+$1.50

various kinds of biological subjects can be found elsewhere.[2,3] The use of lasers and LEDs as light sources was the next step in the technological development of light therapy.

It is clear now that laser therapy cannot be considered separately from physiotherapeutic methods that use such physical factors as low-frequency pulsed electromagnetic fields; microwaves; time-varying, static, and combined magnetic fields; focused ultrasound; direct-current electricity; etc. Some common features of biological responses to physical factors have been briefly analyzed.[5]

As this handbook makes abundantly clear, by the dawn of the 21st century, a certain level of development of (laser) light use in therapy and diagnostics (e.g., photodynamic therapy, optical tomography, etc.) had been achieved. In low-power laser therapy, the question is no longer whether light has biological effects but rather how radiation from therapeutic lasers and LEDs works at the cellular and organism levels and what the optimal light parameters are for different uses of these light sources.

This chapter is organized as follows. First, Section 48.2 briefly reviews clinical applications and considers one of the most topical issues in low-power-laser medicine today, i.e., whether coherent and polarized light has additional benefits in comparison with noncoherent light at the same wavelength and intensity.

Second, direct activation of various types of cells via light absorption in mitochondria is described. Primary photoacceptors and mechanisms of light action on cells as well as mechanisms of cellular signaling are considered (Section 48.3). Section 48.4 describes enhancement of cellular metabolism via activation of nonmitochondrial photoacceptors and possible indirect effects via secondary cellular messengers, which are produced by cells as a result of direct activation. This chapter does not consider systemic effects of low-power laser therapy.

## 48.2 Clinical Applications and Effects of Light Coherence and Polarization

Low-power laser therapy is used by physiotherapists (to treat a wide variety of acute and chronic musculoskeletal aches and pains), by dentists (to treat inflamed oral tissues and to heal diverse ulcerations), by dermatologists (to treat edema, indolent ulcers, burns, and dermatitis), by rheumatologists (to relieve pain and treat chronic inflammations and autoimmune diseases), and by other specialists, as well as general practitioners. Laser therapy is also widely used in veterinary medicine (especially in racehorse-training centers) and in sports-medicine and rehabilitation clinics (to reduce swelling and hematoma, relieve pain, improve mobility, and treat acute soft-tissue injuries). Lasers and LEDs are applied directly to the respective areas (e.g., wounds, sites of injuries) or to various points on the body (acupuncture points, muscle-trigger points). Several books provide details of clinical applications and techniques used.[1,6,7]

Clinical applications of low-power laser therapy are diverse. The field is characterized by a variety of methodologies and uses of various light sources (lasers, LEDs) with different parameters (wavelength, output power, continuous-wave or pulsed operation modes, pulse parameters). Figure 48.1 presents schematically the types of light therapeutic devices, possible wavelengths they can emit, and maximal output power used in therapy. The GaAlAs diodes are used in both diode lasers and LEDs; the difference is whether the device contains the resonator (as the laser does) or not (LED). In recent years, longer wavelengths (~800 to 900 nm) and higher output powers (to 100 mW) have been preferred in therapeutic devices.

One of the most topical and widely discussed issues in the low-power-laser-therapy clinical community is whether the coherence and polarization of laser radiation have additional benefits as compared with monochromatic light from a conventional light source or LED with the same wavelength and intensity.

Two aspects of this problem must be distinguished: the *coherence of light* itself and the *coherence of the interaction* of light with matter (biomolecules, tissues).

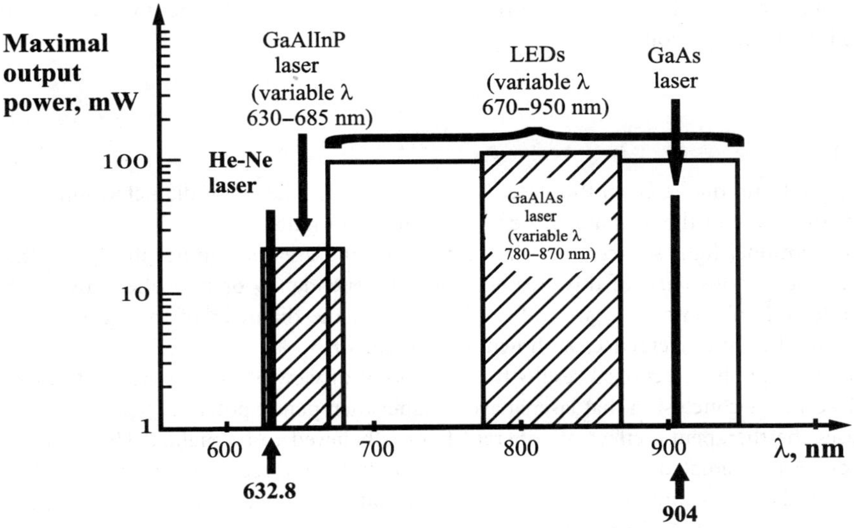


**FIGURE 48.1** Wavelength and maximal output power of lasers and LEDs used in low-power laser therapy.

### 48.2.1 Coherence of Light

The coherent properties of light are described by *temporal* and *spatial* coherence. Temporal coherence of light is determined by the spectral width, $\Delta\nu$, since the coherence time $\tau_{coh}$ during which light oscillates at the point of irradiation has a regular and strongly periodical character:

$$\tau_{coh} \cong \frac{1}{\Delta\nu} \tag{48.1}$$

Here $\Delta\nu$ is the spectral width of the beam in Hz. Since light propagates at the rate $c = 3 \times 10^{10}$ cm/sec, the light oscillations are matched by the phase (i.e., they are coherent) on the length of light propagation $L_{coh}$.

$$L_{coh} = \frac{c}{\Delta\nu[\text{Hz}]} \tag{48.2}$$

or

$$L_{coh} = \frac{1}{\Delta\nu[\text{cm}^{-1}]} \tag{48.2}$$

$L_{coh}$ is called longitudinal coherence. The more monochromatic the light, the longer the length where the light field is coherent in volume. For example, for a multimode He-Ne laser with $\Delta\nu = 500$ MHz, $L_{coh} = 60$ cm. But for a LED emitting at $\lambda = 800$ nm ($= 12{,}500$ cm$^{-1}$), $\Delta\nu = 160$ cm$^{-1}$ (or $\Delta\lambda = 10$ nm), and $L_{coh} = 1/160$ cm$^{-1} \cong 60$ μm, i.e., $L_{coh}$ is longer than the thickness of a cell monolayer (≈10 to 30 μm).

Spatial coherence describes the correlation between the phases of the light field in a lateral direction. For this reason, spatial coherence is also called lateral coherence. The size of the lateral coherence ($\ell_{coh}$) is connected with the divergence ($\varphi$) of the light beam at the point of irradiation:

$$\ell_{coh} \cong \frac{\lambda}{\varphi} \tag{48.3}$$

For example, for a He-Ne laser, which operates in the $TEM_{00}$ mode, the divergence of the beam is determined by the diffraction:

$$\varphi \cong \frac{\lambda}{D} \tag{48.4}$$

where D is the beam diameter. In this case, $\ell_{coh}$ coincides with the beam diameter, since for the $TEM_{00}$ laser mode the phase of the field along the wave front is constant.

With conventional light sources, the size of the emitting area is significantly larger than the light wavelength, and various parts of this area emit light independently or noncoherently. In this case, the size of the lateral coherence $\ell_{coh}$ is significantly less than the diameter of the light beam, and $\ell_{coh}$ is determined by the light divergence, as shown in Equation 48.3.

An analysis of published clinical results from the point of view of various types of radiation sources does not lead to the conclusion that lasers have a higher therapeutic potential than LEDs. But in certain clinical cases the therapeutic effect of coherent light is believed to be higher.[1] However, when human peptic ulcers were irradiated by a He-Ne laser or properly filtered red light was irradiated in a specially designed clinical double-blind study, equally positive results were documented for both types of radiation sources[8] (for a review, see Reference 3).

## 48.2.2 Coherence of Light Interaction with Biomolecules, Cells, and Tissues

The coherent properties of light are not manifested when the beam interacts with a biotissue on the molecular level. This problem was first considered several years ago.[2] The question then arose of whether coherent light was needed for "laser biostimulation" or was it simply a photobiological phenomenon. The conclusion was that under physiological conditions the absorption of low-intensity light by biological systems is of purely noncoherent (i.e., photobiological) nature because the rate of decoherence of excitation is many orders of magnitude higher than the rate of photoexcitation. The time of decoherence of photoexcitation determines the interaction with surrounding molecules (under normal conditions less than $10^{-12}$ sec). The average excitation time depends on the light intensity (at an intensity of 1 $mW/cm^2$ this time is around 1 sec). At 300 K in condensed matter for compounds absorbing monochromatic visible light, the light intensity at which the interactions between coherent light and matter start to occur was estimated to be above the $GW/cm^2$ level.[2] Note that the light intensities used in clinical practice are not higher than tens or hundreds of $mW/cm^2$. Indeed, the stimulative action of various bands of visible light at the level of organisms and cells was known long before the advent of the laser. Also, specially designed experiments at the cellular level have provided evidence that coherent and noncoherent light with the same wavelength, intensity, and irradiation time provide the same biological effect.[9–11] Successful use of LEDs in many areas of clinical practice also confirms this conclusion.

Therefore, it is possible that the effects of light coherence are manifested at the macroscopic (e.g., tissue) level at various depths (L) of irradiated matter. Figure 48.2 presents the coherence volumes ($V_{coh}$) and coherence lengths ($L_{coh}$) for four different light sources. Figure 48.2A presents the data for two coherent-light sources (He-Ne and diode lasers as typical examples of therapeutic devices). Figure 48.2B presents the respective data for noncoherent light (LED and spectrally filtered light from a lamp). Figure 48.2 illustrates how large volumes of tissue are irradiated only by laser sources with monochromatic radiation (Figure 48.2A). For noncoherent-radiation sources (Figure 48.2B) the length of the coherence, $L_{coh}$, is small. This means that only surface layers of an irradiated substance can be achieved by coherent light.

The spatial (lateral) coherence of the light source is unimportant due to strong scattering of light in biotissue when propagated to the depth $L \gg l_{SC}$, where $\ell_{SC}$ is the free pathway of light in relation to scattering. This is because every region in a scattering medium is illuminated by radiation with a wide angle ($\varphi \approx 1$ rd). This means that $\ell_{coh} \cong \lambda$, i.e., the size of spatial coherence $\ell_{coh}$, decreases to the light wavelength (Figure 48.2).

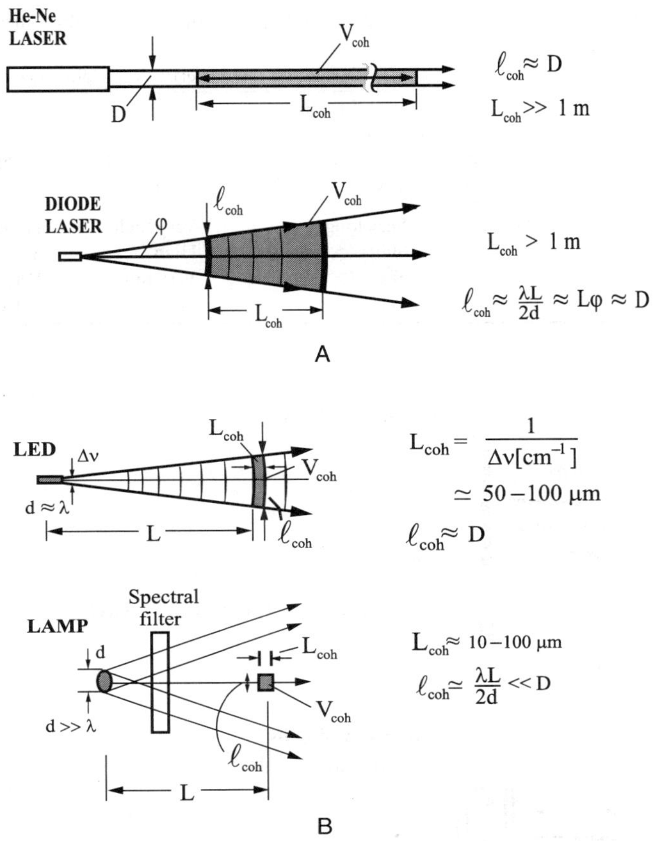


**FIGURE 48.2** Coherence volumes and coherence lengths of light from: (A) laser and (B) conventional sources when a tissue is irradiated. $L_{coh}$ = length of temporal (longitudinal) coherence, $\ell_{coh}$ = size of spatial (lateral) coherence, D = diameter of light beam, d = diameter of noncoherent-light source, φ = beam divergence, Δν = beam spectral width.

Thus, the length of longitudinal coherence ($L_{coh}$) is important when bulk tissue is irradiated because this parameter determines the volume of the irradiated tissue, $V_{coh}$. In this volume, the random interference of scattered light waves and formation of random nonhomogeneities of intensity in space (speckles) occur. For noncoherent-light sources, the coherence length is small (tens to hundreds of microns). For laser sources, this parameter is much higher. Thus, the additional therapeutic effect of coherent radiation, if this indeed exists, depends not only on the length of $L_{coh}$ but also, and even mainly, on the penetration depth into the tissue due to absorption and scattering, i.e., by the depth of attenuation. Table 48.1 summarizes qualitative characteristics of coherence of various light sources, as discussed above.

The difference in the coherence length $L_{coh}$ is unimportant when thin layers are irradiated inasmuch as the longitudinal size of irradiated object $\Delta\ell$ is less than $L_{coh}$ for any source of monochromatic light (filtered lamp light, LED, laser). Examples are the monolayer of cells and optically thin layers of cell suspensions (Figure 48.3A and B). Indeed, experimental results[9–11] on these models provide clear evidence that the biological responses of coherent and noncoherent light with the same parameters are equal. The situation is quite different when a bulk tissue is irradiated (Figure 48.3C). The coherence length $L_{coh}$ is very short for noncoherent-light sources and can play some role only on surface layers of the tissue with thickness $\Delta\ell_{surface}$. For coherent-light sources, the coherence of the radiation is retained along the entire penetration depth L. The random interference of light waves of various directions occurs over this entire distance in bulk tissue ($\Delta\ell_{bulk}$). As a result, a speckle pattern of intensity appears. Maximum values of

**TABLE 48.1** Comparison of Coherence (Temporal and Spatial) of Various Light Sources Used in Clinical Practice and Experimental Work

| | Qualitative Characteristics of Coherence | | | |
|---|---|---|---|---|
| Light Source | Temporal Coherence | Length of Longitudinal (Temporal) Coherence, $L_{coh}$ | Spatial Coherence | Volume of Spatial (Lateral) Coherence, $\ell_{coh}$ |
| Laser | Very high | Very long | Very high | Large |
| LED | Low | Short (>>λ) | High | Small (very thin layer) |
| Lamp with a spectral filter | Low | Short (>>λ) | Very low | Very small |
| Lamp | Very low | Very short (≈λ) | Very low (≈λ) | Extremely small (≈$\lambda^3$) |

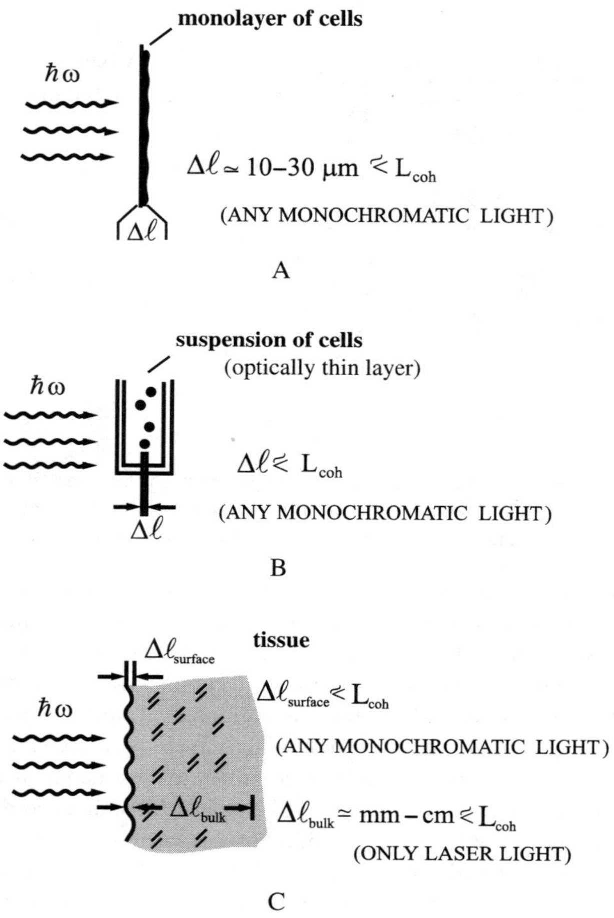


**FIGURE 48.3** Depth ($\Delta\ell$) in which the beam coherency is manifested and coherence length $L_{coh}$ in various irradiated systems: (A) monolayer of cells, (B) optically thin suspension of cells, and (C) surface layer of tissue and bulk tissue.

the intensity appear at the random constructive interference. The minima (i.e., regions of zero intensity) occur at the random destructive interference. The dimensions of these speckles at every occurrence of directed random interference are approximately within the range of the light wavelength, λ. The coherent effects (speckles) appear only at the depth $L_{coh}$. These laser-specific speckles cause a spatially nonhomogeneous deposition of light energy and lead to statistically nonhomogeneous photochemical processes, an increase in temperature, changes in local pressure, deformation of cellular membranes, etc.

For nonpolarized coherent light the random speckles are less pronounced (they have lower contrast) as compared to the speckles caused by coherent polarized light. A special feature of nonpolarized coherent radiation is that the regions with zero intensity appear less often as compared with the action of coherent polarized light. Thus, the polarization of light causes brighter random intensity gradients that can enhance the manifestation of the effects of light coherence when the tissue is irradiated.

Thus, perhaps in scattering biotissue, the main role is played by coherence length (monochromaticity of light) inasmuch as this parameter determines the depth of tissue where the coherent properties of the light beam can potentially be manifested, depending on the attenuation. This is the spatial (lateral) coherence of the beam, i.e., its directivity, which plays the main role in the delivery of light into biotissue. In addition, the direction and orientation of laser radiation could be important factors for some types of tissues (e.g., dental tissue) that have fiber-type structures (filaments). In this case, waveguide propagation effects of light can appear that provide an enhancement of penetration depth.

Considered within the framework of this qualitative picture, some additional (i.e., additional to those effects caused by light absorption by photoacceptor molecules) manifestation of light coherence for deeper tissue is quite possible. This qualitative picture also explains why coherent and noncoherent light with the same parameters produce the same biological effects on cell monolayer,[9] thin layers of cell suspension,[10,11] and tissue surface (e.g., by healing of peptic ulcers[8]). Some additional (therapeutic) effects from the coherent and polarized radiation can appear only in deeper layers of the bulk tissue. To date, no experimental work has been performed to qualitatively and quantitatively study these possible additional effects. In any case, the main therapeutic effects occur due to light absorption by cellular photoacceptors.

## 48.3 Enhancement of Cellular Metabolism via Activation of Respiratory Chain: A Universal Photobiological Action Mechanism

### 48.3.1 Cytochrome *c* Oxidase as the Photoacceptor in the Visible-to-Near-Infrared Spectral Range

Photobiological reactions involve the absorption of a specific wavelength of light by the functioning photoacceptor molecule. The photobiological nature of low-power laser effects[2,3] means that some molecule (photoacceptor) must first absorb the light used for the irradiation. After promotion of electronically excited states, primary molecular processes from these states can lead to a measurable biological effect at the cellular level. The problem is knowing which molecule is the photoacceptor. When considering the cellular effects, this question can be answered by action spectra.

A graph representing photoresponse as a function of wavelength $\lambda$, wave number $\lambda^{-1}$, frequency $\nu$, or photon energy $e$ is called an action spectrum. The action spectrum of a biological response resembles the absorption spectrum of the photoacceptor molecule. The existence of a structured action spectrum is strong evidence that the phenomenon under study is a photobiological one (i.e., primary photoacceptors and cellular signaling pathways exist).[12,13]

The first action spectra in the visible-light region were recorded in the early 1980s for DNA and RNA synthesis rate,[14,15] growth stimulation of *Escherichia coli*,[10,16] and protein synthesis by yeasts[16] for the purpose of investigating the photobiological mechanisms of laser biostimulation. In addition, other action spectra were recorded in various ranges of visible wavelengths: photostimulation of formation of E-rosettes by human lymphocytes, mitosis in L cells, exertion of DNA factor from lymphocytes in the violet–green range,[17] and oxidative phosphorylation by mitochondria in the violet–blue range.[18] All these spectra were recorded for narrow ranges of the optical spectrum and with a limited number of wavelengths, which prevented identification of the photoacceptor molecule.

Full action spectra from 313 to 860 nm for DNA and RNA synthesis rate in both exponentially growing and plateau-phase HeLa cells were also recorded in the early 1980s[19,20] (for a review, see References 2 and 4). The question of the nature of the photoacceptor molecule has since remained open. It was suggested in 1988[21] (see also Reference 4) that the mechanism of low-power laser therapy at the cellular level was based on the absorption of monochromatic visible and NIR radiation by components of the cellular respiratory chain. Absorption and promotion of electronically excited states cause changes in redox properties of these molecules and acceleration of electron transfer (primary reactions). Primary reactions in mitochondria of eukaryotic cells were supposed to be followed by a cascade of secondary reactions (photosignal transduction and amplification chain or cellular signaling) occurring in cell cytoplasm, membrane, and nucleus[21] (for a review, see References 4 and 22). In 1995, an analysis of five action spectra suggested that the primary photoacceptor for the red-NIR range in mammalian cell is a mixed-valence form of cytochrome *c* oxidase[23] (for a review, see Reference 22).

It is remarkable that the five action spectra that were analyzed had very close (within the confidence limits) peak positions in spite of the fact that these processes occurred in different parts of the cells (nucleus and plasma membrane).[19,20,24] However, there were differences in peak intensities. Three of these action spectra only for the red-to-NIR range (wavelengths that are important in low-power laser therapy) are presented in Figure 48.4A, B, and C. Two conclusions were drawn from the action spectra. First, the fact that the peak positions are the same suggests that the primary photoacceptor is the same. Second, the existence of the action spectra implies the existence of cellular signaling pathways inside the cell between photoacceptor and the nucleus as well as between the photoacceptor and cell membrane.

Five action spectra were analyzed, and the bands were identified by analogy with the absorption spectra of the metal-ligand system characteristic of this spectral range[23] (for a review, see References 22 and 25). It was concluded that the ranges 400 to 450 nm and 620 to 680 nm were characterized by the bands pertaining to a complex associated with charge transfer in a metal-ligand system, and within 760 to 830 nm these were d-d transitions in metals, most probably in Cu (II). The range 400 to 420 nm was found to be typical of a $\pi$-$\pi^*$ transition in a porphyrin ring. A comparative analysis of lines of possible d-d transitions and charge-transfer complexes of Cu with our action spectra suggested that the photoacceptor was the terminal enzyme of the mitochondrial respiratory chain cytochrome *c* oxidase. It was suggested that the main contribution to the 825-nm band was made by the oxidized $Cu_A$, to the 760-nm band by the reduced $Cu_B$, to the 680-nm band by the oxidized $Cu_B$, and to the 620-nm band by the reduced $Cu_A$. The 400- to 450-nm band was more likely the envelope of a few absorption bands in the 350- to 500-nm range (i.e., a superposition of several bands). Analysis of the band shapes in the action spectra and the line-intensity ratios also led to the conclusion that cytochrome *c* oxidase cannot be considered a primary photoacceptor when fully oxidized or fully reduced but only when it is in one of the intermediate forms (partially reduced or mixed-valence enzyme)[23] (for a review, see Reference 22) that have not yet been identified.

Taken together, the terminal respiratory-chain oxidases in eukaryotic cells (cytochrome *c* oxidase) and in prokaryotic cells of *E. coli* (cytochrome *bd* complex[26]) are believed to be photoacceptor molecules for red to NIR radiation. In the violet-to-blue spectral range, flavoproteins (e.g., NADH-dehydrogenase[5,21] in the beginning of the respiratory chain) are also among the photoacceptors and terminal oxidases.

One important step in identifying the photoacceptor molecule is to compare the absorption and action spectra. For recording the absorption of a cell monolayer and investigating the changes in absorption under irradiation at various wavelengths of monochromatic light, a sensitive multichannel registration method was developed.[27,28] Figure 48.4D presents an absorption spectrum of a monolayer of HeLa cells dried in air. In these cells, cytochrome *c* oxidase is fully oxidized. A comparison of the peak position of the spectrum in Figure 48.4D and the action spectra in Figure 48.4A, B, and C shows that the peaks near 620, 680, and 820 nm are present in all four spectra, but the peak near 760 nm is practically absent in the absorption spectrum of dry monolayer HeLa cells. Note the suggestion that this peak belongs to $Cu_B$ in reduced state.[23]

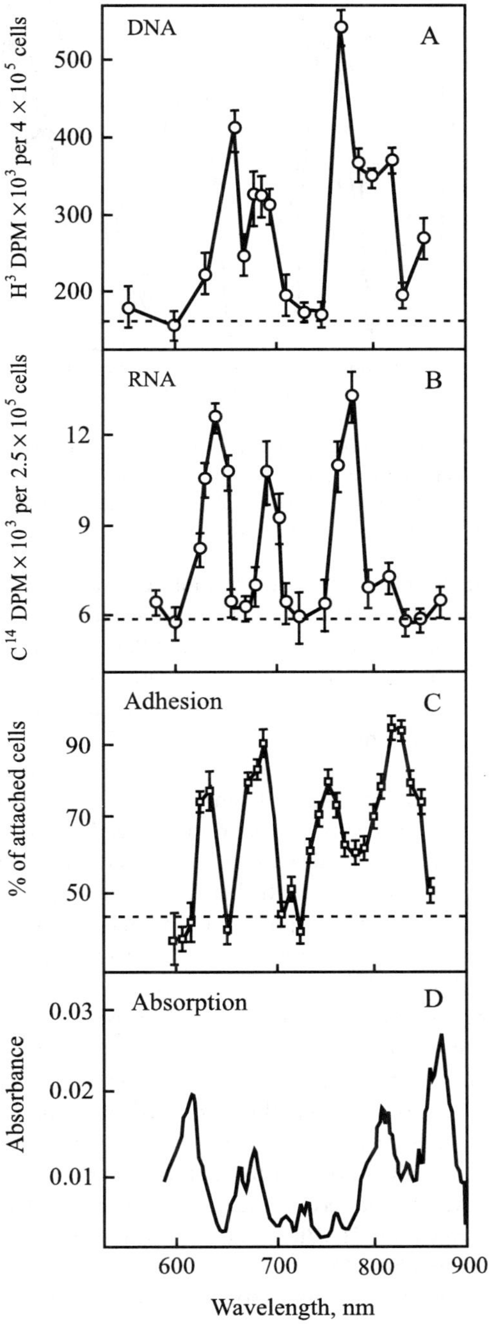


**FIGURE 48.4** Action spectra of (A) DNA and (B) RNA synthesis rate; (C) plasma membrane adhesion of exponentially growing HeLa cells for red to NIR radiation; (D) absorption spectrum of air-dried monolayer of HeLa cells for the same spectral region. (Modified from Karu, T.I. et al., *Nuov. Cim. D*, 3, 309, 1984; Karu, T.I. et al., *Dokl. Akad. Nauk (Moscow)*, 360, 267, 1998; and Karu, T.I. et al., *Lasers Surg. Med.*, 18, 171, 1996.)

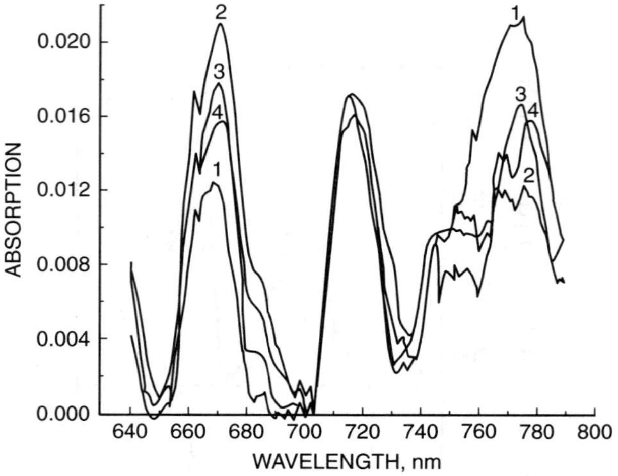


**FIGURE 48.5** Absorption spectrum of monolayer of HeLa cells recorded in open vial immediately after removal of the nutrient medium (curve 1) and following exposure to radiation with $\lambda$ = 820 nm for the first time (curve 2), second time (curve 3), and third time (curve 4), with each exposure lasting 10 sec for a dose $6.3 \times 10^3$ J/m$^2$. (Modified from Karu, T.I. et al., *Dokl. Akad. Nauk (Moscow)*, 360, 267, 1998.)

Later, the absorption spectra were recorded in the monolayer of living HeLa cells, and redox absorbance changes after laser irradiation at different wavelengths were recorded.[27,28] These experiments were performed in open[27] or closed[28] vials. These two conditions differ by, respectively, the partial pressure of oxygen in nutrient medium of cells and by the oxidation state of cytochrome *c* oxidase.

The absorption spectra of a monolayer of living cells in open flasks clearly show the bands at 670 and 775 nm as well as a less distinct band shoulder in the vicinity of 750 nm and band at 718 nm.[27] Exposing the sample for 10 sec to laser radiation with a wavelength of 670 nm and dose of $6.3 \times 10^3$ J/m$^2$ caused changes in its absorption bands around 670, 750, and 775 nm, with the absorption band at 718 nm remaining unchanged. In the action spectra, the band in the neighborhood of 670–680 nm supposedly belongs to the chromophore $Cu_B$ in the oxidized state, while that in the vicinity of 760–770 nm belongs to the chromophore $Cu_B$ in the reduced state.[23] If there is a correspondence between the action spectra bands (Figure 48.4A, B, and C) and the absorption spectra bands recorded in Reference 27, the results are quite natural: as laser irradiation increases absorption in the band at 670 nm — and hence the concentration of the chromophore in the oxidized state, represented by the absorption near 750–770 nm (and the concentration of the reduced chromophore) — decreases.

The exposure of the cellular monolayer to laser light with $\lambda$ = 820 nm[27] was also observed to cause changes in the absorption bands in the vicinity of 670 and 775 nm (Figure 48.5). Note that the action-spectrum band at around 825 nm is supposedly associated with the oxidized chromophore $Cu_A$.[23] Following the first exposure (curve 2), a sharp increase in absorption is observed to occur in the band near 670 nm (and a correspondingly sharp reduction of absorption in the band near 775 nm in comparison with the control, curve 1). The second (curve 3) and the third (curve 4) exposures cause no sharp changes in absorption, which could be due to an equilibrium being established between the oxidized and reduced forms of the chromophore $Cu_B$.

In another set of experiments, the HeLa-cell monolayer was irradiated in the closed vial where the cells had been grown for 72 h.[28] Under these conditions, the respiratory chains are supposedly more reduced as compared with the chains in the previous experiments.[27] The spectrum recorded before the irradiation had strong absorption peaks at 739, 757, and 775 nm and weak maxima at 795, 812, 831, and 873 nm, as well as at 630 nm (Figure 48.6A). A comparison of two sets of spectra[27,28] allows for a rough estimation that the peaks in the red (620 to 680 nm) and NIR regions (812 to 870 nm) are characteristic of the absorption spectra of the more oxidized cytochrome *c* oxidase, and the peaks in the 730- to 775-nm

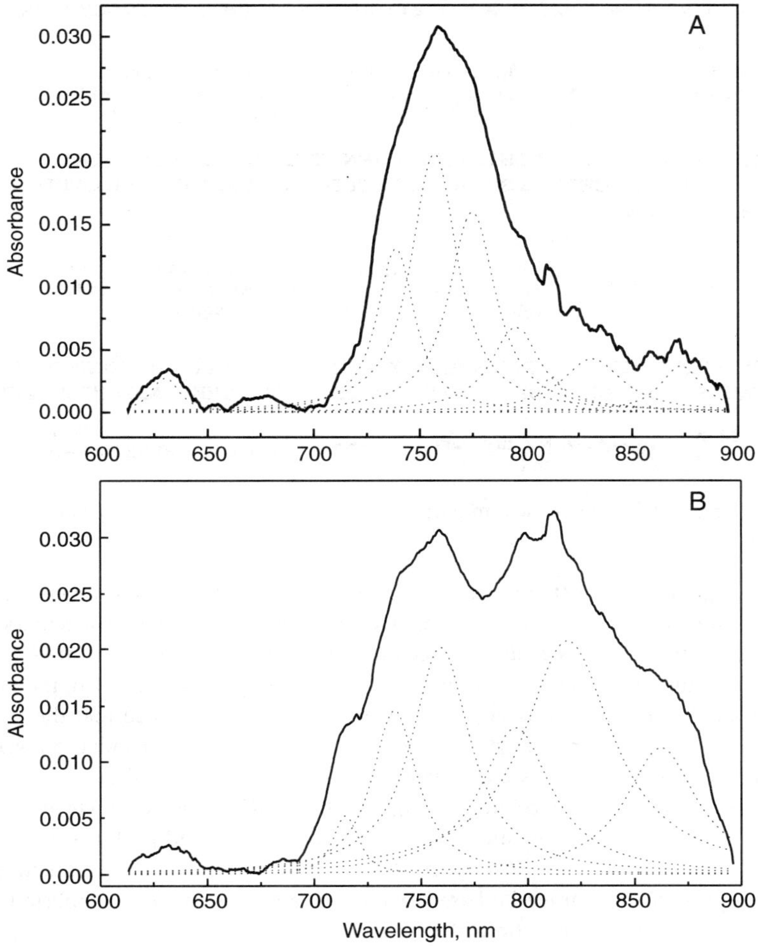


**FIGURE 48.6** Absorption spectra of HeLa monolayer: (A) before and (B) after irradiation at 820, 670, 632.8, and 670 nm in the closed vial (dose at every wavelength $6.3 \times 10^3$ J/m$^2$, irradiation time 10 sec). The dashed lines present the data of Lorentzian fitting of the spectra. (Modified from Karu, T.I. et al., *IEEE J. Sel. Top. Quantum Electron.*, 7, 982, 2001.)

range are characteristic of the spectra of the more reduced cytochrome *c* oxidase. Irradiation of the same HeLa cell monolayer in closed vials at 820, 670, and 632.8 nm and once more at 670 nm caused remarkable changes for the peaks at 739 to 799 nm and at 812 to 873 nm. There were practically no changes in absorption bands in the red region (600 to 700 nm), and a few changes occurred in the green region (peaks at 545 to 581 nm) (Figure 48.6B). It was concluded that cytochrome *c* became more oxidized due to irradiation.[28] The fact that cytochrome *c* oxidase became more oxidized when the tissue or whole cells were irradiated indicates that the oxidative metabolism had been increased.[29] Remarkable redox absorbance changes near 750 to 760 and 820 to 870 nm[28] suggest that irradiation induces structural and functional[30] changes near $Cu_A$ and $Cu_B$ chromophores, respectively. The alteration of peak parameters (width, height, area) at 750 to 760 nm indicates that the structure of the $a_3$-$Cu_B$ site (probably due to ligand-metal interactions) changes.[29,30] Recall that the irradiation of prokaryotic cells *E. coli* with a He-Ne laser also caused partial oxidation of the terminal part of the respiratory chain, cytochrome *bd* complex, while flavoproteins became slightly reduced.[31]

Changes in the absorption of HeLa cells were accompanied by conformational changes in the molecule of cytochrome *c* oxidase (measured by circular dichroism [CD] spectra[32,33]). In the visible spectral range,

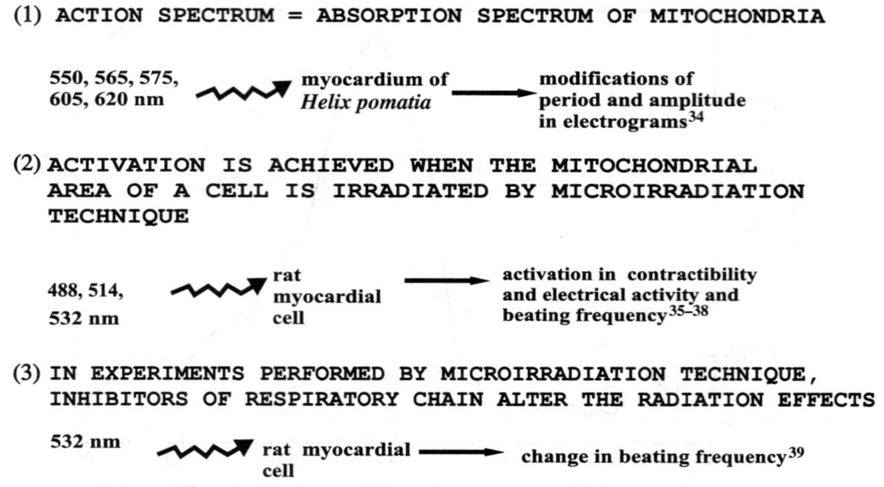
(1) ACTION SPECTRUM = ABSORPTION SPECTRUM OF MITOCHONDRIA
550, 565, 575, 605, 620 nm
myocardium of Helix pomatia
modifications of period and amplitude in electrograms[34]
(2) ACTIVATION IS ACHIEVED WHEN THE MITOCHONDRIAL AREA OF A CELL IS IRRADIATED BY MICROIRRADIATION TECHNIQUE
488, 514, 532 nm
rat myocardial cell
activation in contractibility and electrical activity and beating frequency[35–38]
(3) IN EXPERIMENTS PERFORMED BY MICROIRRADIATION TECHNIQUE, INHIBITORS OF RESPIRATORY CHAIN ALTER THE RADIATION EFFECTS
532 nm
rat myocardial cell
change in beating frequency[39]

**FIGURE 48.7** Experimental data obtained from irradiation of excitable cells indicating photoacceptors are located in the mitochondria.

distinct maxima in CD spectra (the spectra were recorded from 250 to 780 nm) of control cells were found at 566, 634, 680, 712, and 741 nm. After irradiation at 820 nm the most remarkable changes in peak positions as well as in CD signals were recorded in the range 750 to 770 nm — an appearance of a new peak at 767 nm and its shift to 757 nm after the second irradiation. Also, the peaks at 712 and 741 nm disappeared, and a new peak at 601 nm appeared. It was suggested that the changes in degree of oxidation of the chromophores of cytochrome *c* oxidase caused by the irradiation were accompanied by conformational changes in their vicinity. It was further suggested that these changes occured in the environment of $Cu_B$.[33] Even small structural changes in the binuclear site of cytochrome *c* oxidase control both rates of the dioxygen reduction and rates of internal electron- and proton-transfer reactions.[29]

The results of various studies[27,28,32,33] support the suggestion made earlier[21] that the mechanism of low-power laser therapy at the cellular level is based on the increase of oxidative metabolism in mitochondria, which is caused by electronic excitation of components of the respiratory chain (e.g., cytochrome *c* oxidase). Our results also provide evidence that various wavelengths (670, 632.8, and 820 nm) can be used for increasing respiratory activity. The wavelengths that were used in experiments described in References 27, 28, 32, and 33 were chosen in accordance with the maxima in the action spectra (Figure 48.4A and B). Note that 632.8 nm (He-Ne laser) and 820 nm (diode laser or LED) are the most common wavelengths used in therapeutic light sources.

It must be emphasized that when excitable cells (e.g., neurons, cardiomyocytes) are irradiated with monochromatic visible light, photoacceptors are also believed to be the components of the respiratory chain. Since the publication in 1947 of a study by Arvanitaki and Chalazonitis[34] it has been known that mitochondria of excitable cells have photosensitivity. Some of the experimental evidence concerning excitable cells is summarized briefly in Figure 48.7. These experiments were not performed in connection with light therapy. Experimental data[35–39] (see also Reference 40 and Chapter 5 of Reference 5) made it clear that monochromatic visible radiation could cause (via absorption in mitochondria) physiological and morphological changes in nonpigmented excitable cells, which do not contain specialized photoreceptors. Later, similar irradiation experiments were performed with neurons in connection with low-power laser therapy.[5,40] It was shown experimentally in the 1980s that He-Ne laser radiation altered the firing pattern of nerves. In addition, it was found that transcutaneous irradiation with a He-Ne laser mimicked the effect of peripheral stimulation of a behavioral reflex and that dose-related effects existed.[41] And, what is even more important, these findings were found to be connected with pain therapy.[42,43] Later clinical developments of these findings can be found in other publications.[1,6,7]

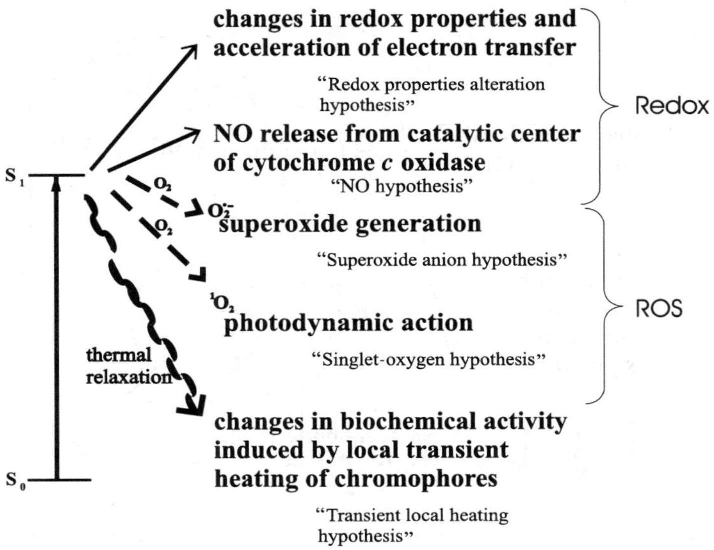


**FIGURE 48.8** Possible primary reactions in photoacceptor molecules after promotion of excited electronic states. ROS = reactive oxygen species.

## 48.3.2 Primary Reactions after Light Absorption

The primary mechanisms of light action after absorption of light quanta and the promotion of electronically excited states have not been established. The suggestions made to date are summarized in Figure 48.8; for simplicity, only singlet states ($S_0$ and $S_1$) are shown. However, triplet states are also involved.

Historically, the first mechanism, proposed in 1981 before recording of the action spectra, was the "singlet-oxygen hypothesis."[44] Certain photoabsorbing molecules like porphyrins and flavoproteins (some respiratory-chain components belong to these classes of compounds) can be reversibly converted to photosensitizers.[45] Based on visible-laser-light action on RNA synthesis rates in HeLa cells and spectroscopic data for porphyrins and flavins, the hypothesis was put forward that the absorption of light quanta by these molecules was responsible for the generation of singlet oxygen $^1O_2$ and, therefore, for stimulation of the RNA-synthesis rate[44] and the DNA synthesis rate.[9] This possibility has been considered for some time as a predominant suppressive reaction when cells are irradiated at higher doses and intensities.[3,5]

The next mechanism proposed was the "redox properties alteration hypothesis" in 1988.[21] Photoexcitation of certain chromophores in the cytochrome *c* oxidase molecule (like $Cu_A$ and $Cu_B$ or hemes *a* and $a_3$[23]) influences the redox state of these centers and, consequently, the rate of electron flow in the molecule.[21]

The latest developments indicate that under physiological conditions the activity of cytochrome *c* oxidase is also regulated by nitric oxide (NO).[46] This regulation occurs via reversible inhibition of mitochondrial respiration. It was hypothesized[47] that laser irradiation and activation of electron flow in the molecule of cytochrome *c* oxidase could reverse the partial inhibition of the catalytic center by NO and in this way increase the $O_2$-binding and respiration rate ("NO hypothesis"). This may be a factor in the increase of the concentration of the oxidized form of $Cu_B$ (Figure 48.5). Recent experimental results on the modification of irradiation effects with donors of NO do not exclude this hypothesis.[48] Note also that under pathological conditions the concentration of NO is increased (mainly due to the activation of macrophages producing NO[49]). This circumstance also increases the probability that the respiration activity of various cells will be inhibited by NO. Under these conditions, light activation of cell respiration may have a beneficial effect.

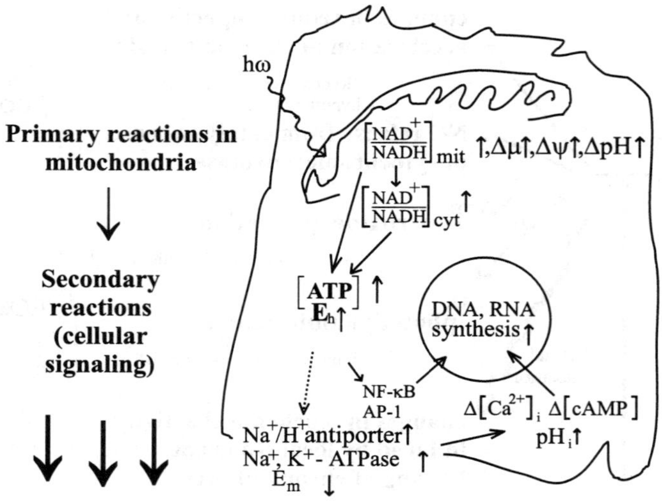


**FIGURE 48.9** Scheme of cellular signaling cascades (secondary reactions) occurring in a mammalian cell after primary reactions in the mitochondria. $E_h$ ↑ = shift of the cellular redox potential to more oxidized direction; the arrows ↑ and ↓ indicate increase or decrease of the respective values, brackets [ ] indicate the intracellular concentration of the respective chemicals.

When electronic states are excited with light, a noticeable fraction of the excitation energy is inevitably converted to heat, which causes a local transient increase in the temperature of absorbing chromophores ("transient local heating hypothesis").[50] Any appreciable time- or space-averaged heating of the sample can be prevented by controlling the irradiation intensity and dose appropriately. The local transient rise in temperature of absorbing biomolecules may cause structural (e.g., conformational) changes and trigger biochemical activity (cellular signaling or secondary dark reactions).[50,51]

In 1993, it was suggested[52] that activation of the respiratory chain by irradiation would also increase production of superoxide anions ("superoxide anion hypothesis"). It has been shown that the production of $O_2^{\bullet-}$ depends primarily on the metabolic state of the mitochondria.[53]

The belief that only one of the reactions discussed above occurs when a cell is irradiated and excited electronic states are produced is groundless. The question is, which mechanism is decisive? It is entirely possible that all the mechanisms discussed above lead to a similar result — a modulation of the redox state of the mitochondria (a shift in the direction of greater oxidation). However, depending on the light dose and intensity used, some of these mechanisms can prevail significantly. Experiments with *E. coli* provided evidence that, at different laser-light doses, different mechanisms were responsible — a photochemical one at low doses and a thermal one at higher doses.[54]

## 48.3.3 Cellular Signaling (Secondary Reactions)

If photoacceptors are located in the mitochondria, how then are the primary reactions that occur under irradiation in the respiratory chain connected with DNA and RNA synthesis in the nucleus (the action spectra in Figure 48.4A and B) or with changes in the plasma membrane (Figure 48.4C)? The principal answer is that between these events are secondary (dark) reactions (cellular signaling cascades or photosignal transduction and amplification chain, Figure 48.9).

Figure 48.9 presents a possible scheme of cellular signaling cascades, which was first proposed to explain the increase in DNA synthesis rate after the irradiation of HeLa cells with monochromatic visible light.[21] New details have been added in recent years,[5,22,25] and the latest version of this scheme is presented in Figure 48.9.

Figure 48.9 suggests three regulation pathways. The first one is the control of the photoacceptor over the level of intracellular ATP. It is known that even small changes in ATP level can significantly alter cellular metabolism (55 for a review, see Reference). However, in many cases the regulative role of redox homeostasis has proved to be more important than that of ATP. For example, the susceptibility of cells to hypoxic injury depends more on the capacity of cells to maintain the redox homeostasis and less on their capacity to maintain the energy status.[56]

The second and third regulation pathways are mediated through the cellular redox state. This may involve redox-sensitive transcription factors (NF-κB and AP-1 in Figure 48.9) or cellular signaling homeostatic cascades from cytoplasm via cell membrane to nucleus (Figure 48.9).[3,21,22] As a whole, the scheme in Figure 48.9 suggests a shift in overall cell redox potential in the direction of greater oxidation.

Recent experimental results of modification of an irradiation effect (increase of plasma-membrane adhesion when HeLa cells are irradiated at 820 nm) with various chemicals support the suggestions presented in Figure 48.9. Among these chemicals were respiratory-chain inhibitors,[57] donors of NO,[58] oxidants and antioxidants,[57] thiol reactive chemicals,[58] and chemicals that modify the activity of enzymes in the plasma membrane.[59] Recall that the overall redox state of a cell represents the net balance between stable and unstable reducing and oxidizing equivalents in dynamic equilibrium and is determined by three couples: NAD/NADH, NADP/NADPH, and GSH/GSSG (GSH = glutathione).

Recent studies have revealed that many cellular signaling pathways are regulated by the intracellular redox state (see References 60 through 63 for reviews). It is believed now that extracellular stimuli elicit cellular responses such as proliferation, differentiation, and even apoptosis through the pathways of cellular signaling. Modulation of the cellular redox state affects gene expression via mechanisms of cellular signaling (via effector molecules like transcription factors and phospholipase $A_2$).[60–62] There are at least two well-defined transcription factors — nuclear factor kappa B (NF-κB) and activator protein (AP)-1 — that have been identified as being regulated by the intracellular redox state (see References 60 and 61 for reviews). As a rule, oxidants stimulate cellular signaling systems, and reductants generally suppress the upstream signaling cascades, resulting in suppression of transcription factors.[64] It is believed now that redox-based regulation of gene expression appears to represent a fundamental mechanism in cell biology.[60,61] It is important to emphasize that in spite of some similar or even identical steps in cellular signaling, the final cellular responses to irradiation can differ due to the existence of different modes of regulation of transcription factors.

It was suggested in 1988 that activation of cellular metabolism by monochromatic visible light was a redox-regulated phenomenon.[21] Specificity of the light action is as follows: the radiation is absorbed by the components of the respiratory chain, and this is the starting point for redox regulation. The experimental data from following years have supported this suggestion.

Dependencies of various biological responses (i.e., secondary reactions) on the irradiation dose, wavelength, pulsation mode, and intensity are available (for reviews, see References 2 through 5). The main features are mentioned here. Dose-biological response curves are usually bell-shaped, characterized by a threshold, a distinct maximum, and a decline phase. In most cases, the photobiological effects depend only on the radiation dose and not on the radiation intensity and exposure time (the reciprocity rule holds true), but in other cases the reciprocity rule proves invalid (the irradiation effects depend on light intensity). Although the biological responses of various cells may be qualitatively similar, they may have essential quantitative differences. The biological effects of irradiation depend on wavelength (action spectra). The biological responses of the same cells to pulsed and continuous-wave (CW) light of the same wavelength, average intensity, and dose can vary. (See Reference 5 for a detailed review).

Figure 48.10 explains magnitudes of low-power laser effects as being dependent on the initial redox status of a cell. The main idea expressed in Figure 48.10 is that cellular response is weak or absent (the dashed arrows on the right side) when the overall redox potential of a cell is optimal or near optimal for the particular growth conditions. The cellular response is stronger when the redox potential of the target cell is initially shifted (the arrows on left side) to a more reduced state (and intracellular pH, $pH_i$, is lowered). This explains why the degrees of cellular responses can differ markedly in different experiments and why they are sometimes nonexistent. A jump in $pH_i$ due to irradiation has been measured experimentally (0.20 units in mammalian cells[65] and 0.32 units in *E. coli*[66]).

Various magnitudes of low-power laser effects (strong effect, weak effect, or no effect at all) have always been one of the most criticized aspects of low-power laser therapy. An attempt was made to quantify the

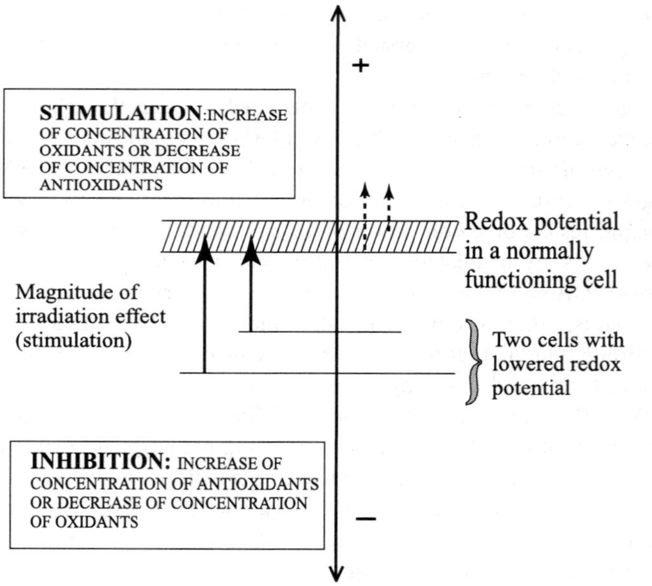


**FIGURE 48.10** Schematic illustration of the action principle of monochromatic visible and NIR radiation on a cell. Irradiation shifts the cellular redox potential in a more oxidized direction. The magnitude of cellular response is determined by the cellular redox potential at the moment of irradiation.

magnitude of irradiation effects as dependent on the metabolic status of *E. coli* cells[67] (for a review, see Reference 26). Recently, the correlation was found between the amount of ATP in irradiated cells and the initial amount of ATP in control cells.[68]

Thus, variations in the magnitude of low-power laser effects at the cellular level are explained by the overall redox state (and $pH_i$) at the moment of irradiation. Cells with a lowered $pH_i$ (in which redox state is shifted to the reduced side) respond stronger than cells with a normal or close-to-normal $pH_i$ value.

### 48.3.4 Partial Derepression of Genome of Human Peripheral Lymphocytes: Biological Limitations of Low-Power Laser Effects

Monochromatic visible light cannot always induce full metabolic activation. One such example is considered in this section. Circulating lymphocytes confronted with an immunological stimulus shift from the resting state ($G_0$-phase of cellular cycle) to one of rapid enlargement, culminating in DNA synthesis and mitosis (blast transformation). The characteristics of biochemical and morphological reactions in lymphocytes under the action of mitogens (agents responsible for blast transformation, e.g., phytohemagglutinin, [PHA]) have been studied for years (for a review, see Reference 69). Cellular responses to a mitogen can be divided into short-term responses without *de novo* protein synthesis and occurring during the first seconds, minutes, and hours after contact with the mitogen starts and long-term ones connected with protein synthesis hours and days after the beginning of stimulation.

Parallel experiments with PHA treatment and He-Ne laser irradiation were carried out, and the results for these two experimental groups were compared with each other and with those of intact control.[70–76] The 10-sec irradiation with a He-Ne laser ($D = 56$ J/m$^2$) induced short-term changes in lymphocytes that were qualitatively similar and quantitatively close to those caused by PHA (which is present in the incubation medium during all experiments). Among short-term responses compared in this set of experiments were $Ca^{2+}$ influx, RNA synthesis, accessibility of chromatin to acridine orange (a test characterizing the chromatin template activity and its transcription function), and steady-state level of *c-myc* mRNA.[70–73] Also, ultrastructural changes of the nucleus[74] and chromatin[75] were found to be similar in

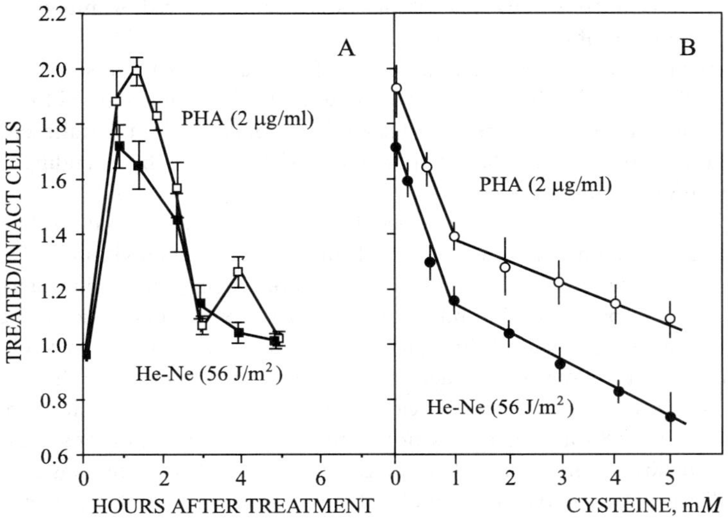


**FIGURE 48.11** Transcription activation (measured by binding of acridine orange to chromatin) of human peripheral lymphocytes: (A) after irradiation with He-Ne laser (10 sec, 56 J/m$^2$) or treatment with phytohemagglutinin (PHA, 2 μg/ml); (B) decrease of transcription activation 1 h after irradiation or PHA treatment depending on concentration of cysteine added immediately after the irradiation or adding PHA. (Modified from Fedoseyeva, G.E. et al., *Lasers Life Sci.*, 2, 197, 1988.)

two experimental groups during the first hours after stimulation. These changes were interpreted as an activation of rRNA metabolism, including its synthesis, processing, and transport.[74]

Two characteristic features of laser-light action were established. First, transcription function was activated in T-lymphocytes but not in B-lymphocytes. At the same time, PHA was stimulative for both types of lymphocytes.[76] Second, despite the similarities in the early responses of lymphocytes to PHA and He-Ne laser radiation, the irradiated lymphocytes did not enter the S-phase of the cell cycle.[71,72] This means that full mitogenic activation did not occur in the irradiated lymphocytes. It is quite possible that the period of irradiation (10 sec in our experiments) was too short to cause the entire cascade of reactions needed for blast transformation. But this time was long enough to cause a boosting effect of blast transformation in PHA-treated lymphocytes.[71,72] The number of blast-transformed cells in the sample, which was irradiated before the beginning of PHA treatment, was 120 to 170% higher depending on PHA concentration.[71,72] It was suggested that the cause of this effect was a partial activation of lymphocytes under irradiation.[71] This may also be a conditioning (priming) effect of certain subpopulations of lymphocytes.[47] It is possible that this is a redox priming effect. Two lines of evidence allow for this suggestion.

First, it is believed that lymphocyte activation under laser radiation starts with mitochondria, as described in Section 48.3.1. This suggestion is supported by experiments showing that ATP extrasynthesis and an increase of energy charge occur in irradiated lymphocytes.[77] Formation of giant mitochondrial structures in the irradiated cells indicated that a higher level of respiration and energy turnover occurred in these cells.[78] But recording of action spectra is needed for further studies of photoacceptors in lymphocytes.

Second, the early transcriptional activation of lymphocytes both by PHA and He-Ne laser radiation (Figure 48.11A) can be eliminated by a reducing agent, cysteine (Figure 48.11B).[70] As seen in Figure 48.11B, the effect depends on the concentration of cysteine. Cysteine also cancels blast transformation of lymphocytes.[79] The activation events in T-lymphocytes and monocytes, which are mediated through translocation of the transcription factor NF-κB, depend on the redox state of these cells.[80] A basal redox equilibrium tending toward oxidation was a prerequisite for the activation of T-lymphocytes and U937 monocytes; both constitutive activation as well as that induced by mitogens were inhibited or even canceled by treatment of cells with reducing agents or antioxidants.[79,80]

Partial mitogenic activation of lymphocytes under He-Ne laser radiation is not the only example of limited activation. For example, silent neurons of *Helix pomatia* did not respond to He-Ne laser irradiation, while the spontaneously active neurons responded strongly under the same experimental conditions.[40] Only 25 to 27% of 3T3 fibroblasts responded to NIR radiation by extending their pseudopodia toward the monochromatic-light source.[81]

The results of experiments with *E. coli* batches showed that they contained a subpopulation that, in response to the irradiation, rapidly began a new cycle of replication and division.[5,26] The number of cells in this subpopulation depended on the cultivation conditions, being smaller in faster-growing populations (e.g., the glucose-grown culture) and larger in slower-growing populations (e.g., the arabinose-grown culture). Presumably, in cells of this light-sensitive subpopulation, the particular metabolic state necessary for the division could be established. Irradiation is what enables the cells to achieve this active state rapidly. Also, this set of experiments[5,26] clearly proved that there is a limit to the specific growth rate of all populations (0.80 $h^{-1}$) that does not depend on growth conditions, and in addition that populations that are already growing at this rate cannot be stimulated. This was also found to be the reason why *E. coli* growth was maximally stimulated not in summer but rather in winter. In autumn and winter, the intact culture featured relatively slow growth. In spring and summer, when the growth of that culture accelerated and the growth rate of the control culture was almost comparable to that of the culture exposed to the optimum dose of red light in the autumn-winter period, irradiation had but little effect.[3]

One conclusion from experiments with cultured cells was that only proliferation of slowly growing subpopulations could be stimulated by irradiation. Also, the experiments with HeLa cells demonstrated that one of the effects of He-Ne laser irradiation of these cells was a decrease of the duration of the $G_1$ period but not other periods of the cell cycle[82] (for a review, see Reference 83).

Taken together, there exist certain biological as well as other limitations connected with the physiological status of an irradiated object.

## 48.4 Enhancement of Cellular Metabolism via Activation of Nonmitochondrial Photoacceptors: Indirect Activation/Suppression

The redox-regulation mechanism cannot occur solely via the respiratory chain (see Section 48.3). Redox chains containing molecules that absorb light in the visible spectral region are usually key structures that can regulate metabolic pathways. One such example is NADPH-oxidase of phagocytic cells, which is responsible for nonmitochondrial respiratory burst. This multicomponent enzyme system is a redox chain that generates reactive oxygen species (ROS) as a response to the microbicidal or other types of activation. Irradiation with the He-Ne[84–87] and semiconductor lasers and LEDs[52,88–91] can activate this chain. The features of radiation-induced nonmitochondrial respiratory burst, which was quantitatively and qualitatively characterized by measurements of luminol-amplified chemiluminescence (CL),[84–91] must be followed. First, nonmitochondrial respiratory burst can be induced both in homogeneous cell populations and cellular systems (blood, spleen cells, and bone marrow) by both CW and pulsed lasers and LEDs. Figure 48.12 presents some examples. Qualitatively, the kinetics of CL enhancement after irradiation is similar to that after treatment of cells with an object of phagocytosis, *Candida albicans*. Quantitatively, the intensity of induced by radiation CL is approximately one order of magnitude lower (Figure 48.12B). This is true for He-Ne laser radiation[84] as well as for radiation of various pulsed LEDs.[88,91] Second, irradiation effects (stimulation or inhibition of CL) on phagocytic cells strongly depend on the health status of the host organism.[85–90] This circumstance can be used for diagnostic purposes. Third, there are complex dependencies on irradiation parameters; irradiation can suppress or activate the nonmitochondrial respiratory burst.[88–91] These problems have been reviewed in detail elsewhere.[5]

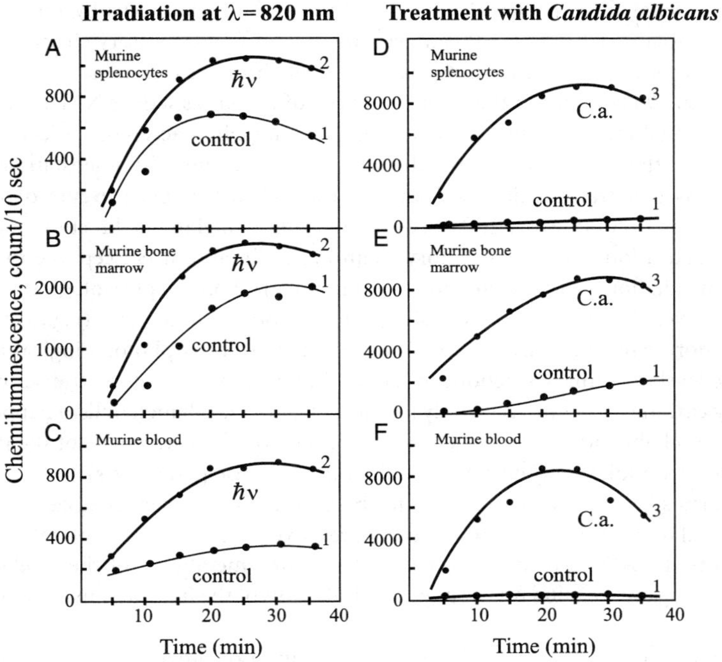


**FIGURE 48.12** Kinetic curves of chemiluminescence of murine splenocytes (A, D), bone marrow (B, E), and blood (C, F) after irradiation (A–C) or treatment (D–F) with *Candida albicans*. Curve 1 denotes everywhere the spontaneous chemiluminescence of control cells; curve 2 is chemiluminescence after irradiation of samples in dose 800 J/m$^2$ (I = 51 W/m$^2$, f = 292 Hz); and curve 3 marks the chemiluminescence induced by the treatment with *C. albicans* (5 × 10$^7$ particles/ml). The measurement error is ≤ ±5%. (From Karu, T.I. et al., *Lasers Surg. Med.*, 21, 485, 1997.)

Finally, reactive-oxygen species, the burst of which is induced by direct irradiation of phagocytes, can activate or deactivate other cells that were not directly irradiated. In this way, indirect activation (or suppression) of metabolic pathways in nonirradiated cells occurs. Cooperative action among various cells via secondary messengers (ROS, lymphokines, cytokines,[92] and NO[93]) requires much more attention when the mechanisms of low-power laser therapy are considered at the organism level.

This chapter did not consider systemic effects of low-power laser therapy at the organism level. The mechanisms of these effects have not yet been established. Perhaps NO plays a role as a secondary messenger for systemic effects of laser irradiation. A possible mechanism connected with the NO-cytochrome *c* oxidase complex was considered earlier (Section 48.3.2). In addition, mechanisms of analgesic effects of laser radiation[94] and systemic therapeutic effects that occur via blood irradiation[95] could be connected with NO.

Recent studies have demonstrated that a number of nonphagocytic cell types, including fibroblasts, osteoblasts, endothelial cells, chondrocytes, kidney mesangial cells, and others, generate ROS (mainly superoxide anion) in low concentrations in response to stimuli.[96] The function of this ROS production is not yet known. It is believed that an NADPH-oxidase (probably different from that in phagocytic cells) is present in nonphagocytic cells as well.[96] To date, the effects of irradiation on this enzyme have not yet been studied.

Another example of important redox chains are NO-synthases, a group of redox-active P450-like flavocytochromes that are responsible for NO generation under physiological conditions.[97] So far, the irradiation effects on these systems have not been investigated.

# 48.5 Conclusion

This chapter considered three principal ways of activating individual cells by monochromatic (laser) light. The photobiological action mechanism via activation of the respiratory chain is a universal mechanism. Primary photoacceptors are terminal oxidases (cytochrome *c* oxidase in eukaryotic cells and, for example, cytochrome *bd* complex in the prokaryotic cell of *E. coli*) as well as NADH-dehydrogenase (for the blue-to-red spectral range). Primary reactions in or with a photoacceptor molecule lead to photobiological responses at the cellular level through cascades of biochemical homeostatic reactions (cellular signaling or photosignal transduction and amplification chain). Crucial events of this type of cell-metabolism activation occur due to a shift of cellular redox potential in the direction of greater oxidation. Cell-metabolism activation via the respiratory chain occurs in all cells susceptible to light irradiation. Susceptibility to irradiation and capability for activation depend on the physiological status of irradiated cells; cells whose overall redox potential is shifted to a more reduced state (e.g., certain pathological conditions) are more sensitive to irradiation. The specificity of final photobiological response is determined not at the level of primary reactions in the respiratory chain but at the transcription level during cellular signaling cascades. In some cases, only partial activation of cell metabolism happens (e.g., priming of lymphocytes). All light-induced biological effects depend on the parameters of the irradiation (wavelength, dose, intensity, radiation time, CW or pulsed mode, pulse parameters).

Other redox chains in cells can also be activated by irradiation. In phagocytic cells irradiation initiates a nonmitochondrial respiratory burst (production of reactive oxygen species, especially superoxide anion) through activation of NADPH-oxidase located in the plasma membrane of these cells. The irradiation effects on phagocyting cells depend on the physiological status of the host organism as well as on radiation parameters.

Direct activation of cells can lead to the indirect activation of other cells. This occurs via secondary messengers released by directly activated cells: reactive oxygen species produced by phagocytes, lymphokines and cytokines produced by various subpopulations of lymphocytes, or NO produced by macrophages or as a result of NO-hemoglobin photolysis of red blood cells.

Coherent properties of laser light are not manifested at the molecular level by light interaction with biotissue. The absorption of low-intensity laser light by biological systems is of a purely noncoherent (i.e., photobiological) nature. At the cellular level, biological responses are determined by absorption of light with photoacceptor molecules. Coherent properties of laser light are unimportant when the cellular monolayer, the thin layer of cell suspension, and the thin layer of tissue surface are irradiated. In these cases, the coherent and noncoherent light with the same wavelength, intensity, and dose provides the same biological response. Some additional (therapeutic) effects from coherent and polarized radiation can occur only in deeper layers of bulk tissue.

## Acknowledgments

I am indebted to Prof. V.S. Letokhov for helpful comments and discussions about the effects of coherent and noncoherent light. Partial financial support from the Ministry of Science, Industry, and Technology of the Russian Federation (grant 108-12(00)) and the Ministry of Science and Technology of the Moscow Region and Russian Foundation of Basic Research (grant 01-02-97025) are acknowledged.

## References

1. Tuner, J. and Hode, L., *Low Level Laser Therapy: Clinical Practice and Scientific Background,* Prima Books, Grängesberg, Sweden, 1999.
2. Karu, T.I., Photobiological fundamentals of low-power laser therapy, *IEEE J. Quantum Electron.*, QE-23, 1703, 1987.
3. Karu, T.I., *Photobiology of Low-Power Laser Therapy*, Harwood Academic, London, 1989.

4. Karu, T.I., Photobiology of low-power laser effects, *Health Phys.*, 56, 691, 1989.
5. Karu, T., *The Science of Low Power Laser Therapy*, Gordon & Breach, London, 1998.
6. Baxter, G.D., *Therapeutic Lasers: Theory and Practice*, Churchill Livingstone, London, 1994.
7. Simunovic, Z., Ed., *Lasers in Medicine and Dentistry*, Vitgraf, Rijeka (Croatia), 2000.
8. Sazonov, A.M., Romanov, G.A., Portnoy, L.M., Odinokova, V.A., Karu, T.I., Lobko, V.V., and Letokhov, V.S., Low-intensity noncoherent red light in complex healing of peptic and duodenal ulcers, *Sov. Med.*, 12, 42, 1985 (in Russian).
9. Karu, T.I., Kalendo, G.S., Letokhov, V.S., and Lobko, V.V., Biostimulation of HeLa cells by low intensity visible light, *Nuov. Cim. D*, 1, 828, 1982.
10. Karu, T.I., Tiphlova, O.A., Letokhov, V.S., and Lobko, V.V., Stimulation of *E. coli* growth by laser and incoherent red light, *Nuov. Cim. D*, 2, 1138, 1983.
11. Bertoloni, G., Sacchetto, R., Baro, E., Ceccherelli, F., and Jori, G., Biochemical and morphological changes in *Escherichia coli* irradiated by coherent and non-coherent 632.8 nm light, *J. Photochem. Photobiol. B Biol.*, 18, 191, 1993.
12. Hartman, K.M., Action spectroscopy, in *Biophysics*, Hoppe, W., Lohmann, W., Marke, H., and Ziegler, H., Eds., Springer-Verlag, Heidelberg, 1983, p. 115.
13. Lipson, E.D., Action spectroscopy: methodology, in *CRC Handbook of Organic Chemistry and Photobiology*, Horspool, W.H. and Song, P.-S., Eds., CRC Press, Boca Raton, FL, 1995, p. 1257.
14. Karu, T.I., Kalendo, G.S., Letokhov, V.S., and Lobko, V.V., Biological action of low-intensity visible light on HeLa cells as a function of the coherence, dose, wavelength, and irradiation dose, *Sov. J. Quantum Electron.*, 12, 1134, 1982.
15. Karu, T.I., Kalendo, G.S., Letokhov, V.S., and Lobko, V.V., Biological action of low-intensity visible light on HeLa cells as a function of the coherence, dose, wavelength, and irradiation regime. II., *Sov. J. Quantum Electron.*, 13, 1169, 1983.
16. Karu, T.I., Tiphlova, O.A., Fedoseyeva, G.E., Kalendo, G.S., Letokhov, V.S. Lobko, V.V., Lyapunova, T.S., Pomoshnikova, N.A., and Meissel, M.N., Biostimulating action of low-intensity monochromatic visible light: is it possible? *Laser Chem.*, 5, 19, 1984.
17. Gamaleya, N.F., Shishko, E.D., and Yanish, G.B., New data about mammalian cells photosensitivity and laser biostimulation, *Dokl. Akad. Nauk S.S.S.R. (Moscow)*, 273, 224, 1983.
18. Vekshin, N.A., Light-dependent ATP synthesis in mitochondria, *Mol. Biol. (Moscow)*, 25, 54, 1991.
19. Karu, T.I., Kalendo, G.S., Letokhov, V.S., and Lobko, V.V., Biostimulation of HeLa cells by low-intensity visible light. II. Stimulation of DNA and RNA synthesis in a wide spectral range, *Nuov. Cim. D*, 3, 309, 1984.
20. Karu, T.I., Kalendo, G.S., Letokhov, V.S., and Lobko, V.V., Biostimulation of HeLa cells by low intensity visible light. III. Stimulation of nucleic acid synthesis in plateau phase cells, *Nuov. Cim. D*, 3, 319, 1984.
21. Karu, T.I., Molecular mechanism of the therapeutic effect of low-intensity laser radiation, *Lasers Life Sci.*, 2, 53, 1988.
22. Karu, T., Primary and secondary mechanisms of action of visible-to-near IR radiation on cells, *J. Photochem. Photobiol. B Biol.*, 49, 1, 1999.
23. Karu, T.I. and Afanasyeva, N.I., Cytochrome oxidase as primary photoacceptor for cultured cells in visible and near IR regions, *Dokl. Akad. Nauk (Moscow)*, 342, 693, 1995.
24. Karu, T.I., Pyatibrat, L.V., Kalendo, G.S., and Esenaliev, R.O., Effects of monochromatic low-intensity light and laser irradiation on adhesion of HeLa cells *in vitro*, *Lasers Surg. Med.*, 18, 171, 1996.
25. Karu, T., Low-power laser effects, in *Lasers in Medicine*, Waynant, R., Ed., CRC Press, Boca Raton, FL, 2002, p. 169.
26. Tiphlova, O. and Karu, T., Action of low-intensity laser radiation on *Escherichia coli*, *CRC Critical Rev. Biomed. Eng.*, 18, 387, 1991.

27. Karu, T.I., Afanasyeva, N.I., Kolyakov, S.F., and Pyatibrat, L.V., Changes in absorption spectra of monolayer of living cells after irradiation with low intensity laser light, *Dokl. Akad. Nauk* (*Moscow*), 360, 267, 1998.
28. Karu, T.I., Afanasyeva, N.I., Kolyakov, S.F., Pyatibrat, L.V., and Welser, L., Changes in absorbance of monolayer of living cells induced by laser radiation at 633, 670, and 820 nm, *IEEE J. Sel. Top. Quantum Electron.*, 7, 982, 2001.
29. Jöbsis-van der Vliet, F.F., Dicovery of the near-infrared window in the body and the early development of near-infrared spectroscopy, *J. Biomed. Opt.*, 4, 392, 1999.
30. Jöbsis-van der Vliet, F.F. and Jöbsis, P.D., Biochemical and physiological basis of medical near-infrared spectroscopy, *J. Biomed. Opt.*, 4, 397, 1999.
31. Dube, A., Gupta, P.K., and Bharti, S., Redox absorbance changes of the respiratory chain components of *E. coli* following He-Ne laser irradiation, *Lasers Life Sci.*, 7, 173, 1997.
32. Kolyakov, S.F., Pyatibrat, L.V., Mikhailov, E.L., Kompanets, O.N., and Karu, T.I., Changes in the spectra of circular dichroism of suspension of living cells after low intensity laser radiation at 820 nm, *Dokl. Akad. Nauk* (*Moscow*), 377, 824, 2001.
33. Karu, T.I., Kolyakov, S.F., Pyatibrat, L.V., Mikhailov, E.L., and Kompanets, O.N., Irradiation with a diode at 820 nm induces changes in circular dichroism spectra (250–750 nm) of living cells, *IEEE J. Sel. Top. Quantum Electron.*, 7, 976, 2001.
34. Arvanitaki, A. and Chalazonitis, N., Reactiones bioelectriques a la photoactivation des cytochromes, *Arch. Sci. Physiol.*, 1, 385, 1947.
35. Berns, M.W., Gross, D.C.L., Cheng, W.K., and Woodring, D., Argon laser microirradiation of mitochondria in rat myocardial cell in tissue culture. II. Correlation of morphology and function in single irradiated cells, *J. Mol. Cell. Cardiol.*, 4, 71, 1972.
36. Berns, M.W. and Salet, C., Laser microbeam for partial cell irradiation, *Int. Rev. Cytol.*, 33, 131, 1972.
37. Salet, C., Acceleration par micro-irradiation laser du rhythme de contraction de cellular cardiaques en culture, *C.R. Acad. Sci. Paris*, 272, 2584, 1971.
38. Salet, C., A study of beating frequency of a single myocardial cell. I. Q-switched laser microirradiation of mitochondria, *Exp. Cell Res.*, 73, 360, 1972.
39. Salet, C., Moreno, G., and Vinzens, F., A study of beating frequency of a single myocardial cell. III. Laser microirradiation of mitochondria in the presence of KCN or ATP, *Exp. Cell Res.*, 120, 25, 1979.
40. Balaban, P., Esenaliev, R., Karu, T., Kutomkina, E., Letokhov, V., Oraevsky, A., and Ovcharenko, N., He-Ne laser irradiation of single identified neurons, *Lasers Surg. Med.*, 12, 329, 1992.
41. Walker, J. and Akhanjee, K., Laser-induced somatosensory evoked potential: evidence of photosensitivity of peripheral nerves, *Brain Res.*, 344, 281, 1985.
42. Walker, J., Relief from chronic pain by laser irradiation, *Neurosci. Lett.*, 43, 339, 1983.
43. Walker, J., Treatment of human neurological problems by laser photostimulation, U.S. patent 4,671,285, 1987.
44. Karu, T.I., Kalendo, G.S., and Letokhov, V.S., Control of RNA synthesis rate in tumor cells HeLa by action of low-intensity visible light of copper laser, *Lett. Nuov. Cim.*, 32, 55, 1981.
45. Giese, A.C., Photosensitization of organisms with special reference to natural photosensitizers, in *Lasers in Biology and Medicine*, Hillenkampf, F., Pratesi, R., and Sacchi, C., Eds., Plenum Press, New York, 1980, p. 299.
46. Brown, G.C., Nitric oxide and mitochondrial respiration, *Biochem. Biophys. Acta*, 1411, 351, 1999.
47. Karu, T., Mechanisms of low-power laser light action on cellular level, in *Lasers in Medicine and Dentistry*, Simunovic, Z., Ed., Vitgraf, Rijeka (Croatia), 2000, p. 97.
48. Karu, T.I., Pyatibrat, L.V., and Kalendo, G.S., Donors of NO and pulsed radiation at $\lambda$=820 nm exert effects on cells attachment to extracellular matrices, *Toxicol. Lett.*, 121, 57, 2001.
49. Hothersall, J.S., Cunha, F.Q., Neild, G.H., and Norohna-Dutra, A., Induction of nitric oxide synthesis in J774 cell lowers intracellular glutathione: effect of oxide modulated glutathione redox status on nitric oxide synthase induction, *Biochem. J.*, 322, 477, 1997.

50. Karu, T.I., Tiphlova, O.A., Matveyets, Yu. A., Yartsev, A.P., and Letokhov, V.S., Comparison of the effects of visible femtosecond laser pulses and continuous wave laser radiation of low average intensity on the clonogenicity of *Escherichia coli*, *J. Photochem. Photobiol. B Biol.*, 10, 339, 1991.
51. Karu, T.I., Local pulsed heating of absorbing chromophores as a possible primary mechanism of low-power laser effects, in *Laser Applications in Medicine and Surgery*, Galletti, G., Bolognani, L., and Ussia, G., Eds., Monduzzi Editore, Bologna, 1992, p. 253.
52. Karu, T., Andreichuk, T., and Ryabykh, T., Changes in oxidative metabolism of murine spleen following diode laser (660–950nm) irradiation: effect of cellular composition and radiation parameters, *Lasers Surg. Med.*, 13, 453, 1993.
53. Forman, N.J. and Boveris, A., Superoxide radical and hydrogen peroxide in mitochondria, in *Free Radicals in Biology*, Vol. 5, Pryor, A., Ed., Academic Press, New York, 1982, p. 65.
54. Karu, T., Tiphlova, O., Esenaliev, R., and Letokhov, V., Two different mechanisms of low-intensity laser photobiological effects on *Escherichia coli*, *J. Photochem. Photobiol. B Biol.*, 24, 155, 1994.
55. Brown, G.C., Control of respiration and ATP synthesis in mammalian mitochondria and cells, *Biochem. J.*, 284, 171, 1992.
56. Chance, B., Cellular oxygen requirements, *Fed. Proc. Fed. Am. Soc. Exp. Biol.*, 16, 671, 1957.
57. Karu, T.I., Pyatibrat, L.V., and Kalendo, G.S., Cell attachment modulation by radiation from a pulsed semiconductor light diode ($\lambda$=820 nm) and various chemicals, *Lasers Surg. Med.*, 28, 227, 2001.
58. Karu, T.I., Pyatibrat, L.V., and Kalendo, G.S., Thiol reactive agents and semiconductor light diode radiation ($\lambda$=820 nm) exert effects on cell attachment to extracellular matrix, *Laser Ther.*, 11, 177, 2001.
59. Karu, T.I., Pyatibrat, L.V., and Kalendo, G.S., Cell attachment to extracellular matrices is modulated by pulsed radiation at 820 nm and chemical that modify the activity of enzymes in the plasma membrane, *Lasers Surg. Med.*, 29, 274, 2001.
60. Gius, D., Boreto, A., Shah, S., and Curry, H.A., Intracellular oxidation reduction status in the regulation of transcription factors NF-$\kappa$B and AP-1, *Toxicol. Lett.*, 106, 93, 1999.
61. Sun, Y. and Oberley, L.W., Redox regulation of transcriptional activators, *Free Rad. Biol. Med.*, 21, 335, 1996.
62. Nakamura, H., Nakamura, K., and Yodoi, J., Redox regulation of cellular activation, *Annu. Rev. Immunol.*, 15, 351, 1997.
63. Kamata, H. and Hirata, H., Redox regulation of cellular signalling, *Cell. Signal.*, 11, 1, 1999.
64. Calkhoven, C.F. and Geert, A.B., Multiple steps in regulation of transcription-factor level and activity, *Biochemistry*, 317, 329, 1996.
65. Chopp, H., Chen, Q., Dereski, M.O., and Hetzel, F.W., Chronic metabolic measurement of normal brain tissue response to photodynamic therapy, *Photochem. Photobiol.*, 52, 1033, 1990.
66. Quickenden, T.R., Daniels, L.L., and Byrne, L.T., Does low-intensity He-Ne radiation affect the intracellular pH of intact *E. coli*? *Proc. SPIE*, 2391, 535,1995.
67. Tiphlova, O. and Karu, T., Dependence of *Escherichia coli* growth rate on irradiation with He-Ne laser and growth substrates, *Lasers Life Sci.*, 4, 161, 1991.
68. Karu, T.I., Pyatibrat, L.V., and Kalendo, G.S., Studies into the action specifics of a pulsed GaAlAs laser ($\lambda$=820 nm) on a cell culture. I. Reduction of the intracellular ATP concentration: dependence on initial ATP amount, *Lasers Life Sci.*, 9, 203, 2001.
69. Ashman, R.F., Lymphocyte activation, in *Fundamental Immunology*, Paul, W.P., Ed., Raven Press, New York, 1984, p. 267.
70. Fedoseyeva, G.E., Smolyaninova, N.K., Karu, T.I., and Zelenin, A.V., Human lymphocyte chromatin changes following irradiation with a He-Ne laser, *Lasers Life Sci.*, 2, 197, 1988.
71. Karu, T.I., Smolyaninova, N.K., and Zelenin, A.V., Long-term and short-term responses of human lymphocytes to He-Ne laser radiation, *Lasers Life Sci.*, 4, 167, 1991.

72. Smolyaninova, N.K., Karu, T.I., Fedoseyeva, G.E., and Zelenin, A.V., Effect of He-Ne laser irradiation on chromatin properties and nucleic acids synthesis of human blood lymphocytes, *Biomed. Sci.*, 2, 121, 1991.
73. Shliakhova, L.N., Itkes, A.V., Manteifel, V.M., and Karu, T.I., Expression of *c-myc* gene in irradiated at 670 nm human lymphocytes: a preliminary report, *Lasers Life Sci.*, 7, 107, 1996.
74. Manteifel, V.M. and Karu, T.I., Ultrastructural changes in human lymphocytes under He-Ne laser radiation, *Lasers Life Sci.*, 4, 235, 1992.
75. Manteifel, V.M., Andreichuk, T.N., and Karu, T.I., Influence of He-Ne laser radiation and phytohemagglutinin on the ultrastructure of chromatin of human lymphocytes, *Lasers Life Sci.*, 6, 1, 1994.
76. Manteifel, V.M. and Karu, T.I., Activation of chromatin in T-lymphocytes nuclei under the He-Ne laser radiation, *Lasers Life Sci.*, 8, 117, 1998.
77. Herbert, K.E., Bhusate, L.L., Scott, D.L., Diamantopolos, C., and Perrett, D., Effect of laser light at 820 nm on adenosine nucleotide levels in human lymphocytes, *Lasers Life Sci.*, 3, 37, 1989.
78. Manteifel, V., Bakeeva, L., and Karu, T., Ultrastructural changes in chondriome of human lympocytes after irradiation with He-Ne laser: appearance of giant mitochondria, *J. Photochem. Photobiol. B Biol.*, 38, 25, 1997.
79. Novogrodski, A., Lymphocyte activation induced by modifications of surface, in *Immune Recognition*, Rosenthal, S., Ed., Academic Press, New York, 1975, p. 43.
80. Israél, N., Gougerot-Pocidalo, M.-A., Aillet, F., and Verelizier, J.-L., Redox status of cells influences constitutive or induced NF-kB translocation and HIV long terminal repeat activity in human T-lymphocytes and monocytic cell lines, *J. Immunol.*, 149, 3386, 1992.
81. Albrecht-Büchler, G., Surface extensions of 3T3 cells towards distant infrared light sources, *J. Cell Biol.*, 114, 494, 1991.
82. Karu, T.I., Pyatibrat, L.V., and Kalendo, G.S., Biostimulation of HeLa cells by low-intensity visible light. V. Stimulation of cell proliferation *in vitro* by He-Ne laser radiation, *Nuov. Cim. D*, 9, 1485, 1987.
83. Karu, T., Effects of visible radiation on cultured cells, *Photochem. Photobiol.*, 52, 1089, 1090.
84. Karu, T.I., Ryabykh, T.P., Fedoseyeva, G.E., and Puchkova, N.I., Induced by He-Ne laser radiation respiratory burst on phagocytic cells, *Lasers Surg. Med.*, 9, 585, 1989.
85. Karu, T.I., Ryabykh, T.P., and Antonov, S.N., Different sensitivity of cells from tumor-bearing organisms to countinuous-wave and pulsed laser radiation ($\lambda$ = 632.8 nm) evaluated by chemiluminescence test. I. Comparison of responses of murine splenocytes: intact mice and mice with transplanted leukemia EL-4, *Lasers Life Sci.*, 7, 91, 1996.
86. Karu, T.I., Ryabykh, T.P., and Antonov, S.N., Different sensitivity of cells from tumor-bearing organisms to countinuous-wave and pulsed laser radiation ($\lambda$ = 632.8 nm) evaluated by chemiluminescence test. II. Comparison of responses of human blood: healthy persons and patients with colon cancer, *Lasers Life Sci.*, 7, 99, 1996.
87. Karu, T.I., Ryabykh, T.P., and Letokhov, V.S., Different sensitivity of cells from tumor-bearing organisms to continuous-wave and pulsed laser radiation ($\lambda$ = 632.8 nm) evaluated by chemiluminescence test. III. Effect of dark period between pulses, *Lasers Life Sci.*, 7, 141, 1997.
88. Karu, T., Andreichuk, T., and Ryabykh, T., Suppression of human blood chemiluminescence by diode laser radiation at wavelengths 660, 820, 880 or 950 nm, *Laser Ther.*, 5, 103, 1993.
89. Karu, T.I., Andreichuk, T.N., and Ryabykh, T.P., On the action of semiconductor laser radiation ($\lambda$ = 820 nm) on the chemiluminescence of blood of clinically healthy humans, *Lasers Life Sci.*, 6, 277, 1995.
90. Karu, T.I., Ryabykh, T.P., Sidorova, T.A., and Dobrynin, Ya.V., The use of a chemiluminescence test to evaluate the sensitivity of blast cells in patients with hemoblastoses to antitumor agents and low-intensity laser radiation, *Lasers Life Sci.*, 7, 1, 1996.
91. Karu, T.I., Pyatibrat, L.V., and Ryabykh, T.P., Nonmonotonic behaviour of the dose dependence of the radiation effect on cells *in vitro* exposed to pulsed laser radiation at $\lambda$ = 820 nm, *Lasers Surg. Med.*, 21, 485, 1997.

92. Funk, J.O., Kruse, A., and Kirchner, H., Cytokine production in cultures of human peripheral blood mononuclear cells, *J. Photochem. Photobiol. B Biol.*, 16, 347, 1992.
93. Naim, J.O., Yu, W. Ippolito, K.M.L., Gowan, M., and Lazafame, R.J., The effect of low level laser irradiation on nitric oxide production by mouse macrophages, *Lasers Surg. Med.*, Suppl. 8, 7 (abstr. 28), 1996.
94. Mrowiec, J., Sieron, A., Plech, A., Cieslar, G., Biniszkiewicz, T., and Brus, R., Analgesic effect of low-power infrared laser radiation in rats, *Proc. SPIE*, 3198, 83, 1997.
95. Vladimirov, Y., Borisenko, G., Boriskina, N., Kazarinov, K., and Osipov, A., NO-hemoglobin may be a light-sensitive source of nitric oxide both in solution and in red blood cells, *J. Photochem. Photobiol. B Biol.*, 59, 115, 2000.
96. Sbarra, A.J. and Strauss, R.R., Eds., *The Respiratory Burst and Its Photobiological Significance*, Plenum Press, New York, 1988.
97. Sharp, R.E. and Chapman, S.K., Mechanisms for regulating electron transfer in multi-centre redox proteins, *Biochem. Biophys. Acta*, 1432, 143, 1999.

# 49

# Image-Guided Surgery




Richard D. Bucholz
*St. Louis University Health Sciences Center*
*St. Louis, Missouri*

Keith A. Laycock
*St. Louis University Health Sciences Center*
*St. Louis, Missouri*


## 49.1 Introduction

Surgery is rooted in a basic inherent conflict. The Hippocratic Oath commands surgeons to abstain from whatever causes harm to the patient, yet the essence of surgery is to invade the body — and therefore to inflict injury. The majority of all surgical research has focused on techniques that alter the balance between improving the lot of the patient suffering from disease and minimizing the mortality and morbidity attendant on the therapeutic act.

Much of the morbidity associated with surgery is incurred during the development of situational knowledge by the surgeon while performing surgery. For a surgeon to reach a lesion the opening into the body must afford both access to the lesion and an appreciation of location. Surgeons use landmarks visualized during dissection to provide course corrections that eventually bring them to the target of choice. Once at the target the opening must be large enough to allow the surgeon to effectively deal with the lesion in an appropriate fashion, which often consists of removing the diseased tissue. The therapeutic ratio between the efficacy and injury of a procedure is therefore a function of the quality of visualization of the surgeon of anatomy during surgery and the efficacy of the instruments, or effectors, at hand to deal with the pathology once visualized. Any technology that holds the promise of improving both visualization and the effectiveness of surgery will therefore have a dramatic effect on the therapeutic ratio and be eagerly embraced by both surgeons and their patients. A prime example of such a technology is image-guided surgery.

Surgeons have always looked to alternative ways of providing visualization without exposing anatomy. The development of medical imaging has revolutionized surgery, enabling many procedures to be performed


0-8493-1116-0/03/$0.00+$1.50

that were hitherto impossible due to safety concerns and localization difficulties. Soon after development of the first imaging modality (radiographic x-ray imaging by Roentgen in 1895) surgeons employed the technology within the operating room (OR) to reduce the invasiveness of surgery. For the first time a surgeon could visualize skeletal injuries and malformations and plan their repair or correction prior to commencing a surgical procedure. Radiography also enabled localization of dense masses or foreign objects inside the body cavity, expediting their removal and minimizing iatrogenic trauma arising from exploratory procedures. In addition, the ability to visualize bony anatomy also enabled the surgeon to use this as a frame of reference when it was not feasible to expose the target site, such as in neurosurgical procedures in which lesions are made within the brain to restore function.

In such neurosurgical procedures, the highest degree of precision is essential to avoid iatrogenic injury to structures adjacent to the target site. The need for accuracy in the insertion of instruments inspired neurosurgeons to develop the field of stereotactic surgery (from *stereos*, meaning solid, as in three-dimensional (3D), and either the Greek *taxis*, meaning ordered or arranged, or the Latin *tactus*, meaning touch), in which the surgical instruments are precisely guided to a selected target within the patient by equipment that uses the patient's image data for navigational reference. Zernov first proposed the concept of a mechanical stereotactic guidance device in 1890,[1] and several clinically employed devices were produced following the invention of radiology.

Stereotactic frames were literally bolted to the patient's head by means of screws driven into the skull, which thus provided a rigid frame of reference within which the exact 3D coordinates of instruments and target could be defined. The surgeon would choose a target and an entry point, and a small calculator would be used to calculate how to set the angles of the moveable arcs attached to the frame. Once these were correctly positioned, an instrument passed through the holder would be restricted to travel only from the entry point to the desired target, with the frame acting to constrain the surgeon to the preset surgical path.

In most of these frame-based coordinate systems, the head ring of the frame defines the plane of origin, with the x and y coordinates describing a unique point within that plane. The z-axis is usually perpendicular to the plane of origin, and any point in the anatomy of the patient can be precisely described using three numbers consisting of the respective $x$, $y$, and $z$ coordinates. Examples of frames that are coordinate-based include the Leksell, Talairach, and Hitchcock frames. The Leksell is an arc-centered system in which navigation during the procedure is confined by a probe carrier held by a series of easily adjustable arcs. The target point in such a system is at the center of the arc system. In non-Cartesian frames, such as the popular Brown-Roberts-Wells frame, targets and trajectories are defined by four angles and a length, and any point within the patient can be reached regardless of its relation to the center of the coordinate system. The latter frame was developed after computers became available, so Cartesian scanner coordinates could be readily converted into spherical coordinates.

However, plain radiographs of the body, consisting of two-dimensional (2D) images depicting the 3D structure of the body, are inherently limited by the loss of information in reducing three dimensions to two. Unfortunately, during the first half of the 20th century, the only image data available for surgical-planning purposes were 2D radiographs obtained preoperatively from several perspectives. A further limitation was that such radiographs did not depict the soft tissue within the skull. Because radiographs do not show the location of the brain without the introduction of a contrast agent, pneumoencephalography was commonly performed prior to a stereotactic procedure to show the location of the ventricles. Fortunately, the third ventricle has two points — the anterior and posterior commissures — that are easily identified on lateral skull radiographs, allowing the position of the third ventricle to be determined accurately. Stereotactic atlases of human cranial anatomy were then developed based on these points. They were adequate for localizing the ventricles, but their accuracy left something to be desired when it came to locating structures such as the cortex that were lateral to the midline. Thus, early stereotactic procedures could be performed with precision only near the midline and in patients with normal anatomy not affected by the presence of space-occupying lesions. Stereotactic surgery was thus of value to neurosurgeons who selected targets close to the midline to restore lost neurological function, but other neurosurgeons still needed to expose large amounts of normal anatomy in order to see the anatomical landmarks required to guide them to their desired targets.

Though stereotactic frames were highly successful, they had several limitations. First, the target and trajectory parameters were usually calculated with respect to preoperative image data and did not necessarily take into account intraoperative changes in anatomy. Also, many surgeons found the frames to be cumbersome and obstructive. These limitations led to the development of devices that could determine position in 3D space without the need for mechanical arcs and calipers. These devices are called 3D digitizers. They assign coordinates to each selected point, enabling coregistration of patients, images, and instruments within the same coordinate system. Finally, the limitation associated with 2D imaging was resolved by the advent of 3D imaging techniques such as computed tomography (CT) and magnetic resonance imaging (MRI). With the advent of 3D imaging and intraoperative localization devices it has become possible to almost completely replace stereotactic frames with frameless navigational systems. These frameless systems are discussed later, but it is first necessary to examine the advances in imaging techniques that have allowed image-guided navigation techniques to achieve their current level of ubiquity.

## 49.2 Advances in Imaging Technology

### 49.2.1 Computed Tomography and Magnetic Resonance Imaging

While radiographs are clearly still invaluable in many medical applications, they have generally been superseded for image-guidance purposes by other imaging modalities, predominantly CT and MRI. CT is excellent for imaging vascular and skull-base lesions and ideal for trauma cases. However, MRI offers superior tissue contrast and is therefore extremely useful for imaging brain tumors in soft tissues and organs, for example. The 2D images acquired with these modalities can be manipulated by relatively inexpensive computer technology into 3D virtual models of the patient's anatomy. Such virtual models can be rotated and examined from all angles to enable the surgeon to determine the best approach to the target. These visualization techniques have thereby markedly increased the range of applications for stereotactic techniques. Preoperative images can also be supplemented by the use of intraoperative imaging, allowing the surgeon to evaluate the progress of the operation and decide if changes to the surgical plan are necessary due to movement of tissue.

### 49.2.2 Functional Imaging

Functional imaging technology, which detects changes in metabolism produced by function and produces images showing the location of such functional activity, has experienced an upsurge in use in recent years. Currently, the main functional imaging modalities are positron emission tomography (PET), single photon emission tomography (SPECT), magnetic source imaging (MSI), functional magnetic resonance imaging (fMRI), and mass spectroscopy. The latest development in this area is magnetoencephalography (MEG), which may be used for mapping normal and continuous abnormal function. It can already identify motor and sensory cortex in the brain, and new algorithms may enable detection of speech activity as well.

In functional imaging, alterations in function, rather than structure, serve as the basis for the imaging. For example, CT develops an image based on the absorption of x-rays by the tissue within the x-ray beam. fMRI detects differences in blood flow while the patient is at rest vs. performing a specific task. Such changes in blood flow are usually induced by the demands of a specific part of the brain performing a specific task. The resultant image is therefore a depiction of the anatomical location of a specific function. Knowledge of the location of function, rather than anatomy, can be particularly useful in procedures either targeting abnormal function for correction (such as in surgery for epilepsy intractable to medical intervention) or in the removal of lesions close to critical areas of the brain such as the motor cortex. As neurosurgery evolves from techniques designed to remove lesions to those designed to restore function it is anticipated that functional imaging will be increasingly used as the imaging modality of choice for image-guided surgery.

While functional imaging has considerable potential, it does have limitations. The images usually demonstrate only function, and the relationship between function and anatomy is not necessarily obvious. Therefore, to detect an abnormality in a specific patient it may be necessary to acquire data from normal controls for comparison purposes or refer to an atlas that delineates the relationship of function to anatomy. Therefore, the use of functional imaging almost always uses a technique called image fusion, in which two different images of the same 3D object are precisely overlaid onto each other to render an image comprised of information from both sources. To perform such an overlay the two images must be related to each other through a process of registration in which points or geometry depicted on both images are used to align the images with each other. The process of registration is key for image-to-image fusion and for image-to-patient fusion and will be covered later in this chapter.

## 49.3 Segmentation and Deformation of Images

When imaging a patient for surgical planning purposes the entire anatomy of part of the patient, such as the head, is normally depicted. Usually, however, only a specific area is of interest for a particular case, and it is desirable to separate this region from the remainder to allow study of the shape and size of the tissue in question. This process is called segmentation and can be accomplished by matching the patient's image to a standardized printed atlas or an electronic atlas that has specific structures already segmented. This process can be rendered very inaccurate due to the significant differences between individual brains. This discrepancy may be further exacerbated by the presence of tumors, abscesses, or accidental trauma that distorts the normal anatomy, making it difficult to identify or orientate the patient's image data relative to the atlas example. A solution is to use an elastic deformation technique to resize and warp the atlas to match the patient's anatomy. This approach constrains the points on the patient image and atlas plate that must be connected but otherwise leaves the structures free to deform to a template. This enables more precise preoperative planning and is valuable if the surgical plan must be updated intraoperatively to reflect brain shift or if ambiguous structures are encountered that must be identified before proceeding further. Several deformable atlases have now been produced for neurosurgical applications,[2,3] though there is still some skepticism in conservative circles about their value.

At first, limited computing capacity meant that digitized data had to be manually segmented. Manual segmentation of images by humans is time-consuming and tedious. Fortunately, initial studies suggest that automated segmentation techniques are as reliable as manual segmentations. Automated techniques are much faster and are robust with respect to diverse disease entities. Use of such automated techniques may avoid the need for invasive monitoring prior to functional neurosurgical procedures. For example, when planning a surgical intervention for epilepsy, it has been necessary to surgically insert depth electrodes to record abnormal activity. Automated segmentation coupled with functional-to-anatomical image fusion may eliminate the need for such complex and costly evaluation prior to surgery. The technique could also accommodate structural lesions such as tumors. After segmenting out the lesion, the brain is deformed to what it would look like without the lesion. First the atlas is deformed to the brain, then the tumor can be grown back into the image dataset. Although this process is very experimental, it might allow for the depiction of the location of pathways altered by the presence of mass lesions.

## 49.4 Registration

The fundamental concept underlying all image-guided surgery is registration. This is the method used for mapping the detailed bits of imaging data of a specific patient in such a way as to relate the bits of data both to one another and to the physical space/anatomy of the patient. Registration has been traditionally performed by identifying and aligning points, called fiducials, seen on the sets of images to their location on the patient in physical space; this process is termed paired-point registration.

In conventional frame-based stereotactic surgery, markers placed on the frame during imaging allowed for the registration of the images to the operative field. Paired-point-based methods have now been supplemented with curve- and surface-matching methods, moment and principal fixed-axes methods, correlation methods, and atlas methods.[4]

## 49.4.1 Paired-Point Methods

The most straightforward method for achieving coregistration of coordinate spaces is to use a set of markers that are visible in the patient's image data set and that can be pointed to by a localization device or digitizer (see below). While only three points (in a nonlinear arrangement) are required to relate two 3D objects to each other, using more points improves the accuracy of the subsequent registration and reduces the risk of failure should one of the markers move relative to the others or become obscured and undetectable by the digitizer.

Some investigators (notably in the radiosurgical field) have used clearly defined anatomical features as reference markers.[5] However, there is a degree of ambiguity associated with feature-based registration, and most neurosurgeons prefer to use artificial markers applied to the patient prior to imaging. Different types and sizes of artificial fiducials are available and may be fixed to the patient's skin using adhesive or embedded directly in the bone. Markers applied to the skin cause no discomfort to the patient but may be dislodged between preoperative imaging and surgery and are more prone to localized movement relative to each other. This can have deleterious effects on the accuracy of both registration and target localization. In contrast, bone-implanted fiducials offer the highest accuracy of any registration technique, including stereotactic frames,[6] because they cannot move relative to the patient's bone and, once implanted, remain in place until the surgery is completed. Implanted markers are most frequently used in procedures involving the skull, spine, or pelvis. However, the trauma and subsequent pain associated with implantation of fiducials in bone militates against their routine use. The inability to fixate fiducials to soft tissue like the liver has essentially prevented any application of image guidance to organs that are not confined to a rigid structure such as the skull.

Regardless of the type of marker used for paired-point registration, the resultant accuracy of the registration process depends not only on the markers but also on how they are distributed over the volume being registered. Regardless of their number, markers arranged within a line will never allow for 3D registration. Further, coplanar markers will allow for registration but will have little accuracy in the z dimension perpendicular to the plane of the markers. For the best possible registration the markers should be spread as widely over the entire 3D volume as possible and as far from each other in each coordinate dimension.

## 49.4.2 Contour and Surface Mapping

As an alternative to matching isolated detectable points, it is also possible to obtain a registration by matching contours obtained from the surface of the patient. Surface contours can be extracted from image data sets using intensity-based segmentation algorithms. Using a digitizer, a random contour from the patient can be matched to the contour from the imaging study. The computer can then align these features with areas of maximum curvature that are the most useful for the process. A further development is to dispense with point matching altogether and simply match large surface areas. In neurosurgical procedures, the surfaces imaged by PET, CT, or MRI can be matched directly with the scalp as imaged in the OR. This permits rapid registration immediately prior to the surgery, obviating the need for repeating preoperative diagnostic imaging with fiducials in place. This saves both time and money, though care must be taken to ensure that the diagnostic images are of sufficient quality to justify relying on them for surgical planning and guidance. Frequently diagnostic images lack the fine spacing needed between images to allow for accurate intraoperative guidance. There is some variation in the degree to which surface-matching techniques can be automated. The best-known surface-matching strategy is that of Pelizarri and Chen,[7] but a number of alternative approaches have been published.

Existing surface-based registration techniques are generally less accurate and less reliable than fiducial marker-based techniques and require considerably more technical support and time. The application accuracy achieved for surface matching is usually in excess of 3.0 mm.[8] Further, it is important to note that contour matching is incapable of registering two perfect spheres using their contours; therefore, points for the registration process should be obtained wherever there is a departure of normal anatomy away from a sphere, such as the area around the face and nose.

A combination of rapid contour mapping with fully automated segmentation would result in a system permitting frequent, rapid, and accurate updates of the registration information at each stage of the surgical procedure, though the technology required for this is still under development.

## 49.5 Frameless Systems

### 49.5.1 Early Approaches to Frameless Localization

As mentioned earlier, stereotactic frames have now been almost completely replaced by frameless navigation systems for most neurosurgical procedures. Among the earliest attempts at 3D localizers were articulated arms with potentiometers at the joints between the links. These sensors determined the angles of the joints by measuring resistance, enabling the orientation of the distal section and location of the tip to be determined. The first reported use of a passive localization arm of this type for neurosurgical guidance was by Watanabe,[9] who used an arm with six joints. A widely used commercial model is the ISG Viewing Wand (ISG Technologies, Toronto, Canada). Guthrie and Adler[10] developed a similar series of arms in which optical encoders replaced the potentiometers to improve accuracy. These arm-based systems are simple and do not require a clear line of sight. However, they can interfere with the surgeon's movements and are difficult to use with surgical instruments.

Another early approach used ultrasound emitters and microphone arrays in a known configuration to localize the position of the instruments. Position was determined on the basis of the time delay between emission of an ultrasonic pulse by the emitter and its detection by each microphone. In 1986, Roberts et al.[11] applied a version of this approach to an operating microscope and developments continue to this day. Barnett et al.[12,13] developed a system in which ultrasonic emitters are fitted to a hand-held interactive localization device (Picker International, Highland Heights, OH), and the detecting microphone arrays are fitted to the side rails of a standard operating table. However, ultrasound-based digitizing systems are limited by their requirement for a clear line of sight between the emitters and detectors, and the systems may be confused by echoes, environmental noise, and the effects of fluctuating air temperature. On the positive side, they are relatively inexpensive, costing no more than half as much as an optical digitizer, can be set up rapidly, and can cover a very large working volume. Also, it is easy to affix the emitters to all kinds of instruments and other equipment.

### 49.5.2 Optical Digitizers

Modern navigation systems make use of light-emitting diodes (LEDs) to track the location of instruments within the operative space and were first developed by Bucholz. The LEDs used in such systems emit near-infrared light that can be detected by at least two charge-coupled-device (CCD) cameras positioned in the OR. Inside the cameras, light from the LEDs is focused onto a layer of several thousand CCD sensor elements, and the infrared energy is converted into electrical impulses. These digital signals may be further processed or converted to analog form. Through triangulation the point of light emission can be calculated with 0.3-mm accuracy.

Optical triangulation techniques are highly accurate and robust. A useful comparison of their properties with those of sonic digitizers has been published.[14] Like sonic digitizers, optical systems require a clear line of sight, but they are unaffected by temperature fluctuations and do not appear to be compromised by surgical light sources. The infrared light is not visible to the surgical team and causes no harmful effects.

A variation on the infrared detection approach uses passive reflectors instead of active emitters, with an infrared source illuminating the operative field. The reflected light is detected as before. Instruments fitted with passive reflectors do not need to have electrical cables attached and are thus easier to sterilize. However, the passive markers cannot be "turned off" and thus appear simultaneously to the CCD cameras at all times. This can confuse the navigation system if multiple instruments are being used and a large number of reflectors are present in the operative field at once. Another passive approach substitutes an ultraviolet (UV) source as the field illuminator, with fluorescent markers reirradiating the UV light at visible frequency levels. However, there is concern that prolonged exposure to UV light in this context may pose a health hazard.

### 49.5.3 Reference Systems

With stereotactic frames the mounting of surgical instruments on the base ring ensured that they were accurately positioned relative to the patient's anatomy. Frameless devices provide no such safeguarded connection between the instruments and the patient's anatomy, which can lead to error if the position of the patient is not accurately known.

An important benefit of surgical navigation systems is the ability to change the patient's position following registration. To avoid the need for reregistration, some means of tracking the head relative to the detection mechanism is required. This is commonly achieved with a reference arc or similar device fixed directly to the patient's head or to the head holder, assuming that the head does not move in relation to the holder. The reference device has at least three emitters that are detectable by the digitizer employed in the design of the system; for optically based systems, a multiplicity of LEDs serves this function, allowing one or more LEDs to be blocked and still allow the position of the head to be determined prior to localization of the surgical instruments.

By monitoring the relative positions of the instrument and the reference arc, the navigation system correlates the position of the instrument to the patient's anatomy on a continuous basis, regardless of the frequent movements of the patient that occur during a standard operation.

### 49.5.4 Effectors

The term effector simply refers to those instruments used by a surgeon to perform surgery on the patient. Although surgical instruments comprise the largest portion of such devices, an effector can also be anything used to treat the patient surgically including surgical microscopes, endoscopes, genetic material, drug polymers, and robots. Such effectors rely on precise positioning to produce the maximal benefit for the patient, so integration of navigational technology is critical to their success.

Essential components of a surgical navigation system are the instruments that, through modifications by the addition of LEDs or reflectors, permit localization. Because the LEDs or spheres must be visible to the camera array at all times, they cannot be mounted on the tip of an instrument, as this will not be visible to the cameras during surgery. Instead, the LEDs or spheres are usually located at a given distance from the tip in the handle of the instrument. If the instrument is essentially linear, such as a forceps, then the minimum requirement for localization is two LEDs mounted in alignment with the tip. Reflective spheres generally do not localize well if they are placed in a linear alignment, as reflective systems experience difficulty when the view of the spheres partially overlap, so reflective arrays usually consist of at least three spheres in a nonlinear arrangement. Given the relatively large size of reflective spheres, LEDs are usually the detector of choice for microsurgical instrumentation.

A bayoneted instrument is commonly employed to allow surgery through a small opening. The offset design of a bayonet instrument allows the handle to be placed off the axis of the line of sight of the surgeon, and its geometry is ideal for use in conjunction with a navigational system. Similarly, for instruments with complex 3D shapes, LEDs or spheres must be placed off axis to allow tracking.

When a specific trajectory into the brain is desired, rigid fixation of instrumentation is preferred to ensure that the instrumentation is aligned along the selected surgical path. Typical situations in which this path is key include tumor biopsy, insertion of depth electrodes for epilepsy, insertion of ventricular catheters, and functional surgery. This function is carried out by a biopsy guide-tube adaptor, which is attached to the reference arc using a standard retractor arm with adjustable tension. The tube is equipped with four LEDs to provide redundancy.

The most popular frameless surgical navigation system is the Trion (formerly called the StealthStation) manufactured by Medtronic. The Trion consists of (1) a UNIX-based workstation that, by communicating with the other components of the system, displays position on a high-resolution monitor or head-mounted display; (2) an infrared optical digitizer with camera array (as described earlier); (3) a reference LED array (e.g., a reference arc) for the patient; and (4) surgical instruments modified by the addition of either passive reflectors or active emitters (usually LEDs). Optional components of the system include a robotically controlled locatable surgical microscope and surgical endoscopes modified by the placement of LEDs.

The Trion utilizes CT or MRI for intraoperative guidance. Images are typically obtained in the axial dimension, and the scanning parameters for each modality are adjusted to achieve roughly cubic voxels: Image files are transmitted from the CT or MRI scanner to the surgical workstation over an Ethernet-based local area network and then converted to a standard file format. During surgery, three standard views (the original axial projection and reconstructed sagittal and coronal images) are displayed on a monitor at all times. A cross-hair pointer superimposed on these images indicates the position of the surgical instrument, endoscope, or microscope focal point. A fourth window can alternatively display a surface-rendered 3D view of the patient's anatomy, real-time video from the endoscope or microscope, or reference data from an anatomical atlas. An alternative view, the navigational view, produces images orthogonal to the surgical instrument rather than the patient and is particularly useful for aligning the instrument with a surgical path.

An operating microscope can also be tracked relative to the surgical field and the position of the focal point displayed on the preoperative images. Certain microscopes can be robotically controlled by the system. By means of motors that move the microscope head, the workstation can drive the scope to focus on a specific point in the surgical field chosen by clicking on the spot as viewed on the workstation display.

A rigid straight fiberscope ("INCLUSIVE" endoscope, Medtronic Sofamor Danek, Memphis, TN) has also been modified to work with the Trion for intraventricular procedures. Four LEDS are attached to the endoscope near the camera mount using a star-shaped adapter, and the geometric configuration of the LEDs is programmed into the surgical navigational system along with the endoscopic dimensions.

## 49.6 Intraoperative Imaging

Images acquired preoperatively obviously depict the situation prior to opening of the cranium. However, the brain is elastic and usually deforms during a surgical procedure, particularly when resecting a large tumor or if a significant volume of CSF is drained. Attempts have been made to model how the brain would respond to surgery, but this is mathematically complex. An alternative solution is to use intraoperative imaging to update and correct preoperative images as the surgery progresses. In its simplest form, the same imaging modality used preoperatively could be employed intraoperatively. Alternatively, a different imaging modality could be used during a procedure and the deformation detected by this imaging modality used to elastically deform the preoperative images. Several different modalities are currently available for intraoperative imaging, and their relative advantages and disadvantages are summarized in Table 49.1.

Ultrasound has been suggested as a low-cost alternative to stereotaxis. Ultrasound data may be acquired as a single 2D image, or a 3D data set may be acquired with newer systems. However, the images are taken at oblique angles to normal anatomical axes and are often difficult to interpret. There is a high signal-to-noise ratio, and some tumors are not visible at all using ultrasound. However, it may be useful if registered to other imaging modalities that enable the recognition of specific structures. Ultrasound-to-MRI registration is accomplished by using a modified probe equipped with LEDs.

**TABLE 49.1** Relative Merits of Intraoperative Imaging Modalities

| Factor/Type | MRI | CT | Ultrasound | Fluoroscopy |
|---|---|---|---|---|
| Cost | – – | – | + | + |
| Continuous duty | + | – – | ++ | - |
| Resolution | + | + | +++ | – – |
| Contrast | ++ | + | + | + |
| Distortion | – | ++ | + | + |
| Signal-to-noise ratio | ++ | + | – – | – |

+ = advantageous, – = disadvantageous.

With the Trion ultrasound images are fed in through a video port and preoperative MR or CT images reformatted to match the ultrasound images. Color Doppler units, such as the Aloka Model 5000 (Aloka Co., Wallingford, CT), can detect the presence of vessels easily and in color. The system has a modified operative display with five active windows that enable the axial, coronal, and sagittal views to be displayed as reformatted MR or CT with or without overlaid ultrasound images.

## 49.7 Intraoperative System Control

As navigational systems become more complex the surgeon must increasingly interact with the system during a procedure. This need will intensify as more complex effectors, such as robots, are brought into the operating room. Furthermore, as these units proliferate in community hospitals, it will be imperative that they be controllable by a small surgical team, without the need for additional technicians to operate the systems.

Several approaches have been developed that allow the surgeon to control the system while scrubbed. One technique involves the use of touch-sensitive flat-panel displays that can be placed in a sterile bag and used for controlling the system as well as indicating position. Another solution is to incorporate voice recognition in the head-mounted display worn by the surgeon during a procedure.

As these systems become more advanced, it will be important to improve the diversity and utility of the information presented. Rather than being limited to simply showing where a surgeon is within a patient's anatomy, these systems can be employed to compare the patient's anatomy to that of previous patient's functional anatomy. This function is served through the use of an atlas of functional anatomy.

## 49.8 Current Applications

Image-guided surgery techniques have proved valuable in diverse procedures including tumor removal and ablation, treatment of arteriovenous malformations, functional neurosurgery for the treatment of epilepsy or tremor-inducing conditions, and implantation of devices to control hydrocephalus (for example). Image guidance also provides significant assistance in various orthopedic procedures including spine surgery, pelvic fixation, and joint replacement.

### 49.8.1 Tumors

Image guidance is of great value for planning tumor-removal procedures. The actual location of certain types of tumor with distinct margins is usually apparent in preoperative imaging, though some forms of tumor are not so easily delineated. In most cases, image guidance facilitates a more complete resection of the tumor during the initial surgery, and subsequent imaging will reveal whether follow-up surgery is necessary to remove any residual material or recurrent growth.

Another application of image guidance for tumor therapy is the placement of radioactive seeds directly inside tumors that are otherwise inoperable or where surgery may result in unwanted side effects. In general, this technique is used to palliate rather than cure a lesion. One area in which this technique has been used successfully is in the management of malignant tumors of the prostate that commonly occur in patients incapable of tolerating extensive resective surgery.

### 49.8.2 Vascular Malformations

Intracerebral vascular malformations normally require surgical intervention to prevent a future hemorrhage and the attendant neurological deficits. Excision of such malformations is particularly difficult when they occur in eloquent areas of the brain such as those responsible for speech or motor-sensory functions. Image-guidance technology enables the malformation to be modeled in three dimensions and thus permits the optimal resection to be planned, taking into account the relationships of the feeding and draining vessels.

### 49.8.3 Ventriculostomy

The most common application of endoscopy within the head is in the performance of third ventriculostomies in patients with obstructive hydrocephalus or aqueductal stenosis.[15–17] Once the endoscope is successfully inserted, it is frequently possible to proceed on the basis of visual guidance. However, image guidance can ensure a straight trajectory from the burr hole through the foramen of Monroe to the floor of the third ventricle and help avoid unseen branches of the basilar artery. Image guidance is also of particular value in cases of abnormal anatomy or where orientation becomes difficult owing to bloody or blurry cerebrospinal fluid.[18] Recent developments in the field of virtual endoscopy promise to aid in the planning and conduct of these procedures.[19]

### 49.8.4 Functional Neurosurgery

Patients with medically intractable epilepsy are now frequently referred for functional neurosurgery to alleviate their condition. To treat the condition surgically the seizure focus must first be identified and localized within the brain. Once localized, the region can be eradicated. These processes require accurate spatial localization during preoperative imaging and electrophysiological investigations, and accurate coregistration of the patient data with the surgical field is essential to prevent destruction of uninvolved tissue.

While the advent of CT and MRI had a significant impact on epilepsy management, the more recent developments of PET, SPECT, fMRI, and magnetic source imaging (MEG) has enabled greater understanding of the underlying neurophysiology and has permitted refinement of treatment protocols. The ability to coregister each of the imaging modalities with the others (and with the patient) is fundamental to the process of determining the nature of the pathology in a given patient.

Video EEG recordings of actual seizure events also provide vital information for determining which cases are suitable for surgery. Invasive recording techniques are used to evaluate more difficult cases, and placement of the required recording hardware can be facilitated by image guidance. A common approach involves the insertion of depth electrodes into critical mesial structures to localize the seizure origins.[20] Another evaluation technique requires the placement of subdural strip electrodes on the cortical surface. Accurate placement of subdural strips is crucial, and frameless image guidance is potentially of great assistance in the process, though many centers still place the strips without such guidance.

Therapeutic procedures for treating seizure disorders associated with structural lesions have been highly successful in producing seizure-free outcomes.[21,22] All such surgical interventions for seizure disorders benefit considerably from image guidance as it is desirable to fully eliminate the causative lesion while minimizing damage to surrounding tissue. It is likely that in the future we will see the development of noninvasive source localization that will obviate the need for surgical implantation of recording electrodes. However, image guidance in therapeutic surgery will still be required. Improvements in noncontact registration should enable the registration of patient image data to be accomplished and updated more rapidly.

### 49.8.5 Stereotactic Radiosurgery

Another area where image guidance has been of considerable assistance is stereotactic radiosurgery in which one or more highly collimated beams of radiation are directed at a tumor. This is usually an outpatient, nonsurgical procedure using the LINAC Scalpel, Gamma Knife, or, more recently, proton-beam therapy.[23,24] For certain types of brain tumor this treatment is an effective alternative to conventional surgery. However, the radiation does not remove the tumor (although it may shrink it), and resulting scar tissue may make any future surgery more difficult. In this technique, the patient first undergoes the application of a stereotactic frame to establish a reference system upon his or her anatomy. The target of interest is then defined by imaging the patient with the frame in place, and the coordinates of the target are calculated using conventional stereotactic techniques. A focused beam of irradiation is then sent through the target point, and the patient or the beam is moved, and the beam repeated, as in LINAC-based stereotactic radiosurgery. Alternatively, the patient's head is precisely placed at the center of numerous intersecting beams such that all of the beams intersect within the target, as in Gamma Knife radiosurgery. Both devices embrace the key concept of image guidance in terms of maximal functionality with minimal invasiveness.

## 49.9 Conclusions

Image guidance directly supports minimally invasive surgery techniques by enabling the surgeon to view the operative site directly without the need for extensive exposure of normal anatomy. In the hands of a well-trained surgeon, the use of image guidance can markedly reduce the risk associated with critical interventions when compared to the situation before navigation was widely adopted.

Based on the advances made in imaging, image guidance has gone from a few applications limited to the midline of the brain to being used in nearly every cranial procedure imaginable, and fusion of intraoperatively acquired ultrasound images to preoperative MRI data may mean that expensive imaging equipment is unnecessary to bridge the gap between the preoperative and intraoperative situations.

As this technology improves, and as more effectors are built that rely on precise placement, image guidance will become the standard of care for all cranial interventions and many other surgical procedures.

### References


1. Zernov, D.N., L'encéphalomètre, *Rev. Gen. Clin. Ther.*, 19, 302, 1890.
2. Nowinski, W.L., Bryan, R.N., and Raghavan, R., *The Electronic Clinical Brain Atlas: Multi-Planar Navigation of the Human Brain*, Thieme, New York, 1997.
3. Nowinski, W.L., Yeo, T.T., and Thirunavuukarasuu, A., Microelectrode-guided functional neurosurgery assisted by Electronic Clinical Brain Atlas CD-ROM, *Comput. Aided Surg.*, 3, 115, 1998.
4. Maurer, C.M. and Fitzpatrick, J.M., A review of medical image registration, in *Interactive Image-Guided Neurosurgery*, Maciunas, R.J., Ed., American Association of Neurological Surgeons, Park Ridge, IL, 1993, p. 17.
5. Adler, J.R., Jr., Image-based frameless stereotactic radiosurgery, in *Interactive Image-Guided Neurosurgery*, Maciunas, R.J., Ed., American Association of Neurological Surgeons, Park Ridge, IL, 1993, pp. 81–89.
6. Maciunas, R.J., Fitzpatrick, J.M., Galloway, R.L., et al., Beyond stereotaxy: extreme levels of application accuracy are provided by implantable fiducial markers for interactive image-guided neurosurgery, in *Interactive Image-Guided Neurosurgery*, Maciunas, R.J., Ed., American Association of Neurological Surgeons, Park Ridge, IL, 1993, p. 259.
7. Pelizzari, C.A., Chen, G.T.Y., Spelbring, D.R., Weichselbaum, R.R., and Chen, C., Accurate three-dimensional registration of CT, PET, and/or MR images of the brain, *J. Comput.-Assisted Tomogr.*, 13, 20, 1989.

8. West, J., Fitzpatrick, J.M., Wang, M.Y., et al., Comparison and evaluation of retrospective intermodality image registration techniques, *J. Comput.-Assisted Tomogr.*, 21, 554, 1997.
9. Watanabe, E., The neuronavigator: a potentiometer-based localization arm system, in *Interactive Image-Guided Neurosurgery*, Maciunas, R.J., Ed., American Association of Neurological Surgeons, Park Ridge, IL, 1993, p. 135.
10. Guthrie, B.L. and Adler, J.R., Jr., Computer-assisted preoperative planning, interactive surgery, and frameless stereotaxy, *Clin. Neurosurg.*, 38, 112, 1992.
11. Roberts, D.W., Strohbehn, J.W., Hatch, J.F., et al., A frameless stereotaxic integration of computerized tomographic imaging and the operating microscope, *J. Neurosurg.*, 64, 545, 1986.
12. Barnett, G.H., Kormos, D.W., Steiner, C.P., et al. Intraoperative localization using an armless, frameless stereotactic wand, *J. Neurosurg.*, 78, 510, 1993.
13. Barnett, G.H., Kormos, D.W., Steiner, C.P., et al. Frameless stereotaxy using a sonic digitizing wand: development and adaptation to the Picker Vistar medical imaging system, in *Interactive Image-Guided Neurosurgery*, Maciunas, R.J., Ed., American Association of Neurological Surgeons, Park Ridge, IL, 1993, p. 113.
14. Bucholz, R.D. and Smith, K.R., A comparison of sonic digitizers versus light-emitting diode-based localization, in *Interactive Image-Guided Neurosurgery*, Maciunas, R.J., Ed., American Association of Neurological Surgeons, Park Ridge, IL, 1993, p. 179.
15. Dalrymple, S.J. and Kelly, P.J., Computer-assisted stereotactic third ventriculostomy in the management of non-communicating hydrocephalus, *Stereotact. Funct. Neurosurg.*, 59, 105, 1992.
16. Drake, J.M., Ventriculostomy for treatment of hydrocephalus, *Neurosurg. Clin. North Am.*, 4, 657, 1993.
17. Kelly, P.J., Stereotactic third ventriculostomy in patients with nontumoral adolescent/adult onset aqueductal stenosis and symptomatic hydrocephalus, *J. Neurosurg.*, 75, 865, 1991.
18. Muacevic, A. and Muller, A., Image-guided endoscopic ventriculostomy with a new frameless armless neuronavigation system, *Comput. Aided Surg.*, 4, 87, 1999.
19. Burtscher, J., Dessl, A., Bale, R., Eisner, W., Auer, A., Twerdy, K., and Felber, S., Virtual endoscopy for planning endoscopic third ventriculostomy procedures, *Pediatr. Neurosurg.*, 32, 77, 2000.
20. McCarthy, G., Spencer, D.D., and Riker, R.J., The stereotaxic placement of depth electrodes in epilepsy, in *Epilepsy Surgery*, Lüders, H., Ed., Raven Press, New York, 1991, p. 371.
21. Engel, J., Jr., van Ness, P.C., Rasmussen, T.B., et al., Outcome with respect to epileptic seizures, in *Surgical Treatment of the Epilepsies*, Engle, J., Jr., Ed., Raven Press, New York, 1993, p. 609.
22. Piepgras, D.G., Sundt, T.M., Jr., Ragoowansi, A.T., et al., Seizure outcome in patients with surgically treated cerebral arteriovenous malformations, *J. Neurosurg.*, 78, 5, 1993.
23. Amin-Hanjani, S., Ogilvy, C.S., Candia, G.J., Lyons, S., and Chapman, P.H., Stereotactic radiosurgery for cavernous malformations: Kjellberg's experience with proton beam therapy in 98 cases at the Harvard Cyclotron, *Neurosurgery*, 42, 1229, 1998; discussion, p. 1236.
24. Seifert, V., Stolke, D., Mehdorn, H.M., and Hoffmann, B., Clinical and radiological evaluation of long-term results of stereotactic proton beam radiosurgery in patients with cerebral arteriovenous malformations, *J. Neurosurg.*, 81, 683, 1994.

# 50

# Optical Methods for Caries Detection, Diagnosis, and Therapeutic Intervention



Daniel Fried
*University of California, San Francisco*
*San Francisco, California*

## 50.1 Optical, Physical, and Thermal Properties of Dental Hard Tissues

### 50.1.1 Optical Properties of Dental Hard Tissue in the Visible and Near-Infrared

Dental enamel is an ordered array of rods of inorganic apatite-like crystals surrounded by a protein/lipid/water matrix. The crystals are approximately 30 to 40 nm in diameter and can be as long as 10 μm. The crystals are clustered together in 4-μm-diameter rods (or prisms), which are roughly perpendicular to

0-8493-1116-0/03/$0.00+$1.50

the tooth surface. Dentin is honeycombed with dentinal tubules of 1 to 3 μm in diameter. Each of these tubules is surrounded by a matrix of needle-shaped, hydroxyapatite (HAP)-like crystals in a protein matrix composed largely of collagen. Because of the complex nature of these materials, the scattering distributions are generally anisotropic and depend on tissue orientation relative to the irradiating light source[1–4] in addition to the polarization of the incident light.

The optical properties of biological tissue can be quantitatively described by defining the optical constants — the absorption ($\mu_a$) and scattering coefficients ($\mu_s$) — that represent the probability that the incident photons will be absorbed or scattered, as well as the scattering phase function $\Phi(\cos\theta)$, which is a mathematical function that describes the directional nature of scattering.[5–8] When these parameters are known, light transport in dental hard tissue can be completely characterized and modeled. The optical parameters for normal enamel and dentin have been reported in the wavelength range of 200 to 700 nm. For enamel, absorption is very weak in the visible range ($\mu_a < 1$ cm$^{-1}$, $\lambda = 400$ to 700 nm) and increases in the ultraviolet (UV) ($\mu_a > 10$ cm$^{-1}$, $\lambda < 240$ nm).[9] For dentin in the 400- to 700-nm wavelength range, the absorption coefficient is essentially wavelength independent, with a value of $\mu_a \sim 4$ cm$^{-1}$.[10] Scattering in enamel is strong in the near-UV and decreases $\sim\lambda^3$ with increasing wavelength to a value of only 2 to 3 cm$^{-1}$ at 1550 nm.[11,12] In contrast, scattering in dentin is strong throughout the near-UV, visible, and near infrared (NIR).[3,10]

Accurate description of light transport in dental hard tissue relies on knowledge of the exact form of the phase function $\Phi(\cos\theta)$ for each tissue scatterer at each wavelength.[8] Therefore, direct measurement of the phase function is necessary. It is important to note that the empirically derived Kubelka-Munk (KM) coefficients that are commonly used in dentistry are not fundamental optical constants and are not appropriate for describing light transport in tissue with forward-directed scattering such as enamel and dentin, particularly in the NIR.[5–8,13] Scattering in most biological tissues can be represented by a Henyey–Greenstein (HG) function with values of $g$ greater than 0.8.[5] The scattering anisotropy ($g$) should be determined within the context of an appropriate phase function based on the nature of the scatterers in the tissue[5,13] and subsequently validated through comparison of simulated scattering distributions with measured distributions of various thickness.[3] Measured angular-resolved scattering distributions could not be represented by a single scattering phase function $\Phi(\cos\theta)$ and required a linear combination of a highly forward-peaked phase function, an HG function, and an isotropic phase function represented by the following equation:[3]

$$\Phi(\cos\theta) = f_d + (1 - f_d)\left(\frac{(1-g^2)}{(1-g^2-2g\cos\theta)^{3/2}}\right) \tag{50.1}$$

The parameter $f_d$ (fraction diffuse) is defined as the fraction of isotropic scatterers. The average value of the cosine of the scattering angle ($\theta$) is called the scattering anisotropy ($g$), and at 1053 nm $g = 0.96$ and $g = 0.93$ for enamel and dentin, respectively.[3] Note that Zijp and ten Bosch[1] and Zijp et al.[4] reported values of $g = 0.4$ for dentin and $g = 0.68$ for enamel, calculated by taking the ratio of the forward- and backward-scattered light. This latter approach is prone to error due to the contribution of surface scattering.[3] At 1053 nm the fraction of isotropic scatterers, ($f_d$), was measured to be 36% for enamel and less than 2% for dentin. The phase function changes with wavelength; therefore, the phase function must be independently determined at each wavelength.

Increased backscatter from the demineralized region of early caries lesions is the basis for the visual appearance of white spot lesions.[14] Increased porosity of the lesion leads to increased scattering at the lesion surface and higher scattering in the body of the lesion, producing an increase in the magnitude of the diffuse reflectance.[15] Attempts at measuring the optical properties of dental caries have been limited to measurements of backscattered light from optically thick, multilayered sections of simulated caries lesions.[15–17] In those measurements, empirical KM coefficients were calculated representing the diffusion

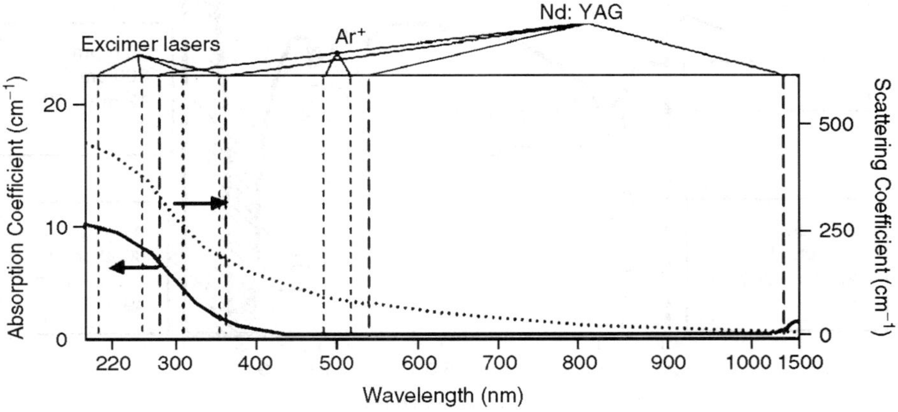


**FIGURE 50.1** The scattering coefficient $\mu_s$ (dotted line) and the absorption coefficient $\mu_a$ (solid line) of dental enamel from the UV-NIR. Laser wavelengths of interest are indicated by the vertical dashed lines. (Compiled from References 9, 11, and 12.)

of light through the tissue. Ko et al.[17] recently proposed that the optical scattering power, the product of the KM scattering parameter and the lesion depth, is a good estimate of enamel demineralization.

Enamel exhibits double refraction or birefringence upon illumination in the visible and NIR.[18] The apatite crystals in dental enamel are highly oriented along the long or c-axis of the enamel prismatic structures. The major source of birefringence in dentin is the highly oriented collagen fibrils in the collagen; the collagen fibers are positively birefringent, while the apatite crystals are negatively birefringent. Demineralized enamel and dentin appear black under observation through crossed polarizers due to depolarization of the incident polarized light. Theuns et al.[19] have related changes in the birefringence of dental enamel during demineralization to mineral content. Polarized-light microscopy measurements over the past several decades on thin sections have demonstrated that caries lesions rapidly depolarize incident polarized light.

## 50.1.2 Optical Properties of Dental Hard Tissue in the Infrared

In the mid-infrared, scattering is negligible, and with accurate knowledge of the absorption coefficient and the reflectivity the light deposition can be described using Beer's law. The magnitude of the absorption coefficient is high in the mid-infrared due to resonant absorption by molecular groups in water, protein, and mineral. The reflectivity can exceed 50% at wavelengths coincident with mineral-absorption bands due to the large increase in the imaginary component of the refractive index (see Section 50.1.3). Coefficients corresponding to carbon dioxide ($CO_2$) laser lines that could be obtained using conventional transmission were determined to be 1168 ± 49 cm$^{-1}$ at 10.3 μm and 819 ± 62 cm$^{-1}$ at 10.6 μm for enamel and 1198 ± 104 cm$^{-1}$ at 10.3 μm and 813 ± 63 cm$^{-1}$ at 10.6 μm for dentin. Enamel-absorption coefficients corresponding to Er:YAG (2.94 μm) and Er:YSGG (2.79 μm) were calculated to be 768 ± 27 cm$^{-1}$ and 451 ± 29 cm$^{-1}$, respectively. The absorption coefficient of dentin at 2.79 μm was calculated to be 988 ± 111 cm$^{-1}$. These absorption coefficients were based on the Beer–Lambert law accounting for Fresnel reflectance losses. Conventional transmission measurements were not possible for the determination of the optical properties of enamel and dentin at 9.3 and 9.6 μm; therefore, alternative methods, such as angular-resolved reflection measurements of polarized light and time-resolved radiometric measurements, were necessary. Duplain et al.[20] measured absorption coefficients of 18,700, 31,300, 6,500, and 5,200 cm$^{-1}$ at 9.3, 9.6, 10.3, and 10.6 μm, respectively, using angular-resolved reflection measurements of polarized light. The magnitude of the coefficients of these researchers in the wavelength range in which

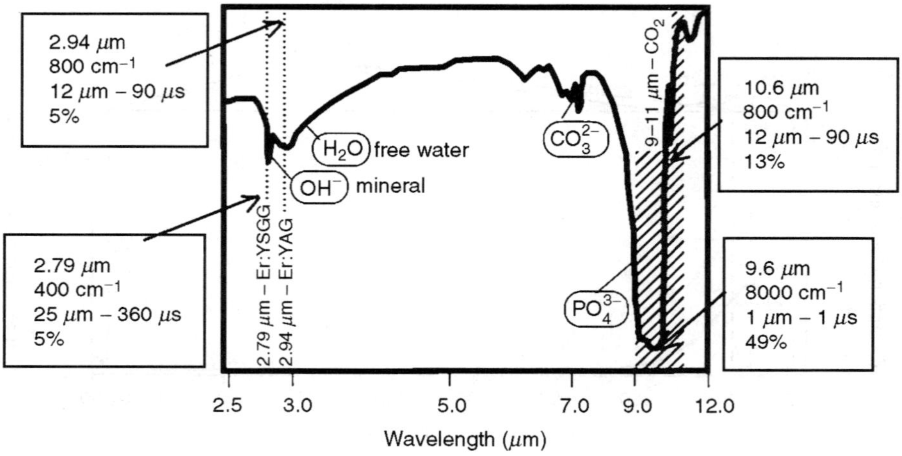


**FIGURE 50.2** An infrared transmission spectrum of dental enamel is shown with the principal molecular absorption groups indicated along with the relevant laser wavelengths. In each box under the indicated laser wavelength is the measured absorption coefficient, depth of absorption, thermal relaxation time calculated from the absorption depth, and the reflectance. (Compiled from References 20, 21, 27, and 28.)

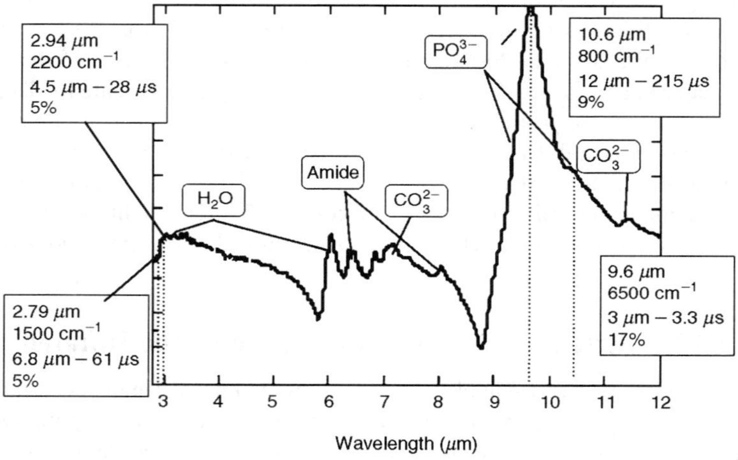


**FIGURE 50.3** An IR transmission spectrum of dentin is shown with the principal molecular absorption groups indicated along with the relevant laser wavelengths. In each box under the indicated laser wavelength is the measured absorption coefficient, depth of absorption, thermal relaxation time calculated from the absorption depth, and the reflectance. (From Fried, D. et al., *Lasers Surg. Med.*, 31, 275, 2002. With permission.)

direct transmission measurements were possible, 10.6 and 10.3 μm, are not consistent with the dental-enamel-absorption coefficients determined using direct-transmission measurements (1168 and 819 $cm^{-1}$ for 10.3 and 10.6 μm, respectively). Moreover, those values are also not consistent with observed surface-modification thresholds based on the known melting range of dental enamel (800 to 1200°C). More recent measurements employing time-resolved radiometry measurements coupled with numerical simulations of thermal relaxation in the tissue[21] are a factor of five to six times lower. Those values are shown in Figures 50.2 and 50.3 for enamel and dentin along with the absorption depth, reflectance, and corresponding thermal relaxation times computed from the absorption depth and thermal property data shown in Table 50.1. These latter values are consistent with time-resolved temperature measurements during laser irradiation and surface-melting thresholds.[26]

**TABLE 50.1** Thermal and Mechanical Properties of Dental Hard Tissues

| Tissue Property | Enamel | Dentin |
|---|---|---|
| Density (g/cm³) | 2.8 | 2.0 |
| Specific heat (J/g °C) | 0.71 | 1.59 |
| Thermal conductivity (W/cm °C) | 0.0093 | 0.0057 |
| Coeff. linear expansion (μm/°C) | 27 | |
| Melting point (°C) | 1280 | |
| Compressive strength (MPa) | 250–550 | 300–380 |
| Tensile strength (MPa) | 10–70 | 50–60 |

Data compiled from References 23 through 25.

## 50.1.3 Infrared Reflectance of Dental Hard Tissue

At $CO_2$ wavelengths, the reflection losses during enamel irradiation are substantial and reduce the laser energy absorbed by the target surface; losses are minimal near 3.0 μm. The fraction of incident laser light reflected at the surface of the tooth is described by the Fresnel reflection formula, $R = [(n_r - 1)^2 + k^2]/[(n_r + 1)^2 + k^2]$, where ($n_r$) is the real component of the refractive index and ($k$) is the attenuation index. The absorption coefficient, ($\mu_\alpha$), ($k$), and the wavelength ($\lambda$) are related by the expression $\mu_\alpha = 4\pi k/\lambda$. Thus, the reflectance of materials can increase markedly and approach 100% in regions of strong absorption, e.g., some metals have absorption coefficients $>10^6$ cm$^{-1}$ and reflectance > 99% in the visible and infrared. The reflectance of dentin and enamel was measured at the $\lambda$ = 2.79, 2.94, 10.6, 10.3, 9.6, and 9.3 μm wavelengths (see Figures 50.3 and 50.4).[20,27,28] The reflectance of enamel is substantially higher at $\lambda$ = 9.6 and 9.3 μm than at $\lambda$= 10.6 μm, near 50%, and must be accounted for when calculating ablation efficiencies, ablation thresholds, and heat deposition in the tooth. Transient and permanent changes in the reflectance of enamel and dentin were observed during and after laser irradiation.[27] These changes resulted in increased energy coupling during irradiation at $\lambda$ = 9.3 and 9.6 μm for fluences that raised the enamel surface temperature several hundred degrees. The reflectance of dentin at $\lambda$ = 9.6 μm permanently increased by as much as 30% as a result of laser irradiation. This change can be attributed to an increase in the mineral density and loss of organics.[27]

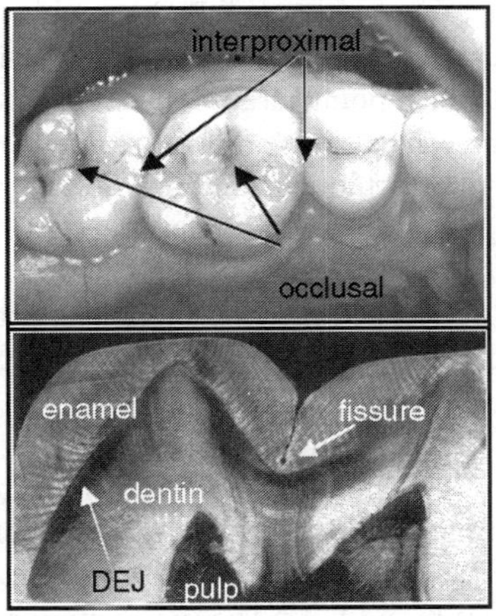


**FIGURE 50.4** Top: The primary (high-risk) sites for dental decay are in the pits and fissures of the posterior teeth and the approximal contact sites between teeth. Bottom: A typical fissure with a diameter of 200 μm in a tooth cross section.

## 50.2 Optical Caries Diagnostics

During the past century the nature of dental decay or dental caries in the United States has changed markedly due to the introduction of fluoride to the drinking water, the advent of fluoride dentifrices and rinses, and improved dental hygiene. In spite of these advances, dental decay continues to be the leading cause of tooth loss in the U.S.[29–31] By age 17, 80% of children have experienced at least one cavity.[32] In addition, two thirds of adults age 35 to 44 have lost at least one permanent tooth to caries. Older adults suffer tooth loss due to the problem of root caries. The nature of the caries problem has changed dramatically, with the majority of newly discovered caries lesions being highly localized to the occlusal pits and fissures of the posterior dentition and the interproximal contact sites between teeth (Figure 50.4). These early carious lesions are often obscured or "hidden" in the complex and convoluted topography of the pits and fissures or are concealed by debris that frequently accumulates in those regions of the posterior teeth. Moreover, such lesions are difficult to detect in the early stages of development. By definition, early caries lesions are those lesions confined to the enamel or incident on the dentinal–enamel junction (DEJ).

In the caries process, demineralization occurs as organic acids generated by bacterial plaque diffuse through the porous enamel of the tooth, dissolving the mineral. If the decay process is not arrested, the demineralization spreads through the enamel and reaches the dentin where it rapidly accelerates due to the markedly higher solubility and permeability of dentin. The lesion spreads throughout the underlying dentin to encompass a large area, resulting in loss of integrity of the tissue and cavitation. Caries lesions are usually not detected until after the lesions have progressed to the point at which surgical intervention and restoration are necessary, often resulting in the loss of healthy tissue structure and weakening of the tooth. Carious lesions also occur adjacent to existing restorations, and diagnostic tools are needed to diagnose the severity of those lesions and determine if an existing restoration needs to be replaced. The diagnostic and treatment paradigms that were developed in the past such as radiography are adequate for large, cavitated lesions; however, they do not have sufficient sensitivity or specificity for the diagnosis of early noncavitating caries, root surface caries, or secondary caries.

New diagnostic tools are needed for the detection and characterization of caries lesions in the early stages of development. In a recent consensus statement released by the NIH entitled *The Diagnosis and Management of Dental Caries throughout Life*, the development of new devices and techniques for caries diagnosis was identified as one of five major areas in which additional research was needed.[32] Caries lesions are routinely detected in the United States using visual/tactile (explorer) methods coupled with radiography. Unfortunately, these methods have numerous shortcomings and are inadequate for the detection of the early stages of the caries process.[33–35] Radiographic methods do not have the sensitivity for early lesions, particularly occlusal lesions, and by the time the lesions are radiolucent they have often progressed well into the dentin, at which point surgical intervention is necessary.[35–37] At that stage in the decay process, it is far too late for preventive and conservative intervention, and a large portion of carious and healthy tissue must be removed, which often compromises the mechanical integrity of the tooth. If left untreated, the decay will eventually infect the pulp, leading to loss of tooth vitality and possibly extraction. The caries process is potentially preventable and curable. If carious lesions are detected early enough, it is likely that they can be arrested/reversed by nonsurgical means through fluoride therapy, antibacterial therapy, dietary changes, or by low-intensity laser irradiation.[32,38] Therefore, one cannot overstate the importance of detecting decay in the early stage of development at which point noninvasive preventive measures can be taken to halt further decay.

In a recent review of conventional methods of caries diagnosis, ten Cate[37] indicated that visual and tactile caries diagnosis was far from ideal and that visual diagnosis of occlusal caries typically has a very low sensitivity. Sensitivities scatter around a value of 0.3, implying that only 20 to 48% of the caries present (usually into the dentin) are found.[37,39,40] The specificity typically exceeds 0.95. The poor sensitivity can be attributed to the "hidden" nature of the majority of occlusal lesions. The area of the lesion accessible for visual and tactile inspection is typically confined to the upper region of the fissure. The bulk of the lesion is not accessible and is most often not detected unless it is so extensive that it is resolvable radiographically.

### 50.2.1 Optical Transillumination (Interproximal Lesions)

Interproximal areas between teeth are generally inaccessible for visual inspection. Bite-wing radiographs are the standard method of diagnosis. Unfortunately, as much as 25% of the interproximal areas of bite-wing x-rays is unresolved due to the overlap with healthy tooth structure on adjoining teeth.[41,42] Values for sensitivity (0.38 and 0.59) and specificity (0.99 and 0.96) are reported for visual and radiological diagnosis, respectively.[43]

Optical transillumination was used extensively before the discovery of x-rays for detection of dental caries. Recently there has been renewed interest in this method with the availability of high-intensity fiber-optic-based illumination systems for the detection of interproximal lesions.[41,44–47] During fiber-optic transillumination (FOTI) a carious lesion appears dark upon transillumination because of decreased transmission due to increased scattering and absorption by the lesion. A digital fiber-optic transillumination system, called DIFOTI, which uses visible light for the detection of caries lesions, has been developed by Electro-Optics Sciences (Irvington, NY)[48] and recently received U.S. Food and Drug Administration (FDA) approval.

### 50.2.2 Fluorescence (DIAGNOdent™ and QLF)

Teeth naturally fluoresce upon irradiation with ultraviolet (UV) and visible light. Alfano et al.[49,50] and Bjelkhagen and Sundstrom[51] demonstrated that laser induced-fluorescence (LIF) of endogenous fluorophores in human teeth could be used as a basis for discrimination between carious and noncarious tissue. Upon illumination with near-UV and visible light and imaging of the emitted fluorescence in the range of 600 to 700 nm, carious/demineralized areas appear dark. The origin of the endogenous fluorescence in teeth in this particular wavelength range has not been established.

Hafstroem-Bjorkman et al. [52] established an experimental relationship between the loss of fluorescence intensity and extent of enamel demineralization. The method was subsequently labeled the QLF method, for quantitative laser fluorescence. An empirical relationship between overall lesion demineralization ($\Delta Z$) vs. loss of fluorescence was established that can be used to monitor lesion progression on smooth surfaces.[53–55] Recent measurements suggest that QLF may be useful for determination of the degree of hydration of the lesion that, in turn, may be indicative of the lesion activity.[56] Unfortunately, it is difficult to apply QLF to occlusal and interproximal lesions, which constitute the majority of carious lesions. Furthermore, the fluorescence method cannot be used to provide information about the subsurface characteristics of the lesion.

Bacteria produce significant amounts of porphyrins, and dental plaque fluoresces upon excitation with red light.[57] A novel caries detection system, the DIAGNOdent™ (KaVo, Biberach, Germany), was recently developed and received FDA approval in the U.S. This device uses a diode laser and a fiber-optic probe designed to detect the NIR fluorescence from porphyrins.[58] This low-cost diagnostic tool is a major step toward better caries detection in occlusal surfaces, greatly aiding in the detection of "hidden" occlusal lesions, i.e., lesions that have progressed into the dentin and are too small to show up on a radiograph.[59] It is important to note that the DIAGNOdent has poor sensitivity (~0.4) for early lesions confined to enamel, and it cannot provide information about the lesion depth and the degree of severity.[60,61]

### 50.2.3 Optical Coherence Tomography for Caries Imaging

Optical coherence tomography (OCT) is a sensitive method for resolving changes in light scattering in early caries lesions. The first OCT images of soft- and hard-tissue structures of the oral cavity were acquired by Colston et al.[62–64] Baumgartner et al.[65,66] presented the first polarization-resolved images or polarization sensitive-OCT images (PS-OCT) of dental caries; however, the penetration depth was limited and the image quality was poor due to limited source power. Feldchtein et al.[67] presented high-resolution dual-wavelength 830- and 1280-nm images of dental hard tissues, enamel and dentin caries, and restorations *in vivo*. Wang et al.[68] measured the birefringence in dentin and enamel and suggested

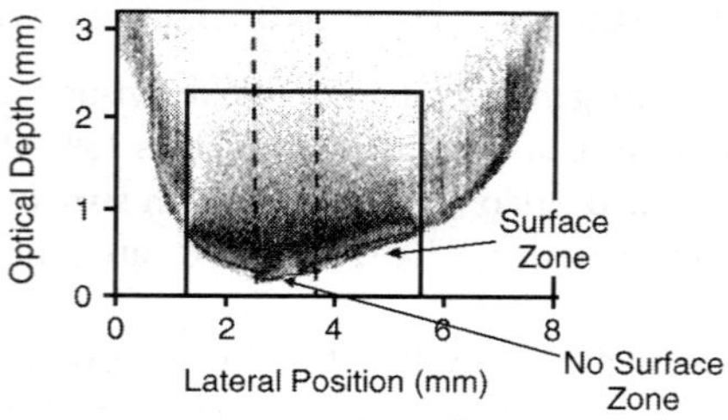


**FIGURE 50.5** Fast-axis PS-OCT scan (orthogonal polarization state) of an obvious smooth-surface interproximal lesion (white spot lesion).[71,72] The scan was taken perpendicular to the long axis of the tooth from the facial to lingual surface. The dark areas on the tooth surface indicate the position of the lesion. The lesion, which is enclosed in the box, appears more active between the dotted lines and lacks a surface zone of reduced scattering. The surface of the lesion contains a thin surface zone of reduced scattering outside of the dotted lines that may indicate remineralization. (From Fried, D. et al., *J. Biomed. Opt.*, 7(4), 618, 2002. With permission.)

that the enamel rods act as waveguides. The following year, Everett et al.[69] presented polarization-resolved images using a high-power 1310-nm broadband source and a bulk optic PS-OCT system. In those images, changes in the mineral density of tooth enamel were resolvable to depths of 2 to 3 mm into the tooth. An important advantage of PS-OCT is that the structure of the carious lesions can be resolved in fine detail. Figure 50.5 shows a fast-axis or S-axis scan — detected polarization orthogonal or perpendicular to incident linear polarization — of an early interproximal lesion. Areas of reduced scattering near the surface may be indicative of remineralization, and those regions of the lesion may be inactive. In the area between the dotted lines, the scattering near the surface of the lesion is very strong. This may be indicative of an active area of the lesion that is progressing. Being able to access the activity of the lesion, i.e., whether it is active and needs intervention or if it has been arrested and can be left alone, is extremely important to the clinician, and no currently available diagnostic technology is capable of such a diagnosis.

According to Kidd,[70] 75% of operative dentistry involves the replacement of existing restorations, and secondary caries is the most common reason given for replacing both amalgam and composite fillings. PS-OCT is well suited for the detection of secondary caries because restorative materials have markedly different scattering properties compared to dental hard tissues. Colston et al.[64] and Feldchtein et al.[67] have shown that OCT can be used to differentiate between restorative materials and enamel and detect decay around the periphery of a restoration. Recent studies have demonstrated that composite can be distinguished from dentin and enamel using PS-OCT and that the degree of demineralization underneath composite restorations can be quantified.[71,72]

Root caries are an increasing problem among our aging population and are the principal cause of tooth loss after the age of 44. The scattering coefficient of dentin is much higher than enamel; therefore, the penetration depth is more limited than for enamel. However, early root caries are localized on the surface of the dentin and cementum. It is difficult to obtain intact thin sections through natural root caries lesions without damaging the fragile, poorly mineralized regions of the lesion. High-resolution x-ray tomography (XTM) can be used to acquire tomograms of tooth mineral density with a resolution of 9 μm.[73,74] Figure 50.6 shows XTM and PS-OCT scans taken from the crown to the root across a small root caries lesion located just below the cementum–enamel junction on a human tooth. The principal advantage of XTM is that tissue slices representing the mineral density vs. position can be extracted from the x-ray tomogram with any desired orientation to match the OCT scan geometry. The XTM image of the area scanned by the PS-OCT system is shown on the right side of Figure 50.6. Taking into account that the high refractive indices of dentin and enamel result in compression of those areas of the image, the agreement between the two imaging modalities is excellent. The entire lesion morphology is reproduced in the fast-axis OCT image, with the characteristic lesion semicircle shape clearly visible. The fast-axis image provides the best match because the confounding influence of the surface reflection and residual coherence are markedly reduced.

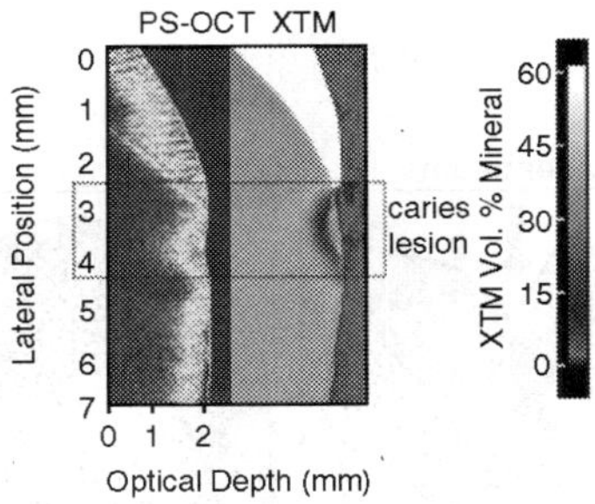


**FIGURE 50.6 (Color figure follows p. 28-30.)** Comparison of the PS-OCT fast-axis image (orthogonal polarization state) with the corresponding x-ray tomogram (XTM) of the mineral density taken from the same region of the tooth (right).[71,72] A small root caries lesion is present just below the cementum–enamel junction shown in the box. The intensity of the OCT images ranges from 12 dB to –45 dB; areas with regions of intensity greater than –5dB are shown in red, and those areas of less than –35 dB are shown in blue. In the XTM image on the right, normal dentin is yellow, enamel is white, the water outside the tooth is indicated in red, and the demineralized area of the lesion is blue (color bar on right). (From Fried, D. et al., *J. Biomed. Opt.*, 7(4), 618, 2002. With permission.)

## 50.3 Therapeutic Applications on Dental Hard Tissues

Since the initial investigations of Stern and Sognnaes[75] more than 30 years ago several unique laser dental applications have evolved for restorative dentistry, i.e., laser ablation of dental hard tissue,[76] caries-inhibition treatments by localized surface heating,[77] and surface conditioning for bonding.[76] For several years medical lasers have been approved by the FDA for soft-tissue vaporization, and even though hard tissue applications have been investigated for over 30 years, only recently have the first lasers been approved for hard-tissue use in the United States. The first lasers to receive approval were the Centuri™ Er:YAG and the Millennium™ Er:YSGG introduced by Premier Laser (Irvine, CA) and Biolase Technology (San Clemente, CA), respectively.

### 50.3.1 New Preventive and Conservative Approaches to Restorative Dentistry

Several developments in dentistry are driving the shift to new more conservative approaches to restorative dentistry than the classical G.V. Black approach that has been the standard for cavity preparation since the 19th century. These include the development of new more effective adhesive restorative materials, a more thorough understanding of the caries process, the now recognized role of fluoride as a remineralizing agent, new and more sensitive methods of caries detection, and the public's demand for more aesthetic restorations. Over the past 50 years, the nature of dental caries has changed, with the majority of new lesions, 90%, occurring in the pits and fissures (occlusal surfaces) of posterior teeth (see Figure 50.4).[78,79]

New caries diagnostic procedures such as those described above are becoming available that enable the dentist to identify early caries lesions before they have spread extensively into the underlying dentin. For lesions of this magnitude, it is too early to use conventional restorations that require the removal of large amounts of healthy tissue. As part of the more conservative approach to restorative dentistry, some experts have revised the G.V. Black cavity-classification scheme to emphasize the importance of early pit and fissure sites.[80] This new conservative approach emphasizes micropreparation with minimal removal of healthy tissue. The laser is ideally suited for this approach for the following reasons.

1. Laser pulses can preferentially ablate carious tissue due to the higher volatility of water and protein that are present in carious tissue at a higher ratio than in normal tissue.
2. They can be tightly focused to drill holes with very high aspect ratios (depth/height) well beyond those obtainable by the dental drill, which is limited by the size of the dental burr (>1 mm). Conservation of enamel structure is paramount for preservation of the natural dentition for retention of sealants and to limit exposure to wear.

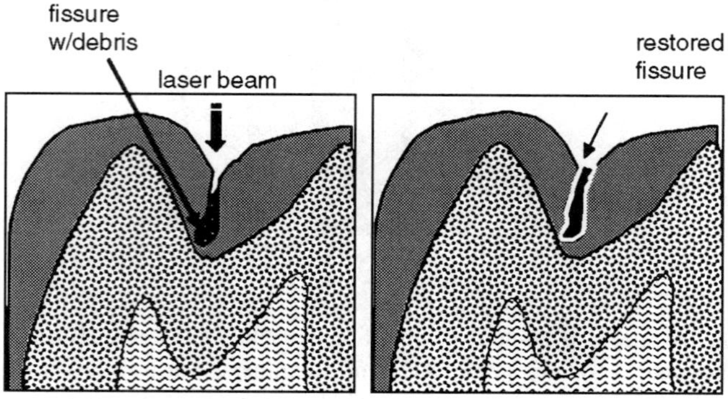


**FIGURE 50.7** Cross section of a typical molar tooth with fissures. (Left) Typical fissures are about 200 μm across and more than a millimeter in depth and may be filled with debris and bacterial plaque. Lasers are ideally suited for removing the debris and ablating the decayed (carious) tissue with minimal removal of healthy enamel. (Right) Afterwards, removal sealants can be applied directly to fill, and if a $CO_2$ laser is used for the preparation, the walls will have an enhanced resistance to further decay (thick white solid line).

3. The laser can be used to open up the neck of the fissures sufficiently to allow the entrance of flowable composite to the base of the fissure.
4. Laser irradiation does not produce a smear layer that must be removed; hence restorative materials can be applied directly to the ablated area without the necessity of further surface preparation and etching.[81] This advantage is of great importance because fissure areas are difficult to etch by conventional means due to the unique enamel morphology. The enamel on the shoulder at the entry to the fissure is often prismless and irregular and may not accept a good etch pattern, so attachment of the resin may be tenuous. Therefore, laser preparation may be superior to the conventional acid etch in pits and fissures.
5. Laser radiation vaporizes water and protein and changes the chemical composition of the remaining mineral of enamel and dentin, thereby decreasing the solubility to acids around the periphery of the restoration site to leave a smooth surface with an enhanced resistance to secondary caries.[77,82–84]
6. Lasers are less likely to require anesthesia and induce less noise and pain.[85–88] Therefore, they are advantageous for treating children and patients with dental phobias.

Thus, lasers are ideally suited for minimally invasive surgical intervention and have the potential for markedly reducing or eliminating the loss of adjacent healthy tissue during the removal of carious tissue from early lesions or the removal of existing composite restorations.

Figure 50.7 illustrates the principle of minimal intervention with lasers.[89] The laser can be precisely focused into the carious fissure of a molar to remove any debris and bacteria, ablate away the carious tissue, and enlarge the narrow neck of the fissure, leaving a smooth crater. If the appropriate irradiation conditions are met, the walls will have enhanced resistance to acid dissolution.[83,84] A flowable composite can subsequently be applied for a highly conservative restoration. By taking this approach the dentist can intervene in a conservative manner before the lesion has progressed to the point that conventional surgical intervention is necessary.

## 50.3.2 Infrared Laser Ablation of Enamel

Enamel is a biological composite containing 12% by volume water, 85% mineral (carbonated hydroxyapatite [CAP]), and 3% protein and lipid. The mineral component is crystalline in nature comprising hexagonal crystallites approximately 40 nm in diameter, which are aligned in enamel rods roughly 5 μm in diameter that run in an S-shaped pattern from the enamel surface to the DEJ. A sheath of protein and water surrounds each rod. Approximately half of the water is bound tightly to the enamel as caged waters of hydration, and the other half is more loosely bound and located between the individual enamel crystals and the rods.[90] The mechanism of interaction of laser light with dental enamel is inherently complex and varies markedly with the laser irradiation wavelength and the nature of the primary absorber in the tissue, water, protein, or mineral. The principal factor limiting the ablation rate of dental hard tissue is the risk of excessive heat deposition in the tooth that may lead to eventual loss of pulpal vitality. The accumulation of heat in the tooth can be minimized through use of a laser wavelength tuned to the maximum absorption coefficient of the tissue irradiated and by judicious selection of the laser-pulse duration to be commensurate with the thermal relaxation time of the deposited energy

Several erbium-laser wavelengths have been investigated for the ablation of dental enamel, i.e., the Er:YAG, 2.94 μm, Er:YLF, 2.81, Er:YSGG, 2.79 μm, Er:YAP, 2.73 μm, and CTE:YAG, 2.69 μm.[91–95] Those wavelengths give access to the water-absorption band from 2.6 to 3.0 μm. These lasers can be operated in either free-running mode, in which the pulse duration can be varied from tens of microseconds to almost a millisecond, or Q-switched mode, in which the pulse duration can vary from tens to hundreds of nanoseconds. Altshuler et al.[92] determined that the ablation threshold in free-running mode increased in the following order: Er:YSGG, Er:YLF, Er:YAP, and Er:YAG. The CTE:YAG ablation threshold is considerably higher than the other erbium systems because absorption in water is weaker, although the $\lambda$ =2.69 μm laser wavelength is still of significant interest because it is readily transmitted through robust, inexpensive low-OH fibers.[96,97] Three studies have investigated variation of the pulse duration during the ablation of enamel. Majaron et al.[98] observed a decrease in the ablation efficiency upon stretching the laser pulse from 200 to 1000 μs. Zeck at al.[99] observed a twofold increase in the ablation efficiency of enamel upon reducing 400-μs Er:YAG laser pulses to 30 μs and a fourfold reduction in the ablation threshold. An even greater reduction in the ablation threshold[100] was observed for Q-switched 200-ns laser pulses, particularly for Er:YSGG laser pulses used in conjunction with water.

$CO_2$ lasers are the most common lasers found in clinics today and have been used for soft-tissue surgical procedures for three decades. The $CO_2$ laser can be designed to operate or "lase" at discrete wavelengths between $\lambda$ = 9 to 11 μm. Those wavelengths correspond to specific rotational–vibrational transitions in the ground state of gas-phase $CO_2$ molecules. Early results with the $CO_2$ laser were discouraging because these studies used continuous-wave (CW) lasers operating at $\lambda$ = 10.6 μm,[101–111] and extensive thermal damage was observed. Recent studies using $CO_2$-laser pulses of submillisecond duration indicate that enamel can be ablated efficiently without generating peripheral damage.[112–114] Even though the pulsed $CO_2$ lasers are a significant improvement over the previous CW systems, the laser parameters are still not well matched to the peak absorption of dental enamel.

Presently, 9.6-μm pulsed $CO_2$ lasers are under development for use in dentistry. One system, manufactured by Argus Photonics Group (APG) (Jupiter, FL), uses transverse excited atmospheric pressure (TEA) technology and uses a pulse duration of 5 to 20 μs.[115,116] Another system from ESC Medical Systems (Yokneam, Israel) uses radio-frequency (RF) technology and a pulse duration of 70 μs.[117] The ESC laser has recently been tested on humans in a limited safety study, and there were no apparent adverse changes to the pulp of the laser-treated teeth.[117]

The laser ablation of dental hard tissue is a water-mediated explosive process at 2.94 and 2.79 μm[93] caused by preferential absorption by water that is localized between the rods of enamel or is intrinsic to the protein/mineral matrix of dentin. During rapid heating the inertially confined water can create enormous subsurface pressures that can lead to the explosive removal of the surrounding mineral matrix.[118] Studies of hard-tissue ablation in the $\lambda$ = 3μm region indicate that large intact particles are ejected at high velocity from the irradiated tissue.[93] Moreover, the normal highly ordered structure of dentin and enamel is conserved during the ablation process (see Figure 50.8). In contrast, the mechanism of ablation of dental hard tissue at the highly absorbed $CO_2$ wavelengths of 9.3 and 9.6 μm is apparently mineral-mediated, and SEM micrographs show that the tissue morphology is markedly changed after irradiation[95] (see Figure 50.8). Surface temperatures at the ablation threshold for enamel indicate that the mechanism of ablation is thermal and occurs at approximately 300 to 400°C for Er:YAG, 800°C for Er:YSGG, and 1200°C for $CO_2$ lasers.[119] Although it is advantageous to ablate enamel at lower surface temperatures to avoid heat deposition, it may also be useful to heat the tissue peripheral to the ablation

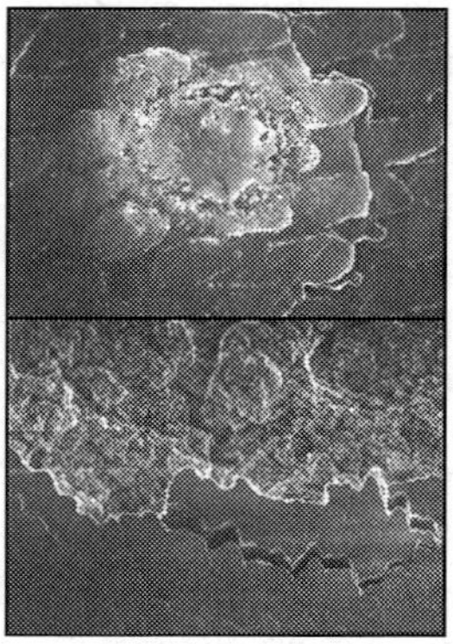

**FIGURE 50.8** Top: SEM image (5000×) of an enamel surface irradiated by a pulsed 9.3-μm $CO_2$ laser near the surface-modification threshold.[28] The enamel prisms are beginning to fuse together as a result of melting of the mineral phase. Bottom: SEM image (2000×) of an enamel surface irradiated by a pulsed 2.94-μm Er:YAG laser just above surface modification threshold. Intact enamel prisms are ejected without melting of the surface.

site to temperatures exceeding 400°C. This creates a zone of increased acid resistance around the restoration site, which may occur to a greater degree after Er:YSGG and $CO_2$ laser irradiation[120] than with Er:YAG laser irradiation.

## 50.3.3 Safety Issues

Heat may accumulate in the tooth to dangerous levels during multiple pulse-laser irradiation. Hence, it is necessary to measure accumulation of heat in the tooth during ablation. The 1965 study by Zach and Cohen on the effect of heat on the pulp of Rhesus monkeys indicated that a temperature rise of 5.5°C in the pulp caused irreversible pulpitis in 15% of the pulps.[121] This 5.5°C temperature rise is typically used as a temperature threshold that should not be exceeded. Higher temperatures may cause thermal damage to the pulp. The accumulation of heat after use of the dental drill for cavity preparations[122] and for the finishing of restorative materials[123] can raise the pulpal temperatures to dangerous levels if air/water cooling is not used. Similarly, the use of multiple pulse irradiation without air/water cooling may also result in the accumulation of heat to levels dangerous to the pulp. Subsurface thermocouple measurements, simulations of heat conduction, and histological examinations during laser irradiation show that the extent of pulpal heating is determined by the rate of deposition of the laser energy in the tooth, the distance from the laser spot to the pulp, and the rate of energy loss from the tooth.[103,105–107,124–127] Miserendino et al.[128] observed that typically the temperature excursions in the pulpal wall occurred 10 to 20 sec after irradiation of the surface of the tooth. Applying air/water cooling for 5 sec after the 5-sec CW ($\lambda$ =10.6 μm) $CO_2$ laser exposures of 10 to 50 J (2 to 10 W and 1-mm spot size) reduced the otherwise excessive temperature rise in the pulp chamber[128] to a safe level.

Strong acoustic waves can be generated during ablation.[129–132] These pressure waves can propagate through the tissue, radiating outward from the site of absorption. Transient stress waves may cause mechanical damage to the surface enamel or dentin surrounding the ablation site, generate cracks, and increase porosity. Generally, biological tissue has a relatively high compressive strength; however; the tensile strength is much weaker, and the tensile forces can cavitate soft tissue and generate cracks in hard tissue. A unipolar compressive stress incident on the interface with another tissue of lower density becomes a bipolar stress wave. The compressive stress wave propagates through the next medium, and a tensile wave or rarefaction wave is reflected back from the interface. It is the reflected tensile-stress wave that has the greatest potential for causing tissue damage. The additional contribution of the explosive release of inertially confined water and $CO_2$ during laser heating and the subsequent ablative recoil may also generate large acoustic transients. The magnitude of the stresses generated in enamel and dentin

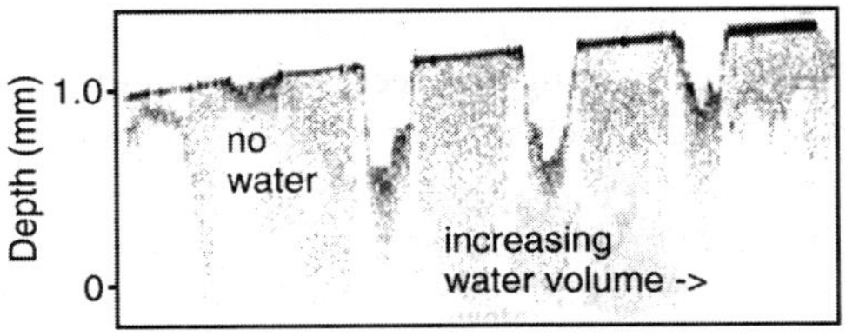


**FIGURE 50.9** Optical cross sections taken with OCT of incisions produced in enamel at a fluence of 100 J/cm$^2$ with a free-running Er:YAG laser with the same number of laser pulses. The depth of the incision varies markedly with the thickness of the applied water layer. The depth of the crater is a minimum with no water and peaks for a critical water thickness and decreases with greater thickness of the water layer.

during $CO_2$ laser irradiation have not been measured, and the effect of varying the pulse duration, fluence, and absorption depth have not been evaluated.

### 50.3.4 Water Augmentation of Ablation

One of the most important and somewhat controversial aspects of hard-tissue ablation is the critical role of water in the ablation process. Initial studies of the role of water focused on preventing tissue dehydration.[133–136] Absorption and diffusion studies in enamel indicate that approximately half of the water in enamel is actually diffusible.[90] Thermal-analysis studies indicate that the tissue must be heated to greater than 200 to 300°C before most of that mobile water is removed.[24] However, the rate for water diffusion is quite slow, on the order of several hours to days; therefore, it is unlikely that sample rehydration has a significant effect during laser irradiation. Several studies have shown that the ablation rate of erbium lasers can be increased by applying a water spray or a layer of water to the surface.[137–139] This phenomena can be seen in Figure 50.9, which shows "optical" cross sections of incisions in enamel produced with a free-running Er:YAG laser with varying volume of water added to the ablation site prior to irradiation.

Several interesting mechanisms have been proposed to explain this phenomenon. One hypothesis is that cavitation bubbles are formed in the water that cut the sound enamel due to the large tensile stresses generated by the collapse of the bubble.[137] Another proposed mechanism suggests that water droplets are rapidly accelerated into the enamel by absorption of the laser beam.[138,140] Other studies suggest that solid particles of ablated material are accelerated against the walls of the crater, resulting in a polishing effect that removes debris and any protruding sharp edges.[141,142]

During high-intensity laser irradiation, marked chemical and physical changes may be induced in the irradiated dental enamel. These changes can have profound effects on the laser ablation/drilling process and may lead to a reduction in the ablation rate and efficiency, increase peripheral thermal damage, and even lead to stalling without further removal of tissue with subsequent laser pulses. Furthermore, thermal decomposition of the mineral can lead to changes in the susceptibility of the modified mineral to organic acids in the oral environment. Morphological changes may result in the formation of loosely attached layers of modified enamel that can delaminate, leading to failure during bonding to restorative materials.[143,144] Therefore, it is important to thoroughly characterize the laser (thermal)-induced chemical and crystalline changes after laser irradiation. The mineral CAP, found in bone and teeth, contains carbonate inclusions that render it highly susceptible to acid dissolution by organic acids generated from bacteria in dental plaque. Upon heating to temperatures in excess of 400°C, the mineral decomposes to form a new mineral phase that has increased resistance to acid dissolution.[77] Recent studies suggest that as a side effect of laser ablation the walls around the periphery of a cavity preparation will be transformed through laser heating into a more acid-resistant phase with an enhanced resistance to future decay.[83,84,145] However, poorly crystalline nonapatite phases of calcium phosphate may have an opposite effect on plaque acid resistance[146] and may increase the quantity of poorly attached grains associated with delamination failures.

Zeck et al.[99] and Rechmann et al.[144] observed that a loosely attached layer of fused enamel was formed during Er:YAG laser ablation if a water spray was not used. Zeck labeled this fused enamel "recrystallites."

Some of the likely sources of these phases are recondensation of vaporized enamel, spallated droplets of ejected melted enamel, and repeated melting and recrystallization of the enamel.[147] Poorly crystalline-fused enamel particles and any surface protrusions or asperities in the ablation crater are likely to inhibit efficient ablation for subsequent laser pulses, leading to stalling and excessive heat accumulation. Forces imparted to the enamel surface by recoiling water particles are not of a sufficient magnitude to ablate the normal intact enamel; however, they may have sufficient strength to cleanse the surface of the crater of these poorly crystalline, loosely adherent mineral phases after the preceding laser pulses. The same recoil forces produce cavitation of the water layer[148–150] and energetically propel the remaining water on the surface several centimeters from the tooth.

## 50.3.5 Laser Ablation of Dentin and Bone

Dentin is composed of an interwoven network of water, small mineral crystals a few nanometers across, and collagen fibers. Intertubular dentin is approximately 47% by volume mineral (CAP), 33% protein (mostly collagen), and 20% water. Bone has a lower volume fraction of mineral and is less dense than dentin. Dentin contains long tubules that are 1 to 3 μm in diameter surrounded by a collar of mineral (peritubular dentin). These tubules extend from the DEJ to the pulp. This uniform orientation of the tubules in dentin facilitates visualization of thermal damage using polarized-light microscopy. In contrast, thermal damage in bone is more difficult to visualize using polarized light due to the lack of uniform orientation of the concentric lamellae. The collagen in dentin and bone is susceptible to thermal denaturation and carbonization (charring). Carbonization and thermal denaturation of dentin may adversely affect the bonding of cements and composite restorative materials and inhibit healing. This is of particular importance for the treatment of root caries, which is an increasingly important problem for our aging population. There have been several studies of bonding restorative materials to laser-treated enamel and dentin.[81,151–154]

Although the emphasis of this chapter is on caries ablation, the optical properties of bone are similar to dentin, and the laser conditions suitable for the ablation of dentin apply equally to bone. Several procedures in oral medicine require the removal and contouring of bone. Thermal damage to bone during laser cutting results in delayed healing time and bone defects.[155,156] Conventional technology often results in mechanical trauma, excessive thermal damage, and hemorrhage.[157,158] Specific areas of application include trepanation for implant placement, precise incisions for distraction osteogenesis for facial reconstruction and correction of craniofacial anomalies, and bone removal and recontouring for periodontology and oral surgery.

Free-running Er:YAG lasers can be used to ablate tissue fairly efficiently in conjunction with a water spray without significant charring of surrounding tissue.[135,159] Absorption in dentin and bone at erbium-laser wavelengths is markedly higher than in enamel because of the higher water content. Absorption is confined to a depth of 4 to 5 μm with a thermal relaxation time of 20 to 30 μs. However, the absorption coefficient of water at 2.94 μm drops by almost an order of magnitude upon heating from room temperature to the critical point at 374°C.[160,161,162] Thus, it is reasonable to assume that absorption occurs to depths greater than 5 μm during laser irradiation; however, that must be confirmed experimentally.

Early studies using CW $CO_2$ lasers operated at 10.6 μm reported extensive cracking and charring of surrounding dentin and bone.[155,156,163–165] These disappointing initial observations caused many laser researchers to overlook the potential of $CO_2$ laser–based systems for hard-tissue ablation. Pulsed $CO_2$ lasers, however, are well suited for use on dental hard tissue due to strong absorption of water, protein, and mineral between 9 and 11 μm. Several researchers[112,166–170] have demonstrated that pulsed $CO_2$ lasers with a pulse duration of less than 20 μs can be used to avoid carbonization of the biopolymer matrix of dentin with the simultaneous application of water cooling. Figure 50.10 shows an incision in dentin with a short-pulsed 9.6-μm $CO_2$ laser. No charring is visible, and the peripheral thermal damage is less than 10 μm. Mineral absorption peaks at 9.6 μm, where the absorption coefficient of dentin is approximately 8000 $cm^{-1}$. This high absorption coefficient indicates that the incident laser light will be absorbed at a depth of under 2 to 3 μm, with a corresponding thermal relaxation time on the order of 5 μs. Ivanenko

**FIGURE 50.10** Top: Incision in dentin produced with a 9.6-μm $CO_2$ laser with a pulse duration of 8 μs. No char, cracks, or discoloration is visible. Bottom: A cross section viewed under polarized light indicates that the zone of thermal damage is less than 10 μm. (From Lee, C., Ragadio, J., and Fried, D., in *Lasers in Dentistry VI,* SPIE, Vol. 3910, 2000, pp. 193–203. With permission.)

and Hering[171] demonstrated that a mechanically Q-switched 10.6-μm $CO_2$ industrial laser could be used to cut bone rapidly at 300 Hz without thermal damage; thus, rapid processing is feasible.

## 50.3.6 Alternative Mechanisms and Wavelengths for Hard-Tissue Ablation

The optical properties of dental hard tissues from $\lambda = 0.4$ to 2.7 μm are characterized by weak absorption and strong scattering and are not well suited for efficient ablation, which requires localization of the energy deposition to the surface of the tissue. Nevertheless, there has been some successful use of lasers emitting within this frequency range for ablation, and they warrant discussion. The absorption of laser light can be increased by adding exogenous chromophores to the surface.[172–174] However, even with the added chromophore, typical ablation thresholds are still high (200 J/cm$^2$) and ablation efficiencies low (0.03 mm$^3$/J).[173] There have been some successes in the removal of dentin using the fiber-coupled neodymium:YAG (Nd:YAG) laser in the contact mode.[175–178] However, this laser wavelength cannot be used to remove sound enamel. Niemz[179] has used picosecond laser pulses to ablate carious and noncarious enamel. The short, highly focused laser pulses generate irradiation in excess of $10^{10}$ W per square centimeter, and the strong electric fields ionize the tissue, forming a plasma.[180,181] Plasma-mediated ablation has several advantages: it produces very clean highly machined craters with minimal peripheral thermal damage, and suitable visible and NIR lasers are available with high spatial beam quality and short pulse durations. Such laser systems, however, are expensive, sophisticated, and yield rather low ablation rates.[182]

UV radiation is highly absorbed by collagen and is efficiently absorbed by dentin.[10,183–187] Hennig et al.[188,189] and Rechmann et al.[190] used the frequency-doubled alexandrite laser (377 nm) to selectively ablate carious dentin. They found at least a fourfold increase in the ablation rate of carious dentin over healthy dentin. There is strong absorption and minimal penetration at $\lambda = 193$ and 248 nm;[184,191] however, the ablation efficiency is very low — 0.001 mm$^3$/J and 0.002 mm$^3$/J for enamel and dentin, respectively.[183,192]

## 50.3.7 Selective Removal of Composite-Secondary Caries

Dental composites are used as adhesives and restorative materials for various dental procedures. They are composed of acrylic resins combined with quartz, zirconia silica, and fused silica filler materials. The most common application of composites is for restorative procedures, i.e., fillings. Composites are color-matched to the tooth, making it difficult for the dentist to differentiate between the enamel and the restorative material. Most restorations will eventually fail and need to be replaced. All of the existing composite from a restoration must be removed to assure proper bonding of the new composite to the tooth structure because new composite does not bond well to residual composite. It is difficult for the clinician to differentiate between the composite of the restoration and the surrounding tooth structure, particularly near the base of the restoration. Hence the dentist frequently removes excessive amounts of healthy tooth structure to ensure complete removal of the composite. The applicability of lasers for the removal of dental composite and bone cement has been investigated for several laser systems including the free-running Er:YAG,[193,194] Nd:YAG,[195,196] XeCl (308 nm),[197] and millisecond-pulsed $CO_2$ lasers,[198] all of which can be used to remove composites, as indicated in the articles listed above; however; they lack selectivity, have undesirable ablation rates, or produce excessive thermal damage. Recently, greater selectivity has been demonstrated using the transverse excited atmospheric pressure (TEA) $CO_2$ laser operating at 10.6 μm[199] and the frequency-tripled Nd:YAG laser operating at 355 nm.[200] In those studies, a TEA $CO_2$ laser with laser pulses of approximately 1-μs duration was used to ablate composite at a rate almost an order of magnitude higher than for enamel. Spectral analysis of the emission plume created by ablating composite and enamel identified a number of spectral lines that could be used to distinguish between the ablation of enamel and composite. However, there was measurable damage to the underlying enamel due to the similar ablation thresholds. Laser pulses of the third harmonic of the Q-switched Nd:YAG (355 nm) can be used to completely remove composite from the surface of the tooth with no discernible damage to the underlying enamel, thereby achieving the desired selectivity.[200]

### 50.3.8 Surface Modification for Caries Prevention — Enamel

The earliest dental laser studies of Yamamoto and Ooya[201] and Stern et al.[202] showed that high-intensity laser radiation rendered enamel mineral more resistant to acid dissolution.[201–203] However, such studies utilized ruby, Nd:YAG (1064 nm), and continuous $CO_2$ lasers at energy levels that produced excessive heat accumulation and extensive peripheral damage, i.e., stress cracking, which prohibited their clinical implementation. More recently, pulsed infrared lasers with wavelength and pulse duration optimally matched to the optical properties of enamel have been used to demonstrate that the transformation of CAP to a more acid-resistant mineral phase is indeed practical with minimal heat deposition in the tooth.[77,203] Several studies over the past 20 years have alluded to the mechanism of inhibition. Thermal-analysis studies of Holcomb and Young[24] and subsequent studies by Fowler and Kuroda[146] indicated that there was substantial loss of carbonate and water at temperatures between 100 and 400°C, which were sufficient to change the crystallinity of the intrinsic mineral. Kuroda and Fowler[204] observed a reduction in the carbonate content of enamel by 66% after irradiation with a CW $CO_2$ laser. Heat treatment between 350 and 650°C reduced the solubility of enamel to acid dissolution,[205–209] even though the permeability was increased. Featherstone et al.[210] correlated the carbonate loss from enamel measured with FTIR with the inhibition of the dissolution rate, providing strong evidence for the transformation of the mineral of enamel to a purer phase. Therefore, the decreased solubility after infrared irradiation can most likely be attributed to the thermal decomposition of the more soluble CAP into the less soluble HAP with corresponding changes in the crystallinity. Other studies have suggested that the production of pyrophosphate, permeability changes, and modification of the protein matrix play a role in inhibition.[205,206,211–213]

Levels of inhibition exceeding 70% have been achieved with $CO_2$ irradiation[214] and levels close to 100% have been achieved by combining laser treatments with fluoride application.[77,172,215] Caries inhibition has also been demonstrated for the erbium-laser wavelengths when a water spray was not employed.[120] The inhibition after Er:YSGG (2.79 μm) laser irradiation (13 J/cm$^2$) was 60%, which approaches the best results obtained for $CO_2$ laser irradiation or daily application of a fluoride dentifrice (70 to 85%).[1–3] Studies of heat-treated and $CO_2$ laser–irradiated enamel have shown that significant chemical and structural changes are induced in enamel at temperatures well below the melting point.[24,146,206,209,215,216] The principal changes are loss of water and carbonate, crystal sintering and fusion, permeability changes, and changes in the crystallinity. Therefore, it may not be necessary to melt and recrystallize the enamel for effective caries-preventive treatments.

### 50.3.9 Surface Modification for Caries Prevention — Dentin

Kantola et al.[217] showed 30 years ago that lasers could be used to increase the mineral content and crystallinity of dentin by preferential removal of the inherent water and protein. Since that time investigators have evaluated the susceptibility of dentin modified by various laser systems to artificial caries-like lesion formation. Nammour et al.[218] and Kinney et al.[219] used CW $\lambda = 10.6$ μm, $CO_2$, and pulsed Nd:YAG lasers operating at very high intensities to irradiate dentin and create a hypermineralized layer of dentin 50 to 100 μm thick with greater than 80% mineral composition on the outer surface. This surface layer was resistant to acid dissolution; however, extensive cracking of the surface was observed, and there was some subsurface demineralization, albeit at a somewhat reduced rate in the study of Nammour et al.[218] Apparently, the laser irradiation did not completely seal the dentinal tubule lumen in those studies; this allowed acid to diffuse through the large surface cracks created during irradiation and the tubules to a depth below the acid-resistant modified layer.

## 50.4 The Future

Dentists are already using lasers and optical diagnostic technology for caries detection and diagnostics and lasers for therapeutic intervention. By combining these methods future dentists may usher in a new era of conservative dentistry in which caries lesions are detected and assessed before they have progressed to the point where the removal of significant amounts of tissue is required. Subsequent to diagnosis lasers will be used to selectively, precisely, and painlessly remove the carious (decayed) tissue while minimizing the loss of healthy tissue. The mineral phase of the surrounding enamel in these high-risk areas will be transformed through laser heating into a more acid-resistant phase to have an enhanced resistance to future decay. Thus, future patients may not have to endure the current standard of treatment that results in the loss of excessive amounts of healthy tissue structure, which thereby affects the appearance and compromises the mechanical integrity of the tooth.

### Acknowledgments

The author acknowledges John D.B. Featherstone, Wolf Seka, Charles Le, John Xie, and Michael Zuerlein for their many contributions to this work, and the NIH/NIDCR for support of the author's dental-photonics research.

### References

1. Zijp, J.R. and ten Bosch, J.J., Angular dependence of HeNe laser light scattering by bovine and human dentine, *Arch. Oral Biol.*, 36, 283, 1991.
2. Zijp, J.R. and ten Bosch, J.J., Theoretical model for the scattering of light by dentin and comparison with measurements, *Appl. Opt.*, 32, 411, 1993.
3. Fried, D., Featherstone, J.D.B., Glena, R.E., and Seka, W., The nature of light scattering in dental enamel and dentin at visible and near-IR wavelengths, *Appl. Opt.*, 34(7), 1278, 1995.
4. Zijp, J.R., ten Bosch, J.J., and Groenhuis, R.A., HeNe laser light scattering by human dental enamel, *J. Dent. Res.*, 74, 1891, 1995.
5. Cheong, W., Prahl, S.A., and Welch, A.J., A review of the optical properties of biological tissues, *IEEE J. Quantum Electron.*, 26, 2166, 1990.
6. Wilson, B.C., Patterson, M.S., and Flock, S.T., Indirect versus direct techniques for the measurement of the optical properties of tissues, *Photochem. Photobiol.*, 46(5), 601, 1987.
7. Van der Zee, P., Methods for measuring the optical properties of tissue samples in the visible and near-infrared wavelength range, in *Medical Optical Tomography: Functional Imaging and Monitoring*, Muller, G., Chance, B., Alfano, R., Arridge, S., Beuthan, J., Gratton, E., Kaschke, M., Masters, B., Svanberg, S., and van der Zee, P., Eds., SPIE, Bellingham, WA, 1993, pp. 450–472.

8. Tuchin, V., *Tissue Optics: Light Scattering Methods and Instruments for Medical Diagnostics*, SPIE, Bellingham, WA, 2000.
9. Spitzer, D. and ten Bosch, J.J., The absorption and scattering of light in bovine and human dental enamel, *Calcif. Tissue Res.*, 17, 129, 1975.
10. ten Bosch, J.J. and Zijp, J.R., Optical properties of dentin, in *Dentine and Dentine Research in the Oral Cavity*, Thylstrup, A., Leach, S.A., and Qvist, V., Eds., IRL Press, Oxford, U.K., 1987, pp. 59–65.
11. Jones, R. and Fried, D., Attenuation of 1310 and 1550-nm laser light through dental enamel, in *Lasers in Dentistry VIII*, Rechmann, P., Fried, D., and Hennig, T., Eds., SPIE, Vol. 4610, San Jose, 2002, pp. 187–190.
12. Fried, D., Featherstone, J.D.B., Glena, R.E., Bordyn, B., and Seka, W., The light scattering properties of dentin and enamel at 543, 632, and 1053 nm, in *Lasers in Orthopedic, Dental, and Veterinary Medicine II*, Dov Gel, D.V.M., O'Brien, S.J., Vangsness, C.T., White, T., and Wiglor, H.A. Eds., SPIE, Vol. 1880, Los Angeles, 1993, pp. 240–245.
13. Jacques, S.L., Alter, C.A., and Prahl, S.A., Angular dependence of HeNe light scattering by human dermis, *Lasers Life Sci.*, 1, 309, 1987.
14. ten Bosch, J.J., van der Mei, H.C., and Borsboom, P.C.F., Optical monitor of *in vitro* caries, *Caries Res.*, 18, 540, 1984.
15. Angmar-Mansson, B. and ten Bosch, J.J., Optical methods for the detection and quantification of caries, *Adv. Dent. Res.*, 1(1), 14, 1987.
16. ten Bosch, J.J. and Coops, J.C., Tooth color and reflectance as related to light scattering and enamel hardness, *J. Dent. Res.*, 74, 374, 1995.
17. Ko, C.C., Tantbirojn, D., Wang, T., and Douglas, W.H., Optical scattering power for characterization of mineral loss, *J. Dent. Res.*, 79(8), 1584, 2000.
18. Schmidt, W.J. and Keil, A., *Polarizing Microscopy of Dental Tissues*, Pergamon Press, New York, 1971.
19. Theuns, H.M., Shellis, R.P., Groeneveld, A., van Dijk, J.W.E., and Poole, D.F.G., Relationships between birefringence and mineral content in artificial caries lesions in enamel, *Caries Res.*, 27, 9, 1993.
20. Duplain, G., Boulay, R., and Belanger, P.A., Complex index of refraction of dental enamel at $CO_2$ wavelengths, *Appl. Opt.*, 26, 4447, 1987.
21. Zuerlein, M., Fried, D., Featherstone, J., and Seka, W., Optical properties of dental enamel at 9–11 μm derived from time-resolved radiometry, *Spec. Top. IEEE J. Quantum Electron.*, 5(4), 1083, 1999.
22. Fried, D., Zuerlein, M.J., Le, C.Q., and Featherstone, J., Thermal and chemical modification of dentin by 9–11 μm $CO_2$ laser pulses of 5–100-μs duration, *Lasers Surg. Med.*, 31, 275, 2002.
23. Duck, F.A., *Physical Properties of Tissue*, Academic Press, San Diego, 1990.
24. Holcomb, D.W. and Young, R.A., Thermal decomposition of human tooth enamel, *Calcif. Tissue Int.*, 31, 189, 1980.
25. Wang, T., Xu, H.C., and Lin, W.Y., Measurements of thermal expansion coefficient of human teeth, *Aust. J. Dent.*, 34(6), 530, 1989.
26. Zuerlein, M. and Fried, D., Depth profile analysis of the chemical and morphological changes in $CO_2$-laser irradiated dental enamel, in *Lasers in Dentistry V*, SPIE, Vol. 3593, San Jose, 1999, pp. 204–210.
27. Fried, D., Glena, R.E., Featherstone, J.D.B., and Seka, W., Permanent and transient changes in the reflectance of $CO_2$ laser irradiated dental hard tissues at $\lambda = 9.3$, 9.6, 10.3, and 10.6 μm and at fluences of 1–20 J/cm$^2$, *Lasers Surg. Med.*, 20, 22, 1997.
28. Fried, D., Zuerlein, M., Featherstone, J.D.B., Seka, W.D., and McCormack, S.M., IR laser ablation of dental enamel:mechanistic dependence on the primary absorber, *Appl. Surf. Sci.*, 128, 852, 1997.
29. Chauncey, H.H., Glass, R.L., and Alman, J.E., Dental caries, principal cause of tooth extraction in a sample of US male adults, *Caries Res.*, 23, 200, 1989.
30. Kaste, L.M., Selwitz, R.H., Oldakowski, R.J., Brunelle, J.A., Winn, D.M., and Brown, L.J., Coronal caries in the primary and permanent dentition of children and adolescents 1–17 years of age: United States, 1988–1991, *J. Dent. Res.*, 75, 631, 1996.

31. Winn, D.M., Brunelle, J. A., Selwitz, R.H., Kaste, L.M., Oldakowski, R.J., Kingman, A., and Brown, L.J., Coronal and root caries in the dentition of adults in the United States, 1988–1991, *J. Dent. Res.*, 75, 642, 1996.
32. NIH Report No. 18, 2001.
33. Lussi, A., Validity of diagnostic and treatment decisions of fissure caries, *Caries Res.*, 25, 296, 1991.
34. Featherstone, J.D.B. and Young, D., The need for new caries detection methods, in *Lasers in Dentistry V*, Featherstone, J.D.B., Rechmann, P., and Fried, D., Eds., SPIE, Vol. 3593, San Jose, 1999, pp. 134–140.
35. Hume, W.R., Need for change in dental caries diagnosis, in *Early Detection of Dental Caries*, Stookey, G.K., Ed., Indiana University, Indianapolis, 1996, pp. 1–10.
36. Featherstone, J.D.B., Clinical implications: new strategies for caries prevention, in *Early Detection of Dental Caries*, Stookey, G.K., Ed., Indiana University, Indianapolis, 1996, pp. 287–296.
37. ten Cate, J.M. and van Amerongen, J. P., Caries diagnosis: conventional methods, in *Early Detection of Dental Caries*, Stookey, G.K., Ed., Indiana University, Indianapolis, 1996, pp. 27–37.
38. Featherstone, J.D.B., Prevention and reversal of dental caries:role of low level fluoride, *Community Dent. Oral Epidemiol.*, 27, 31, 1999.
39. Kidd, E.A.M., Ricketts, D.N.J., and Pitts, N.B., Occlusal caries diagnosis: a changing challenge for clinicians and epidemiologists, *J. Dent. Res.*, 21, 3232, 1993.
40. Wenzel, A., New caries diagnostic methods, *J. Dent. Ed.*, 57, 428, 1993.
41. Pine, C.M., Fiber-optic transillumination (FOTI) in caries diagnosis, in *Early Detection of Dental Caries*, Stookey, G.K., Ed., Indiana University, Indianapolis, 1996, pp. 51–66.
42. Pine, C.M. and ten Bosch, J.J., Dynamics of and diagnostic methods for detecting small carious lesions, *Caries Res.*, 30, 381, 1996.
43. Peers, A., Hill, F.J., Mitropoulos, C.M., and Holloway, P.J., Validity and reproducibility of clinical examination, fibre-optic transillumination, and bite-wing radiology for the diagnossis of small approximal carious lesions, *Caries Res.*, 27, 307, 1993.
44. Peltola, J. and Wolf, J., Fiber optics transillumination in caries diagnosis, *Proc. Finn. Dent. Soc.*, 77, 240, 1981.
45. Barenie, J., Leske, G., and Ripa, L.W., The use of fiber optic transillumination for the detection of proximal caries, *Oral Surg.*, 36, 891, 1973.
46. Holt, R.D. and Azeevedo, M.R., Fiber optic transillumination and radiographs in diagnosis of approximal cariesin primary teeth, *Community Dent. Health*, 6, 239, 1989.
47. Mitropoulis, C.M., The use of fiber optic transillumination in the diagnosis of posterior approximal caries in clinical trials, *Caries Res.*, 19, 379, 1985.
48. Schneiderman, A., Elbaum, M., Schultz, T., Keem, S., Greenebaum, M., and Driller, J., Assessment of dental caries with digital imaging fiber-optic transillumination (DIFOTI): *in vitro* study, *Caries Res.*, 31, 103, 1997.
49. Alfano, R.R. and Yao, S.S., Human teeth with and without dental caries studied by visible luminescent spectroscopy, *J. Dent. Res.*, 60, 120, 1981.
50. Alfano, R.R., Lam, W., Zarrabi, H.J., Alfano, M.A., Cordero, J., and Tata, D.B., Human teeth with and without caries studied by laser scattering, fluorescence and absorption spectroscopy, *IEEE J. Quantum Electron.*, 20, 1512, 1984.
51. Bjelkhagen, H. and Sundstrom, F., A clinically applicable laser luminescence for the early detection of dental caries, *IEEE J. Quantum Electron.*, 17, 266, 1981.
52. Hafstroem-Bjoerkman, U., Sundstroem, F., de Josselin de Jong, E., Oliveby, A., and Angmar-Mansson, B., Comparison of laser fluorescence and longitudinal microradiography for quantitative assessment of *in vitro* enamel caries, *Caries Res.*, 26, 241, 1992.
53. Ando, M., Eckert, A.F., Schemehorn, B.R., Analoui, M., and Stookey, G.K., Relative ability of laser fluorescence techniques to quantitate early mineral loss *in vitro*, *Caries Res.*, 31, 125, 1997.

54. De Jong, E.d.J., Sundstrom, F., Westerling, H., Tranaeus, S., ten Bosch, J.J., and Angmar-Mansson, B., A new method for *in vivo* quantification of changes in initial enamel caries with laser fluorescence, *Caries Res.*, 2, 1995.
55. Lagerweij, M.D., van der Veen, M.H., Ando, M., and Lukantsova, L., The validity and repeatability of three light-induced fluorescence systems: an *in vitro* study, *Caries Res.*, 33, 220, 1999.
56. Van der Veen, M.H., de Josselin de Jong, E., and Al-Kateeb, S., Caries activity detection by dehydration with qualitative light fluorescence, in *Early Detection of Dental Caries II*, Vol. 4, Stookey, G.K., Ed., Indiana University, Indianapolis, 1999, pp. 251–260.
57. Koenig, K., Schneckenburger, H., Hemmer, J., Tromberg, B.J., Steiner, R.W., and Rudolf, W., *In-vivo* fluorescence detection and imaging of porphyrin-producing bacteria in the human skin and in the oral cavity for diagnosis of acne vulgaris, caries, and squamous cell carcinoma, in *Advances in Laser and Light Spectroscopy to Diagnose Cancer and Other Diseases*, Vol. 2135, Alfano, R.R., Ed., SPIE, San Jose, 1994, pp. 129–138.
58. Hibst, R., Paulus, R., and Lussi, A., Detection of occlusal caries by laser fluorescence: basic and clinical investigations, *Med. Laser Appl.*, 16, 205, 2001.
59. Shi, X.Q., Welander, U., and Angmar-Mansson, B., Occlusal caries detection with Kavo DIAGNOdent and radiography: an *in vitro* comparison, *Caries Res.*, 34, 151, 2000.
60. Hibst, R. and Paulus, R., New approach on fluorescence spectroscopy for caries detection, in *Lasers in Dentistry V*, Vol. 3593, Featherstone, R. and Fried, D., Eds., SPIE, San Jose, 1999, pp. 141–148.
61. Lussi, A., Imwinkelreid, S., Pitts, N.B., Longbottom, C., and Reich, E., Performance and reproducibility of a laser fluorescence system for detection of occlusal caries *in vitro*, *Caries Res.*, 33, 261, 1999.
62. Colston, B., Everett, M., Da Silva, L., Otis, L., Stroeve, P., and Nathel, H., Imaging of hard and soft tissue structure in the oral cavity by optical coherence tomography, *Appl. Opt.*, 37(19), 3582, 1998.
63. Colston, B.W., Everett, M.J., Da Silva, L.B., and Otis, L.L., Optical coherence tomography for diagnosis of periodontal diseases, in *Coherence Domain Optical Methods in Biomedical Science and Clinical Applications II*, Tuchin, V.V. and Izatt, V.V., Eds., SPIE, Vol. 3251, San Jose, 1998, pp. 52–58.
64. Colston, B.W., Sathyam, U.S., Da Silva, L.B., Everett, M.J., and Stroeve, P., Dental OCT, *Opt. Express*, 3(3), 230, 1998.
65. Baumgartner, A., Hitzenberger, C.K., Dicht, S., Sattmann, H., Moritz, A., Sperr, W., and Fercher, A.F., Optical coherence tomography for dental structures, in *Lasers in Dentistry IV*, Featherstone, R. and Fried, Eds., SPIE, Vol. 3248, San Jose, 1998, pp. 130–136.
66. Baumgartner, A., Dicht, S., Hitzenberger, C.K., Sattmann, H., Robi, B., Moritz, A., Sperr, W., and Fercher, A.F., Polarization-sensitive optical optical coherence tomography of dental structures, *Caries Res.*, 34, 59, 2000.
67. Feldchtein, F.I., Gelikonov, G.V., Gelikonov, V.M., Iksanov, R.R., Kuranov, R.V., Sergeev, A.M., Gladkova, N.D., Ourutina, M.N., Warren, J.A., and Reitze, D.H., *In vivo* OCT imaging of hard and soft tissue of the oral cavity, *Opt. Express*, 3(3), 239, 1998.
68. Wang, X.J., Zhang, J.Y., Milner, T.E., Boer, J. F.d., Zhang, Y., Pashley, D.H., and Nelson, J.S., Characterization of dentin and enamel by use of optical coherence tomography, *Appl. Opt.*, 38(10), 586, 1999.
69. Everett, M.J., Colston, B.W., Sathyam, U.S., Silva, L.B.D., Fried, D., and Featherstone, J.D.B., Non-invasive diagnosis of early caries with polarization sensitive optical coherence tomography (PS-OCT), in *Lasers in Dentistry V*, Featherstone, J.D.B., Rechmann, P., and Fried, D., Eds., SPIE, Vol. 3593, San Jose, 1999, pp. 177–183.
70. Kidd, E.A.M., Secondary caries, *Int. Dent. J.*, 42, 127, 1992.
71. Fried, D., Xie, J., Shafi, S., Featherstone, J., Breunig, T.M., and Le, C.Q., Imaging caries lesions and lesion progression with polarization optical coherence tomography, in *Lasers in Dentistry VIII*, Rechmann, P., Fried, D., and Hennig, T., Eds., SPIE, Vol. 4610, San Jose, 2002, pp. 113–124.
72. Fried, D., Xie, J., Shafi, S., Featherstone, J.D.B., Breunig, T., and Lee, C.Q., Early detection of dental caries and lesion progression with polarization sensitive optical coherence tomography, *J. Biomed. Opt.*, 7(4), 618, 2002.

73. Kinney, J.H., Balooch, M., Haupt, D.L., Marshall, S.J., and Marshall, G.W., Mineral distribution and dimensional changes in human dentin during demineralization, *J. Dent. Res.*, 74(5), 1179, 1995.
74. Kinney, J.H., Haupt, D.L., Nichols, M.C., Breunig, T.M., Marshall, G.W., and Marshall, S.J., The x-ray tomographic microscope: three-dimensional perspectives of evolving structures, *Nucl. Instrum. Meth. A,* 347, 480, 1994.
75. Stern, R.H. and Sognnaes, R.F., Laser beam effect on hard dental tissues, *J. Dent. Res.*, 43, 873, 1964.
76. Wigdor, H.A., Walsh, J.T., Featherstone, J.D.B., Visuri, S.R., Fried, D., and Waldvogel, J.L., Lasers in dentistry, *Lasers Surg. Med.*, 16, 103, 1995.
77. Featherstone, J.D.B. and Nelson, D.G.A., Laser effects on dental hard tissue, *Adv. Dent. Res.*, 1(1), 21, 1987.
78. Harris, N. and Garcia-Godoy, F., *Primary Preventive Dentistry*, 5th ed., Appleton & Lange, Stamford, CT, 1999.
79. Mertz-Fairhurst, E.J., Pit-and-fissure sealants: a global lack of scientific transfer? *J. Dent. Res.*, 115, 1543, 1992.
80. Mount, G.J. and Hume, W.R., *Preservation and Restoration of Tooth Structure*, Mosby, New York, 1998.
81. Cooper, L.F., Myers, M.L., Nelson, D.G.A., and Mowery, A.S., Shear strength of composite resin bonded to laser pretreated dentin, *J. Prosthet. Dent.*, 60, 45, 1988.
82. Featherstone, J.D.B., Barrett-Vespone, N.A., Kantorowitz, D.F., Lofthouse, J., and Seka, W., Rational choice of $CO_2$ laser conditions for inhibition of caries progression, in *Lasers in Dentistry*, Wigdor, H.A., Featherstone, J.D.B., and White, J.M., Eds., SPIE, Vol. 2394, San Jose, 1995, pp. 57–67.
83. Konishi, N., Fried, D., Featherstone, J.D.B., and Staninec, M., Inhibition of secondary caries by $CO_2$ laser treatment, *Am. J. Dent.*, 12(5), 213, 1999.
84. Young, D.A., Fried, D., and Featherstone, J.D.B., Ablation and caries inhibition of pits and fissures by IR laser irradiation, in *Lasers in Dentistry VI*, SPIE, Vol. 3910, San Jose, 2000, pp. 247–253.
85. Dostalova, T., Jelainkova, H., Kuacerova, H., Krejsa, O., Hamal, K., Kubelka, J., and Prochaazka, S., Noncontact Er:YAG laser ablation: clinical evaluation, *J. Clin. Laser Med. Surg.*, 16, 273, 1998.
86. Keller, U. and Hibst, R., Effects of Er:YAG laser in caries treatment: a clinical pilot study, *Lasers Surg. Med.*, 20, 32, 1997.
87. Pelagalli, J., Gimbel, C.B., Hansen, R.T., Swett, A., and Winn, D.W., Investigational study of the use of Er:YAG laser versus the drill for caries removal and cavity preparation: Phase 1, *J. Clin. Laser Med. Surg.*, 15, 109, 1997.
88. DenBesten, P.K., White, J. M., Pelino, J., Lee, K., and Parkins, F., A randomized prospective parallel controlled study of the safety and effectiveness of Er:YAG laser use in children for caries removal, in *Lasers in Dentistry VI*, SPIE, Vol. 3910, 2000, pp. 171–174.
89. Fried, D., Therapeutic lasers and optical diagnostic technology in dentistry, in *Optics Photonics News*, 1999, pp. 22–26.
90. Dibdin, G.H., The water in human dental enamel and its diffusional exchange measured by clearance of tritiated water from enamel slabs of varying thickness, *Caries Res.*, 27, 81, 1993.
91. Altshuler, G.B., Belikov, A.V., Erofeev, A.V., and Sam, R.C., Optimum regimes of laser destruction of human tooth enamel and dentin, in *Lasers in Orthopedic, Dental, and Veterinary Medicine II*, SPIE, Vol. 1880, Los Angeles, 1993, pp. 101–107.
92. Altshuler, G.B., Belikov, A.V., and Erofeev, A.V., Laser treatment of enamel and dentin by different Er-lasers, in *Lasers in Surgery: Advanced Characterization, Therapeutics, and Systems IV*, Anderson, R.R., Ed., SPIE, Vol. 2128, Los Angeles, 1994, pp. 273–281.
93. Hibst, R. and Keller, U., Mechanism of Er:YAG laser induced ablation of dental hard substances, in *Lasers in Orthopedic, Dental, and Veterinary Medicine II*, Dov Gel, D.V.M., O'Brien, S.J., Vangsness, C.T., White, T., and Wigdor, H.A., Eds., SPIE, Vol. 1880, 1993, pp. 156–162.
94. Hibst, R. and Keller, U., Experimental studies of the application of the Er:YAG laser on dental hard substances: I. Measurement of the ablation rate, *Lasers Surg. Med.*, 9, 338, 1989.
95. Keller, U. and Hibst, R., Experimental studies of the application of the Er:YAG laser on dental hard substances. II. Light microscopic and SEM investigations, *Lasers Surg. Med.*, 9, 345, 1989.

96. Kermani, O., Lubatschowski, H., Asshauer, T., Ertmer, W., Lukin, A., Ermakov, B., and Krieglstein, G.K., Q-switched CTE:YAG laser ablation: basic investigation on soft (corneal) and hard (dental) tissues, *Lasers Surg. Med.*, 13, 537, 1993.
97. Shori, R., Fried, D., Featherstone, J.D.B., and Duhn, C., CTE:YAG applications in dentistry, in *Lasers in Dentistry IV*, Featherstone, J.D.B., Rechmann, P., and Fried, D., Eds., SPIE, Vol. 3248, San, Jose, CA, 1998, pp. 86–91.
98. Majaron, M., Sustercic, D., and Lukac, M., Debris screening and heat diffusion in Er:YAG drilling of hard dental tissues, in *Lasers in Dentistry III*, Wigdor, H.A., Featherstone, J.D.B., and Rechmann, P., Eds., SPIE, Vol. 2973, San Jose, 1997, pp. 11–22.
99. Zeck, M., Benthin, H., Ertl, T., Siebert, G.K., and Muller, G., Scanning ablation of dental hard tissue with erbium laser radiation, in *Medical Applications of Lasers III*, Vol. 2623, 1996, pp. 94–102.
100. Fried, D. and Shori, R., Q-switched Er:YAG Ablation of Dental Hard Tissue, in *Proc. Sixth Int. Congr. Lasers Dent.*, University of Utah, Maui, 1998, pp. 77–79.
101. Lenz, P., Glide, H., and Walz, R., Studies on enamel sealing with the $CO_2$ laser, *Dtsch. Zahnarztl. Z.*, 37, 469, 1982.
102. Kantola, S., Laine, E., and Tarna, T., Laser-induced effects on tooth structure. VI. X-ray diffraction study of dental enamel exposed to a $CO_2$ laser, *Acta Odontol. Scand.*, 31, 369, 1973.
103. Melcer, J., Chaumette, M.T., and Melcer, F., Dental pulp exposed to the $CO_2$ laser beam, *Lasers Surg. Med.*, 7, 347, 1987.
104. Miserendino, L.J., Neiburger, E.J., Walia, H., Luebke, N., and Brantley, W., Thermal effects of continuous wave $CO_2$ laser exposure on human teeth: an *in vitro* study, *J. Endodont.*, 15(7), 302, 1989.
105. Leighty, S.M., Pogrel, M.A., Goodis, H.E., and White, J.M., Thermal effects of the carbon dioxide laser on teeth, *Lasers Life Sci.*, 4(2), 93, 1991.
106. Jeffrey, I.W.M., Lawrenson, B., Saunders, E.M., and Longbottom, C., Dentinal temperature transients caused by exposure to $CO_2$ laser irradiation and possible pulp damage, *J. Dent.*, 18, 31, 1990.
107. Neiburger, E.J. and Miserendino, L., Pulp chamber warming due to $CO_2$ laser exposure, *N.Y. State Dent. J.*, 54(3), 25, 1988.
108. Melcer, J., Latest treatment in dentistry by means of the $CO_2$ laser beam, *Lasers Surg. Med.*, 6, 396, 1986.
109. Palamara, J., Phakey, P.P., Orams, H.J., and Rachinger, W.A., The effect on the ultrastructure of dental enamel of excimer-dye, argon-ion and $CO_2$ lasers, *Scanning Microsc.*, 6(4), 1061, 1992.
110. Pogrel, M.A., Muff, D.F., and Marshall, G.W., Structural changes in dental enamel induced by high energy continuous wave carbon dioxide laser, *Lasers Surg. Med.*, 13, 89, 1993.
111. Ferreira, J.M., Palamara, J., Phakey, P.P., Rachinger, W.A., and Orams, H.J., Effects of continuous-wave $CO_2$ laser on the ultrastructure of human dental enamel, *Arch. Oral Biol.*, 34(7), 551, 1989.
112. Krapchev, V.B., Rabii, C.D., and Harrington, J.A., Novel $CO_2$ laser system for hard tissue ablation, in *Lasers in Surgery: Advanced Characterization, Therapeutics, and Systems IV*, Anderson, R.R., Ed., SPIE, Vol. 2128, 1994, pp. 341–348.
113. Ertl, T. and Muller, G., Hard tissue ablation with pulsed $CO_2$ lasers, in *Lasers in Orthopedic, Dental, and Veterinary Medicine II*, Dov Gel, D.V.M., O'Brien, S.J., Vangsness, C.T., White, T., and Wigdor, H.A., Eds., SPIE, Vol. 1880, 1993, pp. 176–181.
114. Lukac, M., Hocevar, F., Cencic, S., Nemes, K., Keller, U., Hibst, R., Sustercic, D., Gaspirc, B., Skaleric, U., and Funduk, N., Effects of pulsed $CO_2$ and Er:YAG lasers on enamel and dentin, in *Lasers in Orthopedic, Dental, and Veterinary Medicine II*, Dov Gel, D.V.M., O'Brien, S.J., Vangsness, C.T., White, T., and Wigdor, H.A., Eds., SPIE, Vol. 1880, Los Angeles, 1993, pp. 169–175.
115. Fried, D., Murray, M.W., Featherstone, J.D.B., Akrivou, M., Dickenson, K.M., and Duhn, C., Dental hard tissue modification and removal using sealed TEA lasers operating at $\lambda = 9.6$ μm, *J. Biomed. Opt.*, 196, 2001.
116. Fried, D., Murray, M.W., Featherstone, J.D.B., Akrivou, M., Dickenson, K.M., Duhn, C., and Ojeda, O.P., Dental hard tissue modification and removal using sealed TEA lasers operating at $\lambda = 9.6$ μm, in *Lasers in Dentistry V*, Featherstone, J.D.B., Rechmann, P., and Fried, D., Eds., SPIE, Vol. 3593, San Jose, 1999, pp. 196–203.

117. Wigdor, H., Walsh, J.T., and Mostofi, R., The effect of the $CO_2$ laser (9.6 μm) on the dental pulp in humans, in *Lasers in Dentistry VI*, SPIE, Vol. 3910, San Jose, 2000, pp. 158–163.
118. Albagli, D., Perelman, L.T., Janes, van Rosenburg, C., Itzkan, I., and Feld, M.S., Inertially confined ablation of biological tissue, *Lasers Life Sci.*, 6(1), 55, 1994.
119. Fried, D., Visuri, S.R., Featherstone, J.D.B., Seka, W., Glena, R.E., Walsh, J.T., McCormack, S.M., and Wigdor, H.A., Infrared radiometry of dental enamel during Er:YAG and Er:YSGG laser irradiation, *J. Biomed. Opt.*, 1(4), 455, 1996.
120. Fried, D., Featherstone, J.D.B., Visuri, S.R., Seka, W., and Walsh, J.T., The caries inhibition potential of Er:YAG and Er:YSGG laser radiation, in *Lasers in Dentistry II*, Wigdor, H.A., Featherstone, J.D.B., White, J.M., and Neev, J., Eds., SPIE, Vol. 2672, San Jose, 1996, pp. 73–78.
121. Zach, L. and Cohen, G., Pulp response to externally applied heat, *Oral Surg. Oral Med. Oral Pathol.*, 19, 515, 1965.
122. Hamilton, A.I. and Kramer, I.R.H., Cavity preparation with and without waterspray: effects on the human dental pulp and additional effects of further dehydration of the dentine, *Br. Dent. J.*, 126, 281, 1967.
123. Stewart, G.P., Bachman, T.A., and Hatton, J.F., Temperature rise due to finishing of direct restorative materials, *Am. J. Dent.*, 4, 23, 1991.
124. Anic, I., Vidovic, D., Luic, M., and Tudja, M., Laser induced molar tooth pulp chamber temperature changes, *Caries Res.*, 26, 165, 1992.
125. Nammour, S. and Pourtois, M., Pulp temperature increases following caries removal by $CO_2$ laser, *J. Clin. Laser Med. Surg.*, 13(5), 337, 1995.
126. Yu, D., Powell, G.L., Higuchi, W.I., and Fox, J.L., Comparison of three lasers on dental pulp chamber temperature change, *J. Clin. Laser Med. Surg.*, 11(3), 119, 1993.
127. Sandford, M.A. and Walsh, L.J., Differential thermal effects of pulsed vs. continuous $CO_2$ laser radiation on human molar teeth, *J. Clin. Laser Med. Surg.*, 12(3), 139, 1994.
128. Miserendino, L.J., Abt, E., Wigdor, H., and Miserendino, C.A., Evaluation of thermal cooling mechanisms for laser application to teeth, *Laser Surg. Med.*, 13, 83, 1993.
129. Paltauf, G. and Schmidt-Kloiber, H., Modeling and experimental observation of photomechanical effects in tissue-like media, in *Laser-Tissue Interaction VI*, Jacques, S.L., Ed., SPIE, Vol. 2391, San Jose, 1995, pp. 403–412.
130. Motamedi, M. and Rastegar, S., Thermal stress distribution in laser irradiated hard dental tissue: implications for dental applications, in *Laser-Tissue Interaction III*, Jacques, S.L., Ed., SPIE, Vol. 1646, Los Angeles, 1992, pp. 316–321.
131. Perelman, L.T., Albagli, D., Dark, M., Schaffer, J., van Rosenburg, C., Itzkan, I., and Feld, M.S., Physics of laser-induced stress wave propagation, cracking, and cavitation in biological tissue, in *Laser-Tissue Interaction V*, Jacques, S.L., Ed., SPIE, Vol. 2134, Los Angeles, 1994, pp. 144–151.
132. Dingus, R.S. and Scammon, R.J., Grüneisen-stress induced ablation of biological tissue, in *Laser-Tissue Interaction II*, Jacques, S.L., Ed., SPIE, Vol. 1427, Los Angeles, 1991, pp. 45–54.
133. Vickers, V.A., Jacques, S.L., Schwartz, J., Motamedi, M., Rastegar, S., and Martin, J.W., Ablation of hard dental tissues with the Er:YAG laser, in *Laser-Tissue Interaction III*, Jacques, S.L., Ed., SPIE, Vol. 1646, Los Angeles, 1992, pp. 46–55.
134. Walsh, J.T. and Hill, D.A., Erbium laser ablation of bone: effect of water content, in *Laser-Tissue Interaction II*, Jacques, S.L., Ed., SPIE, Vol. 1427, Los Angeles, 1991, pp. 27–33.
135. Wigdor, H.A., Walsh, J. T., and Visuri, S.R., Effect of water on dental material ablation with Er:YAG laser, in *Lasers in Surgery: Advanced Characterization, Therapeutics, and Systems IV*, Anderson, R.R., Ed., SPIE, Vol. 2128, Los Angeles, 1994.
136. Burkes, E.J., Hoke, J., Gomes, E., and Wolbarsht, M., Wet versus dry enamel ablation by Er:YAG laser, *J. Prosthet. Dent.*, 67(6), 847, 1992.
137. Cozean, C., Arcoria, C.J., Pelagalli, J., and Powell, L., Dentistry for the 21st century: Er:YAG laser for teeth, *J. Am. Dent. Assoc.*, 128, 1079, 1997.

138. Rizoiu, I.M. and DeShazer, L.G., New laser-matter interaction concept to enhance hard tissue cutting efficiency, in *Laser-Tissue Interaction V*, Jacques, S.L., Ed., SPIE, Vol. 2134A, San Jose, 1994, pp. 309–317.
139. Majaron, B., Sustercic, D., and Lukac, M., Influence of water spray on Er:YAG ablation of hard dental tissues, in *Medical Applications of Lasers in Dermatology, Opthalmology, Dentistry, and Endoscopy*, SPIE, Vol. 3192, San Remo, 1997, pp. 82–87.
140. Rizoiu, I., Kimmel, A.I., and Eversole, L.R., The effects of an Er, Cr:YSGG laser on canine oral tissues, in *Laser Applications in Medicine and Dentistry*, SPIE, Vol. 2922, Vienna, Austria, 1996, pp. 74–83.
141. Altshuler, G.B., Belikov, A.V., and Erofeev, A.V., Comparative study of contact and noncontact operation mode of hard tooth tissues Er-laser processing, in *Proc. Fifth Int. Congr. Int. Soc. Laser Dent.*, Jerusalem, 1996, pp. 21–25.
142. Majaron, M., Sustercic, D., and Lukac, M., Fiber-tip drilling of hard dental tissues with Er:YAG laser, in *Lasers in Dentistry IV*, Featherstone, J.D.B., Rechmann, P., and Fried, D., Eds., SPIE, Vol. 3248, San Jose, 1998, pp. 69–77.
143. Altshuler, G.B., Belikov, A.V., Erofeev, A.V., and Skrypnik, A.V., Physical aspects of cavity formation of Er-laser radiation, in *Lasers in Dentistry*, Wigdor, H.A., Featherstone, J.D.B., and White, J.M., Eds., SPIE, Vol. 2394, San Jose, 1995, pp. 211–222.
144. Rechmann, P., Glodin, D.S., and Hennig, T., Changes in surface morphology of enamel after Er:YAG irradiation, in *Lasers in Dentistry IV*, Featherstone, J.D.B., Rechmann, P., and Fried, D., Eds., SPIE, Vol. 3248, San Jose, 1998, pp. 62–68.
145. Fried, D., IR ablation of dental enamel, in *Lasers in Dentistry VI*, Featherstone, J.D.B., Rechmann, P., and Fried, D., Eds., SPIE, Vol. 3910, San Jose, 2000, pp. 136–148.
146. Fowler, B. and Kuroda, S., Changes in heated and in laser-irradiated human tooth enamel and their probable effects on solubility, *Calcif. Tissue Int.*, 38, 197, 1986.
147. Fried, D., Ashouri, N., Breunig, T.M., and Shori, R.K., Mechanism of water augmentation during IR laser irradiation of dental enamel, *Lasers Surg. Med.*, 31, 186, 2002.
148. Ith, M., Pratisto, H., Altermatt, H.J., and Weber, H.P., Dynamics of laser-induced channel formation in water and influence of pulse duration on the ablation of biotissue under water with pulsed Er:YAG lasers, *Appl. Phys. B*, 59, 621, 1994.
149. Loertscher, H., Shi, W.Q., and Grundfest, W.S., Tissue ablation through water with erbium:YAG lasers, *IEEE Trans. Biomed. Eng.*, 39(1), 86, 1992.
150. Forrer, M., Ith, M., Frenz, M., Romano, V., Weber, H.P., Silenok, A., and Konov, V.I., Mechanism of channel propagation in water by pulsed erbium laser radiation, in *Laser Interaction with Hard and Soft Tissue*, SPIE, Vol. 2077, Budapest, 1993, pp. 104–112.
151. Visuri, S.R., Gilbert, J.L., and Walsh, J.T., Shear test of composite bonded to dentin: Er:YAG laser vs. dental handpiece preparations., in *Lasers in Dentistry*, Wigdor, H.A., Featherstone, J.D.B., and White, J.M., Eds., SPIE, Vol. 2394, San Jose, 1995, pp. 223–227.
152. Attrill, D.C., Farrar, S.R., King, T.A., Dickinson, M.R., Davies, R.M., and Blinkhorn, A.S., Er:YAG ($\lambda = 2.94$ μm) laser etching of dental enamel as an alternative to acid etching, *Lasers Med. Sci.*, 15, 154, 2000.
153. von Fraunhofer, J.A., Allen, D.J., and Orbell, D.J., Laser etching of enamel for direct bonding, *Angle Orthod.*, 63, 73, 1993.
154. Lieberman, R., Segal, T.H., Nordenberg, D., and Serebro, L.I., Adhesion of composite materials to enamel: comparison between the use of acid and lasing as pretreatment, *Lasers Surg. Med.*, 4, 323, 1984.
155. Clayman, L., Fuller, T., and Beckman, H., Healing of cw and rapid superpulsed, carbon dioxide, laser-induced bone defects, *J. Oral Surg.*, 36, 932, 1978.
156. Friesen, L.R., Cobb, C.M., Rapley, J.W., Forgas, B.L., and Spencer, P., Laser irradiation of bone. II. Healing response following treatment by $CO_2$ and Nd:YAG lasers, *J. Periodontol.*, 68(9), 75, 1997.
157. Spencer, P., Payne, J.T., Cobb, C.M., Peavy, G.M., and Reinisch, L., Effective laser ablation of bone based on the absorption characteristics of water and protein, *J. Periodontol.*, 70(5), 68, 1999.

158. Esposito, M., Hirsh, J.M., Lekholm, U., and Thomsen, P., Review: biological factors contributing to failures of osteointegrated oral implants. I. Success criteria and epidemiology, *Eur. J. Oral Sci.*, 106, 527, 1998.
159. Hibst, R. and Keller, U., Heat effect of pulsed Er:YAG laser radiation, in *Laser Surgery: Advanced Characterization, Therapeutics, and Systems II*, SPIE, Vol. 1200, Los Angeles, 1990, pp. 379–386.
160. Cummings, J.P. and Walsh, J.T., Jr., Erbium laser ablation: the effect of dynamic optical properties, *Appl. Phys. Lett.*, 62(16), 1988, 1993.
161. Vodop'yanov, K.L., Bleaching of water by intense light at the maximum of the $\lambda \sim 3$ µm absorption band, *Sov. Phys. JETP*, 70(1), 114, 1990.
162. Shori, R.K., Walston, A.A., Stafsudd, O.M., Fried, D., and Walsh, J.T., Quantification and modeling of the non-linear changes in absorption coefficient of water at clinically relevant IR laser wavelengths, *Spec. Top. J. Quantum Electron,*, 7(6), 959, 2001.
163. Boehm, R.F., Rich, J., Webster, J., and Janke, S., Thermal stress effects and surface cracking associated with laser use on human teeth, *J. Biomech. Eng.*, 99, 189, 1977.
164. Krause, L.S., Cobb, C.M., Rapley, J.W., Killoy, W.J., and Spencer, P., Laser irradiation of bone. I. An *in vitro* study concerning the effects of the $CO_2$ laser on oral mucosa and subjacent bone, *J. Periodontol.*, 68(9), 872, 1997.
165. Fisher, S.E. and Frame, J.W., The effects of the $CO_2$ surgical laser on oral tissues, *Br. J. Oral Maxillofac. Surg.*, 22, 414, 1984.
166. Koort, H.J. and Frentzen, M., The effect of TEA-$CO_2$-laser on dentine, in *Proc. Third Int. Congr. Lasers Dent.*, Salt Lake City, 1992, abstr. 64.
167. Forrer, M., Frenz, M., Romano, V., Altermatt, H.J., Weber, H.P., Silenok, A., Istomyn, M., and Konov, V.I., Bone-ablation mechanism using $CO_2$ lasers of different pulse duration and wavelength, *Appl. Phys. B*, 56, 104, 1993.
168. Romano, V., Rodriguez, R., Altermatt, H.J., Frenz, M., and Weber, H.P., Bone microsurgery with IR-lasers: a comparative study of the thermal action at different wavelengths, in *Laser Interaction with Hard and Soft Tissue*, SPIE, Vol. 2077, Budapest, 1994, pp. 87–96.
169. Fried, N.M. and Fried, D., Comparison of Er:YAG and 9.6-µm TE $CO_2$ lasers for ablation of skull tissue, *Lasers Surg. Med.*, 28, 335, 2001.
170. Lee, C., Ragadio, J., and Fried, D., Influence of wavelength and pulse duration on peripheral thermal and mechanical damage to dentin and alveolar bone, in *Lasers in Dentistry VI*, Featherstone, J.D.B., Rechmann, P., and Fried, D., Eds., SPIE, Vol. 3910, San Jose, 2000, pp. 193–203.
171. Ivanenko, M.M. and Hering, P., Wet bone ablation with mechanically Q-switched high-repetition-rate $CO_2$ laser, *Appl. Phys. B*, 67, 395, 1998.
172. Tagomori, S. and Morioka, T., Combined effects of laser and fluoride on acid resistance of human dental enamel, *Caries Res.*, 23, 225, 1989.
173. Jennett, E., Motamedi, M., Rastegar, S., Arcoria, C.J., and Fredrickson, C.J., Dye enhanced alexandrite laser for ablation of dental tissue, *Lasers Surg. Med.*, Suppl. 6, 15, 1994.
174. Arcoria, C.J., Fredrickson, C.J., Hayes, D.J., Wallace, D.B., and Judy, M.M., Dye microdrop assisted laser for dentistry, *Lasers Surg. Med.*, Suppl. 5, 17, 1993.
175. White, J.M., Goodis, H.E., Setcos, J.C., Eakle, S., Hulscher, B.E., and Rose, C.L., Effects of pulsed Nd:YAG laser on human teeth: a three-year follow-up study, *J. Am. Dent. Assoc.*, 124, 45, 1993.
176. White, J.M., Goodis, H.E., and Daniels, T.E., Effects of Nd:YAG laser on pulps of extracted teeth, *Lasers Life Sci.*, 4, 3, 191, 1991.
177. White, J.M., Neev, J., Goodis, H.E., and Berns, M.W., Surface temperature and thermal penetration depth of Nd:YAG laser applied to enamel and dentin, in *Laser Surgery: Advanced Characterization, Therapeutics, and Systems III*, Anderson, R.R., Ed., SPIE, Vol. 1643, Los Angeles, 1992, pp. 423–436.
178. White, J.M., Goodis, H.E., Marshall, G.W., and Marshall, S.J., Identification of the physical modification threshold of dentin induced by neodymium and holium YAG lasers using scanning electron microscopy, *Scanning Microsc.*, 7(1), 239, 1993.
179. Niemz, M.H., Cavity preparation with the Nd:YLF picosecond laser, *J. Dent. Res.*, 74, 5, 1194, 1995.

180. Boulnois, J.L., Photophysical processes in recent medical laser developments: a review, *Lasers Med. Sci.*, 1, 47, 1986.
181. Niemz, M.H., Investigation and spectral analysis of the plasma-induced ablation mechanism of dental hydroxyapatite, *Appl. Phys. B*, 58, 273, 1994.
182. Neev, J., Da Silva, L.B., Feit, M.D., Perry, M.D., Rubenchik, A.M., and Stuart, B.C., Ultrashort pulse lasers for hard tissue ablation, *IEEE J. Selected Top. Quantum Electron.*, 2(4), 790, 1996.
183. Neev, J., Liaw, L.L., Raney, D.V., Fujishige, J.T., Ho, P.D., and Berns, M.W., Selectivity, efficiency, and surface characteristics of hard dental tissues ablated with ArF pulsed excimer lasers, *Lasers Surg. Med.*, 11, 499, 1991.
184. Frentzen, M., Koort, H.J., and Thiensiri, I., Excimer lasers in dentistry: future possibilities with advanced technology, *Quintessence Int.*, 23, 117, 1992.
185. Neev, J., Liaw, L.L., Stabholtz, A., Torabinejad, J.T., Fujishige, J.T., and Berns, M.W., Tissue alteration and thermal characteristics of excimer laser interaction with dentin, in *Laser Surgery: Advanced Characterization, Therapeutics, and Systems III,* Anderson, R.R., Ed., SPIE, Vol. 1643, 1992, pp. 386–397.
186. Neev, J., Stabholtz, A., Liaw, L.L., Torabinejad, M., Fujishige, J.T., Ho, P.D., and Berns, M.W., Scanning electron microscopy and thermal characteristics of dentin ablated by a short-pulse XeCl excimer laser, *Lasers Surg. Med.*, 12, 353, 1993.
187. Moss, J.P., Patel, B.C.M., Pearson, G.J., Arthur, G., and Lawes, R.A., Krypton fluoride excimer ablation of tooth tissues: precision tissue machining, *Biomaterials*, 15(12), 1013, 1994.
188. Hennig, T., Rechmann, P., Pilgrim, C., Schwarzmaier, H.-J., and Kaufmann, R., Caries selective ablation by pulsed lasers, in *Lasers in Orthopedic and Veterinary Medicine*, O'Brien, S.J., Dederich, D.N., Wigdor, H., and Trent, A.M., Eds., SPIE, Vol. 1424, Los Angeles, 1991, pp. 99–105.
189. Hennig, T., Rechmann, P., Pilgrim, C., and Kaufmann, R., Basic principles of caries selective ablation by pulsed lasers, in *Proc. Third Int. Congr. Lasers Dent.*, Vol. Salt Lake City, 1992, pp. 119–120.
190. Rechmann, P., Hennig, T., van den Hoff, U., and Kaufmann, R., Caries selective ablation: 377 nm vs. 2.9 μm, in *Lasers in Orthopedic and Veterinary Medicine II*, SPIE, Vol. 1880, Los Angeles, 1994, pp. 235–239.
191. Feuerstein, O., Palanker, D., Fuxbrunner, A., Lewis, A., and Deutsch, D., Effect of the ArF excimer laser on human enamel, *Lasers Surg. Med.*, 12, 471, 1992.
192. Melis, M., Berna, G., Berna, N., and Benvenuti, A., Ablation of hard dental tissues by ArF and XeCl excimer lasers, in *Lasers in Surgery: Advanced Characterization, Therapeutics, and Systems IV*, SPIE, Vol. 2128, 1994, pp. 349–358.
193. Hibst, R. and Keller, U., Removal of dental filling materials by Er:YAG laser radiation, in *Lasers in Orthopedic, Dental, and Veterinary Medicine*, O'Brien, S.J., Dederich, D.N., Wigdor, H., and Trent, A.M., Eds., SPIE, Vol. 1424, 1991, pp. 120–126.
194. Nelson, J.S., Yow, L., Liaw, L.H., Macleay, L., and Zavar, R.B., Ablation of bone and methacrylate by a prototype Er:YAG laser, *Lasers Surg. Med.*, 8(5), 494, 1988.
195. Marshall, S., Marshall, G., Watanabe, L., and White, J., Effects of the Nd:YAG laser on amalgams and composites, *Trans. Acad. Dent. Mater.*, 2, 297, 1989.
196. Thomas, B.W., Hook, C.R., and Draughn, R.A., Laser-aided degradation of composite resin, *Angle Orthod.*, 66(4), 281, 1996.
197. Yow, L., Nelson, J.S., and Berns, M.W., Ablation of bone and PMMA by an XeCl 308 nm excimer laser, *Lasers Surg. Med.*, 9(2), 141, 1989.
198. Sherk, H.H., Lane, G., Rhodes, A., and Black, J., Carbon dioxide laser removal of polymethylmethacrylate, *Clin. Orthopaed. Related Res.*, 310, 67, 1995.
199. Dumore, T. and Fried, D., Selective ablation of orthodontic composite using sub-microsecond IR laser pulses with optical feedback, *Lasers Surg. Med.*, 27(2), 103, 2000.
200. Alexander, R. and Fried, D., Selective removal of dental composite using 355-nm nanosecond laser pulses, *Lasers Surg. Med.*, 30, 240, 2002.

201. Yamamoto, H. and Ooya, K., Potential of yttrium-aluminium-garnet laser in caries prevention, *J. Oral Pathol.*, 38, 7, 1974.
202. Stern, R.H., Sognnaes, R.F., and Goodman, F., Laser effect on *in vitro* enamel permeability and solubility, *J. Am. Dent. Assoc.*, 78, 838, 1966.
203. Featherstone, J.D.B. and Fried, D., Fundamental interactions of lasers with dental hard tissue, *Med. Laser Appl.*, 16, 181, 2001.
204. Kuroda, S. and Fowler, B.O., Compositional, structural and phase changes in *in vitro* laser-irradiated human tooth enamel, *Calcif. Tissue Int.*, 36, 361, 1984.
205. Yamamoto, H. and Sato, K., Prevention of dental caries by Nd: YAG laser irradiation, *J. Dent. Res.*, S9, 1271, 1980.
206. Fox, J.L., Yu, D., Otsuka, M., Higuchi, W.I., Wong, J., and Powell, G.L., Initial dissolution rate studies on dental enamel after $CO_2$ laser irradiation, *J. Dent. Res.*, 71, 1389, 1992.
207. Otsuka, M., Wong, J., Higuchi, W.I., and Fox, J.L., Effects of laser irradiation on the dissolution kinetics of hydroxyapatite preparations, *J. Pharm. Sci.*, 79, 510, 1990.
208. Otsuka, M., Matsuda, Y., Suwa, Y., Wong, J., Fox, J.L., Powell, G.L., and Higuchi, W.I., Effect of carbon dioxide laser irradiation on the dissolution kinetics of self-setting hydroxyapatite cement, *Lasers Life Sci.*, 5(3), 199, 1993.
209. Fox, J.L., Yu, D., Otsuka, M., Higuchi, Wong, J., and Powell, G.L., Initial dissolution rate studies on dental enamel after $CO_2$ laser irradiation, *J. Dent. Res.*, 71(7), 1389, 1992.
210. Featherstone, J.D.B., Fried, D., and Bitten, E.R., Mechanism of laser induced solubility reduction of dental enamel, in *Lasers in Dentistry III*, Wigdor, H.A., Featherstone, J.D.B., and Rechmann, P., Eds., SPIE, Vol. 2973, San Jose, 1997, pp. 112–116.
211. Hsu, J., Fox, J.L., Wang, Z., Powell, G.L., Otzuka, M., and Higuchi, W.I., Combined effects of laser irradiation/solution fluoride on enamel demineralization, *J. Clin. Laser Med. Surg.*, 16, 93, 1998.
212. Higuchi, W.I., Fox, J.L., and Powell, G.L., Methods of preventing tooth decay by laser irradiation and chemical treatment, U.S. patent 4,877,401, 1989.
213. Borggreven, J.M.P.M., van Dijk, J.W.E., and Driessens, F.C.M., Effect of laser irradiation on the permeability of bovine dental enamel, *Arch. Oral Biol.*, 25, 831, 1980.
214. Featherstone, J.D.B., Barrett-Vespone, N.A., Fried, D., Kantorowitz, Z., and Lofthouse, J., $CO_2$ laser inhibition of artificial caries-like lesion progression in dental enamel, *J. Dent. Res.*, 77(6), 1397, 1998.
215. Fox, J.L., Yu, D., Otsuka, M., Higuchi, W.I., Wong, J., and Powell, G.L., The combined effects of laser irradiation and chemical inhibitors on the dissolution of dental enamel, *Caries Res.*, 26, 333, 1992.
216. Fox, J.L., Wong, J., Yu, D., Otsuka, M., Higuchi, W.I., Hsu, J., and Powell, G.L., Carbonate apatite as a model for the effect of laser irradiation on human dental enamel, *J. Dent. Res.*, 73(12), 1848, 1994.
217. Kantola, S., Laine, E., and Tarna, T., Laser-induced effects on tooth structure. VII. X-ray diffraction study of dentine exposed to a $CO_2$ laser, *Acta Odontol. Scand.*, 31, 381, 1973.
218. Nammour, S., Renneboog-Squilbin, C., and Nyssen-Behets, C., Increased resistance to artificial caries-like lesions in dentin treated with $CO_2$ laser, *Caries Res.*, 26, 170, 1992.
219. Kinney, J.H., Haupt, D.L., Balooch, M., White, J.M., Bell, W.L., Marshall, S.J., and Marshall, G.W., The threshold effects of Nd and Ho:YAG laser-Induced surface modification on demineralization of dentin surfaces, *J. Dent. Res.*, 75(6), 1388, 1996.

# VII

# Advanced Biophotonics for Genomics, Proteomics, and Medicine

# 51

# Biochips and Microarrays: Tools for the New Medicine




Tuan Vo-Dinh
*Oak Ridge National Laboratory*
*Oak Ridge, Tennessee*


## 51.1 Introduction

Biochips and microarrays are technologies that will benefit directly from the research advances of the postsequencing era in genomics. Experimental genomics, in combination with the growing body of sequence information, promises to revolutionize the way cells and cellular processes are studied and diseases diagnosed and treated. Information on genomic sequence can be used experimentally with high-density DNA arrays that allow complex mixtures of RNA and DNA to be interrogated in a parallel and quantitative fashion. DNA arrays can be used for many different purposes, especially to measure levels of gene expression (messenger RNA abundance) for tens of thousands of genes simultaneously.[1–4] On the other hand, portable, self-contained biochips with integrated detection microchip systems have great potential for use by the physician at the point of care.[5–11]

As a result, there has been an explosion of interest in the development and applications of biochips and microarrays.[1–5,7,11–29] Biochips are integrated microdevices designed to rapidly and inexpensively perform biochemical procedures for biomedical applications. Because of their miniaturization, low cost, and potential for large-scale automation, biochips can perform more efficiently than currently available laboratory equipment. Biochips with high- and medium-density arrays of oligonucleotides, complementary

DNAs (cDNAs), proteins, and antibodies are among the most powerful and versatile tools in genomics and proteomics for taking full advantage of the large and rapidly increasing body of information on gene structure and function. One of the greatest impacts of microarray and biochip technologies in conjunction with bioinformatics is in enabling an entirely new approach to biological and biomedical research. In the past, researchers studied one or a few genes at a time. With whole-genome sequences and new automated, high-throughput microarray and biochip technologies they can investigate a medical problem systematically and on a large scale. They can study all the genes in a genome or all the gene products in a particular tissue, tumor, or organ. Furthermore, they can investigate how tens of thousands of genes and proteins work together in interconnected networks in a living system. Such knowledge will have a profound impact on the way disorders are diagnosed, treated, and prevented and will bring about revolutionary changes in biomedical research and clinical practice.

This chapter describes the development and applications of biochip and microarray technology in biological research and medical applications. Specific applications of microarrays are described further in Chapter 52 of this handbook.

## 51.2 Biochips and Microarrays: Definition and Classification

### 51.2.1 Biochips and Microarrays: A Definition

The terms *biochip* and *microarray* have often been used indiscriminately. However, these two systems are different in design and concept. The term *biochip* was coined in analogy to the phrase *computer chip*, which, of course, refers to the silicon-based substrate used in the fabrication of miniaturized electronic circuits. Therefore, in the generic sense, the term *biochip* involves the concept of an integrated circuit (IC). However, the term has taken on a variety of meanings over the years. A biochip is often referred to as a material or substrate that has an array of probes for biochemical assays. In general, any device or component incorporating a two-dimensional (2D) array of reaction sites having biological materials on a solid substrate has been referred to as a biochip. Biochips involve both miniaturization, usually in microarray formats, and the possibility of low-cost mass production. In the literature, various terms have been used to describe this new technology: If the probes are nucleic acids, the devices are called DNA biochips, DNA chips, genome chips, DNA microarrays, gene arrays, and genosensor arrays. GeneChip™ is a registered trademark of Affymetrix, Inc., which uses this term to refer to its high-density, oligonucleotide-based DNA arrays. If the probes consist of antibodies or proteins, the devices are referred to as protein chips or protein biochips. A recently developed system with both DNA and antibody probes on the same platform is referred to as multifunctional biochip.[7,9–11]

We define the term *biochip* in the same way that the International Union of Pure and Applied Chemistry (IUPAC) defines *biosensor*.[30] According to IUPAC, a biosensor is "a self-contained integrated device, which is capable of providing specific quantitative or semiquantitative analytical information using a biological recognition element (biochemical receptor), which is retained in direct spatial contact with a transduction element." The interaction of the analyte with the bioreceptor is designed to produce an effect measured by the transducer, which converts the information into a measurable effect, such as an electrical signal. Figure 51.1 illustrates the conceptual principle of the biosensing process. Biochips can be considered an array of individual biosensors that can be individually monitored and generally are used for the analysis of multiple analytes.

We group microarrays and biochips into two general classes: (1) microarray systems, which consist of arrays of probes on a substrate ("chip"), and (2) integrated biochips, which also include detector-array microchips. Microarrays (often called "chips") consist only of arrays of probes (in a chip format) but do not include sensor microchips integrated into the system.[10] Microarrays usually have separate, relatively large detection systems that are suitable for laboratory-based research applications (Figure 51.2). They can have large numbers (tens of thousands) of probes used for identifying multiple biotargets at very high speeds and high throughput by matching with different types of probes via hybridization. Therefore, array plates are very useful for gene-discovery and drug-discovery applications, which often require tens of thousands of assays on a single plate.

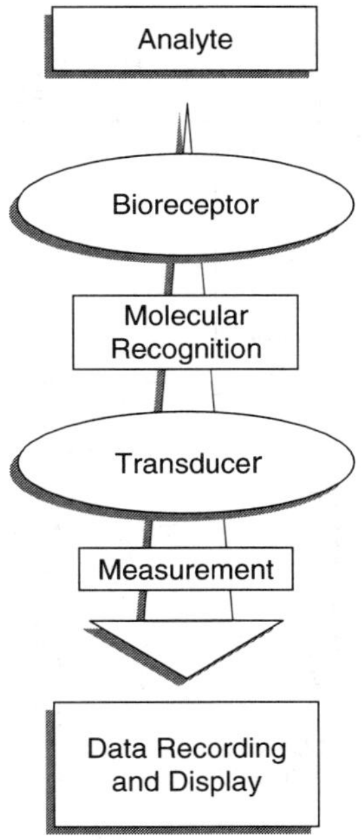


**FIGURE 51.1** Principle of the biosensing process.

On the other hand, integrated biochips include an IC microsensor, which makes these devices very portable and inexpensive (Figure 51.3). These devices generally have medium-density probe arrays (10 to 100 probes) and are most appropriate for medical diagnostics at the physician's office or at the point of care in the field. The characteristic features of microarrays and biochips are summarized in Table 51.1.

### 51.2.2 Biochip Classification

Both microarrays and biochips use probes that consist of biological recognition systems, often called bioreceptors. Biochips can be classified either by the nature of their bioreceptors or by the type of transducer used (Figure 51.4). A detailed description of various types of bioreceptors and transducers is provided in Chapter 20 of this handbook. A bioreceptor is a biological molecular species (e.g., an antibody, an enzyme, a protein, or a nucleic acid) or a living biological system (e.g., cells, tissue, or whole organisms) that uses a biochemical mechanism for recognition. Bioreceptors are the key to specificity for biochip technologies responsible for binding the analyte of interest to the sensor for the measurement. Bioreceptors can take many forms and are as numerous as the different analytes that have been monitored using biosensors. Bioreceptors can generally be classified into five major categories: (1) antibody/antigen, (2) enzymes, (3) nucleic acids/DNA, (4) cellular structures/cells, and (5) biomimetic probes (synthetic probes that mimic receptors of living systems).

Biochips can also be classified by the type of transducers used. The conventional transducer techniques are (1) optical measurements (luminescence, absorption, surface plasmon resonance, etc.), (2) electrochemical measurements, and (3) mass-sensitive measurements (surface acoustic wave, microbalance, etc.).

## 51.3 Microarray Systems

### 51.3.1 Basic Principles

Nucleic acids have been widely used as bioreceptors for microarray and biochip technologies.[31–38] The complementarity of adenine:thymine (A:T) and cytosine:guanine (C:G) pairing in DNA forms the basis

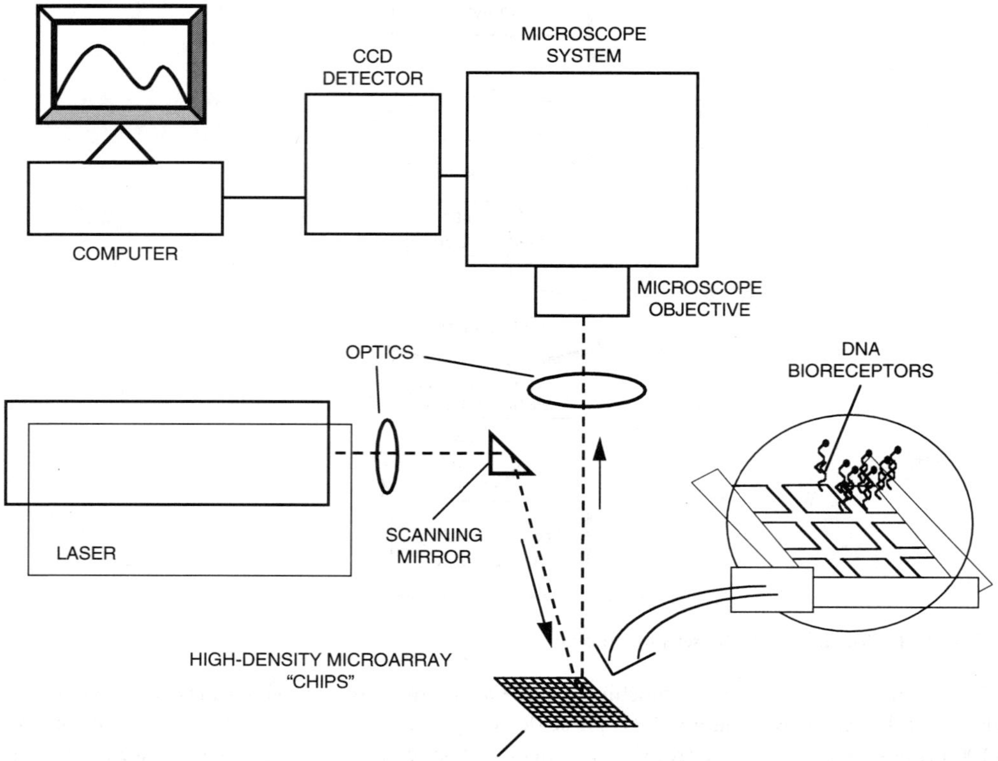


**FIGURE 51.2** Schematic diagram of a microarray system. Microarrays (often called "chips") consist of high-density arrays of probes (in a chip format) but do not include sensor microchips integrated into the system. These microarrays usually have separate detection systems that are relatively large and suitable for laboratory-based research applications.

for the specificity of biorecognition in DNA biochips. If the sequence of bases composing a certain part of the DNA molecule is known, then the complementary sequence can be synthesized and labeled with an optically detectable compound (e.g., a radioactive label or a fluorescent label). When unknown fragments of single-strand DNA (ssDNA) samples, called the target, react (or hybridize) with the probes on the chip, double-strand DNA (dsDNA) fragments are formed only when the target and the probe are complementary according to the base-pairing rule. When the targets contain more than one type of sample, each is labeled with its specific tag.

The microarrays of probes serve as reaction sites, each reaction site containing single strands of a specific sequence of a DNA fragment. The DNA fragments can either be short oligonucleotides — e.g., 20- to 100-nucleotide (nt) sequences — or longer strands of cDNA. The relatively large nucleic-acid components — those longer than 100 nt — are often used in RNA expression analysis,[1] and those with short nucleic acids — with oligonucleotides up to 25 nt — can be used for both RNA expression[39] and sequence analysis.

Probes based on a synthetic biorecognition element, peptide nucleic acid (PNA), have been developed.[34] PNA is an artificial oligoamide that is capable of binding very strongly to complementary oligonucleotide sequences. A surface plasmon resonance sensor has been used to directly detect dsDNA that has been amplified by a polymerase chain reaction (PCR).

Synthetic antibody-like probes produced by molecular-imprint methods are often used as biomimetic probes. Another type of biomimetic probes is aptamers, which are artificial nucleic-acid ligands that can be produced against amino acids, proteins, and other molecules.

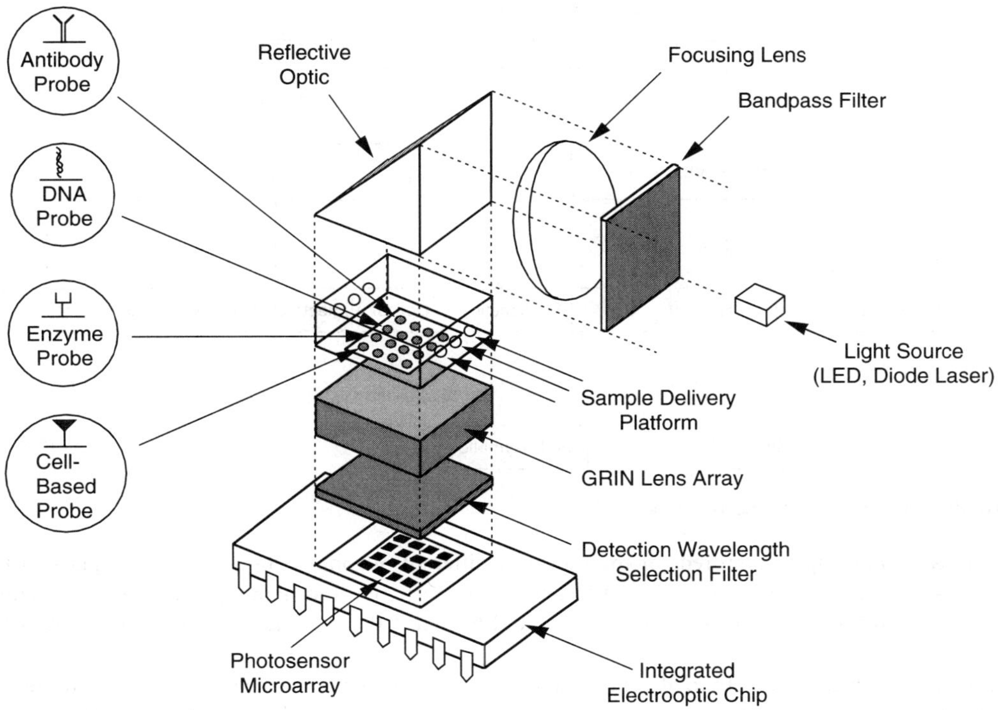


**FIGURE 51.3** Schematic diagram of an integrated biochip. Integrated biochips also include an IC microsensor, which makes these devices very portable and inexpensive. These devices generally have medium-density probe arrays (10 to 100 probes) and are most appropriate for medical diagnostics at the physician's office or at the point of care in the field.

**TABLE 51.I** Characteristic Features of Microarrays and Biochips

| Systems | Bioreceptor | Format | Probe Density | Detection System | Applications |
|---|---|---|---|---|---|
| Microarrays | DNA<br>Antibody<br>Proteins<br>Cells, etc. | Array | High density<br>($10^3$–$10^5$) | Separate system<br>(laboratory system) | Research tools |
| Biochips | DNA<br>Antibody<br>Proteins<br>Cells, etc. | Array | Low-medium density<br>(10–$10^2$) | Integrated system<br>(portable) | Physician's tool<br>Field monitor |

## 51.3.2 Microarray Fabrication

DNA arrays, often called DNA chips, can be fabricated using high-speed robotics on a variety of substrates. The substrates can be thin plates made of silicon, glass, gel, gold, or a polymeric material such as plastic or nylon, or may even be composed of beads at the ends of fiber-optic bundles.[40] Oligonucleotide microarrays are fabricated either by *in situ* light-directed combinatorial synthesis that uses photographic masks for each chip[41] or by conventional synthesis followed by immobilization on glass substrates.[1,2,12]

Arrays with more than 250,000 different oligonucleotide probes or 10,000 different cDNAs per $cm^2$ have been produced.[42] Sequence information is used directly to design high-density, 2D arrays of synthetic oligonucleotides. The probe arrays, made using spatially patterned, light-directed combinatorial chemical

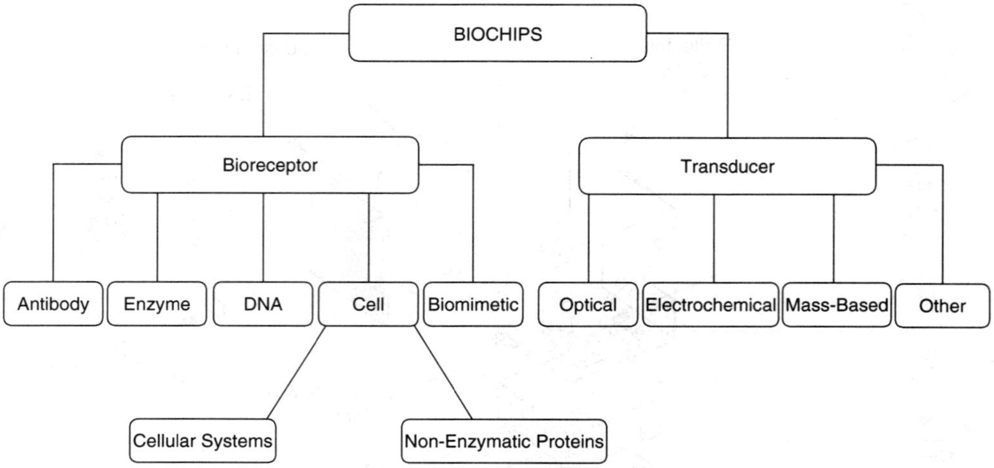


**FIGURE 51.4** Biochip classification schemes.

synthesis, contain up to hundreds of thousands of different oligonucleotides on a small glass surface. The arrays have been designed and used to perform quantitative and highly parallel measurements of gene expression, to discover polymorphic loci, and to detect the presence of thousands of alternative alleles.

A maskless fabrication method of light-directed oligonucleotide microarrays replaces the chrome mask with virtual computer-generated masks. These virtual masks are relayed to a digital micromirror array capable of synthesizing microarrays containing more than 76,000 features measuring 16 $\mu m^2$. A reflective imaging system forms an ultraviolet image of the virtual mask on the active surface of the glass substrate, which is mounted in a flow-cell reaction chamber connected to a DNA synthesizer. Programmed chemical-coupling cycles follow light exposure, and steps are repeated to achieve the desired pattern.[43]

Modified chemistries should allow for other types of arrays to be manufactured using photolithography. Alternative processes, such as ink-jet[44] and spotting techniques, can also be used.[13,14] One method for constructing these arrays using light-directed DNA synthesis with photoactivatable monomers can currently achieve densities on the order of $10^6$ sequences/$cm^2$. One of the challenges facing the developers of this technology is to further increase the volume, complexity, and density of sequence information encoded in these arrays. An approach for synthesizing DNA probe arrays that combines standard solid-phase oligonucleotide synthesis with polymeric photoresist films serving as the photoimageable component opens the way to exploiting high-resolution imaging materials and processes from the microelectronics industry for the fabrication of DNA probe arrays with substantially higher densities.[44]

Activated DNA has been immobilized in aldehyde-containing polyacrylamide gel for use in manufacturing microarrays.[13] In this process, abasic sites were generated in DNA by partial acidic depurination. Amino groups were then introduced into the abasic sites by reaction with ethylenediamine and reduction of the aldimine bonds formed. It was found that DNA could be fragmented at the site of amino-group incorporation or preserved mostly unfragmented. In similar reactions, both amino-DNA and amino-oligonucleotides were attached through their amines to polyacrylamide gel-derivatized with aldehyde groups. ssDNA and dsDNA of 40 to 972 nt or base pairs were immobilized on the gel pads to manufacture DNA microarrays.[13]

Microarrays containing PCR-amplified genomic DNA extracts from mice tumors on a Zetaprobe® membrane have been fabricated using a modified thermal ink-jet printer.[45] This method, using a modified bubble-jet printing system, is a simple and cost-effective procedure for the fabrication of microarrays containing biological samples. Figure 51.5 shows a cross-section view of a generic bubble-jet cartridge illustrating the connection of ink channels to the printhead (Figure 51.5A) and a membrane printed with biological materials extracted from mice tumor cells with fragile histidine triad (FHIT) gene and control

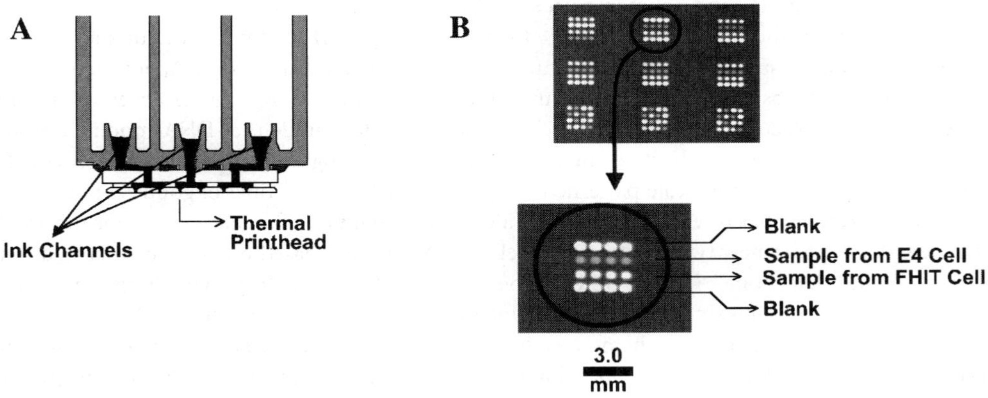


**FIGURE 51.5 (Color figure follows p. 28-30.)** (A) Schematic diagram of a generic bubble-jet cartridge illustrating the connection of ink channels to the printhead. (B) Membrane printed with biological materials using bubble-jet technology. For purposes of better visualization of the spotting, different fluorescent dyes were added for the preparation of this sample. (Adapted from Allain, L.R. et al., *Fresenius J. Anal. Chem.*, 371(2), 146, 2001.)

cells (cells labeled E4). For purposes of better visualization of the spotting, different fluorescent dyes were added for the preparation of this figure.[45] Because of their mass-produced design, ink-jet printers are a much cheaper alternative to conventional spotting techniques.

## 51.3.3 Assay Detection

Because of its inherently sensitive detection capability, fluorescence is the most commonly used technique in DNA hybridization assays. Fluorescent labels can be inserted during the PCR process; subsequently, the sample is tested for hybridization to the microarray. In general, microarray systems are based on fiber-optic probes. Glass and silica plates are used as the probe substrates and are externally connected to a photosensing system, which generally consists of a conventional detection device such as a photomultiplier or a charge-coupled device (CCD). Although the probes on the sampling platform are small, the entire device — containing excitation laser sources and detection systems (often a confocal microscope system) — is relatively large, e.g., tabletop size. The high density of the array and the small size of the probes often require powerful excitation source and high-precision scanning systems, which are also relatively large. These systems have demonstrated their usefulness in genomics research and analysis, but they are laboratory-oriented and involve relatively expensive equipment.

Other optical techniques, such as surface plasmon resonance, has also been used to monitor real-time binding of low-molecular-weight ligands to DNA fragments that were irreversibly bound to the sensor surface via Coulombic interactions. Ellipsometry confirmed that the DNA layer remained stable over a period of several days. Binding rates and equilibrium coverages were determined for various ligands by changing the ligand concentration. In addition, affinity constants and association and dissociation rates were also determined for these various ligands. A label-free optical technique, based on reflectometric interference spectroscopy, has also been used in characterization of low-molecular-weight ligand–DNA interaction.[46]

Ellipsometric and interferometric detection methods have also been used for characterization of DNA probes immobilized on a combinatorial array.[47] The hybridization reaction may be solution based, or it may be driven by an electric field.[48] This technology has been successfully applied to the simultaneous assay of expression of many thousands of genes and to large-scale gene discovery as well as to polymorphism screening and mapping of genomic DNA.

Several researchers have investigated sensing-system-based sandwich-type biosensors formed from liquid-crystalline dispersions formed from DNA–polycation complexes.[49] These sandwich biosensors have been used to detect compounds and physical factors that affect the ability of specific DNA crosslinkers, polycationic molecules, to bind between adjacent DNA molecules.

The application of the principles of atomic force microscopy (AFM) for the use of supported lipid bilayers for anchoring subcellular to the submolecular biomolecules such as DNA, enzymes, and crystalline protein arrays has been studied.[50] Chemical force microscopy (CFM) has also been used to probe the mechanics of molecular recognition between surfaces. The technique uses DNA-modified probes to scan the interactions between short segments of ssDNA containing complementary sequences. These substrates consisted of micron-scale patterned arrays of one or more distinct oligonucleotides.[51]

Raman spectroscopy has recently been used as a detection technique for gene assays. For this method, researchers developed a new type of spectral label for DNA probes based on surface-enhanced Raman scattering (SERS) detection.[29,52–55] The SERS probes have great potential to provide both sensitivity and selectivity via label multiplexing because of the intrinsically narrow bandwidths of Raman peaks. The effectiveness of the new detection scheme was demonstrated using the *gag* gene sequence of the human immunodefficiency (HIV) virus.[55] The development of a multiarray system for DNA diagnostics using visible and near-infrared (NIR) dyes has been reported.[28]

### 51.3.4 Target-Sample Amplification

An important step in DNA microarray assay involves efficient and reproducible messenger RNA (mRNA) and DNA amplification methods. Various amplification technologies are described in detail in Chapter 55 of this handbook. This section briefly discusses two main approaches: (1) a PCR-based approach that has been used to make single-cell cDNA libraries[56,57] and (2) a method that uses multiple rounds of linear amplification based on cDNA synthesis and a template-directed *in vitro* transcription (IVT) reaction. Although the PCR technique is most widely used, the IVT method shows significant promise. Labeled material can be sufficiently produced through multiple-round cDNA/IVT amplification. The IVT technique is highly reproducible and introduces less quantitative bias than PCR-based amplification.[58,59] Messenger RNA from single live neurons has been characterized using the IVT method.[60,61]

The need for only minute amounts of starting material is a significant advantage of the microarray and biochip technologies. This feature, in conjunction with the capability of laser-capture microdissection technology to provide pure samples, has greatly enhanced the reliability and value of the DNA chip technology as a sophisticated genetic-testing technique.[62,63] The combination of microarrays generated with DNA, or RNA extracted from pure cell populations within the tissue, and powerful amplification strategies promises to be especially important for studies that involve human biopsy material from nonhomogeneous tissue, as well as for research in developmental biology, immunology, and neurobiology.

### 51.3.5 DNA-Sequence Analysis

Microarray-based sequence analyses are generally performed using arrays designed to evaluate specific sequences. There are several approaches to analyzing a DNA sequence of interest for all possible changes. The first approach uses probes complementary to a significant subset of sequence changes (such as those described for the cystic fibrosis transmembrane regulator[64]) and measures gain of hybridization signal to these probes relative to reference samples. The approach allows for a partial scan of a DNA segment for all possible sequence variations. Probes are designed such that the location of the interrogated target base is in the centermost position of the potential target/probe duplex and thus provides the best discrimination for hybridization specificity. The probe arrays are made using spatially patterned, light-directed combinatorial chemical synthesis and contain up to hundreds of thousands of different oligonucleotides on a small glass surface. The arrays have been designed and used to perform quantitative and highly parallel measurements of gene expression, to discover polymorphic loci, and to detect the presence of thousands of alternative alleles.[42] In another study, the entire 297-bp HIV-1 protease gene-coding sequence from 167 viral isolates was screened for all single nucleotide changes using this gain-of-signal approach. The DNA sequence of USA HIV-1 clade B proteases was found to be extremely variable, with

47% of the 99 amino-acid positions varied.[65] Naturally occurring mutations in HIV-1-infected patients have important implications for therapy and the outcome of clinical studies.

The second approach to scanning DNA analyses involves the selective loss of hybridization signal to perfect-match probes that are complementary to wild-type sequence in a manner analogous to comparative genomic hybridization studies.[66,67] In the loss-of-signal approach, sequence variations are scored by quantifying relative losses of hybridization signal to perfect-match oligonucleotide probes relative to wild-type reference targets. Loss-of-signal analysis allows for a practical screen to be set up for virtually any sequence variation. A disadvantage of loss-of-signal analysis is that the mutation cannot be discerned; the identity of the sequence change must be established by subsequent dideoxysequencing of the region surrounding the loss-of-signal signature. This approach has been further improved by using internal standards that utilize two-color assays. In this scheme, reference targets with a known sequence can be cohybridized to the arrays along with the test target. Labeling each target with a different fluorophore allows a direct comparison of hybridization signals from the two targets to be made. High-density arrays consisting of over 96,600 oligonucleotides 20 nt in length screen for a wide range of heterozygous mutations in the 3.45-kilobase (kb) exon 11 of the hereditary breast and ovarian cancer gene *BRCA1*.[67] Reference and test samples were cohybridized to these arrays and differences in hybridization patterns quantified by two-color analysis. The ability to scan a large gene rapidly and accurately for all possible heterozygous mutations in large numbers of patient samples will be critical for the future of medicine. DNA minisequencing assays couple target hybridization with enzymatic primer-extension reactions to provide a powerful means of scanning for all possible sequence variations.[22,68,69]

Representational difference analysis (RDA), which compares the results for two tissue samples, is another analysis method used for DNA microarrays and biochips. Subtraction of one set of results from the other shows which genes are active in one sample but less active or completely inactive in the other. The comparison is usually performed between normal and cancer cells, or between metastatic and nonmetastatic cancers. In this way, the evaluation may pinpoint an abnormal cellular process occurring mainly (or even exclusively) in diseased cells. The evaluation may thus serve not only as a diagnostic tool but also as a means to identify therapeutic targets. RDA of cDNA was used for a comparison of the global transcript level of tumor of the larynx and the corresponding normal epithelial tissue for detecting differentially expressed genes. Overall, some 130 gene fragments were identified.[70] After further analysis, these genes could be put in functional groups such as genes whose overexpression was a result of tumor growth or dedifferentiation, genes that played major roles in signal transduction pathways, apoptosis, oncogenes, or tumor-suppressor genes in addition to new, entirely unknown genes.

## 51.3.6 Protein Microarrays

Recently, antibodies have received increasing interest as bioreceptors for research in proteomics. When microarray probes consist of antibodies or proteins, the microarrays are sometimes called protein chips or protein biochips.[45,71–79] The basis for the specificity of immunoassays is the antigen–antibody (Ag–Ab) binding reaction, which is a key mechanism by which the immune system detects and eliminates foreign matter.[80–82] The way in which an antigen and an antigen-specific antibody interact may perhaps be understood as analogous to a lock-and-key fit, in which the specific configurations of a unique key enable it to open a lock. In the same way, an antigen-specific antibody fits its unique antigen in a highly specific manner, so that hollows, protrusions, planes, and ridges on the antigen and the antibody molecules (in a word, the total three-dimensional [3D] structure) are complementary. Due to this 3D shape fitting, and the diversity inherent in individual antibody makeup, it is possible to find an antibody that can recognize and bind to any one of a huge variety of molecular shapes. This unique property of antibodies is the key to their usefulness in immunosensors because this ability to "recognize" molecular structures allows researchers to develop antibodies that bind specifically to chemicals, biomolecules, microorganism components, etc. Such antibodies can then be used as specific detectors to identify and find an analyte of interest that is present, even in extremely small amounts, in a myriad of other chemical substances.

Protein microarrays have received great interest because it is not sufficient just to know the genes involved and their DNA sequences.[71] Proteins formed as a result of the activity of certain genes induce the symptoms of illnesses. To understand and manage disease it is important to investigate those proteins in the body that are responsible for all biological processes (growth, metabolism, disease, etc.). The number of these proteins is estimated at several hundred thousand. For this reason, microarray and biochip technologies are expected to play an important role in the investigation of proteomics.[83]

Proteomics research has two main objectives: (1) quantification of all the proteins expressed in a cell and (2) the functional study of thousands of protein in parallel. The standard methods for identification and quantification are 2D gel separation and mass spectrometry. Microarray-based assays are being used to study protein–protein and protein–ligand interactions because the rapid development of this technology has shortened the analysis process and made it more reliable.[84]

## 51.4 Biochips

### 51.4.1 The Integrated Biochip Concept

Biochips that integrate conventional biotechnology with semiconductor processing, microelectromechanical systems (MEMS), optoelectronics, and digital-signal and image acquisition and processing have received a great deal of interest. There is a strong need for a truly integrated biochip system that comprises probes, samplers, and detectors as well as amplifier and logic circuitry on board. Such a system will be useful in physicians' offices and can be used by relatively unskilled personnel.

An integrated biochip using bioreceptor probes and a detection system in a self-contained microdevice has been developed.[5,7–9,11] As illustrated schematically in Figure 51.3, the biochip concept involves the combination of IC elements, an electro-optics excitation/detection system, and bioreceptor probes in a self-contained and integrated microdevice. A basic biochip includes (1) an excitation light source with related optics, (2) a bioprobe, (3) a sampling platform and delivery system, (4) an optical detector with associated optics and dispersive device, and (5) a signal-amplification/treatment system.

Construction of a biochip involves the integration of several basic elements of very different natures. The basic steps are (1) the selection or development of the bioreceptor, (2) the selection of the excitation source, (3) the selection or development of the transducer, and (4) the integration of the excitation source–bioreceptor–transducer system. The development of the biochip comprises three major elements. The first element involves the development of a bioreceptor probe system: a microarray of bioreceptor probes on a multiarray sampling platform. The second element involves the development of nonradioactive methods for optical detection: the fluorescence technique. The third element involves the development of an integrated electro-optic IC system on a single chip for biosensing: a photodiode-amplifier microchip using complementary metal oxide semiconductor (CMOS) technology.

### 51.4.2 Integrated Chip Fabrication

Most microarray detection systems are very large when the excitation source and detector are considered, making them impractical for anything but laboratory usage. For biochips, however, the sensors, amplifiers, discriminators, and logic circuitry are all built onto the chip. The integrated biochip involves integrated electro-optic sensing photodetectors. Highly integrated biosensors are possible partly because multiple optical sensing elements and microelectronics can be fabricated on a single IC. The biochips include a large-area, *n*-well integrated amplifier-photodiode array that has been designed as a single, custom IC fabricated for the biochip. This IC device is coupled to the multiarray sampling platform and is designed for monitoring very low light levels. The individual photodiodes have 900-$\mu m^2$ sizes and are arrayed on a 1-mm spacing grid. The photodiodes and the accompanying electronic circuitry were fabricated using a standard 1.2-micron *n*-well CMOS process. The use of this standard process allows for the production of photodiodes and phototransistors as well as other numerous types of analog and

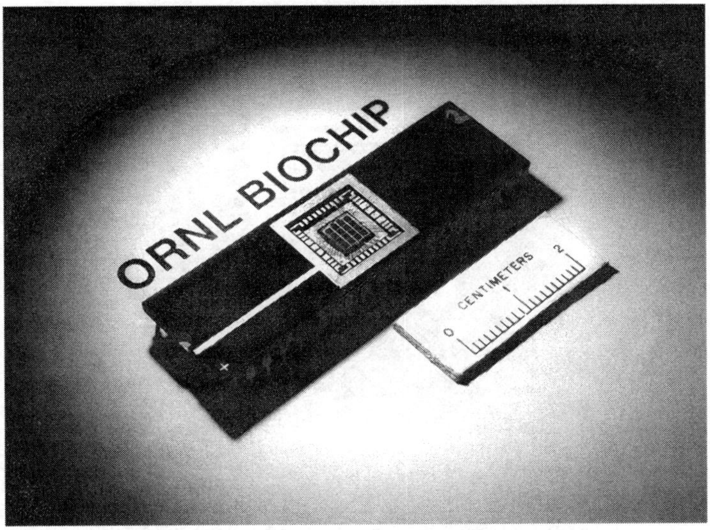


**FIGURE 51.6 (Color figure follows p. 28-30.)** Integrated circuit microchip for a biochip with 4 × 4 sensor array. (From Vo-Dinh, T., *Sensors Actuators*, B51, 52, 1998. With permission.)

digital circuitry in a single IC chip. This feature is the main advantage of the CMOS technology as compared to other detector technologies such as CCDs or charge-injection devices. The photodiodes themselves are produced using the *n*-well structure that is generally used to make resistors or as the body material for transistors. Since the anode of the diode is the p-type substrate material, which is common to every circuit on the IC chip, only the cathode is available for monitoring the photocurrent, and the photodiode is constrained to operate with a reverse bias. Figure 51.6 shows a 16 (4 × 4)-array microchip fabricated for the DNA biochip.

An analog multiplexer was designed to allow any of the elements in the array to be connected to an amplifier. The multiplexer is made from 16 cells for the 4 × 4 array device. Each cell has a CMOS switch controlled by the output of the address decoder cell. The switch is open, connecting the addressed diode to an amplifier. This arrangement makes it possible to connect a 4 × 4 (8 × 8 or 10 × 10) array of light sources (different fluorescent probes, for example) to the photodiode array and read out the signal levels sequentially.

Figure 51.7A shows a photograph of 64 (8 × 8)-array IC microchips. Figure 51.7B shows the design of the different CMOS photodiode regions and signal-processor regions of the 8 × 8 microchip. The CMOS technology makes possible highly integrated biosensors, partly through the capability of fabricating multiple optical sensing elements and microelectronics on a single IC. A 2D array of optical detector-amplifiers was integrated on a single IC chip.

## 51.4.3 Multifunctional Biochip: A Hybrid Chip for Genomics and Proteomics

The development of the first integrated DNA biochip with a phototransistor IC microchip was reported by Vo-Dinh et al.[5–7,85] This work involved the integration of 4 × 4 optical biosensor arrays onto an IC. The usefulness and potential of this DNA biochip was illustrated by the detection of gene fragments of the HIV system.[5]

Nucleic acids are not, however, the only possible substances of interest in medical diagnosis. For example, proteins also need to be analyzed using biochips with antibody probes. These antibody-based biochips are often referred to as protein chips. In general, biochips employ only one type of bioreceptor as probes, i.e., either nucleic acid or antibody probes. A novel integrated biochip system that uses multiple bioreceptors with different functionalities on the same biochip, allowing simultaneous detection of several types of biotargets on a single platform, has been developed.[8,9] This device is referred to as the multifunctional

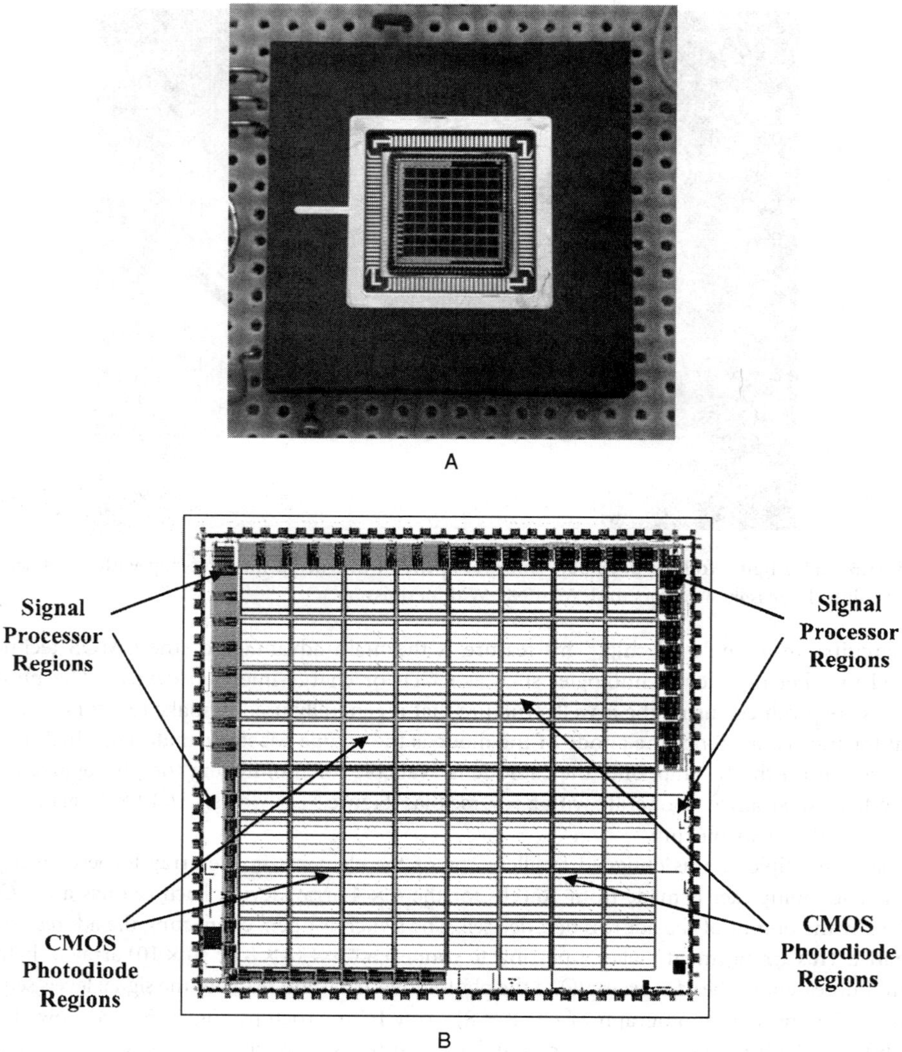


**FIGURE 51.7 (Color figure follows p. 28-30.)** (A) Photograph of the 8 × 8 integrated circuit microchip. (B) Schematic of the electronic design of the 8 × 8 microchip with CMOS photodiode regions and signal processor regions.

biochip (MFB). The integrated electro-optic microchip system developed for this work involved integrated electro-optic sensing photodetectors for the biosensor microchips.

The MFB allows for simultaneous detection of several disease endpoints using different bioreceptors, such as DNA and antibodies, on a single biochip system. The multifunctional capability of the MFB device is illustrated by measurements of different types of bioreceptors using DNA probes specific to gene fragments of the *Mycobacterium tuberculosis* (TB) system and antibody probes targeted to the cancer-related tumor-suppressor gene *p53* on a single biochip platform (Figure 51.8).

## 51.5 Applications in Biology and Medicine

### 51.5.1 Genomics Research

Microarray and biochip technologies provide powerful tools to identify all the genes as well as to understand their functions and how these components work together to coordinate the functions of cells

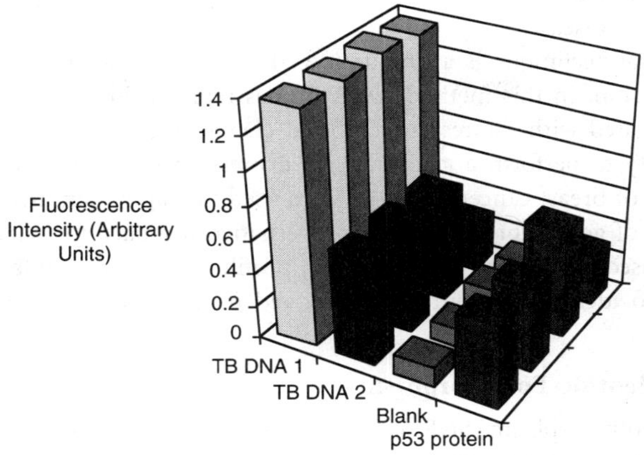


**FIGURE 51.8** The multifunctional biochip used for simultaneous detection of the p53 protein (antibody probe) and the *Microbacterium tuberculosis* gene (DNA probe).

and organisms.[21] Two important applications for the high-density DNA microarray technology involve identification of sequence (gene mutation) and determination and monitoring of expression level, e.g., mRNA abundance (gene expression).

DNA microarray and biochip technologies for genome-scaled screening and the availability of genome information will enable rapid mapping and identification of disease-related genes, which can then be analyzed in more detail. Until recently, functional studies of genes were carried out using the one-gene-one-protein-at-a-time approach. High-density microarray technology can simultaneously analyze the expression of all genes in a given organism containing thousands of different genes.

The collection of genes that are expressed or transcribed from genomic DNA (the expression profile) is a major determinant of cellular phenotype and function that changes rapidly and dramatically in response to various stimulations and during normal cellular events. Microarray hybridization has been applied for a genome-wide transcriptional analysis of cellular mitotic processes.[86] A comprehensive identification of cell-cycle-regulated genes of the yeast *Saccharomyces cerevisiae* has been accomplished.[87] Transcription alterations in DNA replication genes, genes involved with cell-cycle control and chromosomes, and a large collection of genes involved with smooth muscle function, apoptosis, intercellular adhesion, and cell motility have been measured.[88]

With the human-genome-sequencing phase now completed, the postgenomics phase of research involves information analysis and experimental activities initiated as the results of genome information. For instance, the information from open reading frames (ORFs), which indicate the particular parts within the genome encoded for proteins, allows functional assignment of the given genes. Functional assignment is performed using several methods: systematic disruption of the gene to observe the effects of the absence of the functional genes, expression of the genes in transgenic animals, and expression of recombinant proteins.

#### 51.5.1.1 Application of Human Expressed Sequence Tags and cDNA Microarrays

Human expressed sequence tags (ESTs), in combination with high-density cDNA microarray techniques, have been used to map the genome, to explore genes, and to profile downstream gene changes in a host of experiments. EST is a short strand of DNA that is a part of a cDNA molecule and can act as the identifier of a gene. ESTs represent sequences expressed in an organism under particular conditions that are often used in locating and mapping genes. For instance, in lung cancer the effects of overexpression of a tumor suppressor — the phosphatase and tensin homology deleted on chromosome 10 (PTEN) — has been studied.[89] In prostate cancer, ESTs have been used for identification of several potential markers for prostate cancer.[90] Prostate cancer tends to become transformed to androgen-independent disease over time when treated by androgen-deprivation therapy. In one study, two variants of the human prostate cancer cell line LNCaP were used to study gene-expression differences during prostate cancer progression to androgen-independent disease.[90]

The cDNA microarray technique is a unique tool that exploits this wealth of information for the analysis of gene expression. In this method, DNA probes representing cDNA clones are arrayed on a glass slide and interrogated with fluorescently labeled cDNA targets. The power of the DNA biochip technology is the ability to perform a genome-wide expression profile of thousands of genes in one experiment. In a study of breast cancer, cell lines were used for rapid identification and comparison of the differential pattern of gene expression.[91] In an alternative approach for selecting tissue-appropriate cDNAs that could be used to examine the expression profiles of developmental processes and diseases, ESTs were employed to identify 21 neural crest-derived melanocytes in a library of 198 cDNAs for microarray analysis.[92]

### 51.5.1.2 Single-Nucleotide Polymorphism

The single-nucleotide polymorphism (SNP) is the most frequent type of variation in the human genome. Although more than 99% of human DNA sequences are the same across the population, variations in DNA sequence can have a major impact on how humans respond to disease; to such environmental insults as bacteria, viruses, toxins, and chemicals; and to drugs and other therapies. Methods are being developed to detect different types of variation, particularly SNPs, that occur about once every 100 to 300 bases. SNPs provide powerful tools for a variety of medical genetic studies. It is believed that SNP maps will help researchers identify the multiple genes associated with such complex diseases as cancer, diabetes, vascular disease, and some forms of mental illness. These associations are difficult to establish with conventional gene-hunting methods because a single altered gene may make only a small contribution to disease risk. Detection of the presence of particular mutations or polymorphisms using oligonucleotide-array-based analysis of known genomic DNA sequence was first reported in 1989.[93] Genotype was determined nonradioactively by binding of streptavidin-horseradish peroxidase to the biotinylated DNA in a colorimetric assay that detected the signal intensity produced by each allele-specific probe.

In a large-scale survey for SNPs employing a combination of gel-based sequencing and high-density microarray DNA chips, 2.3 megabases of human genomic DNA were examined by a combination of gel-based sequencing and high-density variation-detection DNA chips. The study identified a total of over 3000 candidate SNPs. A genetic map was constructed showing the location of over 2000 of these SNPs. Prototype genotyping chips were developed that allow simultaneous genotyping of 500 SNPs. The study provides a characterization of human diversity at the nucleotide level and demonstrates the possibility of large-scale identification of human SNPs.[22]

Most human cancers are characterized by genomic instability, the accumulation of multiple genetic alterations, and allelic imbalance throughout the genome. Loss of heterozygosity (LOH) is a common form of allelic imbalance, and the detection of LOH has been used to identify genomic regions that harbor tumor-suppressor genes and to characterize tumor stages and progression. Researchers have used high-density oligonucleotide arrays for genome-wide scans for LOH and allelic imbalance in human tumors.[94] The detection of LOH and other chromosomal changes using large numbers of SNP markers should enable identification of patterns of allelic imbalance, with potential prognostic and diagnostic utility. Prototype genotyping chips have been produced to detect 400, 600, and 3000 of these SNPs.[95] Methods have been developed for rapid genotyping of these SNPs using oligonucleotide arrays.[96,97] Each allele of an SNP marker is represented on the array by a set of perfect-match and mismatch probes. High-density DNA-chip-based analysis has also been used to determine the distant history of SNPs, polymorphism detection, and genotyping in current human populations. The studies demonstrate that microarray-based assays allow for rapid comparative sequence analysis of intra- and interspecies genetic variation.[96,97]

### 51.5.1.3 Gene-Expression Studies

DNA microarrays and biochips can also be used to probe gene-expression patterns throughout a genome. For this purpose the probes are selected not to cover the length of an individual gene but to have sequences characteristic of different genes (so-called partial-sequence tags). When a gene is expressed (active), its code is transcribed into single-stranded mRNAs through which the gene transmits its instructions for cellular biosynthesis of a specific protein. Hence, a cell's cytoplasm contains a variety of messengers, depending on which genes are expressed. The messengers' hybridizations with one or another partial-sequence tag on a biochip can serve to reveal the patterns of genes being expressed. Microarray analyses have been successfully applied to identify amplified and overexpressed genes in chromosomal region 17q23, which is known to be augmented in up to 20% of primary breast cancers.[98]

An important objective of current cancer research is to develop a detailed molecular characterization of gene expression in tumor cells and tissues that is linked to clinical information. One study used differential display to identify approximately one quarter of all genes that were aberrantly expressed in a breast cancer cell line.[99] Two clusters of genes, represented by *p53* and maspin, had expression patterns strongly associated with estrogen-receptor status. A third cluster that included HSP-90 tended to be associated with the clinical-tumor stage, whereas a fourth cluster that included keratin 14 tended to be associated with tumor size. The expression levels of these clinically relevant gene clusters allowed breast tumors to be grouped into distinct categories. Gene-expression fingerprints that include these four gene clusters could improve prognostic accuracy and therapeutic outcomes for breast-cancer patients.[99]

High-density oligonucleotide arrays were used to measure messenger RNA levels and to evaluate the transcriptional profiles of fibroblast cultures taken from donors of various ages. mRNA levels were measured in actively dividing fibroblasts isolated from young, middle-aged, and old humans and from humans with progeria, a rare genetic disorder characterized by accelerated aging. The study identified genes whose expression was associated with age-related phenotypes and diseases. The data suggest that an underlying mechanism of the aging process involves increasing errors in the mitotic machinery of dividing cells in the postreproductive stage of life. Results indicated that a central underlying process of aging involves errors in the mitotic machinery that lead to chromosomal pathologies and ultimately to misregulation of essential structural, signaling, and metabolic genes.[100]

Complex changes in patterns of gene expression that accompany the development and progression of cancer and the experimental reversal of tumorigenicity can be investigated by microarray and biochip technologies. The tumorigenic properties of human melanoma cell line can be suppressed by introduction of a normal human chromosome 6. The introduction of this chromosome results in a reduced growth rate, restored contact inhibition, and suppressed soft agar clonogenicity and tumorigenicity in nude mice. A high-density microarray of 1161 DNA elements was used to search for differences in gene expression associated with tumor suppression in this system. Fluorescent probes for hybridization from two sources of cellular mRNA were labeled with different fluorophores to provide a direct and internally controlled comparison of the mRNA levels corresponding to each arrayed gene. The fluorescence signals representing hybridization to each arrayed gene were analyzed to determine the relative abundance in the two samples of mRNAs corresponding to each gene.[3]

High-density cDNA arrays have been used to detect tumorigenesis- and angiogenesis-related alterations in gene-transcript-expression profiles in ovarian cancer.[101] In the ovary, an abundance of specific tumor markers, increased macrophage recruitment mediators, a late-stage angiogenesis profile, and the presence of chemoresistance-related markers distinguished normal and advanced ovarian-cancer-tissue samples. The detection of such parallel changes in pathway- and tissue-specific markers may prove a potential approach ready for application in reproductive-disease diagnostic and therapeutic developments.[101]

Similarly, investigators have applied general approaches to cancer classification based on gene-expression profiles monitored by DNA microarrays to a systematic characterization of gene expression in B-cell malignancies.[102] The molecular classification of tumors on the basis of gene expression can be used to identify previously undetected and clinically significant subtypes of cancer.[102]

A damaging change during cancer progression is the switch from a locally growing tumor to a metastatic killer. This switch is believed to involve numerous alterations that allow tumor cells to complete the complex series of events needed for metastasis. DNA microarrays were used to investigate this metastasis process where a defined pattern of gene expression that correlates with progression from a locally growing tumor to a metastatic phenotype had been discerned.[103] The results demonstrated an enhanced expression of several genes involved in extracellular matrix assembly and of a second set of genes that regulate the actin-based cytoskeleton.[103]

Many genes and signaling pathways controlling cell proliferation, death and differentiation, and genomic integrity are involved in cancer development. Identification of prognostic markers and prognostic parameters for renal cell carcinoma has been performed using cDNA microarray and biochip technologies.[104] Finally, DNA microarray technologies have been applied in genomic analysis and monitoring of the expression profiles of a host of cancerous organs including the glioma,[105] ovary,[101] breast,[106] blood,[107] liver,[108] nasopharyngeal,[109] and lung.[110]

## 51.5.2 Gene-Profile Analysis

High-density DNA microarrays allow for massively parallel genomics profiling and gene-discovery studies and could provide researchers with information on tens of thousand of genes simultaneously. These arrays can be used to test an individual gene throughout all or part of its length for single-base variations from the gene's usual sequence. DNA chips, available from various commercial sources such as Affymetrix (Santa Clara, CA), can be used to detect single-base variations in the human genes *BRCA1* and *BRCA2* (breast-cancer-related), *p53* (a tumor-suppressor gene mutated in many forms of cancer), and *P450* (coding for a key liver enzyme system that metabolizes drugs). Other Affymetrix GeneChips analyze the HIV genome for variations within the code for the viral protease and reverse transcriptase. Predicting drug resistance of a given patient's viral strain is a possible application of these chips.

Measurements of the expression of thousands of genes in a single experiment, revealing many new and potentially important cancer genes, have become possible with techniques such as serial analysis of gene expression and cDNA microarrays. These genome-screening tools are capable of analyzing one tumor at a time. However, techniques to survey hundreds of specimens from patients in different stages of disease are often needed to establish the diagnostic, prognostic, and therapeutic importance of each of the emerging cancer-gene candidates. To facilitate rapid screening for molecular alterations in many different malignancies, a tissue microarray consisting of samples from 17 different tumor types was generated. Altogether, 397 individual tumors — such as breast, colon, kidney, lung, ovary, bladder, colon, stomach, testis head and neck, and endometrial cancer, as well as melanoma — were arrayed in a single paraffin block. The results confirmed and even extended existing data in the literature for the amplification of three extensively studied oncogenes, *CCND1, CMYC,* and *ERBB2.*[111]

A high-throughput microarray technique that facilitates gene expression and copy-number surveys of very large numbers of tumors has been developed.[15] This technique can analyze as many as 1000 cylindrical tissue biopsies from individual tumors distributed in a single tumor-tissue microarray. Sections of the microarray provide targets for parallel *in situ* detection of DNA, RNA, and protein targets in each specimen on the array, and consecutive sections permit rapid analysis of hundreds of molecular markers in the same set of specimens. Detection of six gene amplifications as well as *p53* and estrogen-receptor expression in breast cancer demonstrates the usefulness and potential of this method for identifying new subgroups of tumors.

Gene amplifications and deletions have often been associated with pathogenetic roles in cancer. Thirty thousand radiation-hybrid-mapped cDNAs provide a genomic resource to map these lesions with high resolution. Researchers report the development of a multiplex assay to detect small deletions and insertions by using a modified PCR to evenly amplify each amplicon (PCR/PCR), followed by ligase detection reaction (LDR).[112] This study demonstrates that microarray analysis of PCR/PCR/LDR2 products permits rapid identification of small insertion and deletion mutations in the context of both clinical diagnosis and population studies.

Another group of researchers has developed a cDNA microarray-based comparative genomic hybridization method for analyzing DNA copy-number changes across thousands of genes simultaneously.[26] With this procedure, DNA copy-number alterations of twofold or less could be reliably detected. The researchers have mapped regions of DNA copy-number variation at high resolution in breast cancer cell lines, revealing previously unrecognized genomic amplifications and deletions and new complexities of amplicon structure. They have also identified recurrent regions of DNA amplification that may harbor novel oncogenes and performed comparison and correlations of alterations of DNA copy number and gene expression in parallel analyses. For breast tumors, DNA copy-number information is being compared and correlated with data already collected on *p53* status, microarray gene-expression profiles, and treatment response and clinical outcome. Genome-wide DNA copy-number information on a set of 9 breast cancer cell lines and over 35 primary breast tumors has been collected in the study.

## 51.5.3 Pharmacogenomics and Drug Discovery

High-throughput microarray and biochip technologies have generated increasing interest in the field of pharmacogenomics, which is aimed at finding correlations between therapeutic responses to drugs and the genetic profiles of patients. It is hoped that the discovery of genetic variances that affect drug action will lead to the development of new diagnostic procedures and therapeutic products that enable drugs to be prescribed selectively to patients for whom they will be effective and safe. Many people die each year from adverse responses to medications that are beneficial to others. Others experience serious reactions or fail to respond at all. DNA variants in genes involved in drug metabolism, particularly the cytochrome *P450* multigene family, are the focus of much current research in this area. Enzymes encoded by these genes are responsible for metabolizing most drugs, and their function affects patients' responses to both the drug and the dose. Within the next decade, researchers will begin to correlate DNA variants with individual responses to medical treatments, identify particular subgroups of patients, and develop drugs customized for those populations. This "customized medicine" approach has now come to be recognized as a significant commercial opportunity for genomics.[17]

A number of attempts are ongoing to treat cystic fibrosis, a lethal recessive disease affecting children in the United States and Europe, either by gene therapy or pharmacotherapy. Recently, cDNA-array technology was used in a cystic fibrosis study of human cells concerned with global gene expression and drug effects.[113] The results provided useful pharmacogenomic information with significant relevance to both gene and pharmacological therapy. The functional similarity and specificity of different purine analogs have been determined by comparing the expression profiles of their genome-wide effects on treated yeast, murine, and human cells. Purine libraries have also been shown to be useful tools for analyzing a variety of signaling and regulatory pathways and may lead to the development of new therapeutics.[114]

A method for drug-target validation and identification of secondary drug-target effects based on genome-wide gene-expression patterns using DNA biochip microarrays has been demonstrated in several experiments, which included treatment of mutant yeast strains defective in calcineurin, immunophilins, or other genes with the immunosuppressants cyclosporin A and FK506.[115] The described method permits the direct confirmation of drug targets and recognition of drug-dependent changes in gene expression that are modulated through pathways distinct from the drug's intended target. Such a method may prove useful in improving the efficiency of drug-development programs.

The molecular mechanisms underlying the progression of prostate cancer during hormonal therapy have remained poorly understood. A study of hormone therapy for prostate cancer developed a new strategy, using of a combination of complementary cDNA and tissue-microarray technologies, for the identification of differentially expressed genes in hormone-refractory human prostate cancer.[116]

Microarray and biochip technologies, in conjunction with genomic data and bioinformatics, are expected to make drug development faster, cheaper, and more effective. Most drugs today are based on about 500 molecular targets; genomic research on the genes involved in diseases, disease pathways, and drug-response sites will lead to the discovery of thousands of new targets.

## 51.5.4 Toxicogenomics

The availability of genome-scale DNA-sequence information and reagents has led to the development of a new scientific subdiscipline combining the fields of toxicology and genomics.[23] This subdiscipline, often referred to as toxicogenomics, involves the study of how genomes respond to environmental stressors or toxicants.[117] Toxicogenomics combines genome-wide mRNA-expression profiling with protein-expression patterns using bioinformatics to understand the role of gene–environment interactions in disease and dysfunction. The goal of toxicogenomics is to find correlations between toxic responses to toxicants and changes in the genetic profiles of the objects exposed to such toxicants. For these applications, DNA microarray and biochip technologies, which allow the monitoring of the expression levels of thousands of genes simultaneously, provide an important analytical tool.[27] A general method by which gene expression, as measured by cDNA microarrays, can be used as a highly sensitive and informative marker for toxicity has been described.[23]

Arrays containing expressed-sequence tags for xenobiotic metabolizing enzymes, proteins associated with glutathione regulation, DNA-repair enzymes, heat-shock proteins, and housekeeping genes were used to examine gene expression in response to β-naphthoflavone (β-NF), with results comparable to those of Northern blotting analysis.[118]

## 51.5.5 Genetic Screening of Diseases

An important area of DNA biochip application involves screening genetic diseases. Gene tests can be used to diagnose disease, confirm a diagnosis, provide prognostic information about the course of disease, confirm the existence of a disease, and, potentially, predict the risk of future disease in healthy individuals or their progeny. Currently, several hundred genetic tests are in clinical use, with many more under development. Most current tests detect mutations associated with rare genetic disorders that follow Mendelian inheritance patterns. These diseases include myotonic and Duchenne muscular dystrophies, cystic fibrosis, neurofibromatosis type 1, sickle cell anemia, and Huntington's disease.

These screening procedures are often performed prenatally. DNA biochip technology could be a sensitive tool for the prenatal detection of genetic disorders. Newborn blood cards provide high-quality DNA samples that can reliably support highly multiplexed PCRs. In a single assay, a DNA microarray facilitates the codetection of amplification products diagnostic for several genetic diseases. Initial data utilizing the model systems of sickle cell disease, an autosomal recessive condition caused by a single amino-acid substitution in the beta-globin protein, have been reported.[119]

Genetic alterations and abnormalities could lead to uncontrolled cell division and metathesis, resulting ultimately in cancer. Genetic studies have identified a number of genes that must mutate in order to induce cancer or promote the growth of malignant cells. For example, mutations in *BRCA1,* a gene on chromosome 17, have been found to be related to breast cancer. A high percentage of women with a mutated *BRCA1* gene will develop breast cancer or have an increase in ovarian cancer. A second breast-cancer gene, *BRCA2,* is located in chromosome 13. Another important gene associated with cancer is the *p53* gene, which encodes a nuclear protein that serves as a transcription factor. The *p53* gene is a tumor-suppressor gene that regulates passage of the cell from the G1 into the S phase in the cell cycle. Mutations of *p53* have been associated with a wide variety of cancers, including colon, lung, breast, and bladder cancers. DNA biochips are available for the analysis of human genes including *BRCA1, BRCA2, p53,* and *P450.*

Information on genetic abnormalities that are characteristic of certain tumor types or stages of tumor progression is provided by the rapidly expanding database of comparative genomic hybridization (CGH) publications, which already covers about 1500 tumors. CGH data have led to the discovery of six new gene amplifications as well as a locus for a cancer-predisposition syndrome. The approach based on CGH has now been established as a first-line screening procedure for cancer researchers and will serve as a basis for ongoing efforts to develop the next-generation genome-scanning techniques such as high-resolution microarray technology.[120] Recent microarray studies have involved identification of genes involved in the development or progression of ovarian cancer,[121,122] prostate cancer,[90,123,124] breast cancer,[125] renal cell carcinoma,[104] and urinary bladder cancer.[126]

With the availability of sequence data on the human gnome, it is expected that biochips will soon be used for a wide variety of other genetic tests. The use of DNA microarrays has been suggested for evaluating renal function and disease,[127] analysis of gene expression in multiple sclerosis lesions,[128] quantitative histological analysis of Alzheimer's disease.[129]

### 51.5.6 Pathogen Detection

Another important application of miniaturized biochip devices involves the detection of biological pathogens (e.g., bacteria and viruses) present in the environment, at occupational sites such as hospitals and offices, or in public places. To achieve the required level of sensitivity and specificity in detection, it is often necessary to use a device that is capable of identifying and differentiating a large number of biochemical constituents in complex environmental samples. DNA biochip technologies could offer a unique combination of performance capabilities and analytical features of merit not available in any other current bioanalytical system. Biochip devices that combine automated sample-collection systems and multichannel sensing capability will allow simultaneous detection of multiple pathogens present in complex environmental samples. In this application, the biochip technology could provide an important tool to warn of exposure to pathogenic agents for use in human-health protection and disease prevention.

The list of expected biochip applications could include biochips for diagnosis of infectious diseases such as HIV. Nucleic-acid arrays have been constructed for many different organisms and successfully employed in a host of various experiments.[130–134] In an investigation to obtain a more comprehensive view of the global effects of HIV infection of CD4-positive T cells at the mRNA level, cDNA microarray analysis was performed on approximately 1500 cellular cDNAs at 2 and 3 d postinfection (p.i.) with HIV-1. Host-cell gene expression changed little at 2 d p.i., but at 3 d p.i., 20 cellular genes were identified as differentially expressed. Genes involved in T-cell signaling, subcellular trafficking, and transcriptional regulation, as well as several uncharacterized genes, were among those whose mRNAs were differentially regulated.[135]

Current applications of microarray research involve complete profiling of the genomic sequences of microbial pathogens and hosts. Such studies will ultimately lead to sophisticated new strategies for studying host–pathogen interactions.[136] One such study looked at *Pseudomonas aeruginosa*, an opportunistic pathogen that plays a major role in lung-function deterioration in cystic fibrosis patients. To identify critical host responses during infection, the investigators used high-density DNA microarrays, consisting of 1506 human cDNA clones, to monitor gene expression in the *A549* lung pneumocyte cell line during exposure to *P. aeruginosa.* Identification of differentially regulated genes was analyzed by expression microarray analysis of human cDNAs.[137]

Bacille Calmette-Guerin (BCC) vaccines are live attenuated strains of *Mycobacterium bovis* administered to prevent tuberculosis. To provide for rational approaches to the design of improved diagnostics and vaccines and better understand the differences between *M. tuberculosis, M. bovis,* and the various BCC daughter strains researchers studied their genomic compositions by performing comparative hybridization experiments on a DNA microarray.[138]

### 51.5.7 The Biochip for Rapid Clinical Diagnosis

In general, because high-density microarray technologies involve relatively expensive and bulky detection systems, they are useful for laboratory applications but are not appropriate for clinical use in the physician's office. On the other hand, integrated biochip systems, having medium-density arrays (10 to 100 probes) and a miniaturized detection microchip, are most appropriate for medical diagnostics at the point of care, i.e., the physician's office. Recently, an integrated biochip device with a 16-photodiode array was developed and evaluated for the detection of HIV1 gene fragments.[5,6] This integrated, portable system, described in Section 51.4, is an illustration of the usefulness of the DNA biochip for detection of a specific HIV gene sequence. Actual detection of HIV viruses will require simultaneous detection of multiple gene-sequence regions of the viruses. It has been observed that progression of AIDS causes an increase in the genotype diversity in HIV viruses. HIV viruses appear to defeat the immune system by producing and accumulating these gene mutations as the disease progresses. In this study, a specific DNA-sequence fragment was used as the model system for a feasibility demonstration. Other sequence fragments

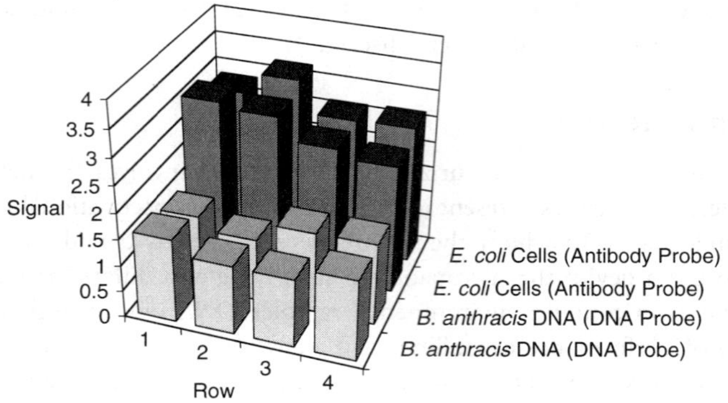


**FIGURE 51.9** Detection of *E. coli* and *B. anthracis* using the biochip system. Two rows of the biochip were used for the detection of $1 \times 10^6$ *E. coli* cells via Cy5-labeled antibody probes, while the other two rows were used for the detection of the 41.2 pmol *B. anthracis* using Cy5-labeled DNA probes.

of HIV viruses or other pathogens of medical interest, such as the *M. tuberculosis* bacterium or the *p53* cancer gene, have been used on the same biochip.[5–8]

As noted earlier, biosensors and biochips usually employ only one type of bioreceptor as probes, i.e., either nucleic-acid or antibody probes. Biochips with DNA probes are often called gene chips, and biochips with antibody probes are often called protein chips. An integrated DNA biochip that uses multiple bioreceptors with different functionalities on the same biochip, allowing simultaneous detection of several types of biotargets on a single platform, has been developed.[8–10] This device is referred to as the MFB.

The unique feature of the MFB is its ability to perform different types of bioassay on a single platform using DNA and antibody probes simultaneously. Figure 51.9 illustrates the detection of *E. coli and B. anthracis* with the MFB system. Each signal bar of the graph represents the signal acquired with an individual detection element of the $4 \times 4$ photosensor array.[9] Sensing arrays on half the biochip show the detection of $1 \times 10^6$ *E. coli* cells via Cy5-labeled antibody probes, while the other half detected the 41.2 pmol *B. anthracis* gene fragment via Cy5-labeled DNA probes. Hybridization of a nucleic-acid probe to DNA biotargets (e.g., gene sequences, bacteria, viral DNA) offers a very high degree of accuracy for identifying DNA sequences complementary to that of the probe. In addition, the MFB's antibody probes take advantage of the specificity of immunological recognition. Figures 51.10 and 51.11 demonstrate the quantitative capability of the biochip in monitoring the immobilization of various biomolecules of medical and environmental interest. Figure 51.10 shows a calibration curve for the detection of a segment of the Bac 816 gene of *B. anthracis* through hybridization of a Cy5-labeled DNA probe. Figure 51.11 shows a calibration curve for the detection of immobilized *E. coli* 157:H7. In contrast to the *B. anthracis* study described above, this screening method involved Ag–Ab interaction, analogous to the Western blot assay. This study demonstrates the feasibility of the MFB for the detection of multiple biotargets of different functionality (DNA, proteins, etc.) on a single biochip platform.

## 51.6 Conclusion

Rapid, simple, cost-effective medical devices for screening multiple diseases and infectious pathogens are essential for early diagnosis and improved treatments of many illnesses. An important factor in medical diagnostics is rapid, selective, and sensitive detection of biochemical substances or biological pathogens at ultratrace levels in biological samples (tissues, blood, and other body fluids).

In research laboratories, microarray technology enables massive parallel mining of biological data, with biological chips providing hybridization-based expression monitoring, polymorphism detection, and genotyping on a genomic scale. High-density microarray biochips containing sequences representative

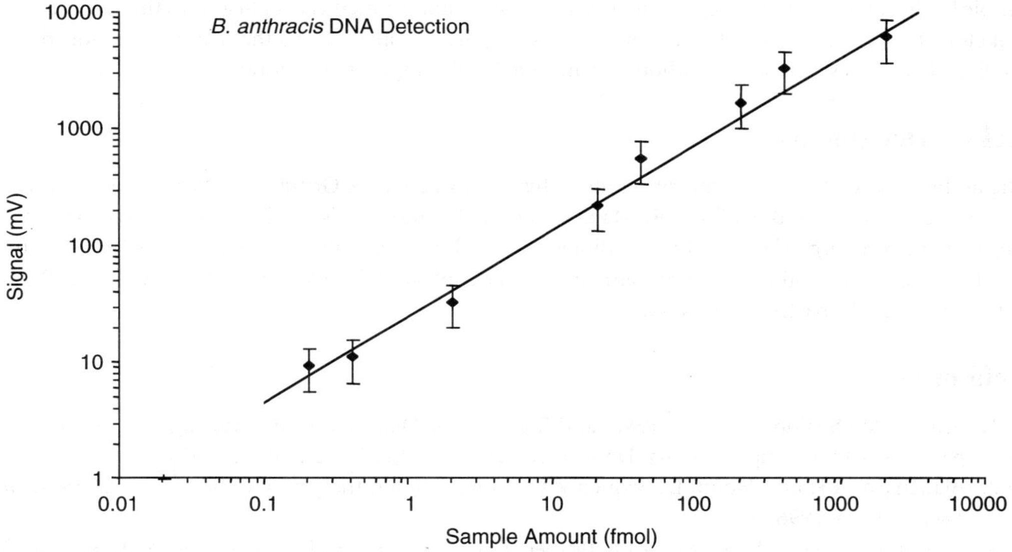


**FIGURE 51.10** Calibration curve for the detection of *B. anthracis* using DNA probes.

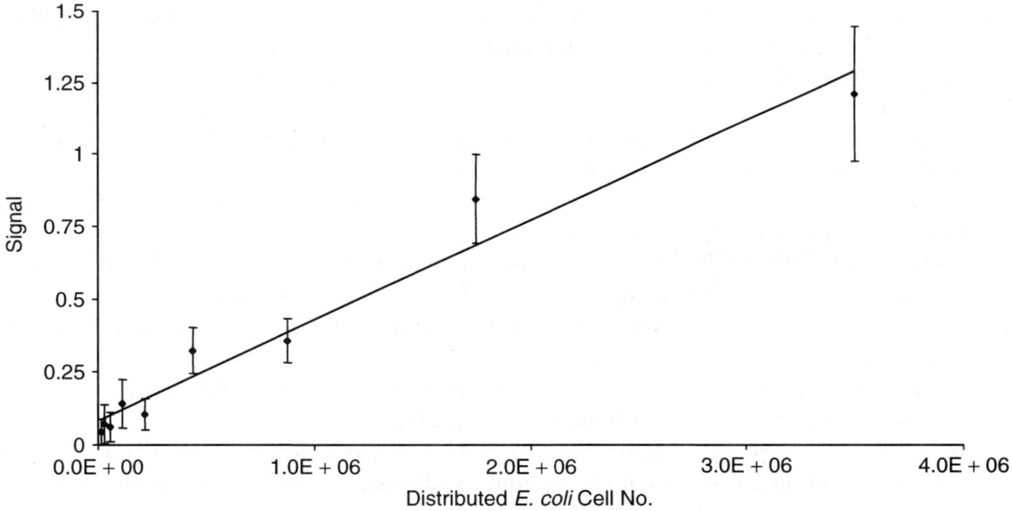


**FIGURE 51.11** Calibration curve for the detection of *E. coli* using antibody probes.

of many human genes may soon permit the expression analysis of the entire human genome in a single reaction. These "genome chips" will provide unprecedented access to key areas of human health, including disease prognosis and diagnosis, drug discovery, toxicology, aging, and mental illness.

At the physician's office, integrated biochip systems offer several advantages in size, performance, analysis capabilities, and cost because of their integrated optical sensing microchip. The small probes (microliter to nanoliter in size) minimize sample requirement and reduce reagent requirements and waste. Highly integrated systems reduce noise and increase signal due to the improved efficiency of sample collection and the reduction of interfaces. The IC technology used in the systems will allow for large-scale production and therefore low cost. The assembly of the various components of the system is made simple by the integration of several elements on a single chip. For medical applications, this cost advantage will allow for the development of extremely low-cost, disposable biochips that can be used for in-home medical diagnostics of diseases without the need to send samples to a laboratory for analysis.

## Acknowledgments

The author acknowledges the contributions of his co-workers G.D. Griffin, A.L. Wintenberg, M. Askari, D.L. Stokes, J. Mobley, B. Cullum, R. Maples, and G.H. Miller. This work was sponsored by the U.S. Department of Energy (DOE) Office of Biological and Environmental Research, by the DOE Chemical and Biological Nonproliferation Program, and under contract DE-AC05-00OR22725 with UT-Battelle, LLC, and by the Department of Navy.

## References

1. Schena, M., Shalon, D., Davis, R.W., and Brown, P.O., Quantitative monitoring of gene-expression patterns with a complementary-DNA microarray, *Science*, 270(5235), 467, 1995.
2. Blanchard, A.P. and Hood, L., Sequence to array: probing the genome's secrets, *Nat. Biotechnol.*, 14(13), 1649, 1996.
3. DeRisi, J., Penland, L., Brown, P.O., Bittner, M.L., Meltzer, P.S., Ray, M., Chen, Y.D., Su, Y.A., and Trent, J.M., Use of a cDNA microarray to analyse gene expression patterns in human cancer, *Nat. Genet.*, 14(4), 457, 1996.
4. Wallace, R.W., DNA on a chip: serving up the genome for diagnostics and research, *Mol. Med. Today*, 3(9), 384, 1997.
5. Vo-Dinh, T., Alarie, J.P., Isola, N., Landis, D., Wintenberg, A.L., and Ericson, M.N., DNA biochip using a phototransistor integrated circuit, *Anal. Chem.*, 71(2), 358, 1999.
6. Vo-Dinh, T., Development of a DNA biochip: principles and applications, *Sensors Actuators*, B51, 52, 1998.
7. Vo-Dinh, T. and Cullum, B., Biosensors and biochips: advances in biological and medical diagnostics, *Fresenius J. Anal. Chem.*, 366(6–7), 540, 2000.
8. Vo-Dinh, T., The multi-functional biochip, in *Sixth Annu. Biochip Technol. Chips Hits '99*, Cambridge Health Institute, Berkeley, CA, 1999.
9. Vo-Dinh, T., Griffin, G.D., Stokes, D.L., and Wintenberg, A.L., A Multi-functional biochip for biomedical diagnostics and pathogen detection, *Sensors Actuators*, in press.
10. Vo-Dinh, T. and Askari, M., Micro arrays and biochips: applications and potential in genomics and proteomics, *Curr. Genomics*, 2, 399, 2001.
11. Vo-Dinh, T., Cullum, B.M., and Stokes, D.L., Nanosensors and biochips: frontiers in biomolecular diagnostics, *Sensors Actuators B Chem.*, 74(1–3), 2, 2001.
12. Blanchard, G.C., Taylor, C.G., Busey, B.R., and Williamson, M.L., Regeneration of immunosorbent surfaces used in clinical, industrial and environmental biosensors: role of covalent and noncovalent interactions, *J. Immunol. Meth.*, 130(2), 263, 1990.
13. Proudnikov, D., Timofeev, E., and Mirzabekov, A., Immobilization of DNA in polyacrylamide gel for the manufacture of DNA and DNA-oligonucleotide microchips, *Anal. Biochem.*, 259(1), 34, 1998.
14. Blanchard, A.P., Kaiser, R.J., and Hood, L.E., High-density oligonucleotide arrays, *Biosensors Bioelectron.*, 11(6–7), 687, 1996.
15. Kononen, J., Bubendorf, L., Kallioniemi, A., Barlund, M., Schraml, P., Leighton, S., Torhorst, J., Mihatsch, M.J., Sauter, G., and Kallioniemi, O.P., Tissue microarrays for high-throughput molecular profiling of tumor specimens, *Nat. Med.*, 4(7), 844, 1998.
16. Marshall, A. and Hodgson, J., DNA chips: an array of possibilities, *Nat. Biotechnol.*, 16(1), 27, 1998.
17. Housman, D. and Ledley, F.D., Why pharmacogenomics? Why now? *Nat. Biotechnol.*, 16(6), 492, 1998.
18. Kricka, L.J., Revolution on a square centimeter, *Nat. Biotechnol.*, 16(6), 513, 1998.

19. Service, R.F., Microchip arrays put DNA on the spot, *Science*, 282(5388), 396, 1998.
20. Service, R.F., Future chips — labs on a chip — coming soon: the pocket DNA sequencer, *Science*, 282(5388), 399, 1998.
21. Schena, M., Heller, R.A., Theriault, T.P., Konrad, K., Lachenmeier, E., and Davis, R.W., Microarrays: biotechnology's discovery platform for functional genomics, *Trends Biotechnol.*, 16(7), 301, 1998.
22. Wang, D.G., Fan, J.B., Siao, C.J., Berno, A., Young, P., Sapolsky, R., Ghandour, G., Perkins, N., Winchester, E., Spencer, J., Kruglyak, L., Stein, L., Hsie, L., Topaloglou, T., Hubbell, E., Robinson, E., Mittmann, M., Morris, M.S., Shen, N.P., Kilburn, D., Rioux, J., Nusbaum, C., Rozen, S., Hudson, T.J., Lipshutz, R., Chee, M., and Lander, E.S., Large-scale identification, mapping, and genotyping of single-nucleotide polymorphisms in the human genome, *Science*, 280(5366), 1077, 1998.
23. Nuwaysir, E.F., Bittner, M., Trent, J., Barrett, J.C., and Afshari, C.A., Microarrays and toxicology: the advent of toxicogenomics, *Mol. Carcinog.*, 24(3), 153, 1999.
24. Ekins, R. and Chu, F.W., Microarrays: their origins and applications, *Trends Biotechnol.*, 17(6), 217, 1999.
25. Khan, J., Saal, L.H., Bittner, M.L., Chen, Y.D., Trent, J.M., and Meltzer, P.S., Expression profiling in cancer using cDNA microarrays, *Electrophoresis*, 20(2), 223, 1999.
26. Pollack, J.R., Perou, C.M., Alizadeh, A.A., Eisen, M.B., Pergamenschikov, A., Williams, C.F., Jeffrey, S.S., Botstein, D., and Brown, P.O., Genome-wide analysis of DNA copy-number changes using cDNA microarrays, *Nat. Genet.*, 23(1), 41, 1999.
27. Afshari, C.A., Nuwaysir, E.F., and Barrett, J.C., Application of complementary DNA microarray technology to carcinogen identification, toxicology, and drug safety evaluation, *Cancer Res.*, 59(19), 4759, 1999.
28. Vo-Dinh, T., Isola, N., Alarie, J.P., Landis, D., Griffin, G.D., and Allison, S., Development of a multiarray biosensor for DNA diagnostics, *Instrum. Sci. Technol.*, 26(5), 503, 1998.
29. Vo-Dinh, T., Allain, L.R., and Stokes, D.L., Cancer gene detection using SERS, *J. Raman Spectrosc.*, 33, 511, 2002.
30. Thevenot, D.R., Toth, K., Durst, R.A., and Wilson, G.S., Electrochemical biosensors: recommended definitions and classification — (technical report), *Pure Appl. Chem.*, 71(12), 2333, 1999.
31. Wang, J., Rivas, G., Fernandes, J.R., Paz, J.L.L., Jiang, M., and Waymire, R., Indicator-free electrochemical DNA hybridization biosensor, *Anal. Chim. Acta*, 375(3), 197, 1998.
32. Barker, S.L.R., Kopelman, R., Meyer, T.E., and Cusanovich, M.A., Fiber-optic nitric oxide-selective biosensors and nanosensors, *Anal. Chem.*, 70(5), 971, 1998.
33. Erdem, A., Kerman, K., Meric, B., Akarca, U.S., and Ozsoz, M., DNA electrochemical biosensor for the detection of short DNA sequences related to the hepatitis B virus, *Electroanalysis*, 11(8), 586, 1999.
34. Sawata, S., Kai, E., Ikebukuro, K., Iida, T., Honda, T., and Karube, I., Application of peptide nucleic acid to the direct detection of deoxyribonucleic acid amplified by polymerase chain reaction, *Biosensors Bioelectron.*, 14(4), 397, 1999.
35. Niemeyer, C.M., Boldt, L., Ceyhan, B., and Blohm, D., DNA-directed immobilization: efficient, reversible, and site-selective surface binding of proteins by means of covalent DNA–streptavidin conjugates, *Anal. Biochem.*, 268(1), 54, 1999.
36. Niemeyer, C.M., Ceyhan, B., and Blohm, D., Functionalization of covalent DNA-streptavidin conjugates by means of biotinylated modulator components, *Bioconjug. Chem.*, 10(5), 708, 1999.
37. Marrazza, G., Chianella, I., and Mascini, M., Disposable DNA electrochemical sensor for hybridization detection, *Biosensors Bioelectron.*, 14(1), 43, 1999.
38. Bardea, A., Patolsky, F., Dagan, A., and Willner, I., Sensing and amplification of oligonucleotide-DNA interactions by means of impedance spectroscopy: a route to a Tay-Sachs sensor, *Chem. Commun.*, 1(21), 1999.
39. Lockhart, D.J., Dong, H.L., Byrne, M.C., Follettie, M.T., Gallo, M.V., Chee, M.S., Mittmann, M., Wang, C.W., Kobayashi, M., Horton, H., and Brown, E.L., Expression monitoring by hybridization to high-density oligonucleotide arrays, *Nat. Biotechnol.*, 14(13), 1675, 1996.

40. Walt, D.R., Biological warfare detection, *Anal. Chem.*, 72(23), 738A, 2000.
41. Hacia, J.G., Woski, S.A., Fidanza, J., Edgemon, K., Hunt, N., McGall, G., Fodor, S.P.A., and Collins, F.S., Enhanced high density oligonucleotide array-based sequence analysis using modified nucleoside triphosphates, *Nucleic Acids Res.*, 26(21), 4975, 1998.
42. Lipshutz, R.J., Fodor, S.P.A., Gingeras, T.R., and Lockhart, D.J., High density synthetic oligonucleotide arrays, *Nat. Genet.*, 21(20), 1999.
43. Singh-Gasson, S., Green, R.D., Yue, Y.J., Nelson, C., Blattner, F., Sussman, M.R., and Cerrina, F., Maskless fabrication of light-directed oligonucleotide microarrays using a digital micromirror array, *Nat. Biotechnol.*, 17(10), 974, 1999.
44. McGall, G., Labadie, J., Brock, P., Wallraff, G., Nguyen, T., and Hinsberg, W., Light-directed synthesis of high-density oligonucleotide arrays using semiconductor photoresists, *Proc. Natl. Acad. Sci. U.S.A.*, 93(24), 13555, 1996.
45. Allain, L.R., Askari, M., Stokes, D.L., and Vo-Dinh, T., Microarray sampling-platform fabrication using bubble-jet technology for a biochip system, *Fresenius J. Anal. Chem.*, 371(2), 146, 2001.
46. Piehler, J., Brecht, A., Gauglitz, G., Maul, C., Grabley, S., and Zerlin, M., Specific binding of low molecular weight ligands with direct optical detection, *Biosensors Bioelectron.*, 12(6), 531, 1997.
47. Gray, D.E., CaseGreen, S.C., Fell, T.S., Dobson, P.J., and Southern, E.M., Ellipsometric and interferometric characterization of DNA probes immobilized on a combinatorial array, *Langmuir*, 13(10), 2833, 1997.
48. Edman, C.F., Raymond, D.E., Wu, D.J., Tu, E.G., Sosnowski, R.G., Butler, W.F., Nerenberg, M., and Heller, M.J., Electric field directed nucleic acid hybridization on microchips, *Nucleic Acids Res.*, 25(24), 4907, 1997.
49. Skuridin, S.G., Yevdokimov, Y.M., Efimov, V.S., Hall, J.M., and Turner, A.P.F., A new approach for creating double-stranded DNA biosensors, *Biosensors Bioelectron.*, 11(9), 903, 1996.
50. Czajkowsky, D.M., Iwamoto, H., and Shao, Z.F., Atomic force microscopy in structural biology: from the subcellular to the submolecular, *J. Electron. Microsc.*, 49(3), 395, 2000.
51. Mazzola, L.T., Frank, C.W., Fodor, S.P.A., Mosher, C., Lartius, R., and Henderson, E., Discrimination of DNA hybridization using chemical force microscopy, *Biophys. J.*, 76(6), 2922, 1999.
52. Vo-Dinh, T., Houck, K., and Stokes, D.L., Surface-enhanced Raman gene probes, *Anal. Chem.*, 66(20), 3379, 1994.
53. Vo-Dinh, T., Surface-enhanced Raman spectroscopy using metallic nanostructures, *Trends Anal. Chem.*, 17(8–9), 557, 1998.
54. Vo-Dinh, T., Stokes, D.L., Griffin, G.D., Volkan, M., Kim, U.J., and Simon, M.I., Surface-enhanced Raman scattering (SERS) method and instrumentation for genomics and biomedical analysis, *J. Raman Spectrosc.*, 30(9), 785, 1999.
55. Isola, N.R., Stokes, D.L., and Vo-Dinh, T., Surface enhanced Raman gene probe for HIV detection, *Anal. Chem.*, 70(7), 1352, 1998.
56. Jena, P.K., Liu, A.H., Smith, D.S., and Wysocki, L.J., Amplification of genes, single transcripts and cDNA libraries from one cell and direct sequence analysis of amplified products derived from one molecule, *J. Immunol. Meth.*, 190(2), 199, 1996.
57. Wang, A.M., Doyle, M.V., and Mark, D.F., Quantitation of messenger-RNA by the polymerase chain-reaction, *Proc. Natl. Acad. Sci. U.S.A.*, 86(24), 9717, 1989.
58. Kwoh, D.Y., Davis, G.R., Whitfield, K.M., Chappelle, H.L., Dimichele, L.J., and Gingeras, T.R., Transcription-based amplification system and detection of amplified human immunodeficiency virus type-1 with a bead-based sandwich hybridization format, *Proc. Natl. Acad. Sci. U.S.A.*, 86(4), 1173, 1989.
59. Guatelli, J.C., Whitfield, K.M., Kwoh, D.Y., Barringer, K.J., Richman, D.D., and Gingeras, T.R., Isothermal, *in vitro* amplification of nucleic-acids by a multienzyme reaction modeled after retroviral replication, *Proc. Natl. Acad. Sci. U.S.A.*, 87(5), 1874, 1990.
60. Luo, L., Salunga, R.C., Guo, H.Q., Bittner, A., Joy, K.C., Galindo, J.E., Xiao, H.N., Rogers, K.E., Wan, J.S., Jackson, M.R., and Erlander, M.G., Gene expression profiles of laser-captured adjacent neuronal subtypes, *Nat. Med.*, 5(1), 117, 1999.

61. Eberwine, J., Yeh, H., Miyashiro, K., Cao, Y.X., Nair, S., Finnell, R., Zettel, M., and Coleman, P., Analysis of gene-expression in single live neurons, *Proc. Natl. Acad. Sci. U.S.A.*, 89(7), 3010, 1992.
62. Bonner, R.F., Emmert Buck, M., Cole, K., Pohida, T., Chuaqui, R., Goldstein, S., and Liotta, L.A., Cell sampling-laser capture microdissection: molecular analysis of tissue, *Science*, 278(5342), 1481, 1997.
63. Ohyama, H., Zhang, X., Kohno, Y., Alevizos, I., Posner, M., Wong, D.T., and Todd, R., Laser capture microdissection-generated target sample for high-density oligonucleotide array hybridization, *Biotechniques*, 29(3), 530, 2000.
64. Cronin, M.T., Fucini, R.V., Kim, S.M., Masino, R.S., Wespi, R.M., and Miyada, C.G., Cystic fibrosis mutation detection by hybridization to light-generated DNA probe arrays, *Hum. Mutat.*, 7(3), 244, 1996.
65. Kozal, M.J., Shah, N., Shen, N.P., Yang, R., Fucini, R., Merigan, T.C., Richman, D.D., Morris, D., Hubbell, E.R., Chee, M., and Gingeras, T.R., Extensive polymorphisms observed in HIV-1 clade B protease gene using high-density oligonucleotide arrays, *Nat. Med.*, 2(7), 753, 1996.
66. Chee, M., Yang, R., Hubbell, E., Berno, A., Huang, X.C., Stern, D., Winkler, J., Lockhart, D.J., Morris, M.S., and Fodor, S.P.A., Accessing genetic information with high-density DNA arrays, *Science*, 274(5287), 610, 1996.
67. Hacia, J.G., Brody, L.C., Chee, M.S., Fodor, S.P.A., and Collins, F.S., Detection of heterozygous mutations in BRCA1 using high density oligonucleotide arrays and two-colour fluorescence analysis, *Nat. Genet.*, 14(4), 441, 1996.
68. Nikiforov, T.T., Rendle, R.B., Goelet, P., Rogers, Y.H., Kotewicz, M.L., Anderson, S., Trainor, G.L., and Knapp, M.R., Genetic bit analysis — a solid-phase method for typing single nucleotide polymorphisms, *Nucleic Acids Res.*, 22(20), 4167, 1994.
69. Pastinen, T., Kurg, A., Metspalu, A., Peltonen, L., and Syvanen, A.C., Minisequencing: a specific tool for DNA analysis and diagnostics on oligonucleotide arrays, *Genome Res.*, 7(6), 606, 1997.
70. Frohme, M., Scharm, B., Delius, H., Knecht, R., and Hoheisel, J.D., Use of representational difference analysis and cDNA arrays for transcriptional profiling of tumor tissue, in *Colorectal Cancer: New Aspects of Molecular Biology and Immunology and Their Clinical Applications*, Buhz, H., Hanski, C., Mann R., Reicken, E.O., and Scherubi, H., Eds., New York Academy of Science, New York, 2000, pp. 85–105.
71. Askari, M., Alarie, J.P., Moreno-Bondi, M., and Vo-Dinh, T., Application of an antibody biochip for p53 detection and cancer diagnosis, *Biotechnol. Progress*, 17(3), 543, 2001.
72. Hancock, W.S., Wu, S.L., and Shieh, P., The challenges of developing a sound proteomics strategy, *Proteomics*, 2(4), 352, 2002.
73. Volinia, S., Francioso, F., Venturoli, L., Tosi, L., Marastoni, M., Carinci, P., Carella, M., Stanziale, P., and Evangelisti, R., Use of peptide microarrays for assaying phospho-dependent protein interactions, *Minerva Biotechnol.*, 13(4), 281, 2001.
74. Service, R.F., Protein chips — searching for recipes for protein chips, *Science*, 294(5549), 2080, 2001.
75. Zhu, H., Klemic, J., Bilgin, M., Hall, D., Bertone, P., Gerstein, M., Reed, M., and Snyder, M., Analysis of yeast proteins using protein chips, *Yeast*, 18, S108, 2001.
76. Borrebaeck, C.A.K., Ekstrom, S., Hager, A.C.M., Nilsson, J., Laurell, T., and Marko-Varga, G., Protein chips based on recombinant antibody fragments: a highly sensitive approach as detected by mass spectrometry, *Biotechniques*, 30(5), 1126, 2001.
77. Voss, D., Protein chips — new microarrays show how proteins interact, *Technol. Rev.*, 104(4), 35, 2001.
78. Zhu, H., Klemic, J.F., Chang, S., Bertone, P., Casamayor, A., Klemic, K.G., Smith, D., Gerstein, M., Reed, M.A., and Snyder, M., Analysis of yeast protein kinases using protein chips, *Nat. Genet.*, 26(3), 283, 2000.
79. Borrebaeck, C.A.K., Antibodies in diagnostics — from immunoassays to protein chips, *Immunol. Today*, 21(8), 379, 2000.
80. Vo-Dinh, T., Tromberg, B.J., Griffin, G.D., Ambrose, K.R., Sepaniak, M.J., and Gardenhire, E.M., Antibody-based fiberoptics biosensor for the carcinogen benzo(a)pyrene, *Appl. Spectrosc.*, 41(5), 735, 1987.

81. Vo-Dinh, T., Nolan, T., Cheng, Y.F., Sepaniak, M.J., and Alarie, J.P., Phase-resolved fiberoptics fluoroimmunosensor, *Appl. Spectrosc.*, 44(1), 128, 1990.
82. Vo-Dinh, T., Alarie, J.P., Cullum, B.M., and Griffin, G.D., Antibody-based nanoprobe for measurement of a fluorescent analyte in a single cell, *Nat. Biotechnol.*, 18(7), 764, 2000.
83. Ziebolz, B., Proteomics: the treasure hunt for diagnosis and treatment, *PharmaChem,* 3(5), 2002.
84. Seeskin, E., Chemical array: the drug discovery superhighway, *Drug Discovery World,* Spring, 43, 2002.
85. Vo-Dinh, T., A hetero-functional biochip for environmental monitoring, *Abstr. Papers Am. Chem. Soc.*, 221, 57-AGRO, 2001.
86. Cho, R.J., Campbell, M.J., Winzeler, E.A., Steinmetz, L., Conway, A., Wodicka, L., Wolfsberg, T.G., Gabrielian, A.E., Landsman, D., Lockhart, D.J., and Davis, R.W., A genome-wide transcriptional analysis of the mitotic cell cycle, *Mol. Cell,* 2(1), 65, 1998.
87. Spellman, P.T., Sherlock, G., Zhang, M.Q., Iyer, V.R., Anders, K., Eisen, M.B., Brown, P.O., Botstein, D., and Futcher, B., Comprehensive identification of cell cycle-regulated genes of the yeast *Saccharomyces cerevisiae* by microarray hybridization, *Mol. Biol. Cell,* 9(12), 3273, 1998.
88. Lockhart, D.J. and Winzeler, E.A., Genomics, gene expression and DNA arrays, *Nature,* 405(6788), 827, 2000.
89. Hong, T.M., Yang, P.C., Peck, K., Chen, J.J.W., Yang, S.C., Chen, Y.C., and Wu, C.W., Profiling the downstream genes of tumor suppressor PTEN in lung cancer cells by complementary DNA microarray, *Am. J. Resp. Cell Mol. Biol.*, 23(3), 355, 2000.
90. Vaarala, M.H., Porvari, K., Kyllonen, A., and Vihko, P., Differentially expressed genes in two LNCaP prostate cancer cell lines reflecting changes during prostate cancer progression, *Lab. Invest.*, 80(8), 1259, 2000.
91. Yang, G.P., Ross, D.T., Kuang, W.W., Brown, P.O., and Weigel, R.J., Combining SSH and cDNA microarrays for rapid identification of differentially expressed genes, *Nucleic Acids Res.*, 27(6), 1517, 1999.
92. Loftus, S.E., Chen, Y., Gooden, G., Ryan, J.F., Birznieks, G., Hilliard, M., Baxevanis, A.D., Bittner, M., Meltzer, P., Trent, J., and Pavan, W., Informatic selection of a neural crest-melanocyte cDNA set for microarray analysis, *Proc. Natl. Acad. Sci. U.S.A.*, 96(16), 9277, 1999.
93. Saiki, R.K., Walsh, P.S., Levenson, C.H., and Erlich, H.A., Genetic-analysis of amplified DNA with immobilized sequence- specific oligonucleotide probes, *Proc. Natl. Acad. Sci. U.S.A.*, 86(16), 6230, 1989.
94. Mei, R., Galipeau, P.C., Prass, C., Berno, A., Ghandour, G., Patil, N., Wolff, R.K., Chee, M.S., Reid, B.J., and Lockhart, D.J., Genome-wide detection of allelic imbalance using human SNPs and high-density DNA arrays, *Genome Res.*, 10(8), 1126, 2000.
95. Sapolsky, R.J., Hsie, L., Berno, A., Ghandour, G., Mittmann, M., and Fan, J.B., High-throughput polymorphism screening and genotyping with high-density oligonucleotide arrays, *Genet. Anal.-Biomol. Eng.*, 14(5–6), 187, 1999.
96. Rose, N.R, Bigazzi, P.E, and Rapid, T.-A., *Methods in Immunodiagnosis*, John Wiley & Sons, New York, 1999.
97. Hacia, J.G., Fan, J.B., Ryder, O., Jin, L., Edgemon, K., Ghandour, G., Mayer, R.A., Sun, B., Hsie, L., Robbins, C.M., Brody, L.C., Wang, D., Lander, E.S., Lipshutz, R., Fodor, S.P.A., and Collins, F.S., Determination of ancestral alleles for human single-nucleotide polymorphisms using high-density oligonucleotide arrays, *Nat. Genet.*, 22(2), 164, 1999.
98. Barlund, M., Forozan, F., Kononen, J., Bubendorf, L., Chen, Y.D., Bittner, M.L., Torhorst, J., Haas, P., Bucher, C., Sauter, G., Kallioniemi, O.P., and Kallioniemi, A., Detecting activation of ribosomal protein S6 kinase by complementary DNA and tissue microarray analysis, *J. Natl. Cancer Inst.*, 92(15), 1252, 2000.
99. Martin, K., Kritzman, D.M., Price, L.M., Koh, B., Kwan, C.P., Zhang, X.H., Mackay, A., O'Hare, M.J., Kaelin, D.M., Mutter, G.L., Pardee, A.B., and Sager, R., Linking gene expression patterns to therapeutic groups in breast cancer, *Cancer Res.*, 60(8), 2232, 2000.

100. Ly, D.H., Lockhart, D.J., Lerner, R.A., and Schultz, P.G., Mitotic misregulation and human aging, *Science*, 287(5462), 2486, 2000.
101. Martoglio, A.M., Tom, B.D.M., Starkey, M., Corps, A.N., Charnock-Jones, D.S., and Smith, S.K., Changes in tumorigenesis- and angiogenesis-related gene transcript abundance profiles in ovarian cancer detected by tailored high density cDNA arrays, *Mol. Med.*, 6(9), 750, 2000.
102. Alizadeh, A.A., Eisen, M.B., Davis, R.E., Ma, C., Lossos, I.S., Rosenwald, A., Boldrick, J.G., Sabet, H., Tran, T., Yu, X., Powell, J.I., Yang, L.M., Marti, G.E., Moore, T., Hudson, J., Lu, L.S., Lewis, D.B., Tibshirani, R., Sherlock, G., Chan, W.C., Greiner, T.C., Weisenburger, D.D., Armitage, J.O., Warnke, R., Levy, R., Wilson, W., Grever, M.R., Byrd, J.C., Botstein, D., Brown, P.O., and Staudt, L.M., Distinct types of diffuse large B-cell lymphoma identified by gene expression profiling, *Nature*, 403(6769), 503, 2000.
103. Clark, E.A., Golub, T.R., Lander, E.S., and Hynes, R.O., Genomic analysis of metastasis reveals an essential role for RhoC, *Nature*, 406(6795), 532, 2000.
104. Moch, H., Schraml, P., Bubendorf, L., Mirlacher, M., Kononen, J., Gasser, T., Mihatsch, M.J., Kallioniemi, O.P., and Sauter, G., High-throughput tissue microarray analysis to evaluate genes uncovered by cDNA microarray screening in renal cell carcinoma, *Am. J. Pathol.*, 154(4), 981, 1999.
105. Galloway, A.M. and Allalunis-Turner, J., cDNA expression array analysis of DNA repair genes in human glioma cells that lack or express DNA-PK, *Radiat. Res.*, 154(6), 609, 2000.
106. Perou, C.M., Sorlie, T., Eisen, M.B., van de Rijn, M., Jeffrey, S.S., Rees, C.A., Pollack, J.R., Ross, D.T., Johnsen, H., Aksien, L.A., Fluge, O., Pergamenschikov, A., Williams, C., Zhu, S.X., Lonning, P.E., Borresen-Dale, A.L., Brown, P.O., and Botstein, D., Molecular portraits of human breast tumours, *Nature*, 406(6797), 747, 2000.
107. Liu, T.X., Zhang, J.W., Tao, J., Zhang, R.B., Zhang, Q.H., Zhao, C.J., Tong, J.H., Lanotte, M., Waxman, S., Chen, S.J., Mao, M., Hu, G.X., Zhu, L., and Chen, Z., Gene expression networks underlying retinoic acid-induced differentiation of acute promyelocytic leukemia cells, *Blood*, 96(4), 1496, 2000.
108. Saurin, J.C., Joly-Pharaboz, M.O., Pernas, P., Henry, L., Ponchon, T., and Madjar, J.J., Detection of Ki-ras gene point mutations in bile specimens for the differential diagnosis of malignant and benign biliary strictures, *Gut*, 47(3), 357, 2000.
109. Fung, L.F., Lo, A.K.F., Yuen, P.W., Liu, Y., Wang, X.H., and Tsao, S.W., Differential gene expression in nasopharyngeal carcinoma cells, *Life Sci.*, 67(8), 923, 2000.
110. Lindblad-Toh, K., Tanenbaum, D.M., Daly, M.J., Winchester, E., Lui, W.O., Villapakkam, A., Stanton, S.E., Larsson, C., Hudson, T.J., Johnson, B.E., Lander, E.S., and Meyerson, M., Loss-of-heterozygosity analysis of small-cell lung carcinomas using single-nucleotide polymorphism arrays, *Nat. Biotechnol.*, 18(9), 1001, 2000.
111. Schraml, P., Kononen, J., Bubendorf, L., Moch, H., Bissig, H., Nocito, A., Mihatsch, M.J., Kallioniemi, O.P., and Sauter, G., Tissue microarrays for gene amplification surveys in many different tumor types, *Clin. Cancer Res.*, 5(8), 1966, 1999.
112. Favis, R., Day, J.P., Gerry, N.P., Phelan, C., Narod, S., and Barany, F., Universal DNA array detection of small insertions and deletions in BRCA1 and BRCA2, *Nat. Biotechnol.*, 18(5), 561, 2000.
113. Srivastava, M., Eidelman, O., and Pollard, H.B., Pharmacogenomics of the cystic fibrosis transmembrane conductance regulator (CFTR) and the cystic fibrosis drug CPX using genome microarray analysis, *Mol. Med.*, 5(11), 753, 1999.
114. Gray, N.S., Wodicka, L., Thunnissen, A., Norman, T.C., Kwon, S.J., Espinoza, F.H., Morgan, D.O., Barnes, G., LeClerc, S., Meijer, L., Kim, S.H., Lockhart, D.J., and Schultz, P.G., Exploiting chemical libraries, structure, and genomics in the search for kinase inhibitors, *Science*, 281(5376), 533, 1998.
115. Marton, M.J., DeRisi, J.L., Bennett, H.A., Iyer, V.R., Meyer, M.R., Roberts, C.J., Stoughton, R., Burchard, J., Slade, D., Dai, H.Y., Bassett, D.E., Hartwell, L.H., Brown, P.O., and Friend, S.H., Drug target validation and identification of secondary drug target effects using DNA microarrays, *Nat. Med.*, 4(11), 1293, 1998.

116. Bubendorf, L., Kolmer, M., Kononen, J., Koivisto, P., Mousses, S., Chen, Y.D., Mahlamaki, E., Schraml, P., Moch, H., Willi, N., Elkahloun, A.G., Pretlow, T.G., Gasser, T.C., Mihatsch, M.J., Sauter, G., and Kallioniemi, O.P., Hormone therapy failure in human prostate cancer: analysis by complementary DNA and issue microarrays, *J. Natl. Cancer Inst.*, 91(20), 1758, 1999.
117. Medlin, J.F., Timely toxicology, *Environ. Health Perspect.*, 107(5), A256, 1999.
118. Bartosiewicz, M., Trounstine, M., Barker, D., Johnston, R., and Buckpitt, A., Development of a toxicological gene array and quantitative assessment of this technology, *Arch. Biochem. Biophys.*, 376(1), 66, 2000.
119. Dobrowolski, S.F., Banas, R.A., Naylor, E.W., Powdrill, T., and Thakkar, D., DNA microarray technology for neonatal screening, *Acta Paediatr.*, 88, 61, 1999.
120. Forozan, F., Karhu, R., Kononen, J., Kallioniemi, A., and Kallioniemi, O.P., Genome screening by comparative genomic hybridization, *Trends Genet.*, 13(10), 405, 1997.
121. Wang, K., Gan, L., Jeffery, E., Gayle, M., Gown, A.M., Skelly, M., Nelson, P.S., Ng, W.V., Schummer, M., Hood, L., and Mulligan, J., Monitoring gene expression profile changes in ovarian carcinomas using cDNA microarray, *Gene*, 229(1–2), 101, 1999.
122. Ono, K., Tanaka, T., Tsunoda, T., Kitahara, O., Kihara, C., Okamoto, A., Ochiai, K., Takagi, T., and Nakamura, Y., Identification by cDNA microarray of genes involved in ovarian carcinogenesis, *Cancer Res.*, 60(18), 5007, 2000.
123. Xu, J.C., Stolk, J.A., Zhang, X.Q., Silva, S.J., Houghton, R.L., Matsumura, M., Vedvick, T.S., Leslie, K.B., Badaro, R., and Reed, S.G., Identification of differentially expressed genes in human prostate cancer using subtraction and microarray, *Cancer Res.*, 60(6), 1677, 2000.
124. Elek, J., Park, K.H., and Narayanan, R., Microarray-based expression profiling in prostate tumors, *In Vivo*, 14(1), 173, 2000.
125. Sgroi, D.C., Teng, S., Robinson, G., LeVangie, R., Hudson, J.R., and Elkahloun, A.G., *In vivo* gene expression profile analysis of human breast cancer progression, *Cancer Res.*, 59(22), 5656, 1999.
126. Richter, J., Wagner, U., Kononen, J., Fijan, A., Bruderer, J., Schmid, U., Ackermann, D., Maurer, R., Alund, G., Knonagel, H., Rist, M., Wilber, K., Anabitarte, R., Hering, F., Hardmeier, T., Schonenberger, A., Flury, R., Jager, P., Fehr, J.L., Schraml, P., Moch, H., Mihatsch, M.J., Gasser, T., Kallioniemi, O.P., and Sauter, G., High-throughput tissue microarray analysis of cyclin E gene amplification and overexpression in urinary bladder cancer, *Am. J. Pathol.*, 157(3), 787, 2000.
127. Hsiao, L.L., Stears, R.L., Hong, R.L., and Gullans, S.R., Prospective use of DNA microarrays for evaluating renal function and disease, *Curr. Opinion Nephrol. Hypertension*, 9(3), 253, 2000.
128. Whitney, L.W., Becker, K.G., Tresser, N.J., Caballero-Ramos, C.I., Munson, P.J., Prabhu, V.V., Trent, J.M., McFarland, H.F., and Biddison, W.E., Analysis of gene expression in multiple sclerosis lesions using cDNA microarrays, *Ann. Neurol.*, 46(3), 425, 1999.
129. Hanzel, D.K., Trojanowski, J.Q., Johnston, R.F., and Loring, J.F., High-throughput quantitative histological analysis of Alzheimer's disease pathology using a confocal digital microscanner, *Nat. Biotechnol.*, 17(1), 53, 1999.
130. DeRisi, J.L., Iyer, V.R., and Brown, P.O., Exploring the metabolic and genetic control of gene expression on a genomic scale, *Science*, 278(5338), 680, 1997.
131. Wodicka, L., Dong, H.L., Mittmann, M., Ho, M.H., and Lockhart, D.J., Genome-wide expression monitoring in *Saccharomyces cerevisiae, Nat. Biotechnol.*, 15(13), 1359, 1997.
132. White, K.P., Rifkin, S.A., Hurban, P., and Hogness, D.S., Microarray analysis of *Drosophila* development during metamorphosis, *Science*, 286(5447), 2179, 1999.
133. Chambers, J., Angulo, A., Amaratunga, D., Guo, H.Q., Jiang, Y., Wan, J.S., Bittner, A., Frueh, K., Jackson, M.R., Peterson, P.A., Erlander, M.G., and Ghazal, P., DNA microarrays of the complex human cytomegalovirus genome: profiling kinetic class with drug sensitivity of viral gene expression, *J. Virol.*, 73(7), 5757, 1999.

134. Gingeras, T.R., Ghandour, G., Wang, E.G., Berno, A., Small, P.M., Drobniewski, F., Alland, D., Desmond, E., Holodniy, M., and Drenkow, J., Simultaneous genotyping and species identification using hybridization pattern recognition analysis of generic *Mycobacterium* DNA arrays, *Genome Res.*, 8(5), 435, 1998.
135. Geiss, G.K., Bumgarner, R.E., An, M.C., Agy, M.B., van't Wout, A.B., Hammersmark, E., Carter, V.S., Upchurch, D., Mullins, J.I., and Katze, M.G., Large-scale monitoring of host cell gene expression during HIV-1 infection using cDNA microarrays, *Virology*, 266(1), 8, 2000.
136. Cummings, C.A. and Relman, D.A., Using DNA microarrays to study host-microbe interactions, *Emerging Infect. Dis.*, 6(5), 513, 2000.
137. Ichikawa, J.K., Norris, A., Bangera, M.G., Geiss, G.K., van't Wout, A.B., Bumgarner, R.E., and Lory, S., Interaction of *Pseudomonas aeruginosa* with epithelial cells: identification of differentially regulated genes by expression microarray analysis of human cDNAs, *Proc. Natl. Acad. Sci. U.S.A.*, 97(17), 9659, 2000.
138. Behr, M.A., Wilson, M.A., Gill, W.P., Salamon, H., Schoolnik, G.K., Rane, S., and Small, P.M., Comparative genomics of BCG vaccines by whole-genome DNA microarray, *Science*, 284(5419), 1520, 1999.

# 52

# Array Technologies and Multiplex Genetic Analysis


Youxiang Wang*
*ACLARA Biosciences, Inc.*
*Mountain View, California*

Carmen Virgos
*ACLARA Biosciences, Inc.*
*Mountain View, California*

Sharat Singh
*ACLARA Biosciences, Inc.*
*Mountain View, California*

Maureen T. Cronin**
*ACLARA Biosciences, Inc.*
*Mountain View, California*

Stephen J. Williams
*ACLARA Biosciences, Inc.*
*Mountain View, California*

Elaine S. Mansfield***
*ACLARA Biosciences, Inc.*
*Mountain View, California*




## 52.1 Introduction

One of the greatest challenges in human genetics is to systematically identify and classify genetic polymorphisms and elucidate their contribution to human health and disease. During the past 20 years, the genetic components of more than 1000 human monogenetic disorders have been identified.[1] However, many fewer genetic factors important to common polygenic diseases have been identified. These complex disorders result from the cumulative effects of many independent susceptibility loci, none of which alone is either necessary or sufficient to cause disease. Because of this, the classical mapping strategies for monogenetic disorders, particularly linkage studies, have been largely unsuccessful in identifying the principal genetic components of complex diseases.

The analysis of DNA-sequence variation is important, particularly in dissecting the genetic basis of disease and in gauging individual xenobiotic responsiveness.[2] Several types of DNA variants exist in the human genome: single nucleotide polymorphisms (SNPs), microsatellite markers, and deletions or insertions of single or multiple base pairs. SNPs are highly abundant (on average one per thousand bases) and generally randomly distributed across the genome. By definition the least frequent allele of an SNP occurs in 1% or more of the population, while disease markers may have a much lower frequency.[3–7] The International SNP Map Working Group has recently introduced a high-density map of publicly available SNPs with the aim of providing a public resource to support human population genetic studies and to help in the identification of diagnostic and therapeutic targets for common diseases.[8]

The human genome is estimated to contain more than 3 million SNPs[3–6] and, as of August 2001, 2.98 million SNPs have been deposited into the public database, dbSNP, at the National Center for Biotechnology Information (http://www.ncbi.nlm.gov/SNP). This collection includes 1.79 million nonredundant


Current affiliations: * Turnerdesigns, Sunnyvale, California
** Genomic Health, Inc., Redwood City, California
*** Department of Nephrology, Stanford University School of Medicine, Stanford, California



0-8493-1116-0/03/$0.00+$1.50

SNPs, making this an extremely useful set of markers for a wide range of genetic studies. In particular, genotyping large numbers of SNPs in linkage or large-scale association studies is expected to provide sufficient power to detect susceptibility loci with weak penetrance in complex disease traits including common human diseases and altered drug responses.[9] For example, high-density SNP analysis would be useful in studies to identify contributions of individual genetic polymorphisms to complex diseases such as diabetes in which many genes have some influence. New drug targets and informative genetic diagnostic tests are expected to emerge from studies of this type. However, improved technologies capable of genotyping large numbers of SNPs across large sample sets accurately, rapidly, and cost-effectively are needed to make these studies feasible.

Short tandem repeat (STR) sequences, also known as microsatellite polymorphisms, are another type of informative DNA-sequence variation. STRs consist of tandemly repeated DNA-sequence elements (2 to 5 bp) that are highly polymorphic and widely distributed throughout the human genome. These characteristics have made them ideal tools for genomic mapping and for individual identification where they have been widely applied in paternity and forensic analysis. A human microsatellite map including more than 9000 STR markers is now available.[10] Similar maps are appearing for other organisms and have become the basis for "quantitative trait locus" (QTL) mapping, a method that can reveal levels of genetic complexity underlying observed morphological and behavioral traits. New generations of genetic-analysis devices, including instruments and methods capable of performing sensitive, high-resolution, high-throughput STR separations, will support these and future STR-based studies.

Traditionally, slab-gel electrophoresis has been widely used as an analytical technique to study genetic variation. However, increasing demand for high-throughput genotyping requires fast, parallel, high-resolution analysis of multiple targets and multiple samples. The performance of slab-gel systems is limited in throughput because they provide relatively slow separations and require significant manual intervention; thus, they do not meet current requirements. By comparison, capillary electrophoresis (CE) eliminates manual gel pouring and sample-loading steps and offers excellent separation speed and resolution. High-throughput CE analyses are possible because sample injection and gel loading can be automated and 96 capillaries can be run simultaneously in an array format.[11] Overall throughput can be further improved by using multiplexed assay formats; for example, dyes with different fluorescence emission wavelengths to discriminate targets can be combined in a single analytical run. In the present paper, we will summarize array-based genotyping technologies and describe in particular plastic microfluidic device technology as it may be applied to achieve high-throughput screening of SNPs and STRs.

## 52.2 Array-Based Technologies

Array technologies are driven primarily by the need to increase throughput and automation while reducing costs by limiting consumption of expensive reagents. System miniaturization and integration are the enabling technologies that provide accurate, fast analysis of large sample sets while providing cost reduction. Array technologies provide increased throughput by increasing either the number of samples that can be analyzed simultaneously (e.g., capillary array electrophoresis) or per-sample information content (e.g., microarrays).

Oligonucleotide microarray technology enables genomic research and potentially diagnostic applications since it provides a means of simultaneously analyzing thousands of DNA sequences. DNA microarrays or DNA chips are libraries of synthetic oligonucleotides or PCR products generated form cDNA attached to a substrate such as coated glass, silicon, or plastic or nylon membranes. Arrays are organized in such a way that each sequence has a defined location. Microarrays can be used in applications that require the sequence-specific detection of nucleic acids such as gene-expression profiling and SNP scoring. There are four general types of commercially available arrays: printed filter arrays, printed glass slide arrays, *in situ* synthesized oligonucleotide arrays, and electronically addressable chip arrays. In addition, equipment and reagents for producing printed arrays are commercially available, and detailed protocols for microarray printing can be found on the Web (e.g., http://cmgm.stanford.edu/pbrown/). For a review of the available instrumentation and associated technology used to produce arrays, see Reference 12.

After hybridization, microarrays are scanned to measure the relative abundance of labeled target hybridized to complementary immobilized probes. Typical protocols for hybridizing filter arrays require that the hybridization targets (DNA or RNA) be labeled with radioisotopes, such as $^{33}P$, which can be detected using a PhosphorImager® (Molecular Dynamics, Sunnyvale, CA) or x-ray film. Targets hybridized to plastic- or glass-slide arrays are typically labeled with one or more fluorescent dyes. Labeling methods incorporating multiple fluorescent dyes, coupled with high-resolution imaging, offer the advantage of measuring comparative signal abundance, allowing relative quantitation measurements to be made more accurately, as spot-to-spot signal variation can be normalized. In this way, more representative quantitation can be assessed for each of the hundreds to thousands of probes represented in the array.

Microarrays are valuable tools for surveying the "big picture" for biological processes by simultaneously looking at the interactions of multiple genes or polymorphisms. However, typically only one sample is processed per array analysis, and therefore large numbers of hybridization arrays are needed for most genetic association studies. In contrast, capillary array electrophoresis (CAE) provides two-dimensional multiplicity, as many measurements may be made in parallel on many samples.

The mobility-based separation mechanism of gel electrophoresis is broadly applicable to the analysis of a wide range of molecules including proteins and nucleic acids. A practical solution to the need for high-throughput analyses using CE was provided by the introduction of CAE in 1992 by Huang et al.[13] In CAE, individual capillaries are loaded independently but bundled together for parallel, simultaneous detection using a scanning detection system that addresses the entire array of capillaries during each analytical run. Generally, small (50- to 100-µm internal diameter, 200-µm outer diameter) gel-filled capillaries are bundled near the array outlet to provide the equivalent of high-density, closely packed "gel lanes." Several CAE systems are now commercially available for medium- and high-throughput genetic analysis. These generally include a high degree of automation (sample loading, electrophoresis, and analysis) and reduced hands-on operator time.[12,14]

Sequenom[15] has developed another array format technology that is based on matrix-assisted laser desorption ionization time-of-flight (MALDI-TOF) mass spectrometry. This DNA MassArray™ Technology integrates sample preparation with signal detection and a low level of multiplexing (3 to 5 SNPs per assay) in a fully automated process to support high-throughput SNP analysis. In addition, readout signals can be accurately quantified, which means that DNA sample pooling can be used for estimating the allele frequencies of polymorphic markers to provide an additional degree of efficiency to the process.

In the postgenomic era, the scientific community's appetite for large volumes of biological information is unlikely to diminish, and continually more advanced approaches to high-throughput genetic analysis will be needed. Since CE was demonstrated in microfabricated devices in the early 1990s,[16–18] the "lab-on-a-chip" devices have increased the speed and throughput of genetic analysis by an order of magnitude. The precise control of injected sample plug volumes and the short separation lengths afforded by microfabricated chips relative to standard CE have significantly reduced separation time.[19] Furthermore, chip-based devices provide a single platform to integrate all processing steps from sample preparation through analysis on the same platform and eventually on the same device.

## 52.3 Plastic Chips and Their Application to Electrophoresis

Many different manufacturing procedures and substrate materials have been investigated for microchip CE-device fabrication. Early devices were constructed using glass or quartz because these substrates are optically transparent and exhibit electroosmotic flow properties similar to those of fused silica.[17,20] Glass chips are generally constructed using modified processes developed to manufacture silicon chips. Disadvantages of glass devices include limitations on the geometry of etched features (imposed by the isotropic acid-etch characteristics of glass), the high temperature bonding required for sealing, and the fragile nature of the final device.

While early microfabricated electrophoretic devices used glass or silica as substrates, a few groups began exploring the use of polymeric microstructures. For example, Eckstrom et al.[21] and Soane and Soane[22] described the use of various polymers including fluoropolymers and silicone rubber. The ease

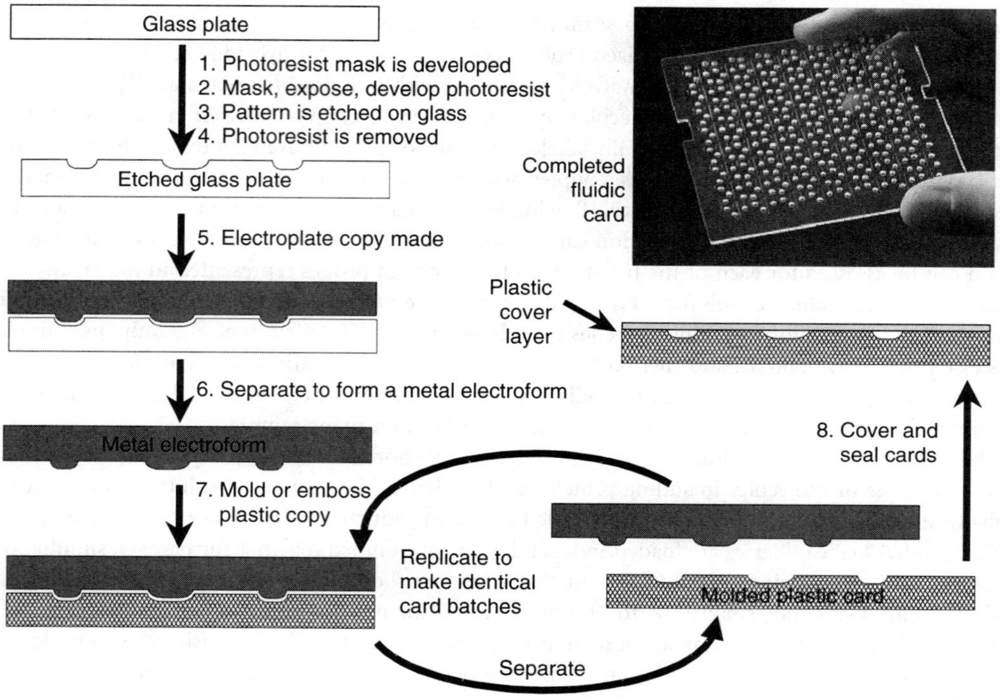


**FIGURE 52.1** Plastic LabCard™ device fabrication process.

of fabrication and relatively low cost of devices manufactured from these materials have made them an attractive alternative to optical-quality glass. In particular, there has been growing interest in the application of poly (dimethylsiloxane)-based (PDMS) microchip CE devices.[23,24] PDMS is a hydrophobic elastomer in which channels can be formed by curing the polymer over a master tool formed with the desired microchip features. Many devices can be replicated from a single master without requiring expensive clean-room facilities. To date, several groups have demonstrated CE-based separations in PDMS microchip devices.[25–27]

ACLARA Biosciences, Inc. (Mountain View, CA) is a leading developer of plastic microfluidic array technologies, often referred to as "lab-on-a-chip" systems. ACLARA's proprietary technology targets genomic applications (RNA and DNA analysis) and high-throughput pharmaceutical drug screening. Their plastic LabCard™ microfluidic array devices are mass produced and disposable. This allows researchers to rapidly perform large numbers of biochemical measurements in a format that is inexpensive, miniaturized, automated, and free from sample cross-contamination worries.

Use of fabrication technology to manufacture miniaturized channel networks for electrophoresis in thin, plastic substrates is an expertise at ACLARA.[28] Microfluidic devices are made of interconnected networks of microchannels and tiny-volume reservoirs. Fluids are transported between reservoirs and through interconnected microchannels using electrokinetic mechanisms. These multichannel plastic LabCard devices are replicated from masters (typically formed in glass, silicon, or nickel) by processes that include injection molding, compression molding, casting, and hot embossing. The master is produced either by direct etching of a material (for example, acid-etching of glass) or, in the case of nickel masters, by electroplating nickel from an aqueous solution directly onto an etched substrate. The latter approach yields a robust master than can be used many thousands of times to produce molded plastic devices. The process flow to generate substrates with high-precision patterns of microsize channel features is illustrated in Figure 52.1.

Confocal laser-induced fluorescence (LIF) is used as a sensitive detection method for separations performed in microchips. The effective rejection of background fluorescence with this approach can enhance signal-to-noise ratios, especially for plastic microfluidic devices, which frequently contain plasticizers that can yield high fluorescent background signals. Experimental procedures and the instrumental setup for scanning devices have been described in detail elsewhere.[29] Briefly, the "breadboard" workstations used include an inverted epifluorescence microscope and custom-built power supplies. A laser or filtered ultraviolet lamp provides the excitation source, while the fluorescent emission is detected by a photomultiplier tube or charge-coupled device (CCD).

## 52.4 Multiplex STR Analysis

STR polymorphisms are most easily scored by PCR amplification, followed by size separation of amplicons using denaturing electrophoresis. To address increased typing demands in biomedical research and individual identification applications, STR analyses have been converted from slab-gel systems to CE and, more recently, to CAE systems.[30] Microfluidic arrays offer additional advantages over traditional CAE analysis of STR markers because separation times can be dramatically reduced. High-throughput, multiplexed STR genotyping has recently been performed in multichannel plastic devices.[31] Figure 52.2 shows the electrophoretic separation of a multiplexed STR allelic ladder (AmpFLSTR® Blue, Applied Biosystems, Foster City, CA) using a multichannel plastic LabCard device. The high-resolution separation (to 300 bp) was complete in less than 10 min, much faster than typically achieved by standard CE, CAE, or slab-gel electrophoresis.

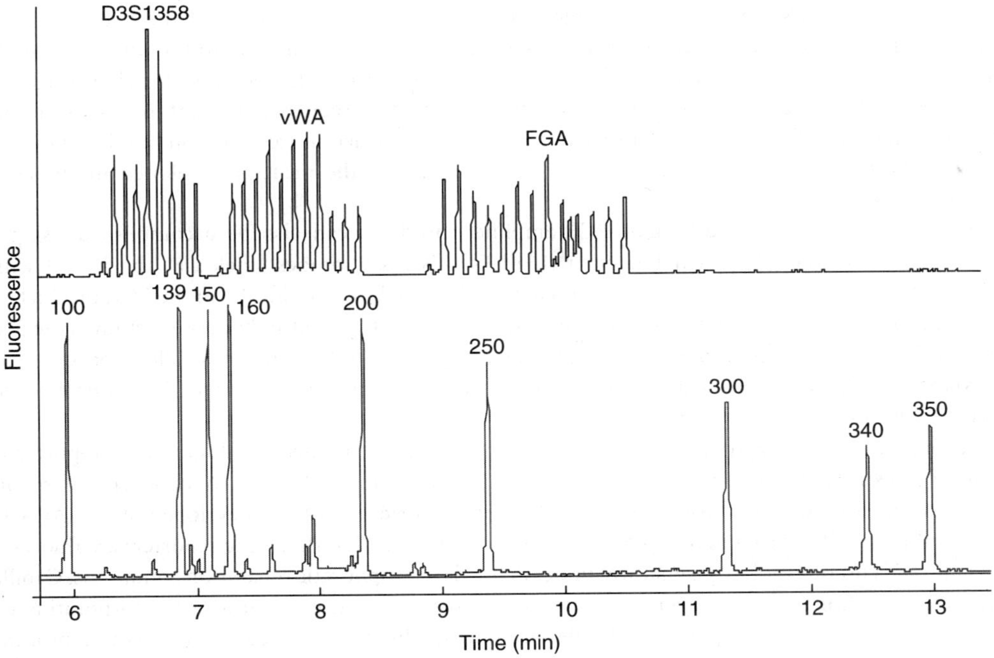


**FIGURE 52.2** Simultaneous, two-color separation of three STR loci and migration standard in a 32-channel plastic LabCard™ device. The AmpFLSTR® Blue allelic ladder standard for D3S1358, vWA, and FGA (upper panel) was combined with a comigrating sizing standard, GeneScan®-350 ROX™ (Applied Biosystems) (lower panel) and analyzed simultaneously using a plastic LabCard device. The separation matrix used was 4% (w/v) linear polyacrylamide with 7 *M* urea in 1 × TTE buffer. Separation was performed at 160 V/cm and 33°C with a separation length to the detector of 4.5 cm. Fluorescent signal was measured with a laser-induced fluorescence detector, as outlined in the text. Electropherograms were plotted and analyzed using DA × 7 (Van Mierlo Software Consultancy, Eindhoven, the Netherlands). The STR sample was separated in less than 10 min. (Data courtesy of Yining Shi, ACLARA Biosciences, Inc., Mountain View, CA.)

## 52.5 Multiplex SNP Detection

A variety of methodologies and platforms currently exist for detecting SNPs present in genomic DNA.[32] Most methods begin with polymerase chain reaction (PCR) for DNA target amplification to provide enough DNA for detection. More recently, significant progress has been made in detecting and measuring the abundance of target sequences using "real-time" PCR.[33] SNP detection using real-time amplification monitors accumulation of individual PCR products as they are synthesized and fluorescently labeled during the amplification reaction. Three basic real-time PCR product-detection methods exist: (1) labeling double stranded DNA (dsDNA) using intercalating dyes, (2) generating fluorescence-energy transfer during amplification (also known as the FRET approach), and (3) releasing quenched fluorescence during amplification (such as the Taqman® approach).

Higuchi et al.[34] first demonstrated the real-time PCR using ethidium bromide as a reporter. A variety of dyes are known to exhibit increased fluorescence upon binding to dsDNA. This detection method is used in conjunction with the specific amplification of wild type or mutant allele by the amplification refractory mutation system (ARMS).[35] The fluorescence of wild-type or mutant PCR products are continuously monitored in parallel, allele-specific reactions using ethidium bromide or SYBR® dyes (Molecular Probes, Eugene, OR), which bind to the accumulating PCR product and generate a proportional fluorescent signal. One drawback of this method is that the nonspecific binding of the dyes to proteins, residual primers, or spurious PCR products can result in false positive results or high backgrounds.

The fluorescence resonance energy transfer (FRET) technique relies on the binding of two differentially labeled fluorescent DNA probes to the target sequence. The close proximity of the two dyes on the adjacent, hybridized probes increases resonance-energy transfer from one dye to the other, leading to a unique fluorescence signal. Mismatches caused by SNPs that disrupt the adjacent binding of either of the probes results in signal loss that can be used to detect mutant sequences present in a DNA sample.[36] Alternatively, signal of a dual-labeled FRET probe may be quenched in free solution but detectable with allele-specific hybridization. This approach to probe design is the principle exploited in "molecular beacon" assays.

Real-time fluorescence detection of PCR-amplified products can also be accomplished using the TaqMan® assay method (Applied Biosystems).[37,38] In this assay, probes labeled at the 5′-end with a fluorescent reporter dye and at the 3′-end with a quencher dye bind specifically to the PCR product. The 5′-3′ exonuclease activity of thermostable polymerases, such as *Taq* DNA polymerase, removes the 5′ dye during PCR, releasing the fluorescence from quenching. For SNP scoring, two allele-specific probes (TaqMan probes), each labeled with a different fluorescent reporter dye, are used to discriminate between alternative alleles.

The TaqMan assay design is difficult to multiplex due to the limited combination of reporter and quencher dyes and the overlapping fluorescent emissions of these dye sets. A new class of reporter molecules called eTag™ (electrophoretic mobility tag) reporters has been developed at ACLARA that allow multiplex SNP scoring in 5′-nuclease-based assays, such as TaqMan. eTag reporters are fluorescent molecules with unique mobility signatures under defined electrophoretic separation conditions. Capillary instruments and microchannel devices can be used as readout platforms for assays incorporating eTag reporter technology. These reporter molecules are released from hybridized oligonucleotide probes by enzymatic cleavage. Subsequently, released eTag reporters are electrophoretically separated and decoded based on migration time to identify the targets present in the original DNA sample.

An example of a multiplexed SNP assay using TaqMan and eTag reporter-labeled, allele-specific probes is illustrated in Figure 52.3. Large families of generic reporters are being developed and are ideally suited to electrophoretic analysis. The eTag reporter strategy for multiplexing assays provides a method for detecting one or more target nucleic-acid sequences in a mixture because the labels give each target a unique electrophoretic signature. This biochemistry can be integrated with microfluidic-based LabCard systems to provide high-throughput, multiplexed SNP genotyping.

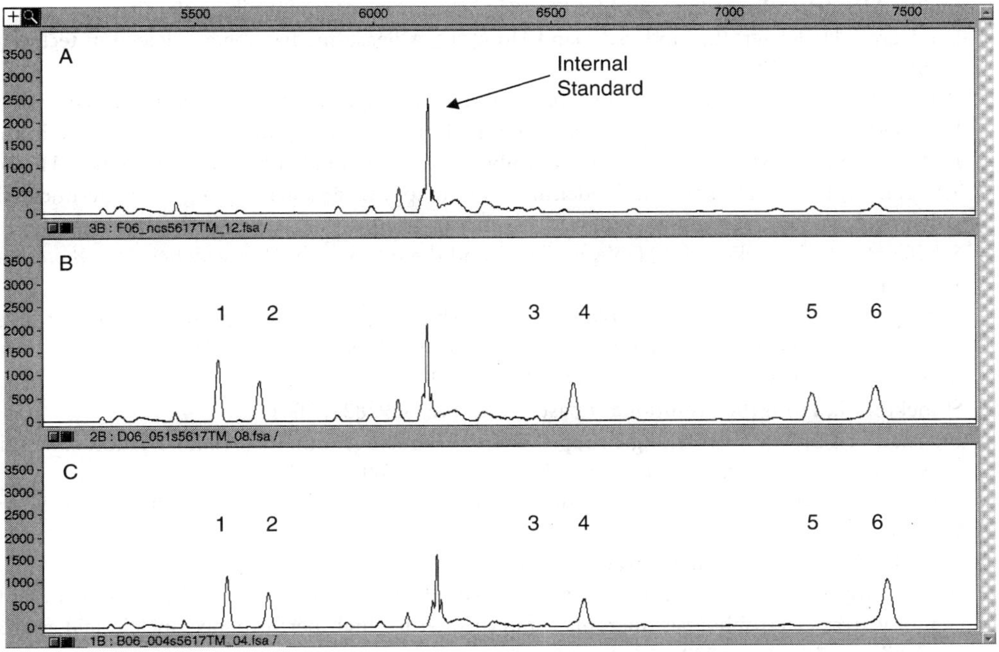


**FIGURE 52.3** Electropherogram illustrating the results of a multiplexed SNP assay using the eTag reporter technology in a TaqMan assay format. A water blank served as a negative control (panel A), and two different human genomic DNA samples were analyzed (panels B and C). The six eTag reporter labels were detected by CAE using an ABI Prism® 3100 genetic analyzer. Migration position of enzymatically released reporters relative to the internal Rox-dye standard permits genotype identification. The polymorphisms evaluated using allele-specific probes were ApoB Thre71Ile, ApoB Arg3500Gln, and CETP Ile405Val, where peak 1 = ApoB Ile71, peak 2 = ApoB Thre71, peak 3 = ApoB Gln3500, peak 4 = ApoB Arg3500, peak 5 = CETP Val405, and peak 6 = CETP Ile405. Sample DNA in panel B was heterozygous for ApoB Thre71Ile (1, 2) and CETP Ile405Val (5, 6) and homozygous for ApoB Arg3500Gln (4, 4). DNA sample in panel C was heterozygous for ApoB Thre71Ile (1, 2) and homozygous for both ApoB Arg3500Gln (4, 4) and CETP Ile405Val (6, 6).

## 52.6 Conclusion

High-throughput genetic analysis can be achieved in plastic microfluidic chips. Aside from SNP and STR scoring, additional applications, including gene-expression analysis, dsDNA fragment separation, and DNA sequencing,[28] as well as protein separation,[39] have also been demonstrated. In the future, fully integrated LabCard devices employing interfacing structures to allow complete sample-to-answer testing, the true "lab-on-a-chip" devices, will be possible using some of the assay formats described in this chapter.

### Acknowledgments

The authors thank Tracy Matray and colleagues for kindly providing eTag reporter-labeled probes and Yining Shi for his work on the microchannel separations of STR allelic ladders.

## References


1. Online Mendelian Inheritance in Man™ URL: http://www.ncbi.nlm.nih.gov/ominm.
2. Landegren, U., Kaiser, R., Sanders, J., and Hood, L., A ligase-mediated gene detection technique, *Science*, 242, 229, 1988.
3. Wang, D.G., Fan, J.B., Siao, C.J., Berno, A., Young, P., Sapolsky, R., Ghandour, G., Perkins, N., Winchester, E., Spencer, J., Kruglyak, L., Stein, L., Hsie, L., Topaloglou, T., Hubbell, E., Robinson, E., Mittmann, M., Morris, M.S., Shen, N., Kilburn, D., Rioux, J., Nusbaum, C., Rozen, S., Hudson, T.J., Lander, E.S., et al., Large-scale identification, mapping, and genotyping of single-nucleotide polymorphisms in the human genome, *Science*, 280, 1077, 1998.
4. Schafer, A.J. and Hawkins, J.R., DNA variation and the future of human genetics, *Nat. Biotechnol.*, 16, 33, 1998.
5. Landegren, U., Nilsson M., and Kwok, P.Y., Reading bits of genetic information: methods for single-nucleotide polymorphism analysis, *Genome Res.*, 8, 769, 1998.
6. Brookes, A.J., The essence of SNPs, *Gene*, 234, 177, 1999.
7. Stonekin, M., From the evolutionary past ... , *Nature*, 409, 821, 2001.
8. International SNP Map Working Group, A map of human genome sequence variation containing 1.42 million single-nucleotide polymorphisms, *Nature*, 409, 928. 2001.
9. Risch, N. and Merikangas, K., The future of genetic studies of complex human diseases, *Science*, 273, 1516, 1996.
10. The CEPH Genotype database available at http://www.cephb.fr/cephdb/.
11. Shi, Y., Simpson, P.C., Scherer, J.R., Wexler, D., Skibola, C., Smith, M.T., and Mathies, R.A., Radial capillary array electrophoresis microplate and scanner for high-performance nucleic acid analysis, *Anal. Chem.*, 71, 5354, 1999.
12. Meldrum, D., Automation for genomics, part two: sequencers, microarrays, and future trends, *Genome Res.*, 10, 1288, 2000.
13. Huang, X.C., Quesada, M.A., and Mathies, R.A., DNA sequencing using capillary array electrophoresis, *Anal. Chem.*, 64, 967, 1992.
14. Graham, C.A. and Hill, A.J.M., Introduction to DNA sequencing, in *DNA Sequencing Protocols*, 2nd ed., Humana Press, Totowa, NJ, 2001, p. 1.
15. Jurinke, C., van den Boom D., and Cantor, C.R., Automated genotyping using the DNA MassArray™ technology, in *DNA Arrays: Methods and Protocols*, 1st ed., Humana Press, Totowa, NJ, 2001, p. 103.
16. Manz, A., Fettinger, J.C., Verpoorte, E., Lüdi, H., Widmer H.M., and Harrison, D.J., Micromachining of monocrystalline silicon and glass for chemical analysis systems: a look into next century's technology or just a fashionable craze? *Trends Anal. Chem.*, 10, 144, 1991.
17. Harrison, D.J., Manz, A., Fan, Z., Luedi, H., and Widmer, H.M., Capillary electrophoresis and sample injection systems integrated on a planar glass chip, *Anal. Chem.*, 64, 1926, 1992.
18. Woolley, A.T. and Mathies, R.A., Ultra-high-speed DNA sequencing using capillary electrophoresis chips, *Anal. Chem.*, 67, 3676, 1995.
19. Salas-Solano, O., Carrilho, E., Kotler, L., Miller, A.W., Goetzinger, W., Sosic, Z., and Karger, B.L., Routine DNA sequencing of 1000 bases in less than one hour by capillary electrophoresis with replaceable linear polyacrylamide solutions, *Anal. Chem.*, 70, 3996, 1998.
20. Effenhauser, C.S., Paulus, A., Manz, A., and Widmer, H.M., High-speed separation of antisense oligonucleotides on a micromachined capillary electrophoresis device, *Anal. Chem.*, 66, 2949, 1994.
21. Eckstrom, B., Jacobson, G., Ohman, O., and Sjodin, H., Microfluidic structure and process for its manufacture, world patent wo9116966, 1991.
22. Soane, D.S. and Soane, Z.M., Method and device for moving molecules by the application of a plurality of electrical fields, U.S. patent 5126022, 1992.
23. McDonald, J. C., Duffy, D.C., Anderson, J.R., Chiu, D.T., Wu, H., Schueller, O.J., and Whitesides, G.M., Fabrication of microfluidic systems in poly(dimethylsiloxane), *Electrophoresis*, 21, 27, 2000.

24. Ocvirk, G., Munroe, M., Tang, T., Oleschuk, R., Westra, K., and Harrison, D.J., Electrokinetic control of fluid flow in native poly(dimethylsiloxane) capillary electrophoresis devices, *Electrophoresis*, 21, 107, 2000.
25. Chan, J.H., Timperman, A.T., Qin, D., and Aebersold, R., Microfabricated polymer devices for automated sample delivery of peptides for analysis by electrospray ionization tandem mass spectrometry, *Anal. Chem.*, 71, 4437, 1999.
26. Duffy, D.C., McDonald, J.C., Schueller, O.J.A., and Whitesides, G.M., Rapid prototyping of microfluidic systems in poly(dimethylsiloxane), *Anal. Chem.*, 70, 4974, 1998.
27. Effenhauser, C.S., Gerard, J.M. Bruin, G.J.M., Paulus, A., and Ehrat, M., Integrated capillary electrophoresis on flexible silicone mircrodevices: analysis of DNA restriction fragments and detection of single DNA molecules on microchips, *Anal. Chem.*, 69, 3451, 1997.
28. Cronin, M.T. and Mansfield, E.S., Microfluidic arrays in genetic analysis, *Expert Rev. Mol. Diagn.*, 1, 109, 2001.
29. McCormick, R.M., Nelson, R.J., Alonso-Amigo, M.G., Benvegnu, D.J., and Hooper, H.H., Microchannel electrophoretic separations of DNA in injection-molded plastic substrates, *Anal. Chem.*, 69, 2626, 1997.
30. Mansfield, E.S., Robertson, J.M., Vainer, M., Isenberg, A.R., Frazier, R.R., Ferguson, K., Chow, S., Harris, D.W., Barker, D.L., Gill, P.D., Budowle, B., and McCord, B.R., Analysis of multiplexed short tandem repeat (STR) systems using capillary array electrophoresis, *Electrophoresis*, 19, 101, 1998.
31. Sassi, A.P., Paulus, A., Cruzado, I.D., Bjornson, T., and Hooper, H.H., Rapid, parallel separations of D1S80 alleles in a plastic microchannel chip, *J. Chromatogr.*, 894, 203, 2000.
32. Shi, M.M., Enabling large-scale pharmacogenetic studies by high-throughput mutation detection and genotyping technologies, *Clin. Chem.*, 47, 164, 2001.
33. Wittwer, C., Herrmann, M.G., Moss, A.A., and Rasmussen, R.P., Continuous fluorescence monitoring of rapid cycle DNA amplification, *Biotechniques*, 22, 130, 1997.
34. Higuchi, R., Fockler, C., Dollinger, G., and Watson, R., Kinetic PCR analysis: real-time monitoring of DNA amplification reactions, *Biotechnology*, 11, 1026, 1993.
35. Newton, C.R., Graham, A., Heptinstall, L.E., Powell, S.J., Summers, C., Kalsheker, N., Smith, J.C., and Markham, A.F., Analysis of any point mutation in DNA: the amplification refractory mutation system (ARMS), *Nucleic Acid Res.*, 17, 2503, 1989.
36. Chen, X., Levine, L., and Kwok, P.Y., Fluorescence polarization in homogeneous nucleic acid analysis, *Genome Res.*, 8, 549, 1999.
37. Holland, P.M., Abramson, R.D., Watson, R., and Gelfand, D.H., Detection of specific polymerase chain reaction product by utilizing the 5′-3′ exonuclease activity of *Thermus aquaticus* DNA polymerase, *Proc. Natl. Acad. Sci. U.S.A.*, 88, 7276, 1991.
38. Livak, K.J., Oligonucleotides with fluorescent dyes at opposite ends provide a quenched probe system useful for detecting PCR product and nucleic acid hybridization, *PCR Meth. Appl.*, 4, 357, 1995.
39. Gottschlich, N., Jacobson, S.C., Culbertson, C.T., and Ramsey, J.M., Two-dimensional electrochromatography/capillary electrophoresis on a microchip, *Anal. Chem.*, 73, 2669, 2001.

# 53

# DNA Sequencing Using Fluorescence Detection


Steven A. Soper
*Louisiana State University*
*Baton Rouge, Louisiana*

Clyde V. Owens
*Louisiana State University*
*Baton Rouge, Louisiana*

Suzanne J. Lassiter
*Louisiana State University*
*Baton Rouge, Louisiana*

Yichuan Xu
*Louisiana State University*
*Baton Rouge, Louisiana*

Emanuel Waddell
*Louisiana State University*
*Baton Rouge, Louisiana*




## 53.1 General Considerations

### 53.1.1 What Is DNA?

#### 53.1.1.1 Organization of Genome

The blueprint for all cellular structures and functions is encoded in the genome of any organism. The genome consists of deoxyribonucleic acid (DNA), which is tightly coiled into narrow threads that, when completely stretched, reach a length of approximately 1.5 m yet possess a width of only ~2.0 nm ($2 \times 10^{-9}$ m). The threads of DNA are typically associated with many different types of proteins and are organized into structures called chromosomes that are housed within the nucleus of cells in most eukaryotic organisms. There are 23 pairs of chromosomes within the human genome.

DNA is composed of several different chemical units: a deoxyribose sugar unit, phosphate group, and one of four different nucleotide bases (adenine [A], guanine [G], cytosine [C], or thymine [T]). At the molecular level, it is the order of these bases that carries the code to build proteins within the cell that inevitably control the function of various cells and also determine an organism's physical characteristics. In the human genome, the 23 pairs of chromosomes contain 3 billion bases. The length of the chromosomes


0-8493-1116-0/03/$0.00+$1.50

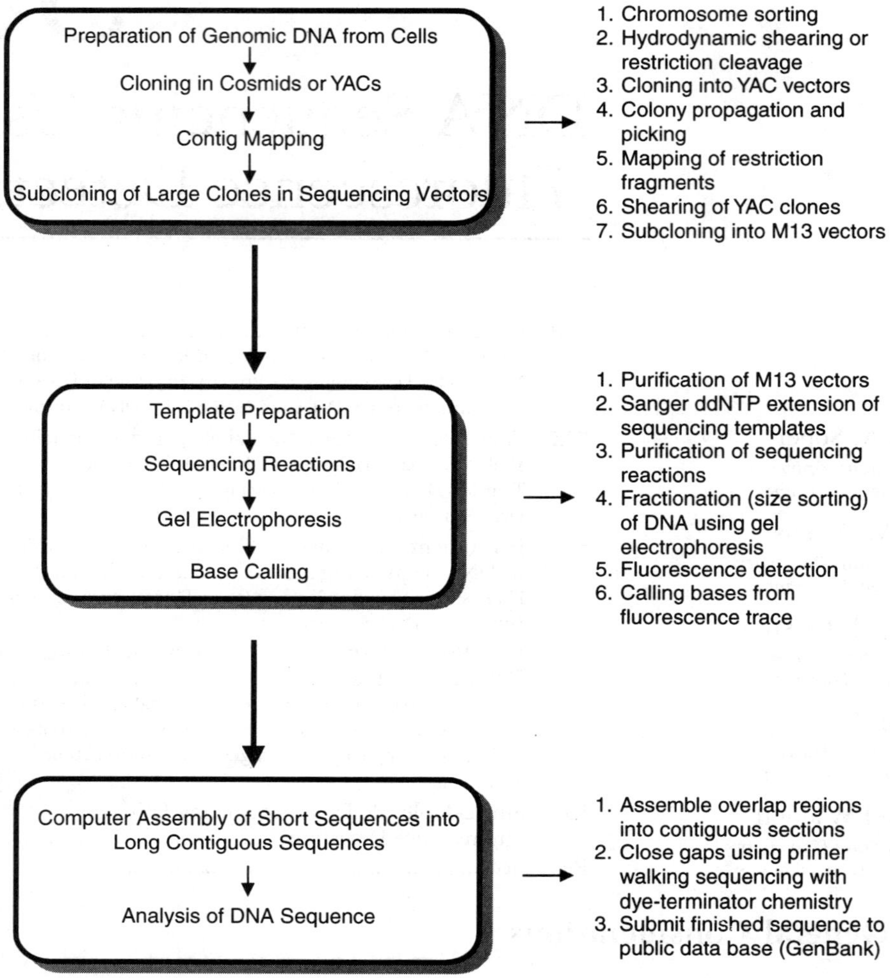


**FIGURE 53.1** Flow chart showing the processing steps typically involved in sequencing DNA starting with isolating genomic DNA from cells.

(in base pairs) varies greatly, with the smallest chromosome containing 50 million bases (chromosome Y) and the longest containing 250 million bases (chromosome 1). It is the primary function of DNA sequencing to determine the order of these nucleotide bases.

#### 53.1.1.2 Functions of Genes

The coding regions (regions that carry the information for the construction of proteins) of the genome are contained within genes. Each gene consists of a specific sequence of nucleotide bases, with three nucleotide bases (codon) directing the cells' protein-synthesizing machinery to add a specific amino acid to the target protein. It is estimated that within the human genome are approximately 30,000 genes, with the length of the genes varying greatly. However, only approximately 10% of the human genome is thought to contain protein-coding sequences.

## 53.1.2 What Is DNA Sequencing?

Figure 53.1 shows a flow chart depicting the important steps involved in the process of DNA sequencing. There are three primary steps — mapping, sequencing, and assembly. Within each of these three general steps are a number of substeps, which include such processes as cloning and subcloning (mapping), template preparation and gel electrophoresis (sequencing), and the computer algorithms required to assemble the small bits of sequencing data into the contiguous strings that comprise the intact chromosome. Each chromosome may consist of hundreds of millions of nucleotide bases; unfortunately, the sequencing phase of the intricate process can handle only pieces of DNA that vary from 1000 to 2000 base pairs (bp) in length. Therefore, the chromosome must typically be broken down into manageable pieces using either restriction enzymes or mechanical shearing and then cloned into bacterium to increase the copy number of the individual pieces of DNA. Following sequencing of each cloned fragment the individual pieces must be reassembled into a contiguous strand representing the entire chromosome. This process typically involves sophisticated computer algorithms to search for commonalties in the small fragments and overlap them to build the sequence of the entire chromosome.

### 53.1.2.1 DNA-Sequencing Factories

To accomplish the lofty goal of sequencing the entire human genome, sequencing factories have been assembled to produce large amounts of data and deposit these data into public databases for easy accessibility by the general scientific and medical communities. One such production-scale sequencing center is the Human Genome Sequencing Center at the Baylor College of Medicine (www.hgsc.bcm.tmc.edu). Some statistics on this particular sequencing center help put the sequencing demands into perspective. As of May 1999, the center had deposited over 26 Mbp ($26 \times 10^6$ bp, 0.7% of the human genome) of sequence data into the public databases; in addition, the center typically runs approximately 14 automated fluorescence-based DNA-sequencing machines 12 h a day. This amounts to 50,000 sequencing reactions a month.

This chapter focuses on the sequencing phase of the genome-processing steps. It is this particular step on which fluorescence — both hardware and probe development — has had a profound impact in terms of augmenting the throughput of acquiring sequencing data. We begin by briefly introducing (or refreshing) the reader to the molecular and geometrical structure of DNA and then review the common schemes used to sequence DNA. We also discuss the various electrophoretic modes for fractionating DNA based on size, an integral component in high-throughput sequencing. We then examine common strategies of fluorescence detection used in many sequencing machines; our discussion includes hardware and probe (labeling-dye) developments.

## 53.1.3 Structure of DNA

### 53.1.3.1 Nucleotide Bases

As stated in the previous section, the basic building blocks of DNA are a deoxyribose sugar unit, a phosphate group, and the nucleotide base, of which there are four; when combined chemically, these building blocks form an individual nucleotide. Figure 53.2 shows the chemical structure of the four nucleotide units that comprise DNA. The bases are grouped into two different classes — the pyrimidines (T, C) and the purines (A, G). In the case of ribonucleic acid (RNA), the structural differences include the incorporation of a ribose sugar (inclusion of a hydroxyl group at the 2′ position) and also the substitution of uracil for thymine.

Since DNA — more specifically, chromosomes — is composed of many of these individual nucleotide units strung together, the nucleotides are covalently attached via phosphodiester linkages, which occur between the phosphate group on the 5′ site of one sugar unit and the 3′ hydroxyl group on another nucleotide (see Figure 53.2). Therefore, DNA exists as a biopolymer, with the repeating units being deoxyribose and phosphate residues that are always linked together by the same type of linkage and form the backbone of the DNA molecule. However, the order of bases along the biopolymer backbone can vary greatly and imparts a high degree of individuality to any particular DNA molecule.

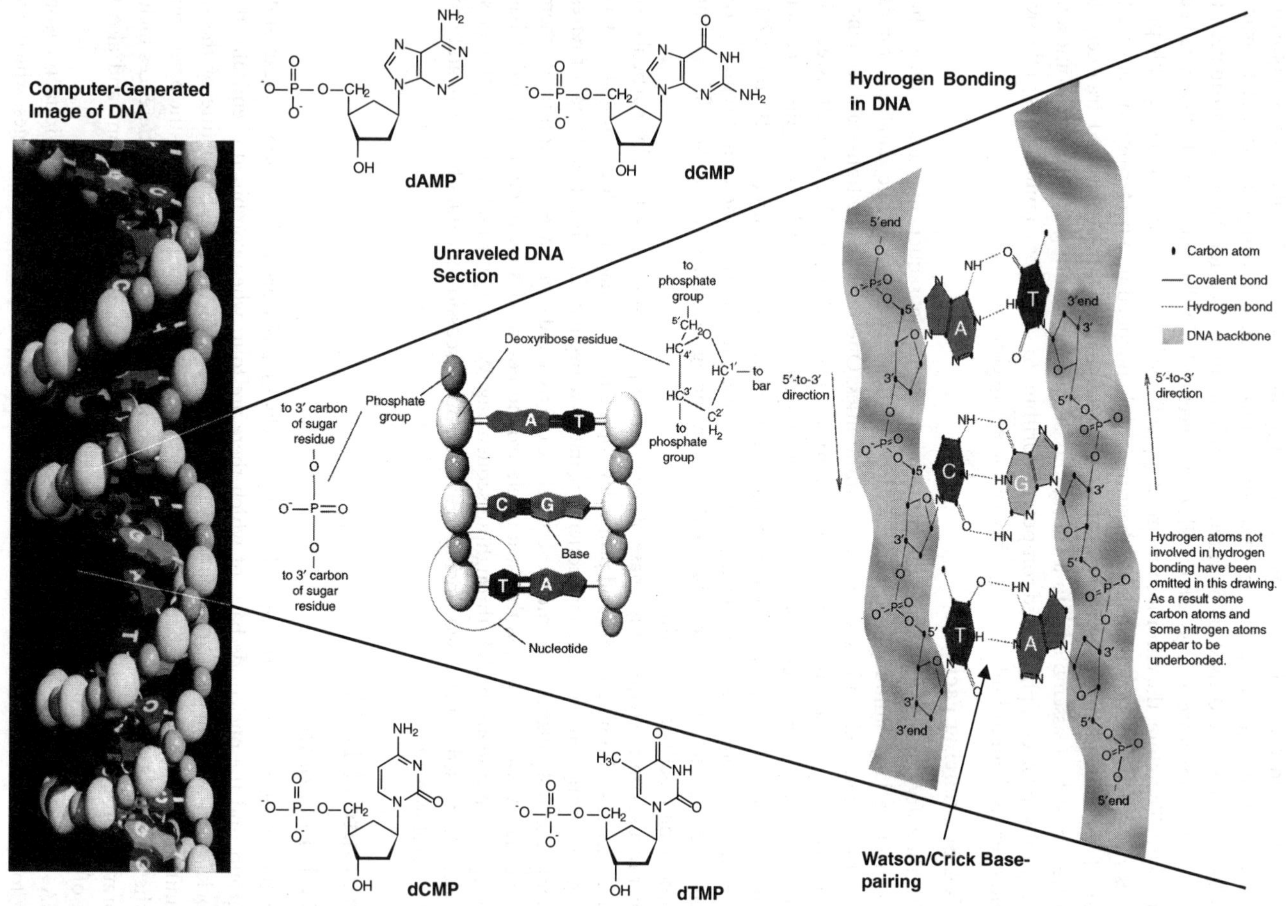


**FIGURE 53.2** Chemical structures of deoxyribonucleic acid (DNA). The computer image shows the double-helical nature of DNA. Also shown are the hydrogen bonding between nucleotide bases (Watson–Crick base pairing) and the structures of the nucleotide building blocks.

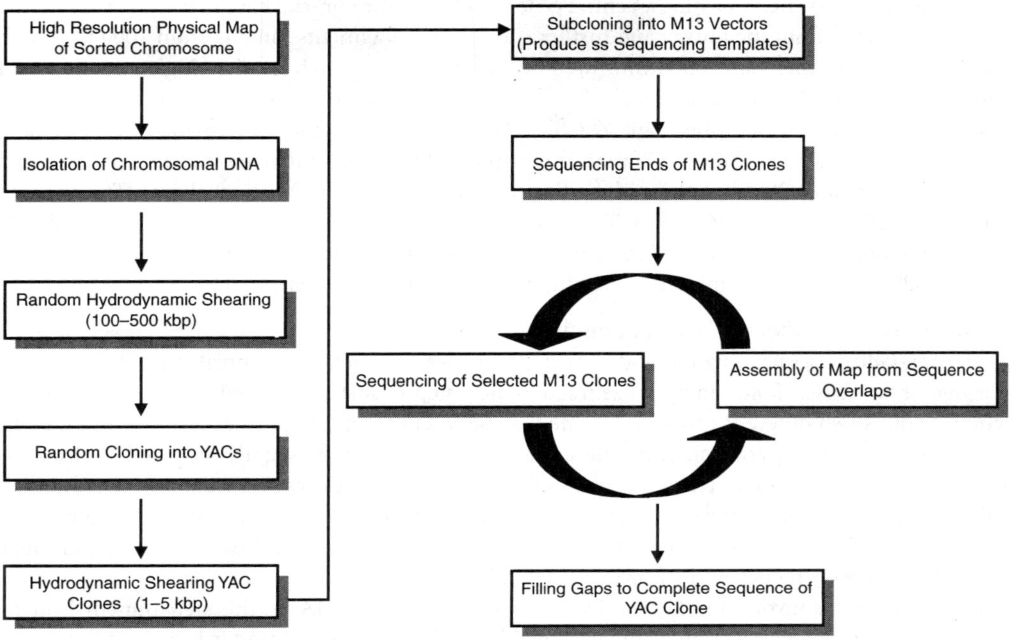


**FIGURE 53.3** Processing flow chart of ordered shotgun sequencing of DNA.

### 53.1.3.2 Watson–Crick Base Pairing

The three-dimensional structure of DNA is known to differ greatly from that of proteins, which are also biopolymers composed of different amino acid residues. In 1953, James Watson and Francis Crick discovered that DNA existed in a double-helical structure (see Figure 53.2), with the sugar-phosphate backbone oriented on the outside of the molecule and the bases positioned on the inside of the double helix.[1] They also surmised that the two strands of the double-stranded molecule were held together via hydrogen bonds between a pair of bases on opposing strands. From modeling, they found that A could pair only with T and C with G (see Figure 53.2). Each of these base pairs possesses a symmetry that permits it to be placed into the double helix in two different ways (A to T and T to A; C to G and G to C). Thus, all possible permutations (four) of sequence can exist for all four bases. In spite of the irregular sequence of bases along each strand, the sugar–phosphate backbone assumes a very regular helical structure, with each turn of the double helix composed of ten nucleotide units.

## 53.1.4 Methods for Determining the Primary Structure of DNA

The important factor to consider when developing a sequencing strategy for whole genomes is to remember that only small sections (1000 to 2000 bp) of DNA can be sequenced and that entire chromosomes are composed of well over $1 \times 10^6$ bp. The actual process of generating DNAs that can be handled by sequencing machines is typically involved and requires a number of cloning and purification steps followed by actual sequencing and then assembling the pieces into contiguous regions of the target chromosome. While this chapter does not cover these specific processes, Figure 53.3 provides a schematic diagram of a typical strategy used in many sequencing laboratories. This strategy is termed an ordered shotgun approach and starts with the mechanical shearing (breaking apart) of intact chromosomes into pieces composed of 100,000 to 200,000 bp.[2] These sheared DNAs are then cloned into yeast artificial chromosomes (YACs), one sheared section per YAC that allows one to amplify the number of copies of the insert. Cloning provides an unlimited amount of material for sequencing and serves as the basis for construction of libraries, which are random sets of cloned DNA fragments. Genomic libraries are sets of overlapping fragments encompassing an entire genome. Once these libraries have been constructed, single inserts are extracted from the library and further sheared into fragments ranging from 1000 to 2000 bp. These fragments are then subcloned into M13 vectors to produce high-quality single-stranded DNAs (ssDNAs) appropriate for actual sequencing. Typically, M13 subclones are sequenced from both ends to allow construction of maps of the single YAC inserts, which are then used to provide a scaffold for the complete sequence analysis of the YAC insert. The important procedures that are the focus of this chapter are those that are actually used to produce the sequence data of the M13 subclones. The two most common procedures are the Maxam–Gilbert chemical degradation method and the Sanger dideoxy-chain-termination method.[3,4] Both of these methods have in common the production of a nested set of fragments, all of which are terminated (cleaved) at a common base or bases.

#### 53.1.4.1 Maxam–Gilbert DNA Sequencing

The Maxam–Gilbert sequencing method uses chemical cleavage methods to break ssDNA molecules at either one or two bases, followed by a size fractionation step to sort the cleaved products. The cleavage reaction involves two different reactions — one that cleaves at G and A (purine) residues and the other that cleaves at C and T (pyrimidine) residues. The first reaction can be slightly modified to cleave at G only and the second at T only. Therefore, one can run four separate cleavage reactions — G only, A + G, T only, T + C — from which the sequence can be deduced. In each cleavage reaction, the general process involves chemically modifying a single base, removing the modified base from its sugar, and, finally, breaking the bond of the exposed sugar in the DNA backbone.

The chemical steps involved in G cleavage are shown in Figure 53.4A. In this step, dimethylsulfate is used to methylate G. After eviction of the modified base via heating the strand is broken at the exposed sugar by subjecting the DNA to alkali conditions. To cleave at both A and G residues, the procedure is identical to the G cleavage reaction except that a dilute acid is added after the methylation step (see Figure 53.4B). The reaction that cleaves at either a C or C and T residue is carried out by subjecting the DNA to hydrazine to remove the base and piperidine to cleave the sugar–phosphate backbone. The extent of each reaction can be carefully limited so that each strand is cleaved at only one site.

Figure 53.4C depicts the entire Maxam–Gilbert process. As shown in the figure, four cleavage reactions are run — G, G + A, T, T + C. Prior to chemical cleavage, the intact DNA strand is labeled (typically with a $^{32}P$ radiolabel for detection at the 5′ end). Following chemical cleavage the reactions are run in an electrophoresis gel (polyacrylamide) and separated based on size. The actual sequence of the strand is then deduced from the generated gel pattern.

#### 53.1.4.2 Sanger Chain-Termination Method

Unlike the Maxam–Gilbert method, the Sanger procedure is an enzymatic method and involves construction of a DNA complement to the template whose sequence is to be determined. The complement is a strand of DNA that is constructed with a polymerase enzyme, which incorporates single nucleotide bases according to Watson–Crick base-pairing rules (A–T, G–C) (see Figure 53.5). The nested set of fragments is produced by interrupting the polymerization by inserting into the reaction cocktail a base that has a structural modification — the lack of a hydroxyl group at the 3′ position of the deoxyribose sugar (dideoxynucleotide, ddNTP). Figure 53.5A shows the chemical structure of a ddNTP. When mixed with the deoxynucleotides (dNTP), the polymerization proceeds until the ddNTP is incorporated.

Figure 53.5C shows the entire Sanger chain-termination protocol. The process is typically carried out by adding to the template DNA a primer that has a known sequence and anneals (binds) to a complementary site on the unknown template. This primer can carry some type of label, e.g., a covalently attached fluorochrome. However, situating the fluorescent label on the ddNTP can be done as well. Following annealing of the primer to the template, the reaction is carried out by the addition of a polymerase enzyme, all four dNTPs, and one particular ddNTP. Therefore, four reactions are carried out, each containing a particular ddNTP. Following polymerization, the reactions are loaded onto an electrophoresis gel and size-fractionated. The final step involves reading the sequence from the gel.

The Sanger method has been the preferred sequencing method for most large-scale sequencing projects because of the ease with which it prepares the nested set of fragments. Many times reactions can be run under standard conditions without the need for chemical additions at timed intervals. In addition, the process is very conducive to automation.

## 53.1.5 Modes of Electrophoresis

Whatever fluorescence detection protocol is used for calling bases, the important analytical technique that is required is a fractionation step in which the DNA molecules are sorted by size. While the focus of this chapter is on fluorescence-based detection in sequencing applications, it is informative to briefly discuss some of the common gel electrophoresis platforms that are used because the fluorescence detector is integrated into the sizing step to provide online detection.

All electrophoresis formats have in common the use of an electric field to shuttle the DNAs through a maze consisting of a polymer, either static or dynamic, with pores of various sizes. In all electrophoresis experiments, the mobility of the molecule ($\mu$, $cm^2/Vs$, defined as the steady-state velocity per unit electric-field strength) in an electric field is determined by:

$$\mu = \frac{q}{f} \tag{53.1}$$

where $q$ is the net charge on the molecule, and $f$ is the frictional property of the molecule and is related to its conformational state as well as the molecular weight (MW) of the molecule. For example, proteins can be considered solid spheres; thus, $f \sim (MW)^{1/3}$. In the case of DNAs, either single-stranded or double-stranded, $f \sim (MW)^1$ because the DNA molecule acts as a free-draining coil. Since $q$ is also related to the length of the DNA molecule, one $\mu \sim N_b^0$, where $N_b^0$ indicates that the mobility (in free solution) is independent of the number of bases comprising the DNA molecule. Because of this property of DNAs, the electrophoresis step must include some type of sieving medium, which can be a polymer consisting of pores with a definitive size.

For any type of analytical separation resolution is a key parameter that is optimized to improve the performance of the separation. The resolution $R$ for electrophoretic separations can be calculated from the simple relationship:

$$R = \frac{1}{4}\frac{\Delta\mu_{app}}{\mu_{app,avg}}N^{1/2} \tag{53.2}$$

where $\Delta\mu_{app}$ is the difference in mobility between two neighboring bands, $\mu_{app,avg}$ is the average electrophoretic mobility for the same neighboring bands, and $N$ is the plate number, which represents the efficiency (bandwidth) for the electrophoresis. This equation shows that the resolution can be improved by increasing the difference in the mobility of the two bands (increase selectivity, gel property) or increasing the plate numbers (narrower bands). As a matter of reference, when $R = 0.75$, two bands are baseline resolved. For DNA sequencing, the accuracy in the base call depends intimately on the resolution obtained during gel fractionation.

A common polymer used for DNA sequencing is linear or crosslinked polyacrylamides or polyethylene oxides, both of which possess the appropriate pore size for sorting ssDNAs. Polyacrylamides are prepared from the acrylamide monomer ($-CH_2 = CHCONH_2$), which is copolymerized with a certain percentage of crosslinker, *N,N′*-methylene-*bis*-acrylamide ($-CH_2(CHCOCH = CH_2)_2^-$), in the presence of a catalyst accelerator-chain-initiator mixture. The porosity of the gel is determined by the relative proportion of acrylamide monomer to crosslinking agent. For reproducible fractionation of DNAs, they must be maintained in a single-strand conformation, which is accomplished by adding denaturants, such as urea or formamide, to the sieving gel.

The existence of a gel network contributes significantly to the electrophoretic migration pattern (i.e., electrophoretic mobility) observed in the sequencing process. Any DNA fragments to be fractionated will inevitably encounter the gel network of polymer threads. This encounter increases the effective

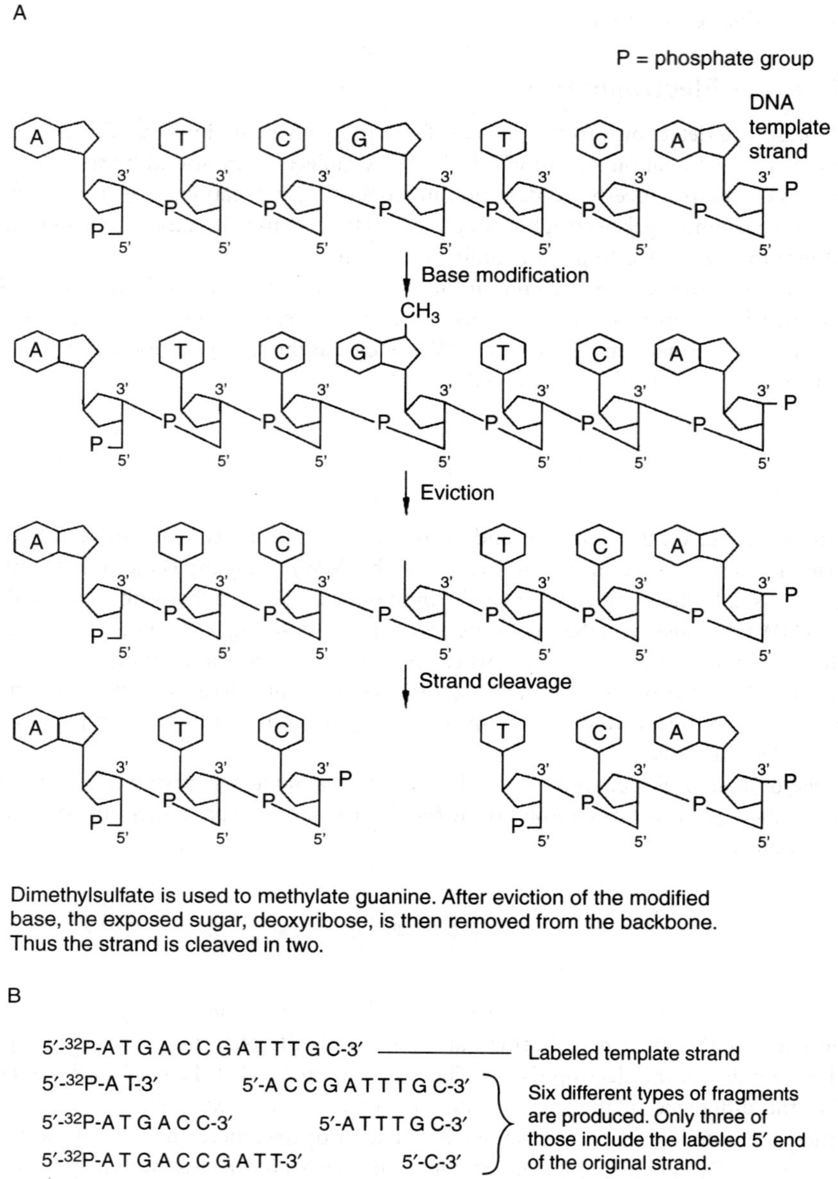


**FIGURE 53.4** Maxam–Gilbert chemical degradation method for DNA sequencing. (A) Chemical cleavage method for a guanine residue using dimethylsulfate to methylate the G residue. (B) Fragments generated from chemical cleavage at only G.

friction and, consequently, lowers the velocity of movement of the molecules. It is obvious that the retardation will be most pronounced when the mean diameter of gel pores is comparable to the size of the DNA fragments. Size therefore plays a critical role in determining the relative electrophoresis mobility and the degree of separation of different DNA fragments. It is this sieving effect that partly determines the resolution obtained with polyacrylamide gels. The gel network also minimizes convection currents caused by small temperature gradients.

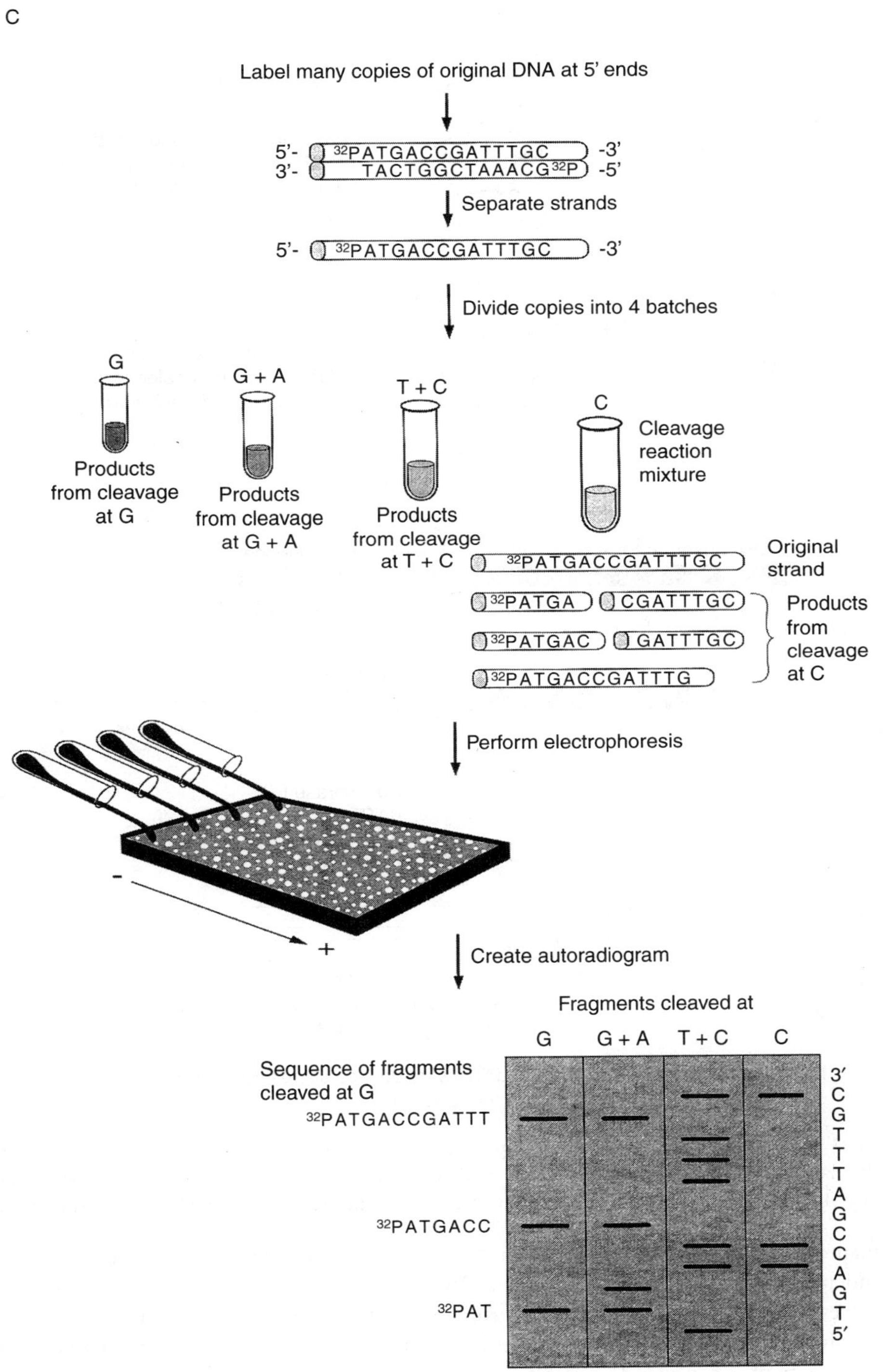


**FIGURE 53.4 (continued)** Maxam–Gilbert chemical degradation method for DNA sequencing. (C) Steps in Maxam–Gilbert sequencing.

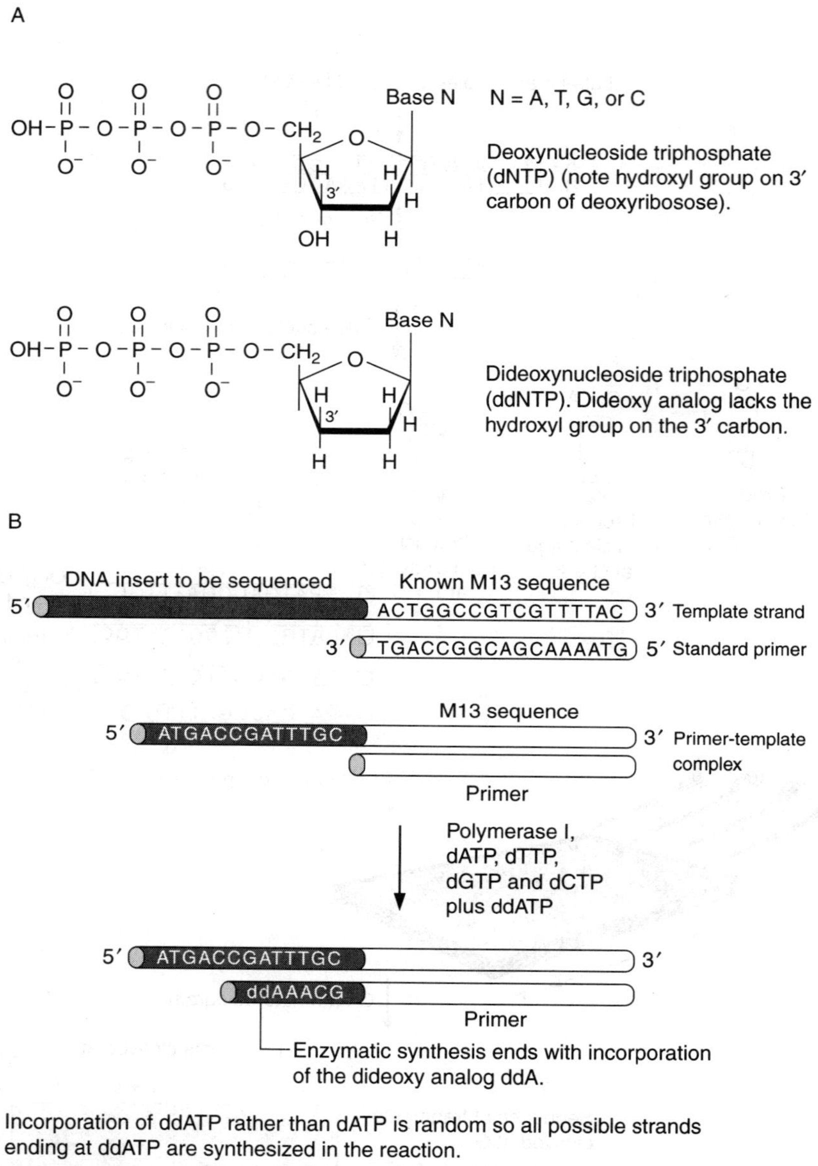


**FIGURE 53.5** Sanger chain-termination DNA-sequencing method. (A) Chemical structures of a dNTP and also a ddNTP that lacks a hydroxyl group on the 3′ site of the deoxyribose sugar. (B) Chain-termination reaction with ddATP. Also shown is primer annealing to the template, which is directed by Watson-Crick base-pairing rules.

The main characteristic property of DNA is that the value of the negative charge on the molecule is almost independent of the pH of the medium. Therefore, their electrophoretic mobility is due mainly to differences in molecular size, not in charge. For a particular gel, the electrophoretic mobility of a DNA fragment is inversely proportional to the logarithm of the number of bases — up to a certain limit. Since each DNA fragment is expected to possess a unique size, its electrophoretic mobility is unique, and it migrates to a unique position within the electric field in a given length of time. Therefore, if a mixture of DNA fragments is subjected to electrophoresis, each of the fragments would be expected to concentrate into a tight migrating band at unique positions in the electric field.

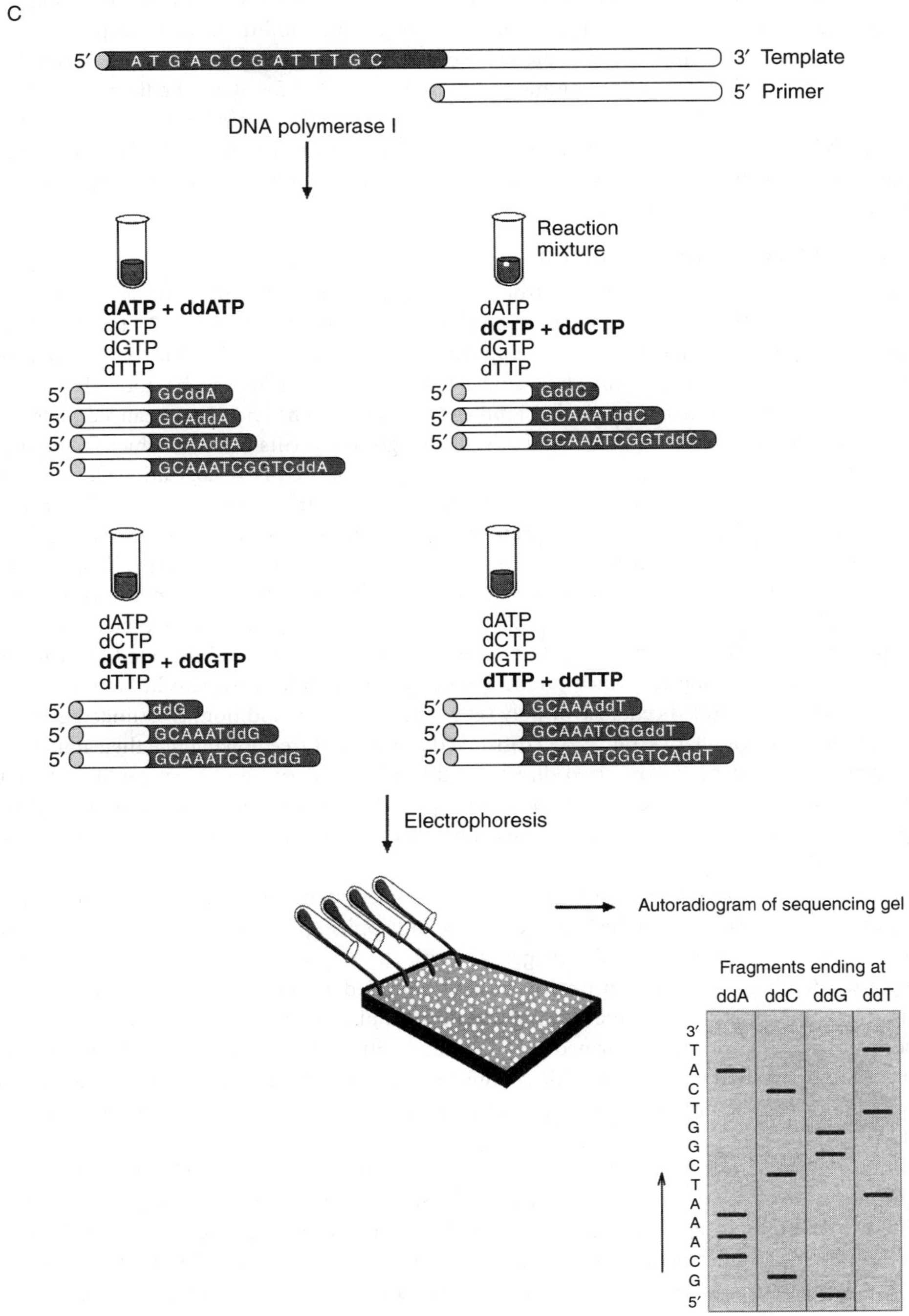


**FIGURE 53.5 (continued)** Sanger chain-termination DNA-sequencing method. (C) Steps in Sanger sequencing.

#### 53.1.5.1 Slab-Gel Electrophoresis

DNA sequencing is usually performed with slab gels. In slab gel electrophoresis (SGE), polymerization of acrylamide as well as the electrophoresis are conducted in a mold formed by two glass plates with a thin spacer. When analytical electrophoresis is performed, several samples are usually run simultaneously in the same slab. Since all samples are present in the same gel, the conditions of electrophoresis are quite constant from sample to sample. In SGE the sample is loaded using a pipette into wells formed into the gel during polymerization. A loading volume of 1 to 10 μl is typical for SGE. The field strength that can be applied to these types of gels ranges from 50 to 80 V/cm, with the upper limit determined by heating (Joule) caused by current flow through the gel. Due to the thick nature of the gel, this heat is not efficiently dissipated, causing convective mixing and thus zone broadening, which limits the upper level on the electric field strength that can be effectively used.

#### 53.1.5.2 Capillary-Gel Electrophoresis

To reduce the development time in the electrophoresis, many sequencing applications now use capillary-gel electrophoresis (CGE), in which a glass tube (i.d. = 50 to 100 μm; length = 30 to 50 cm) contains the gel-sieving matrix, and an electric field is applied to this capillary column. Due to the higher surface-to-volume ratio afforded by the thin glass capillary, large electric fields can be applied (~300 V/cm), which results in shorter electrophoresis development times (2 to 3 h) and also enhanced plate numbers compared to SGE. The glass capillary contains silanol groups on its surface, which at high pH are deprotonated. As such, the glass capillary is negatively charged above pH = 7.0. Cations in the electrolyte solution buildup at the wall of the glass tube, producing an electrical double layer. When a voltage is applied across the capillary, the cations migrate to the wall and exert a force on the surrounding fluid, causing a bulk flow of solution toward the cathode (negative terminal). This electrically induced flow is called an electroosmotic flow and, unfortunately, can interfere with the electrophoresis of the DNAs. Therefore, in DNA separations using glass capillaries, the wall is coated with some type of polymer (e.g., a linear polyacrylamide) to suppress this electroosmostic flow. After the wall is coated, the capillary can then be filled with the sieving gel, which can be a linear polyacrylamide (no cross linking) or some other type of gel, such as a hydroxyl cellulose of poly (ethylene oxide).[5] In addition, crosslinked gels may also be used in these small capillaries, but since crosslinked gels are not free flowing like their noncrosslinked counterparts, they must be polymerized directly within the capillary tube. Most capillary-based DNA sequencers use linear gels because they can be easily replaced using high-pressure pumping, which allows the capillary to be used for multiple sequencing runs. Once polymerized, the crosslinked gels cannot be removed from the capillary.

The electrophoresis is performed after the column is filled with the gel by inserting one end of the capillary into a sample containing the sequencing mix and applying an electric field to the capillary tube. The injection end is typically cathodic (negative), with the opposite end being anodic (positive). By applying a fixed voltage for a certain time period, a controlled amount of sample can be inserted into the column, with the injection volume ranging from 1 to 10 nl. Since the capillary is made of fused silica, fluorescence detection can be performed directly from within the capillary tube. In most cases, the capillary column is coated with a polyimide coating (nontransparent) to give it strength, and the optical detection window can be produced by simply burning off a section of the polymer using a low-temperature flame.

The higher plate numbers (i.e., better resolution) that are obtained in CGE vs. SGE are a direct consequence of the ability to use higher electric fields. Since development time is significantly shorter in CGE due to the ability to apply higher electric fields, band spreading due to longitudinal diffusion is reduced, resulting in higher plate numbers. The ability to use higher electric fields in CGE results from the fact that Joule heating is suppressed because the heat can be effectively dissipated by the high-surface-to-volume-ratio capillary.

## 53.1.6 Detection Methods for DNA Sequencing

### 53.1.6.1 Autoradiographic Detection

Following electrophoresis, the individual DNA bands separated on the gel must be detected and subsequently analyzed. One of the earlier methods implemented to detect DNA bands in gels was autoradiography. In this mode, one of the phosphates of an individual nucleotide is replaced with a radioisotope, typically $^{32}P$ ($\tau_{1/2}$ = 14 d) or $^{35}S$ ($\tau_{1/2}$ = 87 d), both of which are radioprobes that emit β-particles. When Sanger methods are used to prepare the sequencing reactions, either the primer or the dideoxynucleotide can contain the radiolabel. The labeling is done using an enzyme (T4 polynucleotide kinase), which catalyzes the transfer of a γ-phosphate group from ATP to the 5′-hydroxy terminus of a sequencing primer. After the electrophoresis is run, the gel is dried and then situated on an x-ray film. The film is developed (exposure to radiation from radioprobes), and dark bands are produced on the film where the DNA was resident. The sequence is then read manually from the gel.

The primary advantage of this approach is the inexpensive nature of the equipment required to perform the measurement — basically only a gel dryer, film holder, and film. The difficulties associated with this approach are numerous. One important issue is the fact that radioisotopes are used; therefore, waste disposal becomes a difficult problem. Throughput issues (data-production rates) are also a primary concern with autoradiographic detection. For example, radiography can sometimes require several days to expose the film to get strong signals to read the bases from the gel. In addition, the detection is done after electrophoresis, not during. Also, since there is no means of identifying the individual bases using radioprobes, each base must be analyzed in a different lane of the gel. And finally, the bases must be called manually, which often leads to frequent errors in the sequence reconstruction. Therefore, the inability to obtain data-production rates sufficient to accommodate large sequencing projects has made radiographic detection obsolete for high-throughput applications.

### 53.1.6.2 Fluorescence Detection

For most DNA-sequencing applications, irrespective of the separation platform used, fluorescence is the accepted detection protocol, for several important reasons. Fluorescence allows one to perform the base calling and detection in an automated fashion and alleviates the need for manual base calling. In addition, fluorescence can be carried out during the separation, eliminating long film-development times. More importantly, because multiple probes possessing unique spectral properties can be implemented, the four bases comprising the DNA molecule can be identified in a single gel lane, potentially increasing throughput by a factor of four vs. radiographic detection. All of these important advantages associated with fluorescence allow for higher throughputs in DNA-sequencing applications. As such, fluorescence can be considered one of the most important recent technical innovations in DNA sequencing and has made it feasible to consider tackling large genome-sequencing projects such as the human genome.

The first demonstrations on the use of fluorescence in DNA sequencing came with the work of Smith et al.,[6] Prober et al.,[7] and Ansorge et al.[8] Slab gels or large gel tubes were used to fractionate the DNA ladders produced during enzymatic polymerization using Sanger sequencing strategies. The fluorescence detection was accomplished using four spectroscopically unique probes, which allowed the DNA-sequence reconstruction to be done in a single electrophoresis lane of the gel. The chemical structures of the dye labels used in the Smith et al.[6] and Prober et al.[7] experiments are shown in Figures 53.6 and 53.7, respectively. As shown in the figures, the dyes were attached (covalently) either to the sequencing primer or to the dideoxynucleotides. The advantage of using dye-labeled dideoxynucleotides is that the sequencing reactions can be performed in a single reaction tube, whereas the dye-labeled primer reactions must be performed in four separate tubes and pooled prior to electrophoresis. In the case of the dye-labeled terminators, succinylfluorescein analogs were used with slight structural modifications to alter the absorption/emission maxima. The dyes were attached either to the 5 position of the pyrimidine bases or the 7 position of the 7-deazapurines, both of which are nonhydrogen-bonding sites on the nucleotide

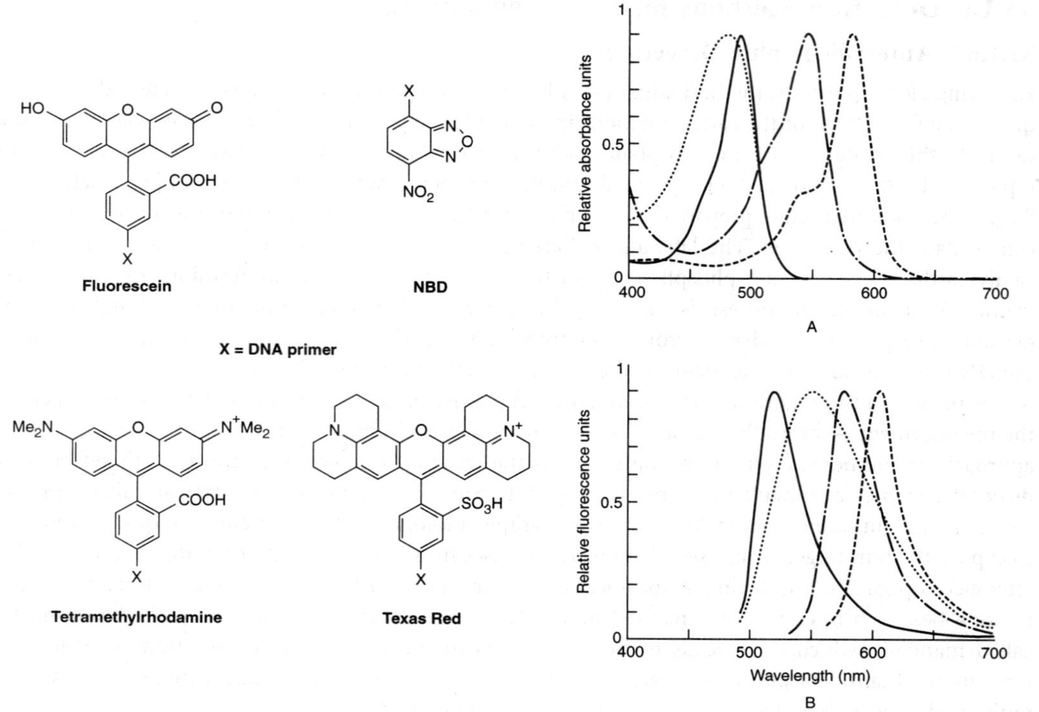


**FIGURE 53.6** Chemical structures and absorbance (A) and emission spectra (B) of the dyes used for labeling primers. (Adapted from Smith, L.M. et al., *Nature*, 321, 674, 1986.)

base. The linker structure is also important, which in this case was a propargylamine, because the presence of the dye on the terminator radically affects its ability to be incorporated by the polymerase enzyme. For dye-labeled primers, the oligonucleotides possessing the appropriate sequence were prepared on a standard DNA synthesizer. For Smith et al.[6] a thymidine derivative was prepared that contained a phosphoramidite at the 3′ carbon and a protected alkyl amino group at the 5′ carbon (typically a 6-carbon linker structure). During the final addition cycle of the oligonucleotide prepared via solid-phase synthesis using phosphoramidite chemistry, the thymidine residue was added, and, following deprotection of the alkyl amino group and cleavage from the support, a free primary amine group resulted, which can be reacted with any amino-reactive fluorescent dye to produce the oligonucleotide derivative.

Figures 53.6 and 53.7 also show the absorption and emission profiles for the dye sets used in these experiments. The major attributes of the dye sets are that they can be efficiently excited with either 488- or 514.5-nm lines of the argon ion laser. In addition, there is minimal separation between the emission maxima of the dyes, which allows processing of the fluorescence on as few detection channels as possible. However, there is significant overlap in the emission spectra of the four dyes, producing severe spectral leakage into other detection channels, which must be corrected by software.

The optical hardware used to process the four-color fluorescence for both systems is depicted in Figure 53.8. For the Smith et al. experiment,[6] a single laser and a single detection channel were used; in this case, the detection channel consisted of a multiline argon ion laser and a conventional photomultiplier tube (PMT). Placed in front of the laser and PMT were filter wheels to select the appropriate excitation wavelength (488 or 514.5 nm) and emission color. The filter pairs used during fluorescence readout were 488/520 nm, 488/550 nm, 514/580 nm, and 514/610 nm. In the case of the Prober et al. experiment,[7] due to the narrow distribution between the excitation maxima of the dyes only excitation using the 488-nm line from the argon ion laser was required as well as two PMT tubes to process the emission from the

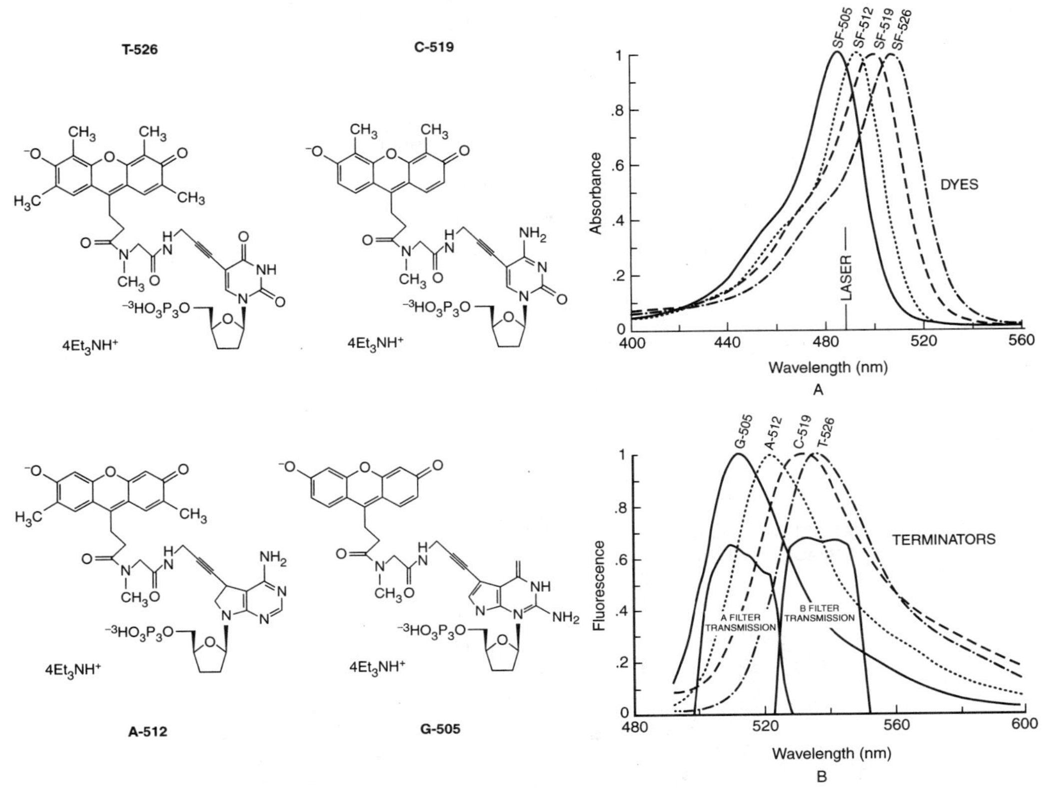


**FIGURE 53.7** Chemical structures and the absorbance (A) and emission spectra (B) of dyes conjugated to ddNTPs. Also shown are the excitation laser wavelength (488 nm) and the filter set used to isolate the fluorescence onto the two detection channels. (Adapted from Prober, J.M. et al., *Science*, 238, 336, 1987.)

four colors. Discrimination of the four colors was accomplished by monitoring the intensity of each dye on both detectors simultaneously. By histograming the ratio of the fluorescence intensity of each dye (produced from an electrophoresis band) on the two detection channels, a discrete value was obtained that allowed facile discrimination of the four different fluorescent dyes (i.e., terminal base). To determine the limit of detection of these fluorescence systems, injections of known concentrations of dye-labeled sequencing primers were electrophoresed. In both cases, the mass detection limit was estimated to be $10^{-17}$ to $10^{-18}$ mol.

Figure 53.9 shows the data output from these systems. Unfortunately, reading the sequence directly from the raw gel data becomes problematic due to several nonidealities. These include signal from a single dye appearing on multiple detection channels due to the broad and closely spaced emission bands, dye-dependent electrophoretic mobility shifts, and nonuniformity in the intensity of the electrophoresis bands due to the enzymatic reaction used to construct the individual DNA size ladders. Thus, several post-electrophoresis-processing steps were required to augment sequence reconstruction of the test template. In the case of the Smith et al. example,[6] these steps involved:

1. High-frequency noise removal using a low-pass Fourier filter.
2. A time delay between measurements at different wavelengths corrected by linear interpolation between successive measurements.
3. A multicomponent analysis performed on each set of four data points, which produced the amount of the four dyes present in the detector as a function of time.

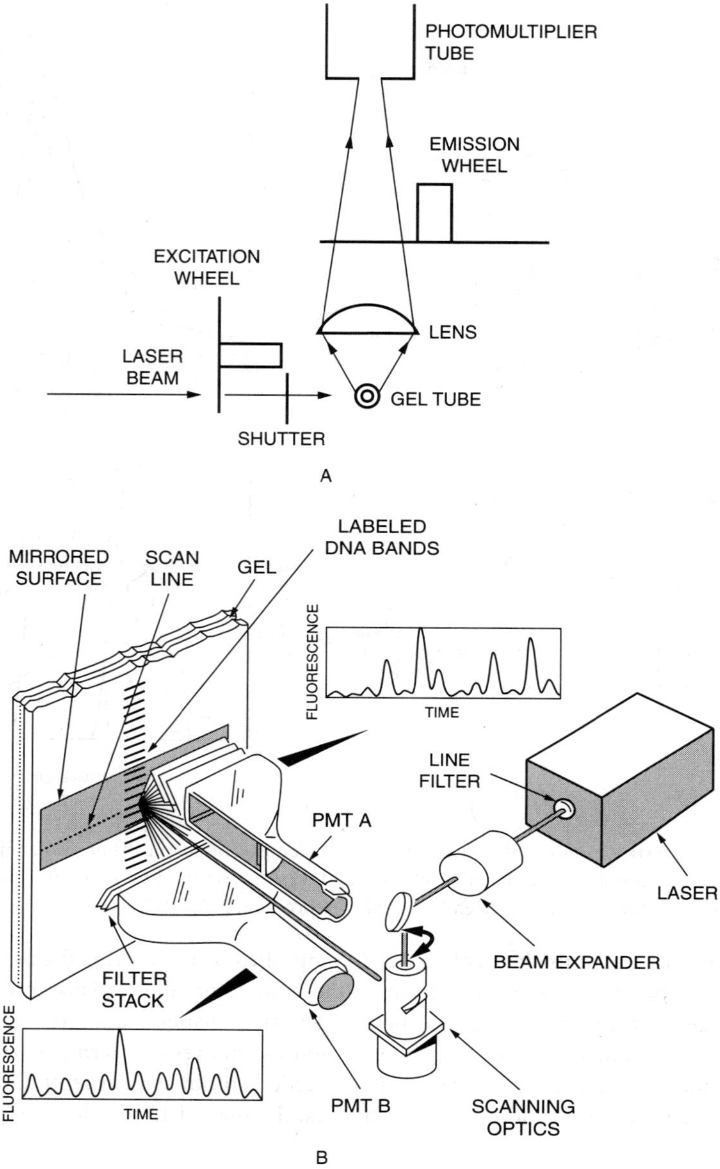


**FIGURE 53.8** Fluorescence detector systems for DNA sequencing. (A) The laser consisted of an argon ion laser operating at 488 and 514 nm. The wavelength was selected using a rotating filter wheel. The emission was collected by a lens and sent through a filter wheel that contained four different filters, one for each of the multicolor dyes used to label the sequencing primers (see Figure 53.6 for their structures and fluorescence properties). (Adapted from Smith et al., *Nature*, 321, 674, 1986.) (B) The laser used for excitation is an argon ion laser operated at 488 nm only. The beam was rastered over the slab-gel plate using scanning optics. The emission was then processed on one of two elongated, stationary photomultiplier tubes (PMTs) (see Figure 53.7 for dye set used with this fluorescence detector). (Adapted from Prober, J.M. et al., *Science*, 238, 336, 1987.)

4. The peaks present in the located data stream.
5. The mobility corrected for the dye attached to each DNA fragment. In this case, it was empirically determined that fluorescein and rhodamine-labeled DNA fragments moved as if they were one base longer than the NBD-labeled fragments, and the Texas Red fragments moved as if they were 1.25 bases longer.

The important performance criteria in any type of automated DNA sequencer are its throughput, the number of bases it can process in a single gel read, and its accuracy in calling bases. In terms of base-calling accuracy, these early instruments demonstrated an error rate of approximately 1%, with a read length approaching 500 bases. The throughput of the instrument described by Prober et al.[7] was estimated to give a raw throughput of 600 bases per hour (12 electrophoresis lanes). Interestingly, many present-day commercial automated sequencers still use similar technology in their machines, and the throughput can be as high as 16,000 bases per hour (96 electrophoresis lanes).

## 53.2 Fluorescent Dyes for DNA Labeling and Sequencing

As stated above, fluorescence detection has had a tremendous impact on DNA sequencing because it possesses a high-speed readout process and has the ability to discriminate among the four nucleotide bases in a single gel lane due to the unique spectral properties of the target dye molecules. A variety of dye sets (typically four dyes per set, one for each nucleotide base) have been developed for DNA-sequencing applications, and regardless of the dye set under consideration certain properties associated with the dyes for DNA-sequencing applications are important. These properties are listed below:

1. *Each dye in the set must be spectroscopically distinct.* The ideal situation from a throughput point of view is to identify each base in a single gel lane instead of four gel lanes. In most cases, the discrimination is based on differences in the emission properties of each dye (distinct emission maxima); however, other fluorescence properties can be used as well such as fluorescence lifetimes.
2. *The dye set should preferably be excited by a single laser source.* It becomes instrumentally difficult to implement multiple excitation sources because lasers are typically used as the source of excitation to improve the limit of detection in the measurement, and the upkeep on multiple lasers becomes problematic.
3. *The dye set should possess high extinction at the excitation frequency and also large quantum yields in the gel matrix used to fractionate the DNAs.* Good photophysical properties are necessary to improve detectability. In addition, the dye set should show reasonable quantum yields in denaturing gels, which consist of high concentrations of urea or formamide.
4. *The dye set should show favorable chemical stability at high temperatures.* A common procedure in most Sanger sequencing strategies is to implement a thermostable polymerase and then subject the sequencing reactions to multiple temperature cycles (55 to 95°C) to amplify the amount of product generated (cycle sequencing). Therefore, the dyes must be able to withstand high-temperature conditions for extended periods of time.
5. *The dyes must induce minimal mobility shifts during the electrophoresis analysis of the sequencing ladders.* Often, the mobility shifts induced by individual dyes cause misordering of the individual bases during sequence reconstruction. As such, post-electrophoresis corrections are often employed to rectify this perturbation in the DNA's mobility.
6. *The dyes must not significantly perturb the activity of the polymerase enzyme.* This is especially true in dye-terminator chemistry because the proximity of the dye to the polymerase enzyme can dramatically influence the enzyme's ability to incorporate that dye–ddNTP conjugate into the polymerized DNA molecule.

### 53.2.1 Visible Fluorescence Dyes for DNA Labeling

The dye set frequently used in many automated, fluorescence-based DNA sequencers is the FAM, JOE, TAMRA, and ROX series (see Figure 53.10), which consists of fluorescein and rhodamine analogs containing a succinimidyl ester for facile conjugation to amine-terminated sequencing primers or terminators. These dyes can be efficiently excited by the 488- and 514.5-nm lines from an argon ion laser; they also possess emission profiles that are fairly well resolved. Many of the Applied Biosystems (Foster City, CA) automated DNA sequencers use this particular dye set (for example, ABI 373 and 377 series, see www.perkin-elmer.com). Figure 53.10 shows emission spectra for this dye set as well as the filter set that

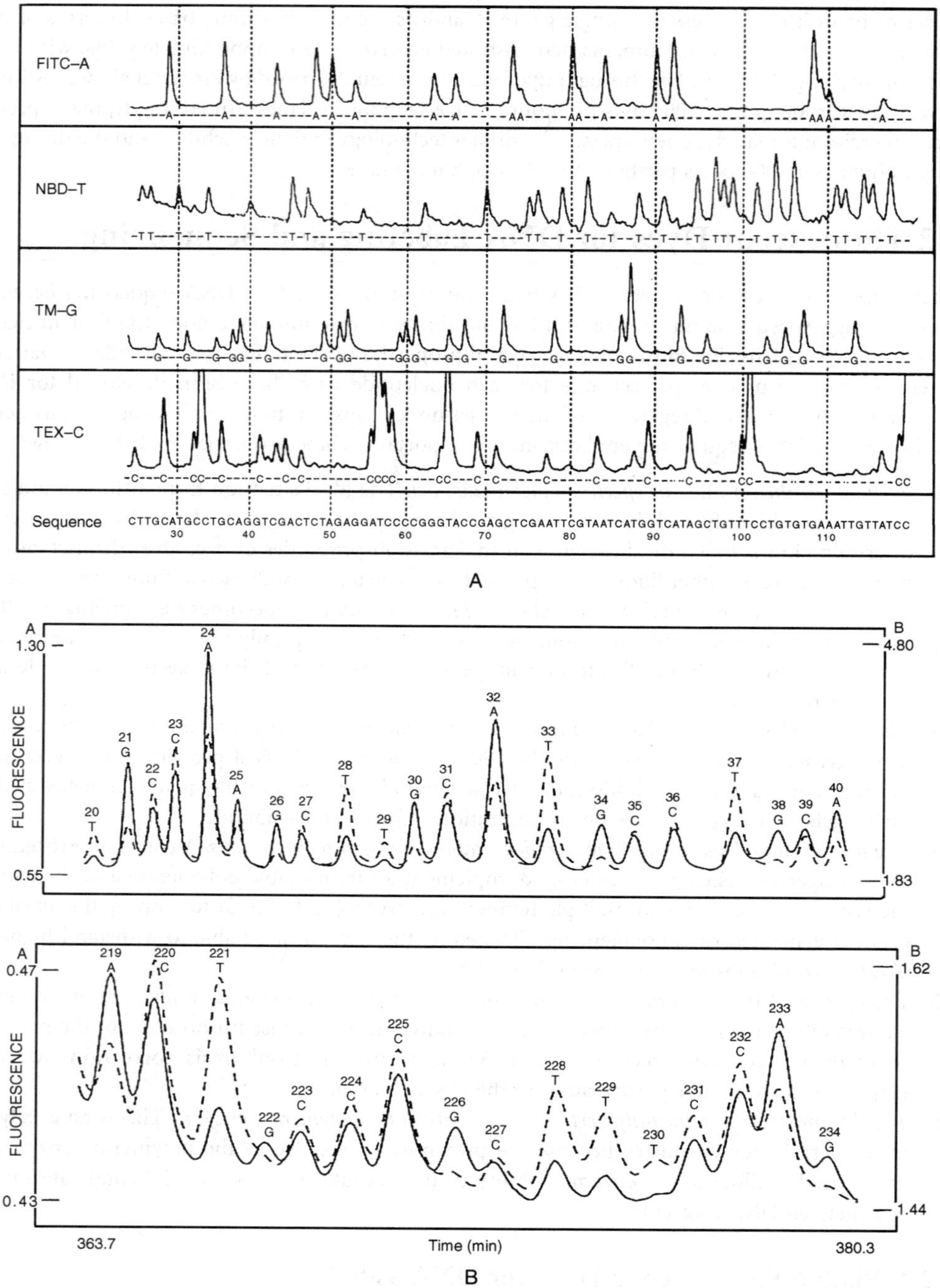


**FIGURE 53.9**

is used to isolate the emission from the dyes onto the appropriate detection channel. While this dye set is fairly robust and works well with typical DNA-cycle-sequencing conditions, this series present some difficulties, i.e., the broad emission profiles, dye-dependent mobility shifts, and inefficient excitation of TAMRA and ROX with the 514.5-nm line from the argon ion laser.

**FIGURE 53.9 (Opposite)** Fluorescence sequencing data obtained from the two fluorescence systems described in Figure 53.8. (A) The dyes (see Figure 53.6) were conjugated to a primer with a sequence, 5′-CCCAGTCACGACGTT-3′, complementary to a site on the M13 phage vector. The reactions were run using Sanger chain-termination methods and run in four different vessels, one for each terminator. The polymerase enzyme was a T7 polymerase, and the reactions were pooled prior to electrophoresis. (Adapted from Smith, et al., *Nature*, 321, 674, 1986.) (B) Sequence of the M13mp8 template using the dye-labeled terminator set shown in Figure 53.7. The base assignment is given as a letter above each electrophoretic peak. The reactions were prepared using Sanger chain-termination methods and prepared in a two-stage fashion. In the first stage, the template (M13mp18) was added to a reaction tube along with the primer, heated to 95°C for 2 min and then cooled in ice water for 5 min. In the second stage, dNTPs and the dye-labeled ddNTPs were added along with an AMV reverse transcriptase enzyme. Following chain extension, the reactions were purified (removal of excess terminators) using gel filtration (Sephadex spin column) and loaded onto the gel. The gel was 8% polyacrylamide containing 7 M urea as the denaturant. In these traces, the solid line represents the fluorescence signal from PMT A, and the dashed line is from PMT B. (Adapted from Prober, J.M. et al., *Science*, 238, 336, 1987.)

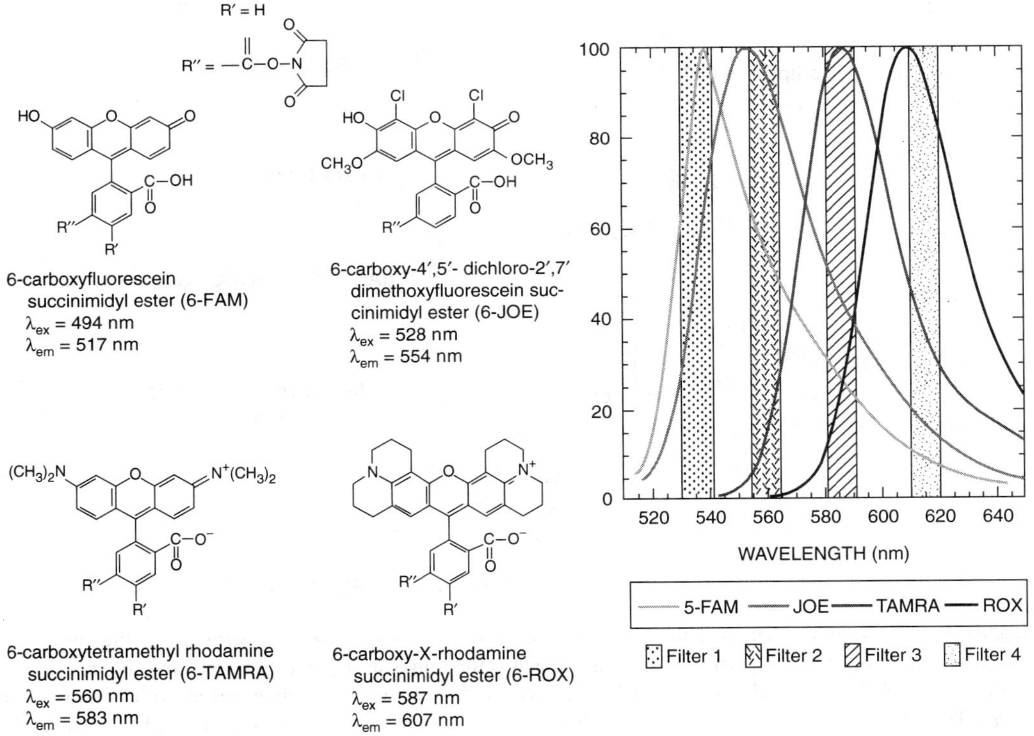


**FIGURE 53.10** Common fluorescence-dye set used for labeling primers for DNA-sequencing applications. The functional group on each dye is a succinimidyl ester, which readily conjugates to primary amine groups. Also shown are the emission spectra of this dye set and the filters used to select the appropriate dye color-processed by each detection channel.

To eliminate dye-dependent mobility shifts and minimize crosstalk between detection channels using four-color processing, a bodipy dye set (4,4-difluoro-4-bora-3α, 4α-diaza-s-indacene-3-proprionic acid) has been used for DNA-sequencing applications.[9] The structures of the dyes are presented in Figure 53.11. Inspection of the fluorescence-emission profiles for this particular dye set indicated that the bandwidths were less than those associated with the FAM/JOE/TAMRA/ROX dye set, which resulted in less spectral leakage between detection channels in the fluorescence readout hardware. Figure 53.11 reveals that all

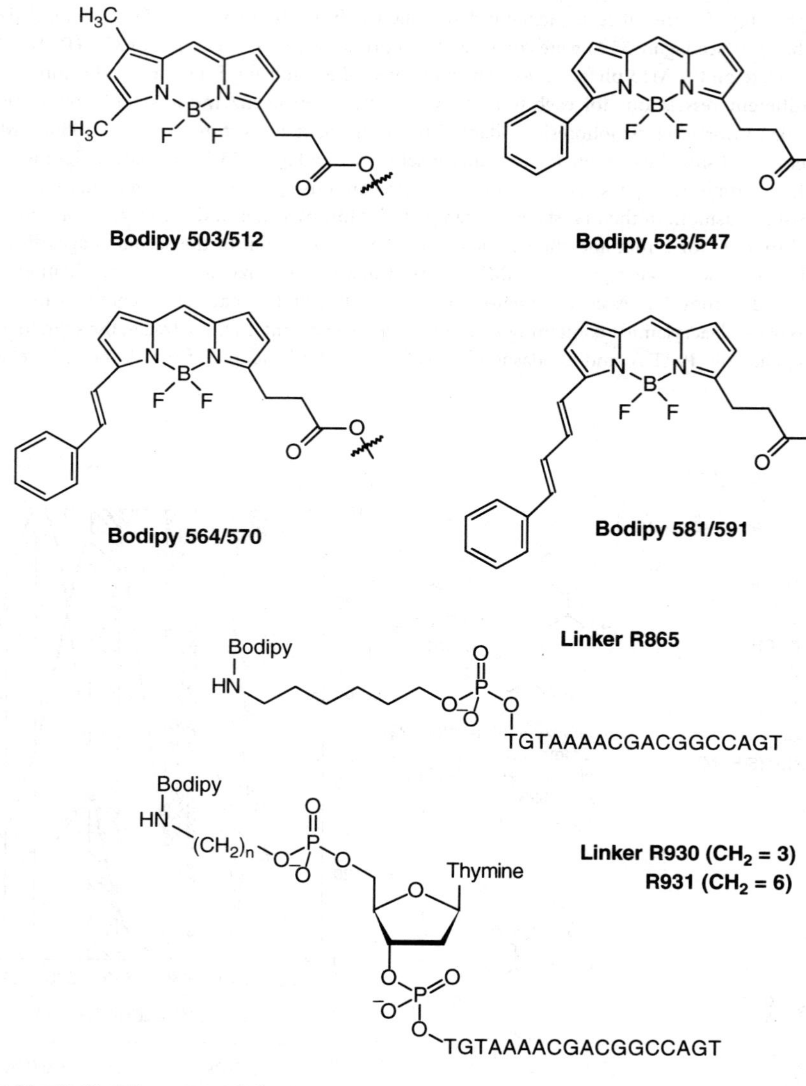


**FIGURE 53.11** Bodipy dye set used for DNA-sequencing applications. Below each structure is the absorption/emission maxima (nm) of the particular dye. Also shown is the universal sequencing primer and the modification required to attach the dye to the primer. The linkers were either three or six carbon linkers with primary amine groups. The linkers are designated as R865 (C-6 linker, U-$C_6$), R930 (C-3 linker, $U^+$-$C_3$), or R931 (C-6 linker, $U^+$-$C_6$). (Adapted from Metzker, M.L. et al., *Science*, 271, 1420, 1996.)

these dyes are neutral and have very similar molecular structures, thereby minimizing dye-dependent mobility shifts. To fully compensate for any mobility-shift differences within the dye set, a modified linker structure was synthesized to account for this difference (see Figure 53.11).

A particular shortcoming associated with this dye set is the chemical instabilities the dyes display when subjected to extended heating at high temperatures. An example of this is shown in Figure 53.12, in which bodipy-labeled primers were used with cycle-sequencing conditions. In cycle sequencing, a linear amplification of the ddNTP-terminated fragments can be generated by subjecting the reaction cocktail to multiple temperature cycles consisting of 95°C (thermal denaturation of the double-stranded DNA molecule), 55°C (annealing the sequencing primer to the DNA template), and 72°C (chain extension using a thermostable polymerase enzyme, such as *Taq* polymerase). The difficulty arises during the 95°C

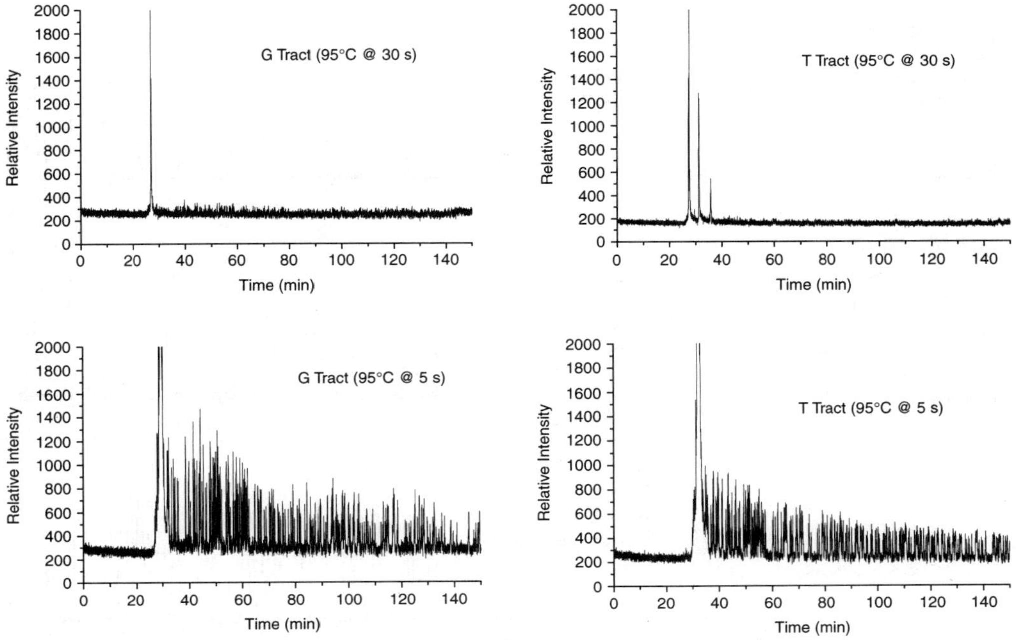


**FIGURE 53.12** Cycle-sequencing effects on bodipy-labeled DNA-sequencing primers. The sequencing reactions were prepared using Sanger chain-termination methods and a *Taq* DNA polymerase. The cycling conditions consisted of 25 cycles with 95°C for 30 or 5 sec; 55°C for 10 sec, and 72°C for 60 sec. The reactions were run with a single terminator (ddGTP, ddTTP) and an M13mp18 template. The sequencing reactions were analyzed in a CGE DNA sequencer using laser excitation at 532 nm. The dyes were bodipy 564/570 (ddGTP) and bodipy 581/591 (ddTTP) (see Figure 53.11 for structures of dyes). The large peak at ~28 min is unextended dye-labeled primer.

step, where significant dye decomposition can result. Figure 53.12 shows fluorescence traces of sequencing ladders prepared using a 95°C step time of either 30 sec or 5 sec. As shown in the figure, there is a significant loss in signal when the cycling time (at 95°C) is 30 sec, but the signal is partially restored when the cycling time is reduced to 5 sec.

Many of the dyes discussed above have been employed in dye-primer sequencing applications, in which the primer used for selecting the DNA polymerization site on the unknown template is determined by Watson–Crick base pairing. An alternative is dye-terminator chemistry, where the fluorescent dyes are covalently attached to the ddNTP used in Sanger sequencing strategies. The advantages of this method are twofold. First, the sequencing reactions can be carried out in a single tube. In dye-primer chemistry, the sequencing reactions are carried out in four separate tubes and then pooled prior to the electrophoresis step. If four spectroscopically unique fluorescent probes attached to the ddNTP are implemented, the reactions can be carried out in a single tube, which reduces reagent consumption and also minimizes sample transfer steps. Second, primers of known sequence need not be synthesized. In primer sequencing, making oligonucleotides of 17 to 23 bases in length can be costly and time-consuming, especially if the dye must be chemically tethered to the primer. The use of dye-terminator chemistry eliminates the need for synthesizing dye-labeled primers. Unfortunately, dye terminators themselves can be quite expensive; in addition, many polymerase enzymes are very sensitive to the type of dye attached to the ddNTP. For example, fluorescein dyes are poor labels for terminators when using the *Taq* polymerase due to its incompatibility with the binding pocket of *Taq*, while the rhodamine dyes are more hydrophobic and are thus more suitable for use with *Taq* polymerase. The result is that the peak heights for the electrophoresis bands can vary tremendously due to differences in incorporation of the dye-modified ddNTPs by the particular polymerase enzyme. In dye-primer chemistry, this disparity is absent due to the large

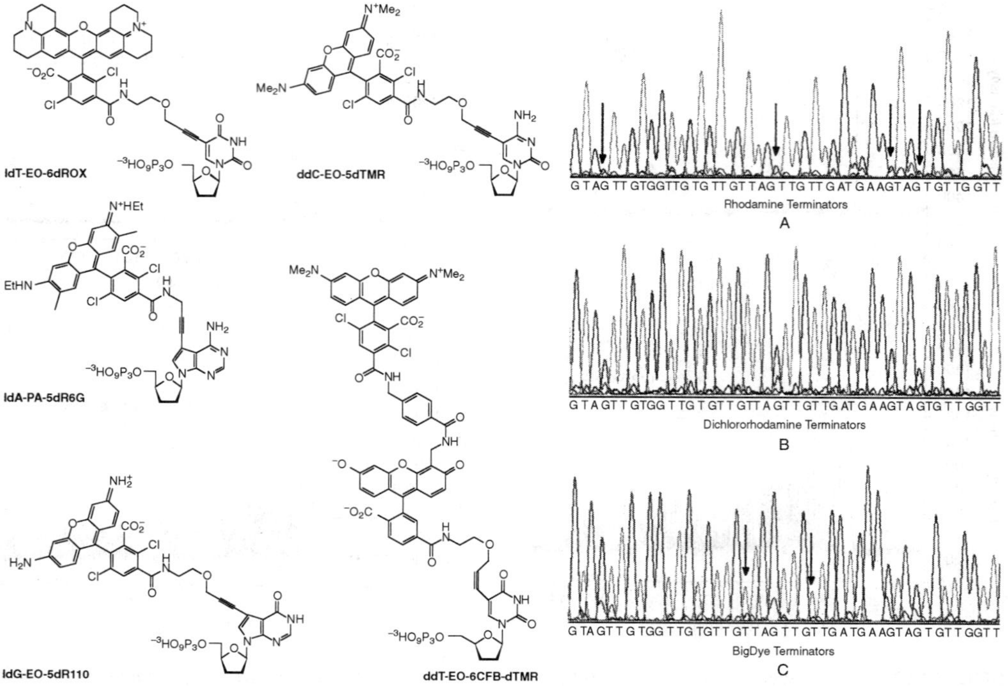


**FIGURE 53.13** Chemical structures of the d-rhodamine dyes used for labeling terminators. Also shown is the structure of an energy transfer terminator (BigDye terminator). The accompanying panels show a comparison of sequencing data obtained using rhodamine- (A), d-rhodamine- (B), and Big Dye- (C) labeled terminators. The arrows in (A) are G-peaks with weak signal. In (C), the arrows indicate weaker T-peaks following G-peaks. The sequencing reactions were run with an AmpliTaq DNA polymerase and cycle-sequencing conditions (30 cycles with a temperature program of 95°C for 30 sec, 55°C for 20 sec, 60°C for 4 min). The DNA template was isolated from a bacterial artificial chromosome. Following chain extension, the reactions were purified using a Centri-Sep spin column to remove excess terminators and then dried in a vacuum and resuspended in 2 to 4 μl of formamide. The reactions were electrophoresed in a slab gel with laser-induced fluorescence detection. (Adapted from Rosenblum, B.B. et al., *Nucleic Acids Res.*, 25, 4500, 1997.)

displacement of the dye from the polymerization site.[10] It is interesting to note that several mutant forms of *Taq* polymerase have been prepared to allow more facile incorporation of dye-labeled terminators.[11] For example, *Taq* Pol I (AmpliTaq FS®) has two modifications in it — a substitution that eliminates the 3′ → 5′ nuclease activity and also a substitution that improves 2′,3′-ddNTP incorporation.

The nature of the dye on the terminator can also influence the mobility of the polymerized DNA fragment as well. For example, rhodamine dyes are typically zwitterionic and thus appear to stabilize hairpin (secondary) structures in the DNA fragment, causing compressions in the electrophoresis data, especially in GC-rich regions of the template. On the other hand, the fluorescein dyes, which are negatively charged, do not suffer from such anomalies.[10]

Slight structural modifications on the base chromophore can influence its incorporation during DNA polymerization. Also, the linker structure can influence the incorporation efficiency as well. Figure 53.13 shows a set of dye-labeled terminators that have been found to give fairly even peak heights in sequencing patterns and produce minimal mobility shifts; the figure also shows the sequencing patterns that were generated with both the rhodamine dyes and the dichlororhodamine (d-rhodamine) analogs.[12] The linker structures chosen for this set were either the propargylamine linker developed by Prober et al.[7] or a propargyl ethoxyamino linker. The choice of linker was based on its ability to accommodate the polymerase enzyme and to minimize the mobility differences within the dye set. For the rhodamine terminators, very

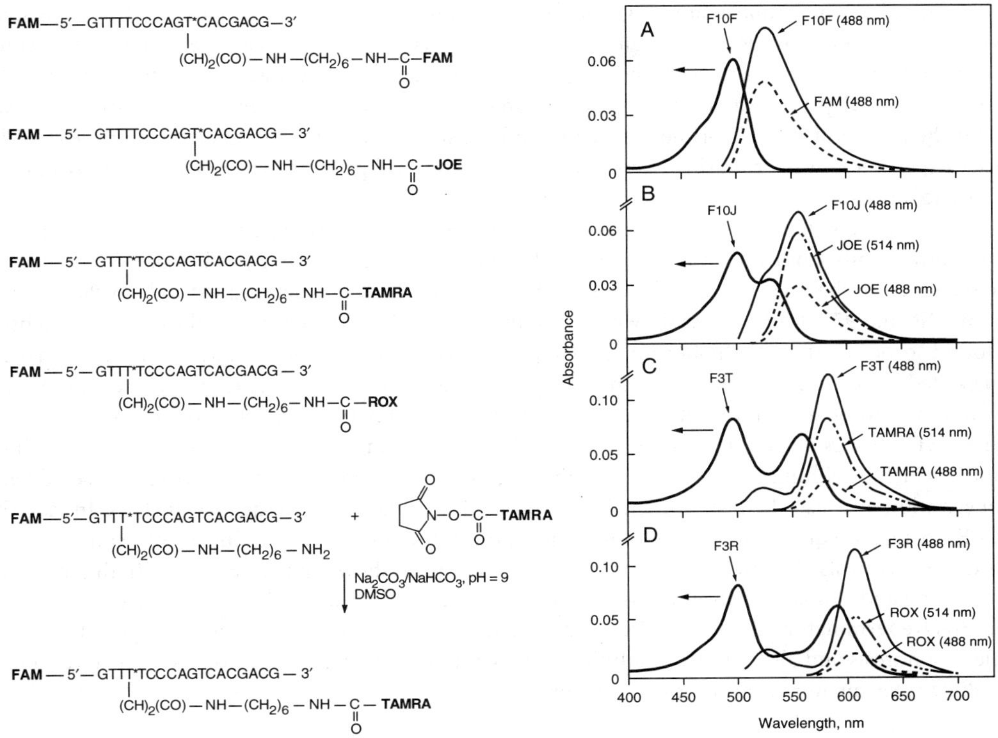


**FIGURE 53.14** Chemical structures and absorbance and fluorescence-emission spectra of energy-transfer (ET)-labeled DNA-sequencing primers. The bottom panel shows the reaction of a dye-labeled FAM primer (5′-end with $-NH_2$ T modified residue) and a succinimidyl TAMRA dye (F3T). The absorption (dark line) and emission spectra for both the ET primers and the single-dye-labeled primers are also shown for comparison purposes. The number in parenthesis is the excitation wavelength used for collecting the emission profile. (Adapted from Ju, J. et al., *Proc. Natl. Acad. Sci. U.S.A.*, 92, 4347, 1995.)

weak G-peaks, which appeared after A-peaks, were observed. However, for the d-rhodamine terminators, this disparity in peak intensity was alleviated, with the peak heights in the pattern being much more uniform.

### 53.2.2 Energy Transfer Dyes for DNA Sequencing

One of the major problems associated with many of the single-dye sets mentioned previously is that their absorption spectra are dispersed over a relatively large spectral range, which provides poor excitation efficiency even for dual-laser (488 and 514 or 543) systems. As such, the red dyes are used at higher concentrations during DNA polymerization to circumvent poor excitation. To overcome this problem without sacrificing spectral dispersion in the emission profiles, the phenomenon of Förster energy transfer has been used to design sequencing primers that can be efficiently excited with a single laser line.[13–16] A brief introduction about these ET (energy transfer) primers in the context of DNA sequencing follows.

The chemical structures of the ET primers developed by Richard Mathies and his group[15] are shown in Figure 53.14. The donor dye in this case was FAM, which could be excited with the 488-nm line of an argon ion laser. The acceptor dyes were FAM, JOE, TAMRA, or ROX. The donor (FAM) was attached to the sequencing primer on the 5′ end during the solid-phase DNA synthetic preparation of the M13 (–40) sequencing primer using phosphoramidite chemistry. The sequencing primer also contained a modified base (T*) that possessed a linker structure with a primary amine. The appropriate acceptor was conjugated to the primary amine group off the modified base following cleavage from the solid support via a succinimidyl ester functional group. The spacer distance between the donor and acceptor was selected by positioning T* within the M13 (–40) primer sequence during solid-phase synthesis. The naming of these ET primers followed the convention donor-spacer (bp)-acceptor. A ten-base spacer was used for the FAM and JOE ET primers, and a three-base spacer was selected for TAMRA and ROX. The choice of spacer size was primarily determined by producing ET primers, which showed uniform electrophoretic mobilities.

Figure 53.14 shows the absorbance and emission profiles of the ET primer series. While the absorption spectra show bands from both the acceptor and donor dyes, the emission profiles are dominated by fluorescence from the acceptor dyes. In fact, the ET efficiency has been determined to be 65% for F10J, 96% for F3R, and 97% for F3T.[15] As shown in the emission profiles in Figure 53.14, the emission intensity was found to be significantly higher for the ET primers compared to their single-dye primer partners when excited with 488-nm laser light from the argon ion laser due to higher efficiency in excitation. This translates into improved fluorescence sensitivity of these ET primers during electrophoresis.

These ET primers do offer some advantages due to their improved detection sensitivities; those advantages include eliminating the need for adjusting the concentrations of the dye primers used during polymerization and also the need for smaller amounts of template in the sequence analysis. In fact, the use of ET primers required about one fourth the amount of template vs. the single-dye primers.

ET dye pairs can also be situated on terminators; an example is shown in Figure 53.13.[12] In this example, d-rhodamine and rhodamine dyes are used with a propargyl ethoxyamino linker. This dye set has been called BigDye terminators. Again, the dyes were selected so as to provide fairly uniform peak heights in the electrophoresis and also uniform mobility shifts within the series. A sample of a sequence run using these BigDye terminators is also shown in Figure 53.13.

### 53.2.3 Near-Infrared Dyes for DNA Sequencing

The attractive feature associated with fluorescence in the near-infrared (NIR) ($\lambda_{ex}$ > 700 nm) includes smaller backgrounds observed during signal collection and the rather simple instrumentation required for carrying out ultrasensitive detection. In most cases, the limit of detection for fluorescence measurements is determined primarily by the magnitude of the background produced from scattering or impurity fluorescence. This is particularly true in DNA sequencing because detection occurs within the gel matrix, which can be a significant contributor of scattering photons. In addition, the use of denaturants in the gel matrix, such as urea (7 M) or formamide, can produce large amounts of background fluorescence. The lower background that is typically observed in the NIR can be attributed to the fact that few species fluoresce in the NIR. In addition, the $1/\lambda^4$ dependence of the Raman cross section also provides a lower scattering contribution at these longer excitation wavelengths.

An added advantage of NIR fluorescence is the fact that the instrumentation required for detection can be rather simple and easy to use. A typical NIR fluorescence-detection apparatus can consist of an inexpensive diode laser and single-photon avalanche diode (SPAD). These components are solid state, which allows the detector to be run for extended periods of time, requiring little maintenance or operator expertise.

NIR fluorescence can be a very attractive detection strategy in gel sequencing because of the highly scattering medium in which the separation must be performed. Due to the intrinsically lower backgrounds that are expected in the NIR vs. the visible, on-column detection can be performed without sacrificing detection sensitivity. To highlight the intrinsic advantages associated with the use of NIR fluorescence detection in capillary-gel DNA-sequencing applications, a direct comparison between laser-induced fluorescence detection at 488-nm excitation and 780-nm excitation has been reported.[17] In this study, a sequencing primer labeled with FAM or a NIR dye were electrophoresed in a capillary-gel column and the detection limits calculated for both systems. The results indicated that the limits of detection for

**TABLE 53.1.** Chemical Structures of Some Typical NIR Fluorescent Dyes and Their Photophysical Properties

| Sulfonated, Heavy-Atom Modified Near-IR Dyes (X = I, Br, Cl, F) | Dye Substitution | Absorbance (nm) | Emission (nm) | $\varepsilon$ ($M^{-1}$ $cm^{-1}$) | $\phi$ | $\tau_f$ (ps) |
|---|---|---|---|---|---|---|
| 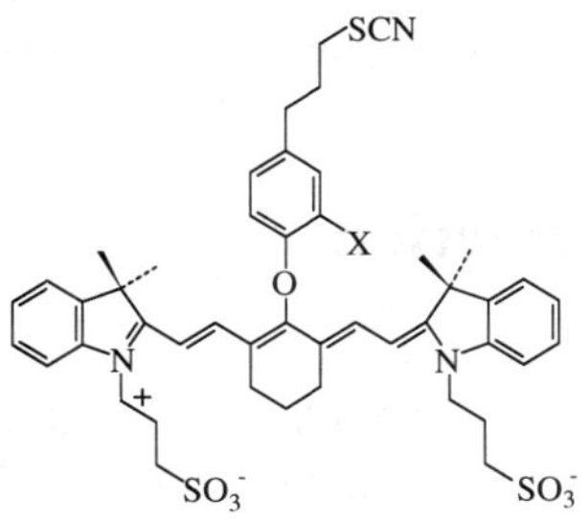 | I | 766 | 796 | 216,000 | 0.15 | 947 |
| | Br | 768 | 798 | 254,000 | 0.14 | 912 |
| | Cl | 768 | 797 | 239,000 | 0.14 | 880 |
| | F | 768 | 796 | 221,000 | 0.14 | 843 |

| IRD 41 | Absorbance (nm) | Emission (nm) | εf | $\varepsilon$ ($M^{-1}$ $cm^{-1}$) |
|---|---|---|---|---|
| 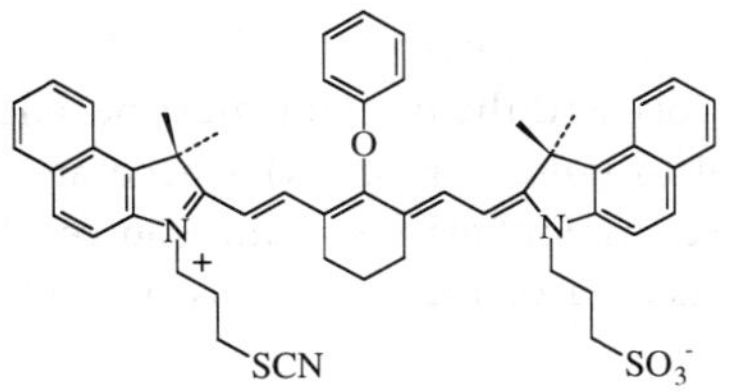 | 787 | 807 | 0.16 | 200,000 |

*Note:* In the top panel of this table, the dyes were modified with an intramolecular heavy atom to produce a set of dyes with four distinct lifetime values. In both cases, the dyes contained an isothiocyanate to allow conjugation to amine-containing molecules.

the NIR case were found to be $3.4 \times 10^{-20}$ mol, while for 488-nm excitation the limit of detection was $1.5 \times 10^{-18}$ mol. The improvement in the limit of detection for the NIR case was observed in spite of the fact that the fluorescence quantum yield associated with the NIR dye was only 0.07, while the quantum yield for the FAM dye was ~0.9. The improved detection limit resulted primarily from the significantly lower background observed in the NIR. NIR has also been demonstrated in sequencing applications using SGE, where the detection sensitivity has been reported to be ~2000 molecules.[18,19]

The fluorophores that are typically used in the NIR include the tricarbocyanine (heptamethine) dyes, which consist of heteroaromatic fragments linked by a polymethine chain (see Table 53.1). The absorption maxima can be altered by changing the length of this polymethine chain or by changing the heteroatom within the heteroaromatic fragments. These NIR dyes typically possess large extinction coefficients and relatively low quantum yields in aqueous solvents.[20] The low-fluorescence quantum yields result primarily from high rates of internal conversion. An additional disadvantage associated with these dyes is their poor water solubility. These dyes show a high propensity toward aggregation, forming aggregates with poor fluorescence properties. This aggregation can be alleviated to a certain degree by inserting charged groups within the molecular framework of the fluorophore, e.g., alkyl-sulfonate groups. In addition, these dyes have short fluorescence lifetimes and are very susceptible to photobleaching vs. the visible dyes typically used in DNA-sequencing applications. In most cases, the NIR chromophore is covalently attached to either a sequencing primer or ddNTP via an isothiocyanate functional group. As shown in Table 53.1, this functional group can be placed on the heteroatom (N in this case) or in the para-position of the bridging phenyl ring. When the isothiocyanate group is situated on the heteroatom, the net charge on the dye becomes neutral and as such has limited water solubility. If the isothiocyanate group is placed on the bridging phenyl ring, this gives a net –1 charge to the dye and improves its water solubility. Unfortunately, the ether linkage is susceptible to nucleophilic attack, especially by dithiothriotol (DTT), which can release the dye from the moiety (sequencing primer or ddNTP) to which it is attached. This can produce a large peak in the electropherogram, which can mask some of the sequencing fragments that comigrate with the free dye. This problem can be alleviated to a certain extent by using an ethanol-precipitation step following DNA polymerization.

## 53.3 Instrumental Formats for Fluorescence Detection in DNA Sequencing

The ability to read the fluorescence during the electrophoresis fractionation of the DNA ladders and also accurately identify the terminal base (Sanger sequencing) is a challenging task due to a number of technical issues. As such, a number of fluorescence-readout devices have been developed for reading such data. When considering the design of a fluorescence detector for sequencing applications, several instrumental constraints must be incorporated into the design including:

1. *High sensitivity.* As pointed out previously, the amount of material (fluorescently-labeled DNA) loaded onto the gel can be in the low attomole range ($10^{-18}$ mol), and the detector must be able to read this fluorescence signature with a reasonably high signal-to-noise ratio (SNR) to accurately call the base. Therefore, in most cases, irrespective of the separation platform, the source of excitation is a laser that is well matched to the excitation maxima of the dye set used in the sequencing experiment.
2. *Spectral-marker base identification.* The instrument must be able to identify one of the four bases terminating the sequencing fragments by accurately processing the fluorescence via spectral discrimination (wavelength) or some other fluorescence property, such as the lifetime. For spectral discrimination, this would require sorting the fluorescence by wavelength using either filters or gratings. In addition, multiple detection channels would be required.
3. *Fluorescence processing from many electrophoresis lanes.* In most sequencing instruments, the fluorescence must be read from multiple gel lanes (SGE) or multiple capillaries (CGE). This can be done by either using a scanning system, in which the relay optic (collection optic) is rastered over the gel lanes or capillaries, or an imaging system, in which the fluorescence from the multiple gel lanes or capillaries are imaged onto some type of multichannel detector, such as a charge-coupled device (CCD) or image-intensified photodiode array.
4. *Robust instrumentation.* Since many sequencing devices are run by novice operators and are also run for extended periods of time, the detector format must be dependable and turnkey in operation.

This short list of requirements for any type of fluorescence-readout device appropriate for sequencing presents itself with many challenges that are often noncomplementary with the sequencing requirements. For example, high sensitivity is particularly demanding because the separation platforms used to fractionate the DNA are becoming smaller; therefore, smaller amounts of material must be inserted into the device. In addition, detecting material directly within the gel matrix (typically a polyacrylamide gel) can be problematic due to the intense scattering photons that it produces. Also, the signal is transitory in that the DNA fragment resides within the probing volume (defined by the laser beam size) for a few seconds. Another issue is that many separation channels or lanes must be interrogated for high-throughput applications. On top of these considerations, a high SNR is required to obtain high accuracy in the base-calling phases of the readout process. As such, significant design considerations go into fabricating a fluorescence detector system for DNA-sequencing applications.

There are two general types of fluorescence detector formats — scanning and imaging-type devices. In most high-throughput sequencing devices, multiple lanes of the slab gel or multiple capillaries are

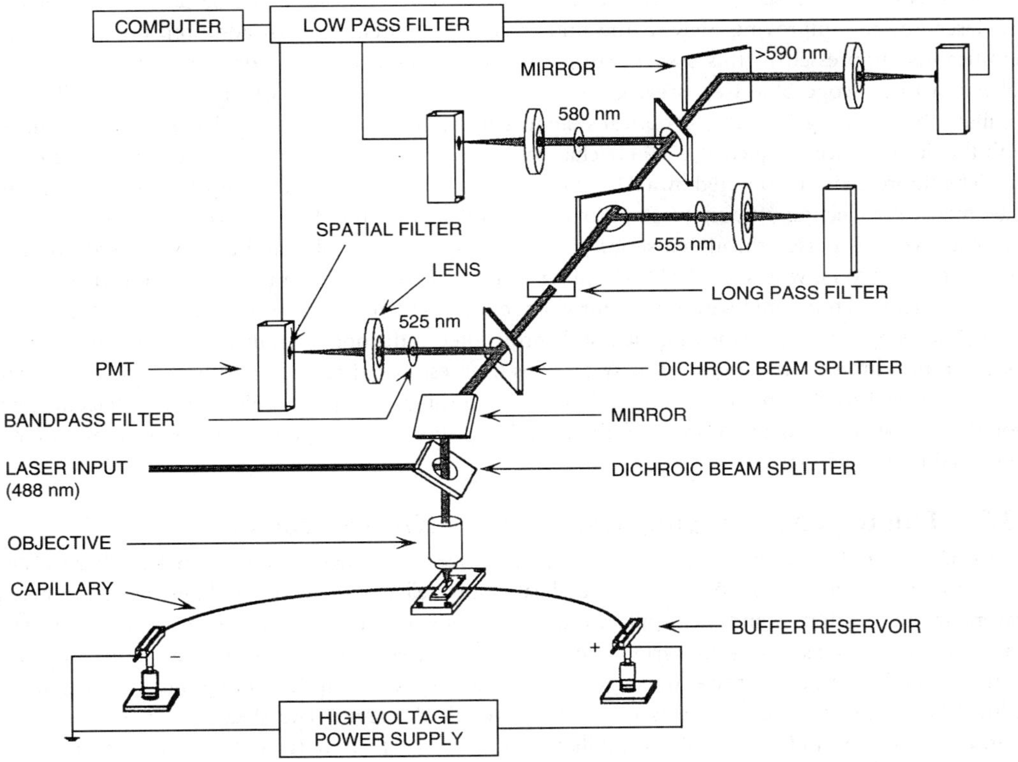


**FIGURE 53.15** Four-color, laser scanning fluorescence system for DNA sequencing applications using capillary gel electrophoresis. The excitation source was an air-cooled Ar ion laser operating at 488 nm (1 mW average power). The collection and focusing optic consisted of a microscope objective (20×, NA = 0.5). The emission was directed onto one of four different PMTs using dichroic filters and further isolated from background photons using a bandpass filter. The fluorescence was sampled at 2 Hz with output filtered using a low-pass filter with a time-constant of 1 sec. (Adapted from Ju, J. et al., *Proc. Natl. Acad. Sci. U.S.A.*, 92, 4347, 1995.)

run in parallel to increase system throughput. For example, many machines run in a 96-lane format because microtiter plates, which are standard plates used to prepare sequencing reactions, come in a 96-well format. In the scanning systems, the excitation beam is tightly focused and irradiates only a single lane, while the relay optic is rastered over the lanes of the gel or the capillary array and the fluorescence from each lane processed sequentially on one set of detection channels. For the imaging systems, all of the electrophoresis lanes are irradiated by a laser or lasers simultaneously, with the fluorescence readout accomplished using a multichannel detector such as a CCD.

## 53.3.1 Fluorescence Scanning Instruments

Figure 53.15 shows a typical scanning system.[21,22] This system uses a confocal geometry with epi-illumination in which the objective used to collect the emitted fluorescence also serves to focus the laser beam into individual capillaries (or lanes of the slab gel) used for the electrophoresis. Following collection the emission is focused onto a spatial filter at the secondary image plan of the collection objective. The laser light (488 nm, argon ion laser, 1 mW) is directed into the objective using a dichroic filter. The capillaries are held into a linear array, and in this particular example the capillary array is translated beneath the microscope objective. As shown in the figure, the laser irradiates only one capillary at a time, but since the beam is tightly focused (diameter = 10 μm) the electronic transition can be saturated at relatively low laser powers, improving SNR in the fluorescence measurement.[23] In addition, the noise can be significantly suppressed in this system, since a pinhole is used in the secondary image plane of the collection microscope objective, preventing scattered out-of-focus light generated at the walls of the capillary from passing through the optical system. The capillary array is scanned at a rate of 20 mm/sec, with the fluorescence sampled at 1500 Hz/channel (color channel) resulting in a pixel image size of 13.3 μm. The fluorescence is collected by a 32× microscope objective (numerical aperture = 0.4), resulting in a geometrical collection efficiency of approximately 12%. Once the fluorescence has been collected by the objective, it is passed through a series of dichroics to sort the color and then processed on one of four different PMTs, with each PMT sampling a different color (spectral discrimination). While the present system is configured with four color channels, the system could easily be configured to do two-color processing as well by removing the last dichroic filter in the optical train and two of the PMTs. The concentration limit of detection of this system has been estimated to be $2 \times 10^{-12}$ M (SNR = 3), which was determined by flowing a solution of fluorescein through an open capillary.[21] The detection limit would be expected to degrade in a gel-filled capillary due to the higher background that would be generated by the gel matrix.

## 53.3.2 Fluorescence-Imaging Systems for DNA Sequencing

Figure 53.16A shows an imaging system for reading fluorescence from multichannel capillary systems for DNA sequencing.[24,25] In this example, a sheath flow cell is used with the laser beam (or beams) traversing the sheath flow and the capillary output dumping into the sheath stream (see Figure 53.18C). This geometry allows simultaneous irradiation of all the material migrating from the capillaries without requiring the laser beam to travel through each capillary, which would cause significant scattering and reduce the intensity of the beam as it traveled through the array. The sheath flow cell also causes contraction of the fluid output of the capillary since the sheath flow runs at a greater linear velocity compared to the sample (capillary) stream. Figure 53.18B shows a fluorescence image of the output from the capillary array, indicating that the sample stream diameter at the probing point was ~0.18 mm and also demonstrating minimal crosstalk between individual capillaries in the array. The laser beams (488 nm from argon ion, 6 mW; 532 nm from frequency-doubled yttrium-aluminum-garnet [YAG], 6 mW) were positioned slightly below the exit end of the capillaries (see Figure 53.18A) with the beams brought colinear using a dichroic filter. The collection optic and focusing optic produced a total magnification of 1 and resulted in a geometrical collection efficiency of 1%. To achieve multicolor processing capabilities, the collected radiation was sent through an image-splitting prism to produce four separated (spectrally) line images on the array detector (one for each dye used to label the terminal bases). In addition, a series of narrow bandpass filters were placed in front of the image-splitting prism to assist in isolating the appropriate colors for data processing. The detector that was used for this system was a two-dimensional CCD camera with a cooled image intensifier. Interestingly, the detection limit reported for this system was found to be $2 \times 10^{-12}$ M when operated in the four-color mode, comparable to that seen for the scanning system discussed above. However, the presence or absence of the gel matrix does not affect the sensitivity of the fluorescence measurement in this case, since the fluorescence interrogation is done off column in the sheath flow. In fact, researchers have reported that the implementation of the sheath flow geometry in gel electrophoresis can offer a significant improvement in limits of detection by minimizing scattering contributions to the background.[26]

When comparing these two fluorescence-readout systems, several issues should be highlighted. One is the duty cycle, which takes into account the loss in signal due to multiple-lane sampling. For any type of scanning system the sampling of the electrophoresis lanes is done in a sequential fashion. For a scanning system sampling 96 capillaries the duty cycle is approximately 1%. However, in the imaging system, all capillaries are sampled continuously and the duty cycle is nearly 100%. Therefore, comparisons of detection limits for any system must include a term for the duty cycle because lower duty cycles will degrade the limit of detection.

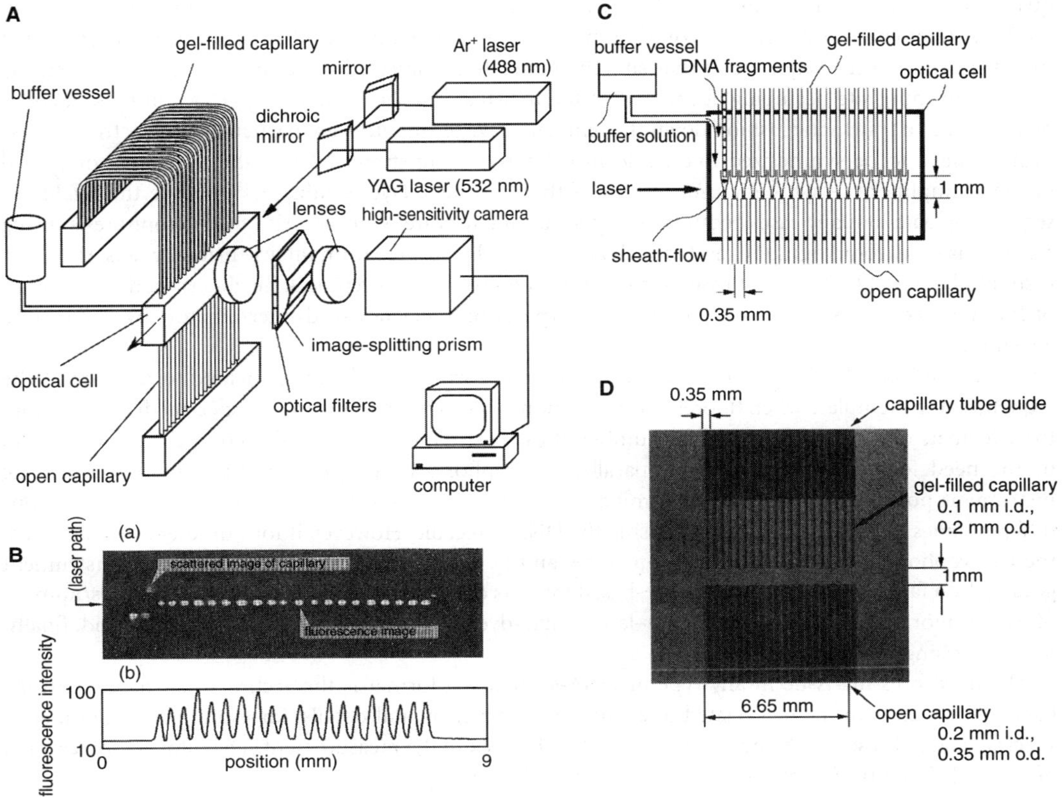


**FIGURE 53.16** (A) Schematic of an imaging laser-induced fluorescence detector for reading four-color fluorescence from capillary-gel arrays. The lasers used for excitation were an argon ion (488 nm) and a YAG laser (532 nm). The lasers were allowed to traverse below the output of the gel columns and the fluorescence collected with a lens system. The fluorescence line image was split into four different color images using a polyhedral image-splitting prism coupled to optical filters. The filtered fluorescence was detected using a two-dimensional CCD. (B) Fluorescence image in the one-color mode across an array of 20 capillaries. The integrated fluorescence intensity is shown in the bottom panel. (C) Schematic view of the multiple sheath flow cell using gravity feed for the sheath flow. Twenty gel-filled capillaries were aligned at a 0.35-mm pitch in an optical cell (26 mm × 26 mm × 4 mm). (D) Photograph of the capillary array aligned in the sheath flow cell. (Adapted from Takahashi, S. et al., *Anal. Chem.*, 66, 1021, 1994.)

## 53.4 DYE Primer/Terminator Chemistry and Fluorescence Detection Formats

In most sequencing applications, dye-labeled primers are used for accumulating sequencing data using automated instruments. This stems from the fact that dye-labeled primers are typically less expensive to use vs. their dye-labeled terminator counterparts. Also, in most applications, small pieces of DNA (1 to 2 kbp in length) are cloned into bacterial vectors for propagation (to increase copy number), such as M13s, which have a known sequence and serve as ideal priming sites. However, dye-labeled primers do present problems; for example, the sequencing reactions must be run in four separate tubes during polymerization and then pooled prior to gel electrophoresis. In addition, unextended primer can result in a large electrophoretic peak (i.e., high intensity), which often masks the ability to call bases close to the primer-annealing site.

Dye-labeled terminators can be appealing to use in certain applications, for example, when high-quality sequencing data is required and in primer-walking strategies. In primer walking, the sequence of the DNA template is initiated at a common priming site using a primer that is complementary to that site. After reading the sequence at that site, the template is subjected to another round of sequencing, with the priming site occurring at the end of the first read. In this way, a long DNA can be sequenced by walking in a systematic fashion down the template. Dye terminators are particularly attractive because primers need to be synthesized frequently in primer-walking strategies, and the need for nonlabeled primers simplifies the synthetic preparation of these primers. Dye terminators improve the quality of sequencing data in many cases because the excess terminators are removed prior to electrophoresis (using size-exclusion chromatography) and as such give clean gel reads free from intense primer peaks. However, it should be noted that in most cases, terminators can produce uneven peak heights (broad distribution of fluorescence intensities) due to the poor incorporation efficiency of dye terminators by polymerase enzymes.

When dye-labeled primers are used, several different formats can be implemented to reconstruct the sequence of the template when fluorescence detection is being used. In the case of spectral discrimination, these formats may vary in terms of the number of dyes used, the number of detection channels required, or the need for running one to four parallel electrophoresis lanes. For example, if the sequencing instrument possesses no spectral-discrimination capabilities, the electrophoresis must be run in four different lanes, one for each base comprising the DNA molecule. However, if four different dyes are used, the electrophoresis can be reduced to one lane, and as a result the production rate of the instrument goes up by a factor of 4. The fluorescence-based formats discussed here include (number of dyes/number of electrophoresis lanes), single-dye/four-lane; single-dye/single-lane; two-color/single-lane, and, finally, four-color/single-lane strategies.

The most pressing issue in any type of DNA-sequencing format is the accuracy associated with the base call, which is intimately related to a number of experimental details, for example, the number of spectral channels used in the instrument as well as the SNR in the measurement. The information content of a signal, $I$, can be determined from the simple relation:[27]

$$I = n \log_2(\mathrm{SNR}) \tag{53.3}$$

where $n$ is the number of spectral channels and SNR is the SNR associated with the measurement. The term $I$ is expressed in bits, and typically two bits are necessary to distinguish between four different signals, but only if there is no spectral overlap between the dyes used for identifying the bases. Unfortunately, in most multicolor systems, the spectra of the dyes used in the sequencing device show significant overlap; thus, many more bits will be required to call bases during the sequencing run.

While the above equation can provide information on how to improve the accuracy of the base call, it does not provide the sequencer with information on the identity of the individual electrophoretic peaks (base call) or the quality of a base call within a single-gel read. For example, if four-color sequencing is used with dye-primer chemistry, how should the data be processed, and what is the confidence with which an electrophoretic peak is called an A, T, G, or C? To provide such information, an algorithm has been developed to not only correct for anomalies associated with fluorescence-based sequencing but also assign a quality score to each called base. The typical algorithm used is called the *Phred* scale, and it employs several steps to process the sequencing data obtained from fluorescence-based, automated DNA sequencers.[28]

The data input into *Phred* consists of a trace, which is electrophoretic data processed into four spectral channels, one for each base. The algorithm consists of four basic steps:

1. *Idealized electrophoretic peak locations (predicted peaks) are determined.* This is based on the premise that most peaks are evenly spaced throughout the gel. In regions where this is not the case (typically during the early and late phases of the electrophoresis), predictions are made as to the number of correct bases and their idealized locations. This step is carried out using Fourier methods as well as the peak-spacing criterion and helps to discriminate noise peaks from true peaks.

2. *Observed peaks are identified in the trace.* Peaks are identified by summing trace values to estimate the area in regions that satisfy the criterion $2 \times v(i) \geq v(i+1) + v(i-1)$, where $v(i)$ is the intensity value at point $i$. If the peak area exceeds 10% of the average area of the preceding ten accepted peaks and 5% of the area of the immediate preceding peak, it is accepted as a true peak.
3. *Observed peaks are matched to the predicted peak locations, omitting some peaks and splitting others.* In this phase of the algorithm, the observed peak arises from one of four spectral channels and thus can be associated with one of the four bases. It is this ordered list of matched observed peaks that determines a base sequence for the DNA template in question.
4. *Observed peaks that are uncalled (unmatched to predicted peaks) are processed.* In this step, an observed peak that did not have a complement in the predicted trace is called and assigned a base and finally inserted into the read sequence.

This algorithm deals mainly with sorting out difficulties associated with the electrophoresis by identifying peaks in the gel traces, especially in areas where the peaks are compressed (poor resolving power) or where multiple peaks are convolved due to significant band broadening produced by diffusional artifacts.

Often, preprocessing of the traces is carried out prior to *Phred* analysis to correct for dye-dependent mobility shifts. In most cases, these mobility shifts are empirically determined by running an electropherogram of a single dye-labeled DNA ladder (for example, T-terminated ladder) and comparing the mobilities to the same ladder labeled with another dye of the set. This type of analysis can be very complex and involved because the mobility shift depends both on the dye and linker structure and on the separation platform used. For example, dyes that show uniform mobility shifts in SGE may not show the same effect in CGE. In addition, these mobility shifts can be dependent on the length of the DNA to which the dye is attached.[29] An example of this phenomenon is shown in Figure 53.17. In this particular example, cyanine dyes were covalently anchored to an M13 (–40) sequencing primer and annealed to an M13 template followed by extension with a single terminator (ddT). The tracts were electrophoresed using CGE. In Figure 53.17A, comparison between a Cy5T7 (–2 charge) and Cy5.5T (–1 charge) tracts indicates that the Cy5T7-labeled fragments migrate faster in the beginning of the run (smaller DNA fragments), but after 300 bp, the two dye-labeled fragments comigrate. In Figure 53.17B, a mobility crossover occurs at ~125 bp for the dyes Sq5T4 (neutral charge) and Cy5.5T12 (–2 charge), with the shorter fragments migrating faster with the Cy5.5T12 label, and after this the Sq5T4-labeled fragments migrate faster. These types of mobility shifts have been ascribed to differences in the net charge of the dye label and to potential dye–DNA base interactions. These interactions, predominantly driven by hydrophobic interactions, may cause loops or hairpin structures on the 5′ end of the dye–DNA complex. These structures would cause a faster migration rate compared to fully extended structures produced by most dye–DNA complexes.

## 53.4.1 Single-Color/Four-Lane

In this processing format, only a single fluorescence detection channel is required to analyze the signal from the labeling dye because only a single dye is used to detect the sequencing fragments produced following chain extension. However, since no color discrimination is implemented, the electrophoresis must be run in four lanes, one for each base, like the format used in traditional autoradiographic detection. While this is a reasonable approach for slab-gel separations, it is not a viable strategy in capillary-gel applications due to the poor run-to-run reproducibility in the migration rates of the fragments traveling through the different capillaries. This is due to differences in the gel from capillary to capillary as well as differences in the integrity of the wall coatings used to suppress the electroosmotic flow. In the slab-gel format, reproducibility in the migration times becomes less of a problem because all of the lanes are run in the same gel matrix.

Figure 53.18 shows the output of a typical single-color/four-lane sequencing device along with the called bases. In this example, each different color trace represents an electropherogram from an individual lane of the slab gel, which is overlaid to allow reconstruction of the sequence of the template. In this case, the device used a single microscope head containing the collection optics, a diode laser, filters, and avalanche photodiode to read the fluorescence from the gel.[18] The microscope scanner is rastered over

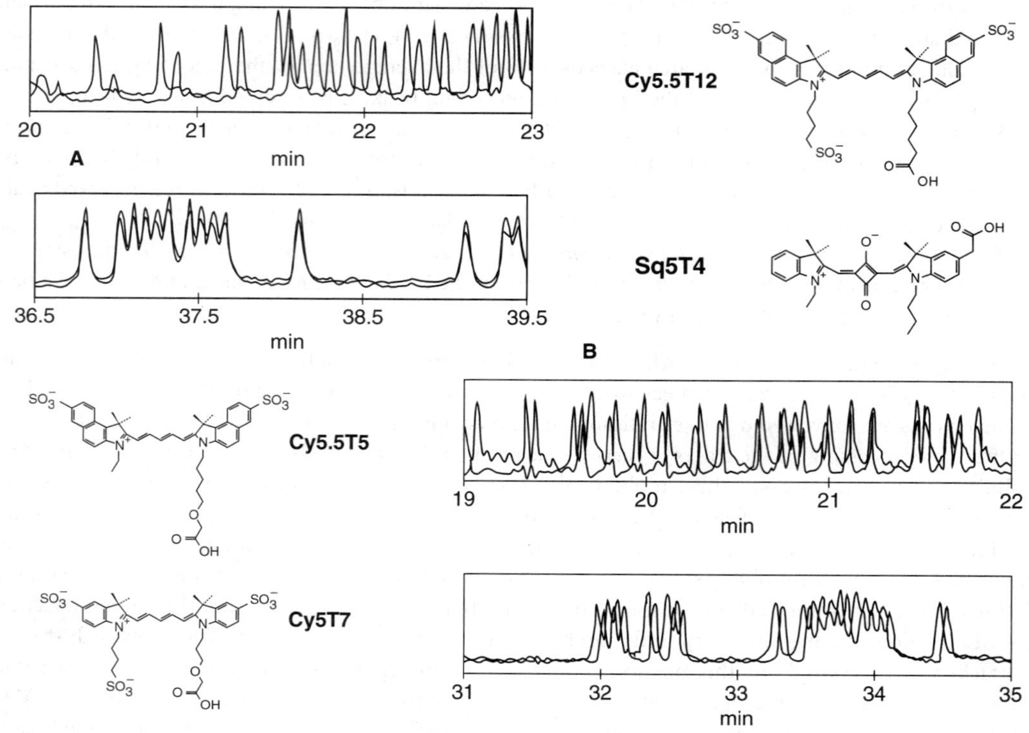


**FIGURE 53.17** Electrophoresis traces and chemical structures of four different cyanine dyes used to end-label DNA sequencing primers. (A) Gel trace of Cy5.5T5 (red trace) and Cy5T7 (blue trace). (B) Gel trace of Cy5.5T12 (red trace) and Sq5T4 (blue trace). In all cases, the ladders were prepared using a single terminator (ddT) and an M13mp18 template. The electrophoresis was carried out in a capillary-gel column (field strength = 185 V/cm) using a hydroxyethyl cellulose sieving buffer. (Adapted from Tu, O. et al., *Nucleic Acids Res.*, 26, 2797, 1998.)

the gel at a rate of ~0.15 cm/sec and monitors fluorescence along a single axis of the gel. The gel is approximately 20 cm in width and 42 cm in length and can accommodate 48 separate lanes. The time required to secure this data was 6 h, with the extended time due primarily to the limited electric field that can be applied to the thick slab gel.

## 53.4.2 Single-Color/Single-Lane

In this sequencing approach, only a single fluorophore is used, and thus only a single laser is required to excite the fluorescence and only a single detection channel is needed to process the fluorescence. The advantage of this approach is that instrumentally it is very simple because the hardware required for detection is simple. In addition, because the sequence is reconstructed from a single electrophoresis lane and not four, the throughput can be substantially higher compared to a single-fluorophore/four-lane method.

The bases are identified by adjusting the concentration ratio of the terminators used during DNA enzymatic polymerization to alter the intensity of the resulting electrophoretic bands.[17,27,30,31] Therefore, if the concentration of the terminators used during DNA polymerization was 4:2:1:0 (A:C:G:T), a series of fluorescence peaks would be generated following electrophoretic sizing with an intensity ratio of 4:2:1:0, and the identification of the terminal bases would be carried out by categorizing the peaks according to their heights. To accomplish this with some degree of accuracy in the base calling, the ability of the DNA polymerase enzyme to incorporate the terminators must be nearly uniform. This can be achieved using a special DNA polymerase, which in this case is a modified T7 DNA polymerase.[32,33] This enzyme has

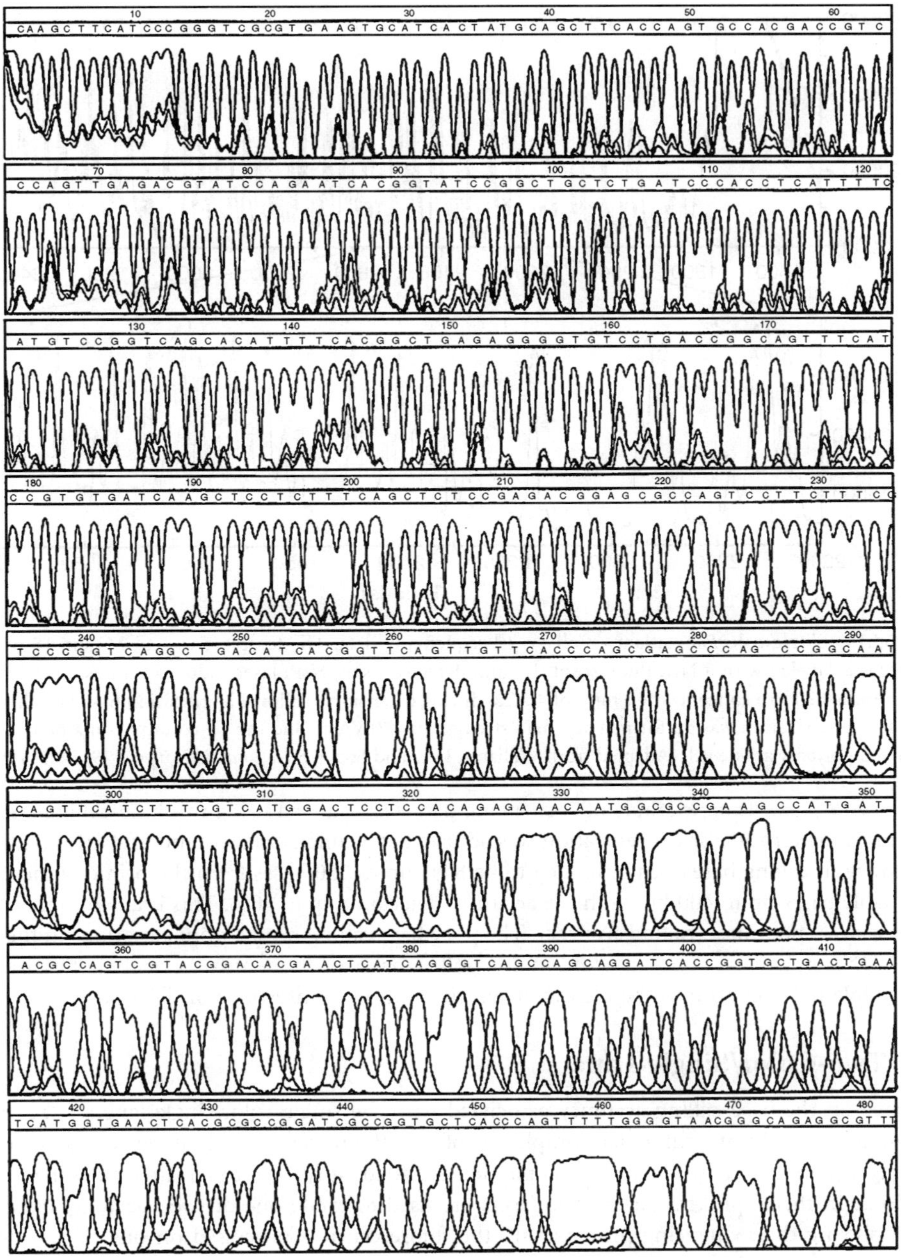

**FIGURE 53.18 (Color figure follows p. 28-30.)** Single-color/four-lane sequencing trace (called bases 1 to 480) of a PCR-amplified λ-bacteriophage template. The sequencing was performed using an IRD-800-labeled primer (21mer) and SGE instrument (Li-COR 4000). The gel consisted of 8% polyacrylamide with 7 *M* urea. The dimensions of the gel were 25 cm (width) by 41 cm (length). The traces from the four lanes were overlaid to reconstruct the sequence of the template.

been modified so as to remove its proofreading capabilities by eliminating its 3′ → 5′ exonuclease activity. Since this method requires uniform incorporation of the terminators, it is restricted to the use of dye-primer chemistry. In addition, since the T7 enzyme is not a thermostable enzyme, cycle sequencing cannot be used.

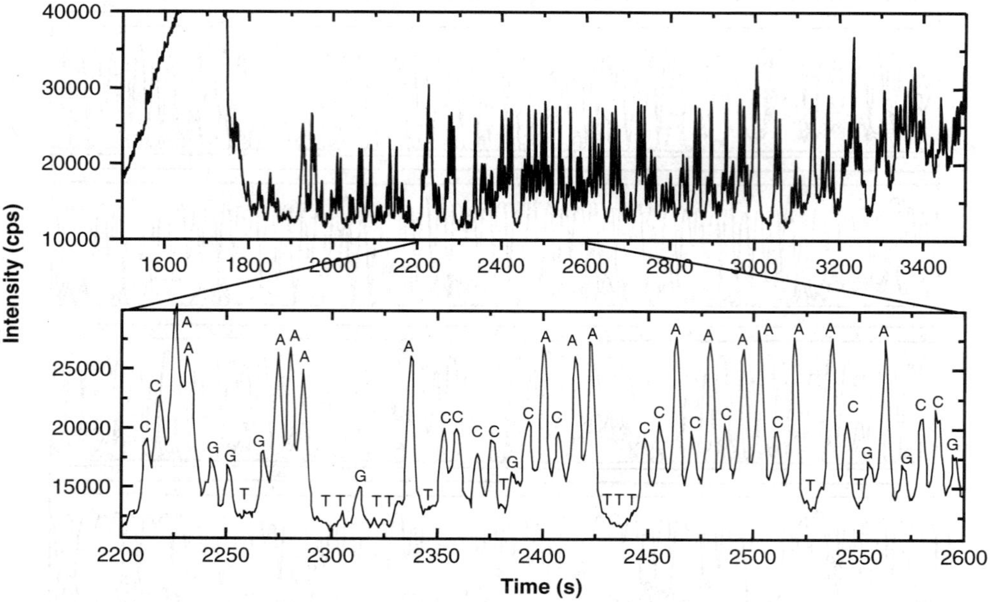


**FIGURE 53.19** Single-color/single-lane capillary-gel sequencing data. The template was an M13mp18 phage, and the primer was labeled with a NIR fluorescent dye and detected using NIR, laser-induced fluorescence. The terminators were adjusted to a concentration ratio of 4:2:1:0 during extension to allow identification based on fluorescence intensities. The separation was performed at a field strength of 250 V/cm, and the gel column contained a 3% T/3% C crosslinked polyacrylamide gel. (Adapted from Williams, D.C. and Soper, S.A., *Anal. Chem.*, 67, 19, 3427, 1995.)

Figure 53.19 presents an example of sequencing data accumulated using this base-calling strategy. In this example, the terminator concentration was adjusted at a concentration ratio of 4:2:1:0 (A:C:G:T).[56] The accuracy in calling bases was estimated to be 84% up to 250 bases from the primer annealing site, with readable bases up to 400 but with the accuracy deteriorating to 60%. This is a common artifact in this approach — poor base-calling accuracy. The poor base calling results from variations in the activity of the T7 DNA polymerase and, in addition, the null signal used to identify Ts. Ambiguities are present when multiple null signals must be identified, which can lead to insertions or deletions.

## 53.4.3 Two-Color/Single-Lane

To improve on the base-calling accuracy associated with the single-color/single-lane strategy without having to increase the instrumental complexity of the fluorescence readout device associated with sequencing instruments, a two-color format may be used to identify the four terminal bases in sequencing applications. In this approach, one or two lasers are used to excite one of two spectrally distinct dyes used for labeling the sequencing primers, and the fluorescence is processed on one of two detection channels, consisting of bandpass filters and photon transducers.

Figure 53.15 shows a schematic of a two-color scanning instrument (last dichroic was not used, and two PMTs were removed). It was used to excite the fluorescence of the labeling dyes, FAM and JOE. Due to their similar absorption maxima, a single laser (488 nm) could efficiently excite the fluorophores. In addition, this dye pair was selected because they produced sequencing fragments that comigrated, and thus no mobility correction was required. The bases were identified using a binary coding scheme, which is shown in Table 53.2.[22] During DNA polymerization, four separate reactions were run, with the A-reaction containing an equimolar mixture of the FAM- and JOE-labeled sequencing primers. In the case of G, only the JOE-labeled primer was present, while for T only the FAM-labeled primer was present, and for C no dye-labeled primer was used. By ratioing the signal in the red (JOE) to green (FAM) channel,

**TABLE 53.2.** Binary Coding Scheme for Two-Color DNA Sequencing

| | FAM | JOE |
|---|---|---|
| A | 1 | 1 |
| C | 0 | 1 |
| G | 1 | 0 |
| T | 0 | 0 |

*Note:* 1 = the presence of the dye-labeled primer during DNA polymerization, and 0 = represents the absence of the dye label.

a value was obtained that could be used to identify the terminal base. The attractive feature of this protocol is that while the absolute intensity of the bands present in the electropherogram may vary by a factor of 20 due to sequence-dependent termination, the ratio varies by a factor of only 1.7. The read length using this binary coding system was up to approximately 350 bases, with the number of errors ~15 (accuracy = 95.7%). The majority of the errors were attributed to C determinations, since a null signal was used to indicate the presence of this base.

To alleviate the errors in the base calling associated with identifying bases using a null signal, a two-dye, two-level approach can be implemented.[34,34] In this method, the bases using a common dye label have the concentration of ddNTPs adjusted during chain extension to alter the intensity of the fluorescence peaks developed during the electrophoresis. Also required in this approach are uniform peak heights, requiring the use of the modified T7 DNA polymerase in the presence of manganese ions. For example, Chen et al.[34] used a FAM-labeled primer for marking Ts and Gs, with the concentration ratio of the ddNTPs adjusted to 2:1 (T:G). Likewise, the As and Cs were identified using a 2:1 concentration ratio of the terminators, with the labeling dye in this case being TAMRA. Sequencing data produced an effective read length of 350 bases, with an accuracy of 97.5%. When the concentration ratios of the terminators sharing a common labeling dye was increased to 3:1, the read length was extended to ~400 bases with a base-calling accuracy >97%. As is evident, the elimination of null signals to identify bases can improve the base-calling accuracy in these types of sequence determinations.

While most fluorescence-labeling strategies for DNA sequencing that depend on differences in intensities of the electrophoretic peaks to identify bases use dye-labeled primers, internal labeling, where the fluorescent dye is situated on the dNTP, can also be used.[36] The advantages associated with using dye-labeled dNTPs are (1) the ability to use a wide range of primers because no dye-labeled primer is required, (2) incorporation of the dye-labeled dNTP can be much more uniform than dye-labeled ddNTPs, and (3) dye-labeled dNTPs are much less expensive than dye-labeled primers and terminators. Using a tetramethylrhodamine-labeled dATP and a fluorescein-labeled dATP, a two-color/single-lane sequencing assay has been reported.[36] For internal labeling, a two-step polymerization reaction was used in which the template was annealed to the sequencing primer (unlabeled) along with the dye-dATP and the four unlabeled dNTPs as well as the polymerase enzyme. The extension reaction was incubated at 37°C for 10 min, after which the appropriate terminator was added and the reaction allowed to proceed for an additional 10 min. The initial extension reaction extended six to eight nucleotides to a quartet of As, with 80 to 90% of the fragments containing a single dye-labeled dATP. Since only two dyes were used in this particular example, the concentration ratio for a pair of terminators sharing a common dye was adjusted (3:1) to allow discrimination based on the intensity of the resulting electrophoretic peaks. Analysis of the sequencing data indicated that the read length was 500 bases with an accuracy of 97%.

## 53.4.4 Four-Color/Single-Lane

The commonly used approach in most commercial DNA-sequencing instruments using fluorescence detection is the four-color/single-lane strategy for identifying the terminal bases in sequencing applications. The primary reasons for using a four-color/single-lane approach are that it provides high accuracy in the base calling, especially for long reads, and the throughput can be high due to the fact that all bases comprising

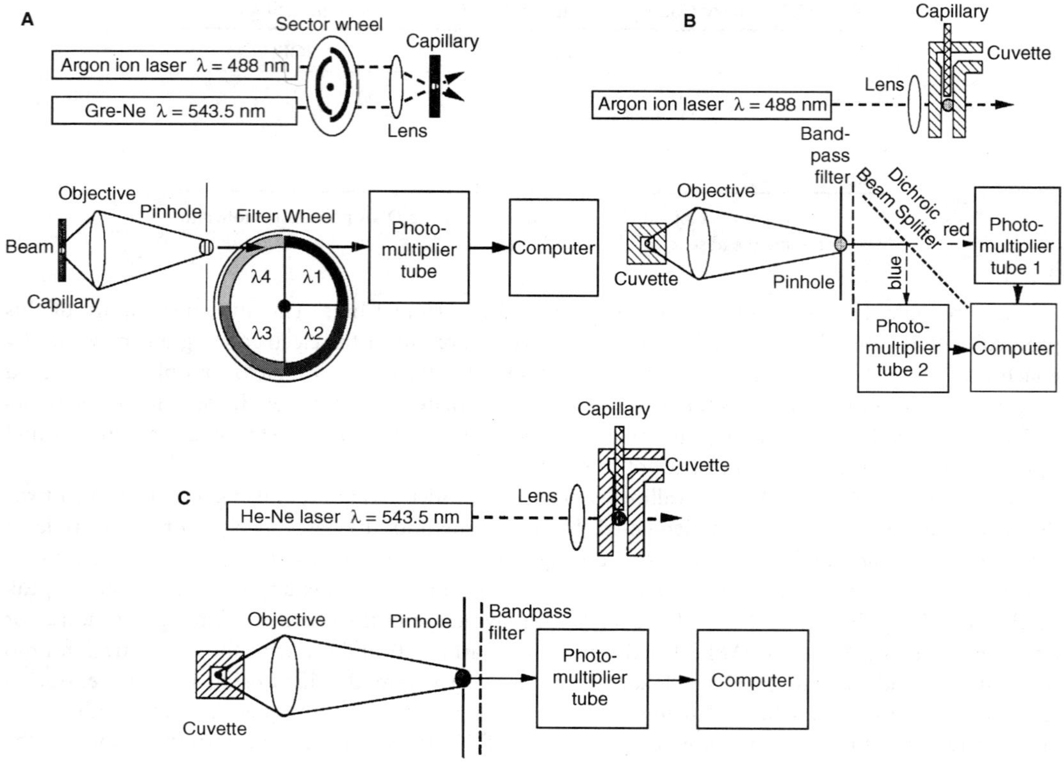


**FIGURE 53.20** Laser-induced fluorescence-detection apparati for processing DNA-sequencing data obtained using CGE: four-color/single-lane (A), two-color/single-lane (B), and one-color/single-lane (C).

the template DNA can be called in a single-gel tract. Unfortunately, a four-color detector requires extensive optical components to sort the fluorescence, and in some cases multiple excitation sources are needed to efficiently excite the fluorophores used to label the individual sequencing ladders. In addition, post-electrophoresis software corrections may be required to account for spectral leakage into detection channels.

Most dye-terminator reads are used with this four-color strategy, since the data analysis (base calling) does not depend on uniform incorporation efficiencies, which are hard to achieve using dye-labeled teminators. The same type of instrumentation that is used for four-color/dye-primer reads can also be used for four-color/dye-terminator reads. The only difference is in terms of the sample preparation protocols and the software corrections in the sequencing data such as different mobility-correction factors. In most cases, a size-exclusion step is used following DNA polymerization to remove excess dye-labeled terminators because they are negatively charged and can mask the sequencing data due to the presence of a large dye-terminator band in the gel.

An example of a four-color detector for capillary-gel electrophoresis is shown in Figure 53.20A, in which two laser sources (argon ion laser, 488 nm, and green helium-neon laser, 543 nm) were used to excite the dye set FAM, JOE, TAMRA, and ROX. To process the emission on a single detection channel, a four-stage filter wheel was synchronized to a sector wheel situated in front of the two lasers. The synchronization was set to pass 488-nm excitation for FAM and JOE and simultaneously place the bandpass filters for FAM and JOE in the optical path. Following this, the 543-nm laser light was passed, and the filters for TAMRA and ROX were situated within the optical path. Figure 53.21 shows typical traces produced from this system in which the sequence of an M13mp18 phage test template was analyzed. A read length exceeding 550 bases was obtained at an accuracy of 97%.

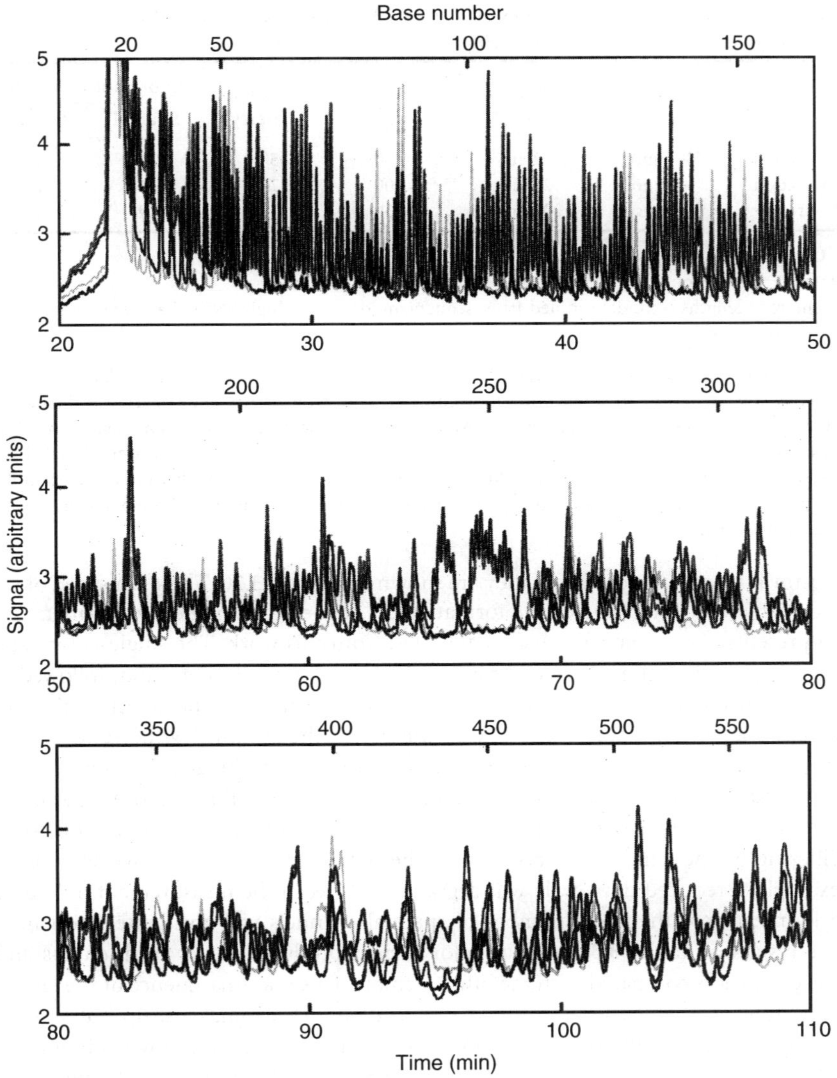


**FIGURE 53.21 (Color figure follows p. 28-30.)** Four-color/single-lane sequencing of an M13mp18 template with a histidine tRNA insert. The numbers along the top of the electropherogram represent cumulative bases from the primer-annealing site. The electrophoresis was performed in a capillary column with a length of 41 cm and an i.d. of 50 μm. The sieving gel consisted of a cross-linked polyacrylamide (6% T/5% C) and was run at a field strength of 150 V/cm. The dyes used for the labeling of the sequencing primers were FAM, JOE, TAMRA, and ROX. Each color represents fluorescence from a different wavelength region (blue = 540 nm, green = 560 nm, yellow = 580 nm, red = 610 nm). (Adapted from Swerdlow, H. et al., *Anal. Chem.*, 63, 2835, 1991.)

## 53.4.5 Choosing the Right Sequencing Format

With a variety of different fluorescence-detection formats available, the question becomes which configuration is better in terms of base calling, both from a read-length and accuracy point of view. In addition, which detection format produces the best SNR in the measurement? Other issues require attention as well, e.g., the complexity of the instrumentation required for detection. A rigorous comparison between the various detector configurations for sequencing applications has been carried out.[27] In this study, three

**TABLE 53.3** Comparison of Figures of Merit for Various Detector Formats for Fluorescence-Based DNA Sequencing

| Detection Mode | Noise (mass)[b] | Detection Limit (mol)[b] | Read Length[c] | Base Calling Accuracy | Information Content[a] 10 amol | 100 zmol |
|---|---|---|---|---|---|---|
| One color | 700 ymol | 2 zmol | 400 | 85% | 14 | 7 |
| Two color | 7 zmol | 20 zmol | 500 | 97% | 21 | 8 |
| Four color | 70 zmol | 200 zmol | 550 | 97% | 29 | 2 |

[a] Calculated from Equation 53.3.
[b] ymol = $10^{-24}$ mol; zmol = $10^{-21}$ mol.
[c] The read lengths were determined from sequencing data with high SNRs that were comparable across the series.

*Note:* The sequencing assays used capillary gel electrophoresis for fractionating the DNA ladders. The gel consisted of a crosslinked polyacrylamide gel with either formamide or urea as the denaturant. In all cases, the template was an M13mp18 phage with dye-primer chemistry used for fluorescence labeling. In the four-color experiment, FAM, JOE, TAMRA, and ROX were the labeling dyes, for the two-color experiment FAM and TAMRA were used, and for the one-color example only TAMRA was used. The sequencing primer was an M13 (–40) primer and the polymerase was a Sequenase enzyme.

different detector formats were used, and they are shown in Figure 53.20. These consisted of a four-color/single-lane format, a two-color/single-lane format, and a single-color/single-lane format.

Table 53.3 presents a summary of the data collected from this work. The single-color experiment, in which there is a single detection channel, provides the lowest limits of detection, followed by the two-color system and the four-color system. The significant improvement in the detection limit for the two-color and single-color systems was partly due to the use of the sheath flow detector that was used in these formats. However, in spite of the use of the sheath flow cell, the general trend is that the lower number of spectral channels typically results in a better SNR in the fluorescence measurement due to the fact that spectral sorting is not required. Spectral sorting causes emission losses due to reflection or inefficient filtering by the bandpass filters used in the optical train. If a filter wheel is used, as in this particular example, a reduced duty cycle will degrade the SNR in the measurement. However, this does not necessarily mean that the lower number of spectral channels will give better sequencing data. As shown by the results of Table 53.3, the four-color format produced better read lengths and favorable base-calling accuracies as compared with the other formats. This is a consequence of the fact that because four spectral channels are being used, the information content in the signal goes up, but only at reasonable loading levels of sample into the sequencing instrument. It is clear that at low loading levels of DNA-sequencing ladders, the one-color or two-color approach may be better due to improved limits of detection.

## 53.4.6 Single-Color/Four-Lifetime Sequencing

While most sequencing applications using fluorescence require spectral discrimination to identify the terminal base during electrophoretic sizing, an alternative approach is to use the fluorescence lifetime of the labeling dye to identify the terminal base. In this method, either time-resolved or phase-resolved techniques can be used to measure the fluorescence lifetime of the labeling dye during the gel electrophoresis separation.

The monitoring and identification of multiple dyes by lifetime discrimination during a gel separation can allow for improved identification efficiency when compared to that of spectral wavelength discrimination. When the identity of the terminal nucleotide base is accomplished through differences in spectral emission wavelengths, errors in the base call can arise from broad, overlapping emission profiles, which results in crosstalk between detection channels. Lifetime discrimination eliminates the problem of crosstalk between detection channels and can also potentially allow processing of the data on a single readout channel. Several other advantages are associated with fluorescence lifetime identification protocols, including:[37]

1. The calculated lifetime is immune to concentration differences.
2. The fluorescence lifetime can be determined with higher precision than fluorescence intensities.
3. Only one excitation source is required to efficiently excite the fluorescent probes, and only one detection channel is needed to process the fluorescence for appropriately selected dye sets.

One potential difficulty associated with this approach is the poor photon statistics (limited number of photocounts) that can result when making such a measurement. This results from the need to make a dynamic measurement (the chromophore is resident in the excitation beam for only 1 to 5 sec) and the low-mass loading levels associated with many DNA electrophoresis formats. Basically, the low number of photocounts acquired to construct the decay profile from which the lifetime is extracted can produce low precision in the measurement, which affects the accuracy in the base call. In addition, the high scattering medium in which the fluorescence is measured (polyacrylamide gel) can produce large backgrounds, again lowering the precision in the measurement. An additional concern with lifetime measurements for calling bases in DNA-sequencing applications is the heavy demand on the instrumentation required for such a measurement. However, the increased availability of pulsed diode lasers and simple avalanche photodiode detectors has had a tremendous impact on the ability to assemble a time-resolved instrument appropriate for sequencing applications.

There are two different formats for measuring fluorescence lifetimes — time-resolved[37–43] and frequency-resolved.[44–48] Since the time-resolved mode is a digital (photon-counting) method, it typically shows a better SNR than a frequency-resolved measurement, making it more attractive for separation platforms that deal with minute amounts of sample. In addition, time-resolved methods allow for the use of time-gated detection in which background photons, which are coincident with the laser pulse (scattered photons), can be gated out electronically, improving the SNR in the measurement.

A typical time-correlated single-photon-counting (TCSPC) device consists of a pulsed excitation source, a fast detector, and timing electronics. A device that has been used for making time-resolved measurements during CGE is shown in Figure 53.22.[39] The light source consisted of an actively pulsed solid-state gallium aluminum arsenide (GaAlAs) diode laser with a repetition rate of 80 MHz and an average power of 5.0 mW at a lasing wavelength of 780 nm. The pulse width of the laser was determined to be ~50 psec (formal width half maximum [FWHM]). The detector selected for this instrument was a SPAD, which has an active area of 150 μm and is actively cooled. In addition, the SPAD has a high single-photon-detection efficiency (>60% above 700 nm). The counting electronics (constant fraction discriminator [CFD], analog-to-digital converter [ADC], time-to-amplitude converter [TAC], and pulse-height analyzer) are situated on a single TCSPC board. The board plugs directly into a PC bus and exhibits a dead time of <260 nsec, allowing efficient processing of single-photon events at counting rates exceeding $2 \times 10^6$ counts/sec. This set of electronics allows for the collection of 128 sequential decay profiles with a timing resolution of 9.77 psec per channel. The instrument possesses a response function of approximately 275 psec (FWHM), adequate for measuring fluorescence lifetimes in the subnanosecond regime.

Probably the most important aspect of lifetime determinations in sequencing applications is the processing or calculation algorithms used to extract the lifetime value from the resulting decay. Since the photon statistics are poor and the accuracy in the base call depends directly on the lifetime differences between fluors in the dye set and the relative precision in the measurement, algorithms that deal with this situation are required as well as those that can be performed online during the electrophoresis. Two algorithms for on-the-fly fluorescence-lifetime determinations have been used — the maximum likelihood estimator (MLE) and the rapid lifetime determination method (RLD). MLE calculates the lifetime via the following relation:[49]

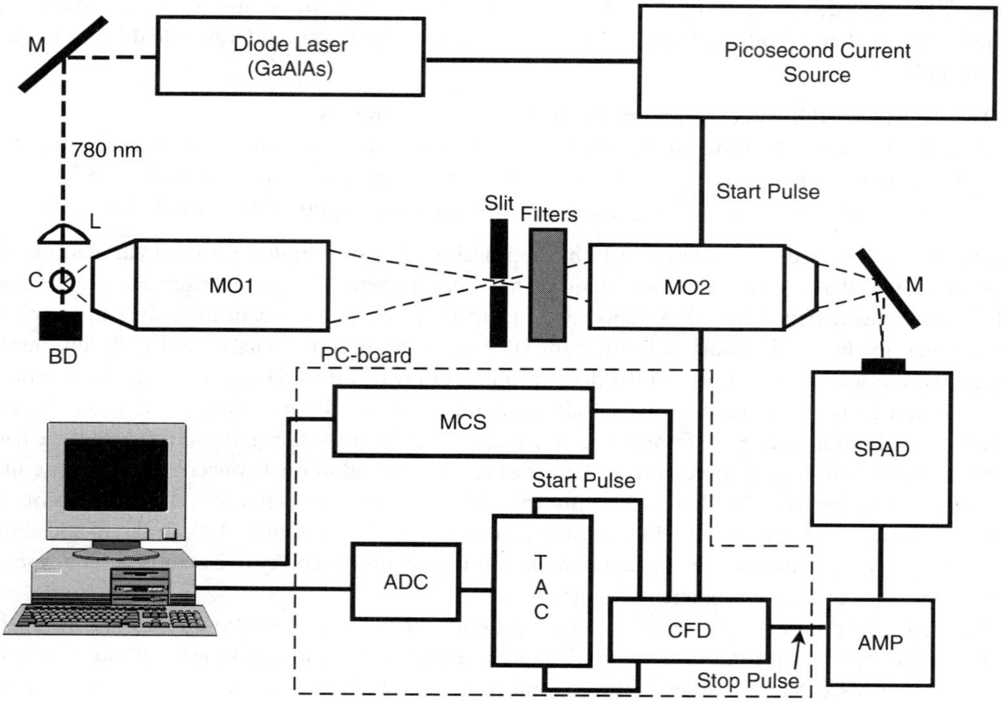


**FIGURE 53.22** Time-correlated single-photon counting detector for CGE. The laser source was a pulsed diode laser that operated at a repetition rate of 80 MHz and lased at 780 nm (average power = 5 mW). The laser was focused onto a capillary-gel column with the emission collected using a 40× microscope objective (NA = 0.85). The fluorescence was imaged onto a slit and then spectrally filtered and finally focused onto the photoactive area of a single-photon-avalanche diode. The electronics for processing the time-resolved data was situated on a single PC board, which was resident on the PC bus. (Adapted from Legendre, B.L. et al., *Rev. Sci. Instrum.*, 67, 3984, 1996.)

$$1+(e^{T/\tau_f}-1)^{-1}-m(e^{mT/\tau_f}-1)^{-1}=N_t^{-1}\sum_{i-1}^{m} iN_i \tag{53.4}$$

where $m$ is the number of time bins within the decay profile, $N_t$ is the number of photocounts in the decay spectrum, $N_i$ is the number of photocounts in time bin $i$, and $T$ is the width of each time bin. A table of values using the left-hand side of the equation is calculated by setting $m$ and $T$ to experimental values and using lifetime values ($\tau_f$) ranging over the anticipated values. The right-hand side of the equation is constructed from CE decay data over the appropriate time range. The fluorescence lifetime may then be determined by matching the value of the right-hand side obtained from the data with the table entry. The relative standard deviation in the MLE may be determined from $N^{-1/2}$.

Fluorescence lifetimes calculated using the RLD method is performed by integrating the number of counts within the decay profile over a specified time interval and using the following relationship:[50]

$$\tau_f = -\Delta t / \ln(D_1/D_0) \tag{53.5}$$

where $\Delta t$ is the time range over which the counts are integrated, $D_0$ is the integrated counts in the early time interval of the decay spectrum, and $D_1$ represents the integrated number of counts in the later time interval. Both the MLE and RLD methods can extract only a single lifetime value from the decay, which in the case of multiexponential profiles would represent a weighted average of the various components comprising the decay.

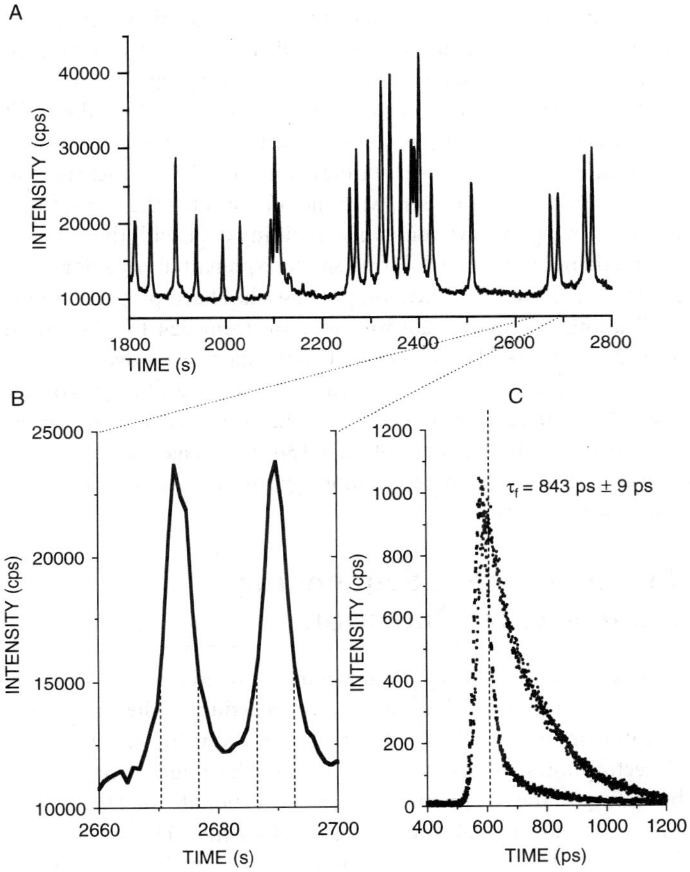


**FIGURE 53.23** CGE of C-terminated fragments produced from an M13mp18 template with time-resolved fluorescence detection (A). The labeling dye (F-substituted, heavy-atom-modified NIR fluorescent label; see Table 53.1) was attached to the 5′ end of a sequencing primer. (B) Expanded view of two peaks selected from the electropherogram showing the time area (dashed lines) over which photoelectrons were used to construct the decay profiles. The decay profiles shown in (C) consist of the gel blank and the dye-labeled DNA fragment. The dashed line shows the start channel in which the calculation was initiated. The lifetime was calculated using Equation 53.3. The electrophoresis was carried at a field strength of 200 V/cm using a cross-linked polyacrylamide gel. (Adapted from Soper, S.A. et al., *Anal. Chem.*, 67, 4358, 1995.)

Wolfrum et al. have also implemented a special pattern recognition technique.[38] Basically, the method involves comparing a pattern to the measured decay and searches for a pattern that best fits the measurement. This algorithm is equivalent to the minimization of a log-likelihood ratio where fluorescent-decay profiles serve as the pattern. Since the pattern-recognition algorithm uses the full amount of information present in the data, it potentially has the lowest error or misclassification probability.

To demonstrate the feasibility of acquiring lifetimes on the fly during the CGE separation of sequencing ladders, C-terminated fragments produced from Sanger chain-terminating protocols and labeled with a NIR fluorophore on the 5′ end of a sequencing primer were electrophoresed and the lifetimes of various components within the electropherogram determined.[37] An example of the data produced from this detection format is shown in Figure 53.23. The average lifetime determined using the MLE method was found to be 843 psec, with a standard deviation of ±9 psec (RSD = 1.9%). The lifetime values calculated here compared favorably to a static measurement performed on the same dye.

Since the base calling is done with lifetime discrimination as opposed to wavelength discrimination, new types of dye sets can and need to be used that suit the identification method. For example, it is not necessary to use dyes with discrete emission maxima, and so structural variations in the dye set can be relaxed. A dye set developed for lifetime discrimination has been prepared and consists of a NIR chromophore that has unique fluorescence lifetimes with the lifetime altered via the addition of an intramolecular heavy atom.[51] Each of these dyes possesses the same absorbance maximum and fluorescence-emission maximum but different fluorescence lifetimes (see Table 53.1). Since these were tricarbocyanine dyes, the lifetimes were found to be <1.0 nsec, with the lifetimes for the dye set ranging from 947 psec to 843 psec when measured in a polyacrylamide gel containing urea, with the observed lifetimes less than what was observed in methanol but still exhibiting single exponential behavior.

A dye set appropriate for lifetime-identification purposes has been prepared and used in four lifetime DNA-sequencing applications.[40] The dyes absorb radiation from 624 to 669 nm and possess lifetimes that range from 1.6 to 3.7 nsec (see Figure 53.24). Unfortunately, the dye set shows multiexponential behavior in sequencing gels containing denaturants (urea, 7 *M*). In addition, to correct for dye-dependent mobility shifts unique linker structures were used. Using this dye set (dye-primer chemistry) and a pulsed diode laser operating at 630 nm, the sequence of an M13mp18 phage was evaluated. The lifetimes were extracted from the decays using a pattern-recognition algorithm. The read length was found to be 660 bases with a calling accuracy of 90%.

## 53.5 Single-Molecule DNA Sequencing Using Fluorescence Detection

The sequencing strategies discussed to this point depend on a fractionation step (electrophoresis) to sort the DNA by size. While much progress has been made in reducing the time required to develop the electropherogram, resulting in increases in the throughput of acquiring sequencing data, many problems arise when using gel electrophoresis. These problems include the ability to sequence DNA pieces that are only 1000 to 2000 bp in length, the requirement of gels to fractionate the DNA, the relatively slow speed associated with the process, and the data-processing requirements. The most pervasive problem is the ability to work only with DNA pieces that are 1000 to 2000 bases in length. Since the chromosomes exceed $1 \times 10^6$ bases, the assembly of the sequence of the entire chromosome with extremely short pieces makes the task daunting. Directly sequencing larger strands of DNA would relieve some the technical challenges associated with assembly. In addition, since YAC clones are ~100,000 bp, this requires the shearing and then subcloning of these DNAs into M13s to produce templates for sequencing. The ability to work with longer DNAs would eliminate the need for this secondary cloning step.

A very attractive approach has been suggested to rapidly sequence DNA strands that are >40,000 bp in length. The process is based on the principle of single-molecule detection using fluorescence photon burst analysis.[52] The process is depicted in Figure 53.25. Basically, the process involves immobilizing a long strand of DNA in a flow stream and then clipping the terminal base using an exonuclease enzyme, releasing the nucleotide from the original strand. The single nucleotide (either fluorescently labeled or nonlabeled) is then carried to a laser beam, which excites the fluorescence with the color used to identify the base clipped from the DNA strand. As shown in the figure, the single-molecule-sequencing approach does not involve a gel-fractionation step, potentially significantly speeding up the sequencing rate. In essence, the sequencing rate is determined by the rate at which the enzyme clips nucleotide bases from the strand, which for many exonuclease enzymes is on the order of 1000 bases per second. There are three key technical challenges associated with this technique: (1) tethering the DNA strand to a bead for holding it stationary within a flow stream, (2) creating a complement of the original DNA strand using dye-labeled dNTPs, and (3) detecting single-dye-labeled nucleotides using laser-induced fluorescence detection.

Since the premise of this sequencing protocol is to analyze single DNA molecules, a single DNA molecule must be selected and then trapped within a flow stream for processing. To accomplish this, a

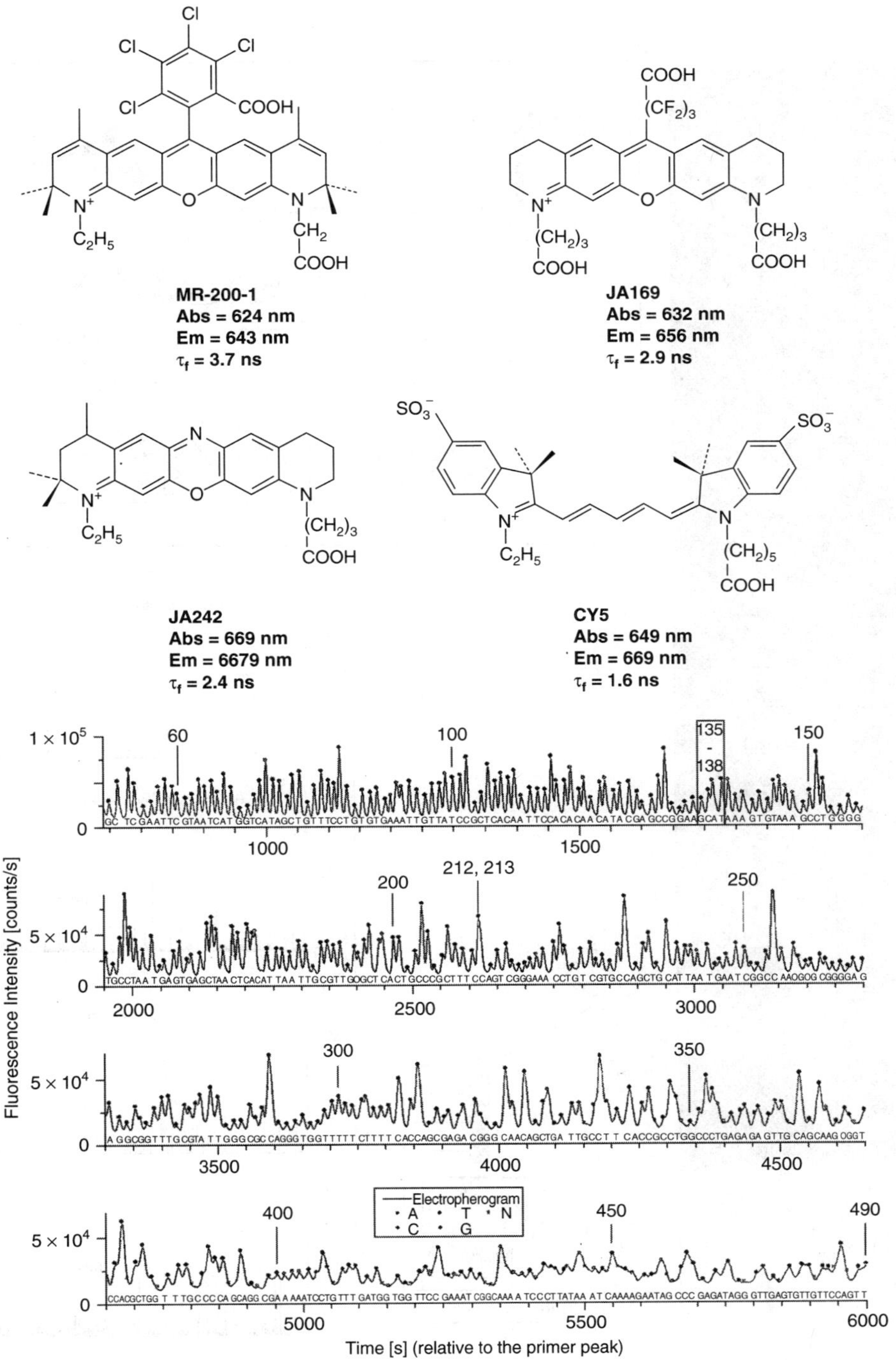


**FIGURE 53.24** Chemical structures of dyes used for multiplex, time-resolved DNA sequencing as well as the fluorescence properties of this dye set. Also shown is the sequencing data obtained when using the dye set shown above conjugated to sequencing primers. The sequencing was performed in a 5% linear polyacrylamide gel containing 7 M urea at a field strength of 160 V/cm. The fluorescence detector consisted of a pulsed diode laser (630 nm) with the optics configured in a confocal geometry. (Adapted from Lieberwirth, U. et al., *Anal. Chem.*, 70, 4771, 1998.)

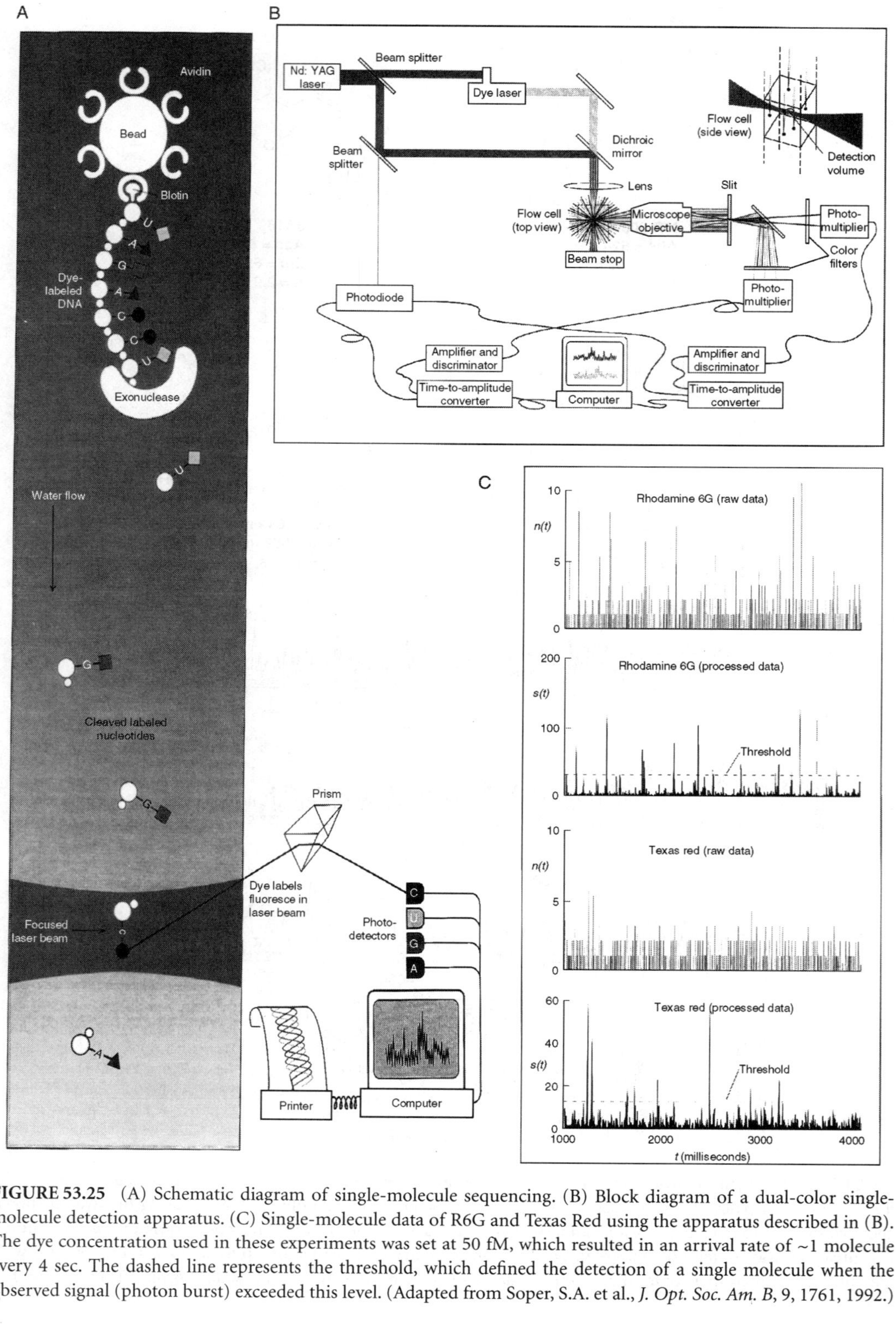


**FIGURE 53.25** (A) Schematic diagram of single-molecule sequencing. (B) Block diagram of a dual-color single-molecule detection apparatus. (C) Single-molecule data of R6G and Texas Red using the apparatus described in (B). The dye concentration used in these experiments was set at 50 fM, which resulted in an arrival rate of ~1 molecule every 4 sec. The dashed line represents the threshold, which defined the detection of a single molecule when the observed signal (photon burst) exceeded this level. (Adapted from Soper, S.A. et al., *J. Opt. Soc. Am. B*, 9, 1761, 1992.)

DNA molecule can be prepared that contains a biotin molecule on its 5′ end. The biotin-modified DNA molecule can then be attached to a microbead (i.d. = 1 to 10 μm) coated with streptavidin. Streptavidin is a protein that contains binding sites for biotin. The association of biotin to streptavidin is very strong ($K_{assoc.} = 10^{15}$ $M^{-1}$) and in addition is stable to both heat and the reagents used in most enzymatic reactions. In this case, a bead that has one DNA molecule attached to it must be selected, as a bead with multiple DNA strands would create a registry problem because the exonucleases work at different rates. Therefore, conditions that produce a sufficient population of beads containing a single DNA molecule must be selected. This can be done statistically by incubating the biotinylated DNA with a large excess of microbeads. This produces a large number of beads with no DNA molecule; therefore, the appropriate beads must be selected by staining the immobilized DNA with a fluorescent intercalating dye and using fluorescence to identify a bead with the appropriate number of DNA molecules attached.

The suspension of the bead containing the DNA molecule in the flow stream can be accomplished using an optical trapping technique. In this case, one or two tightly focused laser beams can be directed onto the bead. This generates an optical trap, which can hold the bead in the flow stream. The size of the trap is determined by the size of the focused laser beam (or beams) and is on the order of 5 μm. Since the trap is generated by a momentum exchange between the photons and the trapped bead, it does not depend on any type of electronic or vibrational transition. As such, the trapping laser can be a far-red or infrared laser that does not bleach the dyes incorporated into the target DNA molecule during polymerization.

The second technical challenge in this scheme is to produce a complement that contains dye-modified oligonucleotides. Since detection is accomplished using fluorescence (typically visible fluorescence) single-molecule detection, it is necessary to covalently label each dNTP with a probe and then, more importantly, build a fluorescent complement of the target DNA using a polymerase enzyme. As with dye-labeled ddNTPs, incorporation of dye-modified dNTPs is a challenge because the incorporation rate and fidelity of modified dNTPs by standard polymerase enzymes is not facile. Therefore, mutants will have to be prepared to accomplish this task. An alternative to using fluorescent dyes is to implement ultraviolet single-molecule detection, which requires unmodified dNTPs. While this strategy is much more forgiving in terms of molecular biology, it places severe challenges on the detection phases of the technique because the nucleotide bases have absorption maxima around 260 nm and the fluorescence quantum yields of the bases are low ($\sim 10^{-4}$) at room temperature in solution.

The final challenge in this approach is the ability to detect single-dye molecules (nucleotides) in solution and identify the single molecules via spectral discrimination. Since a DNA molecule is comprised of four different nucleotide bases, the bases as they are clipped by the exonuclease from the target DNA must be spectrally identified. While several researchers have demonstrated the ability to detect single molecules in flow streams,[53–55] the ability to color-discriminate adds complexity to the instrumentation (multiple lasers, multiple detection channels) and requires careful selection of the dyes. Not only must the dyes be well resolved in terms of their emission maxima, but the photophysics of the dye must be conducive to single-molecule detection, i.e., must have a high quantum yield and favorable photochemical stability. Figure 53.25 shows an example of a dual-color single-molecule-detection apparatus.[56] It consists of a mode-locked neodymium:YAG laser operating at 532 nm (second harmonic) and also a synchronously pumped dye laser ($\lambda_{em}$ = 585 nm). These wavelengths were chosen to match the absorption maxima of the two dyes selected for this experiment, R6G ($\lambda_{ex}$ = 528 nm; $\lambda_{em}$ = 555 nm) and Texas Red ($\lambda_{ex}$ = 578 nm; $\lambda_{em}$ = 605 nm). The fluorescence was processed on one of two photodetectors, which in this case consisted of microchannel plates (MCPs). The fluorescence from each dye was directed onto the appropriate MCP using a dichroic filter, with the emission further isolated from background photons and fluorescence from the other dye using a bandpass filter. The processing electronics consisted of conventional TCSPC electronics, which were used along with the pulsed lasers to allow for implementation of time-gated detection. Time-gated detection reduces the amount of background-scattered photons into the data stream, improving the SNR in the single-molecule measurement. An example of the data output from this dual-color single-molecule detector is shown in Figure 53.25 (raw data and processed data). The raw data were filtered using a weight quadratic sum filter [$S(t)$] given by the following expression:[37]

$$S(t) = \sum_{\tau=0}^{k-1} w(\tau)d(t+\tau)^2 \tag{53.6}$$

where $k$ represents the time range covered by the molecular transit through the laser beam, $w(\tau)$ is weighting factors selected to best distinguish the signal from noise, and $d(t)$ represents the raw data point at time (t). These data make clear that large-amplitude bursts of photons occur with both dyes. The detection efficiency was estimated to be 78% for R6G and 90% for Texas Red. In both cases, the error rate (defined as identifying a molecule when one was not present, false positive) was estimated to be <0.01 per second.

## References

1. Watson, J.D. and Crick, F.H.C., Molecular structure of nucleic acid: a structure of deoxyribonucleic acid, *Nature*, 171, 737, 1953.
2. Chen, E.Y., Schlessinger, D., and Kere, J., Ordered shotgun sequencing, a strategy for integrated mapping and sequencing of YAC clones, *Genomics*, 17, 651, 1993.
3. Maxam, A.M. and Gilbert, W., A new method for sequencing DNA, *Proc. Natl. Acad. Sci. U.S.A.*, 74, 560, 1977.
4. Sanger, F., Nicklen, S., and Coulson, A.R., DNA sequencing with chain-terminating inhibitors, *Proc. Natl. Acad. Sci. U.S.A.*, 74, 5463, 1977.
5. Fung, E.N. and Yeung, E.S., High-speed DNA sequencing by using mixed poly(ethylene oxide) solutions in uncoated capillary columns, *Anal. Chem.*, 67, 1913, 1995.
6. Smith, L.M., Saunders, J.Z., Kaiser, R.J., Hughes, P., Dodd, C.R., Connell, C.R., Heiner, C., Kent, S.B.H., and Hood, L.E., Fluorescence detection in automated DNA sequence analysis, *Nature*, 321, 674, 1986.
7. Prober, J.M., Trainor, G.L., Dam, R.J., Hobbs, F.G., Robertson, C.W., Zagursky, R.J., Cocuzza, A.J., Jensen, M.A., and Baumeister, K., A system for rapid DNA sequencing with fluorescent chain-terminating dideoxynucleotides, *Science*, 238, 336, 1987.
8. Ansorge, W., Sproat, B., Stegeman, J., Schwager, C., and Zenke, M., Automated DNA sequencing: ultrasensitive detection of bands during electrophoresis, *Nucleic Acids Res.*, 15, 4593, 1987.
9. Metzker, M.L., Lu, J., and Gibbs, R.A., Electrophoretically uniform fluorescent dyes for automated DNA sequencing, *Science*, 271, 1420, 1996.
10. Lee, L.G., Connell, C., Woo, S., Cheng, R., McArdle, B., Fuller, C., Halloran, N., and Wilson, R., DNA sequencing with dye-labeled terminators and T7 DNA polymerase: effects of dyes and dNTPs on incorporation of dye-terminators and probability analysis of termination fragments, *Nucleic Acids Res.*, 20, 2471, 1992.
11. Tabor, S. and Richardson, C.C., A single residue in DNA polymerase from *E. coli* DNA polymerase I family is critical for distinguishing between deoxy- and dideoxynucleotides, *PNAS*, 92, 6339, 1995.
12. Rosenblum, B.B., Lee, L.G., Spurgeon, S.L., Khan, S.H., Menchen, S.M., Heiner, C.R., and Chen, S.M., New dye-labeled terminators for improved DNA sequencing patterns, *Nucleic Acids Res.*, 25, 4500, 1997.
13. Hung, S.-C., Mathies, R.A., and Glazer, A.N., Optimization of spectroscopic and electrophoretic properties of energy transfer primers, *Anal. Biochem.*, 252, 78, 1997.
14. Hung, S.-C., Mathies, R.A., and Glazer, A.N., Comparison of fluorescence energy transfer primers with different donor-acceptor dye combinations, *Anal. Biochem.*, 255, 32, 1998.
15. Ju, J., Ruan, C., Fuller, C.W., Glazer, A.N., and Mathies, R.A., Fluorescence energy transfer dye-labeled primers for DNA sequencing and analysis, *Proc. Natl. Acad. Sci. U.S.A.*, 92, 4347, 1995.
16. Lee, L.G., Spurgeon, S.L., Heiner, C.R., Benson, S.C., Rosenblum, B.B., Menchen, S.M., Graham, R.J., Constantinescu, A., Upadhya, K.G., and Cassel, J.M., New energy transfer dyes for DNA sequencing, *Nucleic Acids Res.*, 25, 2816, 1997.

17. Williams, D.C. and Soper, S.A., Ultrasensitive near-IR fluorescence detection for capillary gel electrophoresis and DNA sequencing applications, *Anal. Chem.*, 67, 3427, 1995.
18. Middendorf, L.R., Bruce, J.C., Bruce, R.C., Eckles, R.D., Grone, D.L., Roemer, S.C., Sloiker, G.D., Steffens, D.L., Sutter, S.L., Brumbaugh, J.A., and Patonay, G., Continuous, on-line DNA sequencing using a versatile infrared laser scanner/electrophoresis apparatus, *Electrophoresis*, 13, 487, 1992.
19. Shealy, D.B., Lipowska, M., Lipwoski, J., Narayanan, N., Sutter, S., Strekowski, L., and Patonay, G., Synthesis, chromatographic separation and characterization of near-infrared-labeled DNA oligomers for use in DNA sequencing, *Anal. Chem.*, 67, 247, 1995.
20. Soper, S.A. and Mattingly, Q., Steady-state and picosecond laser fluorescence studies of nonradiative pathways in tricarbocyanine dyes: implications to the design of near-IR fluorochromes with high fluorescence efficiencies, *J. Am. Chem. Soc.*, 116, 3447, 1994.
21. Huang, X.C., Quesada, M.A., and Mathies, R.A., Capillary array electrophoresis using laser-excited confocal fluorescence detection, *Anal. Chem.*, 64, 967, 1992.
22. Huang, X.C., Quesada, M.A., and Mathies, R.A., DNA sequencing using capillary array electrophoresis. *Anal. Chem.*, 64, 2149, 1992.
23. Soper, S.A., Shera, E., Davis, L., Nutter, H., and Keller, R., The photophysical constants of several visible fluorescent dyes and their effects on ultrasensitive fluorescence detection, *Photochem. Photobiol.*, 57, 972, 1993.
24. Kambara, H. and Takahashi, S. Multiple-sheathflow capillary array DNA analyzer, *Nature*, 361, 565, 1993.
25. Takahashi, S., Murakami, K., Anazawa, T., and Kambara, H., Multiple sheath-flow gel capillary-array electrophoresis for multicolor fluorescent DNA detection, *Anal. Chem.*, 66, 1021, 1994.
26. Swerdlow, H., Wu, S., Harke, H., and Dovichi, N., Capillary gel electrophoresis for DNA sequencing: laser-induced fluorescence detection with the sheath flow cuvette, *J. Chromatogr.*, 516, 61, 1990.
27. Swerdlow, H., Zhang, J.Z., Chen, D.Y., Harke, H.R., Grey, R., Wu, S., and Dovichi, N.J., Three DNA sequencing methods using capillary gel electrophoresis and laser-induced fluorescence, *Anal. Chem.*, 63, 2835, 1991.
28. Ewing, B., Hillier, L., Wendl, M., and Green, P., Base-calling of automated sequencer traces using *Phred*. I. Accuracy assessment, *Genomics*, 8, 175, 1998.
29. Tu, O., Knott, T., Marsh, M., Bechtol, K., Harris, D., Barker, D., and Bashkin, J., The influence of fluorescent dye structure on the electrophoretic mobility of end-labeled DNA, *Nucleic Acids Res.*, 26, 2797, 1998.
30. Ansorge, W., Zimmermann, C., Schwager, C., Stegemann, J., Erfle, H., and Voss, H., One label, one tube, Sanger DNA sequencing in one and two lanes on a gel, *Nucleic Acids Res.*, 18, 3419, 1990.
31. Chen, D., Swerdlow, H.P., Harke, H.R., Zhang, J.Z., and Dovichi, N.J., Single-color laser-induced fluorescence detection and capillary gel electrophoresis for DNA sequencing, *Proc. Int. Soc. Opt. Eng.*, 1435, 161, 1991.
32. Tabor, S. and Richardson, C.C., DNA sequence analysis with a modified bacteriophage T7 DNA polymerase, *Proc. Natl. Acad. Sci. U.S.A.*, 84, 4767, 1987.
33. Tabor, S. and Richardson, C.C., DNA sequence analysis with a modified bacteriophage T7 DNA polymerase, *J. Biol. Chem.*, 14, 8322, 1990.
34. Chen, D.Y., Harke, H.R., and Dovichi, N.J., Two-label peak-height encoded DNA sequencing by capillary gel electrophoresis: three examples, *Nucleic Acids Res.*, 20, 4873, 1992.
35. Li, Q. and Yeung, E., Simple two-color base-calling schemes for DNA sequencing based on standard four-label Sanger chemistry, *Appl. Spectrosc.*, 49, 1528, 1995.
36. Starke, H.R., Yan, J., Zhang, Z., Muhlegger, K., Effgen, K., and Dovichi, N., Internal fluorescence labeling with fluorescent deoxynucleotides in two-label peak-height encoded DNA sequencing by capillary electrophoresis, *Nucleic Acids Res.*, 22, 3997, 1994.
37. Soper, S.A., Legendre, B.L., and Williams, D.C., On-line fluorescence lifetime determinations in capillary electrophoresis, *Anal. Chem.*, 67, 4358, 1995.

38. Köllner, M., Fischer, A., Arden-Jacob, J., Drexhage, K.-H., Müller, R., Seeger, S., and Wolfrum, J., Fluorescence pattern recognition for ultrasensitive molecule identification: comparison of experimental data and theoretical approximations, *Chem. Phys. Lett.*, 250, 355, 1996.
39. Legendre, B.L., Williams, D.C., Soper, S.A., Erdmann, R., Ortmann, U., and Enderlein, J., An all solid-state near-infrared time-correlated single photon counting instrument for dynamic lifetime measurements in DNA sequencing applications, *Rev. Sci. Instrum.*, 67, 3984, 1996.
40. Lieberwirth, U., Arden-Jacob, J., Drexhage, K.H., Herten, D.P., Muller, R., Neumann, M., Schulz, A., Siebert, A., Siebert, S., Sagner, G., Klingel, S., Sauer, M., and Wolfrum, J., Multiplexed dye DNA sequencing in capillary gel electrophoresis by diode laser-based time-resolved fluorescence detection, *Anal. Chem.*, 70, 4771, 1998.
41. Sauer, M., Arden-Jacob, J., Drexhage, K.H., Marx, N.J., Karger, A.E., Lieberwirth, U., Muller, R., Neumann, M., Nord, S., Schulz, A., Seeger, S., Zander, C., and Wolfrum, J., On-line diode laser based time-resolved fluorescence detection of labeled oligonucleotides in capillary gel electrophoresis, *Biomed. Chromatogr.*, 11, 81, 1997.
42. Soper, S.A. and Legendre, B.L., Error analysis of simple algorithms for determining fluorescence lifetimes in ultradilute dye solutions, *Appl. Spectrosc.*, 48, 400, 1994.
43. Waddell, E., Stryjewski, W., and Soper, S.A., A fiber-optic-based multichannel time-correlated single photon-counting device with subnanosecond time resolution, *Rev. Sci. Instrum.*, 70, 32, 1999.
44. He, H., Nunnally, B.K., Li, L.C., and McGowen, L.B., On-the-fly fluorescence lifetime detection of dye-labeled primers for multiplexed analysis, *Anal. Chem.*, 70, 3413, 1998.
45. Li, L.C. and McGowen, L.B., On-the-fly frequency-domain fluorescence lifetime detection in capillary electrophoresis, *Anal. Chem.*, 68, 2737, 1996.
46. Li, L.C., He, H., Nunnally, B.K., and McGowen, L.B., On-the-fly fluorescence lifetime detection of labeled DNA primers, *J. Chromatogr.*, 695, 85, 1997.
47. Li, L. and McGowen, L.B., Effects of gel material on fluorescence lifetime detection of dyes and dye-labeled DNA primers in capillary electrophoresis, *J. Chromatogr.*, 841, 95, 1999.
48. Nunnally, B.K., He, H., Li, L.C., Tucker, S.A., and McGowen, L.B., Characterization of visible dyes for four-decay fluorescence detection, *Anal. Chem.*, 69, 2392, 1997.
49. Hall, P. and Sellinger, B., Better estimates of exponential decay parameters, *J. Phys. Chem.*, 85, 2941, 1981.
50. Ballew, R.M. and Demas, J.N., An error analysis of the rapid lifetime determination method for the evaluation of single exponential decays, *Anal. Chem.*, 61, 30, 1989.
51. Flanagan, J.H., Owens, C.V., Romero, S.E., Waddell, E., Kahn, S.H., Hammer, R.P., and Soper, S.A., Near-infrared heavy-atom-modified fluorescent dyes for base-calling in DNA-sequencing applications using temporal discrimination, *Anal. Chem.*, 70, 2676, 1998.
52. Fairfield, E.R., Jett, J., Keller, R., Hahn, J., Krakowski, L., Marrone, B., Martin, J., Ratliff, R., Shera, E., and Soper, S., Rapid DNA sequencing based upon single molecule detection, *Gen. Anal.*, 8, 1, 1991.
53. Shera, E.B., Seitzinger, N., Davis, L., Keller, R., and Soper, S., Detection of single fluorescent molecules, *Chem. Phys. Lett.*, 174, 553, 1990.
54. Soper, S.A., Hahn, J., Nutter, H., Shera, E., Martin, J., Jett, J., and Keller, R., Single molecule detection of R-6G in ethanolic solutions utilizing CW excitation, *Anal. Chem.*, 63, 432, 1991.
55. Soper, S.A., Mattingly, Q.L., and Vegunta, P., Photon burst detection of single near infrared fluorescent dye molecules, *Anal. Chem.*, 65, 740, 1993.
56. Soper, S.A., Davis, L., and Shear, E.B., Detection and identification of single molecules in solution, *J. Opt. Soc. Am. B*, 9, 1761, 1992.

# 54

# Living-Cell Analysis Using Optical Methods




Pierre M. Viallet
*University of Perpignan*
*Perpignan, France*

Tuan Vo-Dinh
*Oak Ridge National Laboratory*
*Oak Ridge, Tennessee*


## 54.1 Introduction: The Green Fluorescent Protein Breakthrough

Previous chapters in this handbook have extensively described new methods in the field of biophotonics and reviewed applications. The scope of this chapter is to present, from the biologist's point of view, a synthesis of the current and potential applications of the previously described techniques. The chapter focuses on techniques that have been or are supposed to be used for studying a type of biomolecular event occurring on a specific intracellular organelle. The advantages and, of course, limitations of each method, as they appear in the light of present knowledge, will be pointed out. An effort will be made to show how some techniques are able to give simultaneous but selective information on biochemical events occurring at different sites of the cell.

The recent breakthroughs in genomics and proteomics have completely changed the way we think about exploring intracellular dynamics. Previously, we would introduce inside the cell hexogenous fluorescent molecules that were selected to monitor some specific biomolecular pathways. We depended entirely on their ability to enter the cell and had few means of directing them to a specific intracellular target. As a consequence, the information obtained could often be biased in various ways. First, the fluorescence signal was often a mixture of signals resulting from different microenvironments due to the nonspecific location of the probe. Second, to be informative the hexogenous fluorescent molecules had to compete with the natural biomolecules, and thus they might induce some imbalance in the intramolecular processes.

The techniques of molecular biology have allowed for the generation of fluorescent chimera that mimic biological molecules and that can be vectorized to or expressed at specific targets (organelles or microcompartments). One can say that cells themselves have been reengineered to give us the requested information, even if potential under- or overexpression may yet undermine this information.

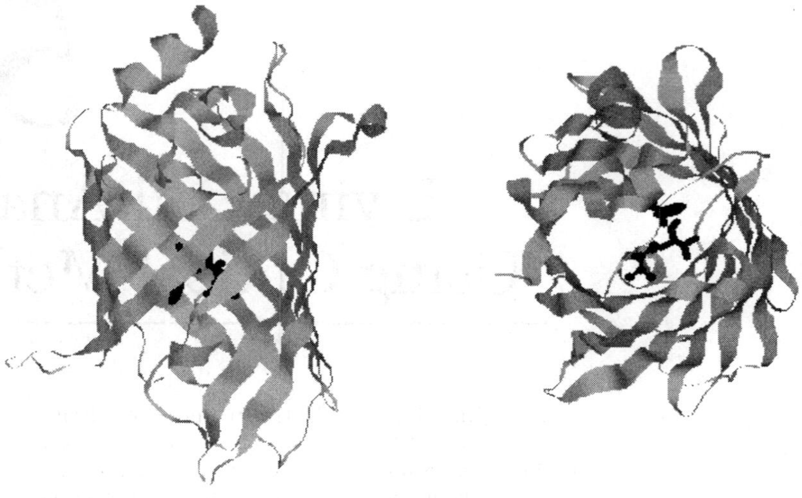

A B

**FIGURE 54.1** Side (A) and top (B) view of the 3D structure of enhanced yellow fluorescent protein (EYFP), showing the 11-stranded β–barrel as ribbon. The three amino-acid residues 64, 65, and 66, which form the chromophore after cyclization and oxygenation, are shown as balls and sticks.

Because these techniques have made it possible to get localized, real-time information on intracellular dynamics, it became necessary to modify and develop optical tools to get dynamic, three-dimensional (3D) microscopic maps of the living cell with nanometric resolution of individual molecules.

The green fluorescence protein (GFP) is a spontaneously fluorescent polypeptide of 27kD (238 amino-acid residues) from the jellyfish *Aequorea victoria* that absorbs ultraviolet (UV) blue light and emits in the green region of the spectrum.[1] Its structure and potential uses for studying living cells have been recently reviewed in detail.[2–7] The GFP chromophore results from a cyclization of three adjacent amino acids (S65, Y66, G67) and the subsequent 1,2dehydrogenation of the tyrosine.[2–7] Although the chromophore by itself is able to absorb light, its fluorescence properties result from the presence of an 11-stranded β-barrel. GFP has been expressed both in bacteria and in eukaryotic cells and has been produced by *in vitro* translation of the GFP messenger RNA (mRNA). Moreover, GFP retains its fluorescence when fused to heterologous proteins on the N- and C-terminals, and these bindings generally do not affect the functionality of the tagged protein.[2,3,6] That leads to the use of GFP as an intracellular reporter. Unfortunately, the fluorescence intensity of the native GFP is not bright enough for most intracellular applications. Therefore, variants have been engineered with different excitation or emission spectra that better match available light sources.[2,3,5–7] Brighter mutants were also found necessary in case of low expression levels in specific cellular microenvironments.[3,5–7] Figure 54.1 shows the structure of a GFP-type variant molecule, the enhanced yellow fluorescent protein (EYFP). Moreover, the time lag between the protein synthesis and fluorescence development, as well as sensitivity for photobleaching, was also a limitation when changes in location or conformation of the protein were sought.[6]

The potential toxicity of GFP has been questioned.[2] The efficiency of all these new mutants has recently been extensively reviewed, as has the incidence of GFP mutations on mRNA transcription and translation rates.[2,6] The only serious limitation to the use of GFP variants seems to be their size, which makes them unusable for tagging small biological molecules such as lipids or small proteins.

## 54.2 Exploring the Protein Factories

All recent studies have demonstrated that both tight control and coordination of gene expression require a high level of organization in the eukaryotic cell nucleus. Both fluorescence *in situ* hybridization (FISH) and immunocytochemical studies have shown that chromosomes occupy distinct territories within the nucleus and that many of the factors involved in transcription and RNA processing are located in nuclear bodies. Active genes in transcription are located at the borders of the condensed chromatin regions, which give them access to the transcription, RNA processing, and transport machinery (see Reference 8 and references therein). Due to its importance for life conservation, it is not surprising that photonics techniques were used to visualize how this special organization is conserved during cell replication, how it is correlated to the cellular functional activities, and how endogenic or exogenic events might disturb it.

### 54.2.1 Information Storage and Conservation: Chromatin Structure and Cellular Division

Advances in the specific fluorescent labeling of chromatin in living cells, in combination with 3D fluorescence microscopy and image analysis, have led to detailed studies of the higher-order architecture of chromatin in the human cell nucleus. Techniques such as time-lapse confocal microscopy (TLCM), fluorescence-resonance energy transfer (FRET), and fluorescence redistribution after photobleaching (FRAP) have opened the way to space-time (4D) studies of the dynamics of this architecture and of its potential role in gene regulation.[9–13] Several features of this architecture are well established; and even if many points remain in the speculative domain, models of the functioning of this super architecture have been proposed (see References 13 and 14 and references therein).

In parallel with the elucidation of the dynamics of chromatin architecture, many biophotonic studies have been devoted to nuclear transformation during mitosis and to the dynamics of postmitotic events. Multiwavelength fluorescence imaging (MWFI) has allowed the observation of the dynamic behavior of chromosomes and of some GFP fusion proteins linked to the nuclear envelope or the centromeres (see References 14 and 15 and references therein).The different pathways of assembly during nuclear-envelope formation of nuclear lamins A and B1 were also investigated by FRAP of lamins tagged with enhanced GFP (EGFP).[16] Time-lapse fluorescence imaging (TLFI) was used for a detailed analysis of mitosis in fission yeast, revealing the consequence of the presence of lagging chromosomes on the rate of anaphase B.[17] The expression of GFP-labeled centrin was used to study the respective contribution of mother and daughter centrioles to centrosome activity and behavior.[18] The dynamics of nucleolar reassembly were monitored in living cells expressing fusions of the processing-related proteins fibrillarin, nucleolin, and B23 with GFP.[19,20] Previously, the effects of mitotic inhibitors on cell-cycle progression were visualized by following the dynamics of chromosomes and microtubules stained with fluorescent chemicals.[21]

### 54.2.2 Information Transmission to Endoplasmic Reticulum Factories

Although many problems still remain unclear, most modern pictures of the nucleus agree with the need for a high level of organization of the chromatin. It is accepted that chromosome territories contain regions in which the degree of chromatin condensation varies. To be transcriptionally active genes must be located at the border of the condensed chromatin, in what Cremer et al. named an "interchromatin domain compartment."[14] This location contains ribonucleoprotein complexes involved in transcription and pre-mRNA splicing that can be detected through multiwavelength fluorescent microscopy.[8,12,14] The use of multifluorescent labeling protocols would make it possible to track gene transcripts in living cells.[22] Figure 54.2 shows a schematic diagram of the main successive steps involved in the synthesis of proteins.

Several approaches based on fluorescence techniques have been developed to detect the different RNAs in living cells and to monitor their dynamic behavior. A discussion of the methods used for introducing fluorescent RNAs inside the nucleus (microinjection of fluorescent RNAs into the nucleus, *in vivo* hybridization of fluorescent oligonucleotides to endogenous RNAs, or expression in cells of fluorescent RNA-binding proteins) is beyond the scope of this chapter. From our point of view, most of these studies can be classified in two sets: those involving only one fluorescent tag and those that require the use of two fluorescent tags. In the first set, when short linear DNA nucleotides (20 to 40 bases) or short antisense fluorescent probes are involved, chemical fluorophores e.g., fluorescein and rhodamine, are used in

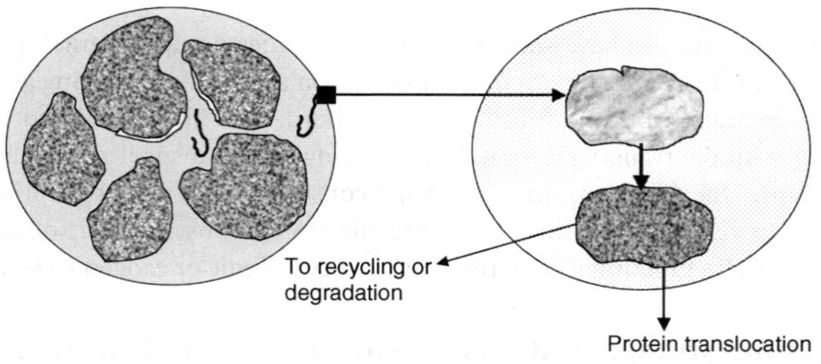


**FIGURE 54.2** Schematic diagram of the main successive steps involved in the synthesis of proteins.

association with TLFl, FRAP, or fluorescence correlation spectroscopy (FCS) (see Reference 8 and references therein). mRNA was also visualized using GFP as a tag on the RNA-loop-binding protein MS2. In the second set, hairpin-shaped oligonucleotide probes named "molecular beacons" are synthesized with fluorescent donor and acceptor chromophores at their 5′ and 3′ ends (see Reference 8 and references therein; Figures 54.3A and B). In the absence of a complementary nucleic-acid strand, the stem–loop conformation induces a FRET signal that prevents the fluorescence emission of the donor fluorophore. On the other hand, hybridization with a complementary sequence opens the stem–loop conformation, increases the distance between the donor–acceptor pair, and decreases the efficacy of FRET, allowing the fluorescence emission of the donor to be detected with confocal laser microscopy. Although the use of such molecular beacons was supposed to increase the signal-to-noise ratio (SNR), this does not always seem to be the case, suggesting that other intracellular mechanisms might sometimes open the stem–loop conformation[8] and included references. But detailed information reported in recent papers provides a model of the movement of mRNAs from transcription site to nuclear pores and of the role played by some nuclear proteins in facilitating or slowing down that complex process.[23,24] The development and applications of molecular beacons are further described in Chapter 57 of this handbook.

### 54.2.3 The "Quality-Control" System

When inserted into endoplasmic reticulum (ER) membranes, newly synthesized proteins encounter the lumenal environment of the ER that contains chaperone proteins. These proteins facilitate the folding reactions necessary for protein oligomerization, maturation, and export from the ER to the Golgi apparatus. Nevertheless, there are few studies that can actually visualize this complex process on living cells using fluorescence techniques.[25] In fact, visualization of this process involves a very impressive task. First, involved chaperones must be tagged to demonstrate their interaction with the target polypeptide. Second, the polypeptide itself must be labeled in a way that allows the properly folded protein to be distinguished from the polypeptide itself and from incompletely or improperly folded proteins. That task could be more demanding than having some GFP mutants at both ends of the target molecule. Nevertheless, interactions between different chaperones have been evidenced in solution.[26]

In fact, it is not certain that such a complex process works properly for each polypeptide entering the ER. On the contrary, there are grounds for suspecting that alteration of the yield of this process may induce some diseases.[27] Details of the mechanisms by which properly folded proteins are sorted from improperly folded ones and from the resident proteins have yet to be clarified and visualized in living cells. Nevertheless, it has been shown that improperly folded proteins accumulate in a novel pre-Golgi compartment, sometimes called ERGIC (for ER-to-Golgi), where they aggregate or are targeted for ER-associated protein degradation.[28,29]

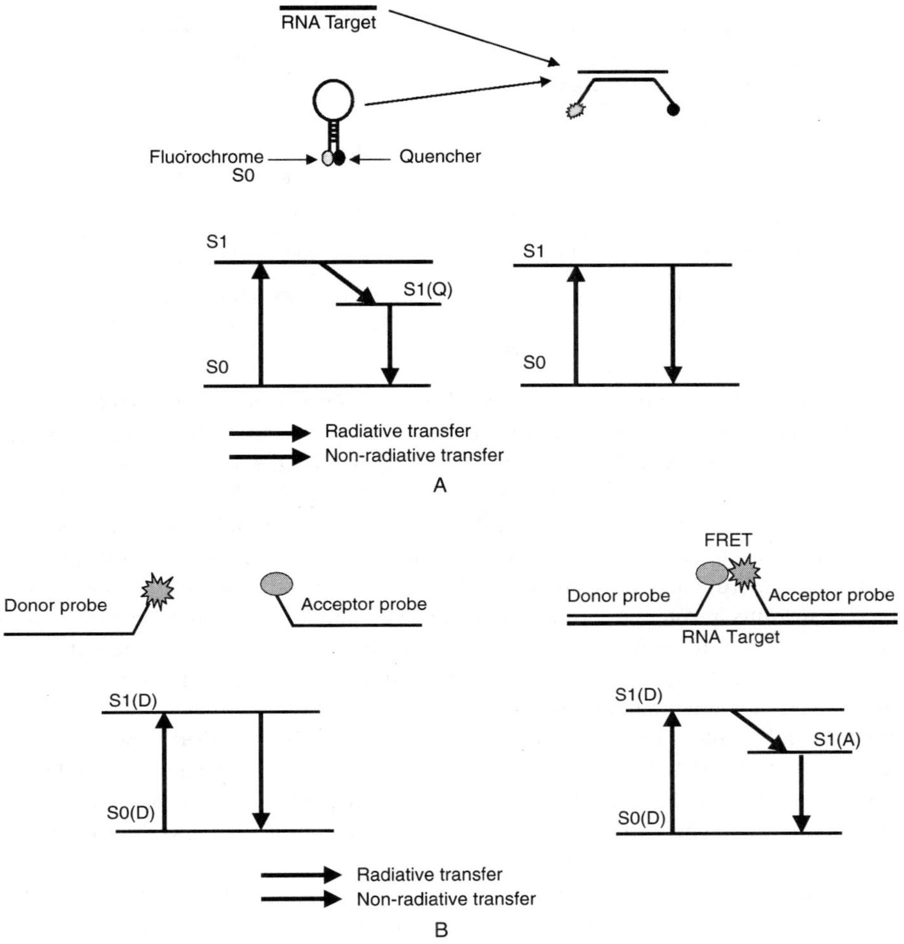


**FIGURE 54.3** (A) General concept of molecular beacons: hybridization with the target-RNA sequence opens the stem-loop conformation that results in emission of fluorescence. (B) Use of double labeling for detecting an RNA sequence: the fluorescence of the acceptor probe is seen only when both the acceptor and donor are properly hybridized.

Recent studies have shown that membranes of some organelles have their own systems for transforming nuclear-encoded preproteins into proteins that can be internalized in these organelles. Although they have not yet been applied to living cells, some methods have been tested in solution, which would allow the study of the conformational changes of these preproteins during the process of internalization in mitochondria.[30]

## 54.3 Monitoring the Location and Activities of Biological Molecules

### 54.3.1 Large Proteins

#### 54.3.1.1 Mapping the Cytoskeleton with GFP-Tagged Proteins

Before the use of GFPs, cytoskeleton research relied primarily on immunofluorescence microscopy techniques that require fixation and hence killing of the specimen before its analysis. The sole method

**TABLE 54.1** Spectral Properties of Some Commercially Available GFP Mutants

| GFP variants | Excitation (nm) | Emission (nm) | QY | Lifetime (ns) τϕ | τM |
|---|---|---|---|---|---|
| GFPwt (jellyfish) | 396 | 508 | 0.8 | nd | |
| GFPbr (Quantum) | 488 | 507 | nd | nd | |
| GFPbr (Clontech) | 488 | 507 | nd | 2.36 ± 0.06 | 2.42 ± 0.03 |
| BFP2br (Clontech) | 380 | 440 | nd | nd | |
| BFPbr (Quantum) | 380 | 440 | nd | nd | |
| YFPbr (Clontech) | 513 | 527 | 0.63 | 2.85 ± 0.12 | 2.88 ± 0.05 |

Data from Sacchetti, A. et al., *Histol. Histopathol.*, 15, 101, 2000; Galigniana, M.D. et al., *J. Biol. Chem.*, 274, 16222, 1999; and Umenishi, F. et al., *Biophys. J.*, 78, 1024, 2000. QY = quantum yield; nd = no determination.

for visualizing cytoskeletal dynamics was the microinjection of purified and fluorescently labeled protein, a protocol of limited use due to technical difficulties. In contrast, the addition of fluorescent tags to cytoskeleton proteins allowed visualization of the reorganization of its spatial structure and the localization of individual cellular components such as proteins and organelles. Furthermore, activities directly associated with the cytoskeleton, such as intracellular transport, cell motility, and morphological plasticity, have been investigated.[2] The studied proteins span a wide range, from proteins incorporated in or bound to microtubules (α and β-tubulins, MAPs, Tau, Kinesin, etc.),[31] to some associated with microfilaments (actins, myosins, myomesin, titin, β1-integrin, etc.), and finally to some incorporated in intermediate filaments (vimentin, lamin A) or septins.[32] The dynamics of microtubules (stabilization, turnover, bending and breaking, 3D organization) have also been investigated in recent studies using time-lapse fluorescence or confocal laser scanning microscopy.[33–39]

The functions of the motor proteins kinesin and dynein have also been studied using GFP tagging.[40,41,42] Total internal reflection (TIR) has been successfully used to image the dynamics of individual GFP-kinesin.[43] Such a method could be used to characterize the dynamics of the rapidly increasing number of identified motor proteins because tagging with GFP is stoichiometric and allows the use of recombinant fusion protein directly from bacterial extracts.

Successful monitoring of microfilaments using GFP-actin has been reported.[44–48] The actin cytoskeleton is an interesting target because it is involved in, and often drives a variety of, cellular processes such as morphogenesis,[49] endocytosis,[50] and cytokinesis, and links with the extracellular matrix.[51,52]

#### 54.3.1.2 Mapping Protein Distribution in Membranes

GFP exhibits fluorescence when it is used to create chimeric GFP proteins that can be easily visualized inside living cells. The main advantage of these GFP chimeras over chemical fluorescent tagging is the ease with which they may be assembled and introduced inside cells.[3,7] Of course, the final steady-state levels of GFP are critical for further applications. As with any other protein, they depend on the kinetics of the transcription and degradation of the GFP–mRNA and on the rate of degradation of the protein.[6,53] Furthermore, GFP chimeras can be used as *in vivo* markers for subcellular structures through the construction of systems that contain the appropriate targeting information.[3] Thus GFP has been successfully used to visualize the localization of membrane proteins.[5,6,54,55] Successful imaging of protein localization in the endoplasmic reticulum or the Golgi complex has also been reported.[3,56,57]

Studying the colocalization of different kinds of proteins requires the use of GFPs that fluoresce at different wavelengths. Besides EGFP, EBFP (Enhanced Blue Fluorescent Protein), EYFP, and ECFP (Enhanced Cyan Fluorescent Protein) have been synthesized; they fluoresce at 440, 527, and 475 nm, respectively. Several investigators have demonstrated the potency of using these spectral variants of GFP to acquire image sets of two or more different organelles or proteins in living cells.[3,5–7,58] In a more sophisticated approach, the lifetimes of these markers can also be used for colocalization studies through fluorescence lifetime imaging (FLIM).[59] Table 54.1 lists spectral properties and lifetimes of some commercially available GFP mutants.

### 53.3.1.3 Monitoring the Intracellular Trafficking of Proteins

Compared with the previous one, this goal is more demanding because it requires that the time necessary to record the fluorescent signal be short enough to allow a visualization of the protein translocation. Cellular expression or microinjection of brighter mutants of GFP allows for the construction of GFP chimeras usable for monitoring the translocation of proteins. However, the stability of the probes against photobleaching becomes a crucial factor for the successive illuminations of the same microscopic field that are necessary to detect any change in localization.

Expression, translocation, and turnover of proteins involve successive transient protein–membrane interactions that may be associated with changes in the membranar morphology. Optical methods used for monitoring the protein traffic have been found useful for monitoring morphological changes involved in the protein trafficking or resulting from the presence of viruses.[60]

Direct observation of nucleocytoplasmic transport (combined with nuclear localization or nuclear export signal) of a GFP-tagged glutathione-S-transferase has been reported,[7,61] while visualization of the transport from reticulum to Golgi has been performed with a GFP-tagged temperature-sensitive mutant protein.[3] Other studies have focused on expression and organellar targeting to test the ability to produce stable "green" transgenic parasites.[62]

Secretion involves shuttling proteins between various intracellular compartments. The transport of secretory vesicles from the trans-Golgi network to the plasma membrane has been monitored using a GFP chimera of the protein chromogranin B and time-lapse imaging. The same imaging technique has been used to demonstrate the direct rapid translocation of MHC class II molecules from the lysosomal compartment to the plasma membrane in cells of the immune system.[3]

Several proteins involved in signal transduction have been tagged with GFP variants to visualize their transport between cytosolic and membrane-bound pools in response to stimulation. The protein kinase B (PKB) (a mitogen-regulated kinase that protects cells from apoptosis) binds to phosphatidylinositol 3,4,5 triphosphate (PIP3), generated and localized in the plasma membrane via its pleckstrin homology (PH) domain, and is subsequently activated. Its translocation upon growth-factor stimulation was monitored in transiently transfecting fibroblasts using a GFP-PKB chimera. Other signaling proteins, such as Bruton's tyrosine kinase, the Ras GTPase-activating protein GAP1(m), and the ARF guanine nucleotide exchange factor ARNO, have also been GFP-tagged and translocated to the plasma membrane via their PH domains upon cell stimulation.[3]

GFP variants with different spectral properties have been used to simultaneously monitor the intracellular movements of several distinct proteins to follow the dynamics of nuclear envelope disassembly and assembly.[3] Fluorescent indicators have been developed to study in detail the cleavage of peptides in living cells as they are transported from the endoplasmic reticulum to the Golgi. When a peptide is labeled with these dyes at specific sites, a FRET is observed that disappears upon proteolytical process inside the cell.[63] Such a study leads the way to a new, elegant, and sophisticated approach to determining the precise mechanisms involved in immune surveillance.

GFP-tagged proteins have also been used to analyze the distribution of chemotactic receptors.[3] Other applications included studies on inhibition of protein shuttling,[64] regulation of membrane diffusion of aquaporin water channels,[65] internal trafficking and surface mobility of a functionally intact 2-adrenergic receptor-GFP conjugate,[5,7] and transient association of proteins with cell contact areas.[66] A quantitative study of fluorescent adenovirus entry has also been reported as a model system for cargo transport from the cell surface to the nucleus.[67]

### 53.3.1.4 Monitoring "Proteins at Work" in Living Cells

Monitoring "proteins at work" is more challenging than protein mapping or trafficking because it requires techniques that allow dormant proteins to be distinguished from activated proteins. Because the physical characteristics of the fluorescence of a molecule strongly depend on its microchemical environment, fluorescence techniques can be used to probe the changes of conformation that occur when a tagged protein is activated. Of course, the labeling of the protein with a fluorescent molecule must obey rules that are somewhat contradictory. The binding must occur at a definite location on the protein. This location must be far enough from the active site of the protein so that the binding does not change the activity of the protein but at the same time close enough to the active site so that it can be sensitive to the 3D conformational change occurring when the protein is activated.

Fluorescent molecules in which changes in chemical microenvironment induce spectral shifts of emission and excitation fluorescence spectra are numerous. Others experience changes in fluorescence intensity. Obtaining unambiguous information on potential changes in conformation for a given protein in living cells generally demands the use of two fluorophores: one of them must report only the total concentration of the protein, while the other must be sensitive to conformational changes. This could be troublesome when the tagged proteins must be microinjected inside the cells.

Another approach to monitoring conformational changes is to take advantage of change in the polarization of the fluorescence emission. Although anisotropy of fluorescence is practically independent of the concentration of the fluorescent dye, this technique has not been widely used in living cells until now, probably due to its relatively low sensitivity.[7]

The lifetime of the first excited state of a molecule is also very sensitive to changes in the chemical microenvironment of the dye while it is insensitive to the dye concentration. It was demonstrated 10 years ago that changes in fluorescence lifetime could in principle be used for FLIM. Until now, the use of this technique has been restricted to a few laboratories, due to technical difficulties.[68–72] Nevertheless, it has recently been demonstrated that FLIM at a single modulation frequency can be used to image and follow the cellular distribution of two, and ultimately three, GFPs simultaneously. FLIM may become more commonly used in the near future due to several advantages over spectral discrimination. Simultaneous readout of data and the need for only one dichroic and long-pass emission filter result in the use of a lower light dose and reduce the risk of photochemical damage to the cells.[59]

GFP chimeras were also found to be invaluable for time-lapse imaging to study the dynamics of proteins in cells of the immune system.[3] The same technique was used to visualize transport of secretory vesicles from the trans-Golgi network to the plasma membrane.[3,6] The dynamics of the distribution of chemotactic receptors on the surface of neutrophils were also monitored. More recently, the availability of GFP variants with different spectral characteristics has made it possible to compare the dynamics of two distinct proteins simultaneously.[3]

GFP chimeras have also demonstrated their potentialities in the field of functional studies. FRAP has been used to show the internalization of a β-adrenergic receptor chimera after stimulation. Its plasma-membrane-diffusion coefficient was also measured.[7] An arrestin chimera was also used to visualize the arrestin-binding kinetics to different G-protein-linked receptor subtypes.[7] GFP chimeras of galactosyltransferase, mannosidase II, and the KDEL receptor were found to have extremely high mobility in the Golgi of live HeLa cells; this implies that protein immobilization is not responsible for the retention of these proteins in the Golgi.[6] FRAP was also used to monitor the mobility of a GFP fusion of E-cadherin at different stages of epithelial cell–cell adhesion.[3]

The intensity of signals obtained with the previously mentioned FRET technique is also independent of the dye concentration, but this technique requires the use of two fluorophores. FRET occurs only when two convenient fluorophores are close enough together, and it falls off with the sixth power of the distance between them. Consequently, this effect is well adapted to monitor 3D conformational changes experienced by a double-labeled macromolecule, but it can also be used for monitoring conformational changes occurring inside a cluster of proteins. This technique has been used recently for monitoring an interaction between a nuclear receptor and its specific activator that is required for transcription of transiently transfected and chromosomally integrated reporter genes.[73] High-resolution FRET microscopy was used for studying the potential role of "lipid rafts" in the interaction between cholera toxin B-subunits and GPI-anchored proteins in the plasma membrane of HeLa cells.[74] Receptor-mediated activation of heterotrimeric GTP-binding proteins was also visualized in living cells by monitoring FRET between α- and β-subunits fused to cyan and yellow fluorescent proteins.[75]

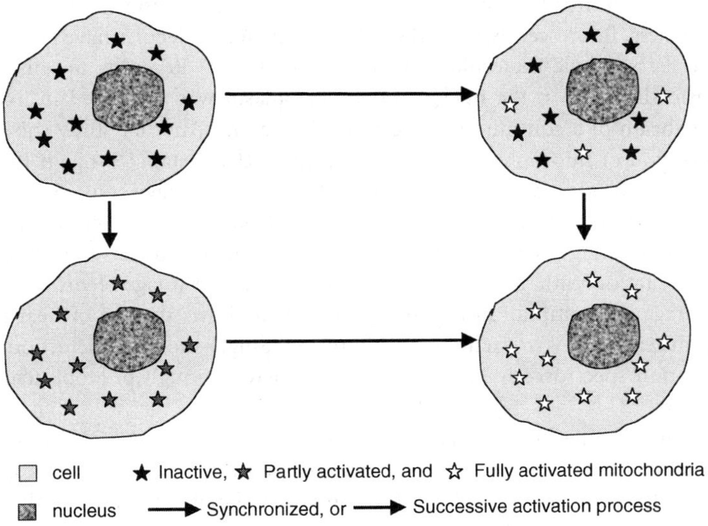


**FIGURE 54.4** Schematic presentation of two potential activation pathways of the mitochondria network.

### 53.3.1.5 Looking at the Single Organelle/Protein Level for an Integrated Approach of Cellular Mechanisms

As discussed in the previous sections, FRET occurs only when the distance between the donor and the acceptor molecule is between 2 and 10 nm. FRET provides us with a way to light up interactions at the molecular level, far below the usual limitation of traditional optics. The interest in getting information at the single-molecule level is obvious. It would improve our basic knowledge of biological events such as signal transduction, gene transcription, and intracellular transport. Furthermore, it would provide invaluable information on the potential synchronization of the response to any stimulus of organelles belonging to the same network (e.g., mitochondriae, ionic channels). Figure 54.4 shows a schematic presentation of two potential activation pathways of the mitochondria network.

Fluorescence microscopy has demonstrated its capability in sensing single molecules in *in vitro* experiments. In particular, it has been shown that single-dye molecules and single-fluorescent proteins emit fluorescent photons in an intermittent on-and-off fashion, a kind of information that cannot be obtained from population-averaged measurements. Although single fluorophores have recently been detected in the membranes of living cells with good SNRs, it is still not clear whether single-molecule fluorescence microcopy can be achieved at the spatial and temporal resolution required for biological experiments. Even if stable and bright fluorophores existed that could be coupled *in vivo* to biological molecules, some crucial challenges would remain. One of them is the development of easy-to-use, affordable equipment that allows high-resolution localization of individual point-like sources in 3D at a rate compatible with that of biological events. Another challenge is quantitative recording of relatively weak signals that must be extracted from the fluorescence background resulting from normal fluorescent cellular components such as flavins or nicotine amine nucleotides. Nevertheless, single GFP molecules have been successfully imaged in living cells using a total internal reflection fluorescence microscope.[76]

A new technique called fluorescence speckle spectroscopy (FSS) has recently been proposed for visualizing the movement, assembly, and turnover of macromolecular assemblies in living cells. A sensitivity of one to seven fluorophores per resolvable unit (270 nm) has been reported using a conventional wide-field epifluorescence light microscope and digital imaging with a low-noise, cooled CCD camera.[77]

Both confocal microscopy and multiphoton microscopy allow the energy required for fluorescence excitation to be concentrated in a tiny volume (femtoliter). Both share the problems linked with the use of far-field optics (spherical aberrations and chromatic aberrations, both in the excitation and in the detection arms) that are difficult to compensate for when deconvolution methods are used for image restoration. Nevertheless, fluorescently labeled lipid and receptor molecules have been monitored on the surface of living cells by single-molecule imaging techniques.[78,79] Recently, the distribution of single molecules of transferrin-TMR in the nucleus or the cytoplasm was visualized in living HeLa cells by tightly focusing the beam of a continuous-wave argon ion laser.[80] Another study showed that ultrahigh resolution colocalization of fluorophores is possible in the FRET distance range if the different fluorophores can be excited at the same wavelength, are bright enough, and have emission properties that allow an unambiguous distinction between the different types.[81] However, a closed-loop piezoscanner must be used that ensures a perfect alignment of each fluorophore on the optical axis to minimize the chromatic aberrations in the excitation path. Nothing is said about the rate of image recording.

The main advantage of multiphoton microscopy is that it allows the use of near-infrared (NIR) or infrared radiation instead of UV to excite fluorophores.[82,83] Single-molecule detection with two-photon fluorescence-correlation spectroscopy has been demonstrated, although photobleaching-resistant fluorophores are required.[82,84]

While limited to membranes, near-field scanning optical microscopy (NSOM) has been found useful for studying interactions between host and malarial skeletal proteins in erythocytes.[85,86] A protocol to extract the near-field fluorescence signal from the composite signal containing both near- and far-field fluorescence has been reported with improved image resolution.[87] An elegant high-resolution, two-photon, near-field scanning spectroscopy has also been described.[82]

Besides these studies on improved methods for detecting single molecules, other studies have been focused on the dynamics of whole-organelle or ionic-channel networks using wide-field spectroscopy and different kinds of image processing.[88–91] Although they do not claim to have reached the single-molecule detection level, these studies contain valuable information on the dynamics of the mitochondrial- or ionic-channel network, strengthening the interest in single-molecule imaging in living cells.

## 53.3.2 Lipids and Small Proteins

As indicated earlier, the direct labeling of proteins with GFP is restricted to relatively large proteins due to the size of GFP. Nevertheless, many elegant methods have been developed to trace membrane lipids due to their importance in the cellular machinery (endocytosis, exocytosis, vesicular trafficking of proteins, transduction of extracellular signals, remodeling of the cytoskeleton, regulation of calcium flux, and apoptosis). In one approach, the membrane receptors of some lipids have been GFP-tagged to visualize their intracellular localization.[92,93] However, these results are difficult to interpret in terms of lipid concentration when the effects of external stimuli are sought. Therefore, fluorescent analogs of these lipids have been synthesized and used to track their intracellular distribution or redistribution.[94–99] The same approach has been used for smaller molecules such as cholesterol or cyclic AMP and GMP.[53,100,101]

Theoretically, the same method can be used for small proteins. As stated previously, the problem consists in finding at least one amino-acid residue — and potentially two if one wants to use FRET techniques that match the following criteria. They must be reactive enough to allow a specific binding of the fluorescent tag and be conveniently located in the protein structure so that tagging does not change the properties of the protein (conformation, binding sites or potential receptors, etc.). Furthermore, tags must experience some changes in their chemical microenvironment upon protein binding (to substrates or receptors) if the intracellular protein activity is to be studied. So far only calmodulin intracellular activity has been studied using these methods.[7]

## 53.3.3 GFP Mutants as Ionic Sensors

Until now, the search for ion-specific, genetically engineered fluorescent probes has been limited mainly to $Ca^{2+}$ and $H^+$ due to their importance in the cell machinery, although a $Cl^-$ probe has recently been reported.[102] Several laboratories reported that some GFP mutants experience pH-sensitive spectral changes.[3,6] Fluorescence intensity was found to decrease with pH, while the shape and position of the

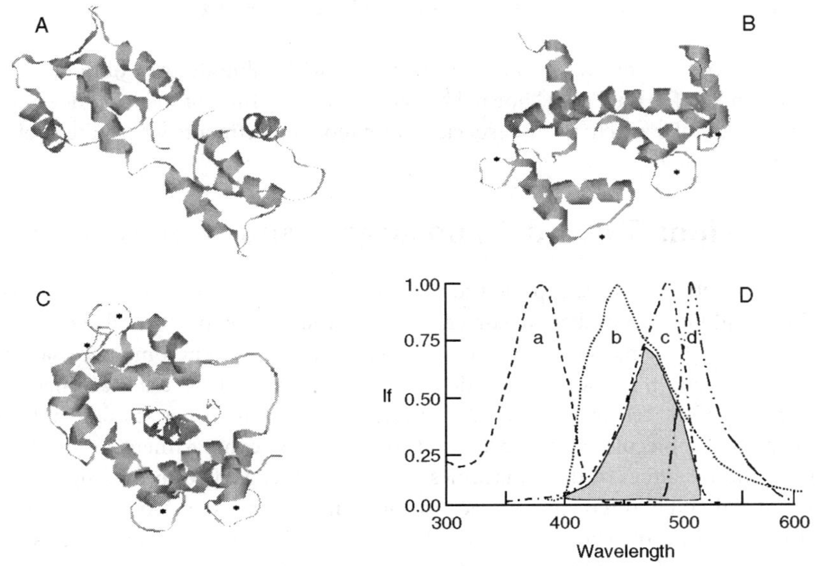


**FIGURE 54.5** Changes in 3D structure of calmodulin upon binding to calcium ions and myosin chain M13. (A) Free calmodulin. (B) In the presence of calcium ions. (C) In the presence of calcium ions and myosin chain M13 (limited to the peptide involved in the interaction for sake of clarity). (D) Respective normalized excitation and emission spectra of GFP mutants Y66H and S65T: (a) excitation spectrum of Y66H, (b) emission spectrum of Y66H, (c) excitation spectrum of S65T, and (d) emission spectrum of S65T. The gray zone shows the wavelength range of overlapping between the emission spectrum of Y66H and the excitation spectrum of S65T.

spectrum were relatively insensitive to pH. Another type of pH-sensitive GFP mutant was developed that experiences a spectral shift upon pH change.[3,6] Both these types of pH sensors have their advantages and limitations. The latter permits the measurement of pH from the ratio of intensity at two wavelengths, a method that is relatively independent of photobleaching and alteration in focal plane. The former allows easier use of confocal microscopy because only one excitation wavelength is required. Recent applications dealt with Golgi and secretory vesicle pH.[101,103,104]

GFPs have also been used for $Ca^{2+}$ measurements. Fluorescent intracellular calcium indicators called "chameleons" have been synthesized that are fusion proteins containing blue fluorescent protein or cyan fluorescent protein; calmodulin, the calmodulin-binding domain of skeletal-muscle myosin light chain kinase (M13); and either GFP or yellow fluorescence protein. In the absence of calcium, this long chain is extended; in the presence of calcium, calmodulin experiences a conformational change. This change favors its binding to M13, which results in a more compact conformation that increases the FRET between the flanking fluorescent proteins.[2,3,5,7,105] Figure 54.5 schematically illustrates the changes in the 3D structure of calmodulin upon binding to calcium ions and myosin chain M13. Mutations inside calmodulin allowed the generation of chameleons with different calcium affinities from 10 n*M* to 10 m*M*. Furthermore, the inclusion of localization signals in the structure makes it possible to target these chameleons to specific intracellular locations for monitoring intraorganellar calcium dynamics. Measurements of rapid $Ca^{2+}$ turnover in the endoplasmic reticulum during signaling have been reported,[106] and the use of yellow fluorescent protein less sensitive to pH and chloride interference has allowed $Ca^{2+}$ measurement in the Golgi.[107] Furthermore, new fluorescent chimeras, dubbed "pericams," have recently been used to monitor free $Ca^{2+}$ dynamics in the cytoplasm, the nucleus, and the mitochondria of HeLa cells.[108]

In another approach to calcium monitoring, a blue fluorescent protein was linked to GFP by a 26-residue spacer containing the calmodulin-binding domain from smooth-muscle myosin chain kinase. Due to the flexibility of that chimera, FRET occurred between the fluorescent proteins that were close in the absence of calcium. This proximity did not exist in the presence of calcium, and FRET was no longer possible.[4]

Since all these probes rely on the interaction of calcium with calmodulin, a ubiquitous protein, they are expected to compete for $Ca^{2+}$ with "normal" proteins. Perhaps this competition does not induce any perturbation in the cellular mechanisms, especially in nanocompartments where the probability of the presence of free $Ca^{2+}$ is low.

## 53.4 Conclusion: Toward Nanosurgery and Nanomedicine

Future biological investigations are expected to focus on monitoring the interactions between the fundamental cellular building blocks that are currently under study. Once the 3D high-resolution structures and biological roles of proteins and RNAs have been elucidated, it will become necessary to determine their precise locations and translocations inside the cell to understand the cell machinery and circuitry. Many vital functions of the cell are performed by modular cellular machines self-assembled from a large number of interacting molecules and translocated from one cell compartment to another. Unraveling the organization and dynamics of these machines requires tools capable of providing *in vivo* 3D microscopic pictures of individual interacting molecules with nanometer resolution and at a time scale compatible with the speed of intracellular processes. The same kinds of tools are also necessary to monitor the cooperative behavior of networks such as the mitochondrial- or ionic-channel networks.

Significant advances in fluorescence techniques have recently improved their spatial resolution beyond the classical diffraction limit of light. Among them, the most promising appear to be wide-field image restoration by computational methods,[109] wide-field single-molecule localization and tracking,[110–112] aperture-[113] and apertureless-type near-field scanning spectroscopy, two-photon excitation microscopies,[114,115] and point-spread-function engineering by stimulated emission depletion.[116] Nevertheless, near-field scanning spectroscopy and total internal reflection fluorescence spectroscopy appear more appropriate for studying the cell surface than for monitoring intact cellular structures.[117] Recently, a new imaging and ultrahigh-resolution colocalization technique has been proposed that can pinpoint the locations of multiple distinguishable probes with nanometer accuracy and is said to better handle the limitations presented by other methods.[81] But to date this technique has not been used *in vivo* and requires probes relatively insensitive to fading.

Among the above-mentioned techniques, multiphoton technology seems to be the most appropriate for delivering a given dose of radiation into a predetermined nanovolume. Not surprisingly, it has been used for both noninvasive optical biopsy and nanosurgery.[82] Due to the use of irradiation wavelengths in the 700- to 1100-nm spectral range, living tissues can be investigated at depths of up to 100 μm. The endogenous fluorophores that can potentially be excited are NADH, NADPH, flavins, and collagen, allowing for the imaging of intracellular structures, the location of the nucleus, and cell morphology in the different skin layers. Imaging of the accumulation of protoporphyrin in tumor cells of mice with solid carcinoma has also been reported, suggesting potential applications of two-photon technology to photodynamic therapy of tumors.

Femtosecond NIR laser pulses have numerous advantages over conventional nanosecond UV laser pulses: high penetration depth, absence of out-of-focus absorption, efficient induction of multiphoton excitation, and absence of plasma shielding and significant heat transfer. Convenient tuning of the laser power makes it possible to confine the energy necessary for material removal to the central part of the beam. Therefore, this technique may provide a noncontact nanoscalpel for surgery inside a cellular organelle that will not have unwanted effects on other cellular compartments.[82] Optical tweezers that have been used for studying the strength of the interaction between fibronectin and its membrane receptor or the microrheology of biopolymer-membrane complexes have also been tested to facilitate *in vitro* fertilization.[118–120]

Nanotechnology is opening up new possibilities for the development of fiber-optics-based nanosensors and nanoprobes of submicron-sized dimensions suitable for intracellular measurements. The ability to examine processes within living cells could provide great advances in our understanding of cellular

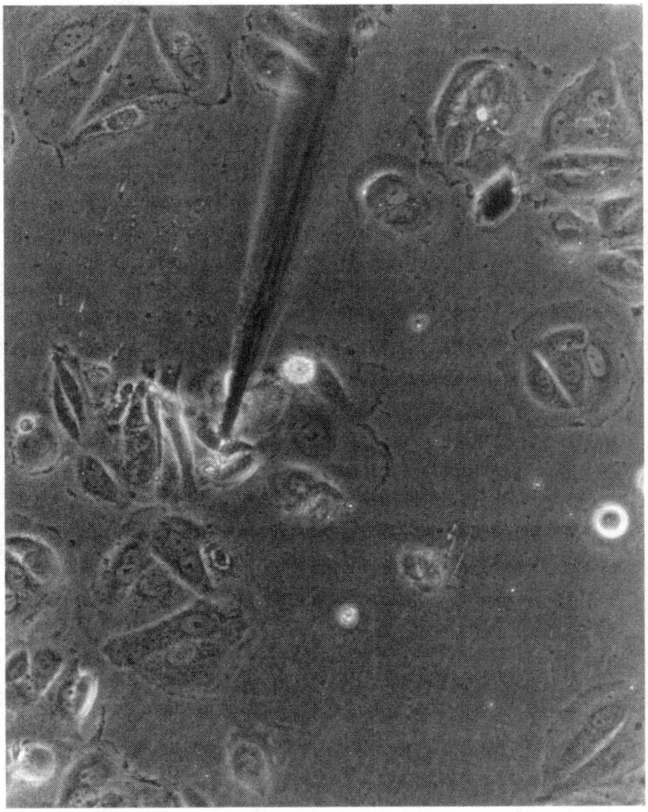

**FIGURE 54.6 (Color figure follows p. 28-30.)** Photograph of an antibody-based nanoprobe used to measure the presence of benzopyrene tetrol in a single cell. The small size of the fiber-optics probe allows manipulation of the nanoprobe at specific locations within the cell. (From Vo-Dinh, T. et al., *Nat. Biotechnol.*, 18, 76, 2000. With permission.)

function, thereby revolutionizing cell biology. Fiber-optic sensors offer important advantages for *in situ* monitoring applications due to the optical nature of the detection signal. These sensors are not subject to electromagnetic interferences from static electricity, strong magnetic fields, or surface potentials. Another advantage of fiber-optic sensors is the small size of optical fibers, which allows for sensing intracellular/intercellular physiological and biological parameters in microenvironments. Figure 54.6 shows a fiber-optic nanosensor with antibody probes for benzopyrene tetrol, a biomarker of human exposure to the carcinogen benzo[a]pyrene that has been recently developed and used for *in vivo* analysis of single cells.[121–125] Nanosensors could provide much needed tools to investigate important biological processes at the cellular level. For instance, the mode of entry of fluorescent proteins or exogenous chemicals into various cells is important, as are subsequent enzymatic processes inside cells. Tracking these species becomes feasible due to their natural fluorescence. Inspection of cells at submicron resolution by optical nanosensors that can interrogate at precise locations has the potential for significant advances in knowledge of cellular structure/function. The development and application of fiber-optic nanosensors are described in detail in Chapter 60 of this handbook.

The methods and techniques reviewed in this chapter have demonstrated that biophotonics may allow imaging of most of the different steps that result in the synthesis of proteins and monitoring of their activity. Thus, perhaps in the near future it will be possible to identify and image which specific steps in this complex process are responsible for any given disease. For instance, such a nanodiagnosis could show that a given disease results from an increase in the misfolding of a protein instead of a decrease in the production of this protein. Will this mean that nanomedicine, involving the delivery of drugs especially designed to repair a dysfunction in a step of any biological pathway, can be expected in the near future and that it will be possible to visualize the benefits through biophotonic technology? The challenge for photonics to elucidate life processes will come from the complexity inherent in biological events, as demonstrated in a recent study.[126] The researchers reported that they have identified 997 messenger RNAs responding to 20 systematic perturbations of the yeast galactose-utilization pathway, providing evidence that approximately 15 to 289 detected proteins are regulated post-transcriptionally.[126]

## Acknowledgments

This research was supported by the French Department of Education and Sciences, by the ORNL Laboratory Directed Research and Development Program (Advanced Nanosystems), and by the Office of Biological and Environmental Research, U.S. Department of Energy, under contract DE-AC05-00OR22725 with UT-Battelle, LLC.

## References

1. Prasher, D.C., Eckenrode, V.K., Ward, W.W., Prendergast, F.G., and Cormier, M.U., Primary structure of the *Aequorea victoria* green-fluorescent protein, *Gene*, 111, 229, 1992.
2. Ludin, B. and Matus, A., GFP illuminates the cytoskeleton, *Trends Cell Biol.*, 8, 72, 1998.
3. Bajno, L. and Grinstein, S., Fluorescent proteins: powerful tools in phagocyte biology, *J. Immunol. Methods*, 232, 67, 1999.
4. Chamberlain, C. and Hahn, K.M., Watching proteins in the wild: fluorescence methods to study dynamics in living cells, *Traffic*, 1, 755, 2000.
5. Latif, R. and Graves, P., Fluorescent probes: looking backward and looking forward, *Thyroid*, 10, 407, 2000.
6. Sacchetti, A., Ciccocioppo, R., and Alberti, S., The molecular determinants of the efficiency of green fluorescent protein mutants, *Histol. Histopathol.*, 15, 101, 2000.
7. Whitaker, M., Fluorescent tags of protein function in living cells, *BioEssays*, 22, 180, 2000.
8. Molenaar, C., Marras, S.A., Slats, J.M.C., Truffert, J. C., Leonaitre, M., Raap, A.P., Dirks, R.W., and Tanke, H.J., Linear 2′o-methyl RNA probes for the visualization of RNA in living cell, *Nucleic Acids Res.*, 29, E89, 2001.
9. Boudonck, K., Dolan, L., and Shaw, P.J., The movement of coiled bodies visualized in living plant cells by the green fluorescent protein, *Mol. Biol. Cell*, 10, 2297, 1999.
10. Haraguchi, T., Ding, D.Q., Yamamoto, A., Kaneda, T., Koujin, T., and Hiraoka, Y., Multiple color fluorescence imaging of chromosomes and microtubules in living cells, *Cell Struct. Function*, 24, 291, 1999.
11. Wachsmuth, M., Waldeck, W., and Langowski, J., Anomalous diffusion of fluorescent probes inside living cell nuclei investigated by spatially-resolved fluorescence correlation spectroscopy, *J. Mol. Biol.*, 298, 677, 2000.
12. Misteli, T., Cell biology of transcription and pre-mRNA splicing: nuclear architecture meets nuclear function, *J. Cell Sci.*, 113, 1841, 2000.
13. Houstmuller, A.B. and Vermeulen, W., Macromolecular dynamics in living cell nuclei revealed by fluorescence redistribution after photobleaching, *Histochem. Cell Biol.*, 115, 13, 2001.
14. Cremer, T., Kreth, G., Koester, H., Fink, R.H.A., Heintzmann, R., Cremer, M., Solovei, I., Zink, D., and Cremer, C., Chromosome territories, interchromatin domain compartments and nuclear matrix: an integrated view of the functional nuclear architecture, *Crit. Rev. Eukaryot. Gene Expression*, 10, 179, 2000.
15. Haraguchi, T., Koujin, T., and Hiraoka, Y., Application of GFP: time-lapse multi-wavelength fluorescence imaging of living mammalian cells, *Acta Histochem. Cytochem.*, 33, 169, 2000.
16. Moir, R.D., Yoon, M., Khuon, S., and Goldman, R.O., Nuclear lamins A and B1: different pathways of assembly during nuclear envelope formation in living cells, *J. Cell Biol.*, 151, 1155, 2000.

17. Pidoux, A.L., Uzawa, S., Perry, P.E., Cande, W.Z., and Allshire, R.C., Live analysis of lagging chromosomes during anaphase and their effect on spindle elongation rate in fission yeast, *J. Cell Sci.*, 113, 4177, 2000.
18. Piel, M., Meyer, P., Khodjakov, A., Reider, C.L., and Bornens, M., The respective contributions of the mother and daugther centrioles to centrosome activity and behavior in vertebrate cells, *J. Cell Biol.*, 149, 317, 2000.
19. Dundr, M., Misteli, T., and Olson, M.O.J., The dynamics of postmitotic reassembly of the nucleolus, *J. Cell Biol.*, 150, 433, 2000.
20. Savino, T.M., Gebrane-Younes, J., De Mey, J., Sibarita, J.B., and Hernandez-Verdun, D., Nucleolar assembly of the rRNA processing machinery in living cells, *J. Cell Biol.*, 153, 1097, 2001.
21. Haraguchi, T., Kaneda, T., and Hiraoka, Y., Dynamics of chromosomes and microtubules visualized by multi-wavelength fluorescence imaging in live mammalian cells: effects on cell cycle progression, *Gene Cells*, 2, 369, 1997.
22. Pederson, T., Fluorescent RNA cytochemistry: tracking gene transcripts in living cells, *Nucleic Acids Res.*, 29, 1013, 2001.
23. Beach, D.L., Salmon, E.D., and Bloom, K., Localization and anchoring of mRNA in budding yeast, *Curr. Biol.*, 9, 569, 1999.
24. Politz, J.C. and Peterson, T., Review: movement of m-RNA from transcription site to nuclear pores, *J. Struct. Biol.*, 129, 252, 2000.
25. Bouvier, M., Oligomerization of G-protein-coupled transmitter receptors, *Nat. Rev. Neurosci.*, 2, 274, 2001.
26. Corbett, E.F., Oikawa, K., Francois, P., Tessier, D.C., Kay, C., Bergeron, J.J.M., Thomas, D.Y., Krause, K.H., and Michalak, M., $Ca^{2+}$ regulation of interactions between endoplasmic reticulum chaperones, *J. Biol. Chem.*, 274, 6203, 1999.
27. Halaban, R., Svedine, S., Cheng, E., Smicum, Y., Aron, R., and Hebert, D.N., Endoplasmic reticulum retention is a common defect associated with tyrosinase-negative albinism, *Proc. Natl. Acad. Sci. U.S.A.*, 97, 5889, 2000.
28. Zhang, Y., Nybroek, G., Sullivan, M.L., McCracken, A.P., Watkins, S.C., Michaelis, S., and Brodsky, J.L., Hsp70 molecular chaperone facilitates endoplasmic reticulum-associated protein degradation of cystic fibrosis transmembrane conductance regulator in yeast, *Mol. Biol. Cell*, 12, 1303, 2001.
29. Kamhi-Nesher, S., Shenkman, M., Tolchinsky, S., Fromm, S.V., Ehrlich, R., and Lederkremer, G.Z., A novel quality control compartment derived from the endoplasmic reticulum, *Mol. Biol. Cell*, 12, 1711, 2001.
30. Stan, T., Ahting, U., Dembowski, M., Kunkele, K.P., Nussberger, S., Neupert, W., and Rapaport, D., Recognition of preproteins by the isolated TOM complex of mitochondria, *EMBO J.*, 19, 4895, 2000.
31. Miller, R.K., Matheos, D., and Rose, M.D., The cortical localization of the microtubule orientation protein, Kar9p, is dependent upon actin and proteins required for polarization, *J. Cell Biol.*, 144, 963, 1999.
32. Cid, V.J., Adamikova, L., Sanchez, M., Molina, M., and Nombela, C., Cell cycle control of septin ring dynamics in the budding yeast, *Microbiology*, 147, 1437, 2001.
33. Vorobjev, I.A., Rodionov, V.I., and Borisy, G.G., Contribution of plus and minus end pathways to microtubule turnover, *J. Cell Sci.*, 112, 2277, 1999.
34. Zhang, C., Hugues, M., and Clarke, P.R., RNA-GTP stabilises microtubule asters and inhibits nuclear assembly in Xenopus egg extracts, *J. Cell Sci.*, 112, 2453, 1999.
35. Odde, D.J., Ma, L., Briggs, A.H., DeMarco, A., and Kirschner, M.W., Microtubule bending and breaking in living fibroblasts, *J. Cell Sci.*, 112, 3283, 1999.
36. Windoffer, R. and Leube, R.E., Detection of cytokeratin dynamics by time-lapse fluorescence microscopy in living cells, *J. Cell Sci.*, 112, 4521, 1999.
37. Mallavarapu, A., Sawin, K., and Mitchison, T., A switch in microtubule dynamics at the onset of anaphase B in the mitotic spindle of *Schizosaccharomyces pombe*, *Curr. Biol.*, 9, 1423, 1999.

38. Granger, C.L. and Cyr, R.J., Microtubule reorganization in tobacco BY-2 cells stably expressing GFP-MBD, *Planta*, 210, 502, 2000.
39. Strohmaier, A.R., Porwol, T., Acker, H., and Spiess, E., Three-dimensional organization of microtubules in tumor cells studied by confocal laser scanning microscopy and computer-assisted deconvolution and image reconstruction, *Cells Tissues Organs*, 167, 1, 2000.
40. Pierce, D.W., Hom-Booher, N., Otsuka, A.J., and Vale, R.D., Single-molecule behavior of monomeric and heteromeric kinesins, *Biochemistry*, 17, 5412, 1999.
41. Xiang, X., Han, G.S., Winkelmann, D.A., Zuo, W.Q., and Morris, N.R., Dynamics of cytoplasmic dynein in living cells and the effect of a mutation in the dynactin complex actin-related protein Arp1, *Curr. Biol.*, 10, 603, 2000.
42. Farkasovsky, M. and Kuntzel, H., Cortical Num1p interacts with the dynein intermediate chain Pac11p and cytoplasmic microtubules in budding yeast, *J. Cell Biol.*, 152, 2, 251, 2001.
43. Inoue, Y., Iwane, A.H., Miyai, T., Muto, E., and Yanagida, T., Motility of single one-headed kinesin molecules along microtubules, *Biophys. J.*, 81, 5, 2838, 2001.
44. Ballestrem, C., Wehrle-Haller, B., and Imhof, B.A., Actin dynamics in living mammalian cells, *J. Cell Sci.*, 111, 1649, 1998.
45. Yumura, S. and Fukui, Y., Spatiotemporal dynamics of actin concentration during cytokinesis and locomotion in Dictyostelium, *J. Cell Sci.*, 111, 2097, 1998.
46. Choidas, A., Jungbluth, A., Sechi, A., Murphy, J., Ullrich, A., and Marriott, G., The suitability and application of a GFP-actin fusion protein for long-term imaging of the organization and dynamics of the cytoskeleton in mammalian cells, *Eur. J. Cell Biol.*, 77, 81, 1998.
47. Evans, R.M. and Simpkins, H., Cisplatin induced intermediate filament reorganization and altered mitochondrial function in 3T3 cells and drug-sensitive and -resistant Walker 256 cells, *Exp. Cell Res.*, 245, 69, 1998.
48. Correia, I., Chu, D., Chou, Y.H., Goldman, R.D., and Matsudaira, P., Integrating the actin and vimentin cytoskeletons: adhesion dependent formation of fimbrin-vimentin complexes in macrophages, *J. Cell Biol.*, 146, 831, 1999.
49. McNiven, M.A., Kim, L., Krueger, E.W., Orth, J.D., Cao, H., and Wong, T.W., Regulated interactions between dynamin and the actin-binding protein cortactin modulate cell shape, *J. Cell Biol.*, 151, 187, 2000.
50. Enqvist-Goldstein, A.E.Y., Kessels, M.M., Chopra, V.S., Hayden, M.R., and Drubin, D.G., An actin-binding protein of he Sla2/Huntingtin interacting protein 1 family is a novel component of clathrin-coated pits and vesicles, *J. Cell Biol.*, 147, 1503, 1999.
51. Kachinsky, A.M., Froehner, S.C., and Milgram, S.L., A PDZ-containing scaffold related to the dystrophin complex at the basolateral membrane of epithelial cells, *J. Cell Biol.*, 145, 391, 1999.
52. Akhtar, N. and Hotchin, N.A., RAC1 regulates adherent junctions through endocytosis of E-cadherin, *Mol. Biol. Cell*, 12, 847, 2001.
53. Wild, N., Herberg, F.W., Hofmann, F., and Dostmann, W.R.G., Expression of a chimeric, cGMP-sensitive regulatory subunit of the cAMP-dependent protein kinase type I$\alpha$, *FEBS Lett.*, 374, 356, 1995.
54. Liu, Z.H., Luo, A.P., Wang, X.Q., and Wu, M., Chromosomal localization of a novel retinoic acid induced gene RA28 and protein distribution of its encoded protein, *Chin. Sci. Bull.*, 45, 1857, 2000.
55. Hettmann, C., Herm, A., Geiter, A., Frank, B., Schwarz, E., Soldati, T., and Soldati, D., A dibasic motif in the tail of a class XIV apicomplexan myosin is an essential determinant of plasma membrane localization, *Mol. Biol. Cell*, 11, 1385, 2000.
56. Reichel, C. and Beachy, R.N., Degradation of tobacco mosaic virus movement protein by the 26S proteasome, *J. Virol.*, 74, 3330, 2000.
57. Giraudo, C.G., Daniotti, J.L., and Maccioni, H.J.F., Physical and functional association of glycolipid *N*-acetyl-galactosaminyl and galactosyl transferases in the Golgi apparatus, *Proc. Natl. Acad. Sci. U.S.A.*, 98, 1625, 2001.

58. Margolin, W., Green fluorescent protein as a reporter for macromolecular localization in bacterial cells, *Methods*, 20, 62, 2000.
59. Pepperkok, R., Squire, A., Geley, S., and Bastiaens, P.I.H., Simultaneous detection of multiple green fluorescent proteins in live cells by fluorescence lifetime imaging microscopy, *Curr. Biol.*, 9, 269, 1999.
60. Riechel, C. and Beachy, R.N., Tobacco mosaic virus infection induces severe morphological changes of the endoplasmic reticulum, *Proc. Natl. Acad. Sci. U.S.A.*, 95, 11169, 1998.
61. Rosorius, O., Heger, P., Stelz, G., Hirschmann, N., Hauber, J., and Stauber, R.H., Direct observation of nucleocytoplasmic transport by microinjection of GFP-tagged proteins in living cells, *Biotechniques*, 27, 350, 1999.
62. Striepen, B., He, C.Y.X., Matrajt, M., Soldati, D., and Roos, D.S., Expression, selection, and organellar targeting of the green fluorescence protein in *Toxoplasma gondii*, *Mol. Biochem. Parasitol.*, 92, 325, 1998.
63. Bark, S.J. and Hahn, K.M., Fluorescent indicators of peptide cleavage in the trafficking compartments of living cells: peptides site-specifically labeled with two dyes, *Methods*, 20, 429, 2000.
64. Galigniana, M.D., Housley, P.R., DeFranco, D.B., and Pratt, W.B., Inhibition of glucocorticoid receptor nucleocytoplasmic shuttling by okadaic acid requires intact cytoskeleton, *J. Biol. Chem.*, 274, 16222, 1999.
65. Umenishi, F., Verbavatz, J.M., and Verkman, A.S., cAMP regulated membrane diffusion of a green fluorescent protein-aquaporin 2 chimera, *Biophys. J.*, 78, 1024, 2000.
66. Rivero, F., Albrecht, R., Dislich, H., Bracco, E., Graciotti, L., Bozzaro, S., and Noegel, A.A., RacF1, a novel member of the Rho protein family in *Dictyostelium discoideum*, associates transiently with cell contact areas, *Mol. Biol. Cell*, 10, 1205, 1999.
67. Nakano, M.Y. and Greber U.F., Quantitative microscopy of fluorescent adenovirus entry, *J. Struct. Biol.*, 129, 57, 2000.
68. Gadella, T.W.J., Jr., Jovin, T.M., and Clegg, R.M., Fluorescence lifetime imaging microscopy (FLIM)-spatial resolution of microstructures on the nanosecond time-scale, *Biophys. Chem.*, 48, 221, 1993.
69. So, P.T.C., French, T., Yu, W.M., Berland, K.M., Dong, C.Y., and Gratton, E., Time-resolved fluorescence spectroscopy using two-photon excitation, *Bioimaging*, 3, 1, 1995.
70. Carlsson, K. and Liljeborg, A., Confocal fluorescence microscopy using spectral and lifetime information to simultaneously record four fluorophores with high channel separation, *J. Microsc.*, 185, 37, 1997.
71. Schneider, P.C. and Clegg, R.M., Rapid acquisition, analysis, and display of fluorescence life-time resolved images for real-time applications, *Rev. Sci. Instrum.*, 68, 4107, 1997.
72. Squire, A. and Bastiaens, P.H.I., Three-dimensional image restoration in fluorescence lifetime imaging microscopy, *J. Microsc.*, 193, 36, 1999.
73. Lliopis, J., Westin, S., Ricote, M., Wang, J., Cho, C.Y., Kurokaw, R., Mullen, T., Rose, D., Rosenfeld, M.G., Tsien, R.Y., and Glass, C.K., Ligand-dependent interactions of coactivators steroid receptor coactivator-1 and peroxisome proliferator-activated receptor binding protein with nuclear hormone receptors can be imaged in living cells and are required for transcription, *Proc. Natl. Acad. Sci. U.S.A.*, 97, 4363, 2000.
74. Kenworthy, A.K., Petranova, N., and Edidin, M., High-resolution FRET microscopy of cholera toxin B-subunit and GPI-anchored proteins in cell plasma membranes, *Mol. Biol. Cell*, 11, 1645, 2000.
75. Janetopoulos, C., Jin, T., and Devreotes, P., Receptor-mediated activation of heteromeric G-proteins in living cells, *Science*, 291, 2408, 2001.
76. Lino, R., Koyama, I., and Kusumi, A., Single molecule imaging of green fluorescent proteins in living cells: E-cadherin forms oligomers on the free cell surface, *Biophys. J.*, 80, 2667, 2001.
77. Waterman-Storer, C.M. and Salmon, E.D., Fluorescent speckle microscopy of microtubules: how low can we go? *FASEB J.*, 13, S225, 1999.

78. Schütz, G.J., Kada, G., Pastushenko, V.P., and Schindler, H., Properties of lipid microdomains in a muscle cell membrane visualized by single molecule microscopy, *EMBO J.*, 19, 892, 2000.
79. Sorkin, A., McClure, M., Huang, F.T., and Carter, R., Interaction of EGF receptor and Grb2 in living cells visualized by fluorescence energy transfer (FRET) microscopy, *Curr. Biol.*, 10, 1395, 2000.
80. Byassee, T.A., Chan, W.C.W., and Nie, S., Probing single molecule in living cells, *Anal. Chem.*, 72, 5606, 2000.
81. Lacoste, T.D., Michalet, X., Pinaud, F., Chemla, D.S., Alivisatos, A.P., and Weiss, S., Ultrahigh-resolution multicolor colocalization of single fluorescent probes, *Proc. Natl. Acad. Sci. U.S.A.*, 97, 9461, 2000.
82. König, K., Multiphoton microscopy in life sciences, *J. Microsc.*, 200, 83, 2000.
83. Diaspro, A. and Robello, M., Two-photon excitation of fluorescence for three-dimensional optical imaging of biological structures, *J. Photochem. Photobiol. B*, 55, 1, 2000.
84. Patterson, G.H. and Piston, D.W., Photobleaching in two-photon excitation microscopy, *Biophys. J.*, 78, 2159, 2000.
85. Enderly, Th., Ha, T., Ogletree, D.F., Chemla, D.S., Magowan, C., and Weiss, S., Membrane specific mapping and colocalization of malarial and host skeletal proteins in the *Plasmodium falciparum* infected erythrocyte by dual color near-field scanning optical microscopy, *Proc. Natl. Acad. Sci. U.S.A.*, 94, 520, 1997.
86. Korchev, Y.E., Raval, M., Lab, M.J., Gorelik, J., Edwards, C.R.W., Raymend, T., and Klenerman, D., Hybrid scanning ion conductance and scanning near-field optical spectroscopy for the study of living cells, *Biophys. J.*, 78, 2675, 2000.
87. Doyle, R.T., Szulzcewski, M.J., and Haydon, P.G., Extraction of near-field fluorescence from composite signals provide high resolution images of glial cells, *Biophys. J.*, 80, 2477, 2001.
88. De Giorgi, F., Lartigue, L., and Ichas, F., Electrical coupling and plasticity of the mitochondrial network, *Cell Calcium*, 28, 365, 2000.
89. Margineantu, D., Capaldi, R.A., and Marcus, A.H., Dynamics of the mitochondrial reticulum in live cells using Fourier imaging correlation spectroscopy and digital video microscopy, *Biophys. J.*, 79, 1833, 2000.
90. Toescu, E.C. and Verkhratsky, A., Assessment of mitochodrial polarization status in living cells based on analysis of the spatial heterogeneity of rhodamine 123 fluorescence staining, *Eur. J. Physiol.*, 440, 941, 2000.
91. Savtchenko, L.P., Gagan, P., Korogod, S.M., and Tye-Dumont, S., Imaging stochastic spatial variability in active channel clusters during excitation of single neurons, *Neurosci. Res.*, 39, 431, 2001.
92. Stauffer, T.P., Ahn, S., and Meyer, T., Receptor-induced transient reduction in plasma membrane PtdIns(4,5)P2 concentration monitored in living cells, *Curr. Biol.*, 8, 343, 1998.
93. Varnai, P. and Balla, T., Visualization of phosphoinositides that bind pleckstrin homology domains: calcium- and agonist-induced dynamic changes and relationship to myo-[$^3$H]inositol-labeled phosphoinositide pools, *J. Cell Biol.*, 143, 501, 1998.
94. Arbuzova, A., Yartushova, K., Tangyas-Mihalyne, B., Morris, A.J., Ozaki, S., Prestwich, G.D., and McLaughlin, S., Fluorescently labeled neomycin as a probe of phosphatdylinositol-4,4-bisphosphate in membranes, *Biochim. Biophys. Acta Biomembranes*, 1464, 35, 2000.
95. Berrie, C.P. and Falasca, M., Patterns within protein/polyphosphoinositide interactions provide specific targets for therapeutic intervention, *FASEB J.*, 14, 2618, 2000.
96. Gillooly, D.J., Morrow, I.C., Lindsay, M., Gould, R., Bryant, N.J., Gaullier, J.M., Parton, R.G., and Stenmark, H., Localization of phosphatidylinositol 3-phosphate in yeast and mammalian cells, *EMBO J.*, 19, 4577, 2000.
97. Oziki, S., De Wald, D.B., Shope, J.C., Chen, J., and Prestwich, G.D., Intracellular delivery of phosphoinositides and inositol phosphates using polyamine carriers, *Proc. Natl. Acad. Sci. U.S.A.*, 97, 11286, 2000.

98. Puri, V., Watanabe, R., Singh, Ro.D., Dominguez, M., Brown, J.D., Wheatley, C.L., Marles, D.L., and Pagano, R.E., Chlathrin-dependent and -independent internalization of plasma membrane sphingolipids initiates two Golgi targeting pathways, *J. Cell Biol.*, 154, 535, 2001.
99. Prattes, S. et al., Intracellular distribution and mobilization of unesterified cholesterol in adipocytes: trigliceride droplets are surrounded by cholesterol-rich ER-like surface layer structures, *J. Cell Sci.*, 113, 2977, 2000.
100. Nagai, Y., Miyazaki, M., Aoki, R., Zama, T., Inouye, S., Hirose, K., Iino, M., and Hagiwara, M., A fluorescent indicator for visualizing cAMP-induced phosphorylation in vivo, *Nat. Biothechnol.*, 18, 313, 2000.
101. Honda, A., Adams, S.R., Sawyer, C.L., Lev-Ram, V., Tsien, R.Y., and Dostmann, W.R.G., Spatiotemporal dynamics of guanosine 3′,5′-cyclic monophosphate revealed by a genetically encoded, fluorescent indicator, *Proc. Natl. Acad. Sci. U.S.A.*, 98, 2437, 2001.
102. Jayaraman, S., Haggie, P., Wachter, R.M., Remington, S.J., and Verkman, A.S., Mechanism and cellular applications of a green fluorescent protein-based halide sensor, *J. Biol. Chem.*, 275, 6047, 2000.
103. Chandy, G., Grave, M., Moore, H.-P.H., and Machen, T.E., Proton leak and CFTR in regulation of Golgi pH in respiratory epithelial cells, *Am. J. Physiol.*, 281, C908, 2001.
104. Blackmore, C.G., Varro, A., Dirnaline, R., Bishop, L., Gallacher, D.V., and Dockray, G.J., Measurement of secretory vesicle pH reveals intravesicular alkalinization by vesicular monoamine transporter type 2 resulting in inhibition of prohormone cleavage, *J. Physiol. London*, 531, 605, 2001.
105. Miyawaki, A., Griesbeck, O., Heim, R., and Tsien, R.Y., Dynamic and quantitative $Ca^{2+}$ measurements using improved cameleons, *Proc. Natl. Acad. Sci. U.S.A.*, 96, 2135, 1999.
106. Yu, R. and Hinkle, P.M., Rapid turnover of calcium in the endoplasmic reticulum during signaling, *J. Biol. Chem.*, 275, 23648, 2000.
107. Griesbeck, O., Baird, G.S., Campbell, R.E., Zacharias, D.A., and Tsien, R.Y., Reducing the environmental sensitivity of yellow fluorescent protein, *J. Biol. Chem.*, 276, 29188, 2001.
108. Nagai, T., Sawano, A., Park, E.S., and Miyawaki, A., Circularly permuted green fluorescent proteins engineered to sense $Ca^{2+}$, *Proc. Natl. Acad. Sci. U.S.A.*, 98, 3197, 2001.
109. Carrington, W.A., Lynch, R.M., Moore, E.D.W., Isenberg, G., Fogarty, K.E., and Fredric, F.S., Superresolution three-dimensional images of fluorescence in cells with minimal light exposure, *Science*, 268, 1483, 1995.
110. Schmidt, T., Schutz, G.J., Baumgartner, W., Gruber, H.J., and Schindler, H., Imaging of single molecule diffusion, *Proc. Natl. Acad. Sci. U.S.A.*, 93, 2926, 1996.
111. Schütz, G.J., Schindler, H., and Schmidt, T., Single-molecule microscopy on model membranes reveals anomalous diffusion, *Biophys. J.*, 73, 1073, 1997.
112. Dickson, R.M., Norris, D.J., Tzeng, Y.L., and Moerner, W.E., Three-dimensional imaging of single molecules solvated in pores of poly(acrylamide) gels, *Science*, 274, 966, 1996.
113. Vickery, S.A. and Dunn, R.C., Scanning near-field fluorescence resonance energy transfer microscopy, *Biophys. J.*, 76, 1812, 1999.
114. Hell, S.W. and Stelzer, E.H.K., Fundamental improvement of resolution with a 4Pi-confocal fluorescence microscope using two-photon excitation, *Opt. Commun.*, 93, 277, 1992.
115. König, K., Two-photon near-infrared excitation in living cells, *J. Near Infrared Spectrosc.*, 5, 27, 1997.
116. Klar, T.A., Jakobs, S., Dyba, M., Egner, A., and Hell, S.W., Fluorescence microscopy with diffraction resolution barrier broken by stimulated emission, *Proc. Natl. Acad. Sci. U.S.A.*, 97, 8206, 2000.
117. Sailer, R., Strauss, W.S.L., Emmert, H., Stock, K., Steiner, R., and Schneckenburger, H., Plasma membrane associated location of sulfonated *meso*-tetra-phenylporphyrins of different hydrophilicity probed by total internal reflection fluorescence spectroscopy, *Photochem. Photobiol.*, 71, 460, 2000.
118. Thoumine, O., Kocian, P., Kottelat, A., and Meister, J.J., Short-term binding of fibroblasts to fibronectin: optical tweezers experiments and probabilistic analysis, *Eur. Biophys. J.*, 29, 398, 2000.

119. Helfer, E., Harlepp, S., Bourdieu, L., Robert, J., MacKintosh, F.C., and Chatenay, D., Microrheology of biopolymer-membrane complexes, *Phys. Rev. Lett.*, 85, 457, 2000.
120. König, K., Robert Feulgen Prize Lecture: laser tweezers and multiphoton microscopes in life sciences, *Histochem. Cell Biol.*, 114, 79, 2000.
121. Vo-Dinh, T., Alarie, J.P., Cullum, B., and Griffin, G.D., Antibody-based nanoprobe for measurements in a single cell, *Nat. Biotechnol.*, 18, 76, 2000.
122. Cullum, B.G.D., Griffin, G.H., Miller, and Vo-Dinh, T., Intracellular measurements in mammary carcinoma cells using fiberoptic nanosensors, *Anal. Biochem.*, 277, 25, 2000.
123. Vo-Dinh, T., Griffin, G.D., Alarie, J.P., Cullum, B., Sumpter, B., and Noid, D., Development of nanosensors and bioprobes, *J. Nanopart. Res.*, 2, 17, 2000.
124. Cullum, B. and Vo-Dinh, T., Development of optical nanosensors for biological measurements, *Trends Biotechnol.*, 18, 388, 2000.
125. Vo-Dinh, T., Cullum, B.M., and Stokes, D.L., Nanosensors and biochips: frontiers in biomolecular diagnostics, *Sensors Actuators*, B74, 2, 2001.
126. Ideker, T., Thorsson, V., Ranish, J.A., Christmas, R., Buhler, J., Eng, J.K., Bumgarner, R., Goodlett, D.R., Aeversold, R., and Hood, L., Integrated genomic and proteomic analyses of a systematically perturbed metabolic network, *Science*, 292, 929, 2001.

# 55

# Amplification Techniques for Optical Detection

Guy D. Griffin
*Oak Ridge National Laboratory*
*Oak Ridge, Tennessee*

M. Wendy Williams
*Oak Ridge National Laboratory*
*Oak Ridge, Tennessee*

Dimitra N. Stratis-Cullum*
*Oak Ridge National Laboratory*
*Oak Ridge, Tennessee*

Tuan Vo-Dinh
*Oak Ridge National Laboratory*
*Oak Ridge, Tennessee*



## 55.1 Introduction

This chapter presents a summary account of various amplification techniques for analysis of biological molecules (species) that use some form of optical detection as the endpoint. The analyte species possibly under consideration include lipids, carbohydrates, proteins, and nucleic acids (and combinations of the same such as glycolipids, glycoproteins, etc.), but this review is restricted to proteins and nucleic acids. Of course, there are many detection strategies for both proteins and nucleic acids, but for the purposes of this review, amplification must be involved. We therefore have limited our discussion to immunoassays and nucleic acid techniques employing amplification.

Immunoassays, which depend fundamentally on antibody recognition of antigens, have proved, over the course of many years, to be powerful analytical techniques for protein detection. They can also be used successfully for the detection of substances of lower molecular weight than the usual size of proteins (e.g., peptides, hormones, etc.), provided that antibodies specific against these molecular structures are developed and specially selected.

Detection of nucleic acids has been revolutionized by the development of the polymerase chain reaction (PCR) and other related techniques that produce large multiplications of specific nucleic acid segments. Even before the advent of these powerful amplification techniques, however, sensitive detection of nucleic acids was achieved using various procedures, some of which relied on the principles of immunoassays.[1] Most nucleic acid detection strategies are based fundamentally on the base-pairing interaction that occurs between complementary strands of polynucleotides (even PCR, in which primers must recognize specific complementary sequences); thus, such complementary base pairing (i.e., hybridization) is a significant component of nucleic acid detection assays.

Finally, some detection strategies seem to wed elements of immunoassays and nucleic acid detection assays (e.g., immuno-PCR; PCR-enzyme-linked immunosorbent assay [ELISA]). In this review, we group such techniques into a third category, referred to as hybrid assays. These admittedly could be classified

*Current affiliation: U.S. Army Research Laboratory, Adelphi, Maryland

as either immunoassays or nucleic acid amplification assays, but, in the hope that it will add clarity, they will be categorized separately.

This review makes no claim to comprehensiveness with regard to every possible assay technique that has been applied to a specific situation. Indeed, given the number of researchers in these fields and the time periods over which these techniques have been developed, it would be an enormous undertaking to be truly comprehensive in such a review. An attempt has been made to cover the more common and, in the authors' perception, more useful analytical techniques, with the further caveat that we have focused almost exclusively on the English language literature. The references given are not meant to be exhaustive with regard to any of the techniques discussed.

In addition, this chapter does not cover radiotracer techniques. In many cases, these procedures formed the basis for great advances in biological knowledge and still demonstrate remarkable sensitivity. However, safety issues and issues of waste disposal have somewhat diminished the attractiveness of radioisotopes for biological applications, hence the increasing reliance on other techniques including a heavy emphasis on optical-based analytical tools. Our goal is to provide a comprehensible introduction to various amplification techniques for optical detection that will allow for further investigation of those techniques that pique the reader's interest. If we have omitted significant analytical techniques, we apologize in advance to those dedicated researchers who have advanced the frontiers of science.

## 55.2 Immunoassays

Immunoassays make use of antibody/antigen recognition, a recognition that has great specificity and the potential for high affinity. Protein detection has been a major focus of immunoassay development, although low-molecular-weight substances (drugs, toxins, hormones, etc.) can also be conveniently detected. In the case of haptens of low molecular weight, these molecules can be conjugated to proteinaceous (or other) materials to form antigenic substances for which hapten-specific antibodies may be developed. It is possible to develop antibodies against specific epitopes (antigenic sites) on high-molecular-weight proteins and against high-molecular-weight carbohydrates. However, the adaptation of immunoassays to carbohydrates and lipids is considerably less rich in associated assay development than the protein/hapten systems, which have received the bulk of the efforts at innovation. Consequently, this review focuses essentially on protein detection by immunoassay. The interaction of biotin and avidin/streptavidin is a rather unique case of molecular recognition that has been extensively used in the development of detection schemes throughout biology. The relatively enormous affinity constant ($K_D = 10^{-15}$ $M^{-1}$) (see Reference 2) of biotin for avidin/streptavidin forms the basis of a recognition reaction of great specificity, and the small size and molecular structure of biotin allow for ready conjugation to a wide variety of molecules. Not surprisingly, this system has been exploited in the development of a wide variety of analytical techniques, which appear in this review.

Immunoassays have been performed in a wide variety of formats. The 96-well microplate-based assays are routinely performed throughout the world but are by no means the only feasible format. Immunoblotting, applied as detection for a wide variety of gel-based separation methods, is also widely used. Similarly, a rich literature has grown up regarding immunoassay techniques applied to histological/cytological specimens. Whether the samples are tissue sections, membrane blots, proteins in physiological fluids, or samples in microplate wells, the immunoassay techniques used for detection show similar basic principles. Thus, we do not, in general, divide the assay strategies according to their application (i.e., histological, immunoblot, etc.) but simply mention that some procedures fit certain applications well. In certain cases, we note an interesting and/or unusual format (such as the use of an unusual solid substrate for assay immobilization), but this is not intended to suggest an exhaustive review of all assay formats. After consideration of the various assay principles used, the reader will likely formulate original modifications suitable for specific applications.

Table 55.1 provides a summary of our review of immunoassays. Immunoassays are used routinely in many areas of biological science including basic research, pharmaceutical development, and clinical and forensic applications. Although the term "immunoassay" encompasses a wide variety of quite different techniques, this review will focus only on techniques involving amplification, many of which fall into the category of enzyme-amplified immunoassays (EIA). This category is in and of itself daunting in the number of variants that have been developed. Thus, an attempt is made here to discuss general principles that govern development of immunoassays of this class, examples of general assay formats, and the suitability of various techniques.

## 53.2.1 Classification of Immunoassays

Immunoassays can basically be classified as heterogeneous or homogeneous, competitive or noncompetitive, and direct or indirect.[3,4] Of course these terms may be used in various combinations (e.g., heterogeneous indirect noncompetitive immunoassay). Although the nomenclature for immunoassays can be quite variable and shortened forms of classification are very commonly used, definitions of the basic classifications are appended below.

Heterogeneous vs. homogeneous refers to whether a separation step is required in the immunoassay. In other words, heterogeneous assays require one or more separation steps, while homogeneous assays occur in solution phase without separation. Heterogeneous assays are commonly associated with a solid phase such as a microtiter plate. Immunoassays can be formatted as either competitive or noncompetitive. In the competitive mode, two pools of antigen, one pool labeled with a marker of some sort and the other consisting of the sample to be analyzed, compete for a limited number of antibody-binding sites. Competitive assays of this sort are, in general, homogeneous assays. In a noncompetitive assay, immobilized antibody binds (captures) antigen, and antigen is detected by a second antibody (detector) that has some label and binds to another epitope on the antigen. Perforce, these types of assays are heterogeneous. However, heterogeneous assays may also be formatted so that the immobilized antibodies capture a mixture of labeled and unlabeled antigen from two pools, i.e., a competitive heterogeneous assay. Alternatively, a limited amount of antigen may be immobilized and limited amounts of labeled antibody added to a mixture of sample antigen in solution. This format is also competitive, as increased antigen in solution produces fewer labeled antibodies binding to the immobilized antigen.

Direct immunoassays are assays in which only primary antibodies, with specificity against the antigen of interest, are used. The antibodies here may be labeled or unlabeled. Indirect assays, on the other hand, involve in some way an immune or other type of complex. For these types of assays, the final complex signals, through some indirect means, the formation of an immunospecific binding event. Examples include the interaction of an antispecies IgG antibody, with attached label, with the primary antigen-specific antibody; or the use of a labeled avidin/streptavidin to bind to a biotinylated antigen-specific antibody. Table 55.1 describes many of the immunoassays as homogeneous/heterogeneous, competitive/noncompetitive, etc. However, every immunoassay procedure is not labeled in this way because many assays have the potential to be applied in multiple fashions.

## 55.2.2 Amplification and Specificity

Table 55.1 presents a summary of various immunoassay techniques that employ amplification as a component of the assay. Amplification has multiple layers of meaning with respect to immunoassays. Fundamentally, in these assays, the desired goal is to detect the event of antibody-to-antigen binding. If the antibody has an attached enzyme, then the unique immunoreaction event is amplified by the processivity of the enzyme acting on the appropriate substrate. On the other hand, when immune complexes (e.g., PAP = peroxidase-antiperoxidase, APAAP = alkaline phosphatase-antialkaline phosphatase) are used, the number of enzyme molecules clustered at the site of a single antibody/antigen interaction can be multiplied.[5,6] These are just two of many possible examples; many immunoassays have more than one multiplication strategy. As a final goal, all these amplifications serve to amplify the detection signal, hopefully lifting it above the noise of the system. Table 55.1 presents a distillation of a large compendium of techniques. For each technique, it presents a brief summary of the principle behind the assay, some comments considered germane to the discussion (including further explanation of amplification and comments about sensitivity), and pertinent references.

Of course, the fundamental question arises of why amplification is needed at all. The simplest answer is that the molecular interaction that one wishes to detect is undetectable without amplification. Of course, as the various immunoassay techniques developed over the years, the natural tendency was to push for greater and greater sensitivity, as the biological questions being asked demanded more powerful techniques. Hence a wide variety of amplification techniques have been developed, some of which admittedly have less applicability than others.

Some of the amplification techniques included in Table 55.1 seem to have minimal amplification associated with them, such as fluoroimmunoassay (FIA), where a fluorescent label is covalently attached to an antibody.[7] In an attempt to be consistent, any technique that results in some amplification of a signal could be legitimately called amplification and will therefore be included. For example, in the simple direct FIA, where fluorophore is attached to antibody, if only one fluorophore is attached per antibody molecule, then the assay would not be considered amplified, since single-molecule antigen/antibody binding would only be detected by one fluorophore. In fact, there are multiple potential sites for fluorophore attachment per antibody molecule, and the usual fluorophore:antibody molar ratio following the labeling reaction is 2:1 to 4:1. Actually, controlling the reaction to insert only one fluorophore would be difficult, if not impossible. Nevertheless, the degree of amplification achieved by fluorophore labeling of antibody is certainly limited to relatively small numbers. Too many fluorophores attached can significantly alter the affinity/specificity of the antibody. In fact, the inherent extreme sensitivity of fluorescence is the major reason that simple FIA has become such a widespread technique. Even with only a few molecules of fluorescent label attached per antibody molecule, the antigen/antibody binding event is still amenable to detection.

In general, in Table 55.1, techniques that give high levels of amplification, as opposed to those that produce only small amounts of amplification, are emphasized. However, the extent of amplification is not necessarily synonymous with sensitivity. For instance, with respect to fluorescence, fluorescence polarization or time-resolved measurements can be highly sensitive detection schemes, even though only a few molecules of dye are attached to an antibody. In spite of the above comments, however, the more powerful amplification schemes, where the potential for manifold multiplication of signal exists, are inherently attractive, if for no other reason than the fact that they offer positive detection even under nonoptimum conditions of assay, instrumentation performance, etc.

The gold standard for amplification in immunoassay is probably the ELISA technique, which is used for protein detection as well as for numerous other biological molecules, notably nucleic acids. Here, the amplification occurs as the enzyme converts substrate to product, a process that generally involves many thousands of molecules per minute.[6,8] This enzymatic process is potentially so powerful that, given the caveat of no denaturization of the enzyme protein and given enough time, it should be possible to detect *one* enzyme molecule bound to an immunoreagent. Seldom if ever is this most stringent detection limit necessary. Colorimetric detection of the ELISA reaction is a commonly used endpoint, and instrumentation for ELISA assays and detection based on this endpoint is readily available. Of course, if enzymatic conversion of a substrate to a colored product (either soluble or insoluble, depending on the endpoint of the amplification) is possible, the logical extension is to apply ELISA amplification to substrates that produce fluorescent or chemiluminescent products. This has been done — and sometimes with remarkable success in terms of sensitivity enhancement (see following discussion).

Simple ELISA, however, does not exhaust the ingenuity of amplification schemes, which have evolved. Numerous coupled-enzyme schemes can enhance enzymatic multiplication. Immune reaction coupled to lysis of liposomes produces release of large numbers of liposomally encapsulated indicator molecules. *In vitro* transcription/translation systems can be used to make multiple copies of a target DNA gene sequence attached to an immunoreagent. In some cases, such as surface-enhanced Raman spectroscopy, the optical detection method itself produces a large part of the signal amplification. Table 55.1 shows a variety of amplification schemes.

At its most fundamental level, the specificity inherent in immunoassays resides in the antibody/antigen interaction.[9] No amount of assay ingenuity can improve an antibody reagent with poor specificity. Limit of detection (LOD) has to do with the specificity and affinity of the antibodies for the analyte and the robustness of the amplification system, as well as with the amount of nonspecific interactions that take place in the assay system. As the LOD is pushed lower, at some point the "noise" due to nonspecific interactions becomes a significant limiting factor. Sensitivity, as opposed to limit of detection (which is the lower limit of analyte that can be detected above background "noise"), refers to the ability of the assay system to discriminate between analyte concentrations that do not vary much from one another. A nonsensitive assay system may have essentially the same LOD as a sensitive assay system, but the sensitive assay system can discriminate among analyte amounts, which the nonsensitive assay could not.

Some simple mathematical considerations demonstrate that the antibodies used in immunoassays constitute the limiting factor in developing highly sensitive, rapid, and powerful immunoassays. A brief demonstration taken from Avrameas[10] illustrates this fact. The fundamental mass-action equation governing antigen/antibody intereactions is:

$$Ab + Ag \underset{K_2}{\overset{K_1}{\Leftrightarrow}} Ab\text{–}Ag \tag{55.1}$$

where $Ab$ = antibody, $Ag$ = antigen, $Ab$–$Ag$ = antigen – antibody complex, and $K_1$ and $K_2$ = rate of forward and reverse reactions, respectively. The association constant ($K_a$) at equilibrium is defined as:

$$K_a = K_1/K_2 = [Ab - Ag]/([Ab]\ [Ag]) \tag{55.2}$$

Generally, $K_1$ is larger than $K_2$, so the [$Ab$–$Ag$] complex, once formed, only dissociates slowly (thus allowing the numerous washings common in EIA procedures). Avrameas cites several examples to illustrate the analytical limitations posed by this equilibrium.[10] These examples serve to indicate that EIAs, like all other assays, have theoretical limitations beyond which they cannot go. Once the inherent limitations are understood, however, EIAs have the potential to detect analytes at extremely low concentrations and with notable precision and accuracy.

### 55.2.3 Comparisons of Immunoassays

Which immunoassays are the most sensitive with the lowest limits of detection? There are no simple answers to this question. Just as there are no simple answers to the question of which is the best immunoassay. The various immunoassays enumerated in Table 55.1 have apparently all been successfully implemented. The choice of a particular immunoassay must be defined by the individual researcher for a particular application. For example, sometimes the speed of analysis may be more important than the extreme sensitivity, or a very low limit of detection may be more valuable than the amount of time it takes to perform the analysis. The following discussion provides only some general comments about comparisons of various techniques; the ultimate choice rests with the reader.

Practically speaking, it is essentially impossible to make meaningful comparisons between one investigator's results and those of another. As mentioned above, a major factor in determining sensitivity is the affinity of the antibodies used for the analyte, and very seldom (if ever) do different investigators use exactly the same antibodies. However, a variety of experiments are described in the literature where investigators compare various immunoassay protocols using the same fundamental antibody reagents. Such studies provide a basis for more meaningful comparisons, although there are still many caveats including such questions as whether the various detection labels attached to the antibodies affect the affinity of the antibody to different extents and whether the composition of the immunoreagent "sandwich," if it becomes highly complex, produces steric hindrance to free access of, say, enzyme substrate.

Comparisons imply some standard to which other quantities are compared. Unfortunately, no universal standard exists for immunoassays. As indicated in Table 55.1, very often comparisons are made with ELISA-based colorimetric assays. In addition to direct comparisons using the same or very similar immunoreagents, one finds more nebulous statements in the literature such as that one technique is fourfold more sensitive than another.

One might argue that a comparison of the LODs observed for various immunoassays (of which there are numerous citations in Table 55.1) should provide a reasonable indication regarding which assays are the most sensitive. However, in addition to the fact that antibody reagents vary from investigator to investigator, the assay protocol itself, particularly where an enzyme step (or steps) is included, can significantly impact limits of detection. For example, usual times of incubation of enzyme-labeled immunoreagent with substrate are 30 min to 1 h. It is almost guaranteed that an overnight incubation of enzyme plus substrate would produce a lower LOD (if the enzyme did not lose activity completely).

Finally, and perhaps most perplexing, is the fact that there can be very large ranges over which sensitivity of one immunoassay is said to be increased compared to another immunoassay (for example, see under fluorescence EIA in Table 55.1, where the sensitivity is said to be 5- to $10^5$-fold increased over conventional EIA). Recall that immunoassays are, by their very nature, complex, and a variety of instrumentation can be used to detect the final endpoint. As more sensitive photon-counting detectors (e.g., charge-coupled device [CCD], avalanche photodiode [APD]) are brought into use, and as more intense excitation sources (lasers) are employed, it should be expected that LODs will continue to decrease. Whether single-molecule detection is needed (or even desirable) is another question altogether.

As mentioned previously, in terms of frequency of use, ELISAs are the gold standard of immunoassays. ELISA formats and techniques are well established, there are choices of multiple substrates, and much experience has been accumulated regarding effective coupling of enzyme to antibody. A good discussion of general ELISA principles and techniques can be found in work published by Tijssen in 1985.[6] During much of the 1970s there was considerable debate as to whether enzyme labels in immunoassays would ever be as useful as radioactive labels in terms of sensitivity of detection.[11] Estimates of detection limits of 5 to 10 attomol for radioactive labels, e.g., $^{125}$I, have been made based on the specific radioactivity of carrier-free $^{125}$I of 4.8 dpm/attomol.[12] Aside from issues of safety, radioactive-waste disposal, and limited half-life, Ishikawa et al.[12] point out that the detection limit of some enzymes is $< 1$ attomol, and using other techniques still based on ELISA (i.e., immune complex transfer immunoassay) the detection limit has been pushed to milliattomoles.[11]

However, Ishikawa et al.[11,12] indicate that to achieve attomole sensitivity by ELISA certain formats and procedures must be used. These include use of a noncompetitive "sandwich" (two-antibody) ELISA format (competitive immunoassay with only one antibody is reportedly unable to achieve attomole sensitivity); the use of thiol groups in the hinge portion of the antibody Fab′ fragment for enzyme coupling, providing better retention of the activities of both antibody and enzyme; and the use of a fluormetric substrate for the enzyme β-galactosidase, which change was said to improve the sensitivity 1000-fold over colorimetric substrates. As for this latter point, Ishikawa et al.[12] state that it is possible to measure 0.02 attomol of β-galactosidase and 0.5 attomol of HRP in a 100-min fluorimetric assay with an assay volume of 0.15 ml.

Shalev et al.[13] also confirm that detection by fluorimetric ELISA could be demonstrated to be more sensitive than radioimmunoassay, and in fact detection levels of 3 to 10 attogram/ml, or 24,000 molecules, could be achieved. The authors discuss essentially three factors that enabled them to increase the total assay sensitivity $\sim 10^5$ to $10^6$ times over a colorimetric ELISA. These factors were use of a fluorogenic substrate, prolongation of incubation time (15 h), and increase in surface area available for antigen binding. Interestingly, while Shalev et al.[13] state that it is generally accepted that fluorometric methods are $10^3$ more sensitive than colorimetric methods, using a fluorogenic substrate instead of a colorimetric one enhanced sensitivity in their ELISA assay by only 16 to 30 times. Ruan et al.[14] also note a $10^2$- to $10^3$-fold increase in sensitivity using a fluorogenic substrate vs. a colorimetric substrate in an ELISA assay for prostaglandin H synthetase.

Chemiluminescent ELISAs are also a useful route to increased assay sensitivity. For example, Brown et al.[15] found that using an enzyme-cascade system and a chemiluminescent endpoint, alkaline phosphatase was detectable at concentrations down to 0.4 attomol. Lewkowich et al.[16] found 12- to 29-fold increases in sensitivity in ELISA assays for interleukins when a chemiluminescent substrate was substituted for a colorimetric substrate. Porakishvili et al.[17] observed increases of 10- to 100-fold in sensitivity when comparing chemiluminescent vs. colorimetric ELISAs. Direct comparison of radiolabel with chemiluminescent label indicates that, on an attomole basis, chemiluminescent labels provide a better possibility for detection (i.e., 1 attomol of carrier-free $^{125}I$ = 5 dpm, while 1 attomol of chemiluminescent label = 50 photon counts).[18]

Other ELISA-based immunoassay methods indicated in Table 55.1 also provide the opportunity for great sensitivity. The cloned enzyme-donor immunoassay, for example, is stated to be the most sensitive of homogeneous enzyme immunoassays.[19] Multienzyme cascade-based assays have demonstrated subattomole (e.g., 0.01 attomol for alkaline phosphatase) sensitivity, even when using a colorimetric substrate.[18] The use of enzyme-mimetic catalysts is possibly a significant step forward, as these small (relative to enzymes) molecules offer higher labeling ratios for antibodies without adversely affecting their affinity and still offer catalytic activity comparable to actual enzymes.[20] The catalyzed reporter deposition immunoassays also offer extremely high sensitivity, with the added potential advantage for certain applications of localization of signal.[21] Finally, hybrid technologies combining elements of immunoassay with DNA-amplification technologies seem to offer the greatest sensitivity of all and approach that "Holy Grail" of detection limits — one molecule. These are discussed later in this chapter.

### 55.2.4 Other Techniques

Here we would like to mention some other techniques that involve some sort of amplification but that, in our opinion, do not fit with other optical detection techniques. This is not to imply that these techniques have no place in immunoassay procedures or that they are somehow inferior to other techniques (the immunogold–silver stain procedure seems exquisitely sensitive in comparison to other histochemical procedures). These techniques are either designed for histological use or in some other way do not lend themselves to rigorous quantitative analysis by optical detection such as with photomultiplier tubes (PMTs), CCDs, spectrometers, etc.

One such technique covers a whole series of assays involving agglutination reactions. Some of these procedures have been used for many decades.[22] Amplification in these assays arises from multiple antigen or antibody molecules immobilized on red blood cells, latex spheres, etc. Reaction with antisera results in agglutination of the immuno-coated particles, which is visually observed. A recent application to substances-of-abuse screening using immunoreagent-coated colored latex spheres to distinguish the various metabolic products in urine by agglutination immunoassay has been reported.[23] Such assays can be very powerful screening tools but are usually read visually. The extent of quantification therefore depends on the judgment and visual acuity of the observer as well other factors such as serial dilutions of antisera used in the assay, extent of antigen labeling of the particles involved in agglutination, etc.

Another intriguing technique, originally developed for electron microscopy and subsequently applied to light microscopy of histological specimens, is the immunogold technique, which has been modified to an ultrasensitive procedure by employing a subsequent silver-precipitation step.[24] In this procedure, gold nanoparticles are used to label an antibody, and this labeled antibody then binds to tissue antigens (as part of a multiantibody "sandwich"). Then a silver precipitate is formed around the gold particles, and since multiple silver atoms precipitate around each gold particle, a large amplification results. This immunogold-silver staining procedure was found to be so sensitive (~200-fold more sensitive than immunoperoxidase staining method[24]) that it was felt necessary to at least bring it to the reader's attention. Several not-so-attractive elements of this technique are the long (overnight) incubation with the immunogold reagent and the requirement for development of the silver stain in the dark. This procedure has been adapted for membrane staining, and we have included an entry toward the end of Table 55.1, as densitometry could be used to quantify the results of the silver stain.

## 55.3 Nucleic Acid Analyses

Although technologies for amplifying and detecting nucleic acids do not have as long a history as immunoassay techniques, they have nonetheless developed in the past few decades into an impressive armamentarium for biological investigations. Many of the caveats that have been mentioned in regard to immunoassay techniques can also be applied to nucleic acid amplification technologies. As mentioned in the immunoassay section, nucleic acid amplification technologies that use radioisotopes for detection have been omitted because they do not strictly fall under the field of optical detection. We recognize that many of the key technologies for nucleic acid detection, such as PCR, reverse-transcription PCR, membrane blotting/hybridization, *in situ* hybridization, etc., as originally described, used radioisotopic detection. Often, in the earlier literature, autoradiographic methods were used to visualize the nucleic acids being detected; thus, strictly speaking, optical detection was used. Radioisotope techniques using such labels as $^{32}P$ have also been found to be very sensitive. Nevertheless, to remain consistent with the immunoassay section and to showcase such techniques as fluorescence and chemiluminescence, we have chosen to omit radioisotopic detection.

As in the immunoassay section, we have had to constrain ourselves somewhat with regard to the definition of amplification. As a number of reviews point out,[25–27] amplification with respect to nucleic acids can refer to target amplification, probe amplification, and/or signal amplification. Thus, PCR, the extremely fertile technology devised in the mid-1980s, amplifies the target DNA (or RNA). Branched-DNA amplification techniques, on the other hand, do not amplify the nucleic acid target but rather amplify the signal of a hybridization event, allowing it to be detected. Probe amplification strategies involve techniques such as ligase chain reaction (LCR). All these techniques, and many others, are explained and referenced in Tables 55.2 and 55.3. In some of these technologies, amplification is more obvious than in others (such as when $10^6$ copies of the target nucleic acid are made and detected, as in PCR). In *in situ* hybridization techniques, on the other hand, where a probe must hybridize to a target sequence on a chromosome, amplification is sometimes only achieved by the presence of multiple detection labels on the probe. Of course, it is experimentally possible to use only one label on the probe, followed by single-molecule detection, but this is very much the exception rather than the rule. The table below, taken from Schweitzer and Kingsmore, summarizes various mainstream technologies discussed in this section on nucleic acid amplification.[26] This is a very useful summary in that it provides in an easy-to-read format an idea of what kind of amplification is involved, the sensitivity, useful dynamic range, and other aspects of the assay systems. Not included in this table are various hybridization technologies that, strictly speaking, may not involve target amplification but that do involve some form of signal amplification to detect the hybridization event.

Properties of Various Nucleic Acid Amplification Technologies

| Property | PCR | LCR | SDA | NASBA | bDNA | Invader | RCA |
|---|---|---|---|---|---|---|---|
| DNA-target amplification | ✓ | ✓ | ✓ | × | × | × | ✓ |
| RNA-target amplification | ✓ | × | ✓ | ✓ | × | × | ✓ |
| DNA-signal amplification | × | × | × | × | ✓ | ✓ | ✓ |
| RNA-signal amplification | × | × | × | × | ✓ | ✓ | ✓ |
| Protein-signal amplification | × | × | × | × | × | × | ✓ |
| Multiplexing | Little | × | Little | Little | × | × | ✓ |
| Mesothermal | × | × | ✓ | ✓ | ✓ | ✓ | ✓ |
| Amplification with cells | ✓ | × | × | × | × | × | ✓ |
| Amplification on microarrays | × | × | ✓ | × | × | × | ✓ |
| Sensitivity (copies) | <10 | 100 | 500 | 100 | 500 | 600 | 1 |
| Range (logs) | 5 | 3 | 4 | 5 | 3 | 4 | 7 |
| Specificity (allele discrimination factor) | 50 | 5000 | 50 | 50 | 10 | 3000 | 100,000 |

Data from Schwitzer, B. and Kingmore, S., *Curr. Opinion Biotechnol.*, 12(1), 21, 2001.

The major focus of this section can be found in Tables 55.2 and 55.3, which provide brief summaries in the same format as Table 55.1 in the immunoassay section of various nucleic acid amplification techniques. Like Table 55.1, Tables 55.2 and 55.3 provide a brief description of the principles behind the assay, commentary, and appropriate references. Table 55.2 discusses PCR-based and non-PCR-based methods for nucleic acid amplification, signal amplification strategies like branched DNA, and various hybridization technologies, *in situ* as well as others. Table 55.3 covers procedures that we call hybrid technologies. Here we include techniques like immuno-PCR and PCR-ELISA, which are in some sense a melding of immunoassay techniques and nucleic acid amplification techniques. For the most part, Tables 55.2 and 55.3 are self-explanatory, and no attempt will be made in this review to systematically discuss the information therein. Only a few comments regarding specific issues follow.

## 55.3.1 Target Amplification

The critical feature of nucleic acids that has allowed the powerful amplification/detection technologies to develop is the complementarity of the nucleotide bases (i.e., base pairing) and the resultant ability to form double-stranded molecular hybrids. From these fundamental chemical properties flow all the varied and ingenious experimental systems that exploit these properties. Due to the base-pairing phenomenon, relatively short primers can be annealed to single strands of DNA, and the succeeding DNA sequence (on the 3′ side) can be replicated (in a complementary sense), giving rise to PCR. Of course, this property of nucleic acids, which has proved to be such a powerful analytical tool, also has the potential to introduce errors because mismatches in base pairing can occur and go undetected, giving rise to spurious amplicons, hybridizations, etc. Issues of specificity, fidelity, and efficiency of PCR have been discussed by various authors and the effects of various reaction conditions on the overall process explained (see, for example, Cha and Thilly[28]). This propensity for incorrect hybridization of probes and primers becomes an even greater problem when various detection labels (fluorophore, biotin, enzymes) appear as decorations on the probe/primer strands.[1] Indeed, one of the powerful advantages of using $^{32}P$ radioactively-labeled nucleic acids is the fact that incorporation of the label has minimal effect on overall hybridization of probe/primer to target.

Even if no overt mispairing occurs, as a result of these decorations, the net effect may be a decrease in thermodynamic stability, which would reduce the effectiveness of hybridization under conditions where some degree of stringency is needed. Therefore, considerable effort has been expended to develop analytical techniques to incorporate detector decorations into probes and primers without affecting (to any significant degree) the part of the probe/primer that must effect recognition of the target nucleic acid.

PCR is, without question, the premier target-amplification technique for nucleic acids, with applications that span the gamut of biological investigations. It has been applied to target nucleic acid amplification in microtubes as well as to tissue/cell sections; in the latter case it is known as *in situ* PCR. Very often when PCR is carried out in microtubes, the amplified product is analyzed by agarose-gel electrophoresis. Commonly ethidium bromide, an intercalating dye, is included in the gel, so that location of the amplified sequence can be visualized by ultraviolet-excited fluorescence. We have included this technique as part of the optical detection technologies because it is certainly optical in nature and comprises the analytical endpoint in many applications of PCR technology. On the other hand, this fluorescence-based analysis of gels is usually not done for quantitative purposes because the information desired from PCR is often in regard to the molecular size of the PCR product, not quantitative information regarding the amplification achieved by PCR.

As stated above, PCR products can be detected by direct staining in gel matrices. However, it is also common to use labeled probes (such as molecular beacons, "scorpions," or "Taq Man,")[26] that bind by hybridization to PCR amplicons and by this binding produce a detection signal (i.e., fluorescence). Another strategy is to use PCR primers with attached labels, so that as the amplification proceeds the amplified products are all labeled. In this type of assay, some separation of labeled amplicons from labeled but unincorporated primers must be carried out.

The enormous amplification power of PCR, coupled with the development of highly efficient instrumentation for the obligatory thermal cycling, has overshadowed all other target-amplification technologies. Table 55.2 also includes such techniques as LCR, strand-displacement amplification, nucleic acid sequence–based amplification, and rolling-circle amplification (RCA). Some of these techniques have real advantages over PCR, such as no requirement for thermal cycling, but none has yet succeeded in unseating PCR from its commanding position.

RCA, a relatively new technology, has the potential for either linear or exponential target amplification and has the additional advantage of isothermal amplification. In this procedure, a primer hybridizes to a circular DNA, and the circular DNA is amplified by polymerase extension. Up to $10^5$ copies of the DNA circle are generated on a concatemerized chain from one primer.[26] The amplification potential of this technique is obviously tremendous. RCA can also be used as a signal amplification technology. In this application, a defined circular DNA segment is amplified, and the concatemerized product is labeled either during the polymerase amplification or subsequently. Again, the larger degree of amplification produced by the rolling-circle technique results in enhanced ability to detect signal.

Of course, PCR and other target amplification techniques not only are useful for detecting DNA but can be applied to RNA as well, since RNA can be converted into a DNA copy using reverse transcriptase. In addition, PCR has been applied to *in situ* hybridization studies, where the detection of a nucleic acid sequence in a tissue sample proves a difficult analytical challenge.[29] In such a case, PCR can be used to make many copies of a primer that is complementary to the target strand of interest. This primer may itself be labeled, or labels may be incorporated during amplification. In the strictest sense, this is not target amplification but probe amplification; however, the net effect remains the same. Oddly, while one might expect diffusion of the amplicons away from the site of the PCR reaction, this does not occur to such an extent as to render the technique unusable.

## 55.3.2 Signal Amplification

Target amplification of nucleic acids is not the only way in which nucleic acid sequences may be detected. Signal amplification technologies can be used in which the signal from a hybridization event where probe binds to target is somehow amplified. Branched-DNA amplification, in which a dendrimer structure is formed upon a probe hybridizing to target and in which the branches of the dendrimer can carry multiple labels, is one example of such a signal amplification technique.[25,26] A powerful advantage of the branched-DNA technique is the fact that the amplification of the signal, rather than the target, permits direct quantification of the amount of target DNA being detected based on the signal. Table 55.2 also includes a variety of other techniques that involve signal amplification. For example, a variety of hybridization techniques involve binding of antibody molecules to haptenylated nucleic acid probes. In these instances, signal amplification can take place in a variety of ways. There can be multiple antibody molecules bound to each probe, each antibody can have multiple fluorophores (for example), or the antibodies can be conjugated with enzymes, thereby allowing the tremendous amplification potential of enzyme processivity to contribute to overall amplification.

For membrane hybridization detection, nucleic acid recognition is quite commonly coupled with enzyme-associated signal amplification to produce highly sensitive analyses. For example, biotinylated probes can be hybridized to nucleic acid targets on membranes, and subsequently avidin/streptavidin conjugated with an enzyme such as alkaline phosphatase can be added. The biotin/avidin linkage immobilizes the enzyme at the specific site on the membrane, and addition of substrate results in signal amplification due to enzyme activity. Either fluorogenic or chemiluminescent substrates may be used, and the resulting assays are quite sensitive.

Assays of this nature are a blend of immunoassay techniques and nucleic acid assay techniques and could thus readily be put into the category of "hybrid assays." Perforce, our division into separate categories is somewhat artificial but may serve a useful purpose in delineating key characteristics of various assay schemes. Another hybridization assay that bears a strong similarity to immunoassay techniques is the use of tyramide signal amplification for *in situ* hybridization studies.[25] This technique, also known as CARD (catalyzed reporter deposition), begins with the specific recognition event of a probe hybridizing to its target *in situ*. Either directly or indirectly, horseradish peroxidase is linked to the hybridization probe, and the hybridization event thus localizes the enzyme at the target site. Addition of a tyramide conjugate results in activation of the tyramide by peroxidase, depositing large numbers of tyramide molecules at the reaction site by binding to electron-rich targets near the site. If the tyramide is conjugated to a fluorophore, the net effect is to deposit many fluorophore molecules at the site of reaction. Further signal amplification can be achieved by using immunochemical techniques, such as adding antifluorophore antibodies conjugated with an enzyme (e.g., horseradish peroxidase), and thereby going through another round of tyramide deposition before detection occurs.

These are just a few examples of signal amplification. A more extensive list of examples can be found in Table 55.2. We end this section by simply mentioning the interesting category of detection molecules termed molecular beacons.[30] These molecules derive their powerful detection capabilities from the fundamental chemical properties of nucleic acids. In their simplest form, beacons consist of a stem–loop structure in which the nucleotides in the loop are complementary to a target sequence, while the nucleotides in the stem portion have a fluorophore and quencher held in close apposition on the two terminal nucleotides of the stem. Hybridization of the loop to its target destabilizes the stem, freeing the fluorophore from the quenching effect of the quencher molecule, thereby resulting in fluorescence. Obviously, the beacon must be carefully designed to accomplish the intended purpose. The molecular-beacon concept can also be easily adapted to FRET-based assays.

# 55.4 Hybrid Technologies

Table 55.3 lists the nucleic acid amplification techniques referred to as "hybrid technologies," i.e., a blending of immunological and nucleic acid detection strategies. We have included basically two technologies in this category — immuno-PCR and PCR-ELISA. Each of these techniques is a mirror image of the other in terms of recognition and amplification technologies employed. In immuno-PCR, recognition/specificity is achieved by antigen/antibody interaction. A nucleic-acid fragment attached (directly or indirectly) to the antibody supplies the target for PCR amplification. Thus, in this case, the nucleic acid serves in place of an enzyme attached to the antibody (e.g., standard ELISA immunoassay) as the amplification vehicle. Obviously, both the specificity of the antibody for the antigen and lack of nonspecific binding must be maximized because otherwise the tremendous amplification power of PCR will serve to amplify not only the immuno-specific reaction but spurious reactions as well.

In contrast, with PCR-ELISA assays the hybridization of primers to target nucleic acid sequences serves to provide specificity, while enzymes conjugated to immunoreagents boost the amplification potential of PCR. In this case, the same pitfalls applicable to PCR analysis again hold, as amplicons arising from spurious hybridizations will be amplified by the ELISA process, thereby contributing to the overall nonspecific background. Of course, the immunological reaction, in and of itself, can also contribute to analytical "noise," since binding of enzyme-conjugated antibody nonspecifically to sites other than antigenic sites (commonly digoxigenin) on the target amplicons will lead to increases in nonspecific background.

Both of these techniques have the potential for large signal amplification. Both also have the potential for high specificity. The PCR-ELISA technology has the advantage of building on many years of experience in both PCR and ELISA development. The immuno-PCR procedure, while seemingly straightforward, has a shorter lifespan of development behind it and thus is in its early stages. Even so, some of the results achieved by immuno-PCR have been impressive. For example, Hendrickson et al.[31] were able to develop a multianalyte immuno-PCR assay for three separate analytes in which sensitivities were found to be three orders of magnitude better than conventional ELISA assays. Table 55.3 explains and lists some examples of these powerful analytical tools.

## Acknowledgment

This research was sponsored by the U.S. Department of Energy, managed by UT-Battelle under contract DE-AC05-00OR22725.

## References


1. Guesdon, J. L., Immunoenzymatic techniques applied to the specific detection of nucleic acids — a review, *J. Immunol. Meth.*, 150(1–2), 33, 1992.
2. Wilchek, M. and Bayer, E.A., The avidin biotin complex in immunology, *Immunol. Today*, 5(2), 39, 1984.
3. Price, C.P. and Newman, D.J., Introduction, in *Principles and Practice of Immunoassay*, 2nd ed., Price, C.P. and Newman, D.J., Eds., Stockton Press, New York, 1997, pp. 3–11.
4. Porstmann, T. and Kiessig, S.T., Enzyme-immunoassay techniques — an overview, *J. Immunol. Meth.*, 150(1–2), 5, 1992.
5. Johnstone, A. and Thorpe, R., Immunocytochemistry, immunohistochemistry and flow cytofluorimetry, in *Immunochemistry in Practice*, 3rd ed., Blackwell Science, Cambridge, 1996, pp. 313–338.
6. Tijssen, P., *Practice and Theory of Enzyme Immunoassays*, Elsevier Science Publishing, New York, 1985.
7. Wood, P. and Barnard, G., Fluoroimmunoassay, in *Principles and Practice of Immunoassay*, 2nd ed., Price, C.P. and Newman, D.J., Eds., Stockton Press, New York, 1997, pp. 391–423.
8. Gosling, J.P., Enzyme immunoassay with and without separation, in *Principles and Practice of Immunoassay*, 2nd ed., Price, C.P. and Newman, D.J., Eds., Stockton Press, New York, 1997, pp. 351–387.
9. Van Regenmortel, M.H.V., The antigen-antibody reaction, in *Principles and Practice of Immunoassay*, 2nd ed., Price, C.P. and Newman, D.J., Eds., Stockton Press, New York, 1997, pp. 15–34.
10. Avrameas, S., Amplification systems in immunoenzymatic techniques, *J. Immunol. Meth.*, 150(1–2), 23, 1992.
11. Ishikawa, E., Hashida, S., Kohno, T., and Hirota, K., Ultrasensitive enzyme-immunoassay, *Clin. Chim. Acta*, 194(1), 51, 1990.
12. Ishikawa, E., Hashida, S., Tanaka, K., and Kohno, T., Development and applications of ultrasensitive enzyme immunoassays for antigens and antibodies, *Clin. Chim. Acta*, 185(3), 223, 1989.
13. Shalev, A., Greenberg, A.H., and McAlpine, P.J., Detection of attograms of antigen by a high-sensitivity enzyme-linked immunoabsorbent assay (HS-ELISA) using a fluorogenic substrate, *J. Immunol. Meth.*, 38(1–2), 125, 1980.
14. Ruan, K.H., Kulmacz, R.J., Wilson, A., and Wu, K.K., Highly sensitive fluorometric enzyme-immunoassay for prostaglandin-H synthase solubilized from cultured-cells, *J. Immunol. Meth.*, 162(1), 23, 1993.
15. Brown, R.C., Weeks, I., Fisher, M., Harbron, S., Taylorson, C.J., and Woodhead, J.S., Employment of a phenoxy-substituted acridinium ester as a long-lived chemiluminescent indicator of glucose oxidase activity and its application in an alkaline phosphatase amplification cascade immunoassay, *Anal. Biochem.*, 259(1), 142, 1998.
16. Lewkowich, I.P., Campbell, J.D., and HayGlass, K.T., Comparison of chemiluminescent assays and colorimetric ELISAs for quantification of murine IL-12, human IL-4 and murine IL-4: chemiluminescent substrates provide markedly enhanced sensitivity, *J. Immunol. Meth.*, 247(1–2), 111, 2001.
17. Porakishvili, N., Fordham, J.L.A., Charrel, M., Delves, P.J., Lund, T., and Roitt, I.M., A low budget luminometer for sensitive chemiluminescent immunoassays, *J. Immunol. Meth.*, 234(1–2), 35, 2000.
18. Johansson, A., Ellis, D.H., Bates, D.L., Plumb, A.M., and Stanley, C.J., Enzyme amplification for immunoassays — detection limit of 100th of an attomole, *J. Immunol. Meth.*, 87(1), 7, 1986.
19. Jenkins, S.H., Homogeneous enzyme-immunoassay, *J. Immunol. Meth.*, 150(1–2), 91, 1992.
20. Genfa, Z. and Dasgupta, P.K., Hematin as a peroxidase substitute in hydrogen-peroxide determinations, *Anal. Chem.*, 64(5), 517, 1992.
21. Bobrow, M.N., Litt, G.J., Shaughnessy, K.J., Mayer, P.C., and Conlon, J., The use of catalyzed reporter deposition as a means of signal amplification in a variety of formats, *J. Immunol. Meth.*, 150(1–2), 145, 1992.

22. Price, C.P. and Newman, D.J., Light-scattering immunoassay, in *Principles and Practice of Immunoassay*, 2nd ed., Price, C.P. and Newman, D.J., Eds., Stockton Press, New York, 1997, pp. 445–480.
23. Aoki, K., Itoh, Y., and Yoshida, T., Simultaneous determination of urinary methamphetamine, cocaine and morphine using a latex agglutination inhibition reaction test with colored latex particles, *Jpn. J. Toxicol. Environ. Health*, 43(5) 285, 1997.
24. Holgate, C.S., Jackson, P., Cowen, P.N., and Bird, C.C., Immunogold silver staining — new method of immunostaining with enhanced sensitivity, *J. Histochem. Cytochem.*, 31(7), 938, 1983.
25. Andras, S.C., Power, J.B., Cocking, E.C., and Davey, M.R., Strategies for signal amplification in nucleic acid detection, *Mol. Biotechnol.*, 19(1), 29, 2001.
26. Schweitzer, B. and Kingsmore, S., Combining nucleic acid amplification and detection, *Curr. Opinion Biotechnol.*, 12(1), 21, 2001.
27. Isaksson, A. and Landegren, U., Accessing genomic information: alternatives to PCR, *Curr. Opinion Biotechnol.*, 10(1), 11, 1999.
28. Cha, R.S. and Thilly, W.G., Specificity, efficiency, and fidelity of PCR, in *PCR Primer: A Laboratory Manual*, Dieffenbach, C.W. and Dveksler, G.S., Eds., Cold Spring Harbor Laboratory Press, Plainview, NY, 1995, pp. 37–51.
29. Nuovo, G.J., Co-labeling using *in situ* PCR: a review, *J. Histochem. Cytochem.*, 49, 11, 1329, 2001.
30. Tyagi, S. and Kramer, F.R., Molecular beacons: probes that fluoresce upon hybridization, *Nat. Biotechnol.*, 14(3), 303, 1996.
31. Hendrickson, E.R., Truby, T.M.H., Joerger, R.D., Majarian, W.R., and Ebersole, R.C., A sensitivity multianalyte immunoassay using covalent DNA-labeled antibodies and polymerase chain-reaction, *Nucleic Acids Res.*, 23(3), 522, 1995.

**TABLE 55.1** Immunoassay Amplification Schemes

| Assay | Principle | Comments | Ref. |
|---|---|---|---|
| | ***Enzyme-Linked Immunoassays [Solid-Phase Heterogeneous]*** | | |
| | *In all enzyme immunoassays (EIA) the final signal detected is produced by the activity of an enzyme, while specificity is obtained by antigen/antibody (Ag/Ab) interaction; addition of enzyme substrate to final immune complex results in accumulation of colored product* | *Large signal amplification because of high rate of processivity of enzymes for substrate molecules* | |
| Direct EIA (ELISA) | Enzyme-labeled Ab recognizes Ag | If capture-Ab system used, enzyme-labeled Ab must recognize different Ag epitope then capture Ab; can be formatted as competitive or noncompetitive assay | 1, 2 |
| Indirect "Sandwich" EIA (Immune Complex) (ELISA) | Antispecies IgG enzyme-labeled Ab binds directly to Ag-recognizing Ab or is bound indirectly via an unlabeled antispecies IgG Ab that recognizes both Ag-specific Ab and enzyme-labeled Ab; both direct and indirect immune complex assays termed ELISA | Enzyme numbers multiplied through multiple epitope sites on IgG Ab | 3–7 |
| | ELISA carried out on cloth [usually plastic surfaces (e.g., microwells)] | Cloth assay described as instantaneous; no incubation time with Ab steps; 5-min incubation with substrate[3] | 3 |
| Bispecific/Chimera Ab Molecules | | | 8–14 |
| | Abs constructed to possess dual specificity for Ag and for enzyme; or chimeric Ab constructed to recognize a hapten (covalently linked to an immunoreagent) and an enzyme; in either case enzyme is indirectly linked to immunocomplex, as opposed to directly linked, as in the case of enzymes covalently coupled to Ab | Bispecific Ab constructed so that it binds to both FITC (used as a label for Ag or Ab) and to reporter (HRP) enzyme; thus, hybrid antibody serves as bridge between FITC-labeled immunoreagent and enzyme molecule; this procedure avoids possible deleterious effect on Ab affinity by covalent binding of enzyme directly to Ab, amplification both in number of FITC on primary Ab, and by enzyme indirectly linked to each FITC molecule[10] | 10 |
| | | or: Bispecific Ab constructed so that it recognizes both specific Ag and HRP; again, advantage in indirect linking of reporter enzyme and specific Ab[11] | 11 |
| | also: Chimeric fusion protein constructed having single-chain variable region of Ab fused to enzyme (alkaline phosphatase); bifunctional immunoreagent thus constructed; other constructs allow expression of enzyme-labeled Fab′ fragments[8,9] | | 8, 9, 12, 13 |

| | | | |
|---|---|---|---|
| | Fusion protein constructed between protein A and enzyme; as immunoreagent sandwich is formed, protein A portion of fusion protein binds to Fc region of Ag-specific Ab, forming, in effect, an enzyme-linked immunoassay | Fusion protein construct of enzyme-protein A avoids random nature of chemically labeling Ab with enzymes; since Ab itself is not labeled in this technique, more likely to have less adverse effect on Ab affinity than if Ab were directly labeled | 14 |
| Enzyme/Antienzyme Immune Complexes | Multiple enzyme molecules localized by antienzyme Ab, the whole complex held at Ag site as above (that is, Ag-specific Ab binds to Ag; antispecies IgG Ab binds to Ag-specific Ab, as well as antienzyme Ab made in same species as Ag-specific Ab; antienzyme Ab binds enzyme molecules; excess enzyme washed away); commonly used complexes include:<br>PAP = horseradish peroxidase complex<br>AP-AAP = alkaline phosphatase complex | Enzyme numbers multiplied; often used in immunohistochemistry; 2- to 50-fold increased sensitivity[15] | 15–21 |
| Antihapten Ab "Sandwich" | Haptens (e.g., FITC, DNP) bound to Ag-specific primary antibody, recognized by enzyme-labeled antihapten Ab; can use bridge antibody (i.e., IgM anti-DNP) and DNP-labeled enzyme for increased sensitivity; can also use DNP-labeled glucose oxidase in addition to primary enzyme (peroxidase) for increased sensitivity | Enzyme numbers multiplied – multiple hapten molecules bound to primary Ab – bridging Ab gives more multiplication; used in immunohistochemistry; fourfold more sensitive than PAP[22] | 22, 23 |
| Cell–ELISA (Enzyme-Linked Immunofilter Assay) | Detect cell surface molecules on cells, using Ag-specific Ab and enzyme-labeled anti-species IgG Ab | Usually applied to cells in microwells; Limitations: may be necessary to use different enzymes (e.g., bacterial β-galactosidase) if endogenous cellular enzyme interferes; if cells are fixed, surface antigens may be destroyed; good correlation with flow cytometric analysis[24] | 2, 24–27 |
| ELISPOT (Spot–ELISA) | Ags secreted by individual cells are captured by Ag-specific capture Ab immobilized on solid surface; cells are removed and Ag-specific reporter Ab is added; either reporter Ab has enzyme label or enzyme-labeled antispecies IgG Ab is added; substrate is added to give insoluble product | Need relatively high Ag density on solid-phase surface; nitrocellulose better than plastic; biotin/avidin systems can be used to advantage in this technique; technique developed to reveal secretion by single cell; does not measure Ag concentration | 26–30 |
| ELIFA (Enzyme-Linked Immunofilter Assay) | Ag (e.g., bacteria) captured on membrane detected by Ag-specific Ab and additional immune complex; antispecies IgG Ab + enzyme-labeled Ab | Detect ~50 to 60 bacterial cells using chemiluminescent or colormetric assay, carried out in ~1 h; very rapid (17 min) EIA reported for airborne antibiotic trapped on cellulose nitrate filter;[31] colorimetric detection | 31–35 |

**TABLE 55.1** Immunoassay Amplification Schemes (continued)

| Assay | Principle | Comments | Ref. |
|---|---|---|---|
| EIA (Enzyme-Labeled Ag) | Ag conjugated with marker enzyme; free Ag in sample competes with enzyme-labeled Ag for binding to immobilized Ab | Heterogeneous, competitive assay format; assay stated not to be as sensitive as radioimmunoassay[36] | 36–38 |
| | Fusion protein made between protein A and enzyme; binding to immobilized Ag-specific Ab results in some of enzyme–Ag fusion protein being immobilized in immune complex | Competitive, heterogeneous assay; Ag in analyte solution competes with enzyme–Ag fusion protein for binding to Ab; amplification due to enzyme; such chimeric constructs for use as Ags require a protein Ag | 37 |
| | or:<br>Enzymes covalently attached to different hapten Ags; incubation with Ag-specific Abs and free Ag in solution results in some enzyme-labeled Ag forming immunocomplex with Ab; addition of second Ab (antispecies IgG) results in precipitation of immune complex; following washing, addition of enzyme substrates results in enzymatic activity and optical signal proportional to amount of labeled Ag bound in original immune complex | Heterogeneous assay format; enzyme-labeled Ag competes with free Ag for binding to limited amount of Ab; use of two different enzymes can allow use of two different enzyme substrates whose products can be distinguished by optical properties;[37] applied to simultaneous assay of two Ags – triiodothyronine and thyroxin | 38 |
| Immune Complex Transfer Immunoassay | Ag is complexed between Ag-specific Ab (Fab′) labeled with enzyme, and Ag-specific Ab labeled with both biotin and DNP; resulting immune complex is trapped on solid phase by anti-DNP Ab bound to solid phase; immune complex eluted from solid phase by DNP-lysine wash; immune complex again trapped on streptavidin-linked solid phase; enzyme substrate is added and product is measured; similar principles applied to noncompetitive heterogeneous assays of small MW haptens | Important step is transfer of immune complex to different solid phases, as this removes nonspecifically bound Ab; lends itself to various assay formats; extreme sensitivity, subattomole, down to 1 zeptomol (600 molecules) in assay of human ferritin[39] | 39–41 |
| Enzyme-Amplified Cascade (RELIA) = Releasable Linker Immunoassay | Ag captured by immobilized capture Ab; second biotin-labeled detector Ab binds to immune complex; addition of alkaline phosphatase-conjugated streptavidin followed by addition of release (i.e., biotin-containing) reagent results in release of enzyme–streptavidin conjugate; enzyme activity detected indirectly by enzyme cascade as follows: alkaline phosphatase converts FADP to FAD, which binds to inactive apoenzyme amino-acid oxidase, resulting in active holoenzyme, which oxidizes proline to produce $H_2O_2$, which further reacts with horseradish peroxidase and other chemicals to form a colored product | Heterogeneous assay format; large multiplication due to involvement of multiple enzymes; sensitivity found to be comparable to fluorometric assays | 42 |

| | | | |
|---|---|---|---|
| Fusion Protein-Based EIA | Genetically fused protein is prepared that has both immunoreactivity and enzymatic activity; this protein can take part in immunoreactions and also serve as detection system by generating enzymatic product that is detected; IA carried out in usual manner; fusion protein is part of immunoreactant complex; (example: firefly luciferase:protein A fusion protein–protein A binds to $F_c$ region of Ig, while luciferase reacts with luciferin to produce luminescence)[43] | Potential for sensitive, specific IA; fusion products may not have same degree of binding affinity/enzyme activity as native proteins | 43 |
| | or: Fusion protein prepared that is Ag-fused to enzyme (i.e., proinsulin-alkalinephosphatase); competitive immunoassay carried out in which fusion protein Ag competes with Ag in sample for binding to Ab[44] | | 44 |
| Nonenzymatic Catalytic Amplification Immunoassay (Mimetic Enzyme EIA) | Immunoreactant "sandwich" set up in usual way (i.e., capture Ab, Ag and detector Ab); detector Ab has conjugated to it a chemical which can serve as a catalyst for a reaction which produces an optically detectable product (e.g., p-hydroxyphenyl acetic acid + $H_2O_2$ in the presence of the catalyst hemin generates a fluorescent product) | Amplification occurs by catalytic activity of reagent conjugated to Ab; also, multiple catalytic reagents can be conjugated to single Ab; small size of catalytic reagent compared to usual size of enzyme should result in less steric inhibition of Ab:Ag binding events; catalytic reagent likely to be more stable than enzyme | 45–48 |
| | ***Enzyme-Linked Immuno Assays [Solution-Homogeneous]*** | | |
| | *Same general principle as for Heterogeneous EIA* | | |
| EMIT = Enzyme-Multiplied Immunoassay Technique | Free Ag in sample competes with enzyme-labeled Ag for binding to Ab; enzyme-labeled Ag bound to Ab results in inhibition of enzyme activity; the more Ag in the sample, the more uninhibited enzyme results, and the larger the resultant signal | Competitive assay; usually limited to low MW Ags – direct proportionality between free analyte concentration and enzyme activity (i.e., the more free Ag present, the less enzyme-labeled Ag is complexed with Ab, and the more enzyme activity is expressed) | 49–51 |
| Substrate-Labeled Immunoassays | Substrate-labeled Ag competes with free Ag; Ab binding blocks substrate from action of enzyme present in solution | Competitive assay; not possible to do assay with excess of substrate, so enzyme kinetics are not zero order; less sensitive than EMIT; direct proportionality between free Ag concentration and enzyme activity | 52, 53 |
| Enzyme-Modulator-Labeled Immunoassay | Enzyme inhibitor bound to Ag; Ab binding blocks enzyme inhibition | Competitive assay; inverse relationship between presence of analyte and enzyme activity; more free Ag, less Ab available to bind to inhibitor–Ag complex; therefore, more enzyme inhibition expressed | 54 |

**TABLE 55.1** Immunoassay Amplification Schemes (continued)

| Assay | Principle | Comments | Ref. |
|---|---|---|---|
| Enzyme Inhibitory Immunoassay | Ab (Fab′ fragments) bound to enzyme (Dextranase); binding of Ag to enzyme-bound Ab results in inhibition of enzyme's ability to bind to insoluble substrate; enzyme substrate is in suspension along with other components; inhibition of enzyme activity produces less product | Designed to allow assay of high MW Ag; signal inversely proportional to analyte Ag concentration | 55, 56 |
| Cofactor-Labeled Immunoassay | Essential enzyme cofactors bound to Ag; free (sample) Ag competes with labeled Ag; when Ab binds to cofactor-labeled Ag, cofactor unable to activate apoenzyme, so enzyme activity is not expressed | Competitive assay; very low background; signal (enzyme activity) directly proportional to free analyte concentration | 57 |
| Cloned Enzyme Donor Immunoassay (CEDIA) | *E. coli* β-galactosidase genetically engineered, so that two inactive subunits, a large polypeptide called enzyme acceptor (EA) and a small polypeptide designated enzyme donor (ED), are present in the assay; EA and ED can spontaneously associate to produce active enzyme; in assay Ag is attached to ED; Ag-specific Ab binds to Ag:ED complex and inhibits reassembly of active enzyme; Ag in solution competes with Ag:ED complex for limited amount of Ab; more free analyte in sample, more active β-galactosidase formed; substrate in solution converted to colored product | Homogeneous, competitive assay format; colored product directly related to amount of analyte in sample; assay simple to perform and rapid; minimal background; high sensitivity | 58–60 |
| Enzyme–Channeling Immunoassay | Coupled enzyme reaction wherein activity is accelerated when two enzymes are brought into close contact; hapten + enzyme 1 co-immobilized on bead; Ab labeled with enzyme 2 binding to immobilized hapten on bead produces coupled reaction; if Ab is complexed to analyte in solution, coupled enzyme reaction does not occur | Inverse proportionality; low background | 61 |
| | ***Biotin: Avidin/Streptavidin Procedures*** | | 62–67 |
| | *Biotin and avidin/streptavidin form integral part of immune complex; due to extremely high association constant of biotin avidin/streptavidin, stable complexes with favorable rates of formation are produced* | *Same general detection methodologies as other immunoassays; biotin:avidin complexes contribute to immune "sandwich"* | |
| Simple Complex | Biotinylated Ab recognizes Ag, and binds to enzyme-labeled avidin/streptavidin | Enzyme number multiplied – multiple biotinylation sites on Ab since biotin is a small molecule; small size of biotin also produces only minimal effects on Ab activity; multiple enzyme molecules on each avidin/streptavidin molecule | 62, 63 |

| | | | |
|---|---|---|---|
| "Sandwich'" Complex | Biotinylated Ag-specific Ab reacts with avidin/streptavidin which also binds biotinylated enzyme – can preform avidin/biotinylated complexes (ABC = avidin-biotin complex) | Enzyme number multiplied (see above) also, avidin has four combining sites for biotin; 2- to 100-fold more sensitive than conventional EIA techniques;[64] ABC likely forms a lattice complex containing several (>3) enzyme molecules | 62–67 |
| | also: Multiple "layers" can be built up to form large immunoreagent–biotin–avidin complexes; example: detector Ab is biotinylated; avidin binds to biotinylated detector Ab; avidin-specific biotinylated Ab binds to avidin; last two steps can be repeated; finally add avidin–enzyme complex followed by enzyme substrate | Amplification results from multiple enzyme molecules; amplification increases with number of "immunolayers" added; enhancement of 10- to 20-fold over standard ELISA for cytokine detection reported[65] | 65 |
| | also: Avidin can be conjugated to either Ag-specific or antispecies IgG Ab; subsequent reaction with biotinylated enzymes results in enzyme-labeled antibody–avidin complexes | Multiplication of enzyme molecules in each immunoreactant complex; five- to eightfold enhancement of sensitivity compared to directly enzyme-labeled Ab and fivefold more sensitive than ABC technique[66] | 66 |
| | ***Coupled Enzyme Procedures*** | | |
| | *Same general idea as for EIA; enzyme linked to immunoreagent produces product that is then indirectly detected* | *Potential for large increase in amplification since secondary enzymes contribute to overall multiplication of molecules responsible for final signal* | |
| Multi-Enzyme Cascades | Product of one enzymatic reaction becomes substrate for another enzyme; also enzyme product (e.g., $NAD^+$) can be recycled many times; example: alkaline phosphatase converts $NADP^+$ to $NAD^+$; alcohol dehydrogenase + ethanol converts $NAD^+$ to NADH; diaphorase oxidizes NADH to $NAD^+$ and reduces a tetrazolium salt to a colored formazan dye | Amplified enzymatic signals; 100-fold increase in sensitivity compared to conventional EIA;[68] 0.01 attomol of alkaline phosphatase was detectable using multi-enzyme cascade[64] | 64, 68 |
| Catalyzed Reporter Deposition (CARD) Immunoassay | Form standard ELISA "sandwich"; enzyme linked to Ab in ELISA sandwich reacts with enzyme substrate and biotinyl tyramide (phenolic structure) to form free radical intermediate of tyramide which binds to receptor on solid surface (i.e., tyrosine residues in close proximity to site of enzyme reaction); addition of enzyme-labeled streptavidin results in binding of streptavidin at biotin sites; addition of enzyme substrate results in final product, which is measured; usually applied to procedures where final product is a precipitate; also feasible to use fluorescein tyramide as substrate for production of "activated" intermediate that is deposited; in this case, fluorescein serves as Ag for subsequent addition of antifluorescein Ab conjugated to enzyme; addition of enzyme substrate follows | Potential for many-fold multiplication of enzyme linked to streptavidin, since many biotinylated tyramides can be deposited; final enzyme does not need to be same as ELISA enzyme; other substrates than biotinyl tyramide can be used; HRP, AP, and β-Gal enzymes can be used; HRP used as the first enzyme to generate the "activated" tyramide; other enzymes may be used for secondary amplification (e.g., AP and β-Gal) | 69–73 |

**TABLE 55.1** Immunoassay Amplification Schemes (continued)

| Assay | Principle | Comments | Ref. |
|---|---|---|---|
| | also: | | |
| | Can use multiple amplification cycles, where sequential additions of tyramide derivative plus ligand which binds to deposited tyramide results in many more deposition sites and deposited secondary enzyme | Can be readily applied to membrane-based immunoassays; CARD procedure improved detection limits 8- to 200-fold over standard ELISA format, depending on cycles of tyramide amplification and enzyme substrate used[69] | |
| | So-called "super" CARD assay; a modified protein with many sites for tyramine immobilization is used in the assay microwells instead of the usual blocking agent; essential point is that more reactive sites are provided for biotinylated tyramide immobilization, thus providing more sites for subsequent enzyme immobilization when avidin-HRP is added | Many more molecules of HRP are eventually immobilized around site of each primary HRP-conjugated antibody than in "standard" CARD; incubation time very significantly reduced from standard CARD assay; "super" CARD assay had limit of detection tenfold lower than ELISA and fivefold lower than regular CARD assay | 70 |
| | Principle is same as "super" CARD; synthetic proteins with many binding sites for tyramide deposition are used as blocking agents around sites of immunoreagents | Applied to immuno-dot type assay; "super" CARD assay found to be ~$1 \times 10^4$-fold more sensitive than standard CARD, using casein as blocking agent in standard CARD vs. modified casein in "super" CARD[70] all other immunoreagents being the same; visual detection of as little as 800 molecules of rabbit IgG reported[70] | |
| Dual Enzyme Cascade | Action of enzyme 1 converts inactive inhibitor into an active inhibitor of enzyme 2; residual activity of enzyme 2 provides measure of amount of enzyme 1 present | Inverse relation between concentration of first enzyme and strength of signal; increase in sensitivity of 125-fold over using single enzyme (i.e., enzyme 1 alone) system observed | 74 |
| Bioluminescence Enhanced Enzyme Immunoassay | Standard immunoassay "sandwich" set up with enzyme attached either to Ab or Ag (competitive format); following this, luciferin derivatives susceptible to hydrolysis (to make free luciferin) by the enzyme in the immune "sandwich" are added; after incubation, some of the product solution transferred to luciferase-containing solution; extent of bioluminescence produced used to quantify immune reaction | Can be formatted for competitive or noncompetitive assay; many assay steps, which contributes to longer assay time; maximal LODs stated to be $10^{-19}$ mol for antigen detection[75] | 75, 76 |
| | ***Fluorescence Immunoassays*** | | |
| | *Two approaches: (1) fluorophores attached to immunoreagent (Ab or Ag); when immunocomplex forms, extent of complex formation can be quantified by fluorescence; (2) EIA carried out, but enzyme produces fluorescent product from substrate* | *Fluorescence inherently more sensitive than absorption spectroscopy; other properties of fluorescence (fluorescence polarization, fluorescence lifetimes) can serve as basis for detection* | |

| | | | |
|---|---|---|---|
| Fluoroimmunoassay (FIA) Direct | Fluorophore-labeled Ab binds to immobilized Ag (Ag can be immobilized by capture Ab); immobilized Ab binds free or fluorophore-labeled Ag | Heterogeneous assay; can be competitive or noncompetitive format; minimal amplification; amplification depends on number of fluorophores bound to either Ab or Ag; requirement that Ag be protein or larger peptide; small hapten that is conjugated to one fluorophore, by definition, does not involve amplification; limitations on fluorophore labeling of Ab, because of loss of affinity for Ag, and for Ag, because of potential loss of antigenic epitopes, self-quenching between adjacent fluorophores on same Ag molecule, etc.; immobilized Ab configuration can be used for competitive assay in which fluorophore-labeled Ag competes with free Ag in analyte for binding to limited number of immobilized Ab sites | 77–91 |
| | also: Can measure either the fluorescence of the immobilized labeled fraction or the fluorescence of the labeled immunoreagent remaining in solution when immobilized Ab concentration is limited | Competitive assay in which fluorophore-labeled Ab binds to Ag in analyte solution; any remaining free labeled Ab is reacted with Ag immobilized on beads; thus, fluorescence is inversely proportional to analyte Ag concentration | |
| FIA (Indirect "Sandwich") | "Sandwich" formed with Ag-specific Ab and fluorescently-labeled antispecies IgG Ab, which binds to Ag-specific Ab | Some amplification, due to multiple fluorophore-labeled Abs bound to primary Ab; heterogeneous assay | 80, 89, 92–95 |
| | also: For better detection sensitivity, immunoreactant "sandwich" can be extended: i.e., after primary Ag-specific Ab binds, fluorophore-labeled antispecies IgG Ab made in another animal species binds to primary Ab; third fluorophore-labeled antispecies IgG (specific for animal species of second Ab) binds to secondary Ab; can be extended further | Multiplication of fluorophore labels with each additional Ab added to "sandwich"; used in histochemistry, cytochemistry | 73 |
| Fluorescent CARD Immunoassay | Same principle as CARD assay (see under Coupled Enzyme Procedures); fluorophore-labeled tyramide is converted by action of enzyme to "activated" tyramide, which binds close to site of activation; thus, a cluster of fluorophores is deposited at the site of immunoreagent localization | Multiplication of fluorophore labels deposited at specific locus due to enzyme activity; used in histochemistry, cytochemistry, and immunoblotting; increase in sensitivity over usual fluorescent Ab techniques in histochemistry stated to be 10- to 100-fold[73] | 73, 96, 97 |
| | also: Applied to detection of Ags on cells and membranes | Enhancements of 4- to 15-fold over standard indirect immunofluorescent staining were observed using CARD[96] | 96 |

**TABLE 55.1** Immunoassay Amplification Schemes (continued)

| Assay | Principle | Comments | Ref. |
|---|---|---|---|
| Fluorescence Immunoassay: Avidin/Streptavidin | Fluorophore-labeled avidin/streptavidin binds to biotinylated Ab | Multiple biotin sites on Ab, multiple fluorophores on avidin; heterogeneous assay | 98 |
| Fluorescence EIA (Enzyme Amplification) | Weakly or nonfluorescent substrate generates fluorescent product upon enzyme action; immune "sandwich" same as for ELISA | Heterogeneous assay; fluorescence inherently more sensitive than absorption (colorimetric); 5- to $10^5$-fold more sensitive than conventional EIA[64] | 64, 100–105 |
| | | Extreme sensitivity with certain fluorogenic substrate/enzyme pairs reported (i.e., single enzyme-molecule detection)[99] and detection of activity in single bacterial cells[100] | 99, 100, 104, 105 |
| FIA/Fluorescence EIA (Magnetic Bead Separation) | Ab bound to magnetic beads captures free Ag in sample; separation of bound from free fractions done by magnet; fluorophore-labeled Ag added to beads; extent of fluorescence inversely related to free Ag | Unique solid-phase separation employed; heterogeneous assay; rapid assay times; ready removal of interferent substances | 106 |
| | or: Immobilized Ab captures Ag to which enzyme-labeled second Ag-specific Ab also binds; after separation, incubation with fluorescence-generating substrate (also variant in which Ag is bound to beads and Ag-specific Ab is assayed by antispecies, enzyme-labeled Ab) | | 107 |
| Microsphere-Based Immunoassay/Flow Cytometry | Microspheres of different sizes are used to provide discrimination for different immunoassays; each size microsphere can be labeled with different immunoreagent and appears as a distinct population by flow cytometry; in particular assay, different Ags immobilized on microspheres react with analyte Abs in sera; addition of fluorophore-labeled antispecies Ab results in fluorescently labeled microspheres if Ab is present in analyte serum; fluorescence detected by flow cytometry | Amplification is minimal since only amplification results from multiple fluorophores on detector Ab; detection can be quite sensitive since laser-induced fluorescence is used; possibility of multiple analyte analysis; flow cytometric assay reported to be fivefold more sensitive than microplate-based EIA[108] | 108 |

| | | | |
|---|---|---|---|
| Total Internal Reflection Spectroscopy with Fluorescence Detection | Immunological reaction between Ag and Ab monitored at interface between optical waveguide and sample solution; detected by fluorescence excited by evanescent wave resulting from total internal reflection; results in detection of Ag-Ab binding events only very near surface at which internal reflection occurs | Capture Ab immobilized on waveguide binds Ag in sample; second Ab with fluorescent label binds to Ag; fluorescence detection close to waveguide (evanescent wave) monitors Ag-fluorophore-labeled Ab binding event; no washing necessary; requires rather elaborate optical setup; free fluorophore-labeled Ab contributes to fluorescence seen at waveguide/solution interface; thus Ag–Ab binding event seen as incremental increase in fluorescence | 109 |
| | | Ag (hapten) immobilized on waveguide surface; fluorophore-labeled Ab binds to immobilized Ag and is detected by evanescent-wave excitation; free Ag in solution competes with immobilized Ag for limited amount of Ab; signal decreases as amount of free Ag in solution increases | 110 |
| | | or: Fluorophore-labeled Ag binds to immobilized Ag on waveguide surface and is detected by evanescent-wave excitation | 111 |
| Fluorescence Polarization Immunoassay (FPIA) | A change (increase) in fluorescence polarization occurs when fluorophore-labeled Ag binds to Ab due to changes in rotary motion attendant upon large increase in size of immune complex compared to Ag alone; competitive assay may be used with free Ag and fluorophore-labeled Ag; can also determine specific Ab levels using fluorophore-labeled Ag; FPIA using fluorophore-labeled Ab to detect Ag does not appear to work due to flexibility of Ab or relatively free rotation of attached fluorophore | Amplification minimal — fluorescent labeling of Ag limited to low MW Ags if fluorophores with nanosecond lifetimes used; competitive, homogeneous assay possible; extent of change in fluorescence polarization inversely proportional to free Ag concentration; no separation of bound vs. free Ag required for assay; sensitivity of assay strongly dependent on Ab affinity; potentially very short assay time; low background; potential for high sensitivity; other proteins in Ag sample may interfere with assay if they bind to fluorophore-labeled Ag; sensitivity sub-ng/ml[112] | 112–115 |
| FPIA with Metal–Ligand Complexes | Same principle as FPIA above, but use of transition-metal (e.g., Re, Ru) ligand complexes allows fluorescence polarization assays for high MW antigens; these complexes display high polarization in absence of rotational diffusion and long lifetimes | Competitive, homogeneous assay; metal–ligand complexes do not show probe–probe interactions, and so higher metal–ligand complex/Ag protein ratios can be used; amplification in terms of fluorophore groups is small; metal–ligand complexes have low extinction coefficients and low quantum yields; long lifetimes allow off-gating of fluorescence interferences | 116–118 |

**TABLE 55.1** Immunoassay Amplification Schemes (continued)

| Assay | Principle | Comments | Ref. |
|---|---|---|---|
| Time-Resolved FIA | Immune "sandwich" formed in usual way, with one of components being biotinylated Ab; the biotinylated Ab serves as recognition sites for streptavidin-containing macromolecular complex that also contains multiple chelator molecules saturated with $Eu^{3+}$; after reaction with immune reagents, unbound macromolecular complex is washed away, and Eu-chelate associated with solid phase is measured by time-resolved fluorescence; homogeneous assay not feasible because of instability of Eu chelate/macromolecular complex; particular chelator used forms highly fluorescent, stable complex with $Eu^{3+}$ and can be covalently linked to proteins | Heterogeneous assay; capture Ab immobilized on solid surface captures Ag; biotinylated Ag-specific Ab also binds to Ag; addition of streptavidin-thyroglobulin-chelator-$Eu^{3+}$ macromolecular complex results in some complex being immobilized at immunoreagent sites, if biotinylated Ab is bound; excess macromolecular complex is washed away, wells are dried, and $Eu^{3+}$ fluorescence determined by time-resolved techniques; large amplification due to hundreds of $Eu^{3+}$ bound in macromolecular complex; use of the macromolecular complex resulted in improved detector limits for a number of antigens, in comparison with smaller streptavidin-thyroglobulin-chelator-$Eu^{3+}$ complexes; laser excitation used; detection limit for α-fetoprotein was 60 attomoles/well;[119] enhancement of Eu fluorescence when multiple chelator–Eu complexes bound to protein, thus, when 160 chelator molecules are bound to thyroglobulin in presence of excess $Eu^{3+}$, fluorescence equivalent to 900 molecules of unconjugated chelator-$Eu^{3+}$ is obtained[119] | 119, 120 |
| | also: Immune complex formed in usual way; same chelator as above is used, but Ab directly labeled with chelator + $Eu^{3+}$; chelator binds to $NH_2$ groups on proteins; after washing and drying, fluorescence of $Eu^{3+}$ measured | Too much labeling of proteins with chelator interferes with biological activity of proteins (e.g., Ab/Ag interaction); extent of labeling tolerated before biological activity decreases depends on particular protein; for one Ab, 18 chelator–$Eu^{3+}$ complexes could be attached per Ab molecule without adverse effect; competitive heterogeneous immunoassay for cortisol demonstrated | |
| Dissociation-Based Time-Resolved Fluoroimmunoassay (DELFIA) | Either Ab or Ag is labeled with stable hydrophilic lanthanide (Eu, Tb) chelate, which is nonfluorescent in aqueous solution; immunoassay takes place in standard way (i.e., immune complex "sandwich"); after all immunoreactions are finished, the lanthanide chelate is dissociated with an enhancement reagent, which contains a diketone that chelates the lanthanide in a micellar environment, resulting in a highly fluorescent solution; separation step required to eliminate labeled reagent not bound in immune complex; new chelate with lipophilic reagent is formed, enhancing lanthanide fluorescence; time-resolved fluorescence (delay time between excitation and emission signal) determined | Can be heterogeneous or homogeneous format; time-resolved fluorescence virtually eliminates background and interferents; generally requires laser excitation; only amplification is in the number of metal-chelates per Ag or Ab; narrow emission bands suggest use of multiple labels (more than one lanthanide chelate) for dual analyte assays | 121–125 |

| | | | |
|---|---|---|---|
| EIA with Time-Resolved Fluorescence Detection | Immunoassay carried out like regular EIA; enzyme substrate consists of a chemical that does not form a chelate with lanthanide ion ($Tb^{3+}$, $Eu^{3+}$), but after enzyme activity a product is formed that chelates to lanthanide ion to form highly fluorescent complex; fluorescence measured by time-resolved methods or second derivative synchronous fluorescence; luminescence (fluorescence) increase directly related to enzyme activity; mainly uses $Tb^{3+}$ | Heterogeneous assay format; time-resolved fluorescence advantages (see above); attomole sensitivity observed for one Ag species;[126] possibility for rapid assay; lanthanide chelates have high quantum yields, large Stokes shifts, narrow emission peaks; potential for several orders of magnitude increase in sensitivity over conventional fluorescence;[127] $Tb^{3+}$ has advantages over $Eu^{3+}$ for this type of assay; observed to be ~30-fold more sensitive than assay where no enzyme amplification was used, but where Eu-chelate served as label for immunoreagent (thus, also time-resolved fluorescence detection);[128] 300-fold more sensitive than colorimetric EIA; ~20 attomole of Ag detected[128] | 126–132 |
| Fluorescence Resonance Energy Transfer (FRET) | Fluorophore donor/Ag conjugate (β-phycoerythrin/thyroxine) competes with free Ag in sample for binding to Ab-fluorophore acceptor conjugate (antithyroxine Ab/Cy5); when labeled Ag binds to labeled Ab, FRET occurs; this transfer is detected by measuring changes in fluorescence lifetime, performed by phase-modulation methodology | Homogeneous, competitive assay; particular pair of dyes chosen allow FRET to occur over greater distances than for most acceptor-donor pairs;[133] eliminates many sources of fluorescent interference; instrumentation rather complex; laser required | 133–137 |
| | or:<br>Two fluorophore-labeled Abs bind to different epitopes on analyte Ag; when immune complex forms, excitation of one fluorophore (donor) results in transfer of energy to second fluorophore (acceptor) because of close apposition; emission from second fluorophore signals formation of immune complex | One Ab labeled with europium (III) cryptate binds to Ag; second Ab labeled with allophycocyanine (acceptor) also binds to other epitope on Ag; excitation by laser of donor results in transfer of excitation energy to donor, if both are bound to the same Ag molecule; signal detected by time-resolved technique | 137 |
| | or:<br>Two fluorophore-labeled Abs used that recognize same Ag; one fluorophore is a $Eu^{3+}$-labeled cryptate; the other fluorophore (on different Ab) is allophycocyanine; formation of immune (Ab/Ag/Ab) complex results in nonradiative energy transfer; excitation of $Eu^{3}$ label, coupled with appropriate filtering and time-resolved detection, results in the detection of fluorescent signal arising from allophycocyanine due to energy transfer in immune complex; fluorescence from $Eu^{3+}$-labeled Ab free in solution or allophycocyanine-labeled Ab free in solution not detected due to elimination by optical filtration or time-resolved measurement | Laser excitation; homogeneous assay format; theoretically, only fluorescence from immune complex detected; amplification arises because of high value of quantum yield from allophycocanine and the high efficiency of the energy transfer; therefore, overall fluorescence yield from immune complex higher than would arise from $Eu^{3+}$ alone; detection limit for prolactin found to be as good as for radioactive assay for this same Ag[134] | 134 |

**TABLE 55.1** Immunoassay Amplification Schemes (continued)

| Assay | Principle | Comments | Ref. |
|---|---|---|---|
| | or: Fluorescamine-labeled Ag reacts with fluorescein-labeled Ab in solution; two fluorophores matched so that excitation of fluorescamine on Ag results in energy transfer so that fluorescin on Ab is excited and fluoresces; addition of unlabeled Ag results in decrease in fluorescence, since fewer labeled Ags are bound in immune complex and participate in FRET | Competitive, homogeneous assay format; amplification due to multiple fluorophores on Ab; approximately equivalent limit of detection compared to ELISA observed, when same Ab reagents used[135] | 135 |
| | or: Ag labeled with chemiluminescent (isoluminol) derivative; binding to fluorescein-labeled Ab sets up conditions for nonradiative energy transfer; addition of microperoxidase and $H_2O_2$ after formation of immune complex results in emission of light due to chemiluminescent reaction and energy transfer to fluorophore-labeled Ab, resulting in fluorescence emission from the Ab-associated fluorophore, since donor/acceptor species are designed to allow FRET | Homogeneous, competitive assay format; amplification due to multiple fluorophores on Ab (4-12 fluoresceins/Ab molecule); since both chemiluminescence and fluorescence occur, ratio of light intensity at two wavelengths used to evaluate extent of energy transfer (and, thus, indirectly to evaluate extent of Ag and Ab binding); assay applied to a variety of Ags, e.g., cAMP, progesterone, IgG; rather complicated experimental apparatus, as chemiluminescence/fluorescence monitored simultaneously; assay for camp found to be comparable in sensitivity to radioimmunoassay for cAMP[136] | 136 |
| Fluorescence Excitation Transfer Immunoassay (FETI) | Fluorophore-labeled Ag binds to Ab labeled with a different fluorophore; immunospecific binding event results in fluorescence excitation transfer, where donor fluorophore emits in same spectral region as acceptor fluorophore absorbs; fluorescence (at donor excitation energies) is quenched when immune complex forms; can also be used for multiepitope Ag, with two different Ag-specific Abs, labeled with donor and acceptor fluorophores | Competitive-type assay; homogeneous; amplification only due to multiple fluorophores on Ag and Ab; requirement for dual labeling | 138 |
| | or: Unlabeled Ag in sample competes with fluorophore-labeled Ag (i.e., "fluorescer" Ag) for binding to Ab that is labeled with a "quencher" (i.e., a fluorophore whose absorption wavelength is closely matched to the fluorescence emission of the "fluorescer" Ag); binding of fluorophore-labeled Ag to quencher-labeled Ab, followed by excitation of the "fluorescer" fluorophore with light of the appropriate wavelength results in energy transfer to quencher, due to short distance between "fluorescer" and quencher in immune complex, and decrease in fluorescence; the more Ag in sample, the less "fluorescer" Ag binds to Ab and therefore the more intense the resulting fluorescence | Homogeneous, competitive assay format; increasing fluorescence obtained with increased concentration of Ag in sample, i.e., signal directly proportional to concentration of Ag in sample;[98] amplification due to multiple fluorophores on "fluorescer" Ag; various pairs of "fluorescer" and quencher pairs may be used; for example, fluorescein as "fluorescer" can be coupled with β-phycoerythrin as quencher or β-phycoerythrin as "fluorescer" may be coupled with Texas Red as quencher[98] | 98 |

| | | | |
|---|---|---|---|
| Fluorescence Modulation Immunoassay | Homogeneous assay; Ag-specific Ab and hydrolytic enzyme are in solution; fluorogenic-labeled hapten Ag is prepared by synthesis; fluorogenic label is a substrate for enzyme listed above, such that action of the enzyme on the fluorogenic-labeled hapten releases a fluorescent product upon enzymatic hydrolysis; when fluorogenic-labeled Ag and unlabeled Ag in solution are added to Ab and enzyme solution, fluorogenic-labeled Ag bound by Ab is unavailable for enzymatic hydrolysis; any free labeled Ag is acted upon by enzyme to produce fluorescent product; only applied to hapten Ag | Competitive assay format; fluorescence, produced by enzymatic action, directly proportional to amount of Ag in analyte solution; amplification by enzyme activity; results obtained with this assay correlated well with EMIT assay;[139] potentially very rapid assay, i.e., 20 min,[140] and simple to perform | 139, 140 |
| Substate-Labeled Fluorescent Immunoassay (SLFIA) | Like substrate-labeled immunoassays where Ab binding to an analyte-substrate conjugate inhibits action of enzyme on substrate; addition of analyte (Ag) in sample results in more sample analyte being bound to Ab, consequently freeing more of analyte-substrate conjugate for breakdown by enzyme; however, analyte (Ag)-substrate conjugate is constructed to contain a fluorophore linked to a quencher via an enzyme-hydrolyzable linkage, while the analyte (Ag) is conjugated to the fluorophore; free analyte-substrate conjugate (not Ab-bound) can be hydrolyzed by enzyme, thereby releasing inhibition of fluorescence due to energy transfer to quencher and resulting in full fluorescence of fluorophore | Homogeneous, competitive assay format; amplification due to enzyme action; fluorescence increases as amount of free Ag (analyte) in sample increases, so signal is directly proportional to free Ag; Li and Burd[141] use AMP bound to FMN as quencher/fluorophore pair, respectively; Ag (theophylline) is bound to AMP; if antitheophylline Ab is unavailable for binding due to amount of theophylline in sample, then nucleotide pyrophosphatase in solution hydrolyzes AMP-FMN linkage, restoring FMN fluorescence | 141 |
| | ***Chemiluminescence Procedures*** | | |
| | *Same concept as fluorescence IA; either an immunoreactant is labeled with a chemiluminescent compound or an enzyme substrate is used that produces a chemiluminescent product upon enzyme action* | *Potentially very sensitive; additional advantage in that background should be very low since no exciting light source is needed (as is the case with fluorescence)* | |
| Chemiluminescence Immunoassay (CIA) | Capture Ab used to immobilize Ag on solid surface as per standard assay; second Ag-specific Ab with chemiluminescent label (acridinium ester) binds to immunocomplex; after washing, addition of appropriate reagents produces chemiluminescence (flash) | Multiple chemiluminescent labels per Ab produce some amplification (see entry under FIA); heterogeneous noncompetitive assay; sensitivity should theoretically increase with increase in specific activity of label (i.e., increased moles of label per mole Ab); calculated detection limit <1 attomol;[142] CIA observed to be more sensitive and with wider working range than radioimmunoassay; chemiluminescent immunoassay using luminol-labeled Abs found to be as sensitive as radioimmunoassay, ~22 luminols per Ab molecule;[143] losses in quantum yield after conjugation of luminol to Ab; sensitivity of chemiluminescence immunoassay for thyrotrophin described as far more sensitive than other immunoassays[144] | 142–149 |

**TABLE 55.1** Immunoassay Amplification Schemes (continued)

| Assay | Principle | Comments | Ref. |
|---|---|---|---|
| | or: Ag labeled with chemiluminescent derivative (isoluminol); Ag-specific Ab immobilized on solid surface; unlabeled Ag in sample competes with labeled Ag for binding to Ab; following binding and washing, addition of appropriate reagents produces chemiluminescence (flash); | Heterogeneous competitive assay; light yield inversely proportional to concentration of analyte; rate of light production not constant, so timing of addition of reagents very important; chemiluminescent materials not usually present in biological samples, so no additional background from this source | |
| | or: Chemiluminescent label chemically attached to Ag; addition of free (unlabeled) Ag in sample sets up competitive assay; Ab binding to chemiluminescent-labeled Ag enhances chemiluminescence significantly; free Ag competes for Ab binding sites — the more Ag in the sample, the less enhanced chemiluminescence is seen | Homogeneous, competitive assay format; Ab concentration has to be carefully adjusted; "burst" of light requires specialized measurement apparatus | |
| Chemiluminescence Enzyme Immunoassay (CLEIA) | Immunoassay done in same way as EIA; enzyme label on one of immunoreagents; enzyme activity on substrate results in product that is chemiluminescent, usually a "glow" chemiluminescence; or action of enzyme on substrate (i.e., hydrogen peroxide) generates product that reacts with other chemicals (luminol) in reaction mixture to produce chemiluminescence; enzymes used include alkaline phosphatase, horseradish peroxidase, and β-galactosidase | Very low background; inherently highly sensitive; enhancers produce stronger, more stable light signal; proportionality over large ranges of analyte concentration; using a derivative of adamantly 1,2-dioxetane phosphate as substrate, $<10^{-20}$ mol of alkaline phosphatase were detected on membrane or in solution;[150] light emission persists for many minutes; more sensitive than colorimetric EIA or radioimmunoassay; sensitivity for CLEIA using alkaline phosphatase as enzyme label 67-fold improved over colorimetric endpoint and ~3-fold improved over time-resolved fluorimetric assay;[151] possibility for very rapid total assay time; photographic film can be used as detector; very rapid response to reach peak light emission (3 min or less);[152] integration of chemiluminescent signal for longer periods of time can be done to increase sensitivity for "glow"-type chemiluminescent reactions[153] | 150–162 |
| | Establishment of sensitivity study; enzyme (alkaline phosphatase) dilutions added to substrate + enhancer; reaction results in chemiluminescence | 1.6 zeptomol of enzyme could be detected with a signal-to-noise ratio of >6 ; luminescence increases with time and reaches plateau by 40 min; proportionality of response over five orders of magnitude concentration of enzyme; enhancer provides 400-fold increase in chemiluminescence efficiency[154] | |

| | | | |
|---|---|---|---|
| | Enzyme-labeled (peroxidase) Ab or Ag used in either noncompetitive or competitive (enzyme-labeled Ag) format in standard immunoassays; in the case of enzyme-labeled Ag assays, labeled and unlabeled (sample) Ag compete for limited number of immobilized Ab molecules; substrate + luminol + enhancer added and chemiluminescence (glow) read 30 to 60 sec later | Heterogeneous assay format; direct comparison of colorimetric, fluorimetric, and chemiluminescent assays using same immunoreagents in same assay format demonstrated that the chemiluminescent assay was 10-fold more sensitive than the colorimetric and ~2.5-fold more sensitive than the fluorimetric assay[155] | |
| | Ag (bacterial) immobilized on membrane; enzyme-conjugated Ab added, followed by addition of substrate, enhancer, and luminescent chemical; chemiluminescence detected | Single-cell luminescence detected using a CCD camera and image processor[156] | |
| | Following electrophoresis, immunoblot analysis on membranes is carried out using immunoreagents in standard way (Ag-specific pimary Ab; antispecies IgG secondary Ab conjugated to enzyme); chemiluminescent substrate added and chemiluminescence measured; can modify immunoreagent "sandwich" by using biotinylated secondary Ab and enzyme-conjugated streptavidin | Chemiluminescent substrates offer advantages over colorimetric substrates including higher sensitivity, shorter incubation times, potential for blot reuse, and ease of imaging; amplification by enzyme turnover and by localization of multiple enzyme molecules in immunoreagent "sandwich" | 163 |
| Electrochemiluminescent Labels | Electrochemiluminescent-active species bound to Ab, Ag, or hapten Ag; chemiluminescence is produced by electrochemical generation of chemical species at an electrode; example of active species reaction: ruthenium (II) tris(bipyridyl)$^{++}$, with tripropylamine; as the electrochemistry occurs, an excited state of the ruthenium complex arises, and this excited state decays with emission of a photon, i.e., chemiluminescence | Cycling of active species through electrochemical oxidation-reduction cycles at electrode produces many photons (multiplication); each electrochemiluminescent label on immunoreactant can therefore emit many photons during this cycling; also, can label Ab or protein Ag with many electrochemiluminescent labels; assay restricted to certain formats (e.g., microbeads, solution) due to requirement for close approach of labeled molecules to electrode; somewhat complicated experimental setup; very large dynamic range (six orders of magnitude); assay can be set up with competitive or noncompetitive, heterogeneous or homogeneous format | 164–166 |
| | or:<br>Ag-specific Ab immobilized on electrode surface; Ag is captured by immobilized Ab, but second detector Ab labeled with terbium chelate also binds to immune complex of Ag:Ab; either excess labeled Ab is washed away or not eliminated; since electrochemiluminescence occurs only very near electrode, removal of labeled Ab not in immune complex may not be necessary; application of voltage produces luminescence, which is measured by time-resolved techniques | Homogeneous (no wash) or heterogeneous assay format; amplification occurs only by multiple labels on Ab; low background | 167 |

**TABLE 55.1** Immunoassay Amplification Schemes (continued)

| Assay | Principle | Comments | Ref. |
|---|---|---|---|
| CLEIA with Coupled Enzyme Partners | Immunoreactions carried out in usual way; enzyme label detected indirectly as follows; primary enzyme (β-galactosidase) bound to detector Ab reacts with substrate to generate product (galactose), which becomes substrate for second enzyme (galactose dehydrogenase); product (NADH) of activity of second enzyme reacts with chemical + $O_2$ + isoluminol and microperoxidase to produce chemiluminescence | Overall procedure is rather complicated and assay time is long; detection limits for β-galactosidase reported to be ~$10^{-20}$ mol for 2-h incubation and $10^{-21}$ mol for 1000-min incubation; immunoassay using this procedure for 17α-hydroxyprogesterone reported to be 43-fold more sensitive than colorimetric ELISA;[168] immunoassays formatted as heterogeneous competitive | 168–172 |
| | also:<br>Detector Ab conjugated to alkaline phosphatase; addition of FADP results in generation of FAD; FAD reacts with glucose oxidase (apoenzyme) to form active holoenzyme; addition of glucose results in production of $H_2O_2$, which reacts with another chemical to produce chemiluminescence | Heterogeneous noncompetitive assay for thyrotrophin demonstrated; LOD of $4 \times 10^{-19}$ mol for alkaline phosphatase;[169] "glow" luminescence | |
| | or:<br>β-Galactosidase activity indirectly coupled to bacterial luciferase activity, using NADH generated from coupled-enzyme system to feed into two enzyme-catalyzed reactions, the final one involving luciferase, to generate light | Complicated procedure with many steps; long assay time; detection limit of β-galactosidase for 2-h incubation is ~$10^{-21}$ mol and ~$10^{-22}$ mol for 1000-min incubation[170] | |
| | ***Liposome Immunoassays*** | | |
| | *Liposome, either directly or indirectly, becomes part of immune complex; release of liposome contents produces signal for detection* | *Potential for large amplification because liposomes can encapsulate many molecules* | |
| Liposome Immune Lysis Assay (LILA) | Liposomes constructed with Ag-specific Ab in liposomal membrane and fluorescent dye encapsulated within; addition of Ag followed by second Ag-specific Ab + complement results in lysis, releasing dye | Homogeneous assay as described; assay time <1 h; some nonspecific lysis observed | 173–176 |
| | or:<br>Ag immobilized indirectly on liposome surface through streptavidin–biotin bridges; addition of complement + Ag-specific Ab results in lysis | Competitive homogeneous assay as described; unique method for immobilizing protein through streptavidin–biotin bridge | 177 |
| | or:<br>Ag immobilized directly on liposome surface reacts with Ag-specific Ab; addition of complement lyses liposome, releasing its contents | Homogeneous, competitive assay; competition for liposome-immobilized Ag and Ag in sample | 178 |

| | | | |
|---|---|---|---|
| Immunoliposome-Based Immunoassay | Liposomes with Ag-specific Ab incorporated in membrane and encapsulating fluorescent dye are formed; incubation with immobilized Ag results in Ab-liposome immobilization; lysis of liposomes with ethanol/detergent releases fluorescent dye; fluorescence proportional to Ag concentration | Competitive, heterogeneous assay, where free Ag concentration decreases number of liposomes subsequently trapped on Ag-treated surface; total assay time somewhat long | 179 |
| | Liposomes formed with encapsulated fluorescent dye and biotin in the liposomal membrane; immobilized Ab captures Ag; second biotinylated Ag-specific Ab binds to immobilized Ag; streptavidin forms bridge between immune complex and liposome; after washing to remove interferents, immobilized liposomes lysed, releasing fluorophore; can also be formatted with streptavidin incorporated in liposomal membrane | Heterogeneous, noncompetitive assay; fluorescence directly proportional to concentration of Ag captured; multiplication due to multiple fluorophores per liposome; stated to be as sensitive as best colorimetric ELISAs for Ag tested[179] | 180, 181 |
| Liposome Lysis-(Substrate)-Linked Immunoassay | Liposomes with Ag in membrane and encapsulating an enzyme substrate are constructed; presence of Ag-specific Ab and complement lyses liposomes, releasing substrate; addition of substrate-specific enzyme results in optically detected product | Could be used to detect specific Ab; potential for homogeneous assay | 182 |
| Liposome Lysis-(Reporter)-Linked Immunoassay | Immunoreaction of Ag-enzyme conjugate with Ab in solution inhibits enzymatic activity (e.g., phospholipase C), which can lyse liposome, thereby releasing liposomal contents, i.e., fluorescent dye molecules; free (sample) Ag competes with enzyme-labeled Ag; increasing free Ag results in increased liposome lysis | Multiplication due to many reporter molecules encapsulated within each liposome; homogeneous competitive assay as described; said to be as sensitive as heterogeneous EIA[183] | 183 |
| | Liposomes formed with encapsulated reporter enzyme; free Ag in sample competes for binding to Ag-specific Ab with Ag conjugated to melittin (a cytolysin); Ab binding to Ag–melittin conjugate inactivates cytolytic activity of melittin; Ag–melittin conjugate not bound by Ab lyses liposomes, releasing enzyme which reacts with substrate in solution | Multiplication due to many reporter enzymes per liposome plus enzyme activity; as described, homogeneous, competitive assay; Ag must be low MW for effective inhibition of melittin by Ab binding | 184, 185 |
| Immunoliposome Assay (Dye-Sensitive Photobleaching) | Immune "sandwich" (Ab-Ag-Ab attached to liposome) formed; 1 Ab has attached erythrosine; liposome has fluorescent compound embedded in membrane; when formed "sandwich" is illuminated by Hg lamp, erythrosine generates singlet $^1O_2$, which photooxidizes fluorescent dye in membrane, decreasing fluorescence; singlet $O_2$ able to oxidize membrane-embedded dye only if site of generation is very close to the liposome surface; can also be set up as competitive Ag assay with erythrosine-labeled Ag competing with unlabeled Ag | Competitive and sandwich-type formats both possible; homogeneous assay; detection limit about the same as, or better than, conventional solid-phase assays[186] | 186 |

**TABLE 55.1** Immunoassay Amplification Schemes (continued)

| Assay | Principle | Comments | Ref. |
|---|---|---|---|
| Immunoliposome Assay (Fluorescence Energy Transfer) | Liposome contains Eu-chelate embedded in liposomal membrane and membrane-attached Ab; free Ag competes with allophycocyanine-labeled Ag; when dye-labeled Ag binds to Ab and long-lived fluorescence measured, excitation of Eu results in energy transfer to allophycocyanine which is detected; energy transfer efficient only when two dyes close together | Competitive, homogeneous assay as described; potential for very low background; only demonstrated with biotin as Ag | 187 |
| | ***Expression Immunoassay*** | | |
| | *Immunoreactant complex formed as usual; DNA attached to immune complex codes for protein that, when expressed, produces signal for detection* | *Potentially large amplification, both because of number of protein molecules synthesized, and the fact that the synthesized protein is an enzyme* | |
| | Immobilized Ag reacted with biotinylated Ag-specific Ab; streptavidin bridge couples biotinylated Ab to biotinylated DNA segment, which codes for α-peptide of β-galactosidase; addition of transcription/translation system, produces α-peptide, which then reacts with inactive M15 protein to reconstitute active enzyme; addition of substrate results in a signal, e.g., chemiluminescence | Assay carried out as solid-phase, heterogeneous assay; transcription/translation system potentially multiplies final active enzyme many-fold over other methodologies; theoretically, background should be zero, since active enzyme should not be present unless immune complex generates component needed for activity; assay procedure more complex (more steps) than usual ELISA and of longer duration; potential for very high sensitivity, depending on chemical (chemiluminescent, fluorescent) or enzyme chosen[188] | 188 |
| | Same general design as above, but DNA codes for firefly luciferase; assay endpoint is bioluminescence | Estimated 12 to 14 molecules of luciferase protein produced from each DNA template; comparison with enzyme amplified, time-resolved fluoroimmunoassay showed expression immunoassay to be considerably more sensitive[189] | 189 |
| | ***Immunoassays/Chromatography*** | | |
| Immunoaffinity Chromatography | Ag-specific Ab immobilized on columns; analytes in sample labeled *in toto* with fluorescent dye; analytes bound on Ag-specific immunoaffinity columns, and eluted sequentially; detection by laser-induced fluorescence | Potentially rapid and sensitive; may be nonuniform labeling of analytes in sample; multiple fluorophores may be attached to single analyte molecules; detection levels at pg/ml for various cytokines | 190 |
| | ***Surface-Enhanced Raman Spectroscopy (SERS) Immunoassays*** | | |
| | *Same principle as usual IA, but SERS-active chemicals are used as labels or substrates; SERS effect produces large signal amplification* | | |

| | | | |
|---|---|---|---|
| Heterogeneous Indirect Immunoassay | Capture Ab adsorbed on rough silver surface captures Ag analyte; Ab labeled with SERS-active molecules binds to Ag; SERS signal or surface-enhanced resonance Raman scattering signal is detected | Multiple SERS labels on each Ab; $10^3$ to $10^6$ (or higher for resonance Raman) enhancement in Raman signal from SERS effect; requires close apposition of SERS label to silver surface; potential for minimal assay steps (e.g., no washes) since only SERS label bound immunologically to silver surface produces signal; marked reduction in fluorescence background; special equipment needed to generate surfaces; laser required for measurements | 191 |
| | As above, but capture Ab adsorbed on gold surface; Ag captured and detected by Ab attached to gold nanoparticles, which in turn are labeled with SERS-active molecules | Simultaneous detection of multiple Ags using different SERS active labels; very narrow Raman bands provide potential for sensitive multianalyte detection; Ab bound to nanoparticles noncovalently; possible problem with Ab dissociation/reassociation leading to "cross talk" | 192 |
| EIA with SERS detection of Enzyme Product | Enzyme immunoassay carried out in standard way on microplates; after enzyme reaction (peroxidase), reaction product (azoaniline) is transferred to silver substrate and the surface-enhanced resonance Raman signal is determined; intensity of signal correlates directly with Ag concentration | See above; advantages over other techniques include low background, high selectivity, amplification from enzyme, as well as SERS effect; disadvantages are the elaborate equipment and the extra detection step at the end of the assay; detection limit stated to be $10^{-15}$ mol/l (for IgG)[193] | 193 |
| | ***Imprinted Polymer-Based Immunoassays*** | | |
| | *Polymer with imprinted molecular sites is the immunorecognition element rather than Ab; otherwise IA is the same* | | |
| Chemiluminescent ELISA | Imprinted polymer serves as recognition element instead of Ab; polymer immobilized in tubes or wells; antigen (hapten) labeled with enzyme competes with Ag in analyte solution for imprinted sites in the polymer; following washing, enzyme activity associated with polymer determined by chemiluminescent assay | Heterogeneous, competitive assay format; signal inversely proportional to concentration of Ag in analyte; detection limit not as low as when usual ELISA assay with AB was used; useful range extends over four orders of magnitude; rapid and potentially robust assay; whether molecularly imprinted polymers can serve as Ab substitutes in the wide range of Ags currently detected by immunoassay still an open question | 194, 195 |
| | ***Miscellaneous Heterogeneous Immunoassays*** | | |
| | *These procedures involve standard immunoreagents but also entail the use of beads, particles, crystals, etc., which serve as solid substrates for immunoreactions* | | |

**TABLE 55.1** Immunoassay Amplification Schemes (continued)

| Assay | Principle | Comments | Ref. |
|---|---|---|---|
| Bacterial Magnetic Particle-Based Immunoassay | Magnetic particles (with lipid membranes) produced by bacteria are used as solid phase; Ag-specific Ab covalently immobilized on particles captures Ag; second (detection) Ab has enzyme label; detection by chemiluminescence (for enzyme-labeled Ab) | "Natural" magnetic particles disperse better than artificial; small size (50 to 100 nm) gives very rapid reaction kinetics; chemiluminescence using enzyme amplification was found to be very sensitive with detection limit of 6.7 zeptomol;[196] detection range extended over 1 to $10^5$ fg/ml; a rapid chemiluminescence assay (~10 min) was also devised, but with detection limits of ~10 ng of IgG/ml | 196 |
| | or: Magnetic particles from bacteria are labeled with fluorescently labeled Ab; addition of Ag results in particle aggregation, which produces decrease in fluorescence due to self quenching by fluorophore | Amplification results from numerous fluorescently labeled Ab molecules bound to each magnetic particle; inverse relationship between Ag concentration and fluorescence; bacterial magnetic-particle assay found to be more sensitive than assay using artificial magnetic particles;[197] applied to detection of bacterial cells (*E. coli*) with limit of detection ~$10^2$ cells/ml[198] | 197, 198 |
| Phosphor Crystals | Immunoreagents (Ab, avidin, protein A) directly linked to phosphor particles; assays carried out on membranes or cells; Ag immobilized on membrane or on cell surface detected by immunoreagent sandwich incorporating phosphor conjugates in one or more layers of the immune "sandwich"; fluorescent detection | Amplification based on formation of immune sandwich (i.e., biotinylated Ab, avidin); fluorescence not affected by environmental factors and does not fade; sensitivity found to be as good as ELISA with colorimetric endpoint[199] | 199 |
| | | Also applications to immunocytochemical analyses using immunoreagents linked to phosphor particles | 200 |
| Microspheres with CARD Amplification | Microspheres with encoding dye (to permit performance of multiple assays simultaneously) were derivatized with Ag on their surface; enzyme-labeled Ab was added to mixture of microspheres + Ag-containing analyte solution; any enzyme-labeled Ab bound to bead-immobilized Ag is detected using CARD technology; fluorescent tyramide is deposited on microspheres and fluorescence read | Competitive, heterogeneous assay format; fluorescent signal decreases as Ag concentration in analyte solution increases; encoding dye on microspheres allows multiplex assays; choice of fluorescent dyes has to be carefully made to allow discrimination between fluorescence of microsphere and tyramide derivative; preparation of Ag-derivatized, dye-encoded microspheres somewhat complex | 201 |

| | | | |
|---|---|---|---|
| Immunomagnetic Separation + Immunoassay | Immunoreagents (e.g., capture Abs) immobilized on magnetic beads, so that analyte can be extracted/isolated from bulk sample; rest of immunoassay can take place in usual manner (i.e., formation of immuno- "sandwich" with or without enzyme conjugate, incubation with substrate and detection) | Total assay time <1 h;[202] fluorescence ELISA using immunomagnetic separation was tenfold more sensitive than standard ELISA[202] | 202, 203 |
| | | Electrochemiluminescence-based assay (see under Chemiluminescence) compared with immunomagnetic separation/fluorescence ELISA; electrochemiluminescence-based assay found to be more sensitive in general, perhaps tenfold;[203] both assays could be completed in <1 h | |
| | ***Immunogold-Based Immunoassays*** | | |
| | *Gold nanoparticles serve as label on immunoreagents and as nuclei for deposition of silver atoms* | | |
| Immunogold/silver Assay | Ag immobilized on membrane or in histological specimen; Ag-specific Ab binds to Ag; secondary antispecies IgG Ab labeled with gold nanoparticles binds to primary Ab; silver subsequently precipitated on gold particles | Addition of silver strongly amplifies gold immunostaining; used for immunoblots/dot blotting; 2-h incubation period; found to be more sensitive than PAP or ABC method (with same time of incubation with detector Ab)[204] | 204, 205 |
| | ***Immunoassays Incorporating Aequorin*** | | |
| | Immune "sandwich" with primary Ab and Ag formed in usual way; secondary (antispecies IgG Ab) Ab, which is biotinylated, is then incorporated into complex; subsequent sequential additions of streptavidin and biotinylated aequorin are made, resulting in immobilization of aequorin at site of immune complex; addition of $Ca^{2+}$ results in luminescent flash, which can be detected | Applied to both microwell and membrane blot assays; heterogeneous format; amplification because of multiple binding sites on streptavidin for biotin; detection may be challenging since light flash decays very significantly by 1 sec; immunoassay for Forsman antigen using aequorin was demonstrated as detecting less than 40 fmol | 206 |

## References


1. Nakane, P.K. and Pierce, G.B., Enzyme-labeled antibodies: preparation and application for the localization of antigens, *J. Histochem. Cytochem.*, 14, 929, 1966.
2. Avrameas, S. and Guilbert, B., A method for quantitative determination of cellular immunoglobulins by enzyme-labeled antibodies, *Eur. J. Immunol.*, 1, 394, 1971.
3. Boyd, S. and Yamazaki, H., Instantaneous cloth-based enzyme immunoassay for the semi-quantitative visual determination of antibodies, *Immunol. Invest.*, 26(3), 313, 1997.
4. Engvall, E. and Perlmann, P., Enzyme-linked immunoadsorbent assay (ELISA): quantitative assay of immunoglobulin G, *Immunochemistry*, 8, 871, 1971.
5. Cleveland, P.H., Richman, D.D., Oxman, M.N., Wickham, M.G., Binder, P.S., and Worthen, D.M., Immobilization of viral-antigens on filter-paper for a I125 staphylococcal protein a immunoassay — rapid and sensitive technique for detection of herpes-simplex virus-antigens and anti-viral antibodies, *J. Immunol. Meth.*, 29(4), 369, 1979.
6. Kemeny, D.M., Titration of antibodies, *J. Immunol. Meth.*, 150(1–2), 57, 1992.
7. Glassy, M.C., Handley, H.H., Cleveland, P.H., and Royston, I., An enzyme immunofiltration assay useful for detecting human monoclonal-antibody, *J. Immunol. Meth.*, 58(1–2), 119, 1983.
8. Kohl, J., Ruker, F., Himmler, G., Razazzi, E., and Katinger, H., Cloning and expression of an HIV-1 specific single-chain Fv region fused to *Escherichia coli* alkaline-phosphatase, *Ann. N.Y. Acad. Sci.*, 646, 106, 1991.
9. Weiss, E. and Orfanoudakis, G., Application of an alkaline-phosphatase fusion protein system suitable for efficient screening and production of Fab-enzyme conjugates in *Escherichia coli*, *J. Biotechnol.*, 33(1), 43, 1994.
10. Karawajew, L., Behrsing, O., Kaiser, G., and Micheel, B., Production and ELISA application of bispecific monoclonal-antibodies against fluorescein isothiocyanate (Fitc) and horseradish-peroxidase (Hrp), *J. Immunol. Meth.*, 111(1), 95, 1988.
11. Milstein, C. and Cuello, A.C., Hybrid hybridomas and their use in immunohistochemistry, *Nature*, 305(5934), 537, 1983.
12. Porstmann, B., Avrameas, S., Ternynck, T., Porstmann, T., Micheel, B., and Guesdon, J.L., An antibody chimera technique applied to enzyme-immunoassay for human α-1-fetoprotein with monoclonal and polyclonal antibodies, *J. Immunol. Meth.*, 66(1), 179, 1984.
13. Guesdon, J.L., Velarde, F.N., and Avrameas, S., Solid-phase immunoassays using chimera antibodies prepared with monoclonal or polyclonal anti-enzyme and anti-erythrocyte antibodies, *An. Immunol.*, C134(2), 265, 1983.
14. Kobatake, E., Nishimori, Y., Ikariyama, Y., Aizawa, M., and Kato, S., Application of a fusion protein, metapyrocatechase protein-a, to an enzyme-immunoassay, *Anal. Biochem.*, 186(1), 14, 1990.
15. Avrameas, S., Indirect immunoenzyme techniques for the intracellular detection of antigens, *Immunochemistry*, 6, 925, 1969.
16. Butler, J.E., McGivern, P.L., and Swanson, P., Amplification of enzyme-linked immunosorbent assay (Elisa) in detection of class-specific antibodies, *J. Immunol. Meth.*, 20, 365, 1978.
17. Clark, C.A., Downs, E.C., and Primus, F.J., An unlabeled antibody method using glucose-oxidase antiglucose oxidase complexes (GAG) — a sensitive alternative to immunoperoxidase for the detection of tissue antigens, *J. Histochem. Cytochem.*, 30(1), 2734, 1982.
18. Cordell, J.L., Falini, B., Erber, W.N., Ghosh, A.K., Abdulaziz, Z., Macdonald, S., Pulford, K.A.F., Stein, H., and Mason, D.Y., Immunoenzymatic labeling of monoclonal-antibodies using immune-complexes of alkaline-phosphatase and monoclonal anti-alkaline phosphatase (APAAP complexes), *J. Histochem. Cytochem.*, 32(2), 219, 1984.
19. Mason, D.Y., Cordell, J.L., Abdulaziz, Z., Naiem, M., and Bordenave, G., Preparation of peroxidase–anti-peroxidase (Pap) complexes for immunohistological labeling of monoclonal-antibodies, *J. Histochem. Cytochem.*, 30(11), 1114, 1982.

20. Sternberger, L.A., Hardy, P.H., Cuculis, J.J., and Meyer, H.G., The unlabeled antibody-enzyme method of immunohistochemistry: preparation and properties of soluble antigen-antibody complex (horseradish peroxidase–anti-horseradish peroxidase) and its use in identification of spirochetis, *J. Histochem. Cytochem.*, 18, 315, 1970.
21. Ternynck, T., Gregoire, J., and Avrameas, S., Enzyme anti-enzyme monoclonal-antibody soluble immune-complexes (EMAC) — their use in quantitative immunoenzymatic assays, *J. Immunol. Meth.*, 58(1–2), 109, 1983.
22. Jasani, B., Thomas, N.D., Navabi, H., Millar, D.M., Newman, G.R., Gee, J., and Williams, E.D., Dinitrophenyl (DNP) hapten sandwich staining (DHSS) procedure — a 10 year review of its principle reagents and applications, *J. Immunol. Meth.*, 150(1–2), 193, 1992.
23. Jasani, B., Newman, G.R., Stanworth, D.R., and Williams, E.D., Immunohistochemical localization of tissue receptors using the dinitrophenyl (DNP) hapten sandwich procedure, *J. Pathol.*, 138(1), 50, 1982.
24. Erdile, L.F., Smith, D., and Berd, D., Whole cell ELISA for detection of tumor antigen expression in tumor samples, *J. Immunol. Meth.*, 258(1–2), 47, 2001.
25. Cobbold, S.P. and Waldmann, H., A rapid solid-phase enzyme-linked binding assay for screening monoclonal-antibodies to cell-surface antigens, *J. Immunol. Meth.*, 44(2), 125, 1981.
26. Sedgwick, J.D. and Czerkinsky, C., Detection of cell-surface molecules, secreted products of single cells and cellular proliferation by enzyme-immunoassay, *J. Immunol. Meth.*, 150(1–2), 159, 1992.
27. Suter, L., Bruggen, J., and Sorg, C., Use of an enzyme-linked immunosorbent-assay (ELISA) for screening of hybridoma antibodies against cell-surface antigens, *J. Immunol. Meth.*, 39, 407, 1980.
28. Sedgwick, J.D. and Holt, P.G., A solid-phase immunoenzymatic technique for the enumeration of specific antibody-secreting cells, *J. Immunol. Meth.*, 57(1–3), 301, 1983.
29. Kalyuzhny, A. and Stark, S., A simple method to reduce the background and improve well-to-well reproducibility of staining in ELISPOT assays, *J. Immunol. Meth.*, 257(1–2), 93, 2001.
30. Czerkinsky, C.C., Nilsson, L.A., Nygren, H., Ouchterlony, O., and Tarkowski, A., A solid-phase enzyme-linked immunospot (ELISPOT) assay for enumeration of specific antibody-secreting cells, *J. Immunol. Meth.*, 65(1–2), 109, 1983.
31. Rowell, F.J., Miao, Z.F., Reeves, R.N., and Cumming, R.H., Direct on-filter immunoassay of some beta-lactam antibiotics for rapid analysis of drug captured from the workplace atmosphere, *Analyst*, 122(12), 1505, 1997.
32. Paffard, S.M., Miles, R.J., Clark, C.R., and Price, R.G., A rapid and sensitive enzyme linked immunofilter assay (ELIFA) for whole bacterial cells, *J. Immunol. Meth.*, 192(1–2), 133, 1996.
33. Paffard, S.M., Miles, R.J., Clark, C.R., and Price, R.G., Amplified enzyme-linked-immunofilter assays enable detection of 50-10(5) bacterial cells within 1 hour, *Anal. Biochem.*, 248(2), 265, 1997.
34. Ijsselmuiden, O.E., Meinardi, M., Vandersluis, J.J., Menke, H.E., Stolz, E., and Vaneijk, R.V.W., Enzyme-linked immunofiltration assay for rapid serodiagnosis of syphilis, *Eur. J. Clin. Microbiol. Infect. Dis.*, 6(3), 281, 1987.
35. Clark, C.R., Hines, K.K., and Mallia, A.K., 96-Well apparatus and method for use in enzyme-linked immunofiltration assay (ELIFA), *Biotechnol. Tech.*, 7(6), 461, 1993.
36. Van Weemen, B.K. and Schuurs, A.H.W.M., Immunoassay using antigen-enzyme conjugate, *FEBS Lett.*, 15(3), 232, 1971.
37. Peterhans, A., Mecklenburg, M., Meussdoerffer, F., and Mosbach, K., A simple competitive enzyme-linked-immunosorbent-assay using antigen-beta-galactosidase fusions, *Anal. Biochem.*, 163(2), 470, 1987.
38. Blake, C., Albassam, M.N., Gould, B.J., Marks, V., Bridges, J.W., and Riley, C., Simultaneous enzyme-immunoassay of two thyroid-hormones, *Clin. Chem.*, 28(7), 1469, 1982.
39. Ishikawa, E., Hashida, S., Kohno, T., and Hirota, K., Ultrasensitive enzyme-immunoassay, *Clin. Chim. Acta*, 194(1), 51, 1990.

40. Hashida, S., Tanaka, K., Kohno, T., and Ishikawa, E., Novel and ultrasensitive sandwich enzyme-immunoassay (sandwich transfer enzyme-immunoassay) for antigens, *Anal. Lett.*, 21(7), 1141, 1988.
41. Hashinaka, K., Hashida, S., Nishikata, I., Adachi, A., Oka, S., and Ishikawa, E., Recombinant p51 as antigen in an immune complex transfer enzyme immunoassay of immunoglobulin G antibody to human immunodeficiency virus type 1, *Clin. Diagn. Lab. Immunol.*, 7(6), 967, 2000.
42. Obzansky, D.M., Rabin, B.R., Simons, D.M., Tseng, S.Y., Severino, D.M., Eggelte, H., Fisher, M., Harbron, S., Stout, R.W., and Dipaolo, M.J., Sensitive, colorimetric enzyme amplification cascade for determination of alkaline-phosphatase and application of the method to an immunoassay of thyrotropin, *Clin. Chem.*, 37(9), 1513, 1991.
43. Kobatake, E., Iwai, T., Ikariyama, Y., and Aizawa, M., Bioluminescent immunoassay with a protein-a luciferase fusion protein, *Anal. Biochem.*, 208(2), 300, 1993.
44. Lindbladh, C., Persson, M., Bulow, L., Stahl, S., and Mosbach, K., The design of a simple competitive Elisa using human proinsulin-alkaline phosphatase conjugates prepared by gene fusion, *Biochem. Biophys. Res. Commun.*, 149(2), 607, 1987.
45. Ci, Y.X., Qin, Y., Chang, W.B., Li, Y.Z., Yao, F.J., and Zhang, W., Fluorometric mimetic enzyme-immunoassay of methotrexate, *Fresenius J. Anal. Chem.*, 349(4), 317, 1994.
46. Ci, Y.X., Qin, Y., Chang, W.B., and Li, Y.Z., Application of a mimetic enzyme for the enzyme-immunoassay for alpha-1-fetoprotein, *Anal. Chim. Acta*, 300(1–3), 273, 1995.
47. Genfa, Z. and Dasgupta, P.K., Hematin as a peroxidase substitute in hydrogen-peroxide determinations, *Anal. Chem.*, 64(5), 517, 1992.
48. Zhu, Q.Z., Zheng, X.Y., Xu, J.G., Li, W.Y., and Liu, F.H., Application of hemin as a labeling reagent in mimetic enzyme immunoassay for hepatitis B surface antigen, *Anal. Lett.*, 31(6), 963, 1998.
49. Engvall, E., Enzyme immunoassay ELISA and EMIT, in *Immunochemical Techniques*, Van Vunakis, H. and Langone, J.J., Eds., Academic Press, New York, 1980, pp. 419–439.
50. Dona, V., Homogeneous colorimetric enzyme-inhibition immunoassay for cortisol in human-serum with Fab anti-glucose 6-phosphate dehydrogenase as a label modulator, *J. Immunol. Meth.*, 82(1), 65, 1985.
51. Rubenstein, K.E., Schneider, R.S., and Ullman, E.F., "Homogeneous" enzyme immunoassay — a new immunochemical technique, *Biochem. Biophys. Res. Commun.*, 47(4), 846, 1972.
52. Burd, J.F., Carrico, R.J., Fetter, M.C., Buckler, R.T., Johnson, R.D., Boguslaski, R.C., and Christner, J.E., Specific protein-binding reactions monitored by enzymatic-hydrolysis of ligand-fluorescent dye conjugates, *Anal. Biochem.*, 77(1), 56, 1977.
53. Burd, J.F., Wong, R.C., Feeney, J.E., Carrico, R.J., and Boguslaski, R.C., Homogeneous reactant-labeled fluorescent immunoassay for therapeutic drugs exemplified by gentamicin determination in human-serum, *Clin. Chem.*, 23(8), 1402, 1977.
54. Place, M.A., Carrico, R.J., Yeager, F.M., Albarella, J.P., and Boguslaski, R.C., A colorimetric immunoassay based on an enzyme-inhibitor method, *J. Immunol. Meth.*, 61(2), 209, 1983.
55. Nishizono, I., Ashihara, Y., Tsuchiya, H., Tanimoto, T., and Kasahara, Y., Enzyme inhibitory homogeneous immunoassay for high molecular-weight antigen (Ii), *J. Clin. Lab. Anal.*, 2(3), 143, 1988.
56. Ashihara, Y., Nishizono, I., Tsuchiya, H., Tanimoto, T., and Kasahara, Y., Homogeneous protein enzyme-immunoassay for macromolecular antigens, *Clin. Chem.*, 31(6), 904, 1985.
57. Carrico, R.J., Christner, J.E., Boguslaski, R.C., and Yeung, K.K., Method for monitoring specific binding reactions with cofactor labeled ligands, *Anal. Biochem.*, 72(1–2), 271, 1976.
58. Engel, W.D. and Khanna, P.L., Cedia Invitro diagnostics with a novel homogeneous immunoassay technique — current status and future prospects, *J. Immunol. Meth.*, 150(1–2), 99, 1992.
59. Henderson, D.R., Friedman, S.B., Harris, J.D., Manning, W.B., and Zoccoli, M.A., Cedia, a new homogeneous immunoassay system, *Clin. Chem.*, 32(9), 1637, 1986.
60. Loor, R., Shindelman, J., Singh, H., and Khanna, P.L., Homogeneous enzyme immunoassay for high molecular weight analytes using recombinant enzyme fragments, *J. Clin. Immunoassay*, 14, 47, 1991.

61. Litman, D.J., Hanlon, T.M., and Ullman, E.F., Enzyme channeling immunoassay — a new homogeneous enzyme-immunoassay technique, *Anal. Biochem.*, 106(1), 223, 1980.
62. Guesdon, J.L., Ternynck, T., and Avrameas, S., Use of avidin-biotin interaction in immunoenzymatic techniques, *J. Histochem. Cytochem.*, 27(8), 1131, 1979.
63. Wilchek, M. and Bayer, E.A., The avidin biotin complex in immunology, *Immunol. Today*, 5(2), 39, 1984.
64. Avrameas, S., Amplification systems in immunoenzymatic techniques, *J. Immunol. Meth.*, 150(1–2), 23, 1992.
65. O'Connor, E., Roberts, E.M., and Davies, J.D., Amplification of cytokine-specific ELISAs increases the sensitivity of detection to 5–20 picograms per milliliter, *J. Immunol. Meth.*, 229(1–2), 155, 1999.
66. Van Gijlswijk, R.P.M., van Gijlswijk-Janssen, D.J., Raap, A.K., Daha, M.R., and Tanke, H.J., Enzyme-labelled antibody avidin conjugates, new flexible and sensitive immunochemical reagents, *J. Immunol. Meth.*, 189(1), 117, 1996.
67. Hsu, S.M., Raine, L., and Fanger, H., Use of avidin-biotin-peroxidase complex (ABC) in immunoperoxidase techniques — a comparison between ABC and unlabeled antibody (PAP) procedures, *J. Histochem. Cytochem.*, 29(4), 577, 1981.
68. Johansson, A., Ellis, D.H., Bates, D.L., Plumb, A.M., and Stanley, C.J., Enzyme amplification for immunoassays — detection limit of 100th of an attomole, *J. Immunol. Meth.*, 87(1), 7, 1986.
69. Bobrow, M.N., Shaughnessy, K.J., and Litt, G.J., Catalyzed reporter deposition, a novel method of signal amplification. 2. Application to membrane immunoassays, *J. Immunol. Meth.*, 137(1), 103, 1991.
70. Bhattacharya, R., Bhattacharya, D., and Dhar, T.K., A novel signal amplification technology based on catalyzed reporter deposition and its application in a Dot-ELISA with ultra high sensitivity, *J. Immunol. Meth.*, 227(1–2), 31, 1999.
71. Bobrow, M.N., Harris, T.D., Shaughnessy, K.J., and Litt, G.J., Catalyzed reporter deposition, a novel method of signal amplification — application to immunoassays, *J. Immunol. Meth.*, 125(1–2), 279, 1989.
72. Bobrow, M.N., Litt, G.J., Shaughnessy, K.J., Mayer, P.C., and Conlon, J., The use of catalyzed reporter deposition as a means of signal amplification in a variety of formats, *J. Immunol. Meth.*, 150(1–2), 145, 1992.
73. Van Gijlswijk, R.P.M., Zijlmans, H., Wiegant, J., Bobrow, M.N., Erickson, T.J., Adler, K.E., Tanke, H.J., and Raap, A.K., Fluorochrome-labeled tyramides: use in immunocytochemistry and fluorescence *in situ* hybridization, *J. Histochem. Cytochem.*, 45(3), 375, 1997.
74. Mize, P.D., Hoke, R.A., Linn, C.P., Reardon, J.E., and Schulte, T.H., Dual-enzyme cascade — an amplified method for the detection of alkaline-phosphatase, *Anal. Biochem.*, 179(2), 229, 1989.
75. Hauber, R. and Geiger, R., A sensitive, bioluminescent-enhanced detection method for DNA dot-hybridization, *Nucleic Acids Res.*, 16(3), 1213, 1988.
76. Geiger, R. and Miska, W., Bioluminescence enhanced enzyme-immunoassay — new ultrasensitive detection systems for enzyme immunoassays. 2., *J. Clin. Chem. Clin. Biochem.*, 25(1), 31, 1987.
77. Aalberse, R.C., Quantitative fluoroimmunoassay, *Clin. Chim. Acta*, 48(1), 109, 1973.
78. Blanchard, G.C. and Gardner, R., Two immunofluorescent methods compared with a radial immunodiffusion method for measurement of serum immunoglobulins, *Clin. Chem.*, 24(5), 808, 1978.
79. Brandtzaeg, P., Evaluation of immunofluorescence with artificial sections of selected antigenicity, *Immunology*, 22, 177, 1972.
80. Capel, P.J.A., Quantitative immunofluorescence method based on covalent coupling of protein to sepharose beads, *J. Immunol. Meth.*, 5(2), 165, 1974.
81. Coons, A.H., Histochemistry with labeled antibody, *Int. Rev. Cytol.*, 5, 1, 1956.
82. Coons, A.H., Creech, H.J., and Jones, R.N., Immunological properties of an antibody containing a fluorescent group, *Proc. Soc. Exp. Biol. Med.*, 47, 200, 1941.
83. Coons, A.H. and Kaplan, M.H., Localization of antigen in tissue cells. II. Improvements in a method for the detection of fluorescent antibody, *J. Exp. Med.*, 91, 1, 1950.

84. Chard, T. and Sykes, A., Fluoroimmunoassay for human choriomammotropin, *Clin. Chem.*, 25(6), 973, 1979.
85. Hebert, G.A., Pittman, B., and Cherry, W.B., Factors affecting the degree of nonspecific staining given by fluorescein isothiocyanate labeled globulins, *J. Immunol.*, 98(6), 1204, 1967.
86. Katsh, S., Leaver, F.W., Reynolds, J.S., and Katsh, G.F., Simple, rapid fluorometric assay for antigens, *J. Immunol. Meth.*, 5(2), 179, 1974.
87. Killander, D., Levin, A., Inoue, M., and Klein, E., Quantification of immunofluorescence on individual erythrocytes coated with varying amounts of antigen, *Immunology*, 19, 151, 1970.
88. Odonnell, C.M. and Suffin, S.C., Fluorescence immunoassays, *Anal. Chem.*, 51(1), A33, 1979.
89. Sainte-Marie, G., A paraffin embedding technique for studies employing immunofluorescence, *J. Histochem. Cytochem.*, 10, 250, 1962.
90. Skurkovich, S.V., Olshansky, A.J., Samoilova, R.S., and Eremkina, E.I., Quantitative-determination of human leukocyte interferon by micro-fluorometric immunoassay with FITC-labeled anti-interferon immunoglobulin, *J. Immunol. Meth.*, 19(2–3), 119, 1978.
91. Sykes, A. and Chard, T., Two-site immunofluorometric assay for pregnancy-specific beta-1-glycoprotein (Sp1), *Clin. Chem.*, 26(8), 1224, 1980.
92. Crawford, H.J., Wood, R.M., and Lessof, M.H., Detection of antibodies by fluorescent-spot techniques, *Lancet*, 2, 1173, 1959.
93. Toussaint, A.J. and Anderson, R.I., Soluble antigen fluorescent-antibody technique, *Appl. Microbiol.*, 13(4), 552, 1965.
94. Strom, R. and Klein, E., Fluorometric quantitation of fluorescein-coupled antibodies attached to the cell membrane, *Proc. Natl. Acad. Sci. U.S.A.*, 63, 1157, 1970.
95. Burgett, M.W., Fairfield, S.J., and Monthony, J.F., Solid-phase fluorescent immunoassay for quantitation of C4 component of human complement, *J. Immunol. Meth.*, 16(3), 211, 1977.
96. Chao, J., DeBiasio, R., Zhu, Z.R., Giuliano, K.A., and Schmidt, B.F., Immunofluorescence signal amplification by the enzyme-catalyzed deposition of a fluorescent reporter substrate (CARD), *Cytometry*, 23(1), 48, 1996.
97. Kerstens, H.M.J., Poddighe, P.J., and Hanselaar, A., A novel *in situ* hybridization signal amplification method based on the deposition of biotinylated tyramine, *J. Histochem. Cytochem.*, 43(4), 347, 1995.
98. Kronick, M.N. and Grossman, P.D., Immunoassay techniques with fluorescent phycobiliprotein conjugates, *Clin. Chem.*, 29(9), 1582, 1983.
99. Rotman, B., Zderic, J.A., and Edelstein, M., Fluorogenic substrates for beta-D-galactosidases and phosphatases derived from fluorescein (3, 6-dihydroxyfluoran) and its monomethyl ether, *Proc. Natl. Acad. Sci. U.S.A.*, 50(1), 1, 1963.
100. Revel, H.R., Luria, S.E., and Rotman, B., Biosynthesis of beta-D-galactosidase controlled by phage-carried genes. I. Induced beta-galactosidase biosynthesis after transduction of gene z+ by phage, *Proc. Natl. Acad. Sci. U.S.A.*, 47(1956), 1961.
101. Kato, K., Hamaguchi, Y., Fukui, H., and Ishikawa, E., Enzyme-linked immunoassay. 2. Simple method for synthesis of rabbit antibody-beta-D-galactosidase complex and its general applicability, *J. Biochem.*, 78(2), 423, 1975.
102. Ruan, K.H., Kulmacz, R.J., Wilson, A., and Wu, K.K., Highly sensitive fluorometric enzyme-immunoassay for prostaglandin-H synthase solubilized from cultured cells, *J. Immunol. Meth.*, 162(1), 23, 1993.
103. Shalev, A., Greenberg, A.H., and McAlpine, P.J., Detection of attograms of antigen by a high-sensitivity enzyme-linked immunoabsorbent assay (HS-ELISA) using a fluorogenic substrate, *J. Immunol. Meth.*, 38(1–2), 125, 1980.
104. Miyai, K., Ishibashi, K., and Kawashima, M., Two-site immunoenzymometric assay for thyrotropin in dried blood samples on filter paper, *Clin. Chem.*, 27(8), 1421, 1981.
105. Ishikawa, E. and Kato, K., Ultrasensitive enzyme immunoassay, *Scand. J. Immunol.*, 8(Suppl. 7), 43, 1981.

106. Nargessi, R.D., Ackland, J., Hassan, M., Forrest, G.C., Smith, D.S., and Landon, J., Magnetizable solid-phase fluorimmunoassay of thyroxine by a sequential addition technique, *Clin. Chem.*, 26(12), 1701, 1980.
107. Birkmeyer, R.C., Diaco, R., Hutson, D.K., Lau, H.P., Miller, W.K., Neelkantan, N.V., Pankratz, T.J., Tseng, S.Y., Vickery, D.K., and Yang, E.K., Application of novel chromium dioxide magnetic particles to immunoassay development, *Clin. Chem.*, 33(9), 1543, 1987.
108. McHugh, T.M., Viele, M.K., Chase, E.S., and Recktenwald, D.J., The sensitive detection and quantitation of antibody to HCV by using a microsphere-based immunoassay and flow cytometry, *Cytometry*, 29(2), 106, 1997.
109. Sutherland, R.M., Dahne, C., Place, J.F., and Ringrose, A.S., Optical detection of antibody antigen reactions at a glass-liquid interface, *Clin. Chem.*, 30(9), 1533, 1984.
110. Kronick, M.N. and Little, W.A., New immunoassay based on fluorescence excitation by internal-reflection spectroscopy, *J. Immunol. Meth.*, 8(3), 235, 1975.
111. Thompson, N.L. and Axelrod, D., Immunoglobulin surface-binding kinetics studied by total internal-reflection with fluorescence correlation spectroscopy, *Biophys. J.*, 43(1), 103, 1983.
112. Dandliker, W.B., Kelly, R.J., Dandliker, J., Farquhar, J., and Levin, J., Fluorescence polarization immunoassay — theory and experimental method, *Immunochemistry*, 10(4), 219, 1973.
113. Dandliker, W.B. and de Saussure, V.A., Fluorescence polarization in immunochemistry, *Immunochemistry*, 7, 799, 1970.
114. Dandliker, W.B. and Feigen, G.A., Quantification of the antigen-antibody reaction by the polarization of fluorescence, *Biochem. Biophys. Res. Commun.*, 5(4), 299, 1961.
115. Jolley, M.E., Fluorescence polarization immunoassay for the determination of therapeutic drug levels in human-plasma, *J. Anal. Toxicol.*, 5(5), 236, 1981.
116. Guo, X.Q., Castellano, F.N., Li, L., and Lakowicz, J.R., Use of a long lifetime Re(I) complex in fluorescence polarization immunoassays of high-molecular weight analytes, *Anal. Chem.*, 70(3), 632, 1998.
117. Terpetschnig, E., Szmacinski, H., and Lakowicz, J.R., Fluorescence polarization immunoassay of a high-molecular-weight antigen based on a long-lifetime Ru–ligand complex, *Anal. Biochem.*, 227(1), 140, 1995.
118. Terpetschnig, E., Szmacinski, H., and Lakowicz, J.R., Long-lifetime metal-ligand complexes as probes in biophysics and clinical chemistry, in *Methods in Enzymology*, Vol. 278, *Fluorescence Spectroscopy*, Brand, L. and Johnson, M.L., Eds., Academic Press, San Diego, 1997, pp. 295–321.
119. Diamandis, E.P., Morton, R.C., Reichstein, E., and Khosravi, M.J., Multiple fluorescence labeling with europium chelators — application to time-resolved fluoroimmunoassays, *Anal. Chem.*, 61(1), 48, 1989.
120. Diamandis, E.P. and Christopoulos, T.K., Europium chelate labels in time-resolved fluorescence immunoassays and DNA hybridization assays, *Anal. Chem.*, 62(22), A1149, 1990.
121. Hemmila, I., Dakubu, S., Mukkala, V.M., Siitari, H., and Lovgren, T., Europium as a label in time-resolved immunofluorometric assays, *Anal. Biochem.*, 137(2), 335, 1984.
122. Hemmila, I., Mukkala, V.M., and Takalo, H., Development of luminescent lanthanide chelate labels for diagnostic assays, *J. Alloys Compounds*, 249(1–2), 158, 1997.
123. Karsilayan, H., Hemmila, I., Takalo, H., Toivonen, A., Pettersson, K., Lovgren, T., and Mukkala, V.M., Influence of coupling method on the luminescence properties, coupling efficiency, and binding affinity of antibodies labeled with europium(III) chelates, *Bioconjug. Chem.*, 8(1), 71, 1997.
124. Mukkala, V.M., Mikola, H., and Hemmila, I., The synthesis and use of activated *N*-benzyl derivatives of diethylenetriaminetetraacetic acids — alternative reagents for labeling of antibodies with metal-ions, *Anal. Biochem.*, 176(2), 319, 1989.
125. Soini, E. and Hemmila, I., Fluoroimmunoassay — present status and key problems, *Clin. Chem.*, 25(3), 353, 1979.
126. Christopoulos, T.K. and Diamandis, E.P., Enzymatically amplified time-resolved fluorescence immunoassay with terbium chelates, *Anal. Chem.*, 64(4), 342, 1992.

127. Soini, E. and Kojola, H., Time-resolved fluorometer for lanthanide chelates — a new generation of non-isotopic immunoassays, *Clin. Chem.*, 29(1), 65, 1983.
128. Evangelista, R.A., Pollak, A., and Templeton, E.F.G., Enzyme-amplified lanthanide luminescence for enzyme detection in bioanalytical assays, *Anal. Biochem.*, 197(1), 213, 1991.
129. Bathrellos, L.M., Lianidou, E.S., and Ioannou, P.C., A highly sensitive enzyme-amplified lanthanide luminescence immunoassay for interleukin 6, *Clin. Chem.*, 44(6), 1351, 1998.
130. Diamandis, E.P., Europium and terbium chelators as candidate substrates for enzyme-labeled time-resolved fluorometric immunoassays, *Analyst*, 117(12), 1879, 1992.
131. Lianidou, E.S., Ioannou, P.C., and Sacharidou, E., Second derivative synchronous scanning fluorescence spectrometry as a sensitive detection technique in immunoassays — application to the determination of alpha-fetoprotein, *Anal. Chim. Acta*, 290(1–2), 159, 1994.
132. Veiopoulou, C.J., Lianidou, E.S., Ioannou, P.C., and Efstathiou, C.E., Comparative study of fluorescent ternary terbium complexes: application in enzyme amplified fluorimetric immunoassay for alpha-fetoprotein, *Anal. Chim. Acta*, 335(1–2), 177, 1996.
133. Ozinskas, A.J., Malak, H., Joshi, J., Szmacinski, H., Britz, J., Thompson, R.B., Koen, P.A., and Lakowicz, J.R., Homogeneous model immunoassay of thyroxine by phase-modulation fluorescence spectroscopy, *Anal. Biochem.*, 213(2), 264, 1993.
134. Mathis, G., Rare-earth cryptates and homogeneous fluoroimmunoassays with human sera, *Clin. Chem.*, 39(9), 1953, 1993.
135. Miller, J.N., Lim, C.S., and Bridges, J.W., Fluorescamine and fluorescein as labels in energy-transfer immunoassay, *Analyst*, 105(1246), 91, 1980.
136. Campbell, A.K. and Patel, A., A homogeneous immunoassay for cyclic-nucleotides based on chemiluminescence energy-transfer, *Biochem. J.*, 216(1), 185, 1983.
137. Zuber, E., Mathis, G., and Flandrois, J.P., Homogeneous two-site immunometric assay kinetics as a theoretical tool for data analysis, *Anal. Biochem.*, 251(1), 79, 1997.
138. Ullman, E.F., Schwarzberg, M., and Rubenstein, K.E., Fluorescent excitation transfer immunoassay — general method for determination of antigens, *J. Biol. Chem.*, 251(14), 4172, 1976.
139. Dean, K.J., Thompson, S.G., Burd, J.F., and Buckler, R.T., Simultaneous determination of phenytoin and phenobarbital in serum or plasma by substrate-labeled fluorescent immunoassay, *Clin. Chem.*, 29(6), 1051, 1983.
140. Wong, R.C., Burd, J.F., Carrico, R.J., Buckler, R.T., Thoma, J., and Boguslaski, R.C., Substrate-labeled fluorescent immunoassay for phenytoin in human serum, *Clin. Chem.*, 25(5), 686, 1979.
141. Li, T.M. and Burd, J.F., Enzymic-hydrolysis of intramolecular complexes for monitoring theophylline in homogeneous competitive-protein-binding reactions, *Biochem. Biophys. Res. Commun.*, 103(4), 1157, 1981.
142. Weeks, I., Beheshti, I., McCapra, F., Campbell, A.K., and Woodhead, J.S., Acridinium esters as high-specific-activity labels in immunoassay, *Clin. Chem.*, 29(8), 1474, 1983.
143. Simpson, J.S.A., Campbell, A.K., Ryall, M.E.T., and Woodhead, J.S., Stable chemiluminescent-labeled antibody for immunological assays, *Nature*, 279(5714), 646, 1979.
144. Weeks, I., Sturgess, M., Siddle, K., Jones, M.K., and Woodhead, J.S., A high-sensitivity immunochemiluminometric assay for human thyrotropin, *Clin. Endocrinol.*, 20(4), 489, 1984.
145. Weeks, I., Campbell, A.K., and Woodhead, J.S., Two-site immunochemiluminometric assay for human alpha-1-fetoprotein, *Clin. Chem.*, 29(8), 1480, 1983.
146. Schroeder, H.R., Hines, C.M., Osborn, D.D., Moore, R.P., Hurtle, R.L., Wogoman, F.F., Rogers, R.W., and Vogelhut, P.O., Immunochemiluminometric assay for hepatitis-B surface-antigen, *Clin. Chem.*, 27(8), 1378, 1981.
147. Kim, J.B., Barnard, G.J., Collins, W.P., Kohen, F., Lindner, H.R., and Eshhar, Z., Measurement of plasma estradiol-17-beta by solid-phase chemi-luminescence immunoassay, *Clin. Chem.*, 28(5), 1120, 1982.
148. Pratt, J.J., Woldring, M.G., and Villerius, L., Chemiluminescence-linked immunoassay, *J. Immunol. Meth.*, 21(1–2), 179, 1978.

149. Kohen, F., Pazzagli, M., Kim, J.B., Lindner, H.R., and Boguslaski, R.C., Assay procedure for plasma progesterone based on antibody-enhanced chemi-luminescence, *FEBS Lett.*, 104(1), 201, 1979.
150. Bronstein, I., Voyta, J.C., Thorpe, G.H.G., Kricka, L.J., and Armstrong, G., Chemi-luminescent assay of alkaline-phosphatase applied in an ultrasensitive enzyme-immunoassay of thyrotropin, *Clin. Chem.*, 35(7), 1441, 1989.
151. Thorpe, G.H.G., Bronstein, I., Kricka, L.J., Edwards, B., and Voyta, J.C., Chemi-luminescent enzyme-immunoassay of alpha-fetoprotein based on an adamantyl dioxetane phenyl phosphate substrate, *Clin. Chem.*, 35(12), 2319, 1989.
152. O'Toole, A., Kricka, L.J., Thorpe, G.H.G., and Whitehead, T.P., Rapid enhanced chemiluminescent enzyme-immunoassay for ferritin monitored using instant photographic film, *Anal. Chim. Acta*, 266(2), 193, 1992.
153. Schmid, J.A. and Billich, A., Simple method for high sensitivity chemiluminescence ELISA using conventional laboratory equipment, *Biotechniques*, 22(2), 278, 1997.
154. Schaap, A.P., Akhavan, H., and Romano, L.J., Chemi-luminescent substrates for alkaline-phosphatase — application to ultrasensitive enzyme-linked immunoassays and DNA probes, *Clin. Chem.*, 35(9), 1863, 1989.
155. Thorpe, G.H.G., Kricka, L.J., Moseley, S.B., and Whitehead, T.P., Phenols as enhancers of the chemiluminescent horseradish-peroxidase luminol hydrogen-peroxide reaction — application in luminescence-monitored enzyme immunoassays, *Clin. Chem.*, 31(8), 1335, 1985.
156. Yasui, T. and Yoda, K., Imaging of *Lactobacillus brevis* single cells and microcolonies without a microscope by an ultrasensitive chemiluminescent enzyme immunoassay with a photon-counting television camera, *Appl. Environ. Microbiol.*, 63(11), 4528, 1997.
157. Whitehead, T.P., Thorpe, G.H.G., Carter, T.J.N., Groucutt, C., and Kricka, L.J., Enhanced luminescence procedure for sensitive determination of peroxidase-labeled conjugates in immunoassay, *Nature*, 305(5930), 158, 1983.
158. Velan, B. and Halmann, M., Chemiluminescence immunoassay — new sensitive method for determination of antigens, *Immunochemistry*, 15(5), 331, 1978.
159. Nishizono, I., Iida, S., Suzuki, N., Kawada, H., Murakami, H., Ashihara, Y., and Okada, M., Rapid and sensitive chemiluminescent enzyme-immunoassay for measuring tumor-markers, *Clin. Chem.*, 37(9), 1639, 1991.
160. Bronstein, I., Voyta, J.C., Vanterve, Y., and Kricka, L.J., Advances in ultrasensitive detection of proteins and nucleic-acids with chemiluminescence — novel derivatized 1,2-dioxetane enzyme substrates, *Clin. Chem.*, 37(9), 1526, 1991.
161. Lewkowich, I.P., Campbell, J.D., and HayGlass, K.T., Comparison of chemiluminescent assays and colorimetric ELISAs for quantification of murine IL-12, human IL-4 and murine IL-4: chemiluminescent substrates provide markedly enhanced sensitivity, *J. Immunol. Meth.*, 247(1–2), 111, 2001.
162. Bronstein, I., Edwards, B., and Voyta, J.C., 1,2-Dioxetanes — novel chemi-luminescent enzyme substrates — applications to immunoassays, *J. Bioluminescence Chemiluminescence*, 4(1), 99, 1989.
163. Olesen, C.E.M., Mosier, J., Voyta, J.C., and Bronstein, I., Chemiluminescent immunodetection protocols with 1,2-dioxetane substrates, in *Methods in Enzymology*, Vol. 305, *Bioluminescence and Chemiluminescence, Part C*, Ziegler, M.M. and Baldwin, T.O., Eds., Academic Press, San Diego, 2000, pp. 417–427.
164. Erler, K., Elecsys (R) immunoassay systems using electrochemiluminescence detection, *Wien. Klin. Wochenschr.*, 110, 5, 1998.
165. Blackburn, G.F., Shah, H.P., Kenten, J.H., Leland, J., Kamin, R.A., Link, J., Peterman, J., Powell, M.J., Shah, A., Talley, D.B., Tyagi, S.K., Wilkins, E., Wu, T.G., and Massey, R.J., Electrochemiluminescence detection for development of immunoassays and DNA probe assays for clinical diagnostics, *Clin. Chem.*, 37(9), 1534, 1991.
166. Lin, J.M. and Yamada, M., Electrogenerated chemiluminescence of methyl-9-(p-formylphenyl) acridinium carboxylate fluorosulfonate and its applications to immunoassay, *Microchem. J.*, 58(1), 105, 1998.

167. Kankare, J., Haapakka, K., Kulmala, S., Nanto, V., Eskola, J., and Takalo, H., Immunoassay by time-resolved electrogenerated luminescence, *Anal. Chim. Acta*, 266(2), 205, 1992.
168. Maeda, M., Shimizu, S., and Tsuji, A., Chemiluminescence assay of beta-D-galactosidase and its application to competitive immunoassay for 17-alpha-hydroxyprogesterone and thyroxine, *Anal. Chim. Acta*, 266(2), 213, 1992.
169. Brown, R.C., Weeks, I., Fisher, M., Harbron, S., Taylorson, C.J., and Woodhead, J.S., Employment of a phenoxy-substituted acridinium ester as a long- lived chemiluminescent indicator of glucose oxidase activity and its application in an alkaline phosphatase amplification cascade immunoassay, *Anal. Biochem.*, 259(1), 142, 1998.
170. Tanaka, K. and Ishikawa, E., A highly sensitive bioluminescent assay of beta-D-galactosidase from *Escherichia coli* using 2-nitrophenyl-beta-D-galactopyranoside as a substrate, *Anal. Lett.*, 19(3–4), 433, 1986.
171. Takayasu, S., Maeda, M., and Tsuji, A., Chemi-luminescent enzyme-immunoassay using beta-D-galactosidase as the label and the bis(2,4,6-trichlorophenyl)oxalate-fluorescent dye system, *J. Immunol. Meth.*, 83(2), 317, 1985.
172. Arakawa, H., Maeda, M., and Tsuji, A., Chemi-luminescence enzyme-immunoassay for thyroxine with use of glucose-oxidase and a bis(2,4,6-trichlorophenyl)oxalate fluorescent dye system, *Clin. Chem.*, 31(3), 430, 1985.
173. Ishimori, Y. and Rokugawa, K., Stable liposomes for assays of human sera, *Clin. Chem.*, 39(7), 1439, 1993.
174. Ishimori, Y., Yasuda, T., Tsumita, T., Notsuki, M., Koyama, M., and Tadakuma, T., Liposome immune lysis assay (LILA) — a simple method to measure anti-protein antibody using protein antigen-bearing liposomes, *J. Immunol. Meth.*, 75(2), 351, 1984.
175. Six, H.R., Young, W.W., Uemura, K.I., and Kinsky, S.C., Effect of antibody-complement on multiple vs. single compartment liposomes — application of a fluorometric assay for following changes in liposomal permeability, *Biochemistry*, 13(19), 4050, 1974.
176. Yasuda, T., Naito, Y., Tsumita, T., and Tadakuma, T., A simple method to measure anti-glycolipid antibody by using complement-mediated immune lysis of fluorescent dye-trapped liposomes, *J. Immunol. Meth.*, 44(2), 153, 1981.
177. Pashkov, V.N., Tsurupa, G.P., Griko, N.B., Skopinskaya, S.N., and Yarkov, S.P., The use of strepta-vidin biotin interaction for preparation of reagents for complement-dependent liposome immunoassay of proteins — detection of latrotoxin, *Anal. Biochem.*, 207(2), 341, 1992.
178. Rongen, H.A.H., Bult, A., and van Bennekom, W.P., Liposomes and immunoassays, *J. Immunol. Meth.*, 204(2), 105, 1997.
179. Kobatake, E., Sasakura, H., Haruyama, T., Laukkanen, M.L., Keinanen, K., and Aizawa, M., A fluoroimmunoassay based on immunoliposomes containing genetically engineered lipid-tagged antibody, *Anal. Chem.*, 69(7), 1295, 1997.
180. Rongen, H.A.H., Vanderhorst, H.M., Hugenholtz, G.W.K., Bult, A., Vanbennekom, W.P., and Vandermeide, P.H., Development of a liposome immunosorbent-assay for human interferon-gamma, *Anal. Chim. Acta*, 287(3), 191, 1994.
181. Rongen, H.A.H., Vannierop, T., Vanderhorst, H.M., Rombouts, R.F.M., Vandermeide, P.H., Bult, A., and Vanbennekom, W.P., Biotinylated and streptavidinylated liposomes as labels in cytokine immunoassays, *Anal. Chim. Acta*, 306(2–3), 333, 1995.
182. Yamaji, H., Nakagawa, M., Tomioka, K., Kondo, A., and Fukuda, H., Use of a substrate as an encapsulated marker for liposome immunoassay, *J. Ferment. Bioeng.*, 83(6), 596, 1997.
183. Lim, S.J. and Kim, C.K., Homogeneous liposome immunoassay for insulin using phospholipase C from *Clostridium perfringens*, *Anal. Biochem.*, 247(1), 89, 1997.
184. Litchfield, W.J., Freytag, J.W., and Adamich, M., Highly sensitive immunoassays based on use of liposomes without complement, *Clin. Chem.*, 30(9), 1441, 1984.
185. Nakamura, T., Hoshino, S., Hazemoto, N., Haga, M., Kato, Y., and Suzuki, Y., A liposome immunoassay based on a chemi-luminescence reaction, *Chem. Pharm. Bull.*, 37(6), 1629, 1989.

186. Bystryak, S., Goldiner, I., Niv, A., Nasser, A.M., and Goldstein, L., A homogeneous immunofluorescence assay based on dye-sensitized photobleaching, *Anal. Biochem.*, 225(1), 127, 1995.
187. Okabayashi, Y. and Ikeuchi, I., Liposome immunoassay by long-lived fluorescence detection, *Analyst*, 123(6), 1329, 1998.
188. White, S.R., Chiu, N.H.L., and Christopoulos, T.K., Expression immunoassay based on antibodies labeled with a deoxyribonucleic acid fragment encoding the α-peptide of β-galactosidase, *Analyst*, 123(6), 1309, 1998.
189. Christopoulos, T.K. and Chiu, N.H.L., Expression immunoassay — antigen quantitation using antibodies labeled with enzyme-coding DNA fragments, *Anal. Chem.*, 67(23), 4290, 1995.
190. Phillips, T.M. and Krum, J.M., Recycling immunoaffinity chromatography for multiple analyte analysis in biological samples, *J. Chromatogr. B*, 715(1), 55, 1998.
191. Rohr, T.E., Cotton, T., Fan, N., and Tarcha, P.J., Immunoassay employing surface-enhanced Raman-spectroscopy, *Anal. Biochem.*, 182(2), 388, 1989.
192. Ni, J., Lipert, R.J., Dawson, G.B., and Porter, M.D., Immunoassay readout method using extrinsic Raman labels adsorbed on immunogold colloids, *Anal. Chem.*, 71(21), 4903, 1999.
193. Dou, X., Takama, T., Yamaguchi, Y., Yamamoto, H., and Ozaki, Y., Enzyme immunoassay utilizing surface-enhanced Raman scattering of the enzyme reaction product, *Anal. Chem.*, 69(8), 1492, 1997.
194. Surugiu, I., Danielsson, B., Ye, L., Mosbach, K., and Haupt, K., Chemiluminescence imaging ELISA using an imprinted polymer as the recognition element instead of an antibody, *Anal. Chem.*, 73(3), 487, 2001.
195. Surugiu, I., Ye, L., Yilmaz, E., Dzgoev, A., Danielsson, B., Mosbach, K., and Haupt, K., An enzyme-linked molecularly imprinted sorbent assay, *Analyst*, 125(1), 13, 2000.
196. Matsunaga, T., Kawasaki, M., Yu, X., Tsujimura, N., and Nakamura, N., Chemiluminescence enzyme immunoassay using bacterial magnetic particles, *Anal. Chem.*, 68(20), 3551, 1996.
197. Nakamura, N., Hashimoto, K., and Matsunaga, T., Immunoassay method for the determination of immunoglobulin-G using bacterial magnetic particles, *Anal. Chem.*, 63(3), 268, 1991.
198. Nakamura, N., Burgess, J.G., Yagiuda, K., Kudo, S., Sakaguchi, T., and Matsunaga, T., Detection and removal of *Escherichia coli* using fluorescein isothiocyanate conjugated monoclonal antibody immobilized on bacterial magnetic particles, *Anal. Chem.*, 65(15), 2036, 1993.
199. Beverloo, H.B., Vanschadewijk, A., Zijlmans, H., and Tanke, H.J., Immunochemical detection of proteins and nucleic acids on filters using small luminescent inorganic crystals as markers, *Anal. Biochem.*, 203(2), 326, 1992.
200. Beverloo, H.B., Vanschadewijk, A., Vangelderenboele, S., and Tanke, H.J., Inorganic phosphors as new luminescent labels for immunocytochemistry and time-resolved microscopy, *Cytometry*, 11(7), 784, 1990.
201. Szurdoki, F., Michael, K.L., and Walt, D.R., A duplexed microsphere-based fluorescent immunoassay, *Anal. Biochem.*, 291(2), 219, 2001.
202. Yu, H., Use of an immunomagnetic separation-fluorescent immunoassay (IMS-FIA) for rapid and high throughput analysis of environmental water samples, *Anal. Chim. Acta*, 376(1), 77, 1998.
203. Yu, H., Comparative studies of magnetic particle-based solid phase fluorogenic and electrochemiluminescent immunoassay, *J. Immunol. Meth.*, 218(1–2), 1, 1998.
204. Moeremans, M., Daneels, G., Vandijck, A., Langanger, G., and Demey, J., Sensitive visualization of antigen-antibody reactions in dot and blot immune overlay assays with immunogold and immunogold silver staining, *J. Immunol. Meth.*, 74(2), 353, 1984.
205. Holgate, C.S., Jackson, P., Cowen, P.N., and Bird, C.C., Immunogold silver staining — new method of immunostaining with enhanced sensitivity, *J. Histochem. Cytochem.*, 31(7), 938, 1983.
206. Stults, N.L., Stocks, N.F., Rivera, H., Gray, J., McCann, R.O., Okane, D., Cummings, R.D., Cormier, M.J., and Smith, D.F., use of recombinant biotinylated aequorin in microtiter and membrane-based assays — purification of recombinant apoaequorin from *Escherichia coli*, *Biochemistry*, 31(5), 1433, 1992.

**TABLE 55.2** Nucleic-Acid Amplification Schemes

| Assay | Principle | Comments | Ref. |
|---|---|---|---|
| | ***PCR-Based Nucleic Acid Amplification*** | | |
| | *Target DNA is exponentially amplified by using a thermostable polymerase to copy complementary strands of DNA using one primer for each strand; RNA can enter PCR cycle via initial reaction with reverse transcriptase; various strategies can be used to incorporate a label or other moieties to which a label can be attached during the course of PCR so that amplicons can be detected* | | |
| Polymerase Chain Reaction (PCR) | Two primers are used in interactive process to amplify specific segment of DNA; first DNA target is denatured by heating, followed by hybridization of both primers to complementary strands of target at about 10°C below melting temperature of primers; next step involves raising temperature to 70–73°C, optimizing temperature for extension of primers by thermostable DNA polymerase; this extension may extend beyond region of interesting target in both directions, but is limited in its total length; denaturation, annealing, and primer extension steps are repeated; in presence of excess primers, after a few cycles copies of region of interest in target sequence begin to dominate product formed; each cycle results in doubling of DNA target sequence (exponential amplification); can also be applied to amplifying RNA if RNA is transcribed into cDNA by reverse transcriptase | PCR originally carried out with Klenow fragment of *E. coli* DNA Polymerase I; thermal instability required addition of fresh enzyme at each denaturation step; use of thermostable enzyme is now essentially universal; 25–30 cycles can produce $10^6$-fold amplification; further amplification can be achieved by diluting PCR product and subjecting it to a second round of PCR using fresh enzyme (producing $10^9$- to $10^{10}$-fold amplification); usually a different set of primers (nested) are used that hybridize downstream from the first set of primers; nonspecificity can arise from primers hybridizing to similar DNA sequences elsewhere in target; fragments as long as 2 Kb can be generated; PCR product can be visualized by ethidium bromide staining on agarose gels; a single target can be amplified and detected, but 10–250 DNA copies are the more usual starting material;[1] efficiency of reaction can vary depending on concentration of polymerase, primers, $Mg^{2+}$, etc.; due to exponential nature of PCR, small changes in efficiency can have large effects on final quantity of PCR product | 1–5 |
| Reverse Transcription (RT) PCR | For amplification of target RNA species (e.g. mRNA), RNA target is first reverse transcribed by one of a variety of reverse transcriptase enzymes, forming DNA copy (cDNA) of the RNA; then, PCR can take place using two primers as in conventional PCR; during first round of PCR, one primer binds to cDNA, and complementary strand to cDNA is formed; during second cycle other primer binds to newly formed strand and forms complement; eventually exponential amplification occurs | Amplification of target RNA by PCR | 6–10 |

| | | | |
|---|---|---|---|
| PCR (Fluorescence Detection) | One or both PCR oligonucleotide primers are labeled on the 5′ end with fluorescent dyes; PCR carried out in usual manner; PCR product can be analyzed in numerous ways, but separation of product from unreacted fluorophore-labeled primers must be carried out | Possibility for simultaneous determination of multiple PCR products if different fluorophores are used for labeling; PCR products analyzed by gel electrophoresis or after separation from unreacted primers by microfiltration; technique can be used to detect point mutations and small deletions, chromosome translocations, infectious agents; demonstrated detection of five different PCR products simultaneously, using different fluorescent dyes | 11 |
| Quantitative PCR with Fluorescence Detection | To assess starting quantity of target DNA in PCR reaction an internal DNA standard was used that was same length as target sequence, but with different central sequence; PCR is carried out in usual manner, but constant amount of internal standard is included in each PCR reaction; central sequence of internal standard allows discrimination between internal standard and target-DNA sequence; PCR products analyzed by hybridization reactions with specific probes; in one analysis, PCR products labeled with digoxigenin were captured in microwells by hybridization to a probe; in other analysis, PCR products were captured in microwells by hybridization probe, denatured, and subsequently hybridized to probes with tails of digoxigenin-dUTP; in both cases, addition of antidigoxigenin Ab conjugated to AP, followed by addition of AP substrate 5′-fluorosalicylphosphate results in release of 5′-fluorosalicylic acid, which forms a highly fluorescent complex with $Tb^{3+}$-EDTA; fluorescence is measured by time-resolved techniques | | 12 |
| | also: mRNA detected and quantified using a synthetic RNA as an internal standard in PCR reactions; in this case internal standard is of different size than mRNA target, so both target and standard can be distinguished; RNAs transcribed into cDNAs by reverse transcriptase before PCR | Products analyzed by ethidium bromide-stained agarose gels | 13 |
| PCR (Liquid Chromatography Detection) | PCR products are separated from other reactants and analyzed by liquid chromatography (LC) | PCR product detected by UV absorption after LC; PCR product is also purified during LC process; detection by LC reported to be more sensitive than staining of PCR product by Hoechst dye 33258 or analysis on ethidium bromide stained agarose gels | 14 |
| PCR (Colorimetric Detection) | PCR performed in usual way; following PCR, 3 to 10 subsequent cycles are performed using a pair of nested primers internal to the PCR product to incorporate biotin and a site for a dsDNA-binding protein; DNA-binding protein immobilized in microwell captures smaller nested-primer-generated PCR product; biotin tag on other end of this product reacts with peroxidase-conjugated avidin; addition of peroxidase substrate produces color that provides detection confirmation (indirect) for original PCR product produced | Nested primers achieve specificity because their PCR products will form only if they hybridize to correct sequence generated by first PCR target amplification; use of these labeled primers allows rest of assay to proceed | 15 |

**TABLE 55.2** Nucleic-Acid Amplification Schemes (continued)

| Assay | Principle | Comments | Ref. |
|---|---|---|---|
| PCR (Modified) | PCR carried out as usual, but one primer has a T7 phage promoter sequence attached to it; after PCR, addition of T7 RNA polymerase generates RNA transcripts; an oligonucleotide primer for reverse transcriptase is then added with a sequence complementary to a sequence within region of interest being amplified; addition of reverse transcriptase generates a cDNA product; this final step improves specificity of PCR, since extraneous DNA sequences amplified during PCR will not be targets for hybridization to the reverse-transcriptase primer | Additive amplification from both DNA and RNA types of amplification; products detected on ethidium bromide-stained agarose gels; in application to sequencing, amount of DNA in 150 diploid cells was sufficient for a readable sequence (using radioactive labeling) | 16 |
| PCR (Oligonucleotide Ligation Assay [OLA]) | PCR carried out in usual manner on sequence of interest, followed by ligation reaction using two labeled ligation probes, one labeled with biotin and the other with digoxigenin; following ligation, ligated product is captured on streptavidin-coated microwell; treatment with AP-conjugated antidigoxigenin Ab results in immobilization of AP in microwell; colorimetric detection accomplished by addition of enzyme substrate | Applied to detection of allelic sequence variants; amplification by PCR, ligation and ELISA enzyme; ligation reaction enhances specificity; colorimetric assay found capable of detecting 3 fmol of ligated product | 17 |
| Scorpion Probes | One of the primers for PCR has attached to it the following probe structure; this probe consists of specific sequence complementary to a sequence internal to the PCR amplicons; the whole probe is held in a hairpin loop structure (loop portion is PCR-specific internal sequence) by complementary stem sequences; at 5′ and 3′ side of stems are attached a fluorophore and quencher molecule, respectively; at lower (subdenaturing) temperature, probe stays in hairpin loop structure and fluorescence is quenched; during PCR, hairpin loop is denatured during heating step, primer binds to target during annealing step and is extended; during the next heat denaturation cycle, double-strand amplicon product is denatured as is hairpin-loop probe structure; now loop sequence can hybridize to internal sequence on same amplicon strand to which it is attached, thus fixing the separation of fluorophore and quencher, and producing fluorescence | Amplification by PCR; can detect PCR amplicons in homogeneous solution without separation step; comparisons between Scorpions, Molecular Beacons, and *Taq* Man indicated Scorpions performed better than the other two;[18] Scorpion probes act almost exclusively by unimolecular mechanism | 18, 19 |
| *Taq* Man | PCR reaction carried out in usual manner, but advantage is taken of 5′ → 3′ exonuclease activity of *Taq* DNA polymerase; oligonucleotide probe labeled on 5′ end with quencher and 3′ end with fluorophore is added at start of PCR; fluorescence is quenched until probe binds to internal sequence of amplicons being generated; then exonuclease activity of *Taq* polymerase cleaves probe, releasing fluorophore and producing fluorescence; increasing PCR cycles result in increased fluorescence | Amplification by PCR; detection in homogeneous solution possible, without separation step; quencher and fluorophore must be close to each other for efficient energy transfer but cannot be too close | 20 |

| | | | |
|---|---|---|---|
| FRET-Based Detection | One or both PCR primers have a stem-loop structure attached to their 5′ end; stem-loop sequence does not have to match PCR amplicon sequence; on 5′ end of stem is fluorophore and on 3′ end of stem is quencher; FRET occurs and no fluorescence is detected; PCR reaction carried out in usual manner; modified primer is incorporated into PCR product; during next round of PCR, polymerase synthesizes complementary strand of stem-loop sequence; as double strands of DNA form, stem-loop sequence is linearized and fluorophore and quencher are separated; fluorescence increases | *Taq* DNA polymerase must be free of 5′ → 3′ exonuclease activity; amplification by PCR; technique similar to molecular beacon approach; fluorescence correlates to amount of incorporated primers; ten molecules of starting target can be detected | 21 |
| | ***Post-PCR Product Detection*** | | |
| | *Amplicons from PCR process are labeled subsequent to PCR, allowing detection of PCR product* | | |
| | DNA is amplified by PCR, using a biotinylated primer; following PCR, product is denatured and hybridized to detector oligonucleotide, which is labeled with electrochemiluminescent label; subsequently, biotinylated DNA strand with hybridized detector oligonucleotide attached is captured by reaction with streptavidin-coated beads and subject to electrochemical analysis<br>or:<br>DNA can be amplified by PCR, using one biotinylated primer, while the other primer is labeled with the electrochemiluminescent label; in this case, PCR product (double stranded) is <u>not</u> denatured, but is captured directly by streptavidin-coated beads | Amplification by PCR; electrochemiluminescent label is tris (2,2′-bipyridine) ruthenium (II) chelate (see Table 55.1, Immunoassays); detection limit found to be ~2 attomol, for direct detection on beads using electrochemiluminescence | 22 |
| Bioluminescence Detection | Reverse transcription (RT) PCR carried out as usual; one primer is biotinylated, so PCR product can be captured on streptavidin-treated surface; digoxigenin-labeled probes are hybridized to immobilized PCR product; reaction with antidigoxigenin Ab conjugated to aequorin immobilizes Ab and aequorin at hybridization site; addition of $Ca^{2+}$ triggers flash of light | This technique found to be 30- to 60-fold more sensitive than radioimaging for detecting cytokine mRNA;[23] also permits detection of amplicons at lower PCR cycle numbers; amplification due to PCR | 23, 24 |
| Electrochemiluminescence Detection | One PCR primer biotin-labeled, while other primer labeled (via amino group linker) with ruthenium (II) tris (bipyridy) label; after PCR takes place, double-stranded product is captured on streptavidin-coated beads and then subjected to electrochemiluminescence analysis (see Table 55.1)<br>Alternatively, PCR carried out as usual but with one biotinylated primer; biotin-labeled amplicon captured on streptavidin-treated beads, and hybridized to complementary probe labeled with electrochemiluminescent label (see above); rest of assay as above | | 25, 26 |

**TABLE 55.2** Nucleic-Acid Amplification Schemes (continued)

| Assay | Principle | Comments | Ref. |
|---|---|---|---|
| Molecular Beacons | DNA is amplified by PCR, and subsequently detected by a molecular beacon which is immobilized on a solid surface by a biotin-modified nucleotide linked to avidin bound to surface; molecular beacon is a stem-loop structure in which the loop is complementary to target nucleic-acid sequence and stem has a fluorophore on its 5′ end and a quencher on its 3′ end; as long as the stem-loop structure remains intact, fluorescence is quenched; when target DNA binds to loop, however, stem structure is disrupted, and resulting fluorescence indicates hybridization has occurred | Amplification by PCR; either DNA or RNA can be detected; limit of detection was found to be attomole to subattomole[27] | 27 |
| | or: Molecular beacon used as a probe to hybridize internally to PCR amplicons; beacons included in PCR reaction, and as amplicons accumulate, fluorescence from hybridization of beacons increases | Amplification by PCR; significant fluorescence signal can be detected in early cycles of PCR if target is at fairly high concentration initially | 28 |
| | ***Non-PCR-Based Nucleic Acid Amplification*** | | |
| | *A variety of techniques are used to achieve amplification of target nucleic acids; all involve enzymes; most use a polymerase to generate complementary copies of target molecule* | | |
| Nucleic Acid Sequence–Based Amplification (NASBA) | A target RNA is hybridized to a primer oligonucleotide (primer 1) that has on its 5′ end a promoter sequence for T7 RNA polymerase; a mixture of T7 RNA polymerase, RNase H, and reverse transcriptase are in reaction solution; following hybridization, reverse transcriptase replicates RNA sequence, making a complementary cDNA strand, resulting in RNA/DNA hybrid; RNase H hydrolyses the RNA strand, leaving single cDNA strand (denatured DNA as a template could enter the process at this point); second primer (primer 2) included in the reaction mix hybridizes to the cDNA strand, and reverse transcriptase replicates cDNA, producing a dsDNA with a double-stranded promoter region for T7 RNA polymerase; this enzyme transcribes RNA copies from double-stranded cDNA template; the RNA copies produced can serve as templates for reverse transcriptase; thus, primer 2 binds to newly generated RNA template and reverse transcriptase extends, producing RNA/DNA hybrid hydrolyzed to a single cDNA strand; primer 1 binds to this strand and a double-stranded cDNA is produced by reverse transcriptase, which again can serve as a template for RNA polymerase, etc. | Process is continuous, homogeneous, and isothermal; described as robust, even though multiple enzymes involved; incubation at one temperature for 1.5 to 2 h sufficient to achieve amplification of about $10^9$; NASBA requires fewer cycles than PCR to achieve similar levels of amplification; samples with as little as ten molecules of target found to produce positive results;[29] can be used to detect mRNA | 29 |

| | | | |
|---|---|---|---|
| | | Ten copies of viral RNA could be detected using NASBA and colorimetric detection method; after amplification, amplicons captured with capture probe on microwell plates, and peroxidase-labeled detector probe was hybridized to amplicons; addition of peroxidase substrate produced color proportional to amplicon concentration[30] | 30 |
| | | NASBA can directly, and selectively, amplify ssRNA if denaturation carried out at 65°C, which avoids denaturing DNA; ten copies of interferon gene detected using chemiluminescence detection on membrane[31] | 31 |
| | | RNA polymerase can produce 10 to $10^3$ copies of RNA/single copy of DNA template;[32] thus four to six cycles of NASBA produces $2 \times 10^6$ amplification; total time ~3 to 4 h | 32 |
| | | Within first 15 min, each of the input RNA targets was increased $10^5$-fold; each cDNA template directs synthesis of minimum of 90 copies of detectable RNA during 3SR reaction; 3SR reaction operates only at 37°C, so no denaturation step for dsDNA, as is used in NASBA; DNA template can be amplified by 3SR if denaturation carried out prior to start; much more rapid kinetics than PCR | 33 |
| Rolling-Circle Amplification (RCA) | Small circular single-stranded oligonucleotides serve as the template for DNA synthesis by several DNA polymerases; progressive copying produces a long linear DNA strand consisting of complementary repeats of circular template | Isothermal amplification; replication can proceed hundreds of times round circle; most efficient DNA polymerases lose activity with time | 34 |
| RCA with Condensation of Amplified Circles after Hybridization of Encoding Tags (CACHET-RCA) | RCA with CACHET; fluorescently labeled oligonucleotides also having dinitrophenyl (DNP) label are hybridized at multiple sites on tandem repeats generated by RCA; addition of anti-DNP IgM antibody causes condensation into small object more readily visualized by CCD<br>or:<br>BUDR can be incorporated into RCA product and detected by (1) reaction with biotinylated anti-BUDR Ab and (2) reaction with fluorescently labeled avidin | Single-molecule (DNA) detection reported | 35 |

**TABLE 55.2** Nucleic-Acid Amplification Schemes (continued)

| Assay | Principle | Comments | Ref. |
|---|---|---|---|
| Padlock Probes with RCA | A linear single-stranded probe with two target-DNA-complementary sequences joined by linker region is hybridized to target DNA; complementary sequences on probe are designed to complement contiguous sequence on target; following ligation by a ligase, probe forms circular strand wrapped around target; labels on padlock probe allow detection; once padlock probe is circularized, it can be amplified by RCA before detection, using suitable primer plus enzymes | High specificity; not as likely to have nonspecific reactions as PCR; reduction in nonspecific background; RCA reaction rate stated to be 53 nucleotides/sec for phage 029 DNA polymerase[35] | 35, 36 |
| | or:<br>Using second primer (binding to a distinct site on padlock probe from first primer) in addition to first primer, a continuous pattern of DNA branches can be formed, connected to the first circle;[35] called Hyperbranched RCA | Possible to initiate hyperbranched RCA with as few as 20 molecules of closed circles; each circle estimated to produce $5 \times 10^9$ copies after 90-min reaction[35] | |
| Ligation Amplification Reaction (LAR); also called Ligase Detection Reaction (LDR)/Ligase Chain Reaction (LCR) | In linear amplification, two oligonucleotide probes used; immediately adjacent to each other on DNA target strand, so action of DNA ligase covalently links two oligonucleotides, if nucleotides at junction are perfectly base-paired to target; in exponential amplification both strands of target DNA are used as templates for ligation of oligonucleotide probes (i.e., four oligonucleotide probes), so exponential amplification can occur, because ligation of one set of probes generates a target for complementary probe set; in both cases multiple rounds of denaturation, annealing, and ligation with thermostable ligase result in amplification | Thermostable ligase allows multiple repetitions of ligation reaction, thus linearly increasing product (termed ligase detection reaction); specificity arises from sequence of two oligonucleotides, requirement for perfect base pairing at ligated junction and performing of ligation at elevated temperatures; powerful technique for discriminating single-base substitution; time required for 20 to 30 cycles ≅ 100 to 150 min; sensitivity of LCR: 200 molecules of target DNA amplified ~$2 \times 10^5$-fold after 30 cycles[37] | 37, 38 |
| | also:<br>Oligonucleotide can be fluorophore labeled on 5′ end; upon ligation with downstream oligonucleotide, ligated probe can be detected by fluorescence, following suitable separation methods (i.e., gel electrophoresis) to separate ligated product from unreacted labeled probe; alternatively, one oligonucleotide can be labeled on appropriate end by biotin and ligation product detected after immobilization on streptavidin/solid support[11] | Generation of target-independent products due to blunt-end ligation can limit sensitivity | 39 |
| GAP-LCR | Four oligonucleotide probes are used as in LCR, but two of the probes have 3′ overhangs (extensions) when hybridized to their DNA targets; after hybridization, a gap of 1 to 3 bases exists between adjacent oligonucleotides, which is filled by a thermostable DNA polymerase; resultant “filled-in” probes joined by DNA ligase | Use of probes with noncomplementary 3′ extensions prevents generation of target-independent ligation products; exponential amplification; amplification detected by fluorescence ELISA assay; specificity found to be better than allele-specific PCR | 40 |

| | | | |
|---|---|---|---|
| Strand Displacement Amplification (SDA) | Target amplification technology in which sequence-specific exponential amplification occurs under isothermal conditions; target amplification occurs by process in which two primers bind to opposite ends of target sequence, as in PCR; with SFA, 3′ ends of the primers recognize the target sequence, while 5′ ends have restriction-enzyme-recognition site; DNA replication by exonuclease-deficient form of DNA polymerase produces double-stranded amplicons with a restriction enzyme recognition site; one of strands is protected from enzyme cleavage because modified nucleotide (dATPαS) incorporated during amplification; restriction enzyme nicks unprotected strand; DNA polymerase extends complementary strand starting at the 3′ end, thus displacing nicked strand; polymerization/displacement step regenerates nickable recognition; nicking and polymerization cycle continuously; a fluorogenic probe is also designed to hybridize to target strand (different probe for each target strand) but downstream from primer site; probe acts as primer to generate extended probe; extended probe displaced as above; extended, displaced probe now has 3′ end complementary to second SDA primer; hybridization occurs, and DNA polymerase extends in both directions; probe can be designed with a loop–stem structure, although not necessary; if loop–stem structure, polymerase activity unfolds this structure during polymerization; in any case, conversion of probe to double-strand structure creates restriction endonuclease site; probe constructed with fluorophore and quencher on opposite sides of restriction site; action of endonuclease releases fluorophore from quenching activity | $10^{10}$-fold amplification occurs in 15 min; detection of 10 target copies within 30 min using fluorogenic assay reported; fluorogenic assay relies essentially on FRET-like process; process is complicated conceptually, but appears to work experimentally | 41 |
| | ***Hybridization*** | | |
| | *A probe (oligonucleotide or nucleic-acid segment) with either a label or labeling potential is hybridized to target DNA, and hybridized complex is detected; alternatively, probe can be amplified by PCR process; hybridization commonly applied to membranes, tissue sections, microarrays, beads* | | 42–77 |
| Membrane Detection | DNA transferred to membrane detected with oligonucleotide probes either labeled with alkaline phosphatase (AP) directly or biotin; if biotin-labeled probes are used, membranes subsequently reacted with avidin-AP conjugate; final detection by use of substrate that produces chemiluminescence by AP activity | Subfemtomole detection reported by Tizard et al.;[43] membranes could be successfully stripped and reprobed; enzyme activity provides amplification | 43 |
| | DNA on membrane hybridized to biotinylated probe, followed by reaction with streptavidin/biotinylated AP complex; incubation with luciferin-o-phosphate releases luciferin which now acts as substrate for luciferase | Bioluminescent detection technique; two enzymes theoretically provide large amplification | 44 |

**TABLE 55.2** Nucleic-Acid Amplification Schemes (continued)

| Assay | Principle | Comments | Ref. |
|---|---|---|---|
| | DNA on membrane hybridized to biotinylated probe, followed by reaction with streptavidin-AP conjugate; incubation with either colorimetric or fluorogenic substrate produces visible color or fluorescence | Colorimetric reaction carried out using 5-bromo-4-chloro-3-indolyl phosphate + nitroblue tetrazolium; fluorometric assay used ATTOPHOS™; fluorescence assay carried out in solution using membrane strips in cuvette; amplification by enzyme activity; fluorescent assay ~100-fold more sensitive than colorimetric assay; ~ $2 \times 10^6$ molecules of DNA detected by colorimetric assay and $2 \times 10^4$ molecules by fluorescent assay | 45 |
| | | also: Use of salicylphosphate esters in conjunction with alkaline phosphatase produces well-localized fluorescent products | 46 |
| | DNA on membrane detected by hybridization to digoxigenin-labeled probes; immunochemical techniques (anti-digoxigenin Ab labeled with AP) plus incubation with chemiluminescent sustrate gives chemiluminescence | 10 to 50 fg of target DNA could be detected[42] | 42 |
| Fluorescence *in Situ* Hybridization (FISH) CARD-Based | Same principle as CARD-based immunoassay; hapten-labeled DNA probe hybridizes to DNA target; then either antihapten Ab labeled with peroxidase or streptavidin labeled with peroxidase (if DNA probe is labeled with biotin) is bound to DNA/DNA probe complex; addition of either biotin tyramide or fluorophore tyramide results in deposition of fluorophore or biotin at site of peroxidase action; in the case of biotin deposition, addition of fluorophore-labeled streptavidin completes the detection "sandwich" | Raap et al.[48] indicate much higher signal intensities (at least tenfold) than with conventional FISH; large amplification due to enzyme activity and multiple layers of detection reagents | 48 |
| | or: As above, but after hybridization avidin/biotin/peroxidase complex (ABC) added, followed by biotinylated tyramine; deposition of biotin tyramine at site is detected by addition of fluorescein-avidin conjugate | CARD-FISH technique provided visual detection, when conventional FISH failed | 49 |
| | | A variety of fluorophores were conjugated to tyramine; sensitivity found to be better than conventional FISH[47] | 47, 50 |
| *In Situ* Hybridization (Colorimetric) CARD-Based | After initial hybridization as above, ABC is added, followed by biotinylated tyramine and $H_2O_2$; this is followed by application of ABC followed by addition of colorimetric substrate for detection | Amplification by enzyme activity — both for tyramide deposition and for processive turnover of colorimetric substrate; Kerstens et al.[49] found strong amplification in comparison to conventional *in situ* hybridization | 49, 51 |

| | | | |
|---|---|---|---|
| *In Situ* Hybridization with Probe Extension or Amplification, i.e., *in Situ* PCR | Probe for hybridization to target DNA consists of a single oligonucleotide primer; *Taq* DNA polymerase used to extend primer based on target sequence; various reporter or haptenated nucleotides in the reaction mix are incorporated into extended probe; these serve to provide direct or indirect (immunochemical) detection signals; technique called primed *in situ* DNA synthesis (PRINS); also cycling PRINS, where 2 oligonucleotide primers are used, and essentially a PCR amplification takes place; reporter/haptenated nucleotides in reaction mix incorporated into PCR product | Amplification due to multiple reporter/hapten nucleotides incorporated during probe DNA synthesis; also, for cycling PRINS, multiple copies of probe DNA made | 52–55 |
| | RT *in situ* PCR, for RNA amplification; as above, but cDNA generated by reverse transcriptase; this serves as target for standard PCR | | 56 |
| *In Situ* Hybridization without Probe Amplification; if fluorescent labels used, termed FISH (Fluorescence *in Situ* Hybridization) | DNA probes are developed for hybridization to specific DNA or RNA targets; these probes are either directly labeled with reporter molecules (e.g., fluorophores) or with hapten moieties, which can participate in immunochemical reactions; positive hybridization events are recorded by presence of reporter molecules or by product of the immunochemical reaction (e.g., conversion of an enzyme substrate to a fluorescent product) | Amplification by multiple reporter/hapten sites in DNA probe; also by immunochemical reactions; probes labeled with biotin-, digoxigenin-, or directly with fluorescein-dUTP by nick translation; in case of fluorescein-labeled DNA, results visualized directly, or indirectly, using an antifluorescein AB and antifirst species IgG Ab conjugated to fluorescein; for biotin-labeled probes, fluorescently labeled avidin was used for visualization; for digoxigenin-labeled probes, fluorescent-labeled Ab to digoxigenin was used; immunochemical "sandwiches" of greater complexity also used; direct labeling technique was found to yield lower backgrounds; sensitivity of detection found to be 50 to 100 Kb for directly labeled probes by visual observation; for immunochemical detection, visual detection sensitivity found to be 1 to 5 Kb | 57–63 |
| | | Biotinylated DNA probe hybridized to target DNA and subsequently reacted with avidin/biotinylated peroxidase; substrate for peroxidase was diaminobenzidine, providing colorimetric detection | 64 |

**TABLE 55.2** Nucleic-Acid Amplification Schemes (continued)

| Assay | Principle | Comments | Ref. |
|---|---|---|---|
| | | also: Chemiluminescent detection after using biotin-labeled DNA probe for hybridization to target, followed by immunochemical sandwich (antibiotin Ab; antispecies IgG Ab labeled with peroxidase); addition of enhanced luminol reagent gave chemiluminescence; found to be same order of sensitivity as $^{35}S$-labeled probes detected by autoradiography;[57] other applications where chemiluminescent substrates were used in immunochemical reactions after hybridization of a biotinylated probe to detect two different viral DNAs in same specimen[58] | 57, 58 |
| | | also: Hybridizations on glass microarray detected by enzyme-catalyzed chemiluminescent reactions; 250 attomol target DNA detection limit; enzyme amplification | 59 |
| *In Situ* Hybridization (Colorimetric) | Biotinylated probe hybridized to DNA, followed by antibiotin Ab, biotinylated Ab specific to species IgG of first Ab, followed by avidin and biotin peroxidase complex (ABC); colorimetric substrate for peroxidase added for final detection | Amplification by enzyme activity; diaminobenzidine used as precipitable peroxidase substrate; this was further amplified by treatment with $CuSO_4$ | 49 |
| FISH (Not CARD) | After hybridization of biotinylated probe to DNA, fluoresceinated avidin was added, followed by biotinylated antiavidin Ab, followed by avidin–fluorescein conjugate for fluorescence detection | | 49 |
| Chemical Modification of Probe | Guanine residues in nucleic acids are modified chemically by incorporation of N-2-acetylaminofluorene (AAF) (or 7-iodo derivative) at C-8 position of guanine; following hybridization of modified probe to target, Ab specific to AAF can be used to bind to modified probe, allowing detection by immunochemical methods; also such modified nucleic acids can be immuno-precipitated | Attachment of Ab to nucleic acids allows amplification of detection signal by immunoassay techniques (e.g., ELISA); sensitivity comparable to autoradiographic methods | 65 |

| | | | |
|---|---|---|---|
| Probes Linked to Gold Nanoparticles | Small oligonucleotides are linked through mercaptoalkyl linkage to gold nanoparticles; part of the oligonucleotide sequence is complementary to the target sequence being detected; two different probe oligonucleotides, which recognize adjacent sites on the target, are linked to nanoparticles (more than one pair of oligonucleotide probes are bound per nanoparticle); upon hybridization to the target, an extended polymeric network is formed in which nanoparticles are interlocked by multiple short duplex segments arising from hybridization of probe with target; this close juxaposition of nanoparticles changes the optical properties of color changes from red to blue, upon hybridization | Amplification due to numerous nanoparticles entrapped in polymeric network; specificity of this two probe system seems extremely good; one mismatched base detected; sensitivity of unoptimized system showed detection of ~10 fmol of target oligonucleotide; technique seems to be more sensitive than detection by means of fluorescein-labeled probe as reported in the literature | 66 |
| Bead Capture; Enzyme Amplification | Five different oligonucleotides used for this assay to detect analyte DNA; two oligonucleotides (A and B) have sequences complementary to analyte DNA as well as single-strand overhangs, which act as recognition sites for other oligonucleotide probes; oligoprobe C is biotinylated at one end and has sequence complementary to probe A; it is bound to avidin-coated beads and serves to capture analyte DNA; oligoprobe D is a chemically cross-linked oligonucleotide complex (multimer), which has a complementary sequence to the single-strand overhang of probe B; probe D binds (via probe B) to the analyte DNA and offers a number of sites (due to multimer nature) for attachment of probe E, which is an oligonucleotide complementary to probe D and which also has peroxidase attached to it; addition of substrates results in detection signal, indicating analyte DNA is present | Amplification occurs by using multiple probes for binding to target DNA, the use of multimer oligonucleotide probes that provide multiple sites for attachment of peroxidase, and the enzyme itself; Urdea et al. report that this technique was as sensitive as most sensitive radiolabeled dot-blots; chemiluminescent substrate used in this assay; applied to detection of sub-pg quantities of viral DNA | 67 |
| Hybridization on Microarrays with Optical Detection | Essentially same strategies as used for other hybridization formats; "capture" oligonucleotides attached to solid surface (glass, gel pad, etc.); labeled (often by incorporation of biotin-labeled primers or fluorescently labeled primers in PCR) "target" nucleic acid is hybridized to microarray and detected by various means; for example, biotin-labeled targets can be detected by standard procedures, such as reaction with streptavidin-alkaline phosphatase, and subsequent development of color, fluorescence, or chemiluminescence by addition of appropriate enzyme substrate | | 68–72 |
| PCR/LDR Mutation Detection on Microarray | Multiplex PCR is used to amplify the regions of interest (for mutation analysis); addition of two oligonucleotides which are complementary to sequences on both sides of mutation sets the stage for ligase-dependent ligation, only if sequence at junction of two oligonucleotides is exact match for mutation; one oligonucleotide is fluorescently labeled and the other has overhanging sequence coded for a specific complementary capture oligonucleotide at specific site on microarray; fluorescence detection at this site indicates presence of mutation | | 73 |

**TABLE 55.2** Nucleic-Acid Amplification Schemes (continued)

| Assay | Principle | Comments | Ref. |
|---|---|---|---|
| Invader Assay | Assay depends on use of thermostable flap endonucleases; two overlapping probes are constructed to target sequence; upon hybridization to target, downstream oligonucleotide is cleaved at site of overlap; cleaved portion of oligonucleotide hybridizes to biotin-labeled template, which has a number of A residues at 5′ overhang; use of DNA polymerase and digoxigenin-labeled dU results in incorporation of digoxigenin-dU into probe fragment; double-stranded template + probe (extended by digoxigenin-dU) captured on streptavidin-coated microwell; probe fragment detected by binding of alkaline phosphate-conjugated antidigoxigenin Ab, followed by addition of fluorogenic substrate for alkaline phosphatase | Amplification by thermostable endonuclease (3000 cleavage events per target molecule) and by alkaline phosphatase enzyme bound to Ab; found detection of DNA targets at subattomole levels | 74 |
| | also:<br>First round of invader assay as described above; cleaved probe fragment binds to a biotin-labeled, streptavidin-immobilized probe that also has bound to it another small oligonucleotide with a fluorophore and quencher in close apposition; hybridization of cleaved probe fragment from first round produces another single-nucleotide overlap, also susceptible to endonuclease cleavage; cleavage releases fluorophore, abolishing FRET quenching | Serial version of invader assay can generate more than $10^7$ reporter molecules for each target molecule in 4-h reaction; sensitivity for detection of less than 1000 target molecules reported | 75 |
| Chromosome Painting | Biotinylated oligonucleotide probes specific for some chromosome or chromosomal site hybridized to target chromosome; alternating layers of fluorophore-labeled avidin and biotinylated anti-avidin Ab added to give amplification for detection | Detection of unique intrachromosomal sequences required multiple (3) layers of fluorophore-labeled avidin | 8 |
| Intensification | Biotinylated probes were reacted with target DNA *in situ*, followed by standard immunocytochemical techniques resulting in immobilization of peroxidase at hybridization sites; following incubation with diaminobenzidine and $H_2O_2$, further staining intensification was achieved by silver staining, using a series of silver and gold solutions | More sensitive than colorimetric stain alone | 76 |
| Eu-Labeled Probes | DNA hybridization probes are labeled with a Eu-chelate by introducing aliphatic amino groups on cytosine residues; following hybridization, positive hybridization events detected by time-resolved fluorometry | Amplification by numerous fluorophores on DNA probe — 4 to 8% of total nucleotides labeled;[77] detection limit was 0.15 attomol of target DNA[77] | 77 |
| | ***Branched DNA Technology*** | | |
| | *Target DNA is not amplified, but a series of oligonucleotides complementary to portions of target are used to hybridize to target and form dendrimer structure with many labels that can be used for detection* | | |

| | | | |
|---|---|---|---|
| Branched DNA Amplification | (a) Target DNA is immobilized to solid surface by series of oligonucleotides (called capture probes and capture extenders [CE]) complementary to segments of target DNA; subsequently, other oligonucleotides (label extenders [LE]) complementary to other segments of target DNA hybridize to target, the LE also bind (by complementarity) to a branched DNA segment (amplifier), which then binds many alkaline phosphatase molecules; alkaline phosphatase is bound to a short oligonucleotide with sequence complementary to repeated triplicate sequence on branched DNA;<br>(b) in other developed assays, LE probes bind preamplifier oligonucleotide probes, which then bind many amplifier segments;<br>Incorporation of nonnatural bases, isocytidine and isoguanosine, into amplifier molecules can reduce nonspecific hybridization, thus improving sensitivity; final detection step involves adding chemiluminescent substrate for alkaline phosphatase and detecting luminescence; branched DNA technology amplifies reporter signal, not the target | Amplification of signal is linear, so signal directly related to number of target molecules present in original sample; first generation assay quantifies targets between $10^4$ and $10^7$ molecules; second generation assay quantifies down to ~500 molecules; detection limit of ~50 molecules//ml reported[78] | 78, 79 |
| | | Detection and quantification of mRNA directly from cells; CE and LE probes about 30 bases long and between them provide complete coverage of the mRNA sequence; branched DNA amplifiers have 15 branches, each of which is labeled subsequently with alkaline phosphatase; thus great amplification; solid surface has synthetic DNA designed to hybridize to CE | 80 |
| | | Kern et al.[81] used branched DNA technique for detection of HIV RNA; use of preamplifier intermediate resulted in binding of eight amplifier molecules; each amplifier molecule had 15 branches and each branch could bind three alkaline phosphatase-labeled probes; thus theoretically each HIV RNA could be decorated with as many as 10,080 alkaline phosphatase probes; result detected by chemiluminescent assay; detection limit was 390 HIV RNA copies/ml | 81 |

**TABLE 55.2** Nucleic-Acid Amplification Schemes (continued)

| Assay | Principle | Comments | Ref. |
|---|---|---|---|
| Signal Amplification Cassettes (SAC) | Signal amplification cassettes (SACs) are attached to DNA dendrimers; SACs are oligonucleotides (hairpin or double strand) with 15 adenine overhangs at one or both ends; polymerase activity of T7 DNA polymerase extends strand to fill in overhang with TTP (thymidine triphosphate); for each TTP incorporated, a $PP_i$ is liberated; this reaction continues to recycle because 3′–5′ exonuclease activity regenerates original SAC, which again becomes substrate for DNA polymerase; $PP_i$ generated is converted enzymatically to ATP and luciferase reacts with luciferin + ATP to produce light; many SACs are attached to each DNA dendrimer; potential for tremendous amplification; hybridization of dendrimer to target produces specificity of reaction | Multiple amplification steps involved; many enzymes (DNA polymerase, ATP sulfurylase, luciferase) involved, all contributing to amplification; also amplification by multiple dendrimer sites; detection of 5 zeptomol of DNA dendrimer demonstrated; SAC consisting of just two polymerase reaction sites sufficient to detect low attomole levels of DNA | 82 |

## References


1. Cahill, P., Foster, K., and Mahan, D.E., Polymerase chain-reaction and Q beta replicase amplification, *Clin. Chem.*, 37(9), 1482, 1991.
2. Crescenzi, M., Seto, M., Herzig, G.P., Weiss, P.D., Griffith, R.C., and Korsmeyer, S.J., Thermostable DNA-polymerase chain amplification of T(14-18) chromosome breakpoints and detection of minimal residual disease, *Proc. Natl. Acad. Sci. U.S.A.*, 85(13), 4869, 1988.
3. Chehab, F.F., Doherty, M., Cai, S., Kan, Y.W., Cooper, S., and Rubin, E.M., Detection of sickle-cell-anemia and thalassemias, *Nature*, 329(6137), 293, 1987.
4. Saiki, R.K., Scharf, S., Faloona, F., Mullis, K.B., Horn, G.T., Erlich, H.A., and Arnheim, N., Enzymatic amplification of beta-globin genomic sequences and restriction site analysis for diagnosis of sickle-cell anemia, *Science*, 230(4732), 1350, 1985.
5. Saiki, R.K., Bugawan, T.L., Horn, G.T., Mullis, K.B., and Erlich, H.A., Analysis of enzymatically amplified beta-globin and Hla-Dq-alpha DNA with allele-specific oligonucleotide probes, *Nature*, 324(6093), 163, 1986.
6. Alard, P., Lantz, O., Sebagh, M., Calvo, C.F., Weill, D., Chavanel, G., Senik, A., and Charpentier, B., A versatile ELISA-PCR assay for messenger-RNA quantitation from a few cells, *Biotechniques*, 15(4), 730, 1993.
7. Myers, T.W. and Gelfand, D.H., Reverse transcription and DNA amplification by a thermus-thermophilus DNA-polymerase, *Biochemistry*, 30(31), 7661, 1991.
8. Pinkel, D., Landegent, J., Collins, C., Fuscoe, J., Segraves, R., Lucas, J., and Gray, J., Fluorescence *in situ* hybridization with human chromosome-specific libraries — detection of trisomy-21 and translocations of chromosome-4, *Proc. Natl. Acad. Sci. U.S.A.*, 85(23), 9138, 1988.
9. Rappolee, D.A., Wang, A., Mark, D., and Werb, Z., Novel method for studying messenger-RNA phenotypes in single or small numbers of cells, *J. Cell. Biochem.*, 39(1), 1, 1989.
10. Razin, E., Leslie, K.B., and Schrader, J.W., Connective-tissue mast-cells in contact with fibroblasts express Il-3 messenger-RNA — analysis of single cells by polymerase chain-reaction, *J. Immunol.*, 146(3), 981, 1991.
11. Chehab, F.F. and Kan, Y.W., Detection of specific DNA-sequences by fluorescence amplification — a color complementation assay, *Proc. Natl. Acad. Sci. U.S.A.*, 86(23), 9178, 1989.
12. Bortolin, S., Christopoulos, T.K., and Verhaegen, M., Quantitative polymerase chain reaction using a recombinant DNA internal standard and time resolved fluorometry, *Anal. Chem.*, 68(5), 834, 1996.
13. Wang, A.M., Doyle, M.V., and Mark, D.F., Quantitation of messenger-RNA by the polymerase chain-reaction, *Proc. Natl. Acad. Sci. U.S.A.*, 86(24), 9717, 1989.
14. Katz, E.D., Haff, L.A., and Eksteen, R., Rapid separation, quantitation and purification of products of polymerase chain-reaction by liquid-chromatography, *J. Chromatogr.*, 512, 433, 1990.
15. Kemp, D.J., Smith, D.B., Foote, S.J., Samaras, N., and Peterson, M.G., Colorimetric detection of specific DNA segments amplified by polymerase chain reactions, *Proc. Natl. Acad. Sci. U.S.A.*, 86(7), 2423, 1989.
16. Stoflet, E.S., Koeberl, D.D., Sarkar, G., and Sommer, S.S., Genomic amplification with transcript sequencing, *Science*, 239(4839), 491, 1988.
17. Nickerson, D.A., Kaiser, R., Lappin, S., Stewart, J., Hood, L., and Landegren, U., Automated DNA diagnostics using an ELISA-based oligonucleotide ligation assay, *Proc. Natl. Acad. Sci. U.S.A.*, 87(22), 8923, 1990.
18. Thelwell, N., Millington, S., Solinas, A., Booth, J., and Brown, T., Mode of action and application of Scorpion primers to mutation detection, *Nucleic Acids Res.*, 28(19), 3752, 2000.
19. Whitcombe, D., Theaker, J., Guy, S.P., Brown, T., and Little, S., Detection of PCR products using self-probing amplicons and fluorescence, *Nat. Biotechnol.*, 17(8), 804, 1999.

20. Holland, P.M., Abramson, R.D., Watson, R., and Gelfand, D.H., Detection of specific polymerase chain-reaction product by utilizing the 5′-3′ exonuclease activity of thermus-aquaticus DNA-polymerase, *Proc. Natl. Acad. Sci. U.S.A.*, 88(16), 7276, 1991.
21. Nazarenko, I.A., Bhatnagar, S.K., and Hohman, R.J., A closed tube format for amplification and detection of DNA based on energy transfer, *Nucleic Acids Res.*, 25(12), 2516, 1997.
22. Dicesare, J., Grossman, B., Katz, E., Picozza, E., Ragusa, R., and Woudenberg, T., A high-sensitivity electrochemiluminescence-based detection system for automated PCR product quantitation, *Biotechniques*, 15(1), 152, 1993.
23. Actor, J.K., Kuffner, T., Dezzutti, C.S., Hunter, R.L., and McNicholl, J.M., A flash-type bioluminescent immunoassay that is more sensitive than radioimaging: quantitative detection of cytokine cDNA in activated and resting human cells, *J. Immunol. Meth.*, 211(1–2), 65, 1998.
24. Actor, J.K., Olsen, M., Boven, L.A., Werner, N., Stults, N.L., Hunter, R.L., and Smith, D.F., A bioluminescent assay using AquaLite for PT-PCR amplified RNA from mouse lung, *J. Natl. Inst. Health Res.*, 8, 62, 1996.
25. Kenten, J.H., Casadei, J., Link, J., Lupold, S., Willey, J., Powell, M., Rees, A., and Massey, R., Rapid electrochemiluminescence assays of polymerase chain-reaction products, *Clin. Chem.*, 37(9), 1626, 1991.
26. Blackburn, G.F., Shah, H.P., Kenten, J.H., Leland, J., Kamin, R.A., Link, J., Peterman, J., Powell, M.J., Shah, A., Talley, D.B., Tyagi, S.K., Wilkins, E., Wu, T.G., and Massey, R.J., Electrochemiluminescence detection for development of immunoassays and DNA probe assays for clinical diagnostics, *Clin. Chem.*, 37(9), 1534, 1991.
27. Liu, X., Farmerie, W., Schuster, S., and Tan, W., Molecular beacons for DNA biosensors with micrometer to submicrometer dimensions, *Anal. Biochem.*, 283, 56, 2000.
28. McKillip, J.L. and Drake, M., Molecular beacon polymerase chain reaction detection of *Escherichia coli* O157 : H7 in milk, *J. Food Prot.*, 63(7), 855, 2000.
29. Compton, J., Nucleic-acid sequence-based amplification, *Nature*, 350(6313), 91, 1991.
30. Voisset, C., Mandrand, B., and Paranhos-Baccala, G., RNA amplification technique, NASBA, also amplifies homologous plasmid DNA in non-denaturing conditions, *Biotechniques*, 29(2), 236, 2000.
31. Heim, A., Grumbach, I.M., Zeuke, S., and Top, B., Highly sensitive detection of gene expression of an intronless gene: amplification of mRNA, but not genomic DNA by nucleic acid sequence based amplification (NASBA), *Nucleic Acids Res.*, 26(9), 2250, 1998.
32. Kwoh, D.Y., Davis, G.R., Whitfield, K.M., Chappelle, H.L., Dimichele, L.J., and Gingeras, T.R., Transcription-based amplification system and detection of amplified human immunodeficiency virus type-1 with a bead-based sandwich hybridization format, *Proc. Natl. Acad. Sci. U.S.A.*, 86(4), 1173, 1989.
33. Guatelli, J.C., Whitfield, K.M., Kwoh, D.Y., Barringer, K.J., Richman, D.D., and Gingeras, T.R., Isothermal, *in vitro* amplification of nucleic-acids by a multienzyme reaction modeled after retroviral replication, *Proc. Natl. Acad. Sci. U.S.A.*, 87(5), 1874, 1990.
34. Liu, D.Y., Daubendiek, S.L., Zillman, M.A., Ryan, K., and Kool, E.T., Rolling circle DNA synthesis: small circular oligonucleotides as efficient templates for DNA polymerases, *J. Am. Chem. Soc.*, 118(7), 1587, 1996.
35. Lizardi, P.M., Huang, X.H., Zhu, Z.R., Bray-Ward, P., Thomas, D.C., and Ward, D.C., Mutation detection and single-molecule counting using isothermal rolling-circle amplification, *Nat. Genet.*, 19(3), 225, 1998.
36. Nilsson, M., Malmgren, H., Samiotaki, M., Kwiatkowski, M., Chowdhary, B.P., and Landegren, U., Padlock probes — circularizing oligonucleotides for localized DNA detection, *Science*, 265(5181), 2085, 1994.
37. Barany, F., Genetic-disease detection and DNA amplification using cloned thermostable ligase, *Proc. Natl. Acad. Sci. U.S.A.*, 88(1), 189, 1991.
38. Wu, D.Y. and Wallace, R.B., The ligation amplification reaction (LAR) — amplification of specific DNA-sequences using sequential rounds of template-dependent ligation, *Genomics*, 4(4), 560, 1989.

39. Landegren, U., Kaiser, R., Sanders, J., and Hood, L., A ligase-mediated gene detection technique, *Science*, 241(4869), 1077, 1988.
40. Abravaya, K., Carrino, J.J., Muldoon, S., and Lee, H.H., Detection of point mutations with a modified ligase chain-reaction (Gap-Lcr), *Nucleic Acids Res.*, 23(4), 675, 1995.
41. Nadeau, J.G., Pitner, J.B., Linn, C.P., Schram, J.L., Dean, C.H., and Nycz, C.M., Real-time, sequence-specific detection of nucleic acids during strand displacement amplification, *Anal. Biochem.*, 276(2), 177, 1999.
42. Musiani, M., Zerbini, M., Gibellini, D., Gentilomi, G., Laplaca, M., Ferri, E., and Girotti, S., Chemiluminescent assay for the detection of viral and plasmid DNA using digoxigenin-labeled probes, *Anal. Biochem.*, 194(2), 394, 1991.
43. Tizard, R., Cate, R.L., Ramachandran, K.L., Wysk, M., Voyta, J.C., Murphy, O.J., and Bronstein, I., Imaging of DNA-sequences with chemiluminescence, *Proc. Natl. Acad. Sci. U.S.A.*, 87(12), 4514, 1990.
44. Hauber, R. and Geiger, R., A sensitive, bioluminescent-enhanced detection method for DNA dot-hybridization, *Nucleic Acids Res.*, 16(3), 1213, 1988.
45. Cano, R.J., Torres, M.J., Klem, R.E., and Palomares, J.C., DNA hybridization assay using Attophostm, a fluorescent substrate for alkaline-phosphatase, *Biotechniques*, 12(2), 264, 1992.
46. Evangelista, R.A., Wong, H.E., Templeton, E.F.G., Granger, T., Allore, B., and Pollak, A., Alkyl-substituted and aryl-substituted salicyl phosphates as detection reagents in enzyme-amplified fluorescence DNA hybridization assays on solid support, *Anal. Biochem.*, 203(2), 218, 1992.
47. Van Gijlswijk, R.P.M., Zijlmans, H., Wiegant, J., Bobrow, M.N., Erickson, T.J., Adler, K.E., Tanke, H.J., and Raap, A.K., Fluorochrome-labeled tyramides: use in immunocytochemistry and fluorescence *in situ* hybridization, *J. Histochem. Cytochem.*, 45(3), 375, 1997.
48. Raap, A.K., Vandecorput, M.P.C., Vervenne, R.A.W., Vangijlswijk, R.P.M., Tanke, H.J., and Wiegant, J., Ultra-sensitive FISH using peroxidase-mediated deposition of biotin-tyramide or fluorochrome-tyramide, *Hum. Mol. Genet.*, 4(4), 529, 1995.
49. Kerstens, H.M.J., Poddighe, P.J., and Hanselaar, A., A novel *in situ* hybridization signal amplification method based on the deposition of biotinylated tyramine, *J. Histochem. Cytochem.*, 43(4), 347, 1995.
50. Schmidt, B.F., Chao, J., Zhu, Z.G., DeBiasio, R.L., and Fisher, G., Signal amplification in the detection of single-copy DNA and RNA by enzyme-catalyzed deposition (CARD) of the novel fluorescent reporter substrate Cy3.29-tyramide, *J. Histochem. Cytochem.*, 45(3), 365, 1997.
51. Adams, J.C., Biotin amplification of biotin and horseradish-peroxidase signals in histochemical stains, *J. Histochem. Cytochem.*, 40(10), 1457, 1992.
52. Gosden, J., Hanratty, D., Starling, J., Fantes, J., Mitchell, A., and Porteous, D., Oligonucleotide-primed *in situ* DNA-synthesis (PRINS) — a method for chromosome mapping, banding, and investigation of sequence organization, *Cytogenet. Cell Genet.*, 57(2–3), 100, 1991.
53. Gosden, J. and Hanratty, D., Comparison of sensitivity of 3 haptens in the PRINS reaction, *Technique*, 3, 159, 1992.
54. Gosden, J. and Hanratty, D., PCR *in situ* — a rapid alternative to *in situ* hybridization for mapping short, low copy number sequences without isotopes, *Biotechniques*, 15(1), 78, 1993.
55. Gosden, J. and Lawson, D., Rapid chromosome identification by oligonucleotide-primed *in situ* DNA-synthesis (PRINS), *Hum. Mol. Genet.*, 3(6), 931, 1994.
56. Nuovo, G.J., Co-labeling using *in situ* PCR: a review, *J. Histochem. Cytochem.*, 49(11), 1329, 2001.
57. Lorimier, P., Lamarcq, L., Negoescu, A., Robert, C., Labat-Moleur, F., Gras-Chappuis, F., Durrant, I., and Brambilla, E., Comparison of S-35 and chemiluminescence for HPV *in situ* hybridization in carcinoma cell lines and on human cervical intraepithelial neoplasia, *J. Histochem. Cytochem.*, 44(7), 665, 1996.
58. Gentilomi, G., Musiani, M., Roda, A., Pasini, P., Zerbini, M., Gallinella, G., Baraldini, M., Venturoli, S., and Manaresi, E., Co-localization of two different viral genomes in the same sample by double-chemiluminescence *in situ* hybridization, *Biotechniques*, 23(6), 1076, 1997.

59. Cheek, B.J., Steel, A.B., Torres, M.P., Yu, Y.-Y., and Yang, H., Chemiluminescence detection for hybridization assays on the flow-thru chip, a three-dimensional microchannel biochip, *Anal. Chem.*, 73(24), 5777, 2001.
60. Raap, A.K., Vanderijke, F.M., Dirks, R.W., Sol, C.J., Boom, R., and Vanderploeg, M., Bicolor fluorescence *in situ* hybridization to intron and exon messenger-RNA sequences, *Exp. Cell Res.*, 197(2), 319, 1991.
61. Tkachuk, D.C., Westbrook, C.A., Andreeff, M., Donlon, T.A., Cleary, M.L., Suryanarayan, K., Homge, M., Redner, A., Gray, J., and Pinkel, D., Detection of Bcr-Abl fusion in chronic myelogeneous leukemia by *in situ* hybridization, *Science*, 250(4980), 559, 1990.
62. Wiegant, J., Ried, T., Nederlof, P.M., Vanderploeg, M., Tanke, H.J., and Raap, A.K., *In situ* hybridization with fluoresceinated DNA, *Nucleic Acids Res.*, 19(12), 3237, 1991.
63. Wachtler, F., Hartung, M., Devictor, M., Wiegant, J., Stahl, A., and Schwarzacher, H.G., Ribosomal DNA is located and transcribed in the dense fibrillar component of human sertoli-cell nucleoli, *Exp. Cell Res.*, 184(1), 61, 1989.
64. Dewit, P.E., Kerstens, H.M.J., Poddighe, P.J., Vanmuijen, G.N.P., and Ruiter, D.J., DNA *in situ* hybridization as a diagnostic-tool in the discrimination of melanoma and spitz nevus, *J. Pathol.*, 173(3), 227, 1994.
65. Tchen, P., Fuchs, R.P.P., Sage, E., and Leng, M., Chemically modified nucleic acids as immunodetectable probes in hybridization experiments, *Proc. Natl. Acad. Sci. U.S.A. Biol. Sci.*, 81(11), 3466, 1984.
66. Elghanian, R., Storhoff, J.J., Mucic, R.C., Letsinger, R.L., and Mirkin, C.A., Selective colorimetric detection of polynucleotides based on the distance-dependent optical properties of gold nanoparticles, *Science*, 277(5329), 1078, 1997.
67. Urdea, M.S., Running, J.A., Horn, T., Clyne, J., Ku, L., and Warner, B.D., A novel method for the rapid detection of specific nucleotide-sequences in crude biological samples without blotting or radioactivity — application to the analysis of hepatitis-B virus in human-serum, *Gene*, 61(3), 253, 1987.
68. Call, D.R., Chandler, D.P., and Brockman, F., Fabrication of DNA microarrays using unmodified oligonucleotide probes, *Biotechniques*, 30(2), 368, 2001.
69. Guo, Z., Guilfoyle, R.A., Thiel, A.J., Wang, R.F., and Smith, L.M., Direct fluorescence analysis of genetic polymorphisms by hybridization with oligonucleotide arrays on glass supports, *Nucleic Acids Res.*, 22(24), 5456, 1994.
70. Pease, A.C., Solas, D., Sullivan, E.J., Cronin, M.T., Holmes, C.P., and Fodor, S.P.A., Light-generated oligonucleotide arrays for rapid DNA-sequence analysis, *Proc. Natl. Acad. Sci. U.S.A.*, 91(11), 5022, 1994.
71. Yershov, G., Barsky, V., Belgovskiy, A., Kirillov, E., Kreindlin, E., Ivanov, I., Parinov, S., Guschin, D., Drobishev, A., Dubiley, S., and Mirzabekov, A., DNA analysis and diagnostics on oligonucleotide microchips, *Proc. Natl. Acad. Sci. U.S.A.*, 93(10), 4913, 1996.
72. Langer, P.R., Waldrop, A.A., and Ward, D.C., Enzymatic-synthesis of biotin-labeled polynucleotides — novel nucleic-acid affinity probes, *Proc. Natl. Acad. Sci. U.S.A. Biol. Sci.*, 78(11), 6633, 1981.
73. Favis, R. and Barany, F., Mutation detection in K-ras, BRCA1, BRCA2, and p53 using PCR/LDR and a universal DNA microarray, *Ann. N.Y. Acad. Sci.*, 39, 2000.
74. Lyamichev, V., Mast, A.L., Hall, J.G., Prudent, J.R., Kaiser, M.W., Takova, T., Kwiatkowski, R.W., Sander, T.J., de Arruda, M., Arco, D.A., Neri, B.P., and Brow, M.A.D., Polymorphism identification and quantitative detection of genomic DNA by invasive cleavage of oligonucleotide probes, *Nat. Biotechnol.*, 17(3), 292, 1999.
75. Hall, J.G., Eis, P.S., Law, S.M., Reynaldo, L.P., Prudent, J.R., Marshall, D.J., Allawi, H.T., Mast, A.L., Dahlberg, J.E., Kwiatkowski, R.W., de Arruda, M., Neri, B.P., and Lyamichev, V.I., Sensitive detection of DNA polymorphisms by the serial invasive signal amplification reaction, *Proc. Natl. Acad. Sci. U.S.A.*, 97(15), 8272, 2000.

76. Mullink, H., Vos, W., Jiwa, M., Horstman, A., Vandervalk, P., Walboomers, J.M.M., and Meijer, C., Application and comparison of silver intensification methods for the diaminobenzidine and diaminobenzidine nickel end-product of the peroxidation reaction in immunohistochemistry and *in situ* hybridization, *J. Histochem. Cytochem.*, 40(4), 495, 1992.
77. Hurskainen, P., Dahlen, P., Ylikoski, J., Kwiatkowski, M., Siitari, H., and Lovgren, T., Preparation of europium-labeled DNA probes and their properties, *Nucleic Acids Res.*, 19(5), 1057, 1991.
78. Collins, M.L., Irvine, B., Tyner, D., Fine, E., Zayati, C., Chang, C.A., Horn, T., Ahle, D., Detmer, J., Shen, L. P., Kolberg, J., Bushnell, S., Urdea, M.S., and Ho, D.D., A branched DNA signal amplification assay for quantification of nucleic acid targets below 100 molecules/ml, *Nucleic Acids Res.*, 25(15), 2979, 1997.
79. Urdea, M.S., Horn, T., Fultz, T., Anderson, M., Running, J.A., Hamren, S., Ahle, D., and Chang, C.A., Branched DNA amplification multimers for the sensitive direct detection of human hepatitis virus, *Nucleic Acids Res.*, 24 (Symp. Ser.), 197, 1991.
80. Warrior, U., Fan, Y.H., David, C.A., Wilkins, J.A., McKeegan, E.M., Kofron, J.L., and Burns, D.J., Application of QuantiGene (TM) nucleic acid quantification technology for high throughput screening, *J. Biomol. Screening*, 5, 5, 343, 2000.
81. Kern, D., Collins, M., Fultz, T., Detmer, J., Hamren, S., Peterkin, J.J., Sheridan, P., Urdea, M., White, R., Yeghiazarian, T., and Todd, J., An enhanced-sensitivity branched-DNA assay for quantification of human immunodeficiency virus type 1 RNA in plasma, *J. Clin. Microbiol.*, 34(12), 3196, 1996.
82. Capaldi, S., Getts, R.C., and Jayasena, S.D., Signal amplification through nucleotide extension and excision on a dendritic DNA platform, *Nucleic Acids Res.*, 28(7), e21(i–viii), 2000.

**TABLE 55.3** Hybrid Technologies

| Assay | Principle | Comments | Ref. |
|---|---|---|---|
| | ***Immuno-PCR*** | | |
| | *Technique combining antibody (Ab) specificity for molecular recognition with amplification potential of PCR* | | |
| Immuno-PCR | Immobilized antigen (Ag) is recognized by Ag-specific Ab, which is linked to specific segment of DNA through some linker molecule (*vide infra*); following immune recognition, PCR reaction is carried out on segment of attached DNA; PCR products are detected as final step in analysis: (a) linker molecule is a streptavidin-protein A chimera; protein A component binds to Ab and streptavidin binds to biotinylated DNA (linear plasmid pUC19) forming a bridge between DNA and Ab;[1] (b) linker molecule is avidin; biotinylated Ab and biotinylated DNA bind to avidin and form complete complex,[2] and (c) primary Ab recognizes Ag; secondary biotinylated Ab recognizes primary Ab; modified avidin (to lessen nonspecific binding) forms bridge between biotinylated Ab and biotinylated DNA | Specificity resides in Ag/Ab recognition and in specificity of PCR for target sequence defined by primers; possibility for quantification of Ag concentrations below level at which PCR reaction saturates; tremendous amplification because of exponential nature of PCR; universal usefulness of single immuno-PCR protocol for detection of variety of Ags<br><br>Because of extreme amplification of PCR, any nonspecific binding of DNA or Ab to other immobilized reagents or solid substrate contributes to background; avidin-/biotin-based linker systems must be carefully titrated to form usable complexes (i.e., Ab linker-DNA) that actually contribute to PCR; may be necessary to preform Ab-linker-DNA complexes; scrupulous washing required during assay steps to eliminate nonspecific binding; standard microtiter plates used for immunoassay are suitable for PCR thermal cycling; DNA target used for PCR amplification should, ideally, be very different from any DNA present in sample to avoid possible interferences in PCR amplification arising from exogenous DNA; total assay time likely to be longer than ELISA-type assays; increasing PCR cycle number increases sensitivity, but also increases nonspecific amplifications; amplification also occurs because more than one copy of biotinylated target DNA can be immobilized at each avidin/streptavidin site, if the stoichiometric ratios are titered carefully | 1–10 |

| | |
|---|---|
| or:<br>Immuno-PCR carried out in usual manner, but during PCR amplification a hapten-labeled (digoxigenin) nucleotide is incorporated into PCR products; one PCR primer also has a biotin label to permit capture of the PCR products; subsequently, the biotinylated, hapten-labeled PCR products are captured on a streptavidin-coated solid surface, enzyme-labeled antihapten Ab is added, and the addition of enzyme substrate results in accumulation of detected product | Amplification also from multiple biotinylation sites on primary detection Ab and multiple hapten sites on each individual PCR product molecule; plus enzyme amplification |
| or:<br>Immuno-PCR procedure carried out, and DNA product produced by amplification is detected by staining with a dsDNA selective intercalating fluorescent dye (Pico Green) | |
| or:<br>Usual immuno-PCR protocol, but capture Ab used to immobilize Ag of interest, followed by reaction with another Ag-specific biotinylated Ab; streptavidin serves as bridge to biotinylated target DNA; PCR primers each have one fluorescein/primer; after PCR, amplified product hybridized to capture oligonucleotide immobilized to solid surface; addition of antifluorescein Ab conjugated to enzyme allows standard ELISA endpoint | Total assay is time-consuming; very extensive washing required for certain steps; more than one biotin bound per detection Ab, resulting in more than one target DNA bound/detection Ab<br><br>Sensitivity:<br>Enhancement of $10^5$ vs. colorimetric ELISA using same immunoreagents and linker; 580 molecules of Ag detected,[1] estimated to be several orders of magnitude greater sensitivity than radioimmunoassay[1]<br>Enhancement of $10^8$ vs. colorimetric ELISA; 41 molecules/ml (or 2 molecules per microwell) of Ag detected, with signal:noise $\cong 4$ (Reference 3)<br>Detection limits on the order of 10 to $10^7$ molecules for various antigens, if primary Ag-specific Abs used at higher concentrations; up to $10^9$ enhancement of sensitivity compared to colorimetric ELISA[4]<br>Enhancement of 16-fold compared to colorimetric ELISA in terms of sensitivity, using the same immunoreagents for both assays[5]<br>Sensitivity enhancement of $\sim 10^4$ compared to colorimetric ELISA[6]<br>Enhancement of $10^5$ compared to colorimetric ELISA; lower limit of Ag detection $\cong 6000$ molecules[8] |

**TABLE 55.3** Hybrid Technologies (continued)

| Assay | Principle | Comments | Ref. |
|---|---|---|---|
| | | Detection limits for mouse IgG as Ab were 10 attomol for direct detection using an intercalating dye; 1 attomol for colorimetric ELISA detection and 0.1 attomol for fluorescent enzyme substrate; fluorescence-based assay allowed quantitative detection over six orders of magnitude; compared to standard ELISA using fluorescent substrate, immuno-PCR was 1000-fold more sensitive[9] | |
| Immuno-PCR with DNA-Streptavidin Nanostructures | Essential principle same as immuno-PCR but an oligomeric biotinylated DNA-streptavidin network is used as attachment to biotinylated Ab, rather than single streptavidin bridge between Ab and DNA; rest of analysis identical | Oligomeric DNA-streptavidin complexes provide superior immuno-PCR performance; better signal intensities than standard immuno-PCR; approx. tenfold increase in sensitivity over standard immuno-PCR, and 100- to 1000-fold increase in sensitivity over ELISA immunoassay | 11 |
| *In Situ* Immuno-PCR | Immuno-PCR carried out as usual, but on tissue sections; after binding DNA-labeled Ab to specific Ag, *in situ* PCR carried out, followed by hybridization to digoxigenin-labeled DNA probe and subsequent immunostaining | Amplification by PCR and immunoenzyme step; sensitivity of *in situ* immuno-PCR compared to immunohistochemical analysis using: (1) avidin-biotin-peroxidase complex, (2) alkaline phosphatase, antialkaline phosphatase complex, (3) tyramide signal amplification (i.e., CARD, see Table 55.1), (4) *in situ* PCR; *in situ* immuno-PCR found to be most sensitive of all amplification systems | 12 |
| Immuno-RCA (Rolling-Circle Amplification) | Oligonucleotide that is primer for RCA is attached to an Ab (either primary Ag-specific Ab or secondary Ab); following formation of immunoreactant complex, RCA is initiated by addition of circular template; hundreds of tandemly linked copies of template generated in a few minutes; amplified DNA detected in a variety of ways, e.g., direct incorporation of hapten- or fluorophore-labeled nucleotides into template copies; hybridization with fluorophore or enzyme-labeled oligonucleotides | RCA can be carried out with linear or geometric kinetics; amplification not only by RCA, but each Ab can have more than one oligonucleotide primer attached per Ab molecule; even in linear kinetic mode, immuno-RCA could detect IgE with 100-fold sensitivity increase over conventional ELISA | 13 |
| | also: Incorporate fluoresceinated nucleotides into RCA product; use alkaline phosphatase-conjugated anti-fluorescein + fluorescent substrate to detect signal | Signal amplification by immuno-RCA applied to protein analysis on glass microarrays; 75 cytokines measured simultaneously with femtomole sensitivity | 14 |

| | | | |
|---|---|---|---|
| Immune Capture + PCR | Ag-specific Ab used to capture analyte Ag, thereby concentrating it; PCR subsequently applied to amplify diagnostic DNA target sequence, followed by detection of PCR product | Applied to detection of bacteria in various samples | 15 |
| Immuno-Detection Amplified by T7 RNA Polymerase (IDAT) | Very similar to immuno-PCR with following differences; double-stranded "reporter" DNA still attached to Ab via a linker, but this DNA contains the T7 promoter sequence; Ab binding to Ag target also immobilizes "reporter" DNA; action of T7 RNA polymerase produces multiple copies of RNA, this multiplication being linear and directly related to number of "reporter" templates; RNA is subsequently detected | Very large amplification from T7 RNA polymerase; authors state that fluorescence-detection technique was developed, but no data are shown (all data based on radiometric technique); authors state that detection of only a few copies of an Ag in a mixture should be possible using IDAT | 16 |
| | ***PCR-ELISA*** | | |
| | *PCR process used to amplify target DNA; incorporation of various haptens or other reagents allows immunological detection using ELISA technique* | | |
| PCR-ELISA | Hybrid technique where specificity and amplification of PCR are coupled to detection by ELISA; different formats, but one example is as follows; target-DNA sequence is amplified by standard PCR techniques; digoxigenin-modified nucleotide incorporated into amplicons during PCR; after denaturation, single-stranded amplicons hybridized to biotinylated capture probe and hybridized product captured on streptavidin-coated surface (solid-phase process necessary for ELISA assay); subsequently, antidigoxigenin Ab conjugated to enzyme (i.e., horseradish peroxidase) added and binds to digoxigenin sites on PCR product; after washing away excess Ab, substrate for enzyme added, and color/fluorescence/chemiluminescence detected as in standard ELISA; technique can be applied to detection of mRNA with reverse transcription step before PCR | Amplification potential is great, since PCR and enzyme amplification both used in different parts of assay; specificity for target DNA only as good as PCR specificity and the specificity of the Abs used in the ELISA assay; blend of established immunoassay technique and PCR; demonstrated detection sensitivity of <10 spores of *B. anthracis* in 100 g soil sample, using a colorimetric assay;[17] Hall et al.[18] measured mRNA expression using biotinylated primers in PCR to enable capture of PCR products on avidin-coated plates; PCR amplicons were labeled with digoxigenin; antidigoxigenin Ab conjugated to alkaline phosphatase; colorimetric assay used for detection | 17–22 |
| | DNA amplified by PCR, using biotinylated primers; PCR product immobilized in microwells by hybridization to internal capture probe; streptavidin–AP conjugate or strepavidin-peroxidase conjugate subsequently added and allowed to react; either colorimetric or fluorogenic substrate added for detection | Amplification by both PCR and enzymatic (AP or peroxidase) means; colorimetric and fluorometric assays compared to detection using ethidium bromide-stained agarose gels; colorimetric assay found to have about the same sensitivity as analysis by gels, while chemiluminescent assay was tenfold more sensitive; colorimetric substrates were tetramethylbenzidine for peroxidase and p-nitrophenylphosphate for AP; chemiluminescent substrate (for AP) was Lumi-Phos 530; approximately five cells of *Salmonella* could be detected using PCR + chemiluminescent method; colorimetric detection limits were 50 cells | 23 |

## References


1. Sano, T., Smith, C.L., and Cantor, C.R., Immuno-PCR — very sensitive antigen-detection by means of specific antibody-DNA conjugates, *Science*, 258(5079), 120, 1992.
2. Ruzicka, V., Marz, W., Russ, A., and Gross, W., Immuno-PCR with a commercially available avidin system, *Science*, 260(5108), 698, 1993.
3. Chang, T.C. and Huang, S.H., A modified immuno-polymerase chain reaction for the detection of β-glucuronidase from *Escherichia coli*, *J. Immunol. Meth.*, 208(1), 35, 1997.
4. Case, M.C., Burt, A.D., Hughes, J., Palmer, J.M., Collier, J.D., Bassendine, M.F., Yeaman, S.J., Hughes, M.A., and Major, G.N., Enhanced ultrasensitive detection of structurally diverse antigens using a single immuno-PCR assay protocol, *J. Immunol. Meth.*, 223(1), 93, 1999.
5. Sanna, P.P., Weiss, F., Samson, M.E., Bloom, F.E., and Pich, E.M., Rapid induction of tumor-necrosis-factor-alpha in the cerebrospinal-fluid after intracerebroventricular injection of lipopolysaccharide revealed by a sensitive capture immuno-PCB assay, *Proc. Natl. Acad. Sci. U.S.A.*, 92(1), 272, 1995.
6. Furuya, D., Yagihashi, A., Yajima, T., Kobayashi, D., Orita, K., Kurimoto, M., and Watanabe, N., An immuno-polymerase chain reaction assay for human interleukin-18, *J. Immunol. Meth.*, 238(1–2), 173, 2000.
7. Sperl, J., Paliwal, V., Ramabhadran, R., Nowak, B., and Askenase, P.W., Soluble T-cell receptors — detection and quantitative assay in fluid-phase via ELISA or immuno-PCR, *J. Immunol. Meth.*, 186(2), 181, 1995.
8. Zhou, H., Fisher, R.J., and Papas, T.S., Universal immuno-PCR for ultra-sensitive target protein detection, *Nucleic Acids Res.*, 21(25), 6038, 1993.
9. Niemeyer, C.M., Adler, M., and Blohm, D., Fluorometric polymerase chain reaction (PCR) enzyme-linked immunosorbent assay for quantification of immuno-PCR products in microplates, *Anal. Biochem.*, 246(1), 140, 1997.
10. Maia, M., Takahashi, H., Adler, K., Garlick, R.K., and Wands, J.R., Development of a 2-site immuno-PCR assay for hepatitis-B surface-antigen, *J. Virol. Meth.*, 52(3), 273, 1995.
11. Niemeyer, C.M., Adler, M., Pignataro, B., Lenhert, S., Gao, S., Chi, L.F., Fuchs, H., and Blohm, D., Self-assembly of DNA-streptavidin nanostructures and their use as reagents in immuno-PCR, *Nucleic Acids Res.*, 27(23), 4553, 1999.
12. Cao, Y., Kopplow, K., and Liu, G.Y., *In situ* immuno-PCR to detect antigens, *Lancet*, 356(9234), 1002, 2000.
13. Schweitzer, B., Wiltshire, S., Lambert, J., O'Malley, S., Kukanskis, K., Zhu, Z.R., Kingsmore, S.F., Lizardi, P.M., and Ward, D.C., Immunoassays with rolling circle DNA amplification: a versatile platform for ultrasensitive antigen detection, *Proc. Natl. Acad. Sci. U.S.A.*, 97(18), 10113, 2000.
14. Schweitzer, B., Roberts, S., Grimwade, B., Shao, W.P., Wang, M.J., Fu, Q., Shu, Q.P., Laroche, I., Zhou, Z.M., Tchernev, V.T., Christiansen, J., Velleca, M., and Kingsmore, S.F., Multiplexed protein profiling on microarrays by rolling-circle amplification, *Nat. Biotechnol.*, 20(4), 359, 2002.
15. Widjojoatmodjo, M.N., Fluit, A.C., Torensma, R., Keller, B.H.I., and Verhoef, J., Evaluation of the magnetic immuno PCR assay for rapid detection of salmonella, *Eur. J. Clin. Microbiol. Infect. Dis.*, 10(11), 935, 1991.
16. Zhang, H.T., Kacharmina, J.E., Miyashiro, K., Greene, M.I., and Eberwine, J., Protein quantification from complex protein mixtures using a proteomics methodology with single-cell resolution, *Proc. Natl. Acad. Sci. U.S.A.*, 98(10), 5497, 2001.
17. Beyer, W., Pocivalsek, S., and Bohm, R., Polymerase chain reaction-ELISA to detect *Bacillus anthracis* from soil samples — limitations of present published primers, *J. Appl. Microbiol.*, 87(2), 229, 1999.
18. Hall, L.L., Bicknell, G.R., Primrose, L., Pringle, J.H., Shaw, J.A., and Furness, P.N., Reproducibility in the quantification of mRNA levels by RT-PCR-ELISA and RT competitive-PCR-ELISA, *Biotechniques*, 24(4), 652, 1998.

19. Alard, P., Lantz, O., Sebagh, M., Calvo, C.F., Weill, D., Chavanel, G., Senik, A., and Charpentier, B., A versatile ELISA-PCR assay for messenger-RNA quantitation from a few cells, *Biotechniques*, 15(4), 730, 1993.
20. Allen, R.D., Pellett, P.E., Stewart, J.A., and Koopmans, M., Nonradioactive PCR enzyme-linked-immunosorbent-assay method for detection of human cytomegalovirus DNA, *J. Clin. Microbiol.*, 33(3), 725, 1995.
21. Muramatsu, Y., Yanase, T., Okabayashi, T., Ueno, H., and Morita, C., Detection of *Coxiella burnettii* in cow's milk by PCR-enzyme-linked immunosorbent assay combined with a novel sample preparation method, *Appl. Environ. Microbiol.*, 63(6), 2142, 1997.
22. Sawant, S.G., Antonacci, R., and Pandita, T., Determination of telomerase activity in HeLa cells after treatment with ionizing radiation by telomerase PCR ELISA, *Biochemica*, 4, 22, 1997.
23. Soumet, C., Ermel, G., Boutin, P., Boscher, E., and Colin, P., Chemiluminescent and colorimetric enzymatic assays for the detection of PCR-amplified *Salmonella* sp. products in microplates, *Biotechniques*, 19(5), 792, 1995.

# 56

# Fluorescent Probes in Biomedical Applications

Darryl J. Bornhop
*Texas Tech University*
*Lubbock, Texas*

Kai Licha
*Schering AG*
*Berlin, Germany*



## 56.1 Introduction

Optical imaging techniques for the assessment of tissue anatomy, physiology, and metabolic and molecular function have emerged as an essential tool for both the basic researcher and the clinical practitioner. One concern of clinical practitioners is that too much harmful radiation is used to detect diseased tissue. The attractiveness of optical imaging techniques arises from the fact that fluorescent dyes can be detected at low concentrations and nonionizing, harmless radiation can be applied repeatedly to the patient. Furthermore, the remarkable progress in the development of optical instrumentation in the last two decades (laser excitation and detection systems) has decisively contributed to the growing applicability of optical imaging techniques, which have the advantage of being cheap, small in size, and, therefore, readily at hand to solve clinical problems. The design of contrast agents for optical *in vivo* imaging of diseased tissues has also emerged and is reflected by an increasing number of publications in this area.[1,2] Novel probes have been synthesized and characterized for their ability to monitor disease-specific anatomical, physiological, and molecular parameters through their optical signals. The multifarious world of chromophores and fluorophores provides various parameters that can be exploited for diagnostic measurement and detection.

This chapter describes the essential principles of optical imaging and current medical opportunities that are related to the use of fluorescent probes as exogenously applied agents; this is followed by a review of recent progress in the design of these probes for biomedical imaging purposes. The agents discussed in this chapter are categorized by their structural families, covering the class of cyanine dyes, tetrapyrrole compounds, lanthanide chelates, and other entities. For selected examples, the physicochemical, biochemical, and pharmacological features constituting the diagnostic efficacy of the compounds are illustrated.

0-8493-1116-0/03/$0.00+$1.50

# 56.2 Principles of Optical Imaging

## 56.2.1 Tissue Optics and Function of Dyes as Contrast Agents

The term optical imaging encompasses a large variety of different disciplines. In general, the method uses light within the ultraviolet (UV) and the near-infrared (NIR) spectral region to obtain information on the optical characteristics of tissue. Generally, the interaction of photons with tissue is based on absorption of light, scattering of light, and emission of fluorescence. These three parameters can be used separately to characterize tissue optical properties.

A fundamental observation for optical diagnostic procedures relates to the fact that the penetration depth of light in living tissue strongly depends on the wavelength used[3] because the number of absorption and scattering events in tissue is a function of wavelength.[4]

For wavelengths below approximately 600 nm the penetration of light into tissue is limited to a depth of hundredths of micrometers up to a few millimeters due to strong absorption of the photons, so that only superficial assessment of tissues in this spectral region is possible. The absorption of light in tissue originates from oxy- and deoxyhemoglobin and several tissue components such as porphyrins, melanin, NADH and flavins, collagen, elastin, and lipopigments. Most of these chromophores that contribute to tissue absorption exhibit characteristic fluorescence spectra throughout the visible (VIS) spectral region up to approximately 700 nm. Fluorescence of these intrinsic fluorescent markers (autofluorescence) has been studied as a source of specific spectral information on tissue structure and pathophysiological states[5,6] and was thoroughly exploited to identify diseased tissue areas, e.g., in endoscopy[7,8] or cardiovascular diagnosis.[9] Direct visual inspection or characterization using microscopic techniques is established practice in medicine. Hence, the modality is capable of generating images of tissue structures with high spatial resolution, as does microscopy.

A primary field of application of optical imaging technology is the examination of tissue surfaces via optical fibers incorporated in endoscopes or laparoscopes, as well as of ocular diseases through ophthalmoscopes and direct assessment of skin diseases or during surgical procedures. The only modality found in daily and widespread clinical use so far is the imaging of ocular diseases in ophthalmology. Fluorescein and indocyanine green (ICG) are established as fluorescent agents to enhance fluorescence angiography.[10] Currently under clinical evaluation is the fluorescence-guided identification of tumor margins during surgery as a tool to improve the accuracy and safety of tumor resection. For this purpose, ICG,[11] fluorescein-based conjugates,[12,13] and 5-ALA[14] have been studied.

Imaging of larger tissue volume requires light within the NIR spectral range (700 to 900 nm) because the absorption coefficient of tissue is relatively small, resulting in penetration depths of up to a few centimeters (Figure 56.1).[15] Thus, the identification of inhomogeneities exhibiting a difference in absorption or fluorescence compared to the bulk tissue is possible. However, due to scattering, photons do not follow straight paths when propagating through tissue limiting the spatial resolution of images obtained (diffuse imaging). Nevertheless, tissue absorption is mainly determined by oxyhemoglobin, deoxyhemoglobin, and water. These exhibit a well-defined minimum in absorption in the NIR spectral region (see Figure 56.1) and provide information that can be used to quantitatively calculate important physiological parameters such as blood concentration (total hemoglobin) and oxygenation (ratio oxy-/deoxyhemoglobin). These absorption data, together with tissue-dependent scattering properties, can be fitted by mathematical models to reconstruct the most probable photon propagation through tissue and generate a spatial map of tissue optical properties for a given illumination and detection geometry. This method has been applied primarily to detect breast tumors and image brain function (see Chapter 57 of this handbook). It is beyond the scope of this review to elucidate the underlying basics in more detail. The interested reader is referred to References 4, 16, and 17.

The exogenously introduced optical contrast agents principally provide the opportunity of engendering disease-specific signals within the tissue. This enables the display of physiological and molecular conditions that are characteristic of certain disease states and progression. The detection of fluorescent contrast agents is comparable to nuclear imaging methods; in both modalities the photon sources (fluorescent

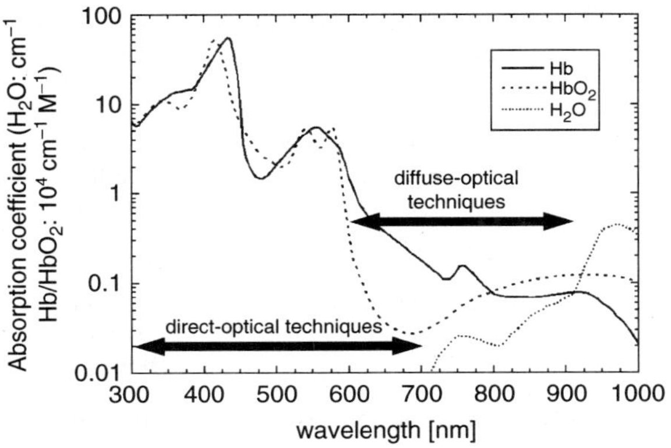


**FIGURE 56.1** Contributions of oxyhemoglobin, deoxyhemoglobin, and water to tissue absorption depending on wavelength. Diffuse optical imaging techniques require NIR light between 700 and 900 nm to achieve maximal penetration into tissue, while for direct optical imaging VIS light of 300 to 700 nm is usually used.

dye or radionuclide) are distributed within the tissue. Generally, the choice of the spectral range of absorption and fluorescence for the dye dictates whether it is detectable on tissue surfaces (UV-VIS dyes) or from deeper tissue areas (NIR dyes). A prerequisite for sensitive detection of an absorbing dye is a high extinction coefficient at the desired excitation wavelength. If fluorescence photons are acquired, the compounds should exhibit large Stokes' shifts (spectral distance between absorption and fluorescence maximum) and high fluorescence quantum yields in physiological media. If fluorescence is recorded within the UV/VIS spectral region, both autofluorescence and the administered contrast agent will contribute to the observed signal, while in the NIR spectral region tissue autofluorescence is negligible due to the absence of endogenous NIR fluorophores. In the latter case, the distribution of the contrast agent is nearly exclusively revealed by the detected signal. The disadvantage of fluorophores over radionuclides is that excitation light must be brought to the dye, so that this factor additionally contributes to penetration-depth limitations. An important advantage, however, is related to the fact that fluorescence can be excited continuously and is not limited to inherent properties as is radioactive decay. A set of photophysical properties is accessible, ranking from changes in fluorescence quantum yield and fluorescence lifetime to alterations of spectral signatures. Moreover, for many dyes these parameters are influenced by local physiological or molecular conditions, such as pH, ions, or oxygen, and have therefore been used for their monitoring and quantification. Most importantly, light applied to fluorescence imaging is nonionizing radiation, rendering it harmless and nontoxic.

## 56.2.2 Imaging Techniques and Medical Applications

Optical imaging techniques and applications may be divided into two groups. On the one hand, imaging of superficial objects acquires directly reflected or scattered photons, while for diffuse imaging techniques photons are recorded after they have passed through relatively thick tissue and optical properties of the tissue are spatially reconstructed using mathematical models. Different technical solutions and instrumental geometries are required for each of these applications. Illumination with light of a desired wavelength and detection with suitable devices (e.g., charge-coupled device, or CCD cameras) generally yields images from superficial structures in reflection geometry. This process is similar to conventional photography. A primary field of application is the examination of tissue surfaces via optical fibers embedded in endoscopes or laparoscopes, as well as of ocular diseases through ophthalmoscopes and direct assessment of skin diseases or during surgical procedures. Several types of superficial diseases in hollow organs have been monitored using fluorescence-guided endoscopy. Tetrapyrrol-based agents have been examined for this purpose,[18] but the most promising results have been obtained with ALA. Clinical studies included the diagnosis of urinary bladder cancer,[19] bronchial cancer,[20] and gastrointestinal diseases.[7] An essential task was

**TABLE 56.1** Medical Applications of Optical Diagnostics

| Modality/Discipline | Signal | Clinical Indication |
|---|---|---|
| | ***Direct-Optical*** | |
| Ophthalmology | Reflected-light fluorescence | Ocular diseases |
| Intraoperative diagnostics | | Detection of tumor boundaries |
| Dermatology | | Skin tumors |
| | ***Via Endoscopic or Catheter Devices*** | |
| Endoscopy | Reflected-light fluorescence | Tumors of the GI tract, lung, bladder, cervix, oral cavity |
| Cardiovascular imaging | | Atherosclerotic plaques |
| | ***Diffuse Imaging and Mathematical Reconstruction*** | |
| Optical mammography | Reflected- or transmitted light fluorescence | Breast tumors |
| Brain imaging | | Brain perfusion, stroke |

to improve the detection of dysplastic changes and carcinoma *in situ*, which escape the endoscopist's eye and require more reliable techniques. Besides ALA, other fluorescent agents such as those presented in Chapter 57 were suggested for these purposes.[21–24]

The assessment of larger tissue volumes has been realized for different transillumination geometries in which tissue areas are illuminated with light, and the transmitted, scattered light is detected at defined positions (e.g., 180° projection geometry or 0°C reflection geometry).[25,26] Diffuse optical tomography (DOT) is based on the detection of photons at multiple positions and the mathematical reconstruction of three-dimensional (3D) optical images.[4,16,17]

The application of DOT for breast-cancer detection and characterization has stimulated a great deal of research in the past few years. Several approaches for the detection of breast cancer using NIR light and utilizing intrinsic tissue optical properties have been followed. The first attempts at detecting breast tissue by means of tissue transilluimnation were reported as far back as 1929,[27] and the development of more powerful light sources and monitoring systems led to a revival of this technique in the 1980s and 1990s. Several clinical studies revealed the limitations of this modality — low spatial resolution and an inability to differentiate between malignant and benign tissue — which limit its clinical usefulness, especially in relation to x-ray mammography.[15,26,28,29] DOT using continuous-wave (CW), modulated, or pulsed light demonstrated advances in the quantification and 2D or 3D display of tissue absorption, scattering, vascularization, and oxygenation. However, only limited patient data have been acquired and published so far with respect to oxygenation.[4,15,30,31] The application of contrast agents was proposed repeatedly and will be the subject of subsequent chapters. The first contrast-enhanced imaging in a clinical setting was reported by Ntziachristos et al., who have demonstrated uptake and localization of ICG in breast lesions using DOT.[32]

Table 56.1 summarizes the various experimental and clinically established modalities and clinical applications that employ exogenously administered optical contrast agents.

# 56.3 Dyes as Contrast Agents for Optical Imaging

## 56.3.1 Cyanine Dyes

### 56.3.1.1 Nonspecific Cyanine Dyes of Classical Contrast-Agent Format

The structural class of cyanine dyes comprises chromophoric structures of more or less arbitrary absorption and fluorescence from the VIS to NIR spectral range (approximately 450 to 900 nm). Cyanine dyes generally exhibit very high molar-extinction coefficients (>150,000 $M^{-1}$ $cm^{-1}$) and good fluorescence quantum yields (up to 50%). They have been originally prepared for the photography industry and were

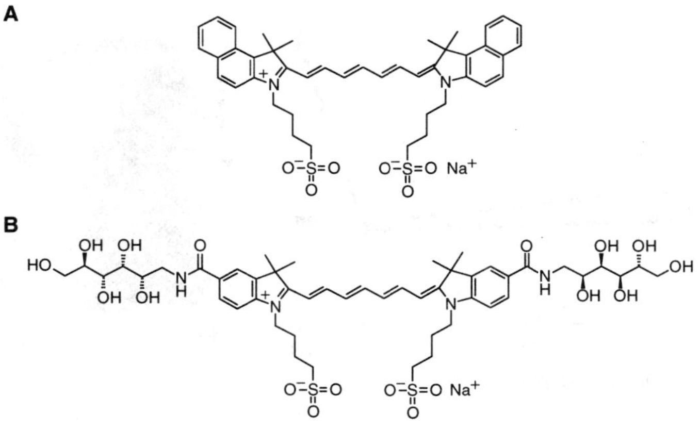


**FIGURE 56.2** Chemical structure of (A) indocyanine green (ICG) and (B) the hydrophilic derivative SIDAG.

adapted to many applications in analytical chemistry, bioanalytics, and biomedicine. (See References 33 and 34 for a comprehensive discussion.)

The first attempts at designing such probes for *in vivo* applications date back to the 1950s, when ICG, a prominent representative of a NIR-absorbing cyanine dye, was first synthesized (see Figure 56.2 for its chemical structure). ICG was clinically applied as a drug for the assessment of hepatic function and cardiac output, for which this compound exhibits favorable pharmacokinetic properties.[35] It has recently been discovered as an imaging agent for ocular diseases to visualize vascular disorders of the retina and choroidea. In a few studies, ICG has been reported as a potential NIR contrast agent for the detection of tumors in animal research[36,37] and at the clinical level.[32]

For tumor-imaging purposes it has been recognized that the high plasma-protein-binding, intravasal distribution and resulting rapid plasma clearance of ICG by the liver leads to a quickly disappearing signal loss, which limits its potential as contrast agent.[37] One possible way to improve tissue retention and differentiation is through the use of nonspecific contrast agents similar to contrast agents for magnetic resonance imaging and computed tomography (CT), which usually are highly hydrophilic structures and achieve contrast enhancement based on morphological and physiological properties of tumor tissue such as increased tumor vasculature, endothelial leakiness, and enlarged extracellular volume. Thus, attempts were made to design structurally related agents of improved properties. The literature describes the synthesis and comparative pharmacokinetic characterization of ICG-related indotricarbocyanine derivatives with carboxy, hydroxyalkyl, and monosaccharide residues.[37,38] Figure 56.2 shows the chemical structures of ICG and the derivative SIDAG, a hydrophilic glucamide-derivatized indotricarbocyanine. Probes such as SIDAG are characterized by a highly increased hydrophilicity and reduced plasma-protein binding. It was assumed that after systemic administration the compounds would be capable of leaving the intravascular space and extravasate into the extracellular compartment, with a certain degree of preference for tumor tissues.

In animal studies, SIDAG showed improved efficacy as an optical contrast agent vs. ICG based on a higher tumor concentration and tumor-to-normal tissue contrast shortly after injection and, in addition, an unexpectedly elevated fluorescence contrast at 24 h after injection (Figure 56.3).[37–39]

Although contrast enhancement with the agents covered in this chapter mainly relies on a more distinct perfusion of tumors compared to normal tissue (as is achieved with ICG) and does not employ target-specific moieties, the value of contrast-enhanced tumor imaging was successfully demonstrated. These agents represent the classical format of a contrast agent.

#### 56.3.1.2 Targeted Cyanine-Dye Conjugates

One way to improve the selectivity of fluorescent dyes for diseased tissues is based on the high instrumental detection sensitivity for optical signals, which makes it possible to monitor signals derived from

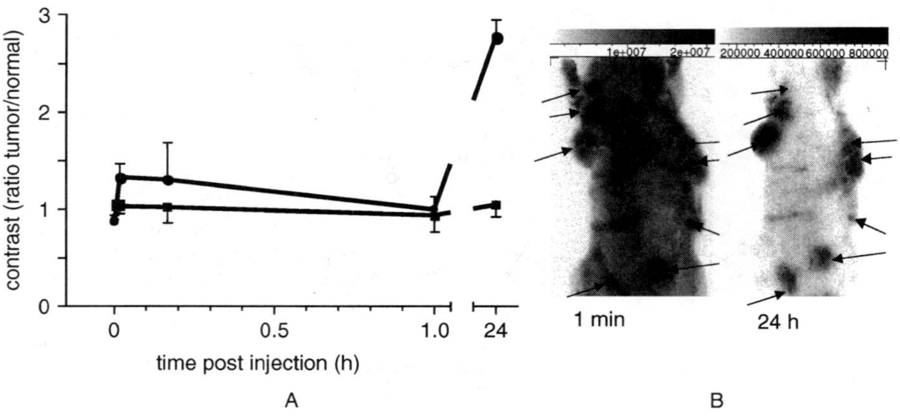


**FIGURE 56.3** Fluorescence-imaging performance of hydrophilic dye SIDAG in rats with chemically induced multiple mammary carcinoma (dose: 2 μmol/kg) in comparison to rats after injection of the same dose of ICG. (A) Time course of SIDAG (squares) and ICG (triangles) and (B) corresponding fluorescence images with SIDAG obtained at 1 min and 24 h after injection. Arrows indicate multiple tumor sites. (From Licha, K. et al., *Acad. Radiol.*, 9(Suppl. 2), 320, 2002. With permission.)

molecular states and events. Optical molecular imaging can therefore be considered the nonradiative counterpart to radionuclide-based methods, of course always limited by the penetration of light into tissue. Hence, novel probes consisting of an efficiently fluorescing dye label and a biological targeting unit (e.g., an antibody, peptide, oligonucleotide, or small molecule) have been reported in recent publications and will be discussed below.

A primary synthetic goal of chemical and physicochemical research in the area of fluorescent probes has been to provide novel fluorescent markers for application in immunoassays, screening assays, or genetic analysis. Many of the structures developed for these purposes contain reactive or activatable groups, e.g., carboxylic acids,[40–42] *N*-hydroxysuccinimidyl esters,[43–45] isothiocyanates,[45] or maleimido[46] functionalities. Figure 56.4 depicts selection of chemical structures. Well established is the commercially available CyDye™-series (e.g., Cy3, Cy5, Cy5.5, Cy7; Amersham-Pharmacia Biotechnology, Piscataway, NJ). Those and other cyanine-dye derivatives were employed for the preparation of various targeted cyanine-dye conjugates.

The strategy of coupling cyanine-dye labels to antibodies for *in vivo* diagnostic purposes was first reported by Folli et al. and Ballou et al.[47–50] The authors demonstrated target-specific uptake in experimental tumor models by planar-reflection imaging of superficial fluorescence patterns and fluorescence-microscopy methods.

Generally, imaging contrast is improved when the probe localizes in the target area with high binding affinity and maintains its concentration level while the circulating fraction is cleared from the blood to the greatest possible extent. Therefore, from the standpoint of background signal, contrast improves when going from full-size antibodies to targeting vehicles of reduced molecular weight such as antibody fragments and peptides.

The use of engineered antibody fragments of reduced molecular weight but still high binding affinities as target-specific carrier molecules for fluorescent dyes[51–54] was reported recently. Neri et al. described antibody single-chain fragments that were identified by phage-display library technology[51] and that exhibited high binding affinities against an extracellular angiogenesis marker, the fibronectin isoform ED-B-FN. This matrix protein is present exclusively in neoplastic blood vessels during angiogenesis and is therefore a promising target for specific fluorescent ligands. After labeling with fluorophores the antibody fragments were succesfully applied to the *in vivo* fluorescence imaging of tumor angiogenesis in animal models.[51–53]

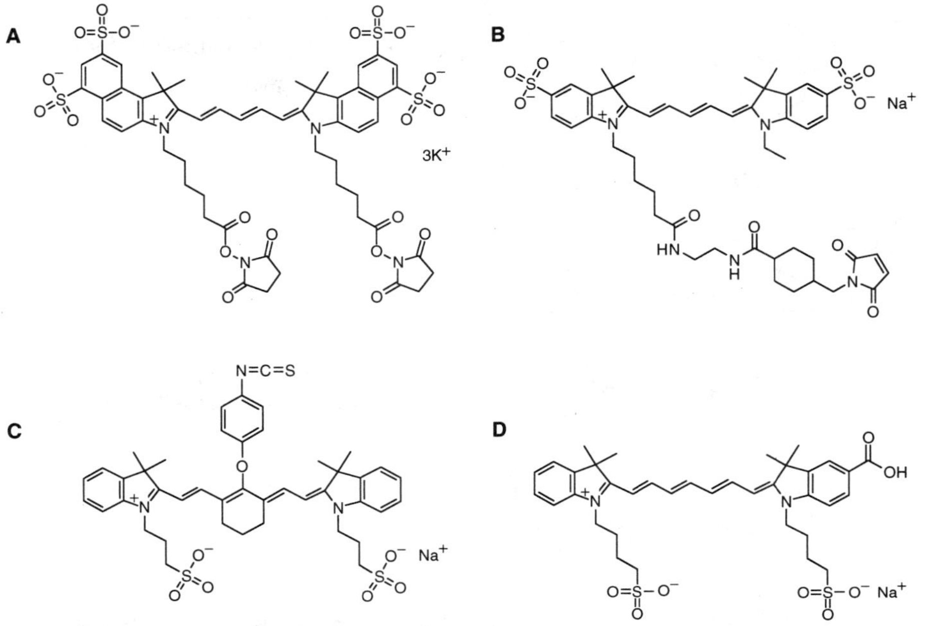

**FIGURE 56.4** Carbocyanine-dye labels with different reactive groups for fluorescence labeling purposes. (A) Commercially available Cy5.5-bis-NHS-ester, (B) maleimido-derivative of Cy5,[46] (C) indotricarbocyanine with isothiocyanate functionality,[45] and (D) derivative with carboxy group at exterior ring position.[40]

Synthetic peptides as vehicles for diagnostic molecules offer a promising strategy for the further reduction of molecular weight. Many tumors are known to overexpress receptors for specific peptide ligands, e.g., somatostatin (SST), bombesin, or vasoactive intestinal peptide (VIP).[56] Pharmacologically optimized derivatives of these natural ligands in radiolabeled form are already clinically established as radiolabeled probes for the receptor scintigraphy of tumors, e.g., OcteoScan,® an $^{111}$In-DTPA-conjugated SST analog.[57] The literature reports some examples in which these principles were successfully adapted to optical receptor imaging by replacing the radiolabel by fluorescent cyanine dyes.[21,22,40,41,58] The synthesis and *in vivo* characterization of different conjugates between cyanine dyes and peptides, such as the SST analog octreotate,[21,22,40] the bombesin-derived analog bombesate,[41,58] and structurally optimized VIP,[59,60] were the subject of studies by various investigators. Figure 56.5 illustrates the structures of selected cyanine dye–peptide conjugates. The authors demonstrated that the conjugates were accessible by standard solid-phase peptide synthesis, which led to structurally well-defined and pure products. The fluorescence properties of the conjugated dyes were less affected only by the peptide, typically leading to fluorescence quantum yields for indotricarbocyanine conjugates in the range of 10%.[40]

Fluorescence-imaging experiments using SST-receptor positive animal-tumor models revealed receptor-mediated uptake resulting in elevated fluorescence at the tumor site compared to surrounding normal tissue areas. Unlike *in vivo* imaging features of nonspecific extracellular dyes or fluorescent single-chain-fragment conjugates, the highest contrast was achieved already within 1 to 2 h after intravenous injection and lasted up to 24 h[22,58] (Figure 56.6). This behavior is based on receptor-mediated accumulation of the peptide conjugates into tumor cells, while nonbound molecules undergo rapid body clearance via the renal pathway.[22]

As an interesting alternative to vehicles of biological origin, the targeting of the LDL (low-density lipoprotein) receptor was demonstrated with "small-molecule" conjugates between different indotricarbocyanines and cholesteryl laurate, which can bind to LDLs and thus mediates internalization of the entire conjugate into LDL-receptor-expressing tumor cells.[61]

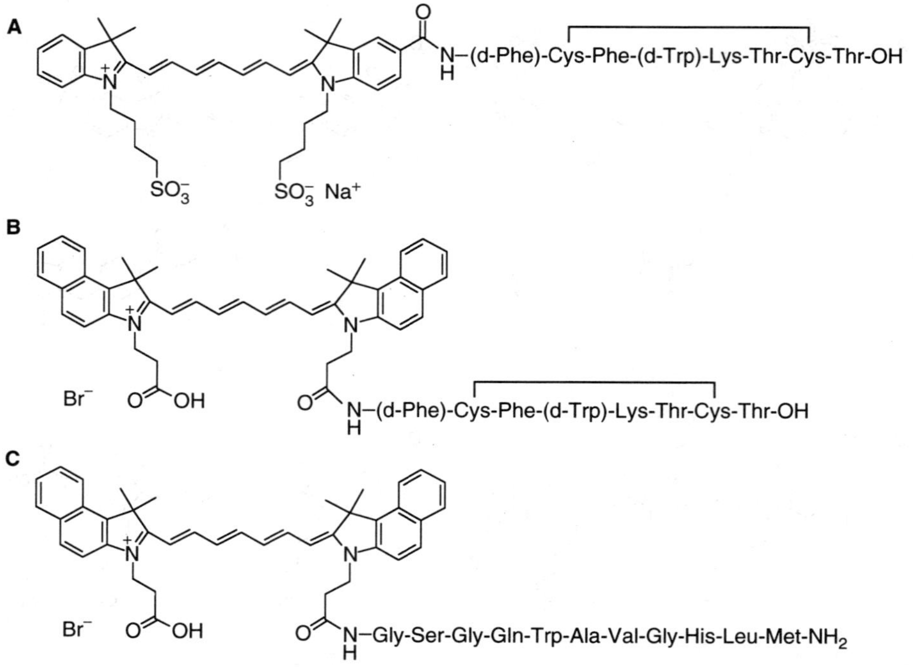


**FIGURE 56.5** Chemical structures of cyanine dye–peptide conjugates designed for receptor-targeted fluorescence imaging of tumors. (A, B) cyanine dye conjugated to the SST-receptor-binding peptide octreotate[21,22,40,41,58] and (C) a bombesin-receptor-avid peptide sequence.[41,58]

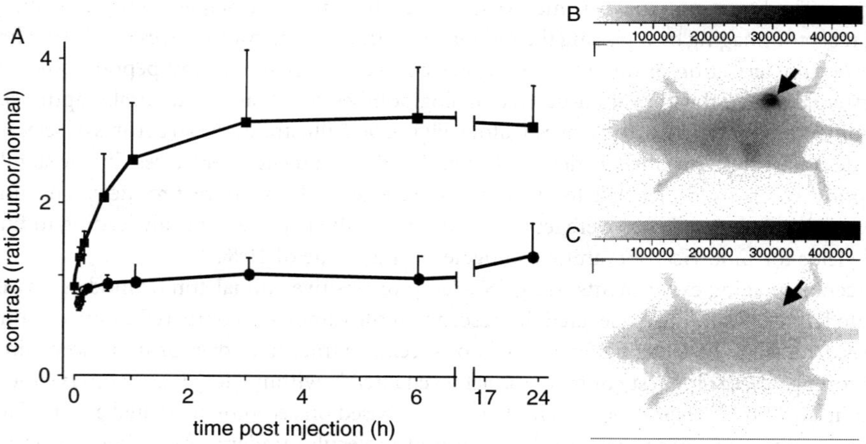


**FIGURE 56.6** Fluorescence-imaging performance of indotricarbocyanine octreotate (see Figure 56.5, compound A) in mice bearing an SSTR2-receptor-expressing pancreatic tumor (dose: 0.02 μmol/kg). (A) Time course of indotricarbocyanine octreotate (squares) vs. a control conjugate where two cysteins in the octreotate sequence are replaced by methionine (circles). Corresponding fluorescence images obtained at 6 h p.i. for the indotricarbocyanine octreotate (B) and the control conjugate (C). (From Licha, K. et al., *Acad. Radiol.*, 9 (Suppl. 2), 320, 2002. With permission.)

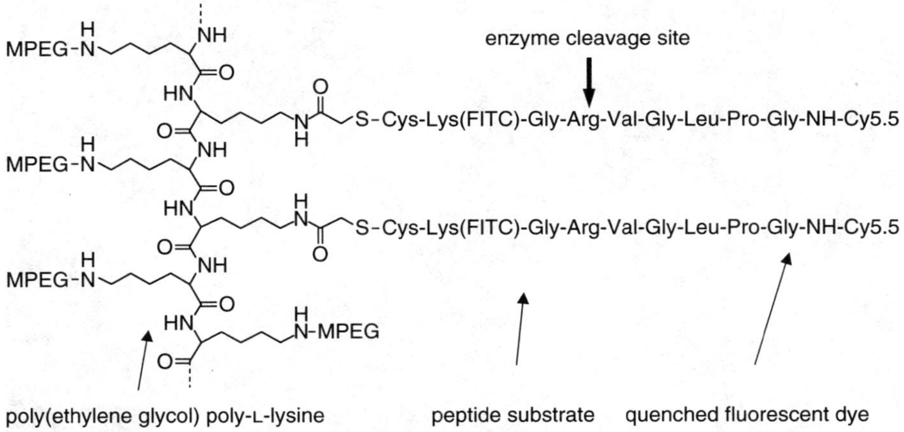


**FIGURE 56.7** Structures of of Cy5.5-labeled poly(ethylene glycol) poly-(L-lysine) graft polymers with a peptide sequence cleavable by the proteolytic enzyme MMP-2. Arrows indicate the enzyme cleavage sites.[65]

### 56.3.1.3 Activatable Cyanine Dye Conjugates ("Smart Probes")

A number of groups have demonstrated enzyme-activatable conjugate complexes where the quenched signal (fluorescence, for example) is turned on by tumor-associated lysosomal protease activity *in vivo*.[23,62–67] These probes carry Cy5.5 molecules (excitation 670 nm, emission 700 nm) that are bound to long circulating graft copolymers consisting of poly-L-lysine and poly(ethylene)glycol. An intratumoral NIR fluorescence signal is generated when tumor-associated proteases cleave the macromolecule, thereby liberating Cy5.5 fragments and affording previously quenched fluorescence.

This approach utilizes the unique opportunity of modulating optical signals through intramolecular fluorescence-quenching effects. Thus, the agents can report information on protein function and molecular conversion (enzymatic activity) and, furthermore, circumvent the necessity of target-specific accumulation, as the circulating fraction, which needs to be cleared for targeted agents, avoids fluorescence detection.

The contrast agent is composed of a poly(ethylene glycol) poly-(L-lysine) graft polymer, which is labeled with Cy5.5 (see Figure 56.4) at the ε-amino groups of free lysines.[23,62,63] This drug exhibits increased fluorescence signals in the presence of the proteolytic enzymes cathepsin B and H by cleavage of the polylysine backbone and leads to a substantial signal increase both in cell culture and animal tumors.[23,63,64]

To broaden the specificity of the probe as a substrate for any other desired enzyme, a cleavable peptide sequence was incorporated in the polymeric backbone. As illustrated in Figure 56.6, the graft polymer was modified in such a way that a desired peptide sequence could be linked to the graft polymer. Fluorophores were conjugated at the N-terminal amino group of the peptide, again being subject to fluorescence quenching (Figure 56.7). These peptide spacers can be easily modified to act as a substrate for tumor- or angiogenesis-specific enzymes with proteolytic activity, e.g., matrix metalloproteinases (MMPs)[65] or the serine protease thrombin.[67]

The particular promise of applying probes for the visualization of protein function relates to the contrast-enhanced imaging of diseased tissues and is expected to have significant impact as a tool for the monitoring of drug response and efficacy at the molecular level. Figure 56.8 shows the results of an experiment in which a probe activatable by MMP-2 was used to monitor treatment response in a mouse model.[65,68]

## 56.3.2 Tetrapyrrole-Based Dyes

### 56.3.2.1 Synthetic Porphyrins, Chlorins, and Related Structures

The main area of application of tetrapyrrole-based compounds, such as porphyrins, chlorins, benzochlorins, phthalocyanines, and expanded porphyrins, has been for photodynamic therapy (PDT).[69–71] While PDT is treated elsewhere in this book (see Chapter 38), it is noteworthy, given the focus of this review,

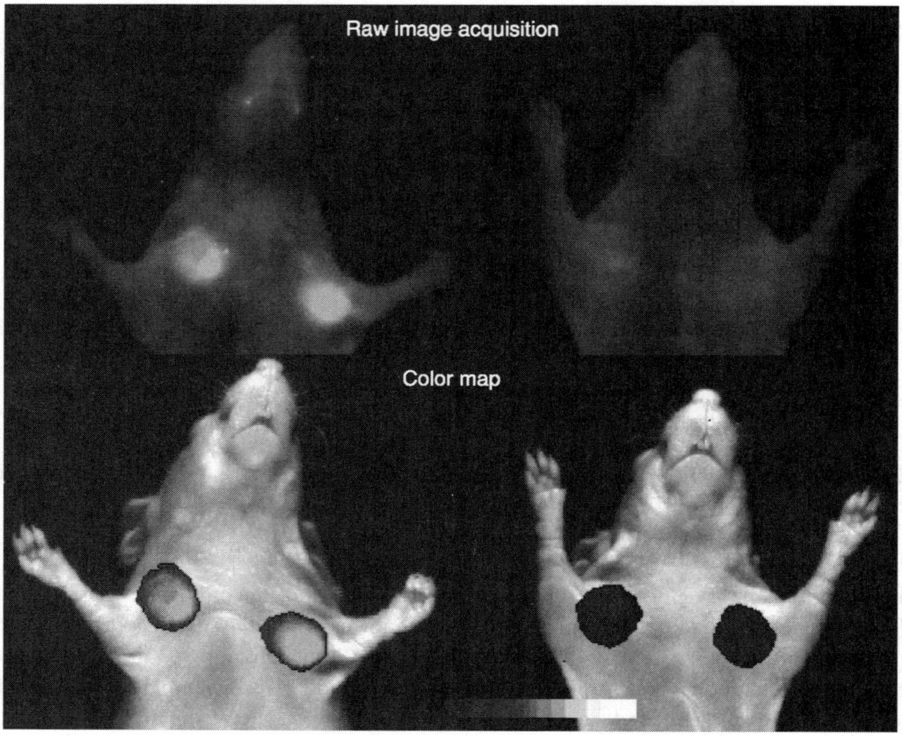


**FIGURE 56.8 (Color figure follows p. 28-30.)** Fluorescence imaging of HT1080 tumor-bearing mice 48 h after injection of Cy5.5-labeled poly(ethylene glycol) poly-(L-lysine) graft polymer (see Figure 56.6) into animals treated with the MMP-inhibitor primomastat vs. untreated animals. The top row shows raw fluorescence images, and the bottom row shows color-coded intensity profiles superimposed onto white-light images. The results show a significantly less fluorescence signal in treated animals relative to the untreated group. (From Bremer, C. et. al., *Nat. Med.*, 7(6), 743, 2001. With permission.)

that many agents originally designed for PDT have shown to be applicable for diagnostic purposes, as these structures exhibit strong fluorescence in the VIS-to-NIR spectral region. The main rationale behind this type of application has been to use the fluorescence emission for real-time assessment of therapy progress and effectiveness during photodynamic treatment.

A variety of PDT compounds have been studied for their diagnostic capabilities, e.g., hematoporphyrin (HpD),[72] meso-tetra-*m*-hydroxyphenylchlorin (*m*-THPC),[73] benzoporphyrin derivatives (BPD),[74] sulfonated phthalocyanines,[75] pheophorbides,[76] derivatives of chlorin e6,[77] and, finally, the expanded porphyrin derivative lutetium texaphyrin (Lu-Tex).[78] Figure 56.9 illustrates selected chemical structures.

Attempts to enhance the selectivity of photosensitizers by conjugating them to target-specific vehicles have been described.[79] Like the approaches based on cyanine dyes, conjugates with antibodies,[80,81] antibody fragments,[53,82] peptides,[83] and the serum proteins albumin and transferrin,[84,85] as well as estradiol[86] and cholesteryl laurate,[87] have been prepared and studied in animals.

#### 56.3.2.2 5-Aminolevulinic Acid (ALA) and Protoporphyrin IX (PpIX)

The underlying mechanism of fluorescence enhancement using ALA differs fundamentally from other approaches where exogenous agents are applied. ALA is not fluorescent per se but represents an essential precursor in the heme biosynthetic pathway. In the last step of the biosynthesis of heme, iron is incorporated into PpIX. After administration of ALA the intracellular synthesis of heme is stimulated with a certain degree of selectivity for tumor cells and engenders a temporarily increased intracellular PpIX

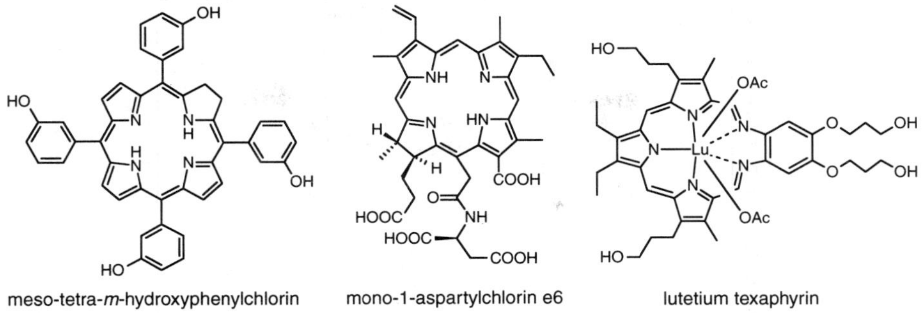


**FIGURE 56.9** Chemical structures of photosensitizers applied as diagnostic agents for the fluorescence detection of tumors.

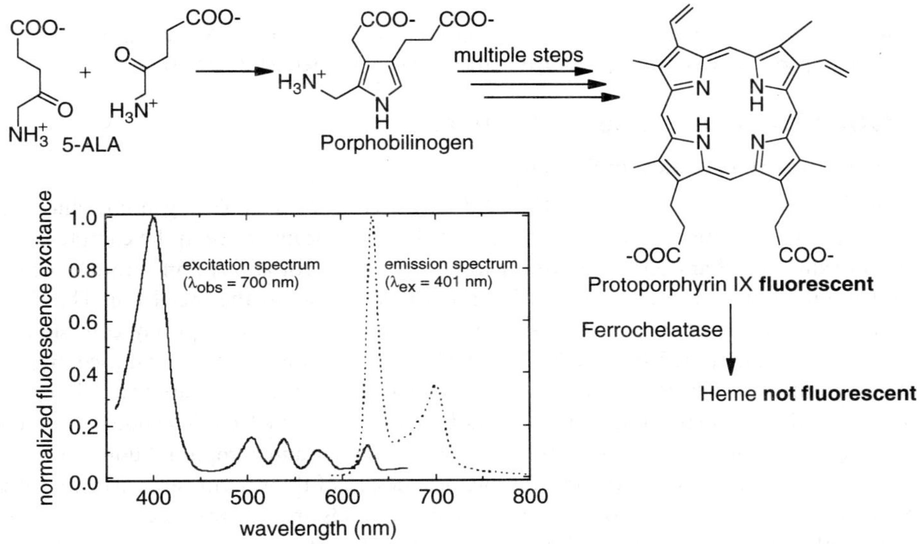


**FIGURE 56.10** Synthetic pathway of heme biosynthesis leading to intracellular formation of fluorescent protoporphyrin IX (PpIX). Fluorescence excitation spectrum and fluorescence emission spectrum of PpIX in methanol. (Figure courtesy of B. Ebert, Physikalisch-Technische Bundesanstalt, Berlin.)

concentration in tumor cells relative to healthy tissue. This phenomenon is based on the finding that the incorporation of iron is catalyzed by the enzyme ferrochelatase, which was suggested to have lower activity in many tumors, thereby leading to a "bottleneck effect" at the level of PpIX formation. Figure 56.10 illustrates the essential steps. Upon irradiation with light, a cell-destructive, photodynamic effect is induced that is of much higher selectivity for malignant cells and accompanied with fewer side effects than is usually achieved with exogenously applied photosensitizers. The principles of PDT using ALA were reviewed, for example, by Peng et al.[88]

While heme is not fluorescent, PpIX exhibits a fluorescence-emission spectrum typical of porphyrins of this structural class (Figure 56.10). Thus, PpIX fluorescence has been used to detect and visualize tumors and other tissue abnormalities in a large variety of clinical applications[7,19,20,89,90] (see also Chapter 2).

Successful attempts to improve ALA delivery to cells were followed by synthesizing ester derivatives of ALA, particularly alkyl esters of different alkyl chain length,[19,91,92] which are converted into free ALA through ester cleavage by esterases. Many of these compounds have proven to penetrate more effectively into tissues and lead to enhanced amounts of accumulated PpIX,[19,91–94] with the highest efficacy described for the *n*-hexyl ester.[91,93,94]

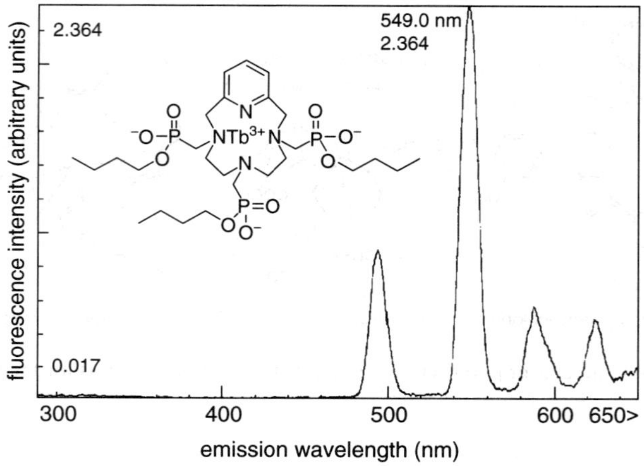


**FIGURE 56.11** Chemical structure and fluorescence emission spectrum of Tb-[*N*-2-pyridylmethyl)-*N′*,*N″*,*N‴*-*tris*(methylenephosphonic acid butyl ester)-1,4,7,10-tetraazacyclododecane] (Tb-PCTMB).

## 56.3.3 Other Dyes and Reporter Systems

### 56.3.3.1 Fluorescent Lanthanide Chelates

Many of the lanthanides, e.g., $La^{3+}$, $Eu^{3+}$, $Tb^{3+}$, $Nd^{3+}$, and $Yb^{3+}$, form stable organometallic complexes with unique optical properties.[95] When an aromatic structure is located close to the complexing moiety, the metal ion exhibits a characteristic emission spectrum upon light absorption through the aromatic system and intramolecular energy transfer of the absorbed energy to the metal ion. This "sensitized luminescence emission" is of long lifetime — up to milliseconds — and typically consists of several emission bands throughout the VIS up to NIR spectral range depending on the metal used. In particular, $Tb^{3+}$ and $Eu^{3+}$ complexes have been synthesized in large structural diversity,[96,97] applied in biotechnology, and used for the development of screening assays based on time-resolved detection techniques.[34,98,99] Terbium yields a bright green fluorescence (major peak at 550 nm), and europium fluoresces in red color (major peak at 600 nm). Another focus for the chemical design of fluorescent rare-earth complexes was driven by the fact that the optical properties of these compounds are sensitive to environmental conditions, such as pH, $pO_2$, glucose, halide ions, and alkali metals, permitting their use as chemical and biological sensors.[100,101]

The application of such compounds for the *in vivo* fluorescence detection of cancer has been proposed by several researchers.[24,102–104] More specifically, a pyclen-based macrocyclic complex with Terbium, Tb-[*N*-2-pyridylmethyl)-*N′*,*N″*,*N*-*tris*(methylenephosphonic acid butyl ester)-1,4,7,10-tetraazacyclododecane] or Tb-PCTMB (Figure 56.11) is extremely luminescent, water soluble at millimolar levels, thermodynamically stable, and nontoxic.[24,102–105] This compound has been detected at the picomolar level in rat intestinal-tissue endoscopies, facilitating an enhanced visual detection of chemically induced colon cancers in the rats.[24,103,105]

Bioconjugatable chelating derivatives of similar structure, consisting of a 1,4,7,10-tetraazacyclododecane system bearing a light-harvesting quinaldine structure, two phosphonic acids, and a single carboxylic acid moiety for fluorescence labeling of target-specific biomolecules, have been described.[106,107]

### 56.3.3.2 Fluorescein, Oxazine, and Thiazine Dyes

Dyes based on the fluorescein chromophore are probably the most frequently applied fluorescent dyes in bioanalytics and biomedical diagnostics. A broad spectrum of reactive derivatives, such as 5-aminofluorescein or fluorescein-isothiocyanate (FITC) and many further differently derivatized analogs,[108] are commercially available. Fluorescein is a drug routinely used for optical diagnosis in ophthalmology in medical imaging.[109]

A recent *in vivo* imaging approach has been the application of fluorescent serum albumin conjugates with 5-aminofluorescein for the intraoperative detection of tumor margins.[12] The synthesis of fluorescent cobalamin derivatives employing FITC and other fluorescein-related fluorophors (Oregon Green, naphthofluorescein) was reported, and the use of these conjugates for the intraoperative visualization of cobalamin receptors was suggested.[110] An oxazin-type dye, ethyl Nile blue A, was used as a marker for the fluorescence-guided identification of premalignant lesions in animal models.[111] Finally mentioned is the phenothiazine dye, toluidine blue, which is established as histological staining dye both in the laboratoty and for clinical examinations.[112]

The Haugland group (see www.MolecularProbes.com) has successfully implemented a host of fluorescein- and oxazine-based dyes into molecular-imaging regimes. Currently, many probes are commercially available for visual signal amplification, organelle probes, fluorescent tracers, and viability kits that extend across many diverse cell lines.

#### 56.3.3.3 Zinc Sensing

Intracellular zinc is thought to be available in a cytosolic pool of free bound Zn(II) ions in the micromolecular to picomolar range.[113] Several Zn(II)-responsive transcription factors are known to mediate zinc homeostasis *in vivo* and are thought to do so by monitoring changes in this hypothetical pool of free Zn(II).[113–119] In research conducted by O'Halloran et al., the response of the mechanism of zinc sensors that control metal uptake or export was calibrated against thermodynamically defined free Zn(II) concentration, and it was shown that whereas the cellular zinc quota is millimolar, free Zn(II) concentrations that trigger zinc uptake or export are femtomolar, thereby suggesting extraordinary intracellular zinc-capacity binding. The cells would then exercise tight control over cytosolic metal concentrations.[113]

#### 56.3.3.4 Connexin Trafficking

Recombinant proteins containing tetracysteine tags have been shown to be successfully labeled in living cells via multicolor and electron-microscopic imaging of connexin trafficking.[120] Tsien et al. have shown that this approach makes discrimination between older and younger protein molecules possible. FlAsH and ReAsH, two fluorescent labeling ligands, label temporally separated pools of Cx43-TC to facilitate the recording of junctional-plaque renewal over time. This approach describes a method for studying the life cycle of proteins including assembly and internalization.[120]

#### 56.3.3.5 PEBBLES

PEBBLES stands for sol-gel probes encapsulated by biologically localized embedding. Kopelman et al. have reported sol-gel-based optical nanosensors that are ratiometrically referenced and can provide real-time measurements of subcellular molecular oxygen[121] and intracellular pH and calcium.[122] The polymer sensors easily incorporate multiple signaling dyes, such as oxygen sensitive $[Ru\text{-}(dpp)_3]^{2+}$ and oxygen-insensitive Oregon green 488-dextran, and are on the order of 50 to 300 nm (radius) in size.

#### 56.3.3.6 Smart DNA Detection

Tan et al. have developed sensitive fluorescent DNA probes that can be used for real-time biomolecular recognition of target DNA sequences.[123] These probes, called "molecular beacons," consist of a hairpin-shaped oligonucleotide that contains both a fluorophore and a quenching moiety. Molecular beacons act like on–off switches that are normally off (no fluorescence). The dye is activated upon hybridization of the stem to complementary DNA. As a result, the stem hybrid unwinds, thereby increasing the distance between the quencher and the fluorescent molecule and selectively generating a quantifiable signal. Two forms of energy transfer that exist in molecular beacons are direct energy transfer and fluorescence resonance energy transfer (FRET). While both forms of energy transfer are distance-dependent, FRET has the added complication of requiring spectral overlap between the donor's (fluorescent dye) emission and the acceptor's (nonfluorescent quencher) absorption spectra. This effect has been further used for the detection of thrombin through specially designed molecular aptamer beacons, employing a fluorescein/Dabcyl quenching system and a thrombin-binding DANN oligonucleotide.[124]

## 56.4 Conclusions

Remarkable optical techniques for the imaging and detection of diseases have emerged in the past few years. The current literature in this area outlines the broad applicability of light-based instrumental solutions for many clinical disciplines. The strength of optical imaging is that it allows for the combination of conventional display of tissues with the promise of highly sensitive detection of molecular signals. Thus, remarkable progress in the design of novel fluorescent probes has been made. In that respect, the chemistry of fluorescent dyes offers various opportunities to the acquisition of optical signals. Several parameters, such as the absorption coefficients, fluorescence quantum yields, fluorescence decay times, and fluorescence quenching/recovery processes, are accessible for reporting physiological states, molecular conditions, and molecular function. Unlike radioactive decay, fluorescence is sensitive to its chemical environment, thereby broadening the applicability to the sensing of chemical analytes.

Progress in biotechnology, from which new biological targets and designated biological vehicles will arise, is credited with having a tremendous impact on the design of specific fluorescent probes. Optical techniques will likely be of increasing importance for sophisticated clinical diagnostic methods, for both laboratory purposes and clinical applications. Yet it remains to be identified which combination of fluorescent probe, biological principle, and instrumental solution will be able to solve the most urgent clinical questions and provide practical assets for the clinician.

### Acknowledgment

Ms. Gretchen Cohenour is acknowledged for her assistance in preparing this chapter.

### References

1. Bornhop, D.J., Contag, C.H., Licha, K., and Murphy, C.J., Advances in contrast agents, reporters, and detection, *J. Biomed. Opt.*, 6, 106, 2001.
2. Weissleder, R., A clearer vision for *in vivo* imaging, *Nat. Biotechnol.*, 19, 316, 2001.
3. Wan, S., Parrish, J.A., Anderson, R.R., and Madden, M., Transmittance of nonionizing radiation in human tissues, *Photochem. Photobiol.*, 34, 679, 1981
4. Tromberg, B.J., Shah, N., Lanning, R., Cerussi, A., Espinoza, J., Pham, T., Svaasand, L., and Butler, J., Non-invasive *in vivo* characterization of breast tumors using photon migration spectroscopy, *Neoplasia*, 2, 26, 2001.
5. Wagnières, G.A., Star, W.M., and Wilson, B.C., *In vivo* fluorescence spectroscopy and imaging for oncological applications, *Photochem. Photobiol.*, 68, 603, 1981.
6. Andersson-Engels, S., Star, W.M., and Wilson, B.C., *In-vivo* fluorescence imaging for tissue diagnostics, *Phys. Med. Biol.*, 42, 815, 1997.
7. Stepp, H., Sroka, R., and Baumgartner, R., Fluorescence endoscopy of gastrointestinal diseases: basic principles, techniques, and clinical experience, *Endoscopy*, 30, 379, 1998.
8. Moesta, K.T., Ebert, B., Handke, T., Nolte, D., Nowak, C., Haensch, W.E., Pandey, R.K., Dougherty, T.J., Rinneberg, Rinneberg, H., and Schlag, P.M., Protoporphyrin IX occurs naturally in colorectal cancers and their metastases, *Cancer Res.*, 61, 991, 2001.
9. Maarek, J.M., Marcu, L., Fishbein, M.C., and Grundfest, W.S., Time-resolved fluorescence of aortic wall: use for improved identification of atherosclerotic lesions, *Lasers Surg. Med.*, 27, 241, 2000.
10. Richards, G., Soubrane, G., and Yanuzzi, L., Eds., *Fluorescein and ICG Angiography*, Thieme, Stuttgart, Germany, 1998.
11. Haglund, M.M., Berger, M.S., and Hochman, D.W., Enhanced optical imaging of human gliomas and tumor margins, *Neurosurgery*, 38, 308, 1996.
12. Kremer, P. et al., Laser-induced fluorescence detection of malignant gliomas using fluorescein-labeled serum albumin: experimental and preliminary clinical results, *Neurol. Res.*, 22, 481, 2000.

13. Kuriowa, T., Kajiamoto, Y., and Ohta, T., Comparison between operative finding on malignant glioma by a fluorescein surgical microscopy and histological findings, *Neurol. Res.*, 21, 130, 1999.
14. Stummer, W. et al., Fluorescence-guided resection of glioblastoma multiforme by using 5-aminolevulinic acid-induced porphyrins: a prospective study in 52 consecutive patients, *J. Neurosurg.*, 93, 1003, 2000.
15. Grosenick, D., Wabnitz, H., Rinneberg, H., Moesta, K.T., and Schlag, P.M., Development of a time domain optical mammograph and roofridge *in vivo* applications, *Appl. Opt.*, 38, 2927, 1999.
16. De Haller, E.B., Time-resolved transillumination and optical tomography, *J. Biomed. Opt.*, 1, 7, 1996.
17. Hawrysz, D.J. and Sevick-Muraca, E.M., Developments toward diagnostic breast cancer imaging using near-infrared optical measurements and fluorescent contrast agents, *Neoplasia*, 2, 388, 2000.
18. Fisher, A.M.R., Murphree, A.L., and Gomer, C.J., Clinical and preclinical photodynamic therapy, *Lasers Surg. Med.*, 17, 2, 1995.
19. Lange, N. et al., Photodetection of early human bladder cancer based on the fluorescence of 5-aminolevulinic acid hexyl ester induced protoporphyrin IX: a pilot study, *Br. J. Cancer*, 80, 185, 1997.
20. Baumgartner, R. et al., Inhalation of 5-aminolevulinic acid: a new technique for fluorescence detection of early stage lung cancer, *J. Photochem. Photobiol. B*, 36, 169, 1996.
21. Achilefu, S., Dorshow, R.B., Bugaj, J.E., and Rajagopalan, R., Novel receptor-targeted fluorescent contrast agents for *in vivo* tumor imaging, *Invest. Radiol.*, 35, 479, 2000.
22. Becker, A. et al., Receptor targeted optical imaging of of tumor with near infrared fluorescent ligands, *Nat. Biotechnol.*, 19, 327, 2001.
23. Weissleder, R. et al., *In vivo* imaging of enzyme activity with activatable near infrared in fluorescent probes, *Nat. Biotechnol.*, 17, 375, 1999.
24. Houlne, M.P., Hubbard, D.S., Kiefer, G.E., and Bornhop, D.J., Imaging and quantitation of tissue selective lanthanide chelates using an endoscopic fluorometer, *J. Biomed. Opt.*, 3, 145, 1998.
25. Jarlman, O., Berg, R., Andersson-Engels, S., Svanberg, S., and Pettersson, H., Laser transillumination of breast tissue phantoms using time-resolved techniques, *Eur. Radiol.*, 6, 387, 1996.
26. Moesta, K.T. , Fantini, S., Jess, H., Totkas, S., Franceschini, M., Kaschke, M., and Schlag, P., Contrast features of breast cancer in frequency-domain laser scanning mammography, *J. Biomed. Opt.*, 3, 129, 1998.
27. Cutler, M., Transillumination as an aid in the diagnosis of breast lesions, *Surg. Gynecol. Obstet.*, 48, 721, 1929.
28. Drexler, B., Davies, J.L., and Schofield, G., Diaphanography in the diagnosis of breast-cancer, *Radiology*, 157, 41, 1985
29. Franceschini, M.A. et al., Frequency-domain techniques enhance optical mammography: initial clinical results, *Proc. Natl. Acad. Sci. U.S.A.*, 94, 6468, 1997.
30. Nioka, S. et al., Optical imaging of human breast cancer, *Adv. Exp. Med. Biol.*, 361, 171, 1994.
31. Pogue, B.W. et al., Hemoglobin imaging of breast tumors with near-infrared tomography, *Radiology*, 218, 261, 2000.
32. Ntziachristos, V., Yodh, A.G., Schnall, M., and Chance, B., Concurrent MRI and diffuse optical tomography of breast after indocyanine green enhancement, *Proc. Natl. Acad. Sci. U.S.A.*, 97, 2767, 2000.
33. Shealy, D.B. et al., Synthesis, chromatographic-separation, and characterization of near-infrared-labeled DNA oligomers for use in DNA-sequencing, *Anal. Chem.*, 67, 247, 1995.
34. Daehne, S., Resch-Genger, U., and Wolfbeis, O.S., Eds., *Near-Infrared Dyes for High Technology Applications*, NATO ASI Series, Kluwer Academic Publishers, London, 1998.
35. Caesar, J., Shaldon, S., Chiandussi, L., Guevara, L., and Sherlock, S., The use of indocyanine green in the measurement of hepatic blood flow and as a test of hepatic function, *Clin. Sci.*, 21, 43, 1961.

36. Gurfinkel, M., Thompson, A.B., Ralston, W., Troy, T.L., Moore, A.L., Moore, T.A., Gust, D., Tatmann, D., Reynolds, J.S., Muggenburg, B., Nikula, K., Pandey, R., Mayer, R.H., Hawrysz, D.J., and Sevick-Muraca, E.M., Pharmacokinetics of ICG and HPPH-car for the detection of normal and tumor tissue using fluorescence, near-infrared reflectance imaging: a case study, *Photochem. Photobiol.*, 72, 94, 2000.
37. Licha, K., Riefke, B., Ntziachristos, V., Becker, A., Chance, B., and Semmler, W., Hydrophilic cyanine dyes as contrast agents for near-infrared tumor imaging: synthesis, photophysical properties and spectroscopic *in vivo* characterization, *Photochem. Photobiol.*, 72, 392, 2000.
38. Licha, K. et al., Cyanine dyes as contrast agents in biomedical optical imaging, *Acad. Radiol.*, 9(Suppl. 2), 320, 2002.
39. Ebert, B., Sukowski, U., Grosenick, D., Wabnitz, H., Moesta, K.T., Licha, K., Becker, A., Semmler, W., Schlag, P.M., and Rinneberg, H., Near-infrared fluorescent dyes for enhanced contrast in optical mammography: phantom experiments, *J. Biomed. Opt.*, 6, 134, 2001.
40. Licha, K., Becker, A., Hessenius, C., Bauer, M., Wisniewski, S., Henklein, P., Wiedenmann, B., and Semmler, W., Synthesis, characterization, and biological properties of cyanine-labeled somatostatin analogues as receptor-targeted fluorescent probes, *Bioconjug. Chem.*, 12, 44, 2001.
41. Achilefu, S. et al., Synthesis, *in vitro* receptor binding, and *in vivo* evaluation of fluorescein and carbocyanine peptide-based optical contrast agents, *J. Med. Chem.*, 45, 2003, 2002.
42. Lin, Y., Weissleder, R., and Tung, C.H., Novel near-infrared cyanine fluorochromes: synthesis, properties, and bioconjugation, *Bioconjug. Chem.*, 13, 605, 2002.
43. Narayanan, N. and Patonay, G., A new method for the sysntesis of heptamethine cyanine dyes — synthesis of new near-infrared fluorescent labels, *J. Org. Chem.*, 60, 2391, 1995.
44. Mujumdar, S.R. et al., Cyanine-labeling reagents: sulfobenzindocyanine succinimidyl esters, *Bioconjug. Chem.*, 7, 356, 1996.
45. Flanagan, J.H., Jr., Khan, S., Menchen, S., Soper, S.A., and Hammer, R.P., Functionalized tricarbocyanine dyes as near-infrared fluorescent probes for biomolecules, *Bioconjug. Chem.*, 8, 751, 1997.
46. Gruber, H.J., Hahn, C.D., Kada, G., Riener, C.K., Harms, G.S., Ahrer, W., Dax, T.G., and Knaus, H.G., Anomalous fluorescence enhancement of Cy3 and Cy3.5 versus anomalous fluorescence loss of Cy5 and Cy7 upon covalent linking to IgG and noncovalent binding to avidin, *Bioconjug. Chem.*, 11, 696, 2000.
47. Folli, S., Westermann, P., Braichotte, D., Pelegrin, A., Wagnieres, G., Van den Berg, H., and Mach, J.P., Antibody-indocyanine conjugates for immunophotodetection of human squamous-cell carcinoma in nude-mice, *Cancer Res.*, 54, 2643, 1994.
48. Fisher, G.W., Ballou, B., Deng, J.S., Hakala, T.R., Srivastava, M., and Farkas, D.L., Three-dimensional imaging of nucleolin trafficking in normal cells, transfectants, and heterokaryons, *Biophys. J.*, 70, 343, 1996.
49. Ballou, B., Fisher, G.W., Hakala, T.R., and Farkas, D.L., Tumor detection and visualization using cyanine fluorochrome-labeled antibodies, *Biotechnol. Prog.*, 13, 649, 1997.
50. Ballou, B., Fisher, G.W., Deng, J.S., Hakala, T.R., Srivastava, M., and Farkas, D.L., Cyanine fluorochrome-labeled antibodies *in vivo*: assessment of tumor imaging using Cy3, Cy5, Cy5.5, and Cy7, *Cancer Detection Prev.*, 22, 251, 1998.
51. Neri, D., Carnemolla, B., Nissim, A., Balza, E., Leprini, A., Querze, G., Pini, A., Tali, L., Halin, C., Neri, P., and Zardi, L., Targeting by affinity-matured recombinant antibody fragments of an angiogenesis associated fibronectin isoform, *Nat. Biotechnol.*, 15, 1271, 1997.
52. Birchler, M., Neri, G., Tarli, L., Halin, C., Viti, F., and Neri, D., Infrared photodetection for the *in vivo* localisation of phage-derived antibodies directed against angiogenic markers, *J. Immunol. Meth.*, 231, 239, 1999.
53. Birchler, M., Viti, F., Zardi, L., Spiess, B., and Neri, D., Selective targeting and photocoagulation of ocular angiogenesis mediated by a phage-derived human antibody fragment, *Nat. Biotechnol.*, 17, 984, 1999.

54. Ramjiawan, B., Pradip, M., Aftanas, A., Kaplan, H., Fast, D., Mantsch, H.H., and Jackson, M., Noninvasive localization of tumors by immunofluorescence imaging using a single chain Fv fragment of a human monoclonal antibody with broad cancer specificity, *Cancer*, 89, 1134, 2000.
55. Nilsson, F., Tarli, L., Viti, F., and Neri, D., The use of phage display for the development of tumour targeting agents, *Adv. Drug Delivery Rev.*, 43, 165, 2000.
56. Goldsmith, S.J., Receptor imaging: competitive or complementary to antibody imaging? *Semin. Nucl. Med.*, 27, 85, 1997.
57. Krenning, E.P., Kwekkeboom, D.J., and Bakker, W.H., Somatostatin receptor scintigraphy with [IN-111-DTPA-D-PHE(1)]- and [I-123-TYR(3)]-octreotide — the Rotterdam experience with more than 1000 patients, *Eur. J. Nucl. Med.*, 20, 716, 1993.
58. Bugaj, J.E., Achilefu, S., Dorshow, R.B., and Rajaogopalan, R., Novel fluorescent contrast agents for optical imaging of *in vivo* tumors based on a receptor-targeted dye-peptide conjugate platform, *J. Biomed. Opt.*, 6, 122, 2001.
59. Licha, K., Bharagava, S., Reinlander, C., Becker, A., Schneider-Mergener, J., and Volkmer-Engert, R., Highly parallel nano-synthesis of cleavable peptide-dye conjugates on cellulose membranes, *Tetrahedron Lett.*, 41, 1711, 2000.
60. Bhargava, S. et al., A complete substitutional analysis of VIP for better tumor imaging properties, *J. Mol. Recogn.*, 15, 145, 2002.
61. Zhang, G. et al., Tricarbocyanine cholesteryl laurates labeled LDL: new near infrared fluorescent probes (NIRFs) for monitoring tumors and gene therapy of familial hypercholesterolemia, *Bioorg. Med. Chem. Lett.*, 12, 1485, 2002.
62. Bogdanov, A., Martin, C., Bogdanova, A.V., Brady, T.J., and Weissleder, R., An adduct of *cis*-diaminedichlorplatinum (II) and poly(ethylene glycol)-poly(L-lysine)-succinate: synthesis and cytotoxic properties, *Bioconjug. Chem.*, 7, 144, 1996.
63. Tung, C.H., Bredow, S., Mahmood, U., and Weissleder, R., Potential cathepsin-D sensitive near infrared fluorescent probe for *in vivo* imaging, *Bioconjug. Chem.*, 10, 892, 1999.
64. Mahmood, U., Tung, C.T., Bogdanov, A., and Weissleder, R., Near-infrared optical imaging of protease activity for tumor detection, *Radiology*, 21, 866, 1999.
65. Bremer, C., Tung, C.H., and Weissleder, R., *In vivo* molecular target assessment of MMP-2 inhibition, *Nat. Med.*, 7, 743, 2001.
66. Bremer, C. et al., Imaging of differential protease expression in breast cancers for detection of aggressive tumor phenotypes, *Radiology*, 222, 814, 2002.
67. Tung, C.H. et al., A novel near-infrared fluorescence sensor for detection of thrombin activation in blood, *Chem. Biochem.*, 3, 2007, 2002.
68. Weissleder, R., Scaling down imaging: molecular mapping of cancer in mice, *Nat. Rev. Cancer*, 2, 1, 2002.
69. Boyle, R.W. and Dolphin, D., Structure and biodistribution relationships of photodynamic sensitizers, *Photochem. Photobiol.*, 64, 469, 1996.
70. Ochsner, M., Photophysical and photobiological processes in the photodynamic therapy of tumours, *J. Photochem. Photobiol. B*, 39, 1, 1997.
71. MacDonald, I.J. and Dougherty, T.J., Basic principles of photodynamic therapy, *J. Porphyrins Phthalocyanines*, 5, 105, 2001.
72. Dougherty, T.J. et al., Energetics and efficiency of photoinactivation of murine tumor cells containing hematoporphyrin, *Cancer Res.*, 30, 1368, 1972.
73. Alian, W., Andersson-Engels, S., Savanberg, K., and Svanberg, S., Laser-induced fluorescence studies of meso-tetra(hydroxyphenyl)chlorin in malignant and normal tissues in rat, *Br. J. Cancer*, 70, 880, 1994.
74. Andersson-Engels, S., Ankerst, J., Johnansson, J., Svanberg, K., and Svanberg, S., Laser-induced fluorescence in malignant and normal tissue of rats injected with benzoporphyrin derivative, *Photochem. Photobiol.*, 57, 978, 1993.

75. Cubeddu, R., Pifferi, A., Taroni, P., Torricelli, A., Valentini, G., Comelli, D., D'Andrea, C., Angelini, V., and Canti, G., Fluorescence imaging during photodynamic therapy of experimental tumors in mice sensitized with disulfonated aluminum phthalocyanine, *Photochem. Photobiol.*, 72, 690, 2000.
76. Tassetti, V., Hajri, A., Sowinska, M., Evrard, S., Heisel, F., Cheng, L.Q., Miehe, J.A., Marescaux, J., and Aprahamian, M., *In vivo* laser-induced fluorescence imaging of a rat pancreatic cancer with pheophorbide-a, *Photochem. Photobiol.*, 65, 997, 1997.
77. Gomi, S., Nishizuka, T., Ushiroda, O., Uchida, N., Takahashi, H., and Sumi, S., The structures of mono-L-aspartyl chlorin e6 and its related compounds, *Heterocycles*, 48, 2231, 1998.
78. Blumenkranz, M.S., Woodburn, K. W., Qing, F., Verdooner, S., Kessel, D., and Miller, R., Lutetium texaphyrin (Lu-Tex): a potential new agent for ocular fundus angiography and photodynamic therapy, *Am. J. Ophthalmol.*, 129, 353, 2000.
79. Rosenkranz, A.A., Jans, D.A., and Sobolev, A.S., Nuclear and nucleolar localization of parathyroid hormone-related protein, *Immunol. Cell Biol.*, 78, 452, 2000.
80. Yarmush, M.L., Thorpe, W.P., Strong, L., Raestraw, S.L., Toner, M., and Tompkins, R.G., Antibody-targeted photolysis, *Crit. Rev. Ther. Drug Carrier Syst.*, 10, 197, 1993.
81. Vrouenraets, M.B., Visser, G.W., Loup, C., Meunier, B., Stigter, M., Oppelaar, H., Stewart, F.A., Snow, G.B., and van Dongen, G.A., Targeting of a hydrophilic photosensitizer by use of internalizing monoclonal antibodies: a new possibility for use in photodynamic therapy, *Int. J. Cancer*, 88, 108, 2000.
82. Duska, L.R., Amblin, M.R., Bamberg, M.P., and Hasan, T., Biodistribution of charged F(ab′)(2) photoimmunoconjugates in a xenograft model of ovarian cancer, *Br. J. Cancer*, 75, 837, 1997.
83. Bisland, S.K., Singh, D., and Gariepy, J., Peptide-based intracellular shuttle able to facilitate gene transfer in mammalian cells, *Bioconjug. Chem.*, 10, 982, 1999.
84. Hamblin, M.R. and Newman, E.L., Photosensitizer targeting in photodynamic therapy. 1. Conjugates of hematoporphyrin with albumin and transferring, *J. Photochem. Photobiol. B*, 26, 45, 1994.
85. Brasseur, N., Langlois, R., La Madeleine, C., Ouellet, R., and van Lier, J.E., Receptor-mediated targeting of phthalocyanines to macrophages via covalent coupling to native or maleylated bovine serum albumin, *Photochem. Photobiol.*, 69, 345, 1999.
86. James, D.A., Swamy, N., Paz, N., Hanson, R.N., and Ray, R., Synthesis and estrogen receptor binding affinity of a porphyrin-estradiol conjugate for targeted photodynamic therapy of cancer, *Bioorg. Med. Chem. Lett.*, 9, 2379, 1999.
87. Zheng, G. et al., Low-protein lipoprotein reconstituted by pyropheophordide cholestrol oleate as target-specific photosensitizer, *Bioconjug. Chem.*, 13, 392, 2002.
88. Peng, Q., Berg, K., Moan, J., and Kongshaug, M., 5-Aminolevulinic acid-based photodynamic therapy: principles and experimental research, *Photochem. Photobiol.*, 65, 235, 1997.
89. Andersson-Engels, S., Berg, R., and Svanberg, S., Multi-colour fluorescence imaging in combination with photodynamic therapy of D-amino levulinic acid (ALA) sensitised skin malignancies, *Bioimaging*, 3, 134, 1995.
90. Hewett, J., Nadeau, V., Ferguson, J., Mosely, H., Ibbotson, S., Allen, J.W., Sibbett, W., and Padgett, M., The application of a compact multispectral imaging system with integrated excitation source to *in vivo* monitoring of fluorescence during topical photodynamic therapy of superficial skin cancers, *Photochem. Photobiol.*, 73, 278, 2001.
91. Kloek, J. and Beijersbergen van Henegouwen, G.M.J., Prodrugs of 5-aminolevulinic acid for photodynamic therapy, *Photochem. Photobiol.*, 64, 994, 1996.
92. Kloek, J., Akkermans, W., and Beijersbergen van Henegouwen, G.M.J., Derivatives of 5-aminolevulinic acid for photodynamic therapy: enzymatic conversion into protoporphyrin, *Photochem. Photobiol.*, 67, 150, 1998.
93. Gaullier, J.M., Berg, K., Peng, Q., Anholt, H., Selbo, P.K., and Moan, J., Use of 5-aminolevulinic acid esters to improve photodynamic therapy on cells in culture, *Cancer Res.*, 57, 1481, 1997.
94. Uehlinger, P., Zellweger, M., Wagnieres, G., Juillerat-Jeanneret, L., van de Bergh, H., and Lange, N., 5-Aminolevulinic acid and its derivatives: physical chemical properties and protoporphyrin IX formation in cultured cells, *J. Photochem. Photobiol.*, 54, 72, 2000.

95. Lamture, J.D. and Wensel, T.G., A novel reagent for labeling macromolecules with intensely luminescent lanthanide complexes, *Tetrahedron Lett.*, 34, 4141, 1993.
96. Chen, J. and Selvin, P.R., Thiol-reactive luminescent chelates of terbium and europium, *Bioconjug. Chem.*, 10, 311, 1999.
97. Werts, M.H.V., Verhoeven, J.W., and Hofstraat, J.W., Efficient visible light sensitisation of water-soluble near-infrared luminescent lanthanide complexes, *J. Chem. Soc. Perkin Trans. 2,* 433, 2000.
98. Mathis, G., Probing molecular-interactions with homogeneous techniques based on rare-earth cryptates and fluorescence energy-transfer, *Clin. Chem.*, 41, 1391, 1995.
99. Lakowicz, J.R., Gryczynski, I., and Gryczynski, Z., Novel fluorescence sensing methods for high throughput screening, *J. Biomol. Screening*, 5, 123, 2000.
100. De Silva, A.P., Gunaratne, H.G.N., and Rice, T.E., Proton-controlled switching of luminescence in lanthanide complexes in aqueous solution: pH sensors based on long-lived emission, *Angew. Chem.*, 108, 2253, 1996.
101. Parker, D., Senanayake, P.K., and Gareth Williams, J.A., Luminescent sensors for pH, $O_2$, halide and hydroxide ions using phenanthridine as a photosensitiser in macrocyclic europium(III) and terbium(III) complexes, *J. Chem. Soc. Perkin Trans. 2,* 10, 2129, 1998.
102. Houlne, M.P., Agent, T.S., Kiefer, G.E., McMillan, K., and Bornhop, D.J., Spectroscopic characterization and tissue imaging using site-selective polyazacyclic terbium(III) chelates, *Appl. Spectrosc.*, 50, 1221, 1996.
103. Bornhop, D.J. et al., Fluorescent tissue site-selective lanthanide chelate, Tb-PCTMB for enhanced imaging of cancer, *Anal. Chem.*, 71, 2607, 1999.
104. Faulkner, S., Beeby, A., Dickins, R.S., Parker, D., and Williams, J.A.G., Generating a warm glow: lanthanide complexes which luminesce in the near-IR, *J. Fluoresc.*, 9, 45, 1999.
105. Hubbard, D.S., Houlne, H.P., Kiefer, G.E., Janseen, H.F., Hacker, C., and Bornhop, D.J., Diagnostic imaging using rare-earth chelates, *Lasers Med. Sci.*, 13, 14, 1998.
106. Griffin, J.M.M., Skwierawska, A.M., Manning, H.C., Marx, J.N., and Bornhop, D.J., Simple, high yielding synthesis of trifunctional fluorescent lanthanide chelates, *Tetrahedron Lett.*, 42, 3823, 2001.
107. Manning, H.C. et al., Facile, efficient conjugation of a trifunctional lanthanide chelate to a peripheral benzodiazepine receptor ligand, *Org. Lett.*, 4, 1075, 2002.
108. Oefner, P.J. et al., High resolution liquid-chromatography of fluorescent dye-labeled nucleic acids, *Anal. Biochem.*, 39, 223, 1994.
109. Hogan, R.N. and Zimmerman, C.F., Sodium fluorescein and other tissue dyes, in *Textbook of Ocular Pharmacology*, Zimmerman, T.J., Ed., Lippincott-Raven, Philadelphia, 1997, p. 849.
110. Smeltzer, C.C., Cannon, M.J., Pinson, P.R., Munger, J.D., Jr., West, F.G., and Grissom, B., Synthesis and characterization of fluorescent cobalamin (CobalaFluor) derivatives for imaging, *Org. Lett.*, 3, 799, 2001.
111. Van Staveren, H.J., Speelman, O.C., Witjes, M.J.H., Cincotta, L., and Star, W.M., Fluorescence imaging and spectroscopy of ethyl Nile blue A in animal models of (pre)malignancies, *Photochem. Photobiol.*, 73, 32, 2001.
112. Takeo, Y. et al., Endoscopic mucosal resection for early esophageal cancer and esophageal dysplasia, *Hepatogastroenterology*, 48, 453, 2001.
113. Outten, C. and O'Halloran, T.V., Femtomolar sensitivity of metalloregulatory proteins controlling zinc homeostasis, *Science*, 292, 2488, 2001.
114. Huckle, J.W. et al., Isolation of a prokaryotic metallothionein locus and analysis of transcriptional control by trace-metal ions, *Mol. Microbiol.*, 7, 177, 1993.
115. Zhao, H. and Eide, D.J., Zap1p: a metalloregulatory protein involved in zinc-responsive transcriptional regulation in *Saccharomyces cerevisiae*, *Mol. Cell. Biol.*, 17, 5044, 1997.
116. Heuchel, R. et. al., The transcription factor MTF-1 is essential for basal and heavy metal-induced metallothionein gene expression, *EMBO J.*, 13, 2870, 1994.
117. Thelwell, C., Robinson, N.J., and Turner-Cavet, J.S., An SmtB-like repressor from Synechocystis PCC 6803 regulates a zinc exporter, *Proc. Natl. Acad. Sci. U.S.A.*, 95, 10728, 1998.

118. Singh, V.K. et al., ZntR is an autoregulatory protein and negatively regulates the chromosomal zinc resistance operon of *S. aureus, Mol. Microbiol.*, 33, 200, 1999
119. Gaballa, A. and Helmann, J.D., Identification of a zinc-specific metalloregulatory protein, zur, controlling zinc transport in *Bacillus subtilis, J. Bacteriol.*, 180, 5815, 1998.
120. Tsien, R.Y. et al., Multicolor and electron microscopic imaging of connexin trafficking, *Science*, 296, 503, 2002.
121. Kopelman, R. et al., A real-time ratiometric method for the determination of molecular oxygen inside living cells using sol-gel-based spherical optical nanosensors with applications to rat C6 glioma, *Anal. Chem.*, 73, 4124, 2001.
122. Clark, H.A., Kopelman, R., Tjalkens, R., and Philbert, M.A., Optical nanosensors for chemical analysis inside single living cells. Part 2: Sensors for pH and calcium and the intracellular application of PEBBLE sensors, *Anal. Chem.*, 71, 4837, 1999.
123. Fang, X., Li, J.J., Perlette, J., and Tan, W., Molecular beacons: novel DNA probes for biomolecular recognition, *Anal. Chem.*, 72, 747A, 2000.
124. Li, J.J., Fan, X., and Tan, W., Molecular aptamer beacons for real-time protein recognition, *Biochem. Biophys. Res. Commun.*, 292, 31, 2002.

Kemin Wang
*Hunan University*
*Changsha, Hunan*
*People's Republic of China*
*and*
*University of Florida*
*Gainesville, Florida*

Jun Li
*Hunan University*
*Changsha, Hunan*
*People's Republic of China*

Xiaohong Fang
*University of Florida*
*Gainesville, Florida*

Sheldon Schuster
*University of Florida*
*Gainesville, Florida*

Marie Vicens
*University of Florida*
*Gainesville, Florida*

Shannon Kelley
*University of Florida*
*Gainesville, Florida*

Hua Lou
*University of Florida*
*Gainesville, Florida*

Jianwei Jeffery Li
*University of Florida*
*Gainesville, Florida*

Terry Beck
*TriLink BioTechnologies, Inc.*
*San Diego, California*

Richard Hogrefe
*TriLink BioTechnologies, Inc.*
*San Diego, California*

Weihong Tan
*University of Florida*
*Gainesville, Florida*

# 57 Novel Fluorescent Molecular Beacon DNA Probes for Biomolecular Recognition



## 57.1 Introduction

Fluorescent probes for biomolecular recognition are of great importance in the fields of chemistry, biology, and medical sciences, as well as in biotechnology. These probes have been used for mechanism studies of biological functions and in ultrasensitive detection of biological species responsible for many diseases.[1] In the post-genome era, quantitative studies of genomic information for disease diagnosis and prevention and drug discovery will be fast growing areas of research and development. This has led to a continued demand for advanced biomolecular recognition probes with high sensitivity and high specificity. The molecular beacon (MB), a recently developed single-stranded DNA (ssDNA) molecule,[2] appears to be a very promising probe for quantitative genomic studies. MBs are hairpin-shaped oligonucleotides that contain both fluorophore and quencher moieties and act like switches that are normally closed to bring the fluorophore/quencher pair together to turn fluorescence "off." When prompted to undergo conformational changes that open the hairpin structure, the fluorophore and the quencher are separated, and fluorescence is turned "on." MBs were first developed in 1996. Since then, they have

0-8493-1116-0/03/$0.00+$1.50

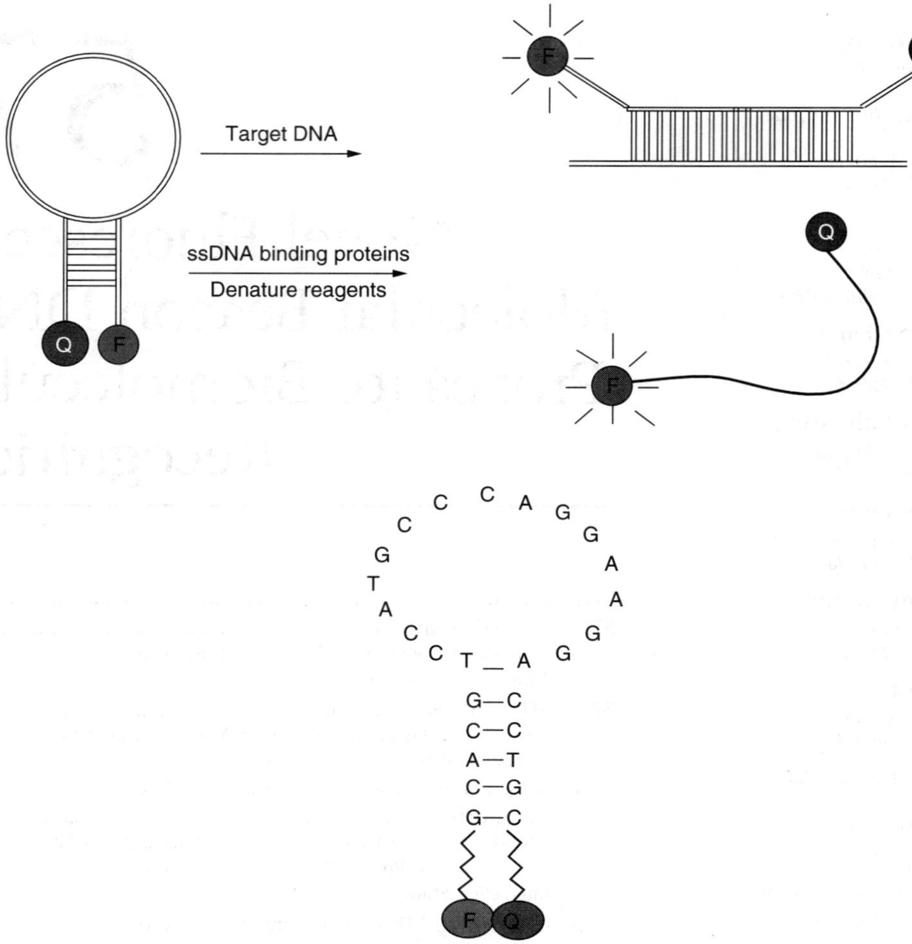


**FIGURE 57.1** Mechanism of operation of molecular beacon DNA probes. On their own these molecules are non-fluorescent because the stem hybrids keep the fluorophores close to the quenchers. Molecular beacons emit intense fluorescence only when the stems are apart through hybridization of DNA molecules with sequences complementary to their loop sequences, unwinding the stem hybrids upon interaction with ssDNA binding protein, if denaturing reagents are used, or under other denaturing conditions.

provided a variety of exciting opportunities in DNA/RNA/protein studies both inside living-cell specimens and in solution.

## 57.2 The MB Principle

### 57.2.1 Molecular Recognition and Fluorescence-Energy Transfer

The hybridization of a nucleic-acid strand to its complement target is one of the most specific well-known molecular recognition events. A MB makes use of this unique feature for DNA/RNA/protein studies. MBs are synthetic short oligonucleotides that possess a loop and stem structure, as schematically shown in Figure 57.1.

Its signal-transduction mechanism for molecular recognition is based on fluorescence-resonance energy transfer (FRET).[1] The loop portion of a MB is a probe sequence complementary to a target nucleic-acid molecule. The arm sequences flanking either side of the probe are complementary to each other but are unrelated to the target sequence. These arm sequences, which have five to seven base pairs,

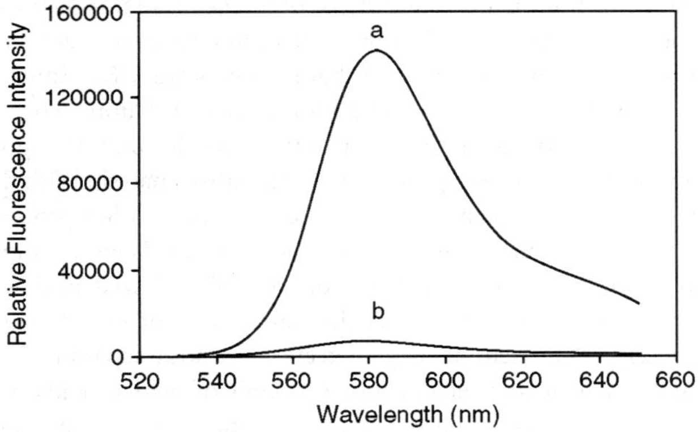


**FIGURE 57.2** Fluorescence spectra of MBs obtained at room temperature. Spectra were taken for MB hybridization with cDNA molecules (a) and with non-cDNA molecules (b).

anneal to form the MB's stem. A fluorescent moiety is covalently attached to the end of one arm, and a quenching moiety is covalently attached to the other end. The fluorescent dye serves as an energy donor, and the nonfluorescent quencher plays the role of an acceptor. The stem keeps these two moieties in close proximity to each other, causing the fluorescence of the fluorophore to be quenched by energy transfer. Thus, the probe is unable to fluoresce. When the probe encounters a target molecule, the loop forms a hybrid that is longer and more stable than the stem. The molecular beacon undergoes a spontaneous conformational change that forces the stem apart. The fluorophore and the quencher are moved away from each other, leading to the restoration of fluorescence. Unhybridized MBs do not fluoresce; thus, it is not necessary to remove them to observe hybridized probes.

Two forms of energy transfer may exist in MBs: direct energy transfer and FRET.[3] Direct energy transfer requires contact between the two moieties in a MB. The collision between the fluorophore and the quencher can distort the energy level of the excited fluorophore, causing the fluorescence of the fluorophore to be quenched by energy transfer. In brief, the quenching moiety dissipates the energy that it receives from the fluorophore as heat, rather than emitting it as light. The other mechanism, FRET, can occur between the two moieties in a MB with a relatively long distance (20 to 100 Å). It requires a spectral overlap between the donor's emission spectrum and the acceptor's absorption spectrum. The rate of FRET is directly proportional to the inverse sixth power of the separation distance of the donor and the acceptor or the quencher.[1] Both forms of the energy transfer are strongly dependent on the distance between the dye moieties. Therefore, the spatial separation of the fluorophore and the quencher in a MB determines the energy-transfer efficiency. When a target DNA hybridizes to a MB, a substantial increase of the fluorescence signal is observed (Figure 57.2) due to larger separation distance between the two moieties in a MB. As DABCYL (dimethylaminophenylazobenzoic acid) has been found to efficiently quench a large variety of fluorophores independent of the spectral overlap, direct energy transfer may be dominant in MBs.[4]

### 57.2.2 Advantages of MB Probes

The hybridization of a nucleic-acid strand to its complement has been widely used in many areas of research and development. Various types of DNA fluorescent probes have been designed for these applications. For example, nucleic-acid blotting techniques have been used to make great strides in our understanding of gene organization and function. In these techniques, a solid support is used to immobilize DNA fragments.[5] A labeled oligonucleotide probe containing the sequence of interest is then used to hybridize with its counterpart. In today's laboratories, nucleic-acid techniques are increasingly being harnessed for use in practical applications such as the molecular diagnosis of disease. But what if one needs to monitor the real-time synthesis of nucleic acids as it is happening? Or what if your research calls for the labeling of nucleic acids within living cells? Methods based on immobilization of hybridized nucleic acids, probes requiring intercalation reagents, or other means requiring the isolation of probe–target complexes cannot be used for these applications. MBs can overcome these difficulties.

The inherent fluorescent-signal-transduction mechanism enables a MB to function as a sensitive probe with high signal-to-background ratio for real-time monitoring. Its fluorescence intensity can increase more than 200 times when it meets the target under optimal conditions.[4] This provides the MBs with a significant advantage over other fluorescent probes in ultrasensitive analysis. With this inherent sensitivity, individual MB DNA molecules have been imaged, and their hybridization process has been monitored on a single-molecule basis.[6] MBs can be used in situations where it is either impossible or undesirable to isolate the probe–target hybrids from an excess of the unhybridized probes, such as in real-time monitoring of polymerase chain reactions (PCRs) in sealed tubes or in detection of messenger RNAs (mRNAs) within living cells. The usefulness of "detection without separation" for these applications cannot be overemphasized. This feature enables the synthesis of nucleic acids to be monitored as it is occurring, in either sealed tubes or in living specimen, without additional manipulations.

Another major advantage of MBs is their molecular-recognition specificity. They are extraordinarily target-specific, ignoring nucleic-acid target sequences that differ by as little as a single nucleotide. While current techniques for routine detection of single-base-pair DNA mutations are often labor-intensive and time-consuming,[7] MBs provide a simple and promising tool for the diagnosis of genetic disease and for gene-therapy study. This specificity of MBs comes from its loop-and-stem structure. The stem hybrid acts as a counterweight for the loop hybrid. Experiments have shown that the range of temperatures in which perfect complementary DNA (cDNA) targets form hybrids but mismatched DNA targets do not is significantly wider for MBs than the corresponding range observed for conventional linear probes.[8] Therefore, MBs can easily discriminate DNA targets that differ from one another by a single nucleotide. Thermodynamic studies revealed that the enhanced specificity is a general feature of structurally constrained DNA probes.

## 57.3 MB Synthesis and Characterization

The synthesis of a MB is similar to that of dual labeling a short oligomer with two dyes. The length of the loop sequence (15 to 40 nucleotides) is chosen so that the probe–target hybrid is stable at the temperature of probing. The stem sequences (five to seven nucleotides) should be strong enough to form the hairpin structure for efficient fluorescence quenching while still weak enough to be dissociated when a cDNA hybridizes with the loop of the MB. Also, the stem sequence must be designed so as not to interfere with the probe sequences. Since DABCYL can serve as a universal quencher for many fluorophores,[4] MBs are generally synthesized using DABCYL-CPG (controlled-pore glass) as starting material. Various fluorescent-dye molecules can be covalently linked to the 5′ end to report fluorescence at various wavelengths. There are carbon chain linkers between the bases and the labeled-dye molecules. The stem and the linker keep the fluorophore and the quencher in close proximity and increase the probability for their direct contact. There are four important steps in this synthesis.[2] First, a CPG solid support is derivatized with DABCYL and used to start the synthesis at the 3′ end of the oligonucleotide. The rest of the nucleotides are added sequentially using standard cyanoethylphosphoramidite chemistry. Second, a primary amine group at the 5′ end is linked to the phosphodiester bond by a six-carbon spacer arm. A trityl protecting group at the ultimate 5′ end protects the amine group. Third, the oligonucleotide is hydrolyzed and removed from the CPG and then purified by reversed-phase HPLC. Fourth, the purified oligonucleotide is removed from the trityl group and labeled with a fluorophore. After labeling, the excess dye is removed by gel-filtration chromatography on Sephadex G-25. The oligonucleotide is then purified again by reversed-phase HPLC. The product peak from the HPLC is collected. The synthesized MB is characterized by ultraviolet (UV) and by mass spectroscopy.[9] The purification of the MB after synthesis is critically important to ensure a high signal-to-noise ratio and to achieve ultrahigh sensitivity. A detailed protocol for MB synthesis can be found on the Internet at http:\\www.molecular-beacons.org. There are also approximately ten commercial companies specializing in custom synthesis of MBs. Today, molecular beacons with specific sequences are readily available at affordable prices and without tedious synthesis by individual investigators.

## 57.3.1 Synthesis and Purification of MBs

MB synthesis is now commercially available. Even though there are many companies involved in MB synthesis, the techniques are similar. Here we give a description of a typical method for MB synthesis (from http://www.molecular-beacons.org). The starting material for the synthesis of MBs is an oligonucleotide that contains a sulfhydryl group at its 5′ end and a primary amino group at its 3′ end. DABCYL is coupled to the primary amino group using an amine-reactive derivative of DABCYL. The oligonucleotides that are coupled to DABCYL are then purified. The protective trityl moiety is then removed from the 5′-sulfhydryl group, and a fluorophore is introduced in its place using an iodoacetamide derivative. Recently, a CPG column that introduces a DABCYL moiety at the 3′ end of an oligonucleotide has become available that enables the synthesis of a molecular beacon completely on a DNA synthesizer. The whole sequence of the molecular beacon used throughout this protocol is shown as the following: fluorescein-5′-GCGAGCTAGGAAACACCA AAGATGATAT TTGCTCGC-3′-dabcyl, where the underlines identify the arm sequences.

### 57.3.1.1 Coupling of DABCYL

1. Dissolve 50 to 250 nmol dry oligonucleotide in 500 μl of 0.1 *M* sodium bicarbonate, pH 8.5. Dissolve about 20 mg of DABCYL (4-(4′-dimethylaminophenylazo)benzoic acid) succinimidyl ester (molecular probes) in 100 μl *N,N*-dimethylformamide and add to a stirring solution of the oligonucleotide in 10-μl aliquots at 20-min intervals. Continue stirring for at least 12 h.
2. Remove particulate material by spinning the mixture in a microcentrifuge for 1 min at 10,000 rpm. To remove unreacted DABCYL, pass the supernatant through a gel-exclusion column. Equilibrate a Sephadex G-25 column (NAP-5, Pharmacia, Peapack, NJ) with buffer A, load the supernatant, and elute with 1 ml buffer A. Filter the elute through a 0.2-μm filter (Centrex MF-0.4, Schleicher & Schuell, Keene, NH) before loading on the HPLC column.
3. Purify the oligonucleotides on a C-18 reverse-phase column (Waters), using a linear elution gradient of 20 to 70% buffer B in buffer A, and run for 25 min at a flow rate of 1 ml/min. Monitor the absorption of the elution stream at 260 nm and 491 nm. A typical chromatogram is shown in Figure 57.3. Collect the peak that absorbs in both wavelengths, which contain oligonucleotides with a protected sulfhydryl group at their 5′ end and DABCYL at their 3′ end (peak D).
4. Precipitate the collected material with ethanol and salt and spin in a centrifuge for 10 min at 10,000 rpm; discard the supernatant, dry the pellet, and dissolve it in 250 μl buffer A. As explained in Figure 57.3, we can collect the desired portion of the synthesized DNA probes.

### 57.3.1.2 Coupling of Fluorophore

1. To remove the trityl moiety, add 10 μl of 0.15 *M* silver nitrate and incubate for 30 min. Add 15 μl of 0.15 *M* dithiothreitol to this mixture and shake for 5 min. Spin for 2 min at 10,000 rpm and transfer the supernatant to a new tube. Dissolve about 40 mg 5-iodoactamidofluorescein (Molecular Probes, Eugene, OR) in 250 μl of 0.2 *M* sodium bicarbonate, pH 9.0, and add it to the supernatant. Incubate the mixture for 90 min. Each of these solutions should be prepared just before use.
2. Remove excess fluorescein from the reaction mixture by gel-exclusion chromatography and purify the oligonucleotides coupled to fluorescein by HPLC, following the instructions in steps 2 and 3 of the previous section. A sample chromatogram is shown in Figure 57.3. Collect the fractions corresponding to peak D, which absorb at wavelengths 260 and 491 nm and are fluorescent when observed with a UV lamp in a dark room. If a different fluorophore is coupled in place of fluorescein, its maximum absorption wavelength should be used instead of 491 nm.
3. Precipitate the collected material and dissolve the pellet in 100 μl TE buffer. Determine the absorbance at 260 nm and estimate the yield (1 OD260 = 33 μg/ml).

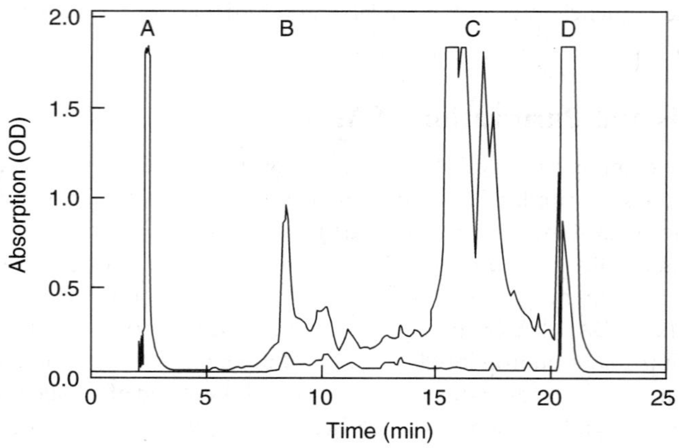


**FIGURE 57.3** Chromatographic separation of oligonucleotides coupled to DABCYL. The top line represents absorption at 260 nm, and the bottom line represents absorption at 491 nm. The oligonucleotides in peaks A and B do not contain trityl moieties, whereas the oligonucleotides in peaks C and D are protected by trityl moieties. The oligonucleotides in peaks B and D are coupled to DABCYL, whereas the oligonucleotides in peaks A and C are not coupled to DABCYL. Peak D should be collected.

### 57.3.1.3 Automated Synthesis

1. Use a CPG column to introduce DABCYL (Glen Research, Sterling, VA) at the 3′ end of the oligonucleotide during automated synthesis. At the 5′ end of the oligonucleotide either a thiol or an amino modifier can be introduced for a subsequent coupling to a fluorophore, or a fluorophore can be directly introduced during automated synthesis using a phosphoramidite. The 5′ modifiers and fluorophores should remain protected with a trityl moiety during the synthesis. Perform post-synthetic steps as recommended by the manufacturer of the DNA synthesizer. Dissolve the oligonucleotide in 600 μl buffer A.
2. When the fluorophore is to be introduced manually, purify the oligonucleotide protected with atrityl moiety. Remove the trityl moiety from the purified oligonucleotide and continue with the coupling of the fluorophore, as described above.
3. When a 5′ fluorophore is introduced via automated synthesis, purify the oligonucleotide protected with the trityl moiety and then remove the trityl moiety from the purified oligonucleotide. Precipitate the molecular beacon with ethanol and salt and dissolve the pellet in 100 μl TE buffer. Determine the absorbance at 260 nm and estimate the yield.

## 57.3.2 MB Characterization

### 57.3.2.1 Characterization of MB by Mass Spectrometry

Once the MBs are synthesized, various methods will be used to evaluate the quality of the product. One of the effective ways of doing this is to run a mass spectrometry and determine the molecular weight. Usually, an aliquot of the MB product collected from HPLC was taken out for analysis by mass spectrometry. Matrix-assisted laser desorption/ionization time-of-flight mass spectrometer was used to confirm the molecular weight of the MB. For example, the calculated molecular weight of a MB used in biosensor development is 10076. As shown in Figure 57.4, the main peak in the mass spectrum was 10085.7 Da. The difference between the measured value and the calculated one is only about 0.1%, which is less than the typical error, 0.2%, in mass-spectrum measurement. Another small peak appeared near a position corresponding to half of the MW of the MB. This was due to the molecules that had lost two electrons in the desorption process during the measurement. The mass-spectrum results support the synthesis of the designed MB.

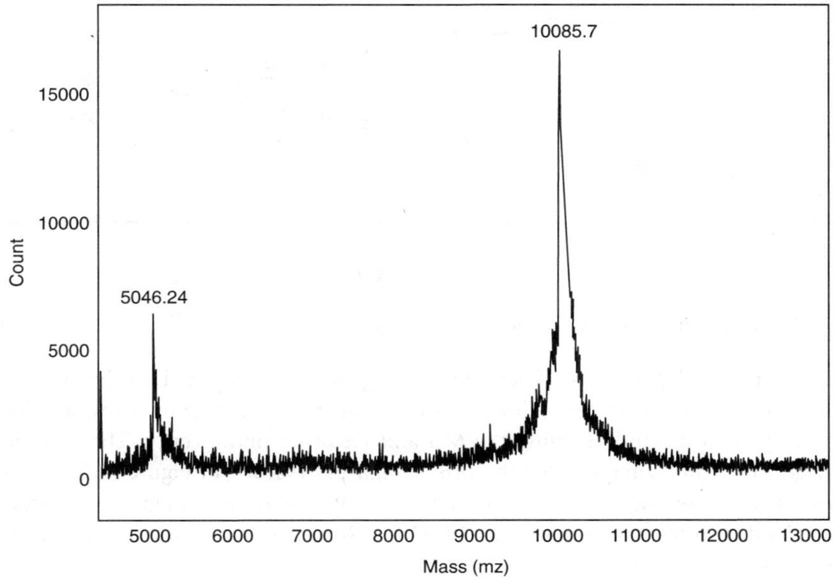


**FIGURE 57.4** Mass spectrum of the synthesized biotinylated molecular beacon.

#### 57.3.2.2 MB Hybridization Activity

MBs can be tested by hybridization to evaluate their activity in DNA/RNA reaction. For example, a newly synthesized MB has been used for DNA hybridization in solution. The MB's hybridization properties were tested using fluorescence measurements performed on a SPEX Industries F-112A spectrophotometer. A submicro quartz cell was used for the hybridization experiment. Two 200-μl sample solutions were prepared: the MB and a fivefold molar excess of its cDNA, and the MB and a fivefold molar excess of a non-cDNA. The concentrations of MB in the two solutions were 50 n*M*. They were incubated for 20 min in the hybridization buffer (20 m*M* Tris-HCl, 50 m*M* KCl, and 5 m*M* $MgCl_2$, pH = 8.0). Emission spectra were recorded at room temperature with excitation at 515 nm for the MB labeled with fluorescein. Hybridization of the MB inside the solution has shown strong fluorescence signal when ssDNA molecules reacted with their cDNA molecules. Theoretically, the enhancement could be as high as more than 200-fold with optimal design of the sequence and under optimal hybridization and optical-detection conditions. The solution with the non-cDNA has no enhancement under the same conditions. Hybridization dynamics of the MB has also been investigated. These experiments clearly show that the MB synthesized is what we have designed and can be used for DNA/RNA studies.

## 57.3.3 MB Lifetime Measurements

Many studies have been undertaken to effectively design and understand the structure of MBs. In addition, many studies have begun to probe how the MB binds to its target. The biggest question in MB research is the opening/closing mechanism of the probe as well as its specificity toward its DNA/RNA targets. Out of many potentially useful techniques, lifetime measurement is ideal for probing the state of the fluorophore in MBs either in closed or open forms. It helps in understanding the opening/closing mechanism of the molecular beacon. This technique is being used to study MBs and their cDNA interaction as well as MB at high pH and different temperatures in expectation of overall comprehension of MBs.

The frequency-domain lifetime instrument was first studied with rhodamine 6G (R6G). The average lifetime on this instrument was found to be 4.06 nsec. Lifetime measurements were taken on the MB at a pH of 12.5. At this pH, the MB exhibited two lifetimes, with average lifetimes of 4.17 and 1.61 nsec. It

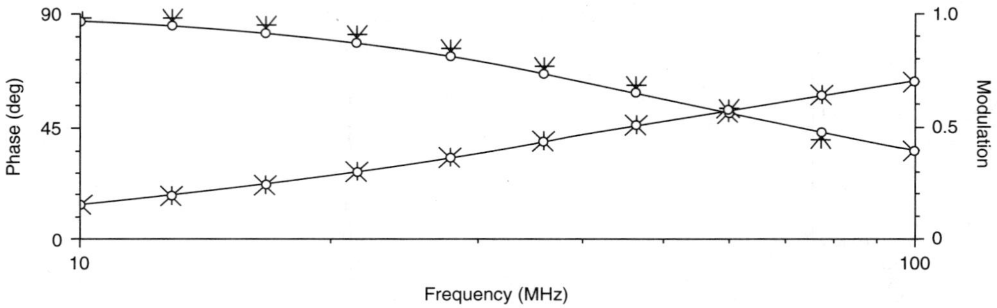


**FIGURE 57.5** Lifetime scan of molecular beacon at high pH.

is a well-known fact that DNA can be affected by a change in pH. At high pH, the bases undergo alkaline hydrolysis, causing denaturation, and a low pH would cause the cleavage of bases. In addition, a high pH causes the stem to unwind, opening the MB and restoring fluorescence. The two lifetime values obtained suggest two conformations for the MB at high pH (shown in Figure 57.5) — the open state (unquenched R6G) and the closed state (quenched R6G). Lifetime measurements are difficult to obtained directly for MBs due to the quenching of DABCYL on R6G. Specific measurements must be taken to enhance the signal.

The effects of temperature on lifetime measurements were studied with the MB. The lifetimes at the melting point $T_m$ of the MB were found to be 3.89 and 0.67 nsec. At a high temperature of 85°C, the lifetime of the MB was found to be 4.01 nsec. Studies were also conducted involving MB targets such as cDNA and single-stranded binding protein (SSB) at varying concentration ratios. At a MB-to-cDNA concentration ratio of 1:1, lifetime values of 4.23 and 0.75 nsec were obtained. For the following concentration ratios of 1:4 and 1:8 of MB and cDNA, lifetime values of 3.91 and 3.97 nsec were found, respectively. A lifetime of 3.87 nsec was found for MB binding to SSB in a 1:2 concentration ratio. It is seen that the open and closed forms can be studied using lifetime measurements due to the distance separating the quencher and the fluorophore. Future detailed experiments to obtain precise lifetime values of MB will help determine the overall mechanism of molecular beacons.

## 57.4 MBs as Fluorescent Probes for DNA/RNA Monitoring

### 57.4.1 DNA/RNA PCR Detection

#### 57.4.1.1 Real-Time Monitoring of PCR

MB probes are suitable tools for real-time monitoring of DNA/RNA amplification during PCRs.[2,4] They can be simply added to a sealed PCR tube. The fluorescence signal is monitored at the annealing step of every cycle. At the annealing temperature, the target amplicons' products bind to the MB to generate fluorescence, while the unbound MB remains in the closed form without fluorescence. The MB–amplicon hybrids dissociate at elevated temperatures, where MBs do not interfere with polymerization. The fluorescence signal increases with the increased cycling, and it directly indicates the concentration of the amplicons in the PCR process. The MB assay is fast, sensitive, and nonradioactive. The PCR tubes are sealed during the entire measurement, thereby avoiding carryover contamination. Compared to TaqMan probes,[10] which are another type of fluorescent probe used in PCR monitoring, MBs give more reliable genotyping results, especially in a GC-rich region. Moreover, MBs allow for sensitive and quantitative detection of low-abundance sequence variants over a wider range. For example, *Chlamdophila felis* infection of cats was detected by a MB to identify the major outer-membrane protein gene. The detection limit was fewer than ten genomic copies.[11]

#### 57.4.1.2 Gene Typing and Mutation Detection for Disease Study and Diagnosis

For the detection of genetic mutations, a method called spectral genotyping has been developed.[7] The principle involves two different MB probes with different loop sequences: one specific for a wild-type allele and the other for a mutant allele. These two MBs also have two different fluorophores. The fluorescence measured at the two emission wavelengths during amplification indicates whether the samples are homozygous wild type, homozygous mutants, or heterozygote. The MB-based PCR mutation-detection method has been used for the study of many diseases,[12–19] for example, acquired immunodeficiency syndrome (AIDS). The polymorphisms in the gene for human CC-chemokine receptors CCR5 and CCR2, which are associates of human immunodeficiency virus type 1 (HIV-1), have also been studied.[12–15] Using MBs researchers are able to investigate the HIV disease mechanism, progression process, and therapeutic effects.

MB-based PCR is also a promising tool for rapid and reliable clinical diagnosis. An example is the method developed to assay four pathogenic retroviruses: HIV-1, HIV-2, and human T-lymphotrophic virus types I and II.[15] The retroviral DNA sequences were amplified by simultaneous PCR reaction, which contained four sets of primers and four MBs, each one being specific for one of the four amplicons and labeled with a different colored fluorophore. The color of the fluorescence generated in the course of amplification identified which retroviruses were present. The number of thermal cycles required for the generation of each color was used to measure the number of copies of each retroviral sequence originally present in the sample. Fewer than ten retroviral genomes were detected. Moreover, ten copies of a rare retrovirus were detected in the presence of 100,000 copies of an abundant retrovirus. There were no false positives for 96 clinical samples. This method will be useful in screening donated blood and transplantable tissue.

It is known that MBs have excellent characteristics for mutation detection. Therefore, sequence analysis of HIV-1 from 74 persons with acute infections identified eight strains with mutations in the reverse transcriptase (RT) gene by real-time nucleic-acid sequence-based amplification assay with molecular beacons.[20] The results illustrate that infection with nucleoside-analog-resistant HIV leads to newly infected individuals with mutants that are sensitive to nucleoside analogs, but it is only a single mutation removed from drug-resistant HIV.

MB-based, real-time PCR allele discrimination has been used to detect all three chemokine receptor mutations that are associated with HIV-1 disease.[21] These spectral genotyping assays were used to genotype 3923 individuals from a globally distributed set of 53 populations. The results showed that CCR2–641 and CCR5–59653T genetic variants were found in almost all populations studied, and their allele frequencies were greatest (similar to 35%) in Africa and Asia but decreased in Northern Europe.

The fast one-tube assay that identifies and distinguishes among all subtypes — A, B, C, and CRFs, AE, and AG — of HIV-1 has been developed by using subtype-specific MBs with multiple fluorophores. The lower detection level of the assay was approximately $10^3$ copies of HIV-1 RNA per reaction. However, the assay in this format would be unsuitable for clinical use but could possibly be used for epidemiological monitoring as well as vaccine-research studies.[22] Meanwhile, all HIV-1 groups, subtypes, and CRFs were detected and quantified with equal efficiency by the addition of molecular beacons to the amplification reaction. The lower level of quantification was 100 copies of HIV-1 RNA with a dynamic range of linear quantification between $10^2$ and $10^7$ RNA molecules.[23]

In conclusion, the real-time monitored HIV-I assay is a fast and sensitive assay with a large dynamic range of quantification and is suitable for quantification of most if not all subtypes and groups of HIV-1.

#### 57.4.1.3 Bacteria Detection

Bacteria exist almost everywhere in the world. Scientists have found bacteria in many different environments. The simplest detection method for bacteria is through observation by microscopy. To identify different kinds of bacteria, a fluorescence-strain method associated with PCR has been developed. Last year, MB PCR was demonstrated to show positive results more rapidly than traditional agarose-gel electrophoresis analysis of PCR products. Use of MBs allows real-time monitoring of PCR reactions, and the closed-tube format allows simultaneous detection and confirmation of target amplicons without the need for agarose-gel electrophoresis or Southern blotting. Therefore, MBs have become a powerful tool for bacteria detection. To accomplish this, the MB that is to hybridize with a target sequence of the bacteria is incorporated into PCRs containing DNA extract from *Escherichia coli* in an artificially contaminated skim milk.[24] The bacteria can then be sensitively analyzed.

Based on the molecular beacon, a rapid and simple homogeneous fluorescence-PCR assay was developed for the clinical diagnosis of infectious diseases. This method could reproducibly detect *Mycobacterium tuberculosis* at the 10 bacteria/mL level with a higher sensitivity than traditional methods. The feasibility of this method was further supported by successful detection of *Neisseria gonorrhoeae* and *Chlamydia trachomatis*.[25] Another bacteria-analysis method based on MBs was also demonstrated by a real-time PCR assay to detect the presence of *E. coli* O157:H7.[24] MBs were designed to recognize a 26-base-pair region of the rfbE gene, encoding for an enzyme necessary for O-antigen biosynthesis. The specificity of the MB-based PCR assay was very high. All *E. coli* serotype O157 tested were positively identified, while all other species, including the closely related O55, were not detected by the assay. Positive detection of *E. coli* O157:H7 was demonstrated when $>10^2$ colony-forming units (CFU)/ml were present in the samples. The ability of the assay to detect *E. coli* O157:H7 in raw milk and apple juice was demonstrated. The sensitivity was as few as 1 CFU/ml after 6 h of enrichment. These assays could be carried out entirely in sealed PCR tubes, enabling rapid and semiautomated detection of *E. coli* O157:H7 in food and environmental samples.[26] Further development was proposed on a real-time PCR assay to detect the presence of *Salmonella* species using MB, which was designed to recognize a 16-base-pair region on the amplicon of a 122-base-pair section of the himA. As few as two colony-forming units per PCR were detected. The high selectivity of MBs made it feasible to detect similar species such as *E. coli* and *Citrobacter freundii* in real-time PCR assays.[27]

### 57.4.2 mRNA Monitoring inside Living Cells

Obtaining knowledge about subcellular localization and cellular transport pathway of RNAs is of crucial importance for our understanding of basic cell biological and developmental processes. It is also of great interest for the study of functional genomics in the post-genome era. Up till now little has been known about the rate at which RNAs are synthesized and processed as well as their pathways in the cell. *In situ* hybridization (ISH) techniques have been successfully used in the past few decades for the microscopic detection of specific mRNA molecules in fixed and pretreated cells and tissues. However, the fixed cell may not really reflect the situation in living cells. Fixation and cell pretreatment also prevent us from studying the dynamics of RNA synthesis and the transport process when they are actually occurring in the cell. MB-based fluorescence *in vivo* hybridization (FIVH) offers a new way to overcome the limitations of traditional ISH for RNA monitoring.[28]

MBs have been reported to visualize the bFGF (basic fibroblast growth factor) mRNA in human trabecular cells,[29] β-actin mRNA in K562 human leukemia cells,[30] and PTK2 kangaroo rat kidney cells.[31] Successful application of MB in mRNA detection and localization depends mainly on the following three factors: rational design of a MB probe for the specific mRNA, efficient introduction of the MB probe into the cell, and optimization of the experimental conditions for fluorescence imaging and detection. Choosing an appropriate sequence on the target mRNA that will be most accessible to the MB probe hybridization and choosing a corresponding MB loop and stem that will avoid the secondary structure of the loop sequence are the major concerns in the MB probe design. A computer analysis of the RNA and MB folding structure can be of help for the optimal design. Microinjection and liposome delivery have been used as effective tools for the transfer of the MB probes into living cells while maintaining the physiological and structural integrity of the cells. However, special care must be taken to introduce an appropriate amount of MB to minimize the disturbance of the cell during the delivery process. Another approach is the use of a biosensor format in which MB probes are immobilized on a solid carrier surface

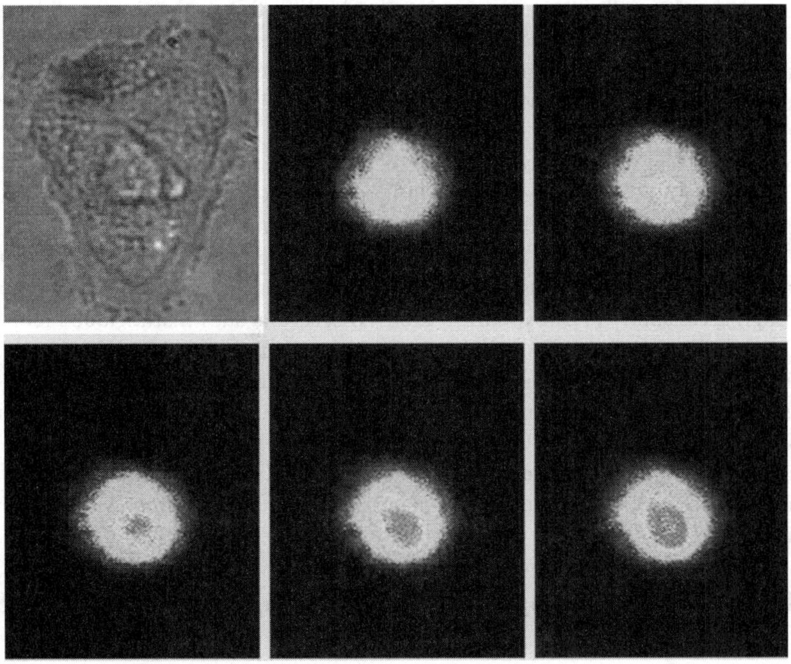

**FIGURE 57.6 (Color figure follows p. 28-30.)** Hybridization of the MB to mRNA in living cells. Clockwise, from top left: An optical image of the kangaroo rat kidney cell and the fluorescence images of the cell at 3, 6, 9, 12, and 15 min after injection of MB to the PtK2 cell.

(see below). A highly sensitive fluorescence-imaging system is usually used to visualize the MB's hybridization with the target mRNA. Control experiments should also be carried out to preclude any signal originating from the opening of the MB by other possible proteins or nuclease cleavage enzymes in cells.

Coupled with the advanced imaging system, MBs have been used for real-time monitoring of the steady-state levels of specific mRNA molecules in the cytoplasm of cultured PtK2 cells.[31] A series of MBs having variable stem–loop sequences with tetramethylrhodamine (TMR) as a fluorophore and DABCYL as a quencher has been designed to serve as hybridization probes and control probes for the PtK2 cell β-actin mRNA detection. Picoliter amounts of MB solution were directly delivered into the cytoplasm of the cells using an Eppendorf microinjection system. Optical images and fluorescence images before and after the MB injection were taken by fluorescence microscopy equipped with an intensified charged-couple device (ICCD) and an argon ion laser excitation system.

Figure 57.6 shows a set of typical images collected for a MB probe hybridized with the β-actin mRNA in one cell. The fluorescence intensity of the cell continuously increased with the hybridization time for the first 15 min and then remained relatively constant for the next 25 min. To determine if the fluorescence-intensity increase represented hybridization of the MB to the target β-actin mRNA, control experiments were conducted as follows. First, control MBs were injected into different PtK2 cells, and all the cells consistently showed no increase in signal intensity over a period of 15 min. Second, the cDNAs of the control MBs were injected into the same respective cells with their corresponding control MBs, and all the cells showed immediate fluorescence enhancement. This clearly shows that the control MBs remained closed after the first injection due to the lack of their specific targets in the cytoplasm of the PtK2 cells. It is also interesting to note that there was a variation in the MBs' relative fluorescence intensity during the same period of hybridization among different cells. The most likely explanation for this difference between cells is different concentrations of β-actin mRNA presented in the cytoplasm of these individual cells.

The results presented in cell mRNA monitoring clearly show that at the single-cell level (1) MBs can be used to successfully monitor, in real time, the hybridization to specific mRNA molecules in the cellular environment and (2) the optical imaging system can differentiate between background fluorescence due to incomplete quenching of control MBs and true signal resulting from MBs designed to hybridize to specific targets.

Using fluorescently labeled DNA probes, researchers are able to investigate the expression of human c-fos mRNA in a living transfected Cos7 cell.[32] Human c-fos mRNA was observed by detecting a hybrid formed with two fluorescently labeled oligodeoxynucleotides and c-fos mRNA in the cytoplasm. Two fluorescent DNAs were prepared, each labeled with a fluorescence molecule different from the other. The FRET occurred when two DNA strands hybridized to an adjacent sequence on the target mRNA. To find sequences of high accessibility of c-fos RNA to the DNAs, several sites that included loop structures on the simulated secondary structure were selected. Each site was examined for the efficiency of hybridization to c-fos RNA by measuring changes in fluorescence spectra when c-fos RNA was added to the pair of the DNAs in solution. A 40mer-specific site was found, and the pair of DNAs for the site was microinjected into Cos7 cells that expressed c-fos mRNA. Hybridization of the pair of the DNAs to c-fos mRNA in the cytoplasm was detected in fluorescence images, indicating FRET. To block the DNAs from accumulating in the nucleus, the DNA probes were bound to a macromolecule (such as streptavidin) to prevent passage to the nuclear pores.

## 57.4.3 MBs for Protein Recognition and Protein/DNA Interaction Study

### 57.4.3.1 Nonspecific Protein-Binding Study

While MB probes were originally designed for nucleic-acid studies, their hairpin structure can also be disturbed to restore fluorescence upon binding to some proteins. The introduction of MBs for protein–DNA interaction studies will be of great interest in understanding many important biological processes and in ultrasensitive protein detection, as proteins play critical roles in many biological processes. This protein-recognition ability was first realized using an *E. coli* SSB.[33] The fluorescence enhancement caused by SSB and by a cDNA target is very comparable (Figure 57.7A). Using MB–SSB binding, it was possible to detect SSB at a concentration as low as $2 \times 10^{-10}$ mol/l using a conventional spectrometer with a mercury lamp. The interaction between the SSB and MB was found to be much faster than that between the cDNA and MB (Figure 57.7B). The fast speed of the protein–DNA binding reaction will provide the basis for rapid protein assays. In addition, there are significant differences in MB binding affinity with different proteins such as albumin, histone, etc. This will lead to the exploration of the potential for selective binding studies of a variety of proteins using designer MBs.

A MB DNA probe has also been used for detailed binding studies of an enzyme, lactate dehydrogenase (LDH).[34] The fluorescence signal of the MB was increased upon interaction with LDH, which was then used for the elucidation of the binding properties and the study of the binding process. Different LDH isoenzymes were found to have different ssDNA binding affinities. The results showed that the stoichiometry of LDH-5/MB binding was 1:1, and the binding constant was $1.9 \times 10^{-7}$ mol/l. Detailed studies of LDH/MB binding, such as salt effects, temperature effects, pH effects, binding sites, binding specificity for different isoenzymes, and competitive binding with different substrates were carried out by a simple fluorescent method using the MB probe. The possibility of using MB probes for quantitative protein detection with ultrasensitivity shows great promise in understanding many important biological processes involving two key biomolecules: nucleic acids and proteins. Although only nonspecific DNA-binding proteins have been investigated so far,[33–35] the method opens the possibility for further development of easily obtainable, modified DNA molecules for real-time specific-protein detection. On the other hand, the study of the nonspecific DNA-binding proteins is also very important. One example is the development of an easy and efficient DNA cleavage enzyme assay using MBs[35] (see below).

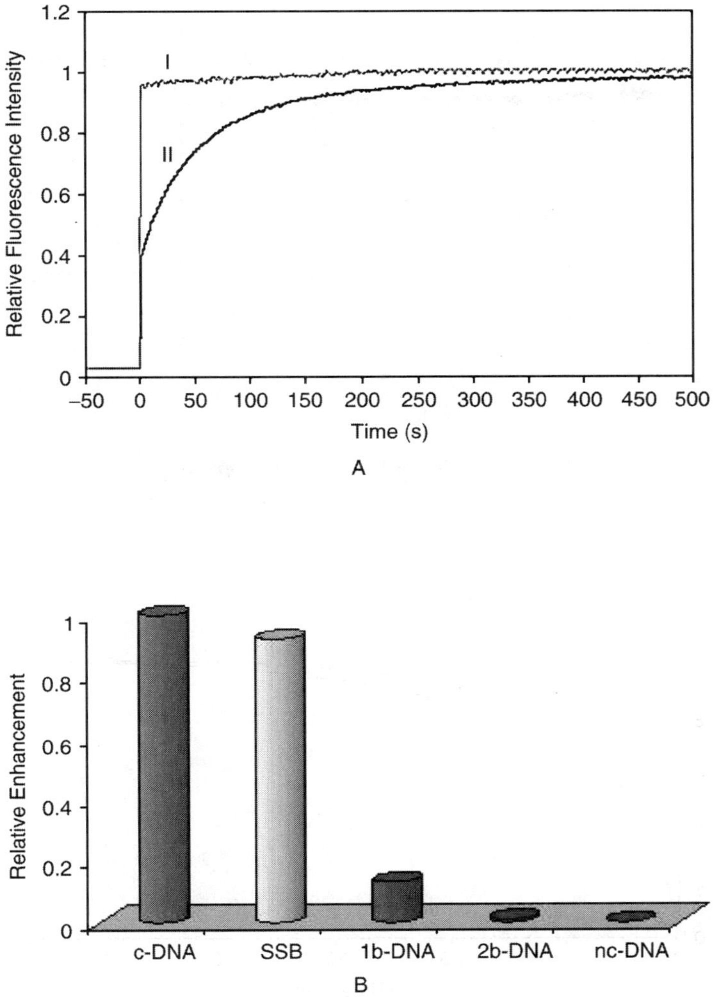


**FIGURE 57.7** (A) The relative fluorescence-intensity time scan of the MB binding reaction with cDNA (I) and SSB (II). The molecular mole ratio is cDNA:MB = 1:1, and SSB:MB is also 1:1. cDNA: cDNA; 1b-DNA: one-base mismatched DNA; 2b-DNA: two-base mismatched DNA; nc-DNA: non-cDNA; SSB: single-stranded binding protein. (B) The relative fluorescence enhancement of the molecular beacon DNA probe when hybridized with different DNA molecules and interaction with SSB. The MB used here is 5′-6FAM-GCTCG TCC ATA CCC AAG AAG GAA G CGAGC-DABCYL-3′.

### 57.4.3.2 Real-Time Enzymatic Cleavage Assay

Traditional methods to assay enzymatic cleavage of ssDNA are discontinuous and time-consuming. The lack of suitable fluorescent probes is an obstacle in the development of fluorescence methods that are continuous and convenient. Based on MB probes, a novel method has been proposed to assay the ssDNA cleavage reaction by single-stranded specific nuclease.[35] The single-stranded nuclease binds to and cleaves the single-stranded loop portion of the MB. The cleavage results in the dissociation of the stem because the five to seven base pairs in the stem are unstable at the cleavage temperature (37°C) when the loop is broken. Consequently, the fluorophore and quencher are completely separated from each other, giving rise to an irreversible fluorescence enhancement that is higher than that caused by the MB's cDNA. Figure 57.8 shows that there is good agreement in the nuclease assay between traditional gel electrophoresis and the fluorescence assay based on MBs. The fluorescence method permits real-time monitoring

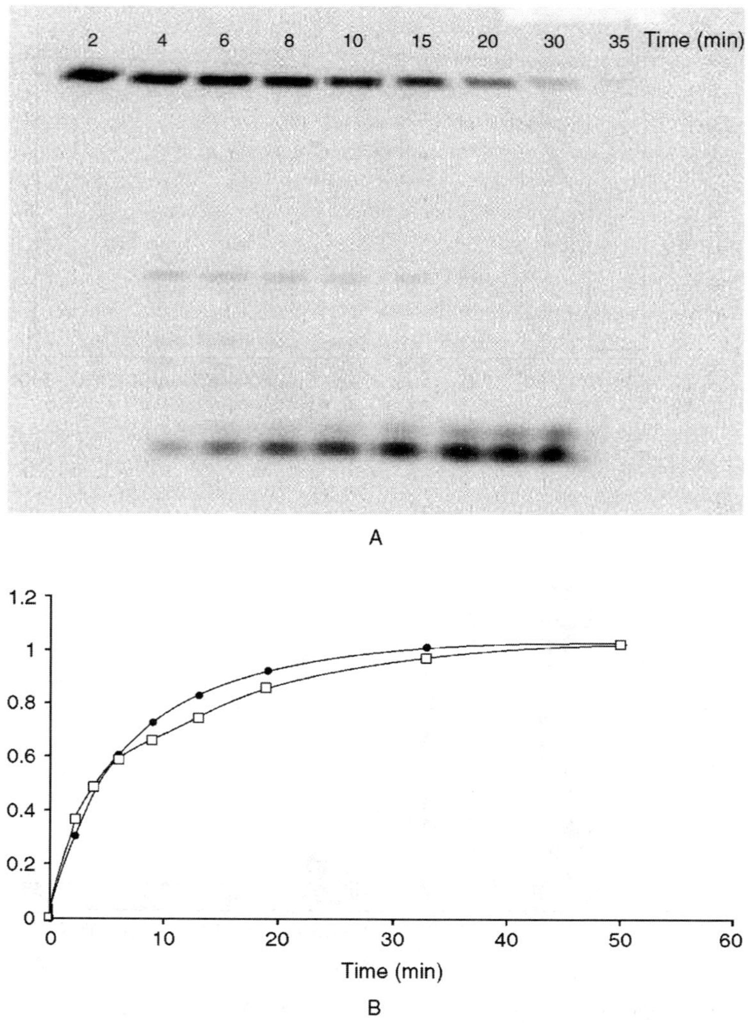


**FIGURE 57.8** The comparison between fluorescence assay and gel electrophoresis assay for the cleavage reaction of the molecular beacon by S1 nuclease. In the fluorescence assay, the time course was recorded for the cleavage reaction. In each scheduled time point, a fixed amount of reaction solution was taken out for gel electrophoresis assay. (A) Polyacrylamide-gel electrophoresis assay. The samples were run on denaturing polyacrylamide gel. The upper bands represent the intact MB, and reaction time points (minutes) are labeled above the bands. (B) Comparing the results on MB cleavage percentage obtained by the two assays. The data for fluorescence assay curve (●) was extracted from the reaction time-course curve; the cleavage percentage in electrophoresis assay (□) at each time point was determined by quantifying the fluorescence decrease of the substrate relative to the total fluorescence.

of the enzymatic cleavage-reaction process, easy characterization of the activity of DNA nucleases, and the study of steady-state cleavage-reaction kinetics. Due to their high sensitivity, reproducibility, and convenience, MBs have been used to observe the study of ssDNA cleavage reactions.

Other cleavage experiments were done as well. For example, efforts have focused on developing imaging probes that can be activated to measure specific enzyme activities *in vivo*. Using cathepsin D as a model target protease, Tung et al. synthesized a long-circulating, synthetic graft copolymer bearing near-infrared (NIR) fluorochromes positioned on cleavable substrate sequences.[36] In its native state, the reporter probe was essentially nonfluorescent at 700 nm due to energy-resonance transfer among the bound fluorochromes (quenching) but became brightly fluorescent when the latter was released by cathepsin D. Using matched rodent tumor models implanted into nude mice expressing or lacking the targeted protease, it could be shown that the former generated sufficient NIR signal to be directly detectable and that the signal was significantly different compared with negative control tumors.

Because MBs carry an appropriate cleavage site within the stem, it has been applied to develop the continuous assay for cleavage of DNA by enediynes. The generality of this approach is demonstrated by using the described assay to directly compare the DNA cleavage by naturally occurring enediynes, nonenediyne small-molecule agents, as well as the restriction endonuclease BamHI.[37] Meanwhile, MBs were used to quantify low levels of type I endonuclease activity.[38] Given the simplicity, speed, and sensitivity of this approach, the described methodology could easily be extended to a high-throughput format and become a new method of choice in modern drug discovery to screen for novel protein-based or small-molecule-derived DNA-cleavage agents.

#### 57.4.3.3 Protein Detection with Specificity

The above nonspecific DNA-binding-protein study opens the possibility for the further development of easily obtainable, modified DNA molecules for real-time specific-protein detection. We have recently developed a new molecular-recognition mechanism by combining the molecular beacon's excellent signal-transduction mechanism with an aptamer's specificity in protein binding for a novel approach to real-time protein detection.[39]

In most biomedical applications, high-affinity recognition and specific-protein recognition are accomplished by using antibodies. Dye-labeled antibodies are often used to detect specific proteins *in vitro*. Although antibodies are highly specific in molecular recognition, they have limitations. Aptamers are a new class of designer DNA/RNA molecules that compete with the antibody in protein recognition.[40,41] Aptamers have many advantages over antibodies: easier synthesis, easier labeling, better reproducibility, easier storage, faster tissue penetration, and shorter blood residence. However, an aptamer itself cannot be used as a fluorescent probe because it lacks the signal-transduction capability to report the binding to a target. The new protein probe, which combines the binding specificity and generality of aptamers with the excellent signal-transduction capability of MBs, has great potential in monitoring protein production in living cells.

The feasibility of this approach has been demonstrated by the successful development of a MB aptamer (MBA) probe for a model protein, thrombin. A MBA probe is constructed by engineering a 17mer thrombin-binding aptamer with a fluorophore and a quencher attached at its two ends. Upon binding to the thrombin, the MBA experiences a significant conformational change, from a loose random coil to a compact unimolecular quadruplex in a solution with low salt concentration. The conformational change brought about by the specific binding is converted to a significant fluorescence signal change of the MBA. This signal change is large enough for sensitive thrombin detection with a detection limit in the sub-nM range.

The MBA–thrombin binding is highly specific, both in target specificity and aptamer-sequence specificity. The aptamer's inherent specificity has been retained in the MBA for the thrombin. To achieve better protein quantitation in living cells, a two-fluorophore MBA based on FRET has been developed for ratiometric imaging.[42] Labeled with two fluorophores, such as coumarin and fluorescein, the MBA "lights up" upon thrombin binding. Comparing this to the direct intensity imaging, the ratio imaging gives a much-improved signal-to-background ratio and hence higher sensitivity and better reliability for precise determination of the proteins desired.

The aptamer beacon has become a sensitive tool for detecting proteins and other chemical compounds. For example, Stanton et al. have designed aptamer beacons[43] for detecting a wide range of ligands. A fluorescence-quenching pair was used to report changes in conformation induced by ligand binding. An antithrombin aptamer was engineered into an aptamer beacon by adding nucleotides to the 5′ end that were complementary to nucleotides at the 3′ end of the aptamer. In the absence of thrombin, the added nucleotides formed a duplex with the 3′ end, forcing the aptamer beacon into a stem–loop structure. In the presence of thrombin, the aptamer beacon forms the ligand-binding structure. This conformational change causes a change in the distance between the fluorophore attached to the 5′ end and a quencher attached to the 3′ end.

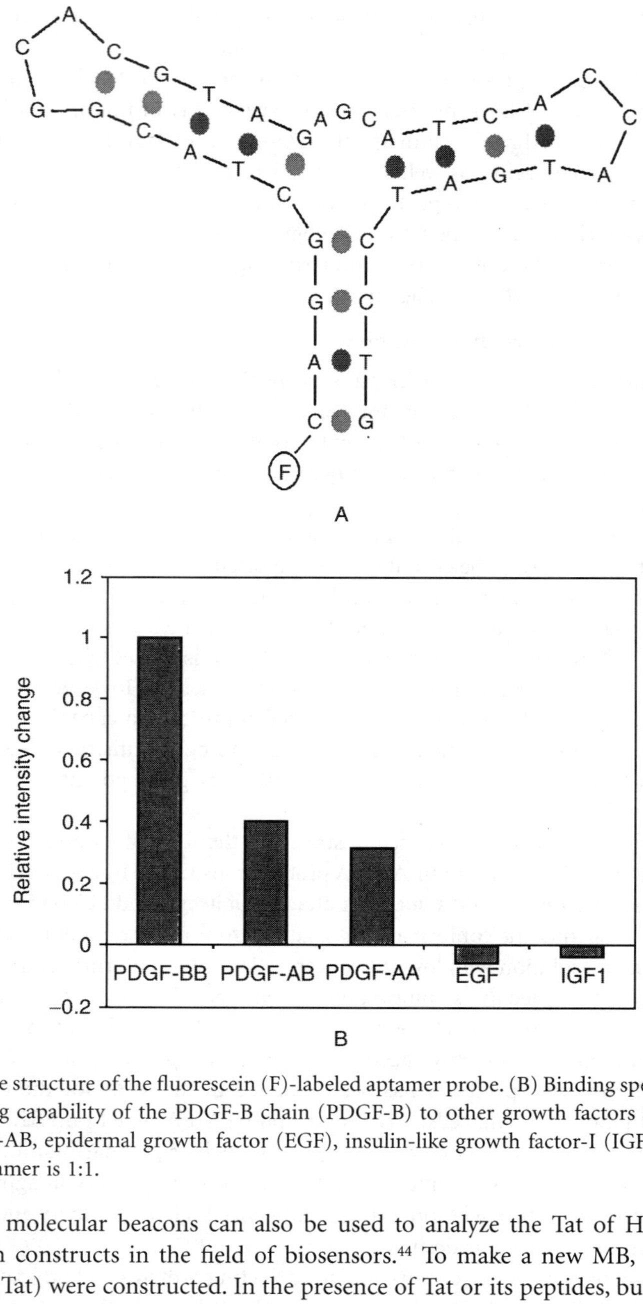


**FIGURE 57.9** (A) The structure of the fluorescein (F)-labeled aptamer probe. (B) Binding specificity of the aptamer, comparing the binding capability of the PDGF-B chain (PDGF-B) to other growth factors such as PDGF isomers PDGF-AA and PDGF-AB, epidermal growth factor (EGF), insulin-like growth factor-I (IGF1). The molar ratio of the protein to the aptamer is 1:1.

Aptamer-derived molecular beacons can also be used to analyze the Tat of HIV and the possible applications of such constructs in the field of biosensors.[44] To make a new MB, two RNA oligomers derived from RNA (Tat) were constructed. In the presence of Tat or its peptides, but not in the presence of other RNA binding proteins, the two oligomers underwent a conformational change to form a duplex that led to the release of fluorophore from the quencher, and thus a significant enhancement of the fluorescence of fluorescein was observed.

Oncoproteins, coded by prooncogene to regulate cell growth and differentiation, play an important role in cancer growth and are often found overexpressed or mutated in malignant tumors. We have designed a new fluorescent probe based on a high-affinity platelet-derived growth factor (PDGF) aptamer (shown in Figure 57.9) for the ultrasensitive detection of oncoprotein PDGF in homogeneous solutions.[45]

The aptamer is labeled with a fluorophore/quencher pair to specifically bind with PDGF. Upon binding, the conformational change of the aptamer results in a FRET-based fluorescence-signal-intensity change of the probe. The aptamer probes are highly sensitive and highly selective. They can detect PDGF in the sub-nM range. Neither the tested extracellular proteins nor other PDGF-related peptide growth factors interfere in the probe binding to PDGF, as shown in Figure 57.9. The new oncoprotein-detection methods are simple and inherit all the advantages of molecular aptamers. They are expected to find wide applications in protein-detection, cancer-diagnosis, and other disease-diagnosis and mechanism studies.

# 57.5 Combining MB and Biosensor Technology

## 57.5.1 Surface-Immobilizable MBs

As the common MB can only be used in homogeneous solution, surface-immobilizable MBs are critical for the development of highly sensitive biosensors for *in vivo* detection and for the study of biomolecular recognition processes at an interface. A biotinylated ssDNA MB has thus been designed.[46] Biotin–avidin binding is one of the most common ways for biomolecule immobilization onto a solid surface and is suitable for DNA hybridization. There are several important considerations regarding the design of biotinylated MB synthesis.[5] Of them, the position of the biotin is very important and should be carefully chosen to minimize the effect of the avidin–biotin bridge on the MB hybridization. It is desirable to add the biotin functional group at the quencher side in the stem of a MB. A spacer between the biotin and the sequence should be added to provide adequate separation to minimize potential interactions between avidin and the DNA sequence. A photostable dye, such as TMR, should be chosen to minimize photobleaching as only a small amount of fluorophores are immobilized on a surface. To immobilize biotinylated MBs, a silica surface is first physically or covalently coated with avidin.[46–49] The biotinylated ssDNA MB then binds to avidin. The binding process is fast and stable. The hybridization properties of MBs on the surface are similar to those in solution.

A method was developed for rapid detection of PCR amplicons based on surface-immobilized PNA–DNA hybrid probes that undergo a fluorescent-linked conformational change in the presence of a cDNA target.[50] Amplicons can be detected by simply adding a PCR to a microtiter-well containing the previously immobilized probe and reading the generated fluorescence. No further transfers or washing steps are involved. The specificity of the method for the detection of ribosomal DNA from *Entamoeba histolytica* was excellent.

## 57.5.2 Optical-Fiber DNA Biosensors Using Surface-Immobilizable MBs

Immobilized MBs have been used for the preparation of optical-fiber DNA biosensors such as submicrometer biosensors and fiber-optic evanescent-wave biosensors.[47,48] The evanescent-wave biosensor was prepared by exposing the core surface of an optical fiber to chemical etching. An evanescent wave generated on the core surface was used for fluorescence excitation in the longitudinal surface of the fiber where the MBs were immobilized (Figure 57.10A).

The microscopic optical-fiber probe was fabricated using either pulling or etching technologies.[51–56] MBs were immobilized only at the small tip (submicromenter in diameter) of the probe. A highly sensitive optical imaging and detection system with an avalanche photon diode or an ICCD[6] was used for the detection of the fluorescence signal from the optical-fiber biosensor. The biosensors were used for the detection of nonlabeled DNA targets in real time and with high sensitivity and one-base-mismatch selectivity. As shown in Figure 57.10B, there was a linear relationship between the initial hybridization reaction rate and the concentration of the cDNA for the submicrometer DNA biosensor. The concentration-detection limit of the target cDNA was 0.3 n*M* for the ultrasmall DNA biosensor. The sensors are stable and reproducible and have remote detection capability. They can be easily regenerated by a 1-min rinse with a 90% formamide solution.[47] They have been applied to the quantitative detection of

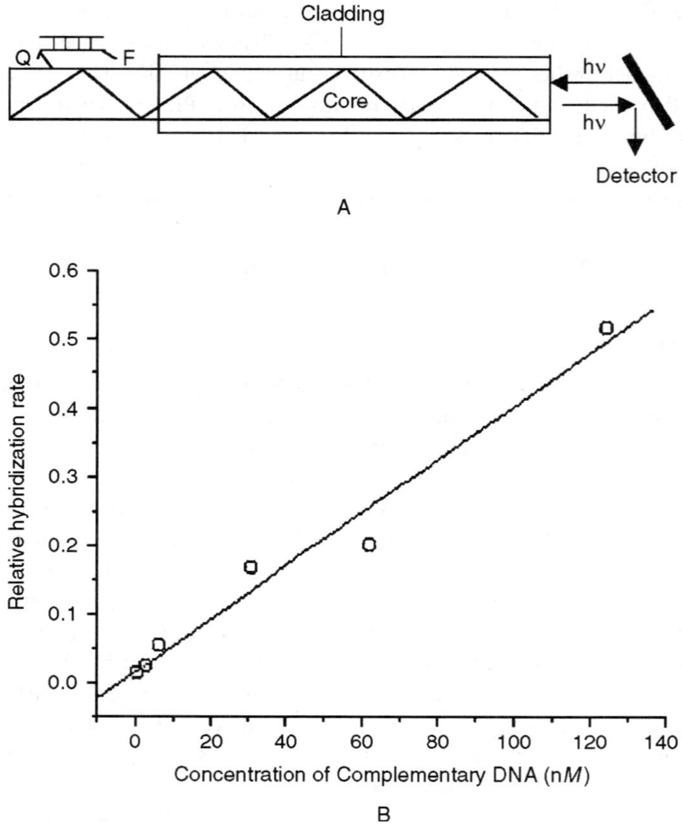


**FIGURE 57.10** (A) Configuration of the evanescent-wave fiber-optical DNA biosensor based on MBs. The MBs are excited by the evanescent wave generated at the core surface of the fiber. The returning fluorescence then travels through the same fiber for detection. (B) Application of the DNA biosensor: linear relationship between the hybridization rate and the concentration of the target DNA.

mRNA sequences and to the study of DNA-hybridization kinetics.[47,48] They hold the potential of direct detection of DNA/RNA targets in living cells without DNA/RNA amplification.

The immobilization method for biotinylated MBs has also enabled the exploration of MB probe arrays for simultaneous multiple-analyte detection.[49,57] MBs with various loop sequences can be immobilized onto the tips of different microscopic-fiber probes.[47] Those biosensors are individually addressable with a spatially resolved imaging system based on the ICCD camera.[6]

We have proposed a bridge structure to immobilize the biotinylated MB.[58] The MB was biotinylated at the quencher side of the stem and linked on a biotinylated glass cover slip through streptavidin, which acted as a bridge between the MB and the glass matrix. An efficient fluorescence-microscope system was constructed to detect the fluorescence change caused by the conformation change of MB in the presence of cDNA target. The proposed biosensor was used to directly detect, in real-time, the target-DNA molecules. The bridge-immobilization method caused the proposed DNA biosensor to have a faster and more stable response. Many efforts were made to develop a more sensitive and stable DNA biosensor. The sol-gel method was shown to be a very efficient approach to immobilize biotinylated MBs.[59] The sol-gel matrix provided a solid support for the immobilized MB, which can be doped on the surface of many materials. This will widen the application of MB in clinical detection and biotechnology. The potential application of DNA biosensors includes the fabrication of submicrometer optic-fiber probes to detect mRNA inside living cells.[60]

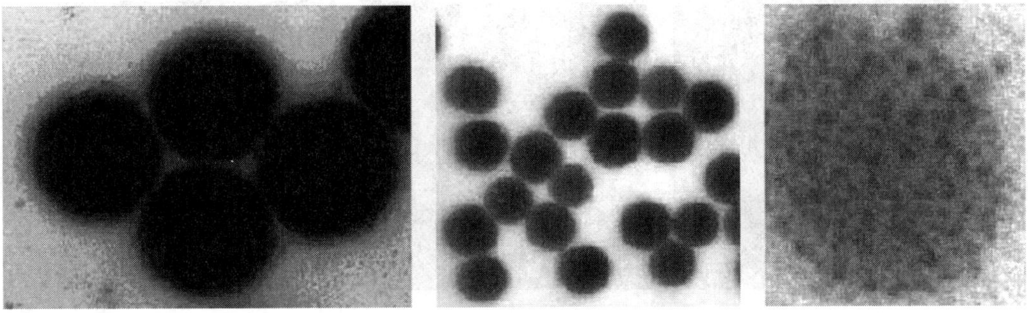

**FIGURE 57.11** TEM images of dye-doped silica-coated nanoparticles: (left) 400 ± 10 nm; (center) 63 ±4 nm; (right) 2 nm. The particles are uniform and have a silica surface for MB immobilization.

### 57.5.3 Nanoparticle MB Biosensor

We have prepared MB sensors based on silica nanoparticles.[61,62] Tens of thousands of uniform nanoparticles have been successfully immobilized with MB. These particles can be attached to the tips of optical fibers for the preparation of a large quantity of fiber probes or introduced into cells as individual sensors. Recently, a gene array has been developed based on the MB using particles and an imaging fiber bundle.[57] Different MB-coated microspheres are randomly distributed in an array of wells etched in a 500-μm diameter optical imaging fiber bundle. To recognize different DNA targets, an optical encoding scheme and an imaging fluorescence-microscope system were used for positional registration and fluorescence-response monitoring. We have also used nanoparticles for MB biosensor development as shown in Figure 57.11. These nanoparticle-based MB biosensors are highly efficient in hybridization as they have a large surface-to-volume ratio.

### 57.5.4 MB Microarray Sensor and Biochip

The multianalyte DNA sensors are expected to provide an easy and fast way for high-throughput gene analysis and disease diagnosis.[57] A methodology for the implementation of multiplexed nucleic-acid hybridization fluorescence assays on microchannel glass substrates. Fluorescence detection was achieved using conventional low-magnification-microscope objective lenses as imaging optics whose depth-of-field characteristics matched the thickness of the microchannel glass chip. The optical properties of microchannel glass were shown, through experimental results and simulations, to be compatible with the quantitative detection of heterogeneous hybridization events taking place along the microchannel sidewalls, with detection limits for oligonucleotide targets in the low-attomole range.[63] A nanoscopic well system for MB biosensors and biochips has also been developed, as shown in Figure 57.12. The system is being used for multiple gene analysis. Overall, a variety of biosensor formats based on MBs have been developed, and their application is just on the horizon.

## 57.6 Prospects

As the human-genome-sequencing project comes to fruition, there will be a resulting change in the focus of quantitative studies of genomic information from the collecting and archiving of genomic data to their analysis and use in prediction and discovery. The key to this new era is research across many disciplinary interfaces and the development and use of new quantitative tools. MBs are ideally suited for and hold great promise in genomics and proteomics. While there are many possibilities and great potentials, we believe that the following three areas are of greatest interest to the bioanalytical sciences and are feasible in the near future. First, more research is expected for MB application in mutation detection for a variety of disease diagnostics and disease-mechanism studies. Efforts will also be made

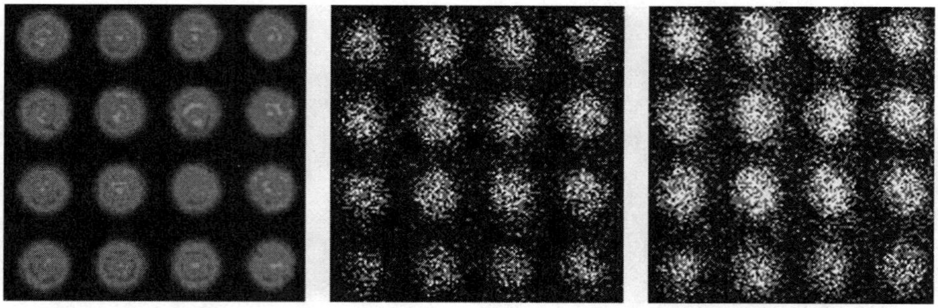

**FIGURE 57.12** The microwell array was made by photolithography. A group of consecutive images was taken as the MB hybridization proceeded. A fixed 1-min interval between two images was used. The figure shows that from the beginning to the end of the hybridization experiment more and more MBs hybridize spontaneously to their targets and undergo a conformational change that results in the generation of fluorescence.

in finding suitable mRNA sequences and real-time studies of RNA processing in living cells and other living specimens without using amplification techniques. This will be expedited with the development of MB biosensors and highly sensitive optical-detection methods.[6,64] Gene expression under different conditions will be studied with precise quantitation. Second, the extraordinarily target-specific capability and the availability of different fluorophore–quencher pairs make MB probes extremely useful for multiple-analyte applications. The detection of many different targets in the same solution can be achieved by the simultaneous use of several MBs, each of which emits light of a different wavelength or which can be excited by light of a different wavelength. These excellent properties also make molecular beacons exquisitely suited for use in the identification of genetic alleles or particular strains of infectious agents. The multiple-MB-probe approach will also be highly useful in combinatorial chemistry for high-throughput-methodology development for drug design and for molecular diagnosis of a variety of genetic diseases, especially with the abundance of genetic information available in the post-genome era. Third, the MB's application as a novel biomolecular recognition agent for proteins will be further explored. With further development, MBs are expected to be useful as intracellular protein-recognition agents to probe proteins in different environments and to monitor protein-DNA/RNA interactions. Preliminary results show significant differences in MB binding affinities by different proteins, which will constitute the basis for highly selective bioassays of a variety of proteins. Approaches to this goal include using designer DNA molecules or aptamer-based[40] MBs for better specificity and using a large array of MB biosensors for pattern recognition. Better understanding of the conformational kinetics of MBs[8] and their fluorescence-energy-transfer mechanism[4,56] will result in optimally designed MB probes for biomolecular recognition with high sensitivity and excellent specificity. All of these developments will open the possibility of using easily obtainable and designer DNA molecules for genomics and proteomics studies, for molecular diagnosis of diseases, and for new drug development.

## Acknowledgments

We thank our colleagues for their help. This work was partially supported by NIH R01-NS39891, NCI R21-CA92581, and NIH NIGMS R01-GM66137.

## References

1. Haugland, R.P., *Handbook of Fluorescent Probes and Research Chemicals*, Molecular Probes, Eugene, OR, 1994.
2. Tyagi, S. and Kramer, F.R., Molecular beacons: probes that fluoresce upon hybridization, *Nat. Biotech.*, 14, 303, 1996.
3. Lakowicz, J.R., *Principles of Fluorescence Spectroscopy*, 2nd ed., Kluwer Academic/Plenum Publishers, New York, 1999.

4. Tyagi, S., Bratu, D.P., and Kramer, F.R., Multicolor molecular beacons for allele discrimination, *Nat. Biotechnol.*, 16, 49, 1996.
5. Meinkoth, J. and Wahl, G., Hybridization of nucleic acids immobilized on solid supports, *Anal. Biochem.*, 138(2), 267, 1984.
6. Fang, X. and Tan, W., Imaging single fluorescent molecules at the interface of an optical fiber probe by evanescent wave excitation, *Anal. Chem.*, 71(15), 3101, 1999.
7. Kostrikis, L.G., Tyagi, S., Mhlanga, M.M., Ho, D.D., and Kramer, F.R., Molecular beacons–spectral genotyping of human alleles, *Science*, 279, 1228, 1998.
8. Bonnet, G., Tyagi, S., Libchaber, A., and Kramer, F.R., Thermodynamic basis of the enhanced specificity of structured DNA probes, *Proc. Natl. Acad. Sci. U.S.A.*, 96(11), 6171, 1999.
9. Matsuo, T., *In situ* visualization of mRNA for basic fibroblast growth factor in living cells, *Biochim. Biophys. Acta*, 1379, 178, 1998.
10. Tapp, I., Malmberg, L., Rennel, E., Wik, M., and Syvanen, A.C., Homogeneous scoring of single-nucleotide polymorphisms: comparison of the 5′-nuclease TaqMan assay and molecular beacon probes, *Biotechniques*, 28(4), 732, 2000.
11. Helps, C., Reeves, N., Tasker, S., and Harbour, D., Use of real-time quantitative PCR to detect *Chlamydophila felis* infection, *J. Clin. Microbiol.*, 39(7), 2675, 2001.
12. Yuan, C.C., Peterson, R.J., Wang, C.D., Goodsaid, F., and Waters, D.J., 5′ Nuclease assays for the loci CCR5- +/Δ32, CCR 2-V64I, and SDF 1-G801A related to pathogenesis of AIDS, *Clin. Chem.*, 46(1), 24, 2000.
13. Lewin, S.R., Vesanen, M., Kostrikis, L., Hurley, A., Duran, M., Zhang, L., Ho, D.D., and Markowitz, M., Use of real-time PCR and molecular beacons to detect virus replication in human immunodeficiency virus type 1-infected individuals on prolonged effective antiretroviral therapy, *J. Virol.*, 73(7), 6099, 1999.
14. Zhang, L.Q., Lewin, S.R., Markowitz, M., Lin, H.H., Skulsky, E., Karanicolas, R., He, Y., Jin, X., Tuttleton, S., Vesanen, M.., Spiegel, H., Kost, R., van Lunzen, J., Stellbrink, H.J., Wolinsky, S., Borkowsky, W., Palumbo, P., Kostrikis, L.G., and Ho, D.D., Measuring recent thymic emigrants in blood of normal and HIV-1–infected individuals before and after effective therapy, *J. Exp. Med.*, 190(5), 725, 1999.
15. Vet, J.A.M., Majithia, A.R., Marras, S.A.E., Dube, S., Poiesz, B.J., and Kramer, F.R., Multiplex detection of four pathogenic retroviruses using molecular beacons, *Proc. Natl. Acad. Sci. U.S.A.*, 96(11), 6394, 1999.
16. Giesendorf, B.A.J, Vet, J.A.M., Tyagi, S., et al., Molecular beacons: a new approach for semiautomated mutation analysis, *Clin. Chem.*, 44(3), 482, 1998.
17. Chen, W. and Martinez, G., Molecular beacons: a real-time polymerase chain reaction assay for detecting salmonella, *Anal. Biochem.*, 280(1), 166, 2000.
18. Schofield, P., Pell, A.N, and Krause, D.O., Molecular beacons: trial of a fluorescence-based solution hybridization technique for ecological studies with ruminal bacteria, *Appl. Environ. Microbiol.*, 63(3), 1143, 1997.
19. Rhee, J.T., Piatek, A.S., Small, P.M, Harris, L.M., Chaparro, S.V., Kramer, F.R., and Alland, D., Molecular epidemiologic evaluation of transmissibility and virulence of *Mycobacterium tuberculosis*, *J. Clin. Microbiol.*, 37(6), 1764, 1999.
20. De Ronde, A., van Dooren, M., van der Hoek, L., Bouwhuis, D., De Rooij, E., van Gemen, B., De Boer, R., and Goudsmit, J., Establishment of new transmissible and drug-sensitive human immunodeficiency virus type 1 wild types due to transmission of nucleoside analogue-resistant virus, *J. Virol.*, 75(2), 595, 2001.
21. Martinson, J.J., Hong, L., Karanicolas, R., Moore, J.P., and Kostrikis, L.G., Global distribution of the CCR2–64I/CCR5–59653T HIV-1 disease-protective haplotype, *AIDS*, 14(5), 483, 2000.
22. De Baar, M.P., Timmer, E.C., Bakker, M., De Rooij, E., van Gemen, B., and Goudsmit, J., One-tube real-time isothermal amplification assay to identify and distinguish human immunodeficiency virus type 1 subtypes A, B, and C and circulating recombinant forms AE and AG, *J. Clin. Microbiol.*, 39(5), 1895, 2001.

23. De Baar, M.P., van Dooren, M.W., De Rooij, E., Bakker, M., van Gemen, B., Goudsmit, J., and De Ronde, A., Single rapid real-time monitored isothermal RNA amplification assay for quantification of human immunodeficiency virus type 1 isolates from groups M, N, and O, *J. Clin. Microbiol.*, 39(4), 1378, 2001.
24. McKillip, J.L. and Drake, M., Molecular beacon polymerase chain reaction detection of *Escherichia coli* O157:H7 in milk, *J. Food Prot.*, 63(7), 855, 2000.
25. Li, Q.G., Liang, J.X., Luan, G.Y., Zhang, Y., and Wang, K., Molecular beacon-based homogeneous fluorescence PCR assay for the diagnosis of infectious diseases, *Anal. Sci.*, 16(2), 245, 2000.
26. Fortin, N.Y., Mulchandani, A., and Chen, W., Use of real-time polymerase chain reaction and molecular beacons for the detection of *Escherichia coli* O157 : H7, *Anal. Biochem.*, 289(2), 281, 2001.
27. Chen, W., Martinez, G., and Mulchandani, A., Molecular beacons: a real-time polymerase chain reaction assay for detecting salmonella similar species such as *Escherichia coli* and *Citrobacter freundii* in real time PCR assays, *Anal. Biochem.*, 280(1), 166, 2000.
28. Fang, X.H., Mi, Y.M., Li, J.J., Terry, B., and Tan, W.H., Real-time imaging of mRNA in single living cells using molecular beacons, *Molecular Beacons: "Lights up" Probes for Living Cell Study, Cell Biochem. Biophys.*, 37, 4, 2002.
29. Dirks, R.W., Molenaar, C., and Tanke, H.J., Methods for visualizing RNA processing and transport pathway in living cells, *Histochem. Cell. Biol.*, 15, 3, 2001.
30. Matsuo, T., *In situ* visualization of messenger RNA for basic fibroblast growth factor in living cells, *Bba-gen Subjects*, 1379(2), 178, 1998.
31. Sokol, D.L., Zhang, X.L., Lu, P.Z., and Gewitz, A.M., Real time detection of DNA RNA hybridization in living cells, *Proc. Natl. Acad. Sci. U.S.A.*, 95, 11538, 1998.
32. Tsuji, A., Koshimoto, H., Sato, Y., Hirano, M., Sei-Iida, Y., Kondo, S., and Ishibashi, K., Direct observation of specific messenger RNA in a single living cell under a fluorescence microscope, *Biophys. J.*, 78(6), 3260, 2000.
33. Li, J., Fang, X., Schuster, S., and Tan, W., Molecular beacons: a novel approach to detect protein-DNA interactions, *Angew. Chem. Int. Ed.*, 39, 1049, 2000.
34. Fang, X., Li, J., and Tan, W., Using molecular beacons to probe the interactions between lactic dehydrogenase and single strand DNA, *Anal. Chem.*, 72(14), 3250, 2000.
35. Li, J., Geyer, R., and Tan, W., Using molecular beacons as a sensitive fluorescence assay for enzymatic cleavage of single-stranded DNA, *Nucleic Acids Res.*, 28(e52), 1, 2000.
36. Tung, C.H., Mahmood, U., Bredow, S., and Weissleder, R., *In vivo* imaging of proteolytic enzyme activity using a novel molecular reporter, *Cancer Res.*, 60(17), 4953, 2000.
37. Biggins, J.B., Prudent, J.R., Marshall, D.J., Ruppen, M., and Thorson, J.S., A continuous assay for DNA cleavage: the application of "break lights" to enediynes, iron-dependent agents, and nucleases, *Proc. Natl. Acad. Sci. U.S.A.*, 97(25), 13537, 2000.
38. Strouse, R.J., Hakki, F.Z., Wang, S.C., DeFusco, A.W., Garrett, J.L., and Schenerman, M.A., Using molecular beacons to quantify low levels of type I endonuclease activity, *Biopharm. Appl. Technol. Biopharm. Dev.*, 13(4), 40, 2000.
39. Li, J.J., Fang, X., and Tan, W., Molecular aptamer beacon: a novel molecular probe for real time protein recognition, *Biochem. Biophys. Res. Commun.*, 292(1), 31, 2002.
40. Osborne, S.E., Matsumura, I., and Ellington, A.D., Aptamers as therapeutic and diagnostic reagents: problems and prospects, *Curr. Opinion Chem. Biol.*, 1, 5, 1997.
41. Jayasena, S.D., Aptamers: an emerging class of molecules that rival antibodies in diagnostics, *Clin. Chem.*, 45, 1628, 1999.
42. Zhang, P., Beck, T., and Tan, W., Design of molecular beacon DNA probe with two fluorophores, *Angew. Chem. Int. Ed.*, 40(2), 402, 2000.
43. Hamaguchi, N., Ellington, A., and Stanton, M., Aptamer beacons for the direct detection of proteins, *Anal. Biochem.*, 294(2), 126, 2001.
44. Yamamoto, R. and Kumar, P.K.R., Molecular beacon aptamer fluoresces in the presence of Tat protein of HIV-1, *Genes Cells*, 5(5), 389, 2000.

45. Fang, X., Vicens, M., Cao, Z., and Tan, W., Signaling oncoprotein by molecular beacon aptamer fluorescent probe, *J. Am. Chem. Soc.*, submitted.
46. Fang, X., Liu, X., Schuster, S., and Tan, W., Designing a novel molecular beacon for surface-immobilized DNA hybridization studies, *J. Am. Chem. Soc.*, 121, 2921, 1999.
47. Liu, X., Farmerie, W., Schuster, S., and Tan, W., Molecular beacons for DNA biosensors with micrometer to submicrometer dimensions, *Anal. Biochem.*, 283, 56, 2000.
48. Liu, X. and Tan, W., A fiber-optic evanescent wave DNA biosensor based on novel molecular beacons, *Anal. Chem.*, 71, 5054, 1999.
49. Fang, X., Liu, X., and Tan, W., Single and multiple molecular beacon probes for DNA by hybridization studies on a silica glass surface, *Proc. SPIE,* 3602, 149, 1999.
50. Ortiz, E., Estrada, G., and Lizardi, P.M., PNA molecular beacons for rapid detection of PCR amplicons, *Mol. Cell. Probes,* 12(4), 219, 1998.
51. Tan, W., Shi, Y., Smith, S., Birnbaum, D., and Kopelman, R., Submicrometer intracellular chemical optical fiber sensors, *Science*, 258, 778, 1992.
52. Zeisel, D., Dutoit, B., Deckert, V., Roth, T., and Zenobi, R., Optical spectroscopy and laser desorption on a nanometer scale, *Anal. Chem.*, 69, 749, 1997.
53. Perlette, J. and Tan, W., Real time detection of molecular beacon mRNA hybridization in single living cell, *Anal. Chem.*, 73, 5631, 2001.
54. Tan, W.H., Optical measurements on the nanometer scale, *Trends Anal. Chem.*, 17, 501, 1998.
55. Tan, W.H., Kopelman, R., Barker, S.L., and Miller, M.T., Ultrasmall optical sensors for cellular measurements, *Anal. Chem.*, 71, 606A, 1999.
56. Zhang, P., Beck, T., and Tan, W., Design of molecular beacon DNA probe with two fluorophores, *Angew. Chem. Int. Ed.*, 40, 402, 2001.
57. Steemers, F.J, Ferguson, J.A., and Walt, D.R., Screening unlabeled DNA targets with randomly ordered fiber-optic gene arrays, *Nat. Biotechnol.*, 18(1), 91, 2000.
58. Li, J., Tan, W., Wang, K., Xiao, D., Yang, X., He, X., and Tang, Z., Ultrasensitive optical DNA biosensor based on surface immobilization of molecular beacon by a bridge structure, *Anal. Sci.*, 17(10), 1149, 2001.
59. Li, J., Tan, W., Wang, K., Yang, X., Tang, Z., and He, X., Optical DNA biosensor based on molecular beacon immobilized on sol-gel membrane, *SPIE Int. Conf. Sensing Unit Sensor Technol.*, Wuhan, China, 2001.
60. Li, J., Tan, W., Wang, K., He, X., Yang, X., and Tang, Z., Cellular application of optical fiber DNA biosensor based on biotinylated molecular beacon by sol-gel immobilization, *Anal. Chem.*, submitted.
61. Santra, S., Tapec, R., Theodoropoulou, N., Dobson, J., Hebard, A., and Tan, W., Synthesis and characterization of silica coated iron oxide nanoparticles in microemulsion: the effect of non-ionic surfactants, *Langmuir,* 17(10), 2900, 2001.
62. Santra, S., Wang, K.M., Tapec, R., and Tan, W., Development of novel dye-doped silica nanoparticles for biomarker application, *J. Biomed. Opt.*, 6(2), 160, 2001.
63. Benoit, V., Steel, A., Torres, M., Lu, Y.Y., Yang, H.J., and Cooper, J., Evaluation of three-dimensional microchannel glass biochips for multiplexed nucleic acid fluorescence hybridization assays, *Anal. Chem.*, 73(11), 2412, 2001.
64. Zhang, P. and Tan, W., Direct observation of single-molecule generation at a solid–liquid interface, *Chem. Eur. J.*, 6, 1087, 2000.

# 58

# Luminescent Quantum Dots as Advanced Biological Labels

Warren C. W. Chan
*University of Toronto*
*Toronto, Canada*

Shuming Nie
*Georgia Tech and Emory University*
*Atlanta, Georgia*



## 58.1 Introduction

The development of optical-detection labels has substantially impacted many areas of biomedical research such as high-throughput drug screening, clinical diagnostics, and *in vivo* monitoring of gene expression and enzyme activity.[1,2] These probes are traditionally based on organic dyes conjugated to biomolecules.[3] Because of their complex molecular structures, however, organic fluorophores often exhibit unfavorable absorption and emission properties such as photobleaching, environmental quenching, broad and asymmetric emission spectra, and the inability to excite more than two to three colors at a single wavelength. These problems can be overcome by exploiting the unique optical properties of metal and semiconductor nanoparticles. In fact, recent research by several groups has linked such nanoparticles to peptides,[4] proteins,[5–7] and DNA[8–13] and has demonstrated their applications in assembling new materials,[14] in homogeneous bioassays,[12–15] and in multicolor fluorescence imaging and detection.[5–10]

Semiconductor nanocrystals, also known as quantum dots, are composed of atoms from group I-VII, II-VI, or III-V elements and are defined as particles with physical dimensions smaller than the exciton Bohr radius.[16,17] For spherical CdSe particles, this occurs when the particle diameter is less than ~10 nm. This quantum-confinement effect gives rise to unique optical and electronic properties that are unavailable in either discrete atoms or in bulk solids. In the last 20 years, significant progress has led to the synthesis of high-quality nanocrystals in large quantities. Prior to 1993, attempts to synthesize quantum dots were conducted in aqueous buffers with added stabilizing agents (e.g., thioglycerol or polyphosphate) or in micelles.[18–22] These procedures yielded quantum dots with low quantum yields ($< 20\%$) and broad size distributions (relative standard deviation or RSD $> 15\%$). In 1993, Bawendi et al. synthesized highly luminescent quantum dots by using a high-temperature organometallic procedure.[23] This method was later improved by three independent research groups.[24–26] These nanocrystals have nearly perfect crystal structures, 30 to 50% quantum yields, and a narrow size range (RSD $< 5\%$). In 1998, Nie, Alivisatos, and their co-workers pioneered the use of these quantum dots as biological labels.[5,6]

0-8493-1116-0/03/$0.00+$1.50

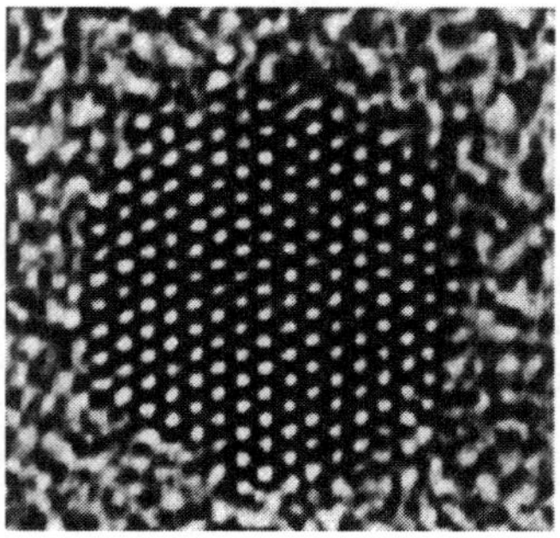

**FIGURE 58.1** High-resolution transmission-electron micrograph of a hexagonally shaped CdSe quantum dot. Note the highly crystalline, ordered layers (rows) of alternating Cd and Se atoms. (Adapted from Alivisatos, A.P., *Science*, 271, 933, 1996.)

This chapter discusses the biological applications of semiconductor nanocrystals. The main sections deal with quantum-dot synthesis, quantum-confinement and optical properties, surface chemistry, and bioconjugation. We conclude with a brief discussion of future challenges and prospects.

## 58.2 Synthesis

Both group II-VI (e.g., CdSe, CdTe, CdS, and ZnSe) and group III-V (e.g., InP and InAs) nanocrystals have been synthesized and characterized.[27–35] In particular, CdSe quantum dots are highly luminescent and have been extensively studied.[23–26] Figure 58.1 shows a high-resolution transmission electron microscopy (TEM) image of a single CdSe quantum dot. Group II-VI nanocrystals can be prepared in either aqueous or organic media. Two aqueous methods have been developed. In the first method, stock solutions consisting of cadmium and sulfur precursors are directly injected into a hot aqueous solution (containing stabilizing agents and micelles).[18–22] Nucleation of nanocrystals occurs within 5 min, as observed by a change in the solution color. Solvent additives (e.g., stabilizing agents and ions) and environmental factors (e.g., pH and temperature) are used to control the nanocrystal size. This method is simple and does not require specialized equipment, but the resulting nanocrystals have broad emission spectra (~50 nm full width at half maximum [FWHM]) and low quantum yields (<10%). In the second method, nanocrystals are grown in yeast cells such as *Candida glabrata* or *Schizosaccharomyces pombe.*[36–38] When cultured in the presence of cadmium or zinc ions, yeasts express proteins with the general structure of $(\gamma\text{-Glu-Cys})_n\text{-Gly}$. Thiol and carboxylic groups from amino-acid residuals bind to the metal ions and induce nucleation. Further addition of cadmium or zinc ions results in the growth of nanoparticles. These nanocrystals have optical properties similar to those synthesized by the first method.

Improved schemes to synthesize nanocrystals have been developed by several groups.[23–26] In one approach, organometallic and chalcogenide compounds are dissolved in tri-*n*-butyl phosphine or tri-*n*-octyl phosphine (TOP), which are injected into a hot coordinating solvent (e.g., tri-*n*-octylphosphine oxide or TOPO) at 340 to 360°C. Rapid nucleation is observed by changes in the reaction mixture's color. A (4,4)-coordination crystal lattice is formed, and the nanoparticles have a Wurtzite-type structure. With this procedure, technical-grade (90 to 92% purity) TOPO is preferred, but the role of impurities such as alkyl phosphonics and alkylphosphonic acids is still unclear. Peng et al. recently used phosphonic or fatty-acid additives in pure TOPO to synthesize high-quality CdS, CdSe, and CdTe nanocrystals.[39,40] The results indicate that CdO and $Cd(CH_3COO)_2$ can be used to replace dimethylcadmium, an expensive organometallic compound. By using organic solvents to synthesize quantum dots, the surfactant molecules TOPO and TOP reside on the particle's surface and act to stabilize the nanocrystals against aggregation. Due to the nonpolar nature of TOPO and TOP, the resulting nanocrystals are soluble in organic solvents such as chloroform and hexane.

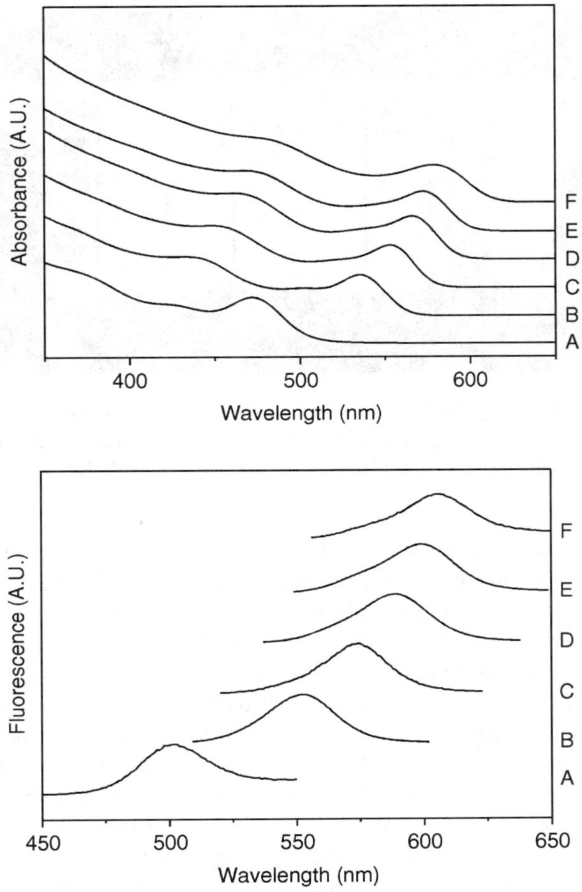


**FIGURE 58.2** (Top) Absorbance profiles of quantum dots after injection of 3 mL $Cd(CH_3)_2$/Se stock solution (time = 0 min) (A), grown for 5 min (B), and after successive injection of $Cd(CH_3)_2$/Se stock solution (C–F). (Bottom) Fluorescence profiles of quantum dots after injection of $Cd(CH_3)_2$/Se stock solution (time = 0 min) (A), grown for 5 min (B), and after successive injection of $Cd(CH_3)_2$/Se stock solution (C–F). (From W.C.W. Chan and S.M. Nie, unpublished data.)

Both the nanocrystal size and shape can be changed by manipulating the kinetics of particle growth. Recently, Peng et al. developed a novel method to control the shape of CdSe nanoclusters.[41,42] Wurtzite-type particles grow in an anisotropic fashion in which growth is favored along the *c*-axis. Manipulation of solvent conditions (e.g., addition of hexyl-phosphonic acid and tetradecylphosphonic acid in pure TOPO) alters the growth rates, leading to the formation and isolation of nanoparticles with discrete aspect ratios (1:1 to 1:30). The size of nanocrystals can be controlled by several procedures. For example, nucleated CdSe quantum dots can be grown at high temperatures (300°C or higher) for an extended period of time (ranging from 2 min to 2 weeks, depending on the desired particle size).[43] In this process (also known as Ostwald ripening), smaller nanocrystals are broken down, and the dissolved atoms are transferred to larger crystals. The rate of this "ripening" process depends on both the temperature and the amount of the limiting reagent.[44,45]

Continuous injection of precursor solutions into the CdSe reaction mixture (in TOPO) at 300°C also produces larger nanocrystals. Figure 58.2 shows the absorption and fluorescence spectra of CdSe quantum dots after successive injections. As the size of nanocrystals increases, there is a shift in the absorbance and fluorescence emission to the red. The fluorescence shift from 500 to 550 nm shows crystal growth

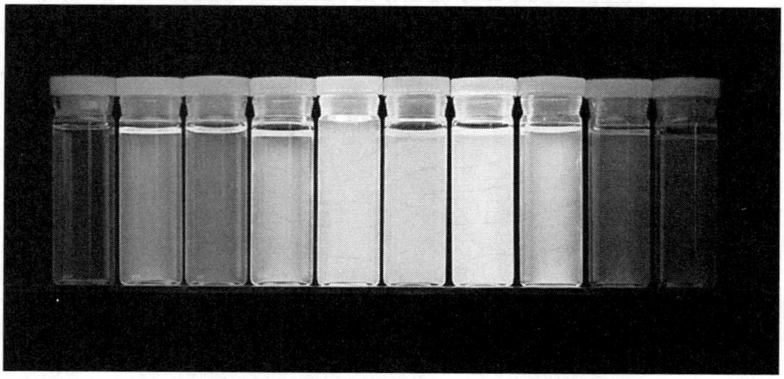

**FIGURE 58.3 (Color figure follows p. 28-30.)** Ten distinguishable emission colors of ZnS-capped CdSe quantum dots excited with a near-UV lamp. From left to right (blue to red), the emission maxima are located at 443, 473, 481, 500, 518, 543, 565, 587, 610, and 655 nm. (Adapted from Han, M.Y. et al., *Nat. Biotechnol.*, 19, 631, 2001. With permission.)

at 300°C for 5 min before the addition of excess precursors. After repeated injections of 1 ml $Cd(CH_3)_2$/Se stock solution, the fluorescence-emission peak shifts from 550 to 575, 590, 600, and 610 nm.[46] Smaller nanocrystals, with fluorescence-emission peaks at 530 nm or less, can be synthesized by slowing the Ostwald ripening process. This is achieved by rapidly decreasing the temperature after initial injection of the $Cd(CH_3)_2$/Se stock solution to prevent formation of large nanocrystals.

The quantum yields of these CdSe nanocrystals are usually 2 to 15% but can be dramatically increased by capping the dots with an inorganic layer such as ZnS or CdS. Zinc sulfide has a wider bandgap than that of CdSe (0.9 eV greater),[47] but the Zn–S bond length is similar to that of Cd–Se (~12% larger).[26] These conditions allow the deposition of a thin ZnS layer on the CdSe core by epitaxial growth.[24–26] In practice, the $Zn(CH_3)_2$/S solution is added slowly in small aliquots to the CdSe/TOPO solution to prevent ZnS nucleation. The quantum yields of the capped CdSe nanocrystals are typically about 30 to 50% and have been reported to be as high as 85% at room temperature.[25] Figure 58.3 depicts a set of different-sized ZnS-capped CdSe quantum dots excited with a hand-held ultraviolet (UV) lamp.

The observed increase in quantum yields can be attributed to the removal of shallow trap states on the surface of quantum dots. These states are defined as discrete electronic energy levels arising from structural defects.[48] For CdSe nanocrystals these defects are mainly located on the surface, where the bonding geometries of chalcogenide atoms are different than those in the interior. These shallow traps lead to a broad emission peak at 700 to 800 nm.[24–26] As the ZnS layer becomes thicker, the quantum efficiency increases to ~50% and then levels off. Several groups have determined that the optimal capping thickness is about one to two monolayers of ZnS.[24,49]

Group III-V nanocrystals can also be prepared by using the TOPO/TOP solvent system. To synthesize InP, a dehalosilylation reaction occurs when $InCl_3$ and $P(Si(CH_3)_3)$ are injected into a TOPO/TOP solution at 240 to 265°C. The size distribution of the InP nanoparticles is narrowed by size-selective precipitation. InP nanocrystals in the size range of 2.5 to 6.0 nm are isolated with a relative standard deviation of 10 to 15%.[28–30] Similar procedures have been used to prepare InAs nanocrystals.[50]

## 58.3 Optical Properties

The optical properties of semicondutor nanoclusters stem from interactions between electrons, holes, and their local environments. Semiconductor quantum dots absorb photons when the energy of excitation exceeds the bandgap energy. During this process, electrons are promoted from the valence band to the conduction band. Measurements of UV–visible spectra reveal a large number of energy states in quantum dots. The lowest excited energy state is shown by the first observable peak (also known as the quantum-confinement

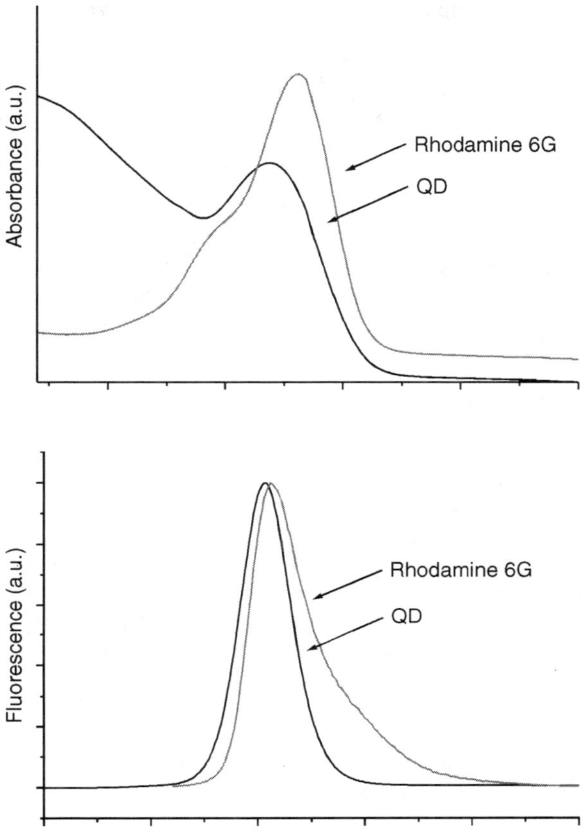


**FIGURE 58.4** Comparison of the excitation (top) and emission (bottom) profiles between rhodamine 6G and CdSe quantum dots. (From Chan, W.C.W., Ph.D. thesis, Department of Chemistry, Indiana University, Bloomington, 2001. With permission.)

peak) at a shorter wavelength than the fluorescence-emission peak. Excitation at shorter wavelengths is possible because multiple electronic states are present at higher energy levels. In fact, the molar-extinction coefficient gradually increases toward shorter wavelengths (Figure 58.4).[24–26] This is an important feature for biological applications because it allows simultaneous excitation of multicolor quantum dots with a single light source.[5,6]

Light emission arises from the recombination of mobile or trapped charge carriers. The emission from mobile carriers is called excitonic fluorescence and is observed as a sharp peak. The emission spectra of single ZnS-capped CdSe quantum dots are as narrow as 13 nm (FWHM) at room temperature.[6] Defect states in the crystal interior or on its surface can trap the mobile charge carriers (electrons or holes), leading to a broad emission peak that is red-shifted from the excitonic peak. Nanocrystals with a large number of trap states generally have low quantum yields, but surface capping or passivation can remove these defect sites and improve the fluorescence quantum yields (as discussed in the previous section).

The excited-state lifetimes of nanocrystals have three exponential components. In bulk measurements, the lifetimes are 5 nsec, 20 to 30 nsec, and 80 to 200 nsec, with the 20- to 30-nsec lifetime dominating.[51,52] These excited-state decay rates are slightly slower than those of organic dyes (1 to 5 nsec) but much faster than those of lanthanide probes (1 μsec to 1 msec). Single-dot measurement further reveals that the excited-state lifetimes depend on the emission intensity, but the exact origins of this multiexponential behavior remain unclear.[51]

The excitonic fluorescence depends on the nanocrystal size. Research by several groups has demonstrated a linear relationship between the particle size and the bandgap energy.[24,43] This quantum-size

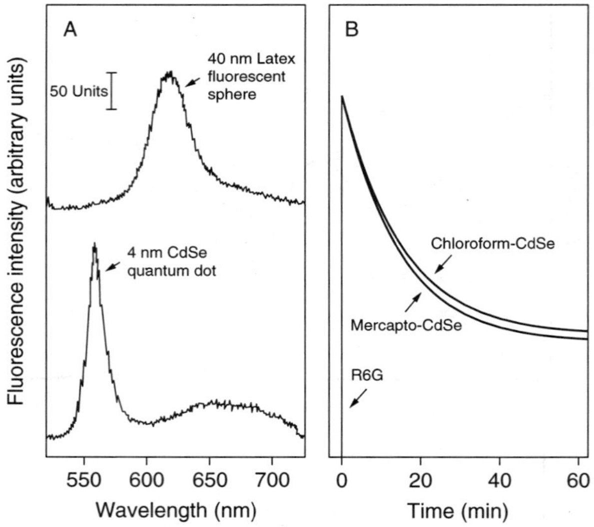


**FIGURE 58.5** Comparison of photophysical properties between luminescent quantum dots and organic dyes. (A) Wavelength-resolved spectra obtained from a single 40-nm fluorescent latex sphere and a single mercapto-quantum dot. The broad emission peak around 660 nm is believed to arise from surface defects on the quantum dot. (B) Time-resolved photobleaching curves for the original quantum dots, the solubilized quantum dots, and the dye rhodamine 6G. (From Chan, W.C.W. and Nie, S.M., *Science*, 281, 2016, 1998. With permission.)

effect is similar to that observed for a "particle in a box." Outside of the box, the potential energy is considered to be infinitely high. Thus, mobile carriers (similar to the particle) are confined within the dimensions of the nanocrystal (similar to the box) with discrete wave functions and energy levels. As the physical dimensions of the box become smaller, the bandgap energy becomes higher. For CdSe nanocrystals, the sizes of 2.5 and 5.5 nm correspond to fluorescence emission at 500 and 620 nm, respectively. In addition to size, the emission wavelength can be varied by changing the semiconductor material. For example, InP and InAs quantum dots usually emit in the far-red and near-infrared,[28–30] while CdS and ZnSe dots often emit in the blue or near-UV.[53] It is also interesting to note that elongated quantum dots (called quantum rods) show linearly polarized emission,[54] whereas the fluorescence emission from spherical CdSe dots is circularly polarized or not polarized.[55,56]

In comparison with organic dyes such as rhodamine 6G and fluorescein, CdSe nanocrystals show similar or slightly lower quantum yields at room temperature. The lower quantum yields of nanocrystals are compensated by their larger absorption cross sections and much-reduced photobleaching rates. Bawendi et al.[23,24] estimated that the molar-extinction coefficients of CdSe quantum dots are about $10^5$ to $10^6$ $M^{-1}$ $cm^{-1}$, depending on the particle size and the excitation wavelength. These values are 10 to 100 times larger than those of organic dyes but similar to the absorption cross sections of phycoerytherin, a multichromophore fluorescent protein. Chan and Nie[6] estimated that single ZnS-capped CdSe quantum dots are ~20 times brighter than single rhodamine 6G molecules. Similarly, phycoerytherin is estimated to be 20 times brighter than fluorescein.[57]

Another attractive feature of using quantum dots as biological labels is their high photostability. Gerion et al. examined the photobleaching rate of silica-coated ZnS-capped CdSe quantum dots against that of rhodamine 6G.[58] The quantum-dot emission stayed constant for 4 h, while rhodamine 6G was photobleached after only 10 min. It has been suggested that capped CdSe nanocrystals are 100 to 200 times more stable than organic dyes and fluorescent proteins (Figure 58.5).[6] Under intense UV excitation, single phycoerytherin molecules were found to photobleach after 70 sec, while the fluorescence emission of quantum dots remained unchanged after 600 sec.[46] The photobleaching of quantum dots is believed to arise from a slow process of photo-induced chemical decomposition. Henglein et al. speculated that CdS decomposition was initiated by the formation of S or SH radicals upon optical excitation.[59,60] These radicals can react with $O_2$ from the air to form a $SO_2$ complex, resulting in slow particle degradation.

Single quantum dots have been shown to emit photons in an intermittent on–off fashion,[61,62] like a "blinking" behavior reported for single fluorescent dye molecules, proteins, polymers, and metal nanoparticles.[63–66] The fluorescence of single quantum dots turns on and off at a rate that depends on the excitation power. This phenomenon has been suggested to arise from a light-induced process involving photoionization and slow charge neutralization of the nanocrystal.[61] When two or more electron–hole pairs are generated in a single nanocrystal, the energy released from the combination of one pair could be transferred to the remaining carriers, one of which is preferentially ejected into the surrounding matrix. Subsequent photogenerated electron–hole pairs transfer their energy to the resident, unpaired carrier, leading to nonradiative delay and dark periods. The luminescence is restored only when the ejected carrier returns to neutralize the particle. Banin et al. believe that thermal trapping of electrons and holes is also a contributing factor because they observed a dependence of the blinking rate on temperature.[67] A further finding is that single dots exhibit random fluctuations in emission wavelength (spectral wandering) over time.[62,68] This effect is attributed to interactions between excitons with optically induced surface changes.

## 58.4 Surface Chemistry

Colloidal-metal nanoparticles have found widespread use in biology and medicine in the last 20 years, but semiconductor nanocrystals have only recently been used for biological labeling.[5,6] The reason is that high-quality quantum dots are often protected with hydrophobic ligands and are not compatible with biomolecules. Murphy et al.[69,70] were the first to demonstrate the use of semiconductor nanoparticles as biosensors in which oligonucleotide molecules were detected by nonspecific interactions with the nanocrystal surface. For other applications, there is a great need to develop simple methods for conjugating biomolecules to quantum dots. TOPO-synthesized nanocrystals are preferred over aqueous-synthesized dots due to their superior optical properties. The surface chemistry is a key factor in determining quantum-dot solubility, biocompatibility, and molecular conjugation.

The complex surface of nanocrystals has been studied by nuclear magnetic resonance (NMR) and x-ray photoelectron spectroscopy (XPS).[71,72] Morphologically, quantum dots are not smooth spherical particles but faceted with many planes and edges. They are generally considered to be negatively charged because of molecules adsorbed on the surface.[73] TOPO strongly coordinates to the surface metal atoms, while TOP or TBP is weakly bound and can easily desorb. The surface properties (such as TOPO molecules interacting with each other and their dimerization at the phosphine oxide end)[74] improve the long-term solubility of quantum dots in organic solvents.

Two methods have been developed to prepare water-soluble quantum dots. Alivisatos et al.[5] reported the use of a silica/siloxane coating for creating water-soluble ZnS-capped CdSe quantum dots. In this procedure, 3-(mercaptopropyl)trimethoxysilane (MPS) is directly adsorbed onto the nanocrystals in which TOPO-molecules are displaced. A silica/siloxane shell is formed on the surface by introduction of a base and hydrolysis of the silanol-groups (from MPS). Polymerizing silanol groups helps stabilize the nanocrystals against flocculation. The quantum dots become soluble in intermediate polar solvents such as methanol and dimethyl sulfoxide. Further reaction with bifunctional methoxy compounds, such as aminopropyl trimethoxysilane or trimethoxysilyl propyl urea, renders the particles soluble in aqueous buffer. The second method involves direct adsorption of bifunctional ligands such as mercaptoacetic acid or dithiothrietol to the quantum-dot surface.[6] Here, a mercapto compound and an organic base are added together to TOPO quantum dots dissolved in chloroform. The base deprotonates both the thiol and the carboxylic groups, leading to favorable electrostatic binding between negatively charged sulfur atoms and surface $Cd^{2+}$ or $Zn^{2+}$ ions. The semiconductor quantum dots precipitate out of the solution but can be redissolved in an aqueous solution ($pH > 5$).

Polymerized siloxane-coated quantum dots are highly stable against flocculation, but only small amounts (in the milligram range) can be prepared per batch. Also, residual silanol groups on the nanocrystal surface often lead to precipitation and gel formation at neutral pH. In comparison, the direct adsorption procedure yields gram-size quantities of water-soluble dots, but the mercapto ligands are not completely stable. Slow desorption of mercaptoacetic-acid molecules often results in aggregation and precipitation of the solubilized quantum dots. The stability of the mercapto-derivatized dots depends on the density of mercapto molecules on the surface, its size, and the number of thiol functional groups per mercapto molecule.[46] When solubilized with mercaptoacetic acid, the quantum dots are stable for 3 to 7 d in neutral phosphate buffered saline (PBS) buffer. Conjugation of complex proteins onto the surface further stabilizes the particles, making them stable for 3 months to 1 year (depending on the protein).

It should be noted that the quantum efficiencies of ZnS-capped CdSe nanocrystals are generally lower in water than in chloroform. The highest quantum yield reported for water-soluble quantum dots is ca. 20 to 30%.[5] Traditionally, it has been held that capping a second semiconductor layer protects and isolates the core from the outside environment.[26] Recent research in our group indicates that the ZnS layer is not uniform and that surface-adsorbed molecules can still influence the properties of the nanocrystal core.[46] In fact, both large and small molecules, such as polymers, proteins, and ions, have been found to influence the optical properties of ZnS-capped CdSe quantum dots.

## 58.5 Bioconjugation

For specific biomolecular binding or recognition, it is important to develop stable and functional quantum-dot bioconjugates. Reactive functional groups that can be placed on the nanocrystal surface include primary amines, carboxylic acids, alcohols, and thiols. Primary amines react with carboxylic acids, catalyzed with a carbodiimide or sulfo *N*-hydroxysuccinimide ester, to form a stable amide bond. A thioether bond is formed when sulfhydryl groups react with maleimides. Other methods for linking fluorescent labels to biomolecules are described in a book by Hermanson.[75] Another approach to linking biomolecules to nanocrystals is to use a thiol-exchange reaction. Here, mercapto-coated quantum dots (by direct adsorption) are mixed with thiolated biomolecules (such as oligonucleotides and proteins).[8,76] After overnight incubation at room temperature, a chemical equilibrium is reached between the adsorbed thiols and the free thiols. A similar approach has recently been used by Bawendi et al.[7] in which engineered proteins with a linear polylysine chain are directly adsorbed onto negatively charge nanocrystals through electrostatic interactions. Figure 58.6 shows a number of common bioconjugation methods.

The surface area of quantum dots is large enough for linking to multiple biomolecules. Two to five protein molecules and 50 or more small molecules (such as oligonucleotides and peptides) may be conjugated to a single 4-nm dot.[75] We envision multifunctional or smart nanostructures being developed by conjugating both DNA and proteins to quantum dots. The attached proteins or peptides could guide the nanoparticles to specific cells (e.g., cancer cells) or to specific locations inside a cell where messenger RNA molecules are detected.

A number of applications have been demonstrated for quantum dots including DNA hybridization, immunoassays, and receptor-mediated endocytosis (Figure 58.7). In addition, Weiss et al.[77,78] showed the ability to measure distances between ZnS-capped CdSe quantum dots and fluorescent beads beyond the diffraction limit. Lifetime measurement was also used for time-gated fluorescence imaging of tissue sections.[52] The background fluorescence lifetimes are similar to those of organic dyes and thus interfere with the target signals. For lanthanide-chelating probes, the excited-state lifetimes are very long, but their quantum efficiencies are relatively low.[79,80] In contrast, semiconductor nanocrystals have excellent emission efficiencies and relatively long decay times, both of which are attractive features for imaging tissue sections.

Quantum dots for mulitplexed optical encoding of biomolecules have recently been demonstrated by Han et al.[81] Polystyrene beads are embedded multicolor CdSe quantum dots at various color and intensity combinations (Figure 58.8). The use of six colors and ten intensity levels can theoretically encode one million protein or nucleic-acid sequences. Specific capturing molecules, such as peptides, proteins, and oligonucleotides, are covalently linked to the beads and encoded by the bead's spectroscopic signature. A single light source is sufficient for reading all the quantum-dot-encoded beads. To determine whether

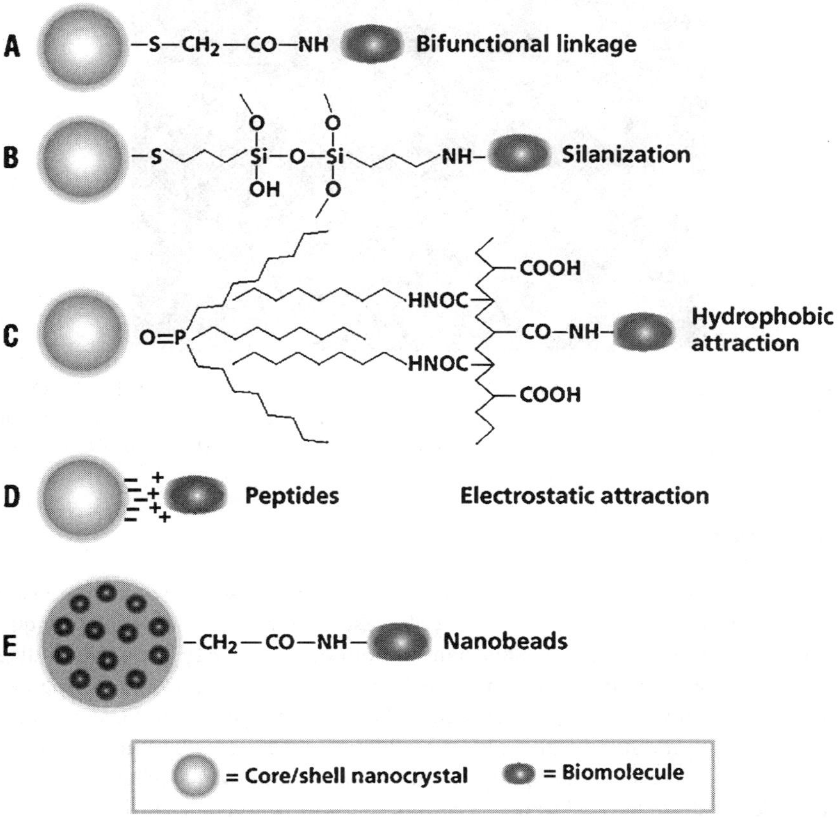


**FIGURE 58.6** Schematic illustration of bioconjugation methods. (Courtesy of Dr. Mingyong Han, University of Singapore.)

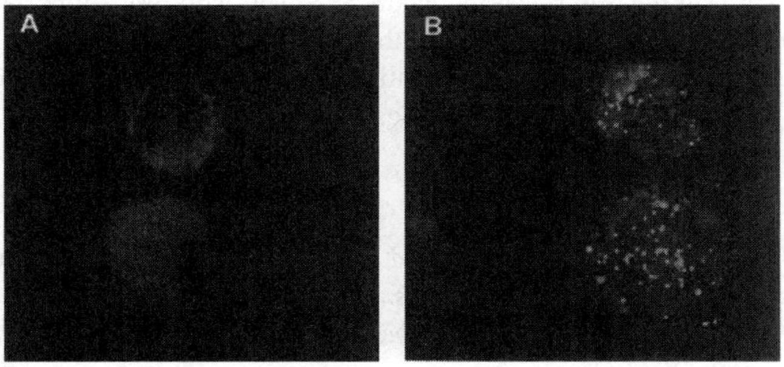


**FIGURE 58.7** Luminescence images of cultured HeLa cells incubated with (A) mercapto-QDs (control) and (B) QD-transferrin conjugates. The QD bioconjugates were transported into the cell by receptor-mediated endocytosis and detected as clusters or aggregates. (From Chan, W.C.W., and Nie, S.M., *Science*, 281, 2016, 1998. With permission.)

an unknown analyte is captured or not, conventional assay methodologies (similar to direct or sandwich immunoassay) can be applied. This so-called "bar-coding technology" can be used for gene profiling and high-throughput drug and disease screening.

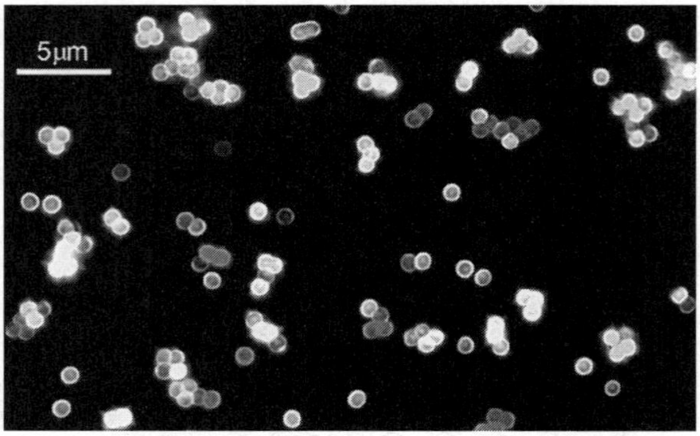


**FIGURE 58.8 (Color figure follows p. 28-30.)** Fluorescence micrograph of a mixture of CdSe/ZnS QD-tagged beads emitting single-color signals at 484, 508, 547, 575, and 611 nm. The beads were spread and immobilized on a polylysine-coated glass slide, which caused a slight clustering effect. (From Han, M.Y. et al., *Nat. Biotechnol.*, 19, 631, 2001. With permission.)

# 58.6 Conclusion

A pervasive trend in biotechnology is the development of multiplexed sensing and ultrasensitive imaging technologies for the rapid molecular profiling of cells, tissues, and organs. The ability to quickly screen a large number of genes and proteins is important to both drug discovery and medical diagnostics. In the next 10 years, we envision novel platforms based on multicolor quantum dots being developed for massively parallel biosensing and analytical detection. These multiplexing technologies will likely combine the advantages of quantum dots with those of microfluidics and microarrays. In addition, noninvasive molecular-imaging technologies could be developed by using luminescent quantum dots. This should allow viral particles to be followed *in vivo*, drug molecules to be analyzed in biological systems, and tumor cells to be tracked in real time.[82]

## Acknowledgments

We thank Dr. Mingyong Han for help preparing Figure 58.6. This work was supported by grants from the National Institutes of Health (R01 GM58173) and the Department of Energy (DOE FG02-98ER14873). Shuming Nie acknowledges the Whitaker Foundation for a Biomedical Engineering Award and the Beckman Foundation for a Beckman Young Investigator Award.

## References

1. Kricka, L.J., *Nonisotopic Probing, Blotting, and Sequencing*, Academic Press, New York, 1995.
2. Diamondis, E.P. and Christopoulos, T.K., Eds., *Immunoassay*, Academic Press, New York, 1996.
3. Haugland, R.P., *Handbook of Fluorescent Probes and Research Chemicals: Molecular Probes*, Molecular Probes, Eugene, OR, 1995.
4. Whaley, S.R., English, D.S., Hu, E.L., Barbara, P.F., and Belcher, A.M., Selection of peptides with semiconductor binding specificity for directed nanocrystal assembly, *Nature*, 405, 665, 2000.
5. Bruchez, M., Jr., Moronne, M., Gin, P., Weiss, S., and Alivisatos, A.P., Semiconductor nanocrystals as fluorescent biological labels, *Science*, 281, 2013, 1998.
6. Chan, W.C.W. and Nie, S.M., Quantum dot bioconjugates for ultrasensitive nonisotopic detection, *Science*, 281, 2016, 1998.
7. Mattoussi, H., Mauro, J.M., Goldman, E.R., Anderson, G.P., Sundar, V.C., Mikulec, F.V., and Bawendi, M.G., Self-assembly of CdSe-ZnS quantum dots bioconjugates using an engineered recombinant protein, *J. Am. Chem. Soc.*, 122, 12142, 2000.

8. Mitchell, G.P., Mirkin, C.A., and Letsinger R.L., Programmed assembly of DNA functionalized quantum dots, *J. Am. Chem. Soc.*, 121, 8122, 1999.
9. Pathak, S., Choi, S.-K., Arnheim, N., and Thompson, M.E., Hydroxylated quantum dots as luminescent probes for *in situ* hybridization, *J. Am. Chem. Soc.*, 123, 4103, 2001.
10. Mirkin, C.A., Letsinger, R.L., Mucic, R.C., and Storhoff, J.J., A DNA-based method for rationally assembling nanoparticles into macroscopic materials, *Nature*, 382, 607, 1996.
11. Alivisatos, A.P., Johnsson, K.P., Peng, X., Wilson, T.E., Loweth, C.J., Bruchez, M.P., Jr., and Schultz, P.G., Organization of nanocrystal molecules using DNA, *Nature*, 382, 609, 1996.
12. Elghanian, R., Storhoff, J.J., Mucic, R.C. Letsinger, R.L., and Mirkin, C.A., Selective colorimetric detection of polynucleotides based on the distance-dependent optical properties of gold nanoparticles, *Science*, 277, 1078, 1997.
13. Dubertret, B., Calame, M., and Libchaber, A.J., Single-mismatch detection using gold-quenched fluorescent oligonucleotides, *Nat. Biotechnol.*, 19, 365, 2001.
14. Storhoff, J.J. and Mirkin, C.A., Programmed materials synthesis with DNA, *Chem. Rev.*, 99, 1849, 1999.
15. Reynolds, R.A., III, Mirkin, C.A., and Letsinger, R.L., Homogeneous, nanoparticle-based quantitative colormetric detection of oligonucleotides, *J. Am. Chem. Soc.*, 122, 3795, 2000.
16. Brus, L.E., A simple model for the ionization potential, electron affinity, and aqueous redox potentials of small semiconductor crystallites, *J. Chem. Phys.*, 79, 5566, 1983.
17. Alivisatos, A.P., Semiconductor clusters, nanocrystals, and quantum dots, *Science*, 271, 933, 1996.
18. Fojtik, A., Weller, H., Koch, U., and Henglein, A., Photochemistry of colloidal metal sulfides: photophysics of extremely small CdS particles: Q-state CdS and magic agglomeration numbers, *Ber. Bunsenges. Phys. Chem.*, 88, 969, 1984.
19. Spanhel, L., Haase, M., Weller, H., and Henglein, A., Photochemistry of colloidal semiconductors: surface modification and stability of strong luminescing CdS particles, *J. Am. Chem. Soc.*, 109, 5649, 1987.
20. Kortan, A.R., Hull, R., Opila, R.L., Bawendi, M.G., Steigerwald, M.L., Carroll, P.J., and Brus, L.E., Nucleation and growth of CdSe on ZnS quantum crystallite seeds, and vice versa, in inverse micelle media, *J. Am. Chem. Soc.*, 112, 1327, 1990.
21. Lianos, P. and Thomas, J.K., Cadmium sulfide of small dimensions produced by inverted micelles, *Chem. Phys. Lett.*, 125, 399, 1986.
22. Dannhauser, T., O'Neil, M., Johansson, K., Whitten, D., and McLendon, G., Photophysics of quantized colloidal semiconductors: dramatic luminescence enhancement by binding of simple amines, *J. Phys. Chem.*, 90, 6074, 1986.
23. Murray, C.B., Norris, D.J., and Bawendi, M.G., Synthesis and characterization of nearly monodisperse CdE (E = S, Se, Te) semiconductor nanocrystallites, *J. Am. Chem. Soc.*, 115, 8706, 1993.
24. Dabbousi, B.O., Rodriguez-Viejo, J., Mikulec, F.V., Heine, J.R., Mattoussi, H., Ober, R., Jensen, K.F., and Bawendi, M.G., (CdSe)ZnS core-shell quantum dots: synthesis and characterization of a size series of highly luminescent nanocrystallites, *J. Phys. Chem. B*, 101, 9463, 1997.
25. Peng, X.G., Schlamp, M.C., Kadavanich, A.V., and Alivisatos, A.P., Epitaxial growth of highly luminescent CdSe/CdS core/shell nanocrystals with photostability and electronic accessibility, *J. Am. Chem. Soc.*, 119, 7019, 1997.
26. Hines, M.A. and Guyot-Sionnest, P., Synthesis of strongly luminescing ZnS-capped CdSe nanocrystals, *J. Phys. Chem. B*, 100, 468, 1996.
27. Weller, H., Quantized semiconductor particles: a novel state of matter for materials science, *Adv. Mat.*, 5, 88, 1993.
28. Guzelian, A.A., Katari, J.E.B., Kadavanich, A.V., Banin, U., Hamad, K., Juban, E., Alivisatos, A.P., Wolters, R.H., Arnold, C.C., and Heath, J.R., Synthesis of size-selected, surface-passivated InP nanocrystals, *J. Phys. Chem.*, 100, 7212, 1996.
29. Prieto, J.A., Armelles, G., Groenen, J., and Cales, R., Size and strain effects in the E-1-like optical transitions of InAs/InP self-assembled quantum dot structures, *Appl. Phys. Lett.*, 74, 99, 1999.

30. Micic, O.I., Cheong, H.M., Fu, H., Zunger, A., Sprague, J.R., Mascarenhas, A., and Nozik, A.J., Size-dependent spectroscopy of InP quantum dots, *J. Phys. Chem. B*, 101, 4904, 1997.
31. Schreder, B., Schmidt, T., Ptatschek, V., Winker, U., Materny, A., Umback, E., Lerch, M., Muller, G., Kiefer, W., and Spanhel, L., CdTe/CdS clusters with "core-shell" structure in colloids and films: the path of formation and thermal breakup, *J. Phys. Chem. B*, 104, 1677, 2000.
32. Kapitonov, A.M., Stupak, A.P., Gaponenko, S.V., Petrov, E.P., Rogach, A.L., and Eychmuller, A., Luminescence properties of thiol-stabilized CdTe nanocrystals, *J. Phys. Chem. B*, 103, 10109, 1999.
33. Hao, E., Sun, H., Zhou, Z., Liu, J. Yang, B., and Shen, J., Synthesis and optical properties of CdSe and CdSe/CdS nanoparticles, *Chem. Mater.*, 11, 3096, 1999.
34. Trindade, T. and O'Brien, P., A single source approach to the synthesis of CdSe nanocrystallites, *Adv. Mater.*, 8, 161, 1996.
35. Rogach, A.L., Kornowski, A., Gao, M., Eychmuller, A., and Weller, H., Synthesis and characterization of a size series of extremely small thiol-stabilized CdSe nanocrystals, *J. Phys. Chem. B*, 103, 3065, 1999.
36. Dameron, C.T., Reese, R.N., Mehra, R.K., Kortan, A.R., Carroll, P.J., Steigerwald, M.L., Brus, L.E., and Winger, D.R., Biosynthesis of cadmium sulfide quantum semiconductor crystallites, *Nature*, 338, 596, 1989.
37. Dameron, C.T., Smith, B.R., and Winge, D.R., Glutathione-coated cadmium-sulfide crystallites in Candida-Glabrata, *J. Biol. Chem.*, 264, 17355, 1989.
38. Bae, W. and Mehra, R.K., Cysteine-capped ZnS nanocrystallites: preparation and characterization, *J. Inorg. Biochem.*, 70, 125, 1998.
39. Qu, L., Peng, Z.A., and Peng, X.G., Alternative routes toward high quality CdSe nanocrystals, *NanoLetters*, 1, 333, 2001.
40. Peng, Z.A. and Peng, X.G., Formation of high-quality CdTe, CdSe, and CdS nanocrystals using CdO as precursor, *J. Am. Chem. Soc.*, 123, 183, 2001.
41. Peng, X.G., Manna, L., Yang, W.D., Wickham, J., Scher, E., Kadavanich, A., and Alivisatos, A.P., Shape control of CdSe nanocrystals, *Nature*, 404, 59, 2000.
42. Peng, Z.A. and Peng, X.G., Mechanisms of the shape evolution of CdSe nanocrystals, *J. Am. Chem. Soc.*, 123, 1389, 2001.
43. Peng, X.G., Wickham, J., and Alivisatos, A.P., Kinetics of II-VI and III-V colloidal semiconductor nanocrystal growth: "focusing" of size distributions, *J. Am. Chem. Soc.*, 120, 5343, 1998.
44. DeSmet, Y., Deriemaeker, L., Parloo, E., and Finsy, R., On the determination of Ostwald ripening rates from dynamic light scattering measurements, *Langmuir*, 15, 2327, 1999.
45. DeSmet, Y., Deriemaeker, L., and Finsy, R., A simple computer simulation of Ostwald ripening, *Langmuir*, 13, 6884, 1997.
46. Chan, W.C.W., *Semiconductor Quantum Dots for Biological Detection and Imaging*, Ph.D. dissertation, Department of Chemistry, Indiana University, Bloomington, 2001.
47. Madelung, O., *Semiconductors — Basic Data*, Springer-Verlag Publishing, New York, 1996.
48. Bawendi, M.G., Steigerwald, M.L., and Brus, L.E., The quantum mechanics of larger semiconductor clusters (quantum dots), *Annu. Rev. Phys. Chem.*, 41, 477, 1990.
49. Talapin, D.V., Rogach, A.L., Kornowski, A., Haase, M., and Weller, H., Highly luminescent monodisperse CdSe and CdSe/ZnS nanocrystals synthesized in a hexadecylamine-triocytlphosphine oxide-trioctylphosphine mixture, *NanoLetters*, 1, 207, 2001.
50. Shi, J.Z., Zhu, K., Zheng, Q., Zhang, L., Ye, L., Wu, J., and Zuo, J., Ultraviolet (340–390 nm), room temperature, photoluminescence from InAs nanocrystals embedded in $SiO_2$ matrix, *Appl. Phys. Lett.*, 70, 2586, 1997.
51. Dahan, M., Laurence, T., Schumacher, A., Chemla, D.S., Alivisatos, A.P., Sauer, M., and Weiss, S., Fluorescence lifetime study of single qdots, *Biophys. J.*, 78, 2270, 2000.
52. Dahan, M., Laurence, T., Pinaud, F., Chemla, D.S., Alivisatos, A.P., Sauer, M., and Weiss, S., Time-gated biological imaging by use of colloidal quantum dots, *Opt. Lett.*, 26, 825, 2001.

53. Hines, M.A. and Guyot-Sionnest, P., Bright UV-blue luminescent colloidal ZnSe nanocrystals, *J. Phys. Chem. B*, 102, 3655, 1998.
54. Hu, J.T., Li, L., Yang, W., Manna, L., Wang, L., and Alivisatos, A.P., Linearly polarized emission from colloidal semiconductor quantum rods, *Science*, 292, 2060, 2001.
55. Efros, A.L., Luminescence polarization of CdSe microcrystals, *Phys. Rev. B*, 46, 7448, 1992.
56. Empedocles, S.A., Neuhauser, R., and Bawendi, M.G., Three-dimensional orientation measurements of symmetric single chromophores using polarization microscopy, *Nature*, 399, 126, 1999.
57. Mathies, R. and Stryer, L., Single-molecule fluorescence detection, in *Applications of Fluorescence in the Biomedical Sciences*, Taylor, D.L., Waggoner, A.S., Lanni, F., Murphy, R.F., and Birge, R., Eds., Alan R. Liss, New York, 1986, p. 129.
58. Gerion, D., Pinaud, F., Willimas, S.C., Parak, W.J., Zanchet, D., Weiss, S., and Alivisatos, A.P., Synthesis and properties of biocompatible water-soluble silica-coated CdSe/ZnS semiconductor quantum dots, *J. Phys. Chem. B*, 105, 8861, 2001.
59. Henglein, A., Photo-degradation and fluroescence of colloidal-cadmium sulfide in aqueous solution, *Ber. Bunsenges. Phys. Chem.*, 86, 301, 1982.
60. Baral, S., Fojtik, A., Weller, H., and Henglein, A., Photochemistry and radiation chemistry of colloidal semiconductors: intermediates of the oxidation of extremely small particles of CdS, ZnS, and $Cd_3P_2$ and size quantization effects (a pulse radiolysis study), *J. Am. Chem. Soc.*, 108, 375, 1986.
61. Nirmal, M., Dabbousi, B.O., Bawendi, M.G., Macklin, J.J., and Brus, L.E., Fluorescence intermittency in single cadmium selenide nanocrystals, *Nature*, 383, 802, 1996.
62. Empedocles, S.A. and Bawendi, M.G., Spectroscopy of single CdSe nanocrystallites, *Acc. Chem. Res.*, 32, 389, 1999.
63. Dickson, R.M., Cubitt, A.B., Tsien, R.Y., and Moerner, W.E., On/off blinking and switching behaviour of single molecules of green fluorescent protein, *Nature*, 388, 355, 1997.
64. VandenBout, D.A., Yip, W.T., Hu, D., Fu, D.K., Swager, T.M., and Barbara, P.F., Discrete intensity jumps and intramolecular electronic energy transfer in the spectroscopy of single conjugated polymer molecules, *Science*, 277, 1074, 1997.
65. Ambrose, W.P., Goodwin, P.M., Martin, J.C., and Keller, R.A., Alterations of single molecule fluorescence lifetimes in near-field optical microscopy, *Science*, 265, 364, 1994.
66. Krug, J.T., II, Wang, G.D., Emory, S.R., and Nie, S., Efficient Raman enhancement and intermittent light emission observed in single gold nanocrystals, *J. Am. Chem. Soc.*, 121, 9208, 1999.
67. Banin, U., Bruchez, M., Alivisatos, A.P., Ha, T., Weiss, S., and Chemla, D.S., Evidence for a thermal contribution to emission intermittency in single CdSe/CdS core/shell nanocrystals, *J. Chem. Phys.*, 110, 1195, 1999.
68. Blanton, S.A., Hines, M.A., and Guyot-Sionnest, P., Photoluminescence wandering in single CdSe nanocrystals, *Appl. Phys. Lett.*, 69, 3905, 1996.
69. Mahtab, R., Rogers, J.P., and Murphy, C.J., Protein-sized quantum dot luminescence can distinguish between straight, bent, and kinked oligonucleotides, *J. Am. Chem. Soc.*, 117, 9099, 1995.
70. Mahtab, R. Rogers, J.P., Singleton, C.P., and Murphy, C.J., Preferential adsorption of a kinked DNA to a neutral curved surface — comparisons to and implications for nonspecific DNA-protein interactions, *J. Am. Chem. Soc.*, 118, 7028, 1996.
71. Becerra, L.R., Murray, C.B., Griffin, R.G., and Bawendi, M.G., Investigation of the surface morphology of capped CdSe nanocrystallites by P-31 nuclear magnetic resonance, *J. Chem. Phys.*, 100, 3297, 1994.
72. Katari, J.E.B., Colvin, V.L., and Alivisatos, A.P., X-ray photoelectron spectroscopy of CdSe nanocrystals with applications to studies of the nanocrystal surface, *J. Phys. Chem.*, 98, 4109, 1994.
73. O'Neil, M., Marohn, J., and McLendon, G., Dynamics of electron-hole pair recombination in semiconductor clusters, *J. Phys. Chem.*, 94, 4356, 1990.

74. Lorenz, J.K. and Ellis, A.B., Surfactant-semiconductor interfaces: perturbation of the photoluminescence of bulk cadmium selenide by adsorption of tri-n-octylphosphine oxide as a probe of solution aggregation with relevance to nanocrystal stabilization, *J. Am. Chem. Soc.*, 120, 10970, 1998.
75. Hermanson, G.T., *Bioconjugate Techniques*, Academic Press, New York, 1996.
76. Willard, D.M., Carillo, L.L., Jung, J., and van Orden, A., CdSe-ZnS quantum dots as resonance energy transfer donors in a model protein–protein binding assay, *NanoLetters*, 1, 469, 2001.
77. Lacoste, T.D., Michalet, X., Pinaud, F., Chemla, D.S., Alivisatos, A.P., and Weiss, S., Ultrahigh-resolution multicolor colocalization of single fluorescent probes, *Proc. Natl. Acad. Sci. U.S.A.*, 97, 9461, 2000.
78. Weiss, S., Shattering the diffraction limit of light: a revolution in fluorescence microscopy, *Proc. Natl. Acad. Sci. U.S.A.*, 97, 8747, 2000.
79. Li, M. and Selvin, P.R., Luminescent polyaminocarboxylate chelates of terbium and europium — the effect of chelate structure, *J. Am. Chem. Soc.*, 117, 8132, 1995.
80. Selvin, P.R., Jancarik, L., Li, M., and Hung, L.W., Crystal structure and spectroscopic characterization of a luminescent europium chelate, *Inorg. Chem.*, 35, 700, 1996.
81. Han, M.Y., Gao, X., Su, J.Z., and Nie, S., Quantum-dot-tagged microbeads for multiplexed optical coding of biomolecules, *Nat. Biotechnol.*, 19, 631, 2001.
82. Mitchell, P., Turning the spotlight on cellular imaging, *Nat. Biotechnol.*, 19, 1013, 2001.

# 59

# PEBBLE Nanosensors for *in Vitro* Bioanalysis


Eric Monson
*University of Michigan*
*Ann Arbor, Michigan*

Murphy Brasuel
*University of Michigan*
*Ann Arbor, Michigan*

Martin A. Philbert
*University of Michigan*
*Ann Arbor, Michigan*

Raoul Kopelman
*University of Michigan*
*Ann Arbor, Michigan*




## 59.1 Introduction

In medical and biochemical research, when a sample domain is reduced to micrometer regimes, e.g., living cells or their subcompartments, the real-time measurement of chemical and physical parameters with high spatial resolution and negligible perturbation of the sample becomes extremely challenging. A traditional strength of chemical sensors (optical, electrochemical, etc.) is the minimization of chemical interference between sensor and sample, achieved with the use of inert, "biofriendly" matrices or interfaces. However, when it comes to penetrating individual live cells, even the introduction of a submicron sensor tip can cause biological damage and resultant biochemical consequences. In contrast, individual molecular probes (free-sensing dyes) are physically small enough but usually suffer from chemical interference between probe and cellular components. Our recently developed PEBBLE sensors (probes encapsulated by biologically localized embedding) are nanoscale spherical devices consisting of sensor molecules entrapped in a chemically inert matrix. This protective coating eliminates interferences such as protein binding and membrane/organelle sequestration, which alter dye response. Conversely, the nanosensor matrix also provides protection to the cellular contents, enabling dyes that would usually be toxic to cells to be used for intracellular sensing. In addition, the inclusion of reference dyes allows quantitative, ratiometric fluorescence techniques to be used. Furthermore, the matrix phase allows the implementation of synergistic sensing schemes. PEBBLEs have been used to measure analytes such as calcium, potassium, nitric oxide, oxygen, chloride, sodium, and glucose.

PEBBLE nanosensors (see Figure 59.1) are submicron-sized optical sensors designed specifically for minimally invasive analyte monitoring in viable, single cells with applications for real-time analysis of drug, toxin, and environmental effects on cell function. PEBBLE is a general term that describes a family


0-8493-1116-0/03/$0.00+$1.50

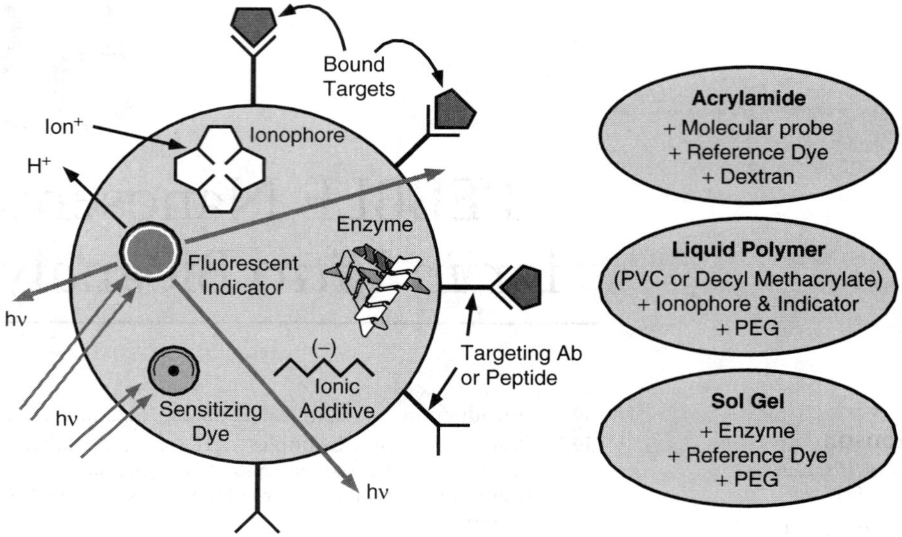


**FIGURE 59.1** Schematic diagram of a PEBBLE nanosensor showing many options available within this flexible, integrated device platform. On the right, current matrix materials are presented with typical constituents.

of matrices and nanofabrication techniques used to miniaturize many existing optode technologies. The main classes of PEBBLE nanosensors are based on matrices of polyacrylamide hydrogel, sol-gel silica, and cross-linked decyl methacrylate. These matrices have been used to fabricate sensors for $H^+$, $Ca^{2+}$, $K^+$, $Na^+$, $Mg^{2+}$, $Zn^{2+}$, $Cl^-$, $NO_2^-$, $O_2$, NO, and glucose that range from 30 to 600 nm in size. A host of delivery techniques have been used to successfully deliver PEBBLE nanosensors into mouse oocytes, rat alveolar macrophages, rat C6 glioma, and human neuroblastoma cells.

PEBBLEs were developed specifically for biological applications and fill a niche that lies between pulled microoptodes and free molecular probes (naked indicator-dye molecules). The strength of the PEBBLE concept lies in two related but distinct roles. First and foremost, the PEBBLE protects the cell from the toxicity inherent in some free molecular dyes and at the same time protects indicator dyes from cellular interferents such as protein binding. The second role, which is possible because the PEBBLE matrix creates a separate sensing phase distinct from the cellular environment, is that multiple dyes, ionophores, and other components can be combined to create complex sensing schemes. These schemes can include reference dyes to allow ratiometric imaging or ionophore/chromoionophore combinations that allow the use of highly selective, nonfluorescent ionophores. Both the protection and the powerful sensing flexibility come in a nanopackage, which, in terms of minimal mechanical and physical perturbation, is closer to "free molecular dyes" than most other sensing platforms. However, the nanosensor preserves the excellent chemical sensing and biocompatibility of macrosensors and surpasses their performance in terms of response time and absolute detection limit.

PEBBLEs are a direct outgrowth of the pulled optical-fiber nanotechnology developed for biosensing by Tan et al.[1,2] and continuing in the work of Rosenzweig,[3,4] Shortreed,[5,6] and Barker.[7,8] In their study, Dourado and Kopelman formalized the specific advantages of having nano-scale dimension sensors.[9] In most instances, there is an explicit functional dependence of optode characteristics on the sensor radius $r$. For instance, the absolute detection limit decreases with $r^3$ (good!) and the response time is reduced as $r^2$ (good!). The signal-to-noise ratio, though, decreases with $r$ (bad!) but not $r^3$ (luckily!) under standard working conditions. Other features that improve, as sensors get smaller, include sample volume, sensitivity, invasiveness, spatial resolution, dissipation of heat in sensor or sample, toxicity, and materials cost. Features that may worsen include fluorophore leaching and photodamage to sensor and sample.

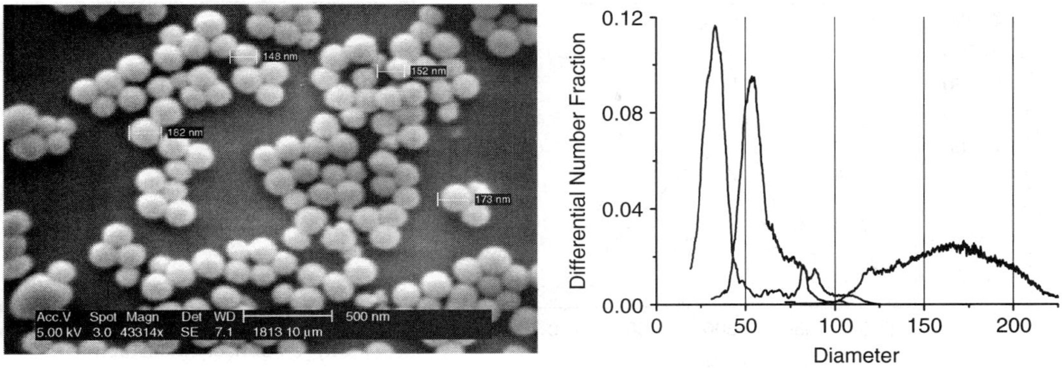


**FIGURE 59.2** Left: Typical scanning electron microscope (SEM) image of sol-gel PEBBLEs. Note the 500-nm scale bar in the image legend used to determine ~160-nm average particle size. Right: Light-scattering results for (left to right) one polyacrylamide and two different sol-gel PEBBLE formulations.

## 59.2 Practical Concept Examples

It is useful to point out concrete examples of the features discussed above before delving into the details of PEBBLE production and application. All PEBBLEs must be well characterized before use, including such measures as nanoparticle size and response calibration. Other essential metrics include tests for constituent leaching, ratiometric stability, response time, and sensitivity to interference from similar analytes and nonspecific protein binding.

### 59.2.1 Ratiometric Sol-Gel Oxygen Sensor: Size, Signal, and Calibration

Depending on the size and matrix material, transmission electron microscropy (TEM), scanning electron microscopy (SEM), and light-scattering measurements are used for PEBBLE-size characterization. The sol-gel, hybrid organic/inorganic silica matrix is typically produced in the 50- to 200-nm size range, as shown by Figure 59.2.

These sol-gel PEBBLEs contain a ruthenium-based dye, $[Ru(dpp)_3]^{2+}$, that has an intensity decrease due to excited state quenching in the presence of molecular oxygen. As a spectrally separated intensity reference, the PEBBLEs also include Oregon Green-488® (Molecular Probes, Eugene, OR), which is insensitive to changes in local oxygen concentrations. Figure 59.3 shows spectra of these PEBBLEs in aqueous solution, in the presence of varying concentrations of oxygen. It is very clear that the Oregon Green reference peak, on the left, remains constant, while the ruthenium peak, on the right, changes in intensity. Also shown (right) is the Stern-Volmer (calibration) plot of fluorescence-intensity ratio vs. oxygen concentration. Although the performance of the sol-gel PEBBLEs is slightly reduced in the aqueous phase as opposed to the gas phase, the sensors still demonstrate good reversibility and reproducibility.[10] The dashed line in Figure 59.3 (right) shows the extent of the biologically relevant oxygen concentrations. The sensors showed at least 95% recovery each time the sensing environments were changed among air-, $O_2$-, or $N_2$-saturated sensor solutions.

### 59.2.2 Ratiometric Zinc PEBBLE Insensitive to Protein Interference

Here, the polyacrylamide (PAA) zinc sensor, based on Newport Green® (Molecular Probes), a zinc-sensitive dye, and Texas Red® (Molecular Probes), a spectrally distinct intensity reference, shows the advantages of PEBBLEs. Quantitative measurements show these sensors to be insensitive to changes in excitation intensity as well as providing protection from nonspecific protein interference. Although Newport Green has good selectivity over intracellular ions, the dye itself is prone to artifacts resulting

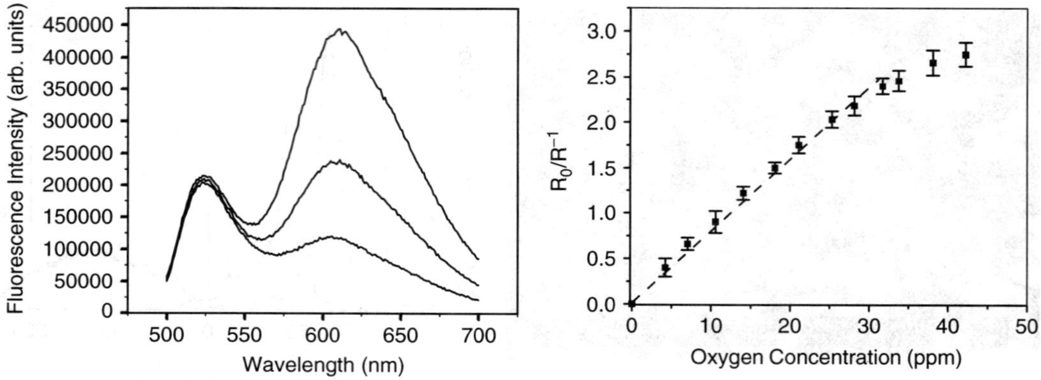


**FIGURE 59.3** Left: Aqueous-phase emission spectra of sol-gel oxygen PEBBLEs excited at 488 nm — top line, PEBBLE solution purged with $N_2$; middle line, PEBBLE solution purged with air; bottom line, PEBBLE solution purged with $O_2$. Right: Stern-Volmer plot of relative-fluorescence-intensity ratios for ratiometric sol-gel oxygen PEBBLEs in aqueous phase. Dashed line denotes biologically relevant range.

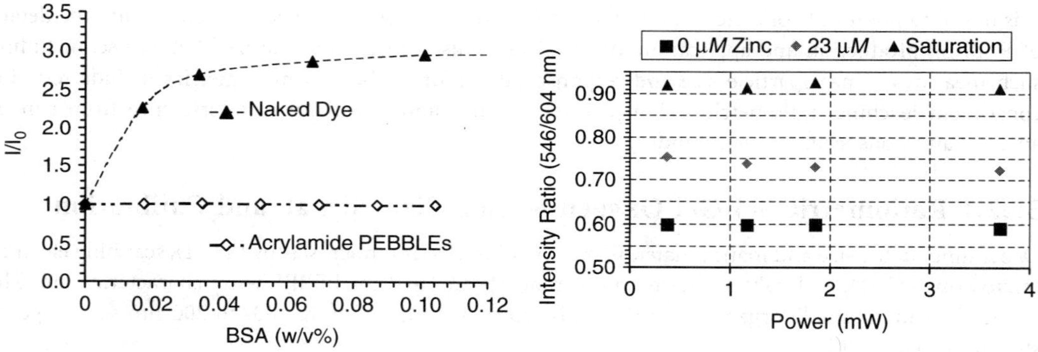


**FIGURE 59.4** Left: Normalized Newport Green emission (530 nm) after addition of successive aliquots of a 10% (w/v) bovine serum albumin (BSA) solution. As little as 0.02% BSA causes a greater than 200% increase in Newport Green-free (naked) dye intensity, but the intensity of the dye embedded in a PAA PEBBLE remains unchanged. Right: Fluorescence-emission-intensity ratio (545 nm/604 nm) from a 10 mg/ml PEBBLE suspension in 10 m*M* Tris buffer monitored using neutral density filters (1.0, 0.5, and 0.3) to attain varied excitation powers at three different zinc concentrations.

from nonspecific binding of proteins, such as bovine serum albumin (BSA), as shown in Figure 59.4 (left). Monitoring of Newport Green's peak at 530 nm reveals a substantial increase in the peak intensity with each successive addition of BSA. The PEBBLEs containing the Newport Green dye, however, are unaffected by the addition of BSA. As little as 0.02% BSA causes an intensity increase of over 200% in the naked Newport Green dye, but the intensity of the Newport Green embedded in the sensor remains unchanged, even at BSA concentrations above 0.10%.[11]

Figure 59.4 (right) demonstrates the advantage of using an integrated ratiometric device over a single-intensity-based dye. Four different excitation-light levels were used for zinc sensing. Although the absolute intensity of fluorescent emission for each dye decreased with decreasing illumination power, the ratio of peak intensities — of Newport Green and Texas Red — remained constant. It is evident from this that fluctuations in the intensity of either a laser or arc lamp would complicate quantitative analysis for intensity-based measurements, while the ratiometric PEBBLEs eliminate the artifacts resulting from power fluctuations. The equivalent would be true as well for insensitivity to fluctuations in the local PEBBLE concentration.

## 59.3 PEBBLE Production Techniques

PEBBLEs represent an advance in nanooptode technology. The science of nanooptode production relies on advances in nanoscale production, using emulsion and dispersion fabrication techniques. The nanoemulsion/dispersion process for preparing PEBBLEs is subtle, and there is no universal method for making hydrophilic, hydrophobic, and amphiphilic nanospheres that contain the right matrix and right chemical components in their proper proportions. Thus, switching from single-dye-containing hydrophilic PAA nanospheres to multicomponent, hydrophobic, liquid-polymer sensors or to inert glass, sol-gel sensors is not yet a routine procedure. However, the production methods, once optimized for a given matrix and its constituents, are based on relatively simple wet-chemistry techniques, as opposed to many complicated physical and chemical nanotechnology schemes. Specific methods for producing sensors from all these matrices as well as the related response mechanisms for each type of sensor are described below.

### 59.3.1 Polyacrylamide Hydrogel

In PAA polymer PEBBLEs, a dye that has a chromometric response to an analyte is entrapped in the matrix pores. Extraction of analyte ions into the hydrogel is not a consideration, though, because water and small ions diffuse freely through the hydrogel. What does occur is the formation of a chromoionophore–analyte complex, similar to the response of the "naked" dye in solution. The dynamic range and selectivity of the PEBBLE depends on the $K_D$ of the dye with respect to the analyte and any interfering ions.

The production of acrylamide PEBBLEs is based on the nanoemulsion techniques studied by Daubresse et al.[12] Some control over particle size and shape can be gained by adjusting surfactant-to-water ratios in the emulsion. The typical polymerization solution consists of 0.4 m*M* fluorescent ionophore (any hydrophilic dye selective for the analyte of interest), 27% acrylamide (monomer), 3% *N,N*-methylenebis(acrylamide) (cross-linker), all in 0.1 *M* phosphate buffer, pH 6.5. One milliliter of this solution is then added to a solution containing 20 ml hexane, 1.8 mmol dioctyl sulfosuccinate sodium salt (surfactant), and 4.24 mmol Brij 30 (surfactant). The solution is stirred under nitrogen for 20 min while cooling in an ice bath. The polymerization is initiated with 24 μl of a 10% ammonium persulfate solution and 12 μl TEMED (initiators); the solution is then allowed to stir at room temperature for 2 h. Hexane is removed by rotary evaporation, and the probes are then rinsed of surfactant with ethanol to give a majority 40-nm probes.[13,14]

### 59.3.2 Sol-Gel Silica/Organic Hybrid

Sol-gel glass has also been used as the matrix for the fabrication of PEBBLE nanosensors because of the superior properties it has for some applications over organic polymers. Sol-gel glass is a porous, high-purity, optically transparent, and homogeneous material,[15] making it an ideal choice as a sensor matrix for quantitative spectrophotometric measurements. Also, it is chemically inert and more photo and thermally stable than polymer matrices. The preparation of sol-gel "glasses" is technically simple, and tailoring the physicochemical properties (i.e., pore size or inner-surface hydrophobicity) of sensor materials can be achieved easily by varying the processing conditions and the concentration or type of reactants used. This enables the pore sizes to be optimized such that the analyte is able to diffuse easily and interact with the sensing molecules, while the latter are prevented from leaking out of the matrix (also true for PAA-based sensors). Furthermore, this "glass" is produced under so-called soft chemical conditions, i.e., low temperatures and relatively mild pH conditions, allowing the inclusion of organic dyes and even biomolecules. It may also be "hybridized" with organic polymers, as shown in the example below.

The reaction solution for the production of oxygen-sensitive sol-gel PEBBLEs consists of the organic, hydrophilic polymer polyethylene glycol (PEG), MW 5000 monomethyl ether (3 g), ethanol (200 proof, 6 ml), Oregon Green-dextran MW 10,000 (0.1 m*M*), $[Ru(dpp)_3]^{2+}$ (0.4 m*M*), and 30% weight ammonia water (3.9 ml), with ammonia serving as catalyst and water being one of the reactants. Upon mixing, the solution becomes transparent and the inorganic "monomer" tetraethyl orthosilicate (TEOS) (0.5 ml) is added dropwise to initiate the hydrolysis of TEOS. The solution is then stirred at room temperature for 1 h to allow the sol-gel reaction (analogous to polymerization) to reach completion. A liberal amount of ethanol is then added to the reaction solution, and the mixture is transferred to an Amicon ultrafiltration cell (Millipore Corp., Bedford, MA). A 100-kDA membrane is used to separate the reacted sol-gel particles (PEBBLEs) from the unreacted monomers, PEG, ammonia, and dye molecules, under a pressure of 10 psi. The PEBBLEs are further rinsed with 500 ml ethanol to ensure that all unreacted chemicals have been removed. The PEBBLE solution is then passed through a suction filtration system (Fisher, Pittsburgh, PA), with a 2-μm filter membrane, to separate the larger particles from the smaller ones. The filtrate (containing the smaller particles) is filtered again, this time with a 0.02-μm filter membrane, to collect the particles, which are then dried to yield a final product consisting of sol-gel PEBBLEs in the size range of 100 to 400 nm in diameter.[10]

### 59.3.3 Decyl Methacrylate Hydrophobic Liquid Polymer

The use of fluorescent indicator molecules in encapsulated form (acrylamide PEBBLEs) has proven valuable in the study of a number of intracellular analytes[11,13,14,16] ($H^+$, $Ca^{2+}$, $Mg^{2+}$, $Zn^{2+}$, $O_2$); however, many ions exist for which no fluorescent indicator dye is sufficiently selective or even available. An alternate class of tandem optical nanosensors is thus required, driving the development of decyl methacrylate (hydrophobic) liquid-polymer PEBBLEs.

A batch of decyl methacrylate PEBBLE sensors is typically made from 210 mg decyl methacrylate, 180 mg hexanedioldimethacrylate, and 300 mg dioctyl sebacate (DOS), with 10 to 30 mmol/kg each of ionophore, chromoionophore, and ionic additives added after spherical-particle synthesis. The spherical particles are prepared by dissolving decyl methacrylate, hexanedioldimethacrylate, and dioctyl sebacate in 2 ml hexane. To a 100-ml round-bottom flask, in a water bath on a hot plate stirrer, 75 ml of pH 2 HCl is added along with 1793 mg of PEG 5000 monomethyl ether and stirred and degassed. The hexane-dissolved monomer cocktail is then added to the reaction flask (under nitrogen) and stirred at full speed, and water bath temperature is raised to 80°C over 30 to 40 min. Potassium peroxodisulfate (6.0 mg) is then added to the reaction, and stirring is reduced to medium speed. The temperature is kept at 80°C for two more hours, after which time the reaction is allowed to return to room temperature and stir for 8 to 12 h. The resulting polymer is suction filtered through a glass microanalysis vacuum filter holder with a Whatman Anodisc filter (0.2-μm pore diameter). The polymer is rinsed three times with water and three times with ethanol to remove excess PEG and unreacted monomer. Tetrahydrofuran (THF) is then used to leach out the DOS and then the PEBBLEs are again filtered and rinsed. They are allowed to dry in a 70°C oven overnight. Dry polymer is then weighed out, and DOS, ionophore, chromoionophore, and ionic additives are added to this dry polymer, so that the resulting polymer will have 40% DOS, 20 mmol/kg ionophore, 10 mmol/kg chromoionophore, and 10 mmol/kg ionic additive. Enough THF is added to this mixture so as to just wet the PEBBLEs. The PEBBLEs are allowed to swell for 8 h, and the THF is then removed by rotary evaporation. The resulting PEBBLE sensors are rinsed with doubly distilled water and allowed to air dry.

## 59.4 Delivery Methods

One of the most important considerations when applying PEBBLE nanosensors to single-cell studies is the (noninvasive) delivery of the PEBBLEs to the cell. The many methods that have been explored include gene gun, picoinjection, liposomal delivery, and sequestration (phagocytosis and pinocytosis) into macrophages. All of these methods are summarized in Figure 59.5.

The method of PEBBLE delivery by gene gun can best be thought of as a shotgun method. PEBBLEs are dried on a plastic (delivery) disk, and this disk is set in front of a rupture disk. Helium pressure is built up behind the rupture disk, which ruptures at a specific helium pressure and propels the PEBBLEs

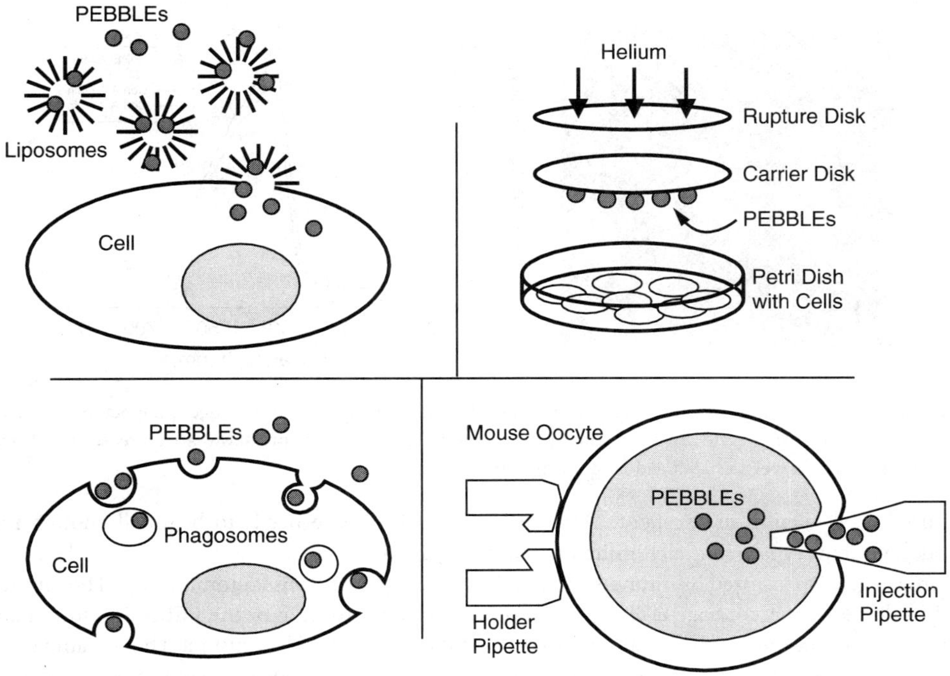


**FIGURE 59.5** Range of delivery methods currently available for PEBBLE nanosensors into single cells for bioanalysis. Clockwise, from upper left, liposomal delivery, gene gun, picoinjection, and phagocytosis.

from the plastic disk into a cell culture. The gene gun can be used to deliver one to thousands of PEBBLEs per cell into a large number of cells very quickly (depending on the concentration of PEBBLEs on the delivery disk).[10,13,16,17] Cell viability is excellent (98% vs. control cells[16]) for small numbers of PEBBLEs and hinges directly on the number of PEBBLEs delivered, the delivery pressure, and the chamber vacuum. The PEBBLE momentum determines whether the PEBBLEs are internalized mainly in the cytoplasm or in the nucleus.

Picoinjection is used to inject picoliter (pl) volumes of PEBBLE-containing solution into single cells (Figure 59.5). This method of delivery depends on the fabrication of pulled capillary "needles" through the use of a pipette-puller and a microforge. The smallest volume deliverable is 10 pl, and the most concentrated PEBBLE solution to work in the pulled capillary syringe is 5 mg/ml PEBBLEs. The maximum number of PEBBLEs one can put in depends on the volume of solution that can be injected without damaging the cell. Picoinjection can give a wide range of PEBBLE concentrations in the cell, and cell viability is good (if done by an expert), but because each cell must be individually injected, the method is time-consuming and tedious.[16]

Commercially available liposomes can also be used to deliver PEBBLEs to cells. The liposomes are prepared in a solution of PEBBLEs and then placed in the cell culture where the liposomes fuse with the cell membranes and empty their contents (the PEBBLE containing solution) into the cell. Three factors play a key role in determining the number of PEBBLEs delivered to each cell with this method — the original concentration of the PEBBLEs, the concentration of liposomes placed in the cell culture, and the length of time the liposomes are left with the cells.[14,16,18] The parameters must be tailored for each cell line used to obtain the desired concentration of PEBBLEs in the cells. While it would be difficult to deliver a single PEBBLE to each cell with this method, it does seem that a low end of 10 to 50 PEBBLEs per cell would be possible, with the high end being the maximum number of PEBBLEs the cell could take without losing viability. Liposomal delivery is useful for delivering PEBBLEs to many cells simultaneously. The challenge is in tailoring the delivery — for the concentrations desirable and for the cell line

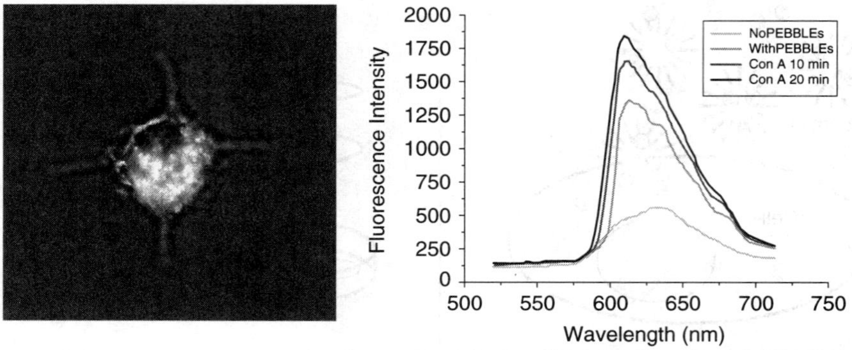


**FIGURE 59.6** Confocal-microscope image (left) of alveolar macrophage with embedded, phagocytosed polyacrylamide PEBBLEs containing Calcium Crimson dye. Fluorescence spectra (right) show an increase in intracellular calcium after cells have been challenged by concanavalin A (Con A).

being used. Cell viability is excellent. Obviously, the PEBBLE size needs to be small enough for this method, and delivery is essentially limited to the cell cytoplasm.

Macrophages (specialized immune-system cells) take up PEBBLEs automatically. The number of PEBBLEs that each macrophage takes up depends on the concentration of the PEBBLE solution and the amount of time the macrophages are allowed to stay in the PEBBLE solution. The advantage of this delivery method is that varying concentrations of PEBBLEs can easily be delivered to macrophages. The disadvantages are that it is useful mainly for macrophages (which are hard to culture) and that PEBBLEs are internalized only into certain cell regions. This method also provides excellent cell viability.[16]

## 59.5 *In Vitro* Bioanalysis

### 59.5.1 Calcium PAA PEBBLEs

The first PEBBLEs produced were acrylamide-based, and one of the first examples of their successful application to cells was with macrophages. Alveolar macrophages were recovered from rat lung lavage using Krebs-Henseleit buffer. Macrophages were maintained in a 5% $CO_2$, 37°C incubator in Dulbecco's modified eagle medium (DMEM) containing 10% fetal bovine serum and 0.3% penicillin, streptomycin, and neomycin. PEBBLE suspensions ranging from 0.3 to 1.0 mg/ml were prepared in DMEM and incubated with alveolar macrophage overnight. Macrophage images were then taken on a confocal microscope, and spectra of the same cells were obtained on the fluorescent microscope (shown in Figure 59.6). Acrylamide PEBBLEs selective for calcium (containing Calcium Crimson™, Molecular Probes, in the acrylamide matrix)[16] were used to monitor calcium in phagosomes within rat alveolar macrophage because of the ease with which macrophage phagocytose particles. This method for delivering the PEBBLEs into cells provided a simple, yet important, test of the PEBBLE sensors in a challenging (acidic) intracellular environment. Macrophages that had phagocytosed 20 nm calcium-selective PEBBLE sensors were challenged with a mitogen, concanavalin A (Con A), inducing a slow increase in intracellular calcium, which was monitored over a period of 20 min. PEBBLE clusters confined to the phagosome enabled correlation of ionic fluxes with stimulation of this organelle.

The calcium PEBBLE in the macrophage experiment clearly demonstrates a time-resolved observation of a biological phenomenon in a single, viable cell. Clearly, relevant time-domain data can be obtained with a fluorescence microscope, spectrograph, and charge-coupled device (CCD). With a confocal-microscope system and the appropriate dye/filter sets, both temporal and spatial resolution can be achieved, as demonstrated below.

Calcium PEBBLEs have also been developed using Calcium Green™-1 (Molecular Probes) dye, in combination with sulforhodamine dye, as sensing components. We note that Calcium Green fluorescence

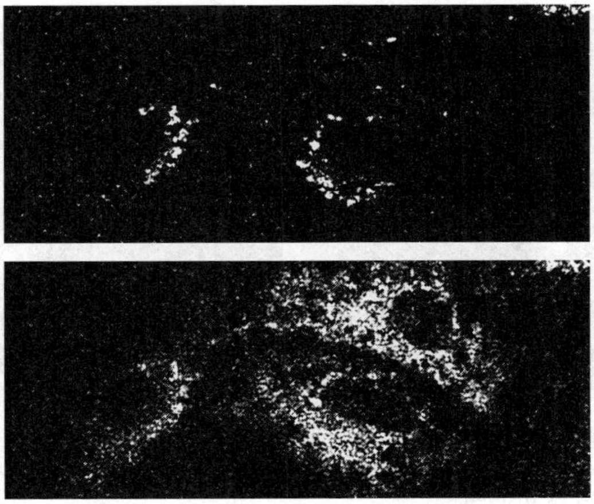

**FIGURE 59.7** Confocal-microscope image, split into green (top) and red (bottom) channels, of human C6 glioma cells containing Calcium Green/sulforhodamine (reference dye) PEBBLEs (toxin diffusing left to right releases mitochondrial calcium into cytosol, as seen by lack of green on right side of image).

increases in intensity with increasing calcium concentrations, while the sulforhodamine fluorescence intensity remains unchanged, regardless of the biologically relevant concentration of ions, pH, or other cellular component; thus, the ratio of the Calcium Green/sulforhodamine intensity gives a good indication of cellular calcium levels, regardless of dye or PEBBLE concentration or fluctuations of light source intensity. Figure 59.7 shows a confocal-microscope image of human C6 glioma cells containing Calcium Green/sulforhodamine PEBBLEs. The top image of the pair is the light intensity from the green (calcium sensitive) fluorescence, and the bottom shows the red intensity (reference), with both dyes confined in the same PEBBLEs. The PEBBLEs were delivered by liposomes to the cytoplasm of the cells. The toxin *m*-dinitrobenzene (DNB) was introduced to the left side of the image and allowed to diffuse to the right. The effect of DNB was the disruption of mitochondrial function, followed by the uncontrolled release of calcium associated with onset of the mitochondrial permeability transition (MPT).[18] Calcium PEBBLEs were used to determine that the half-maximal rate of calcium release ($EC_{50}$) occurred at a tenfold lower concentration of m-DNB in human SY5Y neuroblastoma cells than in human C6 glioma cells.[18]

## 59.5.2 Aqueous Oxygen Sol-Gel PEBBLEs

Sol gel, the newest PEBBLE matrix, gives the flexibility of being able to tailor the properties of the matrix to accept either hydrophilic or hydrophobic dyes. Also, for oxygen their dynamic range is much wider than that of similar acrylamide PEBBLEs.[10] It has also proven to be a matrix compatible with the use of protein-based sensors.[15] Sol-gel PEBBLEs were inserted with a gene gun into rat C6 glioma cells to monitor oxygen. A ratiometric sol-gel PEBBLE sensor ($[Ru(dpp)_3]^{2+}$ oxygen-sensitive dye and Oregon Green™ 488-dextran reference dye) was used.[10] Figure 59.8 shows the confocal images of C6 glioma cells containing sol-gel PEBBLEs under Nomarski illumination overlaid with: (left) the green fluorescence of Oregon Green 488-dextran and (right) the red fluorescence of $[Ru(dpp)_3]^{2+}$. Evidently, the cells still maintained their morphology after the gene-gun injection of PEBBLEs and showed no sign of cell death. The dyes were excited by reflecting, respectively, the 488-nm (Ar-Kr) and the 543-nm (He-Ne) laser lines onto the specimen using a double dichroic mirror. The Oregon Green fluorescence from the PEBBLEs inside the cells (Figure 59.8 left) was detected by passage through a 510-nm long-pass and a 530-nm shortpass filter and the fluorescence of $[Ru(dpp)_3]^{2+}$ (Figure 59.8 right) through a 605-nm (45 nm bandpass) barrier filter. A 40×, 1.4 NA oil-immersion objective was used to image the Oregon Green and $[Ru(dpp)_3]^{2+}$ fluorescence. The distribution of PEBBLEs in overlaid images demonstrated that the green and red fluorescence in Figure 59.8 were truly from PEBBLEs inside cells. Most of the PEBBLEs were loaded into the cytoplasm, but there were also some in the nucleus.

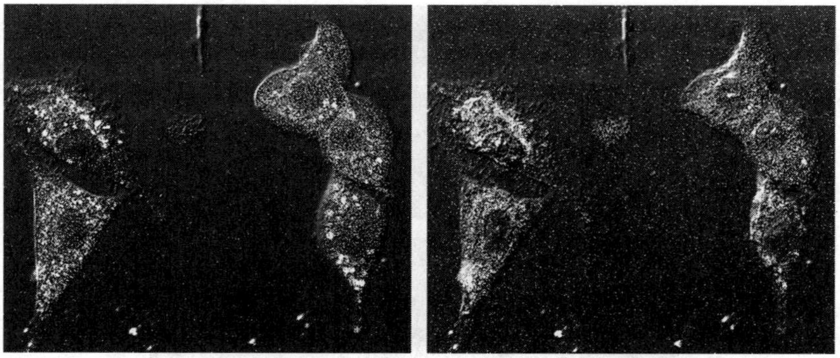

**FIGURE 59.8** Confocal images of rat C6 glioma cells loaded with sol-gel PEBBLEs by gene-gun injection. Nomarski illumination image overlaid with Oregon Green fluorescence (reference, left) and $[Ru(dpp)_3]^{2+}$ fluorescence (right) of the same ratiometric PEBBLEs inside cells.

**TABLE 59.1** Experimental Ratiometric, *in Vitro* Oxygen Results

| Average Intracellular-Oxygen Concentrations (ppm) | Air-Saturated Buffer Solution 8.8 ± 0.8 |
|---|---|
| Cells in air saturated buffer | 7.9 ± 2.1 |
| Cells in $N_2$ saturated buffer (after 25 sec) | 6.5 ± 1.7 |
| Cells in $N_2$ saturated buffer (after 120 sec) | ≤1.5 |

After gene-gun injection, the cells were immersed in DPBS (Dulbecco's phosphate buffered saline), and a spectrum was taken of these cells using 480 ± 10 nm excitation light. The air-saturated DPBS was then replaced by nitrogen-saturated DPBS, causing a decrease in the intracellular-oxygen concentration, and the response of the oxygen PEBBLE sensors inside the cells was monitored during a time period of 2 min. The fluorescence intensity of $[Ru(dpp)_3]^{2+}$ increased successively, indicating that the oxygen level inside the cells had decreased. Average intracellular-oxygen concentrations were determined on the basis of a Stern-Volmer calibration curve, obtained using the fluorescence-microscope Acton spectrometer system,[10] and are summarized in Table 59.1. The comparatively large errors are due to the low resolution of the spectrometer. Note that the measured intracellular-oxygen value (when cells were in air saturated DPBS) is comparable to the value of ~7.1 ppm measured electrochemically inside the much larger islets of Langerhans.[19] These results show that the PEBBLE sensors are responsive when loaded into cells and that they retain their spectral characteristics, enabling a ratiometric measurement to be made.[10]

## 59.5.3 Potassium Decyl Methacrylate PEBBLEs

The acrylamide PEBBLE matrix has been proven to work with any hydrophilic sensing components. However, it is unable to take advantage of the rich history of electrochemical sensors where there exist a host of highly selective, hydrophobic ionophores. In many cases, the selectivity of these ionophores has yet to be matched by hydrophilic dyes (chromoionophores). Highly selective intracellular (and extracellular) hydrophilic indicator dyes are limited to a small set of analytes, such as pH and calcium. While the use of PEBBLEs instead of traditional free "naked" indicators results in protection beneficial to both the cell and the dye, it does not solve the selectivity problems. For instance, hydrophilic potassium indicators will not work in the presence of significantly higher sodium concentrations, and, conversely, sodium indicators will not work in the presence of high potassium concentrations.[20] Obviously, this has serious implications for both intracellular (e.g., high potassium/sodium ion ratios) and extracellular (e.g., high sodium/potassium) applications. Moreover, for many important analyte ions, such as nitrite, no satisfactory color indicators are available. The above problem has been solved in optodes by using *in tandem* an optically silent ionophore (which is highly selective) and a next-door optically visible agent

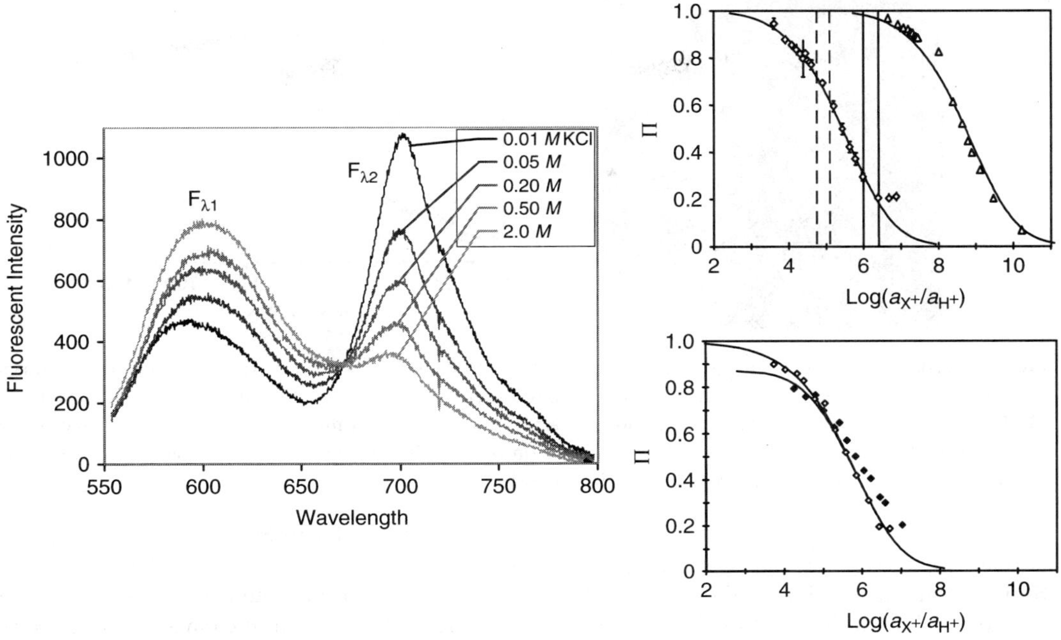


**FIGURE 59.9** Left: Normalized emission spectra from suspended $K^+$ PEBBLE sensors using the pH chromoionophore ETH5350 for ion-correlation spectroscopy in tandem with BME–44. The spectra show response in going from 10 m*M* KCl to 2.0 *M* KCl (well beyond saturation of the sensor), all in 10-m*M* Tris buffer, pH 7.2. Top right: Response of same PEBBLEs to potassium (o) and sodium (Δ), along with theoretical curves. The lines delimit values for log ($a_{K+}/a_{H+}$) typically found in intracellular (solid) and extracellular (dashed) media.[28] Bottom right: Response to additions of KCl in Tris buffer (o) compared to a similar experiment run in a constant background of 0.5 *M* $Na^+$ (□). Solid lines are calculated (not fit) theoretical curves for the $K^+$ response in the presence of 0.5-*M* interfering $Na^+$ using the experimentally determined log-selectivity value of –3.3. $X^+ = K^+$ or $Na^+$.

that plays the role of a *spectator*, or *reporter dye*. While the principles of such tandem sensing schemes were worked out by Bakker and Simon,[21–23] Suzuki,[24,25] and Wolfbeis,[26,27] the first demonstration of such a sensing scheme on the nanoscale occurred with the pulled optodes developed by Shortreed et al.[5,6] The extension of these principles to PEBBLEs required the optimization of a new liquid polymer matrix, decyl methacrylate.[17]

The work described here takes advantage of an indicator with two fluorescence-emission maxima ($\lambda_1$,$\lambda_2$), giving a relative intensity that changes with the degree of protonation (Π). This degree of protonation, Π, can be evaluated in terms of the ratio of the protonated chromoionophore intensity $F_{\lambda 2}$ to the deprotonated chromoionophore intensity $F_{\lambda 1}$ (see Figure 59.9 for spectra) based on an analytically derived relationship.[5]

The degree of protonation (Π) of the indicator spectra obtained from the PEBBLE calibration is related to the analyte concentration by use of the theoretical treatment of ion-exchange sensors developed by Simon, Bakker et al.[5,21–23] For the incorporation of a selective neutral ionophore (BME-44) into a matrix, along with a selective chromoionophore (ETH 5350) for indirect ion monitoring (ion exchange sensors), the metal-ion activity ($a_{K+}$) in solution is a function of the hydrogen-ion activity in solution ($a_{H+}$), the interfering cation activity ($a_{Na+}$), and the constants [$L_{tot}$], [$C_{tot}$], [$R_{tot-}$], which are total ionophore (ligand) concentration, total chromoionophore concentration, and total lipophilic charge-site concentration in the membrane. Note that [CH] is the protonated chromoionophore concentration and [C] is the free-base concentration. The parameter Π has been defined[5–9] as the relative portion of the protonated chromoionophore, $\Pi = [CH]/[C_{tot}]$.

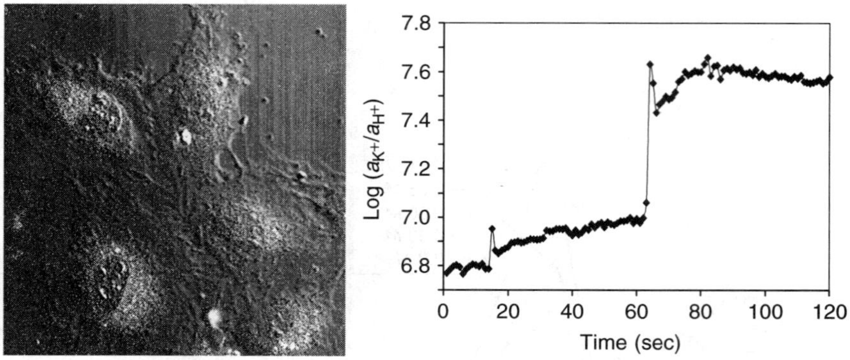


**FIGURE 59.10** Left: Confocal image of decyl methacrylate $K^+$ PEBBLE fluorescence overlaid with Nomarski image of rat C6 glioma cells (488-nm excitation, 580-nm long-pass emission). Right: Ratio data of decyl methacrylate $K^+$ PEBBLEs in C6-glioma cells during the addition of kainic acid (50 μl of 0.4 mg/ml) at 20 sec and at 60 sec. Ratios were converted to log ($a_{K+}/a_{H+}$) using solution calibration of the PEBBLEs. Log ($a_{K+}/a_{H+}$) is seen to increase after kainic acid addition (and subsequent $K^+$ channel openings).

Calibration of a $K^+$ sensor based on these principles is shown in Figure 59.9 (right top) along with normalized spectra (left). For potassium sensing, the chromoionophore is ETH 5350, the ionophore is BME-44, and the lipophilic additive is KTFPB.[6,17] The data points for potassium and sodium responses are plotted along with corresponding theoretical curves. Dashed lines delimit the range of typical extracellular activity ratios, and the solid lines delimit the intracellular levels [log ($a_{K+}/a_{H+}$)].[28]

The response was found to match well with the theory, which was gratifying, considering the small size of the systems. The dynamic range at pH 7.2 extended from 0.63 m*M* to 0.63 *M* $a_{K+}$. The log of the selectivity for potassium vs. sodium, determined by measuring the horizontal separation of the response curves at $\Pi = 0.5$, was –3.3. This selectivity value can be used, along with the mathematical theory for this sensing mechanism, to calculate what the $K^+$ response of the PEBBLEs should be in the presence of 0.5 *M* interfering $Na^+$. Figure 59.9 (lower right) shows these calculated theoretical curves (not fits) along with the corresponding experimental data. This shows a selectivity similar to or better than that obtained for other and larger matrices incorporating BME-44, e.g., –3.1 in PVC-based fiber-optic work, and –3.0 in PVC-based microelectrodes.[6,17] It also exactly matches the value given in the review by Buhlmann et al.[23] for a thin PVC film sensor. This selectivity should be more than sufficient for measurements in intracellular media where potassium concentration[28] is about 100 m*M* and sodium is about 10 m*M*.

The first application of this liquid-polymer class of PEBBLEs was the observation of potassium uptake in rat C6 glioma cells.[17] Decyl methacrylate PEBBLEs were delivered by gene gun using a BioRad (Hercules, CA) Biolistic PDS-1000/He system, with a firing pressure of 650 psi and a vacuum of 15 torr applied to the system. Immediately following PEBBLE delivery, cells were placed on an inverted fluorescent microscope. The gating software for the CCD was set to take continuous spectra at 1.3-sec intervals. After 20 sec and after 60 sec, 50 μl of 0.4 mg/ml kainic acid was injected into the microscope cell. Kainic acid is known to stimulate cells by causing the opening of ion channels. Figure 59.10 (left) shows the confocal fluorescent image of the PEBBLEs overlaid with a Nomarski differential interference contrast image of the cells.[17] Image analysis indicated that the PEBBLE sensors were localized in the cytoplasm of the glioma cells. Figure 59.10 (right) shows the PEBBLE sensors inside the cells responding to the kainic-acid addition. Log ($a_{K+}/a_{H+}$) clearly increases, indicating either an increase in $K^+$ concentration or a decrease in $H^+$ concentration (increase in pH). The amount of kainic acid added is not known to affect the pH of cells in culture, and kainic acid by itself has no effect on the sensors. Thus, the change is likely due to increasing intracellular concentration of $K^+$, which is the expected trend. The membrane of C6 glioma cells can initiate an inward rectifying $K^+$ current, induced by specific $K^+$ channels, a documented role in the control of extracellular potassium.[29] Thus, when stimulated with a channel-opening agonist, the $K^+$ concentration within the glioma cells is indeed expected to increase.

## Acknowledgments

The authors acknowledge the contributions of Dr. Heather Clark, Dr. Jon Aylott, James Sumner, Hao Xu, Dr. Steve Parus, Dr. Ron Tjalkens, Terry Miller, and Dr. Marion Hoyer, and support of National Institutes of Health grant R01-GM-50-300 and Defense Advanced Research Projects Agency grant MDA972-97-1-006.

## References

1. Tan, W., Shi, Z.-Y., Smith, S., Birnbaum, D., and Kopelman, R., Submicrometer intracellular chemical optical fiber sensors, *Science,* 258, 778, 1992.
2. Tan, W., Kopelman, R., Barker, S.L.R., and Miller, M.T., Ultrasmall optical sensors for cellular meaurements, *Anal. Chem.*, 71(17), 606A, 1999.
3. Rosenzweig, Z. and Kopelman, R., Development of a submicrometer optical fiber oxygen sensor, *Anal. Chem.*, 67, 2650, 1995
4. Rosenzweig, Z. and Kopelman, R., Analytical properties and sensor size effects of a micrometer sized optical fiber glucose biosensor, *Anal. Chem.*, 68, 1408, 1996.
5. Shortreed, M., Bakker, E., and Kopelman, R., Miniature sodium-selective ion-exchange optode with fluorescent pH chromoionophores and tunable dynamic range, *Anal. Chem.*, 68(15), 2656, 1996.
6. Shortreed, M.R., Dourado, S., and Kopelman, R., Development of a fluorescent optical potassium-selective ion sensor with ratiometric response for intracellular applications, *Sensors Actuators B*, 38–39, 8, 1997.
7. Barker, S.L.R., Shortreed, M.R., and Kopelman, R., Utilization of lipophilic ionic additives in liquid polymer film optodes for selective anion activity measurements, *Anal. Chem.*, 69(6), 990, 1997.
8. Barker, S.L.R., Thorsrud, B.A., and Kopelman, R., Nitrite- and chloride-selective fluorescent nano-optodes and *in vitro* application to rat conceptuses, *Anal. Chem.*, 70(1), 100, 1998.
9. Dourado, S. and Kopelman, R., Is smaller better? Scaling of characteristics with size of fiber-optic chemical and biochemical sensors, *Proc. SPIE,* 2836, 2, 1996.
10. Xu, H., Aylott, J.W., Kopelman, R., Miller, T.J., and Philbert, M.A., A real-time ratiometric method for the determination of molecular oxygen inside living cells using sol-gel-based spherical optical nanosensors with applications to rat C6 glioma, *Anal. Chem.*, 73(17), 4124, 2001.
11. Sumner, J.P., Aylott, J.W., Monson, E., and Kopelman, R., A fluorescent PEBBLE nanosensor for intracellular free zinc, *Analyst*, 127(1), 11, 2002.
12. Daubresse, C., Granfils, C., Jerome, R., and Teyssie, P., Enzyme immobilization in nanoparticles produced by inverse microemulsion polymerization, *J. Colloid Interface Sci.*, 168, 222, 1994.
13. Clark, H.A., Barker, S.L.R., Kopelman, R., Hoyer, M., and Philbert, M.A., Subcellular optochemical nanobiosensors: probes encapsulated by biologically localised embedding (PEBBLEs), *Sensors Actuators B*, 51, 12, 1998.
14. Clark, H.A., Hoyer, M., Philbert, M.A., and Kopelman, R., Optical nanosensors for chemical analysis inside single living cells. 1. Fabrication, characterization, and methods for intracellular delivery of PEBBLE sensors, *Anal. Chem.*, 71(21), 4831, 1999.
15. Uhlmann, D.R., Teowee, G., and Boulton, J., The future of sol-gel science and technology, *J. Sol-Gel Sci. Technol.*, 8, 1083, 1997.
16. Clark, H.A., Hoyer, M., Parus, S., Philbert, M., and Kopelman, R., Optochemical nanosensors and subcellular applications in living cells, *Mikrochim. Acta*, 131, 121, 1999.
17. Brasuel, M., Kopelman, R., Miller, T.J., Tjalkens, R., and Philbert, M.A., Fluorescent nanosensors for intracellular chemical analysis: decyl methacrylate liquid polymer matrix and ion-exchange-based potassium PEBBLE sensors with real-time application to viable rat C6 glioma cells, *Anal. Chem.*, 73(10), 2221, 2001.

18. Clark, H.A., Kopelman, R., Tjalkens, R., and Philbert, M.A., Optical nanosensors for chemical analysis inside single living cells. 2. Sensors for pH and calcium and the intracellular applications of PEBBLE sensors, *Anal. Chem.*, 71(21), 4837, 1999.
19. Jung, S.-K.G., Waldemar, A., Aspinwall, C.A., and Kauri, L.M., Oxygen microsensor and its application to single cells and mouse pancreatic islets, *Anal. Chem.*, 71(17), 3642, 1999.
20. Haugland, R.P., *Molecular Probes Handbook of Fluorescent Probes and Research Chemicals*, Molecular Probes, Eugene, OR, 1993.
21. Bakker, E. and Simon, W., Selectivity of ion-sensitive bulk optodes, *Anal. Chem.*, 64(17), 1805, 1992.
22. Morf, W.E., Seiler, K., Lehmann, B., Behringer, C., Hartman, K., and Simon, W., Carriers for chemical sensors: design features of optical sensors (optodes) based on selective chromoionophores, *Pure Appl. Chem.*, 61(9), 1613, 1989.
23. Buhlmann, P., Pretsch, E., and Bakker, E., Carrier-based ion-selective electrodes and bulk optodes. 2. Ionophores for potentiometric and optic sensors, *Chem. Rev.*, 98(4), 1593, 1998.
24. Kurihara, K., Ohtsu, M., Yoshida, T., Abe, T., Hisamoto, H., and Suzuki, K., Micrometer-sized sodium ion-selective optodes based on a "tailed" neutral ionophore, *Anal. Chem.*, 71(16), 3558, 1999.
25. Suzuki, K., Ohzora, H., Tohda, K., Miyazaki, K., Watanabe, K., Inoue, H., and Shirai, T., Fiberoptic potassium-ion sensors based on a neutral ionophore and a novel lipophilic anionic dye, *Anal. Chim. Acta*, 237(1), 155, 1990.
26. Mohr, G.J., Lehmann, F., Ostereich, R., Murkovic, I., and Wolfbeis, O.S., Investigation of potential sensitive fluorescent dyes for application in nitrate sensitive polymer membranes, *Fresenius J. Anal. Chem.*, 357(3), 284, 1997.
27. Mohr, G.J., Murkovic, I., Lehmann, F., Haider, C., and Wolfbeis, O.S., Application of potential-sensitive fluorescent dyes in anion- and cation-sensitive polymer membranes, *Sensors Actuators B-Chem.*, 39(1–3), 239, 1997.
28. Ammann, D., *Ion-Selective Microelectrodes*, Springer-Verlag, Berlin, 1986.
29. Emmi, A., Wenzel, H.J., and Schwartzkroin, P.A., Do glia have heart? Expression and functional role for ether-A-go-go currents in hippocampal astrocytes, *J. Neurosci.*, 20(10), 3915, 2000.

# 60

# Nanosensors for Single-Cell Analyses

Brian M. Cullum*
*Oak Ridge National Laboratory*
*Oak Ridge, Tennessee*

Tuan Vo-Dinh
*Oak Ridge National Laboratory*
*Oak Ridge, Tennessee*



## 60.1 Introduction

Over the years, new techniques in chemical and biological sensing have set the stage for great advances in the field of biological research. One of the most recent technological advances in this area has been the development of nanosensors. A nanosensor is a sensor that has dimensions on the nanometer-size scale. Nanosensors of various types (e.g., optical, electrochemical, etc.) have been reported in the literature over the past decade; however, the emphasis of this review will be on optical-based nanosensors.

Optical nanosensors, like larger sensors, can typically be classified into one of two broad categories, depending on the probe used: chemical nanosensors and biological nanosensors.[1–4] Both types of sensors have been used to provide a reliable method of monitoring various chemicals in microscopic environments. They have even been used in the detection of different species within single cells. Nanosensors offer significant improvements over the current methods of cellular analysis involving the uptake of fluorescent indicator dyes because they do not suffer from problems of cellular diffusion. Although the field of nanosensing is relatively new, there are already several excellent reviews on the subject.[5–9] This chapter will review the evolution of optical nanosensors from their beginning (near-field optical microscopy) to the present (biosensors capable of probing subcellular compartments of individual mammalian cells) and discuss their application to biological measurements.

## 60.2 Sensing Principles

A sensor can be generally defined as a device that consists of a recognition element, often called a receptor, and a transducer. Despite the wide variety of receptors and transducers used, all sensors are based on the same principle. First, the sensor is placed in the environment of interest, where the target molecules bind to the receptors in the biosensitive layer of the sensor. These receptors may be ligands for particular chemical species; biological molecules such as antibodies, enzymes, nonenzymatic proteins, and nucleic acids; or even living biological systems such as cells, tissues, or whole organisms. The resulting interaction

*Current affiliation: University of Maryland Baltimore County, Baltimore, Maryland

between the analyte and the receptor then produces an effect that can be measured by the transducer, which converts this effect into some type of measurable signal (e.g., an electrical voltage). Due to this dual-component nature, sensors are typically classified by either the receptor used for molecular recognition or the type of transduction mechanism used for detection. The most common forms of molecular recognition employed in sensors fall into several different categories: (1) nonbiological ligands, (2) antibody/antigen interactions, (3) nucleic-acid interactions, (4) enzymatic interactions, (5) cellular interactions (i.e., microorganisms, proteins), and (6) interactions using biomimetic materials (i.e., synthetic bioreceptors). Common transduction mechanisms used in sensors include optical measurements (e.g., absorption, fluorescence, phosphorescence, Raman, refraction, dispersion, or interference spectrometry, etc.), electrochemical measurements (e.g., amperiometric, potentiometric, etc.), and mass-sensitive measurements (e.g., surface acoustic wave, microbalance, microcantilever, etc.). This chapter will emphasize the principles of operation of sensors that are based on optical transduction and describe some of the most recent applications of nanosensors to single-cell analyses. A more detailed review of the sensing concept and various types of biosensors is provided in greater detail in Chapter 20 of this handbook.

### 60.2.1 Optical-Transduction Techniques

Optical transduction describes a vast array of measurements (e.g., absorption, fluorescence, phosphorescence, Raman, surfaced-enhanced Raman substrate (SERS), refraction, dispersion spectrometry, etc.) that can be performed with optical sensors. In addition to the large number of optical techniques that can be used in the field of sensing, each of them can be used to measure any of many different optical properties. These properties include amplitude, energy, polarization, decay time, and phase. A brief description of the types of information that can be obtained via optical measurements is described below, while more comprehensive information can be found in other chapters of this handbook.

The most commonly measured parameter of the electromagnetic spectrum is amplitude, as it can generally be easily correlated with the concentration of the analyte of interest. Although amplitude is the most commonly measured parameter, the other parameters can also provide a great deal of information about the analyte of interest. For instance, the energy of the electromagnetic radiation can often provide information about changes in the local environment surrounding the analyte, its intramolecular atomic vibrations (i.e., Raman or infrared-absorption spectroscopies), or the formation of new energy levels (i.e., luminescence, ultraviolet-visible absorption). Measurement of the interaction of a free molecule with a fixed surface can often be determined based on polarization measurements. In the case of free molecules in solution, the emitted light is often randomly polarized, whereas when a molecule becomes bound to a fixed surface, the emitted light generally remains polarized. Therefore, the rate of depolarization of a fluorescence signal can provide information on the movement of a molecule. Information can also be gained by measuring the decay time of a specific emission signal (i.e., fluorescence or phosphorescence) because these decay times depend on the excited state of the molecules and their local molecular environment. Another property that can be measured is the phase of the emitted radiation. When electromagnetic radiation interacts with a surface, the speed or phase of that radiation is altered based on the refractive index of that particular medium (i.e., analyte). Therefore, when an analyte binds to the receptor layer of a sensor, a change in the refractive index of the sensing surface could cause a measurable phase shift in the impinging radiation.

### 60.2.2 Molecular Receptors

One of the key factors in determining the usefulness of any chemical or biological sensor is its ability to specifically measure the analyte of interest without having other chemical species interfere with the measurement. It is the selective binding properties of these receptors that are the key to a sensor's specificity. A wide variety of receptors have been used in the development of sensors over the years. In fact, the various receptors that have been used are more numerous than the different analytes that have been monitored using optical sensors. Although the amount and type of different receptors used are continually growing, they can generally be classified into one of six major categories: (1) nonbiological ligands, (2) antibodies, (3) enzymes, (4) nucleic acids, (5) cellular structures/cells, and (6) biomimetic receptors.

#### 60.2.2.1 Nonbiological Ligands

Nonbiological ligands constitute a very broad category of molecules that range from species with only weak molecular interactions (e.g., hydrogen bonding, van der Waals forces, etc.) to molecules capable of binding or complexing ionic species. Optical sensors employing receptors based on weak-force interactions include indicator dyes immobilized in hydrophilic or hydrophobic membranes, which allow only certain chemical species to come into contact with the indicator dye,[10–13] as well as many other forms of interactions.[14–16] Ligands that rely on ionic interactions or analyte binding include species such as ethylene diamine tetra acetic acid (EDTA), Calcium Green™, fluorescein, etc.

Since a large number of these receptor molecules are based simply on the charge state of the analyte, these receptors often experience a great deal of cross reactivity with other similar species. Due to the cross reactivity of these nonbiological receptors and their subsequent lack of specificity when employed in complex environments, such as biological systems (e.g., cells), there has recently been a shift toward biological receptor molecules, or bioreceptors. These bioreceptors generally possess a much greater specificity than their nonbiological counterparts due to a need for shape recognition as well as other forms of interactions (i.e., hydrogen bonding, ionic interactions, etc.).

#### 60.2.2.2 Antibodies

Antibodies are biological molecules that exhibit very selective binding capabilities for specific structures. Two distinctly different classes of antibodies exist, both of which have been employed as bioreceptors for biosensors: monoclonal antibodies and polyclonal antibodies. When measurement of a very specific analyte is desired, monoclonal antibodies are often used, while polyclonal antibodies, which exhibit less specificity, are often used for the measurement of an entire class of compounds (e.g., PAHs). Because most biological systems are complex in their composition, the highly specific binding abilities of antibodies make them one of the most powerful bioreceptors used by biosensors. However, there are also disadvantages to using antibodies for analyte binding. For example, to produce an immune response to a specific molecule, a certain molecular size and complexity are necessary. Therefore, immunogenic proteins typically have molecular weights of greater than 5 kDa.

#### 60.2.2.3 Enzymes

Enzymes represent another class of bioreceptors that are commonly used in biosensors. In addition to the often highly specific binding capabilities of enzymes, their catalytic activity also makes them very useful. With the exception of a small group of catalytic ribonucleic-acid molecules, all enzymes are proteins, and the function of these proteins varies dramatically from enzyme to enzyme. Some enzymes require no additional chemical groups other than their amino-acid residues for activity, while others require an additional chemical component called a cofactor, which may be either one or more inorganic ions (e.g., $Fe^{2+}$, $Mg^{2+}$, $Mn^{2+}$, $Zn^{2+}$, etc.) or a more complex organic or metallo-organic molecule called a coenzyme. Both types of enzymes have been used in the development of biosensors. Sensors employing enzymes that do not require a cofactor or coenzyme are typically used when the analyte of interest is difficult to measure. In such cases, the reaction of the analyte with the enzyme produces a product that is readily measurable by a particular measurement technique. In addition, the catalytic nature of the bioreceptor means that the sensor is often reusable or reversible.

Biosensors employing enzymes that require an additional chemical component can be extremely sensitive and have extremely low detection limits for that additional chemical component. In such a sensor, the indirect detection of the cofactor or coenzyme of interest is typically performed by measuring a product of the enzymatic reaction. Since a single enzyme along with its respective cofactor or coenzyme can catalyze many reactions and thus produce a large number of reaction products, the indirect measurement of very small amounts of cofactor or coenzyme is possible.

#### 60.2.2.4 Nucleic Acids

Another biorecognition mechanism capable of a high degree of specificity involves the hybridization of deoxyribonucleic acid (DNA) or ribonucleic acid (RNA) to a complementary sequence. Within the last decade, the use of nucleic acids as bioreceptors for biosensor and biochip technologies has grown dramatically.[17–22] The specificity of biorecognition in these DNA-based biosensors, which are often referred to as genosensors, is based on the complementarity of the cytosine:guanosine (C:G) and adenine:thymine (A:T) base pairing. If the sequence of bases comprising a certain segment of the DNA molecule is known, then the complementary sequence, often called a probe, can be synthesized and labeled with an optically detectable compound (e.g., a fluorescent label). By unwinding the double-stranded DNA into single strands, adding the probe, and annealing the strands, the unwound DNA will hybridize to its complementary-probe sequence. Such probe strands were used extensively for the rapid testing of various sequences in the human-genome project.

In addition to simple strands of complementary oligonucleotides with some type of optical label (e.g., fluorescent label, etc.), a relatively new class of bioreceptor, known as a molecular beacon (MB), has been developed. A MB is an oligonucleotide strand that has a long sequence that is complementary to the analyte sequence and, on either end, a small region (several nucleic acids long) that is complementary to the other end. In addition, a fluorescent dye is attached to one end, and a quencher molecule is attached to the opposite end. Therefore, when the MB is not in the presence of the analyte, the two ends bind to each other, keeping the fluorophore and quencher molecules in close proximity and resulting in a nonfluorescent complex. However, in the presence of the analyte, the large loop structure binds to the analyte, causing the two ends to separate and thereby preventing the quencher molecule from acting on the fluorophore and producing a fluorescent complex. Much greater detail about these compounds can be found in Chapter 57 of this handbook.

#### 60.2.2.5 Cellular Structures/Cells

Cellular structures and cells comprise the most diverse category of bioreceptors that have been used in the development of biosensors.[23–57] These bioreceptors are based on biorecognition by either an entire cell/microorganism or a specific cellular component capable of specific binding to certain chemical or biochemical species. Because of its great diversity, this category of bioreceptor can be further divided into three major subclasses: (1) cellular systems, (2) enzymes, and (3) nonenzymatic proteins. However, due to the importance and the large number of biosensors that employ enzymes as their biorecognition element, these have been given their own classification and were discussed previously. In addition, since the focus of this chapter is nanosensors, and whole cells or cellular systems are too large to fit on a sensor with nanometer dimensions, we will focus in this section solely on nonenzymatic proteins.

Many proteins within cells often serve as bioreceptors for intracellular reactions that will take place later or in another part of the cell (e.g., carrier proteins, channel proteins on a membrane, etc.). Regardless of the purpose, these proteins provide a means of molecular recognition through one type or another (i.e., active site or potential sensitive site). Because of the various biorecognition properties of these different nonenzymatic proteins, many different types of biosensors have been constructed for one application or another.

#### 60.2.2.6 Biomimetic Receptors

The final major class of receptor used in biosensors, known as biomimetic receptors, is not comprised of true biological molecules. In fact, biomimetic receptors are human-made receptors that have been designed to mimic biological molecules. Since the fabrication of the first biomimetic receptor for biosensing applications, several different methods have been developed for the construction of such receptors.[58–70] These methods include genetic engineering of molecules, artificial membrane fabrication, and molecular imprinting.

Recombinant techniques that allow for the synthesis or modification of a wide variety of binding sites within biological molecules have provided powerful tools for the design and synthesis of bioreceptors

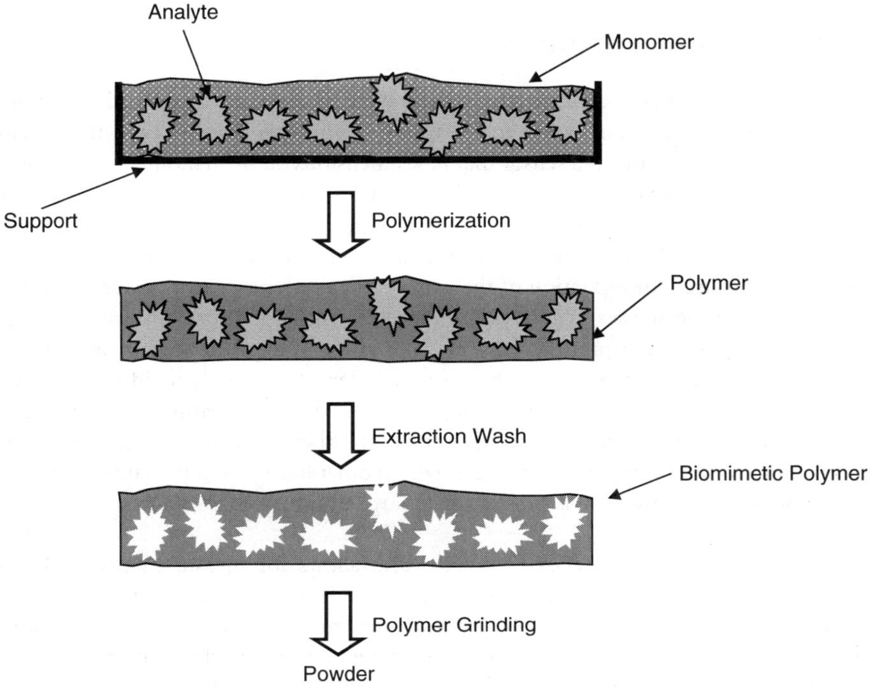


**FIGURE 60.1** Conceptual illustration of the molecular imprinting process.

with desired properties. While biochemists and molecular biologists have been modifying various amino-acid sequences in a variety of proteins for many years, it is only in the past decade that such techniques have begun to be applied to biosensors for the enhancement of a bioreceptor's binding properties (i.e., binding affinity, specificity, etc.).

The second major class of biomimetic receptors — artificial membrane fabrication — has also been used for many different applications over the years. In one such application, Stevens et al. developed an artificial membrane by incorporating gangliosides into a matrix of diacetylenic lipids (5 to 10% of which were derivatized with sialic acid).[71] After the lipids self-assembled into Langmuir-Blodgett layers, they were photopolymerized via ultraviolet irradiation into polydiacetylene membranes. These newly fabricated membranes were then attached to biosensors for the detection of cholera toxins. As the cholera toxins became bound to the membrane they caused the membrane to change color from its original blue color to red, which was subsequently monitored via absorption measurements.

The newest and final class of biomimetic receptors is those created by a technique known as molecular imprinting (see Figure 60.1). This technique consists of mixing analyte molecules with a monomer solution and a large amount of cross linkers. As the monomers begin to polymerize, a solid plastic is created that is full of the particular analyte that is to be measured. Following polymerization, the hard polymer is ground into a fine powder and the analyte molecules are extracted with organic solvents to remove them from the polymer network. After extraction, the resulting polymer powder has molecular holes or binding sites that are complementary to the selected analyte. Therefore, by attaching some of the molecularly imprinted polymer powder on a biosensor, receptor sites for molecules with the exact three-dimensional (3D) structure as the analyte used in the imprinting process are now present. One of the primary reasons for the growing popularity of molecular imprinting is the rugged nature of a polymer relative to a biological sample. Unlike biological molecules, molecularly imprinted polymers (MIPs) can withstand harsh environments such as those experienced in an autoclave or denaturing chemical environment. However, MIPs typically exhibit reception abilities slightly less than antibodies or other biological receptor molecules.

# 60.3 Optical Nanosensors

Over the years, new techniques in sensing have set the stage for great advances in the field of biological research. One of the most recent advances in the field of optical sensors has been the development of optical nanosensors. A nanosensor is a sensor that has dimensions on the nanometer-size scale. Presently, the most common nanosensors employ fiber optics with tip diameters ranging between 20 and 100 nm. These sensors are based on the same basic principles as larger, more conventional optical sensors, except for their excitation process. Because the diameter of the optical fiber's tip is significantly smaller than the wavelength of light used for excitation of the analyte, the photons cannot escape from the tip of the fiber to be absorbed by the species of interest, as is the case in larger fiber-optic sensors. Instead, in a fiber-optic nanosensor, after the photons have traveled as far down the fiber as possible excitons or evanescent fields continue to travel through the remainder of the tip, providing excitation for the fluorescent species of interest present in the sensing layer. An additional feature of this excitation process is that only species that are in extremely close proximity to the fiber's tip can be excited, thereby precluding the excitation of interfering autofluorescent species within other locations of the sample. This is extremely important in biological systems such as cells, where a large number of autofluorescent species are often present.

The nano-scale size of this new class of sensors also allows for measurements in the smallest of environments. One such environment that has evoked a great deal of interest is that of individual cells. These nanosensors make it possible to probe individual chemical species in specific locations within a cell. Previously, such measurements were limited almost entirely to the field of fluorescence microscopy, where a fluorescent dye is inserted into a cell and allowed to diffuse throughout. Depending on the particular fluorescent dye that was chosen, changes in the fluorescence properties of the dye could then be monitored as the dye came into contact with the analyte of interest. By performing these analyses in an imaging format, it is therefore possible to determine the presence of the analyte at various locations. Since this technique relies on imaging the fluorescent dye, it requires the homogenous dispersion of the dye through the various locations in the cell, which is limited by intracellular conditions (i.e., pH, etc.) or often does not even occur due to cellular compartmentalization. Nanosensors therefore offer significant improvements over such methods in many cases as they allow the user to perform analyses at any desired location without a need for homogenous dispersion of the indicator dye.

## 60.3.1 Fabrication of Optical Nanofibers

The construction of fiber-optic probes with nanometer-sized tips is a prerequisite of optical-fiber-based nanosensors. Depending on the specific application and environment in which the nanosensor is to be used, different tip diameters, taper angles, and smoothnesses of the fiber surfaces are required. To achieve the various diameters and tapers on the tips of these fibers, two different fabrication procedures have been developed.

### 60.3.1.1 Heated Pulling Process

The most commonly employed procedure involves the use of a heated pulling instrument, such as a laser-based micropipette puller. These instruments use a $CO_2$ laser to heat the fiber and a tension device that pulls along the major axis of the fiber while it is heated (see Figure 60.2). As the fiber is pulled, the heated region begins to taper down to a point until the fiber has been pulled into two pieces, each having one large end and one tapered end with nanometer-scale dimensions. By varying the parameters associated with the pulling process (i.e., heating temperature, tension applied to the fiber, etc.), it is possible to precisely control the size of the tip, with diameters ranging from less than 20 nm to more than 1000 nm.[72–76] Figure 60.3 shows a scanning electron micrograph (SEM) of an optical fiber that has been pulled in such an instrument. This technique, while requiring accessibility to a relatively expensive micropipette-pulling instrument, makes it possible to reproducibly create optical fibers with nanometer-scale tips in just seconds.

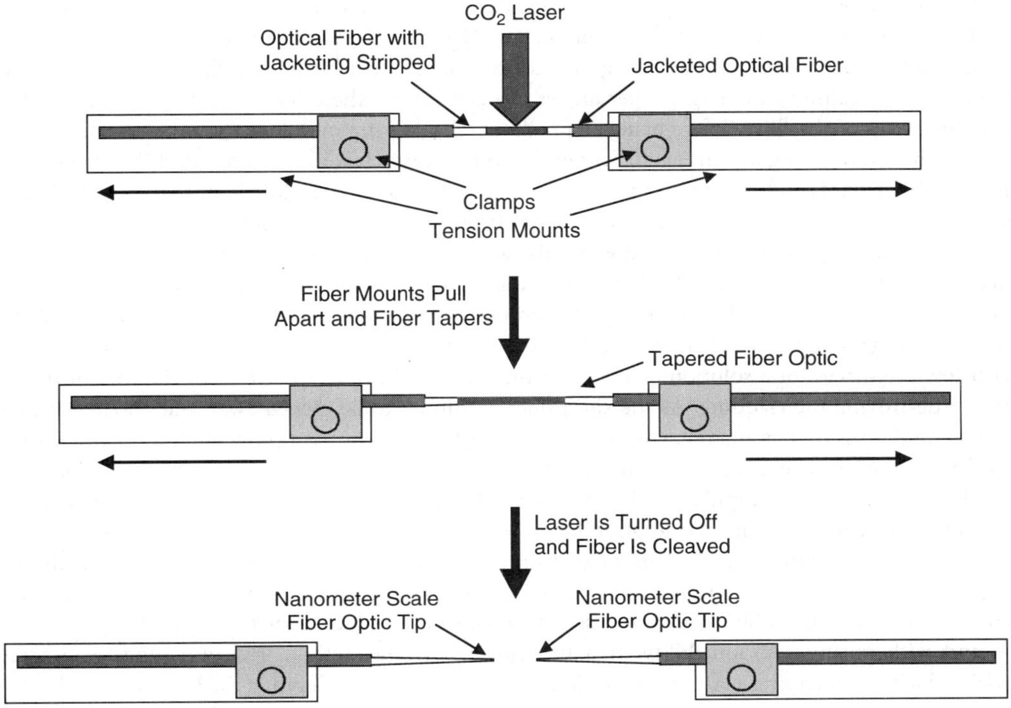


**FIGURE 60.2** Diagram depicting the fiber-pulling procedure for construction of a nanofiber.

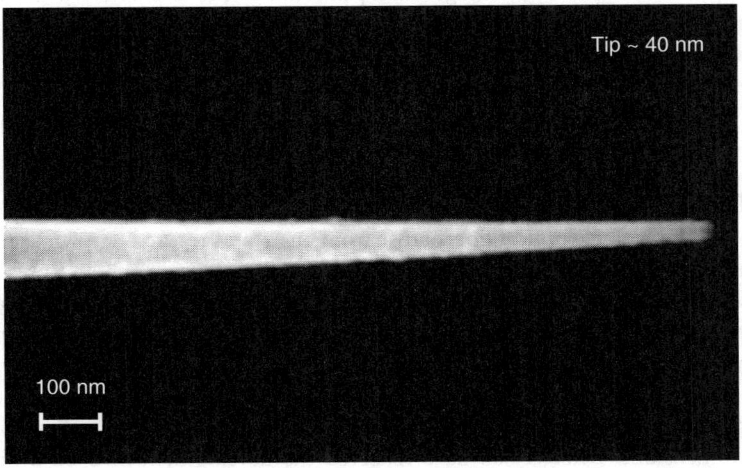


**FIGURE 60.3** A scanning electron micrograph of a nanofiber produced by the micropipette-pulling procedure. The tip of this fiber is approximately 50 nm in diameter.

#### 60.3.1.2 Chemical Etching

The second means of producing optical fibers with nanometer-scale tips is through a chemical-etching process. This process has been found to be an inexpensive and effective alternative to the pulling process. Two different types of etching procedures have been reported in the literature: (1) Turner etching, a method involving the mixture of hydrofluoric acid (HF) with an organic solvent,[77,78] and (2) tube etching, a method developed by Stockle et al.[79] In the Turner method, a fiber is placed in the meniscus between the HF and an organic overlayer. Once in place, the HF dissolves the silica fiber, forming a point. With this technique, fibers with large taper angles and tip diameters comparable to the pulling method have been created. Because of the larger taper angles associated with these fibers, excitation light can travel closer to the tip of the fiber before being trapped, thereby causing more energy to be coupled with the fluorophore and resulting in a more efficient excitation process. However, as a result of the dual chemical nature of the etchant solution, environmental parameters such as temperature fluctuations and vibrations can cause the characteristics of the tip to vary significantly from batch to batch.

To overcome the problems associated with this two-phase chemical-etching process and the batch-to-batch variability, a new form of etching, known as tube etching, has recently been developed. Tube etching involves etching an optical fiber with a single-component solution of HF. In this method, a silica fiber that has an organic/polymer cladding is polished optically flat on both ends, prior to placing one end just below the surface of a solution of HF. Over time, the acid begins to etch away the core of the fiber without destroying the cladding. As the silica fiber continues to be etched away, the polymer cladding acts as a wall, causing any microcurrents in the HF to form a blunted tip on the fiber's core. As more time passes, the silica core material continues to dissolve until it is no longer below the surface of the HF solution. At this point, capillary action draws the HF up the walls of the cladding, allowing it to drain down the silica core back into the solution below. As this happens, it causes the fiber to be etched into the shape of a cone with a large smooth taper. Control over the taper angle of the fiber and the diameter of the tip can be performed simply by varying the depth at which the fiber is placed in the HF and the etching time. Using this tube etching process, it is possible to produce a batch of fibers in approximately 2 h and with greater reproducibility than the Turner etching method. However, due to difficulties associated with submerging a large group of fibers to exactly the same depth in the HF, the reproducibility of this etching method is still less than that obtained by pulling the fibers. In addition, due to the large taper angle associated with these etched fibers as well as those fabricated with the Turner etching process, they are limited in the environments in which they can probe. Because the taper angle is so large, they tend to cause a great deal of physical damage when performing cellular analyses with smaller cells (e.g., mammalian somatic cells, etc.) and are therefore useful only with large cells such as oocytes and certain neurons. It is primarily for this reason that the most common method for creating nanometer-sized tips on fiber-optic nanosensors is performed using the heated pulling technique.

## 60.3.2 Evolution of Optical-Based Nanosensors

### 60.3.2.1 Near-Field Optics and Spectroscopy

Optical sensors using nanoscale optical fibers arose from an area of research known as near-field optical microscopy. Near-field optical microscopy is a relatively new area of research and uses light sources or detectors with emitting or collecting apertures that are smaller than the wavelength of light used for imaging. The most common method for performing such experiments is by placing a pinhole in front of the detector, thereby effectively reducing the size of the detector aperture.[80,81] A variation of this technique is to construct an excitation probe with dimensions that are smaller than the wavelength of light being used for sample interrogation. This small excitation probe can then act as a light source with subwavelength dimensions. Betzig et al. have reported the development of such a probe capable of providing images of a sample with an optimum spatial resolution of approximately 12 nm. This probe was constructed by pulling a single-mode optical fiber with a micropipette puller and then coating the wall of the fiber with ~100 nm of aluminum. By applying an aluminum coating to the walls of the fiber, leakage of light from the tapered region of the fiber, where the cladding was pulled too thin to maintain the total internal reflection, is prevented, thereby confining the excitation radiation to the fiber tip (~20 nm). Using this first nanofiber-based system operating in the illumination mode (i.e., with the probe acting as a localized excitation source) images of a known pattern were reconstructed by raster scanning the probe over it. In addition, signal enhancements of greater than $10^4$ (see References 72 and 82) over previous near-field probe analyses[83–86] were obtained.

Due to the high degree of spatial resolution provided by near-field microscopy, these probes have been used for the detection of single molecules. In particular, one of the more recent applications of these nanofibers to single-molecule detection has been for the detection of dye-labeled DNA molecules using near-field surface-enhanced resonance Raman spectroscopy (NFSERRS).[87,88] This NFSERRS technique made it possible to obtain chemical images of single DNA molecules labeled with the fluorescent dye brilliant cresyl blue (BCB). In that work, the dye-labeled DNA strands were spotted onto a SERS that was prepared by evaporating silver on a nanoparticle-coated substrate.[87,88] NFSERRS spectra were then obtained by illuminating the sample with the nanoprobe and measuring the Raman signal with a charge-coupled device (CCD) mounted on a spectrometer. By raster scanning the nanoprobe across the sample, a two-dimensional image of the DNA molecules was reconstructed and normalized for surface topography based on the intensity of the Rayleigh scatter.

Near-field optical microscopy using taper fiber optics is an area of research that holds a great deal of potential for the mapping of surfaces for individual biological molecules. In addition, since many biological compounds exhibit properties such as luminescence that are typically much stronger than Raman signals, it might be possible in the near future to map out the location of individual molecules of specific chemicals in human tissues such as neurotransmitters in the brain. Single-molecule detection would therefore represent the absolute limit of detection for any compound and could open new horizons in the investigation of the complex chemical reactions and pathways of biological systems. Near-field microscopy is further described in detail in Chapter 12 of this handbook.

#### 60.3.2.2 Chemical Nanosensors

Shortly after their development for near-field microscopy, nanofibers began to be applied to the field of chemical sensing. With the advent of these nanofibers, it became possible to probe for specific chemicals in 3D structures over very short distances (<50 nm). Because of this ability to perform highly spatially localized analyses, the monitoring of concentration gradients and spatial inhomogeneties in submicroscopic environments (e.g., cells) via various spectroscopic techniques has become possible. However, due to the small sampling volume probed by nanofibers, the number of analyte species in the excitation volume is very small, making it important to use a sensitive spectroscopic technique for the analysis (e.g., fluorescence). In addition to using a sensitive spectroscopic technique, it is also important to use a sensitive detection scheme, such as the one shown in Figure 60.4, which is commonly used with fiber-optic nanosensors and nanobiosensors. In this system, an excitation source of some type (typically a laser) is launched into a fiber optic that is attached to the large, proximal end of the nanosensor. The nanosensor, which is held in place by an x-y-z micromanipulator mounted on an inverted microscope, is then used to position the tip of the sensor to the desired location while looking through the eyepiece of the microscope. Once in place, the excitation source is turned on and the dye, which is sensitive to the presence of the analyte, is excited. As the excited dye relaxes to the ground state, the resulting fluorescence emission is collected by the microscope optics and detected with either a CCD or a photomultiplier tube (PMT).

Since the construction and use of the first optical nanosensors by Tan et al. in 1992[73,74] several different optical-fiber-based chemical nanosensors have been reported for the measurement of pH,[89–92] various ion concentrations,[93,94] and other chemical species.[95] In this original work, a micropipette puller was used to taper multimode and single-mode fibers down to diameters of 100 to 1000 nm at the tip.[73,74] These tapered fibers were then used in the production of pH nanosensors via a three-step process. The first step in the process was to apply a thick layer of aluminum to the walls of the fiber using a vacuum evaporator to ensure total internal reflection over the tapered region of the fiber. During this aluminum-deposition process it was important to ensure that the tip of the fiber remained free of metal, thereby providing a silica surface for binding of the probe or receptor molecules to the fiber's tip. To achieve this, the fibers were placed in the evaporator system and rotated with their tips facing away from the source of the evaporating metal. By precisely angling the fibers, their sides would shadow the tips from the evaporating metal.[75,76,96] This coating method is illustrated in Figure 60.5. Once the walls of the fiber were

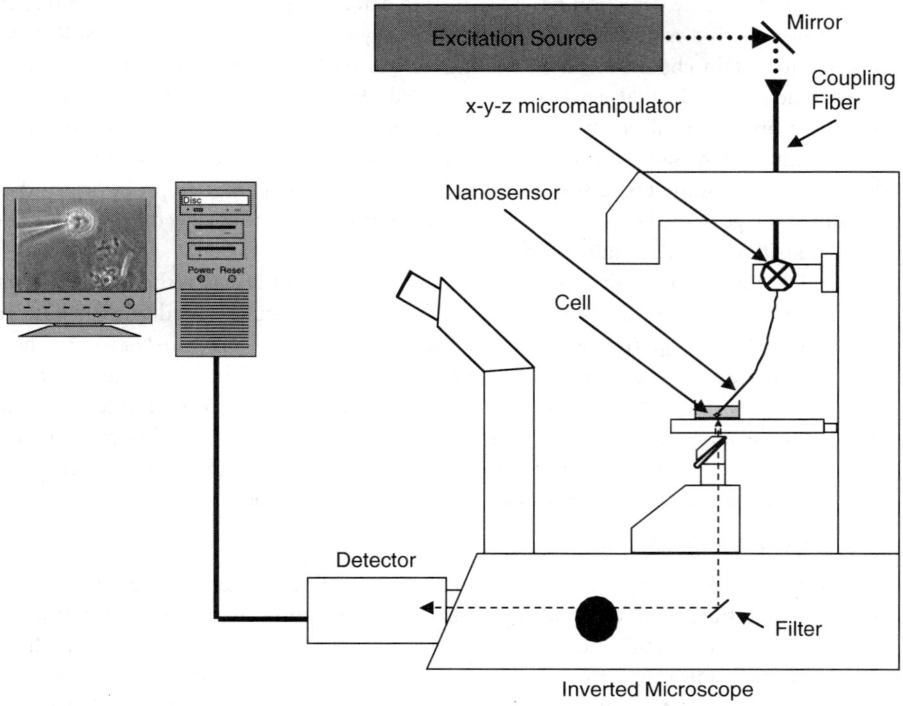


**FIGURE 60.4** Schematic diagram showing a typical experimental setup used for nanosensor measurements within a single cell or microscopic sample. The nanosensor is mounted in an x-y-z micropositioning system mounted on an inverted microscope. The fluorescent light is then collected by the mroscope and monitored with a detector.

coated with aluminum, the next step was to silanize the fiber tip to allow for cross linking to a polymer coating. The final step in the fabrication of this pH nanosensor was to then attach the pH-sensitive dye, acrylofluoresceinamine, to the silanized fiber tip through a variation of a photopolymerization process that has been used often in the construction of larger chemical sensors.[97,98] However, because of the near-field excitation provided by these small fiber probes, the cross linking of the polymer solution was restricted to the near field of the fiber. Following fabrication of this sensor, characterization of its properties was performed. During this characterization, the sampling volume of the sensor was determined to be greater than six orders of magnitude smaller than conventional fiber-optic chemical sensors, making it ideal for subcellular measurements. The reduced sampling volume and response time of the sensor were evaluated by measuring the pH in 10-μm-diameter pores in a polycarbonate membrane. This evaluation found that the response time of the sensor (300 msec) was 100-fold faster than conventional fiber-optic chemical sensors and that it was both stable and reversible with respect to pH changes.

The first reported application of fiber-optic nanosensors for biological measurements was performed on rat embryos.[73] In this study, pH nanosensors like the ones described above were inserted into the extraembryonic space of a rat conceptus, with minimal damage to the surrounding visceral yolk sac, and pH measurements were made. From these measurements, values of the pH in the extraembryonic fluid of rat conceptuses ranging in age from 10 to 12 d old were then compared for any differences. In a similar study using the same pH sensor, indirect measurements of nitrite and chloride levels in the yolk sac of rat conceptuses were also performed.[99] Because of the minimally invasive nature of such measurements, such techniques hold much promise for biological analyses and may aid in furthering our understanding of the effect that environmental factors have on embryonic growth.

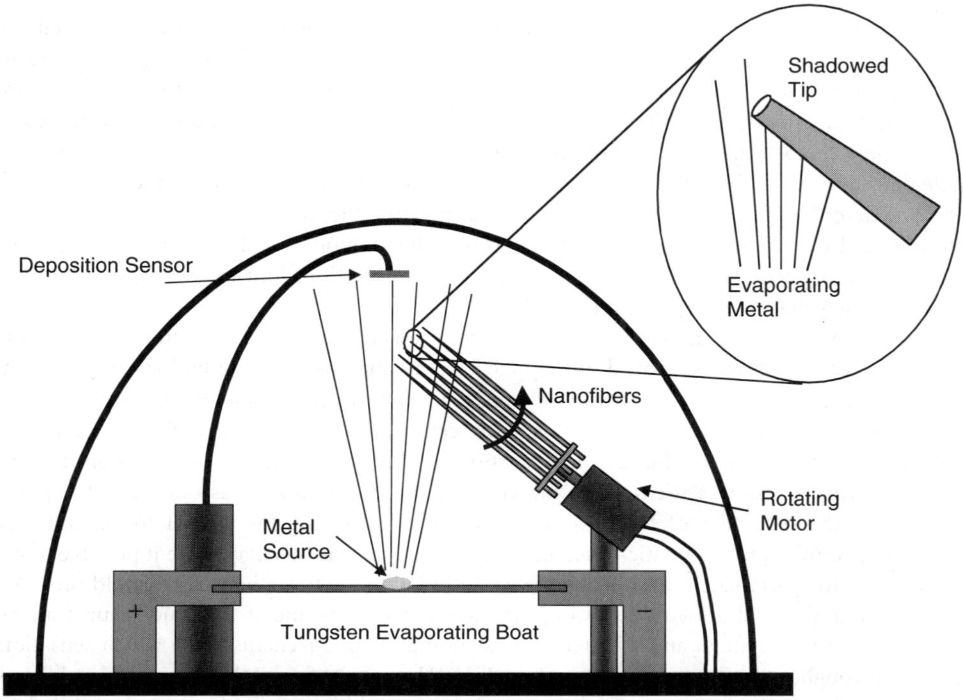


**FIGURE 60.5** Illustration of the procedure for coating the walls of fiber-optic nanosensors with a reflective metal.

As the field of optical-fiber-based nanosensors has evolved, the size of the environments in which they probe continues to shrink. In fact, the measurement of chemical species inside individual cells has even been reported using fiber-optic chemical nanosensors. In the first of these studies, optochemical nanosensors for sodium ions ($Na^+$) were developed using a process similar to the one described above. These sensors were then used for the measurement of sodium-ion ($Na^+$) concentrations in the cytoplasmic space inside a single mouse oocyte, one of the largest mammalian cells (~100 μm in diameter).[8] These sensors allowed for the monitoring of the relative $Na^+$ concentrations while ion channels were opened and closed by the external stimulant, kainic acid. In another application of fiber-optic nanosensors to biological measurements, calcium-ion-sensitive nanosensors were developed and used to measure the calcium-ion fluctuations in vascular smooth muscle cells while the cells were stimulated.[100] These fluctuations were then directly correlated to the stimulation events that were being performed, providing another potential application for these nanometer-sized optical sensors.

### 60.3.2.3 Nanobiosensors

Due to the complexity of biological systems and the number of possible interferences to chemical nanosensors, an obvious need for added specificity arose. This specificity was achieved by the development of nanobiosensors. As with their larger counterparts, the added specificity provided by biological receptor molecules allowed for the analysis of complex environments with minimal affects from other species. Application of nanofibers to the field of biosensors was first reported in 1996 by Vo-Dinh and coworkers.[75] In this work, antibody-based nanobiosensors were developed and characterized for benzo[a]pyrene tetrol (BPT), a DNA adduct of the carcinogen benzo[a]pyrene (BaP). Because it is a biomarker for human exposure to BaP, the measurement of BPT is of great interest in early cancer diagnosis. Fabrication of these first nanobiosensors was performed in a four-step process. The first step in this process was to taper a 600-μm-diameter fiber optic down to 40 nm at the tip. This is followed by application of a thick layer of silver (approximately 200 nm) to the walls of the fiber to prevent light leakage as well as to enhance light delivery to the end of the fiber. Following the silver coating, the fiber tip was silanized to create attachment sites for the antibodies. Once silanized, antibodies were attached to the distal end of the nanofiber via a covalent-binding procedure. The first step in this antibody-binding procedure was to activate the silanized fiber with a solution of carbonyl diimidazole, which was then followed by incubation of the activated fiber with the antibody of interest for 3 d at 4°C. Once the antibodies were bound, the biosensors were characterized in terms of retention of antibody-binding ability as well as sensitivity and absolute detection limits. Tests were performed by placing the nanobiosensors in a measurement system like the one described for chemical nanosensors and inserting the distal end of the sensor into a series of solutions of BPT. These measurements revealed that the antibodies had retained more than 95% of their native binding affinity for BPT and that the absolute detection limit for BPT using these sensors was approximately 300 zeptomoles (zepto = $10^{-21}$).

In the several years since this work, optical nanobiosensors employing many different forms of bioreceptor molecules have been developed and applied to the analysis of many biologically relevant species.[100,101] This has included nanobiosensors for the detection of nitric oxide via fluorescence detection of the nonenzymatic protein cytochrome c or fluorescently labeled cytochrome c[100] as well as an enzymatic-based nanobiosensor for the indirect detection of glutamate.[101] In the latter design, the enzyme glutamate dehydrogenase was used as the bioreceptor. When glutamate was bound to the glutamate dehydrogenase, the reduction of NAD to NADH occurred, allowing for NADH to be measured via fluorescence spectroscopy. Enzymatic-based nanobiosensors such as this may make it possible to continuously monitor the glutamate levels released from an individual cell. Such a sensor could prove to be a significant tool in the field of neurophysiology because glutamate is one of the major neurotransmitters in the central nervous system, and a better understanding of the mechanisms by which sensations are transmitted throughout the human body may be achieved by monitoring of the glutamate levels produced by individual cells. Although enzymes have previously been used in the field of biosensors for their ability to be regenerated, this was the first example of the use of an enzyme in the fabrication of a nanobiosensor.

Several studies have been published of fiber-optic nanobiosensors being applied to *in vitro* measurements of single cells.[9,76,96,102,103] In one such study, nanobiosensors for BPT were prepared as described earlier and used for the measurement of intracellular concentrations of BPT in the cytosol of two different cell lines: (1) human mammary carcinoma cells and (2) rat liver epithelial cells.[76] The cells in both lines were spherical in shape and had diameters of approximately 10 μm. Several hours prior to making a measurement a known amount of BPT was added to the culturing medium of the cells. Then immediately prior to making any measurements, the culturing medium was rinsed several times and replaced with BPT-free medium. This work revealed that the concentration of BPT in each of the different cell lines was the same, suggesting that the means of transport of the BPT into the cells was the same in both cases (probably by diffusion). By performing similar measurements inside various subcellular compartments or organelles, it may be possible to obtain critical information about the location of BPT formation within cells and its transport throughout cells during the process of carcinogenesis. In addition, this work also showed that the insertion of a nanobiosensor into a mammalian somatic cell not only appears to have no effect on the cell membrane but also does not effect the cell's normal function. This was demonstrated by inserting a nanobiosensor into a cell that was just beginning to undergo mitosis and monitoring cell division following a 5-min incubation of the fiber in the cytoplasm and fluorescence measurement. Figure 60.6 contains a series of time-lapse images from this experiment showing the initial stages of mitosis in which the fiber-optic nanobiosensor is interrogating the cell all the way through division of the cell into two identical daughter cells.

### 60.3.2.4 Nanoparticle Chemical Sensors

The development of fiber-optic-based nanosensors and nanobiosensors has already had a large impact on the fields of biological and biomedical research in the short time that they have existed. Nevertheless, significant advances and variations are constantly being made in optical nanosensors, from the use of new, more selective bioreceptors to the development of different types of nanobiosensors for application in various environments. One such advance in the last couple of years has been the development of the first nanoparticle-based optochemical sensors with nanometer-scale sizes in all three dimensions.[104]

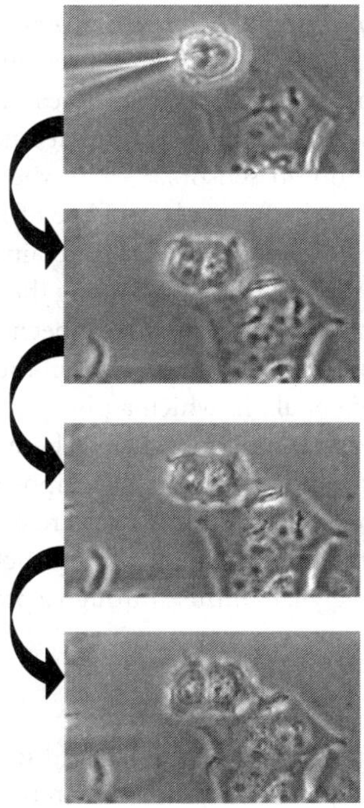


**FIGURE 60.6** Series of time-lapse microscopic images of a cell undergoing mitosis prior to and immediately following interrogation with a fiber-optic nanobiosensor. The cell is approximately 10 μm in diameter.

#### *60.3.2.4.1 PEBBLE Sensors*

Since this first nanoparticle-based sensor was first described, several types of nanoparticle sensors have been developed, each having its own advantages and disadvantages. One such nanoparticle sensor, known as a PEBBLE (probe encapsulated by biologically localized embedding), is comprised of a fluorescent indicator dye embedded in a 20-nm diameter polymer or sol-gel sphere.[105,106] This chapter will provide a brief overview of this topic; a much more extensive description of their development and application can be found in Chapter 59 of this handbook. This encapsulating sphere protects the indicator dye from cellular degradation and the cell from the toxic effects of the dyes while allowing ions to penetrate and interact with the indicator dye. Application of these sensors to the measurement of many different ions within a cell has been demonstrated since their first development. In such analyses, large quantities of these sensors are randomly injected in a cell with a gene gun or similar device to ensure the presence of a sensor at every desired location. Once in place, the cell is illuminated and the resulting fluorescence signal measured over the autofluorescent background of the cell. Presently, PEBBLE sensors have been developed for pH, $Ca^{+2}$, NO, molecular oxygen, and intracellular free $Zn^{+2}$.[105–110]

#### *60.3.2.4.2 Quantum-Dot Biosensors*

Simultaneous with the development of PEBBLEs was the development of quantum-dot-based biosensors capable of monitoring individual chemical species inside a living cell. These nanoparticle biosensors are comprised of ZnS particles capped with Cd-selenide and attached to biological receptors including antibodies, oligonucleotides, and enzymes.[111–115] These quantum dots offer several significant advantages over conventional fluorescent indicator dyes used in chemical sensing. First, they are biocompatible, unlike most indicator dyes, which are often toxic to the cell being investigated. Second, their emission is much more intense than typical indicator dyes. Finally, they are more photostable than dye molecules. These last two properties are very important when trying to monitor changes in chemical or biochemical species concentrations over time because most fluorescent dyes exhibit rapid and significant photobleaching with the small amount used in cellular analyses.

#### *60.3.2.4.3 Biologically Encapsulated Chemical Sensors*

Another class of nanoparticle-based optochemical sensor that has been developed for cellular analyses employs biological encapsulation of fluorescent indicator dyes to ensure biocompatibility with the cell being investigated.[116–121] These biologically encapsulated nanoparticle sensors can be further divided into liposome sensors[120,121] and lipobead sensors.[116,118] In the case of liposome-based sensors, a fluorescent indicator dye that is sensitive to the particular analyte of interest is encapsulated in the internal aqueous compartment of a liposome. This allows the dye to retain its solution-based characteristics (i.e., spectral emission profile, response time, etc.) while preventing the toxic dye molecules from diffusing throughout the cell. Liposome-based nanoparticle sensors have been developed for both pH and molecular oxygen and applied to various cellular analyses. A variation on these liposome-encapsulated sensors has recently been developed by Rosenzweig et al., in which a phospholipid membrane and fluorescent indicator dye are immobilized onto a polystyrene nanoparticle. These sensors, known as lipobeads, are more stable and less susceptible to biological degradation than liposome-based sensors. In addition, by fabricating these sensors so that the fluorescent indicator dyes are embedded in the phospholipid membrane, they can potentially be used for the measurement of larger chemical species than the liposome sensors or PEBBLEs, which require the analyte to diffuse through a protective coating before interacting with the dye.

## 60.4 Conclusion

Due to the increasing interest in nanotechnology and its practical use, application of nanosensors and nanobiosensors to different types of cellular measurements is expanding rapidly. Furthermore, it has been shown that these nanosensors and nanobiosensors often provide more reliable measurements of subcellular chemical species than conventional techniques, such as pretreatment with and subsequent monitoring of fluorescent dyes. The use of nanosensor and nanobiosensor technologies has already begun and will continue to revolutionize the field of cellular biology in the future.

### Acknowledgments

This research was sponsored in part by the ORNL-LDRD project (Advanced Nanosensors Systems Project) as well as the Office of Biological and Environmental Research, U.S. Department of Energy, under contract DE-AC05-00OR22725 managed by UT-Battelle, LLC. B.M. Cullum was also supported by an appointment to the Oak Ridge National Laboratory Postdoctoral Research Associates Program administered jointly by the Oak Ridge National Laboratory and Oak Ridge Institute for Science and Education.

### References

1. Nice, E.C. and Catimel, B., Instrumental biosensors: new perspectives for the analysis of biomolecular interactions, *Bioessays*, 21(4), 339, 1999.
2. Weetall, H.H., Chemical sensors and biosensors, update, what, where, when and how, *Biosensors Bioelectron.*, 14(2), 237, 1999.
3. Tess, M.E. and Cox, J.A., Chemical and biochemical sensors based on advances in materials chemistry, *J. Pharm. Biomed. Anal.*, 19(1–2), 55, 1999.
4. Braguglia, C.M., Biosensors: an outline of general principles and application, *Chem. Biochem. Eng. Q.*, 12(4), 183, 1998.
5. Lu, J.Z. and Rosenzweig, Z., Nanoscale fluorescent sensors for intracellular analysis, *Fresenius J. Anal. Chem.*, 366(6–7), 569, 2000.
6. Vo-Dinh, T. and Cullum, B., Biosensors and biochips: advances in biological and medical diagnostics, *Fresenius J. Anal. Chem.*, 366(6–7), 540, 2000.
7. Vo-Dinh, T., Cullum, B.M., and Stokes, D.L., Nanosensors and biochips: frontiers in biomolecular diagnostics, *Sensors Actuators B Chem.*, 74(1–3), 2, 2001.

8. Tan, W.H., Kopelman, R., Barker, S.L.R., and Miller, M.T., Ultrasmall optical sensors for cellular measurements, *Anal. Chem.*, 71(17), 606A, 1999.
9. Cullum, B.M. and Vo-Dinh, T., The development of optical nanosensors for biological measurements, *Trends Biotechnol.*, 18(9), 388, 2000.
10. Cajlakovic, M., Lobnik, A., and Werner, T., Stability of new optical pH sensing material based on cross-linked poly(vinyl alcohol) copolymer, *Anal. Chim. Acta*, 455(2), 207, 2002.
11. Zhang, Z.J., Zhang, Y.K., Ma, W.B., Russell, R., Grant, C.L., Seitz, W.R., and Sundberg, D.C., Polyvinyl-alcohol as a matrix for immobilizing indicators for fiber optic chemical sensors, *Abstr. Papers Am. Chem. Soc.*, 196, 115, 1988.
12. Lechna, M., Holowacz, I., Ulatowska, A., and Podbielska, H., Optical properties of sol-gel coatings for fiberoptic sensors, *Surf. Coatings Technol.*, 151, 299, 2002.
13. Mohr, G.J., Werner, T., Oehme, I., Preininger, C., Klimant, I., Kovacs, B., and Wolfbeis, O.S., Novel optical sensor materials based on solubilization of polar dyes in apolar polymers, *Adv. Mater.*, 9(14), 1108, 1997.
14. Wolfbeis, O.S., Fiber optic chemical sensors and biosensors, *Anal. Chem.*, 72(12), 81R, 2000.
15. Seitz, W.R., Chemical sensors based on immobilized indicators and fiber optics, *CRC Crit. Rev. Anal. Chem.*, 19(2), 135, 1988.
16. Seitz, W.R., Chemical sensors based on fiber optics, *Anal. Chem.*, 56(1), A16, 1984.
17. Erdem, A., Kerman, K., Meric, B., Akarca, U.S., and Ozsoz, M., DNA electrochemical biosensor for the detection of short DNA sequences related to the hepatitis B virus, *Electroanalysis*, 11(8), 586, 1999.
18. Sawata, S., Kai, E., Ikebukuro, K., Iida, T., Honda, T., and Karube, I., Application of peptide nucleic acid to the direct detection of deoxyribonucleic acid amplified by polymerase chain reaction, *Biosensors Bioelectron.*, 14(4), 397, 1999.
19. Niemeyer, C.M., Boldt, L., Ceyhan, B., and Blohm, D., DNA-directed immobilization: efficient, reversible, and site-selective surface binding of proteins by means of covalent DNA-streptavidin conjugates, *Anal. Biochem.*, 268(1), 54, 1999.
20. Marrazza, G., Chianella, I., and Mascini, M., Disposable DNA electrochemical sensor for hybridization detection, *Biosensors Bioelectron.*, 14(1), 43, 1999.
21. Bardea, A., Patolsky, F., Dagan, A., and Willner, I., Sensing and amplification of oligonucleotide-DNA interactions by means of impedance spectroscopy: a route to a Tay-Sachs sensor, *Chem. Commun.*, 1, 21, 1999.
22. Wang, J., Rivas, G., Fernandes, J.R., Paz, J.L.L., Jiang, M., and Waymire, R., Indicator-free electrochemical DNA hybridization biosensor, *Anal. Chim. Acta*, 375(3), 197, 1998.
23. Gooding, J.J. and Hibbert, D.B., The application of alkanethiol self-assembled monolayers to enzyme electrodes, *Trends Anal. Chem.*, 18(8), 525, 1999.
24. Franchina, J.G., Lackowski, W.M., Dermody, D.L., Crooks, R.M., Bergbreiter, D.E., Sirkar, K., Russell, R.J., and Pishko, M.V., Electrostatic immobilization of glucose oxidase in a weak acid, polyelectrolyte hyperbranched ultrathin film on gold: fabrication, characterization, and enzymatic activity, *Anal. Chem.*, 71(15), 3133, 1999.
25. Patolsky, F., Zayats, M., Katz, E., and Willner, I., Precipitation of an insoluble product on enzyme monolayer electrodes for biosensor applications: characterization by faradaic impedance spectroscopy, cyclic voltammetry, and microgravimetric quartz crystal microbalance analyses, *Anal. Chem.*, 71(15), 3171, 1999.
26. Pemberton, R.M., Hart, J.P., Stoddard, P., and Foulkes, J.A., A comparison of 1-naphthyl phosphate and 4 aminophenyl phosphate as enzyme substrates for use with a screen-printed amperometric immunosensor for progesterone in cows' milk, *Biosensors Bioelectron.*, 14(5), 495, 1999.
27. Serra, B., Reviejo, A.J., Parrado, C., and Pingarron, J.M., Graphite-Teflon composite bienzyme electrodes for the determination of L-lactate: Application to food samples, *Biosensors Bioelectron.*, 14(5), 505, 1999.

28. Cosnier, S., Senillou, A., Gratzel, M., Comte, P., Vlachopoulos, N., Renault, N.J., and Martelet, C., A glucose biosensor based on enzyme entrapment within polypyrrole films electrodeposited on mesoporous titanium dioxide, *J. Electroanal. Chem.*, 469(2), 176, 1999.
29. Blake, R.C., Pavlov, A.R., and Blake, D.A., Automated kinetic exclusion assays to quantify protein binding interactions in homogeneous solution, *Anal. Biochem.*, 272(2), 123, 1999.
30. Hara-Kuge, S., Ohkura, T., Seko, A., and Yamashita, K., Vesicular-integral membrane protein, VIP36, recognizes high- mannose type glycans containing alpha 1–2 mannosyl residues in MDCK cells, *Glycobiology*, 9(8), 833, 1999.
31. Shih, Y.T. and Huang, H.J., A creatinine deiminase modified polyaniline electrode for creatinine analysis, *Anal. Chim. Acta*, 392(2–3), 143, 1999.
32. Nelson, R.W., Jarvik, J.W., Taillon, B.E., and Tubbs, K.A., BIA MS of epitope-tagged peptides directly from *E. coli* lysate: multiplex detection and protein identification at low-femtomole to subfemtomole levels, *Anal. Chem.*, 71(14), 2858, 1999.
33. Piehler, J., Brecht, A., Hehl, K., and Gauglitz, G., Protein interactions in covalently attached dextran layers, *Colloids Surf. B*, 13(6), 325, 1999.
34. Kim, H.J., Hyun, M.S., Chang, I.S., and Kim, B.H., A microbial fuel cell type lactate biosensor using a metal- reducing bacterium, Shewanella putrefaciens, *J. Microbiol. Biotechnol.*, 9(3), 365, 1999.
35. Hu, T., Zhang, X.E., and Zhang, Z.P., Disposable screen-printed enzyme sensor for simultaneous determination of starch and glucose, *Biotechnol. Tech.*, 13(6), 359, 1999.
36. Garjonyte, R. and Malinauskas, A., Amperometric glucose biosensor based on glucose oxidase immobilized in poly(o-phenylenediamine) layer, *Sensors Actuators B Chem.*, 56(1–2), 85, 1999.
37. Lee, Y.C. and Huh, M.H., Development of a biosensor with immobilized L-amino acid oxidase for determination of L-amino acids, *J. Food Biochem.*, 23(2), 173, 1999.
38. Lebron, J.A. and Bjorkman, P.J., The transferrin receptor binding site on HFE, the class I MHC-related protein mutated in hereditary hemochromatosis, *J. Mol. Biol.*, 289, 4(1109), 1999.
39. Gooding, J.J., Situmorang, M., Erokhin, P., and Hibbert, D.B., An assay for the determination of the amount of glucose oxidase immobilised in an enzyme electrode, *Anal. Commun.*, 36(6) 225, 1999.
40. Hu, S.S., Luo, J., and Cui, D., An enzyme-chemically modified carbon paste electrode as a glucose sensor based on glucose oxidase immobilized in a polyaniline film, *Anal. Sci.*, 15(6), 585, 1999.
41. Sergeyeva, T.A., Soldatkin, A.P., Rachkov, A.E., Tereschenko, M.I., Piletsky, S.A., and El'skaya, A.V., Beta-lactamase label-based potentiometric biosensor for alpha-2 interferon detection, *Anal. Chim. Acta*, 390(1–3), 73, 1999.
42. Barker, S.L.R., Zhao, Y.D., Marletta, M.A., and Kopelman, R., Cellular applications of a sensitive and selective fiber optic nitric oxide biosensor based on a dye-labeled heme domain of soluble guanylate cyclase, *Anal. Chem.*, 71(11), 2071, 1999.
43. Huang, T., Warsinke, A., Koroljova-Skovobogat'ko, O.V., Makower, A., Kuwana, T., and Scheller, F.W., A bienzyme carbon paste electrode for the sensitive detection of NADPH and the measurement of glucose-6-phosphate dehydrogenase, *Electroanalysis*, 11(5), 295, 1999.
44. Hall, E.A.H., Gooding, J.J., Hall, C.E., and Martens, N., Acrylate polymer immobilisation of enzymes, *Fresenius J. Anal. Chem.*, 364(1–2), 58, 1999.
45. Anzai, J., Kobayashi, Y., Hoshi, T., and Saiki, H., A layer-by-layer deposition of concanavalin A and native glucose oxidase to form multilayer thin films for biosensor applications, *Chem. Lett.*, 4, 365, 1999.
46. Campbell, T.E., Hodgson, A.J., and Wallace, G.G., Incorporation of erythrocytes into polypyrrole to form the basis of a biosensor to screen for Rhesus (D) blood groups and rhesus (D) antibodies, *Electroanalysis*, 11(4), 215, 1999.
47. Wang, Y.S., Li, G.R., Lu, C.Y., Wan, Z.Y., and Liu, C.X., Preparation of triglyceride enzyme sensor and its application, *Prog. Biochem. Biophys.*, 26(2), 144, 1999.
48. Pancrazio, J.J., Bey, P.P., Cuttino, D.S., Kusel, J.K., Borkholder, D.A., Shaffer, K.M., Kovacs, G.T.A., and Stenger, D.A., Portable cell-based biosensor system for toxin detection, *Sensors Actuators B Chem.*, 53(3), 179, 1998.

49. Zhang, Y.Q., Zhu, J., and Gu, R.A., Improved biosensor for glucose based on glucose oxidase-immobilized silk fibroin membrane, *Appl. Biochem. Biotechnol.*, 75(2–3), 215, 1998.
50. Houshmand, H., Froman, G., and Magnusson, G., Use of bacteriophage T7 displayed peptides for determination of monoclonal antibody specificity and biosensor analysis of the binding reaction, *Anal. Biochem.*, 268(2), 363, 1999.
51. Roos, H., Karlsson, R., Nilshans, H., and Persson, A., Thermodynamic analysis of protein interactions with biosensor technology, *J. Mol. Recognit.*, 11(1–6), 204, 1998.
52. Bin Zhao, Y., Wen, M.L., Liu, S.Q., Liu, Z.H., Zhang, W.D., Yao, Y., and Wang, C.Y., Microbial sensor for determination of tannic acid, *Microchem. J.*, 60(3), 201, 1998.
53. Heim, S., Schnieder, I., Binz, D., Vogel, A., and Bilitewski, U., Development of an automated microbial sensor system, *Biosensors Bioelectron.*, 14(2), 187, 1999.
54. Schmidt, A., StandfussGabisch, C., and Bilitewski, U., Microbial biosensor for free fatty acids using an oxygen electrode based on thick film technology, *Biosensors Bioelectron.*, 11(11), 1139, 1996.
55. Schuler, R., Wittkampf, M., and Chemnitius, G.C., Modified gas-permeable silicone rubber membranes for covalent immobilisation of enzymes and their use in biosensor development, *Analyst*, 124(8), 1181, 1999.
56. Park, I.S. and Kim, N., Simultaneous determination of hypoxanthine, inosine and inosine 5′-monophosphate with serially connected three enzyme reactors, *Anal. Chim. Acta*, 394(2–3), 201, 1999.
57. Situmorang, M., Gooding, J.J., and Hibbert, D.B., Immobilisation of enzyme throughout a polytyramine matrix: a versatile procedure for fabricating biosensors, *Anal. Chim. Acta*, 394(2–3), 211, 1999.
58. Zhang, W.T., Canziani, G., Plugariu, C., Wyatt, R., Sodroski, J., Sweet, R., Kwong, P., Hendrickson, W., and Chaiken, L., Conformational changes of gp120 in epitopes near the CCR5 binding site are induced by CD4 and a CD4 miniprotein mimetic, *Biochemistry*, 38(29), 9405, 1999.
59. Yano, K. and Karube, I., Molecularly imprinted polymers for biosensor applications, *Trends Anal. Chem.*, 18(3), 199, 1999.
60. Cotton, G.J., Ayers, B., Xu, R., and Muir, T.W., Insertion of a synthetic peptide into a recombinant protein framework: a protein biosensor, *J. Am. Chem. Soc.*, 121(5), 1100, 1999.
61. Song, X.D. and Swanson, B.I., Direct, ultrasensitive, and selective optical detection of protein toxins using multivalent interactions, *Anal. Chem.*, 71(11), 2097, 1999.
62. Costello, R.F., Peterson, I.R., Heptinstall, J., and Walton, D.J., Improved gel-protected bilayers, *Biosensors Bioelectron.*, 14(3), 265, 1999.
63. Ramsden, J.J., Biomimetic protein immobilization using lipid bilayers, *Biosensors Bioelectron.*, 13(6), 593, 1998.
64. Girard-Egrot, A.P., Morelis, R.M., and Coulet, P.R., Direct bioelectrochemical monitoring of choline oxidase kinetic behaviour in Langmuir-Blodgett nanostructures, *Bioelectrochem. Bioenerget.*, 46(1), 39, 1998.
65. Wollenberger, U., Neumann, B., and Scheller, F.W., Development of a biomimetic alkane sensor, *Electrochim. Acta*, 43(23), 3581, 1998.
66. Costello, R.F., Peterson, I.P., Heptinstall, J., Byrne, N.G., and Miller, L.S., A robust gel-bilayer channel biosensor, *Adv. Mater. Opt. Electron.*, 8(2), 4752, 1998.
67. Ramstrom, O. and Ansell, R.J., Molecular imprinting technology: Challenges and prospects for the future, *Chirality*, 10(3), 195, 1998.
68. Cornell, B.A., BraachMaksvytis, V.L.B., King, L.G., Osman, P.D.J., Raguse, B., Wieczorek, L., and Pace, R.J., A biosensor that uses ion-channel switches, *Nature*, 387(6633), 580, 1997.
69. Kriz, D.K.M. and Mosbach, K., Introduction of molecularly imprinted polymers as recognition elements in coductometric chemical sensors, *Sensors Actuators B Chem.*, 33(1–3), 178, 1996.
70. Gopel, W. and Heiduschka, P., Interface analysis in biosensor design, *Biosensors Bioelectron.*, 10(9–10), 853, 1995.
71. Charych, D., Cheng, Q., Reichert, A., Kuziemko, G., Stroh, M., Nagy, J.O., Spevak, W., and Stevens, R.C., A 'litmus test' for molecular recognition using artificial membranes, *Chem. Biol.*, 3(2), 113, 1996.

72. Betzig, E., Trautman, J.K., Harris, T.D., Weiner, J.S., and Kostelak, R.L., Breaking the diffraction barrier — optical microscopy on a nanometric scale, *Science*, 251(5000), 1468, 1991.
73. Tan, W.H., Shi, Z.Y., Smith, S., Birnbaum, D., and Kopelman, R., Submicrometer intracellular chemical optical fiber sensors, *Science*, 258(5083), 778, 1992.
74. Tan, W.H., Shi, Z.Y., and Kopelman, R., Development of submicron chemical fiber optic sensors, *Anal. Chem.*, 64(23), 2985, 1992.
75. Alarie, J.P. and Vo-Dinh, T., Antibody-based submicron biosensor for benzo a pyrene DNA adduct, *Polycyclic Aromatic Compounds*, 8(1), 45, 1996.
76. Cullum, B.M., Griffin, G.D., Miller, G.H., and Vo-Dinh, T., Intracellular measurements in mammary carcinoma cells using fiber-optic nanosensors, *Anal. Biochem.*, 277(1), 25, 2000.
77. Hoffmann, P., Dutoit, B., and Salathe, R.P., Comparison of mechanically drawn and protection layer chemically etched optical fiber tips, *Ultramicroscopy*, 61(1–4), 165, 1995.
78. Turner, D.R., U.S. patent 4,469,554, 1984.
79. Stockle, R., Fokas, C., Deckert, V., Zenobi, R., Sick, B., Hecht, B., and Wild, U.P., High-quality near-field optical probes by tube etching, *Appl. Phys. Lett.*, 75(2), 160, 1999.
80. Teague, E.C., *Scanning Microscopy Technology and Applications Society of Photo-Optical Instrumentation Engineering*, SPIE, Bellingham, WA, 1988.
81. Pohl, D.W., Scanning near-field optical microscopy, in *Advances in Optical and Electron Microscopy*, Sheppard, C.J.R. and Mulvey, T., Eds., Academic Press, London, 1984, p. 184.
82. Betzig, E. and Chichester, R.J., Single molecules observed by near-field scanning optical microscopy, *Science*, 262(5138), 1422, 1993.
83. Betzig, E., Lewis, A., Harootunian, A., Isaacson, M., and Kratschmer, E., Near-field scanning optical microscopy (NSOM) — development and biophysical applications, *Biophys. J.*, 49(1), 269, 1986.
84. Durig, U., Pohl, D.W., and Rohner, F., Near-field optical-scanning microscopy, *J. Appl. Phys.*, 59(10), 3318, 1986.
85. Betzig, E., Isaacson, M., and Lewis, A., Collection mode near-field scanning optical microscopy, *Appl. Phys. Lett.*, 51(25), 2088, 1987.
86. Lieberman, K., Harush, S., Lewis, A., and Kopelman, R., A light-source smaller than the optical wavelength, *Science*, 247(4938), 59, 1990.
87. Zeisel, D., Deckert, V., Zenobi, R., and Vo-Dinh, T., Near-field surface-enhanced Raman spectroscopy of dye molecules adsorbed on silver island films, *Chem. Phys. Lett.*, 283(5–6), 381, 1998.
88. Deckert, V., Zeisel, D., Zenobi, R., and Vo-Dinh, T., Near-field surface enhanced Raman imaging of dye-labeled DNA with 100-nm resolution, *Anal. Chem.*, 70(3), 2646, 1998.
89. Samuel, J., Strinkovski, A., Lieberman, K., Ottolenghi, M., Avnir, D., and Lewis, A., Miniaturization of organically doped sol-gel materials — a microns-size fluorescent pH sensor, *Mater. Lett.*, 21(5–6), 431, 1994.
90. McCulloch, S. and Uttamchandoni, D., Sub-micrometre fiber optic chemical sensor, *IEEE Proc. Optoelectron.*, 144, 162, 1995.
91. Tan, W.H., Shi, Z.Y., and Kopelman, R., Miniaturized fiberoptic chemical sensors with fluorescent dye-doped polymers, *Sensors Actuators B Chem.*, 28(2), 157, 1995.
92. Song, A., Parus, S., and Kopelman, R., High-performance fiber optic pH microsensors for practical physiological measurements using a dual-emission sensitive dye, *Anal. Chem.*, 69(5), 863, 1997.
93. Koronczi, I., Reichert, J., Heinzmann, G., and Ache, H.J., Development of a submicron optochemical potassium sensor with enhanced stability due to internal reference, *Sensors Actuators B Chem.*, 51(1–3), 188, 1998.
94. Bui, J.D., Zelles, T., Lou, H.J., Gallion, V.L., Phillips, M.I., and Tan, W.H., Probing intracellular dynamics in living cells with near-field optics, *J. Neurosci. Meth.*, 89(1), 9, 1999.
95. Barker, S.L.R. and Kopelman, R., Development and cellular applications of fiber optic nitric oxide sensors based on a gold-adsorbed fluorophore, *Anal. Chem.*, 70(23), 4902, 1998.
96. Vo-Dinh, T., Alarie, J.P., Cullum, B.M., and Griffin, G.D., Antibody-based nanoprobe for measurement of a fluorescent analyte in a single cell, *Nat. Biotechnol.*, 18(7), 764, 2000.

97. Munkholm, C., Walt, D.R., and Milanovich, F.P., Preparation of $CO_2$ fiber optic chemical sensor, *Abstr. Papers Am. Chem. Soc.*, 193, 183, 1987.
98. Munkholm, C., Parkinson, D.R., and Walt, D.R., Intramolecular fluorescence self-quenching of fluoresceinamine, *J. Am. Chem. Soc.*, 112(7), 2608, 1990.
99. Barker, S.L.R., Thorsrud, B.A., and Kopelman, R., Nitrite- and chloride-selective fluorescent nano-optodes and in in vitro application to rat conceptuses, *Anal. Chem.*, 70(1), 100, 1998.
100. Barker, S.L.R., Kopelman, R., Meyer, T.E., and Cusanovich, M.A., Fiber-optic nitric oxide-selective biosensors and nanosensors, *Anal. Chem.*, 70(5), 971, 1998.
101. Cordek, J., Wang, X.W., and Tan, W.H., Direct immobilization of glutamate dehydrogenase on optical fiber probes for ultrasensitive glutamate detection, *Anal. Chem.*, 71(8), 1529, 1999.
102. Cullum, B.M. and Vo-Dinh, T., Optical nanosensors and biological measurements, *Biofutur*, 2000(205), A1, 2000.
103. Vo-Dinh, T., Griffin, G.D., Alarie, J.P., Cullum, B.M., Sumpter, B., and Noid, D., Development of nanosensors and bioprobes, *J. Nanopart. Res.*, 2, 17, 2000.
104. Sasaki, K., Shi, Z.Y., Kopelman, R., and Masuhara, H., Three-dimensional pH microprobing with an optically-manipulated fluorescent particle, *Chem. Lett.*, 2, 141, 1996.
105. Clark, H.A., Hoyer, M., Philbert, M.A., and Kopelman, R., Optical nanosensors for chemical analysis inside single living cells. 1. Fabrication, characterization, and methods for intracellular delivery of PEBBLE sensors, *Anal. Chem.*, 71(21), 4831, 1999.
106. Clark, H.A., Kopelman, R., Tjalkens, R., and Philbert, M.A., Optical nanosensors for chemical analysis inside single living cells. 2. Sensors for pH and calcium and the intracellular application of PEBBLE sensors, *Anal. Chem.*, 71(21), 4837, 1999.
107. Sumner, J.P., Aylott, J.W., Monson, E., and Kopelman, R., A fluorescent PEBBLE nanosensor for intracellular free zinc, *Analyst*, 127(1), 11, 2002.
108. Clark, H.A., Barker, S.L.R., Brasuel, M., Miller, M.T., Monson, E., Parus, S., Shi, Z.Y., Song, A., Thorsrud, B., Kopelman, R., Ade, A., Meixner, W., Athey, B., Hoyer, M., Hill, D., Lightle, R., and Philbert, M.A., Subcellular optochemical nanobiosensors: probes encapsulated by biologically localised embedding (PEBBLEs), *Sensors Actuators B Chem.*, 51(1–3), 12, 1998.
109. Xu, H., Aylott, J.W., Kopelman, R., Miller, T.J., and Philbert, M.A., A real-time ratiometric method for the determination of molecular oxygen inside living cells using sol-gel-based spherical optical nanosensors with applications to rat C6 glioma, *Anal. Chem.*, 73(17), 4124, 2001.
110. Xu, H., Aylott, J., and Kopelman, R., Sol-gel PEBBLE sensors for biochemical analysis inside living cells, *Abstr. Papers Am. Chem. Soc.*, 219, 97, 2000.
111. Chan, W.C.W. and Nie, S.M., Quantum dot bioconjugates for ultrasensitive nonisotopic detection, *Science*, 281(5385), 2016, 1998.
112. Lyon, W.A. and Nie, S.M., Single molecule methodologies for DNA analysis, *Abstr. Papers Am. Chem. Soc.*, 213, 49, 1997.
113. Taylor, J.R., Fang, M.M., and Nie, S.M., Probing specific sequences on single DNA molecules with bioconjugated fluorescent nanoparticles, *Anal. Chem.*, 72(9), 1979, 2000.
114. Zhang, C.Y., Ma, H., Nie, S.M., Ding, Y., Jin, L., and Chen, D.Y., Quantum dot-labeled trichosanthin, *Analyst*, 125(6), 1029, 2000.
115. Zhang, C.Y., Ma, H., Ding, Y., Jin, L., Chen, D.Y., Miao, Q., and Nie, S.M., Studies on quantum dots-labeled trichosanthin, *Chem. J. Chin. Univ.*, 22(1), 3437, 2001.
116. Ji, J., Rosenzweig, N., Jones, I., and Rosenzweig, Z., Molecular oxygen-sensitive fluorescent lipobeads for intracellular oxygen measurements in murine macrophages, *Anal. Chem.*, 73(15), 3521, 2001.
117. Ji, J., Rosenzweig, N., Griffin, C., and Rosenzweig, Z., Synthesis and application of submicrometer fluorescence sensing particles for lysosomal pH measurements in murine macrophages, *Anal. Chem.*, 72(15), 3497, 2000.
118. McNamara, K.P., Nguyen, T., Dumitrascu, G., Ji, J., Rosenzweig, N., and Rosenzweig, Z., Synthesis, characterization, and application of fluorescence sensing lipobeads for intracellular pH measurements, *Anal. Chem.*, 73(14), 3240, 2001.

119. Nguyen, T., McNamara, K.P., and Rosenzweig, Z., Optochemical sensing by immobilizing fluorophore-encapsulating liposomes in sol-gel thin films, *Anal. Chim. Acta*, 400, 45, 1999.
120. McNamara, K.P., Rosenzweig, N., and Rosenzweig, Z., Liposome-based optochemical nanosensors, *Mikrochim. Acta*, 131(1–2), 57, 1999.
121. McNamara, K.P. and Rosenzweig, Z., Dye-encapsulating liposomes as fluorescence-based oxygen nanosensors, *Anal. Chem.*, 70(22), 4853, 1998.

# 61

# Advanced Photonics: Optical Trapping Techniques in Bioanalysis




Kenji Yasuda
*University of Tokyo*
*Komaba, Meguro-ku, Tokyo, Japan*


## 61.1 Introduction

Knowledge about life processes has expanded dramatically during the 20th century and produced the modern disciplines of genomics and proteomics. However, there remains the great challenge of discovering the integration and regulation of living components in time and space within the cell. As we move into the postgenomic period, the complementarity between genomics and proteomics will become apparent, and the connections between them will be exploited, although genomics, proteomics, or their simple combination will not provide the data necessary to interconnect molecular events in living cells in time and space. The cells in a group are different entities, and differences arise even among cells grown in homogeneous conditions considered to have identical genetic information. These cells respond differently to perturbations.[1] The question is why and how these differences arise. They might be caused by unequal distributions of biomolecules in cells, mutations, interactions between cells, fluctuations of environmental elements that affect the cell, etc. To understand the principle underlying the differences, keeping in mind the possibilities mentioned above, a system is necessary that would allow for the continuous observation of specific cells under fully controlled circumstances such as interactions between cells.


0-8493-1116-0/03/$0.00+$1.50

Conventional systems such as flow cytometry and direct measurement with a microscope have been used for tracking changes in cells. Flow cytometry enables us to obtain the distributions of parameters like concentration, size, shape, DNA content, etc. at the single-cell level in a group.[2,3] The problem with these systems is that they cannot track a specific cell's dynamics continuously because the sample drawn from the culture is discarded after the measurement. In addition, these systems cannot keep cells under isolated conditions or identify a particular cell even after cell division has occurred. Thus, these systems can provide information about the average properties of various single cells, that is, how the group changes, but they cannot provide information about how single cells change. On the other hand, direct measurement of cells in solid media, like agarose-gel plate with microscope, is also used widely.[4–8] This method makes it possible to identify individual cells and track specific cells continuously. Though the cultivation of cells under isolated conditions can begin by first controlling the spread concentration, cells cannot be kept isolated even after cell divisions have occurred; in addition, it is impossible to control the interactions between particular cells because cell positions are fixed from the beginning of cultivation. Thus, these conventional systems are unsatisfactory for understanding the single-cell-level interaction of particular cells.

Thus, techniques were needed to clarify the relationships among genetically identical cells; to meet this need we have developed an on-chip microculture system based on a combination of recent advances in microfabrication techniques and conventional *in vivo* techniques.[9,10] However, the above system lacked a method for controlling the number of cells in a microchamber or for isolating particular cells from the cultured cells. For manipulating cells in microchambers some noncontact forces are required, such as optical tweezers, which have been used as a tool for handling cells, organelles, and biomolecules on a microscope specimen.[11–13] By adding optical tweezers, we have also improved the system for controlling the number of cells in microchambers.[14]

In this section, we describe our on-chip microculture system with optical tweezers. We explain the fabrication setup and examine several examples of practical single-cell cultivation, showing that this system is able to control the environment of cells, especially the effects of isolating cells, which has not been achieved with other systems.

## 61.2 On-Chip Microculture System

### 61.2.1 System Concept and Apparatus Design

As shown in Figure 61.1, the on-chip microculture system enables selective transfer of excess cells from an analysis chamber to a waste chamber (cultivation chamber) through a narrow channel (Figure 61.1A) and selective picking up of a particular cell from the cells in the cultivation chamber (Figure 61.1B).

Figure 61.2 shows the entire on-chip microculture system, which consists of four parts: microchamber-array plate, cover chamber, phase-contrast/fluorescent microscope, and optical tweezers. The cover chamber is a glass cube filled with a buffer medium; it is attached to the array plate to enable the medium in the microchambers to be exchanged through the semipermeable membrane. The volume of the cover chamber is 1 ml, and the maximum flow speed is 10 ml/min. The temperature of the buffer medium is controllable with the help of a Peltier temperature controller. The temperature of the stage is also controlled in the same way. The medium can be exchanged easily at any time during culturing by using stock solutions with different chemical and nutrient concentrations.

Phase-contrast/fluorescent microscopy (Olympus IX-70 inverted microscope with an oil-immersion objective lens, 100×, NA = 1.35) is used to study the growth and division of cells. Phase-contrast and fluorescent images are acquired simultaneously by using a charge-coupled-device (CCD) camera (Olympus, CS230). The cell images are recorded onto videotape and analyzed using a video-capture system on a personal computer (Sony PCV-R73K). The spatial resolution of the images in this system is 0.4 μm when the 100× objective lens is used.

A 1064-nm-wavelength neodymium:yttrium-aluminum-garnet (Nd:YAG) laser (T20-8S, Spectra Physics, Mountain View, CA) is used for noncontact handling as optical tweezers, providing the system with the ability to move cells between microchambers. The cells are trapped at the focal point of the

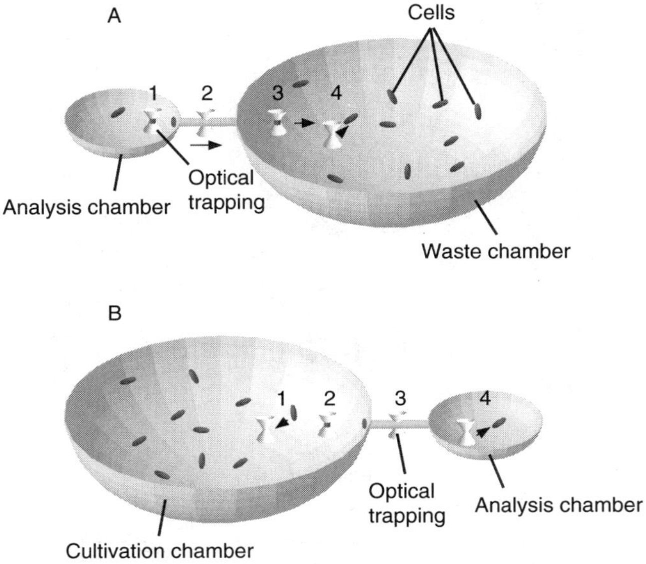


**FIGURE 61.1** Schematic drawing illustrating use of on-chip culture system with optical tweezers. (A) To keep the number of cells in the analysis chamber constant a cell in the chamber is trapped by the optical tweezers and transported into a waste chamber through the channel. (B) A particular cell cultivated in the chamber is transported into an analysis chamber.

laser beam on the microchamber array and transported by moving the focal point by slightly changing the angle of the 1064-nm dichroic mirror. The maximum laser power at the focal point after passing through the 100× objective lens is 40 mW, although we generally used less than 5 mW, which is the minimum laser power for holding bacteria.

## 61.2.2 Microchamber-Array Plate

### 61.2.2.1 Microchamber-Array Glass Slide

As shown in Figure 61.3A, the microchamber-array plate includes an $n \times n$ ($n$ = 20 to 50) array of chambers, where each chamber is 20 to 70 μm in diameter and 5 to 30 μm deep. The biotin-coated microchamber array is covered with a streptavidin-coated cellulose semipermeable membrane (molecular weight [MW] ~25,000) separating the chambers from the nutrient medium circulating through the cover chamber. The semipermeable membrane is fixed on the surface of the glass slide with a streptavidin–biotin attachment, which is strong enough to keep the cells in the microchambers while preventing contamination from the external environment. An example of the spatial arrangement of the microchamber array is shown in Figure 61.3B, where 20-μm (diameter) × 5-μm (depth) microchambers are etched at 100-μm intervals in a 0.17-mm-thick glass slide.

The fabrication process of the microchamber-array plate was as follows. The 0.17-mm-thick glass slide was sonicated in 1 *M* NaOH aqueous solution and washed with water to clean the surface. After cleaning, the dried glass slide was coated with 100-nm-thick chromium by sputtering and next with positive g-line photoresist, OFPR-800 (Tokyo Ohka, Ltd., Kawasaki, Japan) by spinning. Following baking of the slide for 15 min at 85°C, lithography of the microchamber-array patterns was carried out on a contact aligner with a broadband near-ultraviolet (UV) source (G, H, and I lines). The exposed glass slide was then developed and dried again. The exposed region of the glass slide was etched by MPM-E30 solution (Intec Inc., Tokyo, Japan) for chromium and then by hydrofluoric acid (HF) (4.7% [w/v]), $NH_4F$ (36.2% [w/v]) solution (etching velocity = 60 nm/sec at 25°C) for glass. When the shape of the microchambers attained the desired size, the glass slide was washed with water to stop etching and dried again.

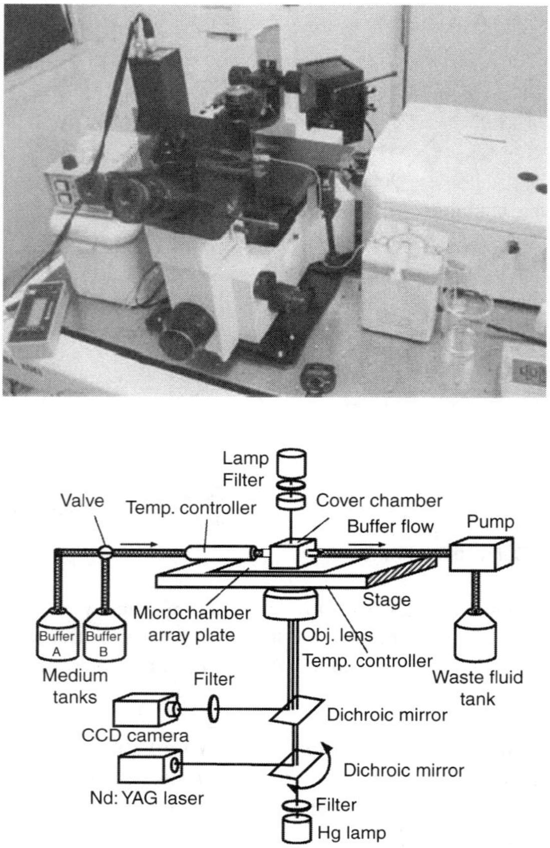


**FIGURE 61.2** Photograph and schematic drawing of on-chip system with four parts: microchamber array plate, cover chamber, phase-contrast/fluorescent microscope, and Nd:YAG laser.

#### 61.2.2.2 Streptavidin Decoration on Cellulose Semipermeable Membrane

First, cellulose semipermeable membrane (Spectora/Por $M_W$ 25,000, Spectrum Laboratories Inc., Dominguez, CA) was washed with pure water and cut in proper sizes adequate for sealing the glass slide. Next, the washed membrane was incubated in 0.2 *M* $NaIO_4$ for 6 h to generate aldehyde functions by oxidation of *cis*-vicinal hydroxyl groups of agarose. Because aldehydic functions react at pH 4 to 6 with primary amines to form Schiff bases, the membrane was washed twice in 0.1 *M* phosphate buffer (pH 6.0) and 15 μg/ml streptavidin hydrazide (Pierce Biotechnology Inc., Rockford, IL) were added. After 2 h incubation, the membrane was washed with water and stocked at 4°C (see Figure 61.4).

#### 61.2.2.3 Biotin Coat on the Surface of Microchamber-Array Glass Slide

The dried microchamber-array glass slide was next dipped in the solution containing 1% 3-(2-amino-ethylaminopropyl) aq. at 25°C to decorate the amino group on the surface of the glass slide. The amino-coated glass slide was dried for 30 min at 120°C. Next, to decorate the biotin on the surface of the glass slide, 1 mg/ml EZ-Link NHS-LC-Biotin (Pierce) aq. was dipped on the surface of the amino glass slide and incubated for 60 min. The biotin-coated glass slide was washed with water, dried at room temperature, and stored (see Figure 61.4).

#### 61.2.2.4 Microchambers Made of Thick Photoresist, SU-8

As shown in Figure 61.5, the microchambers and channels can also be formed on a glass slide lithographically using a thick photoresist, SU-8 (Microlithography Chemical Corp., Watertown, MA) with an aspect ratio of

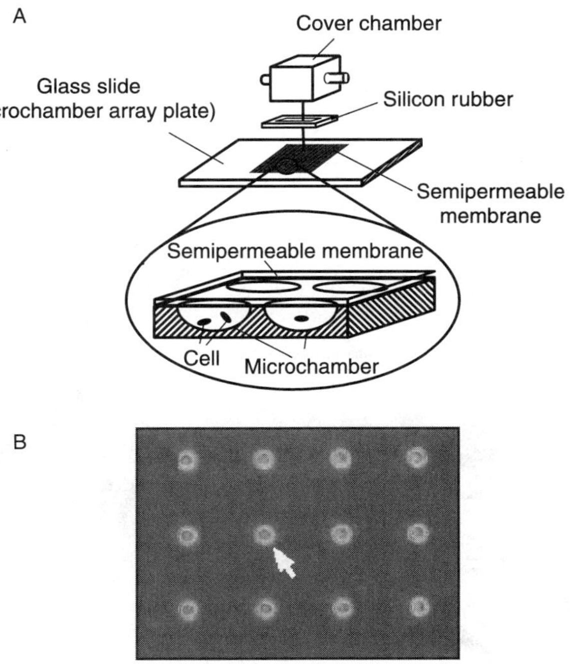


**FIGURE 61.3** Design of microchamber-array plate. (A) Schematic drawing of microchamber array plate. An $n \times n$ ($n$ = 20 to 50) array of microchambers is etched into a 0.17-mm-thick glass slide. Each microchamber is covered with a semipermeable membrane separating the chamber from the nutrient medium circulating through the medium bath in the cover chamber. A single cell or group of cells in the microchamber can thus be isolated from others perfused with the same medium. (B) Optical micrograph of the microchamber-array plate. The arrow indicates the position of one microchamber. The bar indicates a length of 100 μm.

5 μm in height and 5 μm in width, enough to cultivate and transport cells (Figure 61.5A). The process is as follows: (1) spin coat of SU-8 on surface of glass slide at 1400 rpm/sec for 20 sec, (2) soft bake at 125°C on hotplate for 30 min, (3) expose pattern using deep ultraviolet flood at 100 W for 20 sec, (4) bake at 125°C on hotplate for 30 min, and (5) develop and rinse off imaging resist.

Figure 61.5B shows the cross-section view of the microchamber, the core of the microculture system. On a 0.2-mm-thick glass slide, microchambers are fabricated for cell cultivation as described above. The surface of the glass slide is then decorated with biotin and covered with an avidin-decorated cellulose semipermeable membrane (MW ~25,000) to separate the microchambers from the nutrient medium circulating through the cover chamber. The membrane is also fixed to the surface of the glass slide using an avidin–biotin attachment, which is strong enough to keep the cells in the microchambers while preventing contamination from the external environment. Cells are transported between the microchambers with optical tweezers. The buffer medium in the chambers is exchanged with that in the cover chamber through the membrane.

### 61.2.2.5 SU-8 for On-Chip Microculture System

In this experiment, we used a 1064-nm laser for optical trapping. We have chosen the SU-8 photoresist for the following three reasons. First, neither the SU-8 nor the semipermeable membrane has absorbance at 1064 nm. Second, when we used SU-8 for the microchambers, no etching was permitted on the bottom of the glass slide, that is, the bottom of the microchambers was optically flat. In fact, we were able to trap bacteria anywhere in the microchamber array and observed no damage to the microchambers or membrane during our experiments. Third, no nonspecific attachment of cells to the SU-8 surfaces was

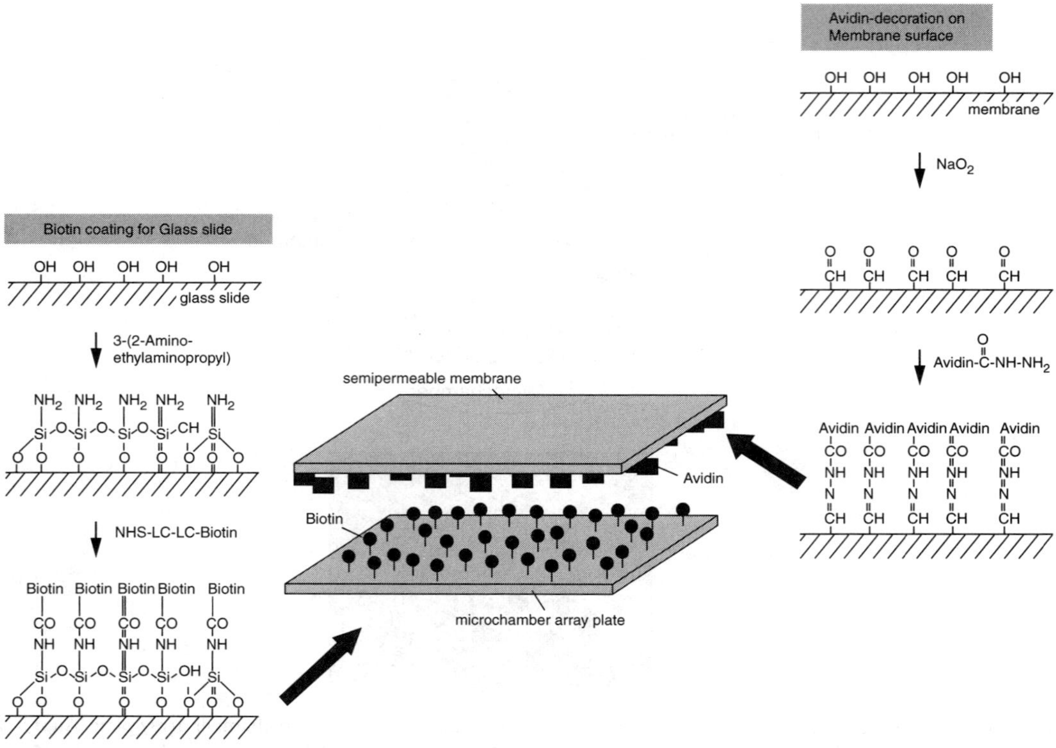


**FIGURE 61.4** Streptavidin decoration and biotin coat procedure.

observed. This was due to the constituents of SU-8: 20 to 25% gamma butyrolactone, 1 to 5% propylene carbonate, 70 to 75% epoxy resin, and mixed triarylsulfonium/hexafluor *o* antimonate salt.

### 61.2.3 Bacterial Assays and Growth Conditions

In our experiments, we used a standard sample of *Escherichia coli* strain JM109 (*end*A1, *rec*A1, *gyr*A96, *thi*, *hsd*R17($r_{k-}$,$m_k^+$), *rel*A1, *sup*E44, λ–, Δ(*lac-pro*AB), [F',*tra*D36, *pro*AB, *lac*I$^q$ZΔM15]; Toyobo Co., Ltd., Tokyo, Japan). Before the on-chip cultivation, the cells were grown overnight in an orbital shaker at a speed of 250 rpm and temperature of 37°C with 100 ml M9LB medium, which consisted of a minimal medium (M9: 4.5 g/l $KH_2PO_4$, 10.5 g/l $K_2HPO_4$, 50 mg/l $MgSO_4.7H_2O$; pH 7.1) containing 1% (v/v) Luria-Bertani medium (LB) and 100 mg/l ampicillin. When the cultured cells reached the stationary phase ($10^8$ to $10^9$ cells/ml), they were fractioned and placed in 500-μl sample tubes and maintained at –80°C as a glycerol stock solution (50% v/v). For each experiment we reactivated the stock cells at 37°C in M9LB medium, using the orbital shaker at 250 rpm. After 20 h of cultivation, the cells returned to the stationary phase and were diluted to $10^5$ cells/ml and cultured again until the concentration reached $10^8$ cells/ml. They were then used as samples for one-cell cultivation. In this cultivation, the number of cells was measured by using a counting glass slide (Erma Optical, Inc., Tokyo, Japan) and the IX-70 optical microscope.

### 61.2.4 Single-Cell Cultivation in On-Chip Microculture Systems

For single-cell cultivation, prepared *E. coli* cells were spread on the microchamber array plate and sealed in with the streptavidin-coated cellulose semipermeable membrane. Figure 61.7 shows the cell number in each microchambers after the lid was attached. As shown in the figure, about 300 (75% of the total) microchambers were vacant and 50 (12.5% of the total) chambers were packed with isolated single cells.

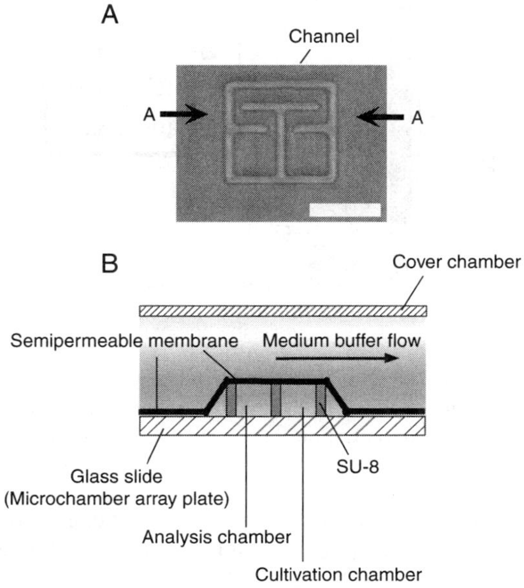


**FIGURE 61.5** Microchamber array using thick photoresist SU-8. There are several microchambers on the microchamber-array plate, including an analysis chamber and a cultivation chamber, connected by a channel. The buffer medium in the chambers is exchanged through a semipermeable membrane. Cells in the microchambers are transported by optical tweezers. (A) Optical micrograph of microchambers and channels constructed lithographically using thick photoresist SU-8. Bar = 50 μm. (B) Schematic drawing of A-A cross-sectional view of microchamber (A).

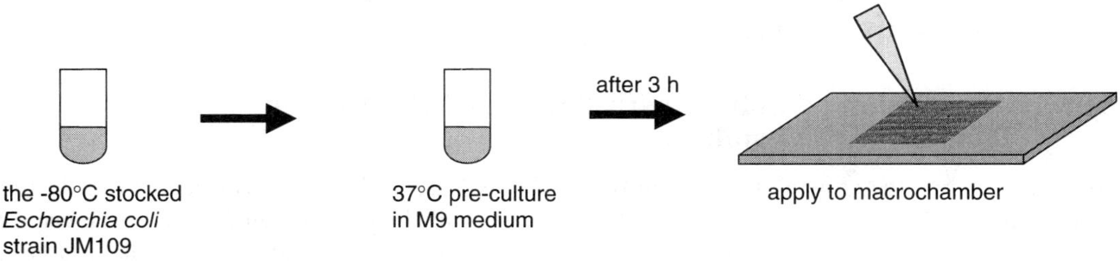


**FIGURE 61.6** Schematic diagram of sample preparation.

Cells entrapped in nonwell regions were sandwiched between the glass slide and the membrane and were no longer moved or grown. When the concentration of cells in the dropping buffer exceeded $10^9$ cells/ml, the batch culture of cells stopped growing as the cells entered the stationary phase. On the other hand, when the concentration was less than $10^7$ cells/ml, cells were in lag phase or early log phase, which was before or just as the number of cells started to increase. Thus, the concentration of *E. coli*, $10^7$ to $10^9$ cells/ml, was appropriate for packing one cell into each microchamber. Concerning how the lid was attached, we generally used one of the following methods: (1) by incubating the membrane on the glass slide until they became attached or (2) by using centrifugation (× 1000 *g*) to remove excess water and attach the membrane on the glass slide. After the membrane was attached to the plate, no *E. coli* trapped in the microchambers could escape. The membrane-sealed microchamber-array plate was then covered with the cover chamber and set on the stage of the microscope. Those microchambers that trapped single *E. coli* cells were chosen for observation. The time-course changes of the chosen isolated *E. coli* were continuously observed and recorded by the CCD camera and VCR (max. 405-min recording).

After recording, the time-course growth of the single *E. coli* cells was analyzed using image-analysis software (1/30 sec and 0.2-μm resolution). We defined the length of an *E. coli* cell as the distance between

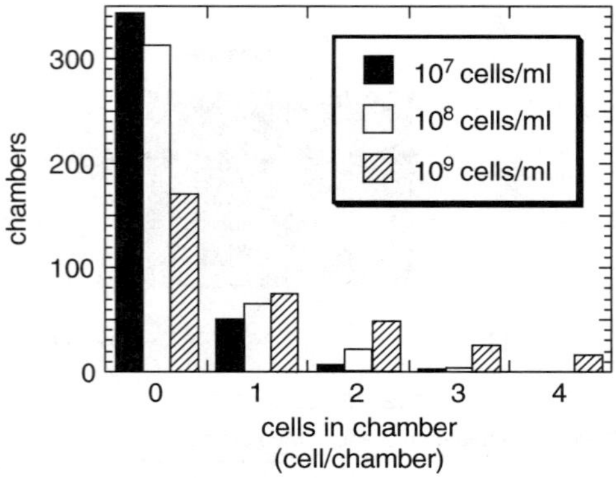


**FIGURE 61.7** Number of cells in each chamber of the microchamber array (400 microchambers) after cells were spread and sealed with membrane. The concentrations of culture cells used for packing were $10^7$, $10^8$, and $10^9$ cells/ml, respectively.

its two ends (see Figure 61.8) and the growth of an *E. coli* cell as its elongation rate. The length measurement included 10 to 15% uncertainty because we could only measure the two-dimensional end-to-end length images of three-dimensionally tumbling *E. coli.*

In this system, we used a 0.17-mm-thick glass slide because the maximum working distance of the X100, N.A. 1.35 objective lens was less than 0.3 mm. As shown in Figure 61.8, no *E. coli* attachment on the glass surface in the microchamber wells was observed during cultivation. Thus, we could measure the time-course change of *E. coli* swimming freely in the microchamber.

# 61.3 Single-Cell Cultivation Using On-Chip Microculture Systems

## 61.3.1 Observation of Cell Growth Rate and Cell-Division Time in Microchambers

We first examined the growth and division of *E. coli* under isolated conditions (Figure 61.8). In this experiment, we used the chamber that was the size of the initial concentration for the isolated single cells — $1.0 \times 10^{10}$ cells/ml. First, in this example, we observed that an isolated *E. coli* cell, 2.9 μm in length, maintained its length for 80 min (Figure 61.8A). After 80 min of sleep (we call this "sleep time"), the cell started growing at a rate of 0.04 μm/min, frequently bending around the septa. When the cell reached a length of 7.4 μm (160 min after inoculation), it divided into two daughter cells (Figure 61.8C and D), each with a length of 3.6 μm. Following the first division, the two daughter cells started growing and tumbling again like the first "mother" cell. The daughter cells divided again after they had grown for 60 to 70 min (Figure 61.8F). The period of cell division after cell growth began increased exponentially, at a rate of 86 min/cell division. Using this method we examined the division time of daughter cells born from isolated mother cells in microchambers. Figure 61.9 shows the distribution of division time of newborn daughter cells ($n$ = 160). As shown in the graph, the cell division of daughter cells started at least 31 min after they were born, and the cell division time was spread from 31 to 223 min. Mean cell-division time, 86.6 min (S.D. 38.1 min), was longer than the peak value of 50 to 70 min. Figure 61.5 explains the initial length dependency of the division time. The samples with an initial length of 3 to 4 μm (filled bars) and 4 to 5 μm (hatched bars) show the same statistical distributions of division time, or a mean time of 72.9 min (SD 31.1 min) for 63 samples and 75.5 min (SD 28.2 min) for 16 samples, respectively, whereas those measuring 2 to 3 μm (open bars) showed a different distribution, or mean division time — 105.1 min (SD 41.0 min) for 64 samples.

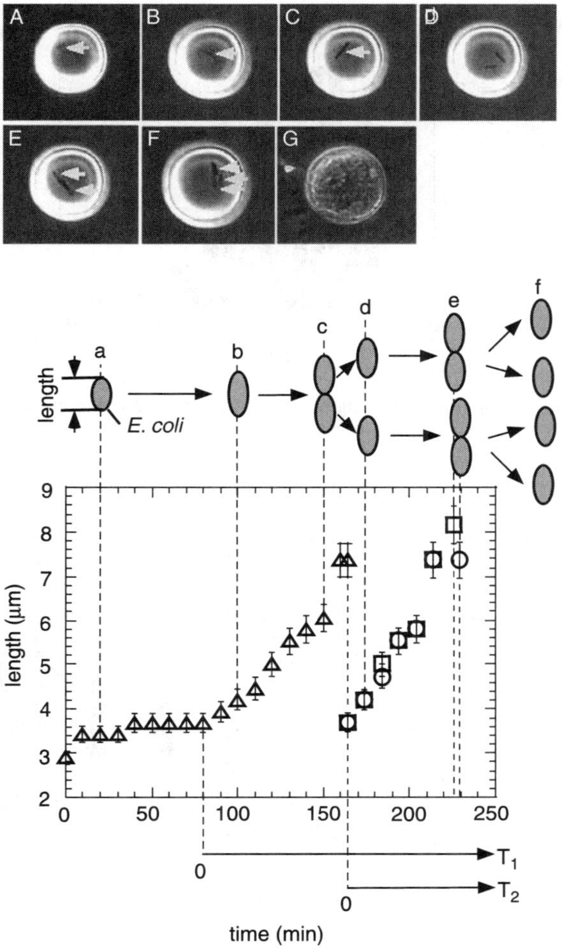


**FIGURE 61.8** Time course of isolated single *E. coli* cell growth in a microchamber. The magnified micrographs at the top (A to G) show the time course of one of the microchambers shown in Figure 61.12 at times of (A) 0 min, (B) 100 min, (C) 150 min, (D) 175 min, (E) 225 min, (F) 230 min, and (G) 15 h after inoculation. The arrows in the micrographs show the positions of *E. coli* cells in the microchamber. The bar indicates a length of 10 μm. The graph at the bottom shows the time-course growth of the individual *E. coli* cells.

This experiment demonstrated several advantages of this system. First, neither contamination from the cover chamber nor escape of *E. coli* cells from the small chamber was observed during the experiment. Second, the direct descendants of single cells could be cultured and compared in the isolated microchambers. Third, the physical properties of the cells in each microchamber could be continuously observed and compared. Finally, the resolution of cell length, 0.2 μm, was an order of magnitude smaller than that of conventional methods like the Coulter Counter (Coulter Electronics, Inc., Hialeah, FL) that determine the sizes of cells electronically.

## 61.3.2 Growth and Cell Cycles of Daughter Cells after Single-Cell Division

We can also compare cell growth and cell division times among isolated single cells and their daughter cells. Daughter cells have the same DNA and cytoplasm as their mother cell, and in this experiment they grew in the same environment, including physical contact with each other. The graph in Figure 61.10

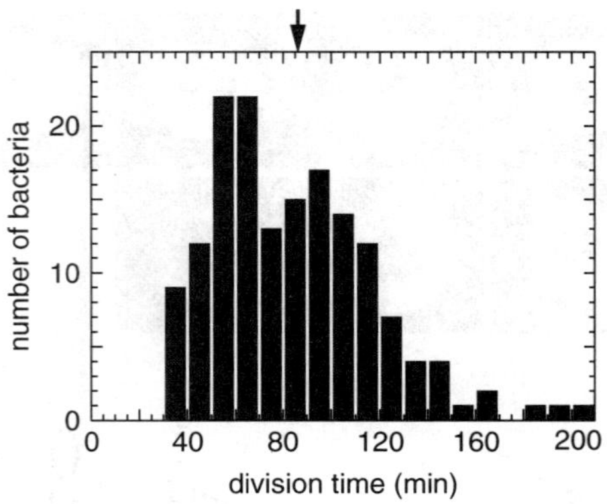


**FIGURE 61.9** Distribution of division time of daughter cells. Arrows indicate the mean time of cell division, 86.6 min (SD = 38.2, $n$ = 160).

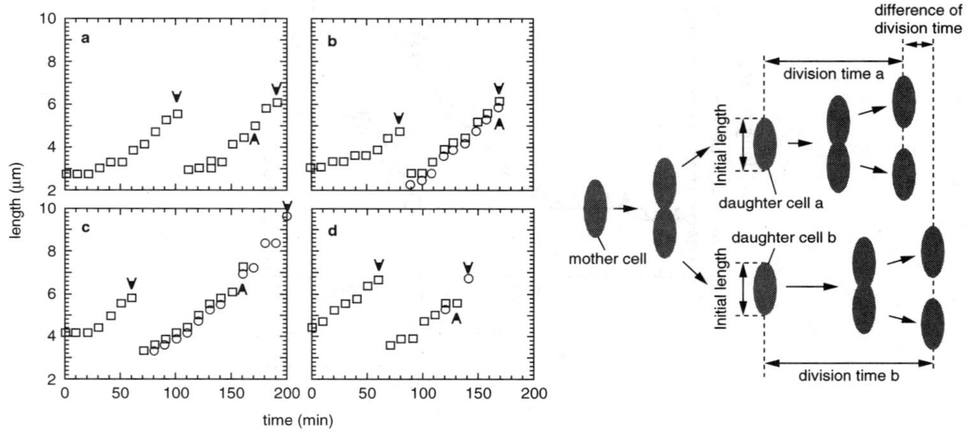


**FIGURE 61.10** Time-course growth of isolated individual *E. coli* cells and two daughters. The arrowheads indicate the cell division occurring in the cells.

shows four examples of such a comparison. An isolated mother cell grew at 0.04 µm/min and divided 85 min after cell growth began, while its daughter cells grew at 0.06 µm/min and divided after 65 to 70 min. In this case, although the growth rate and cell-division time for the mother and daughter cells were different, those for the two daughter cells were almost the same. But none of the eight other comparisons showed strong correlation either between mother and daughter cells or between daughter cells (data not shown). We also observed that mother cells sometimes divided into two daughter cells of obviously unequal lengths (data not shown).

### 61.3.3 Flexible-Medium Buffer Exchange through Semipermeable Membrane

Controlling the medium conditions is important for single-cell analysis. A semipermeable membrane attached on the microchamber-array plate exchanges the medium in microchambers within the time resolution of the VCR system, <1/30 sec, which was checked by a real-time confocal optical microscopy system (CSU21, Yokogawa Electric Corp., Tokyo, Japan). The result indicated no delay of diffusion at 5-µm depth with 1/30-sec resolution. That is, there was no difference of increase of the fluorescent dye's

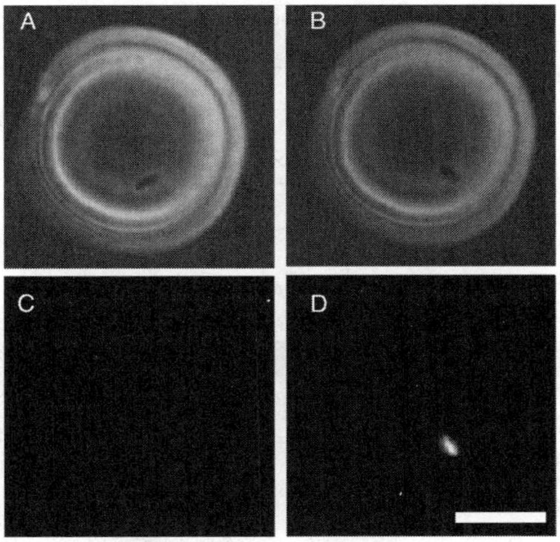


**FIGURE 61.11** Green fluorescent protein expression in *E. coli*, shown (A, B) before IPTG induction and (C, D) 60 min after IPTG induction was begun. (A) and (C) are phase-contrast micrographs; (B) and (D) are fluorescent micrographs. The arrows show the positions of *E. coli*. The bar indicates a length of 10 μm.

concentrations between the top and the bottom of the microchamber after the introduction of the dye into the system. Thus, we concluded the diffusion time was 1/30 sec.

As shown in Figure 61.8, after overnight culturing the number of *E. coli* cells had fully increased in the microchamber (Figure 61.8G). In this case, the concentration of *E. coli* corresponded to $>10^{12}$ cells/ml, which is more than a thousand times larger than the stationary-phase concentration of $>10^9$ cells/ml in batch cultivation with the same sample and medium.

We tested another use of the on-chip culture system — for changing the medium solution in the microchambers through the semipermeable membrane. Single *E. coli* cells were cultured in small chambers with an M9LB medium. To express Green Fluorescent Protein (GFP) in *E. coli* (JM109), we used an M9LB-IPTG medium buffer, which is M9LB with 1 m*M* isopropyl-×-D(-)-thiogalactopyranoside (IPTG; Wako Chemical Co., Tokyo, Japan) and 0.1 m*M* lactose. As shown in Figure 61.11(A) and (B), no GFP expression was observed before IPTG induction. GFP expression was successfully observed 40 min after the medium exchange from M9LB to M9LB-IPTG. Figure 61.11(C) and (D) shows the GFP fluorescence in the same *E. coli* 60 min after the medium exchange, which is almost the same as batch-cultivation data under the same buffer conditions at 37°C.

These results show good system performance for solution exchange, which is difficult for batch- or plate-cultivation methods.

### 61.3.4 Flexible-Chamber-Size Control of Single-Cell Cultivation

This on-chip culture system makes it easy to control the environment for cultivation. For example, cultivation volume can be controlled by using microchambers of various sizes. Figure 61.12 shows one example of volume control by using various microchambers. The size of the chamber shown in Figure 61.12B is 20 (diameter) × 5 μm (depth), which we call a "small chamber," and that in Figure 61.12C is 70 (diameter) × 30 μm (depth), a "large chamber." The volumes of the chambers are $1.0 \times 10^{-10}$ and $7.7 \times 10^{-8}$ ml, respectively. If we culture single cells in these chambers, the estimated concentrations are $1.0 \times 10^{10}$ cells/ml and $1.3 \times 10^{7}$ cells/ml, respectively.

Figure 61.13 shows the chamber-size-growth dependence of single *E. coli* cells. The average growth speeds of five samples were 0.06 μm/min, both in the small chamber and in the large chamber. The mean values of cell-division time were also almost the same for both chambers — around 45 min. The results showed no significant differences in mean values of growth speed and cell-division time, even when the chamber sizes were 1000 times different, i.e., the mean free paths were 10 times different.

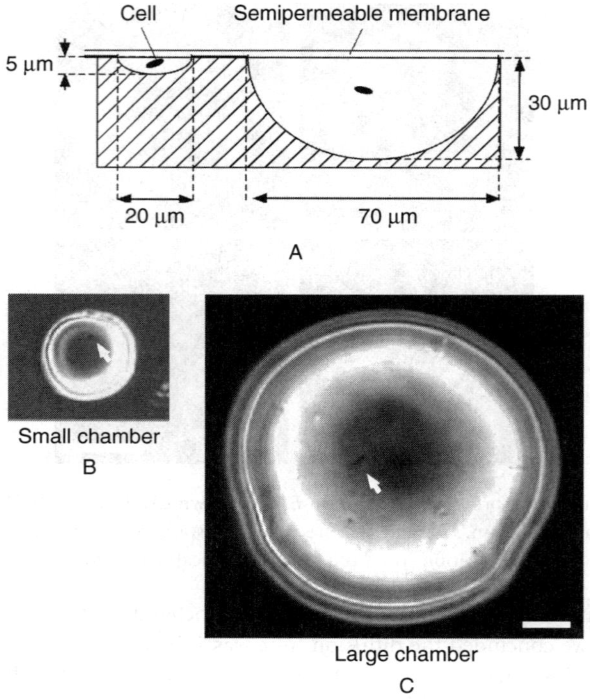


**FIGURE 61.12** (A) Schematic drawing of the two sizes of microchambers The "small chamber" on the left is 20 μm in diameter and 5 μm deep, while the "large chamber" on the right is 70 μm in diameter and 30 μm deep. Optical micrographs of (B) a small chamber and (C) a large chamber. The arrows indicate the positions of *E. coli* cells in the chambers. Bar = 10 μm.

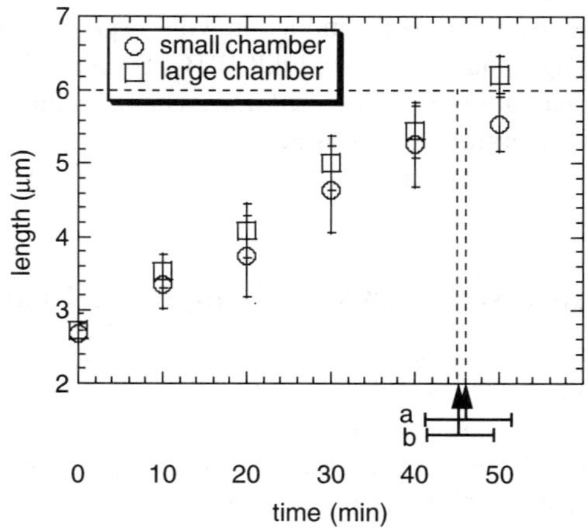


**FIGURE 61.13** Chamber-size dependence of *E. coli* cell growth. The circles indicate a small chamber, 20 μm in diameter and 10 μm deep (Figure 61.12B). The squares indicate a large chamber, 70 μm in diameter and 30 μm deep (Figure 61.12C). The plotted points are mean values of cell length at 10-min intervals, with error bars indicating the SD of five cells. The arrow and bar labeled "a" indicate the mean time and SD of the cell-division time for the small chamber; those labeled "b" indicate these values for the large chamber.

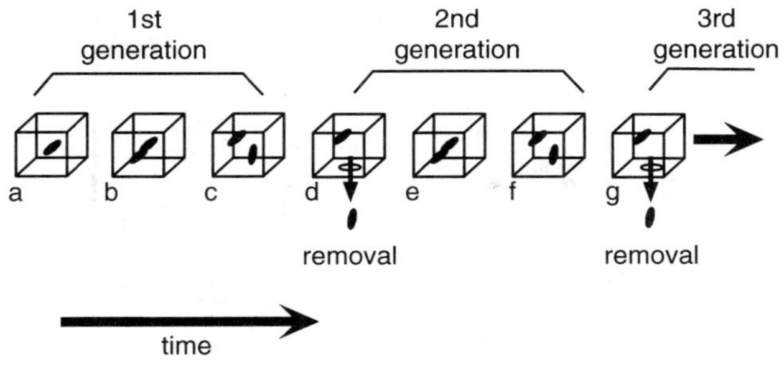


**FIGURE 61.14** Concept and process used to continuously observe direct descendants under isolation conditions.

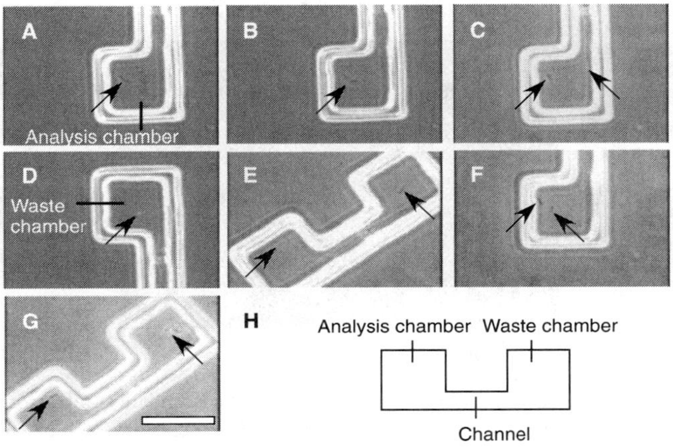


**FIGURE 61.15** Example of continuous observation of direct descendants under isolation conditions in the microchamber array: (A) isolated single cell in the analysis chamber almost divides; (B) the cell has divided into two daughter cells; (C) one of the daughter cells is trapped by optical tweezers and is going to be removed from analysis chamber to waste chamber through the channel; (D) the trapped cell has been removed to the waste chamber; (E) the whole image of two daughter cells after removal; (F) the cell in the third generation divided unequally in length; (G) the whole image of analysis chamber and waste chamber in a fourth-generation cell; (H) schematic drawing of the whole image of the chamber. Bar = 30 μm.

## 61.3.5 Continuous Single-Cell Cultivation with Optical Tweezers

Using this system we also examined whether the direct descendants of an isolated single cell could be observed under the same isolated conditions. For observing the isolated single cells continuously even after cell division occurred, we developed the following method. As shown in Figure 61.14, we first isolated a single cell in the analysis chamber (first generation) and observed its growth (Figure 61.14A). Once cell division occurred, one of the two daughter cells was removed from the chamber quickly before they made physical contact (Figure 61.14C and D). The isolated second-generation cell was observed continuously until it divided again. This process was repeated as many times as possible.

Figure 61.15 shows an example of continuous observation of the direct descendants of a mother cell under isolated conditions, as explained above. First, one mother cell (8.8 μm long) was enclosed in an analysis chamber and observed continuously (Figure 61.15A). Right after the cell divided one of its daughter cells was trapped by optical tweezers and transported to a waste chamber through a narrow channel (cell trap path) (Figure 61.15B through E). The same procedure was done every time the cell in the analysis chamber divided. In the course of observation, the cell in the third generation divided unequally; one was 5.4 μm long, and the other was 3.6 μm (with an error of ±0.4 μm because of limitations

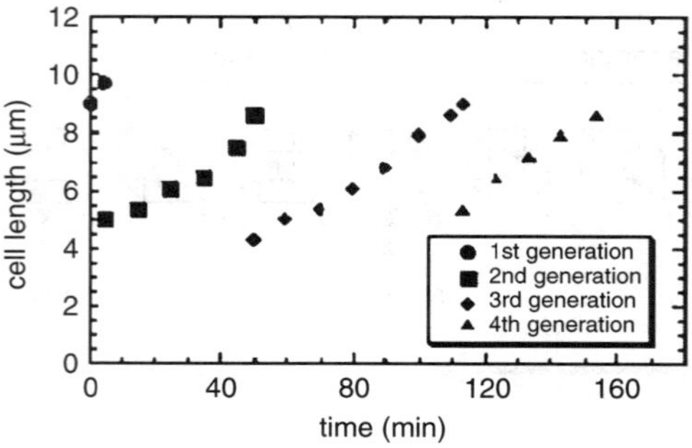


**FIGURE 61.16** Time course of cell growth of isolated single cells shown in Figure 61.15.

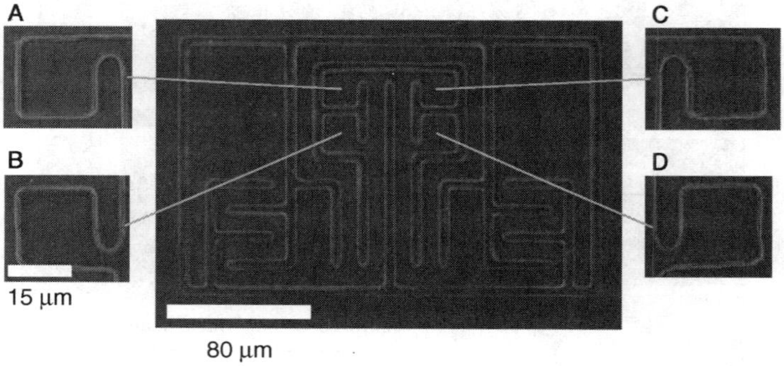


**FIGURE 61.17** Another example of continuous observation of direct descendants under isolated conditions in four-room microchambers for simultaneous observation and comparison. In each of the four chambers (A through D) at the center of the microstructure, isolated single cells are cultivated and compared.

in resolution) (Figure 61.5F). The cell in the fourth generation also divided unequally again; one was 5.3 μm and the other was 3.8 μm. They were quite different from the cell division in the second generation (Figure 61.5B). The lengths of the two daughter cells were almost equal— around 4.9 μm. It has been reported that *E. coli* cells divide equally,[15] but in our system, unequal cell division in length has often been observed.

Figure 61.16 shows the time course of cell growth of four generations of direct descendants in the sample shown in Figure 61.15. As shown in the result, cell cycles for cell division seems independent among generations. In addition, we could not find strong correlation among generations in the elongation velocity or cell-division lengths.

Figure 61.17 shows another example of single-cell cultivation using different chamber designs. In the experiment shown in Figure 61.15, the cells in the waste chamber easily invaded the analysis chamber by swimming because the path (channel) was too simple. Thus, we made the channel pathway in the chambers more labyrinthine. This prevented the escape of the cell from the chambers. Using this chamber array we were able to analyze and compare four cells simultaneously under isolated conditions. The four graphs shown in Figure 61.18 indicate no significant correlation among those four cells during cultivation.

### 61.3.6 Differential Analysis of Single-Cell Cultivation in Microchamber-Array Systems

The results shown in Figures 61.16 and 61.18 illustrate only half of the two daughter cells' properties. We thus improved the method and the microchamber array to enable us to measure all the descendant cells of an isolated mother cell (Figure 61.19).

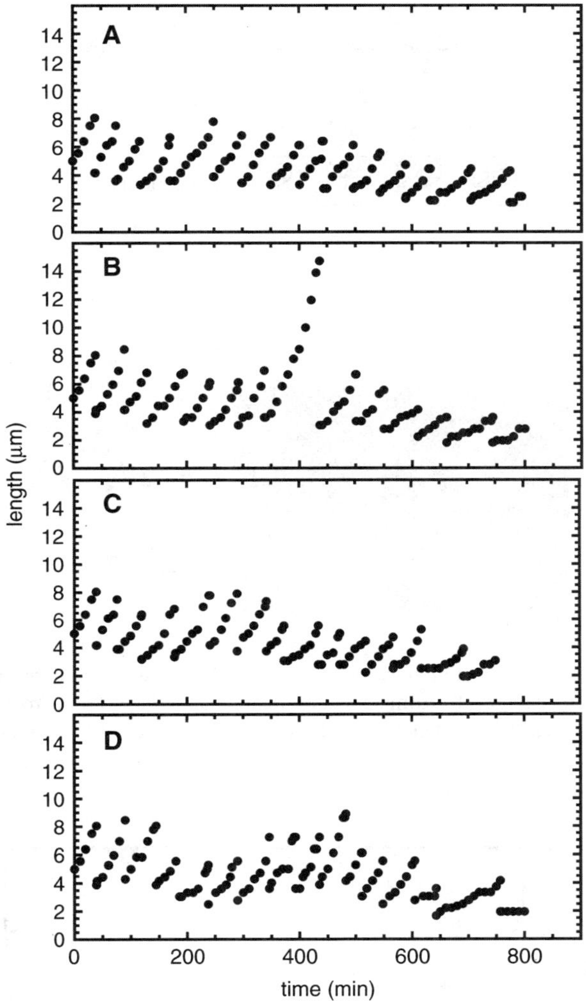


**FIGURE 61.18** Time course of the four cousin cells' growth under isolated conditions. (A) through (D) indicate the samples in chambers A through D, respectively, of Figure 61.17.

Figure 61.19B shows the chamber array we used for complete descendant analysis. The cultivation was done as follows. First, a cell was transferred into chamber 1 (first generation) with optical tweezers. When the cells cultivated in chamber 1 divided into two daughter cells, they were transferred one by one into the second generation's chambers. Next, the isolated daughter cells divided again, and the granddaughter cells were transferred into the third generation's chambers. The same process was followed after the cells of the third generation divided again. Note that all the cells were irradiated during the cell trapping, unlike in the other experiments described above.

As shown in Figure 61.8B, the results indicate the existence of fluctuations of cell length, division time, and division length among direct-descendant cells — even between two sister cells. For example, as shown in the lower graph of Figure 61.20, one of the daughter cells (second generation) was 3.1 μm long at birth and grew as long as 11.9 μm. At this length, it produced two cells, one 9.2 μm long and the other 2.8 μm long, and the ratio of length between newborn cells was almost 3:1. Although the longer cell appeared to be one cell, it might be the size of three cells. If this is true, avoiding several divisions would lead to different patterns of cell reproduction and may affect the proliferation of the cell group.

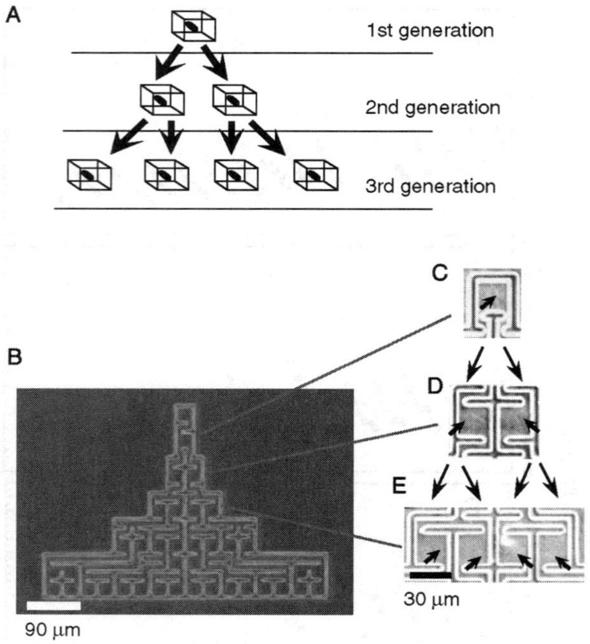


FIGURE 61.19 The process of observation of each descendant cell for generations derived from a single cell (A) and the micrograph of the microchambers for complete descendant analysis (B); (C) microchamber for first generation with isolated single mother cell; (D) two daughters isolated in each chamber; (E) four granddaughters in each of the four microchambers.

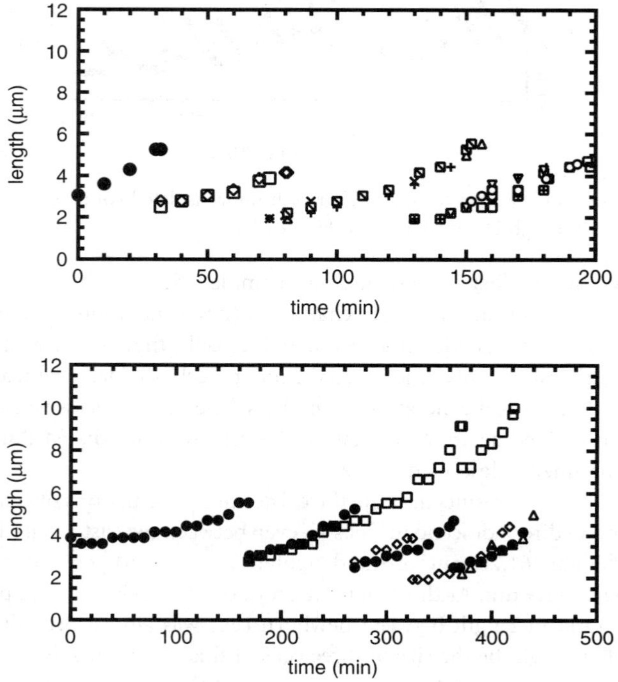


FIGURE 61.20 Two examples of growth of all descendant cells from isolated single cell.

## 61.4 Conclusion

This chapter described the on-chip microcultivation system, which has four advantages; it allows for (1) continuous cultivation of single cells under isolated conditions; (2) control of spatial distribution of cells, interactions of cells, and the number of cells in the chamber; (3) continuous observation of identical specific cells under contamination-free conditions; and (4) control of the environmental conditions surrounding cells such as the cultivation buffer. We also demonstrated two methods for analysis of single-cell differences using the on-chip microculture system and optical tweezers. Although we showed only a few examples of our experimental results, these results demonstrated the unique potential that our system has for detecting several phenomena. Differential analysis should help clarify heterogeneous phenomena such as unequal cell division and cell differentiation, which will become more and more important in the "cellome" era.

### References

1. Spudich, J.L. and Koshland, D.E., Jr., Non-genetic individuality: chance in the single cell, *Nature*, 262, 467, 1976.
2. Åkerlund, T., Nordström, K., and Bernander, R., Analysis of cell size and DNA content in exponentially growing and stationary-phase batch culture of *Escherichia coli*, *J. Bacteriol.*, 177, 6791, 1995.
3. Zhao, R., Natarajan, A., and Srienc, F., A flow injection flow cytometry system for on-line monitoring of bioreactors, *Biotechnol. Bioeng.*, 62, 609, 1999.
4. Donachie, D.D. and Begg, K.J., "Division potential" in *Escherichia coli*, *J. Bacteriol.*, 178, 5971, 1996.
5. Elowitz, M.B. and Leibler, S., A synthetic oscillatory network of transcriptional regulators, *Nature*, 403, 335, 2000.
6. Gardner, T.S., Cantor, C.R., and Collins, J.J., Construction of a genetic toggle switch in *Escherichia coli*, *Nature*, 403, 339, 2000.
7. Panda, A.K., Khan, R.H., Appa Rao, K.B.C., and Totey, S.M., Kinetics of inclusion body production in batch and high cell density fed-batch culture of *Escherichia coli* expressing ovine growth hormone, *J. Bacteriol.*, 75, 161, 1999.
8. Shapiro, J.A. and Hsu, C., *Escherichia coli* K-12 cell-cell interactions seen by time-lapse video, *J. Bacteriol.*, 171, 5963, 1989.
9. Inoue, I., Wakamoto, Y., Moriguchi, H., Okano, K., and Yasuda, K., On-chip culture system for observation of isolated individual cells, *Lab Chip*, 1, 50, 2001.
10. Inoue, I., Wakamoto, Y., and Yasuda, K., Non-genetic variability of division cycle and growth of isolated individual cells in on-chip culture system, *Proc. Jpn. Acad.*, 77, 145, 2001.
11. Ashkin, A., Dziedzic, J.M., Bjorkholm, J.E., and Chu, S., Observation of a single-beam gradient force optical trap for dielectric particles, *Opt. Lett.*, 11, 288, 1986.
12. Ashkin, A., Dziedzic, J.M., and Yamane, T., Optical trapping and manipulation of single cells using infrared laser beams, *Nature*, 330, 769, 1987.
13. Wright, W.H., Sonek, G.J., Tadir, Y., and Berns, M.W., Laser trapping in cell biology, *IEEE J. Quantum Electron.*, 26, 2148, 1990.
14. Wakamoto, Y., Inoue, I., Moriguchi, H., and Yasuda, K., Analysis of single-cell differences by use of an on-chip microculture system and optical trapping, *Fresenius J. Anal. Chem.*, 371, 276, 2001.
15. Marr, A.G., Harvey, R.J., and Trentini, W.C., Growth and division of *Escherichia coli*, *J. Bacteriol.*, 91, 2388, 1996.

# 62

# *In Vivo* Bioluminescence Imaging as a Tool for Drug Development

Christopher H. Contag
*Stanford University School of Medicine*
*Stanford, California*

Pamela R. Contag
*Xenogen Corporation*
*Alameda, California*
*and*
*Stanford University School of Medicine*
*Stanford, California*



## 62.1 Functional Imaging in Target-Directed Paradigms

### 62.1.1 Animal Models in Drug Development

Animal models that are used for biological assessment in drug development typically represent advanced, or even end-stage, disease and often require sacrifice of the animal for analysis. The assays that are subsequently performed on tissue samples can be time-consuming and can only provide a snapshot of the overall disease course, even when performed on large numbers of animals. Therefore, one must consider the limitations of the data that result from conventional animal protocols that employ *ex vivo* assays, which are constrained by sample size, limited to a small number of selected time points, and performed in the absence of intact organ systems. Moreover, in the current paradigm of target-directed drug discovery, the events that occur early in the disease process represent important targets for therapeutic intervention, and therefore development of assays that provide access to information about these targets during therapeutic intervention is essential for evaluating the new classes of compounds that are being developed. Functional imaging that is relatively high-throughput and yields multiparameter data

0-8493-1116-0/03/$0.00+$1.50

will be required to sort out the vast numbers of targets and compounds introduced into drug discovery by this evolving paradigm. Imaging strategies have been designed specifically for laboratory animals, and their use for *in vivo* assessment of therapeutic efficacy has the potential to greatly influence the pharmaceutical industry. One of these imaging modalities, *in vivo* bioluminescent imaging (BLI), and its application to drug discovery and drug development are the subject of this chapter.

### 62.1.2 Optical Imaging

Imaging modalities in the optical regime have the advantages of being sensitive, accessible, relatively low-cost, and rapid. This has led to the development of a number of optical-imaging approaches for *in vivo* analyses.[1–15] Of these, some are particularly well suited for the purpose of accelerating the drug-development process as they address the unmet need of real-time *in vivo* assays that are ideal for imaging small laboratory animals.[5,18–20] From an imaging perspective visible light is safe to use, even in large doses, can penetrate relatively deeply into certain kinds of tissues, does not require a chemical substrate, and is quantifiable under certain circumstances. Light can also be repeatedly introduced into a tissue with safety and is not subject to radioactive decay. As there are few, if any, sources of light in mammalian systems, especially at near-infrared (NIR) wavelengths, background is minimal, if not completely absent, resulting in extraordinary signal-to-noise ratios (SNRs). For tissues and organs like the breast, with low light absorption in the 700- to 850-nm spectral range, optical imaging is clinically feasible;[1,2] however, there may be significantly fewer clinical applications than those developed for animal studies.

### 62.1.3 Imaging in the Preclinical Stage

Increasingly, the subject of conservative use of laboratory animals is resonating with those who see the scientific and economic value of responsible animal-use protocols. To address these issues a number of imaging methods that are based on clinical imaging modalities have been modified for studying animal models. These include magnetic resonance imaging (MRI),[4,5,21–33] positron emission tomography (PET), single-photon-emission tomography (SPECT),[33–42] and x-ray-computed tomography (CT).[43,44] The composite of these and the optical methods represents significant advances in the nascent field of molecular imaging, and many of these imaging methods will likely have a dramatic effect on preclinical studies in laboratory animals.[45]

Incorporating imaging technologies that are rapid and accessible into preclinical studies will yield more and higher-quality experimental data per protocol by increasing the number of times that quantitative data can be collected. By imaging the whole intact live animal at multiple time points researchers can analyze biomolecular processes in the presence of the contextual influences of intact organs. It is an added benefit that with these methods fewer animals can deliver data with greater statistical significance and lower stress. Noninvasive methods can be used to create more predictive animal models that share the characteristics of longitudinal-study design, internal experimental control, molecular information, and quantitative data, and these methods will benefit both scientific inquiry and humane animal use.[45] In addition, imaging can further improve these studies by guiding appropriate endpoint-tissue sampling for histology or biochemical analysis. Imaging of cellular and molecular targets in living-animal models of human biology and disease has been used for animal-research protocols in fields as diverse as cancer biology, microbiology, and gene therapy. Such tools enable noninvasive *in vivo* assessment in individual animals over time, which reveals temporal changes and reduces the number of animals required for a given study.

The advantages of imaging in drug development address many of the limitations at the preclinical stage, and the improved preclinical data sets will lead to better study designs for the clinical trials; in addition, the more user friendly and accessible the methods become the more likely they are to be integrated into these protocols. Optical imaging using bioluminescent reporters for tracking and monitoring cell populations and assessing levels of gene expression or monitoring *in vivo* gene delivery is gaining wide acceptance in the pharmaceutical industry as an accessible and versatile method of approaching the study of animal models of disease.[45] BLI can be incorporated into target validation during the discovery phase of drug development, or it can be used to assess *in vivo* efficacy, safety, and toxicity during preclinical development of specific compounds. With BLI several parameters of drug efficacy and pharmacokinetics can be monitored in the same animal over time, yielding benefits for the discovery and development stages of drug evaluation.

## 62.2 *In Vivo* Bioluminescent Imaging in Drug Discovery

### 62.2.1 *In Vivo* Bioluminescent Imaging

BLI has been used as a noninvasive means of tracking pathogens or tumor cells in animal subjects early in the disease process[4,5,15,18,46–51] and more recently has been used to develop new animal models that incorporate reporter genes into the rodent genome as markers of transcription that reveal developmental changes or response to stress.[19,52,53] These transgenic strategies permit monitoring of a wide variety of biological events noninvasively in living animals and enable relatively high-throughput *in vivo* screens that are likely more predictive of the human response to therapy. Given the accessibility and speed of BLI, researchers will be able to perform *in vivo* efficacy, pharmacokinetics, toxicology, and target validation studies on a larger number of compounds than was possible in the past. As these more physiologically relevant assays are conducted earlier in the drug-development cycle than was previously possible,[20] the decision of whether or not to pursue a compound can be made at a critical time-saving point in the development process.

BLI exploits the light-emitting properties of photoproteins such as luciferase enzymes and is based on the capacity of certain living organisms, including some species of bacteria, algae, coelenterates, beetles, fish, and fireflies, to emit visible light.[20,54,55] Since these chemical reactions, which produce light, can be replicated outside of the organisms to which they are native, bioluminescence has proven extremely useful for research at the cellular and molecular levels and have been used in live-cell assays.[56–59] New developments in detection technology have advanced BLI from biochemical and live-cell assays to include *in vivo* whole-body imaging of pathogens, tumors, and gene-expression patterns in living animals.[20]

Compared to *in vitro* systems or simple transparent organisms, signal attenuation in whole-rodent systems can be severe and varies according to emission wavelength and type and depth of location of the tissues surrounding the cells containing any fluorescent or bioluminescent reporter. Blue-green light (400 to 590 nm) is strongly attenuated by tissue, while red to NIR light (590 to 800 nm) is less affected. Most types of fluorescent and bioluminescent proteins have peak emission at blue to yellow-green wavelengths, although the emission spectrum from firefly luciferase, the most widely used source of bioluminescent light, is broad enough that there is also significant emission at red wavelengths (>600 nm), which penetrate quite deeply into tissue.[60] There are a number of dyes that are fluorescent in the NIR that can be conjugated to biomolecules and are being developed as imaging reagents.[21,61]

Experience with a wide variety of characterized luciferases has yielded methods for *in vivo* discovery using these enzymes as internal sources of light that can be monitored externally as real-time indicators of biological function in living specimens. Ultrasensitive cameras, employing recent technological advances in photon detection, are now being broadly applied to biological questions and can be used successfully to detect the few photons that escape the scattering and absorbing environment of mammalian tissues.[60] The fundamental biochemistry of light production in these organisms consists of the oxidation of a substrate under the catalysis of luciferase enzymes, which produces, as a decay product of the chemical reaction, a heatless light until the excited-state molecules return to the ground state. The luciferin substrates and the structures of the luciferase enzymes vary by organism, as do the mechanisms controlling the speed and intensity of the luminescence. Certain cofactors, such as the nucleotide ATP or other sources of energy, may need to be present for conversion of the substrate to take place.[55,62–66]

Since these reactions can be replicated outside the organisms to which they are native, bioluminescence has proven extremely useful for research at the cellular and molecular levels. The availability of instrumentation that can measure the light emitted in these reactions with great sensitivity and dynamic

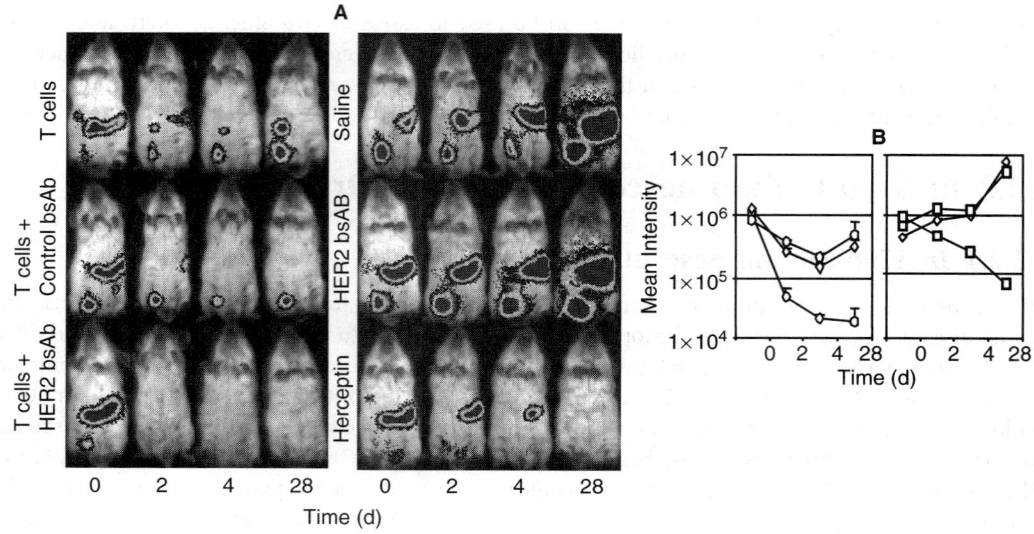


**FIGURE 62.1 (Color figure follows p. 28-30.)** Advancing complex therapeutic strategies for malignancy through imaging. Evaluation of cell-based therapies complexed with molecular therapies may be complicated by having "too many moving parts," and imaging approaches have been used to optimize these therapies by rapidly providing efficacy data without sacrificing the study subjects. In a study by Scheffold et al., the ability to redirect a tumoricidal NK T cell population to a tumor target using a bispecific antibody was revealed using BLI.[68] The control groups in this study consisted of the NK T cells alone, the NK T cells with an irrelevant bispecific antibody, normal saline, the bispecific antibody without additional cells, and herceptin (as a positive control). Temporal analyses for representative animals in each treatment group are shown (A) and data from all animals are plotted (B). Color-coded NK T cells only = light blue, NK T cells and control bispecific antibody = red, NK T cells and bispecific antibody = dark blue, saline = aqua, control bispecific antibody = black, and herceptin = dark green. BLI offers the ability to rapidly study multiple animals in each of six treatment groups and provide accurate whole-body data that are quantitative.

range has made them powerful tools for biochemical and clinical analysis because the optical signatures can be detected at a very low level. Compared to other techniques, such as colorimetric or spectrophotometric indicators, bioluminescent analysis offers the advantages of high sensitivity, wide linear range, low cost per test, and relatively simple and inexpensive equipment. Firefly luciferase is the most commonly used bioluminescent system in research, and bacterial luciferases (Lux) are used in prokaryotic systems.[55,67] The light-producing reaction of the North American firefly, *Photinus pyralis*, is also the most extensively studied bioluminescent system. Firefly bioluminescence, for example, has been used to assay levels of ATP nucleotide, which supplies energy to cells for many important biochemical processes, with extreme sensitivity. Whereas most enzyme assays yield either a product or the disappearance of a substrate, firefly luciferase acts as a quantifiable reactant rather than as a catalyst, whose most significant and easily measured product is light.

## 62.2.2 Linking Correlative Cell-Culture Assays to Animal Models

A significant advantage of *in vivo* luciferase monitoring is that bioluminescent reporters allow for an integrated approach in which the same label can be used in cell-culture correlates of biological processes and then *in vivo* to test predictions made in cultured cells. In this way, a predictable animal model with highly correlative data can offer validation of an established cell-culture assay. An example of this was demonstrated in a study by Scheffold et al. where cytotoxicity assays were modified to use luciferase-labeled target cells, and the efficacy of tumoricidal activity of NK T cells was assessed.[68] The assay was then moved into animal models without changing the reporter, and the same cytotoxic T cells and tumor targets could be studied in *in vivo* models of immune-cell therapies (Figure 62.1).

The data from this study also demonstrated that development and testing of complex treatment regimens could be greatly accelerated using imaging.[68] In this study, the cytotoxicic NK T cells were redirected to the tumor target using a bispecific antibody, and this treatment was compared to herceptin alone and to four other controls including two NK T cell controls and two antibody controls (Figure 62.1). Tumor-burden data were collected twice in the first week and then weekly for 28 d. The number of study groups and data points would not have been possible using conventional assays. Imaging will have an increasing role as therapies are developed that involve immune cells, radiation, DNA-based therapies, and chemotherapies as complementary approaches. The optimization of single drug-treatment regimens can be greatly accelerated using imaging assays that can rapidly measure tumor burden. However, the development of more complex multistep therapeutic strategies such as nonmyeloblative bone-marrow transplantation will depend much more heavily on imaging to assess outcome and determine the mechanisms of action.

# 62.3 Application Areas

## 62.3.1 Cancer Therapies

Effective evaluation of new treatment strategies for malignant disease will require that the animal models closely resemble the human diseases that they are designed to model. The subcutaneous xenograft models of human malignancies are, in some respects, less able to model the human disease than are orthotopic or spontaneous tumor models.[69–71] However, these better models of disease are frequently more difficult to study as the lesions can be located deep within the body and are not accessible for caliper measurements. The disease in these models can involve multiple organ systems and occur in the face of an intact immune system. It is these models that will contribute to our understanding of disease mechanisms as specific genetic elements can be integrated and evaluated in these mice,[69–71] and imaging will provide access to key information on expression of targets, measuring tumor burden at deep-tissue sites and rapidly assessing the degree of metastasis.

### 62.3.1.1 Models of Metastatic and Minimal Disease States

The sensitivity and spatiotemporal nature of BLI-generated data provide an opportunity for detecting and localizing small foci of malignant cells rapidly in laboratory rodents using whole-body assays. Thus, this approach has enabled the study of metastatic and minimal residual disease states in animal models.[46,72] In a study by Wetterwald et al.,[72] a labeled human breast-cancer cell line was used in mouse xenograft model to efficiently evaluate metastatic lesions in bone. Intracardiac injection of tumor cells allowed metastasis to form in a variety of locations including the bone marrow. Micrometastases in the bone marrow can often elude radiographic detection because this detection method requires osteolysis for a signal. BLI, on the other hand, was used to successfully detect the presence and location of such lesions with volumes as small as 0.5 $mm^3$ comprised of approximately $2 \times 10^4$ cells. The microscopic bone-marrow metastases that were detectable by BLI were not associated with osteolytic lesions and hence could not be detected by x-ray imaging. This advancement in the assessment of animal tumor models has greatly accelerated the analyses of these diseases and will improve our ability to develop effective approaches for targeting small numbers of cells present in the early stages of disease progression.

Minimal residual disease remains a major therapeutic hurdle despite the improvements in disease-free survival that have resulted from recent advances in cancer therapies. A majority of cancer patients respond to high-dose chemo- and radiation therapy and enter a state of minimal residual disease. A significant number of these patients will, however, eventually relapse. Effective evaluation of therapies that target small numbers of tumor cells are, therefore, essential for the development of the next generation of cancer therapies that address this critical target. Since protocols that use BLI do not require sacrifice of the animal to obtain sensitive and quantitative data, animals may be followed over time for relapse following the cessation of therapy, and combination therapies that are at first aggressive and secondarily less toxic can be evaluated. The relatively rapid nature of BLI allows for a variety of therapeutic regimens to be measured in the same animal using time for relapse as a criterion by which to measure success. The use of BLI in combination with other imaging modalities will provide greater opportunities for developing strategies to eradicate minimal disease and prevent relapse.

#### 62.3.1.2 Spontaneous Tumor Models

Advancements in subcutaneous and orthotopic models of disease that have been enabled by the application of BLI may be extended to spontaneous-tumor models. The combination of BLI and transgenic technology may yield more predictive animal models of human disease. The introduction of mutations and transgenes in mice has resulted in animals that spontaneously develop malignancies.[69–71] In these mouse models, researchers have been able to study the initial or early events in the initiation of disease, which represent key targets for the treatment of oncogenesis. For example, a conditional transgenic mouse model that develops retinoblastoma-dependent sporadic cancer has been studied.[53] Firefly luciferase was incorporated as a reporter gene in this model, enabling the investigators to follow the animals over the full disease using BLI. The pituitary tumors that developed in these mice could be followed because they were tagged through the expression of the firefly luciferase. The onset of disease and subsequent disease progression and response to therapy in this tumor model were more readily evaluated using a noninvasive approach than could have been accomplished using assays that require sampling of tissues. Coupling spontaneous-tumor models with sensitive imaging modalities will facilitate the analysis of the key processes in the initiation and progression of neoplastic disease, as these models incorporate defined genetic alterations that are linked to oncogenesis.

### 62.3.2 DNA-Based Therapies

The development of therapeutic strategies based on the delivery of therapeutic genes to replace defective genes for the treatment of genetic diseases or in the treatment of malignancy and other diseases has also been constrained by our inability to rapidly analyze our animal model systems noninvasively. Multimodality imaging studies that pair BLI, as a functional imaging modality, with MRI, as a structural imaging modality, demonstrate the strength of such combination approaches in addressing multiple biological questions *in vivo*.[4] In a study by Rehemtulla et al., BLI was used to determine the levels of gene expression following adenoviral-mediated gene delivery in combination with MRI for determining therapeutic efficacy of a therapeutic transgene for cancer-gene therapy.[4] Yeast cytosine deaminase, an enzyme that converts the nontoxic compound 5-fluorocytosine (5FC) into the drug 5-fluorouracil, was the therapeutic gene, and the tumor target was a glioma (rat 9L glioma) in an orthotopic rat model. The ability of prodrug-converting enzymes to facilitate tumor kill was evaluated, and luciferase expression served as the marker of effective delivery of the gene. Diffusion-weighted MRI was used as a surrogate marker of tumor-cell death.[4] The noninvasive assessments of both the extent of gene delivery and the efficacy of the therapy resulted in a more robust model for evaluation of this multicomponent anticancer therapy.

Successful life-long gene replacement was demonstrated in an *in utero* gene-transfer model.[73] In this model, an adeno-associated viral vector carrying a modified luciferase gene was injected into the peritoneal cavity of fetal mice. The luminescent signals from the transduced fetuses were apparent while the animals were *in utero*, indicating that the substrate, luciferin, could cross the placental barrier. After birth these animals continued to express the reporter gene and did so for the 24-month study period. We have developed the ability to diagnose genetic defects in the fetus, and the development of approaches for gene replacement *in utero*, especially for genes encoding secreted proteins where local expression can have a systemic effect, was enabled by this demonstration.

The location and magnitude of expression of therapeutic genes will affect the therapeutic outcome of these experimental therapies. Correlation of gene-delivery parameters with therapeutic outcome is essential for preclinical optimization of such strategies. This is another example of a relatively complex treatment regimen where optimal dosing of the genetic therapy and the prodrug need to be evaluated. Studies that use multimodality imaging should significantly accelerate the decision-making process for the therapeutic indication, drug formulation, and regimen that should move forward to the clinic. Bioluminescent reporter genes may also find clinical use in evaluating the delivery and efficacy of gene therapies and gene vaccines where these genes are intentionally delivered to a tumor or a tissue site.[4,73–76]

## 62.3.3 Infection and Immunity

### 62.3.3.1 Tracking and Monitoring in Infectious Disease

Thousands of potential antimicrobial compounds can be generated using combinatorial chemistries and high-throughput assays, and from this set of compounds potential drug candidates for preclinical testing can be identified using labeled bacteria as targets. Bacterial pathogens that express the bacterial *lux* operon produce light without the need for exogenous substrate addition, and plates of these organisms can be screened rapidly using the same low-light imaging systems that are employed for BLI.[18,20,47,77–81] The bacterial *lux* operon that has been widely used for labeling pathogens was originally derived from *Photorhabdus luminescens* and has since been modified to optimize labeling and use in these assays.[78] The same labeled organisms that are used in the correlative culture assays can be used in the animal models, thereby eliminating the need for different reporters or assays for studies in culture and in animal models. BLI offers opportunities not previously available in the study of infectious disease. The goal in these protocols is integrated studies where a single optical signature can be used to study the pathogen in culture, noninvasively detect the infection throughout the course of infection, and then recover the labeled bacteria, readily distinguishing it from normal flora. A number of published examples reveal how this approach can be used to rapidly assess the efficacy of a drug therapy while using fewer animals than would be required using more traditional assays.[18,47,48,78,79]

The utility of BLI for monitoring bacterial infections *in vivo* was first demonstrated in a mouse model of human typhoid fever.[18] In these early studies, the bacterial pathogens were labeled with a plasmid encoding the *lux* operon; since that time *Salmonella* strains have been labeled by incorporating the *lux* operon into the chromosome of the pathogen by transposon insertion. This eliminates the need for using antibiotics in the animal models to maintain the plasmid and reduces the likelihood of losing the signal due to loss of the plasmid. In these models, the intensity of the luminescent signal is quantitative and correlates with the number of bacteria at oxygenated tissue sites.[18] Because of their small size mice are the most suitable for BLI; however, larger animals such as rats have also been studied using this approach. As the animal increases in size, emitted photons must pass through more tissue that is both absorbing and scattering, and the number of bacteria that is detectable is reduced. Variability is often observed in infectious-disease models, even if the animals are similarly inoculated. The animals in a single-treatment group may have slightly different disease courses. BLI offers the investigator the advantage of identifying these inherent differences prior to initiation of therapy and then assessing the fate of animals with different initial patterns of disease in a given treatment group.

As noninvasive assays permit repeated measurements of the same animal over time, more information can be generated using fewer animals, and each animal becomes its own control, which improves the statistics. In more traditional animal models, data are obtained after serial sacrifices, homogenization of tissue samples, and determination of the number of pathogens present in the tissue. Imaging strategies can be superimposed on these types of studies and also used to direct the *ex vivo* assays by identifying important times and tissues. Where this has been used to validate the imaging approach the intensity of the signal at the sites of infection has been found to correlate well with the number of organisms present.[18,48,78] Whole-body imaging of infected animals often reveals pathogens in tissues not thought to be involved in the disease process, and additional sites of infection are often revealed.[79] These observations may have implications for the molecular mechanisms of disease and routes of infection that had not been previously realized or understood

Although not fully investigated, the use of bioluminescent reporters as indicators of bacterial gene expression *in vivo* will reveal molecular determinants of disease and identify new targets for therapy. Directed gene fusions with *lux* or random integrations of promoterless reporters could not previously be studied in living animals. BLI has opened this door, and spatiotemporal regulation patterns can now be revealed *in vivo*. The patterns of expression would be especially informative in diseases with multiple foci of infection where unique patterns may be apparent at specific times or in specific tissues.

#### 62.3.3.2 Local Delivery of Therapies Using Immune-Cell Homing

BLI is an ideal method for assessing the dynamic changes in immune-cell populations that take place on a whole-body scale. Such data are not available using flow cytometry or immunofluorescence. As the migratory pathways of immune cells within the body relate to the temporal changes of the immune response, they may also relate to treatment approaches. There are several examples of monitoring immune-cell trafficking patterns using BLI and using these data to optimize the delivery therapeutic proteins in specific immune-cell populations.[49,76,82]

In a mouse model for multiple sclerosis in humans, experimental autoimmune encephalomyelitis (EAE), the migratory patterns of autoantigen-reactive CD4+ T cells specific to myelin basic protein (MBP), were monitored and used to deliver an immune modulator. The effector cells of the immune-mediated destruction were transduced with a retrovirus that encoded both a reporter construct (green fluorescent protein [GFP]-*luc* fusion) and interleukin 12 p40 (IL12-p40) as a therapeutic protein.[49] The transduced cells could be selected using a fluorescence-activated cell sorter (FACS), and following adoptive transfer BLI demonstrated that labeled CD4+ T cells expressing IL-12 p40 trafficked to the central nervous system of symptomatic animals. These animals demonstrated a significant reduction in clinical disease. This approach was further substantiated using type II collagen-specific CD4+ T hybridomas and primary CD4+ T cells as vehicles for local delivery of IL-12 p40 in a mouse model of rheumatoid arthritis.[76]

Adoptive immunotherapy, where expression of a proinflammatory antagonist can be localized within the site of tissue destruction, obviates the inherent problems of nonspecific toxicity following systemic administration. Understanding the trafficking patterns of immune cells is a key component of effective development of these types of cell-based therapies, and BLI enabled these studies such that the protocols for local delivery of immunoregulatory proteins could be optimized.

### 62.3.4 Building Transgenic Imaging Markers into Rodent Models of Physiology and Toxicology

Our knowledge of the mouse and human genomes allows transgenic animals to be made that contain a gene modification (insertion, deletion, mutation) in a specific gene that has been identified and characterized. Altered genes can be introduced into the genome by random integration or into specific sites by homologous recombination. As such, transgenic models can be those in which a specific gene has been altered to cause loss of function, overexpression, or anomalous expression of a target gene for increased or altered activity, or they can be those in which reporter genes have been incorporated and coexpressed with an endogenous target gene (Figure 62.2).[19,83] These animals are models for many genetic diseases for which the etiology is known. Conventional transgenic animals with no luciferase reporter must be characterized by traditional methods of genotypic and phenotypic analyses. Phenotypic assays generate data about how a specific gene or mutation may play a role in disease. The assays may be observational, biochemical, or histological or may rely on various whole-body-imaging techniques, as described below. Phenotypic assays identify variations in the transgenic organism relative to the parent strain. With traditional methods, however, these comparisons have been limited to the presence of observable traits or behaviors and by the repertoire of phenotypic assays.

Genetically modified transgenic animals may contain luciferases as markers for gene expression, protein activity, or simply as a tracking molecule to study the effects of genetic mutations on development, immunity, host response, and other metabolic diseases.[20,45,52] These light-producing transgenic animals have been analyzed by BLI, including expression from the HIV-1 promoter[52] and heat-shock protein 32 (ak.a heme oxygenase 1; HO-1)[19] and in the development of spontaneous-tumor models.[53] These types of animal models have demonstrated utility in drug discovery and in toxicity screening (Figure 62.2).

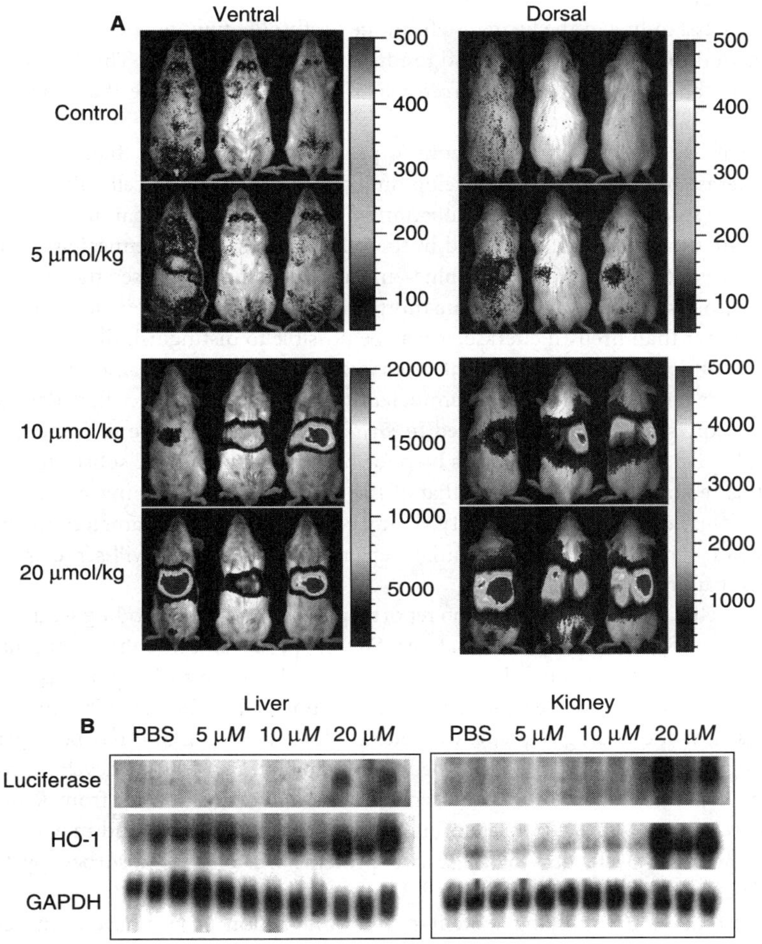


**FIGURE 62.2 (Color figure follows p. 28-30.)** Screening heavy-metal toxicity in a transgenic mouse model.[19] Using a Tg model where the transgene consisted of the heme oxygenase promoter fused to the firefly luciferase coding sequence; dose-dependent increases in luciferase transcription in the liver and kidney following systemic treatment with PBS or three doses of $CdCl_2$ (5, 10, and 20 µmol/kg) were revealed by BLI (A). After imaging the animals were sacrificed and the tissues removed, and total RNA was isolated from the liver and kidneys and analyzed by Northern blot hybridization (B). Levels of mRNA for HO-1, luciferase, and GAPDH were determined for both tissues from three mice at each concentration. Luc signals increased with increasing concentrations of $CdCl_2$ as measured by imaging, and levels of Luc and HO-1 mRNA were elevated in the treatment group that received the highest concentration of $CdCl_2$.

# 62.4 Advances in *in Vivo* Bioluminescence Imaging

## 62.4.1 Dual-Reporter Strategies

Luciferases have been studied for decades and effectively developed as reporter genes for cells and as sensors for use in biochemical assays. Luciferases are used for evaluating gene expression in cultured cells and even in live-cell assays,[56–59,84,85] for revealing the circadian rhythms of small relatively transparent organisms,[86,87] and now for imaging biological processes in living laboratory animals.[20] By virtue of the curious biology conferred upon a variety of organisms that produce luciferases, a detailed understanding of the mechanism of light production and the constraints on light emission is available for some of these enzymes. Our understanding of the luciferase from the firefly (*Photinus pyralis*) is at the level of knowing the relative contribution of single amino-acid residues to light emission.[62,88] This long and active inquiry has provided a broad base of knowledge from which we draw to advance the *in vivo* applications of bioluminescence

Luciferases that use other substrates or emit at peak wavelengths other than that of the commonly used firefly luciferase have been used to develop multiparameter assays in cell cultures.[89] Extending this approach to *in vivo* studies would enable evaluation of two or even several parameters in a given animal. One of the enzymes that has been employed in the cell culture and biochemical assays is the luciferase from the sea pansy (*Renilla reniformis*), a blue-emitting luciferase that uses the substrate coelenterazine.[90,91] Several groups have begun to evaluate this reporter for use *in vivo*.[74,92] Since the *Renilla* luciferase uses a substrate other than firefly luciferase, it may be possible to distinguish the two enzymes biochemically *in vivo*. Distinguishing bioluminescent reporters spectrally is constrained by the wavelengths of emission, which are largely absorbed by mammalian tissue. Despite the fact that *Renilla* luciferase is a blue emitter, its expression has been detected *in vivo*.[74] Use of this enzyme is limited, however, by the absorption of blue light by tissues as well as by poor biodistribution of the substrate. The sensitivity of detection for this reporter will be less than that of the firefly enzyme assuming similar expression levels and specific enzymatic actvities. The ability to image two or more parameters *in vivo* would be a tremendous advance in the field, and the initial steps made in this area[74] will serve as a foundation for continued investigation.

Another approach is to create dual-function reporter genes comprised of coding sequences for reporters that can be detected by two different modalities. Such approaches can enhance the utility of a singe reporter gene, which was demonstrated by Day et al. in a study of expression patterns in fruit flies.[93] The dual-function reporters have enabled the studies of immune-cell trafficking.[49,76,82] In addition, a fusion protein comprised of the coding sequences for *Renilla* luciferase and GFP has been generated for the purpose of using resonance energy transfer between luciferase and GFP to develop protein-proximity assays based on a spectral shift.[94] Bioluminescence resonance energy transfer from *Renilla* luciferase to GFP has been used to study protein–protein interaction in cultured cells[95] and may have utility *in vivo*. However, the emission wavelengths of these two proteins will also be absorbed by tissues, and the sensitivity of detection will not be great.

Two other approaches for studying protein–protein interaction *in vivo* have been described. One is based on principles that were used in the yeast two-hybrid system, and in the mouse counterpart the association of two proteins resulted in transcription of the reporter gene firefly luciferase.[96] The yeast two-hybrid system was developed for the purpose of screening libraries of fusion proteins to identify protein–protein interactions, and the opportunity to perform similar assays *in vivo* may enable similar screening strategies in mammalian cells in the context of the living body. The other approach, first described by Ozawa et al., used the phenomenon of protein–protein splicing that was first described in bacteria.[97–100] This set of approaches may lead to assays that will enhance the development of therapies whose mechanism of action is the disruption of protein–protein associations.

### 62.4.2 Signal-Amplification Strategies

Many promoters that are of interest for drug development may not have high levels of expression and would therefore be difficult to study using reporter-gene strategies. Iyer et al. have addressed this problem by creating transcriptional-amplification methods that enhance the levels of reporter-gene expression in a manner still representative of the regulation of the target gene, in this case for prostate-specific antigen.[101] Nearly a 50-fold amplification was observed for luciferase using this approach, demonstrating that it may be possible to noninvasively assess subtle changes in transcription. Zhang et al. have used such an amplification method in an androgen-responsive DNA-based therapy directed against prostate cancer.[102] Many of the molecular tools developed for cell-based assays can now be extended to *in vivo* studies using BLI, and this is one such application.

Further improvements in the BLI technology's spatial resolution (approaching the millimeter level) may be expected using physics-based diffusion models currently in development. Recent advances in mathematical modeling of light propagation in tissue (photon migration in scattering media and data inversion for image reconstruction), light sources, and detection techniques have improved the quality of measurement analysis and made it possible to array data from diffuse light tomographically.

## 62.5 Summary and Conclusions

For years molecular biologists have been able to study gene regulation in cells and in tissue lysates, and the use of reporter genes has greatly increased our ability to study the tempo of expression. The application of reporter genes in *in vivo* imaging now offers the opportunity to similarly evaluate expression in the living body and interrogate the biology of intact cells noninvasively in living animals where the contextual influences of the intact functional organ systems are retained. Successful sequencing of the human and mouse genomes has formed a foundation that supports functional studies such as BLI, and a logical extension is to study spatiotemporal expression patterns of specific genes in the living body. This will be further complemented by the progress in proteomics, and *in vivo* assays of protein function are among the next steps for BLI. Advances in genetics have made it practical for the first time to approach gene-regulation questions from several directions simultaneously, which will inevitably point the way to many more potential drug targets and, ultimately, the development of new potential therapeutic compounds. The advances in cellular and molecular imaging will allow for more comprehensive analyses of responses to therapy within intact animals. The use of multimodality approaches for structural and functional correlations, and multiple reporter systems to expose related pathways, will enable us to approach questions within the complexity of intact physiology in whole-animal models. Whole-animal imaging has generated high-quality spatiotemporal information and is amenable to highly specific and even multifactorial genetic experiments.

Taken together, the new knowledge from genomics, transgenic methods, and imaging tools is forcing changes in the traditional paradigm of drug development. The animal-testing bottleneck has prompted efforts in many therapeutic areas to develop more accurate animal models that can be evaluated more rapidly using multimodality imaging techniques. Early discovery-phase bottlenecks that develop at the target validation and lead optimization stages also can be addressed by use of new animal models. The goal is to produce more qualified compounds that will progress to the clinic surrounded by complete and relevant preclinical safety and efficacy data. The utility, cost, flexibility, and especially high-quality predictive data generated from whole-animal cellular and molecular imaging define a "value proposition" to the industry. This acknowledged value will, over time, contribute to increased efficiency and effectiveness of drug-discovery and development programs. The gauge that we will use to ultimately evaluate the influence of this technology on drug discovery and development will be the success of compounds in the clinic.

BLI holds great promise for the acceleration of data generation and analysis for the development of new drug candidates by the pharmaceutical industry. This is because 80% of drug candidates fail to satisfy safety and efficacy requirements in clinical trials, and the need to conduct biological assessments in animal models to determine safety and efficacy is the principal obstacle to making the development process more efficient and effective. BLI offers a rapid means of conducting *in vivo* analysis to improve the predictive quality of data used in the assessment of new drug candidates. The technology provides data sets from relevant, intact animal systems, increasing both the efficiency and the effectiveness of selecting new drug candidates for clinical development. Because it is an *in vivo* technology, fewer animals are needed, while time and costs are reduced and more data per protocol are obtained. Overcoming the problems of animal-to-animal variation by using the zero time point as an internal control improves the biostatistics in animal modeling. Rapid animal models, based on internal bioluminescent reporters, are now being developed that will make more predictive and physiologically relevant *in vivo* systems available earlier in the drug-development cycle than was previously possible.[4,5] Given the accessibility and speed of BLI, researchers will be able to perform *in vivo* efficacy, pharmacokinetics, toxicology, and target-validation studies on a larger number of compounds than they were able to do in the past.

Bioluminescent reporter genes may eventually find clinical use in evaluating the delivery and efficacy of gene therapies and gene vaccines where these genes are intentionally delivered to a tumor or a tissue site. Further improvements in the BLI technology's spatial resolution (approaching the millimeter level) may be expected using physics-based diffusion models currently in development. This is an area of investigation that will greatly improve drug development by accelerating the process and providing quantitative *in vivo* data.

## Acknowledgments

This work was supported, in part, by unrestricted gifts from the Mary L. Johnson and Hess Research Funds (CHC) and by grants R33CA88303, R24CA92862, and P20CA86312 (CHC) from the National Institutes of Health.

## References

1. Hintz, S.R., Benaron, D.A., Siegel, A.M., Zourabian, A., Stevenson, D.K., and Boas, D.A., Bedside functional imaging of the premature infant brain during passive motor activation, *J. Perinat. Med.*, 29, 335, 2001.
2. Benaron, D.A., Hintz, S.R., Villringer, A., Boas, D., Kleinschmidt, A., Frahm, J., Hirth, C., Obrig, H., van Houten, J.C., Kermit, E.L., Cheong, W.F., and Stevenson, D.K., Noninvasive functional imaging of human brain using light, *J. Cereb. Blood Flow Metab.*, 20, 469, 2000.
3. Benaron, D.A. and Stevenson, D.K., Optical time-of-flight and absorbance imaging of biologic media, *Science*, 259, 1463, 1993.
4. Rehemtulla, A., Hall, D.E., Stegman, L.D., Chen, G., Bhojani, M.S., Chenevert, T.L., and Ross, B.D., Molecular imaging of gene expression and efficacy following adenoviral-mediated brain tumor gene therapy, *Mol. Imaging*, 1, 43, 2002.
5. Rehemtulla, A., Stegman, L.D., Cardozo, S.J., Gupta, S., Hall, D.E., Contag, C.H., and Ross, B.D., Rapid and quantitative assessment of cancer treatment response using *in vivo* bioluminescence imaging, *Neoplasia*, 2, 491, 2000.
6. Sameni, M., Moin, K., and Sloane, B.F., Imaging proteolysis by living human breast cancer cells, *Neoplasia*, 2, 496, 2000.
7. Jacobs, A., Dubrovin, M., Hewett, J., Sena-Esteves, M., Tan, C.W., Slack, M., Sadelain, M., Breakefield, X.O., and Tjuvajev, J.G., Functional coexpression of HSV-1 thymidine kinase and green fluorescent protein: implications for noninvasive imaging of transgene expression, *Neoplasia*, 1, 154, 1999.
8. Fujimoto, J.G., Pitris, C., Boppart, S.A., and Brezinski, M.E., Optical coherence tomography: an emerging technology for biomedical imaging and optical biopsy, *Neoplasia*, 2, 9, 2000.
9. Tromberg, B.J., Shah, N., Lanning, R., Cerussi, A., Espinoza, J., Pham, T., Svaasand, L., and Butler, J., Non-invasive *in vivo* characterization of breast tumors using photon migration spectroscopy, *Neoplasia*, 2, 26, 2000.
10. Vajkoczy, P., Ullrich, A., and Menger, M.D., Intravital fluorescence videomicroscopy to study tumor angiogenesis and microcirculation, *Neoplasia*, 2, 53, 2000.
11. Ramanujam, N., Fluorescence spectroscopy of neoplastic and non-neoplastic tissues, *Neoplasia*, 2, 89, 2000.
12. Pedersen, M.W., Holm, S., Lund, E.L., Hojgaard, L., and Kristjansen, P.E., Coregulation of glucose uptake and vascular endothelial growth factor (VEGF) in two small-cell lung cancer (SCLC) sublines *in vivo* and *in vitro*, *Neoplasia*, 3, 80, 2001.
13. Kragh, M., Quistorff, B., Lund, E.L., and Kristjansen, P.E., Quantitative estimates of vascularity in solid tumors by non-invasive near-infrared spectroscopy, *Neoplasia*, 3, 324, 2001.

14. Vordermark, D., Shibata, T., and Martin Brown, J., Green fluorescent protein is a suitable reporter of tumor hypoxia despite an oxygen requirement for chromophore formation, *Neoplasia*, 3, 527, 2001.
15. Edinger, M., Sweeney, T.J., Tucker, A.A., Olomu, A.B., Negrin, R.S., and Contag, C.H., Noninvasive assessment of tumor cell proliferation in animal models, *Neoplasia*, 1, 303, 1999.
16. Padera, T.P., Stoll, B.R., So, P.T.C., and Jain, R.K., Conventional and high-speed intravital multiphoton laser scanning microscopy of microvasculature, lymphatics, and leukocyte-endothelial interactions, *Mol. Imaging*, 1, 9, 2002.
17. Hawrysz, D.J. and Sevick-Muraca, E.M., Developments toward diagnostic breast cancer imaging using near-infrared optical measurements and fluorescent contrast agents, *Neoplasia*, 2, 388, 2000.
18. Contag, C.H., Contag, P.R., Mullins, J.I., Spilman, S.D., Stevenson, D.K., and Benaron, D.A., Photonic detection of bacterial pathogens in living hosts, *Mol. Microbiol.*, 18, 593, 1995.
19. Zhang, W., Feng, J.Q., Harris, S.E., Contag, P.R., Stevenson, D.K., and Contag, C.H., Rapid *in vivo* functional analysis of transgenes in mice using whole body imaging of luciferase expression, *Transgenic Res.*, 10, 423, 2001.
20. Contag, P.R., Olomu, I.N., Stevenson, D.K., and Contag, C.H., Bioluminescent indicators in living mammals, *Nat. Med.*, 4, 245, 1998.
21. Bremer, C., Tung, C.H., and Weissleder, R., *In vivo* molecular target assessment of matrix metalloproteinase inhibition, *Nat. Med.*, 7, 743, 2001.
22. Chinnaiyan, A.M., Prasad, U., Shankar, S., Hamstra, D.A., Shanaiah, M., Chenevert, T.L., Ross, B.D., and Rehemtulla, A., Combined effect of tumor necrosis factor-related apoptosis-inducing ligand and ionizing radiation in breast cancer therapy, *Proc. Natl. Acad. Sci. U.S.A.*, 97, 1754, 2000.
23. Evelhoch, J.L., Gillies, R.J., Karczmar, G.S., Koutcher, J.A., Maxwell, R.J., Nalcioglu, O., Raghunand, N., Ronen, S.M., Ross, B.D., and Swartz, H.M., Applications of magnetic resonance in model systems: cancer therapeutics, *Neoplasia*, 2, 152, 2000.
24. Kurhanewicz, J., Vigneron, D.B., and Nelson, S.J., Three-dimensional magnetic resonance spectroscopic imaging of brain and prostate cancer, *Neoplasia*, 2, 166, 2000.
25. Kaplan, O., Firon, M., Vivi, A., Navon, G., and Tsarfaty, I., HGF/SF activates glycolysis and oxidative phosphorylation in DA3 murine mammary cancer cells, *Neoplasia*, 2, 365, 2000.
26. Fleige, G., Nolte, C., Synowitz, M., Seeberger, F., Kettenmann, H., and Zimmer, C., Magnetic labeling of activated microglia in experimental gliomas, *Neoplasia*, 3, 489, 2001.
27. Bogdanov, A., Matuszewski, L., Bremer, C., Petrovsky, A., and Weissleder, R., Oligomerization of paramagnetic substrates result in signal amplification and can be used for MR imaging of molecular targets, *Mol. Imaging*, 1, 16, 2002.
28. Hogemann, D., Josephson, L., Weissleder, R., and Basilion, J.P., Improvement of MRI probes to allow efficient detection of gene expression, *Bioconjug. Chem.*, 11, 941, 2000.
29. Gillies, R.J., Bhujwalla, Z.M., Evelhoch, J., Garwood, M., Neeman, M., Robinson, S.P., Sotak, C.H., and Van Der Sanden, B., Applications of magnetic resonance in model systems: tumor biology and physiology, *Neoplasia*, 2, 139, 2000.
30. Bhujwalla, Z.M., Artemov, D., Natarajan, K., Ackerstaff, E., and Solaiyappan, M., Vascular differences detected by MRI for metastatic versus nonmetastatic breast and prostate cancer xenografts, *Neoplasia*, 3, 143, 2001.
31. Pilatus, U., Ackerstaff, E., Artemov, D., Mori, N., Gillies, R.J., and Bhujwalla, Z.M., Imaging prostate cancer invasion with multi-nuclear magnetic resonance methods: the metabolic Boyden chamber, *Neoplasia*, 2, 273, 2000.
32. Chenevert, T.L.Stegman, L.D., Taylor, J.M., Robertson, P.L., Greenberg, H.S., Rehemtulla, A., and Ross, B.D., Diffusion magnetic resonance imaging: an early surrogate marker of therapeutic efficacy in brain tumors, *J. Natl. Cancer Inst.*, 92, 2029, 2000.
33. Stegman, L.D., Rehemtulla, A., Beattie, B., Kievit, E., Lawrence, T.S., Blasberg, R.G., Tjuvajev, J.G., and Ross, B.D., Noninvasive quantitation of cytosine deaminase transgene expression in human tumor xenografts with *in vivo* magnetic resonance spectroscopy, *Proc. Natl. Acad. Sci. U.S.A.*, 96, 9821, 1999.

34. Ponomarev, V., Doubrovin, M., Lyddane, C., Beresten, T., Balatoni, J., Bornman, W., Finn, R., Akhurst, T., Larson, S., Blasberg, R., Sadelain, M., and Tjuvajev, J.G., Imaging TCR-dependent NFAT-mediated T-cell activation with positron emission tomography *in vivo*, *Neoplasia*, 3, 480, 2001.
35. Gambhir, S.S., Herschman, H.R., Cherry, S.R., Barrio, J.R., Satyamurthy, N., Toyokuni, T., Phelps, M.E., Larson, S.M., Balatoni, J., Finn, R., Sadelain, M., Tjuvajev, J., and Blasberg, R, Imaging transgene expression with radionuclide imaging technologies, *Neoplasia*, 2, 118, 2000.
36. Tjuvajev, J.G., Joshi, A., Callegari, J., Lindsley, L., Joshi, R., Balatoni, J., Finn, R., Larson, S.M., Sadelain, M., and Blasberg, R.G.., A general approach to the non-invasive imaging of transgenes using cis-linked herpes simplex virus thymidine kinase, *Neoplasia*, 1, 315, 1999.
37. Gambhir, S.S., Barrio, J.R., Phelps, M.E., Iyer, M., Namavari, M., Satyamurthy, N., Wu, L., Green, L.A., Bauer, E., MacLaren, D.C., Nguyen, K., Berk, A.J., Cherry, S.R., and Herschman, H.R., Imaging adenoviral-directed reporter gene expression in living animals with positron emission tomography, *Proc. Natl. Acad. Sci. U.S.A.*, 96, 2333, 1999.
38. Mankoff, D.A., Dehdashti, F., and Shields, A.F., Characterizing tumors using metabolic imaging: PET imaging of cellular proliferation and steroid receptors, *Neoplasia*, 2, 71, 2000.
39. Dyszlewski, M., Blake, H.M., Dahlheimer, J.L., Pica, C.M., and Piwnica-Worms, D., Characterization of a novel 99mTc-carbonyl complex as a functional probe of MDR1 P-glycoprotein transport activity, *Mol. Imaging*, 1, 24, 2002.
40. Hackman, T., Doubrovin, M., Balatoni, J., Beresten, T., Ponomarev, V., Beattie, B., Finn, R., Bornmann, W., Blasberg, R., and Tjuvajev, J.G.G., Imaging expression of cytosine deaminase — herpes virus thymidine kinase fusion gene (CD/TK) expression with [124I]FIAU and PET, *Mol. Imaging*, 1, 36, 2002.
41. Hay, R.V., Cao, B., Skinner, R.S., Wang, L.-M., Su, Y., Resau, J.H., Woude, G.F. V., and Gross, M.D., Radioimmunoscintigraphy of tumors autocrine for human met and hepatocyte growth factor/scatter factor, *Mol. Imaging*, 1, 56, 2002.
42. Burt, B.M., Humm, J.L., Kooby, D.A., Squire, O.D., Mastorides, S., Larson, S.M., and Fong, Y., Using positron emission tomography with [(18)F]FDG to predict tumor behavior in experimental colorectal cancer, *Neoplasia*, 3, 189, 2001.
43. Kristensen, C.A., Hamberg, L.M., Hunter, G.J., Roberge, S., Kierstead, D., Wolf, G.L., and Jain, R.K., Changes in vascularization of human breast cancer xenografts responding to antiestrogen therapy, *Neoplasia*, 1, 518, 1999.
44. Paulus, M.J., Gleason, S.S., Kennel, S.J., Hunsicker, P.R., and Johnson, D.K., High resolution X-ray computed tomography: an emerging tool for small animal cancer research, *Neoplasia*, 2, 62, 2000.
45. Contag, P.R., Whole-animal cellular and molecular imaging to accelerate drug development, *Drug Discovery Today*, 7, 555, 2002.
46. Sweeney, T.J., Mailander, V., Tucker, A.A., Olomu, A.B., Zhang, W., Cao, Y., Negrin, R.S., and Contag, C.H., Visualizing the kinetics of tumor-cell clearance in living animals, *Proc. Natl. Acad. Sci. U.S.A.*, 96, 12044, 1999.
47. Francis, K.P., Yu, J., Bellinger-Kawahara, C., Joh, D., Hawkinson, M.J., Xiao, G., Purchio, T.F., Caparon, M.G., Lipsitch, M., and Contag, P.R., Visualizing pneumococcal infections in the lungs of live mice using bioluminescent *Streptococcus pneumoniae* transformed with a novel Gram-positive lux transposon, *Infect. Immunol.*, 69, 3350, 2001.
48. Rocchetta, H.L., Boylan, C.J., Foley, J.W., Iversen, P.W., LeTourneau, D.L, McMillian, C.L., Contag, P.R., Jenkins, D.E., and Parr, T.R., Validation of a noninvasive, real-time imaging technology using bioluminescent *Escherichia coli* in the neutropenic mouse thigh model of infection, *Antimicrob. Agents Chemother.*, 45, 129, 2001.
49. Costa, G.L., Sandora, M.R., Nakajima, A., Nguyen, E.V., Taylor-Edwards, C., Slavin, A.J., Contag, C.H., Fathman, C.G., and Benson, J.M., Adoptive immunotherapy of experimental autoimmune encephalomyelitis via T cell delivery of the IL-12 p40 subunit, *J. Immunol.*, 167, 2379, 2001.

50. Wu, J.C., Sundaresan, G., Iyer, M., and Gambhir, S.S., Noninvasive optical imaging of firefly luciferase reporter gene expression in skeletal muscles of living mice, *Mol. Ther.*, 4, 297, 2001.
51. Wu, J.C., Inubushi, M., Sundaresan, G., Schelbert, H.R., and Gambhir, S.S., Optical imaging of cardiac reporter gene expression in living rats, *Circulation*, 105, 1631, 2002.
52. Contag, C.H., Spilman, S.D., Contag, P.R., Oshiro, M., Eames, B., Dennery, P., Stevenson, D.K., and Benaron, D.A.,Visualizing gene expression in living mammals using a bioluminescent reporter, *Photochem. Photobiol.*, 66, 523, 1997.
53. Vooijs, M., Jonkers, J., Lyons, S., and Berns, A., Noninvasive imaging of spontaneous retinoblastoma pathway-dependent tumors in mice, *Cancer Res.*, 62, 1862, 2002.
54. Hastings, J.W., Chemistries and colors of bioluminescent reactions: a review, *Gene*, 173, 5, 1996.
55. Wilson, T. and Hastings, J.W., Bioluminescence, *Annu. Rev. Cell Dev. Biol.*, 14, 197, 1998.
56. Hooper, C.E., Ansorge, R.E., Browne, H.M., and Tomkins, P., CCD imaging of luciferase gene expression in single mammalian cells, *J. Biolumin. Chemilumin.*, 5, 123, 1990.
57. Hooper, C.E., Ansorge, R.E., and Rushbrooke, J.G., Low-light imaging technology in the life sciences, *J. Biolumin. Chemilumin.*, 9, 113, 1994.
58. White, M.R., Masuko, M., Amet, L., Elliott, G., Braddock, M., Kingsman, A.J., and Kingsman, S.M., Real-time analysis of the transcriptional regulation of HIV and hCMV promoters in single mammalian cells, *J. Cell Sci.*, 108 (Part 2), 441, 1995.
59. White, M.R., Wood, C.D., and Millar, A.J., Real-time imaging of transcription in living cells and tissues, *Biochem. Soc. Trans.*, 24, 411S, 1996.
60. Rice, B.W., Cable, M.D., and Nelson, M.B., *In vivo* imaging of light-emitting probes, *J. Biomed. Opt.*, 6, 432, 2001.
61. Tung, C.H., Mahmood, U., Bredow, S., and Weissleder, R., *In vivo* imaging of proteolytic enzyme activity using a novel molecular reporter, *Cancer Res.*, 60, 4953, 2000.
62. Branchini, B.R., Magyar, R.A., Murtiashaw, M.H., Anderson, S.M., Helgerson, L.C., and Zimmer, M., Site-directed mutagenesis of firefly luciferase active site amino acids: a proposed model for bioluminescence color, *Biochemistry*, 38, 13223, 1999.
63. Baldwin, T.O., Firefly luciferase: the structure is known, but the mystery remains, *Structure*, 4, 223, 1996.
64. Conti, E., Franks, N.P., and Brick, P., Crystal structure of firefly luciferase throws light on a superfamily of adenylate-forming enzymes, *Structure*, 4, 287, 1996.
65. De Wet, J.R., Wood, K.V., DeLuca, M., Helinski, D.R., and Subramani, S., Firefly luciferase gene: structure and expression in mammalian cells, *Mol. Cell Biol.*, 7, 725, 1987.
66. Sandalova, T.P. and Ugarova, N.N., Model of the active site of firefly luciferase, *Biochemistry (Moscow)*, 64, 962, 1999.
67. Sandalova, T. and Lindqvist, Y., Three-dimensional model of the alpha-subunit of bacterial luciferase, *Proteins*, 23, 241, 1995.
68. Scheffold, C., Scheffold, Y., Kornacker, M., Contag, C., and Negrin, R., Real-time kinetics of HER-2/neu targeted cell therapy in living animals, *Cancer Res.*, 62, 5785, 2002.
69. Jacks, T., Remington, L., Williams, B.O., Schmitt, E.M., Halachmi, S., Bronson, R.T., and Weinberg, R.A., Tumor spectrum analysis in p53-mutant mice, *Curr. Biol.*, 4, 1, 1994.
70. Macleod, K.F. and Jacks, T., Insights into cancer from transgenic mouse models, *J. Pathol.*, 187, 43, 1999.
71. Van Dyke, T. and Jacks, T., Cancer modeling in the modern era: progress and challenges, *Cell*, 108, 135, 2002.
72. Wetterwald, A., van der Pluijm, G., Que, I., Sijmons, B., Buijs, J., Karperien, M., Lowik, C.W., Gautschi, E., Thalmann, G.N., and Cecchini, M.G., Optical imaging of cancer metastasis to bone marrow: a mouse model of minimal residual disease. *Am. J. Pathol.*, 160, 1143, 2002.
73. Lipshutz, G.S., Gruber, C.A., Cao, Y., Hardy, J., Contag, C.H., and Gaensler, K.M., *In utero* delivery of adeno-associated viral vectors: intraperitoneal gene transfer produces long-term expression, *Mol. Ther.*, 3, 284, 2001.

74. Bhaumik, S. and Gambhir, S.S., Optical imaging of *Renilla* luciferase reporter gene expression in living mice, *Proc. Natl. Acad. Sci. U.S.A.*, 99, 377, 2002.
75. Gambhir, S.S., Barrio, J.R., Herschman, H.R., and Phelps, M.E., Assays for noninvasive imaging of reporter gene expression, *Nucl. Med. Biol.*, 26, 481, 1999.
76. Nakajima, A., Seroogy, C.M., Sandora, M.R., Tarner, I.H., Costa, G.L., Taylor-Edwards, C., Bachmann, M.H., Contag, C.H., and Fathman, C.G., Antigen-specific T cell-mediated gene therapy in collagen-induced arthritis, *J. Clin. Invest.*, 107, 1293, 2001.
77. Greer, L.F., III and Szalay, A.A., Imaging of light emission from the expression of luciferases in living cells and organisms: a review, *Luminescence*, 17, 43, 2002.
78. Francis, K.P., Joh, D., Bellinger-Kawahara, C., Hawkinson, M.J., Purchio, T.F., and Contag, P.R., Monitoring bioluminescent *Staphylococcus aureus* infections in living mice using a novel luxABCDE construct, *Infect. Immunol.*, 68, 3594, 2000.
79. Burns, S.M., Joh, D., Francis, K.P., Shortliffe, L.D., Gruber, C.A., Contag, P.R., and Contag, C.H., Revealing the spatiotemporal patterns of bacterial infectious diseases using bioluminescent pathogens and whole body imaging, *Contrib. Microbiol.*, 9, 71, 2001.
80. Hamblin, M.R., O'Donnell, D.A., Murthy, N., Contag, C.H., and Hasan, T., Rapid control of wound infections by targeted photodynamic therapy monitored by *in vivo* bioluminescence imaging, *Photochem. Photobiol.*, 75, 51, 2002.
81. Siragusa, G.R., Nawotka, K., Spilman, S.D., Contag, P.R., and Contag, C.H., Real-time monitoring of *Escherichia coli* O157:H7 adherence to beef carcass surface tissues with a bioluminescent reporter, *Appl. Environ. Microbiol.*, 65, 1738, 1999.
82. Hardy, J., Edinger, M., Bachmann, M.H., Negrin, R.S., Fathman, C.G., and Contag, C.H., Bioluminescence imaging of lymphocyte trafficking *in vivo*, *Exp. Hematol.*, 29, 1353, 2001.
83. Zhang, W., Contag, P.R., Madan, A., Stevenson, D.K., and Contag, C.H., Bioluminescence for biological sensing in living mammals, *Adv. Exp. Med. Biol.*, 471, 775, 1999.
84. Frawley, L.S., Faught, W.J., Nicholson, J., and Moomaw, B., Real time measurement of gene expression in living endocrine cells, *Endocrinology*, 135, 468, 1994.
85. Thompson, E.M., Adenot, P., Tsuji, F.I., and Renard, J.P., Real time imaging of transcriptional activity in live mouse preimplantation embryos using a secreted luciferase, *Proc. Natl. Acad. Sci. U.S.A.*, 92, 1317, 1995.
86. Millar, A.J., Short, S.R., Chua, N.H., and Kay, S.A., A novel circadian phenotype based on firefly luciferase expression in transgenic plants, *Plant Cell*, 4, 1075, 1992.
87. Millar, A.J., Carre, I.A., Strayer, C.A., Chua, N.H., and Kay, S.A., Circadian clock mutants in Arabidopsis identified by luciferase imaging, *Science*, 267, 1161, 1995.
88. Kajiyama, N. and Nakano, E., Isolation and characterization of mutants of firefly luciferase which produce different colors of light, *Protein Eng.*, 4, 691, 1991.
89. Grentzmann, G., Ingram, J.A., Kelly, P.J., Gesteland, R.F., and Atkins, J.F., A dual-luciferase reporter system for studying recoding signals, *RNA*, 4, 479, 1998.
90. Lorenz, W.W., McCann, R.O., Longiaru, M., and Cormier, M.J., Isolation and expression of a cDNA encoding *Renilla reniformis* luciferase, *Proc. Natl. Acad. Sci. U.S.A.*, 88, 4438, 1991.
91. Matthews, J.C., Hori, K., and Cormier, M.J., Purification and properties of *Renilla reniformis* luciferase, *Biochemistry*, 16, 85, 1977.
92. Liu, H., Iacono, R.P., and Szalay, A.A., Detection of GDNF secretion in glial cell culture and from transformed cell implants in the brains of live animals, *Mol. Genet. Genomics*, 266, 614, 2001.
93. Day, R.N., Kawecki, M., and Berry, D., Dual-function reporter protein for analysis of gene expression in living cells, *Biotechniques*, 25, 848, 1998.
94. Liu, J., Wang, Y., Szalay, A.A., and Escher, A., Visualizing and quantifying protein secretion using a *Renilla* luciferase-GFP fusion protein, *Luminescence*, 15, 45, 2000.
95. Wang, Y., Wang, G., O'Kane, D.J., and Szalay, A.A., A study of protein-protein interactions in living cells using luminescence resonance energy transfer (LRET) from *Renilla* luciferase to *Aequorea* GFP, *Mol. Gen. Genet.*, 264, 578, 2001.

96. Ray, P., Pimenta, H., Paulmurugan, R., Berger, F., Phelps, M.E., Iyer, M., and Gambhir, S.S., Noninvasive quantitative imaging of protein-protein interactions in living subjects, *Proc. Natl. Acad. Sci. U.S.A.*, 99, 3105, 2002.
97. Ozawa, T., Kaihara, A., Sato, M., Tachihara, K., and Umezawa, Y., Split luciferase as an optical probe for detecting protein-protein interactions in mammalian cells based on protein splicing, *Anal. Chem.*, 73, 2516, 2001.
98. Clarke, N.D., A proposed mechanism for the self-splicing of proteins, *Proc. Natl. Acad. Sci. U.S.A.*, 91, 11084, 1994.
99. Colston, M.J. and Davis, E.O., The ins and outs of protein splicing elements, *Mol. Microbiol.*, 12, 359, 1994.
100. Pietrokovski, S., Conserved sequence features of inteins (protein introns) and their use in identifying new inteins and related proteins, *Protein Sci.*, 3, 2340, 1994.
101. Iyer, M., Wu, L., Carey, M., Wang, Y., Smallwood, A., and Gambhir, S.S., Two-step transcriptional amplification as a method for imaging reporter gene expression using weak promoters, *Proc. Natl. Acad. Sci. U.S.A.*, 98, 14595, 2001.
102. Zhang, L., Adams, J. Y., Billick, E., Ilagan, R., Iyer, M., Le, K., Smallwood, A., Gambhir, S.S., Carey, M., and Wu, L., Molecular engineering of a two-step transcription amplification (TSTA) system for transgene delivery in prostate cancer, *Mol. Ther.*, 5, 223, 2002.

# 63

# Liposome-Based Systems for Biomedical Diagnostics and Therapy

Thuvan Nguyen
*University of New Orleans*
*New Orleans, Louisiana*

Gabriela Dumitrascu
*University of New Orleans*
*New Orleans, Louisiana*

Nitsa Rosenzweig
*Xavier University of Louisiana*
*New Orleans, Louisiana*

Zeev Rosenzweig
*University of New Orleans*
*New Orleans, Louisiana*



## 63.1 Introduction

Phospholipid vesicles, e.g., liposomes, have been used as drug-delivery vesicles in the last three decades. The main advantages of liposomes as drug-delivery carriers include their biocompatibility, ability to effectively encapsulate hydrophilic or hydrophobic drugs, and the sensitivity of their fluid-like membrane to temperature and pH, which enables controlled drug release.[1–6] Several reviews published in recent years describe in detail the use of liposomes as drug-delivery carriers of anticancer drugs.[7–10] In this review, we describe the results of recent studies in our laboratory that focus on the use of liposomes in diagnostics and therapy. We describe the methods used in our laboratory to prepare and characterize liposomes, recent applications of liposomes as fluorescent nanosenors, and extraction and detection tools. We also describe the development of new digital fluorescence-imaging techniques to monitor the release of fluorescent drug analogs from liposomes into single cells.

## 63.2 Basics of Liposomes

Liposomes are spherical phospholipid vesicles that form spontaneously when phospholipids are introduced into aqueous media. The phospholipids are self-assembled to form a bilayer membrane similar to

0-8493-1116-0/03/$0.00+$1.50

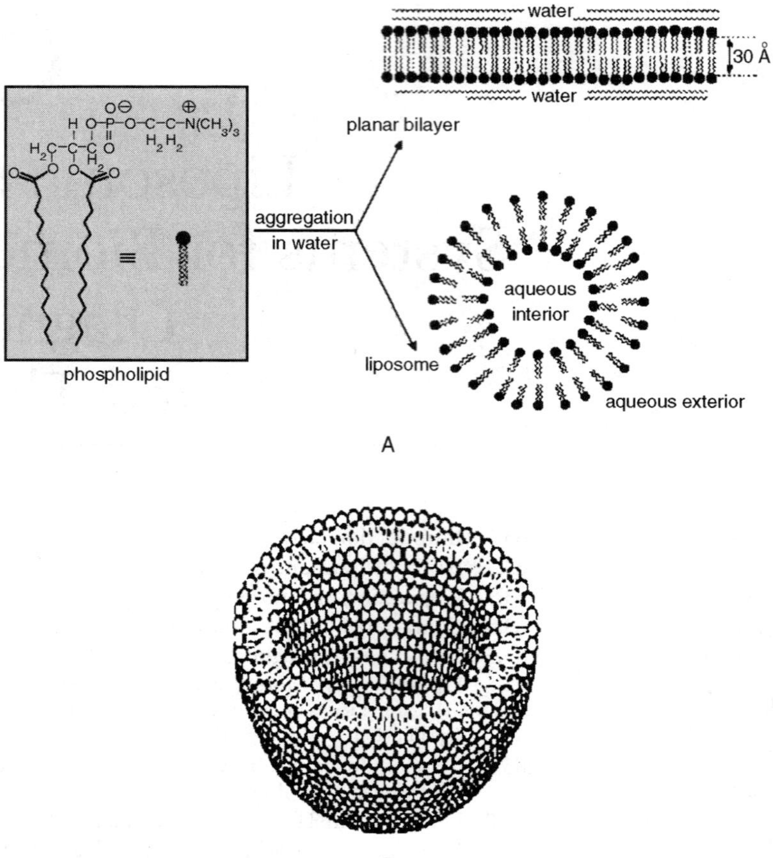


**FIGURE 63.1** (A) Phospholipid self-assembly and (B) lipid vesicles (liposomes).

cell membranes (Figure 63.1). In liposomes, the phospholipid head groups point toward the internal and external aqueous phases of the vesicles. The bilayer membrane may also contain cholesterol and fatty acids to increase its fluidity and flexibility. The morphology of liposomes is classified according to the compartmentalization of aqueous regions between bilayer shells. In unilamellar liposomes, the aqueous compartment is segregated from the external solution by only one membranal bilayer. Unilamellar liposomes are further classified according to their size. Small unilamellar liposomes (SUV) average 100 nm in diameter, while large unilamellar liposomes (LUV) are more than 100 nm in diameter with a maximal size of up to 10 μm. Multilamellar liposomes have more than one bilayer surrounding each aqueous compartment. Multilamellar liposomes typically form large complex honeycomb structures that are difficult to reproduce.

## 63.3 Preparation of Liposomes

There are numerous approaches to preparing liposomes and to modifying their size and shape after formation. The general elements of these approaches involve preparation of the phospholipids for hydration, hydration with agitation, sizing a homogeneous distribution of liposomes, and purification of the products. Typically, the phospholipids are first dissolved in a suitable organic solvent — e.g., chloroform — to get a homogeneous mixture. Then, the solvent is completely removed to yield a dried

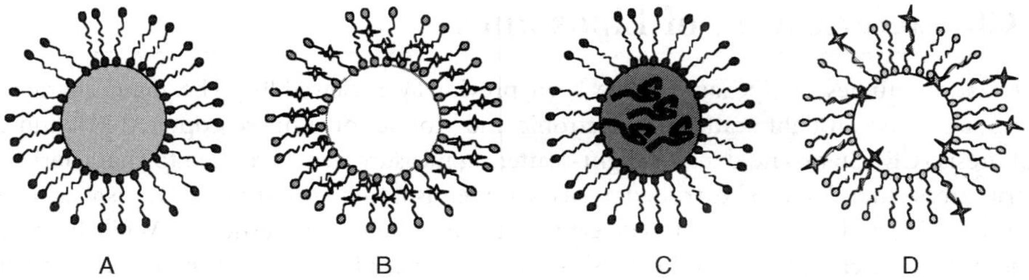


**FIGURE 63.2** Tailoring liposomes. (A) Dyes encapsulated in the internal compartment of the liposomes. (B) Dyes encapsulated in phospholipid membrane of the liposomes. (C)Dextran-conjugated dyes encapsulated in the internal compartment of the liposomes. (D) Dye-DHPE as a part of the phospholipid membrane of the liposomes.

film. Drying the phospholipids from solution volumes of smaller than 1 ml is accomplished using a nitrogen stream for 20 to 30 min. Samples larger than 1 ml are dried by rotary evaporation using a round-bottom flask in a cooling bath. In our laboratory we prepared different fluorescent liposomes using three common methods. The liposomes prepared in our laboratory are illustrated in Figure 63.2. The injection method[11] is used to encapsulate hydrophilic dyes in the internal compartment of liposomes (Figure 63.2A). In this technique, the dried phospholipids are reconstituted with dry isopropanol. The sample is then injected, while vortexing, into a test tube containing an aqueous sample of the molecules (dye, protein) to be encapsulated. As the liposomes form, these molecules are captured in the aqueous compartment of the liposomes with approximately 1 to 10% efficiency. Hydrophobic compounds and dye-labeled phospholipid derivatives are incorporated into the bilayer membrane of liposomes by drying the phospholipids from an organic solvent that already contains the hydrophobic compounds (Figure 63.2B). The dry film is then rehydrated in a buffer solution. This simple method yields liposomes ranging between 100 and 300 nm in diameter. High-energy ultrasonic radiation is useful in preparing small unilamellar liposomes from large multilamellar liposomes.[12] However, as the large liposomes are disrupted, any material previously encapsulated is lost and must be recovered upon formation of the smaller liposomes, normally at lower encapsulation efficiency. Sonication can be performed using either a probe tip or a bath sonicator. Probe sonication is used for reducing the size of larger liposomes to as small as 25 to 50 nm in diameter. However, the high energy used may produce heat capable of degrading the phospholipids.

Another preparation technique is based on dehydration and subsequent rehydration of phospholipids in the presence of the encapsulated material. This method is used primarily to encapsulate water-soluble macromolecules, i.e., fluorescent conjugates of dextrans and proteins (Figure 63.2C).[13,14] Since organic solvents may damage such types of molecules, the entire preparation is carried out in aqueous solution. An aqueous liposome suspension (made by probe sonication) is mixed with the water-soluble macromolecules and dried by rotary evaporation, leaving a multilamellar phospholipid film and the material to be encapsulated. Upon rehydration of the sample and continued incubation in a rotating flask, large liposomes are formed that are capable of encapsulating a high concentration of the solute. Uniformity in liposome size is achieved using an extrusion device, where a preformed liposome sample is passed across a sealed 100-nm-pore-size polycarbonate membrane between two gas-tight glass syringes.[15] When hydrophilic compounds are encapsulated in liposomes, nonencapsulated molecules must be removed from the liposome sample. Passing the liposome sample through a gel-filtration column is a very effective way to trap nonencapsulated molecules in the column while passing the larger liposomes through the column for collection. Using this method we were able to encapsulate dye molecules — dextran-dye conjugates with a molecular weight of up 70 kDA — and even larger protein molecules. Liposomes containing phospholipids labeled with fluorophores are prepared using the injection technique (Figure 63.2D). The fluorophores in these liposomes are attached covalently to the phospholipid headgroup or to the hydrophobic alkyl tail of the phospholipids.

## 63.4 Characterization of Liposomes

Characterization studies of liposomes have been previously reviewed.[16,17] The main characterization techniques are based on light scattering, electronic and atomic force microscopy (AFM), and capillary electrophoresis (CE). Static and dynamic light-scattering measurements are used to characterize the size, size distribution, and shape of extruded vesicles under isotonic conditions.[18] Transmission electron microscopy (TEM) and AFM are used to determine the morphology of liposomes. While light-scattering techniques provide average values of the physical characteristics of liposome suspensions, TEM and AFM provide information on individual liposomes that may or may not represent the entire liposome population. AFM imaging measurements could be conducted on untreated samples in air or in solution, avoiding all the processing, such as fixation, dehydration embedded sectioning, and staining, required in TEM.[19] Since liposomes appear to be stable under conditions of high electric fields like the ones applied in CE, CE can be used to obtain qualitative and quantitative information about the size-to-charge ratio of liposomes.[20,21] The size-to-charge ratio of liposomes is directly related to their mobility in the capillary when a high potential difference is applied. Since an equal distribution of charge on the liposomes is assumed, the electrophoretic distribution observed is due primarily to liposome size. The peak shape of the electropherogram is indicative of liposome size distribution and uniformity. Under normal CE conditions, liposomes produce electropherograms with a smooth broad Gaussian distribution with few spiking events.

## 63.5 Liposomes in Diagnostics

Liposomes have been used as fluorescent labels in homogeneous and heterogeneous assays. Homogeneous assays often rely on the catalytic activity of enzyme-labeled antibodies that are incorporated into the membrane of liposomes.[22] Heterogeneous liposome immunoassays may be performed in a competitive or noncompetitive assay format.[23] They resemble standard enzyme-linked immunosorbent assays, with the second antibody labeled with a fluorescent liposome instead of an enzyme. Single liposomes, which could contain up to 1000 fluorescent molecules, act in these assays as signal amplifiers. Due to the large number of fluorophores in the liposome, its fluorescence is self-quenched. Following several washing steps to remove unbound liposomes that contain the second antibody from the sample, the bound liposomes are lysed with a detergent. The fluorescent molecules are dispersed in the sample, leading to an instantaneous amplification.[24]

Taking advantage of the large signal enhancement obtained when fluorescent liposomes are used in immunoassays, Durst et al.[25] developed an automated flow-injection liposome immunoanalysis (FILIA) system for the measurement of environmental toxicants, using alachlor as a model analyte. In their system, liposomes tagged with alachlor and containing the fluorescent dye sulforhodamine B pass through an immunoreactor column containing antialachlor covalently coupled to glass beads. In the presence of free alachlor, the free alachlor and the alachlor-containing liposomes compete for the binding sites available on the glass beads. The number of liposomes bound to the column is inversely proportional to the concentration of free alachlor. The bound liposomes are lysed, and the level of released dye is quantified and correlated with the free alachlor in the sample.

The successful application of fluorescent liposomes in immunoassays has led to the development of a new research direction in our laboratory that focuses on the fluorescence response of liposomes to environmental stimulation. Our studies in this area are described in the following section.

## 63.6 Liposomes as Fluorescence Nanosensors

Recently we reported the use of fluorescent-dye-encapsulating liposomes for nanoscale sensing of pH, molecular-oxygen, and calcium-ion levels in aqueous solutions.[26,27] The fluorescent liposomes were tested as individual probes and as an assembly in a thin film format. The liposome-containing suspensions and

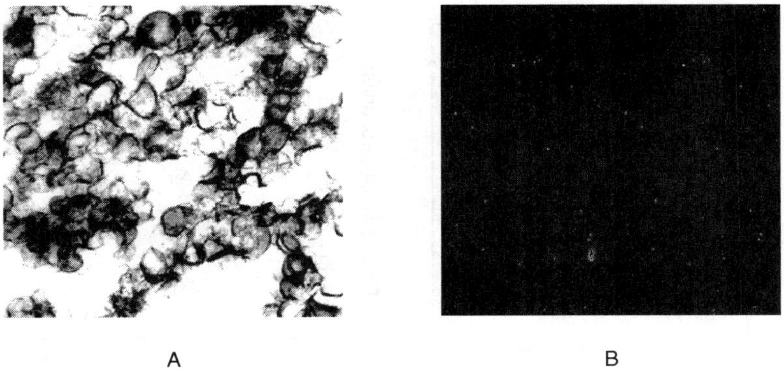


**FIGURE 63.3** (A) An 82,600× TEM of uranyl-acetate-labeled liposomes averaging 70 nm in diameter. (B) Digital fluorescence image of $Ru(phen)_3$-encapsulating liposomes. A 5-μl liposome sample is placed between two microscope cover slips, and fluorescence is detected and imaged by a charge-coupled-device camera. The fluorescence of each liposome is captured on one to ten pixels owing to their lateral diffusion in the sample.

films were stable for up to 14 days when stored under conditions of pH 7.2 and 4°C. Figure 63.3 shows microscopic images of fluorescent liposomes. Figure 63.3A shows a TEM of uranyl-acetate-labeled liposomes. The presence of a relatively homogeneous population of round-shaped unilamellar liposomes averaging about 70 nm in diameter is evident. Figure 63.3B shows a fluorescence image of the dye-encapsulating liposomes. A digital analysis of this image indicates that the ratio between the fluorescence intensity of individual liposomes and the background noise, signal-to-noise, in an air-saturated environment is about 50.

### 63.6.1 pH-Sensing Liposomes

pH-Sensing liposomes were prepared by incorporating fluorescein (a pH-sensitive dye) and tetramethyl rhodamine (pH-insensitive dye) in unilamellar liposomes using the injection method. The liposomes were calibrated against standard solutions of pH 5.0 to 8.0. The pH was determined based on the ratio between the emission peaks of fluorescein at 525 nm and those of tetramethylrhodamine at 575 nm. The pH dependence of the ratio between the fluorescence signals of fluorescein and tetramethylrhodamine, $I_f/I_r$, obtained from 50 pH-sensing liposomes is shown in Figure 63.4A. The dynamic range of the particles was found to be between pH 5.5 and 7.5, with a pH sensitivity of 0.1 pH units.

### 63.6.2 Oxygen-Sensing Liposomes

The fluorescence spectra of nano-sized $Ru(phen)_3$-encapsulating liposomes in oxygenated, air-saturated, and nitrogenated solutions is shown in Figure 63.4B. Due to dynamic quenching by molecular oxygen, the fluorescence intensity of the $Ru(phen)_3$-encapsulating liposome in nitrogenated solution, $I_f(N_2)$, is higher than the fluorescence intensity of the liposome in air-saturated solution, $I_f(\text{air})$, and oxygenated solution, $I_f(O_2)$. The obtained response factor $I_f(N_2)/I_f(O_2) = 7$ is two times higher than the response factor of the same ruthenium diimine complex when trapped in a polymer. In fact, the oxygen response of the $Ru(phen)_3$-encapsulating liposomes is very similar to the oxygen response of a $Ru(phen)_3$ aqueous solution. The increase in the oxygen response over polymer supports indicates that the dye molecules in the liposomes show a free-solution-like behavior. The increased oxygen response partially negates the loss of fluorescence signal due to the miniaturized nanodimensions of the liposomes.

### 63.6.3 Calcium-Ion-Sensing Liposomes

The calcium-ion sensor developed in our laboratory is the first example of a liposome-based biosensor. It employs a calcium-ion-binding protein for recognition and a sensing fluorophore for signal transduction.

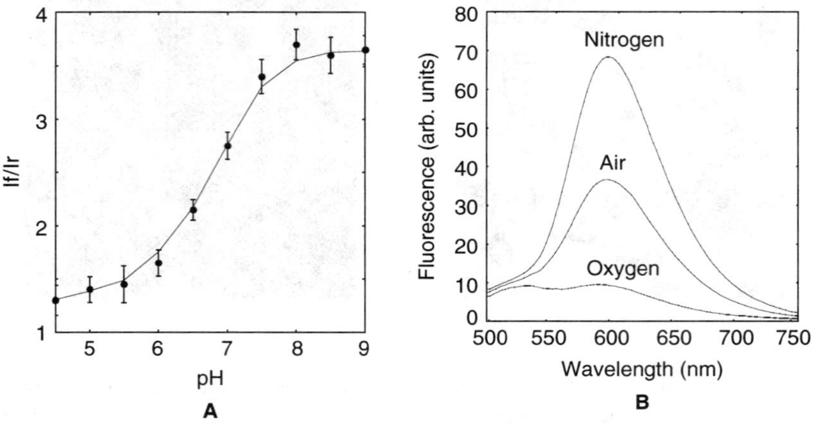


**FIGURE 63.4** (A) pH Response of the fluorescein encapsulating liposomes in solution. (B) Fluorescence response of Ru(phen)$_3$ encapsulating liposomes to oxygenated and nitrogenated solutions in a phosphate-buffered solution at pH 7.2.

To fabricate the sensor, Alexa-488-labeled calmodulin (CaM) molecules were encapsulated in the membrane of liposomes. Upon binding calcium ions, CaM undergoes a conformational change that exposes its hydrophobic cores, which may act as active sites for the interactions with target enzymes or CaM antagonists. Both the carboxy- and amino-terminal domains of CaM then undergo large structural rearrangements from the "closed" conformation (the two helices of each hand are almost antiparallel) to the "open" conformation (the two helices are more perpendicular).[28–35] This conformational change strongly affects the fluorescence intensity of the fluorophore Alexa-488 that is covalently bound to calmodulin. Liposomes containing Alexa-labeled calmodulin show high sensitivity to calcium ions in the micromolar range. The calcium-ion response of Alexa-labeled calmodulin liposomes is fourfold higher than the calcium-ion response of Alexa-labeled calmodulin in solution. This is attributed to the increasing stability of calmodulin when embedded in the liposome membrane. Unlike in commonly used fluorescence indicators, physiological levels of magnesium ions do not interfere with the calcium-ion response of Alexa-CaM-containing liposomes.

Current studies in our laboratory in the area of liposome-based biosensors include the development of a glucose biosensor. Another research direction involves the synthesis of nanometric lipobeads as fluorescence sensors. In lipobeads, the aqueous compartment of liposomes is replaced with a polymer core. While maintaining the biocompatibility of liposomes, lipobeads are expected to show higher mechanical and chemical stability in the cellular environment.

## 63.7 Magnetoliposomes in Extraction and Detection of Antibodies in Biological Samples

Another research direction in our laboratory has been the use of liposomes to extract and detect antibodies in biological samples. We prepared magnetoliposomes following a procedure first developed by Joniau et al.,[36,37] which is based on an entropy-driven exchange of lipids that occurs when fatty-acid-coated magnetic nanoparticles are mixed with a preprepared liposome suspension. Using this technique we synthesized magnetoliposomes containing gold-coated cobalt platinum alloy nanoparticles that were annealed prior to encapsulation. To demonstrate their extraction power, magnetoliposomes labeled with BODIPY fluorescein were used to extract antibodies against BODIPY fluorescein from aqueous solution. Figure 63.5 describes the fluorescence intensity of the sample during three cycles of extraction of antibodies by magnetoliposomes containing BODIPY fluorescein.

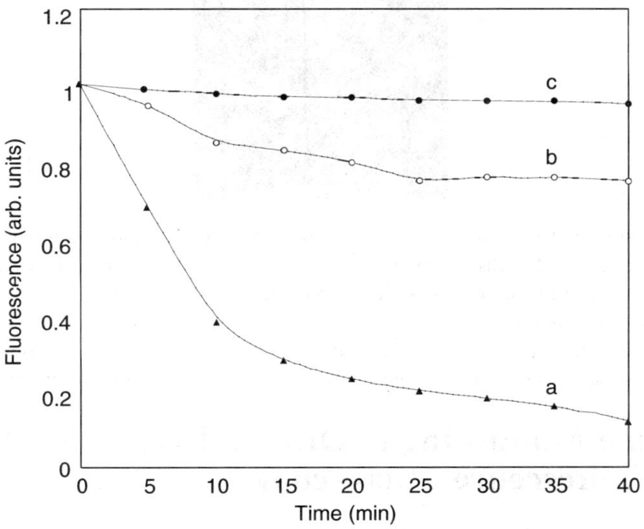


**FIGURE 63.5** Extraction of anti-BODIPY fluorescein from aqueous samples using repeated cycles of application of BODIPY-fluorescein coated magnetoliposomes. A removal of about 90% of the anti-BODIPY-fluorescein molecules is observed.

Curve a shows a 90% decrease in the fluorescence intensity, which is attributed to quenching by antibody molecules of the fluorescence of BODIPY fluorescein when the organic dye molecules bind to the antibody molecules in the solution. Following an incubation period of 40 min at room temperature the particles were separated from the sample solution by passing through a column mounted in the bore of a horseshoe magnet that is used to attract the magnetoliposomes out of the sample. A fresh batch of magnetoliposomes was then added to the antibody solution. Curve b shows that the fluorescence signal decreased by only 20%, indicating a significant reduction in the level of antibody molecules in the solution. This cycle was repeated until a fluorescence decrease of less than 3% was observed over the 40-min incubation time (curve c). Total protein fluorescence-spectroscopy measurements of the solution following three cycles of extraction by magnetoliposomes showed subnanomolar levels of antibodies.

## 63.8 Liposomes in Therapy

The interactions of liposomes with cells have been studied extensively *in vitro* and *in vivo* to explore their possible use as drug carries.[38–40] Significant advances have been made in the use of liposomes in cancer therapy.[41,42] Liposomes tend to localize in solid tumors, and their use increases therapeutic efficacy and decreases nonspecific toxicity. The use of antracycline-containing liposomes to treat stage IV breast cancer has recently become the standard of care.[43] Recent advances in liposome technology include the development of sterically stabilized liposomes to increase their stability *in vivo*[44] and the use of liposome-labeled monoclonal antibodies to selectively target specific cancer cells bearing their corresponding antigens.[45] In our laboratory we developed new fluorescence microscopy techniques to monitor real-time delivery of drugs into single cells. Efficient delivery of drugs from liposomes into cells was expected to increase their efficiency and reduce the effect of the drug on normal cells. The results of this study are described in the following section.

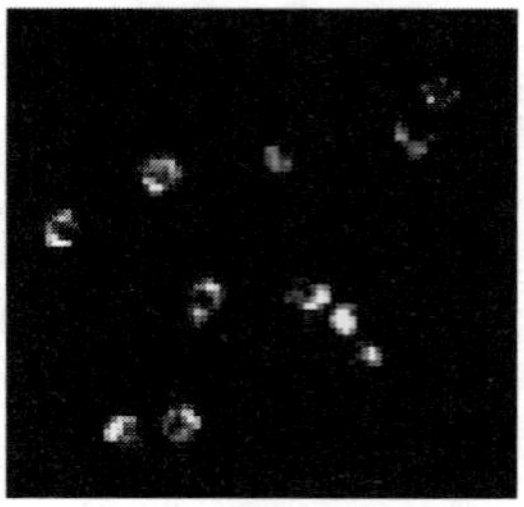

**FIGURE 63.6** A fluorescence of J774 murine macrophages loaded with liposomes containing sulforhodamine. An uneven distribution of the dye in the cells can be seen, which is indicative of compartmentalization of sulforhodamine in the cells. The results of these experiments could be interpreted as limited success. The liposomes were able to deliver sulforhodamine, which is a hydrophilic dye, into the cells. This dye is normally cell impermeable. The use of fusogenic liposomes does not seem to provide an escape route to the dye molecules from the endosomes to the cytoplasm. This suggests that the conditions for endosomal release require further optimization.

## 63.9 Real-Time Monitoring of Drug Delivery from Liposomes Using Fluorescence Microscopy

An efficient liposome-mediated delivery of drugs into cells involves binding the liposomes to the cell membrane, fusion of the liposome with the membrane, and delivery of the liposomal content to the cell cytoplasm. However, in many cellular systems, endocytosis of liposomes takes place rather than fusion. The result of endocytosis is the compartmentalization of the liposomes and the drug in cellular endosomes. In the absence of an escape mechanism, the drug would be ineffective under these conditions. New fusogenic-liposome formulations have been developed to overcome this problem. These liposomes contain active molecules that disrupt the endosomal membrane to facilitate the delivery of the encapsulated drug into the cell cytoplasm.[46–52] Recently introduced fusogenic liposomes are based on the polymorphic-phase transition of dioleoylphosphatidylethanolamine (DOPE).[53–56] DOPE molecules form a bilayer membrane structure at pH 7.2. Under acidic conditions these liposomes undergo a phase transition to adopt a less stable hexagonal-II phase, which enables leakage of the liposomal content to the endosome. The delivery mechanism of the liposomal content from the endosomes to the cytoplasm from these pH-sensitive liposomes is not well understood.[57–62] It was recently suggested that both fusogenic and conventional liposomes fuse with the membrane of the endosome. However, unlike conventional liposomes (e.g., liposomes made of dimyristoyl phosphatidylcholine [DMPC]), fusogenic liposomes also generate pores in the endosome membrane. These pores facilitate the release of the liposomal content to the cytoplasm.

The objective of our study was to find out whether the endosome membrane of cells was destabilized by fusogenic liposomes and to monitor in real time the release of a fluorescent-drug analog from fusogenic liposomes engulfed in the endosome. DOPE/*N*-succinyl DOPE liposomes (7/3, mol/mol ratio), which are liposomes encapsulating sulforhodamine, were used in these experiments. The liposomes were incubated with J 774 murine macrophages because macrophages use endocytosis as the main mechanism of entry for liposomes. Figure 63.6 shows a digital fluorescence image of macrophages loaded with liposomes containing sulforhodamine.

## 63.10 Summary

Studies conducted in our laboratory and by other researchers show that liposomes may be used as highly sensitive analytical probes. Fluorescent liposomes have been previously used as signal enhancers in immunoassays. In our laboratory, we demonstrated for the first time that liposomes could be used as optochemical sensors to monitor the pH, oxygen, and calcium-ion levels of physiological samples. We also showed that liposomes could be used effectively to extract and detect antibodies in biological samples.

Liposomes have been used as effective drug-delivery carriers for the last 30 years. Recent advances in their composition enable a more efficient delivery of drugs. The release of drugs trapped in cellular endosomes remained a challenge. Our studies show that, while liposomes are able to deliver hydrophilic impermeable molecules into cells, their release from endosomes to the cytoplasm is not effective, even when new fusogenic liposomes are used for drug delivery. Further studies are needed to optimize the composition of liposomes and the conditions necessary for endosomal release.

Due to their unique physical and chemical properties and biocompatibility, liposomes will continue to attract the attention of researchers in the foreseeable future. It is expected that recent achievements in the field of liposome research will enable the development of "smart liposomes" that possess targeting capabilities, variety of drug-release properties, and feedback mechanisms based on chemical sensing that will facilitate pulsed delivery of drugs from liposomes when a condition of pathological significance is recorded. To demonstrate this pinciple we are currently developing liposome formulations that release antihistamines in response to the presence of histamines in biological fluids.

## References

1. Uyechi, L.S., Gagne, L., Thurston, G., and Szoka, F.C., Jr., Mechanism of lipoplex gene delivery in mouse lung: binding and internalization of fluorescent lipid and DNA components, *Gene Ther.*, 8(11), 828, 2001.
2. Liang, Y., Belford, S., Tang, F., Prokai, L., Simpkins, J.W., and Hughes, J.A., Membrane fluidity effects of estratrienes, *Brain Res. Bull.*, 54(6), 661, 2001.
3. Wu, J., Lizarzaburu, M.E., Kurth, M.J., Liu, L., Wege, H., Zern, M.A., and Nantz, M.H., Cationic lipid polymerization as a novel approach for constructing new DNA delivery agents, *Bioconjug. Chem.*, 12(2), 251, 2001.
4. Piva, R., del Senno, L., Lambertini, E., Penolazzi, L., and Nastruzzi, C., Modulation of estrogen receptor gene transcription in breast cancer cells by liposome delivered decoy molecules, *J. Steroid Biochem. Mol. Biol.*, 75(2–3), 121, 2000.
5. Chandaroy, P., Sen, A., and Hui, S.W., Temperature-controlled content release from liposomes encapsulating Pluronic F127, *J. Control Release,* 76(1–2), 27, 2001.
6. Toultou, E., Godin, B., Dayan, N., Weiss, C., Piliponsky, A., and Levi-Schaffer, F., Intracellular delivery mediated by an ethosomal carrier, *Biomaterials,* 22(22), 3053, 2001.
7. Dass, C.R. and Su, T., Particle-mediated intravascular delivery of oligonucleotides to tumors: associated biology and lessons from genotherapy, *Drug Delivery,* 8(4), 191, 2001.
8. Meers, P., Enzyme-activated targeting of liposomes, *Adv. Drug Delivery Rev.*, 53(3), 265, 2001.
9. Sihorkar, V. and Vyas, S.P., Potential of polysaccharide anchored liposomes in drug delivery, targeting and immunization, *J. Pharm. Pharm. Sci.*, 4(2), 138, 2001.
10. Oku, N., Tokudome, Y., Asai, T., and Tsukada, H., Evaluation of drug targeting strategies and liposomal trafficking, *Curr. Pharm. Des.*, 6(16), 1669, 2000.
11. Batzri, S. and Korn, E.D., Single bilayer liposomes prepared without sonication, *Biochim. Biophys. Acta*, 298, 1015, 1973.
12. Magalhaes, T., Viotti, A.P., Gomes, R.T., and de Freitas, T.V., Effect of membrane composition and of co-encapsulation of immunostimulants in a liposome-entrapped crotoxin, *Biotechnol. Appl. Biochem.*, 33(Part 2), 61, 2001.
13. Shew, R.L. and Deamer, D.W., A novel method for encapsulation of macromolecules in liposomes, *Biochim. Biophys. Acta*, 816, 1, 1985.
14. Brandl, M. and Gregoriadis, G., Entrapment of haemoglobin into liposomes by the dehydration-rehydration method: vesicle characterization and *in vivo* behaviour, *Biochim. Biophys. Acta*, 1196, 65, 1994.
15. Olson, F., Hunt, P., Szoka, F., Vail, W.J., and Papahadjopolous, D., Preparation of liposomes of defined size distribution by extrusion through polycarbonate membranes, *Biochim. Biophys. Acta*, 557, 9, 1979.

16. Lasic, D.D., Novel applications of liposomes, *Trends Biotechnol.*, 16, 307, 1998.
17. Lundahl, P., Zeng, C.M., Hagglund, C.L., Gottschalk, I., and Greijer, E., Chromatographic approaches to liposomes, proteoliposomes and biomembrane vesicles, *J. Chromatogr. B Biomed. Sci. Appl.*, 722, 103, 1999.
18. Schubert, R. and Stauch, O., Structure of artificial cytoskeleton containing liposomes in aqueous solution studied by static and dynamic light scattering, *Biomacromolecules*, 3, 565, 2002.
19. Paclet, M.H., Coleman, A.W., Vergnaud, S., and Morel, F., P67-phox-mediated NADPH oxidase assembly: imaging of cytochrome b558 liposomes by atomic force microscopy, *Biochemistry*, 39, 9302, 2000.
20. Robert, M.A., Locasscio-Brown, L., MacCrehan, W.A., and Durst, R.A., Liposome behavior in capillary electrophoresis, *Anal. Chem.*, 68, 3434, 1996.
21. Wiedmer, S.K., Hautala, J., Holopainen, J.M., Kinnunen, P.K.J., and Riekkola, M.L., Study on liposomes by capillary electrophoresis, *Electrophoresis*, 22, 1305, 2001.
22. Carmenate, T., Mesa, C., Menendez, T., Falcon, V., and Musacchio, A., Recombinant Opc protein from Neisseria meningitis reconstituted into liposomes elicits opsonic antibodies following immunization, *Biotechnol. Appl. Biochem.*, 34, 63, 2001.
23. Klegerman, M.E., Hamilton, A.J., Huang, S.L., Tiukinhoy, S.D., Khan, A.A., MacDonald, R.C., and McPherson, D.D., Quantitative immunoblot assay for assessment of liposomal antibody conjugation efficiency, *Anal. Biochem.*, 300(1), 46, 2002.
24. Rongen, H.A., Bult, A., and van Bennekom, W.P., Liposomes and immunoassays, *J. Immunol. Meth.*, 204(2), 105, 1997.
25. Reeves, S.G., Rule, G.S., Roberts, M.A., Edwards, A.J., and Durst, R.A., Flow-injection liposome immunoanalysis (FILIA) for alachlor, *Talanta*, 41, 1747, 1994.
26. McNamara, K.P. and Rosenzweig, Z., Dye-encapsulating liposomes as fluorescence-based oxygen nanosensors, *Anal. Chem.*, 70(22), 4853, 1998.
27. McNamara, K.P., Rosenzweig, Z., and Rosenzweig, N., Liposome-based optochemical nanosensors, *Mikrochim. Acta*, 131, 57, 1999.
28. Houdusse, A., Love, M.L., Dominguez, R., Grabarek, Z., and Cohen, C., Structures of four $Ca^{2+}$-bound troponin C at 2.0 Å resolution: further insights into the $Ca^{2+}$-switch in the calmodulin superfamily, *Structure*, 5(12), 1695, 1997.
29. Crivici, A. and Ikura, M., Molecular and structural basis of target recognition by calmodulin, *Annu. Rev. Biophys. Biomol. Struct.*, 24, 85, 1995.
30. Ikura, M., Calcium binding and conformational response in EF-hand proteins, *Trends Biochem. Sci.*, 21, 14, 1996.
31. Evenas, J., Malmendal, A., Thulin, E., Carlstrom, G., and Forsen, S., $Ca^{2+}$ binding and conformational changes in a calmodulin domain, *Biochemistry*, 37, 13744, 1998.
32. Trave, G., Lacombe, P., Pfuhl, M., Saraste, M., and Pastore, A., Molecular mechanism of the calcium-induced conformational change in the spectrin EF-hands, *EMBO J.*, 14(20), 4922, 1995.
33. Nelson, M.R. and Chazin, W.J., An interaction-based analysis of calcium-induced conformational changes in $Ca^{2+}$ sensor proteins, *Protein Sci.*, 7, 270, 1998.
34. Wang, C.A., A note on $Ca^{2+}$ binding to calmodulin, *Biochem. Biophys. Res. Commun.*, 130(1), 426, 1985.
35. Johnson, J.D. and Wittenauer, L.A., A fluorescent calmodulin that reports the binding of hydrophobic inhibitory ligands, *Biochem. J.*, 211, 473, 1983.
36. De Cuyper, M. and Joniau, M., Magnetoliposomes. Formation and structural characterization, *Eur. Biophys. J.*, 15, 311, 1988.
37. Sangregario, C., Weinheman, J., O'Connor, C.J., and Rosenzweig, Z., A new method for the synthesis of magnetoliposomes, *Appl. Phys.*, 85(8), 5699, 1999.
38. Pagano, R.E., Sandra, A., and Takeichi, M., Interactions of phospholipid vesicles with mammaliam cells, *Ann. N.Y. Acad. Sci.*, 308, 185, 1978.

39. Greidziak, M. and Lasch, J., The use of macromolecular protease inhibitors to study liposome-cell interaction, *Biomed. Biochim. Acta*, 50(8), 975, 1991.
40. Leserman, L.D., Weinstein, J.N., Moore, J.J., and Terry, W.D., Specific interaction of myeloma tumor cells with hapten-bearing liposomes containing methotrexate and carboxyfluorescein, *Cancer Res.*, 40(12), 4768, 1980.
41. Kim, S., Liposomes as carriers of cancer chemotherapy. Current status and future prospects, *Drugs*, 46, 618, 1993.
42. Sparano, J.A. and Winer, E.P., Liposomal anthracyclines for breast cancer, *Semin. Oncol.*, 28(4), 32, 2001.
43. Rivera, E., Valero, V., Esteva, F.J., Syrewicz, L., Cristofanilli, M., Rahman, Z., Booser, D.J., and Hortobagyi, G.N., Lack of activity of stealth liposomal doxorubicin in the treatment of patients with anthracycline-resistant breast cancer, *Cancer Chemother. Pharmacol.*, 49(4), 299, 2002.
44. Bakker-Woudenberg, I.A., Long-circulating sterically stabilized liposomes as carriers of agents for treatment of infection or for imaging infectious foci, *Int. J. Antimicrob. Agents*, 19(4), 299, 2002.
45. Kirpotin, D., Park, J.W., Hong, K., Zalipsky, S., Li, W.L., Carter, P., Benz, C.C., and Papahadjopoulos, D., Sterically stabilized anti-HER2 immunoliposomes: design and targeting to human breast cancer cells *in vitro*, *Biochemistry*, 36, 66, 1997.
46. Yatvin, M.B., Krentz, K., Horwitz, B.A., and Shinitzky, M., pH-Sensitive liposomes: possible clinical implications, *Science*, 210, 1253, 1980.
47. Kim, H.S. and Park, Y.S., Gene transfection by quantitatively reconstituted Sendai envelope proteins into liposomes, *Cancer Gene Ther.*, 9(2), 173, 2002.
48. Sagrista, M.L., Bermudez, M., De Madariaga, M.A., and Mora, M., Acylaminophospholipids give negative charge and fusogenic properties to lipid bilayers. Use of *N*-palmitoylphosphatidylethanolamine to obtain long-circulating and fusogenic liposomes to encapsulate tuberculostatic drugs, *Recent Res. Dev. Lipids Res.*, 3, 127, 1999.
49. Ahmad, N., Khan, M.A., and Owais, M., Liposome mediated antigen delivery leads to induction of CD8+ T lymphocyte and antibody responses against the V3 loop region of HIV gp120, *Cell. Immunol.*, 210(1), 49, 2001.
50. Kunisawa, J., Nakanishi, T., Takahashi, I., Okudaira, A., Tsutsumi, Y., Katayama, K., Nakagawa, S., Kiyono, H., and Mayumi, T., Sendai virus fusion protein mediates simultaneous induction of MHC class I/II-dependent mucosal and systemic immune responses via the nasopharyngeal-associated lymphoreticular tissue immune system, *J. Immunol.*, 167(3), 1406, 2001.
51. Hong, M., Lim, S., Oh, Y., and Kim, C., pH-Sensitive, serum-stable and long-circulating liposomes as a new drug delivery system, *J. Pharm. Pharmacol.*, 54(1), 51, 2002.
52. Simoes, S., Slepushkin, V., Duzgunes, N., and Pedroso de Lima, M.C., On the mechanisms of internalization and intracellular delivery mediated by pH-sensitive liposomes, *Biochim. Biophys. Acta*, 1515(1), 23, 2001.
53. Ordeiro, C., Wiseman, D.J., Lutwyche, P., Uh, M., Evans, J.C., Finlay, B.B., and Webb, M.S., Antibacterial efficacy of gentamicin encapsulated in pH-sensitive liposomes against an *in vivo Salmonella enterica* serovar typhimurium intracellular infection model, *Antimicrob. Agents Chemother.*, 44(3), 533, 2000.
54. Webb, M.S, Hui, S.W, and Stepoukus, P.L, Dehydration-induced lamellar-to-hexagonal-II phase transitions in DOPE/DOPC mixture, *Biochim. Biophys. Acta*, 1145, 93, 1993.
55. Morgan, F., Chester, D.C., and Kramer, P.A., The modified stalk mechanism of lamellar/inverted phase transitions and its implications for membrane fusion, *Int. J. Pharm.*, 105, 259, 1994.
56. Fried, D.S., Papahadjopoulos, D., and Debs, R.J., Endocytosis and intracellular processing accompanying transfection mediated by cationic liposomes, *Biochim. Biophys. Acta*, 1278, 41, 1996.
57. Lee, K.D., Hong, K., and Papahadjopoulos, D., Endocytosis and intracellular processing accompanying transfection mediated by cationic liposomes, *Biochim. Biophys. Acta*, 1103, 185, 1992.

58. Van Bambeke, F., Kerkhofs, A., Schanck, A., Remacle, C., Sonveaux, E., Tulkens, P.M., and Mingeot-Leclercq, M., Recognition of liposomes by cells: *in vitro* binding and endocytosis mediated by specific lipid headgroups and surface charge density, *Lipids*, 35(2), 213, 2000.
59. Eue, I., Endocytosis and intracellular fate of liposomes using pyranine as a probe, *Drug Delivery*, 5(4), 265, 1998.
60. Lutwyche, P., Cordeiro, C., Wiseman, D., St.-Louis, M., Uh, M., Hope, M.J., Webb, M.S., and Finlay, B., Intracellular delivery and antibacterial activity of gentamicin encapsulated in pH-sensitive liposomes, *Antimicrob. Agents Chemother.*, 42(10), 2511, 1998.
61. Straubinger, R.M., pH-Sensitive liposomes for delivery of macromolecules into cytoplasm of culturel cells, *Methods Enzymol.*, 221, 361, 1993.
62. Ji, J., Rosenzweig, N., Jones, I., and Rosenzweig, Z., Molecular oxygen-sensitive fluorescent lipobeads for intracellular oxygen measurements in murine macrophages, *Anal. Chem.*, 73(15), 3521, 2001.

# 64

# Surface-Enhanced Raman Scattering (SERS) for Biomedical Diagnostics




Tuan Vo-Dinh
*Oak Ridge National Laboratory*
*Oak Ridge, Tennessee*

David L. Stokes
*Oak Ridge National Laboratory*
*Oak Ridge, Tennessee*


## 64.1 Introduction

Raman spectroscopy is based on vibrational transitions that yield very narrow spectral features characteristic of the investigated sample. Thus, it has long been regarded as a valuable tool for the identification of chemical and biological samples as well as the elucidation of molecular structure, surface processes, and interface reactions. Despite such advantages, Raman scattering suffers from extremely poor efficiency. Compared to luminescence-based processes Raman spectroscopy has an inherently small cross section (e.g., $10^{-30}$ $cm^2$ per molecule), precluding the possibility of analyte detection at low concentration levels without special enhancement processes. Some modes of signal enhancement have included resonance Raman scattering and nonlinear processes such as coherent anti-Stokes Raman scattering. However, the need for high-power, multiple-wavelength excitation sources has limited the widespread use of these techniques.

Nevertheless, there has been a renewed interest in Raman techniques in the past two decades due to the discovery of the surface-enhanced Raman scattering (SERS) effect, which results from the adsorption of molecules on specially textured metallic surfaces. The giant enhancement was first reported in 1974 by Fleischmann et al., who observed the effect for pyridine molecules adsorbed on electrochemically roughened silver electrodes.[1] It was initially believed that the enhancement resulted from the increased surface area produced by the electrochemical roughening, giving rise to increased probed sample density. The teams of Jeanmaire and Van Duyne[2] and Albrecht and Creighton[3] later confirmed the enhancement (up to $10^8$) but attributed the effect to more complex surface-enhancement processes, which continue to be the subject of intense theoretical studies. More recent reports have cited SERS enhancements from $10^{13}$ to $10^{15}$, thereby demonstrating the potential for single-molecule detection with SERS.[4–9]

In the early years following the discovery of SERS (mid-1970s), the focus of SERS studies was on the development of theoretical models to account for the SERS effect.[10–16] However, these studies were limited to a small number of highly polarizable molecules in the 1- to 100-m*M*-concentration range. As a result, despite the great theoretical progress achieved in early SERS studies, interest in SERS as a practical analytical tool was delayed largely because of the common belief that the SERS effect occurred only for a few compounds under specific experimental conditions. Nevertheless, the 1980s marked the beginning of a new phase in SERS evolution, which was characterized by the development of a wide variety of SERS-inducing media and applications to a wide range of chemicals. For example, the use of planar solid-surface-based substrates for SERS-based measurements of a variety of chemicals, including several homocyclic and heterocyclic polyaromatic compounds, was first reported in 1984.[17] These substrates were composed of a planar support material that was covered with dielectric nanoparticles to impart the roughness required for the SERS effect, and then overcoated with a layer of silver. In a similar approach, silver-coated frosted glass was demonstrated to be SERS-active.[18] Silver colloids were investigated for analytical applications.[19–22] Indeed, silver colloids were a popular medium for early SERS-based applications due to the fact that with some skill and patience they could be prepared with chemical reagents available in most common analytical laboratories. In addition, research demonstrated that silver colloids could be used in solid[23,24] as well as free-solution matrices. Researchers also found that through a variety of innovative techniques the reduction process for producing elemental silver colloids could be made to match the chemical and environmental conditions of specific applications. For example, SERS-active "photocolloids" were produced through *in situ* laser-induced photoreduction of silver nitrate solutions.[25] The development of a wide variety of additional SERS-inducing media had been described in the early 1980s; these included electrolyte interfaces, mechanically ground silver, matrix-isolated metal clusters, holographic gratings, silver island films, gold colloids, and tunnel junctions with a silver electrode on a rough surface.[14] Because of the aggressive development of SERS substrates and application to a wide range of chemicals, the potential of SERS as a routine analytical technique was recognized by the mid-1980s.

The SERS technique has since continued to receive increased interest, as is evidenced by the large numbers of papers and review articles.[10–14,26–32] Furthermore, the scope of SERS has been extended to include other surface-enhanced spectroscopies such as surface-enhanced second-harmonic generation[33] and surface-enhanced hyper-Raman scattering.[34]

This chapter provides a brief theoretical background of the SERS effect, followed by a synopsis of the development of SERS-inducing media that have demonstrated repeated success and show great potential for use in biomedical analysis. The chapter also summarizes some of the most significant applications of SERS to date, with a focus on both chemical and biomedical areas. Some highlights of this chapter include reports of various fiber-optic SERS monitors, SERS nanoprobes, near-field SERS probes, SERS-based bioassays, and single-molecule detection.

## 64.2 Theoretical Background

Since the discovery of the SERS effect in the 1970s, extensive fundamental studies have been undertaken to gain a better understanding of the enhancement mechanisms. Nevertheless, it is believed that our current understanding of SERS is still incomplete. For instance, the most established theoretical models predict maximum Raman enhancements on the order of $10^6$ to $10^8$ for the SERS effect. However, recent near-field SERS measurements have shown evidence of SERS enhancement factors as high as $10^{13}$.[8,9] Single-molecule studies have further indicated that enhancement factors on the order of $10^{14}$ to $10^{15}$ may be possible at specific sites referred to as "hot spots."[4–7,35] Although new experimental observations in SERS studies are expected to spur the development of new or refined theoretical models, several current theoretical models could explain some of the more general observations associated with the SERS effect.

A qualitative understanding of the SERS process can be gained through the classical theory of light scattering. Consider an incident-light beam inducing an oscillating dipole, $\Phi$, in a particle that scatters light at the frequency of the dipole oscillation.[36] The dipole moment consists generally of many different harmonic-frequency components; each component may be represented by the following equation:

$$\Phi(t) = \Phi^{o}\cos(2\pi \nu t), \tag{64.1}$$

where $\nu$ is the dipole oscillation (i.e., scattering) frequency and $\Phi^{o}$ is the maximum induced dipole moment for a given frequency component of $\Phi$. When the magnitude of the incident electric field, $E_{inc}$, is not too large, the induced dipole moment can be approximated as:

$$\Phi(t) = P \cdot E_{inc}(t), \tag{64.2}$$

where $P$ is the polarizability of the molecule. The polarizability, which is a tensor quantity, can be qualitatively described as the ease with which molecular orbitals are distorted in the presence of an external field.

Because Raman intensity is proportional to the square of the induced dipole, $\Phi$, it can be inferred from Equation 64.2 that enhancements of either the incident field or the molecular polarizability or both can contribute to Raman enhancements observed via SERS. As a result, theoretical models generally involve two major types of enhancement mechanisms: (1) an "electromagnetic effect" (sometimes referred to as the field effect) in which the molecule experiences large local fields caused by electromagnetic resonances occurring near metal surface structures and (2) a "chemical effect" (also referred to as the molecular effect) in which the molecular polarizability is affected by interactions between the molecule and the metal surface.

The relative contributions of the two mechanisms can vary widely for different molecules largely because of the specificity of the "chemical effect." Electromagnetic interactions, which are more general and relatively long-range (i.e., inversely dependent on the distance cubed between the adsorbed molecule and the metal surface), have been extensively investigated and are reasonably well understood. On the other hand, short-range "chemical effects," which may depend on atomic-scale features of both the metal surface and the adsorbed molecule, are less well known and are currently topics of extensive research. A detailed summary of these studies is far beyond the scope of this chapter but can be found in a number of excellent reviews.[10–15,27,30,37]

## 64.3 Development of SERS Substrates

Roughened metal electrodes, the first SERS-active surfaces to be discovered, served as the basis for extensive theoretical studies. Subsequently, metal colloids gained widespread use in SERS studies because they could easily be prepared chemically with commonly used reagents. While the popularity of metal colloids has persisted since the early 1980s, alternative media — e.g., metal nanoparticle films and nanostructured probes — have since been developed for a wide variety of applications with improved reproducibility. Etched quartz posts coated with silver have also exhibited a great deal of reproducibility. In an effort to simplify fabrication of solid SERS probes, materials have been mechanically roughened to induce the SERS effect. Inherently rough silver membranes, requiring little additional preparation, have also been investigated. Many of these technologies have been the basis of recently developed innovative probes, such as silver-nanoparticle-embedded sol gels and polymers, *in situ* photoreduced sols, and a variety of fiber-optic-based SERS probes.

Some particularly exciting movements in SERS probe development have focused on SERS nanoprobes and near-field probes, which have introduced the possibility of extremely spatially specific SERS analysis as well as single-molecule detection. Another area of intense research has been the development of highly specific solid SERS probes that incorporate coatings of polymers, self-assembled monolayers (SAMs), and bioreceptors. As a result of the extensive development of SERS-based probes, the scope of applications of SERS has been dramatically expanded to include solid, liquid, and airborne samples in many major areas (e.g., environmental, biotechnical, biomedical, and remote-sensing analyses).

## 64.3.1 Metal Electrodes

Electrochemically roughened electrodes were the first medium with which the SERS effect was observed.[1] The observation of this effect resulted in further inaugural studies to confirm it and to establish enhancement factors.[2,3,14] While several metals have been investigated for SERS activity in electrochemical cells,[16,38–41] silver has been most commonly used. During electrochemical preparation, silver at the electrode surface is first oxidized by the reaction $Ag \rightarrow Ag^+ + e^-$; elemental silver is then redeposited in the ensuing reduction process, $Ag^+ + e^- \rightarrow Ag$. This oxidation-reduction procedure generally produces protrusions on the electrode surface in the size range of 25 to 500 nm. Strong SERS signals appear only after several electrochemical oxidation-reduction cycles, often referred to as "activation cycles."

Other metal electrodes roughened by oxidation-reduction cycles have been investigated for use as SERS substrates; these include platinum[39] and copper.[42,43] A distinct advantage of roughened metal electrodes is that they allow selective detection *in situ* by promoting adsorption of particular compounds at specific applied voltages. For example, both SERS and surface-enhanced hyper-Raman spectroscopy (SEHRS) have been reported for pyrazine and pyridine adsorbed on silver electrodes at varying potentials.[44] Oxidized forms of metal electrodes have also been demonstrated to promote enhanced adsorption of compounds otherwise exhibiting low SERS intensity. For example, oxidized silver electrodes have made possible enhanced detection of nerve-agent simulants.[45] In addition, aluminum oxide surfaces have been used for the detection of phthalic-acid isomers and phenylphosphate in the near-infrared (NIR) region.[46,47]

## 64.3.2 Metal Colloids

SERS-active suspensions of elemental metal colloids or nanoparticles of various sizes can be chemically formed in solution. Hence, they can be readily used in suspension for *in situ* solution SERS measurements. Alternatively, they can be immobilized on various solid media for use as surface-based SERS substrates. As with roughened metal electrodes, silver is the most commonly used material. Silver colloids can easily be prepared by reducing a solution of $AgNO_3$ with ice-cold $NaBH_4$,[19–25,48–50] trisodium citrate,[51,52] or hydrogen peroxide under basic conditions.[53] Other more innovative techniques that reduce the need for wet chemistry have been demonstrated. For example, Ahern and Garrell described a unique *in situ* photoreduction method to produce photocolloids in solutions,[25] while another innovative method involved laser ablation of colloids from silver foils into aqueous solutions.[54]

Gold colloids have also been investigated as SERS-active media. Because gold is virtually bioinert, it may prove to be a valuable material for biomedical applications of SERS. Furthermore, gold produces large SERS-enhancement factors when NIR excitation sources are used. NIR excitation radiation is particularly useful in biomedical studies because it allows greater penetration depths in tissues while causing less fluorescence background relative to visible radiation. Gold sols have been used for the SERS detection of various dyes.[55,56] In particular, Kneipp et al. have observed extremely large enhancement factors (up to $10^{14}$) for dyes adsorbed on gold colloids when using NIR excitation.[7] Silver-encapsulated gold nanoparticles[57] and silver oxide sols[58] have also been reported to induce SERS activity.

A primary advantage of colloid hydrosols is that they make SERS an accessible technology. For example, they are easily produced with common chemical reagents. Furthermore, they can be readily characterized by simple UV-absorption spectroscopy. Production of hydrosols does not require the use of expensive, bulky vacuum evaporation chambers, which are common to the production of solid silver-coated nanoparticle-based substrates (see Sections 64.3.3.2 and 64.3.3.3). On the other hand, this wet-chemistry-based procedure is vulnerable to environmental factors as well as experimental error. As a result, the reproducibility in hydrosol sizes can be problematic. Furthermore, the size range can be quite large for particles produced in a single batch. Selection of colloidal particles with optimal sizes requires a tedious procedure.[59] Although some reports have indicated that carefully prepared colloid solutions can last for more than 3 weeks,[20–22,60] they more often tend to be unstable. More specifically, the colloids have a propensity to aggregate into particles that are too large to induce the SERS effect. To offset this disadvantage, stabilizers such as poly(vinylalcohol), poly(vinylpyrrolidone) (PVPL), and sodium dodecyl sulfate have been used to hinder this coagulation process.[61–63]

Metal colloids have also been adapted for use in solid-surface-based substrates. For example, silver colloids have been immobilized on filter-paper supports (cellulose, glass, and quartz fibers) for the SERS detection of various dyes.[23,24] However, the SERS effect was found to depend highly on the specific type of filter paper being used. In a similar approach, filter paper treated with silver colloids has been used for the collection and detection of trace amounts of atmospheric contaminants in aerosols.[64] Extensive studies have been devoted to immobilization of colloids on various substrates pretreated with organic coatings having affinities for both silver and gold. For example, Freeman et al. have demonstrated self-assembly of monodisperse silver and gold-colloid polymers on substrates coated with polymers having cyanide, amine, and thiol functional groups.[65] Self-assembled monolayers (SAMs) of *p*-aminothiophenol[66,67] and *p*-mercaptopyridine[68] have also been used to adsorb gold colloids onto solid gold surfaces. A disadvantage of using SAMs to immobilize colloids is an inherent background SERS signal yielded by the underlying SAM. Nevertheless, such signals have been the basis for fundamental studies regarding the self-assembly of the colloid overlayer. For example, kinetic studies of the assembly of silver and gold colloids on organosilane-polymer-treated surfaces have been pursued.[69,70] Results of these studies have permitted the control and optimization of intercolloid spacing in the production of SAM-based SERS probes. In an innovative method for producing patterns of gold nanoparticles, microcontact printing has been used to deposit corresponding patterns of the underlying poly(dimethyl)siloxane SAMs with amine and thiol functionalities.[71] Finally, SAMs of 3-aminopropyl trimethoxysilane have even been used to immobilize silver colloids on optical fibers for the production of SERS-active fiber-optic sensors.[72]

## 63.3.3 Solid SERS Substrates Based on Metallic Nanostructures

In addition to immobilized colloids, researchers have developed a variety of solid-surface-based SERS substrates that are produced entirely from solid materials, as depicted schematically in Figure 64.1. In contrast to immobilized colloids, the solid-SERS-based probes described in this section exhibit a high degree of reproducibility. Figures 64.2 and 64.3 are scanning electron micrograph (SEM) photographs that illustrate the highly uniform surface features achievable with various solid-surface-based substrate technologies, several of which are described in the following section.

### 64.3.3.1 Metal-Nanoparticle-Island Films

Metallic nanostructured SERS substrates based on metal-island films (Figure 64.1A) are among the most easily prepared surface-based media, granted the availability of a vacuum evaporation system. Such systems are commonly equipped with crystal microbalances for monitoring metal-film thickness. Metal-island films can be produced by depositing a thin (<10 nm) layer of a metal directly onto a smooth solid-base support via sputter-deposition[49] or vacuum evaporation. At such a small thickness, the metal layer forms as aggregated, isolated metal islands, the size and shape of which can be influenced largely by the metal thickness, deposition rate, geometry, and temperature, as well as post-deposition annealing. To optimize the production of SERS-active metal-island films, various diagnostic techniques have been used including optical absorption, atomic force microscopy (AFM), SERS, or combinations thereof.[73,74] More recently, silver-island films have been characterized by combining scanning near-field optical microscopy (SNOM) with Raman spectroscopy, thereby achieving <70-nm resolution.[75] A disadvantage of metal-island films is that they are easily disturbed by solvents encountered in typical biomedical analyses.[76,77] To minimize this disadvantage, buffer metal layers, (3-mercaptopropyl)-trimethoxysilane (MCTMS) layers, and organometallic paint layers have been applied

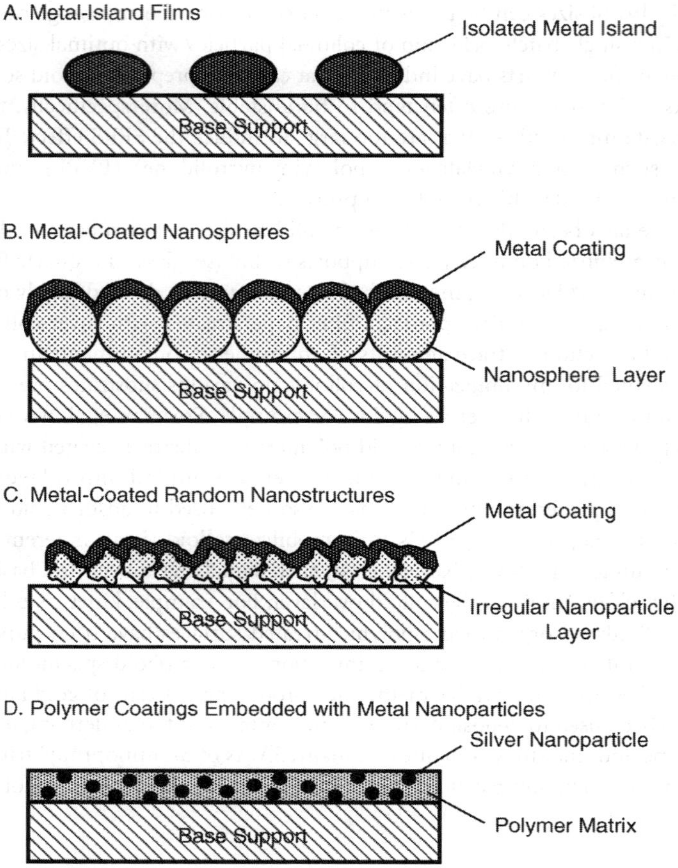


**FIGURE 64.1** Schematic representation of various types of solid-surface-based SERS substrates. (Adapted from From Vo-Dinh T. et al., *J. Raman Spectrosc.*, 30(9), 785, 1999.)

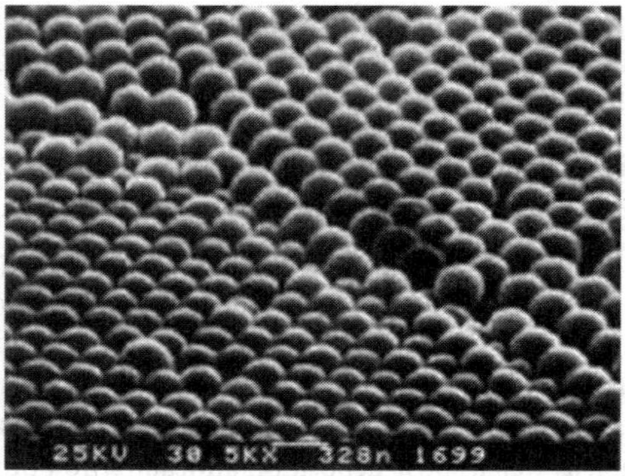


**FIGURE 64.2** Scanning electron micrograph (SEM) of a silver-coated nanosphere-based substrate having 261-nm-radius nanospheres.

to glass supports to stabilize gold island films.[78] Likewise, support media investigated for SERS-active silver islands have included sapphire, carbon, stochastically distributed posts, optical gratings, mica, glass, derivatized glass, and formvar-coated glass.[76,79]

**FIGURE 64.3** SEM photograph of acid-etched quartz posts.

In addition to gold and silver, indium[80,81] and copper[82] have been used in SERS-active metal-island films. Among these metals, silver is the most commonly used. Extensive evaluations of silver-island films have included comparisons to a variety of other silver nanostructured SERS substrates.[83,84] Nevertheless, the less commonly used gold- and copper-island films have been developed for use with NIR (1064.1-nm) excitation.[82] This report is especially promising for biomedical applications because of minimal incidence of fluorescent background in the NIR region. Furthermore, NIR radiation can be propagated through greater distances in tissues than UV or visible light.

#### 64.3.3.2 Metal-Coated Nanosphere Substrates

In our laboratory, we have developed a very dependable solid-surface-based SERS-substrate technology that can be generally described as metal-coated dielectric nanospheres (Figure 64.1B) supported by various planar support media. Nanospheres within a specific size range (e.g., 50 to 500 nm) are spin-coated on a solid support to produce the roughness required to induce the SERS effect. The resulting nanostructured plate is then coated with a layer of silver (50 to 150 nm), which provides the conduction electrons required for the surface plasmon mechanisms. All factors of surface morphology are easily controlled, enabling high batch-to-batch reproducibility. An additional advantage is that the relatively thick layer of silver is less vulnerable to air oxidation than silver islands. Furthermore, the surface is highly resistant to disturbance by sample solvents, making this type of SERS substrate very practical for biomedical applications. Teflon and latex are particularly well suited for SERS substrates because they are commercially available in a wide variety of sizes that can be selected for optimal enhancement.

Preparation of a nanosphere substrate is relatively easy. A 50-μl aliquot of a suspension of nanospheres is deposited evenly over the surface of a dielectric planar support medium, such as filter paper, cellulosic membrane, glass, or quartz.[17,31,85–88] Using a photoresist spinner to produce uniform monolayer coverage, the substrate is then spun at 800 to 2000 rpm for about 20 sec. The spheres adhere to the glass surface, providing uniform coverage. The nanosphere-coated support is then placed in a vacuum evaporator, where silver is deposited on the roughened surface at a rate of 0.15 to 0.2 nm/sec. Figure 64.2 shows a SEM of a silver-coated substrate having 261-nm-radius nanospheres (deposited from a 10% suspension). As demonstrated by the figure, the sphere surfaces exhibit excellent uniformity with respect to nanosphere size, shape, and surface coverage.

Nanosphere-based substrates were among the first to be applied to the detection of a wide variety of compounds of health and environmental interest including organophosphorus agents, chlorinated pesticides, polynuclear aromatic hydrocarbons (PNAs), and DNA adducts.[89–93]

#### 64.3.3.3 Metal-Coated Nanoparticle Subtrates

Nanoparticles with irregular shapes (Figure 64.1C) can also be used in place of regularly shaped nanospheres in the production of dependable, cost-effective SERS substrates. Dieletric nanoparticle materials investigated in our laboratory have included alumina,[94] titanium dioxide,[95] and fumed silica.[86] The production of irregular nanoparticle-based substrates is achieved with an ease equivalent to that for nanosphere-based substrates described above. Generally, 5 to 10% (w/v) aqueous suspensions of the nanoparticles are spin-coated onto solid support media and then coated with 75 to 150 nm of silver via vacuum evaporation. As an alternative to the vacuum-evaporation process, other groups have investigated silver coating via chemical processes.[96,97] Nevertheless, substrates prepared via vacuum evaporation yield exceptional reproducibility.

Alumina-based substrates produced by vacuum evaporation have proven to be among the most dependable, with a typical bath-to-batch variability of <10% RSO for induced signals of selected model compounds. The surface of an alumina-based substrate consists of randomly distributed surface agglomerates and protrusions in the 10- to 100-nm range. These structures produce large electromagnetic fields on the surface when the incident photon energy is in resonance with the localized surface plasmons. Alumina-based substrates have numerous applications.[92,93,96,98–103] In the studies performed in our laboratory, these substrates have been incorporated into a passive vapor dosimeter for the detection of airborne chemicals[98,104] and used for the detection of SERS-active gene probes.[103] Gold coating has also been applied to alumina-coated plates for use in the visible to NIR region,[105] thereby demonstrating greater utility in biomedical applications.

Silver-coated titanium dioxide[95] and fumed silica[86] surfaces also provide efficient SERS-active substrates. Titanium dioxide provides the necessary nano-sized surface roughness for the SERS effect; the nominal particle diameter of titanium dioxide used in our probes is 0.2 μm. When coated with 50 to 100 nm of silver, this type of probe can yield exceptional SERS enhancement with limits of detection of various compounds in the parts-per-billion (ppb) range.[95] Fumed silica substrates have allowed trace detection of PNAs and pesticides.[89,95]

### 64.3.4 SERS Substrates Based on Metal-Coated Quartz Posts

Silver-coated, regularly spaced submicron posts formed in quartz substrates have proven to be a dependable, yet labor-intensive, SERS substrate. Lithographic techniques have been used to control the surface roughness to a degree suitable for testing the electromagnetic model of SERS.[106,107] While these surfaces produce a Raman enhancement on the order of $10^7$, they are difficult to produce with a large surface area. However, an alternative etching procedure for producing quartz posts overcomes this limitation by using an island film as an etching mask on a silica substrate.[17,83,84,107,108] The preparation of silica prolate posts is a multistep operation, as depicted in Figure 64.4. A 500-nm layer of silica is first thermally evaporated onto a fused quartz base support at a rate of 0.1 to 0.2 nm/sec. The resulting thermally deposited crystalline quartz is annealed at 950°C to the fused quartz for 45 min. A 5-nm silver layer is then evaporated onto the thermal silica layer, and the substrate is flash-heated (500°C) for 20 sec, causing the thin silver layer to bead up in small globules. These isolated silver globules act as etch masks when the substrate is subsequently etched for 30 to 60 min in a $CHF_3$ plasma. This etching produces submicron

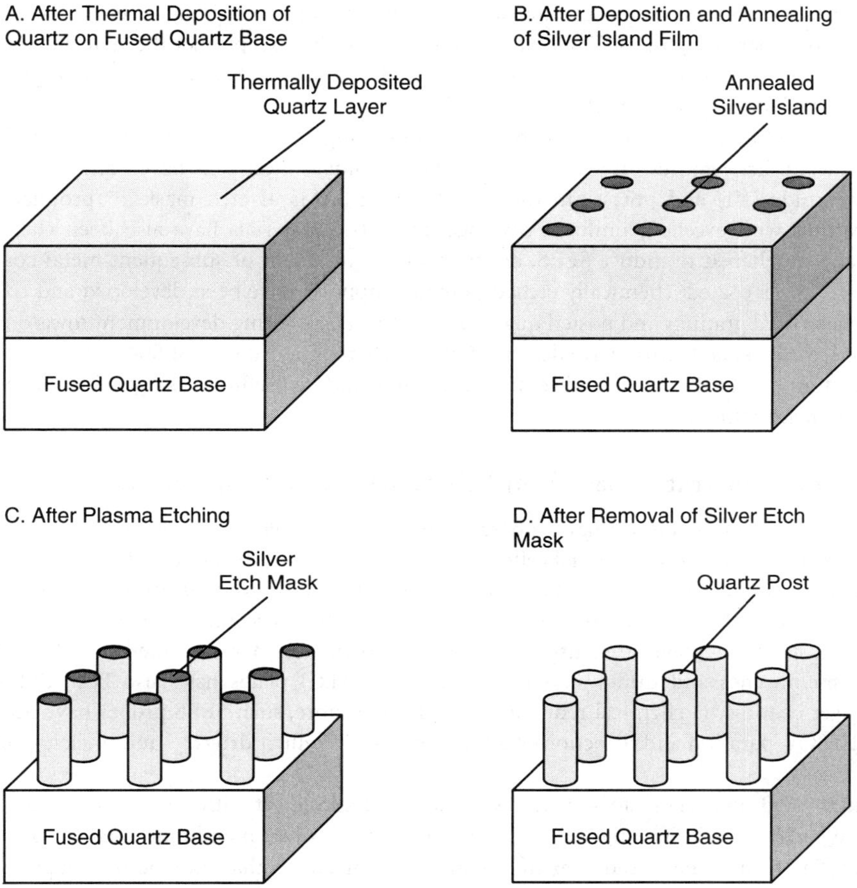


**FIGURE 64.4** Schematic diagram of the production of quartz-post-based SERS-active substrates.

prolate silica posts under the silver globules. Since fused quartz is etched much more slowly than is thermally deposited quartz, the fused quartz base survives the etching process. The posts are then cleaned to remove the silver etch mask and coated with either a continuous 80-nm silver layer[108] or a discontinuous layer with the silver deposition restricted to the tips of the quartz posts.[83,84] Figure 64.3 shows the SEM photograph of the resulting uniformly shaped and distributed quartz posts.

A disadvantage of quartz-post-based substrates is the difficult, time-consuming etching procedure required for post fabrication. Nevertheless, optimized quartz posts can serve as the base for a reusable SERS substrate, provided the silver coating is replaced between uses. It is worthwhile to note, however, that the posts are fragile and hence not amenable to field studies. In contrast, the nanosphere- and nanoparticle-based substrates described above are generally simpler to prepare and very easy to handle. Furthermore, comparative studies indicate that the simple fumed silica-, alumina-, or nanosphere-based probes provide SERS enhancements similar or superior to quartz-post-based probes. Consequently, practical applications involving quartz-post-based SERS probes have been extremely limited.

### 64.3.5 SERS Substrates Based on Chemically and Abrasively Roughened Solid Surfaces

Smooth surfaces can be roughened by chemical and abrasive techniques to impart the required roughness for the SERS effect. In several cases, SERS substrates have been produced through a one-step procedure of roughening metal foils, analogous to redox-roughened metal electrodes. Because the foils act as a source of surface plasmons, no metallic overlayer is required. Copper[109,110] and silver[111–115] foils have been acid-etched to impart roughness. These SERS substrates exhibit exceptional thermal stability, are cost-effective, and can be reused. As a result, the SERS activity of acid-etched silver foils compare favorably with a variety of other types of SERS substrates.[111–115]

Another procedure involves a combination of sandblasting and acid-etching steps to produce 10- to 100-nm-scale surface features. In a more innovative technique, silver foils have been selectively etched with ferrocyanide (III) and potassium thiocyanate using SAMs as etch masks,[116] producing 100-nm surface features with excellent uniform coverage. Dielectric materials have also been chemically and mechanically roughened to induce SERS, but with the requirement of subsequent metal-coating steps. For example, silver-coated, chemically etched polymer supports have been developed and compared to SERS-active crossed gratings and posted quartz wafers.[117] In an exciting development toward biomedical applications, Mullen and Carron have described abrasively roughened optical fibers coated with silver.[118] This integration of a SERS-active surface fiber-optic tip could be useful in a single-fiber SERS probe for low-volume measurements.

## 64.3.6 SERS Substrates Based on Inherently Rough Materials

Certain planar support materials have inherent roughness suitable for inducing the SERS effect. For example, direct metal coating of special cellulosic filter papers produces useful substrates.[27,91,119] SEMs of these cellulosic materials have shown that these surfaces consist of fibrous 10-μm strands with numerous tendrils that provide the necessary protrusions required for the SERS enhancement. For example, silver-coated paper substrates have been used to detect approximately 3 ng/ml methyl red.[120] Solid-phase extraction membranes and thin-layer chromatographic (TLC) plates have also been SERS-activated through silver coating via chemical reduction.[121,122] Furthermore, such SERS probes have been used for combined TLC separation and detection of sulfonamides,[121,122] illicit drugs,[123] and nucleic-purine derivatives.[50]

Inherently rough materials can sometimes induce the SERS effect with little or no additional preparation. One such example investigated in our laboratory is a silver membrane used for air-particulate sampling.[27] The membrane already has micropores and interstices that provide the roughness features required to induce SERS. Furthermore, because it is solid silver, it can be used directly as a SERS-active substrate without applying any additional silver. Another example is black-and-white photographic paper, for which SERS activation is achieved simply through low-level light illumination and development steps.[124] It is important to note, however, that the inherently rough materials suitable for SERS are a very rare find.

## 64.3.7 SERS Substrates Based on Metal-Nanoparticle-Embedded Media

Silver nanoparticles embedded in various solid porous media (Figure 64.1D) have recently been investigated as stable SERS substrates with the potential for selective detection. The *in situ* production of these nanoparticles in solid matrices provides several advantages. For example, a solid matrix not only spatially stabilizes but also physically protects the colloids. In addition, the porosity of such materials as sol gels, cellulose acetate gels, and polycarbonate films permits interaction of analyte compounds with the embedded metal nanoparticles. Furthermore, control of pore size and matrix polarity through simple chemical means (particularly when using the sol-gel technique) can impart selectivity. Finally, the chemical processes generally used to prepare such substrates make the SERS technique accessible to general analytical laboratories.

A silver-nanoparticle-embedded sol-gel substrate produced through the chemical reduction of silver halide particles distributed in the sol-gel matrix has been reported for the detection of neurotransmitters and dopamine.[125,126] *In situ* precipitation of silver chloride nanoparticles via reaction of silver nitrate with trichloroacetic acid throughout the sol-gel matrix was performed before curing. Immediately prior to use, the silver chloride particles were reduced to elemental silver nanoparticles with $FeSO_4 \cdot 7H_20$. An advantage of this technique is that the sol gel can be stored for long periods in the nonreduced form, precluding vulnerability to air oxidation. In a similar approach, chemically reduced silver and gold colloids have been produced in sol-gel-derived xerogel,[127] and silver-doped sol-gel films have been formed through *in situ* chemical reduction of $[Ag(NH_3)_2{}^+]$.[128] Polycarbonate films doped with silver via chemical reduction have been reported as well.[129] As an alternative to chemical reduction, photodeposited gold particles have been produced through UV irradiation of organometallic gold precursors dispersed in sol gels.[130] In a quite different approach, a counter-diffusion technique has been used to embed cellulose acetate films through impregnation with preformed fine silver particles.[131,132]

## 64.3.8 Overcoatings on SERS Substrates

The application of a wide variety of coatings to many of the SERS substrates described in previous sections has been the focus of extensive research in recent years, with the goals of imparting selectivity and robustness to the SERS technology. These coatings have generally been in the form of organic coatings distributed as relatively thick polymer layers or SAMs, ultrathin metal films, dielectric films, and monolayers of bioreceptors. These coatings have improved detection selectivity through enhanced adsorption or selective permeability mechanisms. In some cases, coatings have stabilized SERS-active surface features, thereby extending their shelf lives.

### 64.3.8.1 General Organic, Metallic, and Dielectric Overcoatings

Polymer coatings have been applied to a variety of solid-surface-based SERS substrates. In one study, a relatively thick layer ($\sim 10\ \mu m$) of PVPL was applied to alumina-based silver SERS substrates.[133] In addition to prolonging shelf life, this procedure was demonstrated to preserve surface features vital to the SERS effect by offering resistance to physical disturbance. Furthermore, chemical selectivity was imparted via selective permeability, particularly for compounds having hydrogen-bonding properties. The procedure involved simply dipping the bare silver SERS substrates into a 5% (w/v) methanolic solution of the polymer, followed by room-temperature curing on a level surface for ~30 min. PVPL coatings have also been applied to silver-island-based SERS substrates for selective detection of airborne chemicals.[104,134] In that study, the coating was observed to preserve the delicate silver-island films over a 20-d period.

A combination of polybenzimidazol and mercaptobenzimidazol has been applied to SERS-active, nitric-acid-etched copper foils.[135] This coating inhibited corrosion and thus prolonged the shelf life of the SERS-active substrate. In addition, treatment of silver SERS substrates with various thiols can provide long-term stability, even for substrates stored in water for a month.[136] Propanethiol-coated silver SERS substrates, investigated as potential SERS-based detectors in gas chromatography, have also exhibited enhanced adsorption and detection of hydrophilic compounds.[137] To promote enhanced adsorption of metal ions ($Cu^{+2}$, $Pb^{+2}$, $Cd^{+2}$), 4-(2-pyridylazo) resorcinol coatings have been applied to SERS-active silver.[138] As an example of the enhanced selectivity offered by reactive coatings in the analysis of complex biological matrices, bilirubin and salicylate have been detected at below normal therapeutic levels via direct SERS analysis of whole blood.[139] In that study, application of the coating, diazonium, precluded the need for a time-consuming and possibly risky preparation of a less complex serum sample. A reactive coating has also been proposed for the one-step detection of the illicit drugs amphetamine and methamphetamine.[140]

Extensive efforts have also been devoted to the development of coatings based on inorganic materials. For example, transition metals (Pd, Rh, Pt, Ir, Ru) have been applied to gold substrates as pin-hole-free, ultrathin (2 to 3 monolayers) films to enhance adsorption of CO and NO while preserving the SERS-inducing properties of the underlying gold substrates.[141,142] The ultrathin coatings were achieved through a special electrodeposition process. Silica-coated silver-island-based substrates have also been reported as viable SERS substrates.[143–147] These dielectric-coated substrates have made possible the quantitative detection of dyes adsorbed from solutions. Furthermore, they have been used in fundamental adsorption-kinetics studies. Silica-coated substrates could prove to be especially useful in SERS-based biosensors because of the well-defined hydrophilic and stabilizing properties of the coating. Such substrates could be resistant to oxidation and physical damage, thereby both prolonging shelf life and offering the potential for reuse.

#### 64.3.8.2 SAM Overcoatings

SAMs have proven to be a valuable factor in SERS substrate development and theoretical applications of SERS. For example, they have been used to immobilize SERS-active metal colloids on planar supports in a highly ordered fashion.[67,68,71,129] In fact, specific colloidal patterns have been prepared by microcontact printing of SAMs on planar surfaces before exposure to SERS-active colloids.[71] SAMs have also been used for the production of UV-induced SERS-active photopatterns on silver films coated with *p*-nitrophenol.[148] In more fundamental studies, the inherent SERS signals of SAMs have been used to evaluate the surface uniformity of SERS-probe surfaces including electrochemically roughened gold electrodes and immobilized gold colloids.[149] The application of SAMs to SERS-substrate surfaces also allows selective detection. For example, SAMs of mercaptoalkalinic acids on colloidal-silver substrates have been used for the SERS detection of selectively adsorbed cytochrome *c*[150] and have formed the basis for anchoring DNA capture probes for SERS-active hybridization platforms.[151] Monolayers of cysteamine on silver SERS subtrates have also been reported.[152,153] Such monolayers could promote the selective adsorption of proteins. The effects of electrolytes and pH on the molecular structure and distribution of cysteamine layers have been investigated.[152,153]

#### 64.3.8.3 Bioreceptor Monolayer Overcoatings

The potential use of SERS in biodiagnostic tests has been demonstrated through the use of immobilized monolayers of bioreceptors including oligonucleotides and antibodies. The use of surface-enhanced Raman gene (SERG) probes for medical diagnostics[103,151,154–157] is covered in greater detail in later sections. In the production of SERG probes, SERS-active dye labels can be attached to oligonucleotide primers used in polymerase chain reaction (PCR) amplification of specific target DNA sequences. Following PCR, the resulting labeled target DNA can be denatured, allowing hybridization of specific, single-stranded, labeled DNA to oligonucleotide capture probes, which can be immobilized on solid supports. Contact with SERS-active media permits subsequent detection of the SERG probe. This method combines the high sensitivity of the SERS technique with the inherent molecular specificity offered by DNA-sequence hybridization.

Antibody monolayers have also been the basis of SERS detection with molecular selectivity,[158,159] thus demonstrating the potential for SERS in immunoassays. In one study, Ni et al. demonstrated the simultaneous detection of two antigens in a single sandwich immunoassay using two reporter molecules.[159] Capture antibodies selective for the target antigens were bound to gold colloids. The capture-antibody-coated colloids were then exposed to a mixture of target antigens. Finally, reporter antibodies specific to the immobilized target antigens were immobilized on the colloid via interaction with the target antigens. SERS-active markers on the reporter antibodies permitted extrinsic detection of the immobilized antigens. Each antigen was assigned a different marker, yet both reporters could be detected in a single measurement because of minimal overlap of the respective Raman spectra. In another study, Dou et al. demonstrated the potential for a SERS-based immunoassay with no need for reporter molecules.[158] Instead, they monitored the native SERS signatures of antimouse IgG antibodies adsorbed on gold nanoparticles. Because of the structure-specific nature of Raman scattering, they were able to directly confirm conjugation with antigens through the observation of changes in relative intensities of spectral features of the IgG spectrum. Neither reporter probes nor rinsing steps to remove unbound antigens were required in this assay.

## 64.4 Biomedical Applications of SERS Probes

### 64.4.1 SERS Bioanalysis

The extensive progress in the development of dependable SERS substrates over the past few decades has promoted the application of SERS in the rapidly expanding field of biotechnology, as is demonstrated in several excellent reviews.[154,160–163] The variety of substrates has made these applications far-reaching indeed, spanning liquid- and gas-phase samples from the meso- to nanovolume scale. For example, various SERS-active fiber-optic probes have been developed for analyzing biological samples, as is discussed below. Furthermore, the development of nanoscale fiber-optic probe tips offers the potential for intracellular measurements as well as the integration of SERS with SNOM.[8,9,154,160–163] Reliable substrates have been the basis of SERS-based screening for illicit drugs[123,140,164–168] as well as bioassays based on DNA[103,154,155] and antibody[158,159,169–171] probes. SERS has been applied to the low-level detection of therapeutic drugs[20,172–188] and has even been used to evaluate the interaction between drugs and target species as well as complex matrices,[189–203] sometimes within cells.[204–207] A host of biological compounds (e.g., proteins, amino acids, lipids, fats, fatty acids, DNA, RNA, antibodies, and enzymes) and systems (e.g., lipid bilayers and intracellular environments) have been studied via SERS. Further examples can be found in Table 65.8 of Chapter 65 of this handbook. Such studies could provide valuable insights into cellular processes (some of which promote diseases), thereby aiding in the establishment of the mechanisms of therapeutic drugs and in the development and evaluation of new therapeutic species.

#### 64.4.1.1 SERS-Based Bioassays

Among the most innovative biomedical applications of SERS have been bioassays. DNA-based bioassays are discussed in great detail below. Immunoassays have also received considerable attention for more than a decade. For instance, an early example of a SERS-based immunoassay was reported by Rohr et al. in 1989.[171] The same year, Grabbe and Buck reported on SERS studies of native human immunoglobulin-G.[208] More recently, SERS signatures of native antibodies have been the basis of a simplified immunoassay[158] in which conjugation of the antibody with a target antigen was confirmed through observance of alterations of the SERS spectrum for the native antibody.

As an alternative, extremely sensitive detection can be achieved with reporter-antibody probes tagged with intensely SERS-active compounds or with enzymes that react with substrates to yield SERS-active products. These methods often involve sandwich-immunoassay techniques, which increase the number of required steps but offer the advantages of excellent sensitivity and the potential for "label multiplexing." For example, Ni et al. recently reported the simultaneous detection of two types of antigens in a single assay by using two reporter antibodies.[159] The labels of the reporter probes yielded SERS spectral features with minimal overlap, permitting exploitation of a label multiplex advantage. In a sandwich-assay format, one set of antibodies immobilized the target antigens on gold colloids, while the reporter-antibody probes were subsequently immobilized at allosteric sites of the target antigens.

In other studies, reporter antibodies have been tagged with the enzyme peroxidase.[169,170] Once the peroxidase-labeled antigen was immobilized on the target, it could be exposed to the substrates *o*-phenylenediamine and hydrogen peroxide. A sustained reaction of the enzyme with a multitude of substrate molecules produced an extremely high yield of the SERS-active product, azoaniline. This method has thus far been limited to sandwich-immunoassay formats. For example, in a study to detect membrane-bound enzymes in cells, cells were first exposed to primary antibodies specific to the membrane-bound enzymes, then exposed to a peroxidase-tagged reporter antibody specific to the primary antibody.[169] A format that uses a capture antibody to immobilize a target antigen, and a reporter antibody to bind to an allosteric site of the immobilized target antigen, has also been reported.[170]

#### 64.4.1.2 SERS Detection of Illicit Drugs

Another potential biomedical application of SERS-based analysis is drug screening. Several studies have demonstrated SERS detection of illicit drugs on a variety of substrates.[123,140,164–168] These works have collectively yielded SERS spectra for cocaine–HCl, crack cocaine, amphetamine, methamphetamine, mefenorex, pentylenetetrazole, pemoline, heroin, codeine, and other drugs banned in sports. In one study, stimulant drugs were detected in human urine spiked at μg/ml levels.[168] In other work, SERS was used to distinguish between cocaine and common cutting agents and impurities including benzocaine and lidocaine.[164]

While these studies have demonstrated the potential for direct analysis of complex media with SERS, other efforts have been directed toward the development of reactive coatings to enhance selectivity. For example, Sulk et al. have proposed a reactive coating for the selective detection of amphetamine and methamphetamine.[140] Silver colloids have been most commonly used for this application, but some solid-surface-based media have also been reported, including colloidal silver supported by cellulose paper and ion-exchange paper, as well as acid-etched silver foils.[166] TLC plates coated with silver through vacuum evaporation have also been used for the detection of illicit drugs. TLC plates have the unique advantage of not only providing the roughness needed to induce SERS but also serving as a separation medium. In the TLC study, a mixture of heroin, codeine, and cocaine was separated on bare TLC plates that were subsequently coated with silver to allow detection at approximately 0.2 $\mu g/mm^2$.[123] Separation-based SERS detection of drugs banned in sports has also been achieved through the coupling of SERS with liquid chromatography via a post-column windowless detection flow cell that was supplied with a stream of silver colloids.[167]

#### 64.4.1.3 SERS Studies of Therapeutic Drugs

SERS has been used to detect a large number of therapeutic drugs, as indicated in Table 64.1. A key advantage of SERS spectroscopy is the structural information it provides. In several studies, this advantage has been exploited to confirm interactions between therapeutic drugs and targets. In many cases, such studies have even yielded the establishment of interaction mechanisms, including the identification of specific active sites, complex molecules, and modes of bonding. Observable changes in SERS spectra for drugs and targets before and after interactions have served as the basis for such determinations. Fortunately, interaction between the complexes and the SERS-inducing medium (e.g., silver colloids) often has negligible effects on the drug/target interactions. There have been extensive studies of interactions between a variety of antitumor drugs and DNA including doxorubicin,[200,207] fagaronine,[197] ethoxidine,[197] mitoxanthrone,[184] intoplicine,[201,206] aclacinomycin,[202] saintoplin,[202] acridines,[212] and various ellictipicines.[186,203] Several studies of interactions between topoisomerases and topoisomerase-inhibiting antitumor drugs, including amsacrine[205] and intoplicine,[201,206] have been reported as well. For both drugs, interactions have been confirmed in K562 cancer cells.[205,206] Furthermore, interactions within ternary complexes of topoisomerase-inhibiting drugs, DNA, and topoisomerases have been reported for intoplicine.[201,206] In other *in vivo* studies, the interaction between the antitumor drug doxorubicin and targets has been studied in K562 cancer cells.[207] Interaction between another antitumor drug, dimethylcrocetin, and the retinoic acid receptor, RAR-gamma, has been confirmed in HL60 cancer cells.[204] Both ionic bonds and ring stacking have been determined via SERS as modes of interaction between 9-amino acridine and the metastasis-related protease guanidobenzoatase.[189,190] SERS has been used to identify active sites for the interaction between the antiviral/antiparkinsonian drug amantidine and histidine.[192] The antiviral drugs hypocrellin A[196] and hypericin[198] have been evaluated for interaction with human serum albumin. Another interesting study has demonstrated that SERS spectral changes resulting from the interaction between the antimalarial drug quinacrine and oligonucleotides depend on the nucleotide sequence.[193]

## 64.4.2 SERS Genomics

The last few years have seen a great deal of interest in the development of optical techniques for genomics analysis such as nonradioactive DNA probes for use in biomedical diagnostics, pathogen detection, gene identification, gene mapping, and DNA sequencing. This section presents an overview of SERS methods and instruments, such as DNA mapping and sequencing applications, that can be used in genomics analysis. The hybridization of a nucleic-acid probe to DNA biotargets (e.g., gene sequences, bacteria, viral DNA) permits a very high degree of accuracy in identifying DNA sequences complementary to that probe. The possibility of using SERS for low-level detection of the DNA bases and oligonucleotides was demonstrated in several studies in the 1980s.[50,213–218] More recently, the possibility of using Raman and SERS labels for extremely sensitive detection of DNA has been demonstrated.[103,151,154,155,157] For example, Graham et al. have claimed adequate sensitivity for single-molecule DNA detection via surface-enhanced

**TABLE 64.1** SERS Studies of Therapeutic Drugs

| Drug | Study |
|---|---|
| ***Antitumoral Drugs*** | |
| 9-Aminoacridine | 173, 179, 189, 190 |
| Quinacrine | 173, 193 |
| 6-Mercaptopurine | 175 |
| Hypocrellin A | 196 |
| Fagaronine | 197 |
| Ethoxidine | 197 |
| Dimethylcrocetin | 204 |
| Camptothecins | 180 |
| Amsacrine | 205 |
| Ellipticines | 186, 203, 209, 210 |
| Doxorubicin or adriamycin | 200, 207 |
| Mitoxantrone | 184 |
| Aclacinomycin | 202 |
| Saintopin | 202 |
| Intoplicine | 201, 206 |
| 4′*o*-Tetrahydropyranyl-adriamycin | 207 |
| ***Antiviral Drugs*** | |
| Amantadine | 172, 192 |
| Hypocrellin A | 196 |
| ***Antiretroviral Drugs*** | |
| Hypericin | 181, 198 |
| Emodin | 181 |
| Antiparkinsonian drugs (e.g., amantadine) | 172, 192 |
| Antimalarial drugs (e.g., quinacrine) | 173, 193 |
| Antitubercle bacillus drugs (e.g., pyrazinamide, isoniazid, and isonicotinamide) | 174 |
| Beta blockers (e.g., propranolol, alprenolol, acebutolol, and atenolol) | 182 |
| ***Diuretic Drugs*** | |
| Triamterene | 113 |
| Ameloride | 185, 211 |
| Sulfa drugs | 187, 188 |
| Antimicrobial drugs (e.g., Pefloxacin) | 183, 199 |
| Nitrogen-containing drugs | 20 |
| ***Others*** | |
| Diazepam | 176 |
| Nitrazepam | 176 |
| Amphotericin B | 177 |
| 2-Mercaptopyridine | 211 |
| Pemoline | 211 |
| Triamterene | 211 |

resonance Raman scattering (SERRS).[157] With the use of labeled DNA as gene probes, the SERS technique has been recently applied to the detection of DNA fragments of the human immunodeficiency virus (HIV)[155] as well as oncogenes.[151]

A critical aspect of sequencing the entire human genome involves defining and identifying large insert clones of DNA corresponding to specific regions of the human genome. These maps will be composed of overlapping fragments of human DNA and allow the direct acquisition of DNA fragments that correspond to specific genes.[219–222] An approach that facilitates large-scale genomic sequencing involves developing maps of human chromosomes into maps based on large insert bacterial clones such as bacterial artificial chromosomes (BACs).[220–222] A time-saving method for detecting multiple BAC-clone-labeled probes simultaneously would be especially appealing for the BAC approach to genome sequencing and mapping. The SERS technique can provide this label-multiplex capability. The SERG probes described in this section preclude the need for radioactive labels and have great potential to provide sensitivity, selectivity, and label multiplexing for DNA sequencing as well as clinical assays.

#### 64.4.2.1 Instrumental Systems for SERS Genomics

This section describes two detection systems that allow two different recording modes: (1) spectral recording of individual spots and (2) imaging of the entire two-dimensional (2D) hybridization array plate.

A detection system used for recording the SERS spectrum of individual spots is illustrated in Figure 64.5. An individual spot could correspond to an individual microdot on a hybridization platform. This system can readily be assembled using commercially available or off-the-shelf components. A focused, low-power laser beam is used to excite an individual spot on the hybridization platform. In our studies, a helium-neon laser is used to provide 632.8-nm excitation with approximately 5 mW power. A bandpass filter is used to isolate the 632.8-nm line prior to sample excitation. A signal-collection optical module is used to collect the SERS signal at 180° with respect to propagation of the incident laser beam. The collection module includes a Raman holographic filter, which rejects the Rayleigh scattered radiation before it enters the collection fiber. Finally, the collection fiber is coupled to a spectrograph (ISA, HR-320) equipped with a Princeton Instruments red-enhanced intensified CCD (RE-ICCD)-576S detection system. The signal-collection module can be coupled directly to the spectrograph, bypassing the optical fiber. However, the optical fiber greatly simplifies measurements and improves reproducibility by minimizing critical optical alignment steps.

Since the SERG probe hybridization platform consists of a 2D array of DNA hybridization spots, a method for recording signals from all spots simultaneously would be highly advantageous, reducing analysis time as well as precluding the need for platform scanning. For analysis of SERG-probe-based assays this feat can be accomplished through multispectral imaging (MSI). The concept of MSI is illustrated in Figure 64.6. With conventional imaging, the optical emission from every pixel of an image is without tunable wavelength selection capability (Figure 64.6A). With conventional spectroscopy, the signal at every wavelength within a spectral range can be recorded, but for only a single analyte spot. This condition is the basis for the single-point-analysis system described above. The MSI concept (Figure 64.6C) combines these two recording modalities, thereby allowing the acquisition of a Raman spectrum for every hybridization spot on the assay platform, provided that the entire platform can be included in the instrument field of view. Critical to the success of this concept is an imaging spectrometer. In our studies, a rapid-scanning solid-state device, an acousto-optic tunable filter (AOTF), is used for tunable wavelength selection with image-preserving capability. This compact solid-state device has an effective wavelength range of 450 to 700 nm with a spectral resolution of 2 Å and a diffraction efficiency of 70%. Wavelength tuning is achieved simply by supplying the AOTF with a tunable RF signal.

Figure 64.7 illustrates a system developed to implement the MSI concept. In this system, an expanded, collimated laser beam is used to stimulate a 0.5-cm-diameter region of the assay array platform. In these studies, a krypton ion laser (Coherent, I-70) is used as the excitation source (at 25 to 50 mW). A bandpass filter is used to isolate the 647.1-nm line of the laser. After passing through a bandpass filter (Corion), the laser beam is expanded and collimated using a spatial filter/beam expansion module. An imaging optical module collects and projects the image of the back-illuminated SERG-probe hybridization platform through an AOTF (Brimrose, Model TEAF 10-45-70-S) and onto a charge-coupled-device (CCD) camera (Photometrics, Model CH210). A holographic notch filter (Kaiser, HNPF-647–1.0) is placed in front of the CCD detector for rejection of laser Rayleigh scatter. For these studies, the imaging optical module is a microscope (Nikon, Microphot-SA) equipped with an image port to which the AOTF/CCD detection complex is attached. The image is collected with a 10× objective lens.

Figure 64.8 shows the 2D SERS image of an array pattern of *p*-aminobenzoic acid (PABA), acquired with the MSI system.[154] The diameter of each spot observed in this image is approximately 500 μm. This

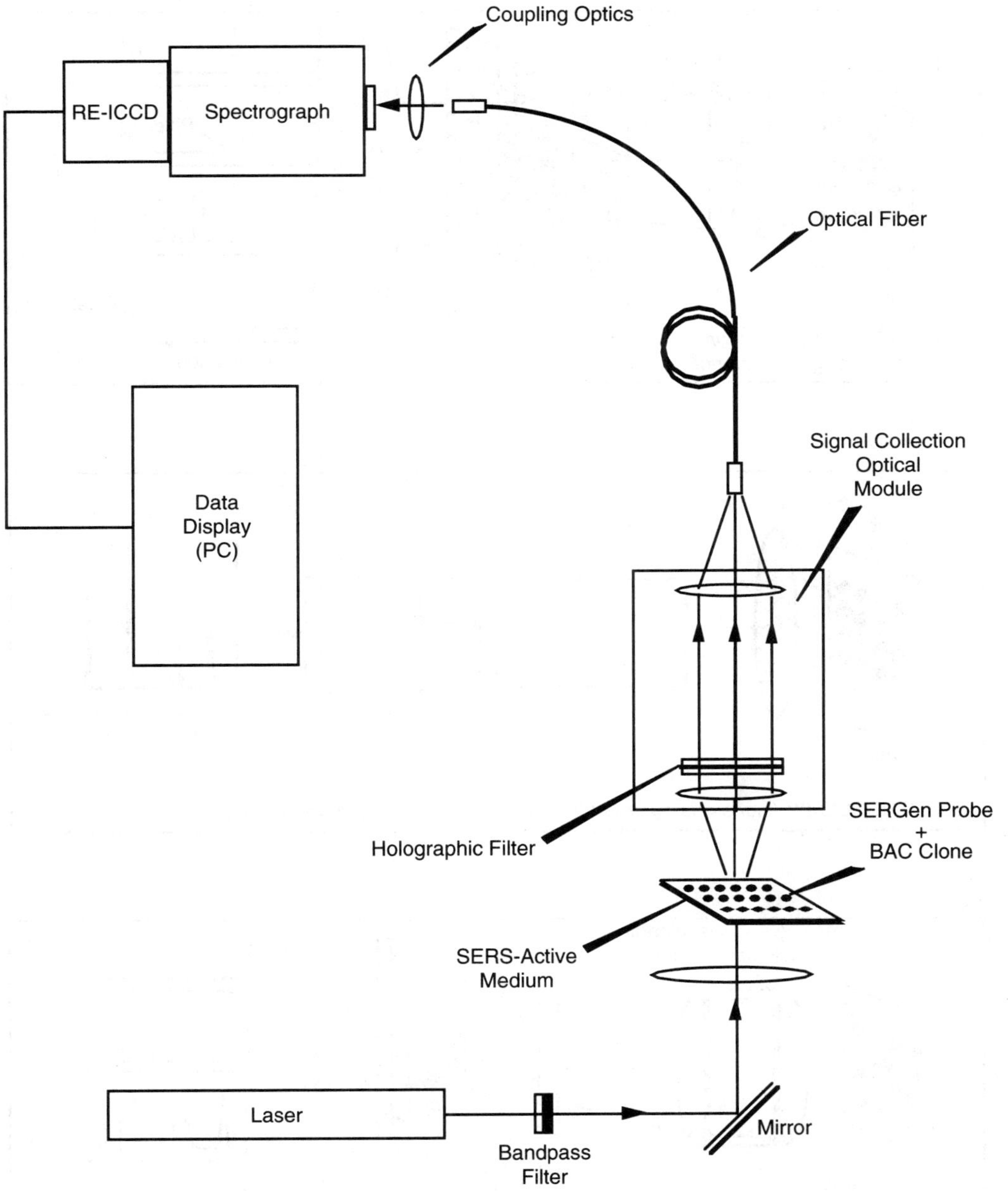


**FIGURE 64.5** Schematic diagram of the instrumental set-up for spectral acquisition from individual spots on a bioassay platform. (Adapted from Vo-Dinh, T. et al., *J. Raman Spectrosc.*, 30(9), 785, 1999.)

result demonstrates the capability of the MSI system to detect a SERS-label spot array deposited on the 5 × 5 mm substrate. The MSI technique has also been used for imaging creasyl fast violet (CFV)-based SERS patterns at a much smaller scale.[154] For example, pattern features with <20-μm diameters and 75-μm spacing were fully resolved when using a 50× objective lens. CFV has been successfully used as a SERS label for the detection of the HIV gene system.[155] These examples, though not based on true hybridizations, illustrate the possibility of performing MSI on DNA-hybridization array platforms with high resolution.

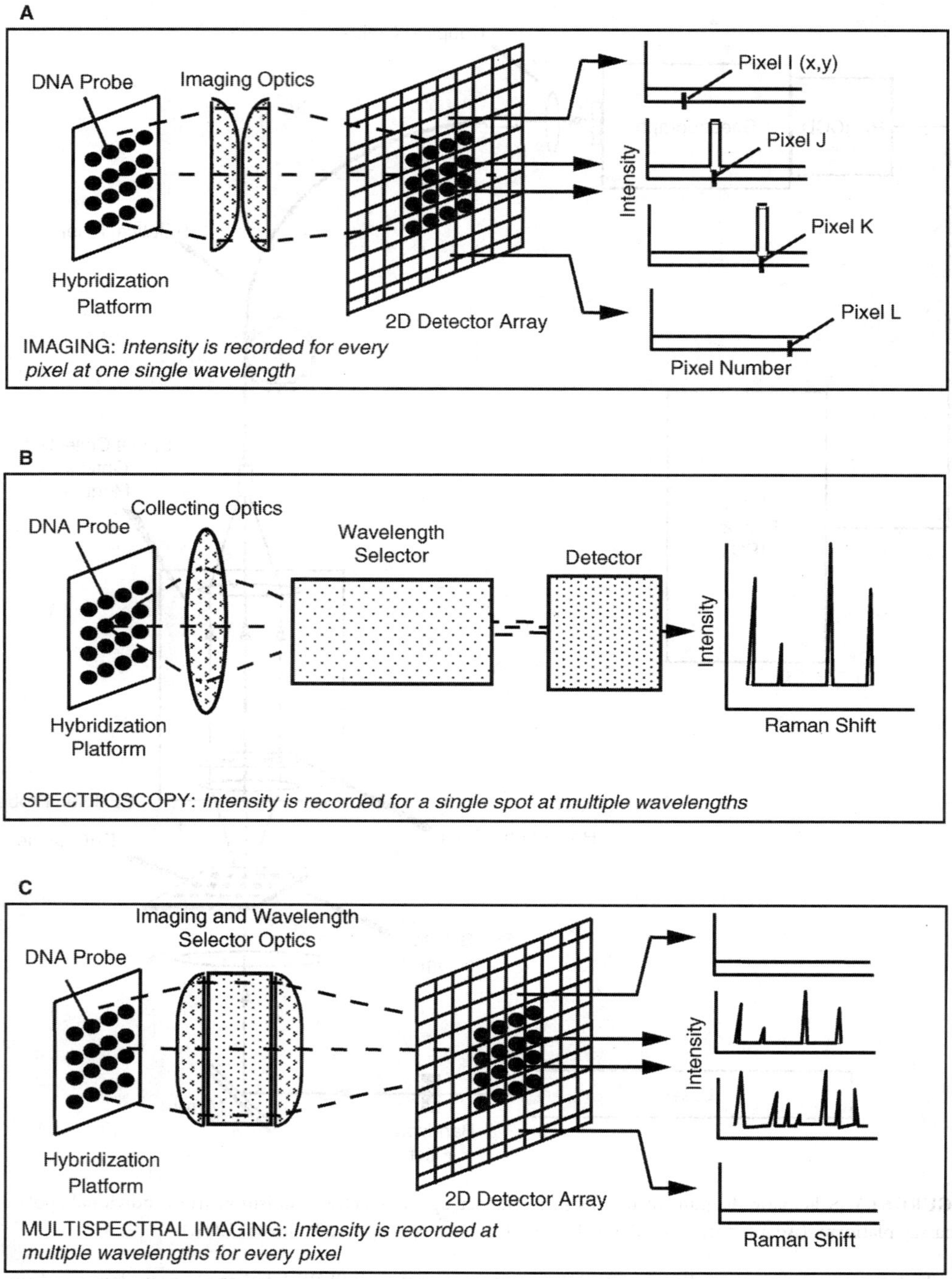


**FIGURE 64.6** Various modes of data acquisition from bioassay platforms: (A) imaging; (B) spectroscopy; (C) multispectral imaging. (Adapted from Vo-Dinh, T. et al., *J. Raman Spectrosc.*, 30(9), 785, 1999.)

#### 64.4.2.2 The SERS Advantage in Label Multiplexing

To fully exploit the label multiplex advantage for DNA-hybridization and SERG detection, several factors must be considered. First, a label must be selected that is SERS-active and compatible with the hybridization platform. The ideal label would exhibit a strong SERS signal when used with the SERS-active

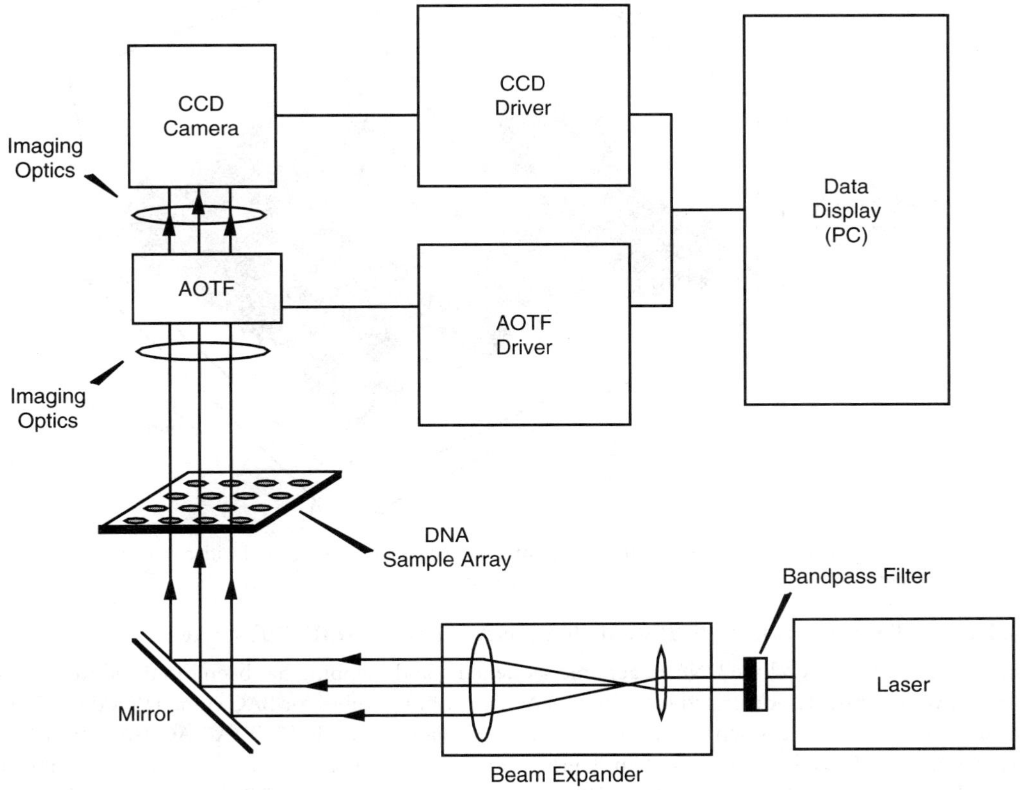


**FIGURE 64.7** Schematic diagram of an instrument for multispectral imaging of bioassay platforms. (Adapted from Vo-Dinh, T. et al., *J. Raman Spectrosc.*, 30(9), 785, 1999.)

substrate of interest (Figure 64.9A). Second, the strong SERS signal should not be affected by being attached to the oligonucleotide selected for the gene probe (Figure 64.9B). Third, the unlabeled oligonucleotide should not exhibit a significant SERS signal relative to the label's signal (Figure 64.9C). Finally, the SERS signal from the labeled probe should not be affected by the hybridization process (Figure 64.9D).

Luminescence labels have been shown to have adequate sensitivity for gene detection.[223] Nevertheless, a more desirable technique would both offer better spectral selectivity than is yielded by broadband luminescence and overcome the need for radioactive labels. The SERG probe is an excellent alternative to the other spectroscopic-based probes. For example, Figure 64.10 compares the fluorescence and SERS spectra of CFV, a SERG-probe label. As shown in the figure, the spectral bandwidth of the CFV label in the fluorescence spectrum is relatively broad (approximately 50 to 60 nm halfwidth), whereas the bandwidth of a characteristic feature (e.g., 585 $cm^{-1}$ band) of the corresponding SERS spectrum of the same CFV label is orders of magnitude narrower (<0.5-nm halfwidth). This observation clearly illustrates the advantage of using SERS for label multiplexing. More specifically, this example demonstrates a 100-factor increase in label multiplexing capacity relative to fluorescence. In a typical Raman spectrum, a 2000-$cm^{-1}$ spectral range can provide approximately 1000 resolvable spectral "intervals" at any given time. Assuming a deduction factor of 10 due to possible spectral overlap, it should be possible to find 100 labels that can be used for the detection of multiple SERG probes simultaneously. This multiplex advantage is particularly useful in high-throughput analyses where multiple gene targets can be screened in a highly parallel multiplex modality.

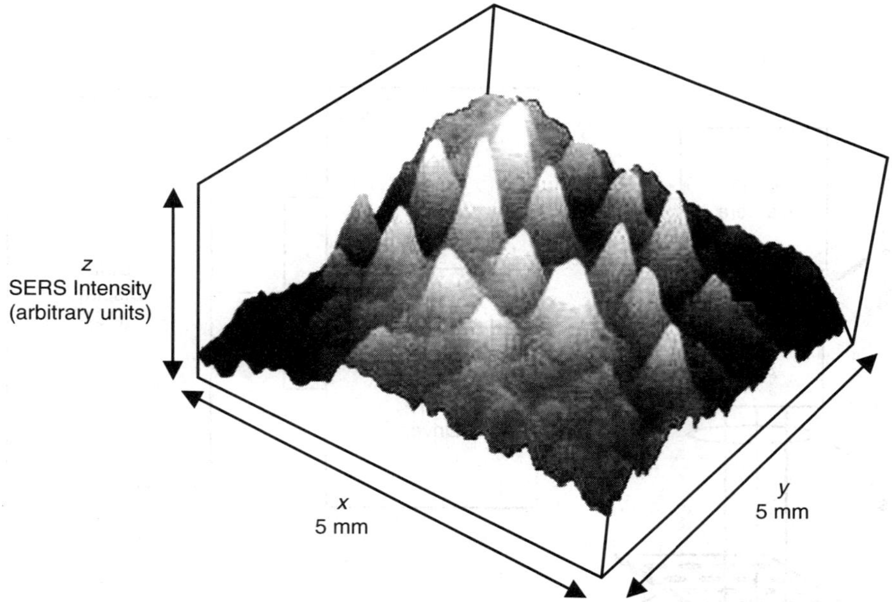


**FIGURE 64.8** Image of a 2D SERS signal pattern for *p*-aminobenzoic acid spots aquired with a multispectral imaging system. (Adapted from Vo-Dinh, T. et al., *J. Raman Spectrosc.*, 30(9), 785, 1999.)

### 64.4.2.3 SERS Detection of SERS-Labeled DNA on Solid SERS Substrates

The importance of BAC clone DNA in genome sequencing and mapping has been discussed in previous sections. We recently demonstrated the potential of using SERG probes for BAC clone DNA detection.[154] In this study, CFV was covalently attached to a 20mer oligonucleotide (5′-TAA-TAC-GAC-TCA-CTA-TAG-GG-3′) of a BAC clone model system. The labeled probe was then spotted on a SERS-active substrate for analysis. Figure 64.11 illustrates the SERS spectrum of this CFV-labeled BAC clone DNA fragment. This spectrum exhibits a series of narrow lines characteristic of CFV, with the strongest at 590 $cm^{-1}$. This intense, sharp line can be attributed to the benzene ring deformation mode. Another less intense but also sharp line at 1195 $cm^{-1}$ could be related to benzene-ring breathing vibrations. Another group of small peaks between 1000 and 1400 $cm^{-1}$ could be associated with aromatic-ring-substitution-sensitive modes. Finally, several peaks that could correspond to benzene-stretch vibrations occur between 1500 and 1650 $cm^{-1}$.

To further demonstrate the applicability of the SERS method in DNA gene-probe technology we have performed hybridization and SERS-detection experiments with BAC clone probes. Hybridization, which involves the joining of a strand of nucleic acid with its corresponding complementary-sequence strand, is a powerful technique for identifying DNA sequences of interest. BAC clone capture probes were immobilized in the wells of an Immuno Maxisorb 96-well plate using Reacti-Bind DNA coating solution (Pierce, Rockford, IL). The capture probes were then incubated with the CFV-labeled BAC clone SERG probes. After stringency washes, silver colloids were added to the wells for SERS detection of the hybridized probes. Hybridization of the labeled probe to BAC DNA could be readily detected via the CFV spectral signature. Selective detection of HIV via hybridization of SERG probes has also been demonstrated.[155]

### 64.4.2.4 Selective Amplification and Detection of SERS Gene Probes via PCR with Labeled Primers

Isola et al. recently demonstrated that primers labeled with SERS-active compounds can be implemented in the PCR process.[155] Furthermore, these researchers hybridized PCR-amplified products to DNA capture probes immobilized on *N*-oxysuccinimide (NOS)-derivatized polystyrene plates (DNA-BIND, Corning-

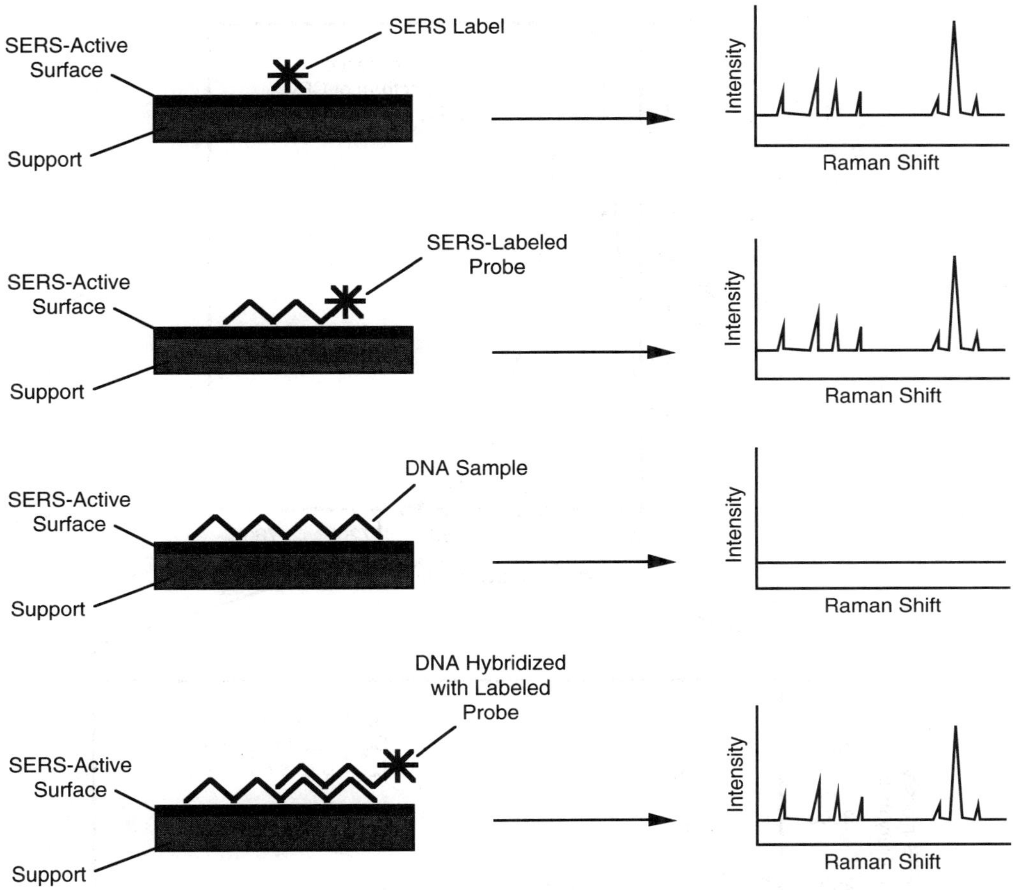


**FIGURE 64.9** Factors considered for development of the SERG probe technology. (Adapted from Vo-Dinh, T. et al., *J. Raman Spectrosc.*, 30(9), 785, 1999.)

Costar). Following hybridization, the plates were SERS-activated through the deposition of a silver-island coating. This method has the potential for combining the spectral selectivity and high sensitivity of SERS with the inherent molecular specificity of PCR and the ensuing DNA-sequence hybridization. The effectiveness of this technique has been demonstrated using the *gag* gene sequence of HIV. This is an especially encouraging result because there is a need for a direct nucleic-acid-based test for the detection of HIV. For example, standard HIV serologic tests, including the enzyme-linked immunoassay and the Western blot assay, do not provide a useful diagnosis of HIV in early infancy because of the overwhelming presence of transplacentally derived maternal antibody in the infant's blood.

Figure 64.12 illustrates the results of the HIV *gag* gene hybridization following PCR amplification with a primer labeled with CFV. Curve 1 shows the SERS spectrum of the hybridized CFV-labeled probe. For comparison, the detection of a nonhybridized CFV-labeled probe (single-stranded DNA, ssDNA) is also shown in curve 2. Curve 3 illustrates a SERS spectrum of free CFV. This study showed that no major alteration in the CFV spectrum occurred as a result of being bound to single- or double-stranded DNA. Furthermore, the researchers observed no CFV spectrum for hybridization evaluation for nonspecific binding between noncomplementary DNA strands. This is demonstrated by the hybridization control spectrum illustrated in curve 4. In this case, the CFV-labeled DNA was a single-stranded primer (designed for PCR). Without the PCR step, there was no complementarity between the CFV-labeled primer and the immobilized capture probe. Curve 5, a blank spectrum for SERS-activated NOS plate with immobilized DNA capture probes, exhibits no major spectral background in the region of interest.

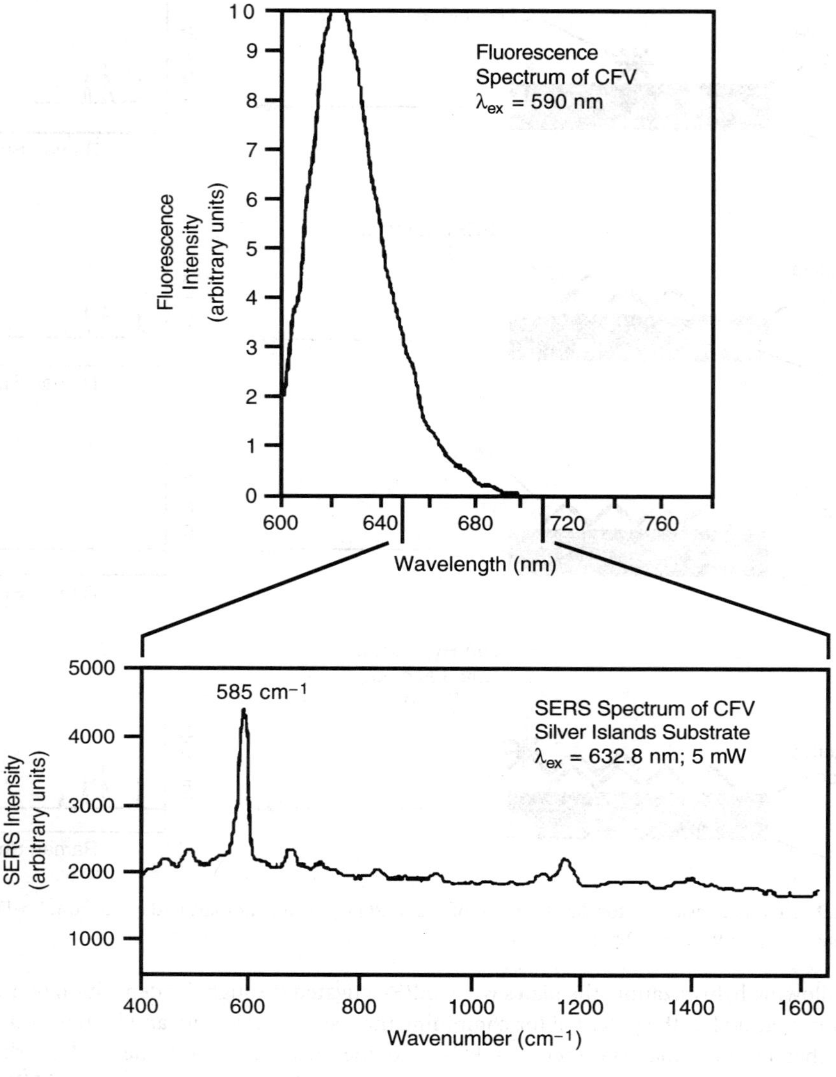


**FIGURE 64.10** Comparison of fluorescence and SERS spectra for the SERG probe label. (Adapted from Vo-Dinh, T. et al., *J. Raman Spectrosc.*, 30(9), 785, 1999.)

### 64.4.2.5 Selective Detection of Labeled DNA Probes via Hybridization on SERS-Active Bioassay Platforms

We have recently developed in our laboratory a SERS-active DNA hybridization platform that greatly enhances the utility of the SERS gene-probe technology.[151] In previous work, hybridization of labeled DNA probes was performed on conventional commercially available platforms (e.g., DNA-bind plates), followed by SERS activation of the platform.[154,155] To make the SERS gene-probe technology more practical, we have developed an inherently SERS-active hybridization platform.[151] No SERS activation of the assay is required following the hybridization step. The new DNA-assay platform is an array of oligonucleotide capture probes immobilized directly on a silver-island-based substrate prepared on glass. The capture probes were anchored directly to the silver surface via reaction with a SAM of alkyl mercaptans, as illustrated in Figure 64.13. The coupling approach involved the esterification under mild conditions of a carboxylic acid (immobilized mercaptoundecanoic acid) with a labile group, an *N*-hydroxysuccinimide (NHS) derivative, and further reaction with a 5′-amine-labeled DNA capture probe,

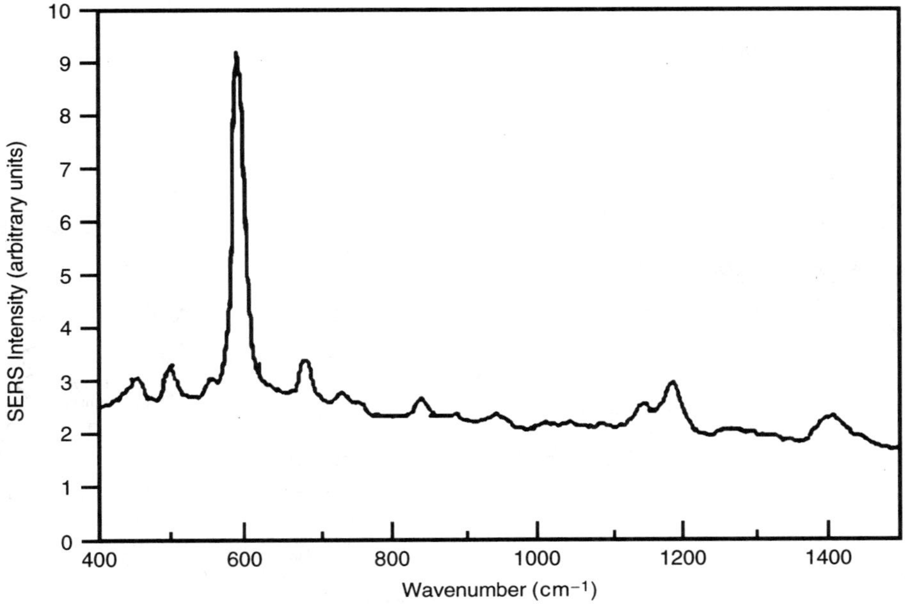


**FIGURE 64.11** SERS spectrum of CFV-labeled BAC clone DNA, krypton laser, 647.1 nm, 10 mW. (Adapted from Vo-Dinh, T. et al., *J. Raman Spectrosc.*, 30(9), 785, 1999.)

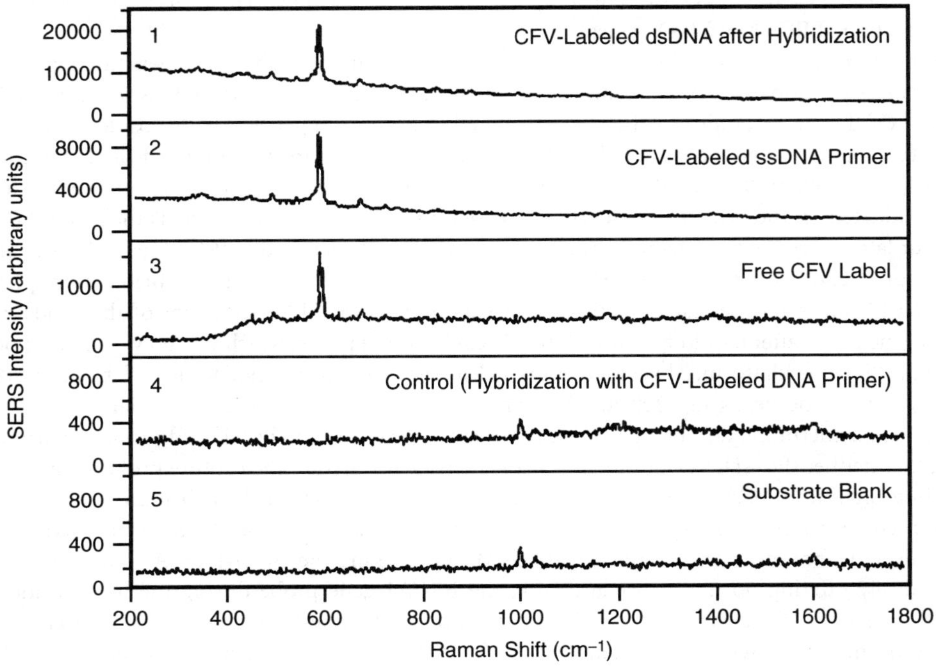


**FIGURE 64.12** Demonstration of the SERG probe technology for the selective detection of the HIV1 *gag* gene using via hybridization after PCR amplification. (Adapted from Isola, N.R. et al., *Anal. Chem.*, 70(7), 1352, 1998.)

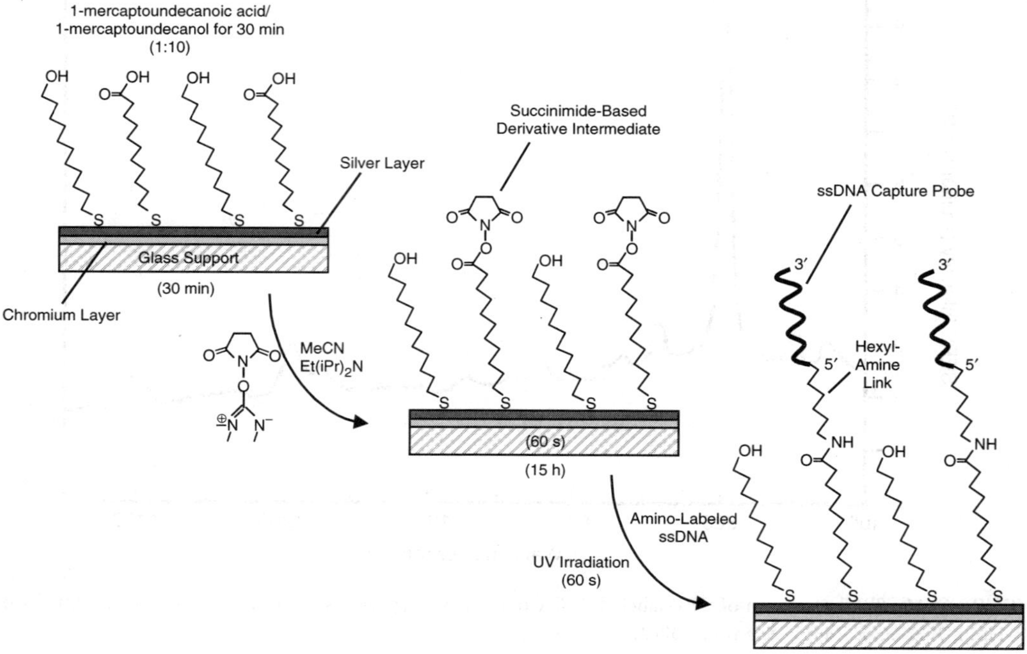


**FIGURE 64.13** Schematic diagram of the production of a SERS-active hybridization platform using a SAM of mercaptoundecanoic acid.

producing a stable amide.[224] Our recent study[151] demonstrated the potential of this technology in cancer-gene detection (BRCA1 and BAX genes).

The breast cancer susceptibility genes BRCA1 and BRCA2 are associated with genetically predisposed breast cancer development. Breast cancer is diagnosed annually in about 180,000 women in the United States.[225] Of these cases, only 5 to 10% are estimated to arise from genetic predisposition. Nevertheless, within the population with a family history of breast cancer, germline mutations in BRCA1 account for a predisposing genetic factor >80% for breast and ovarian cancers.[226] Therefore, intense research has been conducted on the principle of action of BRCA1 and BRCA2 genes, as well as their detection.[227,228]

In our laboratory, we have demonstrated the potential for detecting the BRCA1 gene through hybridization of a rhodamine-B-labeled SERG probe on the new SERS-active hybridization platform. Figure 64.14 illustrates the results of this study. Curve (a) is the SERS spectrum of the rhodamine-B-labeled gene probe after hybridization with the BRCA1 capture probe, which was previously immobilized on the SERS-active platform. This spectrum is a clear signature of the rhodamine-B label. For example, the majority of the peaks in 1200 to 1500 $cm^{-1}$ region can be attributed to Raman-active aromatic vibrations. Furthermore, no new spectral features are observed in the hybridized probe spectrum, indicating that neither the ssDNA capture probe nor the reagents used in the capture probe anchoring steps contribute significant background signals. In contrast, the spectrum for the hybridization control, illustrated in curve (b), shows no spectral features characteristic of rhodamine-B. The hybridization control was a silver surface modified with alkylthiol SAMS but lacking the BRCA1 DNA capture probes, and hence ideally offering no means for attracting the BRCA1 gene probe during the hybridization step. Indeed, the lack of the spectral features characteristic of rhodamine-B indicates that nonspecific binding is insignificant. Similar results have been reported for BAX gene detection using a rhodamine-110-labeled SERG probe.[151]

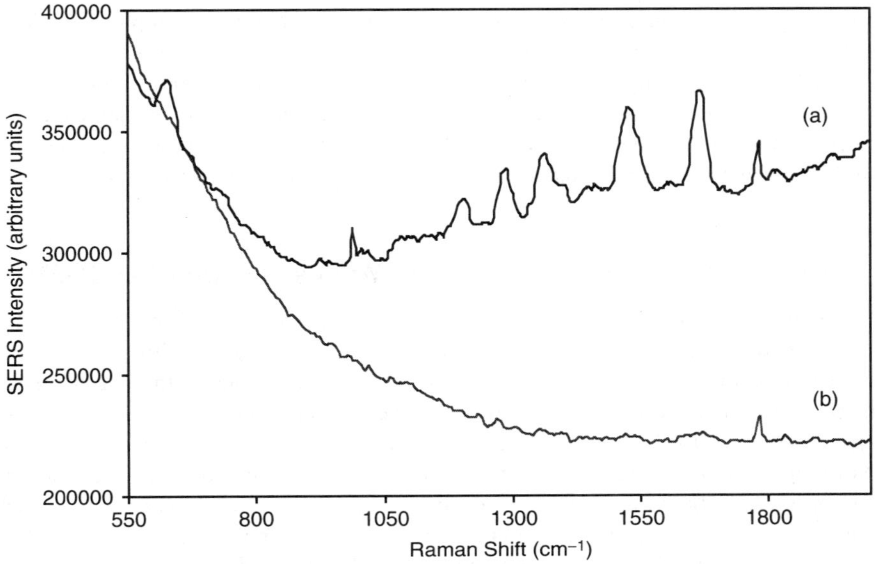


**FIGURE 64.14** Demonstration of the SERG probe technology for the detection of the oncogene BRCA1 via hybridization on a prefabricated SERS-active hybridization platform. (Adapted from Vo-Dinh, T. et al., *J. Raman Spectrosc.*, 33(7), 511, 2002.).

## 64.4.3 Fiber-Optic SERS Monitors for Bioanalysis

SERS-based fiber-optic probes have been developed for remote and *in situ* monitoring for a wide variety of applications.[99,229–231] The general interest in fiber-optic sensors has arisen from the development of SERS substrates that allow direct measurements in liquid samples. The direct measurement of potentially complex samples is afforded by the spectral selectivity of SERS, which can allow discrimination of multiple components detected simultaneously. Such SERS substrates have been incorporated into remote fiber-optic monitors.

The first SERS-based fiber-optic probes developed in our laboratory had a dual-fiber design in which the SERS-inducing substrate was a separate entity to which the fibers were optically coupled.[99,229,230] One optical fiber was used to transmit the laser excitation to the SERS substrate, and a second fiber was used to collect the scattered radiation from the sample.[99] The SERS substrate was a glass-backed, translucent medium based on silver-coated nanoparticles. With this substrate, the excitation and collection fibers could be configured with one of two geometries: either head on (with the fibers positioned on opposite sides of the SERS substrate) or side by side (with the two fibers on the glass-backed side of the substrate). The collection fiber was coupled to a spectrograph. Using a red-enhanced intensified CCD, we were able to detect trace levels of various analytes in milliseconds, even when using excitation and collection fibers as long as 20 m. Dual-fiber systems have also been the basis of several remote sensors based on normal Raman detection, but these systems typically have much lower sensitivity.[232–234]

More recently, several studies have reported use of SERS-based fiber-optic sensors with single-fiber designs in which the SERS-inducing medium is integrated on the tip of the optical fiber.[72,118,231,235,236] In an early report of integrated single-fiber SERS sensors, Mullen and Carron abrasively roughened the optical fiber before applying a metal layer directly to the tip.[118] More recently, we have applied a nanoparticle layer directly to an optical fiber to impart a uniform and reproducible roughness that is critical to the SERS effect. The fiber was simply dipped in an aqueous suspension of 0.1-μm alumina nanoparticles and then coated with a 100-nm layer of silver via vacuum evaporation.[235] Integrated single-fiber SERS sensors have also been fabricated through immobilization of silver-colloid particles on the tips of optical fibers pretreated with (3-aminopropyl)-trimethoxysilane.[72] The simplicity of these integrated single-fiber SERS sensors may enable nanoscale fabrication for more practical biomedical applications (e.g., probing single cells). While nanoscale fibers are not commercially available, it is possible to use a special, commercially available fiber-tapering device (Sutter Instrument Co., Model P-2000) to routinely taper fibers with core diameters as large as 600 μm down to tips of approximately 25 nm in diameter. Alternatively, a tapered tip can be formed more passively through etching with hydrofluoric acid. The tapered end can then be SERS-activated with a dependable SERS-inducing medium (e.g., silver islands or immobilized colloids).

## 64.4.4 SERS-Scanning Near-Field Optical Microscopy (SERS-SNOM)

SNOM has received considerable attention in recent years largely because of the exceptional resolution offered by this new technology. Indeed, SNOM routinely provides optical imaging with resolutions exceeding the diffraction limit.[237,238] Furthermore, SNOM has been coupled with molecular spectroscopy, most commonly through the use of luminescence. The combination of SNOM and SERS has been reported recently.[8,9,237,238] In this study, researchers used a planar silver-island-based SERS substrate to generate SERS signals and a chemically etched, 200-nm optical-fiber tip to deliver the excitation radiation from an argon ion laser (488 nm). The tapered sides of the fiber tip were coated with a thick, opaque layer of metal to limit escape of the excitation radiation to the 200-nm tip, hence permitting extremely localized sample excitation. This factor, combined with a substrate-to-fiber tip spacing of approximately 0.1 nm, enabled the acquisition of spectral and spatial information with subwavelength lateral resolution for CFV and rhodamine-6G molecules distributed on the silver-island substrate. Furthermore, the SERS-SNOM technique demonstrated exceptional sensitivity. Spectra from as few as 300 molecules have been recorded.

## 64.4.5 SERS Single-Molecule Detection

There have been several reports of single-molecule detection using SERS in recent years.[4–7,157,239] A critical factor in these milestone studies has been the development of exceptional SERS substrates. Most of the reports of single-molecule SERS detection have involved the use of metal colloids in suspensions. For example, Kneipp et al. demonstrated single-molecule detection of crystal violet and other dyes in both silver[5,6] and gold colloid suspensions.[7] In these studies, an effective cross section of approximately $10^{-16}$ $cm^2$ per molecule was observed, corresponding to a $10^{14}$ enhancement factor. The gold nanoparticles were commercially available but required proper agglomeration through the addition of NaCl. These results are promising because these dyes could be used as bioassay markers.

In some cases, bioassay markers can be resonance enhanced in addition to benefiting from SERS. For example, a DNA marker, 2,5,1′,3′,7′,9′-hexachloro-6-carboxyfluorescein (HEX) has been used to achieve single-molecule detection via SERRS in a silver-colloid suspension.[157] The HEX signature was observed for $8 \times 10^{-13}$ *M* DNA, which corresponded to less than 1 molecule per probed volume at any time required for measurement. Single-molecule detection has also been demonstrated on planar-surface-based substrates. For example, enhancements to factors of $10^{14}$ to $10^{15}$ have been observed for rhodamine-6G molecules adsorbed on silver colloids immobilized on a polylysine-coated glass surfaces.[4] Similarly, $10^{14}$ to $10^{15}$ factor enhancements have been reported for hemoglobin molecules adsorbed on silver nanoparticles immobilized on a polymer-coated silicon wafer.[239] Researchers in this study reported, however, that single-molecule detection was observed only for hemoglobin molecules situated between and adsorbed to two silver nanoparticles.

In our group, we have progressed toward the goal of single-molecule detection using silver-island-based SERS substrates.[35] We have demonstrated extremely low-level detection of CFV, which is a label we normally use for SERS gene probes. In addition to the giant SERS enhancement produced by the silver islands, a helium-neon laser enabled an additional resonance-enhancement factor at 632.8 nm. Furthermore, the detection system used a spectrograph equipped with a RE-ICCD. A confocal excitation/

collection geometry was implemented with a 100× objective lens. The 0.9 numerical aperture of this lens helped ensure a tight focus and efficient signal collection. Sample preparation for the low-level CFV detection simply involved spotting a 1-μl aliquot of $5 \times 10^{-11}$ M CFV solution (i.e., $3 \times 10^{7}$ molecules) on the silver-island SERS substrate and allowing it to dry after spreading it to a 6.5-mm diameter. Assuming a laser spot diameter of 10 μm and even distribution of CFV molecules within the 6.5-mm diameter spot, we estimated the number of probed molecules to be approximately 70. Nevertheless, signal-to-noise levels on this measurement indicated that this number could be reduced by at least a factor of three. Furthermore, we observed "hot spots" in the area of sample deposition, or micron-scale points of enhanced SERS signal. This phenomenon, along with an improved collection optics and longer signal-integration times, may allow for detection of single CFV molecules on silver-island-based solid substrates.

## 64.5 Conclusions

The development of dependable SERS substrates has spurred renewed interest in Raman scattering as a practical analytical tool in biomedical applications. Until such developments were achieved, there was little interest in Raman-scattering-based analysis because the cross section of normal Raman scattering is miniscule ($10^{-30}$ $cm^2$ per molecule) relative to the cross sections available through other molecular spectroscopies, particularly luminescence. In the past, high laser powers were used to compensate for this shortcoming. However, this procedure had limited effectiveness, largely because it induced photodecomposition on probed molecules, particularly at trace-level concentrations. By contrast, SERS-inducing media enable trace-level detection with relatively low laser powers.

Both liquid- and solid-based SERS-inducing media are now being used for the detection of biomedically significant compounds; moreover, innovative SERS-based biomedical applications are being developed. The development of such substrates and their use in practical analytical applications has required a triumph over the daunting challenge of producing nanoscale structures in reproducible and cost-effective ways. A significant portion of recent efforts has been devoted to investigating SERS coatings to improve selectivity, longevity, and ruggedness. As a result of this work, the scope of applications of SERS has been greatly expanded in the past two decades, evolving from environmental to biomedical applications. Some highlights of SERS-based applications have included single-molecule detection, SERS-based fiber-optic probes and nanoprobes, SERS-based bioassays, and hyphenated techniques such as scanning near-field microscopy (SNOM)/SERS and liquid chromatography (LC)/SERS. As we enter a new century, researchers continue to explore the potential of SERS, opening new horizons in the development of biophotonic probes.

## Acknowledgments

This work was jointly sponsored by the Federal Bureau of Investigation (Project No. 2051-II18-Y1) and the Office of Biological and Environmental Research, U.S. Department of Energy, under contract DE-AC05-00OR22725 with UT-Battelle, LLC, and by the Laboratory Directed Research and Development Program (Advanced Nanosystems Project) at Oak Ridge National Laboratory. David L. Stokes was also supported by an appointment to the Oak Ridge National Laboratory administered by the Oak Ridge Institute for Science and Education.

## References

1. Fleischmann, M., Hendra, P.J., and McQuilla, A.J., Raman-spectra of pyridine adsorbed at a silver electrode, *Chem. Phys. Lett.*, 26(2), 163, 1974.
2. Jeanmaire, D.L. and Van Duyne, R.P., Surface Raman spectroelectrochemistry. 1. Heterocyclic, aromatic, and aliphatic-amines adsorbed on anodized silver electrode, *J. Electroanal. Chem.*, 84(1), 1, 1977.

3. Albrecht, M.G. and Creighton, J.A., Anomalously intense Raman-spectra of pyridine at a silver electrode, *J. Am. Chem. Soc.*, 99(15), 5215, 1977.
4. Nie, S.M. and Emery, S.R., Probing single molecules and single nanoparticles by surface-enhanced Raman scattering, *Science*, 275(5303), 1102, 1997.
5. Kneipp, K., Wang, Y., Kneipp, H., Perelman, L.T., Itzkan, I., Dasari, R., and Feld, M.S., Single molecule detection using surface-enhanced Raman scattering (SERS), *Phys. Rev. Lett.*, 78(9), 1667, 1997.
6. Kneipp, K., Kneipp, H., Deinum, G., Itzkan, I., Dasari, R.R., and Feld, M.S., Single-molecule detection of a cyanine dye in silver colloidal solution using near-infrared surface-enhanced Raman scattering, *Appl. Spectrosc.*, 52(2), 175, 1998.
7. Kneipp, K., Kneipp, H., Manoharan, R., Hanlon, E.B., Itzkan, I., Dasari, R.R., and Feld, M.S., Extremely large enhancement factors in surface-enhanced Raman scattering for molecules on colloidal gold clusters, *Appl. Spectrosc.*, 52(12), 1493, 1998.
8. Deckert, V., Zeisel, D., Zenobi, R., and Vo-Dinh, T., Near-field surface enhanced Raman imaging of dye-labeled DNA with 100-nm resolution, *Anal. Chem.*, 70(13), 2646, 1998.
9. Zeisel, D., Deckert, V., Zenobi, R., and Vo-Dinh, T., Near-field surface-enhanced Raman spectroscopy of dye molecules adsorbed on silver island films, *Chem. Phys. Lett.*, 283(5–6), 381, 1998.
10. Moskovits, M., Surface-enhanced spectroscopy, *Rev. Mod. Phys.*, 57(3), 783, 1985.
11. Wokaun, A., Surface-enhanced electromagnetic processes, *Solid State Phys.–Adv. Res. Appl.*, 38, 223, 1984.
12. Schatz, G.C., Theoretical-studies of surface enhanced Raman-scattering, *Acc. Chem. Res.*, 17(10), 370, 1984.
13. Kerker, M., Electromagnetic model for surface-enhanced Raman-scattering (SERS) on metal colloids, *Acc. Chem. Res.*, 17(8), 271, 1984.
14. Chang, R.K. and Furtak, T.E., *Surface Enhanced Raman Scattering*, Plenum Press, New York, 1982.
15. Pockrand, I., *Surface-Enhanced Raman Vibrational Studies at Solid/Gas Interfaces*, Springer-Verlag, Berlin, 1984.
16. Pemberton, J.E. and Buck, R.P., Detection of low concentrations of a colored adsorbate at silver by surface-enhanced and resonance-enhanced Raman spectrometry, *Anal. Chem.*, 53(14), 2263, 1981.
17. Vo-Dinh, T., Hiromoto, M.Y.K., Begun, G.M., and Moody, R.L., Surface-enhanced Raman spectrometry for trace organic-analysis, *Anal. Chem.*, 56,(9), 1667, 1984.
18. Ni, F. and Cotton, T.M., Chemical procedure for preparing surface-enhanced Raman-scattering active silver films, *Anal. Chem.*, 58(14), 3159, 1986.
19. Sheng, R.S., Zhu, L., and Morris, M.D., Sedimentation classification of silver colloids for surface-enhanced Raman-scattering, *Anal. Chem.*, 58(6), 1116, 1986.
20. Torres, E.L. and Winefordner, J.D., Trace Determination of nitrogen-containing drugs by surface enhanced Raman-scattering spectrometry on silver colloids, *Anal. Chem.*, 59(13), 1626, 1987.
21. Laserna, J.J., Torres, E.L., and Winefordner, J.D., Studies of sample preparation for surface-enhanced Raman spectrometry on silver hydrosols, *Anal. Chim. Acta*, 200(1), 469, 1987.
22. Berthod, A., Laserna, J.J., and Winefordner, J.D., Surface enhanced Raman-spectrometry on silver hydrosols studied by flow-injection analysis, *Appl. Spectrosc.*, 41(7), 1137, 1987.
23. Tran, C.D., Subnanogram detection of dyes on filter-paper by surface-enhanced Raman-scattering spectrometry, *Anal. Chem.*, 56(4), 824, 1984.
24. Tran, C.D., *In situ* identification of paper chromatogram spots by surface enhanced Raman-scattering, *J. Chromatogr.*, 292(2), 432, 1984.
25. Ahern, A.M. and Garrell, R.L., *In situ* photoreduced silver-nitrate as a substrate for surface-enhanced Raman-spectroscopy, *Anal. Chem.*, 59(23), 2813, 1987.
26. Garrell, R.L., Surface-enhanced Raman-spectroscopy, *Anal. Chem.*, 61(6), A401, 1989.
27. Vo-Dinh, T., Surface-enhanced Raman spectrometry, in *Chemical Analysis of Polycyclic Aromatic Compounds*, Vo-Dinh, T., Ed., John Wiley & Sons, New York, 1989.

28. Pemberton, J.E., Surface-enhanced Raman scattering, in *In-situ Studies of Electrochemical Interfaces: A Prospectus*, Abruna, H.D., Ed., VCH Verlag Chemie, Berlin, 1991, pp. 193–265.
29. Cotton, T.M. and Brandt, E.S., Surface-enhanced Raman scattering, in *Physical Methods of Chemistry*, Rossiter, B.W. and Baetzold, R.C., Eds., John Wiley & Sons, New York, 1992.
30. Otto, A., Mrozek, I., Grabhorn, H., and Akemann, W., Surface-enhanced Raman-scattering, *J. Phys.-Condens. Matter*, 4(5), 1143, 1992.
31. Vo-Dinh, T., Surface-enhanced Raman spectroscopy, in *Photonic Probes of Surfaces*, Halevi, P., Ed., Elsevier, New York, 1995.
32. Ruperez, A. and Laserna, J.J., Surface-enhanced Raman spectroscopy, in *Modern Techniques in Raman Spectroscopy*, Laserna, J.J., Ed., John Wiley & Sons, New York, 1996, pp. 227–264.
33. Haller, K.L., Bumm, L.A., Altkorn, R.I., Zeman, E.J., Schatz, G.C., and Vanduyne, R.P., Spatially resolved surface enhanced 2nd harmonic-generation — theoretical and experimental evidence for electromagnetic enhancement in the near-infrared on a laser microfabricated pt surface, *J. Chem. Phys.*, 90(2), 1237, 1989.
34. Golab, J.T., Sprague, J.R., Carron, K.T., Schatz, G.C., and Vanduyne, R.P., A Surface enhanced hyper-Raman scattering study of pyridine adsorbed onto silver: experiment and theory, *J. Chem. Phys.*, 88(2), 7942, 1988.
35. Stokes, D.L., Hueber, D., and Vo-Dinh, T., Towards single-molecule detection with SERS using solid substrates, in Proceedings of Pittsburgh Conference, New Orleans, LA, March 1–5, abstract no. 683, 1998.
36. Stevenson, C.L. and Vo-Dinh, T., Signal expressions in Raman spectroscopy, in *Modern Techniques in Raman Spectroscopy*, Laserna, J.J., Ed., John Wiley & Sons, New York, 1996, pp. 1–39.
37. Kambhampati, P., Child, C.M., Foster, M.C., and Campion, A., On the chemical mechanism of surface enhanced Raman scattering: experiment and theory, *J. Chem. Phys.*, 108(12), 5013, 1998.
38. Pettinger, B., Wenning, U., and Wetzel, H., Surface-plasmon enhanced Raman-scattering frequency and angular resonance of Raman-scattered light from pyridine on Au, Ag and Cu electrodes, *Surf. Sci.*, 101(1–3), 409, 1980.
39. Loo, B.H., Surface-enhanced Raman-spectroscopy of platinum. 2. Enhanced light-scattering of chlorine adsorbed on platinum, *J. Phys. Chem.*, 87(16), 3003, 1983.
40. Fleischmann, M., Graves, P.R., and Robinson, J., The Raman-spectroscopy of the ferricyanide ferrocyanide system at gold, beta-palladium hydride and platinum-electrodes, *J. Electroanal. Chem.*, 182(1), 87, 1985.
41. Carrabba, M.M., Edmonds, R.B., and Rauh, R.D., Feasibility studies for the detection of organic-surface and subsurface water contaminants by surface-enhanced Raman-spectroscopy on silver electrodes, *Anal. Chem.*, 59(21), 2559, 1987.
42. Kudelski, A., Bukowska, J., Janik-Czachor, M., Grochala, W., Szummer, A., and Dolata, M., Characterization of the copper surface optimized for use as a substrate for surface-enhanced Raman scattering, *Vib. Spectrosc.*, 16(1), 21, 1998.
43. Barber, T.E., List, M.S., Haas, J.W., and Wachter, E.A., Determination of nicotine by surface-enhanced Raman-scattering (SERS), *Appl. Spectrosc.*, 48(11), 1423, 1994.
44. Li, W.H., Li, X.Y., and Yu, N.T., Surface-enhanced hyper-Raman spectroscopy (SEHRS) and surface-enhanced Raman spectroscopy (SERS) studies of pyrazine and pyridine adsorbed on silver electrodes, *Chem. Phys. Lett.*, 305(3–4), 303, 1999.
45. Taranenko, N., Alarie, J.P., Stokes, D.L., and VoDinh, T., Surface-enhanced Raman detection of nerve agent simulant (DMMP and DIMP) vapor on electrochemically prepared silver oxide substrates, *J. Raman Spectrosc.*, 27(5), 379, 1996.
46. Haigh, J.A., Hendra, P.J., and Forsling, W., Extension of a novel method for the examination of oxidized aluminum surfaces using near-infrared (NIR) Fourier-transform surface-enhanced Raman-spectroscopy (FT-SERS), *Spectrochim. Acta Part A Mol. Biomol. Spectrosc.*, 50(11), 2027, 1994.

47. Klug, O., Szaraz, I., Forsling, W., and Ranheimer, M., A novel method for investigation of dibasic aromatic acids on oxidized aluminium surfaces by NIR-FT-SERS, *Mikrochim. Acta*(Suppl. 14), 649, 1997.
48. Freeman, R.D., Hammaker, R.M., Meloan, C.E., and Fateley, W.G., A detector for liquid-chromatography and flow-injection analysis using surface-enhanced Raman-spectroscopy, *Appl. Spectrosc.*, 42*(*3), 456, 1988.
49. Ni, F., Sheng, R.S., and Cotton, T.M., Flow-injection analysis and real-time detection of RNA bases by surface-enhanced Raman-spectroscopy, *Anal. Chem.*, 62*(*18), 1958, 1990.
50. Sequaris, J.M.L. and Koglin, E., Direct analysis of high-performance thin-layer chromatography spots of nucleic purine derivatives by surface-enhanced Raman-scattering spectrometry, *Anal. Chem.*, 59*(*3), 525, 1987.
51. Munro, C.H., Smith, W.E., Garner, M., Clarkson, J., and White, P.C., Characterization of the surface of a citrate-reduced colloid optimized for use as a substrate for surface-enhanced resonance Raman-scattering, *Langmuir*, 11*(*10), 3712, 1995.
52. Tarabara, V.V., Nabiev, I.R., and Feofanov, A.V., Surface-enhanced Raman scattering (SERS) study of mercaptoethanol monolayer assemblies on silver citrate hydrosol. Preparation and characterization of modified hydrosol as a SERS-active substrate, *Langmuir*, 14(5), 1092, 1998.
53. Li, Y.S., Cheng, J.C., and Coons, L.B., A silver solution for surface-enhanced Raman scattering, *Spectrochim. Acta Part A Mol. Biomol. Spectrosc.*, 55(6), 1197, 1999.
54. Prochazka, M., Mojzes, P., Stepanek, J., Vlckova, B., and Turpin, P.Y., Probing applications of laser ablated Ag colloids in SERS spectroscopy: improvement of ablation procedure and SERS spectral testing, *Anal. Chem.*, 69*(*24), 5103, 1997.
55. Lee, P.C. and Meisel, D., Adsorption and surface-enhanced Raman of dyes on silver and gold sols, *J. Phys. Chem.*, 86*(*17), 3391, 1982.
56. Hildebrandt, P. and Stockburger, M., Surface-enhanced resonance Raman spectroscopy of rhodamine-6g adsorbed on colloidal silver, *J. Phys. Chem.*, 88*(*24), 5935, 1984.
57. Bright, R.M., Walter, D.G., Musick, M.D., Jackson, M.A., Allison, K.J., and Natan, M.J., Chemical and electrochemical Ag deposition onto preformed Au colloid monolayers: approaches to uniformly-sized surface features with Ag-like optical properties, *Langmuir*, 12*(*3), 810, 1996.
58. Li, Y.S., Surface-enhanced Raman-scattering at colloidal silver-oxide surfaces, *J. Raman Spectrosc.*, 25*(*10), 795, 1994.
59. Emory, S.R. and Nie, S., Screening and enrichment of metal nanoparticles with novel optical properties, *J. Phys. Chem.*, *B*, 102*(*3), 493, 1998.
60. Vo-Dinh, T., Alak, A., and Moody, R.L., Recent advances in surface-enhanced Raman spectrometry for chemical-analysis, *Spectrochim. Acta Part B At. Spectrosc.*, 43*(*4–5), 605, 1988.
61. Siiman, O., Bumm, L.A., Callaghan, R., Blatchford, C.G., and Kerker, M., Surface-enhanced Raman-scattering by citrate on colloidal silver, *J. Phys. Chem.*, 87*(*6), 1014, 1983.
62. Heard, S.M., Grieser, F., and Barraclough, C.G., Surface-enhanced Raman-scattering from amphiphilic and polymer molecules on silver and gold sols, *Chem. Phys. Lett.*, 95, 2, 154, 1983.
63. Lee, P.C. and Meisel, D., Surface-enhanced Raman-scattering of colloid stabilizer systems, *Chem. Phys. Lett.*, 99*(*3), 262, 1983.
64. Ayora, M.J., Ballesteros, L., Perez, R., Ruperez, A., and Laserna, J.J., Detection of atmospheric contaminants in aerosols by surface-enhanced Raman spectrometry, *Anal. Chim. Acta*, 355*(*1), 15, 1997.
65. Freeman, R.G., Grabar, K.C., Allison, K.J., Bright, R.M., Davis, J.A., Guthrie, A.P., Hommer, M.B., Jackson, M.A., Smith, P.C., Walter, D.G., and Natan, M.J., Self-assembled metal colloid monolayers: an approach to SERS substrates, *Science*, 267*(*5204), 1629, 1995.
66. Wang, J., Zhu, T., Fu, X.Y., and Liu, Z.F., Study of chemical enhancement in SERS from Au nanoparticles assembly, *Acta Phys.-Chim. Sinica*, 14*(*6), 485, 1998.
67. Fu, X.Y., Mu, T., Wang, J., Zhu, T., and Liu, Z.F., pH-Dependent assembling of gold nanoparticles on *p*-aminothiophenol modified gold substrate, *Acta Phys.-Chim. Sinica*, 14*(*11), 968, 1998.

68. Zhu, T., Zhang, X., Wang, J., Fu, X.Y., and Liu, Z.F., Assembling colloidal Au nanoparticles with functionalized self-assembled monolayers, *Thin Solid Films*, 329, 595, 1998.
69. Grabar, K.C., Smith, P.C., Musick, M.D., Davis, J.A., Walter, D.G., Jackson, M.A., Guthrie, A.P., and Natan, M.J., Kinetic control of interparticle spacing in Au colloid-based surfaces: rational nanometer-scale architecture, *J. Am. Chem. Soc.*, 118(5), 1148, 1996.
70. Park, S.H., Im, J.H., Im, J.W., Chun, B.H., and Kim, J.H., Adsorption kinetics of Au and Ag nanoparticles on functionalized glass surfaces, *Microchem. J.*, 63(1), 71, 1999.
71. He, H.X., Zhang, H., Li, Q.G., Zhu, T., Li, S.F.Y., and Liu, Z.F., Fabrication of designed architectures of Au nanoparticles on solid substrate with printed self-assembled monolayers as templates, *Langmuir*, 16(8), 3846, 2000.
72. Polwart, E., Keir, R.L., Davidson, C.M., Smith, W.E., and Sadler, D.A., Novel SERS-active optical fibers prepared by the immobilization of silver colloidal particles, *Appl. Spectrosc.*, 54(4), 522, 2000.
73. Vanduyne, R.P., Hulteen, J.C., and Treichel, D.A., Atomic-force microscopy and surface-enhanced Raman-spectroscopy. 1. Ag island films and Ag film over polymer nanosphere surfaces supported on glass, *J. Chem. Phys.*, 99(3), 2101, 1993.
74. Semin, D.J. and Rowlen, K.L., Influence of vapor-deposition parameters on SERS active Ag film morphology and optical-properties, *Anal. Chem.*, 66(23), 4324, 1994.
75. Stockle, R.M., Deckert, V., Fokas, C., Zeisel, D., and Zenobi, R., Sub-wavelength Raman spectroscopy on isolated silver islands, *Vib. Spectrosc.*, 22(1–2), 39, 2000.
76. Roark, S.E. and Rowlen, K.L., Thin Ag films — influence of substrate and postdeposition treatment on morphology and optical-properties, *Anal. Chem.*, 66(2), 261, 1994.
77. Roark, S.E., Semin, D.J., Lo, A., Skodje, R.T., and Rowlen, K.L., Solvent-induced morphology changes in thin silver films, *Anal. Chim. Acta*, 307(2–3), 341, 1995.
78. Mosier-Boss, P.A. and Lieberman, S.H., Comparison of three methods to improve adherence of thin gold films to glass substrates and their effect on the SERS response, *Appl. Spectrosc.*, 53(7), 862, 1999.
79. Mrozek, I. and Otto, A., Quantitative separation of the classical electromagnetic and the chemical contribution to surface enhanced Raman-scattering, *J. Electron Spectrosc. Relat. Phenomena*, 54, 895, 1990.
80. Jennings, C., Aroca, R., Hor, A.M., and Loutfy, R.O., Surface-enhanced Raman-scattering from copper and zinc phthalocyanine complexes by silver and indium island films, *Anal. Chem.*, 56(12), 2033, 1984.
81. Aroca, R. and Martin, F., Trace analysis of tetrasulfonated copper phthalocyanine by surface enhanced Raman-spectroscopy, *J. Raman Spectrosc.*, 17(3), 243, 1986.
82. Jennings, C.A., Kovacs, G.J., and Aroca, R., Fourier-transform surface-enhanced Raman-scattering of langmuir-blodgett monolayers on copper and gold island substrates, *Langmuir*, 9(8), 2151, 1993.
83. Meier, M., Wokaun, A., and Vo-Dinh, T., Silver particles on stochastic quartz substrates providing tenfold increase in Raman enhancement, *J. Phys. Chem.*, 89(10), 1843, 1985.
84. Vodinh, T., Meier, M., and Wokaun, A., Surface-enhanced Raman spectrometry with silver particles on stochastic-post substrates, *Anal. Chim. Acta*, 181, 139, 1986.
85. Goudonnet, J.P., Begun, G.M., and Arakawa, E.T., Surface-enhanced Raman-scattering on silver-coated teflon sphere substrates, *Chem. Phys. Lett.*, 92(2), 197, 1982.
86. Alak, A.M. and Vo-Dinh, T., Silver-coated fumed silica as a substrate material for surface-enhanced Raman-scattering, *Anal. Chem.*, 61(7), 656, 1989.
87. Moody, R.L., Vo-Dinh, T., and Fletcher, W.H., Investigation of experimental parameters for surface-enhanced Raman-scattering (SERS) using silver-coated microsphere substrates, *Appl. Spectrosc.*, 41(6), 966, 1987.
88. Alak, A.M. and Vo-Dinh, T., Surface-enhanced Raman spectrometry of chlorinated pesticides, *Anal. Chim. Acta*, 206(1–2), 333, 1988.
89. Vodinh, T., Miller, G.H., Bello, J., Johnson, R., Moody, R.L., Alak, A., and Fletcher, W.R., Surface-active substrates for Raman and luminescence analysis, *Talanta*, 36(1–2), 227, 1989.

90. Alak, A.M. and Vo-Dinh, T., Surface-enhanced Raman-spectrometry of organophosphorus chemical agents, *Anal. Chem.*, 59(17), 2149, 1987.
91. Vodinh, T., Uziel, M., and Morrison, A.L., Surface-enhanced Raman analysis of benzo *a* pyrene-DNA adducts on silver-coated cellulose substrates, *Appl. Spectrosc.*, 41(4), 605, 1987.
92. Helmenstine, A.M., Li, Y.S., and Vo-Dinh, T., Surface-enhanced Raman-scattering analysis of etheno adducts of adenine, *Vib. Spectrosc.*, 4(3), 359, 1993.
93. Helmenstine, A., Uziel, M., and Vo-Dinh, T., Measurement of DNA-adducts using surface-enhanced Raman spectroscopy, *J. Toxicol. Environ. Health*, 40(2–3), 195, 1993.
94. Bello, J.M., Stokes, D.L., and Vo-Dinh, T., Silver-coated alumina as a new medium for surface-enhanced Raman-scattering analysis, *Appl. Spectrosc.*, 43, 8, 1325, 1989.
95. Bello, J.M., Stokes, D.L., and Vo-Dinh, T., Titanium-dioxide based substrate for optical monitors in surface-enhanced Raman-scattering analysis, *Anal. Chem.*, 61(15), 1779, 1989.
96. Li, Y.S., Vo-Dinh, T., Stokes, D.L., and Yu, W., Surface-enhanced Raman analysis of p-nitroaniline on vacuum evaporation and chemically deposited silver-coated alumina substrates, *Appl. Spectrosc.*, 46(9), 1354, 1992.
97. Li, Y.S. and Wang, Y., Chemically prepared silver alumina substrate for surface-enhanced Raman-scattering, *Appl. Spectrosc.*, 46(1), 142, 1992.
98. Vo-Dinh, T. and Stokes, D.L., Surface-enhanced Raman vapor dosimeter, *Appl. Spectrosc.*, 47(10), 1728, 1993.
99. Alarie, J.P., Stokes, D.L., Sutherland, W.S., Edwards, A.C., and Vo-Dinh, T., Intensified charge coupled device-based fiberoptic monitor for rapid remote surface-enhanced Raman-scattering sensing, *Appl. Spectrosc.*, 46(11), 1608, 1992.
100. Narayanan, V.A., Begun, G.M., Bello, J.M., Stokes, D.L., and Vo-Dinh, T., Analysis of the plant-growth regulator alar (daminozide) and its hydrolysis products using Raman-spectroscopy, *Analysis*, 21(2), 107, 1993.
101. Narayanan, V.A., Begun, G.M., Stump, N.A., Stokes, D.L., and Vo-Dinh, T., Vibrational-spectra of fluvalinate, *J. Raman Spectrosc.*, 24(3), 123, 1993.
102. Narayanan, V.A., Stokes, D.L., and Tuan, V.D., Vibrational spectral-analysis of eosin-Y and erythrosin-B — intensity studies for quantitative detection of the dyes, *J. Raman Spectrosc.*, 25(6), 415, 1994.
103. Vo-Dinh, T., Houck, K., and Stokes, D.L., Surface-enhanced Raman gene probes, *Anal. Chem.*, 66(20), 3379, 1994.
104. Vo-Dinh, T. and Stokes, D.L., Surface-enhanced Raman detection of chemical vapors with the use of personal dosimeters, *Field Anal. Chem. Technol.*, 3(6), 346, 1999.
105. Ibrahim, A., Oldham, P.B., Stokes, D.L., and Vo-Dinh, T., Determination of enhancement factors for surface-enhanced FT- Raman spectroscopy on gold and silver surfaces, *J. Raman Spectrosc.*, 27(12), 887, 1996.
106. Wachter, E.A., Storey, J.M.E., Sharp, S.L., Carron, K.T., and Jiang, Y., Hybrid substrates for real-time SERS-based chemical sensors, *Appl. Spectrosc.*, 49(2), 193, 1995.
107. Liao, P.F., Silver structures produced by microlithography, in *Surface Enhanced Raman Scattering*, Chang, R.K. and Furtak, T.E., Eds., Plenum Press, New York, 1982, pp. 379–390.
108. Enlow, P.D., Buncick, M., Warmack, R.J., and Vo-Dinh, T., Detection of nitro polynuclear aromatic-compounds by surface-enhanced Raman-spectrometry, *Anal. Chem.*, 58(6), 1119, 1986.
109. Xue, G., Dong, J., and Zhang, M.S., Surface-enhanced Raman-scattering (SERS) and surface-enhanced resonance Raman-scattering (SERRS) on $HNO_3$-roughened copper foil, *Appl. Spectrosc.*, 45(5), 756, 1991.
110. Miller, S.K., Baiker, A., Meier, M., and Wokaun, A., Surface-enhanced Raman-scattering and the preparation of copper substrates for catalytic studies, *J. Chem. Soc. Faraday Trans. I*, 80, 1305, 1984.
111. Ruperez, A. and Laserna, J.J., Surface-enhanced Raman-spectrometry on a silver substrate prepared by the nitric-acid etching method, *Anal. Chim. Acta*, 291(1–2), 147, 1994.
112. Ruperez, A. and Laserna, J.J., Surface-enhanced Raman sensor, *Analysis*, 23(2), 91, 1995.

113. Ruperez, A. and Laserna, J.J., Surface-enhanced Raman spectrometry of triamterene on a silver substrate prepared by the nitric acid etching method, *Talanta*, 44(2), 213, 1997.
114. Lu, Y., Xue, G., and Dong, J., $HNO_3$ etched silver foil as an effective substrate for surface-enhanced Raman-scattering (SERS) analysis, *Appl. Surf. Sci.*, 68(4), 485, 1993.
115. Norrod, K.L., Sudnik, L.M., Rousell, D., and Rowlen, K.L., Quantitative comparison of five SERS substrates: sensitivity and limit of detection, *Appl. Spectrosc.*, 51(7), 994, 1997.
116. Cao, Y.H. and Li, Y.S., Constructing surface roughness of silver for surface-enhanced Raman scattering by self-assembled monolayers and selective etching process, *Appl. Spectrosc.*, 53(5), 540, 1999.
117. Szabo, N.J. and Winefordner, J.D., Surface enhanced Raman scattering from an etched polymer substrate, *Anal. Chem.*, 69(13), 2418, 1997.
118. Mullen, K.I. and Carron, K.T., Surface-enhanced Raman-spectroscopy with abrasively modified fiber optic probes, *Anal. Chem.*, 63(19), 2196, 1991.
119. Sutherland, W.S. and Winefordner, J.D., Preparation of substrates for surface-enhanced Raman microprobe spectroscopy, *J. Raman Spectrosc.*, 22(10), 541, 1991.
120. Xue, G., Lu, Y., and Zhang, J.F., Stable SERS substrates used for *in situ* studies of the polymer-metal interface at elevated temperature, *Macromolecules*, 27(3), 809, 1994.
121. Szabo, N.J. and Winefordner, J.D., Evaluation of two commercially available TLC materials as SER substrates, *Appl. Spectrosc.*, 51(7), 965, 1997.
122. Szabo, N.J. and Winefordner, J.D., Evaluation of a solid-phase extraction membrane as a surface-enhanced Raman substrate, *Appl. Spectrosc.*, 52(4), 500, 1998.
123. Horvath, E., Mink, J., and Kristof, J., Surface-enhanced Raman spectroscopy as a technique for drug analysis, *Mikrochim. Acta*, 745, 1997.
124. Gliemann, H., Nickel, U., and Schneider, S., Application of photographic paper as a substrate for surface-enhanced Raman spectroscopy, *J. Raman Spectrosc.*, 29(12), 1041, 1998.
125. Volkan, M., Stokes, D.L., and Vo-Dinh, T., A new surface-enhanced Raman scattering substrate based on silver nanoparticles in sol-gel, *J. Raman Spectrosc.*, 30(2), 1057, 1999.
126. Volkan, M., Stokes, D.L., and Vo-Dinh, T., Surface-enhanced Raman of dopamine and neurotransmitters using sol-gel substrates and polymer-coated fiber-optic probes, *Appl. Spectrosc.*, 54(12), 1842, 2000.
127. Murphy, T., Schmidt, H., and Kronfeldt, H.D., Use of sol-gel techniques in the development of surface- enhanced Raman scattering (SERS) substrates suitable for *in situ* detection of chemicals in sea-water, *Appl. Phys. B-Lasers Opt.*, 69(2), 147, 1999.
128. Lee, Y.H., Dai, S., and Young, J.P., Silver-doped sol-gel films as the substrate for surface-enhanced Raman scattering, *J. Raman Spectrosc.*, 28(8), 635, 1997.
129. Wang, K. and Li, Y.S., Silver doping of polycarbonate films for surface-enhanced Raman scattering, *Vib. Spectrosc.*, 14(2), 183, 1997.
130. Akbarian, F., Dunn, B.S., and Zink, J.I., Surface-enhanced Raman-spectroscopy using photodeposited gold particles in porous sol-gel silicates, *J. Phys. Chem.*, 99(12), 3892, 1995.
131. Imai, Y., Tamai, Y., and Kurokawa, Y., Surface-enhanced Raman scattering of benzoic and thiosalicylic acids adsorbed on fine Ag particle-impregnated cellulose gel films, *J. Sol-Gel Sci. Technol.*, 11(3), 273, 1998.
132. Imai, Y., Kurokawa, Y., Hara, M., and Fukushima, M., Observation of SERS of picolinic acid and nicotinic acid using cellulose acetate films doped with Ag fine particles, *Spectrochim. Acta Part A Mol. Biomol. Spectrosc.*, 53(11), 1697, 1997.
133. Pal, A., Stokes, D.L., Alarie, J.P., and Vo-Dinh, T., Selective surface-enhanced Raman-spectroscopy using a polymer-coated substrate, *Anal. Chem.*, 67, 18, 3154, 1995.
134. Stokes, D.L., Pal, A., Narayanan, V.A., and Vo-Dinh, T., Evaluation of a chemical vapor dosimeter using polymer-coated SERS substrates, *Anal. Chim. Acta*, 399(3), 265, 1999.

135. Carron, K.T., Lewis, M.L., Dong, J.A., Ding, J.F., Xue, G., and Chen, Y., Surface-enhanced Raman-scattering and cyclic voltammetry studies of synergetic effects in the corrosion inhibition of copper by polybenzimidazole and mercaptobenzimidazole at high-temperature, *J. Mater. Sci.*, 28(15), 4099, 1993.
136. Deschaines, T.O. and Carron, K.T., Stability and surface uniformity of selected thiol-coated SERS surfaces, *Appl. Spectrosc.*, 51(9), 1355, 1997.
137. Carron, K.T. and Kennedy, B.J., Molecular-specific chromatographic detector using modified SERS substrates, *Anal. Chem.*, 67(18), 3353, 1995.
138. Crane, L.G., Wang, D.X., Sears, L.M., Heyns, B., and Carron, K., SERS surfaces modified with a 4-(2-pyridylazo)resorcinol disulfide derivative — detection of copper, lead, and cadmium, *Anal. Chem.*, 67(2), 360, 1995.
139. Sulk, R., Chan, C., Guicheteau, J., Gomez, C., Heyns, J.B.B., Corcoran, R., and Carron, K., Surface-enhanced Raman assays (SERA): measurement of bilirubin and salicylate, *J. Raman Spectrosc.*, 30(9), 853, 1999.
140. Sulk, R.A., Corcoran, R.C., and Carron, K.T., Surface enhanced Raman scattering detection of amphetamine and methamphetamine by modification with 2-mercaptonicotinic acid, *Appl. Spectrosc.*, 53(8), 954, 1999.
141. Zou, S.Z. and Weaver, M.J., Surface-enhanced Raman scattering on uniform transition metal films: toward a versatile adsorbate vibrational strategy for solid-nonvacuum interfaces? *Anal. Chem.*, 70(11), 2387, 1998.
142. Wilke, T., Gao, X.P., Takoudis, C.G., and Weaver, M.J., Surface-enhanced Raman-spectroscopy as a probe of adsorption at transition metal-high-pressure gas interfaces — No, Co, and oxygen on platinum-coated gold, rhodium-coated gold, and ruthenium-coated gold, *Langmuir*, 7(4), 714, 1991.
143. Tarcha, P.J., DeSaja-Gonzalez, J., Rodriguez-Llorente, S., and Aroca, R., Surface-enhanced fluorescence on $SiO_2$-coated silver island films, *Appl. Spectrosc.*, 53(1), 43, 1999.
144. Lacy, W.B., Olson, L.G., and Harris, J.M., Quantitative SERS measurements on dielectric-overcoated silver-island films by solution deposition control of surface concentrations, *Anal. Chem.*, 71(13), 2564, 1999.
145. Lacy, W.B., Williams, J.M., Wenzler, L.A., Beebe, T.P., and Harris, J.M., Characterization of $SiO_2$-overcoated silver-island films as substrates for surface-enhanced Raman scattering, *Anal. Chem.*, 68(6), 1003, 1996.
146. Lacy, W.B., Williams, J.M., Wenzler, L.A., Beebe, T.P., and Harris, J.M., Spectroscopic characterization and evaluation of $SiO_2$-overcoated silver islands as a SERS Substrate, *Abstr. Pap. Am. Chem. Soc.*, 209, 50, 1995.
147. Lacy, W.B. and Harris, J.M., Monitoring adsorption-kinetics at oxide surfaces using SiO2-overcoated silver as a SERS substrate, *Abstr. Pap. Am. Chem. Soc.*, 208, 41, 1994.
148. Yang, X.M., Tryk, D.A., Ajito, K., Hashimoto, K., and Fujishima, A., Surface-enhanced Raman scattering imaging of photopatterned self-assembled monolayers, *Langmuir*, 12(23), 5525, 1996.
149. Zhu, T., Yu, H.Z., Wang, J., Wang, Y.Q., Cai, S.M., and Liu, Z.F., Two-dimensional surface enhanced Raman mapping of differently prepared gold substrates with an azobenzene self-assembled monolayer, *Chem. Phys. Lett.*, 265(3–5), 334, 1997.
150. Maeda, Y., Yamamoto, H., and Kitano, H., Self-assembled monolayers as novel biomembrane mimetics. 1. Characterization of cytochrome-*c* bound to self-assembled monolayers on silver by surface-enhanced resonance Raman-spectroscopy, *J. Phys. Chem.*, 99(13), 4837, 1995.
151. Vo-Dinh, T., Allain, L.R., and Stokes, D.L., Cancer gene detection using surface-enhanced Raman scattering (SERS), *J. Raman Spectrosc.*, 33(7), 511, 2002.
152. Michota, A., Kudelski, A., and Bukowska, J., Chemisorption of cysteamine on silver studied by surface-enhanced Raman scattering, *Langmuir*, 16(26), 10236, 2000.

153. Michota, A., Kudelski, A., and Bukowska, J., Influence of electrolytes on the structure of cysteamine monolayer on silver studied by surface-enhanced Raman scattering, *J. Raman Spectrosc.*, 32(5), 345, 2001.
154. Vo-Dinh, T., Stokes, D.L., Griffin, G.D., Volkan, M., Kim, U.J., and Simon, M.I., Surface-enhanced Raman scattering (SERS) method and instrumentation for genomics and biomedical analysis, *J. Raman Spectrosc.*, 30(9), 785, 1999.
155. Isola, N.R., Stokes, D.L., and Vo-Dinh, T., Surface enhanced Raman gene probe for HIV detection, *Anal. Chem.*, 70(7), 1352, 1998.
156. Brown, R., Smith, W.E., and Graham, D., Synthesis of a benzotriazole phosphoramidite for attachment of oligonucleotides to metal surfaces, *Tetrahedron Lett.*, 42(11), 2197, 2001.
157. Graham, D., Smith, W.E., Linacre, A.M.T., Munro, C.H., Watson, N.D., and White, P.C., Selective detection of deoxyribonucleic acid at ultralow concentrations by SERRS, *Anal. Chem.*, 69(22), 4703, 1997.
158. Dou, X., Yamaguchi, Y., Yamamoto, H., Doi, S., and Ozaki, Y., NIR SERS detection of immune reaction on gold colloid particles without bound/free antigen separation, *J. Raman Spectrosc.*, 29(8), 739, 1998.
159. Ni, J., Lipert, R.J., Dawson, G.B., and Porter, M.D., Immunoassay readout method using extrinsic Raman labels adsorbed on immunogold colloids, *Anal. Chem.*, 71(21), 4903, 1999.
160. Nabiev, I. and Manfait, M., Industrial applications of the surface-enhanced Raman-spectroscopy, *Rev. Inst. Fr. Petrole*, 48(3), 261, 1993.
161. Nabiev, I., Chourpa, I., and Manfait, M., Applications of Raman and surface-enhanced Raman-scattering spectroscopy in medicine, *J. Raman Spectrosc.*, 25(1), 13, 1994.
162. Kneipp, K., Kneipp, H., Itzkan, I., Dasari, R.R., and Feld, M.S., Surface-enhanced Raman scattering: a new tool for biomedical spectroscopy, *Curr. Sci.*, 77(7), 915, 1999.
163. Koglin, E. and Sequaris, J.M., Surface enhanced Raman-scattering of biomolecules, *Top. Curr. Chem.*, 134, 1, 1986.
164. Carter, J.C., Brewer, W.E., and Angel, S.M., Raman spectroscopy for the in situ identification of cocaine and selected adulterants, *Appl. Spectrosc.*, 54(12), 1876, 2000.
165. Angel, S.M., Carter, J.C., Stratis, D.N., Marquardt, B.J., and Brewer, W.E., Some new uses for filtered fiber-optic Raman probes: *in situ* drug identification and *in situ* and remote Raman imaging, *J. Raman Spectrosc.*, 30(9), 795, 1999.
166. Perez, R., Ruperez, A., and Laserna, J.J., Evaluation of silver substrates for surface-enhanced Raman detection of drugs banned in sport practices, *Anal. Chim. Acta*, 376(2), 255, 1998.
167. Cabalin, L.M., Ruperez, A., and Laserna, J.J., Surface-enhanced Raman-spectrometry for detection in liquid- chromatography using a windowless flow cell, *Talanta*, 40(11), 1741, 1993.
168. Ruperez, A., Montes, R., and Laserna, J.J., Identification of stimulant-drugs by surface-enhanced Raman-spectrometry on colloidal silver, *Vib. Spectrosc.*, 2(2–3), 145, 1991.
169. Hawi, S.R., Rochanakij, S., Adar, F., Campbell, W.B., and Nithipatikom, K., Detection of membrane-bound enzymes in cells using immunoassay and Raman microspectroscopy, *Anal. Biochem.*, 259(2), 212, 1998.
170. Dou, X., Takama, T., Yamaguchi, Y., Yamamoto, H., and Ozaki, Y., Enzyme immunoassay utilizing surface-enhanced Raman scattering of the enzyme reaction product, *Anal. Chem.*, 69(8), 1492, 1997.
171. Rohr, T.E., Cotton, T., Fan, N., and Tarcha, P.J., Immunoassay employing surface-enhanced Raman-spectroscopy, *Anal. Biochem.*, 182(2), 388, 1989.
172. Rivas, L., Sanchez-Cortes, S., Stanicova, J., Garcia-Ramos, J.V., and Miskovsky, P., FT-Raman, FTIR and surface-enhanced Raman spectroscopy of the antiviral and antiparkinsonian drug amantadine, *Vib. Spectrosc.*, 20(2), 179, 1999.
173. Rivas, L., Murza, A., Sanchez-Cortes, S., and Garcia-Ramos, J.V., Adsorption of acridine drugs on silver: surface-enhanced resonance Raman evidence of the existence of different adsorption sites, *Vib. Spectrosc.*, 25(1), 19, 2001.

174. Wang, Y., Li, Y.S., Wu, J., Zhang, Z.X., and An, D.Q., Surface-enhanced Raman spectra of some anti-tubercle bacillus drugs, *Spectrochim. Acta Part A Mol. Biomol. Spectrosc.*, 56(14), 2637, 2000.
175. Vivoni, A., Chen, S.P., Ejeh, D., and Hosten, C.M., Determination of the orientation of 6-mercaptopurine adsorbed on a silver electrode by surface-enhanced Raman spectroscopy and normal mode calculations, *Langmuir*, 16(7), 3310, 2000.
176. Cinta, S., Iliescu, T., Astilean, S., David, L., Cozar, O., and Kiefer, W., 1,4-benzodiazepine drugs adsorption on the Ag colloidal surface, *J. Mol. Struct.*, 483, 685, 1999.
177. Ridente, Y., Aubard, J., and Bolard, J., Absence in amphotericin B-spiked human plasma of the free monomeric drug, as detected by SERS, *FEBS Lett.*, 446(2–3), 283, 1999.
178. Sockalingum, G.D., Beljebbar, A., Morjani, H., Angiboust, J.F., and Manfait, M., Characterization of island films as surface-enhanced Raman spectroscopy substrates for detecting low antitumor drug concentrations at single cell level, *Biospectroscopy*, 4(5), S71, 1998.
179. Murza, A., Sanchez-Cortes, S., and Garcia-Ramos, J.V., Fluorescence and surface-enhanced Raman study of 9- aminoacridine in relation to its aggregation and excimer emission in aqueous solution and on silver surface, *Biospectroscopy*, 4(5), 327, 1998.
180. Chourpa, I., Beljebbar, A., Sockalingum, G.D., Riou, J.F., and Manfait, M., Structure-activity relation in camptothecin antitumor drugs: Why a detailed molecular characterisation of their lactone and carboxylate forms by Raman and SERS spectroscopies? *Biochim. Biophys. Acta Gen. Subjects*, 1334(2–3), 349, 1997.
181. Sanchez-Cortes, S., Jancura, D., Miskovsky, P., and Bertoluzza, A., Near infrared surface-enhanced Raman spectroscopic study of antiretroviraly drugs hypericin and emodin in aqueous silver colloids, *Spectrochim. Acta Part A Mol. Biomol. Spectrosc.*, 53(5), 769, 1997.
182. Ruperez, A. and Laserna, J.J., Surface-enhanced Raman spectrometry of chiral beta-blocker drugs on colloidal silver, *Anal. Chim. Acta*, 335, 1–2, 87, 1996.
183. Lecomte, S., Moreau, N.J., Manfait, M., Aubard, J., and Baron, M.H., Surface-enhanced Raman spectroscopy investigation of fluoroquinolone DNA/DNA gyrase $Mg^{2+}$ interactions. 1. Adsorption of pefloxacin on colloidal silver — effect of drug concentration, electrolytes, and pH, *Biospectroscopy*, 1(6), 423, 1995.
184. Nabiev, I., Baranov, A., Chourpa, I., Beljebbar, A., Sockalingum, G.D., and Manfait, M., Does adsorption on the surface of a silver colloid perturb drug-DNA interactions — comparative SERS, FT-SERS, and resonance Raman-study of mitoxantrone and its derivatives, *J. Phys. Chem.*, 99(5), 1608, 1995.
185. Calvo, N., Montes, R., and Laserna, J.J., Surface-enhanced Raman-spectrometry of amiloride on colloidal silver, *Anal. Chim. Acta*, 280(2), 263, 1993.
186. Levi, G., Pantigny, J., Marsault, J.P., Christensen, D.H., Nielsen, O.F., and Aubard, J., Surface-enhanced Raman-spectroscopy of ellipticines adsorbed onto silver colloids, *J. Phys. Chem.*, 96(2), 926, 1992.
187. Montes, R. and Laserna, J.J., Fingerprinting and activity of sulpha drugs in surface-enhanced Raman-spectrometry on silver hydrosols, *Analyst*, 115(12), 1601, 1990.
188. Sutherland, W.S., Laserna, J.J., Angebranndt, M.J., and Winefordner, J.D., Surface-enhanced Raman analysis of sulfa drugs on colloidal silver dispersion, *Anal. Chem.*, 62(7), 689, 1990.
189. Murza, A., Sanchez-Cortes, S., Garcia-Ramos, J.V., Guisan, J.M., Alfonso, C., and Rivas, G., Interaction of the antitumor drug 9-aminoacridine with guanidinobenzoatase studied by spectroscopic methods: a possible tumor marker probe based on the fluorescence exciplex emission, *Biochemistry*, 39(34), 10557, 2000.
190. Murza, A., Sanchez-Cortes, S., and Garcia-Ramos, J.V., Surface-enhanced Raman and steady fluorescence study of interaction between antitumoral drug 9-aminoacridine and trypsin-like protease related to metastasis processes, guanidinobenzoatase, *Biopolymers*, 62(2), 85, 2001.
191. Shen, J.K., Ye, Y., Hu, J.M., Shen, H.Y., and Le, Z.F., Surface-enhanced Raman spectra study of metal complexes of *N*-D- glucosamine beta-naphthaldehyde and glycine and their interaction with DNA, *Spectrochim. Acta Part A Mol. Biomol. Spectrosc.*, 57(3), 551, 2001.

192. Stanicova, J., Kovalcik, P., Chinsky, L., and Miskovsky, P., Pre-resonance Raman and surface-enhanced Raman spectroscopy study of the complex of the antiviral and antiparkinsonian drug amantadine with histidine, *Vib. Spectrosc.*, 25(1), 41, 2001.
193. Rivas, L., Murza, A., Sanchez-Cortes, S., and Garcia-Ramos, J.V., Interaction of antimalarial drug quinacrine with nucleic acids of variable sequence studied by spectroscopic methods, *J. Biomol. Struct. Dyn.*, 18(3), 371, 2000.
194. Ye, Y., Hu, J.M., and Zeng, Y.N., Comparison of different spectroscopic methods for the interaction of antitumour compounds with deoxyribonucleic acid, *Chin. J. Anal. Chem.*, 28(7), 798, 2000.
195. Ermishov, M., Sukhanova, A., Kryukov, E., Grokhovsky, S., Zhuze, A., Oleinikov, V., Jardillier, J.C., and Nabiev, I., Raman and surface-enhanced Raman scattering spectroscopy of bis-netropsins and their DNA complexes, *Biopolymers*, 57(5), 272, 2000.
196. Kocisova, E., Jancura, D., Sanchez-Cortes, S., Miskovsky, P., Chinsky, L., and Garcia-Ramos, J.V., Interaction of antiviral and antitumor photoactive drug hypocrellin A with human serum albumin, *J. Biomol. Struct. Dyn.*, 17(1), 111, 1999.
197. Ianoul, A., Fleury, F., Duval, O., Waigh, R., Jardillier, J.C., Alix, A.J.P., and Nabiev, I., DNA binding by fagaronine and ethoxidine, inhibitors of human DNA topoisomerases I and II, probed by SERS and flow linear dichroism spectroscopy, *J. Phys. Chem. B*, 103(11), 2008, 1999.
198. Miskovsky, P., Jancura, D., Sanchez-Cortes, S., Kocisova, E., and Chinsky, L., Antiretrovirally active drug hypericin binds the IIA subdomain of human serum albumin: Resonance Raman and surface-enhanced Raman spectroscopy study, *J. Am. Chem. Soc.*, 120(25), 6374, 1998.
199. Lecomte, S. and Baron, M.H., Surface-enhanced Raman spectroscopy investigation of fluoroquinolones-DNA-DNA gyrase-$Mg^{2+}$ interactions. 2. Interaction of pefloxacin with $Mg^{2+}$ and DNA, *Biospectroscopy*, 3, 1, 31, 1997.
200. Beljebbar, A., Sockalingum, G.D., Angiboust, J.F., and Manfait, M., Comparative FT SERS, resonance Raman and SERRS studies of doxorubicin and its complex with DNA, *Spectrochim. Acta Part A Mol. Biomol. Spectrosc.*, 51(12), 2083, 1995.
201. Nabiev, I., Chourpa, I., Riou, J.F., Nguyen, C.H., Lavelle, F., and Manfait, M., Molecular-interactions of DNA topoisomerase-I and topoisomerase-Ii inhibitor with DNA and topoisomerases and in ternary complexes - binding modes and biological effects for intoplicine derivatives, *Biochemistry*, 33(30), 9013, 1994.
202. Nabiev, I., Chourpa, I., and Manfait, M., Comparative-studies of antitumor DNA intercalating agents, aclacinomycin and saintopin, by means of surface-enhanced Raman-scattering spectroscopy, *J. Phys. Chem.*, 98(4), 1344, 1994.
203. Aubard, J., Schwaller, M.A., Pantigny, J., Marsault, J.P., and Levi, G., Surface-enhanced Raman-spectroscopy of ellipticine, 2-*N*-methylellipticinium and their complexes with DNA, *J. Raman Spectrosc.*, 23(7), 373, 1992.
204. Beljebbar, A., Morjani, H., Angiboust, J.F., Sockalingum, G.D., Polissiou, M., and Manfait, M., Molecular and cellular interaction of the differentiating antitumour agent dimethylcrocetin with nuclear retinoic acid receptor as studied by near-infrared and visible SERS spectroscopy, *J. Raman Spectrosc.*, 28(2–3), 159, 1997.
205. Chourpa, I., Morjani, H., Riou, J.F., and Manfait, M., Intracellular molecular interactions of antitumor drug amsacrine, m-AMSA) as revealed by surface-enhanced Raman spectroscopy, *FEBS Lett.*, 397(1), 61, 1996.
206. Morjani, H., Riou, J.F., Nabiev, I., Lavelle, F., and Manfait, M., Molecular and cellular interactions between intoplicine, DNA, and topoisomerase-Ii studied by surface-enhanced Raman-scattering spectroscopy, *Cancer Res.*, 53(20), 4784, 1993.
207. Nabiev, I.R., Morjani, H., and Manfait, M., Selective analysis of antitumor drug-interaction with living cancer-cells as probed by surface-enhanced Raman-spectroscopy, *Eur. Biophys. J.*, 19(6), 311, 1991.
208. Grabbe, E.S. and Buck, R.P., Surface-enhanced Raman-spectroscopic investigation of human immunoglobulin-G adsorbed on a silver electrode, *J. Am. Chem. Soc.*, 111(22), 8362, 1989.

209. Bernard, S., Schwaller, M.A., Moiroux, J., Bazzaoui, E.A., Levi, G., and Aubard, J., SERS identification of quinone-imine species as oxidation products of antitumor ellipticines, *J. Raman Spectrosc.*, 27(7), 539, 1996.
210. Bernard, S., Schwaller, M.A., Levi, G., and Aubard, J., Metabolism of the antitumor drug N(2)-methyl-9-hydroxy ellipticinium: identification by surface-enhanced Raman spectroscopy of adducts formed with amino acids and nucleic acids, *Biospectroscopy*, 2(6), 377, 1996.
211. Cabalin, L.M., Ruperez, A., and Laserna, J.J., Flow-injection analysis and liquid chromatography: surface-enhanced Raman spectrometry detection by using a windowless flow cell, *Anal. Chim. Acta*, 318(2), 203, 1996.
212. Chourpa, I. and Manfait, M., Specific molecular-interactions of acridine drugs in complexes with topoisomerase-Ii and DNA — SERS and resonance raman-study of M-Amsa in comparison with O-Amsa, *J. Raman Spectrosc.*, 26(8–9), 813, 1995.
213. Koglin, E. and Sequaris, J.M., Interaction of proflavine with DNA studied by colloid surface enhanced resonance Raman-spectroscopy, *J. Mol. Struct.*, 141, 405, 1986.
214. Koglin, E., Sequaris, J.M., Fritz, J.C., and Valenta, P., Surface enhanced Raman-scattering (SERS) of nucleic-acid bases adsorbed on silver colloids, *J. Mol. Struct.*, 114(March), 219, 1984.
215. Koglin, E., Sequaris, J.M., and Valenta, P., Surface Raman-spectra of nucleic-acid components adsorbed at a silver electrode, *J. Mol. Struct.*, 60(January), 421, 1980.
216. Koglin, E., Sequaris, J.M., and Valenta, P., Surface enhanced Raman-spectroscopy of nucleic-acid bases on Ag electrodes, *J. Mol. Struct.*, 79(1–4), 185, 1982.
217. Ervin, K.M., Koglin, E., Sequaris, J.M., Valenta, P., and Nurnberg, H.W., Surface enhanced Raman-spectra of nucleic-acid components adsorbed at a silver electrode, *J. Electroanal. Chem.*, 114(2), 179, 1980.
218. Sequaris, J.M., Fritz, J., Lewinsky, H., and Koglin, E., Surface enhanced Raman-scattering spectroscopy of methylated guanine and DNA, *J. Colloid Interface Sci.*, 105(2), 417, 1985.
219. Foote, S., Vollrath, D., Hilton, A., and Page, D.C., The human y-chromosome — overlapping DNA clones spanning the euchromatic region, *Science*, 258(5079), 60, 1992.
220. Kim, U.J., Shizuya, H., Deaven, L., Chen, X.N., Korenberg, J.R., and Simon, M.I., Selection of a sublibrary enriched for a chromosome from total human bacterial artificial chromosome library using DNA from flow-sorted chromosomes as hybridization probes, *Nucleic Acids Res.*, 23(10), 1838, 1995.
221. Kim, U.J., Birren, B.W., Slepak, T., Mancino, V., Boysen, C., Kang, H.L., Simon, M.I., and Shizuya, H., Construction and characterization of a human bacterial artificial chromosome library, *Genomics*, 34(2), 213, 1996.
222. Kim, U.J., Shizuya, H., Kang, H.L., Choi, S.S., Garrett, C.L., Smink, L.J., Birren, B.W., Korenberg, J.R., Dunham, I., and Simon, M.I., A bacterial artificial chromosome-based framework contig map of human chromosome 22q, *Proc. Natl. Acad. Sci. U.S.A.*, 93(13), 6297, 1996.
223. Richterich, P. and Church, G.M., DNA-Sequencing with direct transfer electrophoresis and non-radioactive detection, *Meth. Enzymol.*, 218, 187, 1993.
224. Boncheva, M., Scheibler, L., Lincoln, P., Vogel, H., and Akerman, B., Design of oligonucleotide arrays at interfaces, *Langmuir*, 15(13), 4317, 1999.
225. Kelsey, J.L. and Hornross, P.L., Breast-cancer — magnitude of the problem and descriptive epidemiology — introduction, *Epidemiol. Rev.*, 15(1), 7, 1993.
226. Martin, A.M. and Weber, B.L., Genetic and hormonal risk factors in breast cancer, *J. Natl. Cancer Inst.*, 92(14), 1126, 2000.
227. Staff, S., Nupponen, N.N., Borg, A., Isola, J.J., and Tanner, M.M., Multiple copies of mutant BRCA1 and BRCA2 alleles in breast tumors from germ-line mutation carriers, *Genes Chromosomes Cancer*, 28(4), 432, 2000.
228. Boyd, J., Sonoda, Y., Federici, M.G., Bogomolniy, F., Rhei, E., Maresco, D.L., Saigo, P.E., Almadrones, L.A., Barakat, R.R., Brown, C.L., Chi, D.S., Curtin, J.P., Poynor, E.A., and Hoskins, W.J., Clinicopathologic features of BRCA-linked and sporadic ovarian cancer, *J. Am. Med. S.*, 283(17), 2260, 2000.

229. Bello, J.M., Narayanan, V.A., Stokes, D.L., and Vo-Dinh, T., Fiberoptic remote sensor for *in situ* surface-enhanced Raman-scattering analysis, *Anal. Chem.*, 62(22), 2437, 1990.
230. Bello, J.M. and Vo-Dinh, T., Surface-enhanced Raman fiberoptic sensor, *Appl. Spectrosc.*, 44(1), 63, 1990.
231. Stokes, D.L., Alarie, J.P., Ananthanarayanan, V., and Vo-Dinh, T., Fiber optics SERS sensors for environmental monitoring, in *SPIE Int. Soc. Opt. Eng.*, 1998, pp. 647–654.
232. McCreery, R.L., Instrumentation for dispersive Raman spectroscopy, in *Modern Techniques in Raman Spectroscopy*, Laserna, J.J., Ed., John Wiley & Sons, New York, 1996, pp. 41–72.
233. Myrick, M.L., Angel, S.M., and Desiderio, R., Comparison of some fiber optic configurations for measurement of luminescence and Raman-scattering, *Appl. Opt.*, 29(9), 1333, 1990.
234. Carrabba, M.M. and Rauh, R.D., U.S. patent no. 5,112,127, 1992.
235. Stokes, D.L. and Vo-Dinh, T., Development of an integrated single-fiber SERS sensor, *Sensors Actuators B Chem.*, 69(1–2), 28, 2000.
236. Stokes, D.L., Alarie, J.P., and Vo-Dinh, T., Surface-enhaced Raman fiberoptic sensors for remote monitoring, in *SPIE Int. Soc. Opt. Eng.*, 1995, pp. 552–558.
237. Pohl, D.W., Denk, W., and Lanz, M., Optical stethoscopy — image recording with resolution lambda/20, *Appl. Phys. Lett.*, 44(7), 651, 1984.
238. Betzig, E., Trautman, J.K., Harris, T.D., Weiner, J.S., and Kostelak, R.L., Breaking the diffraction barrier — optical microscopy on a nanometric scale, *Science*, 251(5000), 1468, 1991.
239. Xu, H.X., Bjerneld, E.J., Kall, M., and Borjesson, L., Spectroscopy of single hemoglobin molecules by surface enhanced Raman scattering, *Phys. Rev. Lett.*, 83(21), 4357, 1999.

# VIII

# Appendix

# 65

# Spectroscopic Data of Biologically and Medically Relevant Species and Samples

Dimitra N. Stratis-Cullum*
*Oak Ridge National Laboratory*
*Oak Ridge, Tennessee*

David L. Stokes
*Oak Ridge National Laboratory*
*Oak Ridge, Tennessee*

Brian M. Cullum**
*Oak Ridge National Laboratory*
*Oak Ridge, Tennessee*

Joon-Myong Song
*Oak Ridge National Laboratory*
*Oak Ridge, Tennessee*

Paul M. Kasili
*Oak Ridge National Laboratory*
*Oak Ridge, Tennessee*

Ramesh Jagannathan
*Oak Ridge National Laboratory*
*Oak Ridge, Tennessee*

Joel Mobley
*Oak Ridge National Laboratory*
*Oak Ridge, Tennessee*

Tuan Vo-Dinh
*Oak Ridge National Laboratory*
*Oak Ridge, Tennessee*



## 65.1 Introduction

This chapter provides several tables summarizing spectroscopic properties of various biologically and medically relevant species and samples. It is divided into two main sections: (1) Fundamental Spectroscopic Properties of Biologically and Medically Relevant Molecules and (2) Spectroscopic Data for Biological and Medical Applications. The spectroscopic data for the first section are provided in the following tables:

- Table 65.1, Characteristic Raman Frequencies
- Table 65.2, Characteristic IR Band Positions of Functional Chemical Groups and Molecules
- Table 65.3, Characteristic IR Absorption Frequencies of Compounds
- Table 65.4. Characteristics of Electronic Transitions between $\sigma$, $\eta$, and $\pi$ Orbitals
- Table 65.5, Absorption Characteristics of Common Fluorophores

For the second section, the spectroscopic data are provided in the following tables, which are grouped into specific spectroscopic methods:

Current affiliations: * U.S. Army Research Laboratory, Adelphi, Maryland
** University of Maryland Baltimore County, Baltimore, Maryland

- Table 65.6, UV-Visible Absorption Data
- Table 65.7, Infrared Absorption Data
- Table 65.8, Raman Data
- Table 65.9A, Fluorescence Data: Biochemical Studies and 65.9B, Fluorescence Data: Cellular Studies, Tissue and Biofluid Studies (*in Vitro*), Animal Studies (*in Vivo*), and Human Clinical Studies (*in Vivo*)
- Table 65.10, Elemental Analysis Data

The spectroscopic information provided in these tables was compiled from a variety of sources: reviews, articles, and book chapters, and almost 700 references, as indicated in the individual tables. We have made special efforts to group the data in each table whenever possible into subclasses that are relevant to biological and biomedical applications. These subclassifications include:

- Biochemical studies (individual compounds)
- Cellular studies (*in vitro* applications)
- Tissue and biofluid studies (*in vitro*)
- Animal studies (*in vivo*)
- Human clinical studies (*in vivo*)

Due to space limitations, the information is not meant to be comprehensive; rather, the aim is to serve as a starting point for a researcher who is unfamiliar with a certain field but would like to obtain the initial information to begin an investigation in a specific area or topic related to biological or medical research. We hope that this chapter will serve as a useful reference source for a broad audience involved in research, teaching, learning, or practice of medical technologies.

## Acknowledgments

This research was sponsored by the U.S. Department of Energy, managed by UT-Battelle under contract DE-AC05-00OR22725. D.N. Stratis-Cullum, D.L. Stokes, B.M. Cullum, R. Jagannathan, J. Mobley, and J.-M. Song were also supported by appointments to the Oak Ridge National Laboratory Postdoctoral Research Associates Program administered jointly by the Oak Ridge National Laboratory and Oak Ridge Institute for Science and Education.

# 65.2 Fundamental Spectroscopic Properties of Biologically and Medically Relevant Species

## 65.2.1 Raman Spectroscopy: Characteristic Frequencies

Table 65.1 provides characteristic Raman frequencies of organic compounds. This table consists of the vibrational frequencies ($cm^{-1}$) for main chemical groups, the type of vibration energy absorbed due to vibrations of polar covalent bonds, the functional groups that cause the vibration, and the kind of compound in which this characteristic vibration would be found.

**TABLE 65.1** Characteristic Raman Frequencies

| Frequency ($cm^{-1}$) | Vibration | Compound |
|---|---|---|
| 3300–3400 | Bonded antisymmetric $NH_2$ stretch | Primary amines |
| 3340–3380 | Bonded OH stretch | Aliphatic alcohols |
| 3374 | CH stretch | Acetylene (gas) |
| 3325–3355 | Bonded antisymmetric $NH_2$ stretch | Primary amides |
| 3300–3350 | Bonded NH stretch | Secondary amines |
| 3300–3335 | ≡CH stretch | Alkyl acetylenes |
| 3250–3300 | Bonded symmetric $NH_2$ stretch | Primary amines |
| 3290–3310 | Bonded NH stretch | Secondary amines |

**TABLE 65.1** Characteristic Raman Frequencies (continued)

| Frequency ($cm^{-1}$) | Vibration | Compound |
|---|---|---|
| 3145–3190 | Bonded symmetric $NH_2$ stretch | Primary amides |
| 3154–3175 | Bonded NH stretch | Pyrazoles |
| 3103 | Antisymmetric $=CH_2$ stretch | Ethylene (gas) |
| 3020–3100 | $CH_2$ stretches | Cyclopropane |
| 3000–3100 | Aromatic CH stretch | Benzene derivatives |
| 3070–3095 | Antisymmetric $=CH_2$ stretch | $C=CH_2$ derivatives |
| 3062 | CH stretch | Benzene |
| 3057 | Aromatic CH stretch | Alkyl benzenes |
| 3000–3040 | CH stretch | C=CHR derivatives |
| 3026 | Symmetric $=CH_2$ stretch | Ethylene (gas) |
| 2980–2990 | Symmetric $=CH_2$ stretch | $C=CH_2$ derivatives |
| 2974–2986 | Symmetric $NH_3^+$ stretch | Alkyl ammonium chlorides (aq. soln.) |
| 2965–2969 | Antisymmetric $CH_3$ stretch | *n*-Alkanes |
| 2912–2929 | Antisymmetric $CH_2$ stretch | *n*-Alkanes |
| 2883–2884 | Symmetric $CH_3$ stretch | *n*-Alkanes |
| 2849–2861 | Symmetric $CH_2$ stretch | *n*-Alkanes |
| 2700–2850 | CHO group (2 bands) | Aliphatic aldehydes |
| 2560–2590 | SH stretch | Thiols |
| 2233–2316 | C≡C stretch (2 bands) | $R–C≡C–CH_3$ |
| 2231–2301 | C≡C stretch (2 bands) | R–C≡C–R′ |
| 2250–2300 | Pseudoantisymmetric N=C=O stretch | Isocyanates |
| 2251–2264 | Symmetric C≡C–C≡C stretch | Alkyl diacetylenes |
| 2259 | C≡N stretch | Cyanamide |
| 2232–2251 | C≡N stretch | Aliphatic nitriles |
| 2220–2100 | Pseudoantisymmetric N=C=S stretch (2 bands) | Alkyl isothiocyanates |
| 2220–2000 | C≡N stretch | Dialkyl cyanamides |
| 2172 | Symmetric C≡C–C≡C stretch | Diacetylene |
| 2134–2161 | $N^+≡C^-$ stretch | Aliphatic isonitriles |
| 2100–2160 | C≡C stretch | Alkyl acetylenes |
| 2140–2156 | C≡N stretch | Alkyl thiocynates |
| 2104 | Antisymmetric N=N=N stretch | $CH_3N_3$ |
| 2094 | C≡N stretch | HCN |
| 2049 | Pseudoantisymmetric C=C=O stretch | Ketene |
| 1974 | C≡C stretch | Acetylene (gas) |
| 1958–1964 | Antisymmetric C=C=C stretch | Allenes |
| 1840–1870 | Symmetric C=O stretch | Saturated five-membered ring cyclic anhydrides |
| 1820 | Symmetric C=O stretch | Acetic anhydride |
| 1788–1810 | C=O stretch | Acid halides |
| 1807 | C=O stretch | Phosgene |
| 1799–1805 | Symmetric C=O stretch | Noncyclic anhydrides |
| 1800 | C=C stretch | $F_2C=CF_2$ (gas) |
| 1795 | C=O stretch | Ethylene carbonate |
| 1792 | C=C stretch | $F_2C=CFCH_3$ |
| 1782 | C=O stretch | Cyclobutanone |
| 1730–1770 | C=O stretch | Halogenated aldehydes |
| 1744 | C=O stretch | Cyclopentanone |
| 1729–1743 | C=O stretch | Cationic *a*-amino acids (aq. soln.) |
| 1734–1741 | C=O stretch | *O*-Alkyl acetates |
| 1720–1740 | C=O stretch | Aliphatic aldehydes |
| 1714–1739 | C=C stretch | $C=CF_2$ derivatives |
| 1736 | C=C stretch | Methylene cyclopropane |
| 1727–1734 | C=O stretch | *O*-Alkyl propionates |
| 1700–1725 | C=O stretch | Aliphatic ketones |
| 1715–1720 | C=O stretch | *O*-Alkyl formates |
| 1694–1712 | C=C stretch | RCF=CFR |
| 1695 | Nonconjugated C=O stretch | Uracil derivatives (aq. soln.) |

**TABLE 65.1** Characteristic Raman Frequencies (continued)

| Frequency ($cm^{-1}$) | Vibration | Compound |
|---|---|---|
| 1644–1689 | C=C stretch | Monofluoroalkenes |
| 1651–1687 | C=C stretch | Alkylidene cyclopentanes |
| 1636–1686 | Amide I band | Primary amides (solids) |
| 1665–1680 | C=C stretch | Tetralkyl ethylenes |
| 1679 | C=C stretch | Methylene cyclobutane |
| 1664–1678 | C=C stretch | Trialkyl ethylenes |
| 1665–1676 | C=C stretch | *trans*-Dialkyl ethylenes |
| 1675 | Symmetric C=O stretch (cyclic dimer) | Acetic acid |
| 1666–1673 | C=N stretch | Aldimines |
| 1672 | Symmetric C=O stretch (cyclic dimer) | Formic acid (aq. soln.) |
| 1655–1670 | Conjugated C=O stretch | Uracil, cytosine, and guanine derivatives (aq. soln.) |
| 1630–1670 | Amide I band | Tertiary amides |
| 1652–1666 | C=N stretch | Ketoximes |
| 1650–1665 | C=N stretch | Semicarbazones (solid) |
| 1636–1663 | Symmetric C=N stretch | Aldazines, ketazines |
| 1654–1660 | C=C stretch | *cis*-Dialkyl ethylenes |
| 1650–1660 | Amide I band | Secondary amides |
| 1649–1660 | C=N stretch | Aldoximes |
| 1610–1660 | C=N stretch | Hydrazones (solids) |
| 1644–1658 | C=C stretch | $R_2C=CH_2$ |
| 1656 | C=C stretch | Cyclohexene, cycloheptene |
| 1649–1654 | Symmetric C=C stretch (cyclic dimer) | Carboxylic acids |
| 1642–1652 | C=N stretch | Thiosemicarbazones (solid) |
| 1590–1650 | $NH_2$ scissors | Primary amines (weak) |
| 1625–1649 | C=C stretch | Allyl derivatives |
| 1640–1648 | N=O stretch | Alkyl nitrites |
| 1638–1648 | C=C stretch | $H_2C=CHR$ |
| 1647 | C=C stretch | Cyclopropene |
| 1638 | C=O stretch | Ethylene dithiocarbonate |
| 1637 | Symmetric C=C stretch | Isoprene |
| 1622–1634 | Antisymmetric $NO_2$ stretch | Alkyl nitrates |
| 1550–1630 | Ring stretches (doublet) | Benzene derivatives |
| 1623 | C=C stretch | Ethylene (gas) |
| 1540–1620 | Three or more coupled C=C stretches | Polyenes |
| 1571–1616 | C=C stretch | Chloroalkenes |
| 1614 | C=C stretch | Cyclopentene |
| 1547–1596 | C=C stretch | Bromoalkenes |
| 1465–1581 | C=C stretch | Iodoalkenes |
| 1575 | Symmetric C=C stretch | 1,3-Cyclohexadiene |
| 1573 | N=N stretch | Azomethane (in soln.) |
| 1566 | C=C stretch | Cyclobutene |
| 1550–1560 | Antisymmetric $NO_2$ stretch | Primary nitroalkanes |
| 1550–1555 | Antisymmetric $NO_2$ stretch | Secondary nitroalkanes |
| 1548 | N=N stretch | 1-Pyrazoline |
| 1535–1545 | Antisymmetric $NO_2$ stretch | Tertiary nitroalkanes |
| 1490–1515 | Ring stretch | 2-Furfuryl group |
| 1500 | Symmetric C=C stretch | Cyclopentadiene |
| 1470–1480 | $OCH_3$, $OCH_2$ deformations | Aliphatic ethers |
| 1460–1480 | Ring stretch | 2-Furfurylidene or 2-furoyl group |
| 1446–1473 | $CH_3$, $CH_2$ deformations | *n*-Alkanes |
| 1465–1466 | $CH_3$ deformation | *n*-Alkanes |
| 1400–1450 | Pseudoantisymmetric N=C=O stretch | Isocyanates |
| 1398–1443 | Ring stretch | 2-Substituted thiophenes |
| 1442 | N=N stretch | Azobenzene |
| 1340–1440 | Symmetric $CO_2$ stretch | Carboxylate ions (aq. soln.) |
| 1400–1415 | Symmetric $CO_2$ stretch | Dipolar and anionic *a*-amino acids (aq. soln.) |

**TABLE 65.1** Characteristic Raman Frequencies (continued)

| Frequency ($cm^{-1}$) | Vibration | Compound |
|---|---|---|
| 1385–1415 | Ring stretch | Anthracenes |
| 1380–1395 | Symmetric $NO_2$ stretch | Primary nitroalkanes |
| 1370–1390 | Ring stretch | Naphthalenes |
| 1368–1385 | $CH_3$ symmetric deformation | *n*-Alkanes |
| 1360–1375 | Symmetric $NO_2$ stretch | Secondary nitroalkanes |
| 1345–1355 | Symmetric $NO_2$ stretch | Tertiary nitroalkanes |
| 1330–1350 | CH deformation | Isopropyl group |
| 1320 | Ring vibration | 1,1-Dialkyl cyclopropanes |
| 1290–1314 | In-plane CH deformation | *trans*-Dialkyl ethylenes |
| 1250–1310 | Amide III band | Secondary amides |
| 1175–1310 | $CH_2$ twist and rock | *n*-Alkanes |
| 1295–1305 | $CH_2$ in-phase twist | *n*-Alkanes |
| 1280–1300 | CC bridge bond stretch | Biphenyls |
| 1275–1282 | Symmetric $NO_2$ stretch | Alkyl nitrates |
| 1240–1280 | Ring stretch | Epoxy derivatives |
| 1276 | Symmetric N=N=N stretch | $CH_3N_3$ |
| 1251–1270 | In-plane CH deformation | *cis*-Dialkyl ethylenes |
| 1266 | Ring "breathing" | Ethylene oxide (oxirane) |
| 1200–1230 | Ring vibration | *para*-Disubstituted benzenes |
| 1200–1220 | Ring vibration | Mono- and 1,2-dialkyl cyclopropanes |
| 1212 | Ring "breathing" | Ethylene imine (aziridine) |
| 1205 | $C_6H_5$ –C vibration | Alkyl benzenes |
| 1188–1196 | Symmetric $SO_2$ stretch | Alkyl sulfates |
| 1188 | Ring "breathing" | Cyclopropane |
| 1165–1172 | Symmetric $SO_2$ stretch | Alkyl sulfonates |
| 950–1150 | CC stretches | *n*-Alkanes |
| 1125–1145 | Symmetric $SO_2$ stretch | Dialkyl sulfones |
| 1144 | Ring "breathing" | Pyrrole |
| 1140 | Ring "breathing" | Furan |
| 1100–1130 | Symmetric C=C=C stretch (2 bands) | Allenes |
| 1130 | Pseudosymmetric C=C=O stretch | Ketene |
| 1112 | Ring "breathing" | Ethylene sulfide |
| 1111 | NN stretch | Hydrazine |
| 1040–1070 | S=O stretch (1 or 2 bands) | Aliphatic sulfoxides |
| 1065 | C=S stretch | Ethylene trithiocarbonate |
| 1020–1060 | Ring vibration | *ortho*-Disubstituted benzenes |
| 990–1040 | Ring vibration | Pyrazoles |
| 1015–1030 | In-plane CH deformation | Monosubstituted benzenes |
| 1010–1030 | Trigonal ring "breathing" | 3-Substituted pyridines |
| 1030 | Trigonal ring "breathing" | Pyridine |
| 1029 | Ring "breathing" | Trimethylene oxide (oxetane) |
| 1026 | Ring "breathing" | Trimethylene imine (azetidine) |
| 990–1010 | Trigonal ring "breathing" | Mono-, *meta*-, and 1,3,5-substituted benzenes |
| 1001 | Ring "breathing" | Cyclobutane |
| 985–1000 | Trigonal ring "breathing" | 2- and 4-Substituted pyridines |
| 992 | Ring "breathing" | Benzene |
| 992 | Ring "breathing" | Pyridine |
| 939 | Ring "breathing" | 1,3-Dialixolane |
| 933 | Ring vibration | Alkyl cyclobutanes |
| 830–930 | Symmetric COC stretch | Aliphatic ethers |
| 914 | Ring "breathing" | Tetrahydrofuran |
| 906 | Symmetric CON stretch | Hydroxylamine |
| 837–905 | CC skeletal stretch | *n*-Alkanes |
| 900–890 | Ring vibration | Alkyl cyclopentanes |
| 850–900 | Symmetric CNC stretch | Secondary amines |
| 899 | Ring "breathing" | Pyrrolidine |

**TABLE 65.1** Characteristic Raman Frequencies (continued)

| Frequency ($cm^{-1}$) | Vibration | Compound |
|---|---|---|
| 886 | Ring "breathing" | Cyclopentane |
| 877 | OO stretch | Hydrogen peroxide |
| 840–851 | Symmetric CON stretch | *O*-Alkyl hydroxylamines |
| 836 | Ring "breathing" | Piperazine |
| 749–835 | $C_4$ skeletal stretch | Isopropyl group |
| 834 | Ring "breathing" | 1,4-Dioxane |
| 832 | Ring "breathing" | Thiophene |
| 832 | Ring "breathing" | Morpholine |
| 720–830 | Ring vibration | *para*-Disubstituted benzenes |
| 820–825 | $C_3O$ symmetrical skeletal stretch | Secondary alcohols |
| 818 | Ring "breathing" | Tetrahydropyran |
| 815 | Ring "breathing" | Piperidine |
| 802 | Ring "breathing" | Cyclohexane (chair form) |
| 700–785 | Ring vibration | Alkyl cyclohexanes |
| 730–760 | $C_4O$ symmetrical skeletal stretch | Tertiary alcohols |
| 650–750 | $C_5$ symmetrical skeletal stretch | *tert*-Butyl group |
| 585–740 | CS stretch (1 or more bands) | Alkyl sulfides |
| 690–735 | "C=S stretch" | Thioamides, thioureas (solid) |
| 733 | Ring "breathing" | Cycloheptane |
| 720–730 | CCl stretch, $P_c$ conformation | Primary chloroalkanes |
| 620–715 | CS stretch (1 or more bands) | Dialkyl sulfides |
| 709 | CCl stretch | $CH_3Cl$ |
| 703 | Ring "breathing" | Cyclooctane |
| 703 | Symmetric $CCl_2$ stretch | $CH_2Cl_2$ |
| 650–690 | Pseudosymmetric N=C=S stretch | Alkyl isothiocyanates |
| 688 | Ring "breathing" | Tetrahydrothiophene |
| 668 | Symmetric $CCl_3$ stretch | $CHCl_3$ |
| 650–660 | CCl stretch, $P_H$ conformation | Primary chloroalkanes |
| 659 | Symmetric CSC stretch | Pentamethylene sulfide |
| 640–655 | CBr stretch, $P_c$ conformation | Primary bromoalkanes |
| 615–630 | Ring deformation | Monosubstituted benzenes |
| 605–615 | CCl stretch, $S_{HH}$ conformation | Secondary chloroalkanes |
| 590–610 | CI stretch, $P_c$ conformation | Primary iodoalkanes |
| 609 | CBr stretch | $CH_3Br$ |
| 577 | Symmetric $CBr_2$ stretch | $CH_2BR_2$ |
| 560–570 | CCl stretch, $T_{HHH}$ conformation | Tertiary chloroalkanes |
| 560–565 | CBr stretch, $P_H$ conformation | Primary bromoalkanes |
| 535–540 | CBr stretch, $S_{HH}$ conformation | Secondary bromoalkanes |
| 539 | Symmetric $CBr_3$ stretch | CHBr |
| 510–525 | SS stretch | Dialkyl sulfides |
| 523 | CI stretch | $CH_3I$ |
| 510–520 | CBr stretch, $T_{HH}$ conformation | Tertiary bromoalkanes |
| 500–510 | CI stretch, $P_H$ conformation | Primary iodoalkanes |
| 480–510 | SS stretch | Dialkyl trisulphides |
| 485–495 | CI stretch, $S_{HH}$ conformation | Secondary iodoalkanes |
| 485–495 | CI stretch, $T_{HHH}$ conformation | Tertiary iodoalkanes |
| 475–484 | Skeletal deformation | Dialkyl diacetylenes |
| 483 | Symmetric $Cl_2$ stretch | $CH_2I_2$ |
| 459 | Symmetric $CCl_4$ stretch | $CCl_4$ |
| 437 | Symmetric $CI_3$ stretch | $CHI_3$ (in soln.) |
| 150–425 | "Chain expansion" | *n*-Alkanes |
| 335–355 | Skeletal deformation | Mono-alkyl acetylenes |
| 267 | Symmetric $CBr_4$ stretch | $CBr_3$ (in soln.) |
| 160–200 | Skeletal deformation | Aliphatic nitriles |
| 178 | Symmetric $CI_4$ stretch | $CI_4$ (solid) |

*Source:* Robinson, J.W., *The CRC Practical Handbook of Spectroscopy*, CRC Press, Boca Raton, FL, 1991.

## 65.2.2 Infrared (IR) Spectroscopy: Characteristic IR Band Positions of Functional Chemical Groups and Molecules

Table 65.2 provides the characteristic IR band positions (frequencies) of organic compounds. This table consists of vibrational frequencies ($cm^{-1}$) of functional groups and the functional groups found within organic molecules that absorb IR energy. These frequencies and assignments have been collected from various references such as http://infrared.als.lbl.gov/IRbands.html, a Web site maintained by the Lawrence Berkeley National Laboratory.

**TABLE 65.2** Characteristic IR Band Positions of Functional Chemical Groups

| Frequency Range ($cm^{-1}$) | Group |
|---|---|
| | ***OH Stretching Vibrations*** |
| 3610–3645 (sharp) | Free OH |
| 3450–3600 (sharp) | Intramolecular H bonds |
| 3200–3550 (broad) | Intermolecular H bonds |
| 2500–3200 (very broad) | Chelate compounds |
| | ***NH Stretching Vibrations*** |
| 3300–3500 | Free NH |
| 3070–3350 | H-bonded NH |
| | ***CH Stretching Vibrations*** |
| 3280–3340 | =C–H |
| 3000–3100 | =C–H |
| 2862–2882, 2652–2972 | C–$CH_3$ |
| 2815–2832 | O–$CH_3$ |
| 2810–2820 | N–$CH_3$ (aromatic) |
| 2780–2805 | N–$CH_3$ (aliphatic) |
| 2843–2863, 2916–2936 | $CH_2$ |
| 2880–2900 | CH |
| | ***SH Stretching Vibrations*** |
| 2550–2600 | Free SH |
| | ***C=N Stretching Vibrations*** |
| 2240–2260 | Nonconjugated |
| 2215–2240 | Conjugated |
| | ***C=C Stretching Vibrations*** |
| 2100–2140 | C=CH (terminal) |
| 2190–2260 | C–C=C–C |
| 2040–2200 | C–C=C–C=CH |
| | ***C=O Stretching Vibrations*** |
| 1700–1900 | Nonconjugated |
| 1590–1750 | Conjugated |
| ~1650 | Amides |
| | ***C=C Stretching Vibrations*** |
| 1620–1680 | Nonconjugated |
| 1585–1625 | Conjugated |
| | ***CH Bending Vibrations*** |
| 1405–1465 | $CH_2$ |
| 1355–1395, 1430–1470 | $CH_3$ |
| | ***C–O–C Vibrations in Esters*** |
| ~1175 | Formates |
| ~1240, 1010–1040 | Acetates |
| ~1275 | Benzoates |

**TABLE 65.2** Characteristic IR Band Positions of Functional Chemical Groups (continued)

| Frequency Range ($cm^{-1}$) | Group |
|---|---|
| | ***C–O–H Stretching Vibrations*** |
| 990–1060 | Secondary cyclic alcohols |
| | ***CH Out-of-Plane Bending Vibrations in Substituted Ethylenic Systems*** |
| 905–915, 985–995 | $–CH=CH_2$ |
| 650–750 | –CH=CH– (*cis*) |
| 960–970 | –CH=CH– (*trans*) |
| 885–895 | $C=CH_2$ |

*Source:* Lawrence Berkeley National Laboratory, http://infrared.als.lbl.gov/IRbands.html.

Table 65.3 provides characteristic vibrational frequencies of main organic bonds. This table consists of the frequency range ($cm^{-1}$), compound type that displays the characteristic absorptions, and organic bonds involved. These frequencies and assignments have been collected from a variety of sources, such as a Web site (http://www.chem.ucla.edu/) maintained by the University of California, Los Angeles.

**TABLE 65.3** Characteristic IR Absorption Frequencies of Compounds

| Frequency Range ($cm^{-1}$) | Compound Type | Bond |
|---|---|---|
| 2960–2850 (Strong) stretch | Alkanes | C–H |
| 1470–1350 (Variable) scissoring and bending | Alkanes | C–H |
| 1380 (Medium-Weak) — doublet — isopropyl, *t*-butyl | $CH_3$ umbrella deformation | C–H |
| 3080–3020 (Medium) stretch | Alkenes | C–H |
| 1000–675 (Strong) bend | Alkenes | C–H |
| 3100–3000 (Medium) stretch | Aromatic rings | C–H |
| 870–675 (Strong) bend | Phenyl ring substitution bands | C–H |
| 2000–1600 (Weak) — fingerprint region | Phenyl ring substitution overtones | C–H |
| 3333–3267 (Strong) stretch | Alkynes | C–H |
| 700–610 (Broad) bend | Alkynes | C–H |
| 1680–1640 (Medium, weak) stretch | Alkenes | C=C |
| 2260–2100 (Weak, strong) stretch | Alkynes | C≡C |
| 1600, 1500 (Weak) stretch | Aromatic rings | C=C |
| 1260–1000 (Strong) stretch | Alcohols, ethers, carboxylic acids, esters | C–O |
| 1760–1670 (Strong) stretch | Aldehydes, ketones, carboxylic acids, esters | C=O |
| 3640–3160 (Strong, broad) stretch | Monomeric — alcohols, phenols | O–H |
| 3600–3200 (Broad) stretch | Hydrogen-bonded — alcohols, phenols | O–H |
| 3000–2500 (Broad) stretch | Carboxylic acids | O–H |
| 3500–3300 (Medium) stretch | Amines | N–H |
| 1650–1580 (Medium) bend | Amines | N–H |
| 1340–1020 (Medium) stretch | Amines | C–N |
| 2260–2220 (Variable) stretch | Nitriles | C≡N |
| 1660–1500 (Strong) asymmetrical stretch | Nitro compounds | $NO_2$ |
| 1390–1260 (Strong) symmetrical stretch | Nitro compounds | $NO_2$ |

*Source:* Robinson, J.W., *The CRC Practical Handbook of Spectroscopy*, CRC Press, Boca Raton, FL, 1991; University of California, Los Angeles Web site (http://www.chem.ucla.edu/).

## 65.2.3 Molecular Spectroscopy: Characteristics of Electronic Transitions between Orbitals

Table 65.4 provides the characteristics of electronic transitions between σ, η, and π orbitals. This table lists the transition type, wavelength (nm), molar absorptivity, ε (L $mol^{-1}$ $cm^{-1}$), and examples of organic compounds with electronic transitions.

**TABLE 65.4** Characteristics of Electronic Transitions between $\sigma$, $\eta$, and $\pi$ Orbitals

| Transition | Wavelength (nm) | $\varepsilon$ (l mol$^{-1}$cm$^{-1}$) | Examples |
|---|---|---|---|
| $\sigma$-$\sigma^*$ | <200 | — | Saturated hydrocarbons |
| $\pi$-$\pi^*$ | 200–500 | $\approx 10^4$ | Alkenes, alkynes, aromatics |
| $\eta$-$\sigma^*$ | 160–260 | $10^2$–$10^3$ | $H_2O$, $CH_3OH$, $CH_3Cl$, $CH_3NH_2$ |
| $\eta$-$\pi^*$ | 250–600 | $10^1$–$10^2$ | Carbonyls, nitro, nitrate, carboxyl |

*Source:* Ingle, J.D. and Crouch, S.R., *Spectrochemical Analysis*, Prentice-Hall, Englewood Cliffs, NJ, 1988.

### 65.2.4 Absorption Characteristics of Common Chromophores

Table 65.5 provides the absorption characteristics of common chromophores. This table consists of groups that are responsible for absorption, wavelength, $\lambda$ (nm), wavenumbers, $\nu$ ($10^3$ cm$^{-1}$), and molar absorptivity, $\varepsilon$ (l mol$^{-1}$cm$^{-1}$).

**TABLE 65.5** Absorption Characteristics of Common Fluorophores

| Group | Example | $\nu$ ($10^3$ cm$^{-1}$) | $\lambda$ (nm) | $\varepsilon$ (l mol$^{-1}$ cm$^{-1}$) |
|---|---|---|---|---|
| C=C | $H_2C{=}CH_2$ | 55 | 182 | 250 |
| | | 57.3 | 174 | 16000 |
| | | 58.6 | 170 | 16500 |
| | | 62 | 162 | 10000 |
| C≡C | H– C≡C–$CH_2$–$CH_3$ | 58 | 172 | 2500 |
| C=O | $H_2CO$ | 34 | 295 | 10 |
| | | 54 | 185 | Strong |
| C=S | $CH_3$–C(=S)– $CH_3$ | 22 | 460 | Weak |
| –$NO_2$ | $CH_3$–$NO_2$ | 36 | 277 | 10 |
| | | 47.5 | 210 | 10000 |
| –N=N– | $CH_3$–N=N– $CH_3$ | 28.8 | 347 | 15 |
| | | >38.5 | <260 | Strong |
| –Cl | $CH_3Cl$ | 58 | 172 | — |
| –Br | $CH_3Br$ | 49 | 204 | 1800 |
| –I | $CH_3I$ | 38.8 | 258 | — |
| | | 49.7 | 201 | 1200 |
| –OH | $CH_3OH$ | 55 | 183 | 200 |
| | | 67 | 150 | 1900 |
| –SH | $C_2H_5SH$ | 43 | 232 | 160 |
| –$NH_2$ | $CH_3NH_2$ | 46.5 | 215 | 580 |
| | | 52.5 | 190 | 3200 |
| –S– | $CH_3$–S–$CH_3$ | 44 | 228 | 620 |
| | | 46.5 | 215 | 700 |
| | | 49.3 | 203 | 2300 |
| C=C–C=C | $H_2C{=}CH{-}CH{=}CH_2$ | 48 | 209 | 25000 |

*Source:* Ingle, J.D. and Crouch, S.R., *Spectrochemical Analysis*, Prentice-Hall, Englewood Cliffs, NJ, 1988.

## 65.3 Spectroscopic Data for Biological and Medical Applications

### 65.3.1 UV-Visible Absorption Spectroscopy

Table 65.6 provides information on application of UV-visible absorption technique to biomedical research. This table covers only representative biochemical species that have been analyzed using UV-visible absorption technique. Absorbance maximum corresponds to the wavelength that was used for the detections of biochemical species in UV-visible absorption technique.

**TABLE 65.6** UV-Visible Absorption Data

| Species | Target Organ or Disease | Absorbance Maximum (nm) | Matrix | Comments | Ref. |
|---|---|---|---|---|---|
| | | ***Biochemical Studies*** | | | |
| 1,8-Bromo-guanosine | | 262 | | | 1 |
| 16α-Hydroxytestosterone, 16β-hydroxytestosterone | | 245 | | | 2 |
| 1-Methylguanosine | | 210, 260 | | | 3 |
| 1-Methylinosine | | 210, 260 | | | 3 |
| 1-Phenylethylamine | | 214 | | | 4 |
| 2-(*S*-alkylthiyl)pyrroline N-oxide | | 258 | | | 5 |
| 2,3-Dihydroxybiphenyl 1,2-dioxygenase-bound DHB | | 299 | | | 6 |
| 2,6-Dimethyldifuro-8-pyrone | | 306 | | Molar extinction coefficient=4700 ($M^{-1}$) | 7 |
| 2α-Hydroxytestosterone, 2β-hydroxytestosterone | | 245 | | | 2 |
| 2-Methoxy-3-isopropylpyrazine | | 307 | | | 8 |
| 2-Methylguanosine | | 210, 260 | | | 3 |
| 3,4-Methylenedioxymethamphetamine | | 214 | | | 4 |
| 3-Acetylmorphine | | 210 | | | 9 |
| 3-Methyluridine+5-methyluridine | | 210, 260 | | | 3 |
| 4-Hydroxyazobenzene-2-carboxylic acid | | 350 | | | 10 |
| 5,10-Methylenetetrahydrofolate | | | | | 11 |
| 5,6-Dihydrouridine | | 210, 260 | | | 3 |
| 6α-Hydroxytestosterone, 6β-hydroxytestosterone | | 245 | | | 2 |
| 6-Acetylmorphine | | 210 | | | 9 |
| 7α-Hydroxytestosterone | | 245 | | | 2 |
| Acetylcodeine | | 210 | | | 12 |
| Adenosine | | 210, 260 | | | 3 |
| α-Human globin chain | | 214 | | | 13 |
| Albonoursin | | 317 | | Molar extinction coefficient=25400 ($M^{-1}$) | 14 |
| Amphetamine | | 214 | | | 4 |
| Anabasine | | 214 | | | 15 |
| Androstenedione | | 245 | | | 2 |
| Asialo-transferrin | | 200 | | | 16 |
| Benoxaprofen | | 305 | | Molar extinction coefficient=25000 ($M^{-1}$) | 17 |
| Benzocaine | | 195 | | | 18 |
| Benzoylecgonine | | 195 | | | 18 |
| β-Human globin chain | | 214 | | | 13 |
| bis(Monoacrylglycerol)phosphate | | 231 | | | 19 |
| bis-*N*-nitroso-caged nitric oxides | | 300 | | | 20 |
| β-Nicotinamine adenine dinucleotide (β-NADH) | | 254 | | | 21 |
| Carprofen | | 240, 262, 300, 328 | | | 17 |
| *cis*-Cinnamoylcocaine, *trans*-cinnamoylcaine | | 195 | | | 18 |
| Cocaine | | 195 | | | 18 |
| Codeine | | 214 | | | 4 |
| Corticosterone | | 245 | | | 2 |
| Cytosine | | 210, 260 | | | 3 |
| Dehydrated β-NADH | | 254 | | | 21 |
| Deoxygenated hemoglobin | | 340, 552 | | | 22 |
| Dexromethorphan | | 220 | | | 9 |
| Dianionic DHB | | 283, 348 | | | 6 |

**TABLE 65.6** UV-Visible Absorption Data (continued)

| Species | Target Organ or Disease | Absorbance Maximum (nm) | Matrix | Comments | Ref. |
|---|---|---|---|---|---|
| Sisialo-transferrin | | 200 | | | 16 |
| Eosin B-protein complex | | 536–544 | | | 23 |
| Ethylmorphine | | 214 | | | 4 |
| Fatty acid | | 195 | | | 24 |
| Ferricytochrome *c* | | 465 | | | 25 |
| Ferulic acid | | 340 | | | 26 |
| Folding acylphosphatase | | 214 | | | 27 |
| Glucose-6-phosphate dehydrogenase | | 214 | | | 28 |
| Green fluorescent protein | | 479 | vacuo | | 29 |
| Guaifenesin | | 220 | | | 9 |
| Guanosine | | 210, 260 | | | 3 |
| Harmaline | | 254 | | | 30 |
| Harmane | | 254 | | | 30 |
| Harmine | | 254 | | | 30 |
| Harmol | | 254 | | | 30 |
| Harmolol | | 254 | | | 30 |
| Hb Aubenas | | 214 | | Type of mutation: β26 Glu-Gly | 13 |
| Hb Chad | | 214 | | Type of mutation: α23 Glu-Lys | 13 |
| Hb Debrousse | | 214 | | Type of mutation: β96 Leu-Pro | 13 |
| Hb Knossos | | 214 | | Type of mutation: β27 Ala-Ser | 13 |
| Hb M-Iwate | | 214 | | Type of mutation: α87 His-Tyr | 13 |
| Hb Olympia | | 214 | | Type of mutation: β20 Val-Met | 13 |
| Hb Ty Gard | | 214 | | Type of mutation: β124 Pro-Gln | 13 |
| Heroin | | 210 | | | 9 |
| Histone H(I) chromosal protein | | 210–230 | | Interaction with daunomycin antibiotic | 31 |
| Inosine | | 210, 260 | | | 3 |
| Leu-enkephalin | | 200 | | | 82 |
| Lidocaine | | 195 | | | 18 |
| Lycopene | | 471 | | | 33 |
| Maitotoxin | | 195 | | | 34 |
| Metamphetamine | | 214 | | | 4 |
| Met-enkephalin | | 200 | | | 32 |
| Methadone | | 214 | | | 4 |
| Monoanionic 2,3-dihydroxybiphenyl (DHB) | | 305 | | | 6 |
| Monsialo-transferrin | | 200 | | | 16 |
| Morphine | | 210 | | | 35 |
| $N^4$-Acetylcytidine | | 210, 260 | | | 3 |
| $N^6$-Methyladenosine | | 210, 260 | | | 3 |
| Naproxen | | 230, 270, 320, 330 | | | 17 |
| Nicotine | | 214 | | | 15 |
| Norharman | | 254 | | | 30 |
| Nornicotine | | 214 | | | 15 |
| Oxygenated hemoglobin | | 415, 542, 577 | | | 22 |
| Pectic polysaccharides | | 235 | | | 36 |
| Phencyclidine | | 214 | | | 4 |
| Pseudoephedrine | | 220 | | | 9 |
| Pseudouridine | | 210, 260 | | | 3 |
| Staurosporine | | 292 | | | 36 |
| Strycnine | | 214 | | | 4 |
| Suprofen | | 300 | | Molar extinction coefficient=14300 ($M^{-1}$) | 17 |
| Testosterone | | 245 | | | 2 |

**TABLE 65.6** UV-Visible Absorption Data (continued)

| Species | Target Organ or Disease | Absorbance Maximum (nm) | Matrix | Comments | Ref. |
|---|---|---|---|---|---|
| Tetrasialo-transferrin | | 200 | | | 16 |
| Thebaine | | 214 | | | 4 |
| Tiaprofenic acid | | 314 | | Molar extinction coefficient=14000 ($M^{-1}$) | 17 |
| Trimebutine maleate | | 214 | | | 38 |
| Trisialo-transferrin | | 200 | | | 16 |
| UDP-galactose | | 262 | | | 1 |
| UDP-glucose | | 262 | | | 1 |
| UDP-*N*-acetylgalactosamine | | 262 | | | 1 |
| UDP-*N*-acetylglucosamine | | 262 | | | 1 |
| Unfolding acylphosphatase | | 214 | | | 27 |
| Uridine | | 210, 260 | | | 3 |
| Xanthosine | | 210, 260 | | | 3 |
| | | ***Tissue/Biofluid Studies (in Vitro)*** | | | |
| (*E*)-5-(2-Bromovinyl)-2′-deoxyuridine | | 254 | | Plasma | 39 |
| (*E*)-5-(2-Bromovinyl)-2′-deoxyuridine | | 254 | | Urine | 39 |
| (*E*)-5-(2-Bromovinyl)-uracil | | 254 | | Plasma | 39 |
| (*E*)-5-(2-Bromovinyl)-uracil | | 254 | | Urine | 39 |
| 2,8-Dihydroxyadenine | | 300 | | Urine | 40 |
| 2-Carboxyibuprofen | | 214 | | Urine | 41 |
| 2-Ethylidene-1,5-dimethyl-3,3-diphenylpyrrolidine | | 195 | | Urine | 42 |
| 2-Hydroxyibuprofen | | 214 | | Urine | 41 |
| 3-Indoxyl sulfate | | 220 | | Urine | 43 |
| 4-Hydroxydebrisoquine | | 195 | | Urine | 44 |
| 4-Hydroxymephenytoin | | 192 | | Urine | 45 |
| 4-Hydroxyphenytoin | | 192 | | Urine | 45 |
| 5-Hydroxyindole-3-acetic acid | | 220 | | Urine | 43 |
| 5-Hydroxytryptophan | | 220 | | Urine | 43 |
| Acetaminophen | | 200 | | Plasma | 45 |
| Acetylcarnitine | | 260 | | Urine | 47 |
| Adenine | | 254 | | Urine | 40 |
| Albendazole | | 280 | | Plasma | 48 |
| Albendazole sulfone | | 280 | | Plasma | 48 |
| Albendazole sulfoxide | | 280 | | Plasma | 48 |
| Allopurinol | | 254 | | Urine | 48 |
| Amphetamine and analogs | | 200 | | Urine | 49 |
| Atropine | | 200 | | Serum | 50 |
| Bambuterol | | 205 | | Plasma | 51 |
| Bunitrolol | | 210 | | Plasma | 52 |
| Carbamazepine | | 210 | | Plasma | 53 |
| Carbamazepine-10,11-epoxide | | 210 | | Plasma | 53 |
| Carnitine | | 260 | | Urine | 47 |
| CGP 6423, CGP 27905 | | 200 | | Urine | 54 |
| Cicletanine | | 214 | | Plasma, urine | 55 |
| Cicletanine glucuronide | | 214 | | Rat urine | 56 |
| Cicletaninesulfate | | 214 | | Rat urine | 56 |
| Clenbuterol | | 214 | | Urine | 57 |
| Creatinine | | 190 | | Urine | 58 |
| Dextrophan | | 200 | | Urine | 59 |
| Dihydrocodeine | | 210 | | Urine | 60 |
| Dihydrocodeine | | 210 | | Plasma | 61 |
| Dihydrocodeine-6-glucuronide | | 210 | | Urine | 60 |
| Ephedrine | | 214 | | Urine | 62 |

**TABLE 65.6** UV-Visible Absorption Data (continued)

| Species | Target Organ or Disease | Absorbance Maximum (nm) | Matrix | Comments | Ref. |
|---|---|---|---|---|---|
| Fluconazole | | 190 | | Plasma | 63 |
| Glucuronides | | 214 | | Urine | 4 |
| Hexanoylcarnitine | | 260 | | Urine | 47 |
| Hexobarbital | | 214 | | Rat plasma | 64 |
| Hippuric acid | | 190 | | Urine | 58 |
| Hippuric acid | | 254 | | Urine | 40 |
| Hypoxanthine | | 254 | | Urine | 40 |
| Ibuprofen | | 220 | | Serum | 65 |
| Isovalerylcarnitine | | 260 | | Urine | 66 |
| Leucovorin, 5-methyl-tetrahydrofolate | | 289 | | Plasma | 47 |
| Levorphanol | | 200 | | Urine | 59 |
| Mepivacaine | | 215 | | Serum | 67 |
| Naproxen | | 220 | | Serum | 43 |
| *N*-Demethyldimethindene | | 200 | | Urine | 68 |
| *N*-Demethyltramadol | | 195 | | Urine | 69 |
| Norcodeine | | 210 | | Plasma | 61 |
| Nordihydrocodeine | | 210 | | Urine | 60 |
| Norverapamil | | 200 | | Plasma | 70 |
| Octanoylcarnitine | | 260 | | Urine | 47 |
| *O*-Demethyl-*N*-demethyltramadol | | 195 | | Urine | 69 |
| *O*-Demethyltramadol | | 214 | | Urine | 71 |
| *O*-Demethyltramadol | | 195 | | Urine | 69 |
| Ondansetron | | 254 | | Serum | 72 |
| Orotic acid | | 280 | | Urine | 73 |
| Oxypurinol | | 254 | | Urine | 40 |
| Palitaxel | | 230 | | Plasma | 74 |
| Palitaxel | | 230 | | Urine | 74 |
| Pentobarbital | | 254 | | Serum | 76 |
| Prilocaine | | 215 | | Serum | 76 |
| Primaquine | | 210 | | Plasma | 51 |
| Propionylcarnitine | | 260 | | Urine | 47 |
| Quinidine | | 220 | | Urine | 43 |
| Quinidine | | 210 | | Plasma | 61 |
| *R*-(−)-Amphetamine | | 195 | | Urine | 77 |
| Reduced haloperidol | | 200 | | Plasma | 78 |
| Salicylate | | 220 | | Urine | 43 |
| Salicylate | | 220 | | Serum | 43 |
| Salicylic acid | | 275 | | Plasma | 46 |
| Salicyluric acid | | 220 | | Urine | 43 |
| Salicyluric acid | | 220 | | Serum | 43 |
| Secobarbital | | 254 | | Serum | 79 |
| Sulfaguanidine | | 260 | | Plasma | 46 |
| Sulfamethoxazole | | 275 | | Plasma | 46 |
| Terbutaline | | 200 | | Urine | 80 |
| Terbutaline | | 210 | | Plasma | 81 |
| Tolbutamide | | 275 | | Plasma | 46 |
| Trimethoprim | | 275 | | Plasma | 46 |
| Tryptophan | | 220 | | Urine | 43 |
| Tryptophan | | 220 | | Serum | 43 |
| Tyrosine | | 220 | | Urine | 43 |
| Tyrosine | | 220 | | Serum | 43 |
| Uracil | | 280 | | Urine | 73 |
| Urate | | 190 | | Urine | 58 |
| Urea | | 190 | | Urine | 58 |
| Uric acid | | 220 | | Serum | 43 |

**TABLE 65.6** UV-Visible Absorption Data (continued)

| Species | Target Organ or Disease | Absorbance Maximum (nm) | Matrix | Comments | Ref. |
|---|---|---|---|---|---|
| Vanillylmandelic acid | | 220 | | Urine | 43 |
| Verapamil | | 215 | | Serum | 82 |
| Warfarin | | 185 | | Plasma | 83 |
| Warfarin | | 310 | | Plasma | 84 |
| Xanthine | | 254 | | Urine | 40 |
| | | ***Animal Studies (in Vivo)*** | | | |
| 8-Core-modified porphyrins, 11-core-modified porphyrins | Colo-26 tumors in BALB/c mice | 694 | | | 85 |
| Aluminium phthalocyanine disulphonic acid | | 685 | | | 86 |
| Disulphonated aluminum phthalocyanine | Tumor-bearing mice | 680–685 | | | 86 |
| Haematoporphyrin derivative | Rat ears | 500, 625 | | | 87 |
| Liposomal zinc(II)-phthalocyanine | Tumor necrosis | 675 | | | 88 |
| *N*-Methyl-pyrrolidone | Tumor necrosis | 671 | | | 89 |
| Photodynamic therapy-induced damage | Rat liver | 625 | | | 90 |
| Photofrin | Mouse tumor | 625 | | | 91 |
| Verteporfin | Canine esophagus | 630 | | | 92 |

## References

1. Lehmann, R., Huber, M., Beck, A., Schindera, T., Rinkler, T., Houdali, B., Weigert, C., Haring, H., Voelter, W., and Schleicher, E.D., Simultaneous, quantitative analysis of UDP-*N*-acetylglucosamine, UDP-*N*-acetylgalactosamine, UDP-glucose and UDP-galactose in human peripheral blood cells, muscle biopsies and cultured mesangial cells by capillary zone electrophoresis, *Electrophoresis,* 21, 3010, 2000.
2. Sanwald, P., Blankson, E.A., Dulery, B.D., Schoun, J., Huebert, N.D., and Dow, J., Isocratic high-performance liquid-chromatographic method for the separation of testosterone metabolites, *J. Chromatogr. B,* 672, 207, 1995.
3. Liebich, H.M., Lehmann, R., Xu, G., Wahl, H.G., and Haring, H., Application of capillary electrophoresis in clinical chemistry: the clinical value of urinary modified nucleosides, *J. Chromatogr. B,* 745, 189, 2000.
4. Bjornsdottir, I. and Hansen, S.H., Fast separation of 16 seizure drug substances using non-aqueous capillary electrophoresis, *J. Biochem. Biophys. Methods,* 38, 155, 1999.
5. Stoyanovsky, D.A., Goldman, R., Jonnalagadda, S.S., Day, B.W., Claycamp, H.G., and Kagan, V.E., Detection and characterization of the electron paramagnetic resonance-silent glutathionyl-5,5-dimethyl-1-pyrroline N-oxide adduct derived from redox cycling of phenoxyl radicals in model systems and HL-60 cells, *Arch. Biochem. Biophys.,* 330, 3, 1996.
6. Vaillancourt, F.H., Barbosa, C.J., Spiro, T.G., Bolin, J.T., Blades, M.W., Turner, R.F., and Eltis, L.D., Definitive evidence for monoanionic binding of 2,3-dihydroxybiphenyl to 2,3-dihydroxybiphenyl 1,2-dioxygenase from UV resonance Raman spectroscopy, UV/Vis absorption spectroscopy, and crystallography, *J. Am. Chem. Soc.,* 124, 2485, 2002.

7. Gahunia, H.K., Lough, A., Vieth, R., and Pritzker, K., A cartilage derived novel compound DDP (2,6-dimethyldifuro-8-pyrone): isolation, purification, and identification, *J. Rheumatol.*, 29, 147, 2002.
8. Cheng, T.B. and Reineccius, G.A., A study of factors influencing 2-methoxy-3-isopropylpyrazine production by *Pseudomonas perolens* using acid trap and UV spectroscopy, *Appl. Microbiol. Biotechnol.*, 36(3), 304, 1991.
9. Xu, X. and Stewart, J.T., MEKC determination of guaifenesin, pseudoephedrine, and dextromethorphan in a capsule dosage form, *J. Liq. Chromatogr. Rel. Technol.*, 23, 1, 2000.
10. Hofstetter, H., Morpurgo, M., Hofstetter, O., Bayer, E.A., and Wilchek, M., A labeling, detection, and purification system based on 4-hydroxyazobenzene-2-carboxylic acid: an extension of avidin-biotin system, *Anal. Biochem.*, 284, 354, 2000.
11. Shin, H.C., Shimoda, M., and Kokue, E., Identification of 5,10-methylenetetrahydrofolate in rat bile, *J. Chromatogr. B*, 661, 237, 1994.
12. Visky, D., Kraszni, M., Hosztafi, S., and Noszal, B., HPCE analysis of hydrolysing morphine derivatives. Quantitation of decomposition rate and mobility, *Chromatographia*, 51, 294, 2000.
13. Saccomani, A., Cecilia, G., Wajcman, H., and Righetti, P.G., Detection of neutral and charged mutations in α- and β-human globin chains by capillary zone electrophoresis in isoelectric, acidic buffers, *J. Chromatogr. A*, 832, 225, 1999.
14. Kanzaki, H., Imura, D., Nitoda, T., and Kawazu, K., Enzymatic dehydrogenation of cyclo(L-Phe-L-Leu) to a bioactive derivative, albonoursin, *J. Mol. Catalysis B-Enzymatic*, 6, 265, 1999.
15. Lochmann, H., Bazzanella, A., Kropsch, S., and Bachmann, K., Determination of tobacco alkaloids in single plant cells by capillary electrophoresis, *J. Chromatogr. A*, 917, 311, 2001.
16. Crivellente, F., Fracasso, G., Valentini, R., Manetto, G., Riviera, A.P., and Tagliaro, F., Improved method for carbohydrate-deficient transferrin determination in human serum by capillary zone electrophoresis, *J. Chromatogr. B*, 739, 81, 2000.
17. Bosca, F., Marin, L., and Miranda, M.A., Photoreactivity of the nonsteroidal anti-inflammatory 2-arylpropionic acids with photosensitizing side effects, *Photochem. Photobiol.*, 74, 637, 2001.
18. Lurie, I.S., Bethea, J.M., Mckibben, T.D., Hays, P.A., Pelligrini, P., Sahai, R., Garcia, A.D., and Weinberger, R., Use of dynamically coated capillaries for the routine analysis of methamphetamine, amphetamine, MDA, MDMA, MDEA, and cocaine using capillary electrophoresis, *J. Forens. Sci.*, 46, 1025, 2001.
19. Luquain, C., Laugier, C., Lagarde, M., and Pageaux, J.F., High-performance liquid chromatography determination of bis(monoacylglycerol) phosphate and other lysophospholipids, *Anal. Biochem.*, 296, 41, 2001.
20. Namiki, S., Kaneda, F., Ikegami, M., Arai, T., Fujimori, K., Asada, S., Hama, H., Kasuya, Y., and Goto, K., bis-*N*-nitroso-caged nitric oxides: photochemistry and biological performance test by rat aorta vasorelaxation, *Bioorg. Med. Chem.*, 7(8), 1695, 1999.
21. Ma, L., Gong, X., Yeung, E.S., Combinational screening of enzyme activity by using multiplexed capillary electrophoresis, *Anal. Chem.*, 72, 3383, 2000.
22. Brown, S.B., *Introduction to Spectroscopy for Biochemistry*, Academic Press, London, 1980.
23. Waheed, A.A. and Gupta, P.D., Application of an eosin B dye method for estimating a wide range of proteins, *J. Biochem. Biophys. Methods*, 33, 187, 1996.
24. Li, Z., Gu, T., Kelder, B., and Kopchick, J.J., Analysis of fatty acids in mouse cells using reversed-phase high-performance liquid chromatography, *Chromatographia*, 54, 463, 2001.
25. Kelm, M., Dahmann, R., Wink, D., and Feelisch, M., The nitric oxide/superoxide assay. Insights into the biological chemistry of the $O_2^{\bullet -}$ interaction, *J. Biol. Chem.*, 272, 9922, 1997.
26. Zupfer, J.M., Churchill, K.E., Rasmusson, D.C., and Fulcher, R.G., Variation in ferulic acid concentration among diverse barley cultivars measured by HPLC and microspectrophotometry, *J. Agric. Food Chem.*, 46, 1350, 1998.
27. Righetti, P.G., Gelfi, C., Bossi, A., Olivieri, E., Castelletti, L., Verzola, B., and Stoyanov, A.V., Capillary electrophoresis of peptides and proteins in isoelectric buffers: an update, *Electrophoresis*, 21, 4046, 2000.

28. St'astna, M., Radko, S.P., and Chrambach, A., Separation efficiency in protein zone electrophoresis performed in capillaries of different diameters, *Electrophoresis,* 21, 985, 2000.
29. Nielsen, S.B., Lapierre, A., Andersen, J.U., Pedersen, U.V., Tomita, S., and Andersen, L.H., Absorption spectrum of the green fluorescent protein chromophore anion *in vacuo, Phys. Rev. Lett.,* 87, 228102, 2001.
30. Cheng, J. and Mitchelson, K.R., Improved separation of six harmane alkaloids by high-performance capillary electrophoresis, *J. Chromatogr. A,* 761, 297, 1997.
31. Zargar, S.J. and Rabbani, A., Interaction of daynomycin antibiotic with histone H-1: ultraviolet spectroscopy and equilibrium dialysis studies, *Int. J. Biol. Macromol.,* 30, 113, 2002.
32. Messana, I., Rossetti, D. V., Cassiano, L., Misiti, F., Giardina, B., and Castagnola, M., Peptide analysis by capillary (zone) electrophoresis, *J. Chromatogr. B,* 699, 149, 1997.
33. Froescheis, O., Moalli, S., Liechti, H., and Bausch, J., Determination of lycopene in tissues and plasma of rats by normal-phase high-performance liquid chromatography with photometric detection, *J. Chromatogr. B,* 739, 291, 2000.
34. Bouaicha, N., Ammar, M., Hennion, M.C., and Sandra, P., A new method for determination of maitotoxin by capillary zone electrophoresis with ultraviolet detection, *Toxicon,* 35, 955, 1997.
35. Proksa, B., Separation of morphine and its oxidation products by capillary zone electrophoresis, *J. Pharm. Biomed. Anal.,* 20, 179, 1999.
36. Hotchkiss, A.T. and Hicks, K.B., Analysis of pectate lyase-generated oligogalacturonic acids by high-performance anion-exchange chromatography with pulsed amperometric detection, *Carbohydrate Res.,* 247, 1, 1993.
37. Gurley, L.R., Umbarger, K.O., Kim, J.M., Bradbury, E.M., and Lehnert, B.E., Development of a high-performance liquid-chromatographic method for the analysis of staurosporine, *J. Chromatogr. B,* 670, 125, 1995.
38. Li, F. and Yu, L., Determination of trimebutine maleate in rat plasma and tissues by using capillary zone electrophoresis, *Biomed. Chromatogr.,* 15, 248, 2001.
39. Olgemoller, J., Hempel, G., Boos, J., and Blaschke, G., Determination of (E)-5-(2-bromovinyl)-2′-deoxyuridine in plasma and urine by capillary electrophoresis, *J. Chromatogr. B,* 726, 261, 1999.
40. Wessel, T., Lanvers, C., Fruend, S., and Hempel, G., Determination of purines including 2,8-dihydroxyadenine in urine using capillary electrophoresis, *J. Chromatogr. A,* 894, 157, 2000.
41. Bjornsdottir, I., Kepp, D.R., Tjornelund, J., and Hansen, S.H., Separation of the enantiomers of ibuprofen and its major phase I metabolites in urine using capillary electrophoresis, *Electrophoresis,* 19, 455, 1998.
42. Lanz, M. and Thormann, W., Characterization of the stereoselective metabolism of methadone and its primary metabolite via cyclodextrin capillary electrophoretic determination of their urinary enantiomers, *Electrophoresis,* 17, 1945, 1996.
43. Caslavska, J., Gassmann, E., and Thormann, W., Modification of a tunable UV-visible capillary electrophoresis detector for simultaneous absorbency and fluorescence detection — profiling of body-fluids for drugs and endogenous compounds, *J. Chromatogr. A,* 709(1), 147, 1995.
44. Lanz, M., Theurillat, R., and Thormann, W., Characterization of stereoselectivity and genetic polymorphism of the debrisoquine hydroxylation in man via analysis of urinary debrisoquine and 4-hydroxydebrisoquine by capillary electrophoresis, *Electrophoresis,* 18, 1875, 1997.
45. Desiderio, C., Fanali, S., Kupfer, A., and Thormann, W., Analysis of mephenytoin, 4-hydroxymephenytoin enantiomers in human urine by cyclodextrin micellar electrokinetic capillary chromatography-simple determination of a hydroxylation polymorphism in man, *Electrophoresis,* 15, 87, 1994.
46. Kunkel, A. and Watzig, H., Micellar electrokinetic capillary chromatography as a powerful tool for pharmacological investigations without sample pretreatment: a precise technique providing cost advantages and limits of detection to the low nanomolar range, *Electrophoresis,* 20, 2379, 1991.
47. Vernez, L., Thormann, W., and Krahenbuhl, S., Analysis of carnitine and acylcarnitines in urine by capillary electrophoresis, *J. Chromatogr. A,* 895, 309, 2000.

48. Prochazkova, A., Chouki, M., Theurillat, R., and Thormann, W., Therapeutic drug monitoring of albendazole: determination of albendazole, albendazole sulfoxide, and albendazole sulfone in human plasma using nonaqueous capillary electrophoresis, *Electrophoresis,* 21, 729, 2000.
49. Varesio, E. and Veuthey, J.-L., Chiral separation of amphetamines by high-performance capillary electrophoresis, *J. Chromatogr. A,* 717, 219, 1995.
50. Jin, L.J., Wang, Y., Xu, R., Go, M.L., Lee, H.K., and Li, S.F.Y., Chiral resolution of atropine, homatropine and eight synthetic tropinyl and piperidinyl esters by capillary zone electrophoresis with cyclodextrin additives, *Electrophoresis,* 20, 198, 1999.
51. Palmarsdottir, S., Mathiasson, L., Jonsson, J.A., and Edholm, L.-E., Determination of a basic drug, bambuterol, in human plasma by capillary electrophoresis using double stacking for large volume injection and supported liquid membranes for sample pretreatment, *J. Chromatogr. B,* 688, 127, 1997.
52. Tanaka, Y., Yanagawa, M., and Terabe, S., Separation of neutral and basic enantiomers by cyclodextrin electrokinetic chromatography using anionic cyclodextrin derivatives as chiral pseudostationary phases, *J. High Resolut. Chromatogr.,* 19, 421, 1996.
53. Kuldvee, R. and Thormann, W., Determination of carbamazepine and carbamzepine-10,11-epoxide in human serum and plasma by micellar electrokinetic capillary chromatography in the absence of electroosmosis, *Electrophoresis,* 22, 1345, 2001.
54. Li, F., Cooper, S.F., and Mikkelsen, S.R., Enantioselective determination of oxprenolol and its metabolites in human urine by cyclodextrin-modified capillary zone electrophoresis, *J. Chromatogr. B,* 674, 227, 1995.
55. Prunonosa, J., Obach, R., Diez-Cascon, A., and Gouesclou, L., Determination of cicletanine enantiomers in plasma by high-performance capillary electrophoresis, *J. Chromatogr.,* 574, 127, 1992.
56. Garay, R.P., Rosati, C., Fanous, K., Allard, M., Morin, E., Lamiable, D., and Vistelle, R., Evidence for (+)-cicletanine sulfate as an active antriuretic metabolite of cicletanine in the rat, *Eur. J. Pharmacol.,* 274, 175, 1995.
57. Altria, K.D., Goodall, D.M., and Rogan, M.M., Quantitative applications and validation of the resolution of enantiomers by capillary electrophoresis, *Electrophoresis,* 15, 824, 1994.
58. Alfazema, L.N., Howells, S., and Perrett, D., Optimised separation of endogenous urinary components using cyclodextrin-modified micellar electrokinetic capillary chromatography, *Electrophoresis,* 21, 2503, 2000.
59. Aumatell, A. and Wells, R.J., Chiral differentiation of the optical isomers of racemethorphan and racemorphan in urine by capillary zone electrophoresis, *J. Chromatogr. Sci.,* 31, 502, 1993.
60. Hufschmid, E., Theurillat, R., Martin, U., and Thormann, W., Exploration of the metabolism of dihydrocodeine via determination of its metabolites in human urine using micellar electrokinetic capillary chromatography, *J. Chromatogr. B,* 668, 159, 1995.
61. Wey, A.B., Zhang, C., and Thormann, W., Head-column field-amplified sample stacking in binary system capillary electrophoresis preparation of extracts for determination of opioids in microliter amounts of body fluids, *J. Chromatogr. A,* 853, 95, 1999.
62. Mazzeo, J.R., Grover, E.R., Swartz, M.E., and Petersen, J.S., Novel chiral surfactant for the separation of enantiomers by micellar electrokinetic capillary chromatography, *J. Chromatogr. A,* 680, 125, 1994.
63. Heeren, F., Tanner, R., Theurillat, R., and Thormann, W., Determination of fluconazole in human plasma by micellar electrokinetic capillary chromatography with detection at 190 nmJ, *J. Chromatogr. A,* 745, 165, 1996.
64. Francotte, E., Cherkaoui, S., and Faupel, M., Separation of the enantiomers of some racemic nonsteroidal aromatase inhibitors and barbiturates by capillary electrophoresis, *Chirality,* 5, 516, 1993.
65. Soini, H., Stefansson, M., Riekkola, M.-L., and Novotny, M.V., Maltooligosaccharides as chiral selectors for the separation of pharmaceuticals by capillary electrophoresis, *Anal. Chem.,* 66, 3477, 1994.

66. Shibukawa, A., Lloyd, D.K., and Wainer, I.W., Simultaneous chiral separation of leucovorin and its major metabolite 5-methyl-tetrahydrofolate by capillary electrophoresis using cyclodextrins as chiral selectors-estimation of the formation constant and mobility of the solute-cyclodextrin complexes, *Chromatographia*, 35, 419, 1993.
67. Siluveru, M. and Stewart, J.T., HPCE determination of R(+) and S(−) mepivacaine in human serum using a derivatized cyclodextrin and ultraviolet detection, *J. Pharm. Biomed. Anal.*, 15, 1751, 1997.
68. Heuermann, M. and Blaschke, G., Simultaneous enantioselective determination and quantification of dimethindene and its metabolite *N*-demethyl-dimethindene in human urine using cyclodextrins as chiral additives in capillary electrophoresis, *J. Pharm. Biomed. Anal.*, 12, 753, 1994.
69. Rudaz, S., Veuthey, J.L., Desiderio, C., and Fanali, S., Simultaneous stereoselective analysis by capillary electrophoresis of tramadol enantiomers and their main phase I metabolites in urine, *J. Chromatogr. A*, 846, 227, 1999.
70. Dethy, J.-M., De Broux, S., Lesne, M., Longstreth, J., and Gilbert, P., Stereoselective determination of verapamil and norverapamil by capillary electrophoresis, *J. Chromatogr. B*, 654, 121, 994.
71. Kurth, B. and Blaschke, G., Achiral and chiral determination of tramadol and its metabolites in urine by capillary electrophoresis, *Electrophoresis*, 20, 555, 1999.
72. Siluveru, M. and Stewart, J.T., Enantioselective determination of S-(+)- and R-(−)-ondansetron in human serum using derivatized cyclodextrin-modified capillary electrophoresis and solid-phase extraction, *J. Chromatogr. B.*, 691, 217, 1997.
73. Salerno, C., D'Eufemia, P., Celli, M., Finocchiaro, R., Crifo, C., and Giardini, O., Determination of urinary orotic acid and uracil by capillary zone electrophoresis, *J. Chromatogr. B*, 734, 175, 1999.
74. Hempel, G., Lehmkuhl, D., Krumpelmann, S., Blaschke, G., and Boos, J., Determination of paclitaxel in biological fluids by micellar electrokinetic chromatography, *J. Chromatogr. A*, 745, 173, 1996.
75. Srinivasan, K. and Bartlett, M.G., Capillary electrophoresis stereoselective determination of R-(+)- and S-(−)-pentobarbital from serum using hydroxypropyl-gamma-cyclodextrin, solid-phase extraction and ultraviolet detection, *J. Chromatogr. B*, 703, 289, 1997.
76. Siluveru, M. and Stewart, J.T., Stereoselective determination of R-(−)- and S-(+)-prilocaine in human serum by capillary electrophoresis using a derivatized cyclodextrin and ultraviolet detection, *J. Chromatogr. B*, 693, 205, 1997.
77. Ramseier, A., Caslavska, J., and Thormann, W., Stereoselective screening for and confirmation of urinary enantiomers of amphetamine, methamphetamine, designer drugs, methadone and selected metabolites by capillary electrophoresis, *Electrophoresis*, 20, 2726, 1999.
78. Wu, S.-M., Ko, W.-K., Wu, H.-L., and Chen, S.-H., Trace analysis of haloperidol and its chiral metabolite in plasma by capillary electrophoresis, *J. Chromatogr. A*, 846, 239, 1999.
79. Srinivasan, K., Zhang, W., and Bartlett, M.G., Rapid simultaneous capillary electrophoretic determination of (R)- and (S)-secobarbital from serum and prediction of hydroxypropyl-gamma-cyclodextrin-secobarbital stereoselective interaction using molecular mechanics simulation, *J. Chromatogr. Sci.*, 36, 85, 1998.
80. Sheppard, R.L., Tong, X., Cai, J., and Henion, J.D., Chiral separation and detection of terbutaline and ephedrine by capillary electrophoresis coupled with ion-spray mass-spectrometry, *Anal. Chem.*, 67, 2054, 1995.
81. Palmarsdottir, S. and Edholm, L.-E., Enhancement of selectivity and concentration sensitivity in capillary zone electrophoresis online coupling with column liquid-chromatography and utilizing a double stacking procedure allowing for microliter injections, *J. Chromatogr. A*, 693, 131, 1995.
82. Soini, H., Riekkola, M.-L., and Novotny, M.V., Chiral separations of basic drugs and quantitation of bupivacaine enantiomers in serum by capillary electrophoresis with modified cyclodextrin buffers, *J. Chromatogr.*, 608, 265, 1992.

83. D'Hulst, A. and Verbeke, N., Separation of the enantiomers of coumarinic anticoagulant drugs by capillary electrophoresis using maltodextrins as chiral modifiers, *Chirality,* 6, 225, 1994.
84. Gareil, P., Gramond, J.P., and Guyon, F., Separation and determination of warfarin enantiomers in human plasma samples by capillary zone electrophoresis using a methylated beta-cyclodextrin-containing electrolyte, *J. Chromatogr.,* 615, 317, 1993.
85. Hilmey, D.G., Abe, M., Nelen, M.I., Stilts, C.E., Baker, G.A., Baker, S.N., Bright, F.V., Davies, S.R., Gollnick, S.O., Oseroff, A.R., Gibson, S.L., Hilf, R., and Detty, M.R., Water-soluble, core-modified porphyrins as novel, longer-wavelength-absorbing sensitizers for photodynamic therapy. II. Effects of core heteroatoms and meso-substituents on biological activity, *J. Med. Chem.,* 45, 449, 2002.
86. Cubeddu, R., Canti, G., Musolino, M., Pifferi, A., Taroni, P., and Valentini, G., *In vivo* absorption spectrum of disulphonated aluminium phthalocyanine in tumor bearing mice, Fifth Biannual Meeting, International Photodynamic Association, Amelia Island, FL, September 21–24, 1994, Abstract 74.
87. Star, W.M., Versteeg, A.A.A., Van Putten, W.L.J., and Marijnissen, J.P.A., Wavelength dependence of hematoporphyrin derivative photodynamic treatment effects on rat ears, *Photochem. Photobiol.,* 52, 547, 1990.
88. Van Leengoed, H.L.L.M., Cuomo, V., Versteeg, A.A.C., Van der Veen, N., Jori, G., and Star, W.M., *In vivo* fluorescence and photodynamic activity of zinc phthalocyanine administered in liposomes, *Br. J. Cancer,* 69, 840, 1994.
89. Schieweck, K., Isele, U., Capraro, H.-G., Ochsner, M., Maurer, Th., Kratz, J., Gentsch, C., Jori, G., Segalla, A., and Biolo, R., Preclinical studies with CGP 55847, liposomal zinc(II)-phthalocyanine, Fifth Biannual Meeting International Photodynamic Association, Amelia Island, FL, September 21–24, 1994, Abstract 8.
90. Farrell, T.J., Olivo, M.C., Patterson, M.S., Wrona, H., and Wilson, B.C., Investigation of the dependence of tissue necrosis on irradiation wavelength and time post injection using a photodynamic threshold dose model, in Spinelli, P., Dal Fante, M., and Marchesini, R., Eds., *Photodynamic Therapy and Biomedical Lasers,* Elsevier, Amsterdam, 1992, p. 830.
91. Potter, W.R., Bellnier, D.A., and Oseroff, A.R., Tissue concentration of sensitizer by *in vivo* reflectance spectroscopy, Fifth Biannual Meeting International Photodynamic Association, Amelia Island, FL, 21–24 September 1994, Abstract 46.
92. Panjehpour, M., DeNovo, R.C., Petersen, M.G., Overholt, B.F., Bower, R., Rubinchik, V., and Kelly, B., Photodynamic therapy using verteporfin (benzoporphyrin derivative monoacid ring A, BPD-NU) and 630 nm laser light in canine esophagus, *Lasers Surg. Med.,* 30, 26, 2002.


## 65.3.2 Infrared Absorption Spectroscopy

Table 65.7 provides information related to applications of FTIR spectroscopy in the field of biomolecular and biomedical research. Extensive literature is available on FTIR of biomolecules; only illustrative examples are given in the Biochemical Studies section, including data on side chains of amino acids (in water), along with a few examples on phospholipids, carbohydrates, and energy metabolites. The subsequent sections — Cellular Studies, Tissue and Biofluid Studies (*in Vitro*), and Animal Studies (*in Vivo*) — are arranged alphabetically based on target organ or disease, beginning on p. 65-24. The following nomenclature is used in this table. The band position of functional groups is given in $cm^{-1}$. Number in parentheses next to band position represents the extinction coefficient (ε) in $M^{-1}$ $cm^{-1}$. Relative band intensities are classified as very strong (VS), strong (S), medium (M), medium to weak (MW), weak (W), and very weak (VW). Table 65.7 is compiled from data from a large cross section of publications, including journals, review articles, and books.

TABLE 65.7 Infrared Absorption Data

| Species | Major IR Bands Functional Group | Shifts ($cm^{-1}$) | Matrix | Comments | Ref. |
|---|---|---|---|---|---|
| | | ***Biochemical Studies*** | | | |
| Acetyl choline receptor | | | | Alpha helical conformation | 1 |
| Alanine | $CH_3$ | 1470 | | In-plane bending (asymmetric) | 2 |
| Alanine | $CH_3$ | 1378 | | In-plane bending (symmetric) | 2 |
| Alanine-based peptides | | | | Random coil and helix | 3 |
| Arginine | $CN_3H_5^+$ | 1672–1673 (420–490) | | Stretching (asymmetric) | 4 |
| Arginine | $CN_3H_5^+$ | 1633–1636 (300–340) | | Stretching (symmetric) | 4 |
| Asparagine | C=O | 1677–1678 (310–330) | | Stretching | 5 |
| Asparagine | $NH_2$ | 1612–1622 (140–160) | | In-plane bending | 5 |
| Aspartic acid | C=O | 1717–1788 (in COOH) | | Stretching | 6 |
| Aspartic acid | COO– | 1574–1579 (290–380) | | Stretching (asymmetric) | 6 |
| Aspartic acid | COO– | 1402 (256) | | Stretching (symmetric) | 6 |
| Aspartic acid | C–O | 1120–1250 (COOH) (100–200) | | Stretching | 6 |
| ATP, GTP, AMP, DAMP, AMP | | | | | 7 |
| Beta-amyloid | | | | Beta sheet structure | 8 |
| Calf thymus DNA | | | | Left-handed Z type | 9 |
| Chymotrypsin | | | | Alpha helical conformation | 10 |
| Collagen IV | | | | Triple helical structure | 11 |
| Cysteine (SH) | SH | 2551 | | Stretching | 12 |
| Cysteine (SH) | $CH_2$ | 1424 | | In-plane bending | 12 |
| Cysteine (SH) | CH(COO–) | 1303 | | In-plane bending | 12 |
| Cysteine (SH) | $CH_2$(COO–) | 1269 | | Wagging | 12 |
| Cytochrome *c* oxidase | | | | Alpha helical conformation | 13 |
| FAD | | | | Characterization | 14 |
| Fibroin | | | | Beta sheet structure | 15 |
| Glutamic acid | COO– | 1556–1560 (450–470) | | Stretching (asymmetric) | 16 |
| Glutamic acid | $CH_2$ | 1452 (SH), 1440 (S) | | In-plane bending | 16 |
| Glutamic acid | $CH_2$ | 1359 (W), 1292 (W) | | Wagging | 16 |
| Glutamic acid | $CH_2$ | 1359 (W) | | Twisting | 16 |
| Glutamic acid | N–C | 1260 (MW) | | Stretching | 16 |
| Glutamic acid | C–H | 1260 (MW) | | In-plane bending | 16 |
| Glutamic acid | $CH_2$ | 1225 (SH), 1187 (MW), 1130 (M) | | Twisting | 16 |
| Glutamic acid | CH | 1225 (SH), 1187 (MW), 1130 (M) | | In-plane bending | 16 |
| Glutamic acid | C–C | 1074 (VW), 1040 (VW), 1018 (VW) | | Stretching | 16 |
| Glutamine | C=O | 1668–1687 (360–380, S) | | Stretching | 17 |
| Glutamine | $NH_2$ | 1586–1611 (220–240, M) | | In-plane bending | 17 |
| Glutamine | $CH_2$ | 1451 (M) | | In-plane bending | 17 |
| Glutamine | C–N | 1410 (S), 1084 (W) | | Stretching | 17 |
| Glutamine | CH | 1410 (S) | | In-plane bending | 17 |
| Glutamine | $CH_2$ | 1315 (M), 1281 (MW) | | Wagging | 17 |
| Glutamine | $CH_2$ | 1256 (MW), 1202 (W) | | Twisting | 17 |
| Glutamine | $NH_2$ | 1410 (S) | | Rocking | 17 |
| Glutamine | C–C | 1052 (M), 999 (MW), 926 (W) | | Stretching | 17 |
| Glycine | $CH_2$ | 1441–1446 | | In-plane bending | 18 |
| Glycine | CH | 1441–1446 | | In-plane bending | 18 |
| Hemoglobin | | | | Apha helical conformation | 19 |
| Histidine | C=C | 1575 (S), 1594 (70, S) | | Stretching | 20 |
| Histidine | C–C | 1575 (S), 1594 (70, S) | | Stretching | 20 |
| Histidine | C=N | 1490 (S) | | Stretching | 20 |
| Histidine | CH | 1490 (S) | | In-plane bending | 20 |

**TABLE 65.7** Infrared Absorption Data (continued)

| Species | Major IR Bands Functional Group | Shifts ($cm^{-1}$) | Matrix | Comments | Ref. |
|---|---|---|---|---|---|
| Histidine | NH | 1423 (M) | | In-plane bending | 20 |
| Histidine | C–N | 1423 (M) | | Stretching | 20 |
| Histidine | $CH_x$ | 1423 (M) | | In-plane bending | 20 |
| Histidine | C=N | 1304 (M) | | Stretching | 20 |
| Histidine | C–N | 1304 (M) | | Stretching | 20 |
| Histidine | C–N | 1265 (S) | | Stretching | 20 |
| Histidine | C–C | 1265 (S) | | Stretching | 20 |
| Histidine | CH | 1229 (S) | | In-plane bending | 20 |
| Histidine | C–N | 1229 (S) | | Stretching | 20 |
| Histidine | NH | 1229 (S) | | | 20 |
| Histidine | C–N | 1153 (M), 1161 (M) | | Stretching | 20 |
| Histidine | NH | 1153 (M), 1161 (M) | | In-plane bending | 20 |
| Histidine | C–N | 1090 (S), 1106 (S) | | Stretching | 20 |
| Histidine | CH | 1090 (S), 1106 (S) | | In-plane bending | 20 |
| Histidine | CH | 995 (M) | | In-plane bending | 20 |
| Histidine | C–C | 995 (M) | | Stretching | 20 |
| Histidine | C–N | 975 (M) | | Stretching | 20 |
| Histidine | $CH_x$ | 975 (M) | | In-plane bending | 20 |
| Histidine | Ring | 975 (M) | | In-plane bending | 20 |
| Histidine | Ring | 941 (S) | | In-plane bending | 20 |
| Histidine | C–C | 941 (S) | | Stretching | 20 |
| Isoleucine | $CH_3$ | 1445 | | In-plane bending (asymmetric) | 21 |
| Isoleucine | $CH_2$ | 1470 | | In-plane bending | 21 |
| Isoleucine | CH | 1320 | | In-plane bending | 21 |
| Leucine | $CH_3$ | 1445 | | In-plane bending (asymmetric) | 2 |
| Leucine | $CH_2$ | 1470 | | In-plane bending | 2 |
| Lysine | $NH_3^+$ | 1626–1629 (60–130) | | In-plane bending (asymmetric) | 4 |
| Lysine | $NH_3^+$ | 1526–1527 (70–100) | | In-plane bending (symmetric) | 4 |
| Lysozyme | | | | Helix and beta sheet | 22 |
| Oligosaccharides | | | | Characterization | 23 |
| Phenylalanine | C–C | 1605 (30), 1585, 1494 (80) | | Stretching | 23 |
| Phospholipids | | | | Fluidity and phase behavior | 24 |
| Phosphatidycholine | | | | Chain length and hydration properties | 25 |
| Phosphatidycholine | | | | Characterization | 26 |
| Phospholipids bilayer | Bilayer | | | Structure and organization | 27 |
| Poly-L-Lysine | | | | Helix-sheet transition studied by FTIR | 28 |
| Polynucleotide | | | | Right-handed B type | 29 |
| Polynucleotide | | | | Triple helical structure | 30 |
| Porin | | | | Beta sheet structure | 31 |
| Proline | $CH_2$ | 1292 (S), 1168 (S) | | Twisting | 32 |
| Proline | $CH_2$ | 1472 (S) | | In-plane bending | 34 |
| Proline | $CH_2$ | 1375 (S), 1317 (S) | | In-plane bending | 34 |
| Proline | C–N | 1435–1465, 979 (M), 945 (M), 911 (M) | | Stretching | 34 |
| Proline | $CH_2$ | 1253 (M),1051 (W), 1033 (M) | | Wagging | 34 |
| Proline | $CH_2$ | 1083 (M) | | Rocking | 34 |
| Proline | C–C | 979 (M), 945 (M), 911 (M) | | Stretching | 34 |
| Rec A | | | | Alpha helical conformation | 33 |
| Rhodopsin | | | | Alpha helical conformation | 34 |
| RNA | | | | Triple helical structure | 35 |
| Serine | $CH_2$ | 1450 | | | 12 |
| Serine | CH | 1364 | | | 12 |
| Serine | C–C | 1364 | | Stretching | 12 |
| Serine | $CH_2$ | 1352 | | Twisting | 12 |

**TABLE 65.7** Infrared Absorption Data (continued)

| Species | Major IR Bands Functional Group | Shifts (cm$^{-1}$) | Matrix | Comments | Ref. |
|---|---|---|---|---|---|
| Serine | COO– | 1352 | | Stretching | 12 |
| Serine | CH | 1352 | | | 12 |
| Serine | $CH_2$ | 1312 | | Wagging | 12 |
| Serine | CH | 1312 | | | 12 |
| Serine | $CH_2$ | 1248 | | Twisting | 12 |
| Serine | COH | 1181 | | | 12 |
| Serine | CO | 1030 | | Stretching | 12 |
| Serine | $NH_3^+$ | 1030 | | Rocking | 12 |
| Serine | CO | 983 | | Stretching | 12 |
| Serum albumin | | | | Alpha helical conformation | 36 |
| Soft cuticle protein | | | | Beta sheet structure | 37 |
| Sugar–metal complexes | | | | Characterization | 38 |
| Sugars | | | | Determination of glucose, fructose, sucrose | 39 |
| Threonine | COH | 1385–1420 (MW), 1225–1330 (MW) | | | 2 |
| Threonine | CH | 1385–1420 (MW), 1225–1330 (MW) | | | 2 |
| Threonine | C–O | 1075–1150 (S), 910–950 | | | 2 |
| Transferrin | | | | Alpha helical conformation | 40 |
| Tryptophan | C–C benzene | 1612 (M), 1576 (W) | | Stretching | 41 |
| Tryptophan | C=C pyrole | 1612 (M), 1576 (W) | | Stretching | 41 |
| Tryptophan | C–C benzene | 1487 (S) | | Stretching | 41 |
| Tryptophan | C–H benzene | 1487 (S) | | In-plane bending | 41 |
| Tryptophan | C–N | 1509 (S) | | Stretching | 41 |
| Tryptophan | C–H pyrole | 1509 (S) | | In-plane bending | 41 |
| Typtophan | N–H | 1509 (S) | | In-plane bending | 41 |
| Tryptophan | C–H benzene | 1455 (VS) | | In-plane bending | 41 |
| Tryptophan | C–C benzene | 1455 (VS) | | Stretching | 41 |
| Tryptophan | C–N | 1455 (VS) | | Stretching | 41 |
| Tryptophan | N–H | 1412 (S) | | In-plane bending | 41 |
| Tryptophan | C–C pyrole | 1412 (S) | | Stretching | 41 |
| Tryptophan | C–H benzene | 1412 (S) | | In-plane bending | 41 |
| Tryptophan | C–C pyrole | 1352 (VS) | | Stretching | 41 |
| Tryptophan | C–N | 1352 (VS) | | Stretching | 41 |
| Tryptophan | C–H benzene | 1352 (VS) | | In-plane bending | 41 |
| Tryptophan | C–C pyrole, C–N | 1334 (VS) | | Stretching | 41 |
| Tryptophan | N–H | 1276 (S) | | In-plane bending | 41 |
| Tryptophan | C–N | 1276 (S) | | Stretching | 41 |
| Tryptophan | C–H benzene | 1245 (S), 970 (SH), 1010 (S), 1147 (W), 1119 (M) | | In-plane bending | 41 |
| Tryptophan | C–H benzene | 1245 (S) | | In-plane bending | 41 |
| Tryptophan | C–C pyrole | 1245 (S), 1119(M), 1010 (S), 970 (SH) | | Stretching | 41 |
| Tryptophan | C–C benzene | 1245 (S) | | Stretching | 41 |
| Tryptophan | C–C pyrole | 1245 (S), 1092 (VS) | | Stretching | 41 |
| Tryptophan | C–H pyrole | 1064 (S) | | In-plane bending | 41 |
| Tryptophan | N–C | 1064 (S) | | Stretching | 41 |
| Tryptophan | C–H pyrole | 1064 (S) | | In-plane bending | 41 |
| Tyrosine | C–C ring | 1614–1621 (85–150) | | Stretching | 42 |
| Tyrosine | C–H ring | 1614–1621 (85–150) | | In-plane bending | 42 |
| Tyrosine | C–C ring | 1516–1518 (340–430) | | Stretching | 42 |
| Tyrosine | C–H ring | 1516–1518 (340–430) | | In-plane bending | 42 |
| Tyrosine | $CH_2$ | 1447–1454 (~80) | | In-plane bending | 42 |
| Tyrosine | $CH_2$ | 1376–1387 (~50) | | Wagging | 42 |

**TABLE 65.7** Infrared Absorption Data (continued)

| Species | Major IR Bands Functional Group | Shifts ($cm^{-1}$) | Matrix | Comments | Ref. |
|---|---|---|---|---|---|
| Tyrosine | C–H | 1326, 1335 (<30), 1290–1295 (~50), 1170–1179 (~50), 1100–1111 (~40) | | In-plane bending | 42 |
| Tyrosine | C–C | 1326, 1335 (<30), 1290–1295 (~50) | | Stretching | 42 |
| Tyrosine | C–O, C–C | 1235–1270 (200) | | Stretching | 42 |
| Tyrosine | COH | 1169–1260 (200), 1083 | | In-plane bending | 42 |
| Unsaturated phospholipids | | | | Conformational studies | 43 |
| Valine | $CH_3$ | 1460 | | In-plane bending (asymmetric) | 2 |
| Valine | CH | 1355, 1320 | | In-plane bending | 2 |

**TABLE 65.7** Infrared Absorption Data (continued)

| Target Organ or Disease | Shifts ($cm^{-1}$) | Matrix | Comments | Ref. |
|---|---|---|---|---|
| | | ***Cellular Studies*** | | |
| Bacteria | | | Differentiation process of bacteria studied by ATR-FTIR (bacteria) | 44 |
| Bacteria | | | Quantification from a binary mixture using FTIR (bacteria) | 45 |
| Bacteria | | | Characterization (bacteria) | 46 |
| Bacteria | | | Effect of antibiotics on bacteria studied by FTIR (bacteria) | 47 |
| Bacteria | | | Identification of bacteria using FTIR (bacteria) | 48 |
| BALB/c | | ZnSe | Transformed by H-ras | 49 |
| Bone marrow cells | | | Detection of hydroxyapatite (rat) | 50 |
| Bronchial | 820–890, O–O stretching | | Detection of damage to airway cells (human) | 51 |
| Cervical cancer | | | Detection of cervical cancer (human) | 52 |
| Cervical cancer | | | Detection of cervical cancer using ATR/FTIR study (human) | 53 |
| Cervical cancer | 1025, Glycogen; 1080, glycogen and nucleic acids; 1240, nucleic acids; 1400, $CH_3$ from lipids and proteins; 1450, $CH_2$ from lipids and proteins | | Detection of cervical cancer (human) | 54 |
| Cervical cancer | Peak at 972 is main indicator of malignancy | | Detection of cervical cancer (human) | 55 |
| *Deinococcus radiodurans* | | | Pharmacological application of FTIR (bacteria) | 56 |
| HL60 | | | Differentiation and apoptosis process studied by FTIR | 57 |
| Keratinocytes | | | Lower organization of lipids compare to human skin (human) | 58 |
| Leukemia | | $BaF_2$ | Detection of leukemia (human) | 59 |
| Leukemia | | | Isolated from thrombocythemic patients (human) | 60 |
| Lung cancer cells | 1030, Glycogen | | FTIR microscopic study (human) | 61 |
| Lung fibroblast | | | Cell cycle and death studied by synchrotron FTIR microscopy (human) | 62 |
| Lymphocytes | | | Histocompatibility matching using FTIR microscopy (human) | 63 |
| Megakaryocytes | | | | 60 |
| NIH3T3 | | ZnSe | Fibroblasts transformed by retrovirus | 64 |
| Normal and malignant fibroblast | | | Characterization using FTIR microscopy (human) | 65 |
| Primary cells | Normal 1082, malignant 1086 | ZnSe | Cells transformed by mouse sarcoma virus (rabbit) | 66 |
| Sperm, erythrocytes | | | FTIR microscopic study (human) | 67 |
| | | ***Tissue and Biofluid Studies (in Vitro)*** | | |
| Alzheimer's disease | | | Brain tissue studied using synchrotron FTIR microscopy (human) | 68 |
| Amniotic fluid | | | Analysis of amniotic fluid using FTIR (human) | 69 |
| Blood | | | Analysis of blood (human) | 70 |
| Brain | | | Enhanced resolution due to deuteration (rat) | 71 |
| Brain | | | Characterization using IR imaging (monkey) | 72 |
| Brain | | | Effect of ARA-C on brain tissue studied using FTIR microscopy (rat) | 73 |
| Brain tumor | | | Characterization of pituitary adenomas (human) | 74 |
| Brain, breast | | | Characterization (human) | 75 |
| Breast | | | Structural variation of proteins from normal to carcinomal states (human) | 76 |

**TABLE 65.7** Infrared Absorption Data (continued)

| Target Organ or Disease | Shifts ($cm^{-1}$) | Matrix | Comments | Ref. |
|---|---|---|---|---|
| Breast | | | Breast cancer detection using FTIR (human) | 77 |
| Breast | | | Grading of tumors using FTIR (human) | 78 |
| Breast cancer | | | Detection of breast cancer using ATR-FTIR (human) | 79 |
| Breast implant biopsies | | | Characterization (human) | 80 |
| Breast implants | | | FTIR imaging (human) | 81 |
| Cartilage | | $CaF_2$ | FTIR imaging studies (bovine) | 82 |
| Cartilage | | | FTIR imaging studies (bovine) | 83 |
| Cells, tissues | | | Detection of trapped $CO_2$ (human) | 84 |
| Cervical cancer | | | Detection of cervical cancer using FTIR microscopy (human) | 85 |
| Cirrhotic liver tissue | | | Liver tissue analysis (human) | 86 |
| Colon cancer | | ZnSe | Identification of polyp and carcinoma (human) | 87 |
| Colon carcinoma | 995–1045, 1195–1245 Phosphate symmetric and asymmetric stretching | | ATR studies (human) | 88 |
| Demineralized bone tissue | | | FTIR imaging studies (bovine) | 89 |
| Dentin biopsies | | | Maturation of organic and mineral contents | 90 |
| Follicle fluid | | | Characterization (human) | 91 |
| Heart | | | Collagen deposition using FTIR microscopy (hamster UM-X 7.1) | 92 |
| Heart | | | Characterization (human) | 93 |
| Heart tissue | | | Detection of myocardial infarction (rat) | 94 |
| Iliac crest biopsies | | | Osteoporosis detected by IR imaging (human) | 95 |
| Iliac crest biopsies | | | FTIR imaging (human) | 96 |
| Implanted tissue | | | Biodiagnostics (dog) | 97 |
| Implants | | | Vascular healing studies using ATR-FTIR (manmade) | 98 |
| Liver | | | Detection of hypercholesterolemicitry using FTIR microscopy (rabbit) | 99 |
| Liver | | | Tissue oxygenation studied in liver transplants using NIR | 100 |
| Liver and heart | 2800–3000, $CH_2$ and $CH_3$ symmetric and asymmetric stretching | | Detection of diabetes | 101 |
| Lung | 1045, Glycogen; 1467, cholesterol | | Lung cancer detection using FTIR microscopy (human) | 102 |
| Lung | | | Detection of mercury poisoning using ATR-FTIR (mouse) | 103 |
| Lymphoid tumors | | | Grading of lymphoid tumors using FTIR microscopy (human) | 104 |
| Mineralized tissue | | | Maturation in horse secondary dentin (horse) | 105 |
| Multiple sclerosis | | | Studies on multiple sclerosis using FTIR microscopy (human) | 106 |
| Multiple sites | | | Glucose sensing using NIR transmission (human) | 107 |
| Myocardial dysfunction | | | Detection of myocardial dysfunction (hamster CMP) | 108 |
| Niemann–Pick type C | | | Detection of Niemann–Pick type C (NPC) (mice BALB/c) | 109 |
| Oral | | | Oral cell carcinoma detection using FTIR (human) | 110 |
| Oral tissue | 1745, Triglycerides, absent in malignant samples | | FTIR-FEWS studies (human) | 111 |
| Oral/oropharyngeal squamous cell carcinoma | | | Oral/oropharyngeal squamous cell carcinoma detection using FTIR microscopy and statistical analysis (human) | 112 |

**TABLE 65.7** Infrared Absorption Data (continued)

| Target Organ or Disease | Shifts ($cm^{-1}$) | Matrix | Comments | Ref. |
|---|---|---|---|---|
| Placenta | | | Detection of gynecological diseases (human) | 113 |
| Plasma | B–H band 2493 | | Detection of boron after boron neutron capture therapy (human) | 114 |
| Polymer implants | | | Biodiagnostics (manmade) | 115 |
| Retinal tissue | | | Studies on oxidative stress using FTIR microscopy (rat) | 116 |
| Saliva, urine, serum | | | Measuring body water (human) | 117 |
| Scrapie infected | | | Scrapie detection in CNS using FTIR microscopy | 118 |
| Serum | | | Quantification of serum components (human) | 119 |
| Serum | | | Quantification of serum components (human) | 120 |
| Serum | | | Quantification of cholesterol (human) | 121 |
| Skin | | | Skin cancer diagnosis using FTIR-FEWS | 122 |
| Skin | | | Basal cell carcinoma detection using FTIR (human) | 123 |
| Stool | | | Quantifying lipid content in biological fluids (human) | 124 |
| Synovial fluid | | | Diagnosis of arthritis using NIR (human) | 125 |
| Urine | | | Quantification of urine components (human) | 126 |
| | | | ***Animal Studies (in Vivo)*** | |
| Brain | | | Noninvasive blood glucose monitoring using ATR-FTIR (human) | 127 |
| Brain | | | Monitoring local cortical blood flow using thermal IR imaging (human) | 128 |
| Brain | | | Thermal IR imaging (monkey) | 129 |
| Lip | | | FTIR-FEWS study (human) | 130 |
| Skin | | | Influence of environmental factors on skin studied using FTIR-FEWS (human) | 131 |
| Skin | | | Cerebrovascular activity studied by NIR (human) | 132 |
| Skin | | | Penetration depth study (human) | 133 |

## References


1. Butler, D.H. and McNamee, M.G., FTIR analysis of nicotinic acetylcholine-receptor secondary structure in reconstituted membranes, *Biochim. Biophys. Acta,* 1150(1), 17, 1993.
2. Colthup, N.B., Daly, L.H., and Wiberley, S.E., *Introduction to Infrared and Raman Spectroscopy,* 2nd ed., Academic Press, New York, 1990.
3. Martinez, G. and Millhauser, G., FTIR spectroscopy of alanine-based peptides — assignment of the amide I′ modes for random coil and helix, *J. Struct. Biol.,* 114, 23, 1995.
4. Venyaminov, S.Y., and Kalnin, N.N., Quantitative IR spectrophotometry of peptide compounds in water ($H_2O$) solutions. I. Spectral parameters of amino acid residue absorption bands, *Biopolymers,* 30, 1243, 1990.
5. Rahmelow, K., Hubner, W., and Ackermann, T., Infrared absorbancies of proteins side chains, *Anal. Biochem.,* 257, 1, 1998.
6. Pinchas, S. and Laulicht, I., *Infrared Spectra of Labelled Compounds,* Academic Press, London, 1971.
7. El Mahdaoui, L., Neault, J.F., and TajmirRiahi, H.A., Carbohydrate-nucleotide interaction. The effects of mono- and disaccharides on the solution structure of AMP, dAMP, ATP, GMP, dGMP, and GTP studied by FTIR difference spectroscopy, *J. Inorg. Biochem.,* 65(2), 123, 1997.
8. Szabo, Z., Klement, E., Jost, K., Zarandi., M, Soos, K., and Penke, B., An FT-IR study of the beta-amyloid conformation: standardization of aggregation grade, *Biochem. Biophys. Res. Commun.,* 265(5), 297, 1999.

9. Tajmirriahi, H.A., Neault, J.F., and Naoui, M., Does DNA acid fixation produce left-handed-z structure, *FEBS Lett.*, 370(1–2), 105, 1995.
10. Chang, Q.L., Liu, H.H., and Chen, J.Y., Fourier-transform infrared-spectra studies of protein in reverse micelles — effect of AOT/Isooctane on the secondary structure of alpha-chymotrypsin, *Biochim. Biophys. Acta,* 1206(2), 247, 1994.
11. Lee, S.M., Lin, S.Y., and Liang, R.C., Secondary conformational structure of type IV collagen in different conditions determined by Fourier transform infrared microscopic spectroscopy, *Artif. Cells Blood. Immobil. Biotechnol.*, 23(2), 193, 1995.
12. Susi, H., Byler, D.M., and Geraaimowicz, W., Vibrational analysis of amino acids: cysteine, serine, beta-chloroalanine, *J. Mol. Struct.*, 102, 63, 1983.
13. Rich, P.R. and Breton, J., FTIR studies of the CO and cyanide adducts of fully reduced bovine cytochrome *c* oxidase, *Biochemistry,* 40(21), 6441, 2001.
14. Birss, V.I., Hinman, A.S., McGarvey, C.E., and Segal, J., *In-situ* FTIR thin-layer reflectance spectroscopy of flavin adenine-dinucleotide at a mercury gold electrode, *Electrochim. Acta,* 39(16), 2449, 1994.
15. Um, I.C., Kweon, H.Y., Park, Y.H., and Hudson, S., Structural characteristics and properties of the regenerated silk fibroin prepared from formic acid, *Int. J. Biol. Macromol.*, 29(2), 91, 2001.
16. Sengupta, P.K. and Krimm, S., Vibrational analysis of peptides, polypeptides and proteins. XXI beta-calcium-poly(L-glutamate), *Biopolymers,* 23, 1565, 1984.
17. Dhamelincourt, P. and Ramirez, F.J., Polarized micro-Raman and FTIR spectra of glutamine, *Appl. Spectrosc.*, 47, 446, 1993.
18. Laulicht, I., Pinchas, S., Samuel, D., and Wasserman, I., The infrared absorption spectrum of oxygen-18 labeled glycine, *J. Phys. Chem.*, 70(9), 2719, 1966.
19. Chen, R.P. and Spiro, T.G., Monitoring the allosteric transition and CO rebinding in hemoglobin with time-resolved FTIR spectroscopy, *J. Phys. Chem. A,* 106(14), 3413, 2002.
20. Hasegawa, K., Ono, T.-A., and Noguchi, T., Vibrational spectra and *ab initio* DFT calculations of 4-methylimidazole and its different protonation forms: infrared and Raman markers of the protonation state of histidine side chain, *J. Phys. Chem. B,* 104, 4253, 2000.
21. Overman, S.A., and Thomas, G.J., Raman markers of nonaromatic side chains in an alpha-helix assembly: ala, asp, glu, gly, ile, leu, lys, ser and val residues of phage fd subunits, *Biochemistry,* 38, 4018, 1999.
22. Perez, C. and Griebenow, K., Fourier-transform infrared spectroscopic investigation of the thermal denaturation of hen egg-white lysozyme dissolved in aqueous buffer and glycerol, *Biotechnol. Lett.*, 22(23), 1899, 2000.
23. Kacurakova, M. and Mathlouthi, M., FTIR and laser-Raman spectra of oligosaccharides in water: characterization of the glycosidic bond, *Carbohydrate Res.*, 284(2), 145, 1996.
24. Lobau, J., Sass, M., Pohle, W., Selle, C., Koch, M.H.J., and Wolfrum, K., Chain fluidity and phase behaviour of phospholipids as revealed by FTIR and sum-frequency spectroscopy, *J. Mol. Struct.*, 481, 407, 1999.
25. Gauger, D.R., Selle, C., Fritzsche, H., and Pohle, W., Chain-length dependence of the hydration properties of saturated phosphatidylcholines as revealed by FTIR spectroscopy, *J. Mol. Struct.*, 565, 25, 2001.
26. Pohle, W., Gauger, D.R., Fritzsche, H., Rattay, B., Selle, C., Binder, H., and Bohlig, H., FTIR-spectroscopic characterization of phosphocholine-headgroup model compounds, *J. Mol. Struct.* 563, 463, 2001.
27. Lewis, R. and McElhaney, R.N., The structure and organization of phospholipid bilayers as revealed by infrared spectroscopy, *Chem. Phys. Lipids,* 96(1–2), 9, 1998.
28. Shibata, A., Yamamoto, M., Yamashita, T., Chiou, J.S., Kamaya, H., and Ueda, I., Biphasic effects of alcohols on the phase transition of poly(L-lysine) between alpha-helix and beta sheet conformations, *Biochemistry,* 31(25), 5733, 1992.

29. Dornberger, U., Spackova, N., Walter, A., Gollmick, F.A., Sponer, J., and Fritzsche, H., Solution structure of the dodecamer d-(CATGGGCC-CATG)(2) is B-DNA. Experimental and molecular dynamics study, *Journal of Biomol. Struct. Dynamics,* 19(1), 159, 2001.
30. Akhebat, A., Dagneaux, C., Liquier, J., and Taillandier, E., Triple helical polynucleotidic structures — an FTIR study of the C+.G.C triplet, *J. Biomol. Struct. Dynamics,* 10(3), 577, 1992.
31. Cabiaux, V., Oberg, K.A., Pancoska, P., Walz, T., Agre, P., and Engel, A., Secondary structures comparison of aquaporin-1 and bacteriorhodopsin: a Fourier transform infrared spectroscopy study of two-dimensional membrane crystals, *Biophy. J.,* 73(1), 406, 1997.
32. Herlinger, A.W. and Long, T.V., Laser-Raman and infrared spectra of amino acids and their metal complexes. 3. Proline and bisprolinato complexes, *J. Am. Chem. Soc.,* 92, 6481, 1970.
33. Butler, B.C., Hanchett, R.H., Rafailov, H., and MacDonald, G., Investigating structural changes induced by nucleotide binding to RecA using difference FTIR, *Biophy. J.,* 82(4), 2198, 2002.
34. Garciaquintana, D., Garriga, P., and Manyosa, J., Quantitative characterization of the structure of rhodopsin in disk membrane by means of Fourier-transform infrared-spectroscopy, *J. Biol. Chem.,* 268(4), 2403, 1993.
35. Klinck, R., Guittet, E., Liquier, J., Taillandier, E., Gouyette, C., and Huynhdinh, T., Spectroscopic evidence for an intramolecular RNA triple-helix, *FEBS Lett.,* 355(3), 297, 1994.
36. Bramanti, E. and Benedetti, E., Determination of the secondary structure of isomeric forms of human serum albumin by a particular frequency deconvolution to Fourier transform IR analysis, *Biopolymers,* 38(5), 639, 1996.
37. Iconomidou, V.A., Chryssikos, G.D., Gionis, V., Willis, J.H., and Hamodrakas, S.J., Soft-cuticle protein secondary structure as revealed by FT-Raman, ATR-FTIR and CD spectroscopy, *Insect. Biochem. Mol. Biol.,* 31(9), 87, 2001.
38. Bandwar, R.P., Sastry, M.D., Kadam, R.M., and Rao, C.P., Transition-metal saccharide chemistry: Synthesis and characterization of D-glucose, D-fructose, D-galactose, D-xylose, D-ribose, and maltose complexes of Co(II), *Carbohydrate Res.,* 297(4), 333, 1997.
39. Sivakesava, S. and Irudayaraj, J., Determination of sugars in aqueous mixtures using mid-infrared spectroscopy, *Appl. Eng. Agric.,* 16(5), 543, 2000.
40. Hadden, J. M., Bloemendal, M., Haris, P.I., Srai, S.K.S., and Chapman, D., Fourier-transform infrared-spectroscopy and differential scanning calorimetry of transferrins — human serum transferrin, rabbit serum transferrin and human lactoferrin, *Biochim. Biophys. Acta,* 1205(1), 59, 1994.
41. Lagant, P., Vergoten, G., and Peticolas, W.L., On the use of ultravolet resonance Raman intensities to elaborate molecular force fields: application to nucleic acid bases and aromatic amino acid residues models, *Biospectroscopy,* 4, 379, 1998.
42. Takeuchi, H. and Harada, I., Normal coordinate analysis of the indole ring, *Spectrochim. Acta,* 42, 1069, 1986.
43. Chia, N.C. and Mendelsohn, R., Conformational disorder in unsaturated phospholipids by FTIR spectroscopy, *Biochim. Biophys. Acta — Biomembranes,* 1283(2), 141, 1996.
44. Gue, M., Dupont, V., Dufour, A., and Sire, O., Bacterial swarming: a biochemical time-resolved FTIR-ATR study of *Proteus mirabilis* swarm-cell differentiation, *Biochemistry,* 40(39), 11938, 2001.
45. Oberreuter, H., Mertens, F., Seiler, H., and Scherer, S., Quantification of micro-organisms in binary mixed populations by Fourier transform infrared (FT-IR) spectroscopy, *Lett. Appl. Microbiol.,* 30(1), 85, 2000.
46. Schultz, C.P., Liu, K., Johnston, J.B., and Mantsch, H.H., Study of chronic lymphocytic leukemia cells by FT-IR spectroscopy and cluster analysis, *Leuk. Res.,* 20(8), 649, 1996.
47. Zeroual, W., Manfait., M, and Choisy, C., FT-IR spectroscopy study of perturbations induced by antibiotic on bacteria (*Escherichia coli*), *Pathol. Biol.* (Paris), 43(4), 300, 1995.
48. van der Mei, H.C., Naumann., D., and Busscher, H.J., Grouping of oral streptococcal species using Fourier-transform infrared spectroscopy in comparison with classical microbiological identification, *Arch. Oral. Biol.,* 38(11), 1013, 1993.

49. Ramesh, J., Salman, A., Hammody, Z., Cohen, B., Gopas, J., Grossman, N., and Mordechai, S., FTIR microscopic studies on normal and H-Ras oncogene transfected cultured mouse fibroblasts, *Eur. Biophy. J. Biophys. Lett.*, 30(4), 250, 2001.
50. Ohgushi, H., Dohi, Y., Katuda, T., Tamai, S., Tabata, S., and Suwa, Y., *In vitro* bone formation by rat marrow cell culture, *J. Biomed. Mater. Res.*, 32(3), 333, 1996.
51. Hemmingsen, A., Allen, J.T., Zhang, S.F., Mortensen, J., and Spiteri, M.A., Early detection of ozone-induced hydroperoxides in epithelial cells by a novel infrared spectroscopic method, *Free Radical Res.*, 31(5), 437, 1999.
52. Lowry, S.R., The analysis of exfoliated cervical cells by infrared microscopy, *Cell. Mol. Biol.*, 44(1), 169, 1998.
53. Wong, P.T.T., Lacelle, S., Fung, M.F.K., Senterman, M., and Mikhael, N.Z., Characterization of exfoliated cells and tissues from human endocervix and ectocervix by FTIR and ATR/FTIR spectroscopy, *Biospectroscopy*, 1(5), 357, 1995.
54. Neviliappan, S., Fang Kan, L., Tiang Lee Walter, T., Arulkumaran, S., and Wong, P.T., Infrared spectral features of exfoliated cervical cells, cervical adenocarcinoma tissue, and an adenocarcinoma cell line (SiSo), *Gynecol. Oncol.*, 85(1), 170, 2002.
55. Morris, B.J., Lee, C., Nightingale, B.N., Molodysky, E., Morris, L.J., Appio, R., Sternhell, S., Cardona, M., Mackerras, D., and Irwig, L.M., Fourier transform infrared spectroscopy of dysplastic, papillomavirus-positive cervicovaginal lavage specimens, *Gynecol. Oncol.*, 56(2), 245, 1995.
56. Melin, A.M., Perromat, A., and Deleris, G., Pharmacologic application of FTIR spectroscopy: effect of ascorbic acid-induced free radicals on *Deinococcus radiodurans*, *Biospectroscopy*, 5(4), 229, 1999.
57. Zhou, J., Wang, Z., Sun, S., Liu, M., and Zhang, H., A rapid method for detecting conformational changes during differentiation and apoptosis of HL60 cells by Fourier-transform infrared spectroscopy, *Biotechnol. Appl. Biochem.*, 33(2), 127, 2001.
58. Pouliot, R., Germain, L., Auger, F.A., Tremblay, N., and Juhasz, J., Physical characterization of the stratum corneum of an *in vitro* human skin equivalent produced by tissue engineering and its comparison with normal human skin by ATR-FTIR spectroscopy and thermal analysis (DSC), *Biochim. Biophys. Acta — Mol. Cell Biol. Lipids*, 1439(3), 341, 1999.
59. Romano, S., Monici, M., Mazzinghi, P., Bernabei, P.A., and Fusi, F., Spectroscopic study of human leukocytes, *Phys. Med.*, 13, 291, 1997.
60. Benedetti, E., Bramanti, E., Papineschi., F., Vergamini, P., and Benedetti, E., An approach to the study of primitive thrombocythemia (PT) megakaryocytes by means of Fourier transform infrared microspectroscopy (FT-IR-M), *Cell Mol. Biol.*, 44(1), 129, 1998.
61. Wang, H.P., Wang, H.C., and Huang, Y.J., Microscopic FTIR studies of lung cancer cells in pleural fluid, *Sci. Total Environ.*, 204(3), 283, 1997.
62. Holman, H.Y., Martin, M.C., Blakely, E.A., Bjornstad, K., and McKinney, W.R., IR spectroscopic characteristics of cell cycle and cell death probed by synchrotron radiation based Fourier transform IR spectromicroscopy, *Biopolymers*, 57(6), 329, 2000.
63. Wood, B.R., Tait, B., and McNaughton, D., Fourier-transform infrared spectroscopy as a tool for detecting early lymphocyte activation: a new approach to histocompatibility matching, *Hum. Immunol.*, 61(12), 1307, 2000.
64. Huleihel, M., Salman, A., Erukhimovitch, V., Ramesh, J., Hammody, Z., and Mordechai, S., Novel spectral method for the study of viral carcinogenesis *in vitro*, *J. Biochem. Biophys. Methods*, 50(2–3), 111, 2002.
65. Lasch, P., Pacifico, A., and Diem, M., Spatially resolved IR microspectroscopy of single cells, *Biopolymers*, 67(4–5), 335, 2002.
66. Huleihel, M., Talyshinsky, M., and Erukhimovitch, V., FTIR microscopy as a method for detection of retrovirally transformed cells, *Spectroscopy*, 15(2), 57, 2001.
67. Wood, B.R., Quinn, M.A., Tait, B., Ashdown, M., Hislop, T., Romeo, M., and McNaughton, D., FTIR microspectroscopic study of cell types and potential confounding variables in screening for cervical malignancies, *Biospectroscopy*, 4(2), 75, 1998.

68. Choo, L.P., Wetzel, D.L., Halliday, W.C., Jackson, M., LeVine, S.M., and Mantsch, H.H., *In situ* characterization of beta-amyloid in Alzheimer's diseased tissue by synchrotron Fourier transform infrared microspectroscopy, *Biophys. J.*, 71(4), 1672, 1996.
69. Liu, K.Z. and Mantsch, H.H., Simultaneous quantitation from infrared spectra of glucose concentrations, lactate concentrations, and lecithin/sphingomyelin ratios in amniotic fluid, *Am. J. Obstet. Gynecol.*, 180, 696, 1999.
70. Krusejarres, J.D., Janatsch, G., Gless, U., Marbach, R., and Heise, H.M., Glucose and other constituents of blood determined by ATR-FTIR-spectroscopy, *Clin. Chem.*, 36(2), 401, 1990.
71. Wetzel, D.L., Slatkin, D.N., and Levine, S.M., FT-IR microspectroscopic detection of metabolically deuterated compounds in the rat cerebellum: a novel approach for the study of brain metabolism, *Cell. Mol. Biol.*, 44(1), 15, 1998.
72. Lewis, E.N., Gorbach, A.M., Marcott, C., and Levin, I.W., High-fidelity Fourier transform infrared spectroscopic imaging of primate brain tissue, *Appl. Spectrosc.*, 50(2), 263, 1996.
73. Lester, D.S., Kidder, L.H., Levin, I.W., and Lewis, E.N., Infrared microspectroscopic imaging of the cerebellum of normal and cytarabine treated rats, *Cell. Mol. Biol.*, 44(1), 29, 1998.
74. Lee, L.S., Lin, S.Y., Chi, C.W., Liu, H.C., and Cheng, C.L., Non-destructive analysis of the protein conformational structure of human pituitary adenomas using reflectance FT-IR microspectroscopy, *Cancer Lett.*, 94(1), 65, 1995.
75. Jackson, M., Choo, L.P., Watson, P.H., Halliday, W.C., and Mantsch, H.H., Beware of connective-tissue proteins — sssignment and implications of collagen absorptions in infrared-spectra of human tissues, *Biochim. Biophys. Acta – Mol. Basis Disease*, 1270(1), 1, 1995.
76. Ci, Y.X., Gao, T.Y., Dong, J.Q., Kan, X., and Guo, Z.Q., FTIR assessment of the secondary structure of proteins in human breast benign and malignant tissues, *Chin. Sci. Bull.*, 44(24), 2215, 1999.
77. Gao, T., Feng, J., and Ci, Y., Human breast carcinomal tissues display distinctive FTIR spectra: implication for the histological characterization of carcinomas, *Anal. Cell. Pathol.*, 18(2), 87, 1999.
78. Jackson, M., Mansfield, J.R., Dolenko, B., Somorjai, R.L., Mantsch, H.H., Watson, P.H., Classification of breast tumors by grade and steroid receptor status using pattern recognition analysis of infrared spectra, *Cancer Detect. Prev.*, 23(3), 245, 1999.
79. Dukor, R.K., Liebman, M.N., and Johnson, B.L., A new, non-destructive method for analysis of clinical samples with FT-IR microspectroscopy. Breast cancer tissue as an example, *Cell. Mol. Biol.*, 44(1), 211, 1998.
80. Ali, S.R., Johnson, F.B., Luke, J.L., and Kalasinsky, V.F., Characterization of silicone breast implant biopsies by Fourier transform infrared mapping, *Cell. Mol. Biol.*, 44(1), 75, 1998.
81. Kidder, L.H., Kalasinsky, V.F., Luke, J.L., Levin, I.W., and Lewis, E.N., Visualization of silicone gel in human breast tissue using new infrared imaging spectroscopy, *Nat. Med.*, 3(2), 235, 1997.
82. Potter, K., Kidder, L.H., Levin, I.W., Lewis, E.N., and Spencer, R.G.S., Imaging of collagen and proteoglycan in cartilage sections using Fourier transform infrared spectral imaging, *Arthritis Rheum.*, 44(4), 846, 2001.
83. Camacho, N.P., West, P., Torzilli, P.A., and Mendelsohn, R., FTIR microscopic imaging of collagen and proteoglycan in bovine cartilage, *Biopolymers*, 62(1), 1, 2001.
84. Schultz, C.P., Eysel, H.H., Mantsch, H.H., and Jackson, M., Carbon dioxide in tissues, cells, and biological fluids detected by FTIR spectroscopy, *J. Phys. Chem.*, 100(16), 6845, 1996.
85. Chiriboga, L., Xie, P., Yee, H., Zarou, D., Zakim, D., and Diem, M., Infrared spectroscopy of human tissue. IV. Detection of dysplastic and neoplastic changes of human cervical tissue via infrared microscopy, *Cell. Mol. Biol.*, 44(1), 219, 1998.
86. Diem, M., Chiriboga, L., and Yee, H., Infrared spectroscopy of human cells and tissue. VIII. Strategies for analysis of infrared tissue mapping data and applications to liver tissue, *Biopolymers*, 57(2), 282, 2000.
87. Argov, S., Ramesh, J., Salman, A., Sinelnikov, I., Goldstein, J., Guterman, H., and Mordechai, S., Diagnostic potential of Fourier-transform infrared microspectroscopy and advanced computational methods in colon cancer patients, *J. Biomed. Opt.*, 7(2), 248, 2002.

88. Bindig, U., Winter, H., Wasche, W., Zelianeos, K., and Muller, G., Fiber-optical and microscopic detection of malignant tissue by use of infrared spectrometry, *J. Biomed. Opt.*, 7(1), 100, 2002.
89. Paschalis, E.P., Verdelis, K., Doty, S.B., Boskey, A.L., Mendelsohn, R., and Yamauchi, M., Spectroscopic characterization of collagen cross-links in bone, *J. Bone Miner. Res.*, 16(10), 1821, 2001.
90. Magne, D., Weiss, P., Bouler, J.M., Laboux, O., and Daculsi, G., Study of the maturation of the organic (type I collagen) and mineral (nonstoichiometric apatite) constituents of a calcified tissue (dentin) as a function of location: a Fourier transform infrared microspectroscopic investigation, *J. Bone Miner. Res.*, 16(4), 750, 2001.
91. Thomas, N., Goodacre, R., Timmins, E.M., Gaudoin, M., and Fleming, R., Fourier transform infrared spectroscopy of follicular fluids from large and small antral follicles, *Hum. Reprod.*, 15(8), 1667, 2000.
92. Liu, K.Z., Dixon, I.M., and Mantsch, H.H., Distribution of collagen deposition in cardiomyopathic hamster hearts determined by infrared microscopy, *Cardiovasc. Pathol.*, 8(1), 41, 1999.
93. Manoharan, R., Baraga, J.J., Rava, R.P., Dasari, R.R., Fitzmaurice, M., and Feld, M.S., Biochemical analysis and mapping of atherosclerotic human artery using FT-IR microspectroscopy, *Atherosclerosis,* 103(2), 18, 1993.
94. Liu, K.Z., Jackson, M., Sowa, M.G., Ju, H.S., Dixon, I.M.C., and Mantsch, H.H., Modification of the extracellular matrix following myocardial infarction monitored by FTIR spectroscopy, *Biochim. Biophys. Acta,* 1315(2), 73, 1996.
95. Mendelsohn, R., Paschalis, E.P., Sherman, P.J., and Boskey, A.L., IR microscopic imaging of pathological states and fracture healing of bone, *Appl. Spectrosc.*, 54(8), 1183, 2000.
96. Mendelsohn, R., Paschalis, E.P., and Boskey, A.L., Infrared spectroscopy, microscopy, and microscopic imaging of mineralizing tissues: spectra-structure correlations from human iliac crest biopsies, *J. Biomed. Opt.*, 4(1), 14, 1999.
97. Lyman, D.J. and Murray-Wijelath, J., Vascular graft healing: I. FTIR analysis of an implant model for studying the healing of a vascular graft, *J. Biomed. Mater. Res.*, 48(2), 17, 1999.
98. Lyman, D.J., Murray-Wijelath, J., Ambrad-Chalela, E., and Wijelath, E.S., Vascular graft healing. II. FTIR analysis of polyester graft samples from implanted bi-grafts, *J. Biomed. Mater. Res.*, 58(3), 221, 2001.
99. Jackson, M., Ramjiawan, B., Hewko, M., and Mantsch, H.H., Infrared microscopic functional group mapping and spectral clustering analysis of hypercholesterolemic rabbit liver, *Cell. Mol. Biol.*, 44(1), 89, 1998.
100. Kitai, T., Tanaka, A., Tokuka, A., Sato, B., Mori, S., Yanabu, N., Inomoto, T., Uemoto, S., Tanaka, K., Yamaoka, Y., Ozawa, K., Someda, H., Fujimoto, M., Moriyasu, F., and Hirao, K., Intraoperative measurement of the graft oxygenation state in living related liver transplantation by near infrared spectroscopy, *Transpl. Int.*, 8(2), 111, 1995.
101. Severcan, F., Toyran, N., Kaptan, N., and Turan, B., Fourier transform infrared study of the effect of diabetes on rat liver and heart tissues in the C-H region, *Talanta,* 53(1), 55, 2000.
102. Yano, K., Ohoshima, S., Gotou, Y., Kumaido, K., Moriguchi, T., and Katayama, H., Direct measurement of human lung cancerous and noncancerous tissues by Fourier transform infrared microscopy: can an infrared microscope be used as a clinical tool? *Anal. Biochem.*, 287(2), 218, 2000.
103. Das, R.M., Ahmed, M.K., Mantsch, H.H., and Scott, J.E., FT-IR spectroscopy of methylmercury-exposed mouse lung, *Mol. Cell. Biochem.*, 145(1), 75, 1995.
104. Andrus, P.G., and Strickland, R.D., Cancer grading by Fourier transform infrared spectroscopy, *Biospectroscopy,* 4(1), 37, 1998.
105. Magne, D., Pilet, P., Weiss, P., and Daculsi, G., Fourier transform infrared microspectroscopic investigation of the maturation of nonstoichiometric apatites in mineralized tissues: a horse dentin study, *Bone,* 29(6), 547, 2001.
106. LeVine, S.M., Wetzel, D.L., Chemical analysis of multiple sclerosis lesions by FT-IR microspectroscopy, *Free Radic. Biol. Med.*, 25(1), 33, 1998.
107. Burmeister, J.J. and Arnold, M.A., Evaluation of measurement sites for noninvasive blood glucose sensing with near-infrared transmission spectroscopy, *Clin. Chem.*, 45(9), 1621, 1999.

108. Bromberg, P.S., Gough, K.M., and Dixon, I.M.C., Collagen remodeling in the extracellular matrix of the cardiomyopathic Syrian hamster heart as assessed by FTIR attenuated total reflectance spectroscopy, *Can. J. Chem.*, 77(11), 1843, 1999.
109. Kidder, L.H., Colarusso, P., Stewart, S.A., Levin, I.W., Appel, N.M., Lester, D.S., Pentchev, P.G., and Lewis, E.N., Infrared spectroscopic imaging of the biochemical modifications induced in the cerebellum of the Niemann–Pick type C mouse, *J. Biomed. Opt.*, 4(1), 7, 1999.
110. Fukuyama, Y., Yoshida, S., Yanagisawa, S., and Shimizu, M., A study on the differences between oral squamous cell carcinomas and normal oral mucosas measured by Fourier transform infrared spectroscopy, *Biospectroscopy,* 5(2), 117, 1999.
111. Wu, J.G., Xu, Y.Z., Sun, C.W., Soloway, R.D., Xu, D.F., Wu, Q.G., Sun, K.H., Weng, S.F., and Xu, G.X., Distinguishing malignant from normal oral tissues using FTIR fiber-optic techniques, *Biopolymers,* 62(4), 185, 2001.
112. Schultz, CP., Liu, K.Z., Kerr, P.D., and Mantsch, H.H., *In situ* infrared histopathology of keratinization in human oral/oropharyngeal squamous cell carcinoma, *Oncol. Res.*, 10(5), 277, 1998.
113. Nishida, Y., Yoshida, S., Li, H.J., Higuchi, Y., Takai, N., and Miyakawa, I., FTIR spectroscopic analyses of human placental membranes, *Biopolymers,* 62(1), 22, 2001.
114. Saini, P., Lai, J.C., and Lu, D.R., FT-IR measurement of mercaptoundecahydrododecaborate in human plasma, *J. Pharm. Biomed. Anal.*, 12(9), 1091, 1994.
115. Afanasyeva, N.I. and Bruch, R.F., Biocompatibility of polymer surfaces interacting with living tissue, *Surface Interface Anal.*, 27(4), 204, 1999.
116. Homan, J.A., Radel, J.D., Wallace, D.D., Wetzel, D.L., and Levine, S.M., Chemical changes in the photoreceptor outer segments due to iron induced oxidative stress: analysis by Fourier transform infrared (FT-IR) microspectroscopy, *Cell. Mol. Biol.*, 46(3), 663, 2000.
117. Jennings, G., Bluck, L., Wright, A., and Elia, M., The use of infrared spectrophotometry for measuring body water spaces, *Clin. Chem.*, 45(7), 1077, 1999.
118. Kneipp, J., Beekes, M., Lasch, P., and Naumann, D., Molecular changes of preclinical scrapie can be detected by infrared spectroscopy, *J. Neurosci.*, 22(8), 2989, 2002.
119. Shaw, R.A., Kotowich, S., Leroux, M., and Mantsch, H.H., Multianalyte serum analysis using mid-infrared spectroscopy, *Ann. Clin. Biochem.*, 35, 624, 1998.
120. Gotshal, Y., Simhi, R., Sela, B.A., and Katzir, A., Blood diagnostics using fiberoptic evanescent wave spectroscopy and neural networks analysis, *Sensors Actuators B,* 42(3), 157, 1997.
121. Liu, K.Z., Shaw, R.A., Man, A., Dembinski., T.C., and Mantsch., H.H., Reagent-free, simultaneous determination of serum cholesterol in HDL and LDL by infrared spectroscopy, *Clin. Chem.*, 48(3), 499, 2002.
122. Sukuta, S. and Bruch, R., Factor analysis of cancer Fourier transform infrared evanescent wave fiberoptical (FTIR-FEW) spectra, *Lasers Surg. Med.*, 24(5), 382, 1999.
123. McIntosh, L.M., Jackson, M., Mantsch, H.H., Stranc, M.F., Pilavdzic, D., and Crowson, A.N., Infrared spectra of basal cell carcinomas are distinct from non-tumor-bearing skin components, *J. Invest. Dermatol.*, 112(6), 951, 1999.
124. Franck, P., Sallerin, J. L., Schroeder, H., Gelot, M.A., and Nabet, P., Rapid determination of fecal fat by Fourier transform infrared analysis (FTIR) with partial least squares regression and an attenuated total reflectance accessory, *Clin. Chem.*, 42(12), 2015, 1996.
125. Shaw, R.A., Kotowich, S., Eysel, H.H., Jackson, M., Thomson, G.T., and Mantsch, H.H., Arthritis diagnosis based upon the near-infrared spectrum of synovial fluid, *Rheumatol. Int.,* 15(4), 159, 165, 1995.
126. Shaw, R.A., Low-Ying, S., Leroux, M., and Mantsch, H.H., Toward reagent-free clinical analysis: quantitation of urine urea, creatinine, and total protein from the mid-infrared spectra of dried urine films, *Clin. Chem.*, 46(9), 1493, 2000.

127. Uemura, T., Nishida, K., Ichinose, K., Shimoda, S., and Shichiri, M., Non-invasive blood glucose measurement by Fourier transform spectroscopic analysis through the mucous membrane of the lip: application of chalcogenide optical fiber system, *Front. Med. Biol. Eng.*, 9(2), 137, 1999.
128. Rausch, M., and Eysel, U.T., Visualization of ICBF changes during cortical infarction thermo encephaloscopy, *Neuroreport*, 7, 2603, 1996.
129. George, J.S., Lewine, J.D., Goggin, A.S., Dyer, R.B., and Flynn, E.R., IR imaging of a monkey's head: local temperature changes in response to somatosensory stimulation, *Adv. Exp. Med. Biol.*, 333, 125, 1993.
130. Brancaleon, L., Bamberg, MP., Sakamaki, T., and Kollias, N., Attenuated total reflection-Fourier transform infrared spectroscopy as a possible method to investigate biophysical parameters of stratum corneum *in vivo*, *J. Invest. Dermatol.*, 116(3), 380, 2001.
131. Brooks, A., Afanasyeva, N.I., Makhine, V., Bruch, R.F., Kolyakov, S.F., Artjushenko, S., and Butvina, L.N., New method for investigations of normal human skin surfaces *in vivo* using fiber-optic evanescent wave Fourier transform infrared spectroscopy (FEW-FTIR), *Surface Interface Anal.*, 27(4), 221, 1999.
132. Smielewski, P., Kirkpatrick, P., Minhas, P., Pickard, J.D., and Czosnyka, M., Can cerebrovascular reactivity be measured with near-infrared spectroscopy? *Stroke*, 26(12), 2285, 1995.
133. Snieder, M. and Hansen, W.G., Crystal effect on penetration depth in attenuated total reflectance Fourier-transform infrared study of human skin, *Mikrochim. Acta*, Suppl. 14, 677, 1999.

### 65.3.3 Raman Spectroscopy

Table 65.8 provides a quick reference for the reader to consult to find Raman-based publications regarding a variety of chemical and biochemical compounds and systems. Works were selected based on both their application and potential significance in the field of biophotonics. As a consequence, this table includes examples of a variety of Raman-based techniques. The following abbreviations are used in this table: NIR = near infrared; IR = infrared; FT = Fourier transform; SERS = surface-enhanced Raman scattering; SERRS = surface-enhanced resonance Raman scattering; UV = ultraviolet; FT SERS = Fourier transform surface-enhanced Raman scattering; LOD = limit of detection; NIR FT = near-infrared Fourier transform; CCD = charge-coupled device; HIV = human immunodeficiency virus; YC-1 = a synthetic benzylindazole derivative or [3-(5′-hydroxymethyl-2′-furyl)-1-benzylindazole; PEO = polyethylene. The instrumental requirements may vary significantly for different techniques; for such details the reader can refer to appropriate chapters of this handbook. Because Raman band assignments can be quite extensive and matrix dependent in Raman scattering, such details are only supplied where reported in great confidence. While the authors hope subsequent editions of this handbook will offer more complete spectral information, this version clearly demonstrates the recently found potential of Raman in biomedical applications. The information in Table 65.8 was compiled using data from more than 200 publications, monographs, and review articles.

**TABLE 65.8** Raman Data

| Species | Target Organ or Disease | Excitation λ (nm) | Major Raman Shifts ($cm^{-1}$) | Comments | Ref. |
|---|---|---|---|---|---|
| | | | ***Biochemical Studies*** | | |
| Aminoglutethimide | Cancer (drug) | | | Examined environmental and solvent effects (e.g., $CCl_4$, $CHCl_3$, and $CH_3CN$) on IR and Raman spectra of aminoglutethimide | 1 |
| Amsacrine | Cancer (drug) | UV | | UV resonance Raman; spectroscopically characterized | 2 |
| Amsacrine | Cancer (drug) | NIR | | FT Raman; spectroscopically characterized | 2 |
| Amsacrine | Leukemia (drug) | UV | 1165, 1265, 1380 | UV resonance; monitored interaction of anticancer drug, amsacrine, with calf thymus DNA; observed spectral changes between free and DNA-bound amsacrine | 3 |
| Amsacrine (m-AMSA) | Topoisomerase II inhibitor/cancer (drug) | | | SERS; monitored intracellular interactions of amsacrine (m-AMSA) within living K562 cancer cells | 4 |
| Antineoplaston-A10 | Cancer (drug) | | | Characterized with vibrational assignments; vibrational modes similar to uracil derivatives, implying that therapy based on resemblance to pyrimidine bases | 5 |
| Avidin | | | | Demonstrated the importance of the side chains of biotin when interacting with avidin | 6 |
| Avidin | | | | Demonstrated the importance of the side chains of biotin when interacting with avidin | 7 |
| Benzo(a)pyrene | Cancer | | | Theoretical study of the oxygenation of benzo(a)pyrene | 8 |
| Biotin | | | | Demonstrated the importance of the side chains of biotin when interacting with avidin | 6 |
| Biotin | | | | Demonstrated the importance of the side chains of biotin when interacting with avidin | 7 |
| Blenoxane (bleomycins A2 and B2) | DNA/cancer (drug) | | | A study in drug/DNA interaction; spectral changes observed between calf thymus DNA and bleomycin/DNA complex | 9 |
| Cinchonine | Malaria (drug) | | 1374 | Phosphate stretch band of cinchonine observed to shift upon interaction with DNA, indicative of coulombic interaction | 10 |
| *cis*-Dichloroplatinum derivatives | Telomerase inhibitor/cancer (drug) | | | Characterized *cis*-dichloroplatinum derivatives with N-donor ligands such as pyridine and 5-substituted quinolines | 11 |
| $CO_2$ | Lung (respiratory quotient test) | | | Inspired, end-tidal, and mixed expired gas compositions analyzed by Raman spectroscopy | 12 |
| Dimethylcrocetin | Retinoic acid nuclear receptor interaction/ cancer (drug) | | 1210, 1165, 1541 | FT Raman; FT SERS; evaluated spectra for free-form dimethylcrocethin (DMCRT) and DMCRT/retinoic acid nuclear receptor in HL60 and K562 cancer cells | 13 |
| DNA | | | | SERS; SERRS; studied adsorption of DNA onto substrates | 14 |
| DNA (via cresyl fast violet) | HIV/AIDS | 633 | | SERS; HIV DNA amplified (PCR) with cresyl fast violet-labeled primers and selectively hybridized to a bioassay platform | 15 |

**TABLE 65.8** Raman Data

| Species | Target Organ or Disease | Excitation λ (nm) | Major Raman Shifts ($cm^{-1}$) | Comments | Ref. |
|---|---|---|---|---|---|
| DNA (via dye label) | | | | SERS; gene probes hybridized to target DNA in solution, spotted on SERS substrate | 16 |
| Gelonin | Type 1 ribosome inactivating protein/AIDS (drug) | | | Determined secondary structure of gelonin to be mainly alpha-helix and beta-sheet with some turn and disordered structure | 17 |
| Gelonin | Type 1 ribosome inactivating protein/cancer (drug) | | | Determined secondary structure of gelonin to be mainly alpha-helix and beta-sheet with some turn and disordered structure. | 17 |
| Guanylate cyclase | Nitric oxide receptor/lung | | 474–492 | Investigated interaction of YC-1 with enzyme soluble guanylate cyclase (sGC) in presence of CO for sGC expressed by baculovirus/Sf9 cell | 18 |
| Guanylate cyclase | Nitric oxide receptor/lung | UV | 478, 495, 525, 567, 572, 1676 | Resonance Raman; characterized the heme domain of rat lung soluble guanylate cyclase | 19 |
| Guanylate cyclase | Nitric oxide receptor/lung | UV | 212, 221 | Resonance Raman; identified histidine 105 in the beta 1 subunit of soluble guanylate cyclase from rat and bovine lung | 20 |
| Guanylate cyclase | Nitric oxide receptor/lung | UV | 497, 520, 1959 | Resonance Raman; spectra illustrated for the alpha(1) beta(1) isoform of lung soluble guanylate cyclase expressed from baculovirus | 21 |
| Guanylate cyclase | Nitric oxide receptor/lung | UV | 1357, 1375, 1473 | Resonance Raman; spectra reported for ferrous and ferric forms of soluble guanylate cyclase from bovine lung | 22 |
| Guanylate cyclase | Nitric oxide receptor/lung | UV | 203, 473, 521, 1681, 1700 | Resonance Raman; spectra reported for reduced, GO-bound, NO-bound, oxidized and oxidized NO-bound forms of soluble guanylate cyclase from bovine lung with/without GTP | 23 |
| $H^+$, $K^+$-ATPase | Stomach | | | Investigated conformational changes induced by replacing $K^+$ ions with $Na^+$ ions; no significant Raman spectral changes observed | 24 |
| $H^+$, $K^+$-ATPase | Stomach | | | Investigated secondary structure of membrane-bound regions of $H^+$, $K^+$-ATPase; determined structure to be largely beta-sheet in character | 25 |
| Heme protein | Heme protein CO activation | | | Investigated mechanism for heme protein activation | 26 |
| Hydroxyapatite | Inclusions (bone) | | | Compared biological hydroxyapatite samples to ox femur | 27 |
| Intoplicine | Topoisomerase II inhibitor/cancer (drug) | | | SERS; intoplicine found to unwind DNA and inhibit calf thymus topisomerase II; intoplicine spectrum eliminated when in the presence of topoisomerase II | 28 |
| Lysozyme | | 632.8 nm | | First interpretable laser-excited spectrum of a native protein | 29 |
| Methotrexate | Dihydrofolate reductase inhibitor/cancer (drug) | | | Ultrasensitive Raman difference spectroscopy illustrated light-induced reaction of methotrexate with NADPH in dihydrofolate active site | 30 |

**TABLE 65.8** Raman Data

| Species | Target Organ or Disease | Excitation λ (nm) | Major Raman Shifts ($cm^{-1}$) | Comments | Ref. |
|---|---|---|---|---|---|
| Methotrexate | Dihydrofolate reductase inhibitor/ rheumatoid arthritis (drug) | | | Ultrasensitive Raman difference spectroscopy illustrated light-induced reaction of methotrexate with NADPH in dihydrofolate active site | 30 |
| Nitrogen | Lung (respiratory quotient test) | | | Inspired, end-tidal, and mixed expired gas compositions analyzed by Raman spectroscopy | 12 |
| Oxygen | Lung (respiratory quotient test) | | | Inspired, end-tidal, and mixed expired gas compositions analyzed by Raman spectroscopy | 12 |
| Proteins | | | | Elucidated kinetics and structures of proteins | 31 |
| Proteins | | | | Elucidated kinetics and structures of proteins | 32 |
| Proteins | | | | Elucidated kinetics and structures of proteins | 33 |
| Proteins | | | | Identified Raman vibrational bands through isotopic substitution | 34 |
| Proteins | | | | Identified Raman vibrational bands through isotopic substitution | 35 |
| Proteins | | | | Identified Raman vibrational bands through isotopic substitution | 36 |
| Proteins | | | | Identified Raman vibrational bands through isotopic substitution | 37 |
| Theraphthal | Cancer (drug-binary catalytic system) | | | Resonance Raman; studied effects of molecular interactions (e.g., with $Ca^+$) and environmental factors (e.g., pH < 6) on spectra of theraphthal | 38 |
| | Drug | | | Studied kinetics and interaction mechanisms between drugs and receptors | 39 |
| | | | ***Cellular Studies*** | | |
| Amsacrine (m-AMSA) | K562 cells/cancer (drug) | | | SERS; monitored intracellular interactions of amsacrine (m-AMSA) within living K562 cancer cells | 4 |
| Bacteriochlorophyll *a* | Photosynthetic bacteria | | | Resonance Raman; substitution of Mg ion by Zn, Ni, Co, and Cu ions studied via Raman scattering; molecule core size shifts caused shifts in resonance | 40 |
| Calcium dipicolinate | Bacteria | UV | | | 41 |
| Carotenoids | Liposomes | | | Egg phosphatidylcholine and egg phosphatidylglycerol; observed β-carotene rapidly taken up *in vitro* and transported to "Gall bodies" | 42 |
| Carotenoids | Lymphocytes | | | Lymphocytes from human blood; observed that carotenoids concentrated in "Gall bodies"; associated carotenoid concentrations with age, decreasing with age | 42 |
| Carotenoids | Lymphocytes/lung cancer | | | Human; compared carotenoid levels in lymphocytes of lung cancer patients and healthy individuals; observed lower carotenoid levels in lung carcinoma patients | 43 |
| Chromatid | Ovary cells/ mutagenesis (photo-induced) | 193 | | Chinese hamster; reported monitoring of mutagenesis and sister chromatin exchange induction via UV irradiation | 44 |

**TABLE 65.8** Raman Data

| Species | Target Organ or Disease | Excitation λ (nm) | Major Raman Shifts ($cm^{-1}$) | Comments | Ref. |
|---|---|---|---|---|---|
| Chromatid | Ovary cells/ mutagenesis (photo-induced) | 308 | | Chinese hamster; reported monitoring of mutagenesis and sister chromatin exchange induction via UV irradiation | 44 |
| Chromatid | Ovary cells/ mutagenesis (photo-induced) | 254 | | Chinese hamster; reported monitoring of mutagenesis and sister chromatin exchange induction via UV irradiation | 44 |
| Demoic acid (neurotoxin) | Phytoplankton | UV | | | 45 |
| Dimethylcrocetin | HL60 and K562 cells/cancer (therapy) | | 1210, 1165, 1541 | FT Raman; FT SERS; evaluated spectra for free-form dimethylcrocethin (DMCRT) and DMCRT/retinoic acid nuclear receptor in HL60 and K562 cancer cells | 13 |
| DNA | FD Virus | UV | | | 46 |
| DNA | Virus | 257 | | | 47 |
| DNA | Virus | 244 | | | 47 |
| DNA | Virus | 238 | | | 47 |
| DNA | Virus | 229 | | | 47 |
| DNA | Virus | | | Resonance Raman; DNA monitored during assembly of viruses | 47 |
| DNA | Virus (T7 bacteriophage) | | | Resonance Raman; monitored DNA spectra for DNA during packaging into viral plasmid with little changes; Mg ions observed to influence packaging | 49 |
| DNA | | Visible | | DNA in chromosomes; investigation of cellular damage at different wavelengths demonstrated that 514.5 nm radiation causes damage while 660 nm radiation does not | 50 |
| DNA | | 514.5; 660 | | DNA in chromosomes | 51 |
| DNA | | Visible | | DNA in chromosomes | 52 |
| DNA | | Tuned | | Resonance Raman; DNA preferentially detected in cells | 53 |
| DNA | | Tuned | | Resonance Raman; DNA preferentially detected in cells | 53 |
| Glucose (via quinoline red) | | | | Attached quinoline red to monitor intake of glucose into cells; glucose intake detected by intensity changes in quinolin red Raman spectrum | 54 |
| Hemoglobin | Rheumatoid arthritis | Visible | | Monitoring of oxygen uptake in human cells; concluded that patients with rheumatoid arthritis uptake oxygen faster but incompletely | 55 |
| Hepatocytes | Liver cancer | Visible | 1040, 1083, 1182, 1241, 1253 | Distinguished between normal and malignant hepatocytes | 56 |
| Hydroxyapatite | C4-2B cell/ prostate cancer | | | Mineral deposition monitored for C4-2B prostate cancer cell line in promineralization media | 57 |
| Intoplicine | K562 cells/cancer (drug) | | | SERS; intoplicine spectrum eliminated when in the presence of topoisomerase II; intoplicine observed as complexed in nucleus and free form in cytoplasm of K562 cells | 28 |

**TABLE 65.8** Raman Data

| Species | Target Organ or Disease | Excitation λ (nm) | Major Raman Shifts ($cm^{-1}$) | Comments | Ref. |
|---|---|---|---|---|---|
| Nucleic acids | Breast cells/cancer | 257 | | Reported spectra for cultured normal and malignant breast cells; differences attributed to relative amounts of nucleic acids and proteins and nuclear hypochromism | 58 |
| Nucleic acids | Cervical cells/ cancer | 257 | | Reported spectra for cultured normal and malignant cervical cells; differences attributed to relative amounts of nucleic acids and proteins and nuclear hypochromism | 58 |
| Nucleic acids | T84 cancer cell | 257 | | | 59 |
| Photosensitizers | Cancer cells | | | Monitored accumulation of photosensitizers in cancer cells | 60 |
| Photosensitizers | Cancer cells | | | Monitored accumulation of photosensitizers in cancer cells | 61 |
| Phthalocyanates | Cancer cells | | 1538 | Determined hydrophilic phthalocyanates accumulate in lysosomes while hydrophobic phthalocyanates accumulate in cytoplasm | 62 |
| Pigments | Lung cancer | Visible | | Pigments in human lymphocytes with high spatial resolution | 63 |
| Pigments | | Visible | | Pigments in human granulocytes with high spatial resolution | 64 |
| Pigments | | Visible | | | 51 |
| Pigments | | Visible | | Pigments in human lymphocytes with high spatial resolution | 66 |
| Protein | Breast cells/cancer | 257 | | Reported spectra for cultured normal and malignant breast cells; differences attributed to relative amounts of nucleic acids and proteins and nuclear hypochromism | 58 |
| Protein | Cervical cells/ cancer | 257 | | Reported spectra for cultured normal and malignant cervical cells; differences attributed to relative amounts of nucleic acids and proteins and nuclear hypochromism | 58 |
| Protein | FD virus | UV | | | 66 |
| Protein | Virus | 257 | | | 47 |
| Protein | Virus | 244 | | | 47 |
| Protein | Virus | 238 | | | 47 |
| Protein | Virus | 229 | | | 47 |
| Protein (active site) | Virus (HAV 3C) | | | Active site structure of HAV 3C monitored as a function of pH | 67 |
| Proteins | Bacteria and viruses | 231 | | UV resonance Raman; obtained fluorescence-free, selective Raman detection of protein aromatic group residues in bacteria and viruses | 68 |
| Proteins | Virus (P22)/ replication | | | Protein side chain and secondary structure monitored as a virus (P22) entered a host cell, replicated components, and assembled new viruses | 69 |
| Protoplasts | Bacteria | UV | | | 41 |
| Purines | Bacteria and viruses | 242 | | UV resonance Raman; obtained fluorescence-free, selective Raman detection of purine aromatic group residues in bacteria and viruses | 68 |

**TABLE 65.8** Raman Data

| Species | Target Organ or Disease | Excitation λ (nm) | Major Raman Shifts ($cm^{-1}$) | Comments | Ref. |
|---|---|---|---|---|---|
| Pyrimidines | Bacteria and viruses | 251 | | UV resonance Raman; obtained fluorescence-free, selective Raman detection of pyrimidine aromatic group residues in bacteria and viruses | 68 |
| Theraphthal | A549 cells/cancer (BCS therapy) | | | Resonance Raman confocal microspectroscopy and imaging; monitored intracellular accumulation, localization, and retention of theraphthal in living cancer cells | 38 |
| | Bacteria | UV | | | 70 |
| | Bacteria | UV | | | 71 |
| | Bacteria | UV | | | 41 |
| | Bacteria | UV | | Examined effects of cultural conditions on deep UV resonance Raman spectra of bacteria | 72 |
| | Bacterial spores | UV | | | 41 |
| | | | ***Tissue/Biofluid Studies (in Vitro)*** | | |
| Acetone | Urine | | 796.5 | Introduced compact, highly sensitive Raman system with integration sphere cell holder. LOD for acetone was 8 mg/dl | 73 |
| Acetone | Urine | | 789 | Human; anti-Stokes Raman; demonstrated quantitative analysis of acetone in urine | 74 |
| Acetone | Urine | NIR | 790 | Human; demonstrated quantitative analysis of acetone in urine with an LOD of 40 mg/dl | 75 |
| Alumina | Bone/implants | | | Changes in alumina spectra investigated after implantation to establish biocompatibility | 76 |
| Ameloride | Urine (drug test) | | | Human; SERS; demonstrated quantitative capability in urine with colloidal silver | 77 |
| Amides | Breast/cancer | 1064 | Normal tissue: 1078, 1300, 1445, 1651; malignant tissue: 1445, 1651; benign tumors: 1240, 1445, 1659 | NIR FT Raman; human | 78 |
| Amides | Skin (stratum corneum) | NIR | | Human; FT Raman; forensic science; demonstrated nondestructive investigation of skin samples from late Neolithic man and contemporary samples | 79 |
| Amides | Various tissues | NIR | | CCD; mammalian tissues; evaluated effects of drying, snap freezing, thawing, and formalin fixation | 80 |
| Amphetamine | Urine (drug test) | | | Human; SERS; first phase of development of 1-step drug test using selective reactive SERS substrate coating; LODs < 20 ppm observed, required amide derivatization | 81 |

**TABLE 65.8** Raman Data

| Species | Target Organ or Disease | Excitation λ (nm) | Major Raman Shifts ($cm^{-1}$) | Comments | Ref. |
|---|---|---|---|---|---|
| Aromatic amino acids | Colon/cancer | 240 | | Human; reported Raman spectra for normal and neoplastic colon mucosa; neoplastic samples showed lower amino-acid-to-nucleotide content ratio | 82 |
| Aromatic amino acids | Colon/cancer | 251 | | Human; reported Raman spectra for normal and neoplastic colon mucosa; neoplastic samples showed lower amino-acid-to-nucleotide content ratio | 82 |
| β-Amylose | Brain/Alzheimer's | | 1628 | Human; post-mortem; distinguished between diseased and healthy tissue | 83 |
| β-Carotene | Breast/cancer | 406.7 | | Human; distinguished between normal and malignant breast tissue, attributing differences to relative concentrations of fatty acids and β-carotene | 84 |
| β-Carotene | Breast/cancer | 457.9 | | Human; distinguished between normal and malignant breast tissue, attributing differences to relative concentrations of fatty acids and β-carotene | 84 |
| β-Carotene | Breast/cancer | 514.5 | | Human; distinguished between normal and malignant breast tissue, attributing differences to relative concentrations of fatty acids and β-carotene | 84 |
| Blood plasma | Cancer | 488 | | Distinguished between cancer patients and healthy individuals by differences in carotenoid spectral regions | 85 |
| Blue particles??? | Cancer tumors | | | Raman microspectroscopy with resolution better than 1 $\mu m^2$ | 86 |
| Calcium oxalates | Kidney stone (calcium oxalate-type) | | 1465, 1492 | Human; quantitative analysis performed on calcium oxalate-type kidney stones | 87 |
| Calcium oxalates | Urine/stones | | | Human; review | 88 |
| Calcium salts | Artery (coronary)/ atherosclerosis | | | Human; combined intravascular ultrasound with Raman spectroscopy | 89 |
| Calculi (assorted) | Urine/stones | NIR | | Human; NIR FT Raman; generated a Raman library and used it to correctly identify composition of various calculi | 90 |
| Carotenoids | Breast/cancer | 488–515.5 | | CCD; human; determined 488–515 nm radiation well suited for monitoring carotenoid features due to resonance enhancement | 91 |
| Carotenoids | Brain/cancer | NIR | | NIR FT Raman; human | 92 |
| Carotenoids | Cardiocvascular tissue/fatty plaques | 514.5 | | | 93 |
| Cholesterol | Artery (coronary)/ atherosclerosis | | | Human; combined intravascular ultrasound with Raman spectroscopy | 89 |
| Cholesterol | Eye (lens)/ cataracts | | | Human; studied changing cholesterol content in lens fibers via Raman scattering; reported cholesterol and protein distribution in eye lens | 94 |
| Cholesterol | Eye (lens)/ cataracts | | | Human; studied changing cholesterol content in lens fibers via Raman scattering; reported cholesterol and protein distribution in eye lens | 95 |

**TABLE 65.8** Raman Data

| Species | Target Organ or Disease | Excitation λ (nm) | Major Raman Shifts ($cm^{-1}$) | Comments | Ref. |
|---|---|---|---|---|---|
| Cholesterol | Ocular lens/ cataracts | Visible | | | 96 |
| Cholesterol | Ocular lens/ cataracts | Visible | | | 97 |
| Collagen | Achilles tendon | Visible | | Bovine | 98 |
| Collagen | Cervix/cancer | NIR | | CCD; human; differentiated between precancerous and normal cervical tissues | 99 |
| Collagen | Cervix/cancer | NIR | | CCD; human; differentiated between precancerous and normal cervical tissues | 100 |
| Collagen | Ocular lens/ cataracts | Visible | | Aggregation and pigmentation monitored | 101 |
| Creatine | Urine | NIR | 692 | Human; demonstrated quantitative analysis of creatine in urine with an LOD of 1.5 mg/dl | 75 |
| Creatinine | Urine | | | Human; SERS needed for creatinine analysis | 102 |
| Cystine | Kidney stone | | 498, 677 | Human; spectra of cystine-type stones almost identical to cystine powder, particularly for S-S and C-S stretches | 103 |
| Cystine | Liver/cystinosis | Visible | 501, 2916–2967 | Human; cystine crystals identified in liver sections | 104 |
| Cystine | Spleen/cystinosis | Visible | 501, 2916–2967 | Human; cystine crystals identified in spleen sections | 104 |
| Cystine | Urine/stones | | | Human; review | 88 |
| Dentin | Teeth | 1064 | | Human; microscope; cross sections of teeth mapped by chemical functionality; dentin and enamel distinguished | 105 |
| Dentin | Teeth | | | Human; studied demineralization of dentin | 106 |
| Dentin | | Visible | | | 107 |
| Elastin | Achilles tendon | Visible | | Bovine | 98 |
| Enamel | Teeth | 1064 | | Human; microscope; cross sections of teeth mapped by chemical functionality; dentin and enamel distinguished | 105 |
| Extracellular proteins | Breast/cancer | 782–830 | | CCD; human; determined that NIR excitation better than blue-green radiation for lipids | 91 |
| Extracellular proteins | Breast/cancer (ductal carcinoma) | 785 | Normal: 1439, 1654; malignant: 1450, 1654 | CCD; human; formalin fixed tissue; observed dramatic spectral differences between normal and malignant tissue samples; diseased tissue dominated by proteins | 108 |
| Extracellular proteins | Breast/cancer (fibrocystic disease) | 785 | | CCD; human; formalin fixed tissue; observed dramatic spectral differences between normal and malignant tissue samples; diseased tissue dominated by proteins | 108 |
| Fatty acids | Breast/cancer | 406.7 | | Human; distinguished between normal and malignant breast tissue, attributing differences to relative concentrations of fatty acids and β-carotene | 84 |
| Fatty acids | Breast/cancer | 457.9 | | Human; distinguished between normal and malignant breast tissue, attributing differences to relative concentrations of fatty acids and β-carotene | 84 |
| Fatty acids | Breast/cancer | 514.5 | | Human; distinguished between normal and malignant breast tissue, attributing differences to relative concentrations of fatty acids and β-carotene | 84 |

**TABLE 65.8** Raman Data

| Species | Target Organ or Disease | Excitation λ (nm) | Major Raman Shifts ($cm^{-1}$) | Comments | Ref. |
|---|---|---|---|---|---|
| Gelatin | Achilles tendon | Visible | | Bovine | 98 |
| Glucose | Blood aqueous humor | | | Human; stimulated Raman scattering; analysis of aqueous humor precluded absorption and scattering complications associated with whole blood | 109 |
| Glucose | Blood plasma | | | Human; anti-Stokes Raman | 110 |
| Glucose | Blood serum | | | Human; anti-Stokes Raman | 110 |
| Glucose | Blood serum | | | Human | 111 |
| Glucose | Urine/diabetes | | 1130.5 | Introduced compact, highly sensitive Raman system with integration sphere cell holder; LOD for glucose was 31 mg/dl | 73 |
| Glucose | Urine/diabetes | | 1130 | Human; anti-Stokes Raman; demonstrated quantitative analysis of glucose in urine | 74 |
| Glucose | Urine/diabetes | NIR | 1130 | Human; demonstrated quantitative analysis of glucose in urine with an LOD of 41 mg/dl | 75 |
| Hepatocytes | Liver/cancer | Visible | 1040, 1083, 1182, 1241, 1253 | Distinguished between normal and malignant hepatocytes | 56 |
| Hydroxyapatites | Bone/implants | | | Changes in hydroxyapatite spectra investigated after implantation to establish biocompatibility | 76 |
| Hydroxyapatites | Bone/titanium implants | NIR | 592–1048 (hydroxy-apatite) 960–1450 (bone) | FT Raman; distinguished between hydroxyapatite and bone when studying hydroxyapatite-treated titanium implants | 112 |
| Hydroxyapatites | Cardiovascular tissue/calcified tissue | 514.5 | | | 113 |
| Hydroxyapatites | Metallic medical implants | NIR | | NIR FT Raman; hydroxyapatite powder spectra compared to spectra for hydroxyapatite coatings from metallic medical implants | 114 |
| Inclusions | Breast | | | Human; studied the effects of silicone breast implants via Raman scattering | 115 |
| L-amphetamine | Urine (Drug test) | | | Human; SERS; vibrational modes used to identify L-amphetamine at concentrations at the mg/ml level with colloidal silver; spectra of drug mixtures in urine reported | 116 |
| Lipids | Colon/cancer | 240 | | Human; reported Raman spectra for normal and neoplastic colon mucosa; neoplastic samples showed lower amino-acid-to-nucleotide content ratio | 82 |
| Lipids | Colon/cancer | 251 | | Human; reported Raman spectra for normal and neoplastic colon mucosa; neoplastic samples showed lower amino-acid-to-nucleotide content ratio | 82 |
| Lipids | Cornea | NIR | | NIR FT Raman; human; investigated thermally induced molecular disorder | 117 |
| Lipids | Eye (lens) | | | Human; studied regional changes in the lens membrane lipid composition corresponding to lens membrane functions | 118 |
| Lipids | Hair | NIR | | NIR FT Raman; human; investigated thermally induced molecular disorder | 119 |

**TABLE 65.8** Raman Data

| Species | Target Organ or Disease | Excitation λ (nm) | Major Raman Shifts ($cm^{-1}$) | Comments | Ref. |
|---|---|---|---|---|---|
| Lipids | Nail | NIR | | NIR FT Raman; human; investigated thermally induced molecular disorder | 119 |
| Lipids | Skin | NIR | | NIR FT Raman; human; investigated thermally induced molecular disorder | 119 |
| Lipids | Skin/cancer | | | Human; examined 11 skin lesion types and noted changes in lipid spectral bands when compared to healthy tissue | 120 |
| Lipids | Skin/cancer (basal cell carcinoma) | 1064 | 1300, 1420–1450 | Human; NIR FT Raman; observed differences between normal skin and basal cell carcinoma attributed to $CH_2$ and $(CH_2)(n)$-in-phase twist | 121 |
| Mefenorex | Urine (drug test) | | | Human; SERS; vibrational modes used to identify mefenorex at concentrations at the mg/ml level with colloidal silver; spectra of drug mixtures in urine reported | 116 |
| Methamphetamine | Urine (drug test) | | | Human; SERS; first phase of development of 1-step drug test using selective reactive SERS substrate coating; LODs < 20 ppm observed, required amide derivatization | 81 |
| Multicomponent | Blood (whole) | NIR | | CCD; human; performed quantitative analysis via partial least squares model | 122 |
| Multicomponent | Blood serum | NIR | | CCD; human; performed quantitative analysis via partial least squares model | 122 |
| Nucleic acids | Cervix/cancer | NIR | | CCD; human; differentiated between precancerous and normal cervical tissues | 99 |
| Nucleic acids | Cervix/cancer | NIR | | CCD; human; differentiated between precancerous and normal cervical tissues | 100 |
| Nucleotides | Colon/cancer | 240 | | Human; reported Raman spectra for normal and neoplastic colon mucosa; neoplastic samples showed lower amino acid to nucleotide content ratio | 82 |
| Nucleotides | Colon/cancer | 251 | | Human; reported Raman spectra for normal and neoplastic colon mucosa; neoplastic samples showed lower amino acid to nucleotide content ratio | 82 |
| Organic enamel | Teeth | | 2941 | Human; archaeological application; samples ranging up to 6000 years old analyzed and dated based on ratio of organic: inorganic band intensities | 123 |
| Pemoline | Urine (drug test) | | | Human; SERS; vibrational modes used to identify pemoline at concentrations at the mg/ml level with colloidal silver; spectra of drug mixtures in urine reported | 116 |
| Pentration enhancers | Skin (stratum corneum) | | | Human; investigated mechanisms of reversible lipid disruption by drug penetration enhancers | 124 |
| Pentration enhancers | Skin (stratum corneum) | | | Human; investigated mechanisms of reversible lipid disruption by drug penetration enhancers | 125 |
| Pentylenetetrazole | Urine (drug test) | | | Human; SERS; vibrational modes used to identify pentylenetetrazole at concentrations at the mg/ml level with colloidal silver; spectra of drug mixtures in urine reported | 110 |

**TABLE 65.8** Raman Data

| Species | Target Organ or Disease | Excitation λ (nm) | Major Raman Shifts ($cm^{-1}$) | Comments | Ref. |
|---|---|---|---|---|---|
| Phosphates | Kidney stone (phosphate-type) | Visible | | Identified phosphate-type kidney stones | 126 |
| Phosphates | Teeth | | 960 | Human; archaeological application; samples ranging up to 6000 years old analyzed and dated based on ratio of organic: inorganic band intensities | 123 |
| Phosphates | Urine/stones | | | Human; review | 88 |
| Phospholipids | Cervix/cancer | NIR | | CCD; human; differentiated between precancerous and normal cervical tissues | 99 |
| Phospholipids | Cervix/cancer | NIR | | CCD; human; differentiated between precancerous and normal cervical tissues | 100 |
| Phospholipids | Ocular lens/ cataracts | Visible | | | 97 |
| Plaque | Artery | | | Human; distinguished areas of plaque in arteries | 127 |
| Plaque | Artery | | | Human; distinguished areas of plaque in arteries | 128 |
| Poly(butylene terephthalate) (PBT) | Implants | | | Goat; confocal Raman microspectroscopy; evaluated influence of PEO on degradation and calcification of dense PEO-PBT copolymer implants | 129 |
| Poly(ethylene oxide) (PEO) | Implants | | | Goat; confocal Raman microspectroscopy; evaluated influence of PEO on degradation and calcification of dense PEO-PBT copolymer implants | 129 |
| Poly(methyl methacrylate) (PMMA) | Bone/implants | | | Changes in PMMA spectra investigated after implantation to establish biocompatibility | 76 |
| Polyethylene | Bone/implants | | | Changes in high-density polyethylene spectra investigated after implantation to establish biocompatibility | 76 |
| Polysaccharides | Skin/cancer (basal cell carcinoma) | 1064 | 840–860 | Human; NIR FT Raman; observed differences between normal skin and basal cell carcinoma attributed to changes in polysaccharide structure | 121 |
| Proteins | Eye (lens)/ cataracts | | | Human; studied changing cholesterol content in lens fibers via Raman scattering; reported cholesterol and protein distribution in eye lens | 94 |
| Proteins | Eye (lens)/ cataracts | | | Human; studied changing cholesterol content in lens fibers via Raman scattering; reported cholesterol and protein distribution in eye lens | 95 |
| Proteins | Hair | NIR | | NIR FT Raman; human; investigated thermally induced molecular disorder | 119 |
| Proteins | Nail | NIR | | NIR FT Raman; human; investigated thermally induced molecular disorder | 119 |
| Proteins | Ocular lens/ Cataracts | Visible | | | 119 |
| Proteins | Skin | NIR | | NIR FT Raman; human; investigated thermally induced molecular disorder | 119 |

**TABLE 65.8** Raman Data

| Species | Target Organ or Disease | Excitation λ (nm) | Major Raman Shifts ($cm^{-1}$) | Comments | Ref. |
|---|---|---|---|---|---|
| Proteins | Skin/cancer | | | Human; examined 11 skin lesion types and noted changes in protein spectral bands when compared to healthy tissue | 120 |
| Proteins | Skin/cancer (basal cell carcinoma) | 1064 | 1640–1680, 1220–1300, 928–940 | Human; NIR FT Raman; observed differences between normal skin and basal cell carcinoma attributed to amide I, amide III, nu (C-C) (valine and proline) vibrational modes | 121 |
| Urates | Urine/stones | | | Human; review | 88 |
| Urea | Urine | | | Human; urea high enough in concentration to monitor with normal Raman | 102 |
| Urea | Urine | | 1016 | Human; anti-Stokes Raman; demonstrated quantitative analysis of urea in urine | 74 |
| Urea | Urine | NIR | 1013 | Human; demonstrated quantitative analysis of urea in urine with an LOD of 4.9 mg/dl | 75 |
| Uric acid | Kidney stone (uric-acid-type) | Visible | | Identified uric-acid-type kidney stones | 130 |
| Uric acid | Urine | | | Human; SERS needed for uric acid analysis | 102 |
| Uric acid | Urine/stones | | | Human; review | 88 |
| | Arterey (coronary) | NIR | | CCD; human | 131 |
| | Artery | NIR | | NIR FT Raman; human | 132 |
| | Artery | NIR | | NIR FT Raman; human | 133 |
| | Artery | NIR | | NIR FT Raman; human | 134 |
| | Artery | NIR | | NIR FT Raman; human | 135 |
| | Artery (aorta) | NIR | | NIR FT Raman; human | 136 |
| | Artery/ atherosclerosis | NIR | | CCD; APOE*3 Leiden transgenic mice | 137 |
| | Blood | NIR | | Human; FT | 138 |
| | Blood serum/ various cancers | | | Human; Raman spectra of blood serum demonstrated to be an indicator for various cancers, each with a unique Raman spectrum | 139 |
| | Bone | NIR | | Investigated sheep and human samples; compared results with pure hydroxy and carbonated apatite; NIR FT Raman | 140 |
| | Bone | NIR | | Investigated sheep and human samples; compared results with pure hydroxy and carbonated apatite; NIR FT Raman | 141 |
| | Brain/cancer | NIR | | NIR FT Raman; rat | 142 |
| | Brain/cancer | NIR | | NIR FT Raman; human | 143 |
| | Brain/Parkinson's disease | 488 | | Monkey; distinguished between white and gray matter | 144 |
| | Breast | NIR | | Human; NIR FT Raman fiber optics; examined normal breast tissues, comparing effectiveness of Koehler laser illumination and the confocal principle | 145 |
| | Breast/cancer | 830 | | CCD; human; demonstrated potential of differentiating between malignant and benign lesions in breast tissues | 146 |
| | Breast/cancer | NIR | | CCD; human | 147 |
| | Breast/cancer | NIR | | CCD; human; spectral imaging of lipids and proteins in thin sections of breast tissue | 148 |

**TABLE 65.8** Raman Data

| Species | Target Organ or Disease | Excitation λ (nm) | Major Raman Shifts ($cm^{-1}$) | Comments | Ref. |
|---|---|---|---|---|---|
| | Cardiovascular tissue/calcified aorta | Visible | | | 149 |
| | Colon/cancer | NIR | | CCD; human; determined Raman has limited utility due to low sensitivity in comparison to fluorescence and absorption; observed excessive interference | 147 |
| | Colon/cancer | UV | | Human | 150 |
| | Connective tissue | NIR | | Human; FT | 138 |
| | Cornea | Visible | | Feline | 151 |
| | Cornea | Visible | | Monitored hydration gradients across the cornea | 152 |
| | Cornea | NIR | | Near IR FT Raman; human. | 153 |
| | Coronary artery/ atherosclerosis | NIR | | CCD; human; classified coronary artery lesions as nonatherosclerotic, noncalcified plaque, and calcified plaque | 154 |
| | Coronary artery/ atherosclerosis | NIR | | CCD; human; classified coronary artery lesions as nonatherosclerotic, noncalcified plaque, and calcified plaque | 155 |
| | Eye | Visible | | Various layers of ocular tissues investigated | 156 |
| | Gallstone | Visible | | | 157 |
| | Gallstone | NIR | | Human; FT Raman; spectral data reported for a range of gallstones | 158 |
| | Gastrointestinal cancer | NIR | | CCD; human | 159 |
| | Gastrointestinal cancer | | | Human; Raman spectroscopy described as yielding the greatest information in comparison to fluorescence, optical coherence tomography, and ultrasound | 160 |
| | Gynecological tissues/cancer | NIR | | Near IR FT Raman; human | 161 |
| | Gynecological tissue/cancer | NIR | | Human; FT; distinguished between normal, benign, and cancerous tissues | 161 |
| | Hair | NIR | | Human; FT Raman spectra reported with band assignments for hair | 162 |
| | Larynx/cancer | 830 | | Human; differentiated between normal, dysplastic, and squamous cell carcinoma of the larynx; salient spectral features observed in 850–950 and 1200–1350 $cm^{-1}$ ranges | 163 |
| | Lung/cancer | 1064 | | Human; totally fluorescence-free Raman spectra obtained for normal and malignant lung tissues | 164 |
| | Mineral deposition disease | Visible | | | 165 |
| | Mineral deposition disease | Visible | | Human; differentiated between mono- and dihydrate oxalates in the myocardium and the pituitary | 166 |
| | Mineral deposition disease/lung | Visible | | | 167 |
| | Mineral deposits/ prosthetic implants | Visible | | | 168 |

**TABLE 65.8** Raman Data

| Species | Target Organ or Disease | Excitation λ (nm) | Major Raman Shifts ($cm^{-1}$) | Comments | Ref. |
|---|---|---|---|---|---|
| | Mineral deposits/ prosthetic implants | Visible | | | 169 |
| | Mineral deposits/ prosthetic implants | Visible | | | 170 |
| | Mineral deposits/ silicone breast implants | Visible | | | 171 |
| | Nail | NIR | | Human; FT Raman spectra reported with band assignments for nails | 162 |
| | Ocular lens | Visible | | | 172 |
| | Ocular lens | Visible | | Monitored disulfide bond formation in eye lens | 173 |
| | Ocular lens | Visible | | | 174 |
| | Ocular lens | Visible | | | 175 |
| | Ocular lens | Visible | | | 176 |
| | Ocular lens | Visible | | Human and rabbit | 177 |
| | Ocular lens | Visible | | Human | 178 |
| | Ocular lens | Visible | | Human | 179 |
| | Ocular lens | NIR | | NIR FT Raman | 180 |
| | Oral tissues/ Cancer | | | Human; distinguished between normal and malignant oral tissues | 181 |
| | Pathological inclusions in breast tissue | Visible | | | 182 |
| | Pathological tissues (inclusions) | Visible | | | 183 |
| | Peripheral artery | NIR | | CCD; human | 184 |
| | Retina/diabetes | NIR | | NIR FT Raman | 185 |
| | Skin (stratum corneum) | NIR | | Human; FT Raman; spectra of human corneum reported with band assignments | 186 |
| | Skin (stratum corneum) | | | Snake; modeled drug diffusion across skin | 187 |
| | Skin (stratum corneum) | | | Pig; modeled drug diffusion across skin | 187 |
| | Skin (stratum corneum) | NIR | | Human; FT Raman; compared normal, hyperkeratotic, and psoriatic stratum corneum | 188 |
| | Skin/cancer | NIR | | NIR FT Raman; human | 189 |
| | Skin/cancer | NIR | | NIR FT Raman; human | 190 |
| | Skin/cancer | NIR | | NIR FT Raman; human | 191 |
| | Skin/cancer (melanoma) | NIR | | FT; human | 138 |
| | Soft tissue/sarcoma | NIR | | CCD; human | 147 |
| | Stomach/cancer | | | Human; malignant tissues observed to yield stringer bands related to OH, NH, C=O stretching and H-O-H bending | 192 |
| | Tooth | NIR | | Human; FT Raman; reported compositional changes in children's teeth | 193 |
| | Tooth dentin | NIR | | Investigated sheep and human samples; compared results with pure hydroxy and carbonated apatite; NIR FT Raman | 140 |

**TABLE 65.8** Raman Data

| Species | Target Organ or Disease | Excitation λ (nm) | Major Raman Shifts ($cm^{-1}$) | Comments | Ref. |
|---|---|---|---|---|---|
| | Tooth enamel | NIR | | Investigated sheep and human samples; compared results with pure hydroxy and carbonated apatite; NIR FT Raman | 140 |
| | Urinary bladder/ cancer | NIR | | CCD; human; determined Raman has limited utility due to low sensitivity in comparison to fluorescence and absorption; observed excessive interference | 147 |
| | Various tissues | UV | | | 147 |
| | | 830 | | CCD; human | 194 |
| | | | ***Animal Studies (in Vivo)*** | | |
| | Cornea | | | Rabbit; confocal Raman; fiber optics; investigated effects of corneal dehydration via observation of changes in stretching modes of OH and CH groups | 195 |
| | Eye (aqueous humor) | | | Rabbit; confocal Raman; compared rabbit and blind human eyes; assessed biochemical properties of eye-specific Raman signatures | 196 |
| | Eye (cornea) | | | Rabbit; confocal Raman; compared rabbit and blind human eyes; assessed biochemical properties of eye-specific Raman signatures | 196 |
| | Eye (lens) | | | Rabbit; confocal Raman; compared rabbit and blind human eyes; assessed biochemical properties of eye-specific Raman signatures | 196 |
| | Ocular lens | | | Rabbit | 175 |
| | Oral tissue/cancer (dysplasia) | | | Rat; *in vivo* spectra obtained for normal and dysplastic rat palate tissue; distinguished between normal tissue, low-grade dysplasia, and high-grade dysplasia/carcinoma | 197 |
| | | | ***Clinical Studies (in Vivo)*** | | |
| Carotenoids | Skin/cancer | | | Human; noninvasive; determined that carotenoid concentration is lower in malignant than healthy skin tissue | 198 |
| | Adipose tissue | 785 | 1077, 1265, 1303, 1444, 1658, 1755 | Human; noninvasive; fiber-optic probe; CCD; distinguished between fat tissue and bone; >200 mW laser power | 199 |
| | Arterey (aorta) | 810 | | CCD; human | 151 |
| | Arterey (aorta) | 810 | | CCD; human | 200 |
| | Bone | 785 | 957, 1032, 1087, 1188, 1248, 1454, 1670 | Human; noninvasive; fiber-optic probe; CCD; distinguished between fat tissue and bone; >200 mW laser power | 199 |
| | Cervix/cancer | NIR | | CCD; human; collected spectra via fiber-optic probe | 201 |
| | Eye (aqueous humor) | | | Confocal Raman; compared rabbit and blind human eyes; assessed biochemical properties of eye-specific Raman signatures | 196 |
| | Eye (cornea) | | | Confocal Raman; compared rabbit and blind human eyes; assessed biochemical properties of eye-specific Raman signatures | 196 |

**TABLE 65.8** Raman Data

| Species | Target Organ or Disease | Excitation λ (nm) | Major Raman Shifts ($cm^{-1}$) | Comments | Ref. |
|---|---|---|---|---|---|
| | Eye (lens) | | | Confocal Raman; compared rabbit and blind human eyes; assessed biochemical properties of eye-specific Raman signatures | 196 |
| | Gastrointestinal tissues | | | Human; Raman spectra collected via fiber-optic probe during routine endoscopy; subtle differences between normal and diseased tissues observed | 202 |
| | Hair | 785 | | Human; noninvasive; fiber-optic probe; CCD; >200 mW laser power | 199 |
| | Mouth | 785 | | Human; noninvasive; fiber-optic probe; CCD; >200 mW laser power | 199 |
| | Nail | NIR | | Human; NIR FT Raman; measured fingernails and discussed feasibility of technique for determination of metabolic disorders, drug use, poisoning, and local infections | 203 |
| | Skin | 785 | | Human; noninvasive; fiber-optic probe; CCD; >200 mW laser power | 199 |
| | Skin/cancer | NIR | | NIR FT Raman; human | 191 |
| | | 830 | | CCD; human | 194 |

## References


1. Glice, M.M., Les, A., and Bajdor, K., IR, Raman and theoretical *ab initio* RHF study of aminoglutethimide — an anticancer drug, *J. Mol. Struct.*, 450, 141, 1998.
2. Buttler, C.A., Cooney, R.P., and Denny, W.A., Raman-spectroscopic studies of amsacrine, *Appl. Spectrosc.*, 46,1540, 1992.
3. Buttler, C.A., Cooney, R.P., and Denny, W.A., Resonance Raman-study of the binding of the anticancer drug amsacrine to DNA, *Appl. Spectrosc.*, 48, 822, 1994.
4. Chourpa, I., Morjani, H., Riou, J.F., and Manfait, M. Intracellular interactions of antitumour drug amsacrine (m-AMSA) as revealed by surface-enhanced Raman spectroscopy, *FEBS Lett.*, 397, 61, 1996.
5. Michalska, D., The Raman and normal coordinate analysis of 3-(*N*-phenylacetylamino)-2,6-piperidinedione, antineoplastin-A10, the new antitumor drug, *Spectrochim. Acta A*, 49, 303, 1993.
6. Torreggiani, A., Fagnano, C., and Fini, G., Involvement of lysine and tryptophan side-chains in the biotin-avidin interaction, *J. Raman Spectrosc.*, 28, 23, 1997.
7. Torreggiani, A. and Fini, G., Raman spectroscopic studies of ligand-protein interactions: the binding of biotin analogues by avidin, *J. Raman Spectrosc.*, 29, 229, 1998.
8. Chiang, H.P., Mou, B., Li, K.P., Chiang, P., Wang, D., Lin, S.J., and Tse, W.S., FT-Raman, FT-IR and normal-mode analysis of carcinogenic polycyclic aromatic hydrocarbons, Part II — a theoretical study of the transition states of oxygenation of benzo(a)pyrene (BaP), *J. Raman. Spectrosc.*, 32, 53, 2001.
9. Rajani, C., Kincaid, J.R., and Petering, D.H., A systematic approach toward the analysis of drug-DNA interactions using Raman spectroscopy: the binding of metal-free bleomycins A(2) and B-2 to calf thymus DNA, *Biopolymers*, 52, 110, 1999.

10. Weselucha-Birczynska, A. and Nakamoto, K., Study of the antimalarial drug cinchonine with nucleic acids by Raman spectroscopy, *J. Raman Spectrosc.*, 27, 915, 1996.
11. Cavigiolio, G., Bendetto, L., Boccaleri, E., Colangelo, D., Viano, I., and Osella, D., Pt(II) complexes with different N-donor ligands for specific inhibition of telomerase, *Inorg. Chim. Acta,* 305, 61, 2000.
12. Hoffman, G.M., Torres, A., and Forster, H.V., Validation of a volumeless breath-by-breath method for measurement of respiratory quotient, *J. Appl. Physiol.*, 75, 1903, 1993.
13. Beljebbar, A., Morjani, H., Angibousat, J.F., Sockalingum, G.D., Polissiou, M., and Manfait, M., Molecular and cellular interaction of the differentiating antitumour agent dimethylcrocetin with nuclear retinoic acid receptor as studied by near-infrared and visible SERS spectroscopy, *J. Raman Spectrosc.*, 28, 159, 1997.
14. Graham, D., Smith, W.E., Linacre, A.M.T., Munro, C.H., Watson, N.D., and White, P.C., Selective detection of deoxyribonucleic acids at ultralow concentrations by SERRS, *Anal. Chem.*, 69, 4703, 1997.
15. Isola, N.R., Stokes, D.L., and Vo-Dinh, T., Surface-enhanced Raman gene probe for HIV detection, *Anal. Chem.*, 70, 1352, 1998.
16. Vo-Dinh, T., Houke, K., and Stokes, D.L., Surface-enhanced Raman gene probes, *Anal. Chem.*, 66, 3379, 1994.
17. Pal, B. and Bajpai, P.K., Spectroscopic characterization of gelonin — assignments secondary structure and thermal denaturation, *Ind. J. Biochem. Biophys.*, 35, 1998, 166.
18. Denninger, J.W., Schelvis, J.P.M., Brandish, P.E., Zhao, Y., Babcock, G.T., and Marletta, M.A., Interaction of soluble guanylate cyclase with YC-1: kinetic and resonance Raman studies, *Biochemistry,* 39, 4191, 2000.
19. Schelvis, J.P.M., Zhao, Y., Marletta, M.A., and Babcock, G.T., Resonance Raman characterization of the heme domain of soluble guanylate cyclase, *Biochemistry,* 37, 16289, 1998.
20. Zhao, Y., Schelvis, J.P.M., Babcock, G.T., and Marletta, M.A., Identification of histidine 105 in the beta 1 subunit of soluble guanylate cyclase as the heme proximal ligand, *Biochemistry,* 37, 4502, 1998.
21. Fan, B.C., Gupta, G., Danziger, R.S., Friedman, J.M., and Rousseau, D.L., Resonance Raman characterization of soluble guanylate cyclase expressed from baculovirus, *Biochemistry,* 37, 1178, 1998.
22. Li, Z.Q., Li, X.Y., Sheu, F.S., Chen, D.M., and Yu, N.T., Isolation, purification and characterization of soluble guanylate cyclase from bovine lung, *Chem. Res. Chin. Univ.*, 13, 111, 1997.
23. Tomita, T., Ogura, T., Tsuyama, S., Imai, Y., and Kitagawa, T., Effects of GTP on CO-bound nitric oxide of soluble guanylate cyclase probed by resonance Raman spectroscopy, *Biochemistry,* 36, 10155, 1997.
24. Raussens, V., Pezolet, M., Ruysschaert, J.M., and Goormaghtigh, E., Structural difference in the $H^+$, $K^+$-ATPase between the E1 and E2 conformations — an attenuated total reflection infrared spectroscopy, UV circular dichroism and Raman spectroscopy study, *Eur. J. Biochem.*, 262, 176, 1999.
25. Rauseens, V., De Jongh, H., Pezelot, H., Ruysschaert, J.M., and Goormaghtigh, E., Secondary structure of the intact $H^+$, $K^+$-ATPase and of its membrane-embedded region — an attenuated total reflection infrared spectroscopy, circular dichroism and Raman spectroscopic study, *Eur. J. Biochem.*, 252, 261, 1998.
26. Vogel, K.M., Spiro, T.G., Shelver, D., Thorsteinsson, M.V., and Roberts, G.P., Resonance Raman evidence for a novel charge relay mechanism of the CO-dependent heme protein transcription factor CooA, *Biochemistry,* 38, 2679, 1999.
27. Walters, M.A., Leung, Y.C., Blumenthal, N.C., LeGeros, R.Z., and Konsker, K.A., A Raman and infrared spectroscopic investigation of biological hydroxyapatite, *J. Inorg. Biochem.*, 39, 193, 1990.

28. Morjani, H., Riou, J.F., Nabiev, I., Lavelle, F., and Manfait, M., Molecular and cellular interactions between intoplicine, DNA, and topoisomerase-II studies by surface-enhanced Raman scattering spectroscopy, *Cancer Res.,* 53, 4784, 1993.
29. Lord, R.C. and Yu, N.T., Laser-excited Raman spectroscopy of biomolecules I. Native lysozyme and its constituent amino acids, *J. Mol. Biol.,* 50, 509, 1970.
30. Chen, Y.Q., Gulotta, M., Cheung, H.T.A., and Callender, R., Light activates reduction of methotrexane by NADPH in the ternary complex with *Escherichia coli* dihydrofolate reductase, *Photochem. Photobiol.,* 69, 77, 1999.
31. Carey, P.R., Raman spectroscopy in enzymology: the first 25 years, *J. Raman Spectrosc.,* 29, 7, 1998.
32. Callender, R., Deng, H., and Gilmanshin, R., Raman difference studies of protein structure and folding, enzymatic catalysis and ligand binding, *J. Raman Spectrosc.,* 29, 15, 1998.
33. Zhao, X., Spiro, T.G., Ultraviolet resonance Raman spectroscopy of hemoglobin with 200 and 2121 nm excitation: H-bonds of tyrosines and prolines, *J. Raman Spectrosc.,* 29, 49, 1998.
34. Overman, S.A., Thomas, G.J., Raman markers of nonaromatic side chains in an alpha-helix assembly: Ala, Asp, Glu, Gly, Ile, Leu, Lys, Ser, and Val residues of phage fd subunits, *Biochemistry,* 38, 4018, 1999.
35. Tsuboi, M., Overman, S.A., and Thomas, G.J., Orientation of tryptophan-26 in coat protein subunits in the filamentous virus Ff by polarized Raman microspectroscopy, *Biochemistry,* 35, 10403, 1996.
36. Overman, S.A. and Thomas, G.J., Amide modes of the alpha-helix: Raman spectroscopy of filamentous virus Fd containing peptid C-13 and H-2 labels in coat protein subunits, *Biochemistry,* 37, 5654, 1998.
37. Overman, S.A. and Thomas, G.J., Novel vibrational assignments for proteins from Raman spectra of viruses, *J. Raman Spectrosc.,* 29, 23, 1998.
38. Feofanov, A.V., Grichine, A.I., Shitova, L.A., Karmakova, T.A., Yakubovskaya, R.I., Egret-Charlier, M., and Vigny, P., Confocal Raman microspectroscopy of thapthal in living cancer cells, *Biophys. J.,* 78, 499, 2000.
39. Torreggiani, A. and Fini, G., Drug-antiserum molecular interactions: a Raman spectroscopic study, *J. Raman Spectrosc.,* 30, 295, 1999.
40. Naveke, A., Lapouge, K., Sturgis, J.N., Hartwich, G., Simonin, J., Scheer, H., and Robert, B., Resonance Raman spectroscopy of metal-substituted bacteriochlorophylls: characterization of Raman bands sensitive to bacteriochlorin conformation, *J. Raman Spectrosc.,* 28, 599, 1997.
41. Manoharan, R., Ghiamati, E., Dalterio, R.A., Britton, K., Nelson, W.H., and Sperry, J.F., UV resonance Raman spectra of bacteria, bacterial spores, protoplasts, and calcium dipicolinate, *J. Microbiol. Methods,* 11, 1, 1990.
42. Ramanauskaite, R.B., Segers Nolten, I.G.M.J., deGrauw, K.J., Sijtsema, N.M., Van der Maas, L., Greve, J., Otto, C., and Figdor, C.G., Carotenoid levels in human lymphocytes, measured by Raman spectroscopy, *Pure Appl. Chem.,* 69, 2131, 1997.
43. Schut, T.C.B., Puppels, G.J., Kraan, Y.M., Greve, J., Van der Maas, L.L.J., and Figdor, C.G., Intracellular carotenoid levels measured by Raman microspectroscopy: comparison of lymphocytes from lung cancer patients and healthy individuals, *Int. J. Cancer,* 74, 20, 1997.
44. Rasmussen, R.E., Hammer-Wilson, M., and Berns, M.W., Mutation and chromatid exchange induction in Chinese hamster ovary (CHO) cells by pulsed excimer laser radiation at 193 nm and 308 nm and continuous UV radiation at 254 nm, *Photochem. Photobiol.,* 49, 413, 1989.
45. Yao, Y., Nelson, W.H., Hargraves, P., and Zhang, J., UV resonance Raman study of demoic acid, a marine neurotoxic amino acid, *Appl. Spectrosc.,* 51, 785, 1997.
46. Wen, Z.Q., Overman, S.A., and Thomas, G.J., Jr., Structure and interactions of the single-stranded DNA genome of filamentous virus fd: investigation by ultraviolet resonance Raman spectroscopy, *Biochemistry,* 36, 7810, 1997.

47. Wen, Z.Q. and Thomas, G.J., Jr., UV resonance Raman spectroscopy of DNA and protein constituents of viruses: assignments and cross sections for excitations at 257, 244, 238, and 229 nm, *Biopolymers,* 45, 247, 1998.
48. Peticolas, W.L., Patapoff, T.W., Thomas, G.A., Postlewait, J., and Powell, J.W., Laser Raman spectroscopy of chromosomes in living eukaryotic cells: DNA polymorphism *in vivo, J. Raman Spectrosc.,* 27, 571, 1996.
49. Overman, S.A., Aubrey, K.L., Reilly, K.E., Osman, O., Hayes, S.J., Serwer, P., and Thomas, G.J., Conformation and interactions of the packaged double-stranded DNA genome of bacteriophage T7, *Biospectroscopy,* 4, S47, 1998.
50. Puppels, G.J., Demul, F.F.M., Otto, C., Greve, J., Robetnicoud, M., Arndtjovin, D.J., and Jovin, T.M., Studying single living cells and chromosomes by confocal Raman microspectroscopy, *Nature,* 347, 301, 1990.
51. Puppels, G.J., Olminkhof, J.H., Segers-Nolten, G.M., Otto, C., de Mul, F.F., and Greve, J., Laser irradiation and Raman spectroscopy of single living cells and chromosomes: sample degradation occurs with 514.5 nm but not with 660 nm laser light, *Exp. Cell. Res.,* 195, 361, 1991.
52. Puppels, G.J., Otto, C., Greve, J., Robert-Nicoud, M., Arndt-Jovin, T.M., Raman microspectroscopy of low pH induced changes in DNA structure of polytene chromosomes, *Biochemistry,* 33, 3386, 1994.
53. Doig, S.J. and Prendergast, F.G., Continuously tunable, quasi-continuous-wave source for ultraviolet resonanace Raman-spectroscopy, *Appl. Spectrosc.,* 49, 247, 1995.
54. Carey, P.R., Resonance Raman labels and Raman labels, *J. Raman Spectrosc.,* 29, 861, 1998.
55. Hoey, S., Brown, D.H., McConnell, A.A., Smith, W.E., Marabani, M., and Sturrock, R.D., Resonance Raman spectroscopy of hemoglobin in intact cells: a probe of oxygen uptake by erythrocytes in rheumatoid arthritis, *J. Inorg. Biochem.,* 34, 189, 1988.
56. Hawi, S.R., Campbell, W.B., Kajdacsy-Balla, A., Murphy, R., Adar, F., and Nithipatikom, K., Characterization of normal and malignant hepatocytes by Raman microspectroscopy, *Cancer Lett.,* 110, 35, 1996.
57. Lin, D.L., Tarnowski, C.P., Zhang, J., Dai, J.L, Rohn, E., Patel, A.H., Morris, M.D., and Keller, E.T., Bone metastatic LNCaP-derivative C4-2B prostate cancer cell line mineralizes *in vitro, Prostate,* 47, 212, 2001.
58. Yazdi, Y., Ramanujam, N., Lotan, R., Mitchell, M.F., Hittelman, W., and Richards-Kortum, R., Resonance Raman spectroscopy at 257 nm excitation of normal and malignant cultured breast and cervical cells, *Appl. Spectrosc.,* 53, 82, 1999.
59. Sureau, F., Chinsky, L., Amirand, C., Ballini, J.P., Duquesne, M., Laigle, A., Turpin, P.Y., and Vigny, P., An ultraviolet micro-Raman spectrometer-resonance Raman spectroscopy within single living cells, *Appl. Spectrosc.,* 44, 1047, 1990.
60. Freeman, T.L., Cope, S.E., Stringer, M.R., Cruse-Sawyer, J.E., Batchelder, D.N., and Brown, S.B., Raman spectroscopy for the determination of photosensitizer localization in cells, *J. Raman Spectrosc.,* 28, 641, 1997.
61. Freeman, T.L., Cope, S.E., Stringer, M.R., Cruse-Sawyer, J.E., Brown, S.B., Batchelder, D.N., and Birbeck, K., Investigation of the subcellular localization of zinc phthalocyanines by Raman mapping, *Appl. Spectrosc.,* 52, 1257, 1998.
62. Arzhantsev, S.Y., Chikishev, A.Y., Koroteev, N.I., Greve, J., Otto, C., and Sijtsema, N.M., Localization study of Co-phthalocyanines in cells by Raman micro(spectro)scopy, *J. Raman Spectrosc.,* 30, 205, 1999.
63. Bakker Schut, T.C., Puppels, G.J., Kraan, Y.M., Greve, J., van der Maas, L.L, and Figdor, C.G., Intracellular carotenoid levels measured by Raman microspectroscopy: comparison of lymphocytes from lung cancer patients and healthy individuals, *Int. J. Cancer,* 74, 20, 1997.
64. Puppels, G.J., Garritsen, H.S.P., Sergersnolten, G.M.J., Demul, F.F.M., and Grefe, J., Raman microscopic approach to the study of human granulocytes, *Biophys. J.,* 60, 1046, 1991.

65. Puppels, G.J., Garritsen, H.S.P., Kummer, J.A., and Greve, J., Carotenoids located in human lymphocyte subpopulations and natural killer cells by Raman microspectroscopy, *Cytometry,* 14, 251, 1993.
66. Grygon, C.A., Perno, J.R., Fodor, S.P.A., and Spiro, T.G., Ultraviolet resonance Raman spectroscopy as a probe of protein structure in the FD virus, *Biotechniques,* 6, 50, 1988.
67. Dinakarpandian, D., Shenoy, B., Pusztai-Carey, M., Malcolm, B.A., and Carey, P.C., Active site properties of the 3C proteinase from hepatitis A virus (a hybrid cisteine/serine protease) probed by Raman spectroscopy, *Biochemistry,* 38, 4943, 1997.
68. Chadha, S., Manoharan, R., Moenne-Loccoz, P., Nelson, W.H., Peticolas, W.L., Sperry, J.F., Comparison of the UV resonance Raman-spectra of bacteria, bacterial-cell walls, and ribosomes excited in the deep UV, *Appl. Spectrosc.,* 47, 38, 1993.
69. Tuma, R. and Thomas, G.J., Mechanisms of virus assembly probed by Raman spectroscopy: the icosahedral bacteriophage P22, *Biophys. Chem.,* 68, 17, 1997.
70. Nelson, W.H., Manoharan, R., and Sperry, J.F., UV resonance Raman studies of bacteria, *Appl. Spectrosc. Rev.,* 27, 67, 1992.
71. Dalterio, R.A., Nelson, W.H., Britt, D., Sperry, J., and Purcell, F.J., A resonance Raman microprobe study of chromobacteria in water, *Appl. Spectrosc.,* 40, 271, 1998.
72. Manoharan, R., Ghiamati, E., Chadha, S., Nelson, W.H., and Sperry, J.F., Effect of cultural conditions on deep UV resonance Raman spectra of bacteria, *Appl. Spectrosc.,* 47, 2145, 1993.
73. Dou, X.M., Yamaguchi, Y., Yamamoto, H., Doi, S., and Ozaki, Y., A highly sensitive compact Raman system without a spectrometer for quantitative analysis of biological samples, *Vibrat. Spectrosc.,* 14, 199, 1997.
74. Dou, X.M., Yamaguchi, Y., Yamamoto, H., Doi, S., and Ozaki, Y., Quantitative analysis of metabolites in urine by anti-Stokes Raman spectroscopy, *Biospectroscopy,* 3, 113, 1997.
75. Dou, X., Yamagouchi, Y., Yamamato, H., Doi, S., and Ozaki, Y., Quantitative analysis of metabolites in urine using a highly precise, compact, near-infrared Raman spectrometer, *Vibrat. Spectrosc.,* 13, 83, 1996.
76. Bertoluzza, A., Fagnano, C., Tinti, A., Morelli, M.A., Tosi, M.R., Maggi, G., and Marchetti, P.G., Raman and infrared spectroscopic study of the molecular characterization of the biocompatibility of prosthetic biomaterials, *J. Raman Spectrosc.,* 25, 109, 1994.
77. Calvo, N., Montres, R., and Laserna, J.J., Surface-enhanced Raman-spectrometry of ameloride on colloidal silver, *Anal. Chim. Acta,* 280, 263, 1993.
78. Alfono, R.R., Liu, C.H., Sha, W.L., Zhu, H.R., Akins, D.L., Cleary, J., Prudente, R., and Clemer, E., Human breast tissue studies by IR Fourier-transform Raman spectroscopy, *Lasers Life Sci.,* 4, 23, 1991.
79. Edwards, H.G.M., Farwell, D.W., Williams, A.C., Barry, B.W., and Rull, F., Novel spectroscopic deconvolution procedure for complex biological systems — vibrational components in the FT-Raman spectra of ice-man and contemporary skin, *J. Chem. Soc. Faraday Trans.,* 91, 3883, 1995.
80. Shim, M.G. and Wilson, B.C., The effects of *ex vivo* handling procedures on the near-infrared Raman spectra of normal mammalian tissues, *Photochem. Photobiol.,* 63, 662, 1996.
81. Sulk, R.A., Corcoran, R.C., and Carron, K.T., Surface enhanced Raman detection of amphetamine and methamphetamine by modification with 2-mercaptonicotinic acid, *Appl. Spectrosc.,* 53, 954, 1999.
82. Boustany, N., Crawford, J.M., Manoharan, R., Dasari, R.R., and Feld, M.S., Analysis of nucleotides and aromatic amino acids in normal and neoplastic colon mucosa by ultraviolet resonance Raman spectroscopy, *Lab. Invest.,* 79, 1201, 1999.
83. Sajid, J., Elhaddoui, A., and Turrell, S., Fourier transform vibrational spectroscopic analysis of human cerebral tissue, *J. Raman Spectrosc.,* 28, 165, 1997.
84. Redd, D.C.B., Feng, Z.C., Yue, K.T., and Gansler, T.S., Raman spectroscopic characterization of human breast tissues: implications for breast cancer diagnosis, *Appl. Spectrosc.,* 47, 787, 1993.

85. Larsson, K. and Hellgren, L., A study of the combined Raman and fluorescence scattering from human blood plasma, *Experientia,* 30, 481, 1974.
86. Huong, P.V., New possibilities of Raman microspectroscopy, *Vibrat. Spectrosc.,* 11, 17, 1996.
87. Kodati, V.R., Tomasi, G.E., Turmin, J.L., and Tu, A.T., Raman-spectroscopic identification of calcium-oxalate-type kidney-stone, *Appl. Spectrosc.,* 44, 1408, 1990.
88. Carmona, P., Bellanato, J., and Escolar, E., Infrared and Raman spectroscopy of urinary calculi: a review, *Biospectroscopy,* 3, 331, 1997.
89. Romer, T.J., Brennan, J.F., Puppels, G.J., Tuinenburg, J., van Duinen, S.G., van der Laarse, A., van der Steen, A.F.W., Bom, N.A., and Bruschke, A.V.C., Intravascular ultrasound combined with Raman spectroscopy to localize and quantify cholesterol and calcium salts in atherosclerotic coronary arteries, *Arterioscl. Throm. Vasc.,* 20, 478, 2000.
90. Hong, T.D.N., Phat, D., Plaza, P., Daudon, M., and Dao, N.Q., Identification of urinary calculi by Raman laser fiber optics spectroscopy, *Clin. Chem.,* 38, 292, 1992.
91. Frank, C.J., Redd, D.C.B., Gansler, T.S., and McCreery, R.L., Characterization of human breast biopsy specimens with near-IR Raman spectroscopy, *Anal. Chem.,* 66, 319, 1994.
92. Mizuno, A., Kitajima, H., Kawauchi, K., Muraishi, S., and Ozaki, Y., Near infrared Fourier transform Raman spectroscopic study of human brain tissues and tumours, *J. Raman Spectrosc.,* 25, 265, 1994.
93. Clarke, R.H., Wang, Q., and Isner, J.M., Laser Raman spectroscopy of atherosclerotic lesions in human coronary artery segments, *Appl. Opt.,* 27, 4799, 1988.
94. Vrensen, G.F.J.M. and Duindam, H.J., Maturation of fiber membranes in the human eye lens — ultrastructural and Raman microspectroscopic observations, *Ophthalmic Res.,* 27(Suppl. 1), 78, 1995.
95. Yaroslavsky, I.V., Yaroslavsky, A.N., Otto, C., Puppels, G.J., Vrensen, G.F.J.M., Duindam, H.J., and Greve, J., Combined elastic and Raman light-scattering of human eye lenses, *Exp. Eye Res.,* 59, 393, 1994.
96. Duindam, J.J., Vrensen, G.F., Otto, C., Puppels, G.J., and Greve, J., New approach to assess the cholesterol distribution in the eye lens: confocal Raman microspectroscopy and filipin cytochemistry, *J. Lipid Res.,* 36, 1139, 1995.
97. Duindam, J.J., Vrensen, G.F., Otto, C., and Greve, J., Cholesterol, phospholipids, and protein changes in focal opacities in the human eye lens, *Invest. Ophthalmol. Vis. Sci.,* 39, 94, 1998.
98. Freshour, B.G. and Koenig, J.L., Raman scattering of collagen, gelatin, and elastine, *Biopolymers,* 14, 379, 1975.
99. Mahadevan-Jensen, A. and Richards-Kortum, R., Raman spectroscopy for the detection of cancers and precancers, *J. Biomed. Opt.,* 1, 31, 1996.
100. Mahadevan-Jensen, A., Mitchell, M.F., Ramanujam, N., Malpica, A., Thompsen, S., Utzinger, U., and Richards-Kortum, R., NIR Raman spectroscopy for *in vitro* detection of cervical precancers, *Photochem. Photobiol.,* 68, 123, 1998.
101. Ozaki, Y., Kaneuchi, F., Iwamoto, T., Yoshiura, M., and Iriyama, K., Nondestructive analysis of biological materials by FT-IR-ATR. 1. Direct evidence for the existence of collagen helix in lens capsule, *Appl. Spectrosc.,* 43, 138, 1989.
102. Premasiri, W.R., Clarke, R.H., and Womble, M.E., Urine analysis by laser Raman spectroscopy, *Lasers Surg. Med.,* 28, 330, 2001.
103. Kodati, V.R. and Tu, A.T., Raman-spectroscopic identification of cystine-type kidney stone, *Appl. Spectrosc.,* 44, 837, 1990.
104. Centeno, J.A., Ishak, K., Mullick, F.G., Gahl, W.A., and O'Leary, T.J., Infrared microspectroscopy and laser Raman microprobe in the diagnosis of cystinosis, *Appl. Spectrosc.,* 48, 569, 1994.
105. Wentrup-Byrne, E., Armstrong, C.A., Armstrong, R.S., and Collins, B.M., Fourier transform Raman microscopic mapping of the molecular components in a human tooth, *J. Raman Spectrosc.,* 28, 151, 1997.
106. Van der Veen, M.H., and ten Bosch, J.J., The influence of mineral loss on the auto-fluorescent behavior of *in vitro* demineralized dentin, *Caries Res.,* 30, 93, 1996.

107. Rippon, W.P., Koenig, J.L., and Walton, A., Laser Raman spectroscopy of biopolymers and proteins, *Agric. Food Chem.*, 19, 692, 1971.
108. Frank, C.J., McCreery, R.L., and Redd, D.C.B., Raman spectroscopy of normal and diseased human breast tissues, *Anal. Chem.*, 67, 777, 1995.
109. Tarr, R.V. and Steffes, P.G., Non-invasive blood glucose measurement system and method using stimulated Raman spectroscopy, U.S. patent 5,243,983.
110. Dou, X.M., Yamaguchi, Y., Yamamoto, H., Uenoyama, H., and Ozaki, Y., Biological applications of anti-Stokes Raman spectroscopy: quantitative analysis of glucose in plasma and serum by a highly sensitive multichannel Raman spectrometer, *Appl. Spectrosc.*, 50, 1301, 1996.
111. Koo, T.-W., Berger, A.J., Itzkan, I., Horowitz, G., and Feld, M.S., Measurement of glucose in blood serum using Raman spectroscopy, *IEEE LEOS Mag.*, 12, 18, 1998.
112. Otto, C., de Grauw, C.J., Duindam, J.J., Sijtsema, N.M., and Greve, J., Applications of micro-Raman imaging in biomedical research, *J. Raman Spectrosc.*, 28, 143, 1997.
113. Clarke, R.H., Hanlon, E.B., Isner, J.M., and Brody, H., Laser Raman spectroscopy of calcified atherosclerotic lesions in cardiovascular tissue, *Appl. Opt.*, 26, 3175, 1987.
114. Tudor, A.M., Melia, C.D., Davies, M.C., Anderson, D., Hastings, G., Morrey, S., Domingos Santos, J., and Monteiro, F., The analysis of biomedical hydroxyapatite powders and hydroxyapatite coatings on metallic medical implants by near-IR Fourier transform Raman-spectroscopy, *Spectrochim. Acta A*, 49, 675, 1993.
115. Frank, C.J., McCreery, R.L., Redd, D.C.B., and Gansler, T.S., Detection of silicone in lymph-node biopsy specimens by near-IR Raman-spectroscopy, *Appl. Spectrosc.*, 47, 387, 1993.
116. Ruperez, A., Montes, R., and Laserna, J.J., Identification of stimulant-drugs by surface-enhanced Raman-spectrometry on colloidal silver, *Vibrat. Spectrosc.*, 2, 145, 1991.
117. Lawson, E.E., Anigbogu, A.N., Williams, A.C., Barry, B.W., and Edwards, H.G., Thermally induced molecular disorder in human stratum corneum lipids compared with a model phospholipid system: FT Raman spectroscopy, *Spectrochim. Acta A*, 54, 543, 1989.
118. Borchman, D., Ozaki, Y., Lamba, O.P., Byrdwell, W.C., Czarnecki, M.A., and Yappert, M.C., Structural characterization of clear human lens lipid-membranes by near-infrared Fourier-transform Raman spectroscopy, *Curr. Eye Res.*, 14, 511, 1995.
119. Gniadecka, M., Faurskov Nielsen, O., Christensen, D.H., and Wulf, H.C., Structure of water, proteins, and lipids in intact human skin, hair and nail, *J. Invest. Dermatol.*, 110, 393, 1998.
120. Gniadecka, M., Wulf, H.C., Nielsen, O.F., Christensen, D.H., and Hercogova, J., Distinctive molecular abnormalities in benign and malignant skin lesions: studies by Raman spectroscopy, *Photochem. Photobiol.*, 66, 418, 1997.
121. Gniadecka, M., Wulf, H.C., Mortensen, N.N., Nielsen, O.F., and Christensen, D.H., Diagnosis of basal cell carcinoma by Raman spectroscopy, *J. Raman Spectrosc.*, 28, 125, 1997.
122. Berger, A.J., Koo, T.-W., Itzkan, I., Horowitz, G., and Feld, M.S., Multicomponent blood analysis by near-infrared Raman spectroscopy, *Appl. Opt.*, 38, 2916, 1999.
123. Bertoluzza, A., Brasili, P., Castri, L., Facchini, F., Fagnano, G., and Tinti, A., Preliminary results in dating human skeletal remains by Raman spectroscopy, *J. Raman Spectrosc.*, 28, 185, 1997.
124. Anigbogu, A.N.C., Williams, A.C., Barry, B.W., and Edwards, H.G.M., Fourier-transform Raman-spectroscopy of interactions between the penetration enhancer dimethyl-sulfoxide and human stratum-corneum, *Int. J. Pharm.*, 125, 265, 1995.
125. Anibogu, A.N.C., Williams, A.C., Barry, B.W., and Edwards, H.G.M., in *Prediction of Percutaneous Penetration: Methods, Measurements and Modelling*, Vol. 3, Brain, K.R., James, V.J., and Walters, K.A., Eds., STS Publishing, Cardiff, 1995, p. 27.
126. Kodati, V.R., Tomasi, G.E., Turmin, J.L., and Tu, A.T., Raman spectroscopic identification of phosphate-type kidney stones, *Appl. Spectrosc.*, 45, 581, 1991.
127. Weinmann, P., Jouan, M., Dao, N.Q., Lacroix, B., Croiselle, C., Bonte, J.P., and Luc, G., Quantitative analysis of cholesterol and cholesteryl esters in human atherosclerotic plaques using near-infrared Raman spectroscopy, *Atherosclerosis*, 140, 81, 1998.

128. Romer, T.J., Brennan, J.F., Schut, T.C.B., Wolthius, R., van der Hoogen, R.C.M., Emeis, J.J., van der Laarse, A., Bruschke, A.V.G., and Puppels, G.J., Raman spectroscopy for quantifying cholesterol in an intact coronary artery wall, *Atherosclerosis*, 141, 117, 1998.
129. Radder, A.M., Van Loon, J.A., Puppels, G.J., Van Bitterswijk, C.A., Degradation and calcification of a PEO/PBT copolymer series, *J. Mater. Sci. Mater. Med.*, 6, 510, 1995.
130. Kodati, V.R., Tu, A.T., and Turmin, J.L., Raman spectroscopic identification of uric-acid-type kidney stone, *Appl. Spectrosc.*, 44, 1134, 1990.
131. Brennan, J.F., Romer, T.J., Lees, R.S., Tercyak, A.M., Kramer, J.R., and Feld, M.S., Determination of human coronary artery composition by Raman spectroscopy, *Circulation*, 96, 99, 1997.
132. Rava, R.P., Baraga, J.J., and Feld, M.S., Near-infrared Fourier-transform Raman spectroscopy of human artery, *Spectrochim. Acta*, 47, 509, 1991.
133. Manoharan, R., Baraga, J.J., Feld, M.S., and Rava, R.P., Quantitative histochemical analysis of human artery using Raman spectroscopy, *J. Photochem. Photobiol. B: Biol.*, 16, 211, 1992.
134. Baraga, J.J., Feld, M.S., and Rava, R.P., *In situ* optical histochemistry of human artery using near infrared Fourier transform Raman spectroscopy, *Proc. Nat. Acad. Sci. U.S.A.*, 89, 3473, 1992.
135. Redd, D.C., Yue, K.T., Martini, L.G., and Kaufman, S.L., Raman spectroscopy of human atherosclerotic plaque: implications for laser angioplasty, *J. Vasc. Interv. Radiol.*, 2, 247, 1991.
136. Liu, C.H., Glassman, W.L.S., Zhu, H.R., Akins, D.L., Deckelbaum, L.I., Stetz, M.L., O'Brien, K., Scott, J., and Alfono, R.R., Near-IR Fourier transform Raman spectroscopy of normal and artherosclerotic human aorta, *Lasers Life Sci.*, 4, 257, 1992.
137. Romer, T.J., Brennan, J.F., Puppels, G.J., Laarse, A.V.D., Princen, H.M.G., Folger, O., Buschman, H.P.J., Jukema, J.W., Havekes, L.M., and Bruschke, A.V.G., Raman spectroscopy provides chemical mappings of atherosclerotic plaques in APOE*3 Leiden transgenic mice, *J. Am. Coll. Cardiol.*, 31, 500A, 1998.
138. Schrader, B., Dippel, B., Frendel, S., Keller, S., Lochte, T., Riedl, M., Sculte, R., and Tatsch, E., NIR FT Raman spectroscopy — a new tool in medical diagnostics, *J. Mol. Struct.*, 408, 247, 1997.
139. Gou, P., Yi, G.H., Xiong, P., Yuan, Y.L., Xie, Q., and Chen, C.X., Raman spectra of the serums from cancerous persons, *Spectrosc. Spectr. Anal.*, 20, 844, 2000.
140. Rehman, I., Smith, R., Hench, L.L., and Bonfield, W., Structural evaluation of human and sheep bone and comparison with synthetic hydroxyapatite by FT Raman spectroscopy, *J. Biomed. Mater. Res.*, 29, 1287, 1995.
141. Smith, R. and Rehman, I., Fourier-transform Raman-spectroscopic studies of human bone, *J. Mater. Sci. Mater. Med.*, 5, 775, 1994.
142. Mizuno, A., Hayashi, T., Tashibu, K., Maraishi, S., Kawauchi, K., and Ozaki, Y., Near-infrared FT-Raman spectra of the rat-brain tissues, *Neurosci. Lett.*, 141, 47, 1992.
143. Keller, S., Schrader, B., Hoffman, A., Schrader, W., Metz, K., Rehlaender, A., Pahnke, J., Ruwe, M., and Budach, W., Application of near-infrared Fourier-transform Raman spectroscopy in medical research, *J. Raman Spectrosc.*, 25, 663, 1994.
144. Ong, C.W., Shen, Z.X., He, Y., Lee, T., and Tang, S.H., Raman microspectroscopy of the brain tissues in the substantia nigra and MPTP-induced Parkinson's disease, *J. Raman Spectrosc.*, 30, 91, 1999.
145. Dipple, B., Tatsch, E., and Schrader, B., Development of an inverted NIR-FT-Raman microscope for biomedical applications, *J. Mol. Struct.*, 408, 247, 1997.
146. Manoharan, R., Shafer, K., Perelman, L., Wu, J., Chen, K., Deinum, G., Fitzmaurice, M., Myles, J., Crowe, J., Dasari, R.R., and Feld, M.S., Raman spectroscopy and fluorescence photon migration for breast cancer diagnosis and imaging, *Photochem. Photobiol.*, 67, 15, 1998.
147. Manoharan, R., Wang, Y., and Feld, M.S., Review: histochemical analysis of biological tissues using Raman spectroscopy, *Spectrochim. Acta A*, 52, 215, 1996.
148. Kline, N. and Treado, P.J., Raman chemical imaging of breast tissue, *J. Raman Spectrosc.*, 28, 119, 1997.
149. Klug, D.D., Singleton, D.L., and Walley, V.M., Laser Raman spectrum of calcified human aorta, *Lasers Surg. Med.*, 12, 13, 1992.

150. Manoharan, R., Wang, Y., Dasari, R.R., Singer, S., Rava, R.P., and Feld, M.S., UV resonance Raman spectroscopy for the detection of colon cancer, *Lasers Life Sci.,* 6, 217, 1995.
151. Goheen, S.C., Lis, L.J., and Jauffman, J.W., Raman spectroscopy of intact feline corneal collagen, *Biochem. Biophys. Acta,* 536, 197, 1978.
152. Bauer, N.J.C., Wicksted, J.P., Jongsma, F.H.M., March, W.F., Hendrikse, F., and Montamedi, M., Non-invasive assessment of the hydration gradient across the cornea using confocal Raman spectroscopy, *Invest. Opthalmol. Vis. Sci.,* 39, 831, 1998.
153. Williams, A.C., Barry, B.W., Edwards, H.G., and Farwell, D.W., A critical comparison of some Raman spectroscopic techniques for studies of human stratum corneum, *Pharm. Res.,* 10, 1642, 1993.
154. Romer, T.J., Brennan, J.F., Fitzmaurice, M., Feldstein, M.L., Deinum, G., Myles, J.L., Kramer, J.R., Less, R.S., and Feld, M.S., Histopathy of human coronary artery atherosclerosis by quantifying its chemical composition by Raman spectroscopy, *Circulation,* 97, 878, 1998.
155. Deinum, G., Rodriguez, D., Romer, T.J., Fitzmaurice, M., Kramer, J.R., and Feld, M.S., Histological classification of Raman spectra of human coronary atherosclerosis using principal component analysis, *Appl. Spectrosc.,* 53, 938, 1999.
156. Jongsma, F.H.M., Erckens, R.J., Wicksted, J.P., Bauer, N.J.C., Hendrikse, F., March, W.F., and Motamedi, M., Confocal Raman spectroscopy system for noncontact scanning of ocular tissues: an *in vitro* study, *Opt. Eng.,* 36, 3193, 1997.
157. Ishida, H., Kamoto, R., Uchida, S., Ishitani, A., Ozaki, Y., Iriyama, K., Tsukie, T., Shibata, K., Ishihara, F., and Kameda, H., Raman microprobe and Fourier transform-infrared microsampling studies of the microstructure of gallstones, *Appl. Spectrosc.,* 41, 407, 1997.
158. Wentrup-Byrne, E., Rintoul, J., Smith, J.L., and Fredericks, P.M., Comparison of vibrational spectroscopic techniques for the characterization of human gallstones, *Appl. Spectrosc.,* 49, 1028, 1995.
159. Bohorfoush, A.G., Tissue spectroscopy for gastrointestinal diseases, *Endoscopy,* 28, 372, 1996.
160. Barr, H., Dix, T., and Stone, N., Optical spectroscopy for the early diagnosis of gastrointestinal malignancy, *Lasers Med. Sci.,* 13, 3, 1998.
161. Liu, C.H., Das, B.B., Glassman, W.L.S., Tang, G.C., Yoo, K.M., Zhu, H.R., Akins, D.L., Lubicz, S.S., Cleary, J., Prudente, R., Celmer, E., Caron, A., and Alfono, R.R., Raman, fluorescence, and time-resolved light scattering as optical diagnostic techniques to separate diseased and normal biomedical media, *J. Photochem. Photobiol. B,* 16, 187, 1992.
162. Williams, A.C., Edwards, H.G.M., and Barry, B.W., Raman-spectra of human keratotic biopolymers — skin, callus, hair and nail, *J. Raman Spectrosc.,* 25, 95, 1994.
163. Stone, N., Stavroulaki, P., Kendall, C., Birchall, M., and Barr, H., Raman spectroscopy for early detection of laryngeal malignancy: preliminary results, *Laryngoscope,* 110, 1756, 2000.
164. Kaminaka, S., Yamazaki, H., Ito, T., Kohda, E., and Hamaguchi, H.O., Near-infrared Raman spectroscopy of human lung tissues: possibility of molecular-level cancer diagnosis, *J. Raman Spectrosc.,* 32, 139, 2001.
165. McGill, N., Dieppe, P.A., Bowden, M., Gardiner, D.J., and Hall, M., Identification of pathological mineral deposits by Raman microspectroscopy, *Lancet,* 337, 77, 1991.
166. Pestaner, J.P., Mullick, F.G., Johnson, F.B., and Centeno, J.A., Calcium oxalate crystals in human pathology: molecular analysis with the laser Raman microprobe, *Arch. Pathol. Lab. Med.,* 120, 537, 1993.
167. Buitveld, H., De Mul, F.F.M., and Greve, J., Identification of inclusions in lung tissues with a Raman microprobe, *Appl. Spectrosc.,* 38, 304, 1984.
168. Hahn, D.W., Wohlfarth, D.L., and Parks, N.L., Analysis of polyethylene wear debris using micro Raman spectroscopy: a report on the presence of beta-carotene, *J. Biomed. Mater. Res.,* 35, 31, 1997.
169. Wolfarth, D.L., Han, D.W., Bushar, G., and Parks, N.L., Separation and characterization of polyethylene wear debris from synovial fluid and tissue samples of revised knee replacements, *J. Biomed. Mater. Res.,* 43, 57, 1997.
170. Kalasinsky, V.F., Johnson, F.B., and Ferwerda, R., Fourier transform IR and Raman microspectroscopy of materials in tissue, *Cell Mol. Biol.,* 44, 141, 1998.

171. Luke, J.L, Kalasinsky, V.F., Turnicky, R.P., Centeno, J.A., Johnson, F.B., and Mullick, F.B., Pathological and biophysical findings associated with silicone breast implants: a study of capsular tissues from 86 cases, *Plast. Reconstr. Surg.*, 100, 1558, 1997.
172. Farrell, R. and McCauley, R., On corneal transparency and its loss with swelling, *J. Opt. Soc. Am.*, 66, 342, 1976.
173. Yu, N.T., DeNagel, D.C., Pruett, P.L., and Kuck, J.F.R., Disulfide bond formation in the eye lens, *Proc. Natl. Acad. Sci. U.S.A.*, 82, 7965, 1985.
174. Mizuno, A., Ozaki, Y., Kamada, Y., Myazaki, H., Itoh, K., and Iriyama, K., Direct measurement of Raman spectra of intact lens in a whole eyeball, *Curr. Eye Res.*, 1, 609, 1981.
175. Yu, N.T, Kuck, J.F.R., and Askren, C.C., Laser Raman spectroscopy of the lens *in situ*, measured in an anesthetized rabbit, *Curr. Eye Res.*, 1, 615, 1982.
176. Yu, N.T., DeNagel, D.C., and Kuck, J.F.R., Ocular lenses, in *Biological Applications of Raman Spectroscopy*, Spiro, T.G., Ed., John Wiley & Sons, New York, 1987, p. 47.
177. Bot, A.C.C., Huizinga, A., de Mul F.F.M., Vrensen, G.F.J.M., and Greve, J., Raman microspectroscopy of fixed rabbit and human lenses and lens slices — new potentials, *Exp. Eye Res.*, 49, 161, 1989.
178. Siebenga, I., Vrensen, G.F., Otto, K., Puppels, G.J., De Mul, F.F., and Greve, J., Aging and changes in protein conformation in the human lens: a Raman microspectroscopy study, *Exp. Eye Res.*, 54, 759, 1992.
179. Smeets, M.H., Vrensen, G.F.J.M., Otto, K., Puppels, G.J., and Greve, J., Local variations in protein structure in the human eye lens — a Raman microspectroscopy study, *Biophys. Biochim. Acta,* 1164, 236, 1993.
180. Nie, S.M., Bergbauer, K.L., Kuck, J.F.R., and Yu, N.T., Near-infrared Fourier-transform Raman spectroscopy in human lens research, *Exp. Eye Res.*, 51, 619, 1990.
181. Venkatakrishna, K., Kurien, J., Pai, K.M., Valiathan, M., Kumar, N.N., Krishna, C.M., Ullas, G., and Kartha, V.B., Optical pathology of oral tissue: a Raman spectroscopy diagnostic method, *Curr. Sci.*, 80, 665, 2001.
182. Schaeberle, M.D., Kalasinsky, V.F., Luke, J.L., Lewis, E.N., Levin, I.W., and Treado, P.J., Raman chemical imaging: histopathology of inclusions in human breast tissue, *Anal. Chem.*, 66, 1829, 1996.
183. Abraham, J.L. and Etz, E., Molecular microanalysis of pathological specimens *in situ* with a laser Raman microprobe, *Science,* 206, 718, 1979.
184. Salenius, J.P., Brennan, J.F., Miller, A., Wang, Y., Aretz, T., Sacks, B., Dasari, R.R., and Feld, M.S., Biochemical composition of human peripheral arteries using near infrared Raman spectroscopy, *J. Vasc. Surg.*, 27, 710, 1998.
185. Sebag, J., Nie, S., Reiser, K., Charles, M.A., and Yu, N.T., Raman spectroscopy of vitreous in proliferative diabetic retinopathy, *Invest. Ophthalmol. Vis. Sci.*, 35, 2976, 1994.
186. Williams, A.C., Edwards, H.G.M., and Barry, B.W., Fourier-transform Raman spectroscopy — a novel application for examining human stratum-corneum, *Int. J. Pharm.*, 81, R11, 1992.
187. Williams, A.C., Barry, B.W., and Edwards, H.G.M., Comparison of Fourier-transform Raman spectra of mammalain and reptilian skin, *Analyst,* 119, 563, 1994.
188. Edwards, H.G.M., Williams, A.C., and Barry, B.W., Potential applications of FT-Raman spectroscopy for dermatological diagnostics, *J. Mol. Struct.*, 347, 379, 1995.
189. Barry, B.W., Edwards, H.G.M., and Williams, A.C., Fourier-transform Raman and infrared vibrational study of human skin — assignment of spectral bands, *J. Raman Spectrosc.*, 23, 641, 1992.
190. Fendel, S. and Schrader, B., Investigation of skin and skin lesions by NIR-FT Raman spectroscopy, *Fresenius J. Anal. Chem.*, 360, 609, 1998.
191. Caspers, P.J., Lucassen, G.W., Wolthuis, R., Bruining, H.A., and Puppels, G.J., *In vitro* and *in vivo* Raman spectroscopy of human skin, *Biospectroscopy,* 4, 569, 1998.
192. Ling, X.F, Li, W.H., Song, Y.Y., Yang, Z.L., Xu, Y.Z., Weng, S.F., Xu, Z., Fu, X.B., Zhou, X.S., and Wu, J.G., FT-Raman spectroscopic investigation on stomach cancer, *Spectrosc. Spectr. Anal.*, 20, 692, 2000.

193. Hendra, P.J., Jones, C., and Warnes, G., *Fourier Transform Raman Spectroscopy: Instrumentation and Chemical Applications*, Ellis Horwood, Chichester, U.K., 1991, p. 209.
194. Brennan, J.F., Wang, Y., Dassari, R.R., and Feld, S.S., Near-infrared Raman spectrometer systems for human tissue studies, *Appl. Spectrosc.*, 51, 201, 1997.
195. Erckens, R.J., Bauer, N.J.C., March, W.F., Jongsma, F.H.M., Hendrikse, F., and Motamedi, M., Non-invasive *in-vivo* assessment of corneal dehydration in the rabbit using confocal Raman spectroscopy, *Invest. Ophthalmol. Vis. Sci.*, 37, 1668, 1996.
196. Bauer, N.J.C., March, W.F., Wicksted, J.P., Hendrikse, F., Jongsma, F.H.M., and Motamedi, M., *In-vivo* confocal Raman spectroscopy of the human eye, *Invest. Ophthalmol. Vis. Sci.*, 37, 3450, 1996.
197. Schut, T.C.B., Witjes, M.J.H., Sterenborg, H.J.C.M., Speelman, O.C., Roodenburg, J.L.N., Marple, E.T., Bruining, H.A., and Puppels, G.J., *In vivo* detection of dysplastic tissue by Raman spectroscopy, *Anal. Chem.*, 72, 6010, 2000.
198. Hata, T.R., Scholz, T.A., Ermakov, I.V., McClane, R.W., Khachik, F., Gellermann, W., and Pershing, L.K., Non-invasive Raman spectroscopic detection of carotenoids in human skin, *J. Invest. Dermatol.*, 115, 441, 2000.
199. Shim, M.G. and Wilson, B.C., Development of an *in vivo* Raman spectroscopic system for diagnostic applications, *J. Raman Spectrosc.*, 28, 131, 1997.
200. Baraga, J.J., Feld, M.S., and Rava, R.P., Rapid near-infrared Raman spectroscopy of human tissue with a spectrograph and CCD detector, *Appl. Spectrosc.*, 46, 187, 1992.
201. Mahadevan-Jensen, A., Mitchell, M.F., Ramanujam, N., Utzinger, U., and Richards-Kortum, R., Development of a fiber optic probe to measure NIR Raman spectra of cervical tissue *in vivo*, *Photochem. Photobiol.*, 68, 427, 1998.
202. Shim, M.G., Song, L.M.W.M., Marcon, N.E., and Wilson, B.C., *In-vivo* near-infrared Raman spectroscopy: demonstration of feasibility during clinical gastrointestinal endoscopy, *Photochem. Photobiol.*, 72, 146, 2000.
203. Schrader, B., Keller, S., Lochte, T., Fendel, S., Moore, D.S., Simon, A., and Sawatski, J., NIR FT Raman spectroscopy in medical diagnosis, *J. Mol. Struct.*, 348, 293, 1995.

### 65.3.4 Fluorescence Spectroscopy

Table 65.9 provides introductory-level information about the various fluorescence properties of many different chemical and biochemical species that have proven useful in the field of biomedical photonics. Due to the overwhelming number of fluorescent chemical species of biomedical importance and the limited space available for compiling this information, this table is far from comprehensive. However, it is the hope of the authors that this table will serve as a useful starting reference for researchers interested in biomedical applications of fluorescence techniques. The information in Table 65.9 was compiled using data from almost 150 publications, review articles, and monographs.

**TABLE 65.9A** Fluorescence Data: Biochemical Studies

| Species | Endogenous/ Exogenous Origin | Absorbance Maxima (nm) | Molar Extinction Coefficient ($M^{-1}$) | Excitation $\lambda_{ex}$ (nm) | Emission $\lambda_{em}$ (nm) | Quantum Yield | Matrix | Lifetime τ1 (rel % τ1) | Lifetime τ2 (rel % τ2) | Lifetime τ3 (rel % τ3) | Comments | Ref. |
|---|---|---|---|---|---|---|---|---|---|---|---|---|
| 3-Hydroxy-kynuremine | Endogenous fluorophore | — | — | 355 | >355 | — | Extracts of human eye (young) | 24000 | 400000 | — | — | 1 |
| 3-Hydroxy-kynuremine | Endogenous fluorophore | — | — | 355 | >355 | — | Extracts of human eye (old) | 0.031–0.061 | — | — | — | 1 |
| 4-Pyridoxic acid (PA) | Endogenous fluorophore | 307 | — | 315 | 425 | — | — | — | — | — | Physiologic pH | 2 |
| Beta-carotene | Endogenous fluorophore | — | — | blue | 520 | — | Powder/solution | 9.6 | 2 | 0.3 | — | 3 |
| Ceroid | Endogenous fluorophore | — | — | 340–395 | 430–460 | — | — | — | — | — | Physiologic pH | 4 |
| Ceroid | Endogenous fluorophore | — | — | 340–395 | 540–640 | — | — | — | — | — | Physiologic pH | 5 |
| Collagen | Endogenous fluorophore | — | — | 280, 265, 330, 450 | 310, 385, 390, 530 | — | Powder | — | — | — | Physiologic pH | 6 |
| Collagen | Endogenous fluorophore | — | — | Ultraviolet | Violet-blue | — | Powder | 2.7 (25) | 8.9 (35) | — | — | 7 |
| Collagen | Endogenous fluorophore | — | — | 340 | 395 | — | Powder/solution | 9.9 | 5.0 | 0.8 | — | 3, 8 |
| Collagen | Endogenous fluorophore | — | — | 270 | 395 | — | Powder/solution | — | — | — | — | 3, 8 |
| Collagen | Endogenous fluorophore | — | — | 285 | 310 | — | Powder/solution | — | — | — | — | 3, 8 |
| Collagen | Endogenous fluorophore | 330 | — | 300–370 | 350–470 | — | — | — | — | — | — | 9 |
| Collagen (Type I) | Endogenous fluorophore | 340, 500 | — | 340 | 410 | — | — | — | — | — | — | 10 |
| Collagen (Type I) | Endogenous fluorophore | 340, 500 | — | 500 | 520 | — | — | — | — | — | — | 10 |
| Collagen, elastin, HP | Endogenous fluorophore | 325 | — | 325 | 400 | — | — | — | — | — | Physiologic pH | 11 |
| Collagen, elastin, LP | Endogenous fluorophore | — | — | 325 | 400 | — | — | — | — | — | Physiologic pH | 12 |

| | | | | | | | | | | | | |
|---|---|---|---|---|---|---|---|---|---|---|---|---|
| Elastin | Endogenous fluorophore | — | — | 350, 410, 450 | 420, 500, 520 | — | Powder | — | — | — | Physiologic pH | 6 |
| Elastin | Endogenous fluorophore | — | — | — | — | — | Powder | 2 (75) | 6.7 (65) | — | — | 7 |
| Elastin | Endogenous fluorophore | 330 | — | 330 | 405 | — | — | — | — | — | — | 10 |
| Elastin | Endogenous fluorophore | — | — | 460 | 520 | — | Powder/solution | 6.7 | 1.4 | — | — | 3, 8 |
| Elastin | Endogenous fluorophore | — | — | 360 | 410 | — | Powder/solution | 7.8 | 2.6 | 0.5 | — | 3, 8 |
| Elastin | Endogenous fluorophore | — | — | 425 | 490 | — | Powder/solution | — | — | — | — | 3, 8 |
| Elastin | Endogenous fluorophore | — | — | 260 | 410 | — | Powder/solution | — | — | — | — | 3, 8 |
| Elastin | Endogenous fluorophore | 348 | — | 300–400 | 330–550 | — | — | — | — | — | — | 9 |
| Endogenous porphyrins | Endogenous fluorophore | — | — | 400 | 610 | — | Powder/solution | — | — | — | — | 13 |
| Endogenous porphyrins | Endogenous fluorophore | — | — | 400 | 675 | — | Powder/solution | — | — | — | — | 13 |
| Eosinophils — circulating | Endogenous fluorophore | — | — | 370, 500 | 440, 550 | — | — | — | — | — | Physiologic pH | 14 |
| Eosinophils — granules | Endogenous fluorophore | — | — | 380 | 520 | — | — | — | — | — | Physiologic pH | 15 |
| Eosinophils — granules | Endogenous fluorophore | — | — | 450 | 520 | — | — | — | — | — | Physiologic pH | 16 |
| Flavin adenine dinucleotide (FAD) | Endogenous fluorophore | — | — | 450 | 515 | — | — | — | — | — | Physiologic pH | 17 |
| Flavin adenine dinucleotide (FAD) | Endogenous fluorophore | — | — | 450 | 515 | — | — | — | — | — | — | 18 |
| Flavin adenine dinucleotide (FAD) | Endogenous fluorophore | — | — | 370 | 535 | — | — | — | — | — | — | 18 |

**TABLE 65.9A** Fluorescence Data: Biochemical Studies (continued)

| Species | Endogenous/ Exogenous Origin | Absorbance Maxima (nm) | Molar Extinction Coefficient ($M^{-1}$) | Excitation $\lambda_{ex}$ (nm) | Emission $\lambda_{em}$ (nm) | Quantum Yield | Matrix | Lifetime τ1 (rel % τ1) | Lifetime τ2 (rel % τ2) | Lifetime τ3 (rel % τ3) | Comments | Ref. |
|---|---|---|---|---|---|---|---|---|---|---|---|---|
| Flavin mononucleotide (FMN) (riboflavin-5′-phosphate) | Endogenous fluorophore | — | — | 440 | 520 | — | Powder/solution | 4.7 | — | — | — | 19 |
| Flavins | Endogenous fluorophore | 215, 260, 388, 455 | — | 200–500 | 500–600 | — | — | — | — | — | — | 9 |
| Flavomono-nucleotide, oxidized (FMN) | Endogenous fluorophore | — | — | 436 | 500 | — | Water | 5.2 (100) | — | — | — | 20 |
| Hematoporphyrins | Endogenous fluorophore | 402 | — | 300–450 | 610–710 | — | — | — | — | — | — | 9 |
| Lipo-pigments | Endogenous fluorophore | 345, 430 | — | 310–470 | 450–670 | — | — | — | — | — | — | 9 |
| Lipofuscin | Endogenous fluorophore | — | — | 340–395 | 430–460, 540–640 | — | — | — | — | — | Physiologic pH | 17 |
| Lipofuscin | Endogenous fluorophore | — | — | 340–395 | 430–460, 540–640 | — | — | — | — | — | Physiologic pH | 4 |
| Lipofuscin | Endogenous fluorophore | — | — | 340–395 | 540–640 | — | — | — | — | — | Physiologic pH | 5 |
| Nicotinamide adenine dinucleotide phosphate (NADPH) | Endogenous fluorophore | — | — | 365 | >410 | — | Water | 0.45–0.60 (75–90) | 1.4–2.2 (10–25) | — | — | 20, 21 |
| Nicotinamide adenine dinucleotide (NAD+) oxidized form | Endogenous fluorophore | 260 | $18 \times 10^{-3}$ | — | — | — | — | — | — | — | Physiologic pH | 22 |
| Nicotinamide adenine dinucleotide (NADH) reduced form | Endogenous fluorophore | — | — | 350 | 460 | — | Powder/solution | 0.6 | 0.2 | — | — | 3, 8 |

| | | | | | | | | | | | | |
|---|---|---|---|---|---|---|---|---|---|---|---|---|
| Nicotinamide adenine dinucleotide (NADH) reduced form | Endogenous fluorophore | 260, 340 | $14.4 \times 10^{-3}$, $6.2 \times 10^{-3}$ | 290 | 440 | — | — | — | — | — | Physiologic pH | 22 |
| Nicotinamide adenine dinucleotide (NADH) reduced form | Endogenous fluorophore | — | — | 340 | 450 | — | — | — | — | — | Physiologic pH | 21 |
| Nicotinamide adenine dinucleotide (NADH) reduced form | Endogenous fluorophore | — | — | 340 | 450 | — | — | — | — | — | Physiologic pH | 18 |
| Nicotinamide adenine dinucleotide (NADH) reduced form | Endogenous fluorophore | 260, 350 | — | 250–280, 320–375 | 425–525 | — | — | — | — | — | — | 9 |
| Phenylalanine | Endogenous fluorophore | 260 | $2.0 \times 10^{-4}$ | 260 | 280 | 0.04 | — | — | — | — | Physiologic pH | 22 |
| Porphyrin | Endogenous fluorophore | — | — | 420 | 640 | — | — | — | — | — | — | 18 |
| Pyridoxal (PL) | Endogenous fluorophore | — | — | 330 | 380 | — | — | — | — | — | Physiologic pH | 2 |
| Pyridoxal 5′-phosphate (PLP) | Endogenous fluorophore | — | — | 330 | 400 | — | — | — | — | — | Physiologic pH | 2 |
| Pyridoxamine (PM) | Endogenous fluorophore | 326 | — | 335 | 400 | — | — | — | — | — | Physiologic pH | 2 |
| Pyridoxine | Endogenous fluorophore | 308 | — | 275–350 | 360–470 | — | — | — | — | — | — | 9 |
| Pyridoxine (PN) | Endogenous fluorophore | 324 | — | 332 | 400 | — | — | — | — | — | Physiologic pH | 2 |
| Tryptophan | Endogenous fluorophore | 280 | $5.6 \times 10^{-3}$ | 280 | 350 | 0.2 | — | — | — | — | Physiologic pH | 22 |
| Tryptophan | Endogenous fluorophore | — | — | 280 | 350 | — | — | — | — | — | Physiologic pH | 21 |
| Tryptophan | Endogenous fluorophore | — | — | 275 | 350 | — | Powder/solution | 2.8 | 1.5 | — | — | 3, 8 |

**TABLE 65.9A** Fluorescence Data: Biochemical Studies (continued)

| Species | Endogenous/ Exogenous Origin | Absorbance Maxima (nm) | Molar Extinction Coefficient ($M^{-1}$) | Excitation $\lambda_{ex}$ (nm) | Emission $\lambda_{em}$ (nm) | Quantum Yield | Matrix | Lifetime τ1 (rel % τ1) | Lifetime τ2 (rel % τ2) | Lifetime τ3 (rel % τ3) | Comments | Ref. |
|---|---|---|---|---|---|---|---|---|---|---|---|---|
| Tryptophan | Endogenous fluorophore | 215 | — | 200–240, 250–280 | 300–380 | — | — | — | — | — | — | 9 |
| Tyrosine | Endogenous fluorophore | 275 | $1.4 \times 10^{-3}$ | 275 | 300 | 0.1 | — | — | — | — | Physiologic pH | 22 |
| Aluminum phthalocyanine | Exogenous fluorophore | — | — | 610 | >645 | — | PBS | 5.0 ± 0.1 | — | — | — | 23 |
| Aluminum phthalocyanine | Exogenous fluorophore | — | — | 610 | >645 | — | PBS + BSA | 5.5 ± 0.2 (92.0 ± 1.0) | 1.0 ± 0.2 (8 ± 1) | — | — | 24 |
| Aluminum phthalocyanine | Exogenous fluorophore | — | — | 610 | >645 | — | 0.01 m*M* CTAB + 0.1 *M* PBS | 6.0 ± 0.1 | — | — | — | 24 |
| Aluminum phthalocyanine | Exogenous fluorophore | — | — | 610 | >645 | — | 1% Triton in 0.1 *M* PBS | 5.9 ± 0.1 | — | — | — | 24 |
| Benzoporphyrin monoacetate (BpD-MA) | Exogenous fluorophore | — | — | 400 | 690 | — | Solution | 5.5 | — | — | Used as a photodynamic therapy agent; optimal time for fluorescence *in vivo* 2–4 h; phototoxic | 25 |
| Coprotoporphyrin IX | Exogenous fluorophore | — | — | 400 | 650–670 | — | DMSO | 20 (100) | — | — | — | 7 |
| Haematoporphyrin (HP) | Exogenous fluorophore | — | — | Visible | 610 | 9 | Solution | 13.5 | — | — | Used as a tumor marking agent; optimal time for fluorescence *in vivo* 2–4 h; phototoxic | 26 |
| Haematoporphyrin derivative (HpD) | Exogenous fluorophore | — | — | 400 | 610 | 2–7 | Solution | 15.5 | 2.5 | — | Used as a photodynamic therapy agent; optimal time for fluorescence *in vivo* 12–72 h; phototoxic | 27, 28 |

| | | | | | | | | | | | | |
|---|---|---|---|---|---|---|---|---|---|---|---|---|
| Mono-L-aspartyl chlorin e6 (MACE) | Exogenous fluorophore | — | — | 410 | 664 | — | Solution | 3.7 | — | — | Used as a photodynamic therapy agent; phototoxic | 29 |
| Nile Blue A (NBA) | Exogenous fluorophore | — | — | 620 | 660 | — | Solution | — | — | — | Used as a photodynamic therapy agent; not phototoxic | 30 |
| Photofrin II | Exogenous fluorophore | — | — | 364 | 615 | — | 0 m*M* CTAB (cetyltrimethyl-ammonium bromide) | 15.4 (100) | — | — | Cationic micelles influence aggregation | 7 |
| Photofrin II | Exogenous fluorophore | — | — | 364 | 630 | — | 0 m*M* CTAB (cetyltrimethyl-ammonium bromide) | 14.6 (64.3) | 2.97 (17.6) | 0.74 (18.1) | Cationic micelles influence aggregation | 31 |
| Photofrin II | Exogenous fluorophore | — | — | 364 | 675 | — | 0 m*M* CTAB (cetyltrimethyl-ammonium bromide) | 14.36 (81.3) | 1.95 (18.7) | — | Cationic micelles influence aggregation | 31 |
| Photofrin II | Exogenous fluorophore | — | — | 364 | 615 | — | 0.005 m*M* CTAB (cetyltrimethyl-ammonium bromide) | 15.04 (94.3) | 2.44 (5.7) | — | Cationic micelles influence aggregation | 31 |
| Photofrin II | Exogenous fluorophore | — | — | 364 | 630 | — | 0.005 m*M* CTAB (cetyltrimethyl-ammonium bromide) | 14.44 (22.4) | 2.37 (14.7) | 0.42 (62.9) | Cationic micelles influence aggregation | 31 |
| Photofrin II | Exogenous fluorophore | — | — | 364 | 675 | — | 0.005 m*M* CTAB (cetyltrimethyl-ammonium bromide) | 14.07 (20.3) | 2.21 (31.2) | 0.67 (48.5) | Cationic micelles influence aggregation | 31 |
| Photofrin II | Exogenous fluorophore | — | — | 364 | 615 | — | 0.01 m*M* CTAB (cetyltrimethyl-ammonium bromide) | 14.99 (80.7) | 2.27 (6.8) | 0.58 (12.5) | Cationic micelles influence aggregation | 31 |
| Photofrin II | Exogenous fluorophore | — | — | 364 | 630 | — | 0.01 m*M* CTAB (cetyltrimethyl-ammonium bromide) | 14.15 (10.4) | 2.2 (14.2) | 0.5 (75.4) | Cationic micelles influence aggregation | 31 |

**TABLE 65.9A** Fluorescence Data: Biochemical Studies (continued)

| Species | Endogenous/ Exogenous Origin | Absorbance Maxima (nm) | Molar Extinction Coefficient ($M^{-1}$) | Excitation $\lambda_{ex}$ (nm) | Emission $\lambda_{em}$ (nm) | Quantum Yield | Matrix | Lifetime τ1 (rel % τ1) | Lifetime τ2 (rel % τ2) | Lifetime τ3 (rel % τ3) | Comments | Ref. |
|---|---|---|---|---|---|---|---|---|---|---|---|---|
| Photofrin II | Exogenous fluorophore | — | — | 364 | 675 | — | 0.01 m*M* CTAB (cetyltrimethyl-ammonium bromide) | 13.84 (7.3) | 2.43 (25.3) | 0.83 (67.4) | Cationic micelles influence aggregation | 31 |
| Photofrin II | Exogenous fluorophore | — | — | 364 | 615 | — | 0.05 m*M* CTAB (cetyltrimethyl-ammonium bromide) | 13.88 (18.2) | 2.47 (28.5) | 0.67 (53.2) | Cationic micelles influence aggregation | 31 |
| Photofrin II | Exogenous fluorophore | — | — | 364 | 630 | — | 0.05 m*M* CTAB (cetyltrimethyl-ammonium bromide) | 12.41 (3.1) | 2.66 (18.9) | 0.62 (78) | Cationic micelles influence aggregation | 31 |
| Photofrin II | Exogenous fluorophore | — | — | 364 | 675 | — | 0.05 m*M* CTAB (cetyltrimethyl-ammonium bromide) | 8.17 (3.2) | 2.67 (34.1) | 1.06 (62.7) | Cationic micelles influence aggregation | 31 |
| Photofrin II | Exogenous fluorophore | — | — | 364 | 630 | — | 0.1 m*M* CTAB (cetyltrimethyl-ammonium bromide) | 10.5 (21.5) | 3.71 (43.2) | 0.96 (35.3) | Cationic micelles influence aggregation | 31 |
| Photofrin II | Exogenous fluorophore | — | — | 364 | 675 | — | 0.1 m*M* CTAB (cetyltrimethyl-ammonium bromide) | 7.55 (49.2) | 3.61 (50.8) | — | Cationic micelles influence aggregation | 31 |
| Photofrin II | Exogenous fluorophore | — | — | 364 | 630 | — | 1.0 m*M* CTAB (cetyltrimethyl-ammonium bromide) | 16.05 (89.6) | 3.38 (10.4) | — | Cationic micelles influence aggregation | 31 |
| Photofrin II | Exogenous fluorophore | — | — | 364 | 675 | — | 1.0 m*M* CTAB (cetyltrimethyl-ammonium bromide) | 15.28 (63.1) | 5.75 (36.9) | — | Cationic micelles influence aggregation | 31 |

| | | | | | | | | | | | | |
|---|---|---|---|---|---|---|---|---|---|---|---|---|
| Protoporphryin IX | Exogenous fluorophore | — | — | 400 | 650–670 | — | DMSO | 17 (89) | 3 (11) | — | — | 7 |
| Protoporphyrin photoproduct | Exogenous fluorophore | — | — | 400 | 650–670 | — | DMSO | 0.7 (38) | 4.5 (62) | — | — | 7 |
| Reduced tin metallo purpurin (Sn.NT2H2) | Exogenous fluorophore | — | — | 410 | 645 | 16 | Solution | 0.65 | — | — | Used as a photodynamic therapy agent; phototoxic | 32 |
| Tetrasulphonated aluminum phthalocyanine (AISPc) | Exogenous fluorophore | — | — | 350 | 675 | 4 | Solution | 5.3 | — | — | Used as a photodynamic therapy agent; optimal time for fluorescence *in vivo* 24–48 h; phototoxic | 33 |
| Zn-protoporphyrin | Exogenous fluorophore | — | — | 400 | 650–670 | — | DMSO | 2 (100) | — | — | — | 7 |

**TABLE 65.9B** Fluorescence Data: Cellular Studies, Tissue and Biofluid Studies (*in Vitro*), Animal Studies (*in Vivo*), and Human Clinical Studies (*in Vivo*)

| Species | Target Organ or Disease | Absorbance Maxima (nm) | Molar Extinction Coefficient ($M^{-1}$) | Excitation $\lambda_{ex}$ (nm) | Emission $\lambda_{em}$ (nm) | Quantum Yield | Matrix | Lifetime $\tau 1$ (rel % $\tau 1$) | Lifetime $\tau 2$ (rel % $\tau 2$) | Lifetime $\tau 3$ (rel % $\tau 3$) | Comments | Ref. |
|---|---|---|---|---|---|---|---|---|---|---|---|---|
| | | | | | | ***Cellular Studies*** | | | | | | |
| 1-Pyreneiso-thiocyanate labeled antimouse IgG | Various cancers | — | — | Blue/visible | Visible | — | Labeled antibodies attached to surface proteins on mouse cells | 20–55 | — | — | Long lifetime allowed for differentiation of diferent types of cells via flow cytometry | 34 |
| 1-Pyrenesulfonyl chloride labeled antimouse IgG | Various cancers | — | — | Blue/visible | Visible | — | Labeled antibodies attached to surface proteins on mouse cells | 20–55 | — | — | Long lifetime allowed for differentiation of diferent types of cells via flow cytometry | 34 |
| 9-Acetoxy-tetra-n-propylporphycene (ATPPn) | Bladder cancer | — | — | Visible | Red/near infrared | — | Human bladder carcinoma cell line | — | — | — | This is a recently developed photosensitizer that has shown promise based on cellular studies with bladder carcinoma cells *in vitro* | 35 |
| Aluminum phthalocyanine | Various cancers | — | — | 610 | >645 | — | Cultured leukemic cells (human erythro-leukemic cells) | 2.2 ± 0.4 (50) | 6.1 ± 0.2 (50) | | — | 24 |
| Aluminum phthalocyanine | Various cancers | — | — | 364 | 630 | — | In L1210 mouse cells, 4 h following i.p. administration | 5.18 (86.33) | 1.52 (13.67) | — | — | 23 |

| | | | | | | | | | | | | |
|---|---|---|---|---|---|---|---|---|---|---|---|---|
| Aluminum phthalocyanine | Various cancers | — | — | 364 | 630 | — | In L1210 mouse cells, 12 h following i.v. administration | 5.9 (39.74) | 2.87 (50.24) | 0.93 (10.01) | — | 23 |
| Chlorophyll-derived photosensitizers | Lung cancer | 760 | — | 760 | Near infrared | — | In OAT 75 small cell lung carcinoma cells | — | — | — | Absorption maximum of the photosensitizer can shift between 20–100 nm depending up local environment within the cell | 36 |
| Flavins | Various diseases | — | — | 365 | 500 | — | Intact yeast cells | 0.2–0.35 (45) | 2.0–3.0 (25) | 6.0–8.0 (30) | — | 20 |
| Flavins | Various diseases | — | — | 365 | 500 | — | Defective yeast cells | 0.30–0.50 (35) | 2.0–3.0 (30) | 6.0–8.0 (35) | τ1 is shorter than τ2 and τ3, but represents the majority of fluorescence signal in intact cells | 20 |
| Flavoproteins, oxidized | Bladder cancer | — | — | 488 | 550–560 | — | Human urothelial cells (normal and malignant) in media | — | — | — | Normal cells exhibited greater than ten times the autofluorescence intensity of tumor cells | 37 |
| Hematoporphyrin derivative (HpD) | Various cancers | — | — | 364 | 630 | — | In L1210 mouse cells, 12 h following i.v. administration | 14.36 (10.83) | 3.12 (20.07) | 0.61 (69.1) | — | 23 |
| Hematoporphyrin | Various cancers | — | — | 364 | 630 | — | In L1210 mouse cells, 4 h following i.p. administration | 15.3 (45.39) | 4.32 (18.16) | 1.11 (36.46) | — | 23 |
| Hematoporphyrin aggregates | Various cancers | — | — | Visible | 675 | — | Various types of cells | — | — | — | Fluorescence emission at 675 nm was assigned to hematoporphyrin monomers | 38 |

**TABLE 65.9B** Fluorescence Data: Cellular Studies, Tissue and Biofluid Studies (*in Vitro*), Animal Studies (*in Vivo*), and Human Clinical Studies (*in Vivo*) (continued)

| Species | Target Organ or Disease | Absorbance Maxima (nm) | Molar Extinction Coefficient ($M^{-1}$) | Excitation $\lambda_{ex}$ (nm) | Emission $\lambda_{em}$ (nm) | Quantum Yield | Matrix | Lifetime $\tau 1$ (rel % $\tau 1$) | Lifetime $\tau 2$ (rel % $\tau 2$) | Lifetime $\tau 3$ (rel % $\tau 3$) | Comments | Ref. |
|---|---|---|---|---|---|---|---|---|---|---|---|---|
| Hematoporphyrin monomers | Various cancers | — | — | Visible | 635 | — | Various types of cells | — | — | — | Fluorescence emission at 635 nm was assigned to hematoporphyrin monomers | 38 |
| Human breast cells (normal and malignant) | Breast cancer | — | — | Ultraviolet/ blue | 340, 440 | — | Human breast cancer cells (normal and malignant) in media | — | — | — | Ratiometric analyses of the fluorescence autofluorescence intensity of the cells at 340 to 440 nm allows for the differentiation of normal and malignant cells | 35 |
| Human semen | Infertility | — | — | 488 | 622 | — | — | — | — | — | — | 39 |
| Human seminal plasma | Infertility | — | — | 488 | 622 | — | — | — | — | — | Sperm motility could be correlated to the autofluorescence emission | 39 |
| Human spermatozoa | Infertility | — | — | 488 | 622 | — | — | — | — | — | Spermatozoa and sperm motility could be correlated to the autofluorescence emission | 39 |

| | | | | | | | | | | | | |
|---|---|---|---|---|---|---|---|---|---|---|---|---|
| Murine fibroblast cells (normal and malignant) | Various cancers | — | — | 290 | Blue-green | — | Murine fibroblast cell lines | — | — | — | Autofluorescence intensity of malignant cells was significantly less than that of normal cells, with tryptophan being the primary component responsible for the difference | 40 |
| *N*-(1-pyrene)-maleimide labeled antimouse IgG | Various cancers | — | — | Blue/visible | Visible | — | Labeled antibodies attached to surface proteins on mouse cells | 20–55 | — | — | Long lifetime allowed for differentiation of diferent types of cells via flow cytometry | 34 |
| Nicotinamide adenine dinucleotide phosphate (NADPH) | Various diseases | — | — | 365 | 450 | — | Intact and defective yeast cells | 0.2–0.3 (30) | 1.4–2.4 (40) | 6.0–8.0 (30) | Deficient cell fluorescence is four times higher | 20 |
| Oral epithelial cells (normal and squamous) | Oral cancer | — | — | 300 | 320–580 | — | Several different culturing media selected to cause cell differentiation | — | — | — | No difference between the differentiated cells could be determined from the resulting spectra | 41 |
| Oral epithelial cells (normal and squamous) | Oral cancer | — | — | 340 | 360–620 | — | Several different culturing media selected to cause cell differentiation | — | — | — | Spectral shifts occurred between the three types of cell differentiation | 41 |
| Oral epithelial cells (normal and squamous) | Oral cancer | — | — | 365 | 400–630 | — | Several different culturing media selected to cause cell differentiation | — | — | — | Small changes in fluorescence intensity and spectral shape occurred between the various differentiated cells | 41 |

**TABLE 65.9B** Fluorescence Data: Cellular Studies, Tissue and Biofluid Studies (*in Vitro*), Animal Studies (*in Vivo*), and Human Clinical Studies (*in Vivo*) (continued)

| Species | Target Organ or Disease | Absorbance Maxima (nm) | Molar Extinction Coefficient ($M^{-1}$) | Excitation $\lambda_{ex}$ (nm) | Emission $\lambda_{em}$ (nm) | Quantum Yield | Matrix | Lifetime τ1 (rel % τ1) | Lifetime τ2 (rel % τ2) | Lifetime τ3 (rel % τ3) | Comments | Ref. |
|---|---|---|---|---|---|---|---|---|---|---|---|---|
| Oral epithelial cells (normal and squamous) | Oral cancer | — | — | 420 | 440–630 | — | Several different culturing media selected to cause cell differentiation | — | — | — | Small changes in fluorescence intensity and spectral shape occurred between the various differentiated cells | 41 |
| Oral epithelial cells (normal and squamous) | Oral cancer | — | — | 200–360 | 380 | — | Several different culturing media selected to cause cell differentiation | — | — | — | Significant changes in intensity and spectral profile occurred between the various differentiated cells | 41 |
| Oral epithelial cells (normal and squamous) | Oral cancer | — | — | 240–415 | 450 | — | Several different culturing media selected to cause cell differentiation | — | — | — | Significant changes in intensity and spectral profile occurred between the various differentiated cells | 41 |
| Oral epithelial cells (normal and squamous) | Oral cancer | — | — | 250–420 | 480 | — | Several different culturing media selected to cause cell differentiation | — | — | — | Significant changes in intensity and spectral profile occurred between the various differentiated cells | 41 |
| Oral epithelial cells (normal and squamous) | Oral cancer | — | — | 270–480 | 520 | — | Several different culturing media selected to cause cell differentiation | — | — | — | Significant changes in intensity and spectral profile occurred between the various differentiated cells | 41 |

| | | | | | | | | | | | | |
|---|---|---|---|---|---|---|---|---|---|---|---|---|
| Ovary cells | Photo-conversion | — | — | 320–400 | 455 | — | Chinese hamster ovary cells | — | — | — | Optical trapping of ovary cells using 730 nm was found to induce an increase in autofluorescnece intensity and a 6-nm red shift in its maxima | 42 |
| Phorbol-13-acetate-12-*N*-methyl-*N*-4-(*N,N′*-di(2-hydroxy-ethyl)amino-7-nitrobenz-2-oxa-1,3-diazole-aminododecanoate (N-C12-Ac(13)) | Cellular signal transduction | 488 | — | 488 | Visible | — | P3HR-1 Burkitt lymphoma cells | — | — | — | Used in the investigation of protein kinase C and its role in signal transduction | 43 |
| Photofrin II | Various cancers | — | — | 364 | 630 | — | In L1210 mouse cells, 4 h following i.p. administration | 14.9 (40.9) | 4.91 (18.32) | 1.18 (41.58) | Cationic micelles influence aggregation | 23 |
| Photofrin II | Various cancers | — | — | 364 | 630 | — | In L1210 mouse cells, 12 h following i.v. administration | 14.81 (7.3) | 4.65 (25.49) | 1.08 (67.27) | Cationic micelles influence aggregation | 23 |
| Protoporphryin | Various cancers | — | — | 420 | 610–690 | — | Incubated meercat kidney cells dorsal skinfold of hamster | 1–2 | 2.0–3.0 | 11.0–14.0 | — | 38 |
| Various types of cells | Cell proliferation | — | — | 320–350 | 450 | — | Various types of cells | — | — | — | Differentiation between slow and rapidly growing cells was possible based on ratiometric analyses | 44 |

**TABLE 65.9B** Fluorescence Data: Cellular Studies, Tissue and Biofluid Studies (*in Vitro*), Animal Studies (*in Vivo*), and Human Clinical Studies (*in Vivo*) (continued)

| Species | Target Organ or Disease | Absorbance Maxima (nm) | Molar Extinction Coefficient ($M^{-1}$) | Excitation $\lambda_{ex}$ (nm) | Emission $\lambda_{em}$ (nm) | Quantum Yield | Matrix | Lifetime τ1 (rel % τ1) | Lifetime τ2 (rel % τ2) | Lifetime τ3 (rel % τ3) | Comments | Ref. |
|---|---|---|---|---|---|---|---|---|---|---|---|---|
| Various types of cells | Cell proliferation | — | — | 340 | 360–660 | — | Various types of cells | — | — | — | Differentiation between slow and rapidly growing cells was possible based on ratiometric analyses | 44 |
| | | | | | ***Tissue and Biofluid Studies (in Vitro)*** | | | | | | | |
| Beta-carotene | Atherosclerosis | — | — | 488 | Visible | — | Arterial tissue incubated with beta-carotene | — | — | — | Beta-carotene was shown to reduce the total autofluorescence intensity of atherosclerotic plaques in arterial tissues | 45 |
| Collagen | Various diseases | — | — | 380 | Visible | — | Cervical tissue | — | — | — | Increases in fluorescence emission due to NADH were found for dysplastic tissue relative to normal tissues | 46 |
| Dentin (healthy and demineralized) | Tooth decay | — | — | 488 | 480–520 | — | Whole tooth | — | — | — | Demineralized dentin exhibited lower fluorescence intensity than healthy dentin at 529 nm | 47 |

| | | | | | | | | | | | | |
|---|---|---|---|---|---|---|---|---|---|---|---|---|
| Dentin (healthy and demineralized) | Tooth decay | — | — | 460–480 | 520 | — | Whole tooth | — | — | — | Demineralized dentin exhibited a more pronounced peak than healthy dentin at 520 nm | 47 |
| Dentin (healthy and carious) | Tooth decay | — | — | 488 | >515 | — | Whole tooth | — | — | — | Autofluorescence correlated well with amount of demineralization | 48 |
| Eosinophils, circulating | Hodgkin's disease/ lymphomas | — | — | 280 | 330 | — | Eosinophils isolated from blood | — | — | — | Fluorescence emission at this wavelength can be attributed to tryptophan | 14 |
| Eosinophils, circulating | Hodgkin's disease/ lymphomas | — | — | 360 | 440 | — | Eosinophils isolated from blood | — | — | — | — | 14 |
| Eosinophils, circulating | Hodgkin's disease/ lymphomas | — | — | 380 | 415 | — | Eosinophils isolated from blood | — | — | — | — | 14 |
| Eosinophils, circulating | Hodgkin's disease/ lymphomas | — | — | 365 | Blue-violet | — | Eosinophils isolated from blood | — | — | — | — | 14 |
| Eosinophils, tissue dwelling | Hodgkin's disease/ lymphomas | — | — | 365 | Amber-gold | — | Eosinophils isolated from various tumor tissues | — | — | — | — | 14 |
| Excised and frozen colon tissue sections (normal and tumor) | Colon cancer | — | — | 351–364 | Visible | — | Human colon tissue | — | — | — | This article characterizes resulting fluorescence emission from various locations in the tissue | 49 |

**TABLE 65.9B** Fluorescence Data: Cellular Studies, Tissue and Biofluid Studies (*in Vitro*), Animal Studies (*in Vivo*), and Human Clinical Studies (*in Vivo*) (continued)

| Species | Target Organ or Disease | Absorbance Maxima (nm) | Molar Extinction Coefficient ($M^{-1}$) | Excitation $\lambda_{ex}$ (nm) | Emission $\lambda_{em}$ (nm) | Quantum Yield | Matrix | Lifetime τ1 (rel % τ1) | Lifetime τ2 (rel % τ2) | Lifetime τ3 (rel % τ3) | Comments | Ref. |
|---|---|---|---|---|---|---|---|---|---|---|---|---|
| Excised and frozen human cervical tissues (normal and malignant) | Cervical cancer | — | — | 365 | 460 | — | Frozen cervical tissue | — | — | — | Autofluorescence emission in the epithelia showed small differences in intensity between severe and mild dysplasia of tissues; epithelial fluorescence is less in malignant tissue than normal tissue | 50 |
| Excised and frozen human cervical tissues (normal and malignant) | Cervical cancer | — | — | 440 | 525 | — | Frozen cervical tissue | — | — | — | Autofluorescence emission in the stroma showed small differences in intensity between inflammatory and severely dysplastic tissues; epithelial fluorescence is less in malignant tissue than normal tissue | 50 |
| Excised arterial tissue | Atherosclerosis | — | — | 458 | Visible | — | Arterial tissue | — | — | — | Good differentiation of normal and atherosclerotic tissue can be seen after excitation with visible wavelengths | 51 |

| | | | | | | | | | | | | |
|---|---|---|---|---|---|---|---|---|---|---|---|---|
| Excised arterial tissue | Atherosclerosis | — | — | 476 | Visible | — | Arterial tissue | — | — | — | Good differentiation of normal and atherosclerotic tissue can be seen after excitation with visible wavelengths | 52 |
| Excised arterial tissue (calcified and noncalcified) | Atherosclerosis | — | — | 495 | 520–800 | — | Arterial tissue | — | — | — | Characteristic spectral and temporal characteristics of 495-nm-induced autofluorescence of normal intima, calcified plaques, and fibro-fatty plaques were determined | 53 |
| Excised arterial tissue (calcified and noncalcified) | Atherosclerosis | — | — | 380 | 400–600 | — | Arterial tissue | — | — | — | Differentiation of atherosclerotic lesions from normal tissue was attempted based on autofluorescence emission (it was found that a better excitation wavelength was 325 nm) | 54 |

**TABLE 65.9B** Fluorescence Data: Cellular Studies, Tissue and Biofluid Studies (*in Vitro*), Animal Studies (*in Vivo*), and Human Clinical Studies (*in Vivo*) (continued)

| Species | Target Organ or Disease | Absorbance Maxima (nm) | Molar Extinction Coefficient ($M^{-1}$) | Excitation $\lambda_{ex}$ (nm) | Emission $\lambda_{em}$ (nm) | Quantum Yield | Matrix | Lifetime τ1 (rel % τ1) | Lifetime τ2 (rel % τ2) | Lifetime τ3 (rel % τ3) | Comments | Ref. |
|---|---|---|---|---|---|---|---|---|---|---|---|---|
| Excised arterial tissue (calcified and noncalcified) | Atherosclerosis | — | — | 450 | 400–600 | — | Arterial tissue | — | — | — | Differentiation of atherosclerotic lesions from normal tissue was attempted based on autofluorescence emission (it was found that a better excitation wavelength was 325 nm) | 54 |
| Excised arterial tissue (calcified and noncalcified) | Atherosclerosis | — | — | 308 | 321–657 | — | Arterial tissue | — | — | — | A XeCl laser was used for the simultaneous ablation and autofluorescence-based diagnosis of atherosclerosis by providing a means of differentiating normal tissue from various plaques | 55 |
| Excised arterial tissue (calcified and noncalcified) | Atherosclerosis | — | — | 351–364 | Visible | — | Arterial tissue | — | — | — | Autofluorescence of collagen, elastin, and ceroid can be used to differentiate normal and atherosclerotic tissues | 56 |

| | | | | | | | | | | | | |
|---|---|---|---|---|---|---|---|---|---|---|---|---|
| Excised arterial tissue (calcified and noncalcified) | Atherosclerosis | — | — | 476 | Visible | — | Arterial tissue | — | — | — | Determination of chemical and morphological characteristics responsible for autofluorescence signals were determined | 57,58 |
| Excised arterial tissue (calcified and noncalcified) | Atherosclerosis | — | — | 325 | 443 | — | Arterial tissue | — | — | — | Normalized fluorescence intensities at 443 nm can be used to differentiate between normal and calcified tissues | 51 |
| Excised arterial tissue (calcified and noncalcified) | Atherosclerosis | — | — | 325 | 375–385 and 435–445 | — | Arterial tissue filled with blood | — | — | — | An autoguidance system for angioplasty was investigated based on ratiometric analyses of autofluorescence at 380 and 440 nm | 59 |
| Excised arterial tissue (calcified) | Atherosclerosis | — | — | 325 | 400–600 (no peak at 480) | — | Arterial tissue | — | — | — | Noncalcified tissue has a distinct peak at 480 nm and calcified tissue shows only a broadband emission | 54 |

**TABLE 65.9B** Fluorescence Data: Cellular Studies, Tissue and Biofluid Studies (*in Vitro*), Animal Studies (*in Vivo*), and Human Clinical Studies (*in Vivo*) (continued)

| Species | Target Organ or Disease | Absorbance Maxima (nm) | Molar Extinction Coefficient ($M^{-1}$) | Excitation $\lambda_{ex}$ (nm) | Emission $\lambda_{em}$ (nm) | Quantum Yield | Matrix | Lifetime τ1 (rel % τ1) | Lifetime τ2 (rel % τ2) | Lifetime τ3 (rel % τ3) | Comments | Ref. |
|---|---|---|---|---|---|---|---|---|---|---|---|---|
| Excised arterial tissue (collagen and elastin components) | Atherosclerosis | — | — | 306–310 | 380 | — | Arterial tissue | — | — | — | By ratioing tryptophan fluorescence intensity to collagen and elastin component, classification of tissue is possible | 60 |
| Excised arterial tissue (noncalcified) | Atherosclerosis | — | — | 248 | 370–460 | — | Arterial tissue | — | — | — | Autofluorescence shows broadband fluorescence emission | 61 |
| Excised arterial tissue (noncalcified) | Atherosclerosis | — | — | 325 | 400–600 (peak at 480) | — | Arterial tissue | — | — | — | Noncalcified tissue has a distinct peak at 480 nm and calcified tissue shows only a broadband emission | 54 |
| Excised arterial tissue (tryptophan component) | Atherosclerosis | — | — | 306–310 | 340 | — | Arterial tissue | — | — | — | By ratioing tryptophan fluorescence intensity to collagen and elastin component, classification of tissue is possible | 60 |

| | | | | | | | | | | | | |
|---|---|---|---|---|---|---|---|---|---|---|---|---|
| Excised bladder tissue (normal mucosa, flat lesions, and papillary tumors) | Bladder cancer | — | — | 337 | 385 and 455 | — | Bladder tissue | — | — | — | Ratiometric measurements of the autofluorescence intensities at 385 and 440 nm were used to distinguish malignant tumors from nonmalignant and inflamed tissue | 62 |
| Excised bladder tissue (normal mucosa, flat lesions, and papillary tumors) | Bladder cancer | — | — | 375–440 (D-light) | Visible | — | Bladder tissue | — | — | — | Autofluorescence imaging could distinguish between normal mucosa, flat lesions, and papillary tumors | 63 |
| Excised brain tissue (normal and Alzheimer's) | Alzheimer's disease | — | — | 647 | 650–850 | — | Human brain tissue (temporal cortex) | — | — | — | Autofluorescence emission from temporal cortex samples was used to correctly diagnose Alzheimer's disease | 64 |
| Excised brain tissue (normal and tumorous) | Brain cancer | — | — | 337 | 460 | — | Human brain tissue | — | — | — | The autofluorescence intensity of normal brain tissue was significantly greater than that of primary brain tumorous regions | 65 |

**TABLE 65.9B** Fluorescence Data: Cellular Studies, Tissue and Biofluid Studies (*in Vitro*), Animal Studies (*in Vivo*), and Human Clinical Studies (*in Vivo*) (continued)

| Species | Target Organ or Disease | Absorbance Maxima (nm) | Molar Extinction Coefficient ($M^{-1}$) | Excitation $\lambda_{ex}$ (nm) | Emission $\lambda_{em}$ (nm) | Quantum Yield | Matrix | Lifetime $\tau 1$ (rel % $\tau 1$) | Lifetime $\tau 2$ (rel % $\tau 2$) | Lifetime $\tau 3$ (rel % $\tau 3$) | Comments | Ref. |
|---|---|---|---|---|---|---|---|---|---|---|---|---|
| Excised breast tissue (normal) | Breast cancer | — | — | 488 | 530, 550, and 590 | — | Normal breast tissue | — | — | — | Presence of three peaks due to absorption by hemoglobin at 420, 542, and 575 nm | 66 |
| Excised breast tissue (tumor) | Breast cancer | — | — | 488 | 530 | — | Tumor breast tissue | — | — | — | Only a single peak is visible in the fluorescence profile | 66 |
| Excised calcified arterial tissue | Atherosclerosis | — | — | 248 | 397, 442, 450, 461, 528, and 558 | — | Arterial tissue | — | — | — | Autofluorescence shows multiple, prominent atomic fluorescence lines | 61 |
| Excised carotid atherosclerotic plaques | Atherosclerosis | — | — | 337 | Visible | — | Carotid arterial tissue | — | — | — | Autofluorescence of carotid atherosclerotic plaques was used to determine the three primary components; fibrous tissue, lipid constituents, and calcified plaque | 67 |
| Excised carotid atherosclerotic plaques | Atherosclerosis | — | — | 476 | Visible | — | Carotid arterial tissue | — | — | — | Autofluorescence of carotid atherosclerotic plaques was used to determine the three primary components; fibrous tissue, lipid constituents, and calcified plaque | 67 |

| | | | | | | | | | | | | |
|---|---|---|---|---|---|---|---|---|---|---|---|---|
| Excised carotid atherosclerotic plaques | Atherosclerosis | — | — | 488 | Visible | — | Carotid arterial tissue | — | — | — | Autofluorescence of carotid atherosclerotic plaques was used to determine the three primary components; fibrous tissue, lipid constituents, and calcified plaque | 67 |
| Excised carotid atherosclerotic plaques | Atherosclerosis | — | — | 458 | Visible | — | Carotid arterial tissue | — | — | — | Autofluorescence of carotid atherosclerotic plaques was used to determine the three primary components; fibrous tissue, lipid constituents, and calcified plaque | 67 |
| Excised cervical tissue (normal and tumor) | Cervical cancer | — | — | 330 | 385 | — | Cervical tissue | — | — | — | Fluorescence intensity of normalized emission spectra is greater for normal tissue than tumor tissue | 68 |
| Excised cervical tissue (normal and tumor) | Cervical cancer | — | — | 365 | 475 | — | Cervical tissue | — | — | — | Fluorescence intensity at 475 nm increases with degree of dysplasia/malignancy | 69 |
| Excised colon tissue (normal and adenomatous) | Colon cancer | — | — | 330, 370, and 430 | 404, 480, and 680 | — | Human colon tissue | — | — | — | — | 70 |

**TABLE 65.9B** Fluorescence Data: Cellular Studies, Tissue and Biofluid Studies (*in Vitro*), Animal Studies (*in Vivo*), and Human Clinical Studies (*in Vivo*) (continued)

| Species | Target Organ or Disease | Absorbance Maxima (nm) | Molar Extinction Coefficient ($M^{-1}$) | Excitation $\lambda_{ex}$ (nm) | Emission $\lambda_{em}$ (nm) | Quantum Yield | Matrix | Lifetime τ1 (rel % τ1) | Lifetime τ2 (rel % τ2) | Lifetime τ3 (rel % τ3) | Comments | Ref. |
|---|---|---|---|---|---|---|---|---|---|---|---|---|
| Excised colon tissue (normal and adenomatous) | Colon cancer | — | — | 325 | 350–600 | — | Human colon tissue | — | — | — | 370 nm was found to be the optimal excitation wavelength for discrimination of normal and adenomatous tissues | 71 |
| Excised colon tissue (normal and tumor) | Colon cancer | — | — | 366 | 480–580 | — | Human colon tissue | — | — | — | Relative fluorescence emission amplitudes of normal and tumor tissue between 480–580 nm varied significantly and tumor tissue displayed more evident shoulder than normal tissue | 72 |
| Excised colon tissue (normal mucosa and adenomatous polyps) | Colon cancer | — | — | 325 | Visible | — | Human colon tissue | — | — | — | Autofluorescence intensity in adenomatous polyps is lower than for normal tissue and small spectral differences can be seen | 73 |

| | | | | | | | | | | | | |
|---|---|---|---|---|---|---|---|---|---|---|---|---|
| Excised heart tissue (nodal conductive and atrial endomyocardial) | Heart arrhythmia | — | — | 308 | 440–500 | — | Heart tissue | — | — | — | Nodal tissue demonstrated a decrease in normalized fluorescence intensity and peak width relative to atrial endomyocardial tissue | 74 |
| Excised heart tissue (nodal conductive and ventricular endocardium) | Heart arrhythmia | — | — | 308 | 430–550 | — | Heart tissue | — | — | — | Nodal tissue demonstrated an increase in fluorescence intensity relative to atrial endomyocardial tissue | 74 |
| Excised human eye lens | Vision problems | — | — | 430–490 | 530–630 | — | Human eye lens | — | — | — | Intrinsic eye lens transmittance can be determined based on autofluorescence analyses | 75 |
| Excised lung tissue (normal) | Lung cancer | — | — | 488 | 530, 550, and 590 | — | Normal lung tissue | — | — | — | Presence of three peaks due to absorption by hemoglogin at 420, 542, and 575 nm | 66 |
| Excised lung tissue (tumor) | Lung cancer | — | — | 488 | 530 | — | Tumor lung tissue | — | — | — | Only a single peak is visible in the fluorescence profile | 66 |

**TABLE 65.9B** Fluorescence Data: Cellular Studies, Tissue and Biofluid Studies (*in Vitro*), Animal Studies (*in Vivo*), and Human Clinical Studies (*in Vivo*) (continued)

| Species | Target Organ or Disease | Absorbance Maxima (nm) | Molar Extinction Coefficient ($M^{-1}$) | Excitation $\lambda_{ex}$ (nm) | Emission $\lambda_{em}$ (nm) | Quantum Yield | Matrix | Lifetime $\tau1$ (rel % $\tau1$) | Lifetime $\tau2$ (rel % $\tau2$) | Lifetime $\tau3$ (rel % $\tau3$) | Comments | Ref. |
|---|---|---|---|---|---|---|---|---|---|---|---|---|
| Excised oral tissue (normal and tumor) | Oral cancer | — | — | 410 | Visible | — | Oral tissue | — | — | — | Autofluorescence emission spectra excited with 410-nm light are capable of differentiating normal and abnormal tissues | 76 |
| Excised oral tissue (normal, malignant, and dysplastic) | Oral cancer | — | — | 410 | 635 | — | Oral tissue | — | — | — | Abnormal tissues exhibited an increase in fluorescence intensity at 635 nm | 76, 77 |
| Excised oral tissue (normal, malignant, and premalignant) | Oral cancer | — | — | 300 | 330–470 | — | Oral tissue | — | — | — | Ratiometric fluorescence measurements at 330 and 470 nm allowed for the differentiation of malignant and premalignant lesions from normal tissues | 78 |
| Excised skin tissue | Skin cancer | — | — | 365 | Visible | — | Skin tissue (147 samples) | — | — | — | Based upon the resulting autofluorescence, it was possible to differentiate non-dysplastic nevi from melanomas and dysplastic nevi (tumor tissue exhibited reduced fluorescence) | 79 |

| | | | | | | | | | | | | |
|---|---|---|---|---|---|---|---|---|---|---|---|---|
| Excised stomach tissue (normal, dysplastic, and tumor) | Stomach cancer | — | — | 325 | 440 and 395 | — | Stomach tissue | — | — | — | Normalized fluorescence intensities at both 440 and 395 nm exhibit significant differences between normal and tumor tissues | 80 |
| Extracts of human eye (soluble and insoluble) | Cataracto-genesis | — | — | Ultraviolet | Blue/green | — | Cortical and nuclear extracts | — | — | — | The green to blue autofluorescence intensity ratio was found to be greater than six for cataractous tissue fractions | 81 |
| Flavins | Various diseases | — | — | 488 | 500–550 | — | Normal rat kidney | 0.357 ± 0.018 (27) | 1.220 ± 0.035 (63) | — | — | 23 |
| Flavins | Various diseases | — | — | 488 | 600–650 | — | Normal rat kidney | 0.204 ± 0.011 (47) | 1.0 ± 0.006 (53) | — | — | 23 |
| Flavins | Various diseases | — | — | 488 | 500–550 | — | Cancerous rat kidney | 0.223 ± 0.015 (59) | 1.966 ± 0.037 (41) | — | — | 23 |
| Flavins | Various diseases | — | — | 488 | 600–650 | — | Cancerous rat kidney | 0.236 ± 0.014 (67) | 1.963 ± 0.037 (33) | — | — | 23 |
| Haematoporphyrin derivative | Various cancers | — | — | 320 | 630 | — | Tumor tissue in DHE-injected rat | 17 | 6 | 0.7 | — | 82 |
| Haematoporphyrin derivative | Various cancers | — | — | 320 | 630 | — | Normal tissue in surrounding muscle | 17 | — | — | — | 82 |
| Melanins derived from 3-hydroxy-anthranilic acid | Colon cancer | — | — | 324 | 413 | — | Human colon tissue | — | — | — | This article characterizes the autofluorescence properties of plasma-soluble melanins | 83 |

**TABLE 65.9B** Fluorescence Data: Cellular Studies, Tissue and Biofluid Studies (*in Vitro*), Animal Studies (*in Vivo*), and Human Clinical Studies (*in Vivo*) (continued)

| Species | Target Organ or Disease | Absorbance Maxima (nm) | Molar Extinction Coefficient ($M^{-1}$) | Excitation $\lambda_{ex}$ (nm) | Emission $\lambda_{em}$ (nm) | Quantum Yield | Matrix | Lifetime τ1 (rel % τ1) | Lifetime τ2 (rel % τ2) | Lifetime τ3 (rel % τ3) | Comments | Ref. |
|---|---|---|---|---|---|---|---|---|---|---|---|---|
| Melanins derived from dopa, catecholamines, catechol, and 3-hydroxy-kynurenine | Colon cancer | — | — | 345 | 445 | — | Human colon tissue | — | — | — | This article characterizes the autofluorescence properties of plasma-soluble melanins | 83 |
| Nicotinamide adenine dinucleotide (NADH) reduced form | Various diseases | — | — | 380 | Visible | — | Cervical tissue | — | — | — | Increases in fluorescence emission due to NADH were found for dysplastic tissue relative to normal tissues | 46 |
| Nicotinamide adenine dinucleotide phosphate (NADPH) | Various diseases | — | — | 320 | 400 | — | Normal arterial wall | 0.3 | 2 | 7 | Calcified plaques have a higher ratio of slow (400 nm) to fast fluorescence (480 nm) | 84 |
| | | | | | ***Animal Studies (in Vivo)*** | | | | | | | |
| 2′,7′-bis-(2 carboxy-ethyl)-5-(and -6)-carboxyfluorescein | Various cancers | — | — | 465 | Visible | — | Grafted tumors on mice | — | — | — | Differentiation of tumors from normal tissue was accomplished based on intracellular pH measurements with this exogenous dye | 85 |

| | | | | | | | | | | | | |
|---|---|---|---|---|---|---|---|---|---|---|---|---|
| Arterial tissue (atherosclerotic) | Atherosclerosis | — | — | Ultraviolet/blue | 410–490 | — | Rabbit arteries | — | — | — | Autofluorescence analyses were capable of monitoring the disruption of atherosclerotic plaques following injection of Russell's viper venom | 86 |
| Benzoporphyrin derivatized-monoacid (BPD-MA) | Various cancers | — | — | 337 | 380–750 | — | Rat tumors (various organs) | — | — | — | BPD-MA exhibited approximately the same demarcation ability as other common photosensitizers | 84 |
| Brain tissue (normal and glioma) | Brain cancer | — | — | 360 | 470 | — | Rat brain | — | — | — | Fluorescence due to NAD(P)H; decreased autofluorescence intensity for gliomas relative to normal tissue | 87 |
| Brain tissue (normal and glioma) | Brain cancer | — | — | 440 | 520 | — | Rat brain | — | — | — | Fluorescence due to flavins; decreased autofluorescence intensity for gliomas relative to normal tissue | 87 |
| Brain tissue (normal and glioma) | Brain cancer | — | — | 490 | 630 | — | Rat brain | — | — | — | Fluorescence due to porphyrins; decreased autofluorescence intensity for gliomas relative to normal tissue | 87 |

**TABLE 65.9B** Fluorescence Data: Cellular Studies, Tissue and Biofluid Studies (*in Vitro*), Animal Studies (*in Vivo*), and Human Clinical Studies (*in Vivo*) (continued)

| Species | Target Organ or Disease | Absorbance Maxima (nm) | Molar Extinction Coefficient ($M^{-1}$) | Excitation $\lambda_{ex}$ (nm) | Emission $\lambda_{em}$ (nm) | Quantum Yield | Matrix | Lifetime τ1 (rel % τ1) | Lifetime τ2 (rel % τ2) | Lifetime τ3 (rel % τ3) | Comments | Ref. |
|---|---|---|---|---|---|---|---|---|---|---|---|---|
| Esophageal multispheroidal tumor (induced by trans-retinoic acid) | Esophageal cancer | — | — | Ultraviolet/ blue | 340, 450, and 520 | — | Rat esophagus | — | — | — | Autofluorescence intensities at 340, 450, and 520 nm demonstrated differences between normal and cancerous tissues | 88 |
| Esophageal tumor (induced by *N*-nitroso-*N*-methylbenzylamine (NMBA)) | Esophageal cancer | — | — | Ultraviolet/ violet | 380 | — | Rat esophagus | — | — | — | Alteration of fluorescence emission correlated to disease progression, from normal to dysplasia to invasive cancer | 89 |
| Foam cell lesions | Atherosclerosis | — | — | 308 | Visible | — | Hyper-cholesterolemic rabbit arteries | — | — | — | Autofluorescence from foam cell lesions exhibited red shifts and spectral broadening similar to oxidized low-density lipoproteins | 90 |
| Heart tissue (native and transplanted) | Heart transplant rejection | — | — | Blue | Visible | — | Rat heart tissue (midtransverse ventricular) | — | — | — | A correlation was found between the severity of tissue rejection and the autofluorescence from the heart | 91 |

| | | | | | | | | | | | | |
|---|---|---|---|---|---|---|---|---|---|---|---|---|
| Heart tissue (normal and hypoxic) | Myocardial hypoxia | — | — | 308 | 350–600 | — | Mouse heart tissue | — | — | — | Hypoxia was found to reduce the autofluorescence between 455–505 nm and remove two spectral peaks at 540 and 580 nm, relative to normal tissue, leaving only a single peak at 555 nm | 92 |
| Hematoporphyrins | Various cancers | — | — | 337 | 630 | — | Rat tumors (various organs) | — | — | — | — | 93 |
| K1735P melanoma | Skin cancer | — | — | Ultraviolet/ violet | 360–700 | — | Melanomas implanted intradermally in the ears of C3H/HeN mice | — | — | — | Resulting autofluorescence spectra showed decreases in the fluorescence intensity over the spectral range of 385–425 nm; however, no spectral differences could be determined between the normal tissue and unpigmented melanomas | 94 |

**TABLE 65.9B** Fluorescence Data: Cellular Studies, Tissue and Biofluid Studies (*in Vitro*), Animal Studies (*in Vivo*), and Human Clinical Studies (*in Vivo*) (continued)

| Species | Target Organ or Disease | Absorbance Maxima (nm) | Molar Extinction Coefficient ($M^{-1}$) | Excitation $\lambda_{ex}$ (nm) | Emission $\lambda_{em}$ (nm) | Quantum Yield | Matrix | Lifetime $\tau 1$ (rel % $\tau 1$) | Lifetime $\tau 2$ (rel % $\tau 2$) | Lifetime $\tau 3$ (rel % $\tau 3$) | Comments | Ref. |
|---|---|---|---|---|---|---|---|---|---|---|---|---|
| Kidney tissue (normal and hypoxic) | Renal hypoxia | — | — | 308 | 350–600 | — | Mouse kidney tissue | — | — | — | Hypoxia was found to reduce the autofluorescence between 455–505 nm and remove two spectral peaks at 540 and 580 nm, relative to normal tissue, leaving only a single peak at 555 nm | 92 |
| MS-2 fibrosarcoma | Various cancers | — | — | 337 | 400–500 | — | Implanted in NALB-CDF1 mice | — | — | — | Autofluorescence intensity was found to be lower in the tumor tissue than in the healthy tissue and the fast component of the biexponential fluorescence decay is significantly lower in the tumor as well | 95 |
| Oral tissue (normal and malignant) | Oral cancer | — | — | 410 | 635 | — | Hamster cheek pouch (tumors induced with 7,12-dimethyl-benz(a)-anthracene) | — | — | — | Neoplastic lesions showed a characteristic fluorescence between 630–640 nm | 96 |

| | | | | | | | | | | | | |
|---|---|---|---|---|---|---|---|---|---|---|---|---|
| Oral tissue (normal and malignant) | Oral cancer | — | — | 350–370 | Visible | — | Hamster cheek pouch | — | — | — | Neoplastic tissue showed an increase in fluorescence intensity relative to normal tissue with excitation between 350–370 nm | 97 |
| Oral tissue (normal and malignant) | Oral cancer | — | — | 400–450 | Visible | — | Hamster cheek pouch | — | — | — | Neoplastic tissue showed an decrease in fluorescence intensity relative to normal tissue with excitation between 400–450 nm, with an optimal wavelength of 410 nm | 97 |
| Oral tissue (normal, hyperplastic, papilloma, and invasive carcinoma) | Oral cancer | — | — | 405 | 430–700 | — | Hamster cheek pouch (tumors induced with 7,12-dimethyl-benz(a)-anthracene) | — | — | — | Ratiometric analyses of the autofluorescence emission at 530/620 nm and 530/630 nm provided a means of differentiating the various stages and types of tumors present | 98 |
| Pancreas tissue (normal and tumor) | Pancreatic cancer | — | — | 355 | 470 and 640 | — | Rat pancreatic tissue (normal and tumor) | — | — | — | — | 99 |

**TABLE 65.9B** Fluorescence Data: Cellular Studies, Tissue and Biofluid Studies (*in Vitro*), Animal Studies (*in Vivo*), and Human Clinical Studies (*in Vivo*) (continued)

| Species | Target Organ or Disease | Absorbance Maxima (nm) | Molar Extinction Coefficient ($M^{-1}$) | Excitation $\lambda_{ex}$ (nm) | Emission $\lambda_{em}$ (nm) | Quantum Yield | Matrix | Lifetime $\tau1$ (rel % $\tau1$) | Lifetime $\tau2$ (rel % $\tau2$) | Lifetime $\tau3$ (rel % $\tau3$) | Comments | Ref. |
|---|---|---|---|---|---|---|---|---|---|---|---|---|
| Pheophorbide-a (Ph-a) | Pancreatic cancer | — | — | 355 | 680 | — | Rat pancreatic tissue (normal and tumor) | — | — | — | Normalization of Ph-a fluorescence intensity was performed via ratiometric analyses with preas autofluorescent signals | 99 |
| Protoporphyrin IX (5-aminolevulinic acid induced) | Liver cancer | — | — | Visible | 635 | — | Heptic tumors in rats | — | — | — | Large quantities of PpIX were found to accumulate in both normal and tumor liver tissue | 100 |
| Protoporphyrin IX (5-aminolevulinic acid induced) | Various cancers | — | — | 405 | 635–705 | — | Rat tumors (various organs) | — | — | — | Maximum build-up of PpIX occurred within 1 h | 101 |
| Trimethoxylated carotenoporphyrin | Various cancers | — | — | 425 | 655 and 720 | — | MS-2 fibrosarcoma in BALB/c mice | — | — | — | The greatest extent of caroteno-porphyrin fluorescence occurred in the liver | 102 |
| Trimethylated carotenoporphyrin | Various cancers | — | — | 425 | 655 and 720 | — | MS-2 fibrosarcoma in BALB/c mice | — | — | — | The greatest extent of caroteno-porphyrin fluorescence occurred in the liver | 102 |

**Human Clinical Studies (in Vivo)**

| | | | | | | | | | | | | |
|---|---|---|---|---|---|---|---|---|---|---|---|---|
| 5,10,15,20-tetra (m-hydroxy-phenyl)chlorin (mTHPC) | Various cancers | — | — | — | — | — | *In vivo* esophageal tissue | 8.5 ± 0.8 | — | — | mTHPC was found to have a fluorescence lifetime of 8.5 ns in esophageal tissue | 103 |
| Bladder tissue (normal and tumor) | Bladder cancer | — | — | 337 | 370–490 | — | *In vivo* bladder tissue | — | — | — | Spectral and temporal differences in the autofluorescence emission were used to differentiate normal and tumor tissues | 103 |
| Bladder tissue (normal and tumor) | Bladder cancer | — | — | 337 | 385–455 | — | *In vivo* bladder tissue | — | — | — | By ratioing the autofluorescence intensity at 385 to the intensity at 455 nm, it was possible to diagnose the tissue with 98% sensitivity | 62 |

**TABLE 65.9B** Fluorescence Data: Cellular Studies, Tissue and Biofluid Studies (*in Vitro*), Animal Studies (*in Vivo*), and Human Clinical Studies (*in Vivo*) (continued)

| Species | Target Organ or Disease | Absorbance Maxima (nm) | Molar Extinction Coefficient ($M^{-1}$) | Excitation $\lambda_{ex}$ (nm) | Emission $\lambda_{em}$ (nm) | Quantum Yield | Matrix | Lifetime τ1 (rel % τ1) | Lifetime τ2 (rel % τ2) | Lifetime τ3 (rel % τ3) | Comments | Ref. |
|---|---|---|---|---|---|---|---|---|---|---|---|---|
| Bladder tissue (normal, inflammatory mucosa, and neoplastic urothelial lesion) | Bladder cancer | — | — | 308 | Visible | — | *In vivo* bladder tissue | — | — | — | The shape of the autofluorescence emission of carcinoma *in situ* tissue was significantly different from both normal and inflamed mucosa, allowing a ratiometric measurement of the intensities at 360 and 440 nm to be used for diagnosis | 104 |
| Bladder tissue (normal, inflammatory mucosa, and neoplastic urothelial lesion) | Bladder cancer | — | — | 337 | Visible | — | *In vivo* bladder tissue | — | — | — | The overall fluorescence intensity of bladder tumor tissue was found to be much less than that of normal tissue, regardless of stage | 104 |
| Bladder tissue (normal, inflammatory mucosa, and neoplastic urothelial lesion) | Bladder cancer | — | — | 480 | Visible | — | *In vivo* bladder tissue | — | — | — | The overall fluorescence intensity of bladder tumor tissue was found to be much less than that of normal tissue, regardless of stage | 104 |

| | | | | | | | | | | | | |
|---|---|---|---|---|---|---|---|---|---|---|---|---|
| Bladder tissue (normal, inflammatory mucosa, and neoplastic urothelial lesion) | Bladder cancer | — | — | 308 | 360 and 440 | — | *In vivo* bladder tissue | — | — | — | Ratiometric analyses of the autofluorescence intensity at 360 to 440 nm were used to differentiate normal or inflamed tissue from neoplastic lesions | 105 |
| Bladder tissue (normal, inflammatory mucosa, and neoplastic urothelial lesion) | Bladder cancer | — | — | 337 | Visible | — | *In vivo* bladder tissue | — | — | — | Autofluorescence intensity of tumors was significantly less than that of normal tissue | 105 |
| Bladder tissue (normal, inflammatory mucosa, and neoplastic urothelial lesion) | Bladder cancer | — | — | 488 | Visible | — | *In vivo* bladder tissue | — | — | — | Autofluorescence intensity of tumors was significantly less than that of normal tissue | 105 |
| Bladder tissue (normal, inflammatory mucosa, and neoplastic urothelial lesion) | Bladder cancer | — | — | 337 | 385 and 455 | — | *In vivo* bladder tissue | — | — | — | Ratiometric analyses based on tissue autofluorescence were used to correctly identify bladder lesions | 106 |
| Brain tissue (normal and malignant gliomas) | Brain cancer | — | — | 460 and 625 | Visible/red/near infrared | — | *In vivo* brain tissue | — | — | — | By combining autofluorescence and diffuse reflectance, differentiation of normal and malignant brain tissue has been shown to be feasible | 107 |

**TABLE 65.9B** Fluorescence Data: Cellular Studies, Tissue and Biofluid Studies (*in Vitro*), Animal Studies (*in Vivo*), and Human Clinical Studies (*in Vivo*) (continued)

| Species | Target Organ or Disease | Absorbance Maxima (nm) | Molar Extinction Coefficient ($M^{-1}$) | Excitation $\lambda_{ex}$ (nm) | Emission $\lambda_{em}$ (nm) | Quantum Yield | Matrix | Lifetime $\tau 1$ (rel % $\tau 1$) | Lifetime $\tau 2$ (rel % $\tau 2$) | Lifetime $\tau 3$ (rel % $\tau 3$) | Comments | Ref. |
|---|---|---|---|---|---|---|---|---|---|---|---|---|
| Bronchial tissue (normal and tumor) | Lung cancer | — | — | 480 | Visible | — | *In vivo* lung tissue | — | — | — | Spectral and temporal differences in the autofluorescence emission were used to differentiate normal and tumor tissues | 103 |
| Bronchial tissue (normal, dyplastic, and carcinoma *in situ* (CIS)) | Lung cancer | — | — | 325 | Green/red | — | *In vivo* lung tissue | — | — | — | Ratiometric measurements of lung tissue over the red and green regions of the spectrum were used to diagnose the tissue as normal, dysplastic, or CIS | 108 |
| Bronchial tissue (normal, metaplastic, and early cancer ) | Lung cancer | — | — | 350–495 | Green-red | — | *In vivo* lung tissue | — | — | — | Absolute autofluorescence measurements allowed for the differentiation of normal, metaplastic, and early cancerous stages of lung tissue | 109 |

| | | | | | | | | | | | | |
|---|---|---|---|---|---|---|---|---|---|---|---|---|
| Bronchial tissue (normal, metaplastic, and early cancer ) | Lung cancer | — | — | 400–480 | 600–800 | — | *In vivo* lung tissue | — | — | — | Autofluorescence differences between normal, metaplastic, and dysplastic tissue were found, with the optimal excitation wavelength for differentiation being 405 nm | 110 |
| Cervical tissue (normal and tumor) | Cervical cancer | — | — | 330 | 385 | — | *In vivo* cervical tissue | — | — | — | The averaged normalized fluorescence intensity was found to be greater for normal tissue as compared to abnormal tissue | 68 |
| Cervical tissue (normal, non-neoplastic normal, and intraepithelial neoplasia (CIN)) | Cervical cancer | — | — | 337 | Visible | — | *In vivo* cervical tissue | — | — | — | Differentiation of tissues was performed based on changes in tissue autofluorescence related to the species collagen, oxyhemoglobin, and NAD(P)H | 112 |
| Collagen | Colon cancer | — | — | 337 | 390 | — | *In vivo* colon tissue | — | — | — | Normal colonic tissue showed a decrease in autofluorescence intensity due to collagen relative to hyperplastic or adenomatous tissues | 113 |

**TABLE 65.9B** Fluorescence Data: Cellular Studies, Tissue and Biofluid Studies (*in Vitro*), Animal Studies (*in Vivo*), and Human Clinical Studies (*in Vivo*) (continued)

| Species | Target Organ or Disease | Absorbance Maxima (nm) | Molar Extinction Coefficient ($M^{-1}$) | Excitation $\lambda_{ex}$ (nm) | Emission $\lambda_{em}$ (nm) | Quantum Yield | Matrix | Lifetime τ1 (rel % τ1) | Lifetime τ2 (rel % τ2) | Lifetime τ3 (rel % τ3) | Comments | Ref. |
|---|---|---|---|---|---|---|---|---|---|---|---|---|
| Endogenous skin protoporphyrins | Skin cancer | — | — | Visible | 600, 620, 640, and 670 | — | *In vivo* skin tissue | 1–5 | — | — | Lipophile skin bacterium, *Propioni-bacterium acnes,* that upon irradiation caused photodynamic activity, was studied | 114 |
| Esophageal tissue (normal and tumor) | Esophageal cancer | — | — | 410 | 450–600 | — | *In vivo* esophageal tissue | — | — | — | Tumor tissue exhibited a decrease in normalized fluorescence intensity at 480 nm relative to normal tissue | 115, 116 |
| Esophageal tissue (normal and tumor) | Esophageal cancer | — | — | Violet-blue | 450–700 | — | *In vivo* esophageal tissue | — | — | — | Specific differences in the autofluorescence spectra of esophageal squamous cell carcinoma, adenocarcinoma of the esphophagus, and adenocarcinoma of the stomach were found | 117, 118 |

| | | | | | | | | | | | | |
|---|---|---|---|---|---|---|---|---|---|---|---|---|
| Esophageal tissue (normal and tumor) | Esophageal cancer | — | — | 337 | 370–490 | — | *In vivo* esophageal tissue | — | — | — | Spectral and temporal differences in the autofluorescence emission were used to differentiate normal and tumor tissues | 103 |
| Flavoproteins in skin tissue | Skin cancer | — | — | 960 (2 photon absorption) | 520 | — | *In vivo* skin tissue | — | — | — | Confocal microscopy was used to determine the spatial location of various autofluorescent species within skin cells | 119 |
| Human cornea | Diabetes | — | — | 360–370 | 532–630 | — | *In vivo* human eyes | — | — | — | Autofluorescence measurements were used to distinguish healthy patients from patients with diabetes mellitus | 120 |
| Human cornea | Diabetes | — | — | 400–410 | 532–630 | — | *In vivo* human eyes | — | — | — | Autofluorescence measurements were used to distinguish healthy patients from patients with diabetes mellitus | 120 |
| Human cornea | Diabetes | — | — | 415–425 | 532–630 | — | *In vivo* human eyes | — | — | — | Autofluorescence measurements were used to distinguish healthy patients from patients with diabetes mellitus | 120 |

**TABLE 65.9B** Fluorescence Data: Cellular Studies, Tissue and Biofluid Studies (*in Vitro*), Animal Studies (*in Vivo*), and Human Clinical Studies (*in Vivo*) (continued)

| Species | Target Organ or Disease | Absorbance Maxima (nm) | Molar Extinction Coefficient ($M^{-1}$) | Excitation $\lambda_{ex}$ (nm) | Emission $\lambda_{em}$ (nm) | Quantum Yield | Matrix | Lifetime $\tau 1$ (rel % $\tau 1$) | Lifetime $\tau 2$ (rel % $\tau 2$) | Lifetime $\tau 3$ (rel % $\tau 3$) | Comments | Ref. |
|---|---|---|---|---|---|---|---|---|---|---|---|---|
| Human cornea | Diabetes | — | — | 425–435 | 532–630 | — | *In vivo* human eyes | — | — | — | Autofluorescence measurements were used to distinguish healthy patients from patients with diabetes mellitus | 120 |
| Human cornea | Diabetes | — | — | 431–441 | 532–630 | — | *In vivo* human eyes | — | — | — | Autofluorescence measurements were used to distinguish healthy patients from patients with diabetes mellitus | 120 |
| Human cornea | Diabetes | — | — | 435–445 | 532–630 | — | *In vivo* human eyes | — | — | — | Autofluorescence measurements were used to distinguish healthy patients from patients with diabetes mellitus | 120 |
| Human cornea | Diabetes | — | — | 445–455 | 532–630 | — | *In vivo* human eyes | — | — | — | Autofluorescence measurements were used to distinguish healthy patients from patients with diabetes mellitus | 120 |

| | | | | | | | | | | | | |
|---|---|---|---|---|---|---|---|---|---|---|---|---|
| Human cornea | Diabetes | — | — | 465–475 | 532–630 | — | *In vivo* human eyes | — | — | — | Autofluorescence measurements were used to distinguish healthy patients from patients with diabetes mellitus | 120 |
| Human cornea | Diabetes | — | — | 475–485 | 532–630 | — | *In vivo* human eyes | — | — | — | Autofluorescence measurements were used to distinguish healthy patients from patients with diabetes mellitus | 120 |
| Human cornea | Diabetes | — | — | 415–491 | 515–630 | — | *In vivo* human eyes | — | — | — | Corneal autofluorescence was found to increase in people with diabetes mellitus | 121 |
| Human eye lens | Vision problems | — | — | Ultraviolet/ blue | 495–520 | — | *In vivo* human eyes | — | — | — | Lens autofluorescence could be correlated to coloration and opalescence of the lens nucleus in humans | 122 |
| Human eye lens | Vision problems | — | — | 430–490 | 530–630 | — | Human eye lens | — | — | — | Intrinsic eye lens transmittance can be determined based on autofluorescence analyses | 75 |

**TABLE 65.9B** Fluorescence Data: Cellular Studies, Tissue and Biofluid Studies (*in Vitro*), Animal Studies (*in Vivo*), and Human Clinical Studies (*in Vivo*) (continued)

| Species | Target Organ or Disease | Absorbance Maxima (nm) | Molar Extinction Coefficient ($M^{-1}$) | Excitation $\lambda_{ex}$ (nm) | Emission $\lambda_{em}$ (nm) | Quantum Yield | Matrix | Lifetime τ1 (rel % τ1) | Lifetime τ2 (rel % τ2) | Lifetime τ3 (rel % τ3) | Comments | Ref. |
|---|---|---|---|---|---|---|---|---|---|---|---|---|
| Laryngeal tissues (normal and tumor) | Laryngeal cancer | — | — | 325 | Visible | — | *In vivo* laryngeal tissue | — | — | — | Images were obtained over several emission bands and ratiometric analyses were performed to differentiate normal tissue from carcinoma tissue | 123 |
| Laryngeal tissues (normal and tumor) | Laryngeal cancer | — | — | 375–440 | Green | — | *In vivo* laryngeal tissue | — | — | — | Tumor autofluorescence intensity was greatly reduced relative to surrounding normal tissue | 124 |
| Laryngeal tissues (normal, dysplastic, carcinoma *in situ*, and microinvasive lesions) | Laryngeal cancer | — | — | 380–460 | Light green | — | *In vivo* laryngeal tissue | — | — | — | Autofluorescence diagnoses were capable of distinguishing normal tissue, dysplastic tissue, carcinoma *in situ*, and microinvasive lesions | 125 |
| Lipofuscin | Vision problems | — | — | Visible | 710 | — | *In vivo* human eyes | — | — | — | Measurements of macular pigment density could be determined based on autofluorescence | 126 |

| | | | | | | | | | | | | |
|---|---|---|---|---|---|---|---|---|---|---|---|---|
| Nicotinamide adenine dinucleotide (NADH) reduced | Colon cancer | — | — | 337 | 460 | — | *In vivo* colon tissue | — | — | — | — | 127 |
| Nicotinamide adenine dinucleotide (NADH) reduced | Colon cancer | — | — | 370 | 460 | — | *In vivo* colon tissue | — | — | — | Ratiometric analyses of autofluorescence emission were used to identify tissues as normal, hyperplastic, or adenomatous | 128 |
| Oral tissues (connective tissues) | Oral cancer | — | — | 365 | Visible | — | *In vivo* oral tissue | — | — | — | Tumor margining was performed using 365-nm light to excite the edges of the tumor | 129 |
| Oral tissues (normal and tumor) | Oral cancer | — | — | 370 | 630–640 | — | *In vivo* oral tissue | — | — | — | Differences between normal and neoplastic tissues were found based on its autofluorescence emission | 130 |
| Oral tissues (normal and tumor) | Oral cancer | — | — | 410 | 630–640 | — | *In vivo* oral tissue | — | — | — | Differences in the autofluorescence emission between normal and neoplastic tissues was found to be optimal following 410-nm excitation | 130 |
| Oral tissues (normal and tumor) | Oral cancer | — | — | 337 | Visible | — | *In vivo* oral tissue | — | — | — | Autofluorescence intensity of contralateral sites was much greater than the abnormal sites | 131 |

**TABLE 65.9B** Fluorescence Data: Cellular Studies, Tissue and Biofluid Studies (*in Vitro*), Animal Studies (*in Vivo*), and Human Clinical Studies (*in Vivo*) (continued)

| Species | Target Organ or Disease | Absorbance Maxima (nm) | Molar Extinction Coefficient ($M^{-1}$) | Excitation $\lambda_{ex}$ (nm) | Emission $\lambda_{em}$ (nm) | Quantum Yield | Matrix | Lifetime τ1 (rel % τ1) | Lifetime τ2 (rel % τ2) | Lifetime τ3 (rel % τ3) | Comments | Ref. |
|---|---|---|---|---|---|---|---|---|---|---|---|---|
| Oral tissues (normal and tumor) | Oral cancer | — | — | 410 | Red/blue | — | *In vivo* oral tissue | — | — | — | The ratio of red fluorescence to blue fluorescence was greater in abnormal tissue than contralateral areas | 131 |
| Oral tissues (normal and tumor) | Oral cancer | — | — | 442 | Red/green | — | *In vivo* oral tissue | — | — | — | Ratiometric images of red autofluorescence and green autofluorescence images were used to identify oral cavity neoplasia | 132 |
| Oral tissues (normal and tumor) | Oral cancer | — | — | 360 | > 480 | — | *In vivo* oral tissue | — | — | — | Fluorescence photography of the tissue autofluorescence could be used to distinguish between benign and malignant oral cavity tumors | 133 |
| Oral tissues (normal and tumor) | Oral cancer | — | — | 350 | 472 | — | *In vivo* oral tissue | — | — | — | Identification of oral cavity neoplasia was demonstrated based on the tissue autofluorescence | 134 |

| | | | | | | | | | | | | |
|---|---|---|---|---|---|---|---|---|---|---|---|---|
| Oral tissues (normal and tumor) | Oral cancer | — | — | 380 | 472 | — | *In vivo* oral tissue | — | — | — | Identification of oral cavity neoplasia was demonstrated based on the tissue autofluorescence | 134 |
| Oral tissues (normal and tumor) | Oral cancer | — | — | 400 | 472 | — | *In vivo* oral tissue | — | — | — | Identification of oral cavity neoplasia was demonstrated based on the tissue autofluorescence | 134 |
| Oral tissues (normal and tumor) | Oral cancer | — | — | 330 | 380 | — | *In vivo* oral tissue | — | — | — | Differentiation of normal and tumor tissues was performed based on a decrease in the 330-nm exciation band in tumor tissue | 135 |
| Oral tissues (normal and tumor) | Oral cancer | — | — | 340 | 390 | — | *In vivo* oral tissue | — | — | — | Diagnosis of the stage of dysplasia was determined based on changes in the autofluorescence emission intensity at 390 nm | 135 |
| Oral tissues (normal and tumor) | Oral cancer | — | — | 370 | Visible | — | *In vivo* oral tissue | — | — | — | — | 130 |
| Oral tissues (normal and tumor) | Oral cancer | — | — | 410 | Visible | — | *In vivo* oral tissue | — | — | — | — | 130 |
| Pharynx tissues (normal and tumor) | Oral cancer | — | — | 370 | Visible | — | *In vivo* pharynx tissue | — | — | — | — | 130 |
| Pharynx tissues (normal and tumor) | Oral cancer | — | — | 410 | Visible | — | *In vivo* pharynx tissue | — | — | — | — | 130 |

**TABLE 65.9B** Fluorescence Data: Cellular Studies, Tissue and Biofluid Studies (*in Vitro*), Animal Studies (*in Vivo*), and Human Clinical Studies (*in Vivo*) (continued)

| Species | Target Organ or Disease | Absorbance Maxima (nm) | Molar Extinction Coefficient ($M^{-1}$) | Excitation $\lambda_{ex}$ (nm) | Emission $\lambda_{em}$ (nm) | Quantum Yield | Matrix | Lifetime $\tau 1$ (rel % $\tau 1$) | Lifetime $\tau 2$ (rel % $\tau 2$) | Lifetime $\tau 3$ (rel % $\tau 3$) | Comments | Ref. |
|---|---|---|---|---|---|---|---|---|---|---|---|---|
| Pharynx tissues (normal and tumor) | Oropharyngeal cancer | — | — | 330 | 380 | — | *In vivo* pharynx tissue | — | — | — | Differentiation of normal and tumor tissues was performed based on a decrease in the 330-nm excitation band in tumor tissue | 135 |
| Pharynx tissues (normal and tumor) | Oropharyngeal cancer | — | — | 340 | 390 | — | *In vivo* pharynx tissue | — | — | — | Diagnosis of the stage of dysplasia was determined based on changes in the autofluorescence emission intensity at 390 nm | 135 |
| Photogem | Various cancers | — | — | 510 | Red/near infrared | — | *In vivo* tissues (lungs, larynx, skin, gastric, esophageal, and gynecological) | — | — | — | Drug accumulation studies were performed and it was found that the accumulation depended dramatically on the tissue type and stage of cancer | 136 |
| Protoporphyrin (5-aminolevulinic acid hexylester hydrochloride-induced) | Various cancers | — | — | Visible | Red | — | *In vivo* bladder tissue | 15.9 ± 1.2 | — | — | PpIX was found to exhibit an mono-exponential decay of 15.9 ns in bladder tissue | 103 |

| | | | | | | | | | | | | |
|---|---|---|---|---|---|---|---|---|---|---|---|---|
| Protoporphyrin (ALA-induced) | Various cancers | — | — | 405 | 550–750 | — | *In vivo* skin tissue | — | — | — | Diagnoses of malignant melanomas based on multiple ratiometric measurements were performed | 137 |
| Protoporphyrin (ALA-induced) | Various cancers | — | — | 435 | 550–750 | — | *In vivo* skin tissue | — | — | — | Diagnoses of malignant melanomas based on multiple ratiometric measurements were performed | 137 |
| Protoporphyrin (ALA-induced) | Various cancers | — | — | 375–440 | Green/red | — | *In vivo* oral tissue | — | — | — | Ratiometric images of both the red (PpIX) fluorescence and green (background) fluorescence were used to differentiate normal and tumor tissues | 138 |
| Protoporphyrin (ALA-induced) | Various cancers | — | — | Visible | Red | — | *In vivo* tumor flank before irradiation | 0.230 (17) | 17.1 (83) | — | — | 139 |
| Protoporphyrin (ALA-induced) | Various cancers | — | — | visible | Red | — | *In vivo* tumor flank after irradiation | 0.270 (5.5) | 5 (45) | — | — | 139 |
| Protoporphyrin (ALA-induced) | Various cancers | — | — | 375–440 | Red | — | *In vivo* laryngeal tissue | — | — | — | PpIX fluorescence allowed for the differentiation of normal tissue from malignant neoplasms | 124 |

**TABLE 65.9B** Fluorescence Data: Cellular Studies, Tissue and Biofluid Studies (*in Vitro*), Animal Studies (*in Vivo*), and Human Clinical Studies (*in Vivo*) (continued)

| Species | Target Organ or Disease | Absorbance Maxima (nm) | Molar Extinction Coefficient ($M^{-1}$) | Excitation $\lambda_{ex}$ (nm) | Emission $\lambda_{em}$ (nm) | Quantum Yield | Matrix | Lifetime $\tau 1$ (rel % $\tau 1$) | Lifetime $\tau 2$ (rel % $\tau 2$) | Lifetime $\tau 3$ (rel % $\tau 3$) | Comments | Ref. |
|---|---|---|---|---|---|---|---|---|---|---|---|---|
| Protoporphyrin (ALA-induced) | Various cancers | — | — | 375–440 | Red | — | *In vivo* oral tissue | — | — | — | PpIX fluorescence allowed for the differentiation of normal tissue from malignant neoplasms and was found to have a 10:1 contrast between malignant and normal tissues | 140 |
| Reduced pyridine nucleotides in skin tissue | Skin cancer | — | — | 730 (2 photon absorption) | Visible | — | *In vivo* skin tissue | — | — | — | Confocal microscopy was used to determine the spatial location of various autofluorescent species within skin cells | 119 |
| Skin tissue | Skin cancer | — | — | 960 (3 photon absorption) | 425 | — | *In vivo* skin tissue | — | — | — | Confocal microscopy was used to determine the spatial location of various autofluorescent species within skin cells | 119 |
| Skin tissue | Skin cancer | — | — | 325 | Visible | — | *In vivo* skin tissue | — | — | — | Autofluorescence emission was found to correlate to photoaging of the skin | 141 |

| | | | | | | | | | | | | |
|---|---|---|---|---|---|---|---|---|---|---|---|---|
| Skin tissue | Skin cancer | — | — | 365 | 440 | — | *In vivo* skin tissue | — | — | — | Results demonstrated that there was no difference between normal and non-melanoma tumors | 142 |
| Skin tissue | Skin cancer | — | — | 375 | 400–700 (peak at 436) | — | *In vivo* skin tissue | — | — | — | Results demonstrated that there was no difference between normal and non-melanoma tumors | 142 |
| Skin tissue | Skin cancer | — | — | 380 | 470 | — | *In vivo* skin tissue | — | — | — | Autofluorescence emission resulting from a skin sample was found to strongly depend on the absorption and scattering properties of the tissue | 143 |
| Tongue tissue (normal and moderately differentiated squamous cell carcinoma) | Oral cancer | — | — | 350 | 390–625 | — | *In vivo* tongue tissue | — | — | — | Abnormal tissue displayed a reduced fluorescence emission between 400 and 450 nm relative to normal tissue | 144 |

**TABLE 65.9B** Fluorescence Data: Cellular Studies, Tissue and Biofluid Studies (*in Vitro*), Animal Studies (*in Vivo*), and Human Clinical Studies (*in Vivo*) (continued)

| Species | Target Organ or Disease | Absorbance Maxima (nm) | Molar Extinction Coefficient ($M^{-1}$) | Excitation $\lambda_{ex}$ (nm) | Emission $\lambda_{em}$ (nm) | Quantum Yield | Matrix | Lifetime τ1 (rel % τ1) | Lifetime τ2 (rel % τ2) | Lifetime τ3 (rel % τ3) | Comments | Ref. |
|---|---|---|---|---|---|---|---|---|---|---|---|---|
| Tongue tissue (normal and moderately differentiated squamous cell carcinoma) | Oral cancer | — | — | 410 | 460–675 | — | *In vivo* tongue tissue | — | — | — | Abnormal tissue displayed a dramatic decrease in fluorescence intensity over the entire specrum, as well as a distinct peak at approximately 630 nm | 144 |
| Tongue tissue (normal and moderately differentiated squamous cell carcinoma) | Oral cancer | — | — | 460 | 490–720 | — | *In vivo* tongue tissue | — | — | — | Abnormal tissue displayed a dramatic decrease in fluorescence intensity over the entire specrum | 144 |

## References


1. Dillon, J. and Atherton, S.J., Time resolved spectroscopic studies on the intact human lens, *Photochem. Photobiol.,* 51, 465, 1990.
2. Bridges, J.W., Davies, D.S., and Williams, R.T., Fluorescence studies on some hydroxypyridines including compounds of vitamin B6 group, *Biochem. J.,* 98, 451, 1966.
3. Andersson-Engels, S., Baert, L., Berg, R., D'Hallewin, M.A., Johansson, J., Stenram, U., Svanberg, K., and Svanberg, S., Fluorescence characteristics of atherosclerotic plaque and malignant tumors, *Proc. SPIE,* 1426, 31, 1991.
4. Eldred, G.E., Miller, G.V., Stark, W.S., and Feeney-Burns, L., Lipofuscin — resolution of discrepant fluorescence data, *Science,* 16, 757, 1982.
5. Sohal, R.S., Assay of lipofuscin ceroid pigment *in vivo* during aging, *Methods Enzymol.,* 105, 484, 1984.
6. Richards-Kortum, R.R., Rava, R.P., Baraga, J., Fitzmaurice, M., Kramer, J. and Feld, M., in *Optronic Techniques in Diagnostic and Therapeutic Medicine,* Pratesi, R., Ed., Plenum Press, New York, 1990.
7. Koenig, K., Schneckenburger, H., Hemmer, J., Tromberg, B.J., and Steiner, R., *In vivo* fluorescence detection and imaging of porphyrin-producing bacteria in the human skin and in the oral cavity for diagnosis of acne vulgaris, caries and squamous cell carcinoma, *Proc. SPIE,* 2135, 68, 1994.
8. Rava, R.P., Richards-Kortum, R., Fitzmaurice, M., Cothren, R., Petras, R., Sivak, M.V., Levin, H., and Feld, M.S., Early detection of dysplasia in colon and bladder tissue using laser-induced fluorescence, *Proc. SPIE,* 1426, 68, 1991.
9. Wagnières, G.A., Star, W.M., and Wilson, B.C., *In vivo* fluorescence spectroscopy and imaging for oncological applications, *Photochem. Photobiol.,* 68(5), 603, 1998.
10. Cheng, S.-H., Master's thesis, University of Texas at Austin, 1992.
11. Fujimoto, D., Isolation and characterization of a fluorescent material in bovine Achilles-tendon collagen, *Biochem. Biophys. Res. Commun.,* 76, 1124, 1977.
12. Eyre, D.R., Paz, M.A., and Gallop, P.M., Cross-linking in collagen and elastin, *Annu. Rev. Biochem.,* 53, 717, 1984.
13. Yang, Y., Ye, T., Li, F., and Ma, P.X., Characteristic autofluorescence for cancer diagnosis and its origin, *Lasers Surg. Med.* 7, 528, 1987.
14. Barnes, D.A., Thomsen, S., Fitzmaurice, M., and Richards-Kortum, R., A characterization of the fluorescent properties of circulating human eosinophils, *Photochem. Photobiol.,* 58(2), 297, 1993.
15. Mayeno, A.N., Hamann, K.J., and Gleich, G.J., Granule-associated flavin adenine-dinucleotide (FAD) is responsible for eosinophil autofluorescence, *J. Leukocyte Biol.,* 51, 172, 1992.
16. Weil, G.J. and Chused, T.M., Eosinophil autofluorescence and its use in isolation and analysis of human eosinophils using flow micro-fluorometry, *Blood,* 57, 1099, 1981.
17. Tsuchida, M., Miura, T., and Aibara, K., Lipofuscin and lipofuscin-like substances, *Chem. Phys. Lipid,* 44(2–4), 297, 1987.
18. Richards-Kortum, R. and Sevick-Muraca, E., Quantitative optical spectroscopy for tissue diagnosis, *Annu. Rev. Phys. Chem.,* 47, 555, 1996.
19. Visser, A.J., Sanetema, J.S., and van Hoek, A., Spectroscopic and dynamic characterization of FMN in reversed micelles entrapped waterpools, *Photochem. Photobiol.,* 39, 11, 1997.
20. Schneckenburger, H. and Konig, K., Fluorescence decay kinetics and imaging of Nad(P)H and flavins as metabolic indicators, *Opt. Eng.,* 31(7), 1447, 1992.
21. Lakowicz, J.R., *Principles of Fluorescence Spectroscopy,* Plenum, New York, 1985.
22. Campbell, I. and Dwek, R., *Biological Spectroscopy,* Benjamin Cummings, Menlo Park, CA, 1984.
23. Cubeddu, R., Ramponi, R., Taroni, P., and Canti, G., Time-gated fluorescence spectroscopy of porphyrin derivatives and aluminium phthalocyanine incorporated *in vivo* in a murine ascitic tumour model, *J. Photochem. Photobiol. B Biol.,* 11, 319, 1991.
24. Ambroz, M., MacRobert, A.J., Morgan, J., Rumbles, G., Foley, M.S.C., and Phillips, D., Time-resolved fluorescence spectroscopy and intracellular imaging of disulphonated aluminium phthalocyanine, *J. Photochem. Photobiol. B Biol.,* 22, 105, 1994.

25. Kessel, D., *In vitro* photosensitization with benzoporphyrin derivative, *Photochem. Photobiol.*, 49, 579, 1989.
26. Kessel, D., Byrne, C.J., and Ward, A.D., Photophysical and photobiological properties of diporphyrin ethers, *Photochem. Photobiol.*, 53, 469, 1991.
27. Andreoni, A. and Cubeddu, R., Photophysical properties of Photofrin II in different solvents, *Chem. Phys. Lett.*, 108, 141, 1984.
28. Moan, J. and Sommer, S., Fluorescence and absorption properties of the components of haematoprophyrin derivative, *Photochem. Photobiophys.*, 3, 93, 1981.
29. Aizawa, K., Okunaka, T., Ohtani, T., Kaabe, H., Yasunaka, Y., O'Hata, S., Ohtomo, N., Nishimiya, K., Conaka, C., Kato, H., Hayata, Y., and Saito, T., Localization of mono-L-aspartl chlorine e6 (NPe6) in mouse tissues, *Photochem. Photobiol.*, 46, 789, 1987.
30. Lin, C.-W., Shulok, J.R, Wong, Y.-K, Schanbacher, C.F., Cinotta, L., and Foley, J.W., Photosensitization, uptake and retention of phenoxazine Nile blue derivatives in human bladder carcinoma cells, *Cancer Res.*, 51, 1109, 1991.
31. Cubeddu, R., Ramponi, R., and Bottiroli, G., Time-resolved fluorescence spectroscopy of hematophorphyrin derivative in micelles, *Chem. Phys. Lett.*, 128, 439, 1986.
32. Kessel, D., Probing the structure of HPD by fluorescence spectroscopy, *Photochem. Photobiol.*, 50(3), 345, 1989.
33. Tralau, C.J., Barr, H., Sandeman, D.R., Barton, T., Lewin, M.R., and Brown, S.G., Aluminum sulfonated phthalocyanine distribution in rodent tumors of the colon, brain and pancreas, *Photochem. Photobiol.*, 46, 777, 1987.
34. Andeoni, A., Bottiroli, G., Colasanti, A., Giangare, M.C., Riccio, P., Roberti, G., and Vaghi, P., Fluorochromes with long-lived fluorescence as potential labels for pulsed laser immunocytofluorometry: photophysical characterization of pyrene derivatives, *J. Biochem. Biophys. Methods*, 29(2), 157, 1994.
35. Aicher, A., Miller, K., Reich, E., and Hautmann, R., Photosensitization of human bladder-carcinoma cells-*in vitro* by 9-acetoxy-tetra-*N*-propylporphycene (Atppn) bound to liposomes from soya phosphatidylcholine, *Opt. Eng.*, 32(2), 342, 1993.
36. Moser, J.G., Ruck, A., Schwarzmaier, H.J., and Westphalfrosch, C., Photodynamic cancer therapy — fluorescence localization and light-absorption spectra of chlorophyll-derived photosensitizers inside cancer cells, *Opt. Eng.*, 31(7), 1441, 1992.
37. Anidjar, M., Ettori, D., Cussenot, O., Meria, P., Desgrandchamps, F., Cortesse, A., Teillac, P., LeDuc, A., and Avrillier, S., Laser induced autofluorescence diagnosis of bladder tumors: dependence on the excitation wavelength, *J. Urol.*, 156(5), 1590, 1996.
38. Seidlitz, H.K., Stettmaier, K., Wessels, J.M., and Schneckenburger, H., Intracellular fluorescence polarization, picosecond kinetics, and light-induced reactions of photosensitizing porphyrins, *Opt. Eng.*, 31(7), 1482, 1992.
39. Amano, T., Kunimi, K., and Ohkawa, M., Fluorescence spectra from human semen and their relationship with sperm parameters, *Arch. Androl.*, 36(1), 9, 1996.
40. Grossman, N., Ilovitz, E., Chaims, O., Salman, A., Jagannathan, R., Mark, S., Cohen, B., Gopas, J., and Mordechai, S., Fluorescence spectroscopy for detection of malignancy: H-ras overexpressing fibroblasts as a model, *J. Biochem. Biophys. Meth.*, 50(1), 53, 2001.
41. Sacks, P.G., Savage, H.E., Levine, J., Kolli, V.R., Alfano, R.R.., and Schantz, S.P., Native cellular fluorescence identifies terminal squamous differentiation of normal oral epithelial cells in culture: a potential chemoprevention biomarker, *Cancer Lett.*, 104(2), 171, 1996.
42. Konig, K., Liu, Y.G., Sonek, G.J., Berns, M.W., and Tromberg, B.J., Autofluorescence spectroscopy of optically trapped cells, *Photochem. Photobiol.*, 62(5), 830, 1995.
43. Balazs, M., Szollosi, J., Lee, W.C., Haugland, R.P., Guzikowski, A.P., Fulwyler, M.J., Damjanovich, S., Feurstein, B.G., and Pershadsingh, H.A., Fluorescent tetradecanoylphorbol acetate: a novel probe of phorbol ester binding domains, *J. Cell Biochem.*, 46(3), 266, 1991.

44. Zhang, J.C., Savage, H.E., Sacks, P.G., Delohery, T., Alfano, R.R., Katz, A., and Schantz, S.P., Innate cellular fluorescence reflects alterations in cellular proliferation, *Lasers Surg. Med.*, 20(3), 319, 1997.
45. Ye, B.Q. and Abela, G.S., Beta-carotene decreases total fluorescence from human arteries, *Opt. Eng.*, 32(2), 326, 1993.
46. Drezek, R., Sokolov, K., Utzinger, U., Boiko, I., Malpica, A., Follen, M., and Richards-Kortum, R., Understanding the contributions of NADH and collagen to cervical tissue fluorescence spectra: modeling, measurements, and implications, *J. Biomed. Opt.*, 6(4), 385, 2001.
47. van der Veen, M.H., ten Bosch, J.J., Autofluorescence of bulk sound and *in vitro* demineralized human root dentin, *Eur. J. Oral Sci.*, 103(6), 375, 1995.
48. Banerjee, A. and Boyde, A., Autofluorescence and mineral content of carious dentin: scanning optical and backscattered electron microscopic studies, *Caries Res.*, 32(3), 219, 1998.
49. Romer, T.J., Fitzmaurice, M., Cothren, R.M., Richards-Kortum, R., Petras, R., Sivak, M.V., and Kramer, J.R., Laser-induced fluorescence microscopy of normal colon and dysplasia in colonic adenomas — implications for spectroscopic diagnosis, *Am. J. Gastroenterol.*, 90(1), 81, 1995.
50. Ramanujam, N., Richards-Kortum, R., Thomsen, S., Mahadevan-Jansen, A., Follen, M., and Chance, B., Low temperature fluorescence imaging of freeze-trapped human cervical tissues, *Opt. Express*, 8(6), 335, 2001.
51. Lucas, A., Radosavljevic, M.J., Lu, E., and Gaffney, E.J., Characterization of human coronary artery atherosclerotic plaque fluorescence emission, *Can. J. Cardiol.*, 6(6), 219, 1990.
52. Richards-Kortum, R., Rava, R.P., Fitzmaurice, M., Kramer, J.R., and Feld, M.S., 476 nm excited laser-induced fluorescence spectroscopy of human coronary arteries — applications in cardiology, *Am. Heart J.*, 122(4), 1141, 1991.
53. Scheu, M., Kagel, H., Zwaan, M., Lebeau, A., and Engelhardt, R., A new concept for a realtime feedback system in angioplasty with a flashlamp pumped dye laser, *Lasers Surg. Med.*, 11(2), 133, 1991.
54. Bosshart, F., Utzinger, U., Hess, O.M., Wyser, J., Mueller, A., Schneider, J., Niederer, P., Anliker, M., and Krayenbuehl, H.P., Fluorescence spectroscopy for identification of atherosclerotic tissue, *Cardiovasc. Res.*, 26(6), 620, 1992.
55. Morguet, A.J., Korber, B., Abel, B., Hippler, H., Wiegand, V., and Kreuzer, H., Autofluorescence spectroscopy using a XeCl excimer-laser system for simultaneous plaque ablation and fluorescence excitation, *Lasers Surg. Med.*, 14(3), 238, 1994.
56. Verbunt, R., Fitzmaurice, M.A., Kramer, J.R., Ratliff, N.B., Kittrell, C., Taroni, P., Cothren, R.M., Baraga, J., and Feld, M., Characterization of ultraviolet laser-induced autofluorescence of ceroid deposits and other structures in atherosclerotic plaques as a potential diagnostic for laser angiosurgery, *Am. Heart J.*, 123(1), 208, 1992.
57. Andersson-Engels, S.J., Svanberg, K., and Svanberg, S., Fluorescence imaging and point measurements of tissue: applications to the demarcation of malignant tumors and atherosclerotic lesions from normal tissue, *Photochem. Photobiol.*, 53(6), 807, 1991.
58. Chaudhry, H.W., Richards-Kortum, R., Kolubayev, T., Kittrell, C., Partovi, F., Kramer, J.R., and Feld, M.S., Alteration of spectral characteristics of human artery wall caused by 476-nm laser irradiation, *Lasers Surg. Med.*, 9(6), 572, 1989.
59. Deckelbaum, L.I., Desai, S.P., Kim, C., and Scott, J.J., Evaluation of a fluorescence feedback-system for guidance of laser angioplasty, *Lasers Surg. Med.*, 16(3), 226, 1995.
60. Baraga, J.J., Rava, R.P., Taroni, P., Kittrell, C., Fitzmaurice, M., and Feld, M.S., Laser induced fluorescence spectroscopy of normal and atherosclerotic human aorta using 306–310 nm excitation, *Lasers Surg. Med.*, 10(3), 245, 1990.
61. Laufer, G.W., Hohla, K., Horvat, R., Henke, K.H., Buchelt, M., Wutzl, G., and Wolner, E., Excimer laser-induced simultaneous ablation and spectral identification of normal and atherosclerotic arterial tissue layers, *Circulation*, 78(4), 1031, 1988.
62. Koenig, F., McGovern, F.J., Althausen, A.F., Deutsch, T.F., and Schomacker, K.T., Laser induced autofluorescence diagnosis of bladder cancer, *J. Urol.*, 156(5), 1597, 1996.

63. Frimberger, D., Zaak, D., Stepp, H., Knuchel, R., Baumgartner, R., Schneede, P., Schmeller, N., and Hofstetter, A., Autofluorescence imaging to optimize 5-ALA-induced fluorescence endoscopy of bladder carcinoma, *Urology*, 58(3), 372, 2001.
64. Hanlon, E.B., Itzkan, I., Dasari, R.R., Feld, M.S., Ferrante, R.J., McKee, A.C., Lathi, D., and Kowall, N.W., Near-infrared fluorescence spectroscopy detects Alzheimer's disease *in vitro*, *Photochem. Photobiol.*, 70(2), 236, 1999.
65. Lin, W.C., Toms, S.A., Motamedi, M., Jansen, E.D., and Mahadevan-Jansen, A., Brain tumor demarcation using optical spectroscopy; an *in vitro* study, *J. Biomed. Opt.*, 5(2), 214, 2000.
66. Alfano, R.R., Tang, G.C., Pradhan, A., Lam, W., Choy, D.S.J., and Opher, E., Fluorescence-spectra from cancerous and normal human-breast and lung tissues, *IEEE J. Quantum Electron.*, 23(10), 1806, 1987.
67. Anastassopoulou, N., Arapoglou, B., Demakakos, P., Makropoulou, M.I., Paphiti, A., and Serafetinides, A.A., Spectroscopic characterisation of carotid atherosclerotic plaque by laser induced fluorescence, *Lasers Surg. Med.*, 28(1), 67, 2001.
68. Richards-Kortum, R., Mitchell, M.F., Ramanujam, N., Mahadevan, A., and Thomsen, S., *In vivo* fluorescence spectroscopy — potential for noninvasive, automated diagnosis of cervical intraepithelial neoplasia and use as a surrogate end-point biomarker, *J. Cell. Biochem.*, Suppl. 19, 111, 1994.
69. Lohmann, W.M., Lohmann, C., and Kunzel, W., Native fluorescence of cervix uteri as a marker for dysplasia and invasive carcinoma, *Eur. J. Obstet. Gynecol. Reprod. Biol.*, 31(3), 249, 1989.
70. Richards-Kortum, R., Rava, R.P., Petras, R.E., Fitzmaurice, M., Sivak, M., and Feld, M.S., Spectroscopic diagnosis of colonic dysplasia, *Photochem. Photobiol.*, 53(6), 777, 1991.
71. Kapadia, C.R., Cutruzzola, F.W., O'Brien, K.M., Stetz, M.L., Enriquez, R., and Deckelbaum, L.I., Laser-induced fluoresence spectroscopy of human colonic mucosa, *Gastroenterology*, 99(1), 150, 1990.
72. Bottiroli, G., Croce, A.C., Locatelli, D., Marchesini, R., Pignoli, E., Tomatis, S., Cuzzoni, C., Dipalma, S., Dalfante, M., and Spinelli, P., Natural fluorescence of normal and neoplastic human colon: a comprehensive *ex vivo* study, *Lasers Surg. Med.*, 16, 48, 1995.
73. Chwirot, B.W., Kowalska, M., Sypniewska, N., Michniewicz, Z., and Gradziel, M., Spectrally resolved fluorescence imaging of human colonic adenomas, *J. Photochem. Photobiol. B*, 50(2–3), 174, 1999.
74. Perk, M., Flynn, G.J., Gulamhusein, S., Wen, Y., Smith, C., Bathgate, B., Tulip, J., Parfrey, N.A., and Lucas, A., Laser induced fluorescence identification of sinoatrial and atrioventricular nodal conduction tissue, *Pacing Clin. Electrophysiol.*, 16(8), 1701, 1993.
75. Larsen, M. and Lund Andersen, H., Lens fluorometry — light-attenuation effects and estimation of total lens transmittance, *Graefes Arch. Clin. Exp. Ophthalmol.*, 229(4), 363, 1991.
76. Ingrams, D.R., Dhingra, J.K., Roy, K., Perrault, D.F., Bottrill, I.D., Kabani, S., Rebeiz, E.E., Pankratov, M.M., Shapshay, S.M., Manoharan, R., Itzkan, I., and Feld, M.S., Autofluorescence characteristics of oral mucosa, *J. Specialties Head Neck*, 19(1), 27, 1997.
77. Roy, K., Bottrill, I.D., Ingrams, D.R., Pankratov, M.M., Rebeiz, E.E., Woo, P., Kabani, S., Shapshay, S., Manoharan, R., Itzkan, I., and Feld, M.S., Diagnostic fluorescence spectroscopy of oral mucosa, *Proc. SPIE*, 2395, 135, 1995.
78. Chen, C.T., Wang, C.Y., Kuo, Y.S., Chiang, H.H., Chow, S.N., Hsiao, I.Y., and Chang, C.P., Light-induced fluorescence spectroscopy: a potential diagnostic tool for oral neoplasia, *Proc. Natl. Sci. Counc. Repub. China B Life Sci.*, 20, 123, 1996.
79. Lohmann, W.N. and Bodeker, R.H., *In situ* differentiation between nevi and malignant melanomas by fluorescence measurements, *Naturwissenschaften*, 78, 456, 1991.
80. Chwirot, B.W., Chwirot, S., Jedrzejczyk, W., Jackowski, M., Raczynska, A.M., Winczakiewicz, J., and Dobber, J., Ultraviolet laser-induced fluorescence of human stomach tissues: detection of cancer tissues by imaging techniques, *Lasers Surg. Med.*, 21(2), 149, 1997.
81. Yappert, M.C., Borchman, D., and Byrdwell, W.C., Comparison of specific blue and green fluorescence in cataractous versus normal human lens fractions, *Invest. Ophthalmol. Vis. Sci.*, 34(3), 630, 1993.

82. Andersson-Engels, S., Johansson, J., Stenram, U., Svanberg, K., and Svanberg, S., Malignant-tumor and atherosclerotic plaque diagnosis using laser-induced fluorescence, *IEEE J. Quantum Electron.*, 26(12), 2207, 1990.
83. Hegedus, Z.L. and Nayak, U., Relative fluorescence intensities of human plasma soluble melanins in normal adults, *Arch. Int. Physiol. Biochim. Biophys.*, 102(6), 311, 1994.
84. Andersson-Engels, S., Ankerst, J., Johansson, J., Svanberg, K., and Svanberg, S., Laser-induced fluorescence in malignant and normal tissue of rats injected with benzoporphyrin derivative, *Photochem. Photobiol.*, 57(6), 978, 1993.
85. Devoisselle, J.M., Maunoury, V., Mordon, S., and Coustaut, D., Measurement of *in vivo* tumorous normal tissue pH by localized spectroscopy using a fluorescent marker, *Opt. Eng.*, 32(2), 239, 1993.
86. Christov, A., Dai, E., Drangova, M., Liu, L.Y., Abela, G.S., Nash, P., McFadden, G., and Lucas, A., Optical detection of triggered atherosclerotic plaque disruption by fluorescence emission analysis, *Photochem. Photobiol.*, 722, 242, 2000.
87. Chung, Y.G., Schwartz, J.A., Gardner, C.M., Sawaya, R.E., and Jacques, S.L., Diagnostic potential of laser-induced autofluorescence emission in brain tissue, *J. Korean Med. Sci.*, 12(2), 135, 1997.
88. Schantz, S.P. and Alfano, R.R., Tissue autofluorescence as an intermediate end-point in cancer chemoprevention trials, *J. Cell. Biochem.*, 48, 199, 1993.
89. Glasgold, R., Glasgold, M., Savage, H., Pinto, J., Alfano, R., and Schantz, S., Tissue autofluorescence as an intermediate end-point in NMBA-induced esophageal carcinogenesis, *Cancer Lett.*, 82(1), 33, 1994.
90. Oraevsky, A.A., Jacques, S.L., Pettit, G.H., Sauerbrey, R.A., Tittel, F.K., Nguy, J.H., and Henry, P.D., XeCl laser-induced fluorescence of atherosclerotic arteries. spectral similarities between lipid-rich lesions and peroxidized lipoproteins, *Circ. Res.*, 72(1), 84, 1993.
91. Morgan, D.C., Wilson, J.E., MacAulay, C.E., MacKinnon, N.B., Kenyon, J.A., Gerla, P.S., Dong, C.M., Zeng, H.S., Whitehead, P.D., Thompson, C.R., and McManus, B.M., New method for detection of heart allograft rejection — validation of sensitivity and reliability in a rat heterotopic allograft model, *Circulation*, 100(11), 1236, 1999.
92. Shehada, R.E.N., Marmarelis, V.Z., Mansour, H.N., and Grundfest, W.S., Laser induced fluorescence attenuation spectroscopy: detection of hypoxia, *IEEE Trans. Biomed. Eng.*, 47(3), 301, 2000.
93. Svanberg, K., Kjellen, E., Ankerst, J., Montan, S., Sjoholm, E., and Svanberg, S., Fluorescence studies of hematoporphyrin derivative in normal and malignant rat-tissue, *Cancer Res.*, 46(8), 3803, 1986.
94. Sterenborg, H., Thomsen, S., Jacques, S.L., and Motamedi, M., *In-vivo* autofluorescence of an unpigmented melanoma in mice — correlation of spectroscopic properties to microscopic structure, *Melanoma Res.*, 5(4), 211, 1995.
95. Colasanti, A., Kisslinger, A., Fabbrocini, G., Liuzzi, R., Quarto, M., Riccio, P., Roberti, G., and Villani, F., MS-2 fibrosarcoma characterization by laser induced autofluorescence, *Lasers Surg. Med.*, 26(5), 441, 2000.
96. Dhingra, J.K., Zhang, X., McMillan, K., Kabani, S., Manoharan, R., Itzkan, I., Feld, M.S., and Shapshay, S.M., Diagnosis of head and neck precancerous lesions in an animal model using fluorescence spectroscopy, *Laryngoscope*, 108(4), 471, 1998.
97. Coghlan, L., Utzinger, U., Drezek, R., Heintzelman, D., Zuluaga, A., Brookner, C., Richards-Kortum, R., Gimenez-Conti, I., and Follen, M., Optimal fluorescence excitation wavelengths for detection of squamous intra-epithelial neoplasia: results from an animal model, *Opt. Express*, 7(12), 436, 2000.
98. Vengadesan, N., Aruna, P., and Ganesan, S., Characterization of native fluorescence from dmba-treated hamster cheek pouch buccal mucosa for measuring tissue transformation, *Br. J. Cancer*, 77(3), 391, 1998.
99. Tassetti, V., Hajri, A., Sowinska, M., Evrard S., Heisel, F., Cheng, L.Q., Miehe, J.A., Marescaux, J., and Aprahamian M., *In vivo* laser-induced fluorescence imaging of a rat pancreatic cancer with pheophorbide-a, *Photochem. Photobiol.*, 65, 997, 1997.

100. Svanberg, K., Liu, D.L., Wang, I., Andersson-Engels, S., Stenram, U., and Svanberg, S., Photodynamic therapy using intravenous delta-aminolaevulinic acid-induced protoporphyrin IX sensitisation in experimental hepatic tumours in rats, *Br. J. Cancer,* 74(10), 1526, 1996.
101. Johansson, J., Berg, R., Svanberg, K., and Svanberg, S., Laser-induced fluorescence studies of normal and malignant tumour tissue of rat following intravenous injection of delta-amino levulinic acid, *Lasers Surg. Med.,* 20(3), 272, 1997.
102. Nilsson, H., Johansson, J., Svanberg, K., Svanberg, S., Jori, G., Reddi, E., Segalla, A., Gust, D., and Moore, T.A., Laser induced fluorescence studies of the biodistribution of carotenoporphyrins in mice, *Br. J. Cancer,* 76(3), 355, 1997.
103. Glanzmann, T., Ballini, J.P., van den Bergh, H., and Wagnières, G., Time-resolved spectrofluorometer for clinical tissue characterization during endoscopy, *Rev. Sci. Instrum.,* 70(10), 4067, 1999.
104. Anidjar, M., Ettori, D., Cussenot, O., Meria, P., Desgrandchamps, F., Cortesse, A., Teillac, P., LeDuc, A., and Avrillier, S., Laser induced autofluorescence diagnosis of bladder tumors: dependence on the excitation wavelength, *J. Urol.,* 156(5), 1590, 1996.
105. Avrillier, S., Tinet, E., Ettori, D., and Anidjar, M., Laser-induced autofluorescence diagnosis of tumors, *Phys. Scripta,* T72, 87, 1997.
106. Koenig, F., McGovern, F.J., Enquist, H., Larne, R., Deutsch, T.F., and Schomacker, K.T., Autofluorescence guided biopsy for the early diagnosis of bladder carcinoma, *J. Urol.,* 159(6), 1871, 1998.
107. Lin, W.C., Toms, S.A., Johnson, M., Jansen, E.D., and Mahadevan-Jansen, A., *In vivo* brain tumor demarcation using optical spectroscopy, *Photochem. Photobiol.,* 73(4), 396, 2001.
108. Lam, S., Hung, J.Y.C., Kennedy, S.M., Leriche, J.C., Vedal, S., Nelems, B., Macaulay, C.E., and Palcic, B., Detection of dysplasia and carcinoma *in situ* by ratio fluorometry, *Am. Rev. Resp. Dis.,* 146(6), 1458, 1992.
109. Zellweger, M., Goujon, D., Conde, R., Forrer, M., van den Bergh, H., and Wagnières, G., Absolute autofluorescence spectra of human healthy, metaplastic, and early cancerous bronchial tissue *in vivo, Appl. Opt.,* 40(22), 3784, 2001.
110. Zellweger, M., Grosjean, P., Goujon, D., Monnier, P., van den Bergh, H., and Wagnières, G., *In vivo* autofluorescence spectroscopy of human bronchial tissue to optimize the detection and imaging of early cancers, *J. Biomed. Opt.,* 6(1), 41, 2001.
111. Ramanujam, N., Mitchell, M.F., Mahadevan, A., Thomsen, S., Silva, E., and Richards-Kortum, R., Fluorescence spectroscopy: a diagnostic tool for cervical intra epithelial neoplasia (CIN), *Gynecol. Oncol.,* 52(1), 31, 1994.
112. Ramanujam, N., Mitchell, M.F., Mahadevan, A., Warren, S., Thomsen, S., Silva, E., and Richards-Kortum, R., *In vivo* diagnosis of cervical intra epithelial neoplasia using 337-nm-excited laser-induced fluorescence, *Proc. Natl. Acad. Sci. U.S.A.,* 91(21), 10193, 1994.
113. Schomacker, K.T., Frisoli, J.K., Compton, C.C., Flotte, T.J., Richter, J.M., Nishioka, N.S., and Deutsch, T.F., ultraviolet laser-induced fluorescence of colonic tissue: basic biology and diagnostic potential, *Lasers Surg. Med.,* 12(1), 63, 1992.
114. Konig, K., Ruck, A., and Schneckenburger, H., Fluorescence detection and photodynamic activity of endogenous protoporphyrin in human skin, *Opt. Eng.,* 31(7), 1470, 1992.
115. Vo-Dinh, T., Panjehpour, M., and Overholt, B.F., Laser-induced fluorescence for esophageal cancer and dysplasia diagnosis, *Proc. N.Y. Acad. Sci.,* 838, 116, 1998.
116. Vo-Dinh, T., Panjehpour, M., Overholt, B.F., and Buckley, P., Laser-induced differential fluorescence for cancer diagnosis without biopsy, *Appl. Spectrosc.,* 51(1), 58, 1997.
117. Mayinger, B., Neidhardt, S., Reh, H., Martus, P., and Hahn, E.G., Fluorescence induced with 5-aminolevulinic acid for the endoscopic detection and follow-up of esophageal lesions, *Gastrointest. Endosc.,* 54(5), 572, 2001.
118. Mayinger, B., Horner, P., Jordan, M., Gerlach, C., Horbach, T., Hohenberger, W., and Hahn, E.G., Endoscopic fluorescence, spectroscopy in the upper GI tract for the detection of GI cancer: initial experience, *Am. J. Gastroenterol.,* 96(9), 2616, 2001.

119. Masters, B.R., So, P.T., and Gratton, E., Multiphoton excitation fluorescence microscopy and spectroscopy of *in vivo* human skin, *Biophys. J.*, 72(6), 2405, 1997.
120. van Schaik, H.J., Alkemade, C., Swart, W., and Van Best, J.A., Autofluorescence of the diabetic and healthy human cornea *in vivo* at different excitation wavelengths, *Exper. Eye Res.*, 68(1), 1, 1999.
121. van Schaik, H.J., Coppens, J., van den Berg, T., and van Best, J.A., Autofluorescence distribution along the corneal axis in diabetic and healthy humans, *Exp. Eye Res.*, 69(5), 505, 1999.
122. Siik, S., Chylack, L.T., Friend, J., Wolfe, J., Teikari, J., Nieminen, H., and Airaksinen, P.J., Lens autofluorescence and light scatter in relation to the lens opacities classification system, LOCS III, *Acta Ophthalmol. Scand.*, 77(5), 509, 1999.
123. Zargi, M., Smid, L., Fajdiga, I., Bubnic, B., Lenarcic, J., and Oblak, P., Detection and localization of early laryngeal cancer with laser-induced fluorescence: preliminary report, *Eur. Arch. Otorhinolaryngol.*, Suppl. 1, S113, 1997.
124. Mehlmann, N., Betz, C.S., Stepp, H., Arbogast, S., Baumgartner, R., Grevers, G., and Leunig, A., Fluorescence staining of laryngeal neoplasms after topical application of 5-aminolevulinic acid: preliminary results, *Lasers Surg. Med.*, 25(5), 414, 1999.
125. Arens, C., Malzahn, K., Dias, O., Andrea, M., and Glanz, H., endoscopic imaging techniques in the diagnosis of laryngeal cancer and its precursor lesions, *Laryngo-Rhino-Otologie* 78(12), 685, 1999.
126. De Leon, H., Ollerenshaw, J.D., Griendling, K.K., and Wilcox, J.N., Adventitial cells do not contribute to neointimal mass after balloon angioplasty of the rat common carotid artery, *Circulation*, 104(14), 1591, 2001.
127. Schomacker, K.T., Frisoli, J.K., Compton, C.C., Flotte, T.J., Richter, J.M., Nishioka, N.S., and Deutsch, T.F., Ultraviolet laser-induced fluorescence of colonic tissue: basic biology and diagnostic potential, *Lasers Surg. Med.*, 12(1), 63, 1992.
128. Cothren, R.M., Sivak, M.V., VanDam, J., Petras, R.E., Fitzmaurice, M., Crawford, J.M., Wu, J., Brennan, J.F., Rava, R.P., Manoharan, R., and Feld, M.S., Detection of dysplasia at colonoscopy using laser-induced fluorescence: a blinded study, *Gastrointest. Endosc.*, 44(2), 168, 1996.
129. Fryen, A., Glanz, H., Lohmann, W., Dreyer, T., and Bohle, R.M., Significance of autofluorescence for the optical demarcation of field cancerisation in the upper aerodigestive tract, *Acta Otolaryngol. (Stockholm)*, 117(2), 316, 1997.
130. Dhingra, J.K., Perrault, D.F., McMillan, K., Rebeiz, E.E., Kabani, S., Manoharan, R., Itzkan, I., Feld, M.S., and Shapshay, S.M., Early diagnosis of upper aerodigestive tract cancer by autofluorescence, *Arch. Otolaryngol. Head Neck Surg.*, 122(11), 1181, 1996.
131. Gillenwater, A., Jacob, R., Ganeshappa, R., Kemp, B., El-Naggar, A.K., Palmer, J.L., Clayman, G., Mitchell, M.F., and Richards-Kortum, R., Noninvasive diagnosis of oral neoplasia based on fluorescence spectroscopy and native tissue autofluorescence, *Arch. Otolaryngol. Head Neck Surg.*, 124(11), 1251, 1998.
132. Kulapaditharom, B. and Boonkitticharoen, V., Laser-induced fluorescence imaging in localization of head and neck cancers, *Ann. Otol. Rhinol. Laryngol.*, 107(3), 241, 1998.
133. Onizawa, K., Saginoya, H., Furuya, Y., and Yoshida, H., Fluorescence photography as a diagnostic method for oral cancer, *Cancer Lett.*, 108, 61, 1996.
134. Heintzelman, D.L., Utzinger, U., Fuchs, H., Zuluaga, A., Gossage, K., Gillenwater, A.M., Jacob, R., Kemp, B., and Richards-Kortum, R.R., Optimal excitation wavelengths for *in vivo* detection of oral neoplasia using fluorescence spectroscopy, *Photochem. Photobiol.*, 72(1), 103, 2000.
135. Kolli, V.R., Shaha, A.R., Savage, H.E., Sacks, P.G., Casale, M.A., and Schantz, S.P., Native cellular fluorescence can identify changes in epithelial thickness *in vivo* in the upper aerodigestive tract, *Am. J. Surg.*, 170(5), 495, 1995.
136. Chissov, V.I., Sokolov, V.V., Filonenko, E.V., Menenkov, V.D., Zharkova, N.N., Kozlov, D.N., Polivanov, I.N., Prokhorov, A.M., Pyhov, R.L., and Smirnov, V.V., Clinical fluorescent diagnosis of tumors using photosensitizer photogem, *Khirurgiia (Moskva)*, 5, 37, 1995.

137. Sterenborg, H., Saarnak, A.E., Frank, R., and Motamedi, M., Evaluation of spectral correction techniques for fluorescence measurements on pigmented lesions *in vivo*, *J. Photochem. Photobiol. B*, 35(3), 159, 1996.
138. Heyerdahl, H., Wang, I., Liu, D.L., Berg, R., Andersson-Engels, S., Peng, Q., Moan, J., Svanberg, S., and Svanberg, K., Pharmacokinetic studies on 5-aminolevulinic acid-induced protoporphyrin IX accumulation in tumours and normal tissues, *Cancer Lett.*, 112(2), 225, 1997.
139. Konig, K., Schneckenburger, H., Ruck, A., and Steiner, R., *In vivo* photoproduct formation during PDT with ALA-induced endogenous phorphyrins, *J. Photochem. Photobiol.*, 18(2–3), 287, 1993.
140. Leunig, A., Rick, K., Stepp, H., Goetz, A., Baumgartner, R., and Feyh, J., Fluorescence photodetection of neoplastic lesions in the oral cavity following topical application of 5-aminolevulinic acid, *Laryngo-Rhino-Otologie*, 75(8), 459, 1996.
141. Leffell, D.J., Stetz, M.L., Milstone, L.M., and Deckelbaum, L.I., *In vivo* fluorescence of human skin, a potential marker of photoaging, *Arch. Dermatol.*, 124(10), 1514, 1988.
142. Sterenborg, H., Motamedi, M., Wagner, R.F., Duvic, M., Thomsen, S., and Jacques, S.L., *In-vivo* fluorescence spectroscopy and imaging of human skin tumors, *Lasers Med. Sci.*, 9(3), 191, 1994.
143. Zeng, H.S., Macaulay, C., Palcic, B., and McLean, D.I., A computerized autofluorescence and diffuse reflectance spectroanalyzer system for *in vivo* skin studies, *Phys. Med. Biol.*, 38(2), 231, 1993.
144. Zuluaga, A.F., Utzinger, U., Durkin, A., Fuchs, H., Gillenwater, A., Jacob, R., Kemp, B., Fan, J., and Richards-Kortum, R., Fluorescence excitation emission matrices of human tissue: a system for *in vivo* measurement and method of data analysis, *Appl. Spectrosc.*, 53(3), 302, 1999.


### 65.3.5 Elemental Analysis

Table 65.10 provides a quick reference of publications pertaining to the elemental analysis using a variety of techniques for biomedical applications. The techniques used to perform the analyses are abbreviated as follows: AAS = atomic absorption spectroscopy; AES = atomic emission spectroscopy; XRF = x-ray fluorescence; ICP = inductively coupled plasma; LIPS = laser-induced plasma spectroscopy; PIXE = particle-induced x-ray emission; MP = microwave-induced plasma-atomic emission spectroscopy; ICP-AES = inductively coupled plasma-atomic emission spectroscopy; MP-AES = microwave plasma atomic emission spectroscopy. QS = Quackenbush Special (mice); $p$ = probability. Future research will undoubtedly provide a more comprehensive compilation, but this version clearly demonstrates the importance of elemental analyses in various biomedical applications. The information in Table 65.10 was compiled using data from almost 120 publications.

**TABLE 65.10** Elemental Analysis Data

| Species | Target Organ or Disease | Technique | Matrix | Comments | Ref. |
|---|---|---|---|---|---|
| | | | ***Biochemical Studies*** | | |
| K | Rat myocardium and skeletal muscle | AAS | Deonized water | | 1 |
| Mg | Rat myocardium and skeletal muscle | AAS | Deonized water | | 1 |
| | | | ***Cellular Studies*** | | |
| B | | AES | Cell suspension | Procedure is applicable to the analysis of boron in the ppm range with a high degree of precision and accuracy | 2 |
| Ca | Rat hepatocytes | AAS | | Effects of hypothermia on cytosolic free calcium concentration ([Ca2+](I)) and total cellular calcium content | 3 |
| Zn | | AAS | | | 4 |
| | | ***Tissue and Biofluid Studies (in Vitro)*** | | | |
| Ab | | AAS | Blood, liver tissue | | 5 |
| Ag | | AAS | Tissue | | 6 |
| Ag | Liver, kidney cortex, 5 brain regions: gray matter of cerebrum, white matter of cerebrum, nucleus lentiformis, cerebellum, brain stem | AAS | Tissue | | 7 |
| Al | Brain | XRF | Brain | | 8 |
| Al | Human organs/tissue (lung, kidney, liver, hair, blood) | ICP-AES, AAS | Human organs/ tissue (lung, kidney, liver, hair, blood) | Reference values human tissues/ organs | 9 |
| Al | | AAS | Fetal serum, amniotic fluid, and organs | Transplacental passage | 10 |
| Al | | ICP-AES | Blood | | 11 |
| Al | | ICP-AES | Human organs | Reference values study | 12 |
| Al | Alzheimer's disease | AAS | Tissue | | 13 |
| Al | Liver | AAS | Tissue | | 14 |
| Al | Liver | AAS | Tissue | Investigation of possible absorption and deposition of bismuth or aluminum from agents used in the treatment of peptic ulcers | 15 |
| Al | Spinal cord, brain stem, cerebellum, forebrain | AAS | | Tissue distribution of Al did not follow that of essential cations as examined in this study | 16 |
| Al | Alzheimer's disease | AAS | Bone | | 17 |
| Al | | LIPS, AAS | Teeth, bone | | 18 |
| Al | Brain | XRF | | | 8 |
| As | Urine | AAS | Urine | | 19 |
| As | Rat kidney | AES | Rat kidney | Results indicate that arsenic accumulates in the kidney cortex synchroneously over time; arsenic also accumulated in liver and red blood cells | 20 |

**TABLE 65.10** Elemental Analysis Data (continued)

| Species | Target Organ or Disease | Technique | Matrix | Comments | Ref. |
|---|---|---|---|---|---|
| As | | AAS | Tissue | | 6 |
| As | | ICP-AES, AAS | | Various biomedical applications | 21 |
| As | Urine | AAS | | | 19 |
| As | Breast milk | AAS | | Environmental toxic exposure of very young children to As by determining As in breast milk; As found not to be excreted into breast milk to significant extent | 22 |
| B | Human organs/tissue (lung, kidney, liver, hair, blood) | ICP-AES, AAS | Human organs/ tissue (lung, kidney, liver, hair, blood) | Reference values human tissue/ organs | 9 |
| B | Blood urine | AE, ICP-AES | | Pharmacokinetics of a compound used for boron neutron capture therapy were studied | 23 |
| B | Cancer (skin, liver) | AES | Tumor, tissue, liver, skin | Procedure is applicable to the analysis of boron in the ppm range with a high degree of precision and accuracy | 2 |
| B | | ICP-AES | Blood | Boron neutron capture therapy (BNCT) | 24 |
| B | | AES | Blood | Boron neutron capture therapy (BNCT) | 25 |
| B | Melanoma | ICP-AES | | | 26 |
| Ba | Human organs/tissue (lung, kidney, liver, hair, blood) | ICP-AES, AAS | Human organs/ tissue (lung, kidney, liver, hair, blood) | Reference values human tissues/ organs | 9 |
| Ba | | ICP-AES | Human organs | Reference values study | 12 |
| Ba | | ICP-AES, AAS | | Various biomedical applications | 21 |
| Ba | Liver | ICP-AES | Tissue | | 27 |
| Be | Human organs/tissue (lung, kidney, liver, hair, blood) | ICP-AES, AAS | Human organs/ tissue (lung, kidney, liver, hair, blood) | Reference values human tissues/ organs | 9 |
| Be | Liver, kidney | AAS | Liver, kidney | Investigation of effects of two chelating agents on toxicity and distribution of Be | 28 |
| Bi | Human organs/tissue (lung, kidney, liver, hair, blood) | ICP-AES, AAS | Human organs/ tissue (lung, kidney, liver, hair, blood) | Reference values human tissues/ organs | 9 |
| Bi | | AAS | Tissue | | 6 |
| Bi | Liver | AAS | Tissue | Investigation of possible absorption and deposition of bismuth or aluminum from agents used in the treatment of peptic ulcers | 15 |
| Ca | Atherosclerotic and normal tissue | LIPS | | Laser angioplasty tissue characterization | 29 |
| Ca | Rat eye tissue | XRF | Rat eye tissue | Hereditary retinal degeneration | 30 |
| Ca | Heart disease | AAS | Aortic valve tissue | Surface structure of decalcified aortic valve tissue | 31 |
| Ca | Heart disease | AAS | Aortic valve tissue | | 32 |
| Ca | | AAS | Stomach, kidneys, bone, liver | | 33 |

**TABLE 65.10** Elemental Analysis Data (continued)

| Species | Target Organ or Disease | Technique | Matrix | Comments | Ref. |
|---|---|---|---|---|---|
| Ca | | ICP-AES | Hair | | 34 |
| Ca | Osteoporosis | AAS | Bone | Steroid osteoporosis treatment | 35 |
| Ca | Wilson's disease | ICP-AES | Brain tissue | | 36 |
| Ca | Liver | ICP-AES | Tissue | | 27 |
| Ca | Spinal cord, brain stem, cerebellum, forebrain | ICP-AES | | | 16 |
| Ca | Meniscal degeneration | ICP-AES | Mensisci | Relationship between meniscal degeneration and element contents | 37 |
| Ca | Heart disease | AAS | Arterial wall | | 38 |
| Ca | | LIPS | Cornea | Plasma emission spectra exhibited significant dependence on sample hydration; this dependence can be used for estimation of water content of irradiated model material and real cornea | 39 |
| Ca | | AES | Cerebrospinal fluid | | 40 |
| Ca | Osteoporosis | LIPS | Hair | | 41 |
| Cd | Human organs/tissue (lung, kidney, liver, hair, blood) | ICP-AES, AAS | Human organs/ tissue (lung, kidney, liver, hair, blood) | Reference values human tissues/ organs | 9 |
| Cd | Aorta | AAS | Aorta | Cadmium accumulation in aortas of smokers | 42 |
| Cd | Prostatic cancer | AAS | Tissue | | 43 |
| Cd | | MP-AES | Blood | | 44 |
| Cd | | | Urine, tissue | | 45 |
| Cd | | ICP-AES | Human organs | Reference values study | 12 |
| Cd | | AAS | Tissue | | 6 |
| Cd | | ICP-AES, AAS | | Various biomedical applications | 21 |
| Cd | Lung | AAS | Tissue | | 46 |
| Cd | Liver | ICP-AES | Tissue | | 27 |
| Co | Human organs/tissue (lung, kidney, liver, hair, blood) | ICP-AES, AAS | Human organs/ tissue (lung, kidney, liver, hair, blood) | Reference values human tissues/ organs | 9 |
| Co | | ICP-AES, AAS | | Various biomedical applications | 21 |
| Co | Lung | AAS | Tissue | | 46 |
| Co | | AAS | Serum | Purpose of this study was to measure the serum cobalt levels and their correlation with clinical and radiological findings in patients with metal-on-metal hip articulating surfaces | 47 |
| Cr | Human organs/tissue (lung, kidney, liver, hair, blood) | ICP-AES, AAS | Human organs/ tissue (lung, kidney, liver, hair, blood) | Reference values human tissues/ organs | 9 |
| Cr | Rat liver, rat kidney | AAS | Tissue | | 48 |
| Cr | Rat tissue, blood | AAS | Rat tissue, blood | | 49 |
| Cr | | ICP-AES | Human organs | Reference values study | 12 |
| Cr | Lung | AAS | Tissue | | 46 |
| Cr | Liver | ICP-AES | Tissue | | 26 |

**TABLE 65.10** Elemental Analysis Data (continued)

| Species | Target Organ or Disease | Technique | Matrix | Comments | Ref. |
|---|---|---|---|---|---|
| Cr | | AAS | Tissue | Objective of this study was to determine whether low plasma chromium concentrations (less than or equal to 3 nmol/l) are associated with altered glucose, insulin, or lipid concentrations during pregnancy | 50 |
| Cu | Human organs/tissue (lung, kidney, liver, hair, blood) | ICP-AES, AAS | Human organs/ tissue (lung, kidney, liver, hair, blood) | Reference values human tissues/ organs | 12 |
| Cu | Rat liver | AAS | Rat liver | Effect of chronic exposure to excess dietary metal supplementation on liver specimens from rats | 51 |
| Cu | Rat kidney | AES | Rat kidney | Results indicate that arsenic and copper accumulate in the kidney cortex synchroneously over time | 52 |
| Cu | Colitis | AAS | Serum | The serum concentrations of copper remained unaltered during colitis | 53 |
| Cu | Colitis/rat colon | AAS | Tissue | | 53 |
| Cu | Chronic alcohol abuse | AAS | Embryo, liver | Maternal hepatic, endometrial, and embryonic levels following alcohol consumption during pregnancy in QS mice | 54 |
| Cu | Rat tissue, blood | AAS | Rat tissue, blood | | 48 |
| Cu | Liver disease | AAS | Tissue, plasma | | 55 |
| Cu | Blood plasma | ICP-AES | Plasma | | 56 |
| Cu | Extrahepatic biliary atresia | AAS | Tissue | | 57 |
| Cu | | ICP-AES | Blood | Reference values for some bulk and trace elements in blood plasma and whole blood (Pb) of mothers and their newborn infants | 58 |
| Cu | | ICP-AES | Liver | | 59 |
| Cu | | ICP-AES | Blood | | 60 |
| Cu | | ICP-AES | Human organs | Reference values study | 12 |
| Cu | Wilson's disease | AAS | Kidney, liver | | 61 |
| Cu | | AAS | Stomach, kidneys, bone, liver | | 33 |
| Cu | | ICP-AES | Hair | | 34 |
| Cu | | AES | Human breast milk | | 62 |
| Cu | Rat small intestine | AAS | Tissue | | 63 |
| Cu | Wilson's disease | ICP-AES | Brain tissue | | 36 |
| Cu | Wilson's disease | ICP-AES | Brain tissue | | 64 |
| Cu | Lung | AAS | Tissue | | 46 |
| Cu | Liver | ICP-AES | Tissue | | 27 |
| Cu | Cataracts | AAS | Tissue | These data support the hypothesis that transition metal–mediated HO production may play a role in the etiology of age-related nuclear cataract | 65 |
| Cu | Liver disease | AAS | Liver, serum | | 66 |

**TABLE 65.10** Elemental Analysis Data (continued)

| Species | Target Organ or Disease | Technique | Matrix | Comments | Ref. |
|---|---|---|---|---|---|
| Cu | Human kidney tumor | XRF, ICP-AES | | | 67 |
| Cu | Breast cancer, stomach cancer, colon cancer | AAS | Serum, tissue | Results suggest that movements of copper out from, and of zinc into, tissues occur, and it is proposed that these changes could be a response to enhanced cytokine production rather than a feature of any mechanism(s) for the onset of cancer | 68 |
| Cu | Heart disease | AAS | Arterial wall | | 38 |
| Cu | Aortoiliacal occlusive disease | AAS | Tissue | | 69 |
| Cu | | AES | Cerebrospinal fluid | | 40 |
| Cu | Hypoglycemia | AAS | Cerebrospinal fluid | Newborns | 70 |
| Fe | Breast milk | AAS | Breast milk | | 71 |
| Fe | Rat eye tissue | XRF | Rat eye tissue | Hereditary retinal degeneration | 31 |
| Fe | Beta-thalassemia | | | | 72 |
| Fe | Chronic alcohol abuse | AAS | Embryo, liver | Maternal hepatic, endometrial, and embryonic levels following alcohol consumption during pregnancy in QS mice | 54 |
| Fe | Rat tissue, blood | AAS | Rat tissue, blood | | 49 |
| Fe | | ICP-AES | Blood | Reference values for some bulk and trace elements in blood plasma and whole blood (Pb) of mothers and their newborn infants | 58 |
| Fe | | AAS | Adipose tissue | | 73 |
| Fe | | ICP-AES | Hair | | 34 |
| Fe | | AES | Human breast milk | | 62 |
| Fe | Rat small intestine | AAS | Tissue | | 63 |
| Fe | Wilson's disease | ICP-AES | Brain tissue | | 36 |
| Fe | Liver | ICP-AES | Tissue | | 27 |
| Fe | Cataracts | AAS | Tissue | These data support the hypothesis that transition metal–mediated HO production may play a role in the etiology of age-related nuclear cataract | 65 |
| Fe | Spinal cord, brain stem, cerebellum, forebrain | ICP-AES | | | 16 |
| Fe | Liver disease | AAS | Liver, serum | | 66 |
| Fe | | AES | Cerebrospinal fluid | | 40 |
| Ga | Lungs, brain, testes, and ovaries | ICP-AES | Lungs, brain, testes, ovaries | Anticancer therapy drug study, organo-metallic gallium (Ga) complex | 74 |
| Hg | Tissues | AAS | Tissues | Improved preparation of small biological samples for mercury analysis | 75 |
| Hg | Methylmercury exposure | AAS | Toenail, hair, blood | Toenails, an easily accessible tissue for estimation of methylmercury exposure, have been shown to be closely correlated with the well-established samples for biomarkers, viz. blood and hair mercury | 76 |

**TABLE 65.10** Elemental Analysis Data (continued)

| Species | Target Organ or Disease | Technique | Matrix | Comments | Ref. |
|---|---|---|---|---|---|
| Hg | Connective tissue diseases | | Urine | | 77 |
| Hg | | ICP-AES, AAS | Hair | | 78 |
| Hg | | AAS | Tissue, blood | | 79 |
| Hg | | ICP-AES, AAS | | Various biomedical applications | 21 |
| I | | ICP-AES | Urine, plasma | | 80 |
| K | Rat eye tissue | XRF | Rat eye tissue | Hereditary retinal degeneration | 30 |
| K | | AES | Blood | | 81 |
| K | Liver | ICP-AES | Tissue | | 27 |
| K | | AES | Cerebrospinal fluid | | 40 |
| Li | Human organs/tissue (lung, kidney, liver, hair, blood) | ICP-AES, AAS | Human organs/ tissue (lung, kidney, liver, hair, blood) | Reference values human tissues/ organs | 9 |
| Li | | AES | Blood | | 81 |
| Li | | ICP-AES | Human organs | Reference values study | 12 |
| Li | | AES | Serum | | 82 |
| Li | | AAS | Blood | | 83 |
| Mg | Human organs/tissue (lung, kidney, liver, hair, blood) | ICP-AES, AAS | Human organs/ tissue (lung, kidney, liver, hair, blood) | Reference values human tissues/ organs | 9 |
| Mg | Chronic alcohol abuse | AAS | Embryo, liver | Maternal hepatic, endometrial, and embryonic levels following alcohol consumption during pregnancy in QS mice | 54 |
| Mg | | AES | Blood | | 81 |
| Mg | | ICP-AES | Human organs | Reference values study | 12 |
| Mg | | ICP-AES, AAS | | Various biomedical applications | 21 |
| Mg | | AAS | Stomach, kidneys, bone, liver | | 33 |
| Mg | | ICP-AES | Hair | | 34 |
| Mg | Wilson's disease | ICP-AES | Brain tissue | | 36 |
| Mg | Liver | ICP-AES | Tissue | | 27 |
| Mg | Spinal cord, brain stem, cerebellum, forebrain | ICP-AES | | | 16 |
| Mg | Meniscal degeneration | ICP-AES | Mensisci | Relationship between meniscal degeneration and element contents | 37 |
| Mg | Breast cancer, stomach cancer, colon cancer | AAS | Serum, tissue | Results suggest that movements of copper out from, and of zinc into, tissues occur, and it is proposed that these changes could be a response to enhanced cytokine production rather than a feature of any mechanism(s) for the onset of cancer | 68 |
| Mg | Heart disease | AAS | Arterial wall | | 38 |
| Mg | | AES | Cerebrospinal fluid | | 40 |
| Mg | Hypoglycemia | AAS | Cerebrospinal fluid | Newborns | 70 |
| Mn | Human organs/tissue (lung, kidney, liver, hair, blood) | ICP-AES, AAS | Human organs/ tissue (lung, kidney, liver, hair, blood) | Reference values human tissues/ organs | 9 |

**TABLE 65.10** Elemental Analysis Data (continued)

| Species | Target Organ or Disease | Technique | Matrix | Comments | Ref. |
|---|---|---|---|---|---|
| Mn | Extrahepatic biliary atresia | AAS | Tissue | | 57 |
| Mn | | AES | Blood | | 81 |
| Mn | | ICP-AES | Human organs | Reference values study | 12 |
| Mn | | ICP-AES | Hair | | 34 |
| Mn | Liver | ICP-AES | Tissue | | 27 |
| Mn | Spinal cord, brain stem, cerebellum, forebrain | AAS | | | 16 |
| Mo | | AAS | Tissue | | 84 |
| Multi-element | Beta-thalassemia-major | PIXE | | | 85 |
| Na | | AES | Blood | | 81 |
| Na | Liver | ICP-AES | Tissue | | 27 |
| Na | | LIPS | Skin | Na line = 589 nm, ablation wavelengths = 1064, 532, 266, 213 nm | 86 |
| Na | | AES | Cerebrospinal fluid | | 40 |
| Ni | | AAS | Blood | | 87 |
| Ni | | ICP-AES | Human organs | Reference values study | 12 |
| Ni | | ICP-AES, AAS | | Various biomedical applications | 21 |
| Ni | | AAS | Blood, liver tissue | | 5 |
| Ni | Lung | AAS | Tissue | | 46 |
| Ni | Liver | ICP-AES | Tissue | | 27 |
| P | Wilson's disease | ICP-AES | Brain tissue | | 36 |
| P | Liver | ICP-AES | Tissue | | 27 |
| P | Meniscal degeneration | ICP-AES | Mensisci | Relationship between meniscal degeneration and element contents | 37 |
| Pb | Human organs/tissue (lung, kidney, liver, hair, blood) | ICP-AES, AAS | Human organs/ tissue (lung, kidney, liver, hair, blood) | Reference values human tissues/ organs | 9 |
| Pb | | ICP-AES | Blood | | 88 |
| Pb | Rat tissue, blood | AAS | Rat tissue, blood | Results show that at 5 ppm Pb exposure, brain and kidney accumulate Pb significantly, confirming that at these exposure levels, blood Pb is not a good index of tissue burden | 89 |
| Pb | | MP-AES | Blood | | 81 |
| Pb | Osteoporosis | AAS | Bone | | 90 |
| Pb | Liver disease | | Blood, tissue | Increased levels of lead were found in the blood of patients who consumed alcohol and those with alcoholic liver disease | 91 |
| Pb | | ICP-AES | Human organs | Reference values study | 12 |
| Pb | | ICP-AES, AAS | | Various biomedical applications | 21 |
| Pb | | PIXE | Bone | | 92 |
| Pb | | ICP-AES | Hair | | 34 |
| Pb | Liver, lung, kidney, brain | AAS | Tissue | Concentrations of lead in liver, lung, kidney, brain, hair, and nails were determined in 32 deceased, long-term exposed male lead smelter workers, and compared with those of 10 male controls | 93 |
| Pb | | XRF | Hair, nails | | 93 |

**TABLE 65.10** Elemental Analysis Data (continued)

| Species | Target Organ or Disease | Technique | Matrix | Comments | Ref. |
|---|---|---|---|---|---|
| Pb | Presenile dementia | AAS | Brain tissue | | 94 |
| Pb | Teeth | AAS | Hard tissue | | 95 |
| Pb | | LIPS, AAS | Teeth, bone | | 18 |
| Pt | | ICP-AES | Blood | | 96 |
| Rb | | AAS | Urine | | 97 |
| S | Wilson's disease | ICP-AES | Brain tissue | | 36 |
| S | Liver | ICP-AES | Tissue | | 27 |
| S | Meniscal degeneration | ICP-AES | Mensisci | Relationship between meniscal degeneration and element contents | 37 |
| Sb | | AAS | Tissue | | 6 |
| Se | Rat lymphoid tissue | AAS | Rat lymphoid tissue | Effect of aging on levels of selenium in the lymphoid tissues of rats | 98 |
| Se | Rat liver | AAS | Rat liver | Effect of chronic exposure to excess dietary metal supplementation on liver specimens from rats | 51 |
| Se | Colitis | AAS | Serum | Serum concentrations of selenium also remained unaltered during colitis | 53 |
| Se | Colitis/rat colon | AAS | Tissue | | 53 |
| Se | Stomach cancer | AAS | Gastric tissue | Se concentration in the biopsies of patients with gastric ulceration and cancer were significantly lower than that in patients with gastritis ($p < 0.05$) and the other conditions ($p < 0.0001$) | 99 |
| Se | Liver | ICP-AES | Tissue | | 27 |
| Se | Human kidney tumor | XRF, ICP-AES | | | 67 |
| Se | Cataracts | AAS | Serum, lens, and aqueous humor | Decreased Se in aqueous humor and sera of patients with senile cataract may reflect defective antioxidative defense systems which may lead to the formation of cataract | 100 |
| Se | | AES | Cerebrospinal fluid | | 40 |
| Si | | AAS | Body fluids, Tissue | | 101 |
| Si | Breast | AAS | Tissue, serum | | 102 |
| Sr | Human organs/tissue (lung, kidney, liver, hair, blood) | ICP-AES, AAS | Human organs/ tissue (lung, kidney, liver, hair, blood) | Reference values human tissues/ organs | 9 |
| Sr | | ICP-AES | Human organs | Reference values study | 12 |
| Sr | | AAS | Serum, urine, bone | | 103 |
| Sr | | AAS | Blood | | 83 |
| Sr | | LIPS, AAS | Teeth, bone | | 18 |
| Ti | | AAS | Tissue | | 103 |
| Tl | Brain | AAS | Tissue | Effects on distribution and lipid peroxidation in brain regions | 105 |
| V | Human organs/tissue (lung, kidney, liver, hair, blood) | ICP-AES, AAS | Human organs/ tissue (lung, kidney, liver, hair, blood) | Reference values human tissues/ organs | 9 |
| V | | AAS | Vaginal tissue | | 106 |

**TABLE 65.10** Elemental Analysis Data (continued)

| Species | Target Organ or Disease | Technique | Matrix | Comments | Ref. |
|---|---|---|---|---|---|
| V | | AAS | Vaginal tissue | | 106 |
| V | | ICP-AES, AAS | | Various biomedical applications | 21 |
| V | Liver | ICP-AES | Tissue | | 28 |
| V | | AAS | Cerum | | 107 |
| W | | ICP-AES, AAS | | Various biomedical applications | 21 |
| Zn | Human organs/tissue (lung, kidney, liver, hair, blood) | ICP-AES, AAS | Human organs/ tissue (lung, kidney, liver, hair, blood) | Reference values human tissues/ organs | 9 |
| Zn | Colitis | AAS | Serum | The serum concentrations of zinc also remained unaltered during colitis | 53 |
| Zn | Colitis/rat colon | AAS | Tissue | | 53 |
| Zn | Chronic alcohol abuse | AAS | Embryo, liver | Maternal hepatic, endometrial, and embryonic levels following alcohol consumption during pregnancy in QS mice | 54 |
| Zn | Rat tissue, blood | AAS | Rat tissue, blood | | 49 |
| Zn | Beta-thalassemia major | | Urine | | 108 |
| Zn | Thoracic empyema | AAS | Serum | | 109 |
| Zn | Extrahepatic biliary atresia | AAS | Tissue | | 58 |
| Zn | | ICP-AES | Blood | Reference values for some bulk and trace elements in blood plasma and whole blood (Pb) of mothers and their newborn infants | 58 |
| Zn | | AES | Blood | | 81 |
| Zn | Prostatic cancer | AAS | Tissue | | 43 |
| Zn | | AAS | Adipose tissue | | 110 |
| Zn | | ICP-AES | Blood | | 61 |
| Zn | | ICP-AES | Human organs | Reference values study | 12 |
| Zn | | AAS | Newborn liver | | 110 |
| Zn | | AAS | Stomach, kidneys, bone, liver | | 33 |
| Zn | | ICP-AES | Hair | | 34 |
| Zn | Wilson's disease | ICP-AES | Brain tissue | | 36 |
| Zn | Liver | AAS | Tissue | | 111 |
| Zn | Liver | ICP-AES | Tissue | | 27 |
| Zn | Spinal cord, brain stem, cerebellum, forebrain | ICP-AES | | | 16 |
| Zn | Liver disease | AAS | Liver, serum | | 65 |
| Zn | Human kidney tumor | XRF, ICP-AES | | | 67 |
| Zn | Breast cancer, stomach cancer, colon cancer | AAS | Serum, tissue | The results suggest that movements of copper out from, and of zinc into, tissues occur and it is proposed that these changes could be a response to enhanced cytokine production rather than being a feature of any mechanism(s) for the onset of cancer | 68 |
| Zn | Heart disease | AAS | Arterial wall | | 38 |
| Zn | Aortoiliacal occlusive disease | AAS | Tissue | | 69 |
| Zn | Nasopharyngeal cancer | AAS | Tissue | | 112 |

**TABLE 65.10** Elemental Analysis Data (continued)

| Species | Target Organ or Disease | Technique | Matrix | Comments | Ref. |
|---|---|---|---|---|---|
| Zn | Breast cancer | AAS | Tissue | | 113 |
| Zn | | AES | Cerebrospinal fluid | | 40 |
| Zn | Hypoglycemia | AAS | Cerebrospinal fluid | Newborns | 70 |
| Zr | | ICP-AES, AAS | | Various biomedical applications | 21 |
| | | ***Animal Studies (in Vivo)*** | | | |
| As | Rat kidney | | | | 114 |
| Cd | Tissues | XRF | Tissues | | 115 |
| Cu | Rat kidney | | | | 114 |
| Hg | Tissues | XRF | Tissues | | 115 |
| Pb | Tissues | XRF | Tissues | | 115 |

## References


1. Alkhamis, K.I., Alhadiyah, B.M., Bawazir, S.A., Ibrahim, O.M., and Alyamani, M.J., Quantification of muscle-tissue magnesium and potassium using atomic-absorption spectrometry, *Anal. Lett.*, 28(6), 103, 1995.
2. Barth, R.F., Adams, D.M., Soloway, A.H., Mechetner, E.B., Alam, F., and Anisuzzaman, A.K.M., Determination of boron in tissues and cells using direct-current plasma atomic emission-spectroscopy, *Anal. Chem.*, 63(9), 890, 1991.
3. Kim, J.S. and Southard, J.H., Alteration in cellular calcium and mitochondrial functions in the rat liver during cold preservation, *Transplantation*, 65(3), 369, 1998.
4. Bax, C.M.R. and Bloxam, D.L., Major pathways of zinc(II) acquisition by human placental syncytiotrophoblast, *J. Cell. Physiol.*, 164(3), 546, 1995.
5. de Pena, Y.P., Vielma, O., Burguera, J.L., Burguera, M., Rondon, C., and Carrero, P., On-line determination of antimony(III) and antimony(V) in liver tissue and whole blood by flow injection-hydride generation-atomic absorption spectrometry, *Talanta*, 55, 2001.
6. da Silva, J.B.B., Bertilia, M., Giacomelli, O., de Souza, I.G., and Curtius, A.J., Iridium and rhodium as permanent chemical modifiers for the determination of Ag, As, Bi, Cd, and Sb by electrothermal atomic absorption spectrometry, *Microchem. J.*, 60(3), 249, 1998.
7. Drasch, G., Gath, H. J., Heissler, E., Schupp, I., and Roider, G., Silver concentrations in human tissues, their dependence on dental amalgam and other factors, *J. Trace Elements Med. Biol.*, 9(2), 82, 1995.
8. Ishihara, R., Ektossabi, A.M., Hanaichi, T., Takeuchi, T., Fujita, Y., Ishihara, Y., and Ohta, T., Aluminum accumulation in human brain tissue, *Int. J. PIXE*, 9(3–4), 259, 1999.
9. Coni, E., Alimonti, A., Fornarelli, L., Beccaloni, E., Sabiioni, E., Pietra, R., Bolis, G.B., Cristallini, E., Stacchini, A., and Caroli, S., Reference valves for elements in human organs — criteria and methods, *Acta Chim. Hung.*, 128(4–5), 563, 1991.
10. Anane, R., Bonini, M., and Creppy, E.E., Transplacental passage of aluminium from pregnant mice to fetus organs after maternal transcutaneous exposure, *Hum. Exp. Toxicol.*, 16(9), 501, 1997.
11. Chappuis, P., Poupon, J., and Rousselet, F., A sequential and simple determination of zinc, copper andaluminum in blood-samples by inductively coupled plasma atomic emission-spectrometry, *Clin. Chim. Acta*, 206(3), 155, 1992.

12. Coni, E., Alimonti, A., Bolis, G.B., Cristallini, E., and Caroli, S., An experimental approach to the assessment of reference values for trace-elements in human organs, *Trace Elements Electrolytes*, 11(2), 84, 1994.
13. DiPaolo, N., Masti, A., Comparini, I. B., Garosi, G., DiPaolo, M., Centini, F., Brardi, S., Monaci, G., and Finato, V., Uremia, dialysis and aluminium, *Int. J. Artif. Organs*, 20(10), 547, 1997.
14. Fiejka, M., Fiejka, E., and Dlugaszek, M., Effect of aluminium hydroxide administration on normal mice: tissue distribution and ultrastructural localization of aluminium in liver, *Pharmacol. Toxicol.*, 78(3), 123, 1996.
15. Gane, E., Sutton, M.M., Pybus, J., and Hamilton, I., Hepatic and cerebrospinal fluid accumulation of aluminium and bismuth in volunteers taking short course anti-ulcer therapy, *J. Gastroenterol. Hepatol.*, 11(10), 911, 1996.
16. Golub, M.S., Han, B., and Keen, C.L., Developmental patterns of aluminum and five essential mineral elements in the central nervous system of the fetal and infant guinea pig, *Biol. Trace Element Res.*, 55(3), 241, 1996.
17. Omahony, D., Denton, J., Templar, J., Ohara, M., Day, J.P., Murphy, S., Walsh, J.B., and Coakley, D., Bone aluminum content in Alzheimer's disease, *Dementia*, 6(2), 69, 1995.
18. Samek, O., Beddows, D.C.S., Telle, H.H., Kaiser, J., Liska, M., Caceres, J.O., and Urena, A.G., Quantitative laser-induced breakdown spectroscopy analysis of calcified tissue samples, *Spectrochim. Acta B*, 56(6), 865, 2001.
19. Becker-Ross, H., Florek, S., and Heitmann, U., A scanning echell monochromator for ICP-OES with dynamic wavelength stabilization and CCD detection, *J. Anal. Atomic Spectrom.*, 57(2), 137, 2000.
20. Ademuyiwa, O. and Elsenhans, B., Time course of arsenite-induced copper accumulation in rat kidney, *Biol. Trace Element Res.*, 74(1), 81, 2000.
21. De Aza, P.N., Guitian, F., De Aza, S., and Valle, F.J., Analytical control of wollastonite for biomedical applications by use of atomic absorption spectrometry and inductively coupled plasma atomic emission spectrometry, *Analyst*, 123(4), 681, 1998.
22. Concha, G., Vogler, G., Nermell, B., and Vahter, M., Low-level arsenic excretion in breast milk of native Andean women exposed to high levels of arsenic in the drinking water, *Int. Arch. Occup. Environ. Health*, 71(1), 42, 1998.
23. Svantesson, E., Capala, J. Markides, K.E., Pettersson, J., Determination of boron-containing compounds in urine and blood plasma from boron neutron capture therapy patients. The importance of using coupled techniques, *Anal. Chem.*, 74(20), 5358, 2002.
24. Brooke, S.L., Green, S., Charles, M.W., and Beddoe, A.H., The measurement of thermal neutron flux depression for determining the concentration of boron in blood, *Phy. Med. Biol.*, 46(3), 707, 2001.
25. Buchar, E., Bednarova, S., Gruner, B., Walder, P., Strouf, O., and Janku, I., Dose-dependent disposition kinetics and tissue accumulation of boron after intravenous injections of sodium mercaptoundecahydrododecaborate in rabbits, *Cancer Chemother. Pharm.*, 29(6), 450, 1992.
26. Iratsuka, J., Yoshino, K., Kondoh, H., Imajo, Y., and Mishima, Y., Biodistribution of boron concentration on melanoma-bearing hamsters after administration of *p*-, *m*-, *o*-boronophenylalanine, *Jpn. J. Cancer Res.*, 91(4), 446, 2000.
27. Galyean, M.L., Ralphs, M.H., Reif, M.N., Graham, J.D., and Braselton, W.E., Effects of previous grazing treatment and consumption of locoweed on liver mineral concentrations in beef steers, *J. Anim. Sci.*, 74(4), 827, 1996.
28. Shukla, S., Sharma, P., Johri, S., and Mathur, R., Influence of chelating agents on the toxicity and distribution of beryllium in rats, *J. Appl. Toxicol.*, 18(5), 331, 1998.
29. Deckelbaum, L.I., Scott, J.J., Stetz, M.L., O'Brien, M.O., and Baker, G., Detection of calcified atherosclerotic plaque by laser-induced plasma emission, *Lasers Surg. Med.*, 12(1), 18, 1992.
30. Sergeant, C., Gouget, B., Llabador, Y., Simonoff, M., Yefimova, M., Courtois, Y., and Jeanny, J.C., Iron in hereditary retinal degeneration: PIXE microanalysis — preliminary results, *Nucl. Instrum. Meth. Phys. Res.*, 158(1–4), 344, 1999.

31. Dahm, M., Dohmen, G., Groh, E., Krummenauer, F., Hafner, G., Mayer, E., Hake, U., and Oelert, H., Decalcification of the aortic valve does not prevent early recalcification, *J. Heart Valve Dis.*, 9(1), 21, 2000.
32. Dahm, M., Prufer, D., Mayer, E., Groh, E., Choi, Y.H., and Oelert, H., Early failure of an autologous pericardium aortic heart valve (ATCV) prosthesis, *J. Heart Valve Dis.*, 7(1), 30, 1998.
33. Dlugaszek, M., Fiejka, M.A., Graczyk, A., Aleksandrowicz, J.C., and Slowikowska, M., Effects of various aluminium compounds given orally to mice on Al tissue distribution anal tissue concentrations of essential elements, *Pharmacol. Toxicol.*, 86(3), 135, 2000.
34. Dombovari, J. and Papp, L., Comparison of sample preparation methods for elemental analysis of human hair, *Microchem. J.*, 59(2), 187, 1998.
35. Emelyanov, A.V., Shevelev, S.E., Murzin, B.A., and Amosov, V.I., Efficiency of calcium and vitamin D-3 in the treatment of steroid osteoporosis in patients with hormone-dependent bronchial asthma, *Ter. Ark.*, 71(11), 68, 1999.
36. Faa, G., Lisci, M., Caria, M. P., Ambu, R., Sciot, R., Nurchi, V.M., Silvagni, R., Diaz, A., and Crisponi, G., Brain copper, iron, magnesium, zinc, calcium, sulfur and phosphorus storage in Wilson's disease, *J. Trace Elements Med. Biol.*, 15(2–3), 155, 2001.
37. Habata, T., Ohgushi, H., Takakura, Y., Tohno, Y., Moriwake, Y., Minami, T., and Fujisawa, Y., Relationship between meniscal degeneration and element contents, *Biol. Trace Element Res.*, 79(3), 247, 2001.
38. Iskra, M., Patelski, J., and Majewski, W., Relationship of calcium, magnesium, zinc and copper concentrations in the arterial wall and serum in atherosclerosis obliterans and aneurysm, *J. Trace Elements Med. Biol.*, 11(4), 248, 1999.
39. Pallikaris, I.G., Ginis, H.S., Kounis, G. A., Anglos, D., Papazoglou, T.G., and Naoumidis, L.P., Corneal hydration monitored by laser-induced breakdown spectroscopy, *J. Refractive Surg.*, 14(6), 655, 1998.
40. Walther, L.E., Streck, S., Winnefeld, K., Walther, B.W., Kolmel, H.W., and Beleites, E., Reference values for electrolytes (Na, K, Ca, Mg) and trace elements (Fe, Cu, Zn, Se) in cerebrospinal fluid, *Trace Elements Electrolytes*, 15(4), 177, 1998.
41. Ohmi, M., Nakamura, M., Morimoto, S., and Haruna, M., Nanosecond time-gated spectroscopy of laser-ablation plume of human hair to detect calcium for potential diagnoses, *Opt. Rev.*, 7(4), 353, 2000.
42. Abu-Hayyeh, S., Sian, M., Jones, K.G., Manuel, A., and Powell, J.T., Cadmium accumulation in aortas of smokers, *Arteriosclerosis Thrombosis Vasc. Biol.*, 21(5), 863, 2001.
43. Brys, M., Nawrocka, A.D., Miekos, E., Zydek, C., Foksinski, M., Barecki, A., and Krajewska, W.M., Zinc and cadmium analysis in human prostate neoplasms, *Biol. Trace Element Res.*, 59(1–3), 145, 1997.
44. Bulska, E., Emteborg, H., Baxter, D.C., Frech, W., Ellingsen, D., and Thomassen, Y., Speciation of mercury in human whole-blood by capillary gas-chromatography with a microwave-induced plasma emission detector system following complexometric extraction and butylation, *Analyst*, 117(3), 657, 1992.
45. Chakraborty, R., Das, A.K., Cervera, M.L., and Delaguardia, M., The atomization of cadmium in graphite furnaces, *Anal. Proc.*, 32(7), 245, 1995.
46. Fortoul, T.I., Osorio, L.S., Tovar, A.T., Salazar, D., Castilla, M.E., and Olaiz Fernandez, G., Metals in lung tissue from autopsy cases in Mexico City residents: comparison of cases from the 1950s and the 1980s, *Environ. Health Perspect.*, 104(6), 186, 1983.
47. Gleizes, V., Poupon, J., Lazennec, J.Y., Chamberlin, B., and Saillant, G., Advantages and limits of determinating serum cobalt levels in patients with metal on metal articulating surfaces, *Rev. Chir. Orthop.*, 85(3), 217, 1999.
48. Anderson, R.A., Bryden, N.A., and Polansky, M.M., Lack of toxicity of chromium chloride and chromium picolinate in rats, *J. Am. Coll. Nutr.*, 16(3), 273, 1997.

49. Anderson, R.A., Bryden, N.A., Polansky, M.M., and Gautschi, K., Dietary chromium effects on tissue chromium concentrations and chromium absorption in rats, *J. Trace Elements Exp. Med.,* 9(1), 11, 1996.
50. Gunton, J.E., Hams, G., Hitchman, R., and McElduff, A., Serum chromium does not predict glucose tolerance in late pregnancy, *Am. J. Clin. Nutr.,* 73(1), 99, 2001.
51. Aburto, E.M., Cribb, A.E., and Fuentealba, C., Effect of chronic exposure to excess dietary copper and dietary selenium supplementation on liver specimens from rats, *Am. J. Vet. Res.,* 62(9), 1423, 2001.
52. Ademuyiwa, O. and Elsenhans, B., Time course of arsenite-induced copper accumulation in rat kidney, *Biol. Trace Element Res.,* 74(1), 81, 2000.
53. Al-Awadi, F.M., Khan, I., Dashti, H.M., and Srikumar, T.S., Colitis-induced changes in the level of trace elements in rat colon and other tissues, *Ann. Nutr. Metab.,* 42(5), 304, 1998.
54. Amini, S.A., Walsh, K., Dunstan, R., Dunkley, P.R., and Murdoch, R.N., Maternal hepatic, endometrial, and embryonic levels of Zn, Mg, Cu, and Fe following alcohol consumption during pregnancy in QS mice, *Res. Commun. Alcohol Subst.,* 16(4), 207, 1995.
55. Baker, A., Gormally, S., Saxena, R., Baldwin, D., Drumm, B., Bonham, J., Portmann, B., and Mowat, A.P., Copper-associated liver-disease in childhood, *J. Hepatol.,* 23(5), 538, 1995.
56. Bamiro, F.O., Littlejohn, D., and Marshall, J., Determination of copper in blood-serum by direct-current plasma and inductively coupled plasma atomic emission-spectrometry, *J. Anal. Atomic Spectrom.,* 3(1), 279, 1988.
57. Bayliss, E.A., Hambidge, K.M., Sokol, R.J., Stewart, B., and Lilly, J.R., Hepatic concentrations of zinc, copper and manganese in infants with extrahepatic biliary atresia, *J. Trace Elements Med. Biol.,* 9(1), 40, 1995.
58. Bertram, C., Bertram, H.P., Schussler, M., and Pfeiffer, M., Element pattern in blood plasma and whole blood from healthy pregnant women and their newborn infants, *Trace Elements Electrolytes,* 15(4), 190, 1998.
59. Braselton, W.E., Stuart, K.J., Mullaney, T.P., and Herdt, T.H., Biopsy mineral analysis by inductively coupled plasma-atomic emission spectroscopy with ultrasonic nebulization, *J. Vet. Diagn. Investig.,* 9(4), 395, 1997.
60. Chappuis, P., Poupon, J., and Rousselet, F., A sequential and simple determination of zinc, copper and aluminum in blood-samples by inductively coupled plasma atomic emission-spectrometry, *Clin. Chim. Acta,* 206(3), 155, 1992.
61. Deng, D.X., Ono, S., Koropatnick, J., and Cherian, M.G., Metallothionein and apoptosis in the toxic milk mutant mouse, *Lab. Invest.,* 78(2), 175, 1998.
62. Dorea, J.G., Iron and copper in human milk, *Nutrition,* 16(3), 209, 2000.
63. During, A., Fields, M., Lewis, C.G., and Smith, J.C., Beta-carotene 15,15′-dioxygenase activity is responsive to copper and iron concentrations in rat small intestine, *J. Am. Coll. Nutr.,* 18(4), 309, 1999.
64. Faa, G., Nurchi, V., Demelia, L., Ambu, R., Parodo, G., Congiu, T., Sciot, R., Vaneyken, P., Silvagni, R., and Crisponi, G., Uneven hepatic copper distribution in Wilson's disease, *J. Hepatol.,* 22(3), 303, 1995.
65. Garner, B., Davies, M.J., and Truscott, R.J.W., Formation of hydroxyl radicals in the human lens is related to the severity of nuclear cataract, *Exp. Eye Res.,* 70(1), 81, 2000.
66. Hatano, R., Ebara, M., Fukuda, H., Yoshikawa, M., Sugiura, N., Kondo, F., Yukawa, M., and Saisho, H., Accumulation of copper in the liver and hepatic injury in chronic hepatitis C, *J. Gastroenterol. Hepatol.,* 15(7), 786, 2000.
67. Homma, S., Sasaki, A., Nakai, I., Sagai, M., Koiso, K., and Shimojo, N., Distribution of copper, selenium, and zinc in human kidney tumors by nondestructive synchrotron-radiation x-ray-fluorescence imaging, *J. Trace Elements Exp. Med.,* 6(4), 163, 1993.

68. Sbir, T., Tamer, L., Erkisi, M., Kekec, Y., Doran, F., Varinli, S., and Taylor, A., Copper, zinc and magnesium in serum and tissues from patients with carcinoma of breast, stomach and colon, *Trace Elements Electrolytes,* 12(3), 113, 1995.
69. Jaakkola, P., Hippelainen, M., and Kantola, M., Copper and zinc concentrations of abdominal-aorta and liver in patients with infrarenal abdominal aortic-aneurysm or aortoiliacal occlusive disease, *Ann. Chir. Gynaecol.,* 83(4), 304, 1994.
70. Aral, Y.Z., Gucuyener, K., Atalay, Y., Hasanoglu, A., Turkyilmaz, C., Sayal, A., and Biberoglu, G., Role of excitatory amino acids in neonatal hypoglycemia, *Acta Paediatr. Jpn.,* 40(4), 303, 1998.
71. Bermejo, P., Pena, E., Dominguez, R., Brmejo, A., Fraga, J.M., and Cocho, J.A., Speciation of iron in breast milk and infant formulas whey by size exclusion chromatography-high performance liquid chromatography and electrothermal atomic absorption spectrometry, *Talanta,* 2000, 50, 1211.
72. Ambu, R., Crisponi, G., Sciot, R., Vaneyken, P., Parodo, G., Iannelli, S., Marongiu, F., Silvagni, R., Nurchi, V., Costa, V., Faa, G., and Desmet, V.J., Uneven hepatic iron and phosphorus distribution in beta-thalassemia, *J. Hepatol.,* 23(5), 544, 1995.
73. Burguera, J.L., Burguera, M., Carrero, P., Rivas, C., Gallignani, M., and Brunetto, M.R., Determination of iron and zinc in adipose-tissue by online microwave-assisted mineralization and flow-injection graphite-furnace atomic-absorption spectrometry, *Anal. Chim. Acta,* 308(1–3), 349, 1995.
74. Collery, P., Domingo, J.L., and Keppler, B.K., Preclinical toxicology and tissue gallium distribution of a novel antitumour gallium compound: tris(8-quinolinolato)gallium(III), *Anticancer Res.,* 16(2), 687, 1996.
75. Adair, B.M. and Cobb, G.P., Improved preparation of small biological samples for mercury analysis using cold vapor atomic absorption spectroscopy, *Chemosphere,* 38(12), 2951, 1999.
76. Alfthan, G.V., Toenail mercury concentration as a biomarker of methylmercury exposure, *Biomarkers,* 2(4), 233, 1997.
77. Arnett, F.C., Fritzler, M.J., Ahn, C., and Holian, A., Urinary mercury levels in patients with autoantibodies to U3-RNP (fibrillarin), *J. Rheumatol.,* 27(2), 405, 2000.
78. Boaventura, G.R., Barbosa, A.C., and East, G.A., Multivessel system for cold-vapor mercury generation — determination of mercury in hair and fish, *Biol. Trace Element Res.,* 60(1–2), 153, 1997.
79. Cominos, X., Athanaselis, S., Dona, A., and Koutselinis, A., Analysis of total mercury in human tissues prepared by microwave decomposition using a hydride generator system coupled to an atomic absorption spectrometer, *Forensic Sci. Int.,* 118(1), 43, 2001.
80. Agut, A., Laredo, F.G., Sanchezvalverde, M.A., Murciano, J., and Tovar, M.D., Plasma levels and urinary excretion of iodine after oral administration of iohexol in dogs and cats, *Invest. Radiol.,* 30(5), 296, 1995.
81. Besteman, A.D., Bryan, G.K., Lau, N., and Winefordner, J.D., Multielement analysis of whole blood using a capacitively coupled microwave plasma atomic emission spectrometer, *Microchem. J.,* 61(3), 240, 1999.
82. Dol, I., Knochen, M., and Vieras, E., Determination of lithium at ultratrace levels in biological fluids by flame atomic emission-spectrometry — use of first-derivative spectrometry, *Analyst,* 117(8), 1373, 1992.
83. Matusiewicz, H., Determination of natural levels of lithium and strontium in human-blood serum by discrete injection and atomic emission-spectrometry with a nitrous-oxide acetylene flame, *Anal. Chim. Acta,* 136, 215, 1982.
84. Marczenko, Z., Lobinski, R., Griepink, B., Wells, D.E., Biemann, K., Gries, W.H., Jackwerth, E., Leroy, M., Lamotte, A., Westmoreland, D.G., Zolotov, Y.A., Ballschmiter, K., Dams, R., Fuwa, K., Grasserbauer, M., Linscheid, M.W., Morita, M., and Huntau, M., Determination of molybdenum in biological materials, *Pure Appl. Chem.,* 63(11), 1627, 1991.
85. Afarideh, H., Amirabadi, A., HadjiSaeid, S.M., Mansourian, N., Kaviani, K., and Zibafar, E., Biomedical studies by PIXE, *Nuclear Instrum. Methods Phys. Res. B — Beam Interactions with Materials and Atoms,* 109, 270, 1996.

86. Hu, X.H., Fang, Q.Y., Cariveau, M.J., Pan, X.N., and Kalmus, G.W., Mechanism study of porcine skin ablation by nanosecond laser pulses at 1064, 532, 266, and 213 nm, *IEEE J. Quantum Electron.*, 37(3), 322, 2001.
87. Chakraborty, R., Das, A.K., Cervera, M.L., and de la Guardia, M., A generalized method for the determination of nickel in different samples by ETAAS after rapid microwave-assisted digestion, *Anal. Lett.*, 30(2), 283, 1997.
88. Alvarado, J., Cavalli, P., Omenetto, N., Rossi, G., Ottaway, J. M., and Littlejohn, D., Direct determination of lead in whole blood using electrothermal vaporization inductively coupled plasma atomic emission spectrometry, *Anal. Lett.*, 22(15), 2975, 1989.
89. Areola, O.O., Jadhav, A.L., and Williams-Johnson, M., Relationship between lead accumulation in blood and soft tissues of rats subchronically exposed to low levels of lead, *Toxic Subst. Mech.*, 18(3), 149, 1999.
90. Bjora, R., Falch, J.A., Staaland, H., Nordsletten, L., and Gjengedal, E., Osteoporosis in the Norwegian moose, *Bone,* 29(1), 70, 2001.
91. Castilla, L., Castro, M., Grilo, A., Guerrero, P., Lopezartiguez, M., Soria, M.L., and Martinezparra, D., Hepatic and blood lead levels in patients with chronic liver disease, *Eur. J. Gastroenterol. Hepatol.*, 7(3), 243, 1995.
92. Deibel, M.A., Savage, J.M., Robertson, J.D., Ehmann, W.D., and Markesbery, W.R., Lead determinations in human bone by particle-induced x-ray-emission (PIXE) and graphite-furnace atomic-absorption spectrometry (GFAAS), *J. Radioanal. Nucl. Chem.*, 195(1), 83, 1995.
93. Gerhardsson, L., Englyst, V., Lundstrom, N.G., Nordberg, G., Sandberg, S., and Steinvall, F., Lead in tissues of deceased lead smelter workers, *J. Trace Elements Med. Biol.,* 9(3), 136, 1995.
94. Haraguchi, T., Ishizu, H., Takehisa, Y., Kawai, K., Yokota, O., Terada, S., Tsuchiya, K., Ikeda, K., Morita, K., Horike, T., Kira, S., and Kuroda, S., Lead content of brain tissue in diffuse neurofibrillary tangles with calcification (DNTC): the possibility of leads neurotoxicity, *Neuroreport* 13(1), cover3–cover3, 2002.
95. Keating, A.D., Keating, J.L., Halls, D.J., and Fell, G.S., Determination of lead in teeth by atomic-absorption spectrometry with electrothermal atomization, *Analyst,* 112(10), 1381, 1987.
96. Dinoto, V., Ni, D., Via, L.D., Scomazzon, F., and Vidali, M., Determination of platinum in human blood using inductively coupled plasma atomic emission-spectrometry with an ultrasonic nebulizer, *Analyst,* 120(6), 1669, 1995.
97. Canavese, C., DeCostanzi, E., Branciforte, L., Caropreso, A., Nonnato, A., Pietra, R., Fortaner, S., Jacono, F., Angelini, G., Gallieni, M., Fop, F., and Sabbioni, E., Rubidium deficiency in dialysis patients, *J. Nephrol.,* 14(3), 169, 2001.
98. Abulaban, F.S., Saadeddin, S.M., Al Sawaf, H.A., and AlBekairi, A.M., Effect of ageing on levels of selenium in the lymphoid tissues of rats, *Med. Sci. Res.*, 25(5), 303, 1997.
99. Burguera, J.L., Villasmil, L.M., Burguera, M., Carrero, P., Rondon, C., Delacruz, A., Brunetto, M.R., and Gallignani, M., Gastric tissue selenium levels in healthy persons, cancer and noncancer patients with different kinds of mucosal damage, *J. Trace Elem. Med. Biol.,* 9(3), 160, 1995.
100. Karakucuk, S., Mirza, G.E., Ekinciler, O.F., Saraymen, R., Karakucuk, I., and Ustdal, M., Selenium concentrations in serum, lens and aqueous humor of patients with senile cataract, *Acta Ophthalmol. Scand.,* 73(4), 329, 1995.
101. Centeno, J.A., Offiah, O.O., Rastogi, T., Dewitt, T.J., and Luke, J.L., Electrothermal atomic absorption determination of silicon in body fluids and tissue specimens, *Abst. Papers Am. Chem. Soc.,* 208, 228, 1994.
102. Leung, F.Y. and Edmond, P., Determination of silicon in serum and breast tissue by electrothermal atomic absorption spectrometry, *Clin. Chem.,* 42(6), 845, 1996.
103. Dhaese, P.C., Van Landeghem, G.F., Lamberts, L.V., Bekaert, V.A., Schrooten, I., and DeBroe, M.E., Measurement of strontium in serum, urine, bone, and soft tissues by Zeeman atomic absorption spectrometry, *Clin. Chem.,* 43(1), 121, 1997.

104. Jorgenson, D.S., Mayer, M.H., Ellenbogen, R.G., Centeno, J.A., Johnson, F.B., Mullick, F.G., and Manson, P.N., Detection of titanium in human tissues after craniofacial surgery, *Plastic Reconst. Surg.*, 99(4), 976, 1997.
105. Galvan-Arzate, S., Martinez, A., Medina, E., Santamaria, A., and Rios, C., Subchronic administration of sublethal doses of thallium to rats, *Toxicol. Lett.*, 116(1–2), 37, 2000.
106. D'Cruz, O.J., Waurzyniak, B., and Uckun, F.A., Subchronic (13-week) toxicity studies of intravaginal administration of spermicidal vanadocene dithiocarbarnate in mice, *Contraception*, 64(3), 177, 2001.
107. Heinemann, G. and Vogt, W., Quantification of vanadium in serum by electrothermal atomic absorption spectrometry, *Clin. Chem.*, 42(8), 1275, 1996.
108. Aydinok, Y., Coker, C., Kavakli, K., Polat, A., Nisli, G., Cetiner, N., Kantar, M., and Cetingul, N., Urinary zinc excretion and zinc status of patients with beta-thalassemia major, *Biol. Trace Element Res.*, 70(2), 165, 1999.
109. Balkan, M.E. and Ozgunes, H., Serum protein and zinc levels in patients with thoracic empyema, *Biol. Trace Element Res.*, 54(2), 105, 1996.
110. Coni, P., Ravarino, A., Farci, A.M.G., Callea, F., Van Eyken, P., Sciot, R., Ambu, R., Marras, A., Costa, V., Faa, G., and Desmet, V.J., Zinc content and distribution in the newborn liver, *J. Pediat. Gastroenterol. Nutr.*, 23(2), 125, 1996.
111. Gabrielson, K.L., Remillard, R.L., and Huso, D.L., Zinc toxicity with pancreatic acinar necrosis in piglets receiving total parenteral nutrition, *Vet. Pathol.*, 33(6), 692, 1996.
112. Jayasurya, A., Bay, B.H., Yap, W.M., Tan, N.G., and Tan, B.K.H., Proliferative potential in nasopharyngeal carcinoma: correlations with metallothionein expression and tissue zinc levels, *Carcinogenesis*, 21(10), 1809, 2000.
113. Jin, R.X., Bay, B.H., Tan, P.H., and Tan, B.K.H., Metallothionein expression and zinc levels in invasive ductal breast carcinoma, *Oncol. Rep.*, 6(4), 871, 1999.
114. Ademuyiwa, O., Elsenhans, B., Nguyen, P.T., and Forth, W., Arsenic copper interaction in the kidney of the rat: influence of arsenic metabolites, *Pharmacol. Toxicol.*, 78(3), 154, 1996.
115. McNeill, F.E., and O'Meara, J.M., *In vivo* measurement of trace toxic metals by K x-ray fluorescence, *Adv. X Ray Anal.*, 41, 910, 1999.

# Index

## A

# B

# C

# D

# E

# F

# H

# I

# J



# K



# L

# M

# N

# O

# P

# Q



# R

# S

# T

# U



# V

## W



## X



## Y



## Z